D1712508

# Instrumental Data for Drug Analysis

Second Edition

Volume 2

Terry Mills III and J. Conrad Roberson

**CRC Press**
**Boca Raton New York**

**Library of Congress Cataloging-in-Publication Data**

Instrumental data for drug analysis / Terry Mills III and J. Conrad Roberson. — 2nd ed.
p. cm. — (CRC series in forensic and police science)
Vol. 5 by Terry Mills III ... [et al.].
Originally published: New York : Elsevier, c1987–c1992, in series:
Elsevier series in forensic and police science.
Includes index in v. 5.
ISBN 0-8493-9522-4 (v. 2)
1. Drugs—Analysis—Standards. 2. Drugs—Spectra—Standards.
3. Instrumental analysis—Standards. I. Mills, Terry.
II. Roberson, J. Conrad. III. Series.
RS189.M54 1993
615′.1901—dc20

93-13539
CIP



Direct all inquiries to CRC Press LLC, 2000 Corporate Blvd., N.W., Boca Raton, Florida 33431.


International Standard Book Number 0-8493-9522-4
Library of Congress Card Number 93-13539
Printed in the United States of America 6 7 8 9 0
Printed on acid-free paper

# INSTRUMENTAL DATA
## FOR
# DRUG ANALYSIS

*CRC Series in Forensic and Police Science*

**BARRY A.J. FISHER,** *Editor*

**TECHNIQUES OF CRIME SCENE INVESTIGATION, Fourth Edition**
Barry A.J. Fisher, Arne Svensson, and Otto Wendel

**SCIENTIFIC EXAMINATION OF QUESTIONED DOCUMENTS**
Ordway Hilton

**INSTRUMENTAL DATA FOR DRUG ANALYSIS, Second Edition**
Terry Mills III and J. Conrad Roberson

# CONTENTS

**VOLUME I**

**PREFACE** vii
**ACKNOWLEDGMENTS** ix
**INTRODUCTION** xi
**DRUG DATA** 1
Acebutolol–Doxapram

**VOLUME 2**

**DRUG DATA** 786
Doxepin–Naltrexone

**VOLUME 3**

**DRUG DATA** 1578
Nandrolone Phenpropionate–Zoxazolamine

**VOLUME 4**

**APPENDIX A** **Standard KBr Infrared Spectra and Standard NMR Solvent Spectra** 2403
**APPENDIX B** **Supplemental Infrared Spectra** 2411
**APPENDIX C** **Supplemental NMR Spectra** 2491
**APPENDIX D** **Ultraviolet Absorption Maxima** 2517
**APPENDIX E** **Mass Spectral Index** 2567
**APPENDIX F** **Infrared Index** 2585
**APPENDIX G** **Gas Chromatographic Data** 2615
**APPENDIX H** **Molecular Formula Index** 2623
**INDEX TO SPECIFIC COMPOUNDS** 2635

# Drug Data
# Doxepin-Naltrexone

# DOXEPIN

$C_{19}H_{21}NO$

Molecular weight: 279.38 (279.16)
Synonyms: N,N-Dimethyl-3-dibenz[b,e]oxepin-11-(6H)-ylidene-1-propanamine
Trade names: Adapin, Sinequan

Use: Antidepressant
HPLC: Si-10; 2A:98B; 4.5
GC: 2278; 250°C

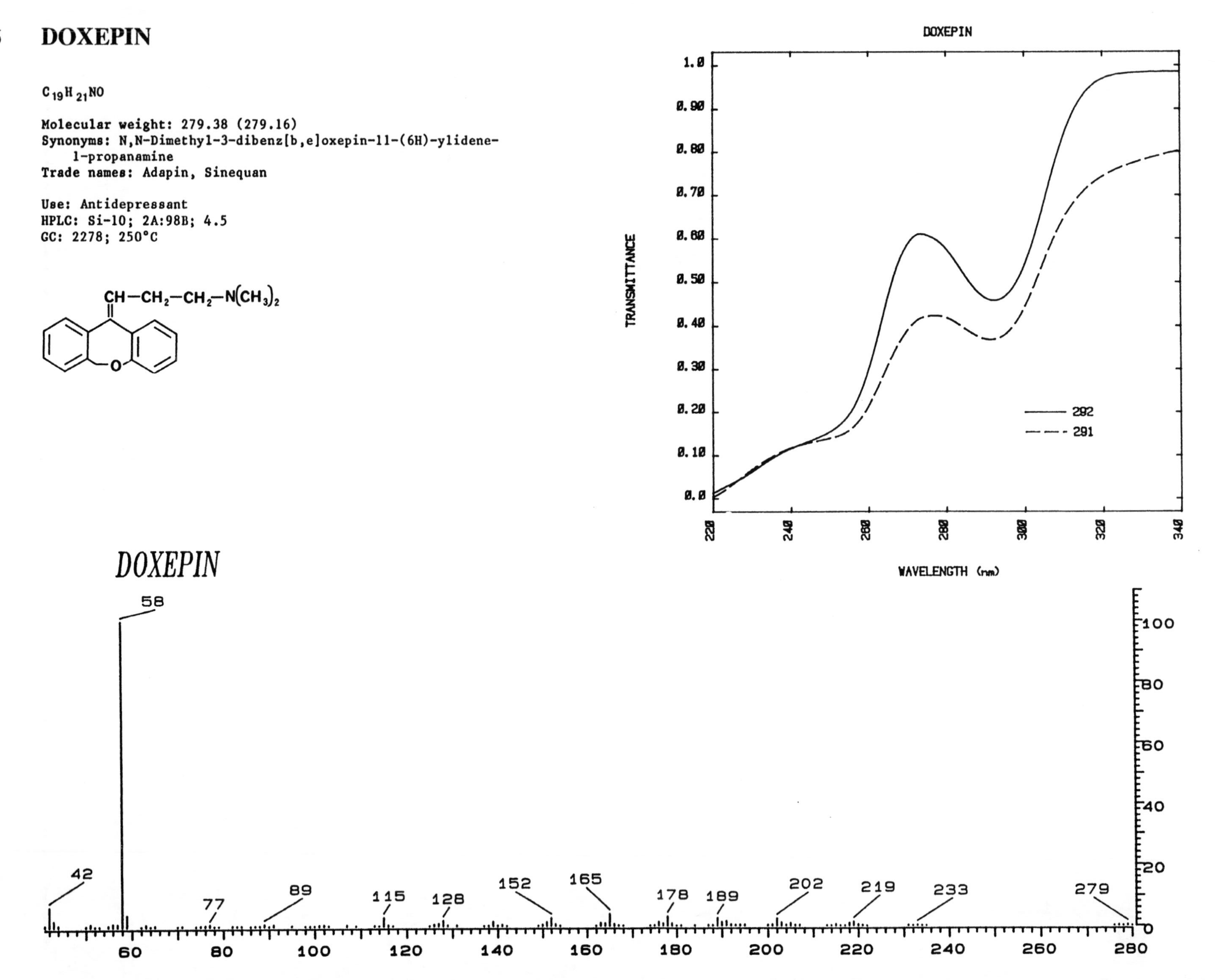

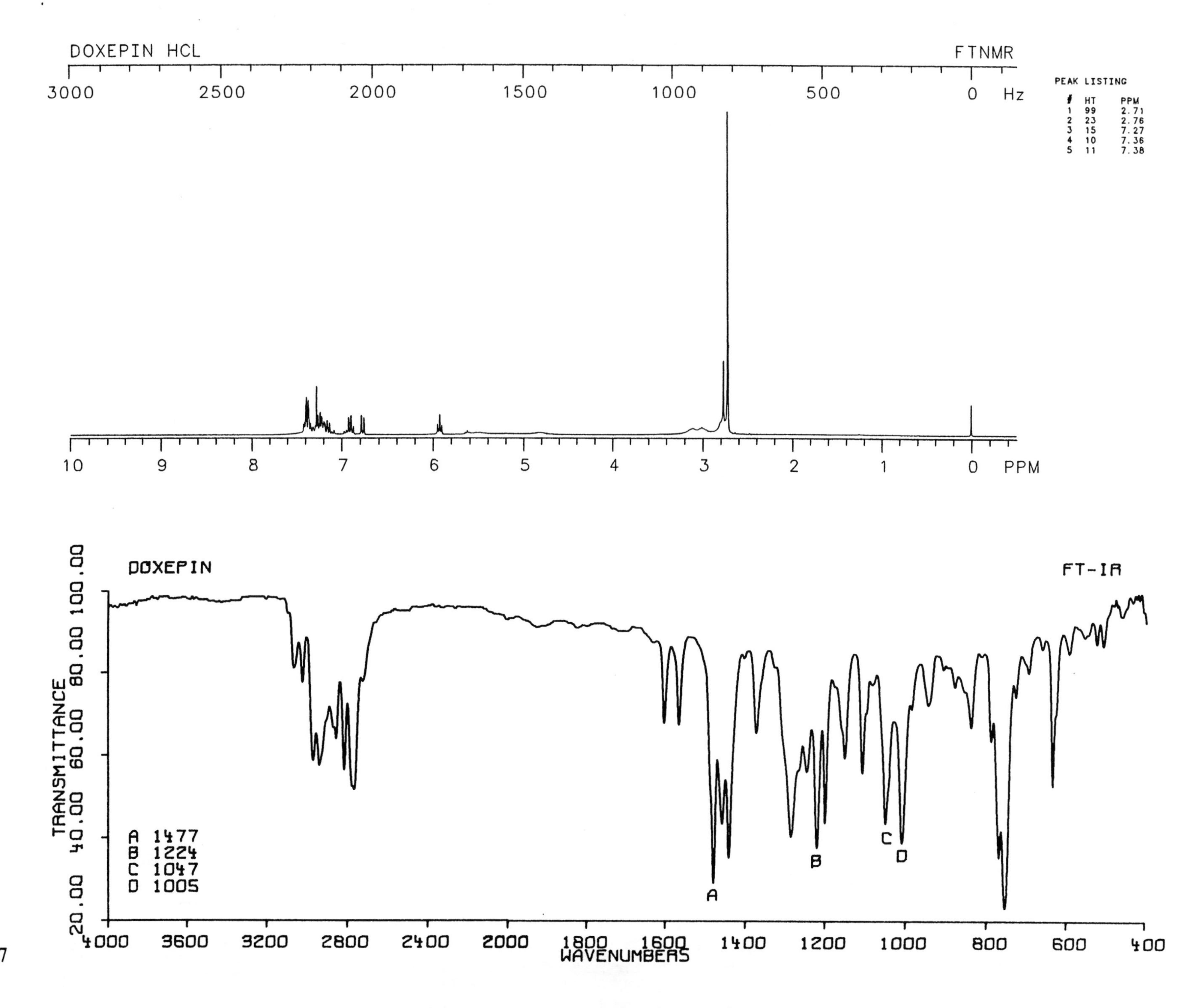
DOXEPIN HCL
FTNMR
3000
2500
2000
1500
1000
500
0 Hz
PEAK LISTING
# HT PPM
1 99 2.71
2 23 2.76
3 15 7.27
4 10 7.36
5 11 7.38
10
9
8
7
6
5
4
3
2
1
0 PPM
DOXEPIN
FT-IR
TRANSMITTANCE
20.00
40.00
60.00
80.00
100.00
A 1477
B 1224
C 1047
D 1005
A
B
C
D
4000
3600
3200
2800
2400
2000
1800
1600
1400
1200
1000
800
600
400
WAVENUMBERS

# DOXORUBICIN

$C_{27}H_{29}NO_{11}$

Molecular weight: 543.54 (543.17)
Synonyms: 10-[(3-Amino-2,3,6-trideoxy-α-L-6,8,11-trihydroxy-8-(hydroxyacetyl)-1-methoxy-5,12-naphthacenedione; 14-hydroxydaunomycin;
Trade names: Adriacin, Adriablastina, Adriamycin

Use: Antibiotic, antineoplastic
HPLC: Si-10; 10A:90B; 4.6
GC:

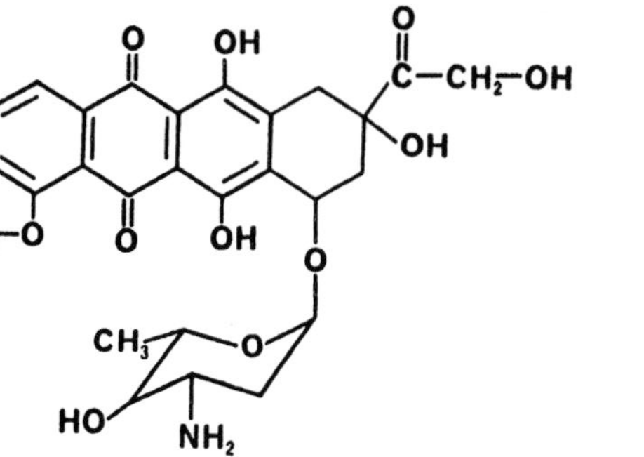


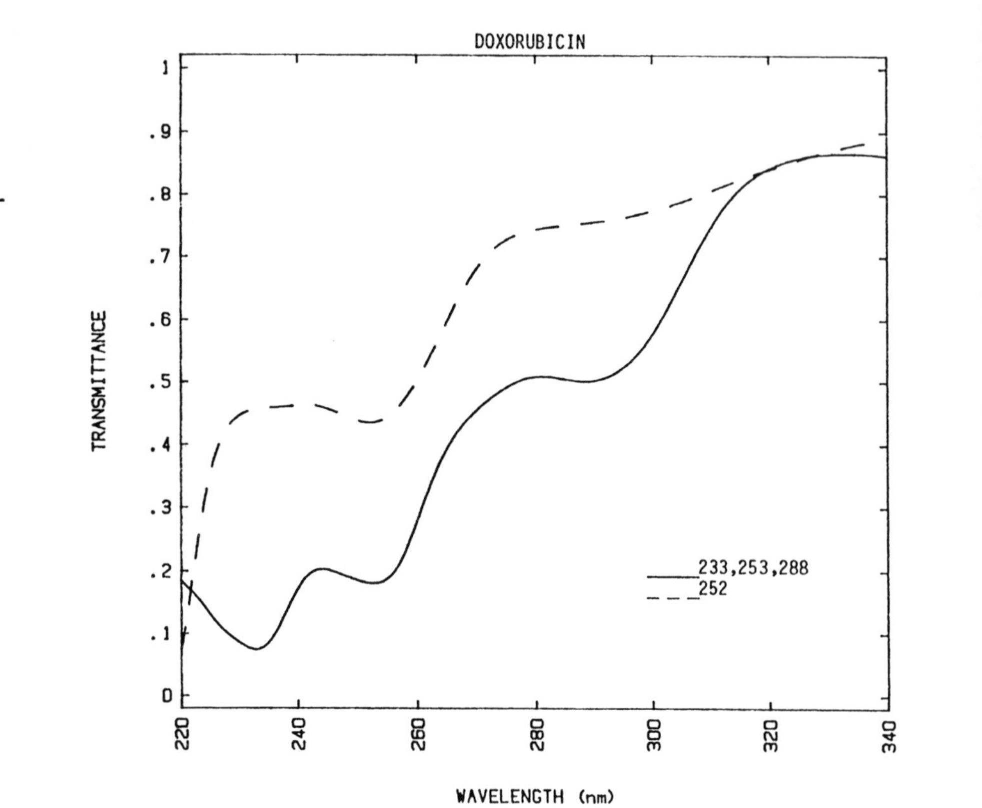


*NO USEFUL MASS SPECTRUM WAS OBTAINED*

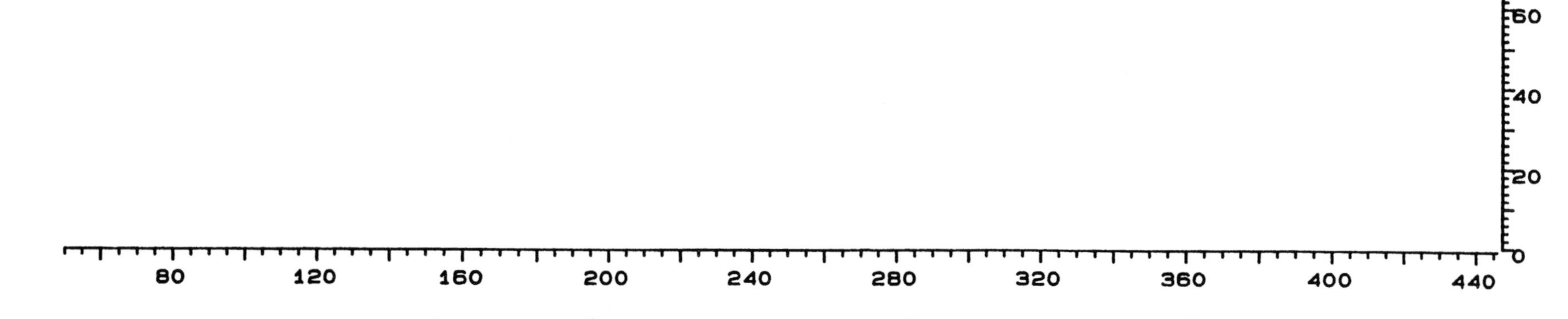

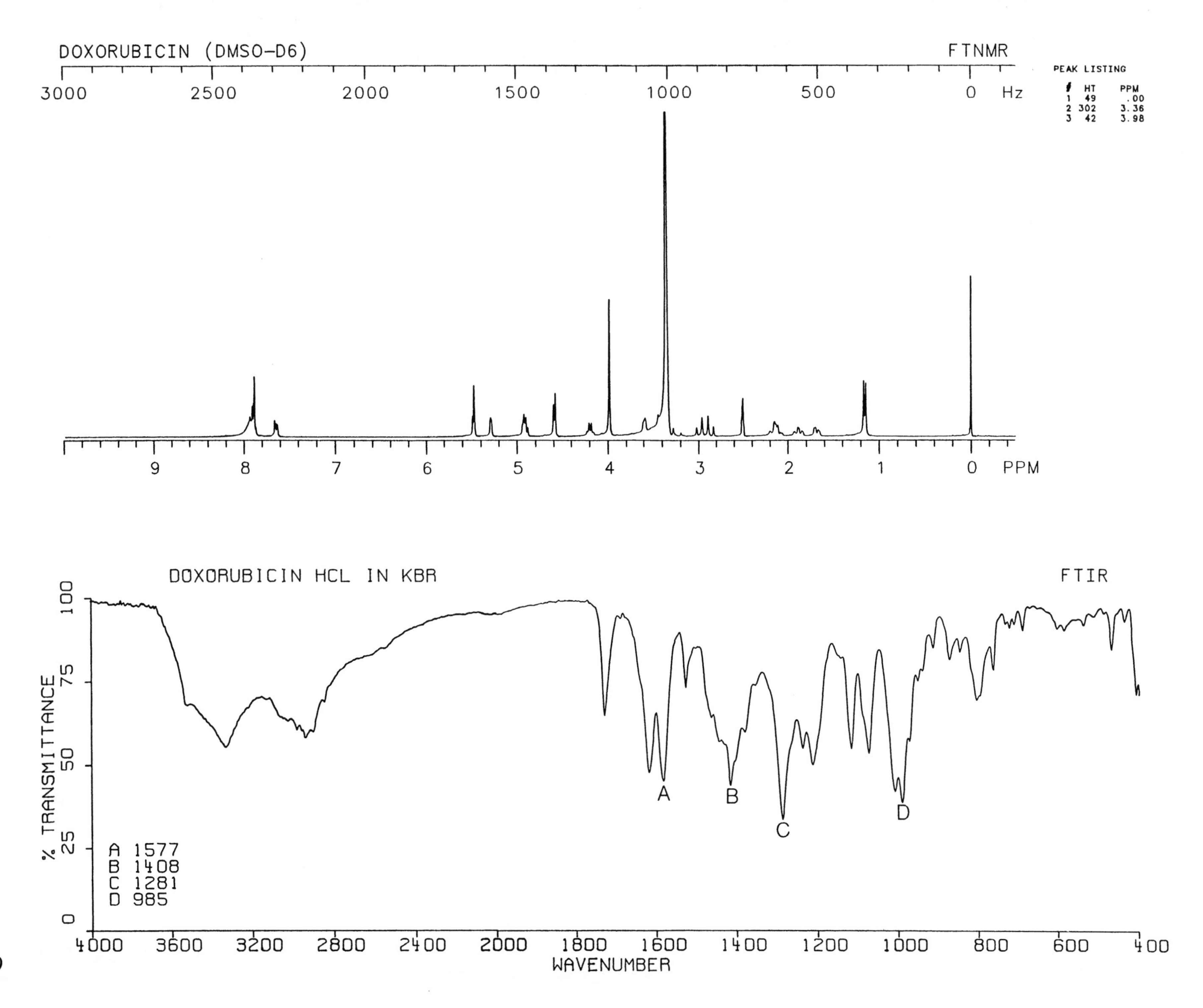
DOXORUBICIN (DMSO-D6)
FTNMR
3000
2500
2000
1500
1000
500
0
Hz
PEAK LISTING
# HT PPM
1 49 .00
2 302 3.36
3 42 3.98
9
8
7
6
5
4
3
2
1
0
PPM
DOXORUBICIN HCL IN KBR
FTIR
% TRANSMITTANCE
100
75
50
25
0
A 1577
B 1408
C 1281
D 985
A
B
C
D
4000
3600
3200
2800
2400
2000
1800
1600
1400
1200
1000
800
600
400
WAVENUMBER

# DOXYCYCLINE

$C_{22}H_{24}N_2O_8$

**Molecular weight:** 444.44 (444.15)
**Synonyms:** 4αS-(Dimethylamino)-1,4,4aα,5,5aα,6,11,12a-octahydro-3,5α,-10,12,12aα-pentahydroxy-6α-methyl-1,11-dioxo-2-naphthacene-carboxamide
**Trade names:** Doxycycline Hyclate, Vibramycin Hyclate, Vibra-Tabs

**Use:** Antibiotic
**HPLC:** Si-10; 5A:95B; 4.0
**GC:**

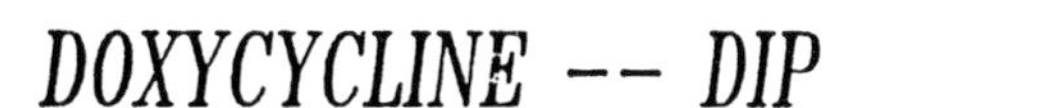


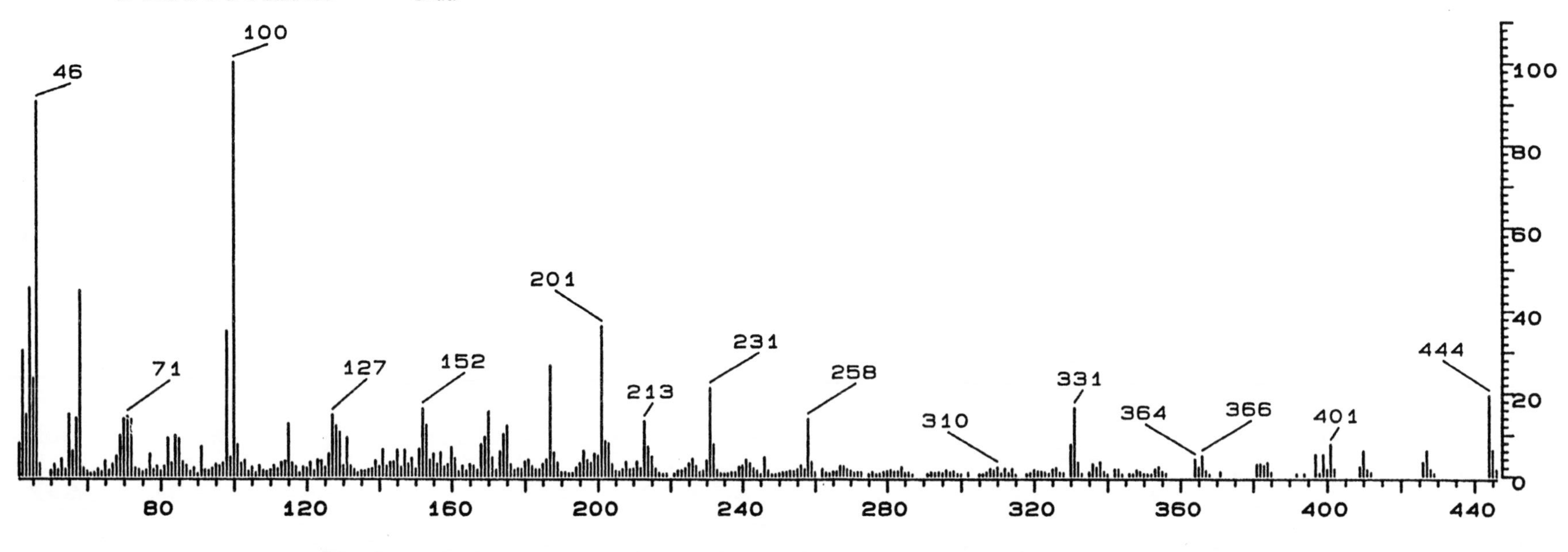

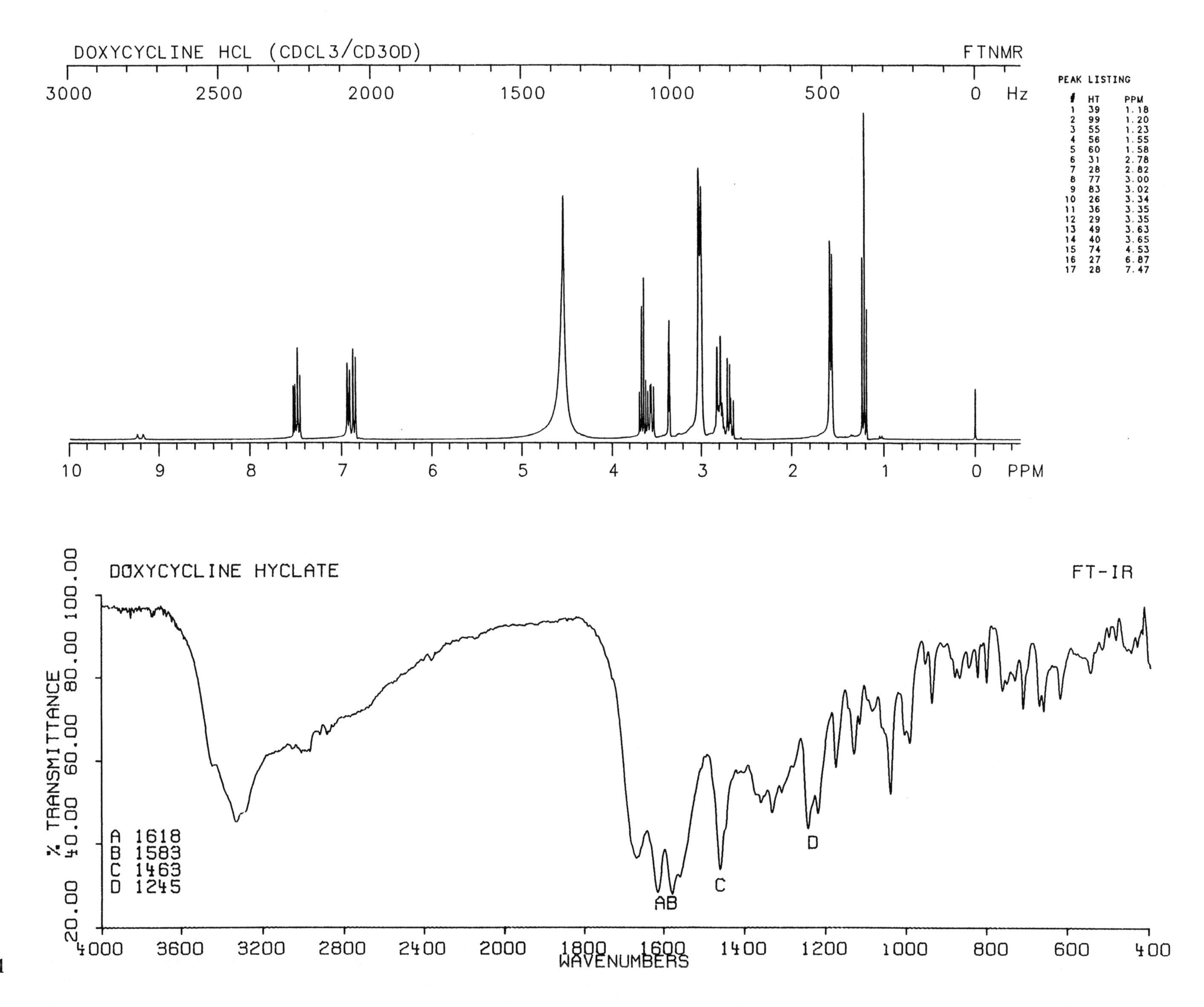
DOXYCYCLINE HCL (CDCL3/CD3OD)
FTNMR
3000 2500 2000 1500 1000 500 0 Hz
PEAK LISTING
# HT PPM
1 39 1.18
2 99 1.20
3 55 1.23
4 56 1.55
5 60 1.58
6 31 2.78
7 28 2.82
8 77 3.00
9 83 3.02
10 26 3.34
11 36 3.35
12 29 3.35
13 49 3.63
14 40 3.65
15 74 4.53
16 27 6.87
17 28 7.47
10 9 8 7 6 5 4 3 2 1 0 PPM
DOXYCYCLINE HYCLATE
FT-IR
% TRANSMITTANCE
20.00 40.00 60.00 80.00 100.00
A 1618
B 1583
C 1463
D 1245
AB
C
D
4000 3600 3200 2800 2400 2000 1800 1600 1400 1200 1000 800 600 400
WAVENUMBERS

# DOXYLAMINE

$C_{17}H_{22}N_2O$

Molecular weight: 270.37 (270.17)
Synonyms: N,N-Dimethyl-2-[1-phenyl-1-(2-pyridinyl)ethoxy]-ethanamine; 2-[α-(2-dimethylaminoethoxy)-α-methylbenzyl]pyridine
Trade names: Bendectin, Doxine, Mereprine, Unisom

Use: Antihistaminic
HPLC: Si-10; 20A:80B; 4.0
GC: 1917; 200°C

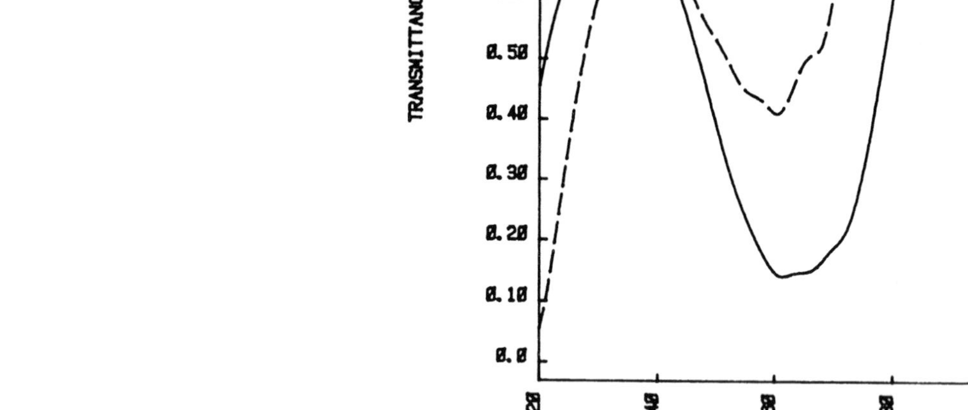


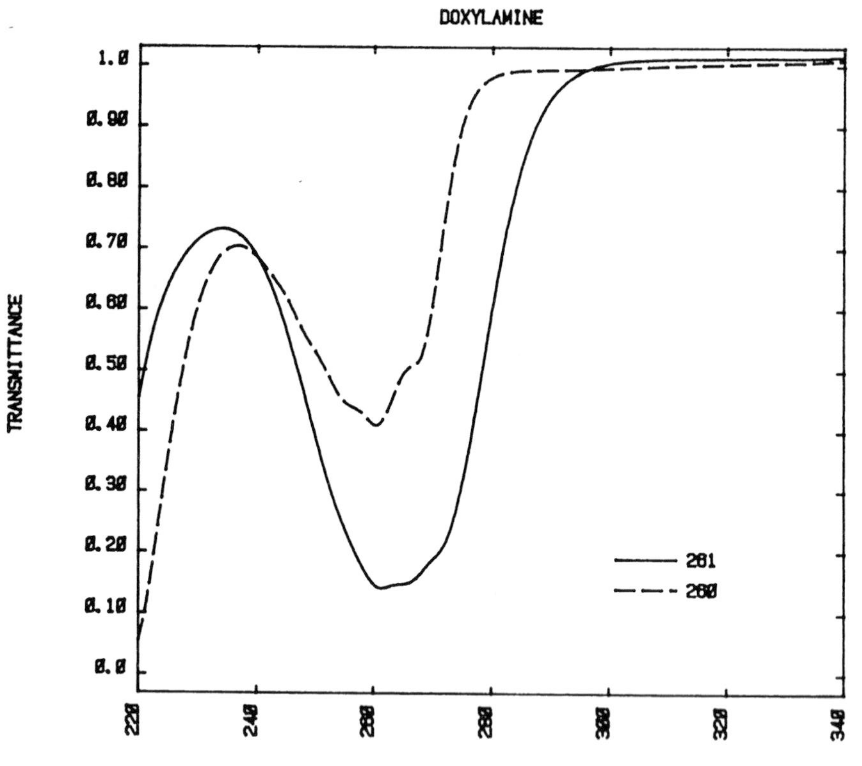


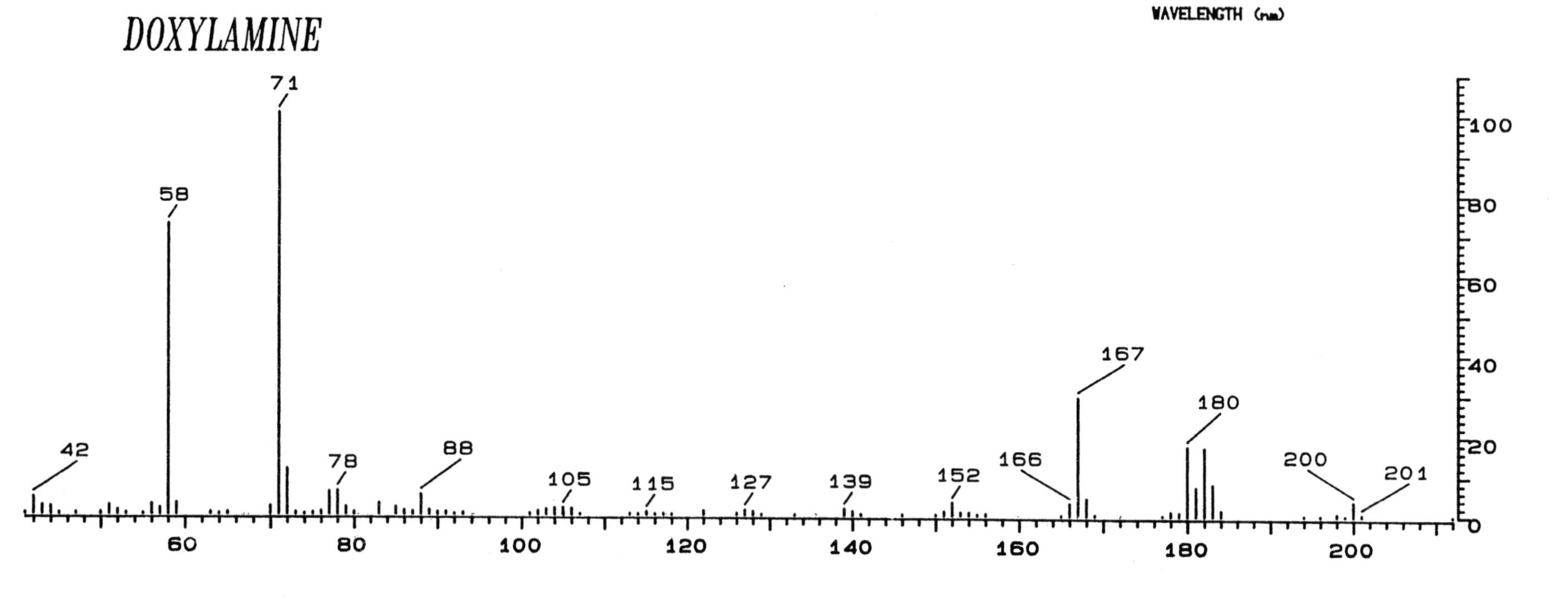

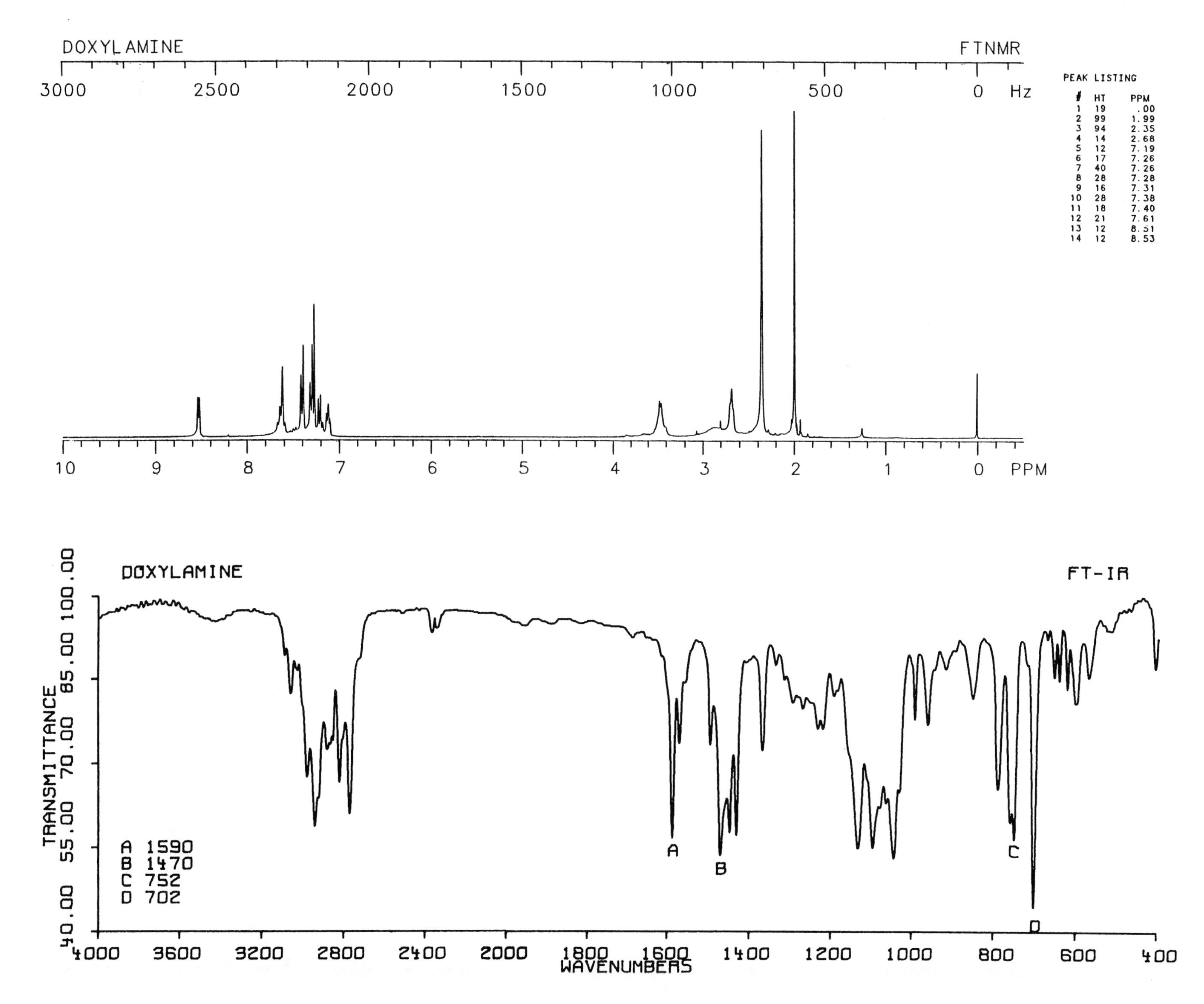
DOXYLAMINE
FTNMR
3000
2500
2000
1500
1000
500
0
Hz
10
9
8
7
6
5
4
3
2
1
0
PPM
PEAK LISTING
# HT PPM
1 19 .00
2 99 1.99
3 94 2.35
4 14 2.68
5 12 7.19
6 17 7.26
7 40 7.26
8 28 7.28
9 16 7.31
10 28 7.38
11 18 7.40
12 21 7.61
13 12 8.51
14 12 8.53
DOXYLAMINE
FT-IR
TRANSMITTANCE
40.00
55.00
70.00
85.00
100.00
A 1590
B 1470
C 752
D 702
A
B
C
D
4000
3600
3200
2800
2400
2000
1800
1600
1400
1200
1000
800
600
400
WAVENUMBERS

# DROMOSTANOLONE PROPIONATE

$C_{23}H_{36}O$

**Molecular weight:** 360.52 (360.27)
**Synonyms:** 2α-Methyl-17β-(1-oxopropoxy)-5α-androstan-3-one; drostandone propionate
**Trade names:** Drolban, Emdisterone, Masterid, Masteril, Masterone, Permastril
**Use:** Antineoplastic
**HPLC:**
**GC:** 2963; 280°C

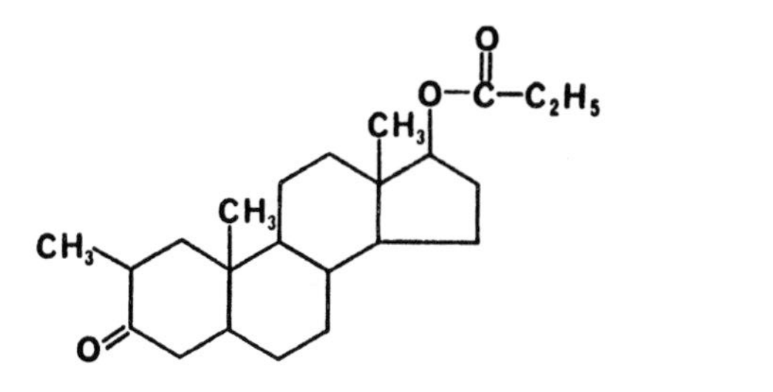


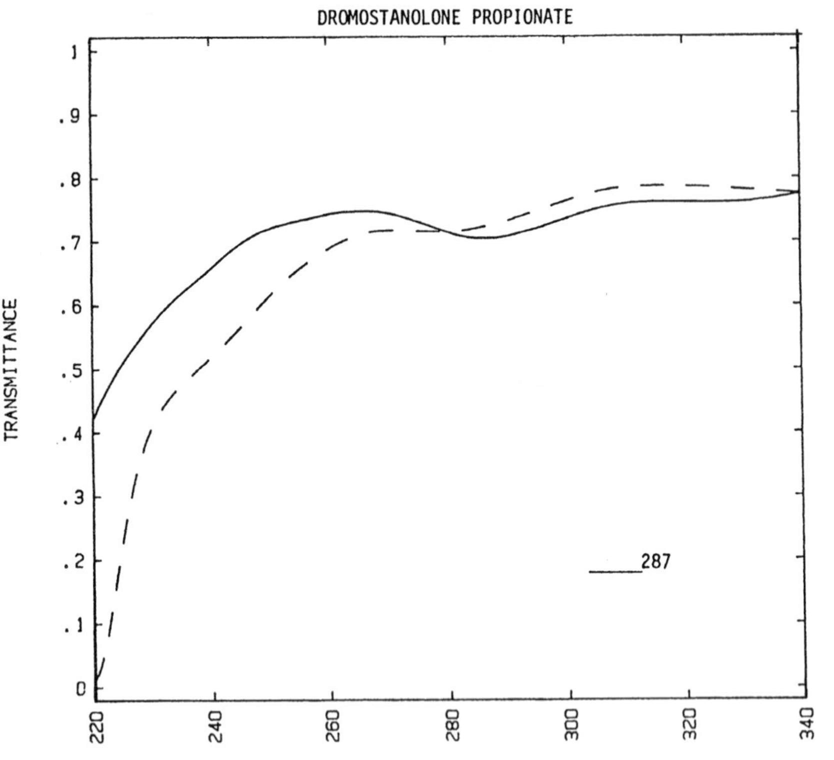

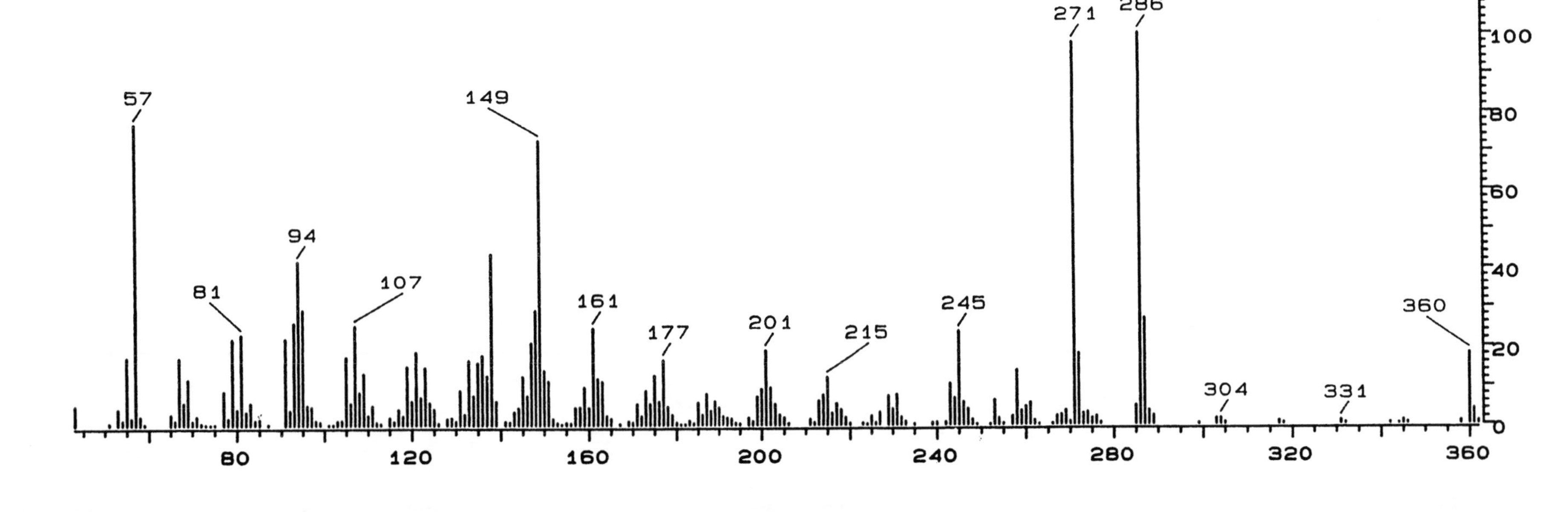

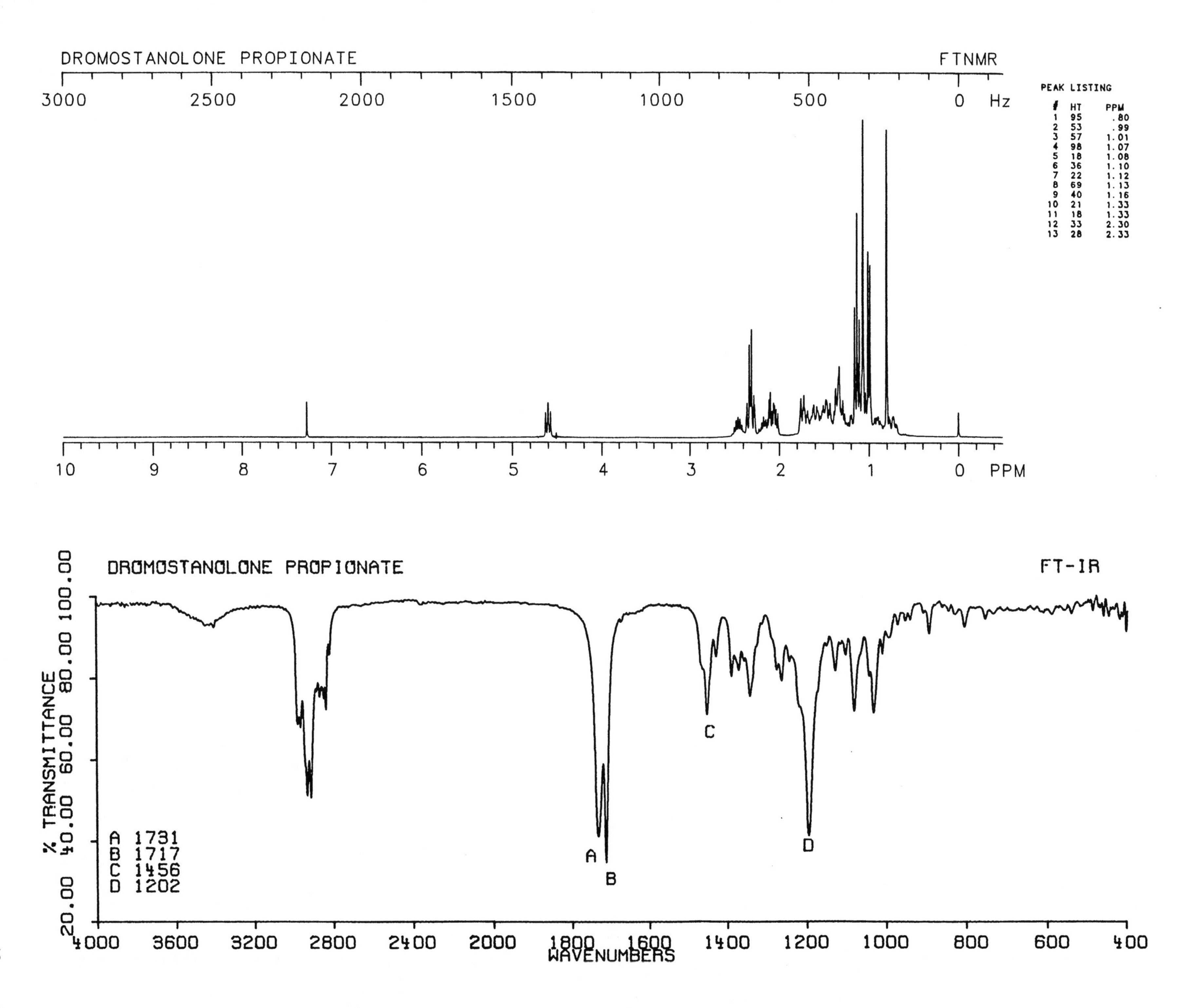
DROMOSTANOLONE PROPIONATE
FTNMR
3000 2500 2000 1500 1000 500 0 Hz
PEAK LISTING
# HT PPM
1 95 .80
2 53 .99
3 57 1.01
4 98 1.07
5 18 1.08
6 36 1.10
7 22 1.12
8 69 1.13
9 40 1.16
10 21 1.33
11 18 1.33
12 33 2.30
13 28 2.33
10 9 8 7 6 5 4 3 2 1 0 PPM
DROMOSTANOLONE PROPIONATE
FT-IR
% TRANSMITTANCE
20.00 40.00 60.00 80.00 100.00
A 1731
B 1717
C 1456
D 1202
A B C D
4000 3600 3200 2800 2400 2000 1800 1600 1400 1200 1000 800 600 400
WAVENUMBERS

# DROPERIDOL

$C_{22}H_{22}FN_3O_2$

Molecular weight: 397.44 (379.17)
Synonyms: 1-[1-[4-(4-Fluorophenyl)-4-oxobutyl]-1,2,3,6-tetrahydro-4-pyridinyl]-1,3-dihydro-2H-benzimidazol-2-one; dehydrobenzperidol
Trade names: Dridol, Droleptan, Inapsine, Innovar

Use: Tranquilizer
HPLC: Si-10; 2A:98B; 7.0
GC: 3022; 280°C

DROPERIDOL

TRANSMITTANCE

WAVELENGTH (nm)

247,276
244,287

DROPERIDOL -- DIP

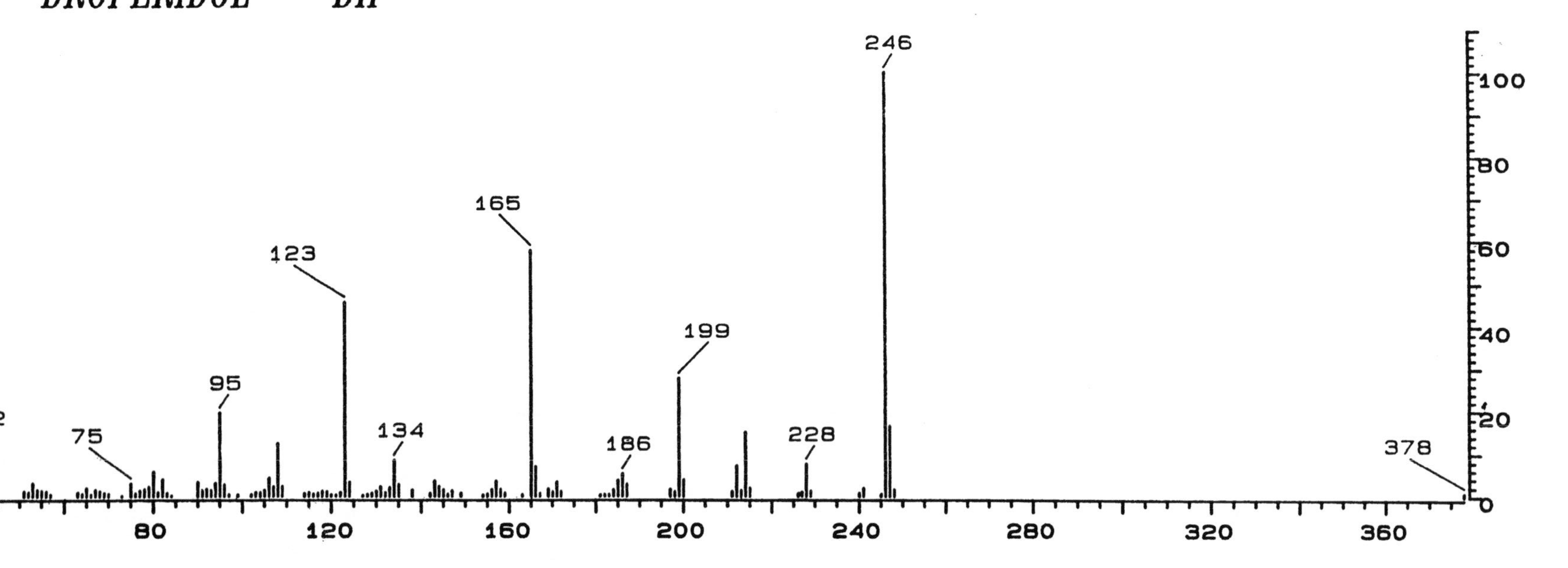

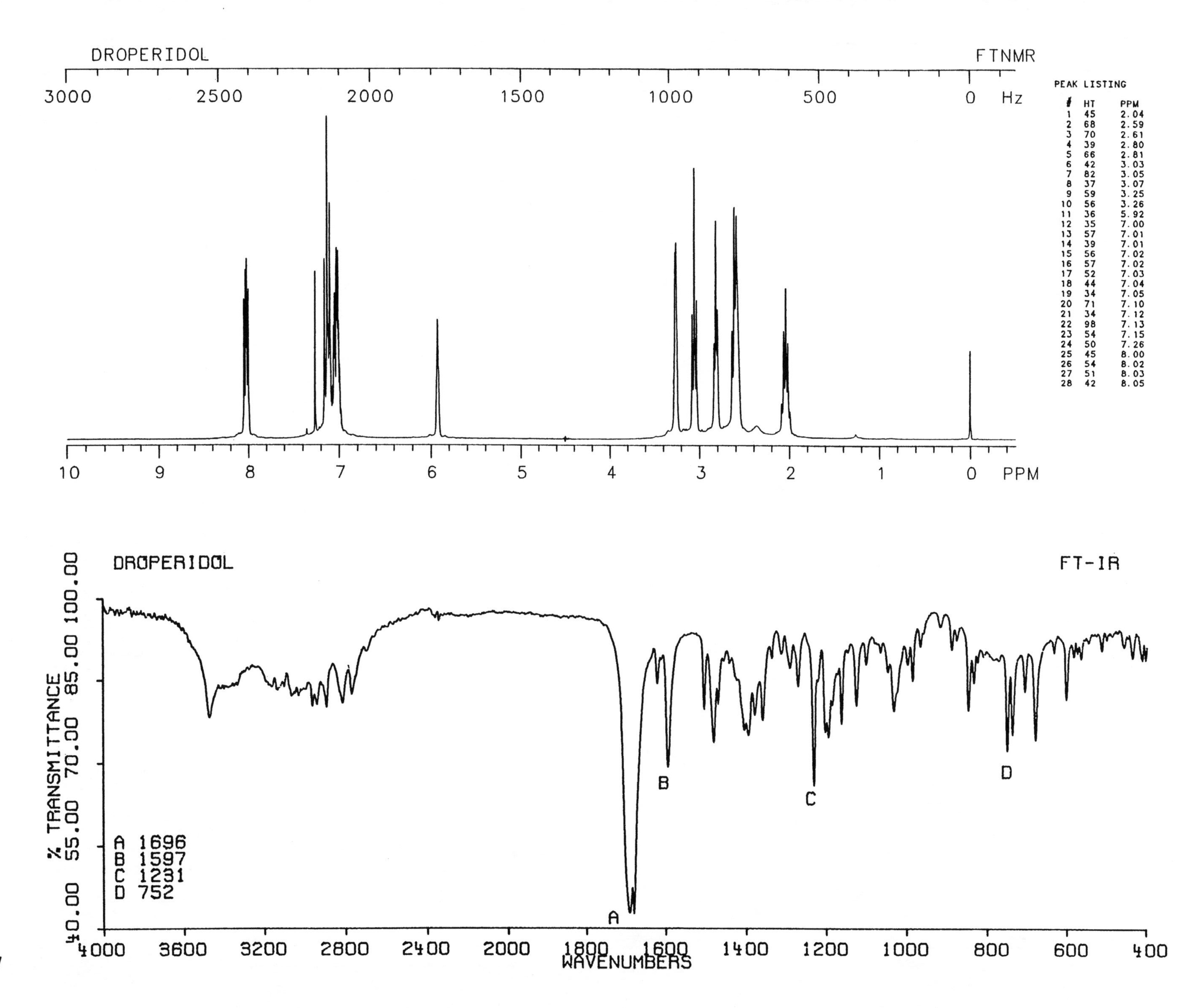
DROPERIDOL
FTNMR
Hz
PPM
PEAK LISTING
# HT PPM
1 45 2.04
2 68 2.59
3 70 2.61
4 39 2.80
5 66 2.81
6 42 3.03
7 82 3.05
8 37 3.07
9 59 3.25
10 56 3.26
11 36 5.92
12 35 7.00
13 57 7.01
14 39 7.01
15 56 7.02
16 57 7.02
17 52 7.03
18 44 7.04
19 34 7.05
20 71 7.10
21 34 7.12
22 98 7.13
23 54 7.15
24 50 7.26
25 45 8.00
26 54 8.02
27 51 8.03
28 42 8.05
DROPERIDOL
FT-IR
% TRANSMITTANCE
WAVENUMBERS
A 1696
B 1597
C 1231
D 752

# DROTEBANOL

$C_{19}H_{27}NO_4$

**Molecular weight:** 333.41 (333.19)
**Synonyms:** 3,4-Dimethoxy-17-methylmorphinan-6-14-diol; oxymethebanol; 6β-14-dihydroxy-3,4-dimethoxy-N-methylmorphinan
**Trade names:** Metebanyl

**Use:** Antitussive
**HPLC:** Si-10; 5A:95B; 5.7
**GC:** 2624; 280°C

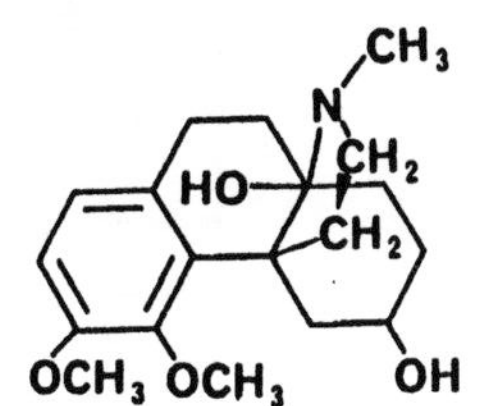


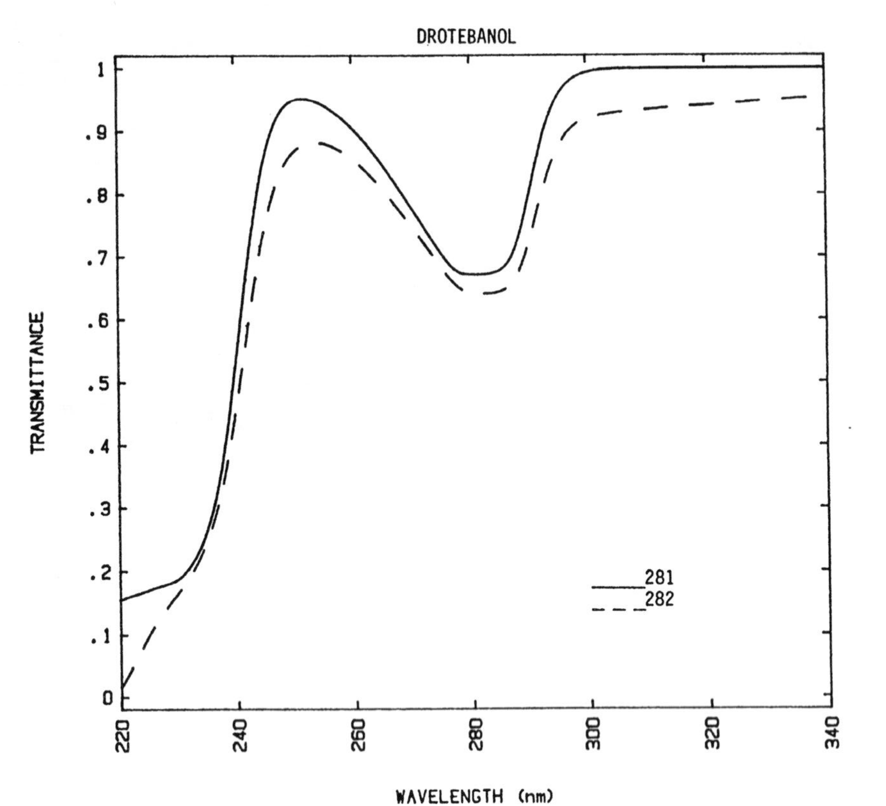

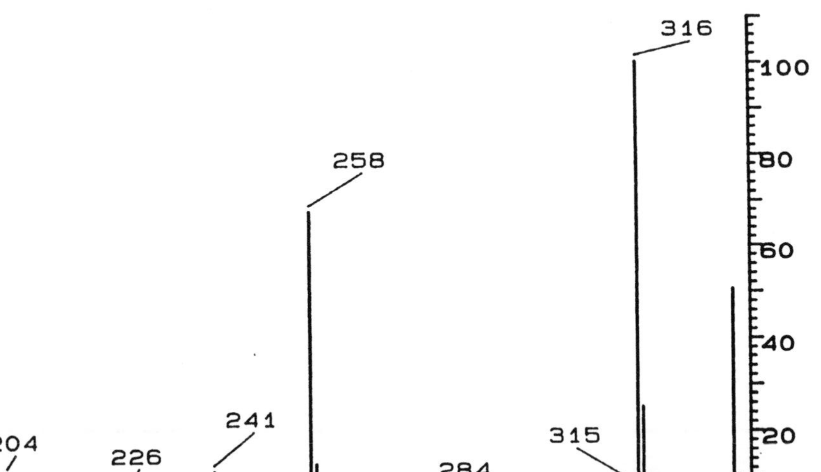

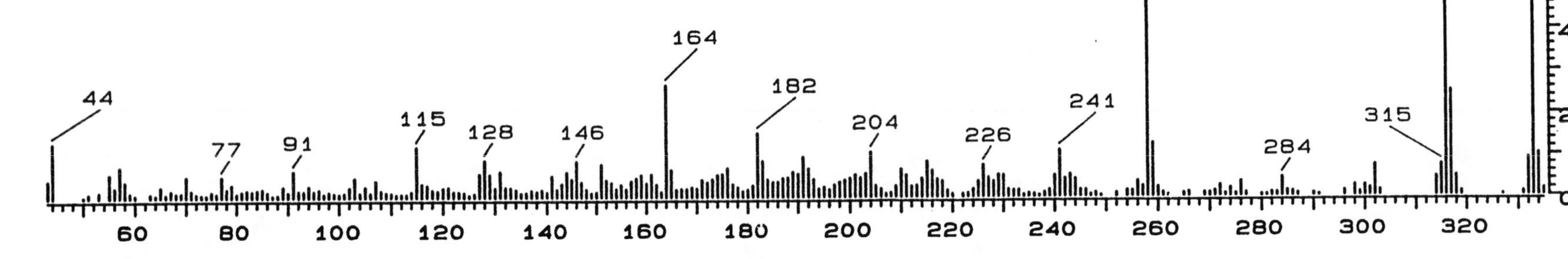

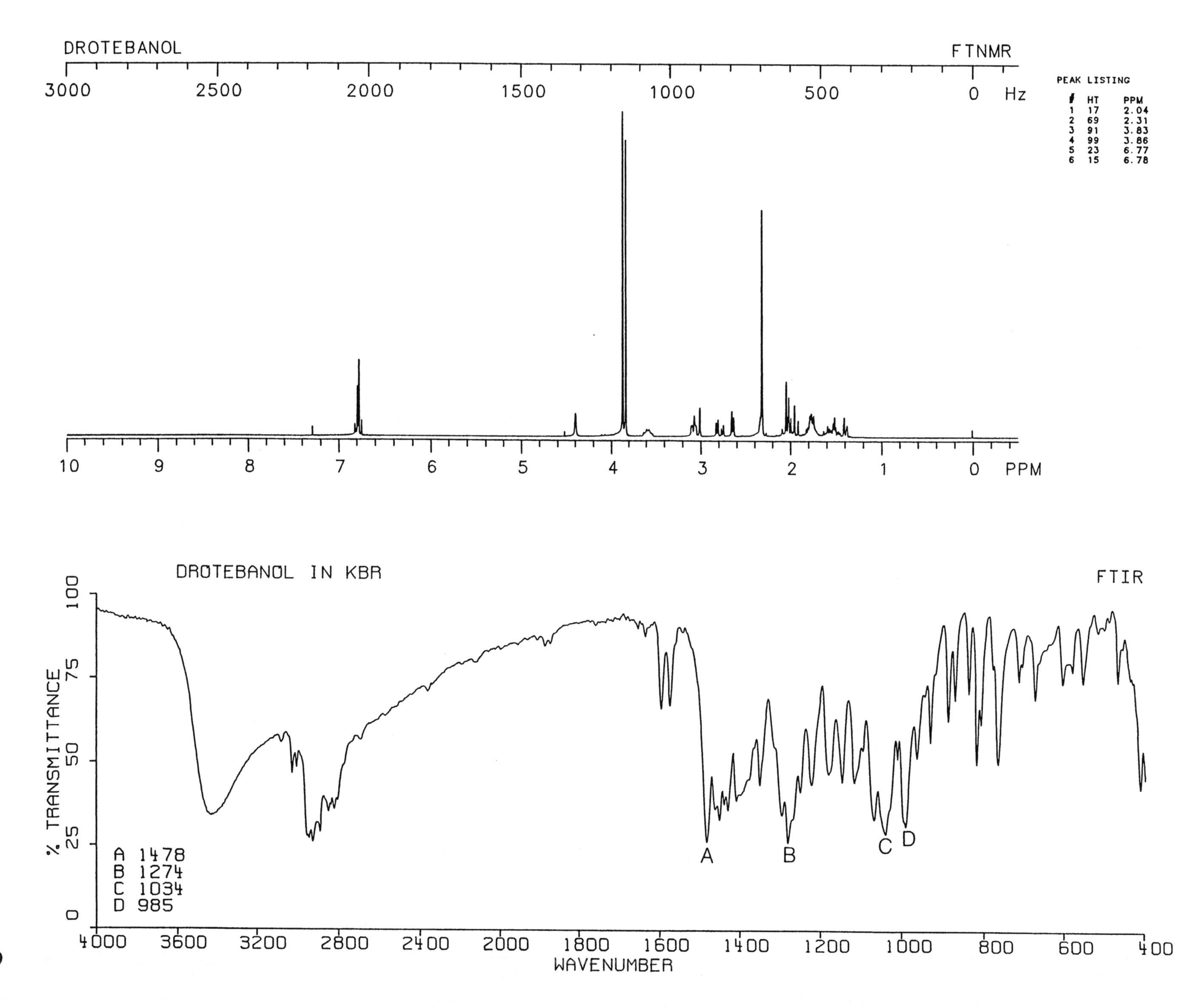
DROTEBANOL
FTNMR
3000 2500 2000 1500 1000 500 0 Hz
10 9 8 7 6 5 4 3 2 1 0 PPM
PEAK LISTING
# HT PPM
1 17 2.04
2 69 2.31
3 91 3.83
4 99 3.86
5 23 6.77
6 15 6.78
DROTEBANOL IN KBR
FTIR
% TRANSMITTANCE
0 25 50 75 100
A 1478
B 1274
C 1034
D 985
A B C D
4000 3600 3200 2800 2400 2000 1800 1600 1400 1200 1000 800 600 400
WAVENUMBER

# DULCITOL

$C_6H_{14}O_6$

Molecular weight: 182.17 (182.08)
Synonyms: Dulcite, Dulcose, Euonymit, Galactitol, Melampyrite, Melampyrum Melampyrin
Trade names: Nitrodulcitol

Use:
HPLC:
GC:

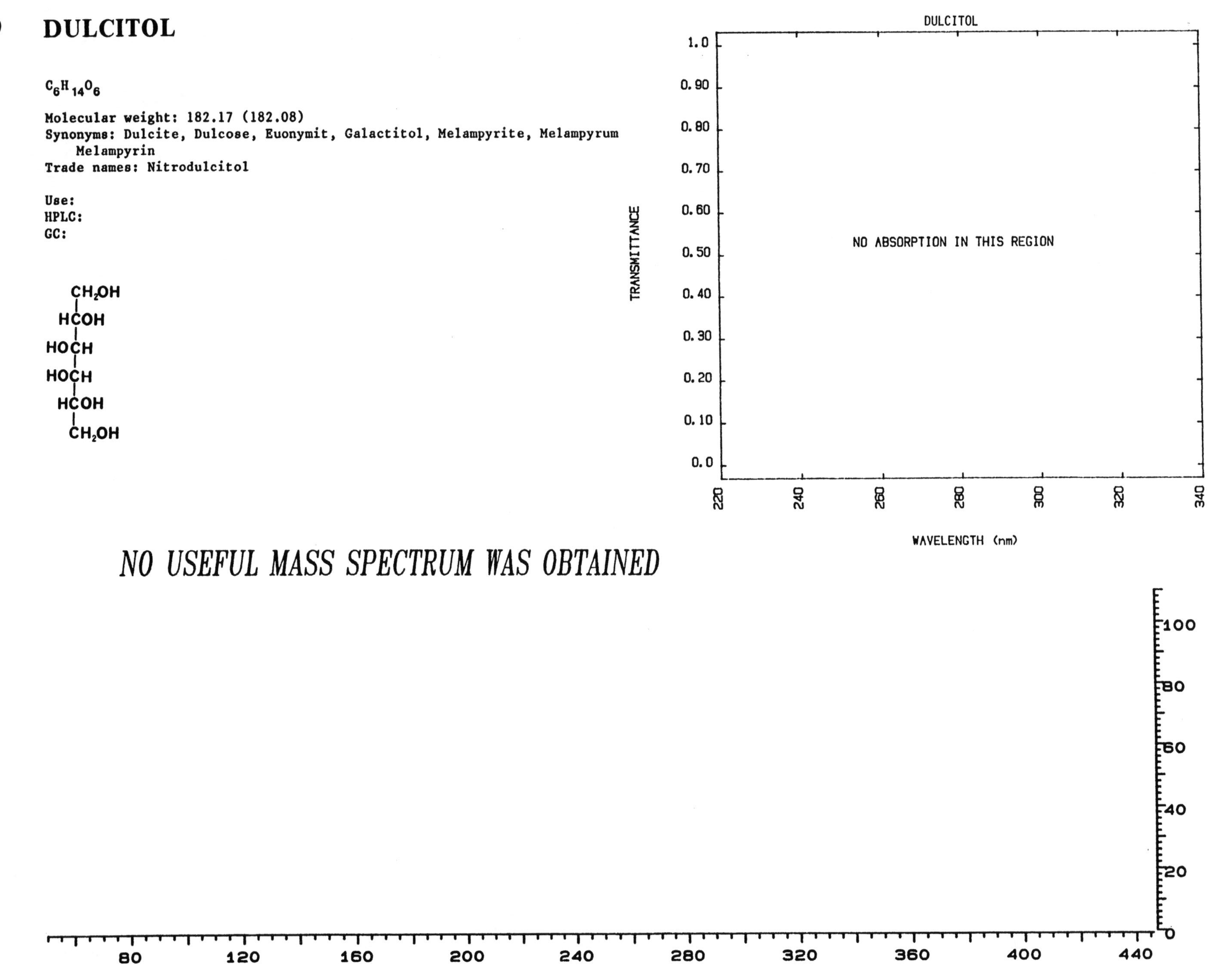

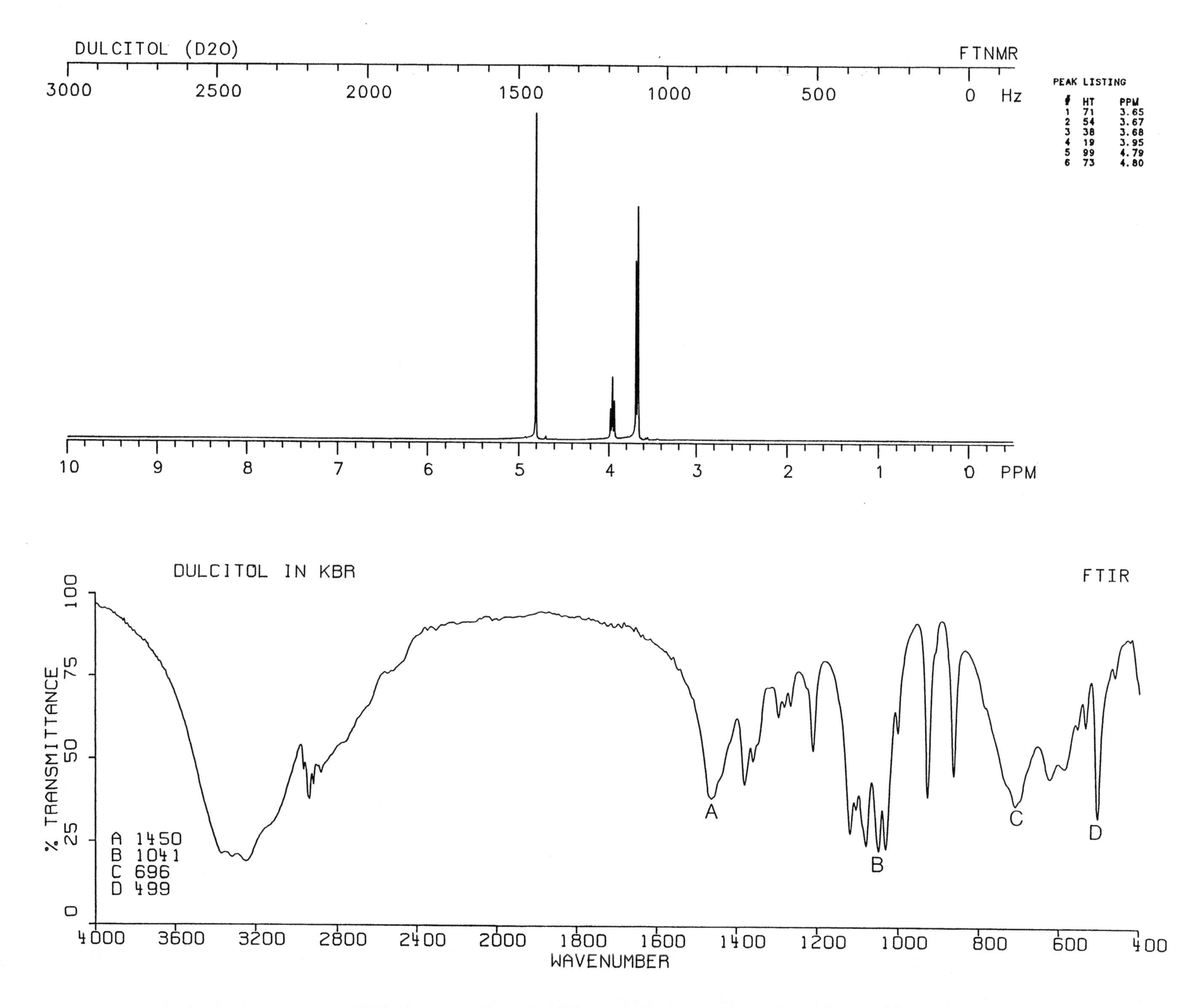
DULCITOL (D2O)
FTNMR
3000 2500 2000 1500 1000 500 0 Hz
PEAK LISTING
# HT PPM
1 71 3.65
2 54 3.67
3 38 3.68
4 19 3.95
5 99 4.79
6 73 4.80
10 9 8 7 6 5 4 3 2 1 0 PPM
DULCITOL IN KBR
FTIR
A 1450
B 1041
C 696
D 499
% TRANSMITTANCE
0 25 50 75 100
4000 3600 3200 2800 2400 2000 1800 1600 1400 1200 1000 800 600 400
WAVENUMBER
A B C D

# DYCLONINE

$C_{18}H_{27}NO_2$

**Molecular weight:** 289.42 (289.20)
**Synonyms:** 1-(4-Butoxyphenyl)-3-(1-piperidinyl)-1-propanone; 4-butoxyphenyl piperidineethyl ketone
**Trade names:** Dyclone

**Use:** Topical anesthetic
**HPLC:** Si-10; 5A:95B; 4.5
**GC:**

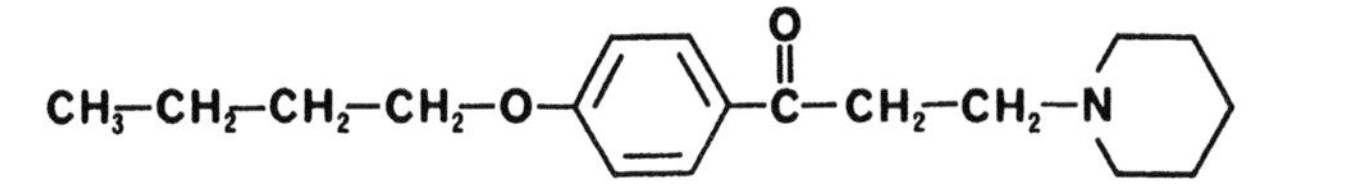


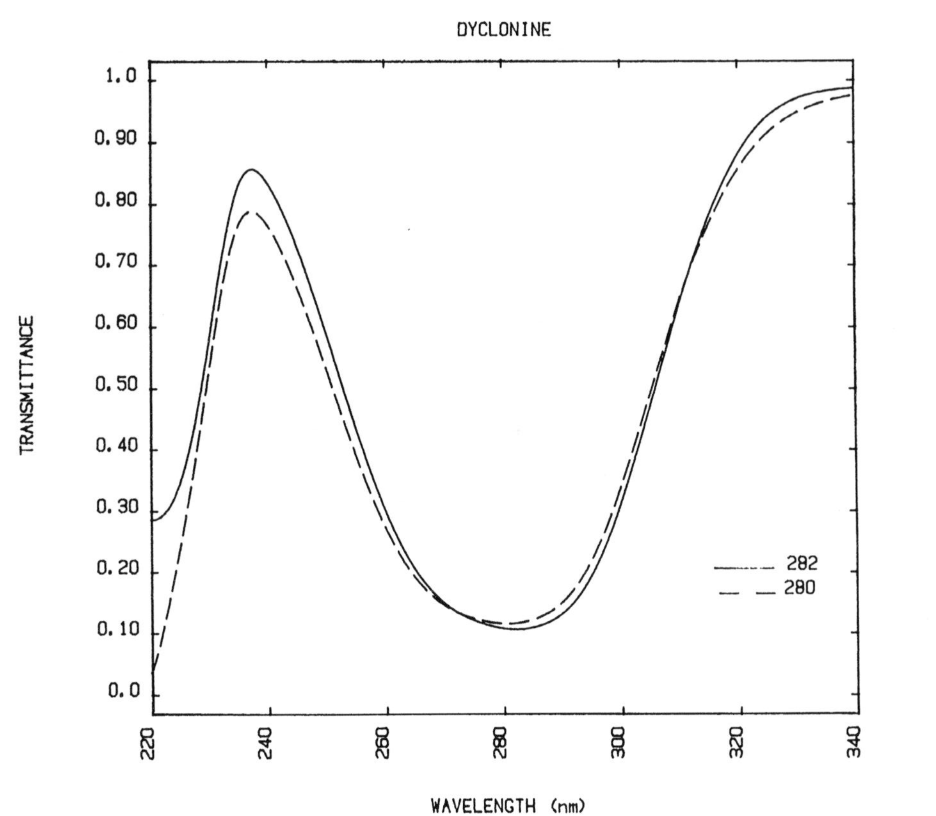

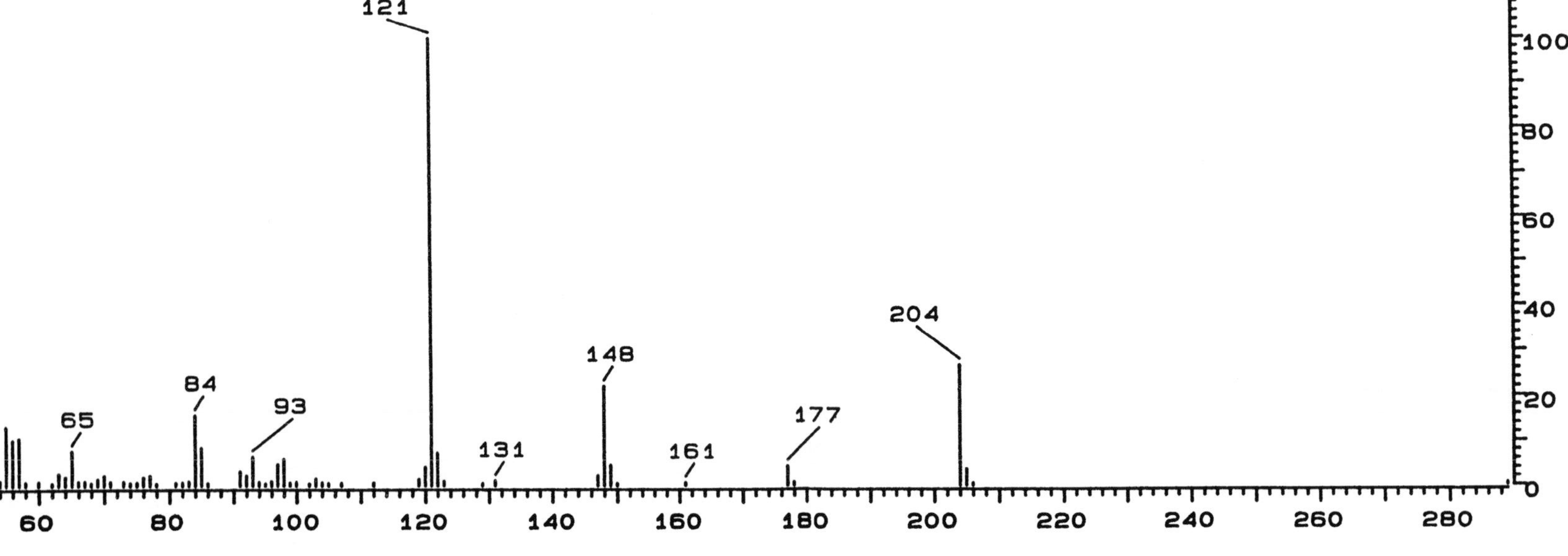

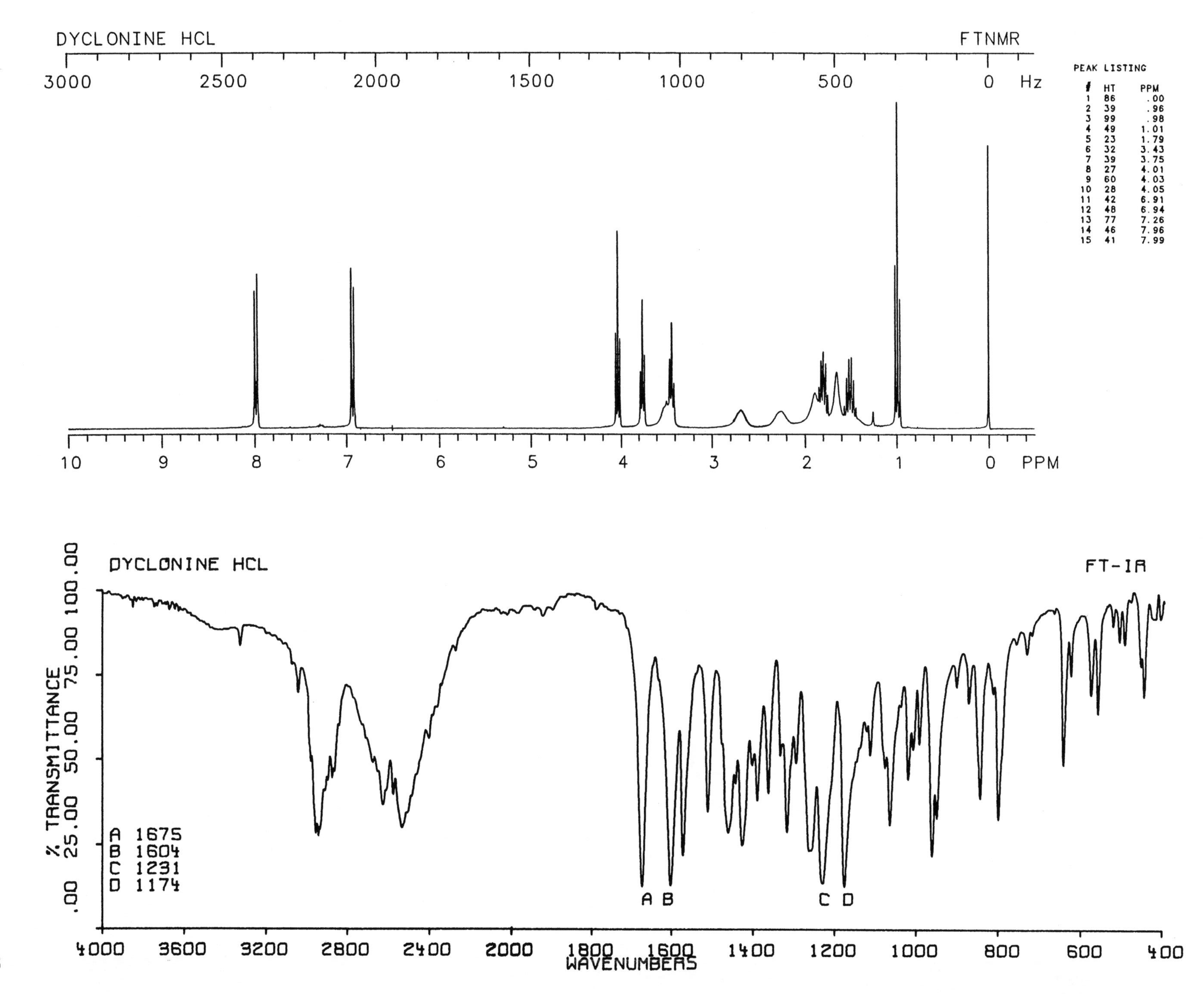
DYCLONINE HCL
FTNMR
3000
2500
2000
1500
1000
500
0
Hz
10
9
8
7
6
5
4
3
2
1
0
PPM
PEAK LISTING
# HT PPM
1 86 .00
2 39 .96
3 99 .98
4 49 1.01
5 23 1.79
6 32 3.43
7 39 3.75
8 27 4.01
9 60 4.03
10 28 4.05
11 42 6.91
12 48 6.94
13 77 7.26
14 46 7.96
15 41 7.99
DYCLONINE HCL
FT-IR
% TRANSMITTANCE
.00
25.00
50.00
75.00
100.00
A 1675
B 1604
C 1231
D 1174
A B
C D
4000
3600
3200
2800
2400
2000
1800
1600
1400
1200
1000
800
600
400
WAVENUMBERS

# DYPHYLLINE

$C_{10}H_{14}N_4O_4$

Molecular weight: 254.25 (254.10)
Synonyms: 7-(2,3-Dihydroxypropyl)-3,7-dihydro-1,3-dimethyl-1H-purine-2,6-dione; (1,2-dihydroxy-3-propyl)theophylline; diprophylline
Trade names: Dilor, Dyflex, Emfaseem, Lufyllin, Neothylline

Use: Muscle relaxant
HPLC: Si-10; 20A:80B; 6.2
GC: 2421; 250°C

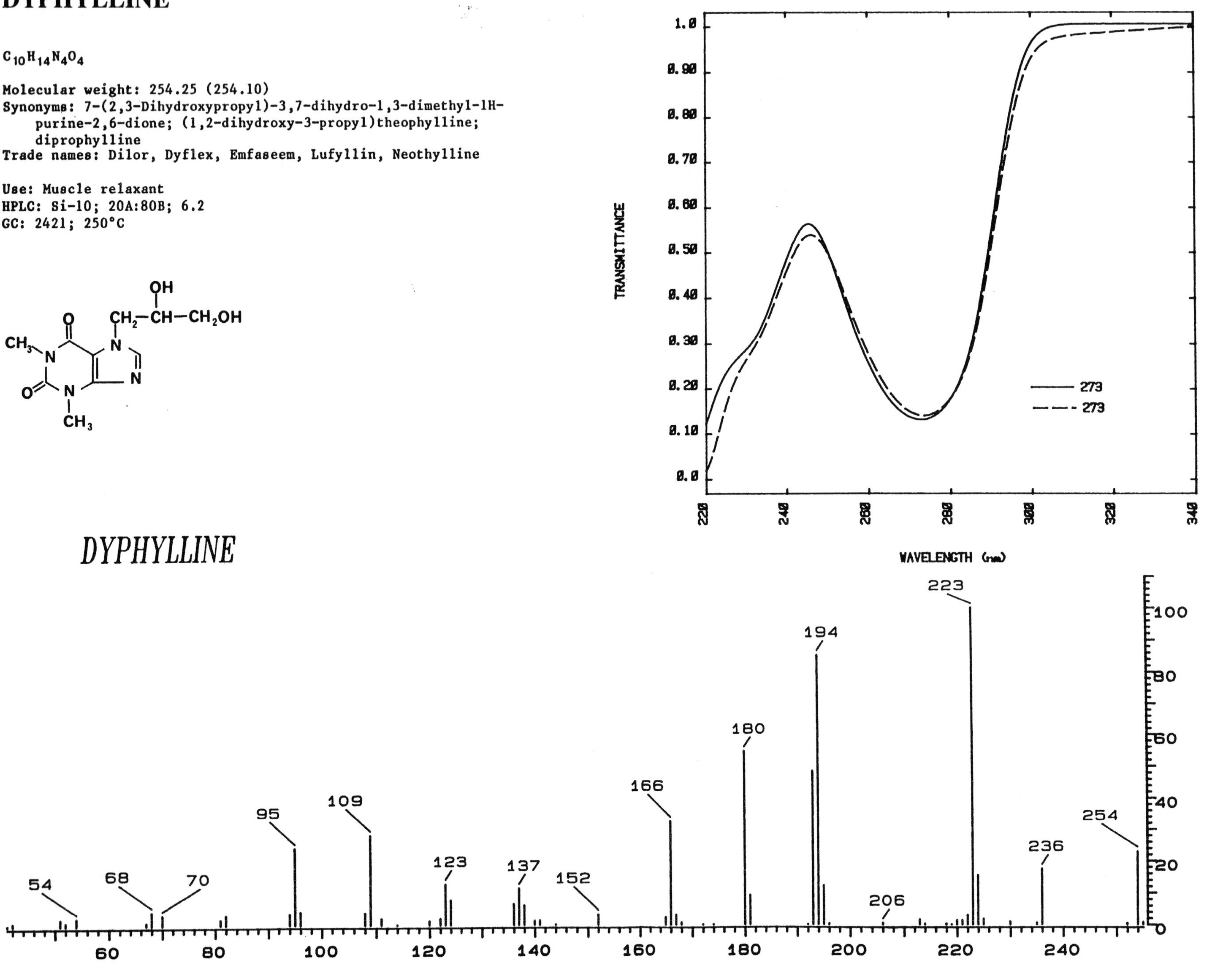

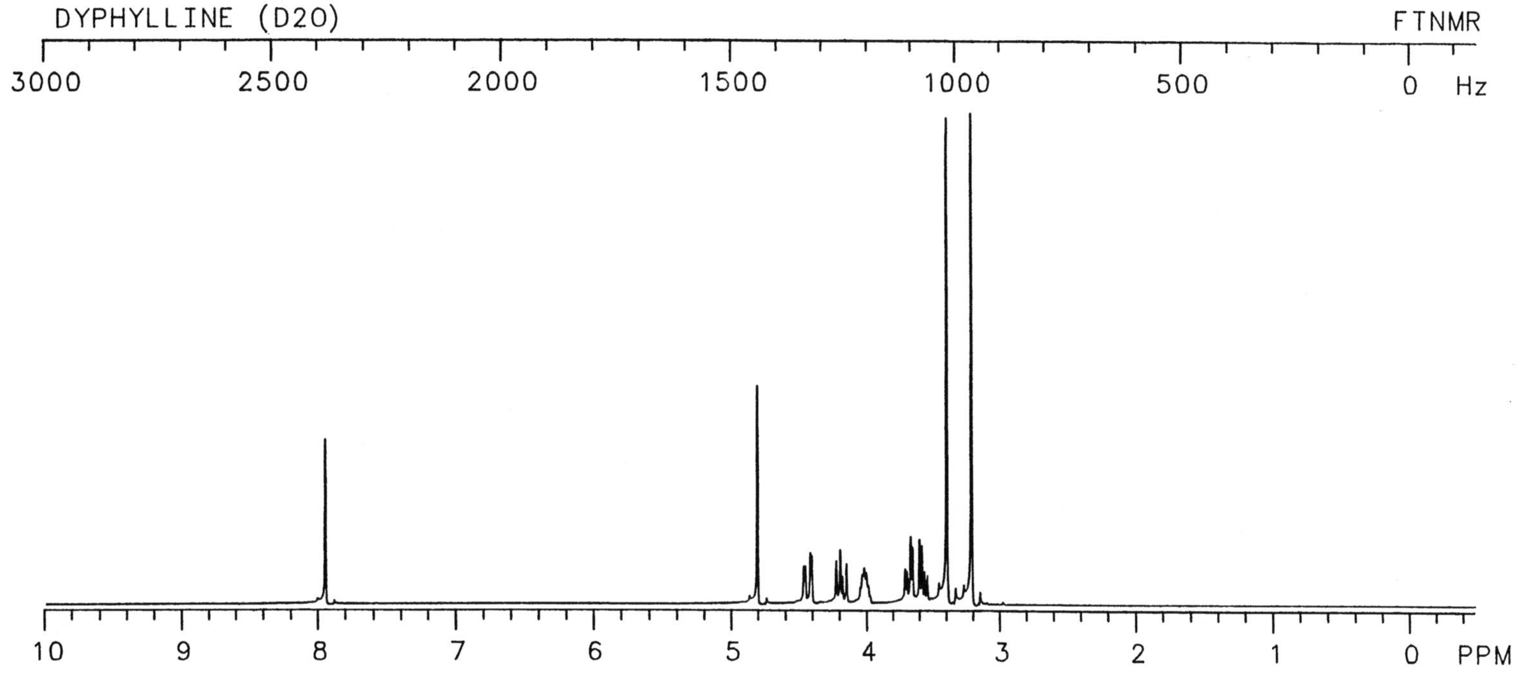

DYPHYLLINE (D2O)

FTNMR

3000 2500 2000 1500 1000 500 0 Hz

10 9 8 7 6 5 4 3 2 1 0 PPM

PEAK LISTING

| # | HT | PPM |
|---|---|---|
| 1 | 99 | 3.21 |
| 2 | 98 | 3.39 |
| 3 | 11 | 3.57 |
| 4 | 13 | 3.59 |
| 5 | 11 | 3.64 |
| 6 | 13 | 3.66 |
| 7 | 8 | 4.14 |
| 8 | 11 | 4.18 |
| 9 | 8 | 4.21 |
| 10 | 9 | 4.39 |
| 11 | 10 | 4.41 |
| 12 | 44 | 4.80 |
| 13 | 33 | 7.94 |

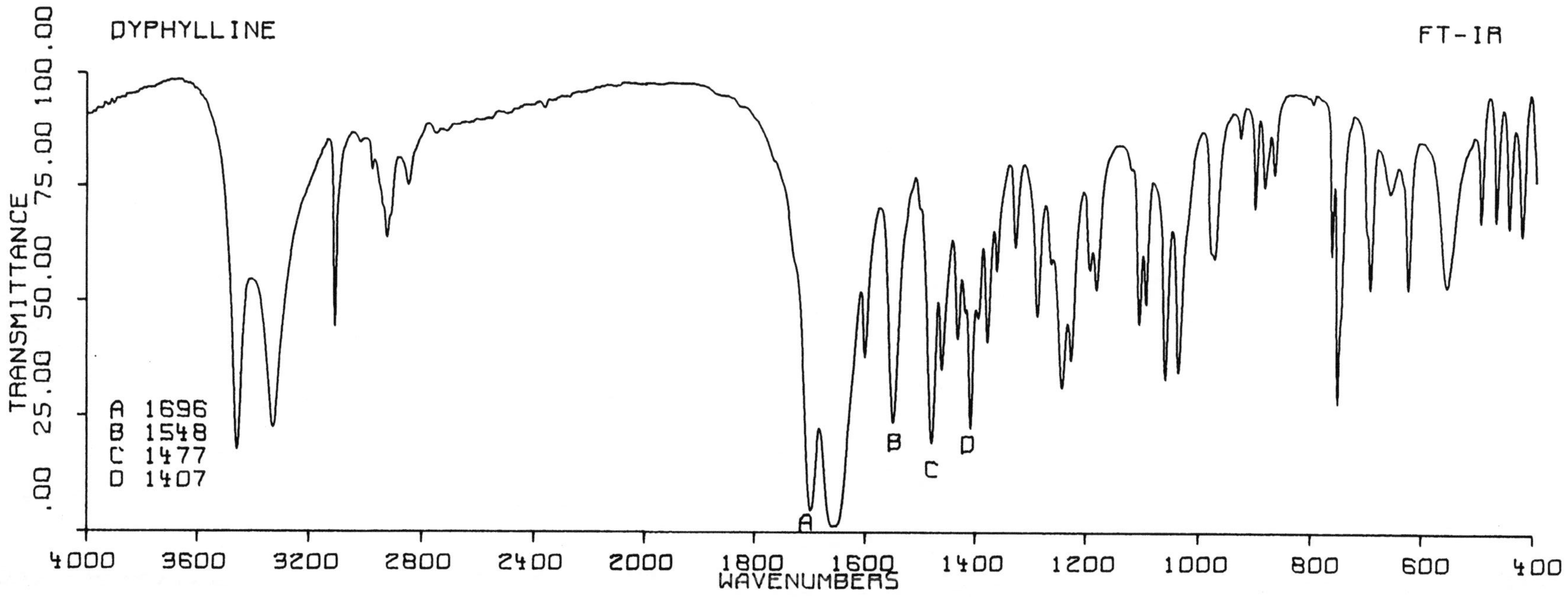

DYPHYLLINE

FT-IR

A 1696
B 1548
C 1477
D 1407

TRANSMITTANCE

.00 25.00 50.00 75.00 100.00

4000 3600 3200 2800 2400 2000 1800 1600 1400 1200 1000 800 600 400

WAVENUMBERS

# ECGONINE

$C_9H_{15}NO_3$

**Molecular weight: 185.22 (185.11)**
**Synonyms: [1R-(exo,exo)]-3-Hydroxy-8-methyl-8-azabicyclo[3.2.1]-octane-2-carboxylic acid**
**Trade names:**

**Use: Topical anesthetic**
**HPLC:**
**GC:**





*ECGONINE -- DIP*

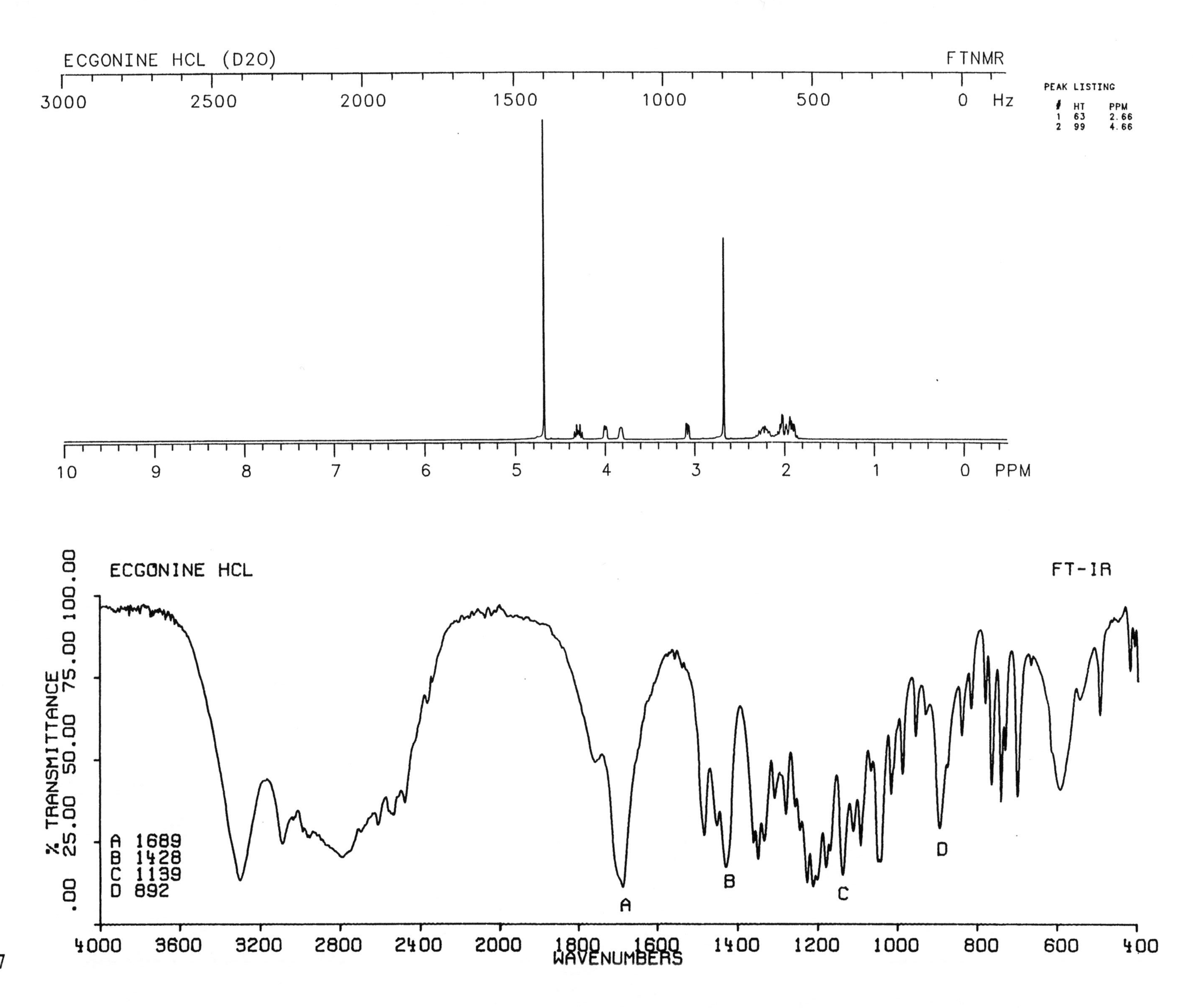

ECGONINE HCL (D2O)
FTNMR
3000 2500 2000 1500 1000 500 0 Hz
PEAK LISTING
# HT PPM
1 63 2.66
2 99 4.66
10 9 8 7 6 5 4 3 2 1 0 PPM
ECGONINE HCL
FT-IR
% TRANSMITTANCE
.00 25.00 50.00 75.00 100.00
A 1689
B 1428
C 1139
D 892
A
B
C
D
4000 3600 3200 2800 2400 2000 1800 1600 1400 1200 1000 800 600 400
WAVENUMBERS

# ECHOTHIOPHATE IODIDE

$C_9H_{23}INO_3PS$

Molecular weight: 383.22 (383.06)
Synonyms: 2-[(Diethoxyphosphinyl)-thio]-N,N,N-trimethylethanaminium iodide; ecothipate iodide; ecostigmine iodide
Trade names: Echodide, Phospholine Iodide, Phospholinjodid

Use: Cholinergic
HPLC:
GC:

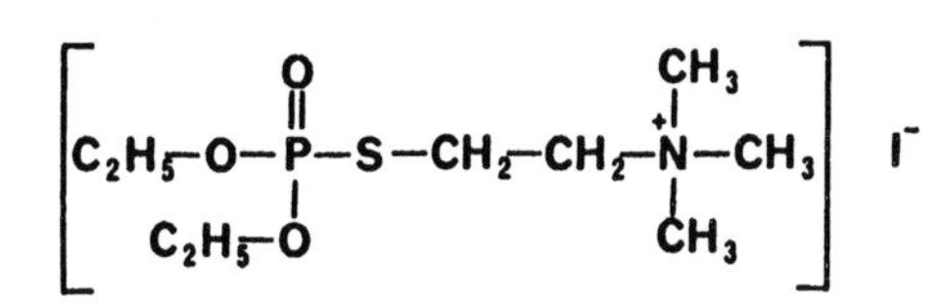


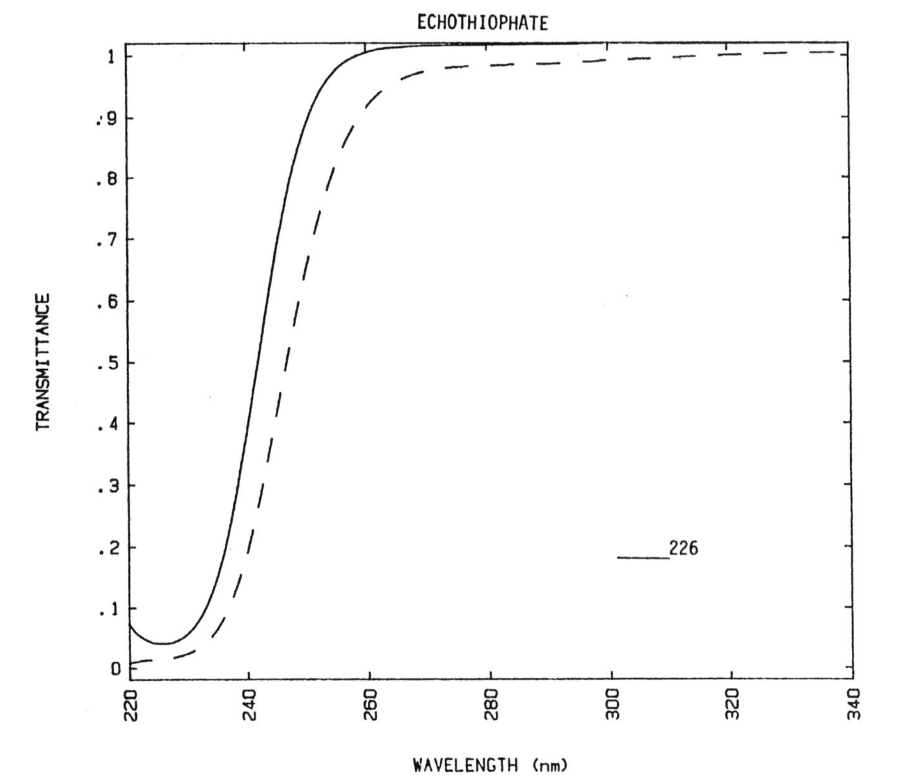


NO USEFUL MASS SPECTRUM WAS OBTAINED

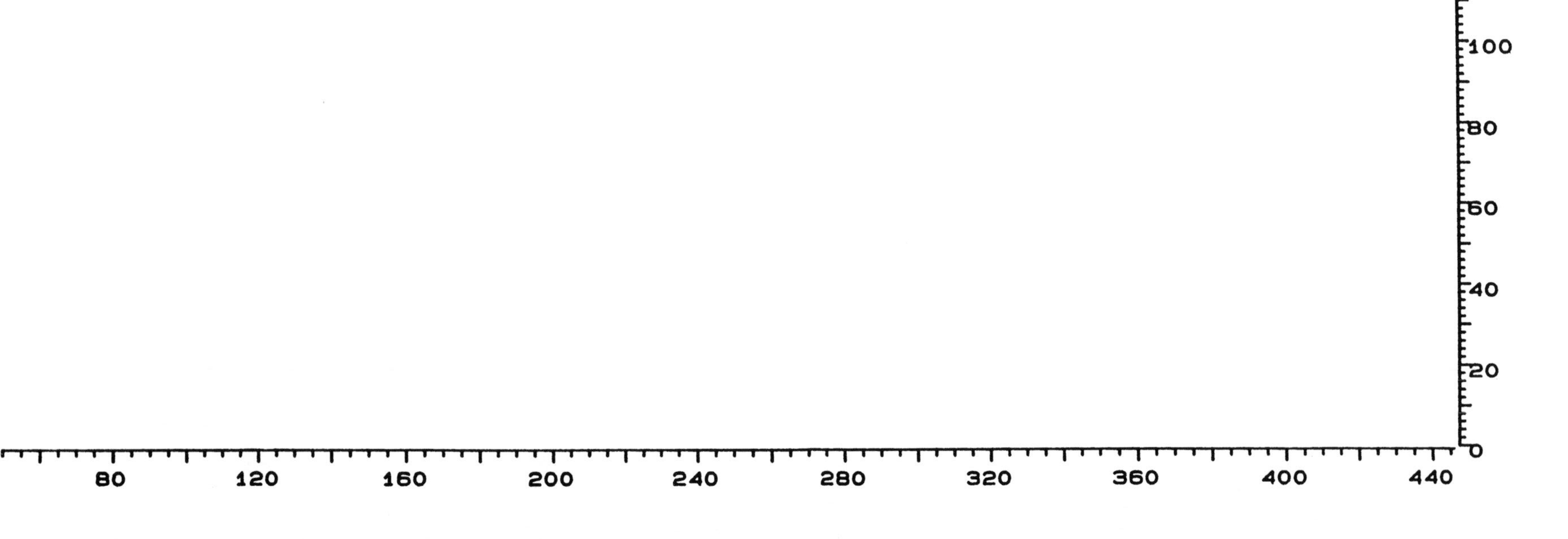

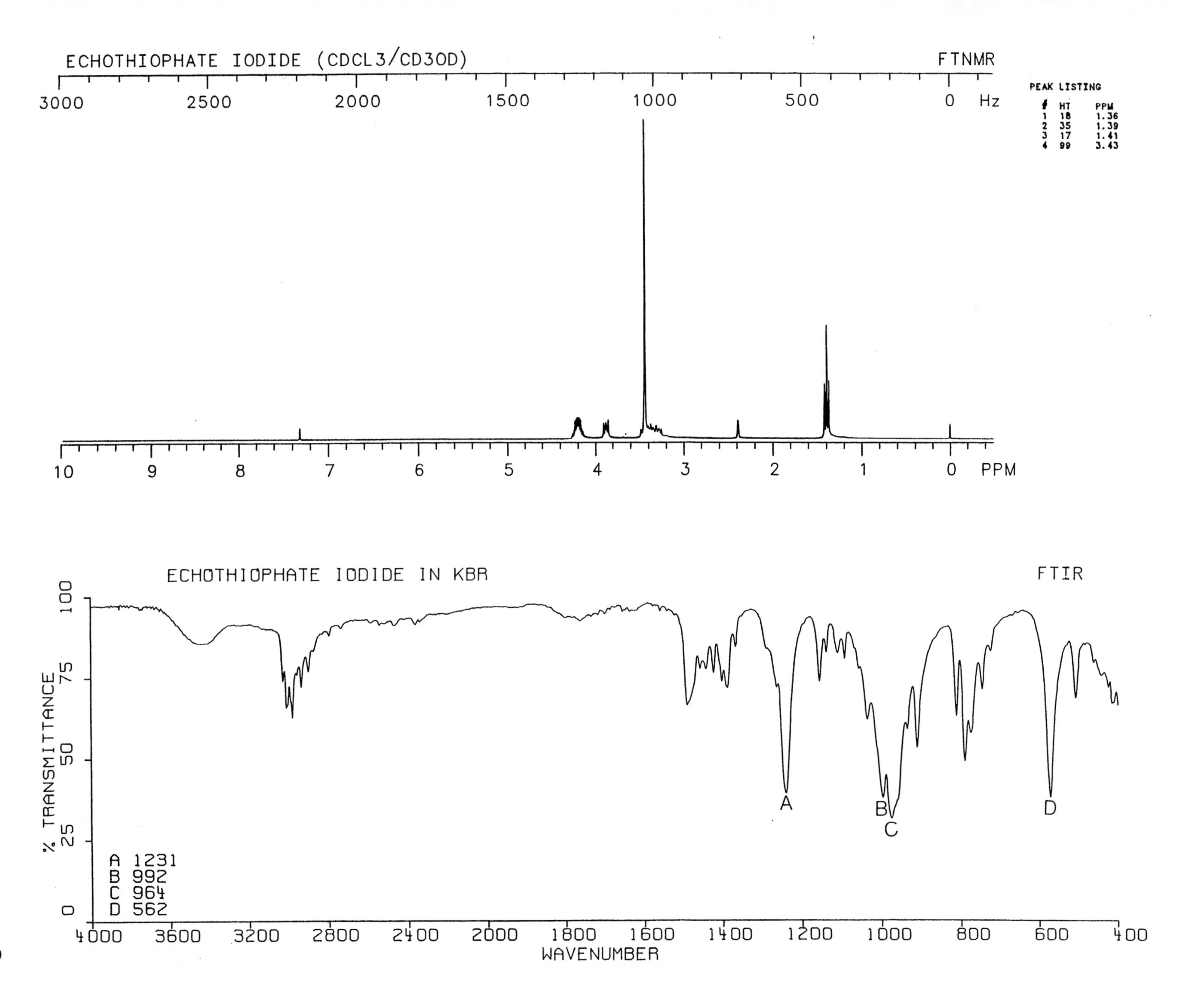
ECHOTHIOPHATE IODIDE (CDCL3/CD3OD)
FTNMR
3000
2500
2000
1500
1000
500
0
Hz
PEAK LISTING
# HT PPM
1 18 1.36
2 35 1.39
3 17 1.41
4 99 3.43
10
9
8
7
6
5
4
3
2
1
0
PPM
ECHOTHIOPHATE IODIDE IN KBR
FTIR
% TRANSMITTANCE
100
75
50
25
0
A
B
C
D
A 1231
B 992
C 964
D 562
4000
3600
3200
2800
2400
2000
1800
1600
1400
1200
1000
800
600
400
WAVENUMBER

# ECONAZOLE NITRATE

$C_{18}H_{15}Cl_3N_2O$

Molecular weight: 381.68 (380.03)
Synonyms: 1-[2-[(4-Chlorophenyl)methoxy]-2-(2,4-dichlorophenyl)-ethyl]-1H-imidazole
Trade names: Ecostatin, Epi-Pevaryl, Econacort, Gyno-Pevaryl, Ifenec, Palavale, Pevaryl, Spectazole
Use: Antifungal
HPLC:
GC: 2910; 280°C

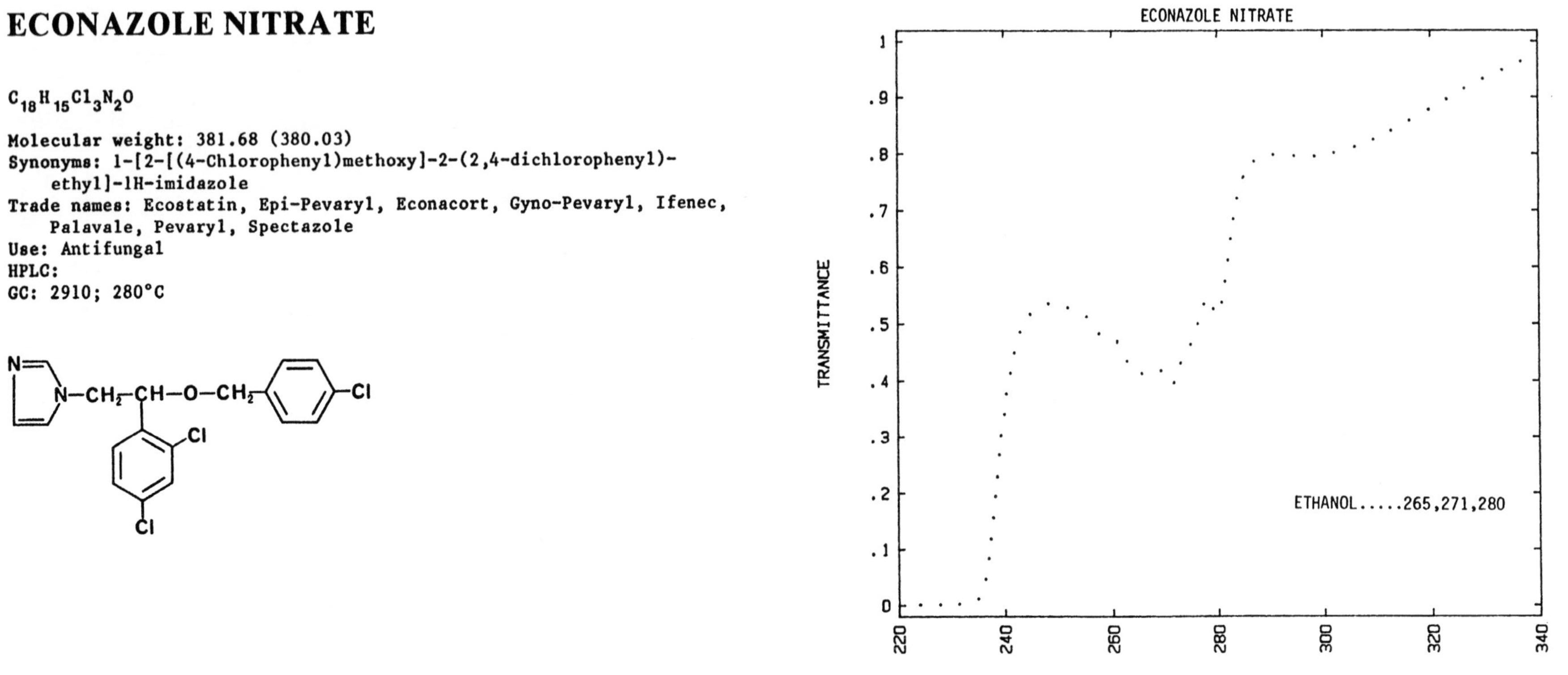


## ECONAZOLE

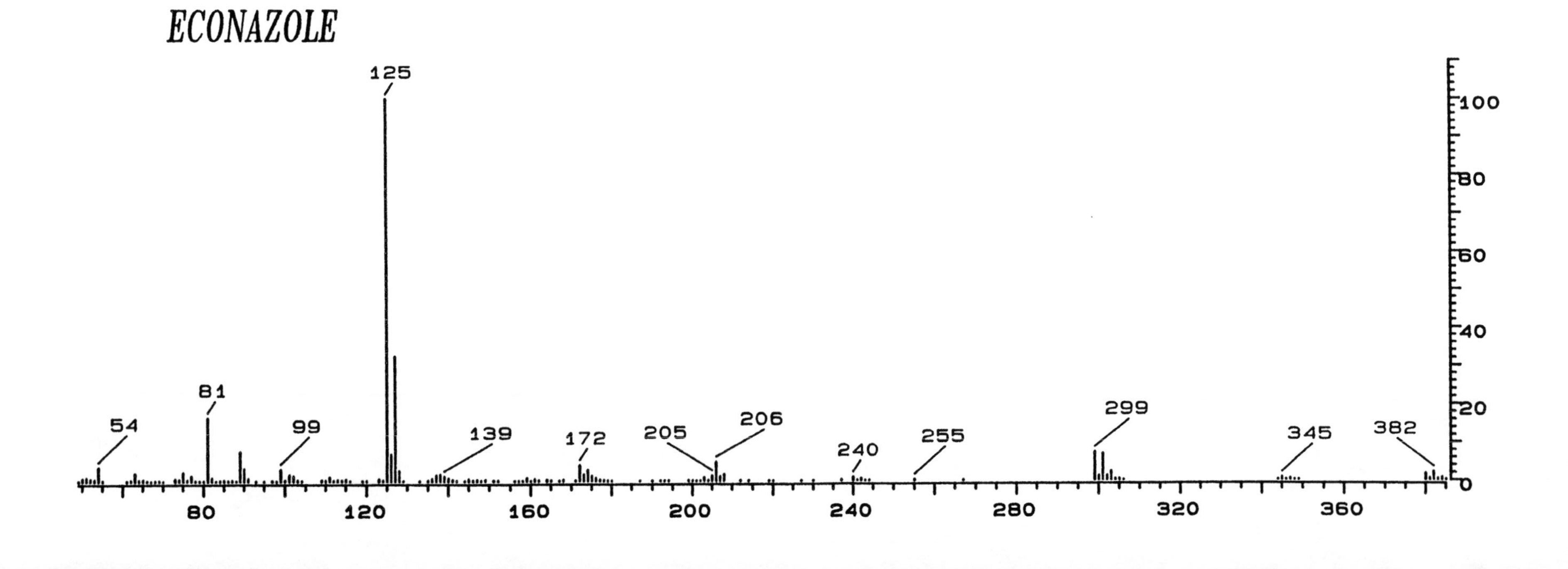

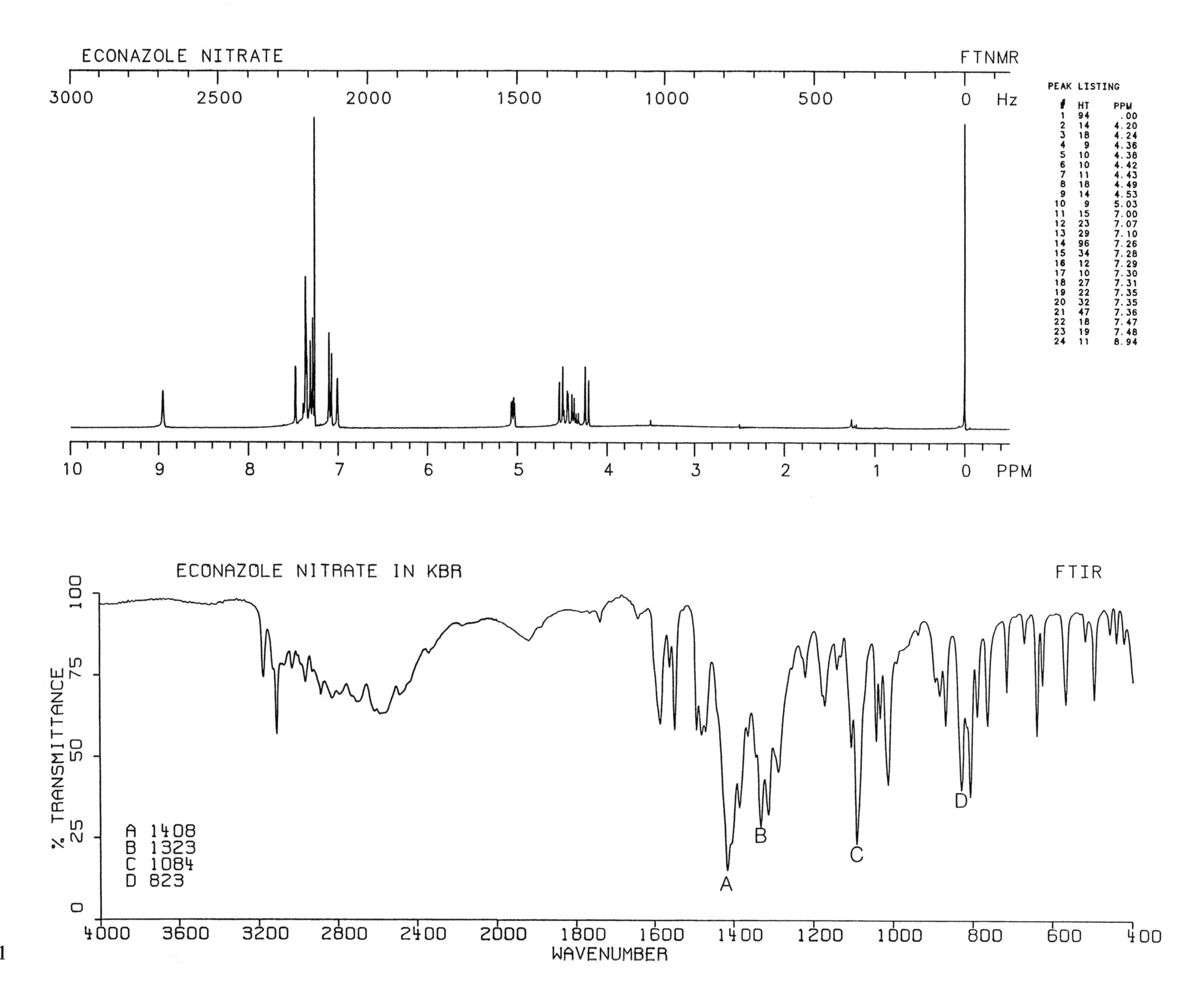
ECONAZOLE NITRATE
FTNMR
3000 2500 2000 1500 1000 500 0 Hz
PEAK LISTING
| # | HT | PPM |
|---|---|---|
| 1 | 94 | .00 |
| 2 | 14 | 4.20 |
| 3 | 18 | 4.24 |
| 4 | 9 | 4.36 |
| 5 | 10 | 4.38 |
| 6 | 10 | 4.42 |
| 7 | 11 | 4.43 |
| 8 | 18 | 4.49 |
| 9 | 14 | 4.53 |
| 10 | 9 | 5.03 |
| 11 | 15 | 7.00 |
| 12 | 23 | 7.07 |
| 13 | 29 | 7.10 |
| 14 | 96 | 7.26 |
| 15 | 34 | 7.28 |
| 16 | 12 | 7.29 |
| 17 | 10 | 7.30 |
| 18 | 27 | 7.31 |
| 19 | 22 | 7.35 |
| 20 | 32 | 7.35 |
| 21 | 47 | 7.36 |
| 22 | 18 | 7.47 |
| 23 | 19 | 7.48 |
| 24 | 11 | 8.94 |
10 9 8 7 6 5 4 3 2 1 0 PPM
ECONAZOLE NITRATE IN KBR
FTIR
% TRANSMITTANCE
100 75 50 25 0
A 1408
B 1323
C 1084
D 823
A
B
C
D
4000 3600 3200 2800 2400 2000 1800 1600 1400 1200 1000 800 600 400
WAVENUMBER

# ECTYLUREA

$C_7H_{12}N_2O_2$

**Molecular weight:** 156.18 (156.09)
**Synonyms:** (Z)-N-(Aminocarbonyl)-2-ethyl-2-butenamide; (α-ethylcrotonyl)urea
**Trade names:** Astyn, Ektyl, Levanil, Nostyn, Cronil, Neuriprocin, Pacetyn

**Use:** Sedative
**HPLC:**
**GC:**

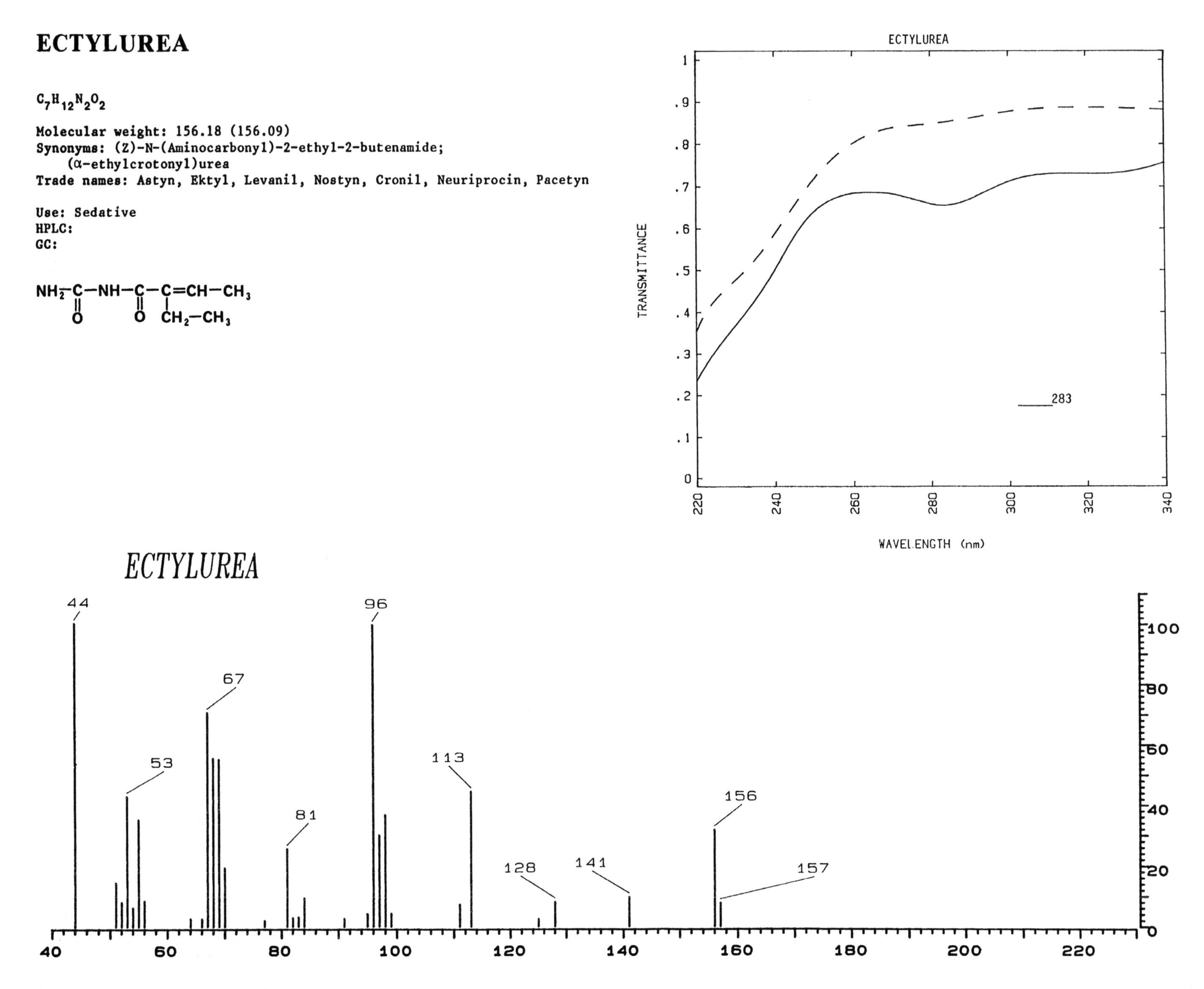

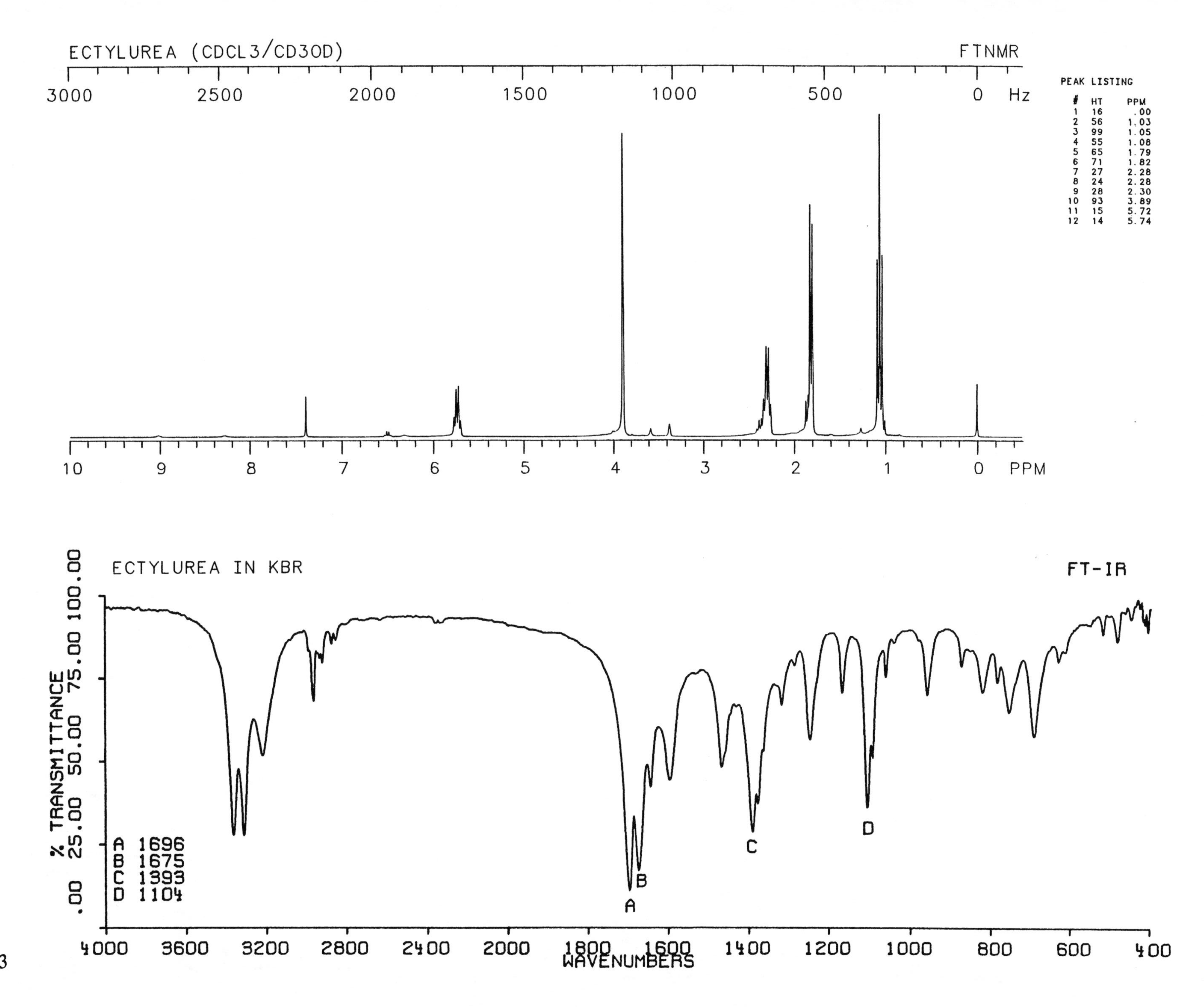
ECTYLUREA (CDCL3/CD3OD)
FTNMR
3000 2500 2000 1500 1000 500 0 Hz
PEAK LISTING
# HT PPM
1 16 .00
2 56 1.03
3 99 1.05
4 55 1.08
5 65 1.79
6 71 1.82
7 27 2.28
8 24 2.28
9 28 2.30
10 93 3.89
11 15 5.72
12 14 5.74
10 9 8 7 6 5 4 3 2 1 0 PPM
ECTYLUREA IN KBR
FT-IR
% TRANSMITTANCE
.00 25.00 50.00 75.00 100.00
A 1696
B 1675
C 1393
D 1104
A
B
C
D
4000 3600 3200 2800 2400 2000 1800 1600 1400 1200 1000 800 600 400
WAVENUMBERS

# EDROPHONIUM CHLORIDE

$C_{10}H_{16}ClNO$

Molecular weight: 201.69 (201.09)
Synonyms: N-Ethyl-3-hydroxy-N,N-dimethylbenzenaminium chloride;
    edrophone chloride
Trade names: Antirex, Tensilon

Use: Cholinergic, muscle stimulant
HPLC:
GC: 1456; 200°C

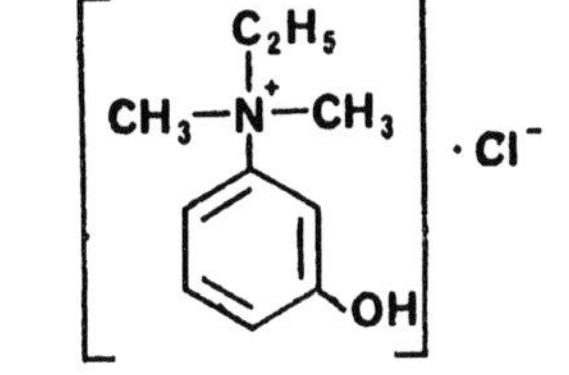


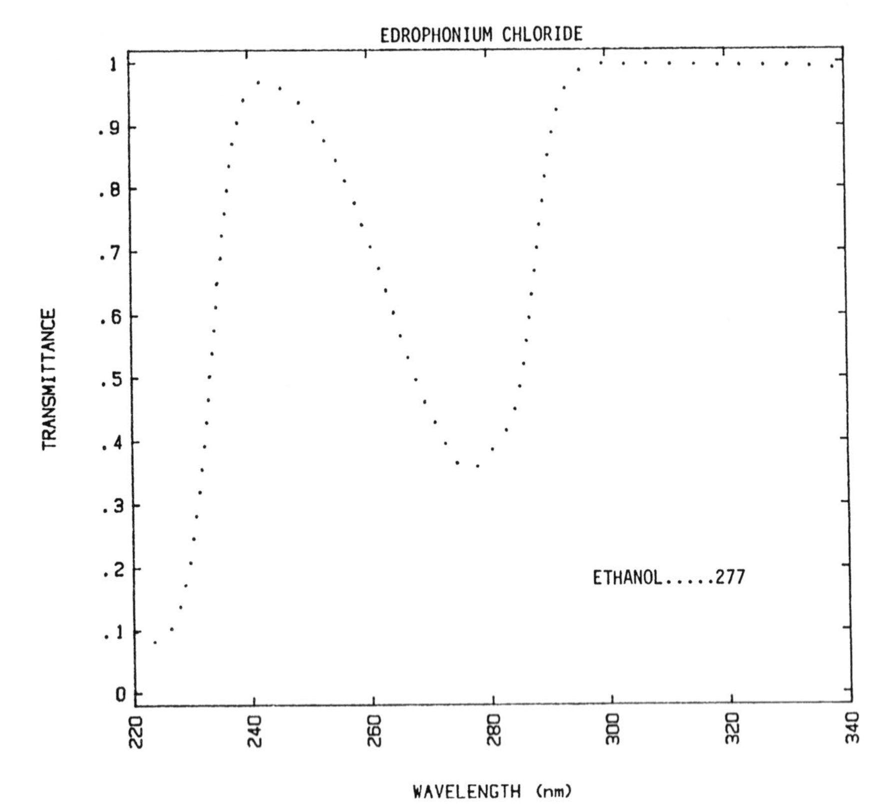


NO USEFUL MASS SPECTRUM WAS OBTAINED

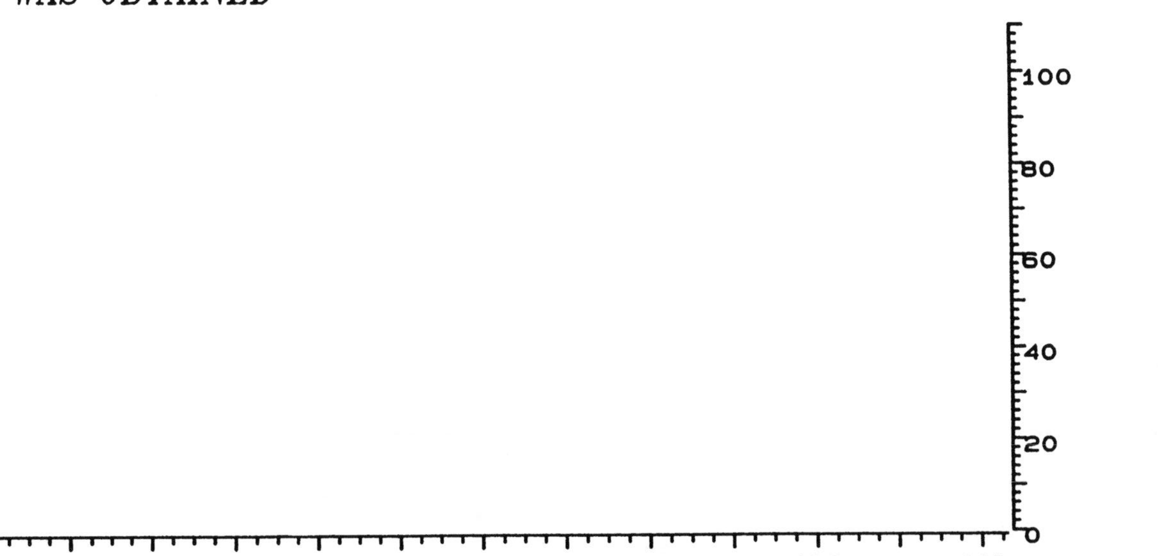

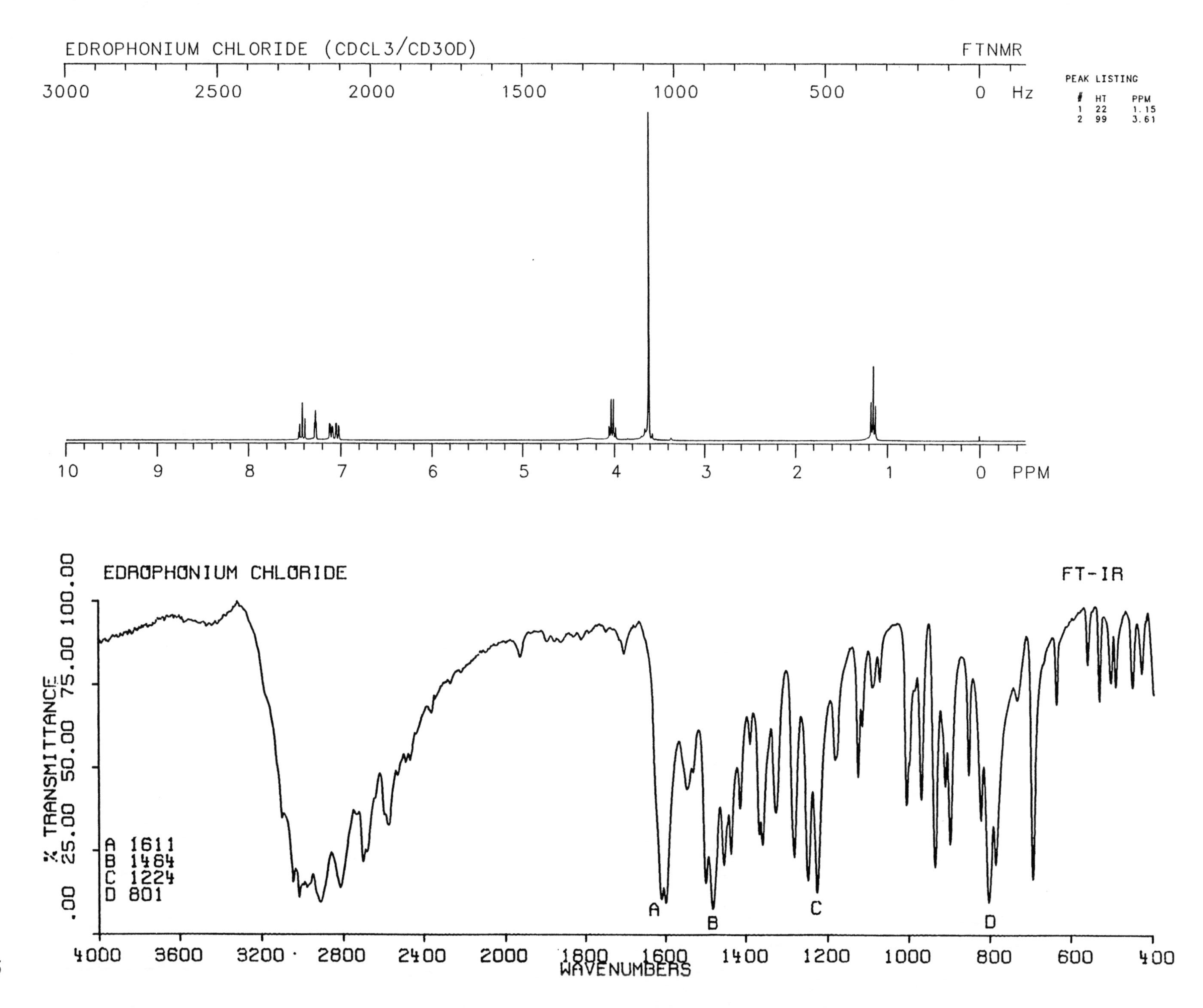
EDROPHONIUM CHLORIDE (CDCL3/CD3OD)
FTNMR
3000 2500 2000 1500 1000 500 0 Hz
PEAK LISTING
# HT PPM
1 22 1.15
2 99 3.61
10 9 8 7 6 5 4 3 2 1 0 PPM
EDROPHONIUM CHLORIDE
FT-IR
% TRANSMITTANCE
.00 25.00 50.00 75.00 100.00
A 1611
B 1484
C 1224
D 801
A B C D
4000 3600 3200 2800 2400 2000 1800 1600 1400 1200 1000 800 600 400
WAVENUMBERS

# EPHEDRINE

$C_{10}H_{15}NO$

Molecular weight: 165.24 (165.12)
Synonyms: α-[1-(Methylamino)ethyl]benzenmethanol; 1-phenyl-2-methylaminopropanol
Trade names: Tedral, Bronkotabs

Use: Bronchodilator
HPLC: Si-10; 20A:80B; 9.0
GC: 1361; 140°C

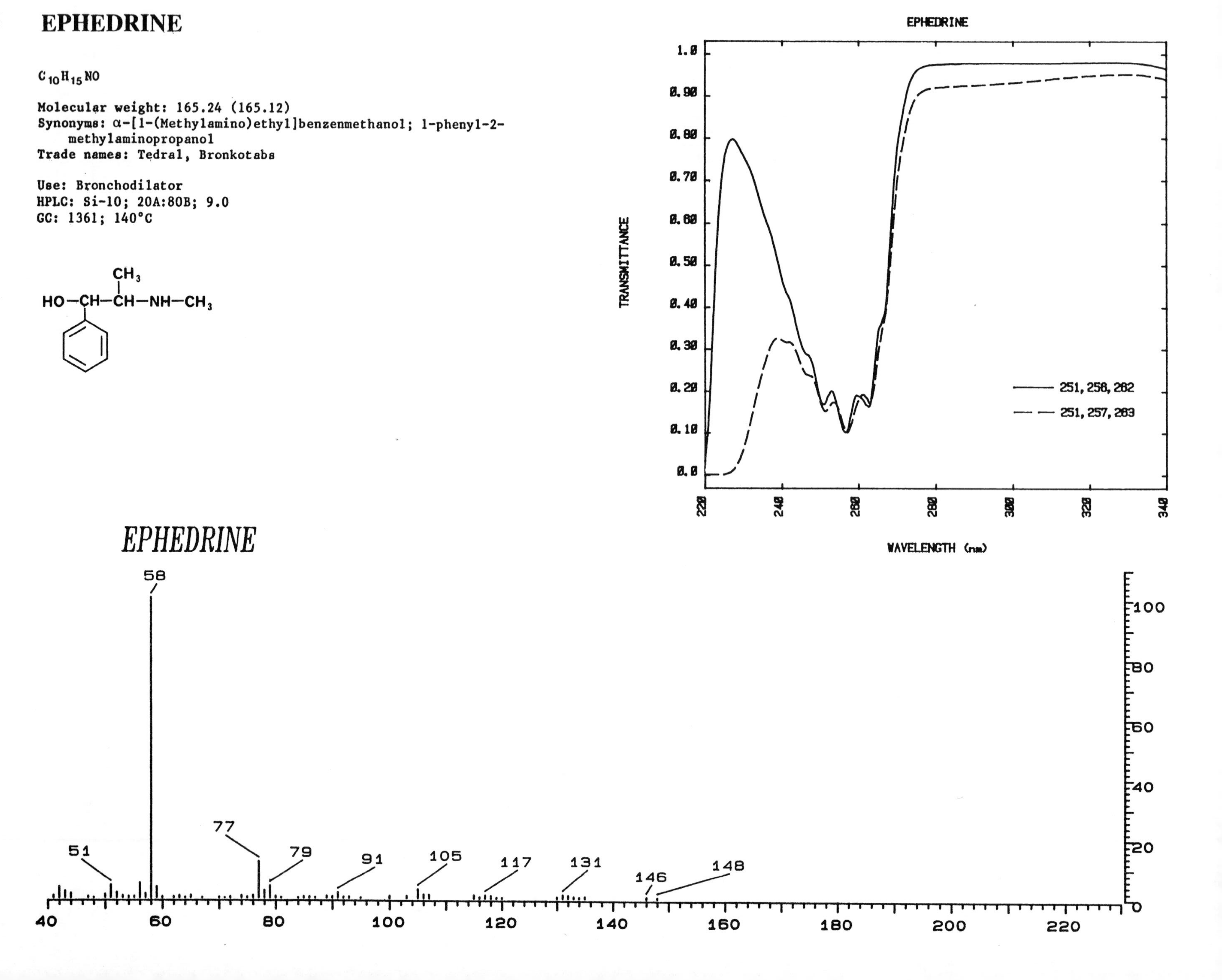

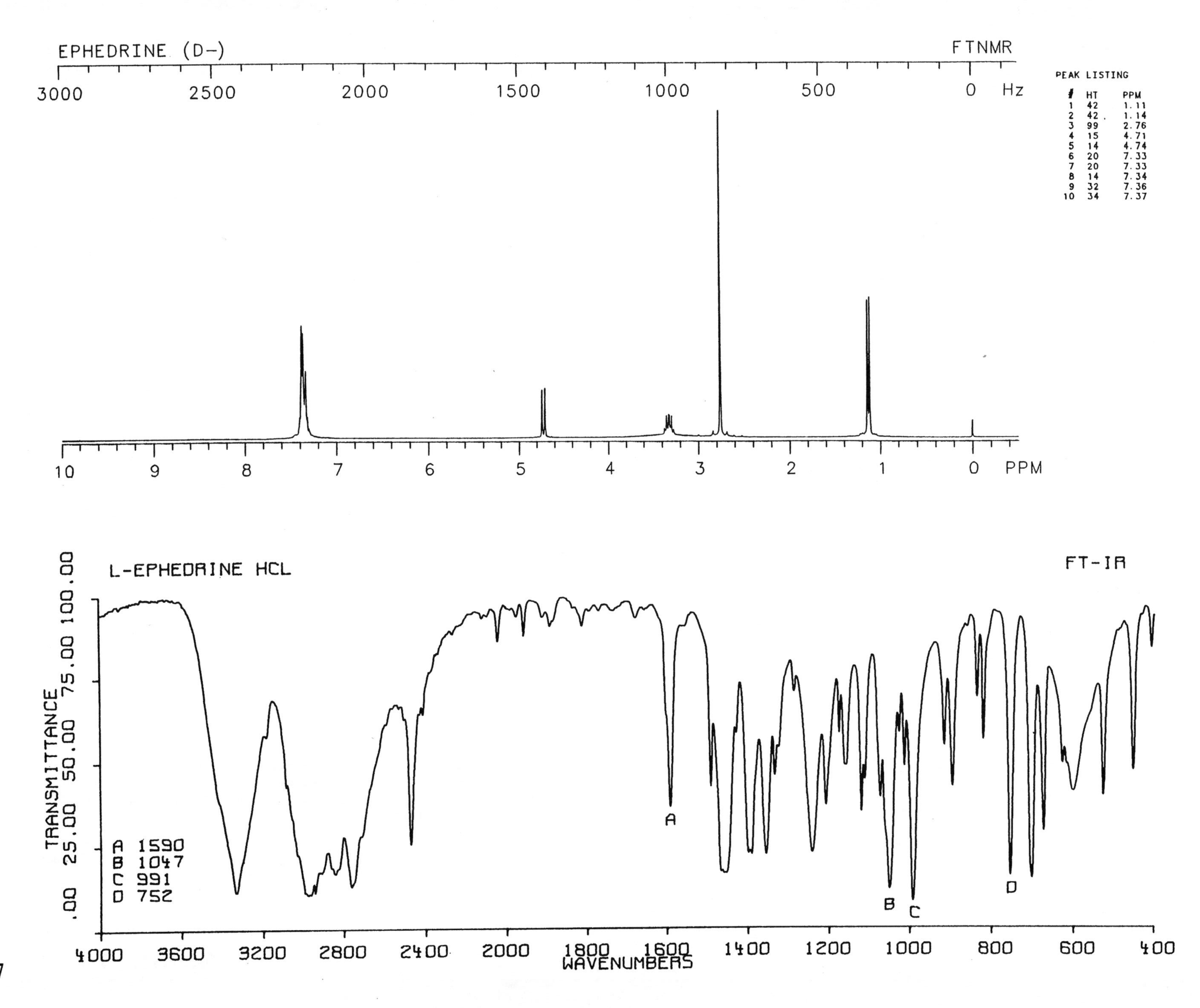
EPHEDRINE (D-)
FTNMR
PEAK LISTING
# HT PPM
1 42 1.11
2 42 1.14
3 99 2.76
4 15 4.71
5 14 4.74
6 20 7.33
7 20 7.33
8 14 7.34
9 32 7.36
10 34 7.37
Hz
PPM
L-EPHEDRINE HCL
FT-IR
A 1590
B 1047
C 991
D 752
TRANSMITTANCE
WAVENUMBERS

# EPINEPHRINE

$C_9H_{13}NO_3$

**Molecular weight:** 183.20 (183.09)
**Synonyms:** 4-[1-Hydroxy-2-(methylamino)ethyl]-1,2-benzenediol; methylaminoethanol catechol

**Trade names:** Adrenalin, Adrenal, Adrephrine, Adrin, Ayerst-Epitrate, Epinephrine, Epinephran, Epirenan, Epipen, Hemisine, Levorenine, Liodocaine, Marcaine, Nephridine, Primatene Mist, Renoform, Surrenine, Supronephrane, Sus-phrine, Takamina, Vaponefrin, Vasotonin
**Use:** Adrenergic
**HPLC:**
**GC:**

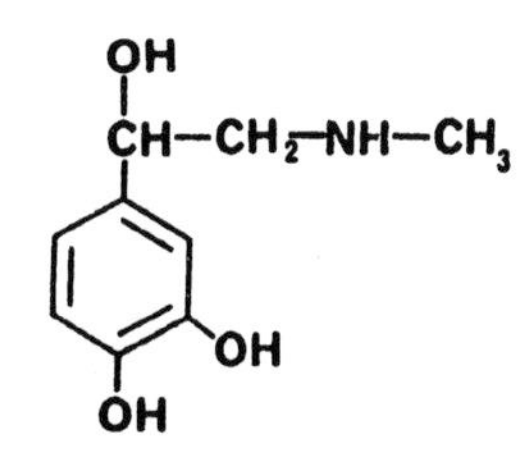


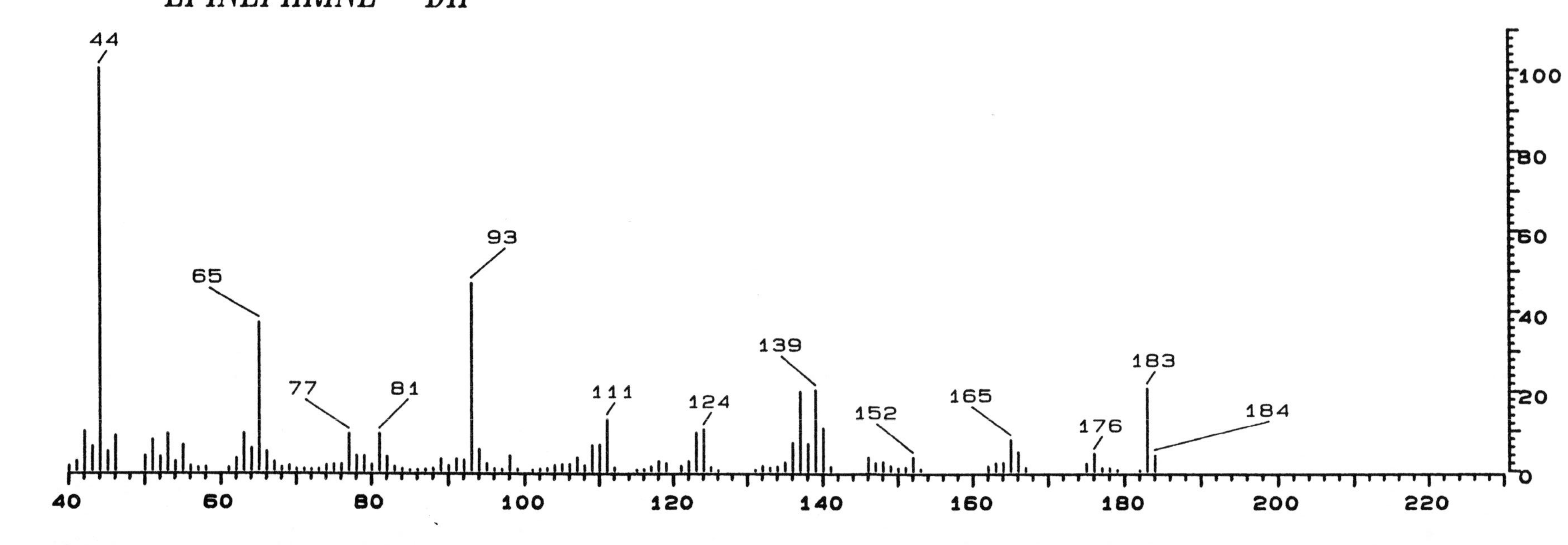

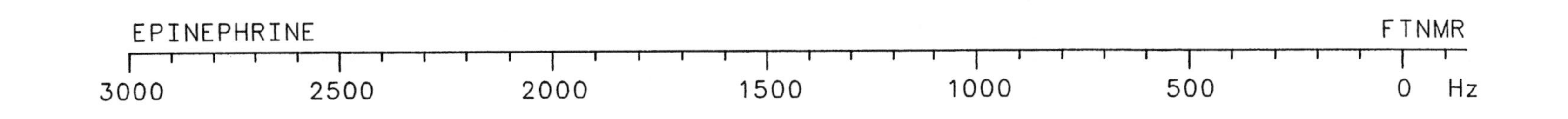


INSUFFICIENT SOLUBILITY

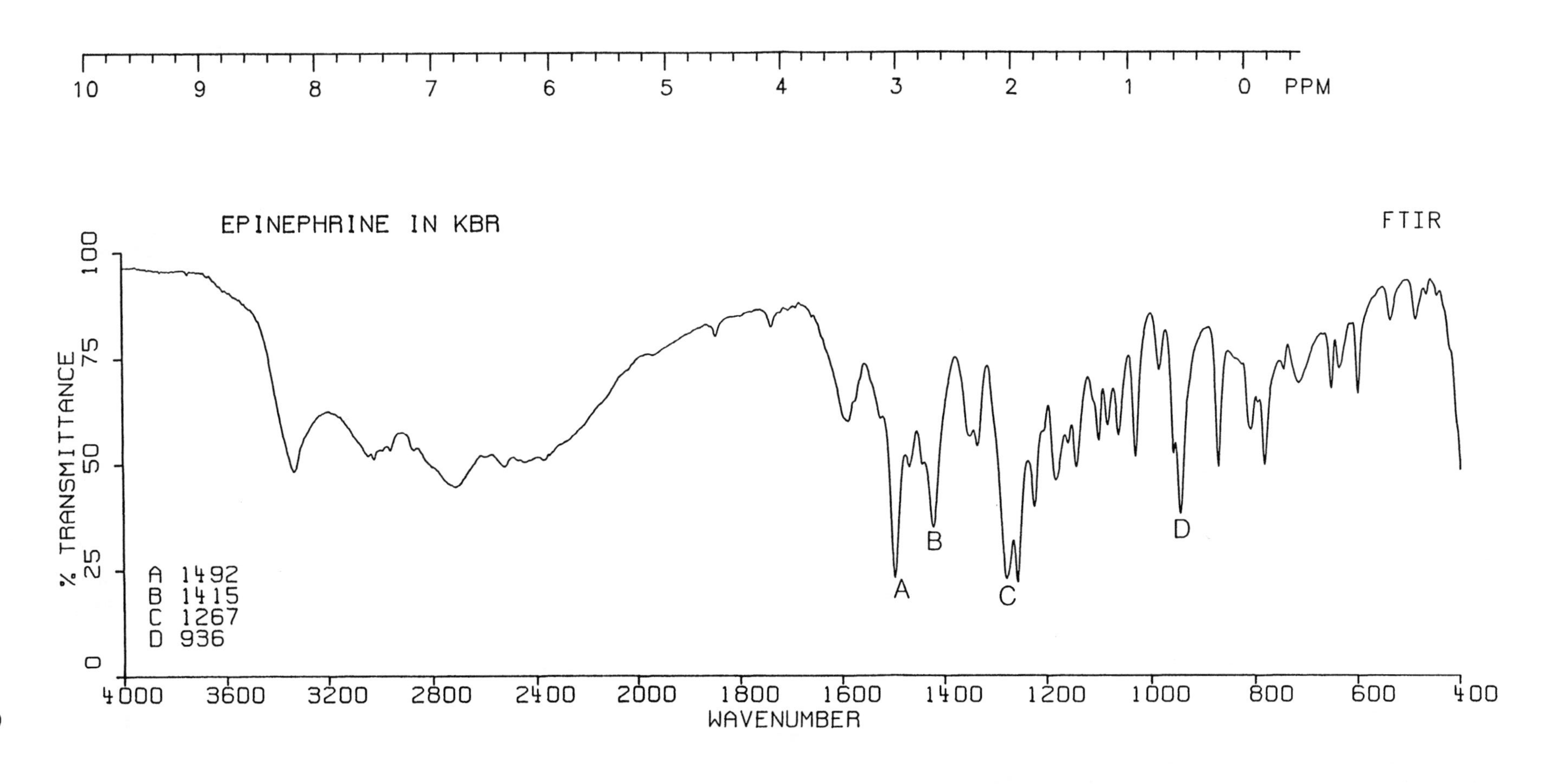

# EQUILIN

$C_{18}H_{20}O_2$

Molecular weight: 268.34 (268.15)
Synonyms: 3-Hydroxyestra-1,2,5(10),7-tetraen-17-one;
1,3,5,7-estratetraen-3-ol-17-one
Trade names:

Use: Estrogen
HPLC:
GC: 2610; 250°C

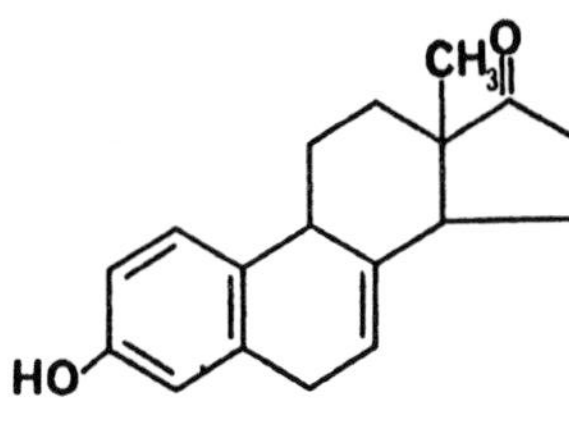


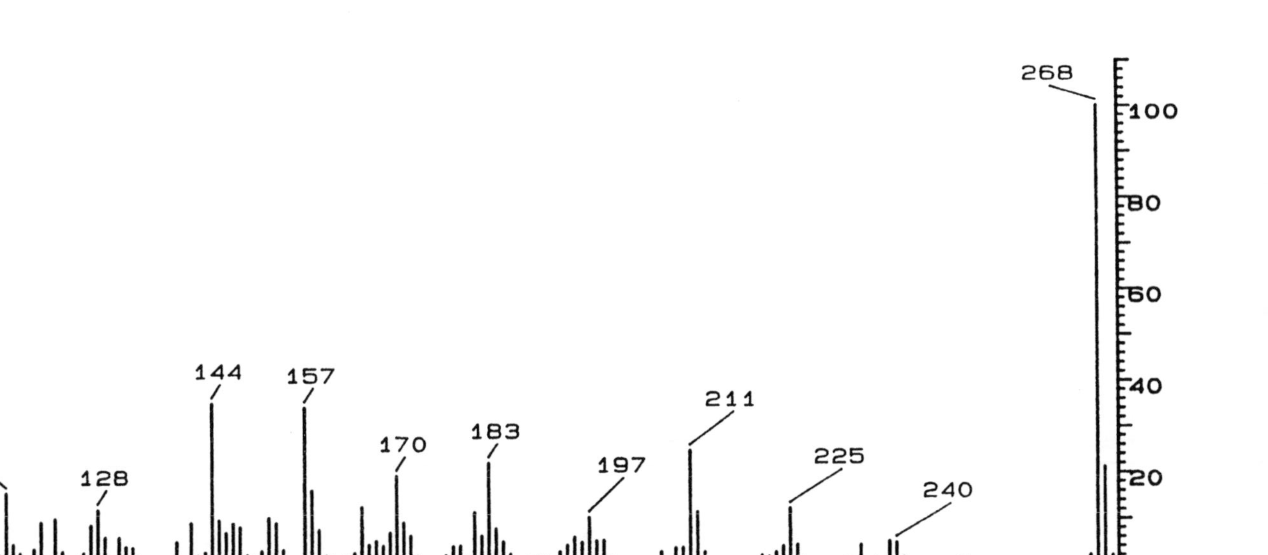

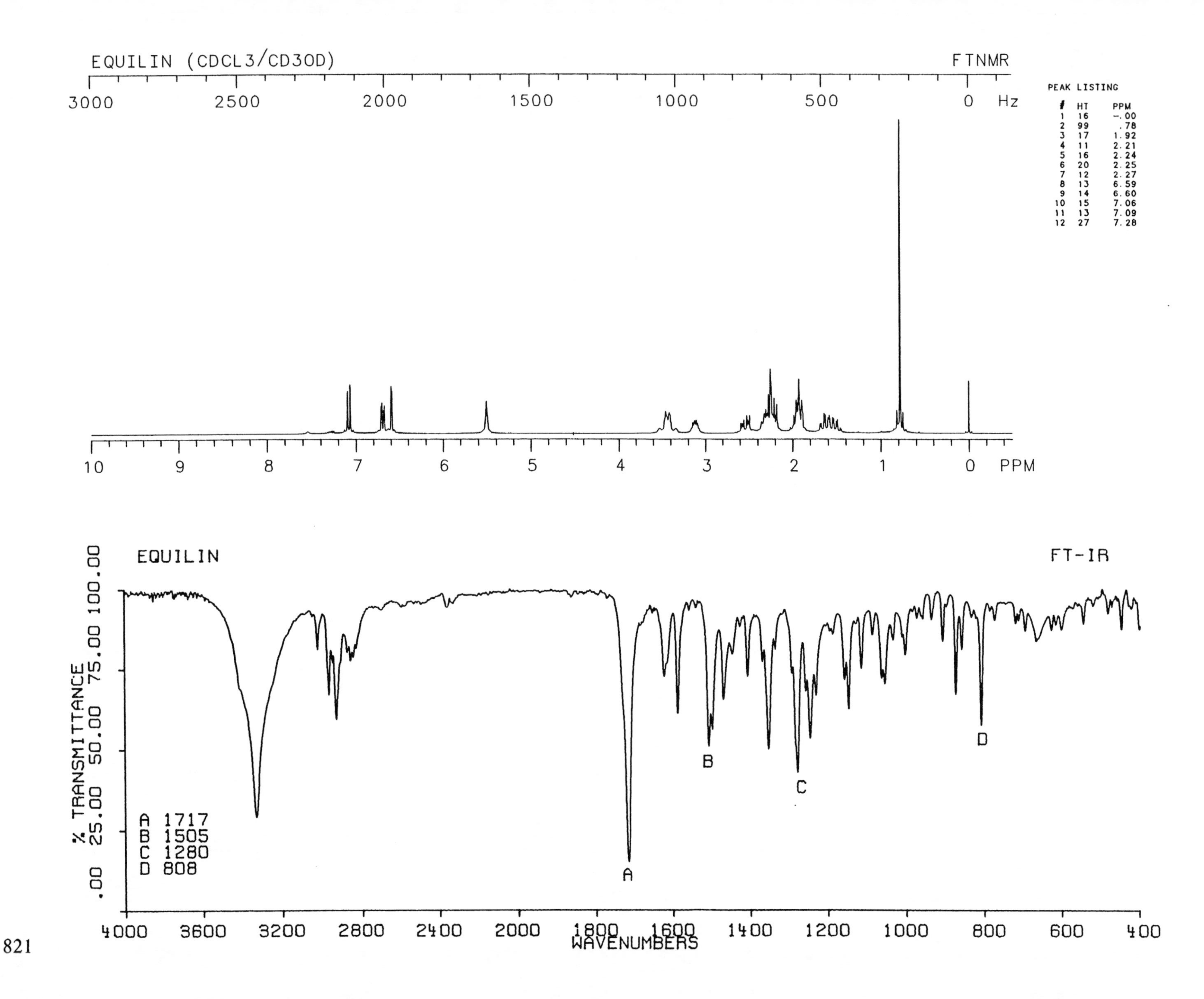
EQUILIN (CDCL3/CD3OD)
FTNMR
3000
2500
2000
1500
1000
500
0 Hz
PEAK LISTING
# HT PPM
1 16 -.00
2 99 .78
3 17 1.92
4 11 2.21
5 16 2.24
6 20 2.25
7 12 2.27
8 13 6.59
9 14 6.60
10 15 7.06
11 13 7.09
12 27 7.28
10 9 8 7 6 5 4 3 2 1 0 PPM
EQUILIN
FT-IR
% TRANSMITTANCE
.00 25.00 50.00 75.00 100.00
A 1717
B 1505
C 1280
D 808
A
B
C
D
4000 3600 3200 2800 2400 2000 1800 1600 1400 1200 1000 800 600 400
WAVENUMBERS

# ERGONOVINE

$C_{19}H_{23}N_3O_2$

**Molecular weight:** 325.39 (325.18)
**Synonyms:** 9,10-Dihydro-N-(2-hydroxy-1-methylethyl)-6-methylergoline-8-carboxamide; D-lysergic acid L-2-propanolamide; ergometrine
**Trade names:** Basergin, Cornocentin, Ergobasine, Ergotocine, Ergostetrine, Ergotrate Maleate, Neofemergen, Syntometrine
**Use:** Oxytocic, ergot alkaloid
**HPLC:** Si-10; 5A:95B; 7.4
**GC:**

ERGONOVINE MALEATE

ETHANOL.....239,315

TRANSMITTANCE

WAVELENGTH (nm)

*ERGONOVINE -- DIP*

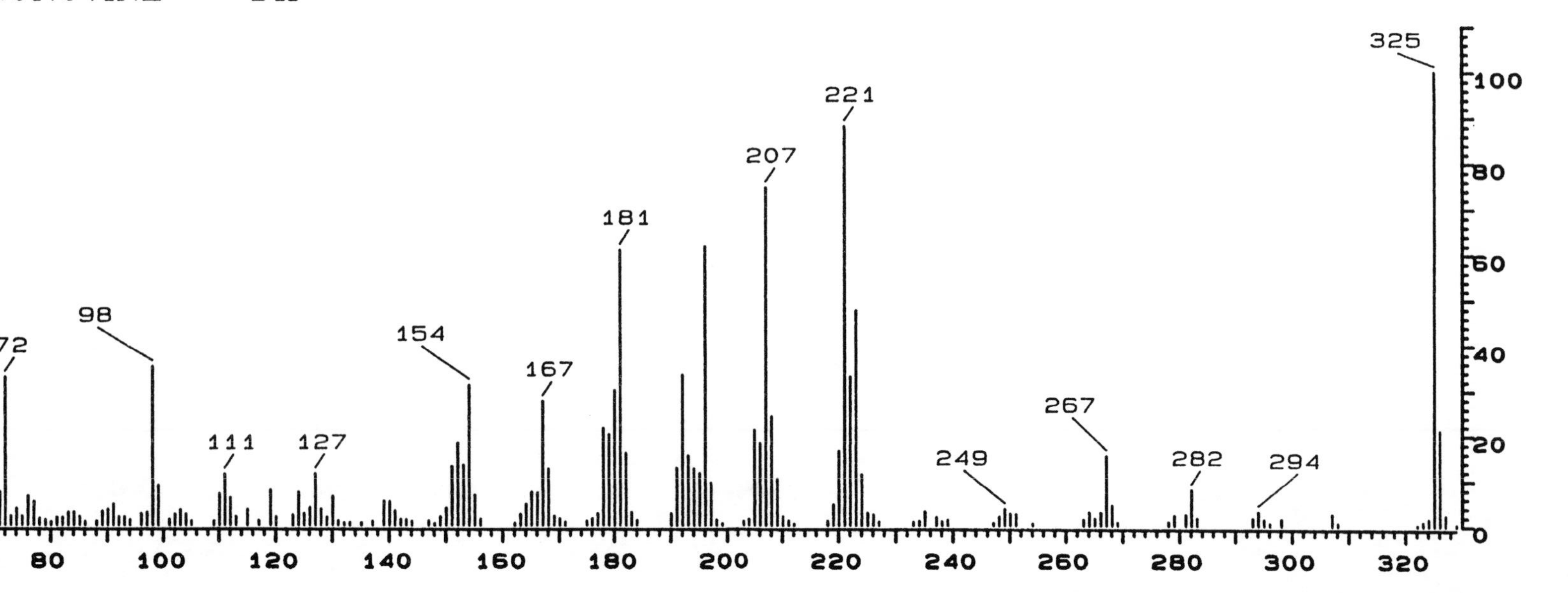

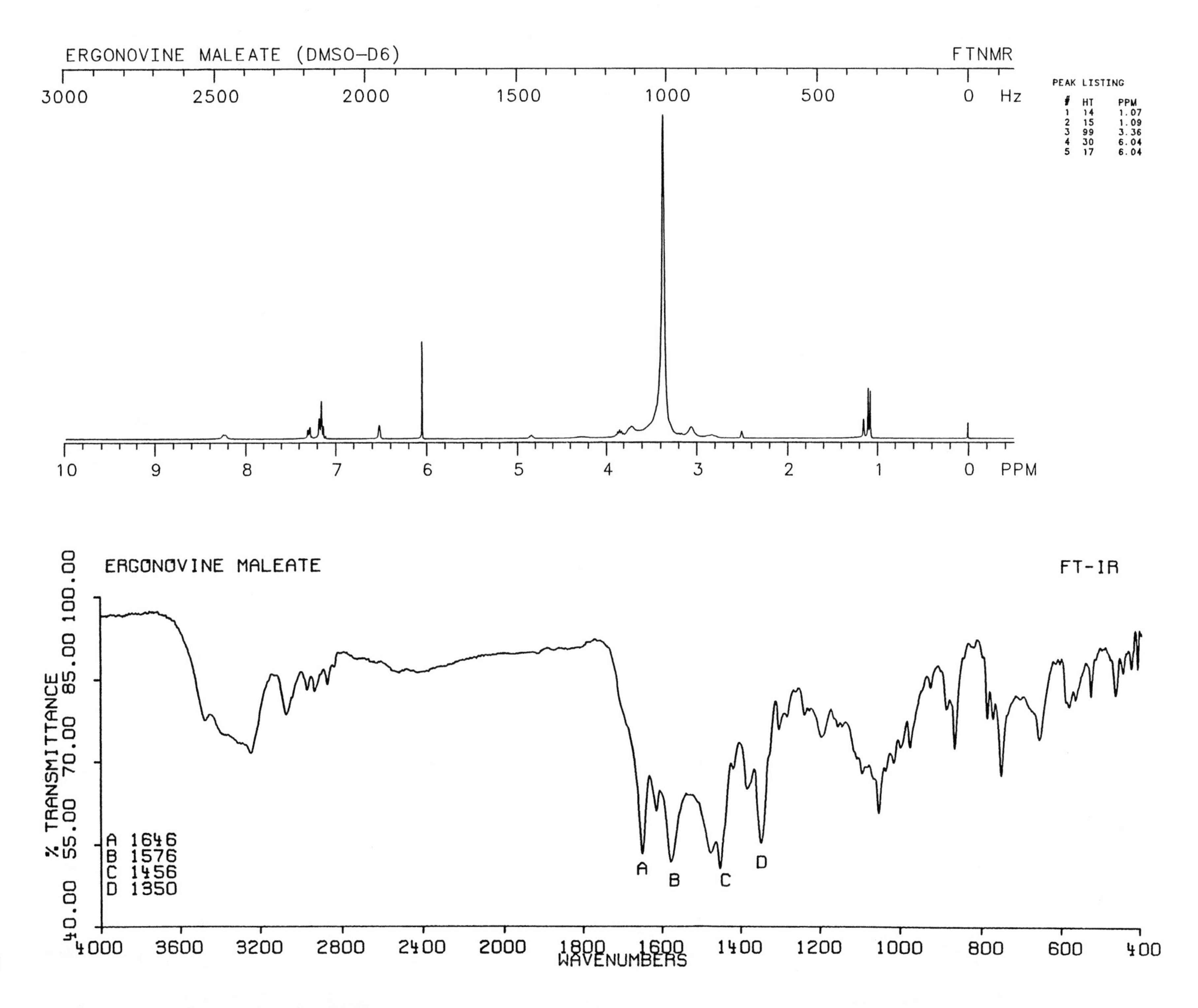
ERGONOVINE MALEATE (DMSO-D6)
FTNMR
3000
2500
2000
1500
1000
500
0 Hz
PEAK LISTING
# HT PPM
1 14 1.07
2 15 1.09
3 99 3.36
4 30 6.04
5 17 6.04
10
9
8
7
6
5
4
3
2
1
0 PPM
ERGONOVINE MALEATE
FT-IR
% TRANSMITTANCE
40.00
55.00
70.00
85.00
100.00
A 1646
B 1576
C 1456
D 1350
A
B
C
D
4000
3600
3200
2800
2400
2000
1800
1600
1400
1200
1000
800
600
400
WAVENUMBERS

# ERGOSTEROL

$C_{28}H_{44}O$

**Molecular weight:** 396.63 (396.34)
**Synonyms:** Ergosta-5,7,22-trien-3β-ol; ergosterin

**Trade names:**

**Use:** Vitamin
**HPLC:**
**GC:**

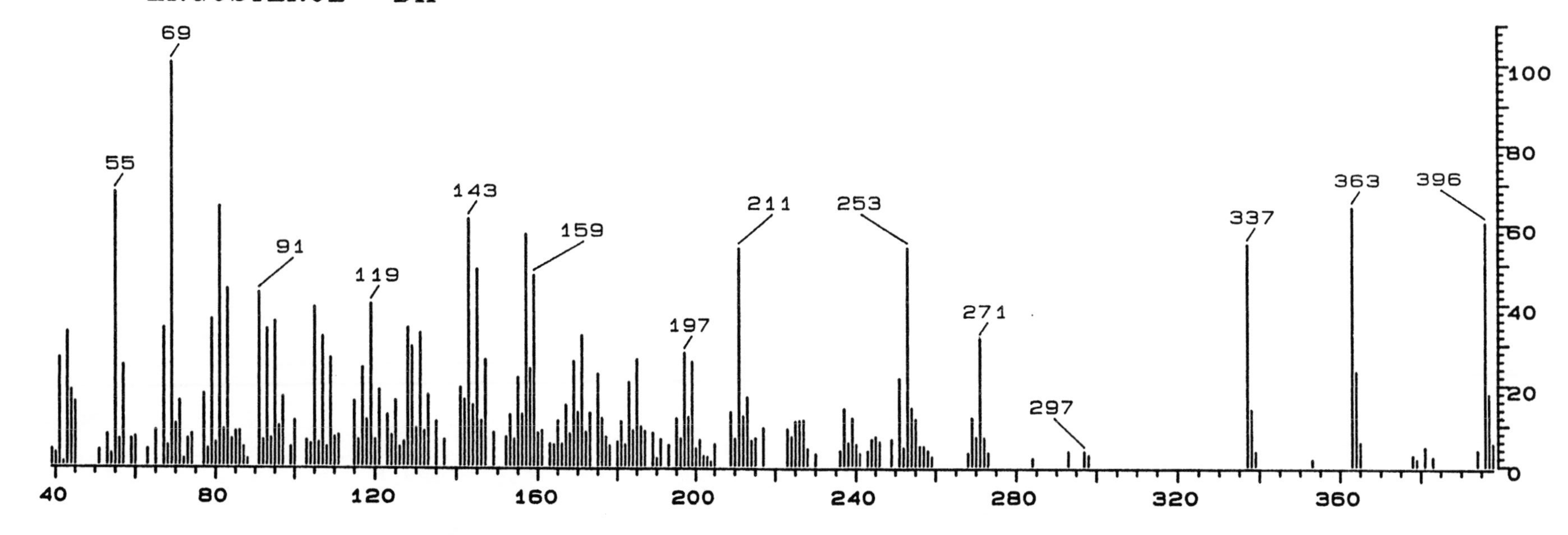

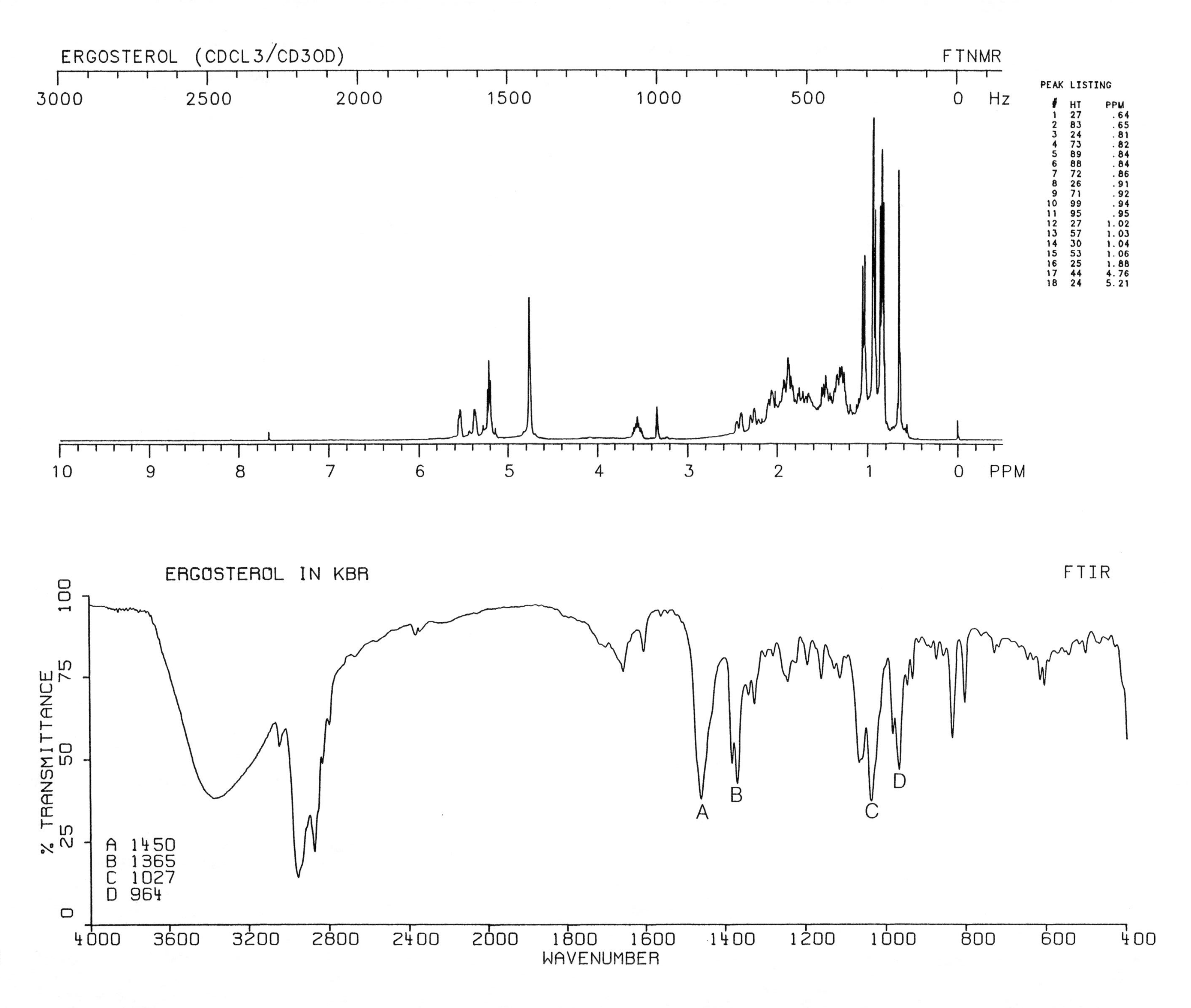

ERGOSTEROL (CDCL3/CD3OD)
FTNMR
3000 2500 2000 1500 1000 500 0 Hz
PEAK LISTING
# HT PPM
1 27 .64
2 83 .65
3 24 .81
4 73 .82
5 89 .84
6 88 .84
7 72 .86
8 26 .91
9 71 .92
10 99 .94
11 95 .95
12 27 1.02
13 57 1.03
14 30 1.04
15 53 1.06
16 25 1.88
17 44 4.76
18 24 5.21
10 9 8 7 6 5 4 3 2 1 0 PPM
ERGOSTEROL IN KBR
FTIR
% TRANSMITTANCE
100 75 50 25 0
A 1450
B 1365
C 1027
D 964
A
B
C
D
4000 3600 3200 2800 2400 2000 1800 1600 1400 1200 1000 800 600 400
WAVENUMBER

# ERGOTAMINE

$C_{33}H_{35}N_5O_5$

**Molecular weight:** 581.65 (581.26)
**Synonyms:** 12'-Hydroxy-2'-methyl-5'α-(phenylmethyl)ergotaman-3',6',18-trione
**Trade names:** Bellergal, Bellergal-S, Bellermine-OD, Cafergot, Ergomar, Ergostat, Gynergen, Medihaler, Migral, Wigraine
**Use:** Vasoconstrictor, ergot alkaloid
**HPLC:**
**GC:** 2394; 250°C

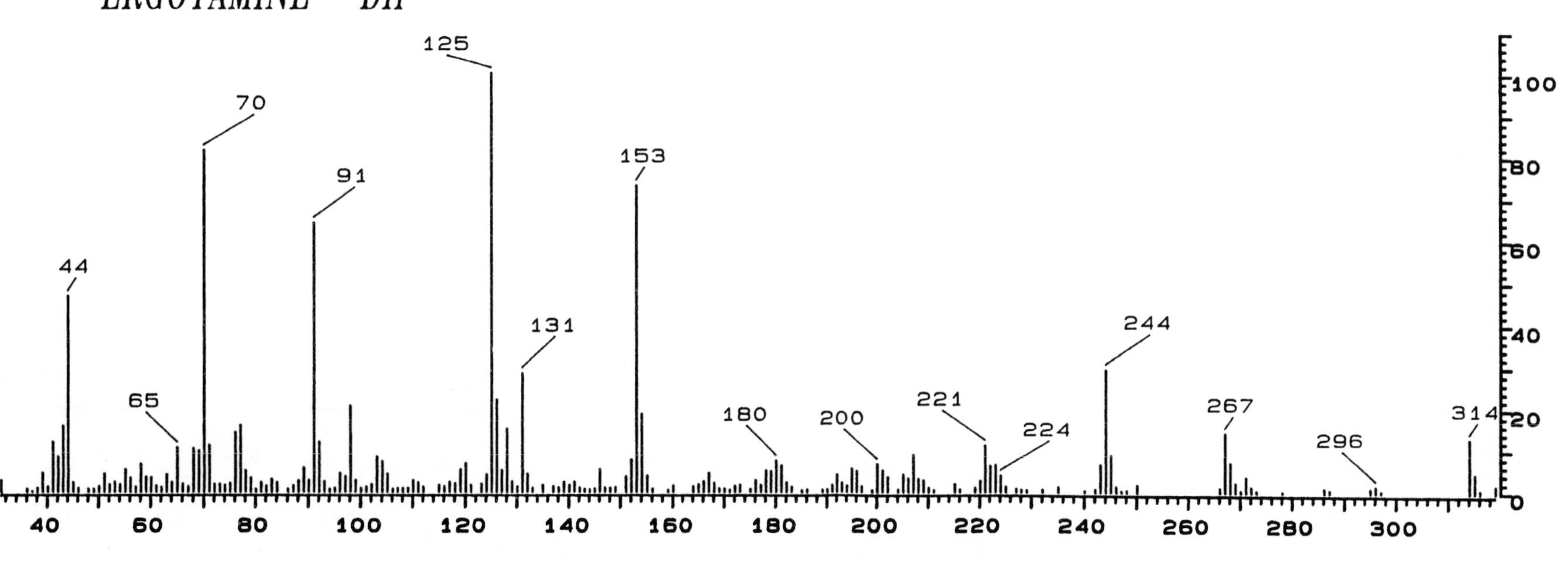

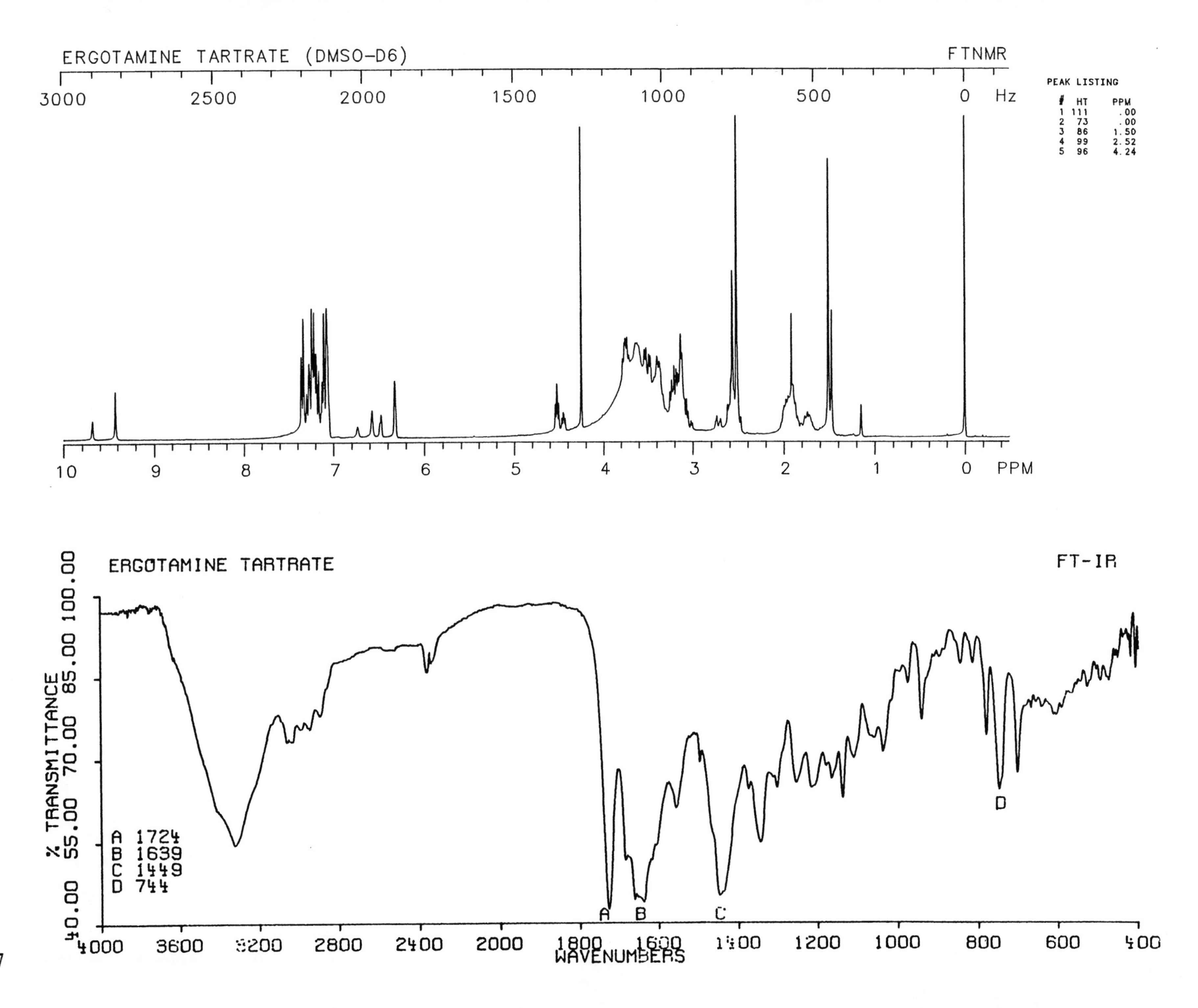
ERGOTAMINE TARTRATE (DMSO-D6)
FTNMR
3000
2500
2000
1500
1000
500
0
Hz
PEAK LISTING
# HT PPM
1 111 .00
2 73 .00
3 86 1.50
4 99 2.52
5 96 4.24
10
9
8
7
6
5
4
3
2
1
0
PPM
ERGOTAMINE TARTRATE
FT-IR
% TRANSMITTANCE
40.00 55.00 70.00 85.00 100.00
A 1724
B 1639
C 1449
D 744
A
B
C
D
4000 3600 3200 2800 2400 2000 1800 1600 1400 1200 1000 800 600 400
WAVENUMBERS

# ERYTHRITYL TETRANITRATE

$C_4H_6N_4O_{12}$

**Molecular weight:** 302.12 (302.00)
**Synonyms:** Erythritol tetranitrate; erythrol tetranitrate; tetranitrol; tetranitrin; nitroerythrite
**Trade names:** Cardilate, Cardiloid

**Use:** Vasodilator
**HPLC:**
**GC:**

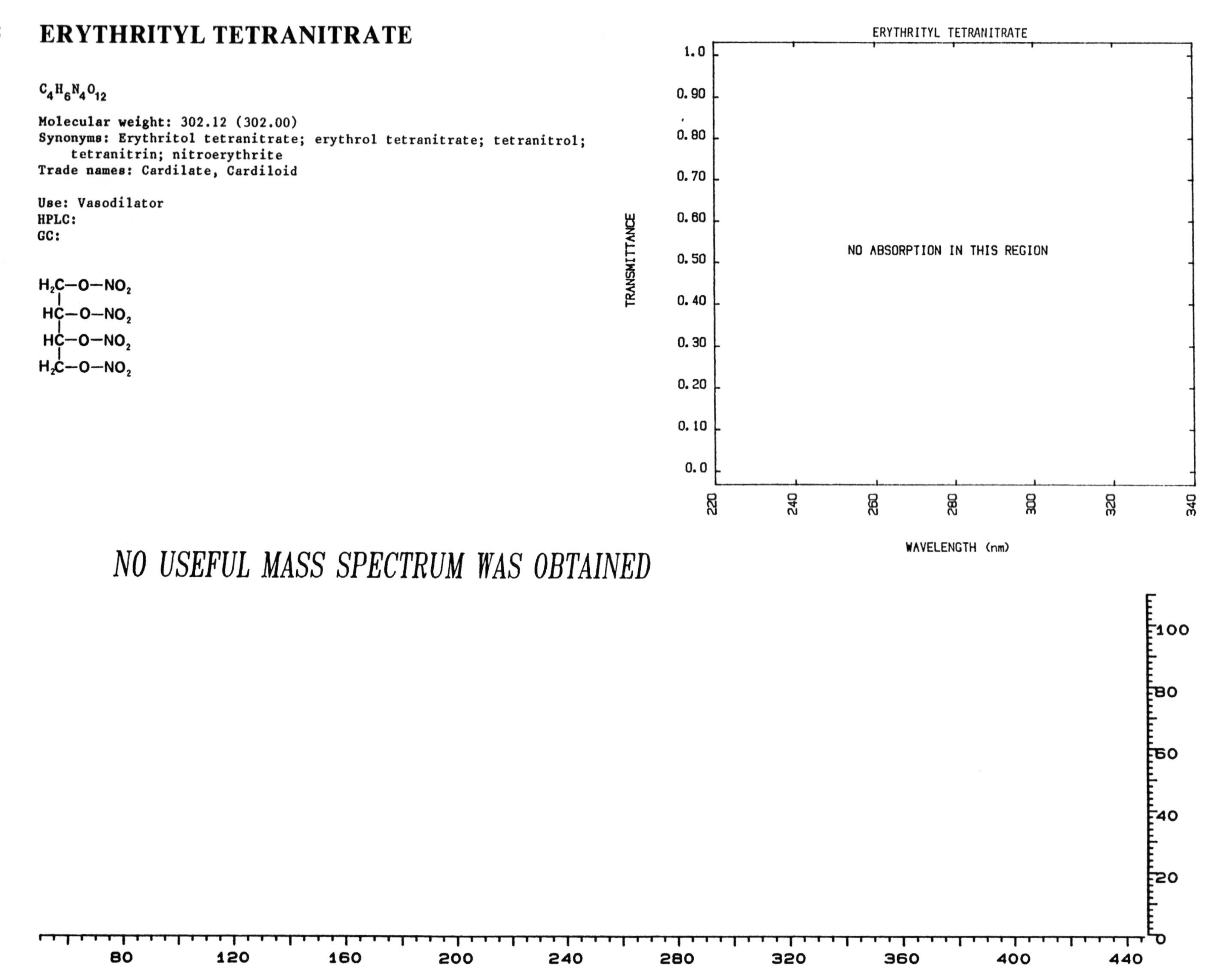

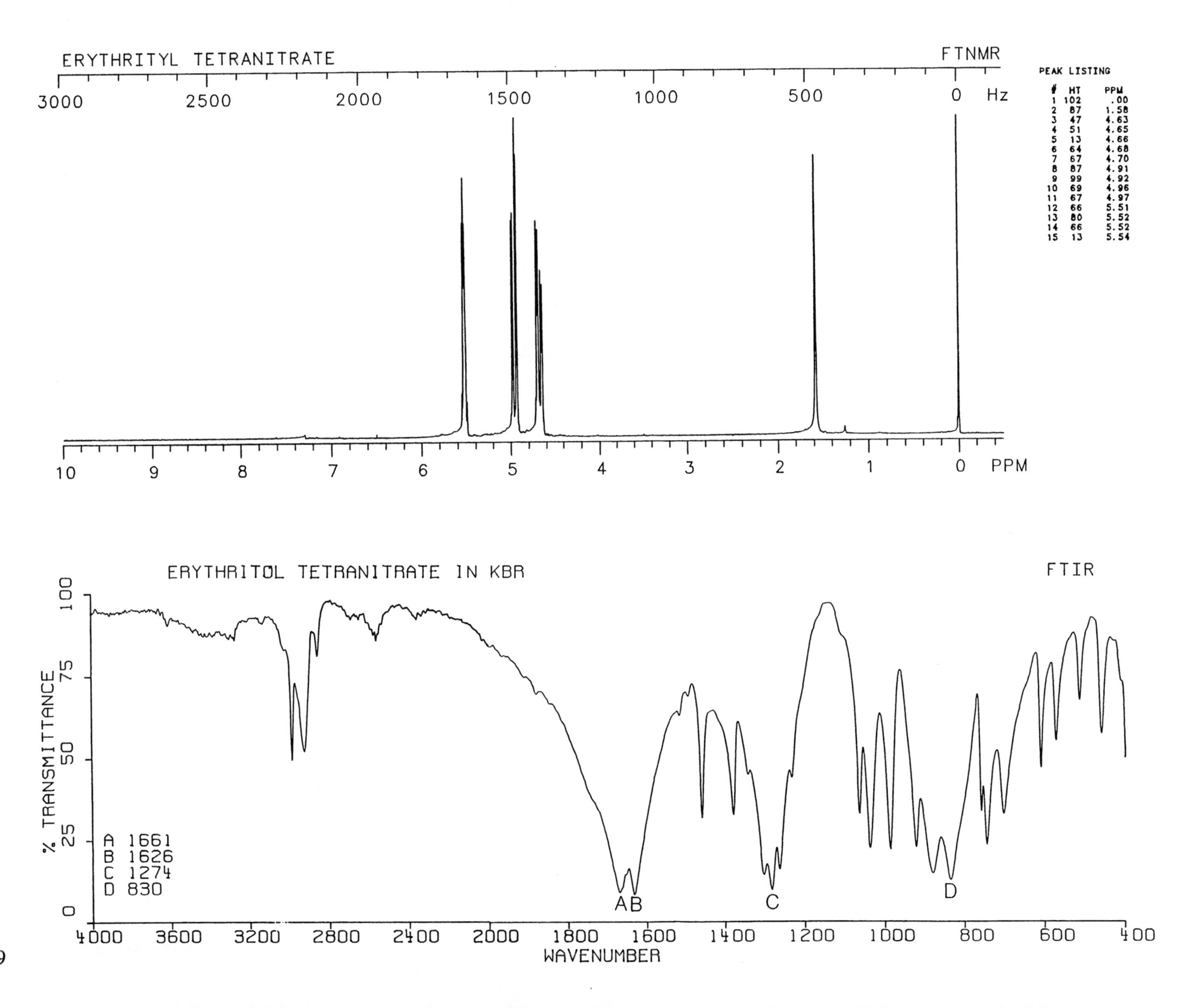
ERYTHRITYL TETRANITRATE
FTNMR
3000 2500 2000 1500 1000 500 0 Hz
10 9 8 7 6 5 4 3 2 1 0 PPM
PEAK LISTING
# HT PPM
1 102 .00
2 87 1.58
3 47 4.63
4 51 4.65
5 13 4.66
6 64 4.68
7 67 4.70
8 87 4.91
9 99 4.92
10 69 4.96
11 67 4.97
12 66 5.51
13 80 5.52
14 66 5.52
15 13 5.54
ERYTHRITOL TETRANITRATE IN KBR
FTIR
% TRANSMITTANCE
100 75 50 25 0
A 1661
B 1626
C 1274
D 830
A B C D
4000 3600 3200 2800 2400 2000 1800 1600 1400 1200 1000 800 600 400
WAVENUMBER

# ERYTHROMYCIN

$C_{37}H_{67}NO_{13}$

**Molecular weight:** 733.94 (733.46)
**Synonyms:** Erythromycin A

**Trade names:** A/T/S, Bristamycin, E.E.S., E-Mycin, Eryc, EryDerm, EryPed, Erythrocin, Ilotycin, Pediamycin, Pediazole, Pfizer-E, Robimycin, SK-Erythromycin, Staticin, Wyamycin
**Use:** Antibacterial
**HPLC:**
**GC:**

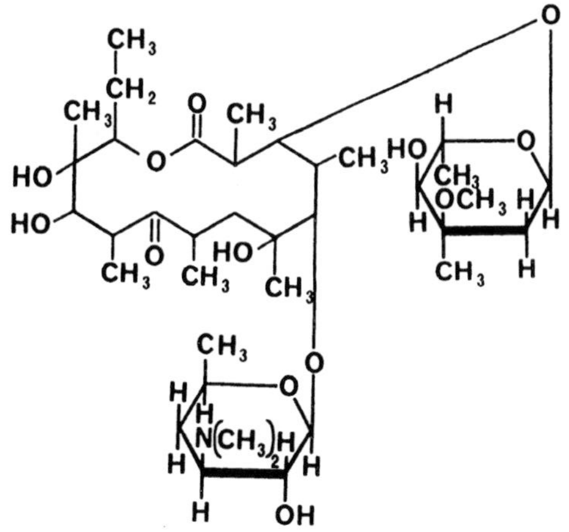

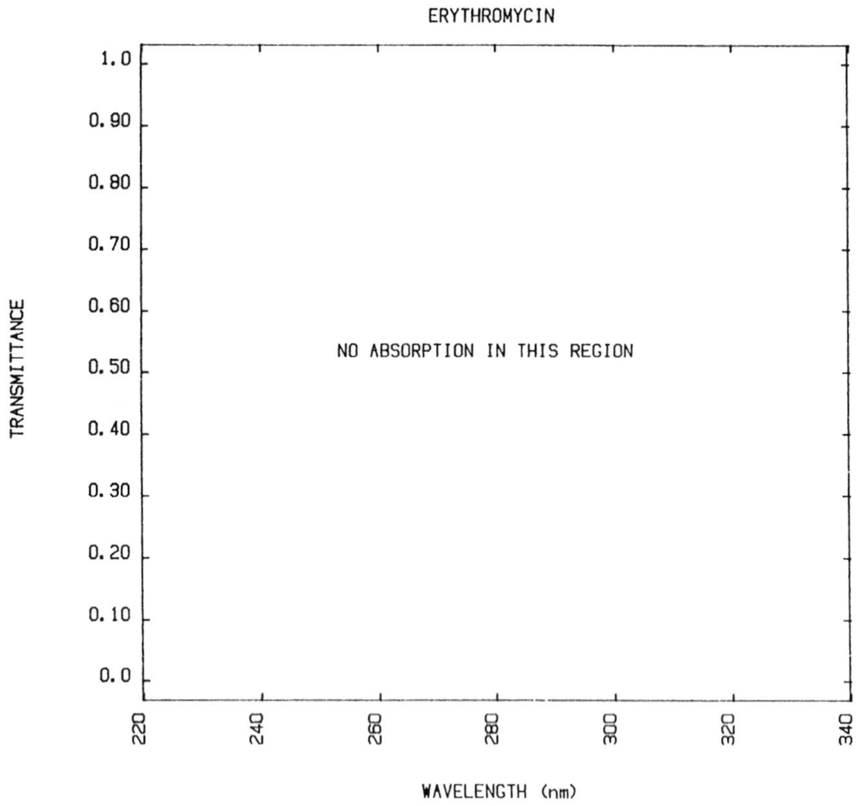

ERYTHROMYCIN

TRANSMITTANCE

NO ABSORPTION IN THIS REGION

WAVELENGTH (nm)

*NO USEFUL MASS SPECTRUM WAS OBTAINED*

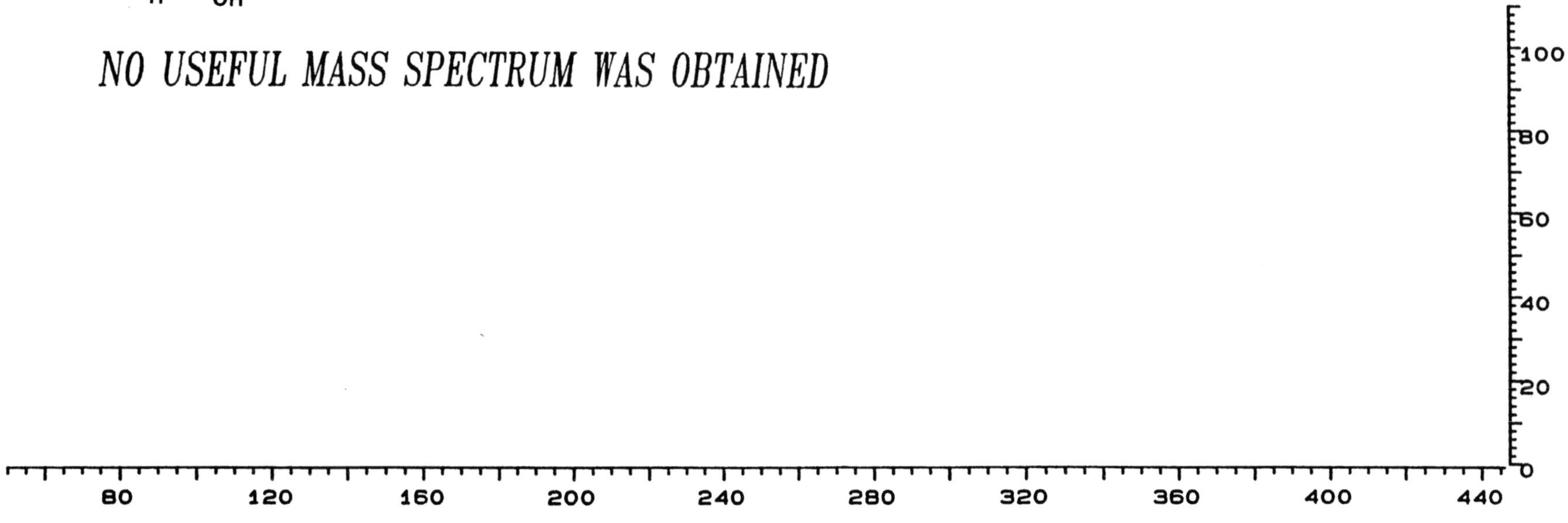

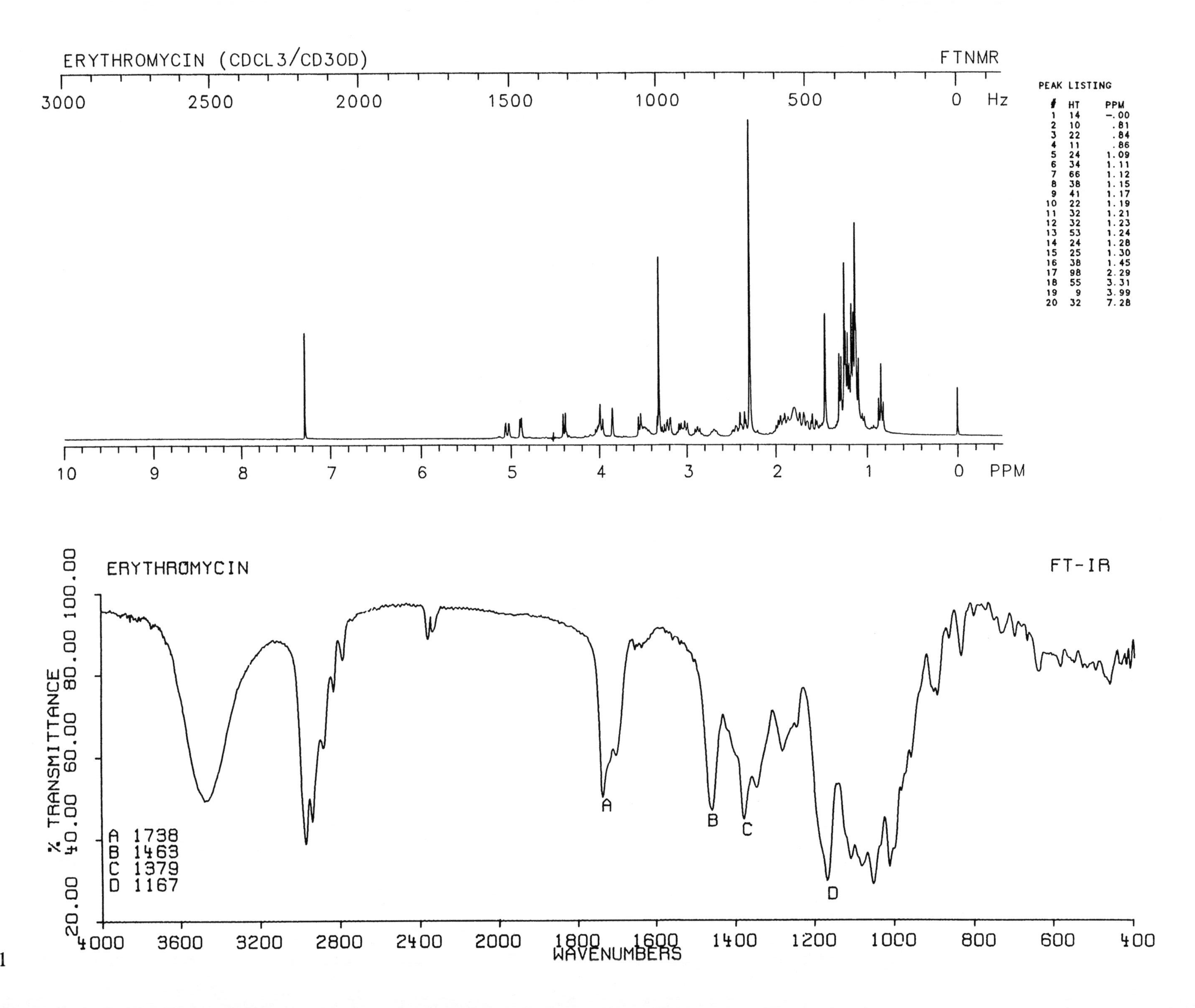
ERYTHROMYCIN (CDCL3/CD3OD)
FTNMR
PEAK LISTING
# HT PPM
1 14 -.00
2 10 .81
3 22 .84
4 11 .86
5 24 1.09
6 34 1.11
7 66 1.12
8 38 1.15
9 41 1.17
10 22 1.19
11 32 1.21
12 32 1.23
13 53 1.24
14 24 1.28
15 25 1.30
16 38 1.45
17 98 2.29
18 55 3.31
19 9 3.99
20 32 7.28
ERYTHROMYCIN
FT-IR
A 1738
B 1463
C 1379
D 1167
% TRANSMITTANCE
WAVENUMBERS

# ESERINE

$C_{15}H_{21}N_3O_2$

Molecular weight: 275.34 (275.16)
Synonyms: 1,2,3,3a,8,8a-Hexahydro-1,3a,8-trimethylpyrrolo-[2,3-b]indol-5-ol methylcarbamate ester; physostigmine; physostigmina
Trade names: Eserine, Physostol

Use: Cholinergic
HPLC: Si-10; 5A:95B; 4.4
GC: 2268; 250°C

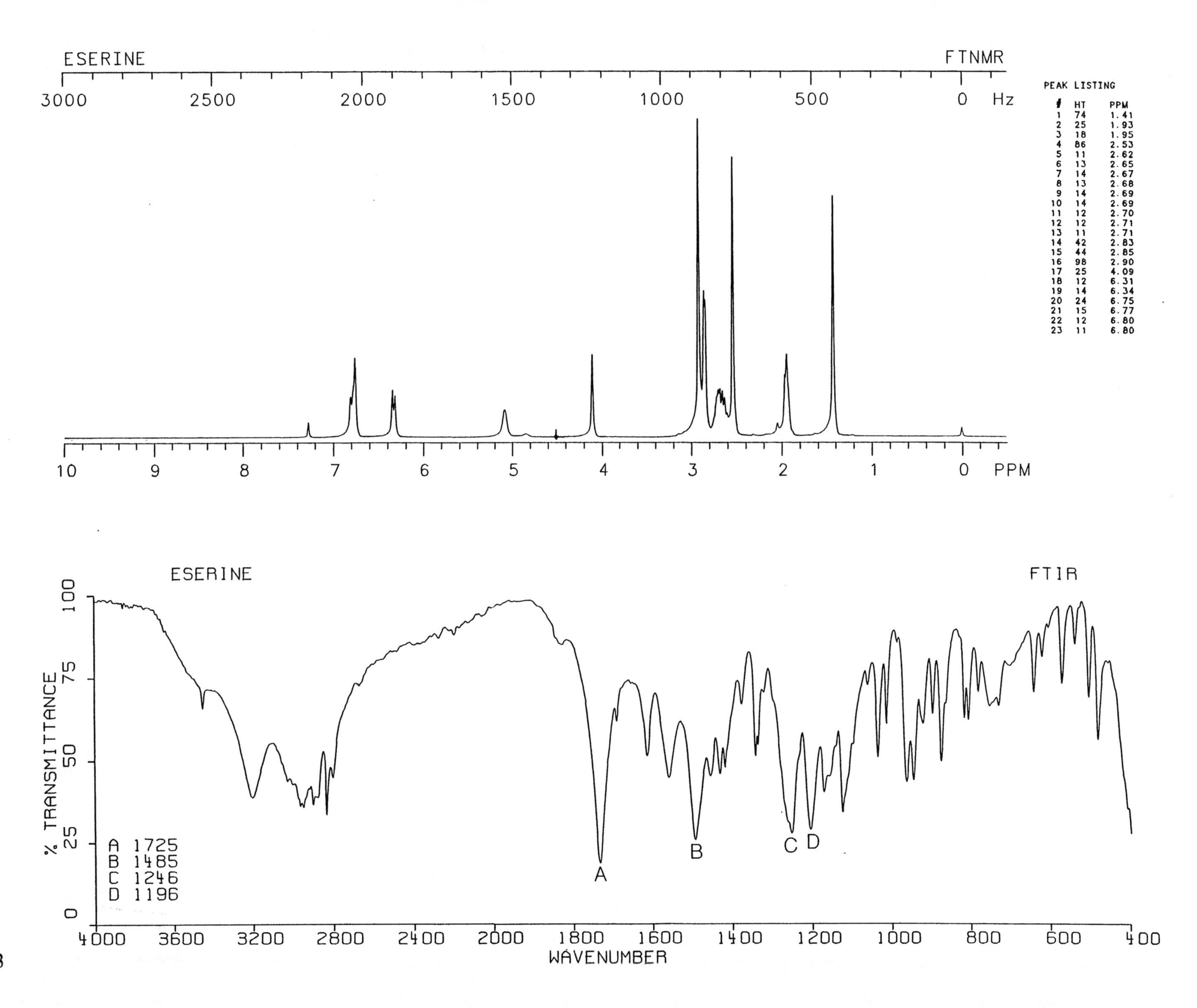
ESERINE
FTNMR
3000 2500 2000 1500 1000 500 0 Hz
10 9 8 7 6 5 4 3 2 1 0 PPM
PEAK LISTING
# HT PPM
1 74 1.41
2 25 1.93
3 18 1.95
4 86 2.53
5 11 2.62
6 13 2.65
7 14 2.67
8 13 2.68
9 14 2.69
10 14 2.69
11 12 2.70
12 12 2.71
13 11 2.71
14 42 2.83
15 44 2.85
16 98 2.90
17 25 4.09
18 12 6.31
19 14 6.34
20 24 6.75
21 15 6.77
22 12 6.80
23 11 6.80
ESERINE
FTIR
% TRANSMITTANCE
100 75 50 25 0
A 1725
B 1485
C 1246
D 1196
A
B
C
D
4000 3600 3200 2800 2400 2000 1800 1600 1400 1200 1000 800 600 400
WAVENUMBER

# ESTRADIOL

$C_{18}H_{24}O_2$

**Molecular weight:** 272.37 (272.18)
**Synonyms:** Estra-1,3,5(10)-triene-3,17-diol; dihydroxyestrin; dihydrotheelin; oestradiol
**Trade names:** Dimenformon, Diogyn, Estrace, Estrovite, Ovasterol, Ovahorman, Ovocyclin, Perlatanol, Provest, Primofol, Profoliol
**Use:** Estrogen
**HPLC:**
**GC:** 2619; 250°C

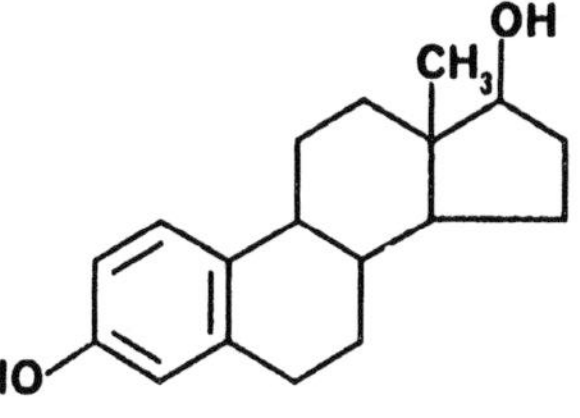


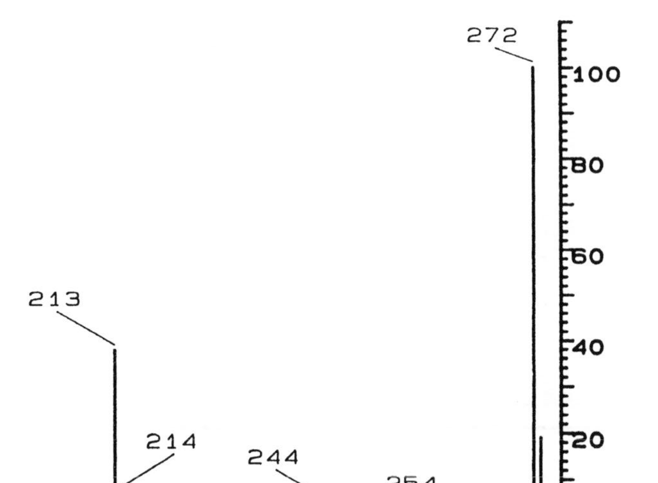


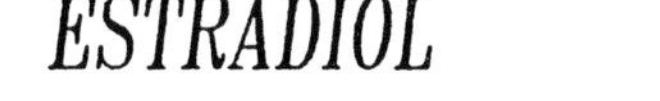


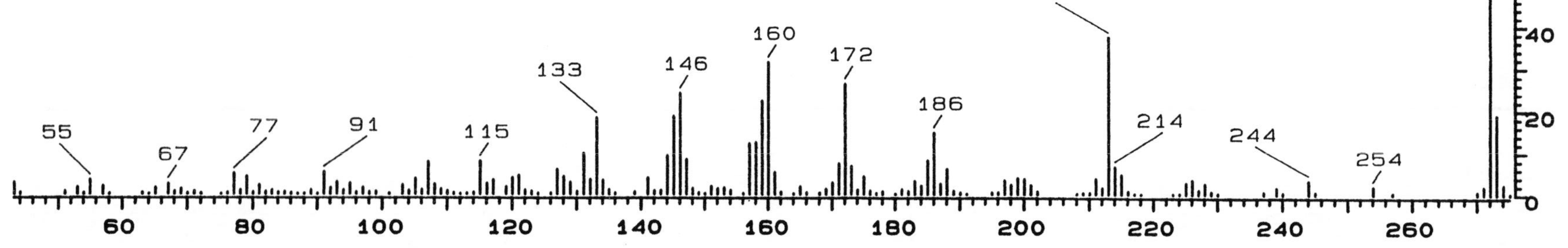


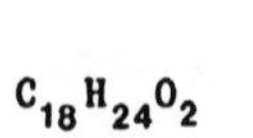

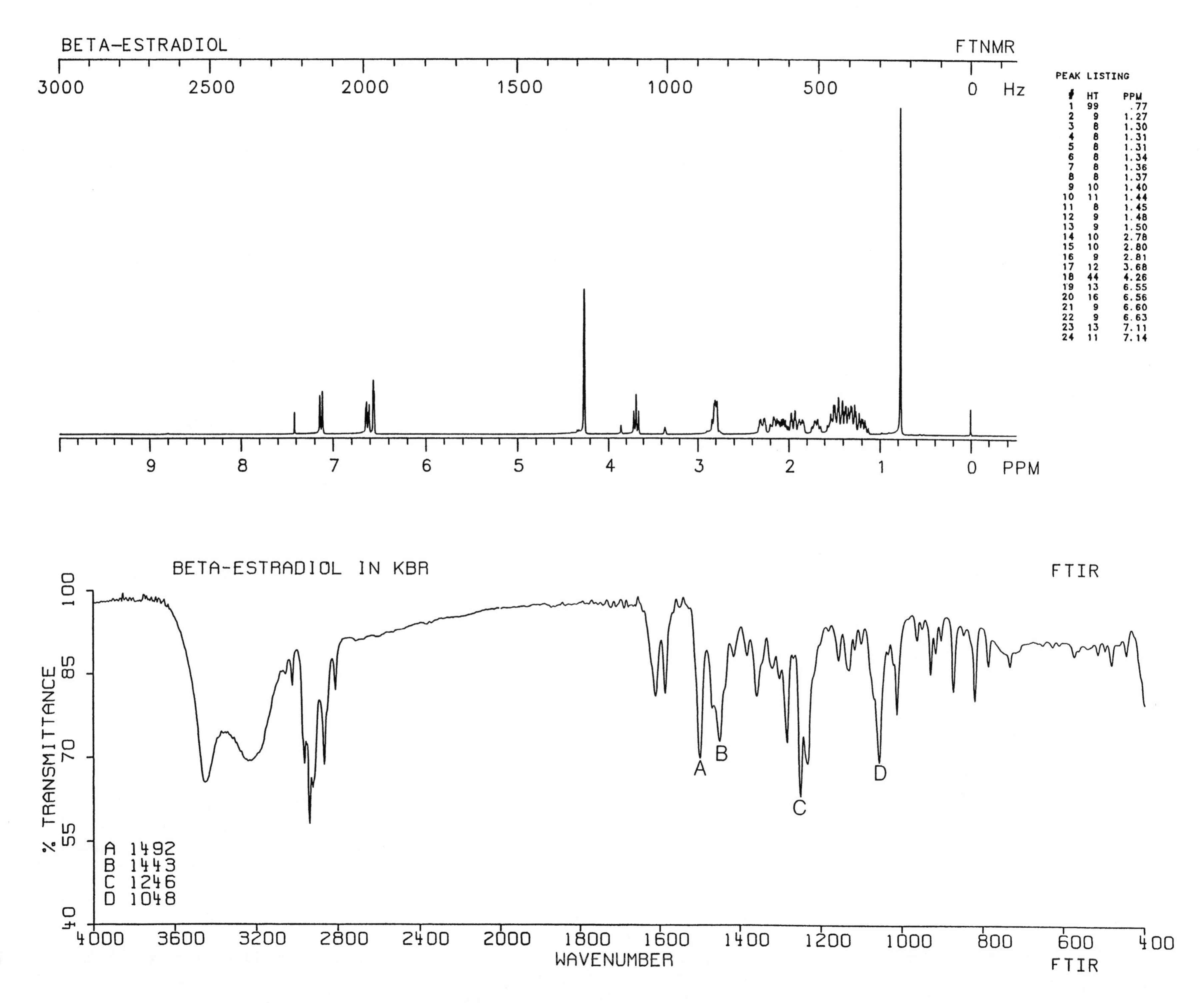
BETA-ESTRADIOL
FTNMR
3000 2500 2000 1500 1000 500 0 Hz
9 8 7 6 5 4 3 2 1 0 PPM
PEAK LISTING
# HT PPM
1 99 .77
2 9 1.27
3 8 1.30
4 8 1.31
5 8 1.31
6 8 1.34
7 8 1.36
8 8 1.37
9 10 1.40
10 11 1.44
11 8 1.45
12 9 1.48
13 9 1.50
14 10 2.78
15 10 2.80
16 9 2.81
17 12 3.68
18 44 4.26
19 13 6.55
20 16 6.56
21 9 6.60
22 9 6.63
23 13 7.11
24 11 7.14
BETA-ESTRADIOL IN KBR
FTIR
% TRANSMITTANCE
100 85 70 55 40
A 1492
B 1443
C 1246
D 1048
4000 3600 3200 2800 2400 2000 1800 1600 1400 1200 1000 800 600 400
WAVENUMBER
FTIR

# ESTRADIOL-3-BENZOATE

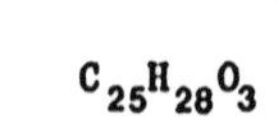

$C_{25}H_{28}O_3$

**Molecular weight:** 376.50 (376.20)
**Synonyms:** β-Estradiol-3-benzoate; oestradiol monobenzoate; 3-benzoyloxyoestra-1,3,5(N)-trien-17β-ol
**Trade names:** Benovocylin, Benzofoline, Difolliculine, Diogyn-B, Estron-B, Femestrone, Hidroestron, Hormogynon, Oestroform, Ovex, Recthormone,
**Use:**
**HPLC:** Si-10; 100B; 4.2
**GC:** 3347; 280°C

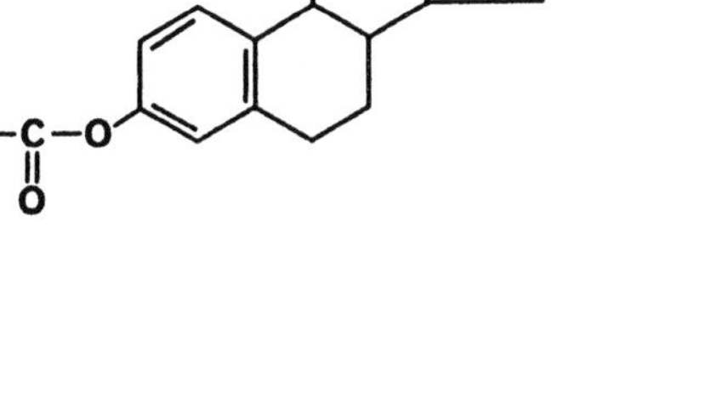


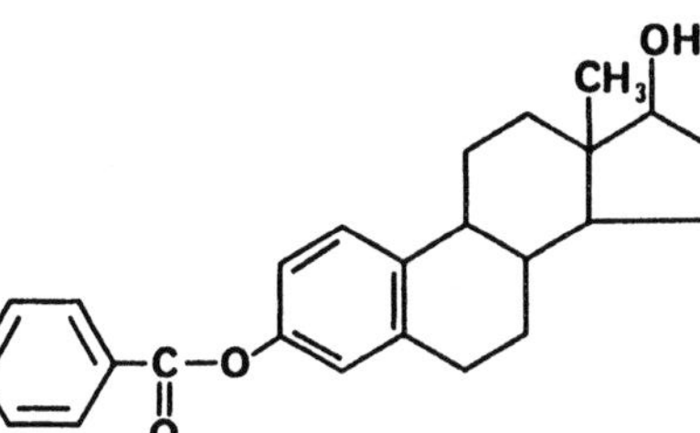

BETA-ESTRADIOL-3-BENZOATE

ETHANOL.....233,267,274

TRANSMITTANCE

WAVELENGTH (nm)

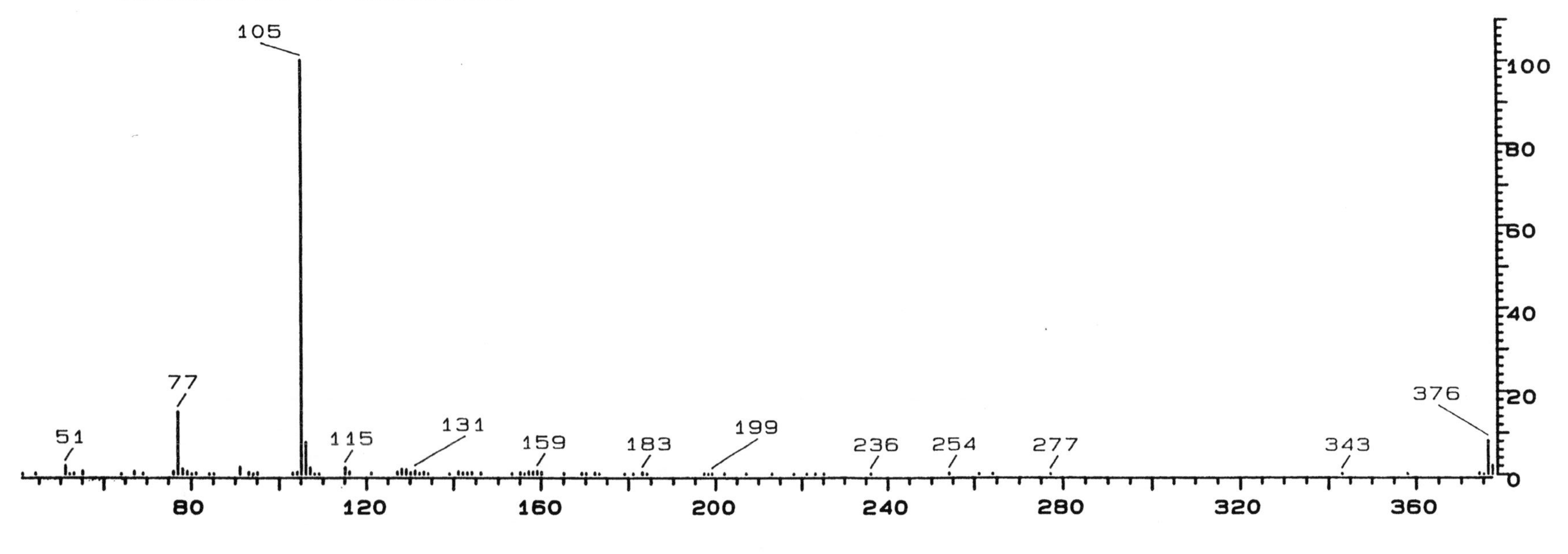

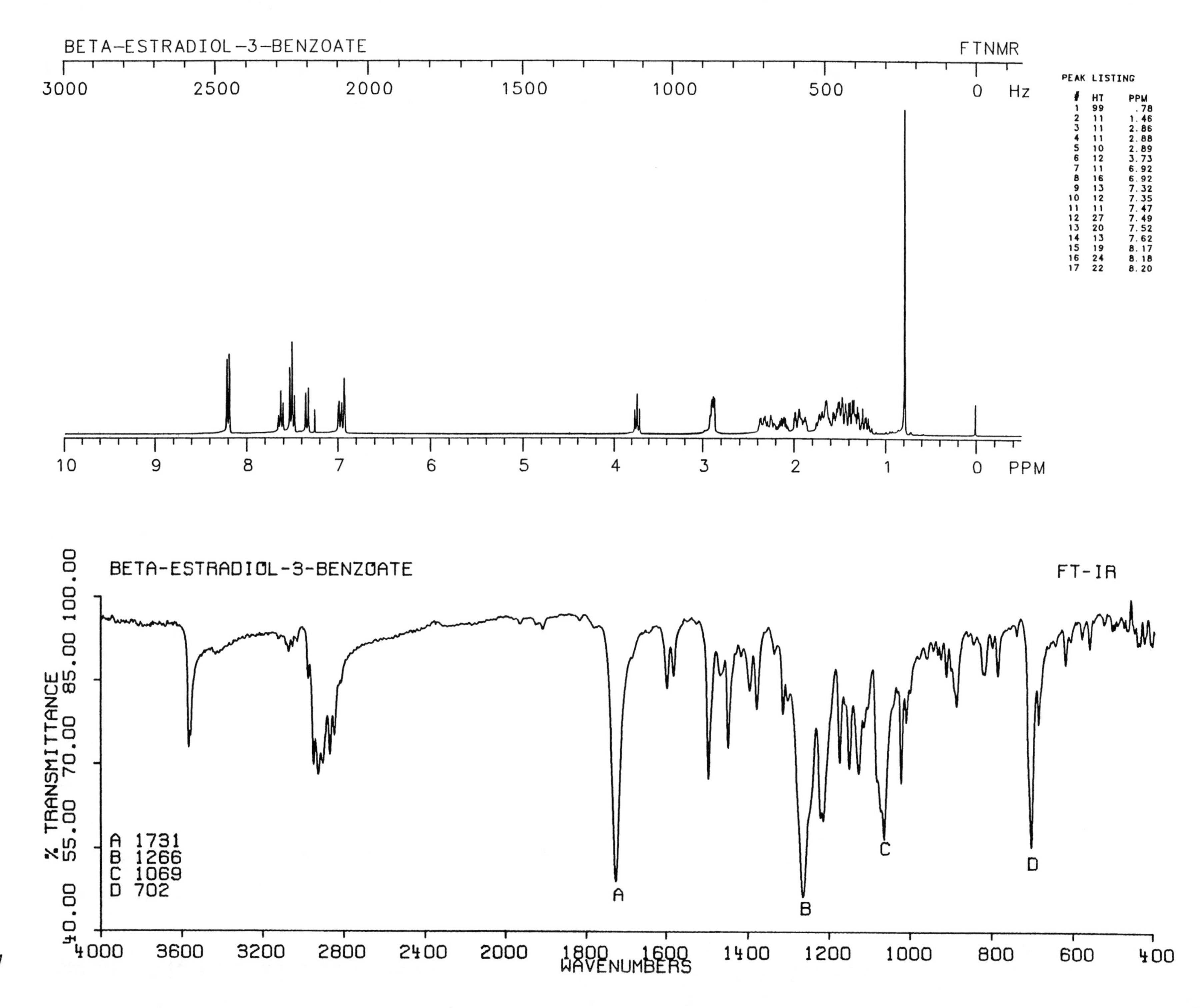
BETA-ESTRADIOL-3-BENZOATE
FTNMR
3000 2500 2000 1500 1000 500 0 Hz
10 9 8 7 6 5 4 3 2 1 0 PPM
PEAK LISTING
# HT PPM
1 99 .78
2 11 1.46
3 11 2.86
4 11 2.88
5 10 2.89
6 12 3.73
7 11 6.92
8 16 6.92
9 13 7.32
10 12 7.35
11 11 7.47
12 27 7.49
13 20 7.52
14 13 7.62
15 19 8.17
16 24 8.18
17 22 8.20
BETA-ESTRADIOL-3-BENZOATE
FT-IR
A 1731
B 1266
C 1069
D 702
% TRANSMITTANCE
40.00 55.00 70.00 85.00 100.00
4000 3600 3200 2800 2400 2000 1800 1600 1400 1200 1000 800 600 400
WAVENUMBERS
A
B
C
D

# ESTRADIOL CYPIONATE

$C_{26}H_{36}O_3$

**Molecular weight:** 396.55 (396.27)
**Synonyms:** Estradiol-17β-cyclopentanepropionate; oestradiol cyclopentyl-propionate; 17β-(3-cyclopentyl)propionyloxyoestra-1,3,5(N)-trien-3-ol
**Trade names:** E-Cypionate, T-E-Cypionate

**Use:** Estrogen
**HPLC:** Si-10; 100B; 4.1
**GC:** 3330; 280°C

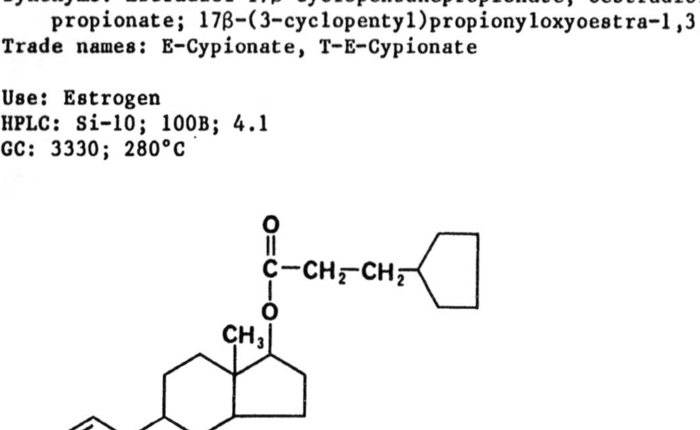


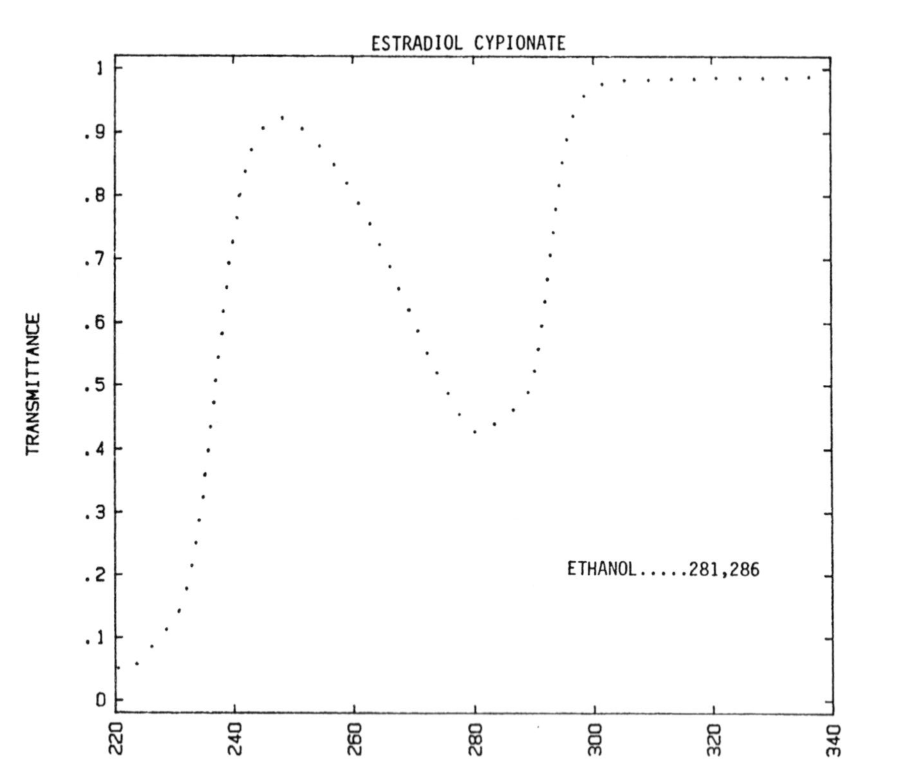


*ESTRADIOL CYPIONATE*

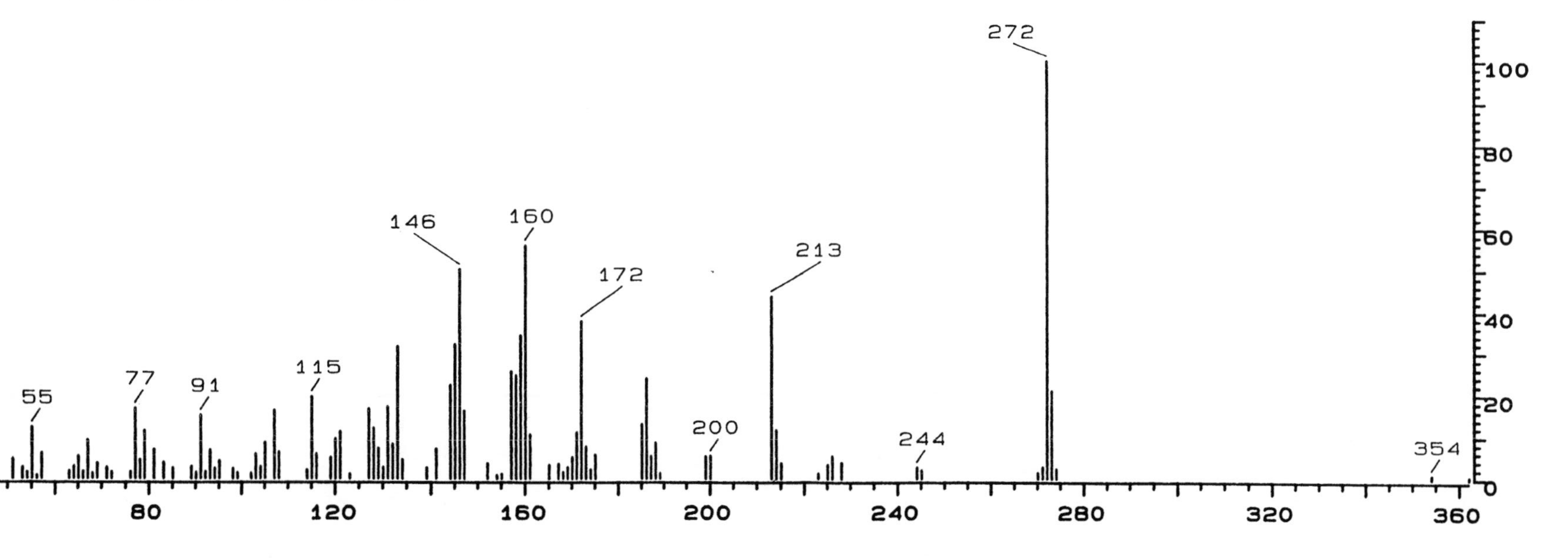

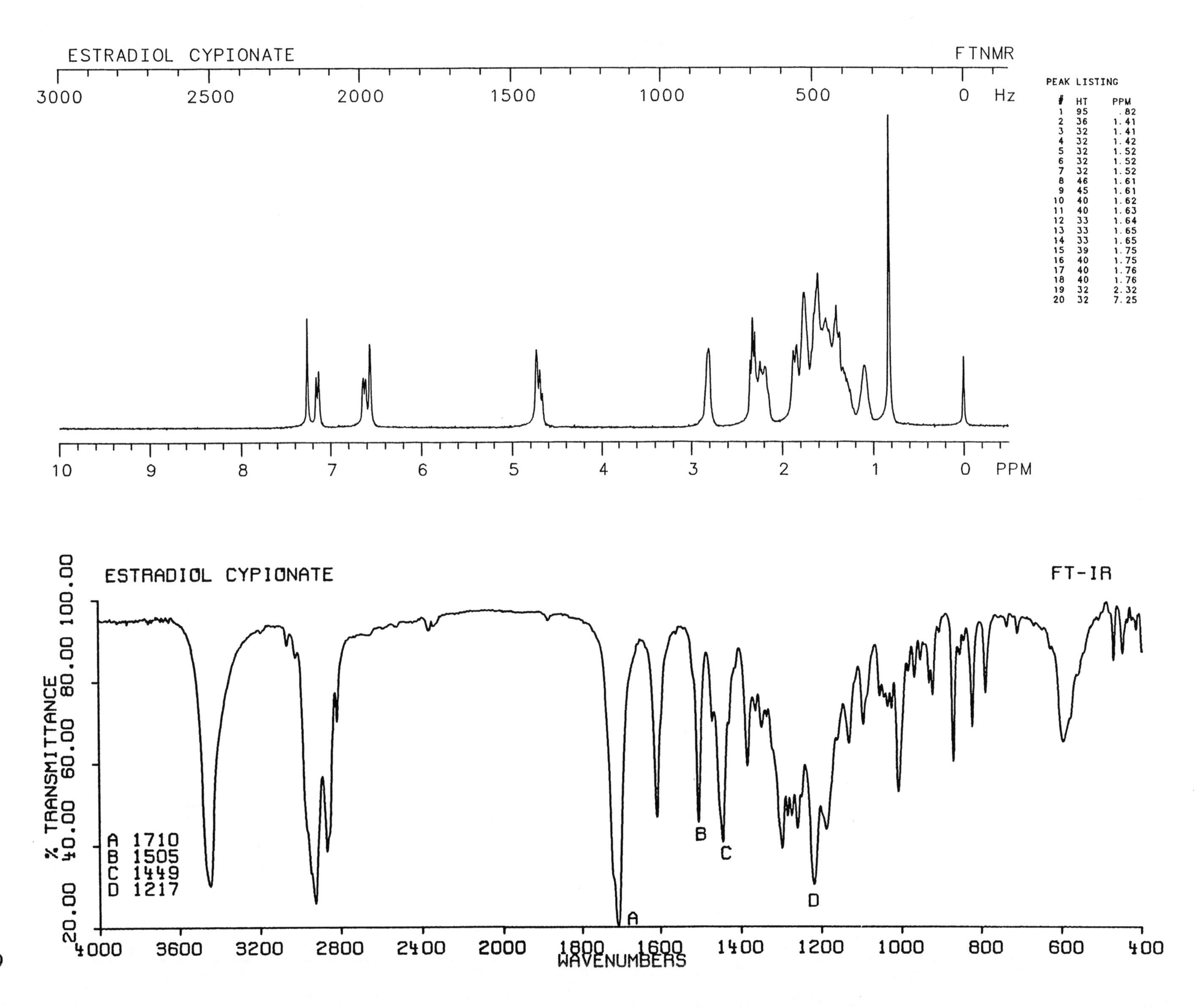
ESTRADIOL CYPIONATE
FTNMR
3000 2500 2000 1500 1000 500 0 Hz
10 9 8 7 6 5 4 3 2 1 0 PPM
PEAK LISTING
# HT PPM
1 95 .82
2 36 1.41
3 32 1.41
4 32 1.42
5 32 1.52
6 32 1.52
7 32 1.52
8 46 1.61
9 45 1.61
10 40 1.62
11 40 1.63
12 33 1.64
13 33 1.65
14 33 1.65
15 39 1.75
16 40 1.75
17 40 1.76
18 40 1.76
19 32 2.32
20 32 7.25
ESTRADIOL CYPIONATE
FT-IR
% TRANSMITTANCE
20.00 40.00 60.00 80.00 100.00
A 1710
B 1505
C 1449
D 1217
A
B
C
D
4000 3600 3200 2800 2400 2000 1800 1600 1400 1200 1000 800 600 400
WAVENUMBERS

# ESTRADIOLDIPROPIONATE

$C_{24}H_{32}O_4$

**Molecular weight:** 384.49 (384.23)
**Synonyms:** Estradiol-3,17β-dipropionate; 3,17β-dipropionyloxyoestra-1,3,5(10)-triene
**Trade names:** Agofollin, Dimenformon Dipropionate, Diovocylin, Ovocyclin, Ovocyclin-P, Progynon-DP
**Use:** Estrogen
**HPLC:**
**GC:** 3101; 280°C

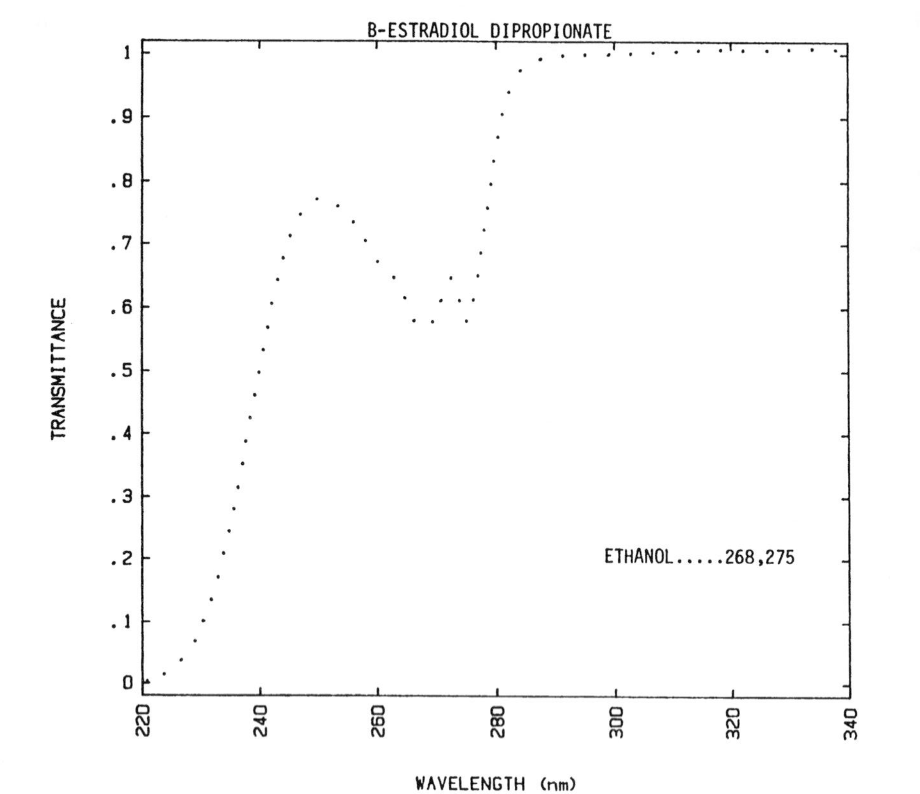

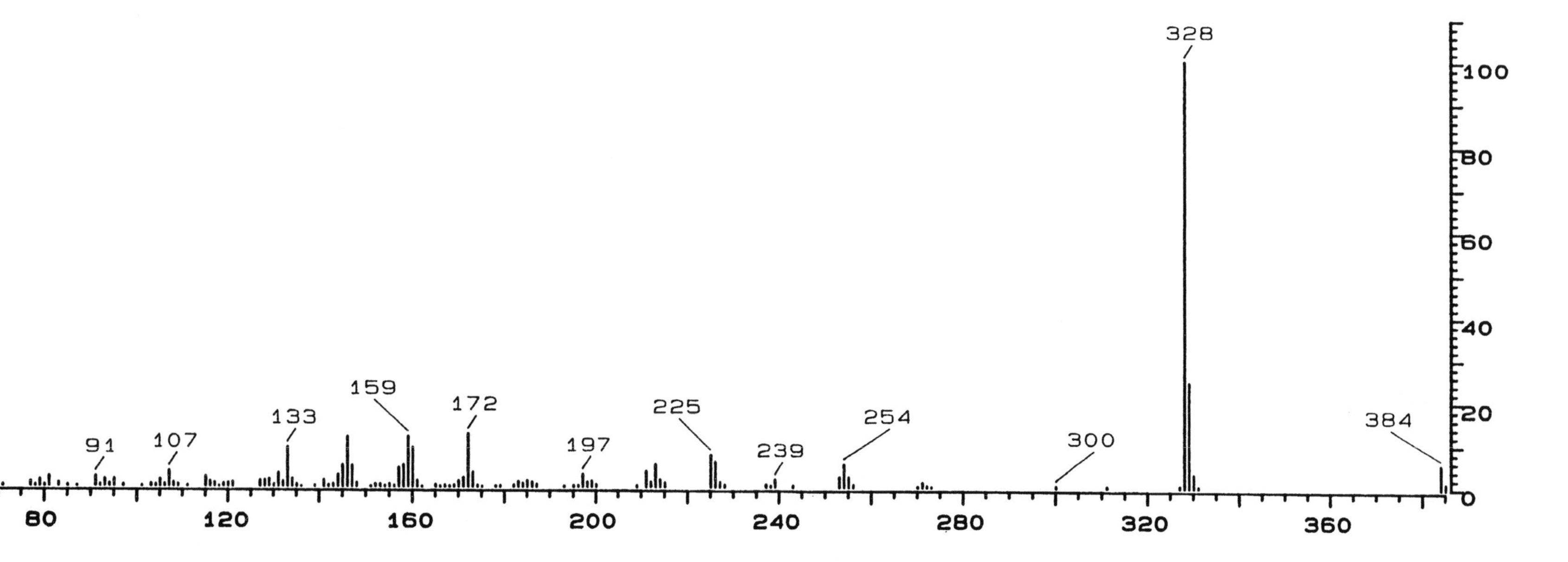

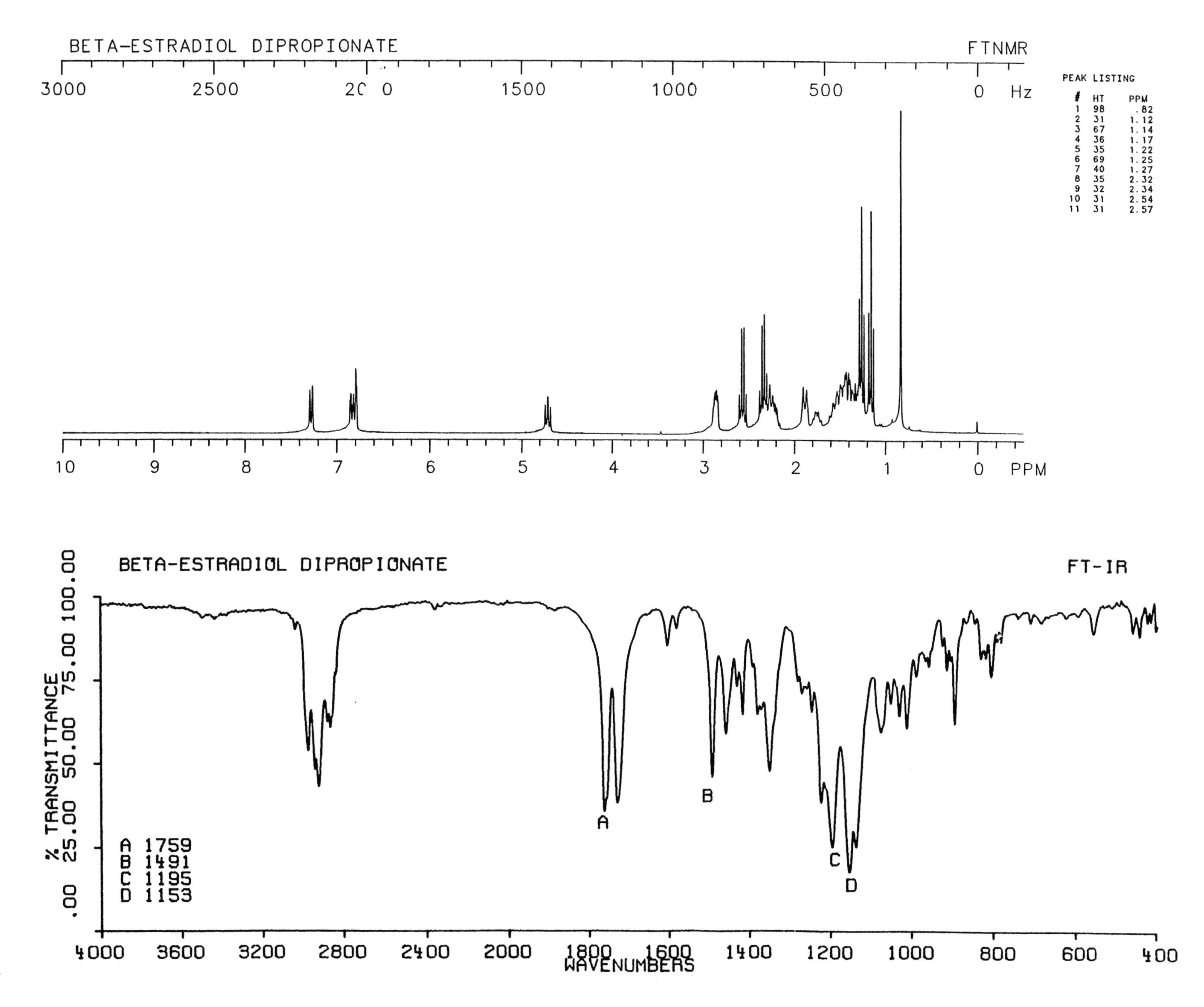
BETA-ESTRADIOL DIPROPIONATE
FTNMR
3000 2500 2000 1500 1000 500 0 Hz
PEAK LISTING
# HT PPM
1 98 .82
2 31 1.12
3 67 1.14
4 36 1.17
5 35 1.22
6 69 1.25
7 40 1.27
8 35 2.32
9 32 2.34
10 31 2.54
11 31 2.57
10 9 8 7 6 5 4 3 2 1 0 PPM
BETA-ESTRADIOL DIPROPIONATE
FT-IR
% TRANSMITTANCE
.00 25.00 50.00 75.00 100.00
A 1759
B 1491
C 1195
D 1153
4000 3600 3200 2800 2400 2000 1800 1600 1400 1200 1000 800 600 400
WAVENUMBERS

# ESTRAMUSTINE

$C_{23}H_{31}Cl_2NO_3$

Molecular weight: 440.41 (439.17)
Synonyms: Estra-1,3,5(10)-triene-3,17-diol-3-[bis(2-chloroethyl)-carbamate]
Trade names: Emcyt

Use: Antineoplastic
HPLC:
GC:

(Cl—CH₂—CH₂)₂N—C(=O)—O— ; CH₃ ; OH

ESTRAMUSTINE PHOSPHATE

TRANSMITTANCE: 0, .1, .2, .3, .4, .5, .6, .7, .8, .9, 1

WAVELENGTH (nm): 220, 240, 260, 280, 300, 320, 340

ETHANOL.....268,275

*ESTRAMUSTINE -- DIP*

63, 70, 115, 128, 168, 183, 213, 272, 439

80, 120, 160, 200, 240, 280, 320, 360, 400, 440

0, 20, 40, 60, 80, 100

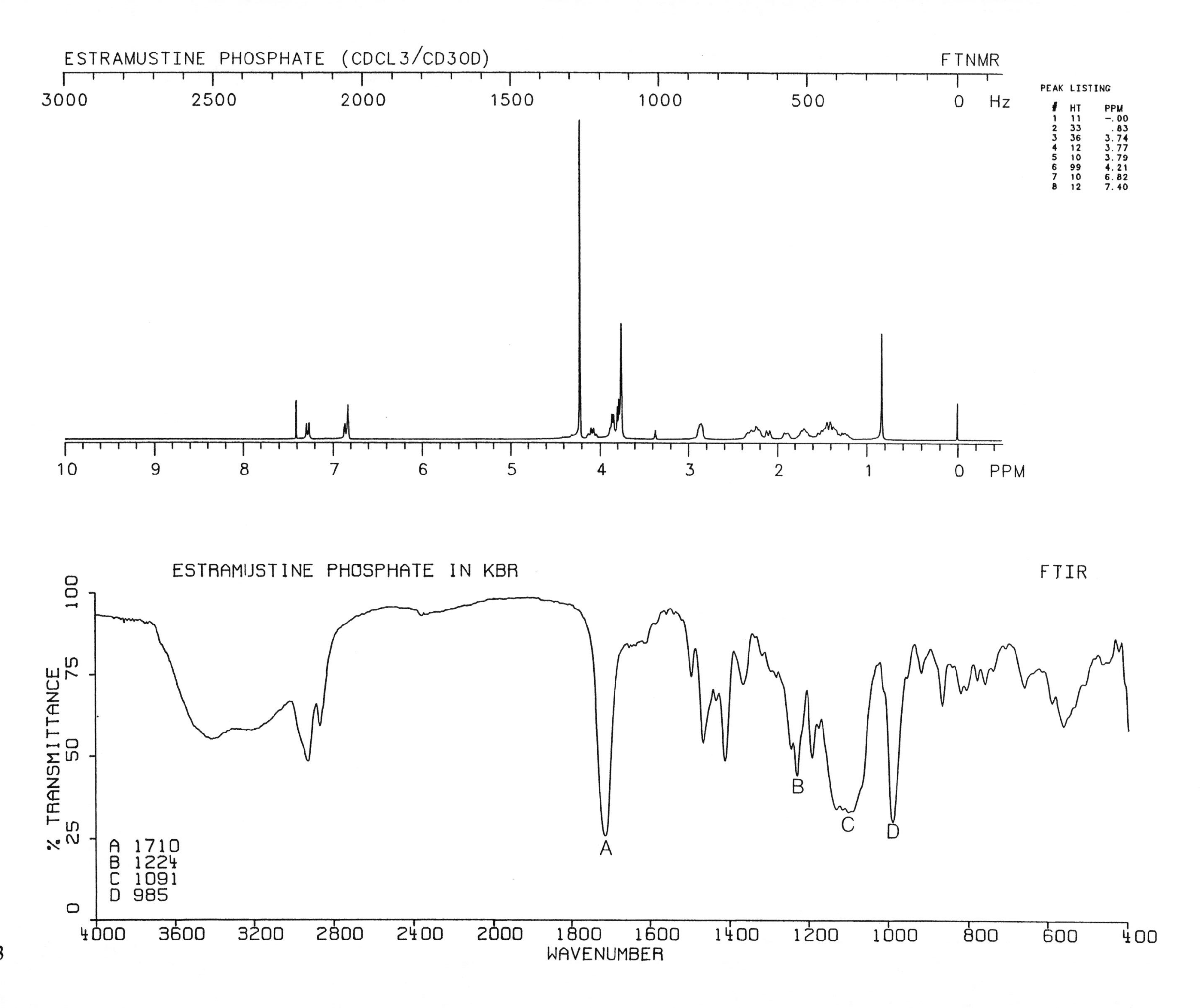

ESTRAMUSTINE PHOSPHATE (CDCL3/CD3OD)
FTNMR
3000 2500 2000 1500 1000 500 0 Hz
PEAK LISTING
# HT PPM
1 11 -.00
2 33 .83
3 36 3.74
4 12 3.77
5 10 3.79
6 99 4.21
7 10 6.82
8 12 7.40
10 9 8 7 6 5 4 3 2 1 0 PPM
ESTRAMUSTINE PHOSPHATE IN KBR
FTIR
100 75 50 25 0
% TRANSMITTANCE
A 1710
B 1224
C 1091
D 985
A B C D
4000 3600 3200 2800 2400 2000 1800 1600 1400 1200 1000 800 600 400
WAVENUMBER

# ESTRIOL

$C_{18}H_{24}O_3$

**Molecular weight:** 288.37 (288.17)
**Synonyms:** Estra-1,3,5(10)-triene-3,16α,17β-triol; oestriol; trihydroxyestrin
**Trade names:** Aacifemine, Destriol, Hormomed, Klimoral, Orgastyptin, Ovesterin, Ovestin, Stiptanon, Theelol, Tridestrin, Triovex
**Use:** Estrogen
**HPLC:** Si-10; 2A:98B; 9.6
**GC:** 2903; 280°C

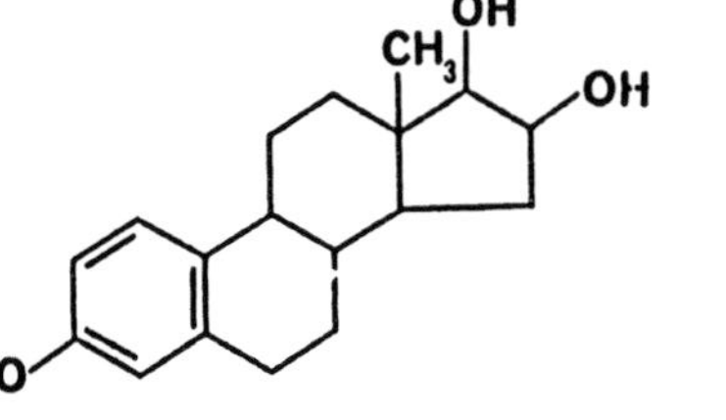


ESTRIOL

TRANSMITTANCE

WAVELENGTH (nm)

277
236,295

*ESTRIOL*

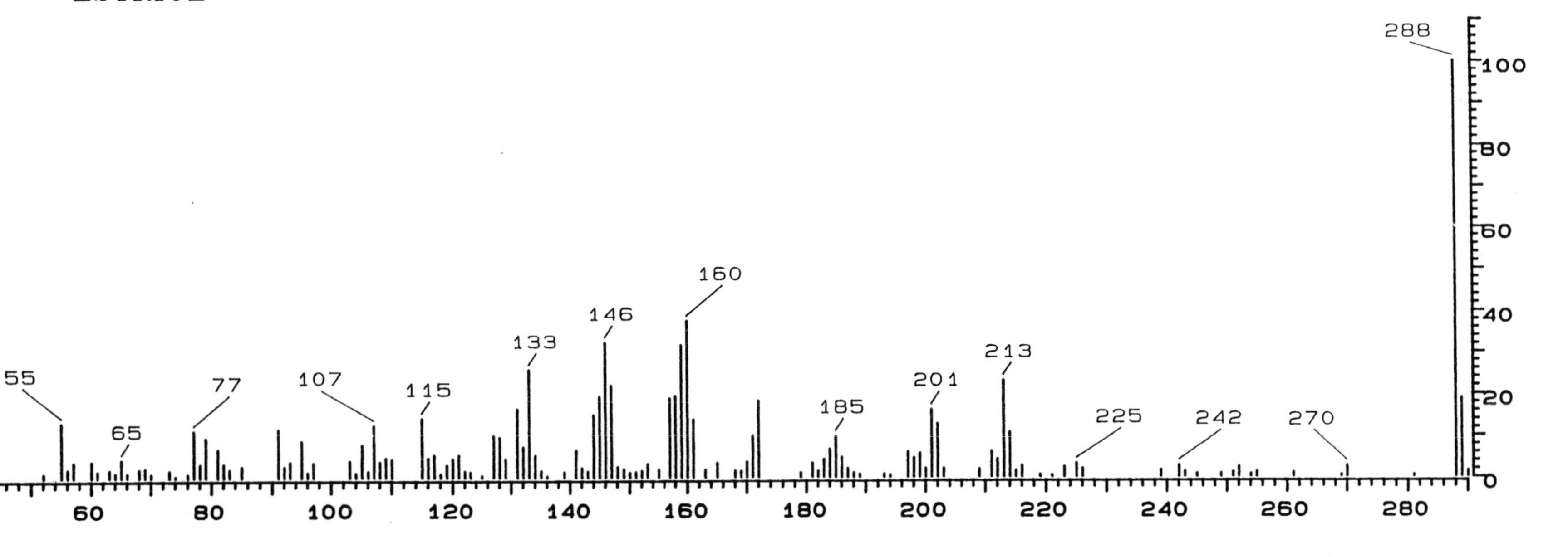

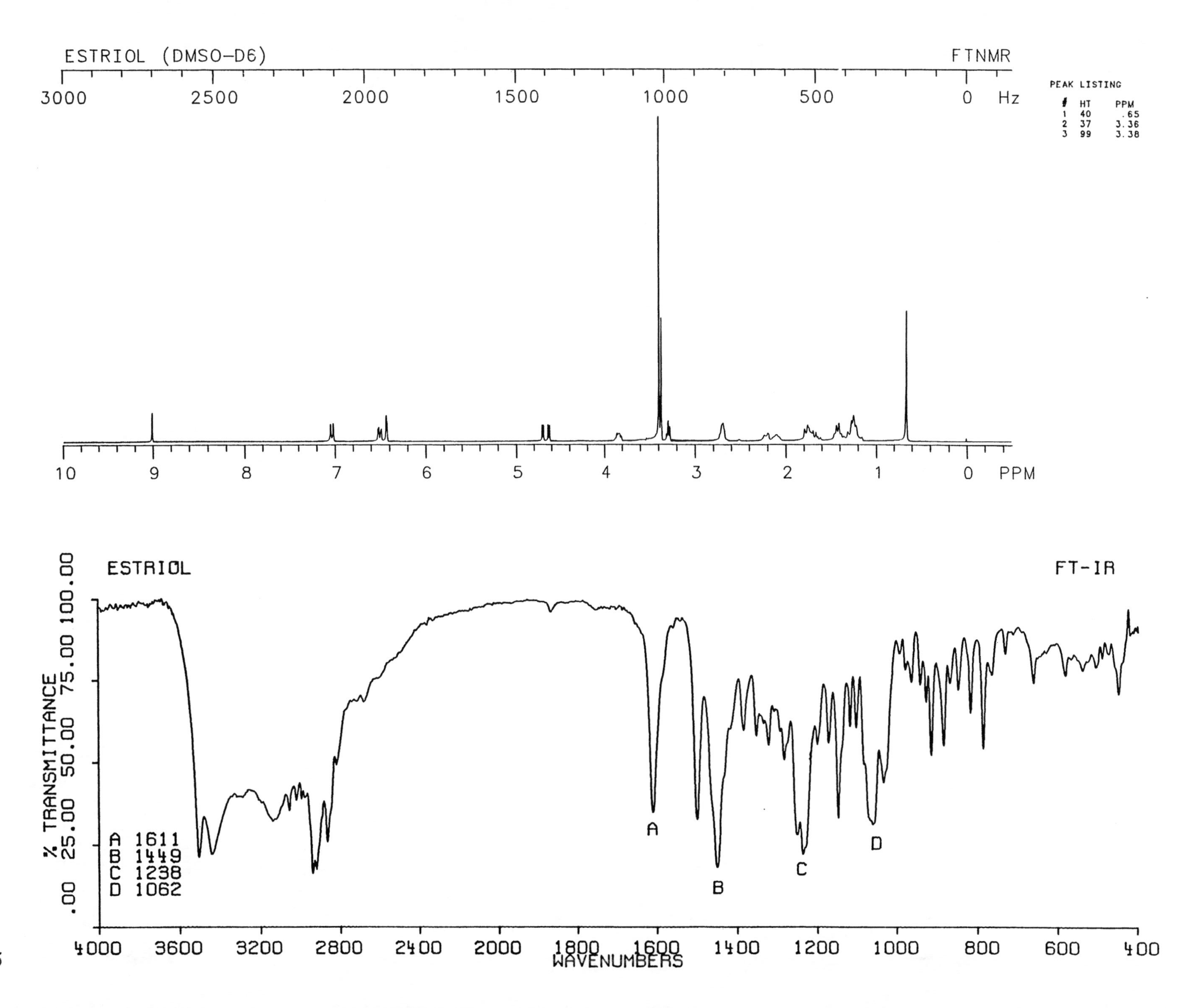
ESTRIOL (DMSO-D6)
FTNMR
3000
2500
2000
1500
1000
500
0 Hz
PEAK LISTING
# HT PPM
1 40 .65
2 37 3.36
3 99 3.38
10
9
8
7
6
5
4
3
2
1
0 PPM
ESTRIOL
FT-IR
% TRANSMITTANCE
.00
25.00
50.00
75.00
100.00
A 1611
B 1449
C 1238
D 1062
A
B
C
D
4000
3600
3200
2800
2400
2000
1800
1600
1400
1200
1000
800
600
400
WAVENUMBERS

# ESTRONE

$C_{18}H_{22}O_2$

Molecular weight: 270.36 (270.16)
Synonyms: 3-Hydroxyestra-1,3,5(10)-trien-17-one; oestrone; folliculin; ketohydroxyoestrin
Trade names: Estrone, Femogen, Hormonin, Kolpon, Natural Estrogenic Substance, Ogen, Oestrilin
Use: Estrogen
HPLC:
GC: 2607; 250°C

ESTRONE

ETHANOL.....281,287

TRANSMITTANCE

WAVELENGTH (nm)

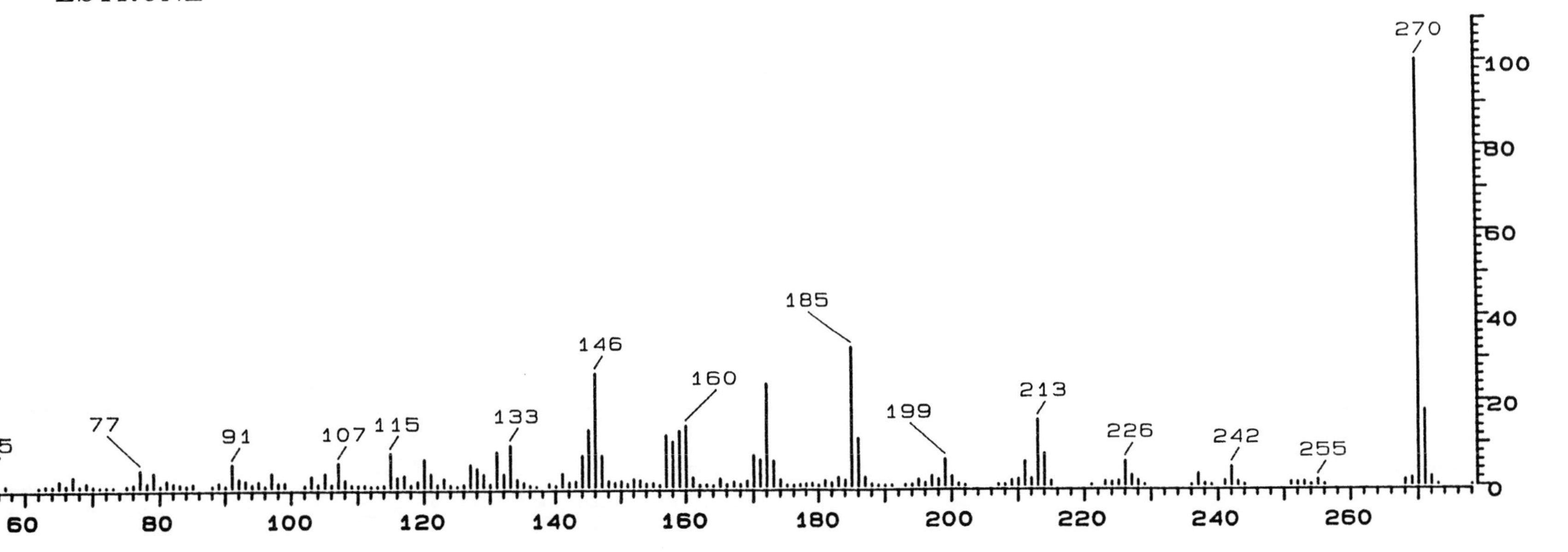

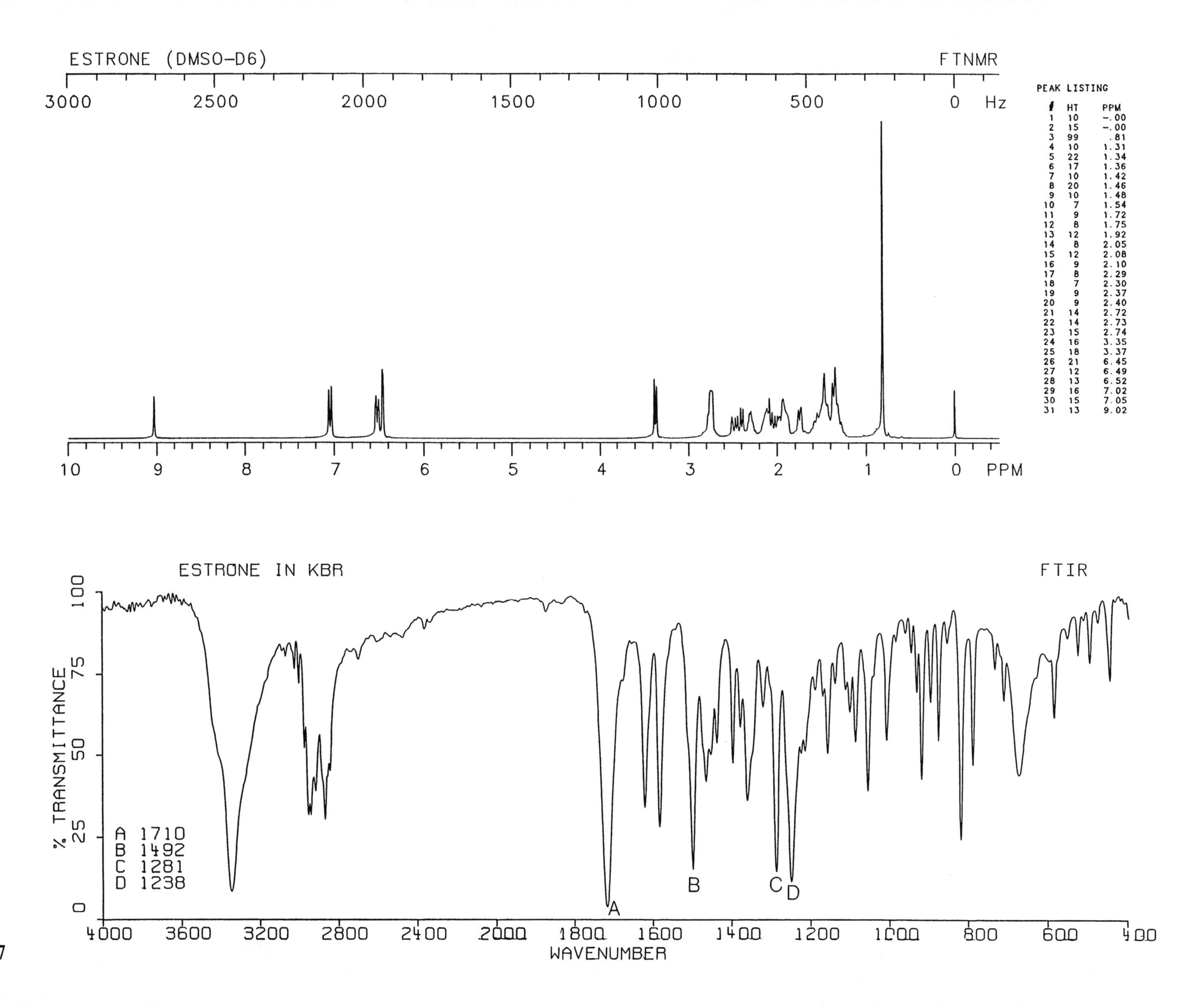
ESTRONE (DMSO-D6)
FTNMR
3000 2500 2000 1500 1000 500 0 Hz
10 9 8 7 6 5 4 3 2 1 0 PPM
PEAK LISTING
# HT PPM
1 10 -.00
2 15 -.00
3 99 .81
4 10 1.31
5 22 1.34
6 17 1.36
7 10 1.42
8 20 1.46
9 10 1.48
10 7 1.54
11 9 1.72
12 8 1.75
13 12 1.92
14 8 2.05
15 12 2.08
16 9 2.10
17 8 2.29
18 7 2.30
19 9 2.37
20 9 2.40
21 14 2.72
22 14 2.73
23 15 2.74
24 16 3.35
25 18 3.37
26 21 6.45
27 12 6.49
28 13 6.52
29 16 7.02
30 15 7.05
31 13 9.02
ESTRONE IN KBR
FTIR
% TRANSMITTANCE
100 75 50 25 0
A 1710
B 1492
C 1281
D 1238
A B C D
4000 3600 3200 2800 2400 2000 1800 1600 1400 1200 1000 800 600 400
WAVENUMBER

# ETHACRYNIC ACID

$C_{13}H_{12}Cl_2O_4$

**Molecular weight:** 303.14 (302.01)
**Synonyms:** [2,3-Dichloro-4-(2-methylene-1-oxobutyl)phenoxy]-acetic acid
**Trade names:** Edecrin

**Use:** Diuretic
**HPLC:** Si-10; 10A:90B; 7.0
**GC:** 2218; 250°C

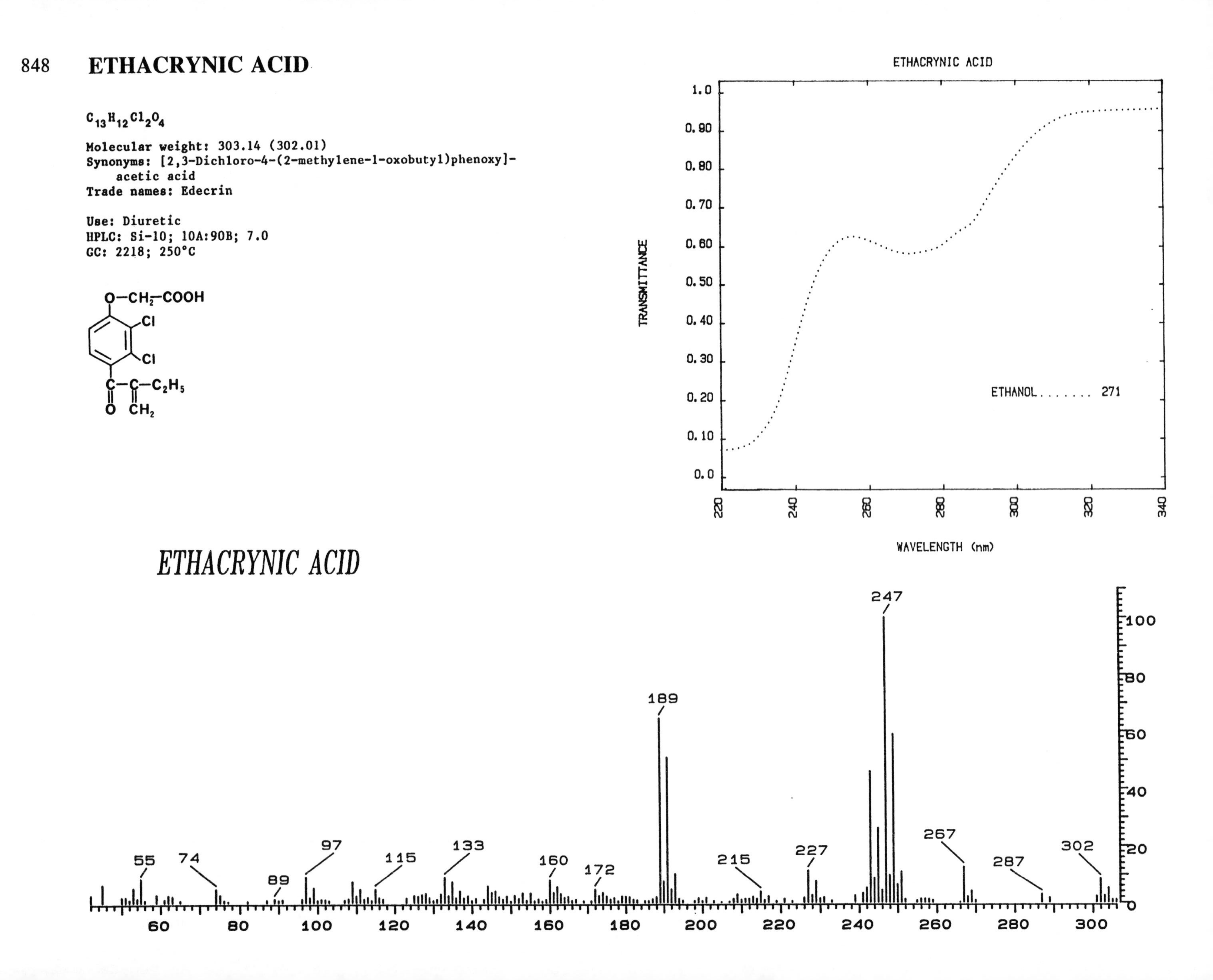

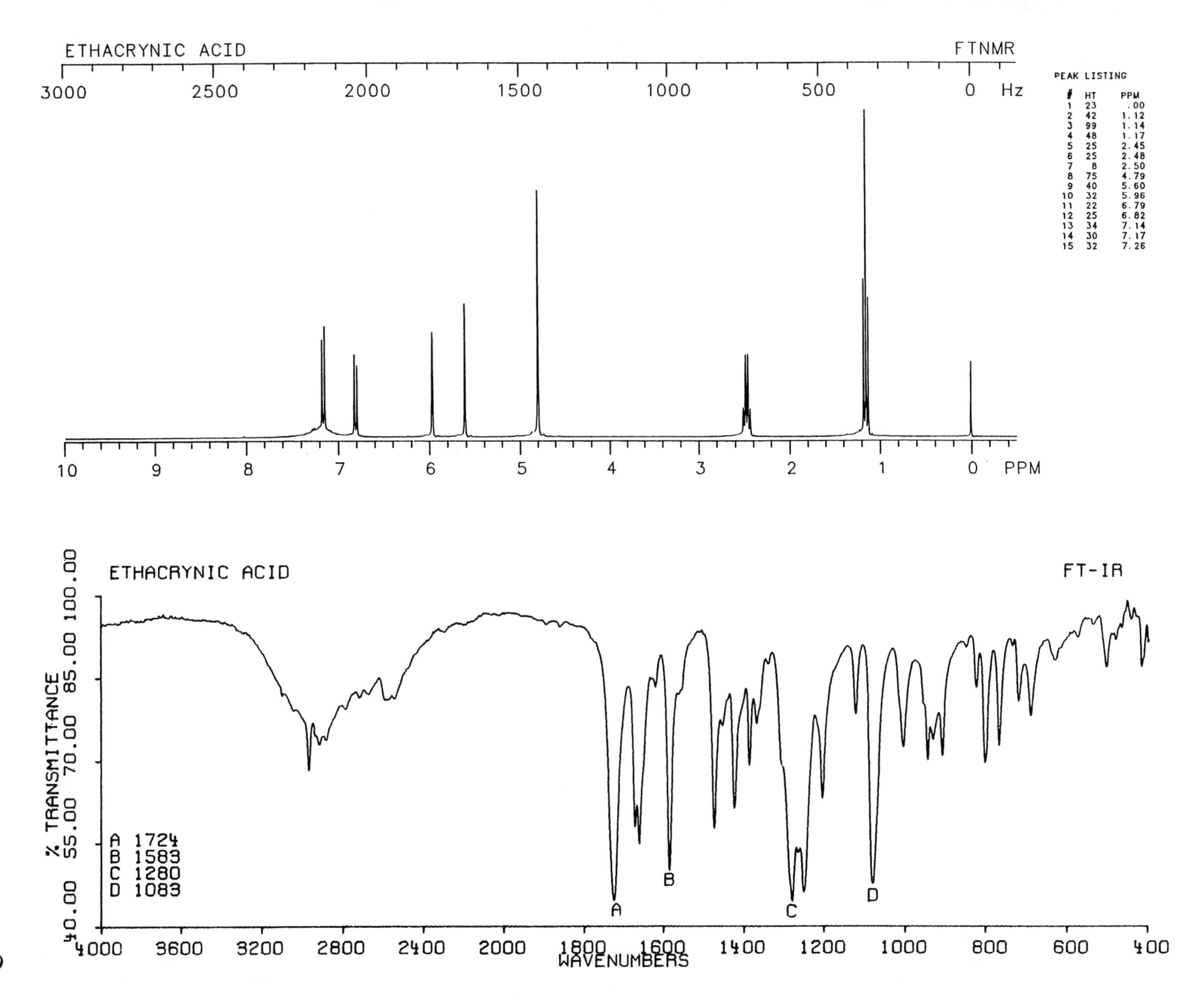
ETHACRYNIC ACID
FTNMR
3000 2500 2000 1500 1000 500 0 Hz
PEAK LISTING
# HT PPM
1 23 .00
2 42 1.12
3 99 1.14
4 48 1.17
5 25 2.45
6 25 2.48
7 8 2.50
8 75 4.79
9 40 5.60
10 32 5.96
11 22 6.79
12 25 6.82
13 34 7.14
14 30 7.17
15 32 7.26
10 9 8 7 6 5 4 3 2 1 0 PPM
ETHACRYNIC ACID
FT-IR
% TRANSMITTANCE
40.00 55.00 70.00 85.00 100.00
A 1724
B 1583
C 1280
D 1083
A B C D
4000 3600 3200 2800 2400 2000 1800 1600 1400 1200 1000 800 600 400
WAVENUMBERS

# ETHALLOBARBITAL

$C_{12}H_{16}N_2O_3$

**Molecular weight:** 236.25 (236.12)
**Synonyms:** 5-(2-Ethylpropenyl)-5-(2-propenyl)-2,4,6(1H,3H,5H)-pyrimidinetrione
**Trade names:**

**Use:** Barbiturate
**HPLC:** Si-10; 1A:99B; 3.4
**GC:** 1588; 200°C

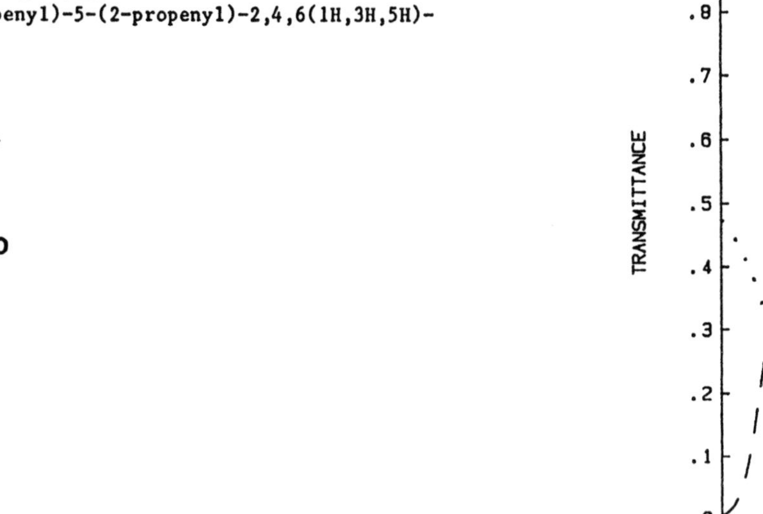

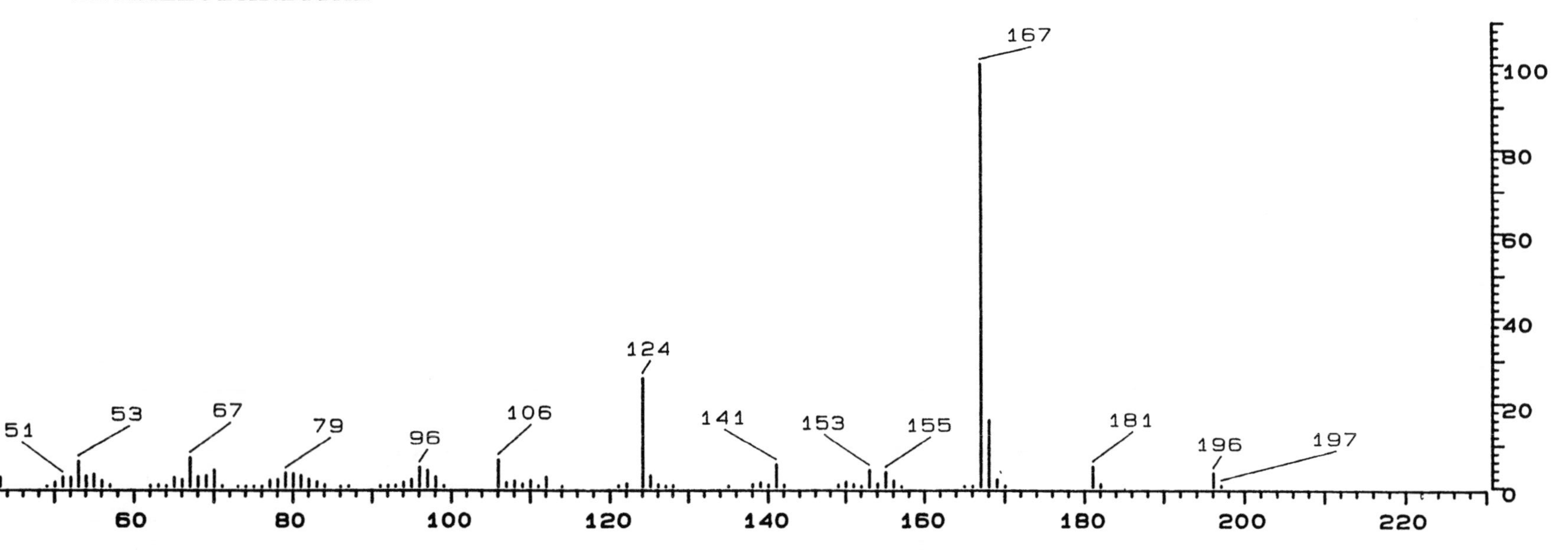

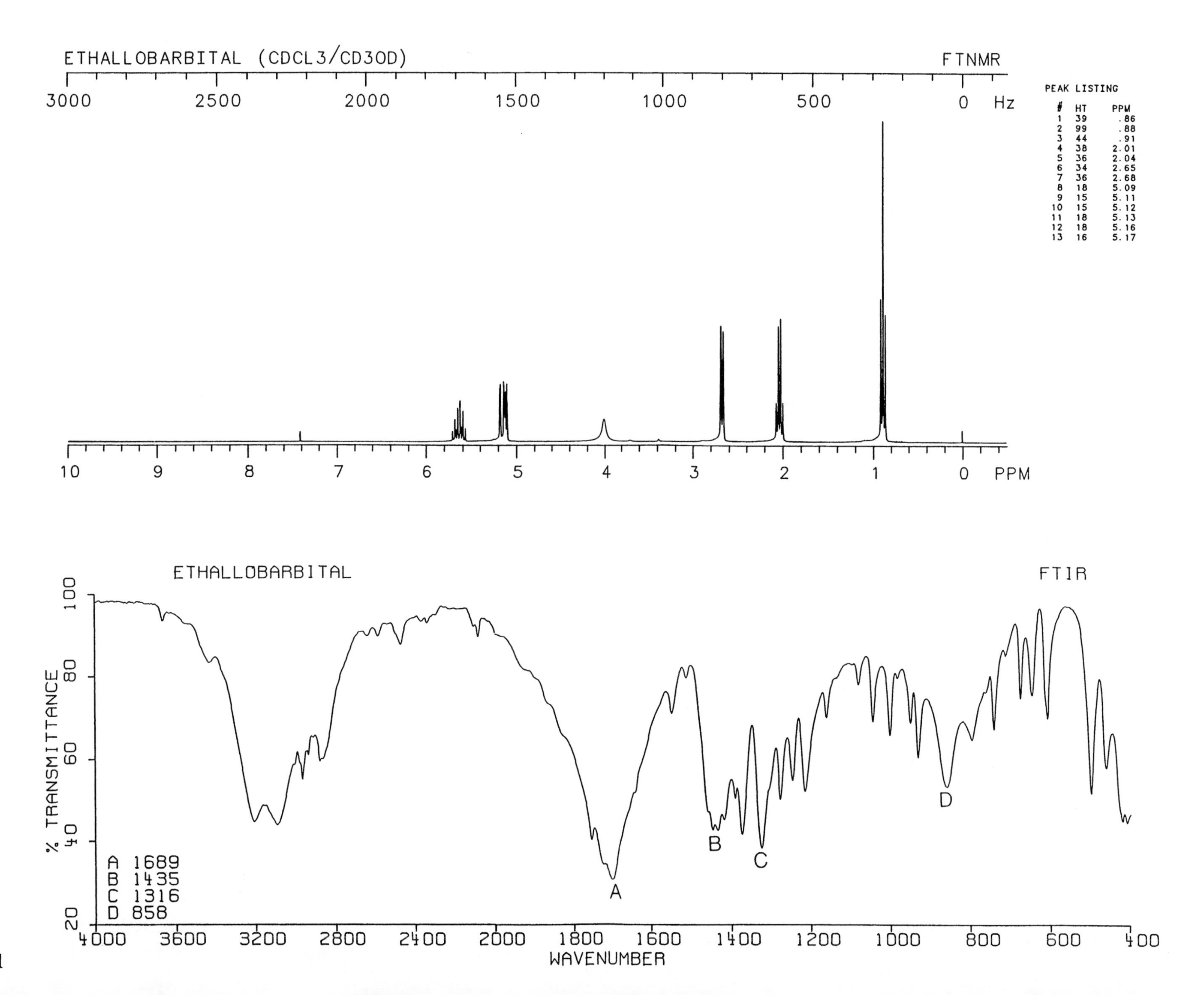
ETHALLOBARBITAL (CDCL3/CD3OD)
FTNMR
3000 2500 2000 1500 1000 500 0 Hz
10 9 8 7 6 5 4 3 2 1 0 PPM
PEAK LISTING
# HT PPM
1 39 .86
2 99 .88
3 44 .91
4 38 2.01
5 36 2.04
6 34 2.65
7 36 2.68
8 18 5.09
9 15 5.11
10 15 5.12
11 18 5.13
12 18 5.16
13 16 5.17
ETHALLOBARBITAL
FTIR
% TRANSMITTANCE
100 80 60 40 20
A 1689
B 1435
C 1316
D 858
A B C D
4000 3600 3200 2800 2400 2000 1800 1600 1400 1200 1000 800 600 400
WAVENUMBER

# ETHAMBUTOL

$C_{10}H_{24}N_2O_2$

**Molecular weight:** 204.31 (204.18)
**Synonyms:** (R)-2,2'-(1,2-Ethanediyldimino)bis-1-butanol

**Trade names:** Myambutol

**Use:** Antibacterial
**HPLC:**
**GC:**

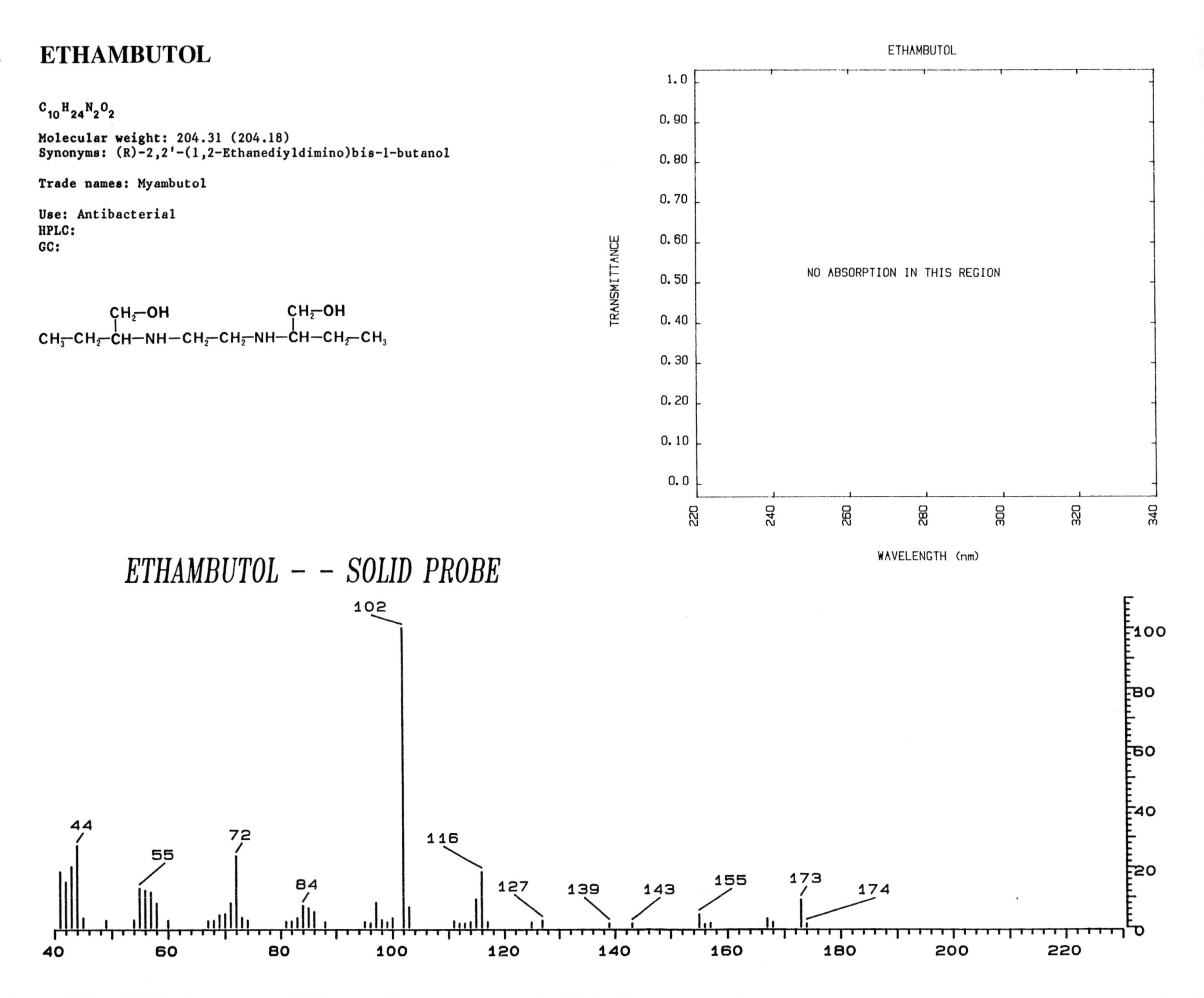

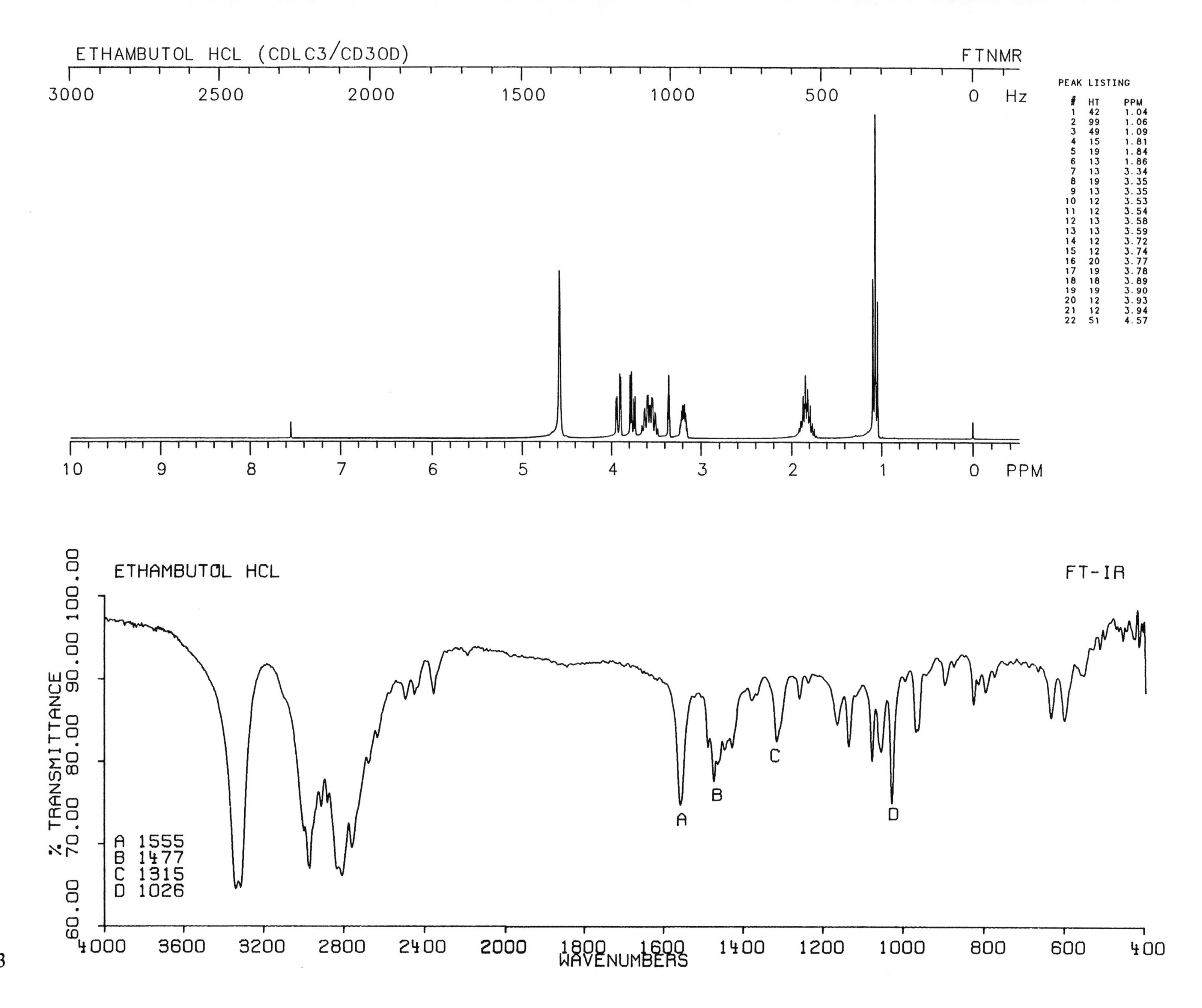
ETHAMBUTOL HCL (CDLC3/CD3OD)
FTNMR
Hz
PPM
PEAK LISTING
# HT PPM
1 42 1.04
2 99 1.06
3 49 1.09
4 15 1.81
5 19 1.84
6 13 1.86
7 13 3.34
8 19 3.35
9 13 3.35
10 12 3.53
11 12 3.54
12 13 3.58
13 13 3.59
14 12 3.72
15 12 3.74
16 20 3.77
17 19 3.78
18 18 3.89
19 19 3.90
20 12 3.93
21 12 3.94
22 51 4.57
ETHAMBUTOL HCL
FT-IR
% TRANSMITTANCE
WAVENUMBERS
A 1555
B 1477
C 1315
D 1026

# ETHAVERINE

$C_{24}H_{29}NO_4$

Molecular weight: 395.50 (395.21)
Synonyms: 1-[(3,4-Diethoxyphenyl)methyl]-6,7-diethoxyisoquinoline; ethylpapaverine
Trade names: Circubid, Ethaquin, Eta-Lent, Ethatab, Isovex-100, Rothan-500
Use: Antispasmodic
HPLC:
GC: 2981; 280°C

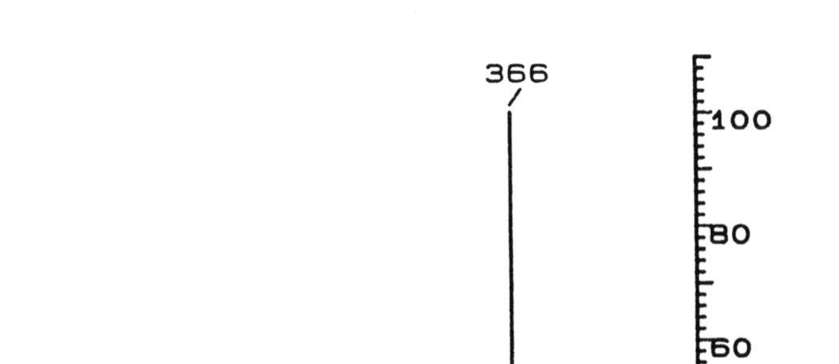


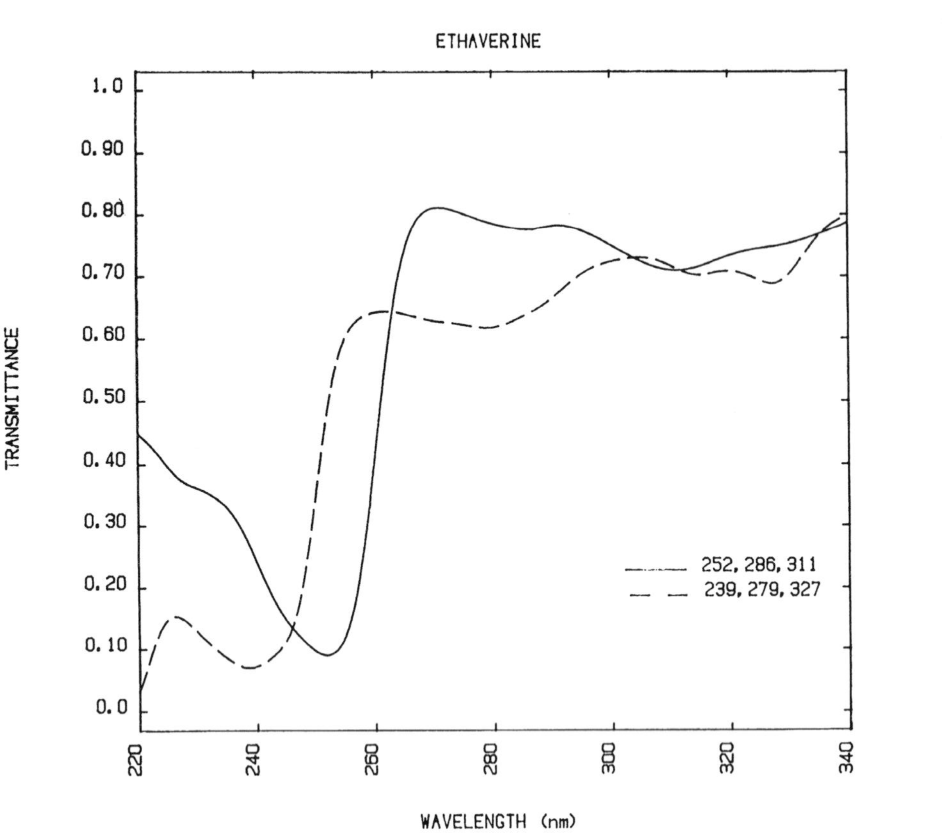


*ETHAVERINE*

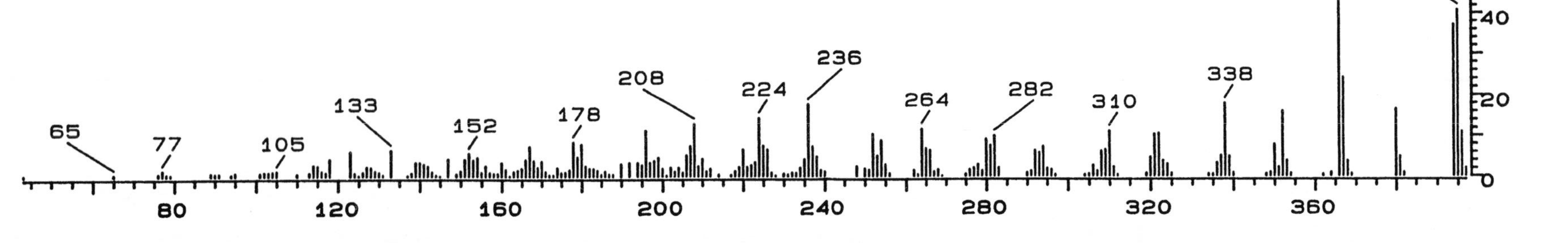

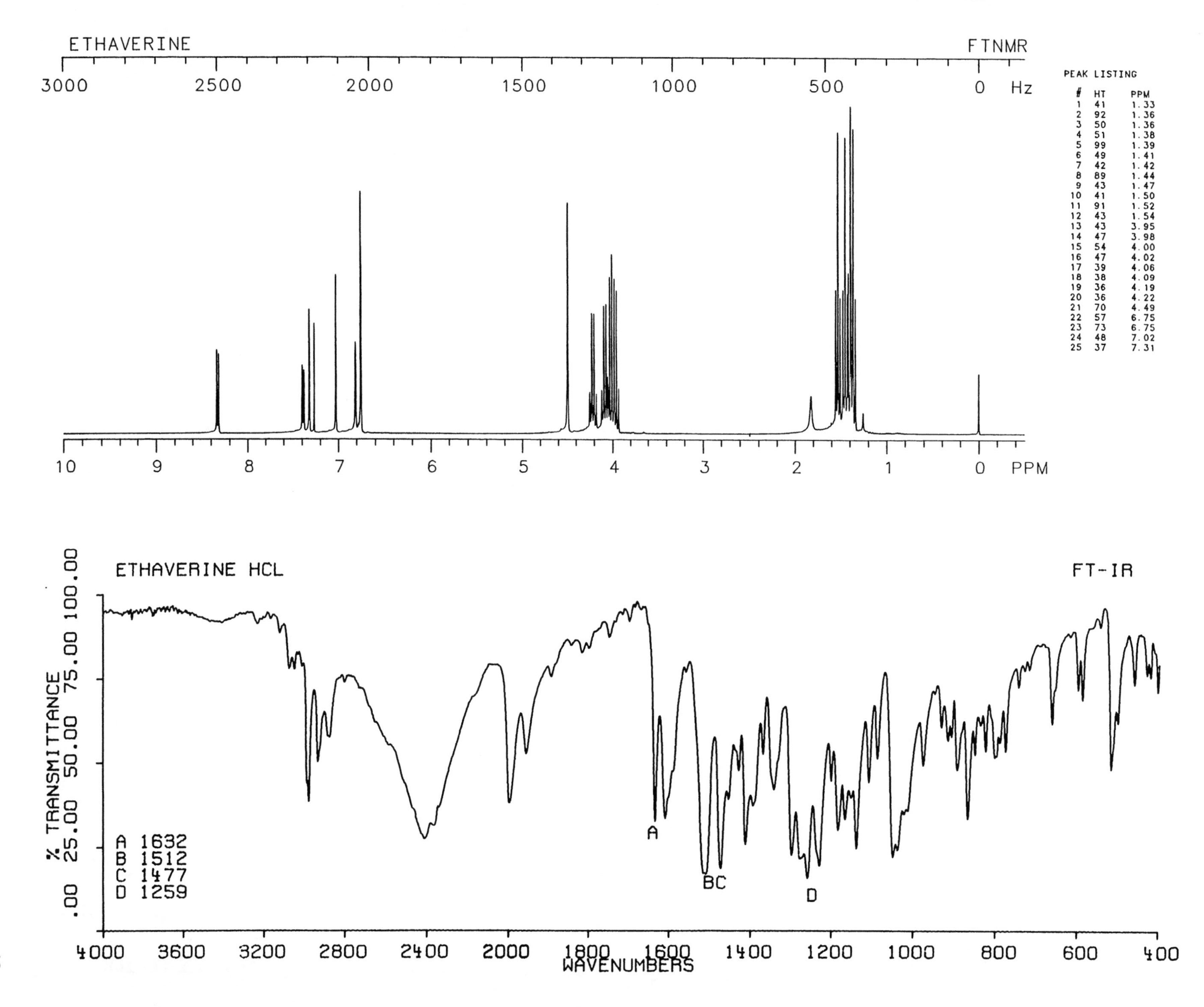

ETHAVERINE
FTNMR
3000 2500 2000 1500 1000 500 0 Hz
10 9 8 7 6 5 4 3 2 1 0 PPM
PEAK LISTING
# HT PPM
1 41 1.33
2 92 1.36
3 50 1.36
4 51 1.38
5 99 1.39
6 49 1.41
7 42 1.42
8 89 1.44
9 43 1.47
10 41 1.50
11 91 1.52
12 43 1.54
13 43 3.95
14 47 3.98
15 54 4.00
16 47 4.02
17 39 4.06
18 38 4.09
19 36 4.19
20 36 4.22
21 70 4.49
22 57 6.75
23 73 6.75
24 48 7.02
25 37 7.31
ETHAVERINE HCL
FT-IR
% TRANSMITTANCE
.00 25.00 50.00 75.00 100.00
A 1632
B 1512
C 1477
D 1259
A
BC
D
4000 3600 3200 2800 2400 2000 1800 1600 1400 1200 1000 800 600 400
WAVENUMBERS

# ETHCHLORVYNOL

$C_7H_9ClO$

**Molecular weight:** 144.60 (144.03)
**Synonyms:** 1-Chloro-3-ethyl-1-penten-4-yl-3-ol; ethyl-β-chlorovinyl ethynyl carbinol
**Trade names:** Placidyl

**Use:** Sedative, hypnotic
**HPLC:** Si-10; 1A:99B; 6.0
**GC:** 982; 80°C

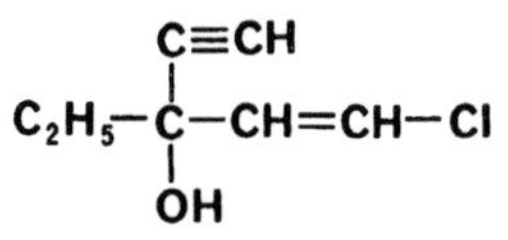

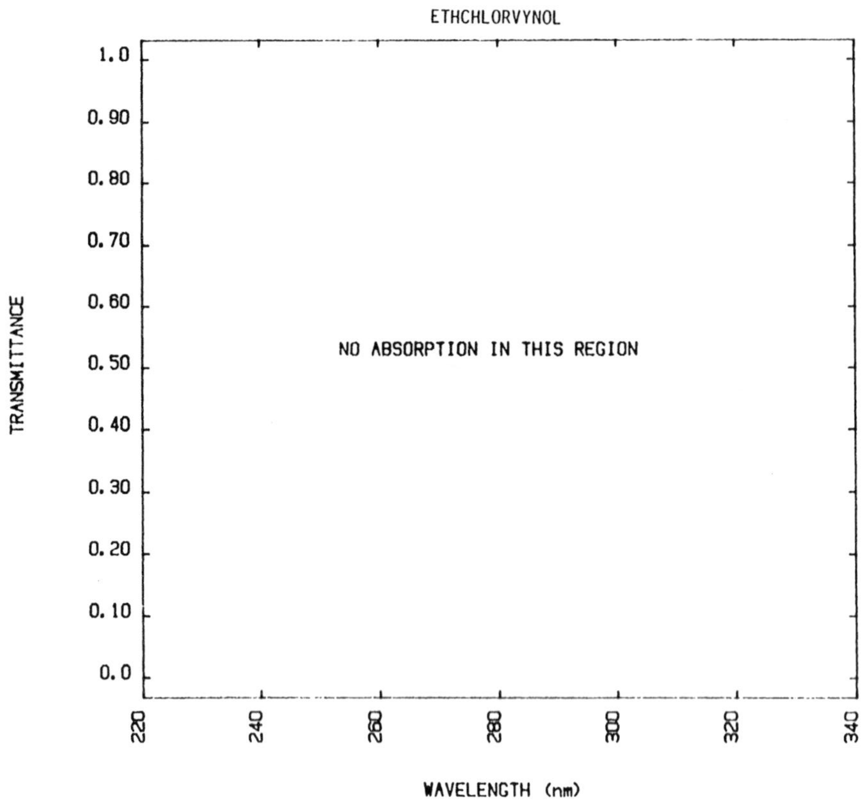

$$C_2H_5-\underset{\substack{|\\ OH}}{\overset{\substack{C\equiv CH\\ |}}{C}}-CH=CH-Cl$$

ETHCHLORVYNOL

TRANSMITTANCE 1.0 0.90 0.80 0.70 0.60 0.50 0.40 0.30 0.20 0.10 0.0

NO ABSORPTION IN THIS REGION

220 240 260 280 300 320 340

WAVELENGTH (nm)

*ETHCHLORVYNOL*

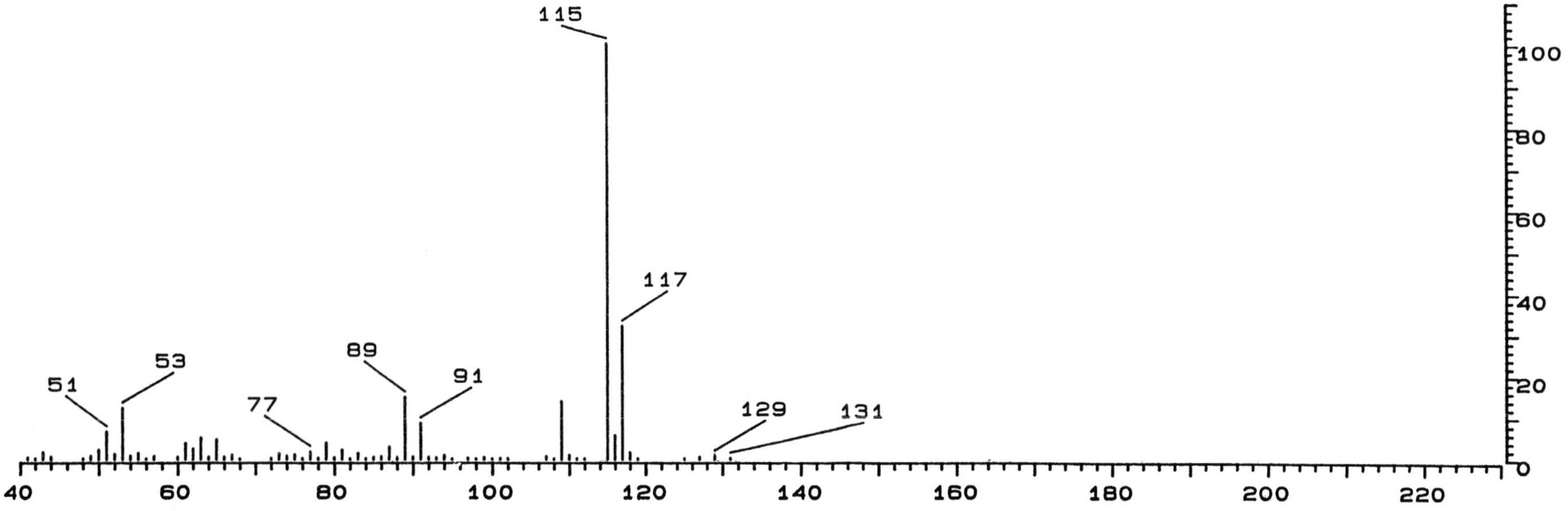

115 117 51 53 77 89 91 129 131

40 60 80 100 120 140 160 180 200 220

0 20 40 60 80 100

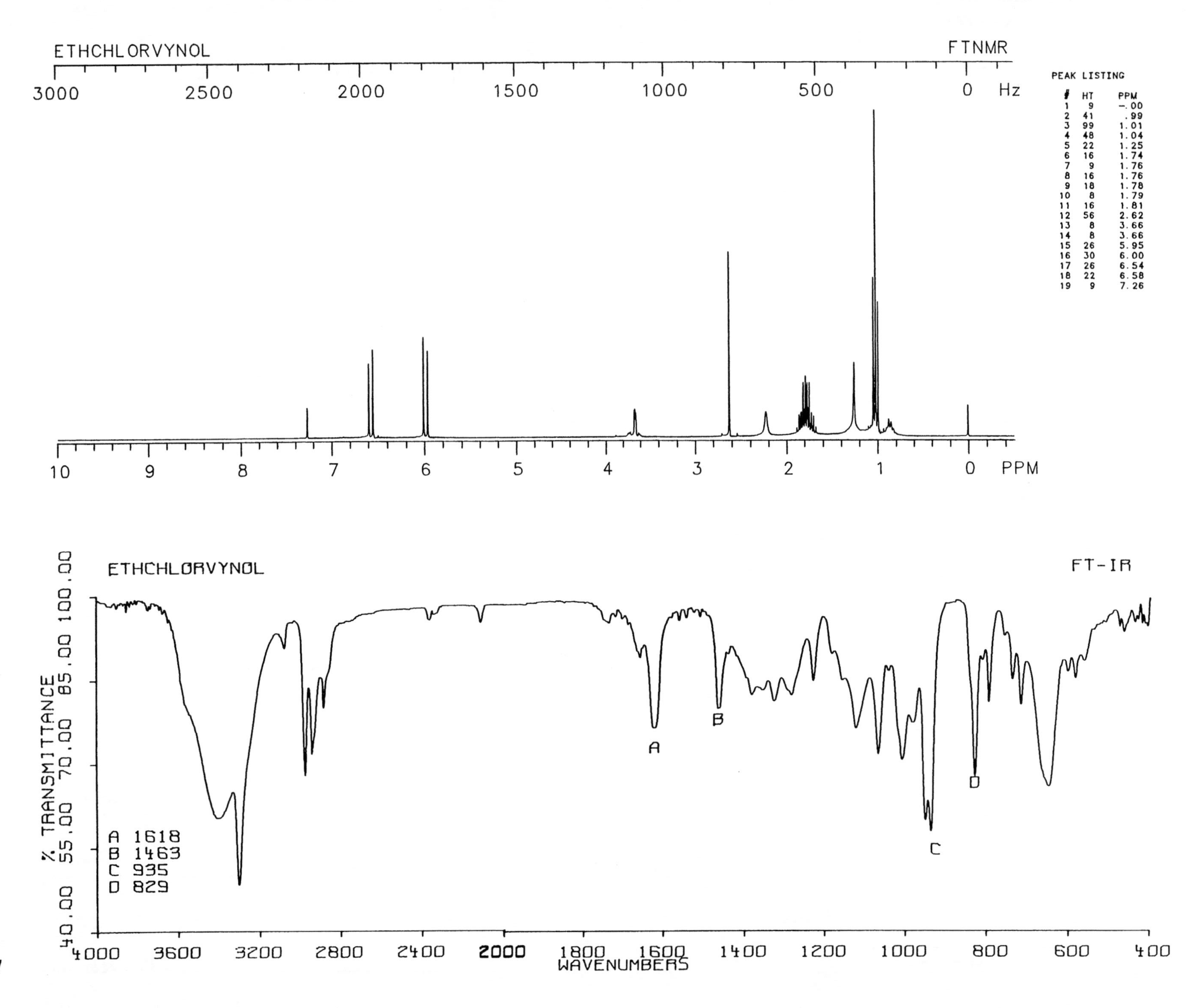
ETHCHLORVYNOL
FTNMR
3000 2500 2000 1500 1000 500 0 Hz
PEAK LISTING
# HT PPM
1 9 -.00
2 41 .99
3 99 1.01
4 48 1.04
5 22 1.25
6 16 1.74
7 9 1.76
8 16 1.76
9 18 1.78
10 8 1.79
11 16 1.81
12 56 2.62
13 8 3.66
14 8 3.66
15 26 5.95
16 30 6.00
17 26 6.54
18 22 6.58
19 9 7.26
10 9 8 7 6 5 4 3 2 1 0 PPM
ETHCHLORVYNOL
FT-IR
% TRANSMITTANCE
40.00 55.00 70.00 85.00 100.00
A 1618
B 1463
C 935
D 829
A
B
C
D
4000 3600 3200 2800 2400 2000 1800 1600 1400 1200 1000 800 600 400
WAVENUMBERS

# ETHINAMATE

$C_9H_{13}NO_2$

Molecular weight: 167.21 (167.10)
Synonyms: 1-Ethynylcyclohexanol carbamate

Trade names: Valmid

Use: Sedative
HPLC: Si-10; 1A:99B; 5.0
GC: 1372; 140°C

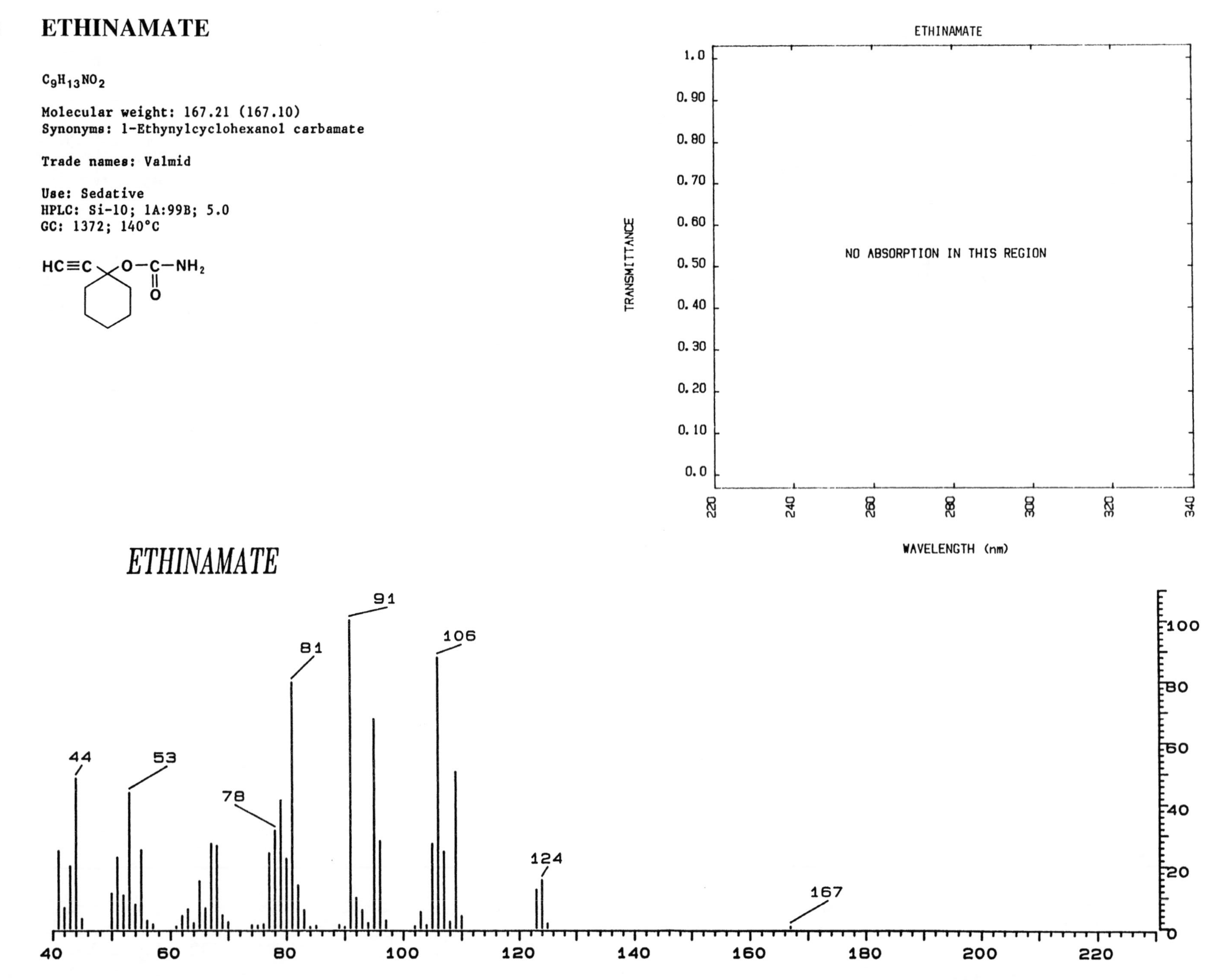

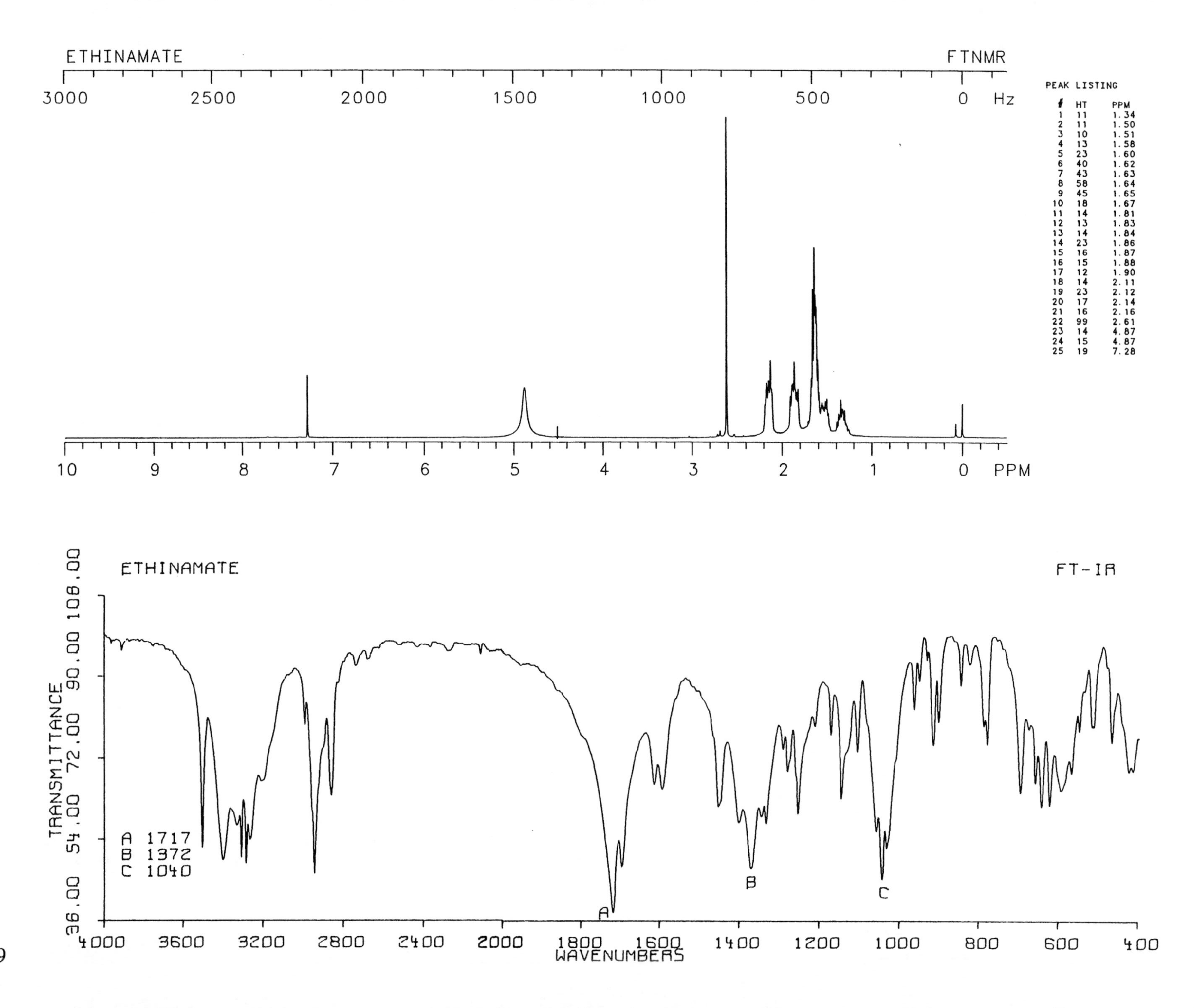
ETHINAMATE
FTNMR
3000
2500
2000
1500
1000
500
0
Hz
PEAK LISTING
# HT PPM
1 11 1.34
2 11 1.50
3 10 1.51
4 13 1.58
5 23 1.60
6 40 1.62
7 43 1.63
8 58 1.64
9 45 1.65
10 18 1.67
11 14 1.81
12 13 1.83
13 14 1.84
14 23 1.86
15 16 1.87
16 15 1.88
17 12 1.90
18 14 2.11
19 23 2.12
20 17 2.14
21 16 2.16
22 99 2.61
23 14 4.87
24 15 4.87
25 19 7.28
10
9
8
7
6
5
4
3
2
1
0
PPM
ETHINAMATE
FT-IR
TRANSMITTANCE
36.00
54.00
72.00
90.00
108.00
A 1717
B 1372
C 1040
A
B
C
4000
3600
3200
2800
2400
2000
1800
1600
1400
1200
1000
800
600
400
WAVENUMBERS

# ETHINAZONE

$C_{17}H_{16}N_2O$

Molecular weight: 264.33 (264.13)
Synonyms: 3-(o-Ethylphenyl)-2-methyl-4(3H)-quinazolinone; etaqualone
Trade names:

Use: Sedative, hypnotic
HPLC: Si-10; 1A:99B; 4.1
GC: 2276; 250°C

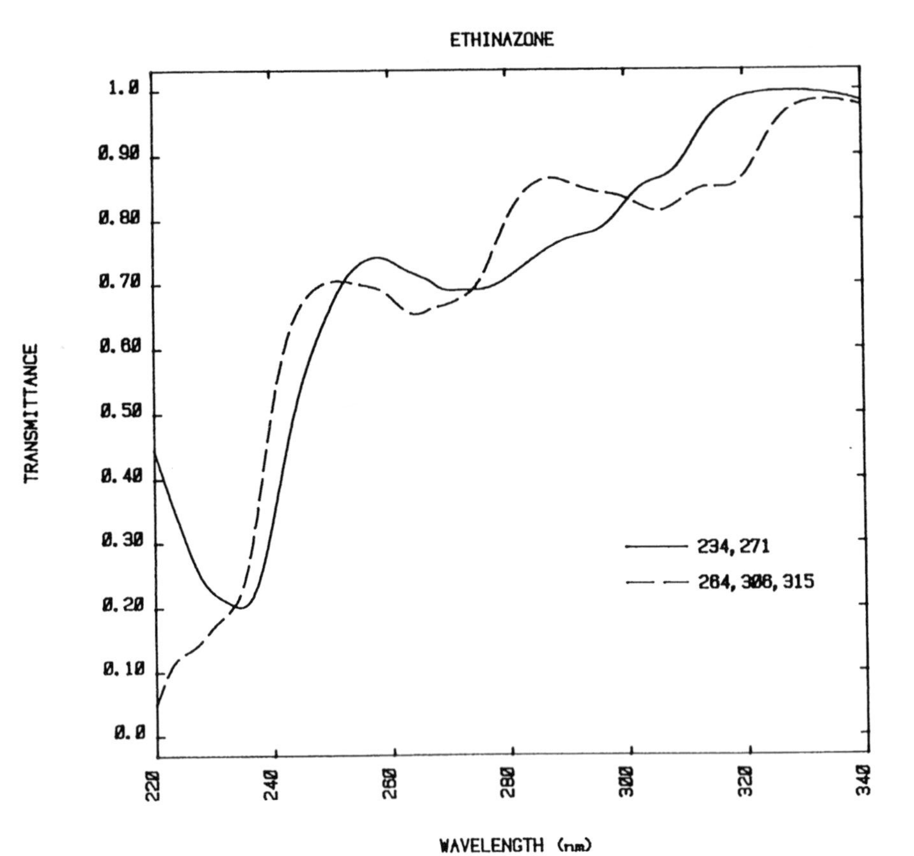


*ETHINAZONE*

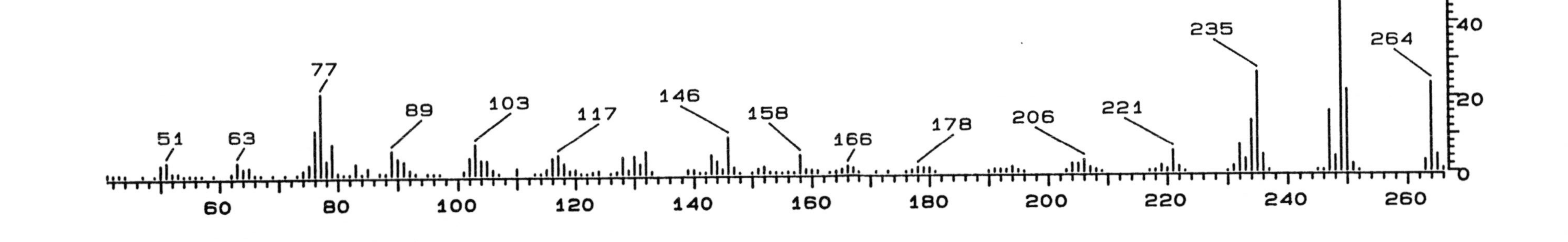

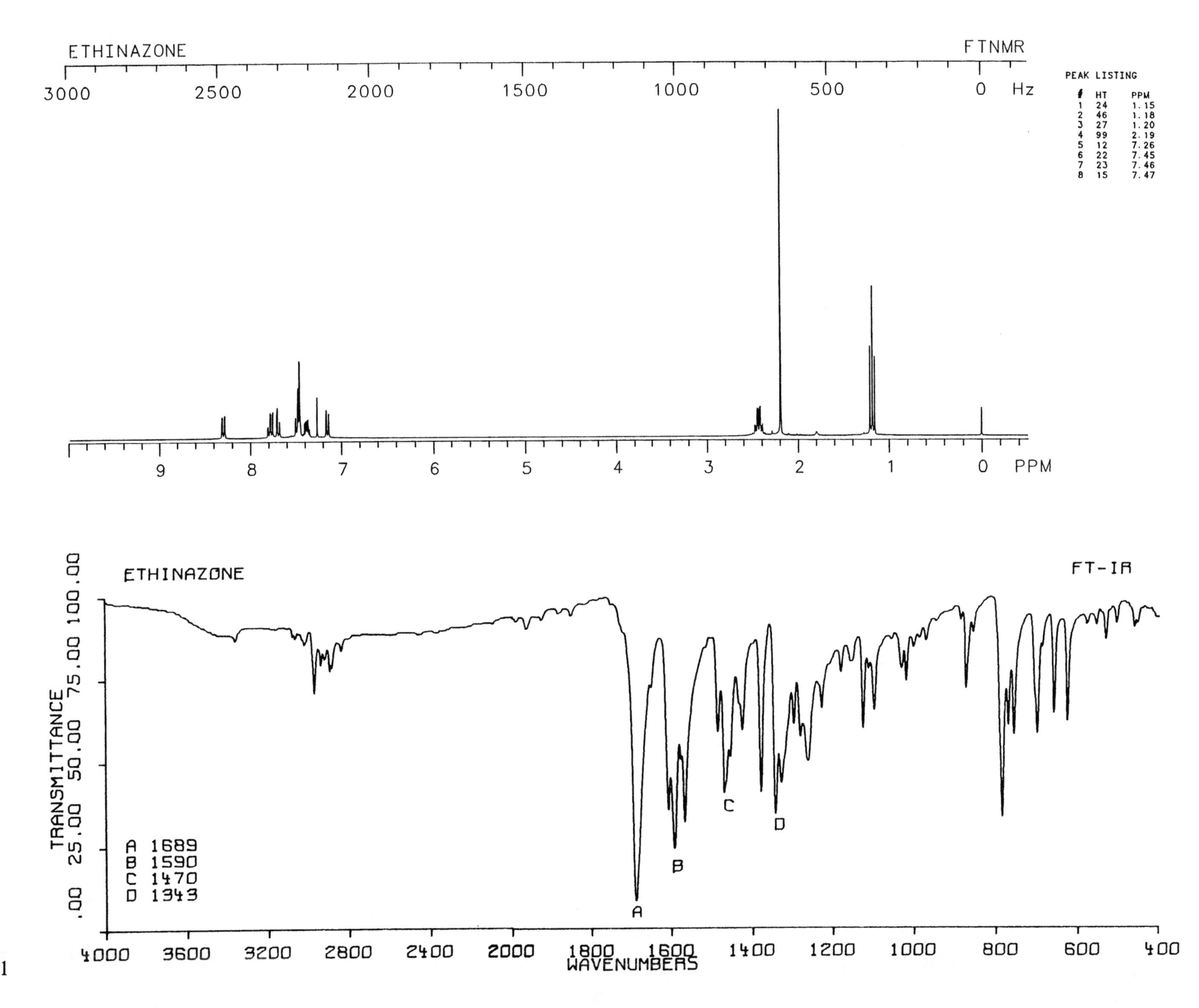
ETHINAZONE
FTNMR
3000
2500
2000
1500
1000
500
0
Hz
PEAK LISTING
# HT PPM
1 24 1.15
2 46 1.18
3 27 1.20
4 99 2.19
5 12 7.26
6 22 7.45
7 23 7.46
8 15 7.47
9
8
7
6
5
4
3
2
1
0
PPM
ETHINAZONE
FT-IR
TRANSMITTANCE
.00
25.00
50.00
75.00
100.00
A 1689
B 1590
C 1470
D 1343
A
B
C
D
4000
3600
3200
2800
2400
2000
1800
1600
1400
1200
1000
800
600
400
WAVENUMBERS

# ETHIONAMIDE

$C_8H_{10}N_2S$

**Molecular weight:** 166.24 (166.06)
**Synonyms:** 2-Ethyl-4-pyridinecarbothioamide; amidazine; ethioniamide; etionamide
**Trade names:** Trecator-SC, Trescatyl

**Use:** Antibacterial
**HPLC:** Si-10; 2A:98B; 5.1
**GC:** 1765; 200°C

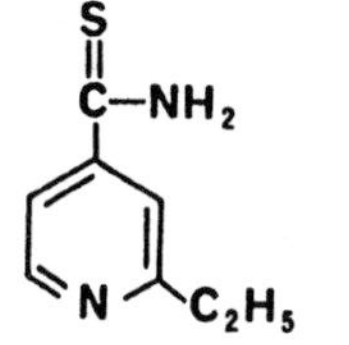


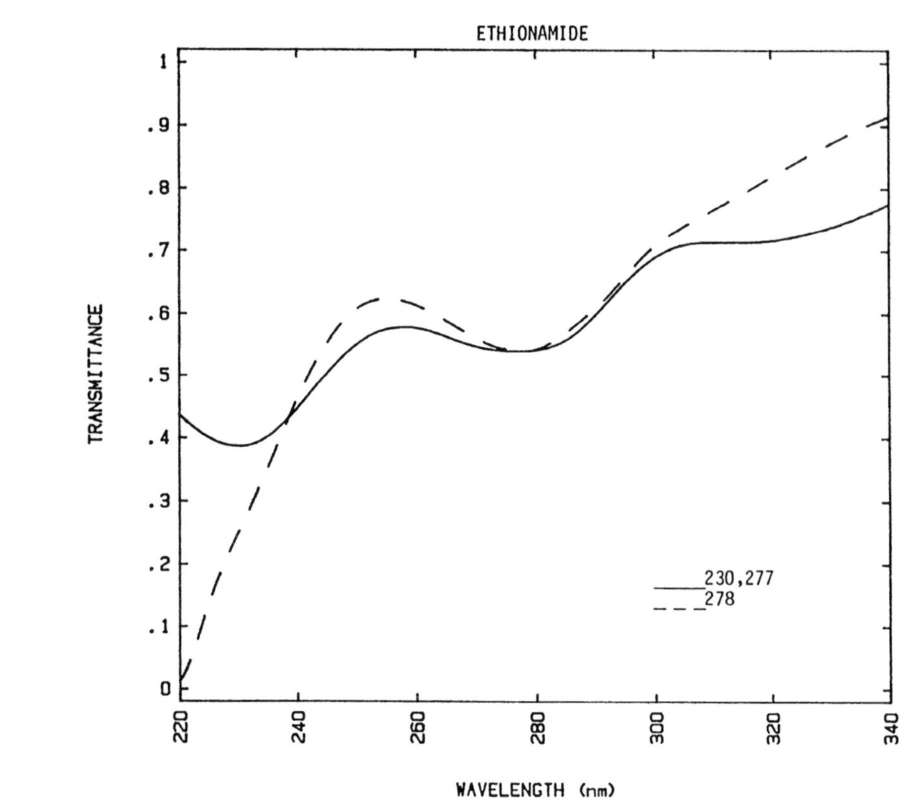

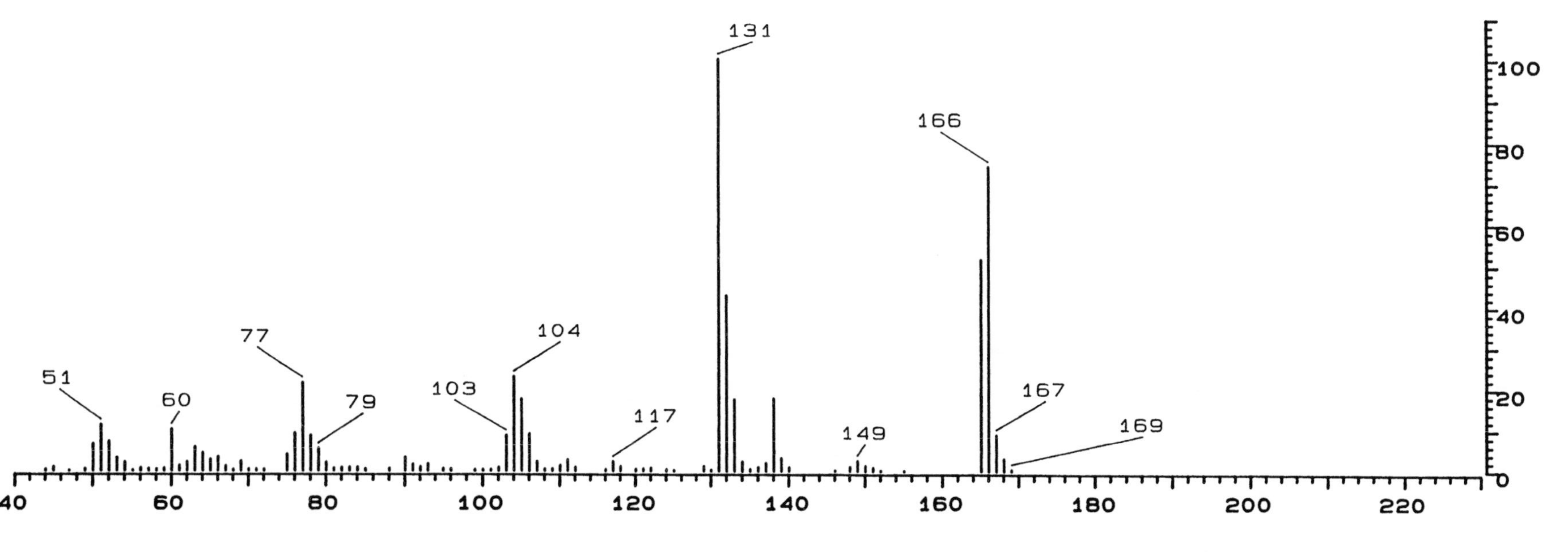

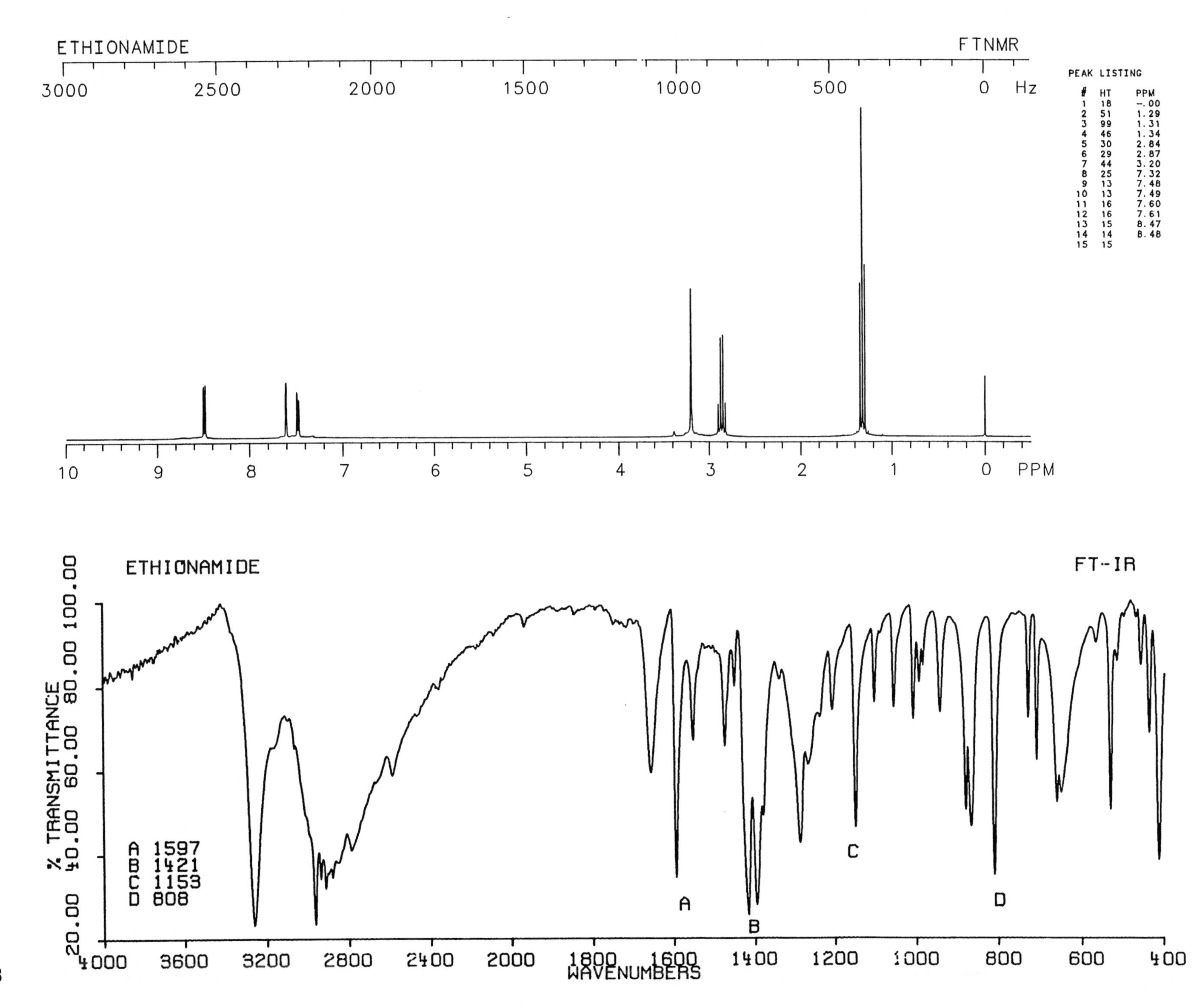
ETHIONAMIDE
FTNMR
3000
2500
2000
1500
1000
500
0 Hz
PEAK LISTING
# HT PPM
1 18 -.00
2 51 1.29
3 99 1.31
4 46 1.34
5 30 2.84
6 29 2.87
7 44 3.20
8 25 7.32
9 13 7.48
10 13 7.49
11 16 7.60
12 16 7.61
13 15 8.47
14 14 8.48
15 15
10
9
8
7
6
5
4
3
2
1
0 PPM
ETHIONAMIDE
FT-IR
% TRANSMITTANCE
20.00
40.00
60.00
80.00
100.00
A 1597
B 1421
C 1153
D 808
A
B
C
D
4000
3600
3200
2800
2400
2000
1800
1600
1400
1200
1000
800
600
400
WAVENUMBERS

# ETHISTERONE

$C_{21}H_{28}O_2$

Molecular weight: 312.44 (312.21)
Synonyms: 17α-Hydroxypregn-4-en-20-yn-3-one; 17α-ethynyltestosterone; aethisteron; pregneninolone; pregnin; praegnin
Trade names: Etherone, Gestoral, Lutidon, Lutogyl, Lutoral, Oral, Ora-Lutin, Pranone, Primolut, Progestoral, Prolidon
Use: Progestin
HPLC:
GC: 2701; 250

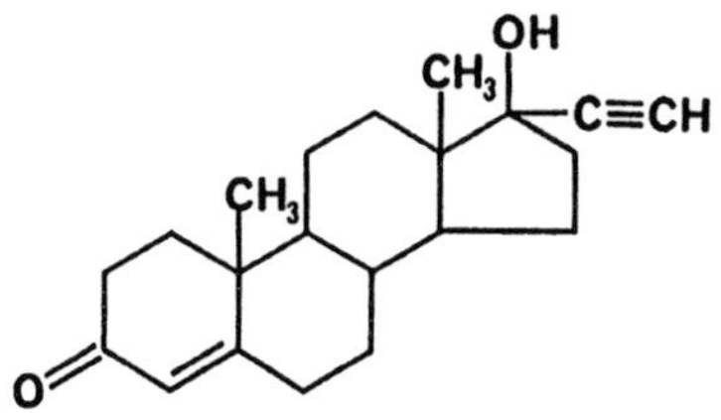

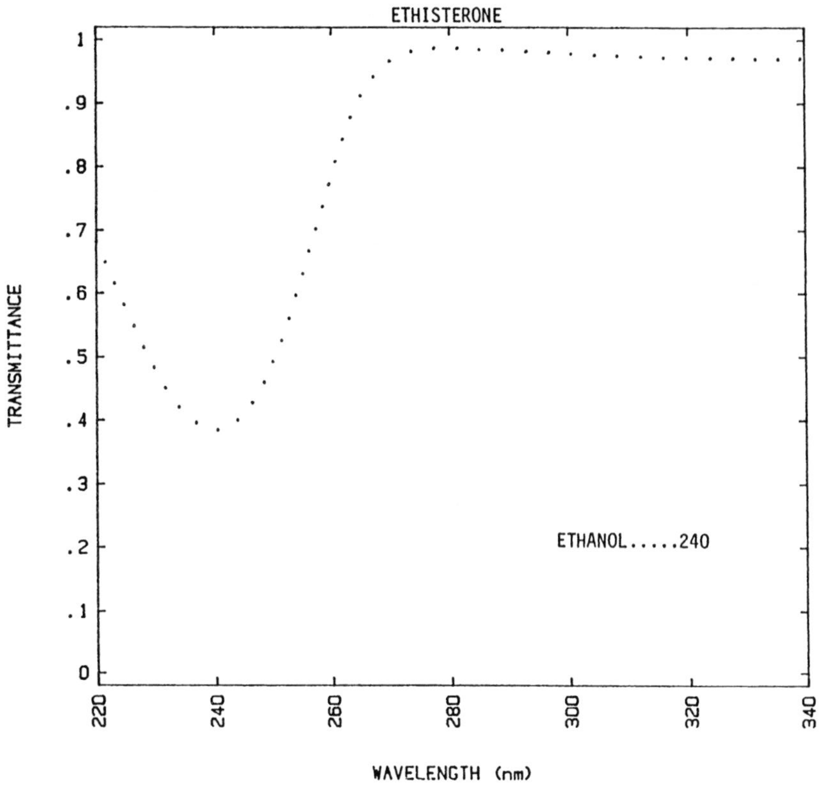

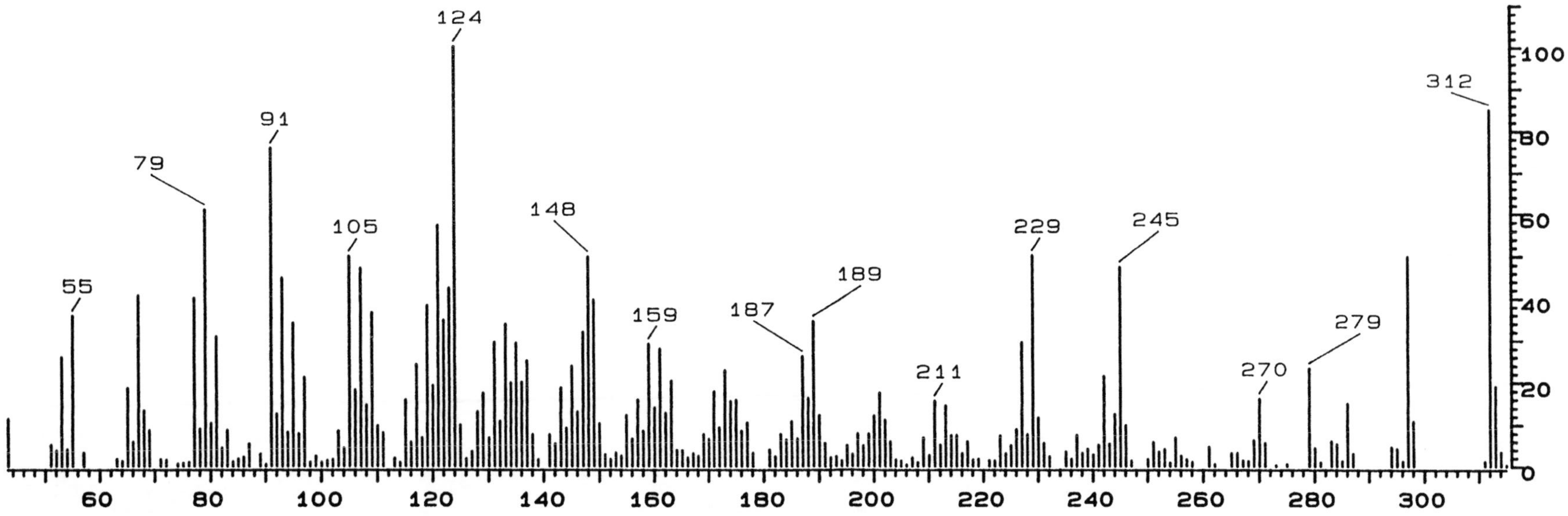

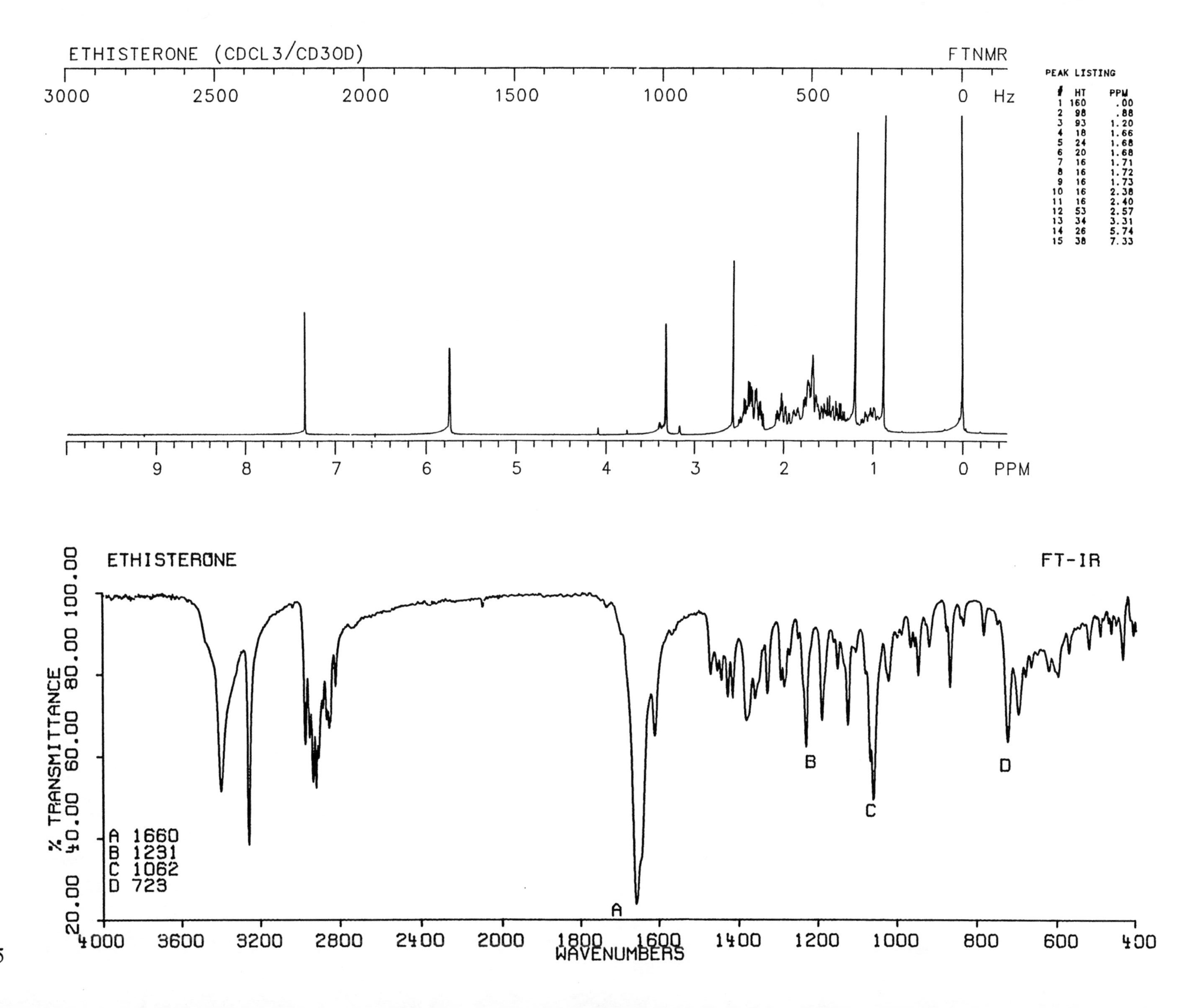
ETHISTERONE (CDCL3/CD3OD)
FTNMR
Hz
PPM
PEAK LISTING
# HT PPM
1 160 .00
2 98 .88
3 93 1.20
4 18 1.66
5 24 1.68
6 20 1.68
7 16 1.71
8 16 1.72
9 16 1.73
10 16 2.38
11 16 2.40
12 53 2.57
13 34 3.31
14 26 5.74
15 38 7.33
ETHISTERONE
FT-IR
% TRANSMITTANCE
WAVENUMBERS
A 1660
B 1231
C 1062
D 723

# ETHOHEPTAZINE

$C_{16}H_{23}NO_2$

Molecular weight: 261.36 (261.17)
Synonyms: Hexahydro-1-methyl-4-phenylazepine-4-carboxylic acid ethyl ester; ethyl heptazine
Trade names: Equagesic, Meprogesic, Zactirin

Use: Analgesic
HPLC: Si-10; 20A:80B; 6.5
GC: 1869; 200°C

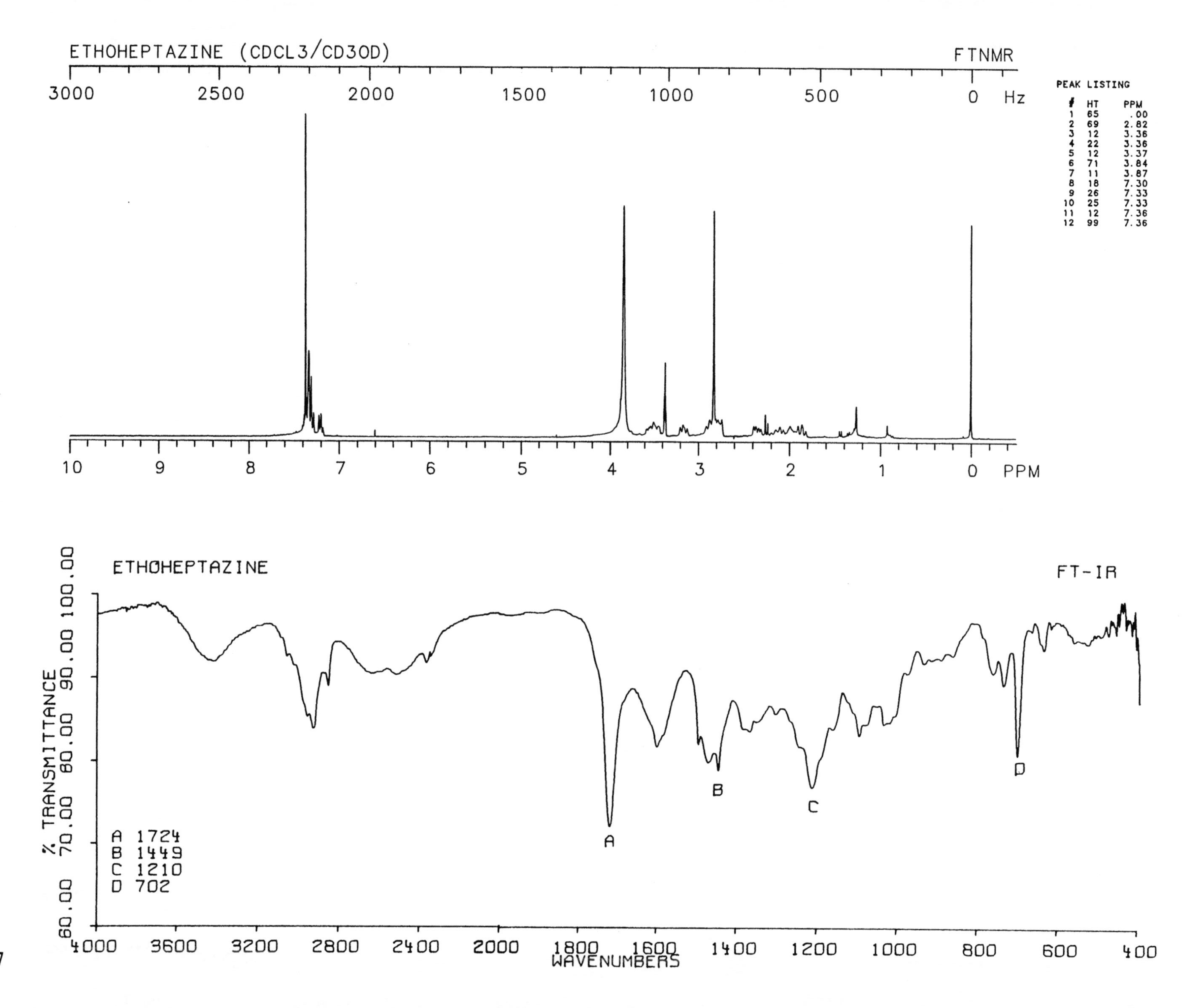
ETHOHEPTAZINE (CDCL3/CD3OD)
FTNMR
3000 2500 2000 1500 1000 500 0 Hz
PEAK LISTING
# HT PPM
1 65 .00
2 69 2.82
3 12 3.36
4 22 3.36
5 12 3.37
6 71 3.84
7 11 3.87
8 18 7.30
9 26 7.33
10 25 7.33
11 12 7.36
12 99 7.36
10 9 8 7 6 5 4 3 2 1 0 PPM
ETHOHEPTAZINE
FT-IR
% TRANSMITTANCE
60.00 70.00 80.00 90.00 100.00
A 1724
B 1449
C 1210
D 702
A
B
C
D
4000 3600 3200 2800 2400 2000 1800 1600 1400 1200 1000 800 600 400
WAVENUMBERS

# ETHOPROPAZINE

$C_{19}H_{24}N_2S$

Molecular weight: 312.48 (312.17)
Synonyms: N,N-Diethyl-α-methyl-10H-phenothiazine-10-ethanamine; profenamine; phenopropazine
Trade names: Parsidol

Use: Antiparkinsonian, anticholinergic
HPLC: Si-10; 5A:95B; 3.8
GC: 2420; 250°C

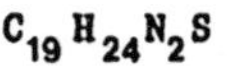

ETHOPROPAZINE

TRANSMITTANCE

1.0
0.90
0.80
0.70
0.60
0.50
0.40
0.30
0.20
0.10
0.0

220 240 260 280 300 320 340

WAVELENGTH (nm)

249, 298
268, 316

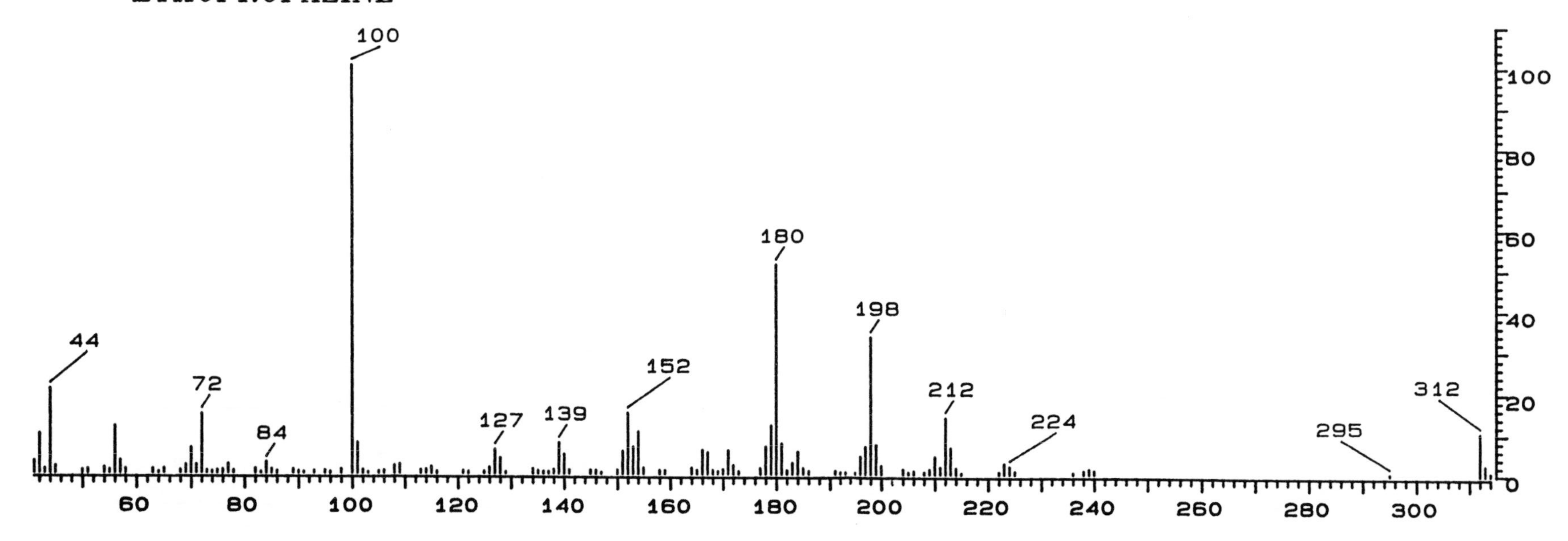

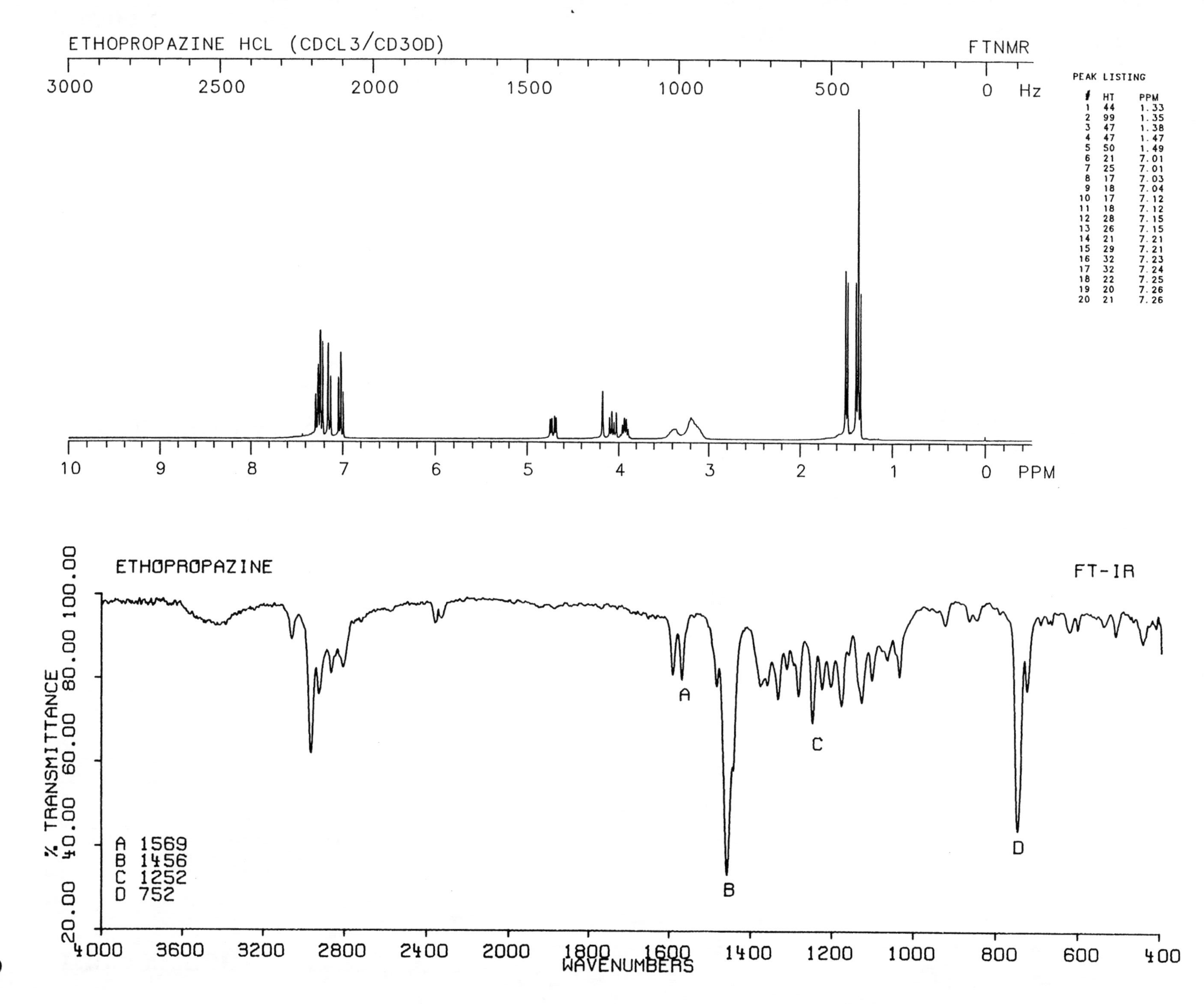
ETHOPROPAZINE HCL (CDCL3/CD3OD)
FTNMR
3000
2500
2000
1500
1000
500
0 Hz
PEAK LISTING
# HT PPM
1 44 1.33
2 99 1.35
3 47 1.38
4 47 1.47
5 50 1.49
6 21 7.01
7 25 7.01
8 17 7.03
9 18 7.04
10 17 7.12
11 18 7.12
12 28 7.15
13 26 7.15
14 21 7.21
15 29 7.21
16 32 7.23
17 32 7.24
18 22 7.25
19 20 7.26
20 21 7.26
10
9
8
7
6
5
4
3
2
1
0 PPM
ETHOPROPAZINE
FT-IR
% TRANSMITTANCE
20.00
40.00
60.00
80.00
100.00
A 1569
B 1456
C 1252
D 752
A
B
C
D
4000
3600
3200
2800
2400
2000
1800
1600
1400
1200
1000
800
600
400
WAVENUMBERS

# ETHOSUXIMIDE

$C_7H_{11}NO_2$

Molecular weight: 141.17 (141.08)
Synonyms: 3-Ethyl-3-methyl-2,5-pyrrolidine-dione; 2-ethyl-2-methylsuccinimide; α-ethyl-α-methylsuccinimide
Trade names: Zarontin

Use: Anticonvulsant
HPLC: Si-10; 1A:99B; 6.0
GC: 1209; 140°C





*ETHOSUXIMIDE*

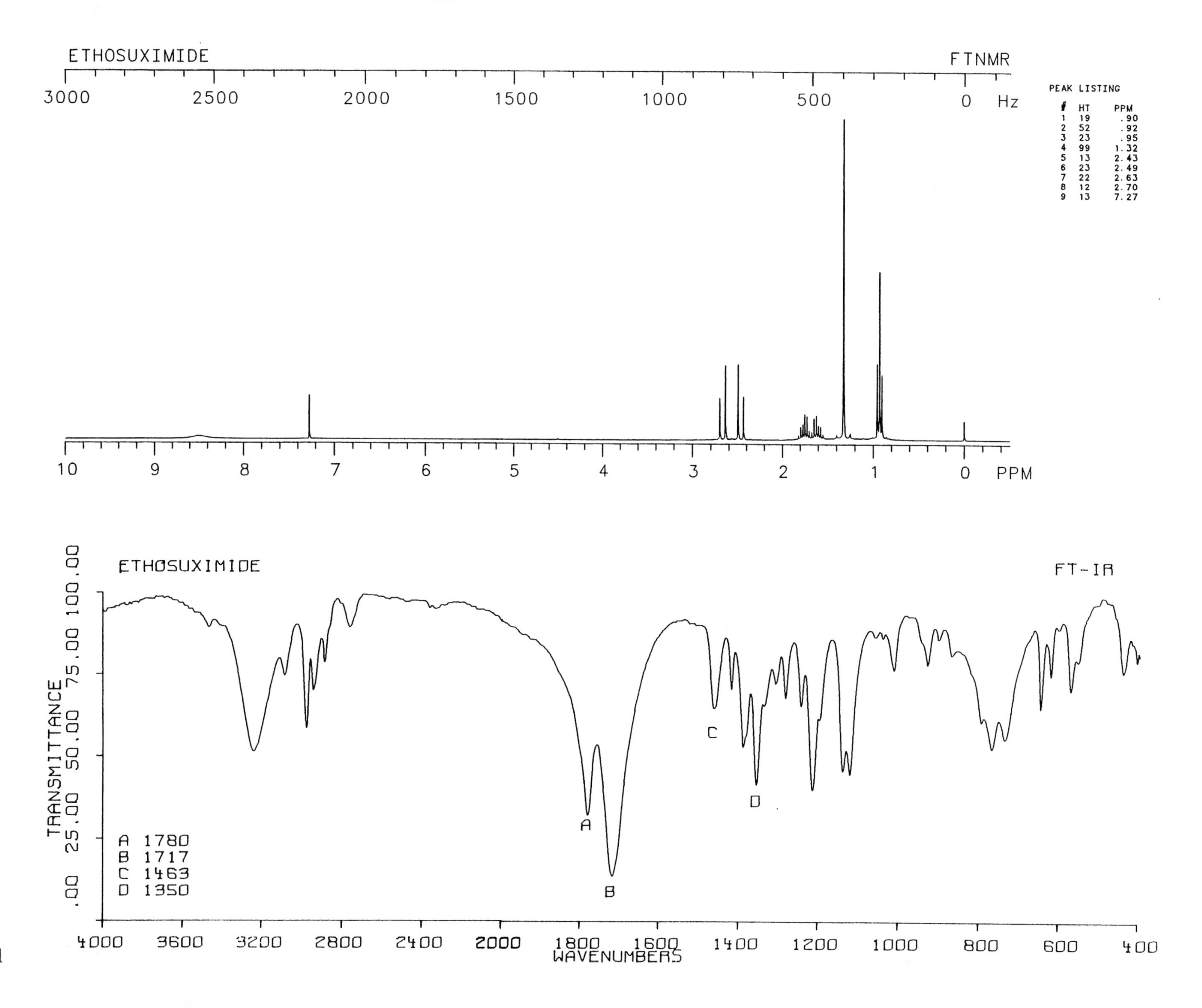
ETHOSUXIMIDE
FTNMR
3000 2500 2000 1500 1000 500 0 Hz
PEAK LISTING
# HT PPM
1 19 .90
2 52 .92
3 23 .95
4 99 1.32
5 13 2.43
6 23 2.49
7 22 2.63
8 12 2.70
9 13 7.27
10 9 8 7 6 5 4 3 2 1 0 PPM
ETHØSUXIMIDE
FT-IR
TRANSMITTANCE
.00 25.00 50.00 75.00 100.00
A 1780
B 1717
C 1463
D 1350
A B C D
4000 3600 3200 2800 2400 2000 1800 1600 1400 1200 1000 800 600 400
WAVENUMBERS

# ETHOTOIN

$C_{11}H_{12}N_2O_2$

**Molecular weight:** 204.23 (204.09)
**Synonyms:** 3-Ethyl-5-phenyl-2,4-imidazolidinedione

**Trade names:** Peganone

**Use:** Anticonvulsant
**HPLC:** Si-10; 1A:99B; 4.0
**GC:** 1618; 200°C

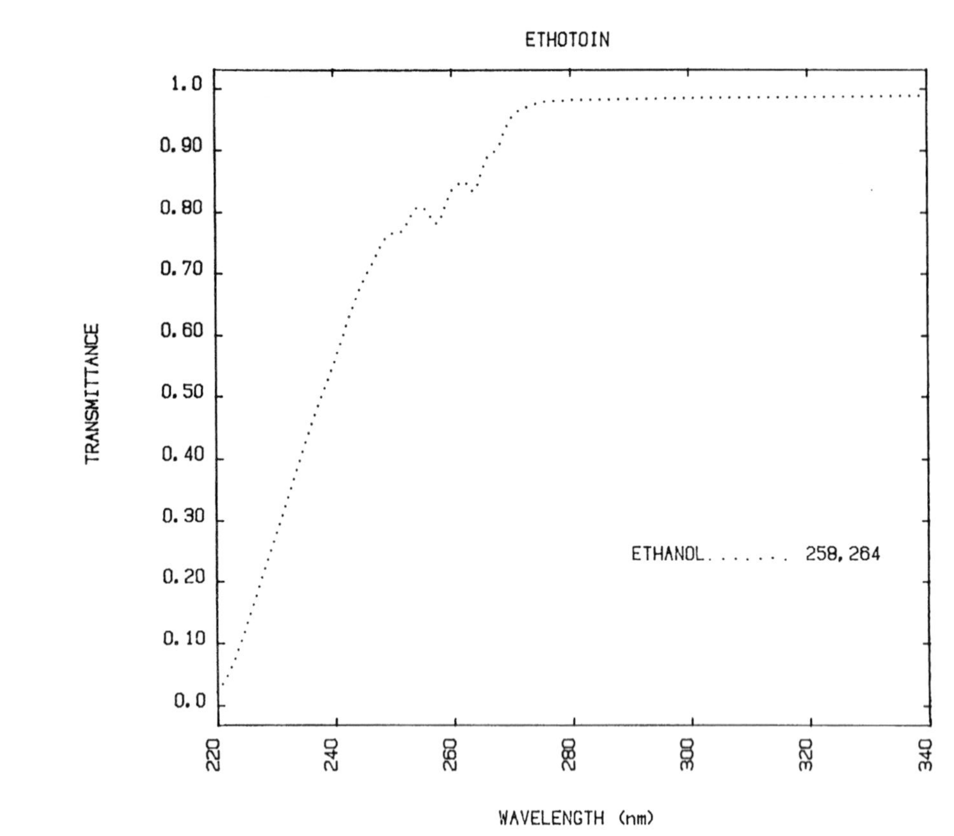


*ETHOTOIN*

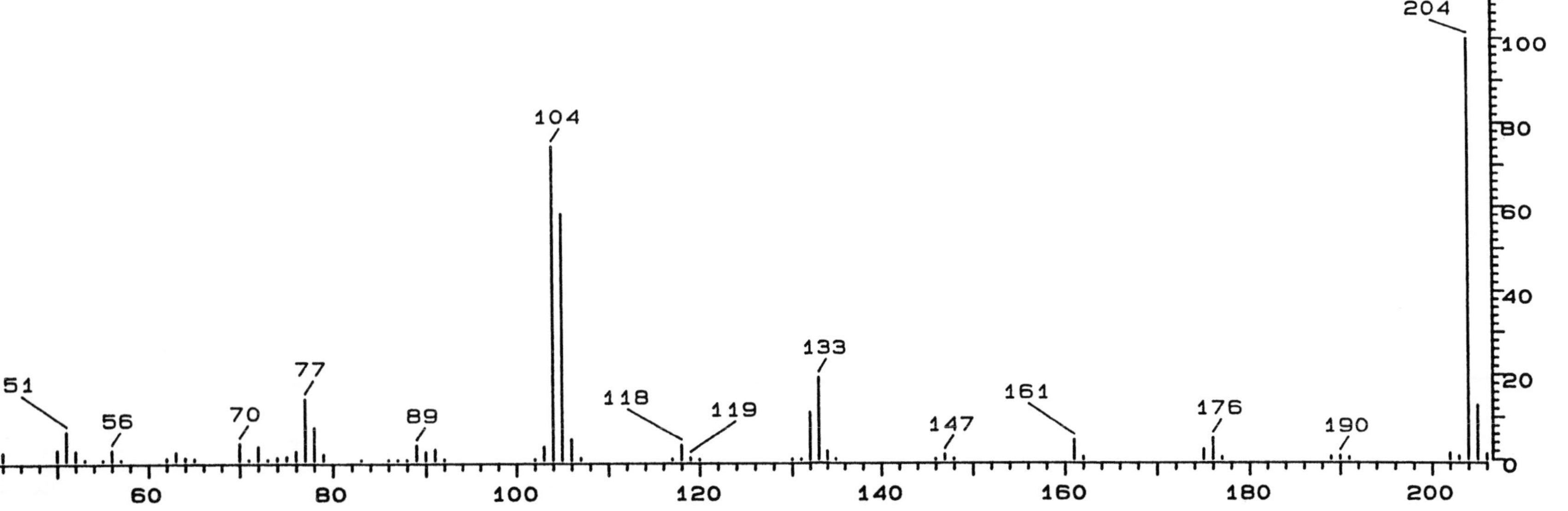


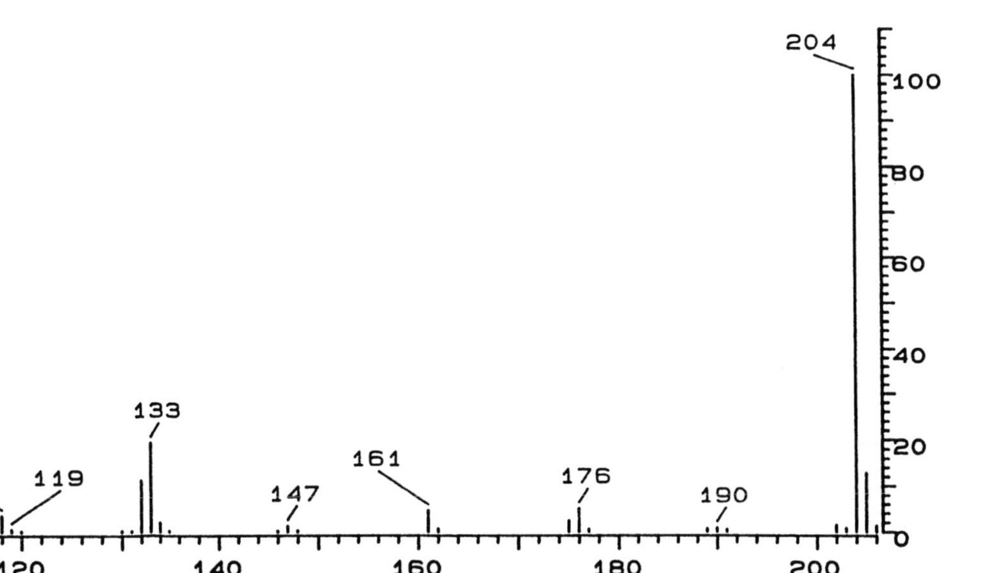

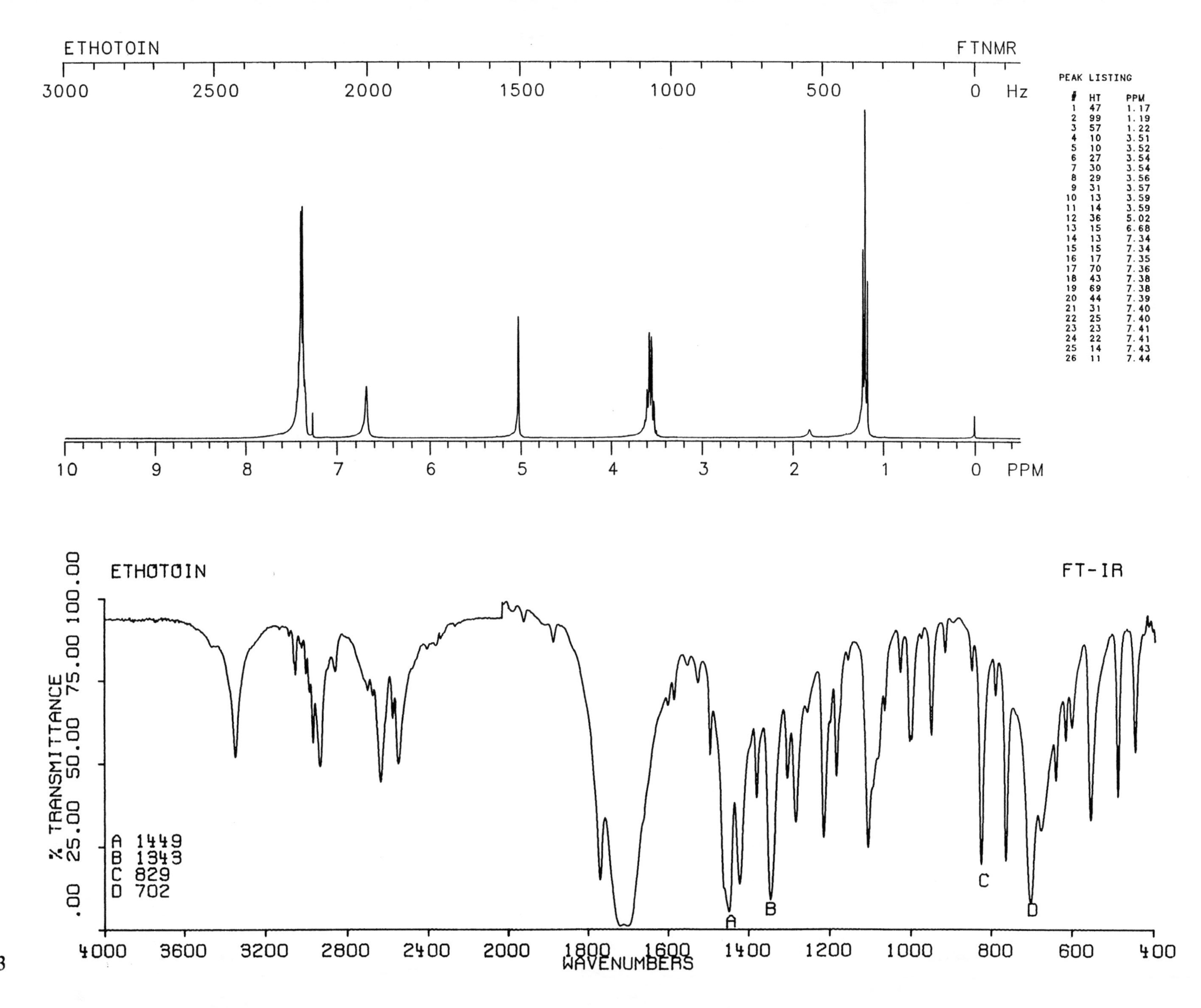
ETHOTOIN
FTNMR
3000
2500
2000
1500
1000
500
0 Hz
10
9
8
7
6
5
4
3
2
1
0 PPM
PEAK LISTING
# HT PPM
1 47 1.17
2 99 1.19
3 57 1.22
4 10 3.51
5 10 3.52
6 27 3.54
7 30 3.54
8 29 3.56
9 31 3.57
10 13 3.59
11 14 3.59
12 36 5.02
13 15 6.68
14 13 7.34
15 15 7.34
16 17 7.35
17 70 7.36
18 43 7.38
19 69 7.38
20 44 7.39
21 31 7.40
22 25 7.40
23 23 7.41
24 22 7.41
25 14 7.43
26 11 7.44
ETHOTOIN
FT-IR
% TRANSMITTANCE
.00
25.00
50.00
75.00
100.00
A 1449
B 1343
C 829
D 702
A
B
C
D
4000
3600
3200
2800
2400
2000
1800
1600
1400
1200
1000
800
600
400
WAVENUMBERS

# ETHOXZOLAMIDE

$C_9H_{10}N_2O_3S_2$

Molecular weight: 258.33 (258.01)
Synonyms: 6-Ethoxy-2-benzothiazolesulfonamide; ethoxyzolamide

Trade names: Cardrase, Ethamide, Glaucotensil, Redupresin

Use: Diuretic
HPLC: Si-10; 1A:99B; 3.4
GC: 2600; 280°C

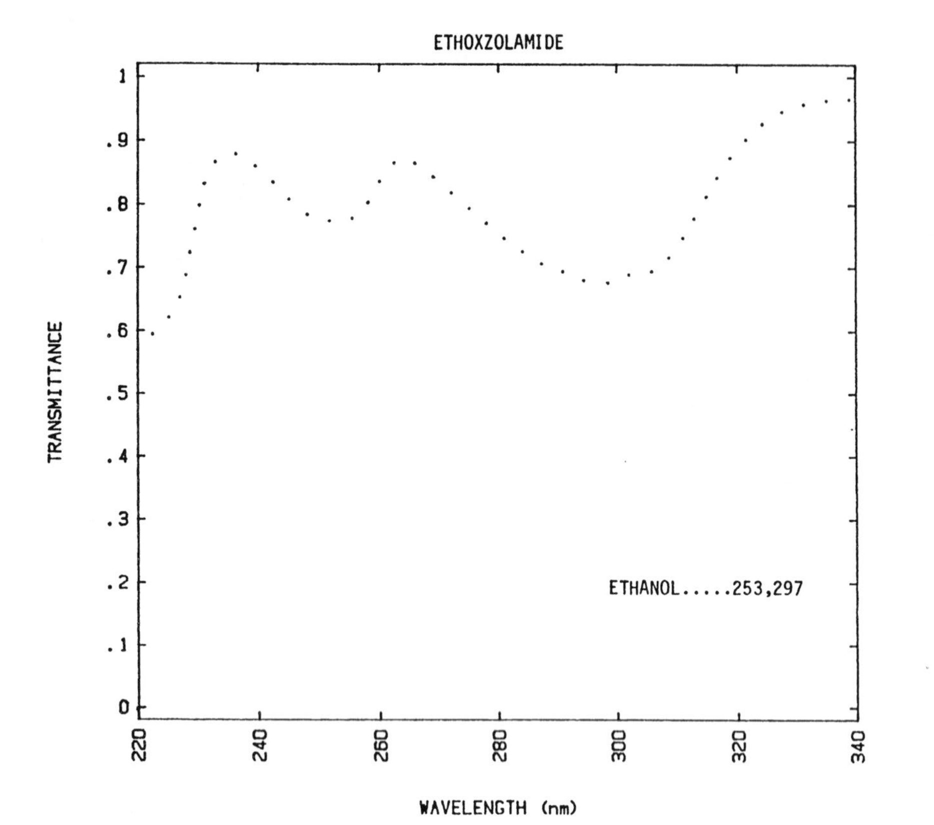


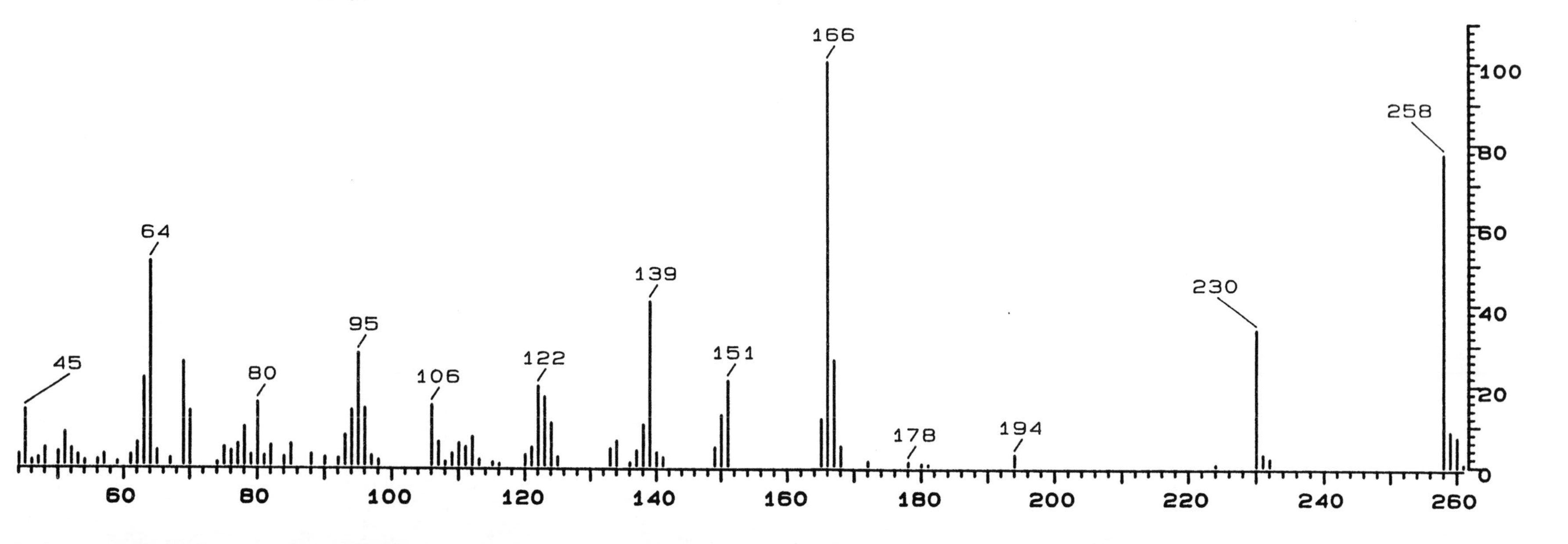

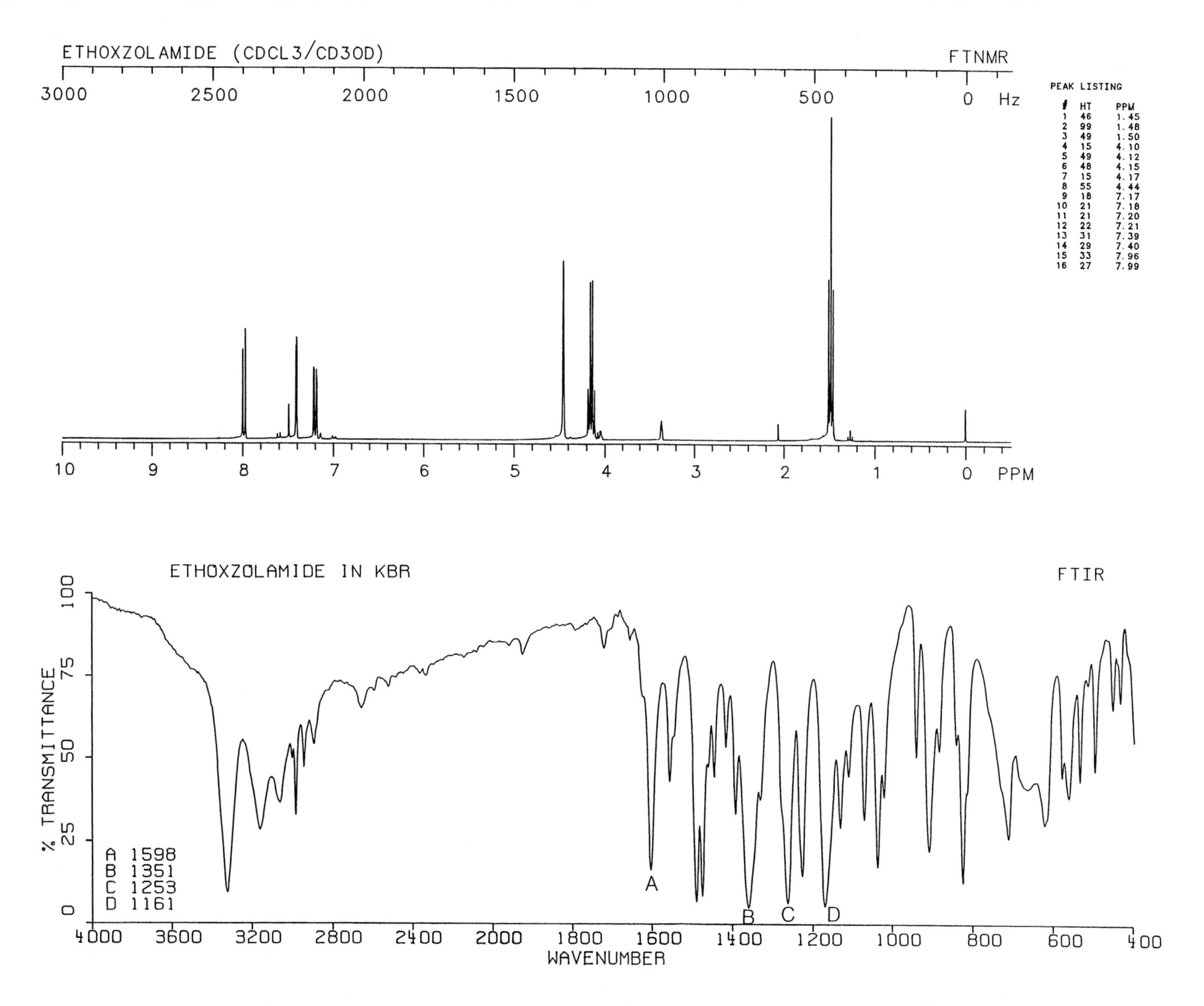
ETHOXZOLAMIDE (CDCL3/CD3OD)
FTNMR
3000 2500 2000 1500 1000 500 0 Hz
10 9 8 7 6 5 4 3 2 1 0 PPM
PEAK LISTING
# HT PPM
1 46 1.45
2 99 1.48
3 49 1.50
4 15 4.10
5 49 4.12
6 48 4.15
7 15 4.17
8 55 4.44
9 18 7.17
10 21 7.18
11 21 7.20
12 22 7.21
13 31 7.39
14 29 7.40
15 33 7.96
16 27 7.99
ETHOXZOLAMIDE IN KBR
FTIR
% TRANSMITTANCE
100 75 50 25 0
A 1598
B 1351
C 1253
D 1161
A B C D
4000 3600 3200 2800 2400 2000 1800 1600 1400 1200 1000 800 600 400
WAVENUMBER

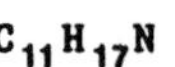

# N-ETHYLAMPHETAMINE

$C_{11}H_{17}N$

Molecular weight: 163.26 (163.14)
Synonyms: N-Ethyl-α-methylbenzenethanamine; 2-ethylamino-1-phenylpropane, NEA
Trade names: Adiparthrol, Apetinil

Use: Anorexic
HPLC: Si-10; 20A:80B; 5.0
GC: 1231; 140°C

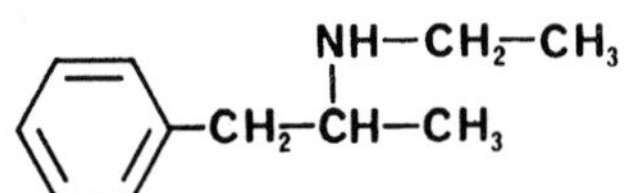


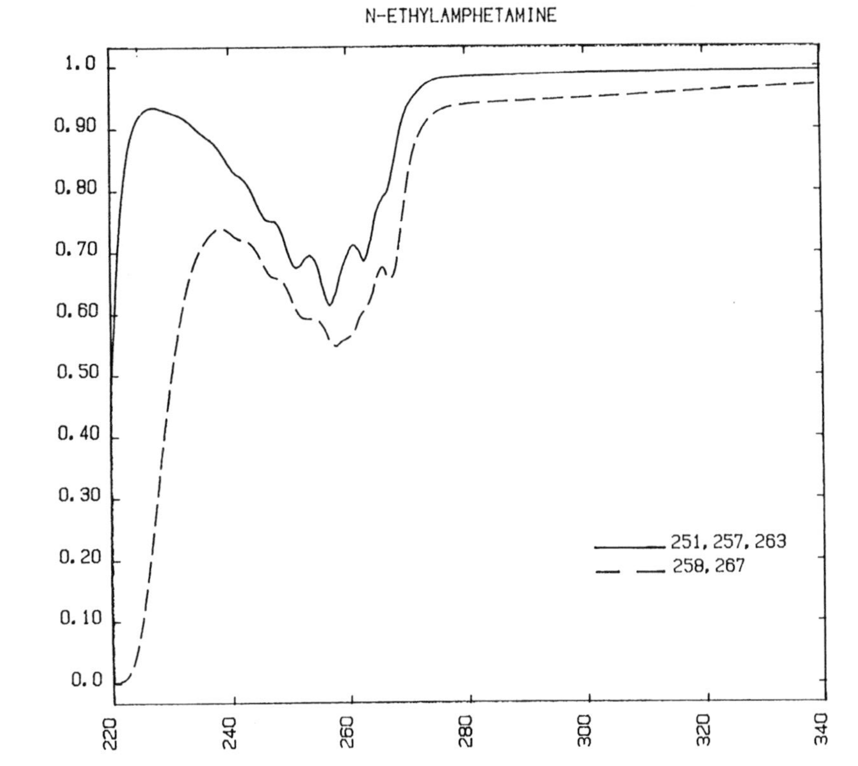

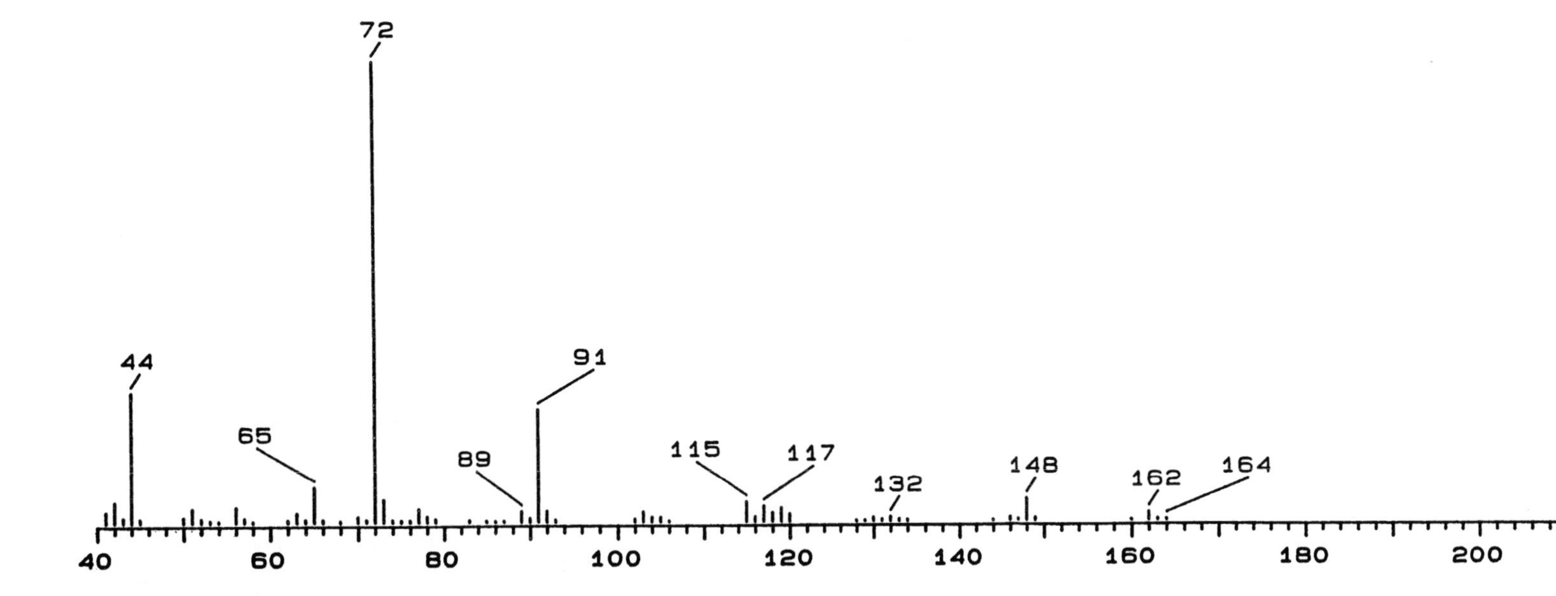

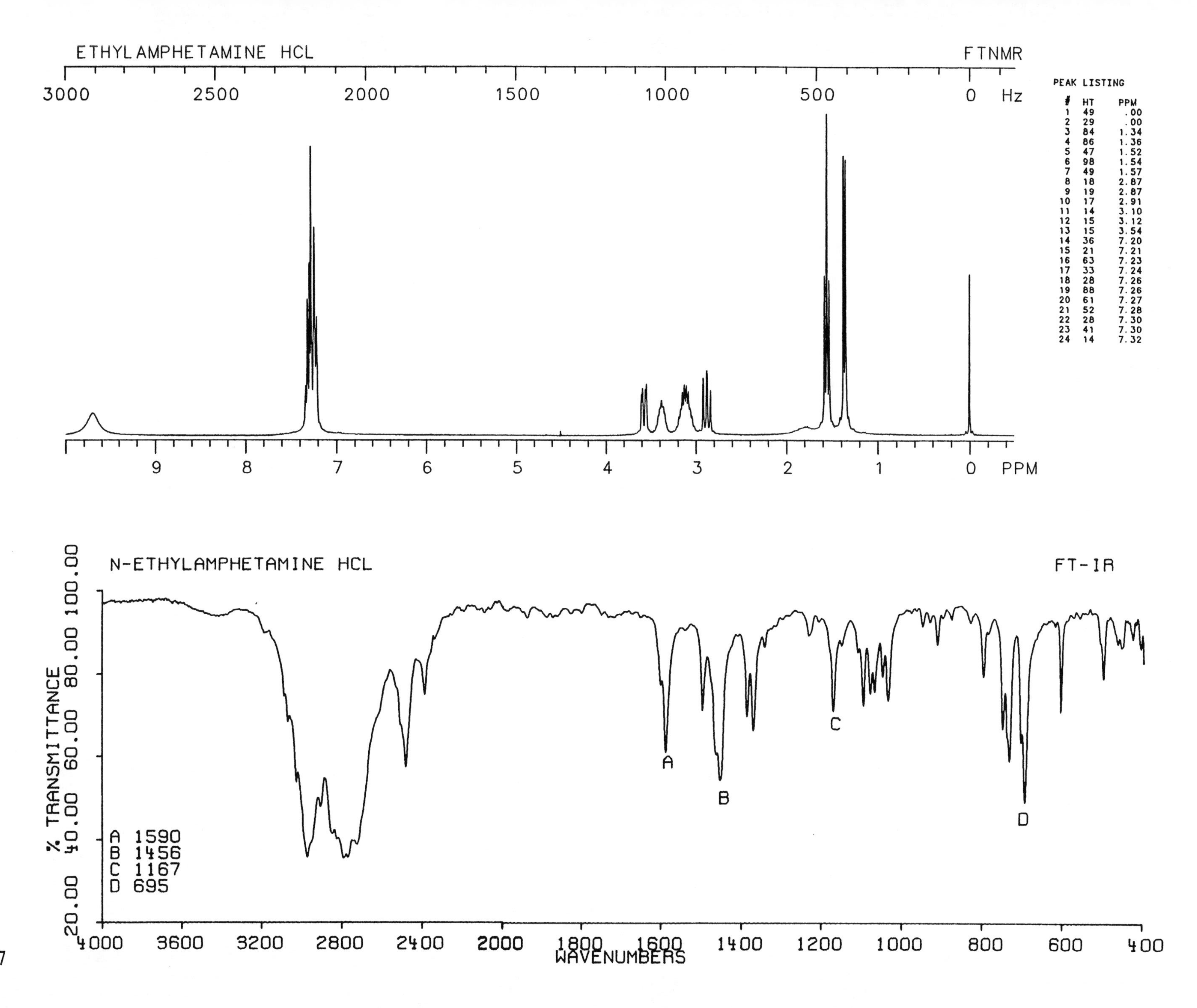

ETHYLAMPHETAMINE HCL
FTNMR
Hz
PPM
PEAK LISTING
# HT PPM
1 49 .00
2 29 .00
3 84 1.34
4 86 1.36
5 47 1.52
6 98 1.54
7 49 1.57
8 18 2.87
9 19 2.87
10 17 2.91
11 14 3.10
12 15 3.12
13 15 3.54
14 36 7.20
15 21 7.21
16 63 7.23
17 33 7.24
18 28 7.26
19 88 7.26
20 61 7.27
21 52 7.28
22 28 7.30
23 41 7.30
24 14 7.32
N-ETHYLAMPHETAMINE HCL
FT-IR
% TRANSMITTANCE
WAVENUMBERS
A 1590
B 1456
C 1167
D 695

# ETHYL CLORAZEPATE

$C_{18}H_{17}ClN_2O_4$

**Molecular weight:** 360.80 (360.09)
**Synonyms:** 7-Chloro-2,3-dihydro-2,2-dihydroxy-5-phenyl-1H-1,4-benzodiazepine-3-ethyl ester
**Trade names:**

**Use:** Tranquilizer
**HPLC:** Si-10; 1A:99B; 10.5
**GC:**

ETHYL CLORAZEPATE

TRANSMITTANCE

1.0
0.90
0.80
0.70
0.60
0.50
0.40
0.30
0.20
0.10
0.0

235, 282

220 240 260 280 300 320 340

WAVELENGTH (nm)

*ETHYL CLORAZEPATE--DIP*

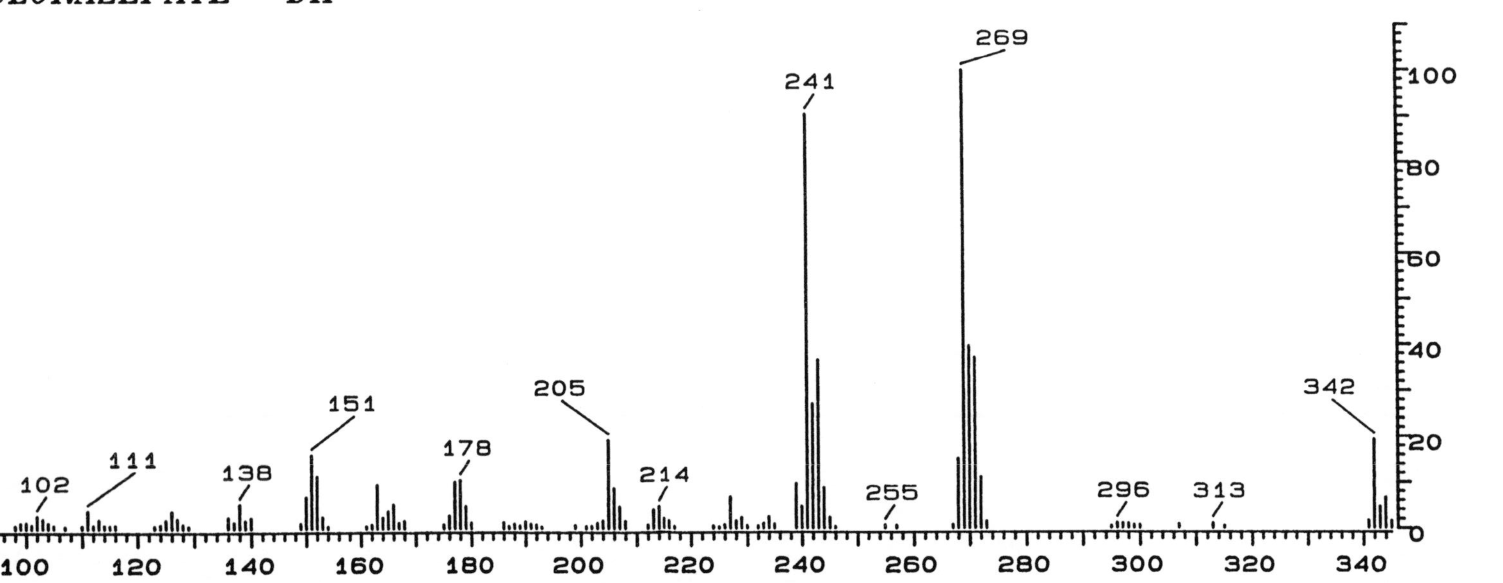

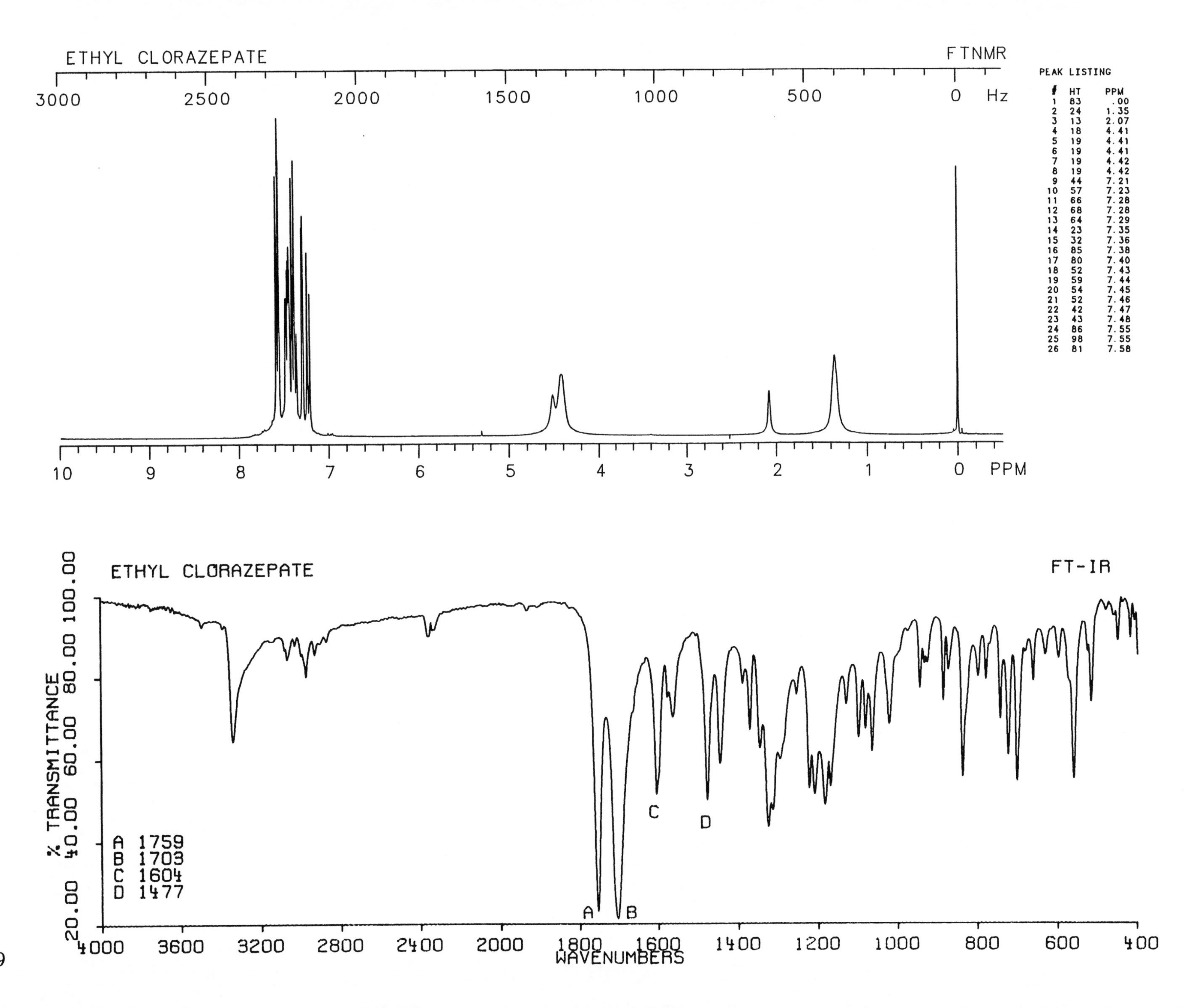
ETHYL CLORAZEPATE
FTNMR
3000 2500 2000 1500 1000 500 0 Hz
10 9 8 7 6 5 4 3 2 1 0 PPM
PEAK LISTING
# HT PPM
1 83 .00
2 24 1.35
3 13 2.07
4 18 4.41
5 19 4.41
6 19 4.41
7 19 4.42
8 19 4.42
9 44 7.21
10 57 7.23
11 66 7.28
12 68 7.28
13 64 7.29
14 23 7.35
15 32 7.36
16 85 7.38
17 80 7.40
18 52 7.43
19 59 7.44
20 54 7.45
21 52 7.46
22 42 7.47
23 43 7.48
24 86 7.55
25 98 7.55
26 81 7.58
ETHYL CLORAZEPATE
FT-IR
% TRANSMITTANCE
20.00 40.00 60.00 80.00 100.00
A 1759
B 1703
C 1604
D 1477
A B C D
4000 3600 3200 2800 2400 2000 1800 1600 1400 1200 1000 800 600 400
WAVENUMBERS

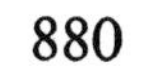

# ETHYLESTRENOL

$C_{20}H_{32}O$

**Molecular weight:** 288.47 (288.25)
**Synonyms:** 19-Nor-17α-pregn-4-en-17-ol; 17β-hydroxy-17α-ethyl-19-nor-4-androstene
**Trade names:** Orabolin

**Use:** Anabolic
**HPLC:**
**GC:** 2402; 280°C

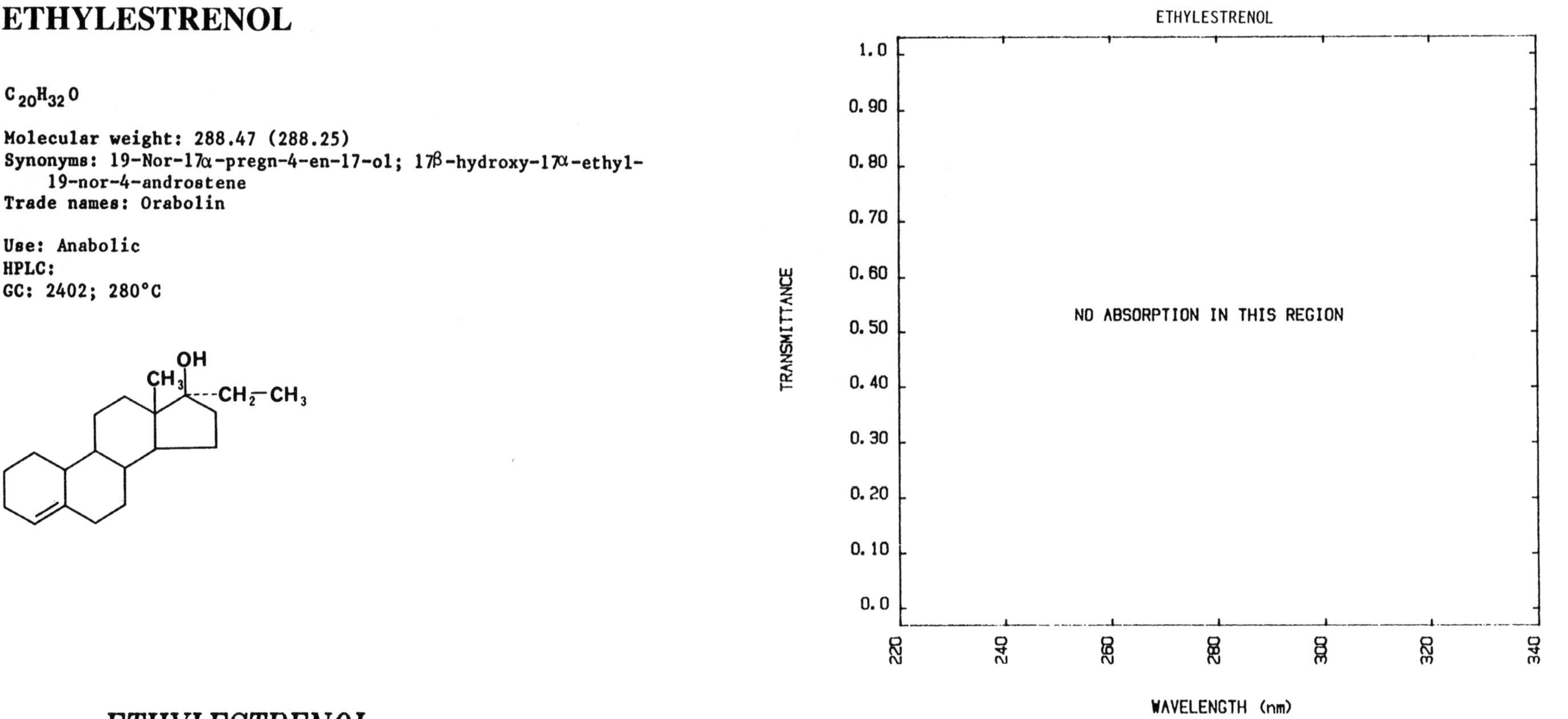


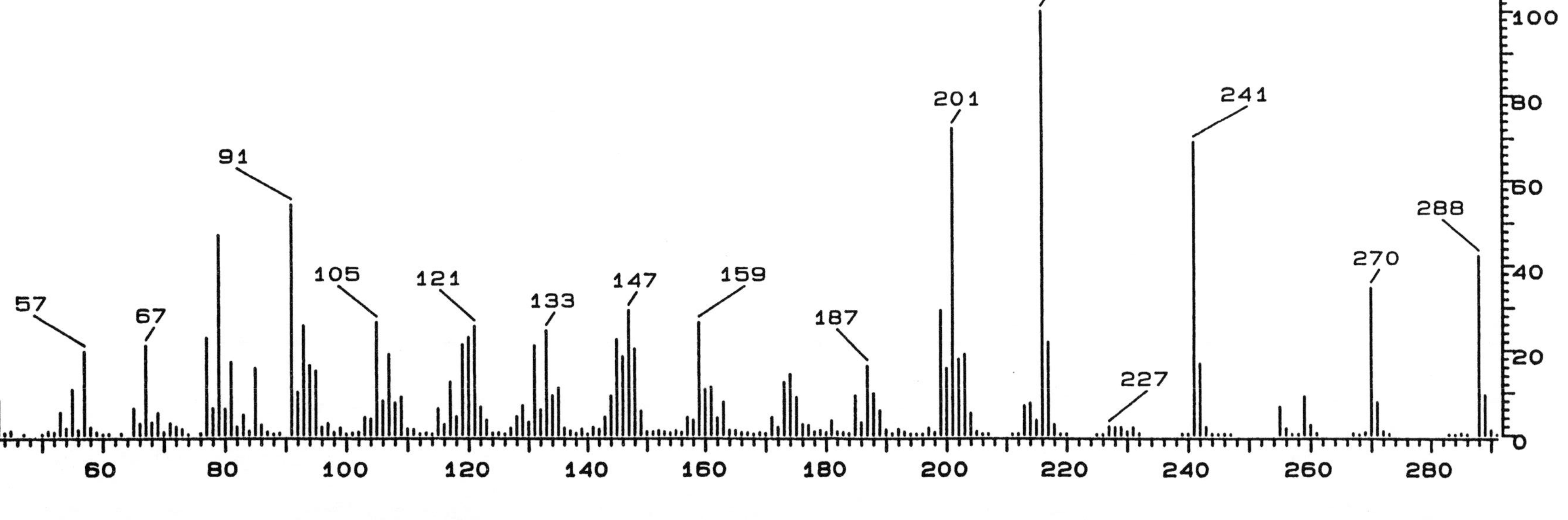

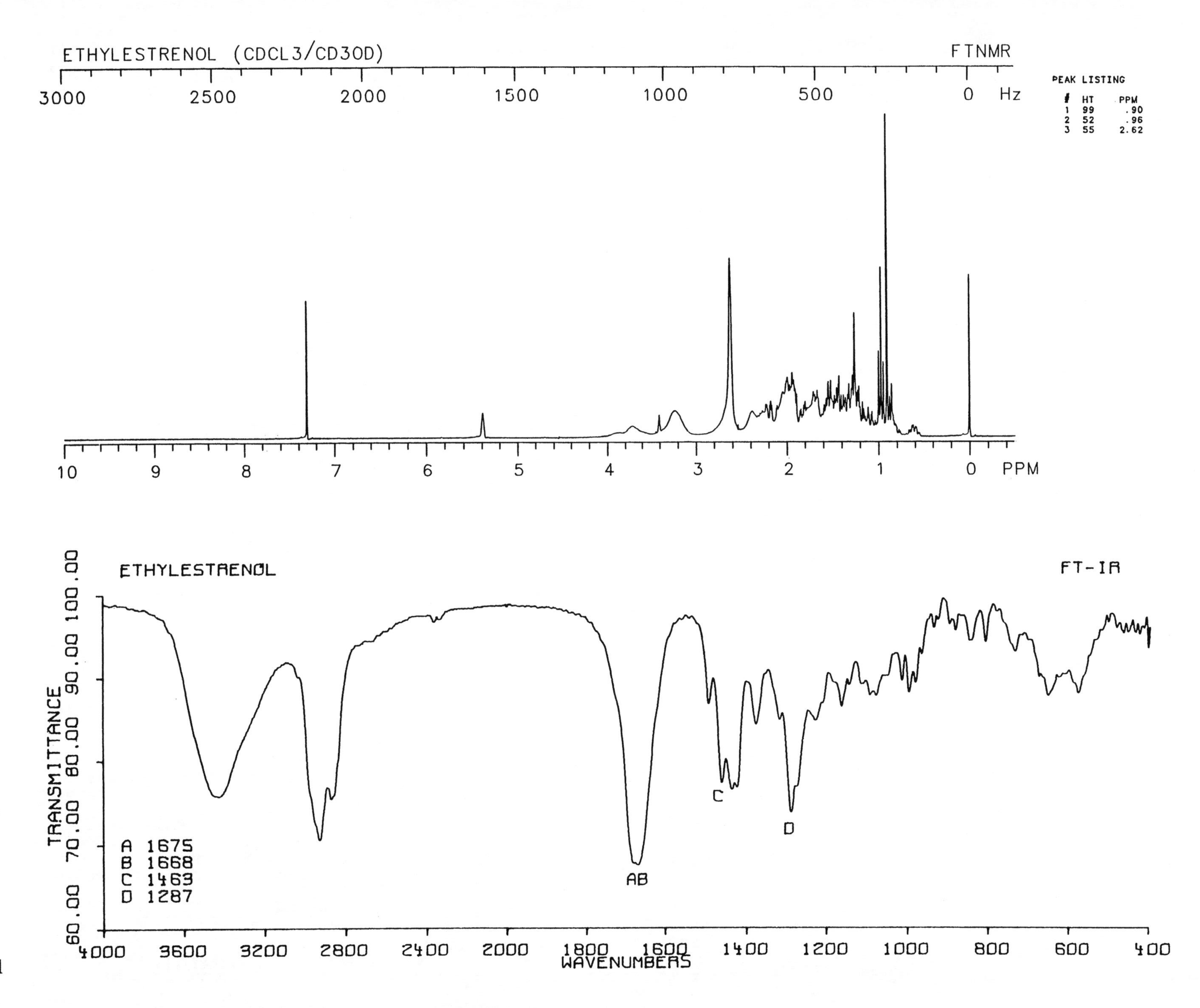
ETHYLESTRENOL (CDCL3/CD3OD)
FTNMR
3000
2500
2000
1500
1000
500
0
Hz
PEAK LISTING
# HT PPM
1 99 .90
2 52 .96
3 55 2.62
10
9
8
7
6
5
4
3
2
1
0
PPM
ETHYLESTRENOL
FT-IR
TRANSMITTANCE
60.00
70.00
80.00
90.00
100.00
A 1675
B 1668
C 1463
D 1287
AB
C
D
4000
3600
3200
2800
2400
2000
1800
1600
1400
1200
1000
800
600
400
WAVENUMBERS

# ETHYL-3-HYDROXYPIPERIDINE

$C_7H_{15}NO$

Molecular weight: 129.20 (129.12)
Synonyms: 1-Ethyl-3-piperidinol

Trade names: Dactilake

Use: Antispasmodic
HPLC: Si-10; 20A:80B; 3.5
GC:

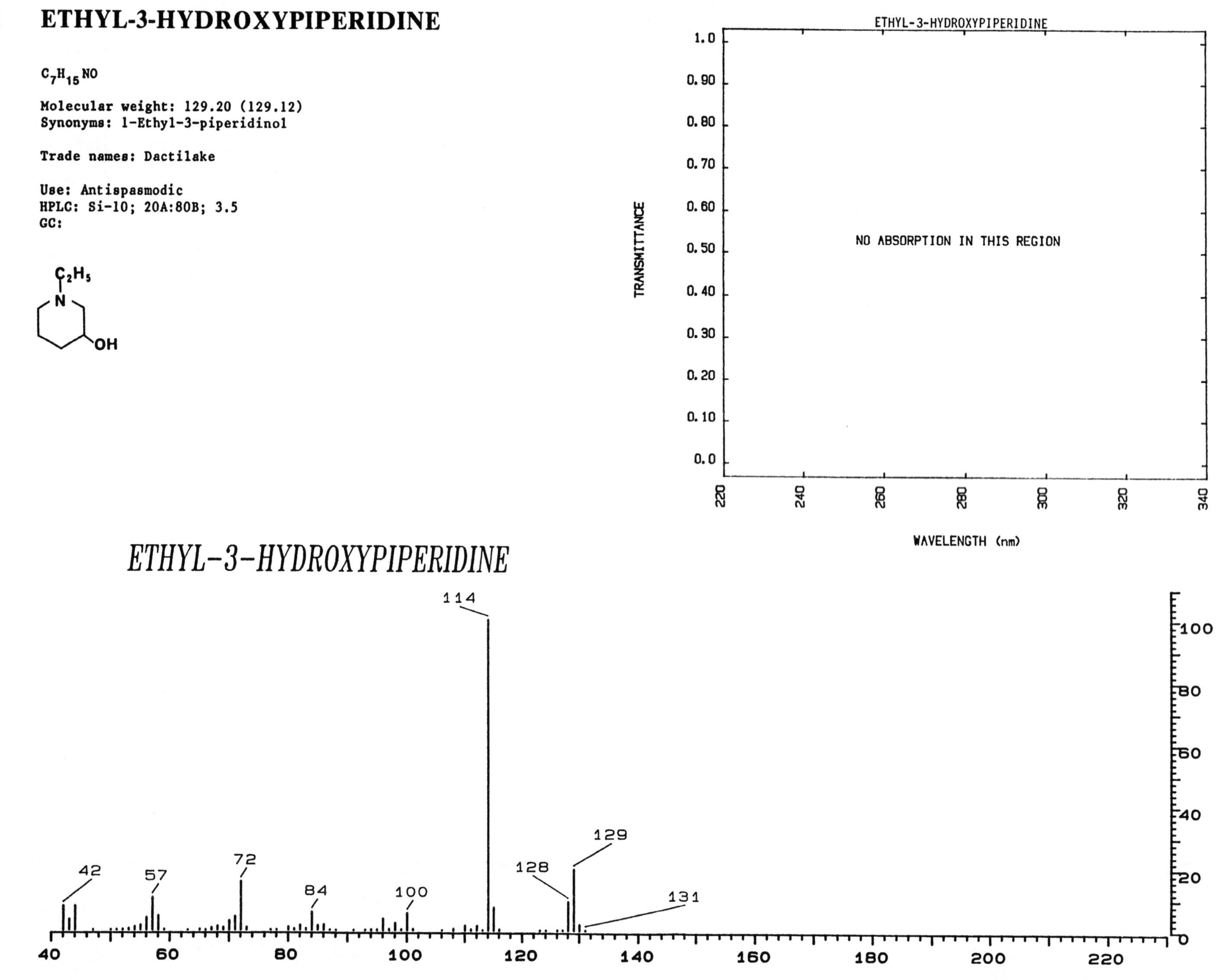

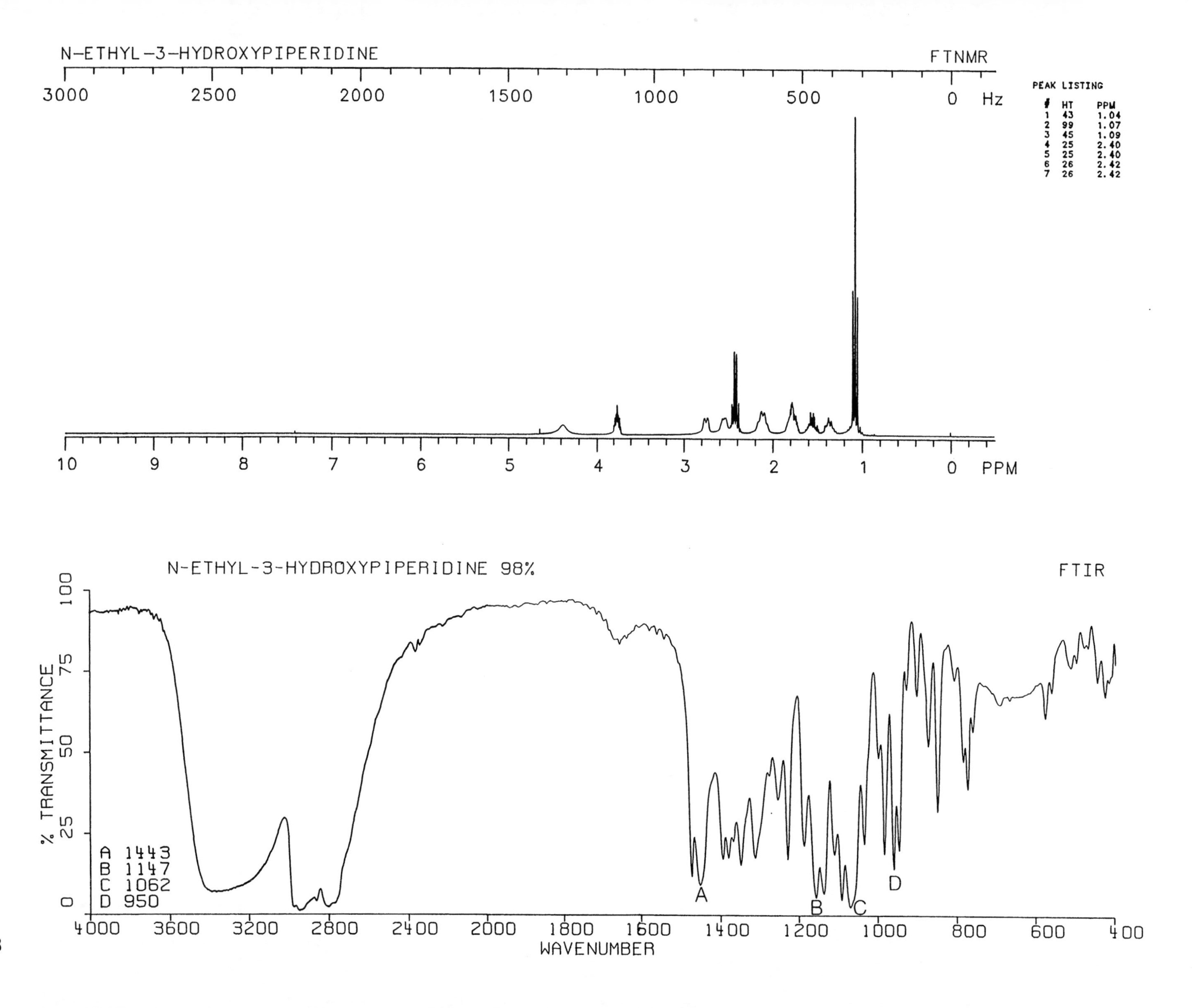
N-ETHYL-3-HYDROXYPIPERIDINE
FTNMR
3000
2500
2000
1500
1000
500
0 Hz
PEAK LISTING
# HT PPM
1 43 1.04
2 99 1.07
3 45 1.09
4 25 2.40
5 25 2.40
6 26 2.42
7 26 2.42
10
9
8
7
6
5
4
3
2
1
0 PPM
N-ETHYL-3-HYDROXYPIPERIDINE 98%
FTIR
% TRANSMITTANCE
100
75
50
25
0
A 1443
B 1147
C 1062
D 950
A
B
C
D
4000
3600
3200
2800
2400
2000
1800
1600
1400
1200
1000
800
600
400
WAVENUMBER

# 2-ETHYLIDENE-1,5-DIMETHYL-3,3-DI-PHENYLPYRROLIDINE

$C_{20}H_{23}N$

**Molecular weight:** 277.41 (277.18)
**Synonyms:** EDDP

**Trade names:**

**Use:** Primary methadone metabolite
**HPLC:**
**GC:** 2081; 200°C

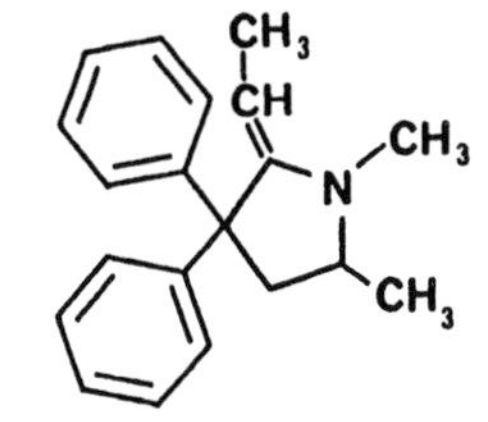


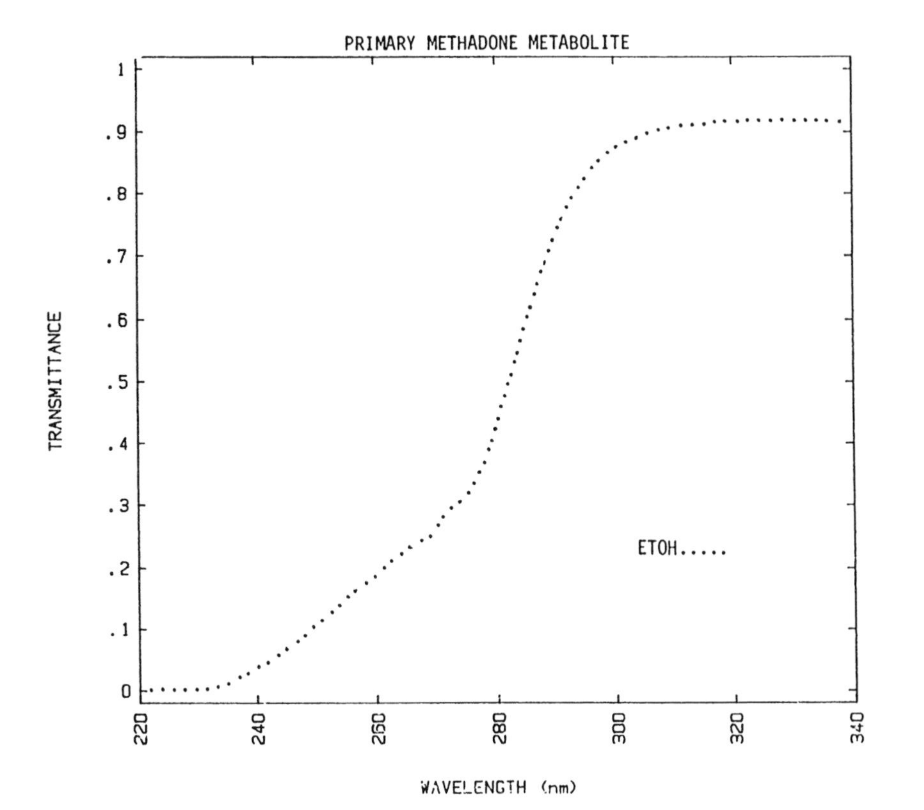

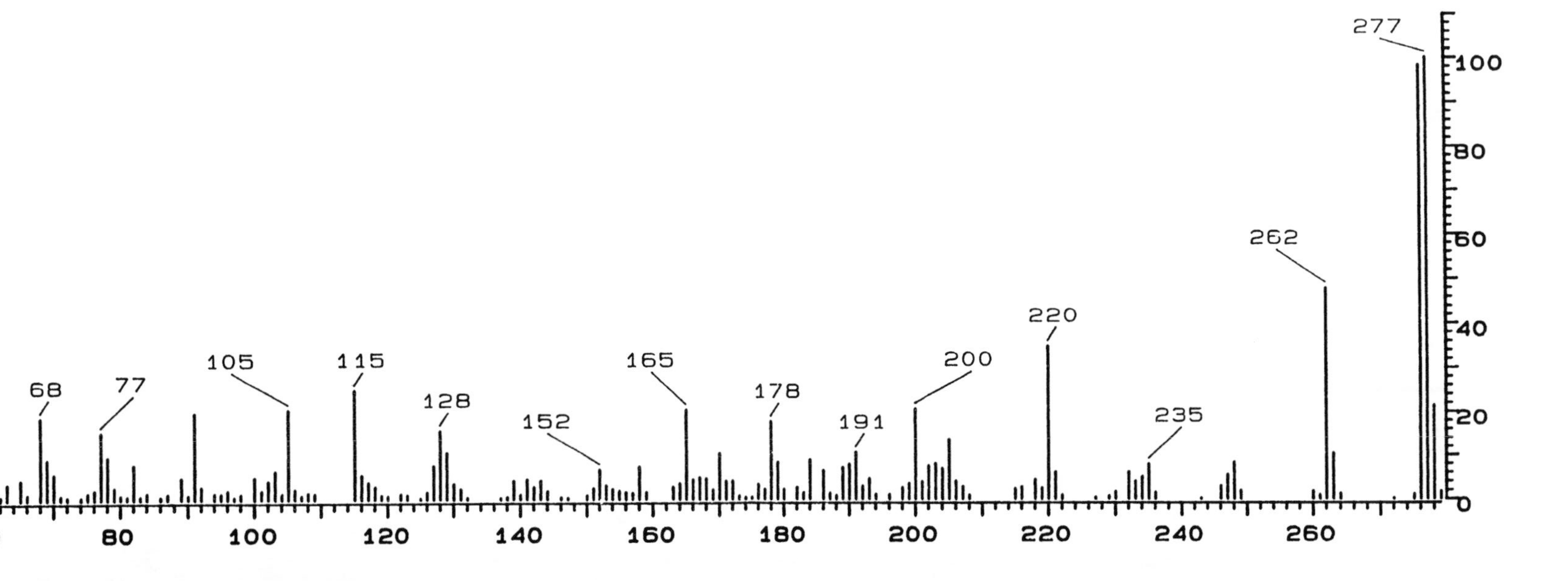

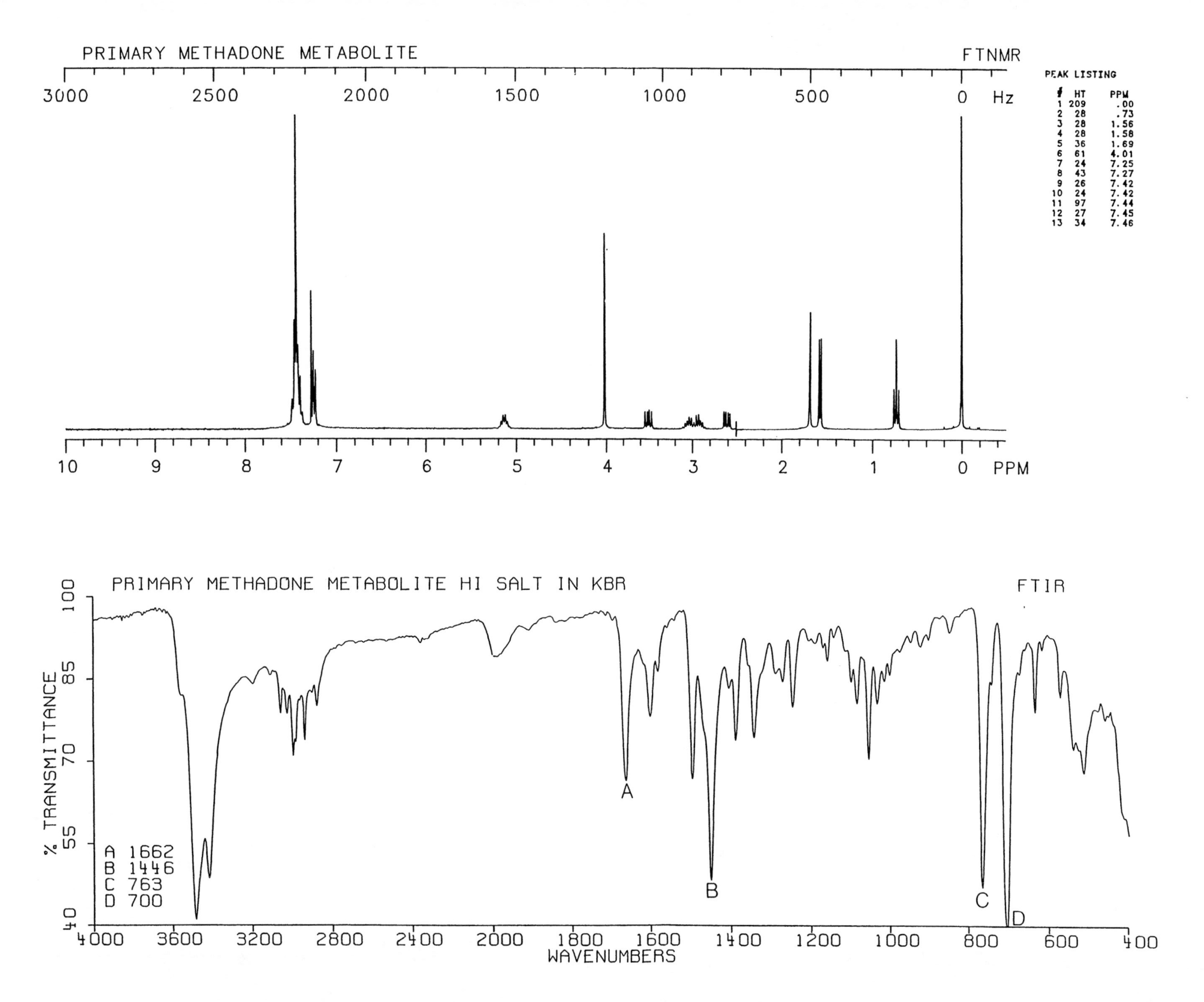
PRIMARY METHADONE METABOLITE
FTNMR
3000 2500 2000 1500 1000 500 0 Hz
10 9 8 7 6 5 4 3 2 1 0 PPM
PEAK LISTING
# HT PPM
1 209 .00
2 28 .73
3 28 1.56
4 28 1.58
5 36 1.69
6 61 4.01
7 24 7.25
8 43 7.27
9 26 7.42
10 24 7.42
11 97 7.44
12 27 7.45
13 34 7.46
PRIMARY METHADONE METABOLITE HI SALT IN KBR
FTIR
% TRANSMITTANCE
100 85 70 55 40
4000 3600 3200 2800 2400 2000 1800 1600 1400 1200 1000 800 600 400
WAVENUMBERS
A 1662
B 1446
C 763
D 700
A
B
C
D

# 2-ETHYL-5-METHYL-3,3-DIPHENYLPYRROLINE

$C_{19}H_{21}N$

**Molecular weight:** 263.38 (263.17)
**Synonyms:** EMDP

**Trade names:**

**Use:** Secondary methadone metabolite
**HPLC:**
**GC:** 1946; 250°C

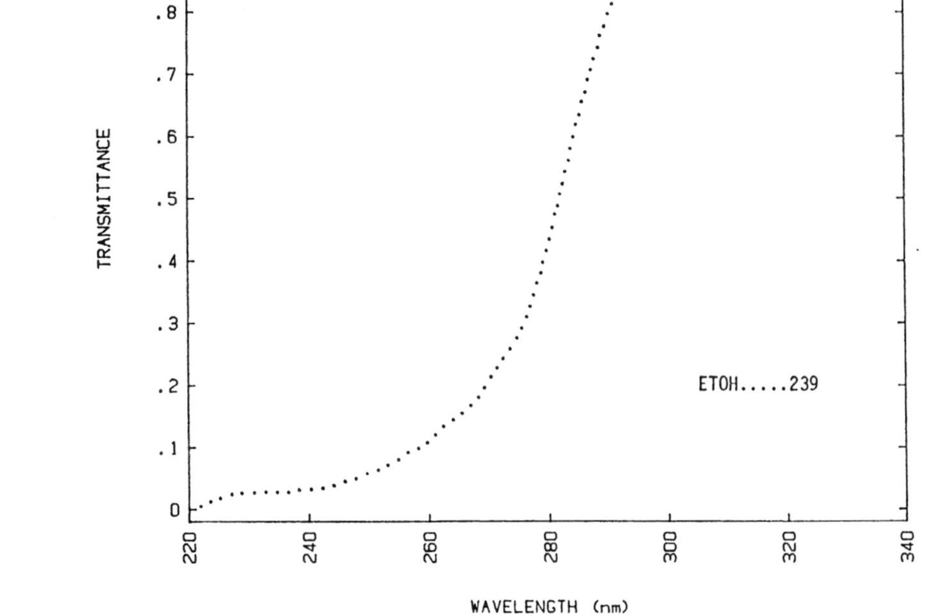


SECONDARY METHADONE METABOLITE

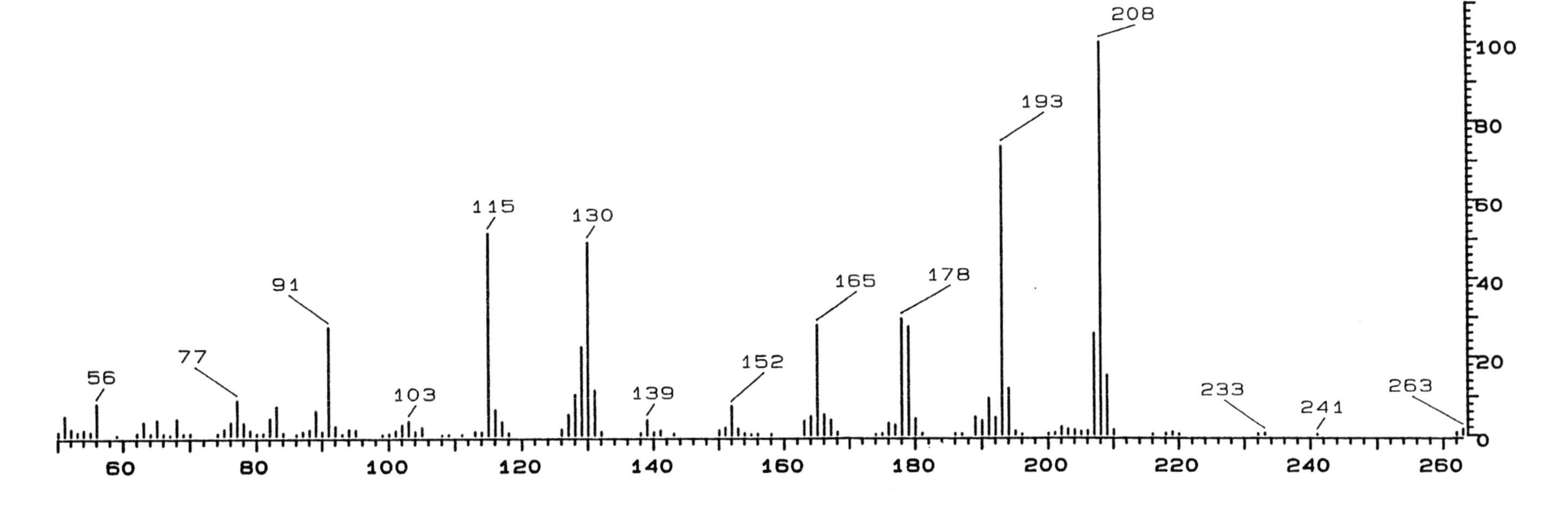

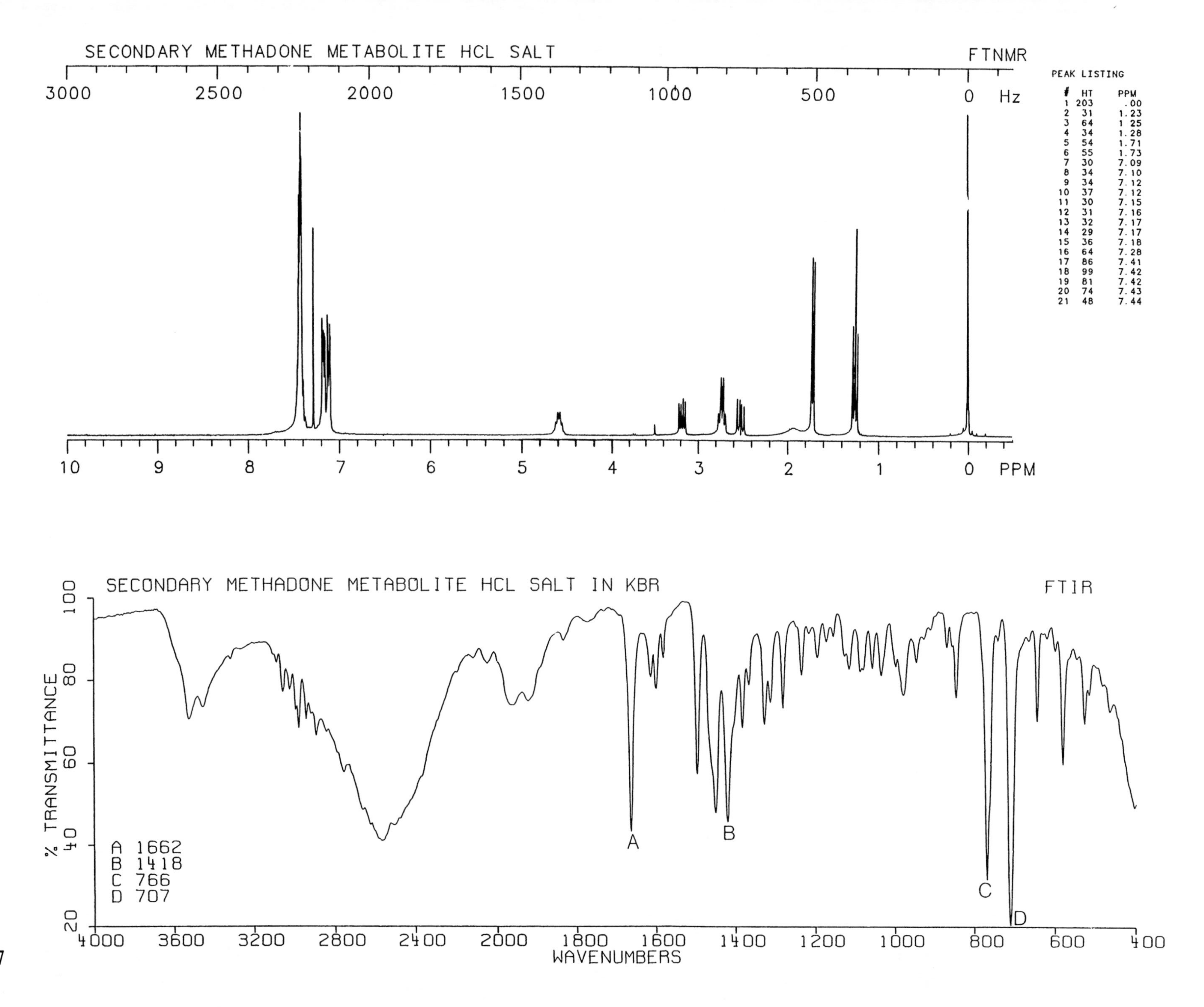

SECONDARY METHADONE METABOLITE HCL SALT
FTNMR
3000 2500 2000 1500 1000 500 0 Hz
10 9 8 7 6 5 4 3 2 1 0 PPM
PEAK LISTING
# HT PPM
1 203 .00
2 31 1.23
3 64 1.25
4 34 1.28
5 54 1.71
6 55 1.73
7 30 7.09
8 34 7.10
9 34 7.12
10 37 7.12
11 30 7.15
12 31 7.16
13 32 7.17
14 29 7.17
15 36 7.18
16 64 7.28
17 86 7.41
18 99 7.42
19 81 7.42
20 74 7.43
21 48 7.44
SECONDARY METHADONE METABOLITE HCL SALT IN KBR
FTIR
% TRANSMITTANCE
100 80 60 40 20
A 1662
B 1418
C 766
D 707
4000 3600 3200 2800 2400 2000 1800 1600 1400 1200 1000 800 600 400
WAVENUMBERS

# ETHYLMORPHINE

$C_{19}H_{23}NO_3$

Molecular weight: 313.40 (313.17)
Synonyms: 7,8-Didehydro-4,5-epoxy-3-ethoxy-17-methylmorphinan-6-ol

Trade names: Dionin

Use: Chemotic, analgesic, antitussive
HPLC: Si-10; 10A:90B; 4.0
GC: 2481; 250°C

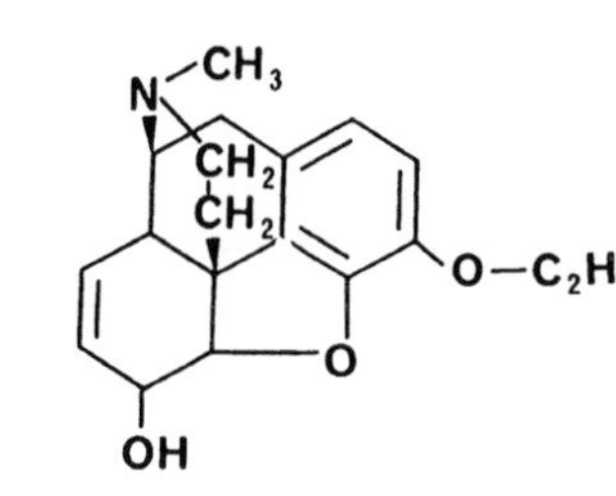


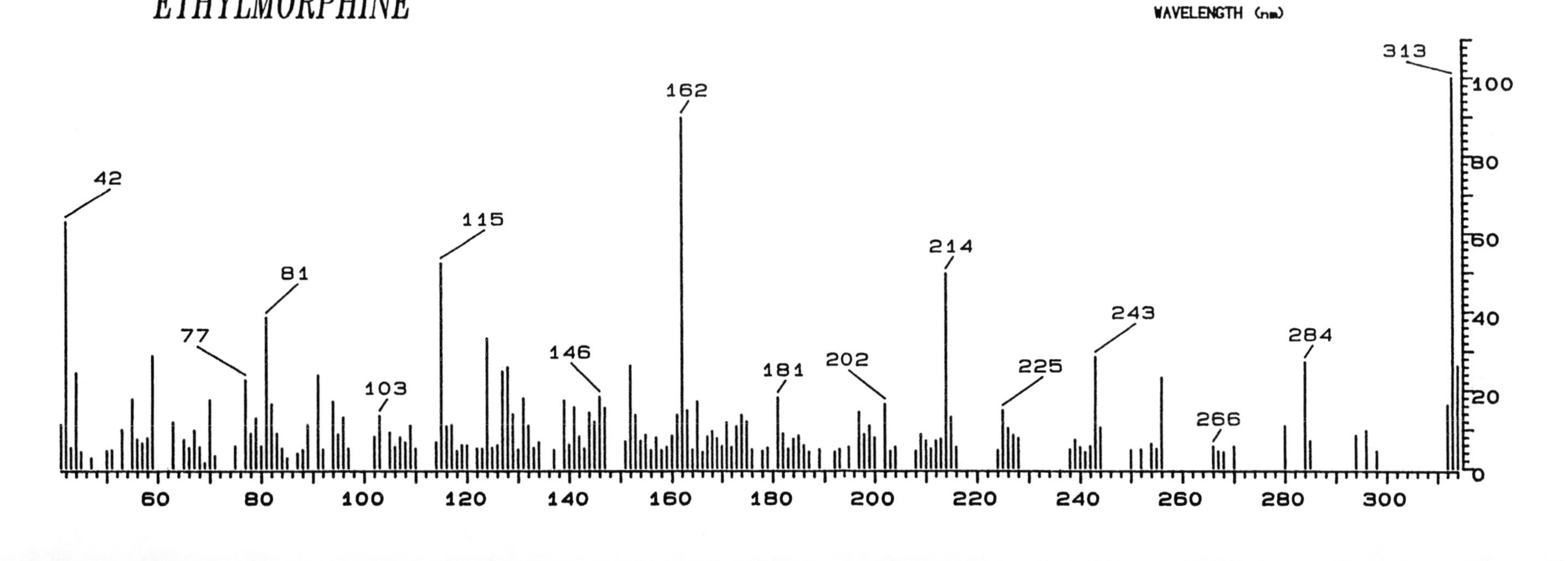

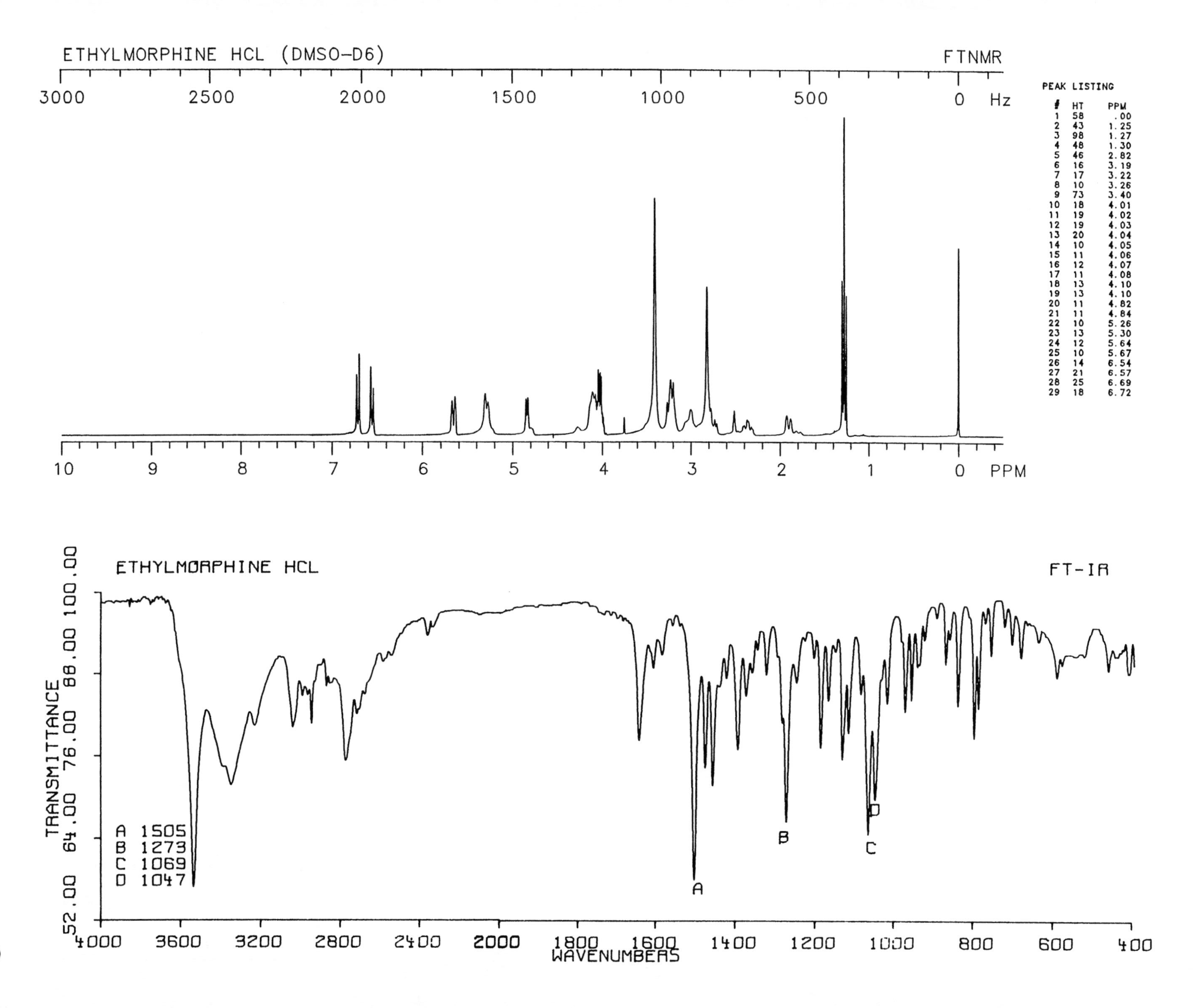
ETHYLMORPHINE HCL (DMSO-D6)
FTNMR
3000 2500 2000 1500 1000 500 0 Hz
10 9 8 7 6 5 4 3 2 1 0 PPM
PEAK LISTING
# HT PPM
1 58 .00
2 43 1.25
3 98 1.27
4 48 1.30
5 46 2.82
6 16 3.19
7 17 3.22
8 10 3.26
9 73 3.40
10 18 4.01
11 19 4.02
12 19 4.03
13 20 4.04
14 10 4.05
15 11 4.06
16 12 4.07
17 11 4.08
18 13 4.10
19 13 4.10
20 11 4.82
21 11 4.84
22 10 5.26
23 13 5.30
24 12 5.64
25 10 5.67
26 14 6.54
27 21 6.57
28 25 6.69
29 18 6.72
ETHYLMORPHINE HCL
FT-IR
TRANSMITTANCE
52.00 64.00 76.00 88.00 100.00
A 1505
B 1273
C 1069
D 1047
4000 3600 3200 2800 2400 2000 1800 1600 1400 1200 1000 800 600 400
WAVENUMBERS

# ETHYLNOREPINEPHRINE

$C_{10}H_{15}NO_3$

**Molecular weight:** 197.24 (197.11)
**Synonyms:** 4-(2-Amino-1-hydroxybutyl)-1,2-benzenediol; ethylnoradrenaline; ethylnorsuprarenin
**Trade names:** Bronkephrine

**Use:** Adrenergic
**HPLC:** Si-10; 20A:80B; 4.8
**GC:**

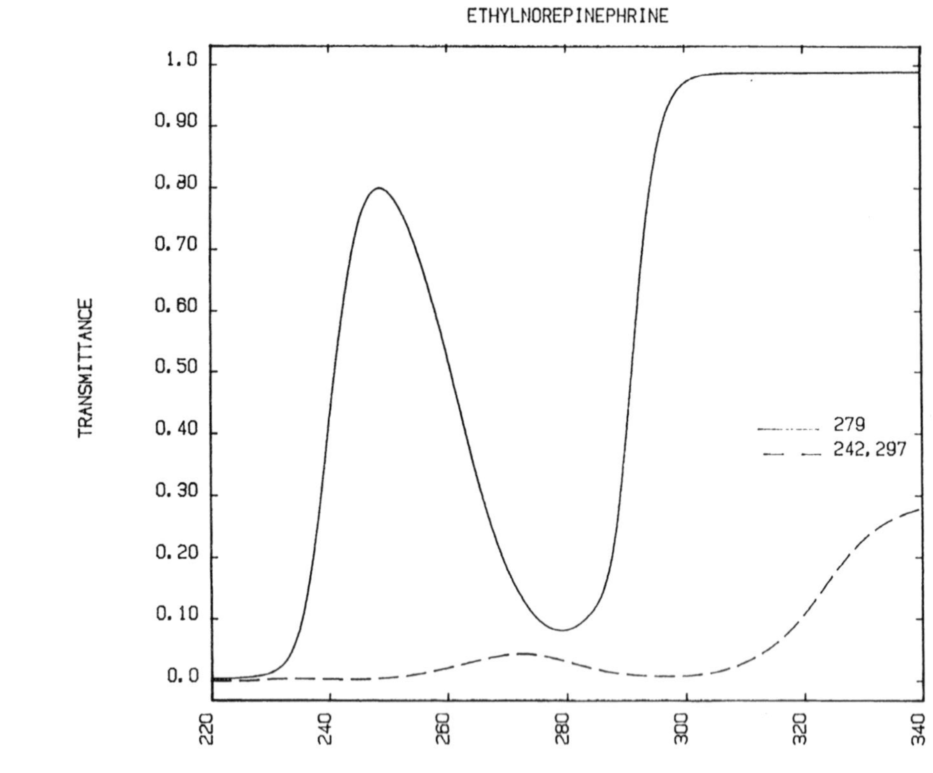

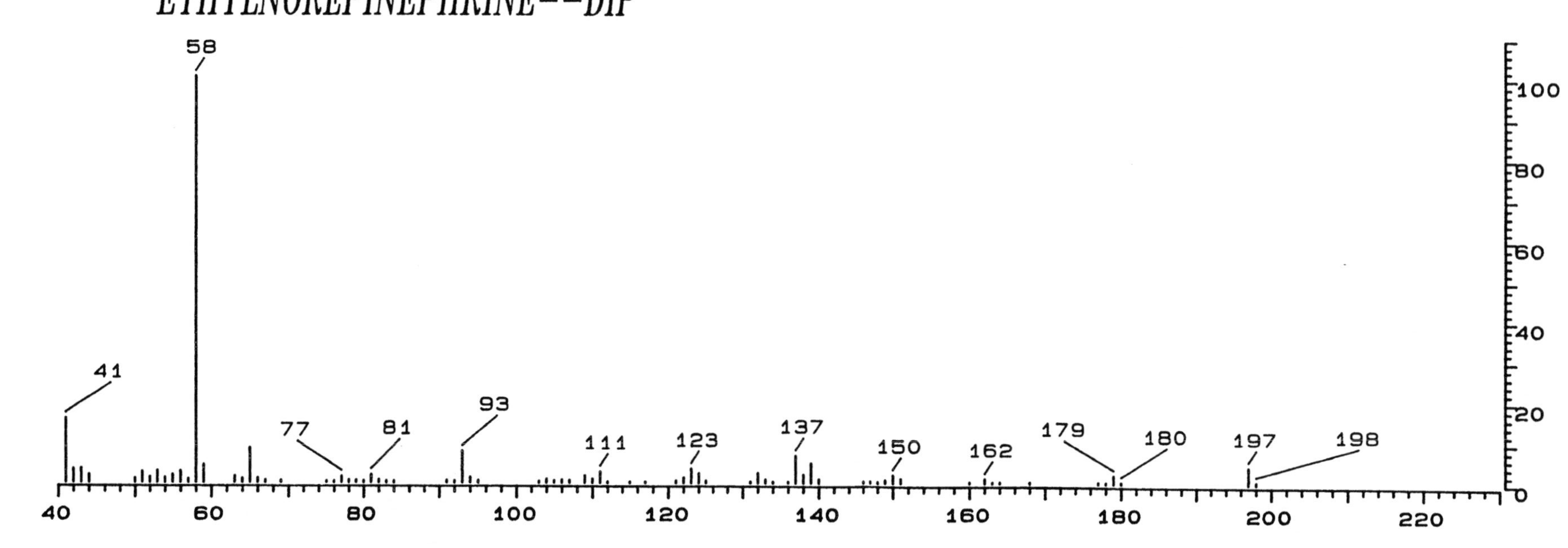

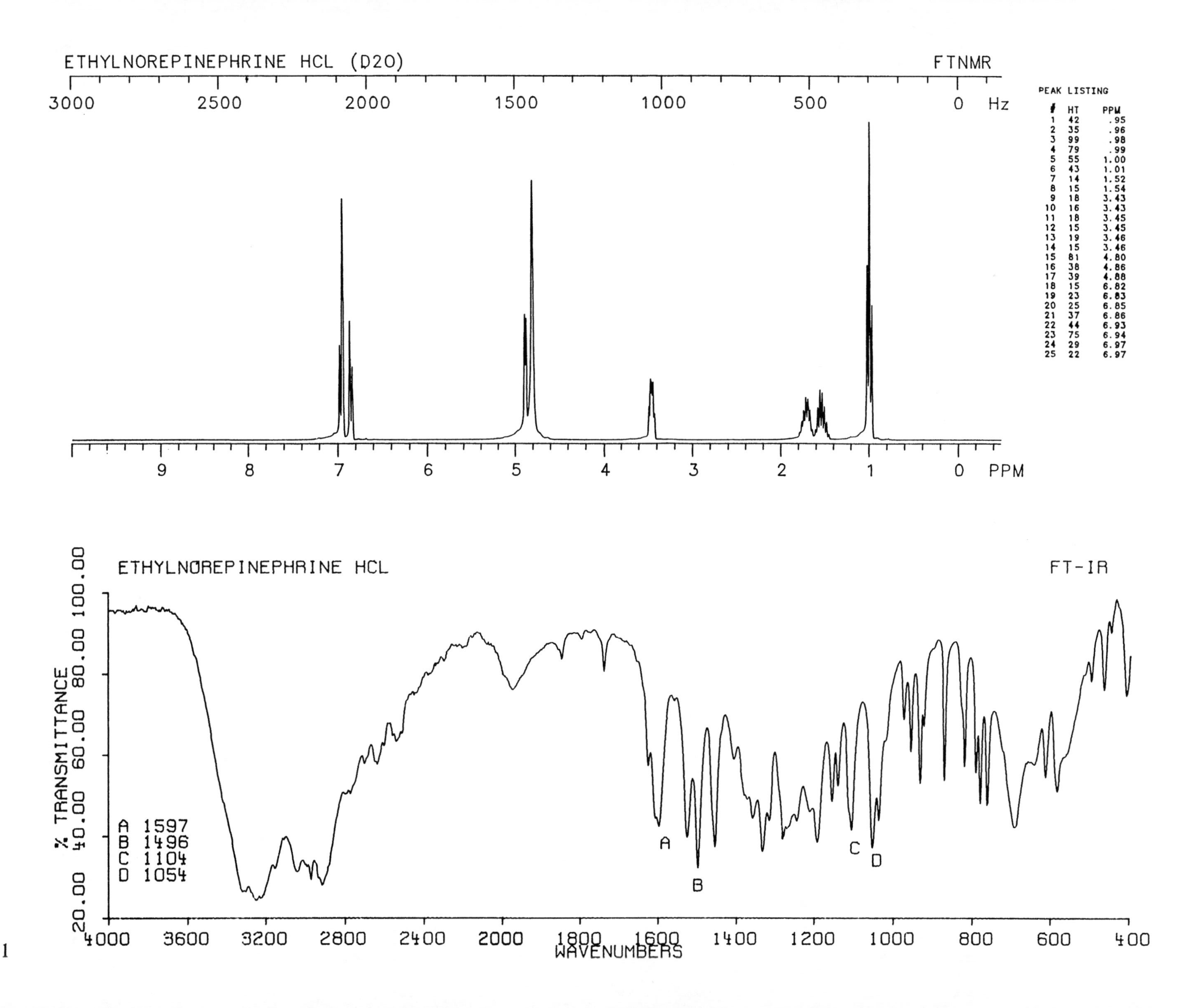
ETHYLNOREPINEPHRINE HCL (D2O)
FTNMR
3000 2500 2000 1500 1000 500 0 Hz
9 8 7 6 5 4 3 2 1 0 PPM
PEAK LISTING
# HT PPM
1 42 .95
2 35 .96
3 99 .98
4 79 .99
5 55 1.00
6 43 1.01
7 14 1.52
8 15 1.54
9 18 3.43
10 16 3.43
11 18 3.45
12 15 3.45
13 19 3.46
14 15 3.46
15 81 4.80
16 38 4.86
17 39 4.88
18 15 6.82
19 23 6.83
20 25 6.85
21 37 6.86
22 44 6.93
23 75 6.94
24 29 6.97
25 22 6.97
ETHYLNOREPINEPHRINE HCL
FT-IR
% TRANSMITTANCE
20.00 40.00 60.00 80.00 100.00
4000 3600 3200 2800 2400 2000 1800 1600 1400 1200 1000 800 600 400
WAVENUMBERS
A 1597
B 1496
C 1104
D 1054
A B C D

# 2-ETHYL-2-PHENYLMALONAMIDE

$C_{11}H_{14}N_2O_2$

Molecular weight: 206.23 (206.11)
Synonyms: PEMA; malonodiamide

Trade names:

Use: Urinary metabolite of primidone
HPLC: Si-10; 2A:98B; 3.8
GC: 2000; 200°C

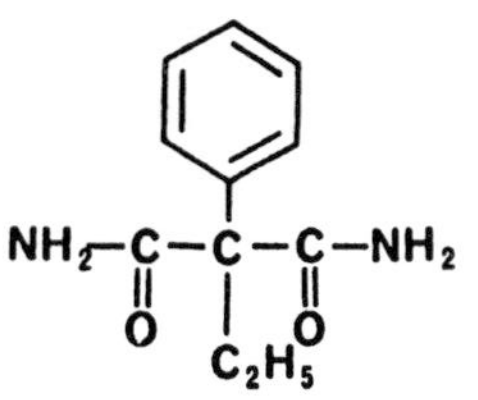


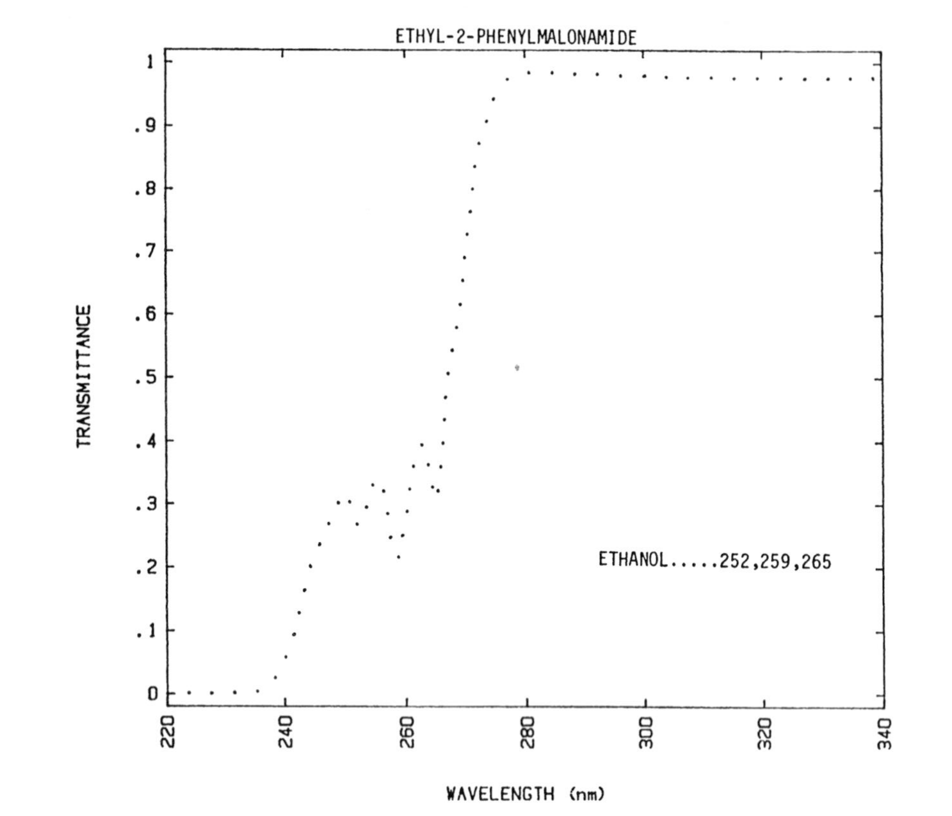

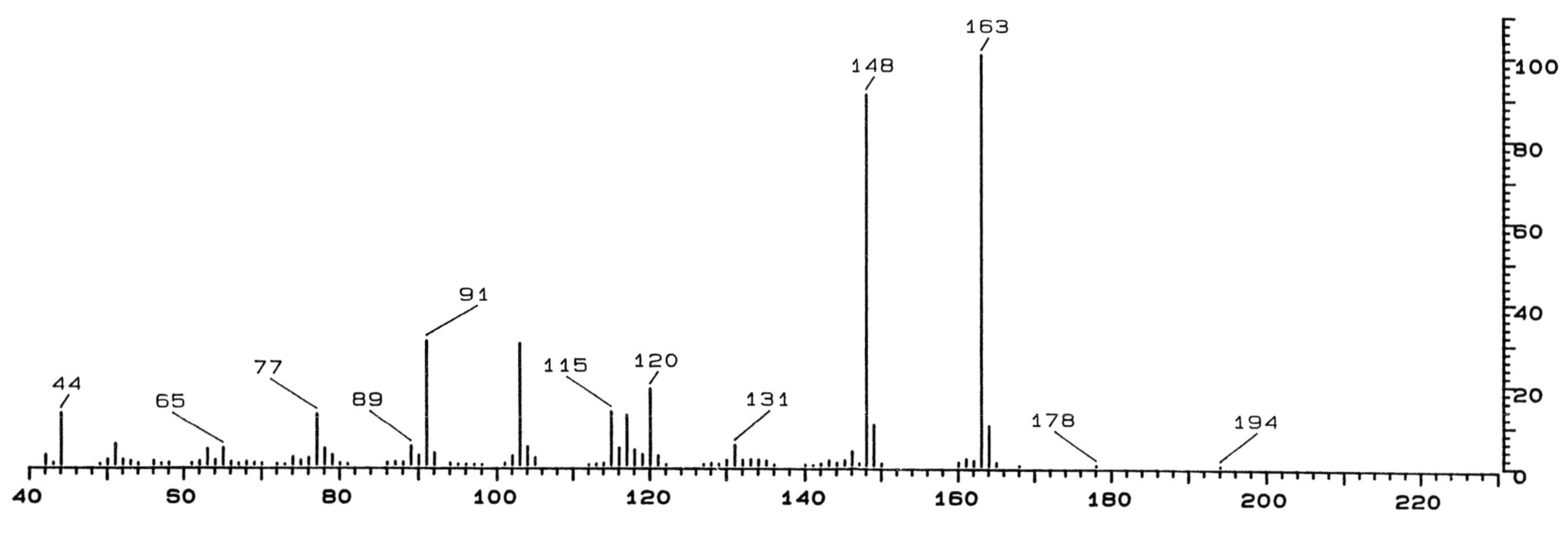

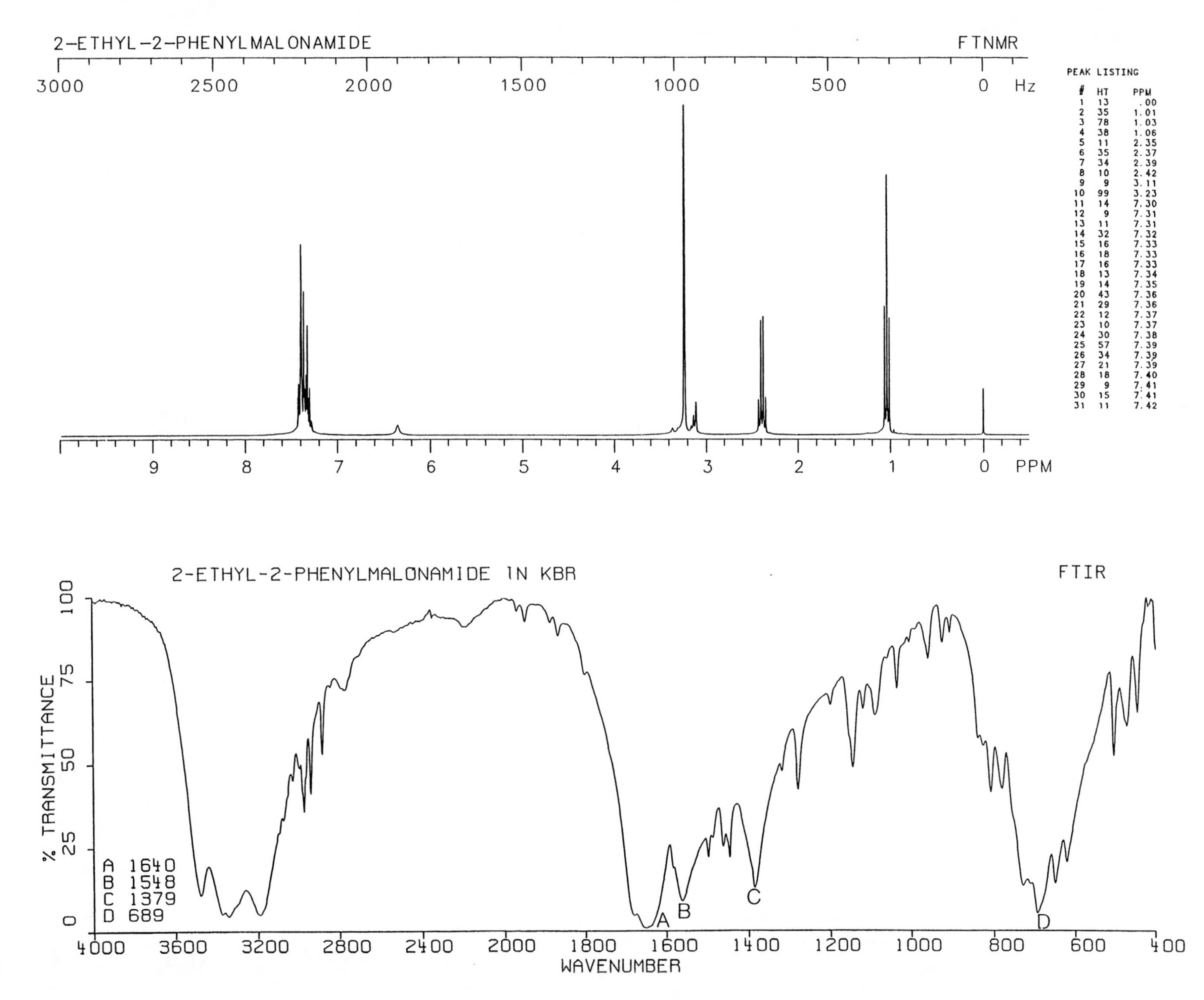
2-ETHYL-2-PHENYLMALONAMIDE
FTNMR
Hz
PPM
PEAK LISTING
# HT PPM
1 13 .00
2 35 1.01
3 78 1.03
4 38 1.06
5 11 2.35
6 35 2.37
7 34 2.39
8 10 2.42
9 9 3.11
10 99 3.23
11 14 7.30
12 9 7.31
13 11 7.31
14 32 7.32
15 16 7.33
16 18 7.33
17 16 7.33
18 13 7.34
19 14 7.35
20 43 7.36
21 29 7.36
22 12 7.37
23 10 7.37
24 30 7.38
25 57 7.39
26 34 7.39
27 21 7.39
28 18 7.40
29 9 7.41
30 15 7.41
31 11 7.42
2-ETHYL-2-PHENYLMALONAMIDE IN KBR
FTIR
% TRANSMITTANCE
WAVENUMBER
A 1640
B 1548
C 1379
D 689

# 5-ETHYL-5-PHENYLHYDANTOIN

$C_{11}H_{12}N_2O_2$

**Molecular weight:** 204.23 (204.09)
**Synonyms:** 5-Ethyl-5-phenyl-2,4-imidazolidinedione

**Trade names:**

**Use:** Anticonvulsant
**HPLC:** Si-10; 1A:99B; 8.2
**GC:** 1965; 200°C

$CH_3$–$CH_2$, $C_6H_5$ (5-ethyl-5-phenyl-2,4-imidazolidinedione ring: NH, C=O, NH, C=O)

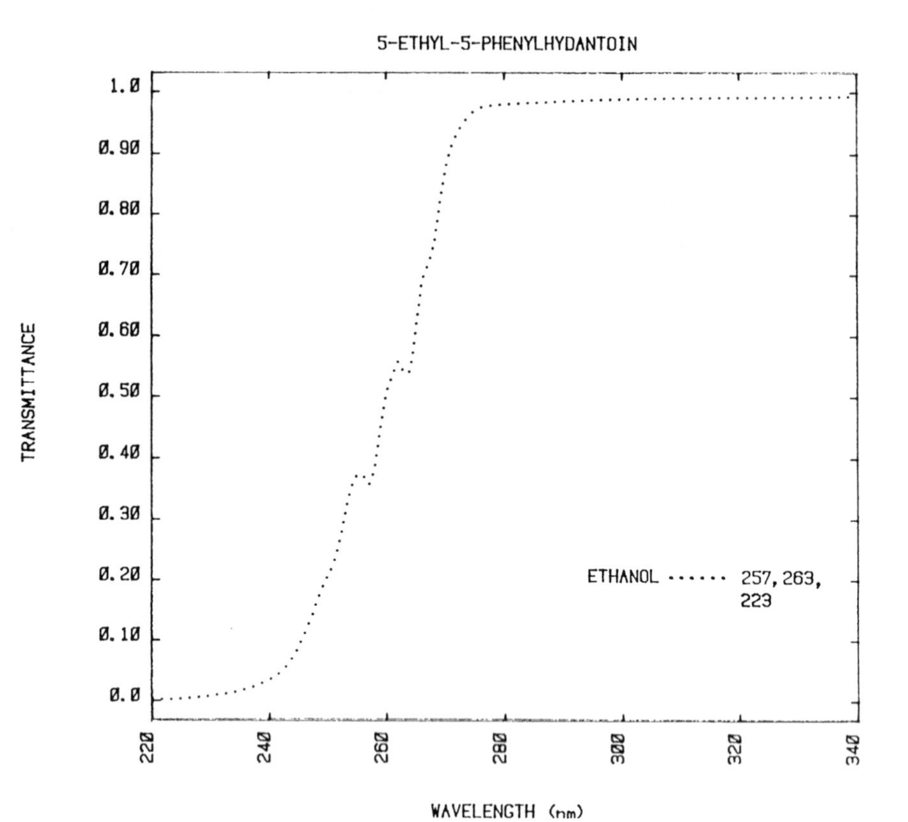


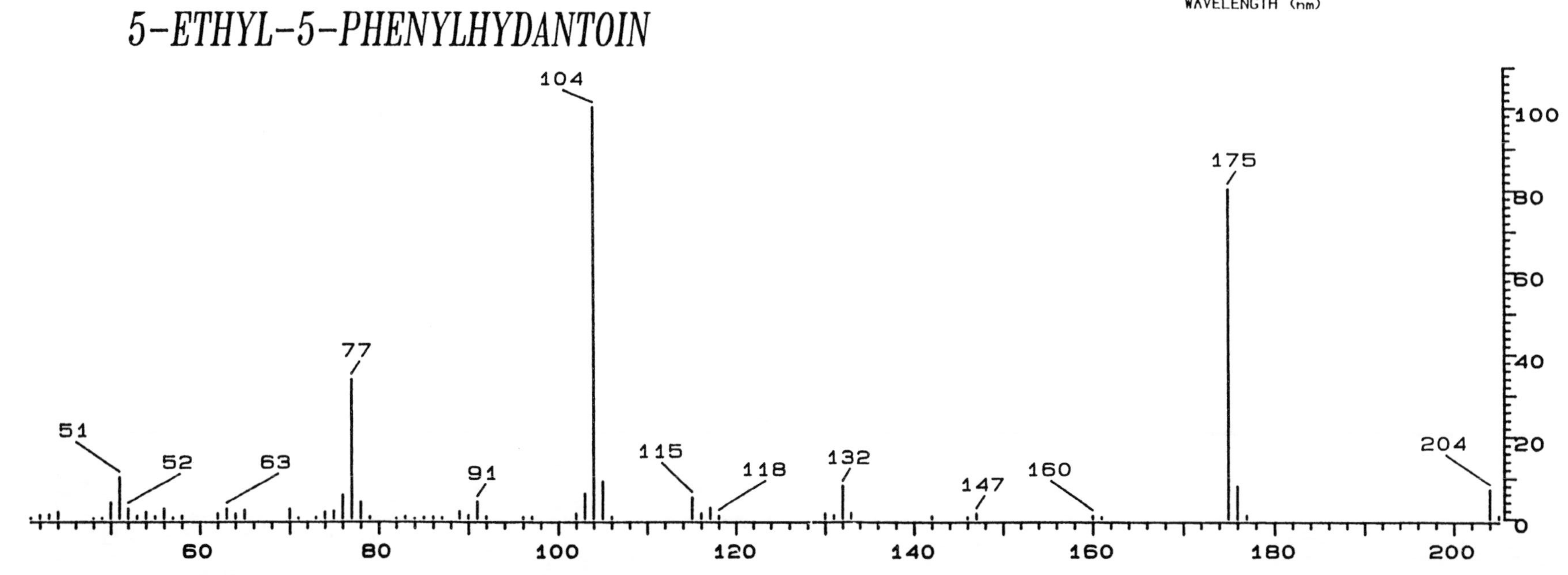

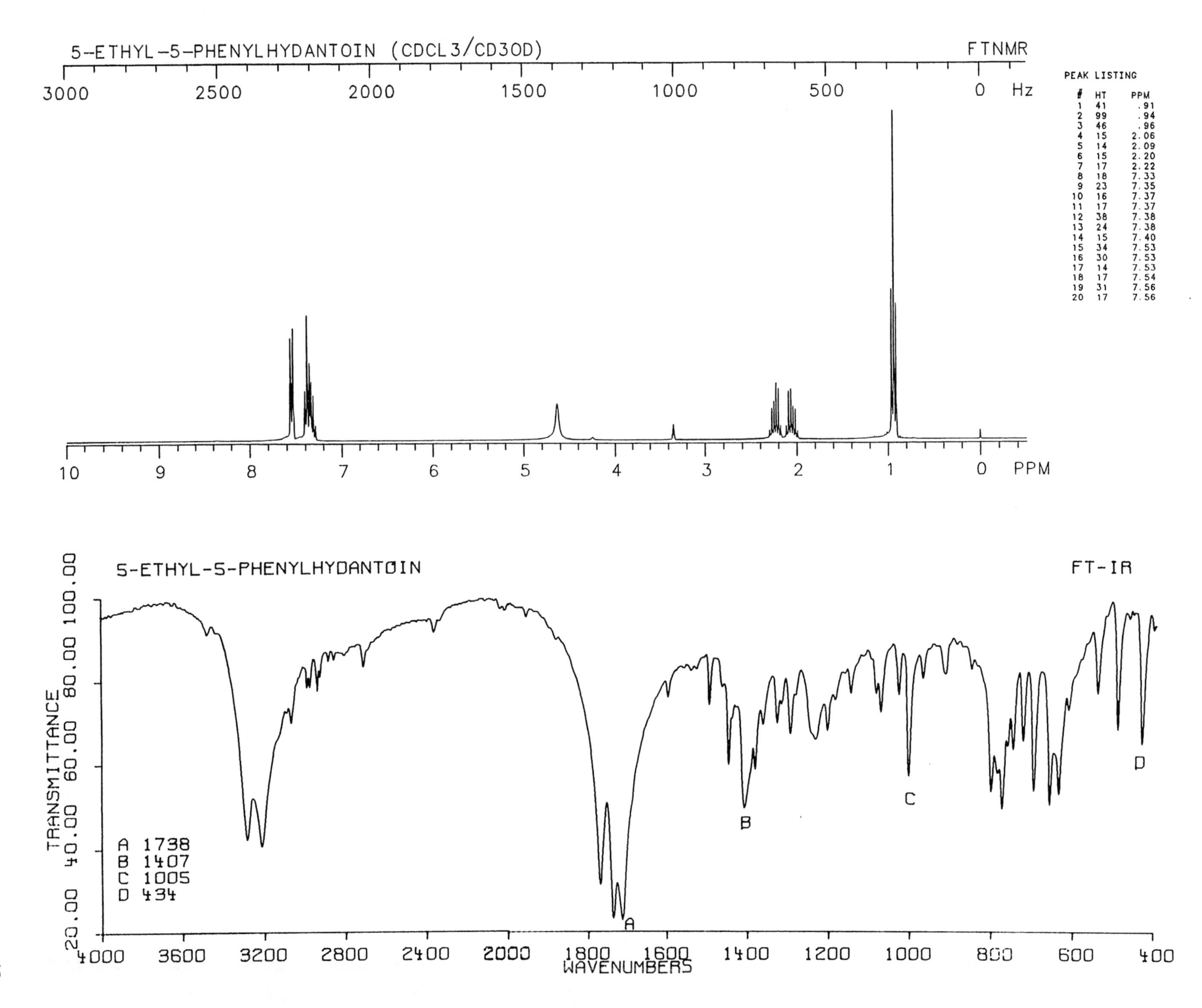
5-ETHYL-5-PHENYLHYDANTOIN (CDCL3/CD3OD)
FTNMR
3000 2500 2000 1500 1000 500 0 Hz
10 9 8 7 6 5 4 3 2 1 0 PPM
PEAK LISTING
# HT PPM
1 41 .91
2 99 .94
3 46 .96
4 15 2.06
5 14 2.09
6 15 2.20
7 17 2.22
8 18 7.33
9 23 7.35
10 16 7.37
11 17 7.37
12 38 7.38
13 24 7.38
14 15 7.40
15 34 7.53
16 30 7.53
17 14 7.53
18 17 7.54
19 31 7.56
20 17 7.56
5-ETHYL-5-PHENYLHYDANTØIN
FT-IR
TRANSMITTANCE
20.00 40.00 60.00 80.00 100.00
A 1738
B 1407
C 1005
D 434
4000 3600 3200 2800 2400 2000 1800 1600 1400 1200 1000 800 600 400
WAVENUMBERS
A
B
C
D

# ETHYL-3-PIPERIDYLBENZILATE

$C_{21}H_{25}NO_3$

Molecular weight: 339.40 (339.18)
Synonyms: 1-Ethylpiperid-3-yl benzilate, JB318

Trade names:

Use: Hallucinogen
HPLC:
GC:

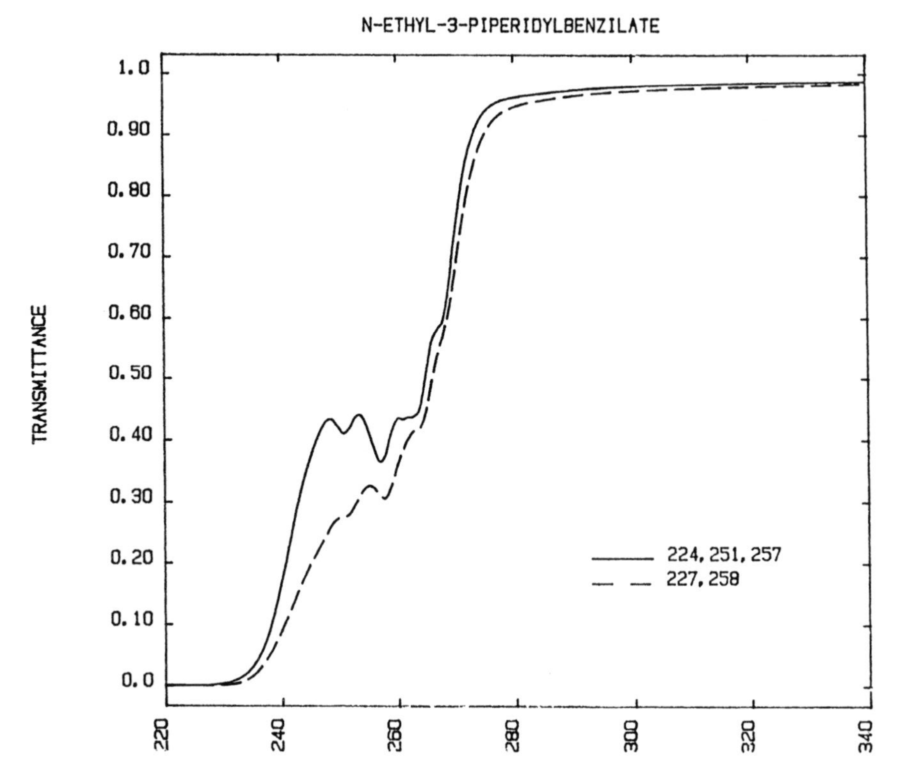


NO USEFUL MASS SPECTRUM WAS OBTAINED

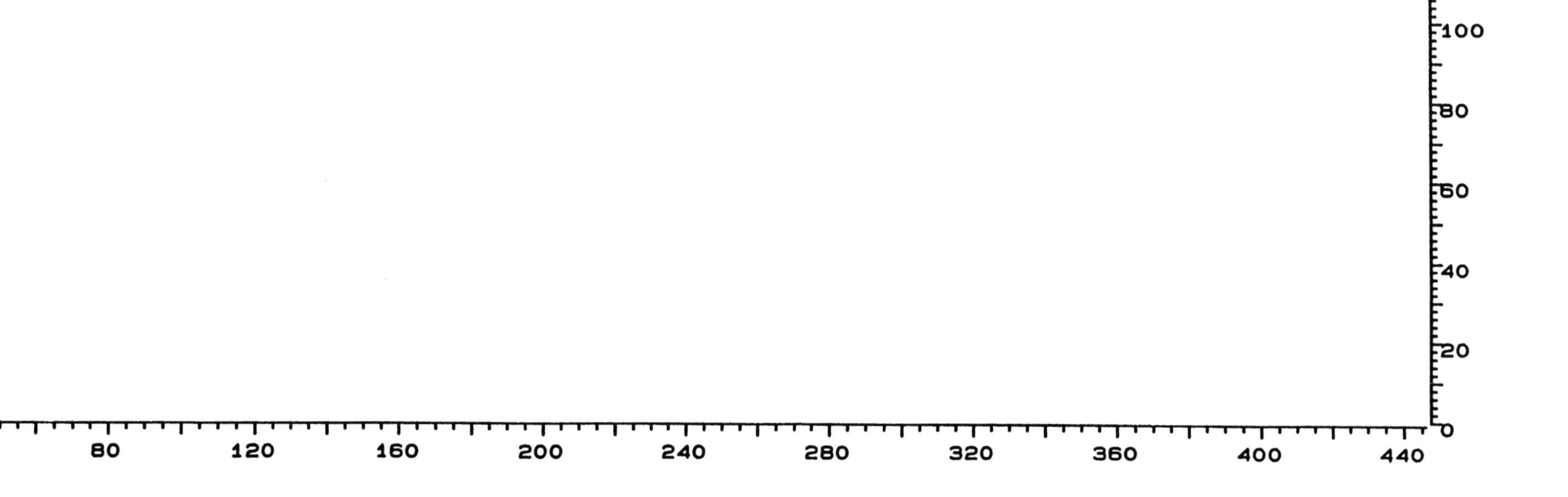

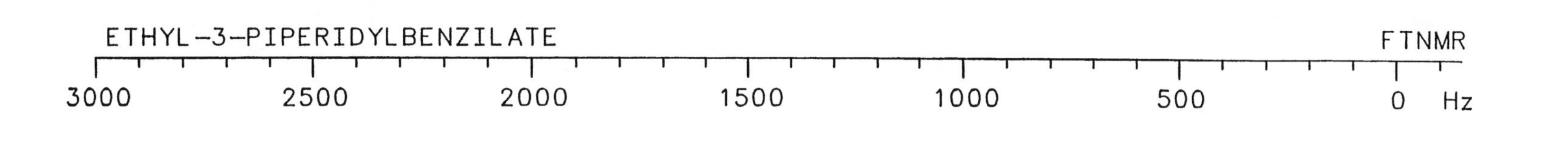
ETHYL-3-PIPERIDYLBENZILATE
FTNMR
3000
2500
2000
1500
1000
500
0 Hz
INSUFFICIENT SAMPLE
10
9
8
7
6
5
4
3
2
1
0 PPM

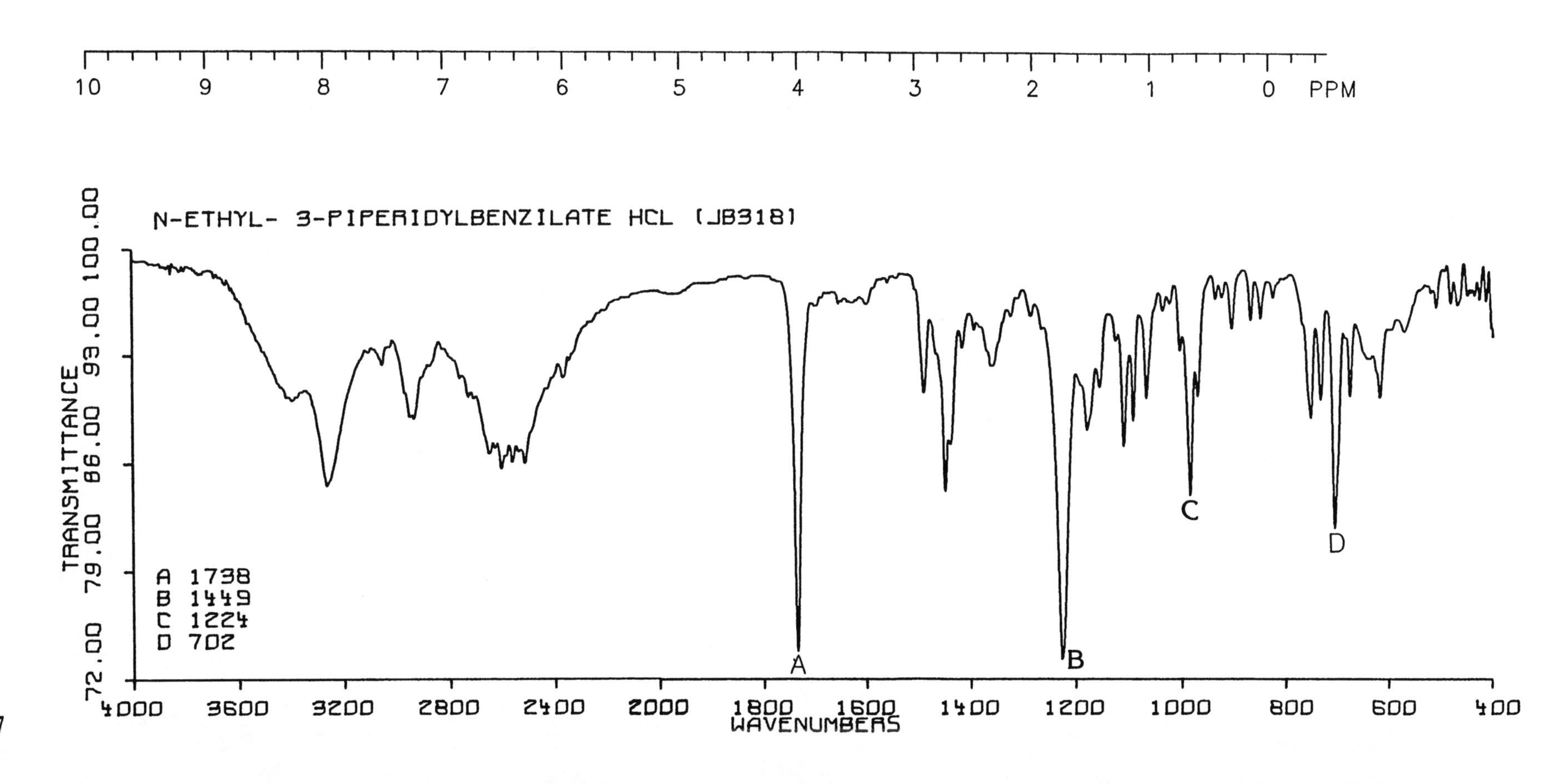
N-ETHYL- 3-PIPERIDYLBENZILATE HCL (JB318)
TRANSMITTANCE
72.00
79.00
86.00
93.00
100.00
A 1738
B 1449
C 1224
D 702
A
B
C
D
4000
3600
3200
2800
2400
2000
1800
1600
1400
1200
1000
800
600
400
WAVENUMBERS

# 5-ETHYL-5-*P*-TOLYBARBITURATE

$C_{13}H_{14}N_2O_3$

Molecular weight: 246.25 (246.10)
Synonyms: 5-(p-Tolyl)-5-(ethyl)-2,4,6(1H,3H,5H)-pyrimidinetrione; tolylbarb
Trade names:

Use: Barbiturate
HPLC: Si-10; 1A:99B; 3.4
GC: 2111; 200°C

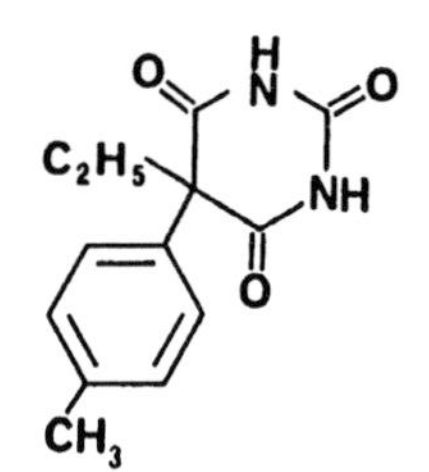


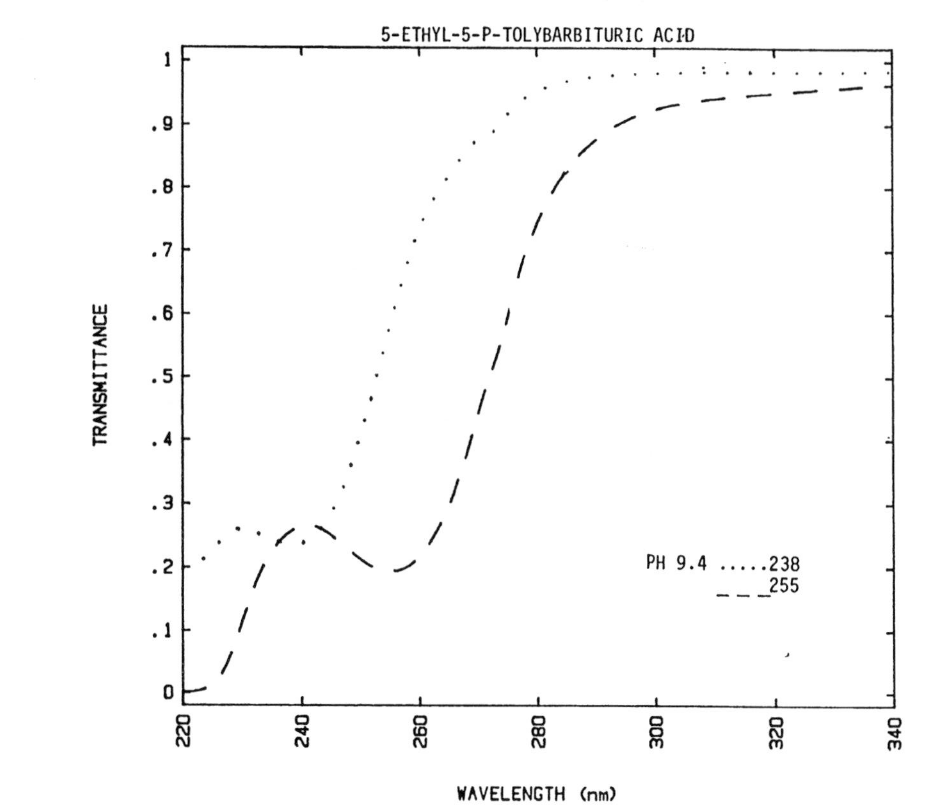


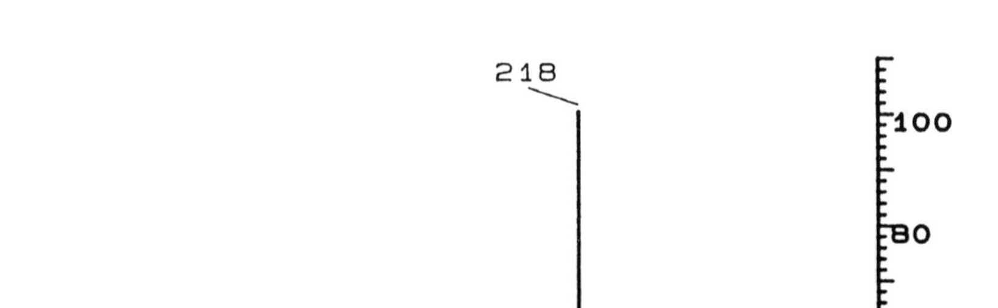


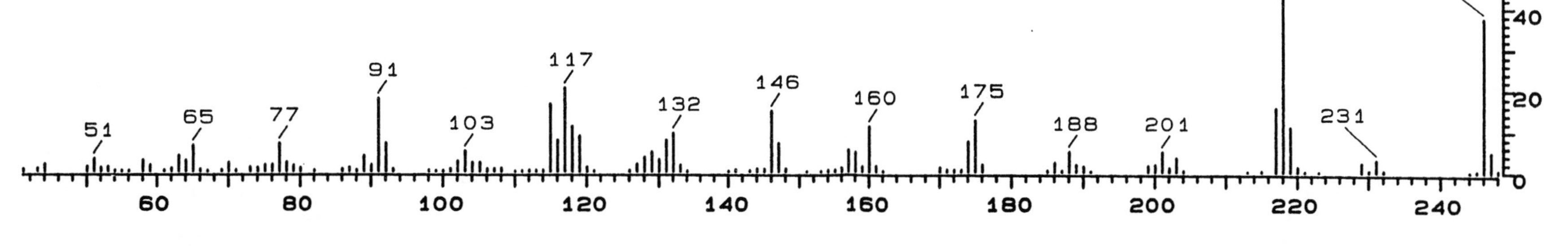

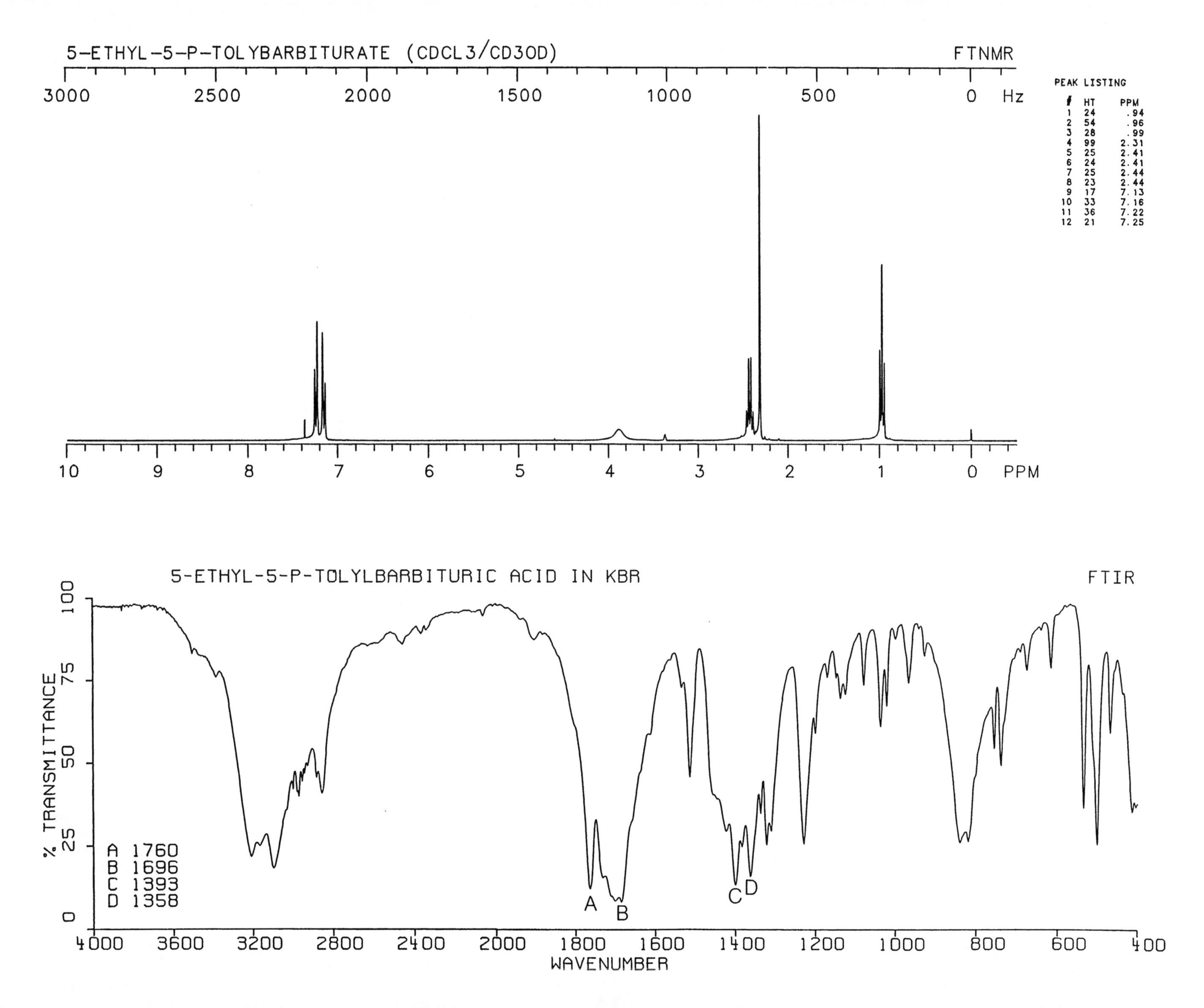
5-ETHYL-5-P-TOLYBARBITURATE (CDCL3/CD3OD)
FTNMR
3000 2500 2000 1500 1000 500 0 Hz
PEAK LISTING
# HT PPM
1 24 .94
2 54 .96
3 28 .99
4 99 2.31
5 25 2.41
6 24 2.41
7 25 2.44
8 23 2.44
9 17 7.13
10 33 7.16
11 36 7.22
12 21 7.25
10 9 8 7 6 5 4 3 2 1 0 PPM
5-ETHYL-5-P-TOLYLBARBITURIC ACID IN KBR
FTIR
% TRANSMITTANCE
100 75 50 25 0
A 1760
B 1696
C 1393
D 1358
A B C D
4000 3600 3200 2800 2400 2000 1800 1600 1400 1200 1000 800 600 400
WAVENUMBER

# ETHYLTRYPTAMINE

$C_{12}H_{16}N_2$

**Molecular weight:** 188.26 (188.13)
**Synonyms:** α-Ethyl-1H-indole-3-ethanamine; α-ethyltryptamine; 3-(2-aminobutyl)indole; etryptamine
**Trade names:** Monase

**Use:** Central stimulant
**HPLC:** Si-10; 10A:90B; 5.1
**GC:** 1859; 200°C

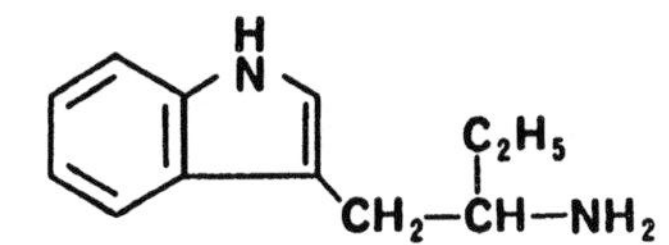


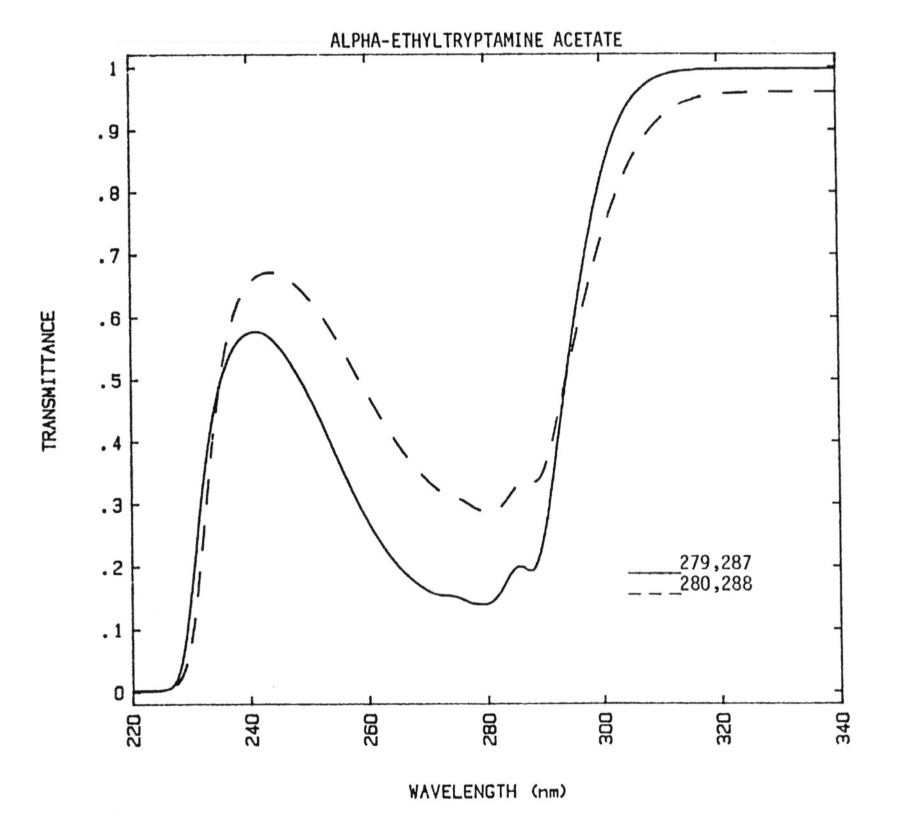

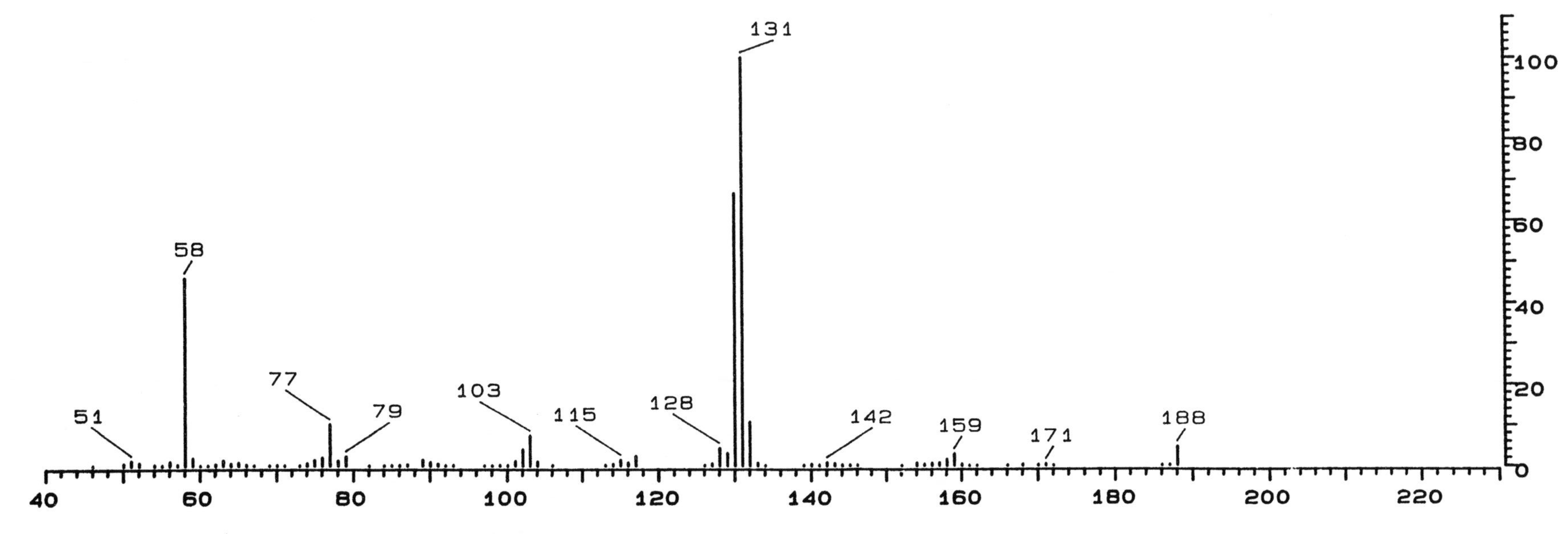

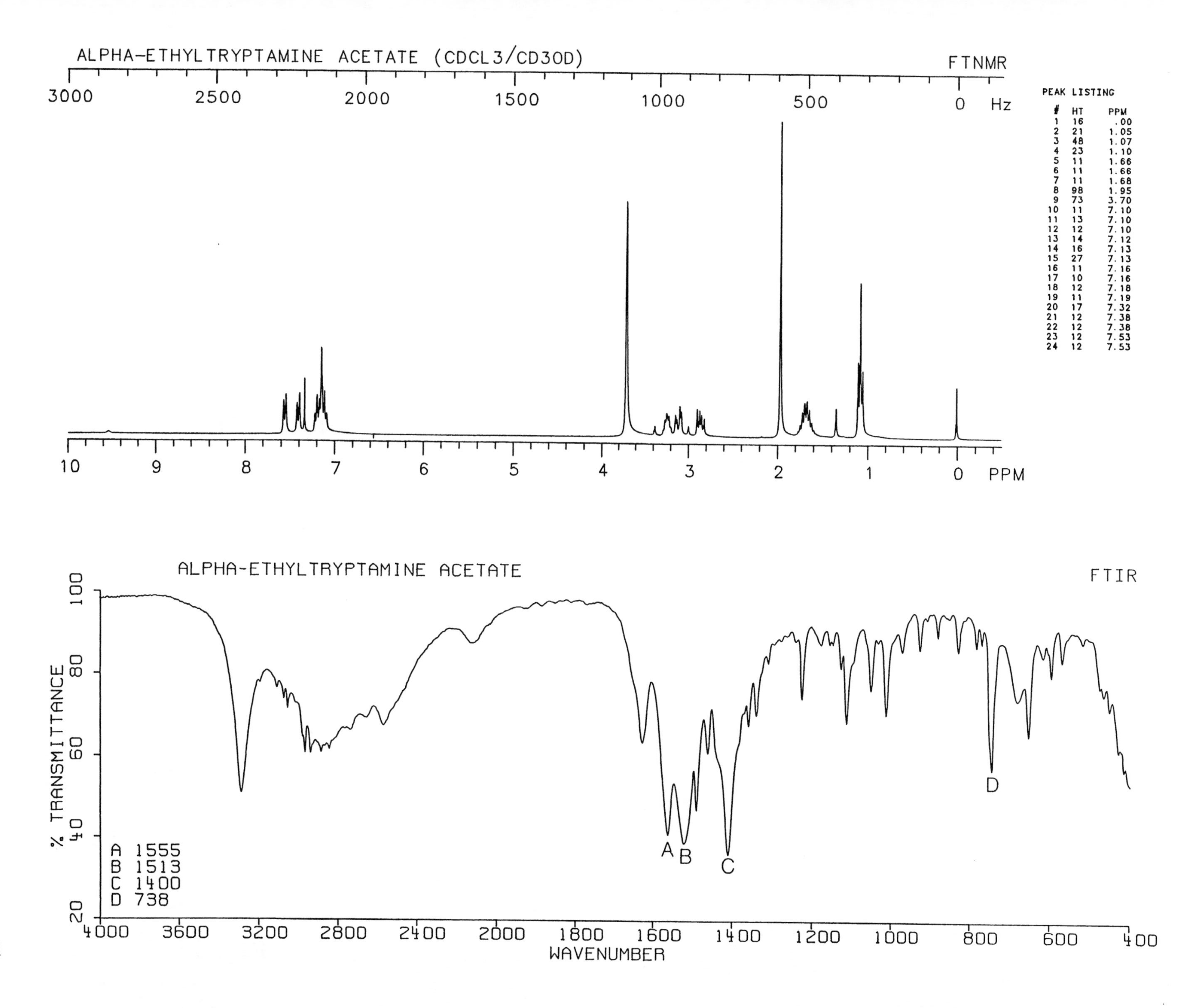
ALPHA-ETHYLTRYPTAMINE ACETATE (CDCL3/CD3OD)
FTNMR
3000 2500 2000 1500 1000 500 0 Hz
10 9 8 7 6 5 4 3 2 1 0 PPM
PEAK LISTING
# HT PPM
1 16 .00
2 21 1.05
3 48 1.07
4 23 1.10
5 11 1.66
6 11 1.66
7 11 1.68
8 98 1.95
9 73 3.70
10 11 7.10
11 13 7.10
12 12 7.10
13 14 7.12
14 16 7.13
15 27 7.13
16 11 7.16
17 10 7.16
18 12 7.18
19 11 7.19
20 17 7.32
21 12 7.38
22 12 7.38
23 12 7.53
24 12 7.53
ALPHA-ETHYLTRYPTAMINE ACETATE
FTIR
% TRANSMITTANCE
100 80 60 40 20
A 1555
B 1513
C 1400
D 738
A B C D
4000 3600 3200 2800 2400 2000 1800 1600 1400 1200 1000 800 600 400
WAVENUMBER

# ETHYNYLESTRADIOL

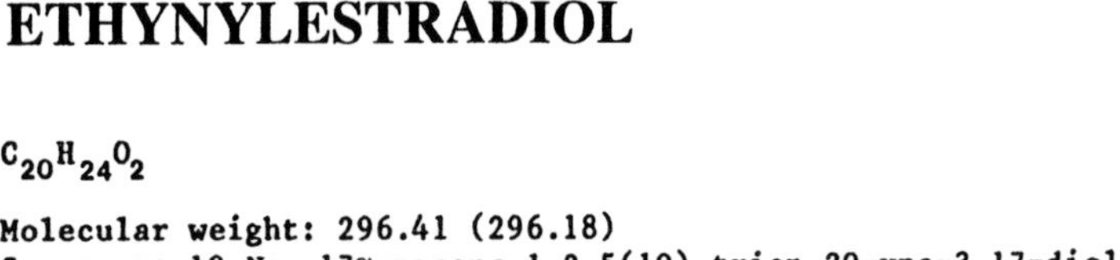

$C_{20}H_{24}O_2$

**Molecular weight:** 296.41 (296.18)
**Synonyms:** 10-Nor-17α-pregna-1,2,5(10)-trien-20-yne-3,17-diol; 17-ethinylestradiol
**Trade names:** Brevicon, Demulen, Estinyl, Loestrin, Modicon, Norinyl, Norlestrin, Ortho-Novum, Ovcon, Ovral
**Use:** Estrogen
**HPLC:** Si-10; 1A:99B; 4.5
**GC:** 2704; 280°C

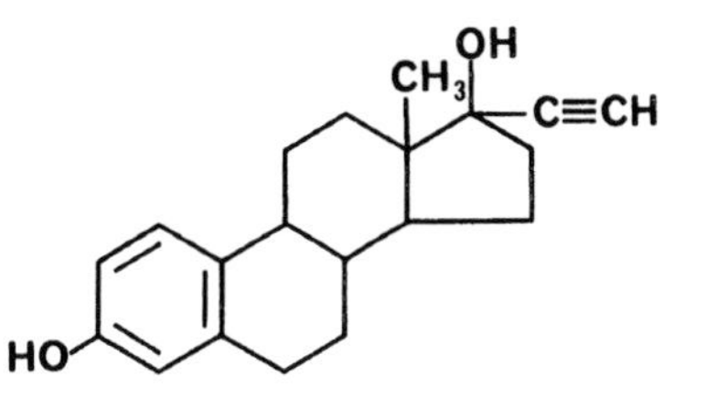


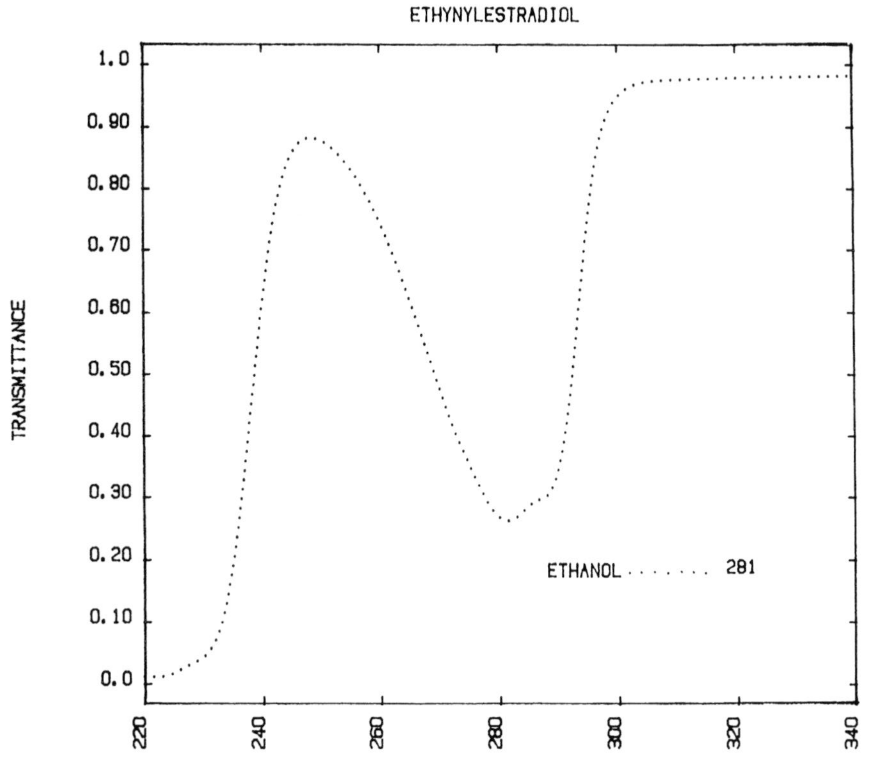


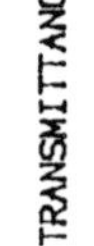


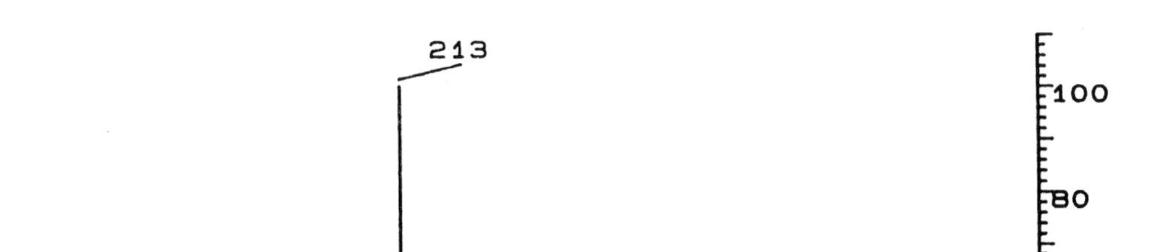


*ETHYNYLESTRADIOL*

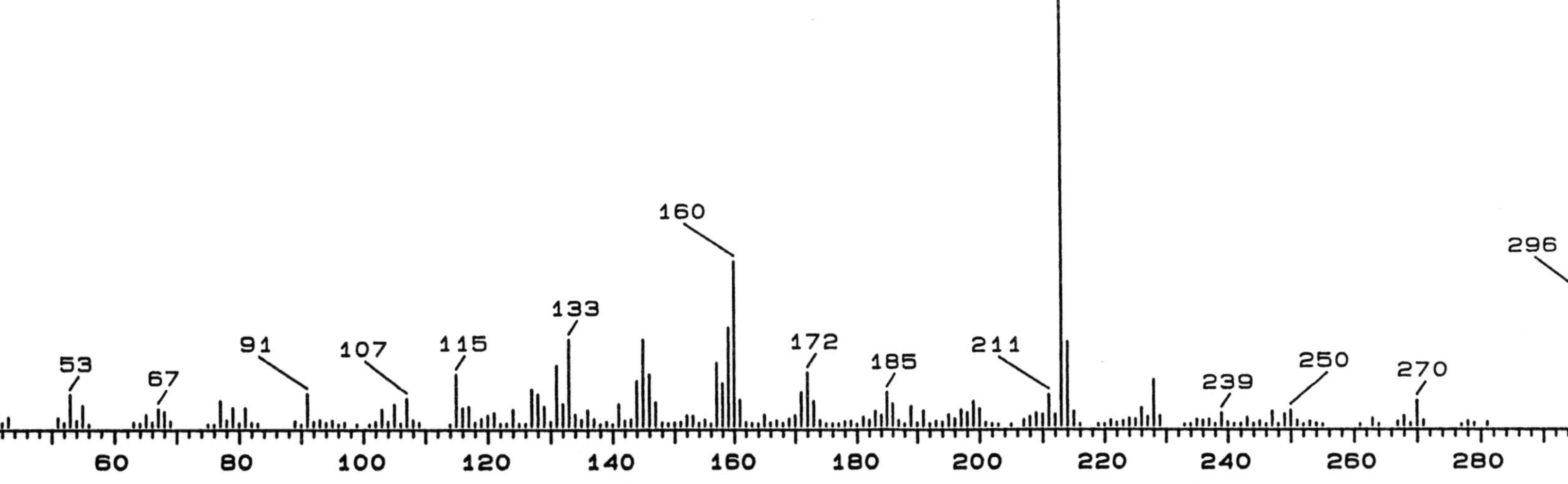

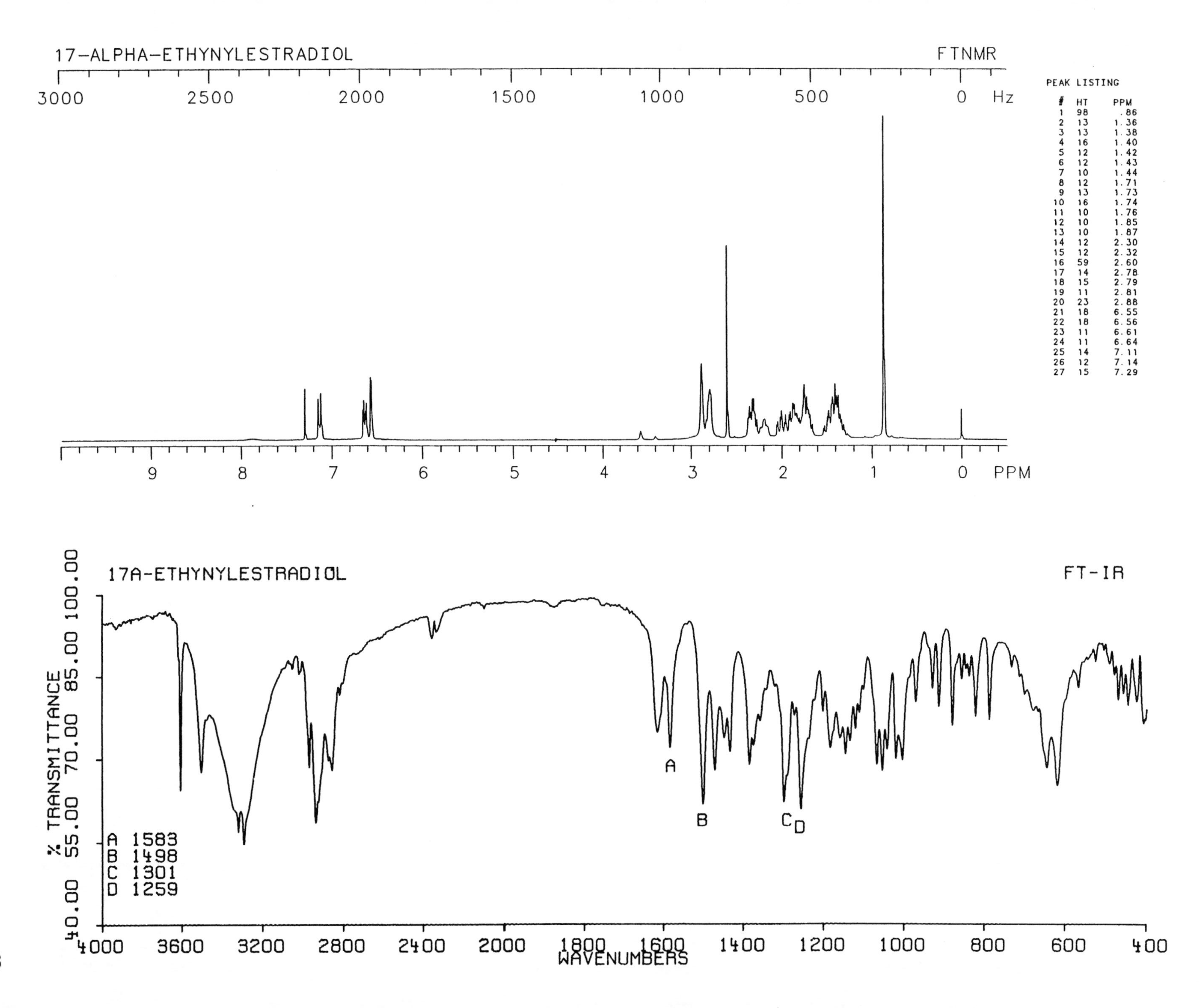
17-ALPHA-ETHYNYLESTRADIOL
FTNMR
Hz
PPM
PEAK LISTING
# HT PPM
1 98 .86
2 13 1.36
3 13 1.38
4 16 1.40
5 12 1.42
6 12 1.43
7 10 1.44
8 12 1.71
9 13 1.73
10 16 1.74
11 10 1.76
12 10 1.85
13 10 1.87
14 12 2.30
15 12 2.32
16 59 2.60
17 14 2.78
18 15 2.79
19 11 2.81
20 23 2.88
21 18 6.55
22 18 6.56
23 11 6.61
24 11 6.64
25 14 7.11
26 12 7.14
27 15 7.29
17A-ETHYNYLESTRADIOL
FT-IR
% TRANSMITTANCE
WAVENUMBERS
A 1583
B 1498
C 1301
D 1259

# ETIDOCAINE

$C_{17}H_{28}N_2O$

**Molecular weight:** 276.43 (276.22)
**Synonyms:** N-(2,6-Dimethylphenyl)-2-(ethylpropylamino)butanamide

**Trade names:** Duranest

**Use:** Local anesthetic
**HPLC:** Si-10; 100B; 4.3
**GC:** 2090; 250°C

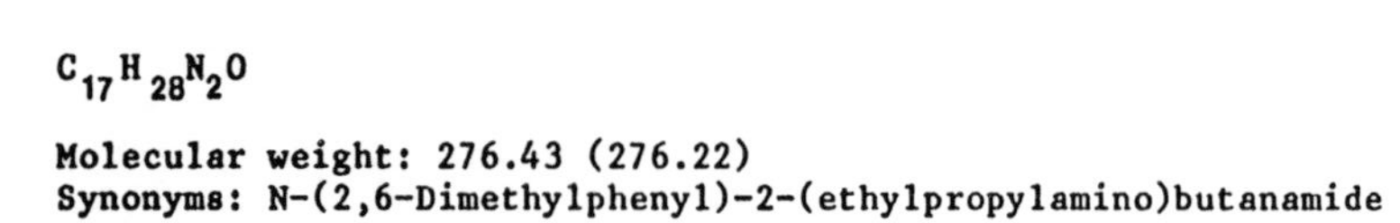

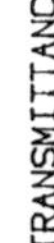

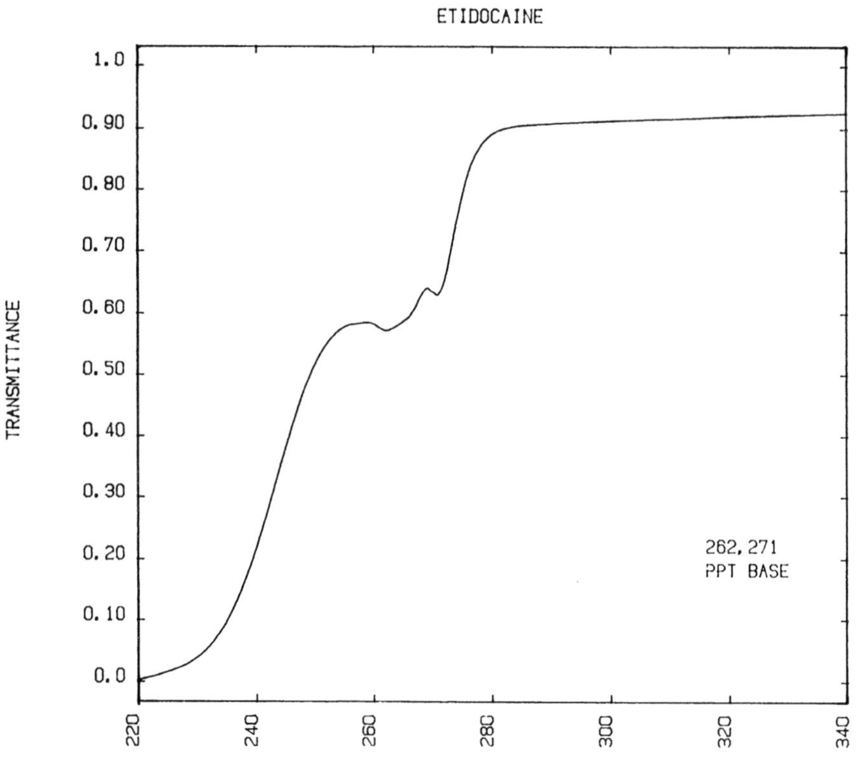


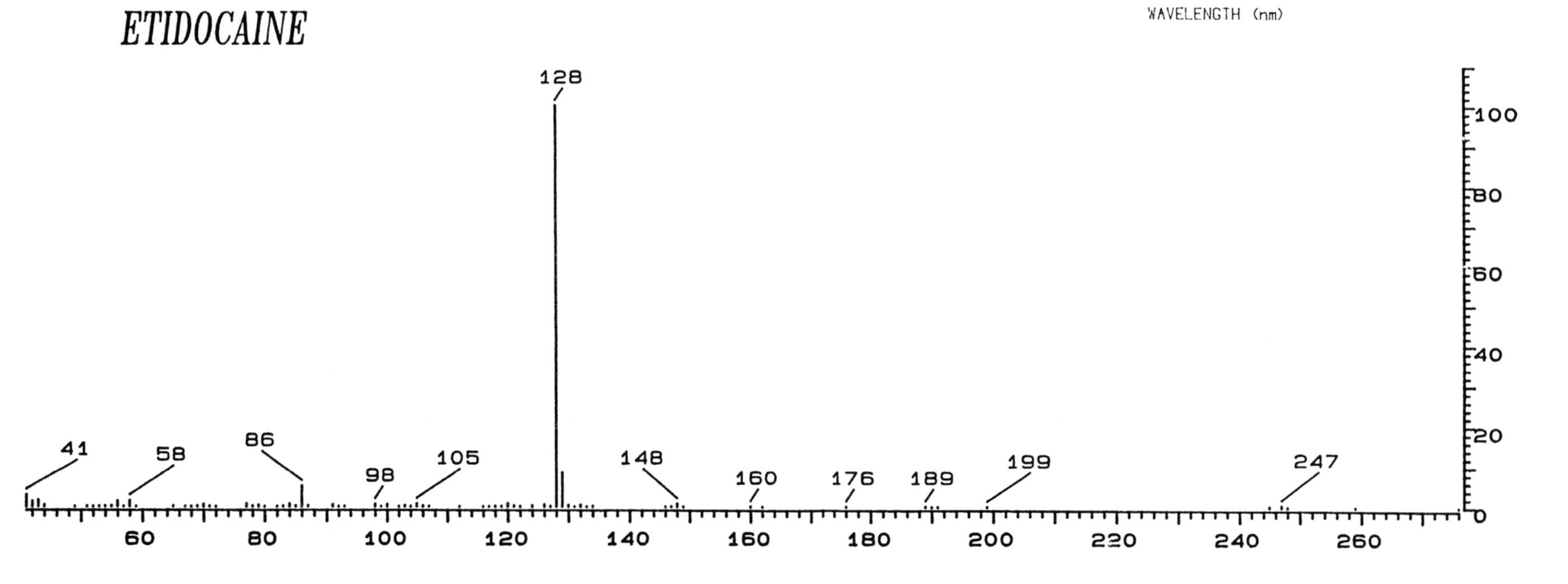

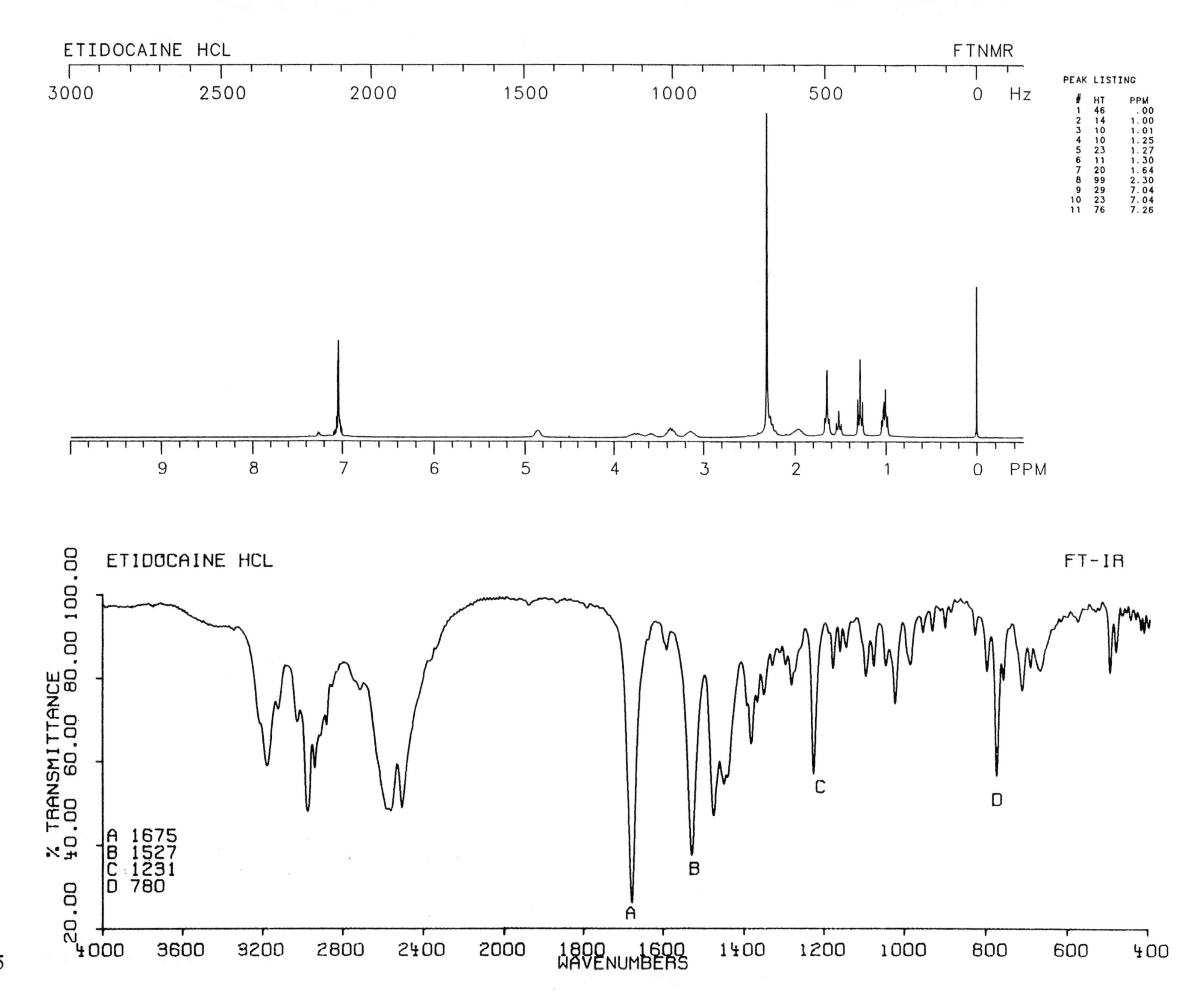
ETIDOCAINE HCL
FTNMR
3000 2500 2000 1500 1000 500 0 Hz
PEAK LISTING
# HT PPM
1 46 .00
2 14 1.00
3 10 1.01
4 10 1.25
5 23 1.27
6 11 1.30
7 20 1.64
8 99 2.30
9 29 7.04
10 23 7.04
11 76 7.26
9 8 7 6 5 4 3 2 1 0 PPM
ETIDOCAINE HCL
FT-IR
% TRANSMITTANCE
20.00 40.00 60.00 80.00 100.00
A 1675
B 1527
C 1231
D 780
A B C D
4000 3600 3200 2800 2400 2000 1800 1600 1400 1200 1000 800 600 400
WAVENUMBERS

# ETONITAZENE

$C_{22}H_{28}N_4O_3$

Molecular weight: 396.49 (396.22)
Synonyms: 2-[(4-Ethoxyphenyl)methyl]-N,N-diethyl-5-nitro-1H-benzimidazole-1-ethanamine
Trade names:

Use: Analgesic
HPLC: Si-10; 2A:98B; 4.2
GC:

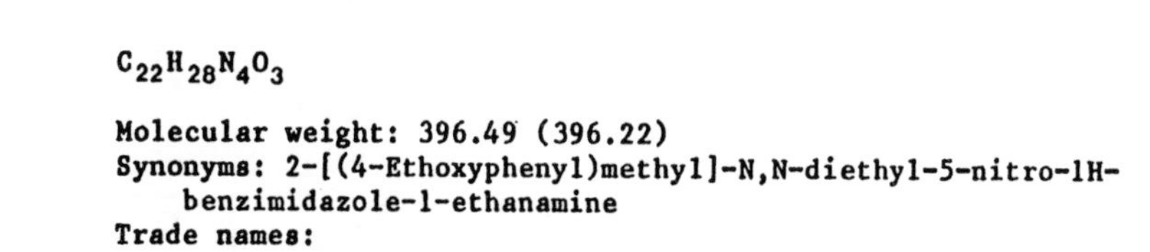


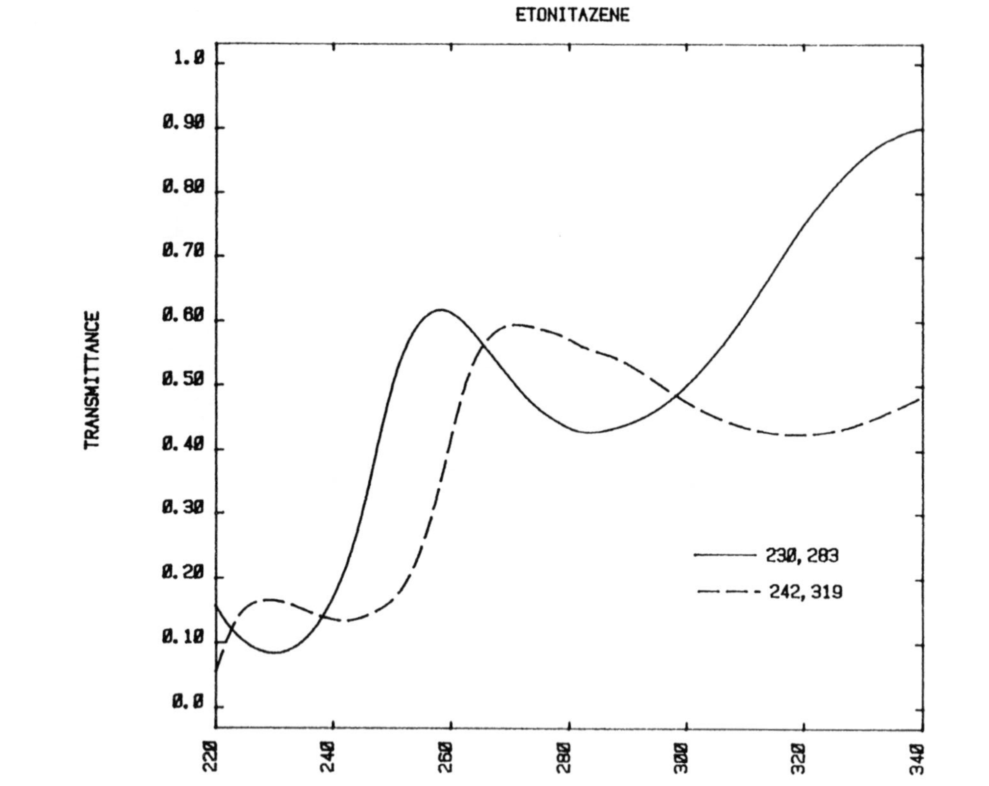


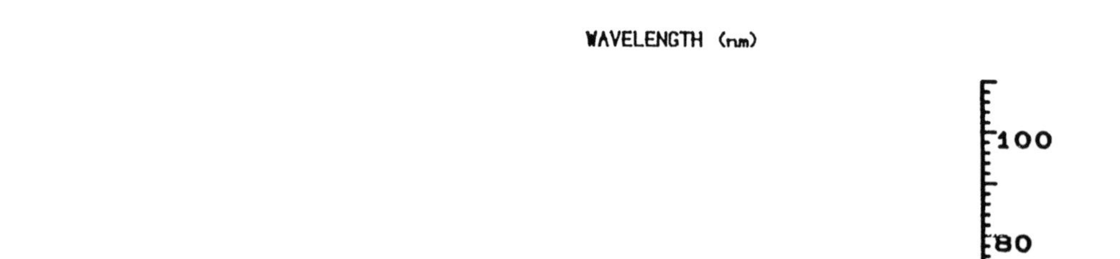

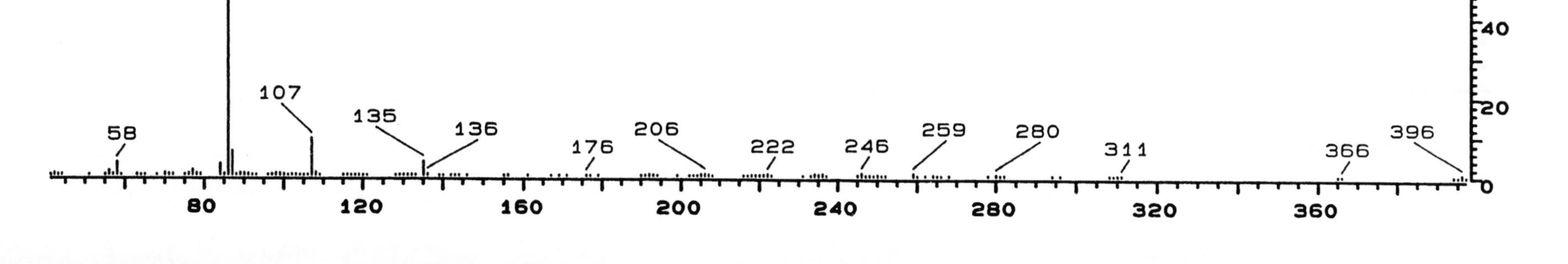

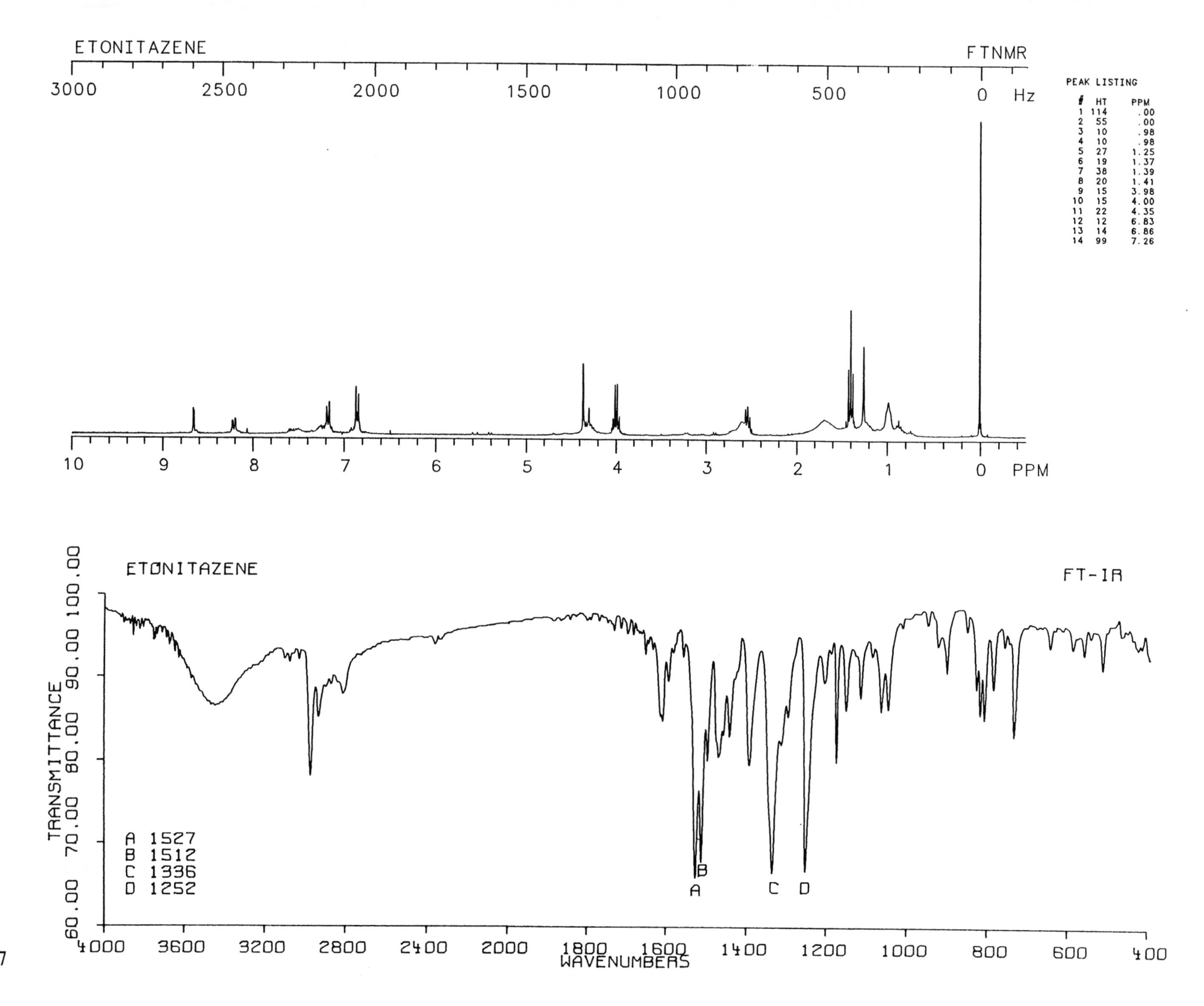
ETONITAZENE
FTNMR
3000 2500 2000 1500 1000 500 0 Hz
PEAK LISTING
# HT PPM
1 114 .00
2 55 .00
3 10 .98
4 10 .98
5 27 1.25
6 19 1.37
7 38 1.39
8 20 1.41
9 15 3.98
10 15 4.00
11 22 4.35
12 12 6.83
13 14 6.86
14 99 7.26
10 9 8 7 6 5 4 3 2 1 0 PPM
ETONITAZENE
FT-IR
TRANSMITTANCE
60.00 70.00 80.00 90.00 100.00
A 1527
B 1512
C 1336
D 1252
A B C D
4000 3600 3200 2800 2400 2000 1800 1600 1400 1200 1000 800 600 400
WAVENUMBERS

# ETOPOSIDE

$C_{29}H_{32}O_{13}$

**Molecular weight:** 588.58 (588.18)
**Synonyms:** 9-[(4,6-O-Ethylidene-β-D-glucopyranosyl)oxy]-5,8,8a,9-tetrahydro-5-(4-hydroxy-3,5-dimethoxyphenyl)furo[3',4':6,7]naphthol[2,3-d]-1,3-dioxol-6(5aH)-one
**Trade names:** Vepesid

**Use:** Antineoplastic
**HPLC:** Si-10; 2A:98B; 4.6
**GC:**

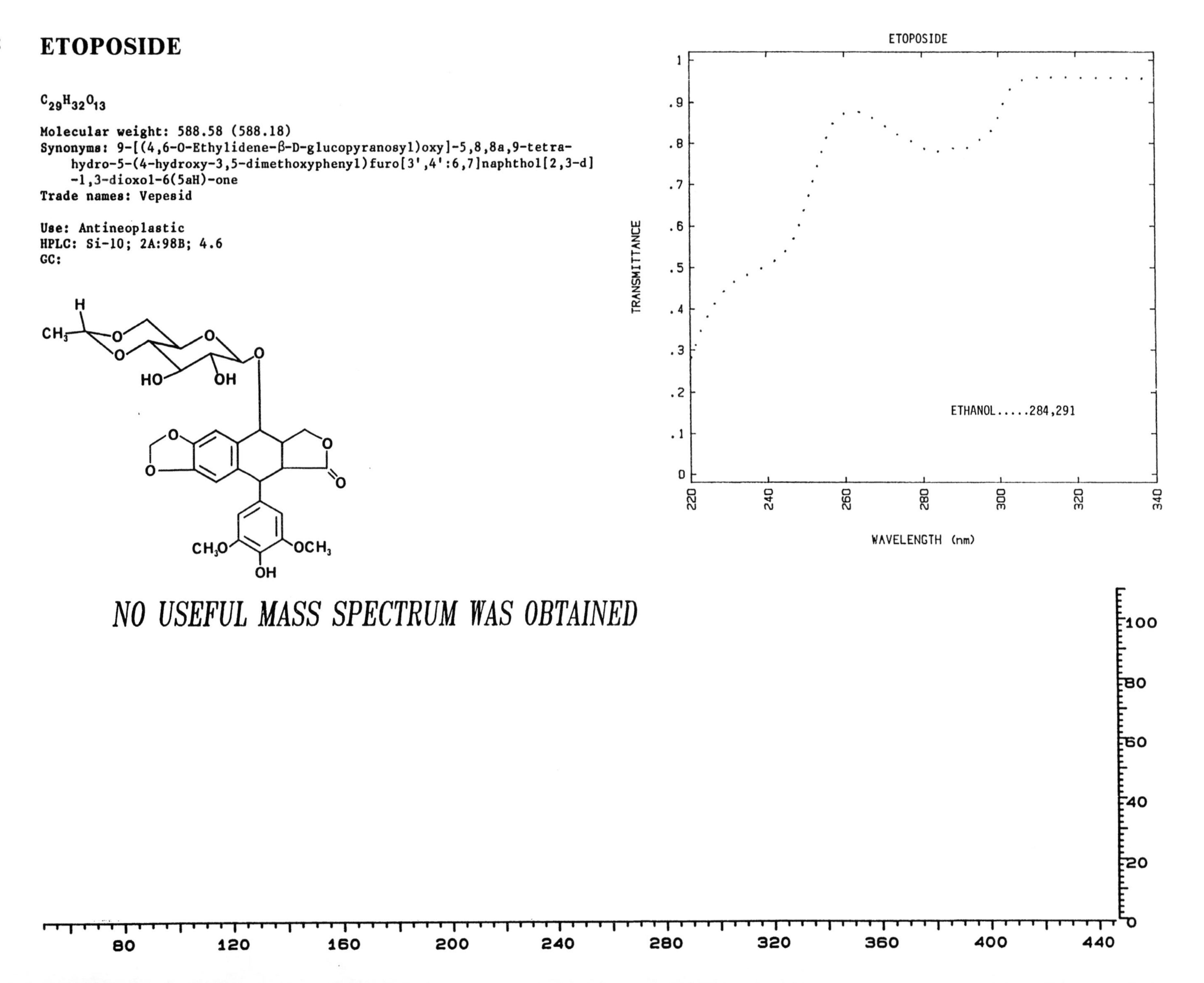

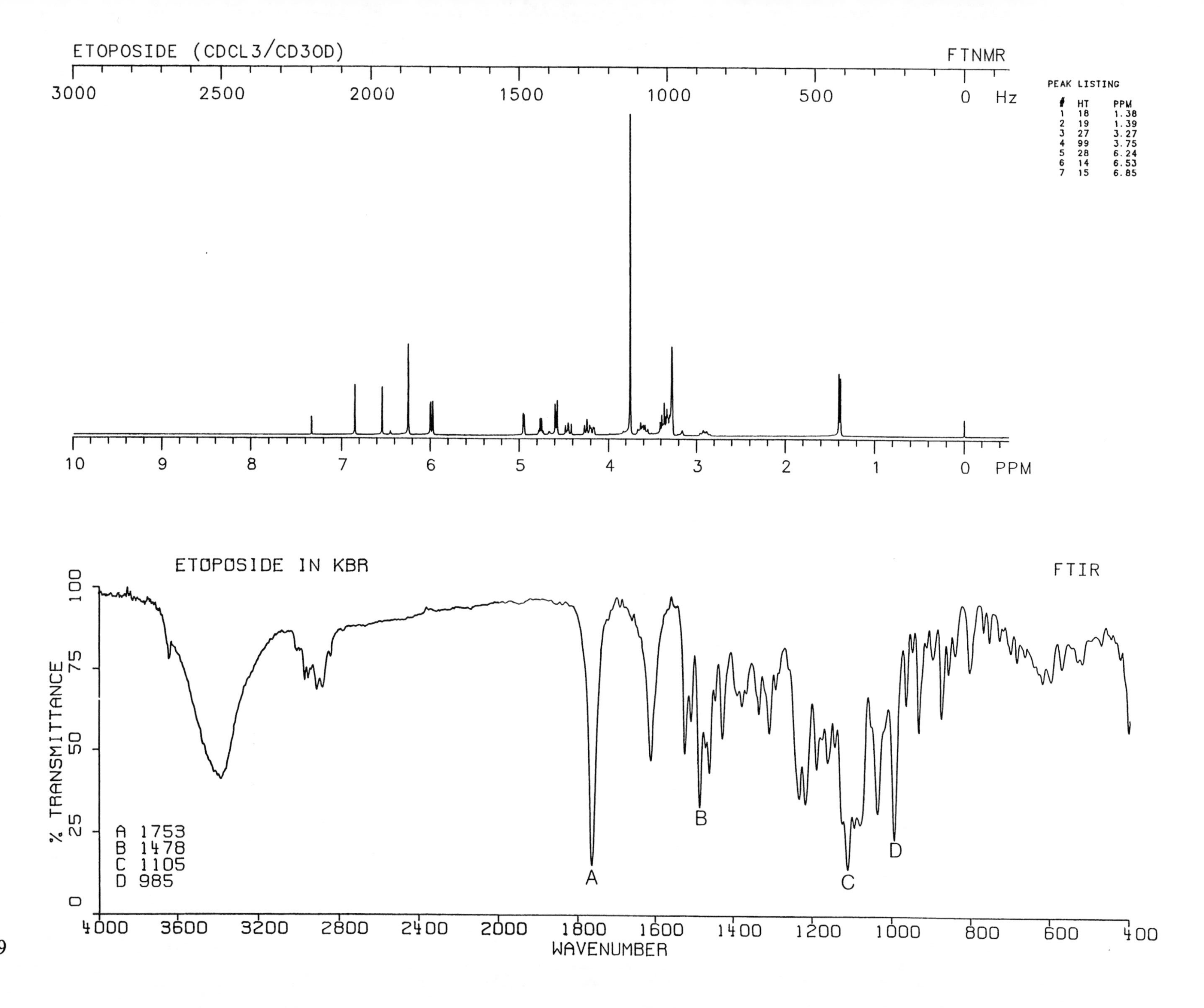

ETOPOSIDE (CDCL3/CD3OD)
FTNMR
3000
2500
2000
1500
1000
500
0 Hz
PEAK LISTING
# HT PPM
1 18 1.38
2 19 1.39
3 27 3.27
4 99 3.75
5 28 6.24
6 14 6.53
7 15 6.85
10
9
8
7
6
5
4
3
2
1
0 PPM
ETOPOSIDE IN KBR
FTIR
% TRANSMITTANCE
100
75
50
25
0
A 1753
B 1478
C 1105
D 985
A
B
C
D
4000
3600
3200
2800
2400
2000
1800
1600
1400
1200
1000
800
600
400
WAVENUMBER

# EUCATROPINE

$C_{17}H_{25}NO_3$

**Molecular weight:** 291.38 (291.18)
**Synonyms:** α-Hydroxybenzeneacetic acid 1,2,2,6-tetramethyl-4-piperidinyl ester
**Trade names:** Euphthalamine

**Use:** Anticholinergic
**HPLC:** Si-10; 5A:95B; 7.1
**GC:** 2120; 250°C

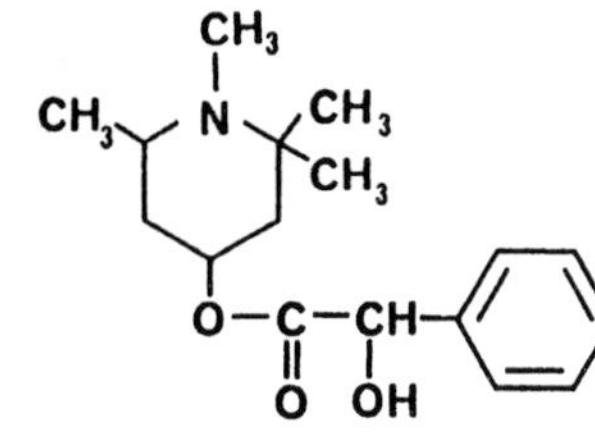


EUCATROPINE

TRANSMITTANCE

WAVELENGTH (nm)

251,257,263
PPT BASE

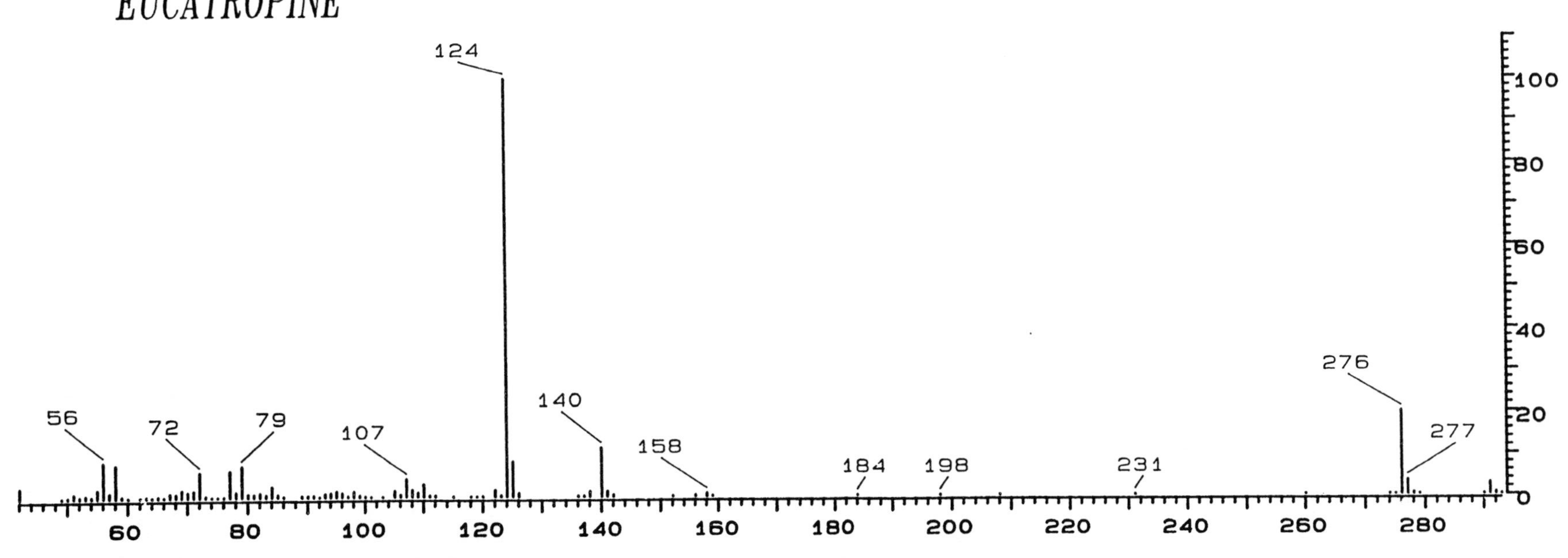

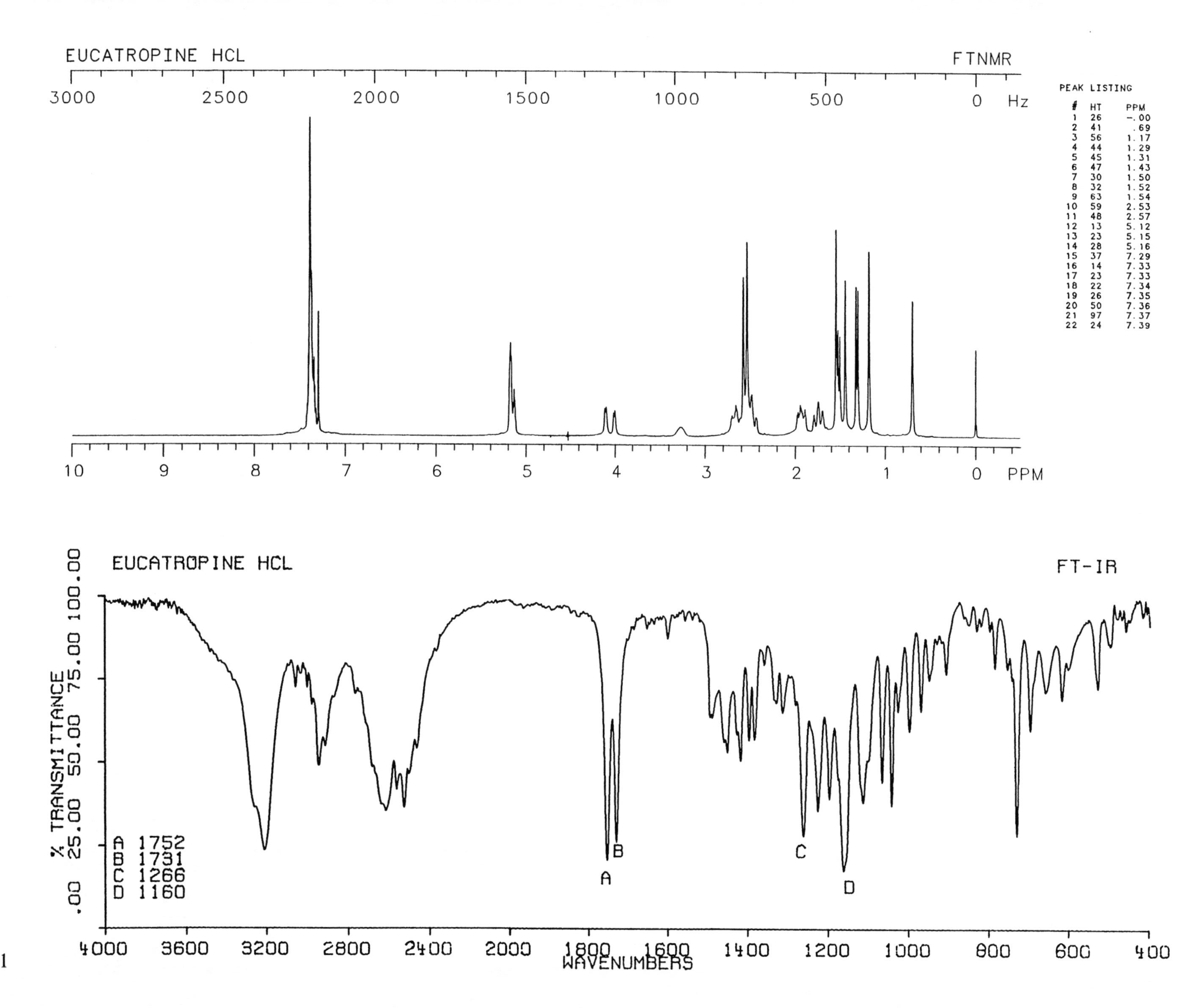
EUCATROPINE HCL
FTNMR
3000 2500 2000 1500 1000 500 0 Hz
PEAK LISTING
# HT PPM
1 26 -.00
2 41 .69
3 56 1.17
4 44 1.29
5 45 1.31
6 47 1.43
7 30 1.50
8 32 1.52
9 63 1.54
10 59 2.53
11 48 2.57
12 13 5.12
13 23 5.15
14 28 5.16
15 37 7.29
16 14 7.33
17 23 7.33
18 22 7.34
19 26 7.35
20 50 7.36
21 97 7.37
22 24 7.39
10 9 8 7 6 5 4 3 2 1 0 PPM
EUCATROPINE HCL
FT-IR
% TRANSMITTANCE
.00 25.00 50.00 75.00 100.00
A 1752
B 1731
C 1266
D 1160
A B C D
4000 3600 3200 2800 2400 2000 1800 1600 1400 1200 1000 800 600 400
WAVENUMBERS

# EUGENOL

$C_{10}H_{12}O_2$

Molecular weight: 164.20 (164.08)
Synonyms: 2-Methoxy-4-(2-propenyl)phenol; allylguaiacol; eugenic acid; caryophyllic acid
Trade names: O-Benzoyleugenol

Use: Analgesic
HPLC: Si-10; 20B:80C; 3.3
GC: 1371; 140°C

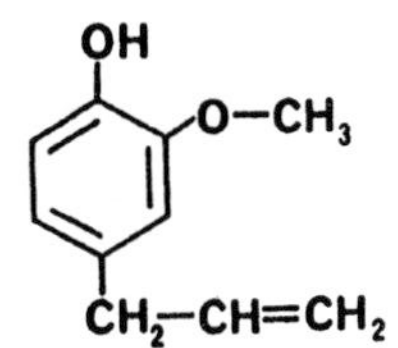


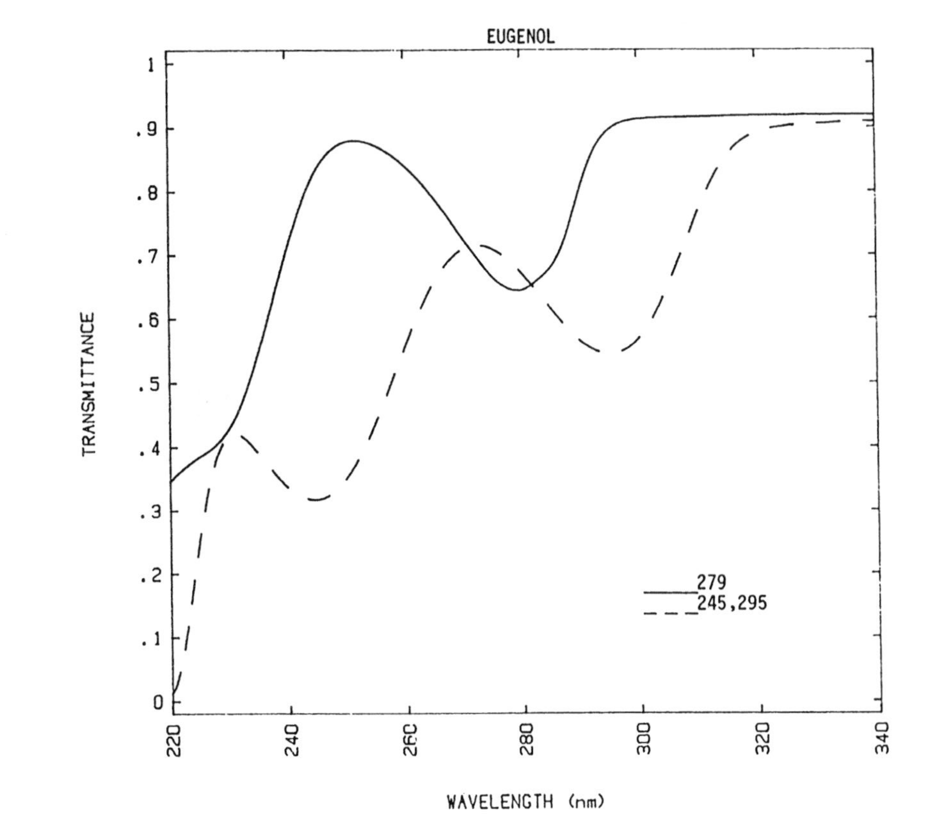


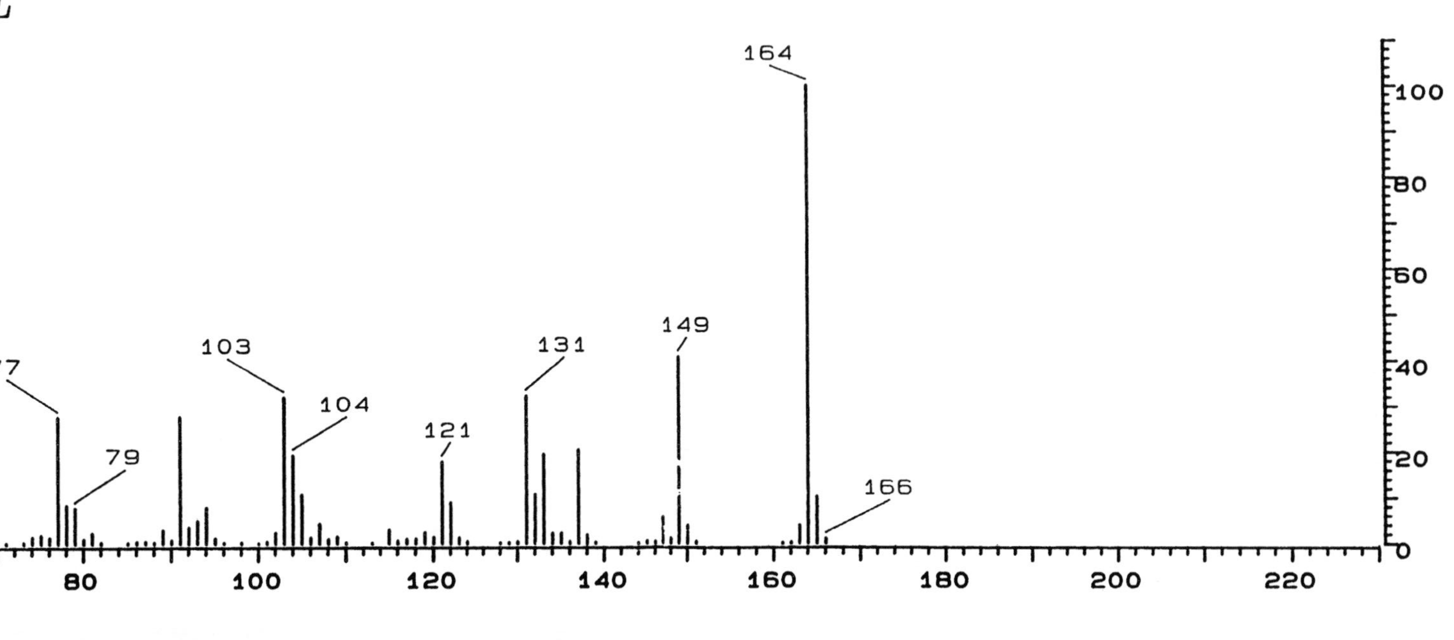

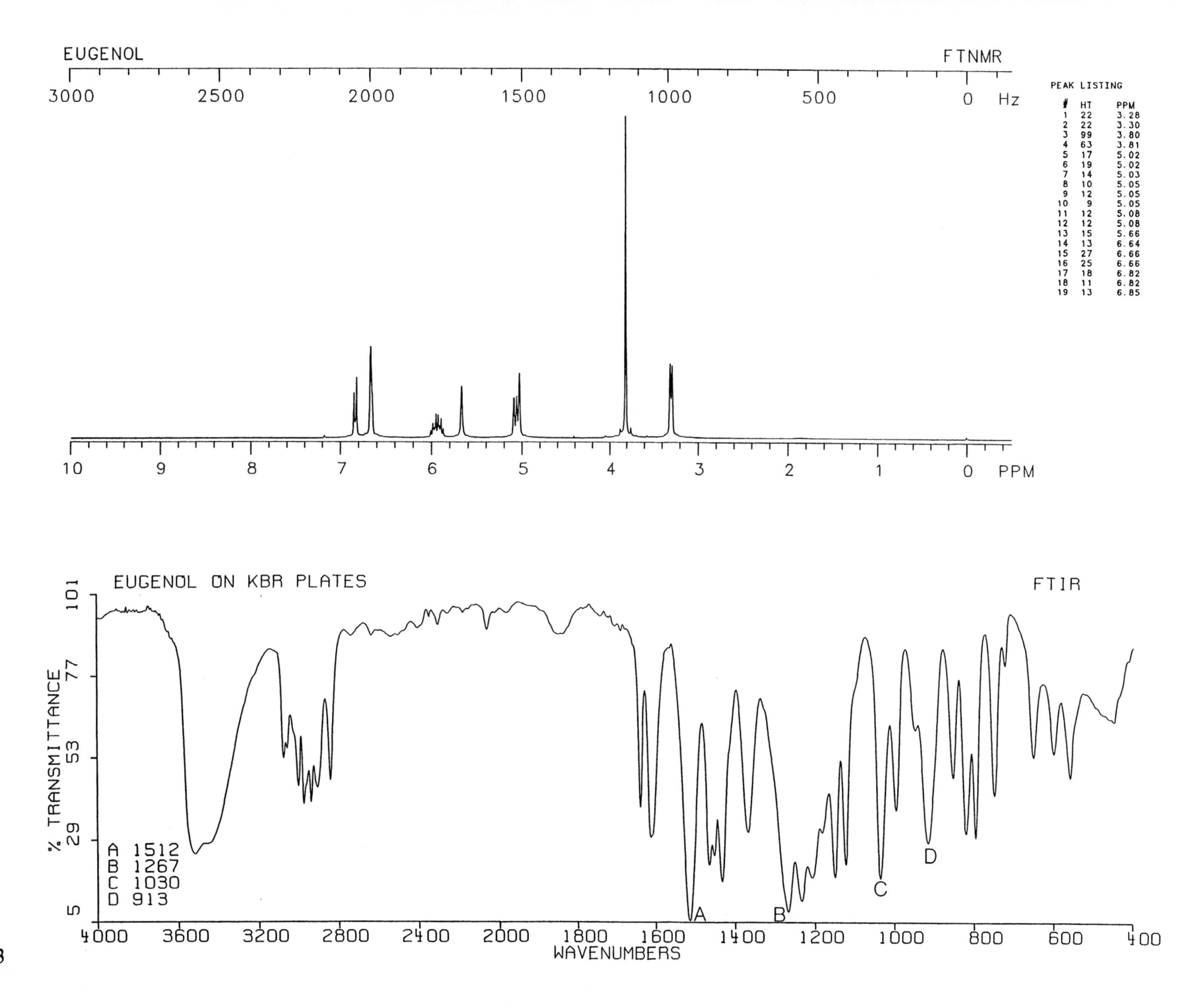
EUGENOL
FTNMR
3000
2500
2000
1500
1000
500
0 Hz
PEAK LISTING
# HT PPM
1 22 3.28
2 22 3.30
3 99 3.80
4 63 3.81
5 17 5.02
6 19 5.02
7 14 5.03
8 10 5.05
9 12 5.05
10 9 5.05
11 12 5.08
12 12 5.08
13 15 5.66
14 13 6.64
15 27 6.66
16 25 6.66
17 18 6.82
18 11 6.82
19 13 6.85
10
9
8
7
6
5
4
3
2
1
0 PPM
EUGENOL ON KBR PLATES
FTIR
101
77
53
29
5
% TRANSMITTANCE
A 1512
B 1267
C 1030
D 913
A
B
C
D
4000
3600
3200
2800
2400
2000
1800
1600
1400
1200
1000
800
600
400
WAVENUMBERS

# FENCAMFAMINE

$C_{15}H_{21}N$

**Molecular weight:** 215.33 (215.17)
**Synonyms:** N-Ethyl-3-phenylbicyclo[2.2.1]-heptan-2-amine;
2-ethylamino-3-phenylnorbornane
**Trade names:** Euvitol, Norcamphane, Reactivan

**Use:** Central stimulant
**HPLC:** Si-10; 5A:95B; 3.3
**GC:** 1711; 200°C

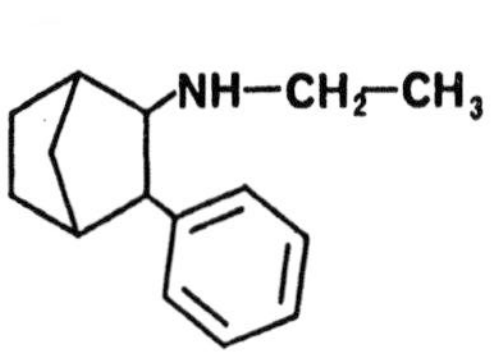


FENCAMFAMINE

TRANSMITTANCE

WAVELENGTH (nm)

252,258,267
262

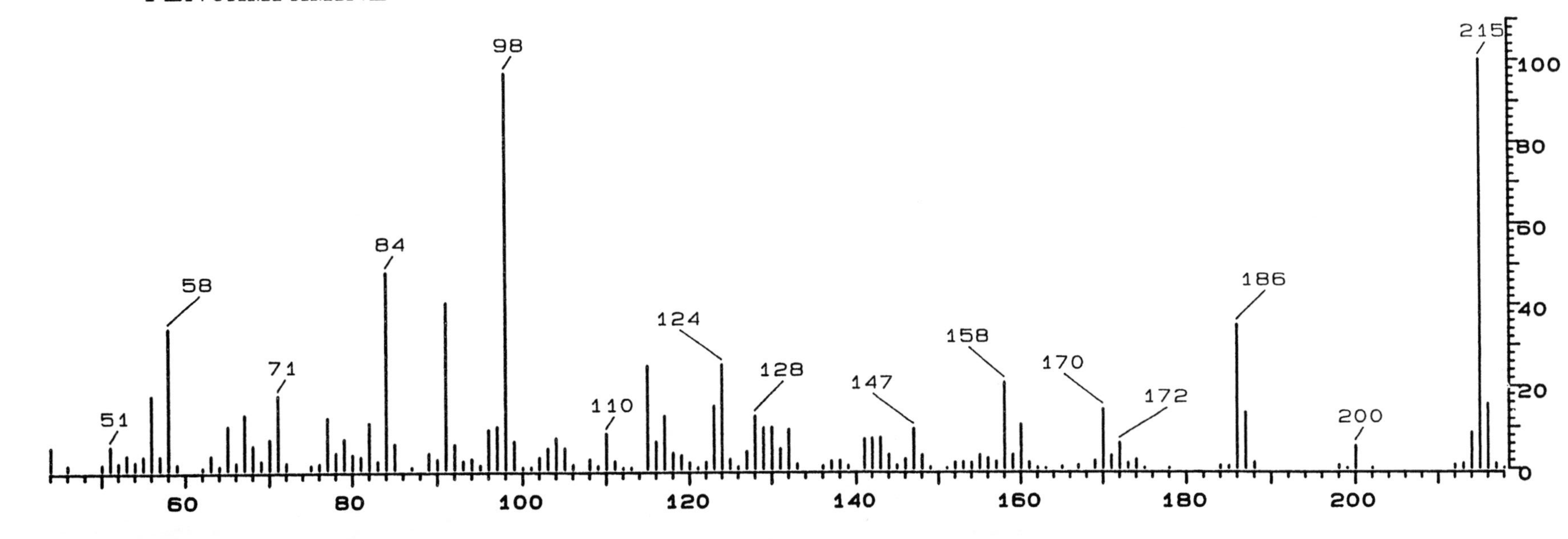

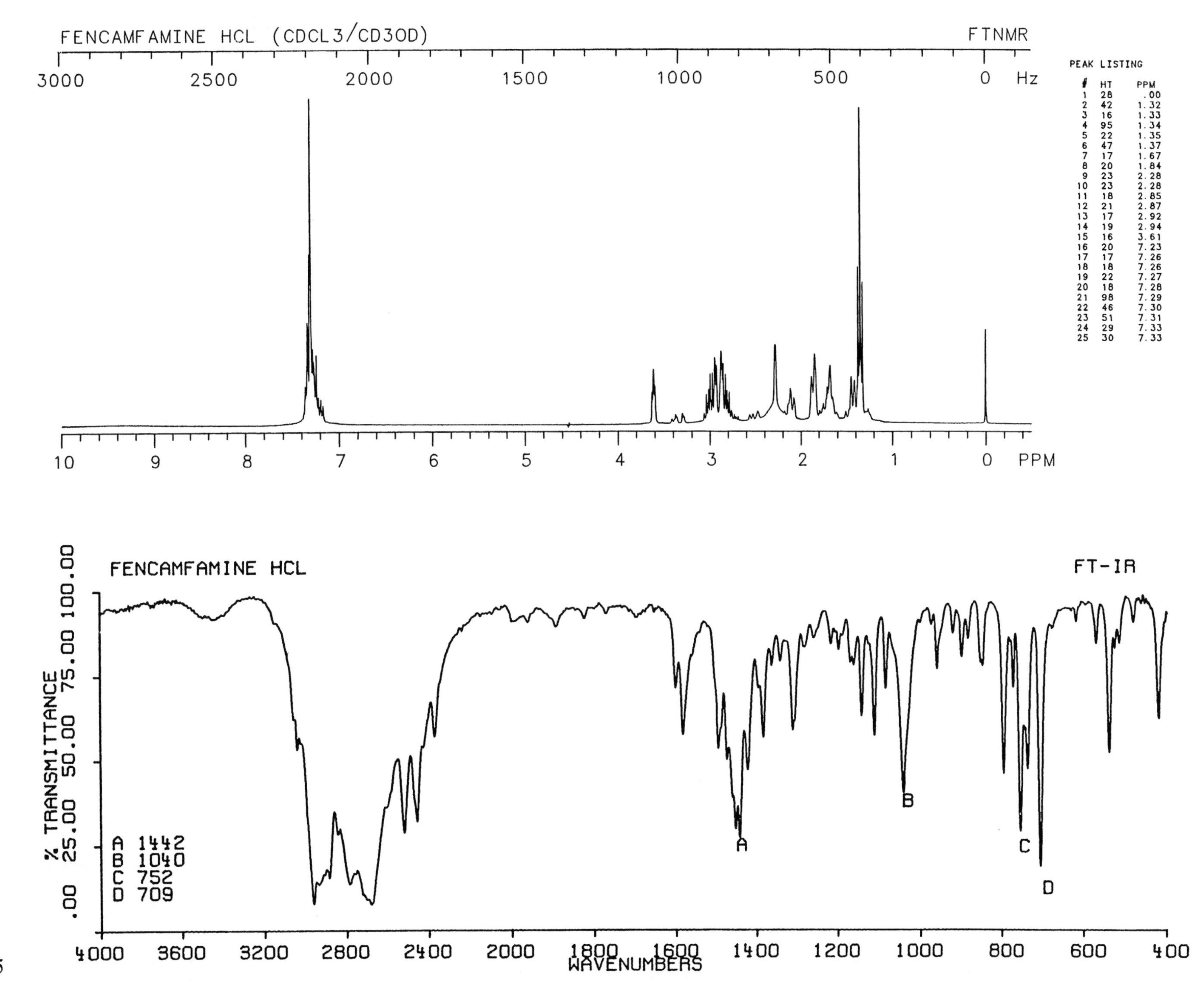
FENCAMFAMINE HCL (CDCL3/CD3OD)
FTNMR
3000
2500
2000
1500
1000
500
0 Hz
PEAK LISTING
# HT PPM
1 28 .00
2 42 1.32
3 16 1.33
4 95 1.34
5 22 1.35
6 47 1.37
7 17 1.67
8 20 1.84
9 23 2.28
10 23 2.28
11 18 2.85
12 21 2.87
13 17 2.92
14 19 2.94
15 16 3.61
16 20 7.23
17 17 7.26
18 18 7.26
19 22 7.27
20 18 7.28
21 98 7.29
22 46 7.30
23 51 7.31
24 29 7.33
25 30 7.33
10
9
8
7
6
5
4
3
2
1
0 PPM
FENCAMFAMINE HCL
FT-IR
% TRANSMITTANCE
.00
25.00
50.00
75.00
100.00
A 1442
B 1040
C 752
D 709
A
B
C
D
4000
3600
3200
2800
2400
2000
1800
1600
1400
1200
1000
800
600
400
WAVENUMBERS

# FENETHYLLINE

$C_{18}H_{23}N_5O_2$

**Molecular weight:** 341.41 (341.19)
**Synonyms:** 3,7-Dihydro-1,3-dimethyl-7-[2-[(1-methyl-2-phenylethyl)-amino]ethyl]-1H-purine-2,6-dione; 7-(3-phenyl-2-propylamino-ethyl)theophylline
**Trade names:** Captagon

**Use:** Stimulant
**HPLC:** Si-10; 2A:98B; 6.5
**GC:** 2894; 280°C

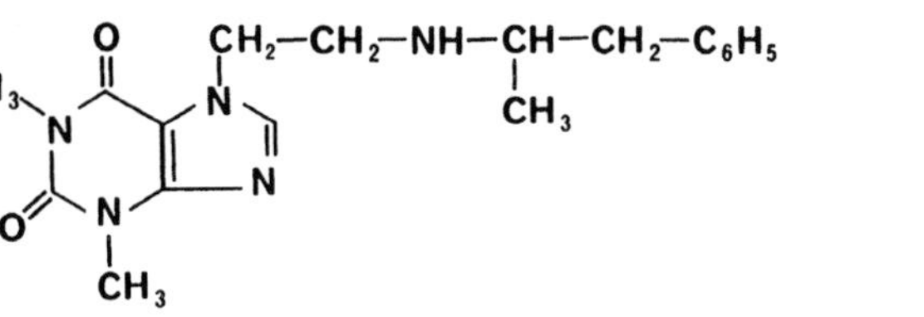


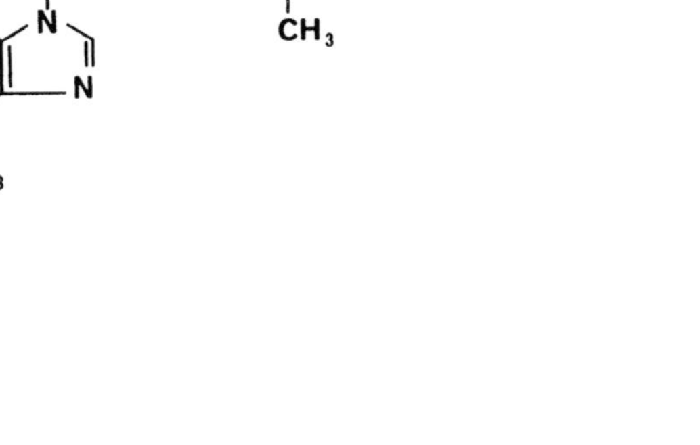


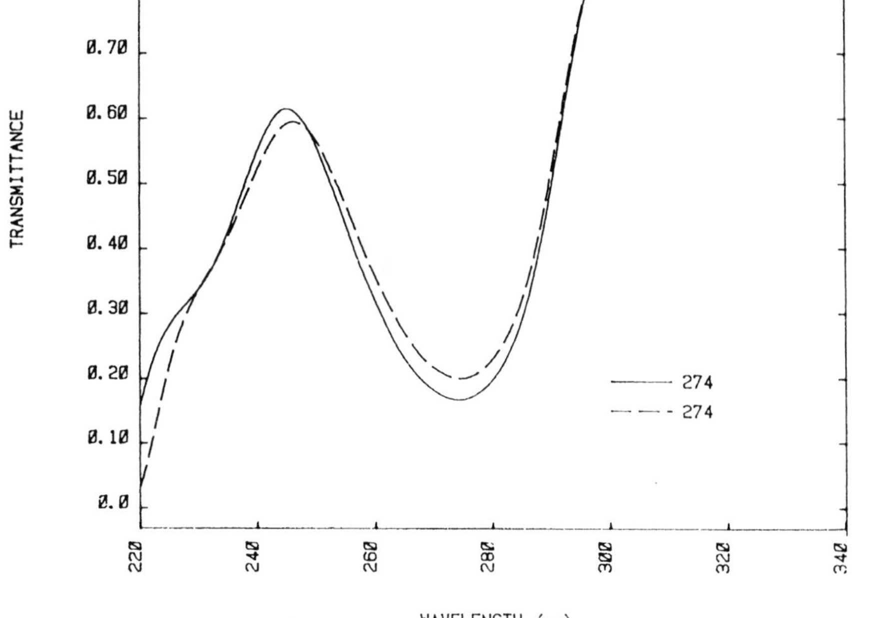


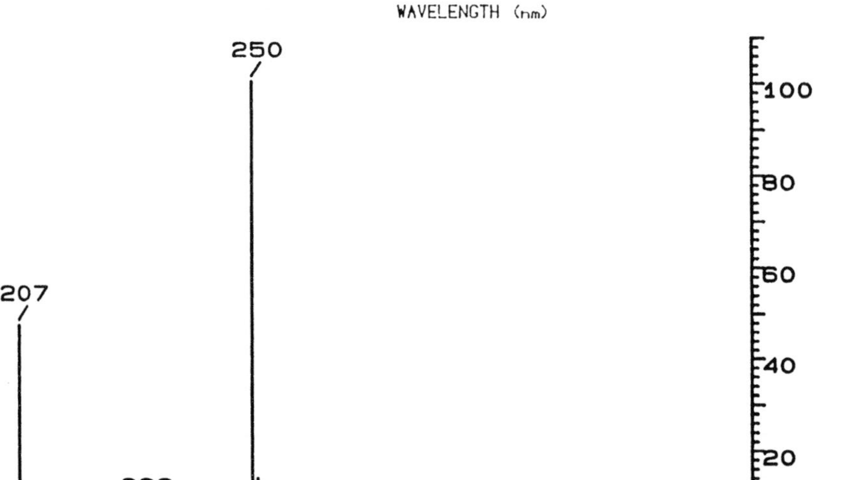

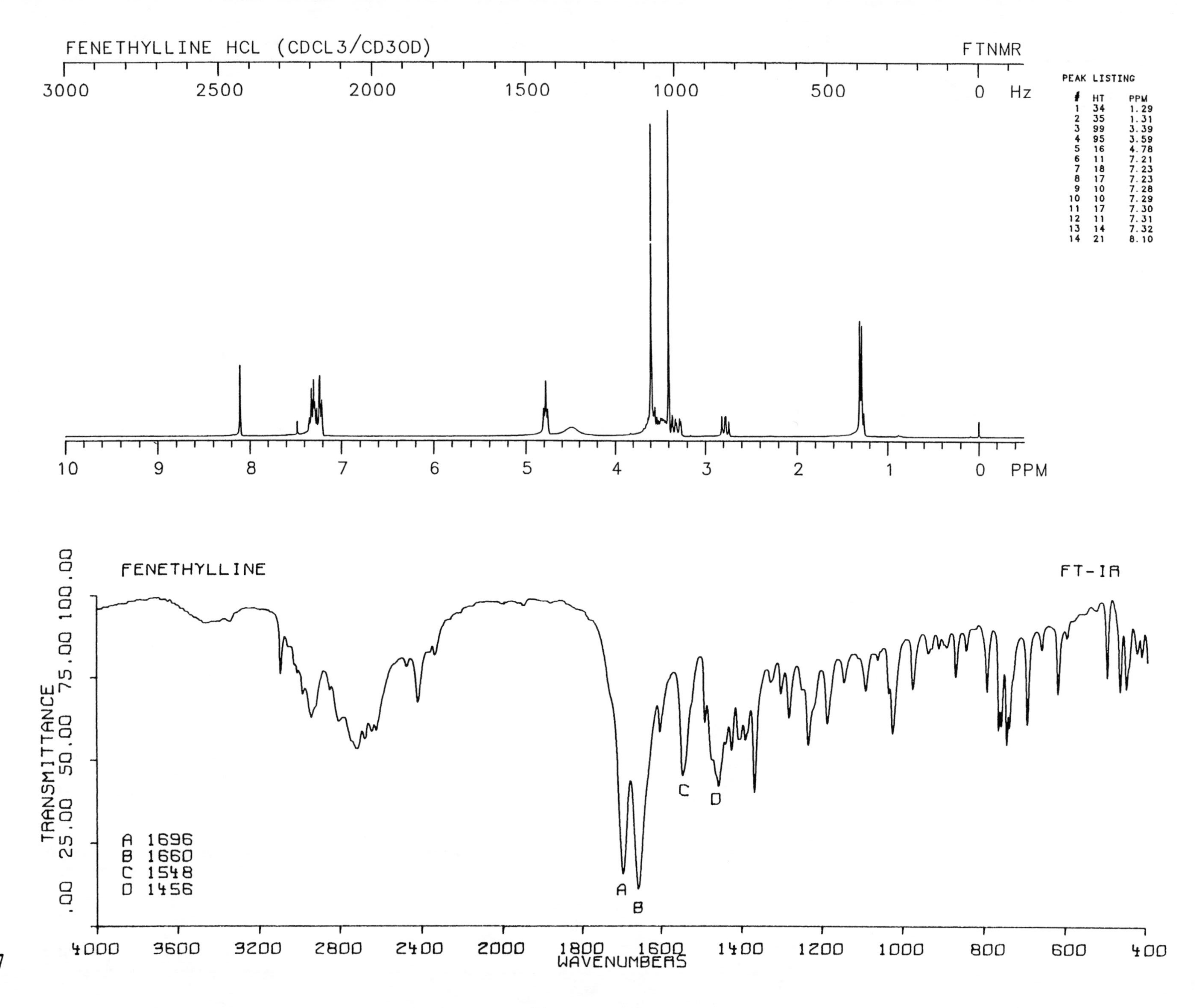
FENETHYLLINE HCL (CDCL3/CD3OD)
FTNMR
3000 2500 2000 1500 1000 500 0 Hz
PEAK LISTING
# HT PPM
1 34 1.29
2 35 1.31
3 99 3.39
4 95 3.59
5 16 4.78
6 11 7.21
7 18 7.23
8 17 7.23
9 10 7.28
10 10 7.29
11 17 7.30
12 11 7.31
13 14 7.32
14 21 8.10
10 9 8 7 6 5 4 3 2 1 0 PPM
FENETHYLLINE
FT-IR
TRANSMITTANCE
.00 25.00 50.00 75.00 100.00
A 1696
B 1660
C 1548
D 1456
A
B
C
D
4000 3600 3200 2800 2400 2000 1800 1600 1400 1200 1000 800 600 400
WAVENUMBERS

# FENFLURAMINE

$C_{12}H_{16}F_3N$

Molecular weight: 231.26 (231.12)
Synonyms: N-Ethyl-α-methyl-3-(trifluoromethyl)benzenethanamine;
N-ethyl-α-methyl-m-(trifluoromethyl)phenethylamine
Trade names: Pondimin

Use: Anorexic
HPLC: Si-10; 2A:98B; 5.0
GC: 1231; 140°C

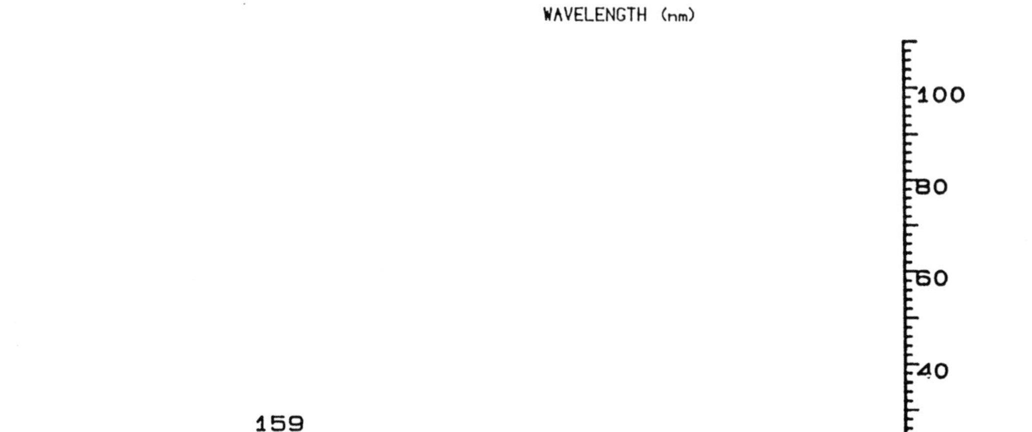


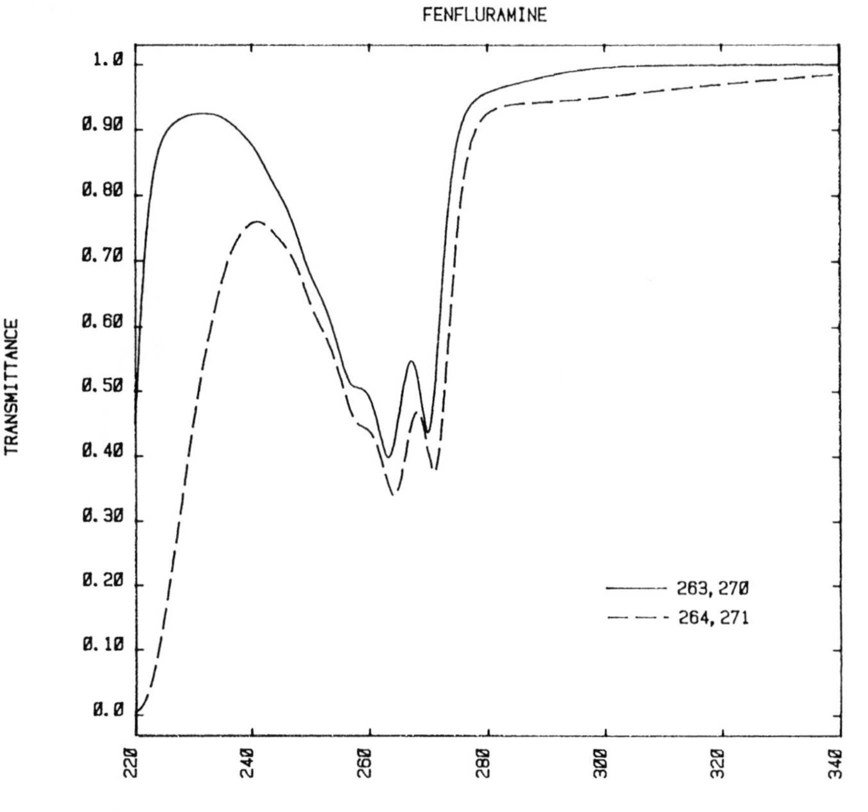


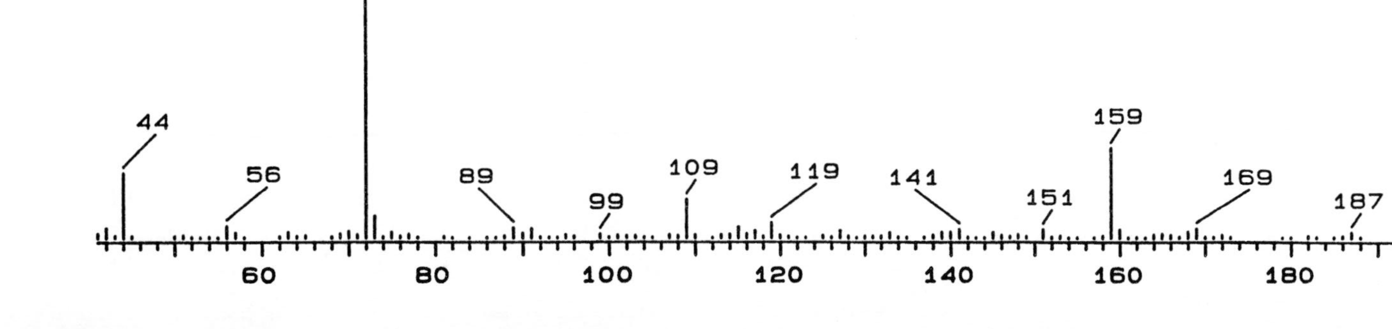

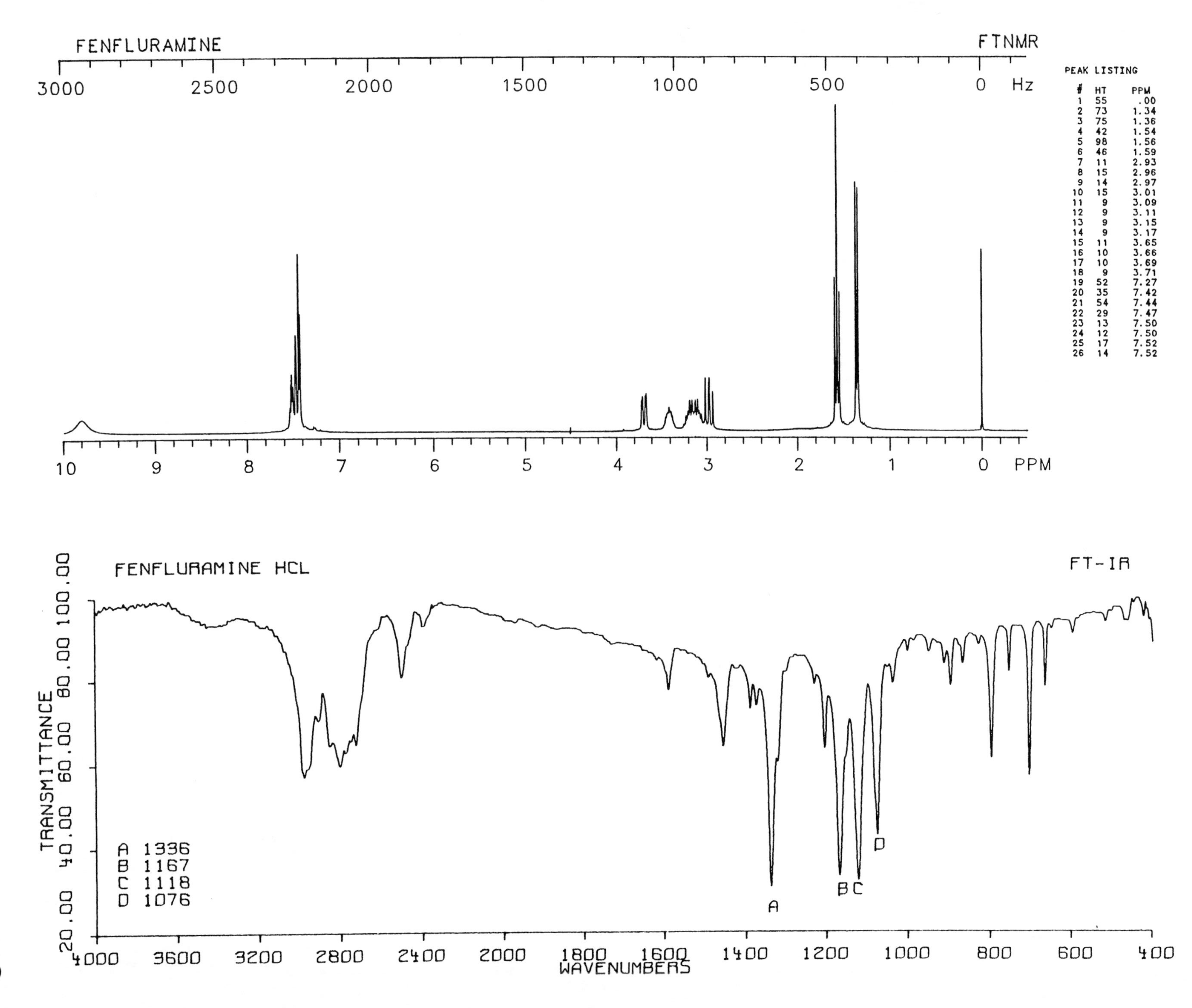
FENFLURAMINE
FTNMR
3000
2500
2000
1500
1000
500
0 Hz
PEAK LISTING
# HT PPM
1 55 .00
2 73 1.34
3 75 1.36
4 42 1.54
5 98 1.56
6 46 1.59
7 11 2.93
8 15 2.96
9 14 2.97
10 15 3.01
11 9 3.09
12 9 3.11
13 9 3.15
14 9 3.17
15 11 3.65
16 10 3.66
17 10 3.69
18 9 3.71
19 52 7.27
20 35 7.42
21 54 7.44
22 29 7.47
23 13 7.50
24 12 7.50
25 17 7.52
26 14 7.52
10
9
8
7
6
5
4
3
2
1
0 PPM
FENFLURAMINE HCL
FT-IR
TRANSMITTANCE
20.00
40.00
60.00
80.00
100.00
A 1336
B 1167
C 1118
D 1076
A
BC
D
4000
3600
3200
2800
2400
2000
1800
1600
1400
1200
1000
800
600
400
WAVENUMBERS

# FENOPROFEN

$C_{15}H_{14}O_3$

Molecular weight: 242.27 (242.09)
Synonyms: α-Methyl-3-phenoxybenzenacetic acid; α-dl-2-(3-phenoxyphenyl)propionic acid
Trade names: Nalfon

Use: Anti-inflammatory
HPLC: Si-10; 20A:80B; 4.0
GC: 2042; 250°C

$C_6H_5-O-C_6H_4-CH(CH_3)-COOH$



*FENOPROFEN*

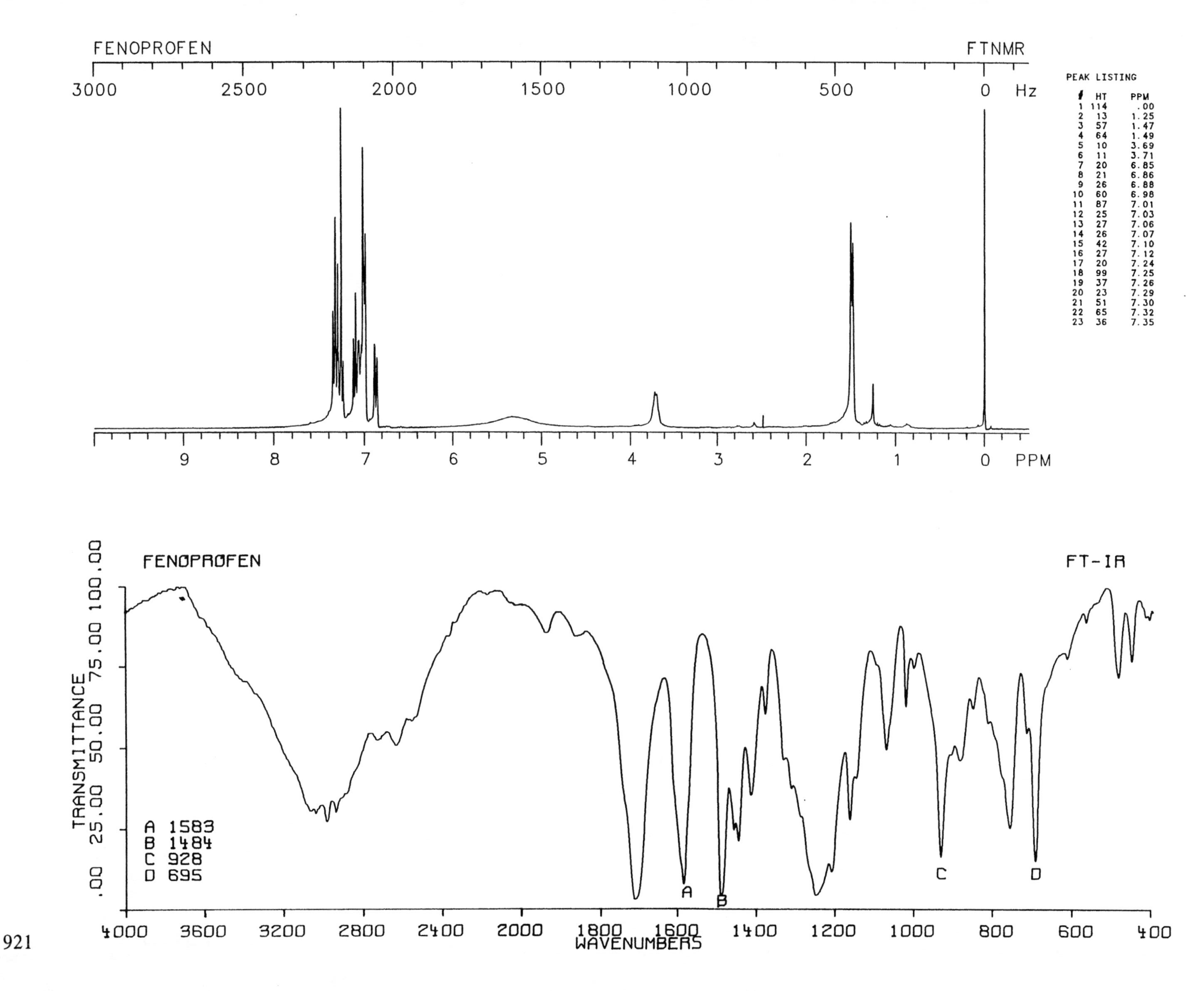
FENOPROFEN
FTNMR
3000 2500 2000 1500 1000 500 0 Hz
PEAK LISTING
# HT PPM
1 114 .00
2 13 1.25
3 57 1.47
4 64 1.49
5 10 3.69
6 11 3.71
7 20 6.85
8 21 6.86
9 26 6.88
10 60 6.98
11 87 7.01
12 25 7.03
13 27 7.06
14 26 7.07
15 42 7.10
16 27 7.12
17 20 7.24
18 99 7.25
19 37 7.26
20 23 7.29
21 51 7.30
22 65 7.32
23 36 7.35
9 8 7 6 5 4 3 2 1 0 PPM
FENOPROFEN
FT-IR
TRANSMITTANCE
.00 25.00 50.00 75.00 100.00
A 1583
B 1484
C 928
D 695
A B C D
4000 3600 3200 2800 2400 2000 1800 1600 1400 1200 1000 800 600 400
WAVENUMBERS

# FENOTEROL

$C_{17}H_{21}NO_4$

Molecular weight: 303.37 (303.15)
Synonyms: 5-[1-Hydroxy-2-[[2-(4-Hydroxyphenyl)-1-methylethyl]
amino]ethyl]-1,3-benzenediol
Trade names: Berotec, Duovent, Partusisten

Use: Bronchodilator
HPLC: Si-10; 10A:90B; 9.8
GC:

HO, HO, OH, CH–CH₂–NH–CH(CH₃)–CH₂, OH

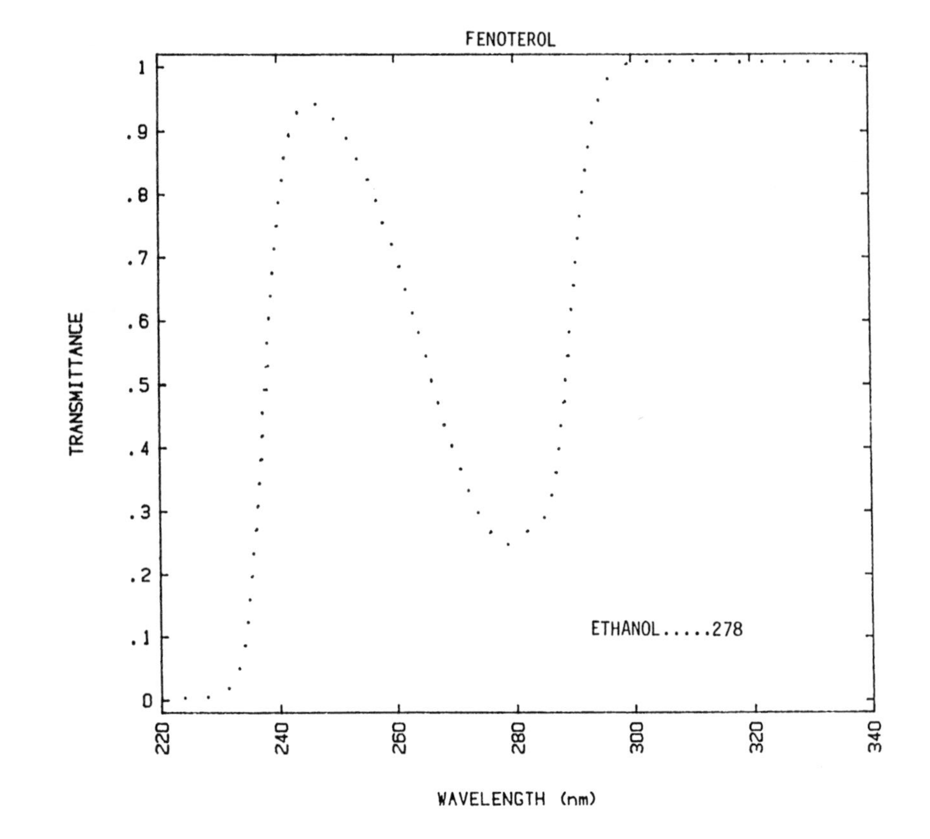


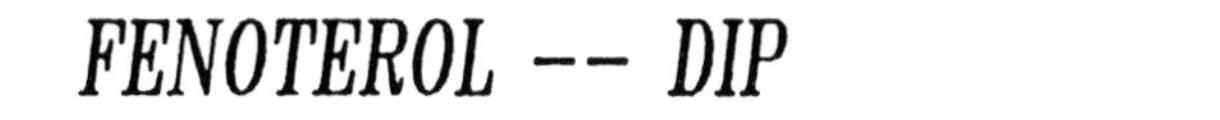

FENOTEROL -- DIP

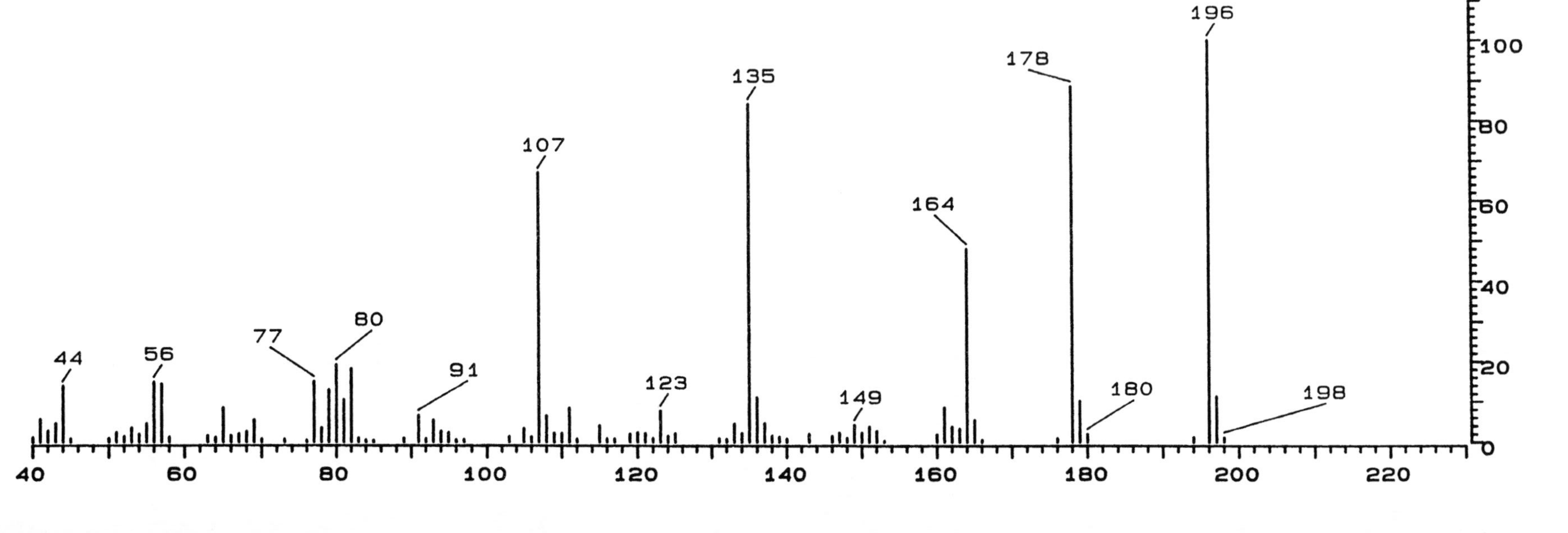

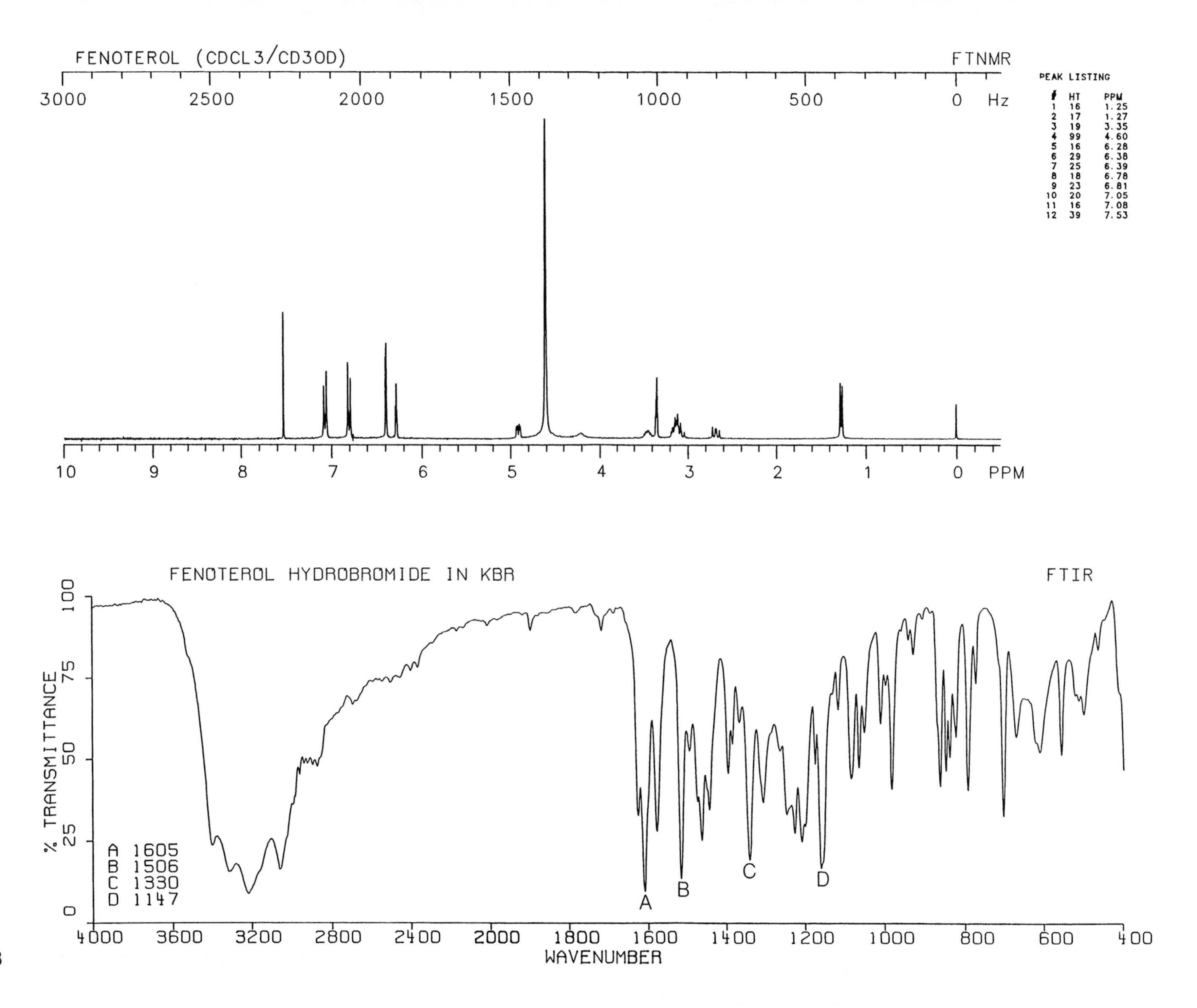
FENOTEROL (CDCL3/CD3OD)
FTNMR
3000 2500 2000 1500 1000 500 0 Hz
10 9 8 7 6 5 4 3 2 1 0 PPM
PEAK LISTING
# HT PPM
1 16 1.25
2 17 1.27
3 19 3.35
4 99 4.60
5 16 6.28
6 29 6.38
7 25 6.39
8 18 6.78
9 23 6.81
10 20 7.05
11 16 7.08
12 39 7.53
FENOTEROL HYDROBROMIDE IN KBR
FTIR
% TRANSMITTANCE
100 75 50 25 0
A 1605
B 1506
C 1330
D 1147
A B C D
4000 3600 3200 2800 2400 2000 1800 1600 1400 1200 1000 800 600 400
WAVENUMBER

# FENPIVERINIUM BROMIDE

$C_{22}H_{29}BrN_2O$

Molecular weight: 417.41 (416.15)
Synonyms: 1-(4-Amino-4-oxo-3,3-diphenylbutyl)-1-methylpiperidinium bromide; fenpipramide methobromide
Trade names: Resantin

Use: Antispasmodic
HPLC: Si-10; 10A:90B; 4.3
GC:

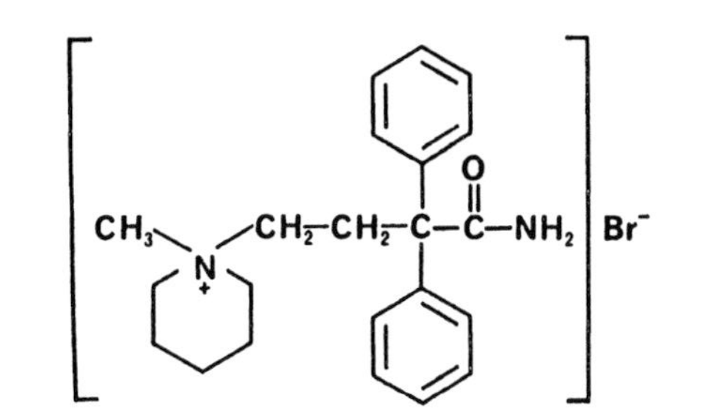


FENPIVERINIUM BROMIDE

TRANSMITTANCE

1
.9
.8
.7
.6
.5
.4
.3
.2
.1
0

220
240
260
280
300
320
340

WAVELENGTH (nm)

252,258,264
252,258

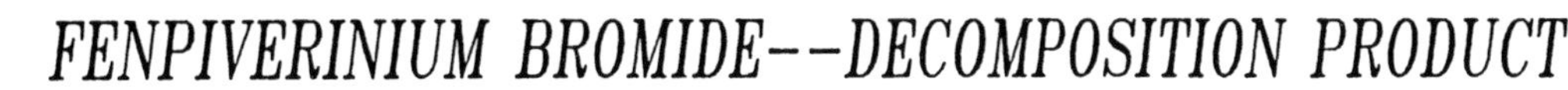


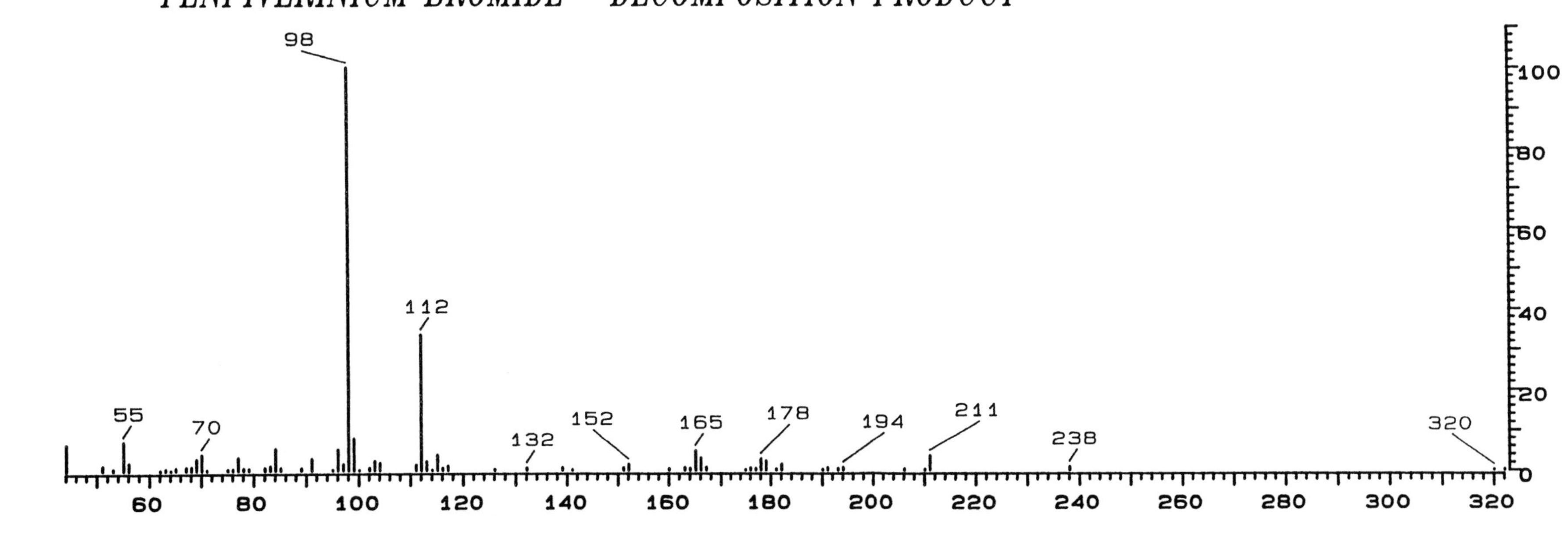

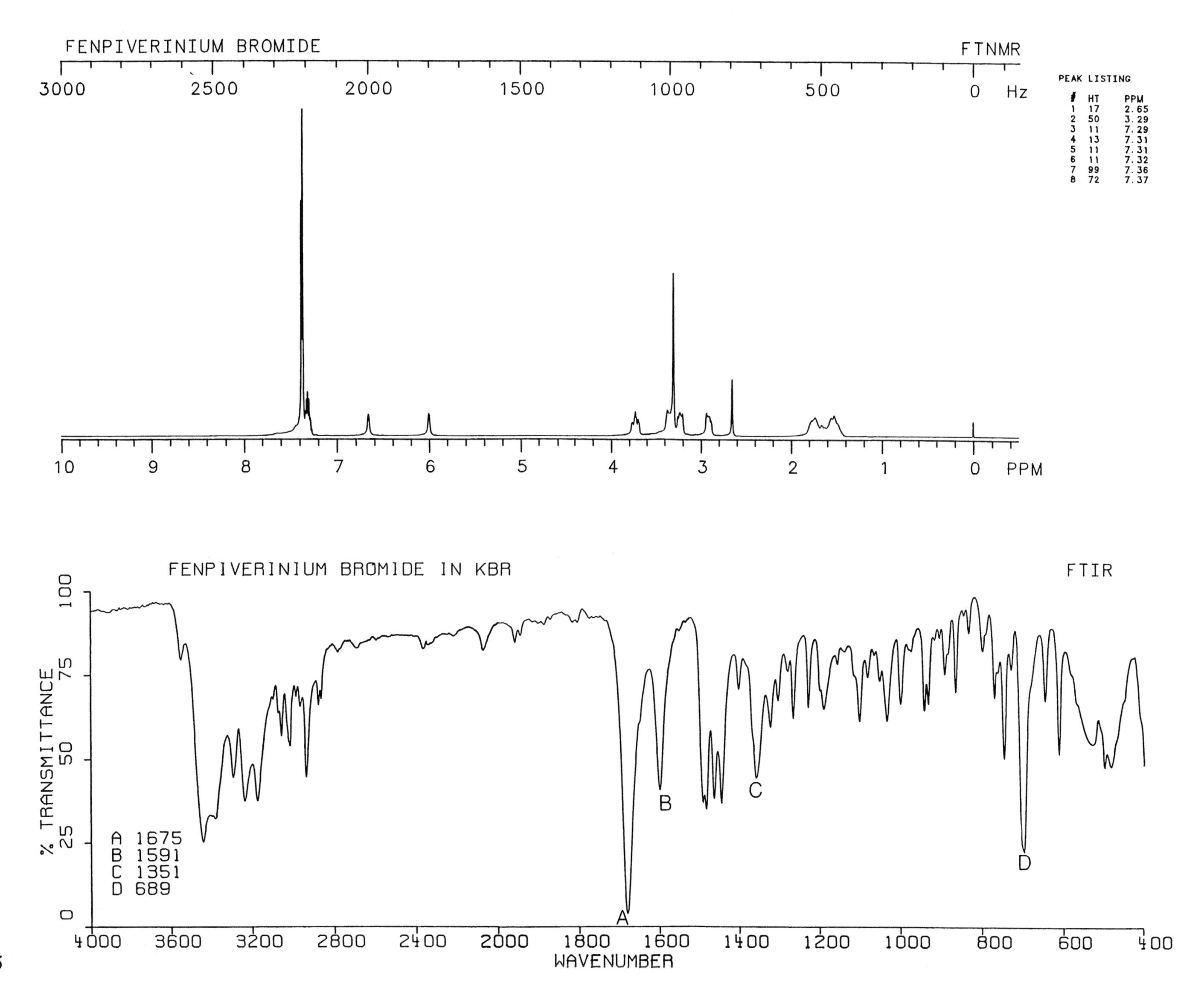
FENPIVERINIUM BROMIDE
FTNMR
PEAK LISTING
| # | HT | PPM |
|---|---|---|
| 1 | 17 | 2.65 |
| 2 | 50 | 3.29 |
| 3 | 11 | 7.29 |
| 4 | 13 | 7.31 |
| 5 | 11 | 7.31 |
| 6 | 11 | 7.32 |
| 7 | 99 | 7.36 |
| 8 | 72 | 7.37 |
Hz
PPM
FENPIVERINIUM BROMIDE IN KBR
FTIR
% TRANSMITTANCE
WAVENUMBER
A 1675
B 1591
C 1351
D 689

# FENTANYL

$C_{22}H_{28}N_2O$

**Molecular weight:** 336.48 (336.22)
**Synonyms:** N-Phenyl-N-[1-(2-phenylethyl)-4-piperidinyl]propanamide

**Trade names:** Innovar, Leptanal, Sublimaze

**Use:** Narcotic analgesic
**HPLC:** Si-10; 2A:98B; 4.8
**GC:** 2734; 250°C

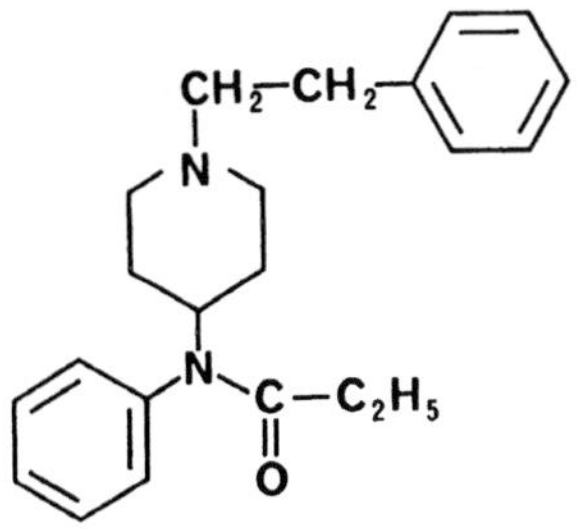

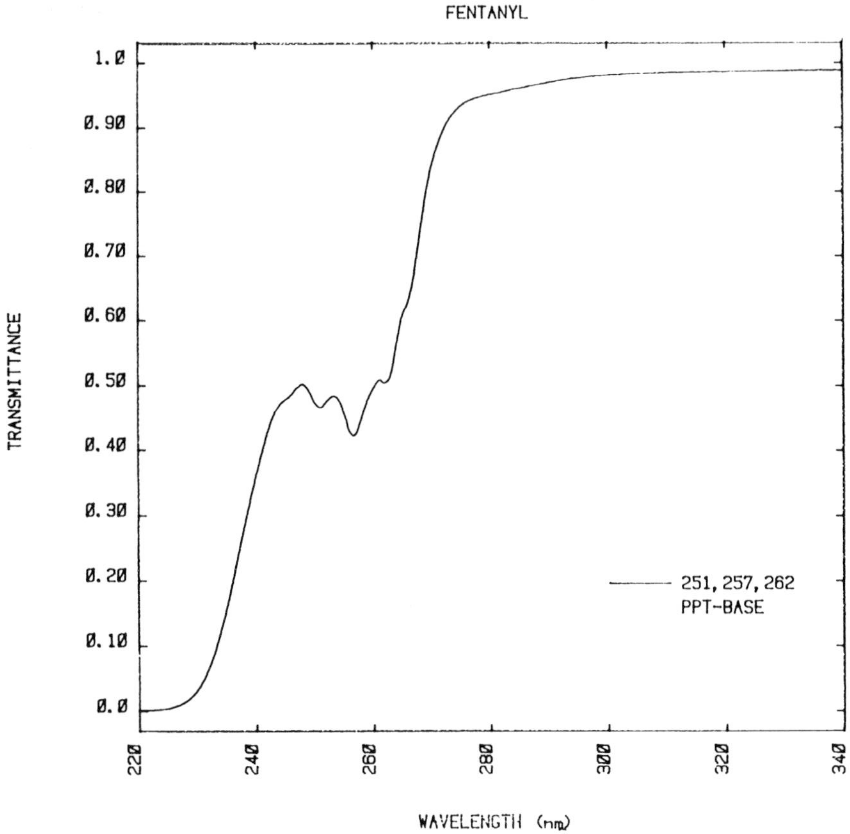

*FENTANYL*

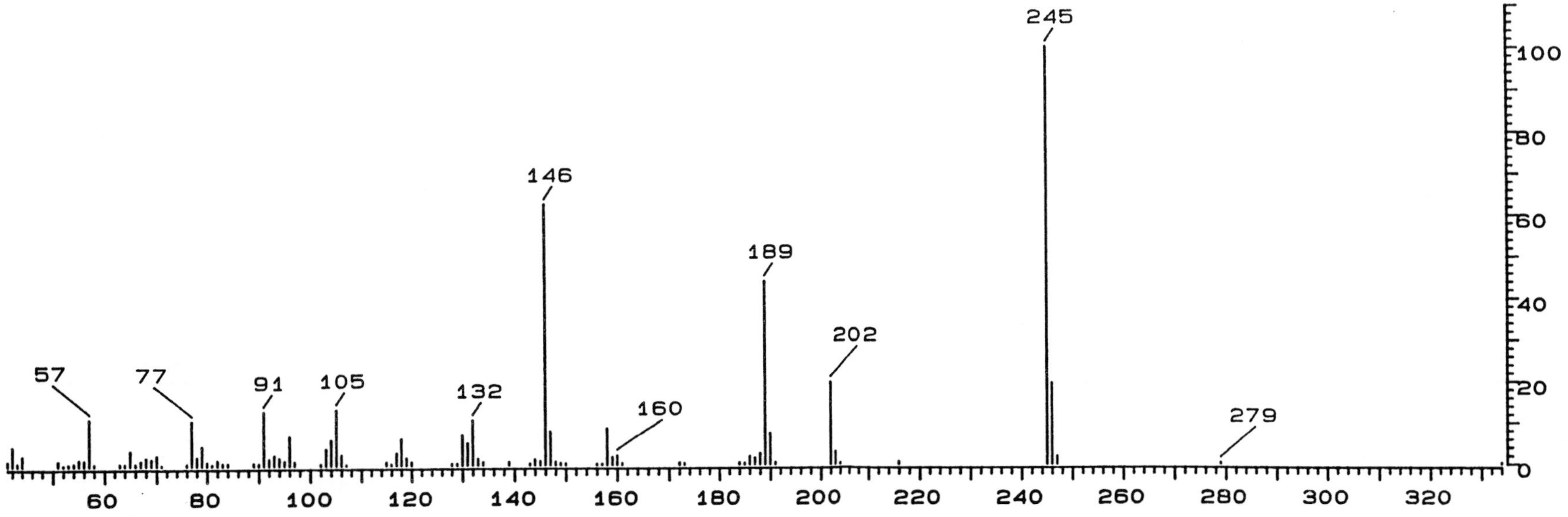

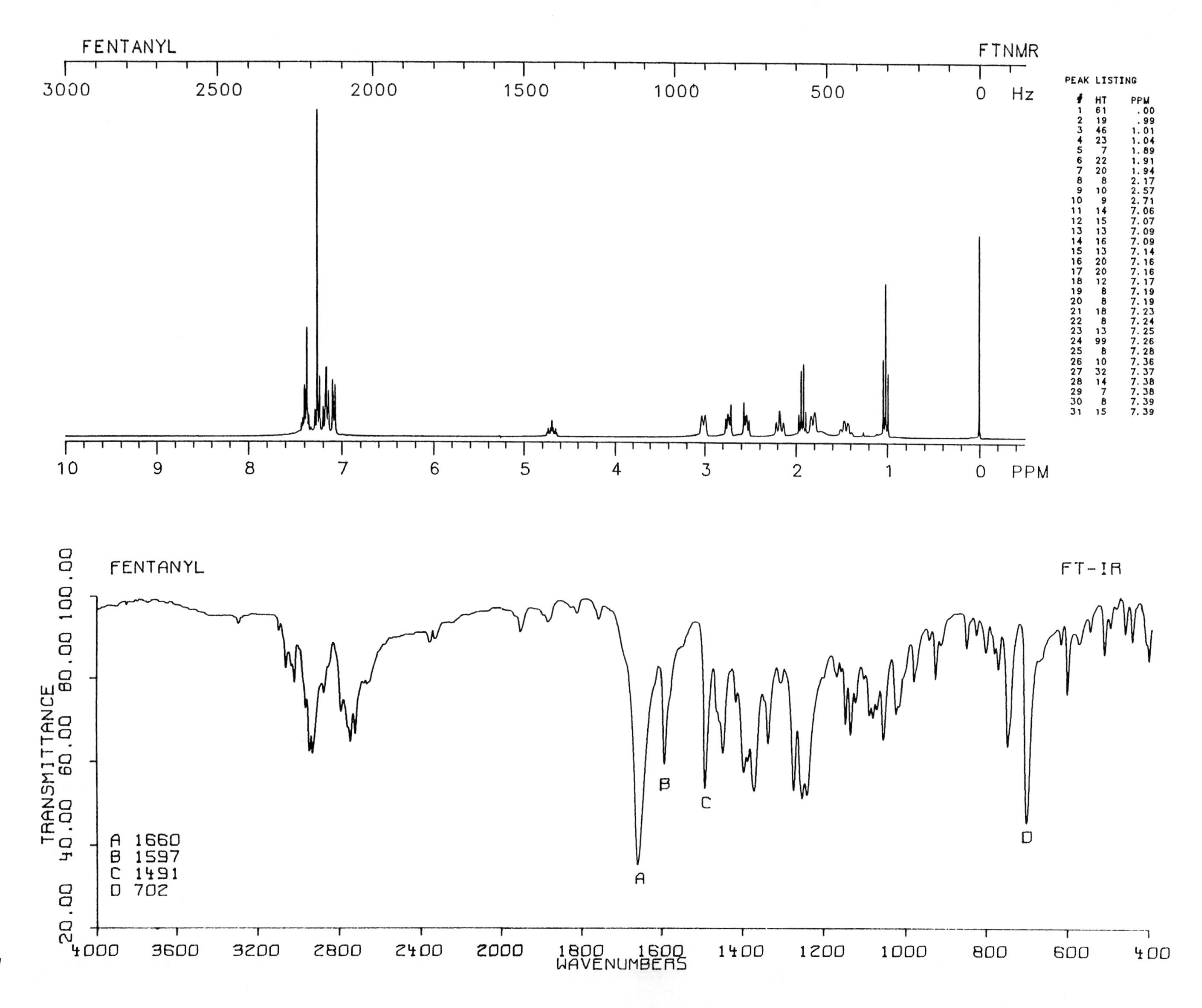
FENTANYL
FTNMR
3000
2500
2000
1500
1000
500
0 Hz
PEAK LISTING
# HT PPM
1 61 .00
2 19 .99
3 46 1.01
4 23 1.04
5 7 1.89
6 22 1.91
7 20 1.94
8 8 2.17
9 10 2.57
10 9 2.71
11 14 7.06
12 15 7.07
13 13 7.09
14 16 7.09
15 13 7.14
16 20 7.16
17 20 7.16
18 12 7.17
19 8 7.19
20 8 7.19
21 18 7.23
22 8 7.24
23 13 7.25
24 99 7.26
25 8 7.28
26 10 7.36
27 32 7.37
28 14 7.38
29 7 7.38
30 8 7.39
31 15 7.39
10
9
8
7
6
5
4
3
2
1
0 PPM
FENTANYL
FT-IR
TRANSMITTANCE
100.00
80.00
60.00
40.00
20.00
A 1660
B 1597
C 1491
D 702
A
B
C
D
4000
3600
3200
2800
2400
2000
1800
1600
1400
1200
1000
800
600
400
WAVENUMBERS

# FENTANYL ACETYL ANALOG

$C_{21}H_{26}N_2O$

Molecular weight: 322.45 (322.21)
Synonyms: N-Phenyl-N[1-(2-phenylethyl)-4-piperidinyl]acetamide

Trade names:

Use:
HPLC: Si-10; 2A:98B; 3.7
GC: 2693; 280°C

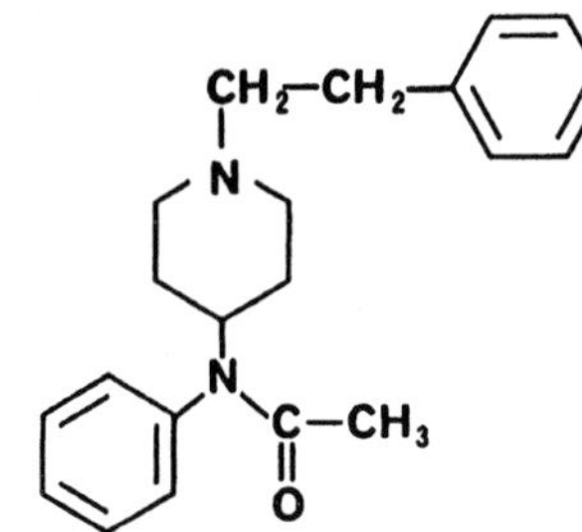


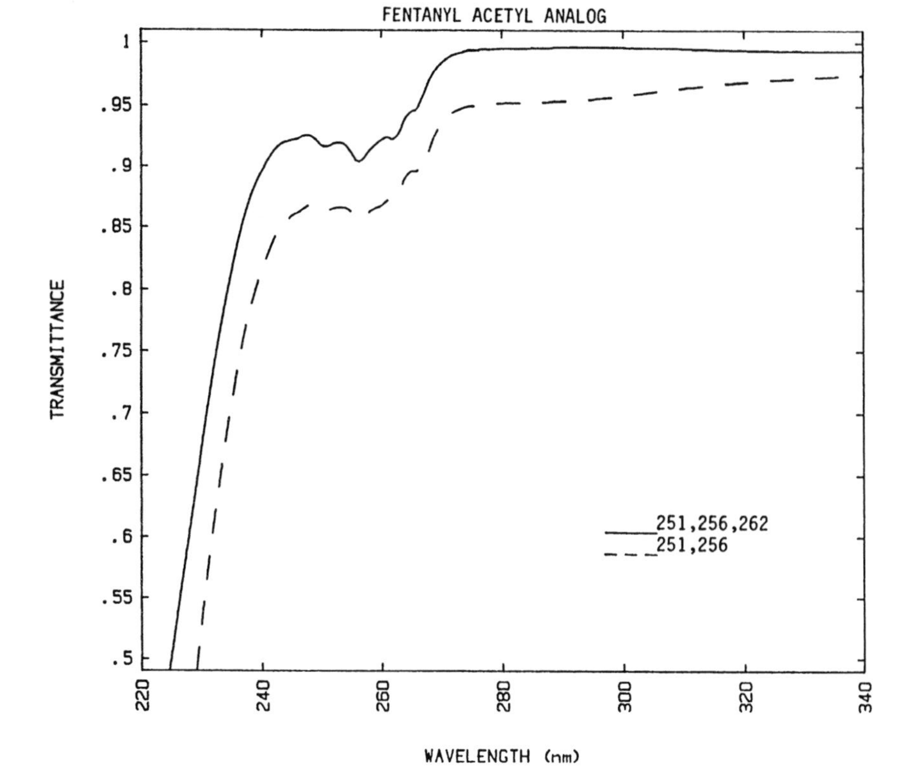

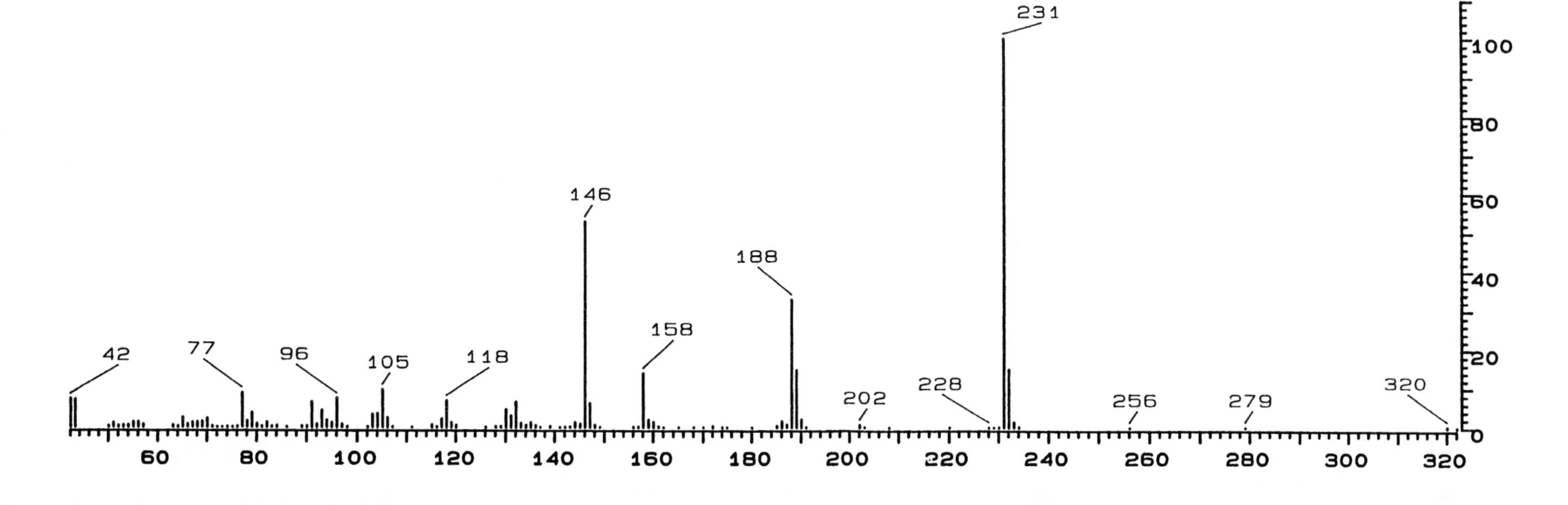

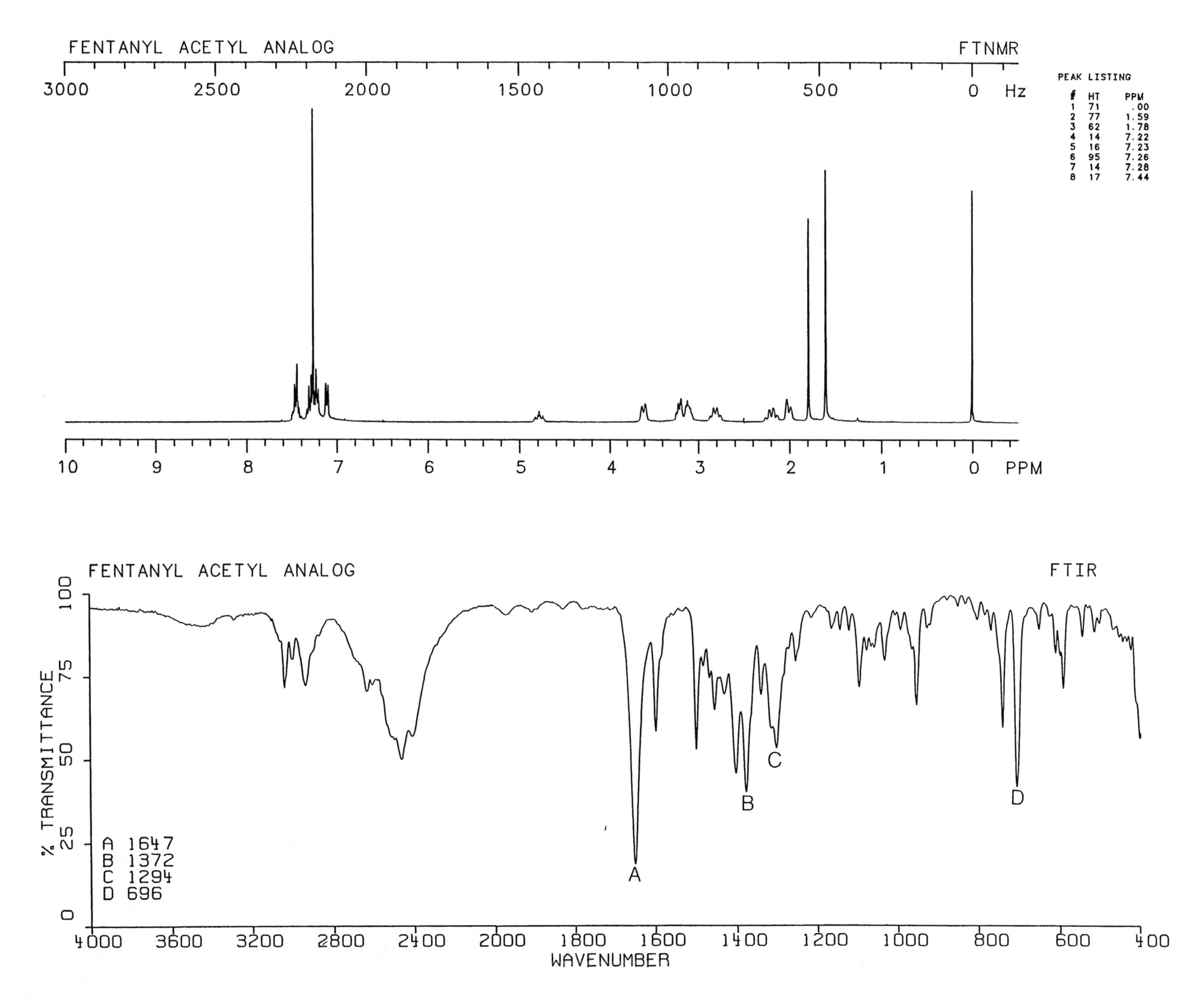
FENTANYL ACETYL ANALOG
FTNMR
3000 2500 2000 1500 1000 500 0 Hz
PEAK LISTING
# HT PPM
1 71 .00
2 77 1.59
3 62 1.78
4 14 7.22
5 16 7.23
6 95 7.26
7 14 7.28
8 17 7.44
10 9 8 7 6 5 4 3 2 1 0 PPM
FENTANYL ACETYL ANALOG
FTIR
% TRANSMITTANCE
100 75 50 25 0
A 1647
B 1372
C 1294
D 696
A
B
C
D
4000 3600 3200 2800 2400 2000 1800 1600 1400 1200 1000 800 600 400
WAVENUMBER

# FENTANYL METHYL ANALOG

$C_{21}H_{26}N_2O$

Molecular weight: 322.45 (322.21)
Synonyms: N-Penyl-N-[(1-phenylmethyl)-4-piperidinyl]propanamide

Trade names:

Use:
HPLC: Si-10; 2A:98B; 4.2
GC: 2632; 280°C

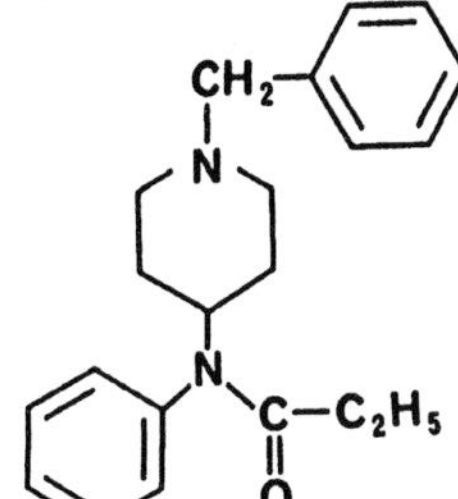


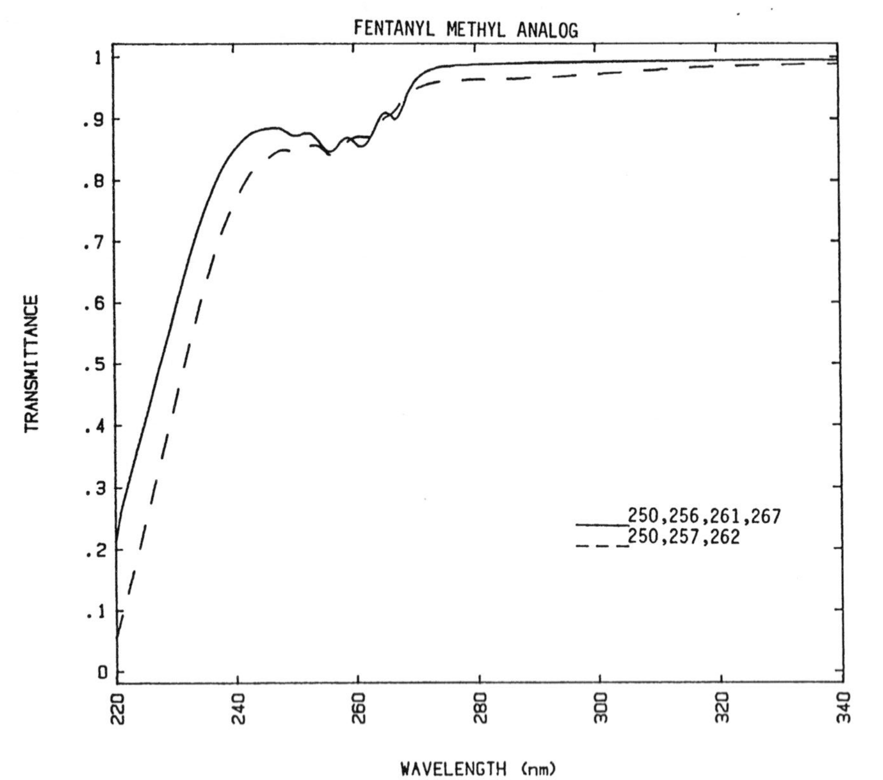

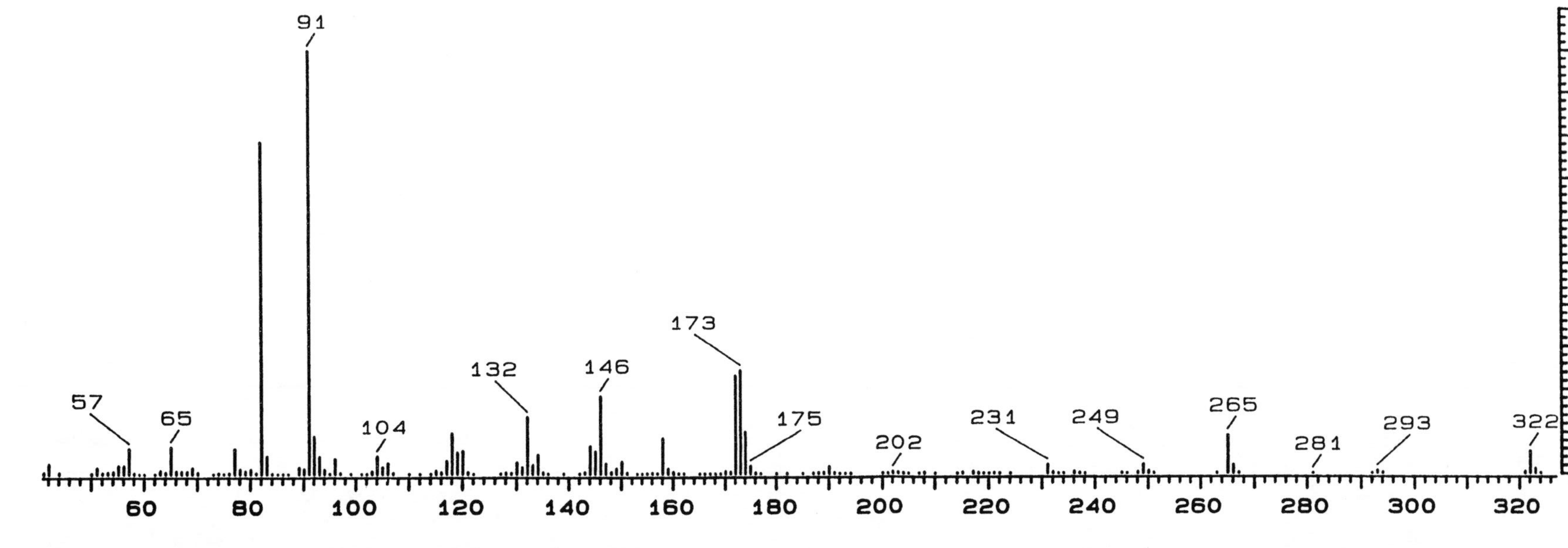

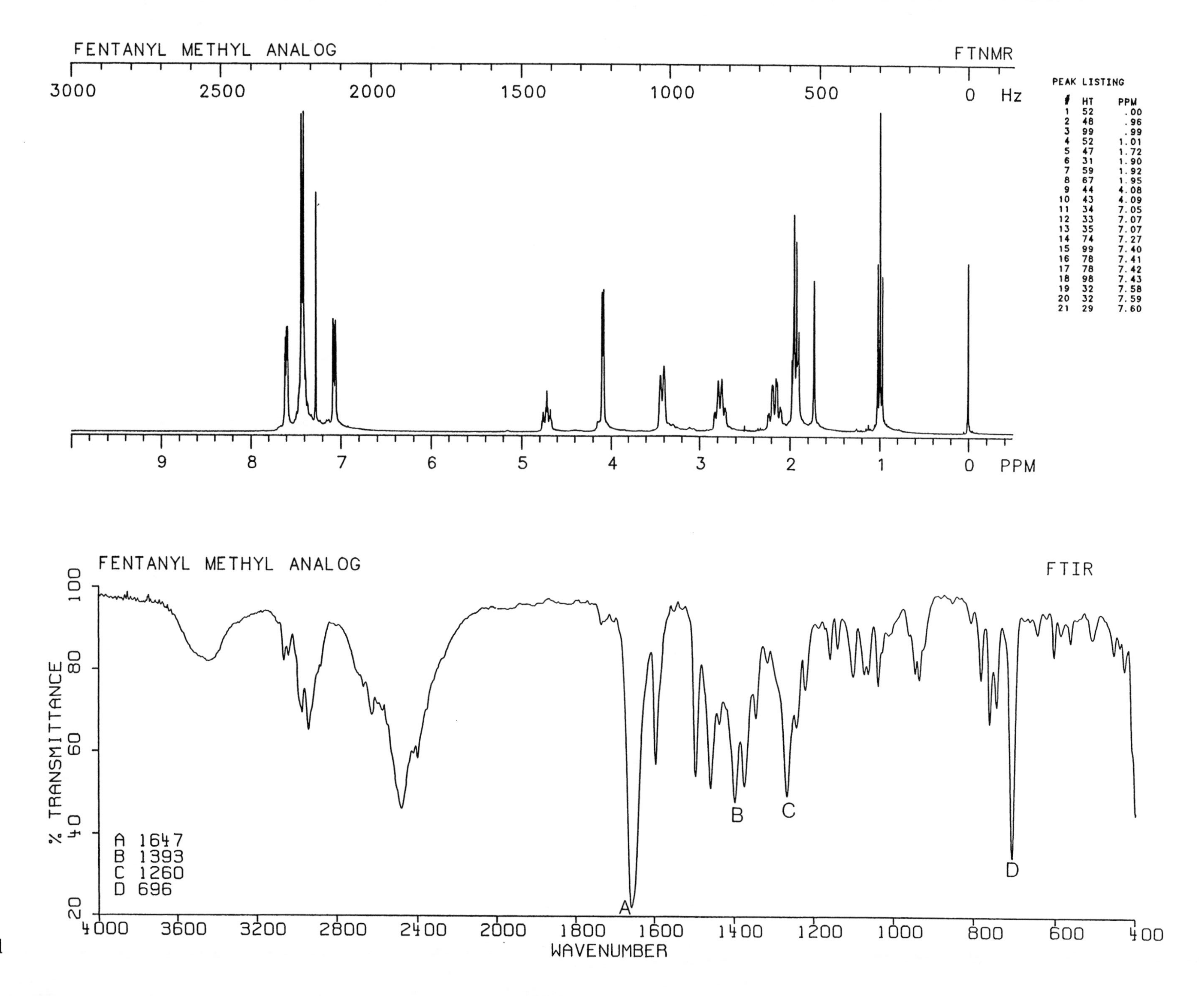
FENTANYL METHYL ANALOG
FTNMR
3000 2500 2000 1500 1000 500 0 Hz
9 8 7 6 5 4 3 2 1 0 PPM
PEAK LISTING
# HT PPM
1 52 .00
2 48 .96
3 99 .99
4 52 1.01
5 47 1.72
6 31 1.90
7 59 1.92
8 67 1.95
9 44 4.08
10 43 4.09
11 34 7.05
12 33 7.07
13 35 7.07
14 74 7.27
15 99 7.40
16 78 7.41
17 78 7.42
18 98 7.43
19 32 7.58
20 32 7.59
21 29 7.60
FENTANYL METHYL ANALOG
FTIR
% TRANSMITTANCE
100 80 60 40 20
A 1647
B 1393
C 1260
D 696
A B C D
4000 3600 3200 2800 2400 2000 1800 1600 1400 1200 1000 800 600 400
WAVENUMBER

# FENTANYL METHYL ACETYL ANALOG

$C_{20}H_{24}N_2O$

**Molecular weight:** 308.42 (308.19)
**Synonyms:** N-Phenyl-N-[(1-phenylmethyl)-4-piperidinyl]acetamide

**Trade names:**

**Use:**
**HPLC:** Si-10; 2A:98B; 4.7
**GC:**

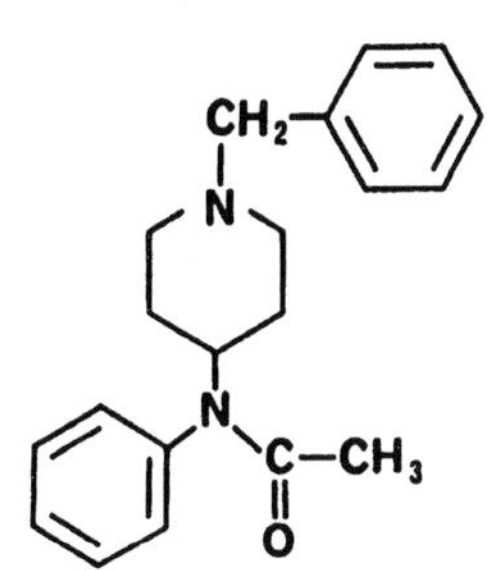


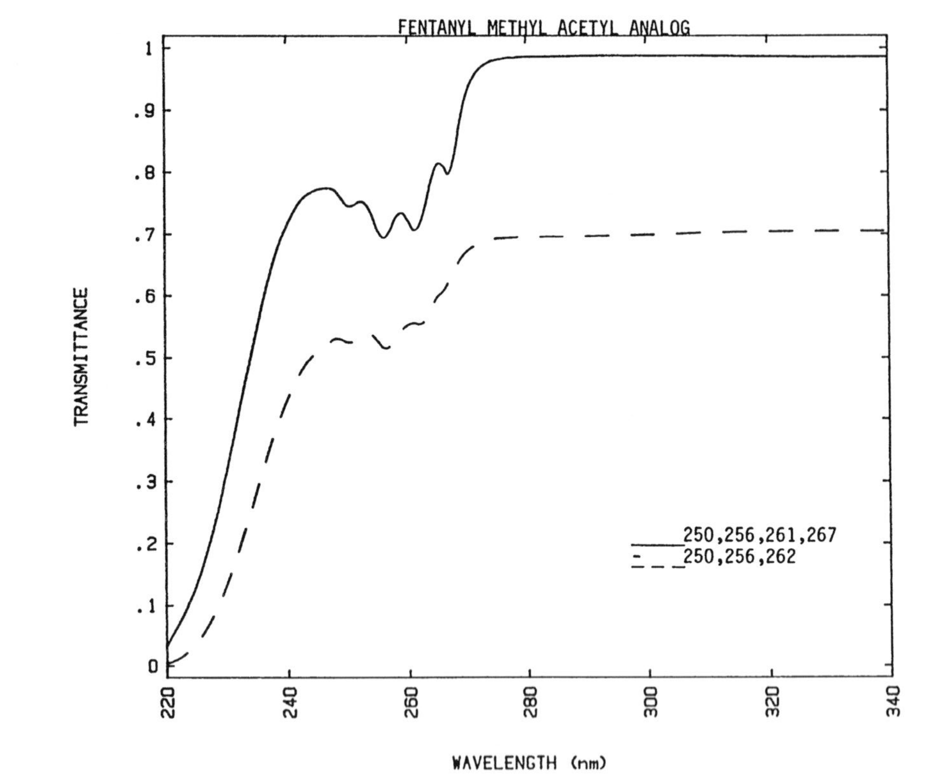

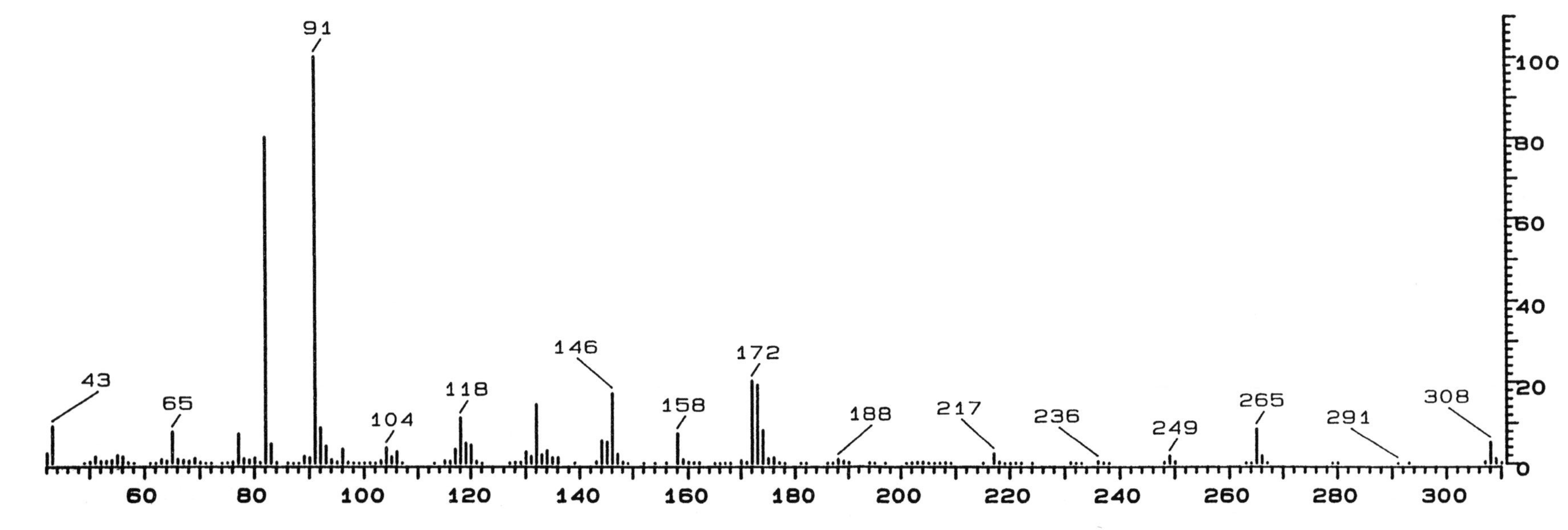

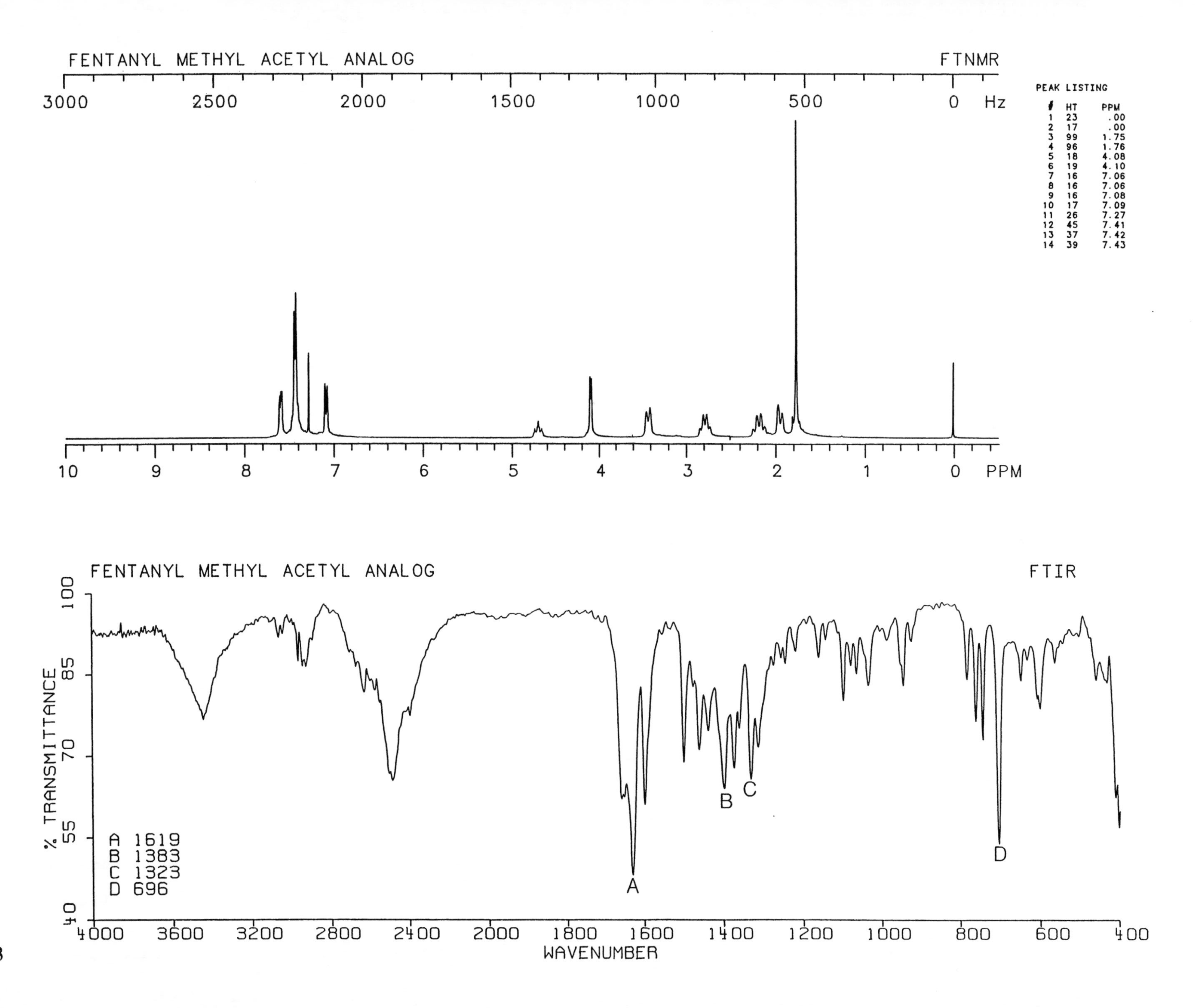
FENTANYL METHYL ACETYL ANALOG
FTNMR
3000 2500 2000 1500 1000 500 0 Hz
PEAK LISTING
# HT PPM
1 23 .00
2 17 .00
3 99 1.75
4 96 1.76
5 18 4.08
6 19 4.10
7 16 7.06
8 16 7.06
9 16 7.08
10 17 7.09
11 26 7.27
12 45 7.41
13 37 7.42
14 39 7.43
10 9 8 7 6 5 4 3 2 1 0 PPM
FENTANYL METHYL ACETYL ANALOG
FTIR
% TRANSMITTANCE
100 85 70 55 40
A 1619
B 1383
C 1323
D 696
A
B
C
D
4000 3600 3200 2800 2400 2000 1800 1600 1400 1200 1000 800 600 400
WAVENUMBER

# FENTANYL 1-META METHYLPHENYL ANALOG

$C_{23}H_{30}N_2O$

**Molecular weight:** 350.50 (350.24)
**Synonyms:** N-Phenyl-N-[1-(2-(3-methylphenyl)ethyl)-4-piperidinyl]-propanamide
**Trade names:**

**Use:**
**HPLC:** Si-10; 2A:98B; 4.0
**GC:** 2848; 280°C

FENTANYL 1-META METHYL PHENYL ANALOG

TRANSMITTANCE

WAVELENGTH (nm)

251,257,263,271
ppt base

FENTANYL 1-META METHYLPHENYL ANALOG

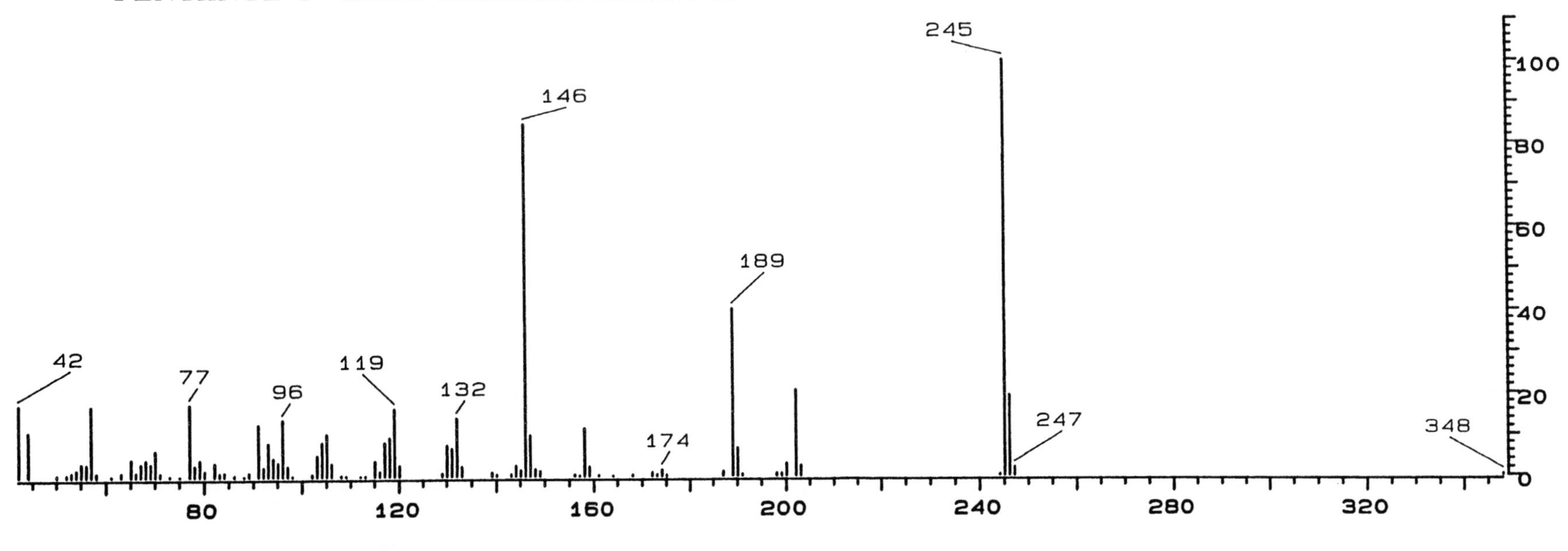

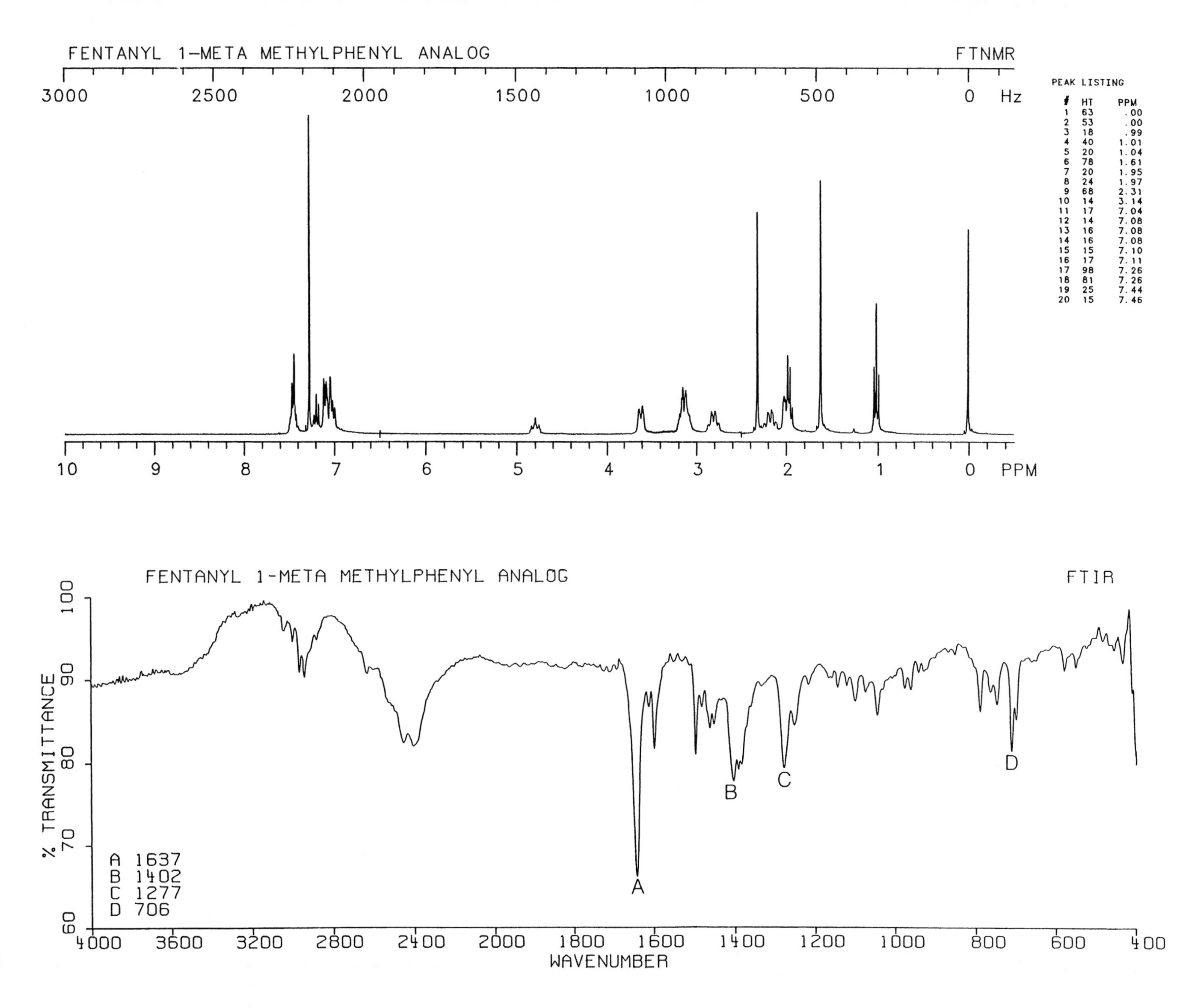

FENTANYL 1-META METHYLPHENYL ANALOG
FTNMR
3000 2500 2000 1500 1000 500 0 Hz
10 9 8 7 6 5 4 3 2 1 0 PPM
PEAK LISTING
# HT PPM
1 63 .00
2 53 .00
3 18 .99
4 40 1.01
5 20 1.04
6 78 1.61
7 20 1.95
8 24 1.97
9 68 2.31
10 14 3.14
11 17 7.04
12 14 7.08
13 16 7.08
14 16 7.08
15 15 7.10
16 17 7.11
17 98 7.26
18 81 7.26
19 25 7.44
20 15 7.46
FENTANYL 1-META METHYLPHENYL ANALOG
FTIR
% TRANSMITTANCE
100 90 80 70 60
A 1637
B 1402
C 1277
D 706
A B C D
4000 3600 3200 2800 2400 2000 1800 1600 1400 1200 1000 800 600 400
WAVENUMBER

# FENTANYL 1-META METHYLPHENYL ACETYL ANALOG

$C_{22}H_{28}N_2O$

**Molecular weight:** 336.47 (336.22)
**Synonyms:** N-Phenyl-N-[1-(2-(3-methylphenyl)ethyl)-4-piperidinyl]-acetamide
**Trade names:**

**Use:**
**HPLC:** Si-10; 2A:98B; 3.5
**GC:** 2809; 280°C

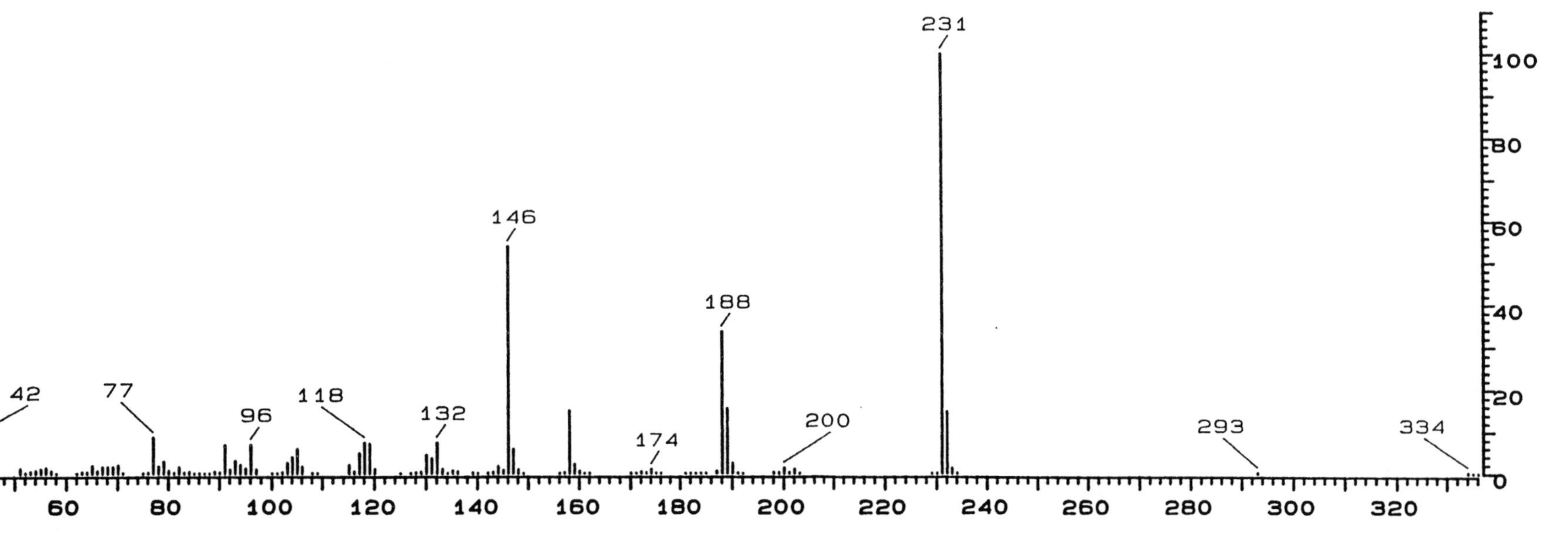

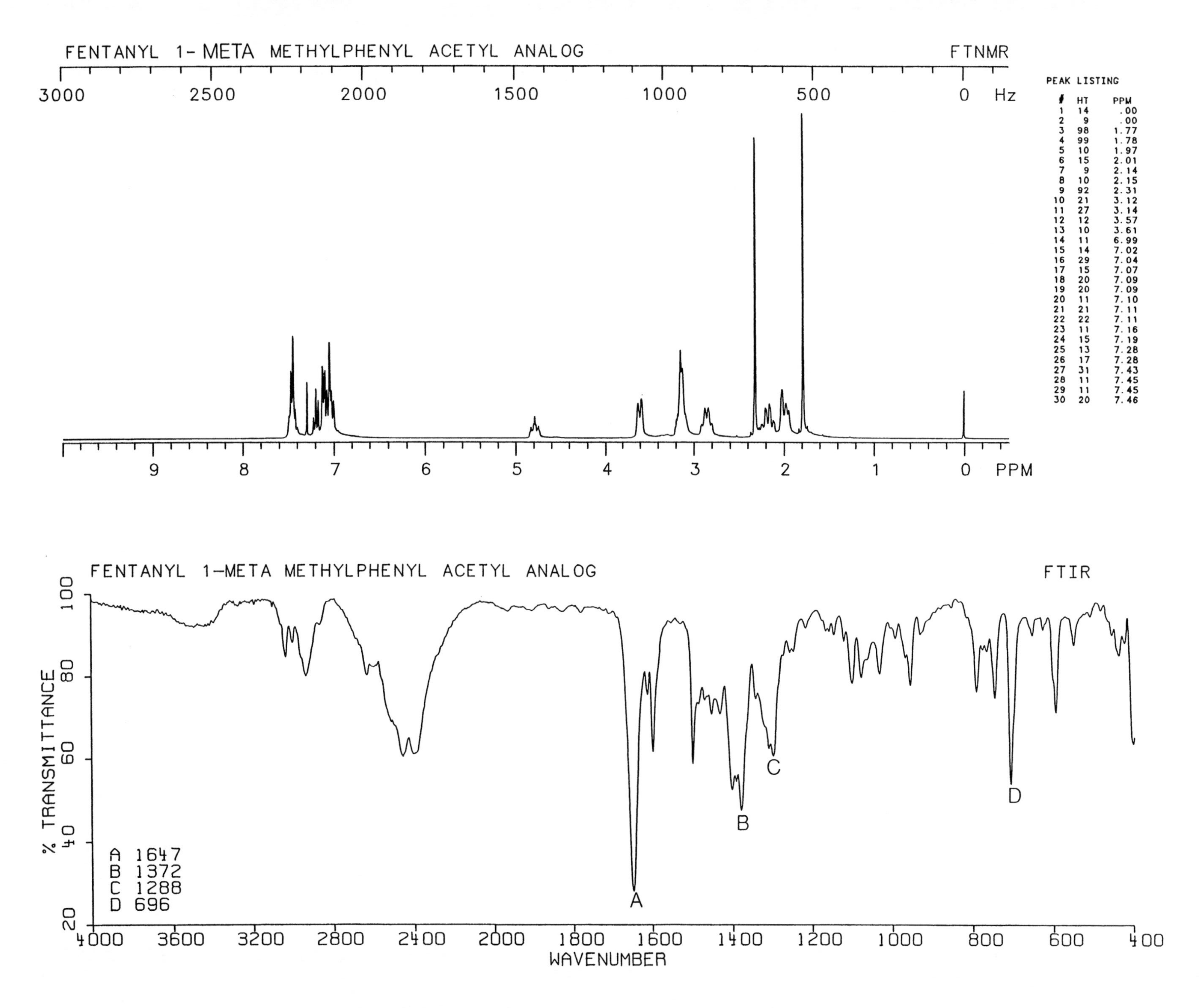
FENTANYL 1-META METHYLPHENYL ACETYL ANALOG
FTNMR
3000
2500
2000
1500
1000
500
0 Hz
9
8
7
6
5
4
3
2
1
0 PPM
PEAK LISTING
# HT PPM
1 14 .00
2 9 .00
3 98 1.77
4 99 1.78
5 10 1.97
6 15 2.01
7 9 2.14
8 10 2.15
9 92 2.31
10 21 3.12
11 27 3.14
12 12 3.57
13 10 3.61
14 11 6.99
15 14 7.02
16 29 7.04
17 15 7.07
18 20 7.09
19 20 7.09
20 11 7.10
21 21 7.11
22 22 7.11
23 11 7.16
24 15 7.19
25 13 7.28
26 17 7.28
27 31 7.43
28 11 7.45
29 11 7.45
30 20 7.46
FENTANYL 1-META METHYLPHENYL ACETYL ANALOG
FTIR
% TRANSMITTANCE
100
80
60
40
20
A 1647
B 1372
C 1288
D 696
A
B
C
D
4000
3600
3200
2800
2400
2000
1800
1600
1400
1200
1000
800
600
400
WAVENUMBER

# FENTANYL 1-ORTHO METHYLPHENYL ACETYL ANALOG

$C_{22}H_{28}N_2O$

**Molecular weight:** 336.47 (336.22)
**Synonyms:** N-Phenyl-N-[1-(2-(2-methylphenyl)ethyl)-4-piperidinyl]-acetamide
**Trade names:**

**Use:**
**HPLC:** Si-10; 2A:98B; 3.6
**GC:** 2832; 280°C

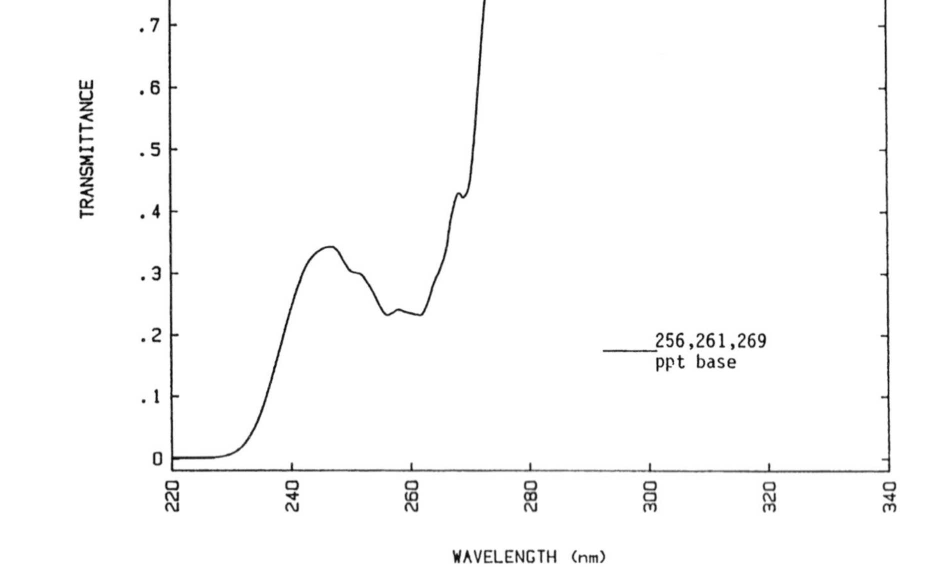

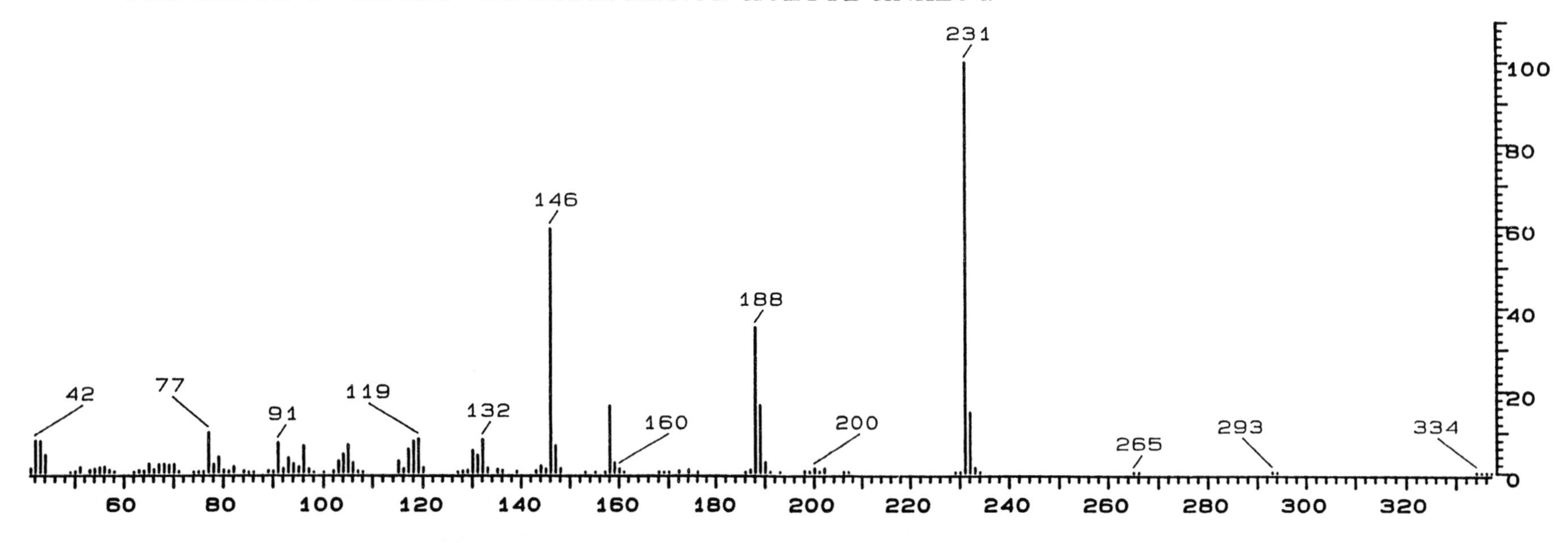

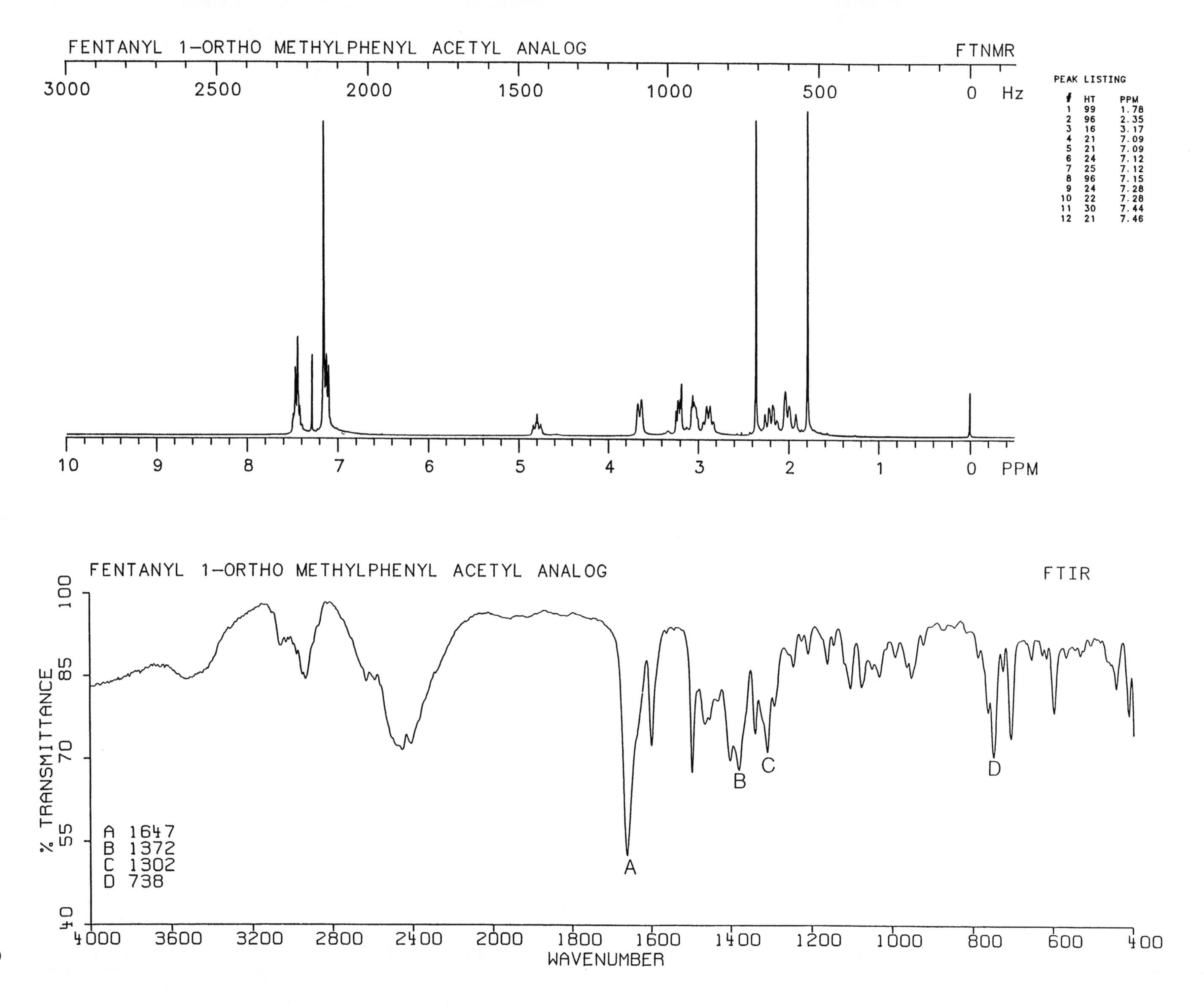
FENTANYL 1-ORTHO METHYLPHENYL ACETYL ANALOG
FTNMR
3000 2500 2000 1500 1000 500 0 Hz
PEAK LISTING
# HT PPM
1 99 1.78
2 96 2.35
3 16 3.17
4 21 7.09
5 21 7.09
6 24 7.12
7 25 7.12
8 96 7.15
9 24 7.28
10 22 7.28
11 30 7.44
12 21 7.46
10 9 8 7 6 5 4 3 2 1 0 PPM
FENTANYL 1-ORTHO METHYLPHENYL ACETYL ANALOG
FTIR
% TRANSMITTANCE
100 85 70 55 40
A 1647
B 1372
C 1302
D 738
A B C D
4000 3600 3200 2800 2400 2000 1800 1600 1400 1200 1000 800 600 400
WAVENUMBER

# FENTANYL 1-PARA METHYLPHENYL ANALOG

$C_{23}H_{30}N_2O$

**Molecular weight:** 350.50 (350.24)
**Synonyms:** N-Phenyl-N-[1-(2-(4-methylphenyl)ethyl)-4-piperidinyl]-propanamide
**Trade names:**

**Use:**
**HPLC:** Si-10; 2A:98B; 3.4
**GC:** 2867; 280°C

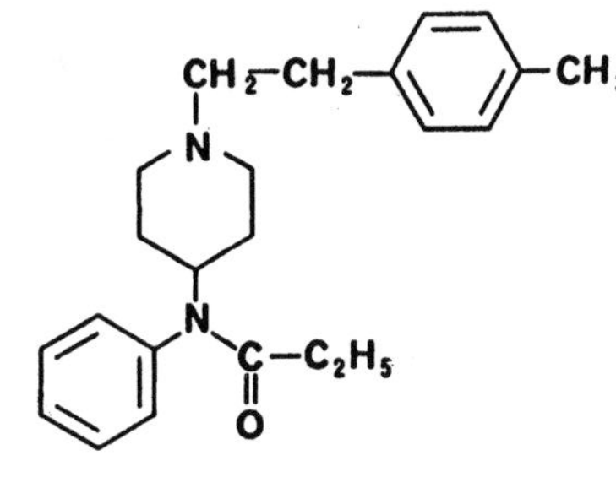


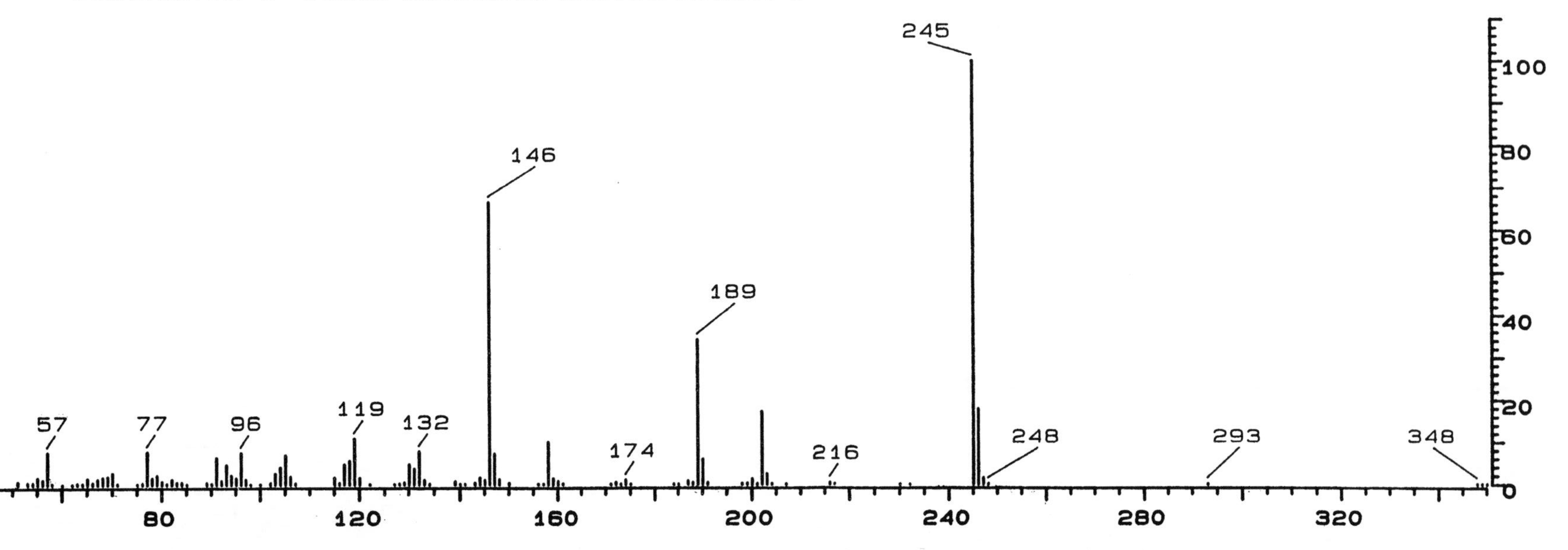

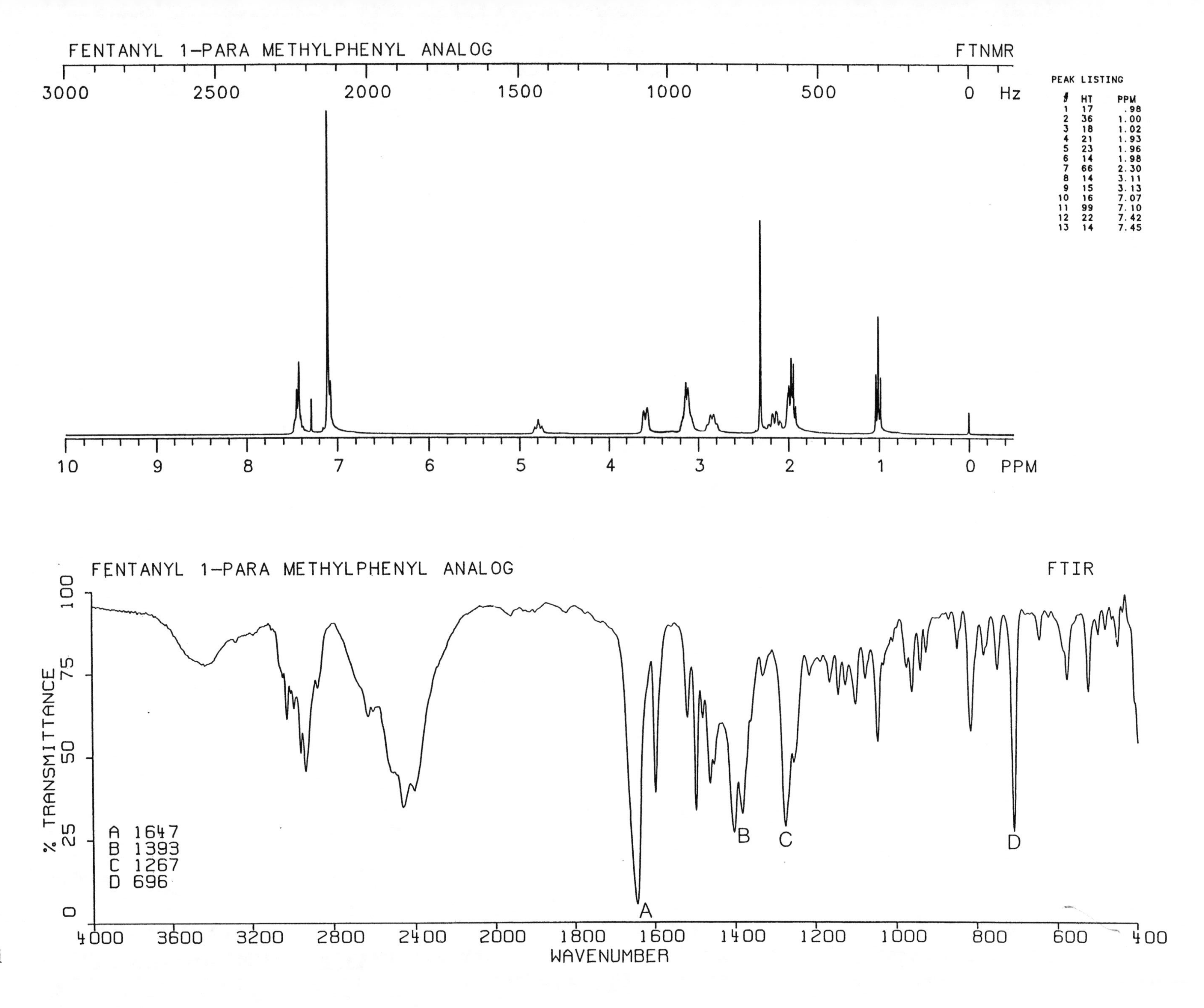

FENTANYL 1-PARA METHYLPHENYL ANALOG
FTNMR
3000 2500 2000 1500 1000 500 0 Hz
PEAK LISTING
# HT PPM
1 17 .98
2 36 1.00
3 18 1.02
4 21 1.93
5 23 1.96
6 14 1.98
7 66 2.30
8 14 3.11
9 15 3.13
10 16 7.07
11 99 7.10
12 22 7.42
13 14 7.45
10 9 8 7 6 5 4 3 2 1 0 PPM
FENTANYL 1-PARA METHYLPHENYL ANALOG
FTIR
% TRANSMITTANCE
100 75 50 25 0
A 1647
B 1393
C 1267
D 696
A B C D
4000 3600 3200 2800 2400 2000 1800 1600 1400 1200 1000 800 600 400
WAVENUMBER

# FENTANYL 1-PARA METHYLPHENYL ACETYL ANALOG

$C_{22}H_{28}N_2O$

**Molecular weight:** 336.47 (336.22)
**Synonyms:** N-Phenyl-N-[1-(2-(4-methylphenyl)ethyl)-4-piperidinyl]-acetamide
**Trade names:**

**Use:**
**HPLC:** Si-10; 2A:98B; 4.4
**GC:** 2823; 280°C

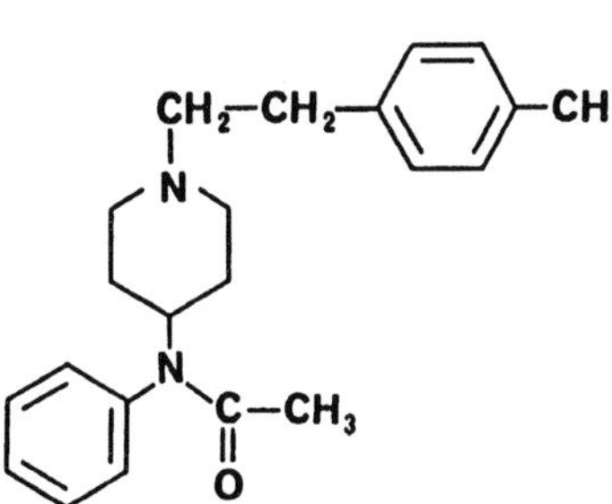


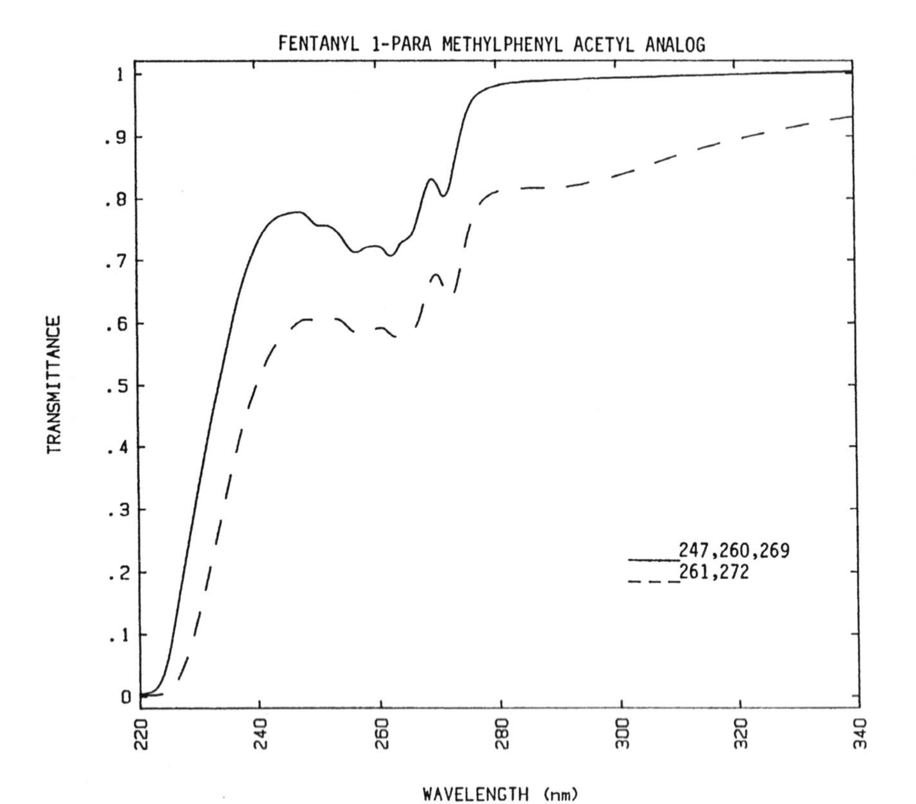


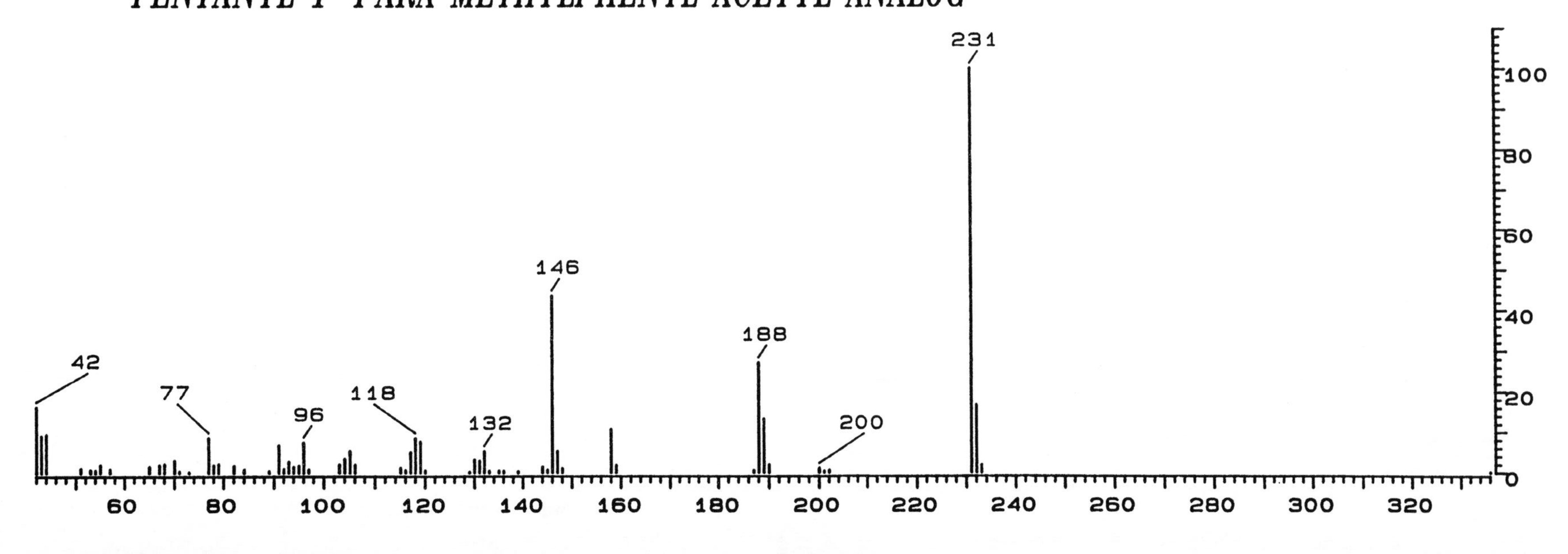

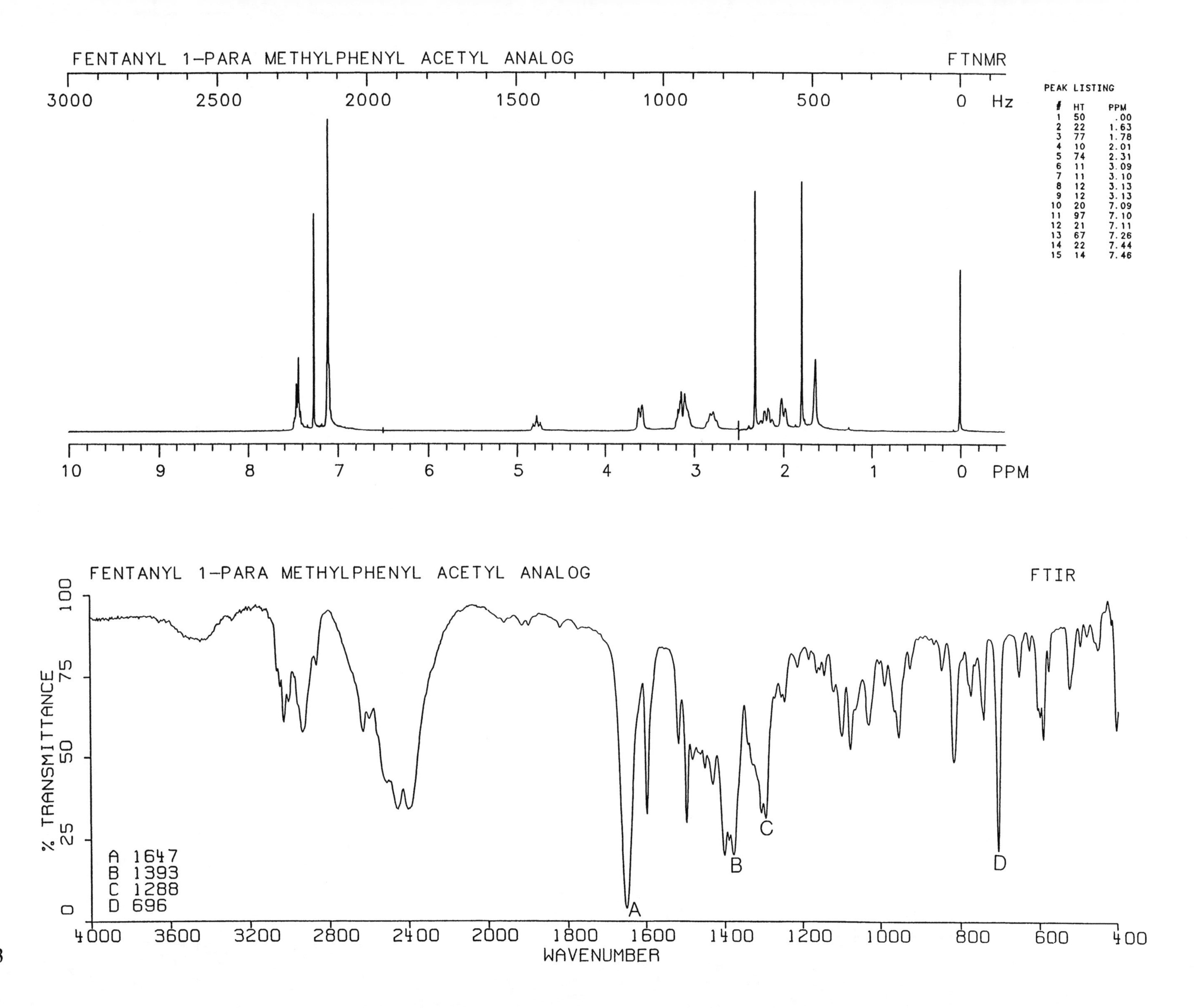
FENTANYL 1-PARA METHYLPHENYL ACETYL ANALOG
FTNMR
3000 2500 2000 1500 1000 500 0 Hz
PEAK LISTING
# HT PPM
1 50 .00
2 22 1.63
3 77 1.78
4 10 2.01
5 74 2.31
6 11 3.09
7 11 3.10
8 12 3.13
9 12 3.13
10 20 7.09
11 97 7.10
12 21 7.11
13 67 7.26
14 22 7.44
15 14 7.46
10 9 8 7 6 5 4 3 2 1 0 PPM
FENTANYL 1-PARA METHYLPHENYL ACETYL ANALOG
FTIR
% TRANSMITTANCE
0 25 50 75 100
A 1647
B 1393
C 1288
D 696
A
B
C
D
4000 3600 3200 2800 2400 2000 1800 1600 1400 1200 1000 800 600 400
WAVENUMBER

# FENTANYL PROPYL ANALOG

$C_{23}H_{30}N_2O$

Molecular weight: 350.50 (350.24)
Synonyms: N-Phenyl-N-[1-(3-phenylpropyl)-4-piperidinyl]-propanamide
Trade names:

Use:
HPLC: Si-10; 2A:98B; 5.5
GC: 2867; 280°C

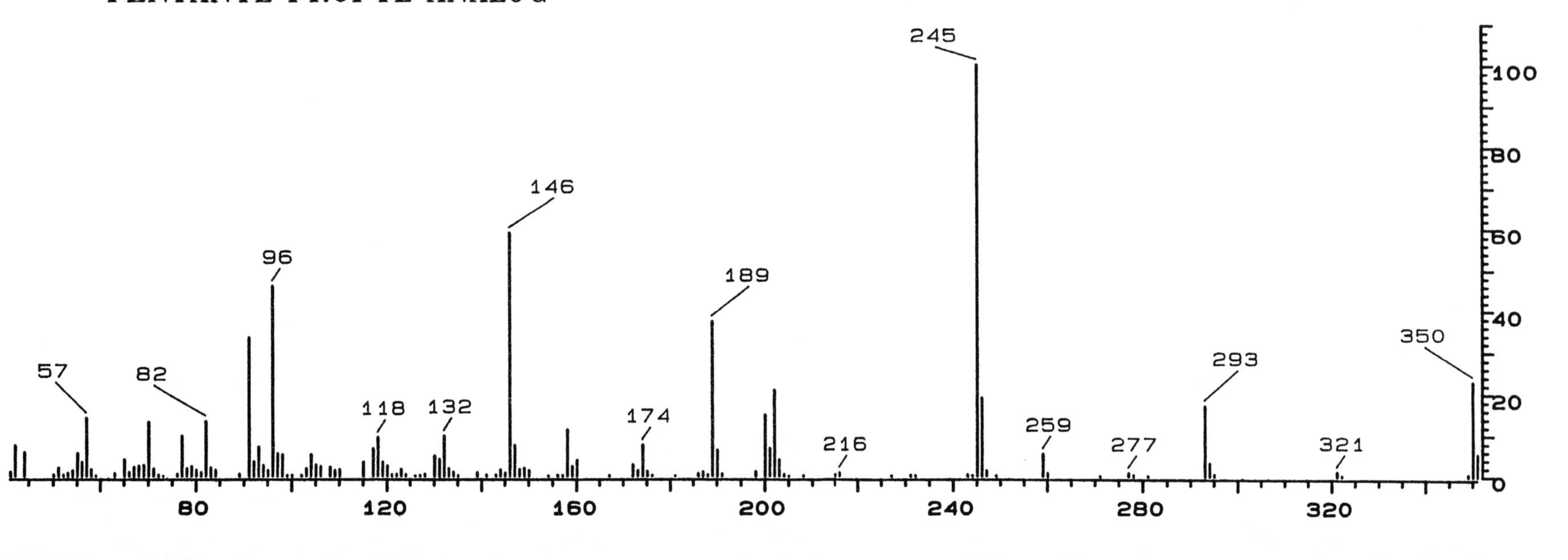

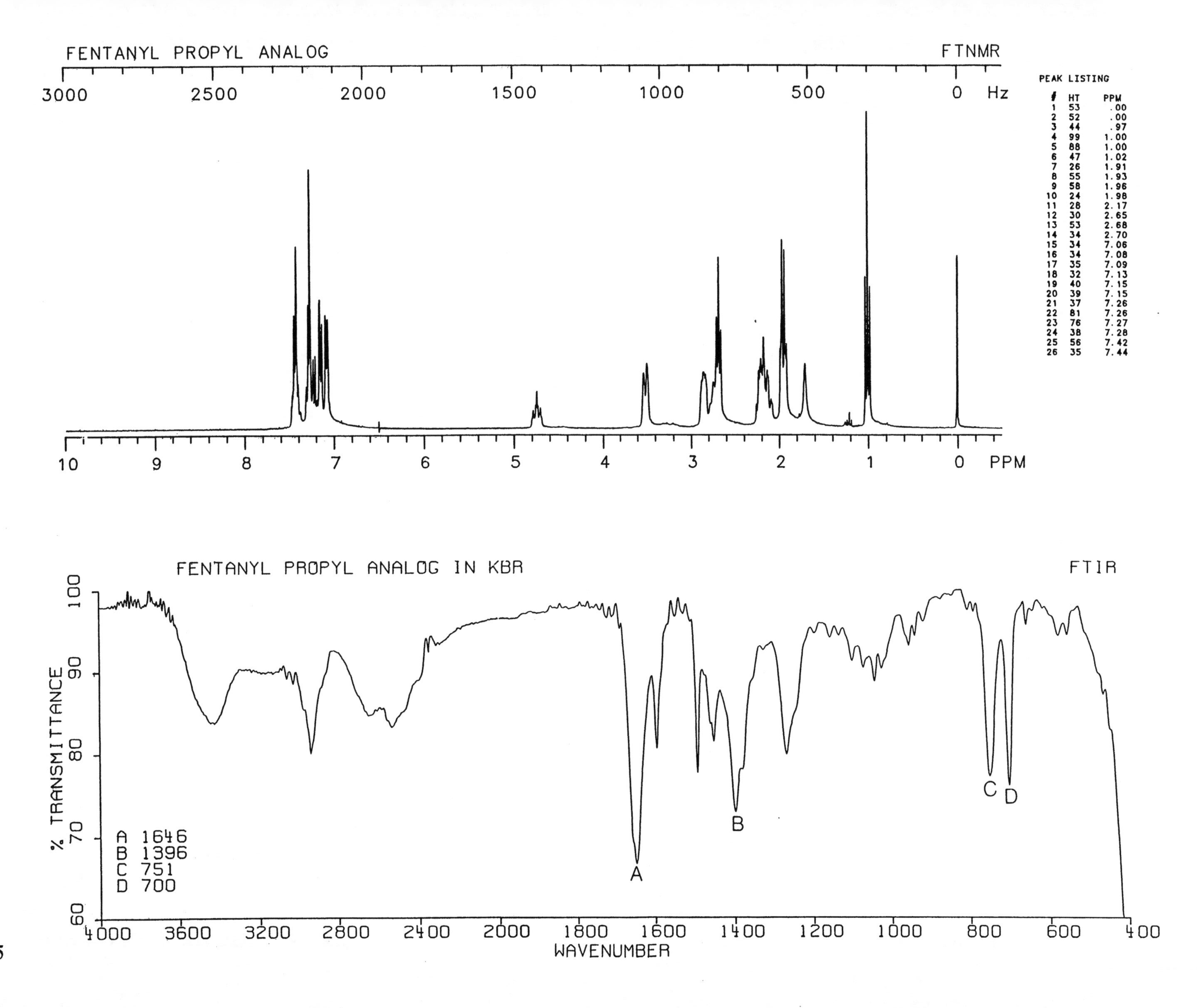

FENTANYL PROPYL ANALOG
FTNMR
3000 2500 2000 1500 1000 500 0 Hz
10 9 8 7 6 5 4 3 2 1 0 PPM
PEAK LISTING
# HT PPM
1 53 .00
2 52 .00
3 44 .97
4 99 1.00
5 88 1.00
6 47 1.02
7 26 1.91
8 55 1.93
9 58 1.96
10 24 1.98
11 28 2.17
12 30 2.65
13 53 2.68
14 34 2.70
15 34 7.06
16 34 7.08
17 35 7.09
18 32 7.13
19 40 7.15
20 39 7.15
21 37 7.26
22 81 7.26
23 76 7.27
24 38 7.28
25 56 7.42
26 35 7.44
FENTANYL PROPYL ANALOG IN KBR
FTIR
% TRANSMITTANCE
100 90 80 70 60
A 1646
B 1396
C 751
D 700
A
B
C
D
4000 3600 3200 2800 2400 2000 1800 1600 1400 1200 1000 800 600 400
WAVENUMBER

# FENTANYL PROPYL ACETYL ANALOG

$C_{22}H_{28}N_2O$

**Molecular weight:** 336.47 (336.22)
**Synonyms:** N-Phenyl-N-[1-(3-phenylpropyl)-4-piperidinyl]acetamide

**Trade names:**

**Use:**
HPLC: Si-10; 2A:98B; 5.0
GC: 2829; 280°C

FENTANYL PROPYL ACETYL ANALOG

TRANSMITTANCE

WAVELENGTH (nm)

251,256
ppt base

*FENTANYL PROPYL ACETYL ANALOG*

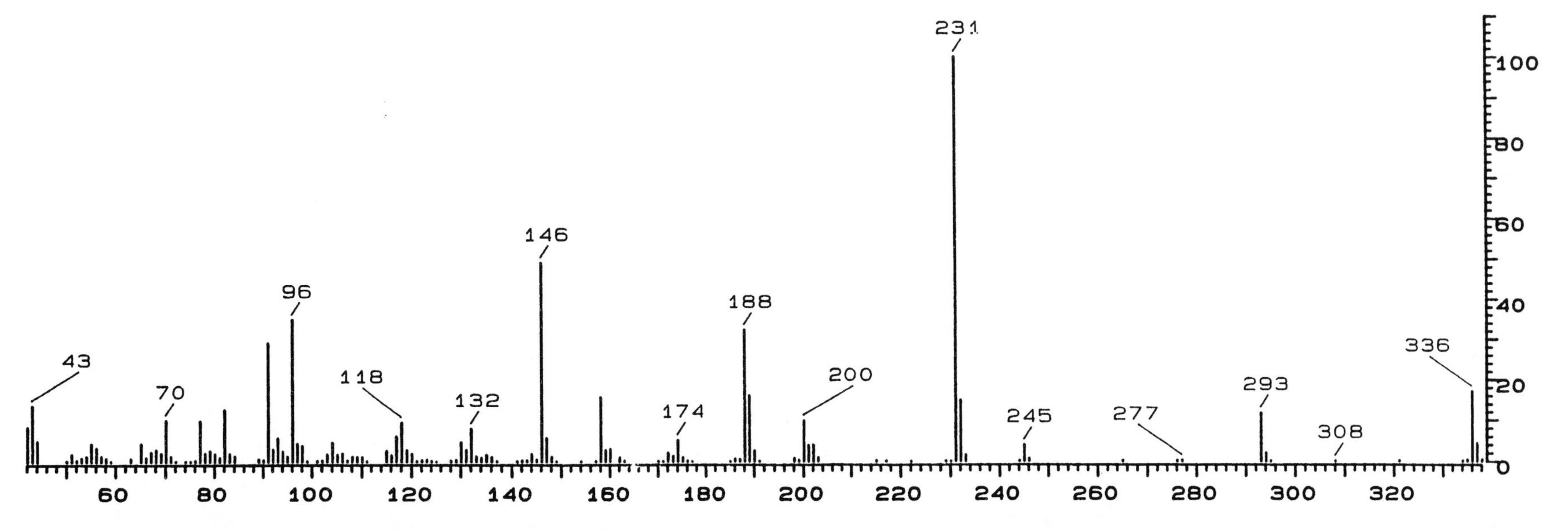

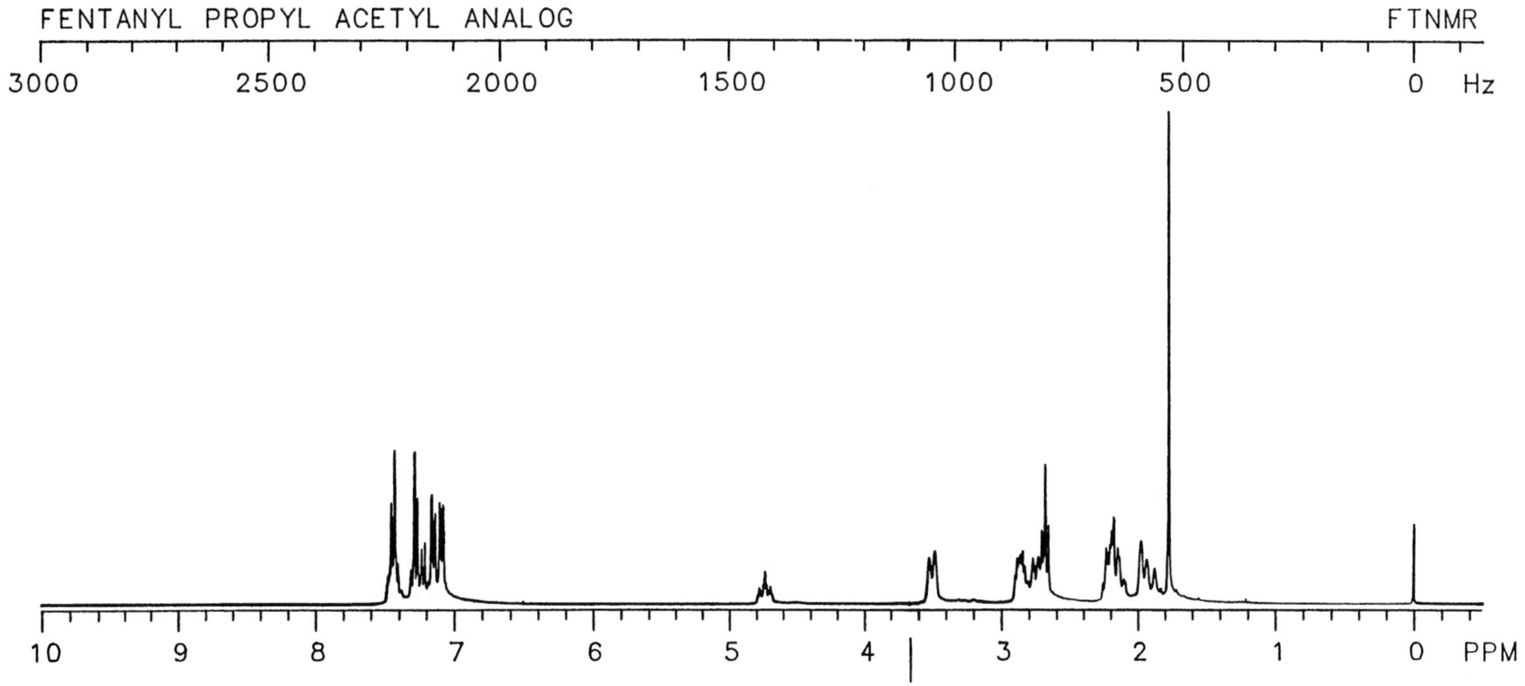

FENTANYL PROPYL ACETYL ANALOG
FTNMR
Hz
PPM


PEAK LISTING

| # | HT | PPM |
|---|---|---|
| 1 | 16 | -.00 |
| 2 | 13 | .00 |
| 3 | 99 | 1.76 |
| 4 | 12 | 1.97 |
| 5 | 17 | 2.17 |
| 6 | 17 | 2.17 |
| 7 | 14 | 2.18 |
| 8 | 12 | 2.19 |
| 9 | 15 | 2.65 |
| 10 | 28 | 2.68 |
| 11 | 14 | 2.70 |
| 12 | 15 | 7.07 |
| 13 | 19 | 7.07 |
| 14 | 20 | 7.07 |
| 15 | 19 | 7.09 |
| 16 | 20 | 7.10 |
| 17 | 18 | 7.13 |
| 18 | 22 | 7.15 |
| 19 | 21 | 7.16 |
| 20 | 12 | 7.20 |
| 21 | 21 | 7.26 |
| 22 | 30 | 7.27 |
| 23 | 30 | 7.28 |
| 24 | 23 | 7.28 |
| 25 | 31 | 7.42 |
| 26 | 20 | 7.44 |

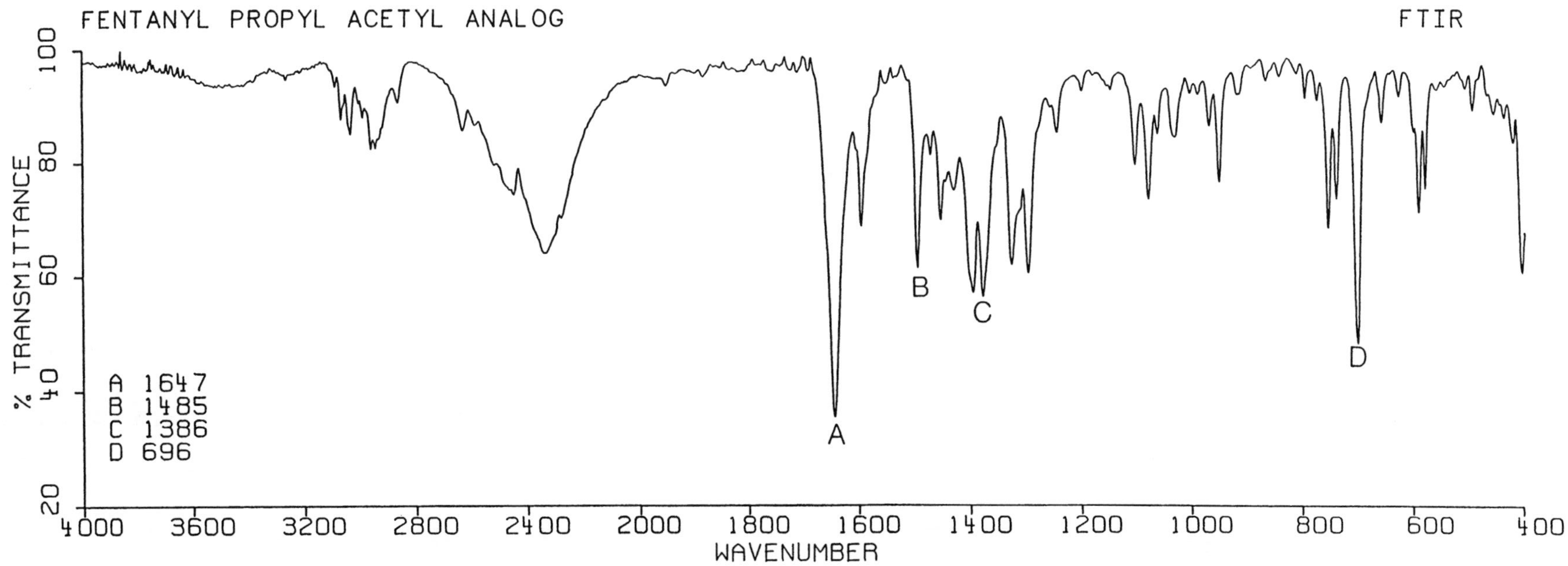

FENTANYL PROPYL ACETYL ANALOG
FTIR
% TRANSMITTANCE
WAVENUMBER
A 1647
B 1485
C 1386
D 696

# FENTANYL ORTHO TOLYL ANALOG

$C_{23}H_{30}N_2O$

**Molecular weight:** 350.50 (350.24)
**Synonyms:** Ortho-Methylphenyl-N-[1-(2-phenylethyl)-4-piperidinyl]-propanamide
**Trade names:**

**Use:**
HPLC: Si-10; 2A:98B; 4.0
GC: 2837; 280°C

FENTANYL ORTHO TOLYL ANALOG

257,263,270
ppt base

TRANSMITTANCE

WAVELENGTH (nm)

*FENTANYL ORTHO TOLYL ANALOG*

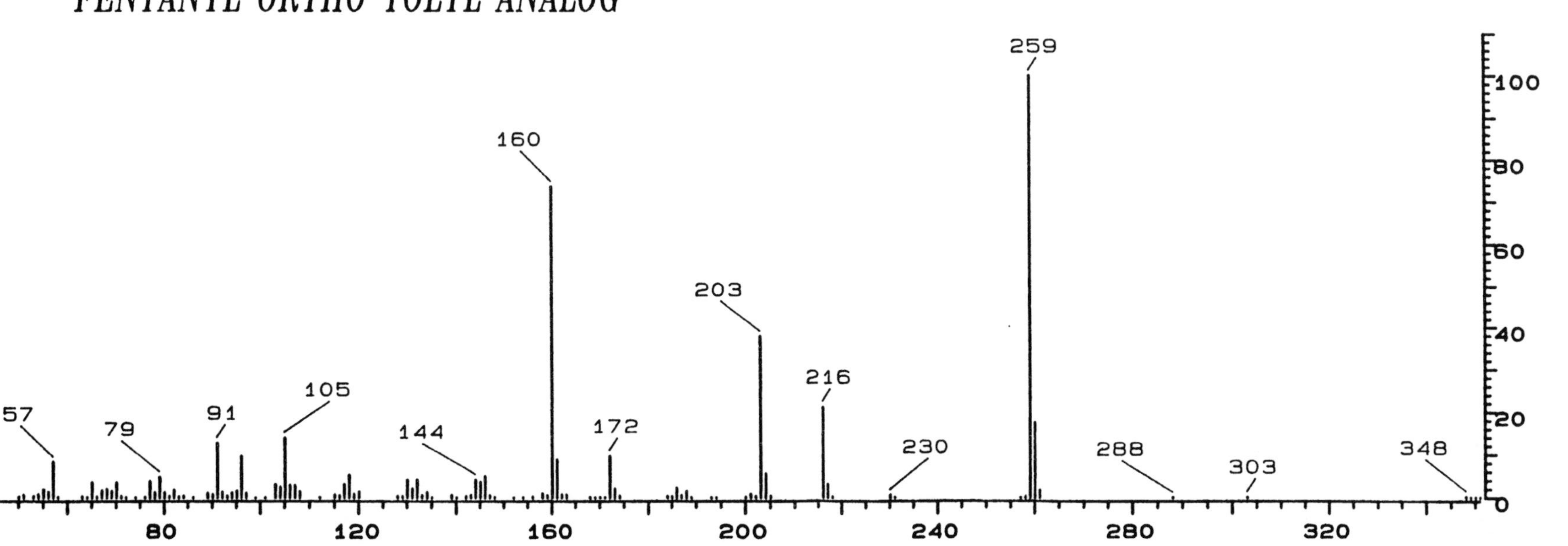

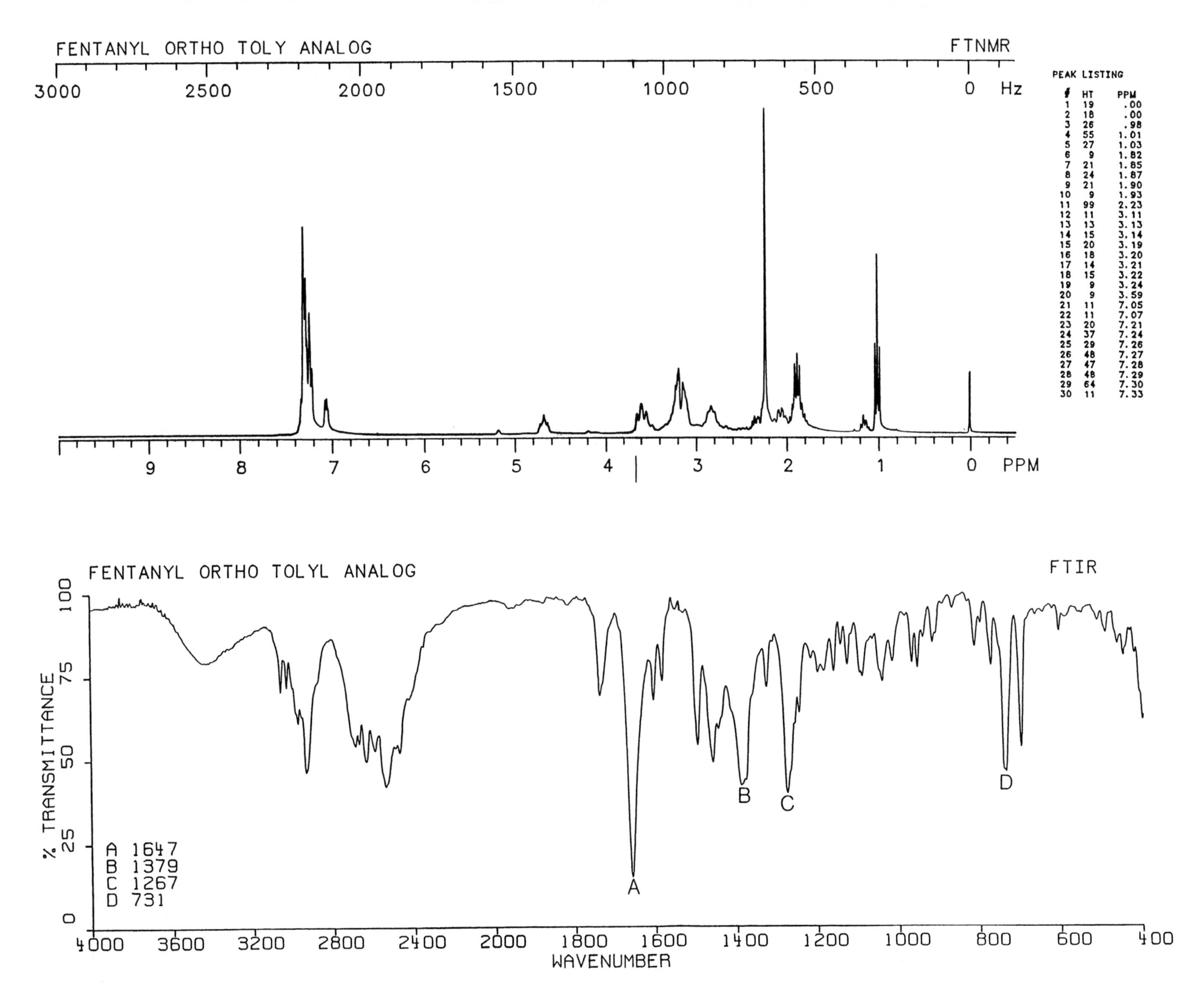
FENTANYL ORTHO TOLY ANALOG
FTNMR
Hz
PPM
PEAK LISTING
| # | HT | PPM |
|---|---|---|
| 1 | 19 | .00 |
| 2 | 18 | .00 |
| 3 | 26 | .98 |
| 4 | 55 | 1.01 |
| 5 | 27 | 1.03 |
| 6 | 9 | 1.82 |
| 7 | 21 | 1.85 |
| 8 | 24 | 1.87 |
| 9 | 21 | 1.90 |
| 10 | 9 | 1.93 |
| 11 | 99 | 2.23 |
| 12 | 11 | 3.11 |
| 13 | 13 | 3.13 |
| 14 | 15 | 3.14 |
| 15 | 20 | 3.19 |
| 16 | 18 | 3.20 |
| 17 | 14 | 3.21 |
| 18 | 15 | 3.22 |
| 19 | 9 | 3.24 |
| 20 | 9 | 3.59 |
| 21 | 11 | 7.05 |
| 22 | 11 | 7.07 |
| 23 | 20 | 7.21 |
| 24 | 37 | 7.24 |
| 25 | 29 | 7.26 |
| 26 | 48 | 7.27 |
| 27 | 47 | 7.28 |
| 28 | 48 | 7.29 |
| 29 | 64 | 7.30 |
| 30 | 11 | 7.33 |
FENTANYL ORTHO TOLYL ANALOG
FTIR
% TRANSMITTANCE
WAVENUMBER
A 1647
B 1379
C 1267
D 731

# FENTANYL META TOLYL ANALOG

$C_{23}H_{30}N_2O$

**Molecular weight:** 350.50 (350.24)
**Synonyms:** Meta-Methylphenyl-N-[1-(2-phenylethyl)-4-piperidinyl]-propanamide
**Trade names:**

**Use:**
**HPLC:** Si-10; 2A:98B; 3.6
**GC:** 2823; 280°C

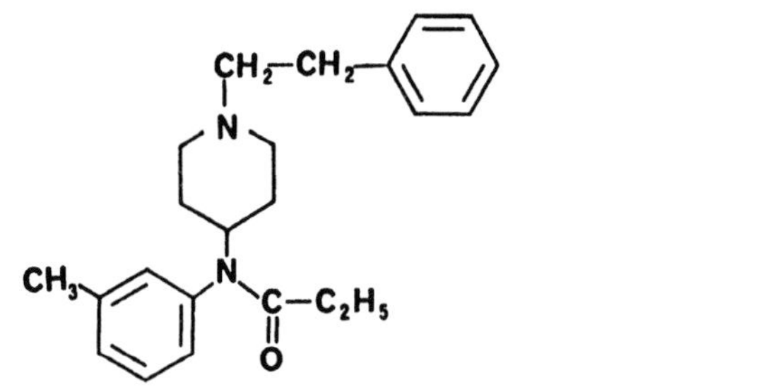


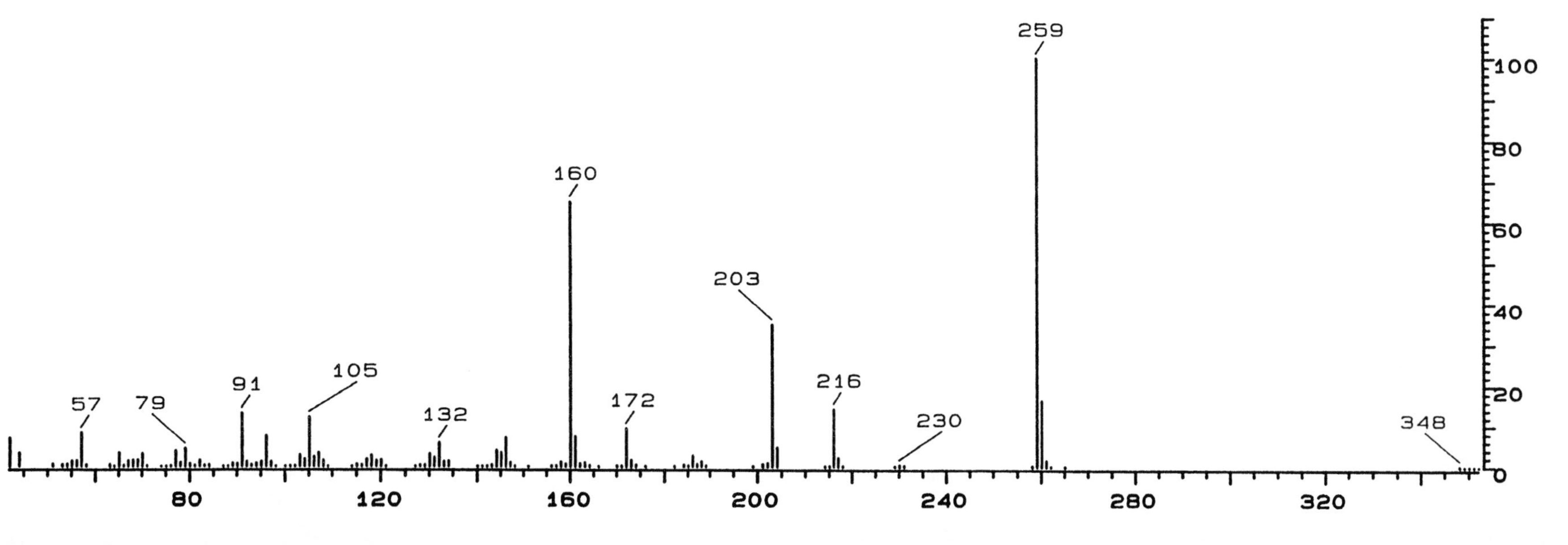

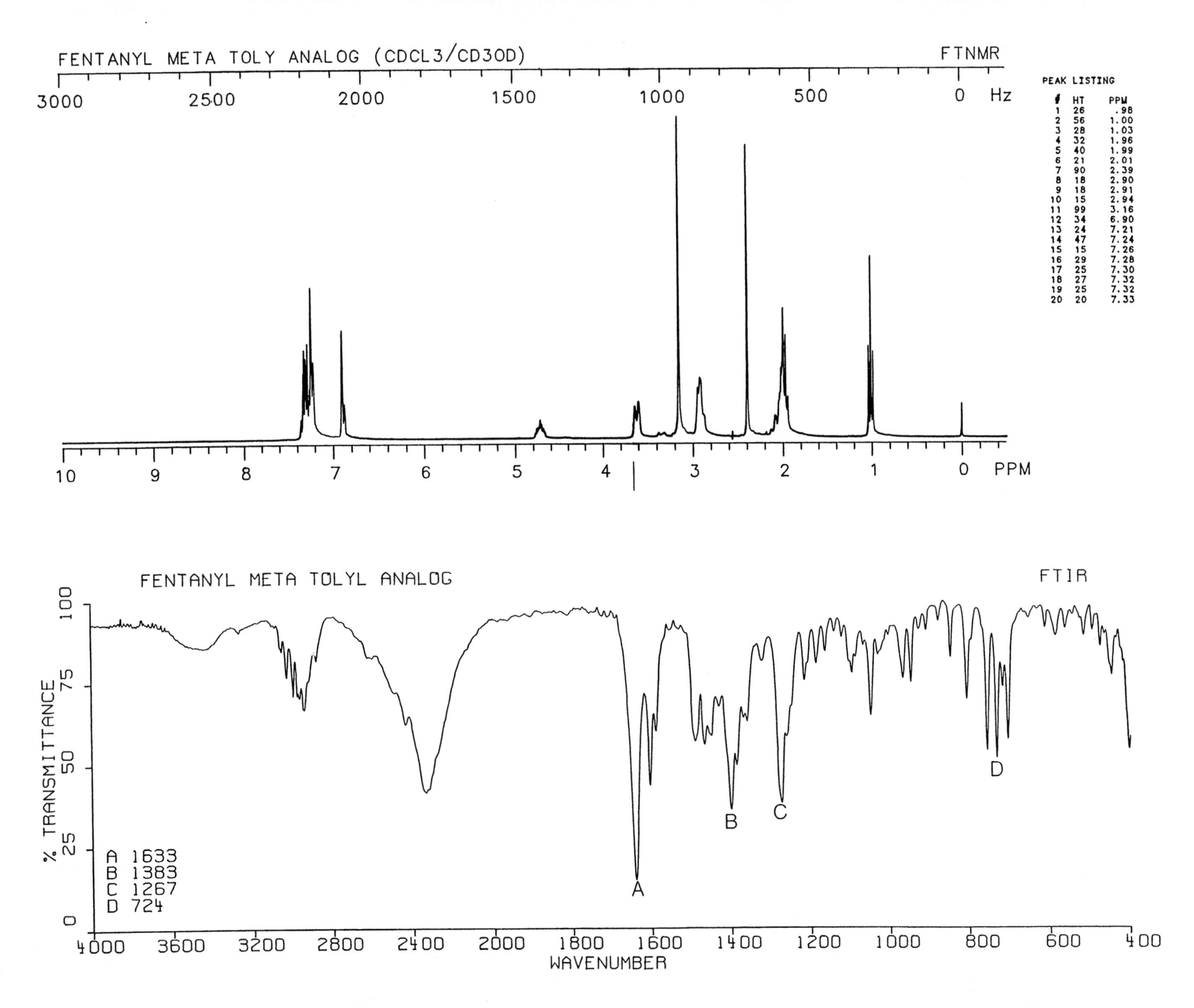

FENTANYL META TOLY ANALOG (CDCL3/CD3OD)
FTNMR
Hz
PPM
PEAK LISTING
# HT PPM
1 26 .98
2 56 1.00
3 28 1.03
4 32 1.96
5 40 1.99
6 21 2.01
7 90 2.39
8 18 2.90
9 18 2.91
10 15 2.94
11 99 3.16
12 34 6.90
13 24 7.21
14 47 7.24
15 15 7.26
16 29 7.28
17 25 7.30
18 27 7.32
19 25 7.32
20 20 7.33
FENTANYL META TOLYL ANALOG
FTIR
% TRANSMITTANCE
WAVENUMBER
A 1633
B 1383
C 1267
D 724

# FENTANYL PARA TOLYL ANALOG

$C_{23}H_{30}N_2O$

**Molecular weight:** 350.50 (350.24)
**Synonyms:** Para-Methylphenyl-N-[1-(2-phenylethyl)-4-piperidinyl]-propanamide
**Trade names:**

**Use:**
**HPLC:** Si-10; 2A:98B; 3.6
**GC:** 2848; 280°C

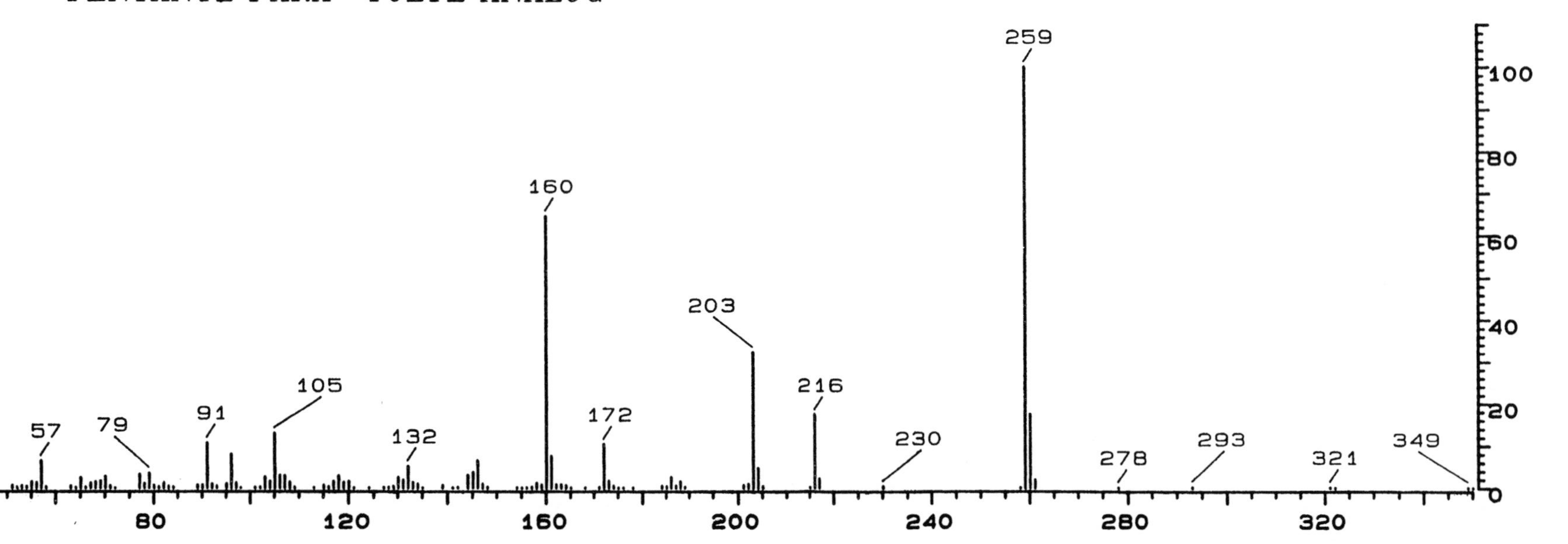

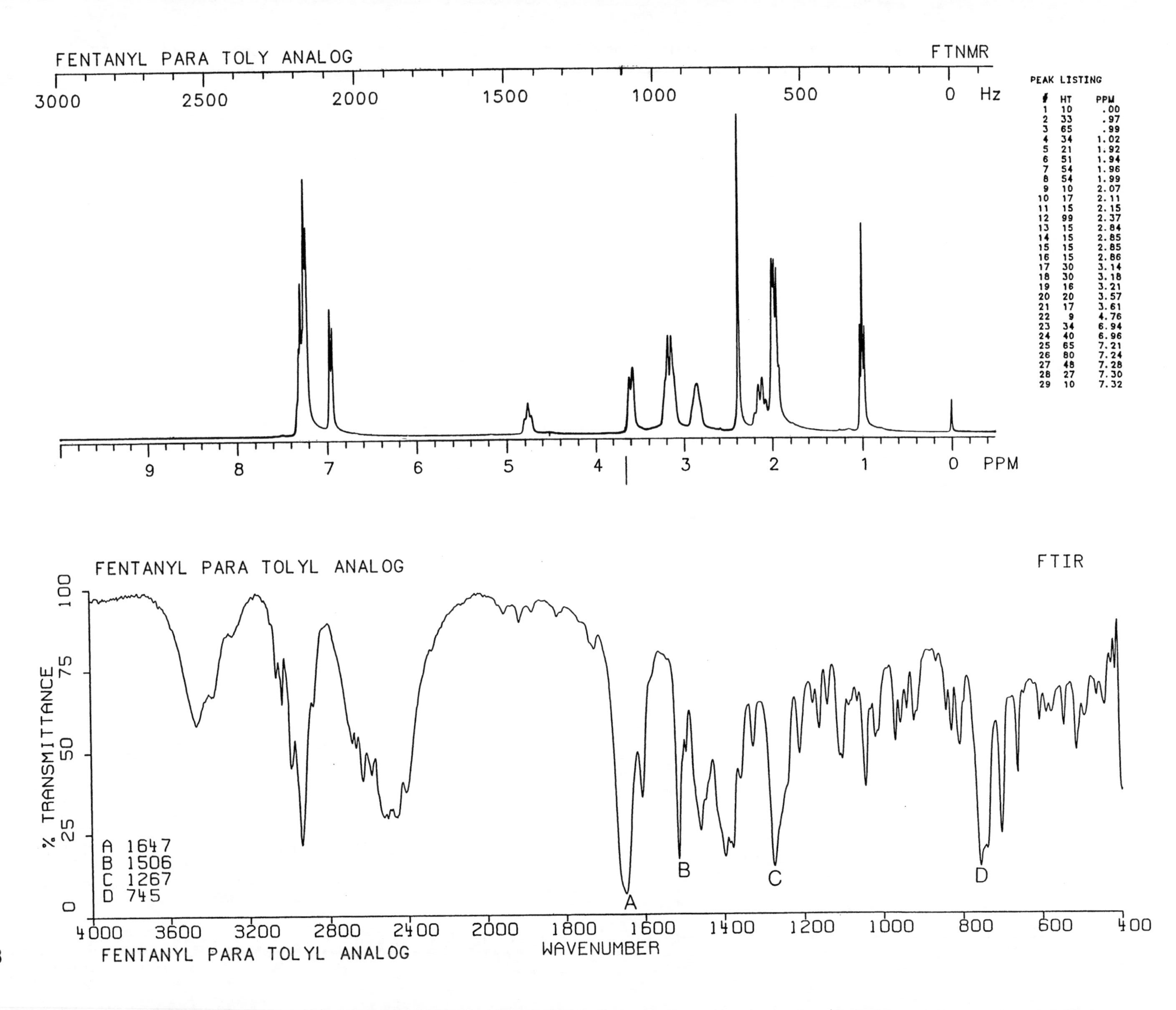
FENTANYL PARA TOLY ANALOG
FTNMR
3000 2500 2000 1500 1000 500 0 Hz
9 8 7 6 5 4 3 2 1 0 PPM
PEAK LISTING
# HT PPM
1 10 .00
2 33 .97
3 65 .99
4 34 1.02
5 21 1.92
6 51 1.94
7 54 1.96
8 54 1.99
9 10 2.07
10 17 2.11
11 15 2.15
12 99 2.37
13 15 2.84
14 15 2.85
15 15 2.85
16 15 2.86
17 30 3.14
18 30 3.18
19 16 3.21
20 20 3.57
21 17 3.61
22 9 4.76
23 34 6.94
24 40 6.96
25 65 7.21
26 80 7.24
27 48 7.28
28 27 7.30
29 10 7.32
FENTANYL PARA TOLYL ANALOG
FTIR
% TRANSMITTANCE
100 75 50 25 0
A 1647
B 1506
C 1267
D 745
A B C D
4000 3600 3200 2800 2400 2000 1800 1600 1400 1200 1000 800 600 400
WAVENUMBER
FENTANYL PARA TOLYL ANALOG

# FENTANYL ORTHO TOLYL ACETYL ANALOG

$C_{22}H_{28}N_2O$

Molecular weight: 336.47 (336.22)
Synonyms: Ortho-Methylphenyl-N-[1-(2-phenylethyl)-4-piperidinyl]-acetamide
Trade names:

Use:
HPLC: Si-10; 2A:98B; 4.6
GC: 2791; 280°C

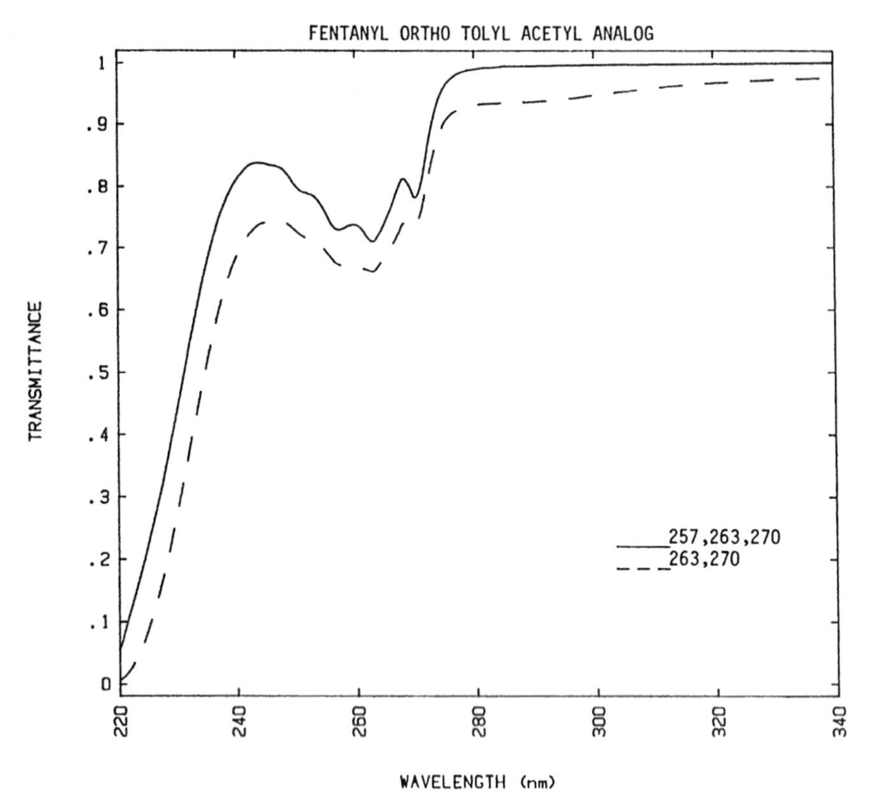


*FENTANYL ORTHO TOLYL ACETYL ANALOG*

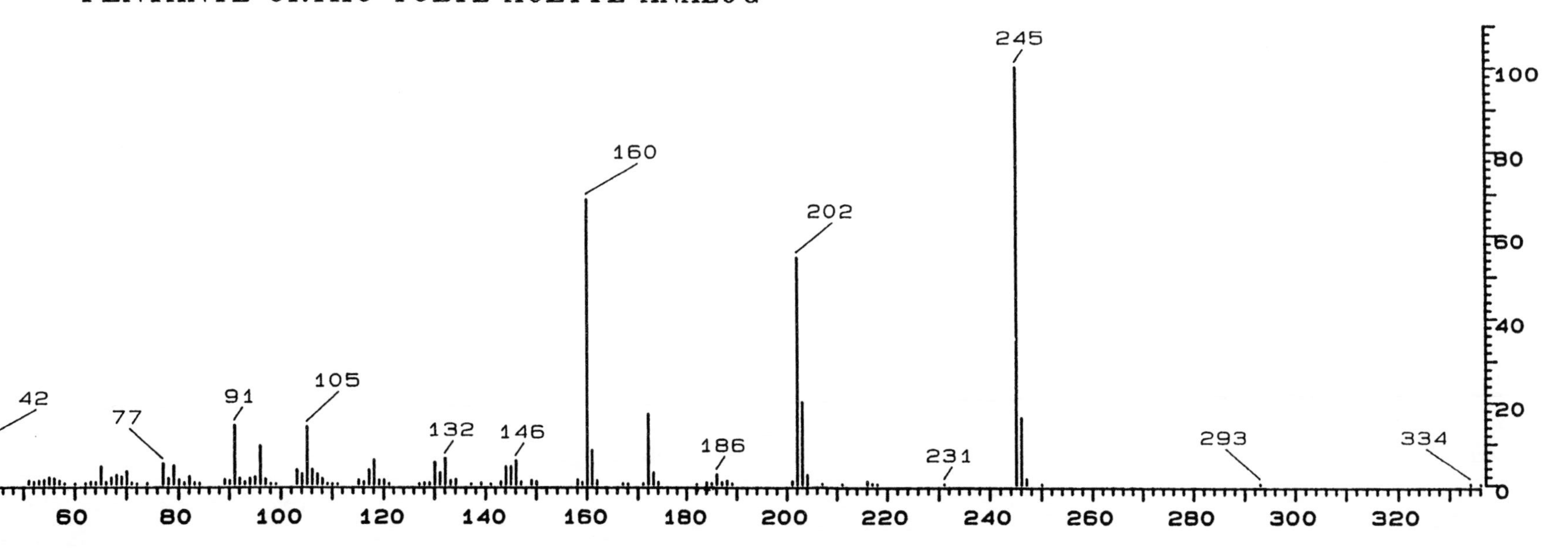

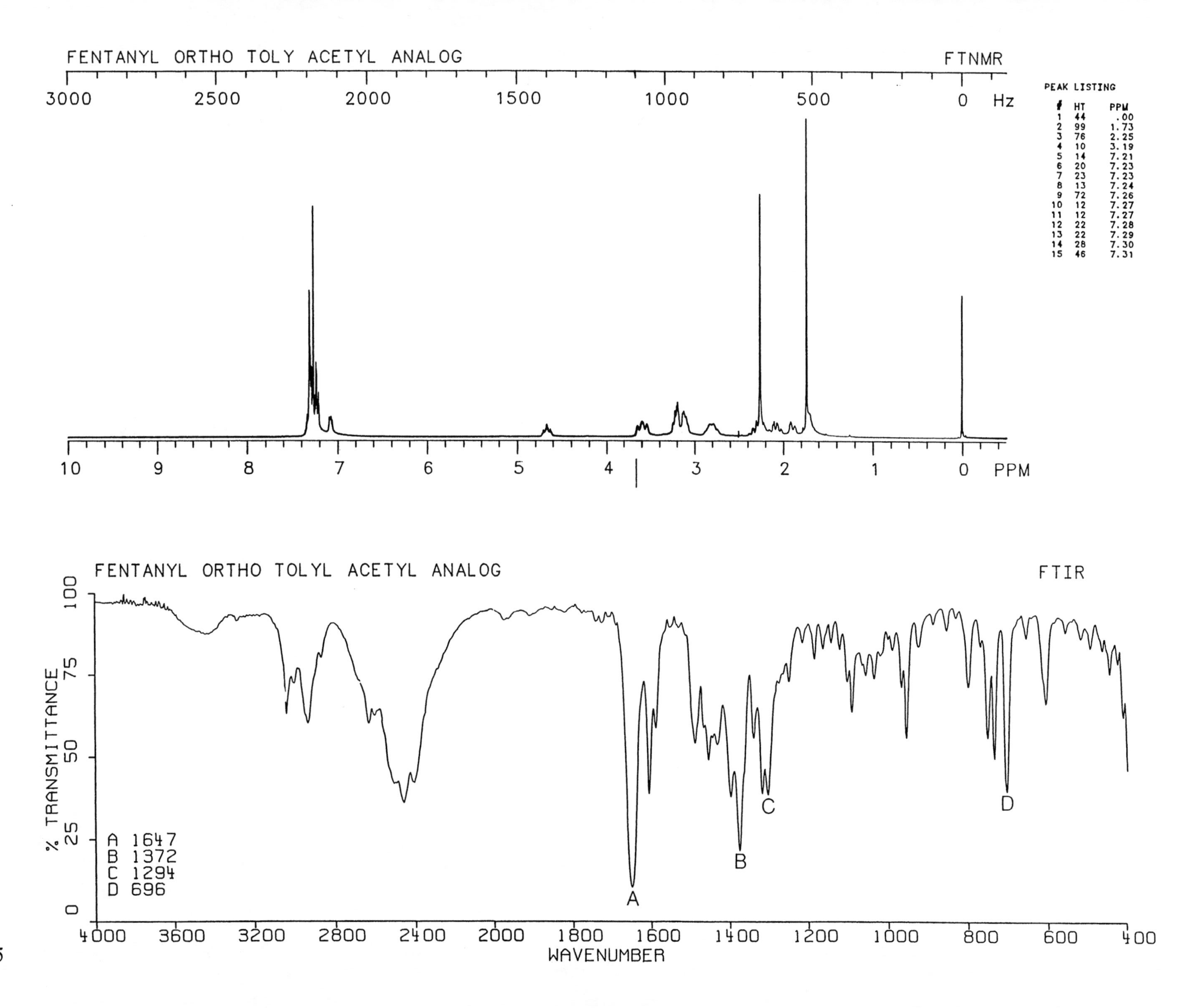
FENTANYL ORTHO TOLY ACETYL ANALOG
FTNMR
3000 2500 2000 1500 1000 500 0 Hz
10 9 8 7 6 5 4 3 2 1 0 PPM
PEAK LISTING
# HT PPM
1 44 .00
2 99 1.73
3 76 2.25
4 10 3.19
5 14 7.21
6 20 7.23
7 23 7.23
8 13 7.24
9 72 7.26
10 12 7.27
11 12 7.27
12 22 7.28
13 22 7.29
14 28 7.30
15 46 7.31
FENTANYL ORTHO TOLYL ACETYL ANALOG
FTIR
% TRANSMITTANCE
100 75 50 25 0
A 1647
B 1372
C 1294
D 696
A
B
C
D
4000 3600 3200 2800 2400 2000 1800 1600 1400 1200 1000 800 600 400
WAVENUMBER

# FENTANYL META TOLYL ACETYL ANALOG

$C_{22}H_{28}N_2O$

**Molecular weight:** 336.47 (336.22)
**Synonyms:** Meta-Methylphenyl-N-[1-(2-phenyllethyl)-4-piperidinyl]-acetamide
**Trade names:**

**Use:**
**HPLC:** Si-10; 2A:98B; 4.1
**GC:** 2770; 280°C

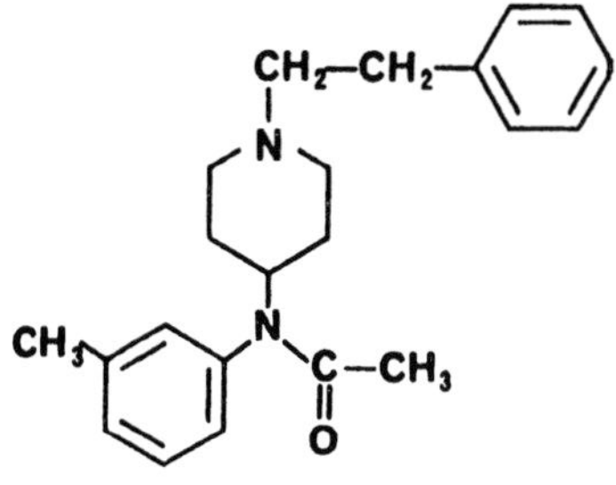


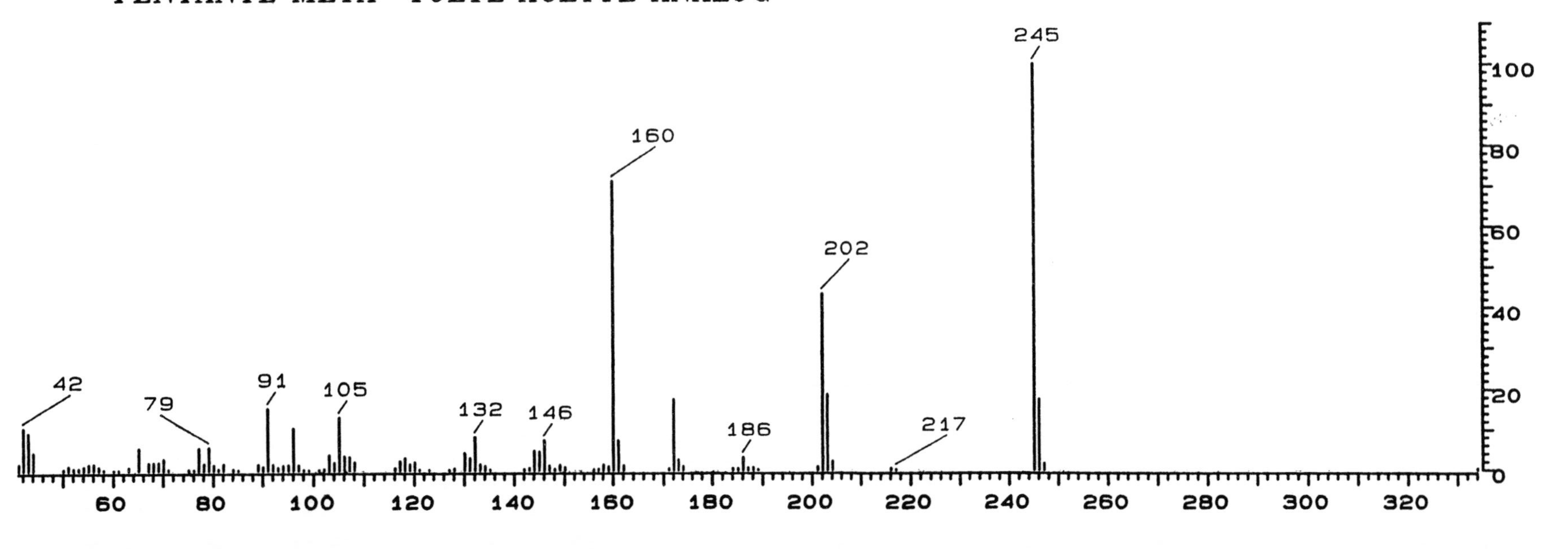

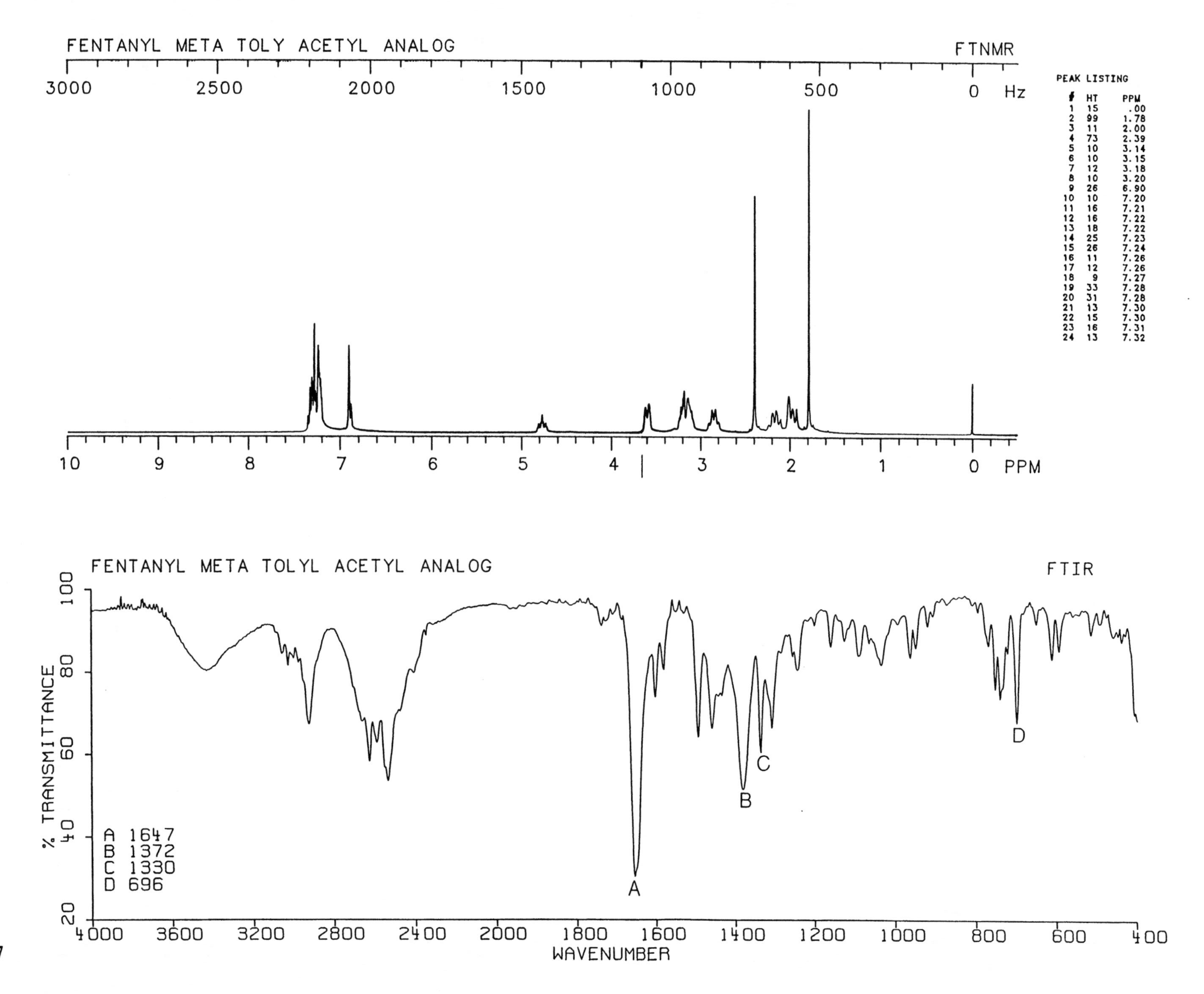
FENTANYL META TOLY ACETYL ANALOG
FTNMR
3000 2500 2000 1500 1000 500 0 Hz
10 9 8 7 6 5 4 3 2 1 0 PPM
PEAK LISTING
# HT PPM
1 15 .00
2 99 1.78
3 11 2.00
4 73 2.39
5 10 3.14
6 10 3.15
7 12 3.18
8 10 3.20
9 26 6.90
10 10 7.20
11 16 7.21
12 16 7.22
13 18 7.22
14 25 7.23
15 26 7.24
16 11 7.26
17 12 7.26
18 9 7.27
19 33 7.28
20 31 7.28
21 13 7.30
22 15 7.30
23 16 7.31
24 13 7.32
FENTANYL META TOLYL ACETYL ANALOG
FTIR
% TRANSMITTANCE
100 80 60 40 20
A 1647
B 1372
C 1330
D 696
A
B
C
D
4000 3600 3200 2800 2400 2000 1800 1600 1400 1200 1000 800 600 400
WAVENUMBER

# FENTANYL PARA TOLYL ACETYL ANALOG

$C_{22}H_{28}N_2O$

**Molecular weight:** 336.47 (336.22)
**Synonyms:** Para-Methylphenyl-N-[1-(2-phenylethyl)-4-piperidinyl]-acetamide
**Trade names:**

**Use:**
**HPLC:** Si-10; 2A:98B; 6.0
**GC:** 2809; 280°C

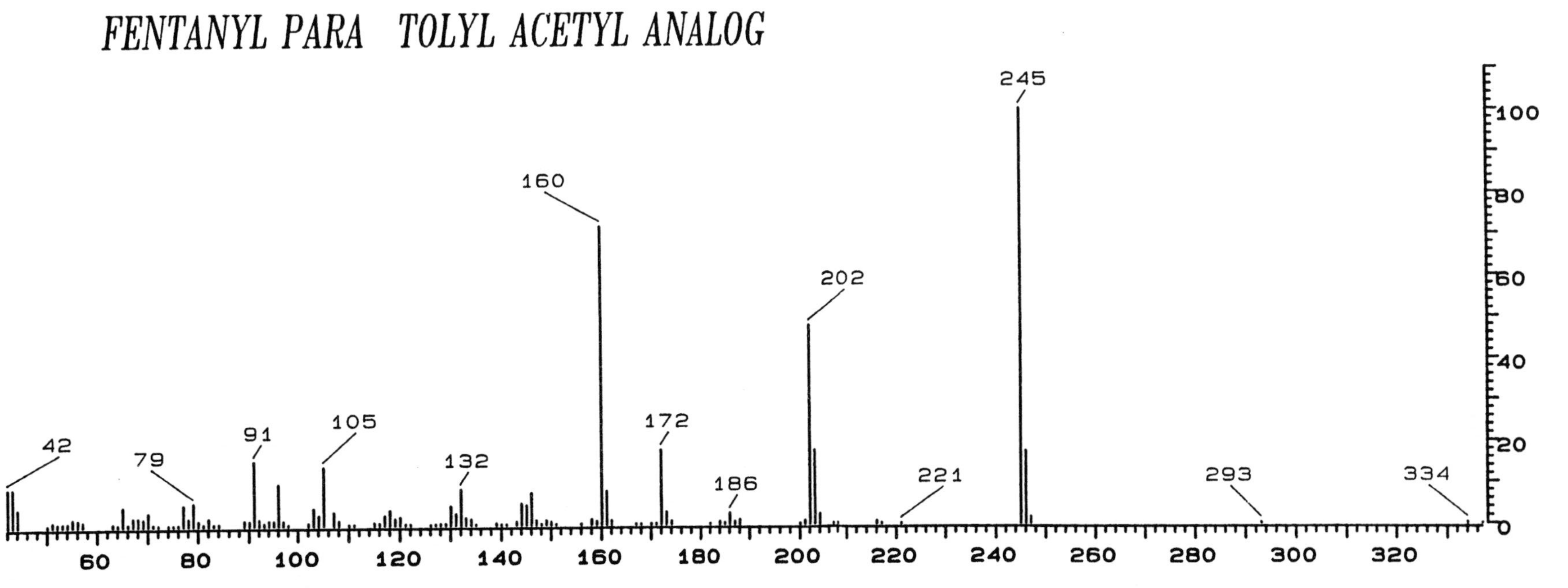

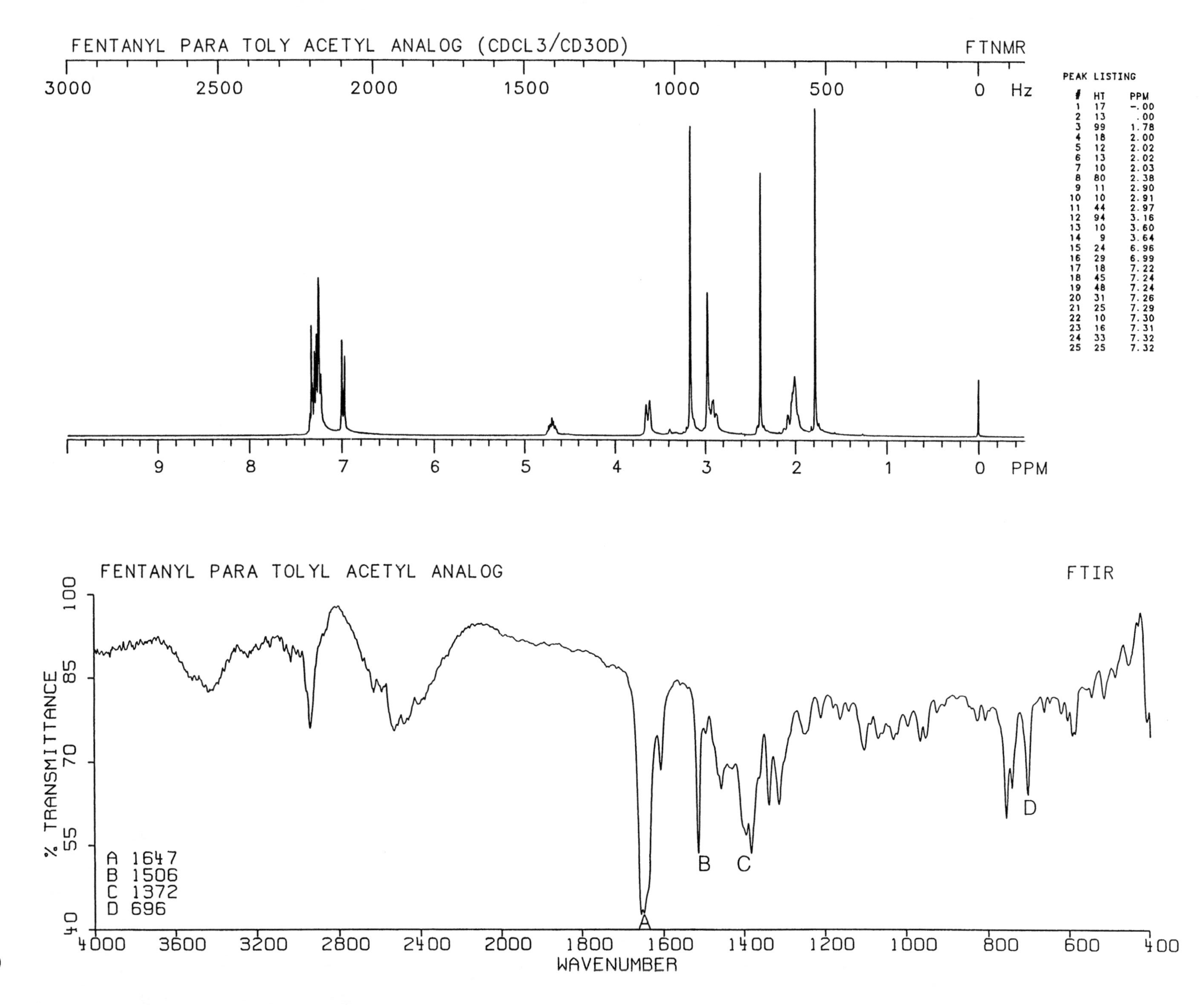

FENTANYL PARA TOLY ACETYL ANALOG (CDCL3/CD3OD)
FTNMR
3000 2500 2000 1500 1000 500 0 Hz
9 8 7 6 5 4 3 2 1 0 PPM
PEAK LISTING
# HT PPM
1 17 -.00
2 13 .00
3 99 1.78
4 18 2.00
5 12 2.02
6 13 2.02
7 10 2.03
8 80 2.38
9 11 2.90
10 10 2.91
11 44 2.97
12 94 3.16
13 10 3.60
14 9 3.64
15 24 6.96
16 29 6.99
17 18 7.22
18 45 7.24
19 48 7.24
20 31 7.26
21 25 7.29
22 10 7.30
23 16 7.31
24 33 7.32
25 25 7.32
FENTANYL PARA TOLYL ACETYL ANALOG
FTIR
% TRANSMITTANCE
100 85 70 55 40
A 1647
B 1506
C 1372
D 696
4000 3600 3200 2800 2400 2000 1800 1600 1400 1200 1000 800 600 400
WAVENUMBER

# FLAVOXATE

$C_{24}H_{25}NO_4$

Molecular weight: 391.47 (391.18)
Synonyms: 3-Methyl-4-oxo-2-phenyl-4H-1-benzopyran-8-carboxylic acid 2-piperidinoethyl ester
Trade names: Urispas

Use: Muscle relaxant
HPLC: Si-10; 2A:98B; 6.0
GC:

FLAVOXATE

TRANSMITTANCE

WAVELENGTH (nm)

240, 293, 318
244, 315

FLAVOXATE

98 111 55 70 133 147 178 189 209 243 247 267 302 307

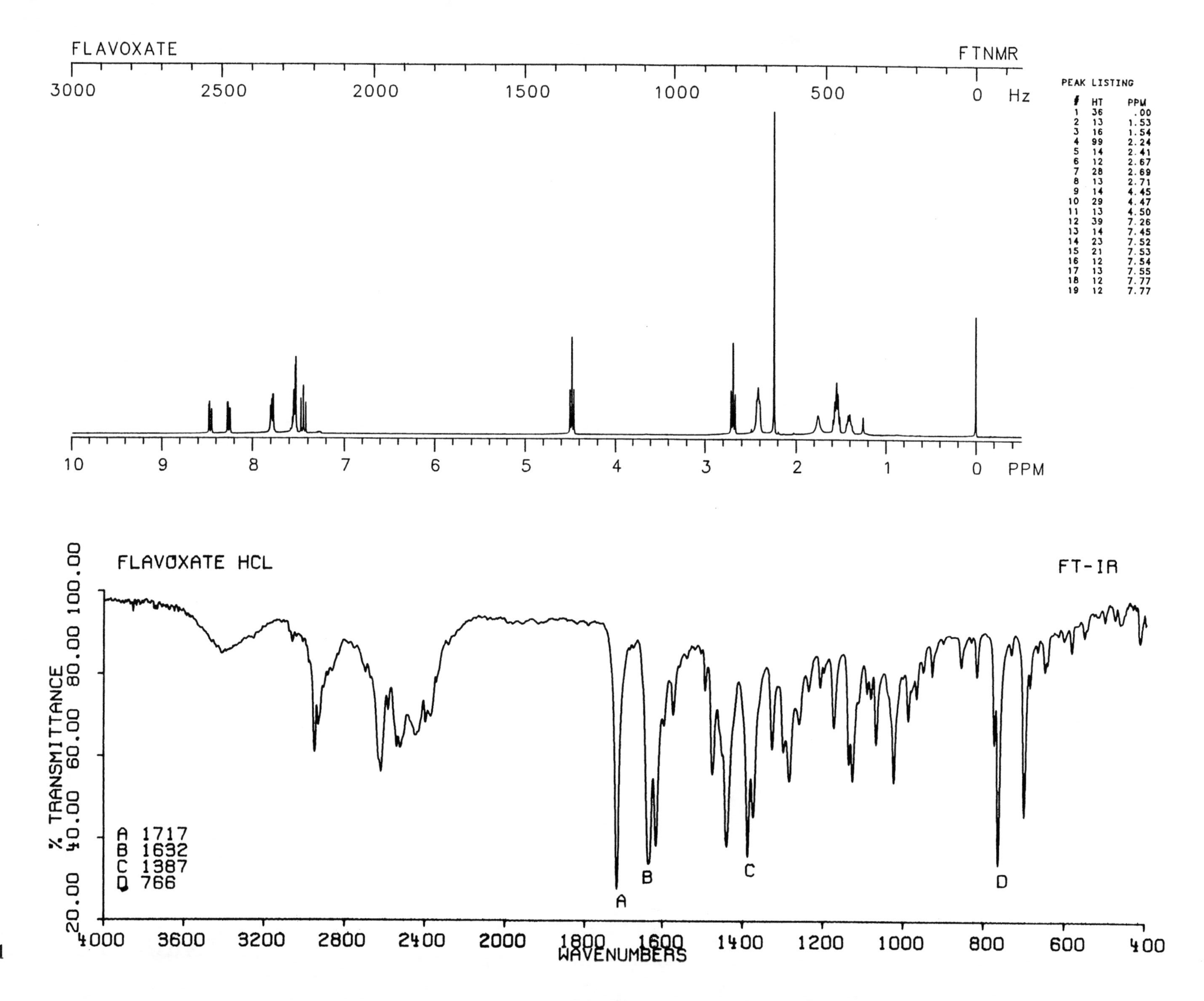
FLAVOXATE
FTNMR
3000 2500 2000 1500 1000 500 0 Hz
10 9 8 7 6 5 4 3 2 1 0 PPM
PEAK LISTING
# HT PPM
1 36 .00
2 13 1.53
3 16 1.54
4 99 2.24
5 14 2.41
6 12 2.67
7 28 2.69
8 13 2.71
9 14 4.45
10 29 4.47
11 13 4.50
12 39 7.26
13 14 7.45
14 23 7.52
15 21 7.53
16 12 7.54
17 13 7.55
18 12 7.77
19 12 7.77
FLAVOXATE HCL
FT-IR
% TRANSMITTANCE
20.00 40.00 60.00 80.00 100.00
A 1717
B 1632
C 1387
D 766
A B C D
4000 3600 3200 2800 2400 2000 1800 1600 1400 1200 1000 800 600 400
WAVENUMBERS

# FLOXURIDINE

$C_9H_{11}FN_2O_5$

**Molecular weight:** 246.21 (246.07)
**Synonyms:** 2'-Deoxy-5-fluorouridine; 5-fluoro-2'-deoxy-β-uridine

**Trade names:** FUDR

**Use:** Antiviral
**HPLC:** Si-10; 5A:95B; 10.0
**GC:**

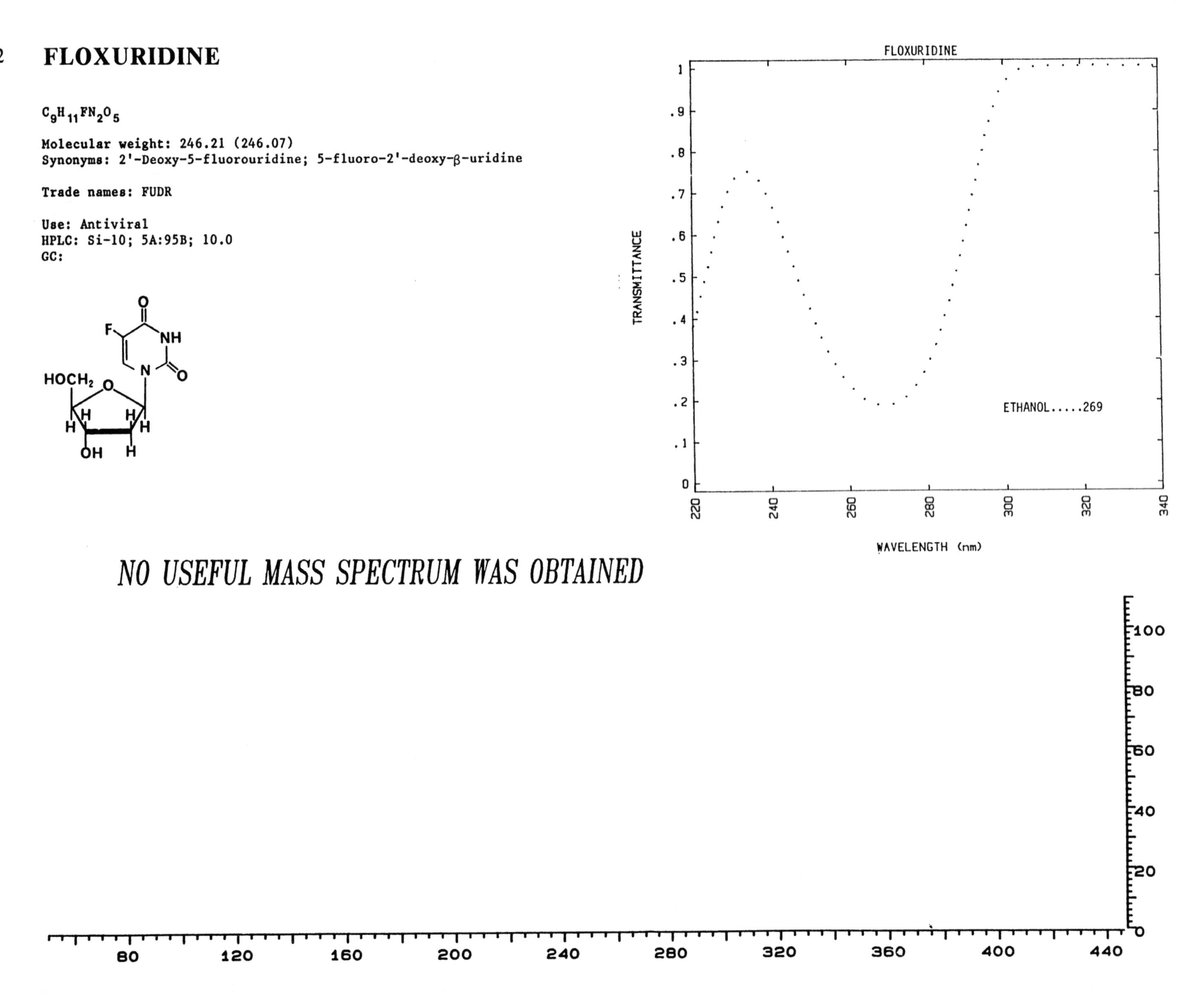

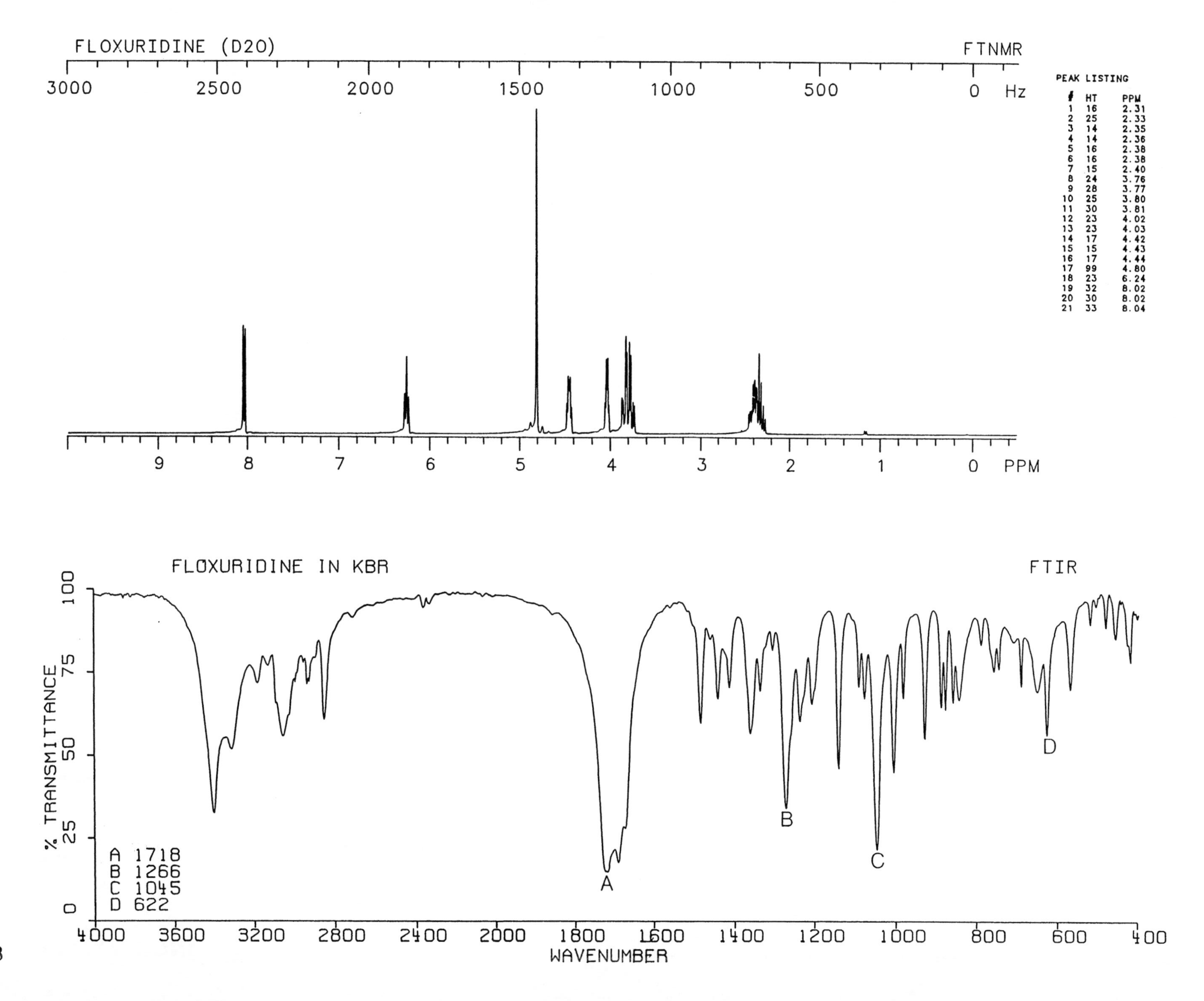

FLOXURIDINE (D20)
FTNMR
3000 2500 2000 1500 1000 500 0 Hz
9 8 7 6 5 4 3 2 1 0 PPM
PEAK LISTING
# HT PPM
1 16 2.31
2 25 2.33
3 14 2.35
4 14 2.36
5 16 2.38
6 16 2.38
7 15 2.40
8 24 3.76
9 28 3.77
10 25 3.80
11 30 3.81
12 23 4.02
13 23 4.03
14 17 4.42
15 15 4.43
16 17 4.44
17 99 4.80
18 23 6.24
19 32 8.02
20 30 8.02
21 33 8.04
FLOXURIDINE IN KBR
FTIR
% TRANSMITTANCE
0 25 50 75 100
A 1718
B 1266
C 1045
D 622
4000 3600 3200 2800 2400 2000 1800 1600 1400 1200 1000 800 600 400
WAVENUMBER

# FLUANISONE

$C_{21}H_{25}FN_2O_2$

Molecular weight: 356.45 (356.19)
Synonyms: 1-(4-Fluorophenyl)-4-[4-(2-methoxyphenyl)-1-piperazinyl]-1-butanone; haloanisone
Trade names: Sedalande

Use: Neuroleptic
HPLC:
GC: 2800; 280°C

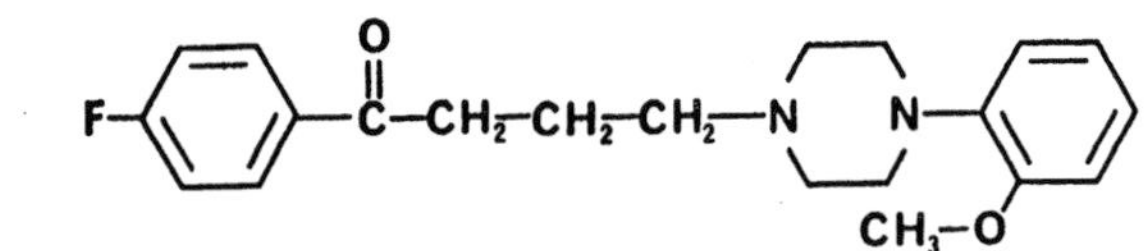


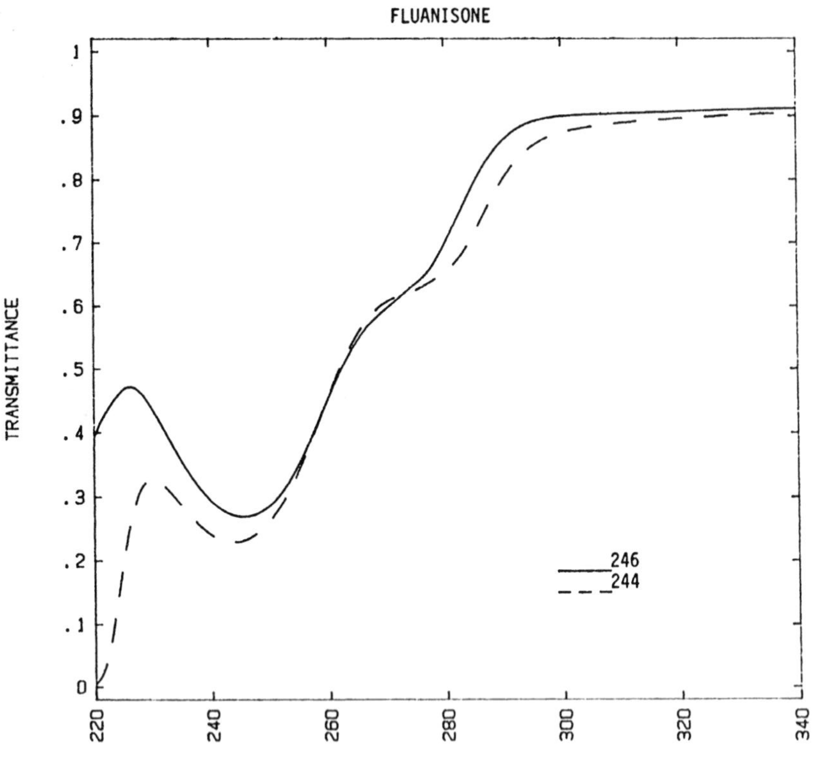

*FLUANISONE*

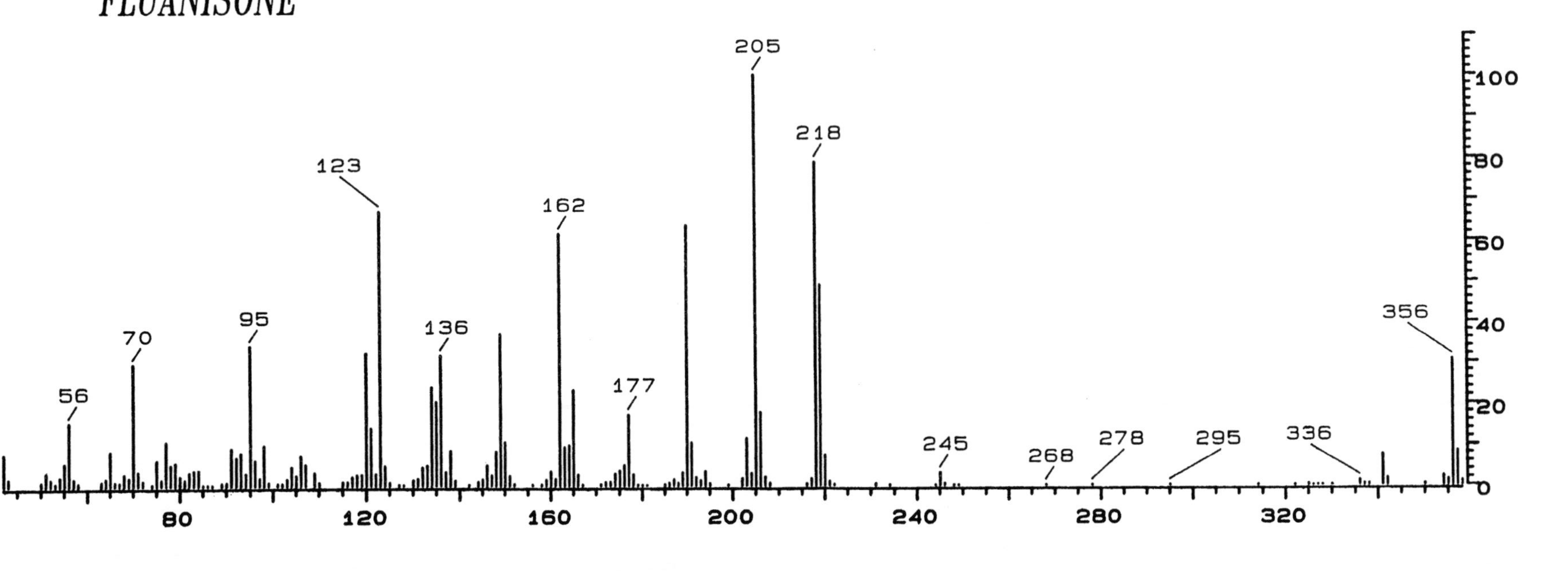

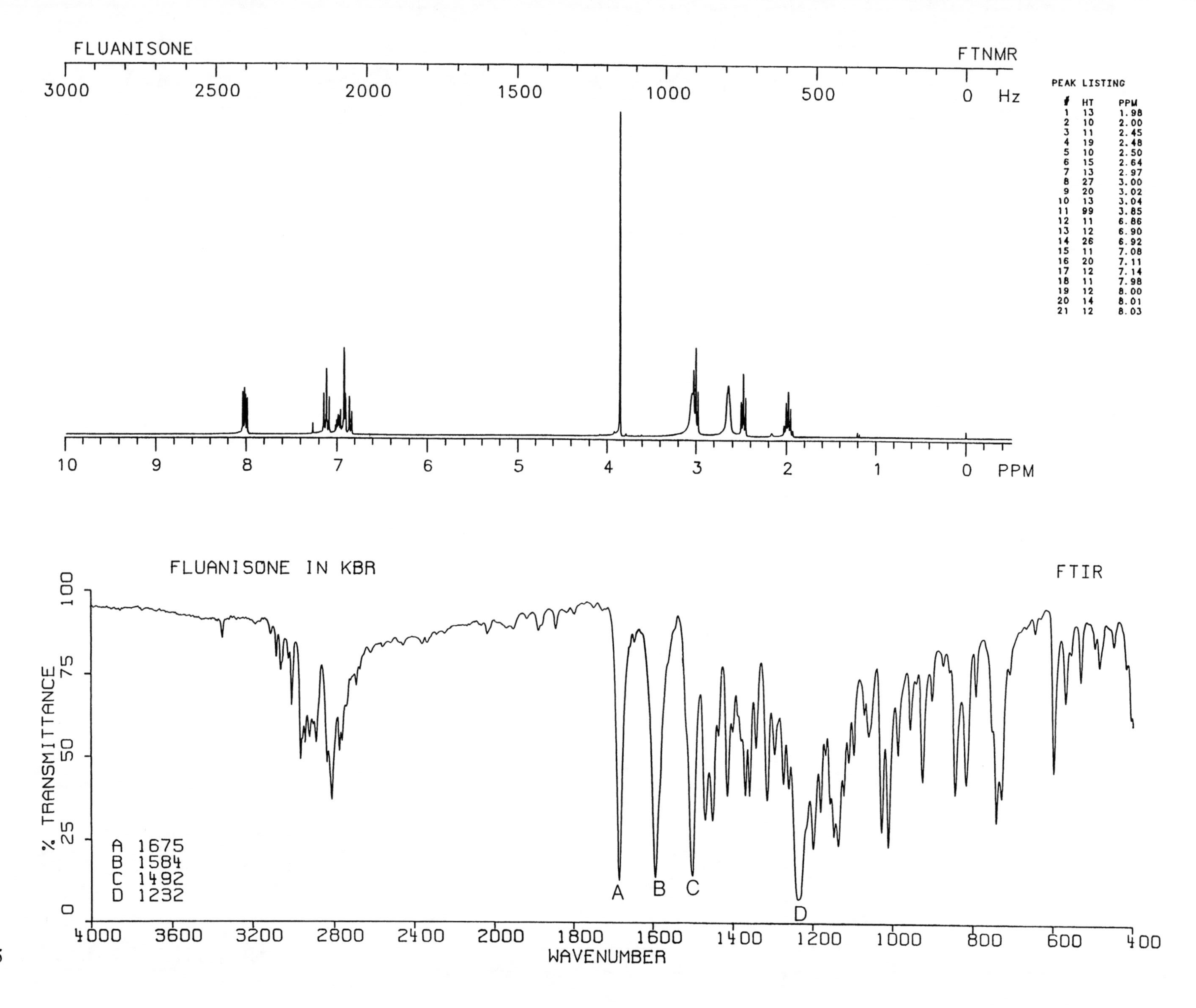


PEAK LISTING

| # | HT | PPM |
|---|---|---|
| 1 | 13 | 1.98 |
| 2 | 10 | 2.00 |
| 3 | 11 | 2.45 |
| 4 | 19 | 2.48 |
| 5 | 10 | 2.50 |
| 6 | 15 | 2.64 |
| 7 | 13 | 2.97 |
| 8 | 27 | 3.00 |
| 9 | 20 | 3.02 |
| 10 | 13 | 3.04 |
| 11 | 99 | 3.85 |
| 12 | 11 | 6.86 |
| 13 | 12 | 6.90 |
| 14 | 26 | 6.92 |
| 15 | 11 | 7.08 |
| 16 | 20 | 7.11 |
| 17 | 12 | 7.14 |
| 18 | 11 | 7.98 |
| 19 | 12 | 8.00 |
| 20 | 14 | 8.01 |
| 21 | 12 | 8.03 |

# FLUDROCORTISONE

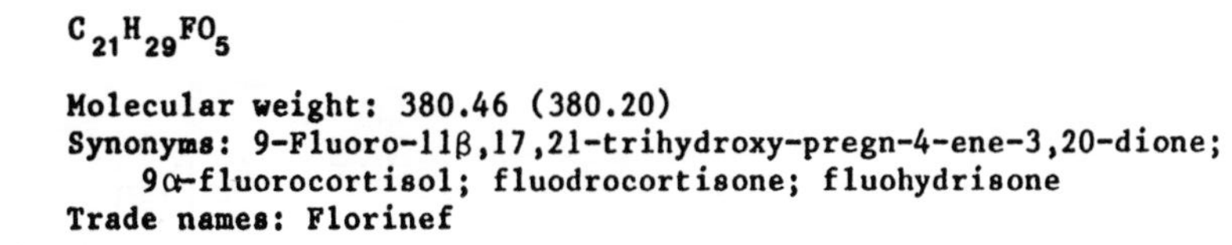

$C_{21}H_{29}FO_5$

**Molecular weight:** 380.46 (380.20)
**Synonyms:** 9-Fluoro-11β,17,21-trihydroxy-pregn-4-ene-3,20-dione; 9α-fluorocortisol; fluodrocortisone; fluohydrisone
**Trade names:** Florinef

**Use:** Anti-inflammatory
**HPLC:**
**GC:**

FLUDROCORTISONE--DIP

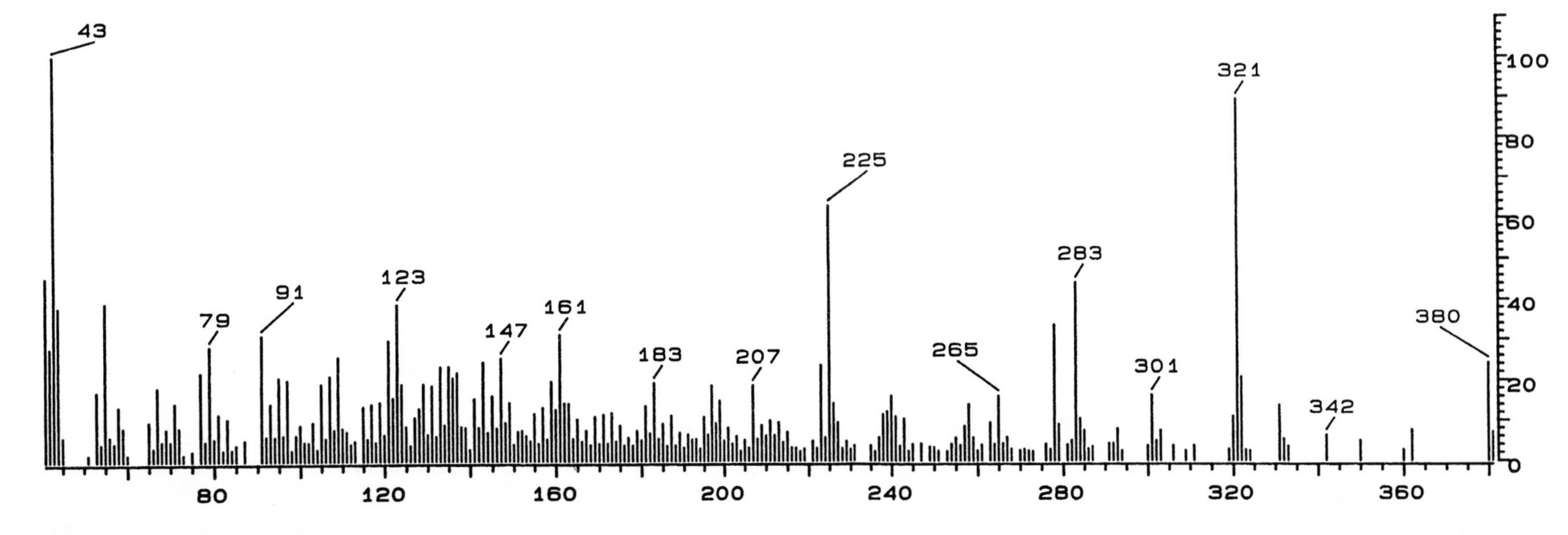

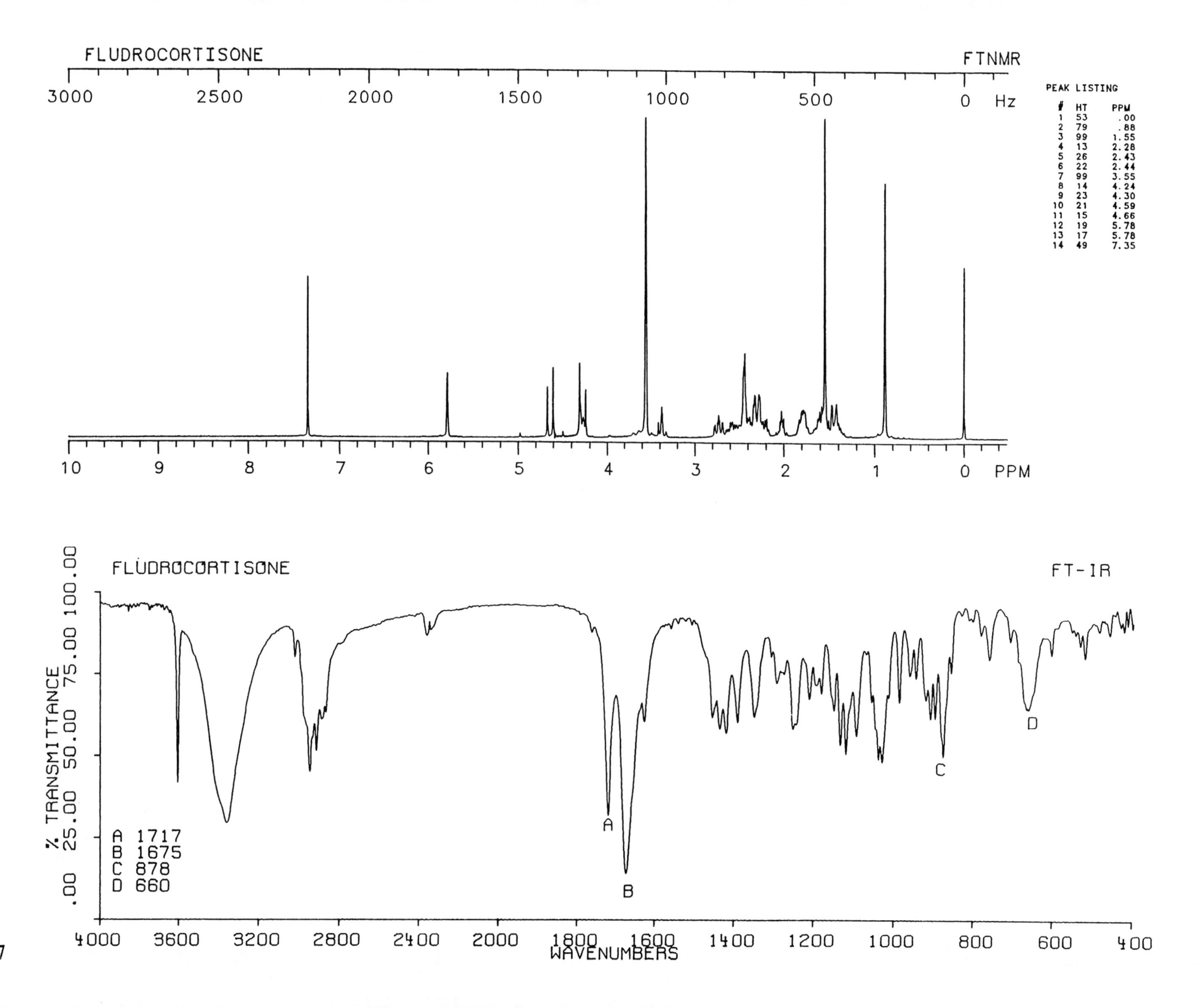
FLUDROCORTISONE
FTNMR
3000 2500 2000 1500 1000 500 0 Hz
PEAK LISTING
# HT PPM
1 53 .00
2 79 .88
3 99 1.55
4 13 2.28
5 26 2.43
6 22 2.44
7 99 3.55
8 14 4.24
9 23 4.30
10 21 4.59
11 15 4.66
12 19 5.78
13 17 5.78
14 49 7.35
10 9 8 7 6 5 4 3 2 1 0 PPM
FLUDROCORTISONE
FT-IR
% TRANSMITTANCE
.00 25.00 50.00 75.00 100.00
A 1717
B 1675
C 878
D 660
A B C D
4000 3600 3200 2800 2400 2000 1800 1600 1400 1200 1000 800 600 400
WAVENUMBERS

# FLUFENAMIC ACID

$C_{14}H_{10}F_3NO_2$

**Molecular weight:** 281.24 (281.07)
**Synonyms:** 2-[[3-(Trifluoromethyl)phenyl]amino]benzoic acid; N-(α,α,α-trifluoro-m-tolyl)anthranilic acid
**Trade names:** Achless, Alfenamin, Ansatin, Arlef, Meralen, Opyrin, Parlef, Paraflu, Ristogen, Surika, Tecramine
**Use:** Anti-inflammatory, analgesic
**HPLC:** Si-10; 10A:90B; 4.7
**GC:** 1549; 200°C

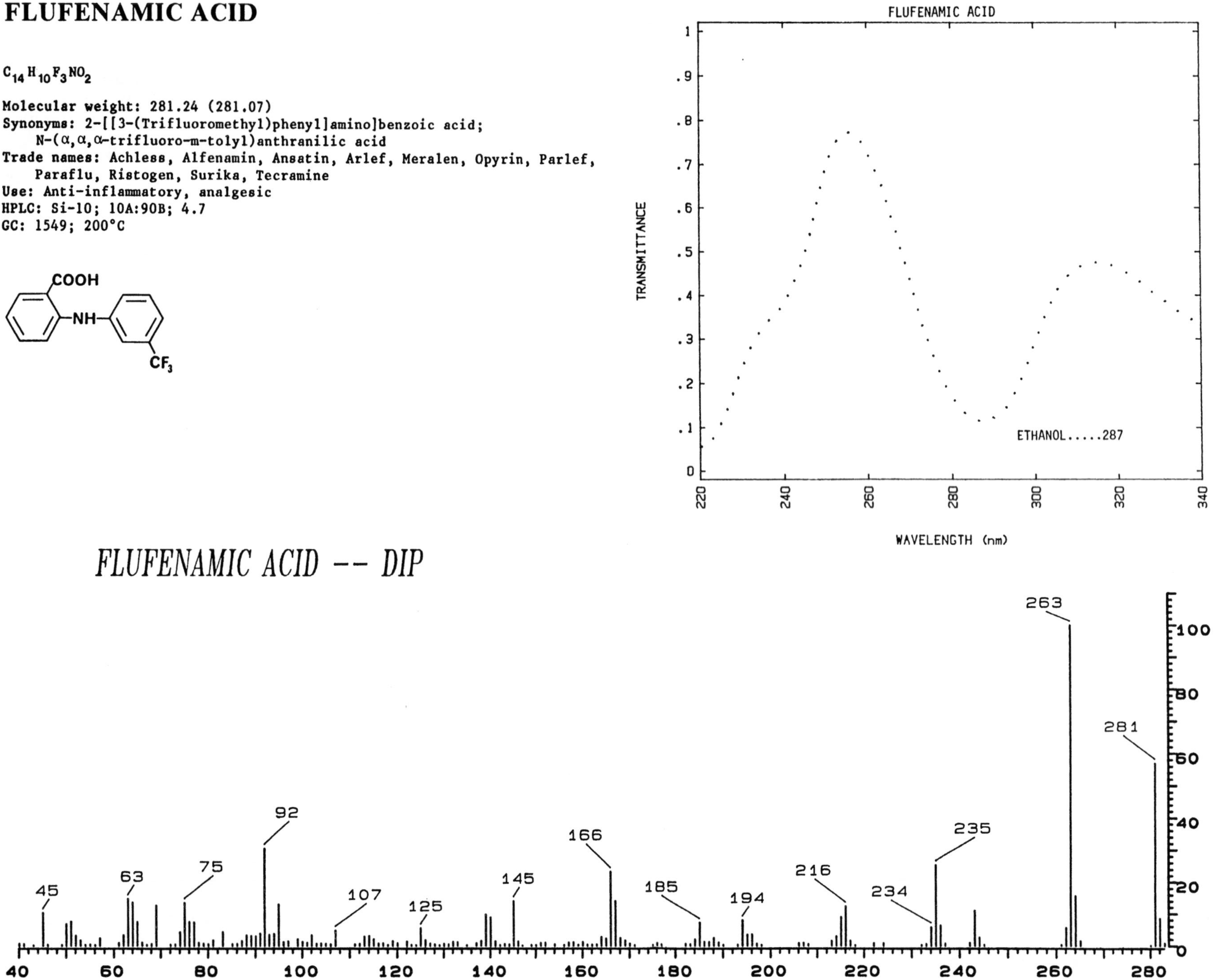

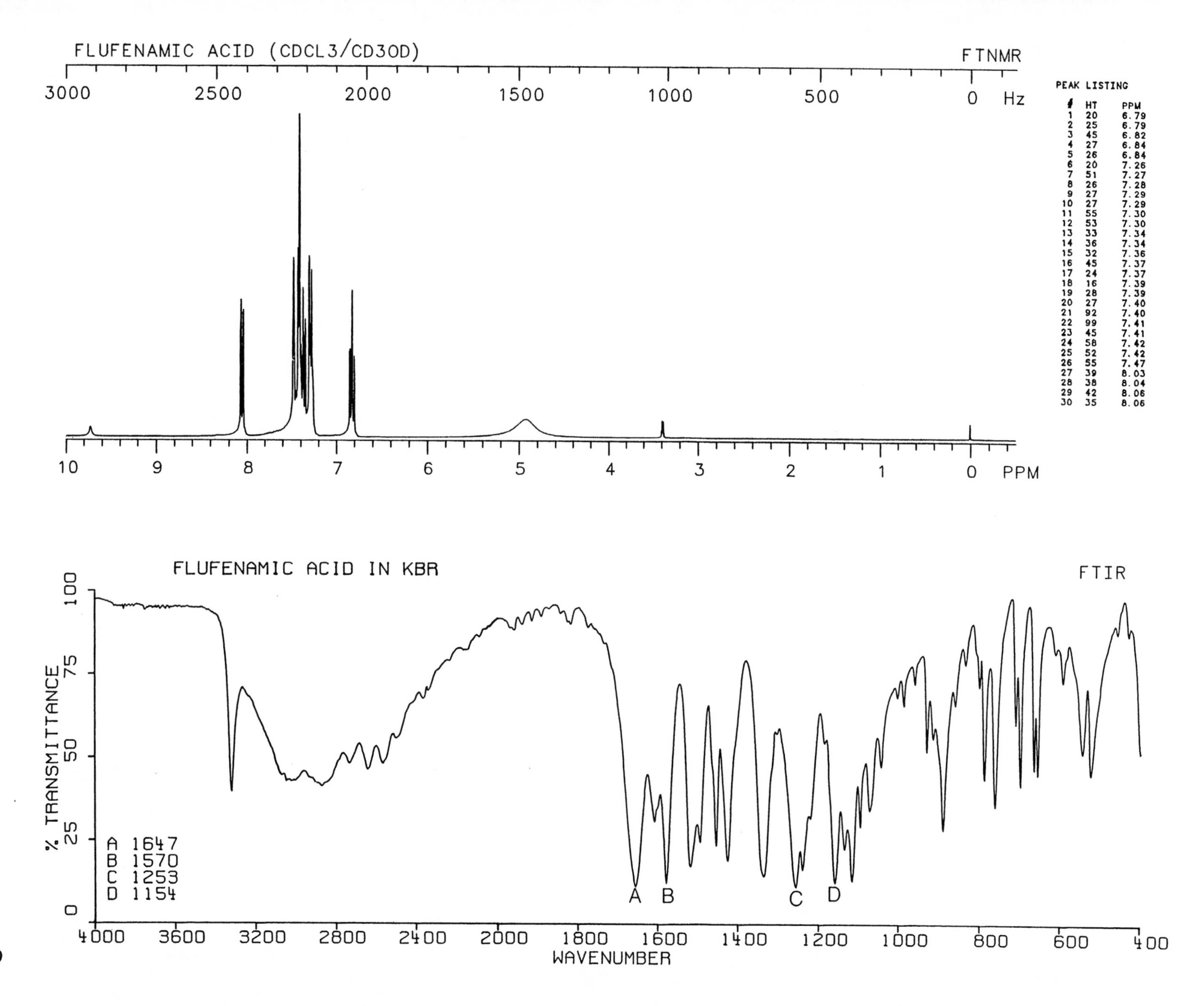

FLUFENAMIC ACID (CDCL3/CD3OD)
FTNMR
3000
2500
2000
1500
1000
500
0
Hz
PEAK LISTING
# HT PPM
1 20 6.79
2 25 6.79
3 45 6.82
4 27 6.84
5 26 6.84
6 20 7.26
7 51 7.27
8 26 7.28
9 27 7.29
10 27 7.29
11 55 7.30
12 53 7.30
13 33 7.34
14 36 7.34
15 32 7.36
16 45 7.37
17 24 7.37
18 16 7.39
19 28 7.39
20 27 7.40
21 92 7.40
22 99 7.41
23 45 7.41
24 58 7.42
25 52 7.42
26 55 7.47
27 39 8.03
28 38 8.04
29 42 8.06
30 35 8.06
10
9
8
7
6
5
4
3
2
1
0
PPM
FLUFENAMIC ACID IN KBR
FTIR
100
75
50
25
0
% TRANSMITTANCE
A 1647
B 1570
C 1253
D 1154
A
B
C
D
4000
3600
3200
2800
2400
2000
1800
1600
1400
1200
1000
800
600
400
WAVENUMBER

# FLUMETHASONE

$C_{22}H_{28}F_2O_5$

**Molecular weight:** 410.46 (410.19)
**Synonyms:** 6 ,9-Difluoro-11β,17α,21-trihydroxy-16α-methylpregna-1,4-diene-3,20-dione; 6α-fluorodexamethasone
**Trade names:** Locorten

**Use:** Anti-inflammatory, glucocorticoid
**HPLC:** Si-10; 100B; 4.0
**GC:**

FLUMETHASONE

TRANSMITTANCE

ETHANOL.....235

WAVELENGTH (nm)

*FLUMETHASONE--DIP*

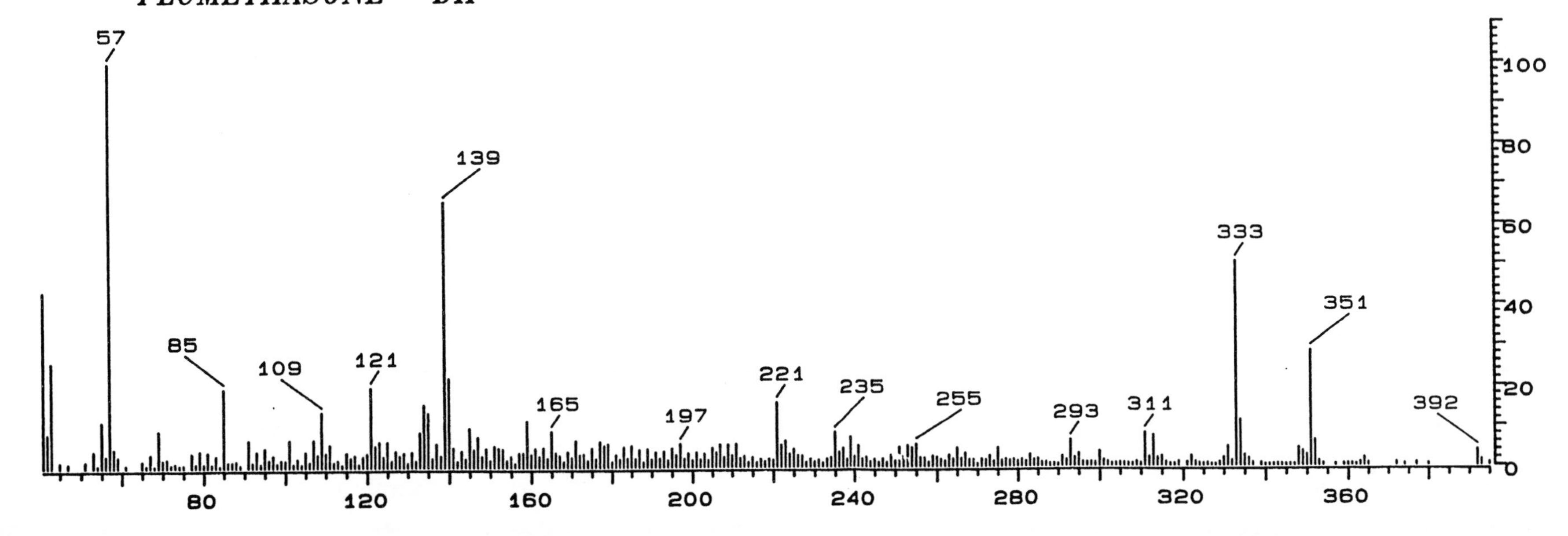

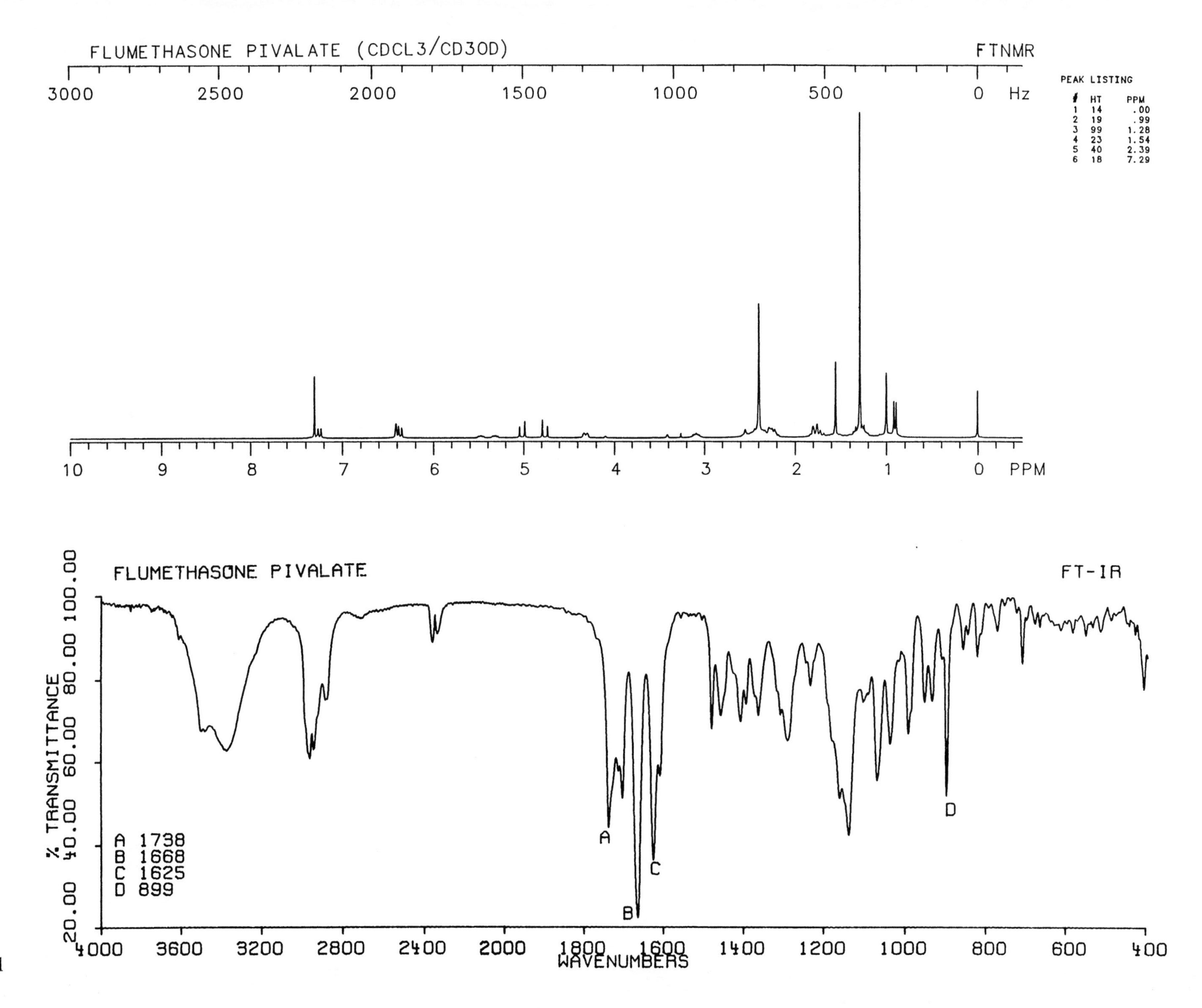

FLUMETHASONE PIVALATE (CDCL3/CD3OD)
FTNMR
3000
2500
2000
1500
1000
500
0 Hz
PEAK LISTING
# HT PPM
1 14 .00
2 19 .99
3 99 1.28
4 23 1.54
5 40 2.39
6 18 7.29
10
9
8
7
6
5
4
3
2
1
0 PPM
FLUMETHASONE PIVALATE
FT-IR
% TRANSMITTANCE
20.00 40.00 60.00 80.00 100.00
A 1738
B 1668
C 1625
D 899
A
B
C
D
4000 3600 3200 2800 2400 2000 1800 1600 1400 1200 1000 800 600 400
WAVENUMBERS

# FLUNARIZINE

$C_{26}H_{26}F_2N_2$

**Molecular weight:** 404.51 (404.21)
**Synonyms:** 1-[Bis(4-fluorophenyl)methyl]-4-(3-phenyl-2-propenyl)piperazine
1-cinnamyl-4-(di-p-fluorobenzhydryl)piperazine
**Trade names:** Flugeral, Gradient, Issium, Sibelium

**Use:** Vasodilator
**HPLC:**
**GC:** 3079; 280°C

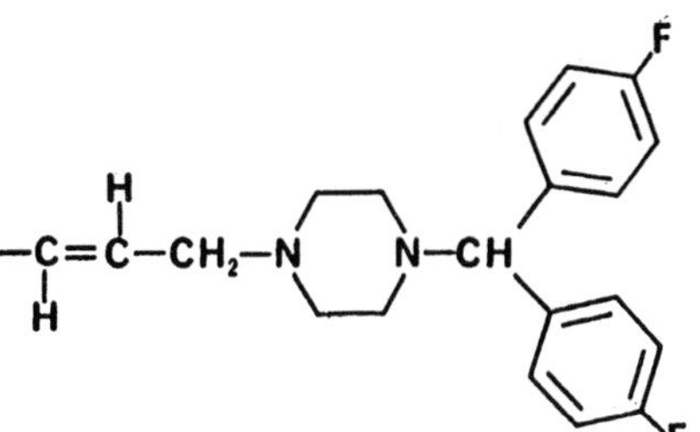


FLUNARIZINE

TRANSMITTANCE

1 .9 .8 .7 .6 .5 .4 .3 .2 .1 0

220 240 260 280 300 320 340

WAVELENGTH (nm)

225,253,291
268

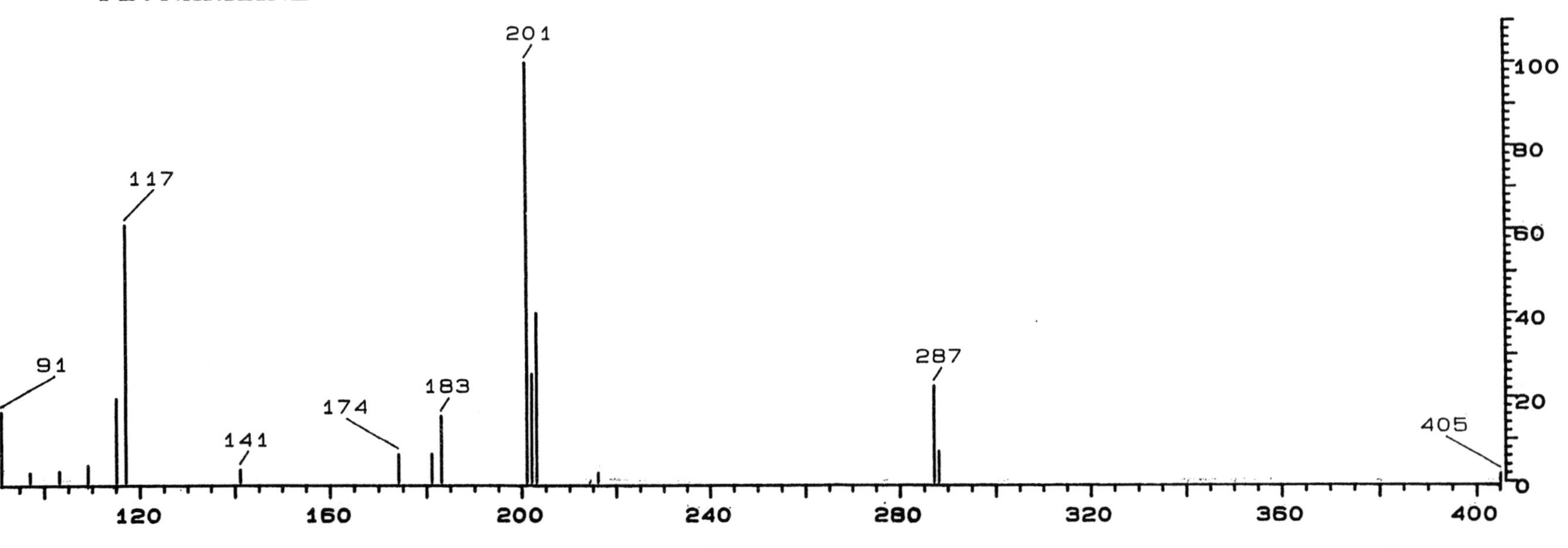

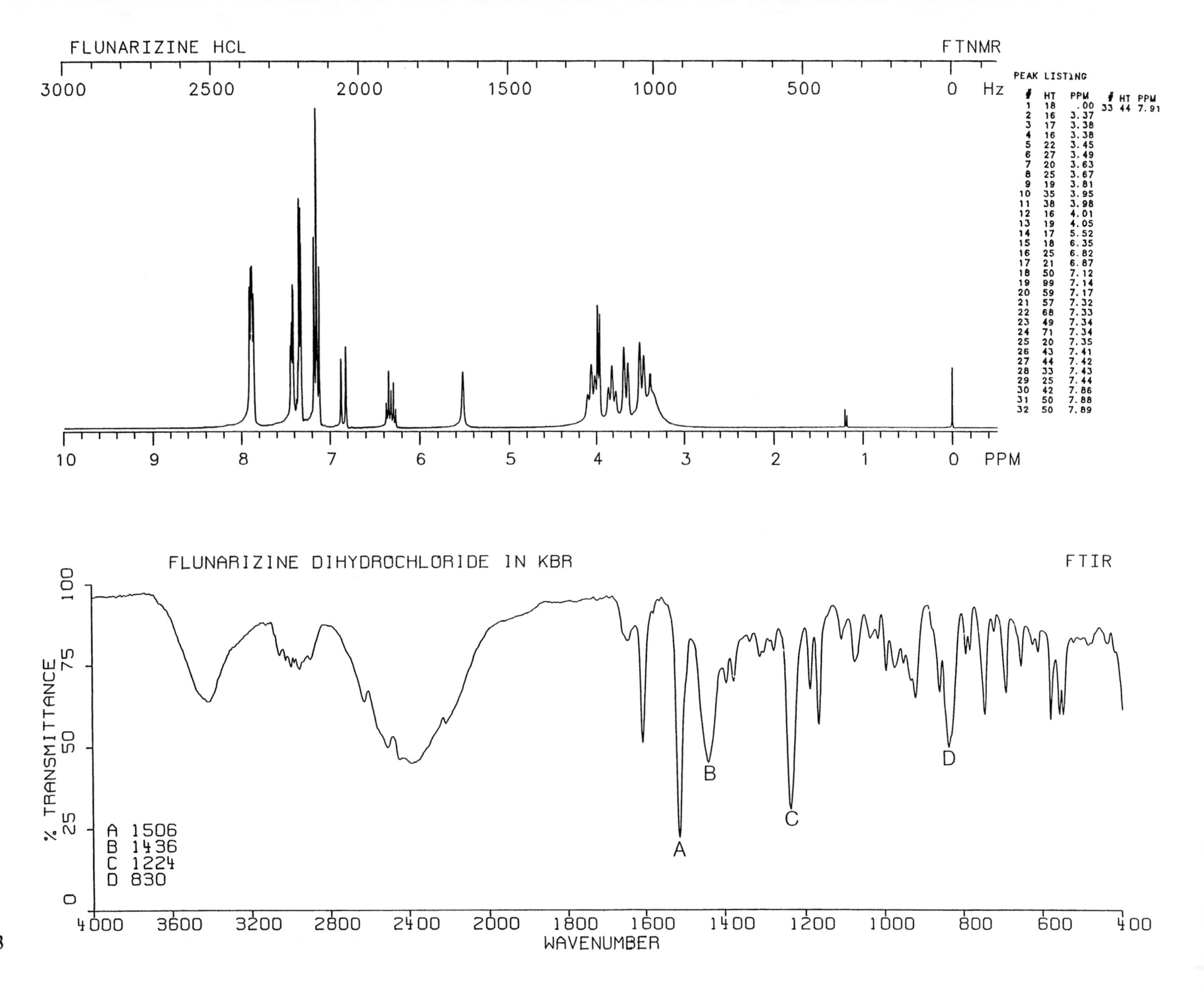

FLUNARIZINE HCL
FTNMR
3000 2500 2000 1500 1000 500 0 Hz
10 9 8 7 6 5 4 3 2 1 0 PPM
PEAK LISTING
# HT PPM
1 18 .00
2 16 3.37
3 17 3.38
4 16 3.38
5 22 3.45
6 27 3.49
7 20 3.63
8 25 3.67
9 19 3.81
10 35 3.95
11 38 3.98
12 16 4.01
13 19 4.05
14 17 5.52
15 18 6.35
16 25 6.82
17 21 6.87
18 50 7.12
19 99 7.14
20 59 7.17
21 57 7.32
22 68 7.33
23 49 7.34
24 71 7.34
25 20 7.35
26 43 7.41
27 44 7.42
28 33 7.43
29 25 7.44
30 42 7.86
31 50 7.88
32 50 7.89
33 44 7.91
FLUNARIZINE DIHYDROCHLORIDE IN KBR
FTIR
% TRANSMITTANCE
100 75 50 25 0
A 1506
B 1436
C 1224
D 830
A B C D
4000 3600 3200 2800 2400 2000 1800 1600 1400 1200 1000 800 600 400
WAVENUMBER

# FLUOCINOLONE ACETONIDE

$C_{24}H_{30}F_2O_6$

**Molecular weight:** 452.50 (452.20)
**Synonyms:** 6α,9-Difluoro-11β,21-dihydroxy-16α,17-[(1-methylethylidine)bis(oxy)]pregna-1,4-diene-3,20-dione
**Trade names:** Derma-Smoothe/FS, Fluonid, Neo-Synalar, Synalar, Synemol
**Use:** Anti-inflammatory, glucocorticoid
**HPLC:** Si-10; 2A:98B; 4.3
**GC:**

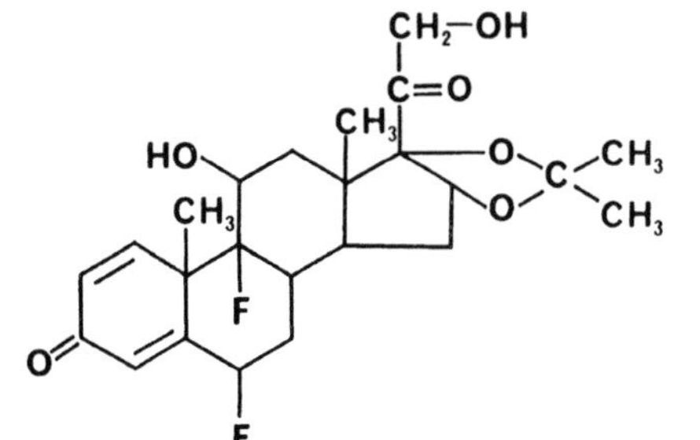


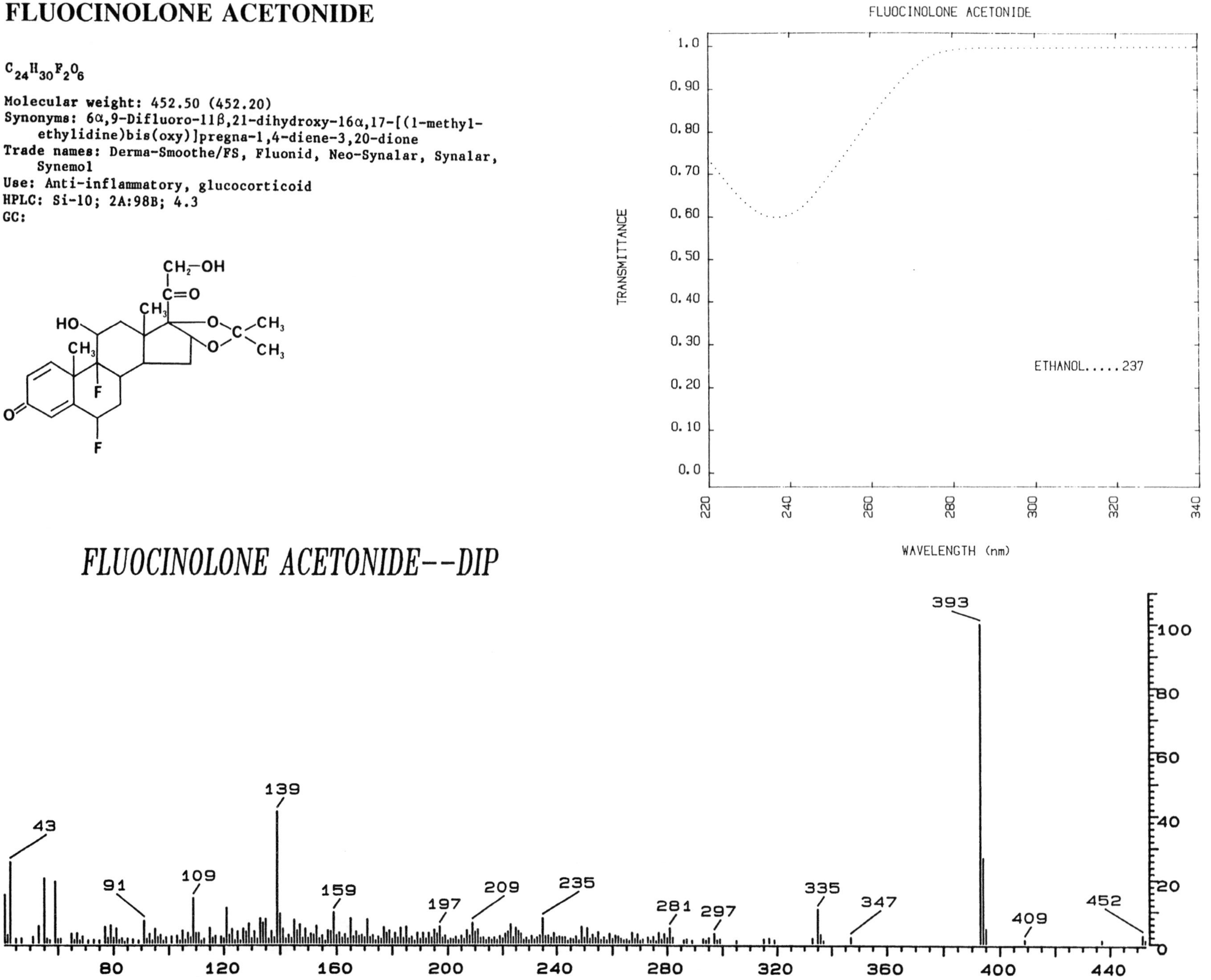

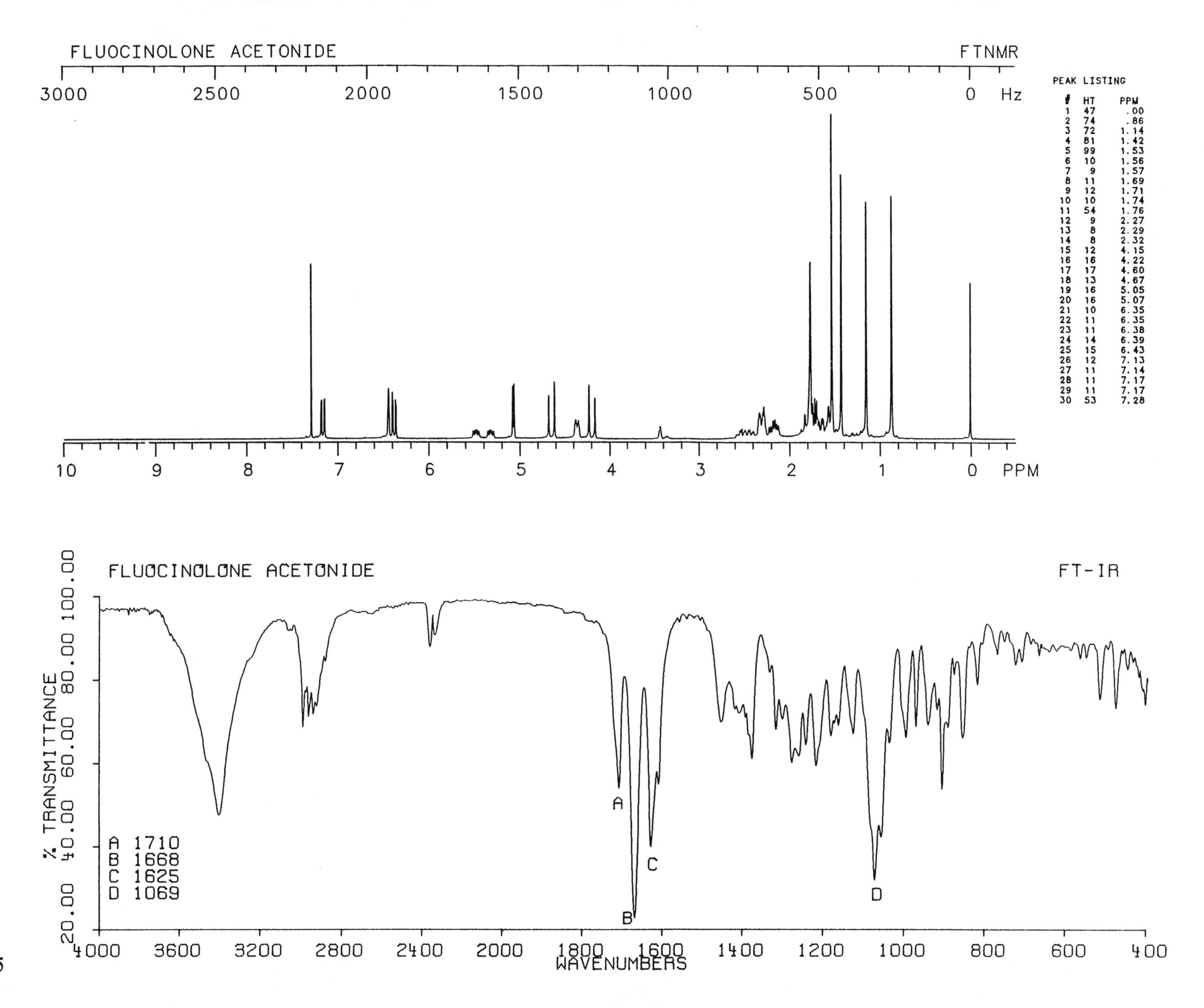
FLUOCINOLONE ACETONIDE
FTNMR
3000 2500 2000 1500 1000 500 0 Hz
10 9 8 7 6 5 4 3 2 1 0 PPM
PEAK LISTING
# HT PPM
1 47 .00
2 74 .86
3 72 1.14
4 81 1.42
5 99 1.53
6 10 1.56
7 9 1.57
8 11 1.69
9 12 1.71
10 10 1.74
11 54 1.76
12 9 2.27
13 8 2.29
14 8 2.32
15 12 4.15
16 16 4.22
17 17 4.60
18 13 4.67
19 16 5.05
20 16 5.07
21 10 6.35
22 11 6.35
23 11 6.38
24 14 6.39
25 15 6.43
26 12 7.13
27 11 7.14
28 11 7.17
29 11 7.17
30 53 7.28
FLUOCINOLONE ACETONIDE
FT-IR
% TRANSMITTANCE
20.00 40.00 60.00 80.00 100.00
A 1710
B 1668
C 1625
D 1069
A
B
C
D
4000 3600 3200 2800 2400 2000 1800 1600 1400 1200 1000 800 600 400
WAVENUMBERS

# FLUOROBENZOYLPROPIONIC ACID

$C_{10}H_9FO_3$

**Molecular weight:** 196.16 (196.05)
**Synonyms:** 3-(p-Fluorobenzoyl)propionic acid

**Trade names:**

**Use:**
**HPLC:** Si-10; 10A:90B; 10.5
**GC:**

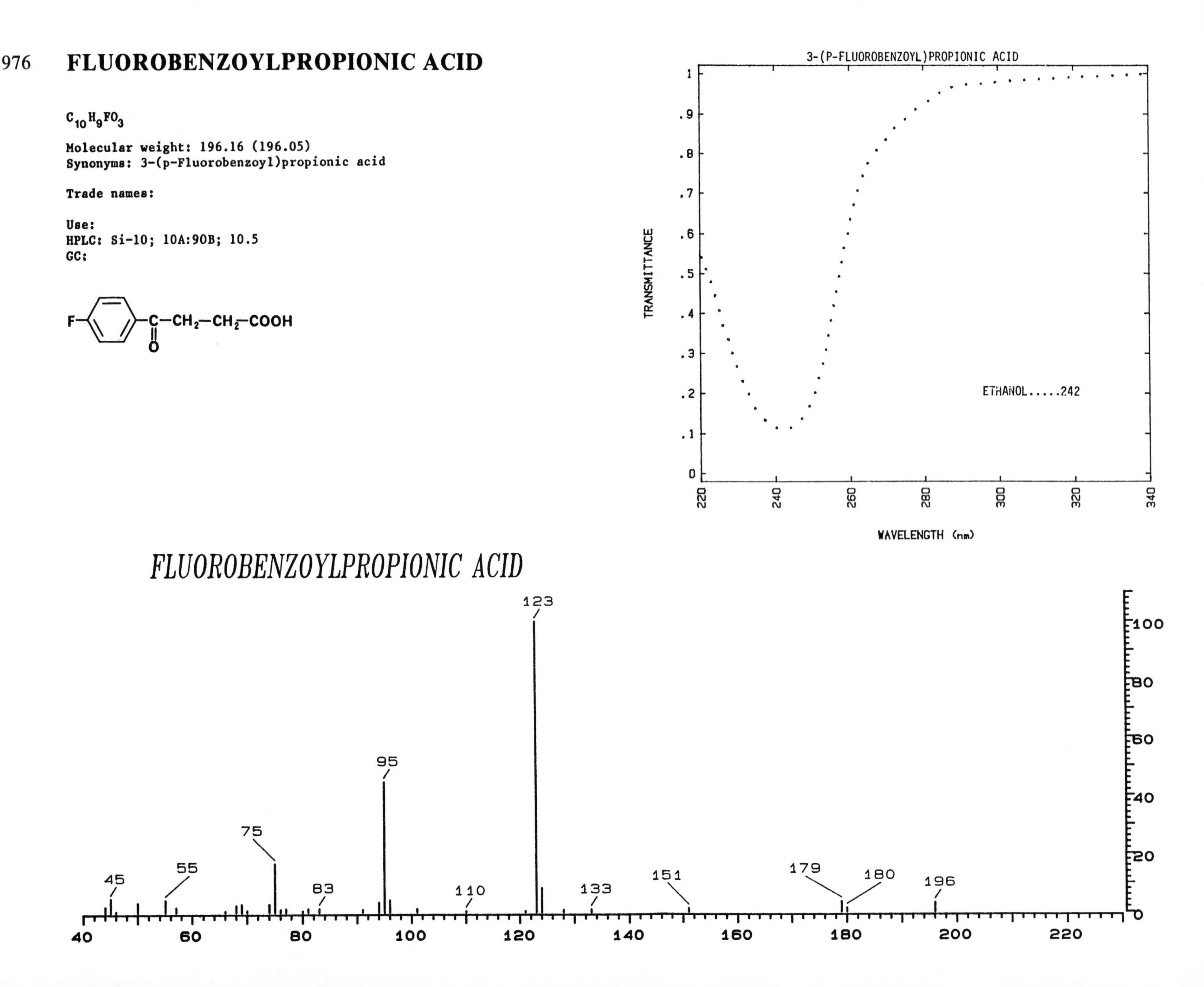

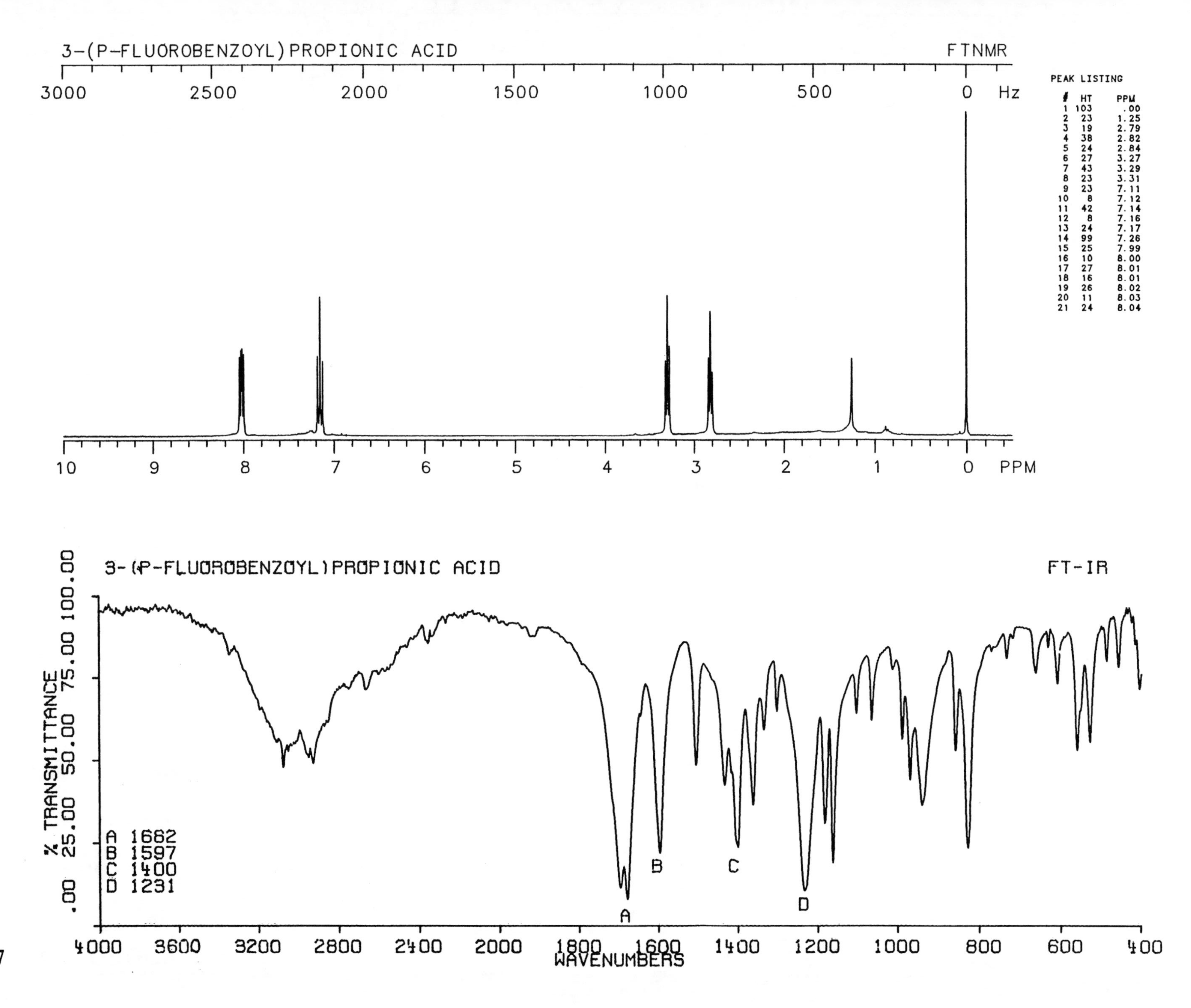
3-(P-FLUOROBENZOYL)PROPIONIC ACID
FTNMR
3000
2500
2000
1500
1000
500
0 Hz
PEAK LISTING
# HT PPM
1 103 .00
2 23 1.25
3 19 2.79
4 38 2.82
5 24 2.84
6 27 3.27
7 43 3.29
8 23 3.31
9 23 7.11
10 8 7.12
11 42 7.14
12 8 7.16
13 24 7.17
14 99 7.26
15 25 7.99
16 10 8.00
17 27 8.01
18 16 8.01
19 26 8.02
20 11 8.03
21 24 8.04
10
9
8
7
6
5
4
3
2
1
0 PPM
3-(P-FLUOROBENZOYL)PROPIONIC ACID
FT-IR
% TRANSMITTANCE
.00
25.00
50.00
75.00
100.00
A 1662
B 1597
C 1400
D 1231
A
B
C
D
4000
3600
3200
2800
2400
2000
1800
1600
1400
1200
1000
800
600
400
WAVENUMBERS

# FLUOROCYTOSINE

$C_4H_4FN_3O$

**Molecular weight:** 129.09 (129.03)
**Synonyms:** 5-Fluorocytosine; 4-amino-5-fluoro-2(1H)-pyrimidinone; flucytosine
**Trade names:** Ancobon

**Use:** Antifungal
**HPLC:** Si-10; 10A:90B; 5.5
**GC:**

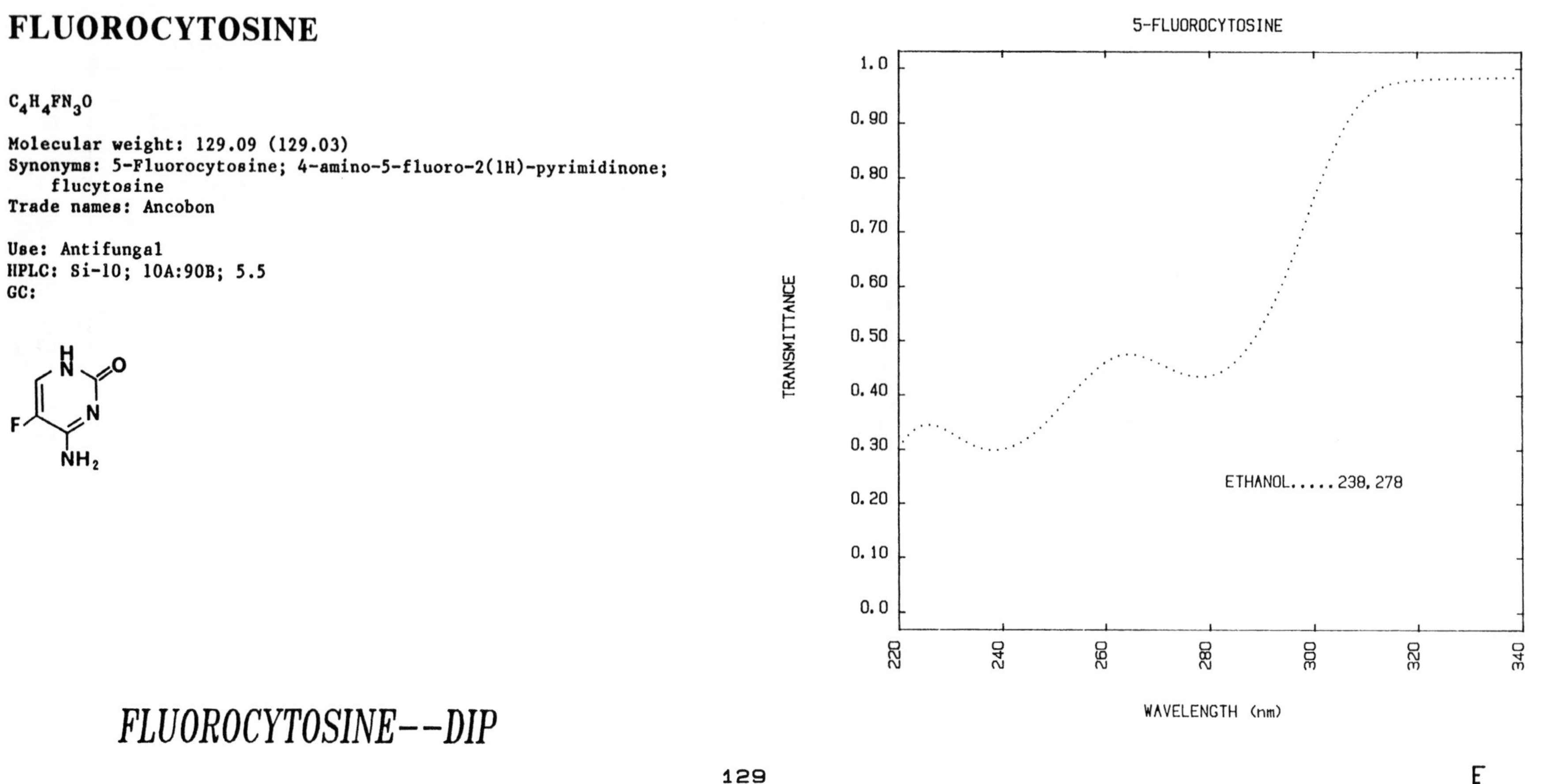


*FLUOROCYTOSINE--DIP*

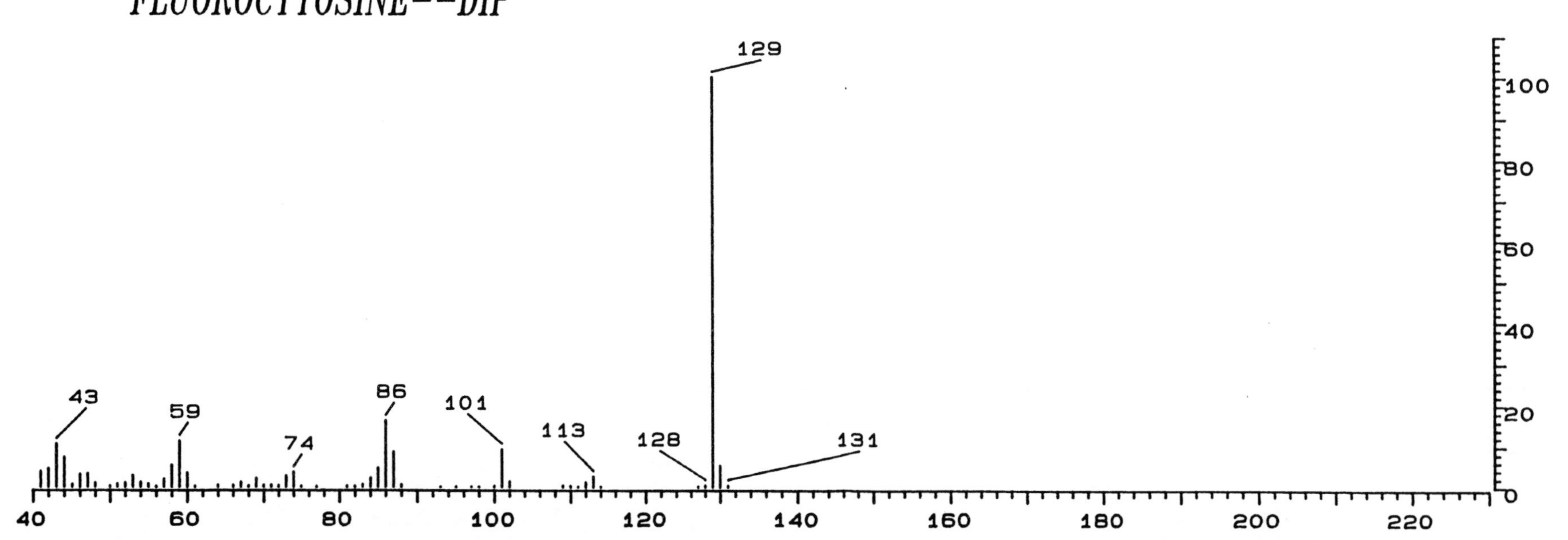

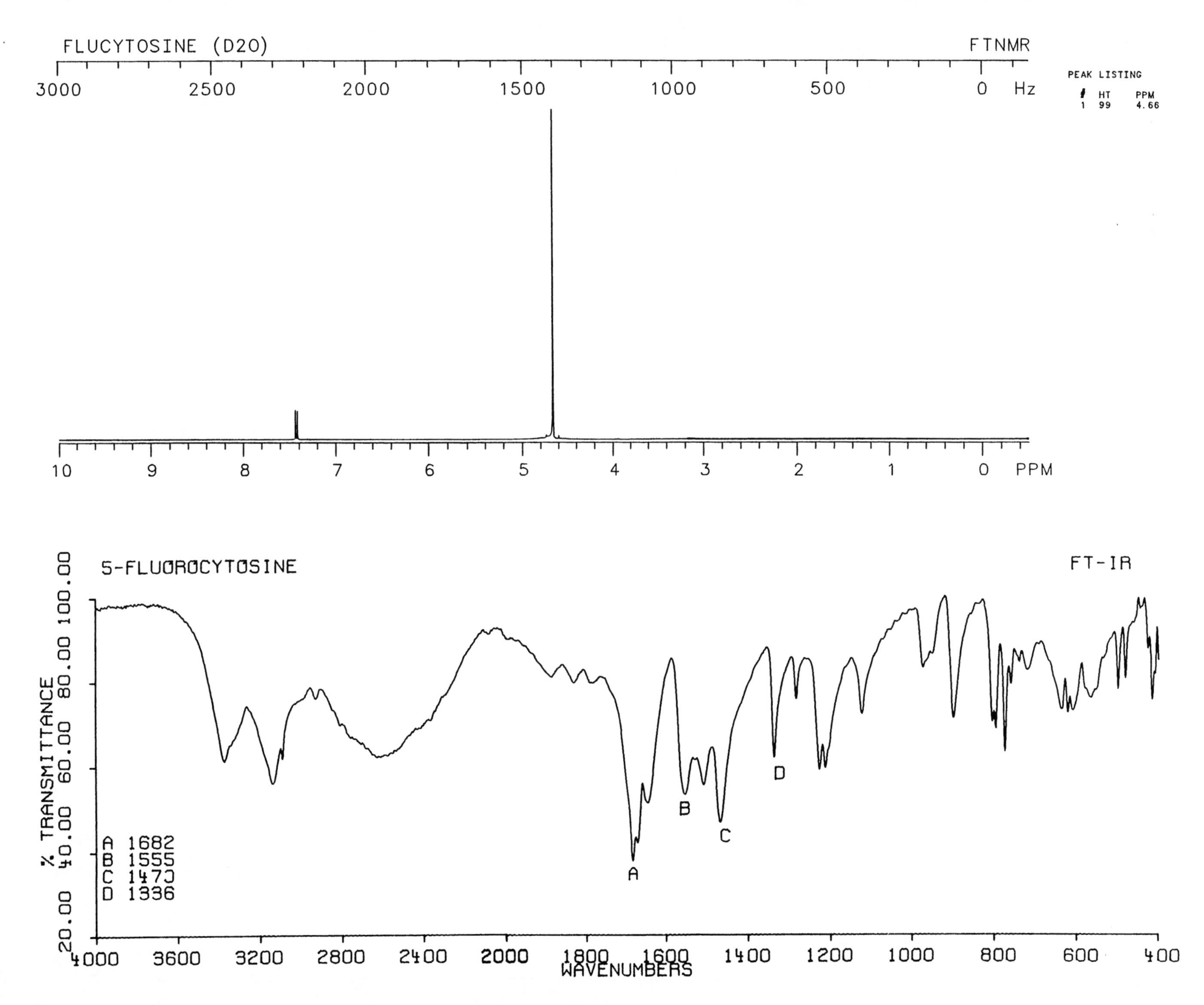

FLUCYTOSINE (D2O)
FTNMR
3000
2500
2000
1500
1000
500
0
Hz
PEAK LISTING
# HT PPM
1 99 4.66
10
9
8
7
6
5
4
3
2
1
0
PPM
5-FLUOROCYTOSINE
FT-IR
% TRANSMITTANCE
20.00
40.00
60.00
80.00
100.00
A 1682
B 1555
C 1470
D 1336
A
B
C
D
4000
3600
3200
2800
2400
2000
1800
1600
1400
1200
1000
800
600
400
WAVENUMBERS

# FLUORODEOXYURIDINE

$C_9H_{11}FN_2O_5$

Molecular weight: 246.20 (246.06)
Synonyms: 2'-Deoxy-5-fluorouridine; 5-fluoro-2'-deoxy-β-uridine;
floxuridine
Trade names: FUDR

Use: Antiviral
HPLC: Si-10; 5A:95B; 4.5
GC:

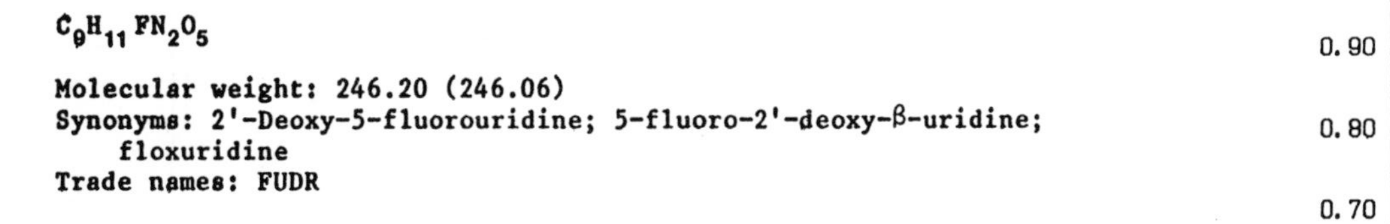




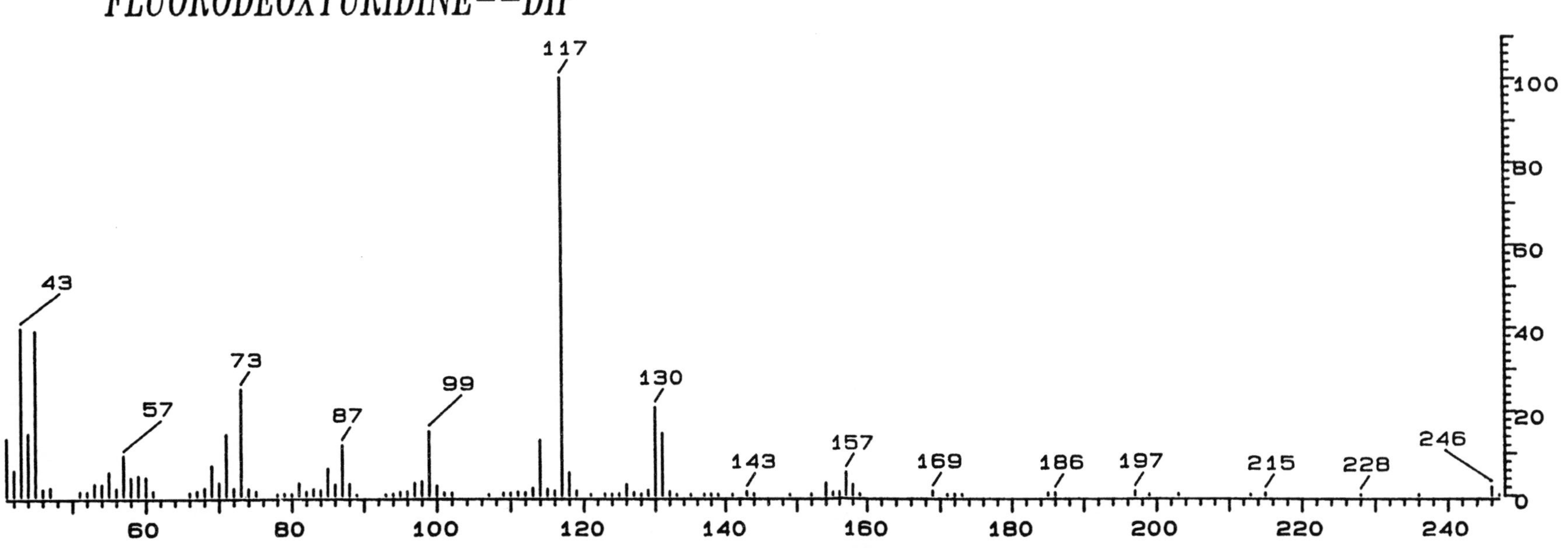

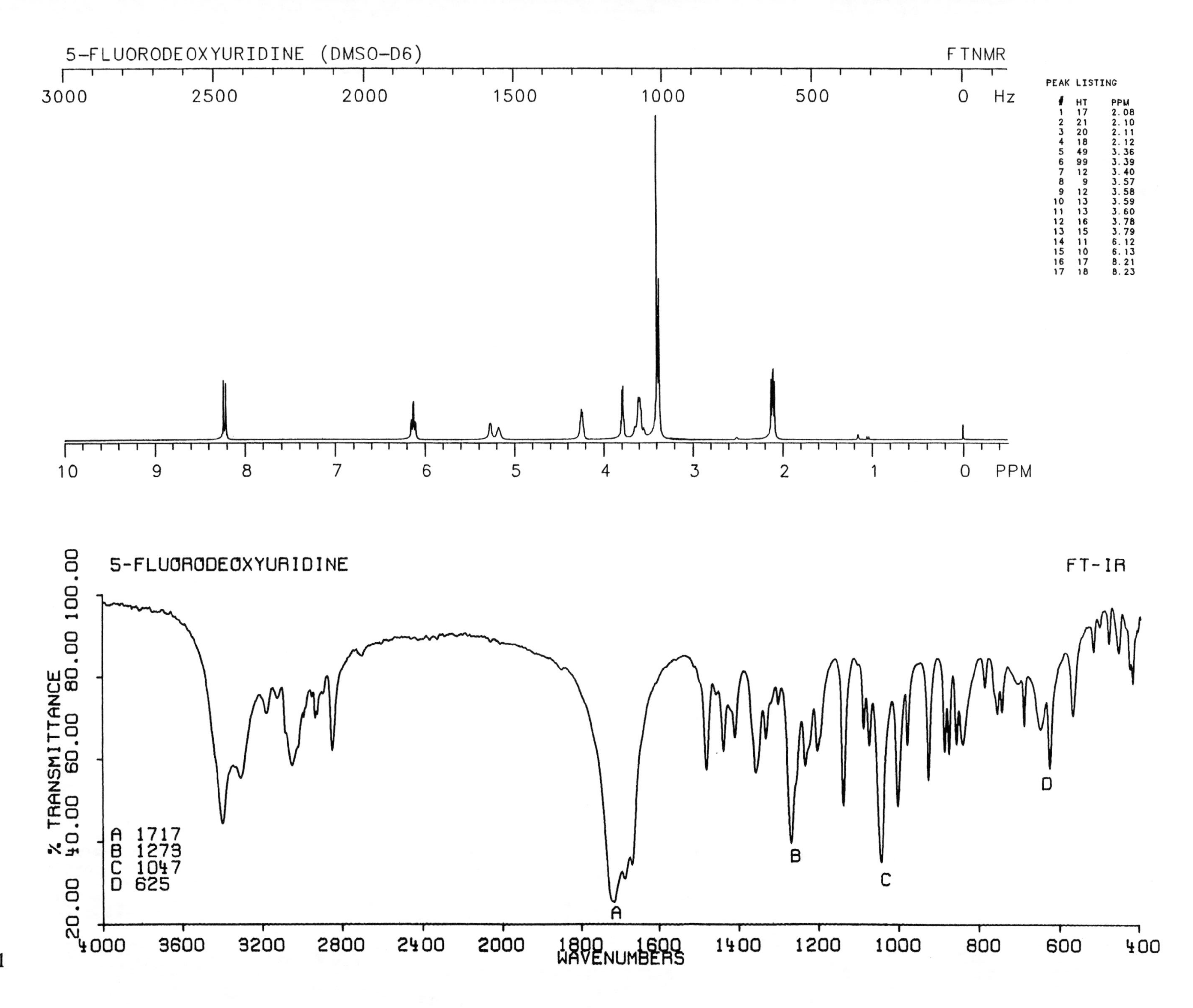

5-FLUORODEOXYURIDINE (DMSO-D6)
FTNMR
3000 2500 2000 1500 1000 500 0 Hz
10 9 8 7 6 5 4 3 2 1 0 PPM
PEAK LISTING
# HT PPM
1 17 2.08
2 21 2.10
3 20 2.11
4 18 2.12
5 49 3.36
6 99 3.39
7 12 3.40
8 9 3.57
9 12 3.58
10 13 3.59
11 13 3.60
12 16 3.78
13 15 3.79
14 11 6.12
15 10 6.13
16 17 8.21
17 18 8.23
5-FLUORODEOXYURIDINE
FT-IR
% TRANSMITTANCE
20.00 40.00 60.00 80.00 100.00
4000 3600 3200 2800 2400 2000 1800 1600 1400 1200 1000 800 600 400
WAVENUMBERS
A 1717
B 1273
C 1047
D 625
A
B
C
D

# 3-FLUOROFENTANYL ANALOG

$C_{22}H_{27}N_2FO$

**Molecular weight:** 354.45 (354.21)
**Synonyms:** Meta-Fluorophenyl-N-[1-(2-phenylethyl)-4-piperidinyl]-propanamide
**Trade names:**

**Use:**
**HPLC:**
**GC:** 2683; 280°C

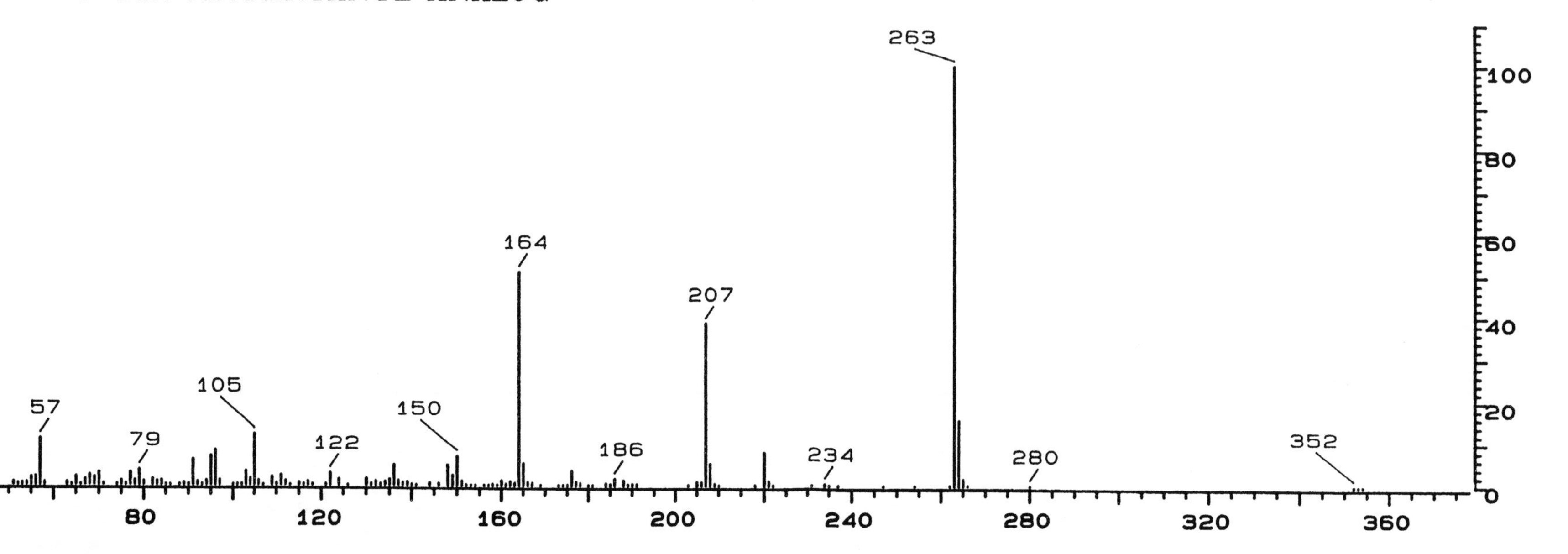

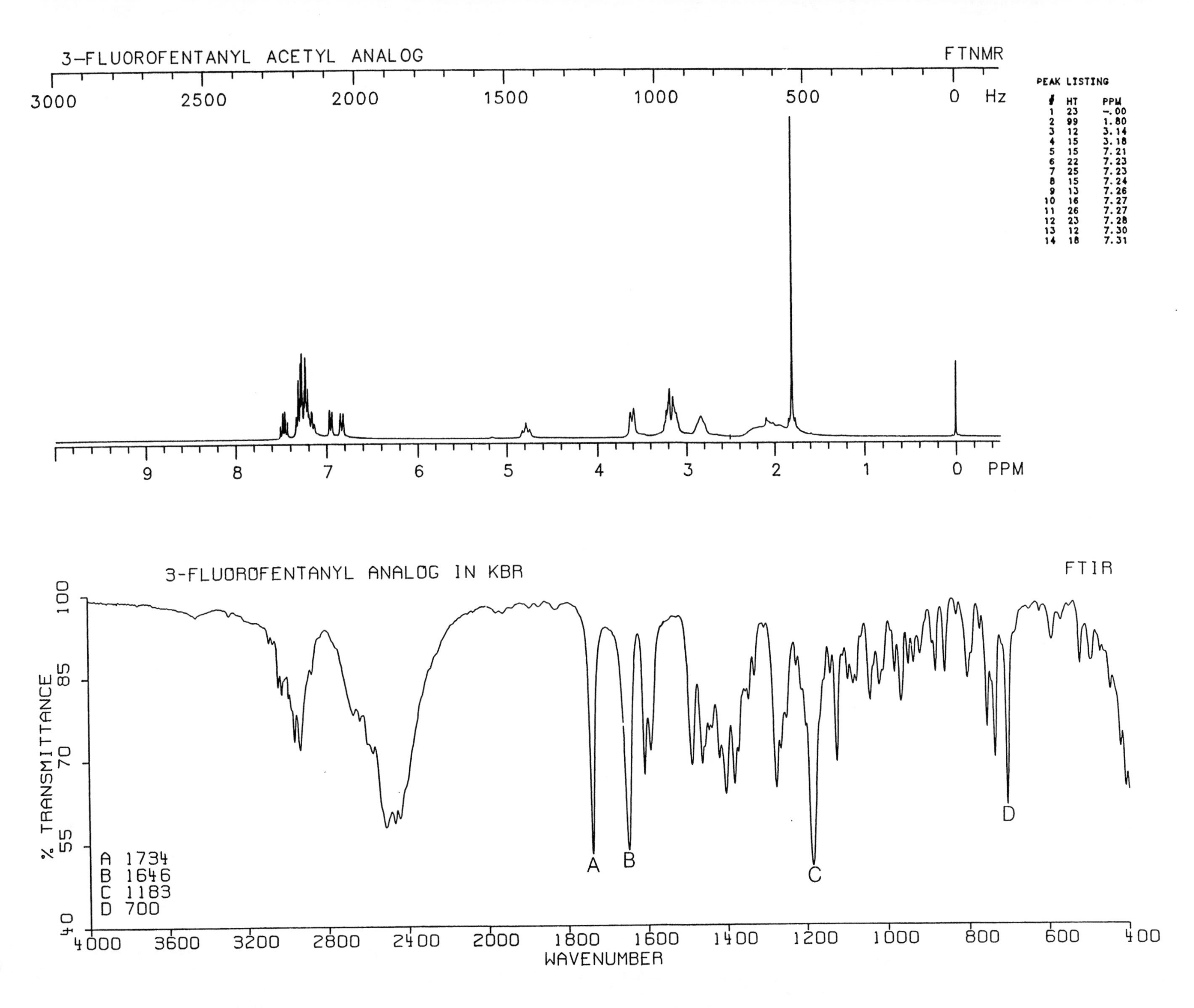

3-FLUOROFENTANYL ACETYL ANALOG
FTNMR
3000 2500 2000 1500 1000 500 0 Hz
PEAK LISTING
# HT PPM
1 23 -.00
2 99 1.80
3 12 3.14
4 15 3.18
5 15 7.21
6 22 7.23
7 25 7.23
8 15 7.24
9 13 7.26
10 16 7.27
11 26 7.27
12 23 7.28
13 12 7.30
14 18 7.31
9 8 7 6 5 4 3 2 1 0 PPM
3-FLUOROFENTANYL ANALOG IN KBR
FTIR
% TRANSMITTANCE
100 85 70 55 40
A 1734
B 1646
C 1183
D 700
A B C D
4000 3600 3200 2800 2400 2000 1800 1600 1400 1200 1000 800 600 400
WAVENUMBER

# PARA-FLUOROFENTANYL ANALOG

$C_{22}H_{27}FN_2O$

**Molecular weight:** 354.48 (354.21)
**Synonyms:** N-(4-Fluorophenyl)-N-[1-(2-phenylethyl)-4-piperidinyl]-propanamide; PFF
**Trade names:**

**Use:** Hallucinogen
**HPLC:** Si-10; 2A:98B; 6.6
**GC:** 2741; 280°C

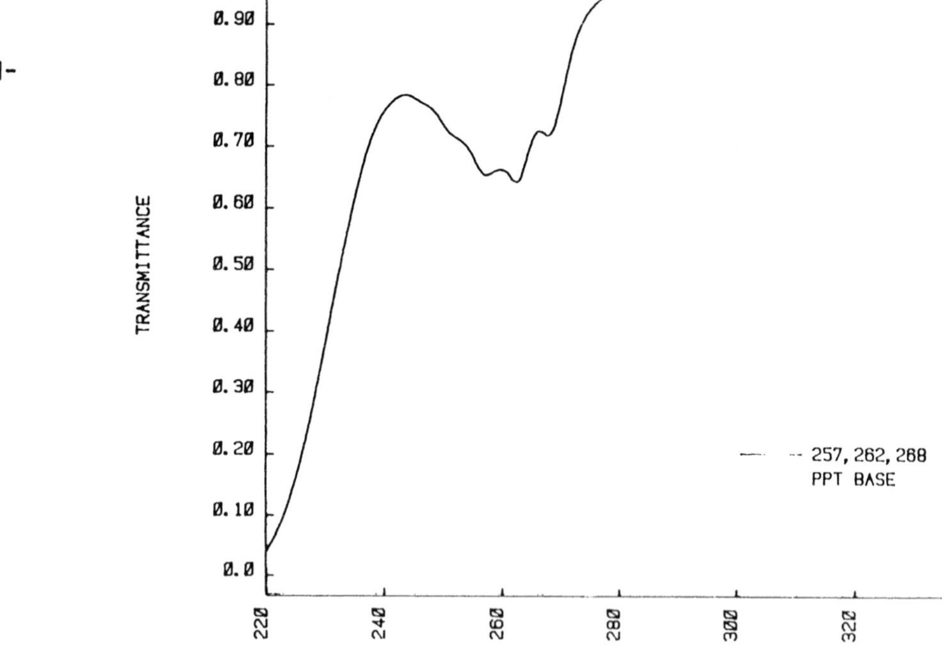


*PARA-FLUOROFENTANYL*

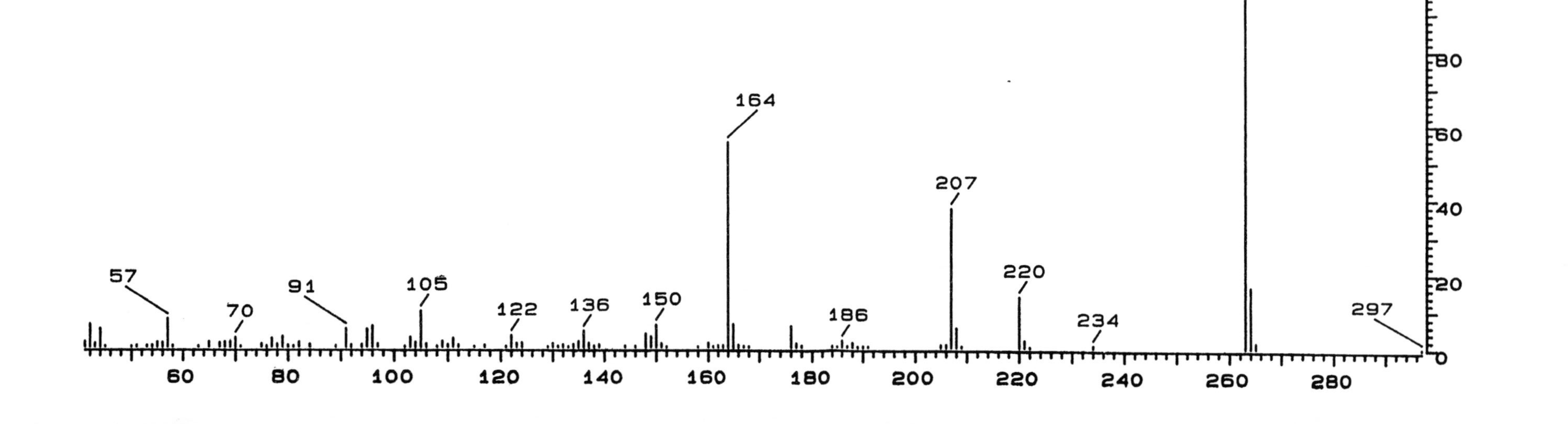

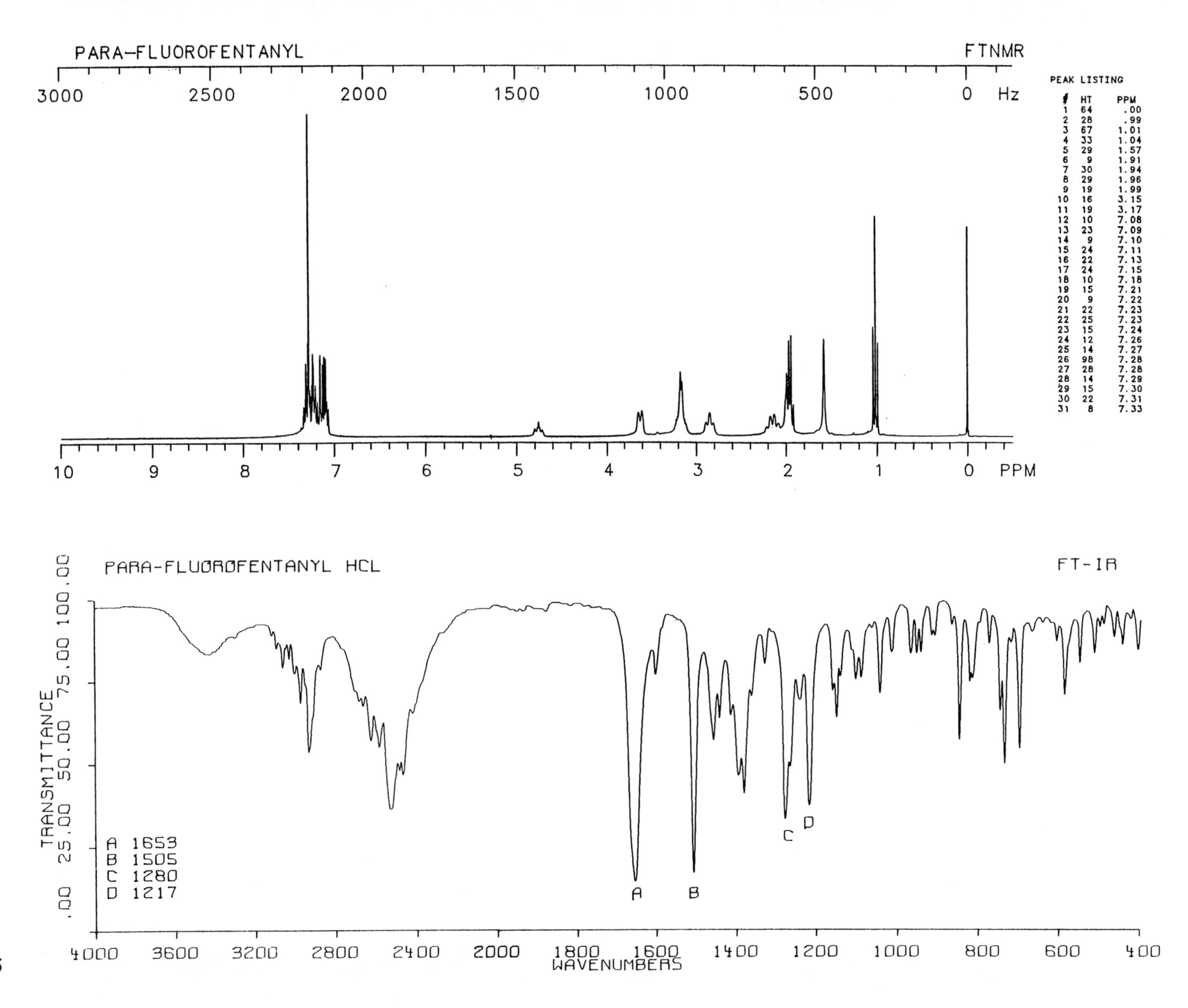

PARA–FLUOROFENTANYL
FTNMR
3000 2500 2000 1500 1000 500 0 Hz
PEAK LISTING
# HT PPM
1 64 .00
2 28 .99
3 67 1.01
4 33 1.04
5 29 1.57
6 9 1.91
7 30 1.94
8 29 1.96
9 19 1.99
10 16 3.15
11 19 3.17
12 10 7.08
13 23 7.09
14 9 7.10
15 24 7.11
16 22 7.13
17 24 7.15
18 10 7.18
19 15 7.21
20 9 7.22
21 22 7.23
22 25 7.23
23 15 7.24
24 12 7.26
25 14 7.27
26 98 7.28
27 28 7.28
28 14 7.29
29 15 7.30
30 22 7.31
31 8 7.33
10 9 8 7 6 5 4 3 2 1 0 PPM
PARA-FLUOROFENTANYL HCL
FT-IR
TRANSMITTANCE
.00 25.00 50.00 75.00 100.00
A 1653
B 1505
C 1280
D 1217
A B C D
4000 3600 3200 2800 2400 2000 1800 1600 1400 1200 1000 800 600 400
WAVENUMBERS

# 2-FLUOROFENTANYL ACETYL ANALOG

$C_{21}H_{25}N_2FO$

Molecular weight: 340.29 (340.20)
Synonyms: Ortho-Fluorophenyl-N-[1-(2-phenylethyl)-4-piperidinyl]-acetamide
Trade names:

Use:
HPLC: Si-10; 2A:98B; 4.7
GC: 2660; 280°C

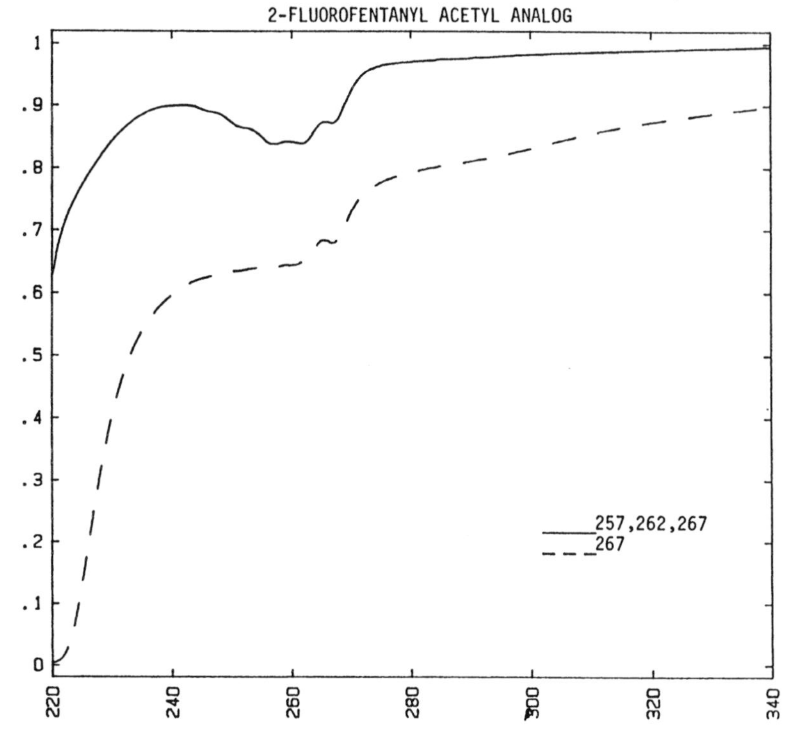
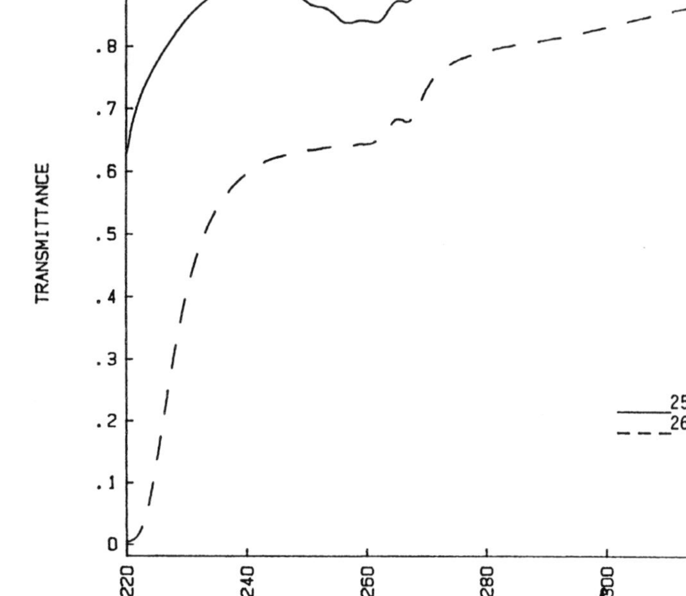

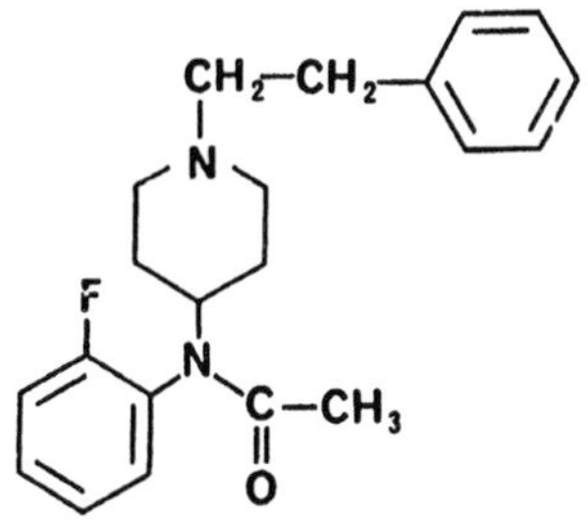

2-FLUOROFENTANYL ACETYL ANALOG

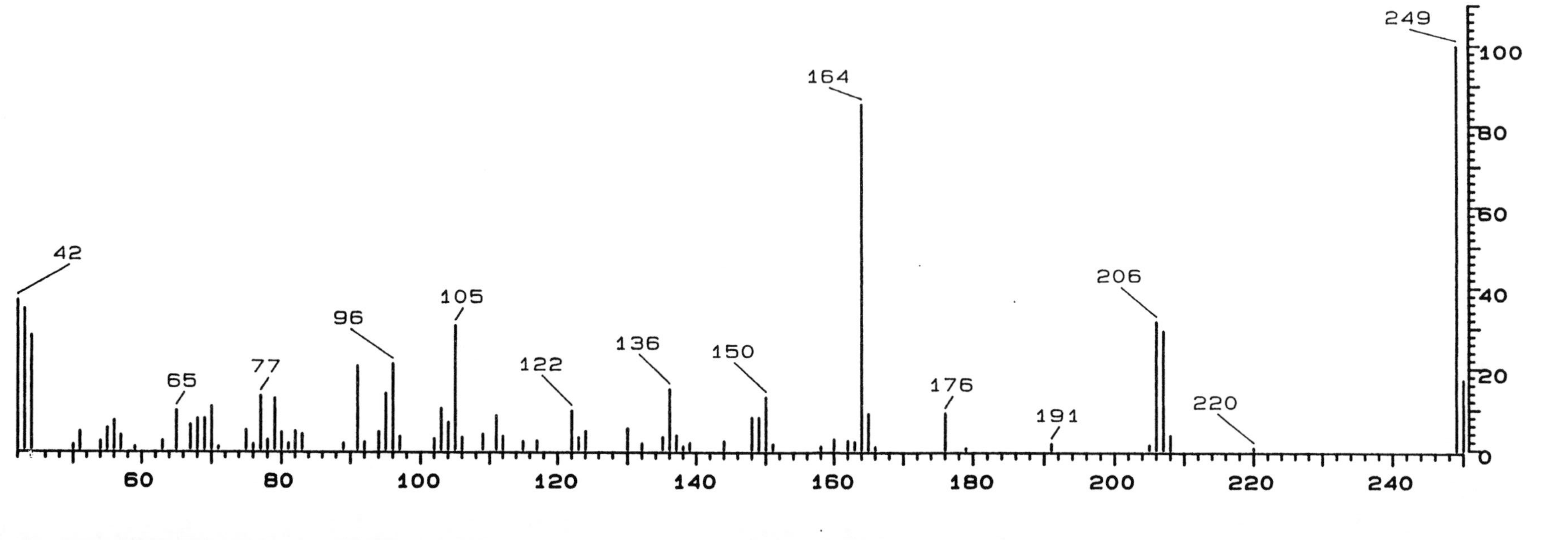

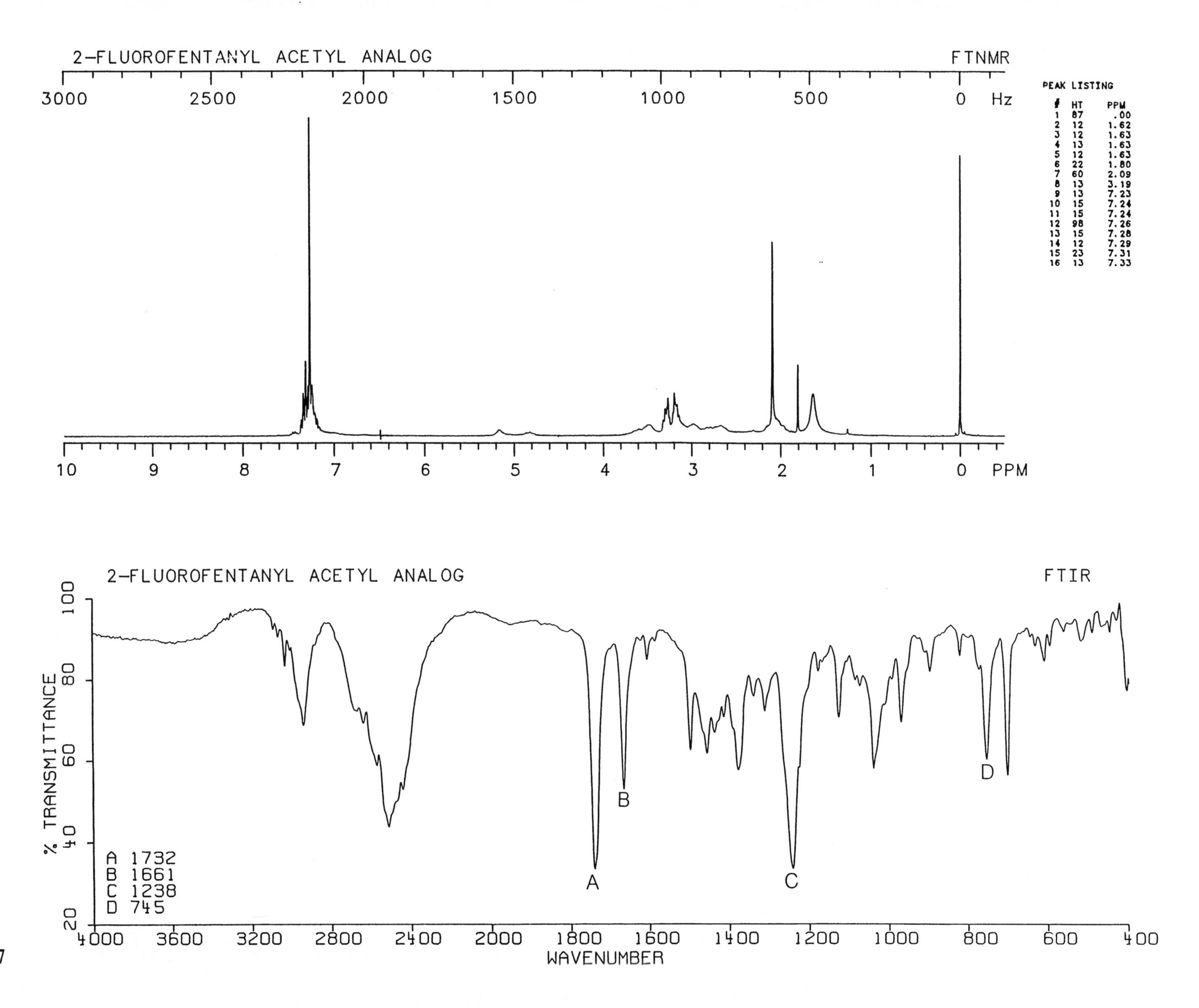
2-FLUOROFENTANYL ACETYL ANALOG
FTNMR
3000
2500
2000
1500
1000
500
0 Hz
PEAK LISTING
# HT PPM
1 87 .00
2 12 1.62
3 12 1.63
4 13 1.63
5 12 1.63
6 22 1.80
7 60 2.09
8 13 3.19
9 13 7.23
10 15 7.24
11 15 7.24
12 98 7.26
13 15 7.28
14 12 7.29
15 23 7.31
16 13 7.33
10 9 8 7 6 5 4 3 2 1 0 PPM
2-FLUOROFENTANYL ACETYL ANALOG
FTIR
% TRANSMITTANCE
100 80 60 40 20
A 1732
B 1661
C 1238
D 745
A
B
C
D
4000 3600 3200 2800 2400 2000 1800 1600 1400 1200 1000 800 600 400
WAVENUMBER

# 3-FLUOROFENTANYL ACETYL ANALOG

$C_{21}H_{25}N_2FO$

**Molecular weight:** 340.24 (340.20)
**Synonyms:** Meta-Fluorophenyl-N-[1-(2-phenylethyl)-4-piperidinyl]-acetamide
**Trade names:**

**Use:**
**HPLC:** Si-10; 2A:98B; 3.8
**GC:** 2612; 280°C

3-FLUOROFENTANYL ACETYL ANALOG

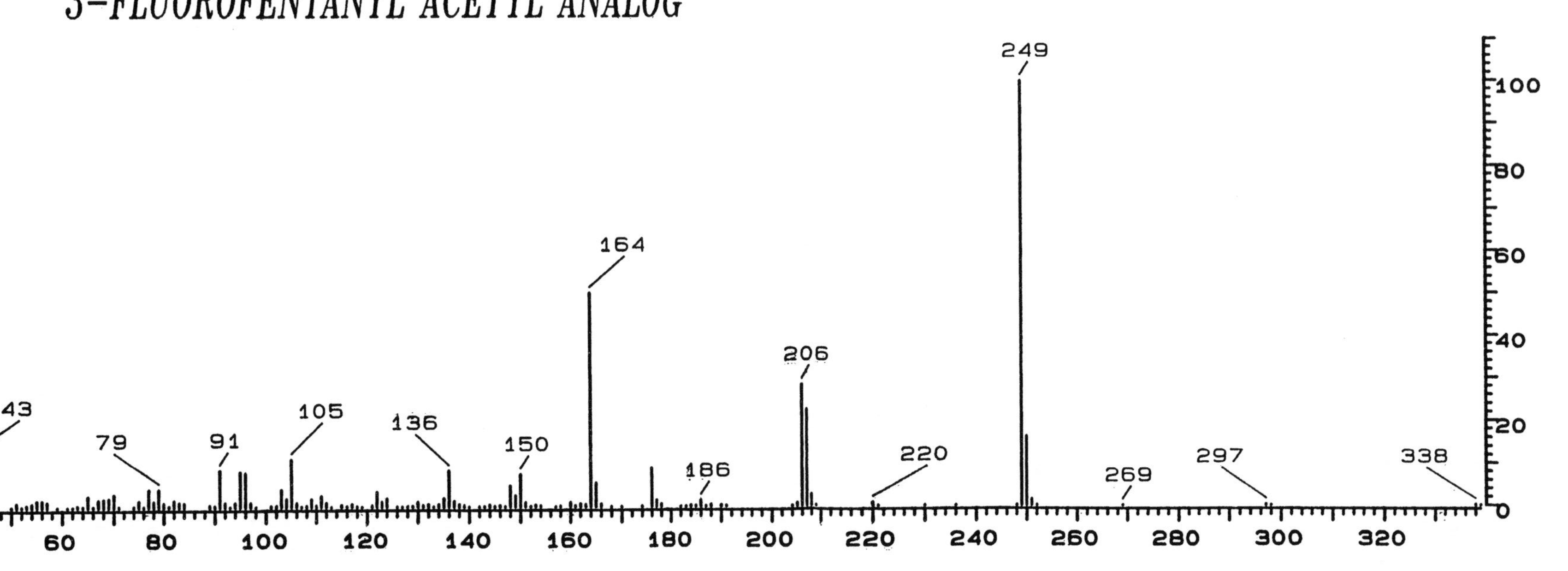

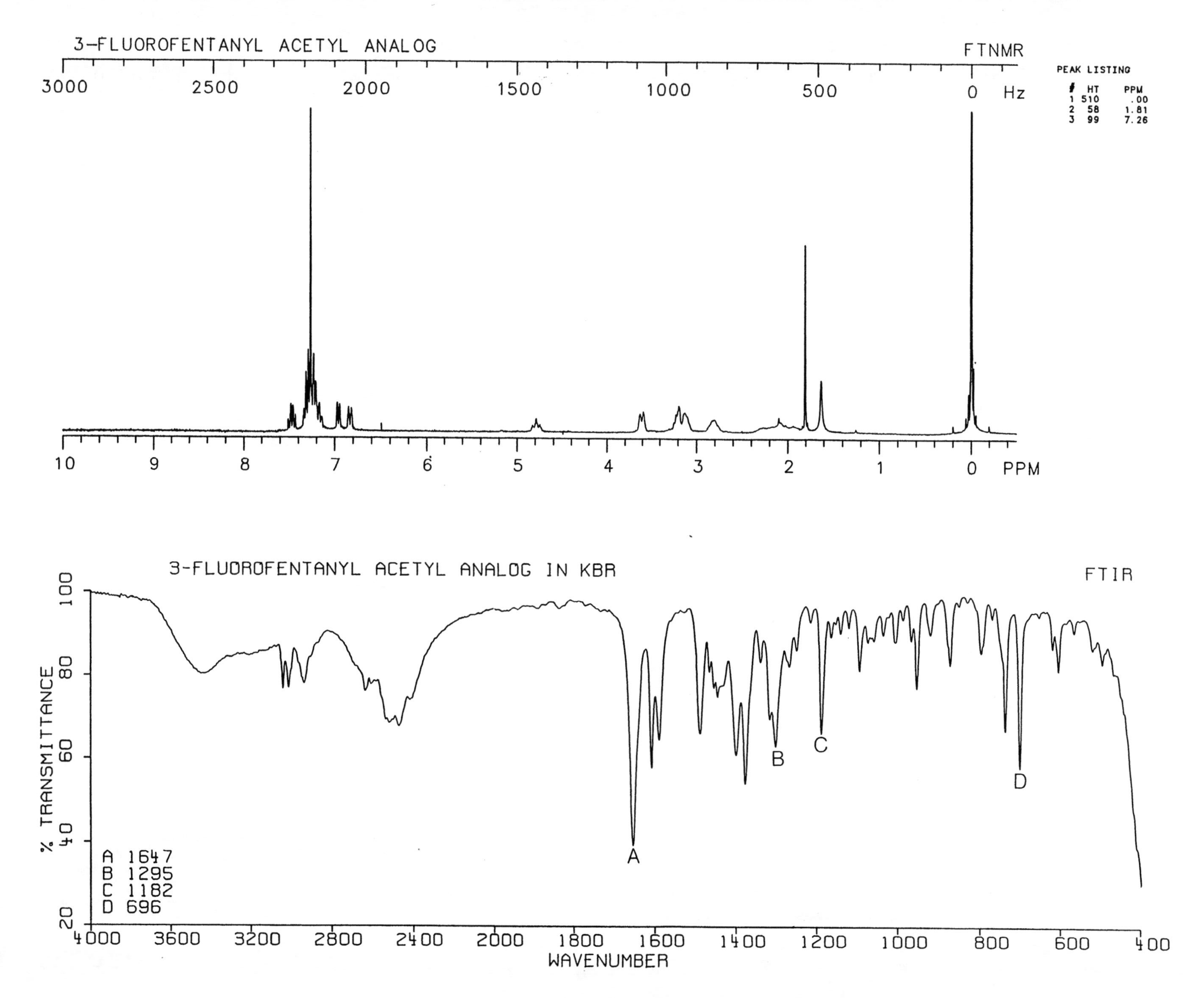
3-FLUOROFENTANYL ACETYL ANALOG
FTNMR
PEAK LISTING
# HT PPM
1 510 .00
2 58 1.81
3 99 7.26
Hz
PPM
3-FLUOROFENTANYL ACETYL ANALOG IN KBR
FTIR
% TRANSMITTANCE
WAVENUMBER
A 1647
B 1295
C 1182
D 696

# FLUOROMETHOLONE

$C_{22}H_{29}FO_4$

Molecular weight: 376.47 (376.21)
Synonyms: 9-Fluoro-11β,17-dihydroxy-6α-methylpregna-1,4-diene-3,20-dione; fluormetholon
Trade names: Oxylane

Use: Glucocorticoid, anti-inflammatory
HPLC: Si-10; 100B; 4.0
GC:

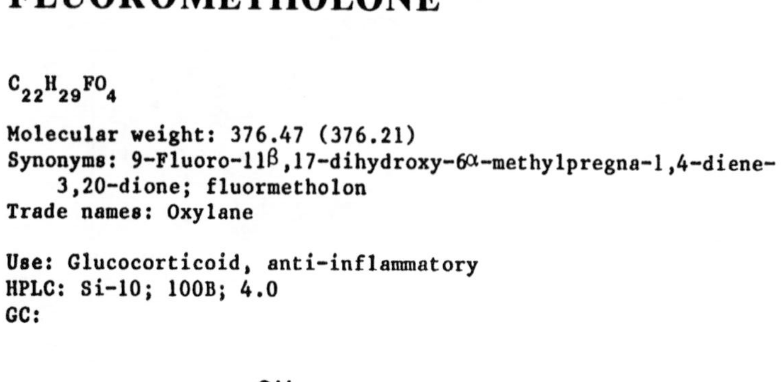


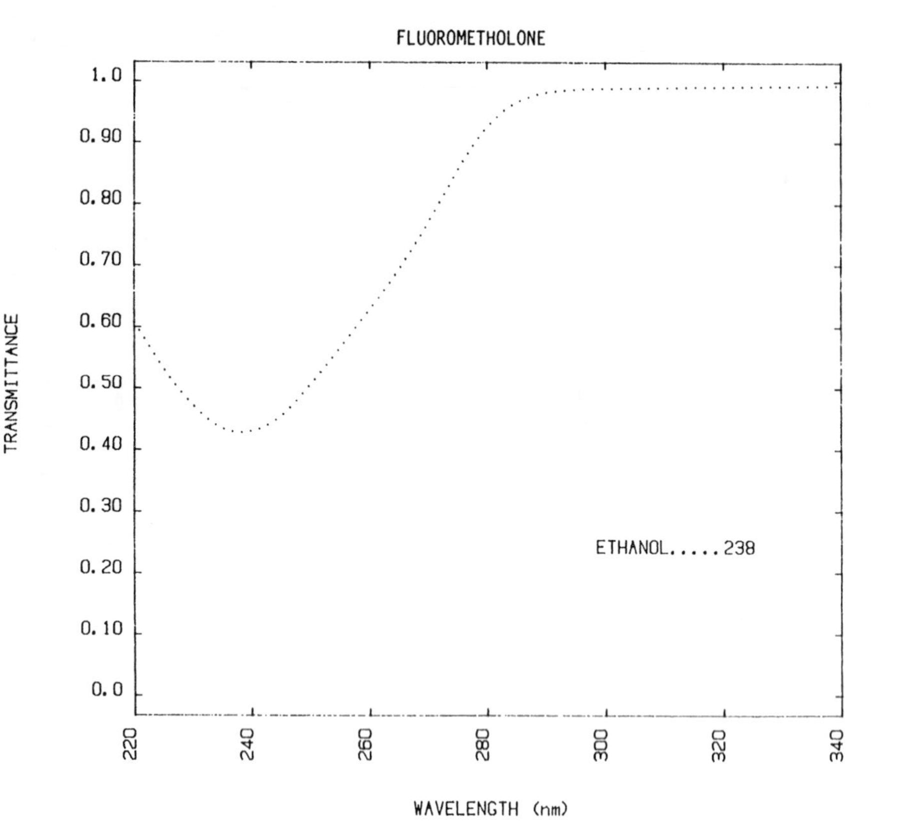


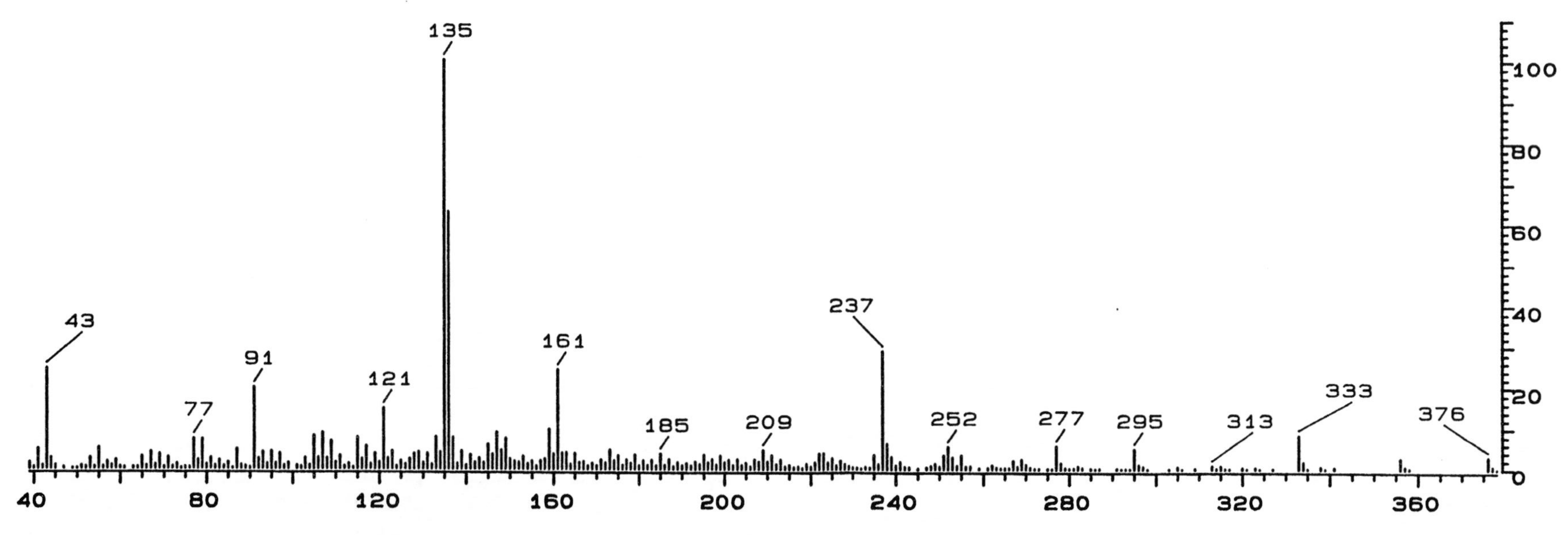

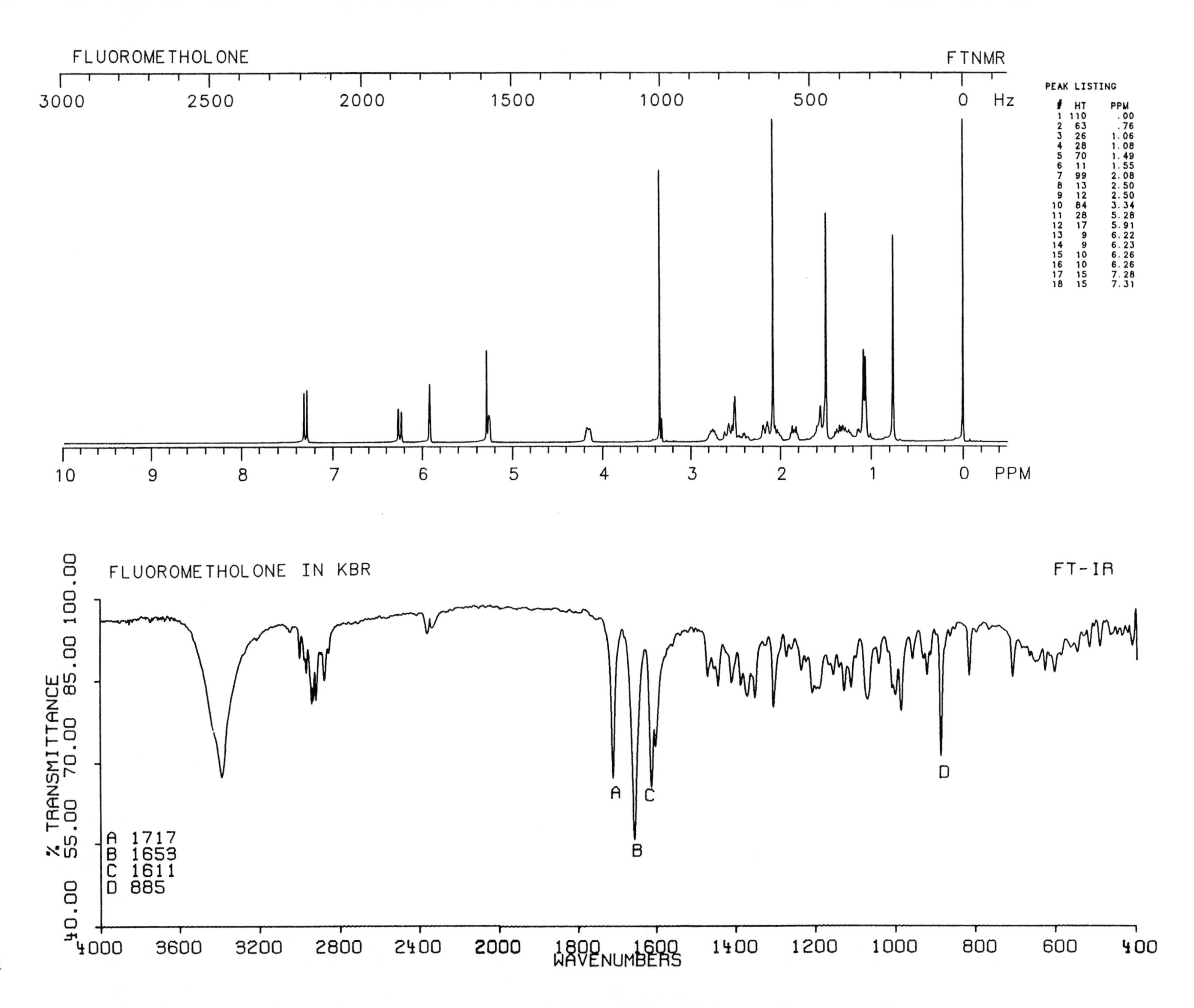
FLUOROMETHOLONE
FTNMR
3000 2500 2000 1500 1000 500 0 Hz
PEAK LISTING
# HT PPM
1 110 .00
2 63 .76
3 26 1.06
4 28 1.08
5 70 1.49
6 11 1.55
7 99 2.08
8 13 2.50
9 12 2.50
10 84 3.34
11 28 5.28
12 17 5.91
13 9 6.22
14 9 6.23
15 10 6.26
16 10 6.26
17 15 7.28
18 15 7.31
10 9 8 7 6 5 4 3 2 1 0 PPM
FLUOROMETHOLONE IN KBR
FT-IR
% TRANSMITTANCE
40.00 55.00 70.00 85.00 100.00
A 1717
B 1653
C 1611
D 885
A
B
C
D
4000 3600 3200 2800 2400 2000 1800 1600 1400 1200 1000 800 600 400
WAVENUMBERS

# FLUOROPHENYL ACETYL GLYCINE

$C_{10}H_{10}FNO_3$

**Molecular weight:** 211.18 (211.06)
**Synonyms:** N-[(4-Fluorophenyl)acetyl]glycine

**Trade names:**

**Use:**
HPLC: Si-10; 20A:80B; 3.3
GC: 2809; 280°C

F–C₆H₄–CH₂–C(=O)–NH–CH₂–COOH

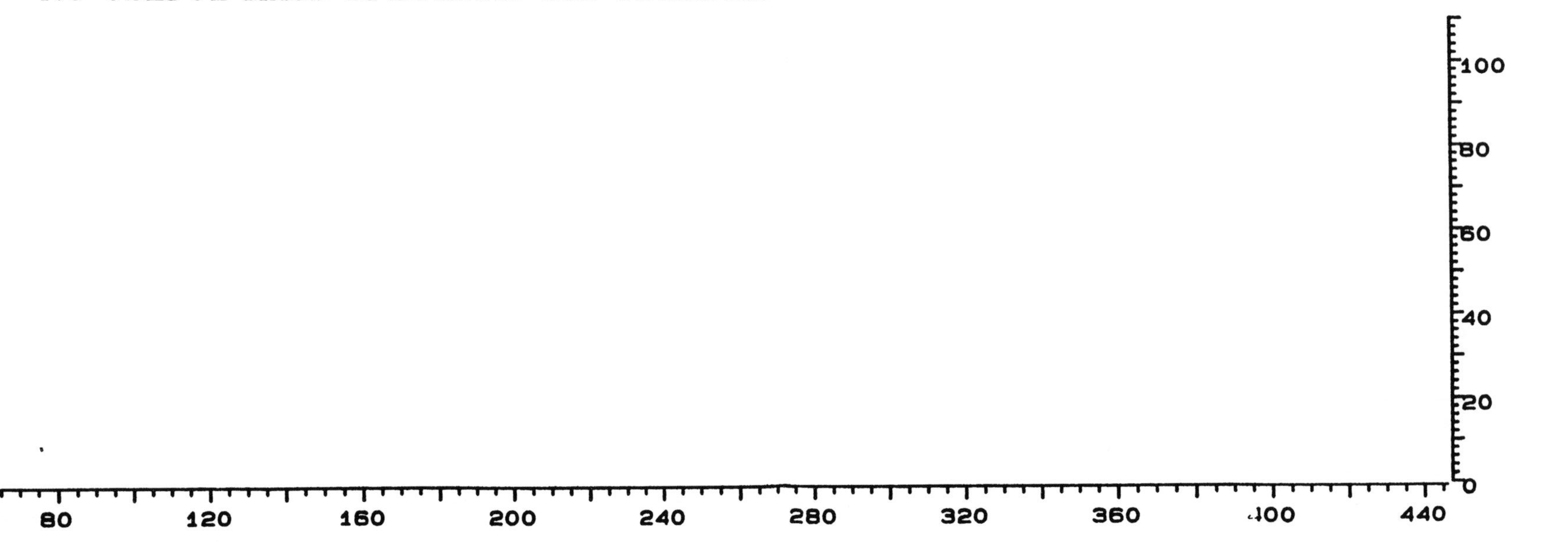


NO USEFUL MASS SPECTRUM WAS OBTAINED

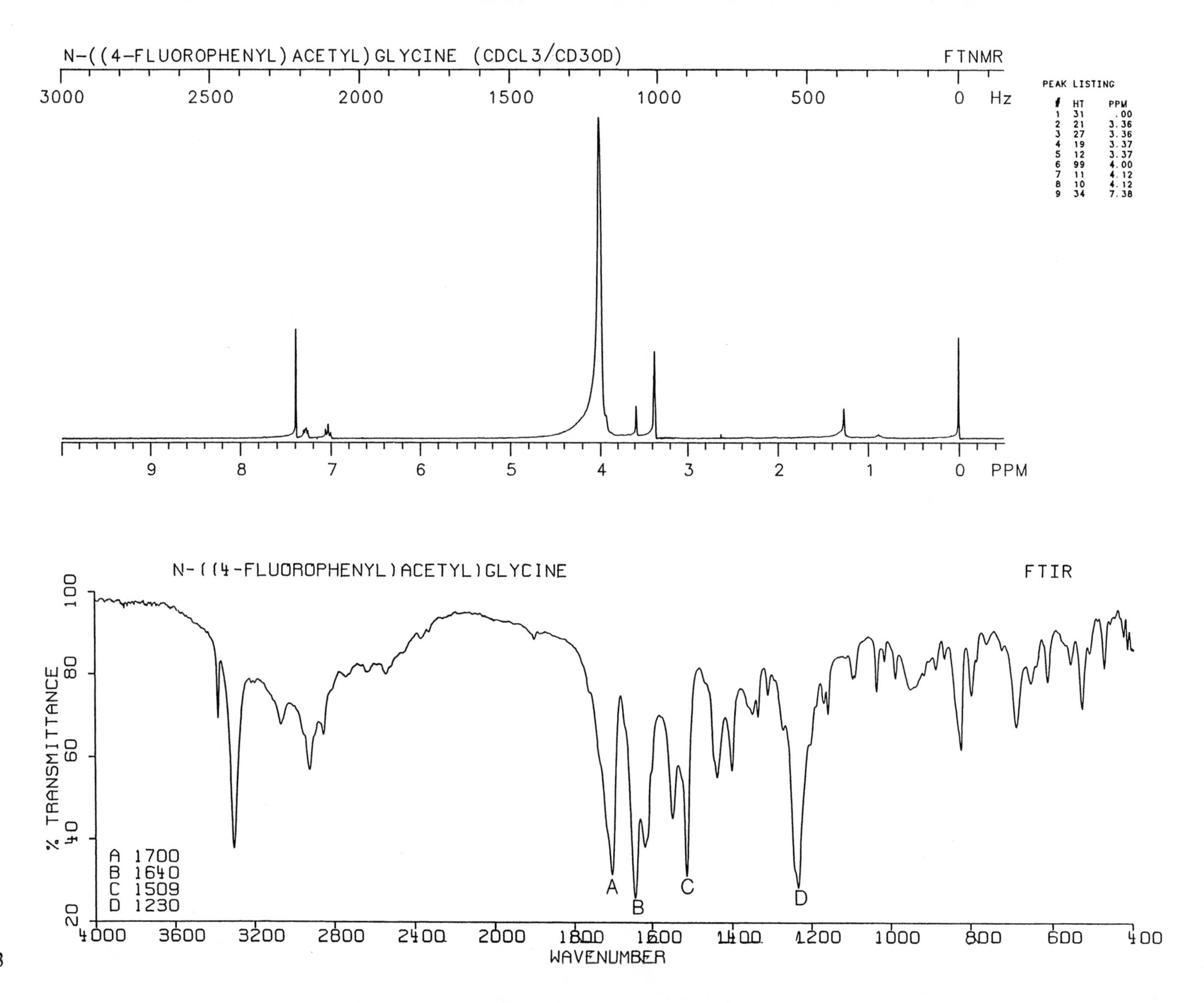
N-((4-FLUOROPHENYL)ACETYL)GLYCINE (CDCL3/CD3OD)
FTNMR
3000 2500 2000 1500 1000 500 0 Hz
9 8 7 6 5 4 3 2 1 0 PPM
PEAK LISTING
# HT PPM
1 31 .00
2 21 3.36
3 27 3.36
4 19 3.37
5 12 3.37
6 99 4.00
7 11 4.12
8 10 4.12
9 34 7.38
N-((4-FLUOROPHENYL)ACETYL)GLYCINE
FTIR
% TRANSMITTANCE
100 80 60 40 20
A 1700
B 1640
C 1509
D 1230
4000 3600 3200 2800 2400 2000 1800 1600 1400 1200 1000 800 600 400
WAVENUMBER

# FLUOROURACIL

$C_4H_3FN_2O_2$

**Molecular weight:** 130.08 (130.02)
**Synonyms:** 5-Fluoro-2,4(1H,3H)-pyrimidinedione; 2,4-dioxo-5-fluoropyrimidine
**Trade names:** Adrucil, Efudex, Fluoroplex, Fluorouracil

**Use:** Antineoplastic
**HPLC:** Si-10; 5A:95B; 4.0
**GC:**

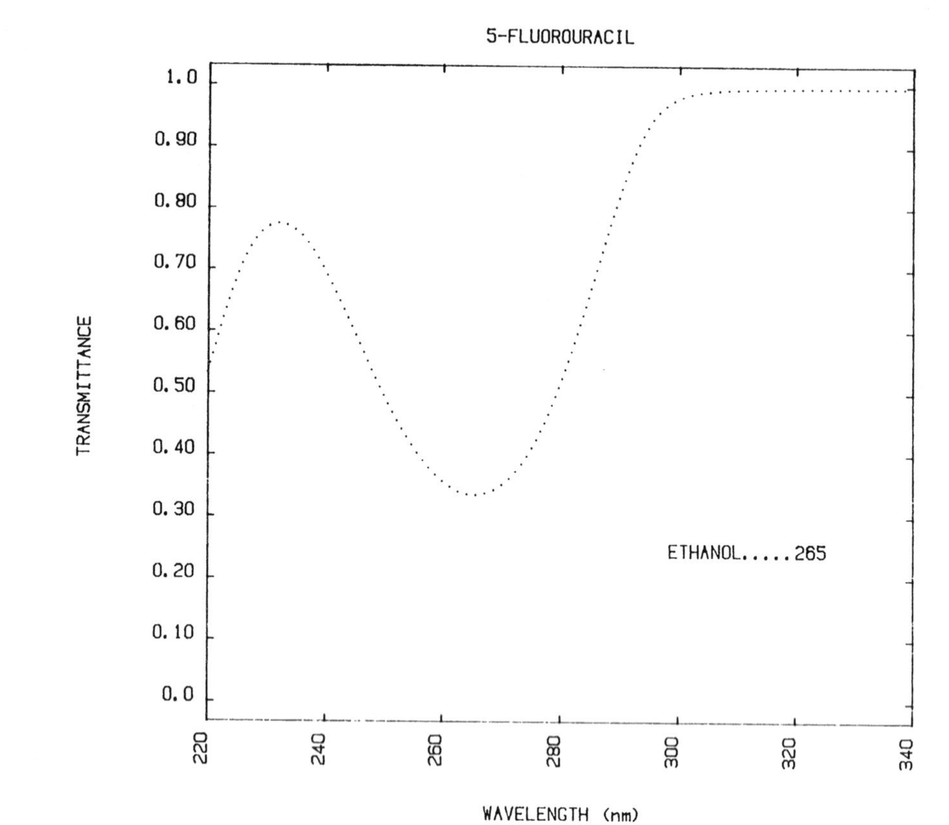


*FLUOROURACIL--DIP*

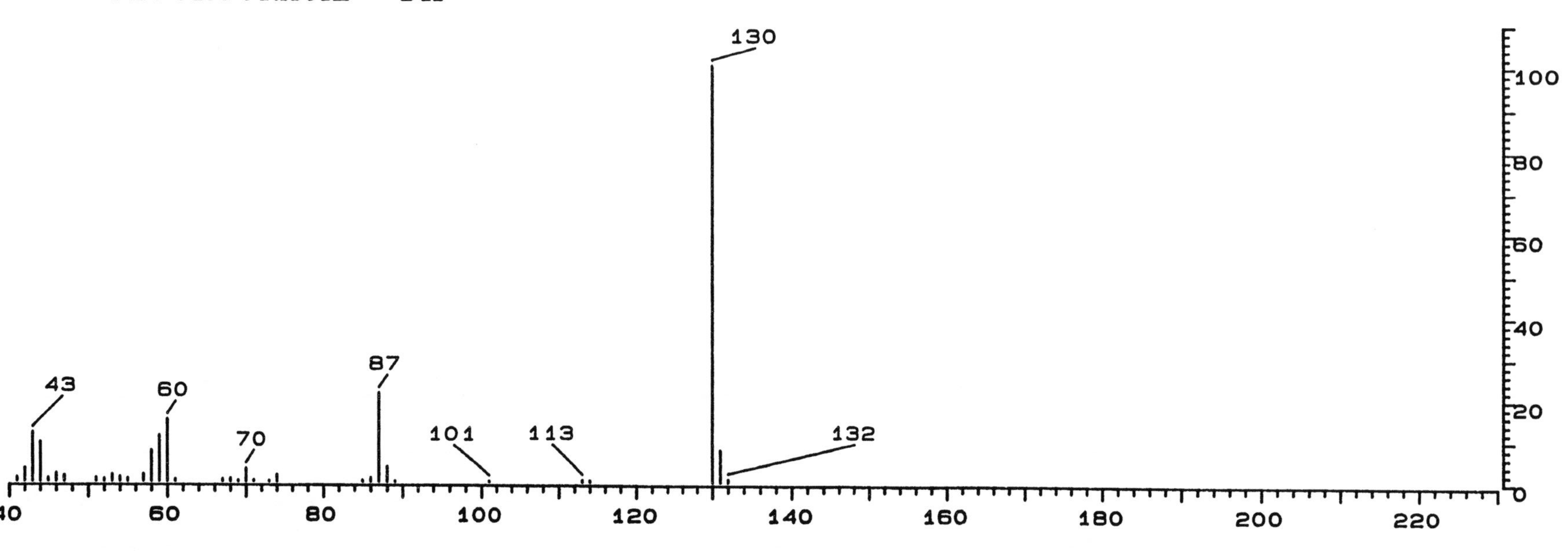

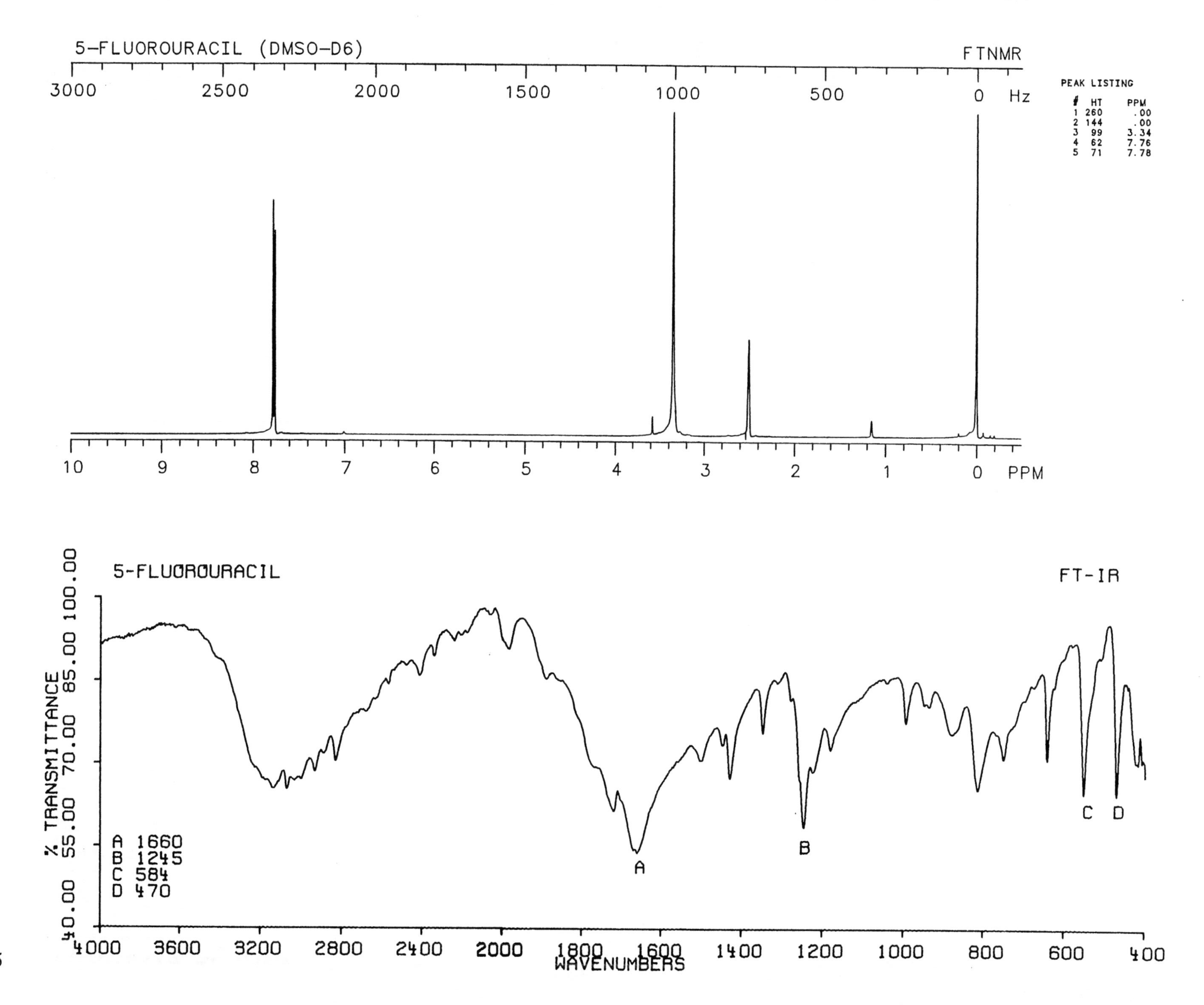

5-FLUOROURACIL (DMSO-D6)
FTNMR
Hz
PPM
PEAK LISTING
# HT PPM
1 260 .00
2 144 .00
3 99 3.34
4 62 7.76
5 71 7.78
5-FLUOROURACIL
FT-IR
% TRANSMITTANCE
WAVENUMBERS
A 1660
B 1245
C 584
D 470

# FLUOXYMESTERONE

$C_{20}H_{29}FO_3$

**Molecular weight:** 336.45 (336.21)
**Synonyms:** 9-Fluoro-11β,17β-dihydroxy-17-methylandrost-4-en-3-one

**Trade names:** Android-F, Halotestin

**Use:** Androgen
**HPLC:** Si-10; 2A:98B; 5.0
**GC:** 2954; 280°C

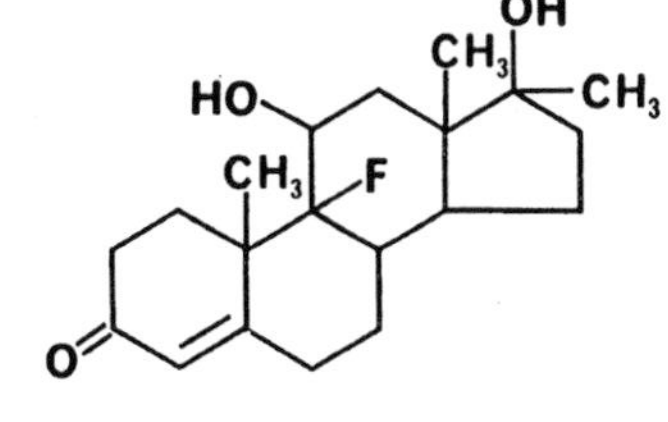


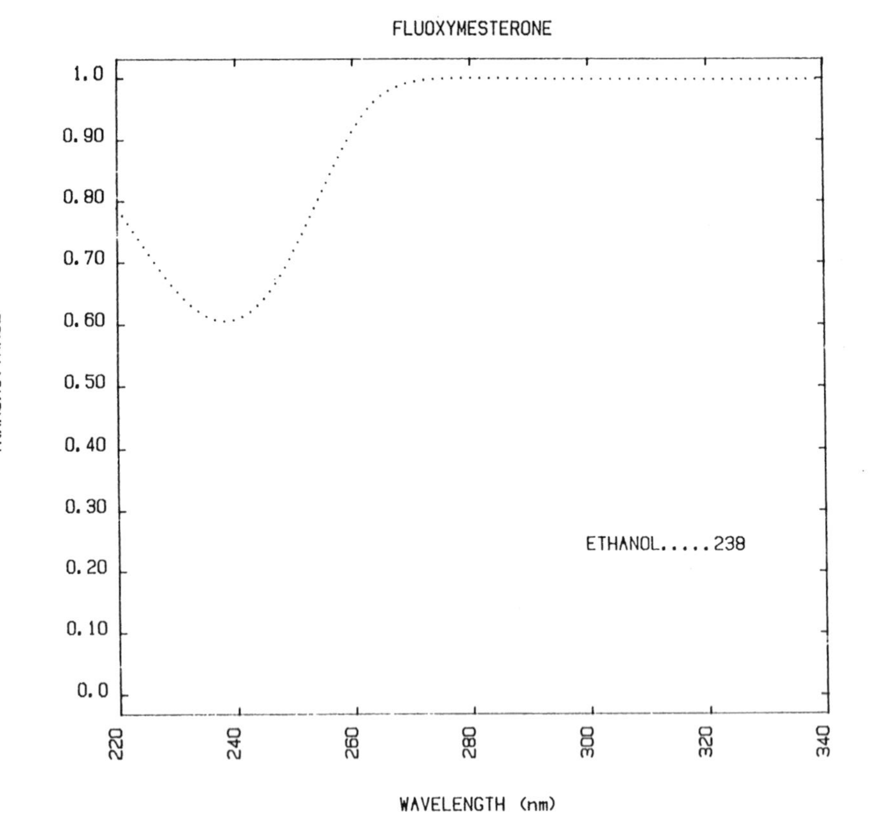


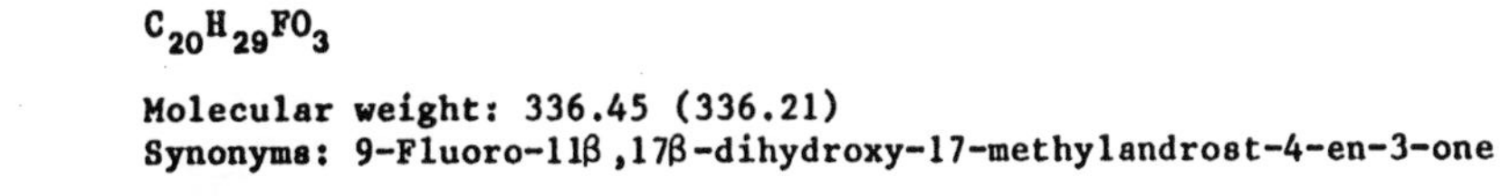


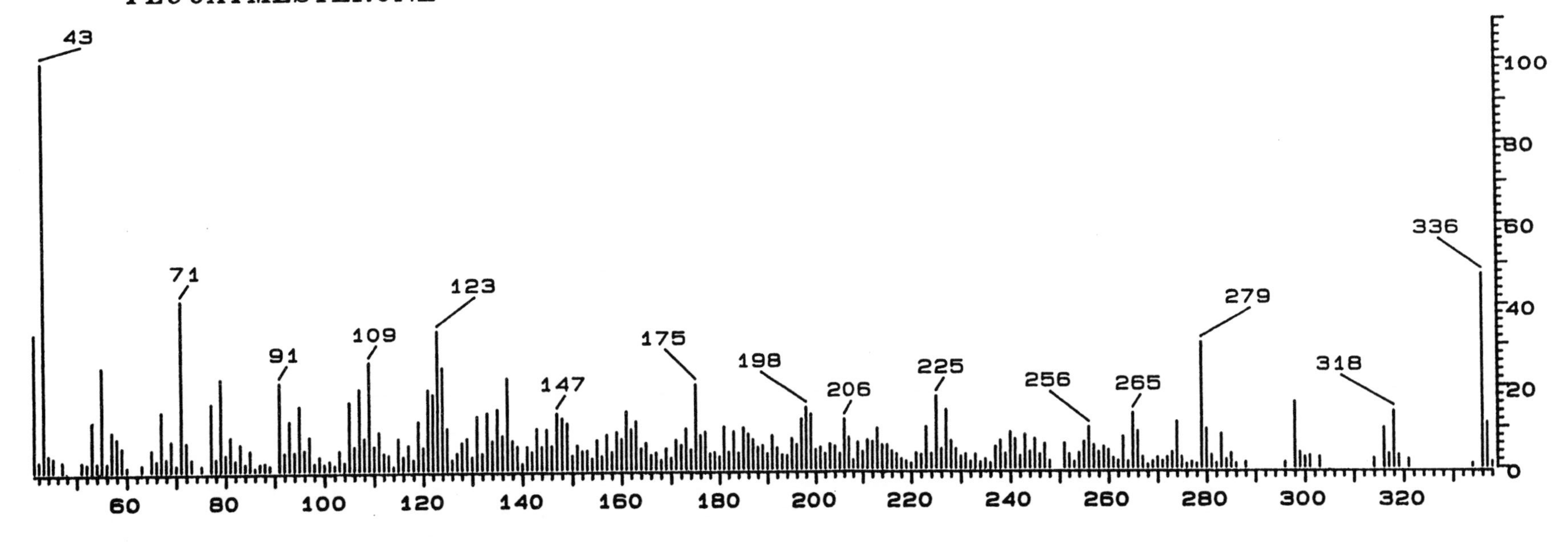

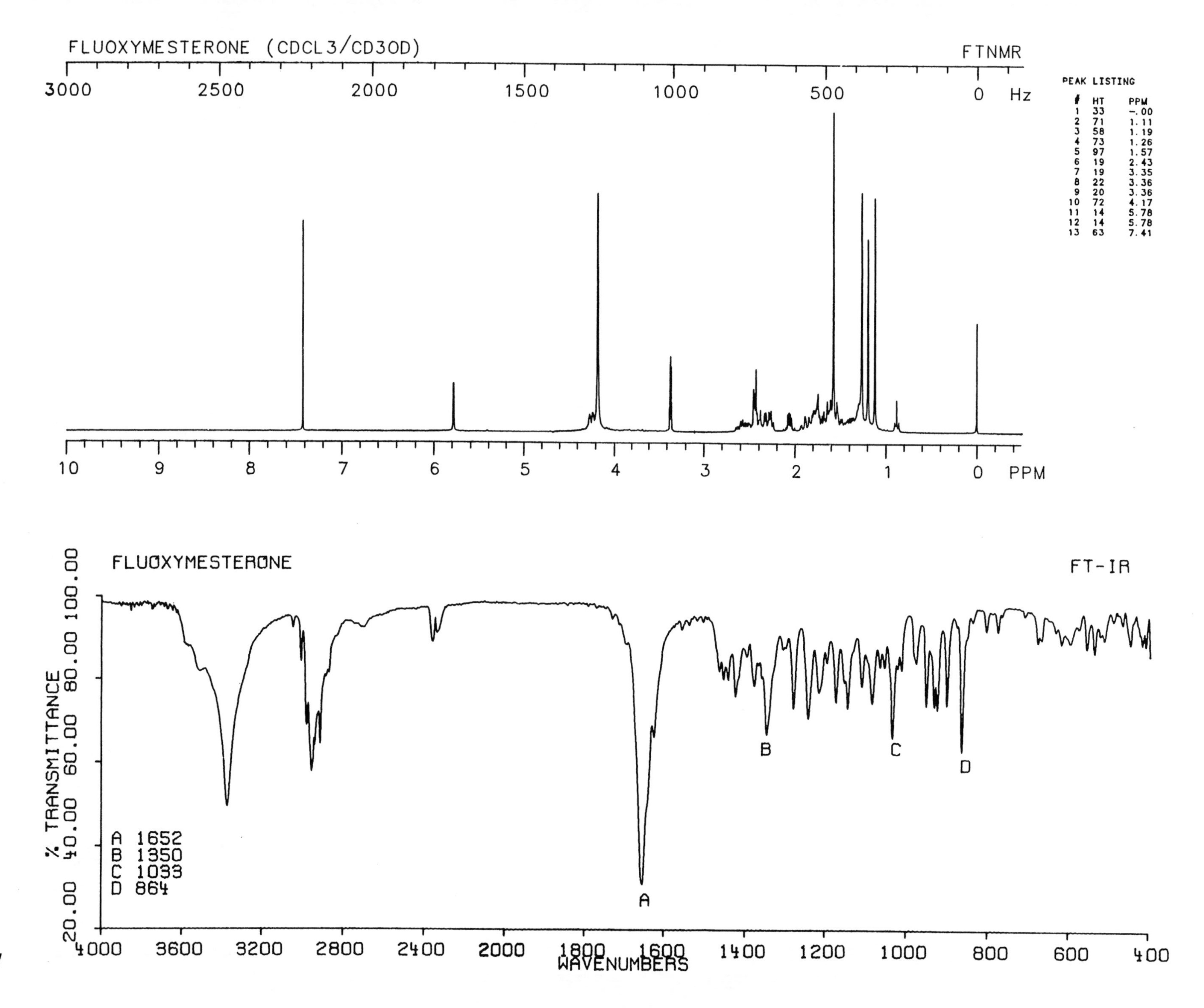
FLUOXYMESTERONE (CDCL3/CD3OD)
FTNMR
3000 2500 2000 1500 1000 500 0 Hz
PEAK LISTING
# HT PPM
1 33 -.00
2 71 1.11
3 58 1.19
4 73 1.26
5 97 1.57
6 19 2.43
7 19 3.35
8 22 3.36
9 20 3.36
10 72 4.17
11 14 5.78
12 14 5.78
13 63 7.41
10 9 8 7 6 5 4 3 2 1 0 PPM
FLUOXYMESTERONE
FT-IR
% TRANSMITTANCE
20.00 40.00 60.00 80.00 100.00
A 1652
B 1350
C 1033
D 864
A
B
C
D
4000 3600 3200 2800 2400 2000 1800 1600 1400 1200 1000 800 600 400
WAVENUMBERS

# FLUPHENAZINE

$C_{22}H_{26}F_3N_3OS$

**Molecular weight:** 437.52 (437.18)
**Synonyms:** 4-[3-[2-(Trifluoromethyl)-10H-phenothiazin-10-yl]-propyl]-1-piperazineethanol
**Trade names:** Permitil, Prolixin, Vespazine

**Use:** Tranquilizer
**HPLC:** Si-10; 20A:80B; 4.5
**GC:** 2102; 280°C

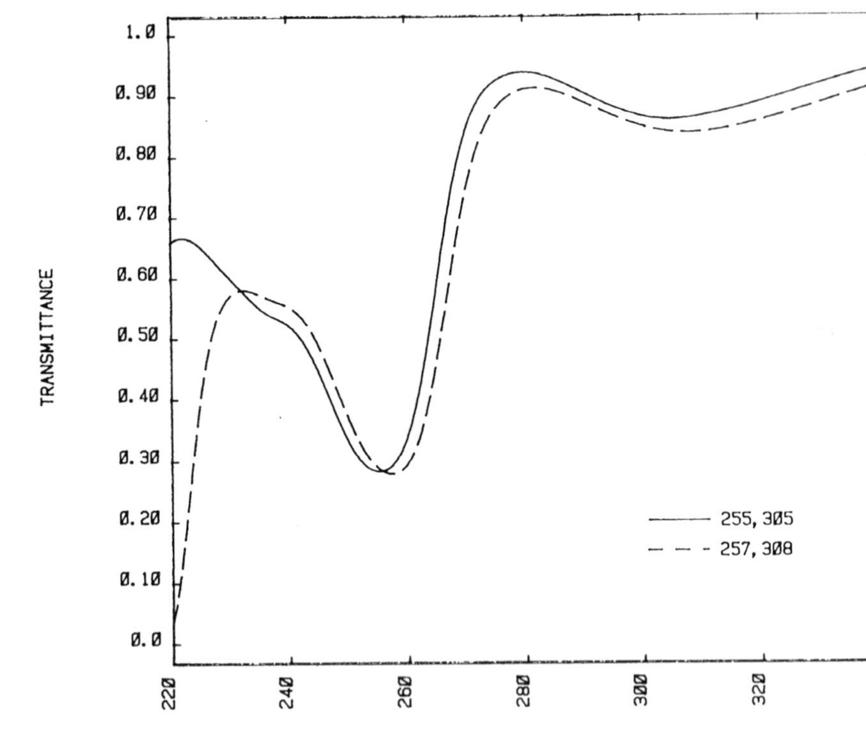


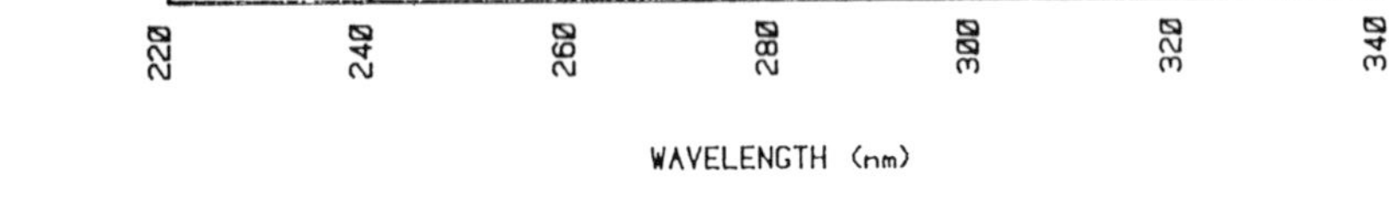

FLUPHENAZINE

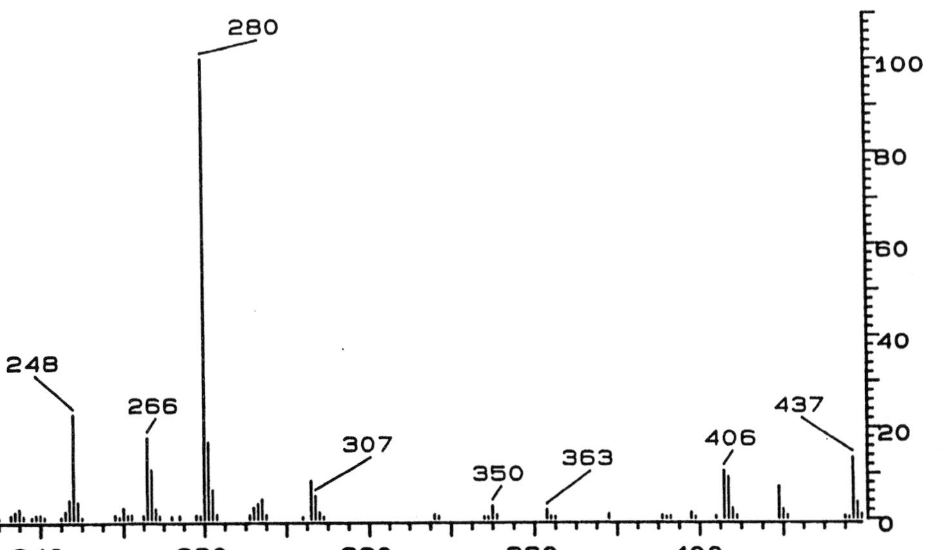


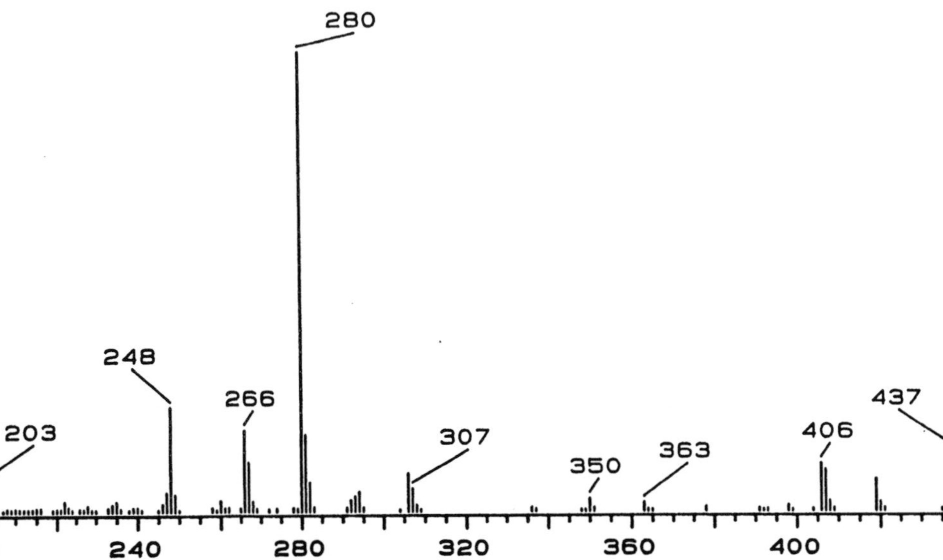

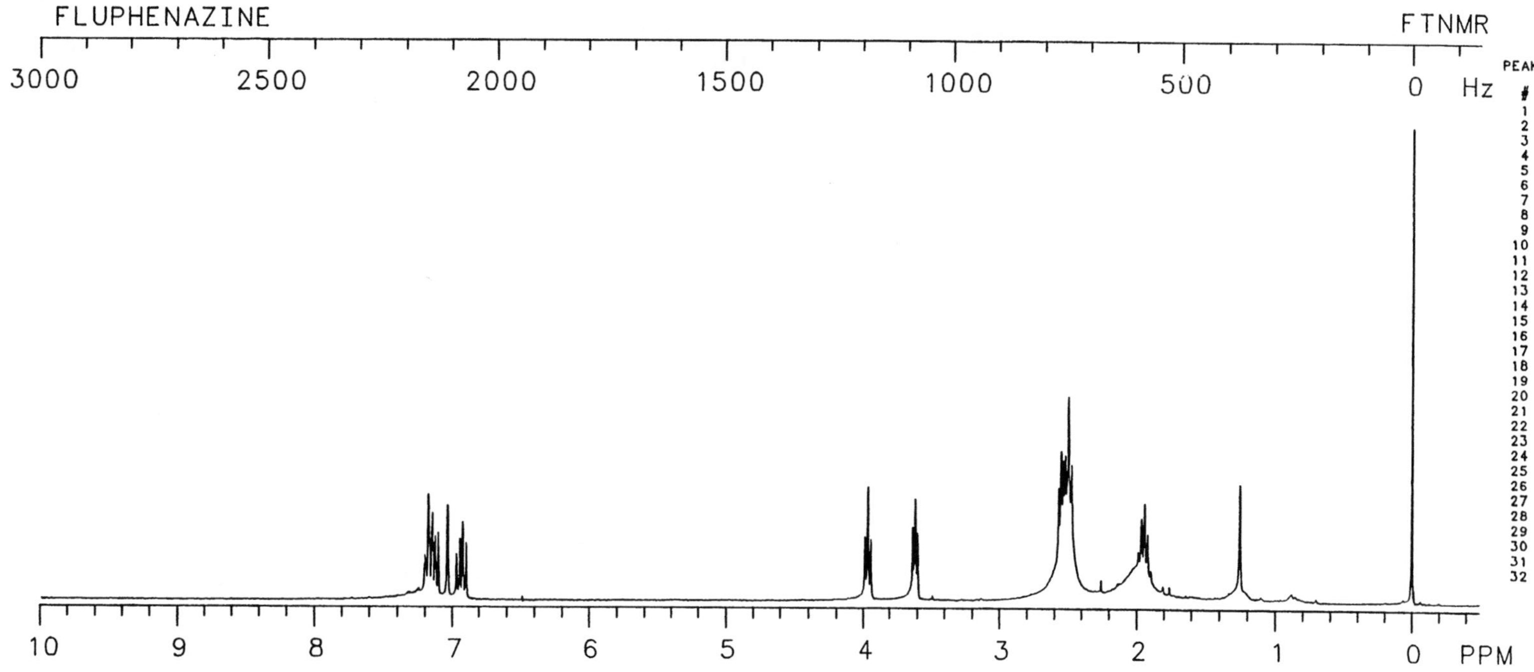

FLUPHENAZINE

FTNMR

3000 2500 2000 1500 1000 500 0 Hz

10 9 8 7 6 5 4 3 2 1 0 PPM

PEAK LISTING

| # | HT | PPM |
|---|---|---|
| 1 | 96 | .00 |
| 2 | 62 | .00 |
| 3 | 23 | 1.25 |
| 4 | 13 | 1.92 |
| 5 | 19 | 1.94 |
| 6 | 16 | 1.96 |
| 7 | 27 | 2.47 |
| 8 | 41 | 2.49 |
| 9 | 29 | 2.52 |
| 10 | 28 | 2.53 |
| 11 | 30 | 2.55 |
| 12 | 23 | 2.57 |
| 13 | 13 | 3.60 |
| 14 | 20 | 3.61 |
| 15 | 15 | 3.63 |
| 16 | 12 | 3.94 |
| 17 | 23 | 3.96 |
| 18 | 13 | 3.98 |
| 19 | 11 | 6.90 |
| 20 | 15 | 6.92 |
| 21 | 12 | 6.94 |
| 22 | 12 | 6.95 |
| 23 | 19 | 7.04 |
| 24 | 13 | 7.11 |
| 25 | 12 | 7.13 |
| 26 | 11 | 7.13 |
| 27 | 17 | 7.15 |
| 28 | 12 | 7.17 |
| 29 | 11 | 7.17 |
| 30 | 14 | 7.17 |
| 31 | 21 | 7.18 |
| 32 | 99 | 7.26 |
| 33 | 64 | 7.26 |

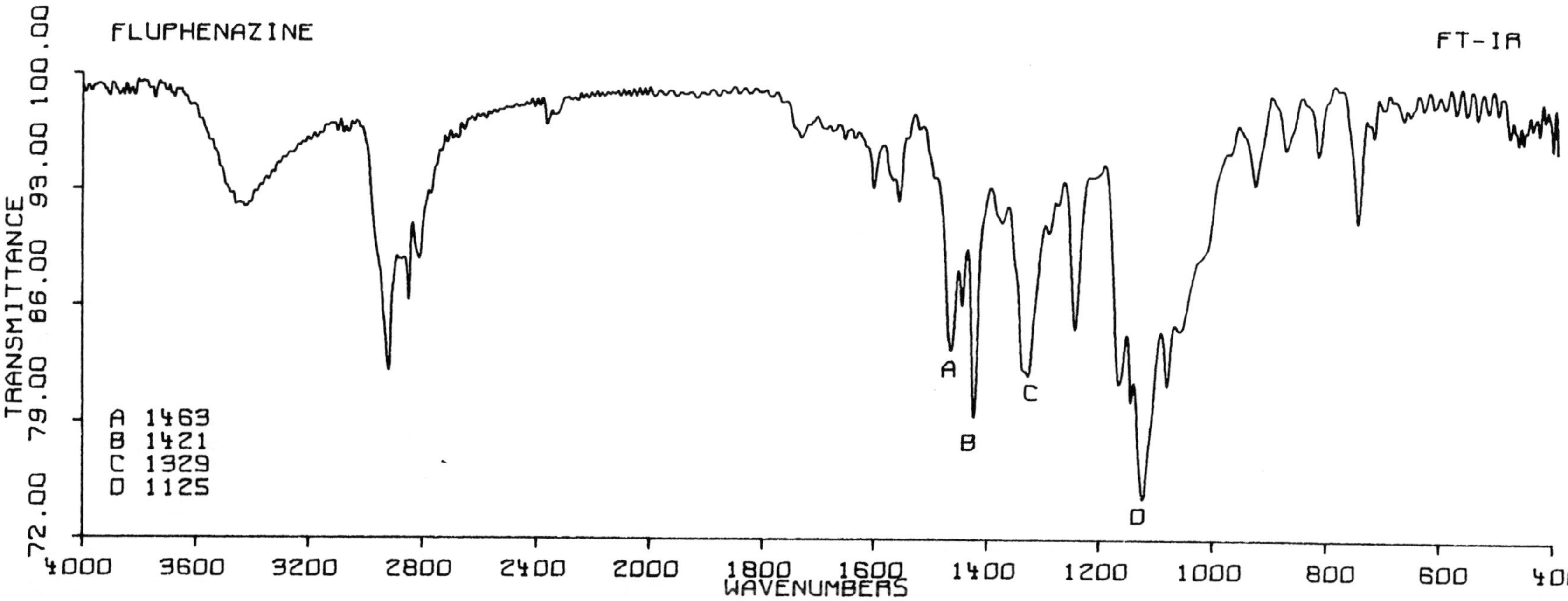

FLUPHENAZINE

FT-IR

TRANSMITTANCE

72.00 79.00 86.00 93.00 100.00

4000 3600 3200 2800 2400 2000 1800 1600 1400 1200 1000 800 600 400

WAVENUMBERS

A 1463
B 1421
C 1329
D 1125

# FLUPREDNISOLONE

$C_{21}H_{27}FO_5$

**Molecular weight:** 378.44 (378.18)
**Synonyms:** 6α-Fluoro-11β,17α,21-trihydroxypregna-1,4-diene-3,20-dione; 6α-fluoroprednisolone
**Trade names:** Alphadrol, Etadrol

**Use:** Anti-inflammatory
**HPLC:**
**GC:**

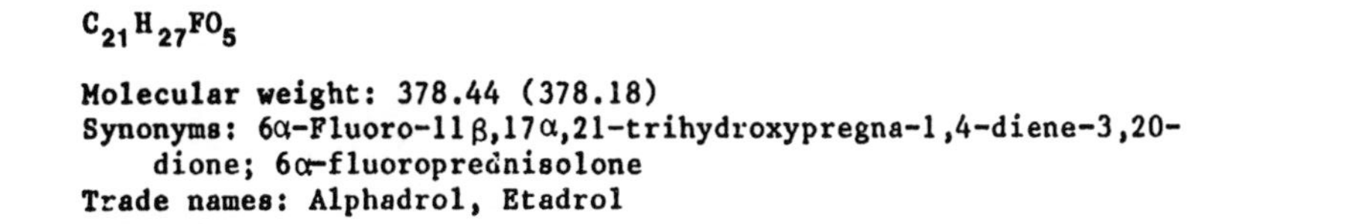


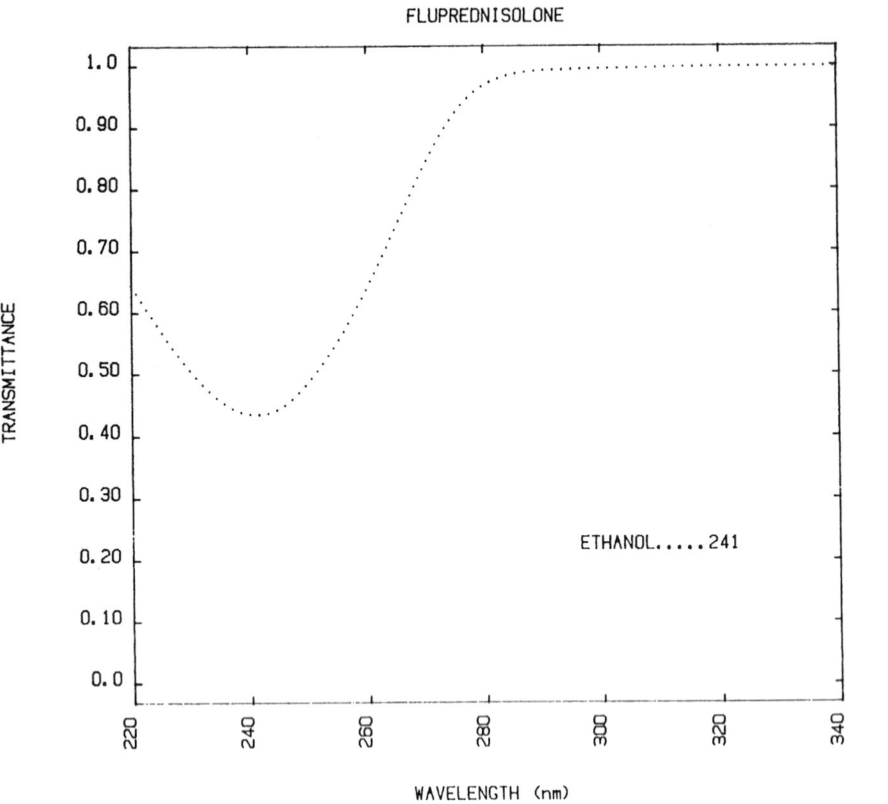


FLUPREDNISOLONE--DIP

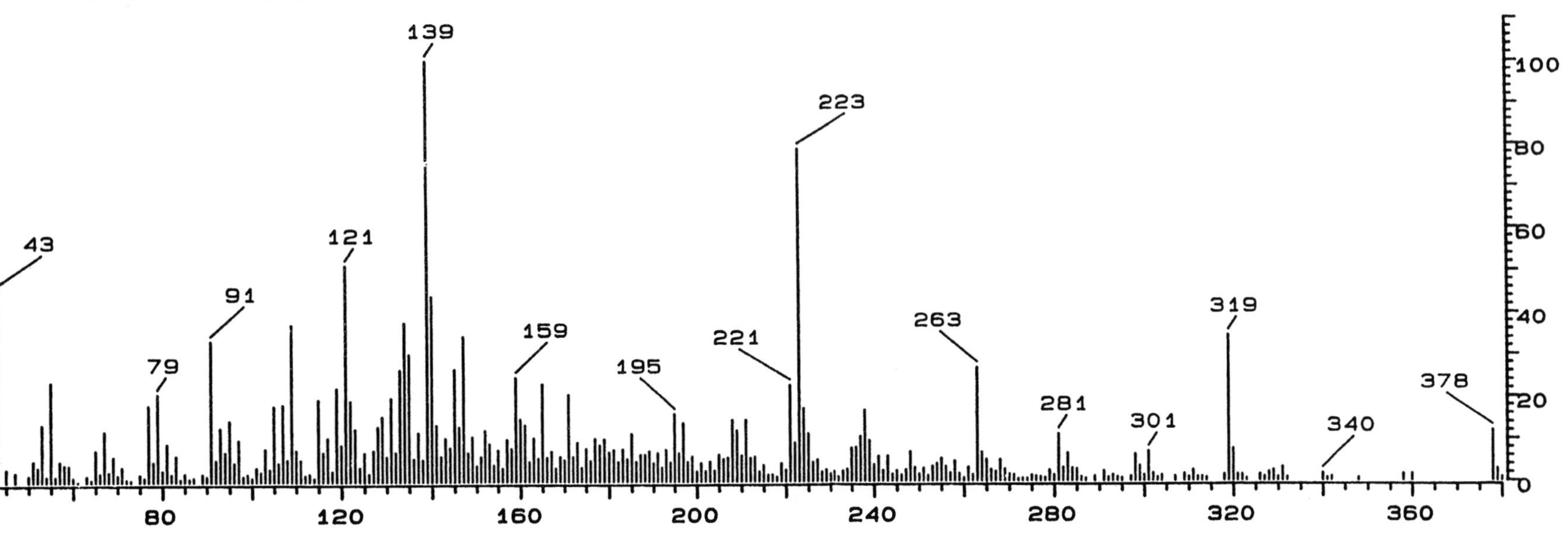

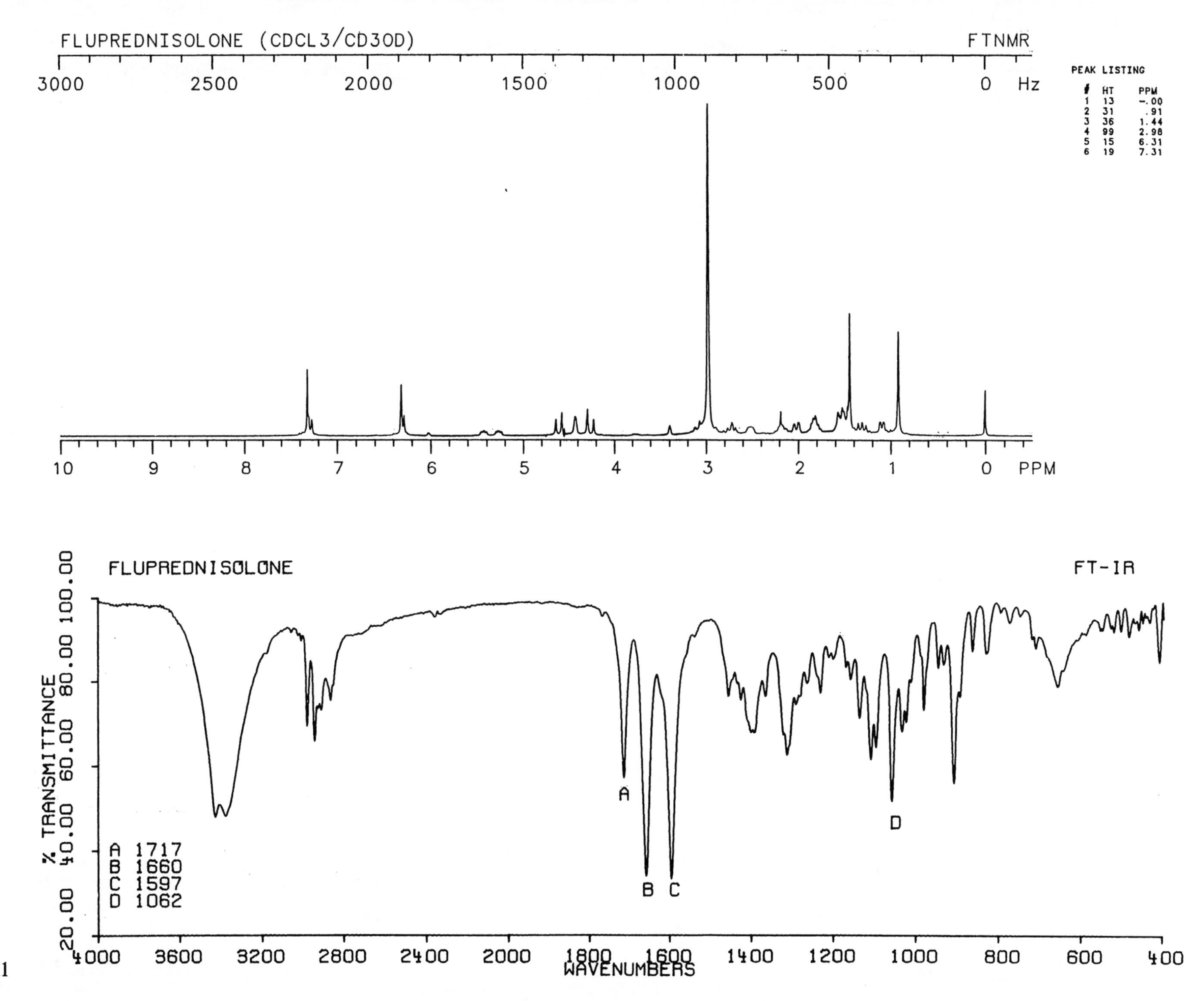

FLUPREDNISOLONE (CDCL3/CD3OD)
FTNMR
3000 2500 2000 1500 1000 500 0 Hz
PEAK LISTING
# HT PPM
1 13 -.00
2 31 .91
3 36 1.44
4 99 2.98
5 15 6.31
6 19 7.31
10 9 8 7 6 5 4 3 2 1 0 PPM
FLUPREDNISOLONE
FT-IR
% TRANSMITTANCE
20.00 40.00 60.00 80.00 100.00
A 1717
B 1660
C 1597
D 1062
4000 3600 3200 2800 2400 2000 1800 1600 1400 1200 1000 800 600 400
WAVENUMBERS

# FLURANDRENOLIDE

$C_{24}H_{33}FO_6$

**Molecular weight:** 436.52 (436.23)
**Synonyms:** 6α-Fluoro-11β,21-dihydroxy-16α,17-[(1-methylethylidene)-bis(oxy)]pregn-4-ene-3,20-dione; flurandrenolone
**Trade names:** Cordran

**Use:** Anti-inflammatory
**HPLC:** Si-10; 1A:99B; 5.8
**GC:**

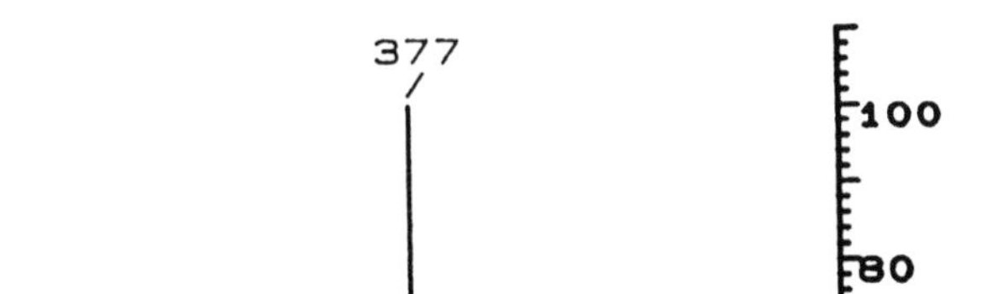

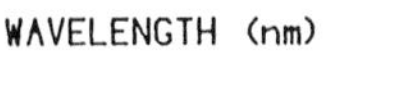

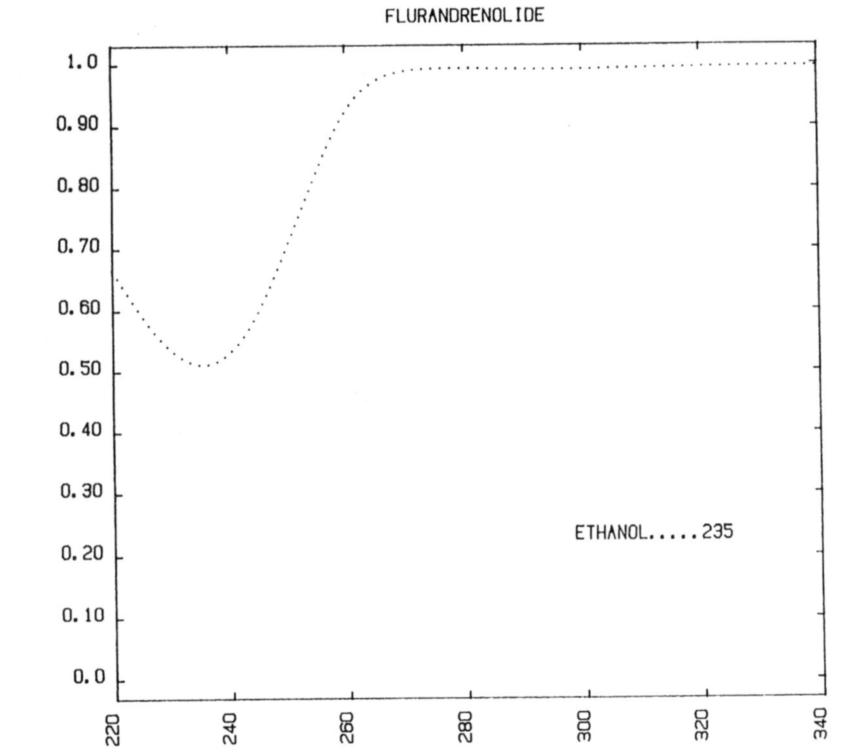

FLURANDRENOLIDE -- DIP

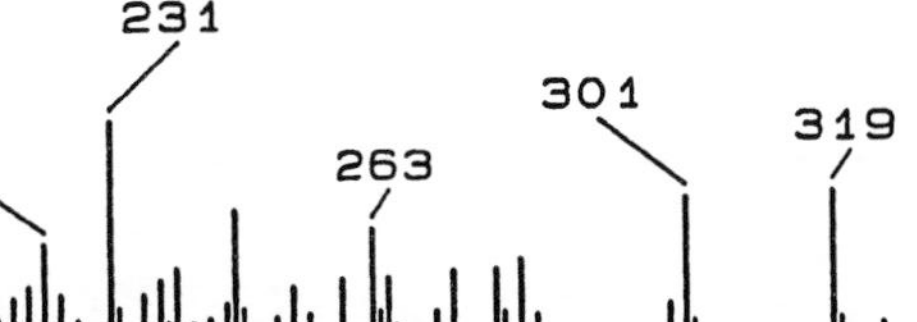

FLURANDRENOLIDE

FTNMR

PEAK LISTING

| # | HT | PPM |
|---|---|---|
| 1 | 33 | .00 |
| 2 | 69 | .86 |
| 3 | 68 | 1.15 |
| 4 | 12 | 1.25 |
| 5 | 13 | 1.26 |
| 6 | 10 | 1.34 |
| 7 | 99 | 1.42 |
| 8 | 78 | 1.45 |
| 9 | 43 | 1.64 |
| 10 | 16 | 1.65 |
| 11 | 19 | 1.66 |
| 12 | 11 | 2.41 |
| 13 | 23 | 3.03 |
| 14 | 11 | 4.20 |
| 15 | 11 | 4.63 |
| 16 | 12 | 4.65 |
| 17 | 16 | 5.06 |
| 18 | 13 | 5.08 |
| 19 | 14 | 6.03 |
| 20 | 15 | 6.03 |
| 21 | 14 | 6.03 |
| 22 | 72 | 7.26 |

3000 2500 2000 1500 1000 500 0 Hz

10 9 8 7 6 5 4 3 2 1 0 PPM

FLURANDRENOLIDE

FT-IR

A 1710
B 1682
C 1090
D 1047

% TRANSMITTANCE

20.00 40.00 60.00 80.00 100.00

4000 3600 3200 2800 2400 2000 1800 1600 1400 1200 1000 800 600 400

WAVENUMBERS

# FLURAZEPAM

$C_{21}H_{23}ClFN_3O$

Molecular weight: 387.88 (387.15)
Synonyms: 7-Chloro-1-[2-(diethylamino)ethyl]-5-(o-fluorophenyl)-1,3-dihydro-2H-1,4-benzodiazepin-2-one
Trade names: Dalmane

Use: Tranquilizer
HPLC: Si-10; 2A:98B; 7.5
GC: 2800; 250°C

$CH_2-CH_2-N(C_2H_5)_2$
N
O
N
Cl
F

FLURAZEPAM
TRANSMITTANCE
1.Ø
Ø.9Ø
Ø.8Ø
Ø.7Ø
Ø.6Ø
Ø.5Ø
Ø.4Ø
Ø.3Ø
Ø.2Ø
Ø.1Ø
Ø.Ø
236, 277
22Ø
24Ø
26Ø
28Ø
3ØØ
32Ø
34Ø
WAVELENGTH (nm)

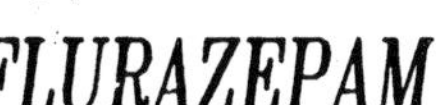


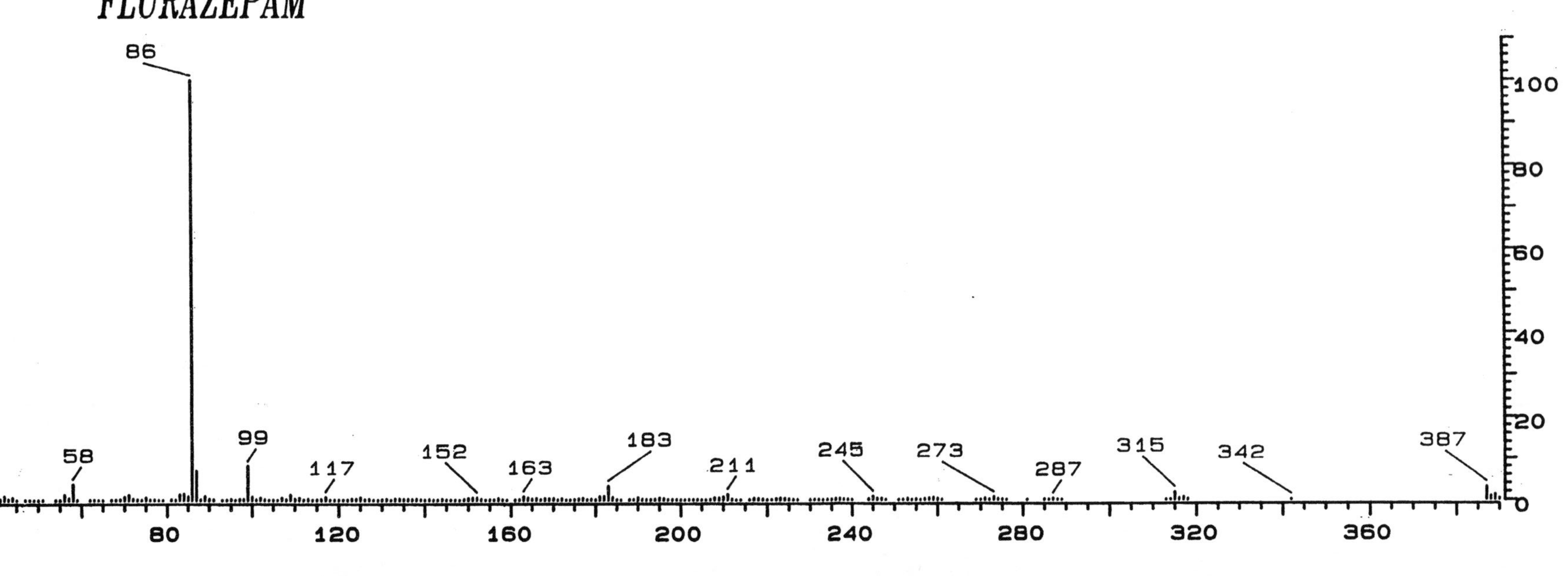

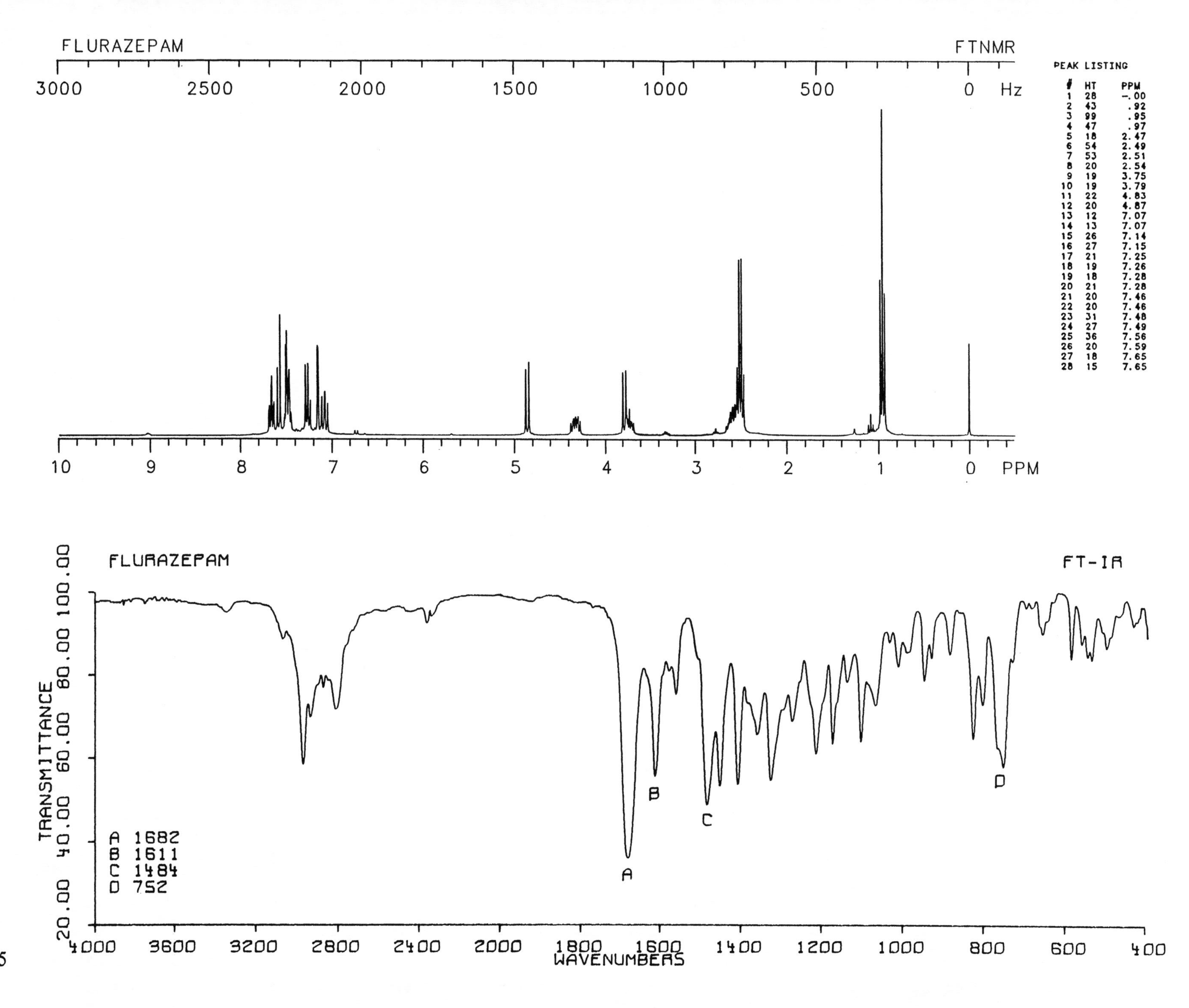
FLURAZEPAM
FTNMR
3000 2500 2000 1500 1000 500 0 Hz
10 9 8 7 6 5 4 3 2 1 0 PPM
PEAK LISTING
# HT PPM
1 28 -.00
2 43 .92
3 99 .95
4 47 .97
5 18 2.47
6 54 2.49
7 53 2.51
8 20 2.54
9 19 3.75
10 19 3.79
11 22 4.83
12 20 4.87
13 12 7.07
14 13 7.07
15 26 7.14
16 27 7.15
17 21 7.25
18 19 7.26
19 18 7.28
20 21 7.28
21 20 7.46
22 20 7.46
23 31 7.48
24 27 7.49
25 36 7.56
26 20 7.59
27 18 7.65
28 15 7.65
FLURAZEPAM
FT-IR
TRANSMITTANCE
20.00 40.00 60.00 80.00 100.00
4000 3600 3200 2800 2400 2000 1800 1600 1400 1200 1000 800 600 400
WAVENUMBERS
A 1682
B 1611
C 1484
D 752
A
B
C
D

# FLUSPIRILENE

$C_{29}H_{31}F_2N_3O$

Molecular weight: 475.59 (475.24)
Synonyms: 8-[4,4-Bis(p-Fluorophenyl)butyl]-1-phenyl-1,3,8-triazaspiro[4.5]decan-4-one
Trade names: Imap, Redeptin

Use: Antipsychotic
HPLC: Si-10; 2A:98B; 4.0
GC:

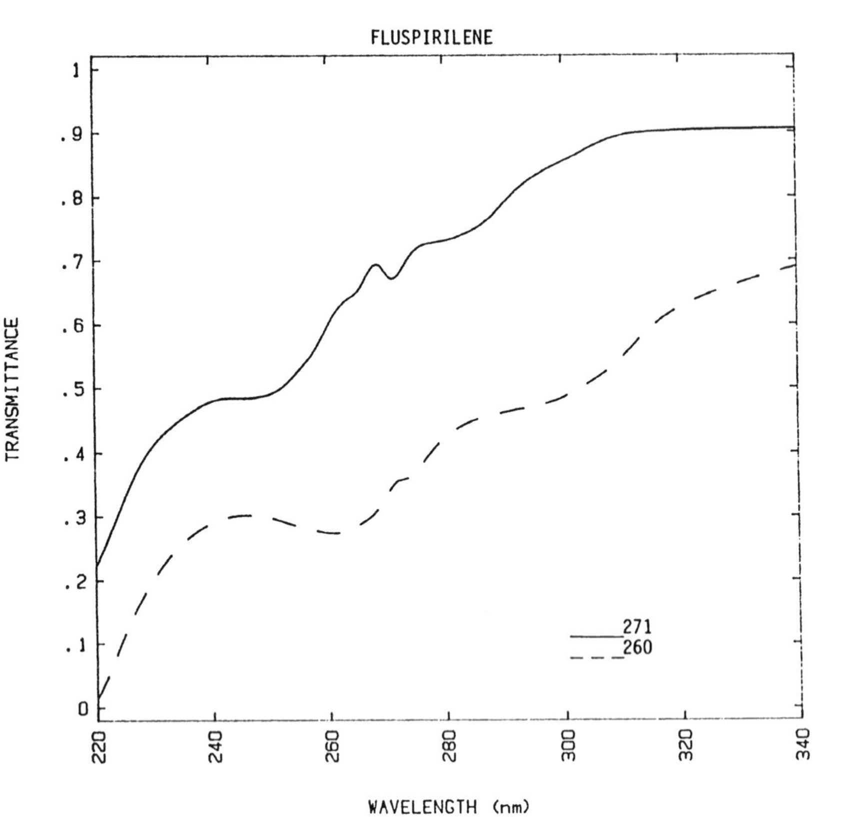

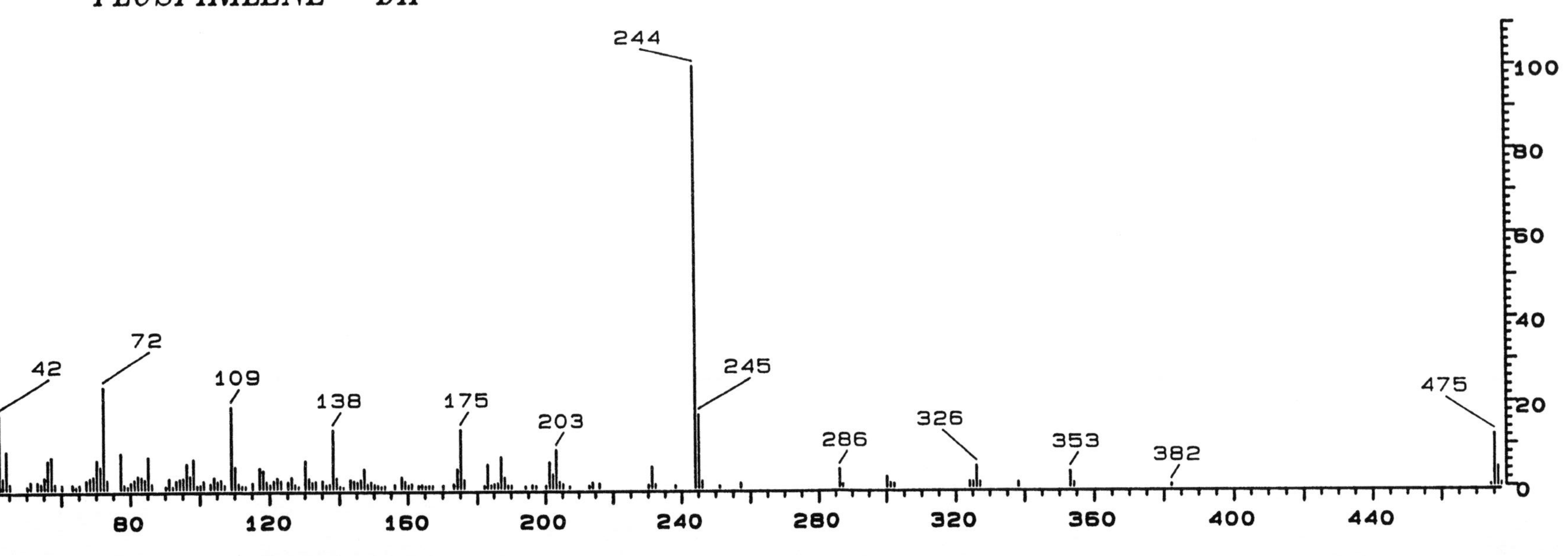

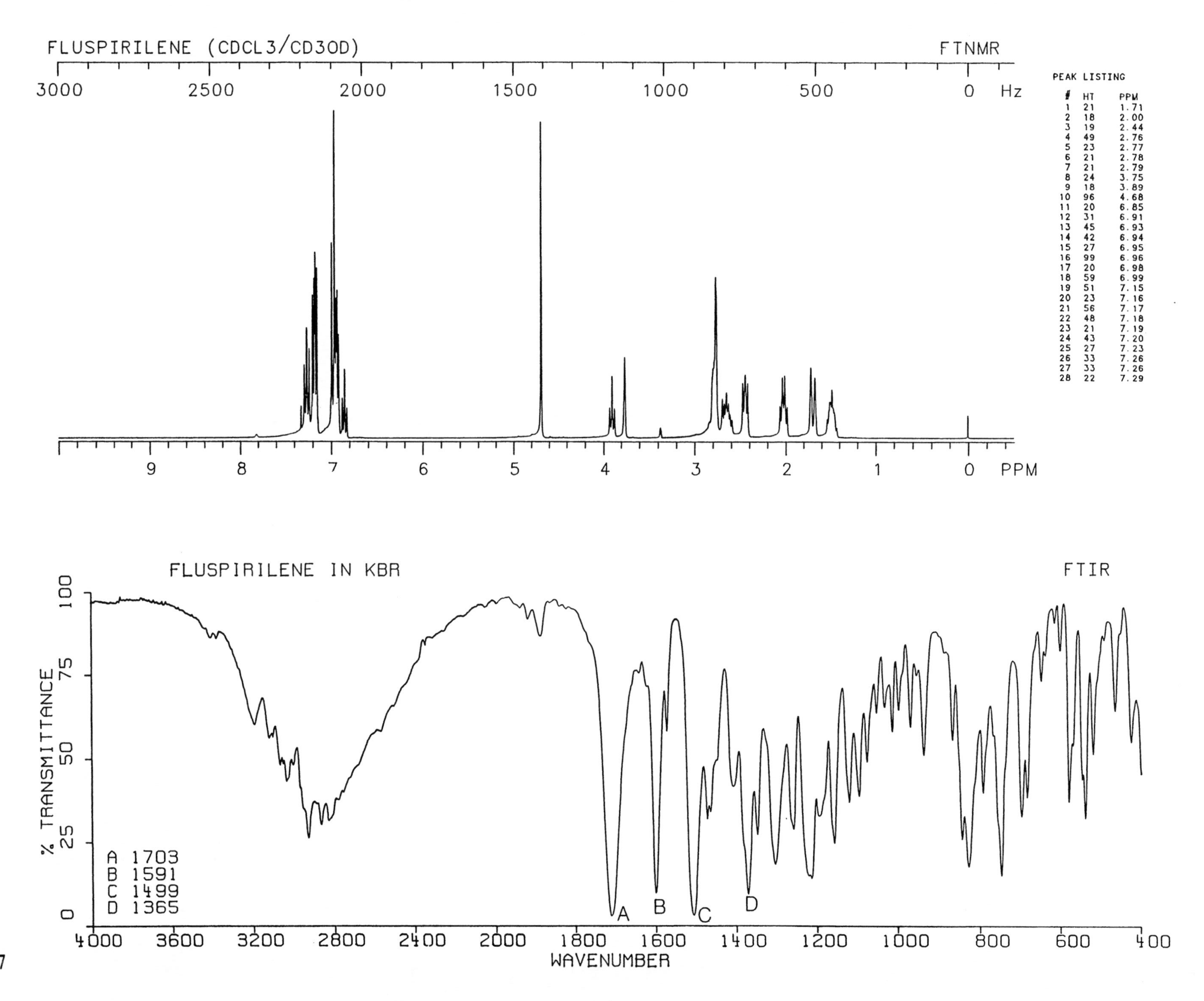

FLUSPIRILENE (CDCL3/CD3OD)
FTNMR
3000 2500 2000 1500 1000 500 0 Hz
9 8 7 6 5 4 3 2 1 0 PPM
PEAK LISTING
# HT PPM
1 21 1.71
2 18 2.00
3 19 2.44
4 49 2.76
5 23 2.77
6 21 2.78
7 21 2.79
8 24 3.75
9 18 3.89
10 96 4.68
11 20 6.85
12 31 6.91
13 45 6.93
14 42 6.94
15 27 6.95
16 99 6.96
17 20 6.98
18 59 6.99
19 51 7.15
20 23 7.16
21 56 7.17
22 48 7.18
23 21 7.19
24 43 7.20
25 27 7.23
26 33 7.26
27 33 7.26
28 22 7.29
FLUSPIRILENE IN KBR
FTIR
100 75 50 25 0
% TRANSMITTANCE
A 1703
B 1591
C 1499
D 1365
A B C D
4000 3600 3200 2800 2400 2000 1800 1600 1400 1200 1000 800 600 400
WAVENUMBER

# FOLIC ACID

$C_{19}H_{19}N_7O_6$

**Molecular weight:** 441.40 (441.14)
**Synonyms:** N-[4-[[(2-Amino-1,4-dihydro-4-oxo-6-pteridinyl)methyl]amino]-benzoyl]-L-glutamic acid; folacin; pteroylglutamic acid; folinsyre
**Trade names:** Al-Vite, Cefol Filmtab, Cevi-Fer, Dayalets, Eldec, Eldercaps, Enivro-Stress, Fero-Folic, Filibon, Folic Acid, Fulvite, Hemocyte-F, Hemo-Vite, Iberet-Folic, Ircon-Fa, Iromin-G, Mega-B, Megadose, Mevanin-C, Nifenix-150, Nu-Iron, Orabex-Tf, Pramet FA, Pramilet-FA, Prenate, Pronemia, Stuartnatal, Trinsicon, Viconforte, Via-Bec, VitaFol, Zenate, Zincvit, Protection III Multi-Vitamin
**Use:** Vitamin
**HPLC:**
**GC:**

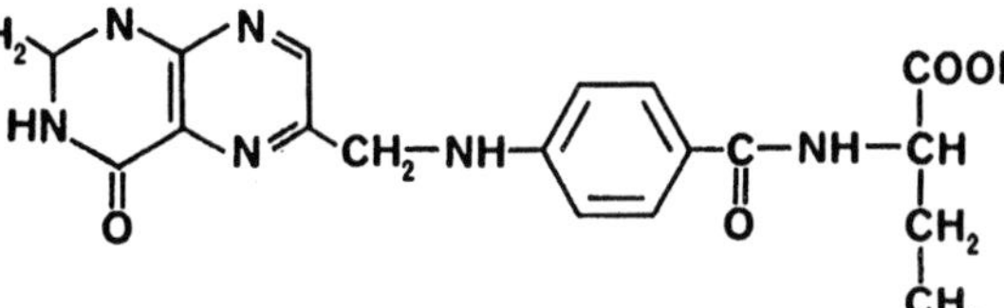


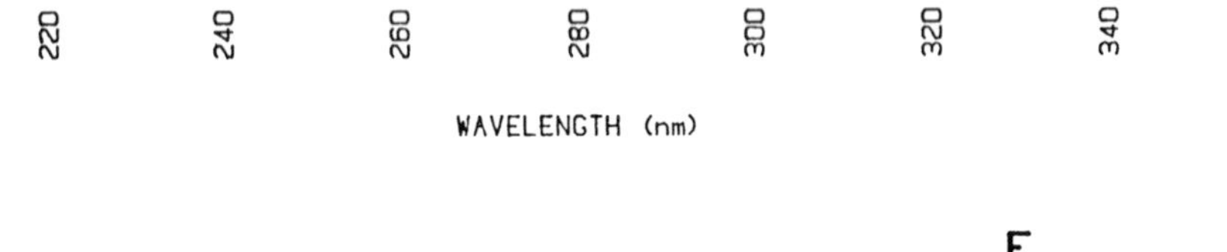


NO USEFUL MASS SPECTRUM WAS OBTAINED

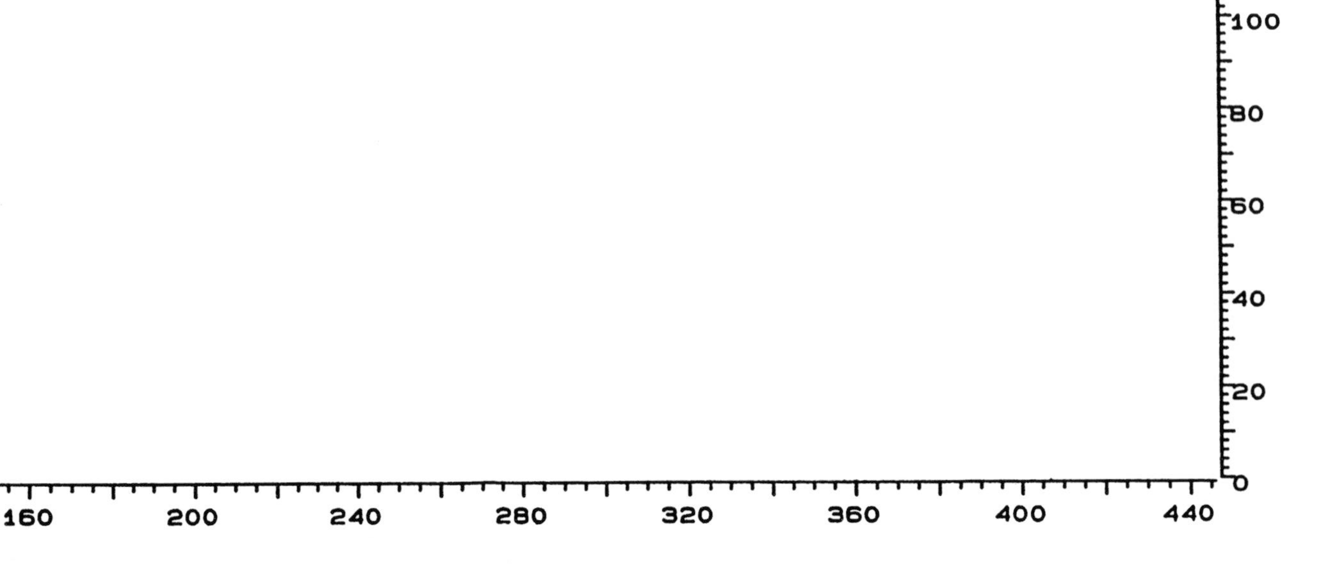

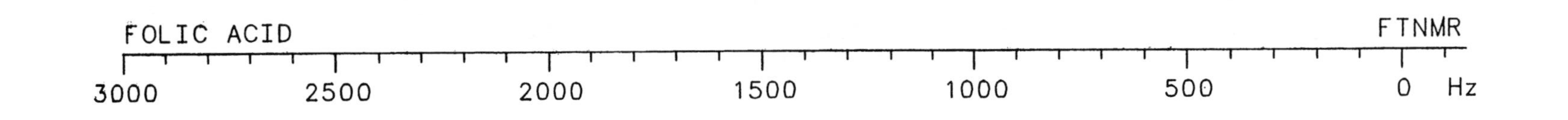


INSUFFICIENT SOLUBILITY

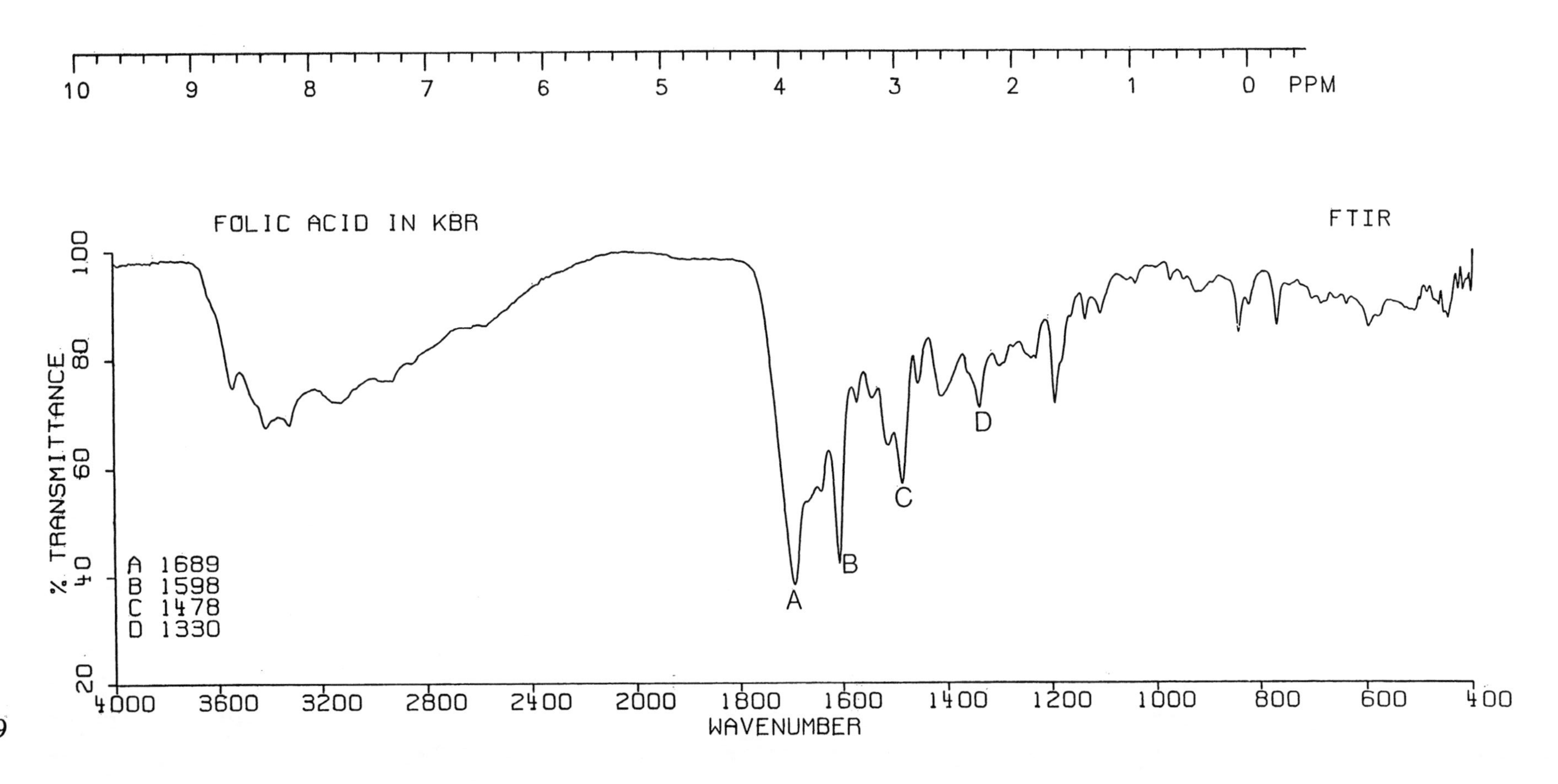

# FOLINIC ACID

$C_{20}H_{23}N_7O_7$

Molecular weight: 473.44 (473.17)
Synonyms: N-[4-[[(2-Amino-5-formyl-1,4,5,6,7,8-hexahydro-4-oxo-6-pteridin yl)methyl]amino]benzoyl]-L-glutamic acid; citrovorum; 5-formyl-5,6,7-8-tetrahydrofolic acid
Trade names: Calcium Folinate, Leucovorin

Use: Antidote for folic acid antagonist
HPLC:
GC:

FOLINIC ACID

TRANSMITTANCE

WAVELENGTH (nm)

286
281

*NO USEFUL MASS SPECTRUM WAS OBTAINED*

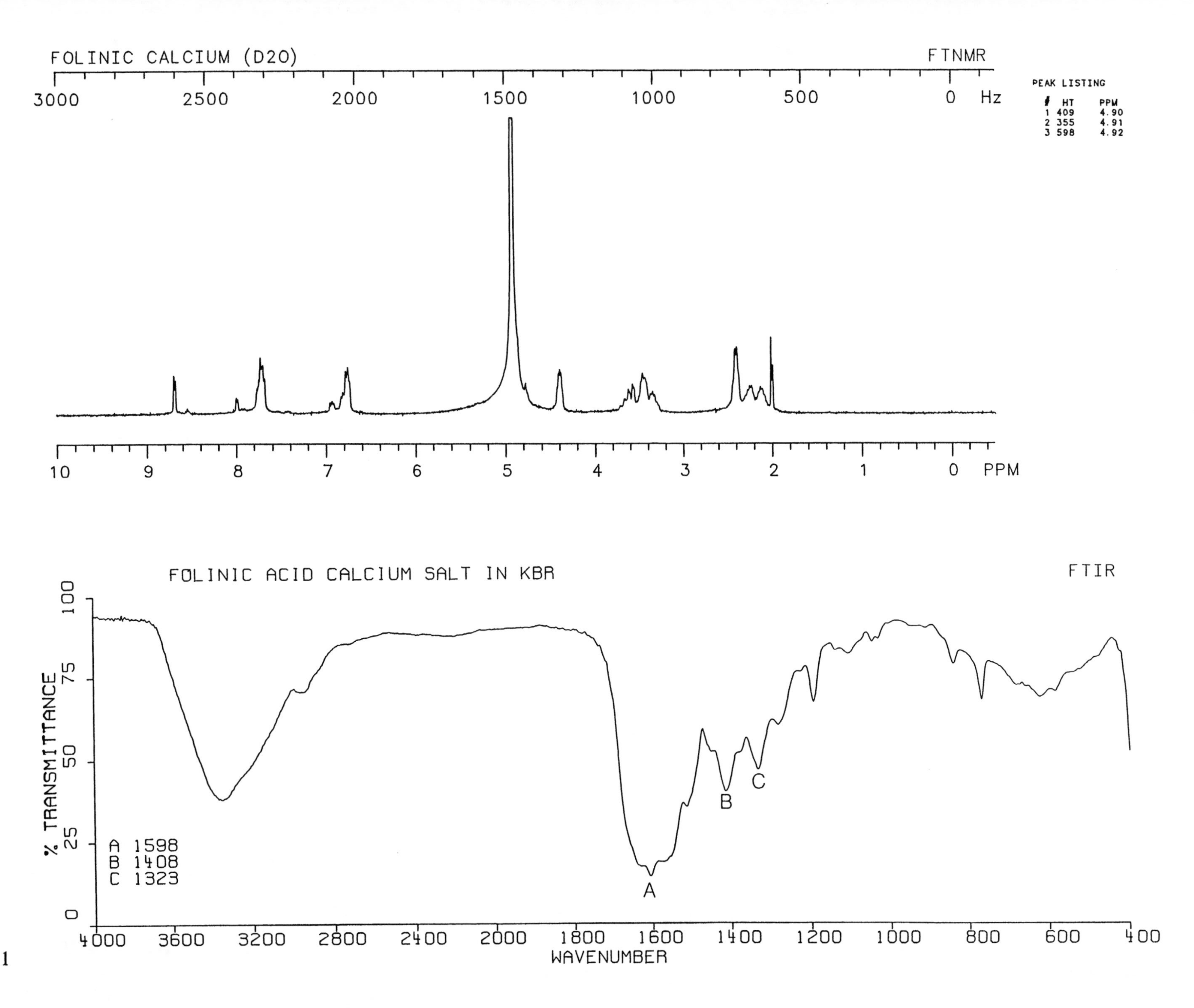
FOLINIC CALCIUM (D2O)
FTNMR
PEAK LISTING
# HT PPM
1 409 4.90
2 355 4.91
3 598 4.92
Hz
PPM
FOLINIC ACID CALCIUM SALT IN KBR
FTIR
% TRANSMITTANCE
WAVENUMBER
A 1598
B 1408
C 1323

# FORMALDEHYDE

$CH_2O$

Molecular weight: 30.03 (30.01)
Synonyms: Methanal; oxomethane; oxymethylene; methylene oxide; formic aldehyde; methyl aldehyde
Trade names: Pedi-DriFoot powder

Use: Preservative, disinfectant
HPLC:
GC:

$O{=}CH_2$

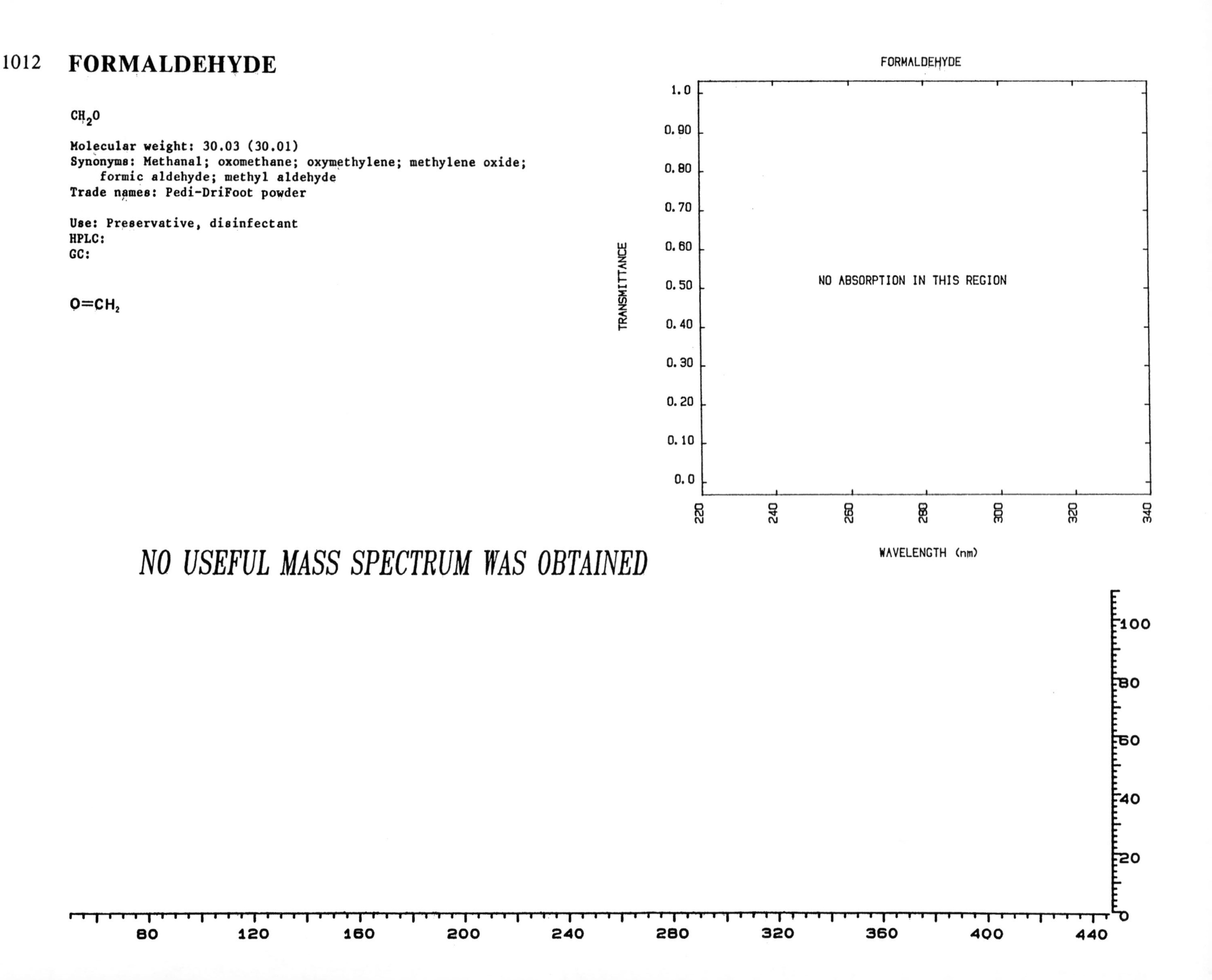

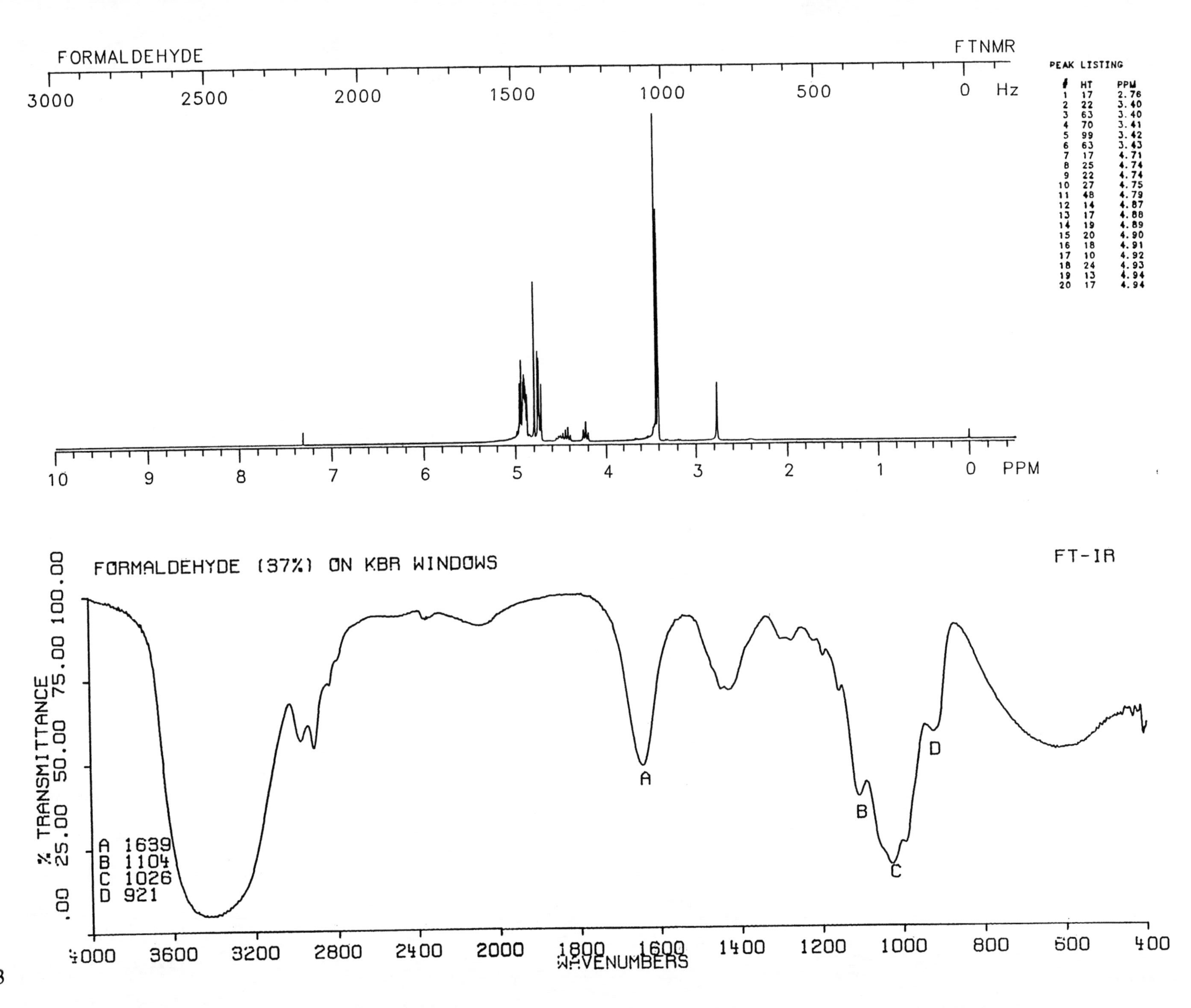

FORMALDEHYDE
FTNMR
3000 2500 2000 1500 1000 500 0 Hz
10 9 8 7 6 5 4 3 2 1 0 PPM
PEAK LISTING
# HT PPM
1 17 2.76
2 22 3.40
3 63 3.40
4 70 3.41
5 99 3.42
6 63 3.43
7 17 4.71
8 25 4.74
9 22 4.74
10 27 4.75
11 48 4.79
12 14 4.87
13 17 4.88
14 19 4.89
15 20 4.90
16 18 4.91
17 10 4.92
18 24 4.93
19 13 4.94
20 17 4.94
FORMALDEHYDE (37%) ON KBR WINDOWS
FT-IR
% TRANSMITTANCE
.00 25.00 50.00 75.00 100.00
A 1639
B 1104
C 1026
D 921
A B C D
4000 3600 3200 2800 2400 2000 1800 1600 1400 1200 1000 800 600 400
WAVENUMBERS

# FORMAMIDE

$CH_3NO$

Molecular weight: 45.04 (45.02)
Synonyms:

Trade names:

Use: Synthesis
HPLC:
GC:

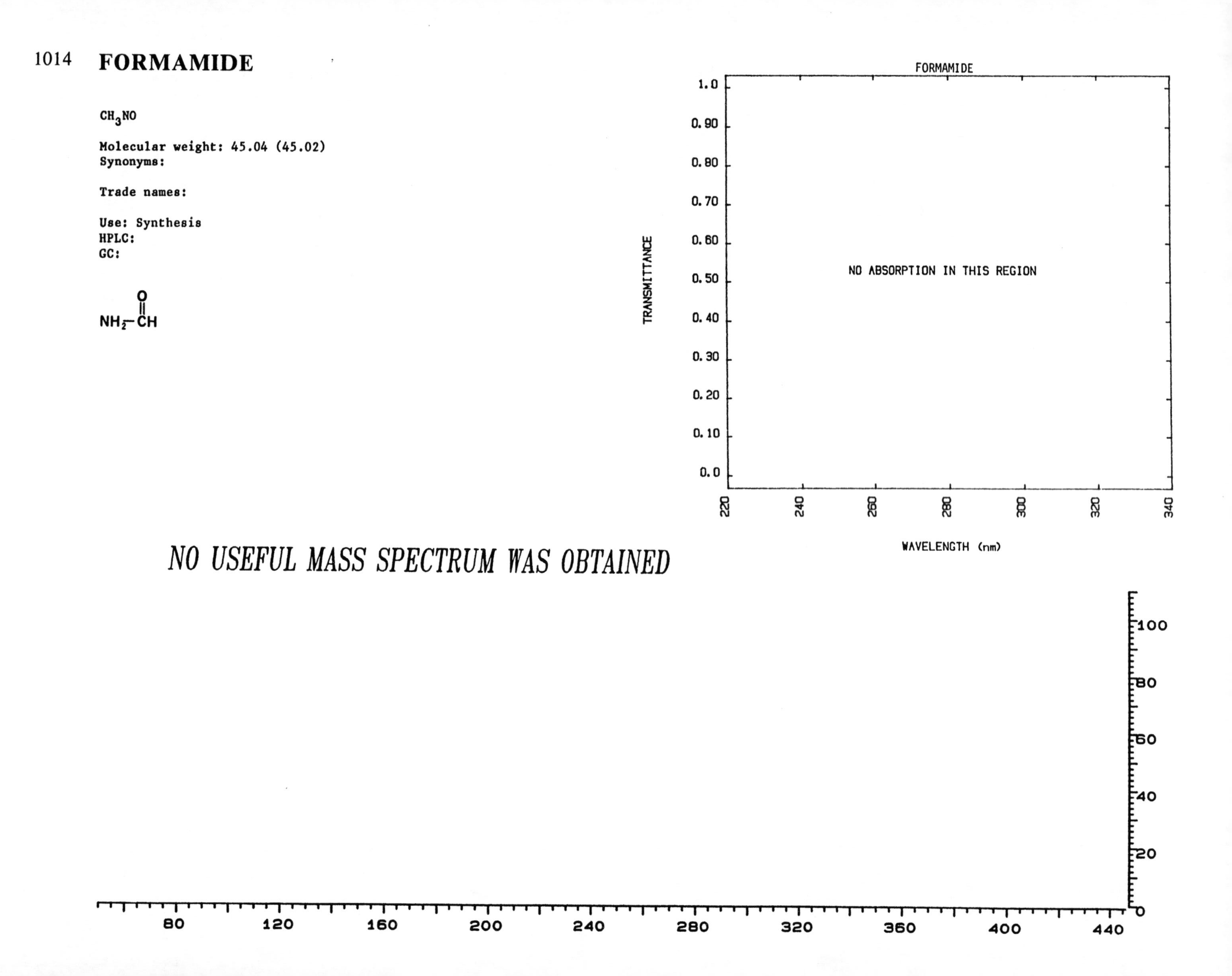

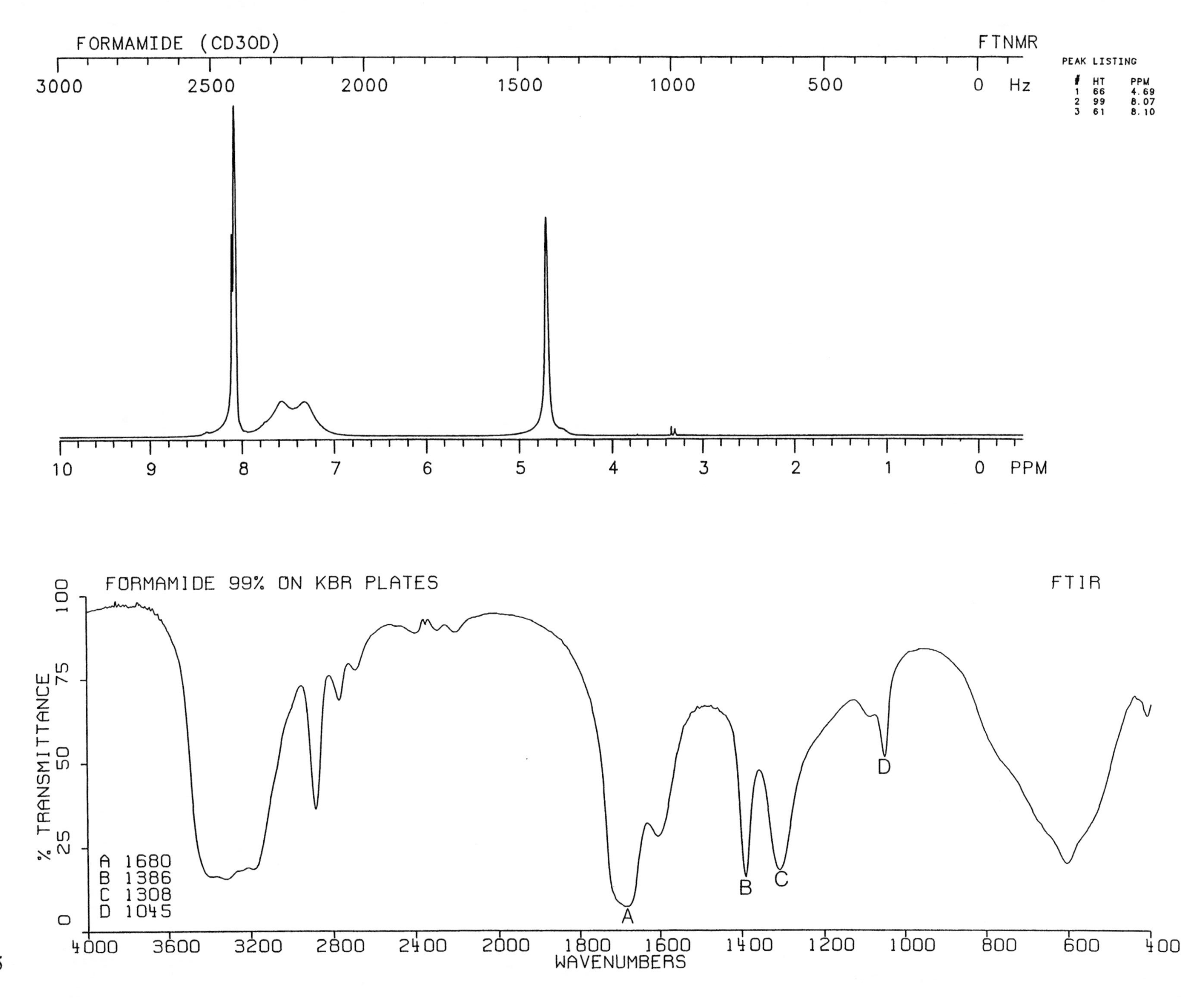
FORMAMIDE (CD3OD)
FTNMR
3000 2500 2000 1500 1000 500 0 Hz
PEAK LISTING
# HT PPM
1 66 4.69
2 99 8.07
3 61 8.10
10 9 8 7 6 5 4 3 2 1 0 PPM
FORMAMIDE 99% ON KBR PLATES
FTIR
100 75 50 25 0
% TRANSMITTANCE
A 1680
B 1386
C 1308
D 1045
A B C D
4000 3600 3200 2800 2400 2000 1800 1600 1400 1200 1000 800 600 400
WAVENUMBERS

# FRUCTOSE

$C_6H_{12}O_6$

**Molecular weight:** 180.16 (180.06)
**Synonyms:** D-Fructose; β-D-fructose; levulose; fruit sugar

**Trade names:**

**Use:** Sugar
**HPLC:**
**GC:**

FRUCTOSE

TRANSMITTANCE

NO ABSORPTION IN THIS REGION

WAVELENGTH (nm)

*NO USEFUL MASS SPECTRUM WAS OBTAINED*

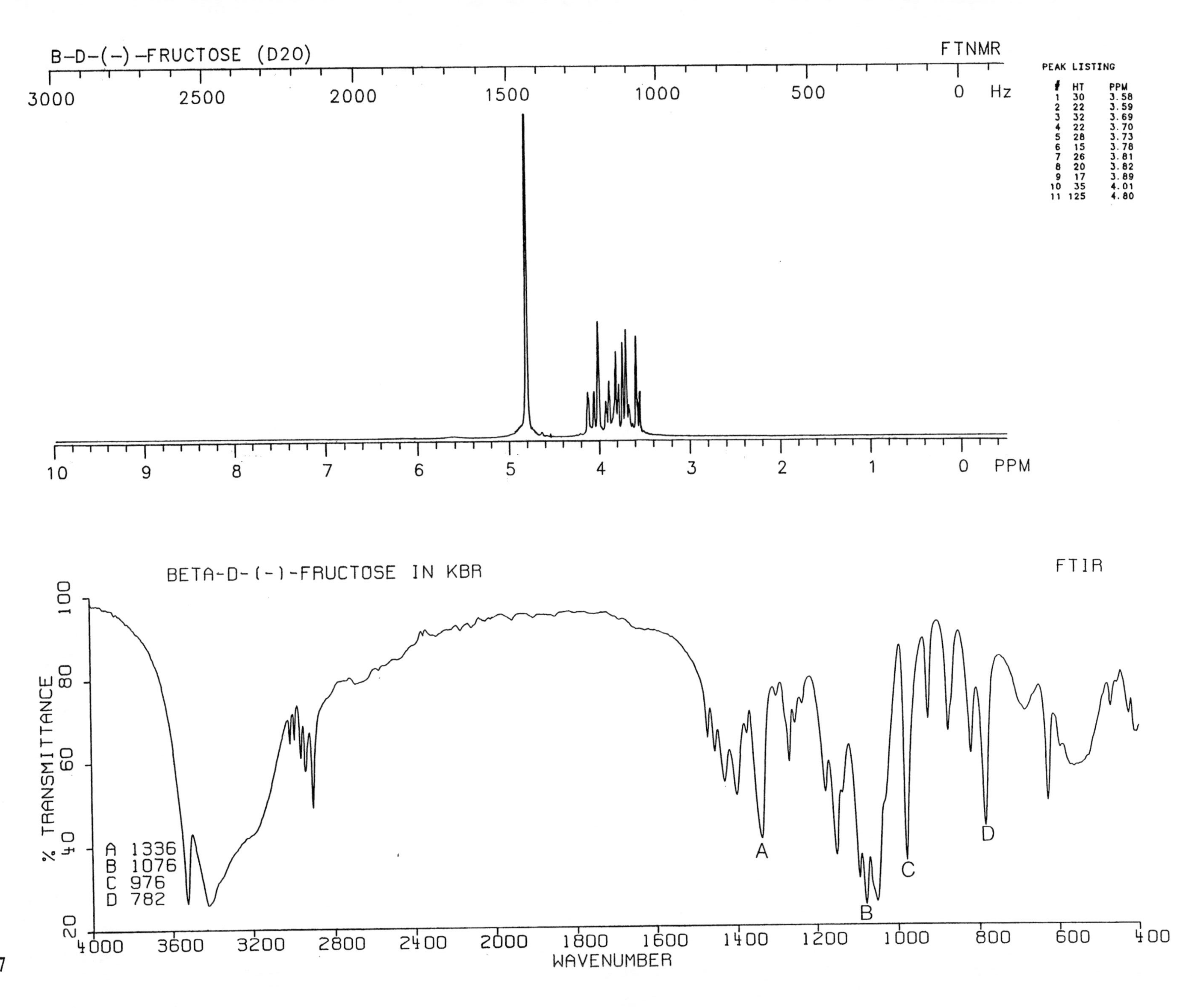
B-D-(-)-FRUCTOSE (D20)
FTNMR
PEAK LISTING
# HT PPM
1 30 3.58
2 22 3.59
3 32 3.69
4 22 3.70
5 28 3.73
6 15 3.78
7 26 3.81
8 20 3.82
9 17 3.89
10 35 4.01
11 125 4.80
Hz
PPM
BETA-D-(-)-FRUCTOSE IN KBR
FTIR
% TRANSMITTANCE
WAVENUMBER
A 1336
B 1076
C 976
D 782

# FUCOSE

$C_6H_{12}O_5$

Molecular weight: 164.16 (164.07)
Synonyms: 6-Deoxy-D-galactose; D-galactomethylose; D-fucose; rhodeose

Trade names:

Use: Sugar
HPLC:
GC:

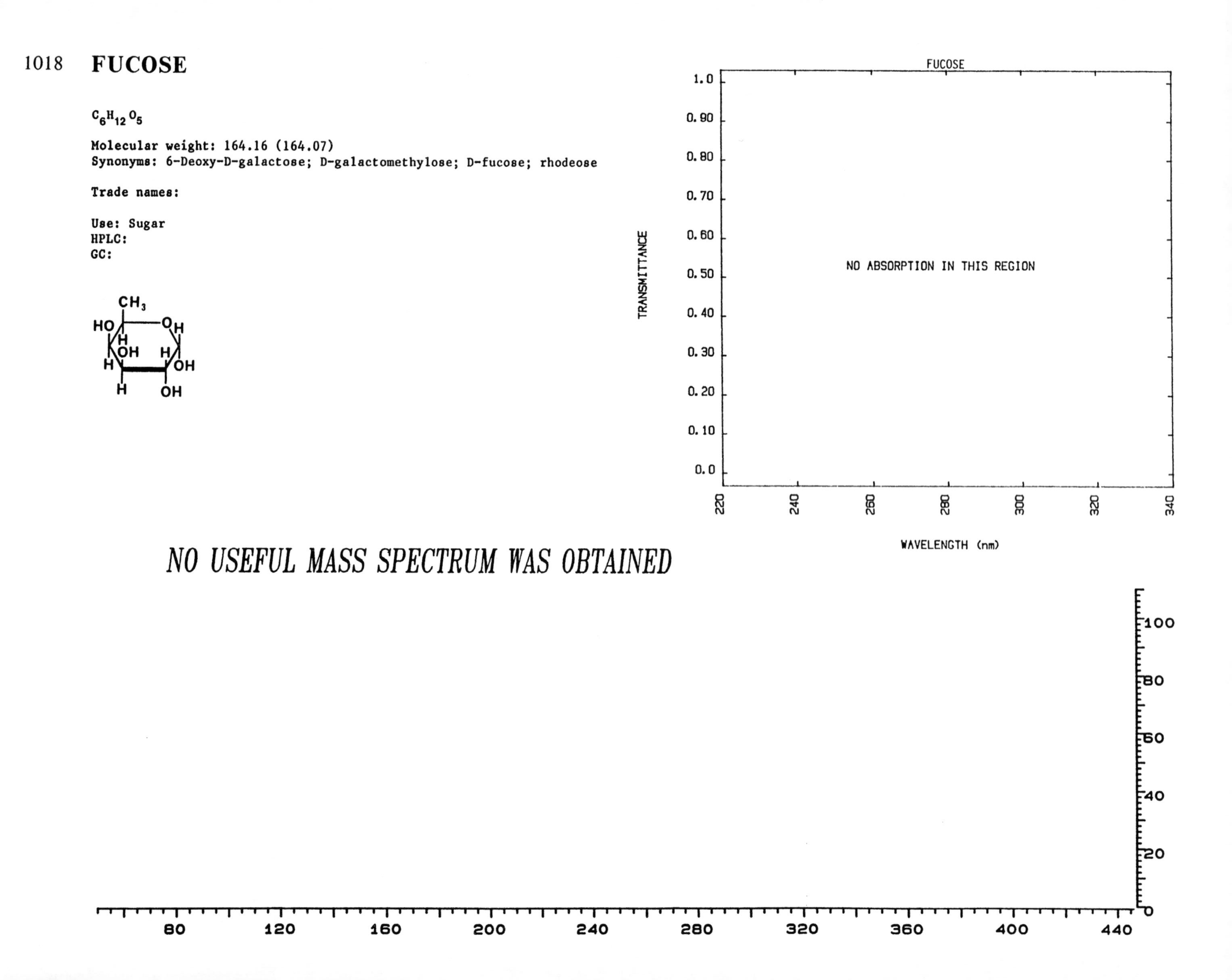

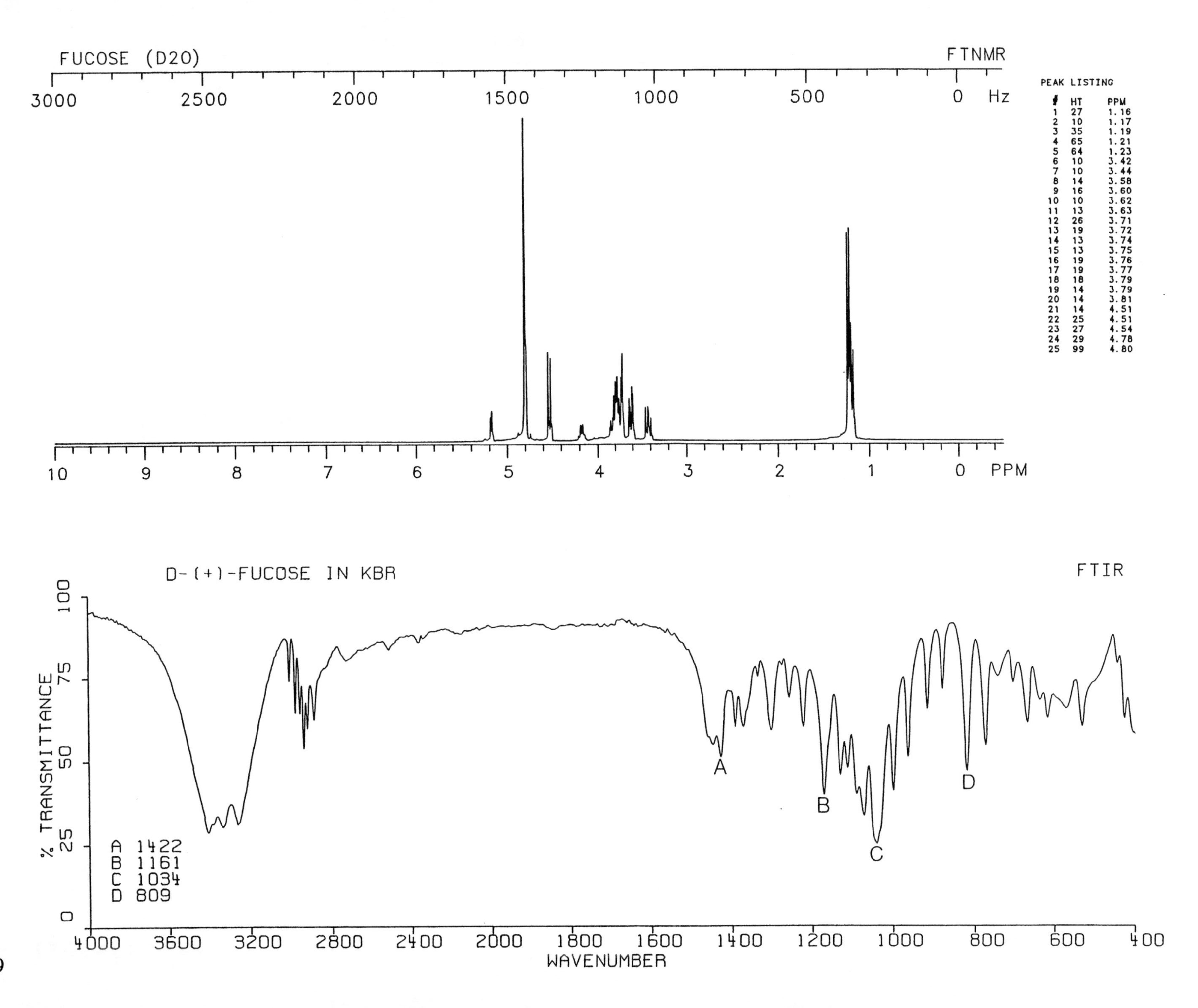
FUCOSE (D2O)
FTNMR
Hz
PPM
PEAK LISTING
# HT PPM
1 27 1.16
2 10 1.17
3 35 1.19
4 65 1.21
5 64 1.23
6 10 3.42
7 10 3.44
8 14 3.58
9 16 3.60
10 10 3.62
11 13 3.63
12 26 3.71
13 19 3.72
14 13 3.74
15 13 3.75
16 19 3.76
17 19 3.77
18 18 3.79
19 14 3.79
20 14 3.81
21 14 4.51
22 25 4.51
23 27 4.54
24 29 4.78
25 99 4.80
D-(+)-FUCOSE IN KBR
FTIR
% TRANSMITTANCE
WAVENUMBER
A 1422
B 1161
C 1034
D 809

# FUMARIC ACID

$C_4H_4O_4$

**Molecular weight:** 116.07 (116.01)
**Synonyms:** Butenedioic acid; trans-1,2-ethylenedicarboxylic acid; allomaleic acid; boletic acid
**Trade names:**

**Use:** Antioxidant
**HPLC:**
**GC:**

HOOC—CH
‖
HC—COOH

FUMARIC ACID

NO ABSORPTION IN THIS REGION

TRANSMITTANCE: 1.0, 0.90, 0.80, 0.70, 0.60, 0.50, 0.40, 0.30, 0.20, 0.10, 0.0

WAVELENGTH (nm): 220, 240, 260, 280, 300, 320, 340

*FUMARIC ACID--DIP*

Labelled peaks: 45, 53, 71, 88, 98, 116, 117

m/z axis: 40, 60, 80, 100, 120, 140, 160, 180, 200, 220

Intensity axis: 0, 20, 40, 60, 80, 100

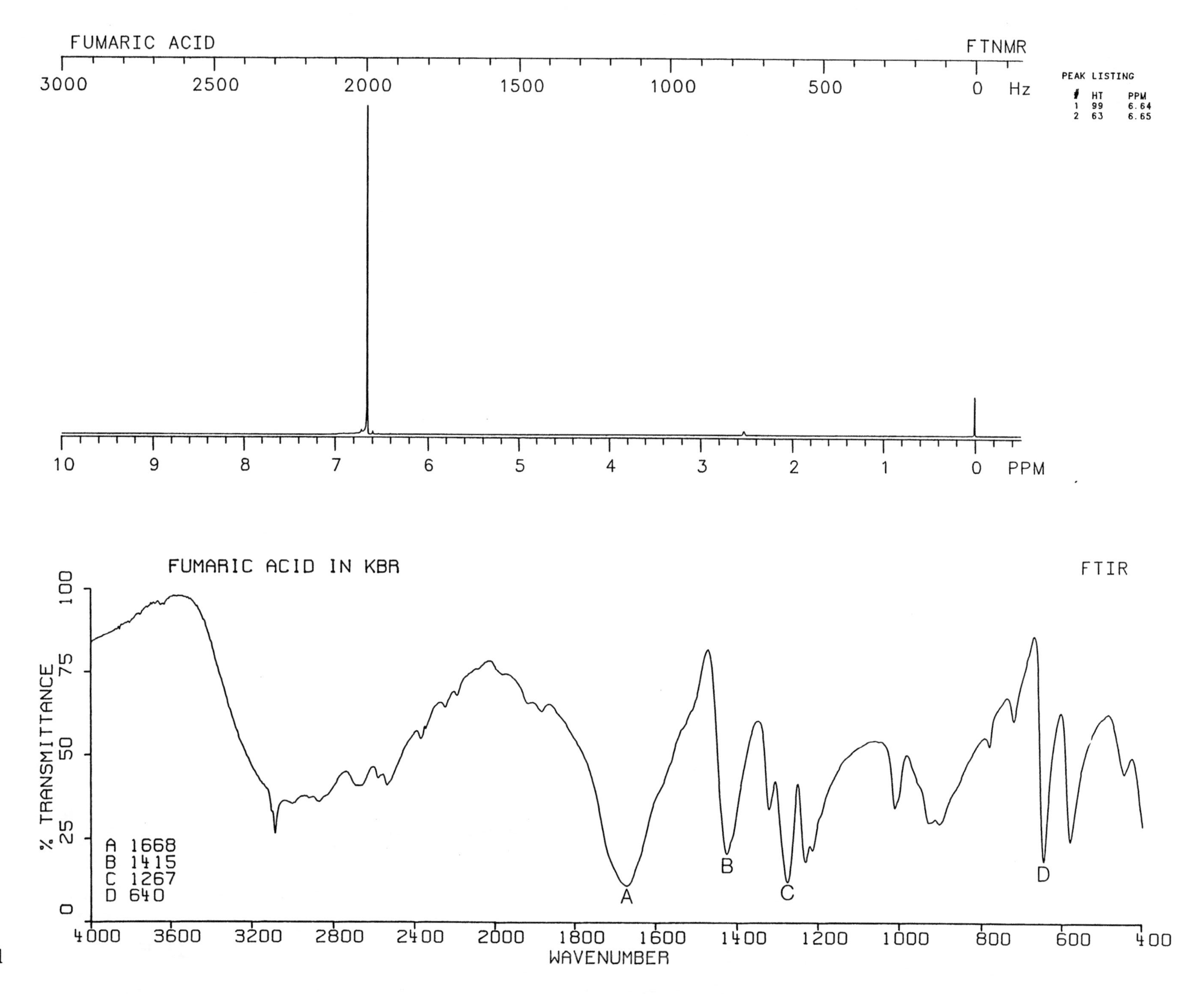
FUMARIC ACID
FTNMR
3000 2500 2000 1500 1000 500 0 Hz
PEAK LISTING
# HT PPM
1 99 6.64
2 63 6.65
10 9 8 7 6 5 4 3 2 1 0 PPM
FUMARIC ACID IN KBR
FTIR
% TRANSMITTANCE
0 25 50 75 100
A 1668
B 1415
C 1267
D 640
A
B
C
D
4000 3600 3200 2800 2400 2000 1800 1600 1400 1200 1000 800 600 400
WAVENUMBER

# FURALTADONE

$C_{13}H_{16}N_4O_6$

Molecular weight: 324.29 (324.11)
Synonyms: 5-(4-Morpholinylmethyl)-3-[[(5-nitro-2-furanyl)methylene]amino]-2-oxazolidinone; nitrofurmethone; furmethonol; nitrofurmethonum
Trade names: Altafur, Altabactina, Furazolin, Ibifur, Medifuran, Unifur, Nitraldone, Otifuril, Sepsinol, Ultrafur, Valsyn
Use: Antibacterial
HPLC: Si-10; 1A:99B; 5.9
GC:

FURALTADONE

TRANSMITTANCE

255

WAVELENGTH (nm)

FURALTADONE

42 70 79 100 128 184 202 319

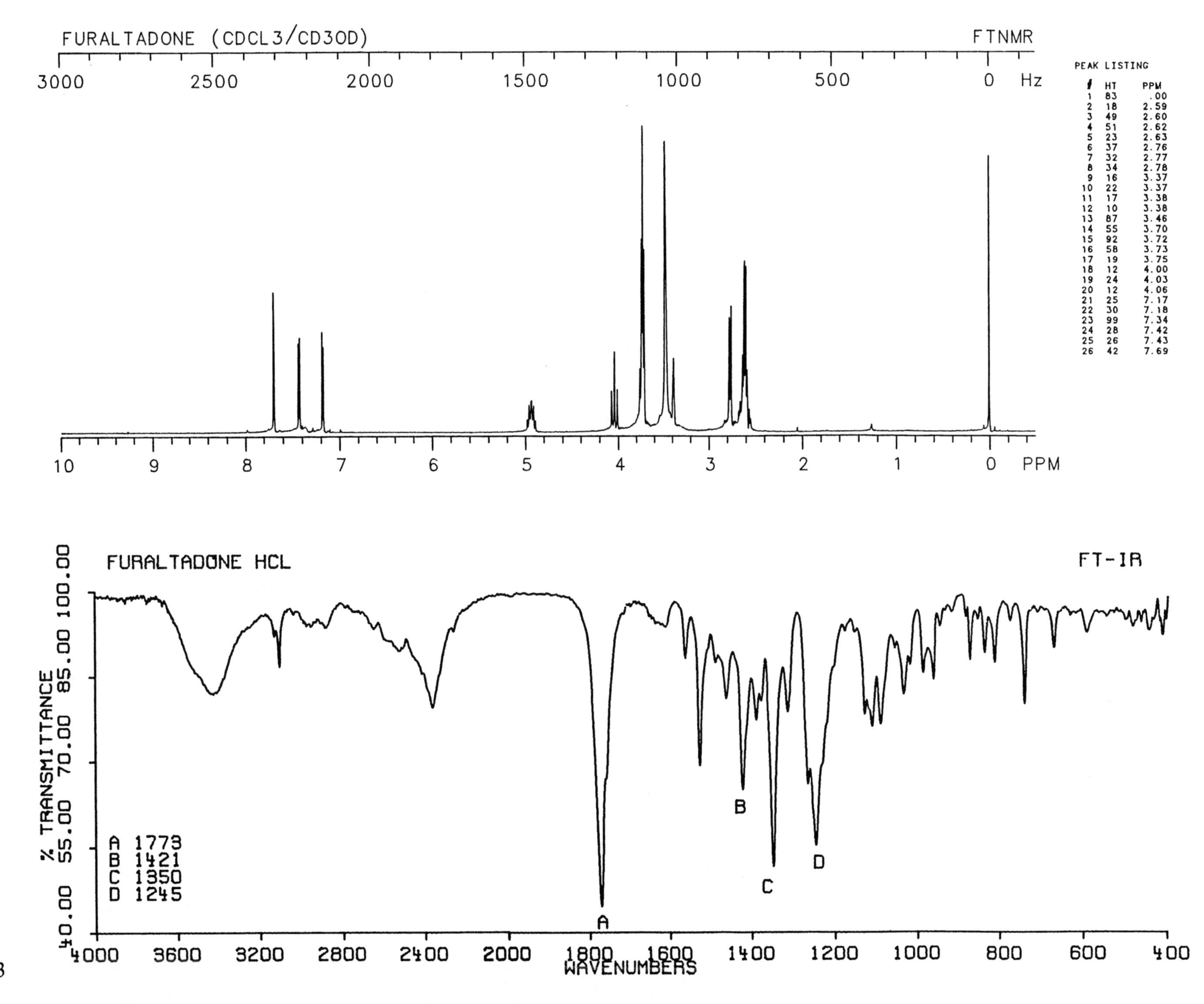
FURALTADONE (CDCL3/CD3OD)
FTNMR
3000
2500
2000
1500
1000
500
0
Hz
PEAK LISTING
# HT PPM
1 83 .00
2 18 2.59
3 49 2.60
4 51 2.62
5 23 2.63
6 37 2.76
7 32 2.77
8 34 2.78
9 16 3.37
10 22 3.37
11 17 3.38
12 10 3.38
13 87 3.46
14 55 3.70
15 92 3.72
16 58 3.73
17 19 3.75
18 12 4.00
19 24 4.03
20 12 4.06
21 25 7.17
22 30 7.18
23 99 7.34
24 28 7.42
25 26 7.43
26 42 7.69
10
9
8
7
6
5
4
3
2
1
0
PPM
FURALTADONE HCL
FT-IR
% TRANSMITTANCE
40.00
55.00
70.00
85.00
100.00
A 1773
B 1421
C 1350
D 1245
A
B
C
D
4000
3600
3200
2800
2400
2000
1800
1600
1400
1200
1000
800
600
400
WAVENUMBERS

# FURAN

$C_4H_4O$

Molecular weight: 68.07 (68.03)
Synonyms: Furfuran; oxole; tetrole; divinylene oxide

Trade names:

Use: Narcotic
HPLC:
GC:

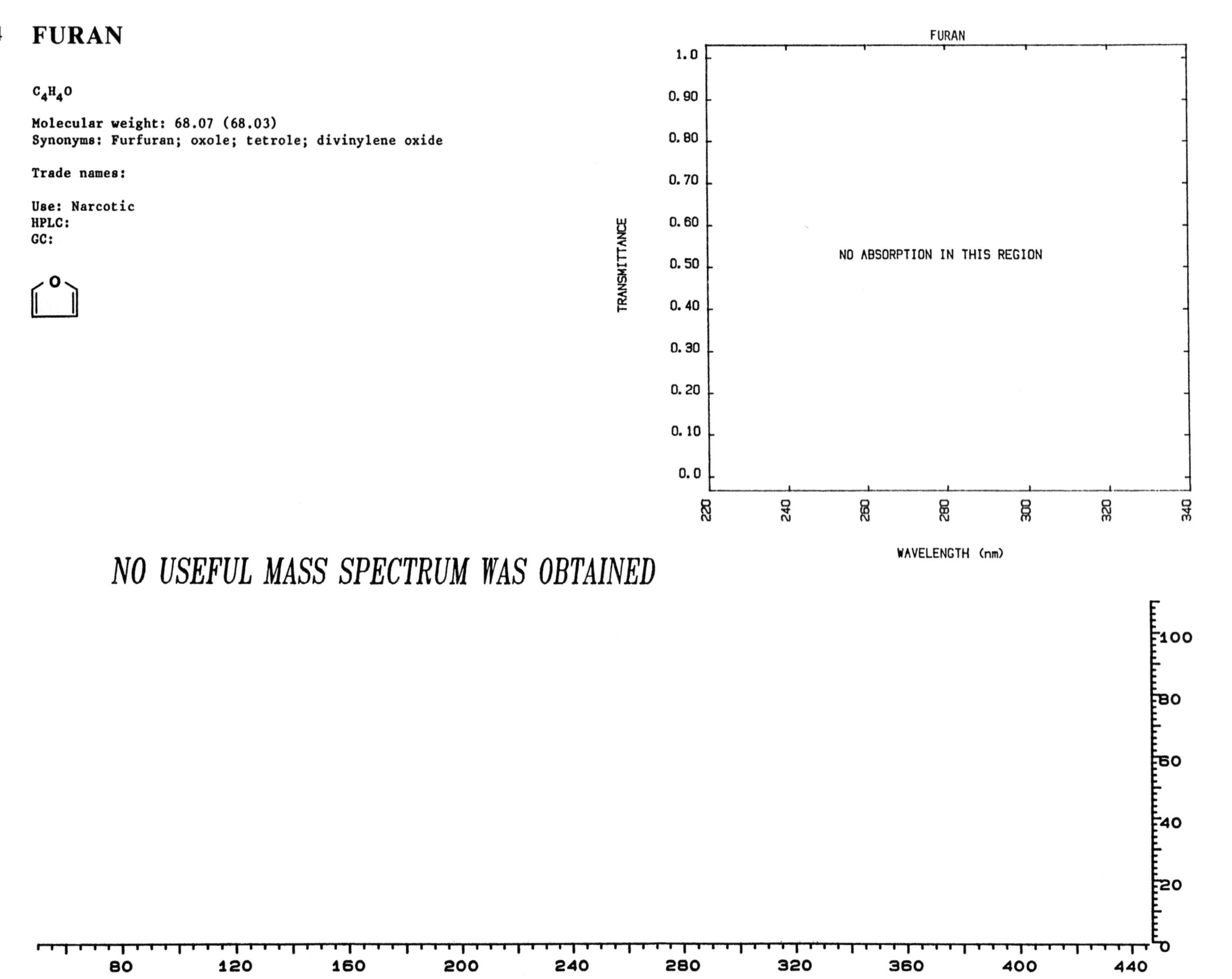

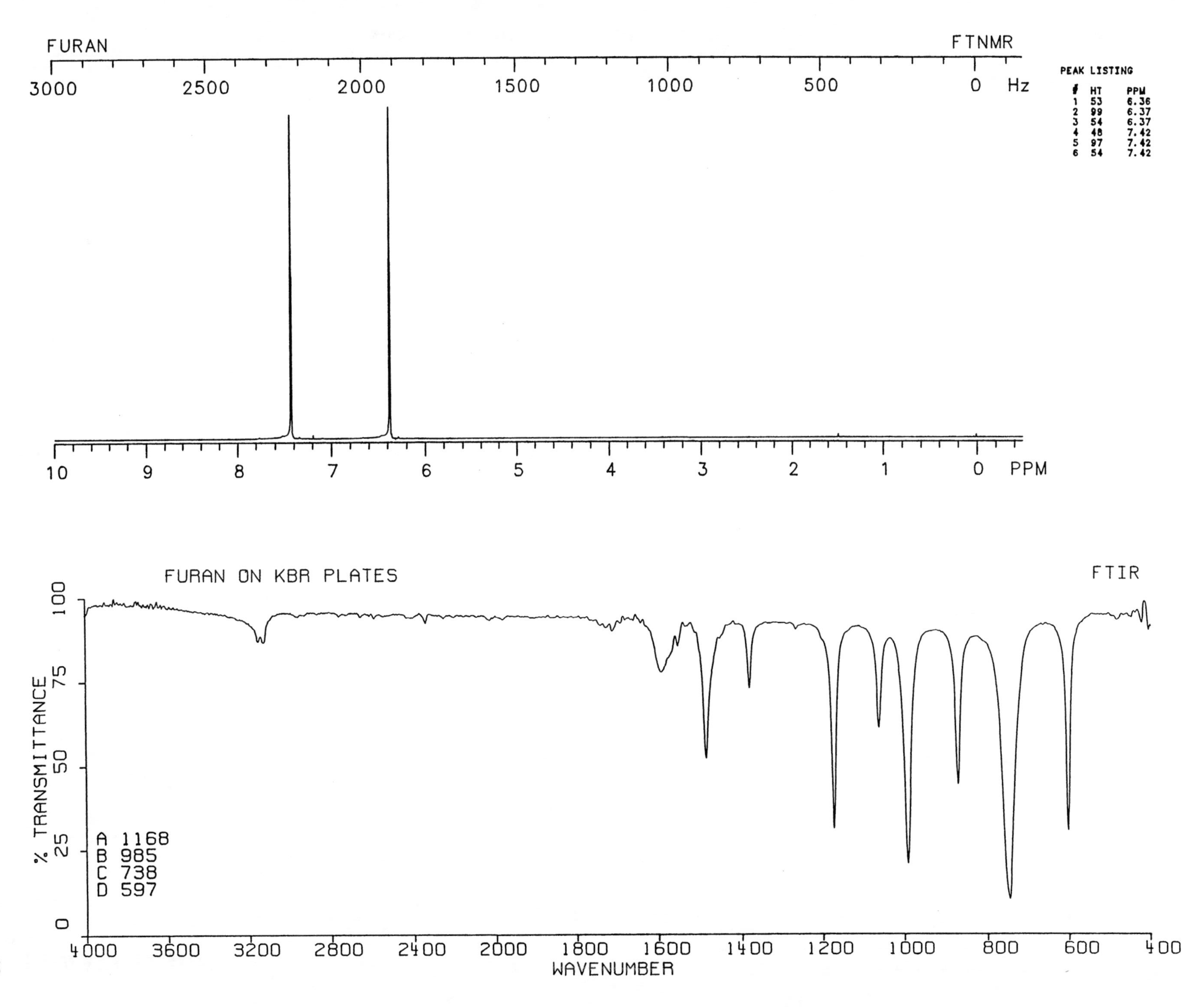
FURAN
FTNMR
3000
2500
2000
1500
1000
500
0 Hz
PEAK LISTING
# HT PPM
1 53 6.36
2 99 6.37
3 54 6.37
4 48 7.42
5 97 7.42
6 54 7.42
10
9
8
7
6
5
4
3
2
1
0 PPM
FURAN ON KBR PLATES
FTIR
% TRANSMITTANCE
100
75
50
25
0
A 1168
B 985
C 738
D 597
4000
3600
3200
2800
2400
2000
1800
1600
1400
1200
1000
800
600
400
WAVENUMBER

# FURAZOLIDONE

$C_8H_7N_3O_5$

**Molecular weight:** 225.16 (225.04)
**Synonyms:** 3-[[(5-Nitro-2-furanyl)methylene]amino]-2-oxazolidinone; nifurazolidonum
**Trade names:** Furoxone

**Use:** Topical anti-infective
**HPLC:** Si-10; 100B; 5.7
**GC:**

$O_2N$ O CH=N N O O

FURAZOLIDONE

TRANSMITTANCE

WAVELENGTH (nm)

258

*FURAZOLIDONE*

51 64 78 79 95 113 139 207 225

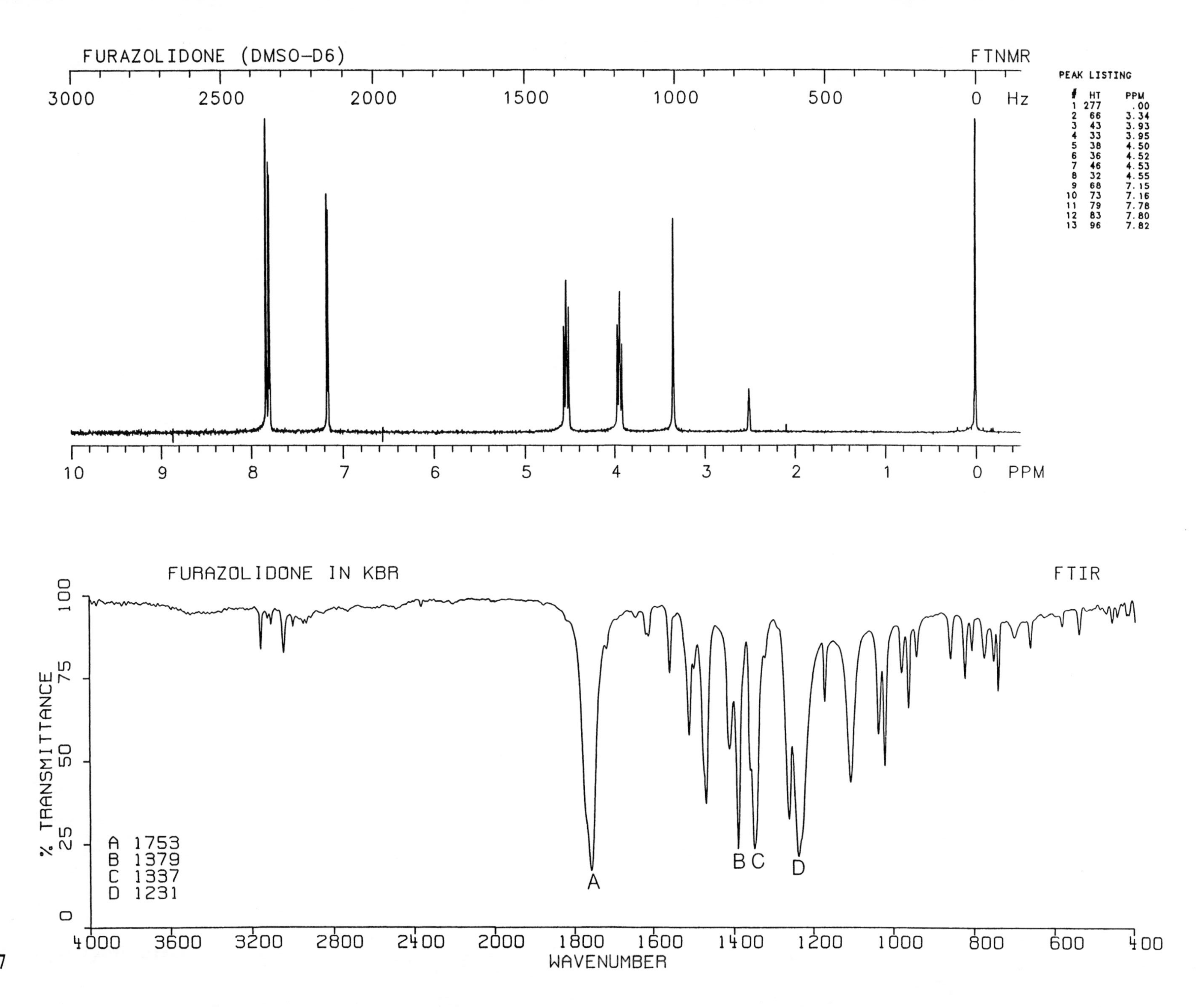
FURAZOLIDONE (DMSO-D6)
FTNMR
Hz
PPM
PEAK LISTING
# HT PPM
1 277 .00
2 66 3.34
3 43 3.93
4 33 3.95
5 38 4.50
6 36 4.52
7 46 4.53
8 32 4.55
9 68 7.15
10 73 7.16
11 79 7.78
12 83 7.80
13 96 7.82
FURAZOLIDONE IN KBR
FTIR
% TRANSMITTANCE
WAVENUMBER
A 1753
B 1379
C 1337
D 1231

# FUROSEMIDE

$C_{12}H_{11}ClN_2O_5S$

**Molecular weight:** 330.74 (330.01)
**Synonyms:** 5-(Aminosulfonyl)-4-chloro-2-[(2-furanylmethyl)amino]-benzoic acid
**Trade names:** Lasix

**Use:** Diuretic, antihypertensive
**HPLC:**
**GC:**

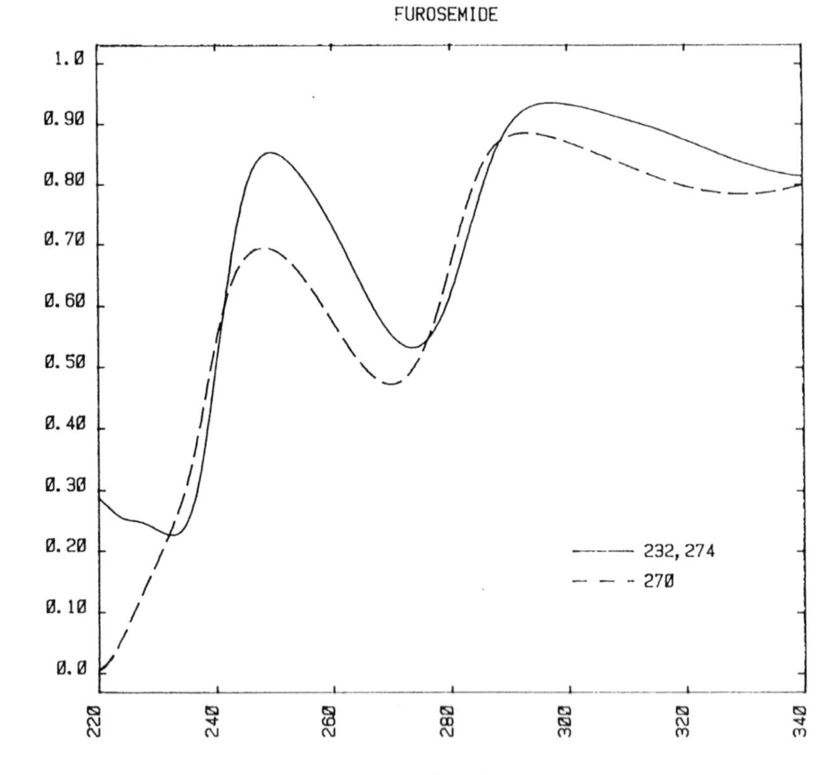


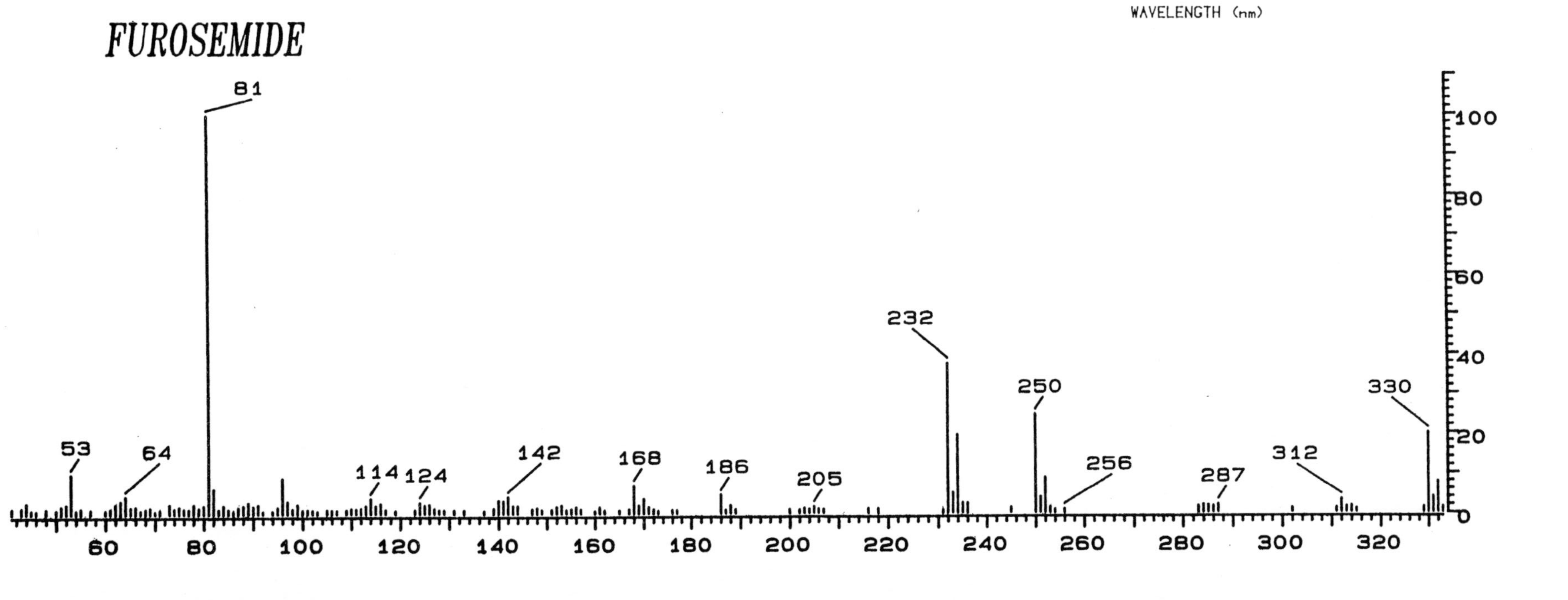

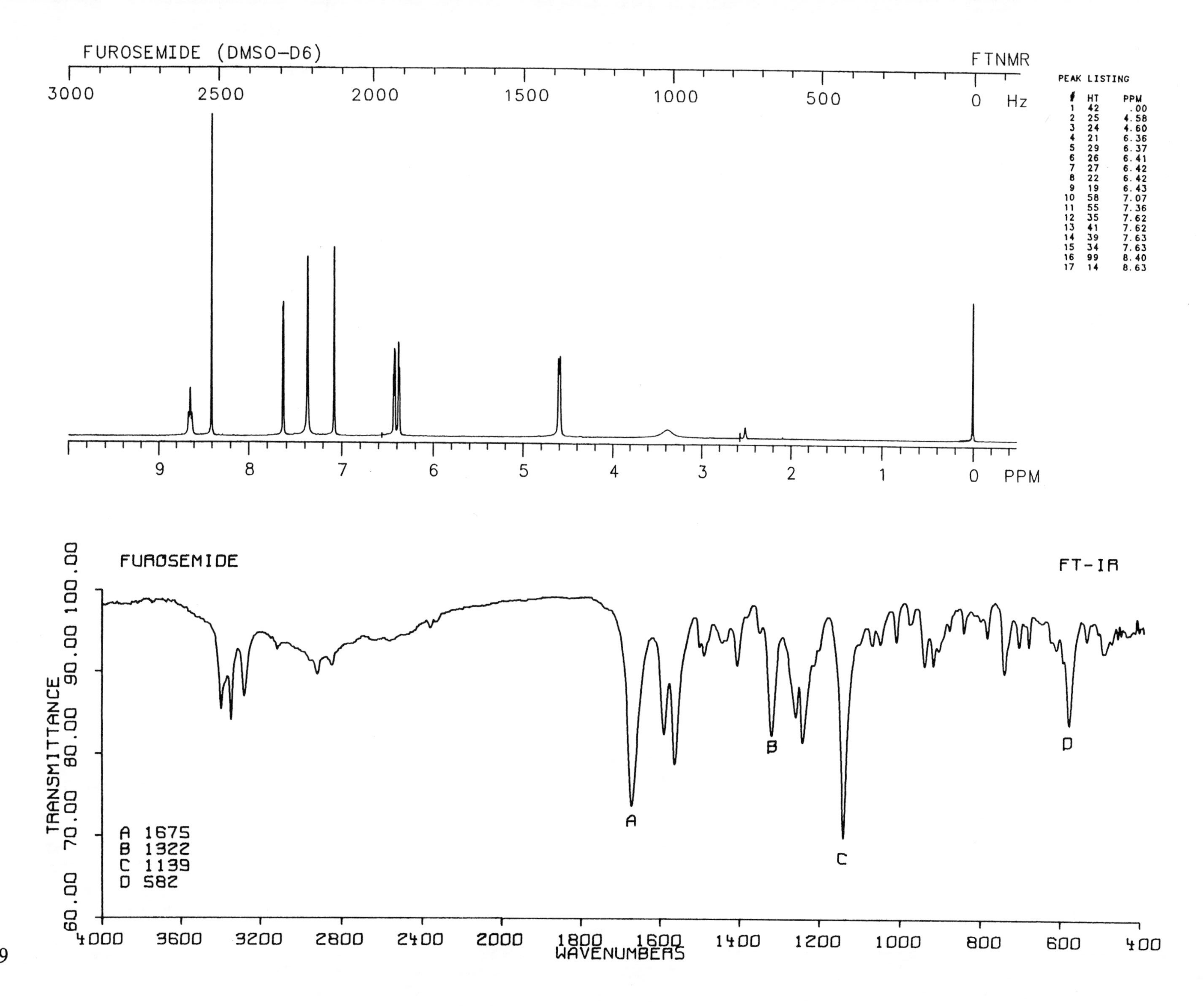
FUROSEMIDE (DMSO-D6)
FTNMR
PEAK LISTING
# HT PPM
1 42 .00
2 25 4.58
3 24 4.60
4 21 6.36
5 29 6.37
6 26 6.41
7 27 6.42
8 22 6.42
9 19 6.43
10 58 7.07
11 55 7.36
12 35 7.62
13 41 7.62
14 39 7.63
15 34 7.63
16 99 8.40
17 14 8.63
3000 2500 2000 1500 1000 500 0 Hz
9 8 7 6 5 4 3 2 1 0 PPM
FUROSEMIDE
FT-IR
TRANSMITTANCE
60.00 70.00 80.00 90.00 100.00
A 1675
B 1322
C 1139
D 582
A
B
C
D
4000 3600 3200 2800 2400 2000 1800 1600 1400 1200 1000 800 600 400
WAVENUMBERS

# 1030 GALACTOSE

$C_6H_{12}O_6$

Molecular weight: 180.16 (180.06)
Synonyms: Cerebrose; brain sugar

Trade names:

Use: Diagnostic aid
HPLC:
GC:

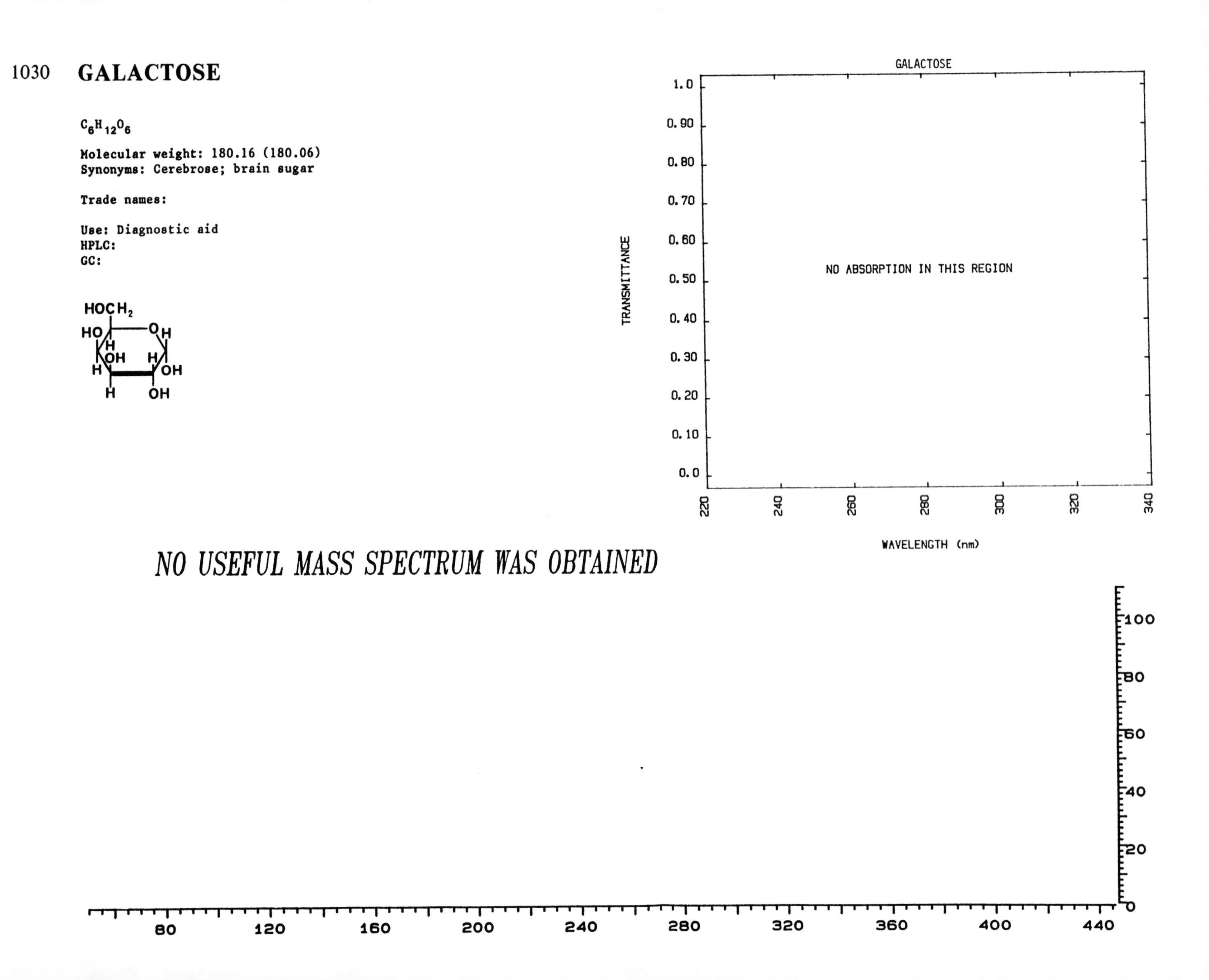

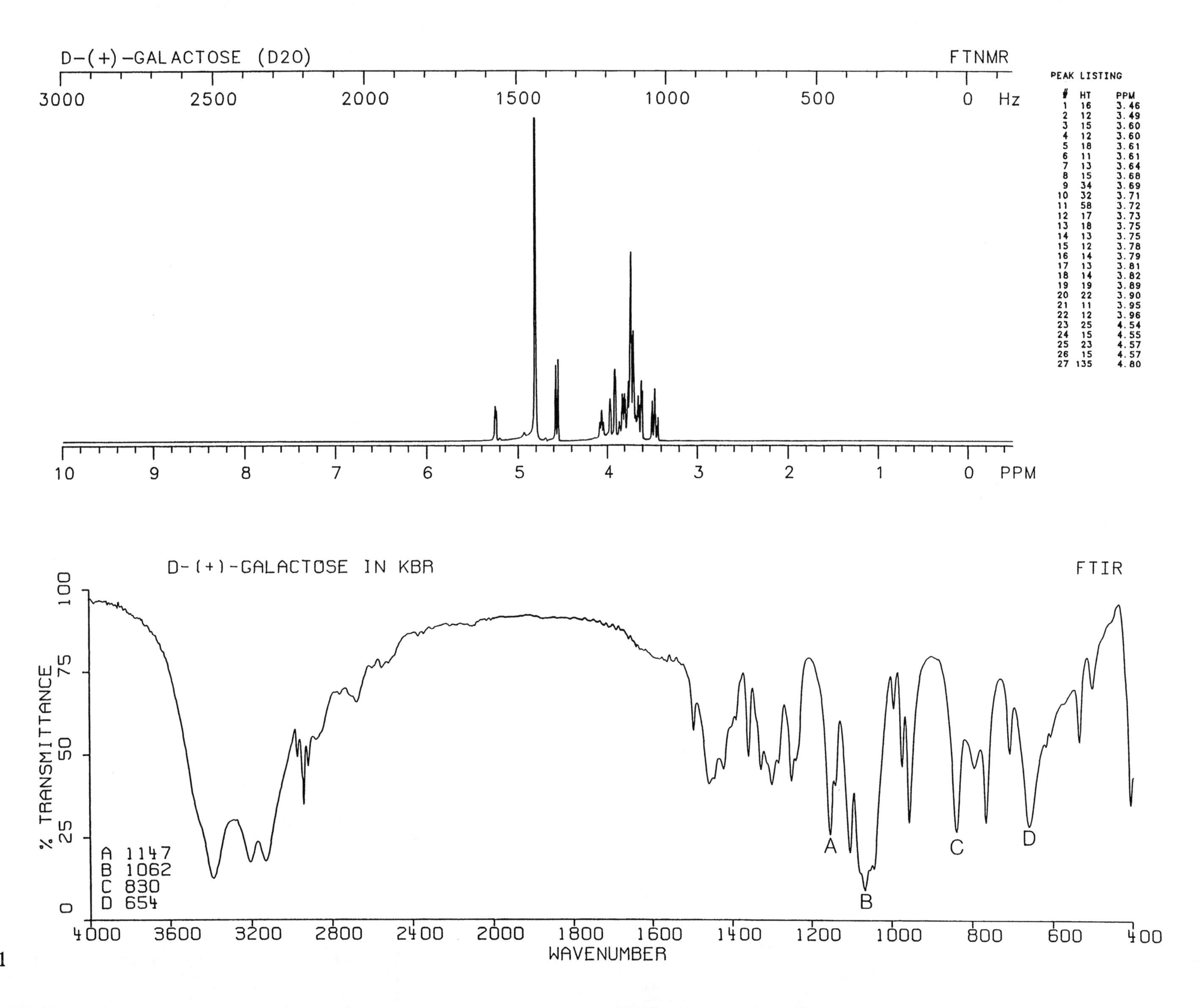
D-(+)-GALACTOSE (D20)
FTNMR
3000 2500 2000 1500 1000 500 0 Hz
10 9 8 7 6 5 4 3 2 1 0 PPM
PEAK LISTING
# HT PPM
1 16 3.46
2 12 3.49
3 15 3.60
4 12 3.60
5 18 3.61
6 11 3.61
7 13 3.64
8 15 3.68
9 34 3.69
10 32 3.71
11 58 3.72
12 17 3.73
13 18 3.75
14 13 3.75
15 12 3.78
16 14 3.79
17 13 3.81
18 14 3.82
19 19 3.89
20 22 3.90
21 11 3.95
22 12 3.96
23 25 4.54
24 15 4.55
25 23 4.57
26 15 4.57
27 135 4.80
D-(+)-GALACTOSE IN KBR
FTIR
% TRANSMITTANCE
0 25 50 75 100
A 1147
B 1062
C 830
D 654
A B C D
4000 3600 3200 2800 2400 2000 1800 1600 1400 1200 1000 800 600 400
WAVENUMBER

# GALLAMINE TRIETHIODIDE

$C_{30}H_{60}I_3N_3O_3$

Molecular weight: 891.56 (891.18)
Synonyms: 2,2',2''-[1,2,3-Benzenetriyltris(oxy)]tris[N,N,N-triethyl-ethanaminium]triiodide; benzcurine iodide; bencurine iodide
Trade names: Flaxedil, Relaxan, Retensin, Tricuran

Use: Skeletal muscle relaxant
HPLC:
GC: 2657; 250°C

$$\left[ C_6H_3(O{-}CH_2{-}CH_2{-}\overset{+}{N}(C_2H_5)_3)_3 \right] \cdot 3I^-$$

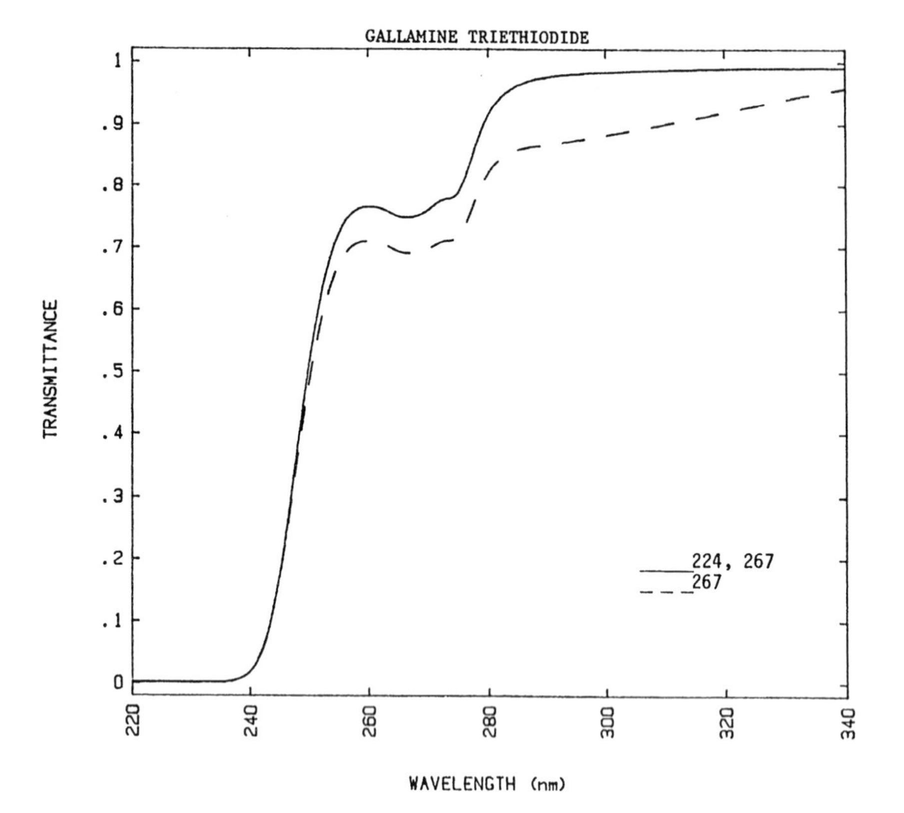

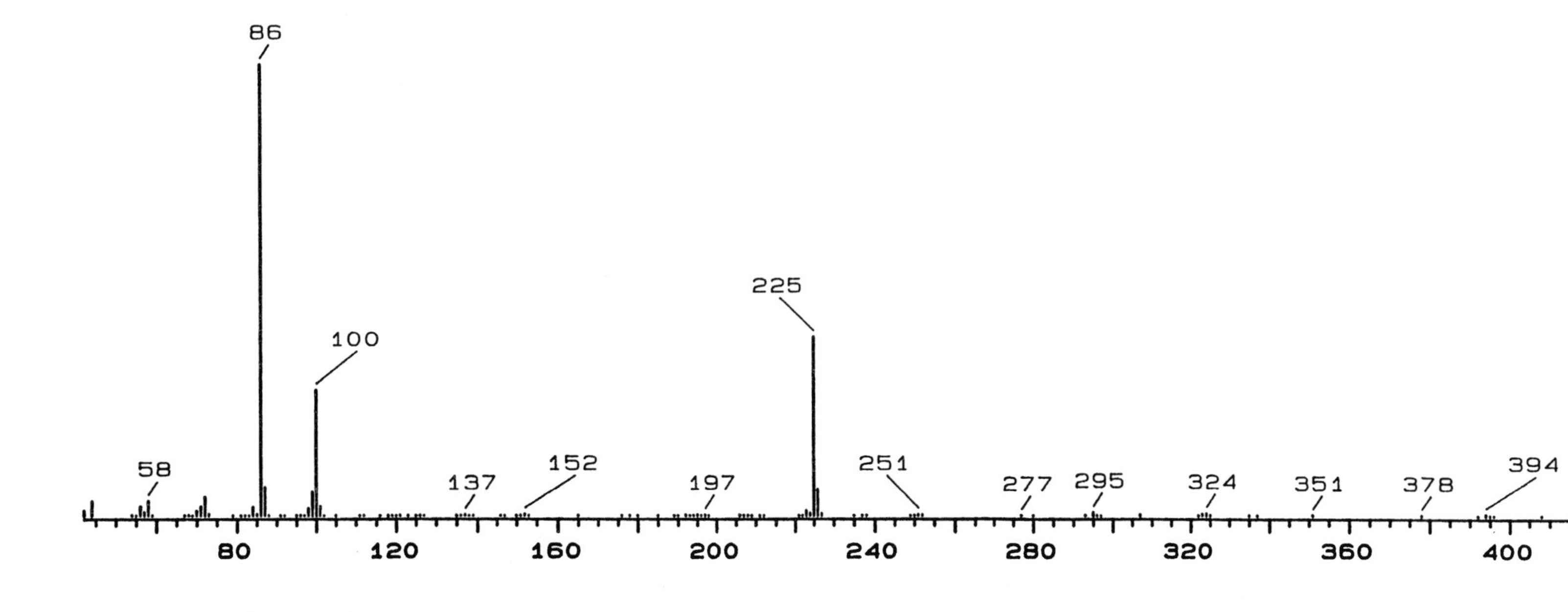

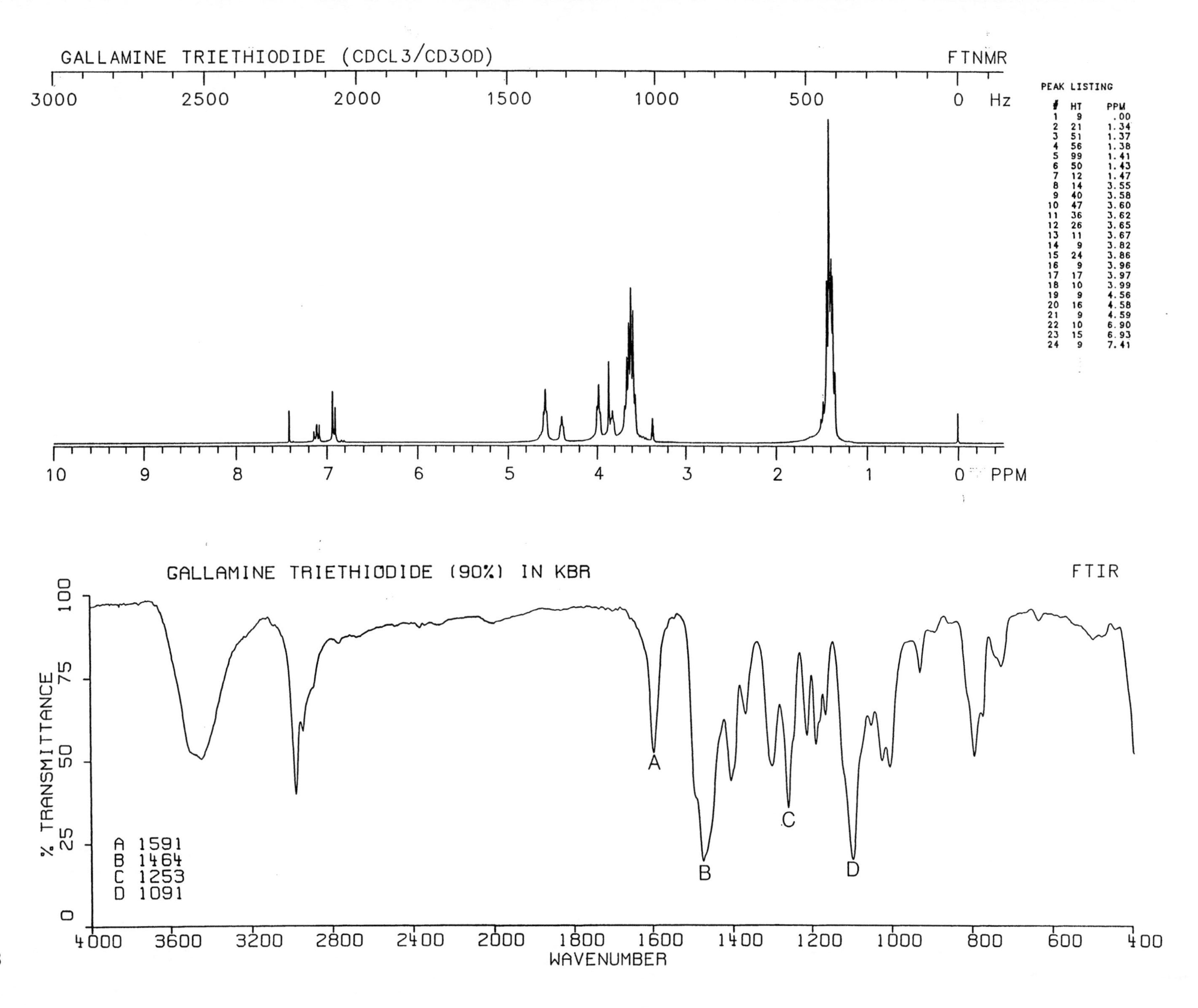

GALLAMINE TRIETHIODIDE (CDCL3/CD3OD)
FTNMR
3000
2500
2000
1500
1000
500
0 Hz
10
9
8
7
6
5
4
3
2
1
0 PPM
PEAK LISTING
# HT PPM
1 9 .00
2 21 1.34
3 51 1.37
4 56 1.38
5 99 1.41
6 50 1.43
7 12 1.47
8 14 3.55
9 40 3.58
10 47 3.60
11 36 3.62
12 26 3.65
13 11 3.67
14 9 3.82
15 24 3.86
16 9 3.96
17 17 3.97
18 10 3.99
19 9 4.56
20 16 4.58
21 9 4.59
22 10 6.90
23 15 6.93
24 9 7.41
GALLAMINE TRIETHIODIDE (90%) IN KBR
FTIR
% TRANSMITTANCE
100
75
50
25
0
A 1591
B 1464
C 1253
D 1091
A
B
C
D
4000
3600
3200
2800
2400
2000
1800
1600
1400
1200
1000
800
600
400
WAVENUMBER

# GALLIC ACID

$C_7H_6O_5$

Molecular weight: 170.12 (170.02)
Synonyms: 3,4,5-Trihydroxybenzoic acid

Trade names:

Use: Antioxidant, astringent
HPLC:
GC:

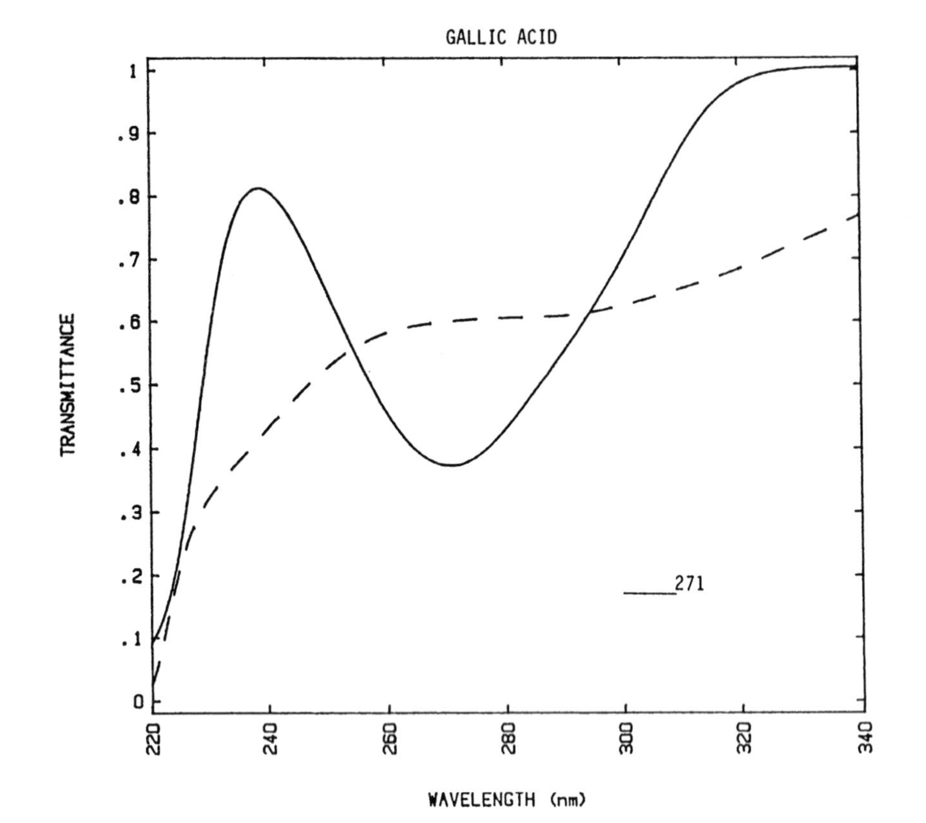


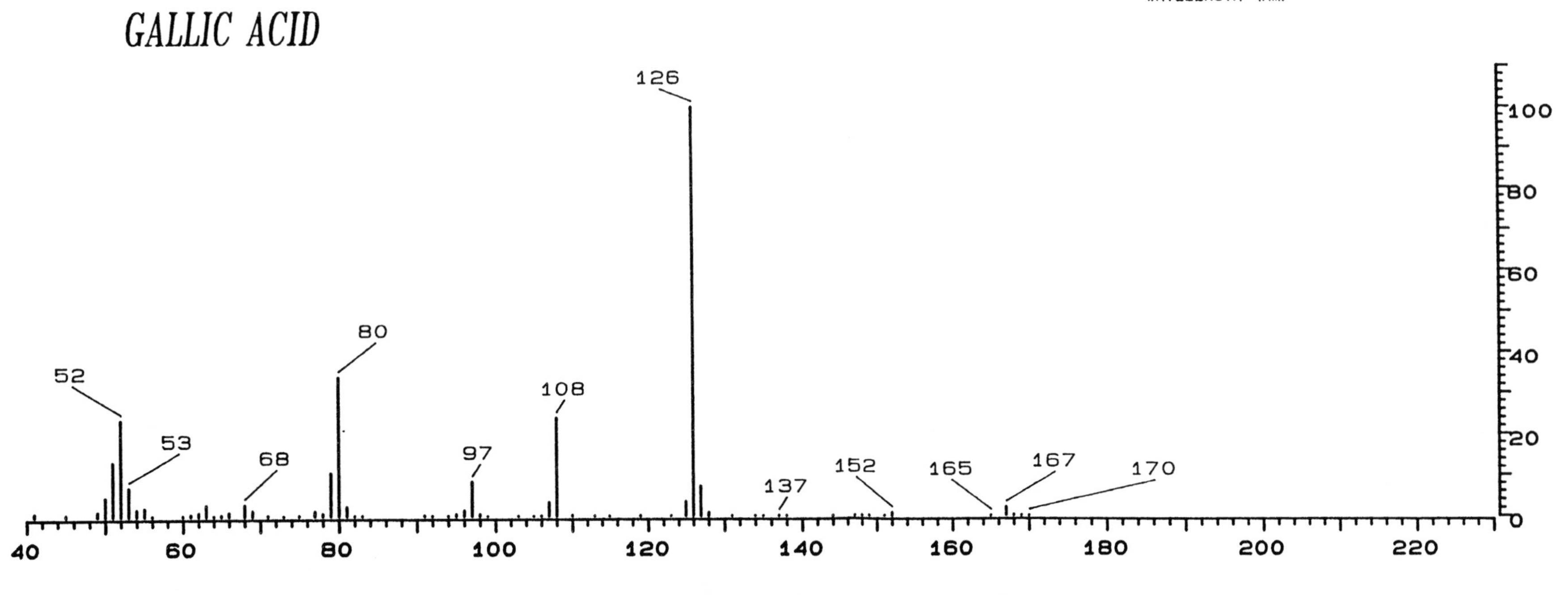

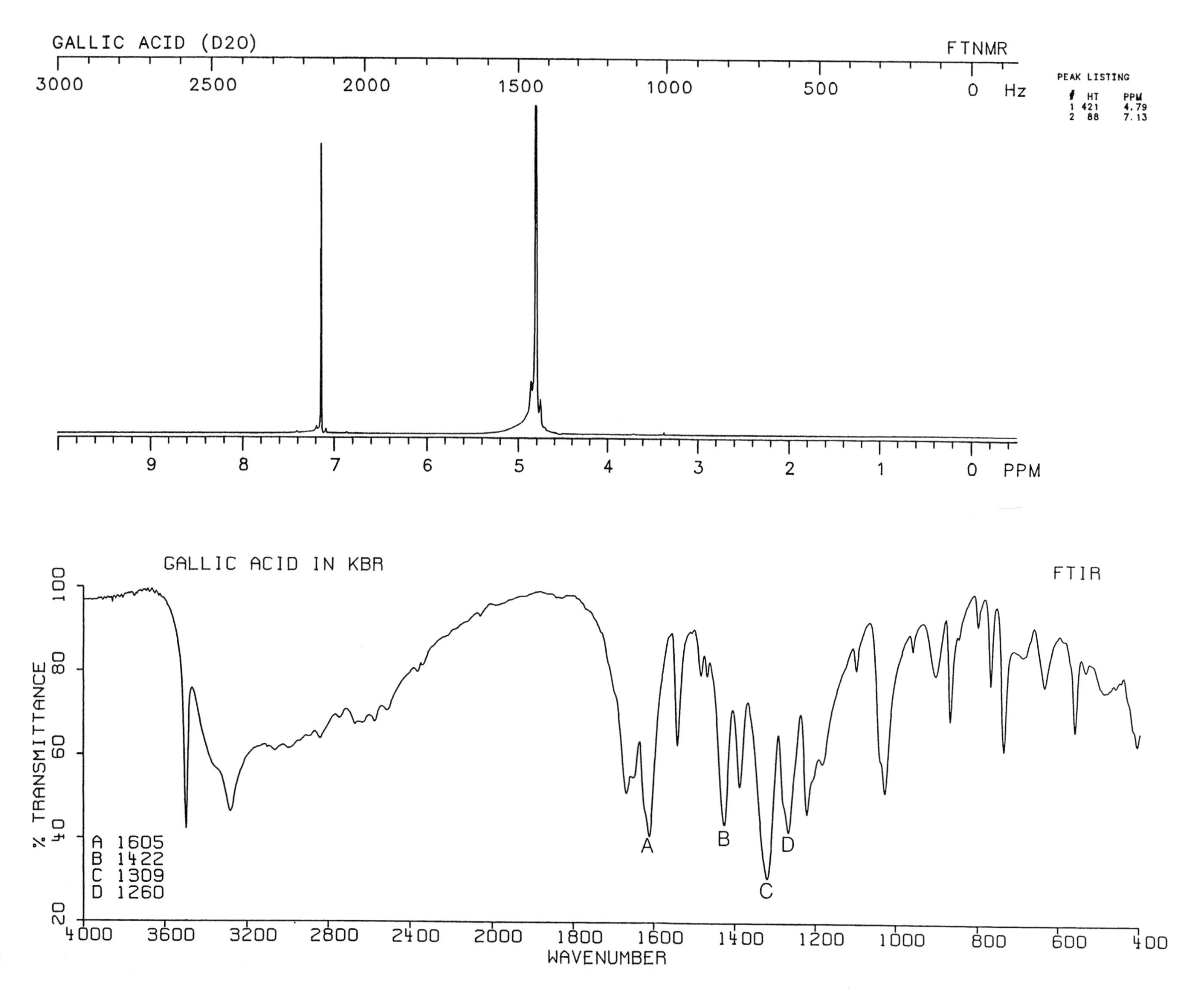
GALLIC ACID (D2O)
FTNMR
3000 2500 2000 1500 1000 500 0 Hz
PEAK LISTING
# HT PPM
1 421 4.79
2 88 7.13
9 8 7 6 5 4 3 2 1 0 PPM
GALLIC ACID IN KBR
FTIR
% TRANSMITTANCE
100 80 60 40 20
A 1605
B 1422
C 1309
D 1260
A
B
C
D
4000 3600 3200 2800 2400 2000 1800 1600 1400 1200 1000 800 600 400
WAVENUMBER

# GELSEMINE

$C_{20}H_{22}N_2O_2$

Molecular weight: 322.40 (322.17)
Synonyms:

Trade names:

Use: Central stimulant, poison
HPLC: Si-10; 5A:95B; 5.0
GC: 2828; 280°C

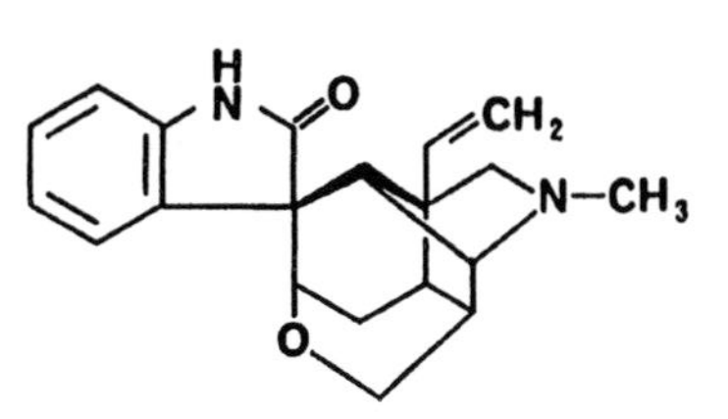


*GELSEMINE*

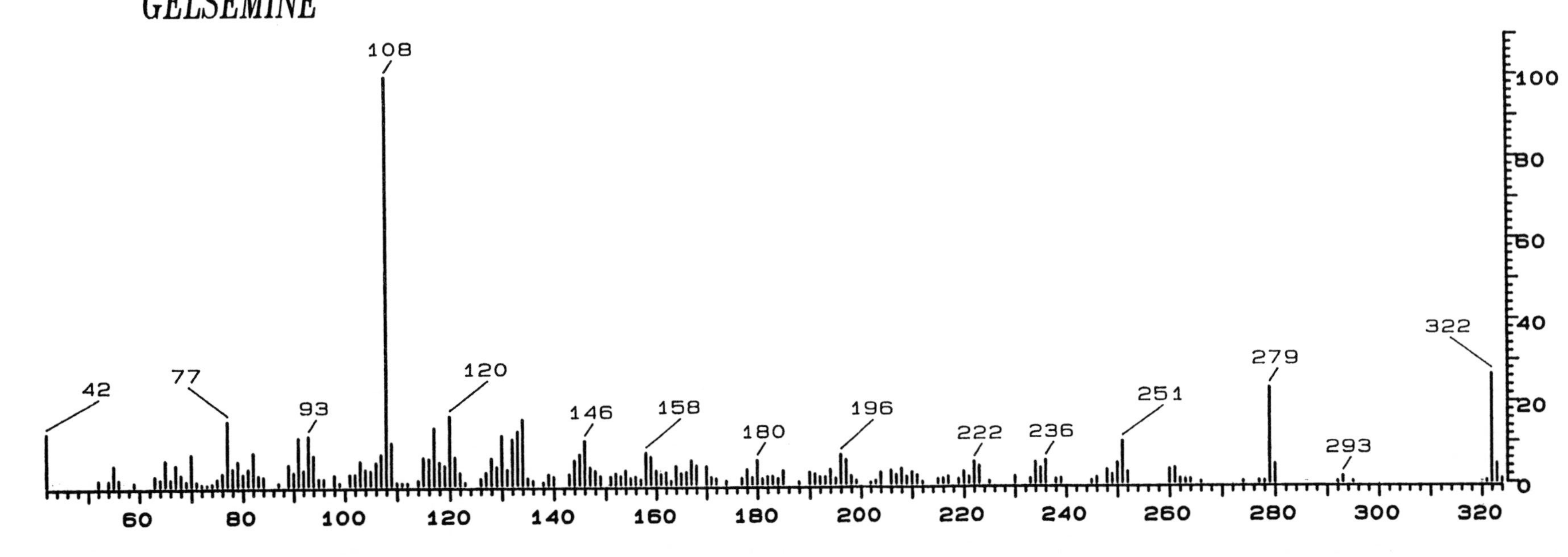

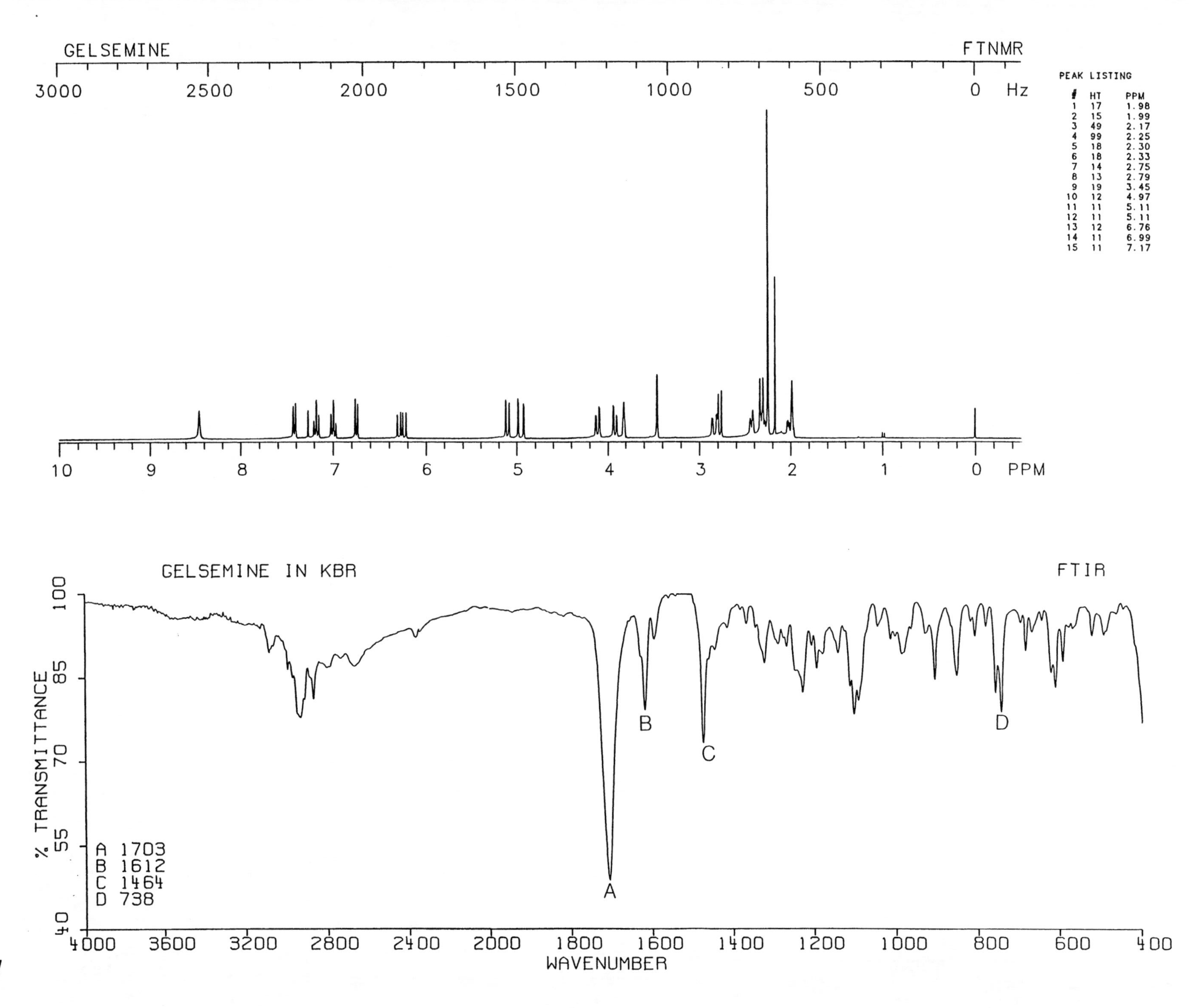
GELSEMINE
FTNMR
3000
2500
2000
1500
1000
500
0
Hz
PEAK LISTING
# HT PPM
1 17 1.98
2 15 1.99
3 49 2.17
4 99 2.25
5 18 2.30
6 18 2.33
7 14 2.75
8 13 2.79
9 19 3.45
10 12 4.97
11 11 5.11
12 11 5.11
13 12 6.76
14 11 6.99
15 11 7.17
10
9
8
7
6
5
4
3
2
1
0
PPM
GELSEMINE IN KBR
FTIR
100
85
70
55
40
% TRANSMITTANCE
A 1703
B 1612
C 1464
D 738
A
B
C
D
4000
3600
3200
2800
2400
2000
1800
1600
1400
1200
1000
800
600
400
WAVENUMBER

# GEMFIBROZIL

$C_{15}H_{22}O_3$

Molecular weight: 250.35 (250.16)
Synonyms: 5-(2,5-Dimethylphenoxy)-2,2-dimethylpentanoic acid

Trade names: Lopid

Use: Lipid regulating agent
HPLC: Si-10; 5A:95B; 12.0
GC: 1939; 200°C

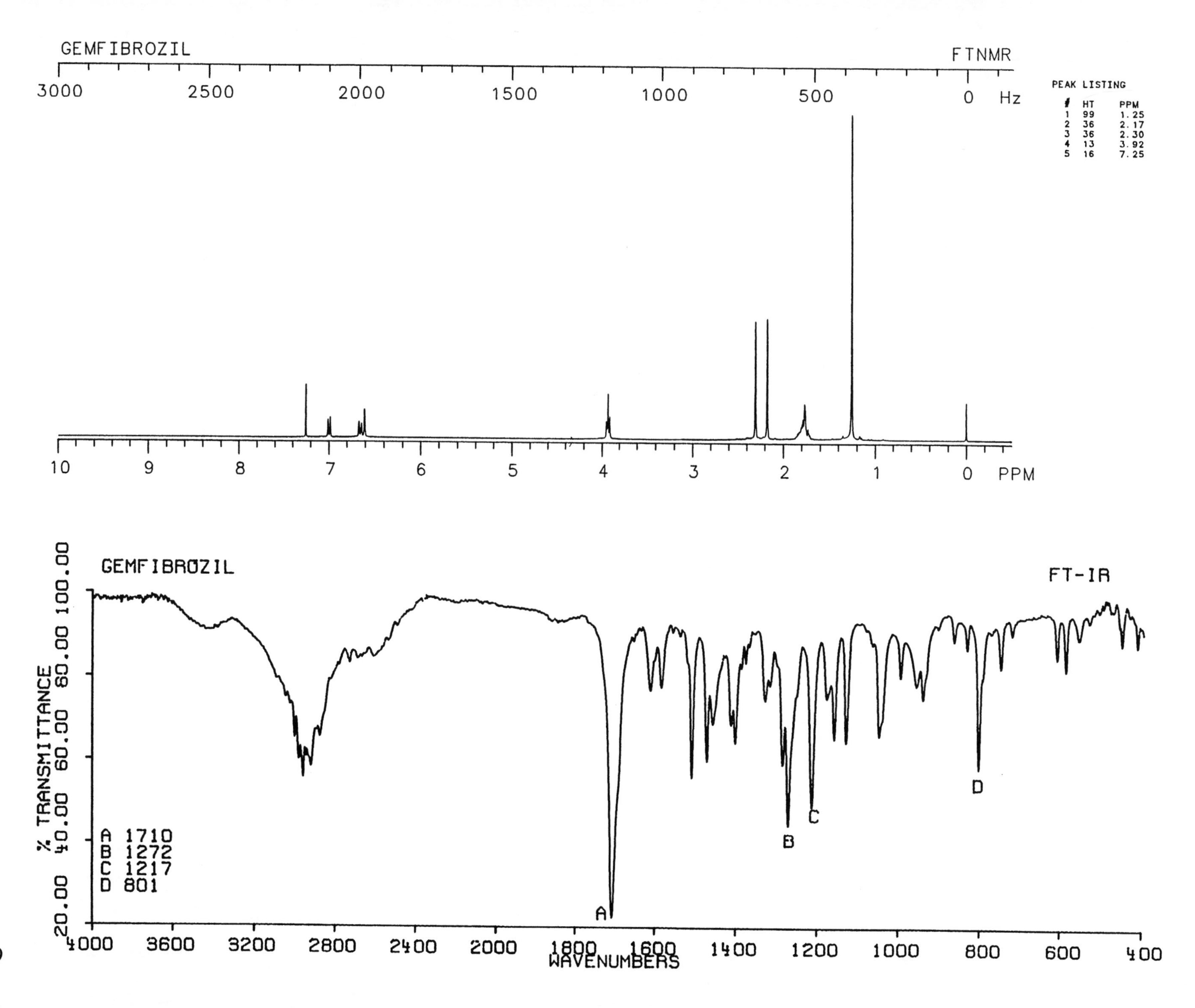

GEMFIBROZIL
FTNMR
3000
2500
2000
1500
1000
500
0 Hz
PEAK LISTING
# HT PPM
1 99 1.25
2 36 2.17
3 36 2.30
4 13 3.92
5 16 7.25
10
9
8
7
6
5
4
3
2
1
0 PPM
GEMFIBROZIL
FT-IR
% TRANSMITTANCE
20.00 40.00 60.00 80.00 100.00
A 1710
B 1272
C 1217
D 801
A
B
C
D
4000 3600 3200 2800 2400 2000 1800 1600 1400 1200 1000 800 600 400
WAVENUMBERS

# GENTAMICIN

**Molecular weight:**
**Synonyms:** Gentamycin

**Trade names:** Garamycin, Gentamicin

**Use:** Antibacterial
**HPLC:**
**GC:**

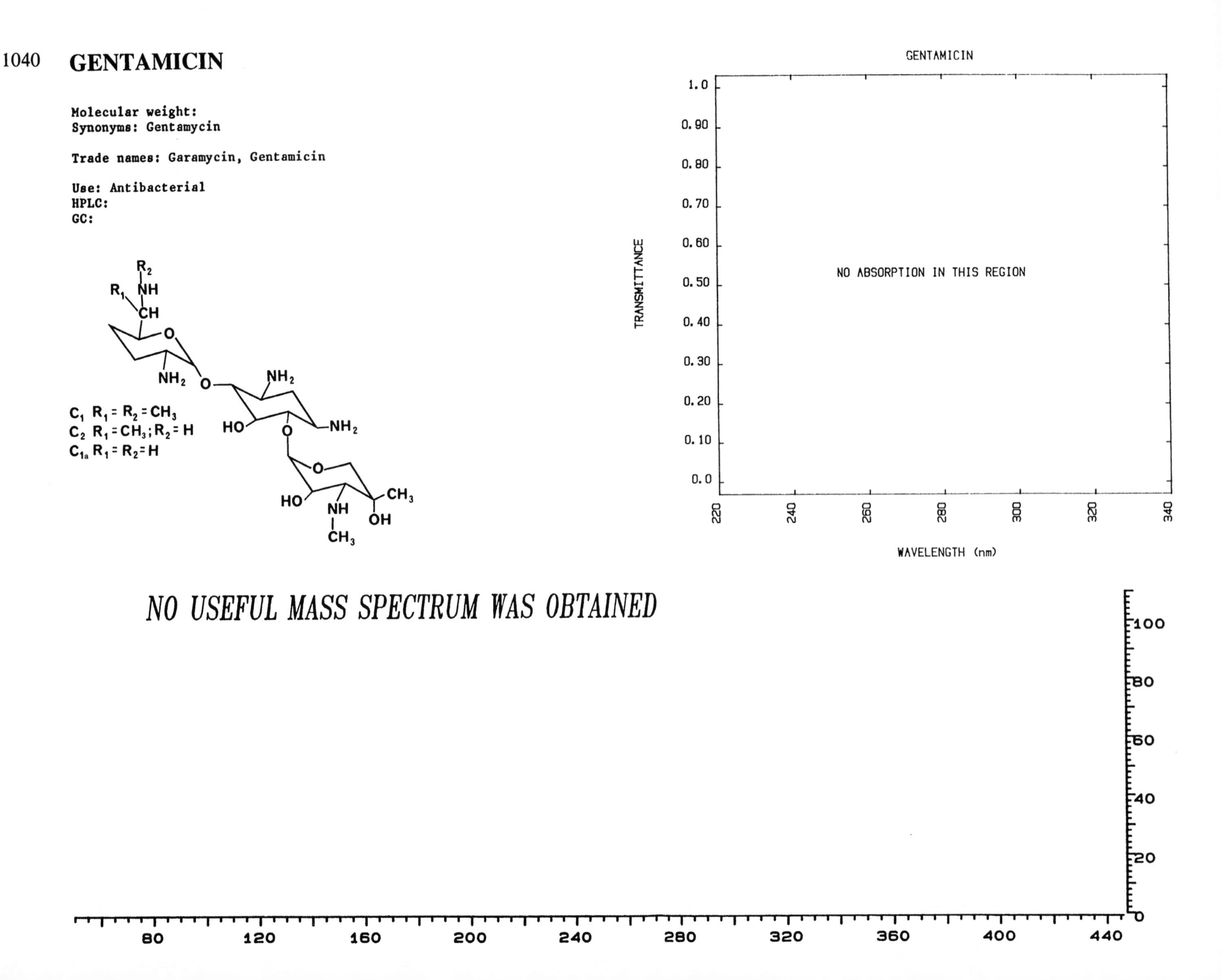

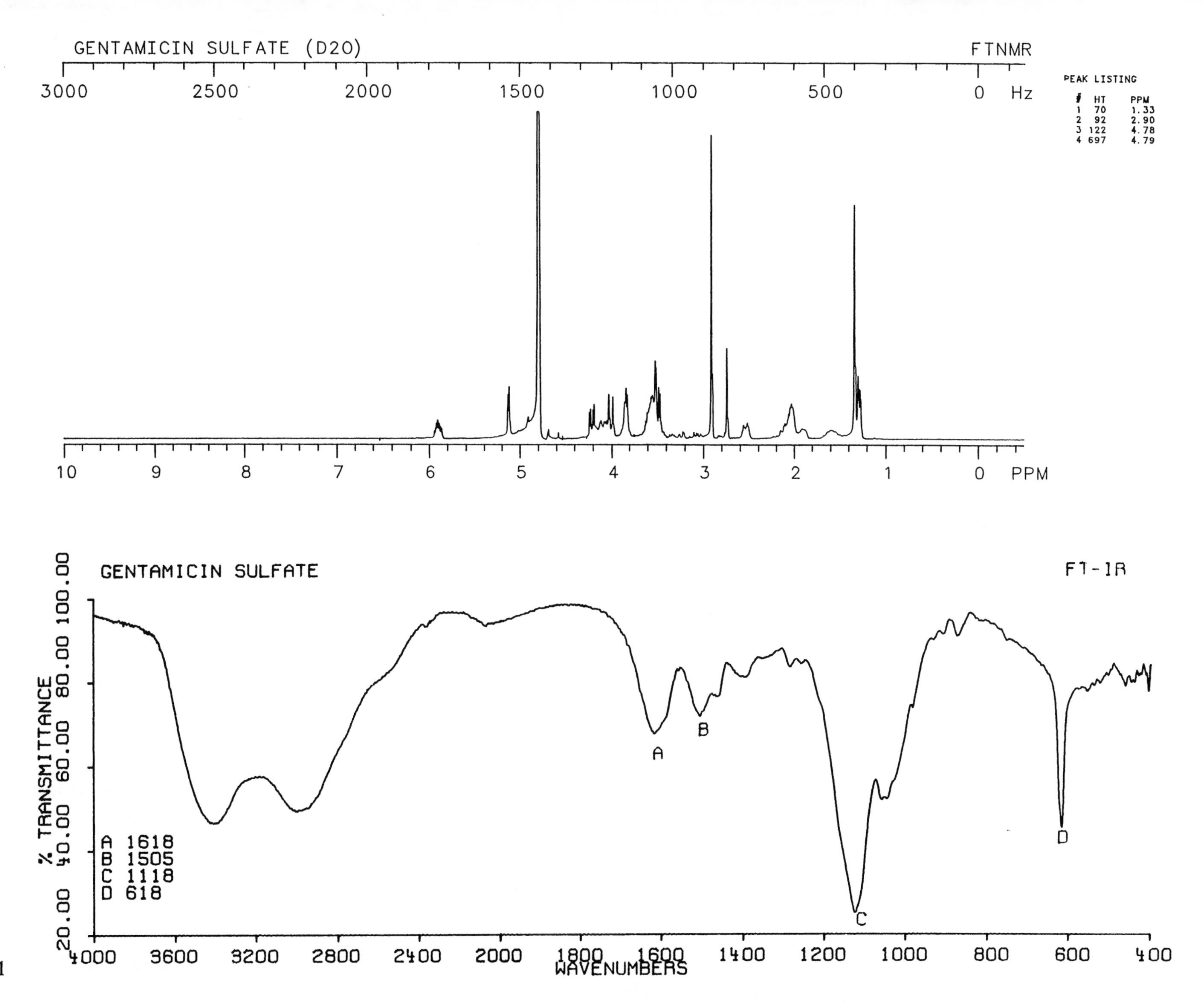
GENTAMICIN SULFATE (D2O)
FTNMR
3000 2500 2000 1500 1000 500 0 Hz
PEAK LISTING
# HT PPM
1 70 1.33
2 92 2.90
3 122 4.78
4 697 4.79
10 9 8 7 6 5 4 3 2 1 0 PPM
GENTAMICIN SULFATE
FT-IR
% TRANSMITTANCE
20.00 40.00 60.00 80.00 100.00
A 1618
B 1505
C 1118
D 618
4000 3600 3200 2800 2400 2000 1800 1600 1400 1200 1000 800 600 400
WAVENUMBERS

# GENTISIC ACID

$C_7H_6O_4$

Molecular weight: 154.12 (154.03)
Synonyms: 2,5-Dihydroxybenzoic acid; 5-hydroxysalicylic acid

Trade names: Gentinatre, Gentisod, Legential, Gentisine

Use: Analgesic, antirheumatic
HPLC:
GC:

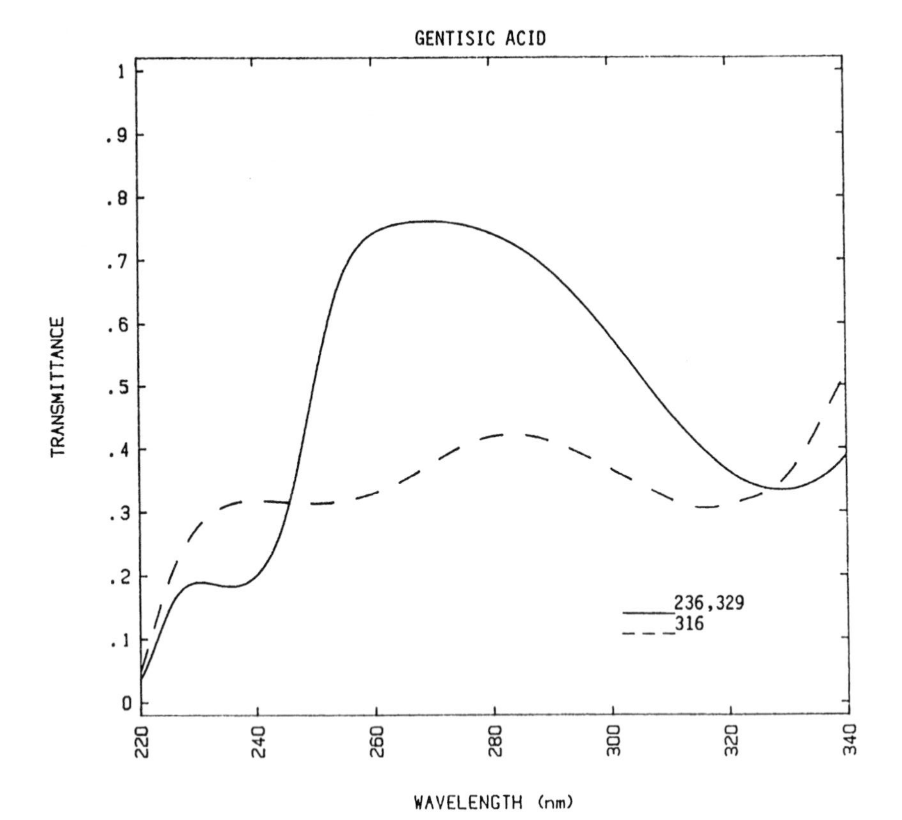


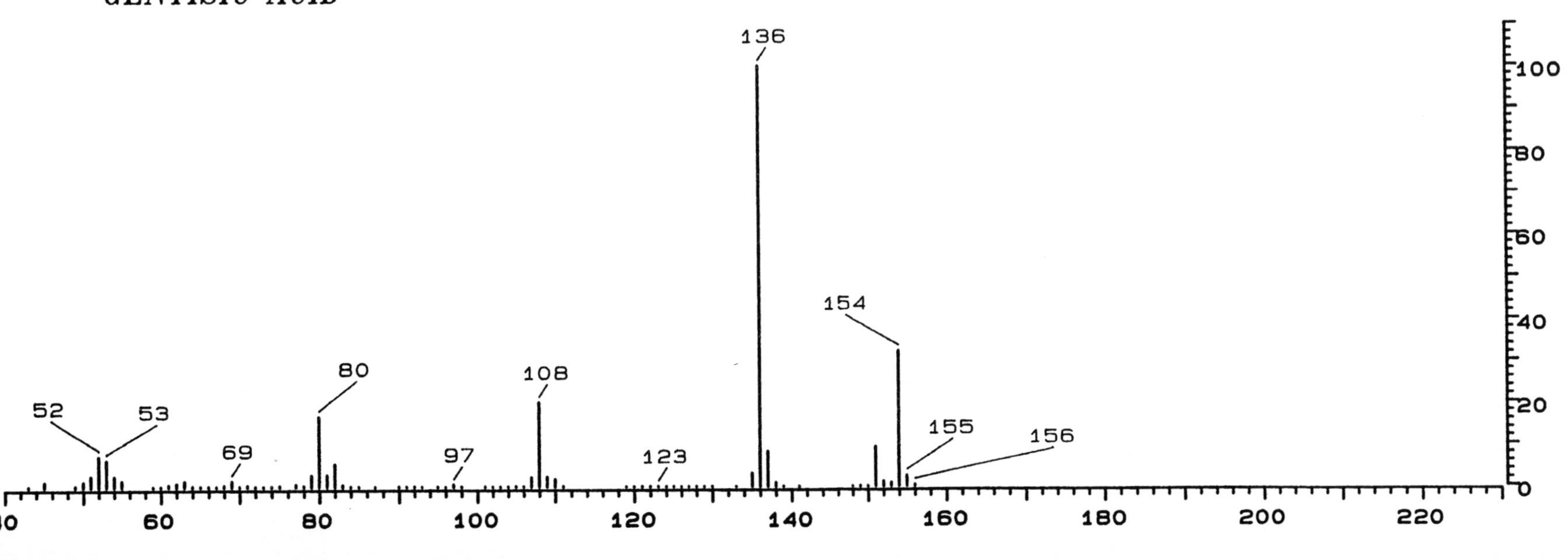

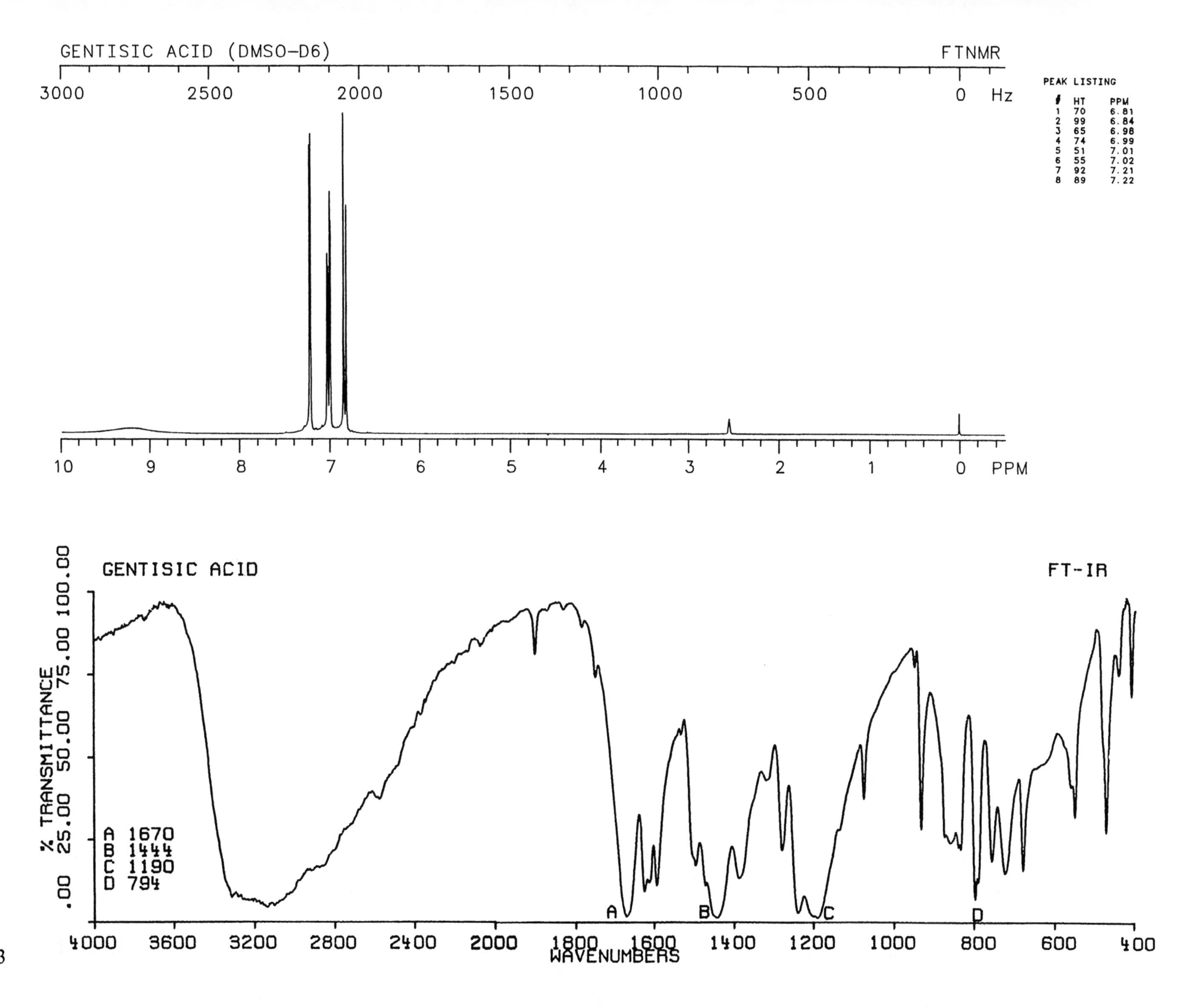
GENTISIC ACID (DMSO-D6)
FTNMR
3000 2500 2000 1500 1000 500 0 Hz
PEAK LISTING
# HT PPM
1 70 6.81
2 99 6.84
3 65 6.98
4 74 6.99
5 51 7.01
6 55 7.02
7 92 7.21
8 89 7.22
10 9 8 7 6 5 4 3 2 1 0 PPM
GENTISIC ACID
FT-IR
% TRANSMITTANCE
.00 25.00 50.00 75.00 100.00
A 1670
B 1444
C 1190
D 794
A B C D
4000 3600 3200 2800 2400 2000 1800 1600 1400 1200 1000 800 600 400
WAVENUMBERS

# GITOXIGENIN

$C_{23}H_{34}O_5$

**Molecular weight:** 390.50 (390.24)
**Synonyms:** 3β,14,16β-Trihydroxy-5β-card-20(22)-enolide

**Trade names:**

**Use:** Cardiotonic
**HPLC:**
**GC:**

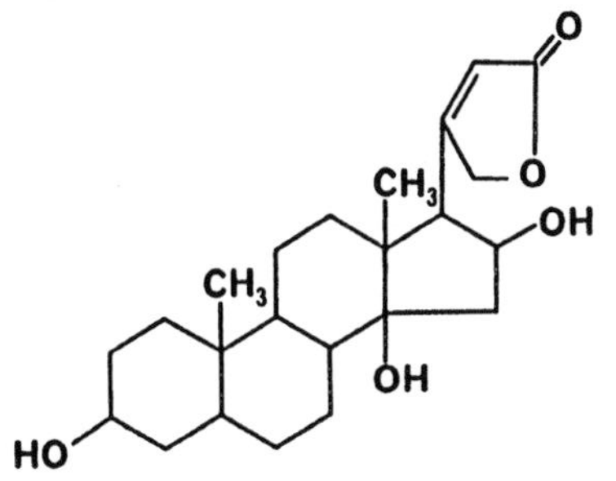

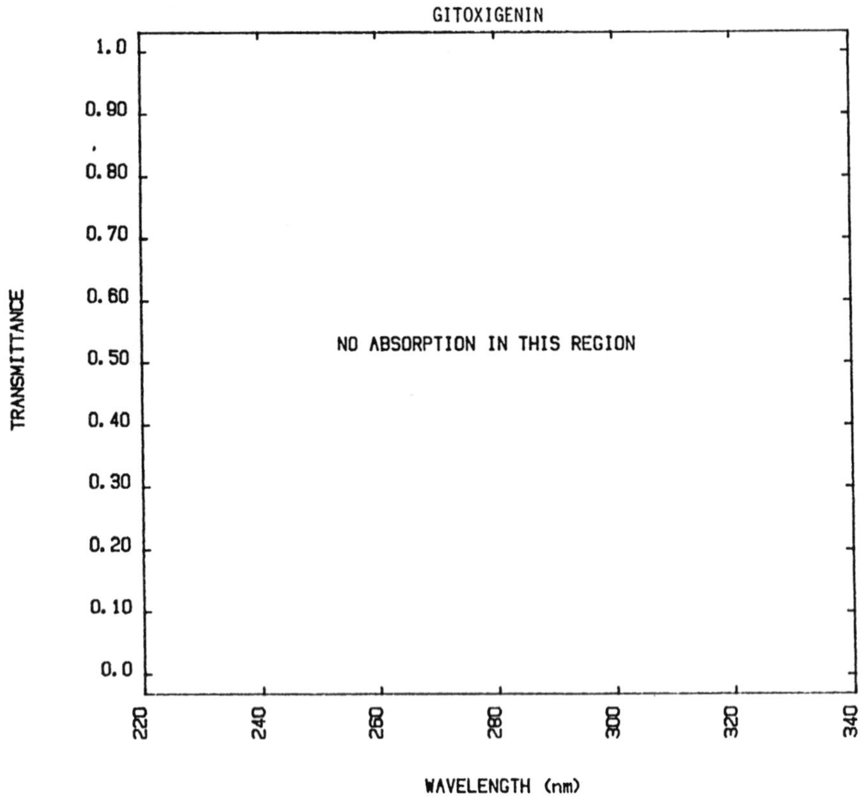

O
O
OH
CH3
CH3
OH
HO

GITOXIGENIN

TRANSMITTANCE
1.0
0.90
0.80
0.70
0.60
0.50
0.40
0.30
0.20
0.10
0.0

NO ABSORPTION IN THIS REGION

220
240
260
280
300
320
340

WAVELENGTH (nm)

*GITOXIGENIN--DIP*

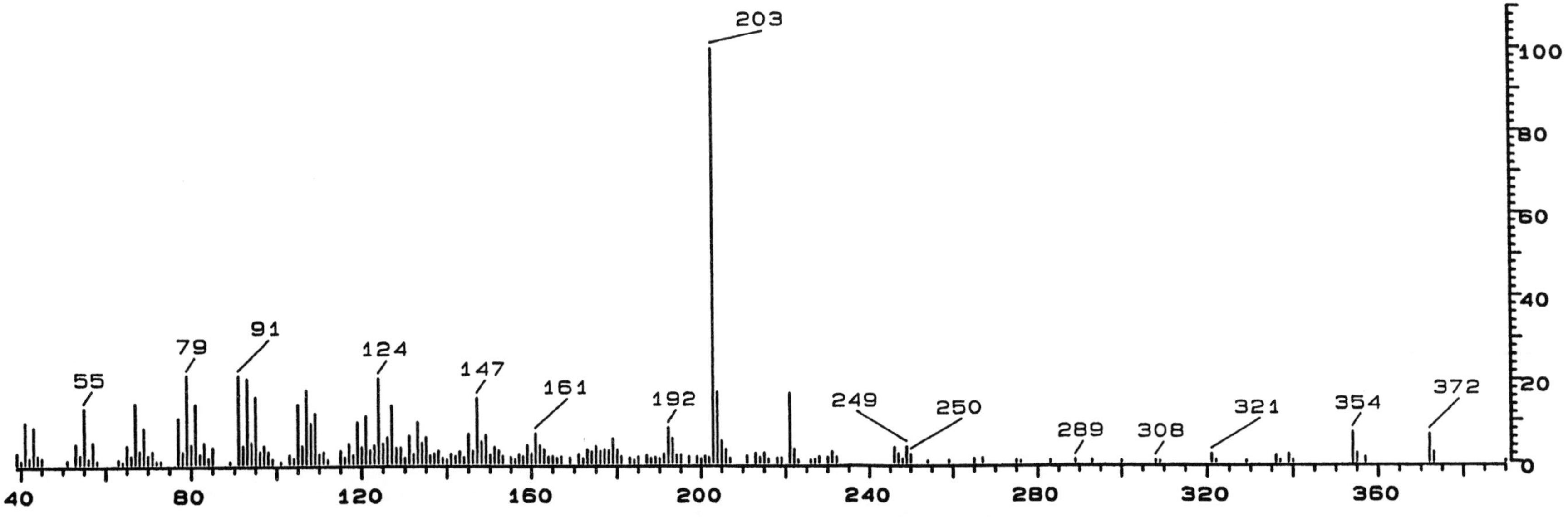

203
55
79
91
124
147
161
192
249
250
289
308
321
354
372

40
80
120
160
200
240
280
320
360

100
80
60
40
20
0

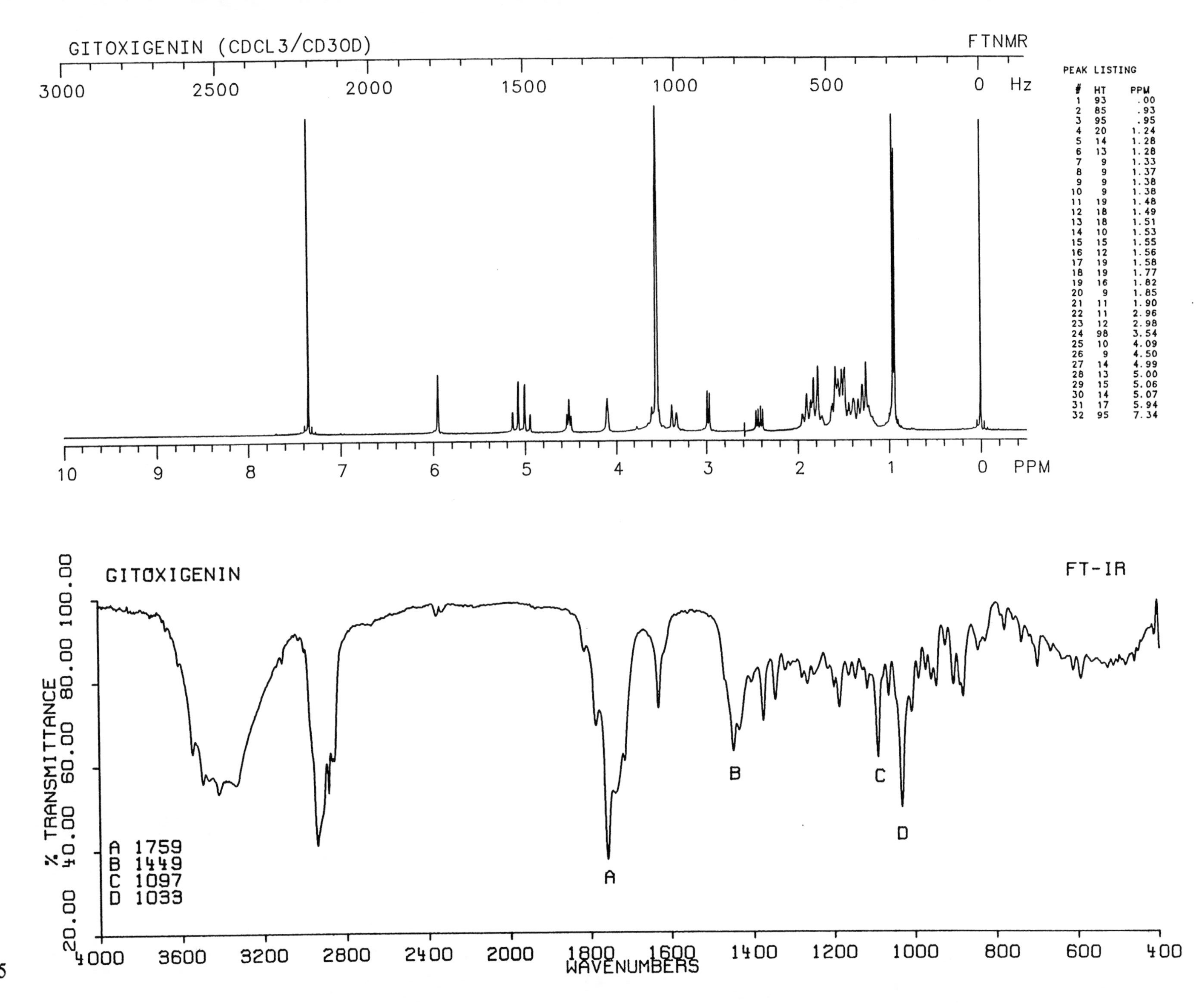
GITOXIGENIN (CDCL3/CD3OD)
FTNMR
PEAK LISTING
# HT PPM
1 93 .00
2 85 .93
3 95 .95
4 20 1.24
5 14 1.28
6 13 1.28
7 9 1.33
8 9 1.37
9 9 1.38
10 9 1.38
11 19 1.48
12 18 1.49
13 18 1.51
14 10 1.53
15 15 1.55
16 12 1.56
17 19 1.58
18 19 1.77
19 16 1.82
20 9 1.85
21 11 1.90
22 11 2.96
23 12 2.98
24 98 3.54
25 10 4.09
26 9 4.50
27 14 4.99
28 13 5.00
29 15 5.06
30 14 5.07
31 17 5.94
32 95 7.34
Hz
PPM
GITOXIGENIN
FT-IR
A 1759
B 1449
C 1097
D 1033
% TRANSMITTANCE
WAVENUMBERS

# GLIPIZIDE

$C_{21}H_{27}N_5O_4S$

**Molecular weight:** 445.55 (445.18)
**Synonyms:** N-[2-[4-[[[(Cyclohexylamino)carbonyl]amino]sulfonyl]phenyl]-ethyl]-5-methylpyrazinecarboxamide; glydiazinamide
**Trade names:** Glibenese, Glucotrol, Minidiab, Minodiab

**Use:** Antidiabetic
**HPLC:** Si-10; 5A:95B; 9.5
**GC:**

$CH_3$–(pyrazine)–C(=O)–NH–$CH_2$–$CH_2$–(phenyl)–$SO_2$–NH–C(=O)–NH–(cyclohexyl)

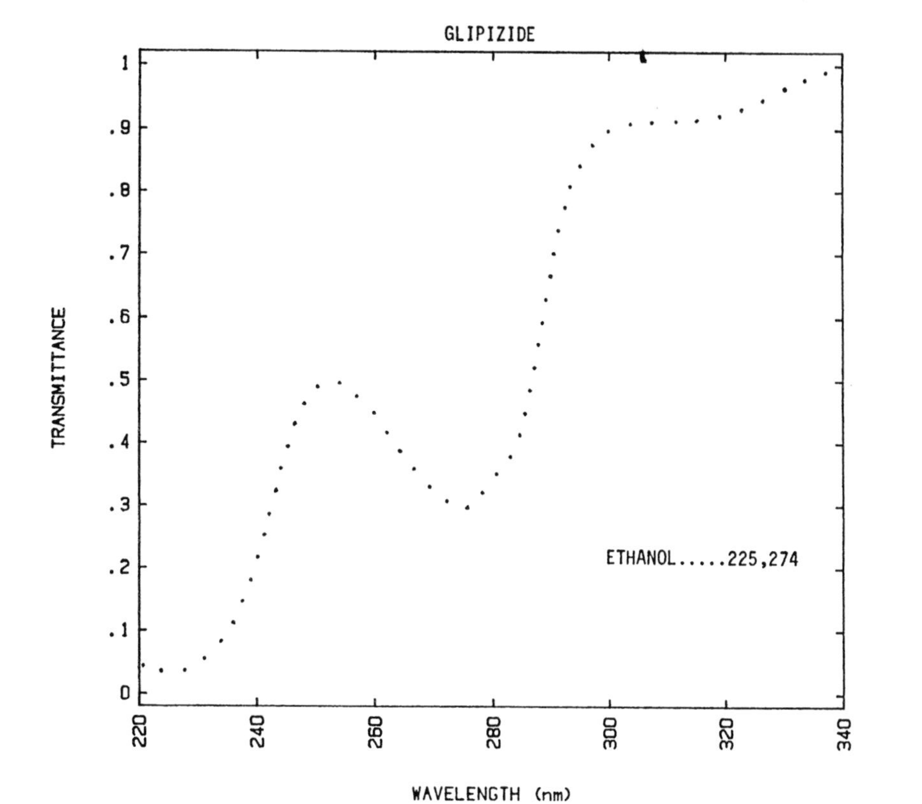


GLIPIZIDE--DIP

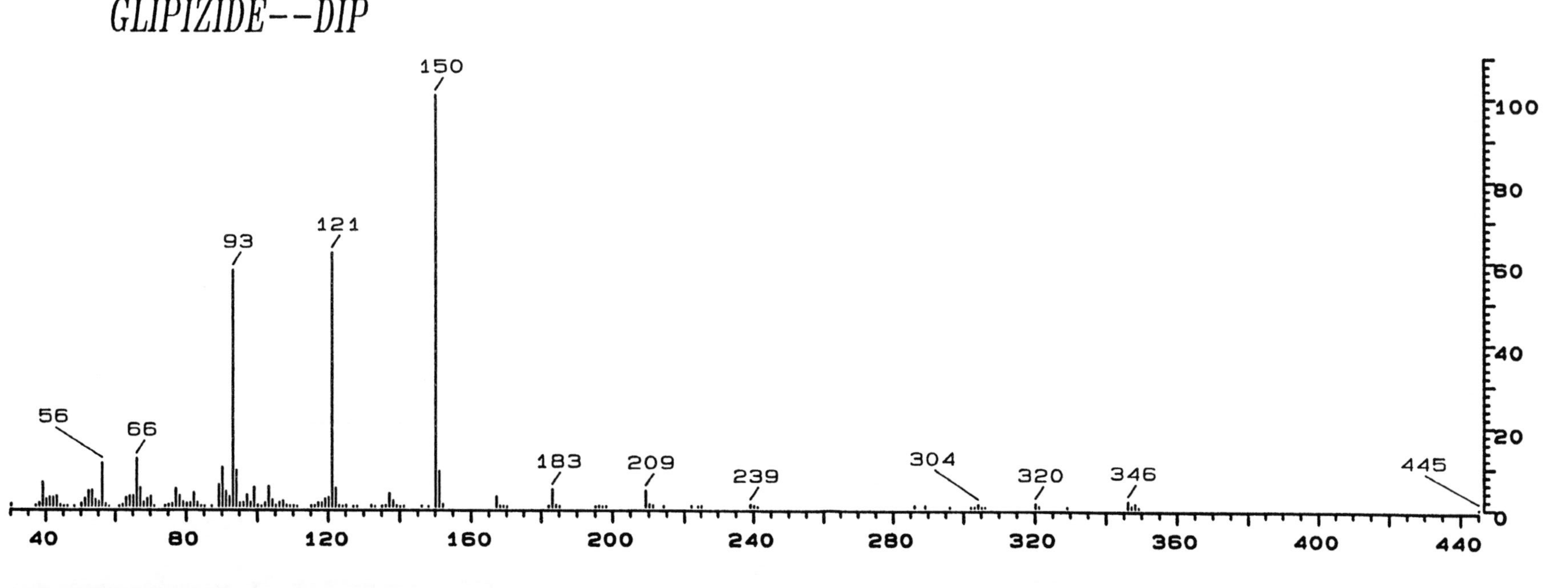

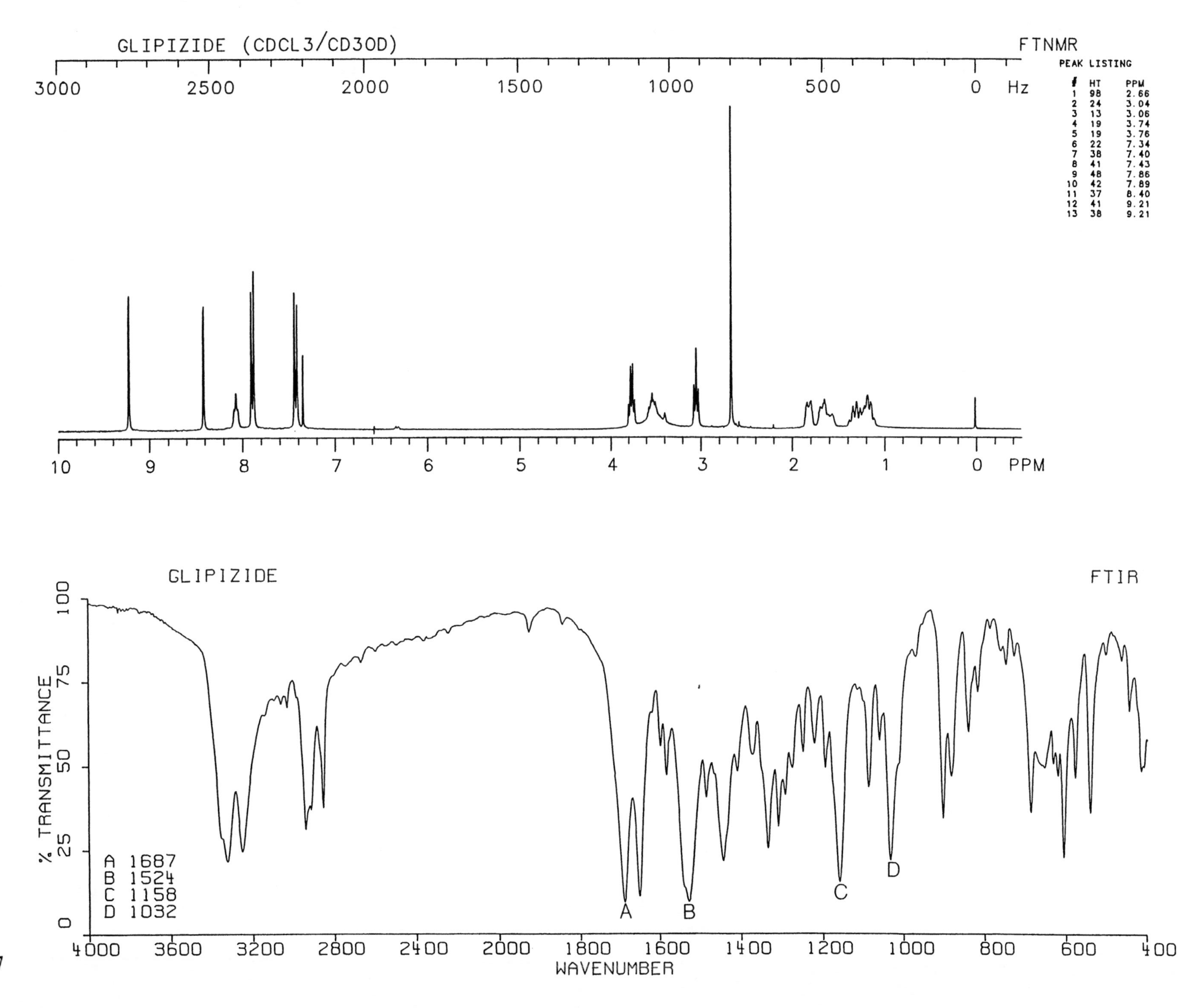

GLIPIZIDE (CDCL3/CD3OD)
FTNMR
PEAK LISTING
# HT PPM
1 98 2.66
2 24 3.04
3 13 3.06
4 19 3.74
5 19 3.76
6 22 7.34
7 38 7.40
8 41 7.43
9 48 7.86
10 42 7.89
11 37 8.40
12 41 9.21
13 38 9.21
3000 2500 2000 1500 1000 500 0 Hz
10 9 8 7 6 5 4 3 2 1 0 PPM
GLIPIZIDE
FTIR
% TRANSMITTANCE
100 75 50 25 0
A 1687
B 1524
C 1158
D 1032
A B C D
4000 3600 3200 2800 2400 2000 1800 1600 1400 1200 1000 800 600 400
WAVENUMBER

# GLUTETHIMIDE

$C_{13}H_{15}NO_2$

Molecular weight: 217.27 (217.11)
Synonyms: 3-Ethyl-3-phenyl-2,6-piperidinedione; α-ethyl-α-phenylglutarimide
Trade names: Doriden

Use: Hypnotic
HPLC: Si-10; 1A:99B; 3.7
GC: 1863; 200°C

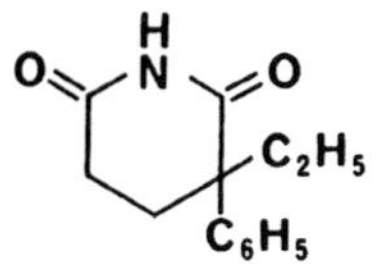

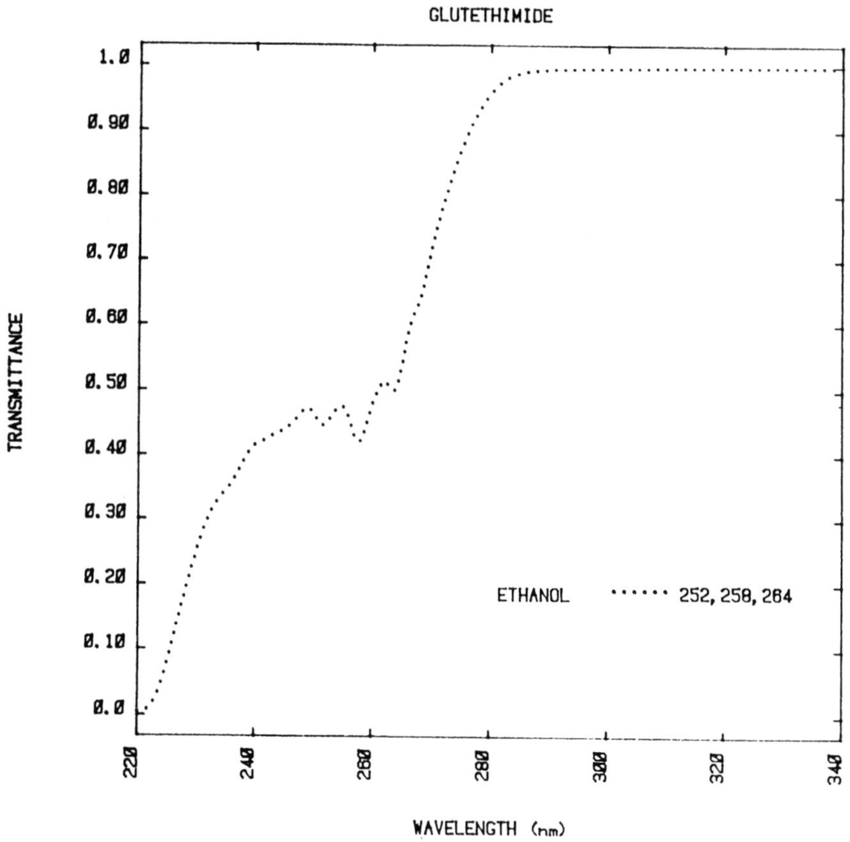

GLUTETHIMIDE

TRANSMITTANCE

1.0, 0.90, 0.80, 0.70, 0.60, 0.50, 0.40, 0.30, 0.20, 0.10, 0.0

ETHANOL ······ 252, 258, 264

220, 240, 260, 280, 300, 320, 340

WAVELENGTH (nm)

*GLUTETHIMIDE*

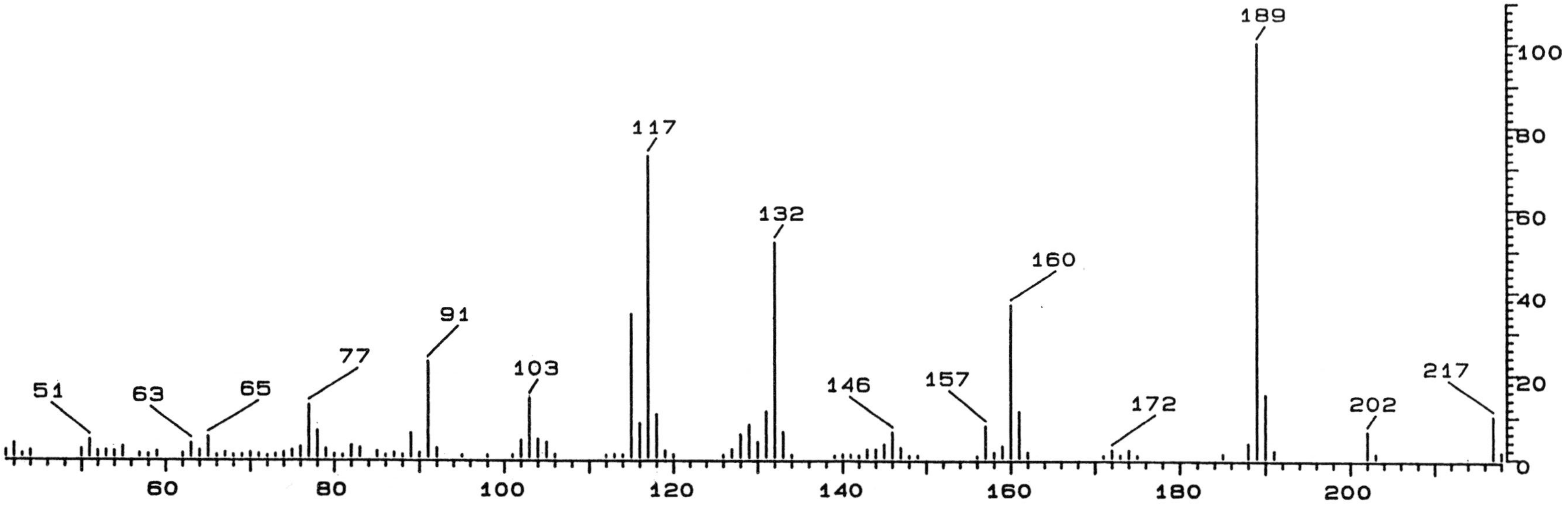

51, 63, 65, 77, 91, 103, 117, 132, 146, 157, 160, 172, 189, 202, 217

60, 80, 100, 120, 140, 160, 180, 200

0, 20, 40, 60, 80, 100

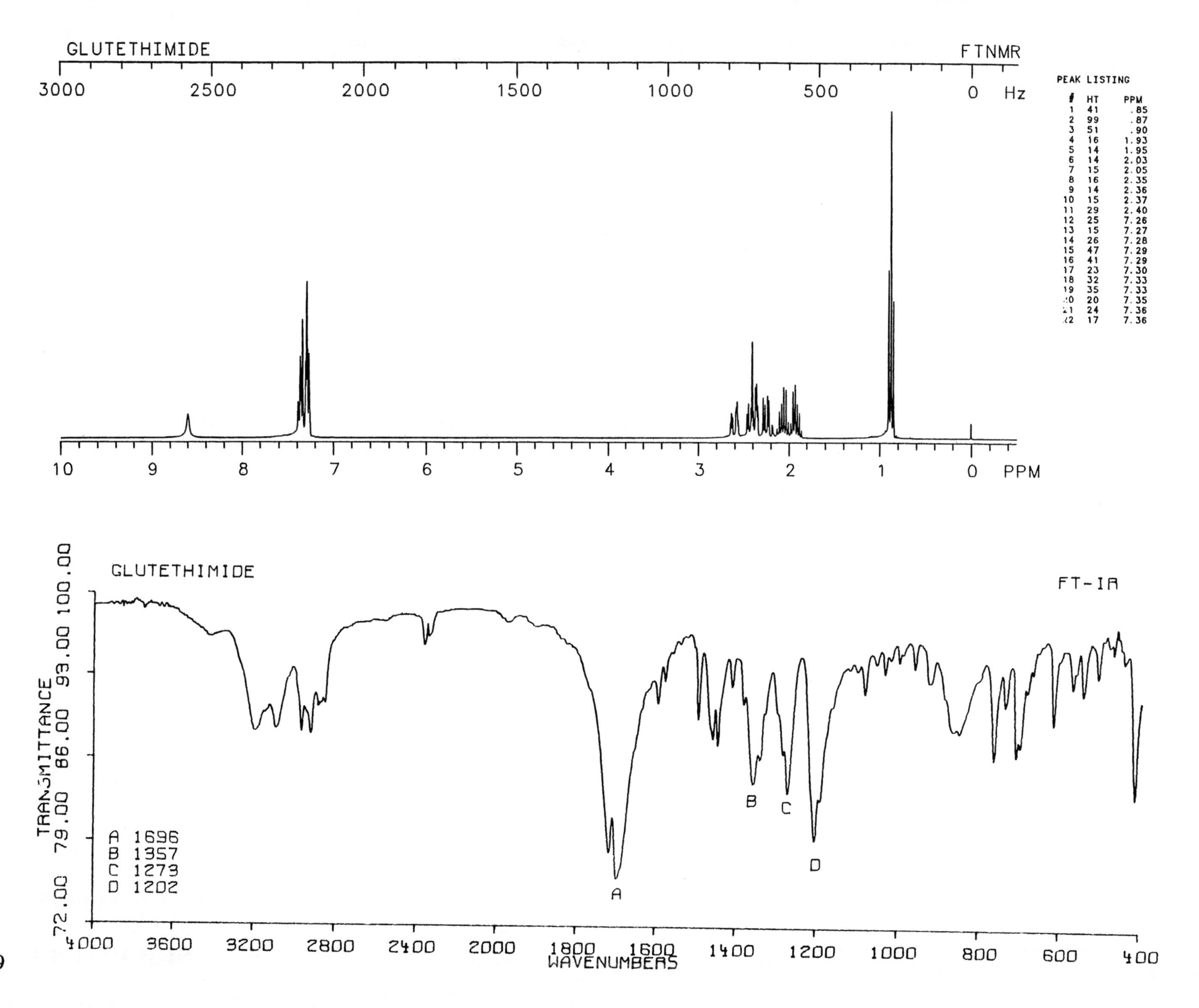
GLUTETHIMIDE
FTNMR
3000 2500 2000 1500 1000 500 0 Hz
10 9 8 7 6 5 4 3 2 1 0 PPM
PEAK LISTING
# HT PPM
1 41 .85
2 99 .87
3 51 .90
4 16 1.93
5 14 1.95
6 14 2.03
7 15 2.05
8 16 2.35
9 14 2.36
10 15 2.37
11 29 2.40
12 25 7.26
13 15 7.27
14 26 7.28
15 47 7.29
16 41 7.29
17 23 7.30
18 32 7.33
19 35 7.33
20 20 7.35
21 24 7.36
22 17 7.36
GLUTETHIMIDE
FT-IR
TRANSMITTANCE
72.00 79.00 86.00 93.00 100.00
A 1696
B 1357
C 1273
D 1202
A B C D
4000 3600 3200 2800 2400 2000 1800 1600 1400 1200 1000 800 600 400
WAVENUMBERS

# GRAMICIDIN

**Polypeptide cyclic antibiotic complex**

**Molecular weight:**
**Synonyms: Gramicidin D; Linear gramicidins**

**Trade names: Mycolog, Myco-Triacet, Mytrex, Neosporin, Nyst-olone, Tri-Thalmic**
**Use: Antibacterial**
**HPLC:**
**GC:**

HCO-Val-Gly-Ala-Leu-Ala-Val-Val-Val-[-Trp-Leu-]$_3$-Trp-$NHCH_2CH_2OH$
(L) (L) (D) (L) (D) (L) (L) (L) (D) (L)

valine-gramicidin A

GRAMICIDIN

TRANSMITTANCE

1 .9 .8 .7 .6 .5 .4 .3 .2 .1 0

ETHANOL.....220,282,290

220 240 260 280 300 320 340

WAVELENGTH (nm)

*NO USEFUL MASS SPECTRUM WAS OBTAINED*

100 80 60 40 20 0

80 120 160 200 240 280 320 360 400 440

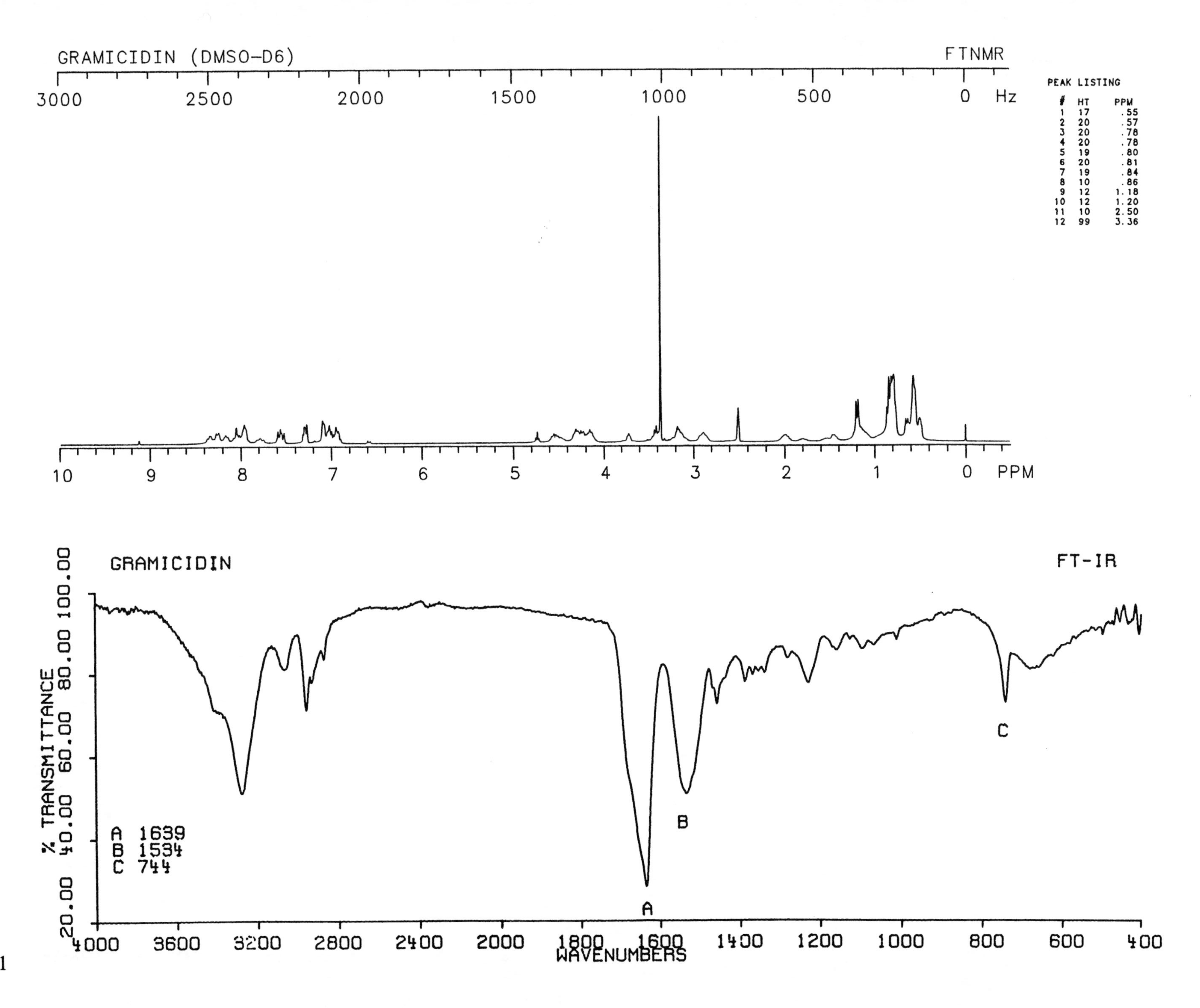
GRAMICIDIN (DMSO-D6)
FTNMR
3000 2500 2000 1500 1000 500 0 Hz
PEAK LISTING
# HT PPM
1 17 .55
2 20 .57
3 20 .78
4 20 .78
5 19 .80
6 20 .81
7 19 .84
8 10 .86
9 12 1.18
10 12 1.20
11 10 2.50
12 99 3.36
10 9 8 7 6 5 4 3 2 1 0 PPM
GRAMICIDIN
FT-IR
% TRANSMITTANCE
20.00 40.00 60.00 80.00 100.00
A 1639
B 1534
C 744
A
B
C
4000 3600 3200 2800 2400 2000 1800 1600 1400 1200 1000 800 600 400
WAVENUMBERS

# GRISEOFULVIN

$C_{17}H_{17}ClO_6$

**Molecular weight:** 352.77 (352.07)
**Synonyms:** (2S-trans)-7-Chloro-2',4,6-trimethoxy-6'-methylspiro-[benzofuran-2(3H),1'-[2]cyclohexene]-3,4'-dione
**Trade names:** Fulvicin, Grifulvin, Grisactin, Gris-PEG

**Use:** Antibiotic, antifungal
**HPLC:** Si-10; 1A:99B; 5.0
**GC:** 2887, 280°C

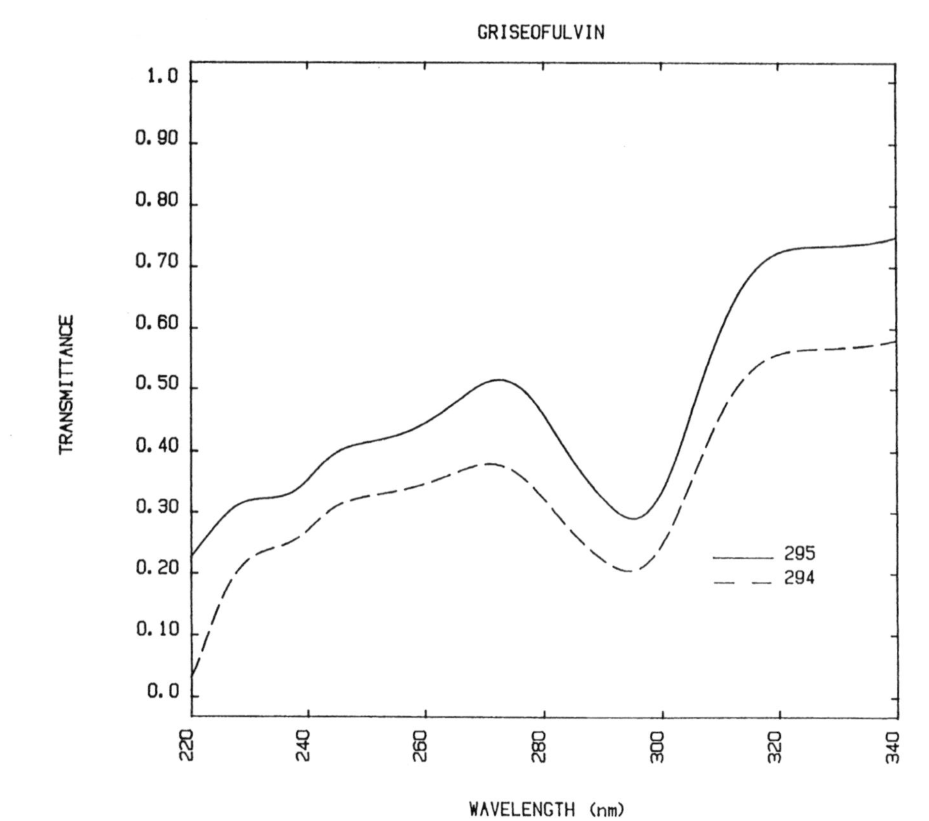


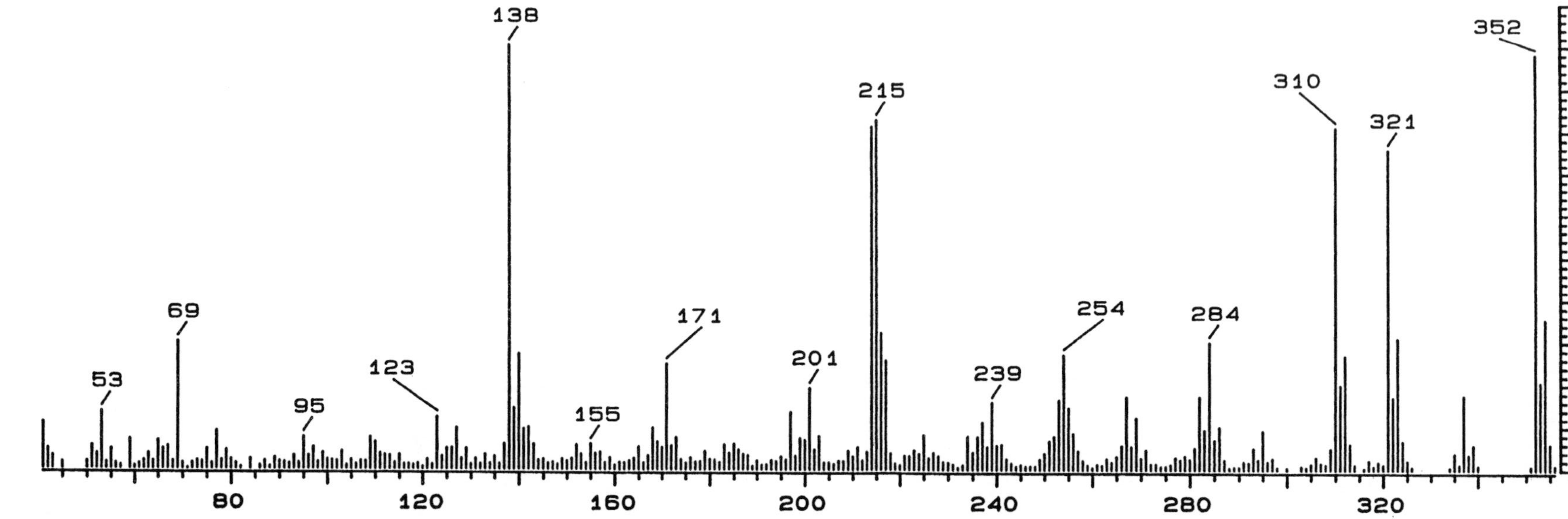


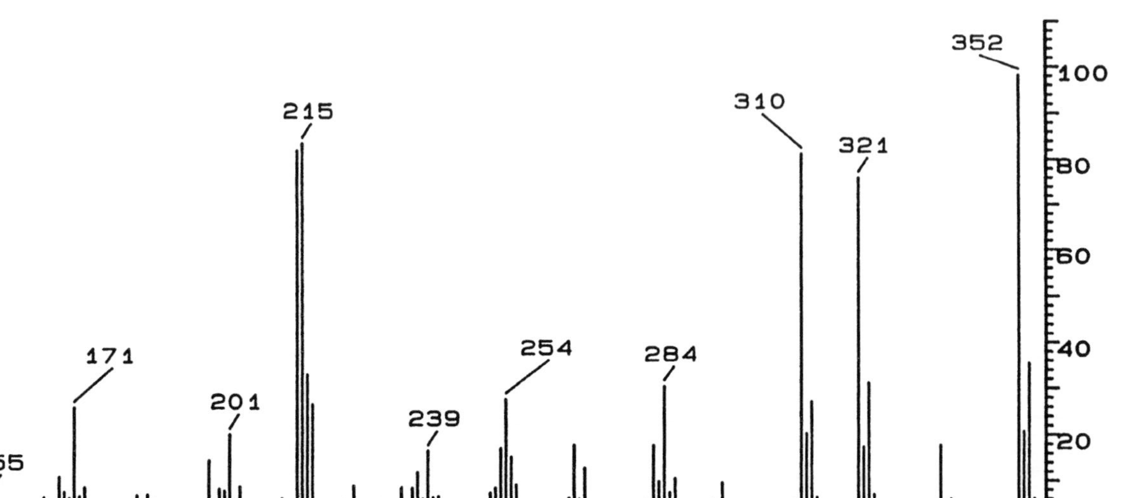

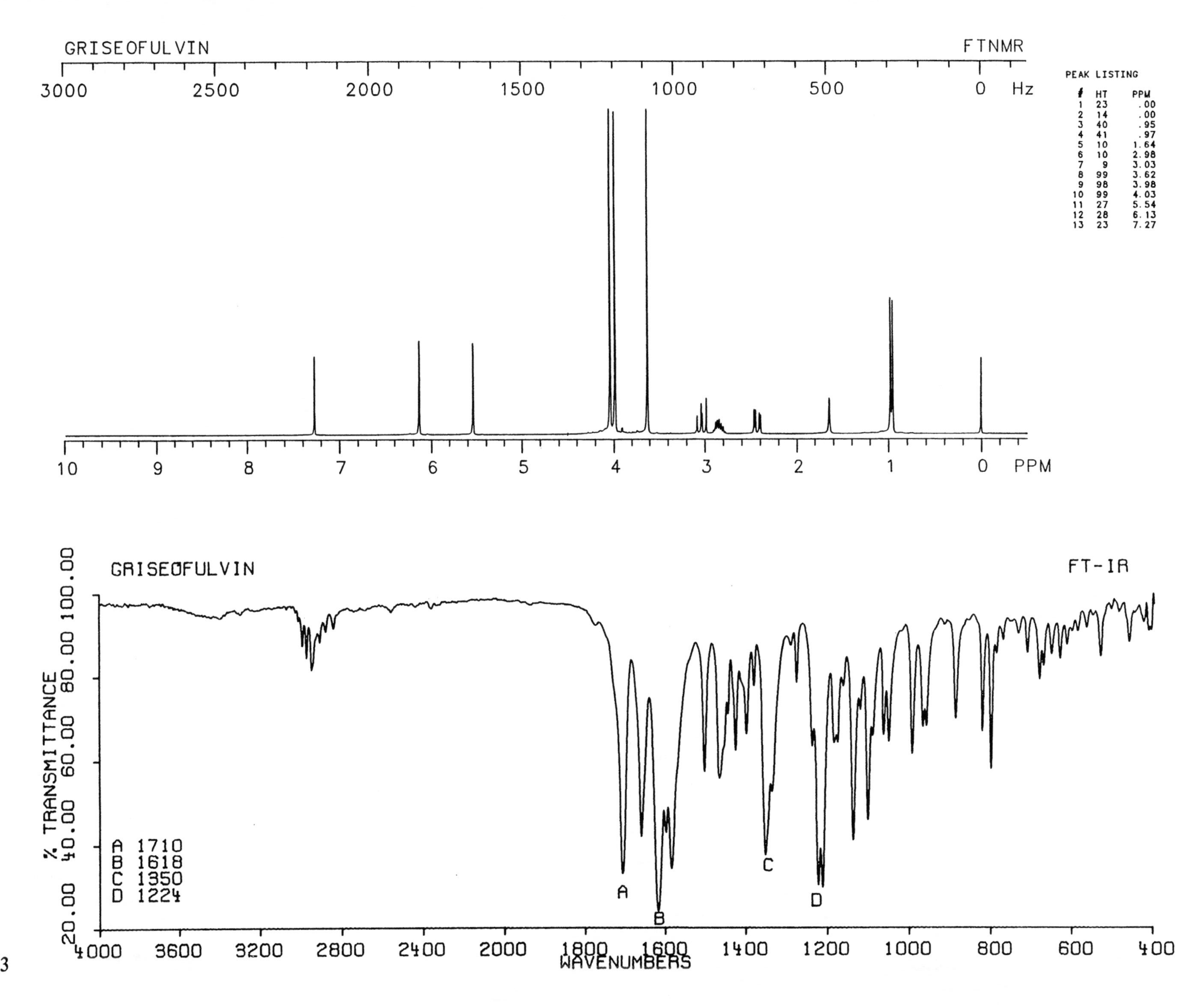
GRISEOFULVIN
FTNMR
PEAK LISTING
| # | HT | PPM |
|---|---|---|
| 1 | 23 | .00 |
| 2 | 14 | .00 |
| 3 | 40 | .95 |
| 4 | 41 | .97 |
| 5 | 10 | 1.64 |
| 6 | 10 | 2.98 |
| 7 | 9 | 3.03 |
| 8 | 99 | 3.62 |
| 9 | 98 | 3.98 |
| 10 | 99 | 4.03 |
| 11 | 27 | 5.54 |
| 12 | 28 | 6.13 |
| 13 | 23 | 7.27 |
GRISEOFULVIN
FT-IR
A 1710
B 1618
C 1350
D 1224
% TRANSMITTANCE
WAVENUMBERS

# GUAIACOL

$C_7H_8O_2$

Molecular weight: 124.13 (124.05)
Synonyms: O-Methoxyphenol; methylcatechol; O-hydroxyanisole; 1-hydroxy-2-methoxybenzene
Trade names:

Use: Expectorant
HPLC: Si-10; 20B:80C; 3.6
GC: 1076; 140°C

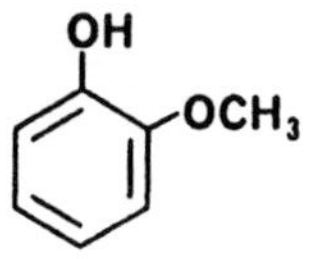

OH
OCH3

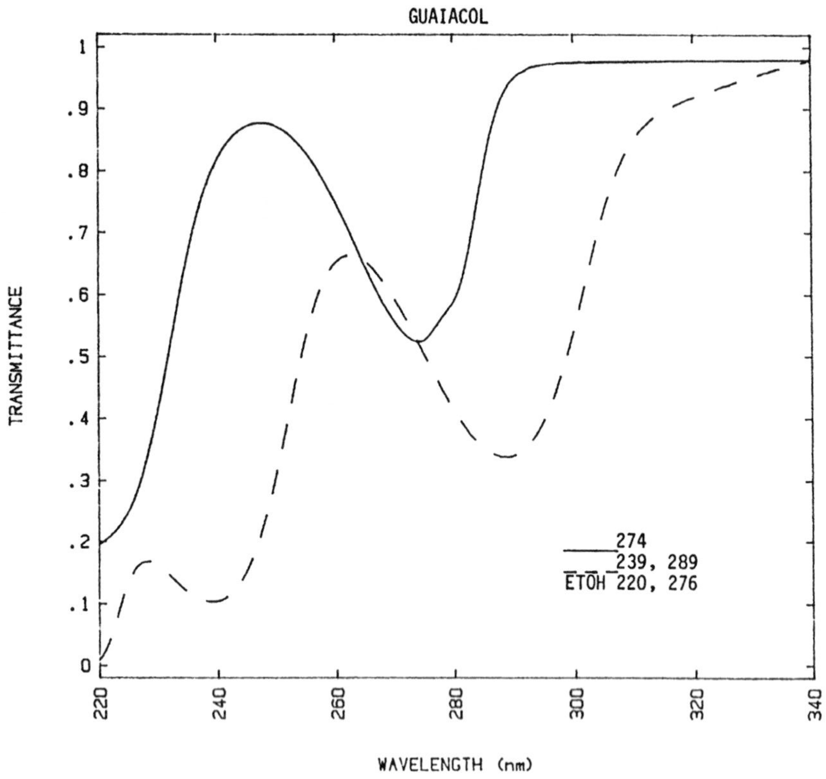

GUAIACOL

TRANSMITTANCE

WAVELENGTH (nm)

274
239, 289
ETOH 220, 276

*GUAIACOL*

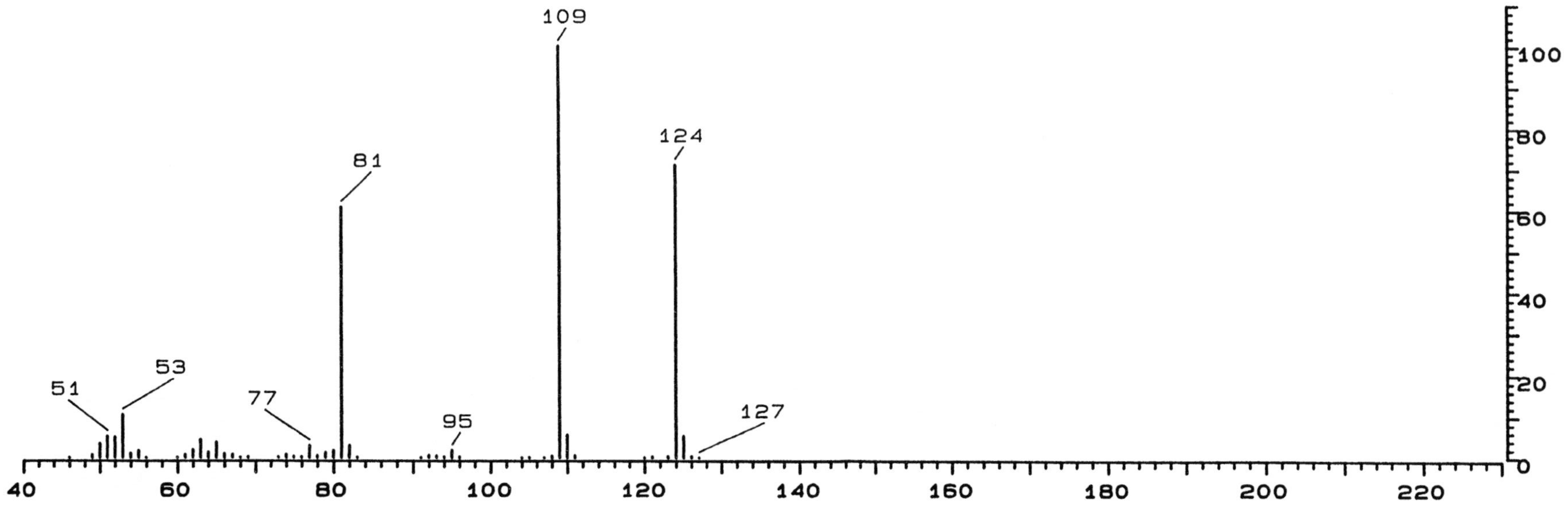

51 53 77 81 95 109 124 127

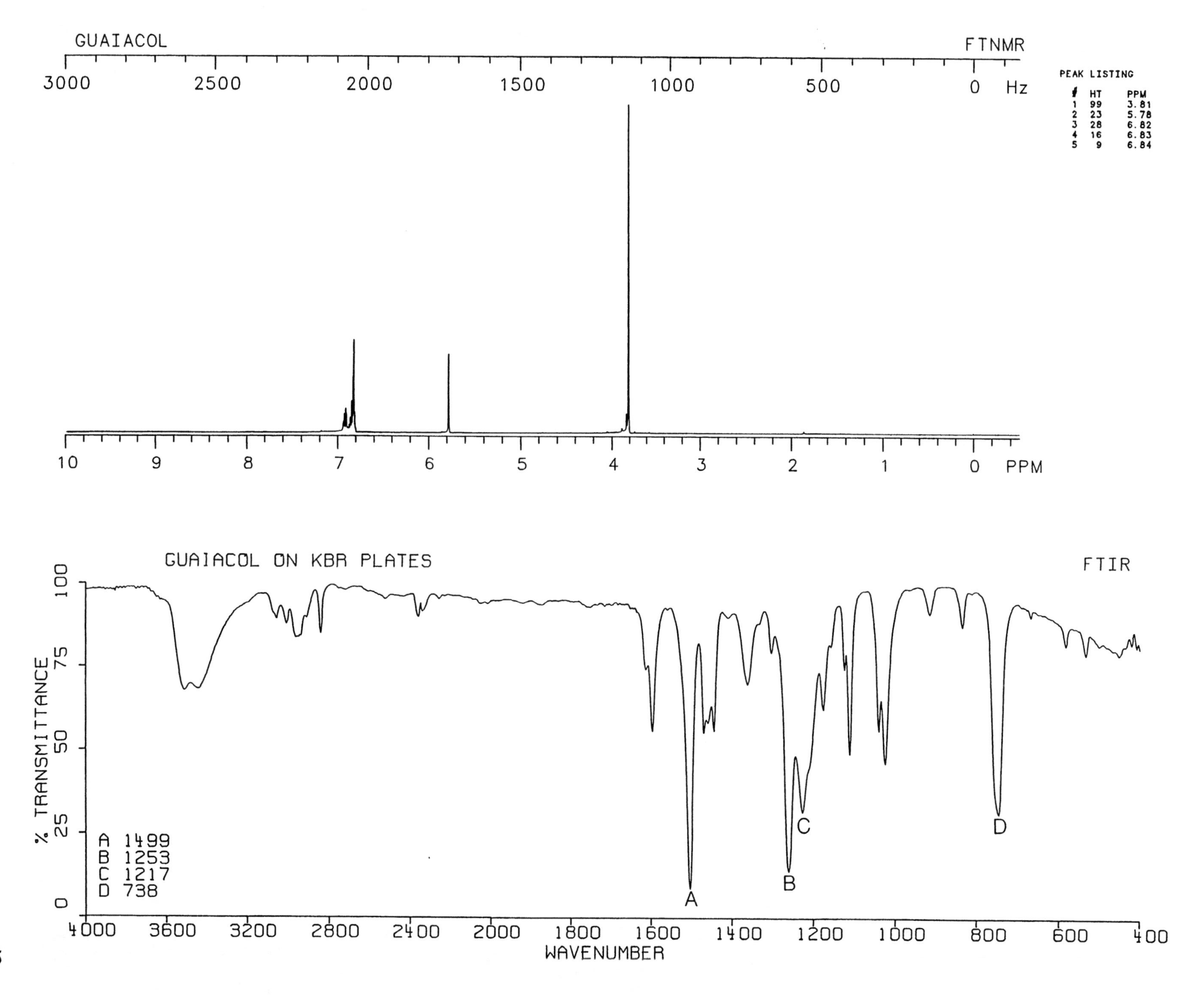
GUAIACOL
FTNMR
3000
2500
2000
1500
1000
500
0 Hz
PEAK LISTING
# HT PPM
1 99 3.81
2 23 5.78
3 28 6.82
4 16 6.83
5 9 6.84
10
9
8
7
6
5
4
3
2
1
0 PPM
GUAIACOL ON KBR PLATES
FTIR
100
75
50
25
0
% TRANSMITTANCE
A 1499
B 1253
C 1217
D 738
A
B
C
D
4000
3600
3200
2800
2400
2000
1800
1600
1400
1200
1000
800
600
400
WAVENUMBER

# GUAIFENESIN

$C_{10}H_{14}O_4$

Molecular weight: 198.22 (198.09)
Synonyms: 3-(o-Methoxyphenoxy)-1,2-propanediol; guaiacol glyceryl ether; glycerol guaiacolate
Trade names: Ingredient in numerous cough preparations

Use: Expectorant
HPLC: Si-10; 2A:98B; 5.0
GC: 1678; 200°C

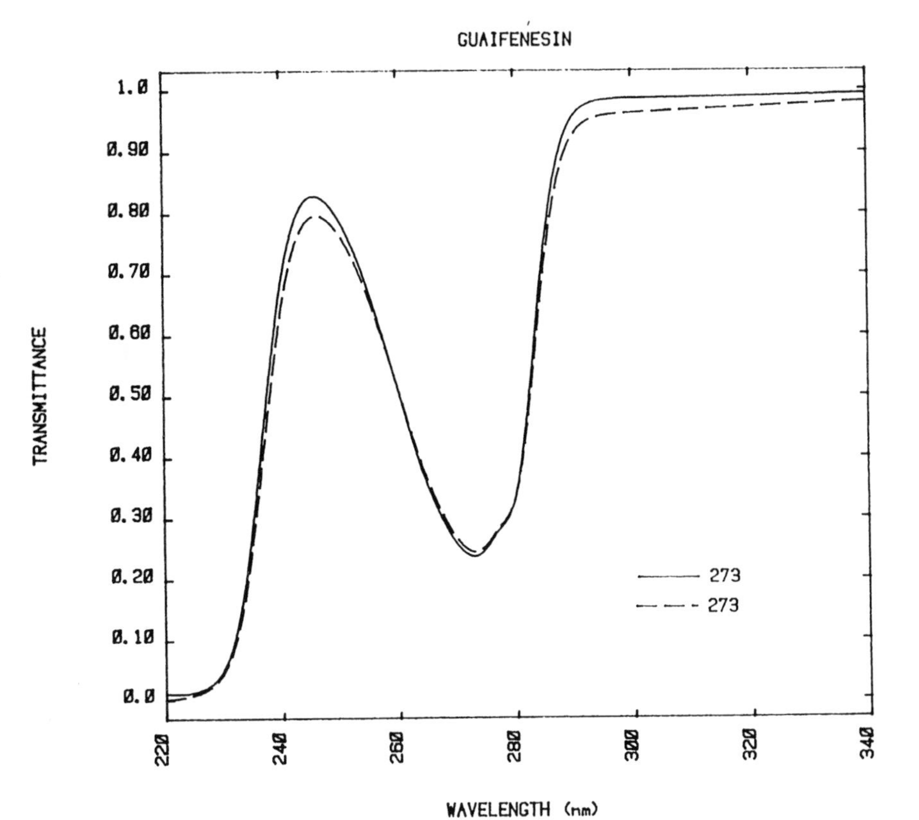


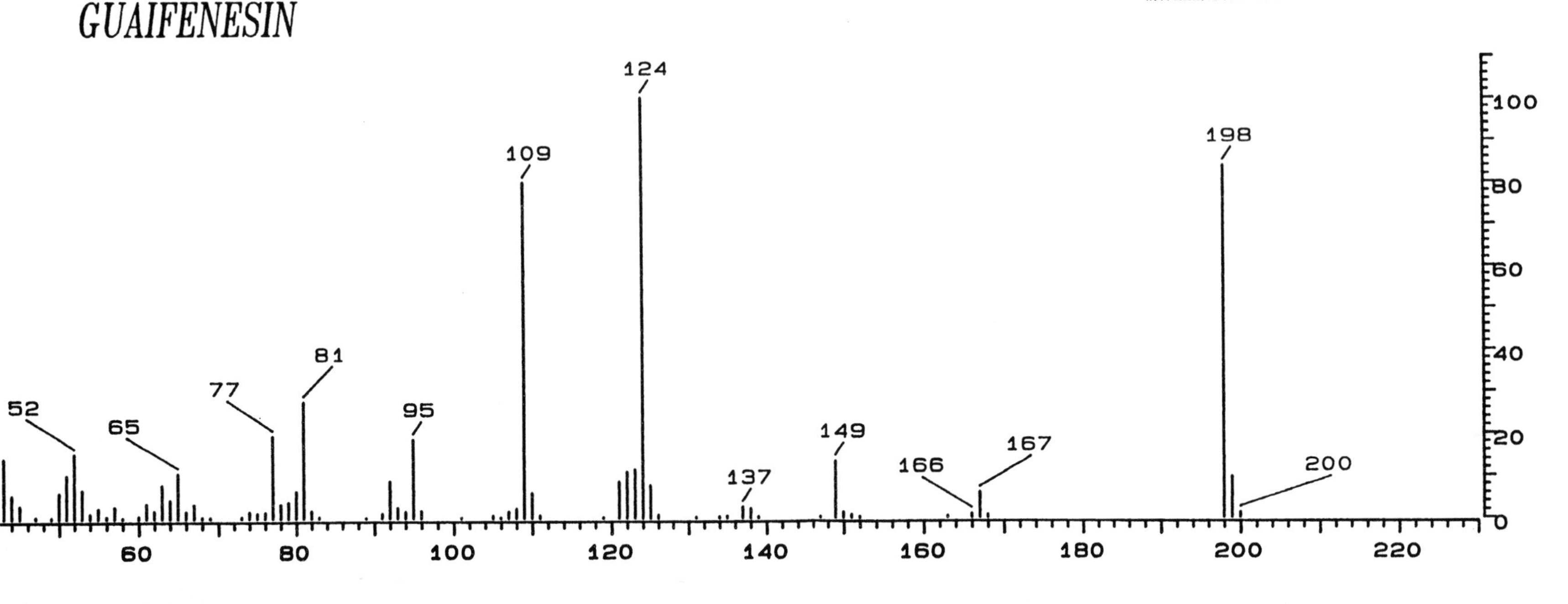

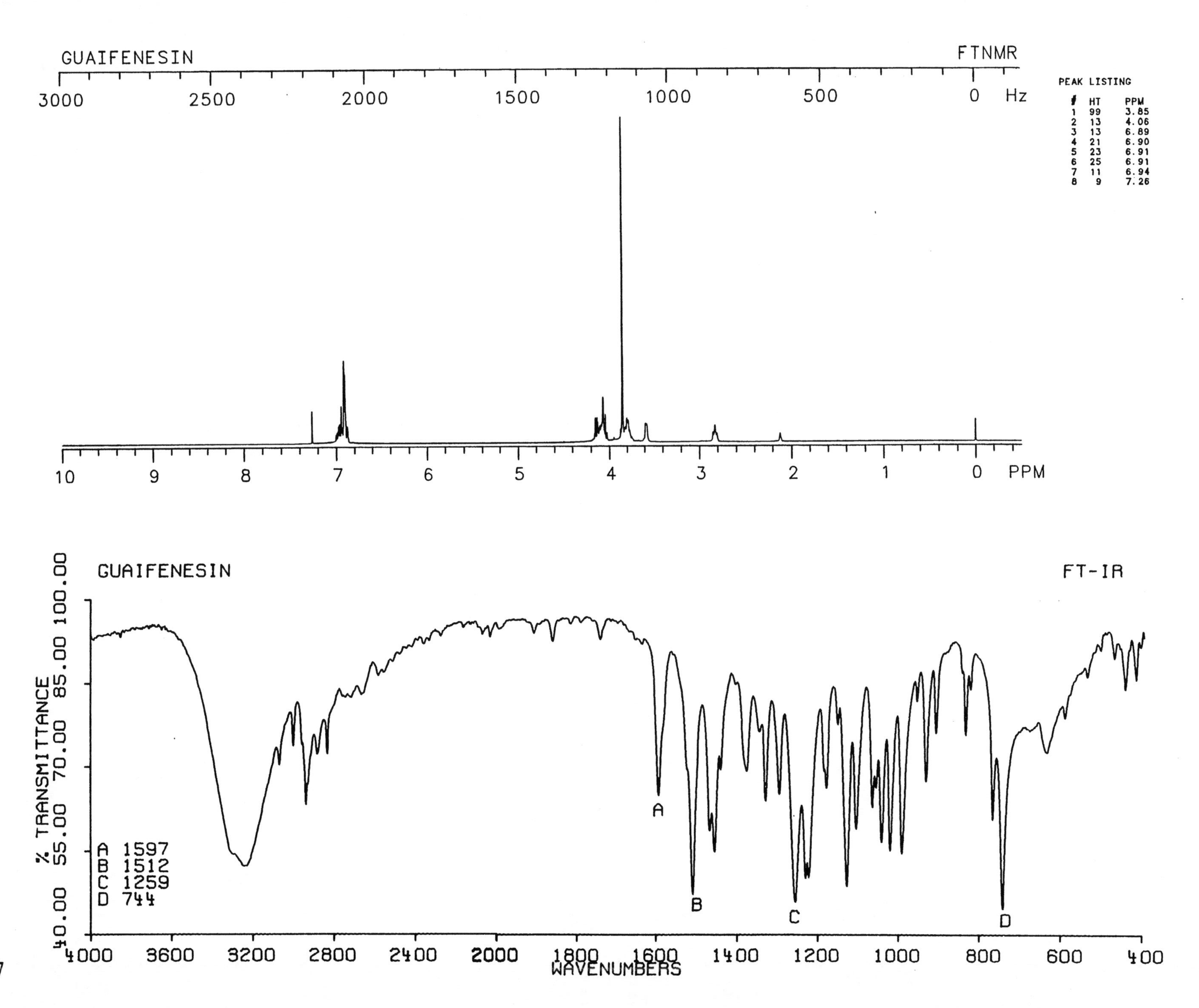
GUAIFENESIN
FTNMR
3000 2500 2000 1500 1000 500 0 Hz
PEAK LISTING
# HT PPM
1 99 3.85
2 13 4.06
3 13 6.89
4 21 6.90
5 23 6.91
6 25 6.91
7 11 6.94
8 9 7.26
10 9 8 7 6 5 4 3 2 1 0 PPM
GUAIFENESIN
FT-IR
% TRANSMITTANCE
40.00 55.00 70.00 85.00 100.00
A 1597
B 1512
C 1259
D 744
A
B
C
D
4000 3600 3200 2800 2400 2000 1800 1600 1400 1200 1000 800 600 400
WAVENUMBERS

# GUANABENZ ACETATE

$C_{10}H_{12}Cl_2N_4O_2$

**Molecular weight:** 291.14 (290.03)
**Synonyms:** 2,6-Dichlorobenzylideneaminoguanidine

**Trade names:** Wytensin

**Use:** Antihypertensive
**HPLC:** Si-10; 5A:95B; 8.9
**GC:** 2228; 250°C

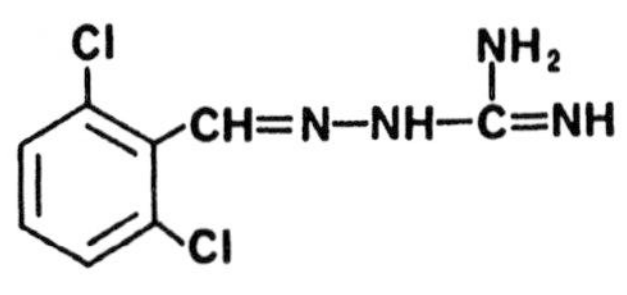


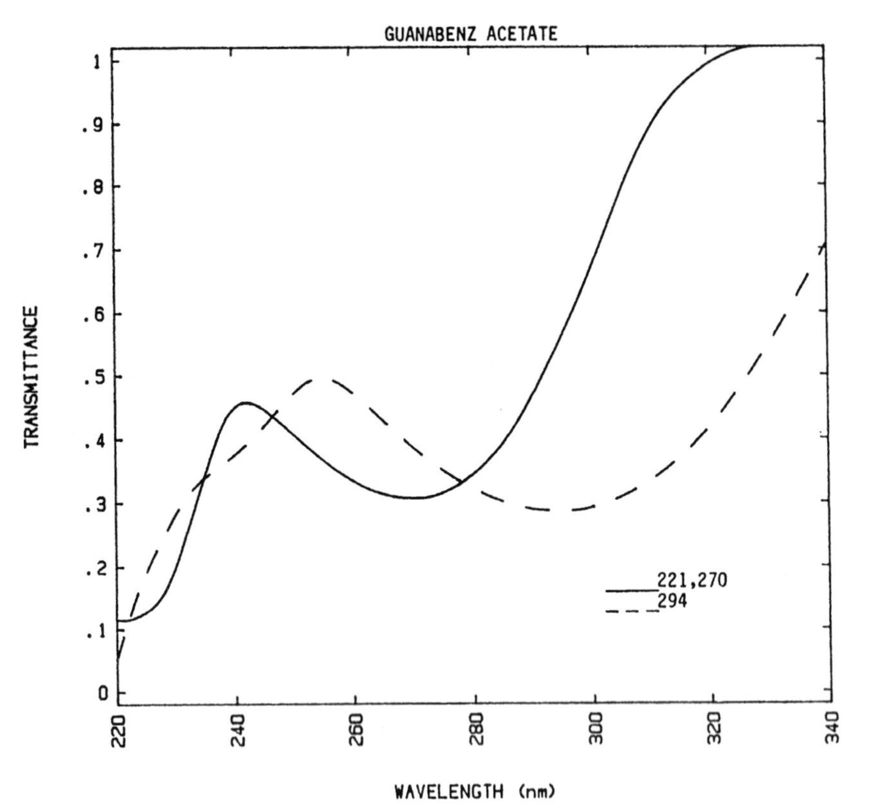

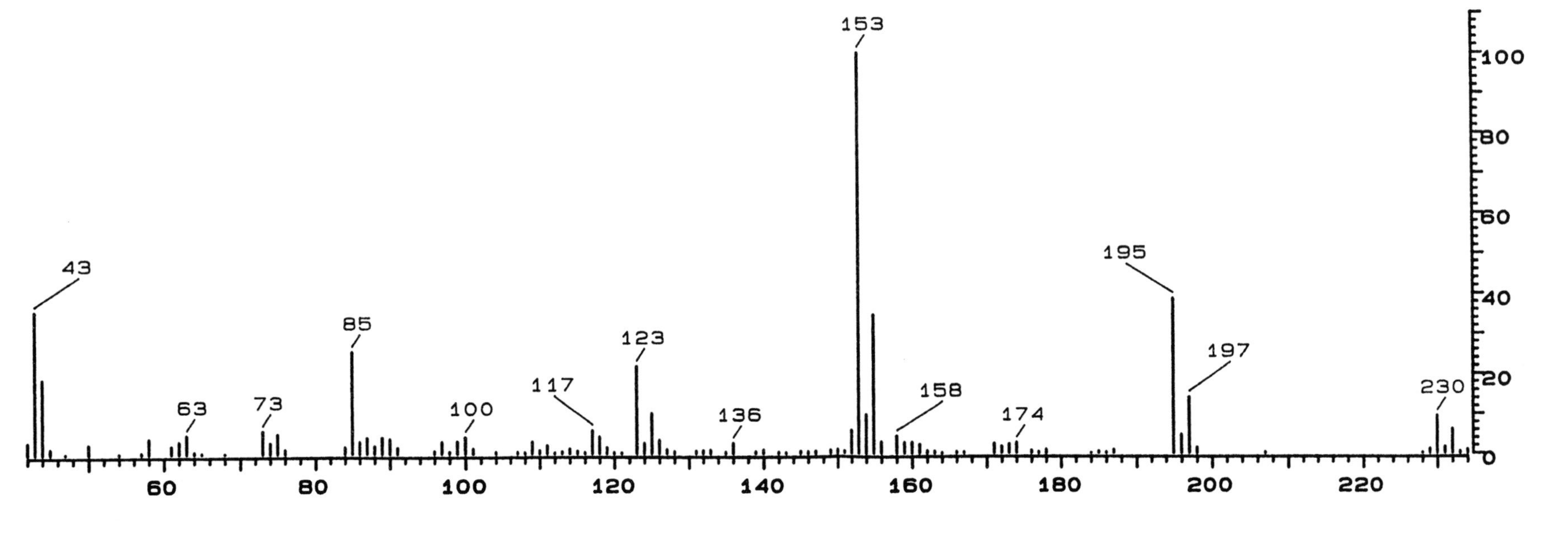

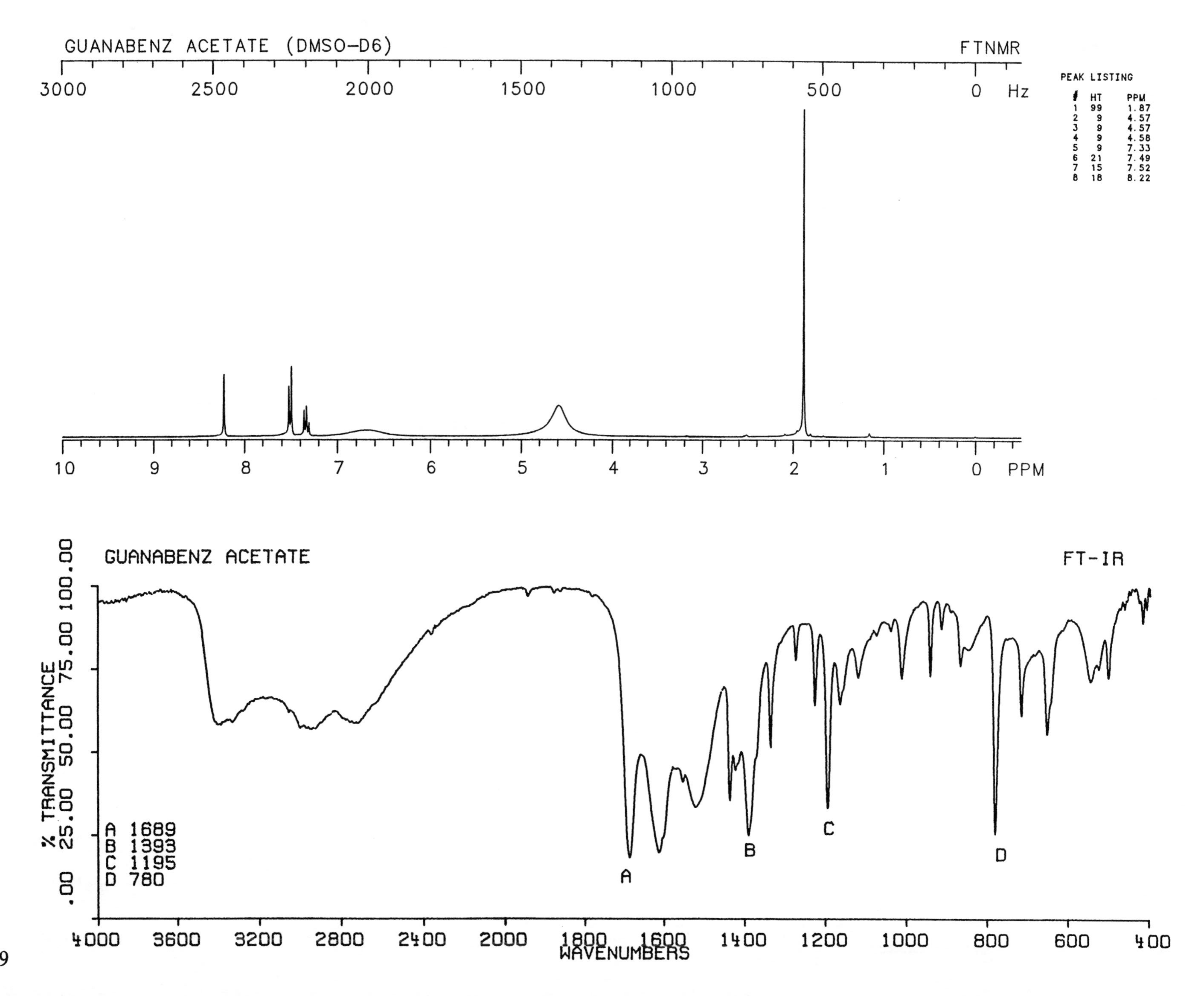
GUANABENZ ACETATE (DMSO-D6)
FTNMR
3000
2500
2000
1500
1000
500
0 Hz
PEAK LISTING
# HT PPM
1 99 1.87
2 9 4.57
3 9 4.57
4 9 4.58
5 9 7.33
6 21 7.49
7 15 7.52
8 18 8.22
10
9
8
7
6
5
4
3
2
1
0 PPM
GUANABENZ ACETATE
FT-IR
% TRANSMITTANCE
.00
25.00
50.00
75.00
100.00
A 1689
B 1393
C 1195
D 780
A
B
C
D
4000
3600
3200
2800
2400
2000
1800
1600
1400
1200
1000
800
600
400
WAVENUMBERS

# GUANADREL

$C_{10}H_{19}N_3O_2$

Molecular weight: 213.28 (213.15)
Synonyms: (1,4-Dioxaspiro[4.5]dec-2-yl-methyl)guanidine

Trade names: Anarel, Hylorel

Use: Antihypertensive
HPLC:
GC:

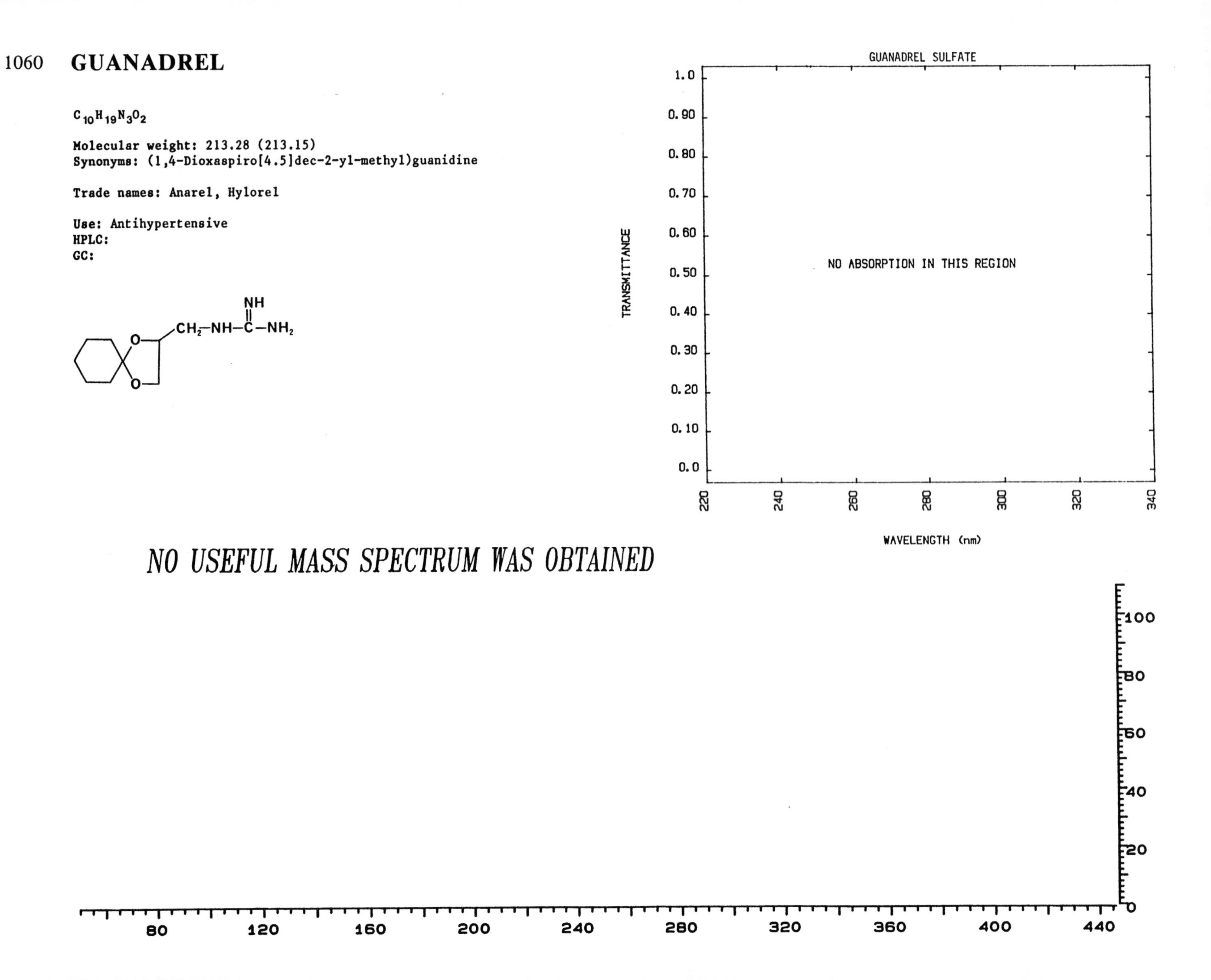

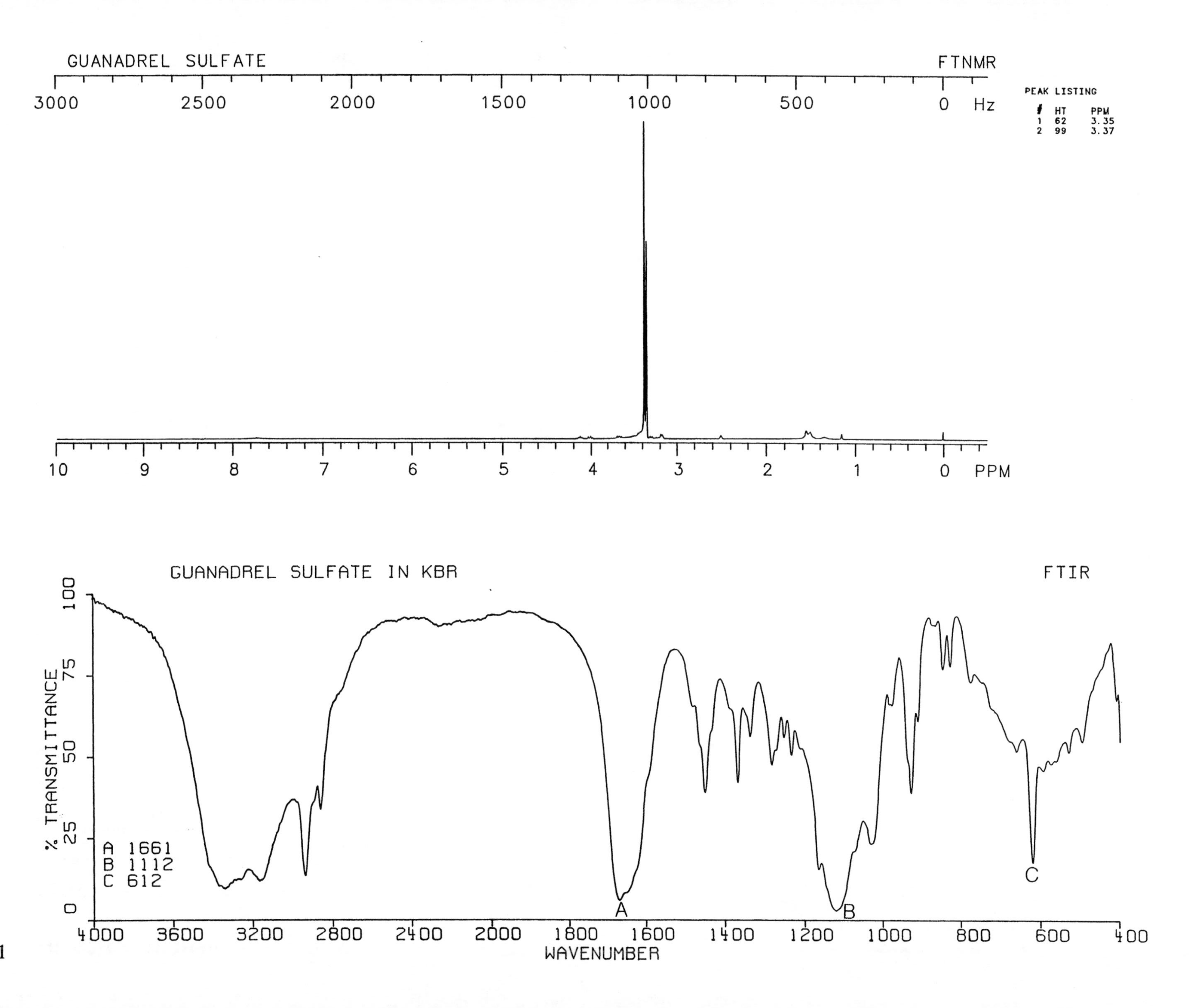
GUANADREL SULFATE
FTNMR
3000
2500
2000
1500
1000
500
0 Hz
PEAK LISTING
# HT PPM
1 62 3.35
2 99 3.37
10
9
8
7
6
5
4
3
2
1
0 PPM
GUANADREL SULFATE IN KBR
FTIR
100
75
50
25
0
% TRANSMITTANCE
A 1661
B 1112
C 612
A
B
C
4000
3600
3200
2800
2400
2000
1800
1600
1400
1200
1000
800
600
400
WAVENUMBER

# GUANETHIDINE

$C_{10}H_{22}N_4$

**Molecular weight:** 198.31 (198.18)
**Synonyms:** [2-(Hexahydro-1(2H)-azocinyl)ethyl]guanidine; oktadin; oktatenzin
**Trade names:** Esimil, Ismelin

**Use:** Antihypertensive
**HPLC:**
**GC:**

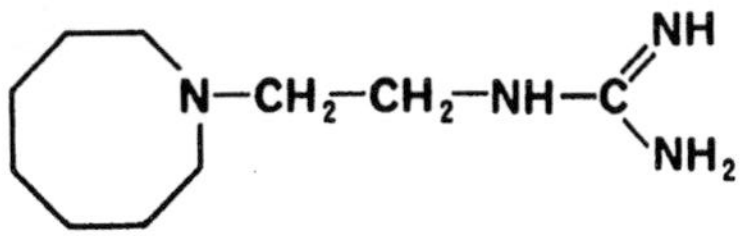

N—CH₂—CH₂—NH—C(=NH)—NH₂

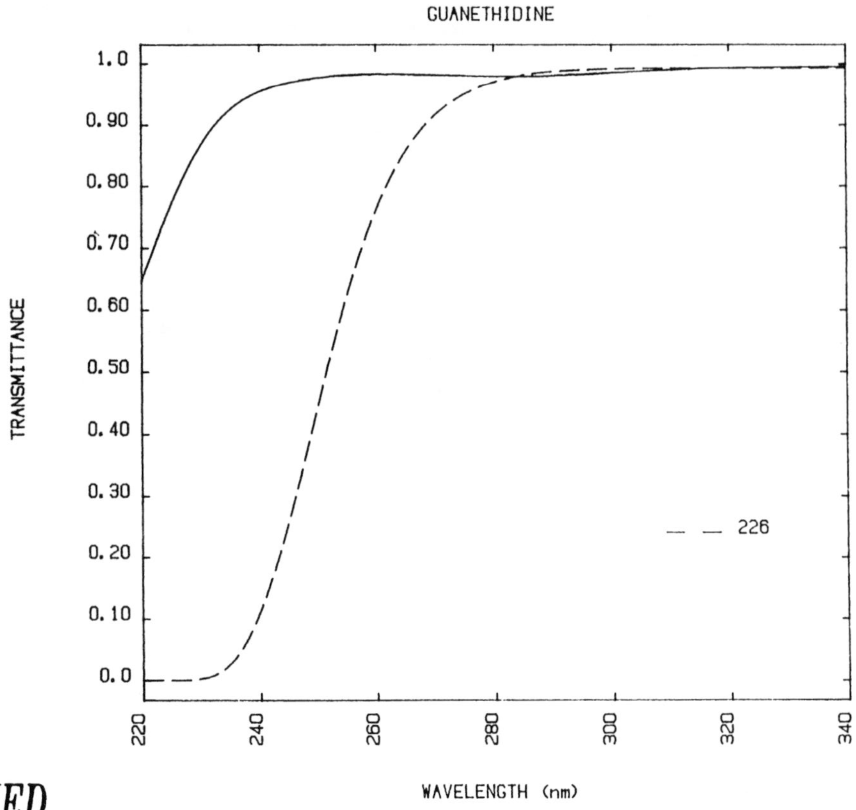

GUANETHIDINE

TRANSMITTANCE

1.0
0.90
0.80
0.70
0.60
0.50
0.40
0.30
0.20
0.10
0.0

— — 226

220 240 260 280 300 320 340

WAVELENGTH (nm)

*NO USEFUL MASS SPECTRUM WAS OBTAINED*

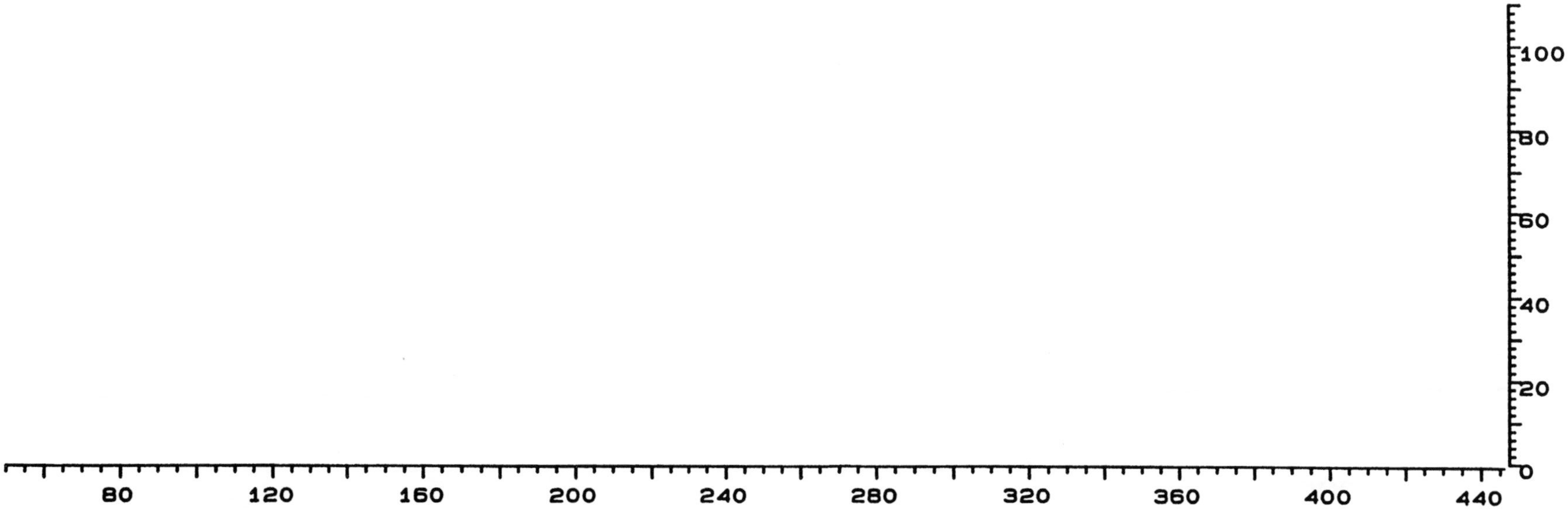

100
80
60
40
20
0

80 120 160 200 240 280 320 360 400 440

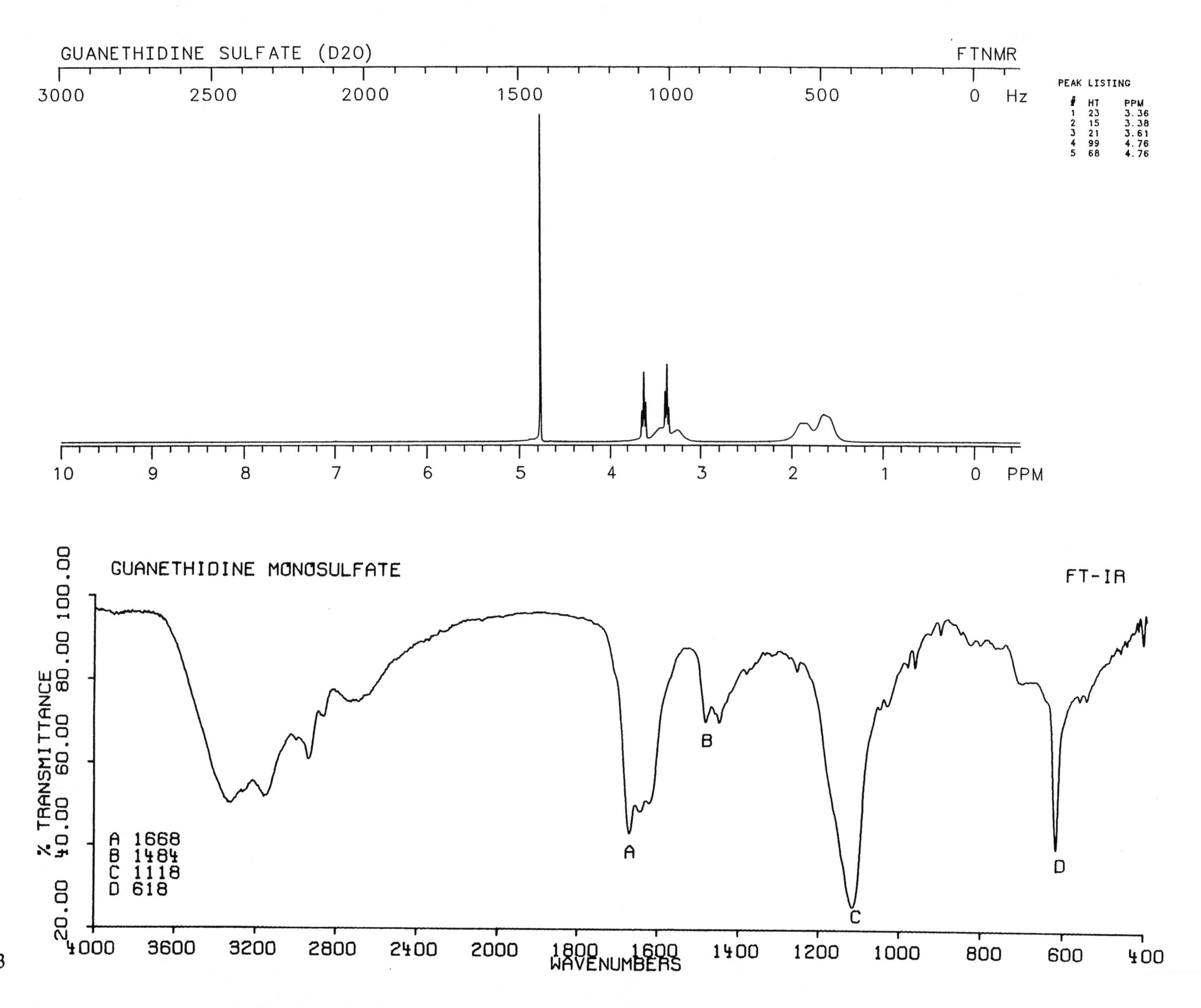
GUANETHIDINE SULFATE (D2O)
FTNMR
3000 2500 2000 1500 1000 500 0 Hz
PEAK LISTING
# HT PPM
1 23 3.36
2 15 3.38
3 21 3.61
4 99 4.76
5 68 4.76
10 9 8 7 6 5 4 3 2 1 0 PPM
GUANETHIDINE MONOSULFATE
FT-IR
% TRANSMITTANCE
20.00 40.00 60.00 80.00 100.00
A 1668
B 1484
C 1118
D 618
A
B
C
D
4000 3600 3200 2800 2400 2000 1800 1600 1400 1200 1000 800 600 400
WAVENUMBERS

# HALAZEPAM

$C_{17}H_{12}ClN_2OF_3$

Molecular weight: 352.75 (352.06)
Synonyms: 7-Chloro-1,3-dihydro-5-phenyl-1-(2,2,2-trifluoroethyl)-2H-1,4-benzodiazepin-2-one
Trade names: Paxipam

Use: Tranquilizer
HPLC: Si-10; 1A:99B; 4.0
GC: 2296; 250°C

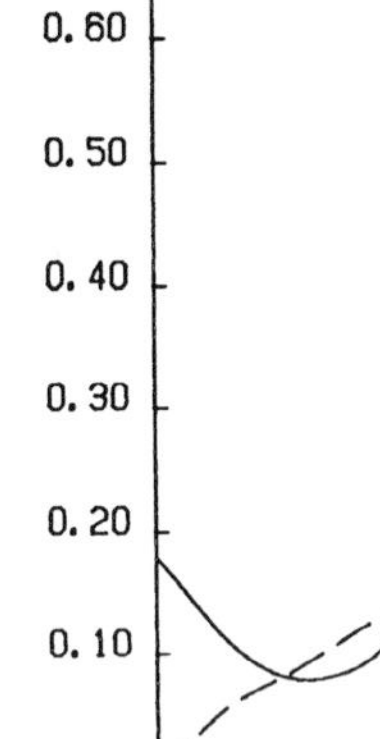


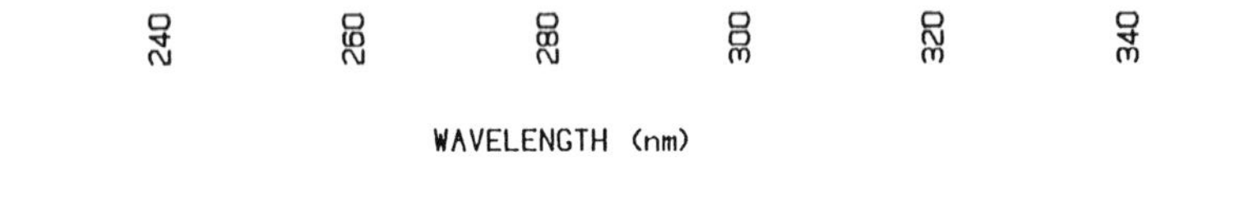


*HALAZEPAM*

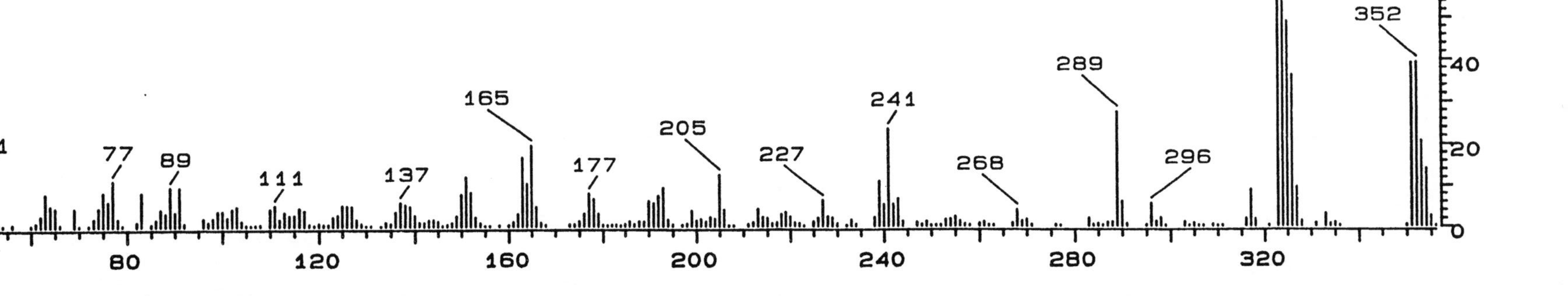

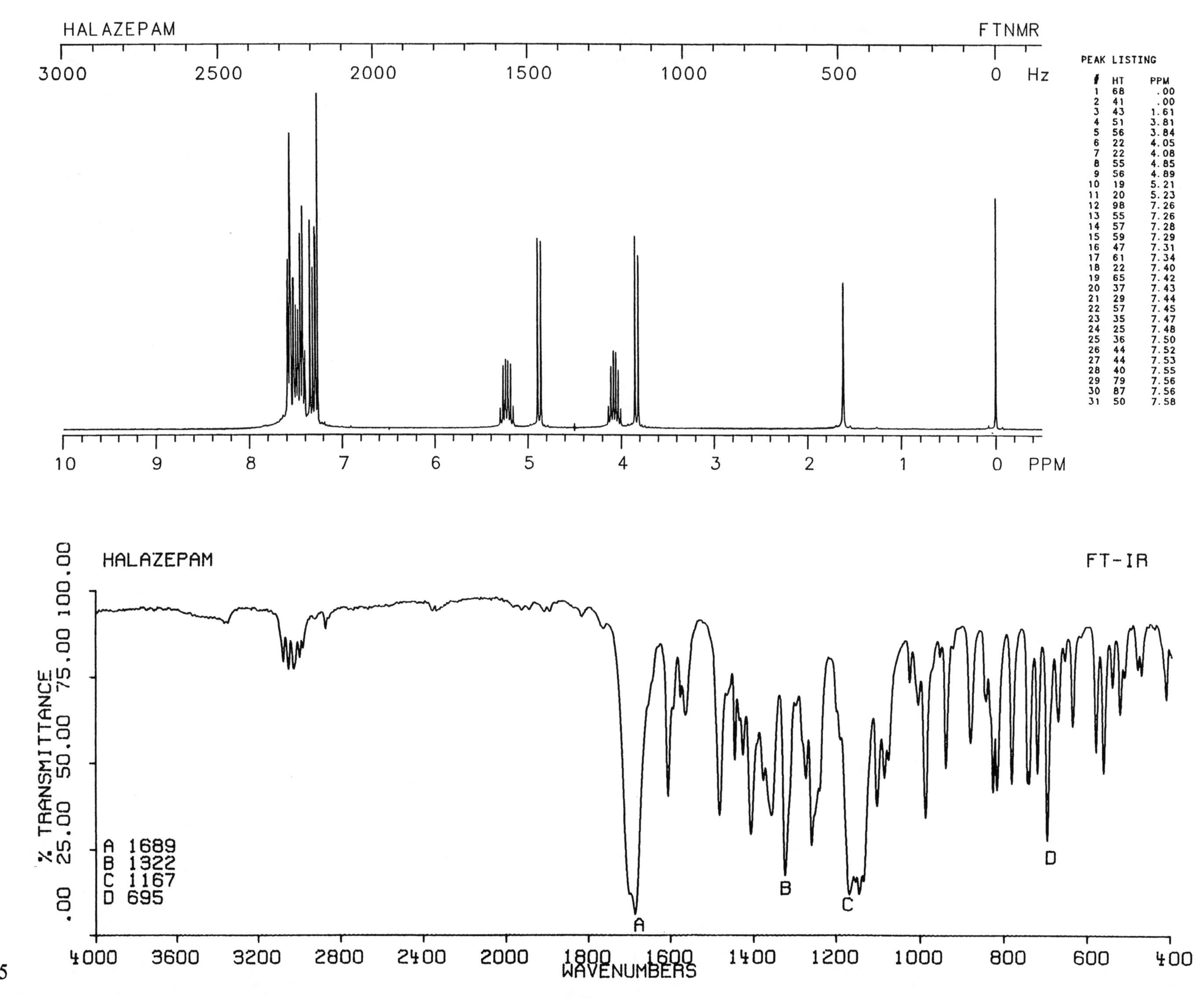

HALAZEPAM
FTNMR
3000
2500
2000
1500
1000
500
0 Hz
10
9
8
7
6
5
4
3
2
1
0 PPM
PEAK LISTING
# HT PPM
1 68 .00
2 41 .00
3 43 1.61
4 51 3.81
5 56 3.84
6 22 4.05
7 22 4.08
8 55 4.85
9 56 4.89
10 19 5.21
11 20 5.23
12 98 7.26
13 55 7.26
14 57 7.28
15 59 7.29
16 47 7.31
17 61 7.34
18 22 7.40
19 65 7.42
20 37 7.43
21 29 7.44
22 57 7.45
23 35 7.47
24 25 7.48
25 36 7.50
26 44 7.52
27 44 7.53
28 40 7.55
29 79 7.56
30 87 7.56
31 50 7.58
HALAZEPAM
FT-IR
% TRANSMITTANCE
.00
25.00
50.00
75.00
100.00
A 1689
B 1322
C 1167
D 695
A
B
C
D
4000
3600
3200
2800
2400
2000
1800
1600
1400
1200
1000
800
600
400
WAVENUMBERS

# HALCINONIDE

$C_{24}H_{32}ClFO_5$

**Molecular weight:** 454.97 (454.19)
**Synonyms:** 21-Chloro-9-fluoro-11β-hydroxy-16α-17-[(1-methylethylidene)bis-(oxy)]pregn-4-ene-3,20-dione
**Trade names:** Halcidern, Halcimat, Halog, Halog-E

**Use:** Topical anti-inflammatory
**HPLC:**
**GC:** 3323; 280°C

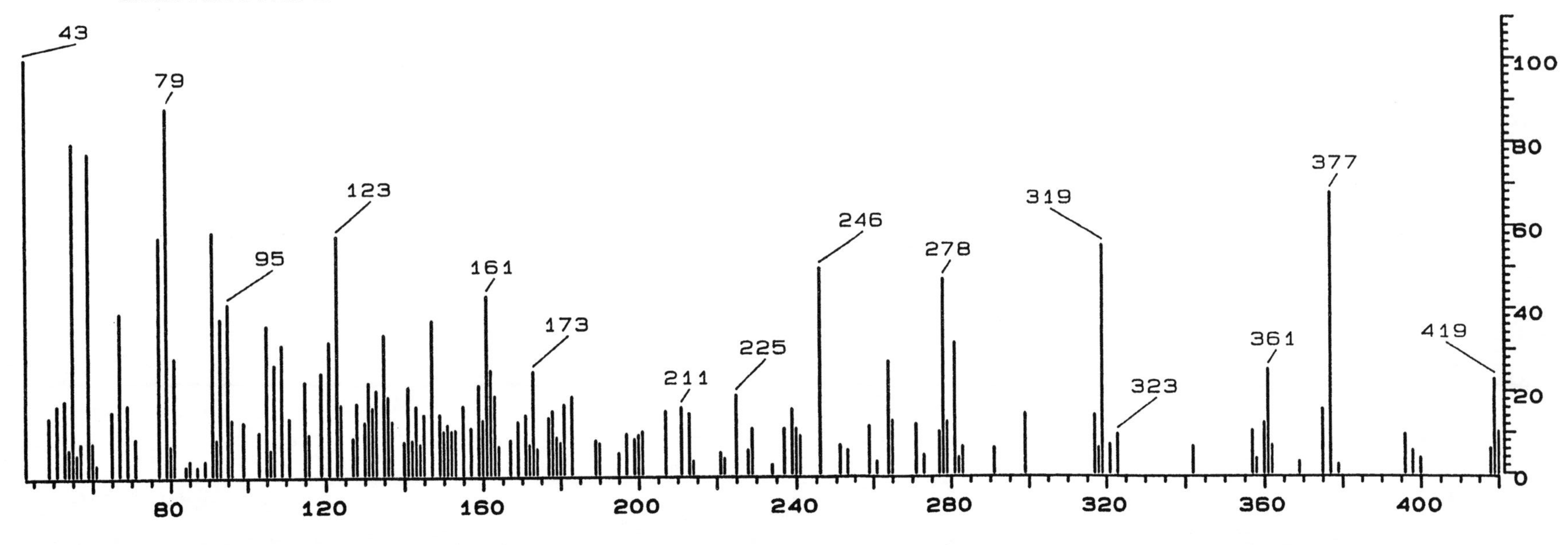

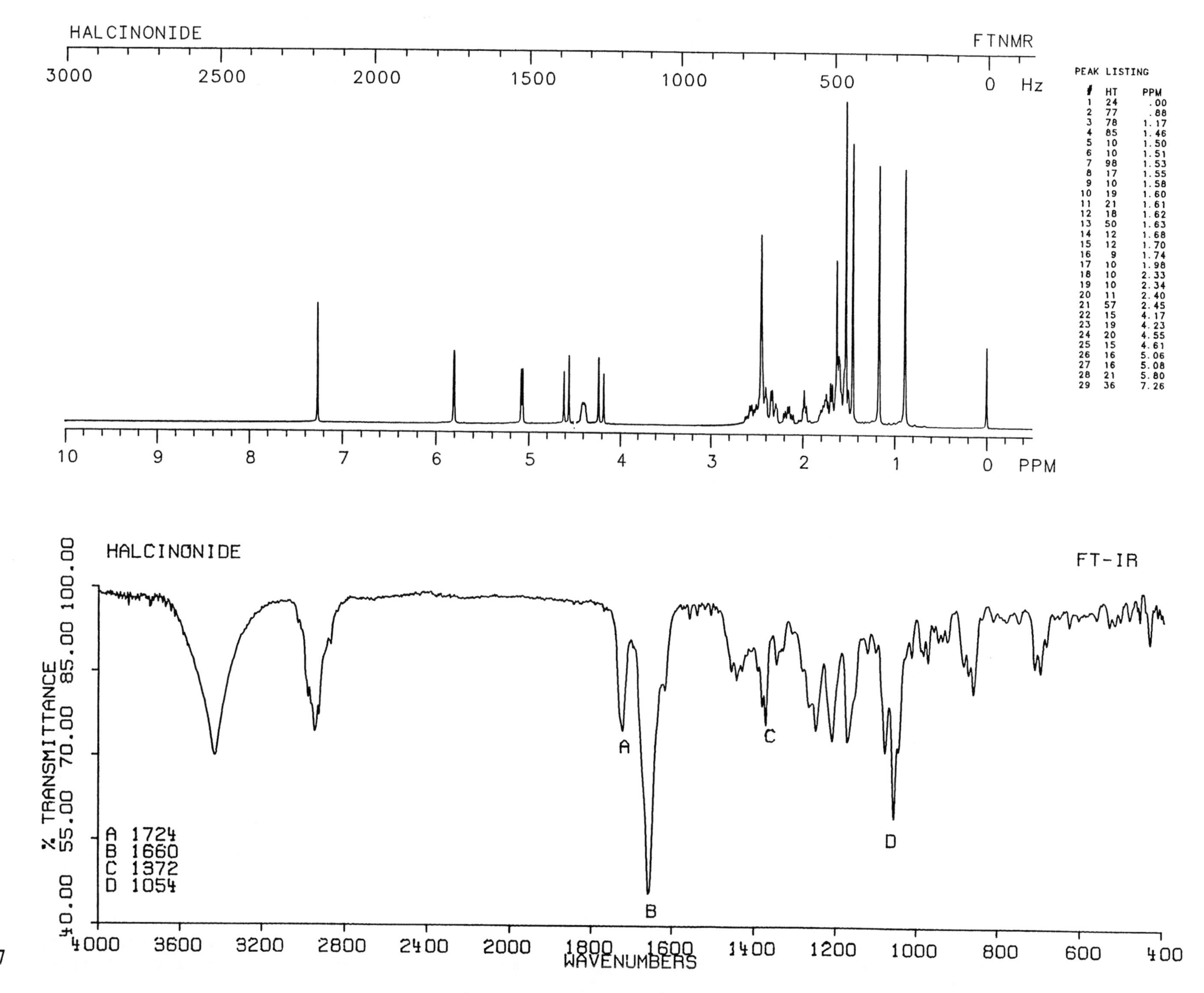
HALCINONIDE
FTNMR
3000 2500 2000 1500 1000 500 0 Hz
10 9 8 7 6 5 4 3 2 1 0 PPM
PEAK LISTING
# HT PPM
1 24 .00
2 77 .88
3 78 1.17
4 85 1.46
5 10 1.50
6 10 1.51
7 98 1.53
8 17 1.55
9 10 1.58
10 19 1.60
11 21 1.61
12 18 1.62
13 50 1.63
14 12 1.68
15 12 1.70
16 9 1.74
17 10 1.98
18 10 2.33
19 10 2.34
20 11 2.40
21 57 2.45
22 15 4.17
23 19 4.23
24 20 4.55
25 15 4.61
26 16 5.06
27 16 5.08
28 21 5.80
29 36 7.26
HALCINONIDE
FT-IR
% TRANSMITTANCE
40.00 55.00 70.00 85.00 100.00
A 1724
B 1660
C 1372
D 1054
A B C D
4000 3600 3200 2800 2400 2000 1800 1600 1400 1200 1000 800 600 400
WAVENUMBERS

# HALOPERIDOL

$C_{21}H_{23}ClFNO_2$

**Molecular weight:** 375.87 (375.14)
**Synonyms:** 4-[4-(4-Chlorophenyl)-4-hydroxy-1-piperidinyl]-1-(4-fluorophenyl)-1-butanone
**Trade names:** Haldol

**Use:** Tranquilizer
**HPLC:** Si-10; 5A:95B; 5.0
**GC:** 3018; 280°C

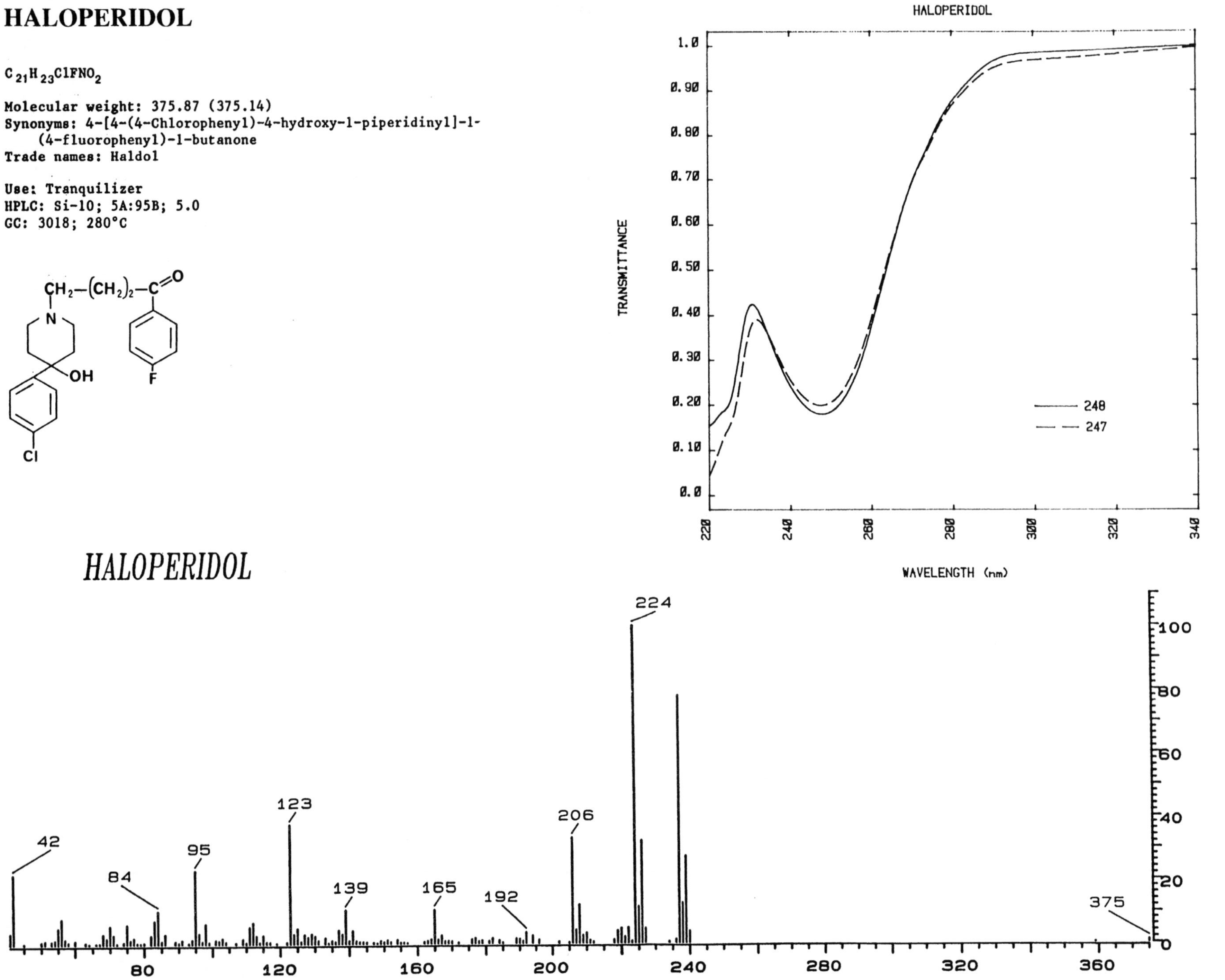

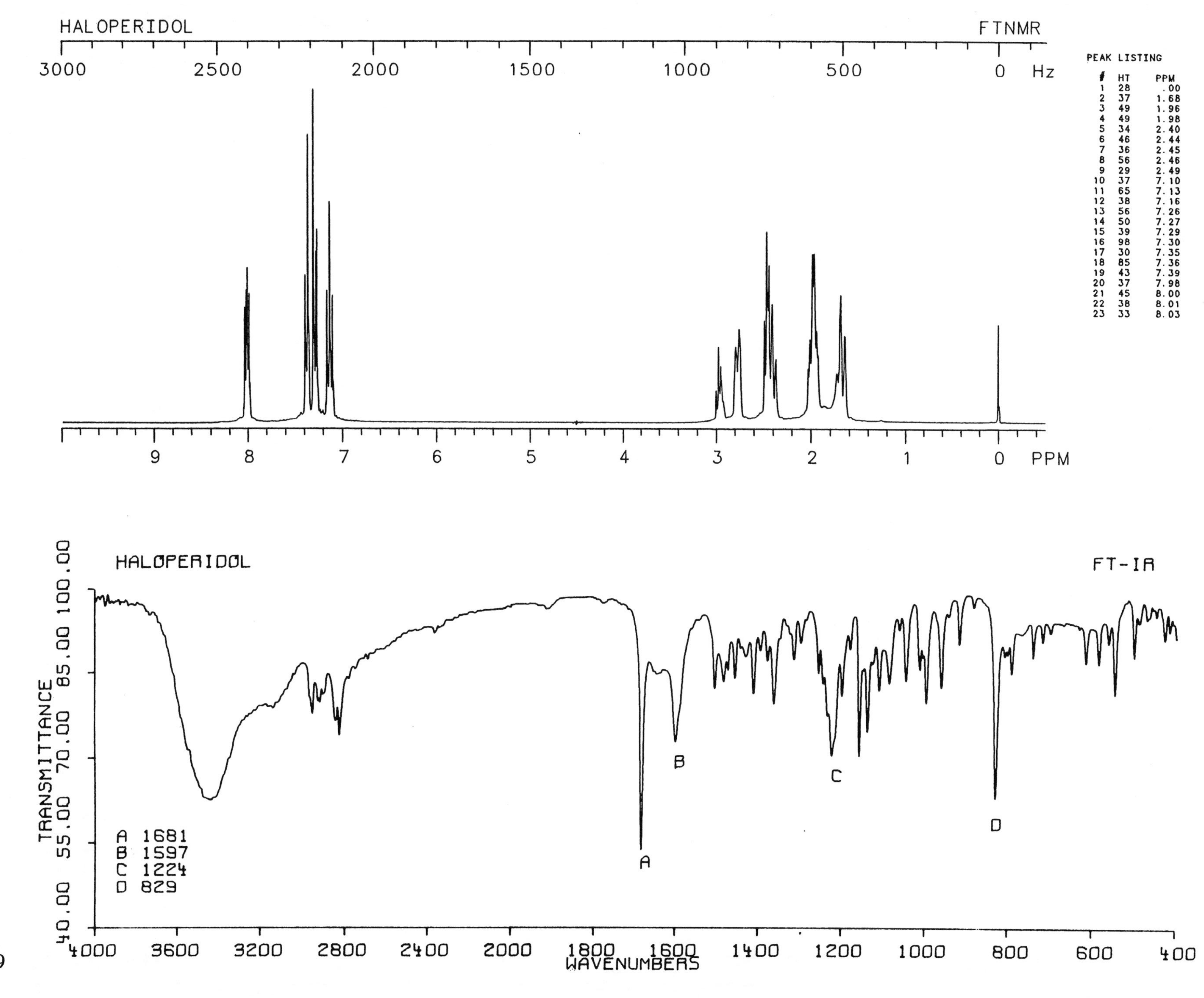

HALOPERIDOL
FTNMR
3000 2500 2000 1500 1000 500 0 Hz
9 8 7 6 5 4 3 2 1 0 PPM
PEAK LISTING
# HT PPM
1 28 .00
2 37 1.68
3 49 1.96
4 49 1.98
5 34 2.40
6 46 2.44
7 36 2.45
8 56 2.46
9 29 2.49
10 37 7.10
11 65 7.13
12 38 7.16
13 56 7.26
14 50 7.27
15 39 7.29
16 98 7.30
17 30 7.35
18 85 7.36
19 43 7.39
20 37 7.98
21 45 8.00
22 38 8.01
23 33 8.03
HALOPERIDOL
FT-IR
TRANSMITTANCE
40.00 55.00 70.00 85.00 100.00
A 1681
B 1597
C 1224
D 829
A
B
C
D
4000 3600 3200 2800 2400 2000 1800 1600 1400 1200 1000 800 600 400
WAVENUMBERS

# HALOPROGIN

$C_9H_4Cl_3IO$

Molecular weight: 361.41 (359.84)
Synonyms: 3-Iodo-2-propynyl-2,4,5-trichlorophenyl ether

Trade names: Halotex

Use: Antibacterial
HPLC:
GC:

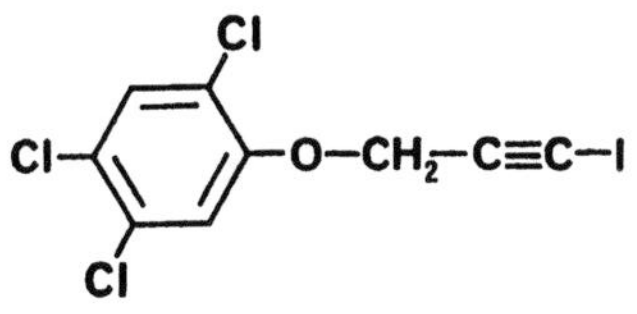

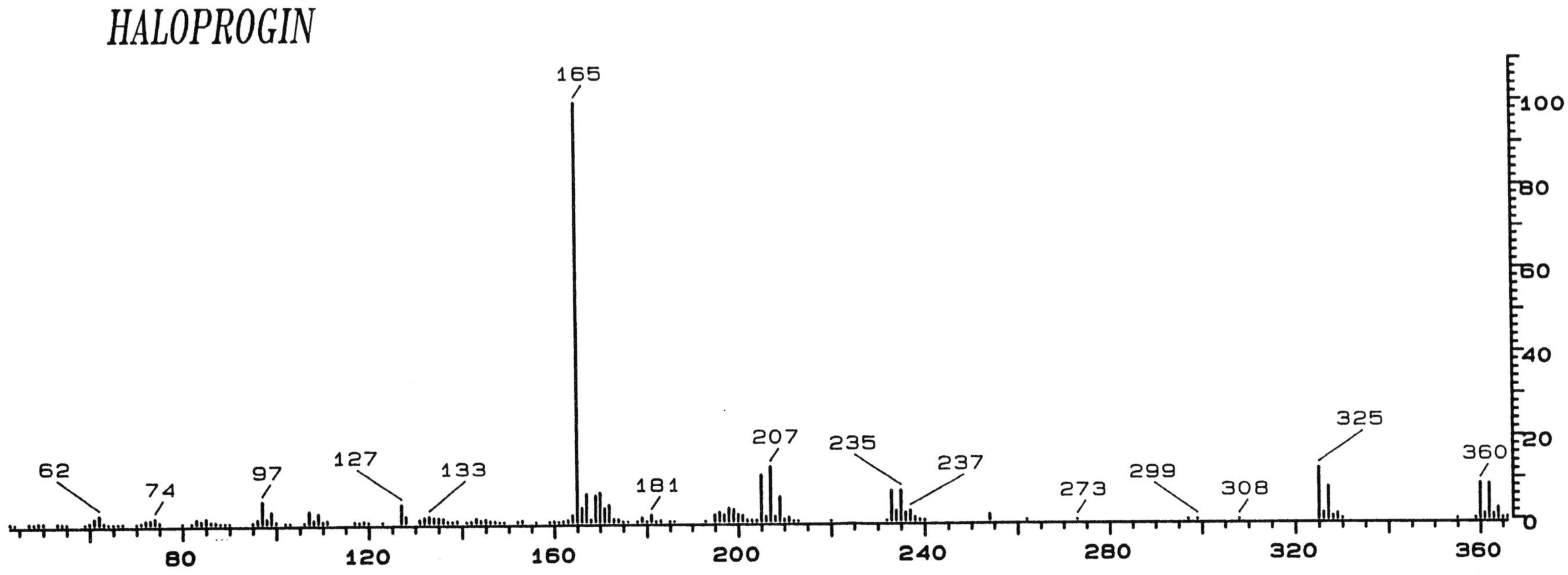

Cl
Cl
Cl
O—CH$_2$—C≡C—I

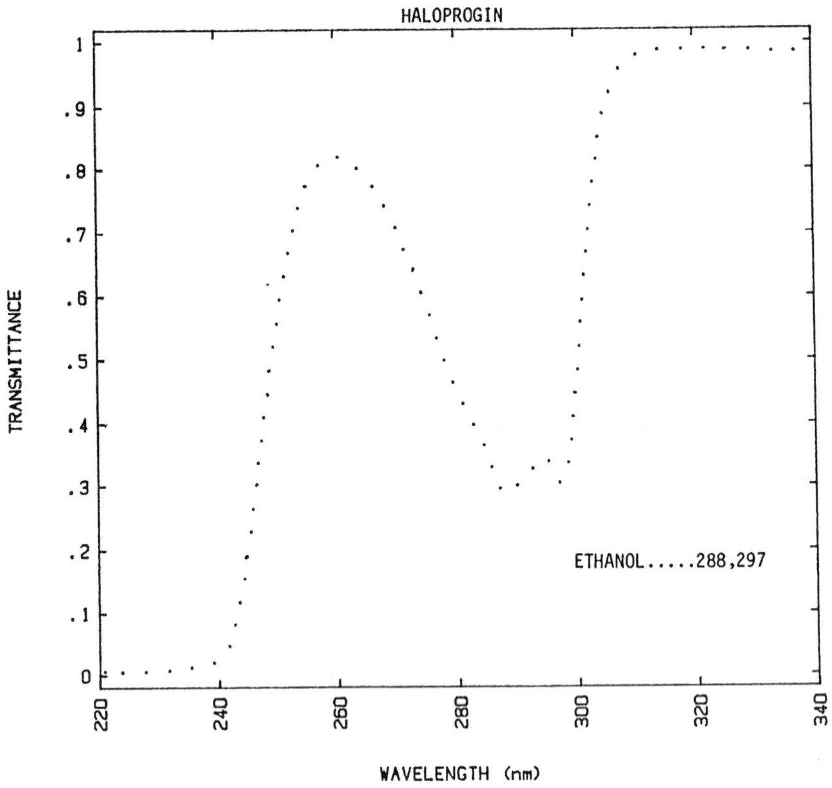

HALOPROGIN
1
.9
.8
.7
.6
.5
.4
.3
.2
.1
0
TRANSMITTANCE
ETHANOL.....288,297
220
240
260
280
300
320
340
WAVELENGTH (nm)

*HALOPROGIN*

165
62
74
97
127
133
181
207
235
237
273
299
308
325
360
80
120
160
200
240
280
320
360
100
80
60
40
20
0

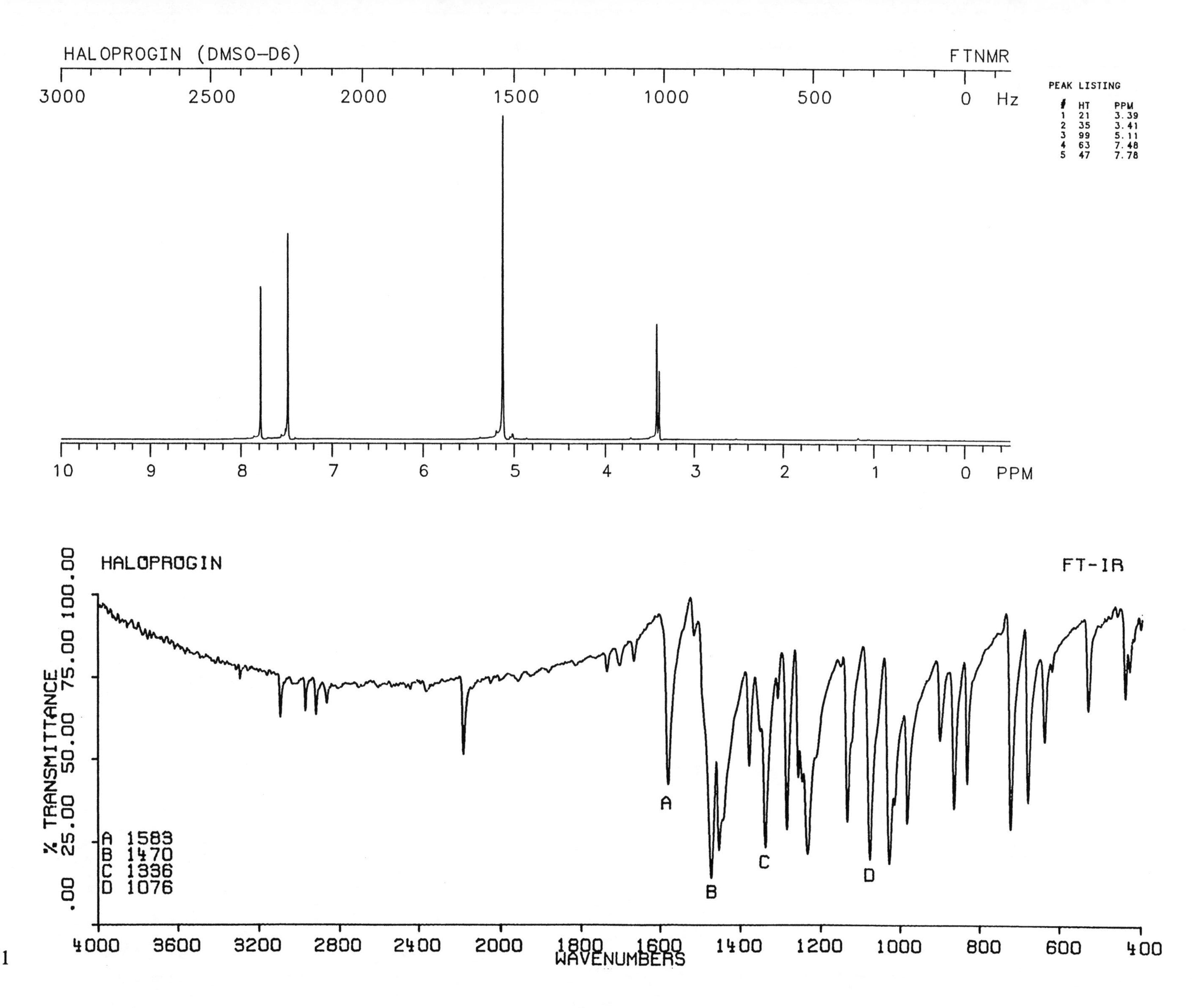
HALOPROGIN (DMSO-D6)
FTNMR
PEAK LISTING
# HT PPM
1 21 3.39
2 35 3.41
3 99 5.11
4 63 7.48
5 47 7.78
HALOPROGIN
FT-IR
% TRANSMITTANCE
WAVENUMBERS
A 1583
B 1470
C 1336
D 1076

# HARMALINE

$C_{13}H_{14}N_2O$

**Molecular weight:** 214.26 (214.11)
**Synonyms:** 4,9-Dihdryo-7-methoxy-1-methyl-3H-pyrido[3,4-b]indole; harmidine; 3,4-dihydroharmine
**Trade names:**

**Use:** Central stimulant, hallucinogen
**HPLC:** Si-10; 20A:80B; 4.4
**GC:** 2287; 250°C

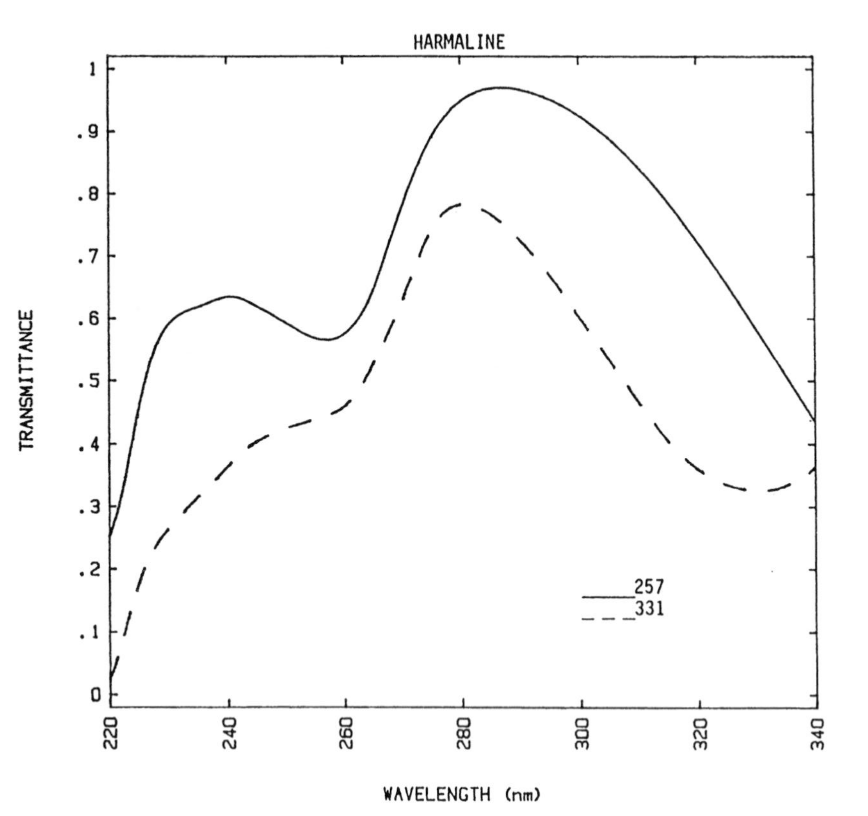

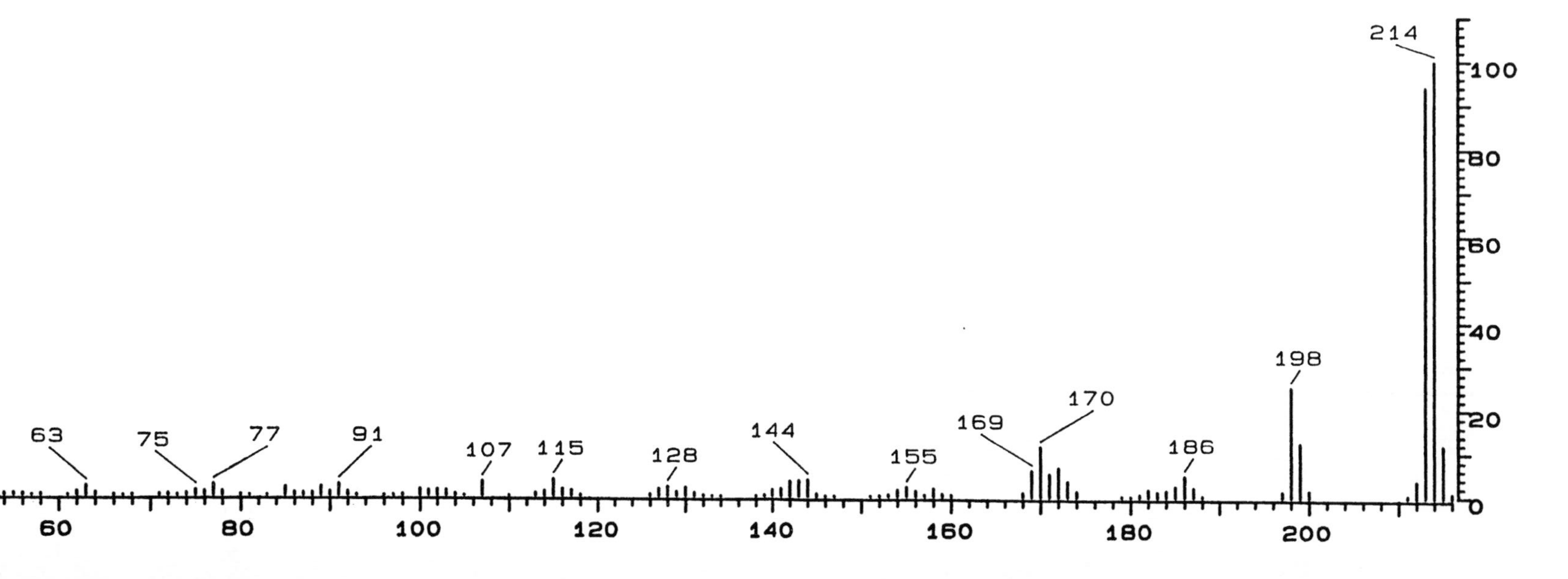

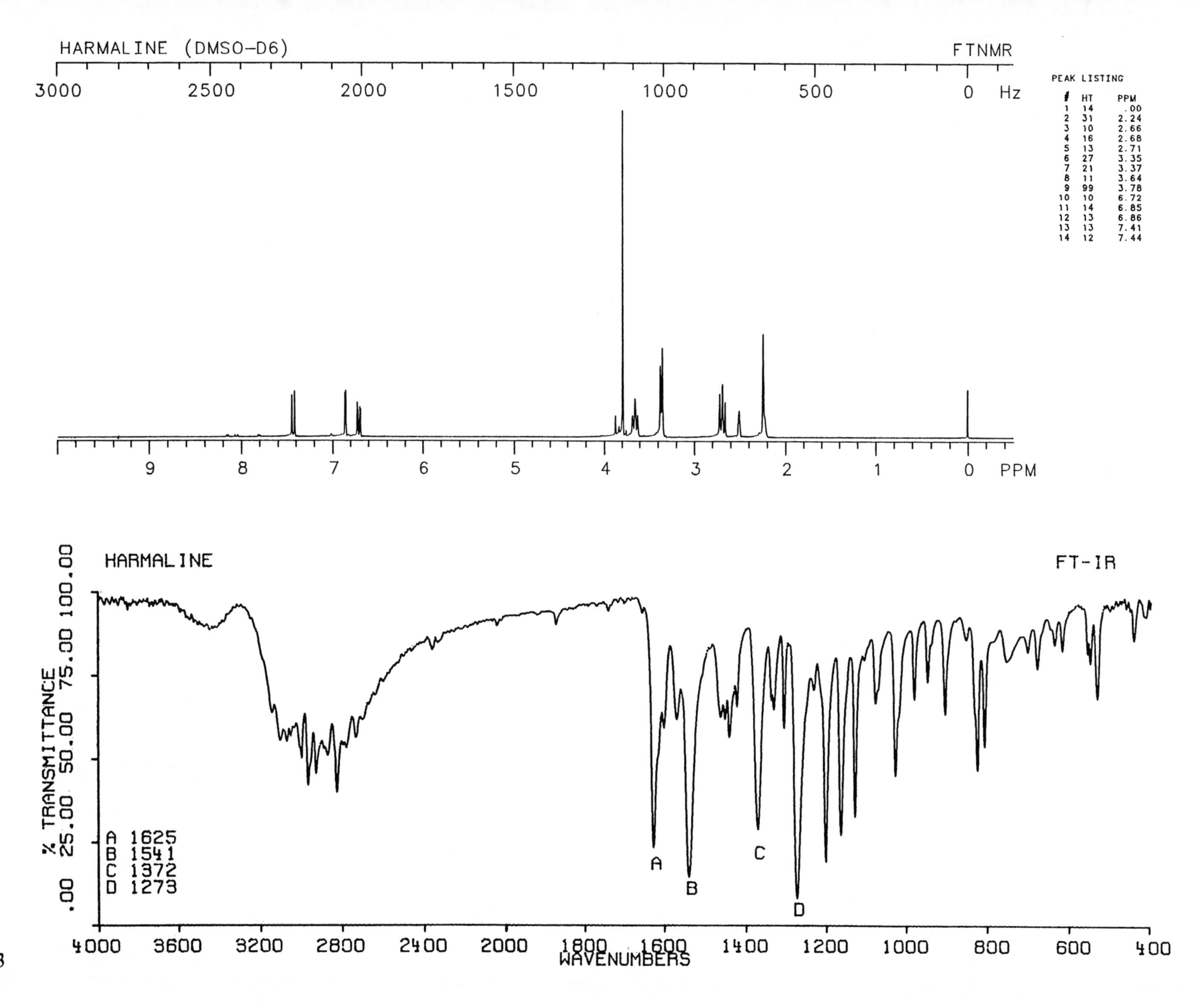
HARMALINE (DMSO-D6)
FTNMR
3000 2500 2000 1500 1000 500 0 Hz
PEAK LISTING
| # | HT | PPM |
|---|---|---|
| 1 | 14 | .00 |
| 2 | 31 | 2.24 |
| 3 | 10 | 2.66 |
| 4 | 16 | 2.68 |
| 5 | 13 | 2.71 |
| 6 | 27 | 3.35 |
| 7 | 21 | 3.37 |
| 8 | 11 | 3.64 |
| 9 | 99 | 3.78 |
| 10 | 10 | 6.72 |
| 11 | 14 | 6.85 |
| 12 | 13 | 6.86 |
| 13 | 13 | 7.41 |
| 14 | 12 | 7.44 |
9 8 7 6 5 4 3 2 1 0 PPM
HARMALINE
FT-IR
% TRANSMITTANCE
.00 25.00 50.00 75.00 100.00
A 1625
B 1541
C 1372
D 1273
A B C D
4000 3600 3200 2800 2400 2000 1800 1600 1400 1200 1000 800 600 400
WAVENUMBERS

# HARMALOL

$C_{12}H_{12}N_2O$

**Molecular weight: 200.24 (200.10)**
**Synonyms: 4,9-Dihdyro-1-methyl-3H-pyrido[3,4-b]indol-7-ol**

**Trade names:**

**Use: Narcotic, anthelmintic**
**HPLC: Si-10; 20A:80B; 8.7**
**GC:**

HO N H CH₃ N

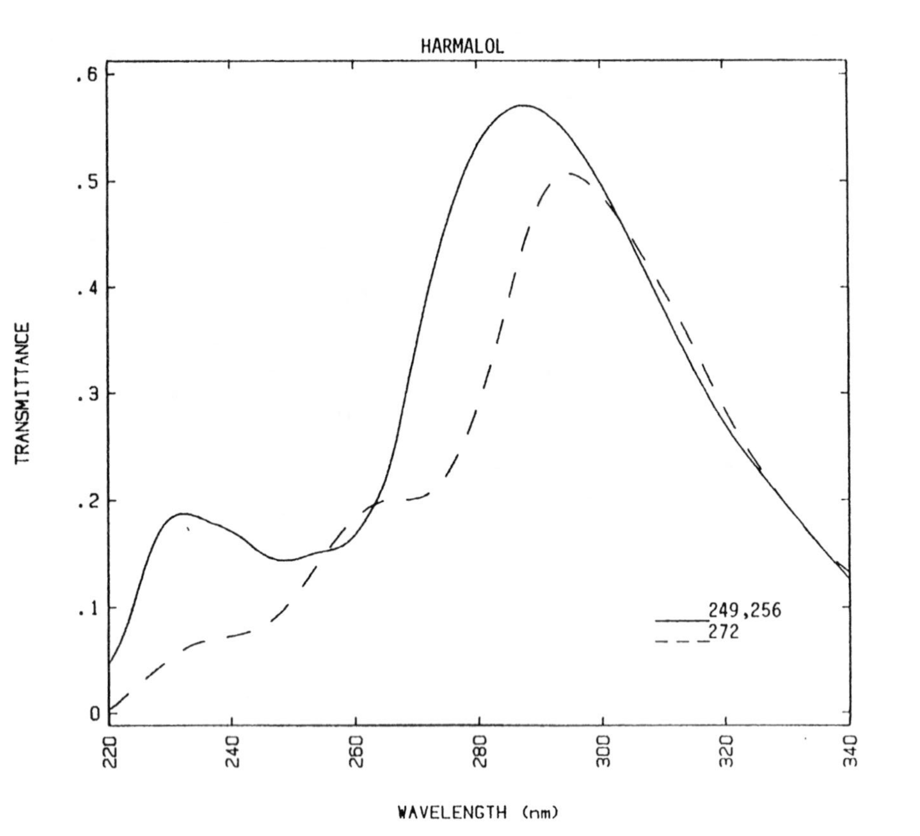

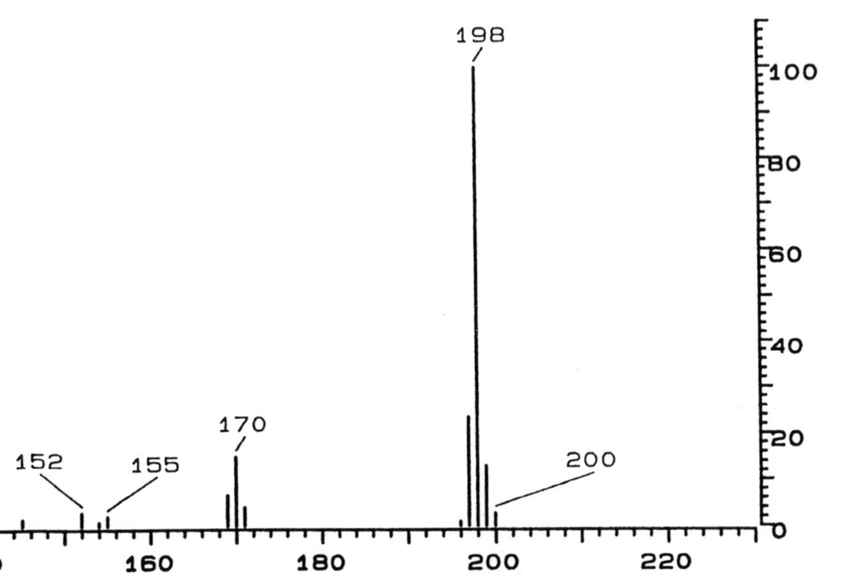
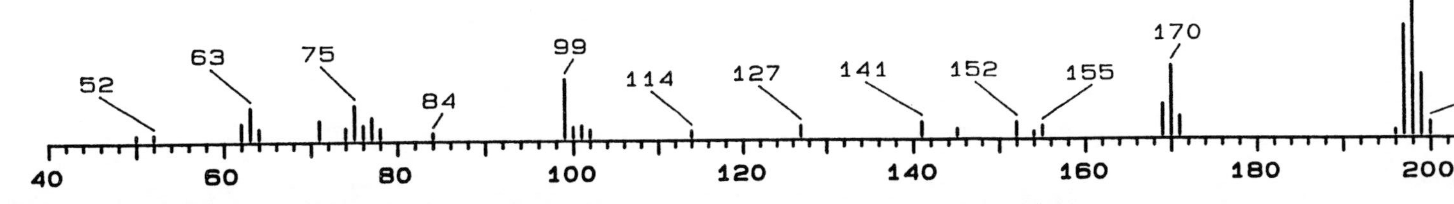

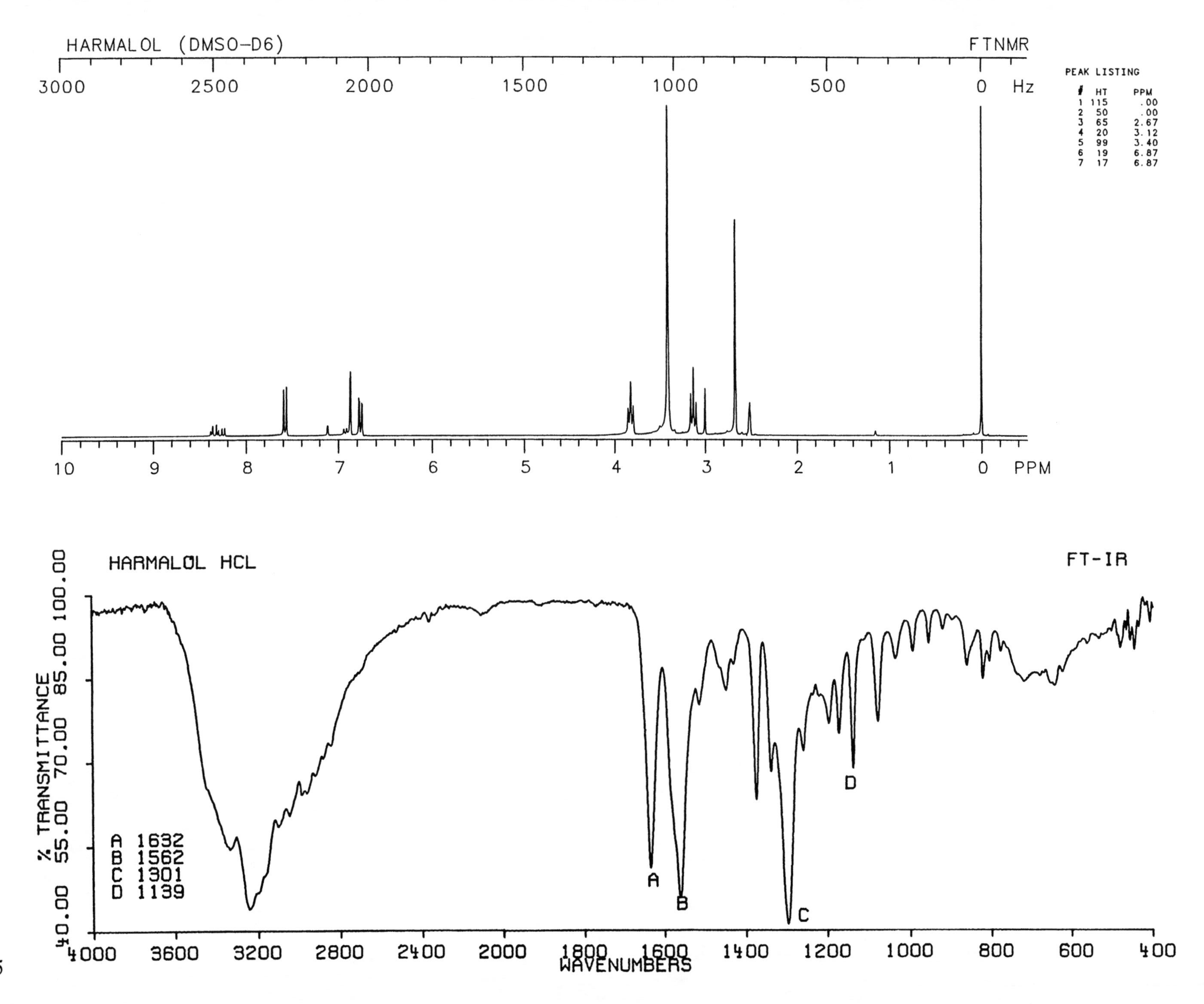
HARMALOL (DMSO-D6)
FTNMR
3000
2500
2000
1500
1000
500
0 Hz
PEAK LISTING
# HT PPM
1 115 .00
2 50 .00
3 65 2.67
4 20 3.12
5 99 3.40
6 19 6.87
7 17 6.87
10
9
8
7
6
5
4
3
2
1
0 PPM
HARMALOL HCL
FT-IR
% TRANSMITTANCE
40.00
55.00
70.00
85.00
100.00
A 1632
B 1562
C 1301
D 1139
A
B
C
D
4000
3600
3200
2800
2400
2000
1800
1600
1400
1200
1000
800
600
400
WAVENUMBERS

# HARMINE

$C_{13}H_{12}N_2O$

Molecular weight: 212.25 (212.10)
Synonyms: 7-Methoxy-1-methyl-9H-pyrido[3,4-b]indole; bainisterine; yageine; telepathine; leucoharmine
Trade names:

Use: Hallucinogen
HPLC: Si-10; 2A:98B; 11.0
GC: 2284; 250°C

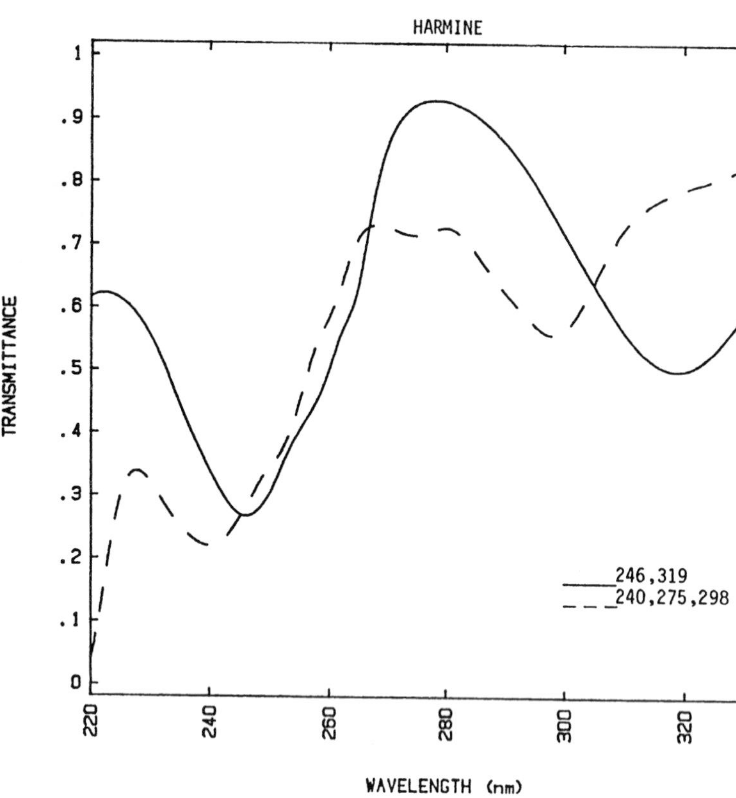


*HARMINE*

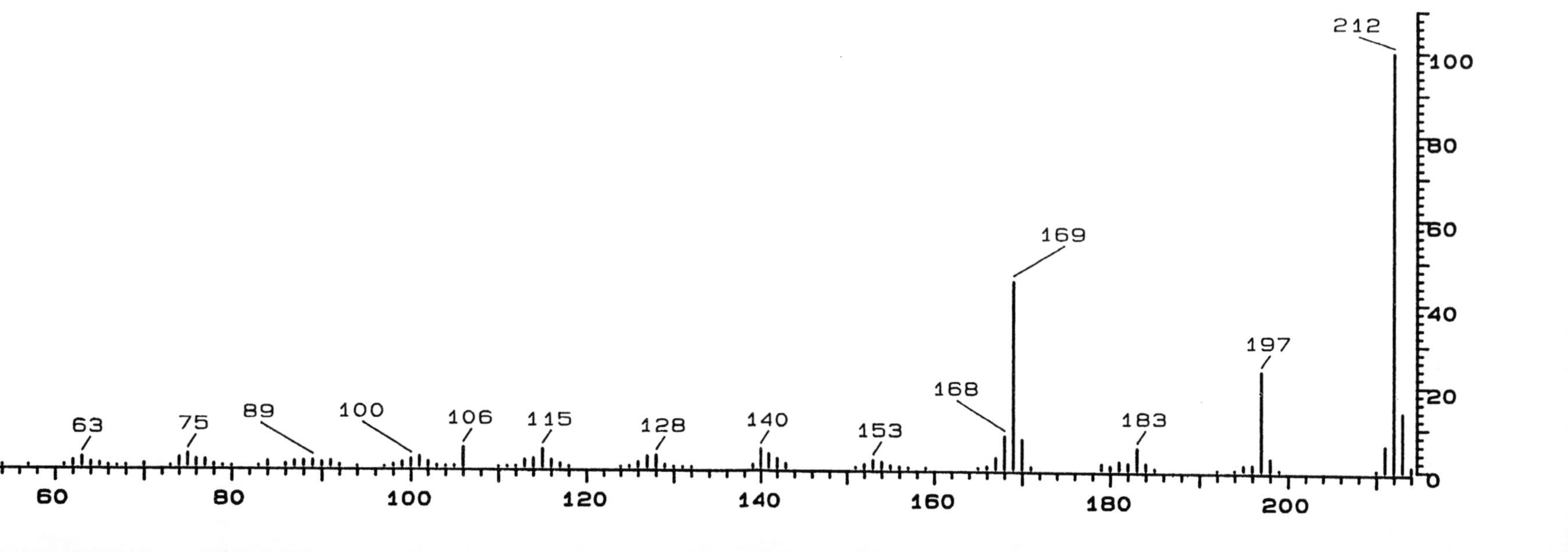

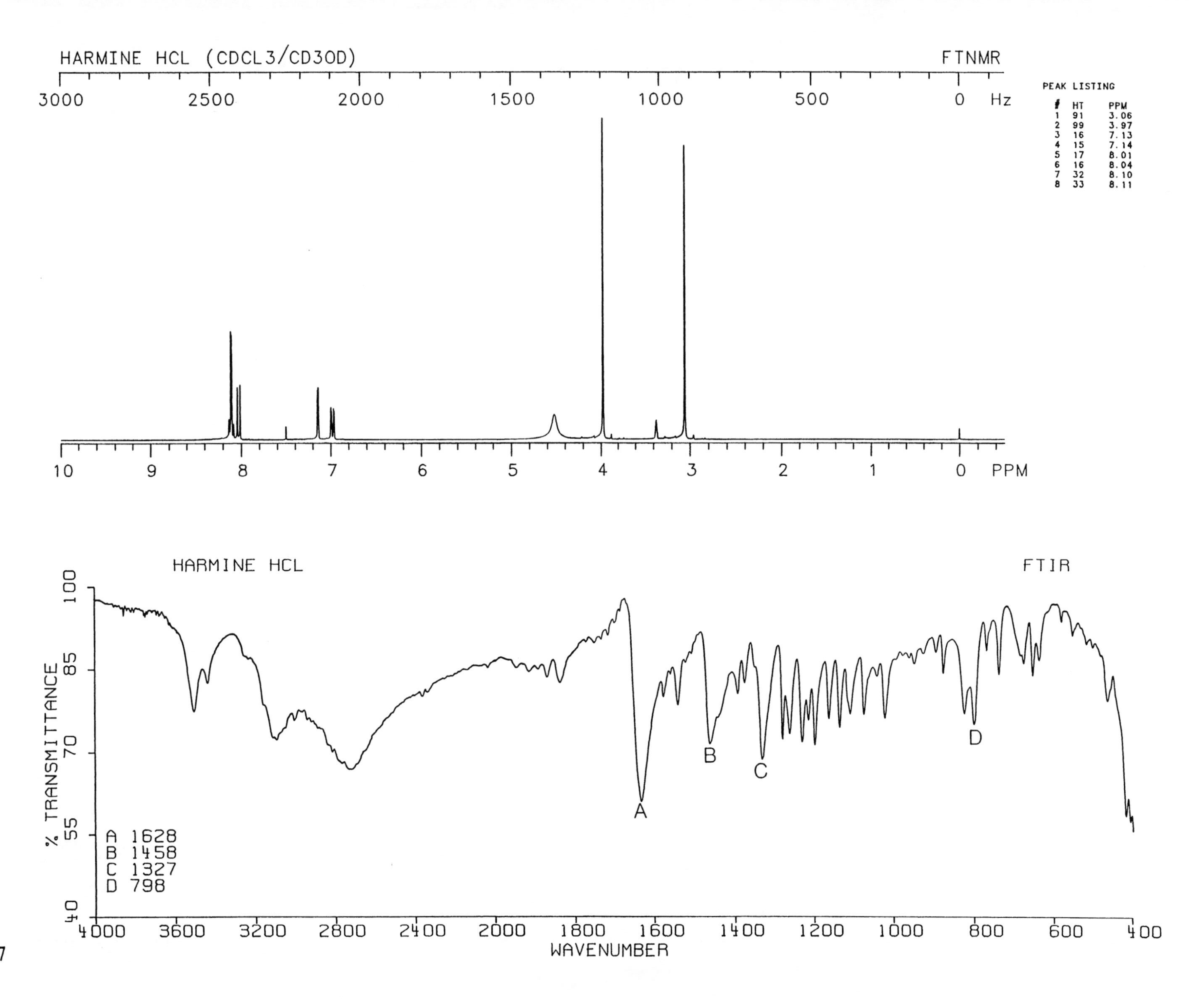
HARMINE HCL (CDCL3/CD3OD)
FTNMR
3000 2500 2000 1500 1000 500 0 Hz
PEAK LISTING
# HT PPM
1 91 3.06
2 99 3.97
3 16 7.13
4 15 7.14
5 17 8.01
6 16 8.04
7 32 8.10
8 33 8.11
10 9 8 7 6 5 4 3 2 1 0 PPM
HARMINE HCL
FTIR
100 85 70 55 40
% TRANSMITTANCE
A 1628
B 1458
C 1327
D 798
A
B
C
D
4000 3600 3200 2800 2400 2000 1800 1600 1400 1200 1000 800 600 400
WAVENUMBER

# HECOGENIN

$C_{27}H_{42}O_4$

Molecular weight: 430.61 (430.31)
Synonyms: (25R)-3β-Hydroxy-5α-spirostan-12-one

Trade names:

Use: Steroidal hormone
HPLC:
GC: 3350; 280°C

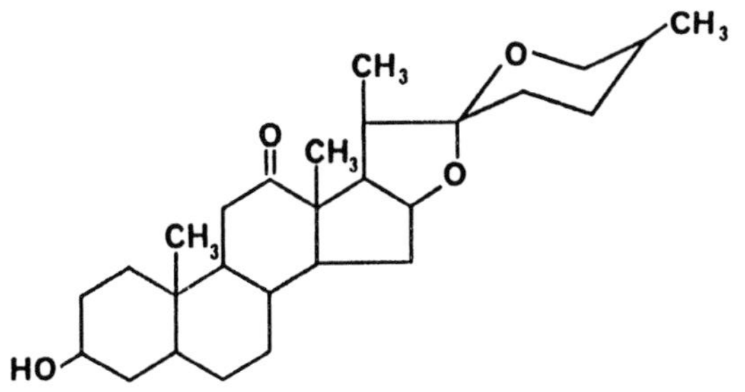

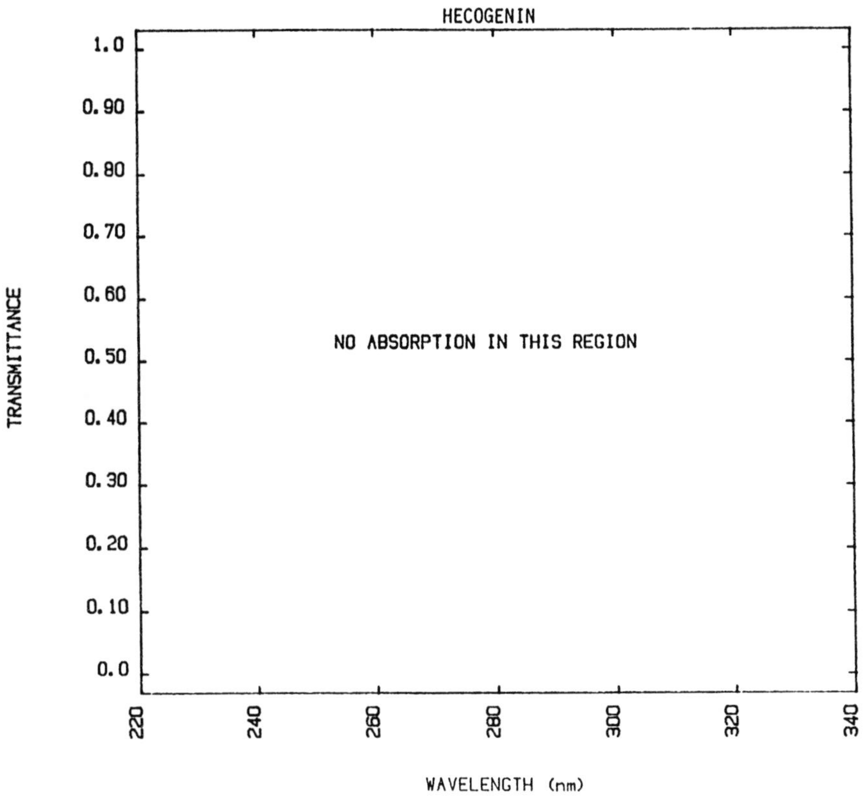

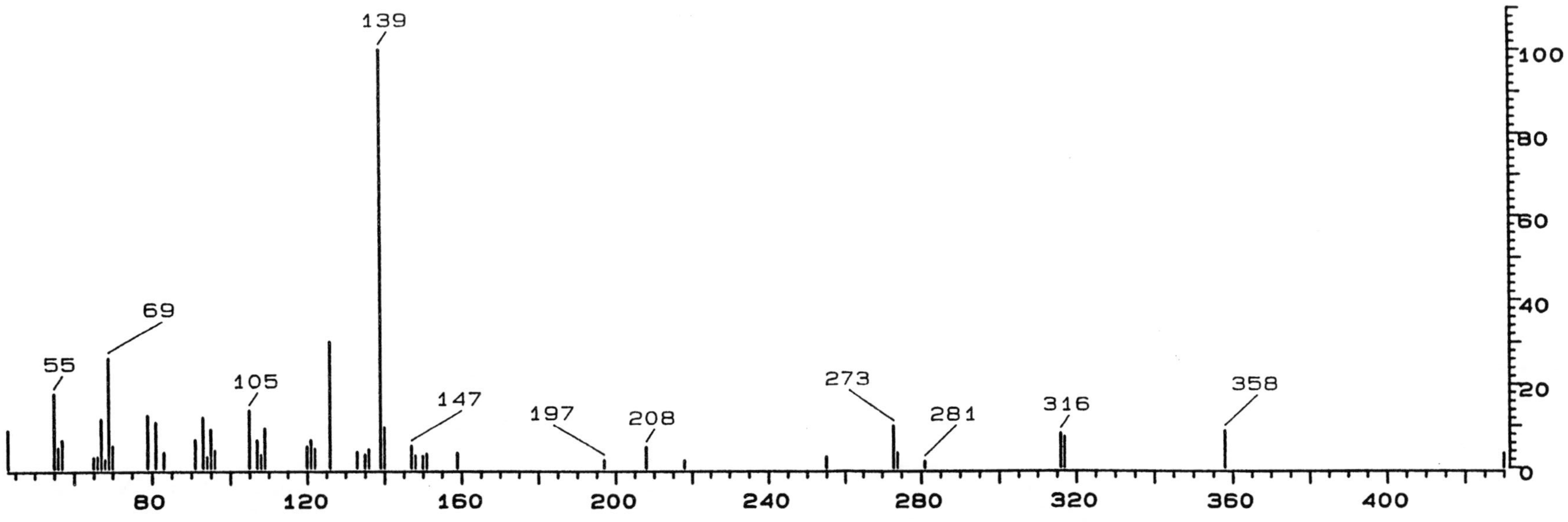

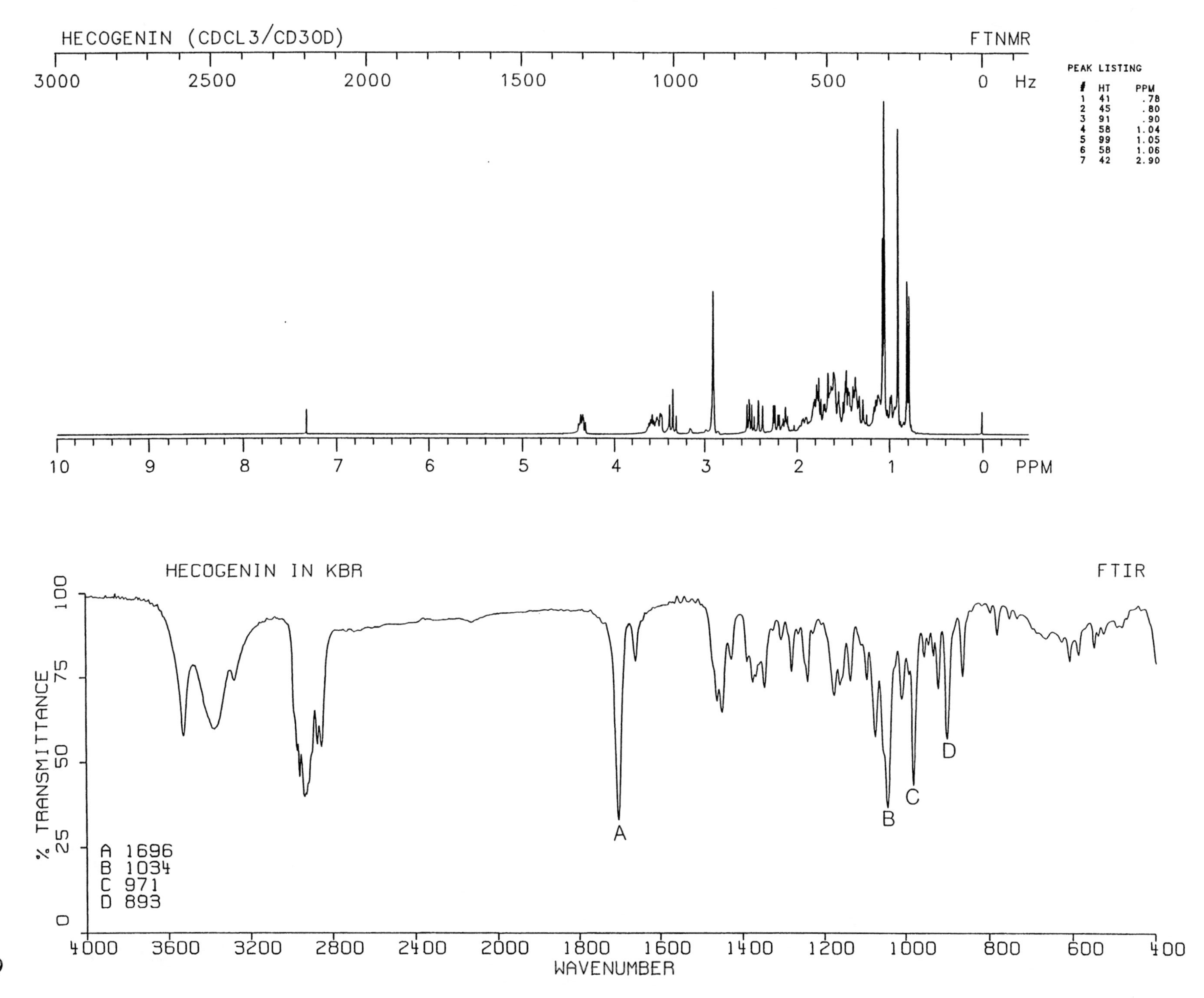
HECOGENIN (CDCL3/CD3OD)
FTNMR
3000 2500 2000 1500 1000 500 0 Hz
PEAK LISTING
# HT PPM
1 41 .78
2 45 .80
3 91 .90
4 58 1.04
5 99 1.05
6 58 1.06
7 42 2.90
10 9 8 7 6 5 4 3 2 1 0 PPM
HECOGENIN IN KBR
FTIR
% TRANSMITTANCE
100 75 50 25 0
A 1696
B 1034
C 971
D 893
A B C D
4000 3600 3200 2800 2400 2000 1800 1600 1400 1200 1000 800 600 400
WAVENUMBER

# HEPTABARBITAL

$C_{13}H_{18}N_2O_3$

**Molecular weight:** 250.30 (250.13)
**Synonyms:** 5-(1-Cyclohepten-1-yl)-5-ethyl-2,4,6(1H,3H,5H)-pyrimidinetrione; 5-ethyl-5-cycloheptenylbarbituric acid
**Trade names:** Heptadorm, Medomin

**Use:** Hypnotic, sedative
**HPLC:** Si-10; 1A:99B; 5.0
**GC:** 2122; 250°C

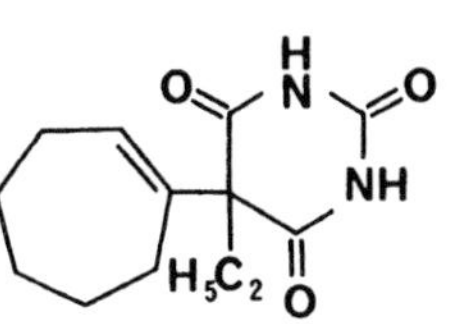


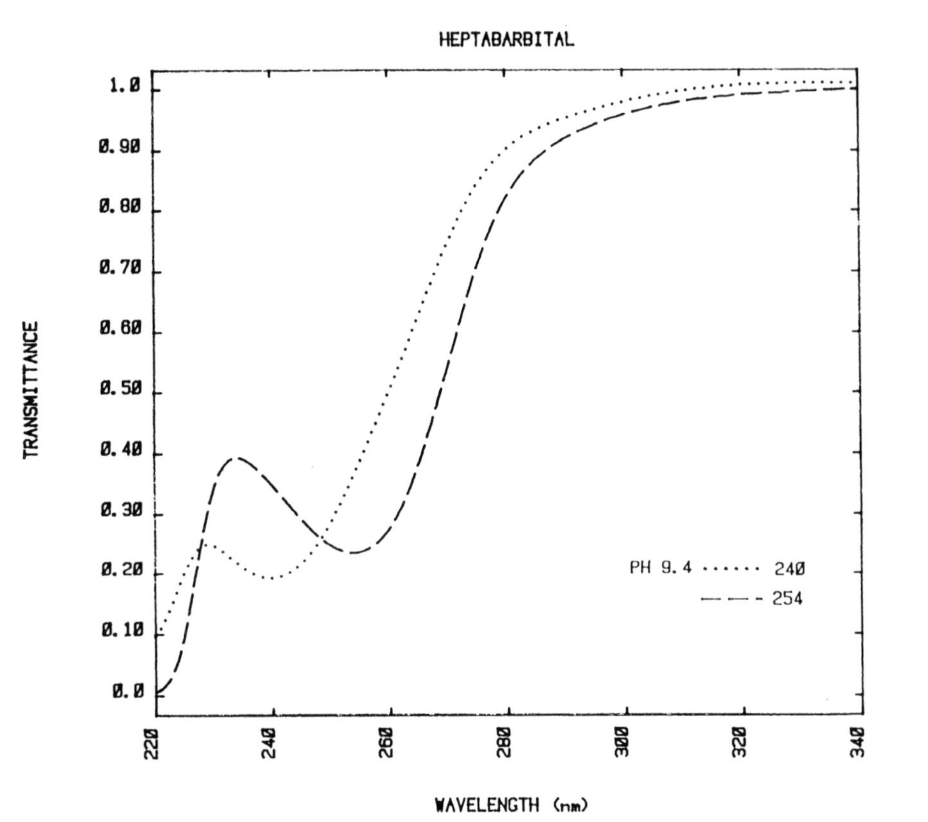

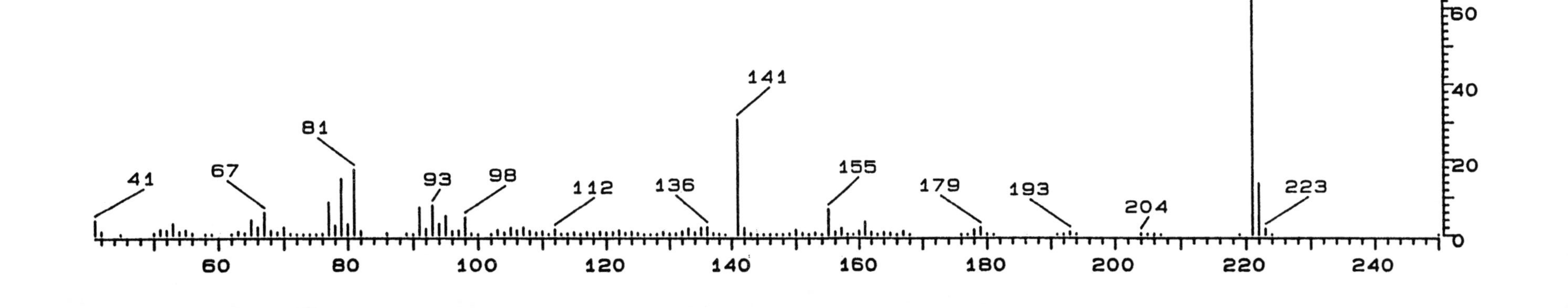

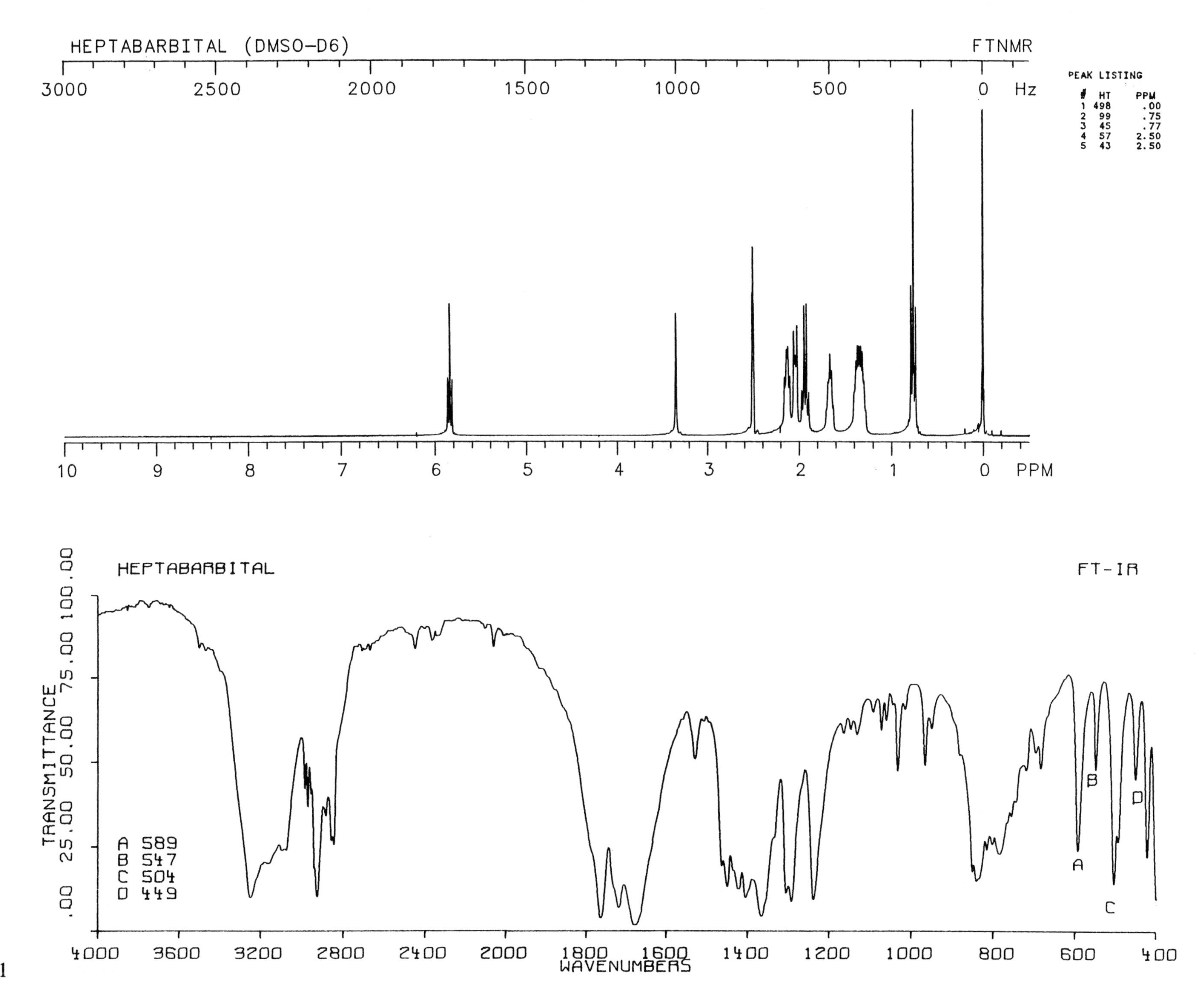
HEPTABARBITAL (DMSO-D6)
FTNMR
3000
2500
2000
1500
1000
500
0 Hz
PEAK LISTING
# HT PPM
1 498 .00
2 99 .75
3 45 .77
4 57 2.50
5 43 2.50
10
9
8
7
6
5
4
3
2
1
0 PPM
HEPTABARBITAL
FT-IR
TRANSMITTANCE
.00
25.00
50.00
75.00
100.00
A 589
B 547
C 504
D 449
A
B
C
D
4000
3600
3200
2800
2400
2000
1800
1600
1400
1200
1000
800
600
400
WAVENUMBERS

# HEROIN

$C_{21}H_{23}NO_5$

Molecular weight: 369.42 (369.16)
Synonyms: 7,8-Didehydro-4,5α-epoxy-17-methylmorphinan-3,6α-diol diacetate; diamorphine; acetomorphine; diacetylmorphine
Trade names:

Use: Narcotic analgesic
HPLC: Si-10; 2A:98B; 8.5
GC: 2666; 250°C

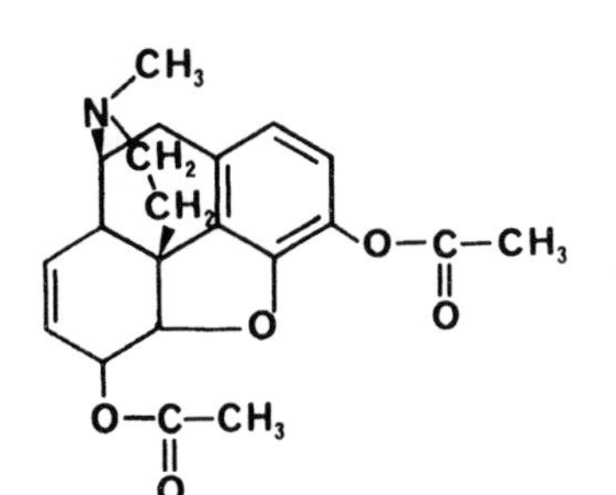


*HEROIN*

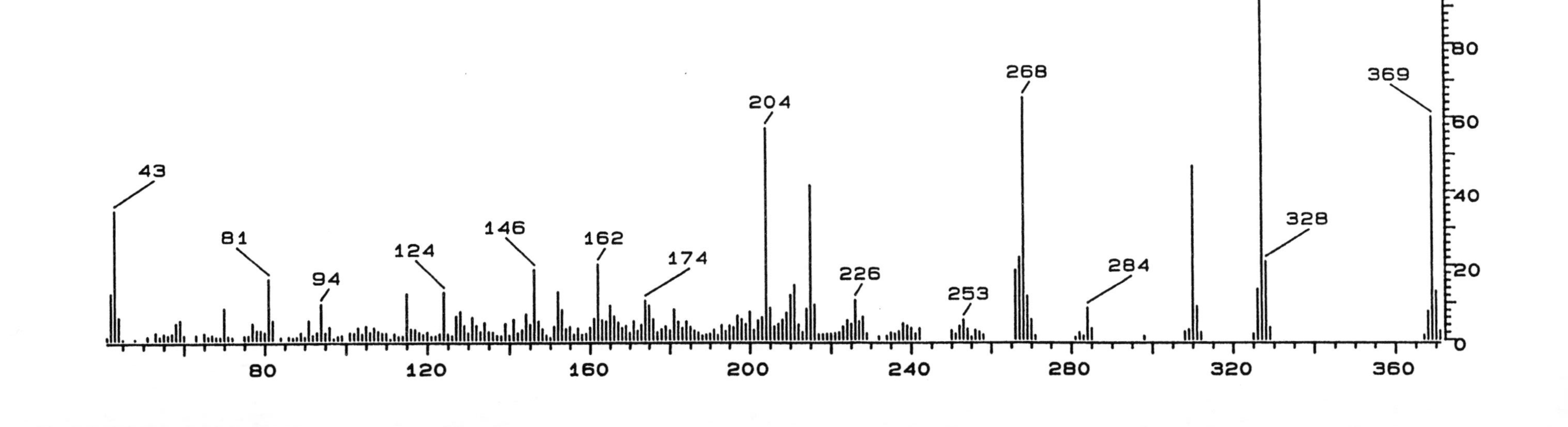

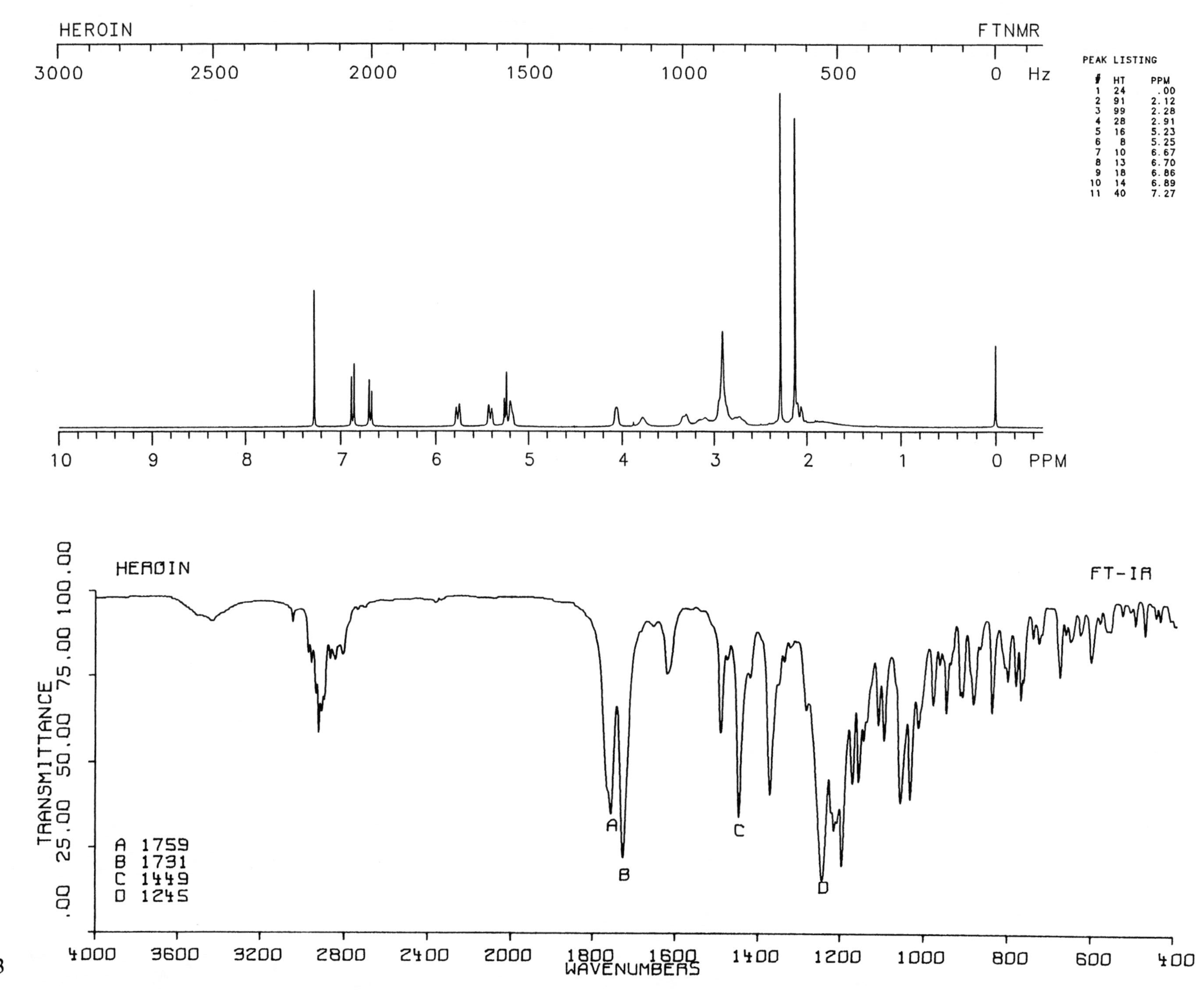
HEROIN
FTNMR
3000
2500
2000
1500
1000
500
0 Hz
PEAK LISTING
# HT PPM
1 24 .00
2 91 2.12
3 99 2.28
4 28 2.91
5 16 5.23
6 8 5.25
7 10 6.67
8 13 6.70
9 18 6.86
10 14 6.89
11 40 7.27
10
9
8
7
6
5
4
3
2
1
0 PPM
HEROIN
FT-IR
TRANSMITTANCE
.00
25.00
50.00
75.00
100.00
A 1759
B 1731
C 1449
D 1245
A
B
C
D
4000
3600
3200
2800
2400
2000
1800
1600
1400
1200
1000
800
600
400
WAVENUMBERS

# HEXACHLOROPHENE

$C_{13}H_6Cl_6O_2$

Molecular weight: 406.92 (403.85)
Synonyms: 2,2'-Methylenebis[3,4,6-trichlorophenol]; hexachlorophane
Trade names: Anacal, Hepadist, Hexaphenyl, Phaisohex, Phisoscrub, pHiso-Hex, Sterzac
Use: Topical anti-infective
HPLC:
GC: 2868; 280°C

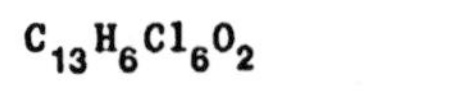


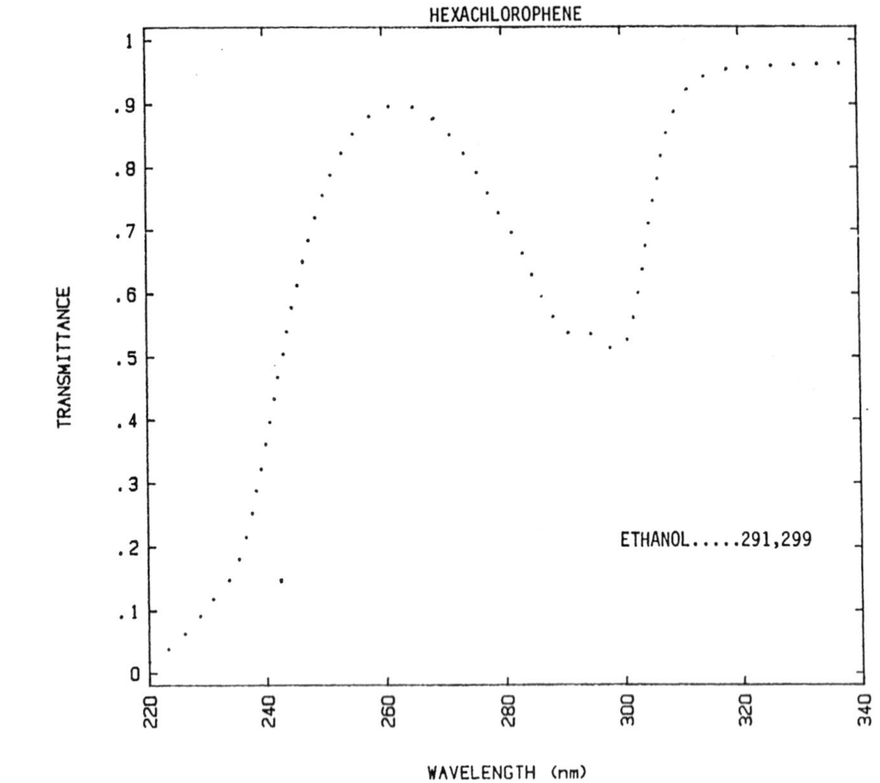

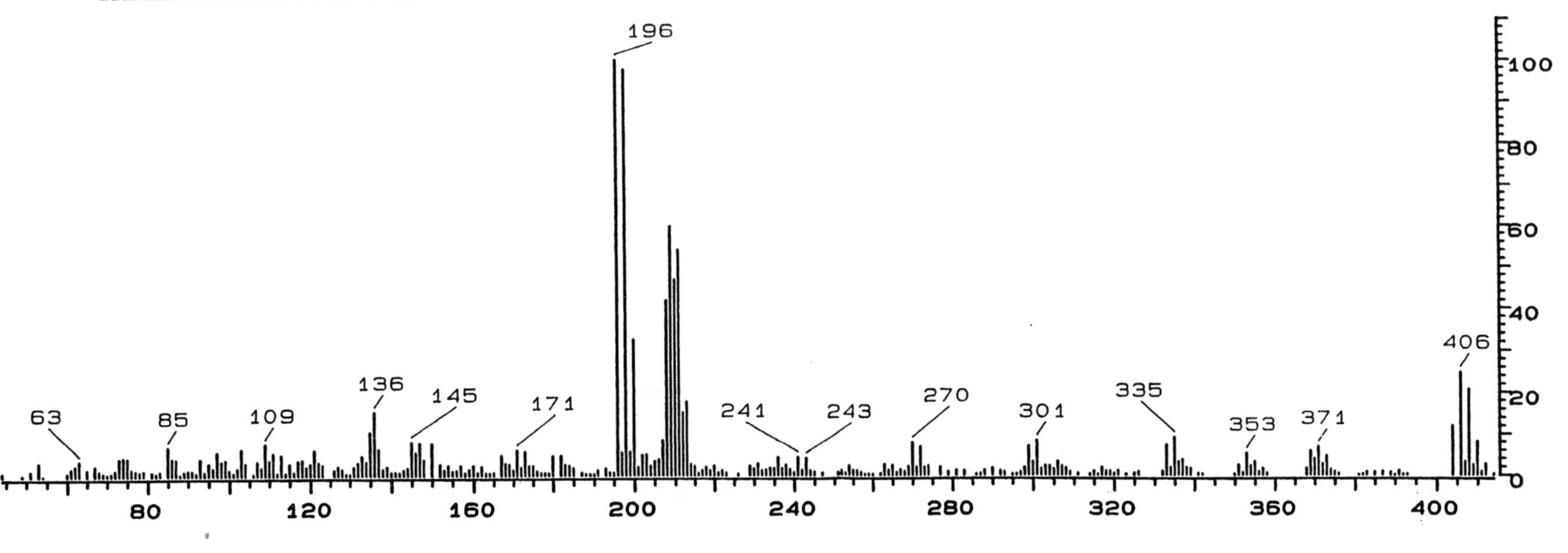

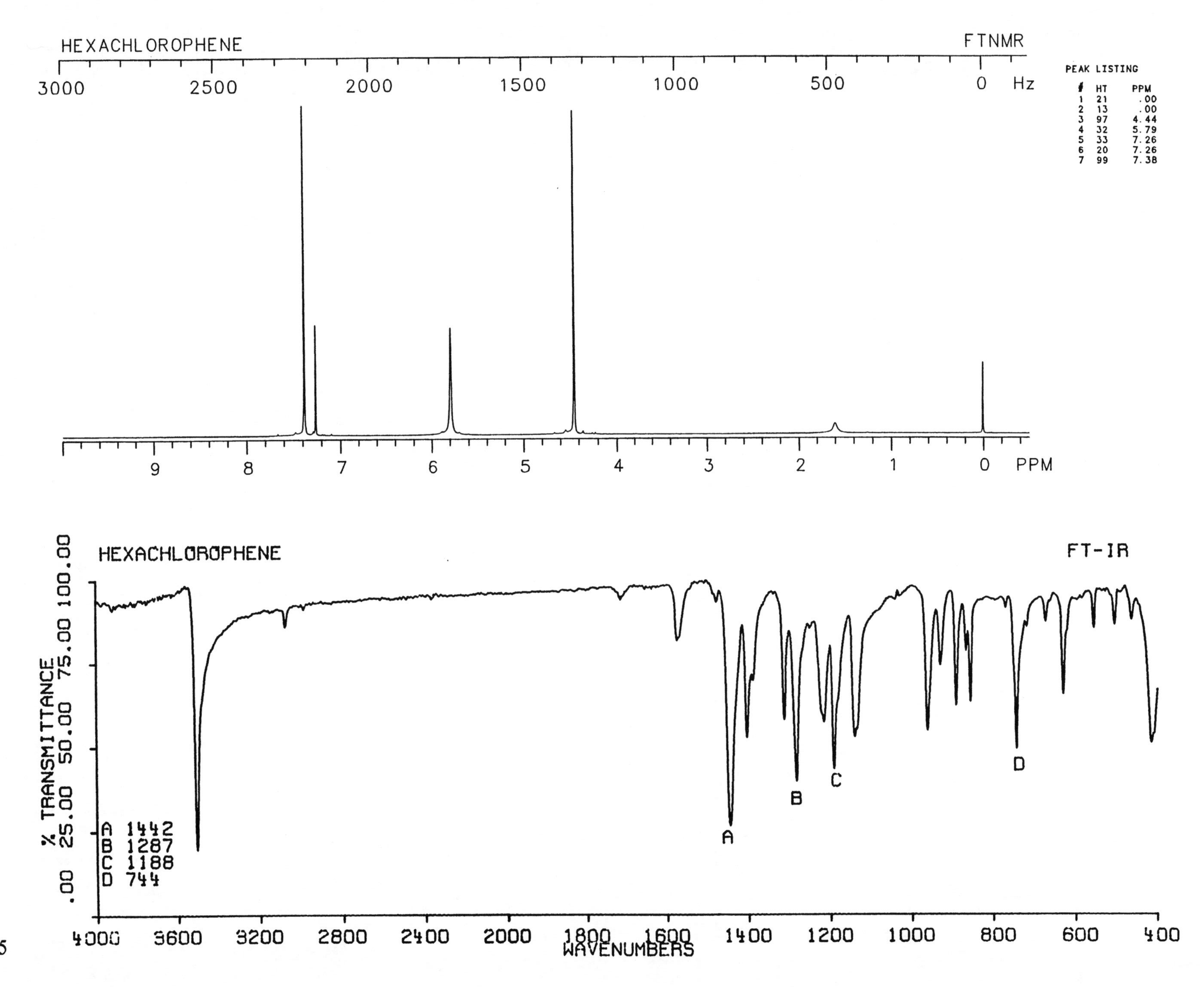
HEXACHLOROPHENE
FTNMR
3000 2500 2000 1500 1000 500 0 Hz
PEAK LISTING
# HT PPM
1 21 .00
2 13 .00
3 97 4.44
4 32 5.79
5 33 7.26
6 20 7.26
7 99 7.38
9 8 7 6 5 4 3 2 1 0 PPM
HEXACHLOROPHENE
FT-IR
% TRANSMITTANCE
.00 25.00 50.00 75.00 100.00
A 1442
B 1287
C 1188
D 744
A B C D
4000 3600 3200 2800 2400 2000 1800 1600 1400 1200 1000 800 600 400
WAVENUMBERS

# HEXAHYDROCANNABINOL

$C_{23}H_{30}O_2$

Molecular weight: 338.49 (338.22)
Synonyms: Hexahydro-6,6,9-trimethyl-3-pentyl-6H-dibenzo[b,d]-pyran-1-ol
Trade names:

Use:
HPLC:
GC: 2437; 250°C

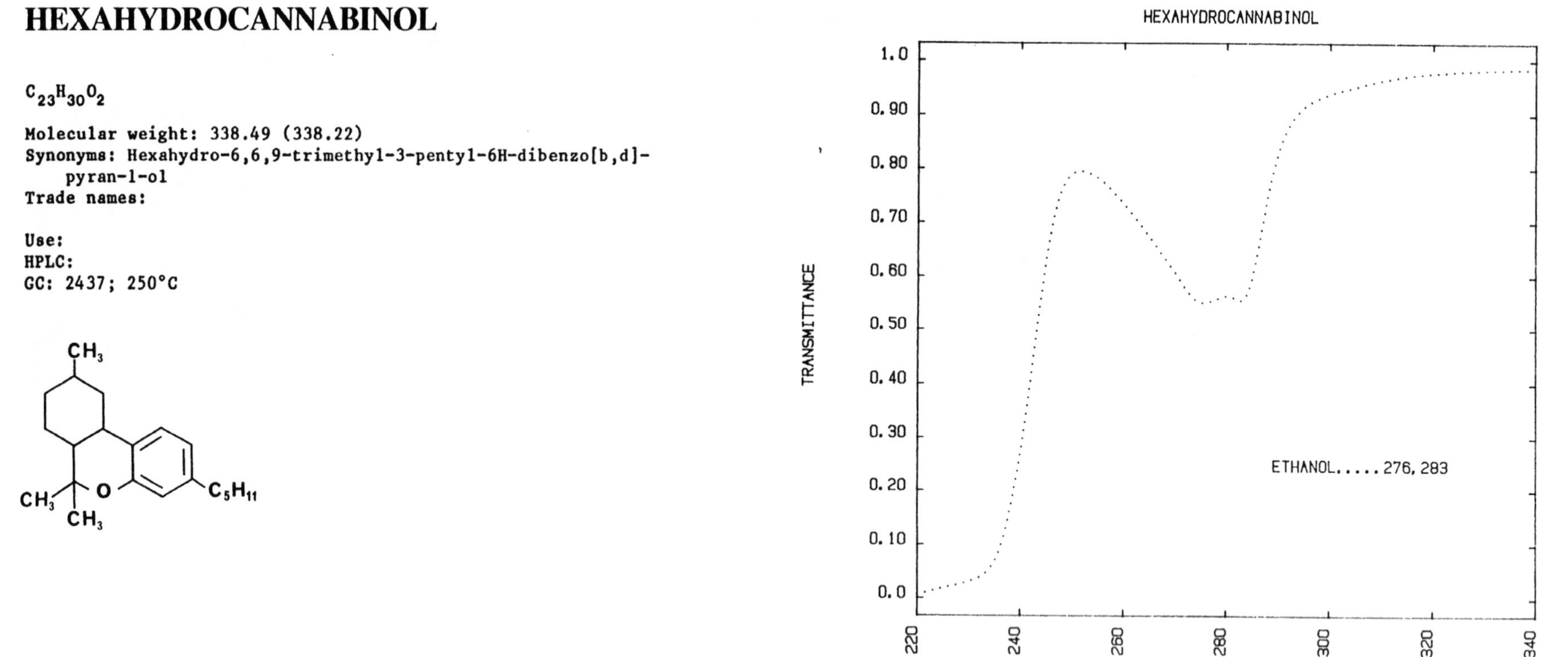


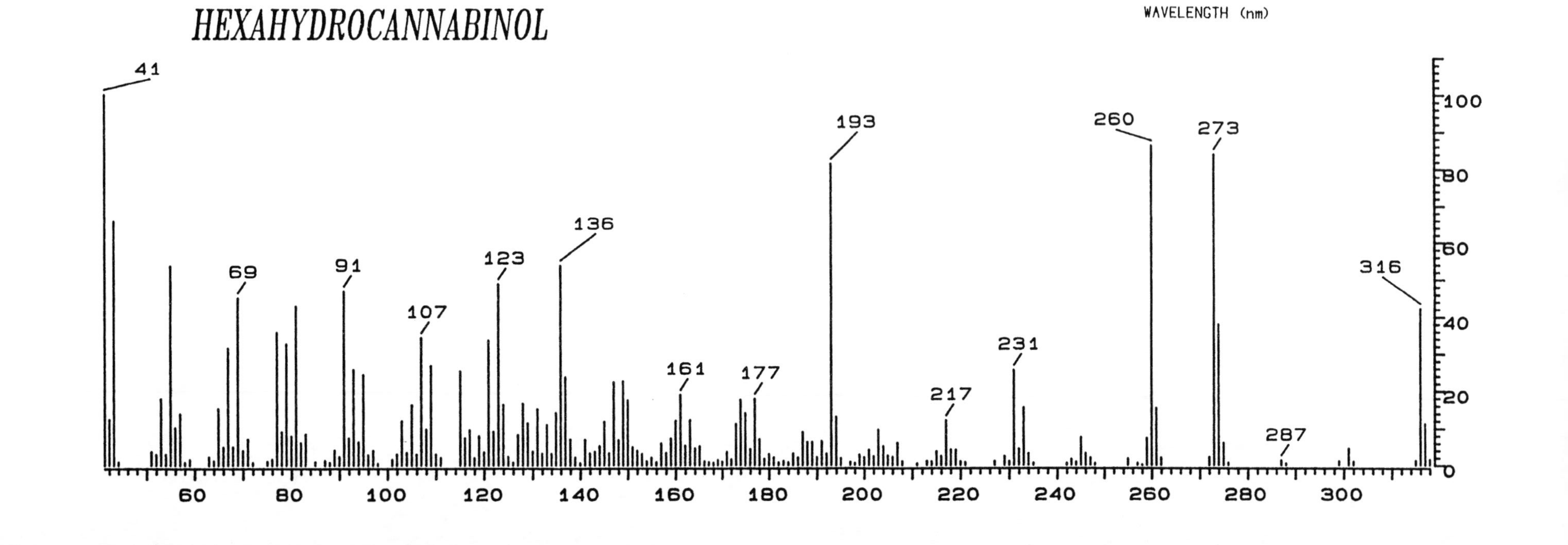

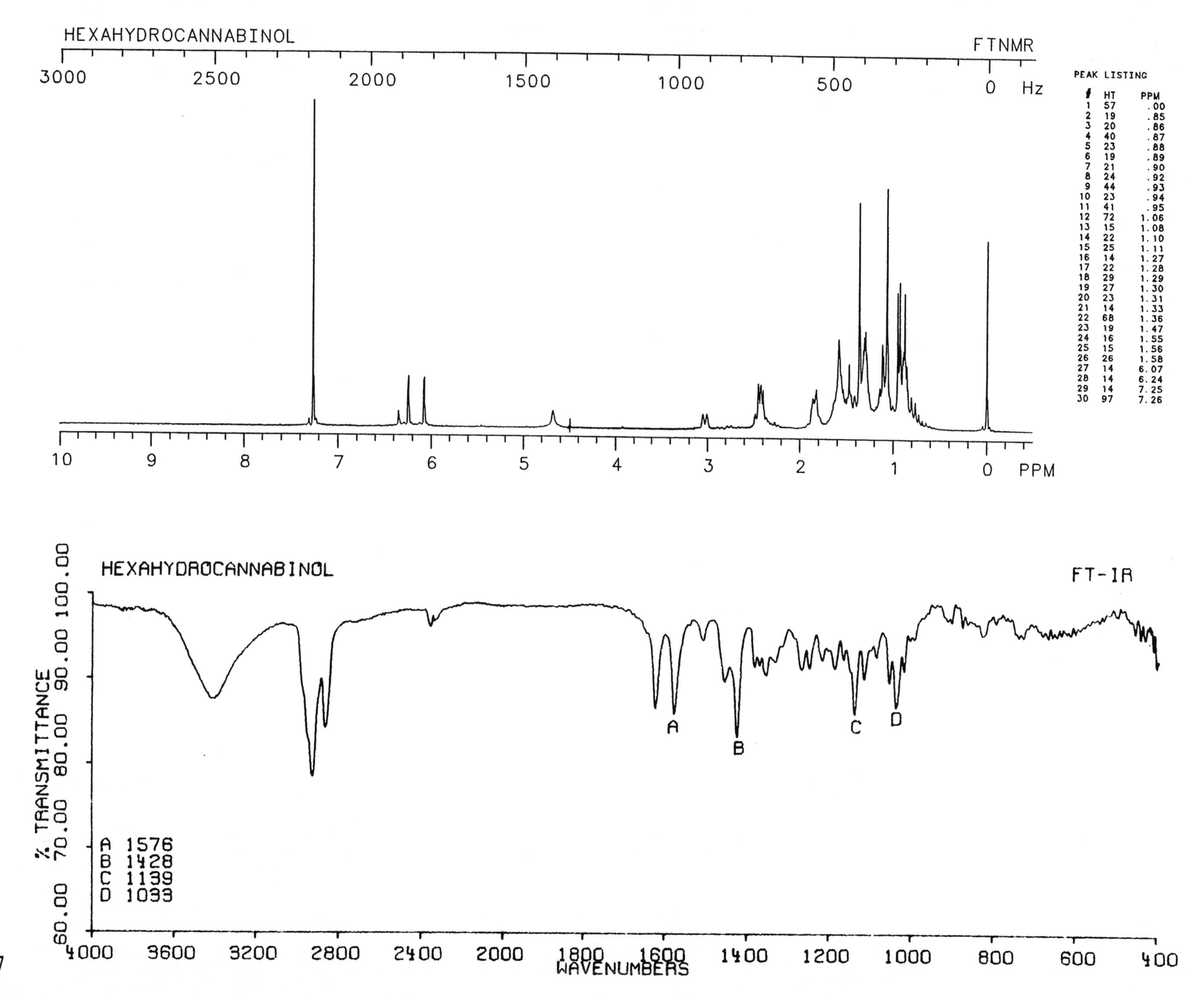
HEXAHYDROCANNABINOL
FTNMR
3000 2500 2000 1500 1000 500 0 Hz
10 9 8 7 6 5 4 3 2 1 0 PPM
PEAK LISTING
# HT PPM
1 57 .00
2 19 .85
3 20 .86
4 40 .87
5 23 .88
6 19 .89
7 21 .90
8 24 .92
9 44 .93
10 23 .94
11 41 .95
12 72 1.06
13 15 1.08
14 22 1.10
15 25 1.11
16 14 1.27
17 22 1.28
18 29 1.29
19 27 1.30
20 23 1.31
21 14 1.33
22 68 1.36
23 19 1.47
24 16 1.55
25 15 1.56
26 26 1.58
27 14 6.07
28 14 6.24
29 14 7.25
30 97 7.26
HEXAHYDROCANNABINOL
FT-IR
% TRANSMITTANCE
60.00 70.00 80.00 90.00 100.00
A B C D
A 1576
B 1428
C 1139
D 1033
4000 3600 3200 2800 2400 2000 1800 1600 1400 1200 1000 800 600 400
WAVENUMBERS

# HEXAMETHONIUM BROMIDE

$C_{12}H_{30}Br_2N_2$

**Molecular weight:** 362.21 (360.08)
**Synonyms:** N,N,N,N',N',N'-Hexamethyl-1,6-hexanediaminium dibromide; hexamethone bromide; hexonium bromide
**Trade names:** Esametina, Gangliostat, Vegolysen, Gastrometonio

**Use:** Antihypertensive
**HPLC:**
**GC:**

$$\left[(CH_3)_3\overset{+}{N}-(CH_2)_6-\overset{+}{N}(CH_3)_3\right]\cdot 2Br^-$$



HEXAMETHONIUM BROMIDE--DIP

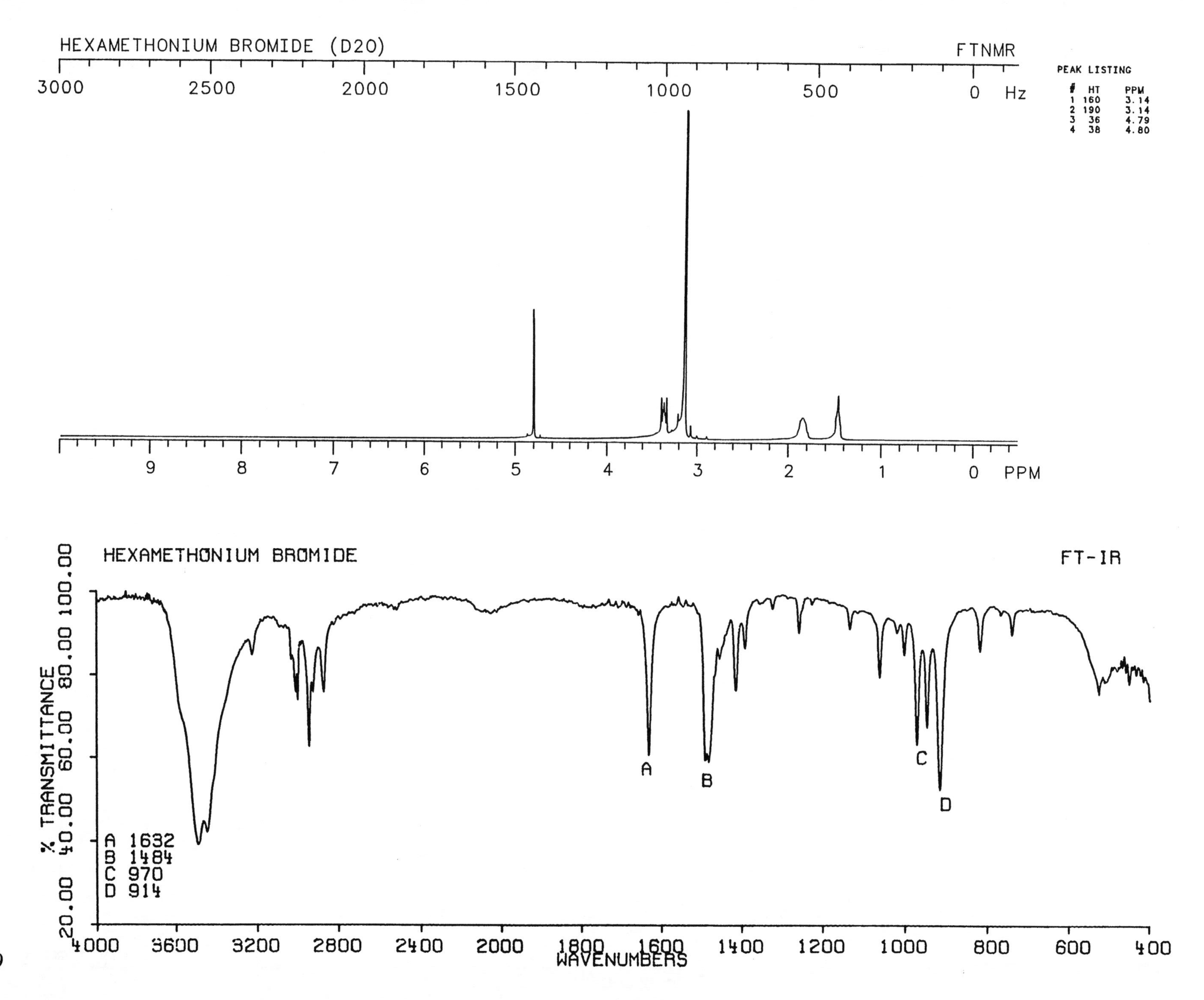
HEXAMETHONIUM BROMIDE (D20)
FTNMR
3000 2500 2000 1500 1000 500 0 Hz
PEAK LISTING
# HT PPM
1 160 3.14
2 190 3.14
3 36 4.79
4 38 4.80
9 8 7 6 5 4 3 2 1 0 PPM
HEXAMETHONIUM BROMIDE
FT-IR
% TRANSMITTANCE
20.00 40.00 60.00 80.00 100.00
A 1632
B 1484
C 970
D 914
A
B
C
D
4000 3600 3200 2800 2400 2000 1800 1600 1400 1200 1000 800 600 400
WAVENUMBERS

# HEXESTROL

$C_{18}H_{22}O_2$

Molecular weight: 270.36 (270.16)
Synonyms: 4,4'-(1,2-Diethyl-1,2-ethanediyl)bisphenol; hexoestrol; dihydrodiethylstilbestrol; hexanoestrol; synestrol; synoestrol
Trade names: Cycloestrol, Hexanoestrol, Hormoestrol, Retalon, Synthovo

Use: Estrogen
HPLC:
GC: 2346; 250°C

HO—C6H4—CH(C2H5)—CH(C2H5)—C6H4—OH

HEXESTROL

TRANSMITTANCE

WAVELENGTH (nm)

ETHANOL.....229,279

HEXESTROL

55 77 91 107 120 135 152 165 181 197 212 213 241 270

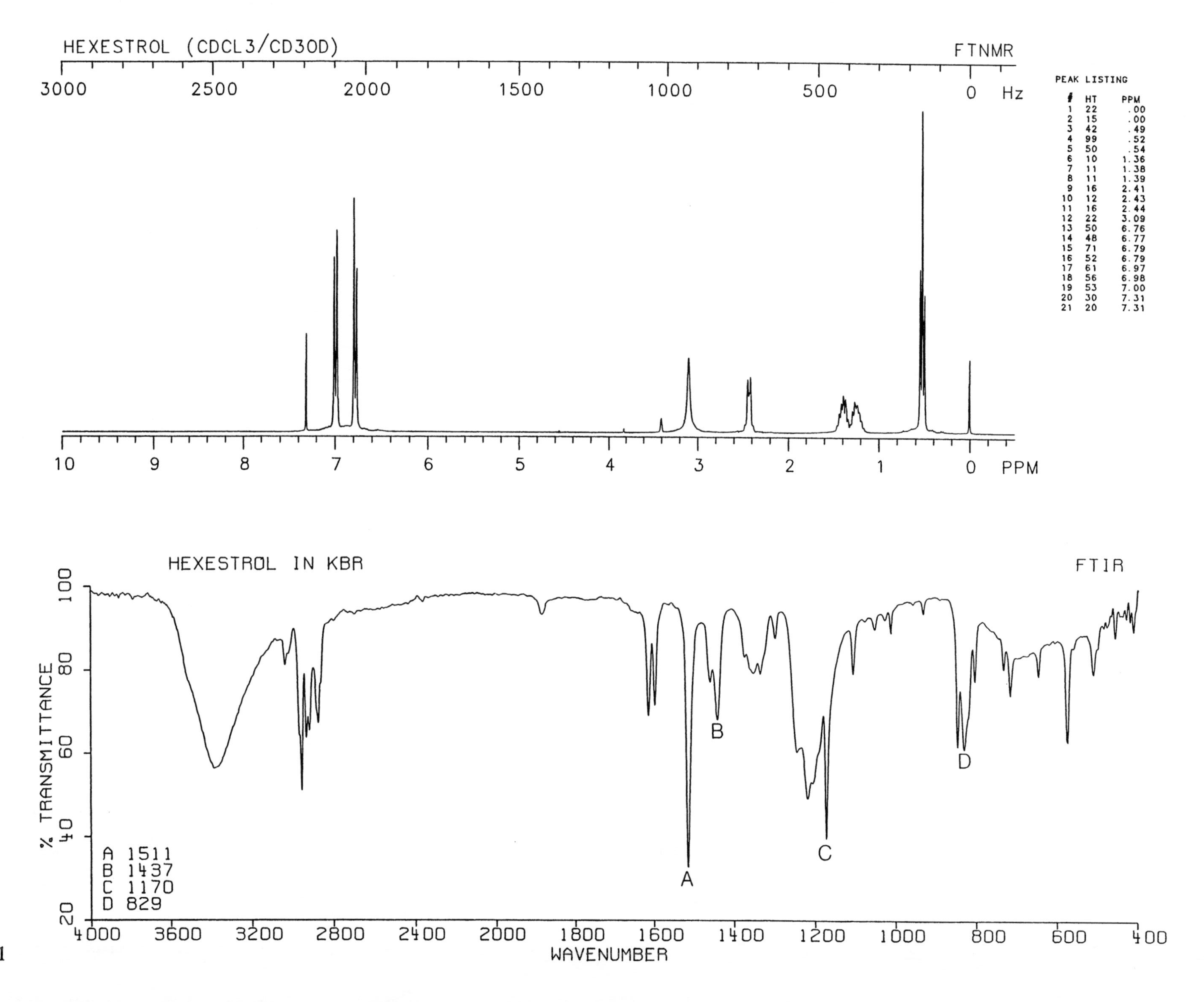
HEXESTROL (CDCL3/CD3OD)
FTNMR
3000
2500
2000
1500
1000
500
0 Hz
PEAK LISTING
# HT PPM
1 22 .00
2 15 .00
3 42 .49
4 99 .52
5 50 .54
6 10 1.36
7 11 1.38
8 11 1.39
9 16 2.41
10 12 2.43
11 16 2.44
12 22 3.09
13 50 6.76
14 48 6.77
15 71 6.79
16 52 6.79
17 61 6.97
18 56 6.98
19 53 7.00
20 30 7.31
21 20 7.31
10
9
8
7
6
5
4
3
2
1
0 PPM
HEXESTROL IN KBR
FTIR
100
80
60
40
20
% TRANSMITTANCE
A 1511
B 1437
C 1170
D 829
A
B
C
D
4000
3600
3200
2800
2400
2000
1800
1600
1400
1200
1000
800
600
400
WAVENUMBER

# HEXETIDINE

$C_{21}H_{45}N_3$

Molecular weight: 339.59 (339.36)
Synonyms: 1,3-Bis(2-ethylhexyl)hexahydro-5-methyl-5-pyrimidinamine

Trade names: Glypesin, Hexoral, Hextril, Oraldene, Sterisil, Sterilate, Sterisol, Triocil
Use: Antifungal
HPLC:
GC: 2157; 250°C

HEXETIDINE

TRANSMITTANCE

NO ABSORPTION IN THIS REGION

WAVELENGTH (nm)

HEXETIDINE

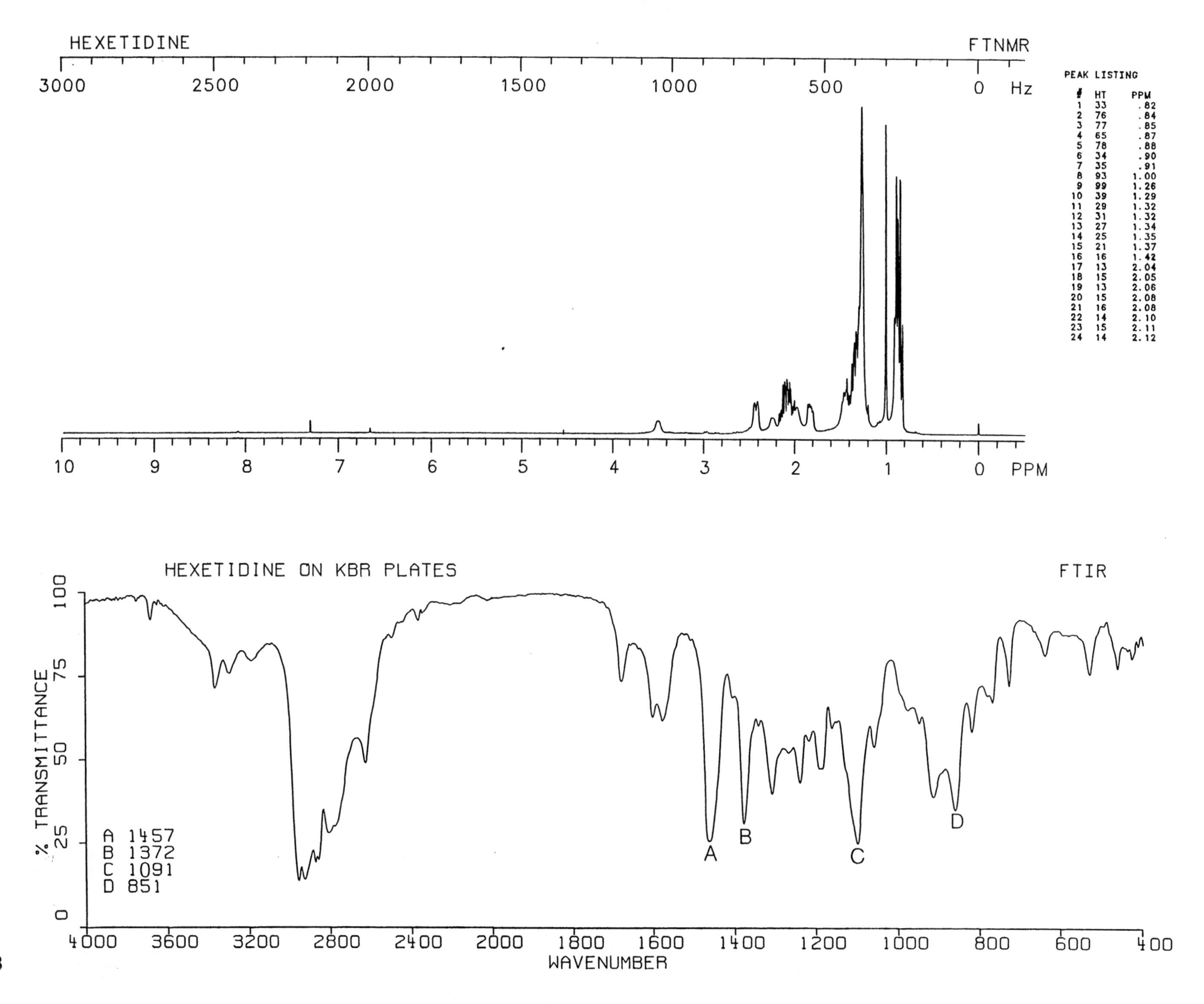

HEXETIDINE
FTNMR
3000
2500
2000
1500
1000
500
0
Hz
PEAK LISTING
| # | HT | PPM |
|---|---|---|
| 1 | 33 | .82 |
| 2 | 76 | .84 |
| 3 | 77 | .85 |
| 4 | 65 | .87 |
| 5 | 78 | .88 |
| 6 | 34 | .90 |
| 7 | 35 | .91 |
| 8 | 93 | 1.00 |
| 9 | 99 | 1.26 |
| 10 | 39 | 1.29 |
| 11 | 29 | 1.32 |
| 12 | 31 | 1.32 |
| 13 | 27 | 1.34 |
| 14 | 25 | 1.35 |
| 15 | 21 | 1.37 |
| 16 | 16 | 1.42 |
| 17 | 13 | 2.04 |
| 18 | 15 | 2.05 |
| 19 | 13 | 2.06 |
| 20 | 15 | 2.08 |
| 21 | 16 | 2.08 |
| 22 | 14 | 2.10 |
| 23 | 15 | 2.11 |
| 24 | 14 | 2.12 |
10
9
8
7
6
5
4
3
2
1
0
PPM
HEXETIDINE ON KBR PLATES
FTIR
100
75
50
25
0
% TRANSMITTANCE
A 1457
B 1372
C 1091
D 851
A
B
C
D
4000
3600
3200
2800
2400
2000
1800
1600
1400
1200
1000
800
600
400
WAVENUMBER

# HEXOBARBITAL

$C_{12}H_{16}N_2O_3$

**Molecular weight:** 236.27 (236.12)
**Synonyms:** 5-(1-Cyclohexen-1-yl)-1,5-dimethyl-2,4,6(1H,3H,5H)-pyrimidinetrione; 5-(1-cyclohexen-1-yl)-1,5-dimethylbarbituric acid; hexobarbitone
**Trade names:** Cyclonal, Hexenal

**Use:** Sedative, hypnotic
**HPLC:** Si-10; 1A:99B; 5.0
**GC:** 1866; 200°C

CH3 O N O CH3 NH O

HEXOBARBITAL

TRANSMITTANCE

1.0 0.90 0.80 0.70 0.60 0.50 0.40 0.30 0.20 0.10 0.0

PH 9.4 ······ 244
— — 244

220 240 260 280 300 320 340

WAVELENGTH (nm)

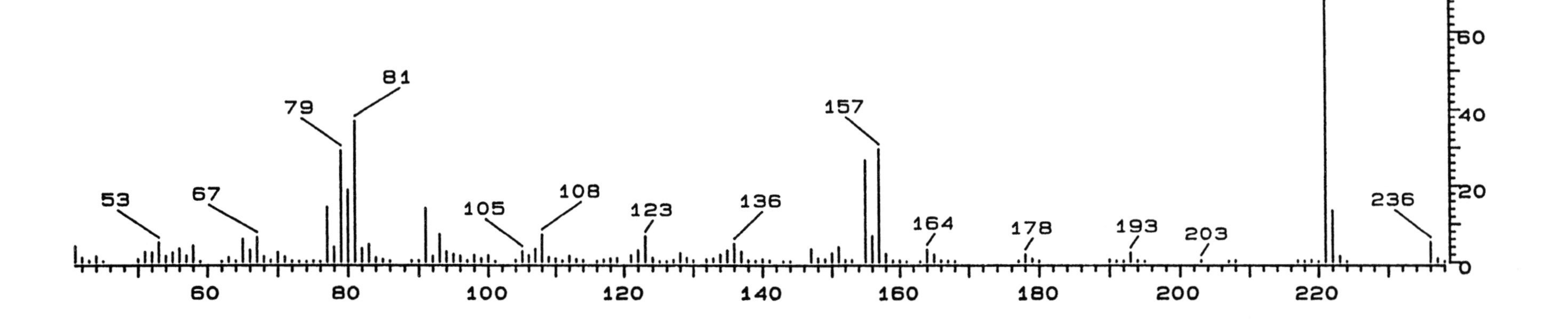

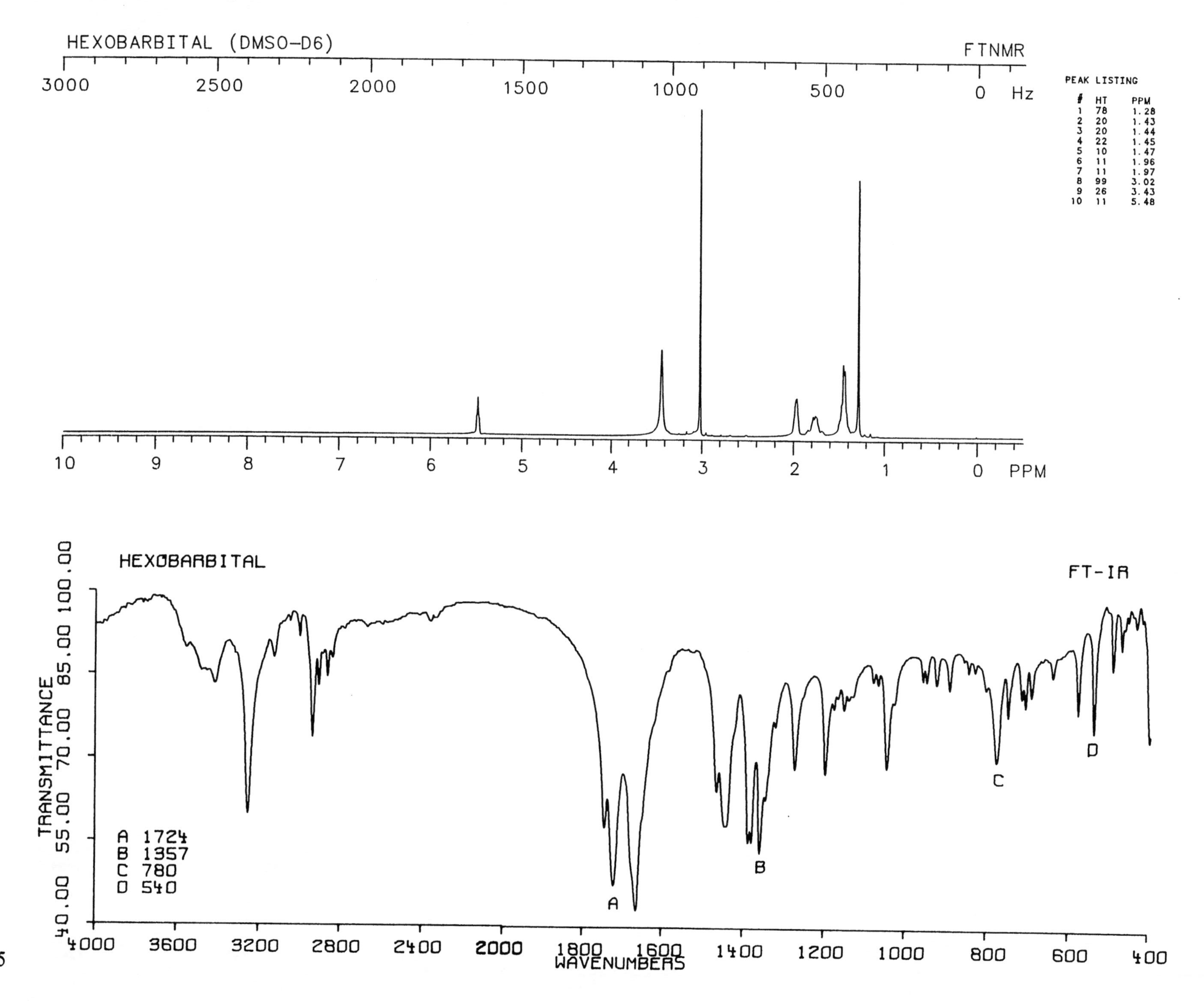
HEXOBARBITAL (DMSO-D6)
FTNMR
Hz
PPM
PEAK LISTING
# HT PPM
1 78 1.28
2 20 1.43
3 20 1.44
4 22 1.45
5 10 1.47
6 11 1.96
7 11 1.97
8 99 3.02
9 26 3.43
10 11 5.48
HEXOBARBITAL
FT-IR
TRANSMITTANCE
WAVENUMBERS
A 1724
B 1357
C 780
D 540

# HEXOCYCLIUM METHYLSULFATE

$C_{21}H_{36}N_2O_5S$

**Molecular weight:** 428.60 (428.23)
**Synonyms:** 4-(2-Cyclohexyl-2-hydroxy-2-phenylethyl)-1,1-dimethylpiperazinium methyl sulfate
**Trade names:** Tral

**Use:** Anticholinergic
**HPLC:**
**GC:**

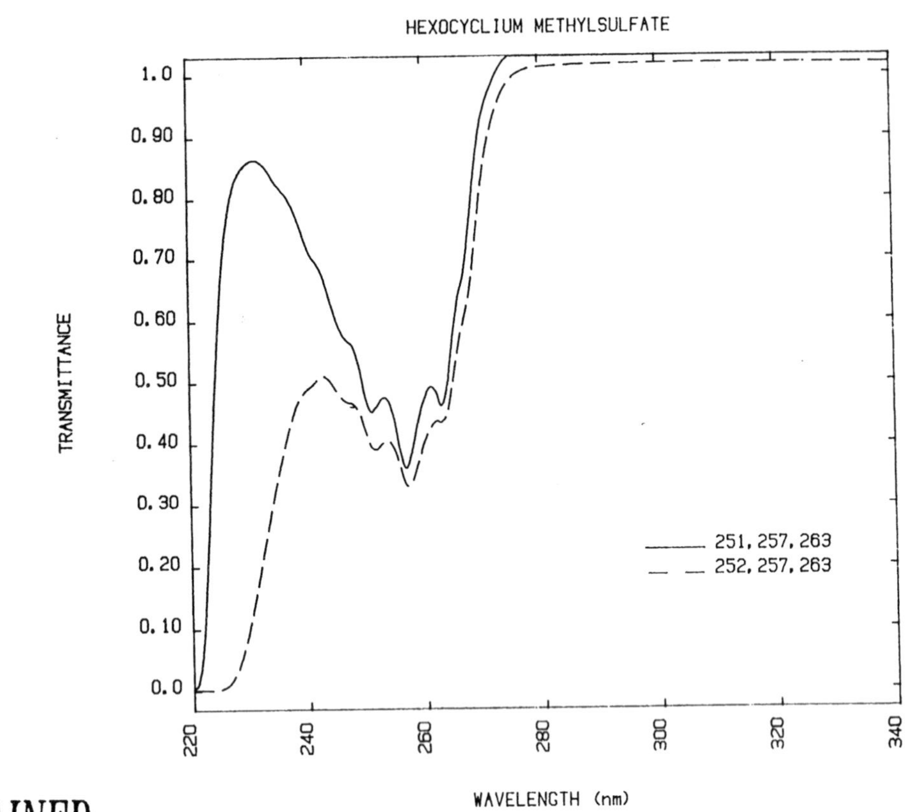


NO USEFUL MASS SPECTRUM WAS OBTAINED

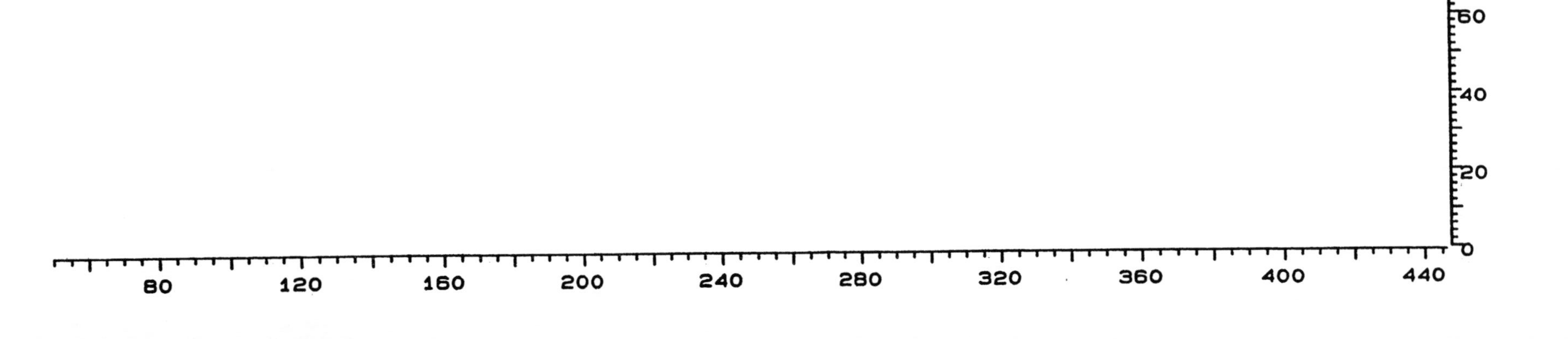

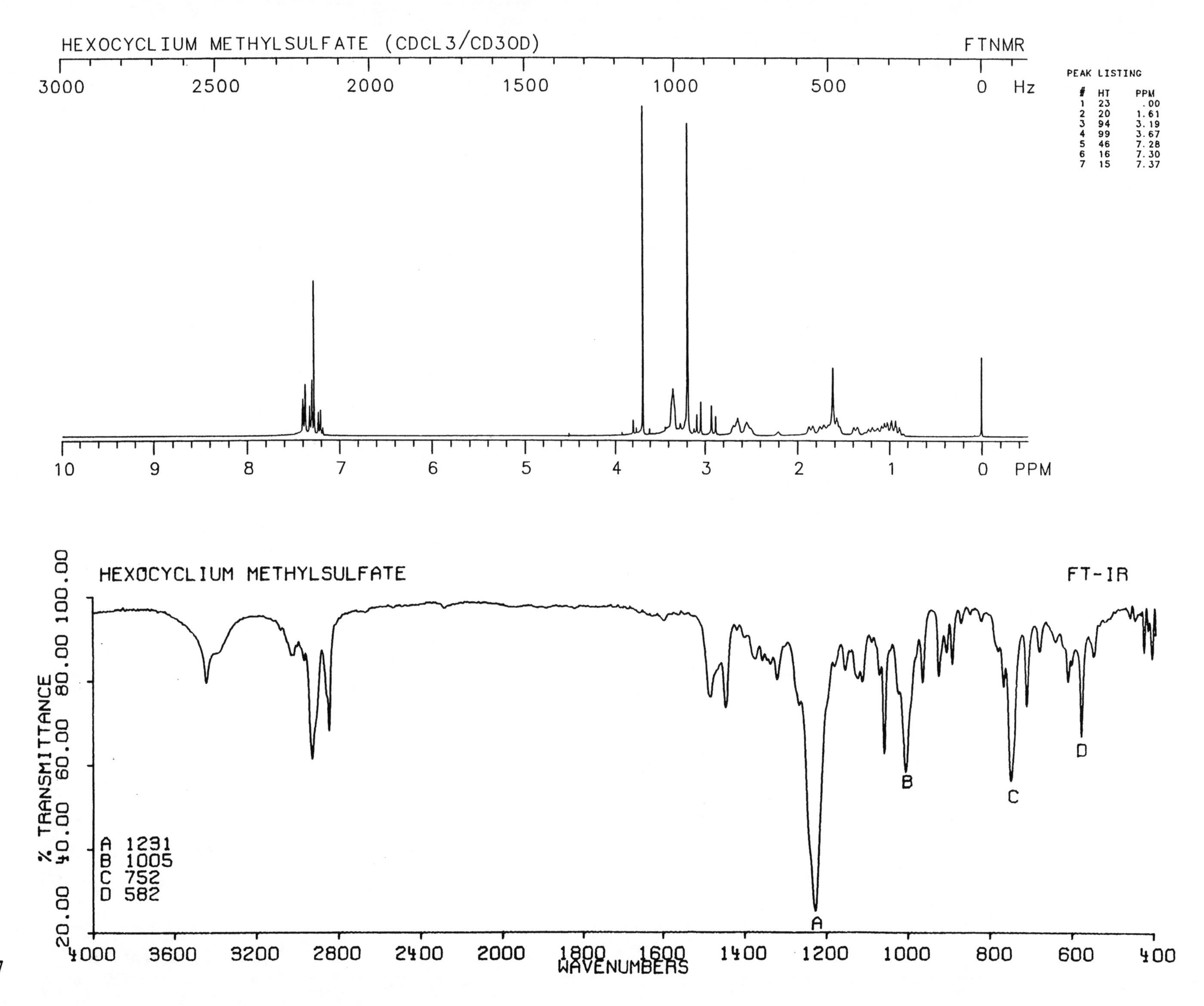

HEXOCYCLIUM METHYLSULFATE (CDCL3/CD3OD)
FTNMR
3000
2500
2000
1500
1000
500
0 Hz
PEAK LISTING
# HT PPM
1 23 .00
2 20 1.61
3 94 3.19
4 99 3.67
5 46 7.28
6 16 7.30
7 15 7.37
10
9
8
7
6
5
4
3
2
1
0 PPM
HEXOCYCLIUM METHYLSULFATE
FT-IR
% TRANSMITTANCE
20.00 40.00 60.00 80.00 100.00
A 1231
B 1005
C 752
D 582
4000 3600 3200 2800 2400 2000 1800 1600 1400 1200 1000 800 600 400
WAVENUMBERS
A
B
C
D

# HEXYL NICOTINATE

$C_{12}H_{17}NO_2$

**Molecular weight:** 207.30 (207.13)
**Synonyms:** N-Hexylnicotinate; nicotinic acid hexylester

**Trade names:** Transvasin

**Use:** Topical vasodilator
**HPLC:**
**GC:** 1651; 200°C

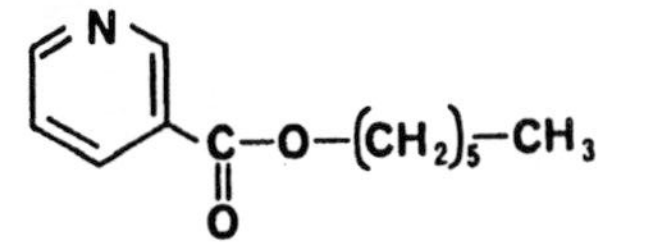


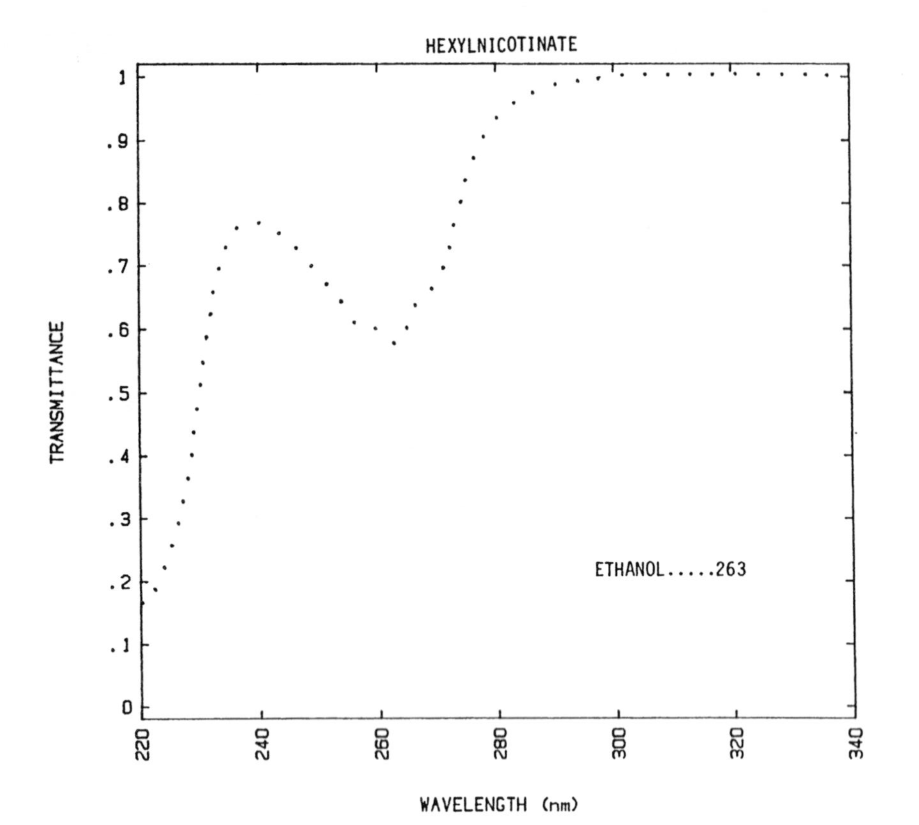

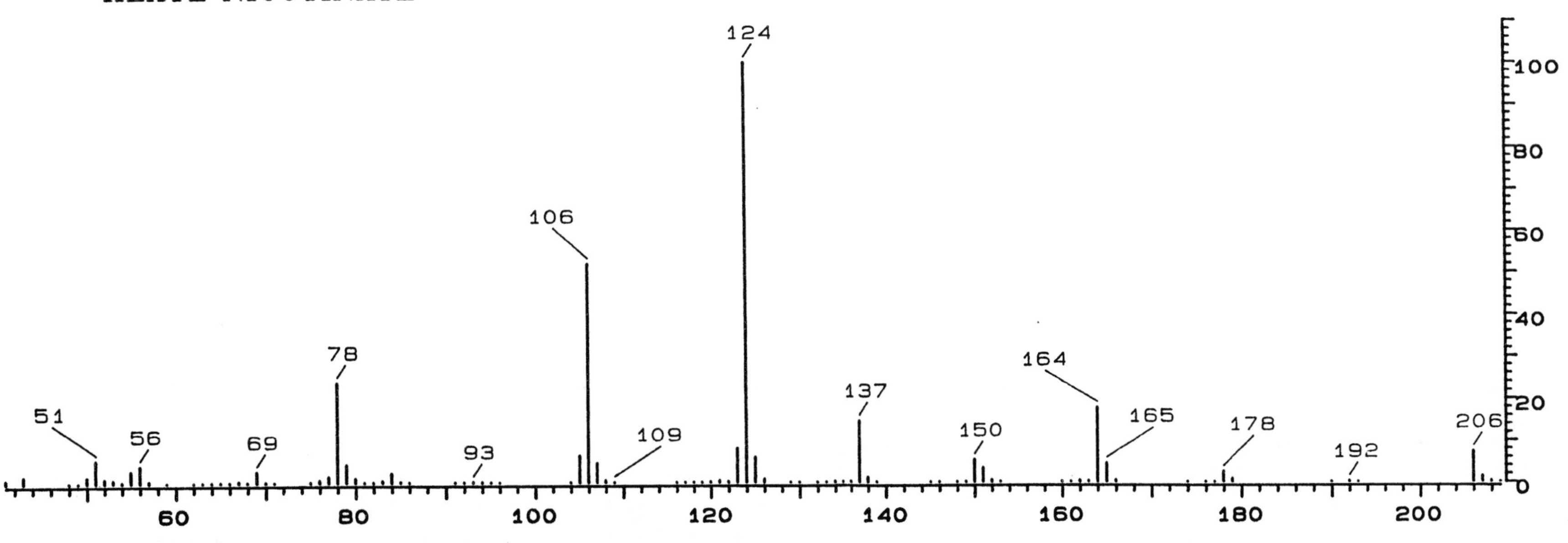

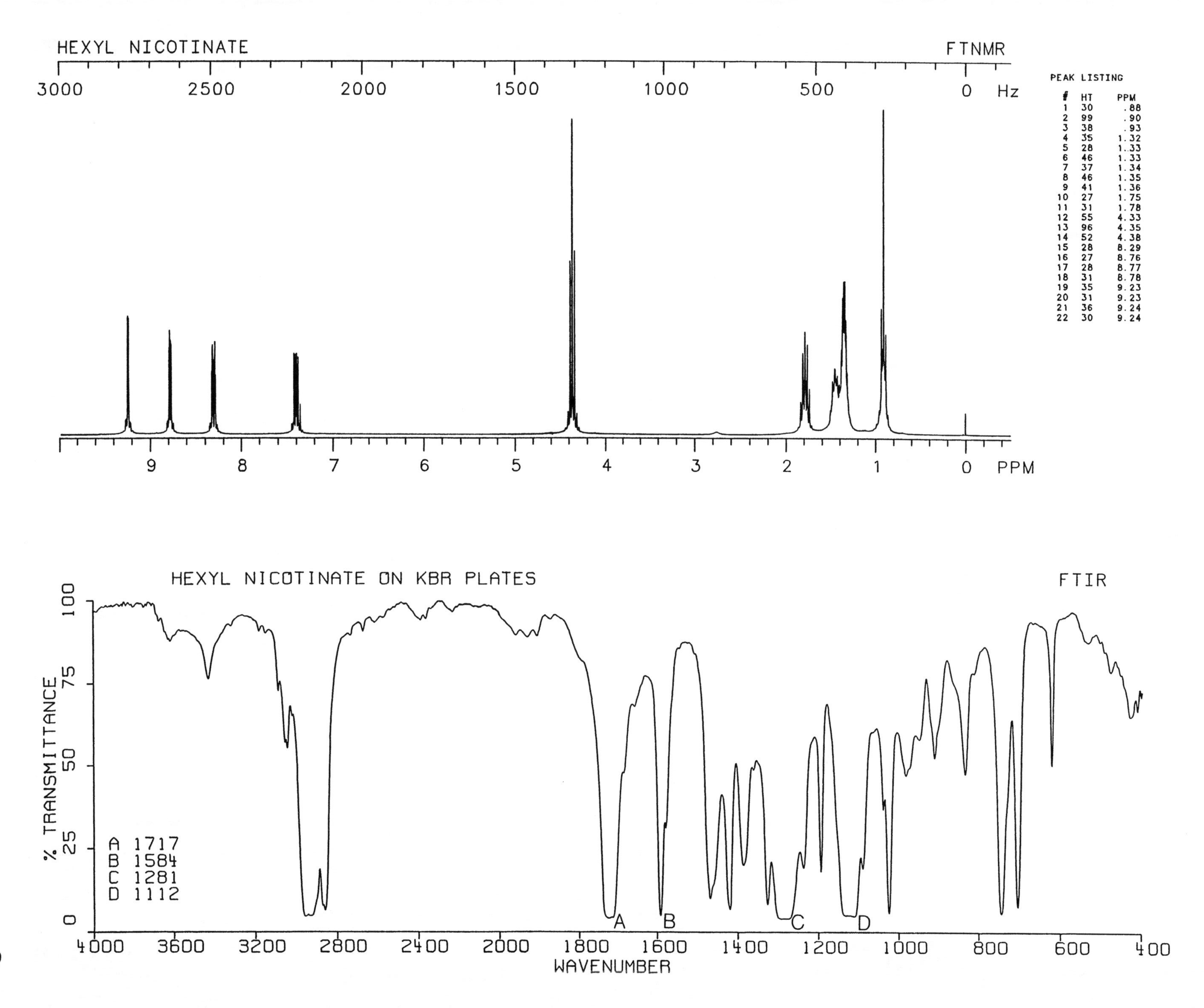

HEXYL NICOTINATE
FTNMR
3000 2500 2000 1500 1000 500 0 Hz
9 8 7 6 5 4 3 2 1 0 PPM
PEAK LISTING
| # | HT | PPM |
|---|---|---|
| 1 | 30 | .88 |
| 2 | 99 | .90 |
| 3 | 38 | .93 |
| 4 | 35 | 1.32 |
| 5 | 28 | 1.33 |
| 6 | 46 | 1.33 |
| 7 | 37 | 1.34 |
| 8 | 46 | 1.35 |
| 9 | 41 | 1.36 |
| 10 | 27 | 1.75 |
| 11 | 31 | 1.78 |
| 12 | 55 | 4.33 |
| 13 | 96 | 4.35 |
| 14 | 52 | 4.38 |
| 15 | 28 | 8.29 |
| 16 | 27 | 8.76 |
| 17 | 28 | 8.77 |
| 18 | 31 | 8.78 |
| 19 | 35 | 9.23 |
| 20 | 31 | 9.23 |
| 21 | 36 | 9.24 |
| 22 | 30 | 9.24 |
HEXYL NICOTINATE ON KBR PLATES
FTIR
% TRANSMITTANCE
100 75 50 25 0
A 1717
B 1584
C 1281
D 1112
4000 3600 3200 2800 2400 2000 1800 1600 1400 1200 1000 800 600 400
WAVENUMBER

# HEXYLRESORCINOL

$C_{12}H_{18}O_2$

**Molecular weight:** 194.26 (194.13)
**Synonyms:** 4-Hexyl-1,3-benzenediol; 4-hexyl-1,3-dihydroxybenzene

**Trade names:** Caprokol, Crystoids, Gelovermin, Oxana, Sucrets

**Use:** Topical antiseptic, anthelmintic
**HPLC:** Si-10; 1A:99B; 6.2
**GC:** 1822; 200°C

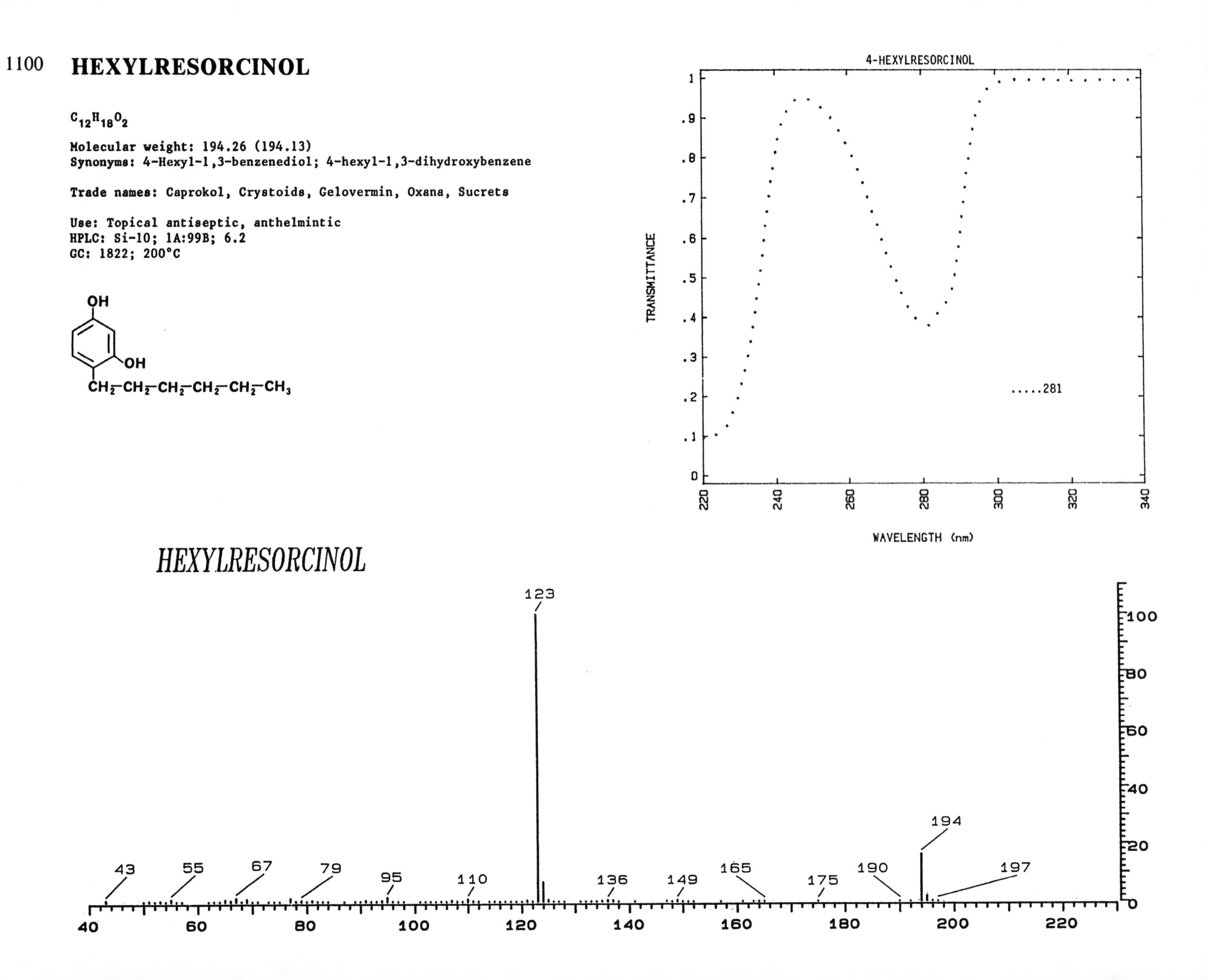

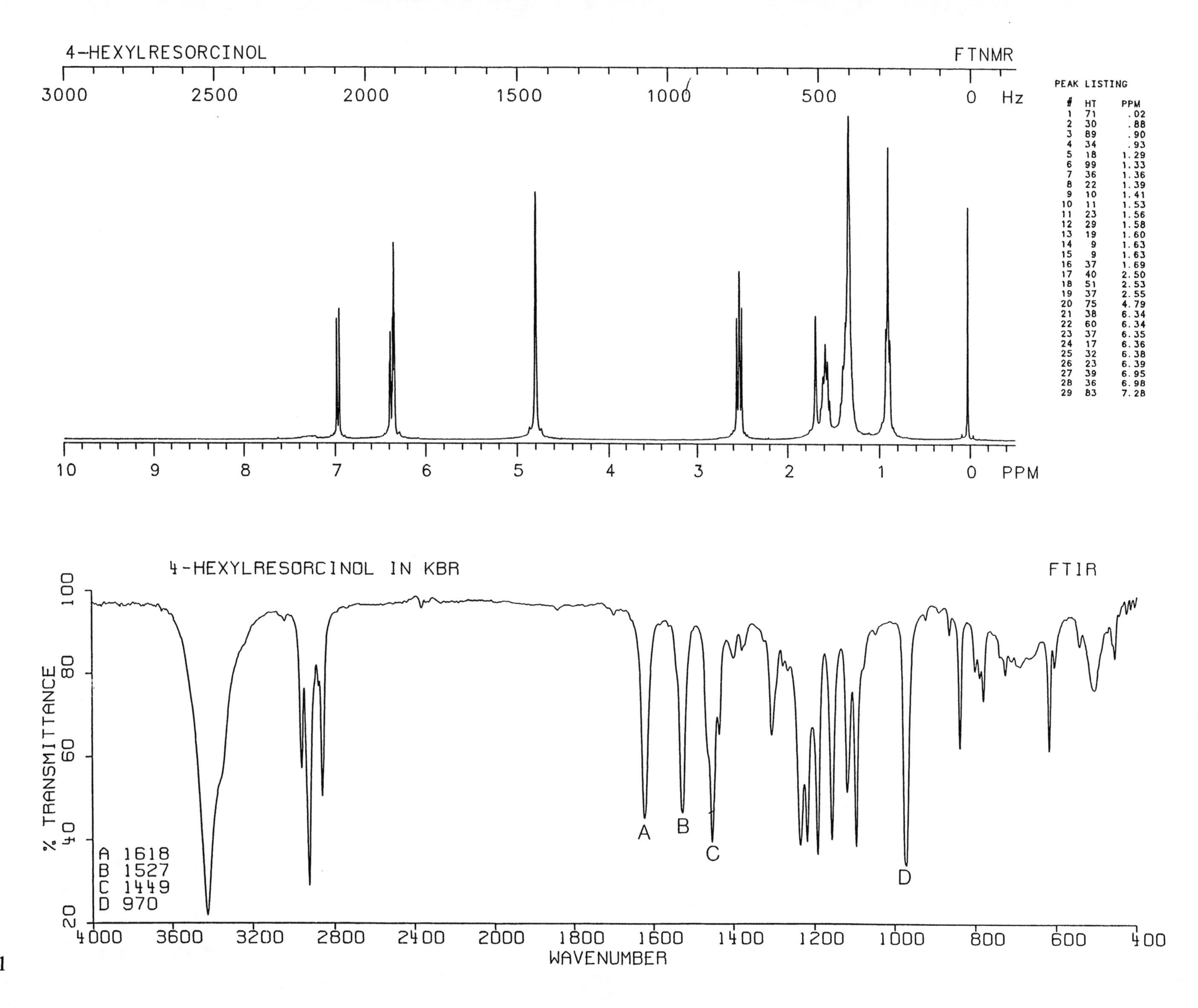
4-HEXYLRESORCINOL
FTNMR
3000 2500 2000 1500 1000 500 0 Hz
10 9 8 7 6 5 4 3 2 1 0 PPM
PEAK LISTING
# HT PPM
1 71 .02
2 30 .88
3 89 .90
4 34 .93
5 18 1.29
6 99 1.33
7 36 1.36
8 22 1.39
9 10 1.41
10 11 1.53
11 23 1.56
12 29 1.58
13 19 1.60
14 9 1.63
15 9 1.63
16 37 1.69
17 40 2.50
18 51 2.53
19 37 2.55
20 75 4.79
21 38 6.34
22 60 6.34
23 37 6.35
24 17 6.36
25 32 6.38
26 23 6.39
27 39 6.95
28 36 6.98
29 83 7.28
4-HEXYLRESORCINOL IN KBR
FTIR
% TRANSMITTANCE
100 80 60 40 20
4000 3600 3200 2800 2400 2000 1800 1600 1400 1200 1000 800 600 400
WAVENUMBER
A B C D
A 1618
B 1527
C 1449
D 970

# HIPPURIC ACID

$C_9H_9NO_3$

Molecular weight: 179.17 (179.06)
Synonyms: N-Benzoylglycine; benzoylaminoacetic acid;
benzamidoacetic acid
Trade names:

Use: Present in human urine
HPLC:
GC:

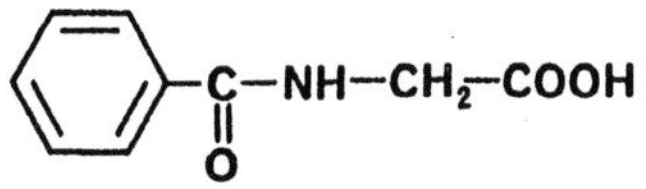

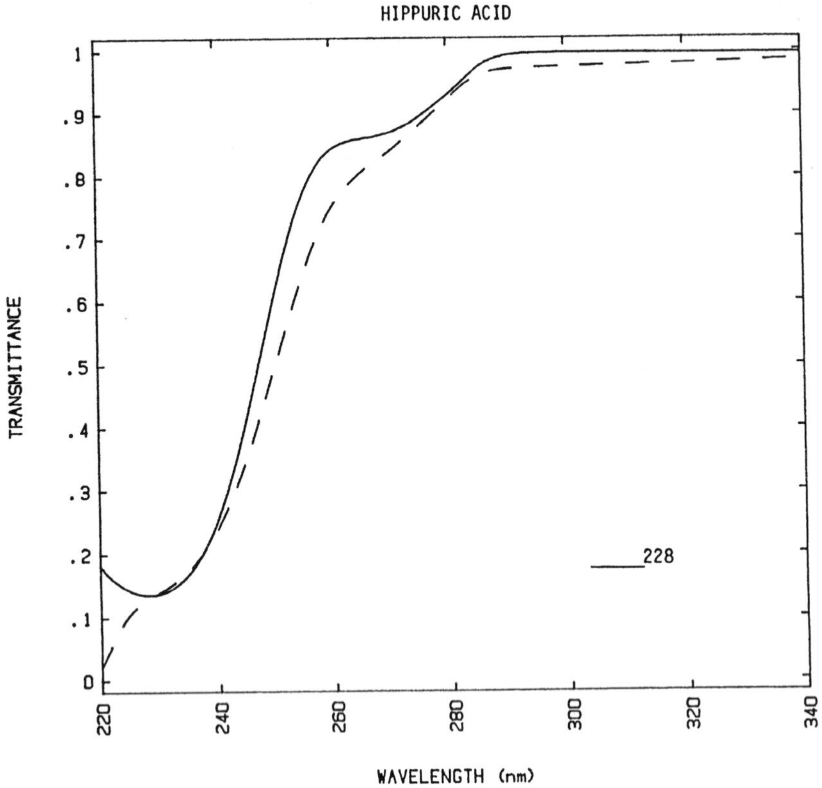

$C_6H_5$–C(=O)–NH–$CH_2$–COOH

HIPPURIC ACID

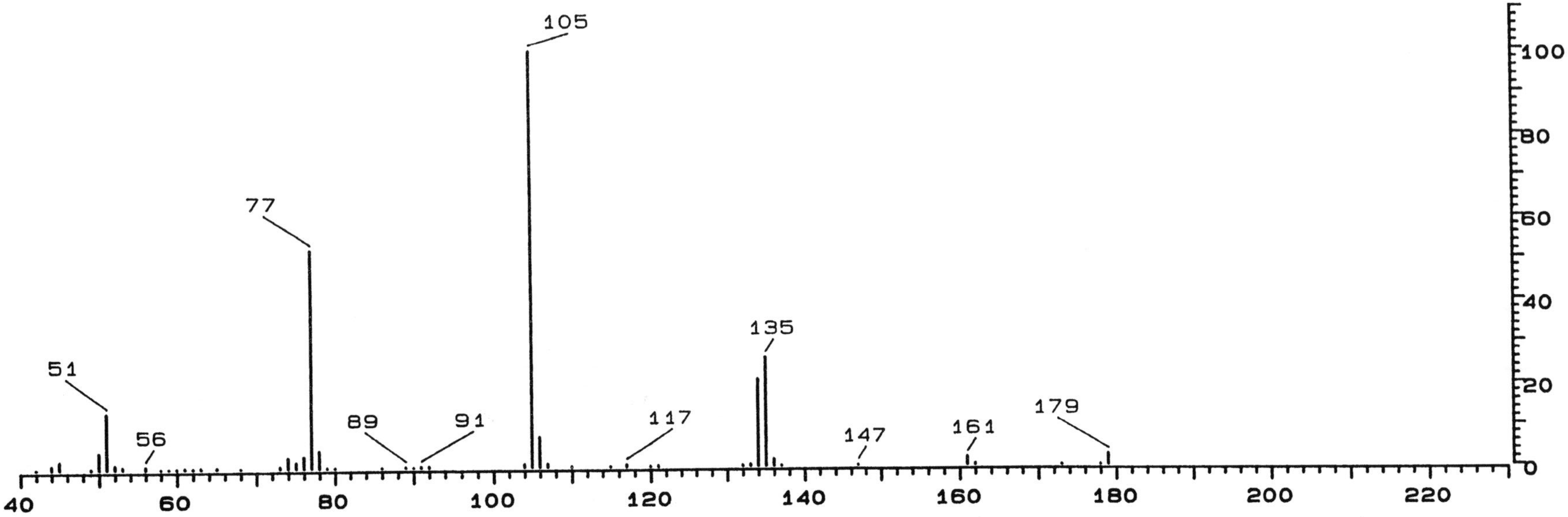

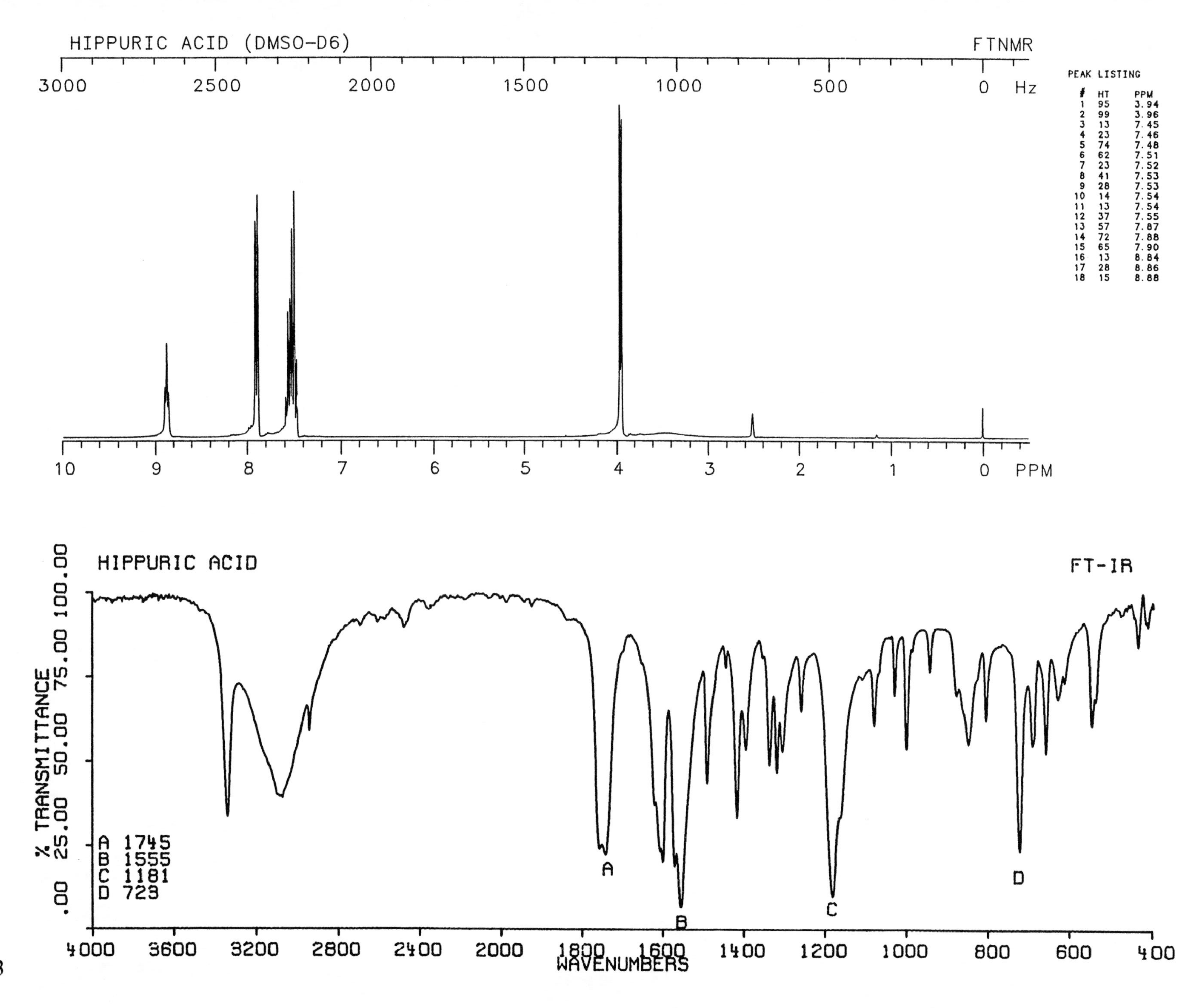
HIPPURIC ACID (DMSO-D6)
FTNMR
3000
2500
2000
1500
1000
500
0 Hz
PEAK LISTING
# HT PPM
1 95 3.94
2 99 3.96
3 13 7.45
4 23 7.46
5 74 7.48
6 62 7.51
7 23 7.52
8 41 7.53
9 28 7.53
10 14 7.54
11 13 7.54
12 37 7.55
13 57 7.87
14 72 7.88
15 65 7.90
16 13 8.84
17 28 8.86
18 15 8.88
10
9
8
7
6
5
4
3
2
1
0 PPM
HIPPURIC ACID
FT-IR
% TRANSMITTANCE
.00
25.00
50.00
75.00
100.00
A 1745
B 1555
C 1181
D 723
A
B
C
D
4000
3600
3200
2800
2400
2000
1800
1600
1400
1200
1000
800
600
400
WAVENUMBERS

# HISTAMINE

$C_5H_9N_3$

Molecular weight: 111.15 (111.08)
Synonyms: 2-(4-Imidazolyl)ethylamine; ergamine; ergotidine

Trade names: Histamine phosphate

Use: Diagnostic aid
HPLC: Si-10; 50A:50B; 3.4
GC: 1548; 200°C

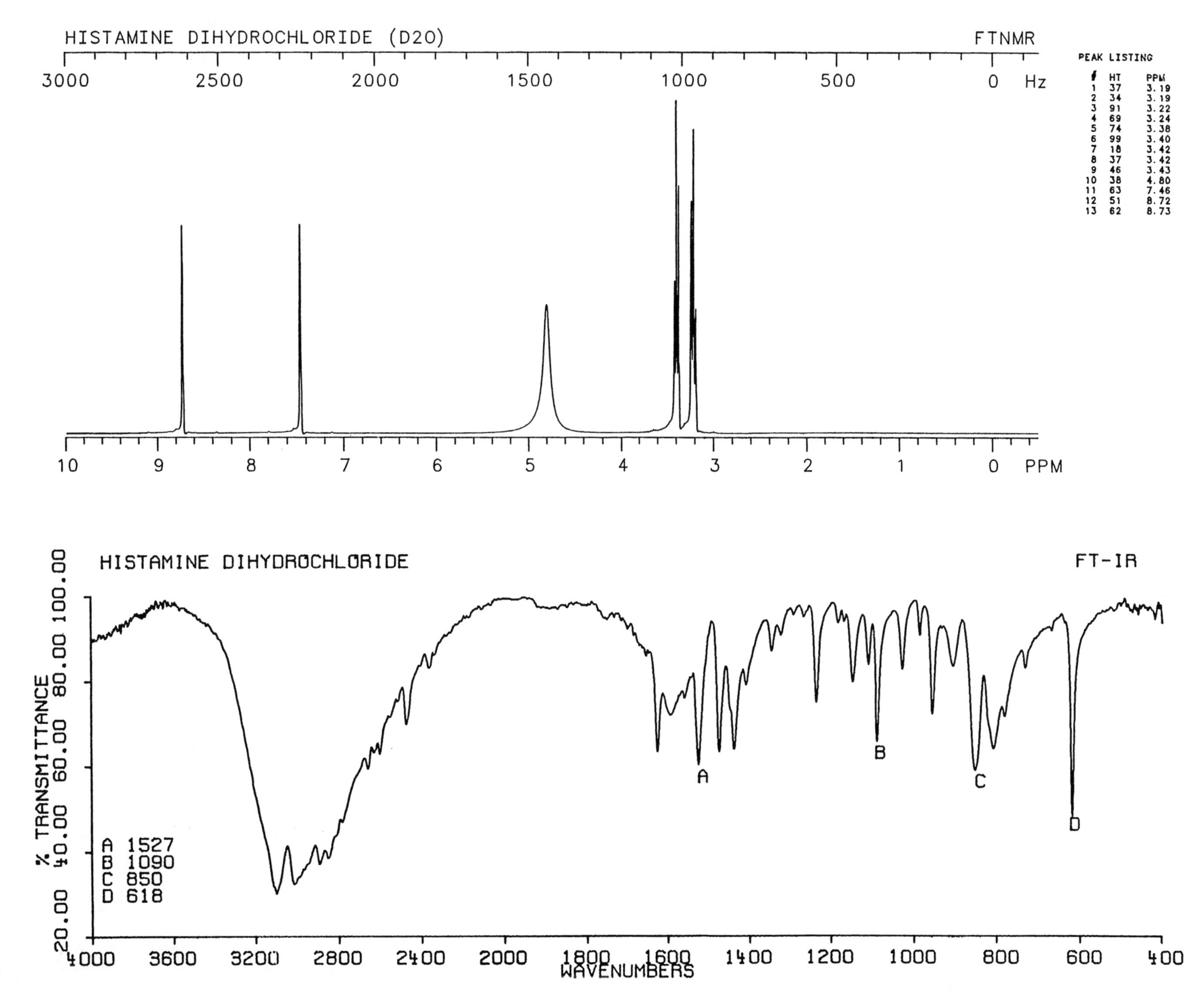
HISTAMINE DIHYDROCHLORIDE (D2O)
FTNMR
3000 2500 2000 1500 1000 500 0 Hz
10 9 8 7 6 5 4 3 2 1 0 PPM
PEAK LISTING
# HT PPM
1 37 3.19
2 34 3.19
3 91 3.22
4 69 3.24
5 74 3.38
6 99 3.40
7 18 3.42
8 37 3.42
9 46 3.43
10 38 4.80
11 63 7.46
12 51 8.72
13 62 8.73
HISTAMINE DIHYDROCHLORIDE
FT-IR
% TRANSMITTANCE
20.00 40.00 60.00 80.00 100.00
4000 3600 3200 2800 2400 2000 1800 1600 1400 1200 1000 800 600 400
WAVENUMBERS
A 1527
B 1090
C 850
D 618
A
B
C
D

# HOMATROPINE

$C_{16}H_{21}NO_3$

Molecular weight: 275.35 (275.15)
Synonyms: endo-α-Hydroxybenzeneacetic acid 8-methyl-8-azabicyclo[3.2.1]oct-3-yl ester
Trade names: Dia-Quel, Gustase-Plus, Homapin, Sinulin

Use: Anticholinergic
HPLC: Si-10; 10A:90B; 11.0
GC: 2155; 250°C

N—CH3
OH
O—C—CH
O C6H5

HOMATROPINE
TRANSMITTANCE
1.0
0.90
0.80
0.70
0.60
0.50
0.40
0.30
0.20
0.10
0.0
220
240
260
280
300
320
340
WAVELENGTH (nm)
251, 257, 263
258

HOMATROPINE
124
82
94
42
67
107
142
201
218
275
60
80
100
120
140
160
180
200
220
240
260
100
80
60
40
20
0

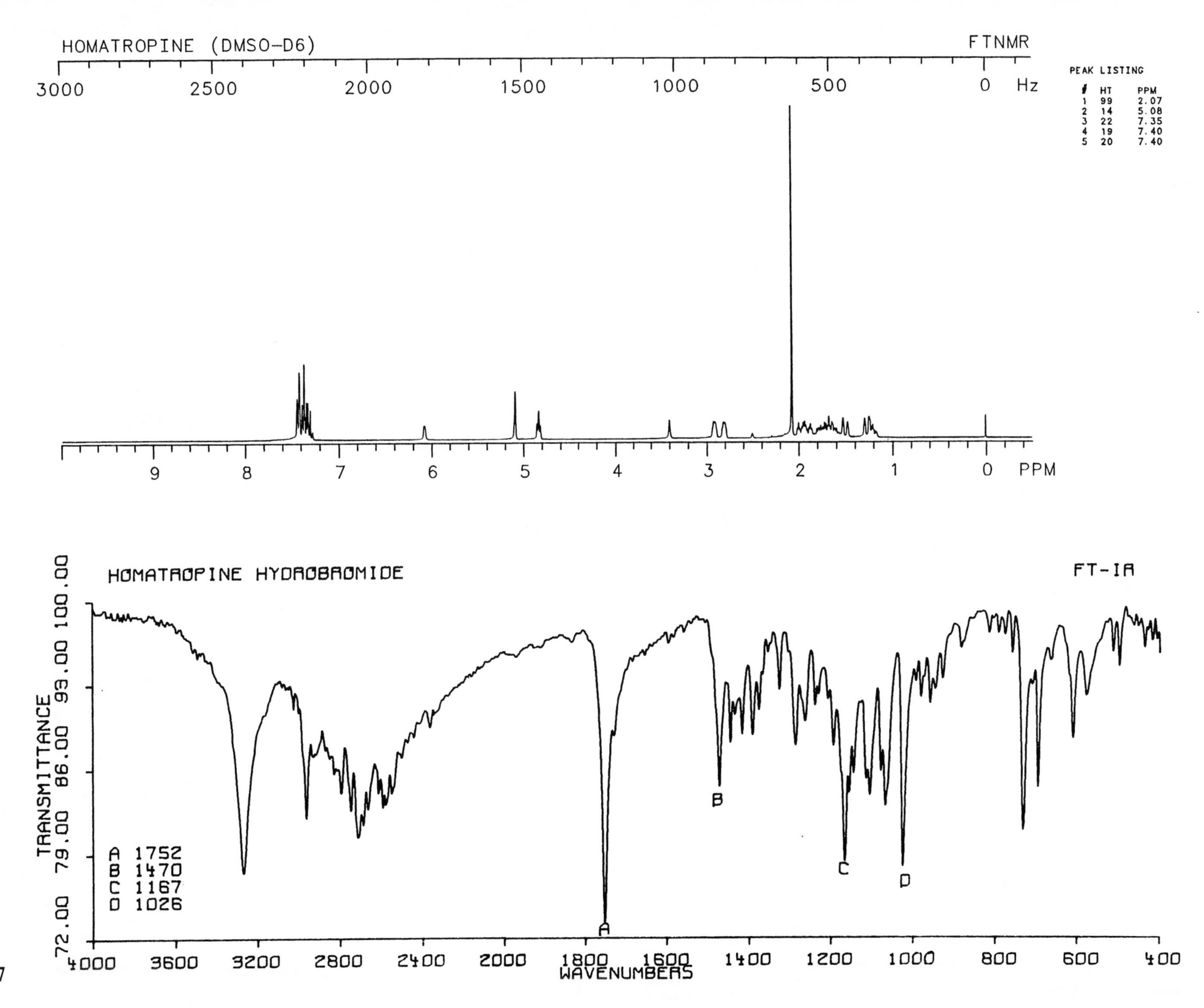
HOMATROPINE (DMSO-D6)
FTNMR
3000 2500 2000 1500 1000 500 0 Hz
PEAK LISTING
# HT PPM
1 99 2.07
2 14 5.08
3 22 7.35
4 19 7.40
5 20 7.40
9 8 7 6 5 4 3 2 1 0 PPM
HOMATROPINE HYDROBROMIDE
FT-IR
TRANSMITTANCE
72.00 79.00 86.00 93.00 100.00
A 1752
B 1470
C 1167
D 1026
A
B
C
D
4000 3600 3200 2800 2400 2000 1800 1600 1400 1200 1000 800 600 400
WAVENUMBERS

# HOMOVANILLIC ACID

$C_9H_{10}O_4$

**Molecular weight:** 182.17 (182.06)
**Synonyms:** 4-Hydroxy-3-methoxybenzeneacetic acid; 4-Hydroxy-3-methoxyphenylacetic acid
**Trade names:**

**Use:** Metabolite in human urine
**HPLC:** Si-10; 20A:80B; 4.4
**GC:** 1701; 200°C

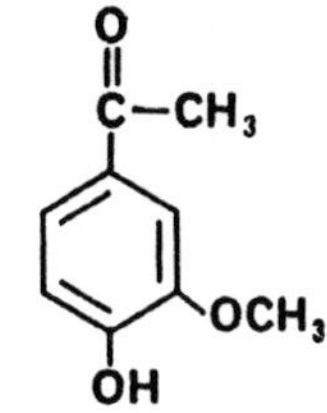


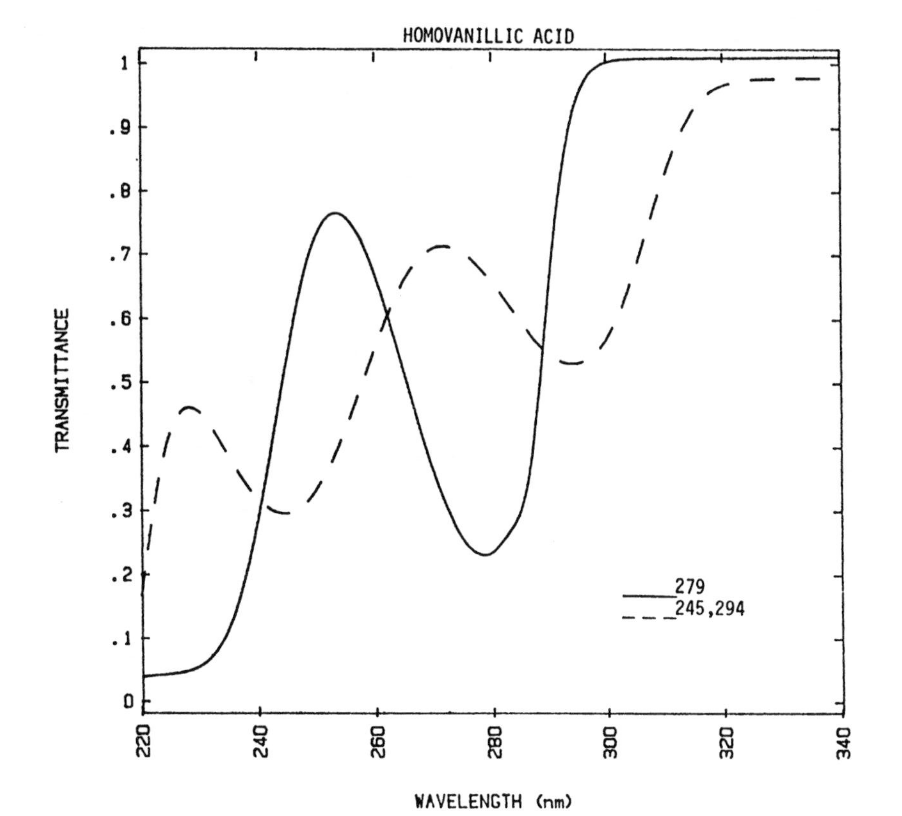

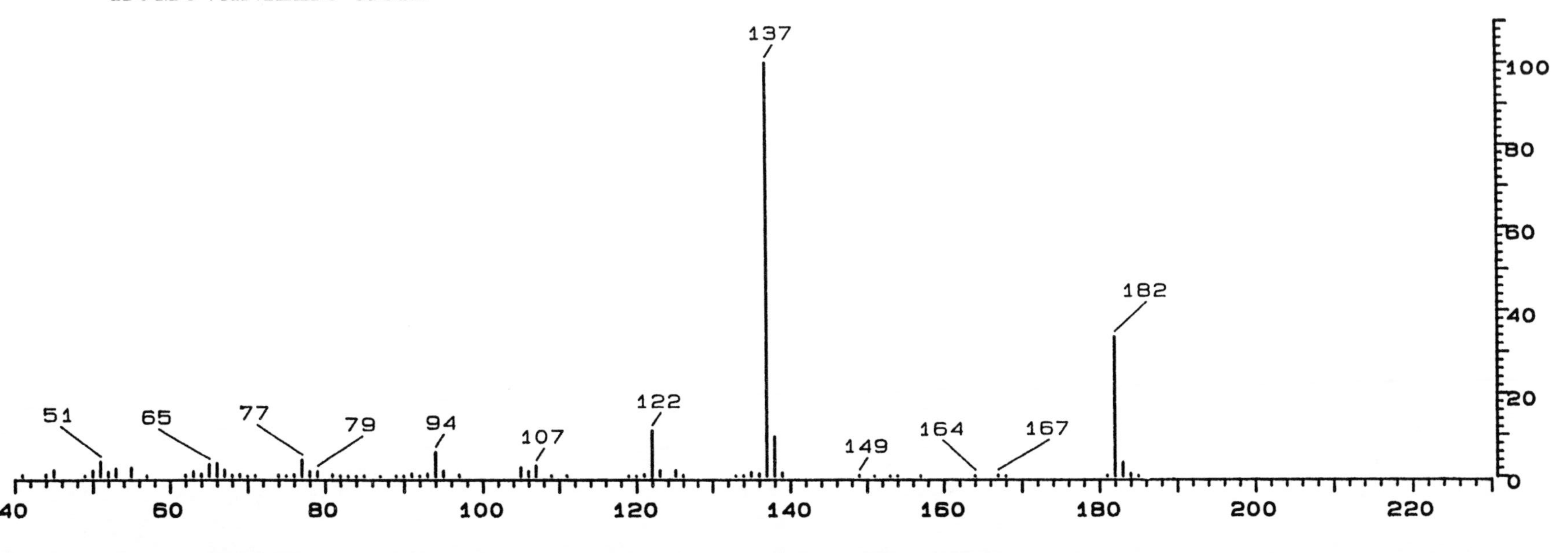

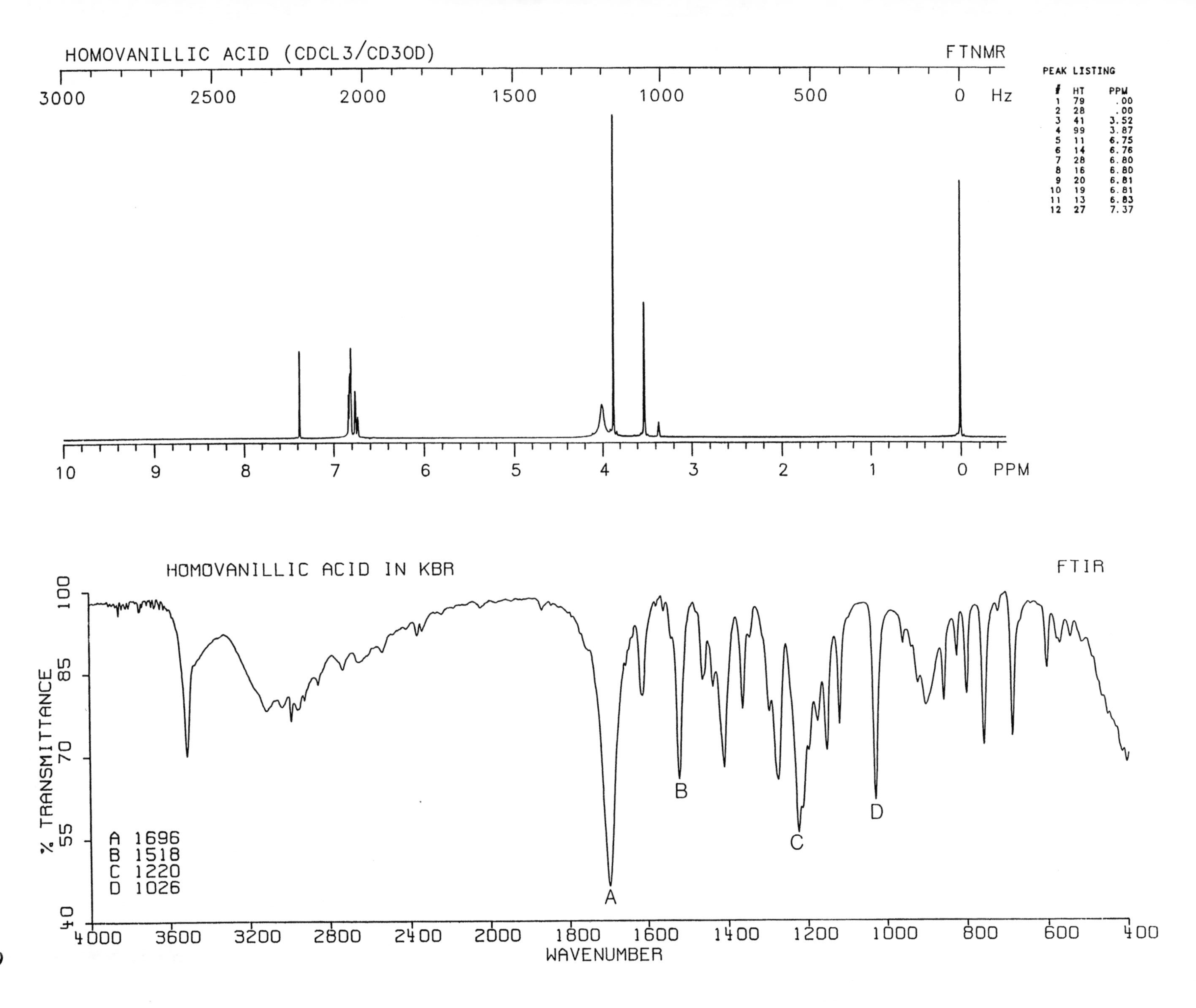
HOMOVANILLIC ACID (CDCL3/CD3OD)
FTNMR
Hz
PPM
PEAK LISTING
# HT PPM
1 79 .00
2 28 .00
3 41 3.52
4 99 3.87
5 11 6.75
6 14 6.76
7 28 6.80
8 16 6.80
9 20 6.81
10 19 6.81
11 13 6.83
12 27 7.37
HOMOVANILLIC ACID IN KBR
FTIR
% TRANSMITTANCE
WAVENUMBER
A 1696
B 1518
C 1220
D 1026

# HYDANTOIN

$C_3H_4N_2O_2$

Molecular weight: 100.08 (100.03)
Synonyms: 2,4-Imidazolidinedione; 2,4-(3H,5H)-imidazoledione; glycolylurea
Trade names:

Use:
HPLC: Si-10; 2A:98B; 5.9
GC: 1581; 200°C

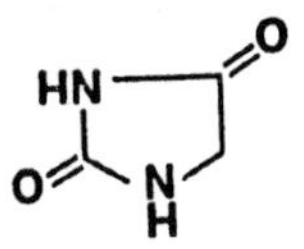

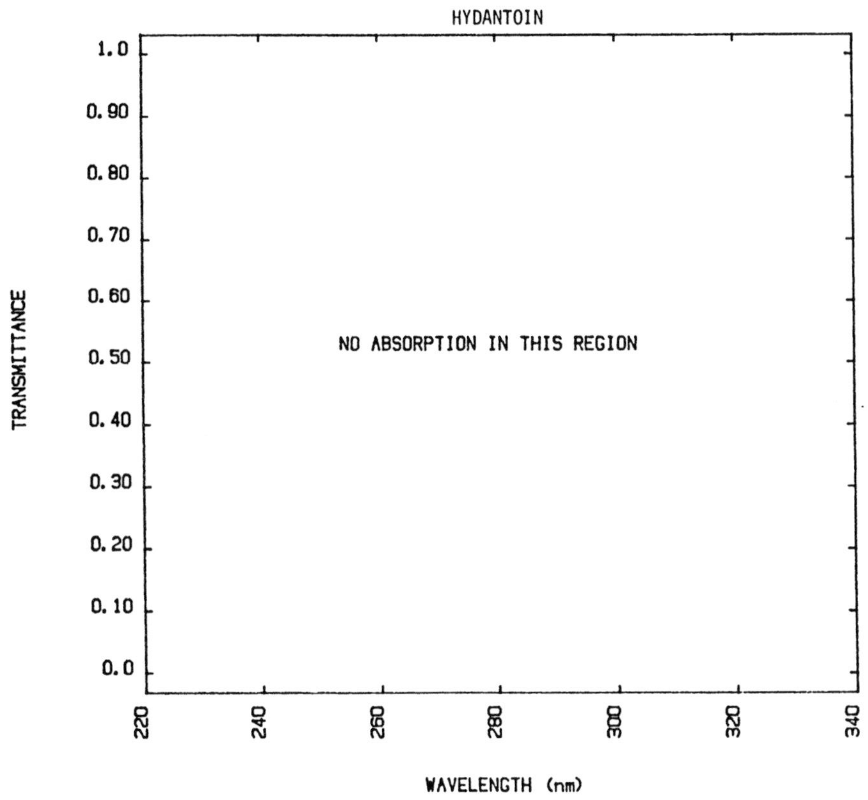

HN O N H O

HYDANTOIN

TRANSMITTANCE

1.0 0.90 0.80 0.70 0.60 0.50 0.40 0.30 0.20 0.10 0.0

NO ABSORPTION IN THIS REGION

220 240 260 280 300 320 340

WAVELENGTH (nm)

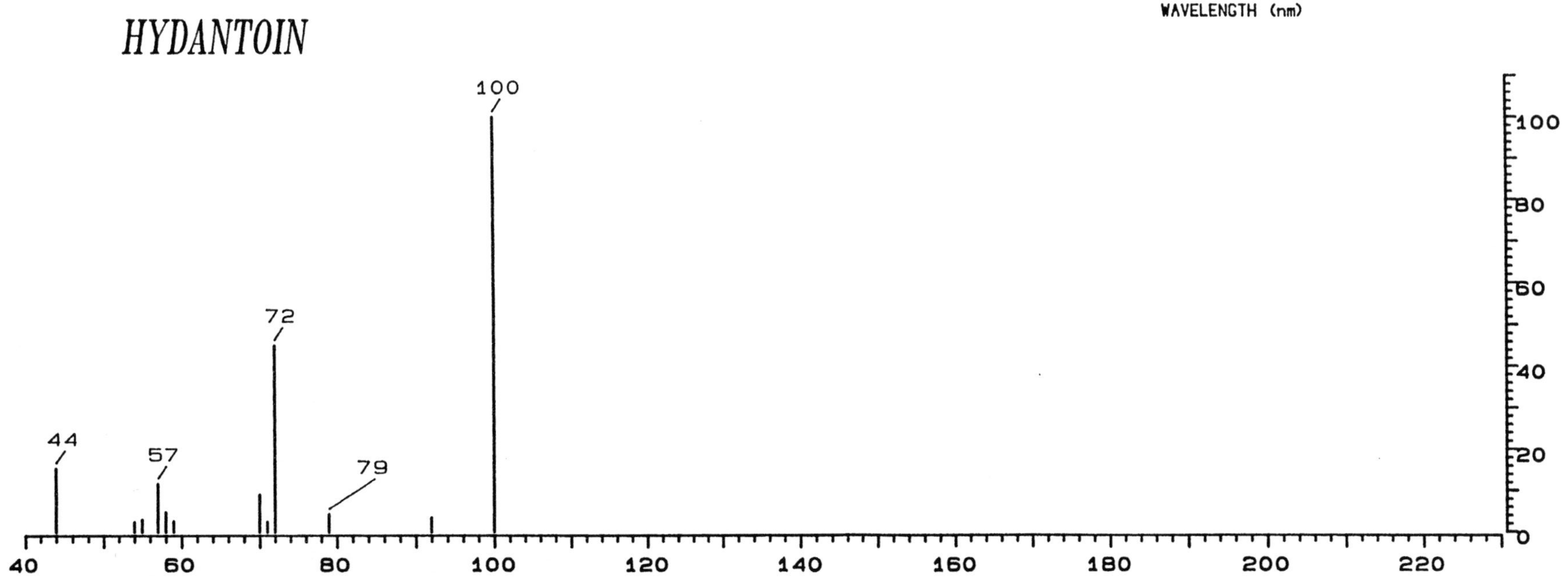

*HYDANTOIN*

100 72 44 57 79

100 80 60 40 20 0

40 60 80 100 120 140 160 180 200 220

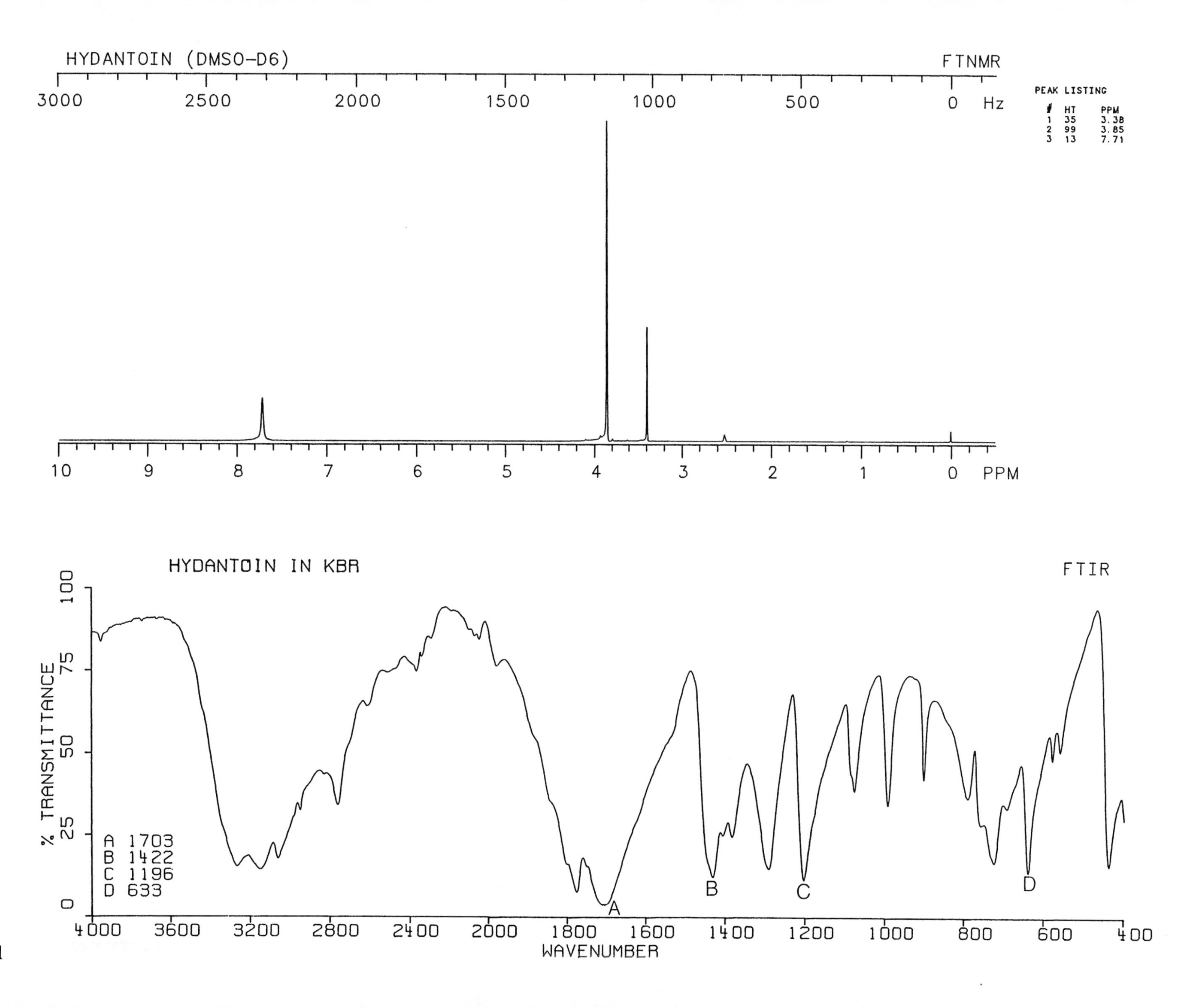
HYDANTOIN (DMSO-D6)
FTNMR
3000 2500 2000 1500 1000 500 0 Hz
PEAK LISTING
# HT PPM
1 35 3.38
2 99 3.85
3 13 7.71
10 9 8 7 6 5 4 3 2 1 0 PPM
HYDANTOIN IN KBR
FTIR
% TRANSMITTANCE
100 75 50 25 0
A 1703
B 1422
C 1196
D 633
4000 3600 3200 2800 2400 2000 1800 1600 1400 1200 1000 800 600 400
WAVENUMBER

# HYDRALAZINE

$C_8H_8N_4$

Molecular weight: 160.18 (160.08)
Synonyms: 1(2H)-Phthalazinone hydrazone; hydrallazine

Trade names: Apresazide, Apresoline, Serpasil, Unipres

Use: Antihypertensive
HPLC: Si-10; 20A:80B; 4.0
GC:

NH—NH₂ (phthalazine structure)

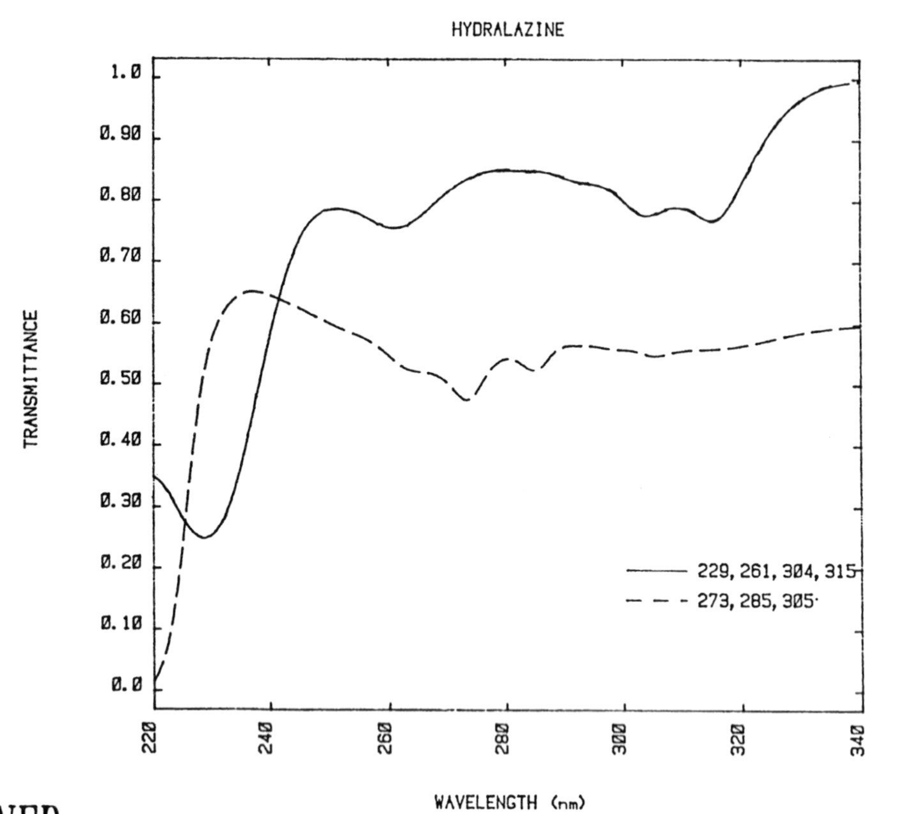


*NO USEFUL MASS SPECTRUM WAS OBTAINED*

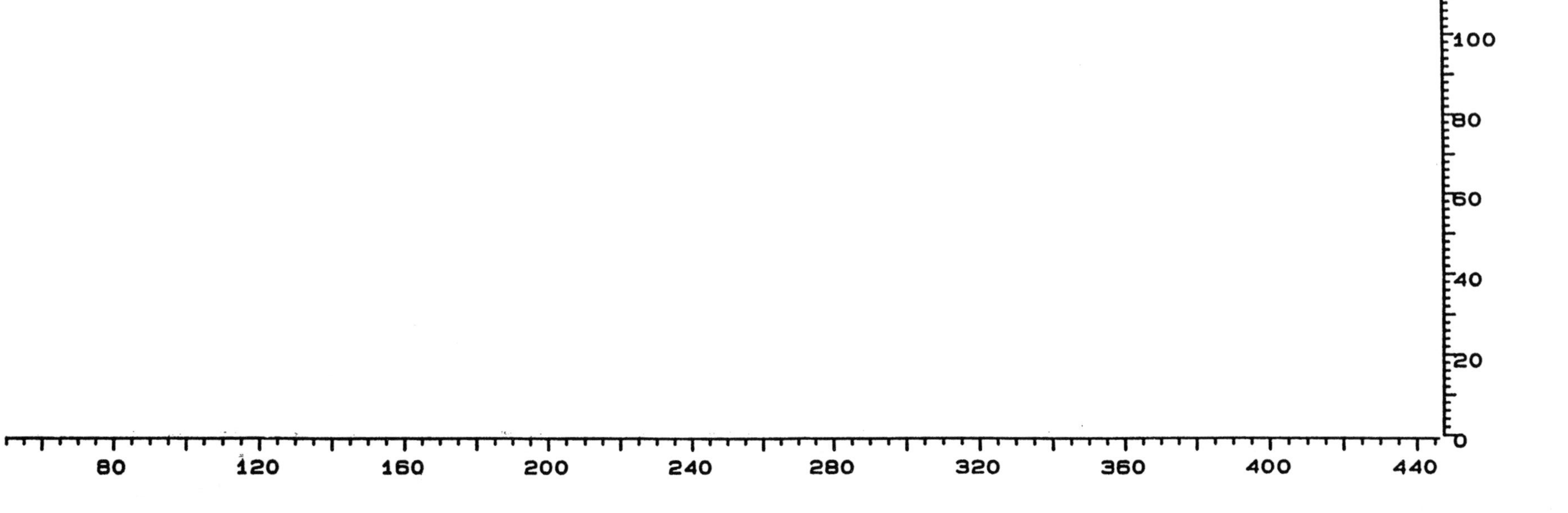

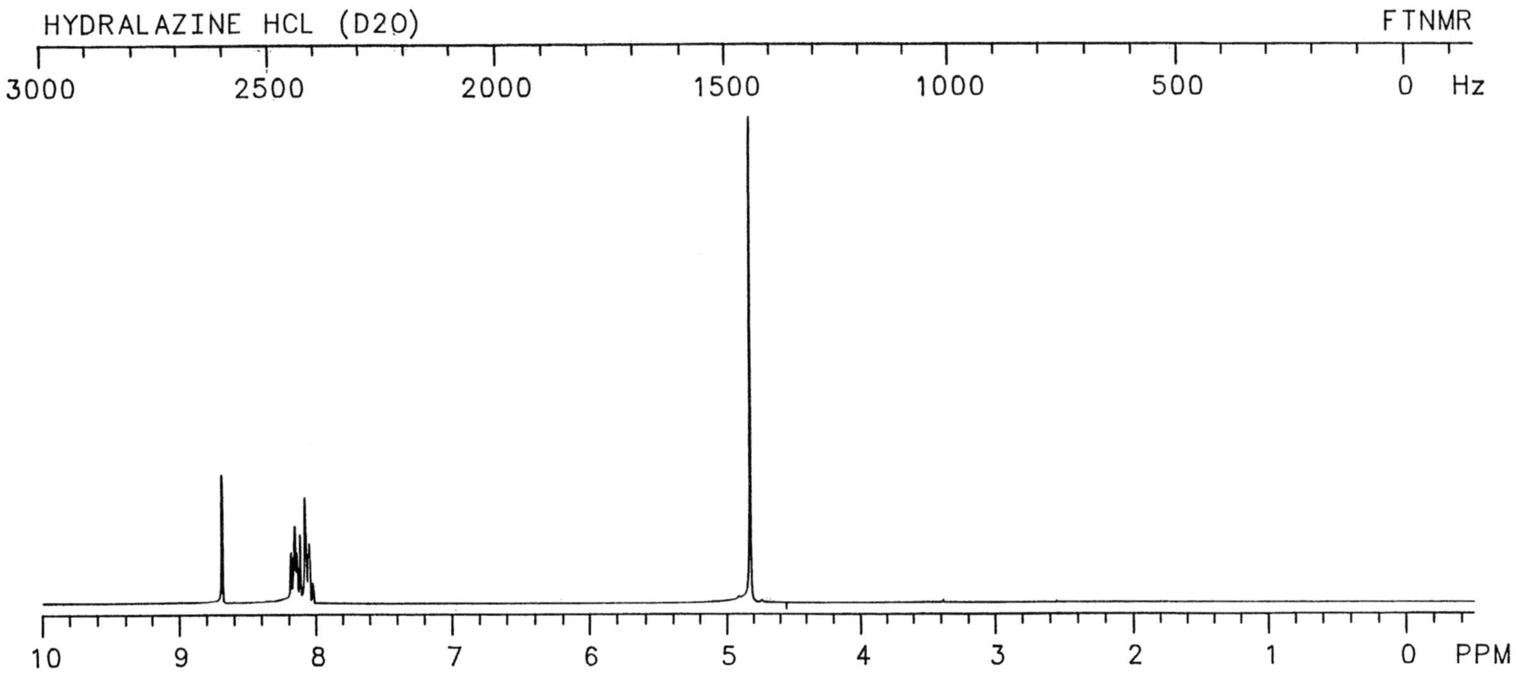

HYDRALAZINE HCL (D2O)

FTNMR

3000 2500 2000 1500 1000 500 0 Hz

PEAK LISTING

| # | HT | PPM |
|---|---|---|
| 1 | 98 | 4.81 |
| 2 | 12 | 8.04 |
| 3 | 9 | 8.05 |
| 4 | 14 | 8.06 |
| 5 | 21 | 8.07 |
| 6 | 14 | 8.11 |
| 7 | 10 | 8.13 |
| 8 | 9 | 8.13 |
| 9 | 15 | 8.15 |
| 10 | 10 | 8.17 |
| 11 | 26 | 8.68 |

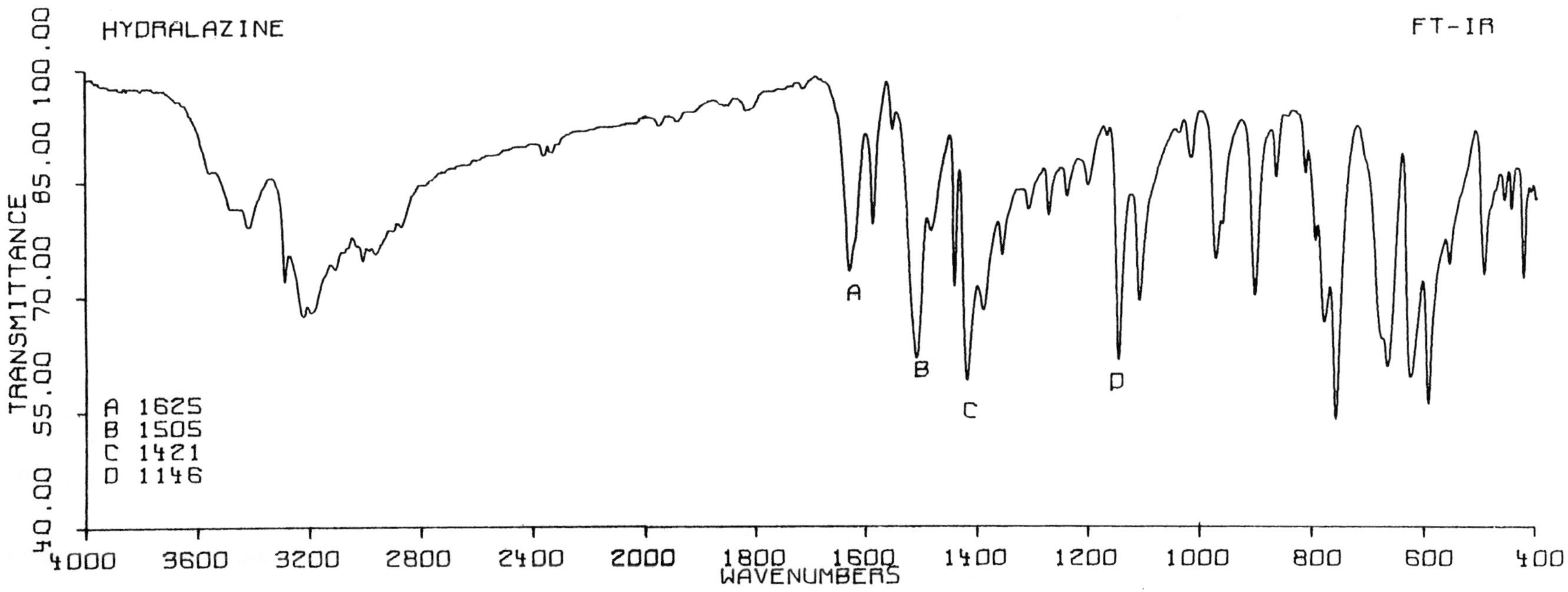

10 9 8 7 6 5 4 3 2 1 0 PPM

HYDRALAZINE

FT-IR

TRANSMITTANCE 40.00 55.00 70.00 85.00 100.00

A 1625
B 1505
C 1421
D 1146

4000 3600 3200 2800 2400 2000 1800 1600 1400 1200 1000 800 600 400
WAVENUMBERS

# HYDRASTININE

$C_{11}H_{13}NO_3$

**Molecular weight:** 207.22 (207.09)
**Synonyms:** 5,6,7,8-Tetrahydro-6-methyl-1,3-dioxolo[4,5-g]isoquinolin-5-ol
**Trade names:**

**Use:** Cardiotonic
**HPLC:** Si-10; 30A:70B; 6.9
**GC:** 1669; 200°C

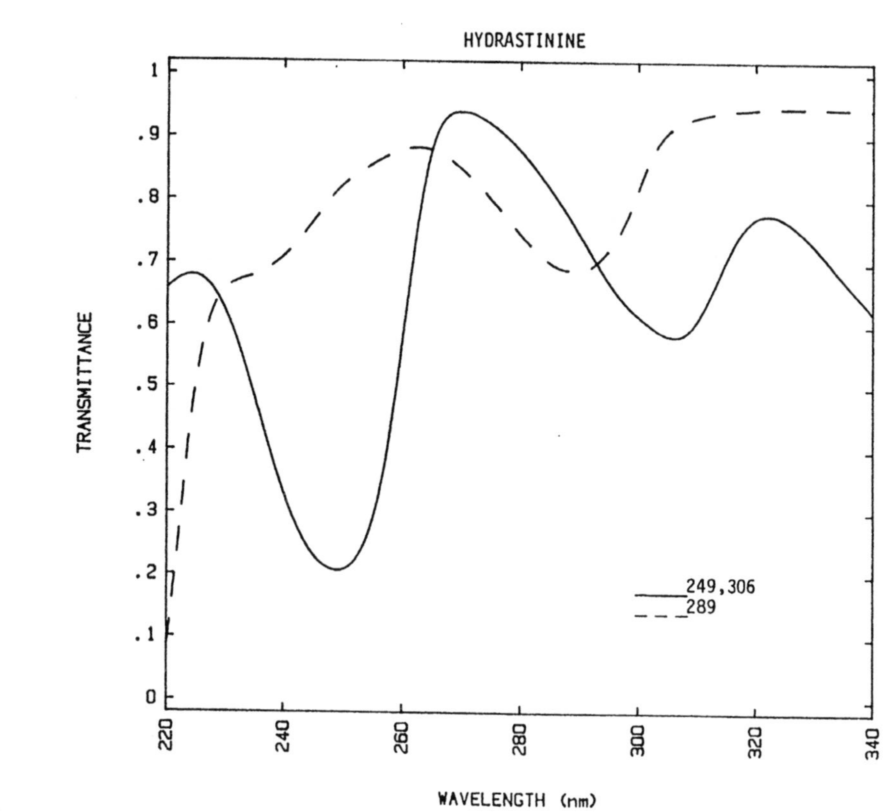


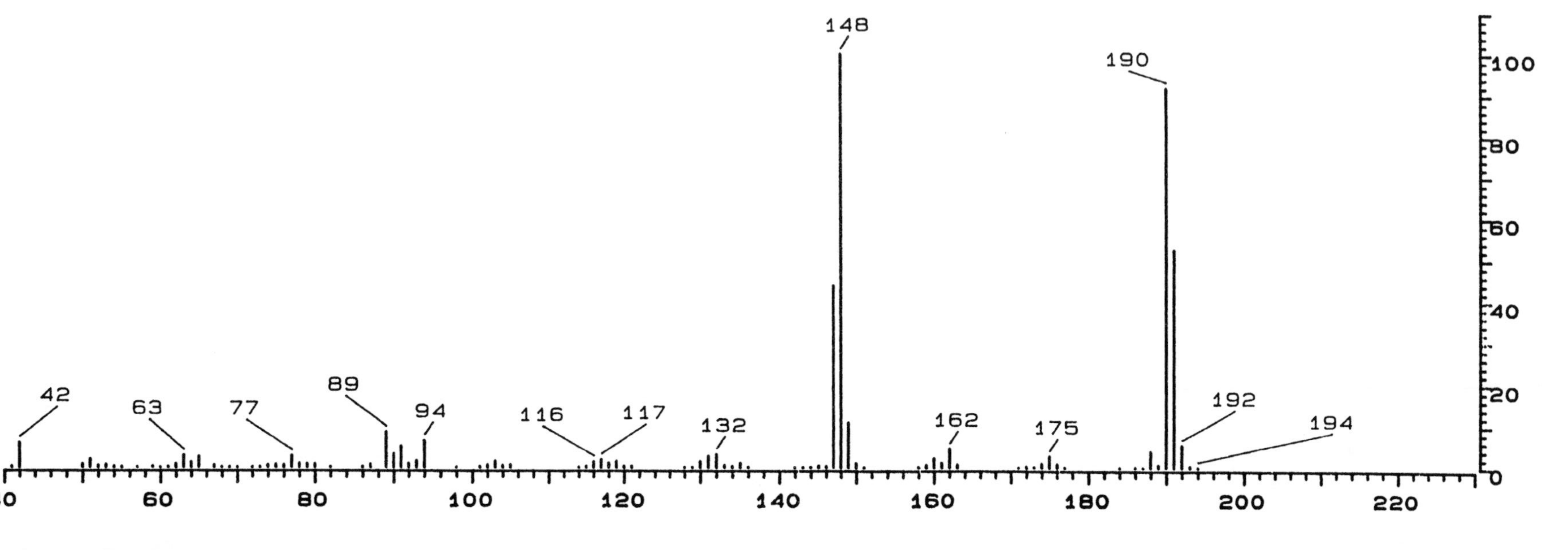

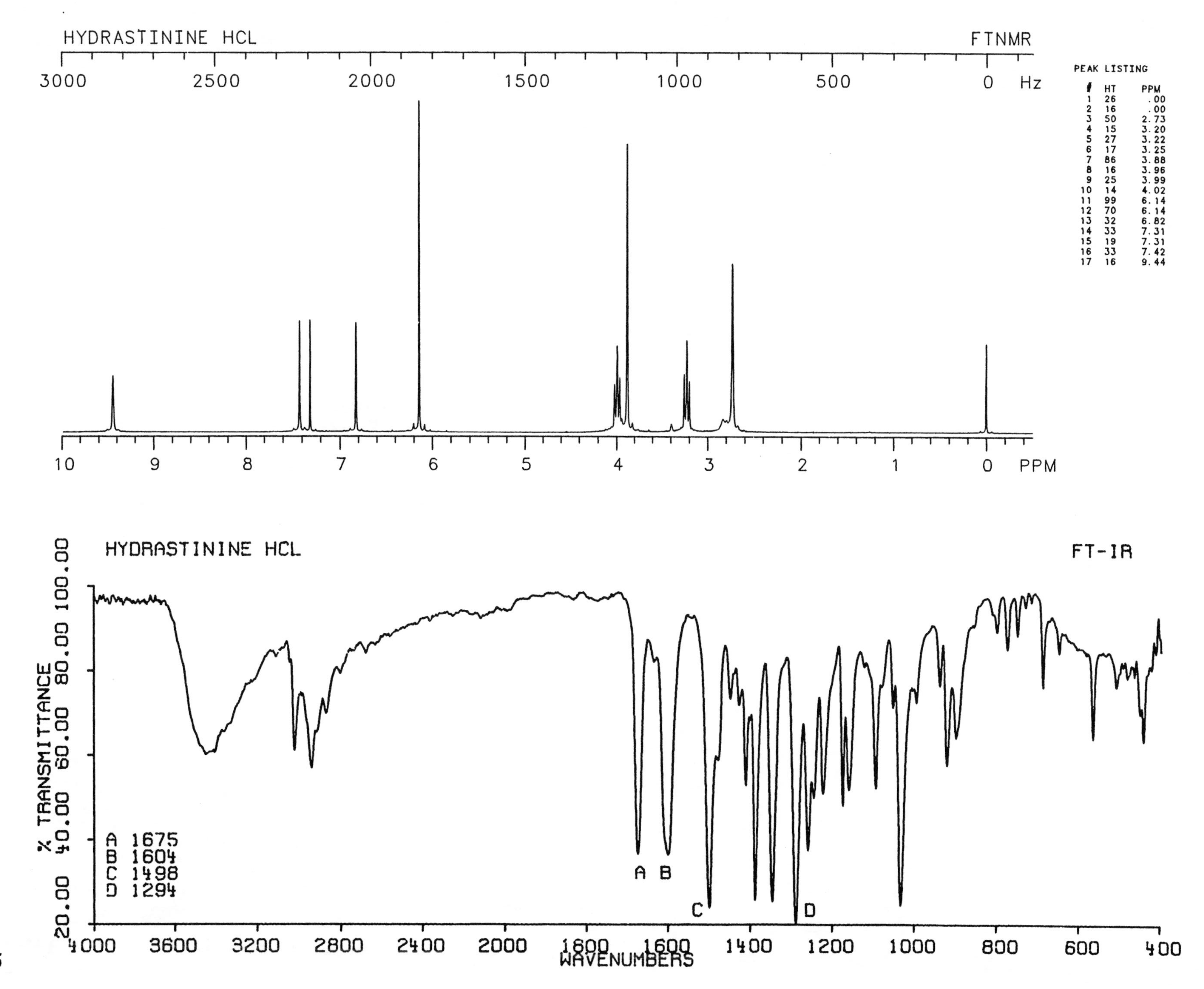
HYDRASTININE HCL
FTNMR
PEAK LISTING
# HT PPM
1 26 .00
2 16 .00
3 50 2.73
4 15 3.20
5 27 3.22
6 17 3.25
7 86 3.88
8 16 3.96
9 25 3.99
10 14 4.02
11 99 6.14
12 70 6.14
13 32 6.82
14 33 7.31
15 19 7.31
16 33 7.42
17 16 9.44
Hz
PPM
HYDRASTININE HCL
FT-IR
% TRANSMITTANCE
WAVENUMBERS
A 1675
B 1604
C 1498
D 1294

# HYDROCHLOROTHIAZIDE

$C_7H_8ClN_3O_4S_2$

Molecular weight: 297.73 (296.97)
Synonyms: 6-Chloro-3,4-dihydro-2H-1,2,4-benzothiadiazine-7-sulfonamide 1,1-dioxide
Trade names: Aldactazide, Aldoril, Apresazide, Dyazide, Esimil, HydroDiuril, Hydropres, Inderide, Oretic, Unipres
Use: Diuretic
HPLC: Si-10; 10A:90B; 4.5
GC:

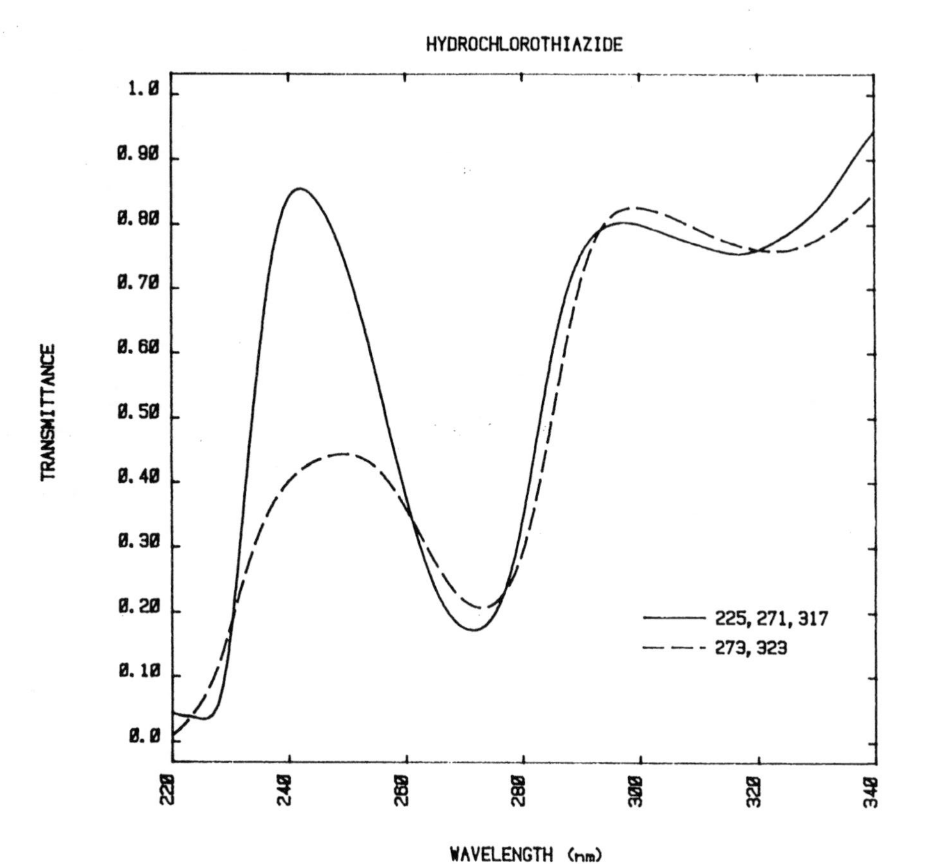

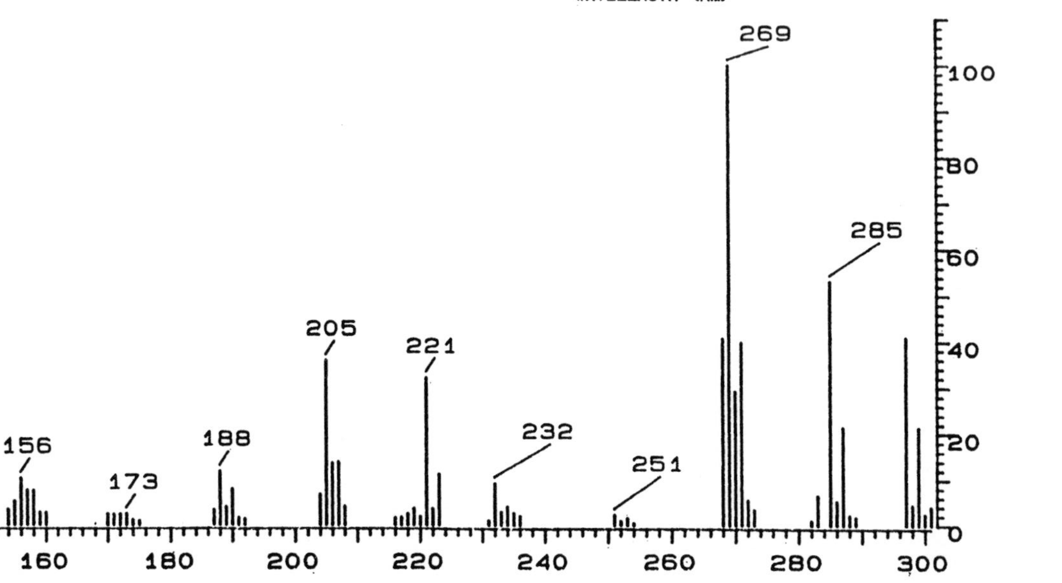

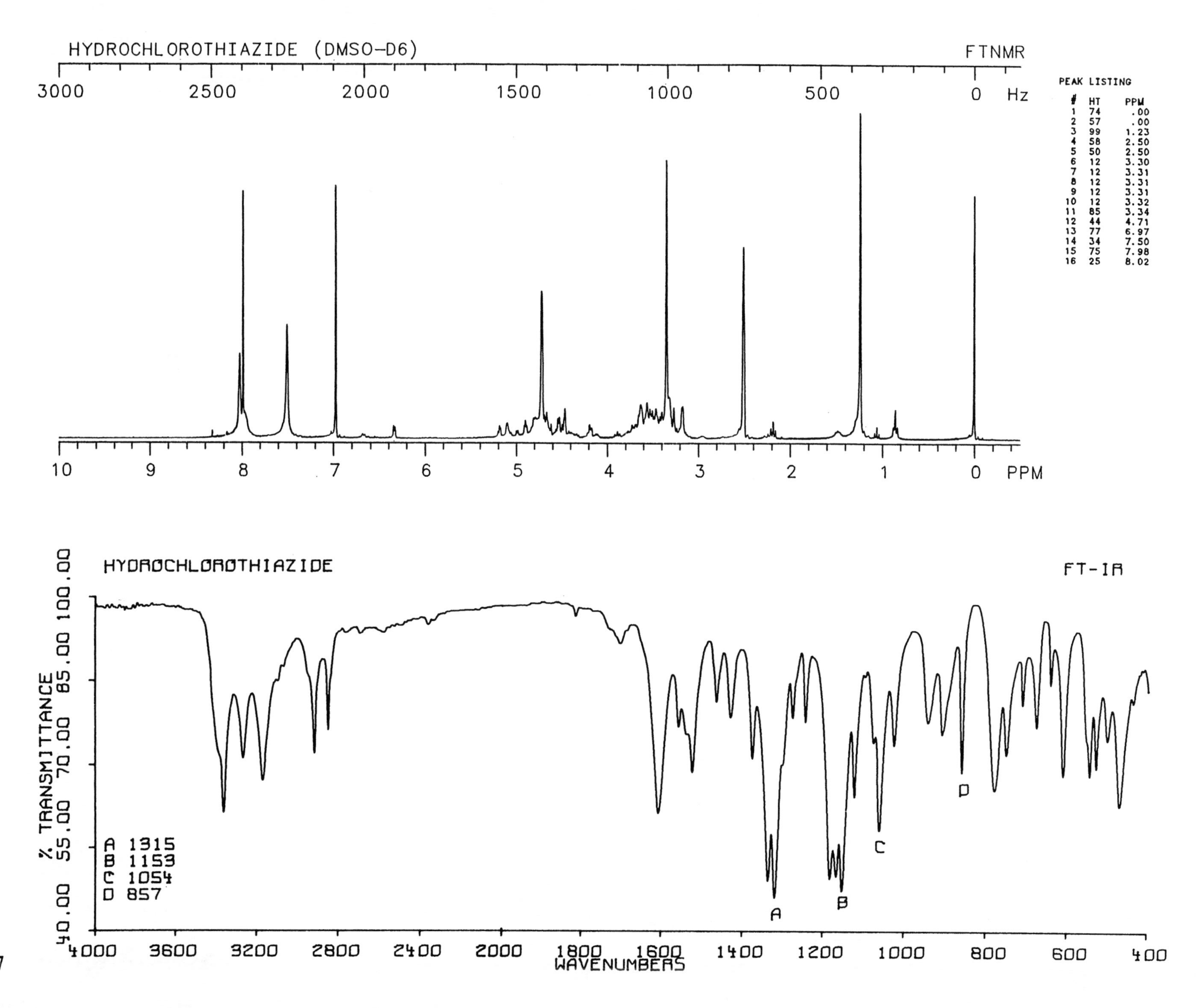

HYDROCHLOROTHIAZIDE (DMSO-D6)
FTNMR
3000 2500 2000 1500 1000 500 0 Hz
10 9 8 7 6 5 4 3 2 1 0 PPM
PEAK LISTING
# HT PPM
1 74 .00
2 57 .00
3 99 1.23
4 58 2.50
5 50 2.50
6 12 3.30
7 12 3.31
8 12 3.31
9 12 3.31
10 12 3.32
11 85 3.34
12 44 4.71
13 77 6.97
14 34 7.50
15 75 7.98
16 25 8.02
HYDROCHLOROTHIAZIDE
FT-IR
% TRANSMITTANCE
40.00 55.00 70.00 85.00 100.00
A 1315
B 1153
C 1054
D 857
A B C D
4000 3600 3200 2800 2400 2000 1800 1600 1400 1200 1000 800 600 400
WAVENUMBERS

# HYDROCODONE

$C_{18}H_{21}NO_3$

Molecular weight: 299.37 (299.15)
Synonyms: 4,5α-Epoxy-3-methoxy-17-methylmorphinan-6-one; dihydrocodeinone
Trade names: Dicodid, Hycodan, Tussend, Triaminic, Tussionex

Use: Narcotic analgesic, antitussive
HPLC: Si-10; 5A:95B; 8.0
GC: 2506; 250°C

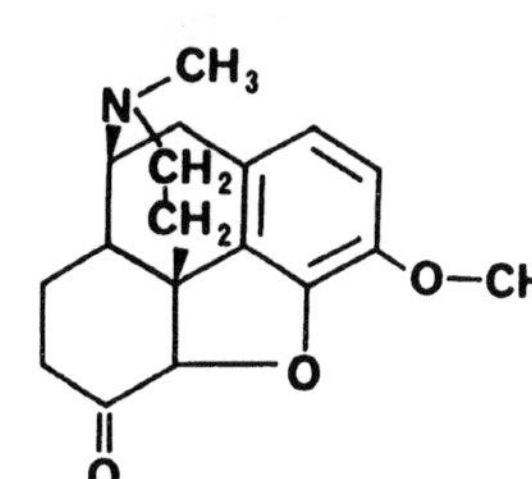


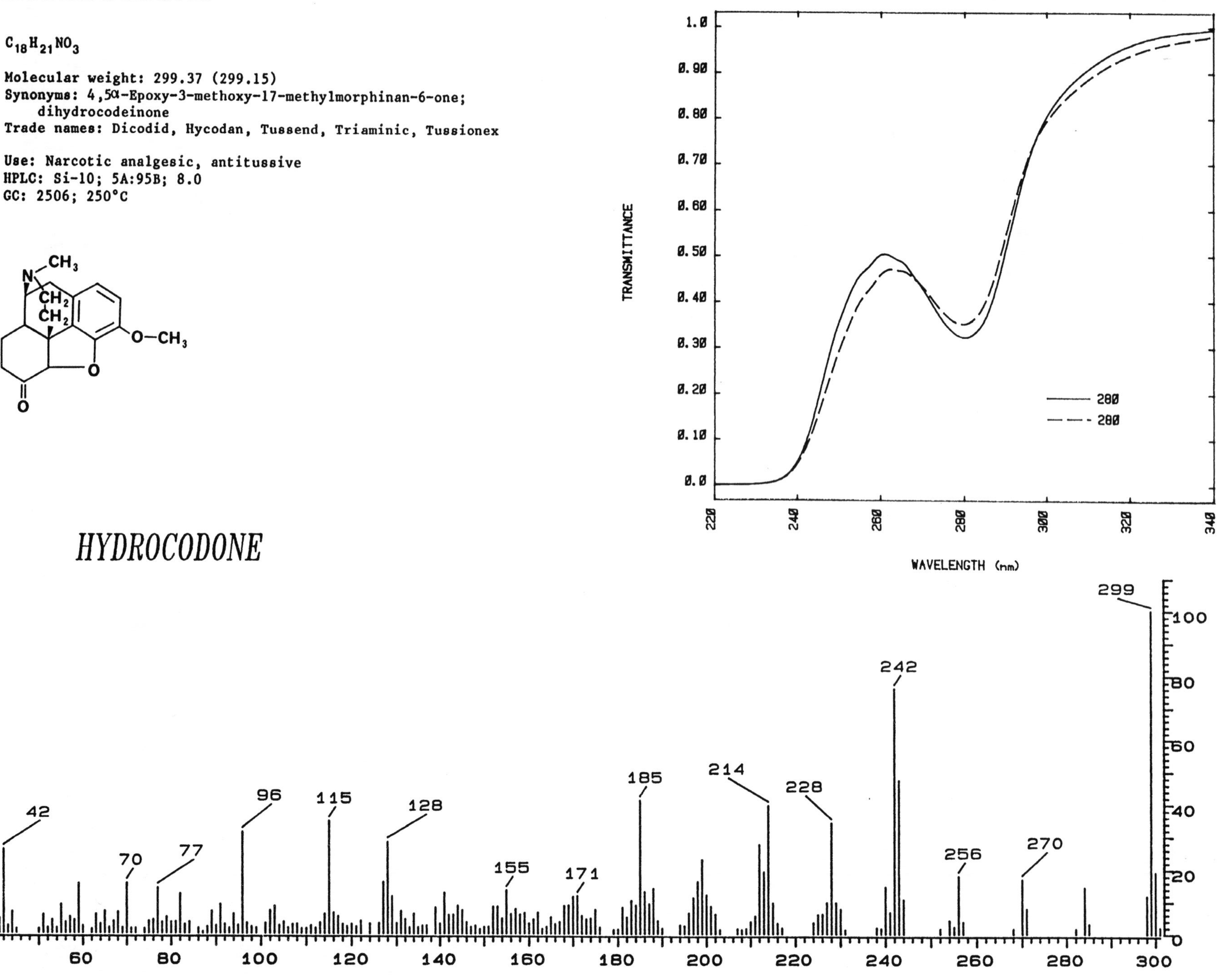

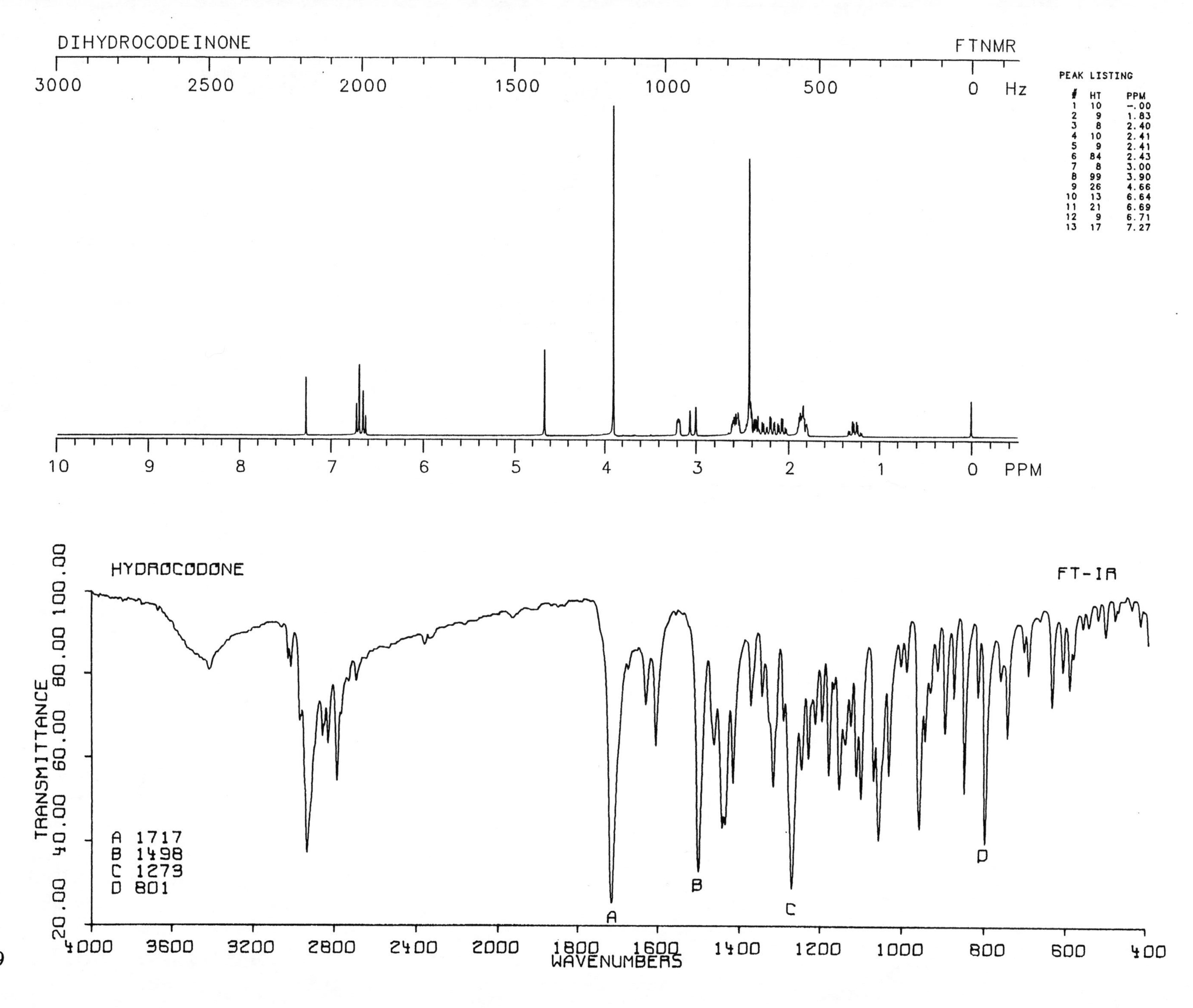
DIHYDROCODEINONE
FTNMR
3000
2500
2000
1500
1000
500
0
Hz
10
9
8
7
6
5
4
3
2
1
0
PPM
PEAK LISTING

| # | HT | PPM |
|---|---|---|
| 1 | 10 | -.00 |
| 2 | 9 | 1.83 |
| 3 | 8 | 2.40 |
| 4 | 10 | 2.41 |
| 5 | 9 | 2.41 |
| 6 | 84 | 2.43 |
| 7 | 8 | 3.00 |
| 8 | 99 | 3.90 |
| 9 | 26 | 4.66 |
| 10 | 13 | 6.64 |
| 11 | 21 | 6.69 |
| 12 | 9 | 6.71 |
| 13 | 17 | 7.27 |

HYDROCODONE
FT-IR
TRANSMITTANCE
20.00
40.00
60.00
80.00
100.00
A 1717
B 1498
C 1273
D 801
A
B
C
D
4000
3600
3200
2800
2400
2000
1800
1600
1400
1200
1000
800
600
400
WAVENUMBERS

# HYDROCORTISONE

$C_{21}H_{30}O_5$

Molecular weight: 362.47 (362.21)
Synonyms: 11β,17,21-Trihydroxypregn-4-ene-3,20-dione;
17-hydroxycorticosterone
Trade names: Acticort, Corticaine, Dermacort, Hydrocortisone

Use: Steroid, glucocorticoid
HPLC: Si-10; 5A:95B; 4.5
GC:





HYDROCORTISONE

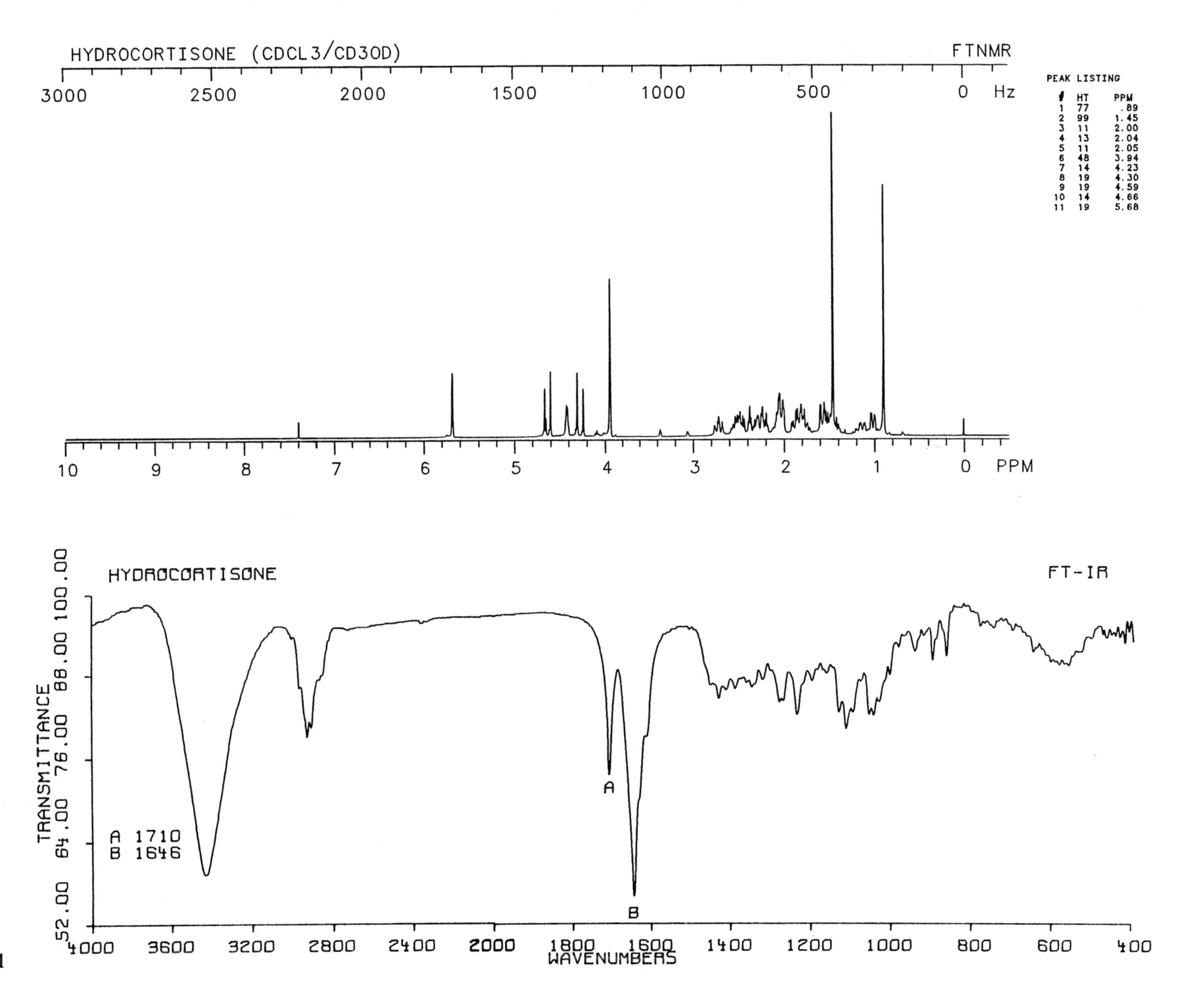

HYDROCORTISONE (CDCL3/CD3OD)
FTNMR
Hz
PPM
PEAK LISTING
# HT PPM
1 77 .89
2 99 1.45
3 11 2.00
4 13 2.04
5 11 2.05
6 48 3.94
7 14 4.23
8 19 4.30
9 19 4.59
10 14 4.66
11 19 5.68
HYDROCORTISONE
FT-IR
TRANSMITTANCE
WAVENUMBERS
A 1710
B 1646
A
B

# HYDROFLUMETHIAZIDE

$C_8H_8F_3N_3O_4S_2$

Molecular weight: 331.28 (330.99)
Synonyms: 3,4-Dihydro-6-(trifluoromethyl)-2H-1,2,4-benzothiadiazine-7-sulfonamide 1,1-dioxide; dihydroflumethiazide; trifluoromethylhydrothiazide
Trade names: Diucardin, Hydrenox, Hydro-Fluserpine, Leodrine, Rautrax, Rontyl, Saluron, Salutensin
Use: Antihypertensive, diuretic
HPLC:
GC:

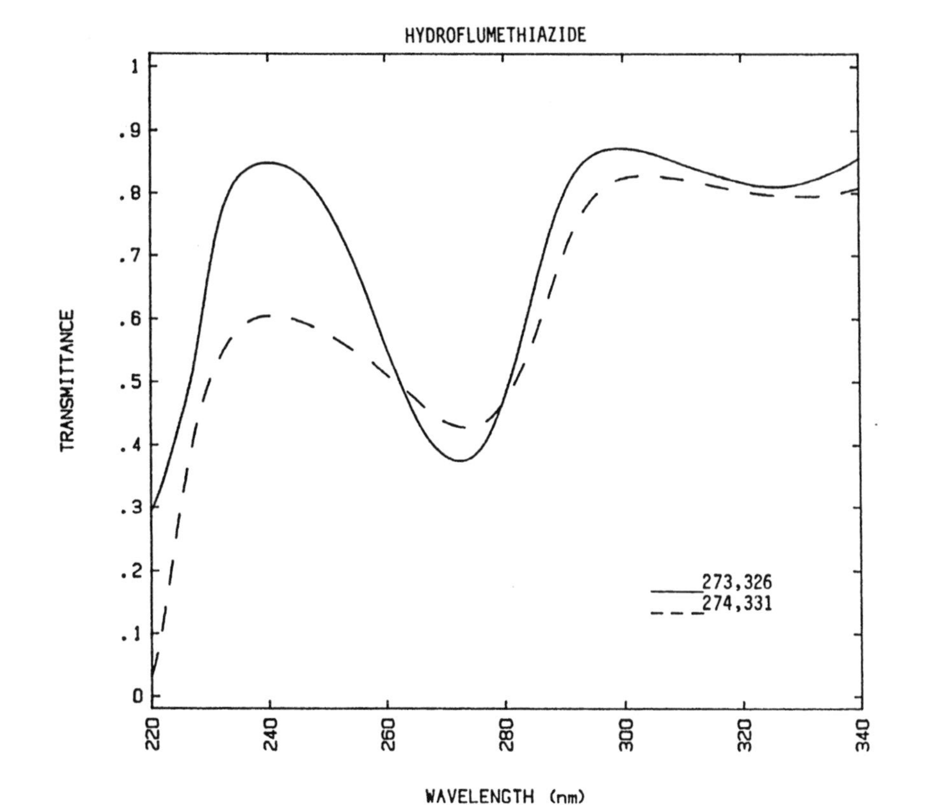

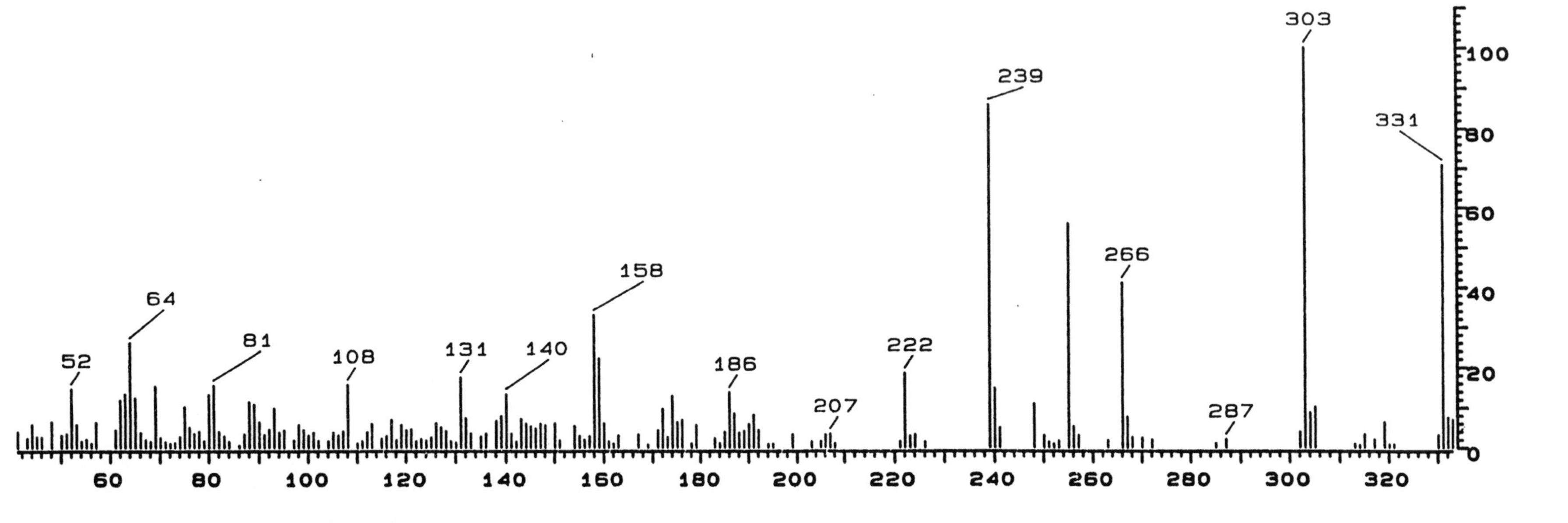

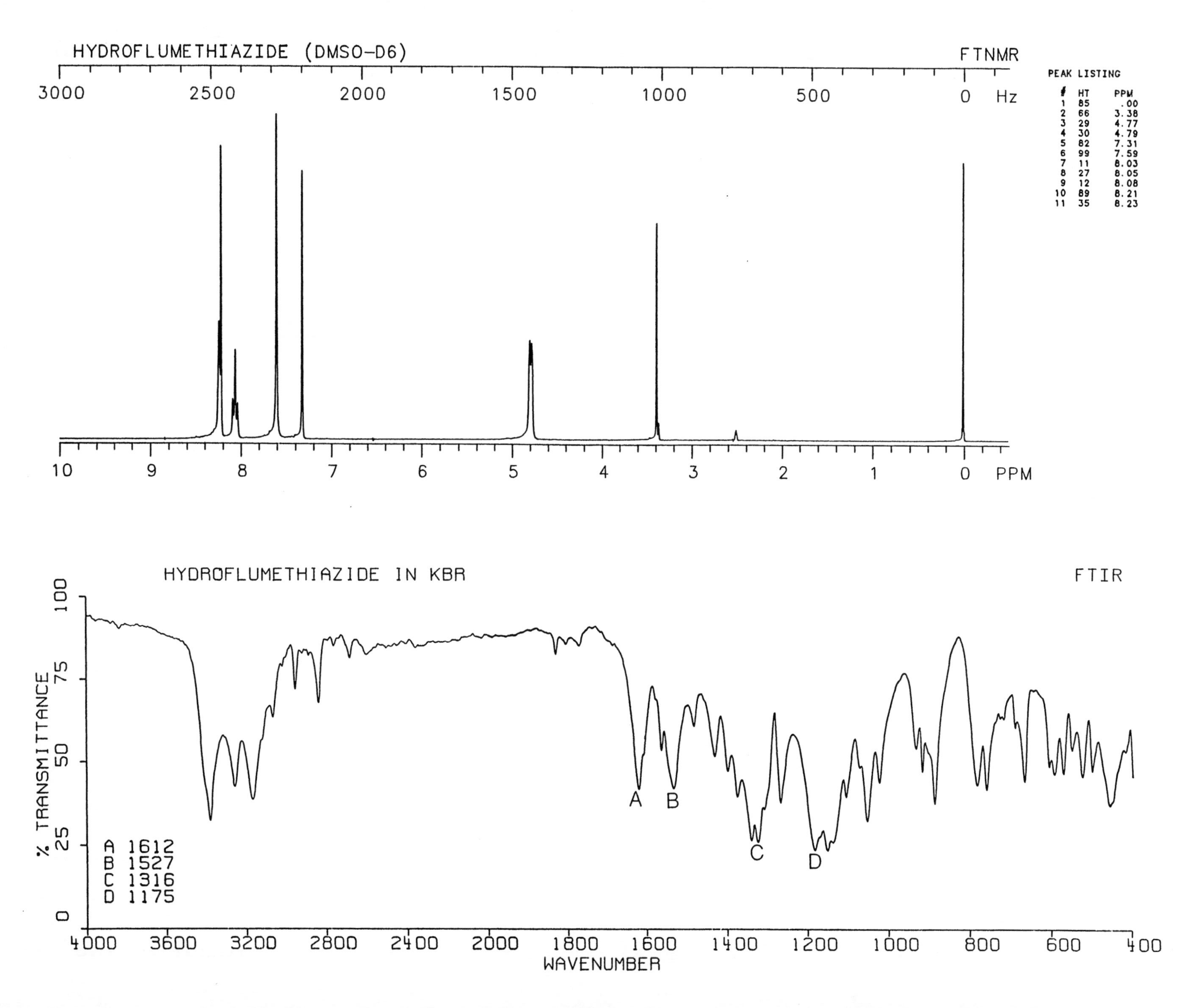

HYDROFLUMETHIAZIDE (DMSO-D6)
FTNMR
3000
2500
2000
1500
1000
500
0 Hz
PEAK LISTING
# HT PPM
1 85 .00
2 66 3.38
3 29 4.77
4 30 4.79
5 82 7.31
6 99 7.59
7 11 8.03
8 27 8.05
9 12 8.08
10 89 8.21
11 35 8.23
10 9 8 7 6 5 4 3 2 1 0 PPM
HYDROFLUMETHIAZIDE IN KBR
FTIR
% TRANSMITTANCE
100 75 50 25 0
A B C D
A 1612
B 1527
C 1316
D 1175
4000 3600 3200 2800 2400 2000 1800 1600 1400 1200 1000 800 600 400
WAVENUMBER

# HYDROMORPHONE

$C_{17}H_{19}NO_3$

Molecular weight: 285.34 (285.14)
Synonyms: 4,5α-Epoxy-3-hydroxy-17-methylmorphinan-6-one; dihydromorphinone
Trade names: Dilaudid

Use: Narcotic analgesic
HPLC: Si-10; 20A:80B; 4.2
GC: 2562; 250°C

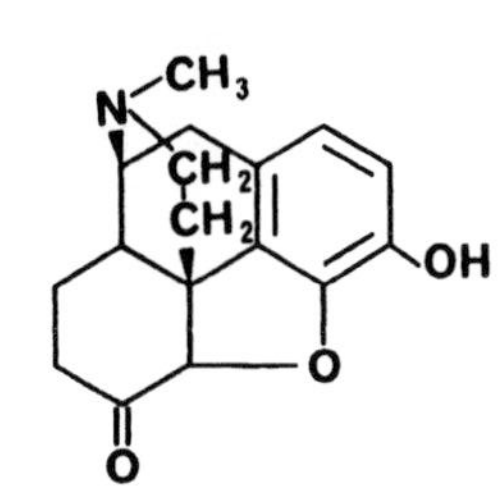


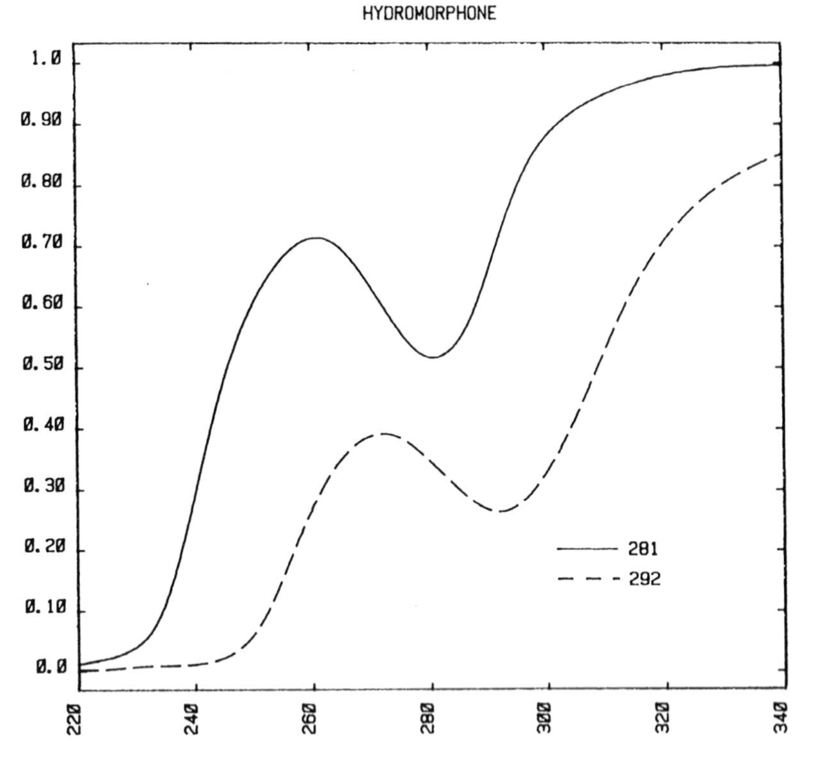

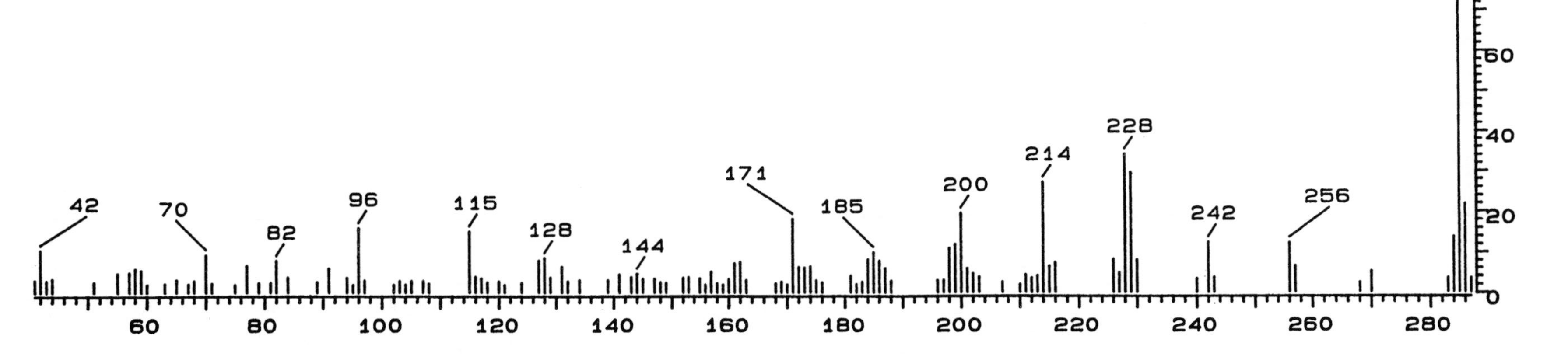

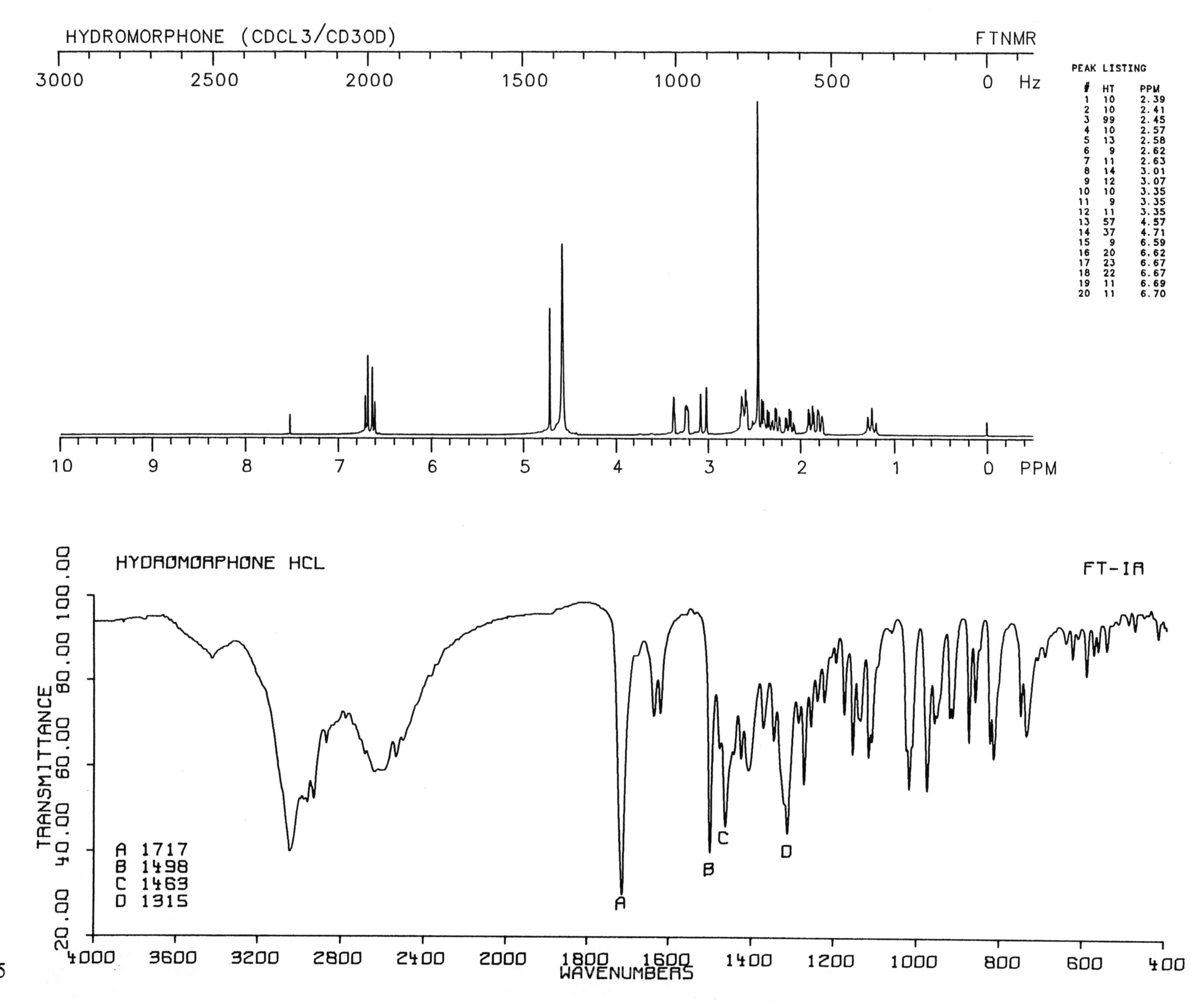
HYDROMORPHONE (CDCL3/CD3OD)
FTNMR
3000 2500 2000 1500 1000 500 0 Hz
10 9 8 7 6 5 4 3 2 1 0 PPM
PEAK LISTING
# HT PPM
1 10 2.39
2 10 2.41
3 99 2.45
4 10 2.57
5 13 2.58
6 9 2.62
7 11 2.63
8 14 3.01
9 12 3.07
10 10 3.35
11 9 3.35
12 11 3.35
13 57 4.57
14 37 4.71
15 9 6.59
16 20 6.62
17 23 6.67
18 22 6.67
19 11 6.69
20 11 6.70
HYDROMORPHONE HCL
FT-IR
TRANSMITTANCE
20.00 40.00 60.00 80.00 100.00
A 1717
B 1498
C 1463
D 1315
4000 3600 3200 2800 2400 2000 1800 1600 1400 1200 1000 800 600 400
WAVENUMBERS
A B C D

# HYDROQUINONE

$C_6H_6O_2$

Molecular weight: 110.11 (110.04)
Synonyms: 1,4-Benzenediol; hydroquinol; quinol

Trade names: Banquin, Eldopaque, Eldoquin, Melanex, Phiaquin, Solaquin

Use: Bleaching agent
HPLC: Si-10; 12A:98B; 4.6
GC: 1264; 140°C

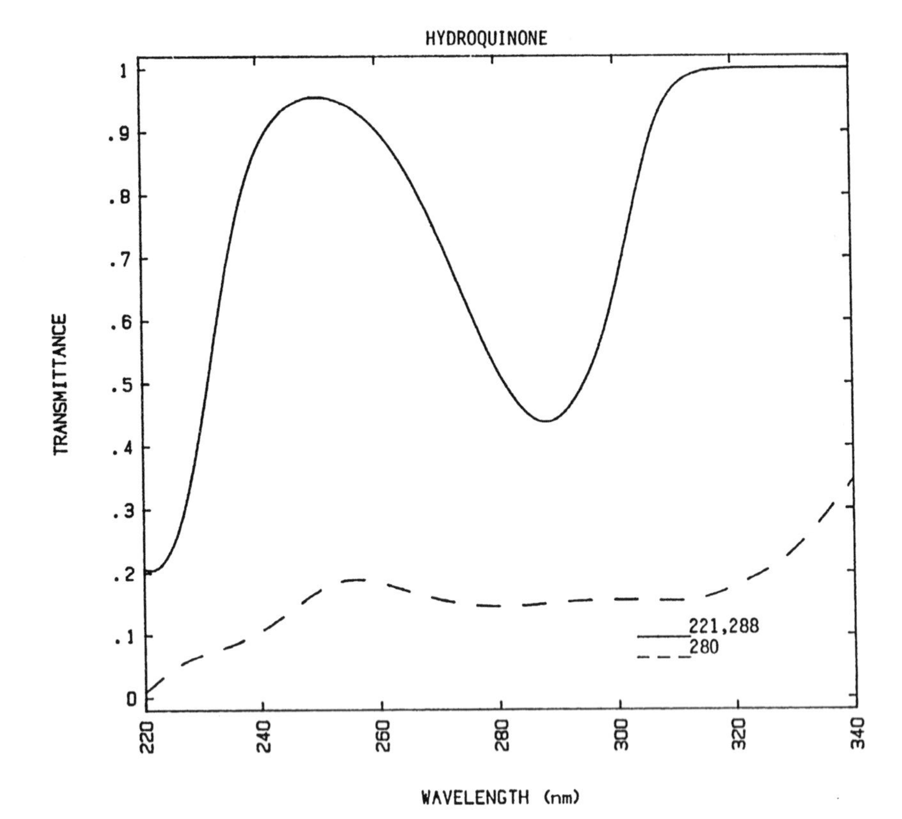


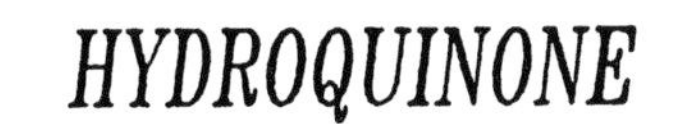


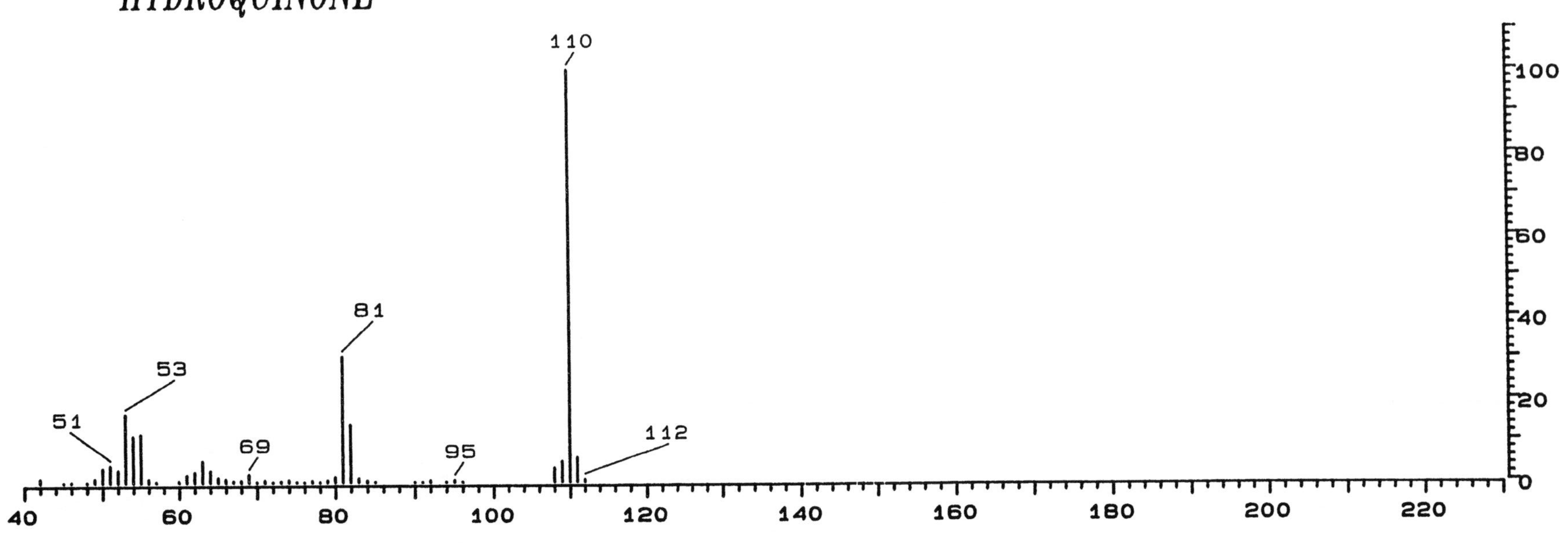

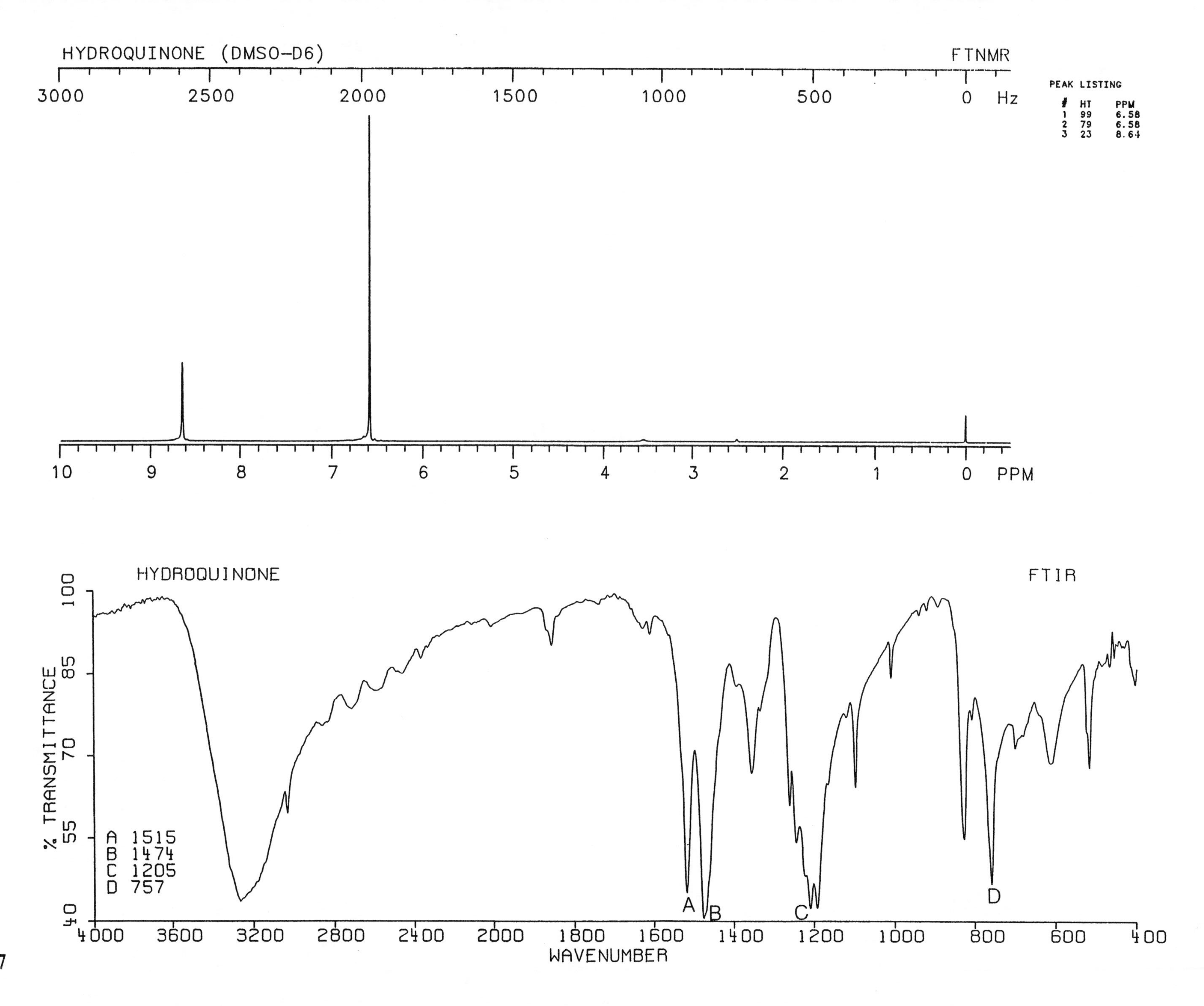
HYDROQUINONE (DMSO-D6)
FTNMR
3000 2500 2000 1500 1000 500 0 Hz
PEAK LISTING
# HT PPM
1 99 6.58
2 79 6.58
3 23 8.64
10 9 8 7 6 5 4 3 2 1 0 PPM
HYDROQUINONE
FTIR
% TRANSMITTANCE
100 85 70 55 40
A 1515
B 1474
C 1205
D 757
A B C D
4000 3600 3200 2800 2400 2000 1800 1600 1400 1200 1000 800 600 400
WAVENUMBER

# HYDROXOCOBALAMINE

$C_{62}H_{89}CoN_{13}O_{15}P$

Molecular weight: 1346.41 (1345.57)
Synonyms: Cobinamide hydroxide phosphate 3'-ester with 5,6-dimethyl-1-α-D-ribofuranosylbenzimidazole inner salt; vitamin $B_{12a}$, hydroxocobemine
Trade names: Axion, Axlon, Cobalex, Depogamma, Docelan, Docevita, Duradoce, Hydrogrisevit, Hydrovit, Hyxobamine, Idrogriseovit, Oxobemin, Redisol, Sytobex, Vitadurin
Use: Vitamin
HPLC: Si-10; 50A:50B; 4.3
GC:

$$[HO-CoX] \underset{}{\overset{H_2O}{\rightleftharpoons}} [H_2O \cdot CoX]^+ OH^-$$

CoX = Vitamin $B_{12}$ minus CN

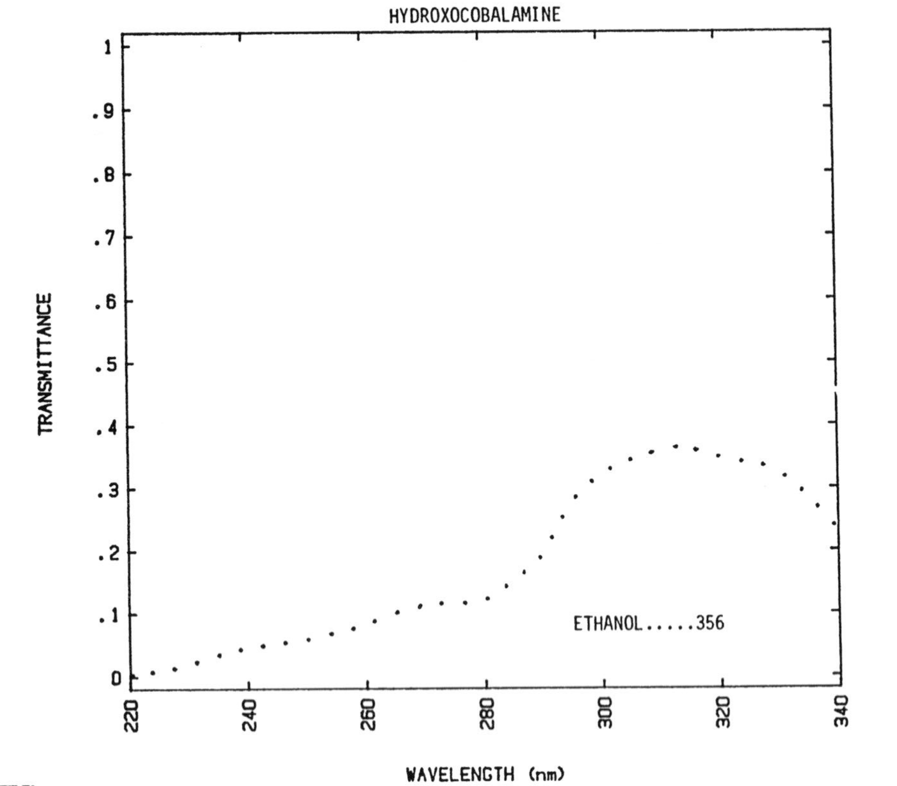


NO USEFUL MASS SPECTRUM WAS OBTAINED

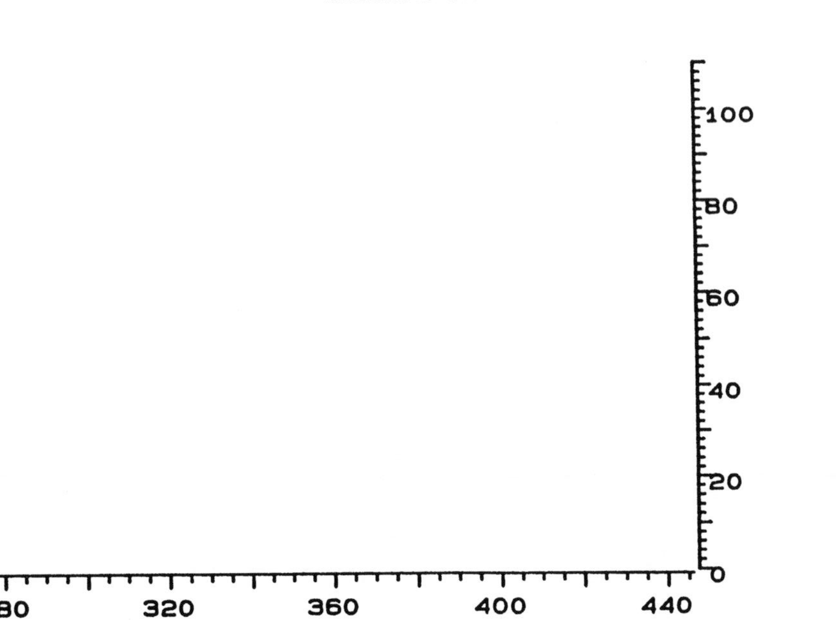

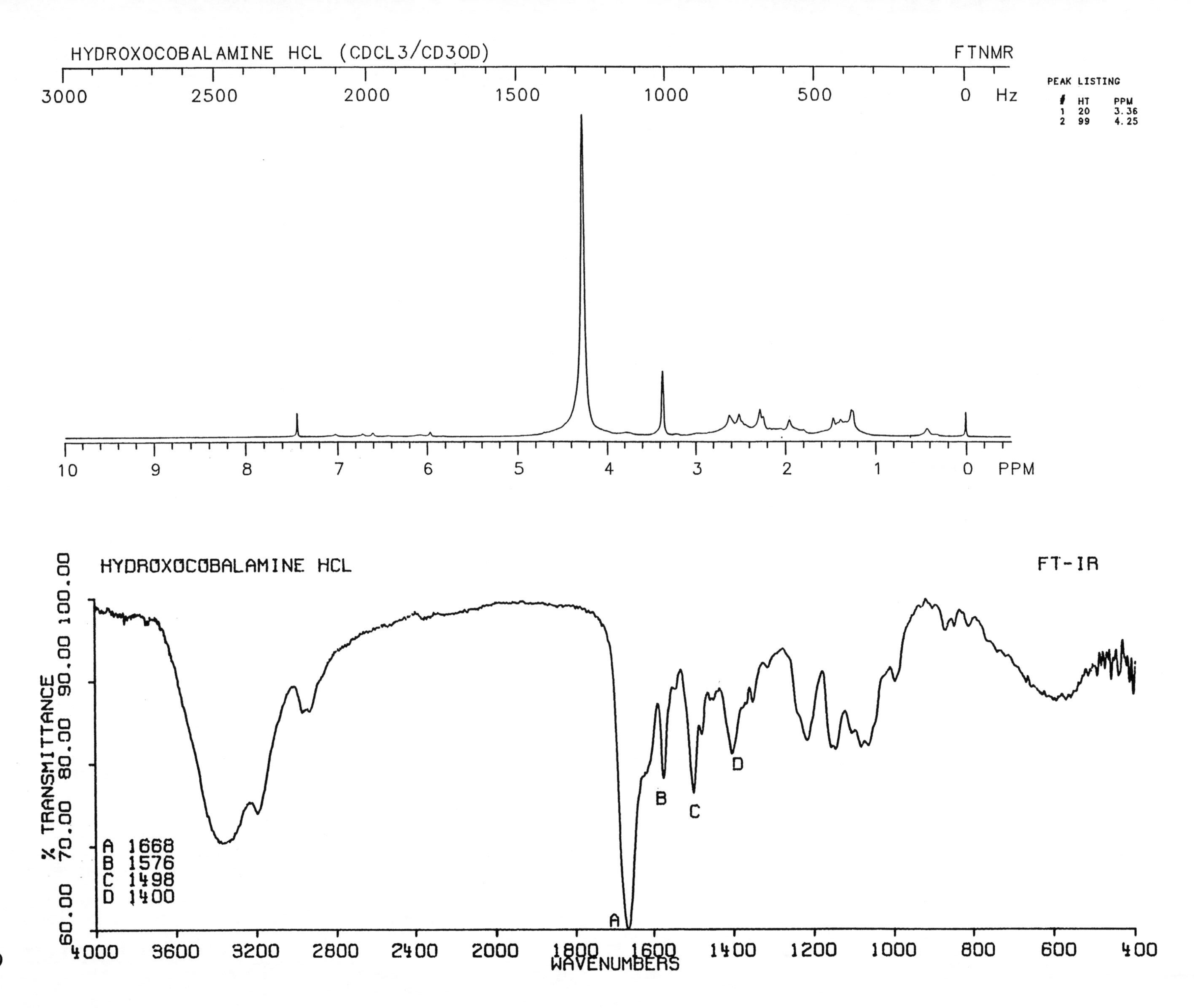
HYDROXOCOBALAMINE HCL (CDCL3/CD3OD)
FTNMR
3000
2500
2000
1500
1000
500
0
Hz
PEAK LISTING
# HT PPM
1 20 3.36
2 99 4.25
10
9
8
7
6
5
4
3
2
1
0
PPM
HYDROXOCOBALAMINE HCL
FT-IR
% TRANSMITTANCE
60.00
70.00
80.00
90.00
100.00
A 1668
B 1576
C 1498
D 1400
A
B
C
D
4000
3600
3200
2800
2400
2000
1800
1600
1400
1200
1000
800
600
400
WAVENUMBERS

# HYDROXYAMOXAPINE

$C_{17}H_{16}ClN_3O_2$

Molecular weight: 329.78 (329.09)
Synonyms: 7-Hydroxy-2-chloro-11-(1-piperazinyl)dibenz-[b,f]-[1,4]-oxazepine
Trade names:

Use: Metabolite
HPLC: Si-10; 5A:95B; 8.0
GC: 2966; 280°C

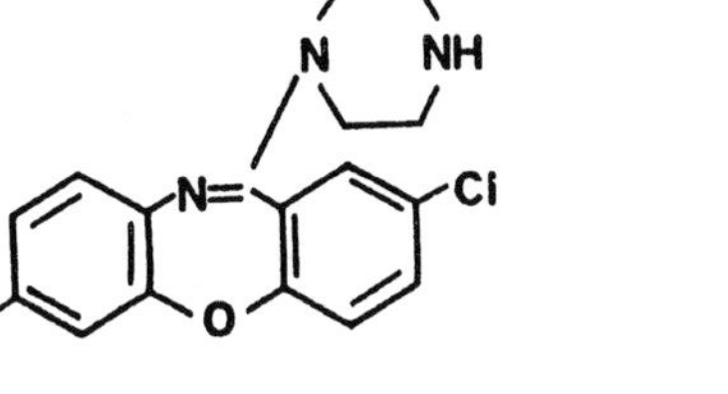


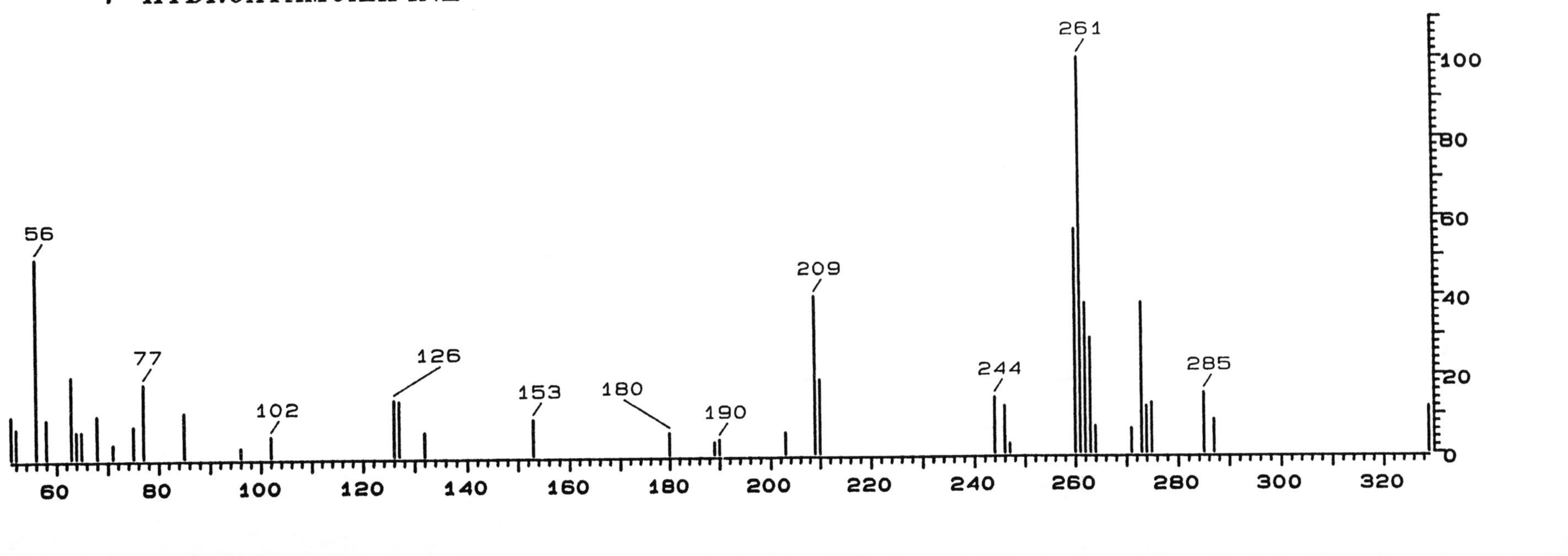

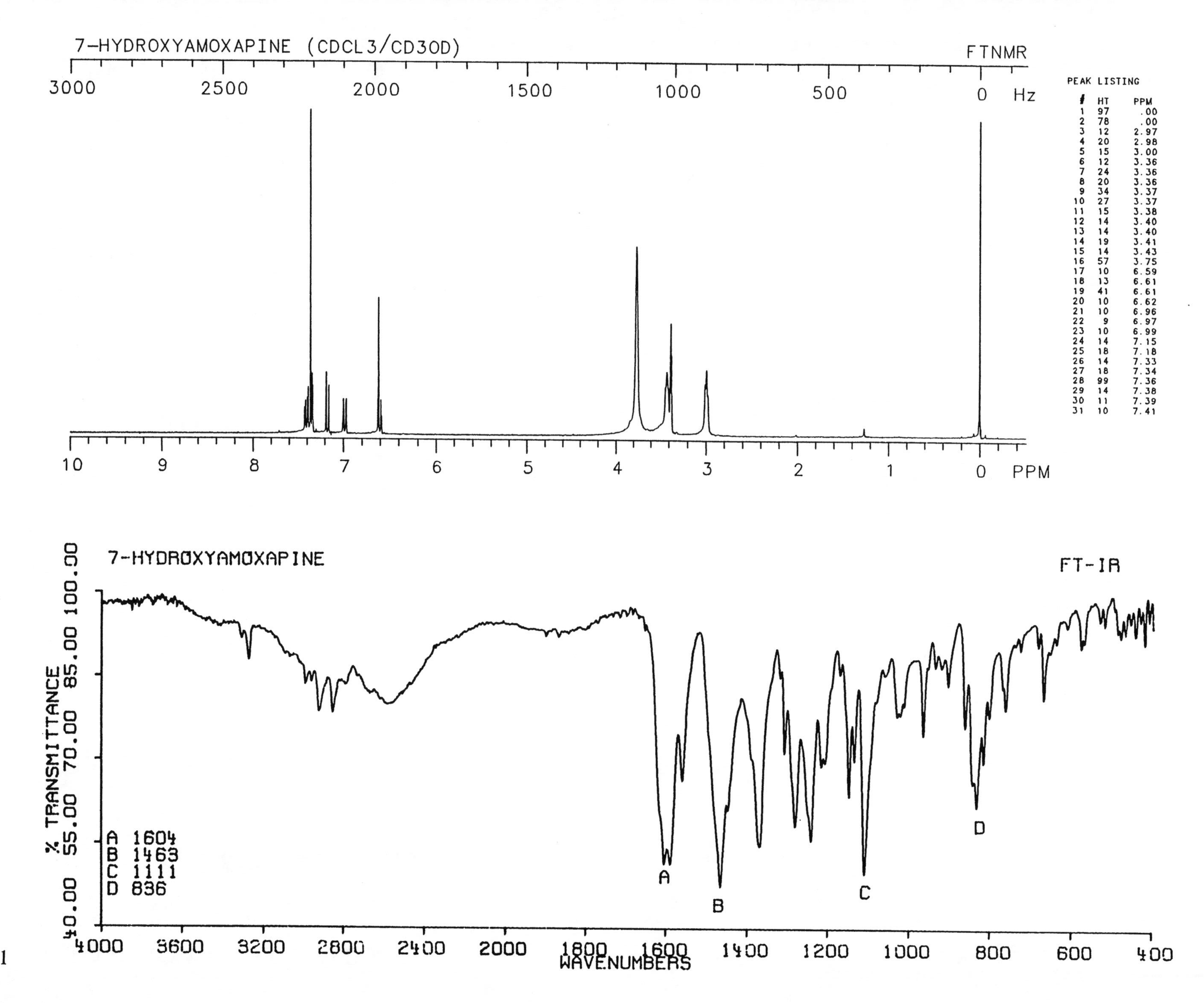
7-HYDROXYAMOXAPINE (CDCL3/CD3OD)
FTNMR
3000 2500 2000 1500 1000 500 0 Hz
10 9 8 7 6 5 4 3 2 1 0 PPM
PEAK LISTING
# HT PPM
1 97 .00
2 78 .00
3 12 2.97
4 20 2.98
5 15 3.00
6 12 3.36
7 24 3.36
8 20 3.36
9 34 3.37
10 27 3.37
11 15 3.38
12 14 3.40
13 14 3.40
14 19 3.41
15 14 3.43
16 57 3.75
17 10 6.59
18 13 6.61
19 41 6.61
20 10 6.62
21 10 6.96
22 9 6.97
23 10 6.99
24 14 7.15
25 18 7.18
26 14 7.33
27 18 7.34
28 99 7.36
29 14 7.38
30 11 7.39
31 10 7.41
7-HYDROXYAMOXAPINE
FT-IR
% TRANSMITTANCE
40.00 55.00 70.00 85.00 100.00
A 1604
B 1463
C 1111
D 836
A B C D
4000 3600 3200 2800 2400 2000 1800 1600 1400 1200 1000 800 600 400
WAVENUMBERS

# 8-HYDROXYAMOXAPINE

$C_{17}H_{16}ClN_3O_2$

Molecular weight: 329.78 (329.09)
Synonyms: 8-Hydroxy-2-chloro-11-(1-piperazinyl)dibenz-[b,f]-[1,4]-oxazepine
Trade names:

Use: Metabolite
HPLC: Si-10; 5A:95B; 9.4
GC: 3026; 280°C

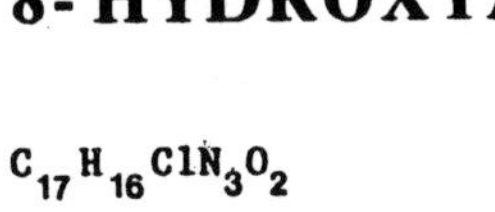


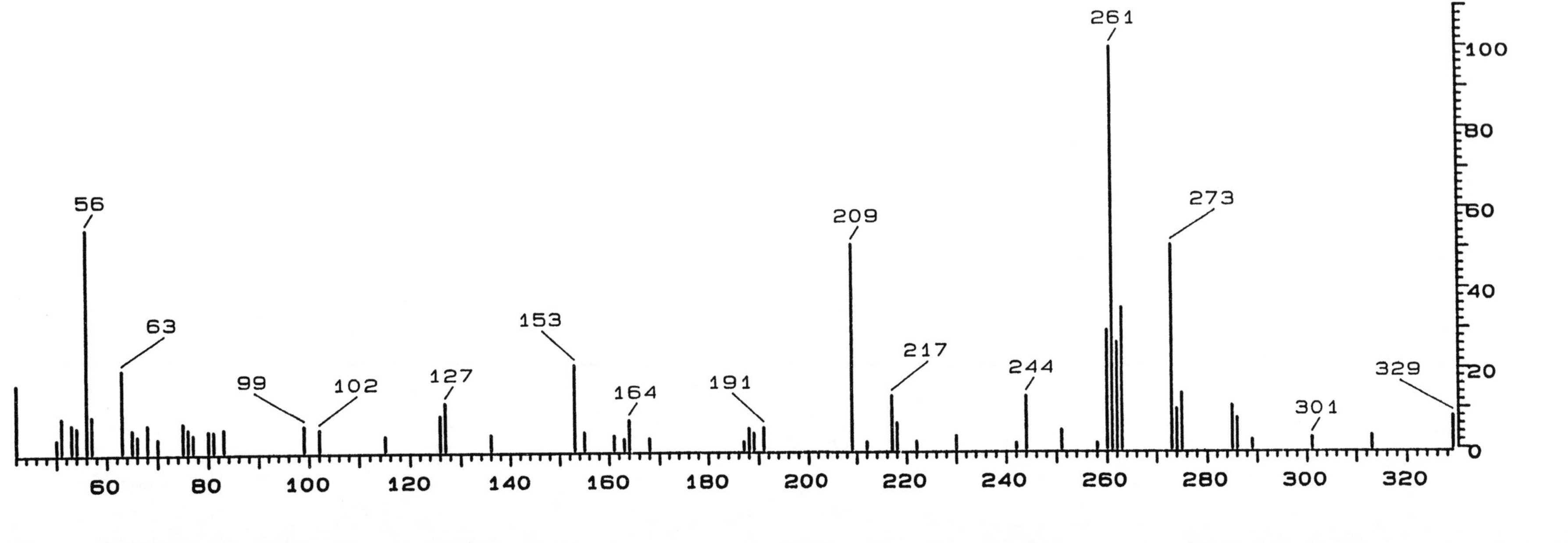

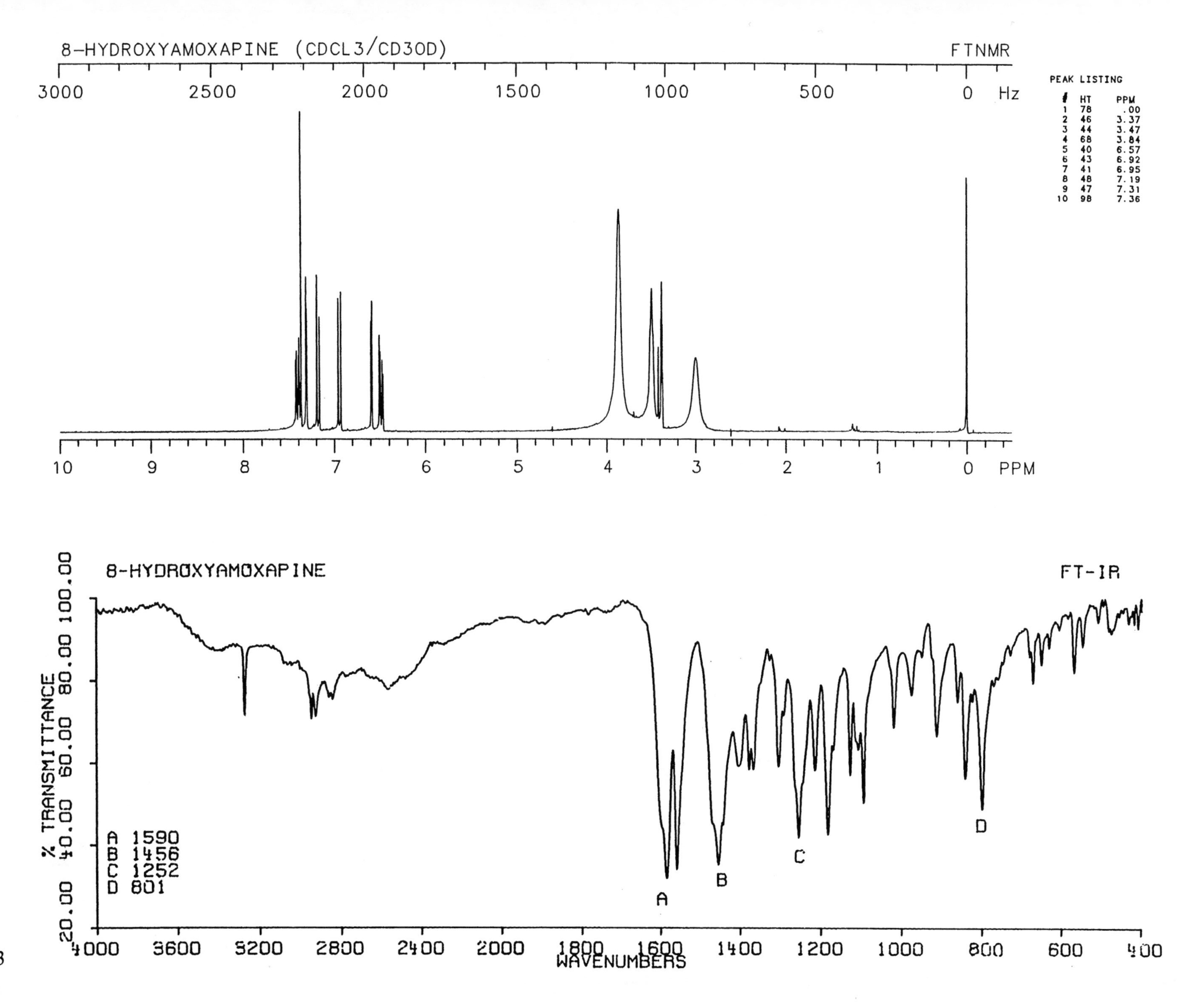
8-HYDROXYAMOXAPINE (CDCL3/CD3OD)
FTNMR
Hz
PPM
PEAK LISTING
# HT PPM
1 78 .00
2 46 3.37
3 44 3.47
4 68 3.84
5 40 6.57
6 43 6.92
7 41 6.95
8 48 7.19
9 47 7.31
10 98 7.36
8-HYDROXYAMOXAPINE
FT-IR
% TRANSMITTANCE
WAVENUMBERS
A 1590
B 1456
C 1252
D 801

# HYDROXYAMPHETAMINE

$C_9H_{13}NO$

**Molecular weight:** 151.21 (151.10)
**Synonyms:** p-(2-Aminopropyl)phenol; dl-p-hydroxy-α-methyl-phenethylamine
**Trade names:** Paredrine

**Use:** Adrenergic
**HPLC:** Si-10; 20A:80B; 5.0
**GC:** 1501; 200°C

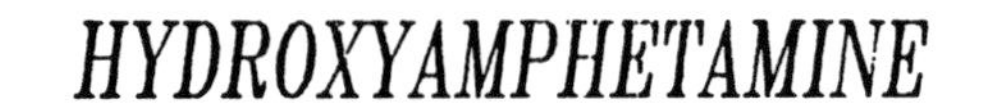


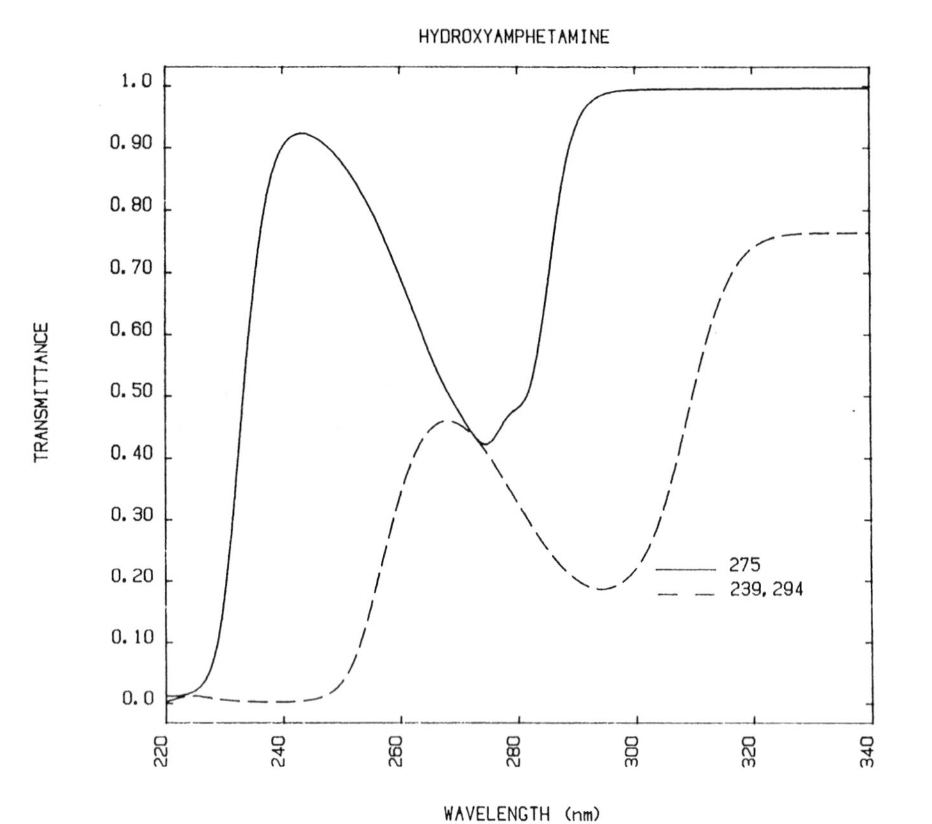


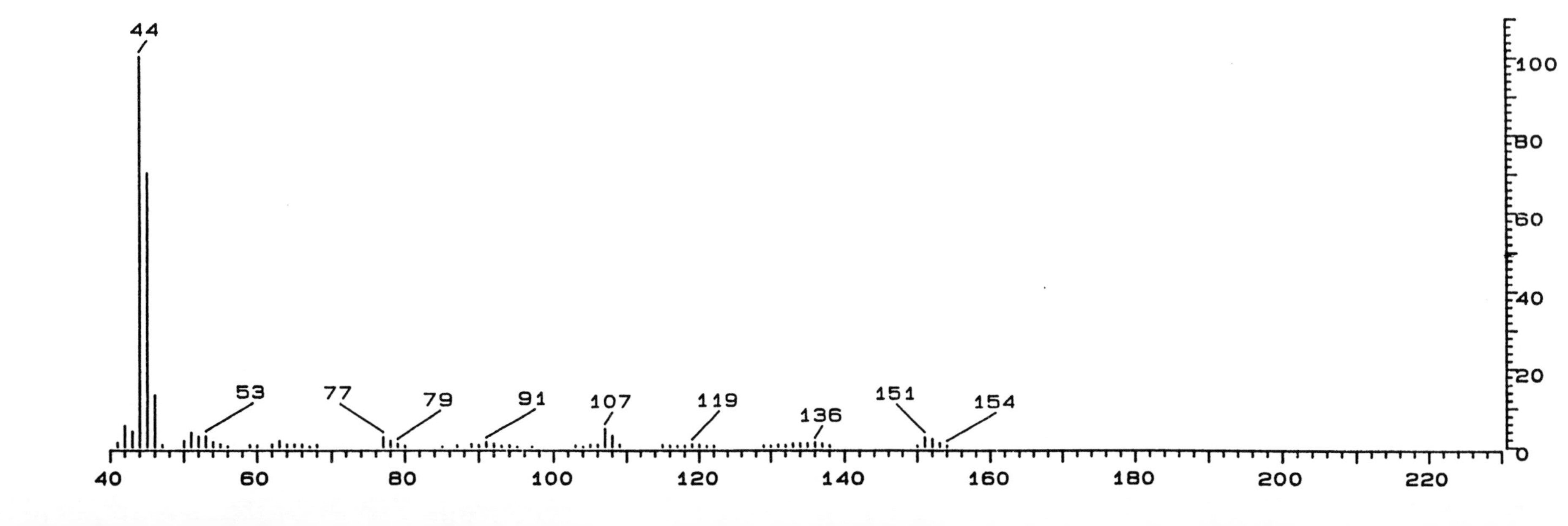

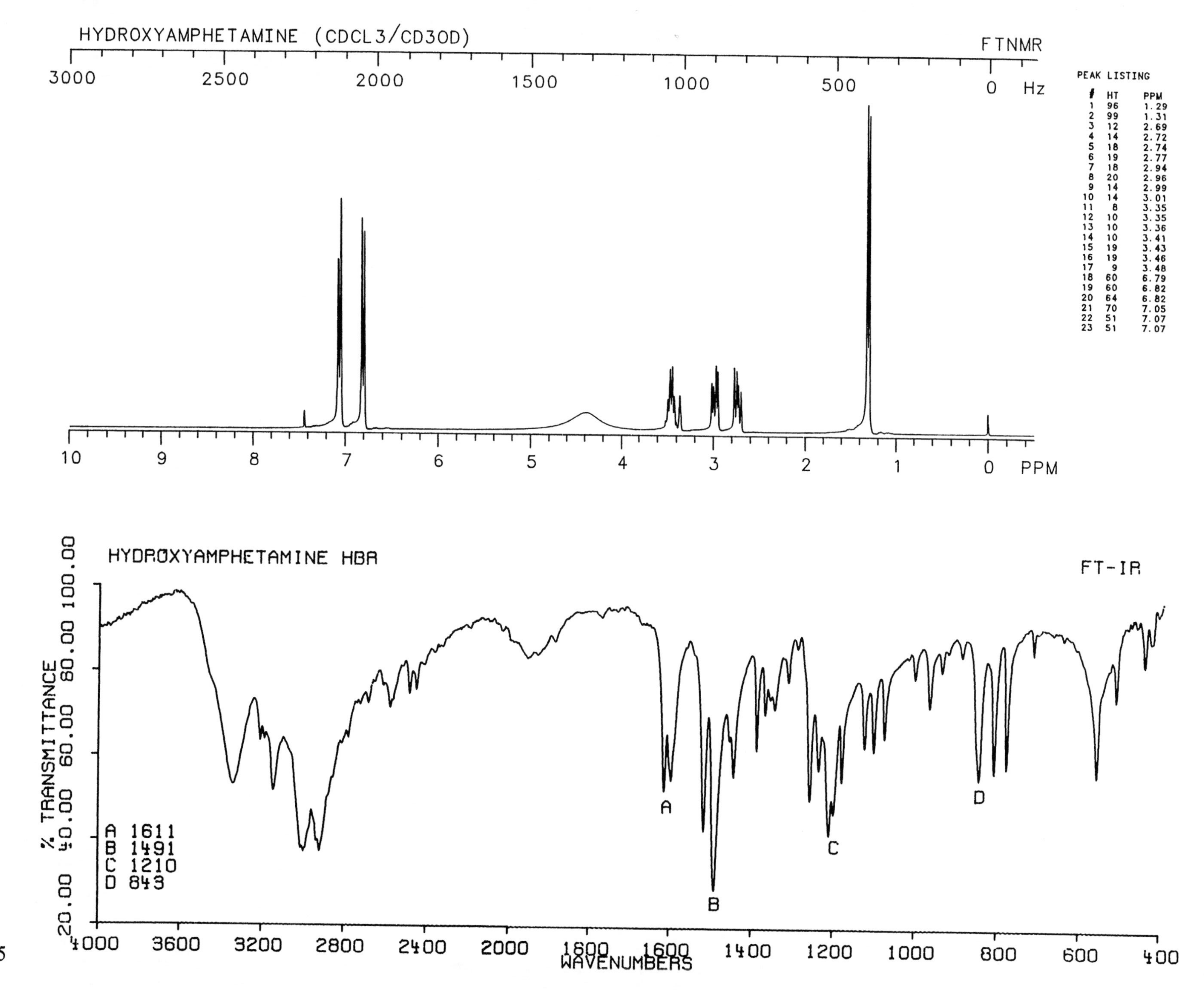
HYDROXYAMPHETAMINE (CDCL3/CD3OD)
FTNMR
Hz
PPM
PEAK LISTING
# HT PPM
1 96 1.29
2 99 1.31
3 12 2.69
4 14 2.72
5 18 2.74
6 19 2.77
7 18 2.94
8 20 2.96
9 14 2.99
10 14 3.01
11 8 3.35
12 10 3.35
13 10 3.36
14 10 3.41
15 19 3.43
16 19 3.46
17 9 3.48
18 60 6.79
19 60 6.82
20 64 6.82
21 70 7.05
22 51 7.07
23 51 7.07
HYDROXYAMPHETAMINE HBR
FT-IR
% TRANSMITTANCE
WAVENUMBERS
A 1611
B 1491
C 1210
D 843

# HYDROXYCHLOROQUINE

$C_{18}H_{26}ClN_3O$

**Molecular weight:** 335.88 (335.17)
**Synonyms:** 2-[[4-[(7-Chloro-4-quinolinyl)amino]penyl]ethylamino]-ethanol; oxychloroquine; oxichlorochine
**Trade names:** Plaquenil

**Use:** Antimalarial
**HPLC:** Si-10; 20A:80B; 6.0
**GC:**

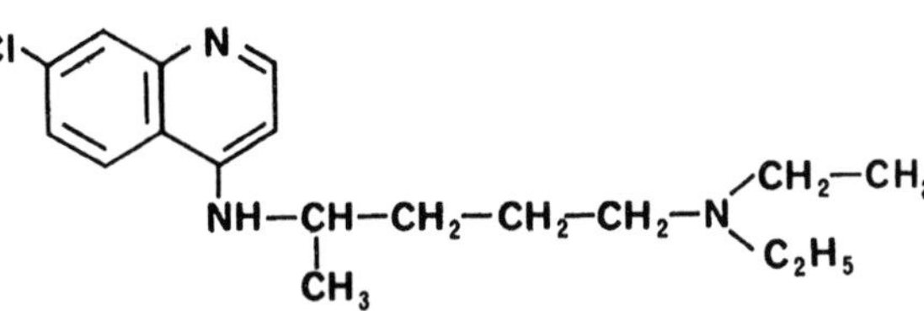


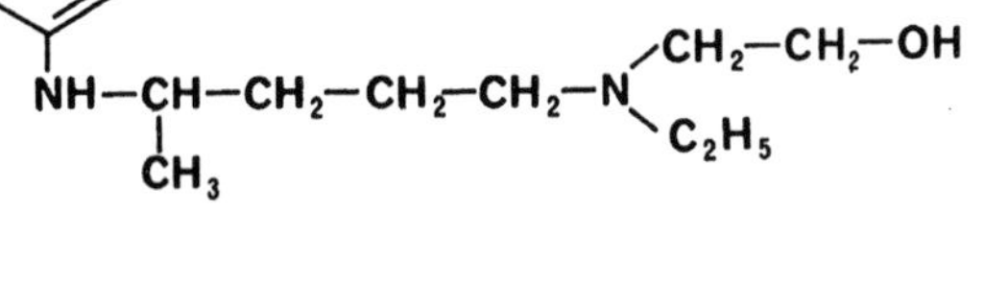


HYDROXYCHLOROQUINE

TRANSMITTANCE

1.0
0.90
0.80
0.70
0.60
0.50
0.40
0.30
0.20
0.10
0.0

220 240 260 280 300 320 340

WAVELENGTH (nm)

—— 235, 256, 330
— — 254, 331

HYDROXYCHLOROQUINE

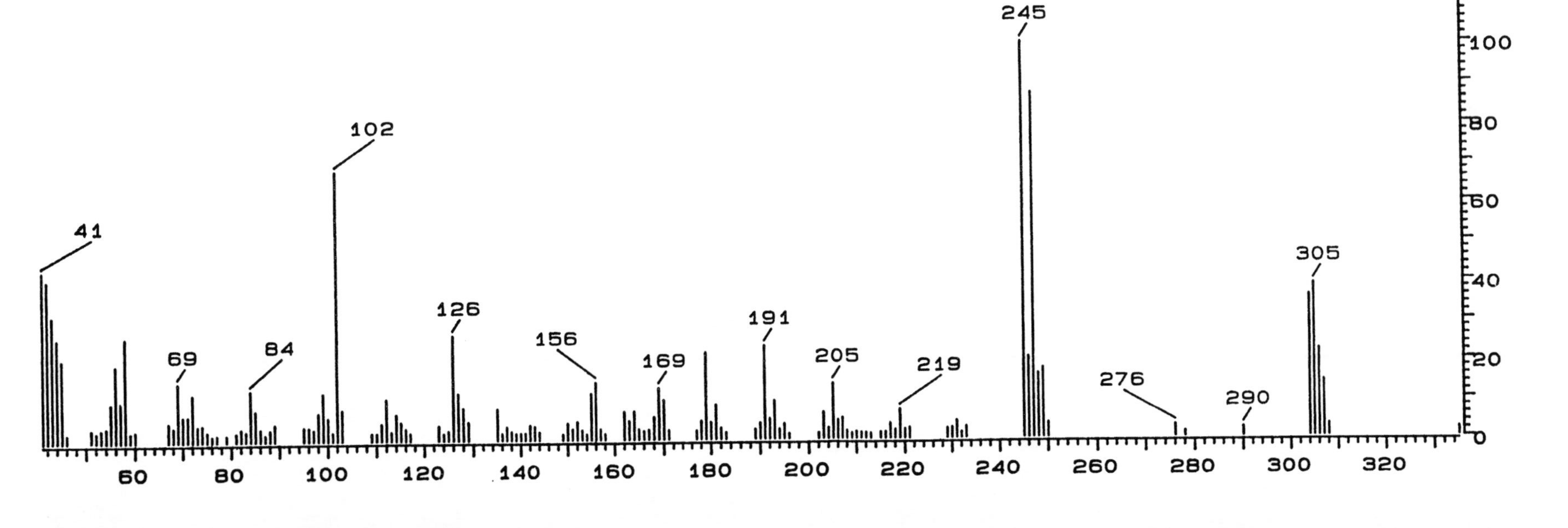

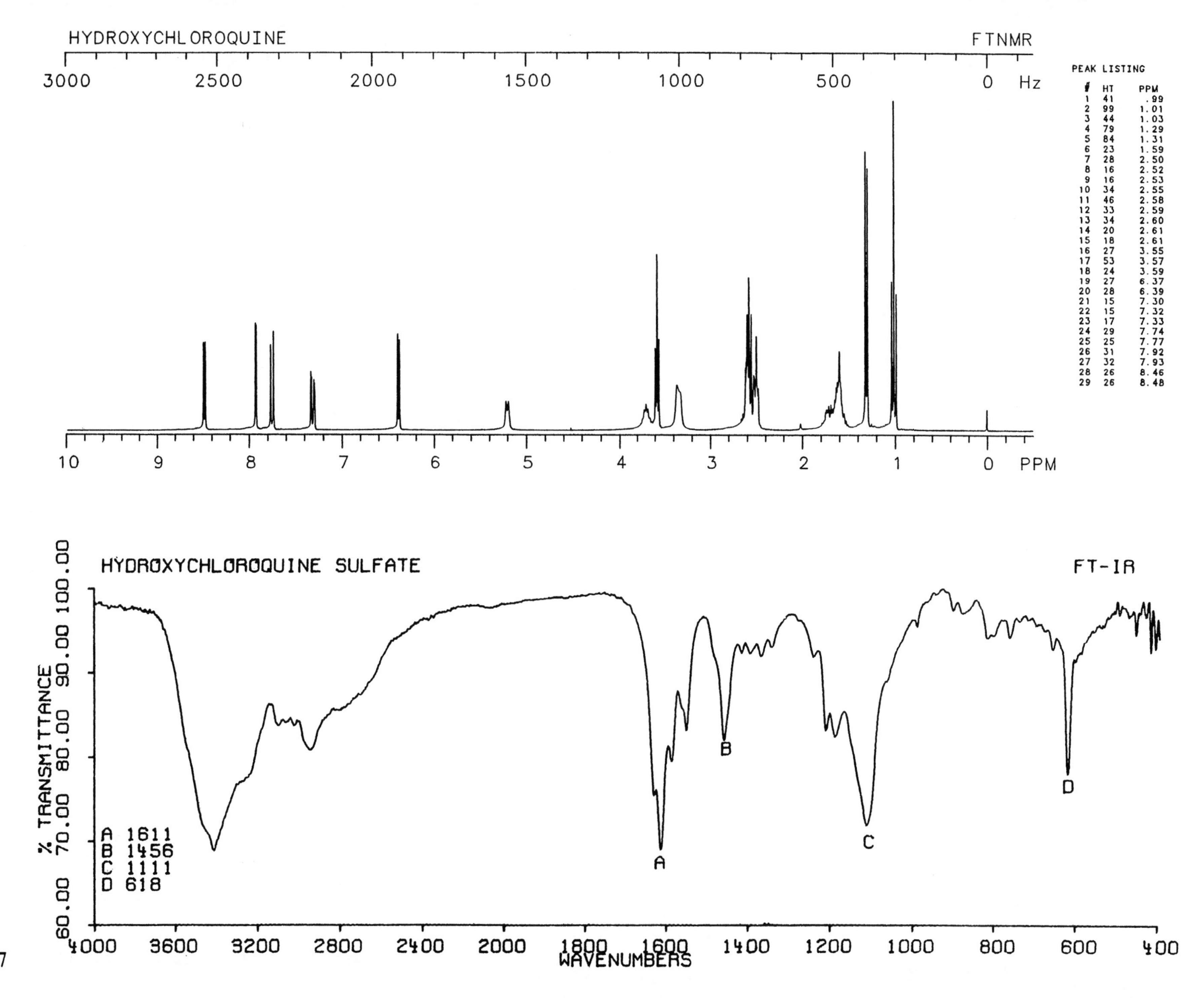
HYDROXYCHLOROQUINE
FTNMR
Hz
PPM
PEAK LISTING
# HT PPM
1 41 .99
2 99 1.01
3 44 1.03
4 79 1.29
5 84 1.31
6 23 1.59
7 28 2.50
8 16 2.52
9 16 2.53
10 34 2.55
11 46 2.58
12 33 2.59
13 34 2.60
14 20 2.61
15 18 2.61
16 27 3.55
17 53 3.57
18 24 3.59
19 27 6.37
20 28 6.39
21 15 7.30
22 15 7.32
23 17 7.33
24 29 7.74
25 25 7.77
26 31 7.92
27 32 7.93
28 26 8.46
29 26 8.48
HYDROXYCHLOROQUINE SULFATE
FT-IR
% TRANSMITTANCE
WAVENUMBERS
A 1611
B 1456
C 1111
D 618

# *N*-1-HYDROXYETHYLFLURAZEPAM

$C_{17}H_{14}ClFN_2O_2$

Molecular weight: 332.76 (332.07)
Synonyms: 7-Chloro-1-(2-hydroxyethyl)-5-(o-fluorophenyl)-1,3-dihydro-2H-1,4-benzodiazepin-2-one
Trade names:

Use:
HPLC: Si-10; 2A:98B; 5.0
GC: 2724; 250°C

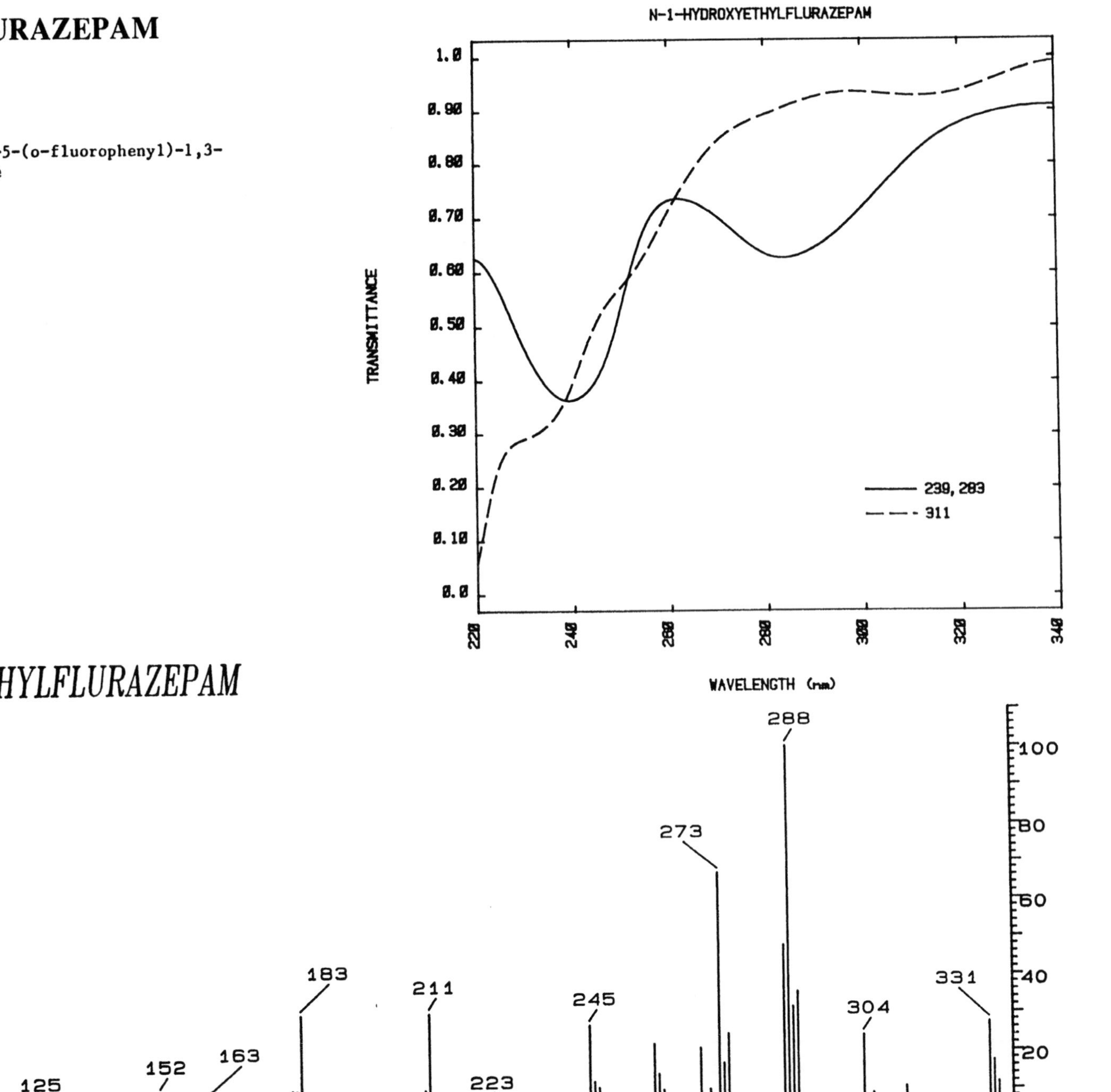

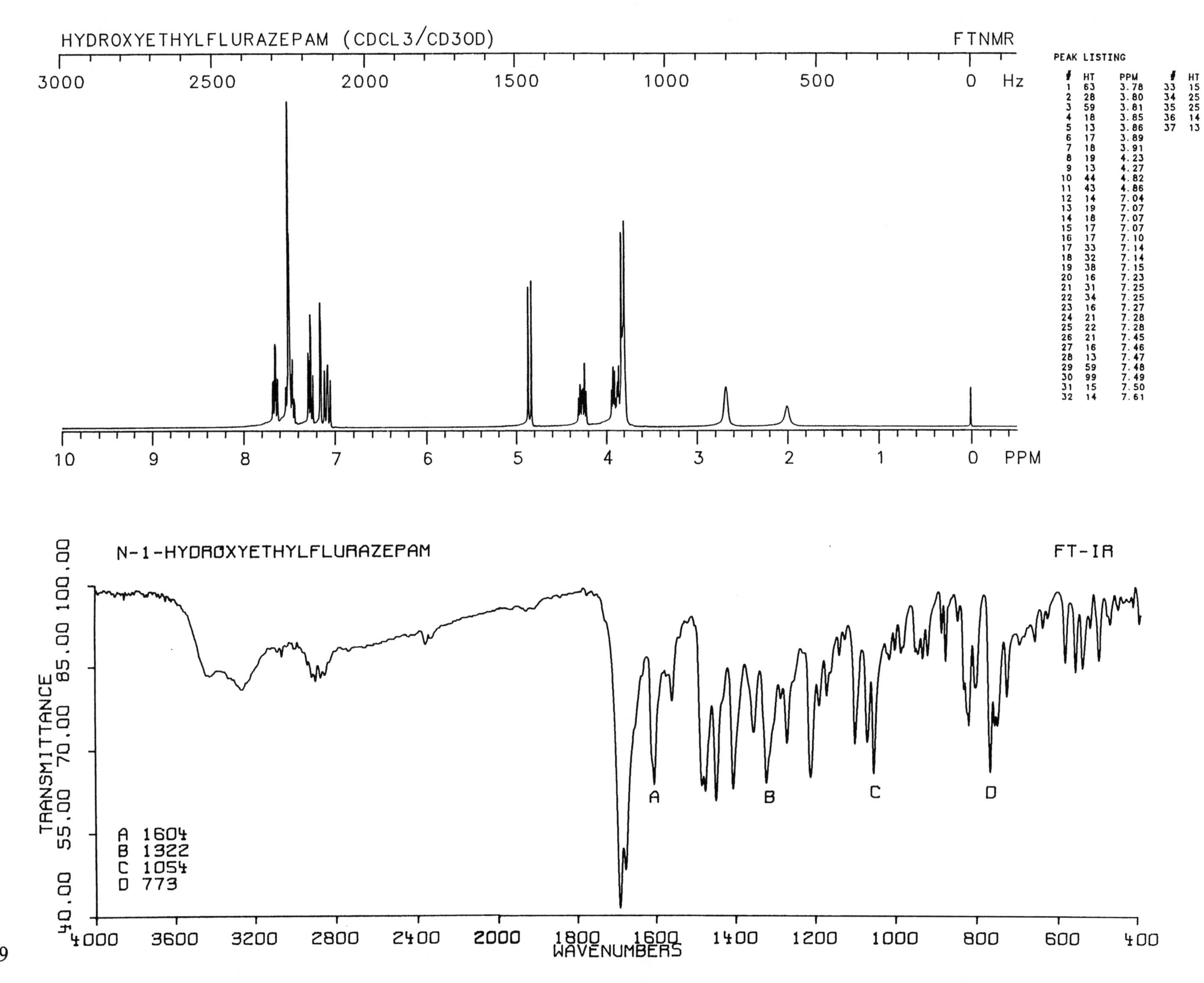
HYDROXYETHYLFLURAZEPAM (CDCL3/CD3OD)
FTNMR
3000 2500 2000 1500 1000 500 0 Hz
10 9 8 7 6 5 4 3 2 1 0 PPM
PEAK LISTING
# HT PPM # HT
1 63 3.78 33 15
2 28 3.80 34 25
3 59 3.81 35 25
4 18 3.85 36 14
5 13 3.86 37 13
6 17 3.89
7 18 3.91
8 19 4.23
9 13 4.27
10 44 4.82
11 43 4.86
12 14 7.04
13 19 7.07
14 18 7.07
15 17 7.07
16 17 7.10
17 33 7.14
18 32 7.14
19 38 7.15
20 16 7.23
21 31 7.25
22 34 7.25
23 16 7.27
24 21 7.28
25 22 7.28
26 21 7.45
27 16 7.46
28 13 7.47
29 59 7.48
30 99 7.49
31 15 7.50
32 14 7.61
N-1-HYDROXYETHYLFLURAZEPAM
FT-IR
TRANSMITTANCE
40.00 55.00 70.00 85.00 100.00
4000 3600 3200 2800 2400 2000 1800 1600 1400 1200 1000 800 600 400
WAVENUMBERS
A B C D
A 1604
B 1322
C 1054
D 773

# HYDROXYETHYLTHEOPHYLLINE

$C_9H_{12}N_4O_3$

**Molecular weight:** 224.22 (224.09)
**Synonyms:** 3,7-Dihydro-1,3-dimethyl-8-(2-hydroxyethyl)-1H-purine-2,6-dione
**Trade names:**

**Use:**
**HPLC:** Si-10; 2A:98B; 7.0
**GC:** 2137; 250°C

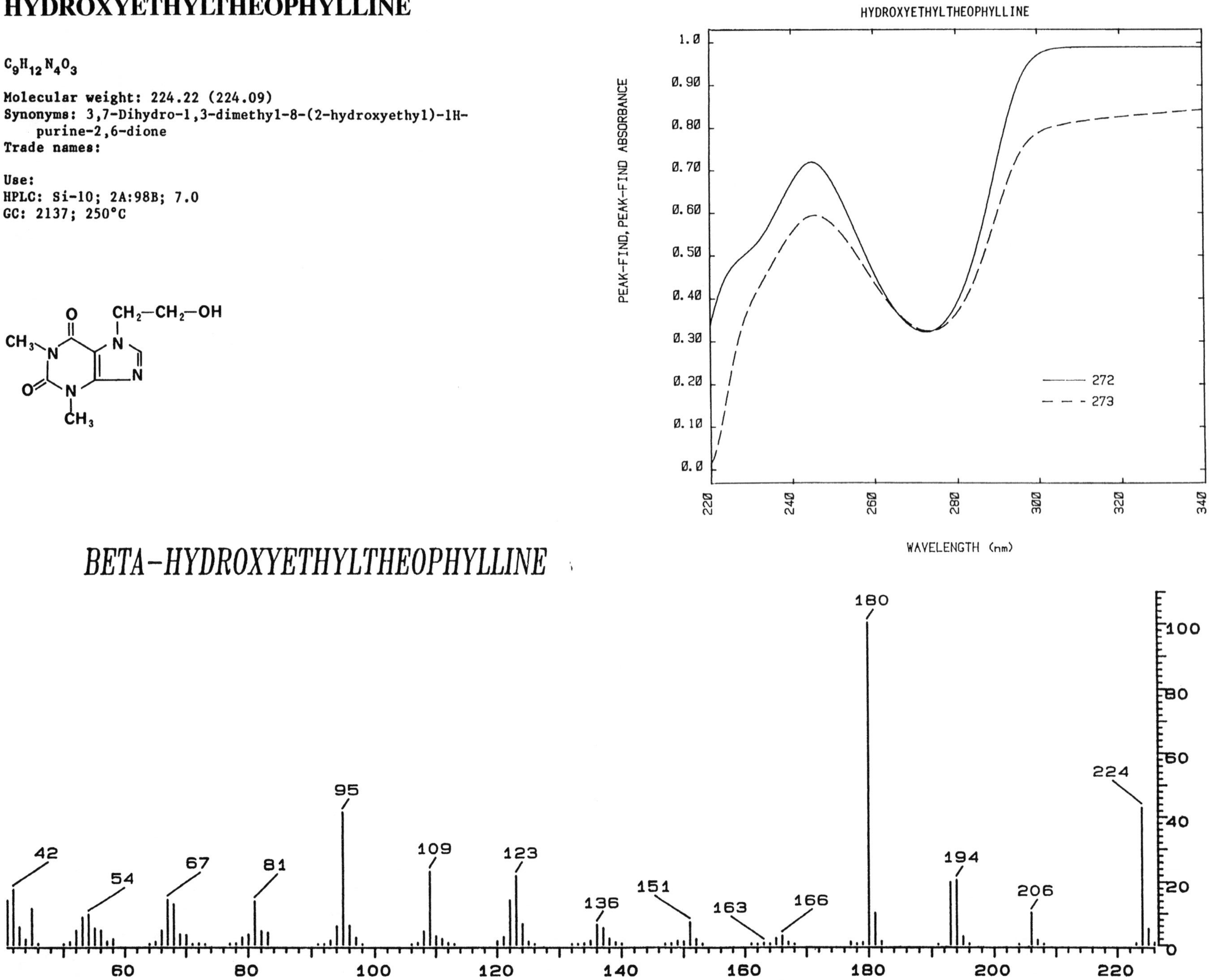

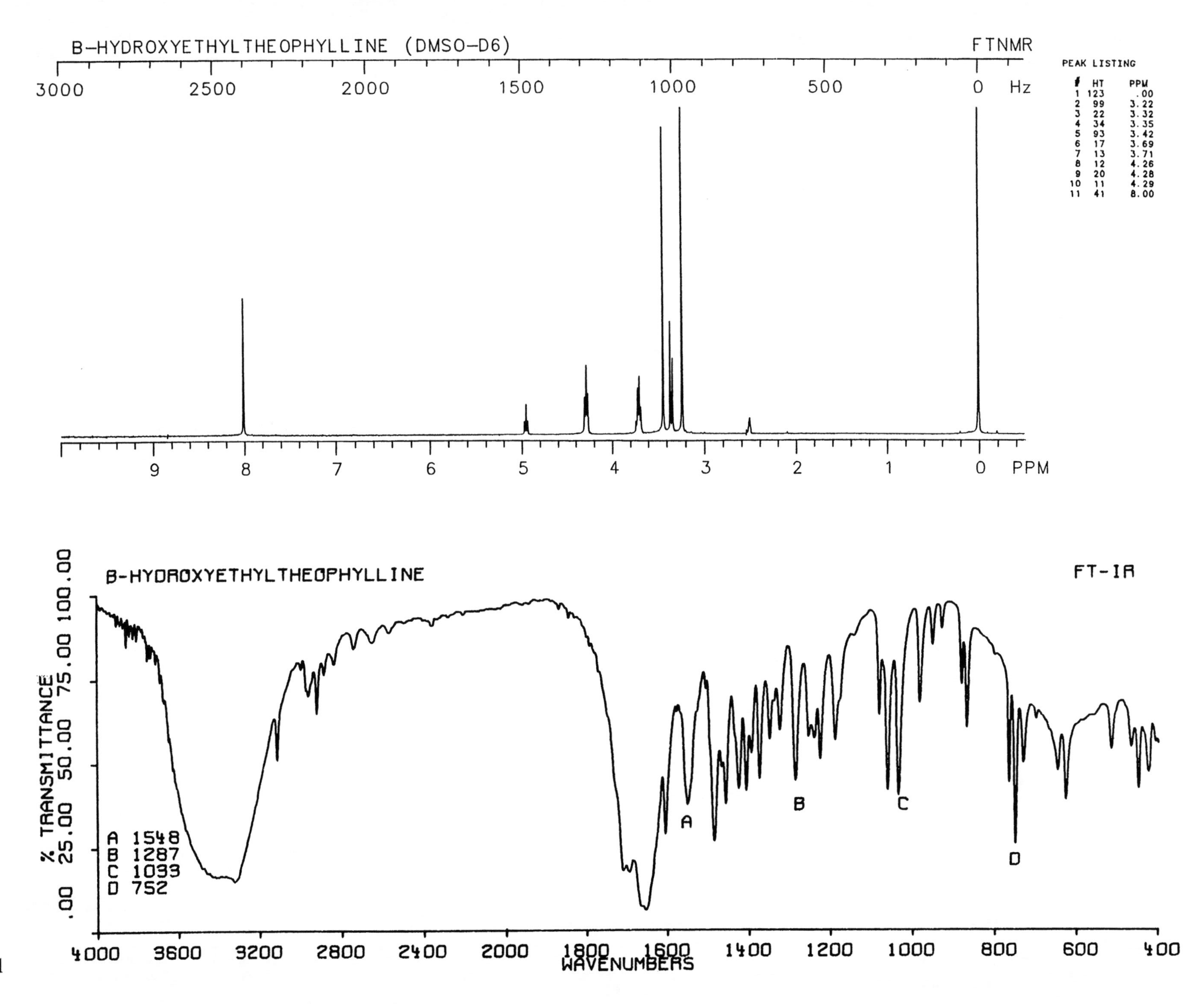

B-HYDROXYETHYLTHEOPHYLLINE (DMSO-D6)
FTNMR
3000 2500 2000 1500 1000 500 0 Hz
9 8 7 6 5 4 3 2 1 0 PPM
PEAK LISTING
# HT PPM
1 123 .00
2 99 3.22
3 22 3.32
4 34 3.35
5 93 3.42
6 17 3.69
7 13 3.71
8 12 4.26
9 20 4.28
10 11 4.29
11 41 8.00
B-HYDROXYETHYLTHEOPHYLLINE
FT-IR
% TRANSMITTANCE
.00 25.00 50.00 75.00 100.00
A 1548
B 1287
C 1033
D 752
A
B
C
D
4000 3600 3200 2800 2400 2000 1800 1600 1400 1200 1000 800 600 400
WAVENUMBERS

# 5-HYDROXYINDOLE-3-ACETIC ACID

$C_{10}H_9NO_3$

Molecular weight: 191.19 (191.06)
Synonyms: 5-Hydroxyindole-3-acetic acid

Trade names:

Use: Metabolite
HPLC: Si-10; 20A:80B; 6.1
GC:

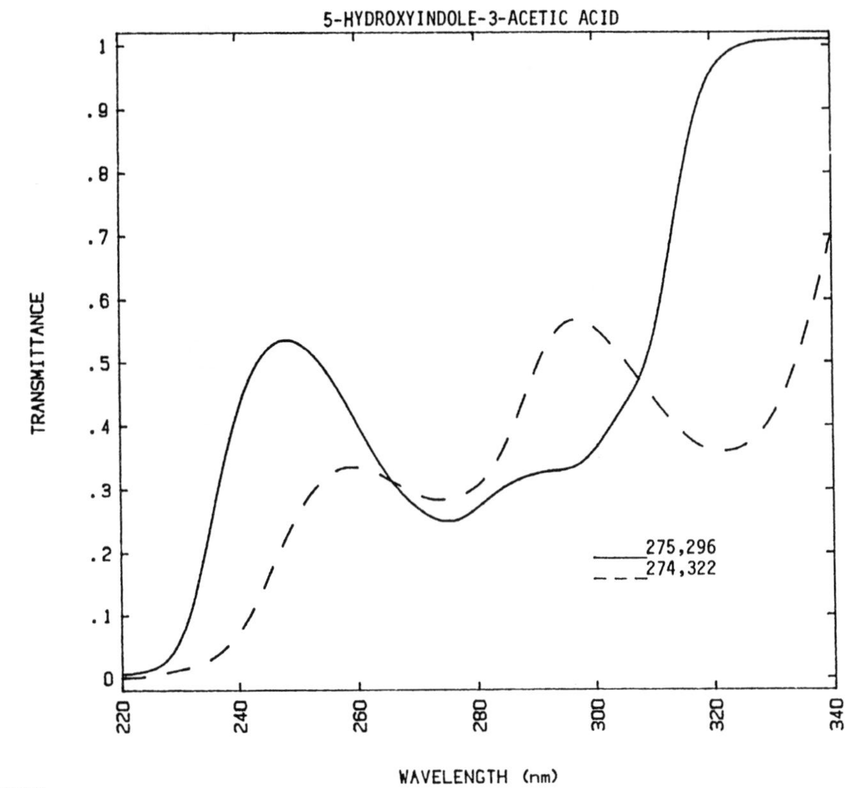


*NO USEFUL MASS SPECTRUM WAS OBTAINED*

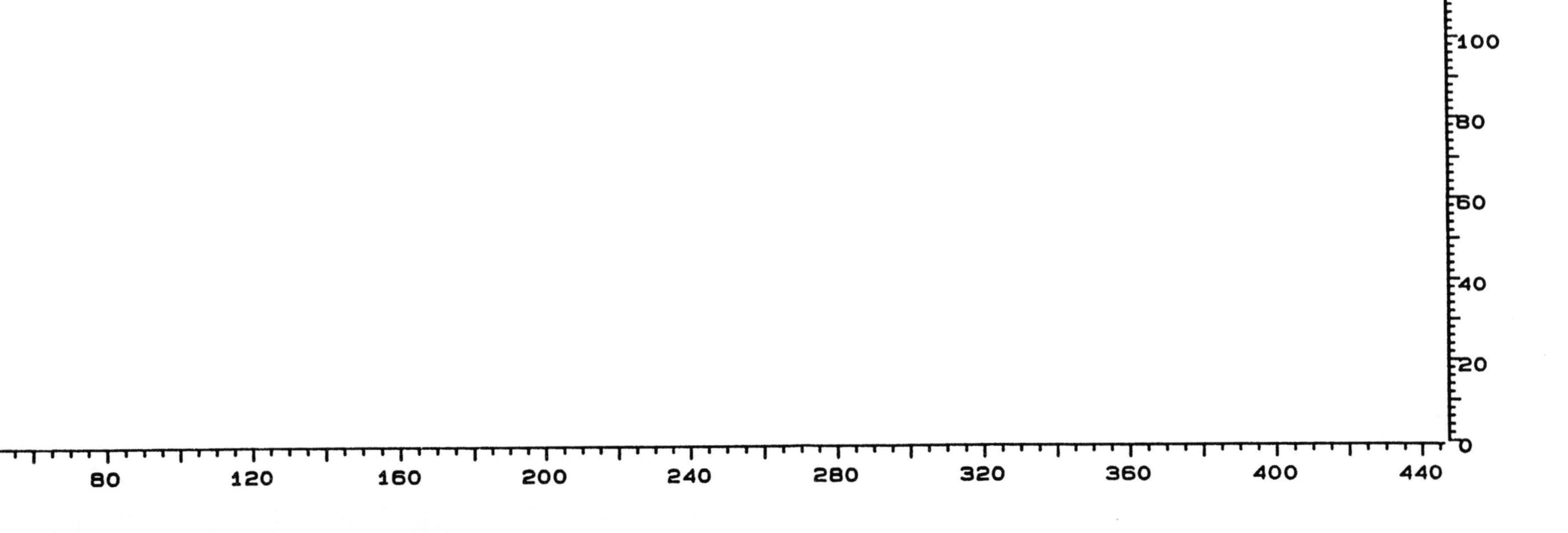

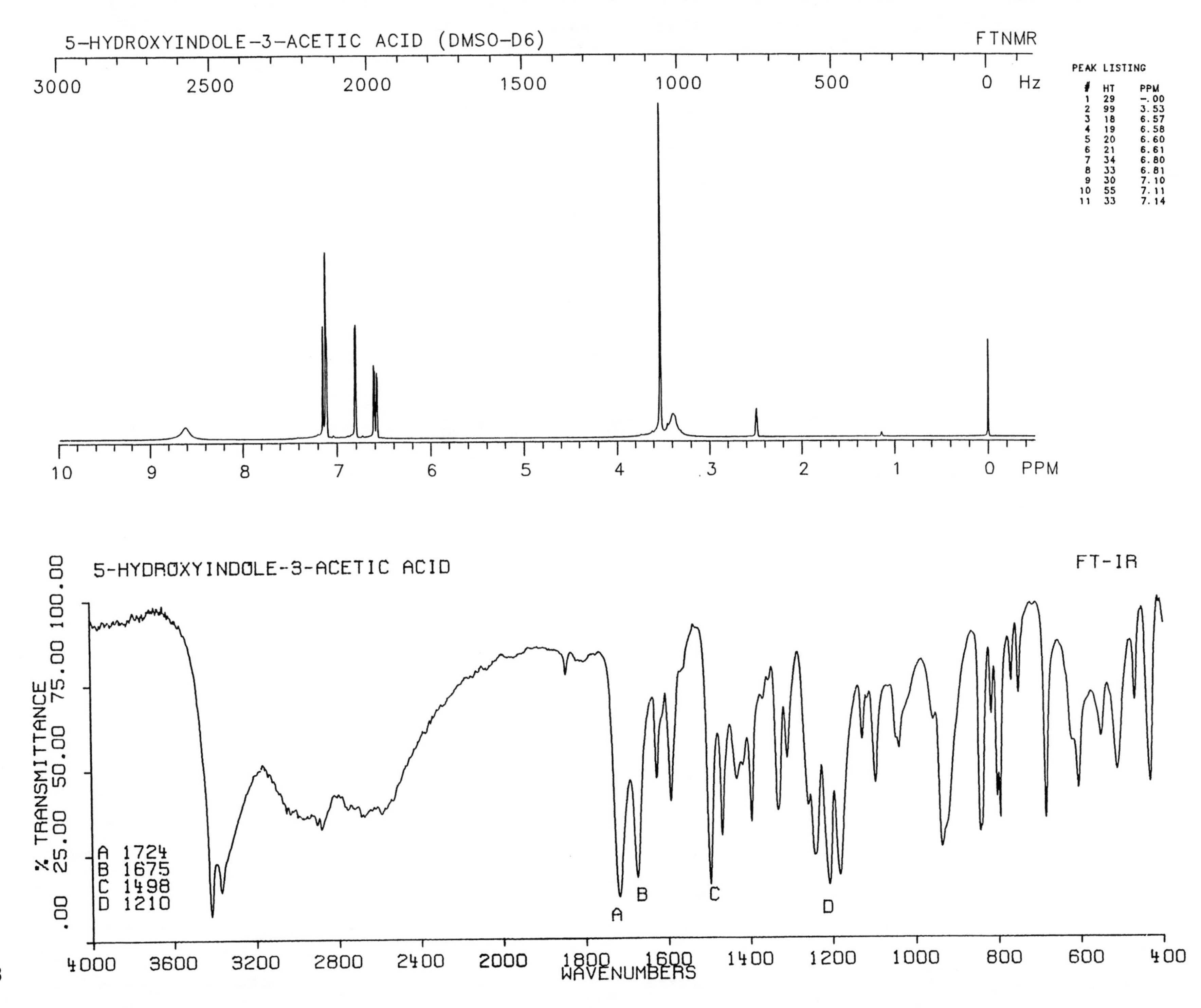
5-HYDROXYINDOLE-3-ACETIC ACID (DMSO-D6)
FTNMR
PEAK LISTING
# HT PPM
1 29 -.00
2 99 3.53
3 18 6.57
4 19 6.58
5 20 6.60
6 21 6.61
7 34 6.80
8 33 6.81
9 30 7.10
10 55 7.11
11 33 7.14
5-HYDROXYINDOLE-3-ACETIC ACID
FT-IR
A 1724
B 1675
C 1498
D 1210
% TRANSMITTANCE
WAVENUMBERS

# HYDROXYLAMINE

$H_3NO$

**Molecular weight:** 33.03 (33.02)
**Synonyms:**

**Trade names:**

**Use:** Synthesis
**HPLC:** Si-10; 20A:80B; 6.1
**GC:**

$NH_2$—OH

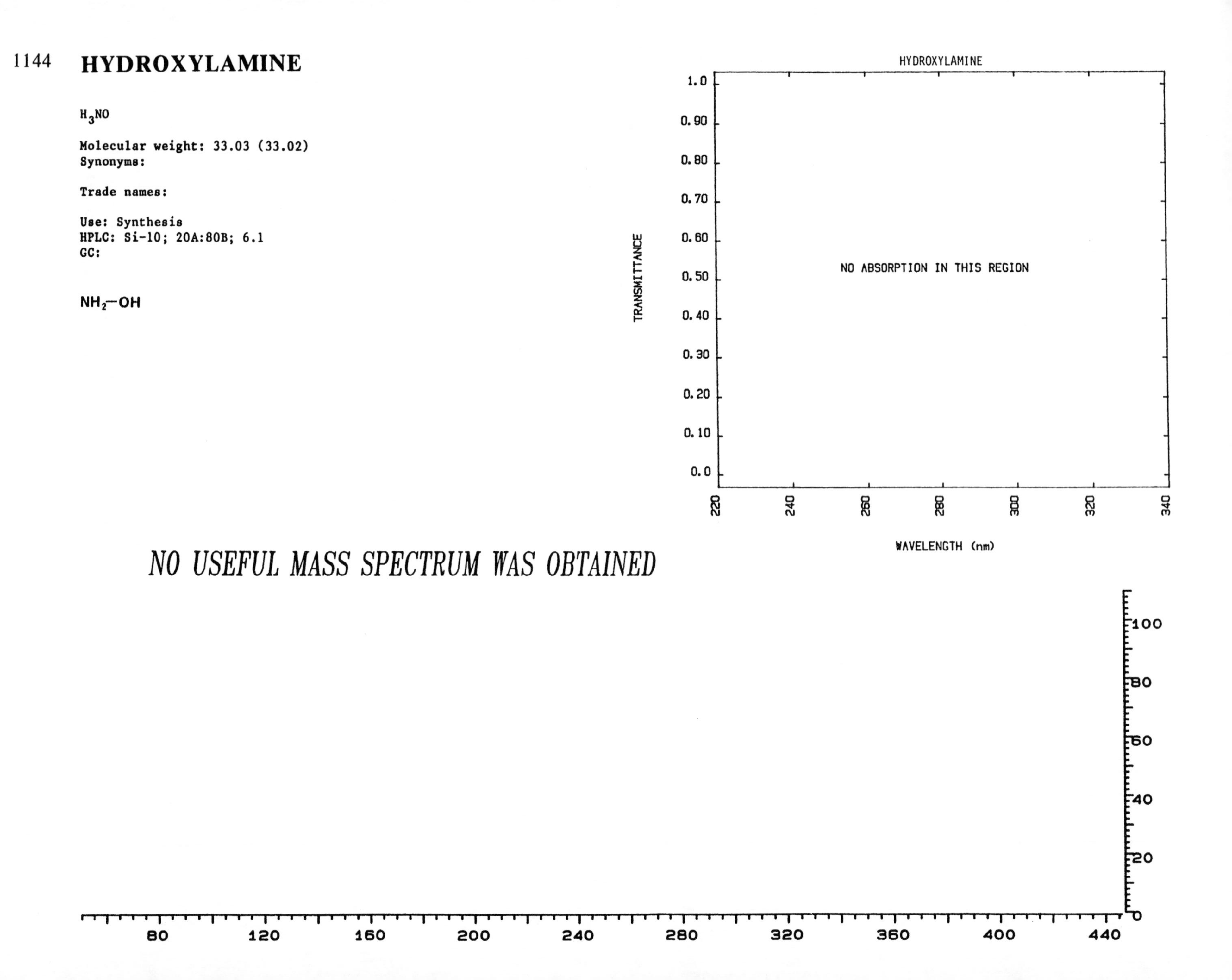

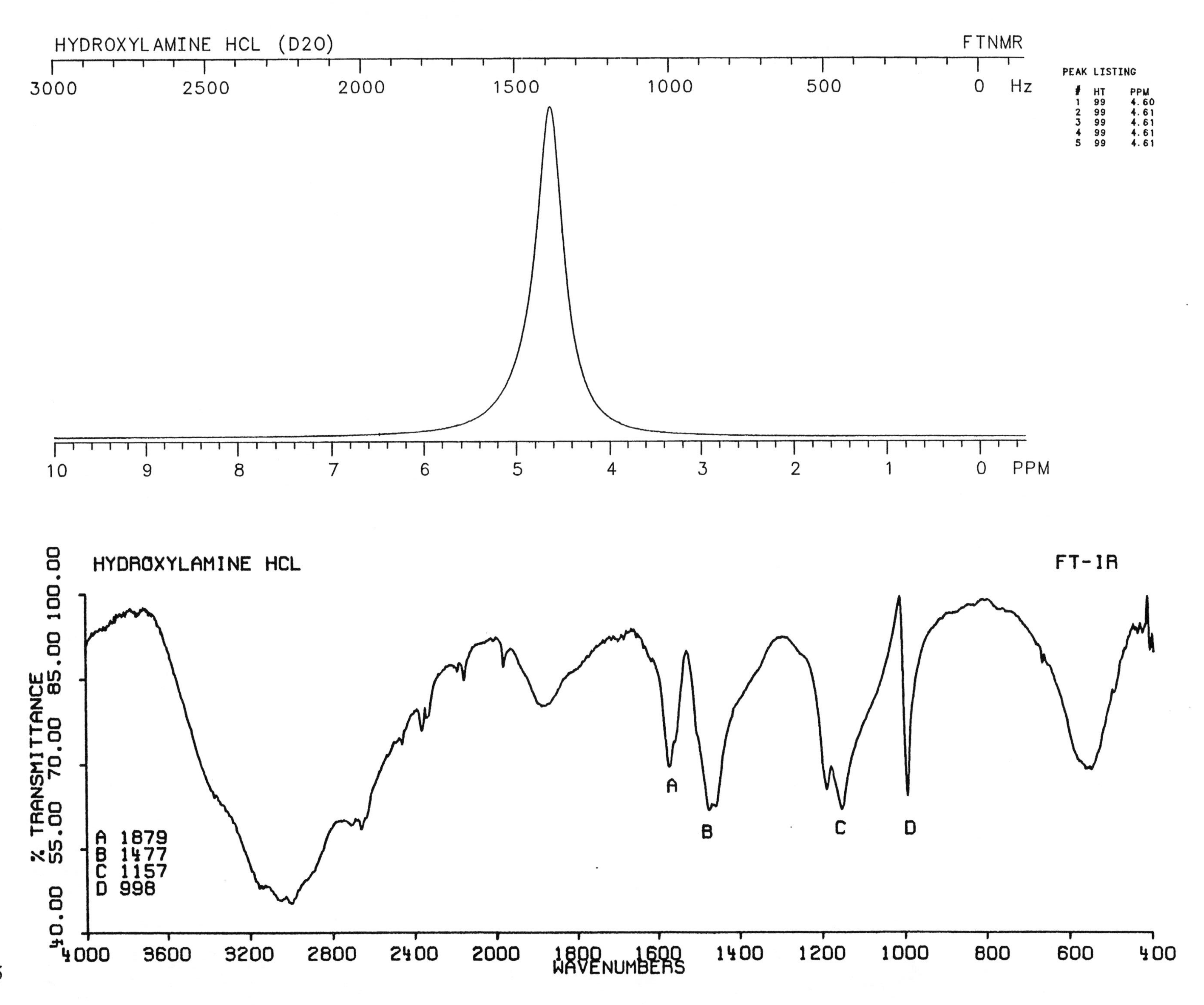
HYDROXYLAMINE HCL (D2O)
FTNMR
3000
2500
2000
1500
1000
500
0 Hz
PEAK LISTING
# HT PPM
1 99 4.60
2 99 4.61
3 99 4.61
4 99 4.61
5 99 4.61
10
9
8
7
6
5
4
3
2
1
0 PPM
HYDROXYLAMINE HCL
FT-IR
% TRANSMITTANCE
40.00
55.00
70.00
85.00
100.00
A 1879
B 1477
C 1157
D 998
A
B
C
D
4000
3600
3200
2800
2400
2000
1800
1600
1400
1200
1000
800
600
400
WAVENUMBERS

# 7- HYDROXYLOXAPINE

$C_{18}H_{18}ClN_3O_2$

**Molecular weight:** 343.80 (343.11)
**Synonyms:** 7-Hydroxy-2-chloro-11-(4-methyl-1-piperazinyl)-dibenz[b,f]-[1,4]-oxazepine
**Trade names:**

**Use:** Metabolite
**HPLC:**
**GC:** 3034; 280°C

7-HYDROXYLOXAPINE

TRANSMITTANCE

ETHANOL.....253,298

WAVELENGTH (nm)

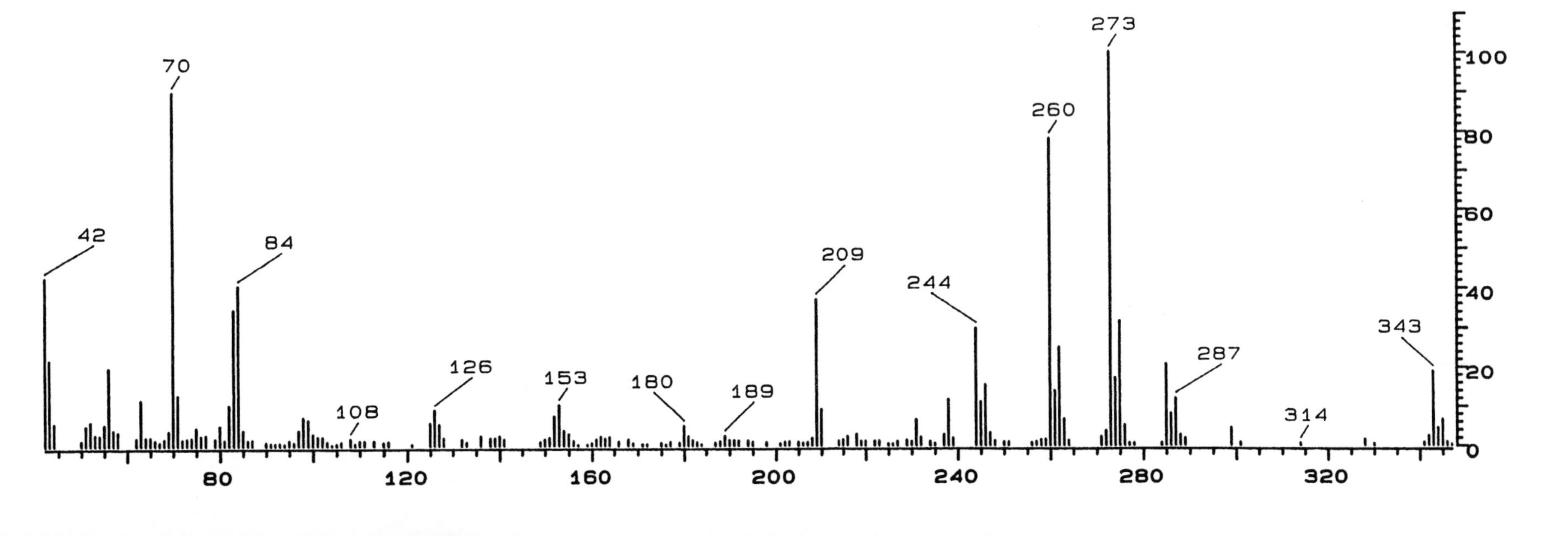

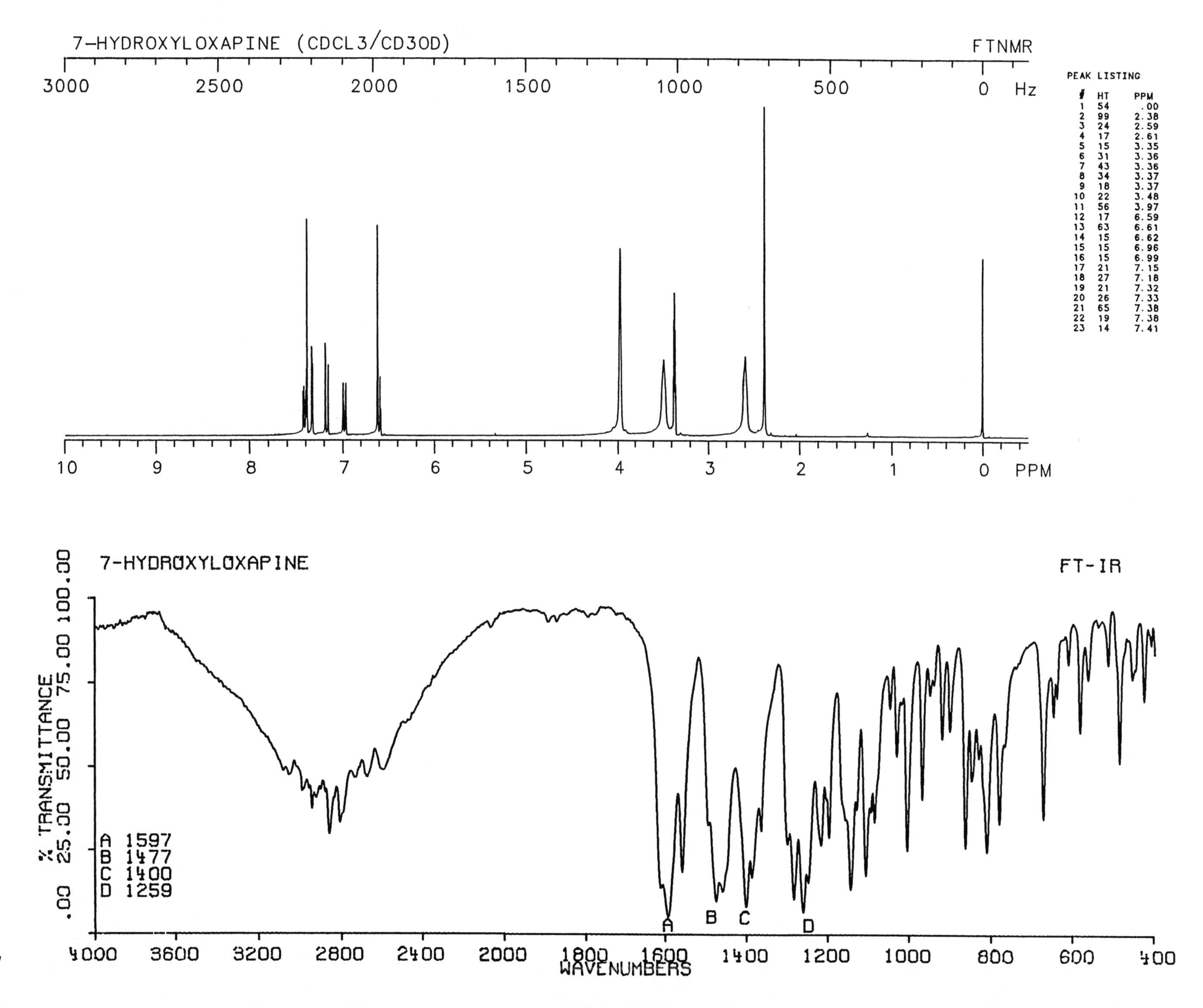
7-HYDROXYLOXAPINE (CDCL3/CD3OD)
FTNMR
3000 2500 2000 1500 1000 500 0 Hz
10 9 8 7 6 5 4 3 2 1 0 PPM
PEAK LISTING
# HT PPM
1 54 .00
2 99 2.38
3 24 2.59
4 17 2.61
5 15 3.35
6 31 3.36
7 43 3.36
8 34 3.37
9 18 3.37
10 22 3.48
11 56 3.97
12 17 6.59
13 63 6.61
14 15 6.62
15 15 6.96
16 15 6.99
17 21 7.15
18 27 7.18
19 21 7.32
20 26 7.33
21 65 7.38
22 19 7.38
23 14 7.41
7-HYDROXYLOXAPINE
FT-IR
% TRANSMITTANCE
.00 25.00 50.00 75.00 100.00
A 1597
B 1477
C 1400
D 1259
A B C D
4000 3600 3200 2800 2400 2000 1800 1600 1400 1200 1000 800 600 400
WAVENUMBERS

# 8-HYDROXYLOXAPINE

$C_{18}H_{18}ClN_3O_2$

Molecular weight: 343.80 (343.11)
Synonyms: 8-Hydroxy-2-chloro-11-(4-Methyl-1-piperazinyl)-dibenz[b,f]-[1,4]-oxazepine
Trade names:

Use: Metabolite
HPLC:
GC: 2979; 280°C

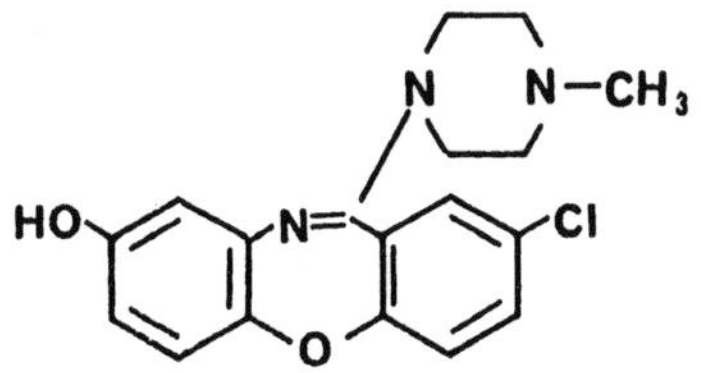

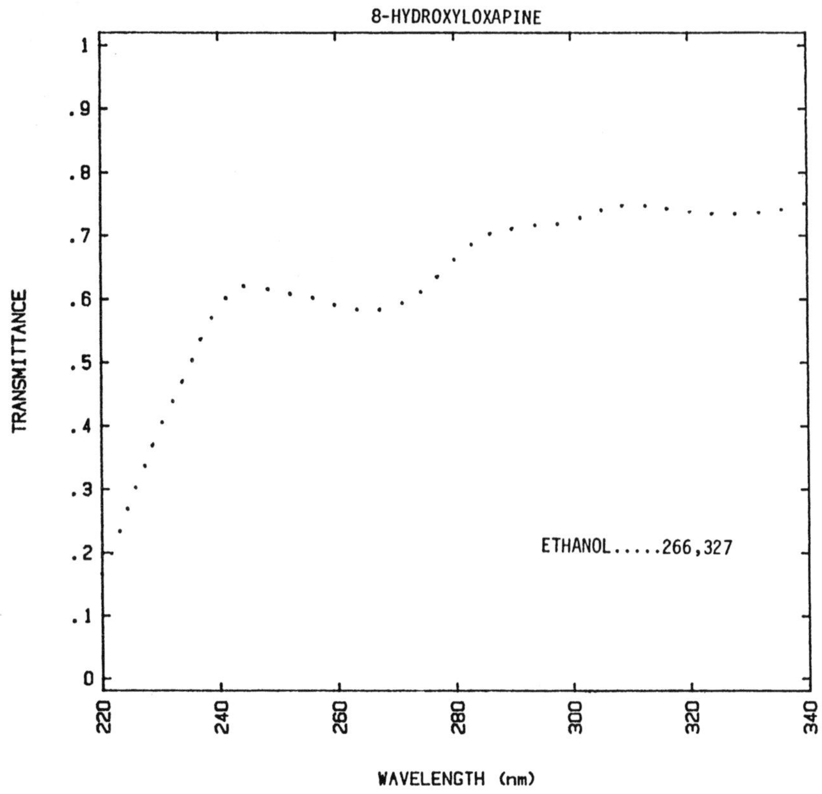

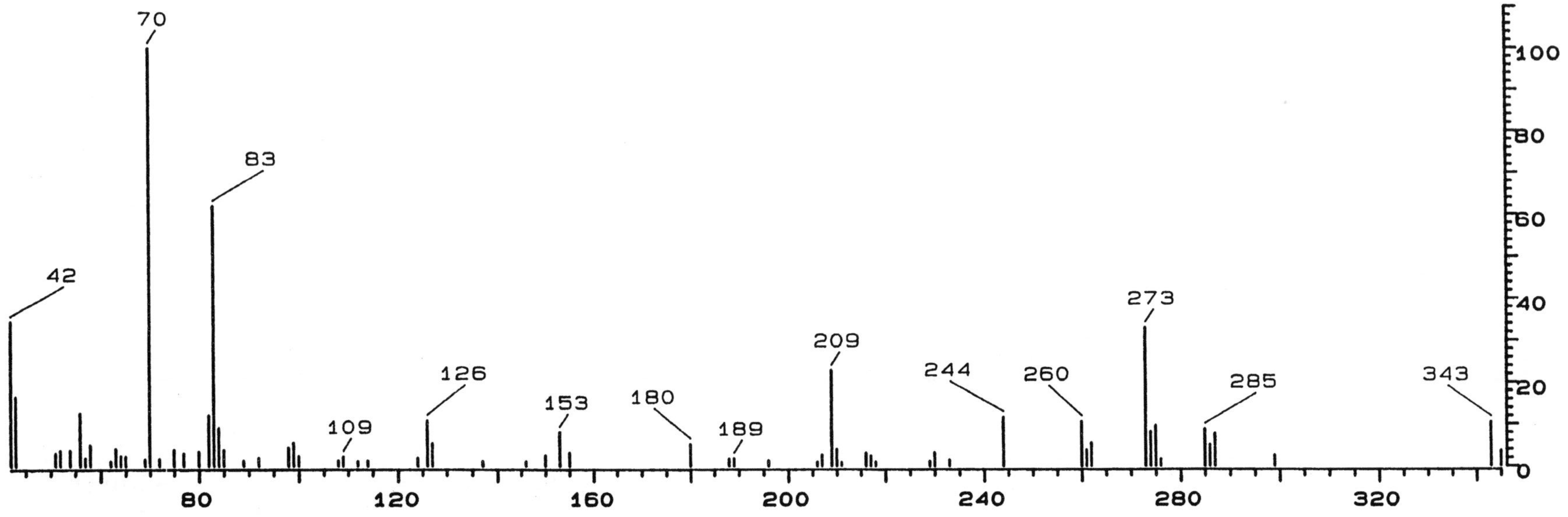

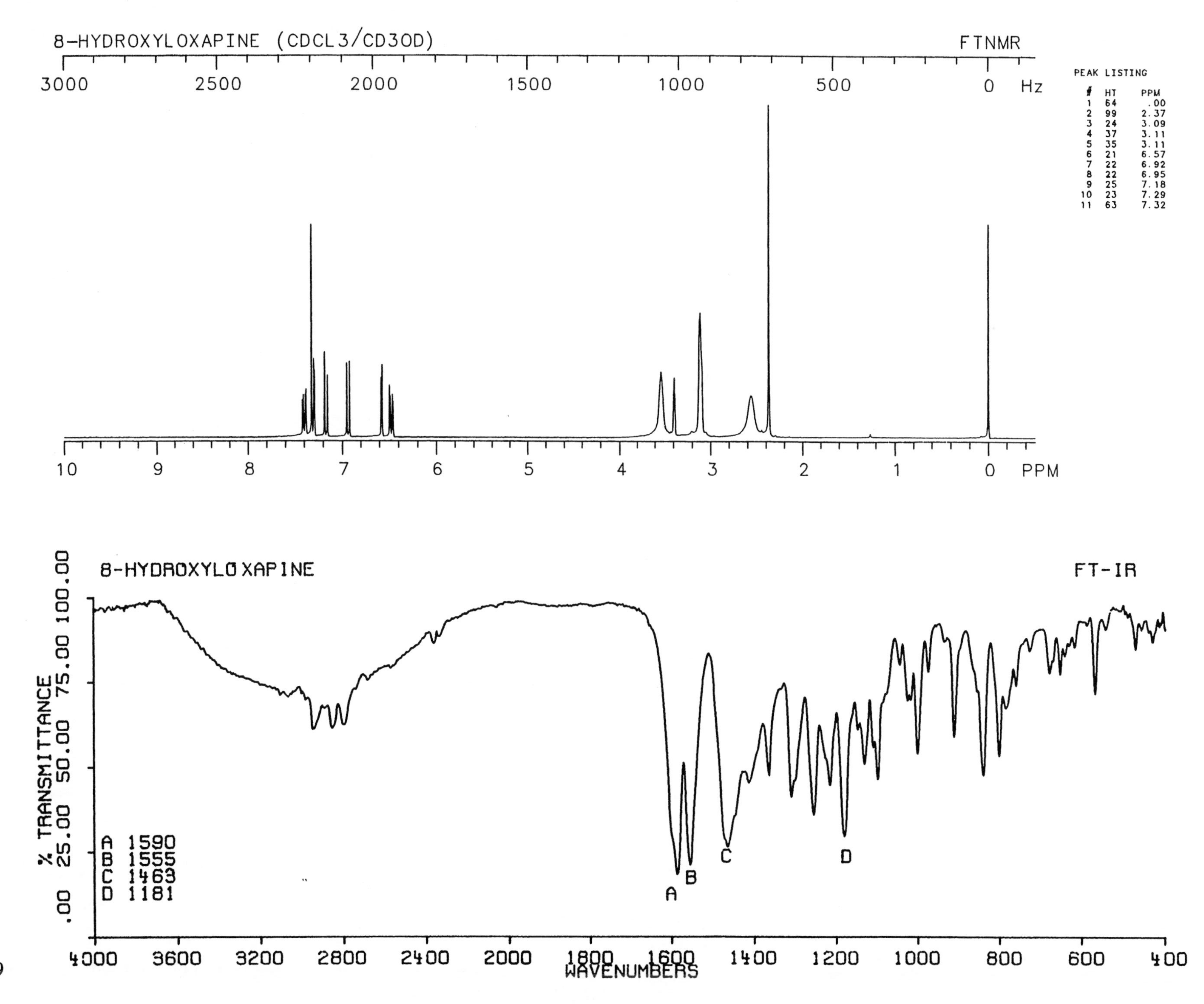
8-HYDROXYLOXAPINE (CDCL3/CD3OD)
FTNMR
Hz
PPM
PEAK LISTING
# HT PPM
1 64 .00
2 99 2.37
3 24 3.09
4 37 3.11
5 35 3.11
6 21 6.57
7 22 6.92
8 22 6.95
9 25 7.18
10 23 7.29
11 63 7.32
8-HYDROXYLOXAPINE
FT-IR
% TRANSMITTANCE
WAVENUMBERS
A 1590
B 1555
C 1463
D 1181

# HYDROXYMEPROBAMATE

$C_9H_{18}N_2O_5$

**Molecular weight:** 234.22 (234.12)
**Synonyms:** Hydroxy-2-methyl-2-propyl-1,3-propanediol dicarbamate

**Trade names:**

**Use:** Metabolite
**HPLC:**
**GC:** 1706; 200°C

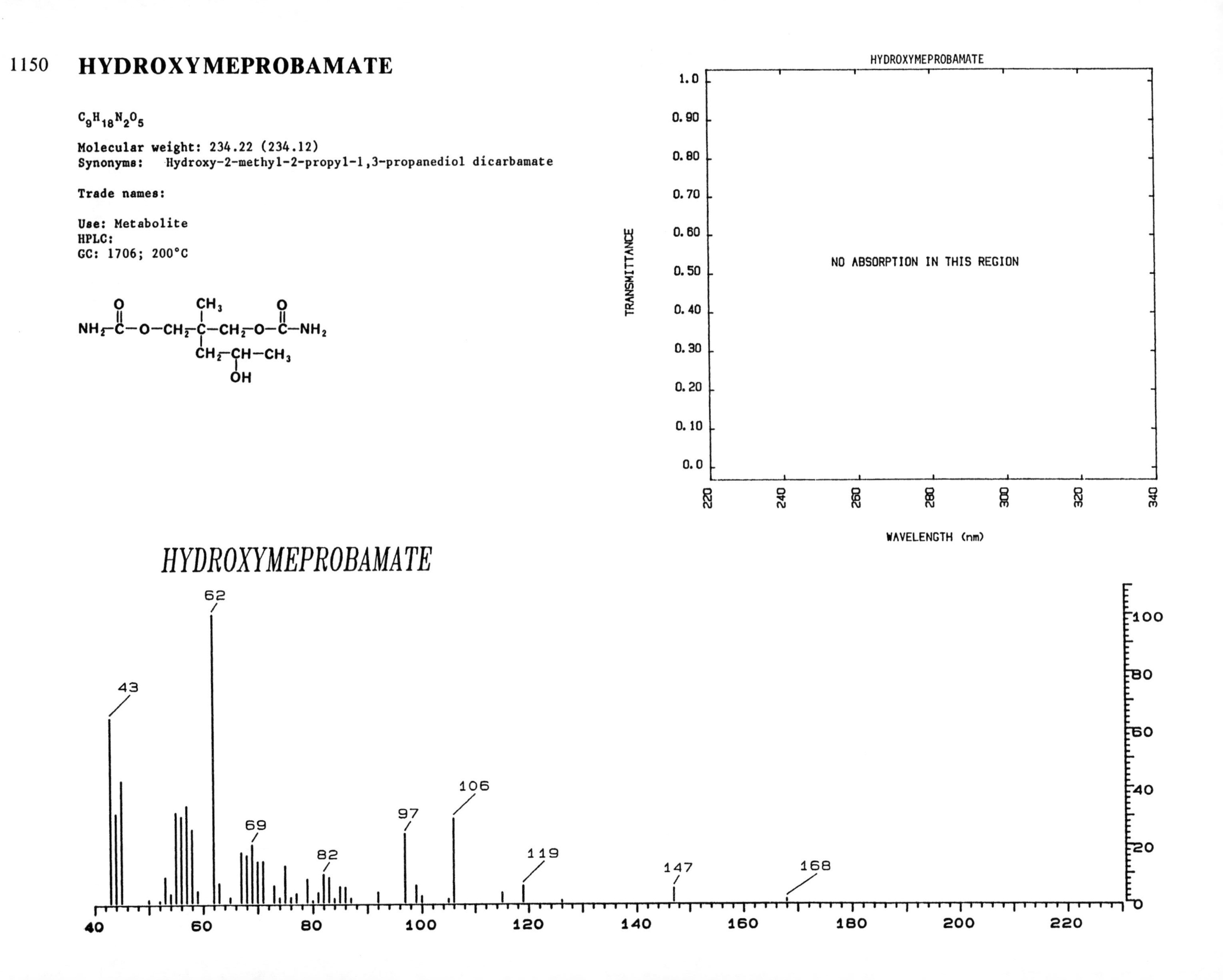

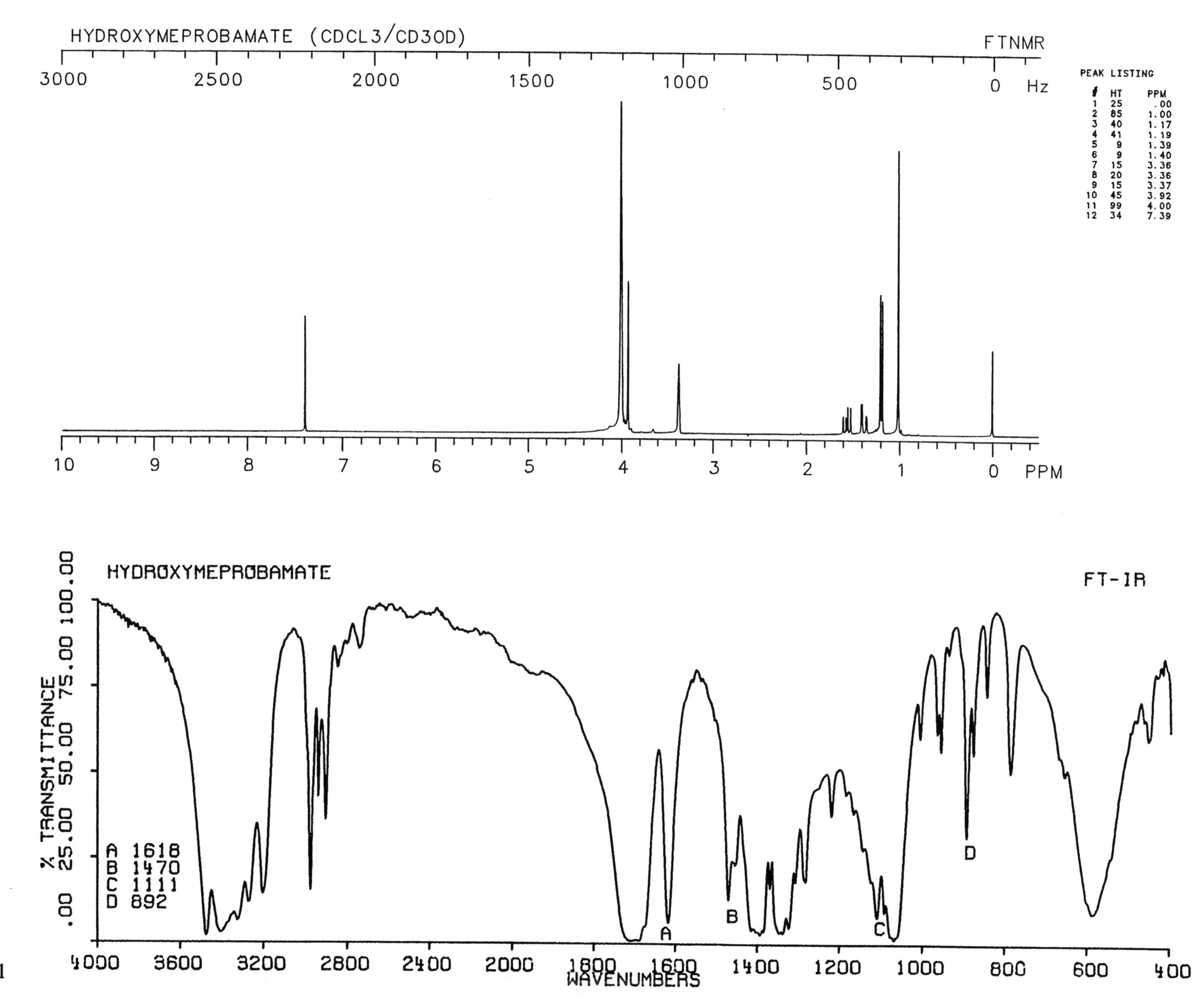
HYDROXYMEPROBAMATE (CDCL3/CD3OD)
FTNMR
3000 2500 2000 1500 1000 500 0 Hz
10 9 8 7 6 5 4 3 2 1 0 PPM
PEAK LISTING
# HT PPM
1 25 .00
2 85 1.00
3 40 1.17
4 41 1.19
5 9 1.39
6 9 1.40
7 15 3.36
8 20 3.36
9 15 3.37
10 45 3.92
11 99 4.00
12 34 7.39
HYDROXYMEPROBAMATE
FT-IR
% TRANSMITTANCE
.00 25.00 50.00 75.00 100.00
A 1618
B 1470
C 1111
D 892
A
B
C
D
4000 3600 3200 2800 2400 2000 1800 1600 1400 1200 1000 800 600 400
WAVENUMBERS

# HYDROXYMETHOXYMANDELIC ACID

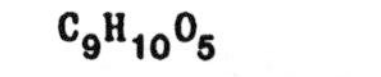

$C_9H_{10}O_5$

**Molecular weight:** 198.17 (198.05)
**Synonyms:** α,3-Dihydroxy-4-methoxybenzeneacetic acid; 3-hydroxy-4-methoxymandelic acid
**Trade names:**

**Use:** Metabolite
**HPLC:** Si-10; 20A:80B; 4.3
**GC:** 1518; 200°C

HYDROXY-4-METHOXYMANDELIC ACID

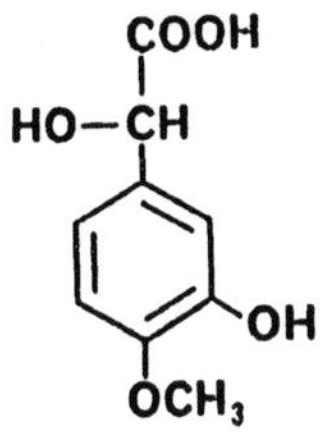

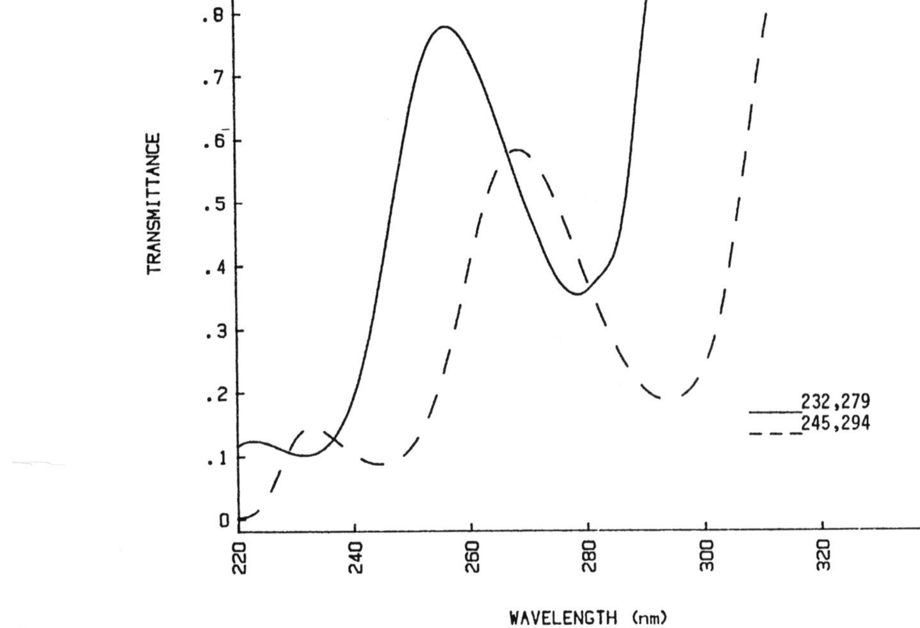

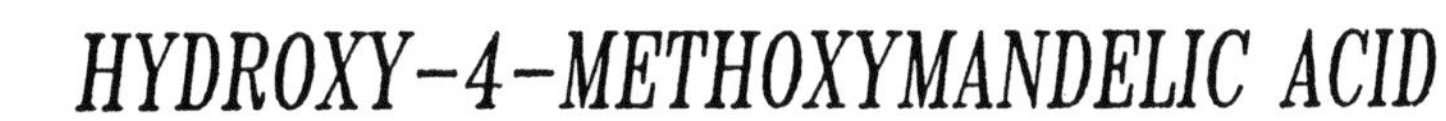

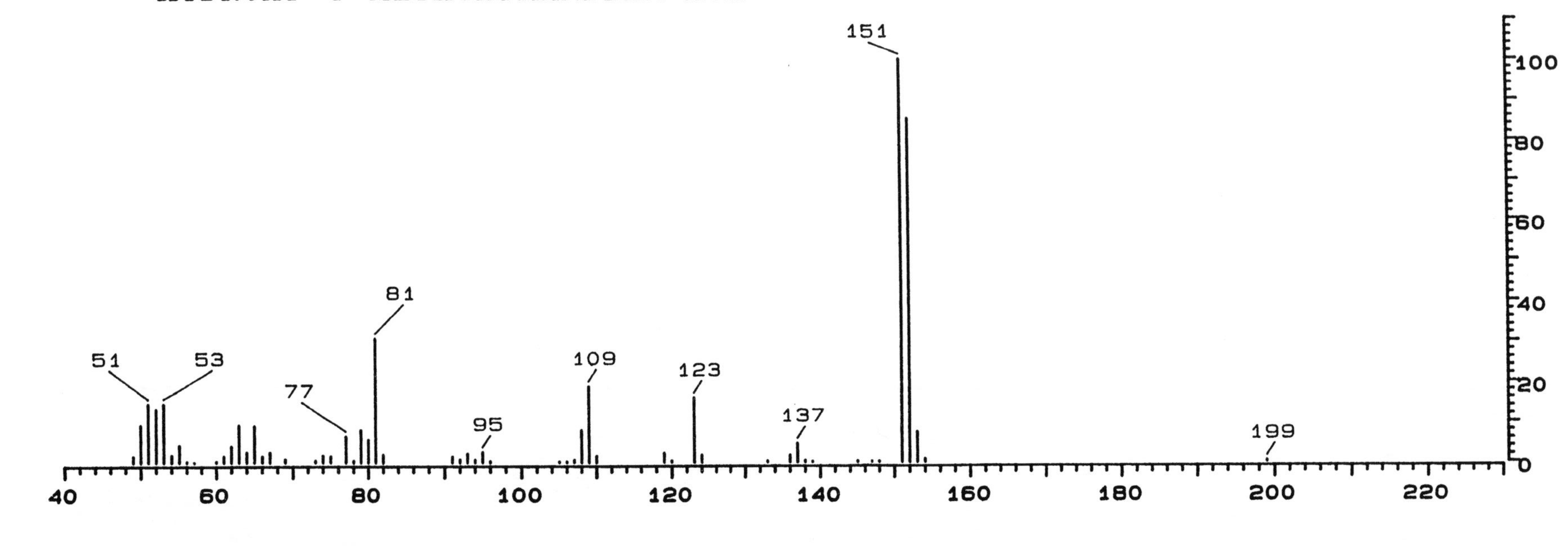

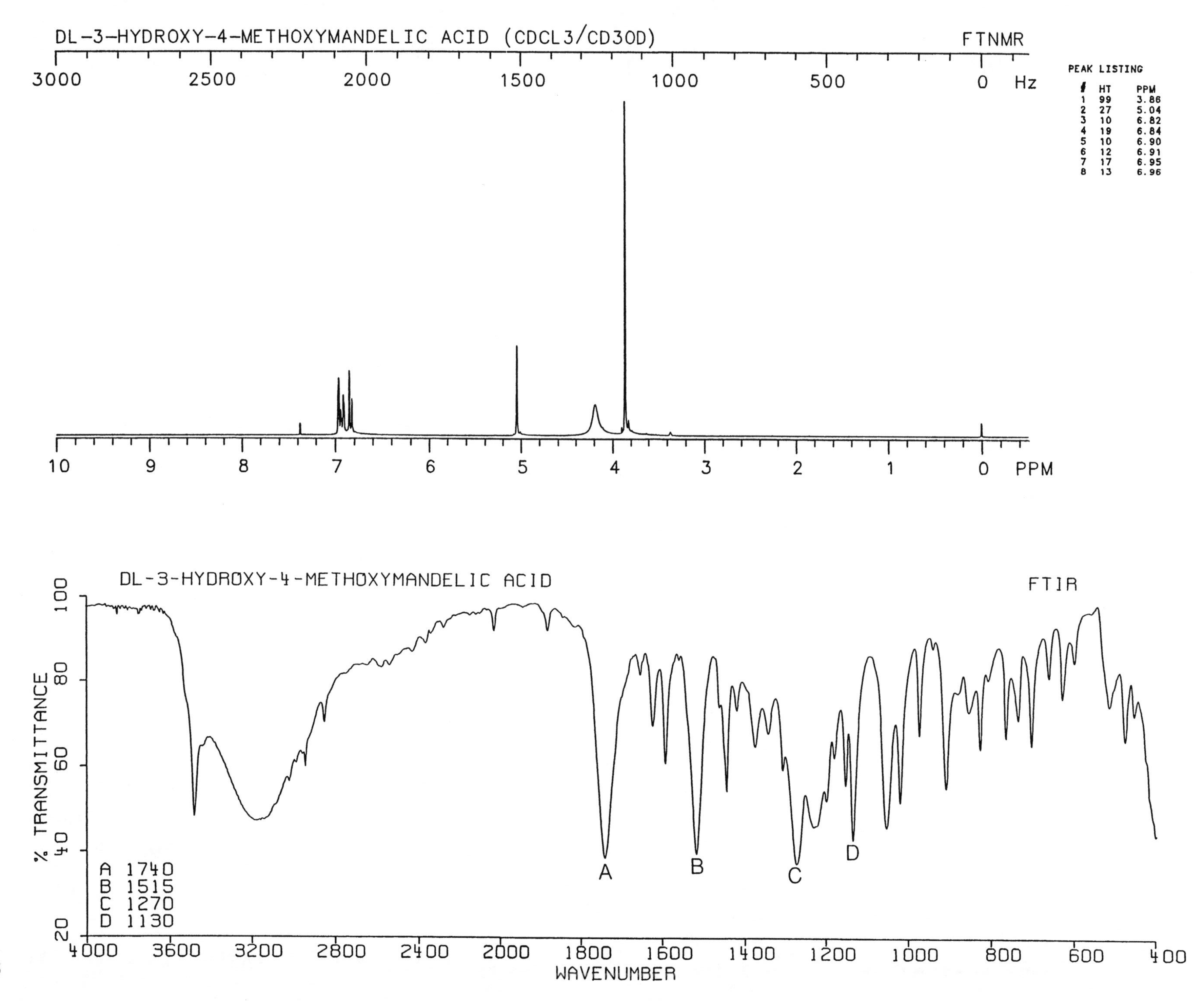
DL-3-HYDROXY-4-METHOXYMANDELIC ACID (CDCL3/CD3OD)
FTNMR
3000 2500 2000 1500 1000 500 0 Hz
10 9 8 7 6 5 4 3 2 1 0 PPM
PEAK LISTING
# HT PPM
1 99 3.86
2 27 5.04
3 10 6.82
4 19 6.84
5 10 6.90
6 12 6.91
7 17 6.95
8 13 6.96
DL-3-HYDROXY-4-METHOXYMANDELIC ACID
FTIR
% TRANSMITTANCE
100 80 60 40 20
A 1740
B 1515
C 1270
D 1130
A B C D
4000 3600 3200 2800 2400 2000 1800 1600 1400 1200 1000 800 600 400
WAVENUMBER

# HYDROXYMETHYLMETHAQUALONE

$C_{16}H_{14}N_2O_2$

**Molecular weight:** 266.29 (266.11)
**Synonyms:** 2-Hydroxymethyl-2-methyl-3-O-tolyl-4(3H)-quinazolinone

**Trade names:**

**Use:** Metabolite
**HPLC:** Si-10; 2A:98B; 3.1
**GC:** 2429; 250°C

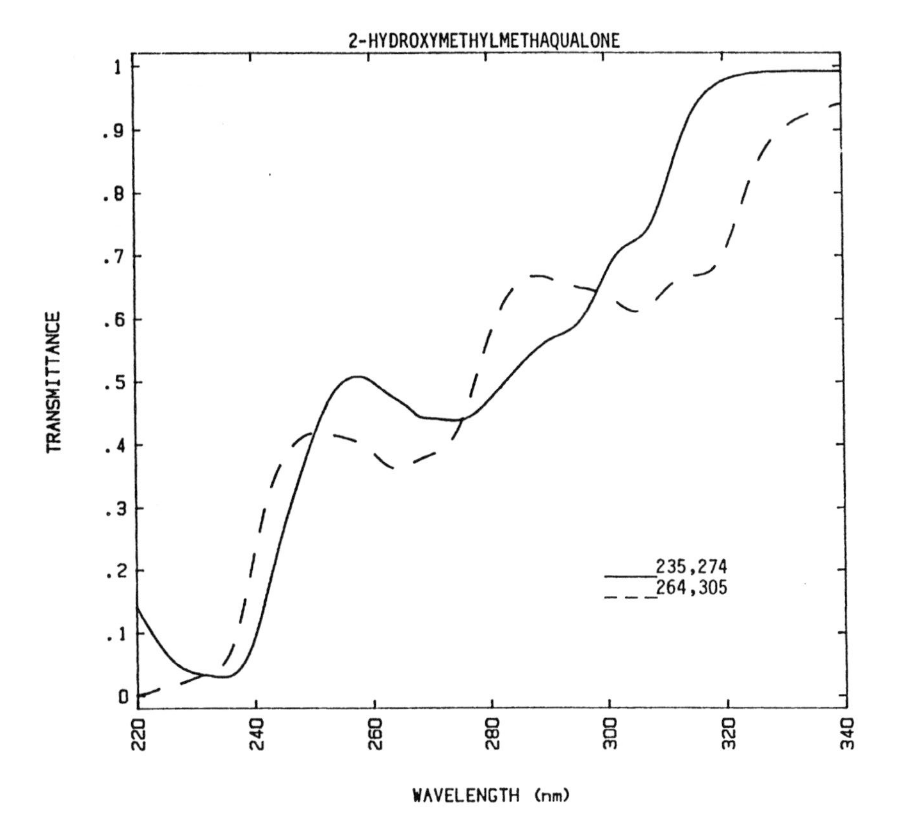

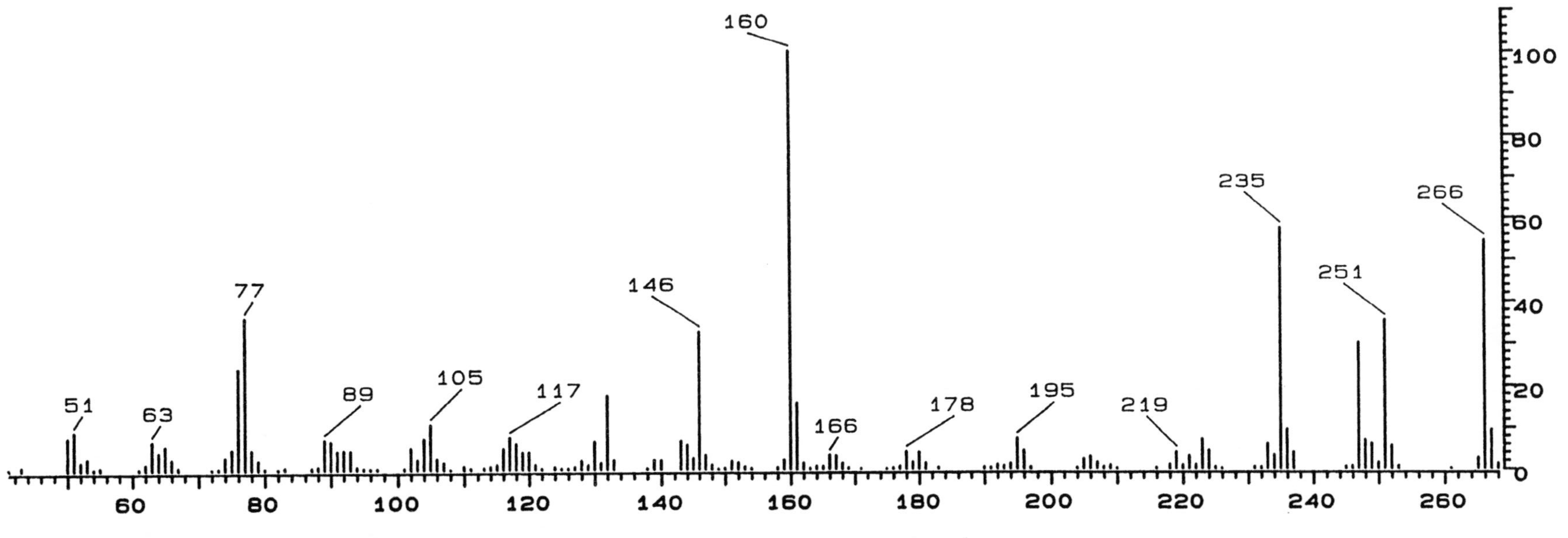

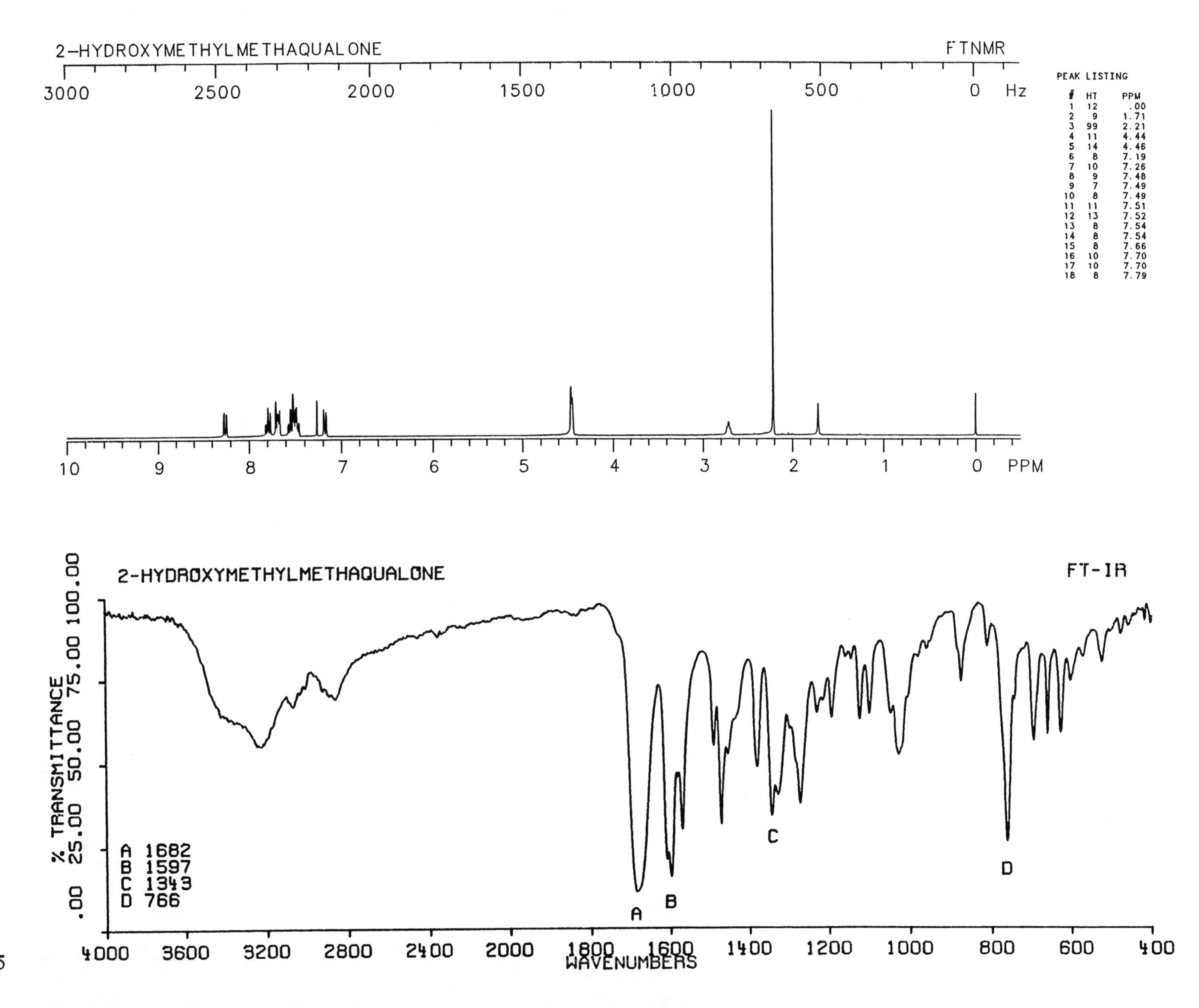

2-HYDROXYMETHYLMETHAQUALONE
FTNMR
3000 2500 2000 1500 1000 500 0 Hz
PEAK LISTING
# HT PPM
1 12 .00
2 9 1.71
3 99 2.21
4 11 4.44
5 14 4.46
6 8 7.19
7 10 7.26
8 9 7.48
9 7 7.49
10 8 7.49
11 11 7.51
12 13 7.52
13 8 7.54
14 8 7.54
15 8 7.66
16 10 7.70
17 10 7.70
18 8 7.79
10 9 8 7 6 5 4 3 2 1 0 PPM
2-HYDROXYMETHYLMETHAQUALONE
FT-IR
% TRANSMITTANCE
.00 25.00 50.00 75.00 100.00
A 1682
B 1597
C 1343
D 766
A B C D
4000 3600 3200 2800 2400 2000 1800 1600 1400 1200 1000 800 600 400
WAVENUMBERS

# HYDROXYMETHYLTRYPTAMINE

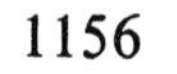

$C_{11}H_{14}N_2O$

**Molecular weight:** 190.25 (190.11)
**Synonyms:** 5-Hydroxymethyl-1H-indol-3-ethanamine

**Trade names:**

**Use:**
**HPLC:**
**GC:** 2170; 250°C

HO—$CH_2$ (indole) $CH_2$—$CH_2$—$NH_2 \cdot H_2C_2O_4$

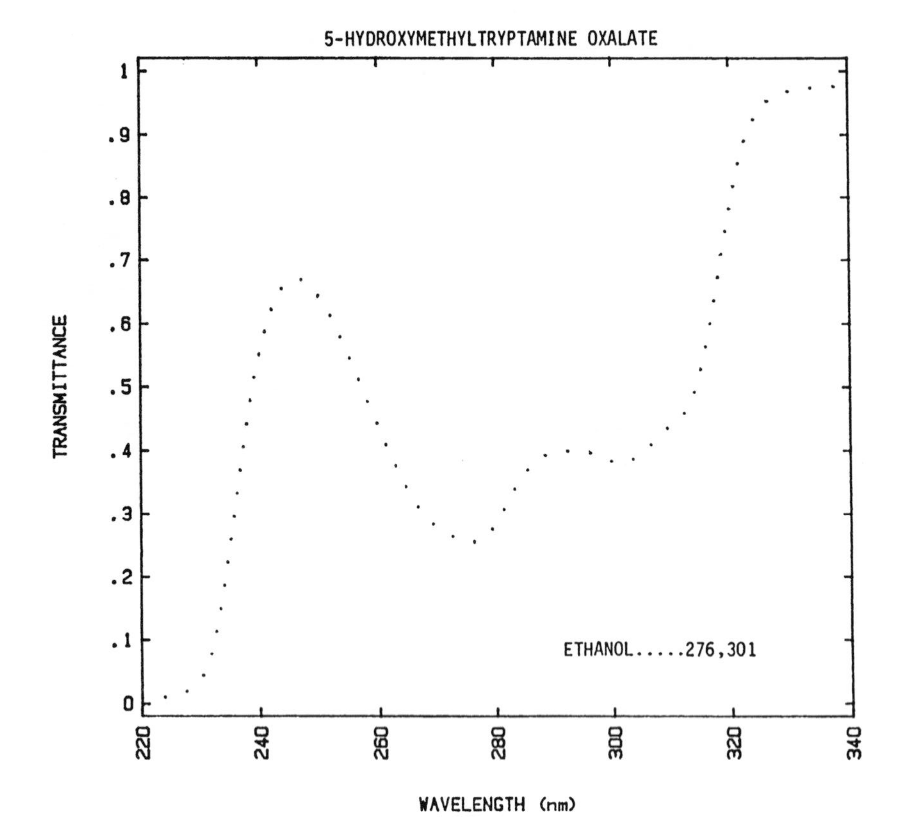


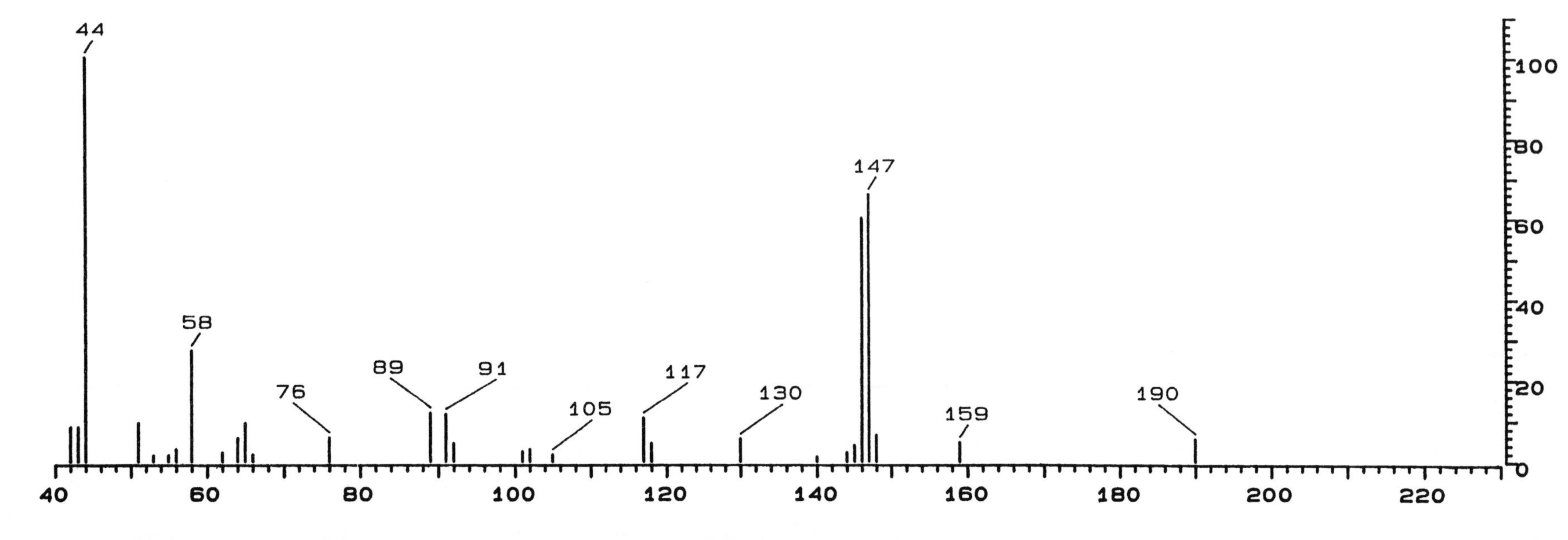

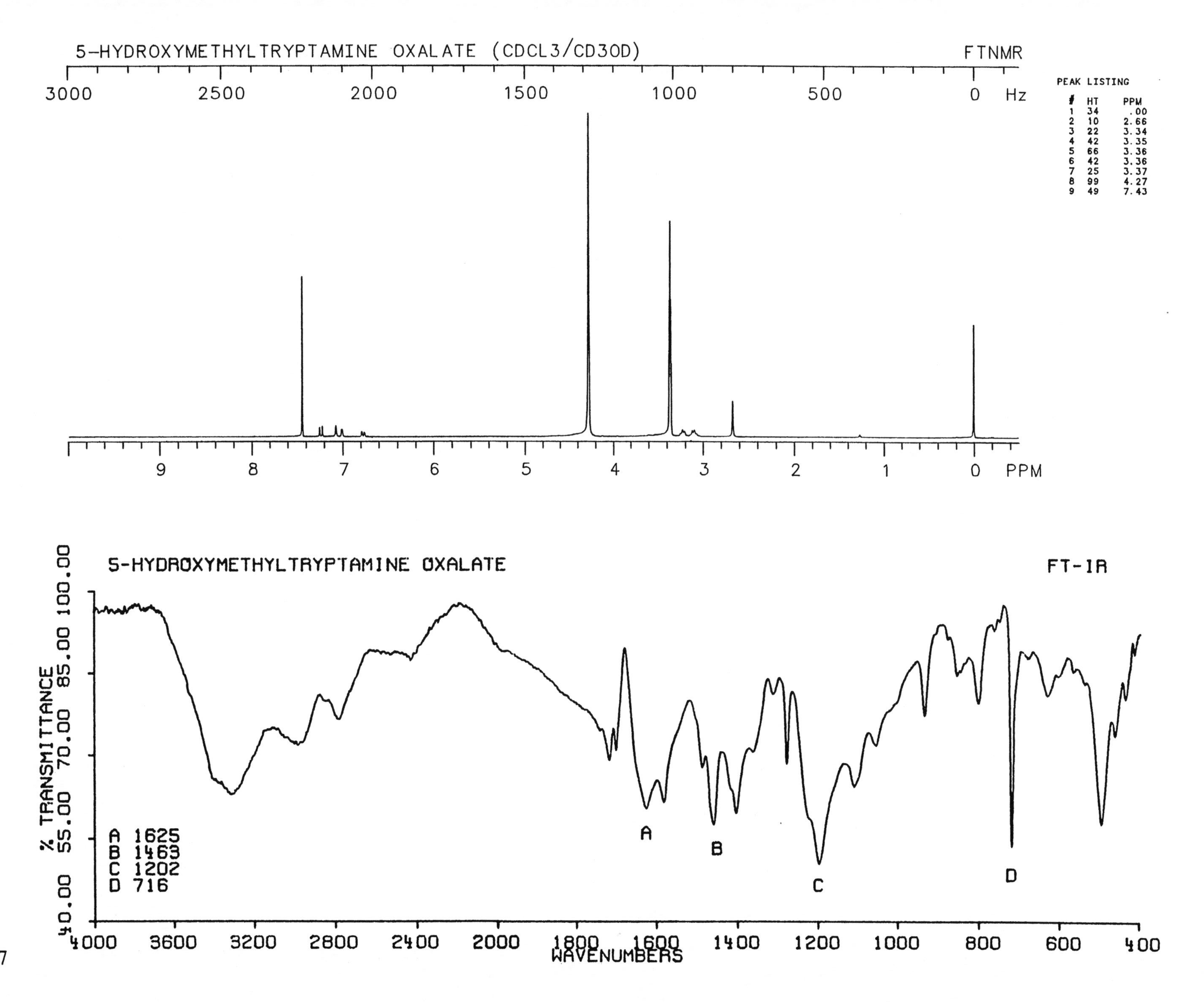

5-HYDROXYMETHYLTRYPTAMINE OXALATE (CDCL3/CD3OD)
FTNMR
3000 2500 2000 1500 1000 500 0 Hz
9 8 7 6 5 4 3 2 1 0 PPM
PEAK LISTING
# HT PPM
1 34 .00
2 10 2.66
3 22 3.34
4 42 3.35
5 66 3.36
6 42 3.36
7 25 3.37
8 99 4.27
9 49 7.43
5-HYDROXYMETHYLTRYPTAMINE OXALATE
FT-IR
% TRANSMITTANCE
40.00 55.00 70.00 85.00 100.00
A 1625
B 1463
C 1202
D 716
A
B
C
D
4000 3600 3200 2800 2400 2000 1800 1600 1400 1200 1000 800 600 400
WAVENUMBERS

# HYDROXY-1,4-NAPHTHOQUINONE

$C_{10}H_6O_3$

**Molecular weight:** 174.15 (174.03)
**Synonyms:** 2-Hydroxy-1,4-naphthalenedone; Lawsone

**Trade names:**

**Use:** Sunscreen agent
**HPLC:** Si-10; 10A:90B; 8.8
**GC:** 1526; 200°C

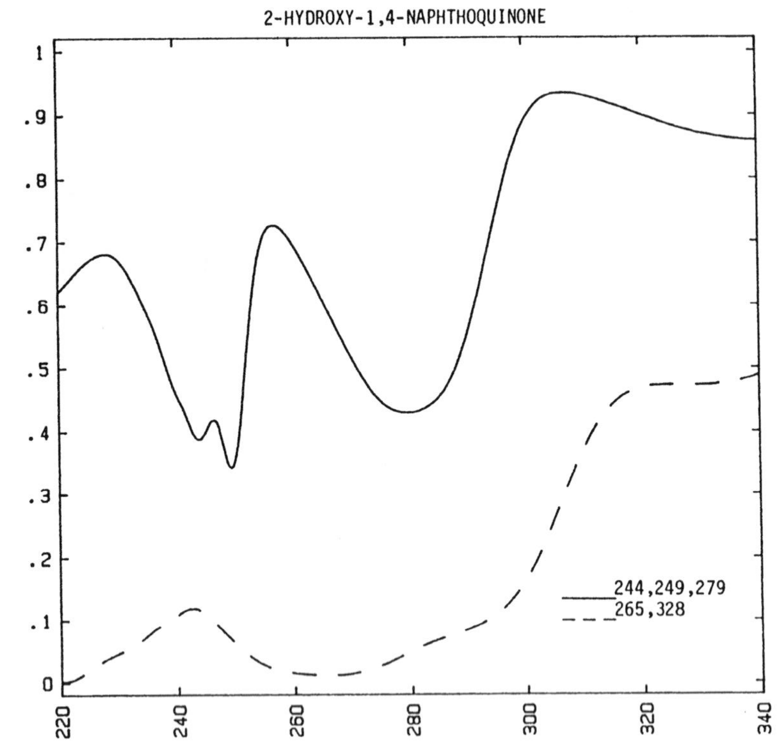
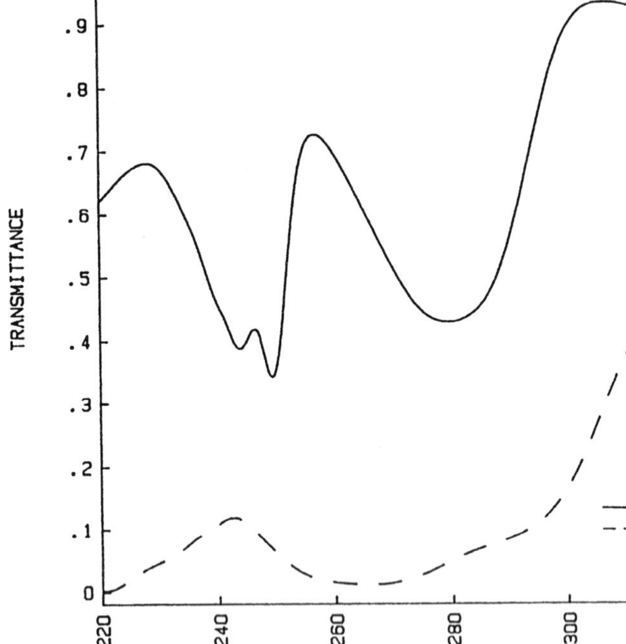


*HYDROXYNAPHTHOQUINONE*

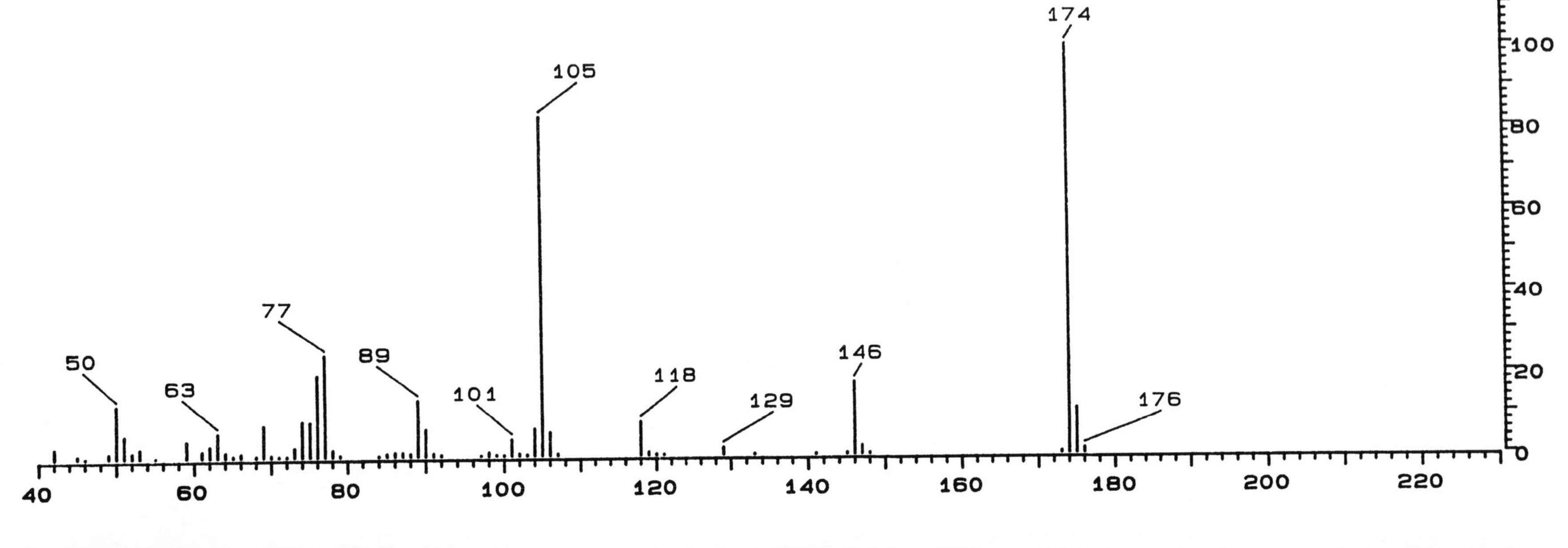

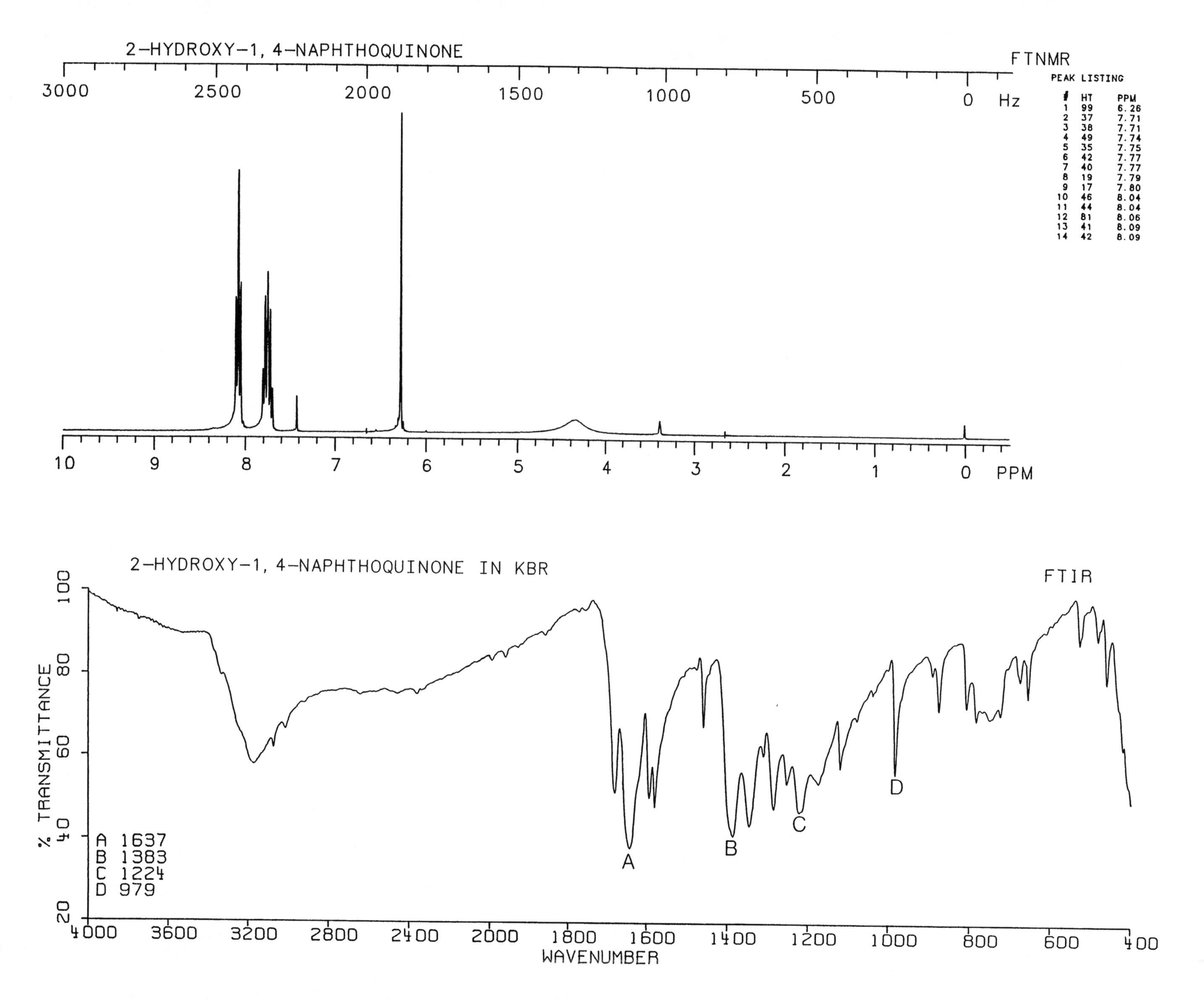
2-HYDROXY-1, 4-NAPHTHOQUINONE
FTNMR
PEAK LISTING
# HT PPM
1 99 6.26
2 37 7.71
3 38 7.71
4 49 7.74
5 35 7.75
6 42 7.77
7 40 7.77
8 19 7.79
9 17 7.80
10 46 8.04
11 44 8.04
12 81 8.06
13 41 8.09
14 42 8.09
3000 2500 2000 1500 1000 500 0 Hz
10 9 8 7 6 5 4 3 2 1 0 PPM
2-HYDROXY-1, 4-NAPHTHOQUINONE IN KBR
FTIR
% TRANSMITTANCE
100 80 60 40 20
A 1637
B 1383
C 1224
D 979
A B C D
4000 3600 3200 2800 2400 2000 1800 1600 1400 1200 1000 800 600 400
WAVENUMBER

# HYDROXYNICOTINIC ACID

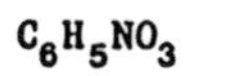

$C_6H_5NO_3$

**Molecular weight:** 139.09 (139.03)
**Synonyms:** 6-Hydroxy-3-pyridinecarboxylic acid

**Trade names:**

**Use:** Metabolite
**HPLC:** Si-10; 20A:80B; 5.1
**GC:**

HO, N, COOH

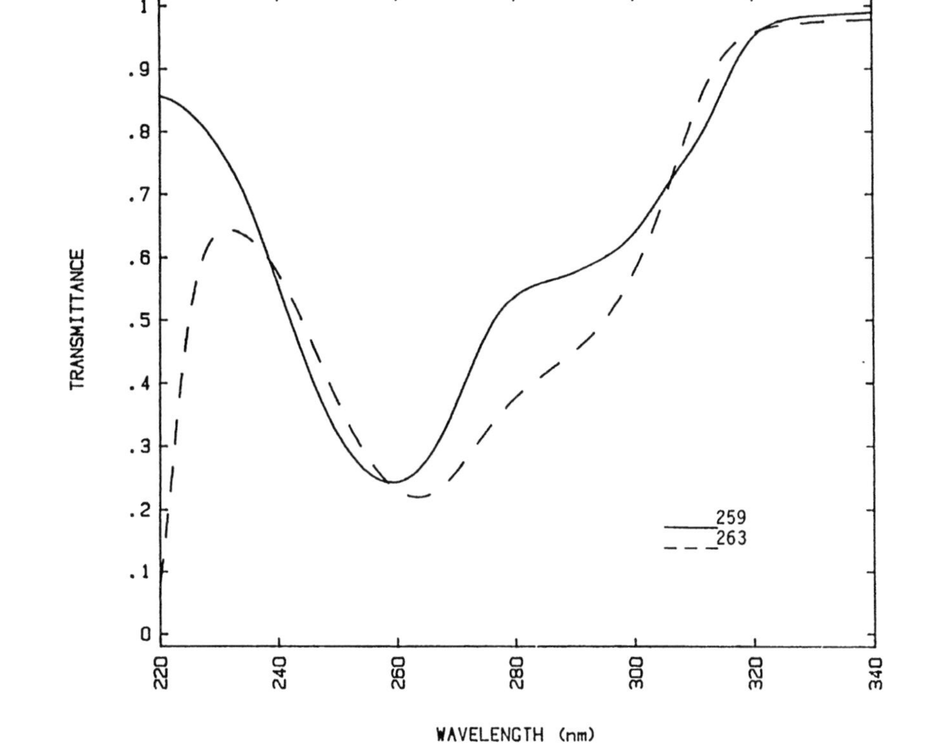

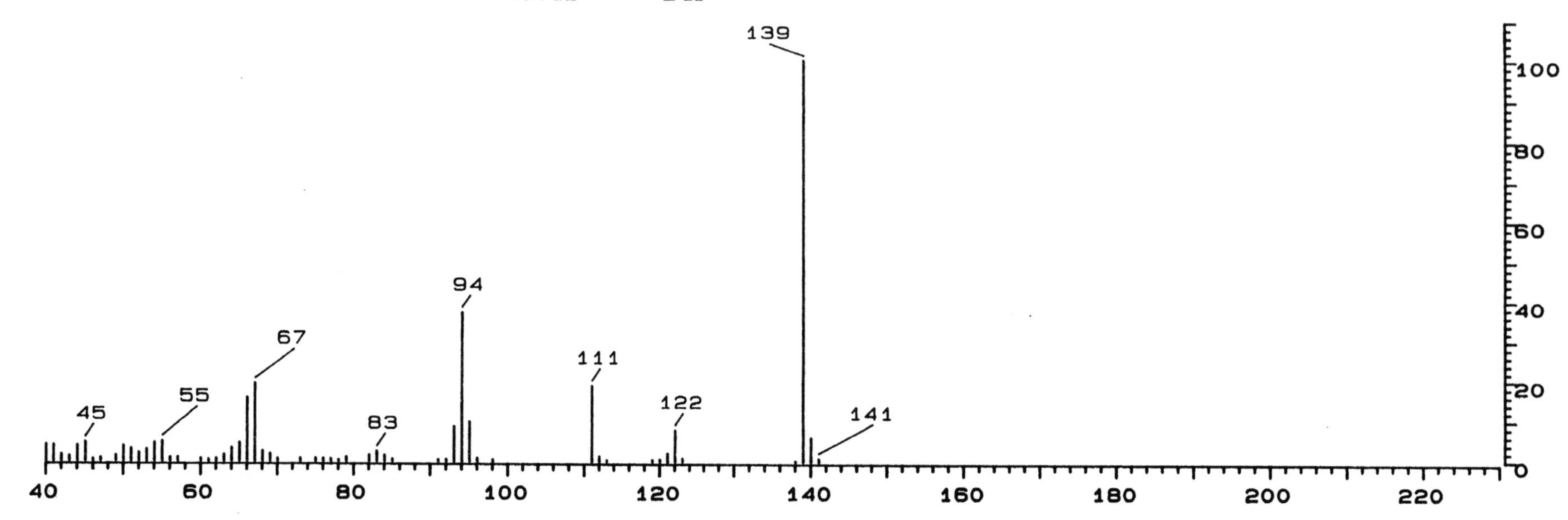

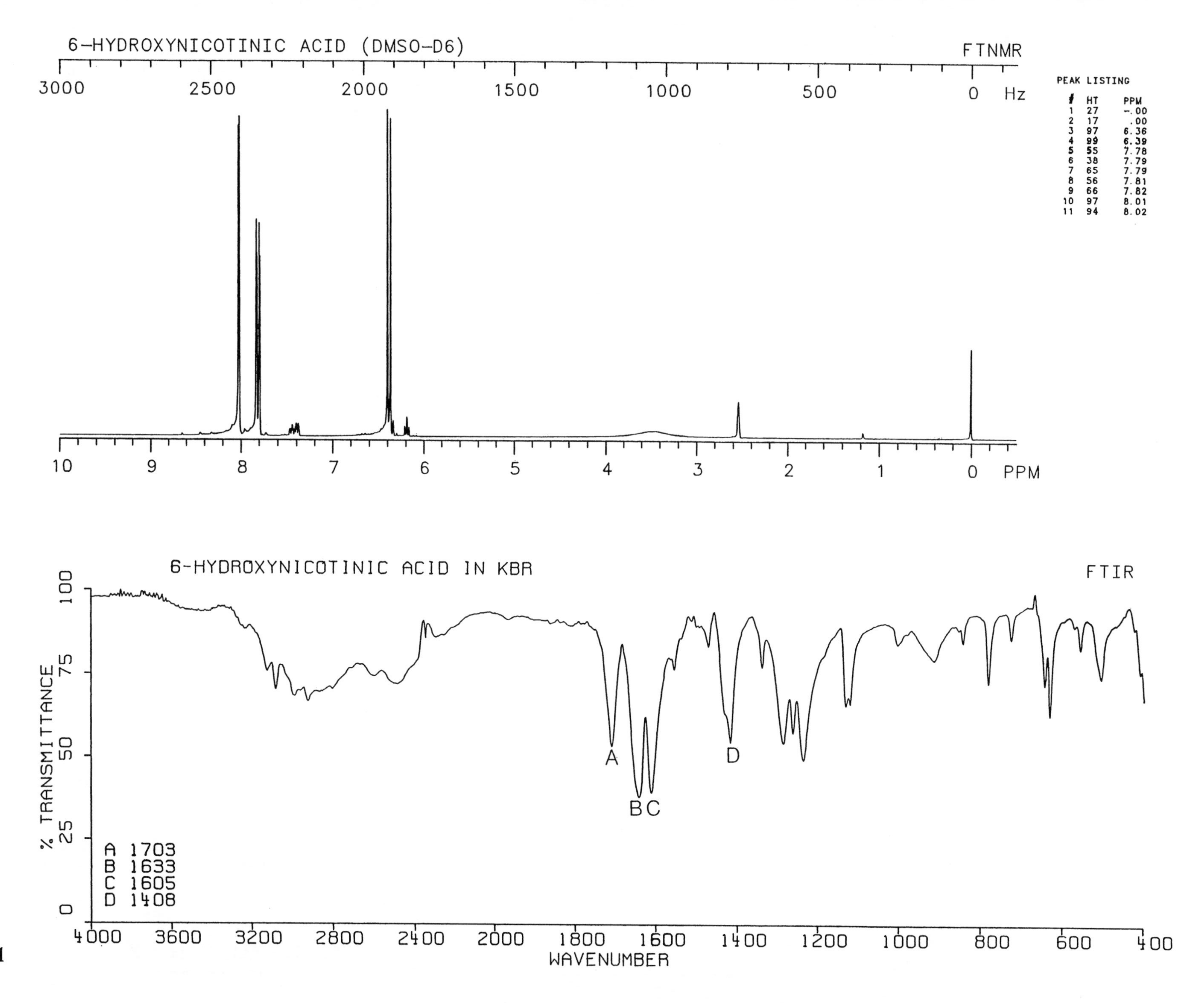
6-HYDROXYNICOTINIC ACID (DMSO-D6)
FTNMR
3000 2500 2000 1500 1000 500 0 Hz
PEAK LISTING
# HT PPM
1 27 -.00
2 17 .00
3 97 6.36
4 99 6.39
5 55 7.78
6 38 7.79
7 65 7.79
8 56 7.81
9 66 7.82
10 97 8.01
11 94 8.02
10 9 8 7 6 5 4 3 2 1 0 PPM
6-HYDROXYNICOTINIC ACID IN KBR
FTIR
% TRANSMITTANCE
100 75 50 25 0
A B C D
A 1703
B 1633
C 1605
D 1408
4000 3600 3200 2800 2400 2000 1800 1600 1400 1200 1000 800 600 400
WAVENUMBER

# HYDROXYPHENOBARBITAL

$C_{12}H_{12}N_2O_4$

**Molecular weight:** 248.22 (248.08)
**Synonyms:** 4-Hydroxy-5-ethyl-5-phenyl-2,3,6(1H,3H,5H)-pyrimidinetrione; 5-ethyl-5-(p-hydroxyphenyl)barbituric acid
**Trade names:**

**Use:** Urinary metabolite
**HPLC:** Si-10; 5A:95B; 4.3
**GC:** 2410; 250°C

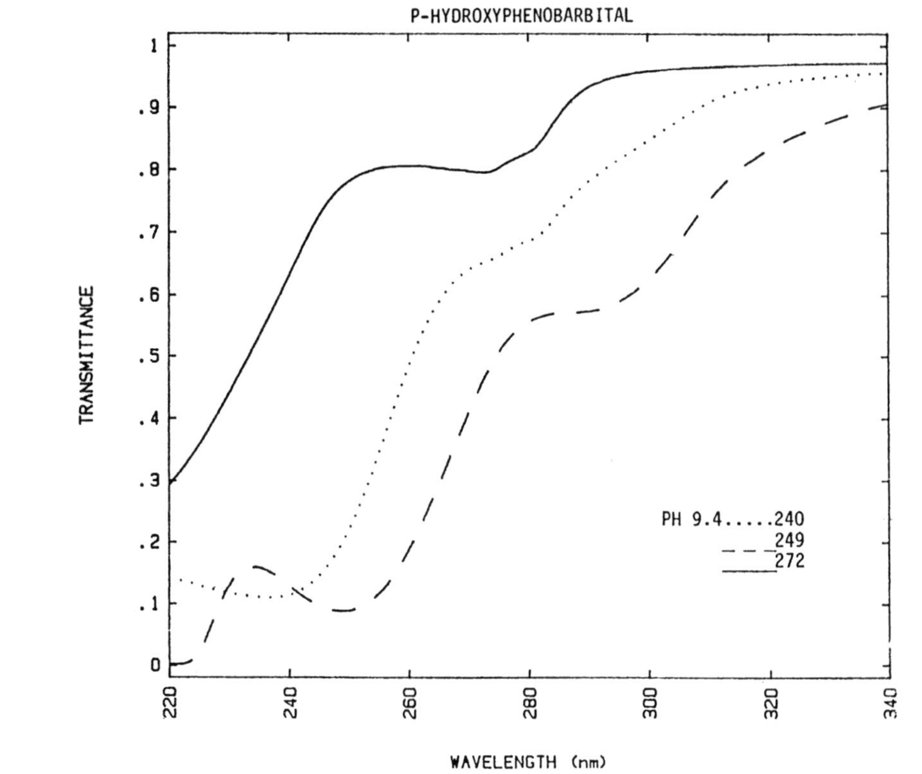

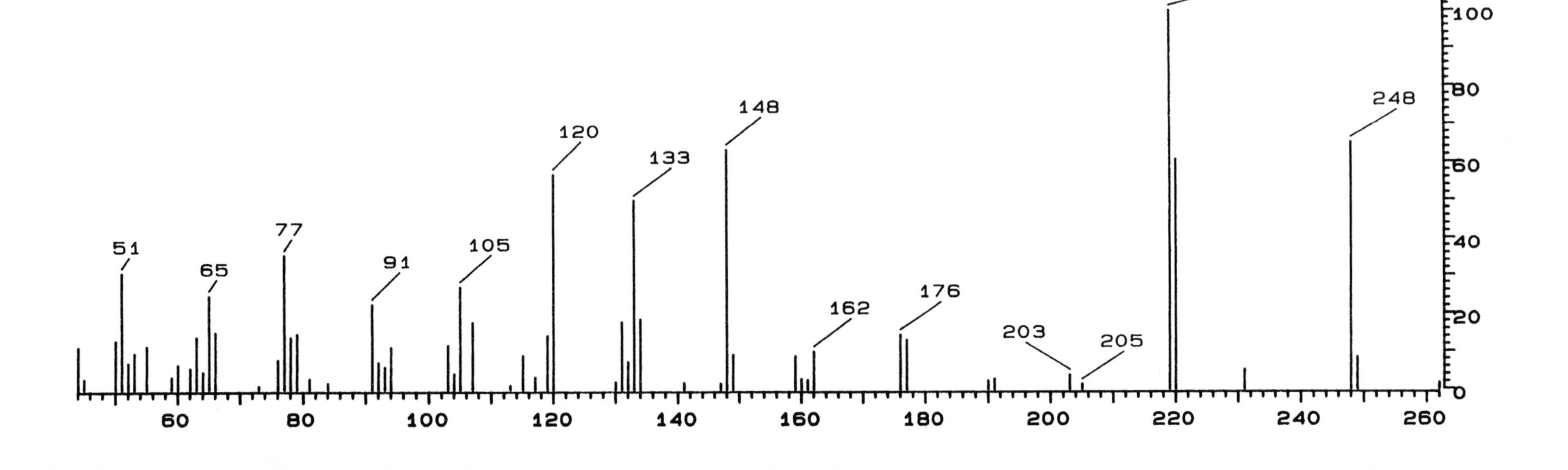

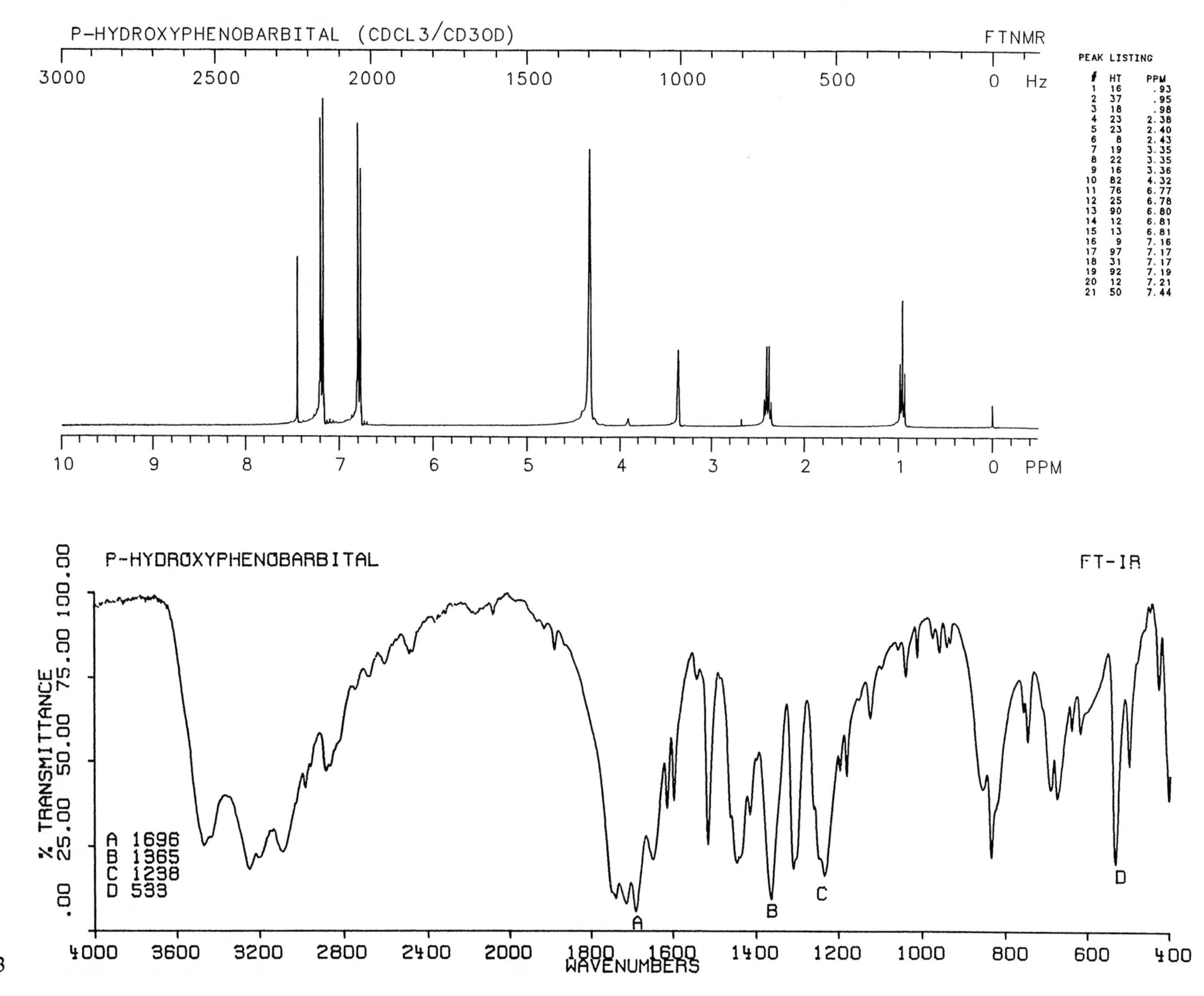

P-HYDROXYPHENOBARBITAL (CDCL3/CD3OD)
FTNMR
3000
2500
2000
1500
1000
500
0 Hz
10
9
8
7
6
5
4
3
2
1
0 PPM
PEAK LISTING
# HT PPM
1 16 .93
2 37 .95
3 18 .98
4 23 2.38
5 23 2.40
6 8 2.43
7 19 3.35
8 22 3.35
9 16 3.36
10 82 4.32
11 76 6.77
12 25 6.78
13 90 6.80
14 12 6.81
15 13 6.81
16 9 7.16
17 97 7.17
18 31 7.17
19 92 7.19
20 12 7.21
21 50 7.44
P-HYDROXYPHENOBARBITAL
FT-IR
% TRANSMITTANCE
.00
25.00
50.00
75.00
100.00
A 1696
B 1365
C 1238
D 533
A
B
C
D
4000
3600
3200
2800
2400
2000
1800
1600
1400
1200
1000
800
600
400
WAVENUMBERS

# 5-M-HYDROXYPHENYL-5-PHENYLHYDANTOIN

$C_{15}H_{12}N_2O_3$

Molecular weight: 268.28 (268.09)
Synonyms: 5-(3-Hydroxyphenyl)-5-phenyl-2,4-imidazolidinedione

Trade names:

Use:
HPLC:
GC: 2915; 280°C

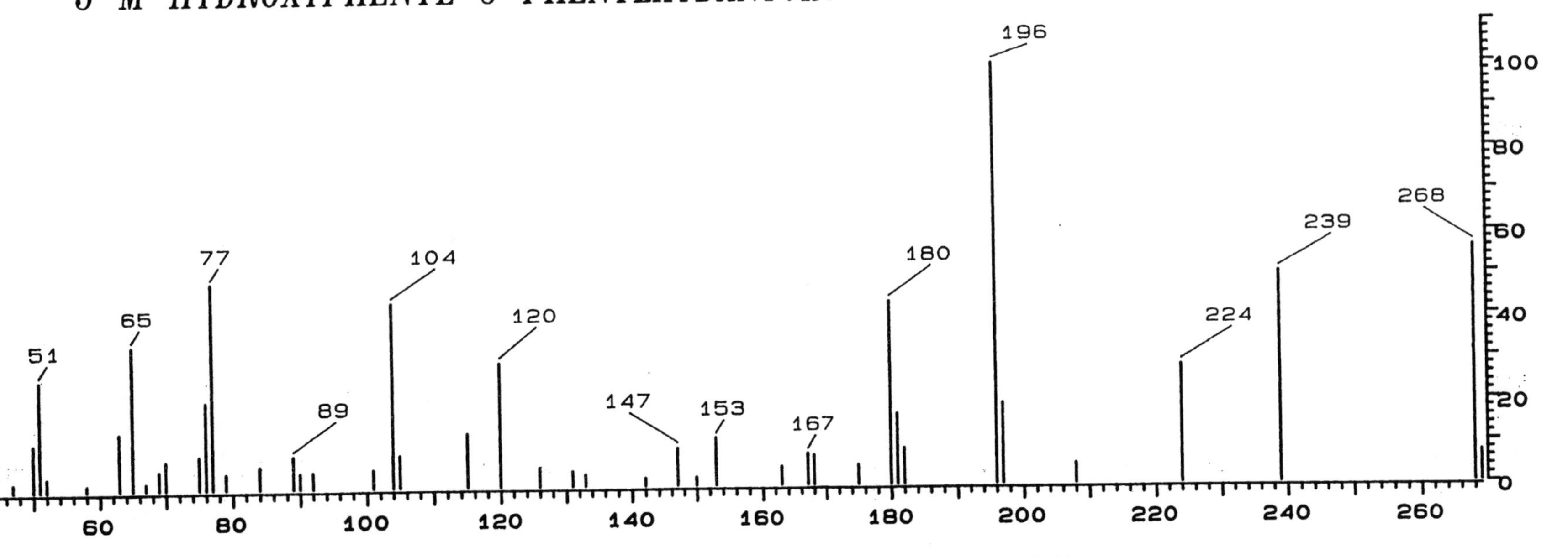

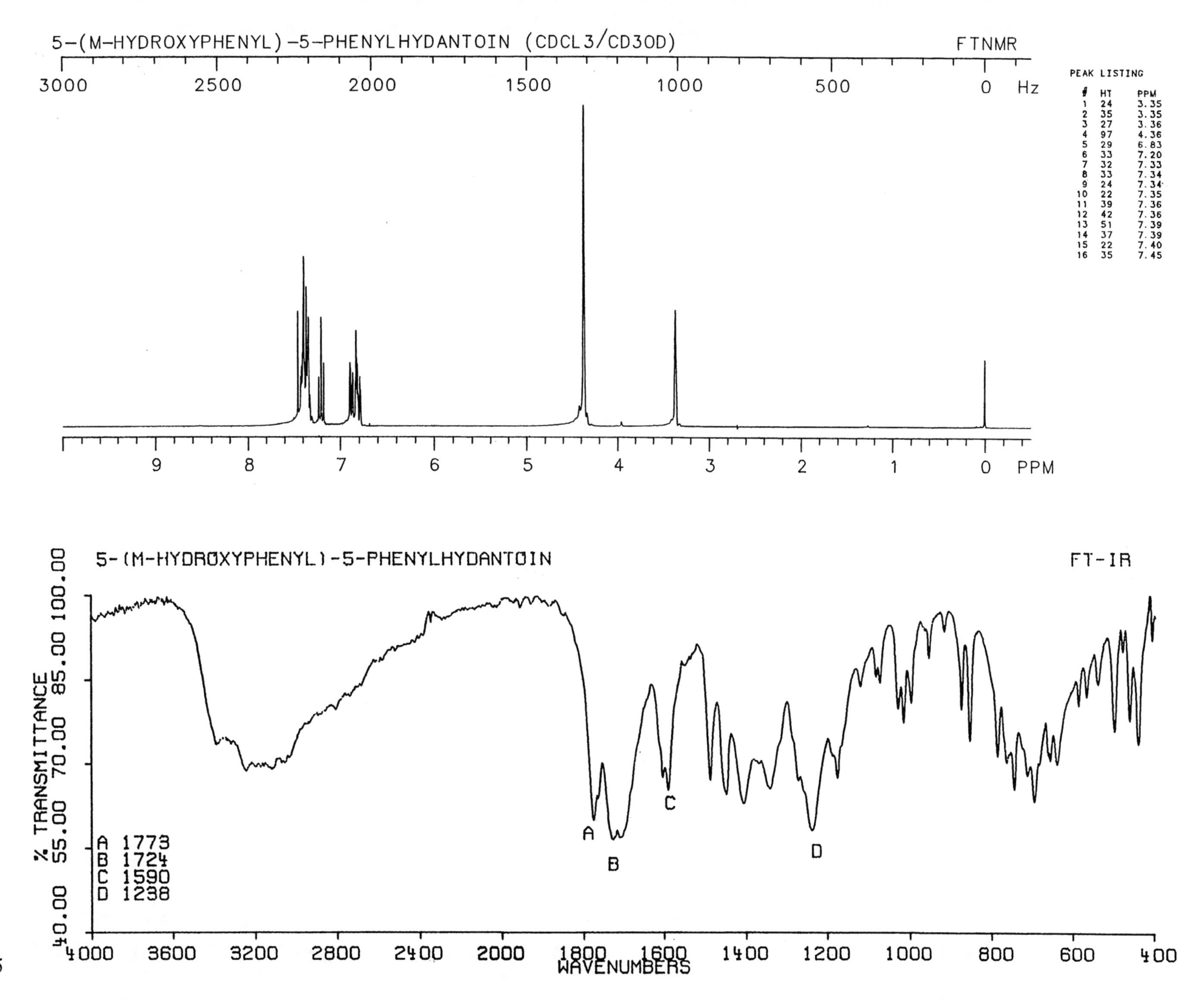
5-(M-HYDROXYPHENYL)-5-PHENYLHYDANTOIN (CDCL3/CD3OD)
FTNMR
3000 2500 2000 1500 1000 500 0 Hz
PEAK LISTING
# HT PPM
1 24 3.35
2 35 3.35
3 27 3.36
4 97 4.36
5 29 6.83
6 33 7.20
7 32 7.33
8 33 7.34
9 24 7.34
10 22 7.35
11 39 7.36
12 42 7.36
13 51 7.39
14 37 7.39
15 22 7.40
16 35 7.45
9 8 7 6 5 4 3 2 1 0 PPM
5-(M-HYDROXYPHENYL)-5-PHENYLHYDANTOIN
FT-IR
% TRANSMITTANCE
40.00 55.00 70.00 85.00 100.00
A 1773
B 1724
C 1590
D 1238
4000 3600 3200 2800 2400 2000 1800 1600 1400 1200 1000 800 600 400
WAVENUMBERS

# 5-*P*-HYDROXYPHENYL-5-PHENYLHYDANTOIN

$C_{15}H_{12}N_2O_3$

**Molecular weight:** 268.28 (268.09)
**Synonyms:** 5-(4-Hydroxyphenyl)-5-phenyl-2,4-imidazolidinedione

**Trade names:**

**Use:**
**HPLC:**
**GC:** 2946; 280°C

5-P-HYDROXYPHENYL-5-PHENYLHYDANTOIN

TRANSMITTANCE

WAVELENGTH (nm)

273
241,289
ETOH 277

5-P-HYDROXYPHENYL-5-PHENYLHYDANTOIN

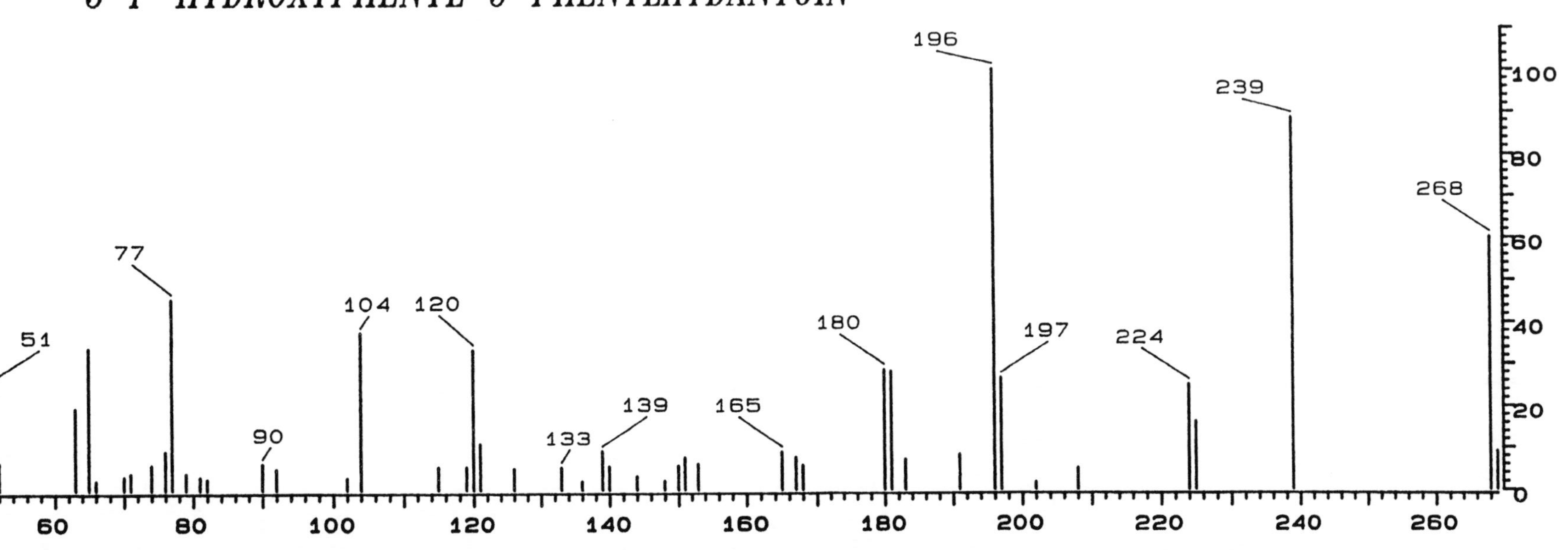

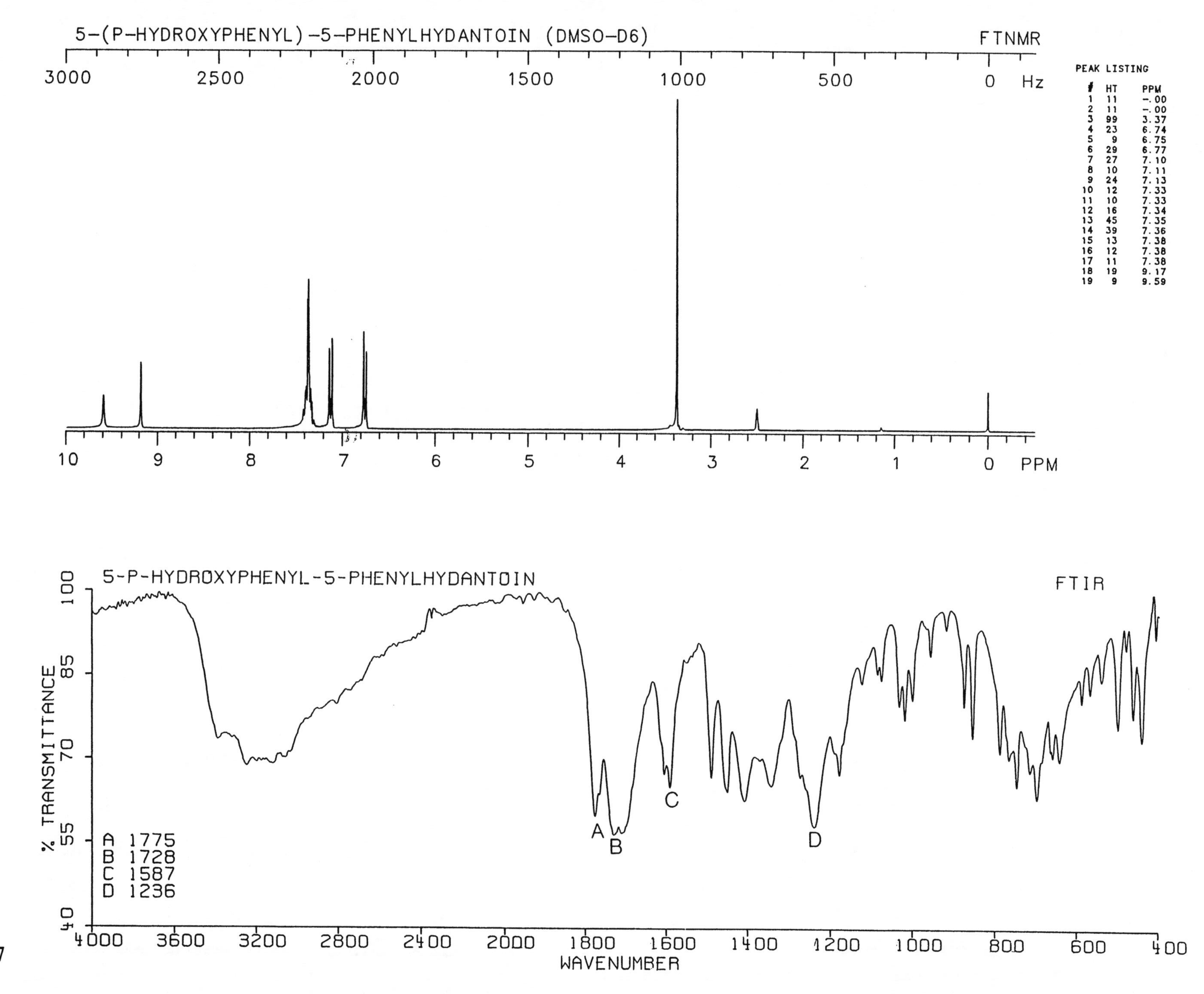
5-(P-HYDROXYPHENYL)-5-PHENYLHYDANTOIN (DMSO-D6)
FTNMR
3000 2500 2000 1500 1000 500 0 Hz
10 9 8 7 6 5 4 3 2 1 0 PPM
PEAK LISTING
# HT PPM
1 11 -.00
2 11 -.00
3 99 3.37
4 23 6.74
5 9 6.75
6 29 6.77
7 27 7.10
8 10 7.11
9 24 7.13
10 12 7.33
11 10 7.33
12 16 7.34
13 45 7.35
14 39 7.36
15 13 7.38
16 12 7.38
17 11 7.38
18 19 9.17
19 9 9.59
5-P-HYDROXYPHENYL-5-PHENYLHYDANTOIN
FTIR
% TRANSMITTANCE
100 85 70 55 40
A 1775
B 1728
C 1587
D 1236
A B C D
4000 3600 3200 2800 2400 2000 1800 1600 1400 1200 1000 800 600 400
WAVENUMBER

# HYDROXYPHENYLPYRUVIC ACID

$C_9H_8O_4$

Molecular weight: 180.16 (180.04)
Synonyms: 4-Hydroxyphenyl-2-oxopropanoic acid

Trade names:

Use:
HPLC: Si-10; 2A:98B; 4.9
GC:

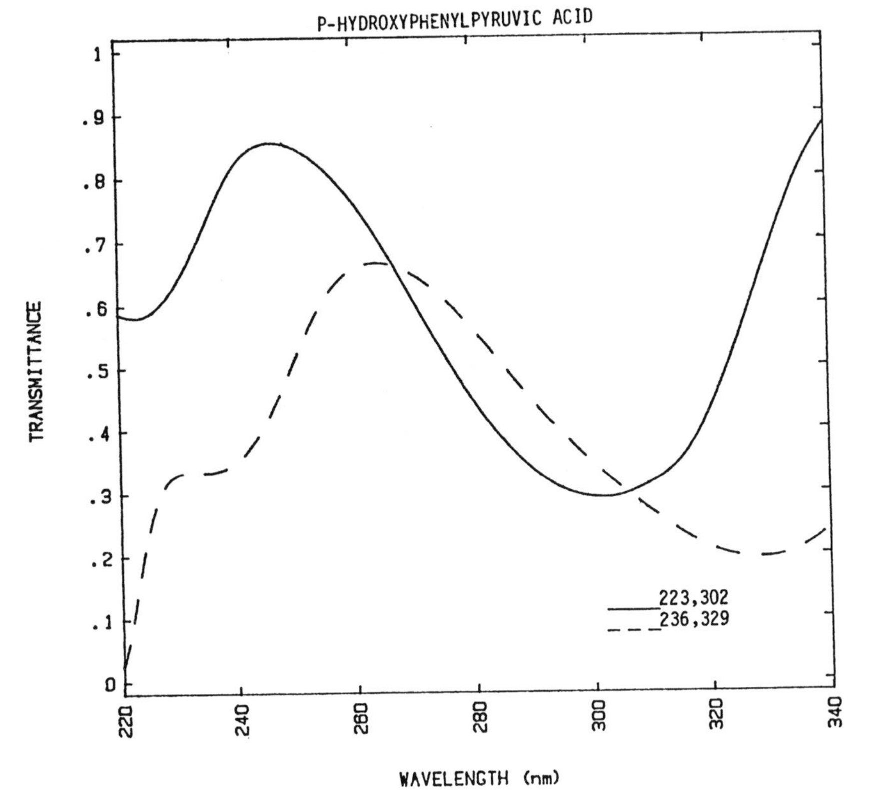


HYDROXYPHENYLPYRUVIC ACID--DIP

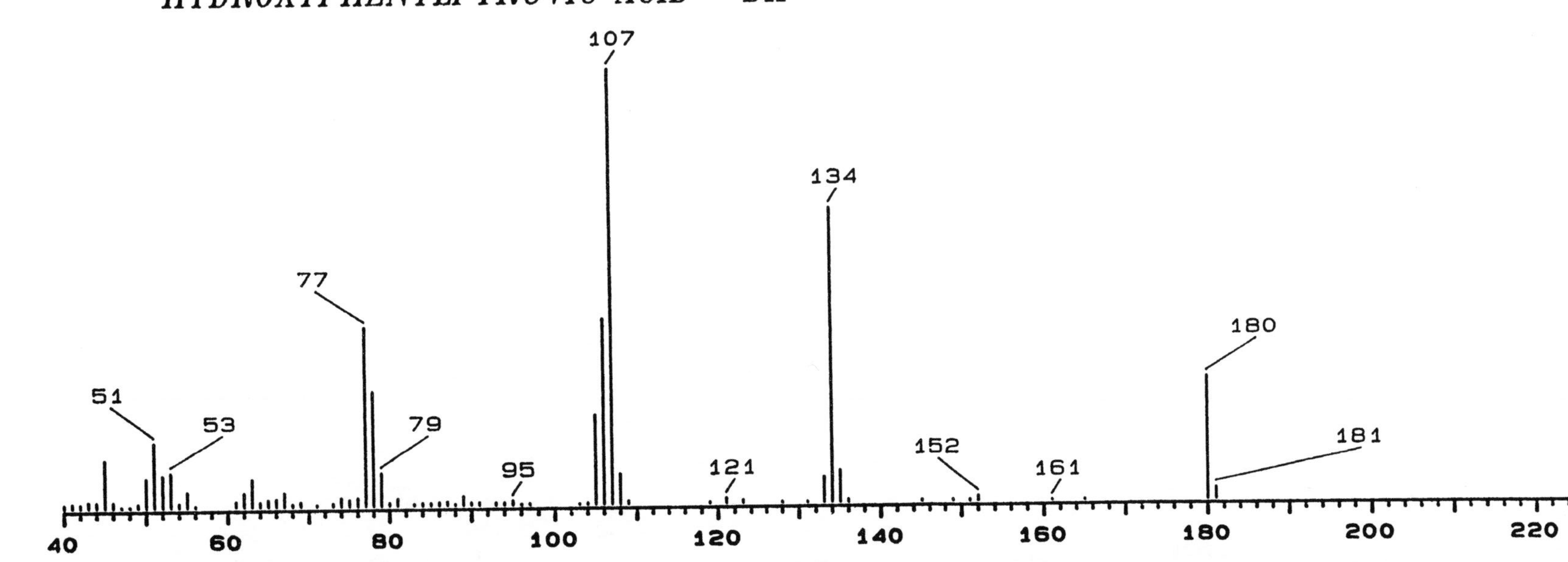

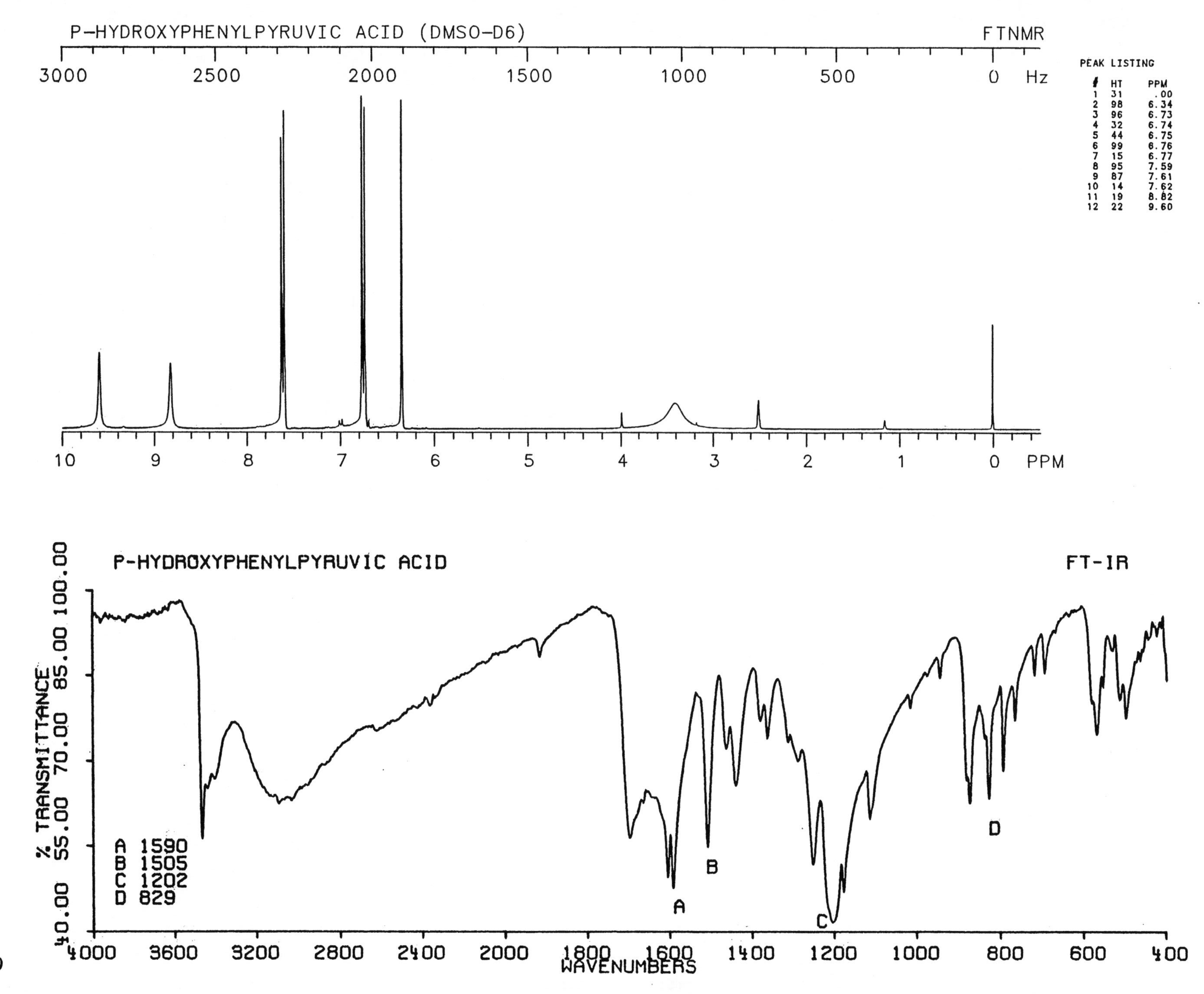
P-HYDROXYPHENYLPYRUVIC ACID (DMSO-D6)
FTNMR
3000
2500
2000
1500
1000
500
0 Hz
PEAK LISTING
# HT PPM
1 31 .00
2 98 6.34
3 96 6.73
4 32 6.74
5 44 6.75
6 99 6.76
7 15 6.77
8 95 7.59
9 87 7.61
10 14 7.62
11 19 8.82
12 22 9.60
10
9
8
7
6
5
4
3
2
1
0 PPM
P-HYDROXYPHENYLPYRUVIC ACID
FT-IR
% TRANSMITTANCE
40.00 55.00 70.00 85.00 100.00
A 1590
B 1505
C 1202
D 829
A
B
C
D
4000 3600 3200 2800 2400 2000 1800 1600 1400 1200 1000 800 600 400
WAVENUMBERS

# 11-α-HYDROXYPROGESTERONE

$C_{21}H_{30}O_3$

**Molecular weight:** 330.45 (330.22)
**Synonyms:** 11-Hydroxypregn-4-ene-3,20-dione; 4-pregnen-11-ol-3,20-dione

**Trade names:** Prodrox

**Use:** Steroid
**HPLC:**
**GC:** 3096; 280°C

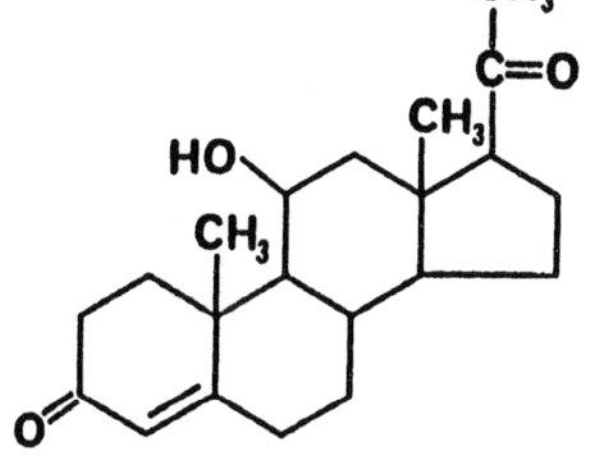


11-HYDROXYPROGESTERONE

TRANSMITTANCE

ETHANOL.....241

WAVELENGTH (nm)

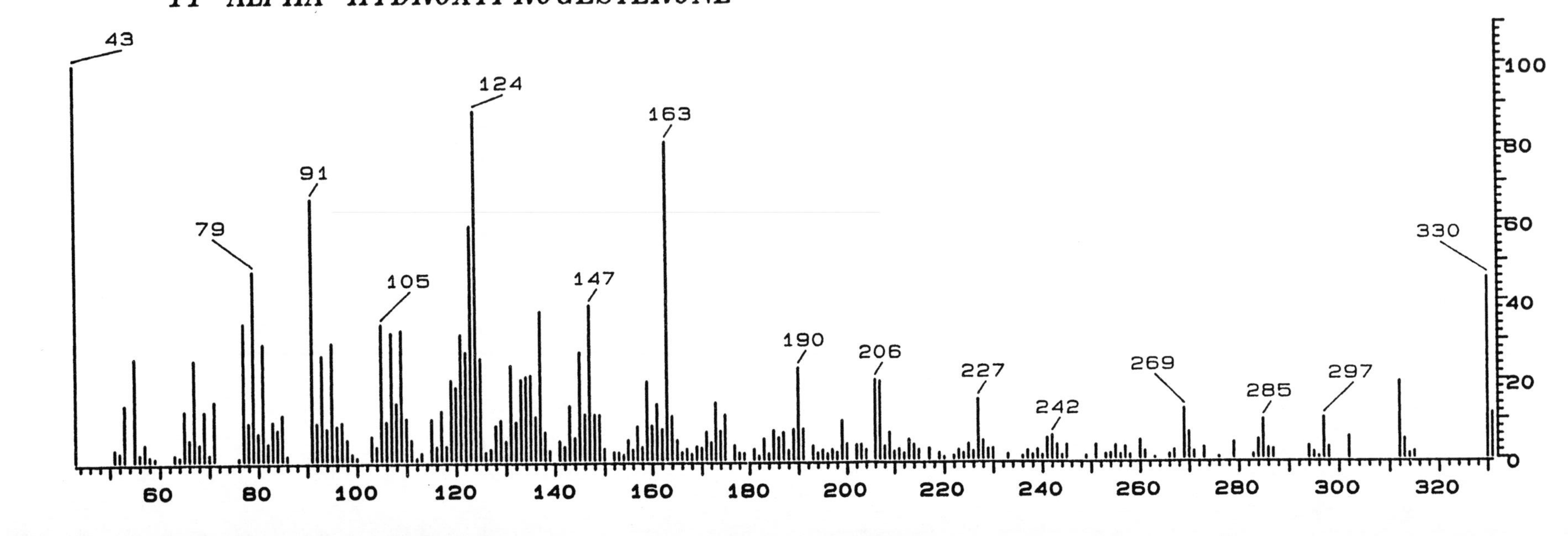

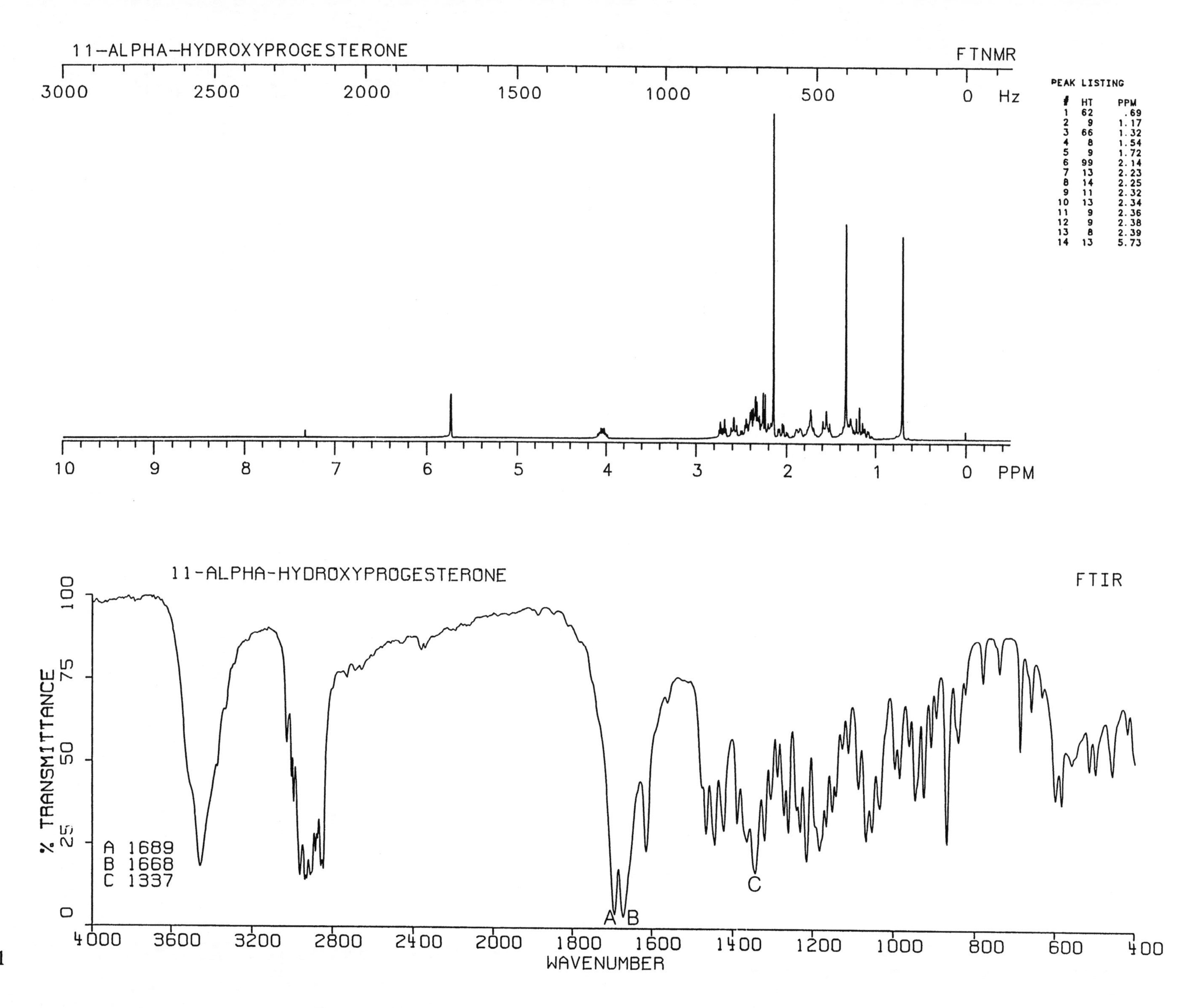
11-ALPHA-HYDROXYPROGESTERONE
FTNMR
3000 2500 2000 1500 1000 500 0 Hz
10 9 8 7 6 5 4 3 2 1 0 PPM
PEAK LISTING
# HT PPM
1 62 .69
2 9 1.17
3 66 1.32
4 8 1.54
5 9 1.72
6 99 2.14
7 13 2.23
8 14 2.25
9 11 2.32
10 13 2.34
11 9 2.36
12 9 2.38
13 8 2.39
14 13 5.73
11-ALPHA-HYDROXYPROGESTERONE
FTIR
% TRANSMITTANCE
100 75 50 25 0
A 1689
B 1668
C 1337
4000 3600 3200 2800 2400 2000 1800 1600 1400 1200 1000 800 600 400
WAVENUMBER

# 17-α-HYDROXYPROGESTERONE (17-α)

$C_{21}H_{30}O_3$

**Molecular weight:** 330.45 (330.22)
**Synonyms:** 17-Hydroxypregn-4-ene-3,20-dione; 4-pregnen-17α-ol-3,20-dione

**Trade names:** Delalutin, Gestageno, Proluton, Prodrox, Relutin

**Use:** Progestin, steroid
**HPLC:**
**GC:** 3002; 280°C

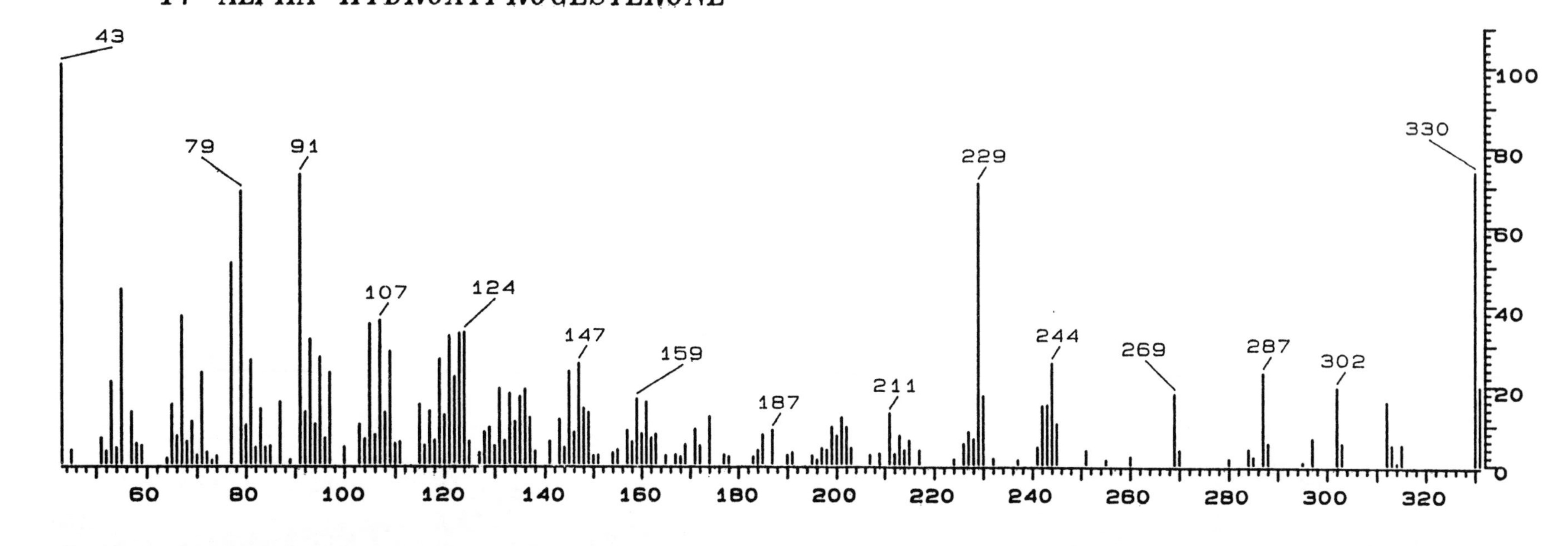

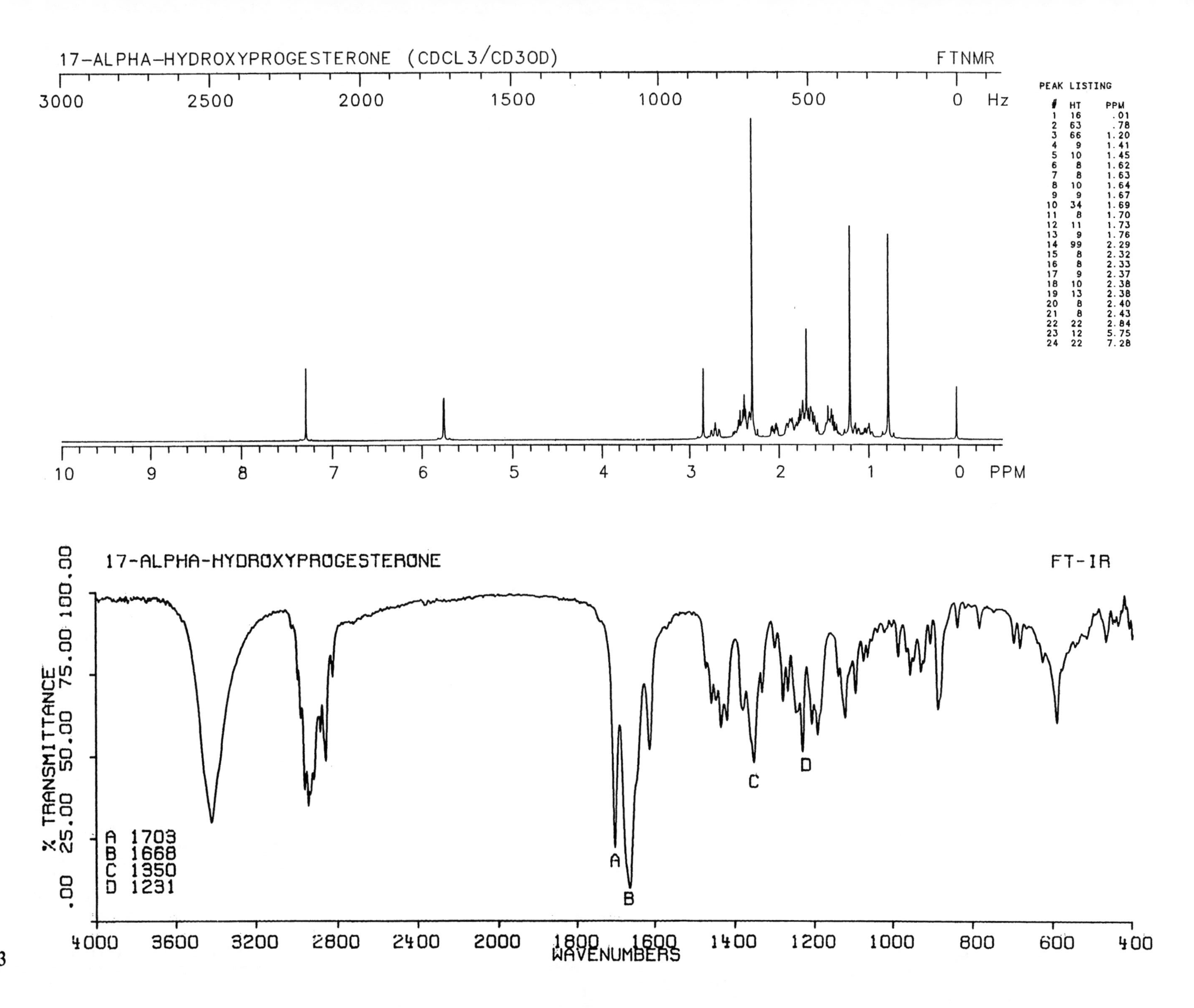
17-ALPHA-HYDROXYPROGESTERONE (CDCL3/CD3OD)
FTNMR
3000 2500 2000 1500 1000 500 0 Hz
10 9 8 7 6 5 4 3 2 1 0 PPM
PEAK LISTING
# HT PPM
1 16 .01
2 63 .78
3 66 1.20
4 9 1.41
5 10 1.45
6 8 1.62
7 8 1.63
8 10 1.64
9 9 1.67
10 34 1.69
11 8 1.70
12 11 1.73
13 9 1.76
14 99 2.29
15 8 2.32
16 8 2.33
17 9 2.37
18 10 2.38
19 13 2.38
20 8 2.40
21 8 2.43
22 22 2.84
23 12 5.75
24 22 7.28
17-ALPHA-HYDROXYPROGESTERONE
FT-IR
% TRANSMITTANCE
.00 25.00 50.00 75.00 100.00
A 1703
B 1668
C 1350
D 1231
A B C D
4000 3600 3200 2800 2400 2000 1800 1600 1400 1200 1000 800 600 400
WAVENUMBERS

# HYDROXYQUINOLINE

$C_9H_7NO$

Molecular weight: 145.15 (145.05)
Synonyms: 8-Quinolinol; oxyquinoline; oxybenzopyridine; phenopyridine; oxine
Trade names: Derma Medicone-HC, Chinosolum, Oxyquinol, Rectal Medicone-HC

Use: Disinfectant
HPLC: Si-10; 1A:99B; 4.6
GC: 1408; 200°C

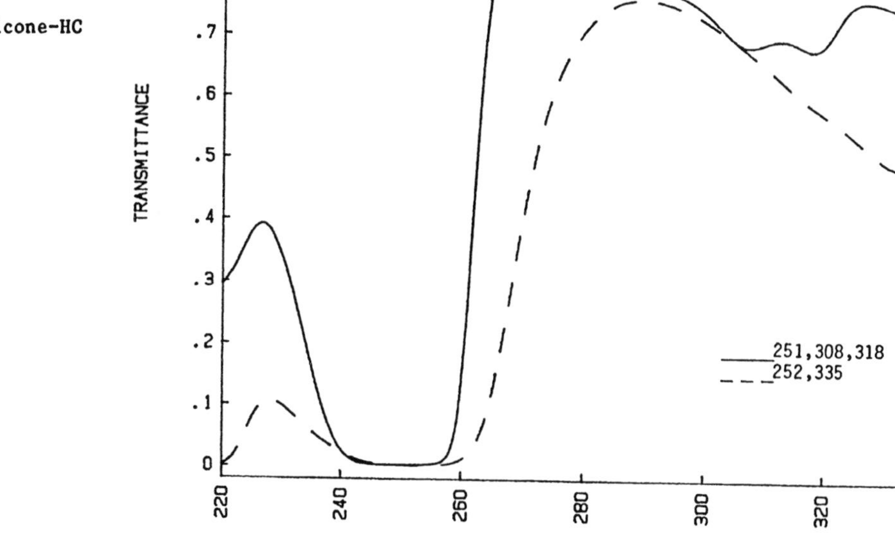


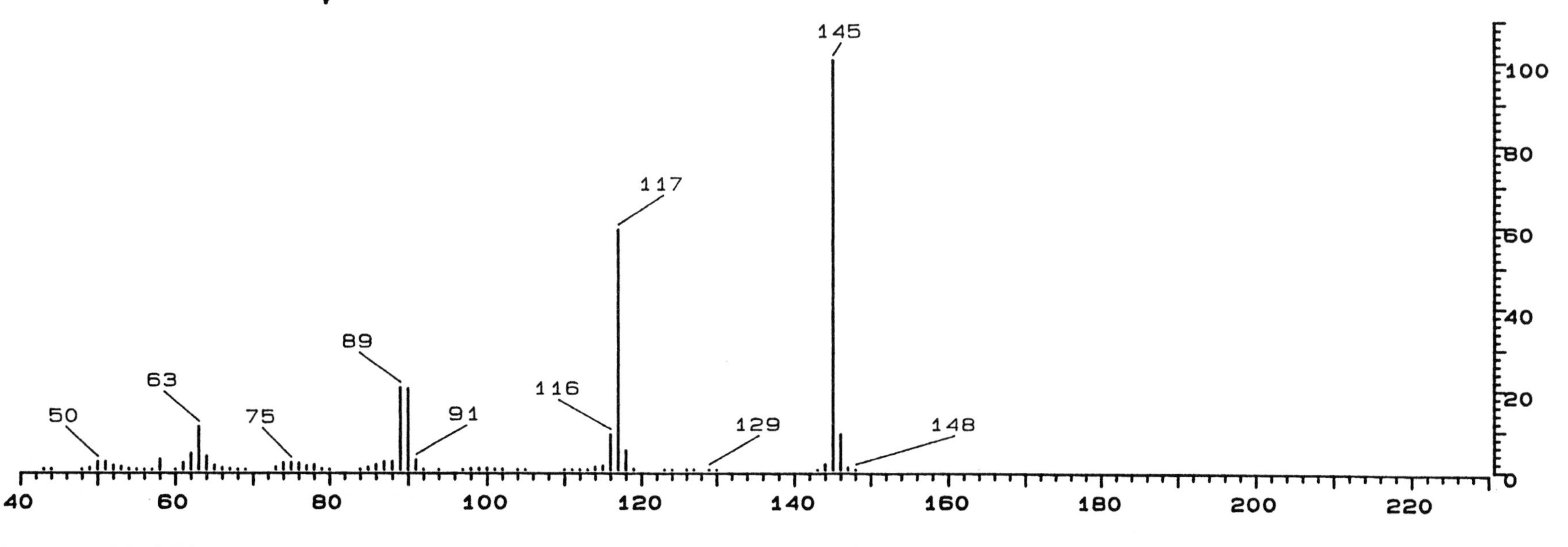

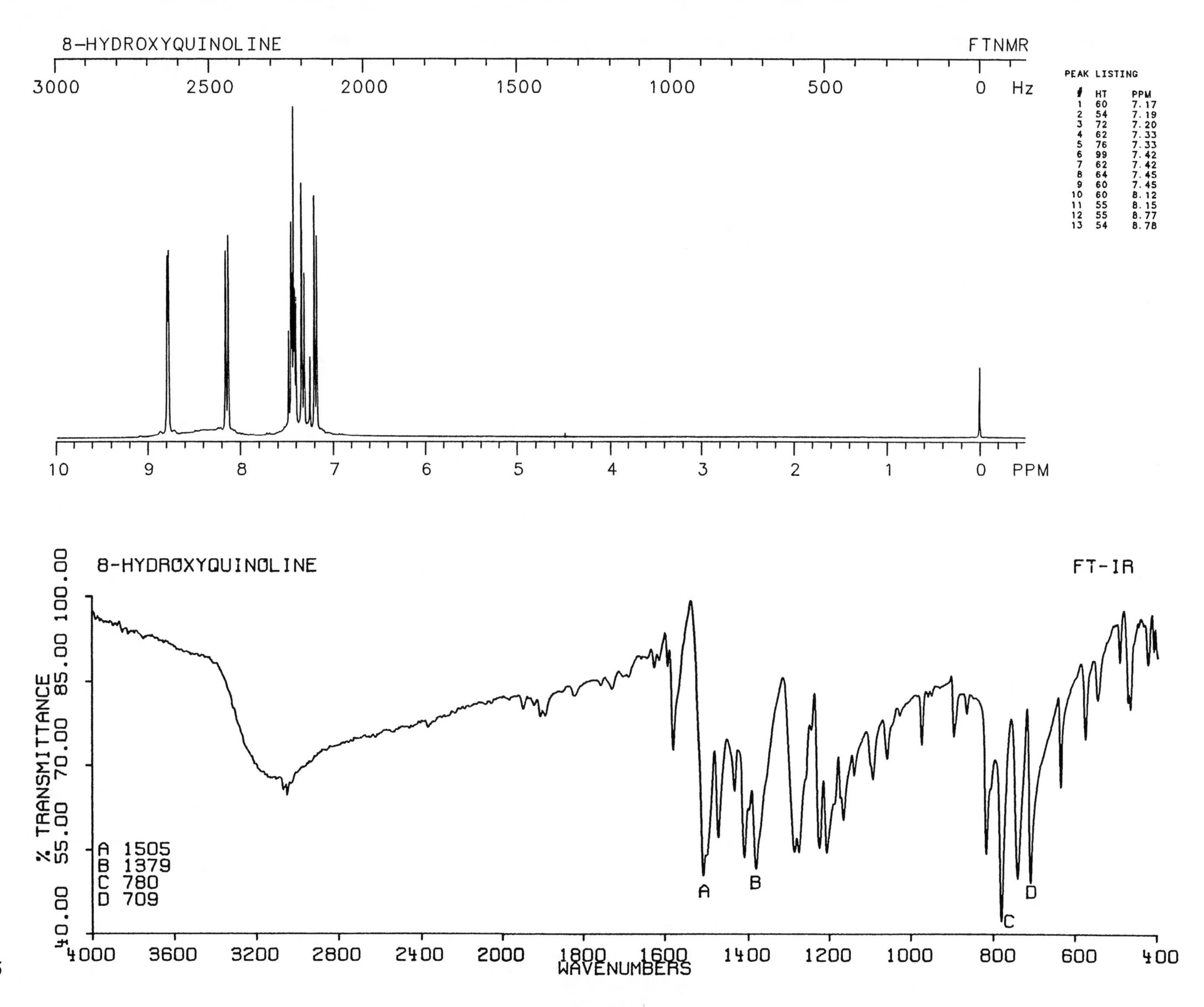
8-HYDROXYQUINOLINE
FTNMR
Hz
PPM
PEAK LISTING
# HT PPM
1 60 7.17
2 54 7.19
3 72 7.20
4 62 7.33
5 76 7.33
6 99 7.42
7 62 7.42
8 64 7.45
9 60 7.45
10 60 8.12
11 55 8.15
12 55 8.77
13 54 8.78
8-HYDROXYQUINOLINE
FT-IR
% TRANSMITTANCE
WAVENUMBERS
A 1505
B 1379
C 780
D 709

# 11-HYDROXY-DELTA-9-TETRAHYDROCANNABINOL

$C_{20}H_{28}O_3$

**Molecular weight:** 316.42 (316.20)
**Synonyms:** 11-Hydroxy-tetrahydro-6,6,9-trimethyl-3-pentyl-6H-dibenzo[b,d]pyran-1-ol
**Trade names:**

**Use:** Metabolite
**HPLC:** Si-10; 1A:99B; 4.6
**GC:** 2826; 280°C

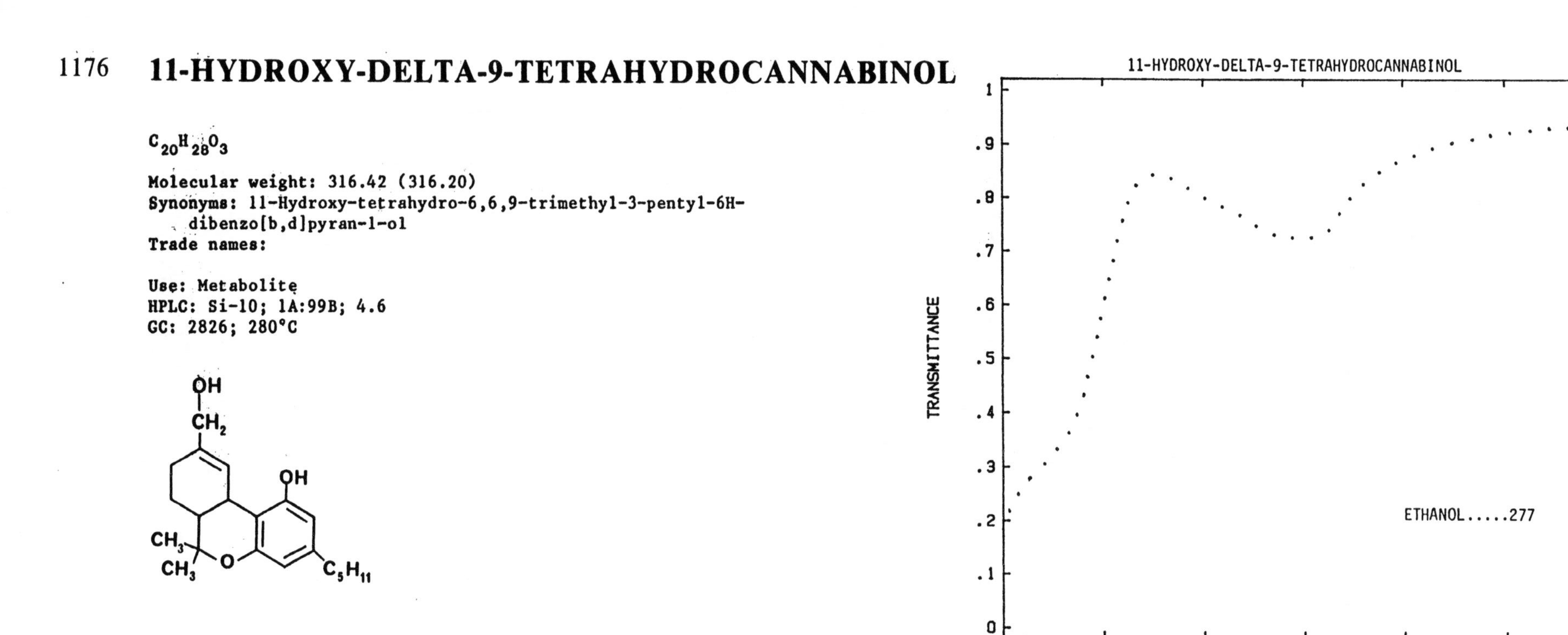

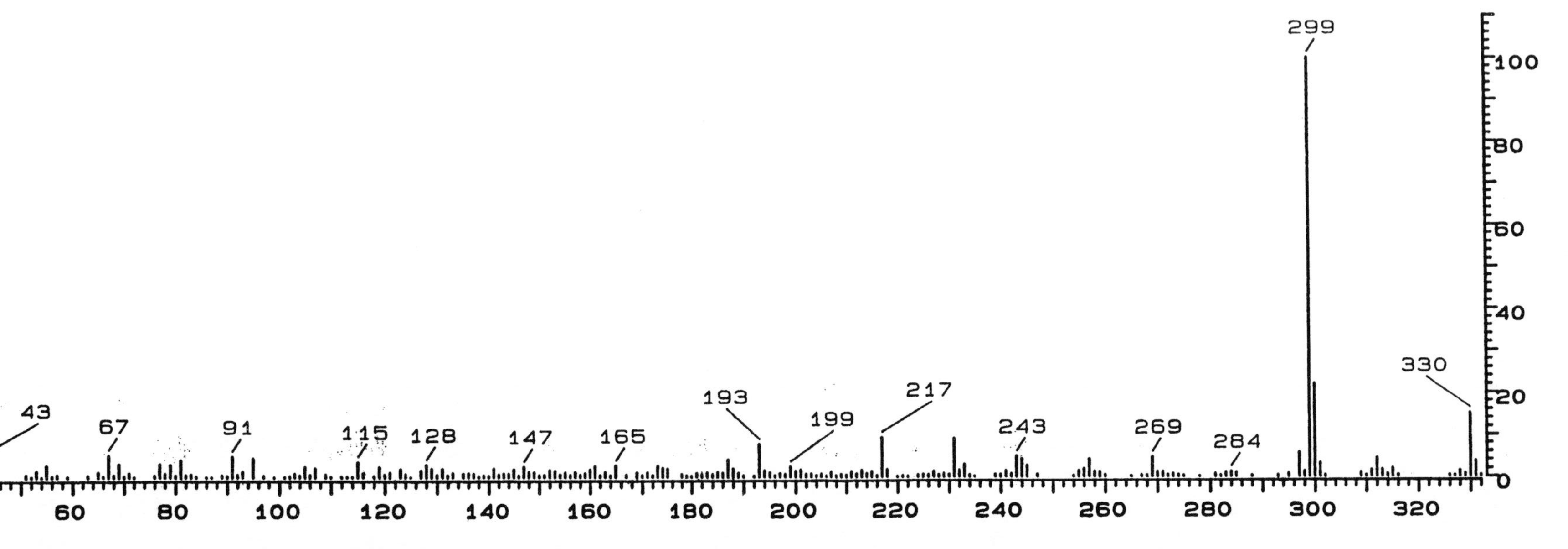

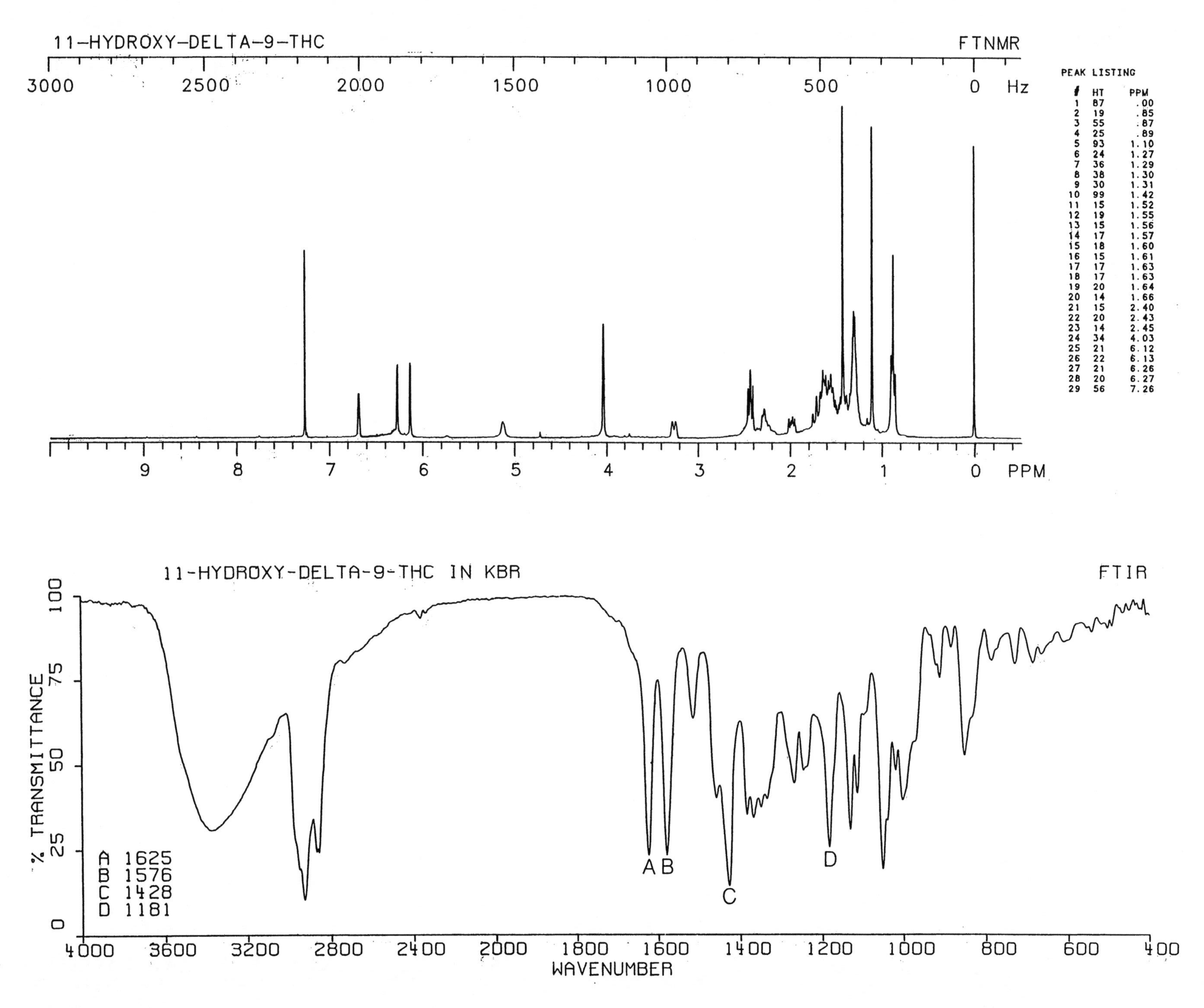
11-HYDROXY-DELTA-9-THC
FTNMR
3000 2500 2000 1500 1000 500 0 Hz
9 8 7 6 5 4 3 2 1 0 PPM
PEAK LISTING
# HT PPM
1 87 .00
2 19 .85
3 55 .87
4 25 .89
5 93 1.10
6 24 1.27
7 36 1.29
8 38 1.30
9 30 1.31
10 99 1.42
11 15 1.52
12 19 1.55
13 15 1.56
14 17 1.57
15 18 1.60
16 15 1.61
17 17 1.63
18 17 1.63
19 20 1.64
20 14 1.66
21 15 2.40
22 20 2.43
23 14 2.45
24 34 4.03
25 21 6.12
26 22 6.13
27 21 6.26
28 20 6.27
29 56 7.26
11-HYDROXY-DELTA-9-THC IN KBR
FTIR
% TRANSMITTANCE
100 75 50 25 0
A 1625
B 1576
C 1428
D 1181
A B C D
4000 3600 3200 2800 2400 2000 1800 1600 1400 1200 1000 800 600 400
WAVENUMBER

# 5-HYDROXYTRYPTAMINE CREATININE

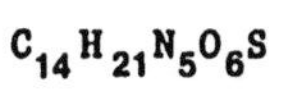

$C_{14}H_{21}N_5O_6S$

Molecular weight: 387.38 (387.12)
Synonyms: 3-(2-Aminoethyl)-1H-indol-5-ol creatinine sulfate;
Serotonin, enteramine, thrombotonin
Trade names: Antemovis

Use:
HPLC: Si-10; 20A:80B; 7.1
GC:

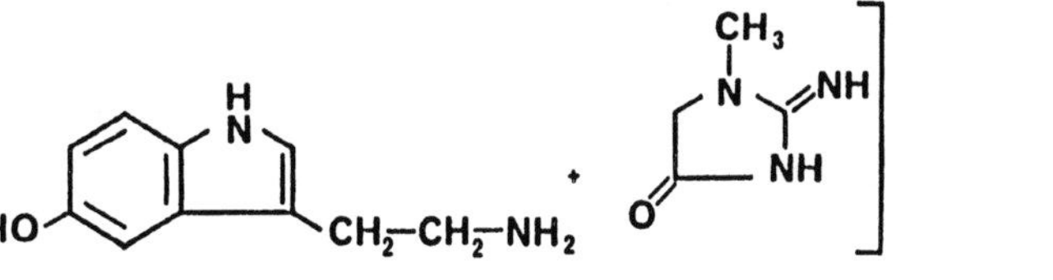


5-HYDROXYTRYPTAMINE

TRANSMITTANCE

ETHANOL.....276

WAVELENGTH (nm)

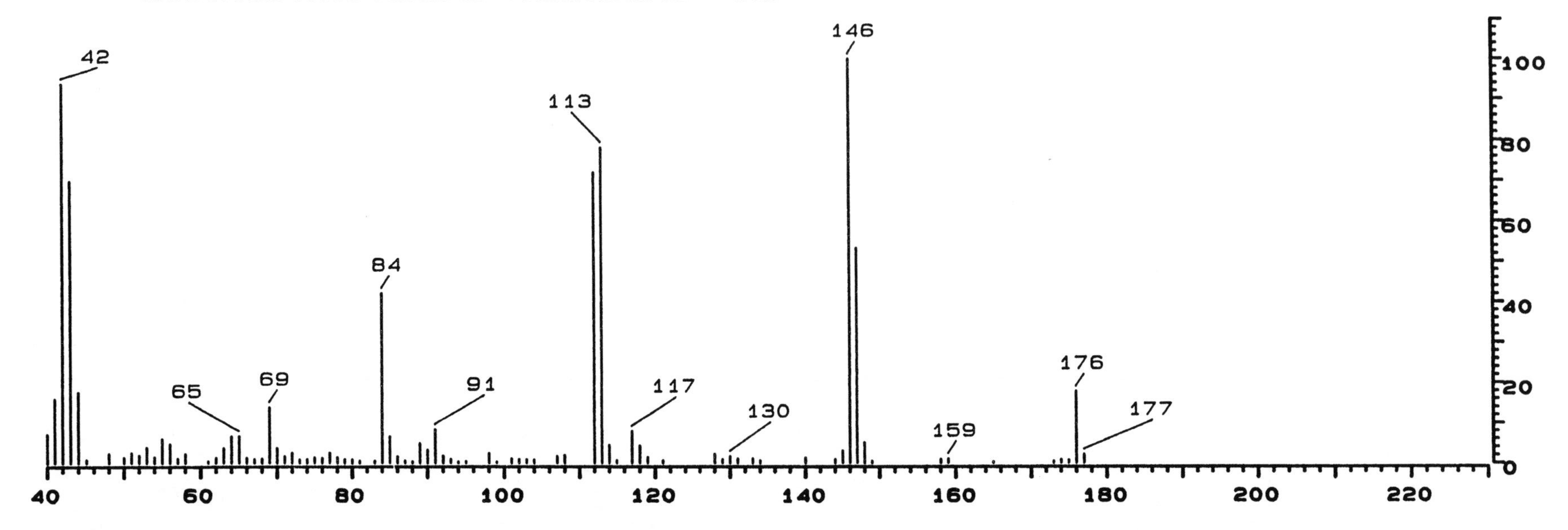

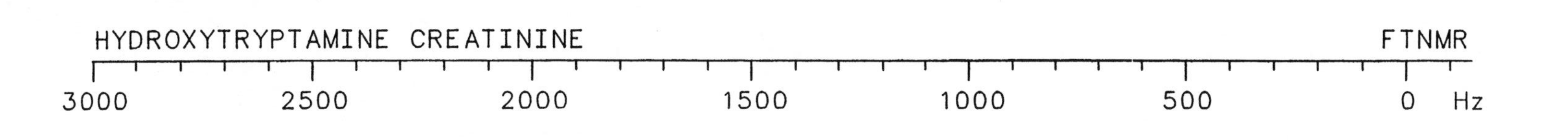


INSUFFICIENT SOLUBILITY

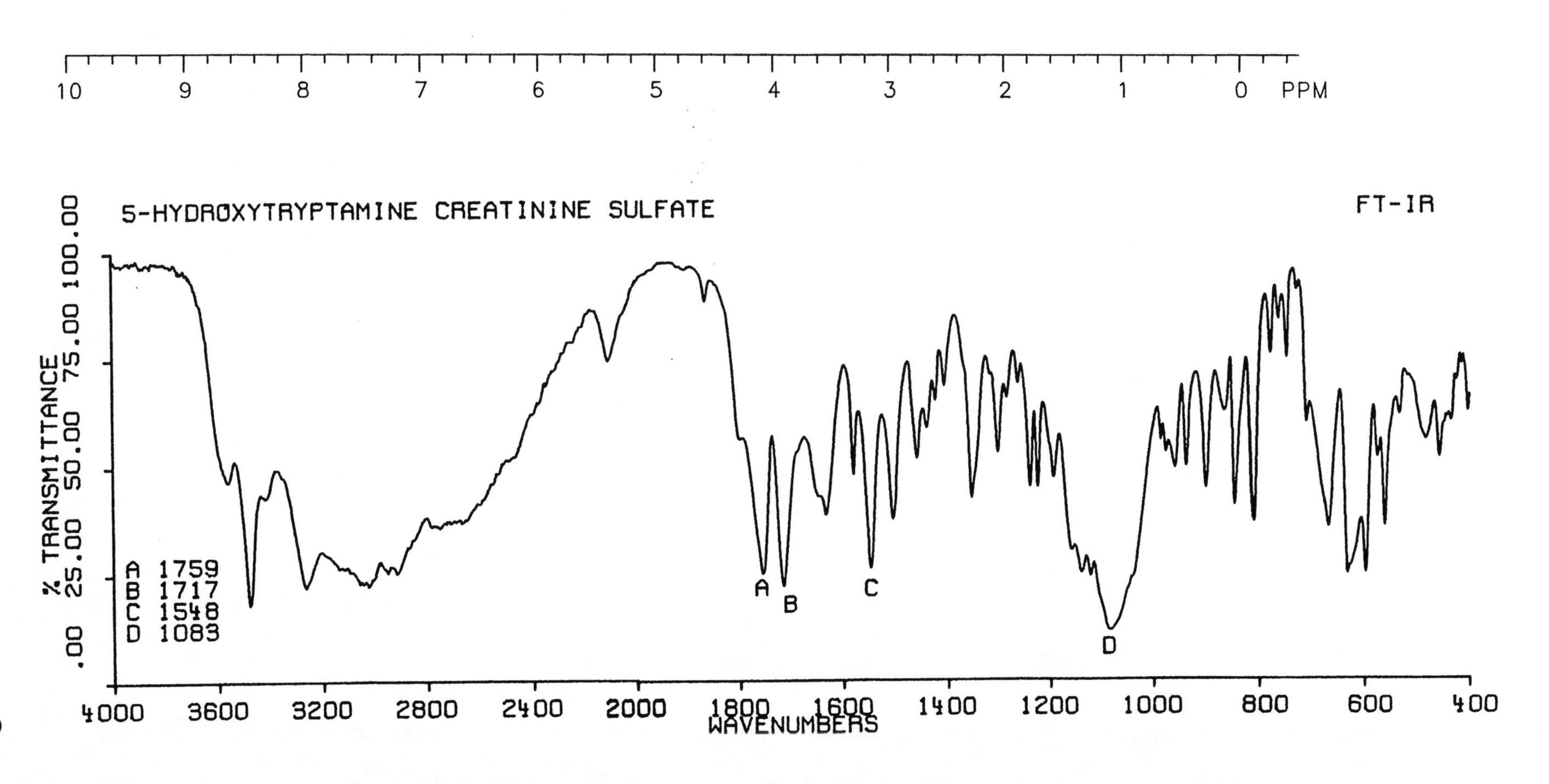

# HYDROXYUREA

$CH_4N_2O_2$

**Molecular weight: 76.06 (76.03)**
**Synonyms: Hydroxycarbamide**

**Trade names: Hydrea**

**Use: Antineoplastic**
**HPLC:**
**GC:**

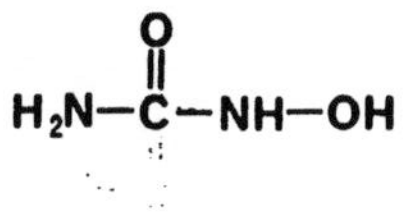

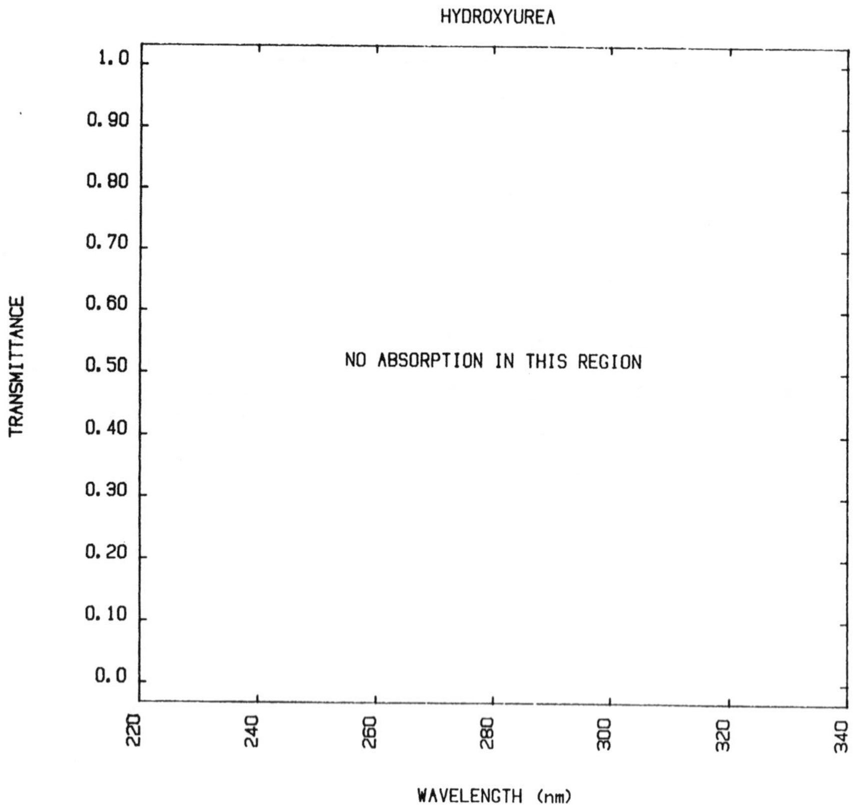

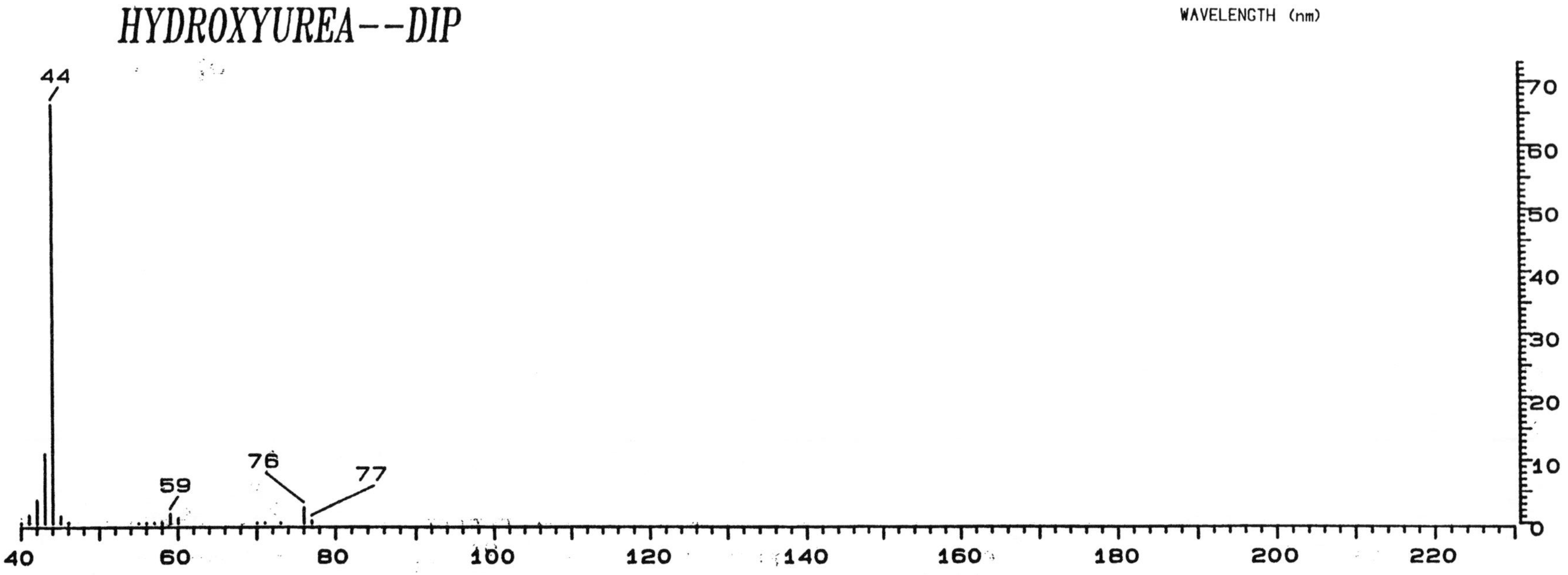

$H_2N-C(=O)-NH-OH$

HYDROXYUREA

NO ABSORPTION IN THIS REGION

TRANSMITTANCE

WAVELENGTH (nm)

HYDROXYUREA--DIP

44 59 76 77

HYDROXYUREA (DMSO-D6)

FTNMR

3000 2500 2000 1500 1000 500 0 Hz

PEAK LISTING

| # | HT | PPM |
|---|---|---|
| 1 | 52 | 3.44 |
| 2 | 78 | 6.24 |
| 3 | 75 | 8.32 |
| 4 | 99 | 8.65 |

10 9 8 7 6 5 4 3 2 1 0 PPM

HYDROXYUREA

FT-IR

A 1646
B 1111
C 822
D 561

% TRANSMITTANCE 20.00 40.00 60.00 80.00 100.00

4000 3600 3200 2800 2400 2000 1800 1600 1400 1200 1000 800 600 400

WAVENUMBERS

# HYDROXYZINE

$C_{21}H_{27}ClN_2O_2$

Molecular weight: 374.91 (374.18)
Synonyms: 2-[2-[4-[(4-Chlorophenyl)phenylmethyl]-1-piperazinyl]-ethoxy]ethanol
Trade names: Atarax, Cartrax, Marax, Sedaril, Vistrax, Vistaril

Use: Tranquilizer
HPLC: Si-10; 2A:98B; 10.0
GC: 2934; 280°C

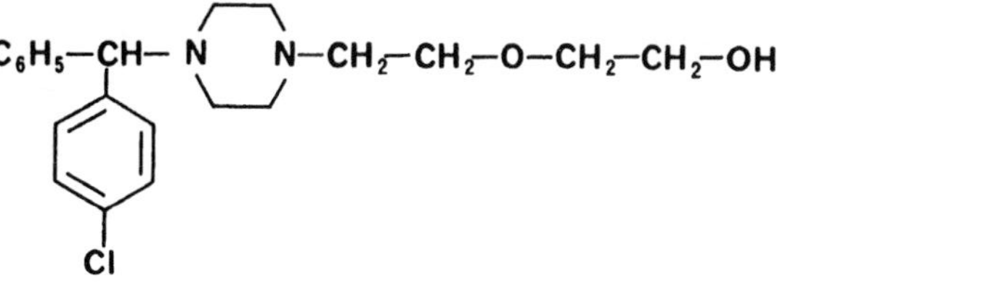


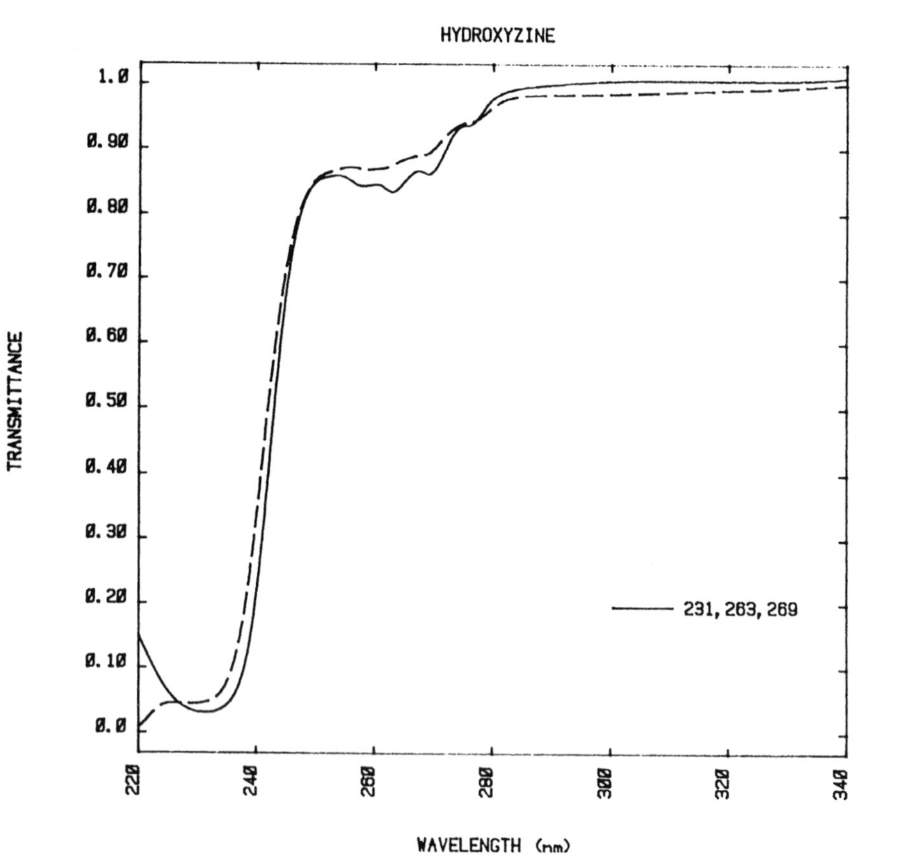


*HYDROXYZINE*

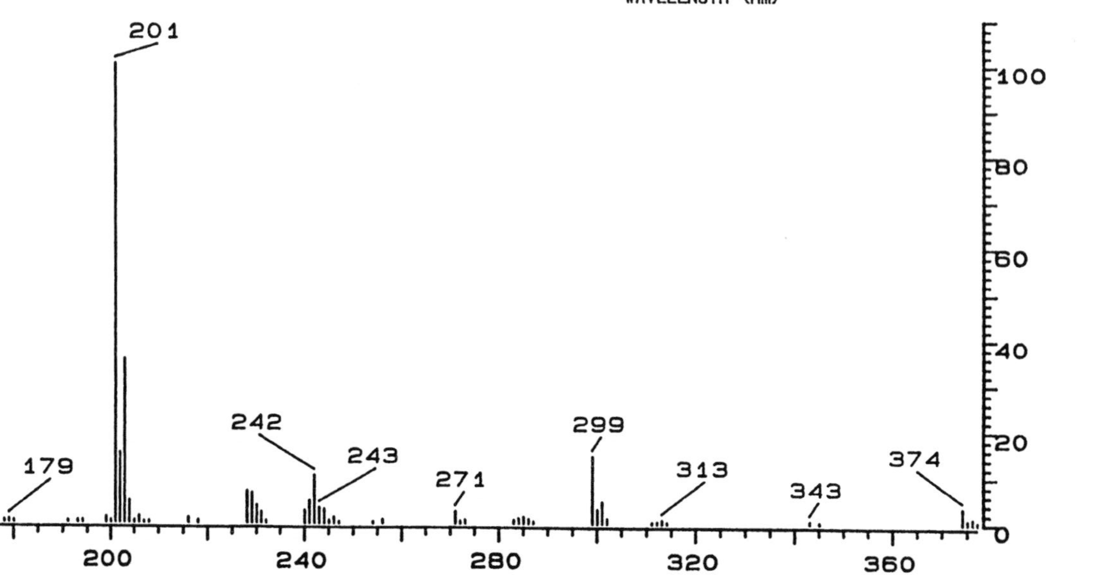

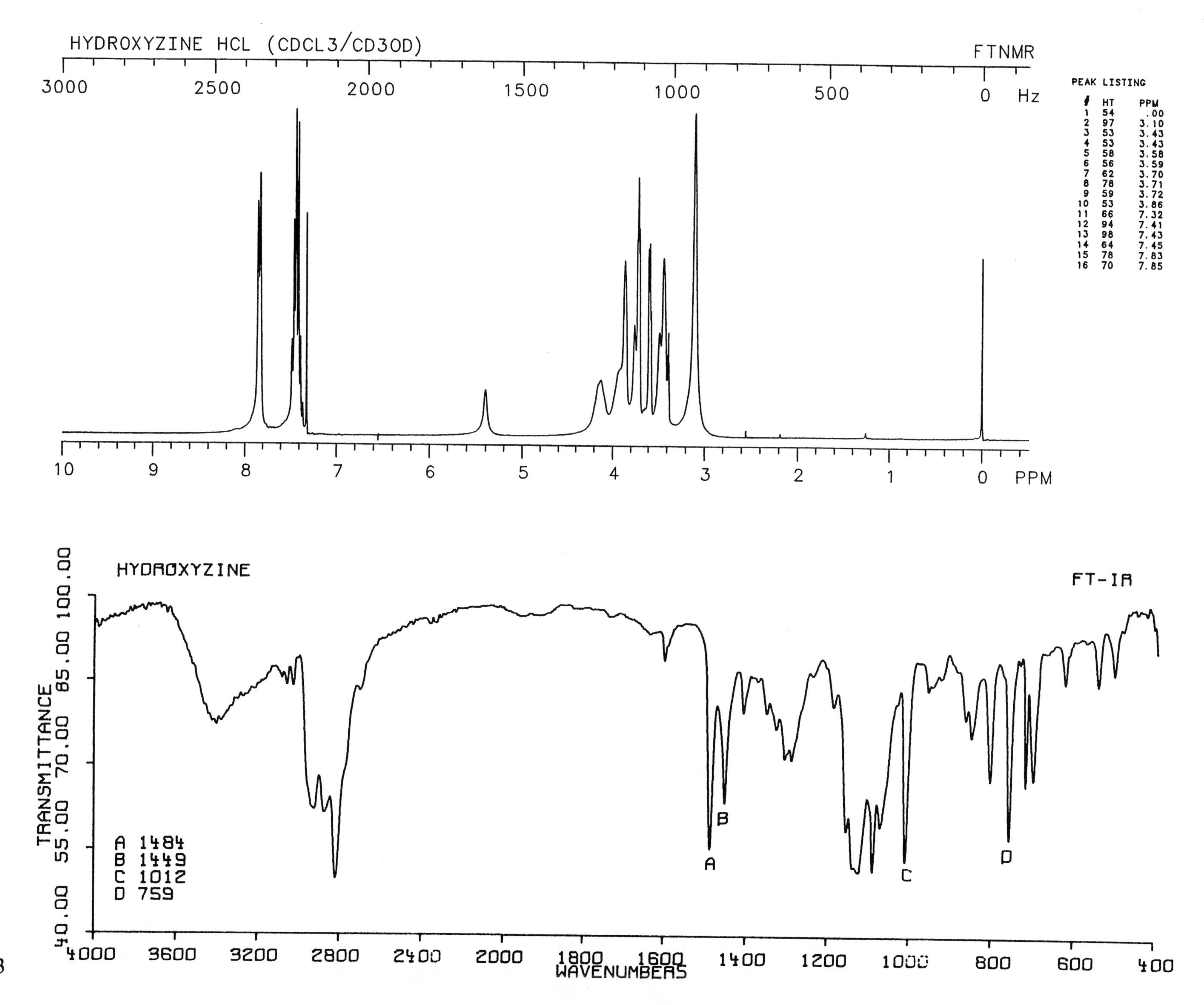
HYDROXYZINE HCL (CDCL3/CD3OD)
FTNMR
3000
2500
2000
1500
1000
500
0
Hz
PEAK LISTING
# HT PPM
1 54 .00
2 97 3.10
3 53 3.43
4 53 3.43
5 58 3.58
6 56 3.59
7 62 3.70
8 78 3.71
9 59 3.72
10 53 3.86
11 66 7.32
12 94 7.41
13 98 7.43
14 64 7.45
15 78 7.83
16 70 7.85
10
9
8
7
6
5
4
3
2
1
0
PPM
HYDROXYZINE
FT-IR
TRANSMITTANCE
40.00
55.00
70.00
85.00
100.00
A 1484
B 1449
C 1012
D 759
A
B
C
D
4000
3600
3200
2800
2400
2000
1800
1600
1400
1200
1000
800
600
400
WAVENUMBERS

# HYOSCYAMINE

$C_{17}H_{23}NO_3$

**Molecular weight:** 289.38 (289.17)
**Synonyms:** [3(S)-endo]-α(Hydroxymethyl)-benzeneacetic acid-8-methyl-8-azabicyclo[3.2.1]oct-3-yl ester; ℓ-hyoscyamine; duboisine
**Trade names:** Kutrase, Levsin, Levsinex, Prosed, Urised

**Use:** Anticholinergic
**HPLC:**
**GC:** 2256; 250°C

CH3
N
O
CH2—OH
O—C—CH
C6H5

HYOSCYAMINE

1.0
0.90
0.80
0.70
0.60
0.50
0.40
0.30
0.20
0.10
0.0
TRANSMITTANCE
220
240
260
280
300
320
340
WAVELENGTH (nm)
—— 246, 251, 257, 263
– – 251, 257, 263

*HYOSCYAMINE*

124
42
67
82
94
140
141
193
272
289
60
80
100
120
140
160
180
200
220
240
260
280
100
80
60
40
20
0

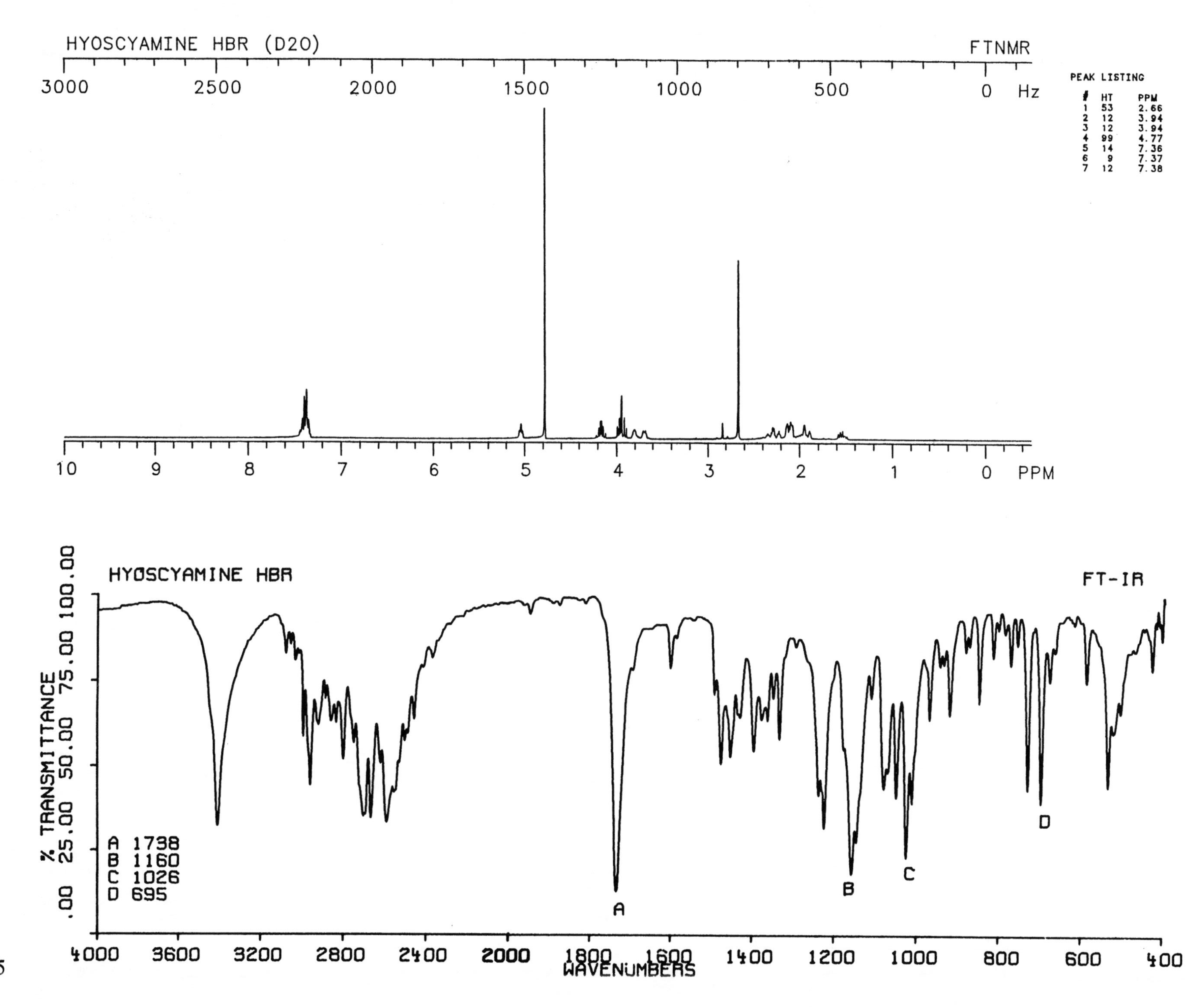
HYOSCYAMINE HBR (D2O)
FTNMR
3000 2500 2000 1500 1000 500 0 Hz
PEAK LISTING
# HT PPM
1 53 2.66
2 12 3.94
3 12 3.94
4 99 4.77
5 14 7.36
6 9 7.37
7 12 7.38
10 9 8 7 6 5 4 3 2 1 0 PPM
HYOSCYAMINE HBR
FT-IR
% TRANSMITTANCE
.00 25.00 50.00 75.00 100.00
A 1738
B 1160
C 1026
D 695
A
B
C
D
4000 3600 3200 2800 2400 2000 1800 1600 1400 1200 1000 800 600 400
WAVENUMBERS

# IBOGAINE

$C_{20}H_{26}N_2O$

Molecular weight: 310.44 (310.20)
Synonyms: 7-Ethyl-6,6a,7,8,9,10,12,13-octahydro-2-methoxy-6,9-methano-5H-pyrido[1,2-a]azepino-[4.5-b]indole
Trade names:

Use: Antidepressant
HPLC: Si-10; 50A:50B; 4.0
GC: 2958; 280°C

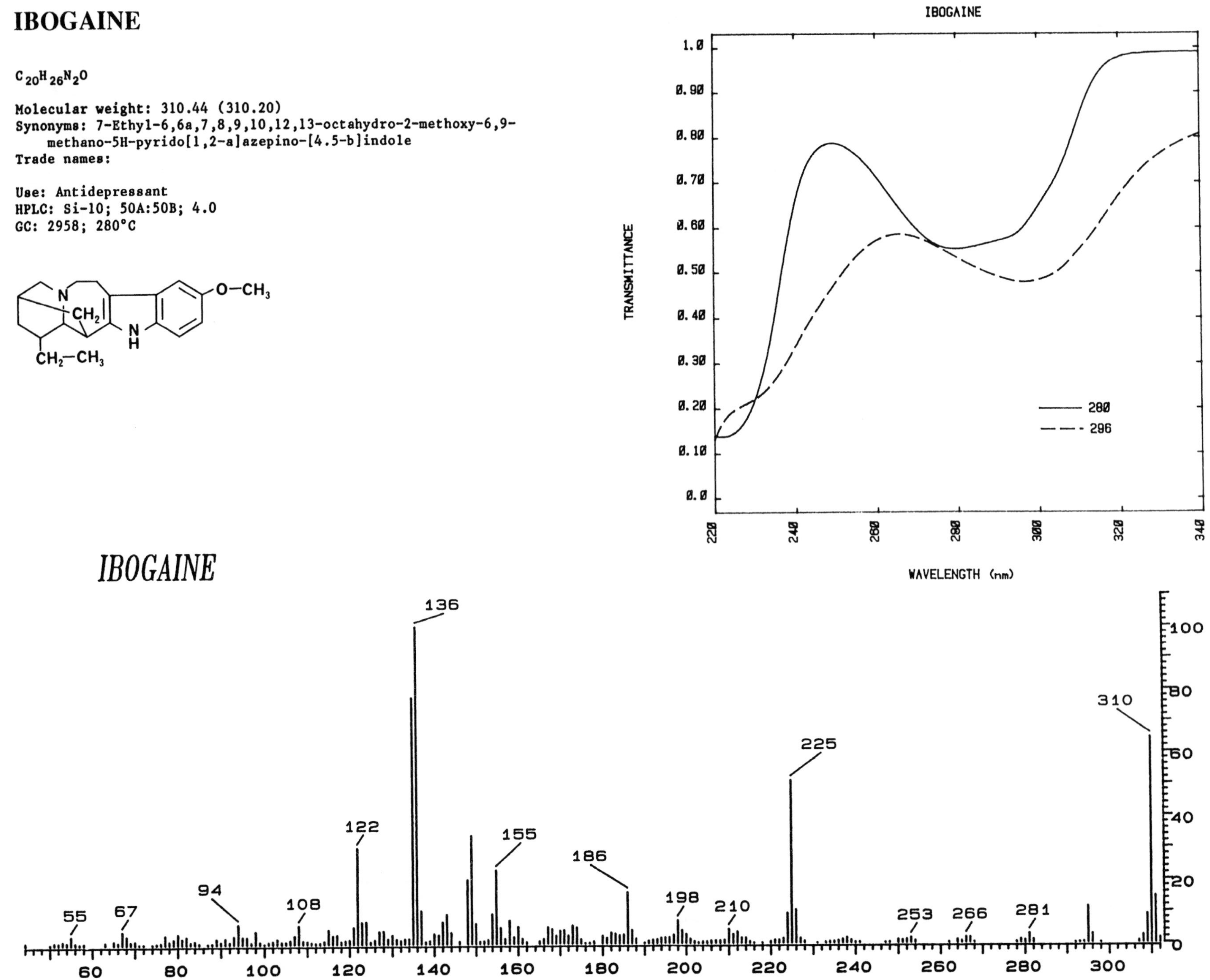

IBOGAINE

FTNMR

Hz: 3000 2500 2000 1500 1000 500 0

PPM: 10 9 8 7 6 5 4 3 2 1 0

PEAK LISTING

| # | HT | PPM |
|---|---|---|
| 1 | 22 | .89 |
| 2 | 33 | .90 |
| 3 | 28 | .91 |
| 4 | 47 | .93 |
| 5 | 27 | .94 |
| 6 | 28 | .94 |
| 7 | 33 | .95 |
| 8 | 22 | 3.71 |
| 9 | 34 | 3.72 |
| 10 | 21 | 3.73 |
| 11 | 28 | 3.74 |
| 12 | 22 | 3.74 |
| 13 | 84 | 3.76 |
| 14 | 63 | 3.77 |
| 15 | 43 | 3.83 |
| 16 | 46 | 3.83 |
| 17 | 99 | 7.28 |

IBOGAINE HCL

FT-IR

TRANSMITTANCE: 20.00 40.00 60.00 80.00 100.00

A 1627
B 1592
C 1486
D 1217

WAVENUMBERS: 4000 3600 3200 2800 2400 2000 1800 1600 1400 1200 1000 800 600 400

# IBUPROFEN

$C_{13}H_{18}O_2$

**Molecular weight:** 206.26 (206.13)
**Synonyms:** α-Methyl-4-(2-methylpropyl)benzeneacetic acid; 2-(4-isobutylphenyl)propionic acid
**Trade names:** Motrin

**Use:** Anti-inflammatory
**HPLC:** Si-10; 20A:80B; 4.0
**GC:** 1621; 200°C

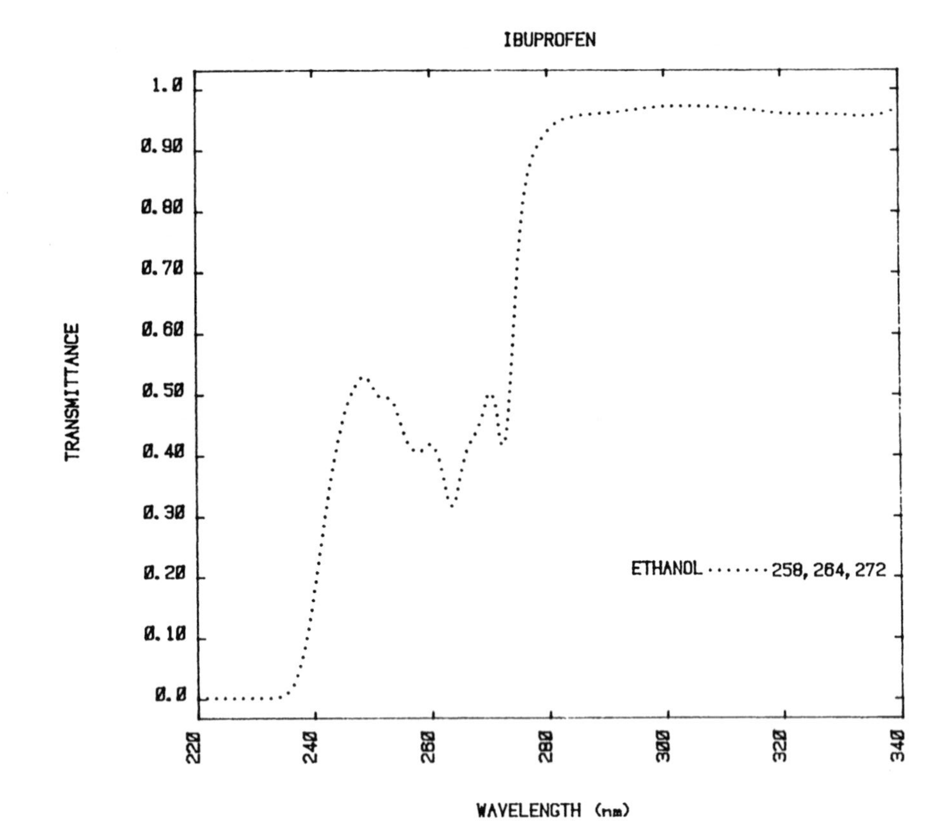


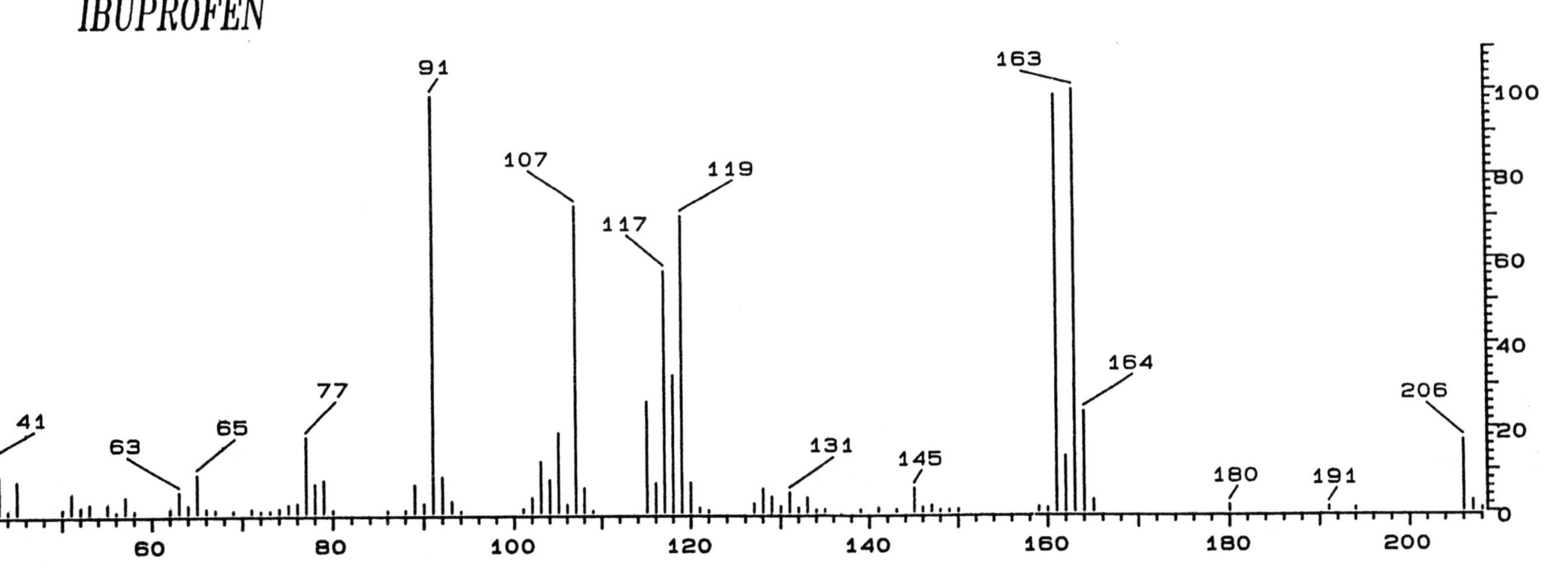

IBUPROFEN

FTNMR

3000 2500 2000 1500 1000 500 0 Hz

PEAK LISTING

| # | HT | PPM |
|---|---|---|
| 1 | 90 | .88 |
| 2 | 99 | .90 |
| 3 | 42 | 1.47 |
| 4 | 45 | 1.50 |
| 5 | 27 | 2.42 |
| 6 | 26 | 2.45 |
| 7 | 28 | 7.10 |
| 8 | 28 | 7.20 |

9 8 7 6 5 4 3 2 1 0 PPM

IBUPROFEN

FT-IR

TRANSMITTANCE

20.00 40.00 60.00 80.00 100.00

A 1724
B 1505
C 1421
D 1231

4000 3600 3200 2800 2400 2000 1800 1600 1400 1200 1000 800 600 400

WAVENUMBERS

# IMIDAZOLE

$C_3H_4N_2$

Molecular weight: 68.08 (68.04)
Synonyms: Glyoxaline; 1,3-diazole; iminazole; miazole; pyrro[b]monazole; 1,3-diaza-2,4-cyclopentadiene
Trade names:

Use: Synthesis
HPLC: Si-10; 5A:95B; 6.1
GC: 1020; 140°C

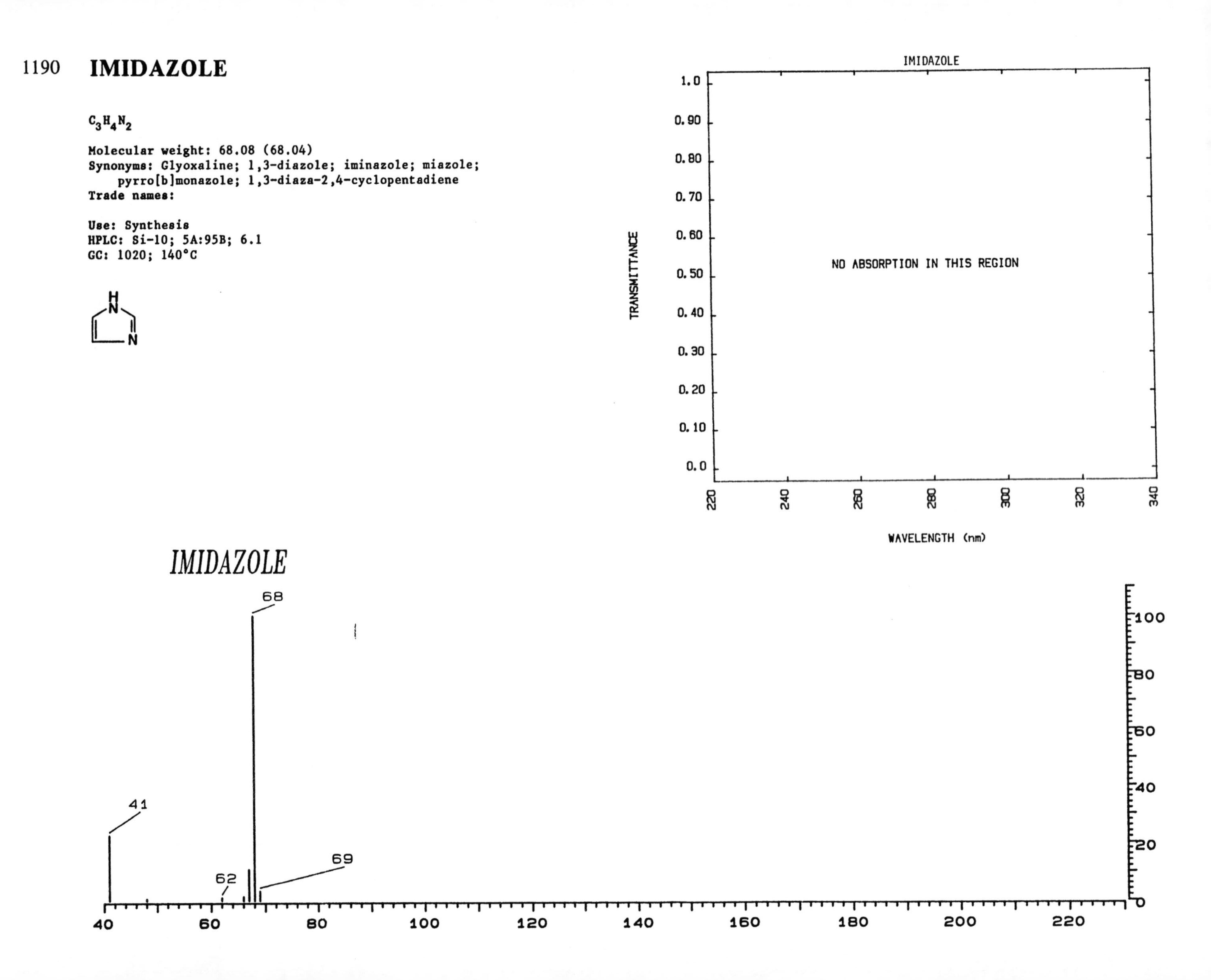

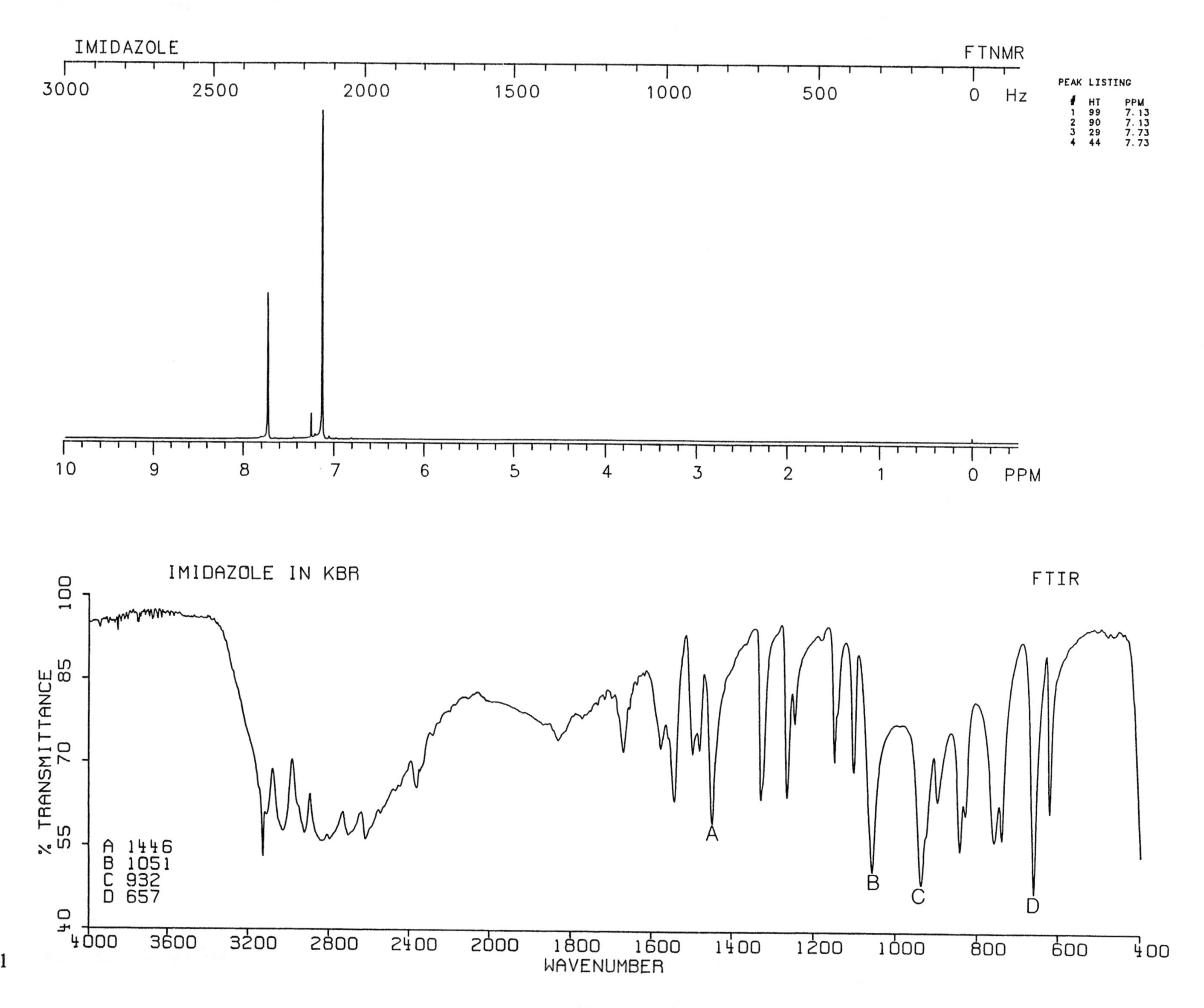
IMIDAZOLE
FTNMR
3000 2500 2000 1500 1000 500 0 Hz
PEAK LISTING
# HT PPM
1 99 7.13
2 90 7.13
3 29 7.73
4 44 7.73
10 9 8 7 6 5 4 3 2 1 0 PPM
IMIDAZOLE IN KBR
FTIR
% TRANSMITTANCE
100 85 70 55 40
A 1446
B 1051
C 932
D 657
A
B
C
D
4000 3600 3200 2800 2400 2000 1800 1600 1400 1200 1000 800 600 400
WAVENUMBER

# IMINOSTILBENE

$C_{14}H_{11}N$

Molecular weight: 193.25 (193.09)
Synonyms: 5H-dibenz[b,f]azepine

Trade names:

Use: Metabolite of carbamazepine
HPLC:
GC: 2026; 200°C

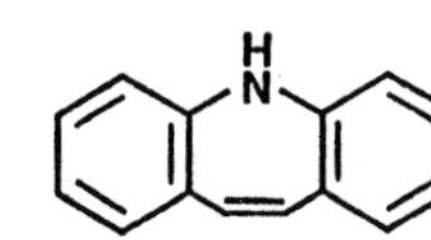


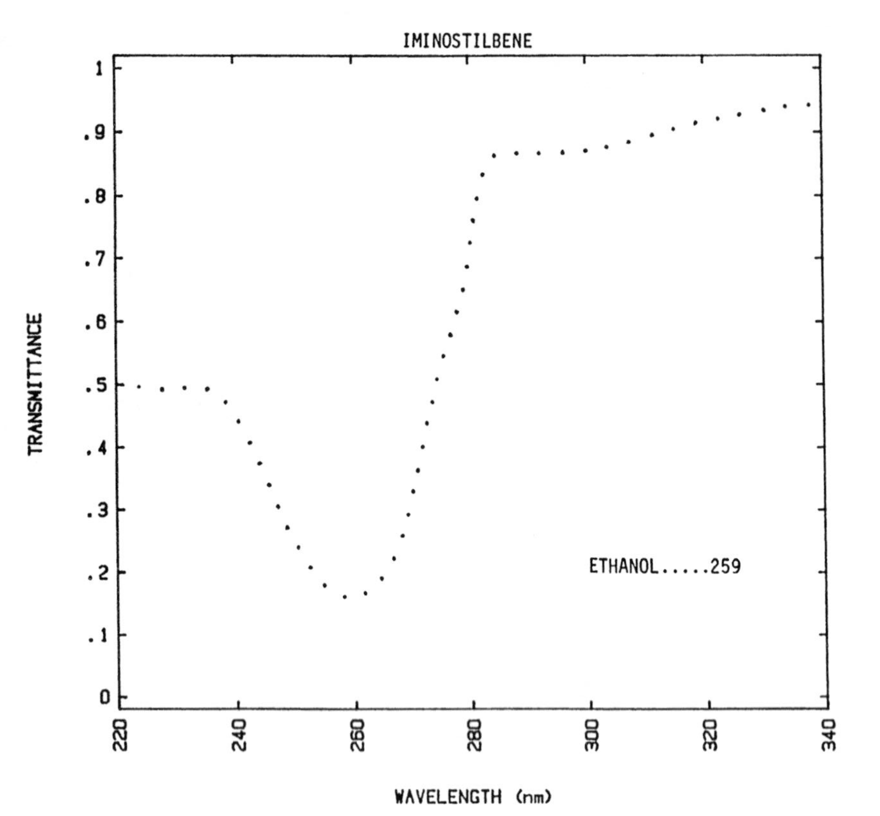


*IMINOSTILBENE*

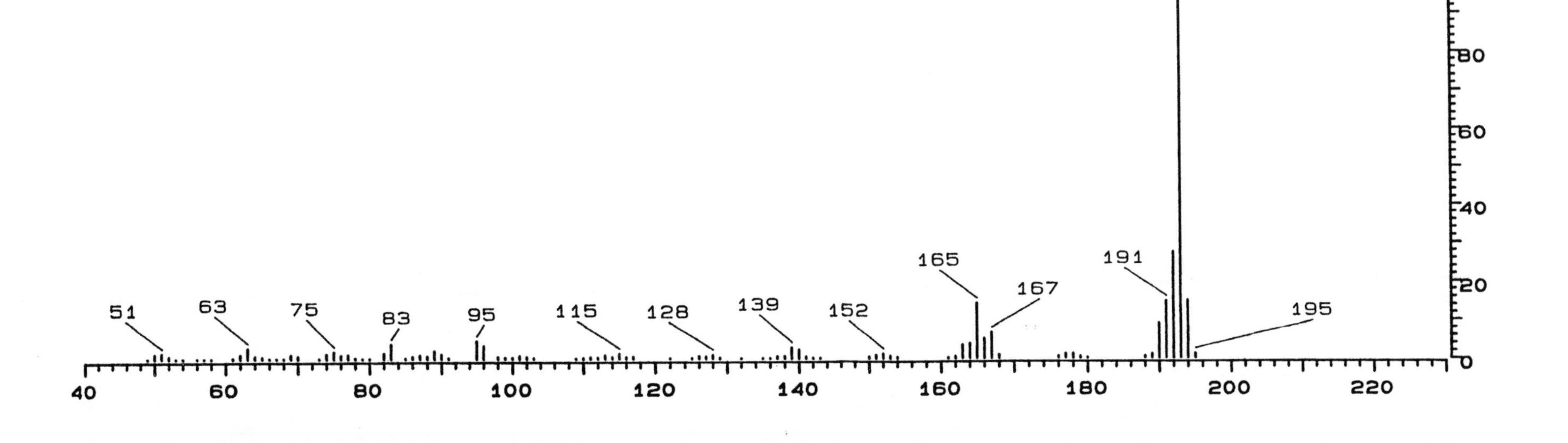

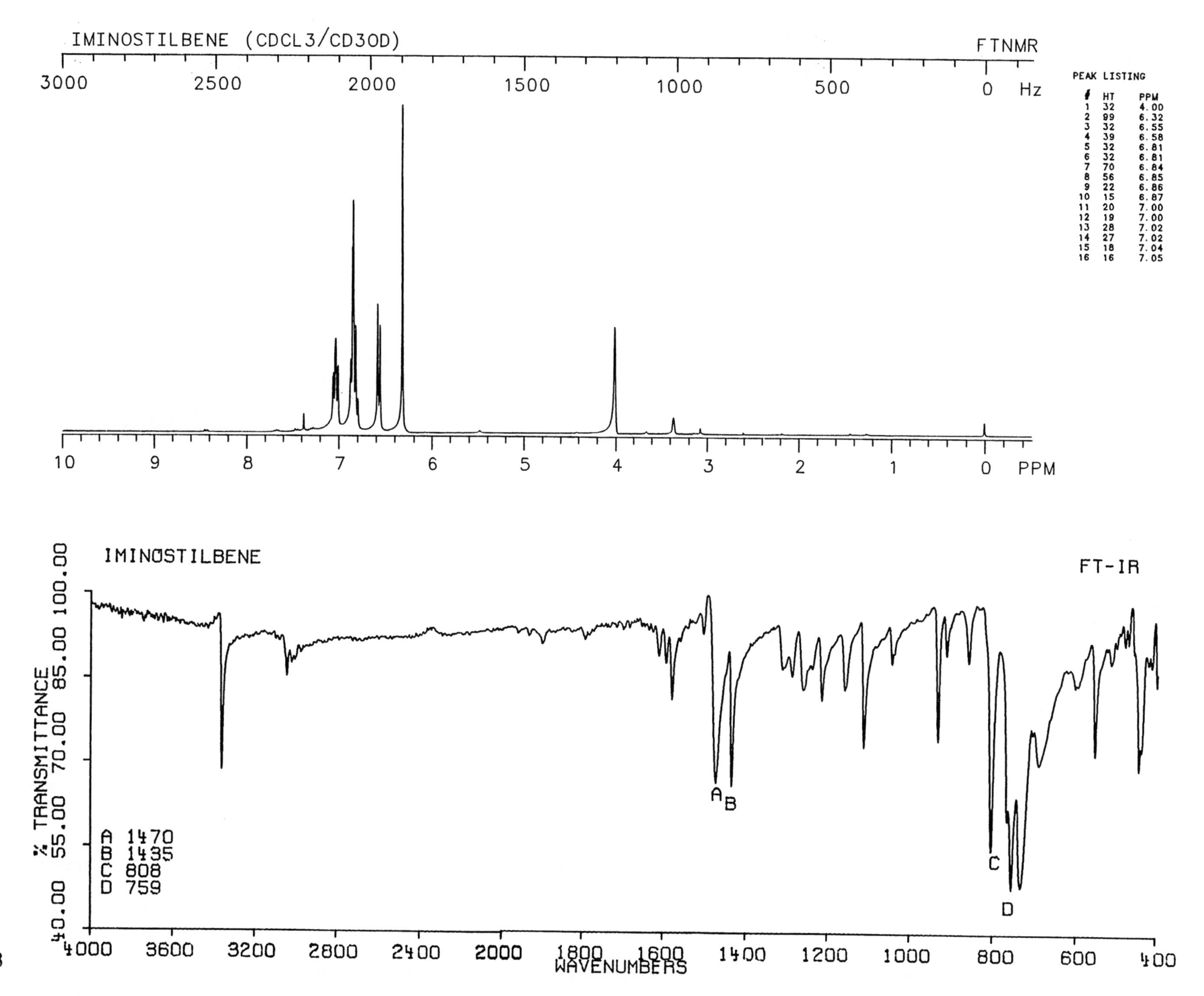
IMINOSTILBENE (CDCL3/CD3OD)
FTNMR
PEAK LISTING
# HT PPM
1 32 4.00
2 99 6.32
3 32 6.55
4 39 6.58
5 32 6.81
6 32 6.81
7 70 6.84
8 56 6.85
9 22 6.86
10 15 6.87
11 20 7.00
12 19 7.00
13 28 7.02
14 27 7.02
15 18 7.04
16 16 7.05
Hz
PPM
IMINOSTILBENE
FT-IR
% TRANSMITTANCE
WAVENUMBERS
A 1470
B 1435
C 808
D 759

# IMIPRAMINE

$C_{19}H_{24}N_2$

**Molecular weight:** 280.41 (280.19)
**Synonyms:** 10,11-Dihydro-N,N-dimethyl-5H-dibenz[b,f]azepine-5-propanamine
**Trade names:** Janimine, Tofranil

**Use:** Antidepressant
**HPLC:** Si-10; 5A:95B; 5.5
**GC:** 2276; 250°C

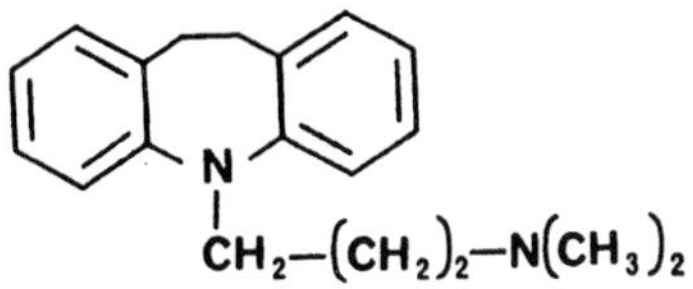

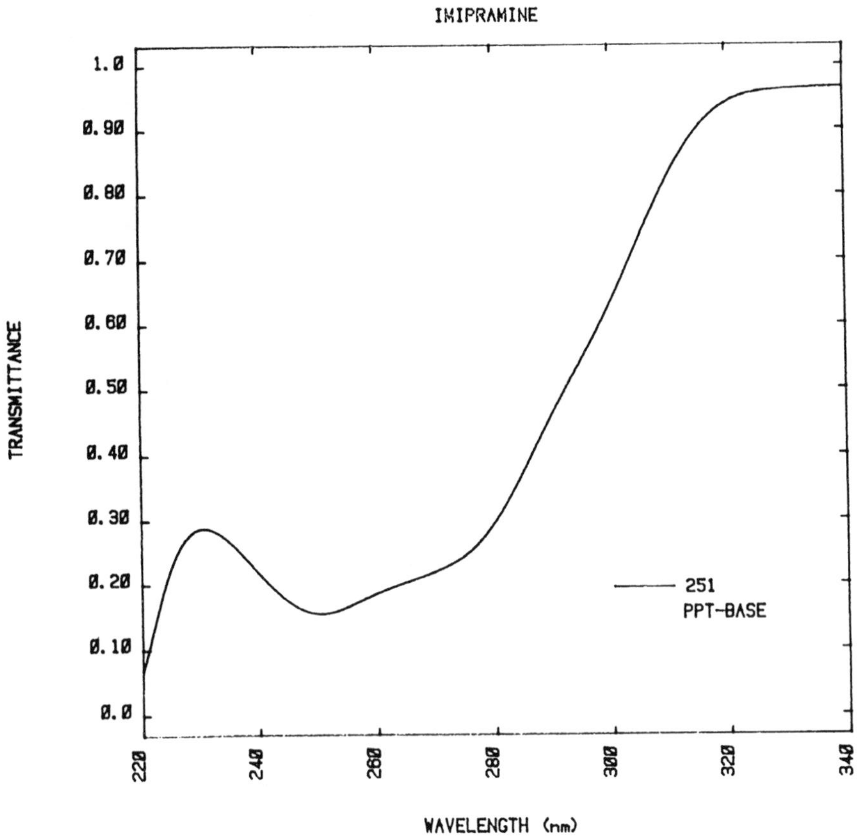

$CH_2—(CH_2)_2—N(CH_3)_2$

IMIPRAMINE

TRANSMITTANCE

1.0 0.90 0.80 0.70 0.60 0.50 0.40 0.30 0.20 0.10 0.0

251 PPT-BASE

220 240 260 280 300 320 340

WAVELENGTH (nm)

*IMIPRAMINE*

234 58 193 220 208 280 85 130 91 42 152 165 179 117 265

60 80 100 120 140 160 180 200 220 240 260 280

100 80 60 40 20 0

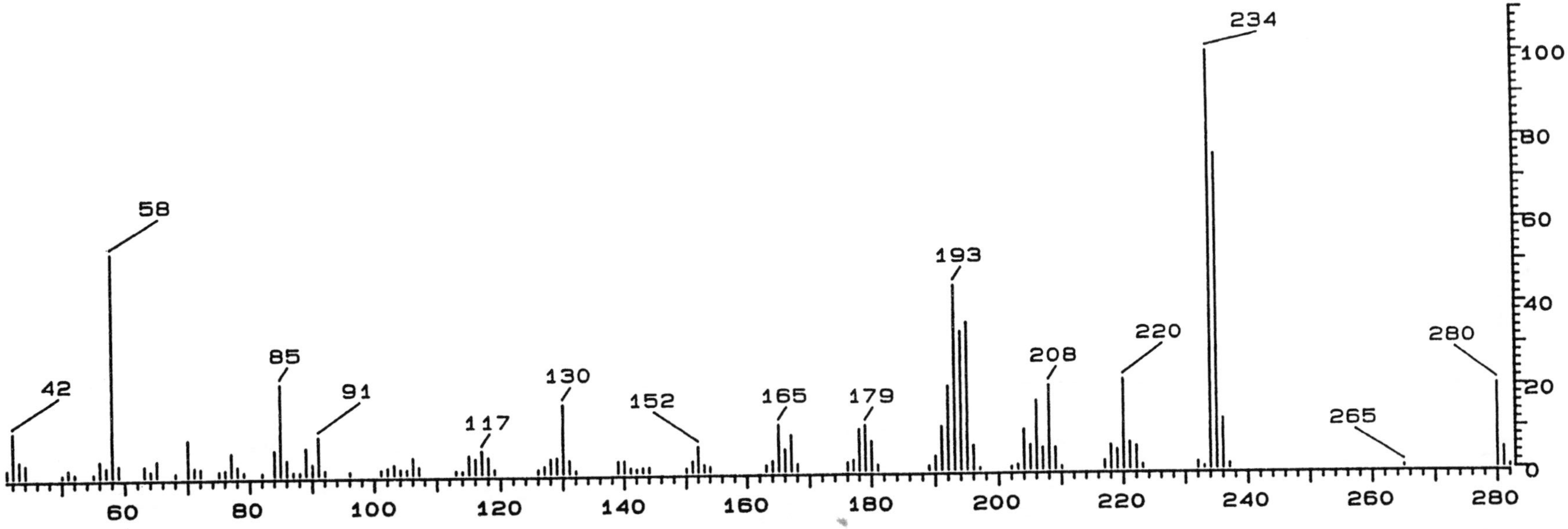

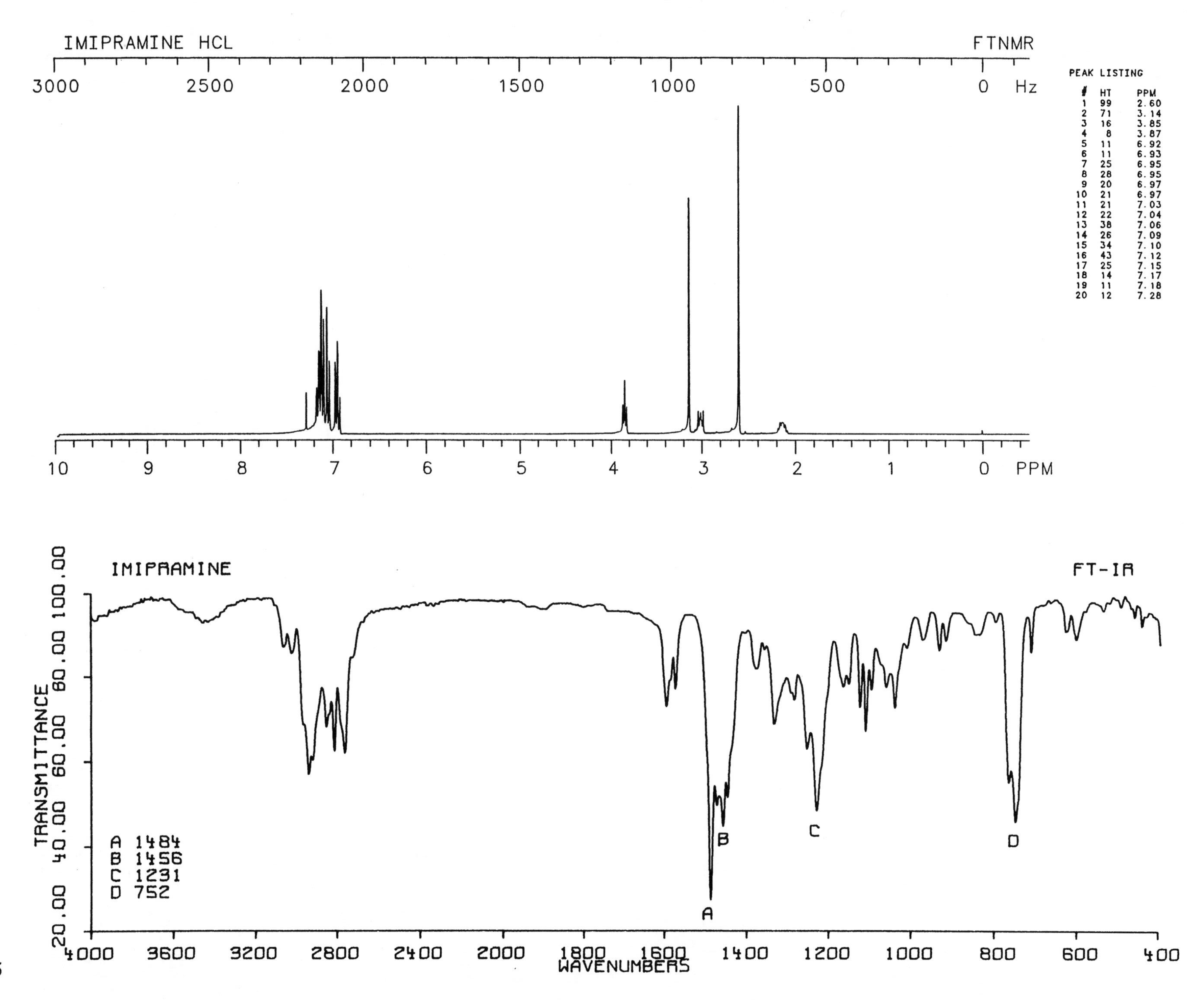
IMIPRAMINE HCL
FTNMR
3000
2500
2000
1500
1000
500
0 Hz
PEAK LISTING
# HT PPM
1 99 2.60
2 71 3.14
3 16 3.85
4 8 3.87
5 11 6.92
6 11 6.93
7 25 6.95
8 28 6.95
9 20 6.97
10 21 6.97
11 21 7.03
12 22 7.04
13 38 7.06
14 26 7.09
15 34 7.10
16 43 7.12
17 25 7.15
18 14 7.17
19 11 7.18
20 12 7.28
10 9 8 7 6 5 4 3 2 1 0 PPM
IMIPRAMINE
FT-IR
TRANSMITTANCE
20.00 40.00 60.00 80.00 100.00
A 1484
B 1456
C 1231
D 752
A
B
C
D
4000 3600 3200 2800 2400 2000 1800 1600 1400 1200 1000 800 600 400
WAVENUMBERS

# INDAPAMIDE

$C_{16}H_{16}ClN_3O_3S$

Molecular weight: 365.84 (365.06)
Synonyms: 3-(Aminosulfonyl)-4-chloro-N-(2,3-dihydro-2-methyl-1H-indol-1-yl)benzamide
Trade names: Fludex, Ipamix, Lozol, Natrilix

Use: Diuretic
HPLC: Si-10; 1A;99B; 5.0
GC:

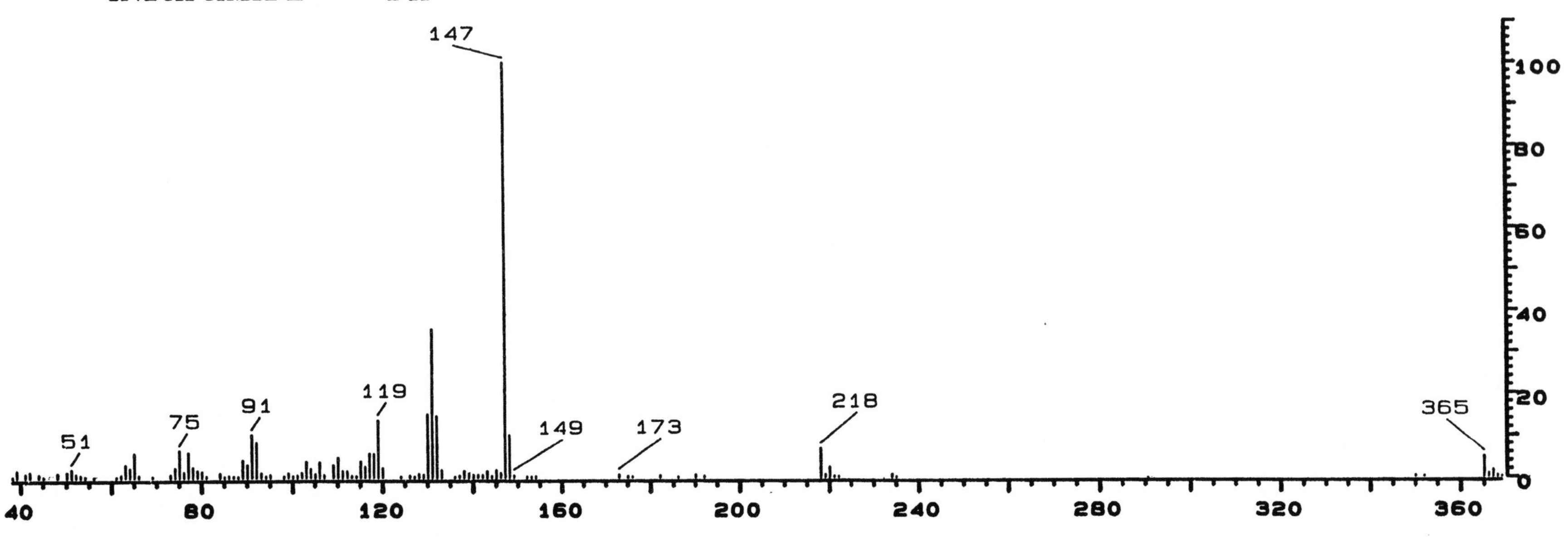

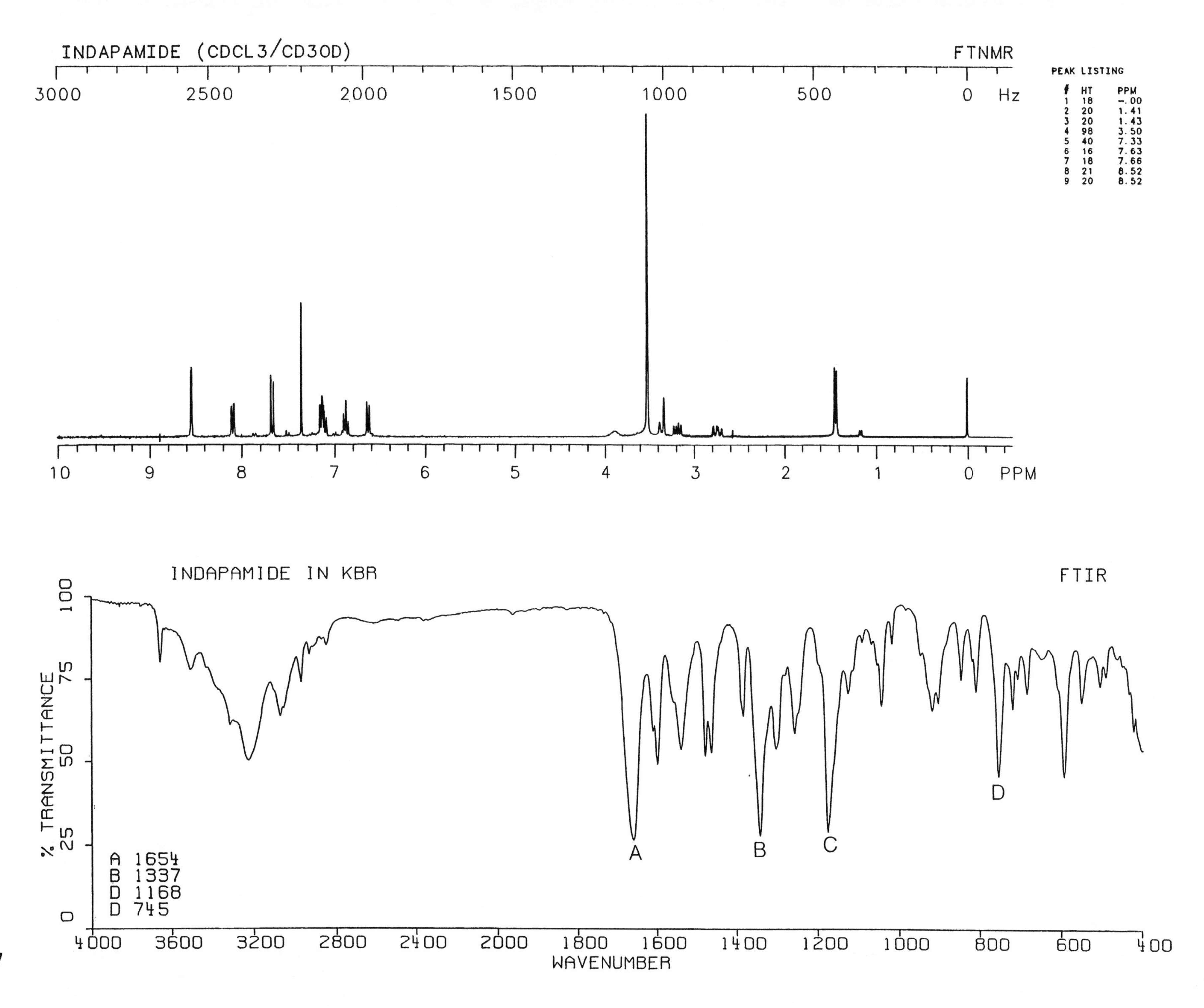

INDAPAMIDE (CDCL3/CD3OD)
FTNMR
3000 2500 2000 1500 1000 500 0 Hz
PEAK LISTING
# HT PPM
1 18 -.00
2 20 1.41
3 20 1.43
4 98 3.50
5 40 7.33
6 16 7.63
7 18 7.66
8 21 8.52
9 20 8.52
10 9 8 7 6 5 4 3 2 1 0 PPM
INDAPAMIDE IN KBR
FTIR
% TRANSMITTANCE
100 75 50 25 0
A 1654
B 1337
D 1168
D 745
A B C D
4000 3600 3200 2800 2400 2000 1800 1600 1400 1200 1000 800 600 400
WAVENUMBER

# INDOLE-3-ACETIC ACID

$C_{10}H_9NO_2$

Molecular weight: 175.18 (175.06)
Synonyms: Indoleacetic acid; heteroauxin

Trade names:

Use: Plant hormone, synthesis
HPLC: Si-10; 20A:80B; 5.3
GC: 1916; 200°C

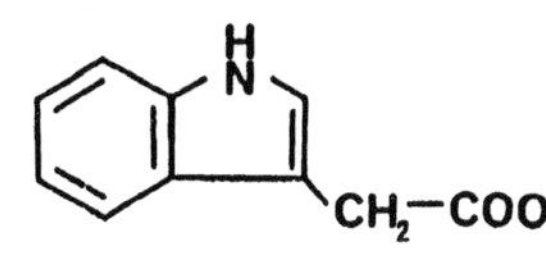


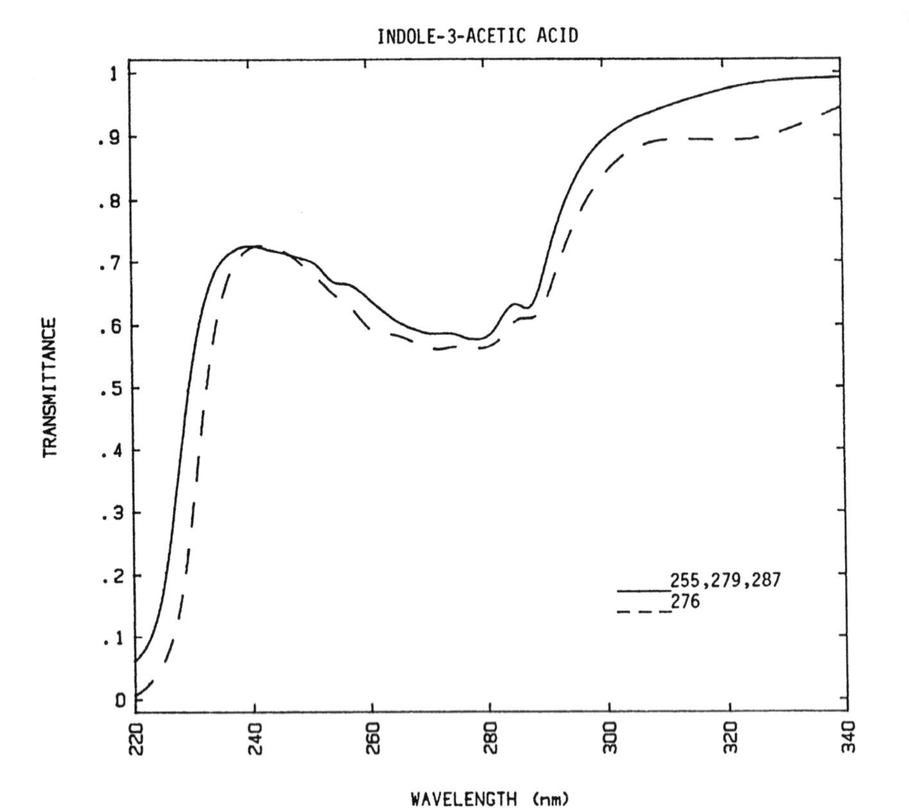


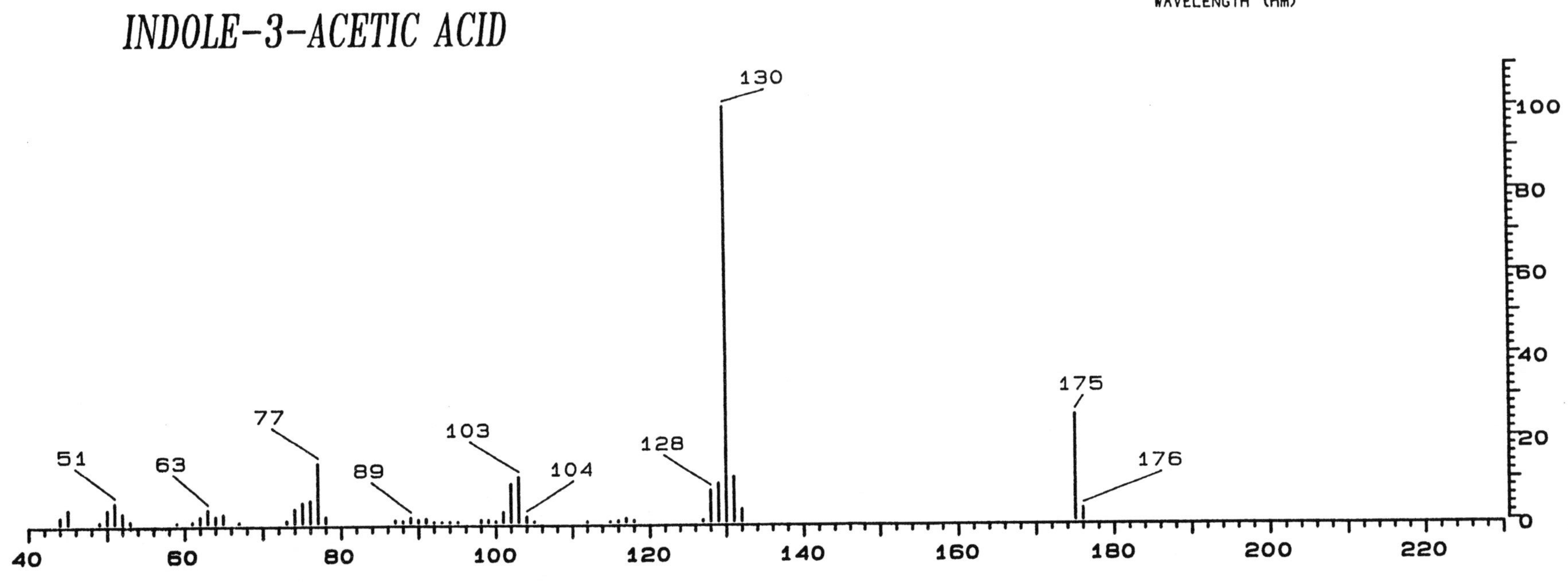

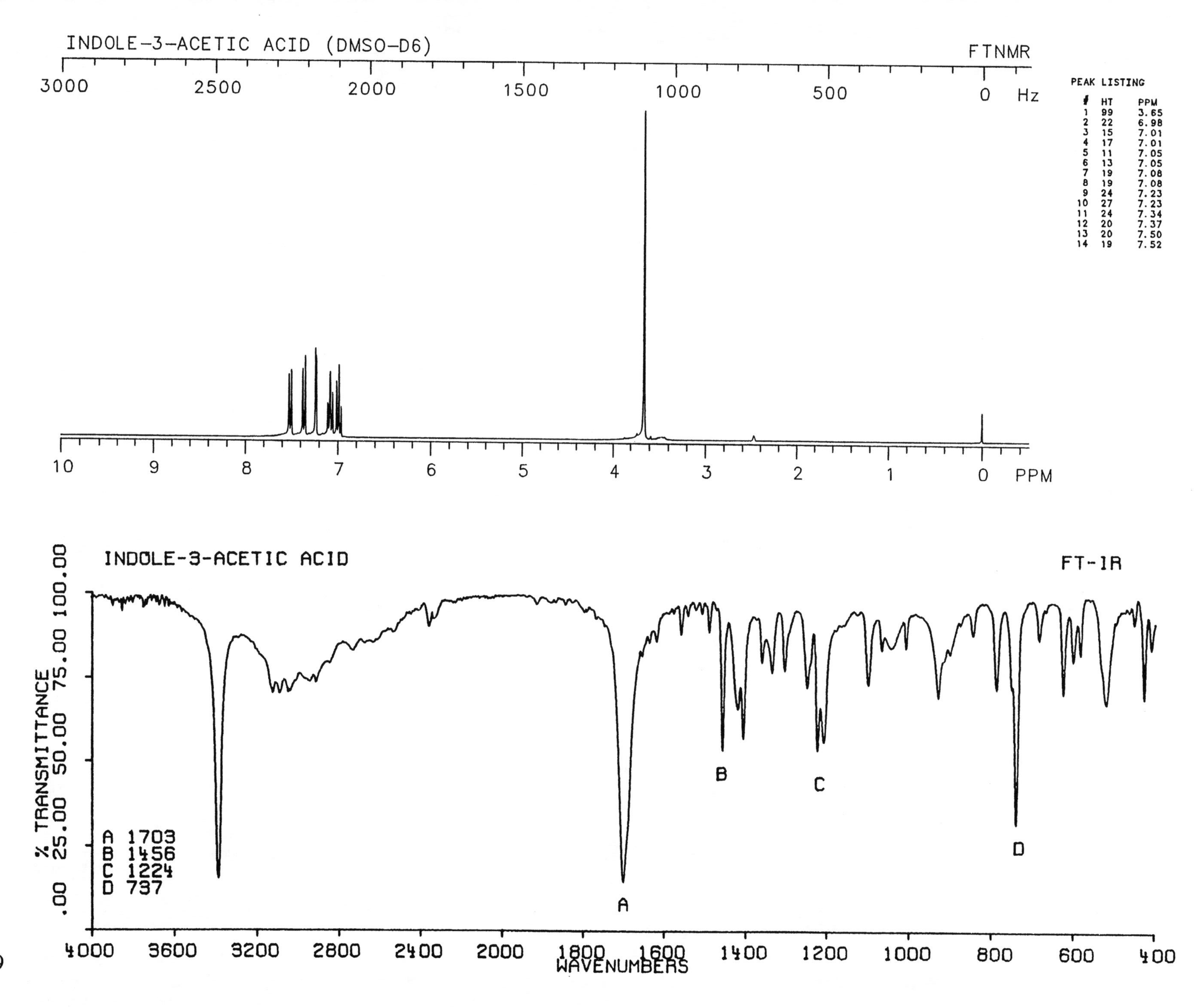

INDOLE-3-ACETIC ACID (DMSO-D6)
FTNMR
3000 2500 2000 1500 1000 500 0 Hz
PEAK LISTING
# HT PPM
1 99 3.65
2 22 6.98
3 15 7.01
4 17 7.01
5 11 7.05
6 13 7.05
7 19 7.08
8 19 7.08
9 24 7.23
10 27 7.23
11 24 7.34
12 20 7.37
13 20 7.50
14 19 7.52
10 9 8 7 6 5 4 3 2 1 0 PPM
INDOLE-3-ACETIC ACID
FT-IR
% TRANSMITTANCE
.00 25.00 50.00 75.00 100.00
A 1703
B 1456
C 1224
D 737
A
B
C
D
4000 3600 3200 2800 2400 2000 1800 1600 1400 1200 1000 800 600 400
WAVENUMBERS

# INDOMETHACIN

$C_{19}H_{16}ClNO_4$

**Molecular weight:** 357.79 (357.08)
**Synonyms:** 1-(p-Chlorobenzoyl)-5-methoxy-2-methylindole-3-acetic acid

**Trade names:** Indocin

**Use:** Anti-inflammatory
**HPLC:** Si-10; 20A:80B; 4.5
**GC:** 2641; 250°C

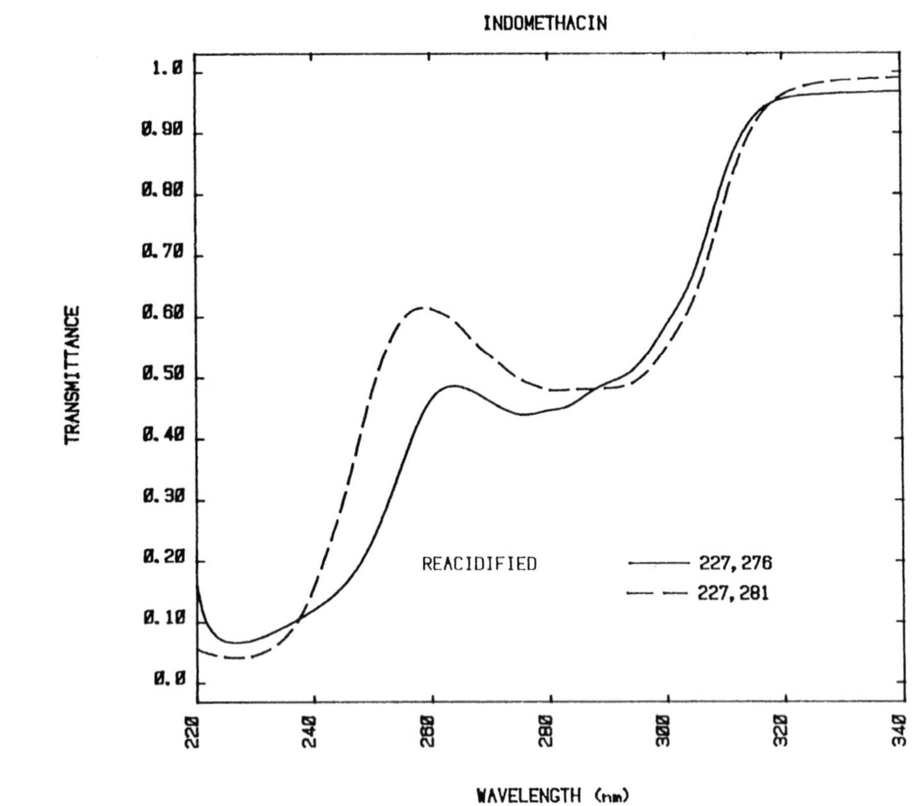


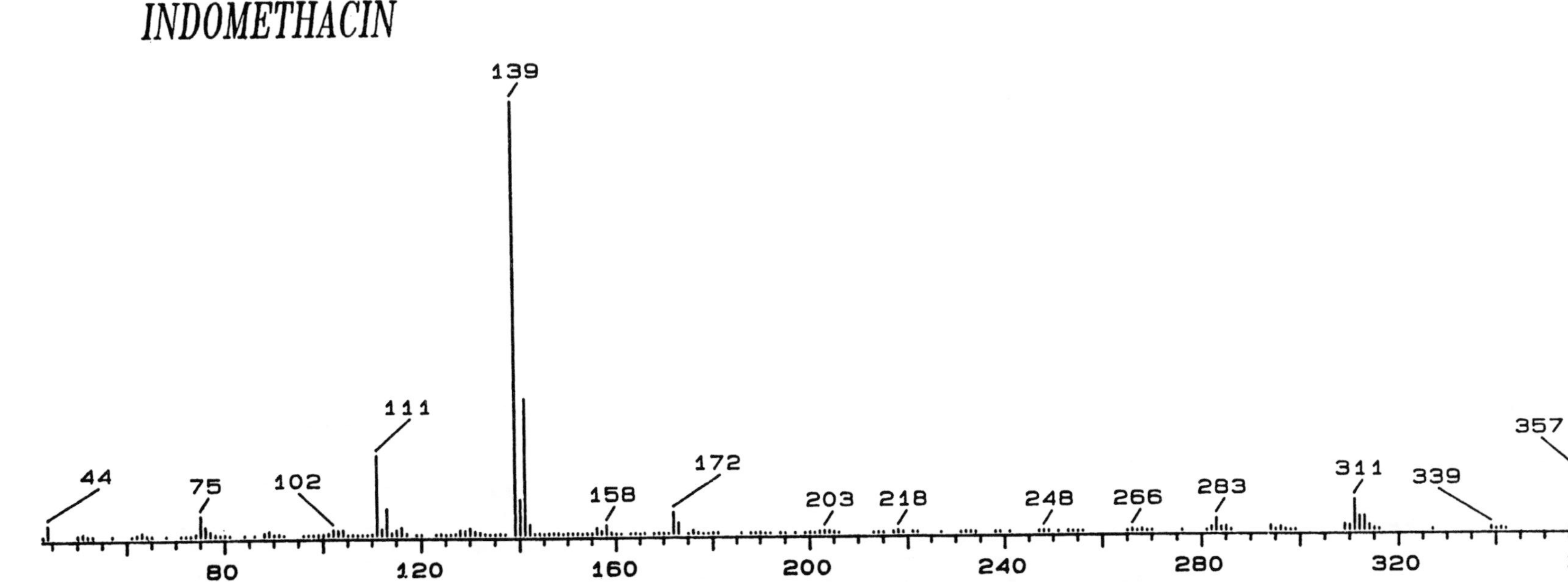

INDOMETHACIN

FTNMR

3000 2500 2000 1500 1000 500 0 Hz

PEAK LISTING

| # | HT | PPM |
|---|---|---|
| 1 | 66 | 2.37 |
| 2 | 37 | 3.66 |
| 3 | 99 | 3.82 |
| 4 | 15 | 6.84 |
| 5 | 13 | 6.95 |
| 6 | 12 | 6.95 |
| 7 | 18 | 7.44 |
| 8 | 23 | 7.47 |
| 9 | 23 | 7.64 |
| 10 | 20 | 7.66 |

10 9 8 7 6 5 4 3 2 1 0 PPM

INDOMETHACIN

FT-IR

A 1696
B 1675
C 1357
D 759

AB C D

TRANSMITTANCE 20.00 40.00 60.00 80.00 100.00

4000 3600 3200 2800 2400 2000 1800 1600 1400 1200 1000 800 600 400
WAVENUMBERS

# INOSITOL

$C_6H_{12}O_6$

Molecular weight: 180.16 (180.06)
Synonyms: myo-Inositol; meso-inositol; hexahydroxycyclohexane; cyclohexitol; meat sugar
Trade names: Amino-Cerv, B-50, Cardenz, Lecithin, Mega-B, Megadose, Sclerex
Use: Lipotropic
HPLC: $LiNH_2$; 70D:30E; 9.0
GC:

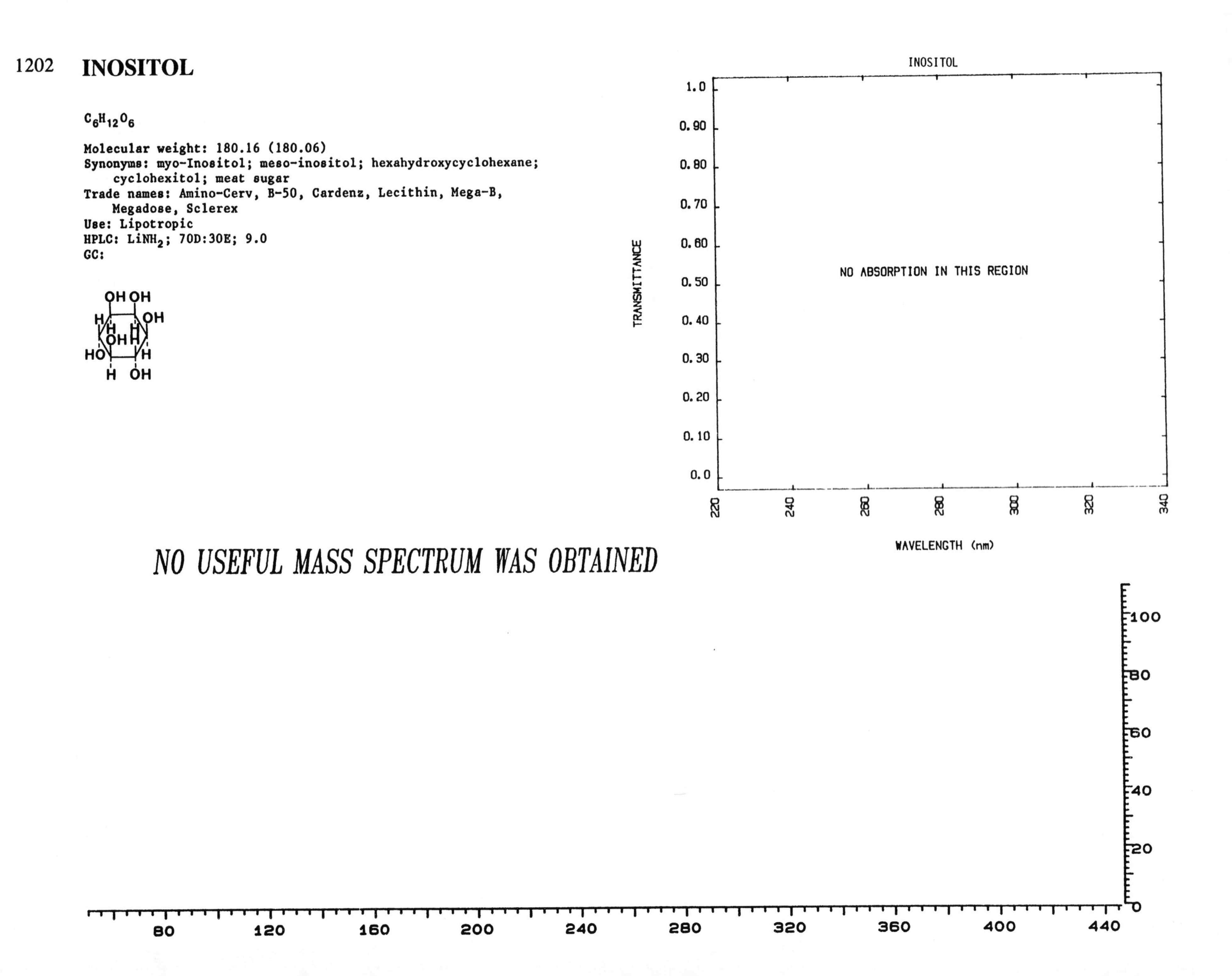

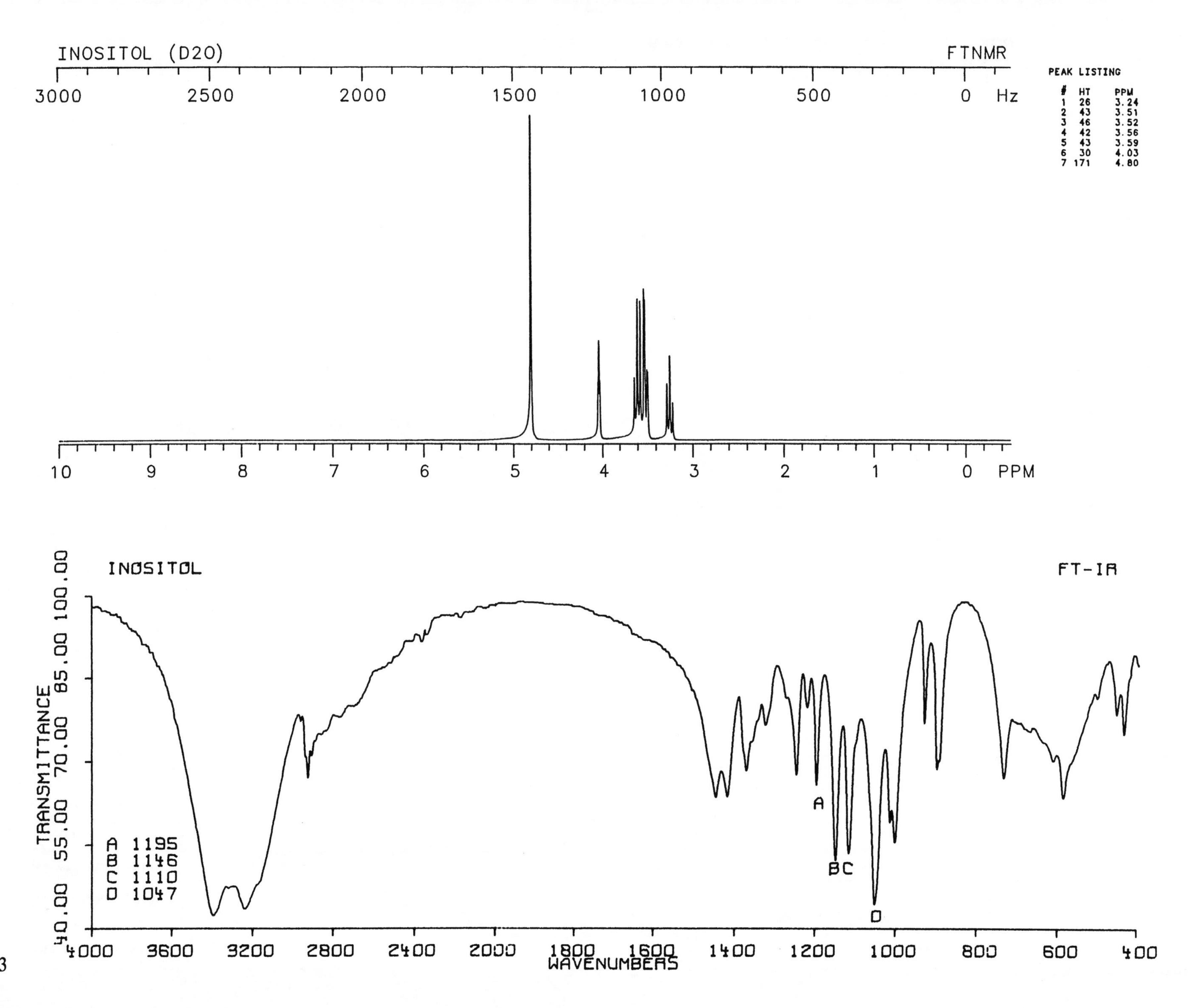
INOSITOL (D2O)
FTNMR
3000 2500 2000 1500 1000 500 0 Hz
PEAK LISTING
# HT PPM
1 26 3.24
2 43 3.51
3 46 3.52
4 42 3.56
5 43 3.59
6 30 4.03
7 171 4.80
10 9 8 7 6 5 4 3 2 1 0 PPM
INOSITOL
FT-IR
TRANSMITTANCE
40.00 55.00 70.00 85.00 100.00
A 1195
B 1146
C 1110
D 1047
A
BC
D
4000 3600 3200 2800 2400 2000 1800 1600 1400 1200 1000 800 600 400
WAVENUMBERS

# IODIPAMIDE

$C_{20}H_{14}I_6N_2O_6$

**Molecular weight:** 1139.41 (1139.51)
**Synonyms:** 3,3'-[(1,6-Dioxo-1,6-hexanediyl)diimino]bis[2,4,6-triiodobenzoic acid]; adipiodone
**Trade names:** Biligrafin, Cholegrafin, Cholografin, Cholospect, Endografin Intrabilix
**Use:** Diagnostic aid
**HPLC:**
**GC:**

COOH COOH
I I I I
NH—C—(CH₂)₄—C—NH
O O
I I

IODIPAMIDE

TRANSMITTANCE

1 .9 .8 .7 .6 .5 .4 .3 .2 .1 0

220 240 260 280 300 320 340

WAVELENGTH (nm)

ETHANOL.....239

NO USEFUL MASS SPECTRUM WAS OBTAINED

100 80 60 40 20 0

80 120 160 200 240 280 320 360 400 440

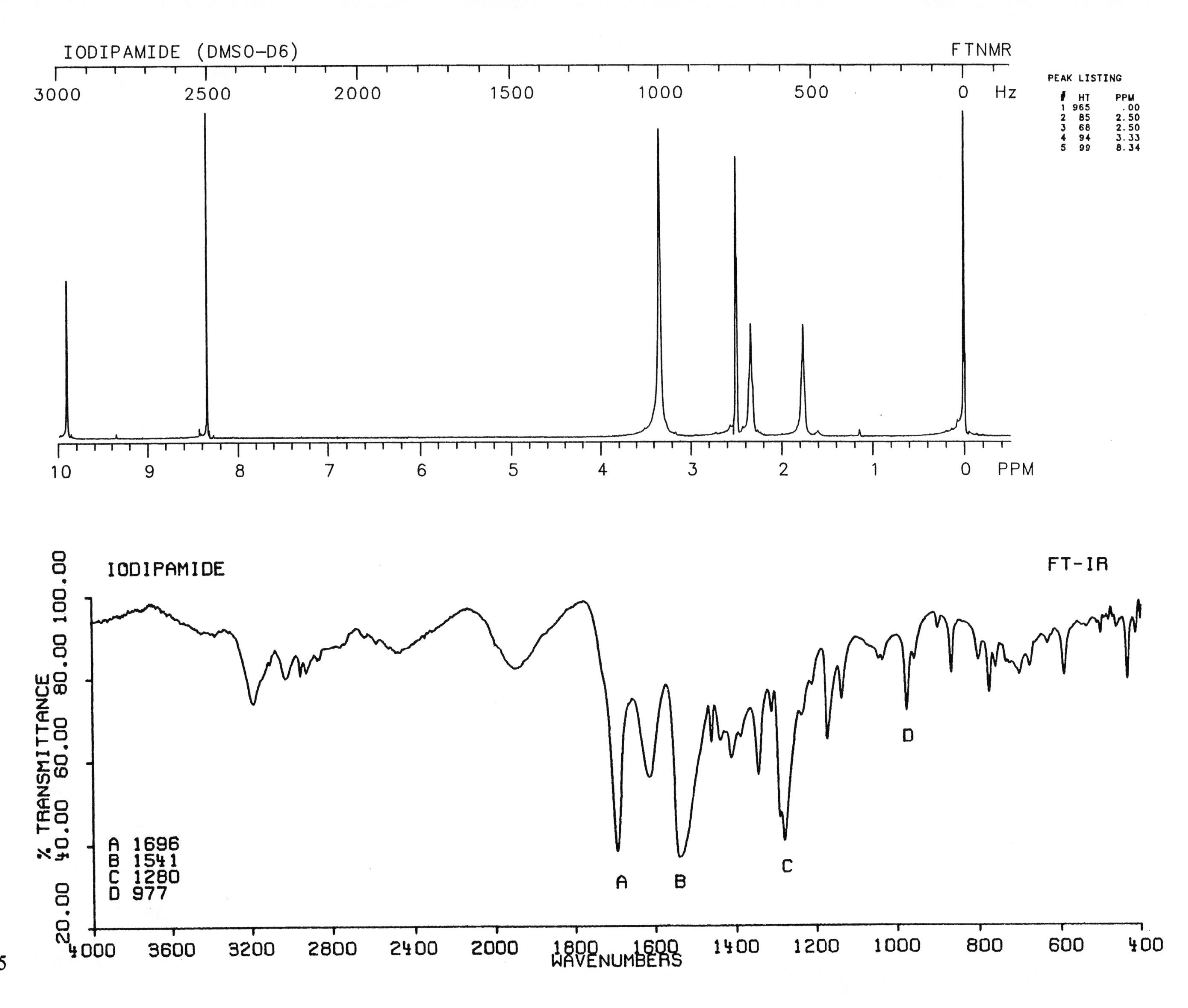
IODIPAMIDE (DMSO-D6)
FTNMR
3000
2500
2000
1500
1000
500
0
Hz
PEAK LISTING
# HT PPM
1 965 .00
2 85 2.50
3 68 2.50
4 94 3.33
5 99 8.34
10
9
8
7
6
5
4
3
2
1
0
PPM
IODIPAMIDE
FT-IR
% TRANSMITTANCE
20.00 40.00 60.00 80.00 100.00
A 1696
B 1541
C 1280
D 977
A
B
C
D
4000 3600 3200 2800 2400 2000 1800 1600 1400 1200 1000 800 600 400
WAVENUMBERS

# IODOCHLORHYDROXYQUIN

$C_9H_5ClINO$

**Molecular weight:** 305.50 (304.91)
**Synonyms:** 5-Chloro-7-iodo-8-quinolinol; chloroiodoquin; iodochlorohydroxyquinoline
**Trade names:** F-E-P, HCV, Nystaform, Pedi-Cort V, Racet, Vioform

**Use:** Topical anti-infective
**HPLC:** Si-10; 100B; 10.0
**GC:** 1967; 200°C

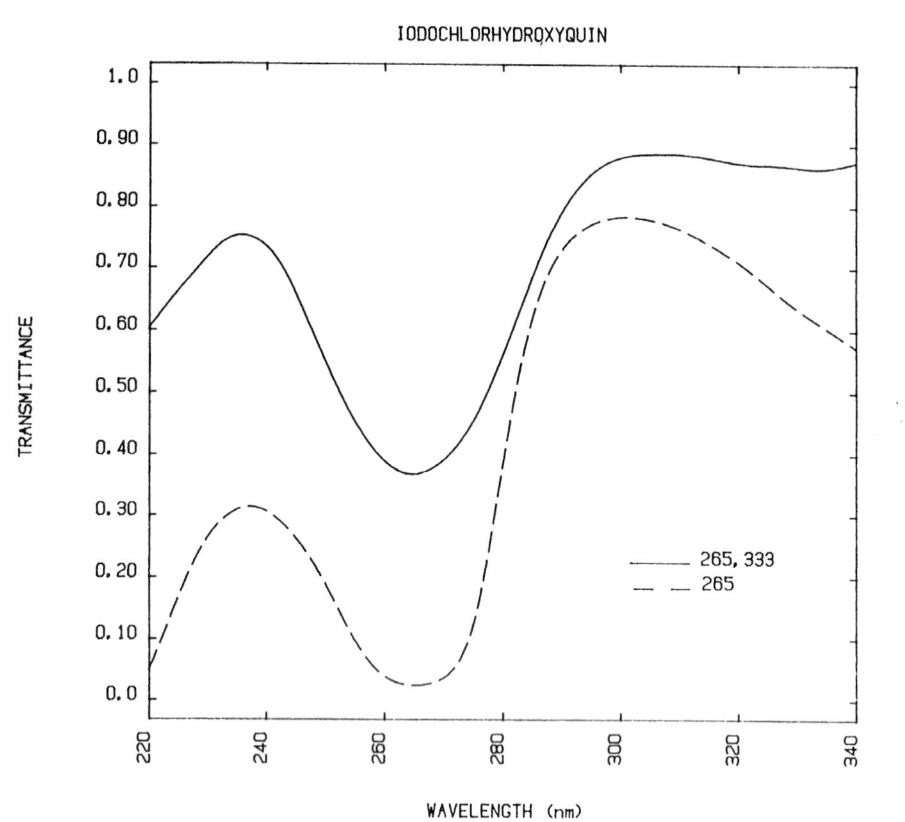


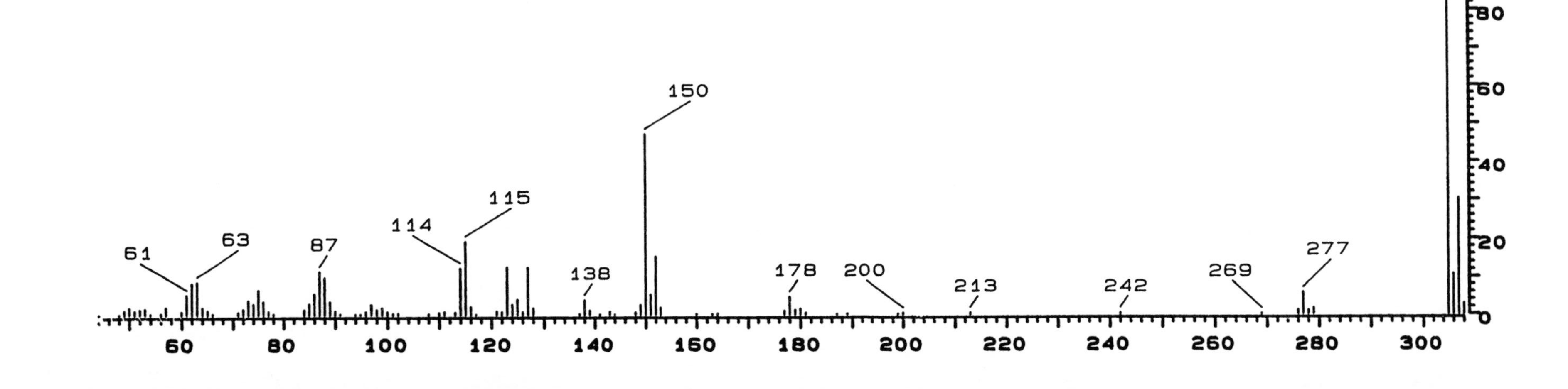

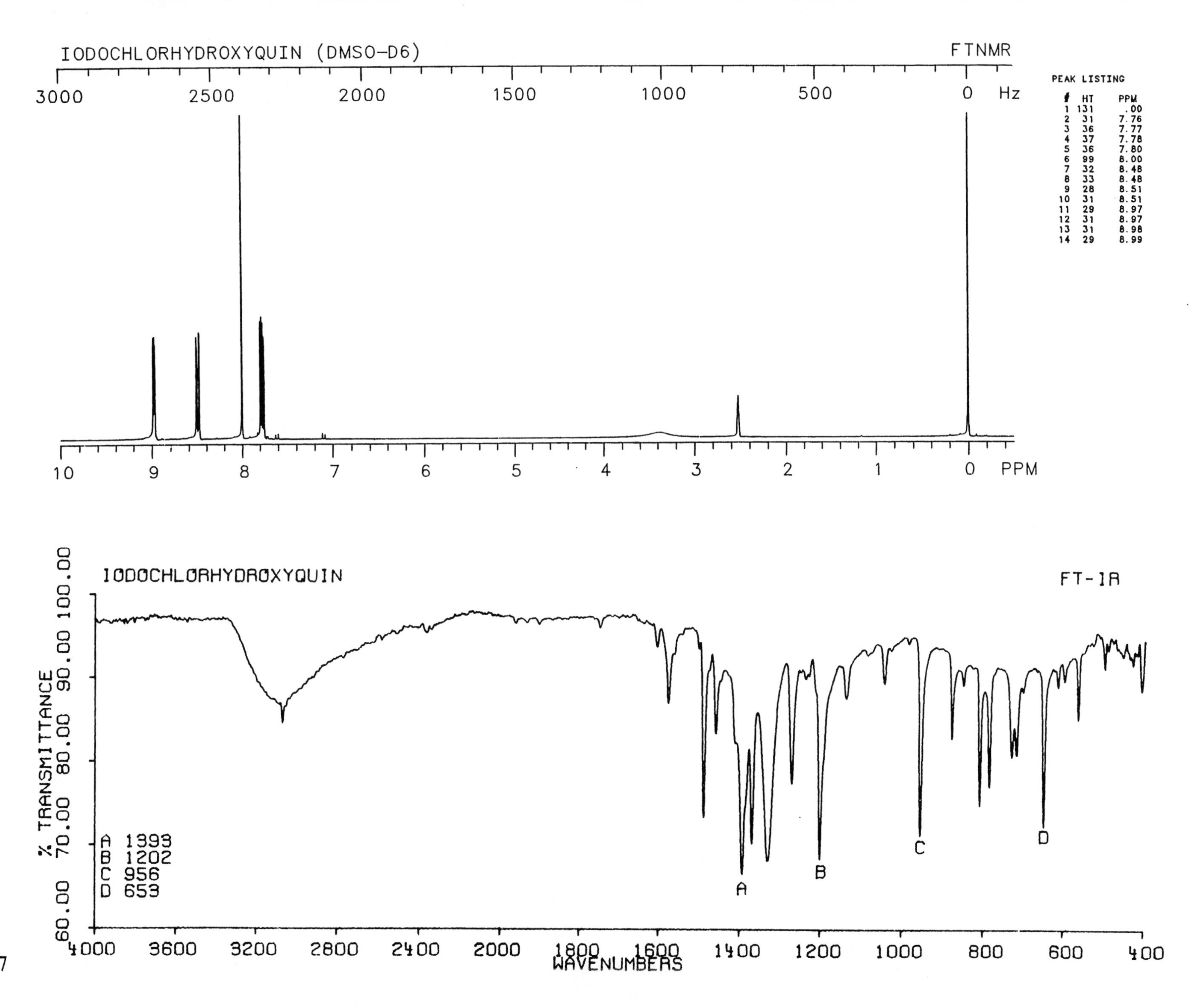

IODOCHLORHYDROXYQUIN (DMSO-D6)
FTNMR
PEAK LISTING
# HT PPM
1 131 .00
2 31 7.76
3 36 7.77
4 37 7.78
5 36 7.80
6 99 8.00
7 32 8.48
8 33 8.48
9 28 8.51
10 31 8.51
11 29 8.97
12 31 8.97
13 31 8.98
14 29 8.99
IODOCHLORHYDROXYQUIN
FT-IR
A 1393
B 1202
C 956
D 653
% TRANSMITTANCE
WAVENUMBERS

# IOPANOIC ACID

$C_{11}H_{12}I_3NO_2$

**Molecular weight:** 570.93 (570.80)
**Synonyms:** 3-Amino- α-ethyl-2,4,6-triiodobenzenepropanic acid; iodopanoic acid
**Trade names:** Telepaque

**Use:** Diagnostic aid
**HPLC:** Si-10; 10A:90B; 5.7
**GC:** 2555; 250°C

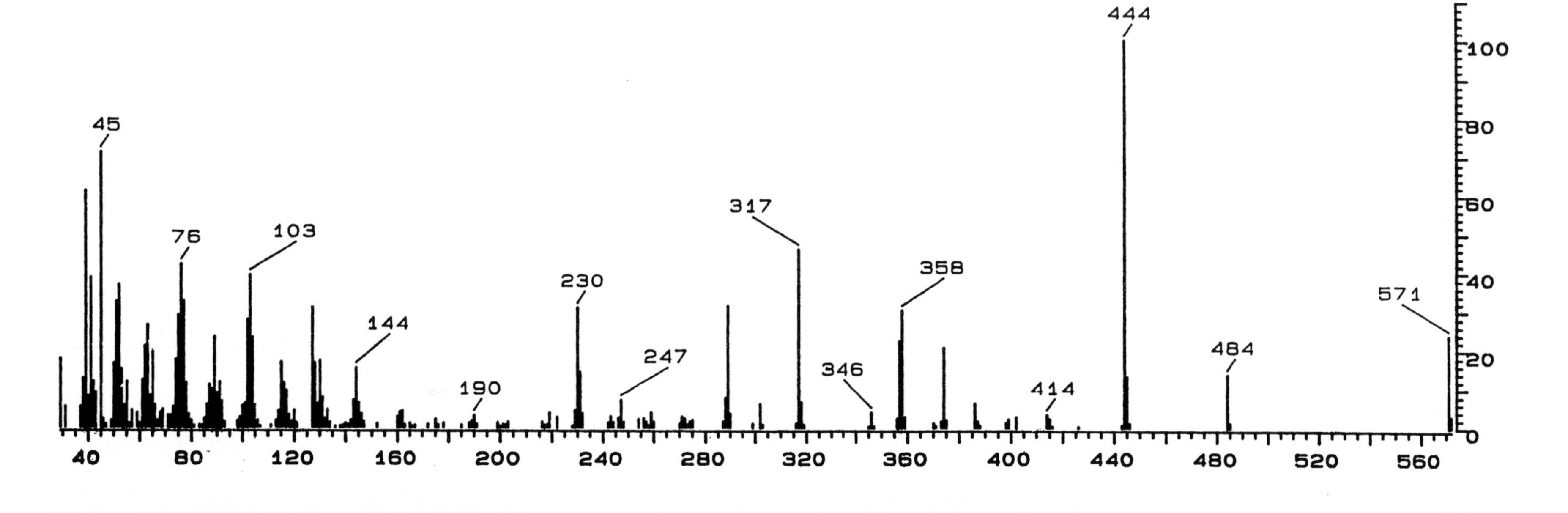

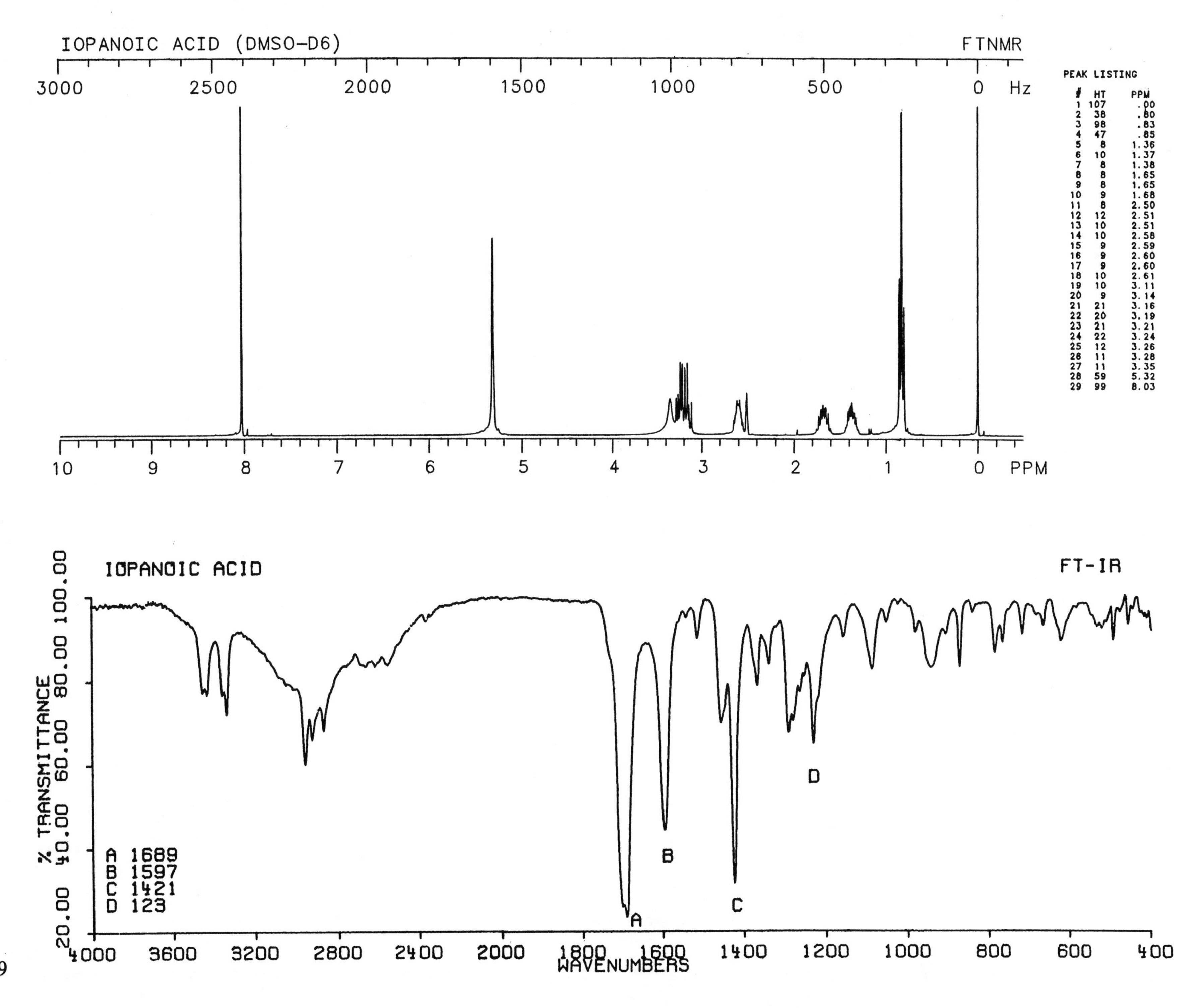
IOPANOIC ACID (DMSO-D6)
FTNMR
3000 2500 2000 1500 1000 500 0 Hz
10 9 8 7 6 5 4 3 2 1 0 PPM
PEAK LISTING
# HT PPM
1 107 .00
2 38 .80
3 98 .83
4 47 .85
5 8 1.36
6 10 1.37
7 8 1.38
8 8 1.65
9 8 1.65
10 9 1.68
11 8 2.50
12 12 2.51
13 10 2.51
14 10 2.58
15 9 2.59
16 9 2.60
17 9 2.60
18 10 2.61
19 10 3.11
20 9 3.14
21 21 3.16
22 20 3.19
23 21 3.21
24 22 3.24
25 12 3.26
26 11 3.28
27 11 3.35
28 59 5.32
29 99 8.03
IOPANOIC ACID
FT-IR
% TRANSMITTANCE
20.00 40.00 60.00 80.00 100.00
A 1689
B 1597
C 1421
D 123
A
B
C
D
4000 3600 3200 2800 2400 2000 1800 1600 1400 1200 1000 800 600 400
WAVENUMBERS

# IPODATE

$C_{12}H_{13}I_3N_2O_2$

Molecular weight: 597.96 (597.81)
Synonyms: 3-[[(Dimethylamino)methylene]amino]-2,4,6-triiodobenzenepropanoic acid
Trade names: Oragrafin

Use: Diagnostic aid
HPLC: Si-10; 5A:95B; 9.5
GC:

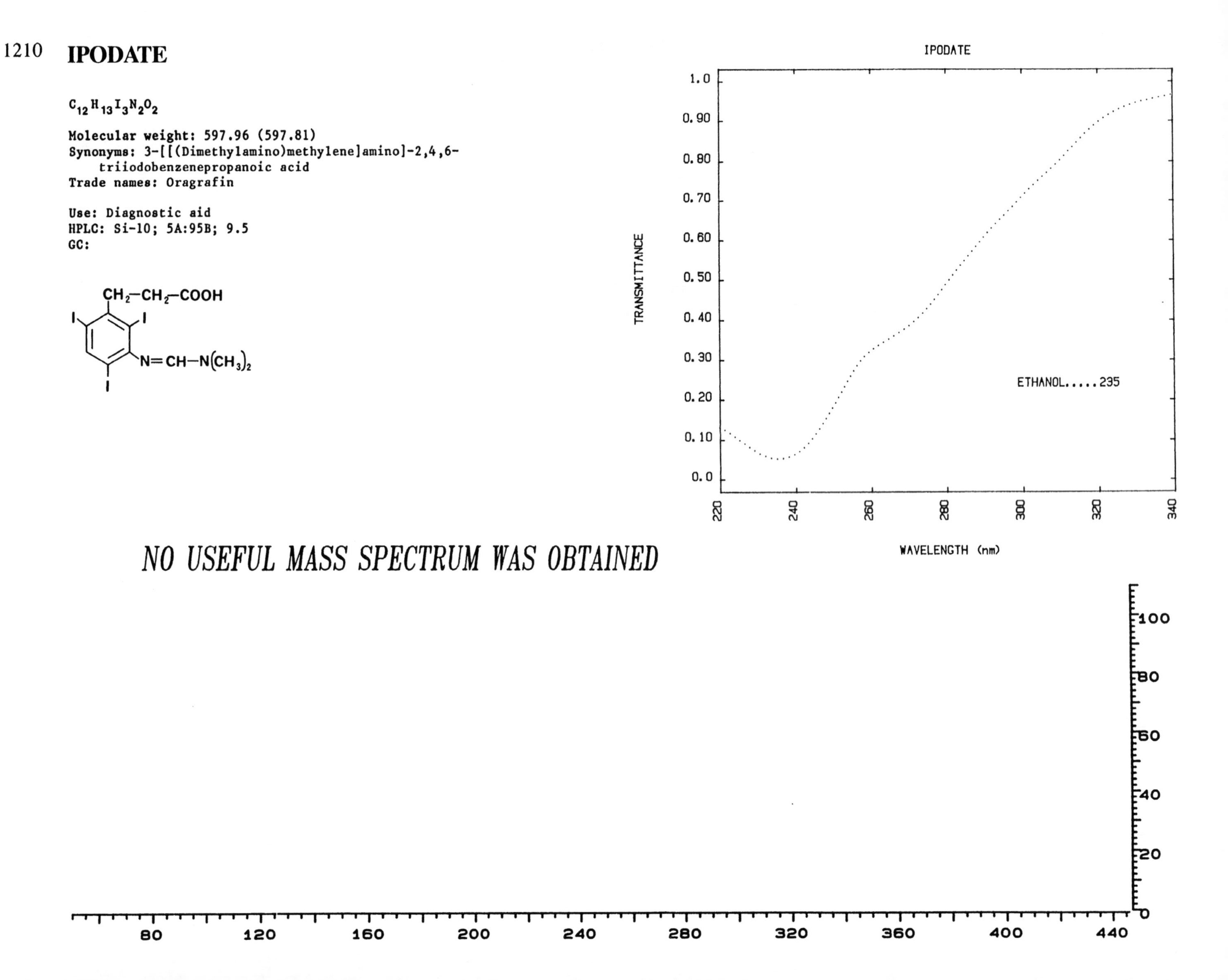

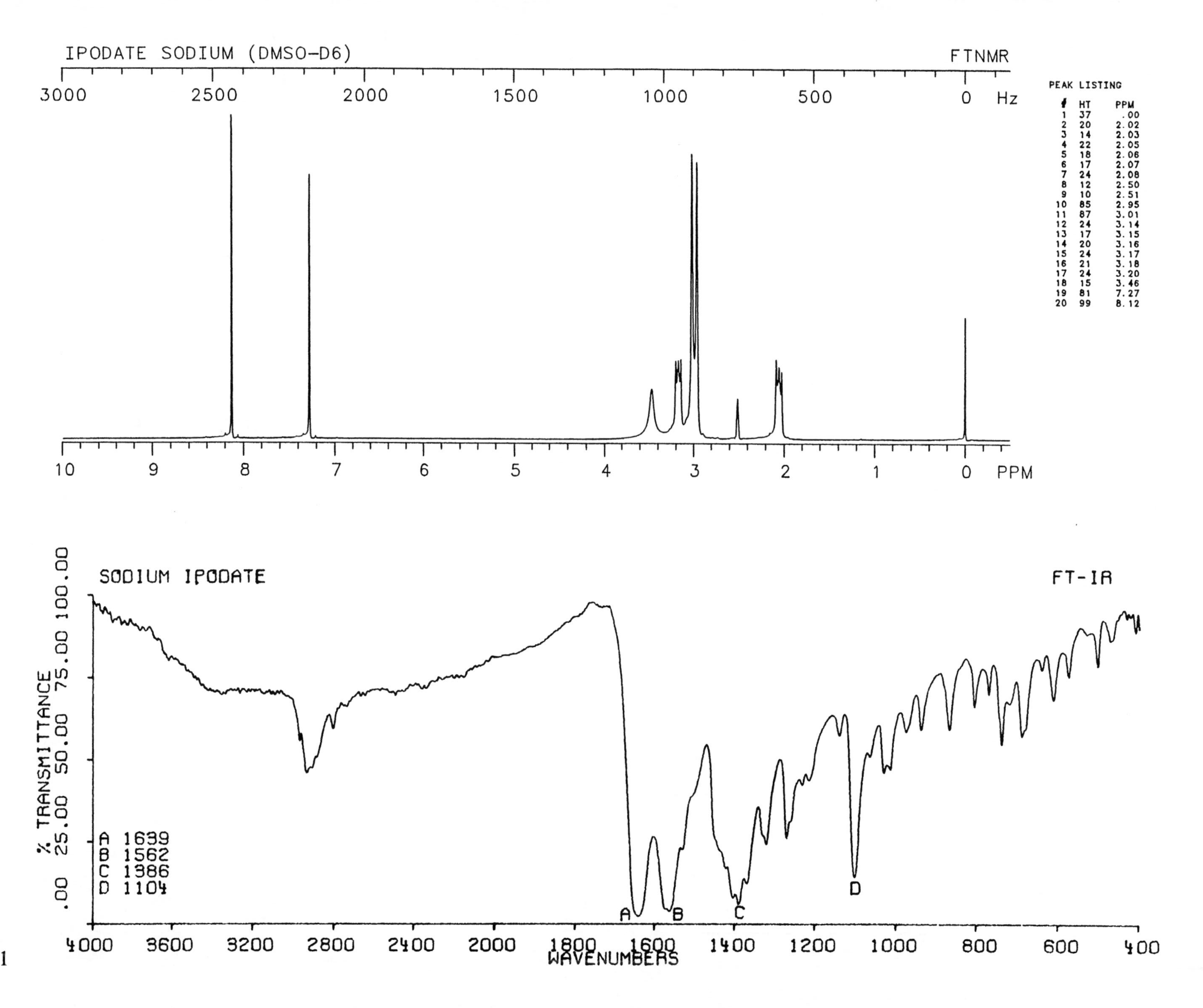
IPODATE SODIUM (DMSO-D6)
FTNMR
3000
2500
2000
1500
1000
500
0 Hz
PEAK LISTING
# HT PPM
1 37 .00
2 20 2.02
3 14 2.03
4 22 2.05
5 18 2.06
6 17 2.07
7 24 2.08
8 12 2.50
9 10 2.51
10 85 2.95
11 87 3.01
12 24 3.14
13 17 3.15
14 20 3.16
15 24 3.17
16 21 3.18
17 24 3.20
18 15 3.46
19 81 7.27
20 99 8.12
10
9
8
7
6
5
4
3
2
1
0 PPM
SODIUM IPODATE
FT-IR
% TRANSMITTANCE
.00
25.00
50.00
75.00
100.00
A 1639
B 1562
C 1386
D 1104
4000
3600
3200
2800
2400
2000
1800
1600
1400
1200
1000
800
600
400
WAVENUMBERS

# IPRONIAZID

$C_9H_{13}N_3O$

Molecular weight: 179.22 (179.11)
Synonyms: 4-Pyridinecarboxylic acid 2-(1-methylethyl)hydrazide; 1-isonicotinyl-2-isopropylhydrazine
Trade names: Euphozid, Marsilid

Use: Antidepressant, antitubercular
HPLC: Si-10; 2A:98B; 8.1
GC: 1604; 200°C





IPRONIAZID

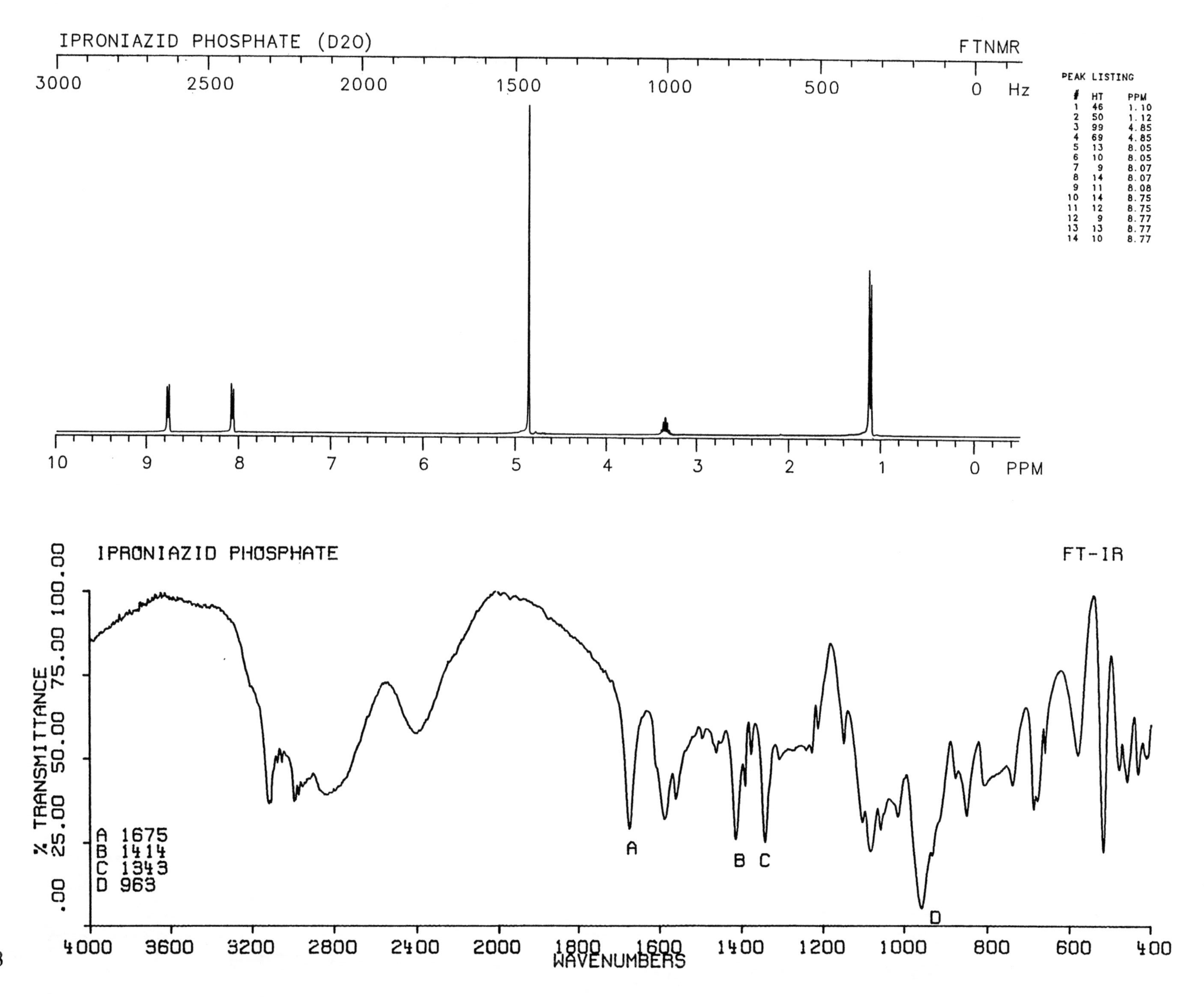

IPRONIAZID PHOSPHATE (D2O)
FTNMR
3000 2500 2000 1500 1000 500 0 Hz
PEAK LISTING
# HT PPM
1 46 1.10
2 50 1.12
3 99 4.85
4 69 4.85
5 13 8.05
6 10 8.05
7 9 8.07
8 14 8.07
9 11 8.08
10 14 8.75
11 12 8.75
12 9 8.77
13 13 8.77
14 10 8.77
10 9 8 7 6 5 4 3 2 1 0 PPM
IPRONIAZID PHOSPHATE
FT-IR
% TRANSMITTANCE
.00 25.00 50.00 75.00 100.00
A 1675
B 1414
C 1343
D 963
A
B C
D
4000 3600 3200 2800 2400 2000 1800 1600 1400 1200 1000 800 600 400
WAVENUMBERS

# ISOAMYL NITRITE

$C_5H_{11}NO_2$

**Molecular weight:** 117.15 (117.08)
**Synonyms:** 3-Methylbutyl nitrous acid; isopentyl nitrite

**Trade names:**

**Use:** Coronary vasodilator
**HPLC:**
**GC:** 676; 80°C

$(CH_3)_2CH{-}CH_2{-}CH_2{-}O{-}N{=}O$

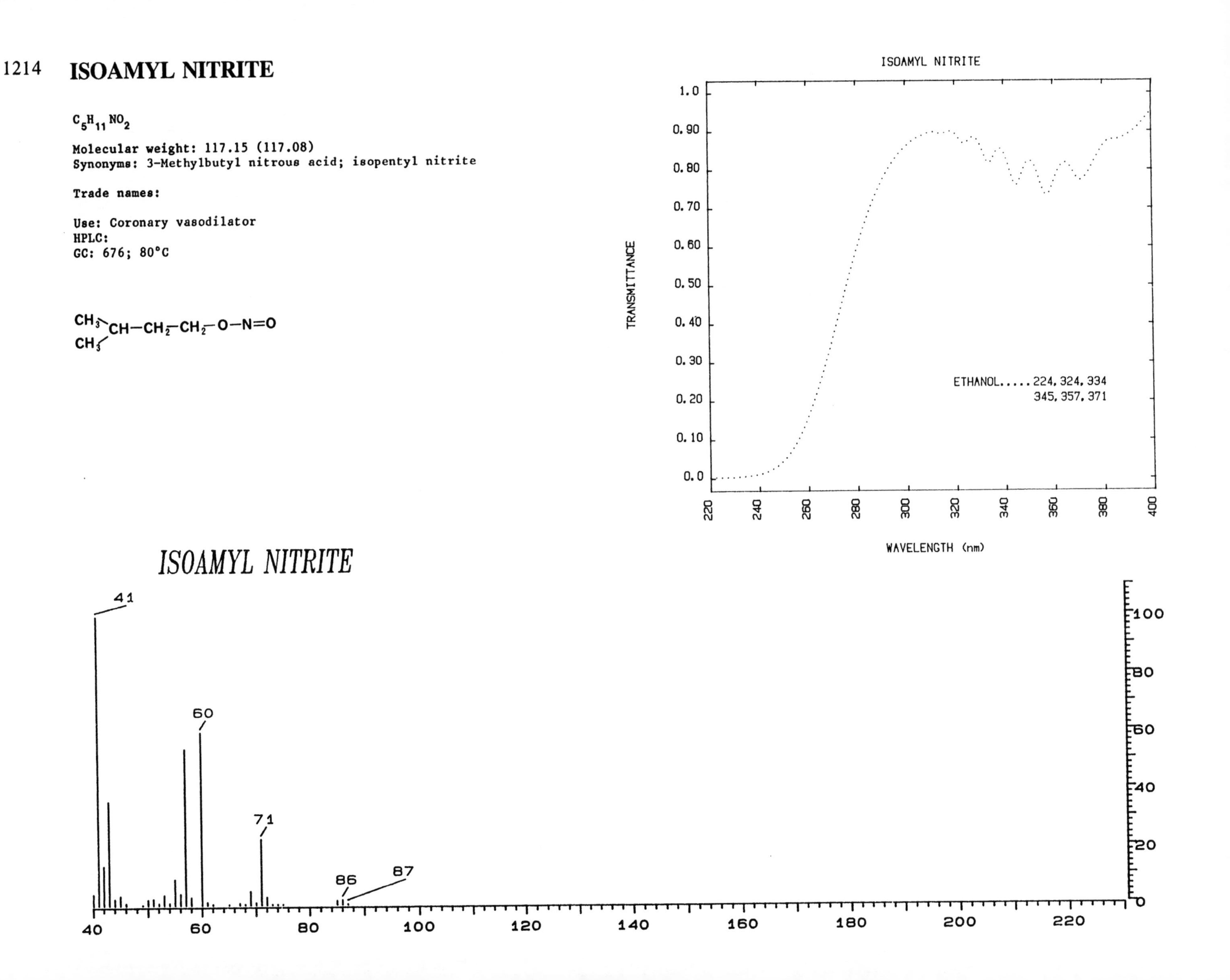

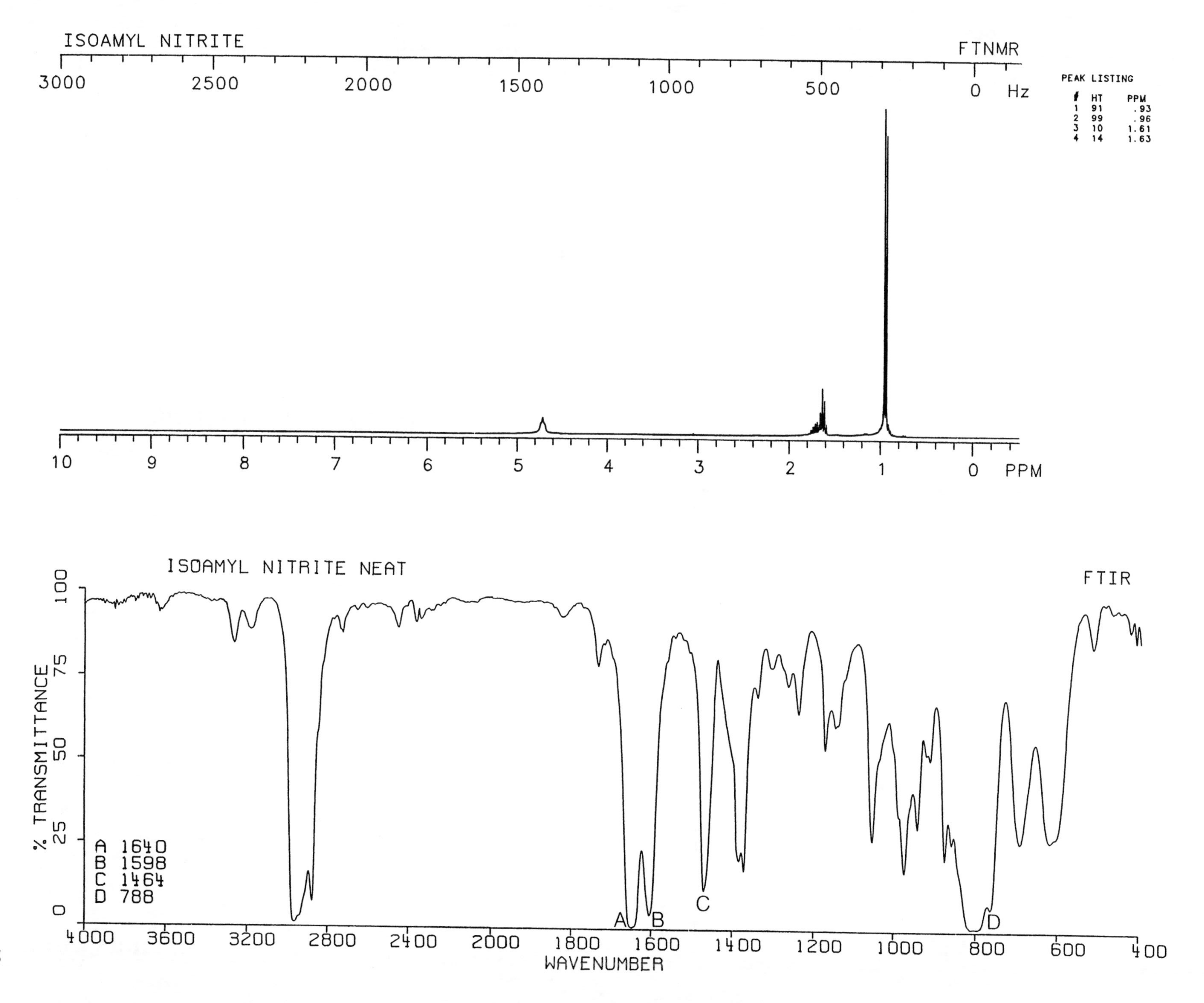
ISOAMYL NITRITE
FTNMR
3000 2500 2000 1500 1000 500 0 Hz
PEAK LISTING
# HT PPM
1 91 .93
2 99 .96
3 10 1.61
4 14 1.63
10 9 8 7 6 5 4 3 2 1 0 PPM
ISOAMYL NITRITE NEAT
FTIR
% TRANSMITTANCE
100 75 50 25 0
A 1640
B 1598
C 1464
D 788
4000 3600 3200 2800 2400 2000 1800 1600 1400 1200 1000 800 600 400
WAVENUMBER

# ISOBARBITURIC ACID

$C_4H_4N_2O_3$

Molecular weight: 128.07 (128.02)
Synonyms: 2,4,5-Trihydroxypyrimidine

Trade names:

Use: Barbiturate
HPLC: Si-10; 50A:50B; 4.5
GC:





ISOBARBITURIC ACID--DIP

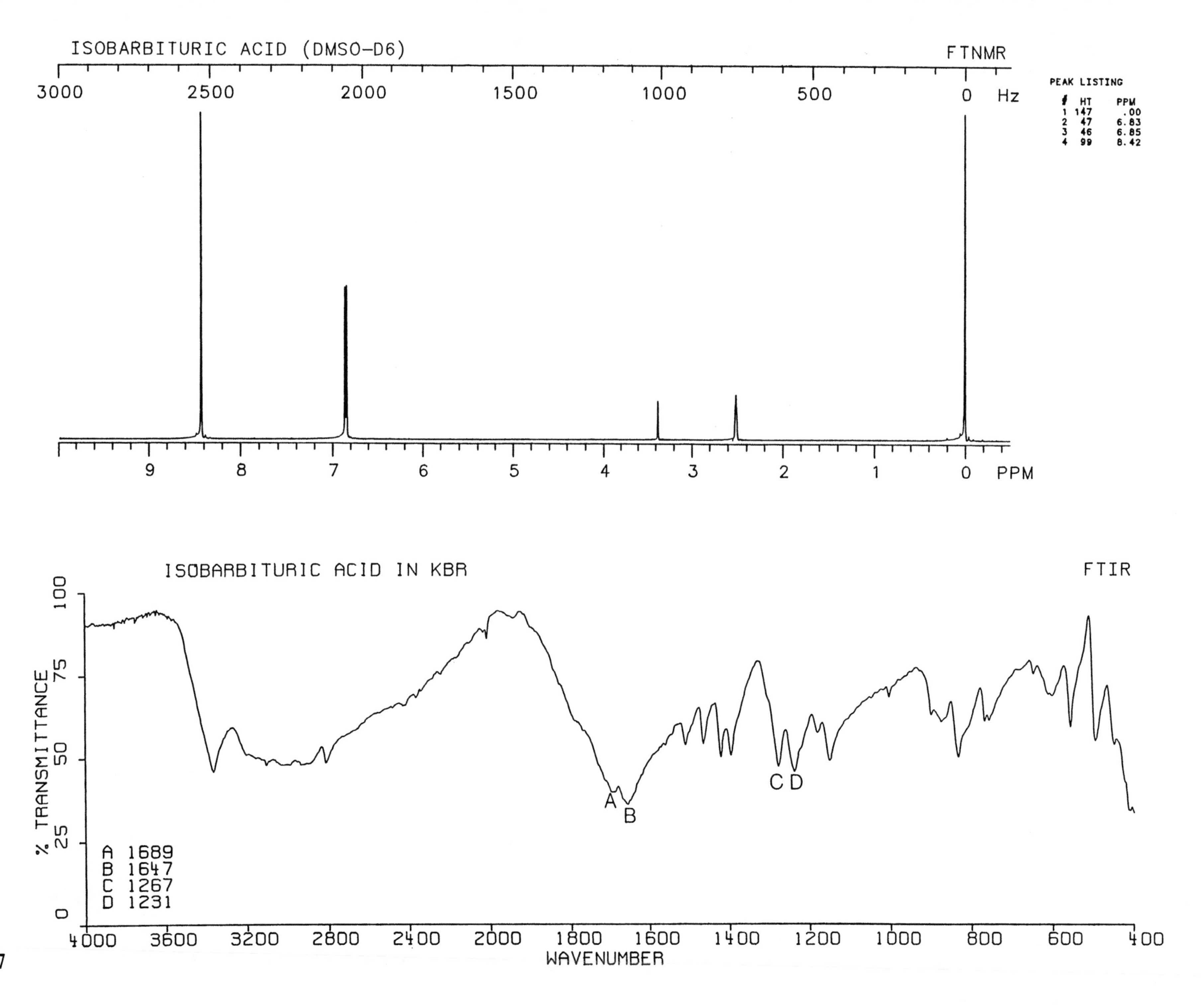
ISOBARBITURIC ACID (DMSO-D6)
FTNMR
Hz
PPM
PEAK LISTING
# HT PPM
1 147 .00
2 47 6.83
3 46 6.85
4 99 8.42
ISOBARBITURIC ACID IN KBR
FTIR
% TRANSMITTANCE
WAVENUMBER
A 1689
B 1647
C 1267
D 1231

# 3-ISOBUTYL-1-METHYLXANTHINE

$C_{10}H_{14}N_4O_2$

Molecular weight: 222.25 (222.11)
Synonyms: 3,7-Dihydro-3-(isobutyl)-1-methyl-1H-purine-2,6-dione

Trade names:

Use:
HPLC: Si-10; 2A:98B; 6.0
GC: 2145; 250°C

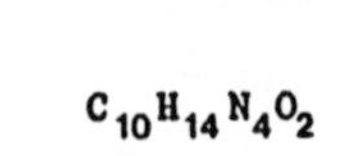


3-ISOBUTYL-1-METHYLXANTHINE

TRANSMITTANCE

1.0
0.90
0.80
0.70
0.60
0.50
0.40
0.30
0.20
0.10
0.0

220 240 260 280 300 320 340

WAVELENGTH (nm)

—— 270
- - - 276

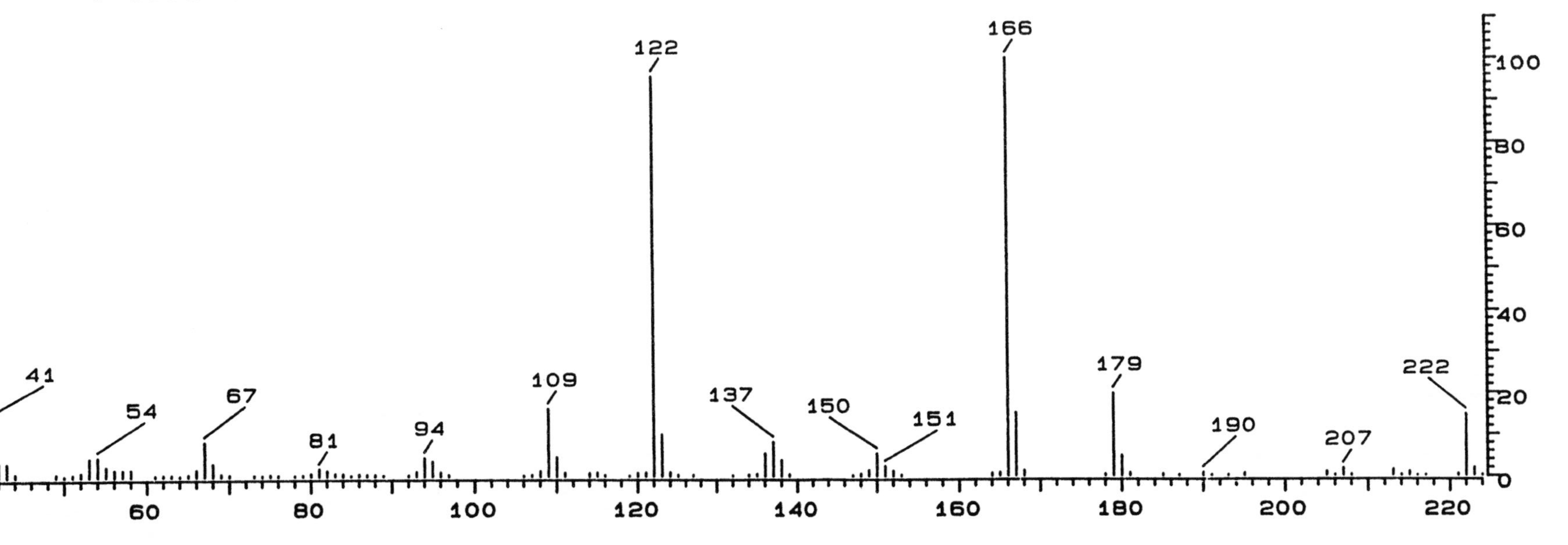

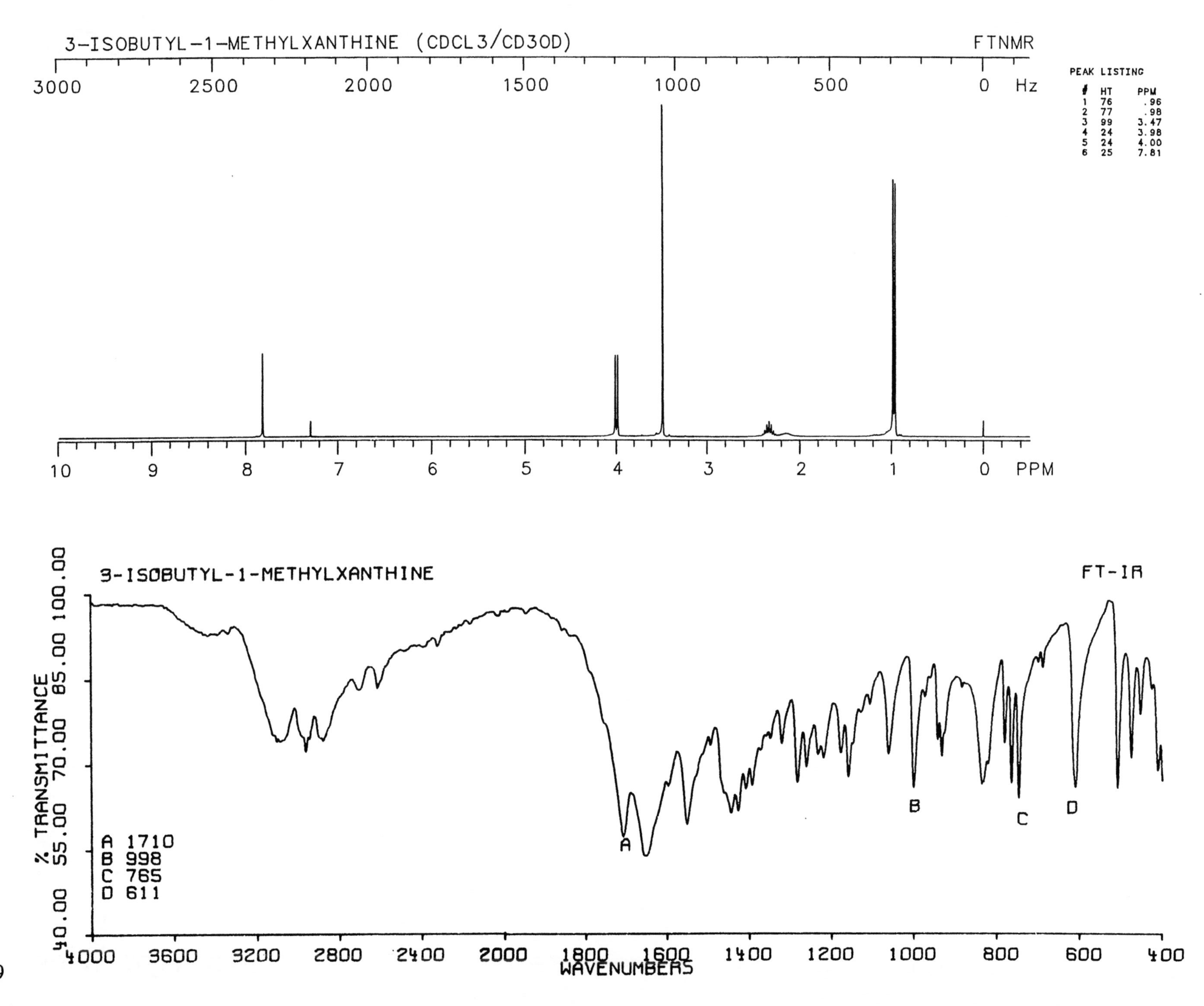
3-ISOBUTYL-1-METHYLXANTHINE (CDCL3/CD3OD)
FTNMR
3000
2500
2000
1500
1000
500
0 Hz
PEAK LISTING
# HT PPM
1 76 .96
2 77 .98
3 99 3.47
4 24 3.98
5 24 4.00
6 25 7.81
10
9
8
7
6
5
4
3
2
1
0 PPM
3-ISOBUTYL-1-METHYLXANTHINE
FT-IR
% TRANSMITTANCE
40.00
55.00
70.00
85.00
100.00
A 1710
B 998
C 765
D 611
A
B
C
D
4000
3600
3200
2800
2400
2000
1800
1600
1400
1200
1000
800
600
400
WAVENUMBERS

# ISOBUTYL NITRITE

$C_4H_9NO_2$

**Molecular weight:** 103.12 (103.06)
**Synonyms:**

**Trade names:**

**Use:** Vasodilator
**HPLC:**
**GC:** 600; 80°C

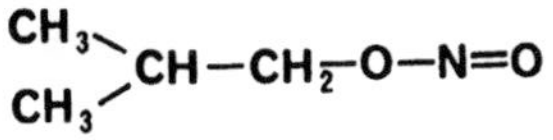

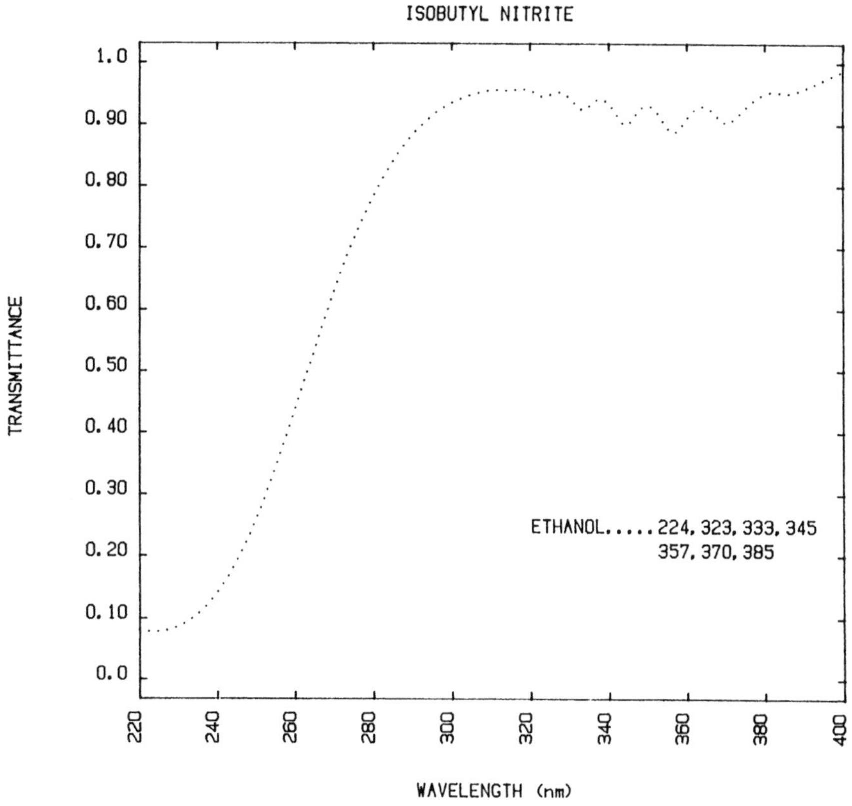

$CH_3(CH_3)CH{-}CH_2{-}O{-}N{=}O$

ISOBUTYL NITRITE

TRANSMITTANCE

1.0 0.90 0.80 0.70 0.60 0.50 0.40 0.30 0.20 0.10 0.0

ETHANOL.....224, 323, 333, 345
357, 370, 385

220 240 260 280 300 320 340 360 380 400

WAVELENGTH (nm)

*ISOBUTYL NITRITE*

41 57 71 102

80 60 40 20 0

40 60 80 100 120 140 160 180 200 220

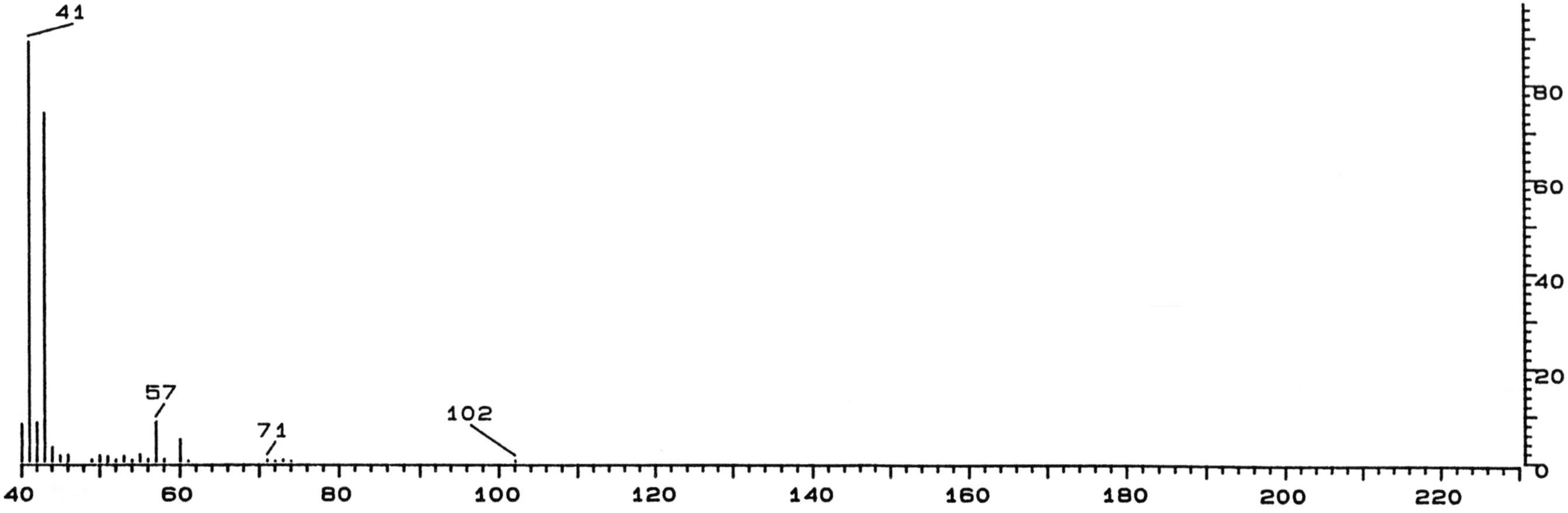

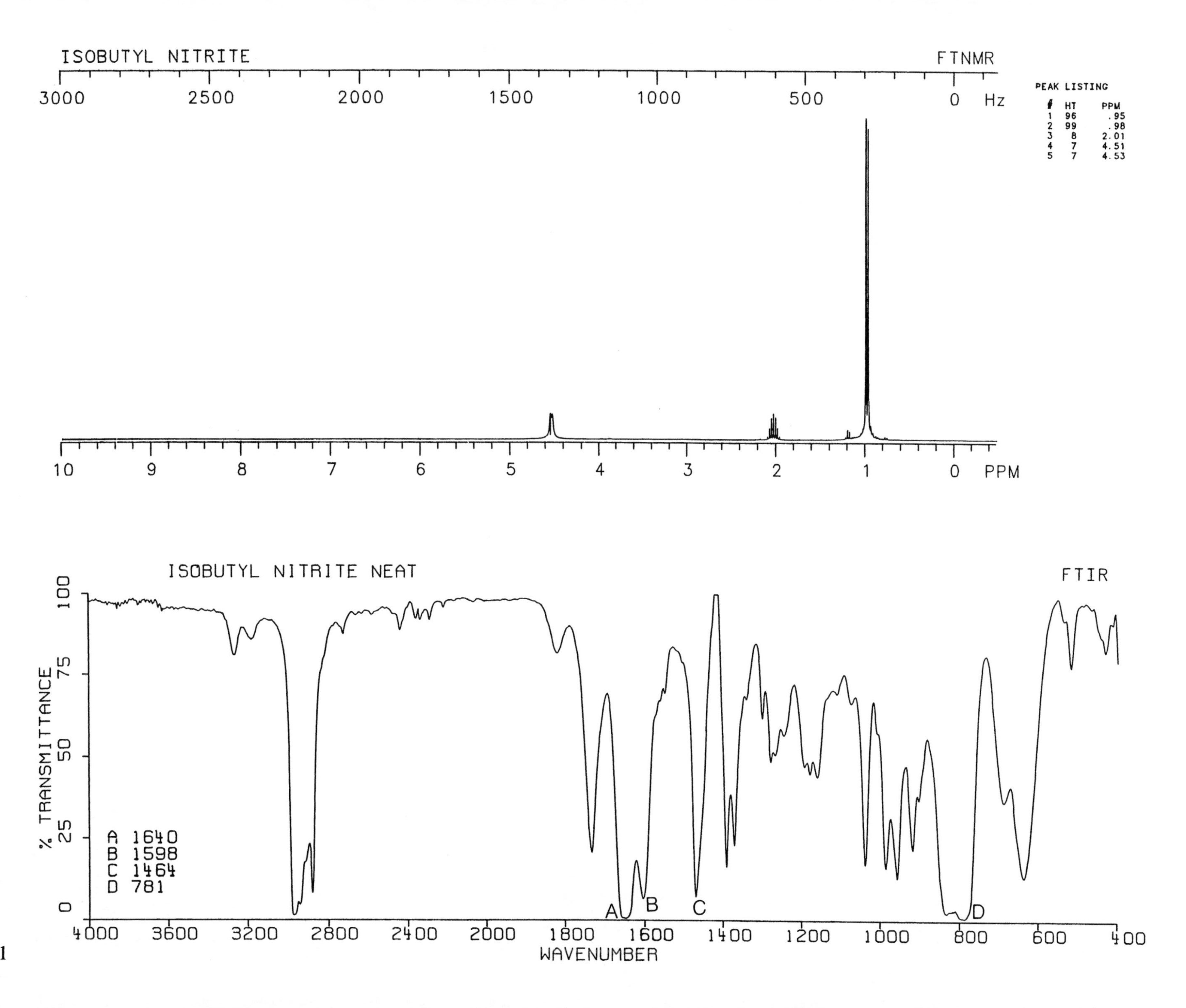
ISOBUTYL NITRITE
FTNMR
3000
2500
2000
1500
1000
500
0 Hz
PEAK LISTING
# HT PPM
1 96 .95
2 99 .98
3 8 2.01
4 7 4.51
5 7 4.53
10
9
8
7
6
5
4
3
2
1
0 PPM
ISOBUTYL NITRITE NEAT
FTIR
% TRANSMITTANCE
100
75
50
25
0
A 1640
B 1598
C 1464
D 781
A
B
C
D
4000
3600
3200
2800
2400
2000
1800
1600
1400
1200
1000
800
600
400
WAVENUMBER

# ISOCARBOXAZID

$C_{12}H_{13}N_3O_2$

**Molecular weight:** 231.26 (231.10)
**Synonyms:** 5-Methyl-3-isoxazolecarboxylic acid 2-benzylhydrazide

**Trade names:** Marplan

**Use:** Antidepressant
HPLC: Si-10; 1A:99B; 4.0
GC:

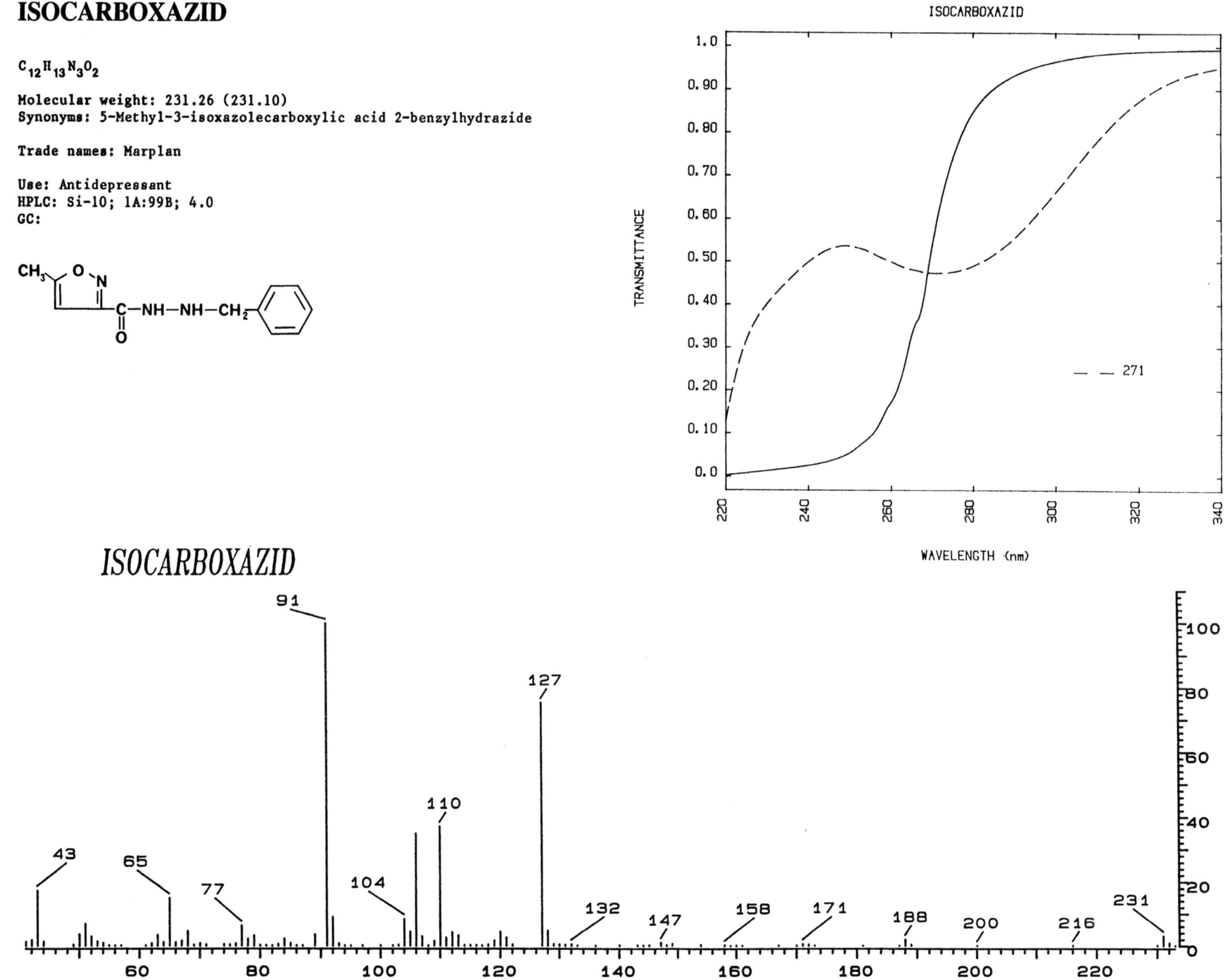

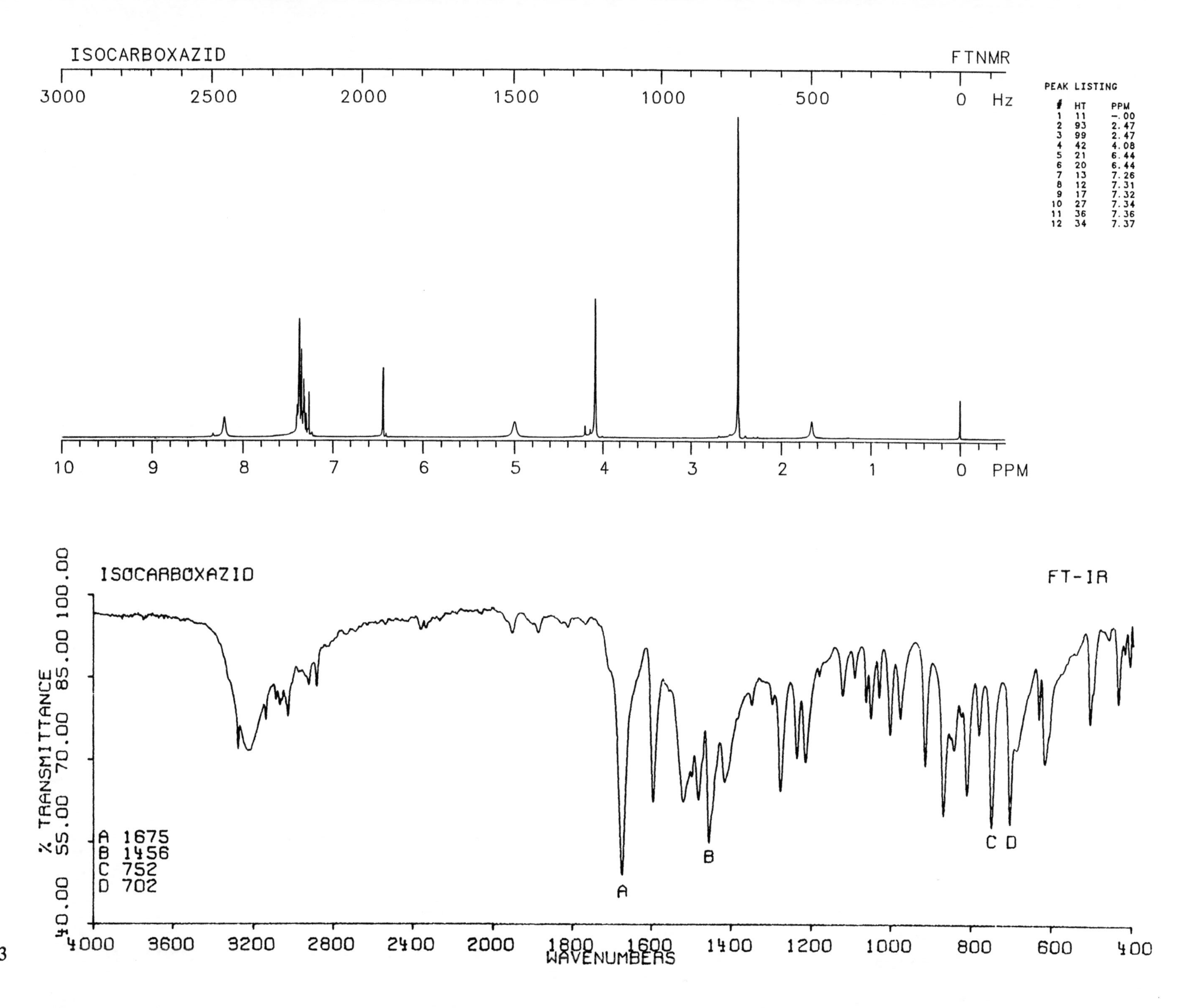
ISOCARBOXAZID
FTNMR
3000
2500
2000
1500
1000
500
0 Hz
PEAK LISTING
# HT PPM
1 11 -.00
2 93 2.47
3 99 2.47
4 42 4.08
5 21 6.44
6 20 6.44
7 13 7.26
8 12 7.31
9 17 7.32
10 27 7.34
11 36 7.36
12 34 7.37
10
9
8
7
6
5
4
3
2
1
0 PPM
ISOCARBOXAZID
FT-IR
% TRANSMITTANCE
100.00
85.00
70.00
55.00
40.00
A 1675
B 1456
C 752
D 702
A
B
C D
4000
3600
3200
2800
2400
2000
1800
1600
1400
1200
1000
800
600
400
WAVENUMBERS

# ISOETHARINE

$C_{13}H_{21}NO_3$

Molecular weight: 239.32 (239.15)
Synonyms: 3,4-Dihydroxy-α-[1-(isopropylamino)propyl]benzyl alcohol; N-isopropylethylnorepinephrine; isoetarine
Trade names: Bronkometer, Bronkosol

Use: Bronchodilator
HPLC: Si-10; 20A:80B; 4.5
GC:

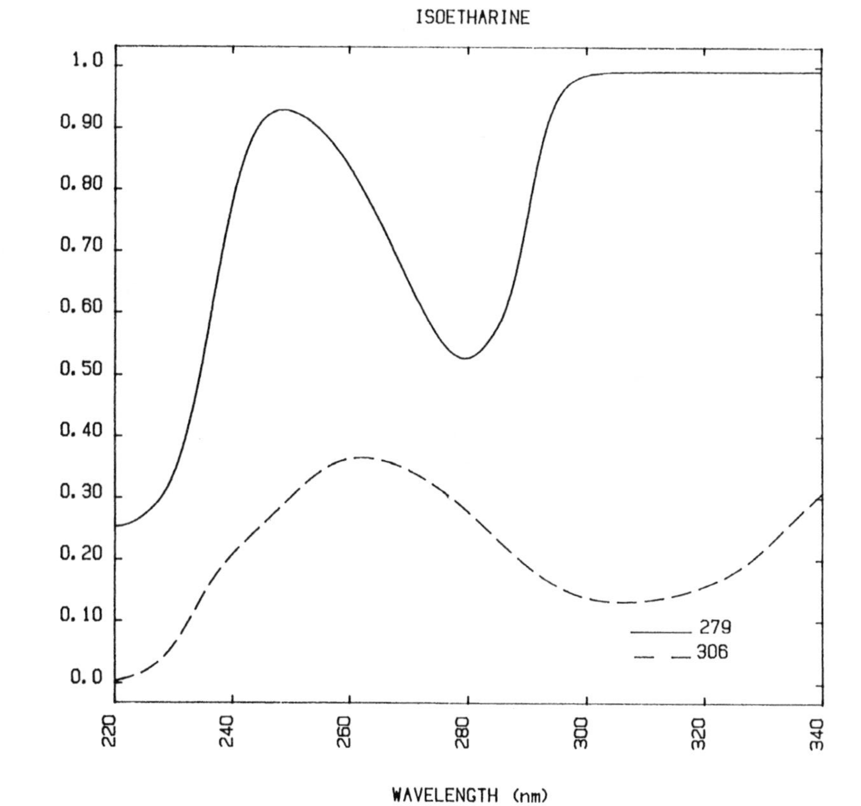

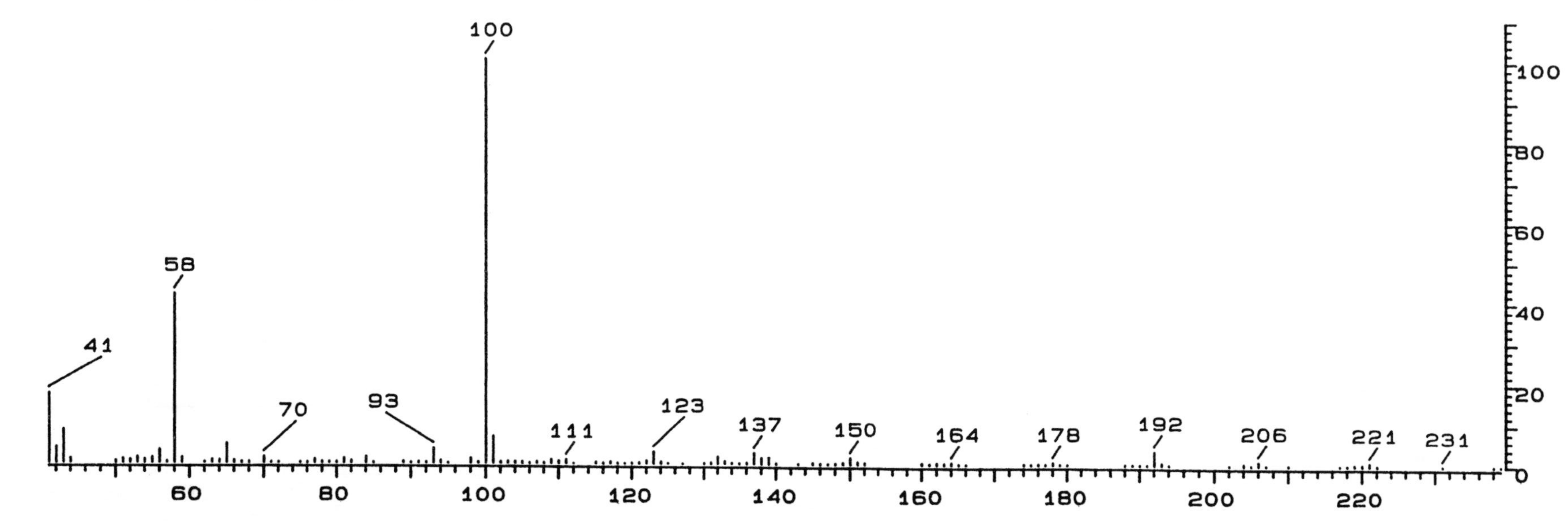

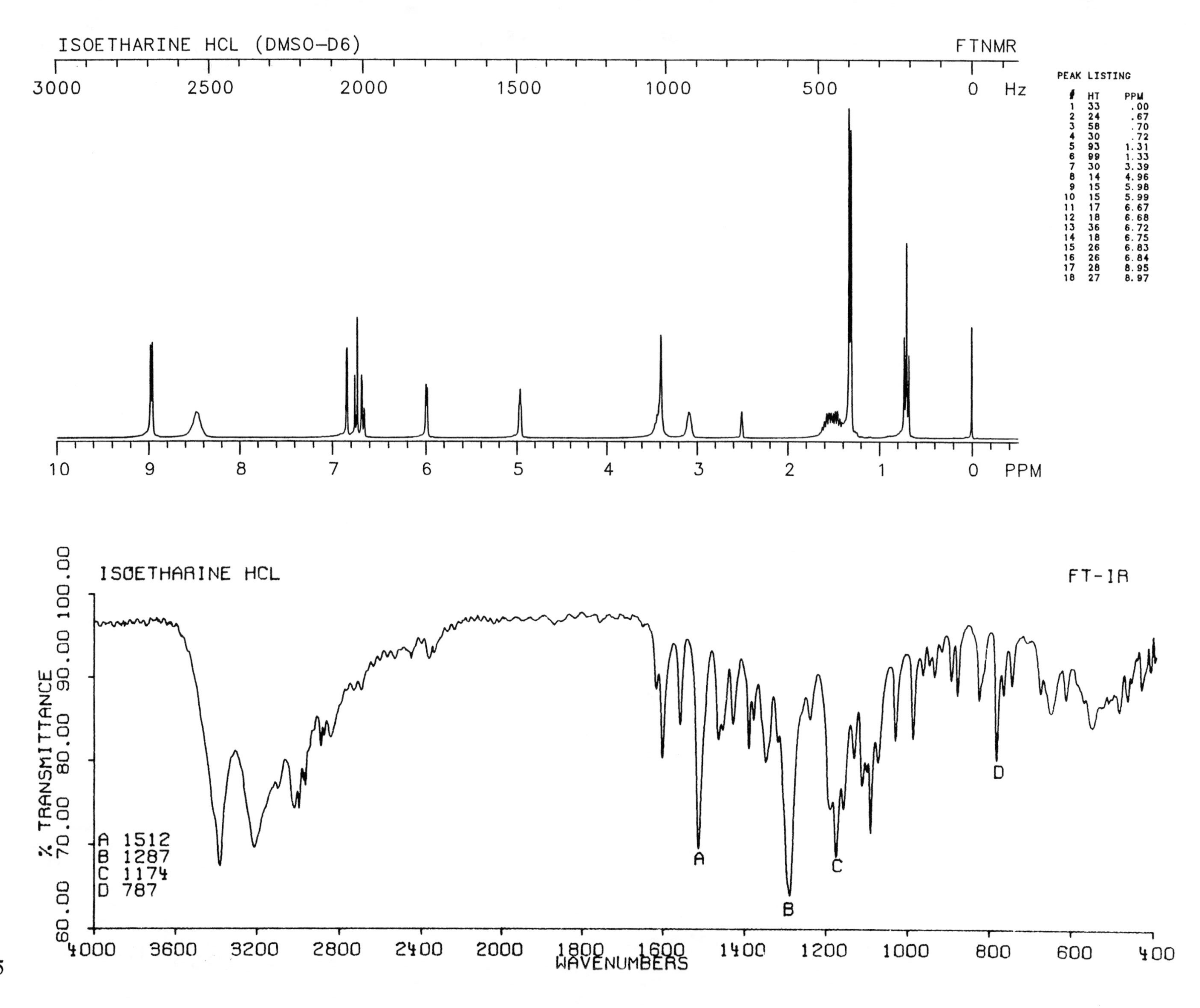

ISOETHARINE HCL (DMSO-D6)
FTNMR
3000
2500
2000
1500
1000
500
0 Hz
PEAK LISTING
# HT PPM
1 33 .00
2 24 .67
3 58 .70
4 30 .72
5 93 1.31
6 99 1.33
7 30 3.39
8 14 4.96
9 15 5.98
10 15 5.99
11 17 6.67
12 18 6.68
13 36 6.72
14 18 6.75
15 26 6.83
16 26 6.84
17 28 8.95
18 27 8.97
10 9 8 7 6 5 4 3 2 1 0 PPM
ISOETHARINE HCL
FT-IR
% TRANSMITTANCE
60.00 70.00 80.00 90.00 100.00
A 1512
B 1287
C 1174
D 787
A
B
C
D
4000 3600 3200 2800 2400 2000 1800 1600 1400 1200 1000 800 600 400
WAVENUMBERS

# ISOMETHADONE

$C_{21}H_{27}NO$

**Molecular weight:** 309.43 (309.21)
**Synonyms:** 6-(Dimethylamino)-5-methyl-4,4-diphenyl-3-hexanone; isoamidone
**Trade names:** Isoadanone, Liden

**Use:** Narcotic analgesic
**HPLC:** Si-10; 2A:98B; 5.2
**GC:** 2195; 250°C

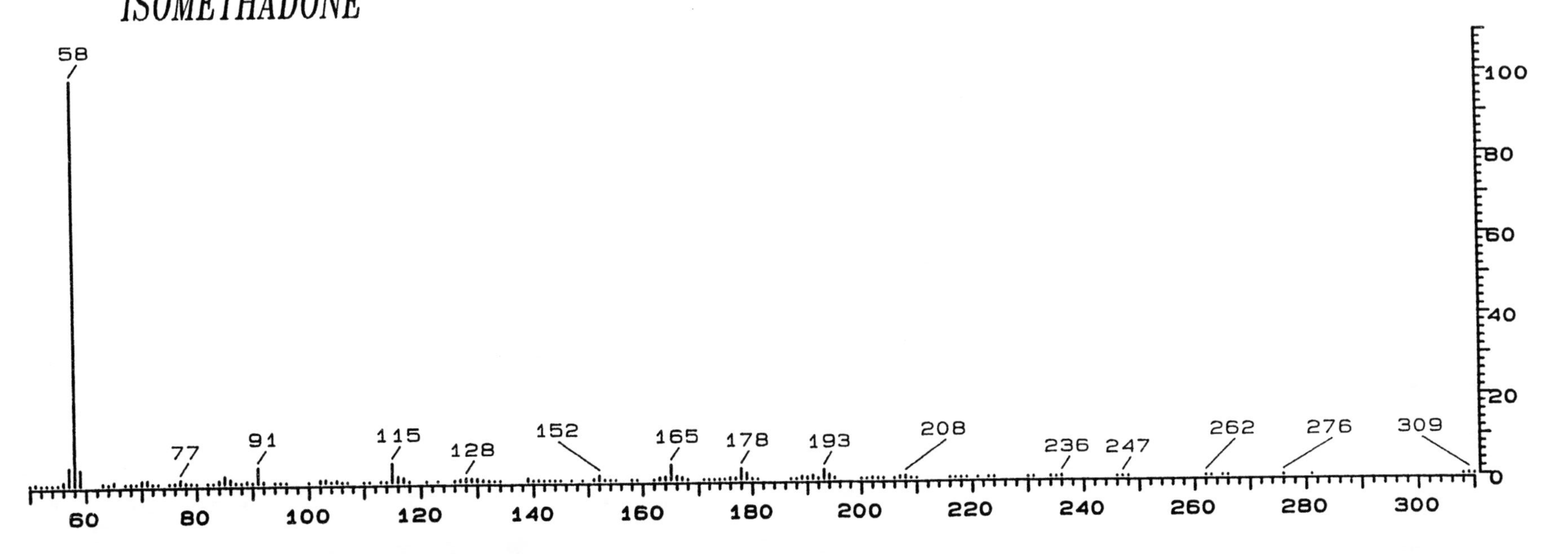

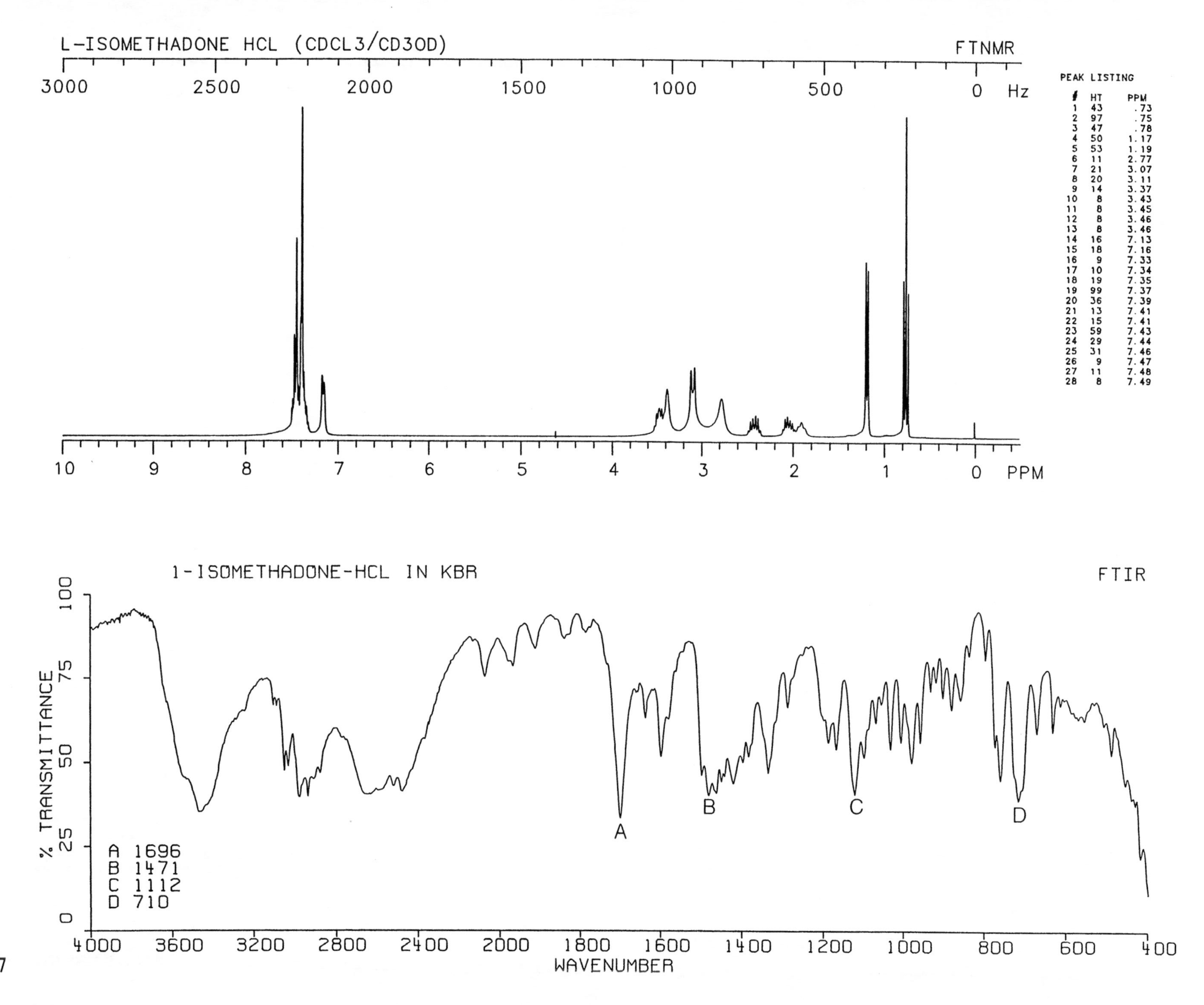

L-ISOMETHADONE HCL (CDCL3/CD3OD)
FTNMR
3000 2500 2000 1500 1000 500 0 Hz
10 9 8 7 6 5 4 3 2 1 0 PPM
PEAK LISTING
# HT PPM
1 43 .73
2 97 .75
3 47 .78
4 50 1.17
5 53 1.19
6 11 2.77
7 21 3.07
8 20 3.11
9 14 3.37
10 8 3.43
11 8 3.45
12 8 3.46
13 8 3.46
14 16 7.13
15 18 7.16
16 9 7.33
17 10 7.34
18 19 7.35
19 99 7.37
20 36 7.39
21 13 7.41
22 15 7.41
23 59 7.43
24 29 7.44
25 31 7.46
26 9 7.47
27 11 7.48
28 8 7.49
1-ISOMETHADONE-HCL IN KBR
FTIR
% TRANSMITTANCE
100 75 50 25 0
A B C D
A 1696
B 1471
C 1112
D 710
4000 3600 3200 2800 2400 2000 1800 1600 1400 1200 1000 800 600 400
WAVENUMBER

# ISONIAZID

$C_6H_7N_3O$

**Molecular weight:** 137.14 (137.06)
**Synonyms:** 4-Pyridinecarboxylic acid hydrazide; isonicotinic acid hydrazide
**Trade names:** INH, Nydrazid, Rifamate

**Use:** Antibacterial, tuberculostatic
**HPLC:** Si-10; 5A:95B; 7.7
**GC:** 1602; 200°C

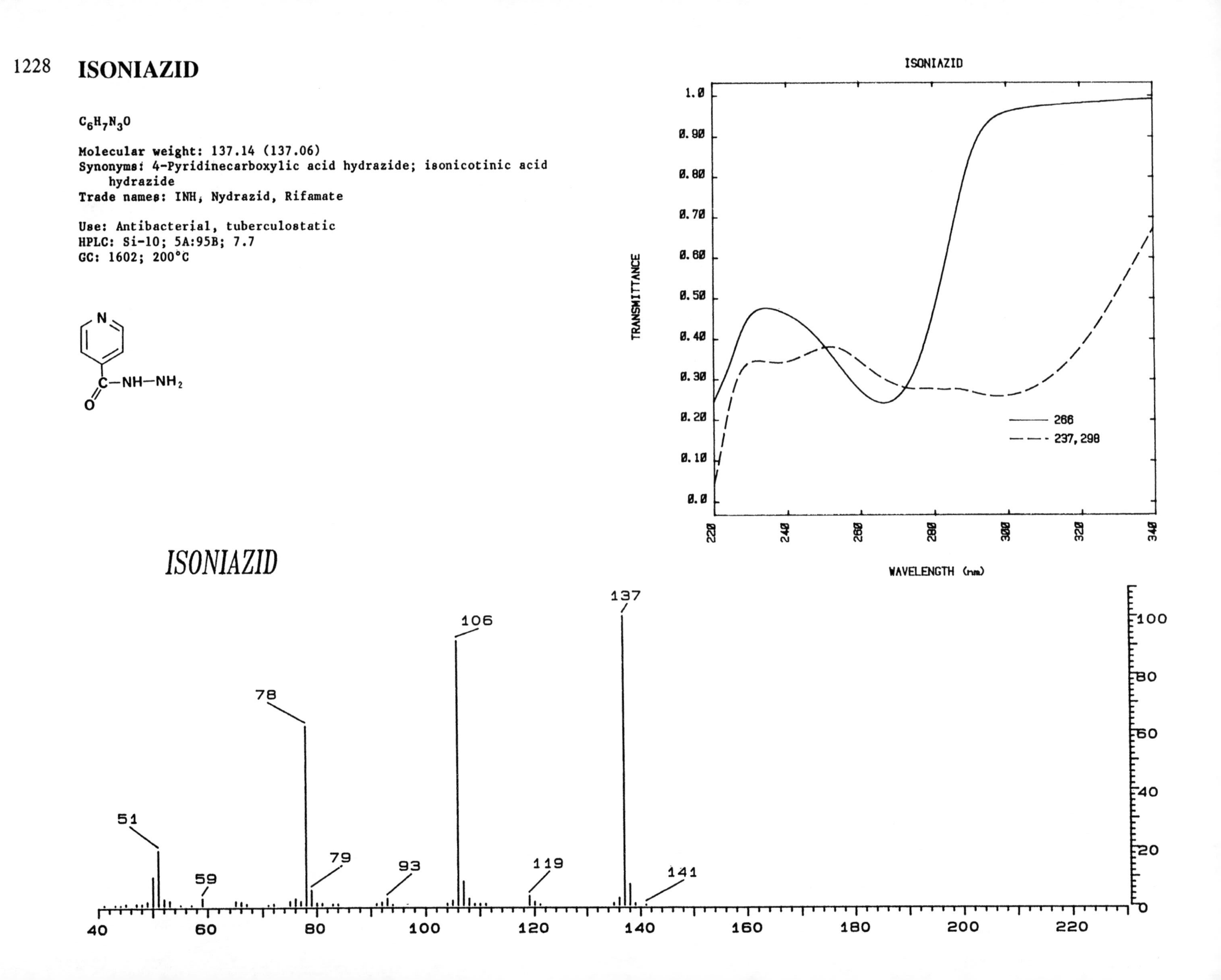

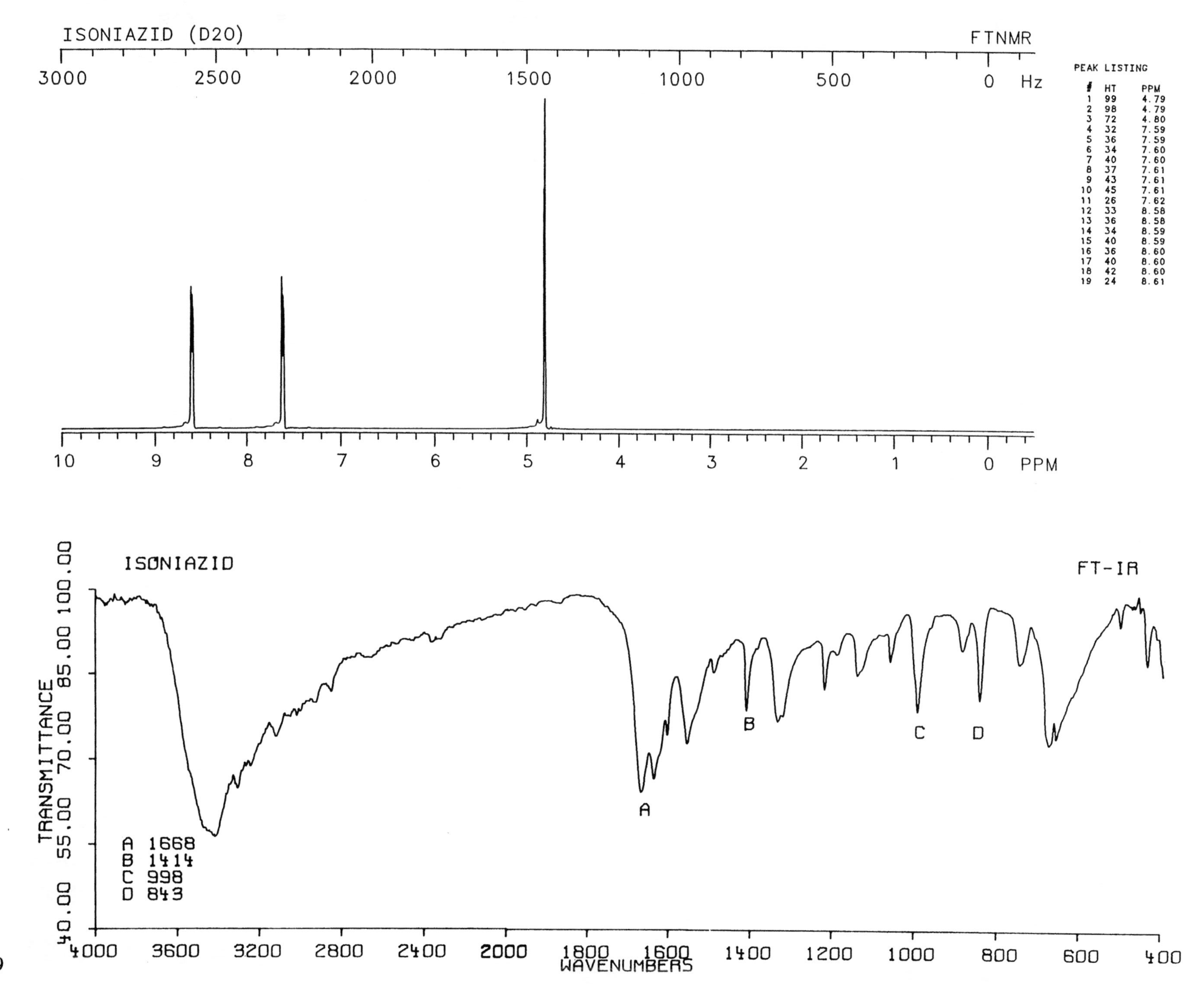
ISONIAZID (D2O)
FTNMR
3000
2500
2000
1500
1000
500
0 Hz
PEAK LISTING
# HT PPM
1 99 4.79
2 98 4.79
3 72 4.80
4 32 7.59
5 36 7.59
6 34 7.60
7 40 7.60
8 37 7.61
9 43 7.61
10 45 7.61
11 26 7.62
12 33 8.58
13 36 8.58
14 34 8.59
15 40 8.59
16 36 8.60
17 40 8.60
18 42 8.60
19 24 8.61
10
9
8
7
6
5
4
3
2
1
0 PPM
ISONIAZID
FT-IR
TRANSMITTANCE
40.00
55.00
70.00
85.00
100.00
A 1668
B 1414
C 998
D 843
A
B
C
D
4000
3600
3200
2800
2400
2000
1800
1600
1400
1200
1000
800
600
400
WAVENUMBERS

# ISONICOTINAMIDE

$C_6H_6N_2O$

Molecular weight: 122.13 (122.05)
Synonyms: 4-Pyridinecarboxylic acid amide; isonicotinic acid amide

Trade names:

Use:
HPLC: Si-10; 5A:95B; 6.0
GC: 1458; 200°C

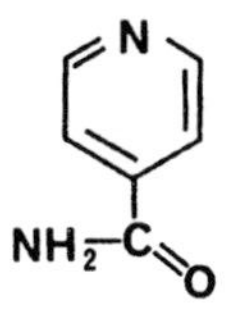

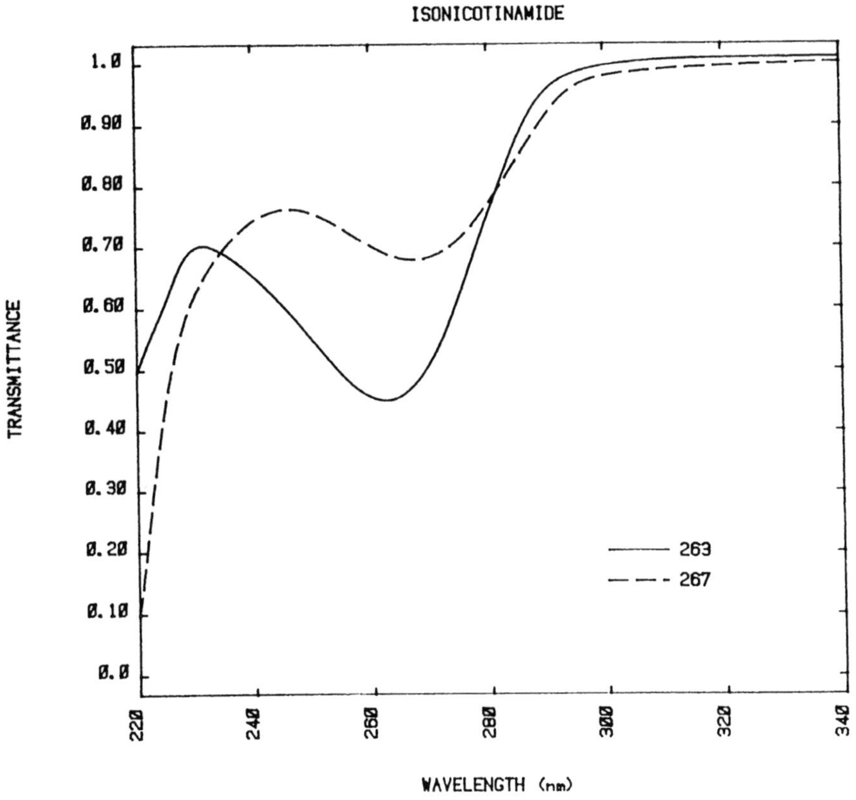

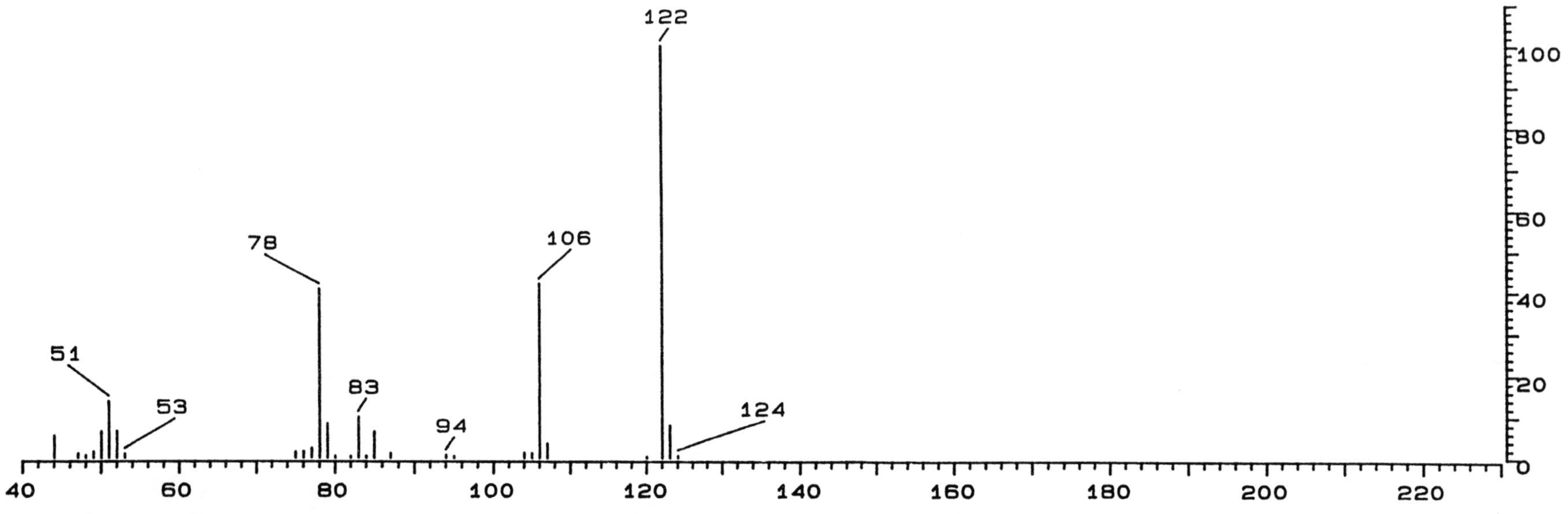

ISONICOTINAMIDE

ISONICOTINAMIDE

ISONICOTINAMIDE (DMSO-D6)

FTNMR

PEAK LISTING

| # | HT | PPM |
|---|---|---|
| 1 | 40 | .00 |
| 2 | 23 | .00 |
| 3 | 23 | 3.40 |
| 4 | 15 | 7.77 |
| 5 | 82 | 7.78 |
| 6 | 63 | 7.78 |
| 7 | 53 | 7.79 |
| 8 | 99 | 7.80 |
| 9 | 15 | 8.28 |
| 10 | 74 | 8.72 |
| 11 | 46 | 8.73 |
| 12 | 45 | 8.74 |
| 13 | 75 | 8.74 |

ISONICOTINAMIDE

FT-IR

A 1668
B 1625
C 1597
D 1548

# ISONICOTINIC ACID

$C_6H_5NO_2$

Molecular weight: 123.11 (123.03)
Synonyms: 4-Pyridinecarboxylic acid; γ-picolinic acid

Trade names:

Use:
HPLC: Si-10; 20A:80B; 6.3
GC:

N
COOH

ISONICOTINIC ACID

TRANSMITTANCE

1.0
0.90
0.80
0.70
0.60
0.50
0.40
0.30
0.20
0.10
0.0

220 240 260 280 300 320 340

WAVELENGTH (nm)

270
266

ISONICOTINIC ACID -- DIP

123
106
78
51
53
79
94
125

40 60 80 100 120 140 160 180 200 220

0 20 40 60 80 100

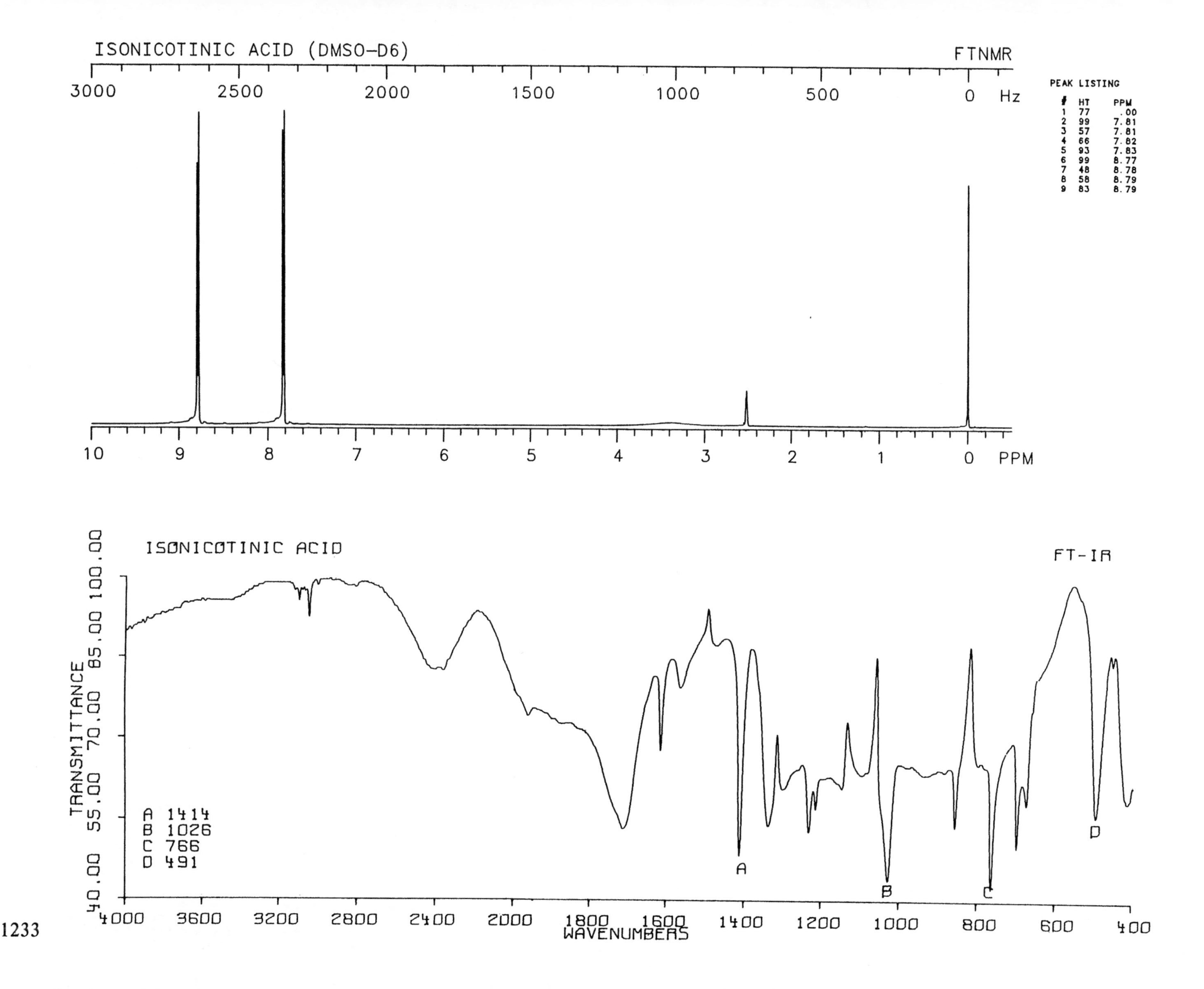
ISONICOTINIC ACID (DMSO-D6)
FTNMR
3000 2500 2000 1500 1000 500 0 Hz
PEAK LISTING
# HT PPM
1 77 .00
2 99 7.81
3 57 7.81
4 66 7.82
5 93 7.83
6 99 8.77
7 48 8.78
8 58 8.79
9 83 8.79
10 9 8 7 6 5 4 3 2 1 0 PPM
ISONICOTINIC ACID
FT-IR
TRANSMITTANCE
40.00 55.00 70.00 85.00 100.00
A 1414
B 1026
C 766
D 491
A
B
C
D
4000 3600 3200 2800 2400 2000 1800 1600 1400 1200 1000 800 600 400
WAVENUMBERS

# ISOPROPAMIDE IODIDE

$C_{23}H_{33}IN_2O$

**Molecular weight:** 480.42 (480.16)
**Synonyms:** γ-(Aminocarbonyl)-N-methyl-N,N-bis(1-methylethyl)-γ-phenylbenzenepropanaminium iodide
**Trade names:** Combid, Darbid, Ornade, Priamide, Prochlor-Iso, Pro-Iso, Stelabid, Tyrimade
**Use:** Anticholinergic
**HPLC:**
**GC:** 2269; 250°C

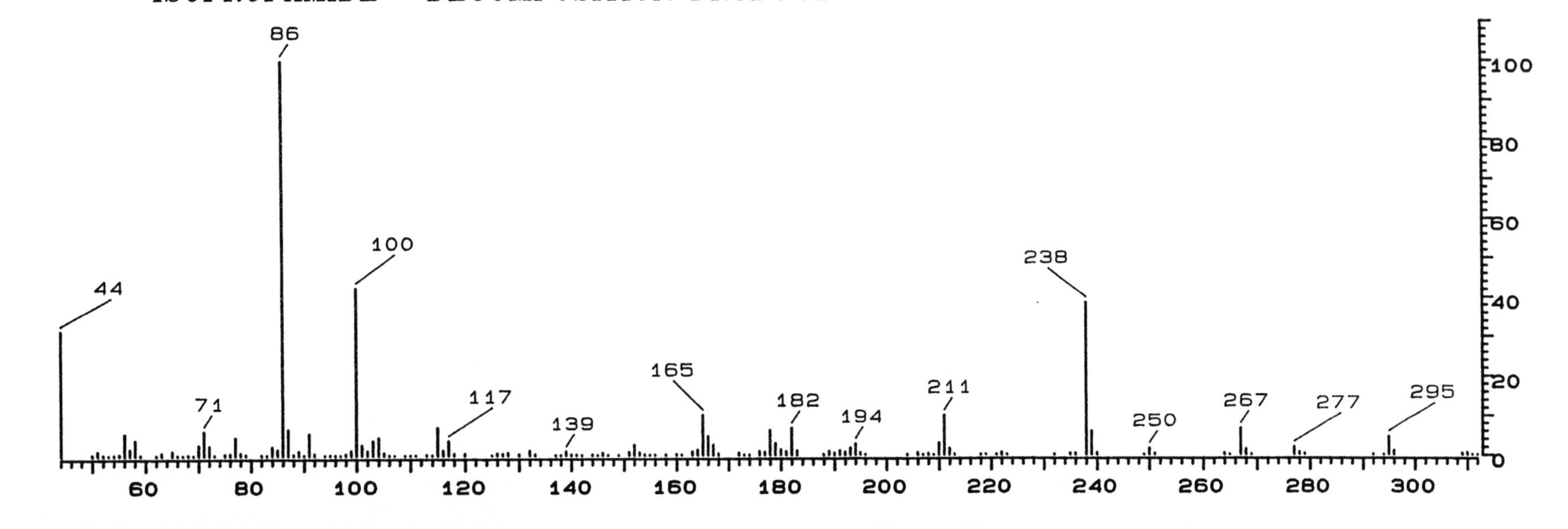

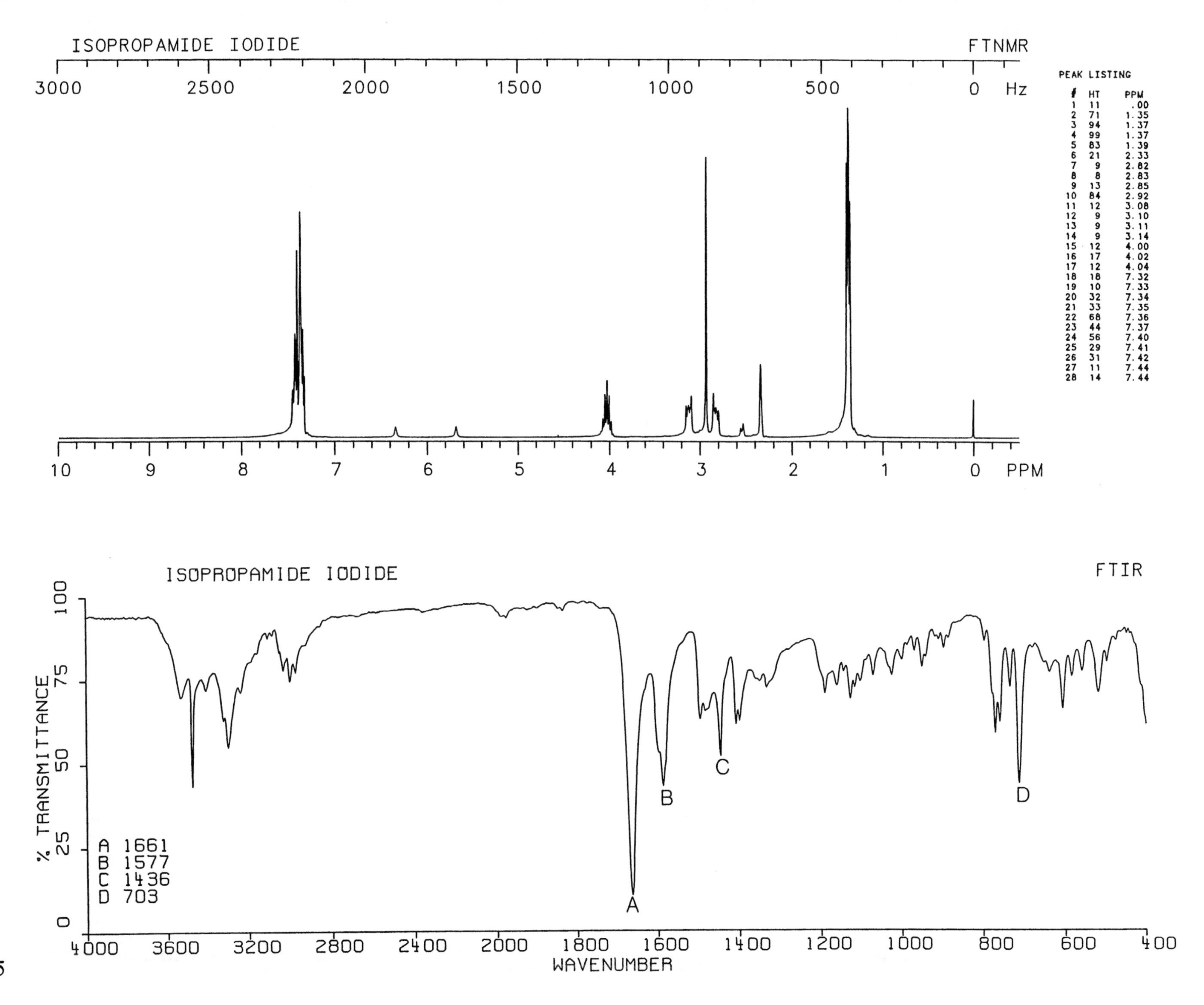
ISOPROPAMIDE IODIDE
FTNMR
3000
2500
2000
1500
1000
500
0 Hz
10
9
8
7
6
5
4
3
2
1
0 PPM
PEAK LISTING
# HT PPM
1 11 .00
2 71 1.35
3 94 1.37
4 99 1.37
5 83 1.39
6 21 2.33
7 9 2.82
8 8 2.83
9 13 2.85
10 84 2.92
11 12 3.08
12 9 3.10
13 9 3.11
14 9 3.14
15 12 4.00
16 17 4.02
17 12 4.04
18 18 7.32
19 10 7.33
20 32 7.34
21 33 7.35
22 68 7.36
23 44 7.37
24 56 7.40
25 29 7.41
26 31 7.42
27 11 7.44
28 14 7.44
ISOPROPAMIDE IODIDE
FTIR
% TRANSMITTANCE
100
75
50
25
0
A 1661
B 1577
C 1436
D 703
A
B
C
D
4000
3600
3200
2800
2400
2000
1800
1600
1400
1200
1000
800
600
400
WAVENUMBER

# *N*-(ISOPROPYL)-1-PHENYLCYCLOHEXYLAMINE

$C_{15}H_{23}N$

**Molecular weight:** 217.35 (217.18)
**Synonyms:** N-(Isopropyl)-1-phenylcyclohexanamine; N-(isopropyl)-1-phenylcyclohexamine
**Trade names:**

**Use:** Hallucinogen
**HPLC:** Si-10; 5A:95B; 5.0
**GC:** 1620; 200°C

N-[ISOPROPYL]-1-PHENYLCYCLOHEXYLAMINE

257, 261, 268
PPT BASE

TRANSMITTANCE

WAVELENGTH (nm)

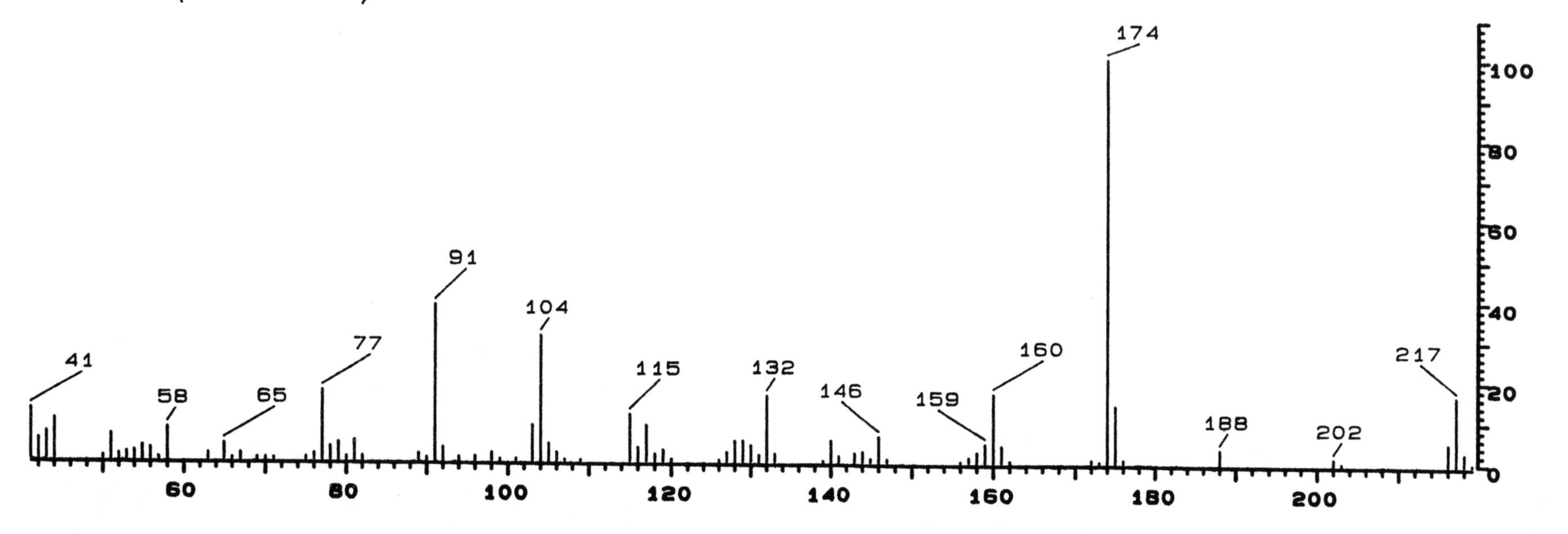

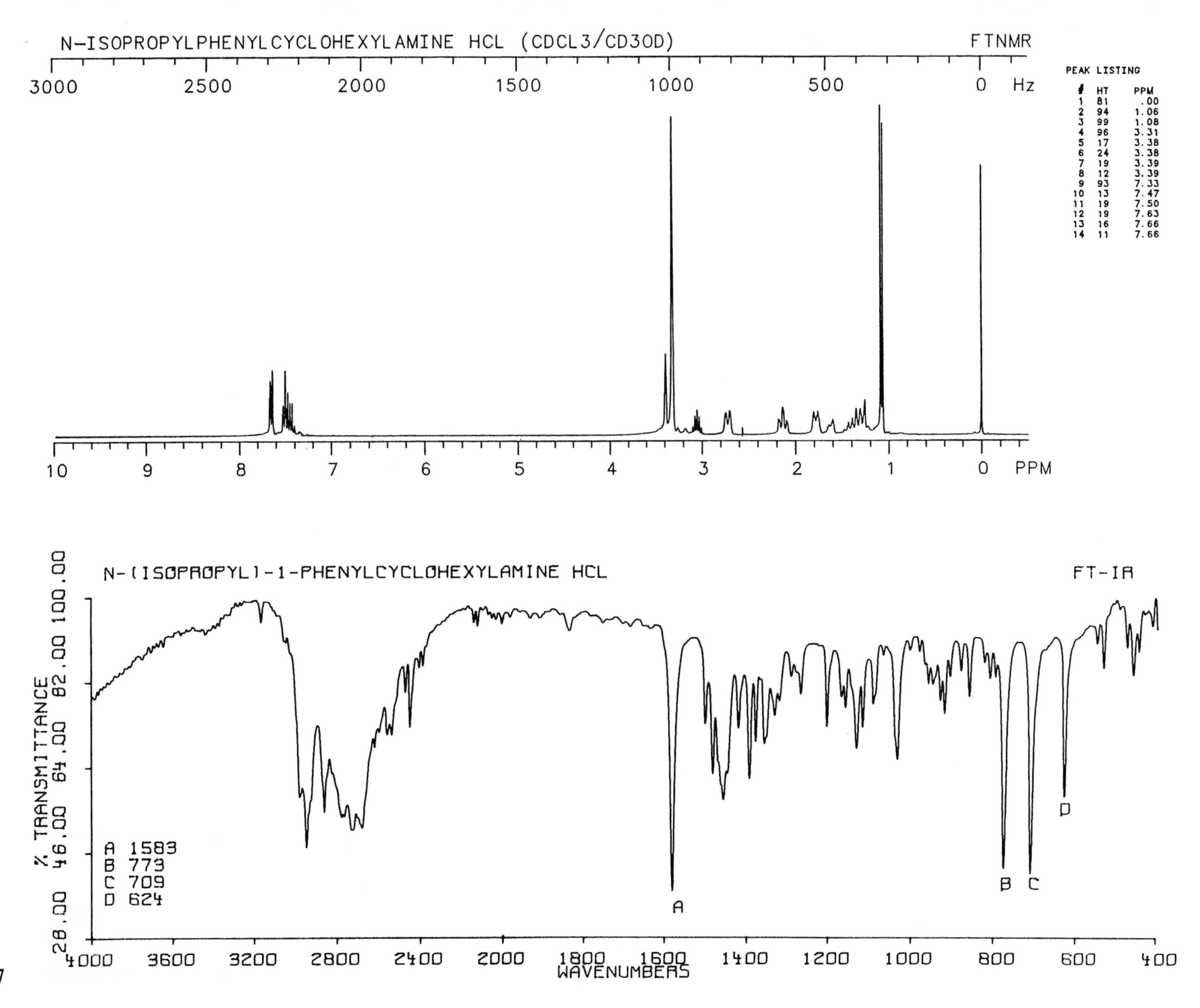
N-ISOPROPYLPHENYLCYCLOHEXYLAMINE HCL (CDCL3/CD3OD)
FTNMR
3000 2500 2000 1500 1000 500 0 Hz
PEAK LISTING
# HT PPM
1 81 .00
2 94 1.06
3 99 1.08
4 96 3.31
5 17 3.38
6 24 3.38
7 19 3.39
8 12 3.39
9 93 7.33
10 13 7.47
11 19 7.50
12 19 7.63
13 16 7.66
14 11 7.66
10 9 8 7 6 5 4 3 2 1 0 PPM
N-(ISOPROPYL)-1-PHENYLCYCLOHEXYLAMINE HCL
FT-IR
% TRANSMITTANCE
28.00 46.00 64.00 82.00 100.00
A 1583
B 773
C 709
D 624
A
B C
D
4000 3600 3200 2800 2400 2000 1800 1600 1400 1200 1000 800 600 400
WAVENUMBERS

# ISOPROTERENOL

$C_{11}H_{17}NO_3$

Molecular weight: 211.26 (211.12)
Synonyms: 4-[1-Hydroxy-2-[(1-methylethyl)amino]ethyl]-1,2-benzenediol; N-isopropylnoradrenaline
Trade names: Isoproterenol, Mucomyst, Vapo-Iso

Use: Sympathomimetic, bronchodilator
HPLC: Si-10; 20A:80B; 7.0
GC:

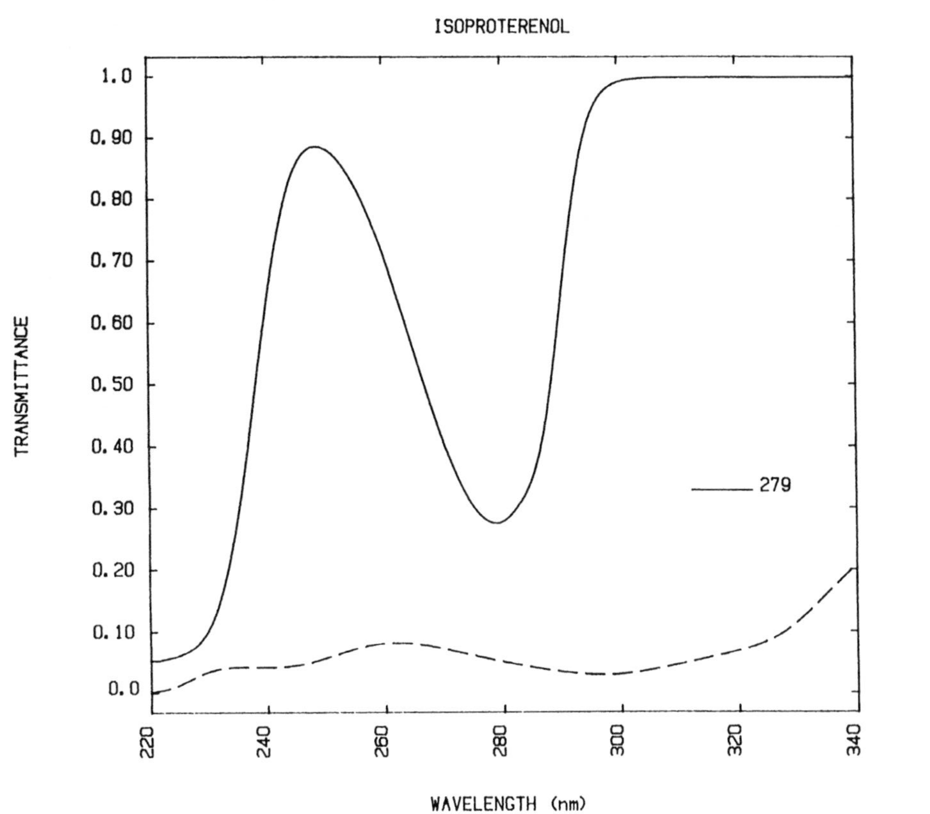


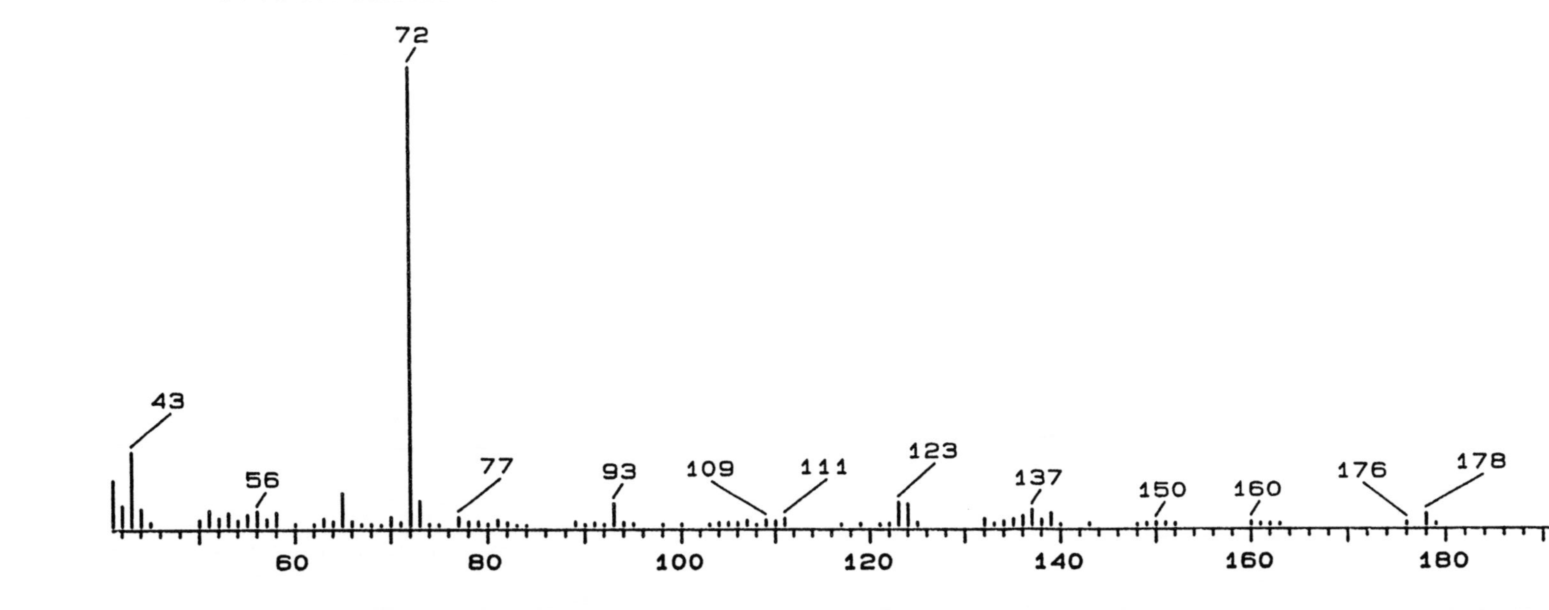

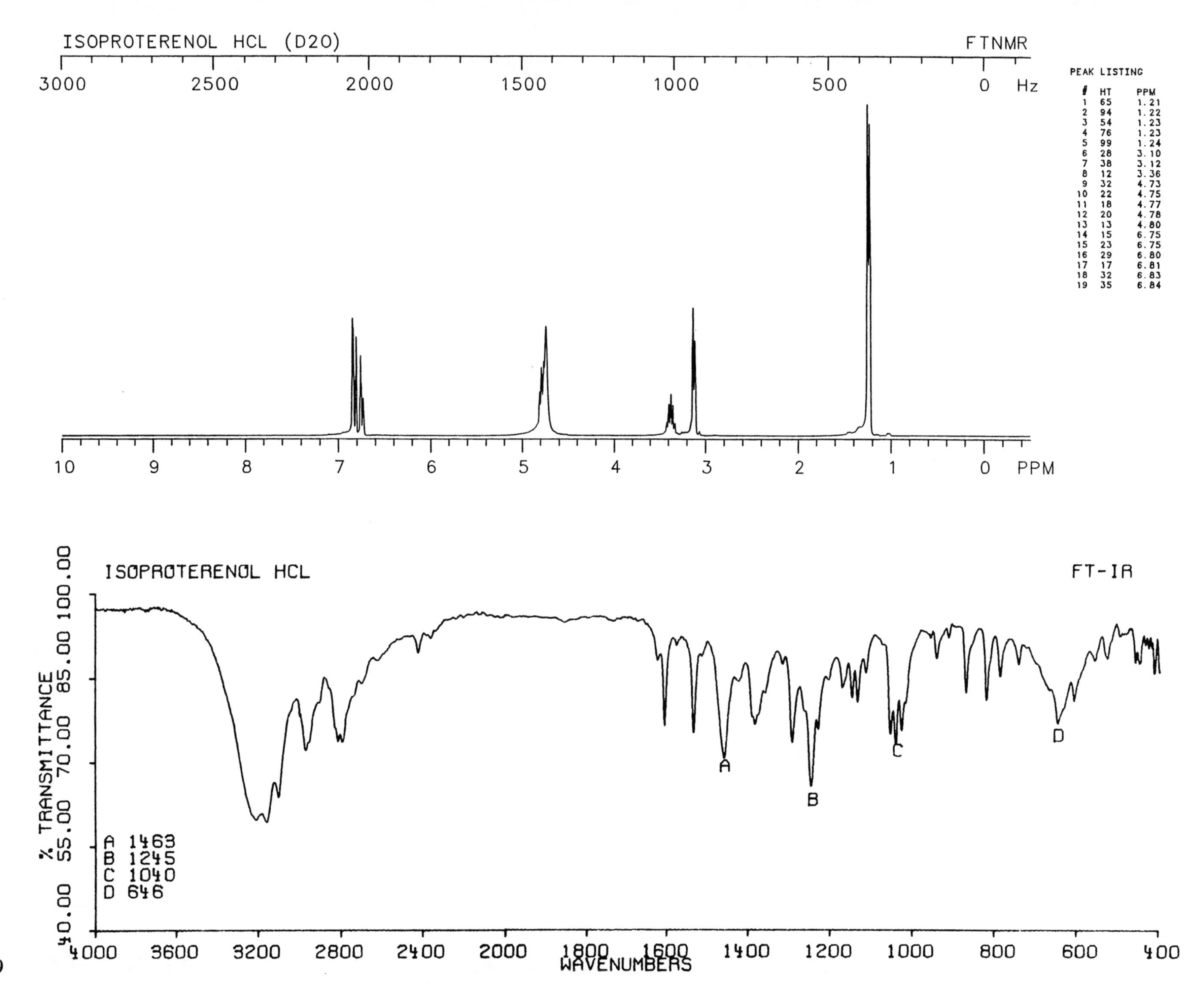

ISOPROTERENOL HCL (D2O)
FTNMR
3000
2500
2000
1500
1000
500
0 Hz
PEAK LISTING
# HT PPM
1 65 1.21
2 94 1.22
3 54 1.23
4 76 1.23
5 99 1.24
6 28 3.10
7 38 3.12
8 12 3.36
9 32 4.73
10 22 4.75
11 18 4.77
12 20 4.78
13 13 4.80
14 15 6.75
15 23 6.75
16 29 6.80
17 17 6.81
18 32 6.83
19 35 6.84
10
9
8
7
6
5
4
3
2
1
0 PPM
ISOPROTERENOL HCL
FT-IR
% TRANSMITTANCE
40.00
55.00
70.00
85.00
100.00
A 1463
B 1245
C 1040
D 646
A
B
C
D
4000
3600
3200
2800
2400
2000
1800
1600
1400
1200
1000
800
600
400
WAVENUMBERS

# ISOSAFROLE

$C_{10}H_{10}O_2$

Molecular weight: 162.18 (162.07)
Synonyms: 5-(1-Propenyl)-1,3-benzodioxole

Trade names:

Use: Synthesis
HPLC:
GC: 1365; 140°C

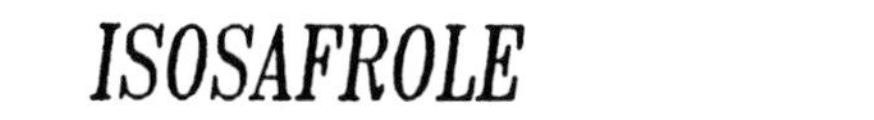


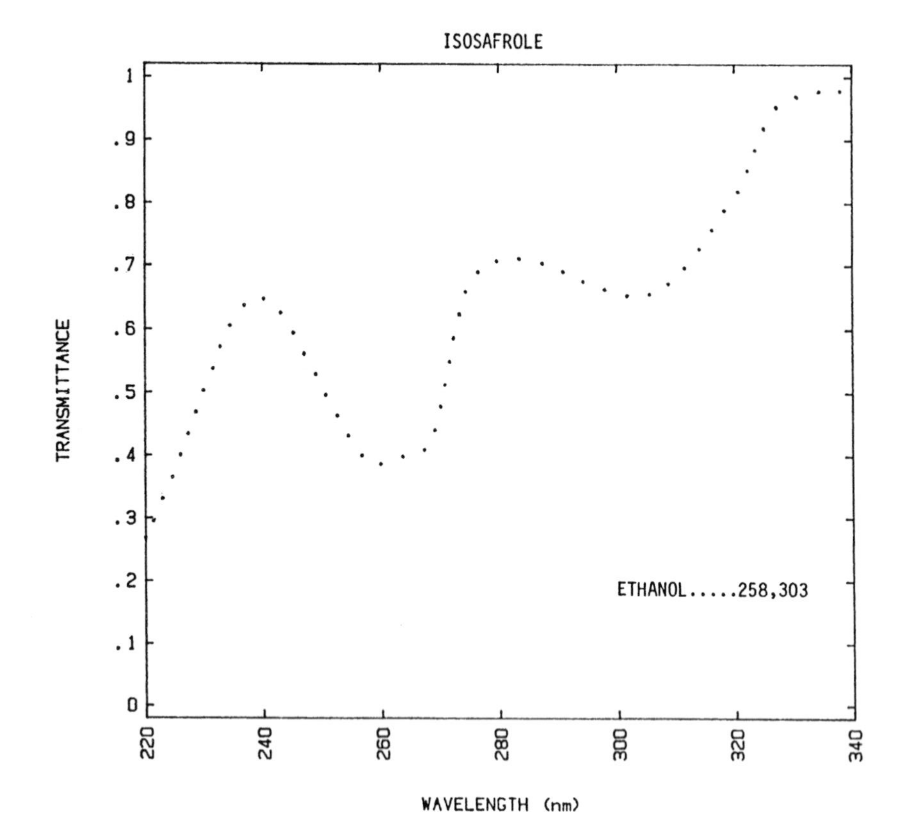


*ISOSAFROLE*

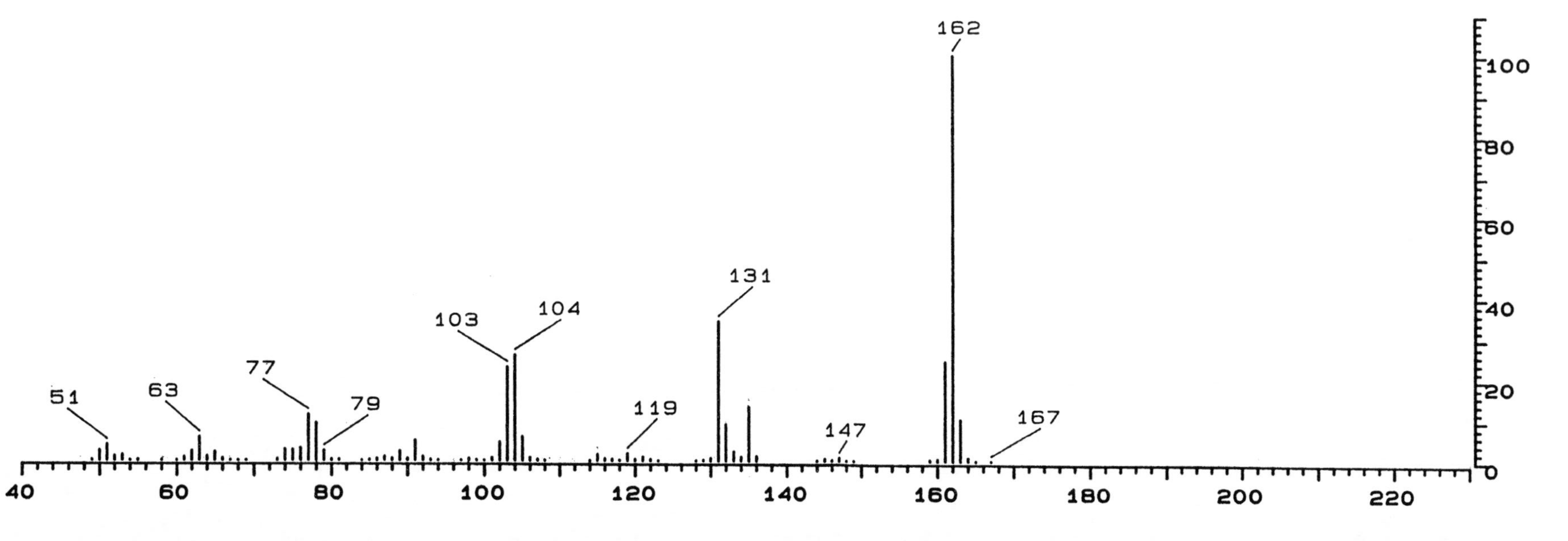

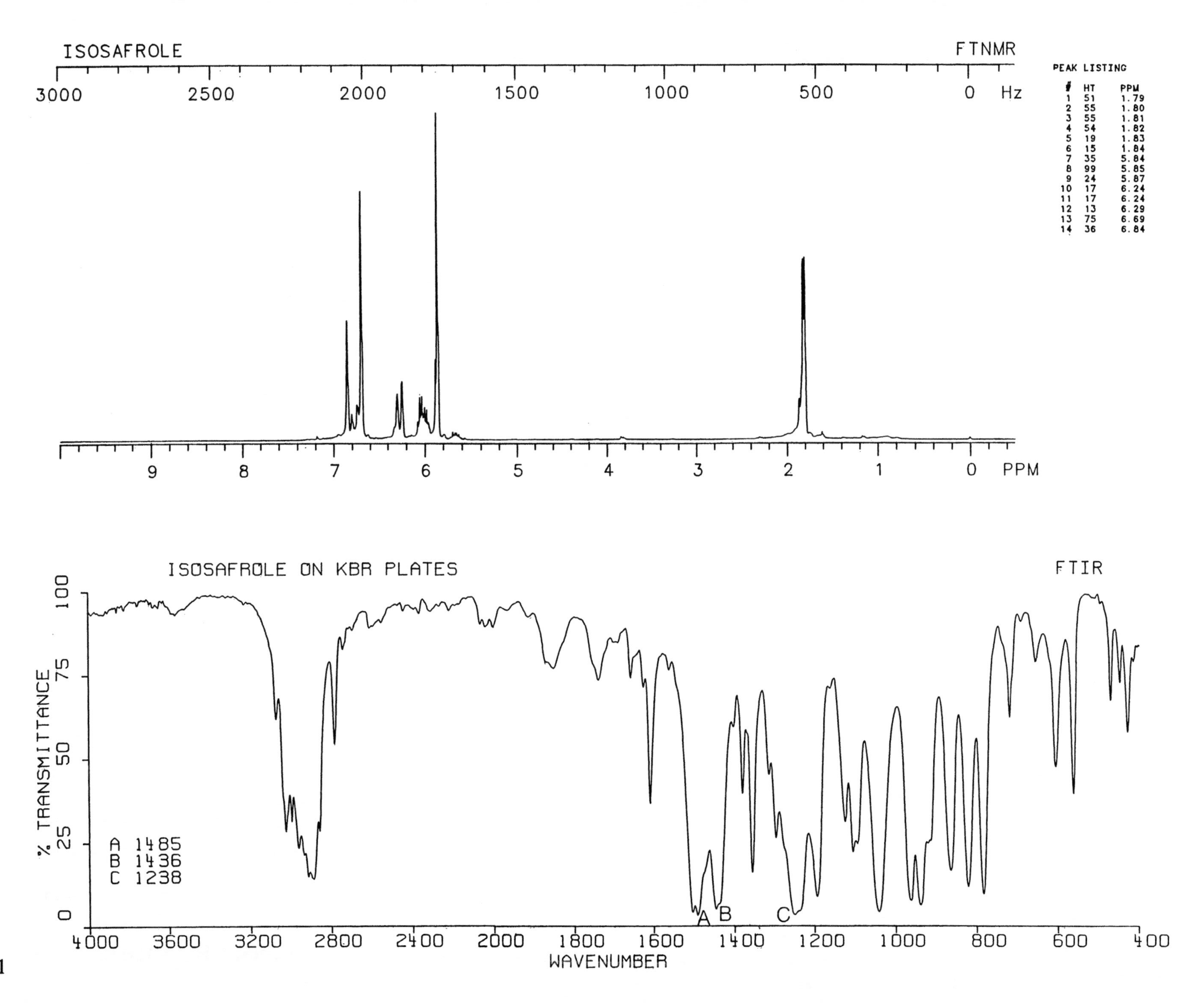
ISOSAFROLE
FTNMR
3000 2500 2000 1500 1000 500 0 Hz
9 8 7 6 5 4 3 2 1 0 PPM
PEAK LISTING
# HT PPM
1 51 1.79
2 55 1.80
3 55 1.81
4 54 1.82
5 19 1.83
6 15 1.84
7 35 5.84
8 99 5.85
9 24 5.87
10 17 6.24
11 17 6.24
12 13 6.29
13 75 6.69
14 36 6.84
ISOSAFROLE ON KBR PLATES
FTIR
% TRANSMITTANCE
0 25 50 75 100
A 1485
B 1436
C 1238
A B C
4000 3600 3200 2800 2400 2000 1800 1600 1400 1200 1000 800 600 400
WAVENUMBER

# ISOSORBIDE DINITRATE

$C_6H_8N_2O_8$

**Molecular weight:** 236.14 (236.03)
**Synonyms:** 1,4:3,6-Dianhydro-D-glucitol dinitrate; dinitrosorbide

**Trade names:** Dilatrate-SR, Iso-Bid, Isordil, Isosorbide Dinitrate, Isotrate, Sorate, Sorbide, Sorbitrate
**Use:** Coronary vasodilator
**HPLC:**
**GC:**

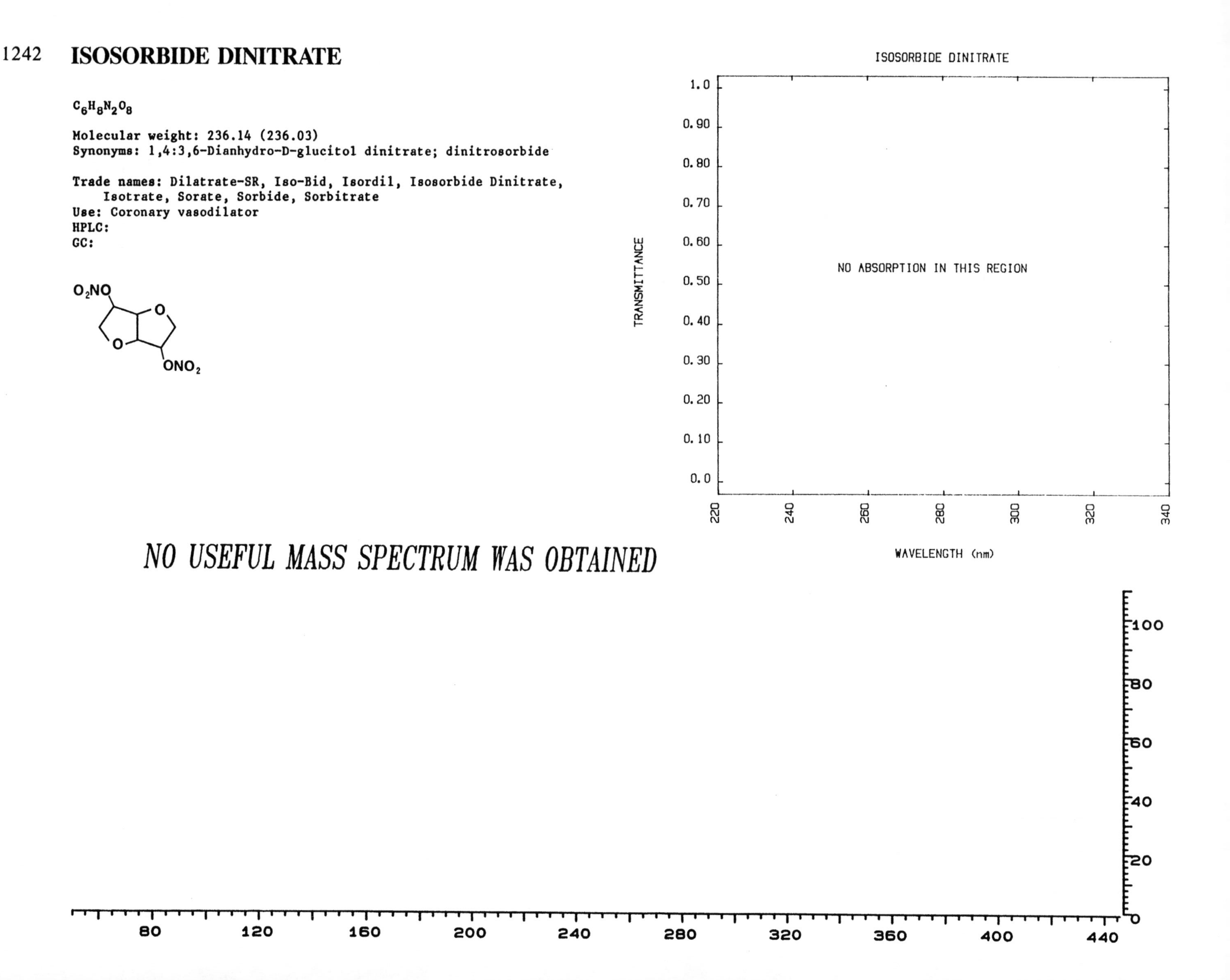

ISOSORBIDE DINITRATE

FTNMR

3000 2500 2000 1500 1000 500 0 Hz

9 8 7 6 5 4 3 2 1 0 PPM

PEAK LISTING

| # | HT | PPM |
|---|---|---|
| 1 | 24 | .00 |
| 2 | 35 | 1.58 |
| 3 | 37 | 3.91 |
| 4 | 37 | 3.93 |
| 5 | 63 | 3.95 |
| 6 | 53 | 3.97 |
| 7 | 52 | 4.06 |
| 8 | 47 | 4.07 |
| 9 | 28 | 4.09 |
| 10 | 39 | 4.10 |
| 11 | 35 | 4.11 |
| 12 | 96 | 4.13 |
| 13 | 99 | 4.14 |
| 14 | 76 | 4.14 |
| 15 | 88 | 4.15 |
| 16 | 18 | 4.18 |
| 17 | 13 | 4.56 |
| 18 | 13 | 4.57 |
| 19 | 60 | 4.58 |
| 20 | 54 | 4.59 |
| 21 | 40 | 4.99 |
| 22 | 62 | 5.01 |
| 23 | 30 | 5.03 |
| 24 | 34 | 5.37 |
| 25 | 48 | 5.38 |
| 26 | 91 | 5.38 |
| 27 | 82 | 5.39 |
| 28 | 21 | 5.41 |
| 29 | 41 | 7.26 |

ISOSORBIDE DINITRATE

FT-IR

% TRANSMITTANCE .00 25.00 50.00 75.00 100.00

A 1668
B 1632
C 1090
D 864

4000 3600 3200 2800 2400 2000 1800 1600 1400 1200 1000 800 600 400
WAVENUMBERS

# ISOXSUPRINE

$C_{18}H_{23}NO_3$

**Molecular weight:** 301.39 (301.17)
**Synonyms:** 4-Hydroxy-α-[1-[(1-methyl-2-phenoxyethyl)amino]ethyl]-benzenemethanol
**Trade names:** Isosuprine, Vasodilan

**Use:** Vasodilator
**HPLC:**
**GC:** 2350; 250°C

HO–C6H4–CH(OH)–CH(CH3)–NH–CH(CH3)–CH2–O–C6H5

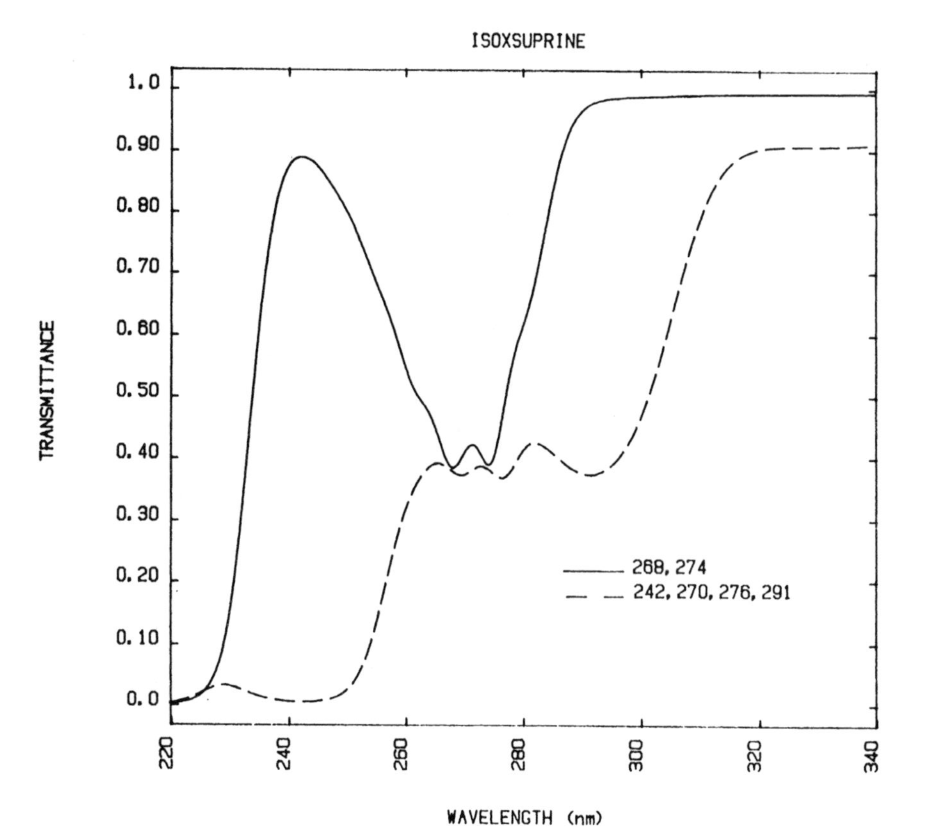

*ISOXSUPRINE--DIP*

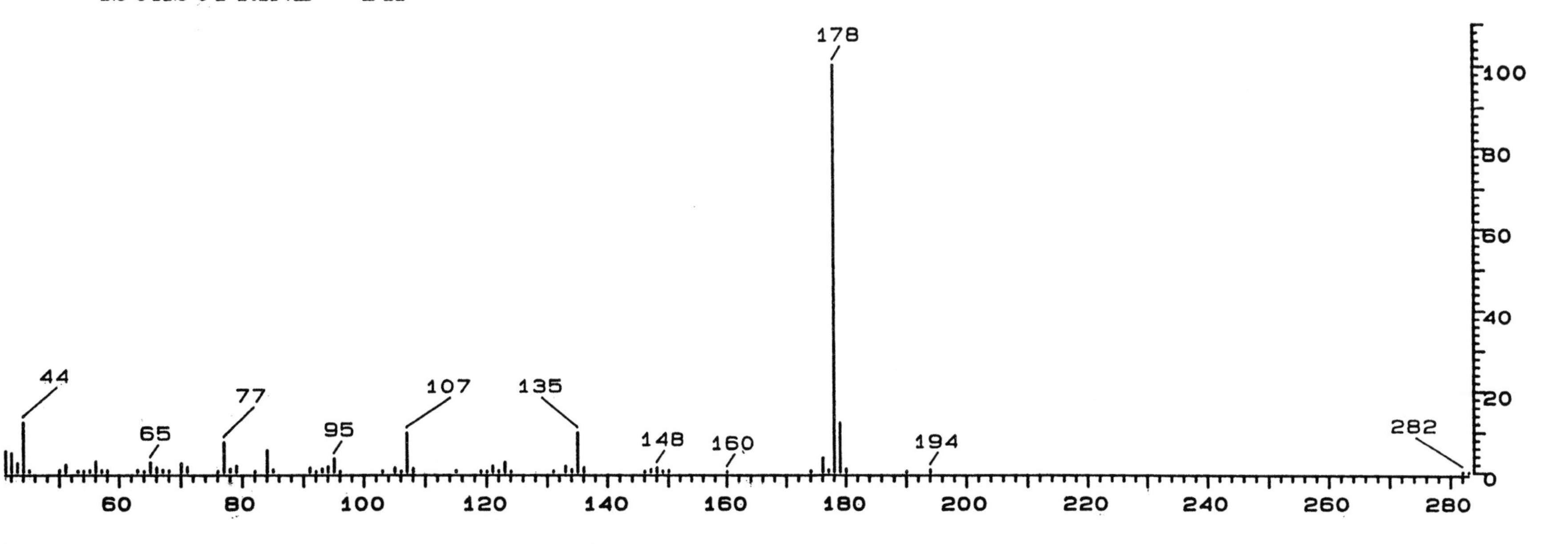

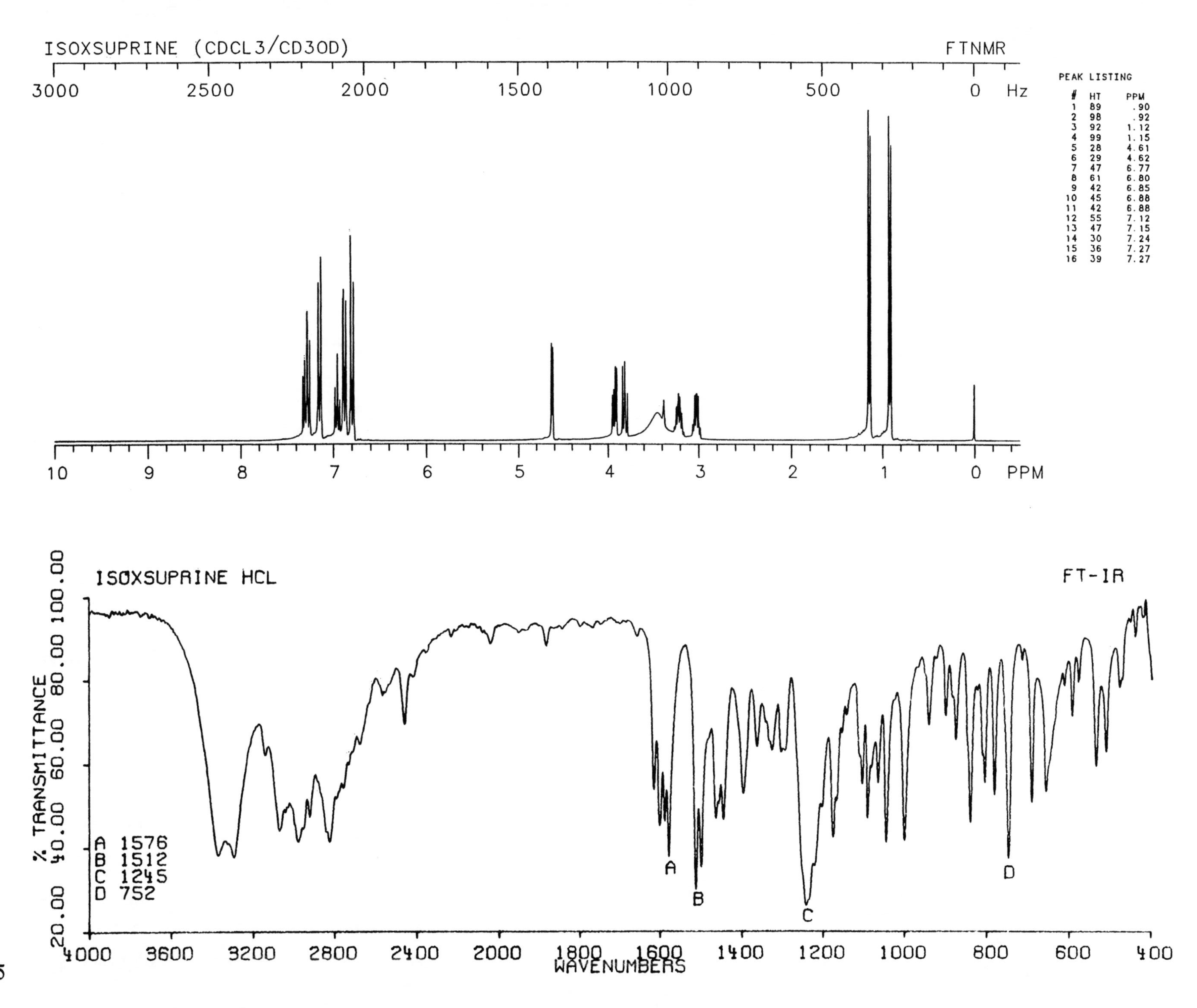
ISOXSUPRINE (CDCL3/CD3OD)
FTNMR
3000 2500 2000 1500 1000 500 0 Hz
10 9 8 7 6 5 4 3 2 1 0 PPM
PEAK LISTING
# HT PPM
1 89 .90
2 98 .92
3 92 1.12
4 99 1.15
5 28 4.61
6 29 4.62
7 47 6.77
8 61 6.80
9 42 6.85
10 45 6.88
11 42 6.88
12 55 7.12
13 47 7.15
14 30 7.24
15 36 7.27
16 39 7.27
ISOXSUPRINE HCL
FT-IR
% TRANSMITTANCE
20.00 40.00 60.00 80.00 100.00
A 1576
B 1512
C 1245
D 752
A
B
C
D
4000 3600 3200 2800 2400 2000 1800 1600 1400 1200 1000 800 600 400
WAVENUMBERS

# KANAMYCIN

$C_{18}H_{36}N_4O_{11}$

**Molecular weight:** 484.51 (484.24)
**Synonyms:** O-3-Amino-3-deoxy-α-D-glucopyranosyl-(1→6)-O-[6-amino-6-deoxy-α-D-glucopyranosyl-(1→4)]-2-deoxy-α-D-streptamine
**Trade names:** Kantrex

**Use:** Antibacterial
**HPLC:**
**GC:**

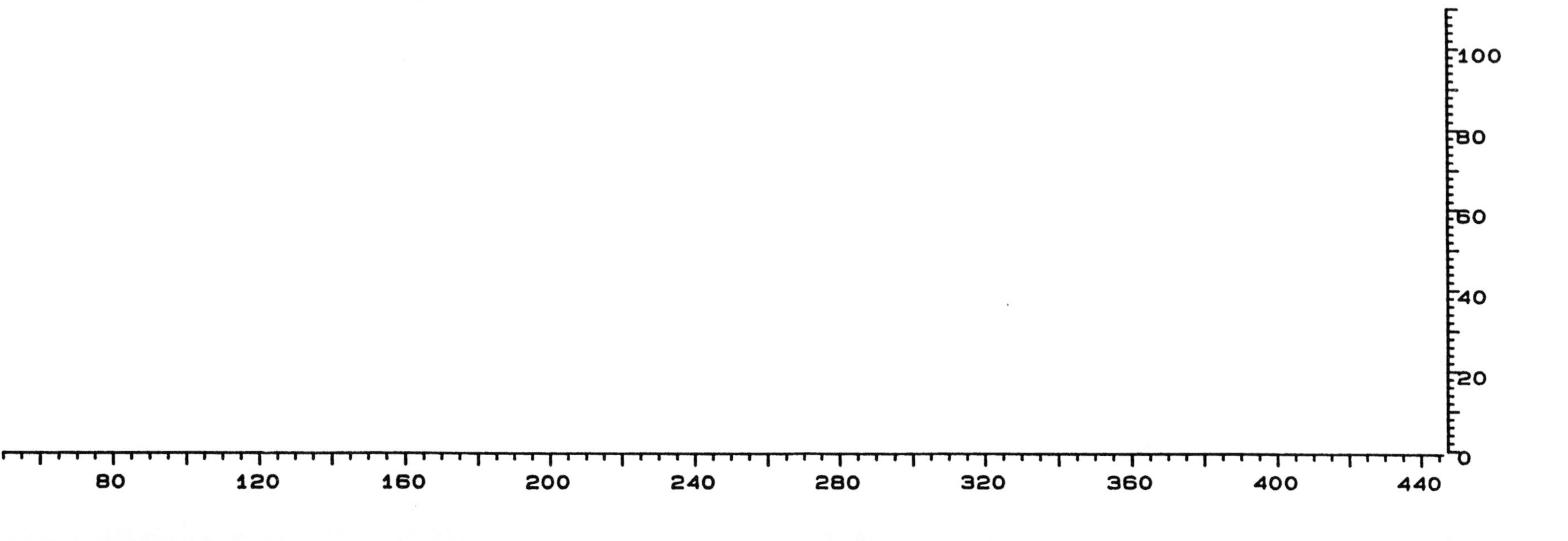

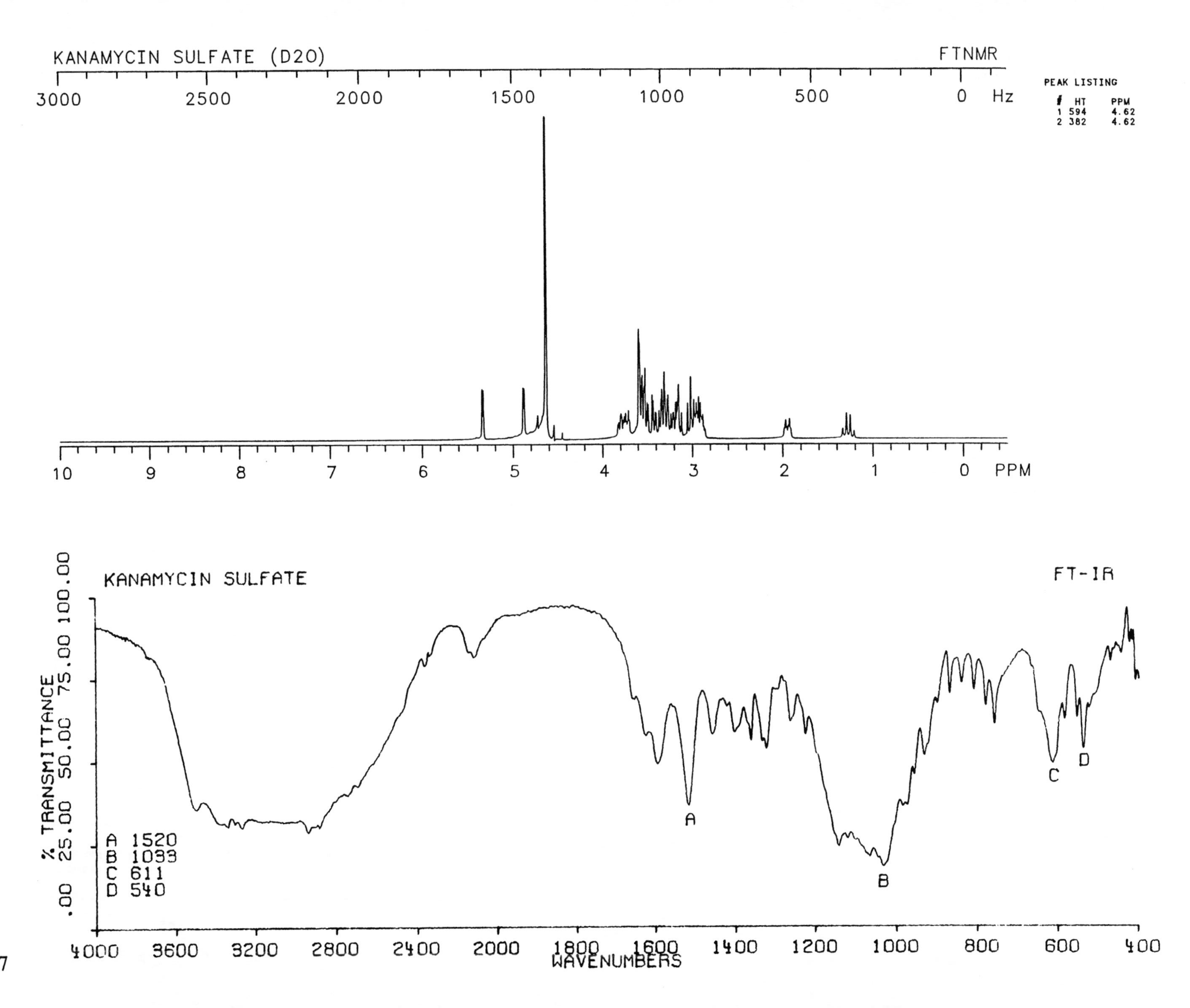
KANAMYCIN SULFATE (D2O)
FTNMR
3000
2500
2000
1500
1000
500
0
Hz
PEAK LISTING
# HT PPM
1 594 4.62
2 382 4.62
10
9
8
7
6
5
4
3
2
1
0
PPM
KANAMYCIN SULFATE
FT-IR
% TRANSMITTANCE
.00
25.00
50.00
75.00
100.00
A 1520
B 1033
C 611
D 540
A
B
C
D
4000
3600
3200
2800
2400
2000
1800
1600
1400
1200
1000
800
600
400
WAVENUMBERS

# KETAMINE

$C_{13}H_{16}ClNO$

**Molecular weight:** 237.73 (237.09)
**Synonyms:** 2-(o-Chlorophenyl)-2-(methylamino)cyclohexanone

**Trade names:** Ketaject, Ketalar

**Use:** Anesthetic
**HPLC:** Si-10; 2A:98B; 4.0
**GC:** 1864; 200°C

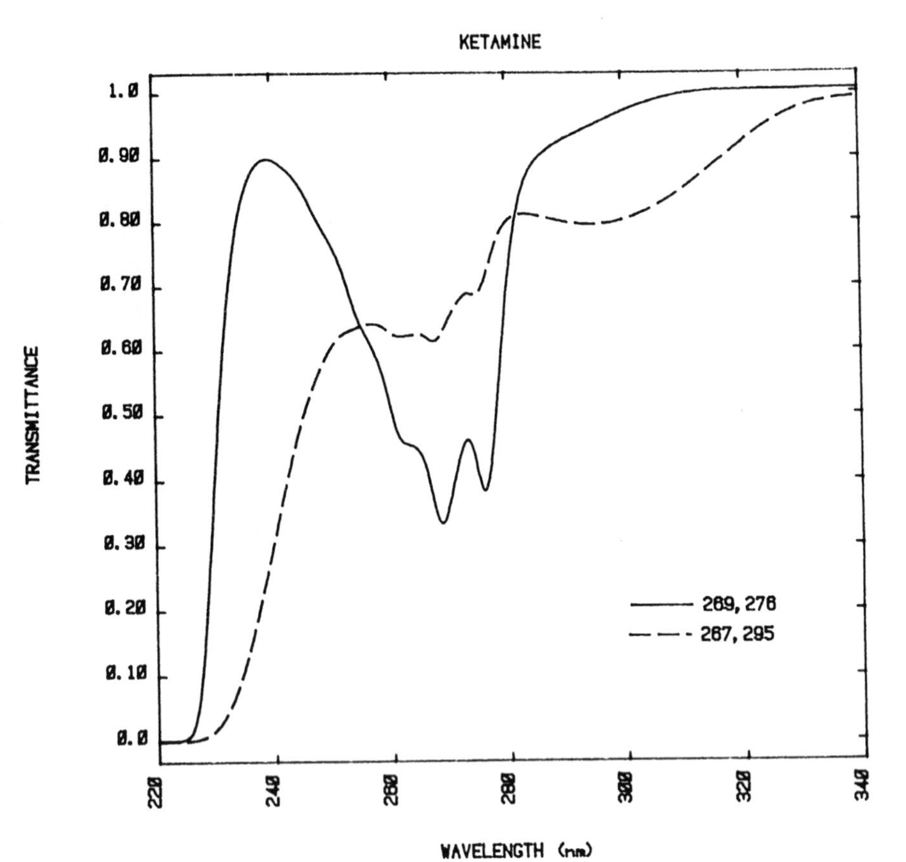


*KETAMINE*

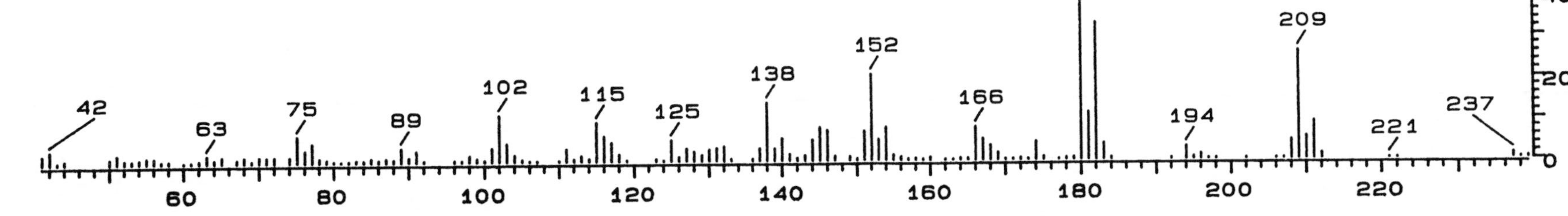
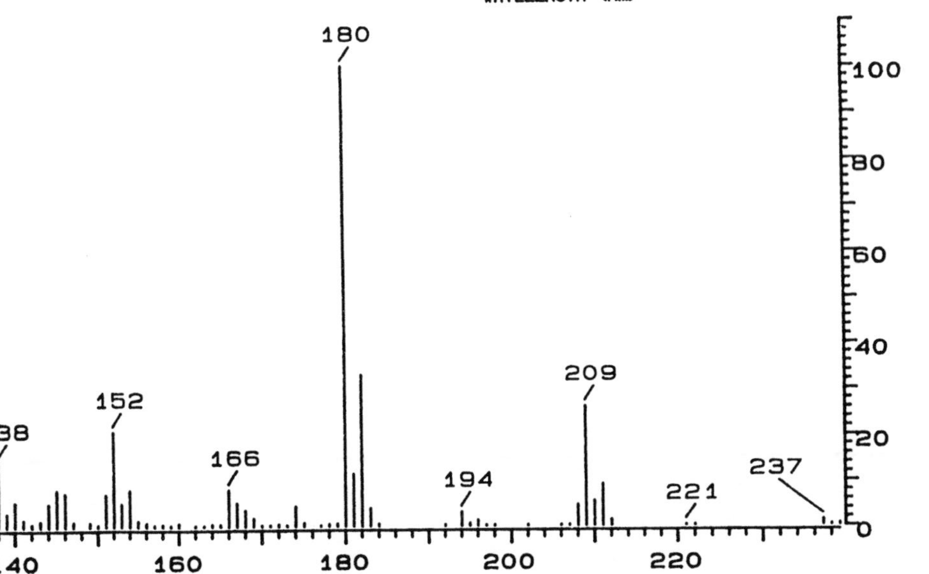

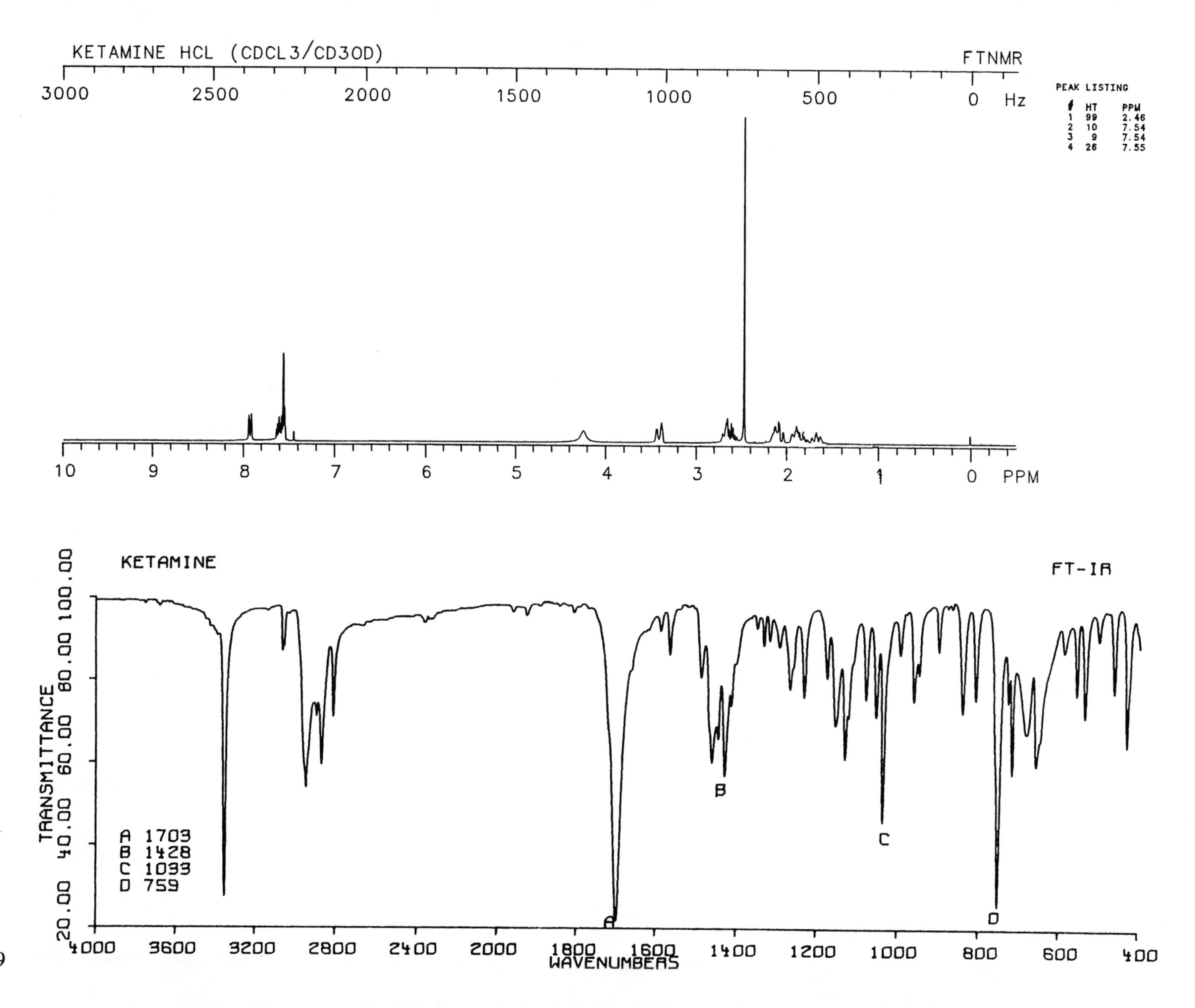
KETAMINE HCL (CDCL3/CD3OD)
FTNMR
3000 2500 2000 1500 1000 500 0 Hz
PEAK LISTING
# HT PPM
1 99 2.46
2 10 7.54
3 9 7.54
4 26 7.55
10 9 8 7 6 5 4 3 2 1 0 PPM
KETAMINE
FT-IR
TRANSMITTANCE
20.00 40.00 60.00 80.00 100.00
A 1703
B 1428
C 1033
D 759
4000 3600 3200 2800 2400 2000 1800 1600 1400 1200 1000 800 600 400
WAVENUMBERS

# KETOCONAZOLE

$C_{26}H_{28}O_4N_4Cl_2$

**Molecular weight:** 531.44 (530.15)
**Synonyms:** cis-1-Acetyl-4-[4-[[2-(2,4-dichlorophenyl)-2-(1H-imidazol-1-ylmethyl)-1,3-dioxolan-4-yl]methoxy]phenyl]piperazine
**Trade names:** Nizoral

**Use:** Antifungal agent
**HPLC:** Si-10; 5A:95B; 3.7
**GC:**

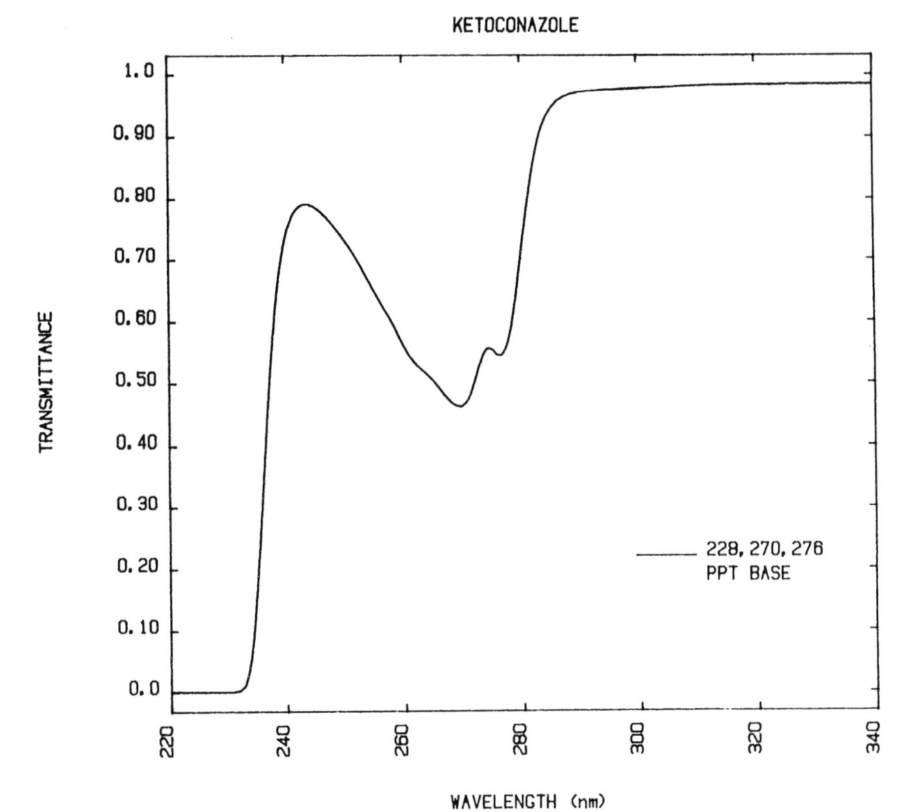


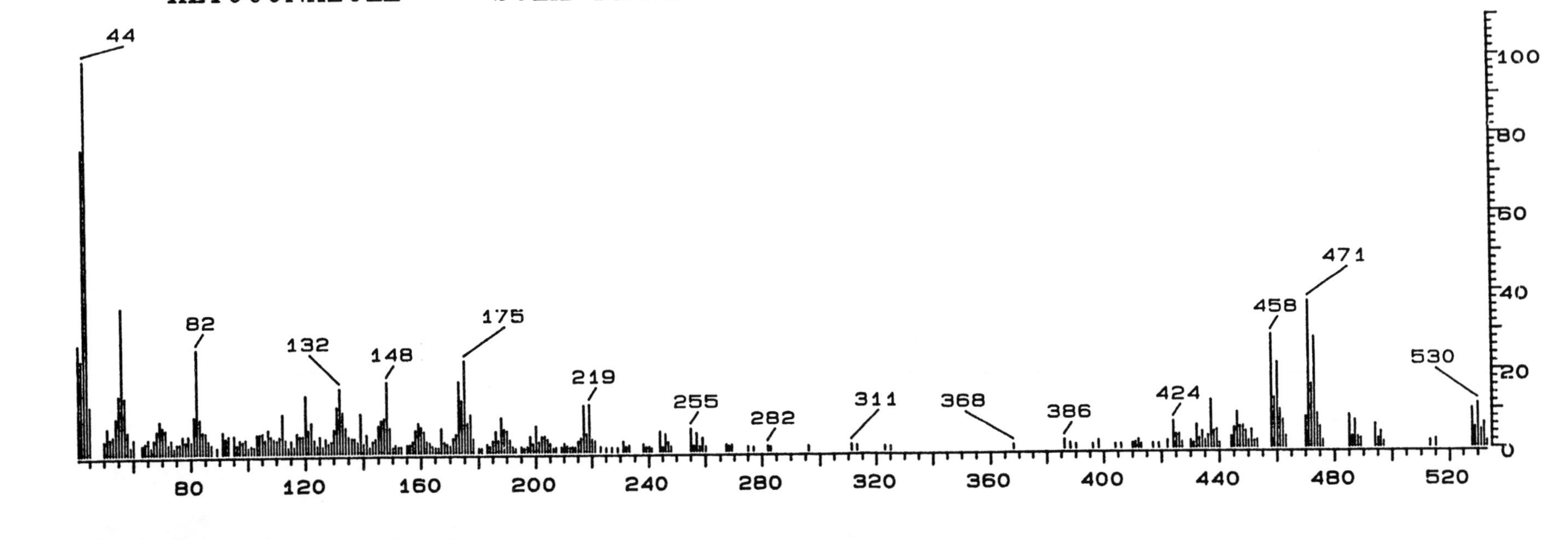

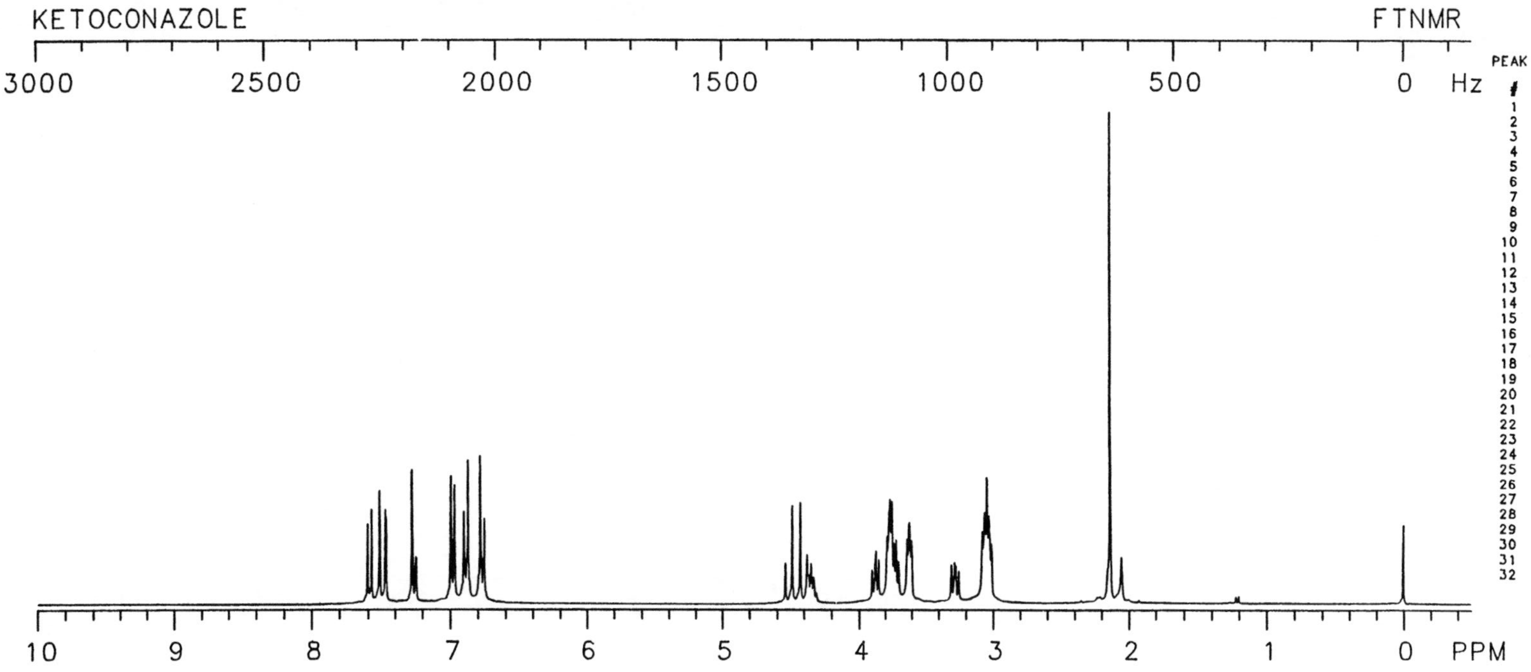

KETOCONAZOLE

FTNMR

3000 2500 2000 1500 1000 500 0 Hz

10 9 8 7 6 5 4 3 2 1 0 PPM

PEAK LISTING

| # | HT | PPM | # | HT | PPM |
|---|---|---|---|---|---|
| 1 | 15 | .00 | 33 | 23 | 6.97 |
| 2 | 11 | .00 | 34 | 25 | 6.99 |
| 3 | 9 | 2.05 | 35 | 9 | 7.24 |
| 4 | 99 | 2.13 | 36 | 9 | 7.24 |
| 5 | 12 | 3.01 | 37 | 13 | 7.27 |
| 6 | 17 | 3.02 | 38 | 27 | 7.27 |
| 7 | 25 | 3.04 | 39 | 26 | 7.27 |
| 8 | 18 | 3.06 | 40 | 17 | 7.46 |
| 9 | 14 | 3.07 | 41 | 16 | 7.46 |
| 10 | 8 | 3.27 | 42 | 18 | 7.47 |
| 11 | 12 | 3.59 | 43 | 22 | 7.51 |
| 12 | 16 | 3.61 | 44 | 21 | 7.51 |
| 13 | 13 | 3.63 | 45 | 18 | 7.57 |
| 14 | 8 | 3.69 | 46 | 15 | 7.57 |
| 15 | 12 | 3.71 | 47 | 16 | 7.59 |
| 16 | 12 | 3.72 | 48 | 13 | 7.60 |
| 17 | 20 | 3.74 | | | |
| 18 | 21 | 3.76 | | | |
| 19 | 18 | 3.76 | | | |
| 20 | 13 | 3.78 | | | |
| 21 | 8 | 3.84 | | | |
| 22 | 10 | 3.86 | | | |
| 23 | 8 | 4.34 | | | |
| 24 | 9 | 4.37 | | | |
| 25 | 20 | 4.42 | | | |
| 26 | 19 | 4.48 | | | |
| 27 | 8 | 4.53 | | | |
| 28 | 17 | 6.75 | | | |
| 29 | 29 | 6.78 | | | |
| 30 | 28 | 6.87 | | | |
| 31 | 18 | 6.90 | | | |
| 32 | 23 | 6.96 | | | |

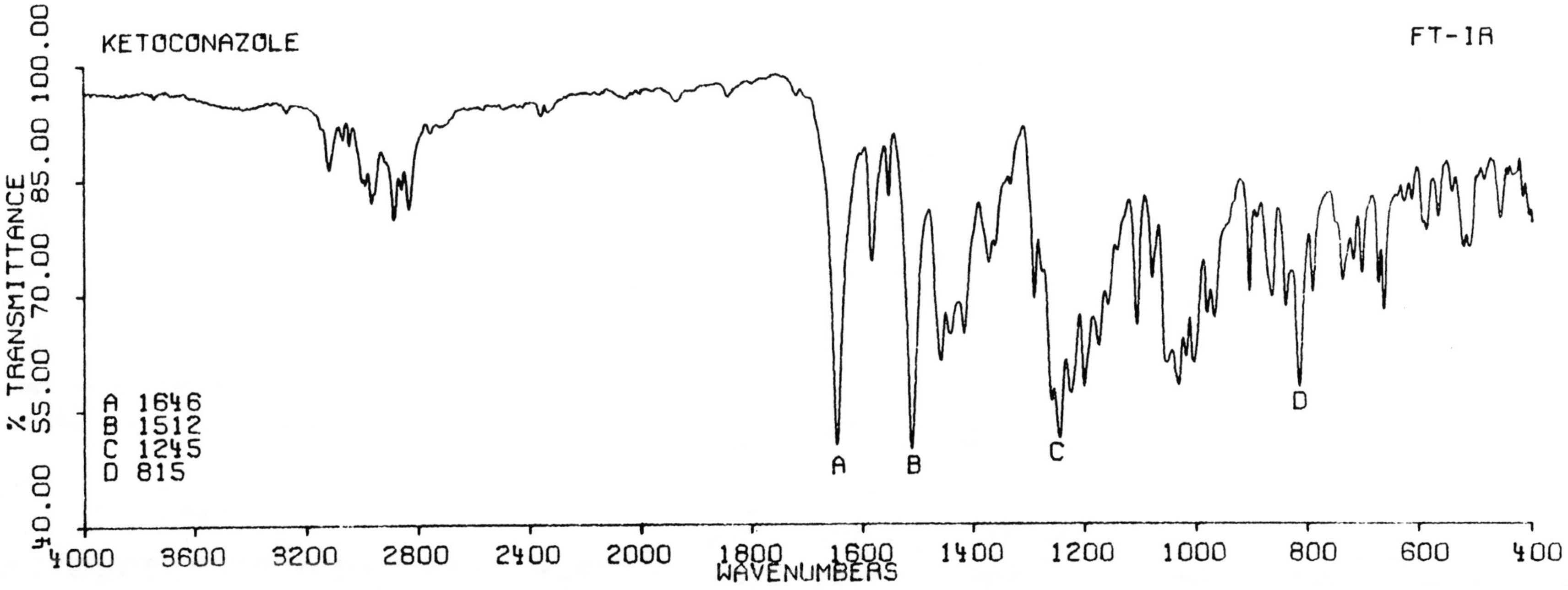

KETOCONAZOLE

FT-IR

A 1646
B 1512
C 1245
D 815

% TRANSMITTANCE

40.00 55.00 70.00 85.00 100.00

4000 3600 3200 2800 2400 2000 1800 1600 1400 1200 1000 800 600 400

WAVENUMBERS

# KETOGLUTARIC ACID

$C_5H_6O_5$

**Molecular weight:** 146.10 (146.02)
**Synonyms:** 2-Oxopentanedioic acid; 2-oxoglutaric acid; 2-oxo-1,5-pentanedioic acid; α-ketoglutaric acid
**Trade names:** Ornicetil, Ornithine-α-ketoglutarate

**Use:** Amino acid metabolism
**HPLC:**
**GC:**

$HOOC-CH_2-C(=O)-CH_2-COOH$

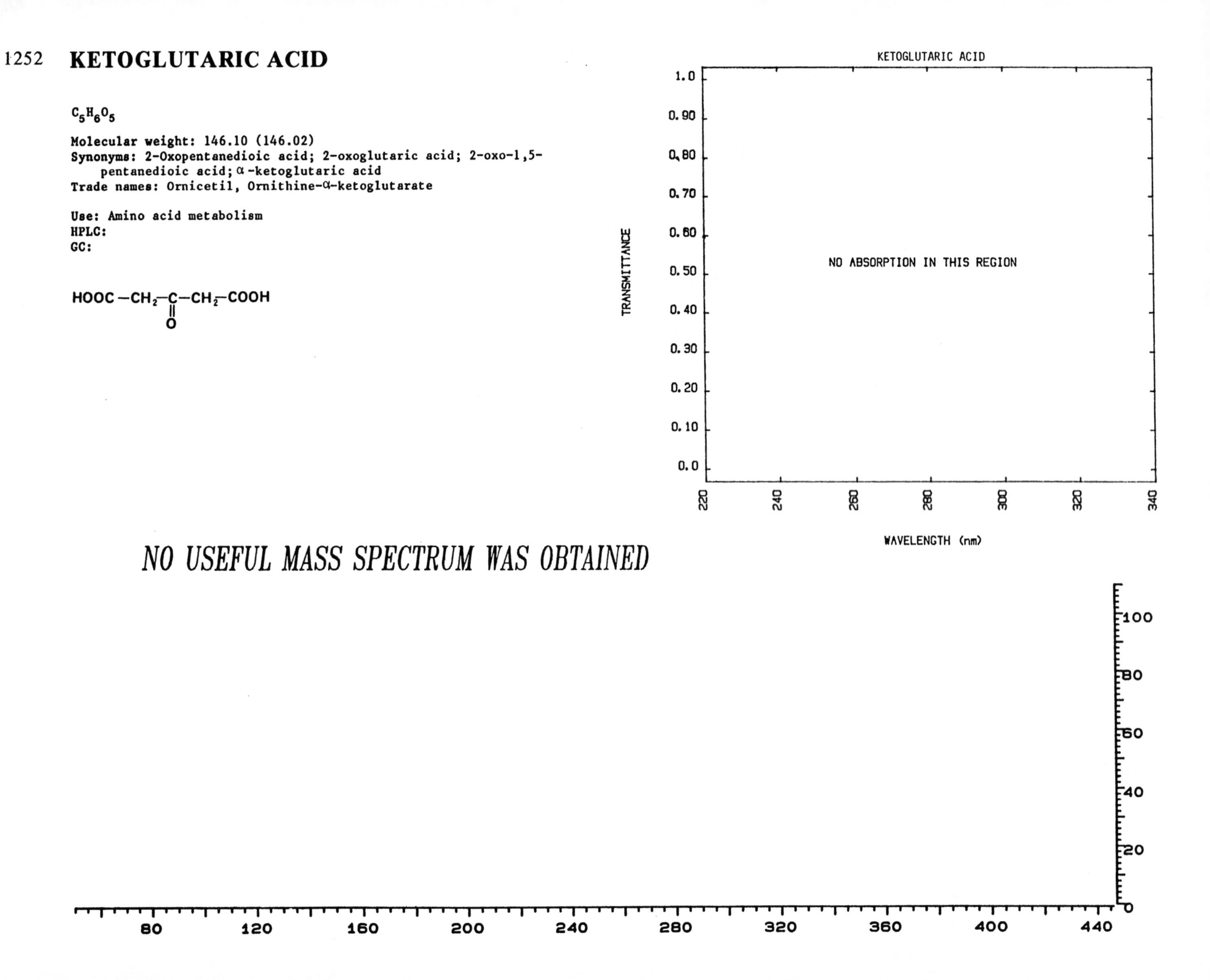

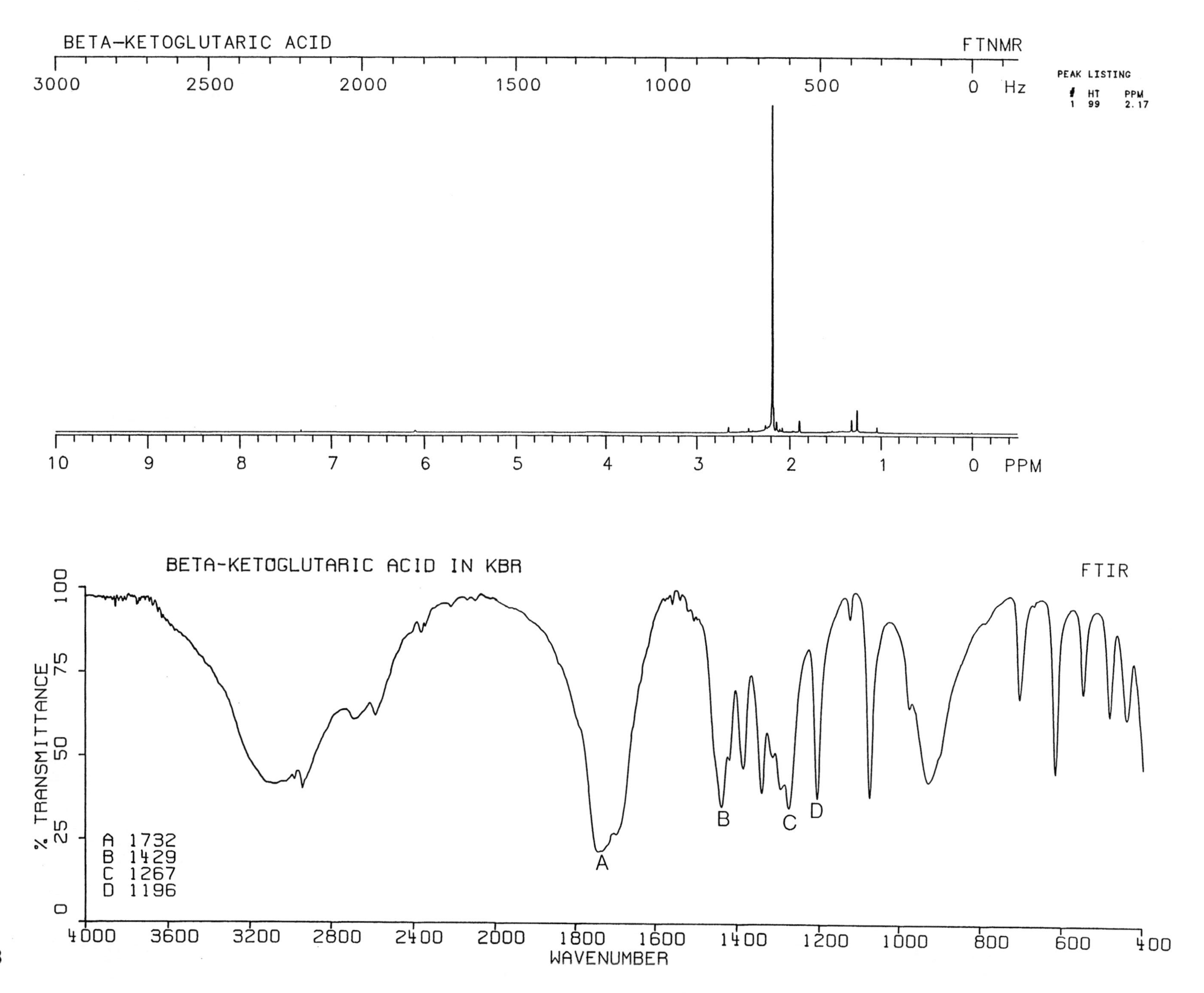
BETA-KETOGLUTARIC ACID
FTNMR
3000
2500
2000
1500
1000
500
0 Hz
PEAK LISTING
# HT PPM
1 99 2.17
10
9
8
7
6
5
4
3
2
1
0 PPM
BETA-KETOGLUTARIC ACID IN KBR
FTIR
% TRANSMITTANCE
100
75
50
25
0
A 1732
B 1429
C 1267
D 1196
A
B
C
D
4000
3600
3200
2800
2400
2000
1800
1600
1400
1200
1000
800
600
400
WAVENUMBER

# KETOPROFEN

$C_{16}H_{14}O_3$

**Molecular weight:** 254.29 (254.09)
**Synonyms:** 3-Benzoyl-α-methylbenzeneacetic acid; m-benzoyl-hydratropic acid
**Trade names:** Alrheumart, Alrheumun, Capisten, Fastum, Ketopron, Kefenid, Lertus, Iso-K, Meprofen, Orudis, Oruvail
**Use:** Anti-inflammatory, analgesic
**HPLC:** Si-10; 10A:90B; 7.1
**GC:** 2440; 250°C

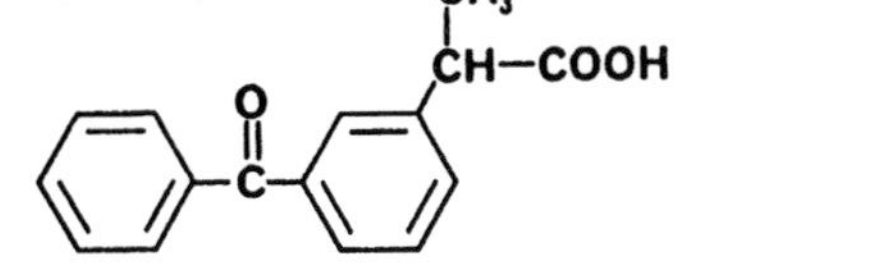
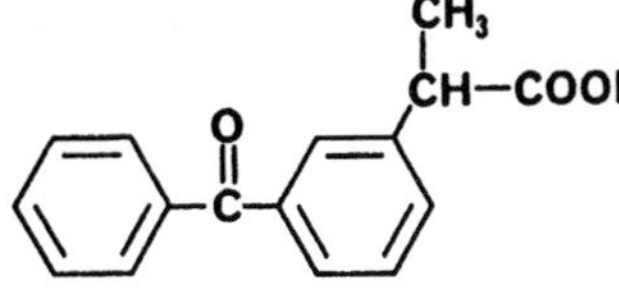


KETOPROFEN

TRANSMITTANCE

WAVELENGTH (nm)

259
260

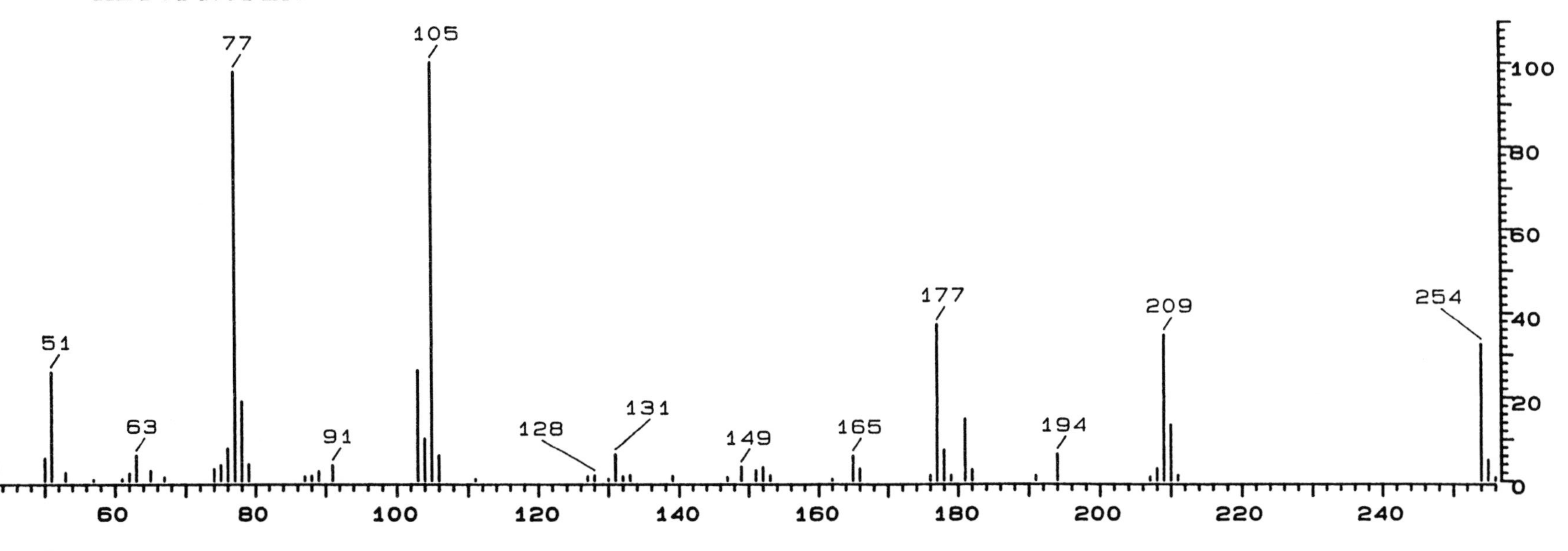

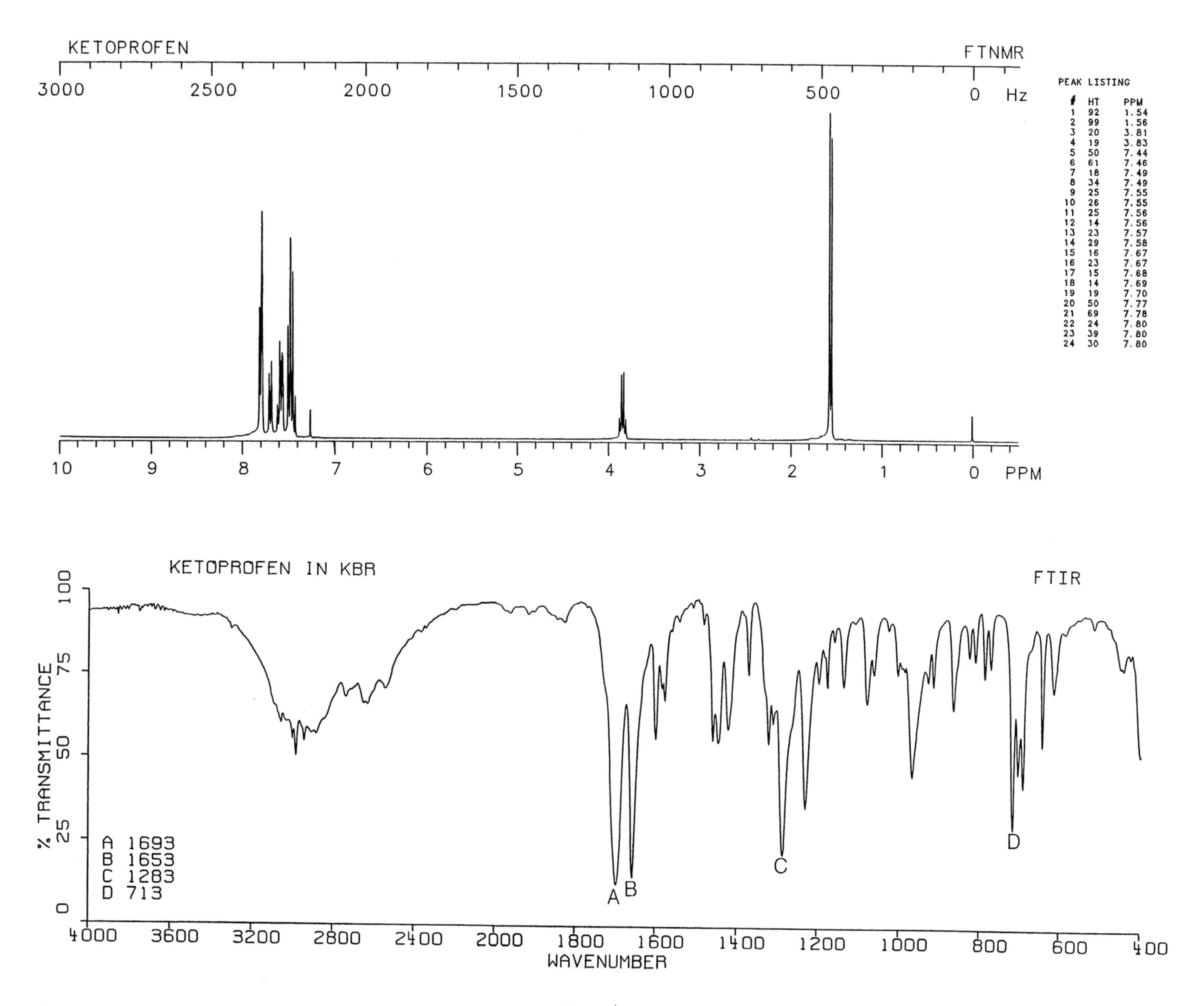
KETOPROFEN
FTNMR
3000 2500 2000 1500 1000 500 0 Hz
10 9 8 7 6 5 4 3 2 1 0 PPM
PEAK LISTING
# HT PPM
1 92 1.54
2 99 1.56
3 20 3.81
4 19 3.83
5 50 7.44
6 61 7.46
7 18 7.49
8 34 7.49
9 25 7.55
10 26 7.55
11 25 7.56
12 14 7.56
13 23 7.57
14 29 7.58
15 16 7.67
16 23 7.67
17 15 7.68
18 14 7.69
19 19 7.70
20 50 7.77
21 69 7.78
22 24 7.80
23 39 7.80
24 30 7.80
KETOPROFEN IN KBR
FTIR
% TRANSMITTANCE
100 75 50 25 0
A 1693
B 1653
C 1283
D 713
A B C D
4000 3600 3200 2800 2400 2000 1800 1600 1400 1200 1000 800 600 400
WAVENUMBER

# LABETALOL

$C_{19}H_{24}N_2O_3$

Molecular weight: 328.41 (328.18)
Synonyms: 2-Hydroxy-5-[1-hydroxy-2-[(1-methyl-3-phenylpropyl)amino]-ethyl]benzamide; ibidomide
Trade names: Normodyne, Presdate, Trandate, Vescal

Use: Adrenergic acid
HPLC: Si-10; 10A:90B; 4.4
GC: 1268; 140°C

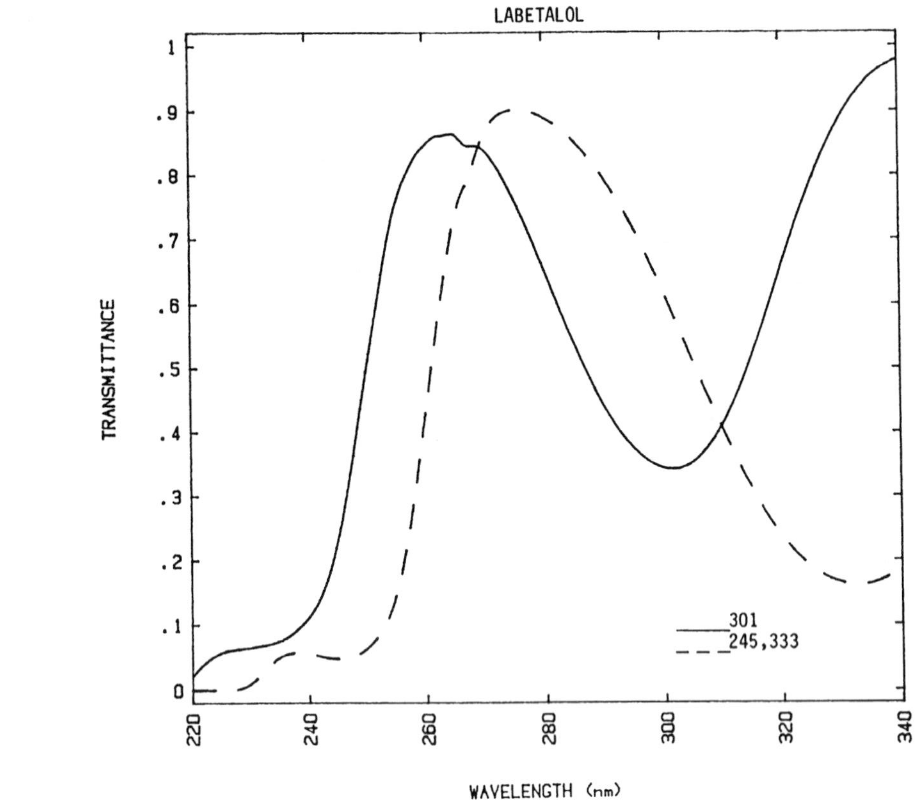

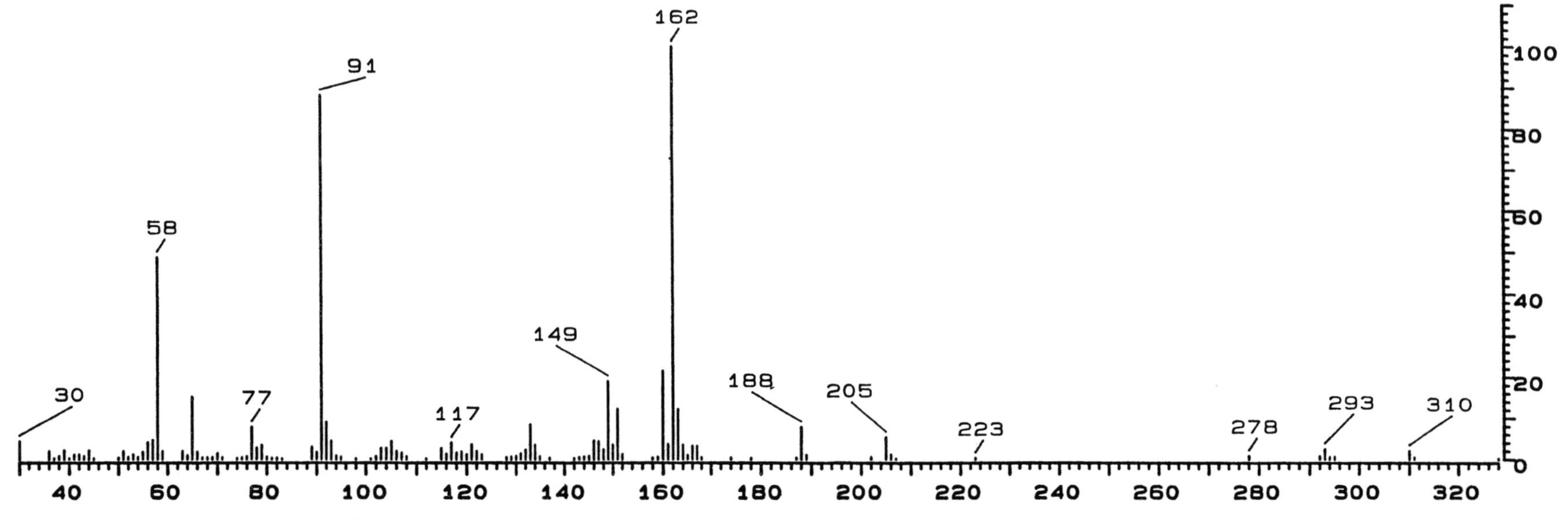

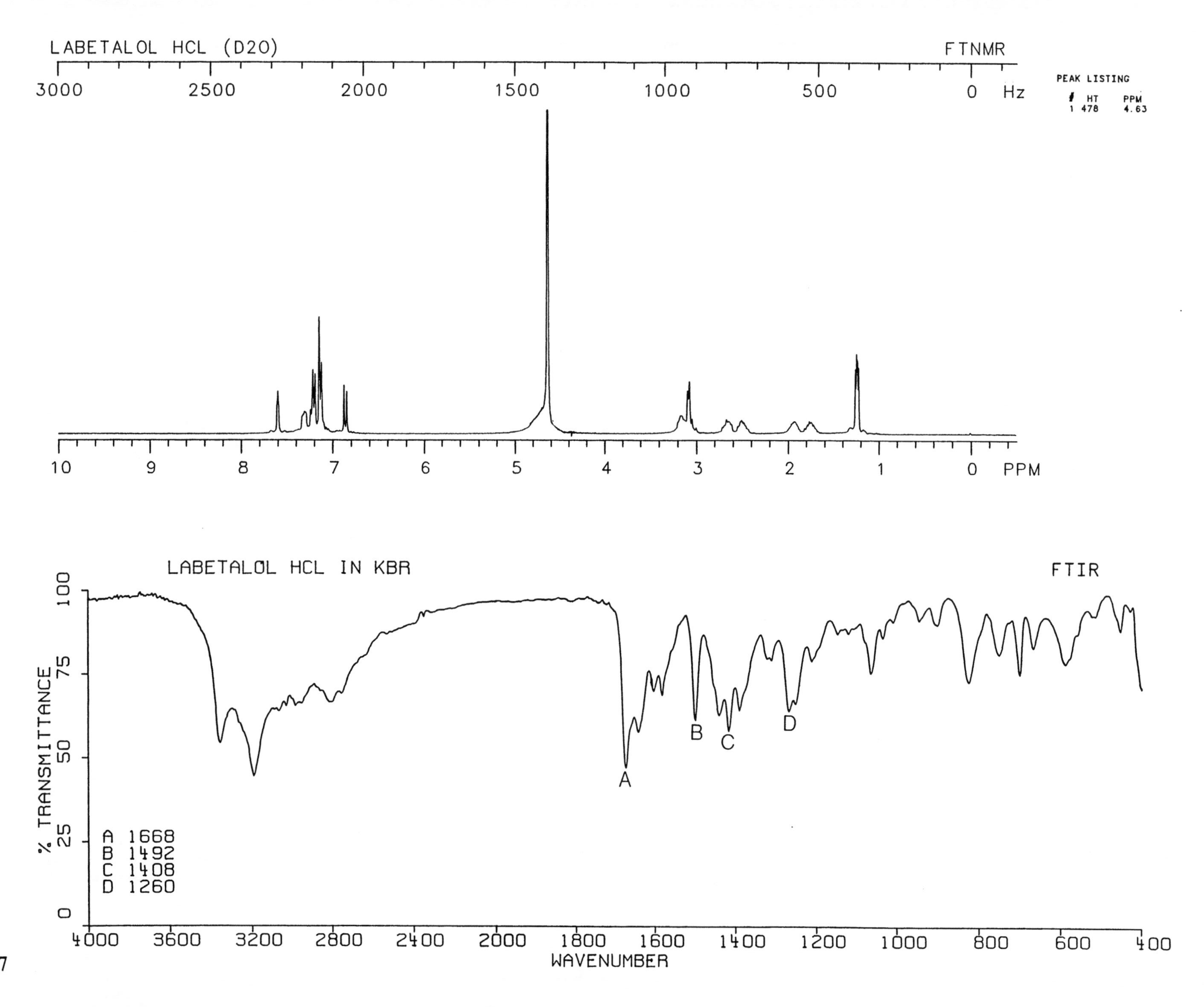

LABETALOL HCL (D2O)
FTNMR
3000 2500 2000 1500 1000 500 0 Hz
PEAK LISTING
# HT PPM
1 478 4.63
10 9 8 7 6 5 4 3 2 1 0 PPM
LABETALOL HCL IN KBR
FTIR
% TRANSMITTANCE
100 75 50 25 0
A 1668
B 1492
C 1408
D 1260
A
B
C
D
4000 3600 3200 2800 2400 2000 1800 1600 1400 1200 1000 800 600 400
WAVENUMBER

# LACTOSE

$C_{12}H_{22}O_{11}$

**Molecular weight:** 342.30 (342.12)
**Synonyms:** 4-O-β-d-Galactopyranosyl-d-glucose; milk sugar

**Trade names:**

**Use:** Nutrient
**HPLC:** Li-$NH_2$; 70D:30E; 10.1
**GC:**

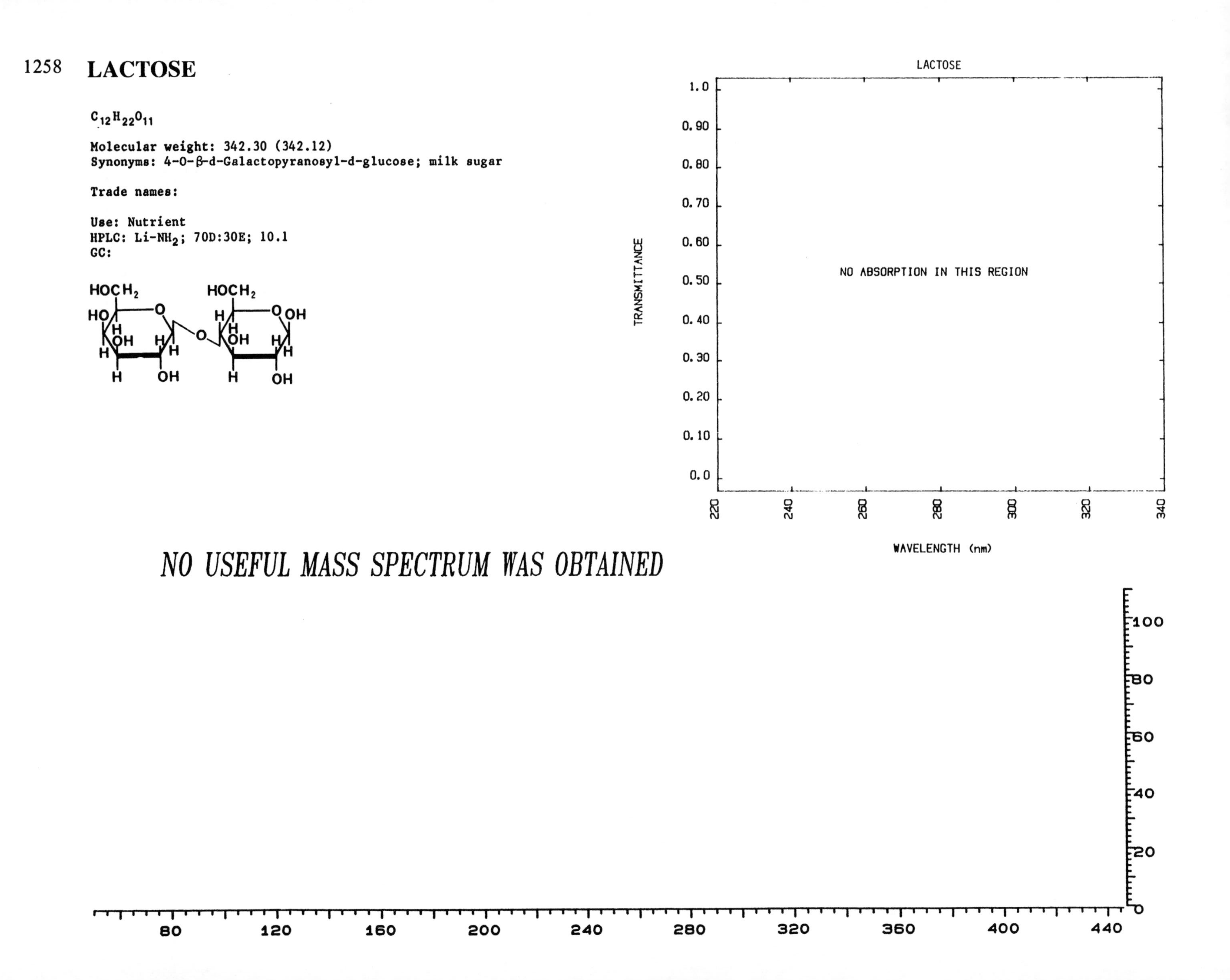

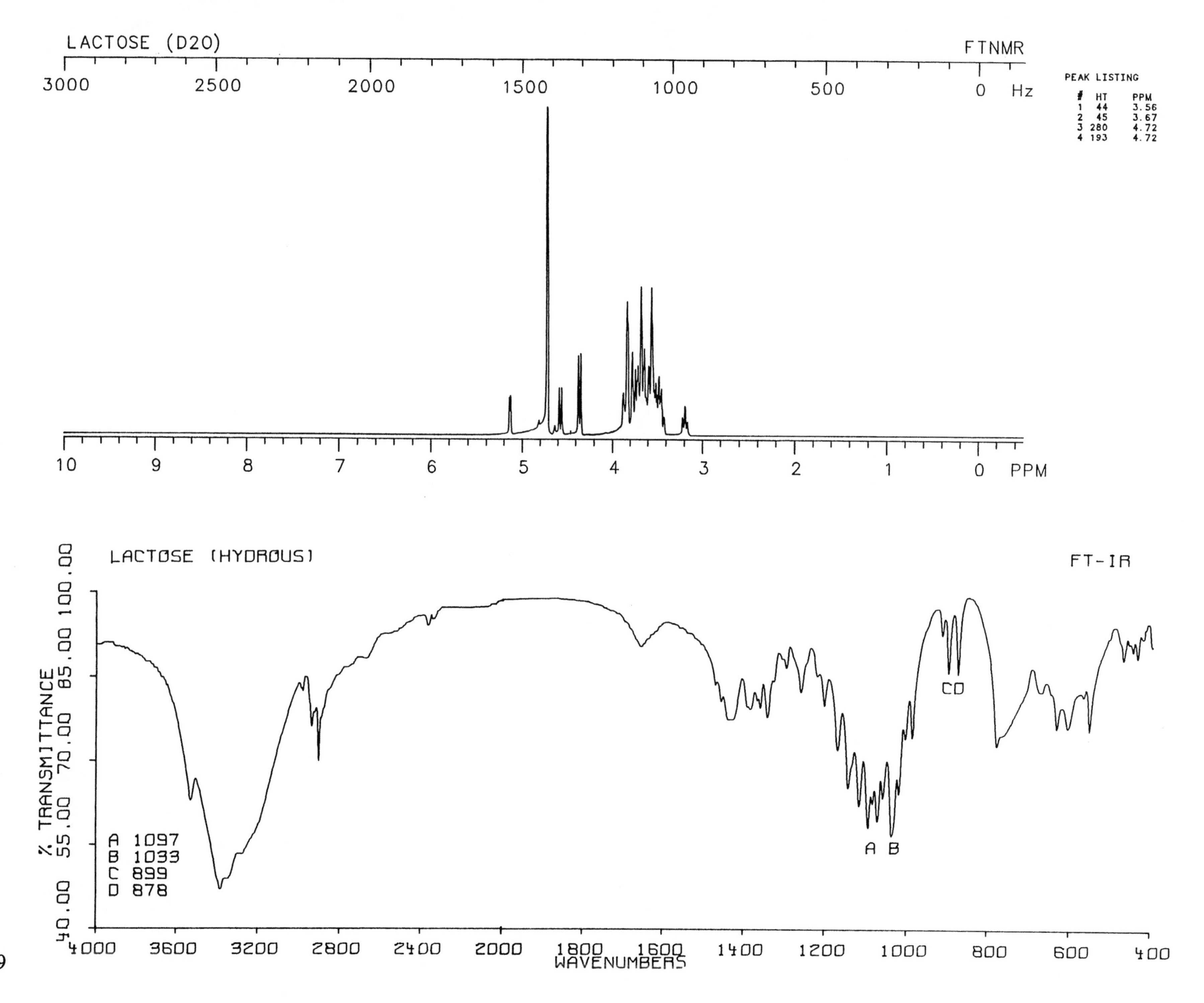

LACTOSE (D20)
FTNMR
3000
2500
2000
1500
1000
500
0 Hz
PEAK LISTING
# HT PPM
1 44 3.56
2 45 3.67
3 280 4.72
4 193 4.72
10
9
8
7
6
5
4
3
2
1
0 PPM
LACTOSE (HYDROUS)
FT-IR
% TRANSMITTANCE
40.00
55.00
70.00
85.00
100.00
A 1097
B 1033
C 899
D 878
A B
CD
4000
3600
3200
2800
2400
2000
1800
1600
1400
1200
1000
800
600
400
WAVENUMBERS

# LACTULOSE

$C_{12}H_{22}O_{11}$

**Molecular weight:** 342.30 (342.12)
**Synonyms:** 4-O-β-D-Galactopyranosyl-D-fructose

**Trade names:** Cephulac, Chronulac

**Use:** Laxative
**HPLC:**
**GC:**

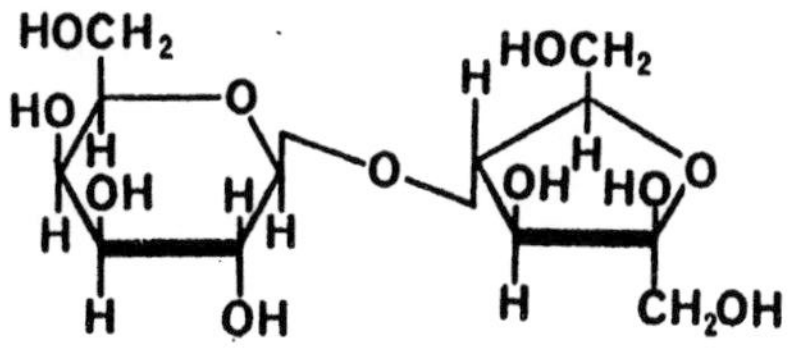

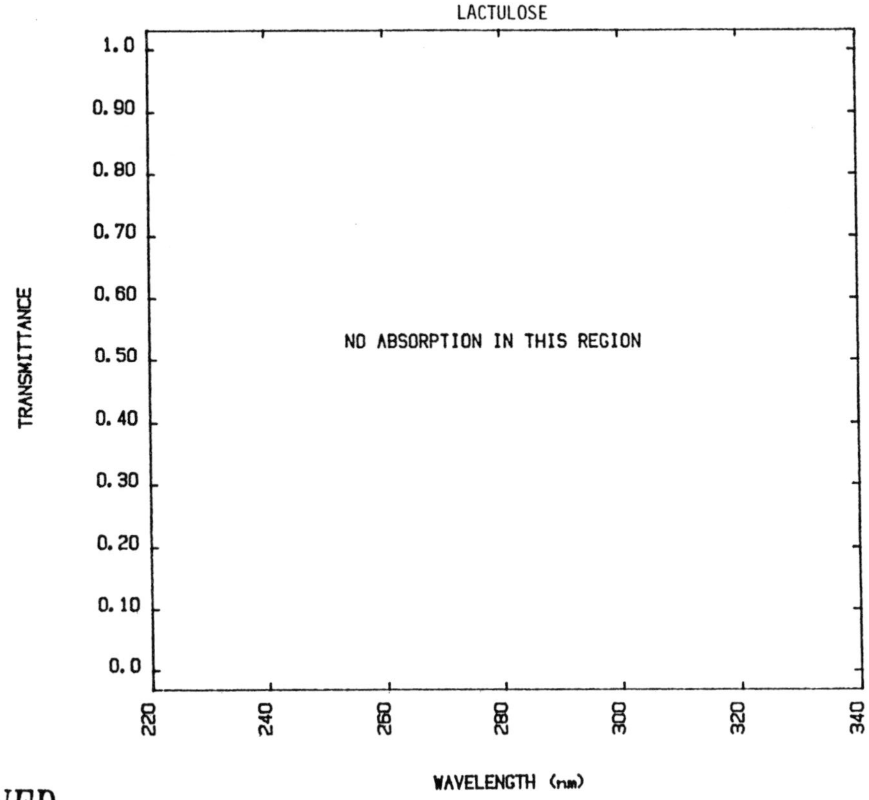

NO USEFUL MASS SPECTRUM WAS OBTAINED

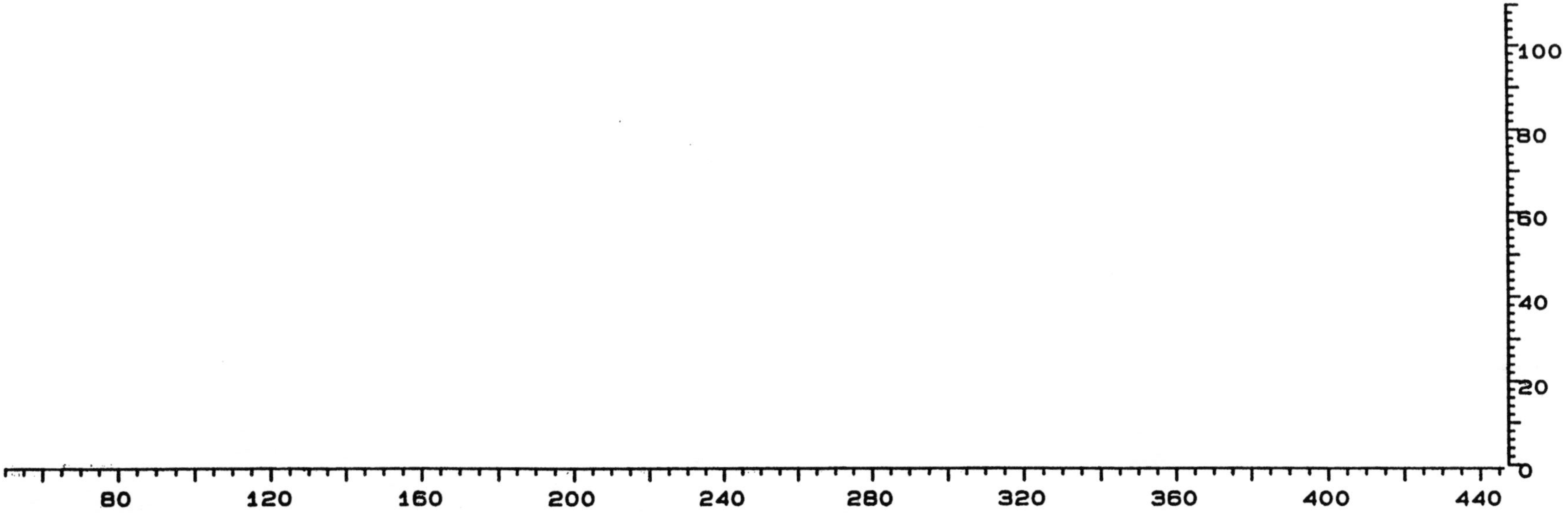

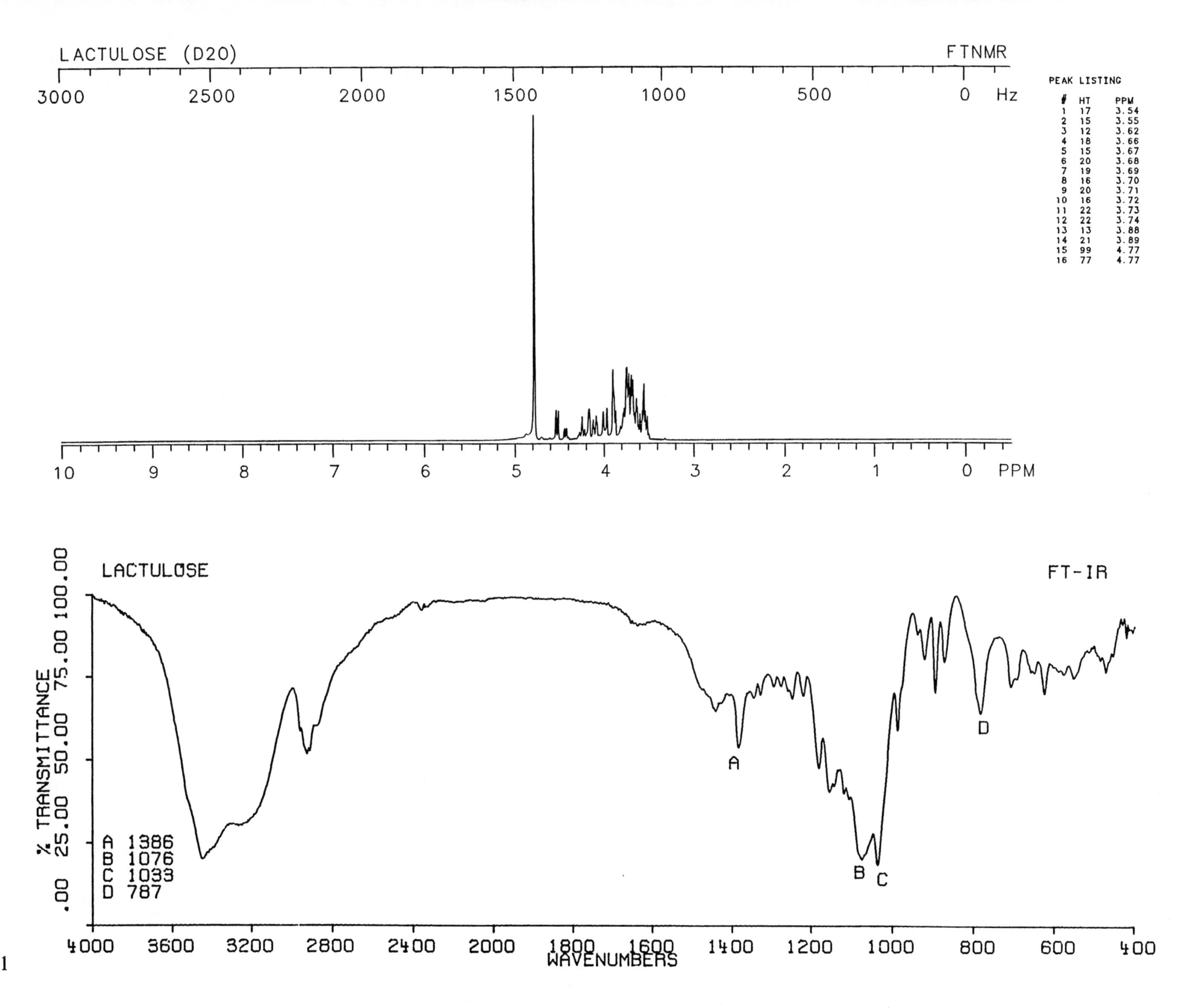
LACTULOSE (D20)
FTNMR
Hz
PPM
PEAK LISTING
# HT PPM
1 17 3.54
2 15 3.55
3 12 3.62
4 18 3.66
5 15 3.67
6 20 3.68
7 19 3.69
8 16 3.70
9 20 3.71
10 16 3.72
11 22 3.73
12 22 3.74
13 13 3.88
14 21 3.89
15 99 4.77
16 77 4.77
LACTULOSE
FT-IR
% TRANSMITTANCE
WAVENUMBERS
A 1386
B 1076
C 1033
D 787

# LEVALLORPHAN

$C_{19}H_{25}NO$

**Molecular weight:** 283.42 (283.19)
**Synonyms:** 17-Allylmorphinan-3-lol; ℓ,N-allyl-3-hydroxymorphinan

**Trade names:** Lorfan

**Use:** Narcotic antagonist
**HPLC:** Si-10; 10A:90B; 4.5
**GC:** 2400; 250°C

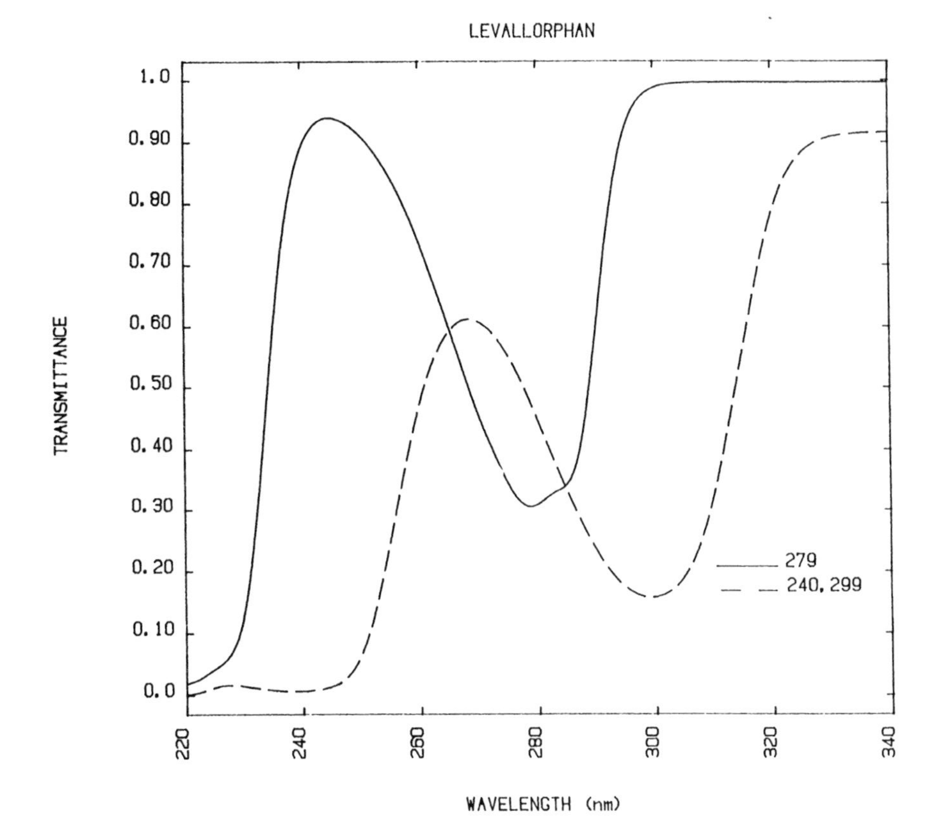


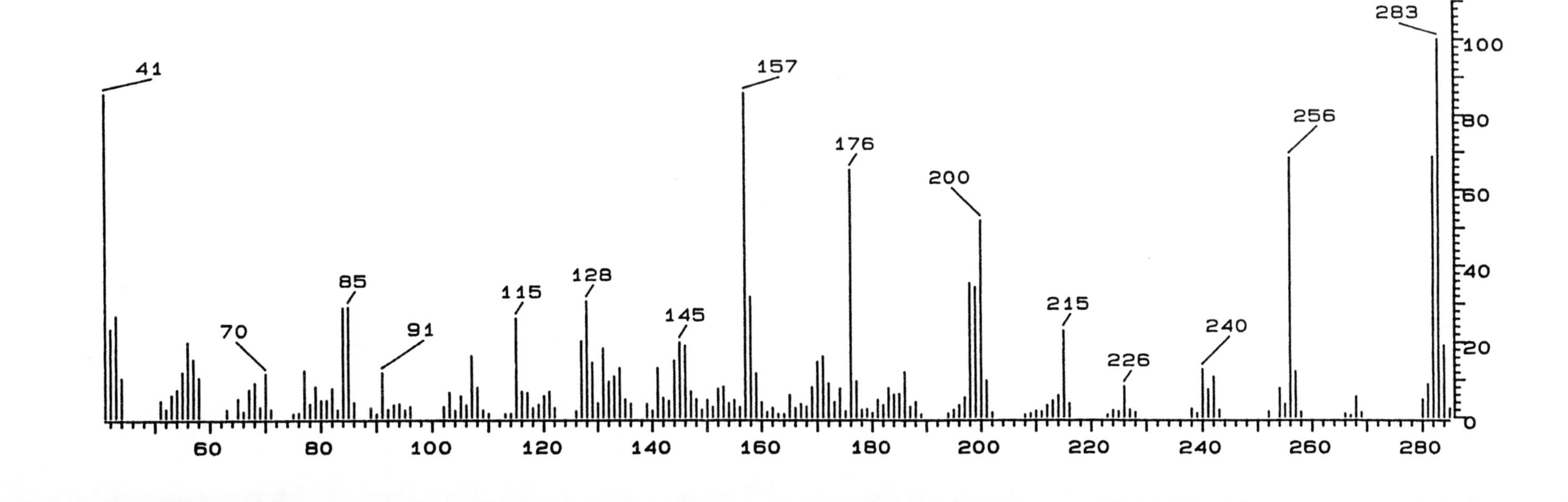

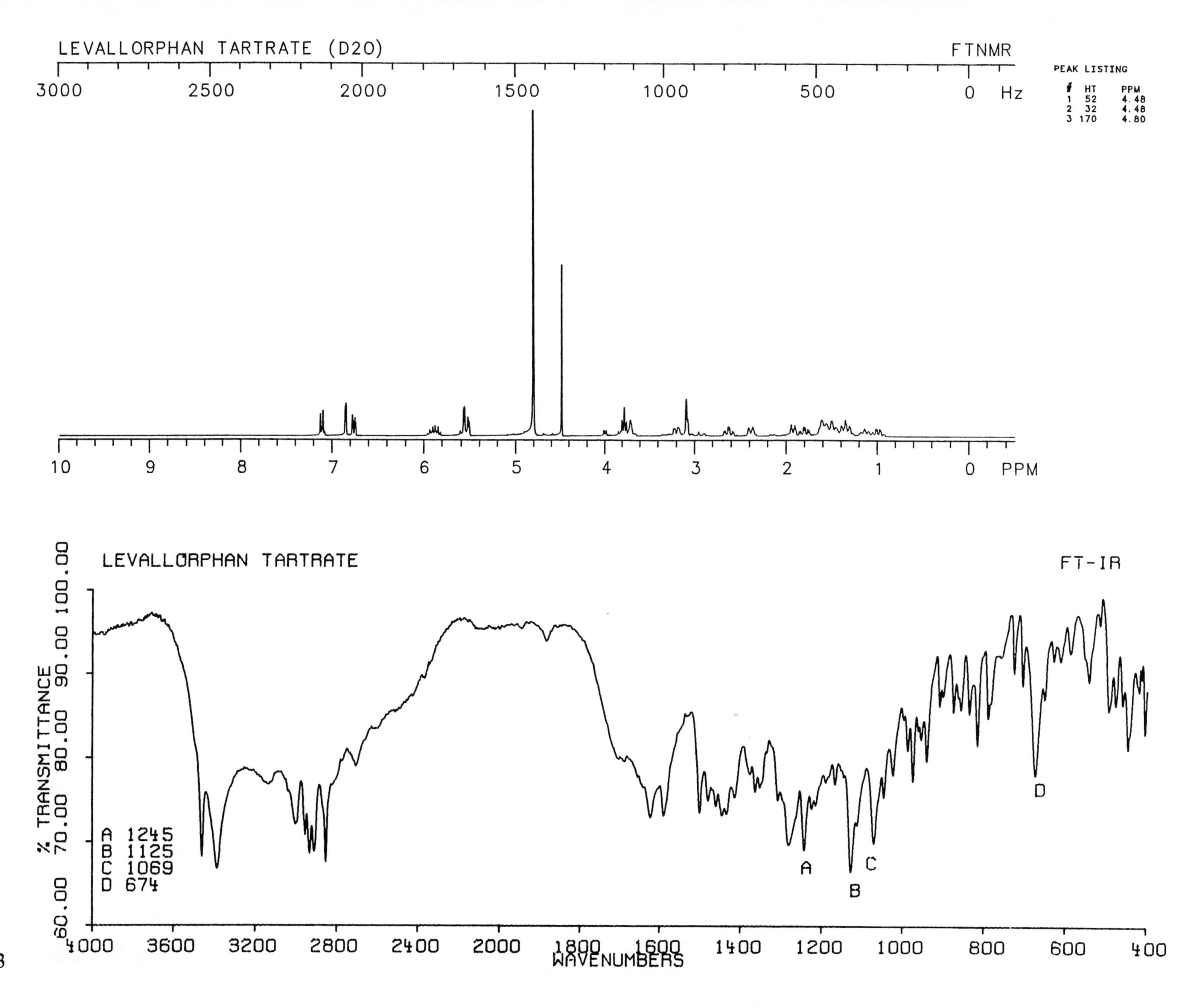

LEVALLORPHAN TARTRATE (D2O)
FTNMR
3000
2500
2000
1500
1000
500
0 Hz
PEAK LISTING
# HT PPM
1 52 4.48
2 32 4.48
3 170 4.80
10
9
8
7
6
5
4
3
2
1
0 PPM
LEVALLORPHAN TARTRATE
FT-IR
% TRANSMITTANCE
60.00
70.00
80.00
90.00
100.00
A 1245
B 1125
C 1069
D 674
A
B
C
D
4000
3600
3200
2800
2400
2000
1800
1600
1400
1200
1000
800
600
400
WAVENUMBERS

# LEVAMISOLE

$C_{11}H_{12}N_2S$

Molecular weight: 204.31 (204.07)
Synonyms: (-)-2,3,5,6-Tetrahydro-6-phenylimidazo[22,1-b]thiazole; ℓ-tetramisole
Trade names: Levasole, Nemicide, Solaskil, Tramisol, Worm-Chek

Use: Antidepressant, anthelmintic
HPLC: Si-10; 1A:99B; 5.5
GC: 2026; 250°C

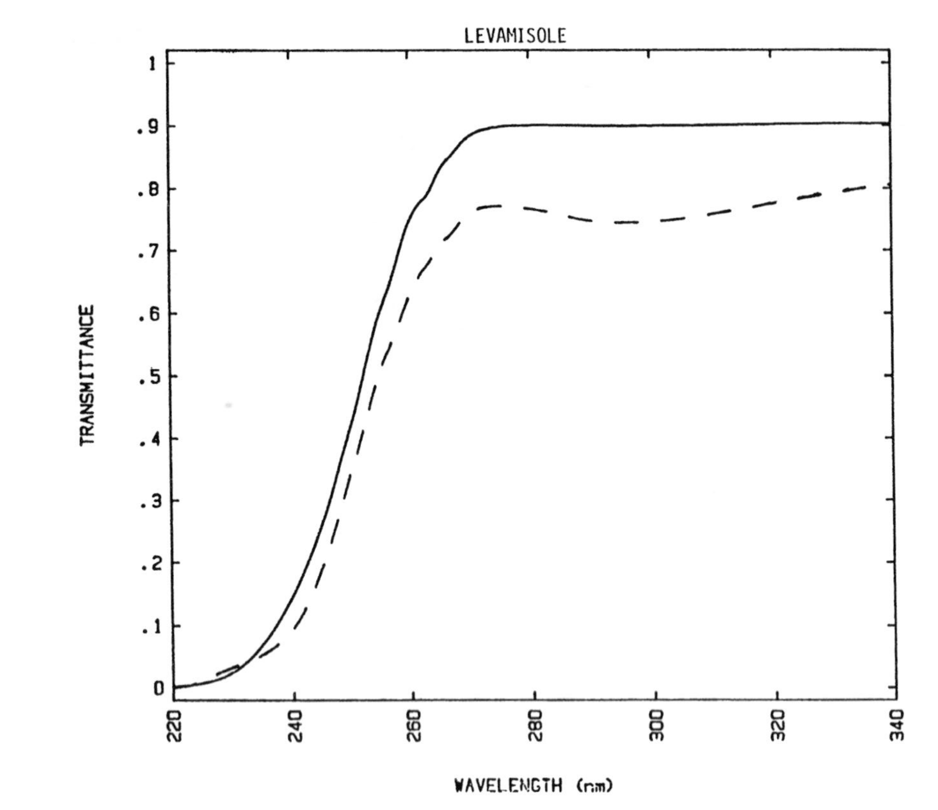

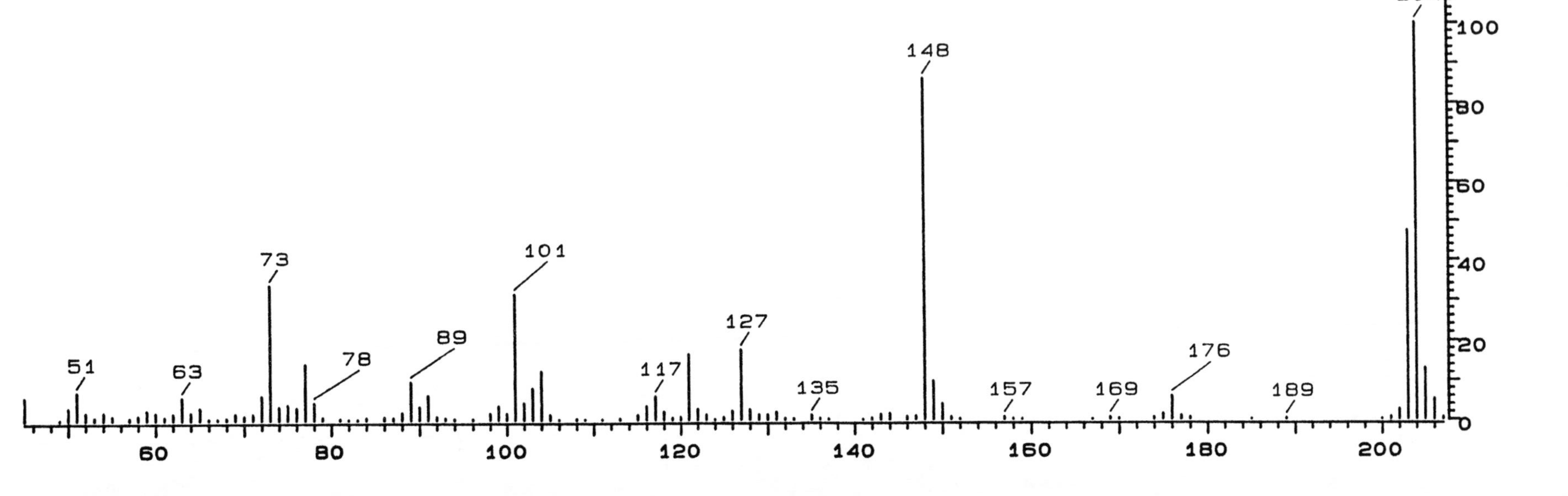

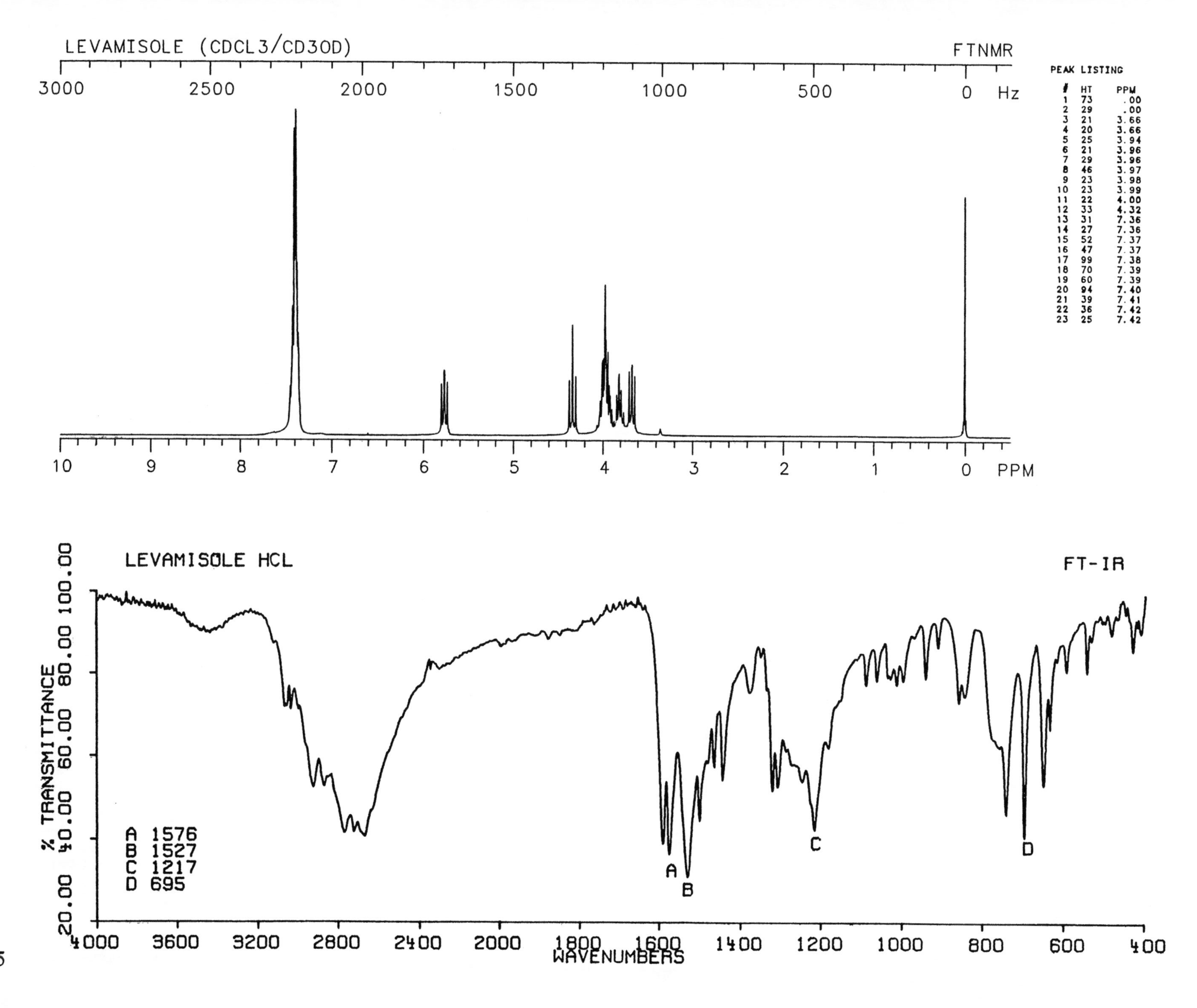
LEVAMISOLE (CDCL3/CD3OD)
FTNMR
Hz
PPM
PEAK LISTING
# HT PPM
1 73 .00
2 29 .00
3 21 3.66
4 20 3.66
5 25 3.94
6 21 3.96
7 29 3.96
8 46 3.97
9 23 3.98
10 23 3.99
11 22 4.00
12 33 4.32
13 31 7.36
14 27 7.36
15 52 7.37
16 47 7.37
17 99 7.38
18 70 7.39
19 60 7.39
20 94 7.40
21 39 7.41
22 36 7.42
23 25 7.42
LEVAMISOLE HCL
FT-IR
% TRANSMITTANCE
WAVENUMBERS
A 1576
B 1527
C 1217
D 695

# LEVODOPA

$C_9H_{11}NO_4$

**Molecular weight:** 197.19 (197.07)
**Synonyms:** 3-Hydroxy-1-tyrosine; L-dopa

**Trade names:** Larodopa, Sinemet

**Use:** Anticholinergic, antiparkinsonian
**HPLC:**
**GC:**

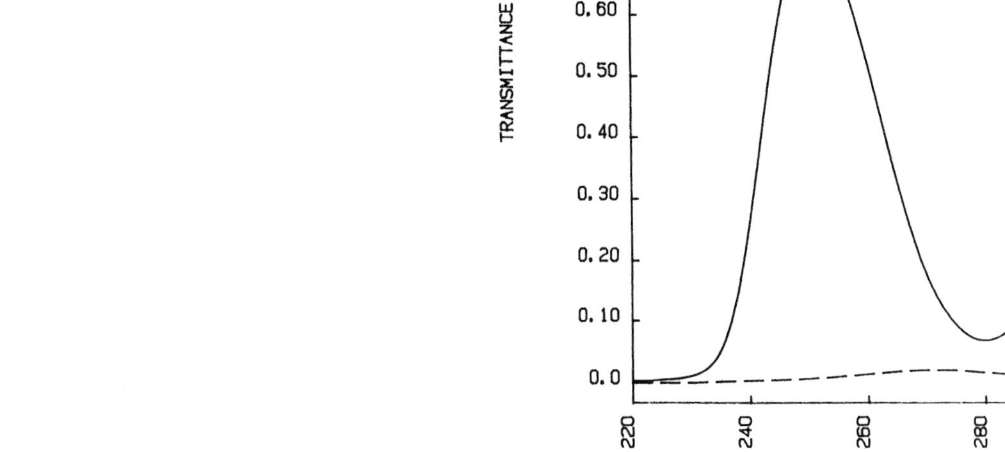


## *LEVODOPA - - SOLID PROBE*

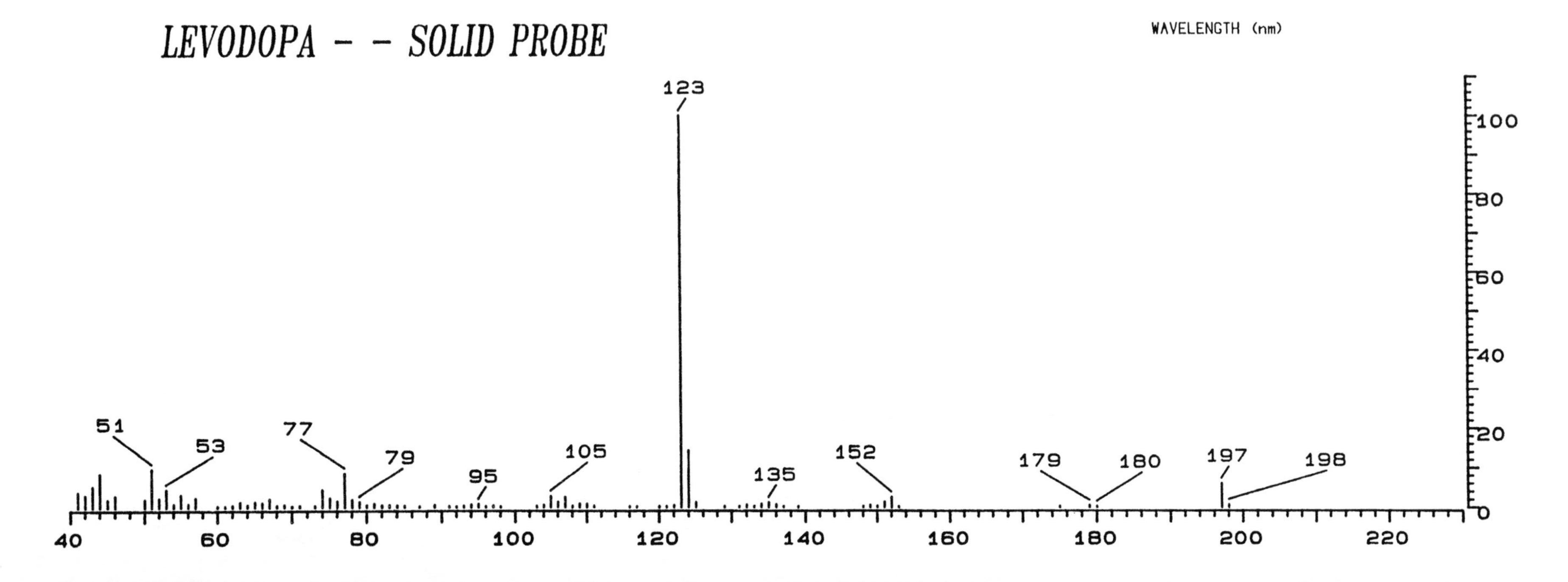

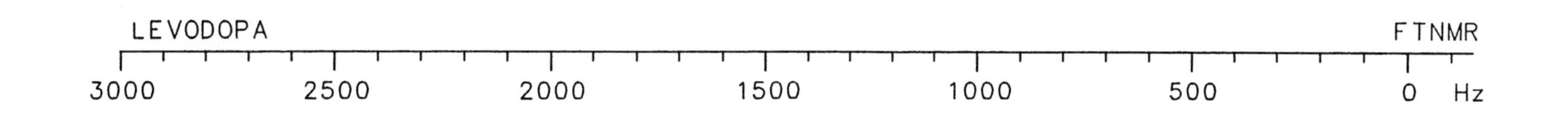


INSUFFICIENT SOLUBILITY

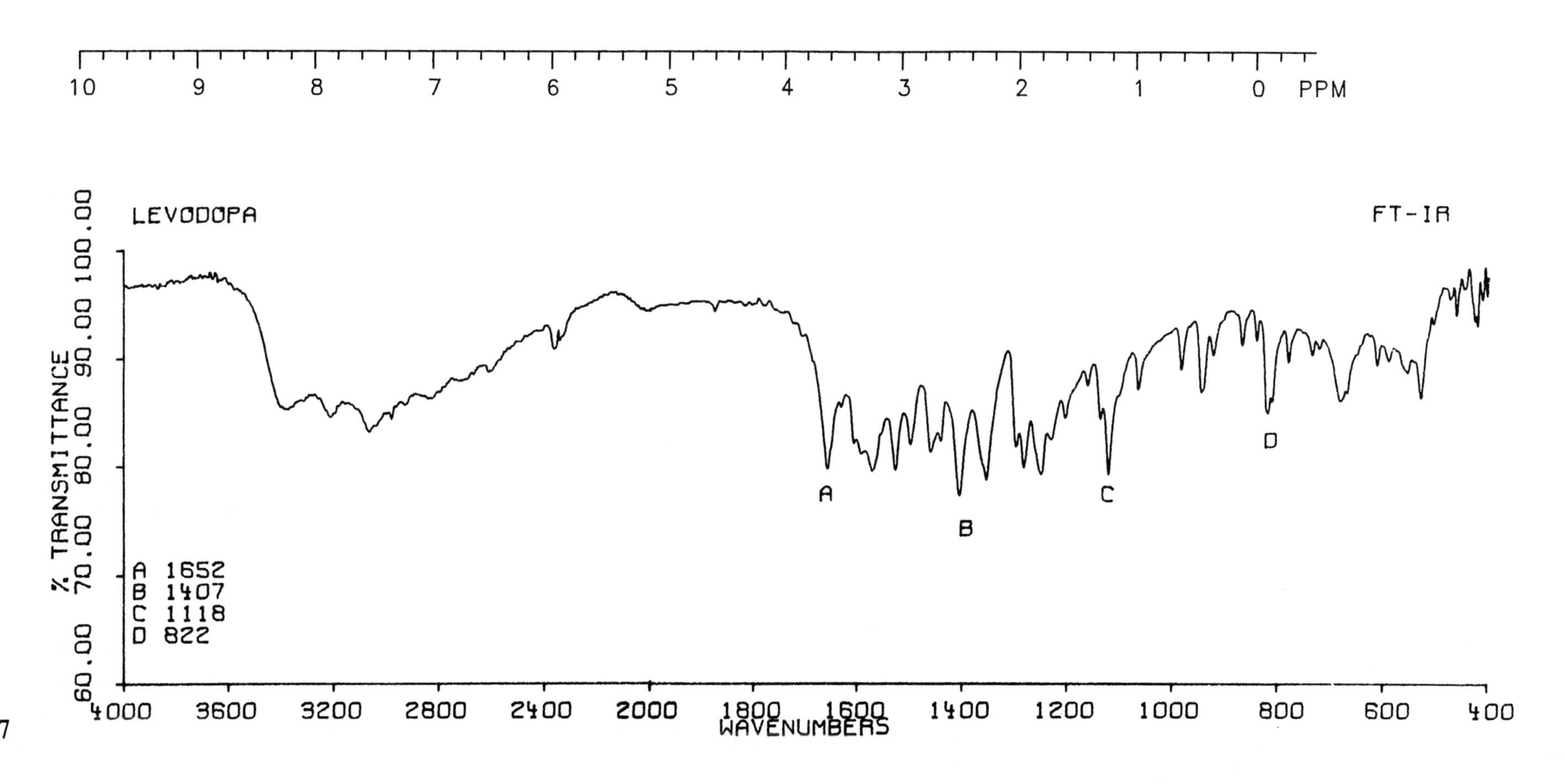

# LEVORPHANOL

$C_{17}H_{23}NO$

**Molecular weight:** 257.38 (257.18)
**Synonyms:** 17-Methylmorphinan-3-ol; 1,3-hydroxy-N-methylmorphinan

**Trade names:** Levo-Dromoran

**Use:** Narcotic analgesic
**HPLC:** Si-10; 50A:50B; 6.5
**GC:** 2303; 250°C

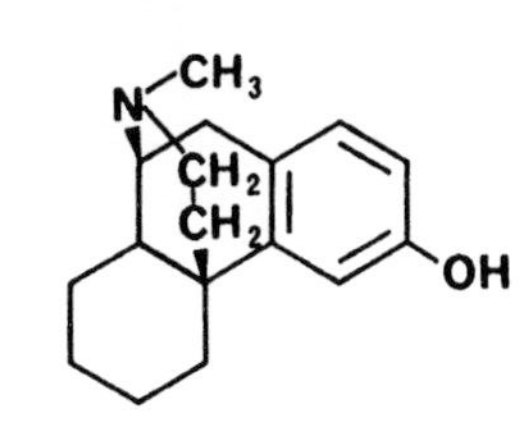


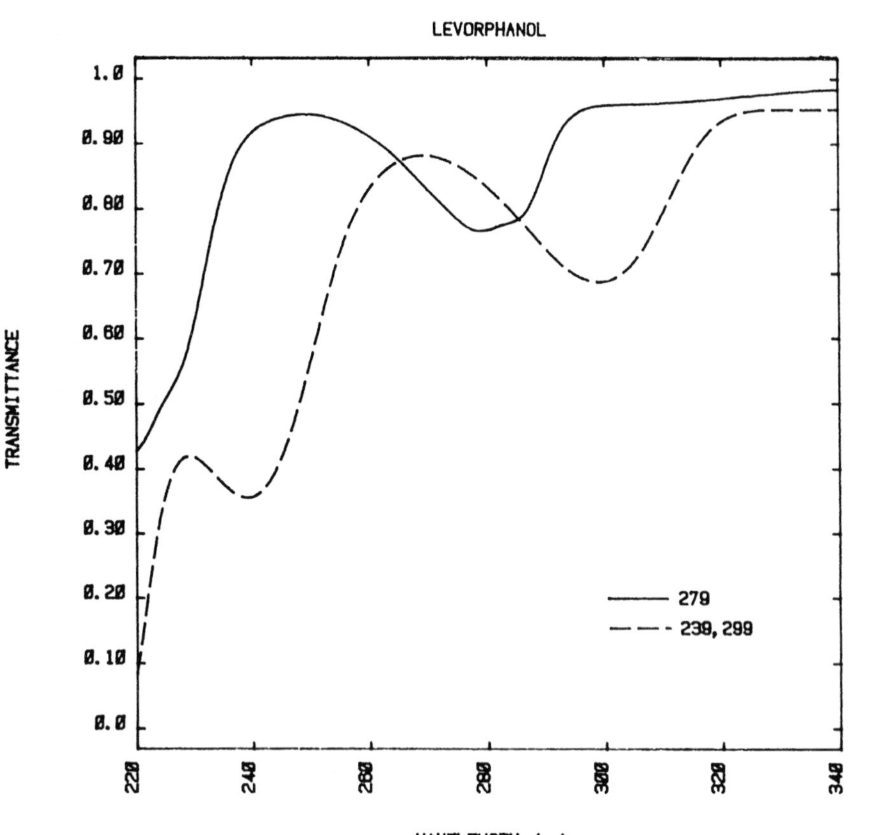

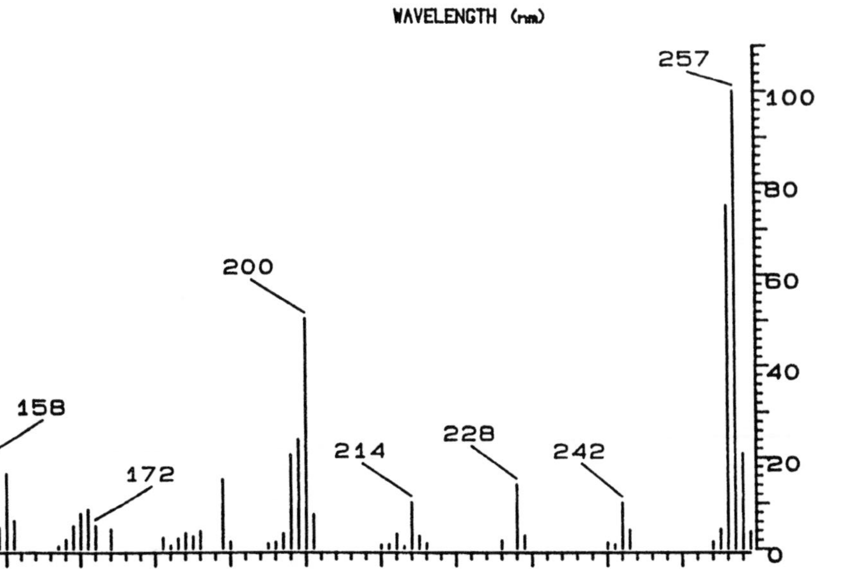

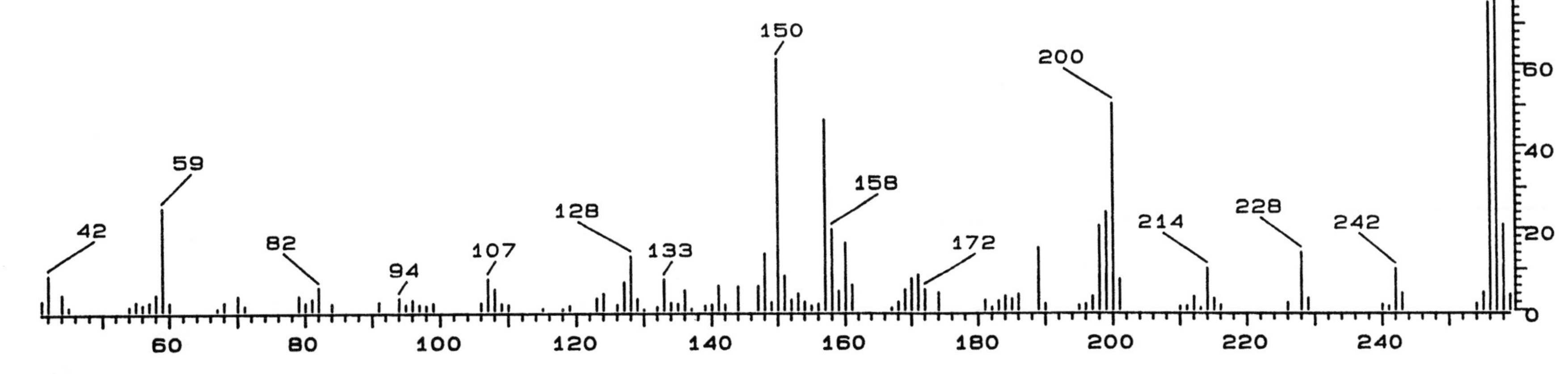

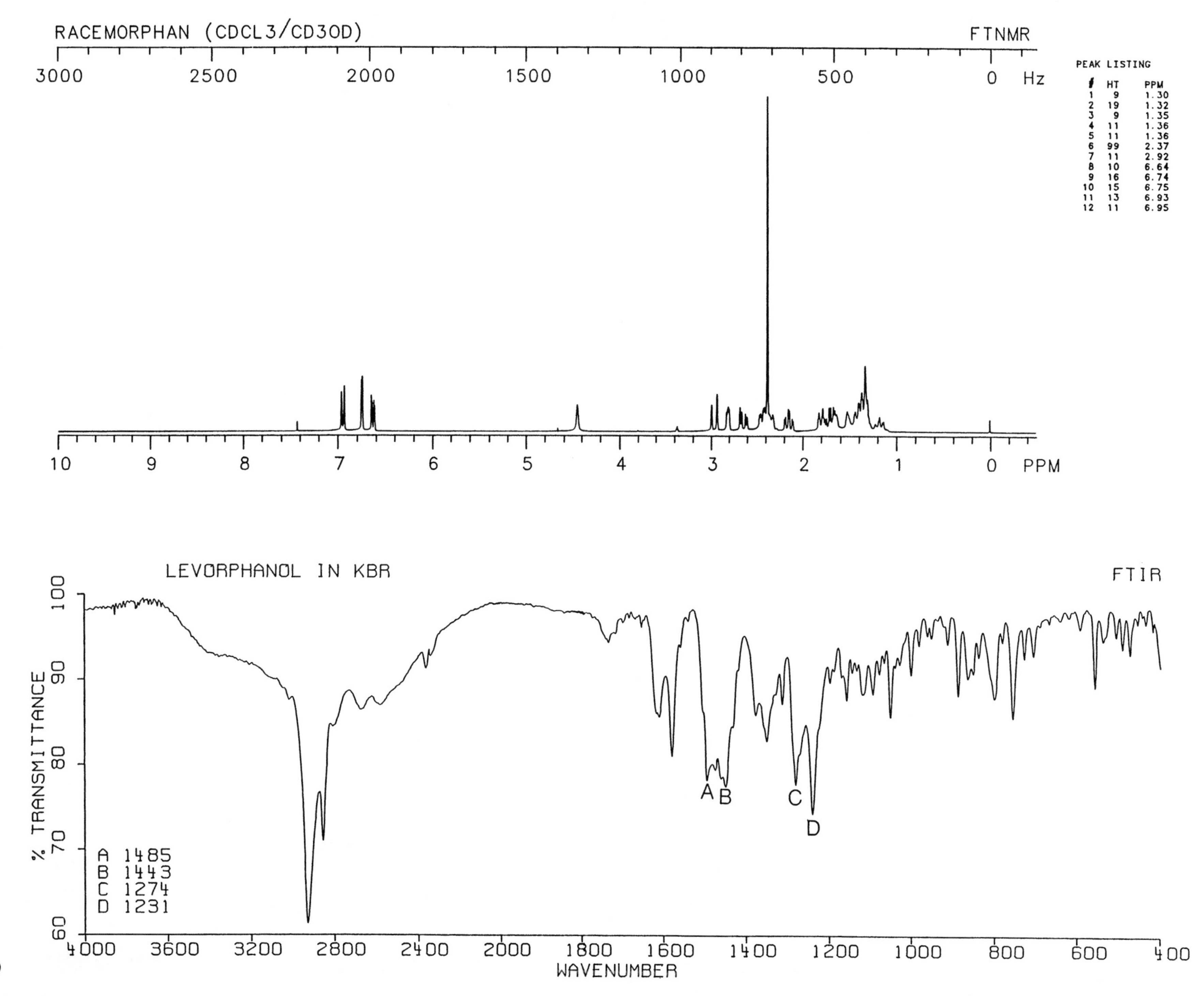

RACEMORPHAN (CDCL3/CD3OD)
FTNMR
3000 2500 2000 1500 1000 500 0 Hz
10 9 8 7 6 5 4 3 2 1 0 PPM
PEAK LISTING
# HT PPM
1 9 1.30
2 19 1.32
3 9 1.35
4 11 1.36
5 11 1.36
6 99 2.37
7 11 2.92
8 10 6.64
9 16 6.74
10 15 6.75
11 13 6.93
12 11 6.95
LEVORPHANOL IN KBR
FTIR
% TRANSMITTANCE
100 90 80 70 60
A 1485
B 1443
C 1274
D 1231
A B C D
4000 3600 3200 2800 2400 2000 1800 1600 1400 1200 1000 800 600 400
WAVENUMBER

# LIDOCAINE

$C_{14}H_{22}N_2O$

**Molecular weight:** 234.34 (234.17)
**Synonyms:** 2-(Diethylamino)-N-(2,6-dimethylphenyl)acetamide; lignocaine
**Trade names:** Anestacon, Lidosporin, Xylocaine

**Use:** Local anesthetic
**HPLC:** Si-10; 1A:99B; 4.0
**GC:** 1895; 200°C

$NH-C(=O)-CH_2-N(C_2H_5)_2$ (on 2,6-dimethylphenyl ring, $CH_3$, $CH_3$)

LIDOCAINE

TRANSMITTANCE

— 262, 271

WAVELENGTH (nm)

*LIDOCAINE*

86 42 58 72 105 118 120 134 148 160 176 191 209 221 234

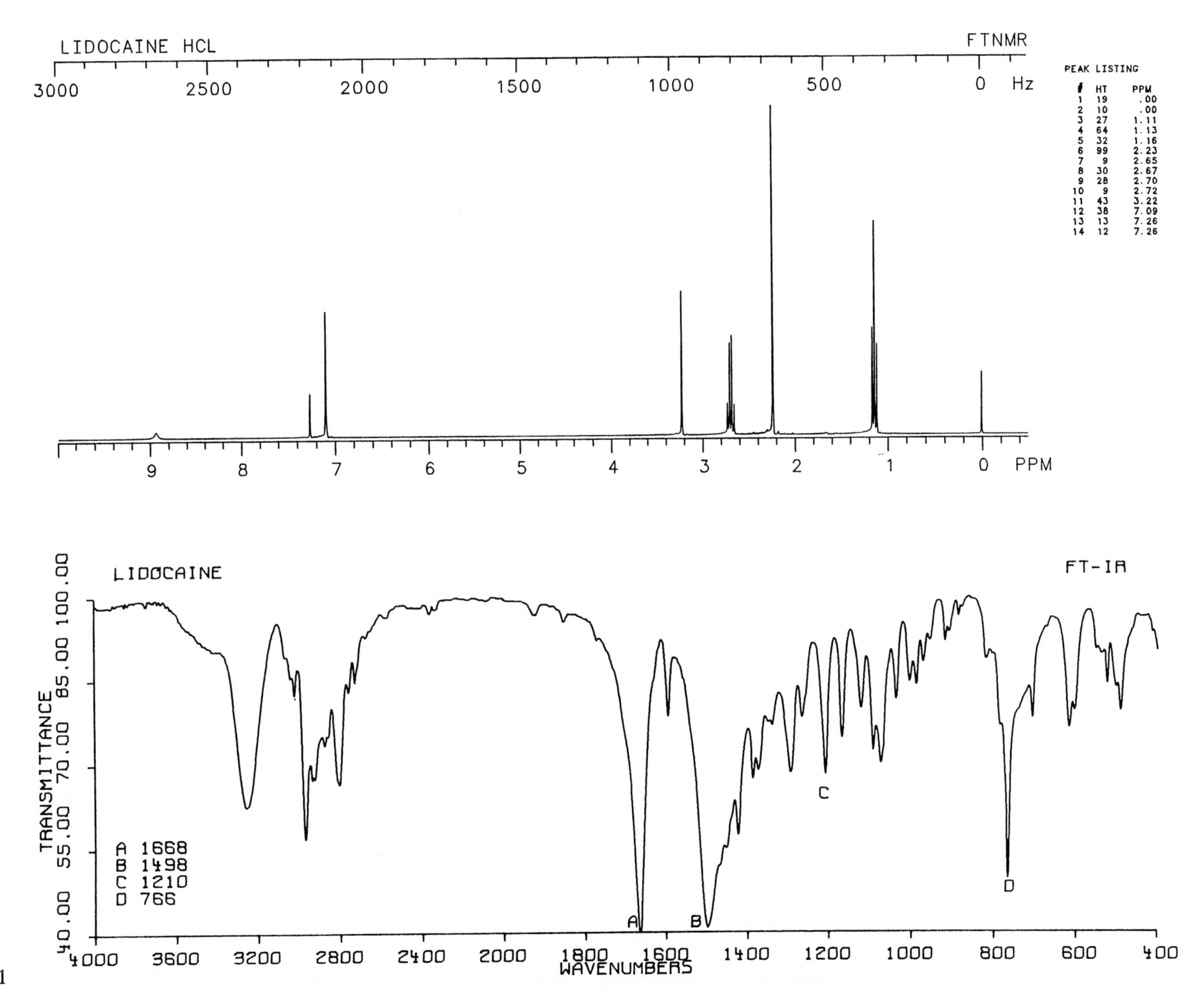
LIDOCAINE HCL
FTNMR
3000
2500
2000
1500
1000
500
0 Hz
PEAK LISTING
# HT PPM
1 19 .00
2 10 .00
3 27 1.11
4 64 1.13
5 32 1.16
6 99 2.23
7 9 2.65
8 30 2.67
9 28 2.70
10 9 2.72
11 43 3.22
12 38 7.09
13 13 7.26
14 12 7.26
9
8
7
6
5
4
3
2
1
0 PPM
LIDOCAINE
FT-IR
TRANSMITTANCE
40.00
55.00
70.00
85.00
100.00
A 1668
B 1498
C 1210
D 766
A
B
C
D
4000
3600
3200
2800
2400
2000
1800
1600
1400
1200
1000
800
600
400
WAVENUMBERS

# LIDOFLAZINE

$C_{30}H_{35}F_2N_3O$

Molecular weight: 491.63 (491.28)
Synonyms: 4-[4,4-Bis(4-fluorophenyl)butyl]-N-(2,6-dimethylphenyl)-1-piperazineacetamide; ordiflazine
Trade names: Angex, Clinium, Corflazine, Klinium

Use: Vasodilator
HPLC: Si-10; 2A:98B; 4.3
GC:

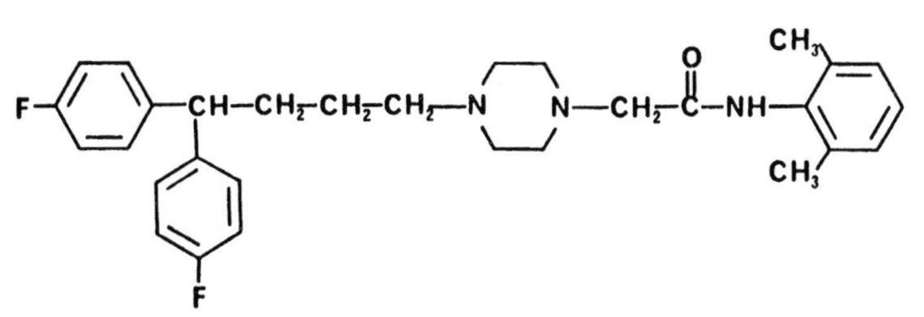


LIDOFLAZINE

TRANSMITTANCE

WAVELENGTH (nm)

265,271
PPT BASE

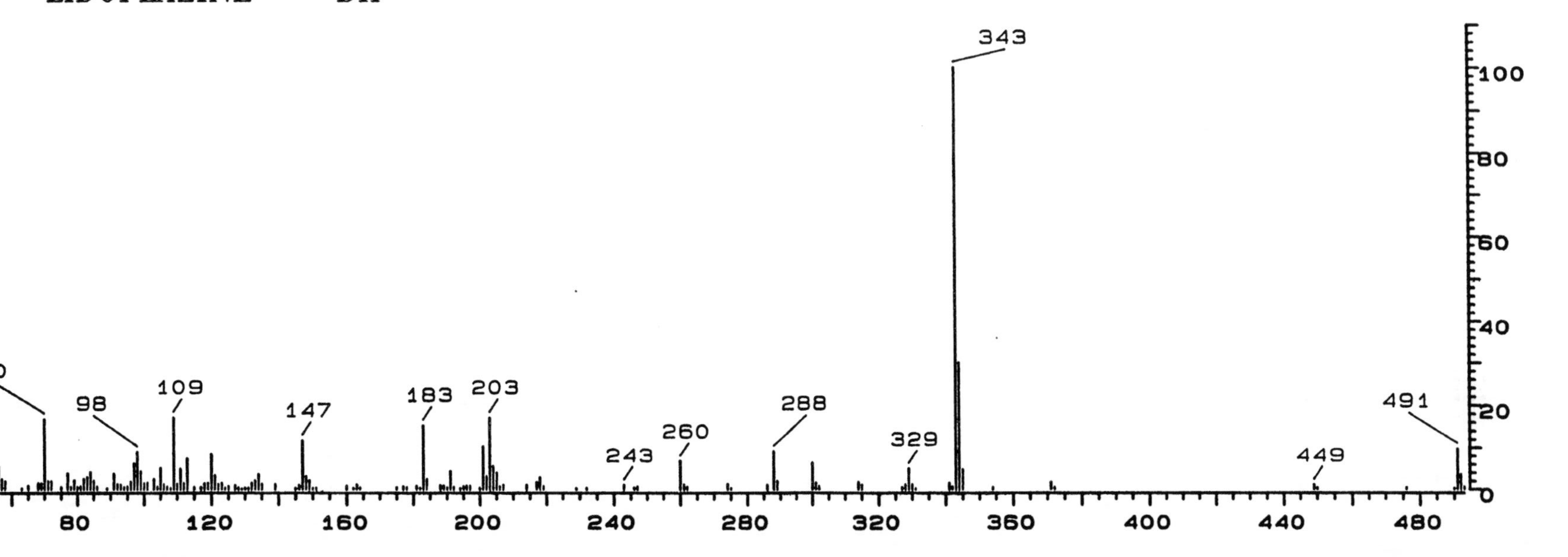

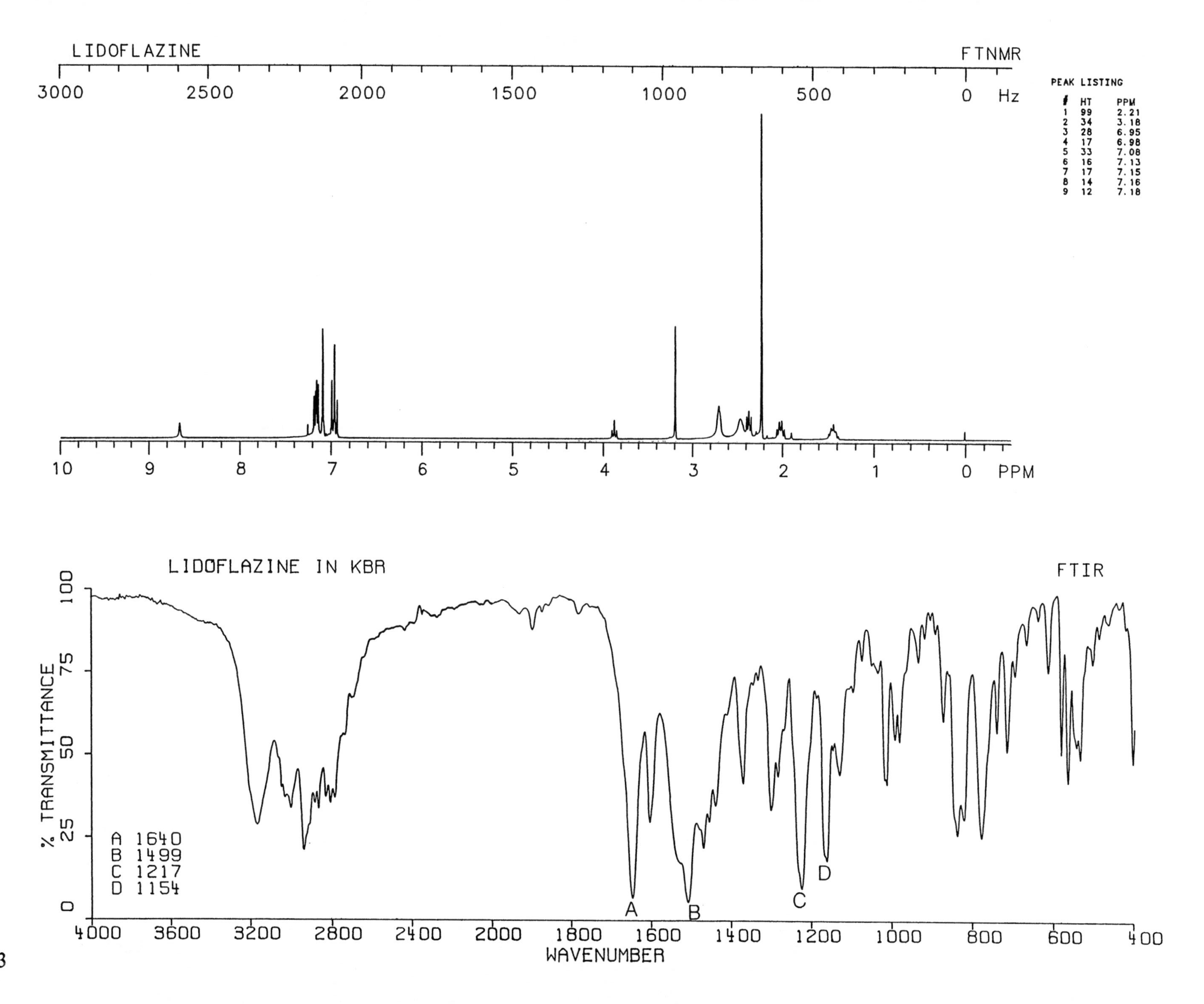
LIDOFLAZINE
FTNMR
3000 2500 2000 1500 1000 500 0 Hz
PEAK LISTING
# HT PPM
1 99 2.21
2 34 3.18
3 28 6.95
4 17 6.98
5 33 7.08
6 16 7.13
7 17 7.15
8 14 7.16
9 12 7.18
10 9 8 7 6 5 4 3 2 1 0 PPM
LIDOFLAZINE IN KBR
FTIR
% TRANSMITTANCE
100 75 50 25 0
A 1640
B 1499
C 1217
D 1154
A B C D
4000 3600 3200 2800 2400 2000 1800 1600 1400 1200 1000 800 600 400
WAVENUMBER

# LINCOMYCIN

$C_{18}H_{34}N_2O_6S$

Molecular weight: 406.55 (406.21)
Synonyms: Methyl-6,8-dideoxy-6-(1-methyl-4-propyl-2-pyrrolidine-carboxamido)-1-thio-D-erythro-α-D-galacto-octopyranoside (α-form); lincolnensin
Trade names: Lincocin

Use: Antibacterial
HPLC:
GC:

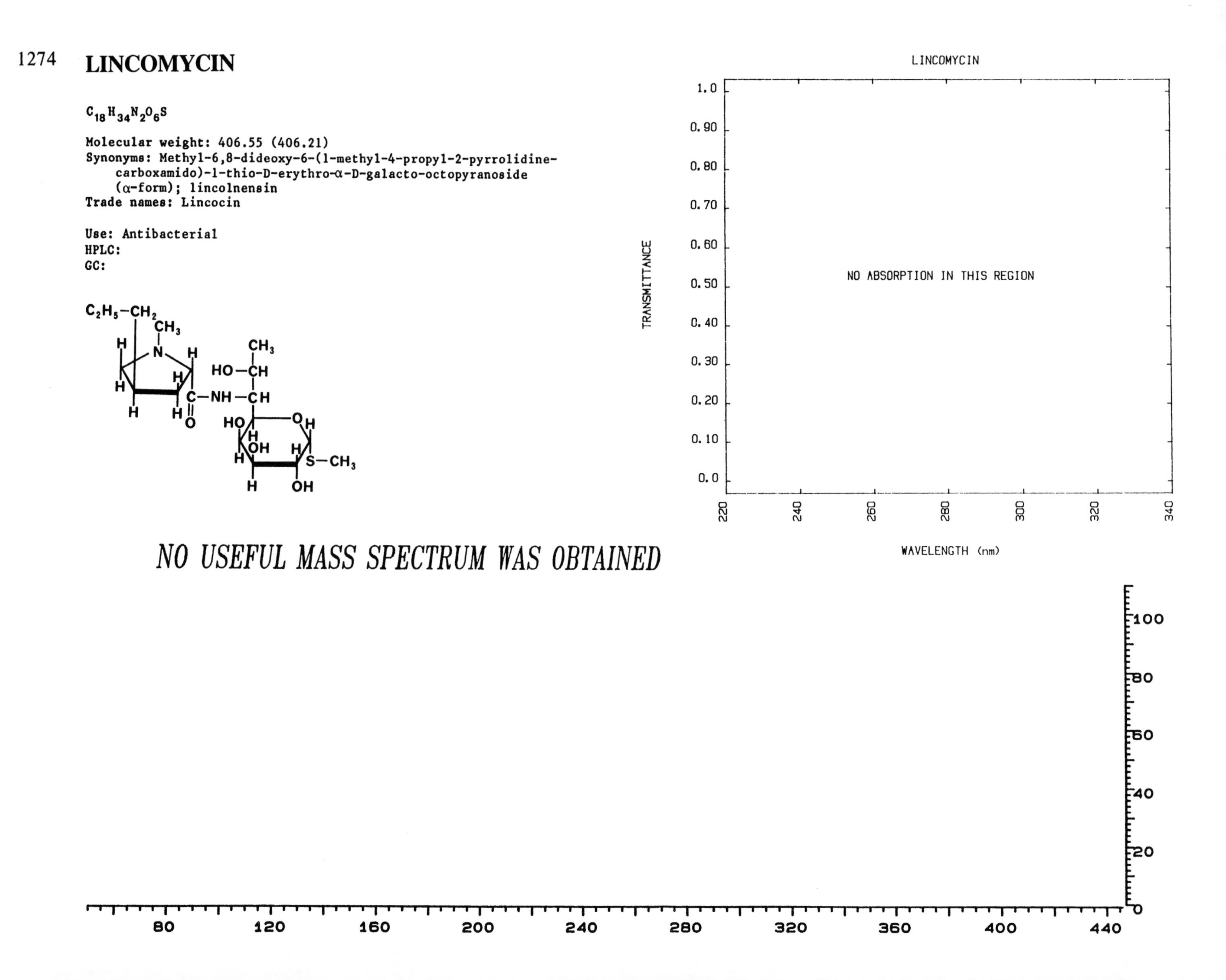

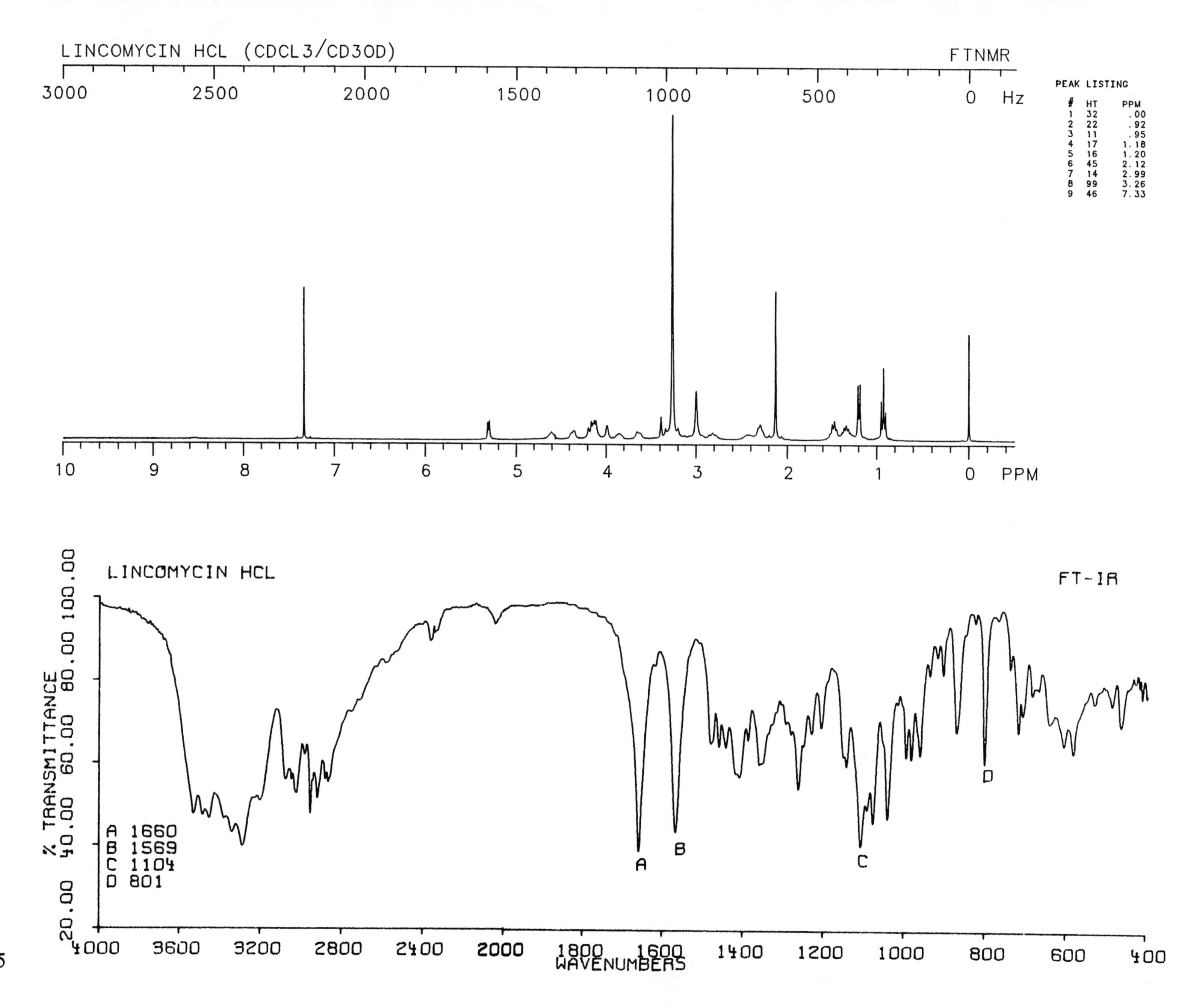

LINCOMYCIN HCL (CDCL3/CD3OD)
FTNMR
Hz
PPM
PEAK LISTING
# HT PPM
1 32 .00
2 22 .92
3 11 .95
4 17 1.18
5 16 1.20
6 45 2.12
7 14 2.99
8 99 3.26
9 46 7.33
LINCOMYCIN HCL
FT-IR
% TRANSMITTANCE
WAVENUMBERS
A 1660
B 1569
C 1104
D 801

# LINOLEIC ACID

$C_{18}H_{32}O_2$

**Molecular weight:** 280.44 (280.24)
**Synonyms:** 9,12-Octadecadienoic acid; 9,12-linoleic acid; linolic acid
**Trade names:** Clinolamide, Linolexamide

**Use:** Nutrient (essential fatty acid)
**HPLC:**
**GC:**

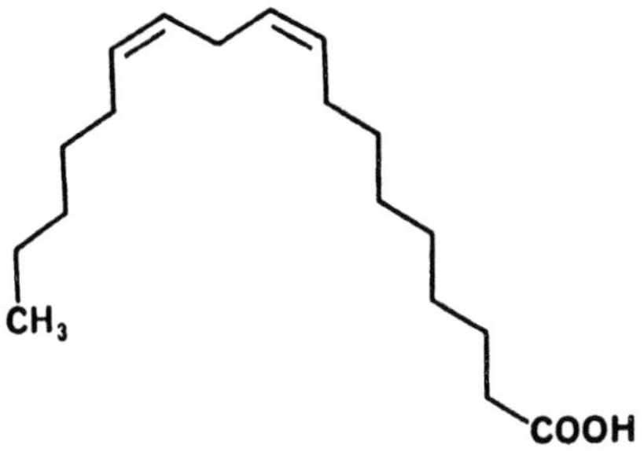

CH$_3$ COOH

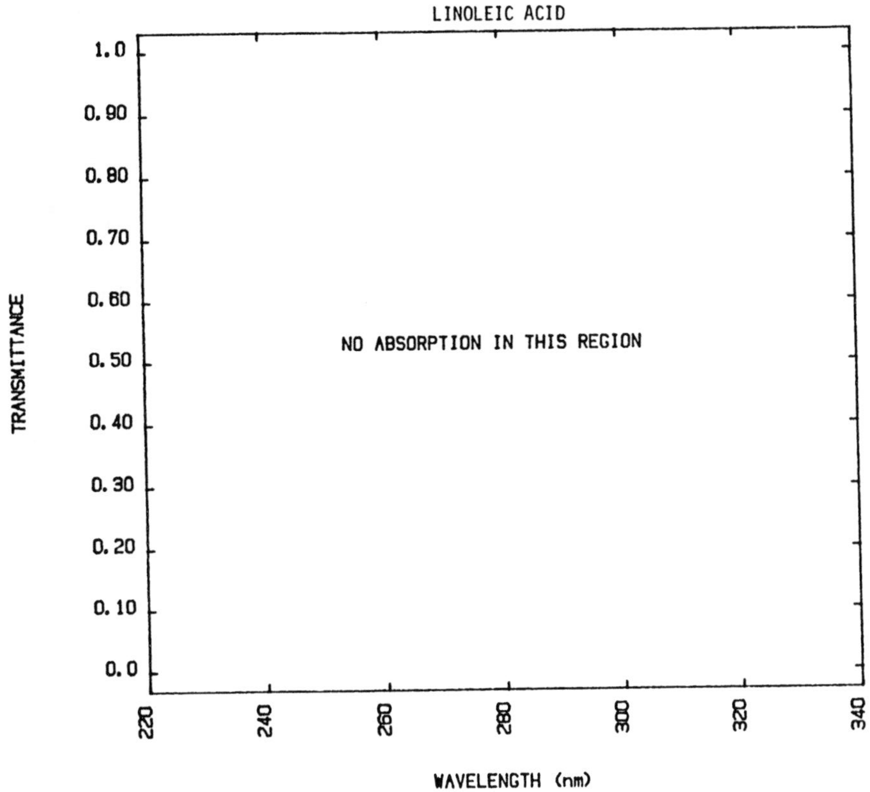

LINOLEIC ACID

TRANSMITTANCE

1.0 0.90 0.80 0.70 0.60 0.50 0.40 0.30 0.20 0.10 0.0

NO ABSORPTION IN THIS REGION

220 240 260 280 300 320 340

WAVELENGTH (nm)

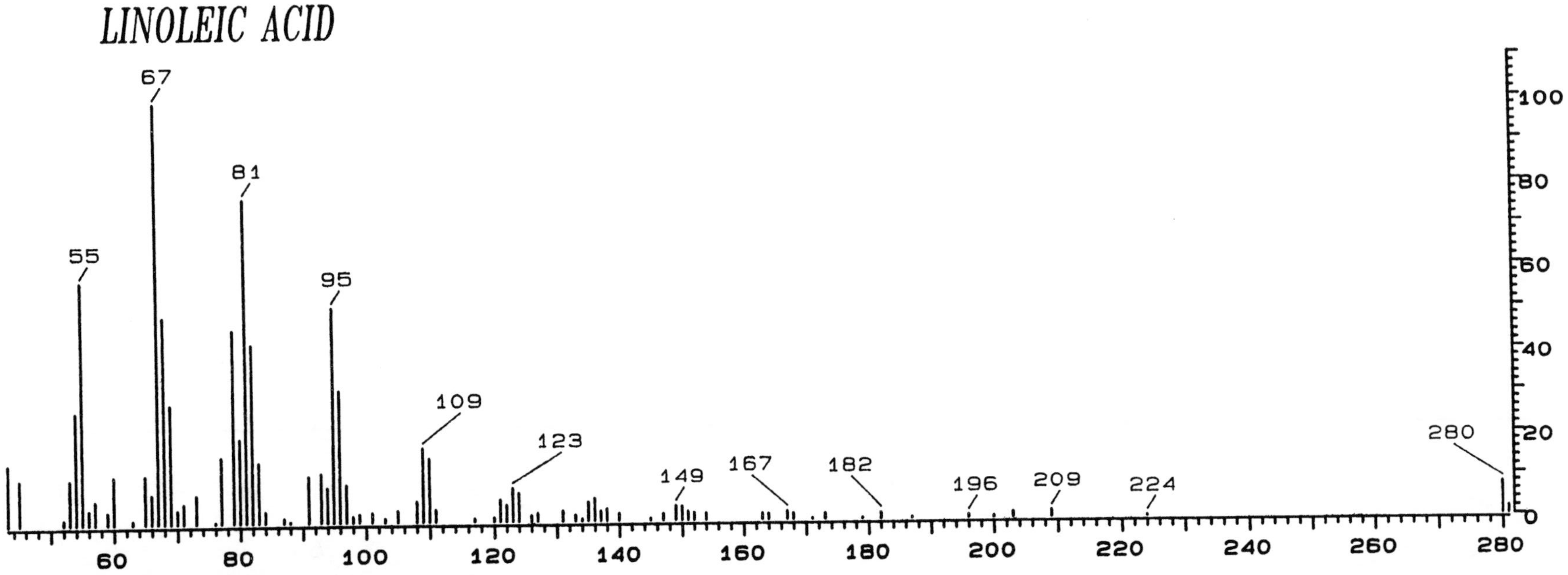

*LINOLEIC ACID*

67 81 55 95 109 123 149 167 182 196 209 224 280

60 80 100 120 140 160 180 200 220 240 260 280

0 20 40 60 80 100

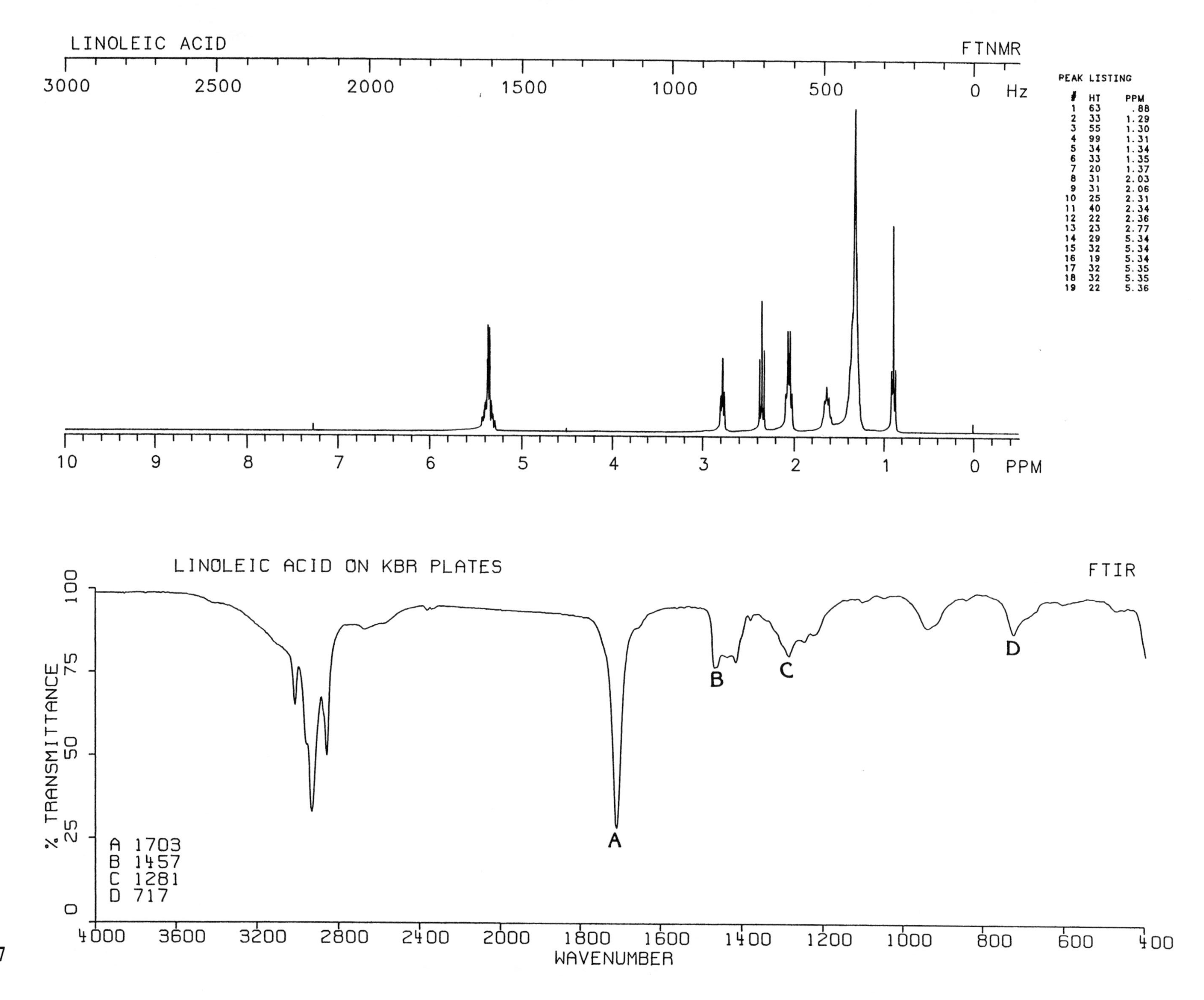

LINOLEIC ACID
FTNMR
3000 2500 2000 1500 1000 500 0 Hz
PEAK LISTING
# HT PPM
1 63 .88
2 33 1.29
3 55 1.30
4 99 1.31
5 34 1.34
6 33 1.35
7 20 1.37
8 31 2.03
9 31 2.06
10 25 2.31
11 40 2.34
12 22 2.36
13 23 2.77
14 29 5.34
15 32 5.34
16 19 5.34
17 32 5.35
18 32 5.35
19 22 5.36
10 9 8 7 6 5 4 3 2 1 0 PPM
LINOLEIC ACID ON KBR PLATES
FTIR
% TRANSMITTANCE
100 75 50 25 0
A 1703
B 1457
C 1281
D 717
A B C D
4000 3600 3200 2800 2400 2000 1800 1600 1400 1200 1000 800 600 400
WAVENUMBER

# LOBELINE

$C_{22}H_{27}NO_2$

**Molecular weight:** 337.47 (337.20)
**Synonyms:** 2-[6-(2-Hydroxy-2-phenylethyl)-1-methyl-2-piperidinyl]-1-phenylethanone
**Trade names:** Bantron, Inflatine, Lobron, Lobidan, Lobeton, Lobelidine, Toban, Unilobin, Zoolobelin
**Use:** Respiratory stimulant
**HPLC:** Si-10; 2A:98B; 3.2
**GC:** 1069; 140°C

O CH3 OH
C—CH2 N CH2—CH

L-LOBELINE

TRANSMITTANCE

WAVELENGTH (nm)

248
247

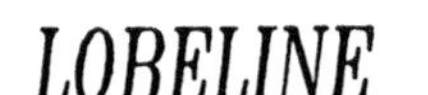


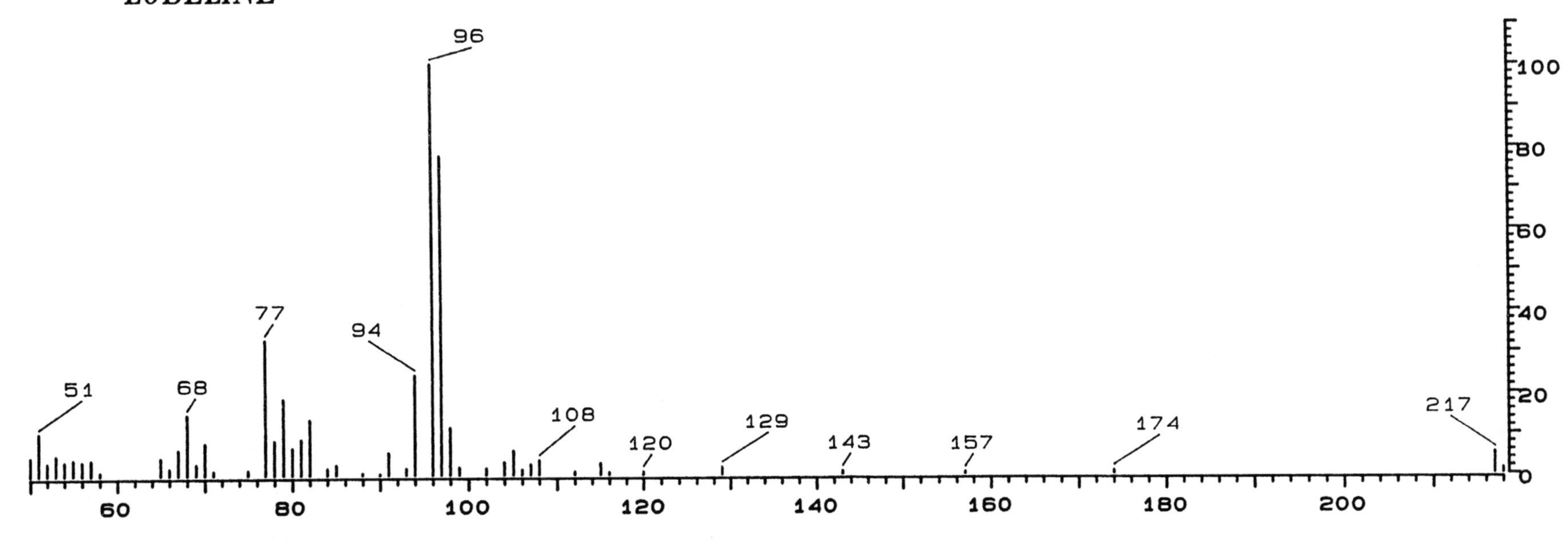

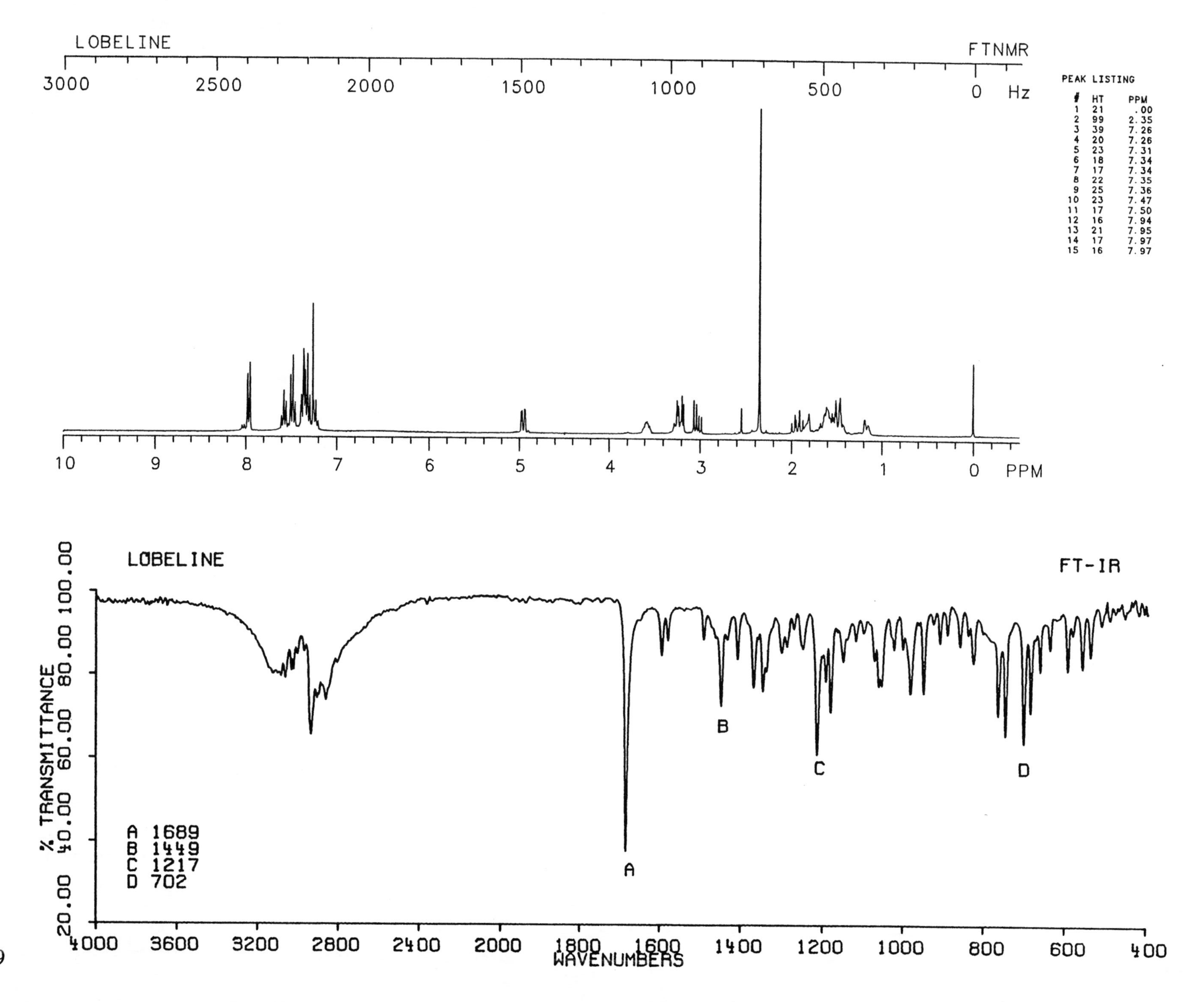
LOBELINE
FTNMR
3000 2500 2000 1500 1000 500 0 Hz
PEAK LISTING
# HT PPM
1 21 .00
2 99 2.35
3 39 7.26
4 20 7.26
5 23 7.31
6 18 7.34
7 17 7.34
8 22 7.35
9 25 7.36
10 23 7.47
11 17 7.50
12 16 7.94
13 21 7.95
14 17 7.97
15 16 7.97
10 9 8 7 6 5 4 3 2 1 0 PPM
LOBELINE
FT-IR
% TRANSMITTANCE
20.00 40.00 60.00 80.00 100.00
A 1689
B 1449
C 1217
D 702
A
B
C
D
4000 3600 3200 2800 2400 2000 1800 1600 1400 1200 1000 800 600 400
WAVENUMBERS

# LOMUSTINE

$C_9H_{16}ClN_3O_2$

**Molecular weight:** 233.70 (233.09)
**Synonyms:** 1-(2-Chloroethyl)-3-cyclohexyl-1-nitrosourea

**Trade names:** CeeNU

**Use:** Antineoplastic
**HPLC:** Si-10; 50B:50C; 3.5
**GC:**

NO
Cl—CH$_2$—CH$_2$—N—C(=O)—NH—(cyclohexyl)

LOMUSTINE

TRANSMITTANCE: 1.0, 0.90, 0.80, 0.70, 0.60, 0.50, 0.40, 0.30, 0.20, 0.10, 0.0

ETHANOL.....230

WAVELENGTH (nm): 220, 240, 260, 280, 300, 320, 340

## LOMUSTINE – – SOLID PROBE

Peaks: 41, 55, 67, 83, 97, 108, 126, 153, 233

Axis: 60, 80, 100, 120, 140, 160, 180, 200, 220; 0, 20, 40, 60, 80, 100

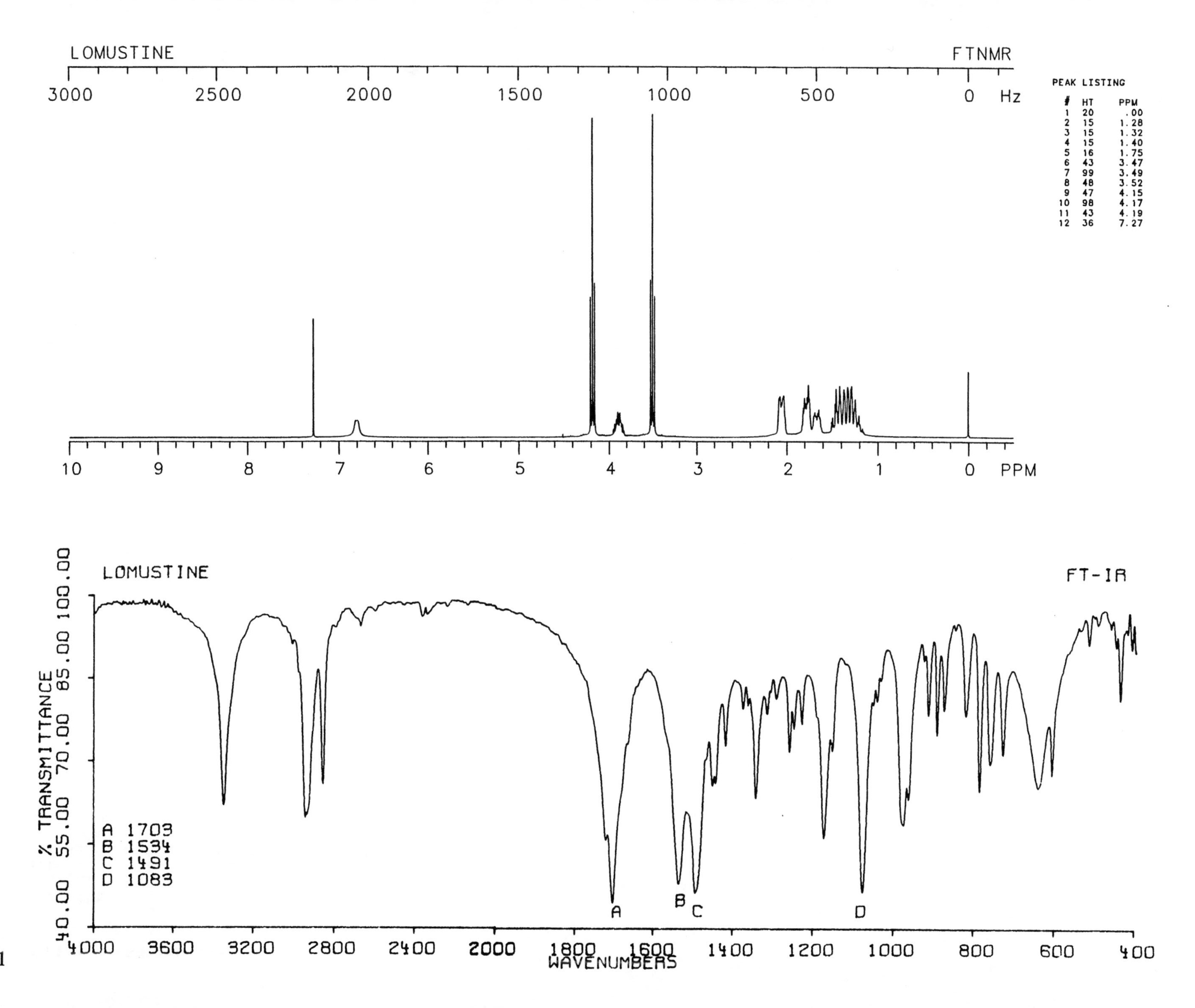
LOMUSTINE
FTNMR
3000
2500
2000
1500
1000
500
0
Hz
PEAK LISTING
# HT PPM
1 20 .00
2 15 1.28
3 15 1.32
4 15 1.40
5 16 1.75
6 43 3.47
7 99 3.49
8 48 3.52
9 47 4.15
10 98 4.17
11 43 4.19
12 36 7.27
10
9
8
7
6
5
4
3
2
1
0
PPM
LOMUSTINE
FT-IR
% TRANSMITTANCE
40.00
55.00
70.00
85.00
100.00
A 1703
B 1534
C 1491
D 1083
A
B
C
D
4000
3600
3200
2800
2400
2000
1800
1600
1400
1200
1000
800
600
400
WAVENUMBERS

# LOPERAMIDE

$C_{29}H_{33}ClN_2O_2$

Molecular weight: 477.05 (476.22)
Synonyms: 4-(4-Chlorophenyl)-4-hydroxy-N,N-dimethyl-α,α-diphenyl-1-piperidinebutanamide
Trade names: Imodium

Use: Antidiarrheal
HPLC: Si-10; 10A:90B; 4.0
GC:

O
C—N(CH3)2
C6H5—C—CH2—CH2—N
C6H5
OH
Cl

LOPERAMIDE

TRANSMITTANCE

1.0
0.90
0.80
0.70
0.60
0.50
0.40
0.30
0.20
0.10
0.0

225, 252, 258, 264
PPT BASE

220
240
260
280
300
320
340

WAVELENGTH (nm)

*LOPERAMIDE - - SOLID PROBE*

42
72
115
139
165
206
238
266
402
432

80
120
160
200
240
280
320
360
400
440

100
80
60
40
20
0

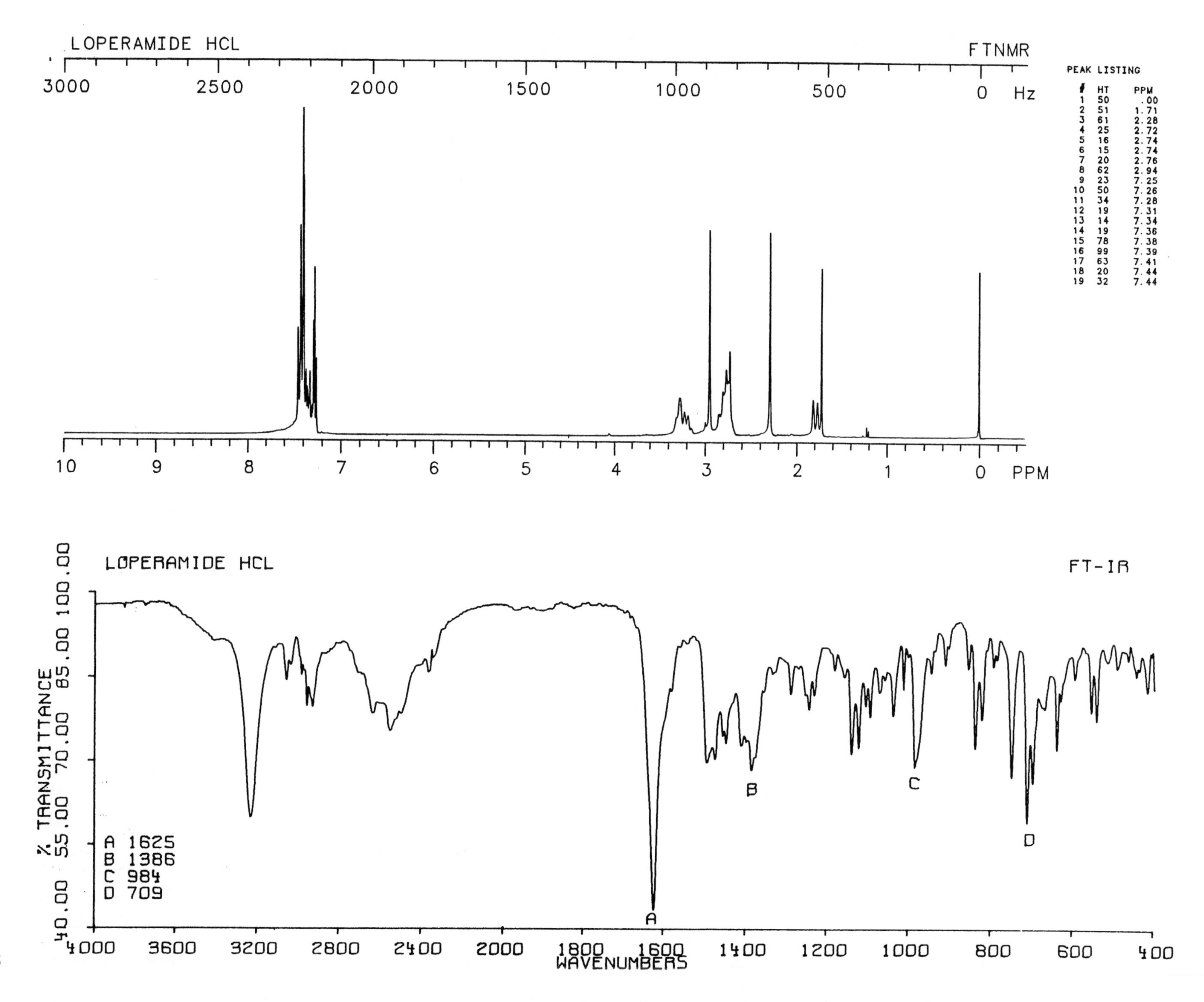
LOPERAMIDE HCL
FTNMR
3000
2500
2000
1500
1000
500
0 Hz
PEAK LISTING
# HT PPM
1 50 .00
2 51 1.71
3 61 2.28
4 25 2.72
5 16 2.74
6 15 2.74
7 20 2.76
8 62 2.94
9 23 7.25
10 50 7.26
11 34 7.28
12 19 7.31
13 14 7.34
14 19 7.36
15 78 7.38
16 99 7.39
17 63 7.41
18 20 7.44
19 32 7.44
10
9
8
7
6
5
4
3
2
1
0 PPM
LOPERAMIDE HCL
FT-IR
% TRANSMITTANCE
40.00
55.00
70.00
85.00
100.00
A 1625
B 1386
C 984
D 709
A
B
C
D
4000
3600
3200
2800
2400
2000
1800
1600
1400
1200
1000
800
600
400
WAVENUMBERS

# LORAZEPAM

$C_{15}H_{10}Cl_2N_2O_2$

**Molecular weight:** 321.16 (320.01)
**Synonyms:** 7-Chloro-5-(o-chlorophenyl)-1,3-dihydro-3-hydroxy-2H-1,4-benzodiazepin-2-one
**Trade names:** Ativan

**Use:** Tranquilizer
**HPLC:** Si-10; 5A:95B; 4.0
**GC:** 2465; 250°C

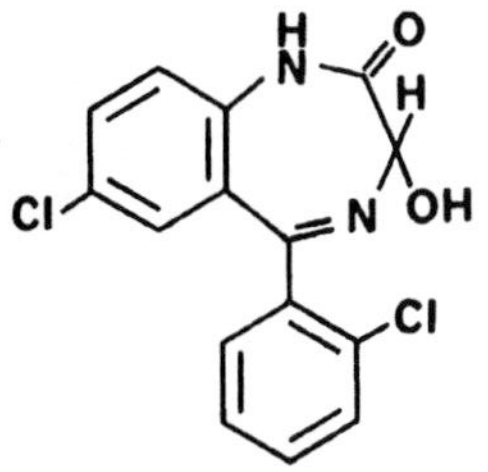

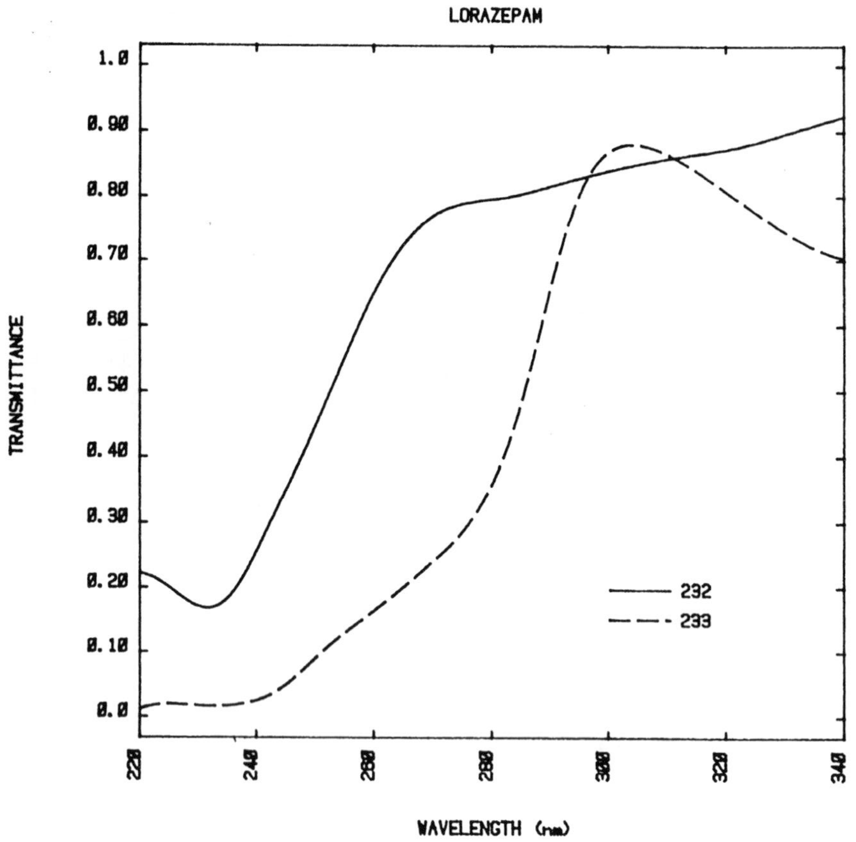

*LORAZEPAM*

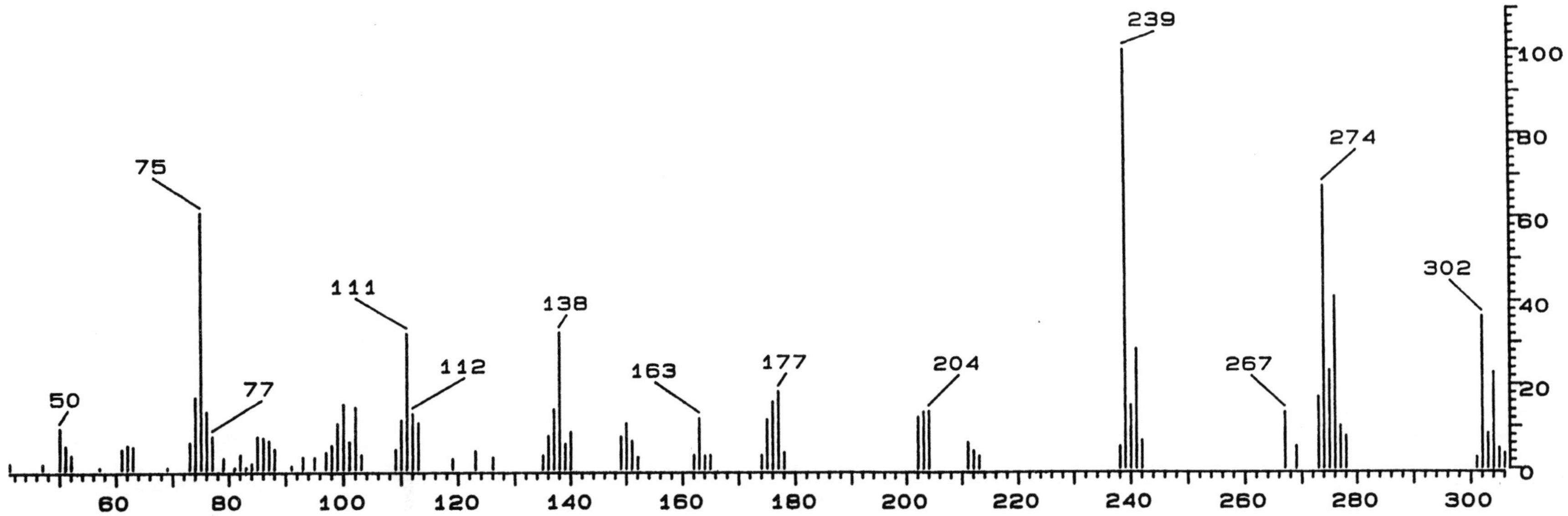

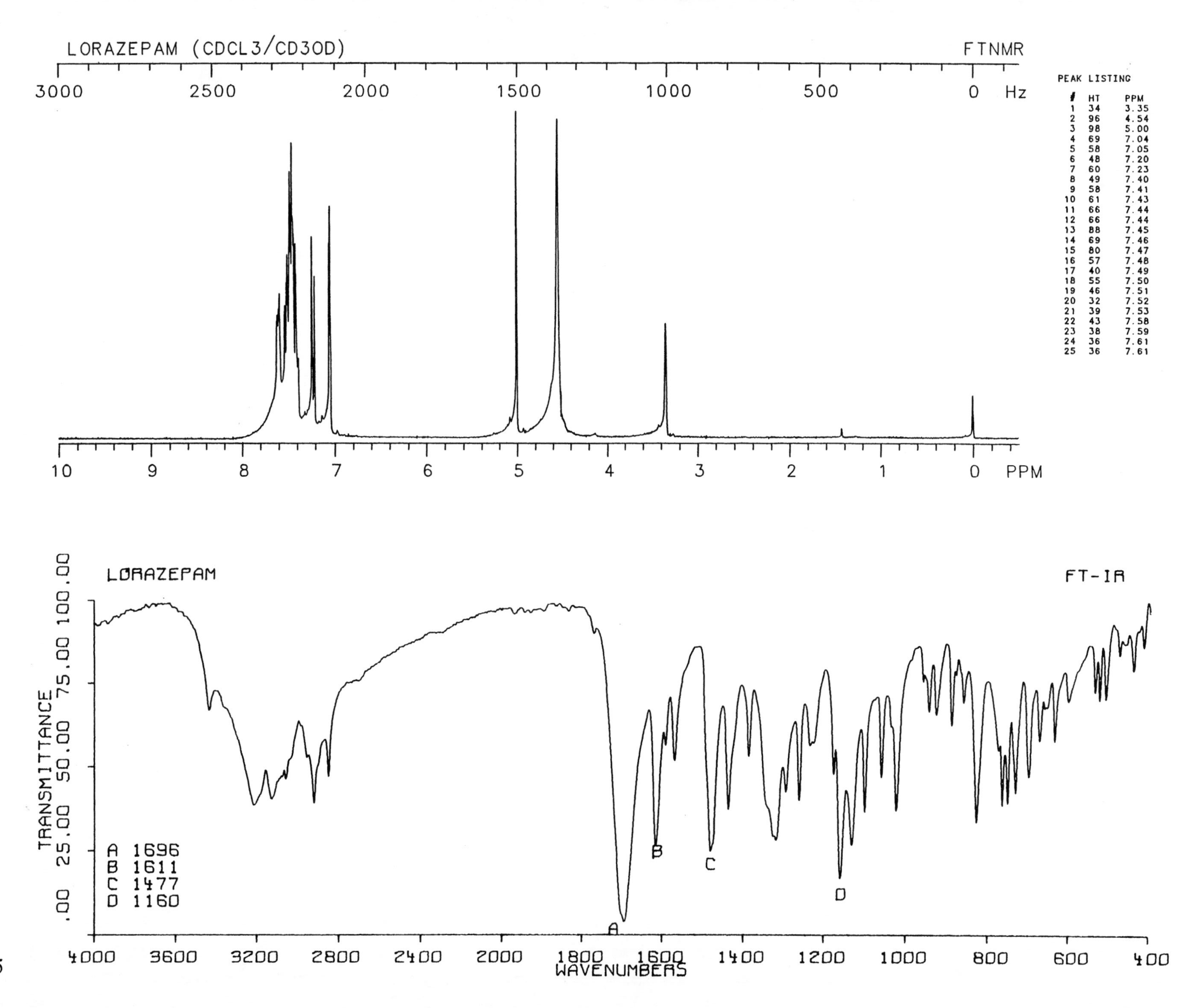
LORAZEPAM (CDCL3/CD3OD)
FTNMR
3000
2500
2000
1500
1000
500
0
Hz
PEAK LISTING
# HT PPM
1 34 3.35
2 96 4.54
3 98 5.00
4 69 7.04
5 58 7.05
6 48 7.20
7 60 7.23
8 49 7.40
9 58 7.41
10 61 7.43
11 66 7.44
12 66 7.44
13 88 7.45
14 69 7.46
15 80 7.47
16 57 7.48
17 40 7.49
18 55 7.50
19 46 7.51
20 32 7.52
21 39 7.53
22 43 7.58
23 38 7.59
24 36 7.61
25 36 7.61
10
9
8
7
6
5
4
3
2
1
0
PPM
LORAZEPAM
FT-IR
TRANSMITTANCE
.00
25.00
50.00
75.00
100.00
A 1696
B 1611
C 1477
D 1160
A
B
C
D
4000
3600
3200
2800
2400
2000
1800
1600
1400
1200
1000
800
600
400
WAVENUMBERS

# LOXAPINE

$C_{18}H_{18}ClN_3O$

**Molecular weight:** 327.81 (327.11)
**Synonyms:** 2-Chloro-11-(4-methyl-1-piperazinyl)-dibenz[b,f]-[1,4]oxazepine
**Trade names:** Loxitane

**Use:** Tranquilizer
**HPLC:** Si-10; 2A:98B; 8.5
**GC:** 2621; 250°C

LOXAPINE

TRANSMITTANCE

1.0 0.90 0.80 0.70 0.60 0.50 0.40 0.30 0.20 0.10 0.0

220 240 260 280 300 320 340

WAVELENGTH (nm)

252, 293
249, 298

*LOXAPINE*

42 70 83 99 137 139 164 193 202 228 245 257 283 297 327

60 80 100 120 140 160 180 200 220 240 260 280 300 320

0 20 40 60 80 100

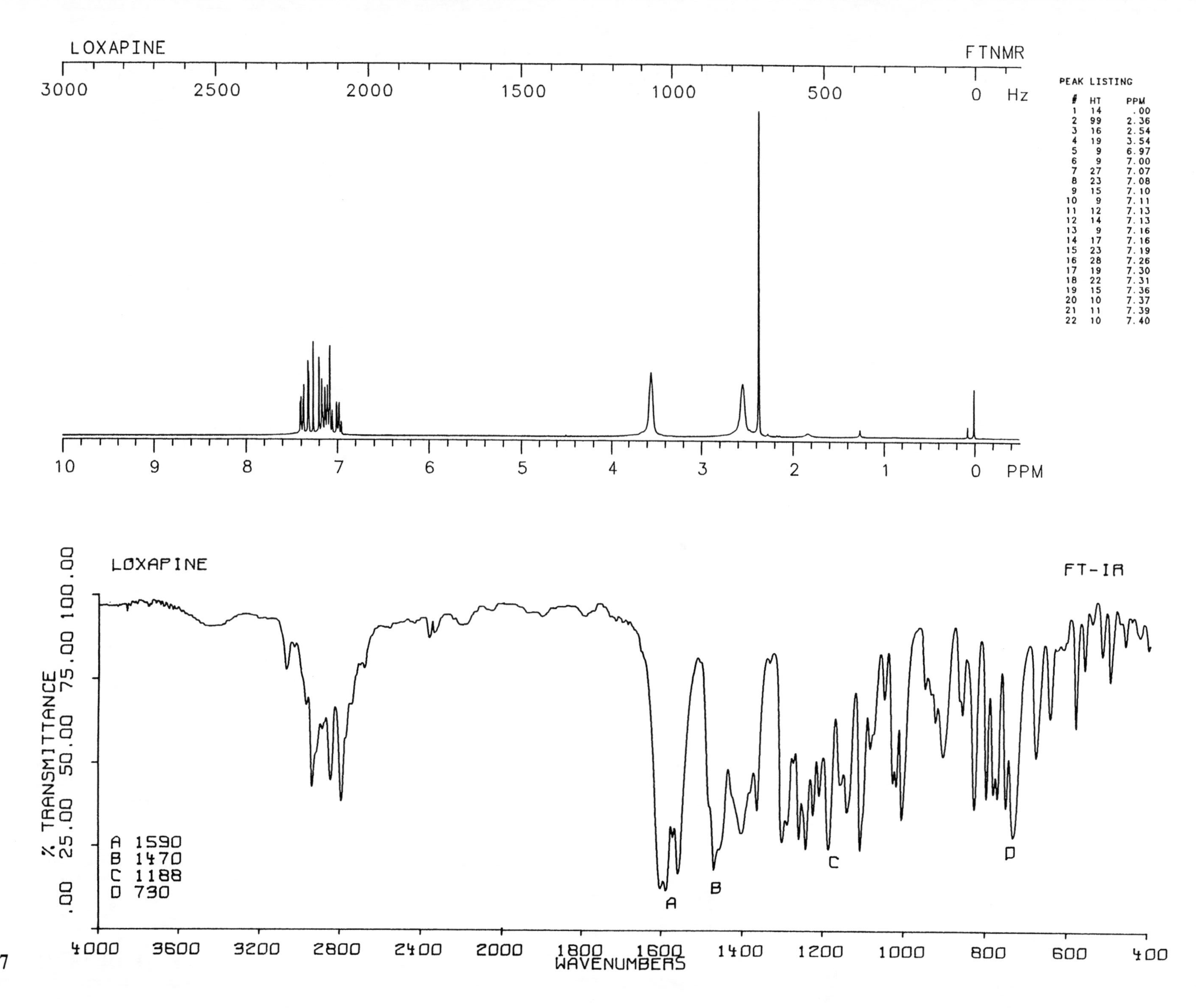
LOXAPINE
FTNMR
3000
2500
2000
1500
1000
500
0
Hz
PEAK LISTING
| # | HT | PPM |
|---|---|---|
| 1 | 14 | .00 |
| 2 | 99 | 2.36 |
| 3 | 16 | 2.54 |
| 4 | 19 | 3.54 |
| 5 | 9 | 6.97 |
| 6 | 9 | 7.00 |
| 7 | 27 | 7.07 |
| 8 | 23 | 7.08 |
| 9 | 15 | 7.10 |
| 10 | 9 | 7.11 |
| 11 | 12 | 7.13 |
| 12 | 14 | 7.13 |
| 13 | 9 | 7.16 |
| 14 | 17 | 7.16 |
| 15 | 23 | 7.19 |
| 16 | 28 | 7.26 |
| 17 | 19 | 7.30 |
| 18 | 22 | 7.31 |
| 19 | 15 | 7.36 |
| 20 | 10 | 7.37 |
| 21 | 11 | 7.39 |
| 22 | 10 | 7.40 |
10
9
8
7
6
5
4
3
2
1
0
PPM
LOXAPINE
FT-IR
% TRANSMITTANCE
.00
25.00
50.00
75.00
100.00
A 1590
B 1470
C 1188
D 730
A
B
C
D
4000
3600
3200
2800
2400
2000
1800
1600
1400
1200
1000
800
600
400
WAVENUMBERS

# LYSERGIC ACID

$C_{16}H_{16}N_2O_2$

Molecular weight: 268.32 (268.12)
Synonyms: 9,10-Didehydro-6-methylergoline-8β-carboxylic acid

Trade names:

Use: Hallucinogen
HPLC: Si-10; 30A:70B; 6.0
GC:

HN COOH N CH3

LYSERGIC ACID

TRANSMITTANCE

1.0 0.90 0.80 0.70 0.60 0.50 0.40 0.30 0.20 0.10 0.0

220 240 260 280 300 320 340

WAVELENGTH (nm)

224, 310
239, 308

*LYSERGIC ACID*

44 69 83 97 111 127 139 154 167 192 207 224 225 250 268

60 80 100 120 140 160 180 200 220 240 260

0 20 40 60 80 100

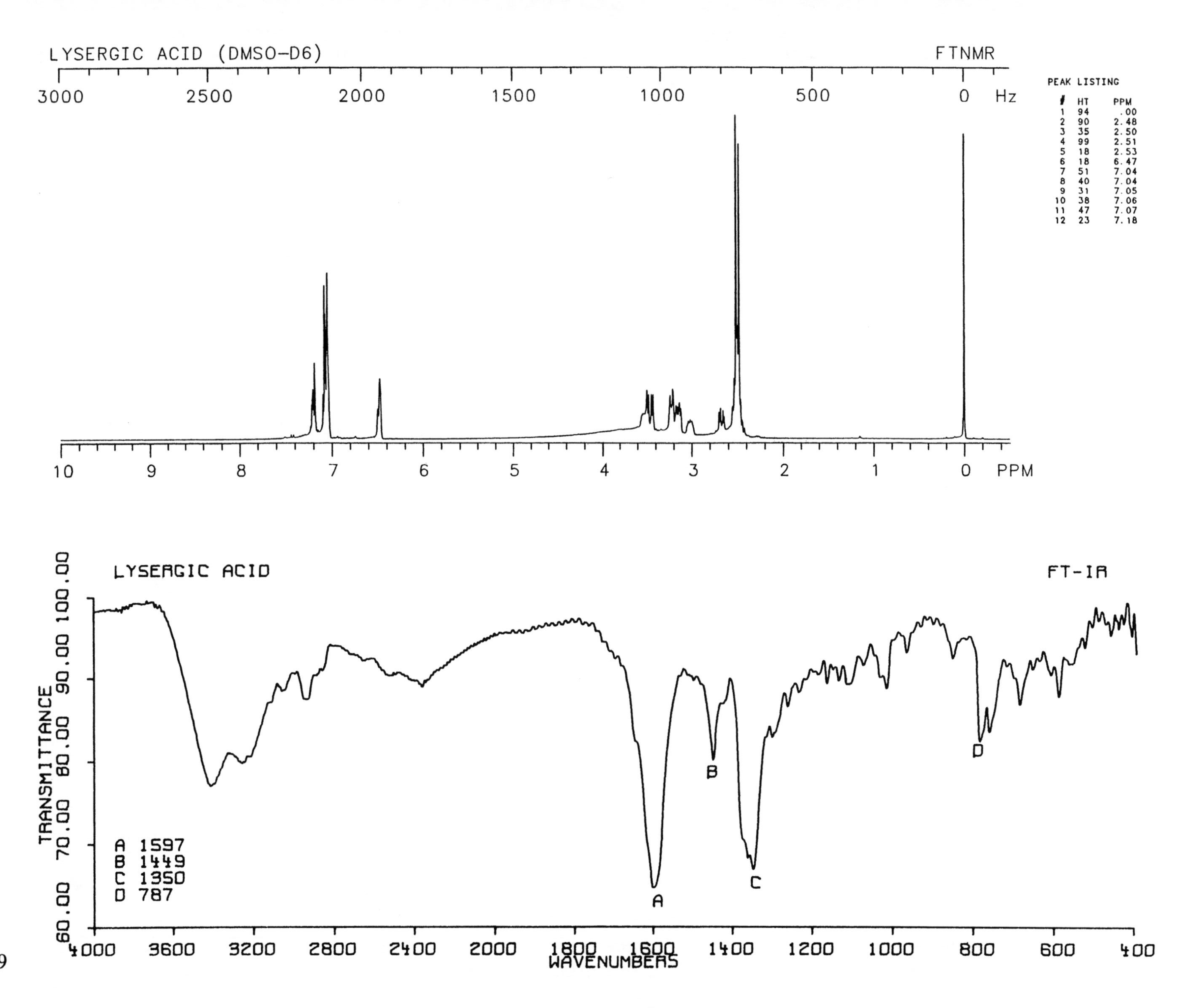
LYSERGIC ACID (DMSO-D6)
FTNMR
3000
2500
2000
1500
1000
500
0 Hz
PEAK LISTING
# HT PPM
1 94 .00
2 90 2.48
3 35 2.50
4 99 2.51
5 18 2.53
6 18 6.47
7 51 7.04
8 40 7.04
9 31 7.05
10 38 7.06
11 47 7.07
12 23 7.18
10
9
8
7
6
5
4
3
2
1
0 PPM
LYSERGIC ACID
FT-IR
TRANSMITTANCE
60.00
70.00
80.00
90.00
100.00
A 1597
B 1449
C 1350
D 787
A
B
C
D
4000
3600
3200
2800
2400
2000
1800
1600
1400
1200
1000
800
600
400
WAVENUMBERS

# LYSERGIC ACID DIETHYLAMIDE

$C_{20}H_{25}N_3O$

**Molecular weight:** 323.44 (323.20)
**Synonyms:** 9,10-Didehydro-N,N-diethyl-6-methylergoline-8β-carboxamide; N,N-diethyl-d-lysergamide; d-lysergic acid diethylamide; LSD; LSD-25; lysergide
**Trade names:**

**Use:** Hallucinogen
**HPLC:** Si-10; 5A:95B; 4.0
**GC:**

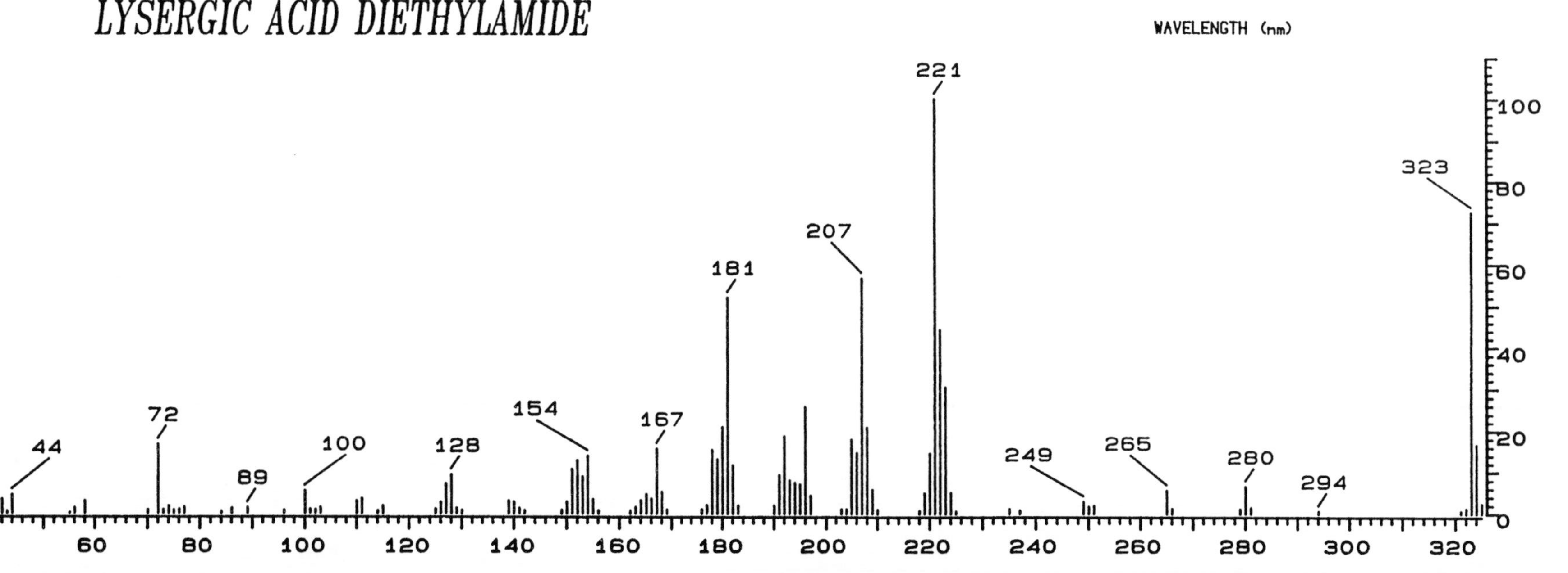

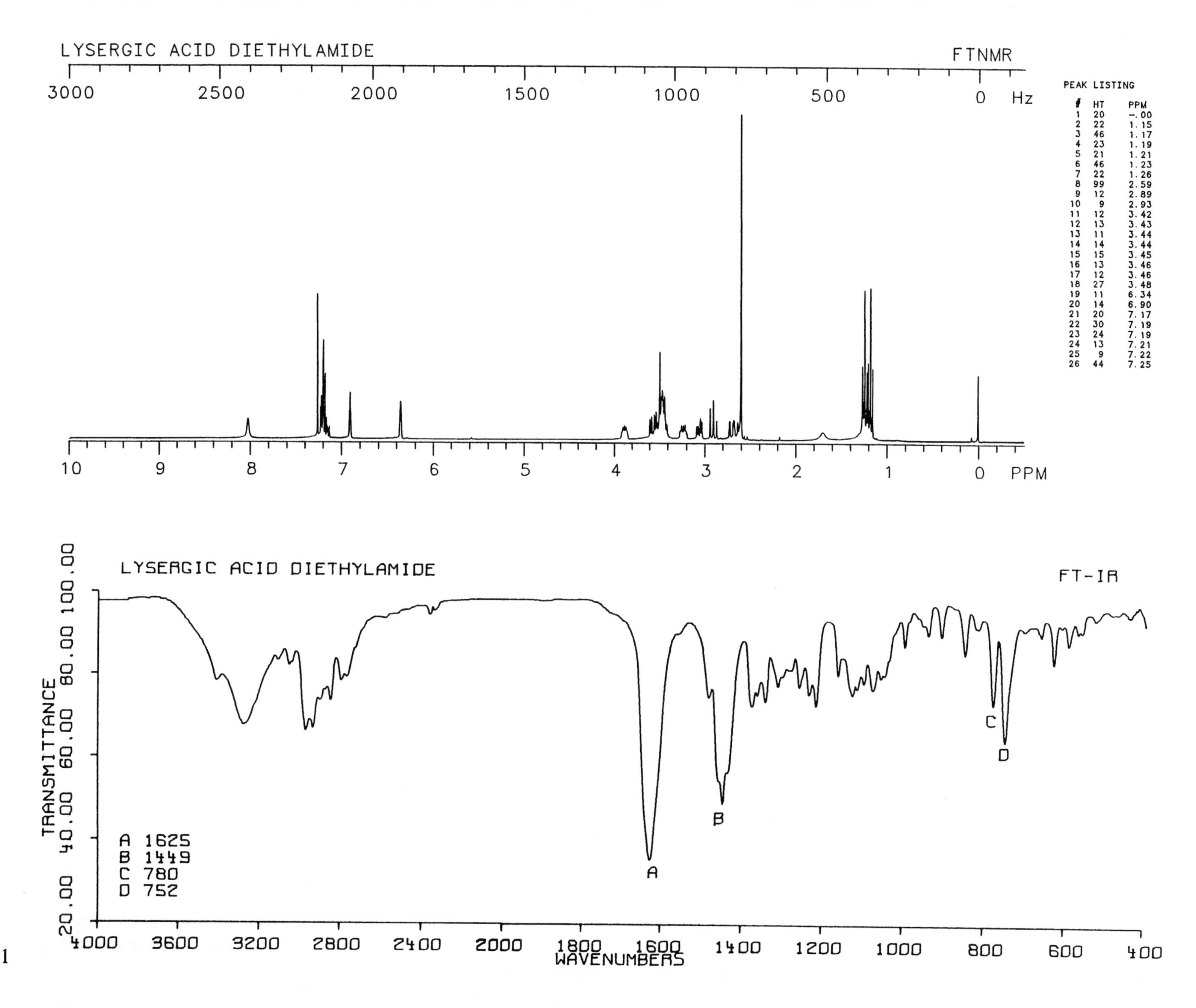
LYSERGIC ACID DIETHYLAMIDE
FTNMR
3000 2500 2000 1500 1000 500 0 Hz
10 9 8 7 6 5 4 3 2 1 0 PPM
PEAK LISTING
# HT PPM
1 20 -.00
2 22 1.15
3 46 1.17
4 23 1.19
5 21 1.21
6 46 1.23
7 22 1.26
8 99 2.59
9 12 2.89
10 9 2.93
11 12 3.42
12 13 3.43
13 11 3.44
14 14 3.44
15 15 3.45
16 13 3.46
17 12 3.46
18 27 3.48
19 11 6.34
20 14 6.90
21 20 7.17
22 30 7.19
23 24 7.19
24 13 7.21
25 9 7.22
26 44 7.25
LYSERGIC ACID DIETHYLAMIDE
FT-IR
TRANSMITTANCE
20.00 40.00 60.00 80.00 100.00
4000 3600 3200 2800 2400 2000 1800 1600 1400 1200 1000 800 600 400
WAVENUMBERS
A 1625
B 1449
C 780
D 752
A
B
C
D

# LYSERGIC ACID N-(METHYLPROPYL)AMIDE

$C_{20}H_{25}N_3O$

**Molecular weight:** 323.44 (323.20)
**Synonyms:** 9,10-Didehydro-N-methyl-N-propyl-6-methylergoline-8β-carboxamide; LAMPA
**Trade names:**

**Use:** Hallucinogen
**HPLC:** Si-10; 5A:95B; 4.0
**GC:**

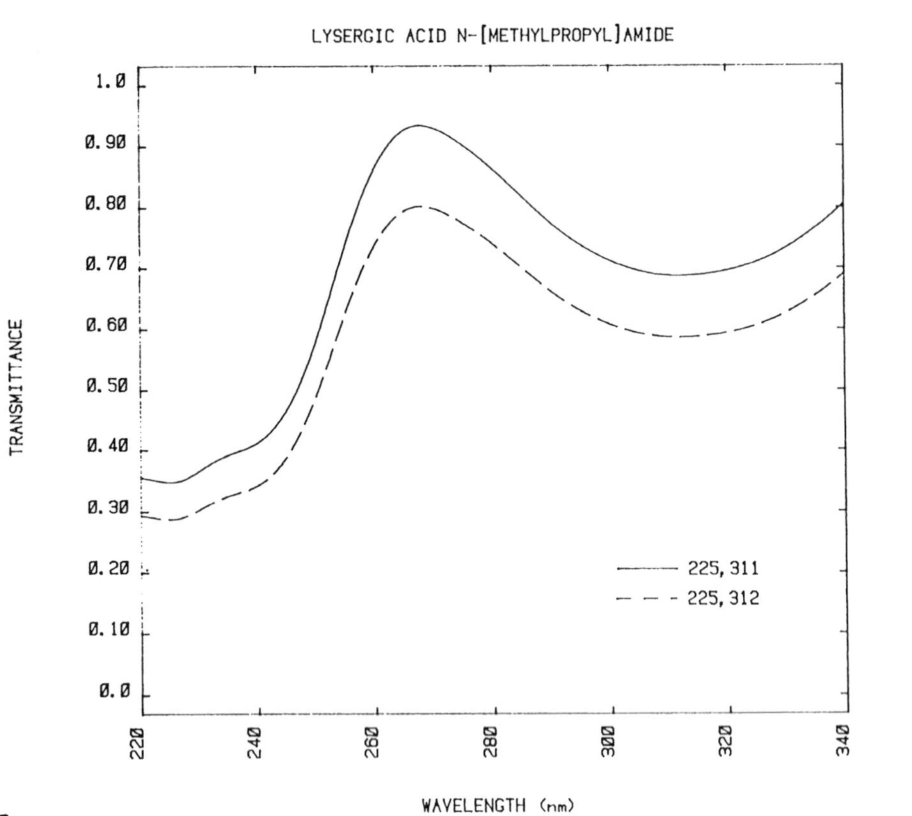


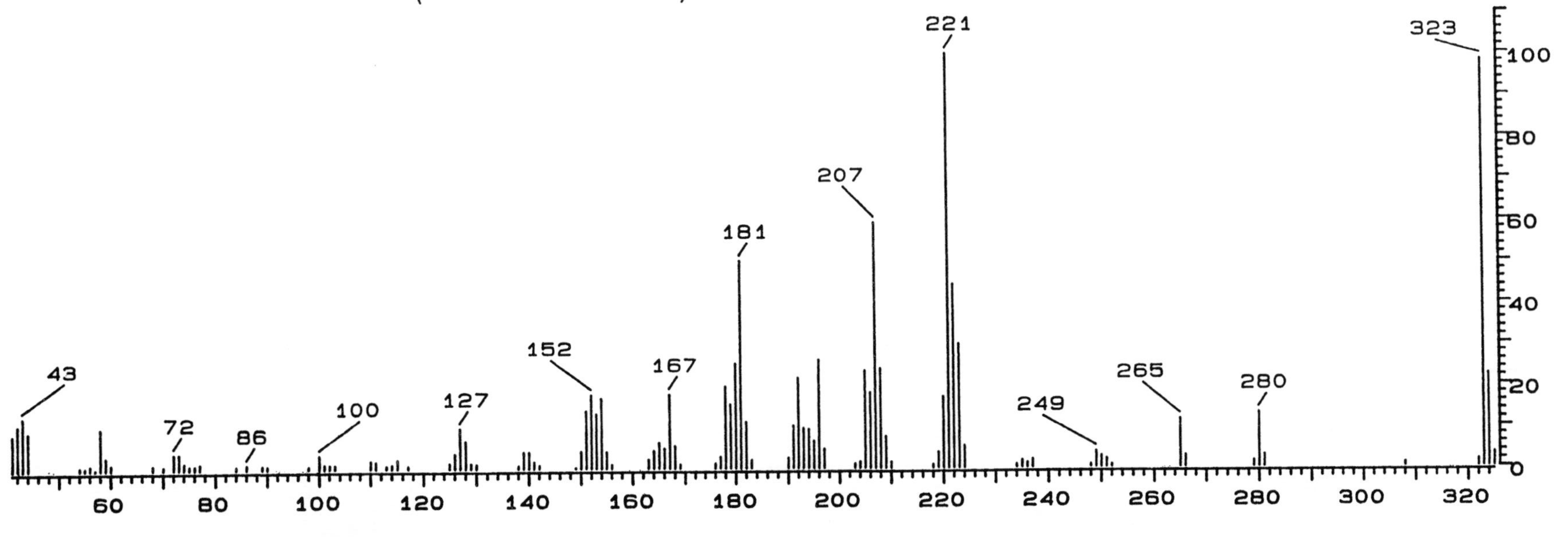

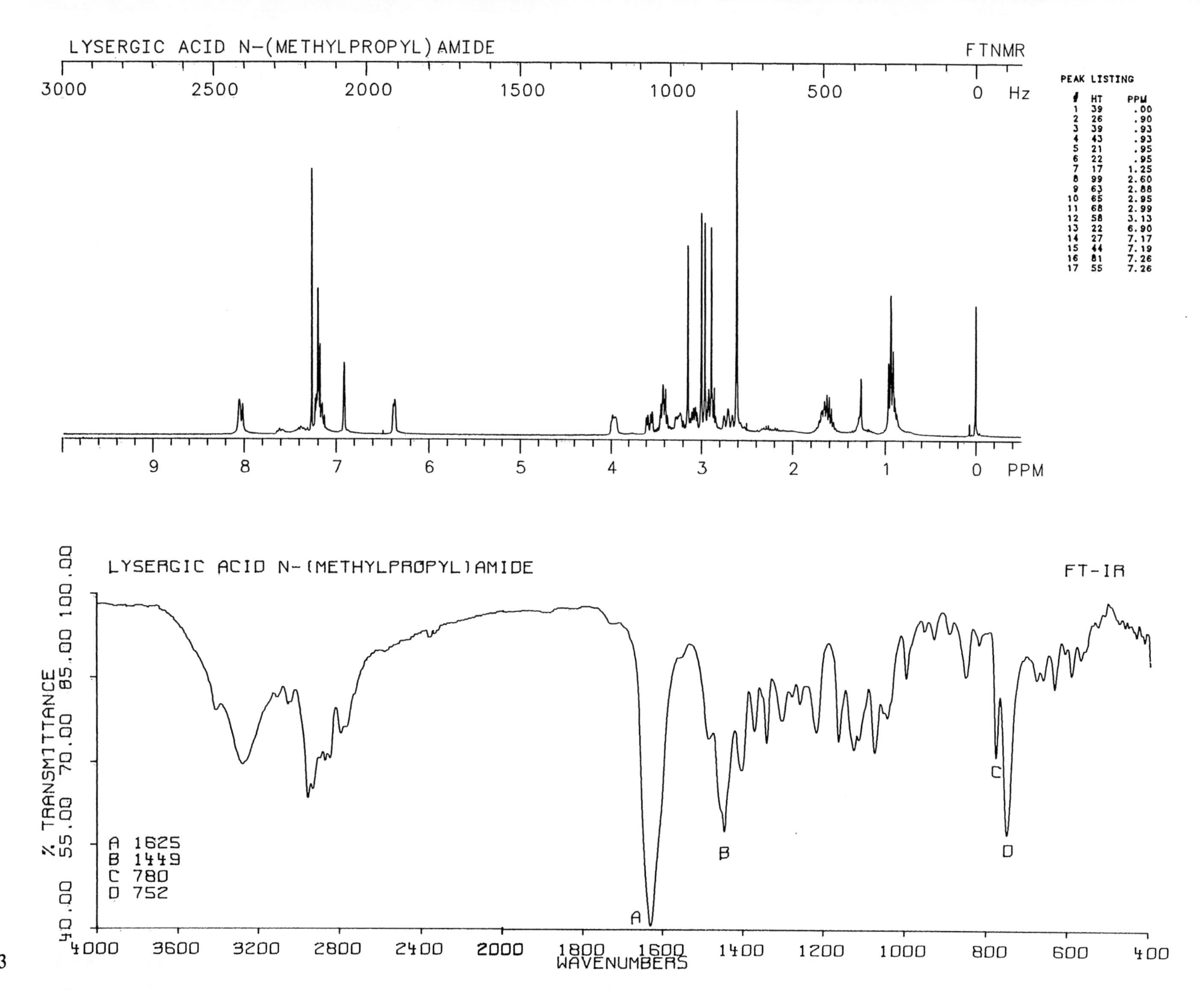

LYSERGIC ACID N-(METHYLPROPYL) AMIDE
FTNMR
3000
2500
2000
1500
1000
500
0 Hz
PEAK LISTING
# HT PPM
1 39 .00
2 26 .90
3 39 .93
4 43 .93
5 21 .95
6 22 .95
7 17 1.25
8 99 2.60
9 63 2.88
10 65 2.95
11 68 2.99
12 58 3.13
13 22 6.90
14 27 7.17
15 44 7.19
16 81 7.26
17 55 7.26
9
8
7
6
5
4
3
2
1
0 PPM
LYSERGIC ACID N-(METHYLPROPYL)AMIDE
FT-IR
% TRANSMITTANCE
40.00
55.00
70.00
85.00
100.00
A 1625
B 1449
C 780
D 752
A
B
C
D
4000
3600
3200
2800
2400
2000
1800
1600
1400
1200
1000
800
600
400
WAVENUMBERS

# LYSERGOL

$C_{16}H_{18}N_2O$

Molecular weight: 254.33 (254.14)
Synonyms: 9,10-Didehydro-6-methylergoline-6-β methanol

Trade names:

Use: Hallucinogen
HPLC:
GC:

HOH₂C, NH, N, CH₃

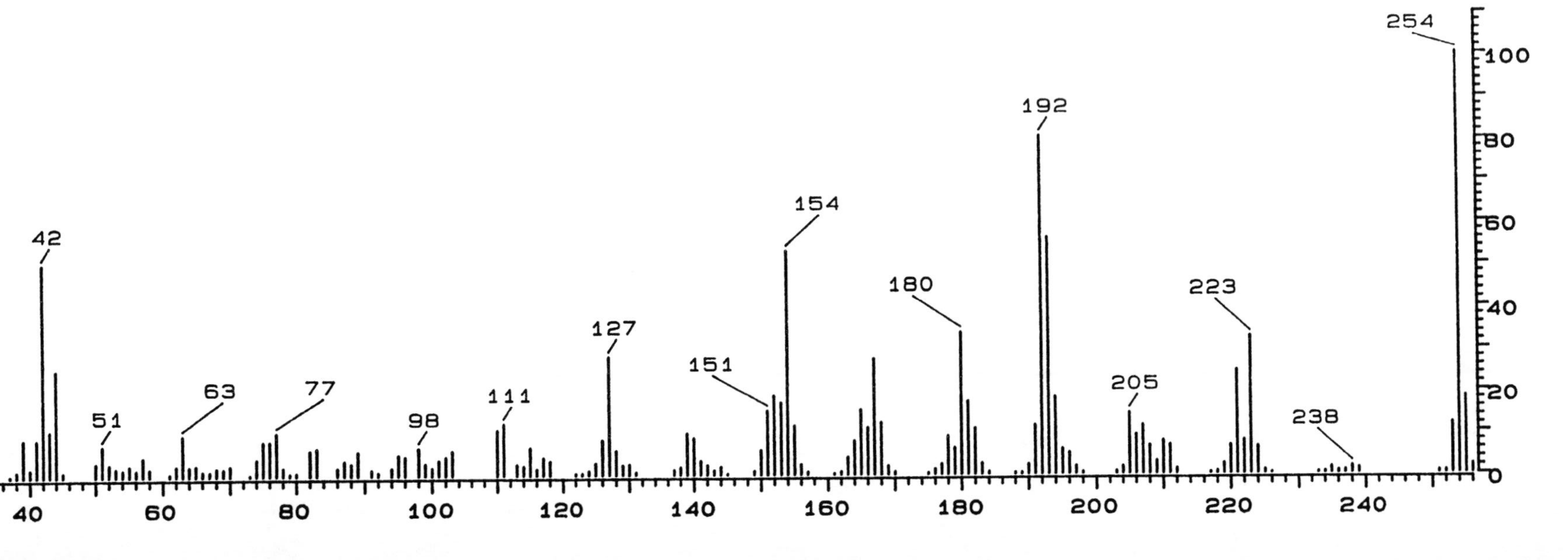

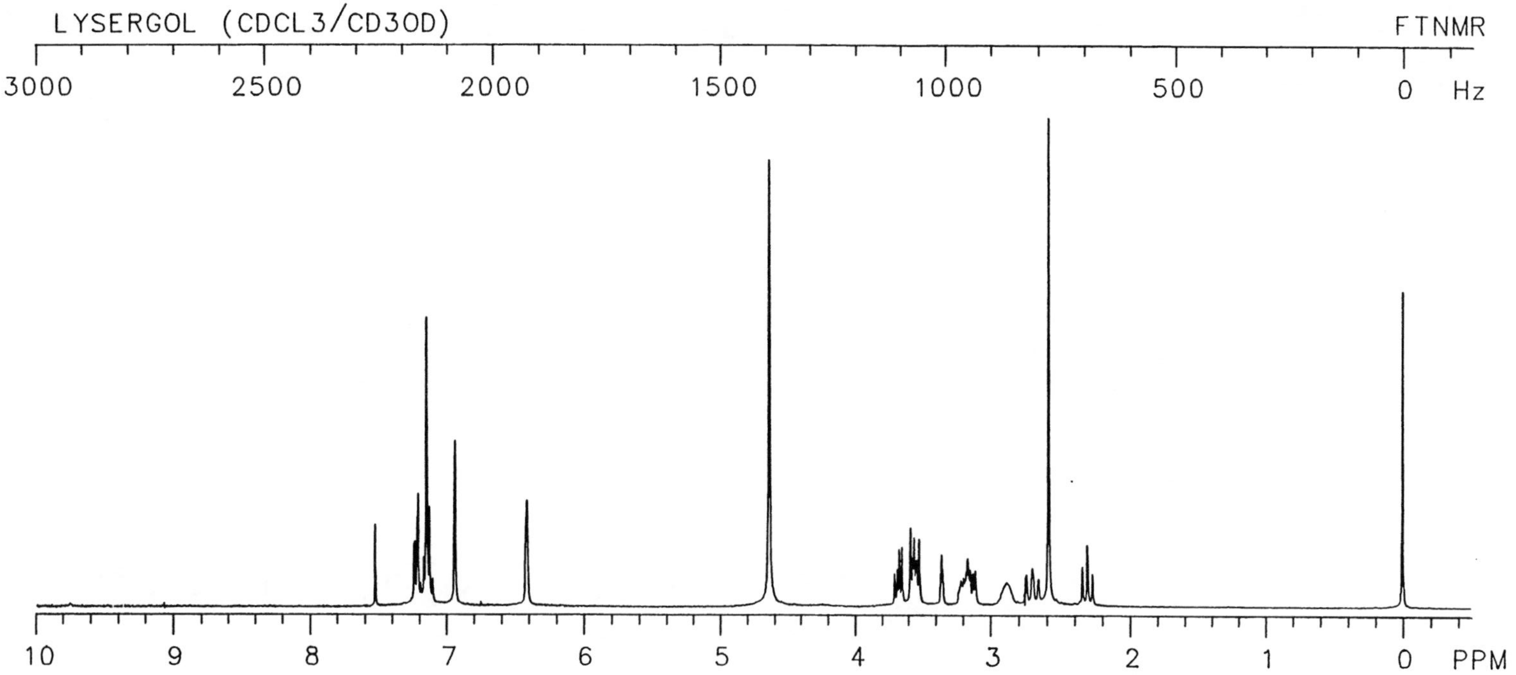
LYSERGOL (CDCL3/CD3OD)
FTNMR
3000
2500
2000
1500
1000
500
0 Hz
10
9
8
7
6
5
4
3
2
1
0 PPM

PEAK LISTING

| # | HT | PPM |
|---|---|---|
| 1 | 64 | .00 |
| 2 | 12 | 2.30 |
| 3 | 8 | 2.34 |
| 4 | 99 | 2.58 |
| 5 | 8 | 2.70 |
| 6 | 7 | 3.14 |
| 7 | 10 | 3.16 |
| 8 | 8 | 3.34 |
| 9 | 10 | 3.35 |
| 10 | 13 | 3.51 |
| 11 | 9 | 3.53 |
| 12 | 9 | 3.54 |
| 13 | 14 | 3.55 |
| 14 | 10 | 3.56 |
| 15 | 16 | 3.57 |
| 16 | 12 | 3.64 |
| 17 | 11 | 3.66 |
| 18 | 7 | 3.67 |
| 19 | 90 | 4.63 |
| 20 | 21 | 6.41 |
| 21 | 33 | 6.93 |
| 22 | 30 | 6.94 |
| 23 | 20 | 7.12 |
| 24 | 58 | 7.14 |
| 25 | 10 | 7.16 |
| 26 | 23 | 7.20 |
| 27 | 18 | 7.21 |
| 28 | 13 | 7.23 |
| 29 | 12 | 7.23 |
| 30 | 16 | 7.52 |

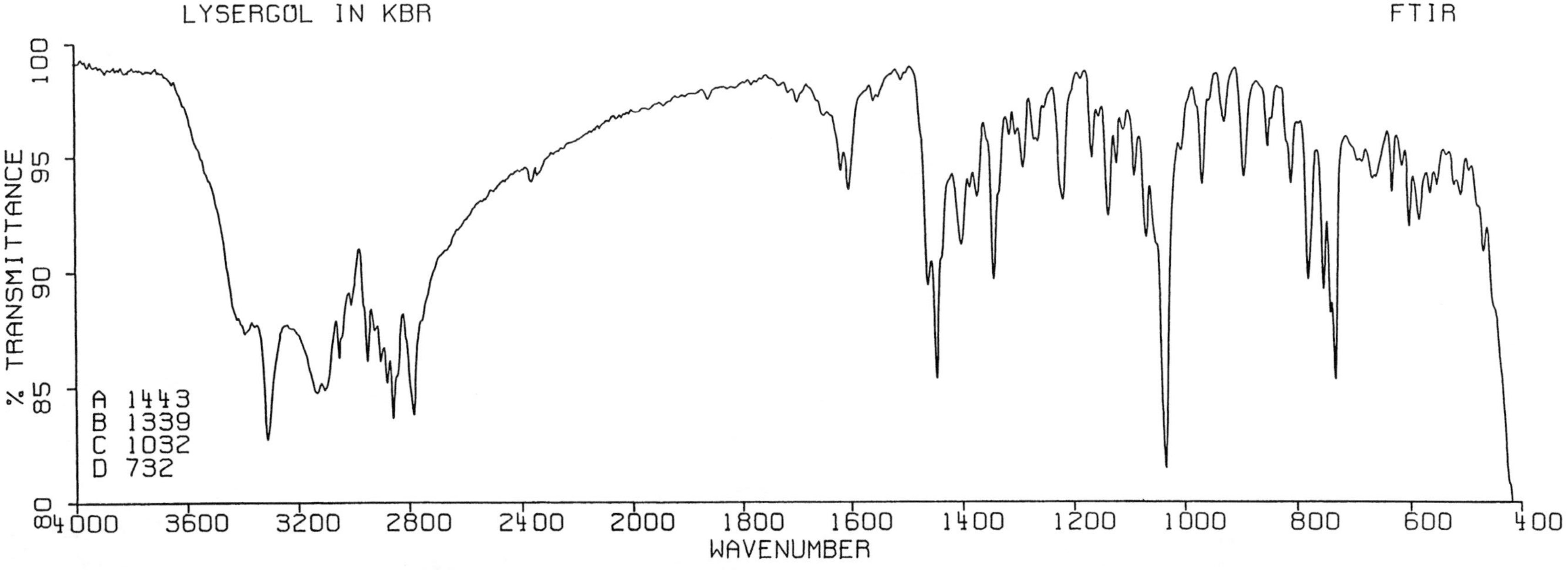
LYSERGOL IN KBR
FTIR
% TRANSMITTANCE
100
95
90
85
80
A 1443
B 1339
C 1032
D 732
4000
3600
3200
2800
2400
2000
1800
1600
1400
1200
1000
800
600
400
WAVENUMBER

# *D*-LYXOSE

$C_5H_{10}O_5$

Molecular weight: 150.13 (150.05)
Synonyms: α-D-Lyxose

Trade names:

Use: Sugar
HPLC:
GC:

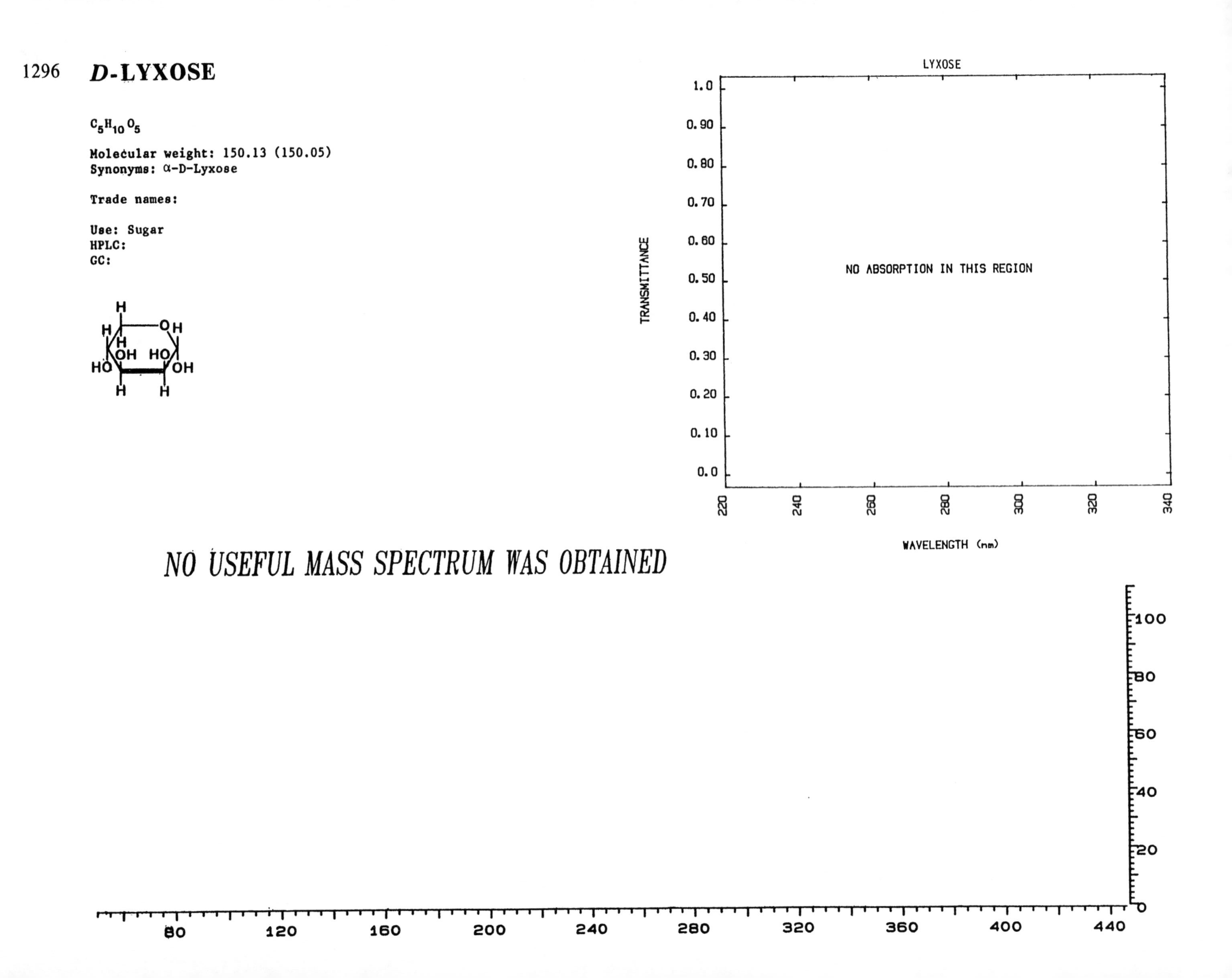

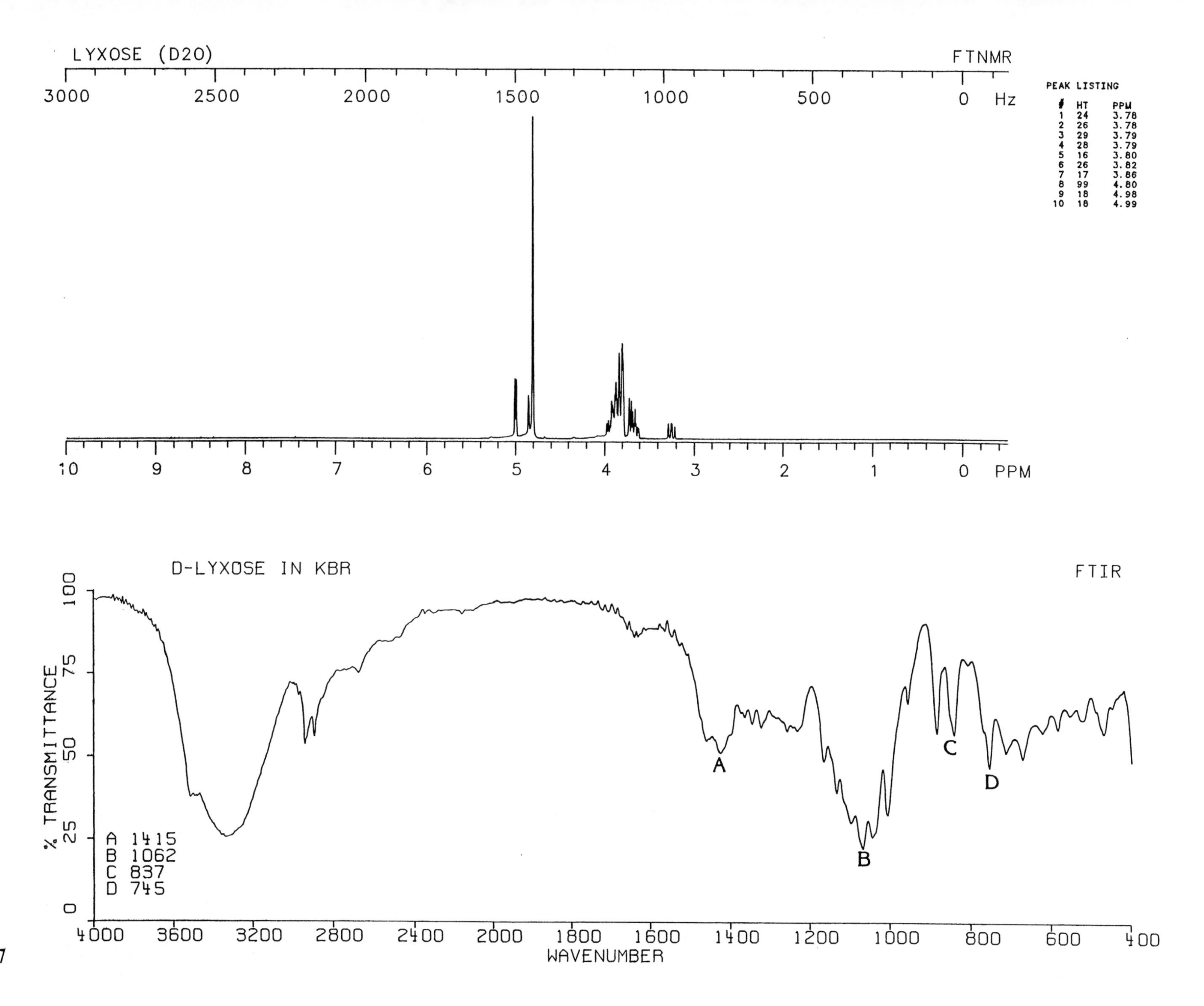
LYXOSE (D2O)
FTNMR
3000
2500
2000
1500
1000
500
0 Hz
PEAK LISTING
# HT PPM
1 24 3.78
2 26 3.78
3 29 3.79
4 28 3.79
5 16 3.80
6 26 3.82
7 17 3.86
8 99 4.80
9 18 4.98
10 18 4.99
10
9
8
7
6
5
4
3
2
1
0 PPM
D-LYXOSE IN KBR
FTIR
% TRANSMITTANCE
100
75
50
25
0
A 1415
B 1062
C 837
D 745
A
B
C
D
4000
3600
3200
2800
2400
2000
1800
1600
1400
1200
1000
800
600
400
WAVENUMBER

# MAFENIDE

$C_7H_{10}N_2O_2S$

**Molecular weight:** 186.25 (186.05)
**Synonyms:** α-Amino-p-toluenesulfonamide; 4-homosulfanilamide; maphenide; bensulfamide; sulfabenzamine; p-aminomethylbenzene sulfonamide
**Trade names:** Sulfamylon

**Use:** Antibacterial
**HPLC:**
**GC:** 2076; 250°C

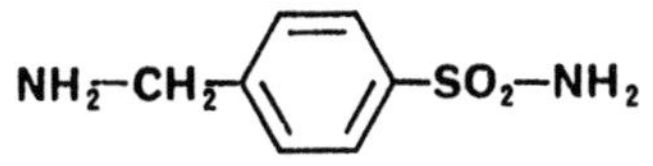

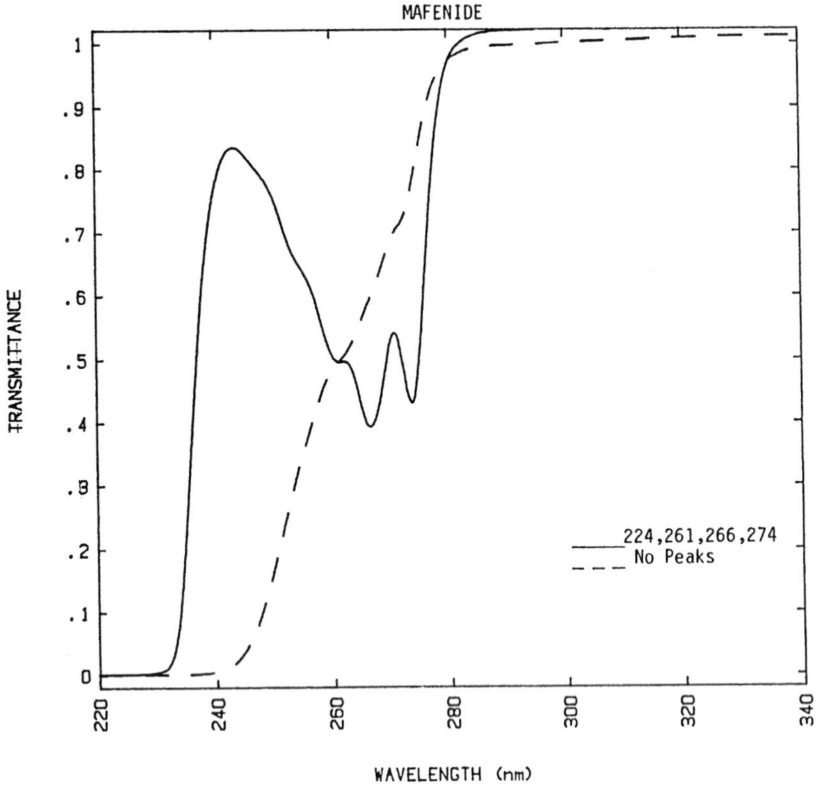

$NH_2$—$CH_2$—$C_6H_4$—$SO_2$—$NH_2$

MAFENIDE

TRANSMITTANCE

1 .9 .8 .7 .6 .5 .4 .3 .2 .1 0

220 240 260 280 300 320 340

WAVELENGTH (nm)

224,261,266,274
No Peaks

*MAFENIDE*

51 63 77 89 94 106 141 158 185 187

40 60 80 100 120 140 160 180 200 220

0 20 40 60 80 100

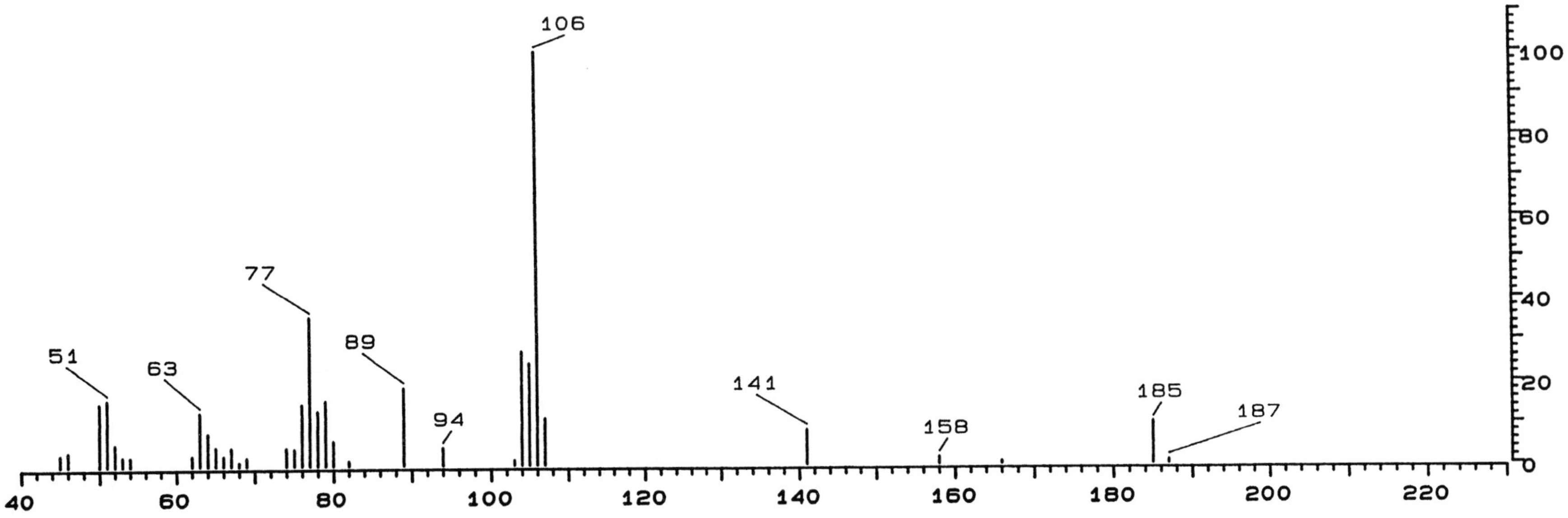

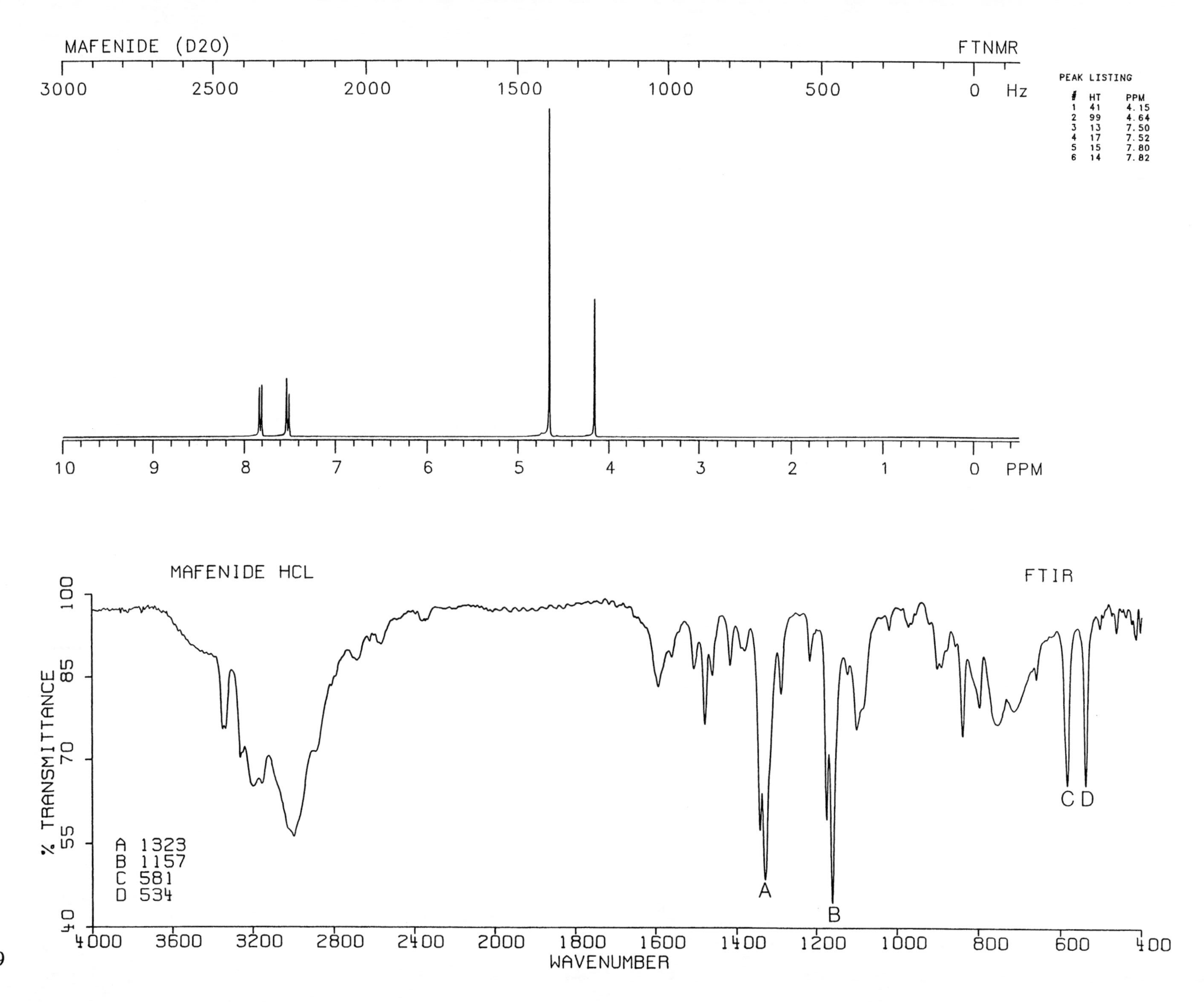

MAFENIDE (D2O)
FTNMR
3000
2500
2000
1500
1000
500
0 Hz
PEAK LISTING
# HT PPM
1 41 4.15
2 99 4.64
3 13 7.50
4 17 7.52
5 15 7.80
6 14 7.82
10
9
8
7
6
5
4
3
2
1
0 PPM
MAFENIDE HCL
FTIR
100
85
70
55
40
% TRANSMITTANCE
A 1323
B 1157
C 581
D 534
A
B
C D
4000
3600
3200
2800
2400
2000
1800
1600
1400
1200
1000
800
600
400
WAVENUMBER

# MALTOSE

$C_{12}H_{22}O_{11}$

**Molecular weight:** 342.31 (342.17)
**Synonyms:** 4-O-α-D-Glucopyranosyl-D-glucose; malt sugar; maltobiose; 4-(α-D-glucosido)-D-glucose
**Trade names:** Maltos

**Use:** Nutrient, sweetener, pharmaceutical dispensing agent
**HPLC:**
**GC:**

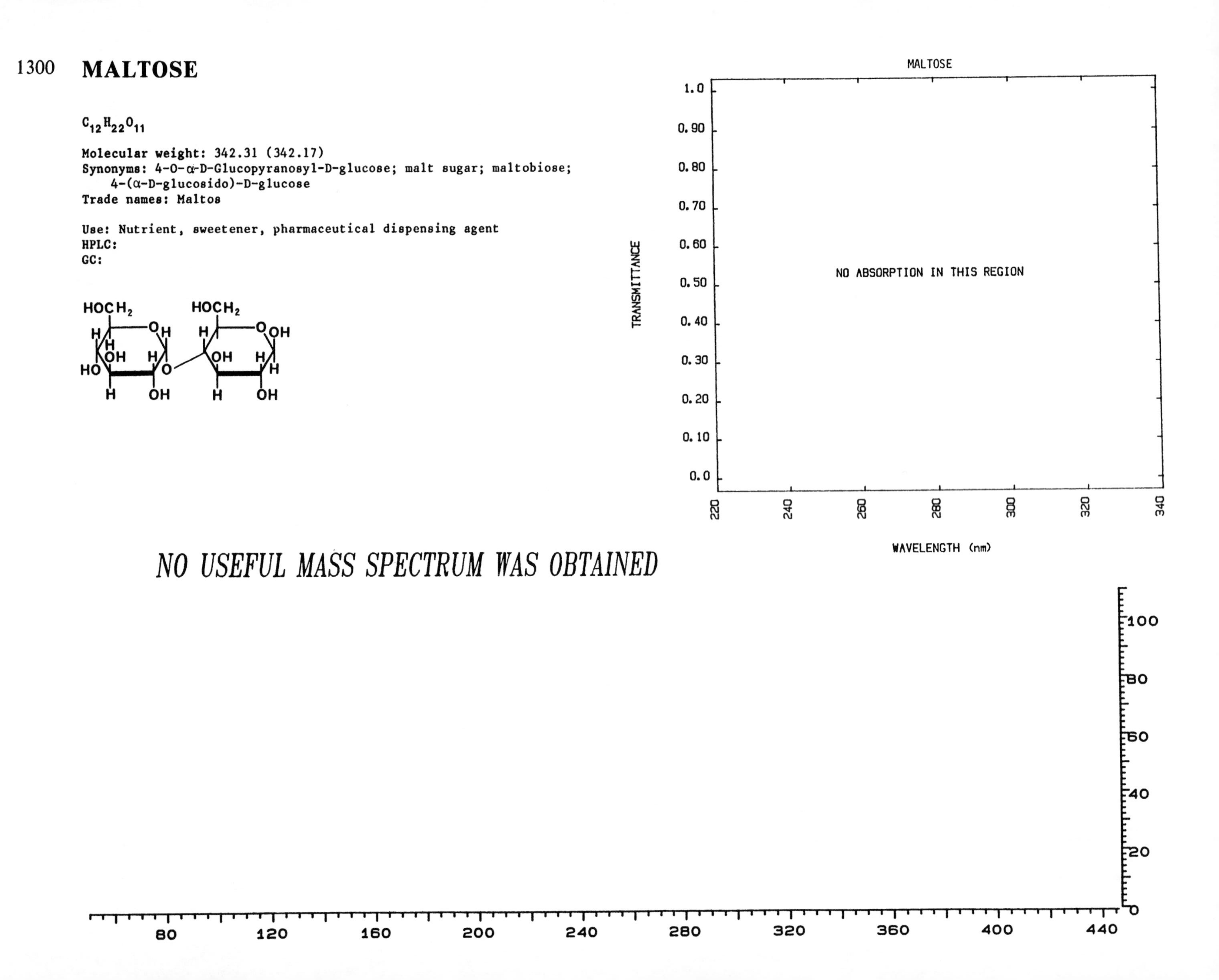

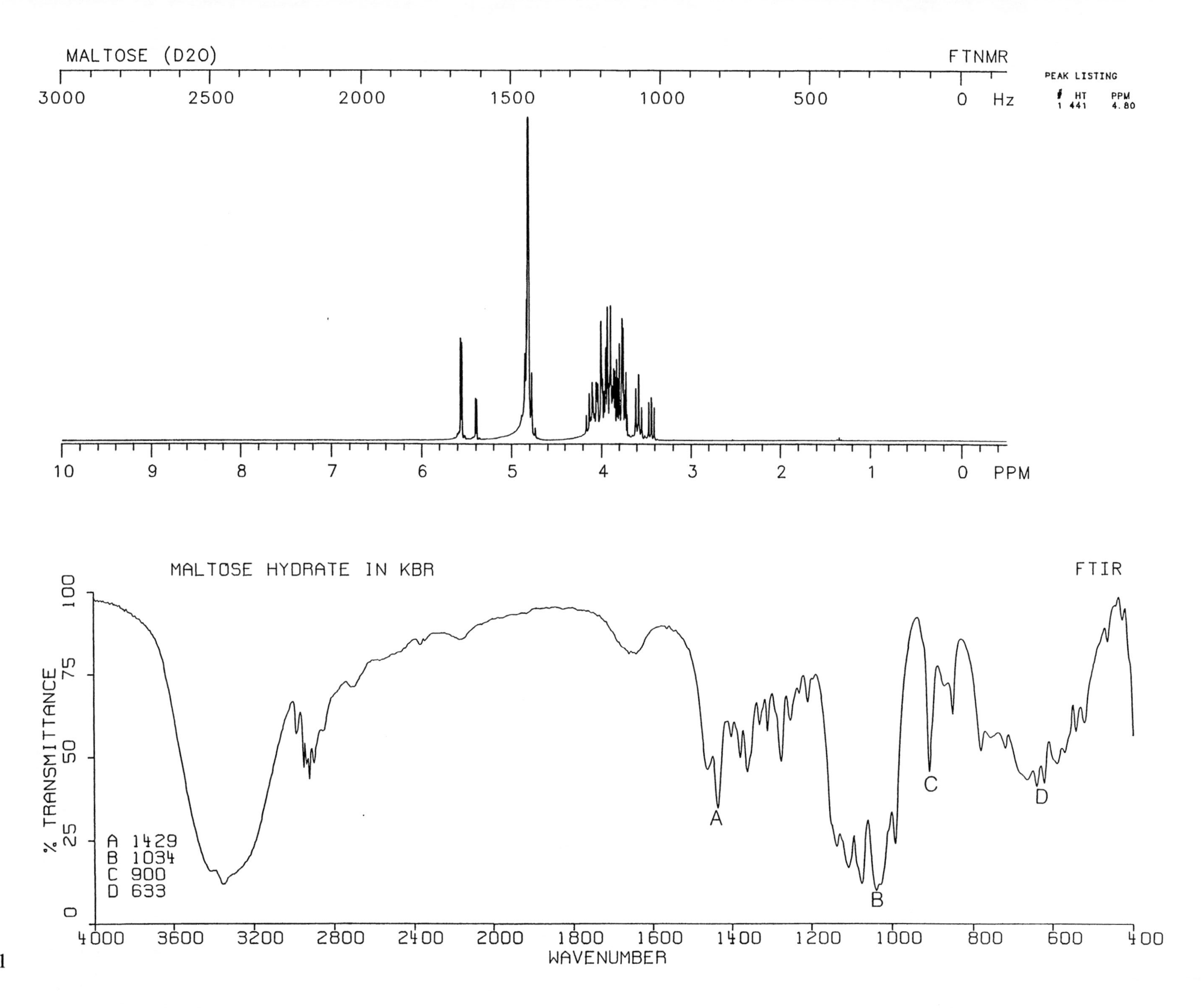
MALTOSE (D2O)
FTNMR
PEAK LISTING
# HT PPM
1 441 4.80
3000 2500 2000 1500 1000 500 0 Hz
10 9 8 7 6 5 4 3 2 1 0 PPM
MALTOSE HYDRATE IN KBR
FTIR
% TRANSMITTANCE
0 25 50 75 100
A 1429
B 1034
C 900
D 633
A
B
C
D
4000 3600 3200 2800 2400 2000 1800 1600 1400 1200 1000 800 600 400
WAVENUMBER

# MANDELIC ACID

$C_8H_8O_3$

**Molecular weight:** 152.40 (152.05)
**Synonyms:** α-Hydroxybenzeneacetic acid; α-hydroxy-α-toluic acid; α-hydroxy phenylacetic acid; phenylhydroxyacetic acid; phenylglycolic acid; amylgdalic acid; amygdalinic acid; paramandelic acid
**Trade names:** Camdelate, Uromaline

**Use:** Urinary antiseptic
**HPLC:** Si-10; 20A:80B; 3.6
**GC:** 1496; 200°C

OH
|
$C_6H_5$–CH–COOH

MANDELIC ACID

TRANSMITTANCE

WAVELENGTH (nm)

——— 251,257,262
– – – 252,257,263

MANDELIC ACID

51 63 77 79 91 107 118 135 152 154

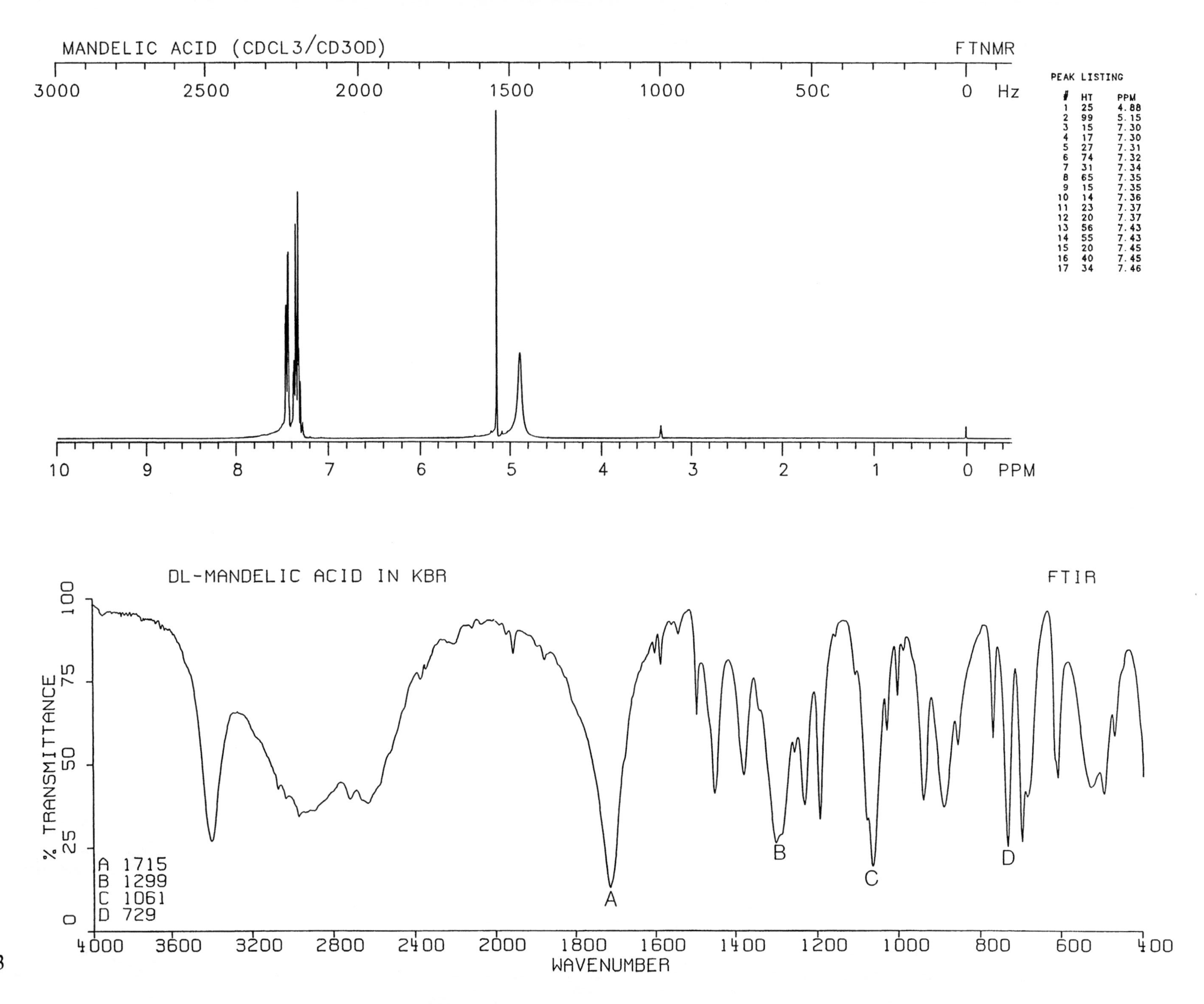

MANDELIC ACID (CDCL3/CD3OD)
FTNMR
3000
2500
2000
1500
1000
50C
0
Hz
PEAK LISTING
# HT PPM
1 25 4.88
2 99 5.15
3 15 7.30
4 17 7.30
5 27 7.31
6 74 7.32
7 31 7.34
8 65 7.35
9 15 7.35
10 14 7.36
11 23 7.37
12 20 7.37
13 56 7.43
14 55 7.43
15 20 7.45
16 40 7.45
17 34 7.46
10
9
8
7
6
5
4
3
2
1
0
PPM
DL-MANDELIC ACID IN KBR
FTIR
% TRANSMITTANCE
100
75
50
25
0
A 1715
B 1299
C 1061
D 729
A
B
C
D
4000
3600
3200
2800
2400
2000
1800
1600
1400
1200
1000
800
600
400
WAVENUMBER

# MANNITOL

$C_6H_{14}O_6$

Molecular weight: 182.17 (182.08)
Synonyms: d-Mannitol; mannite; manna sugar

Trade names: Manicol, Mannidex, Diosmol, Osmitrol, Osmosal

Use: Diuretic
HPLC: Li-$NH_2$; 70D:30E; 6.8
GC:

$CH_2OH$
HOCH
HOCH
HCOH
HCOH
$CH_2OH$

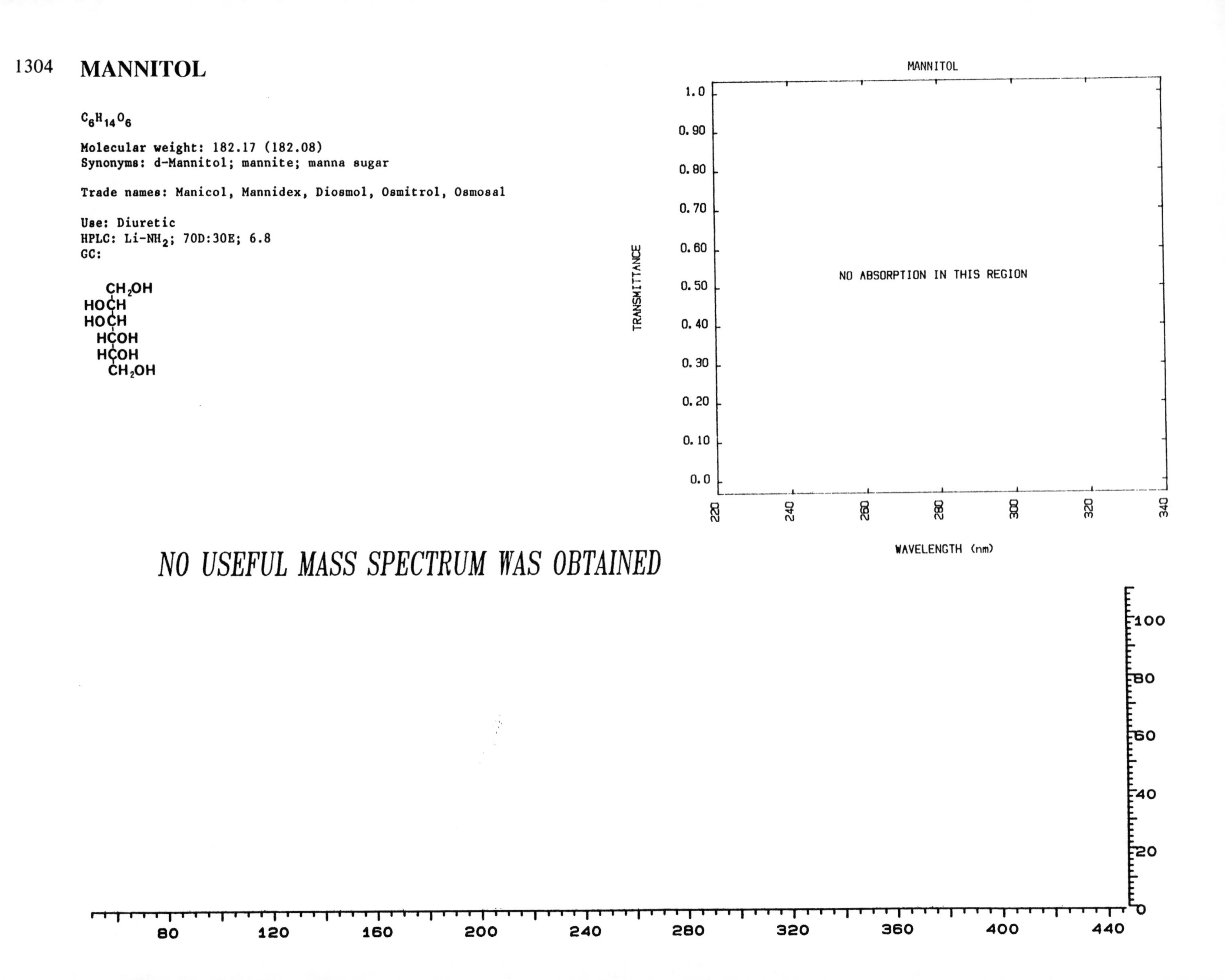

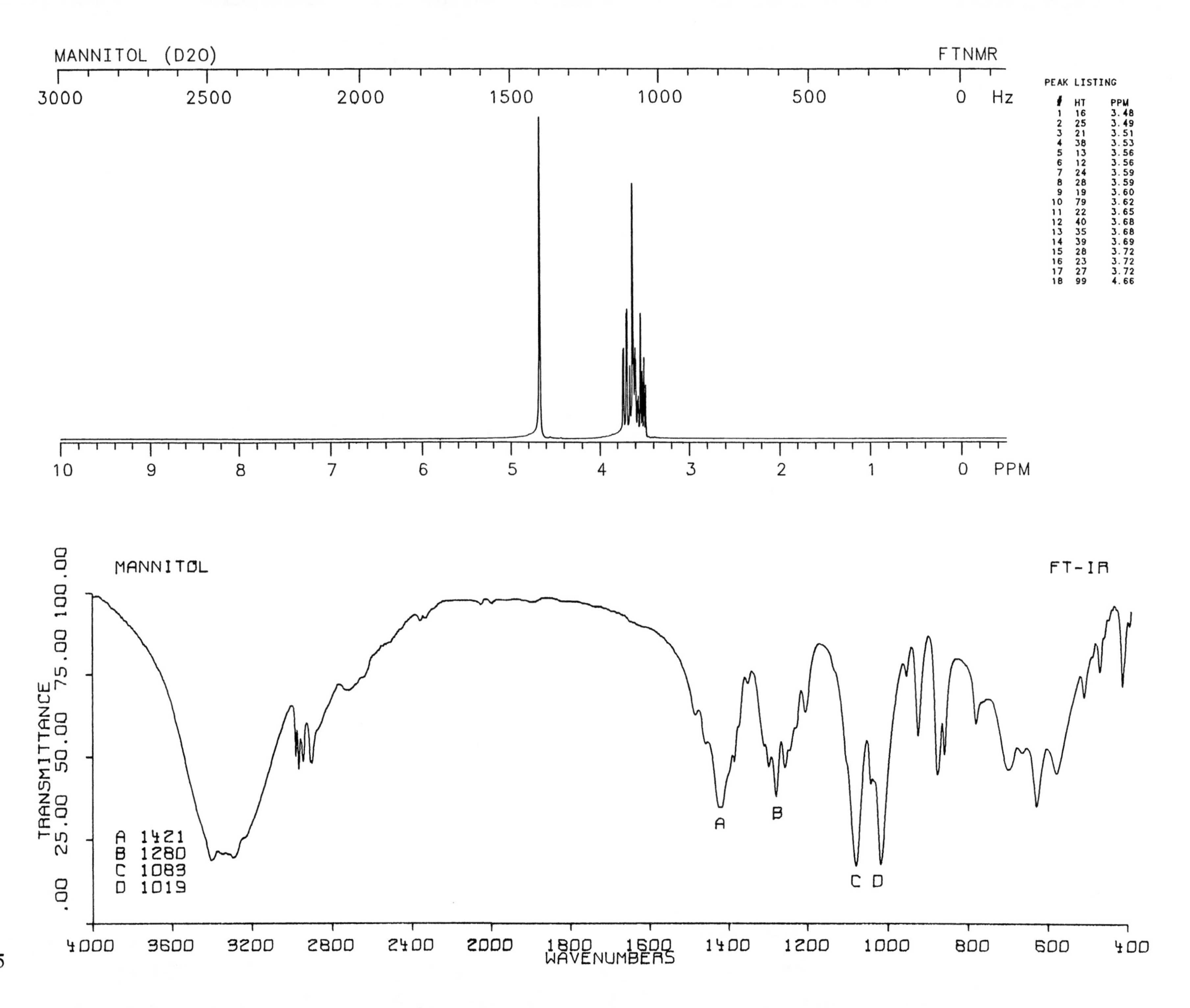

MANNITOL (D2O)
FTNMR
3000 2500 2000 1500 1000 500 0 Hz
PEAK LISTING
# HT PPM
1 16 3.48
2 25 3.49
3 21 3.51
4 38 3.53
5 13 3.56
6 12 3.56
7 24 3.59
8 28 3.59
9 19 3.60
10 79 3.62
11 22 3.65
12 40 3.68
13 35 3.68
14 39 3.69
15 28 3.72
16 23 3.72
17 27 3.72
18 99 4.66
10 9 8 7 6 5 4 3 2 1 0 PPM
MANNITOL
FT-IR
TRANSMITTANCE
.00 25.00 50.00 75.00 100.00
A 1421
B 1280
C 1083
D 1019
A B C D
4000 3600 3200 2800 2400 2000 1800 1600 1400 1200 1000 800 600 400
WAVENUMBERS

# *D*-MANNOSE

$C_6H_{12}O_6$

Molecular weight: 180.16 (180.06)
Synonyms: Seminose; carubinose

Trade names:

Use: Sugar
HPLC:
GC:

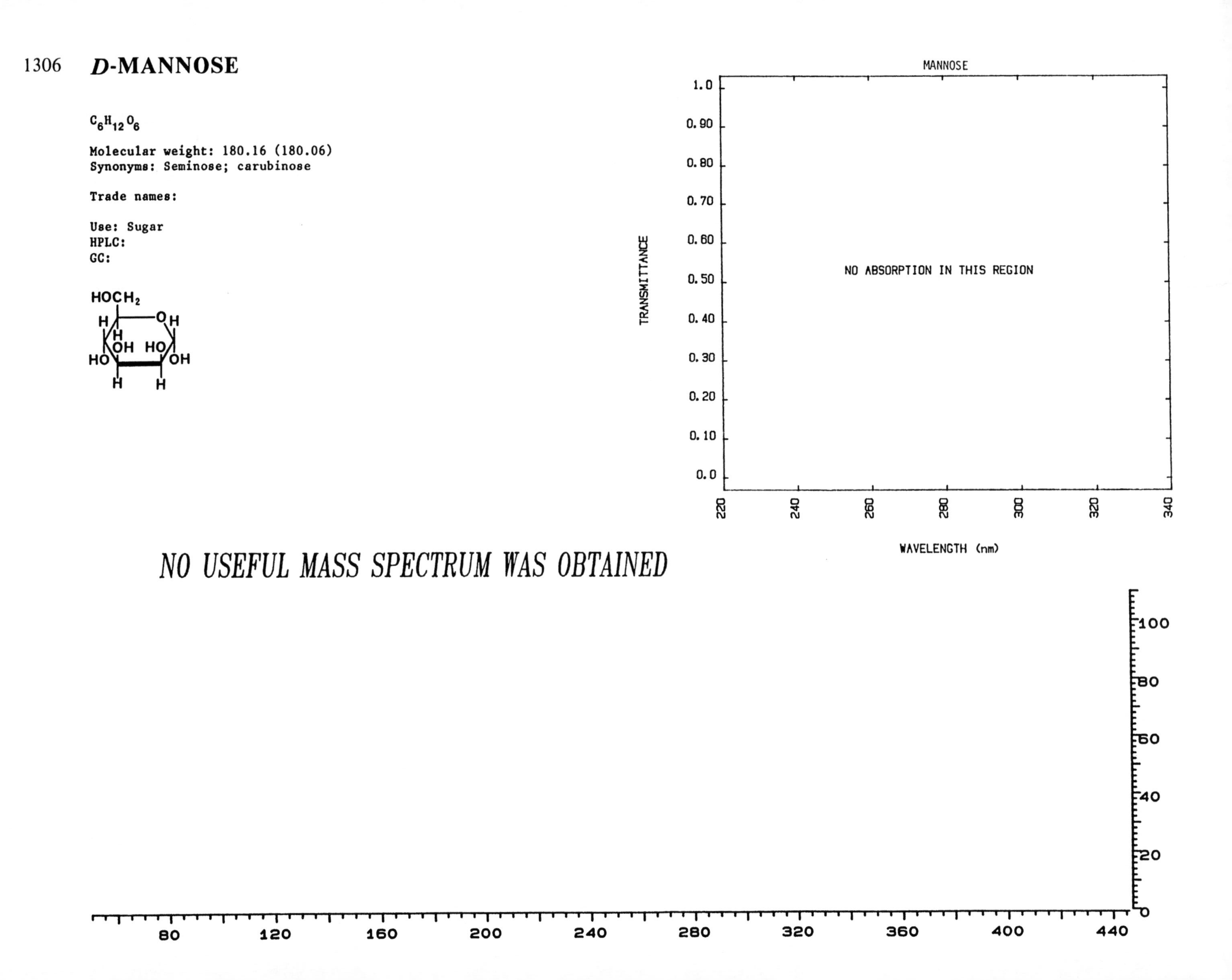


NO USEFUL MASS SPECTRUM WAS OBTAINED

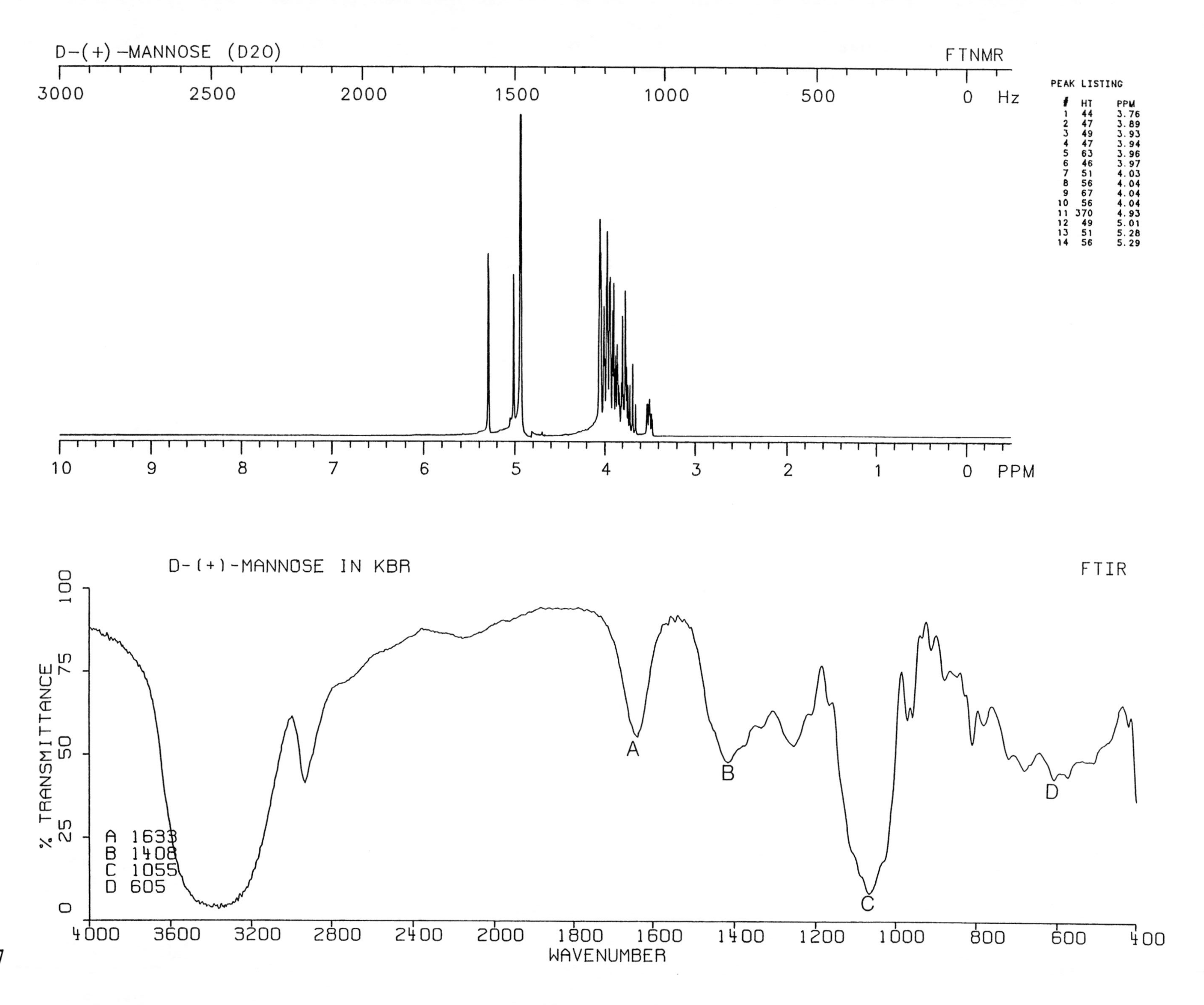
D-(+)-MANNOSE (D2O)
FTNMR
3000 2500 2000 1500 1000 500 0 Hz
10 9 8 7 6 5 4 3 2 1 0 PPM
PEAK LISTING
# HT PPM
1 44 3.76
2 47 3.89
3 49 3.93
4 47 3.94
5 63 3.96
6 46 3.97
7 51 4.03
8 56 4.04
9 67 4.04
10 56 4.04
11 370 4.93
12 49 5.01
13 51 5.28
14 56 5.29
D-(+)-MANNOSE IN KBR
FTIR
% TRANSMITTANCE
0 25 50 75 100
A 1633
B 1408
C 1055
D 605
4000 3600 3200 2800 2400 2000 1800 1600 1400 1200 1000 800 600 400
WAVENUMBER

# MAPROTILINE

$C_{20}H_{23}N$

**Molecular weight:** 277.41 (277.18)
**Synonyms:** N-Methyl-9,10-ethanoanthracene-9(10H)-propanamine

**Trade names:** Ludiomil

**Use:** Antidepressant
**HPLC:** Si-10; 20A:80B; 6.0
**GC:** 2380; 250°C

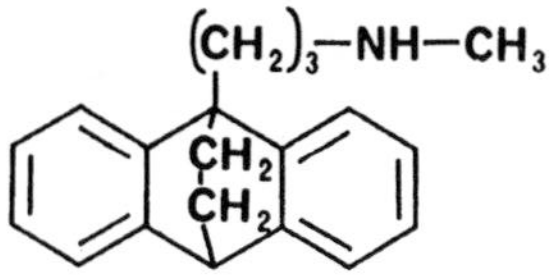

(CH2)3—NH—CH3
CH2
CH2

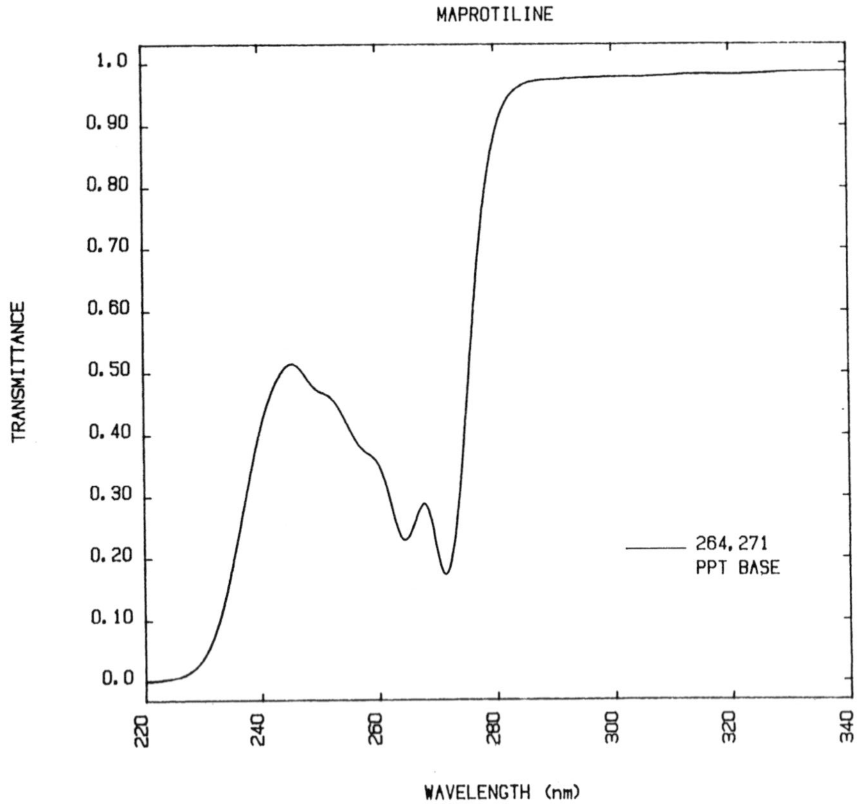

MAPROTILINE

TRANSMITTANCE

1.0 0.90 0.80 0.70 0.60 0.50 0.40 0.30 0.20 0.10 0.0

264, 271
PPT BASE

220 240 260 280 300 320 340

WAVELENGTH (nm)

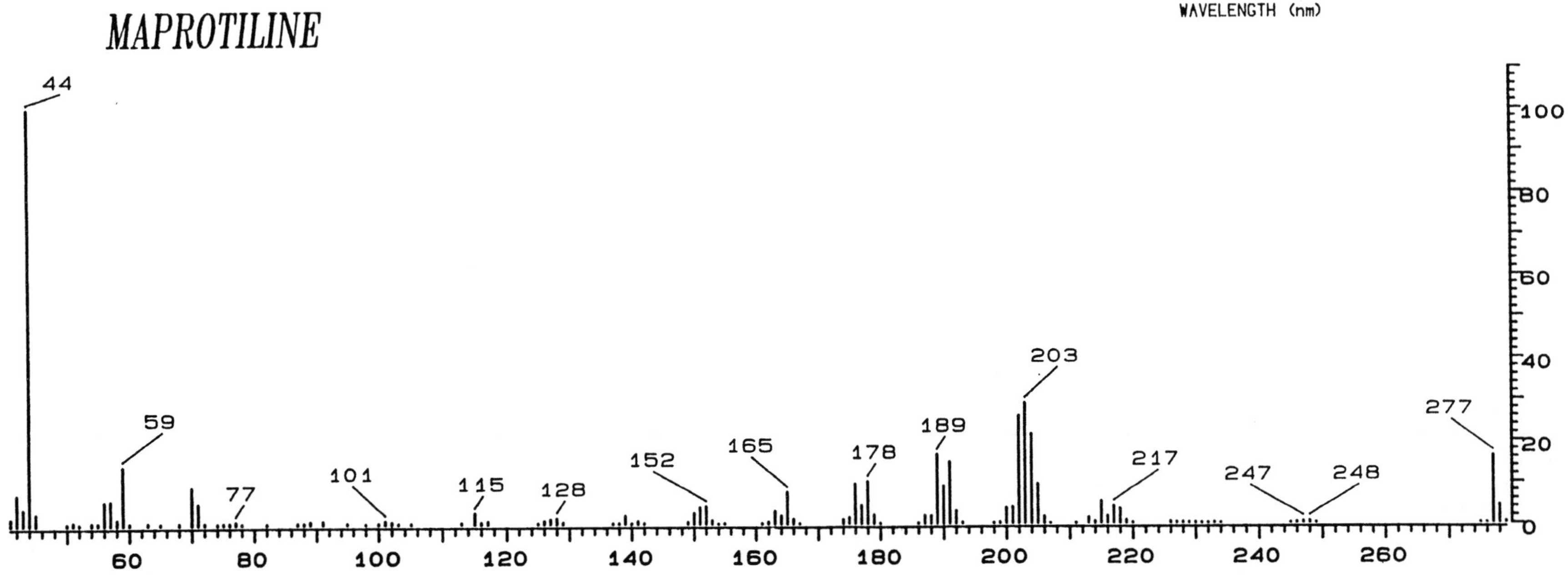

*MAPROTILINE*

44 59 77 101 115 128 152 165 178 189 203 217 247 248 277

60 80 100 120 140 160 180 200 220 240 260

100 80 60 40 20 0

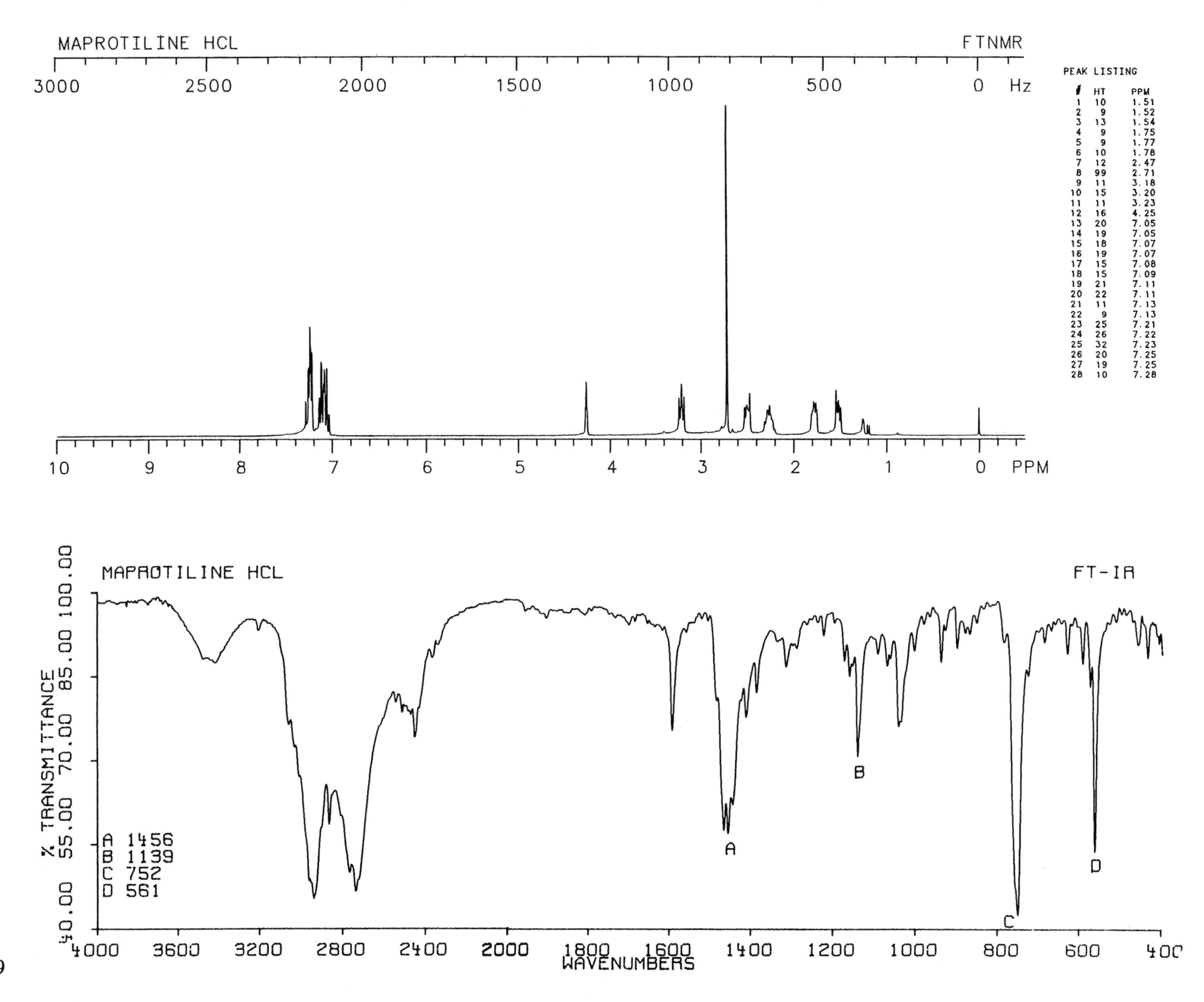
MAPROTILINE HCL
FTNMR
3000 2500 2000 1500 1000 500 0 Hz
PEAK LISTING
# HT PPM
1 10 1.51
2 9 1.52
3 13 1.54
4 9 1.75
5 9 1.77
6 10 1.78
7 12 2.47
8 99 2.71
9 11 3.18
10 15 3.20
11 11 3.23
12 16 4.25
13 20 7.05
14 19 7.05
15 18 7.07
16 19 7.07
17 15 7.08
18 15 7.09
19 21 7.11
20 22 7.11
21 11 7.13
22 9 7.13
23 25 7.21
24 26 7.22
25 32 7.23
26 20 7.25
27 19 7.25
28 10 7.28
10 9 8 7 6 5 4 3 2 1 0 PPM
MAPROTILINE HCL
FT-IR
% TRANSMITTANCE
40.00 55.00 70.00 85.00 100.00
A 1456
B 1139
C 752
D 561
A B C D
4000 3600 3200 2800 2400 2000 1800 1600 1400 1200 1000 800 600 400
WAVENUMBERS

# MAZINDOL

$C_{16}H_{13}ClN_2O$

**Molecular weight:** 284.74 (284.07)
**Synonyms:** 5-(4-Chlorophenyl)-2,5-dihydro-3H-imidazo[2,1-a]-isoindol-5-ol
**Trade names:** Sanorex

**Use:** Anorexic
**HPLC:** Si-10; 5A:95B; 4.7
**GC:** 2424; 250°C

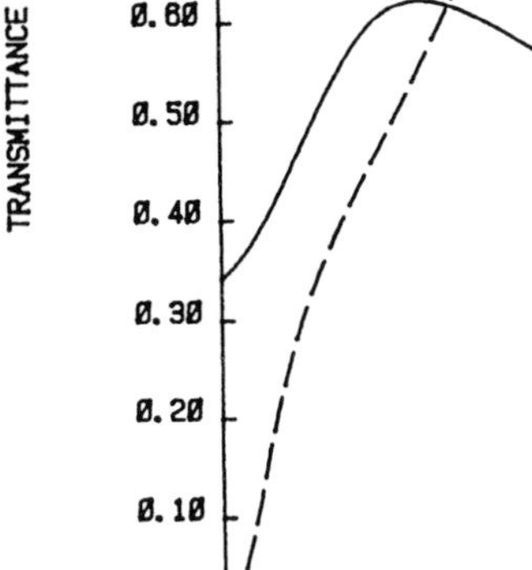

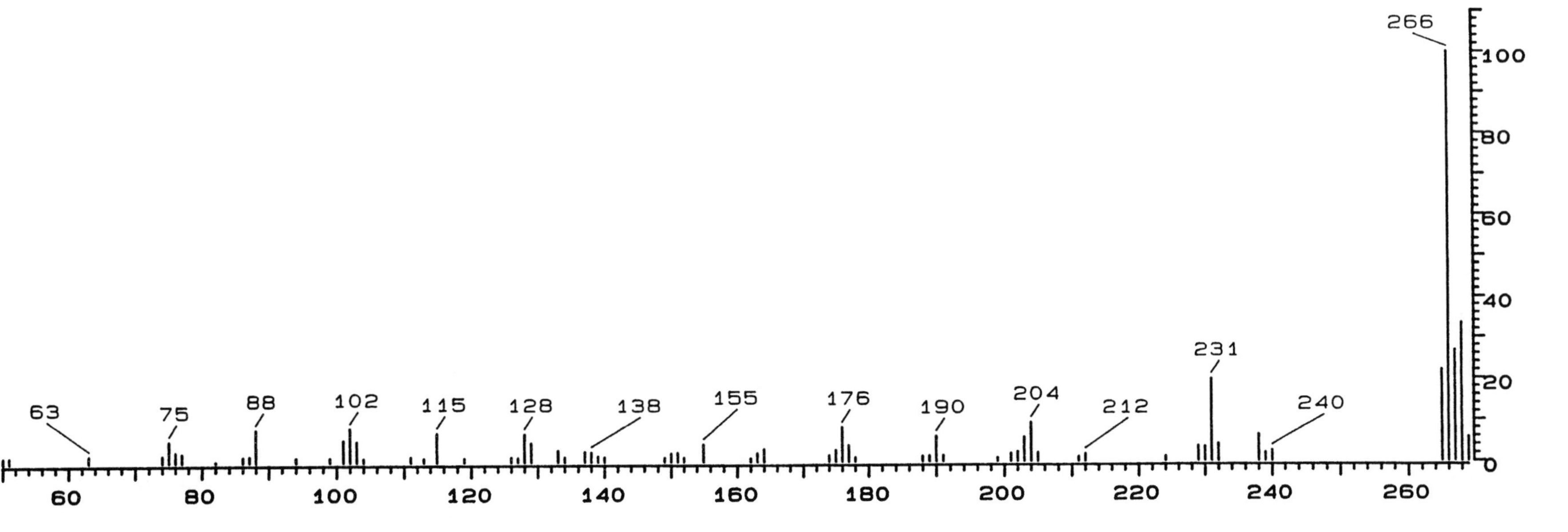

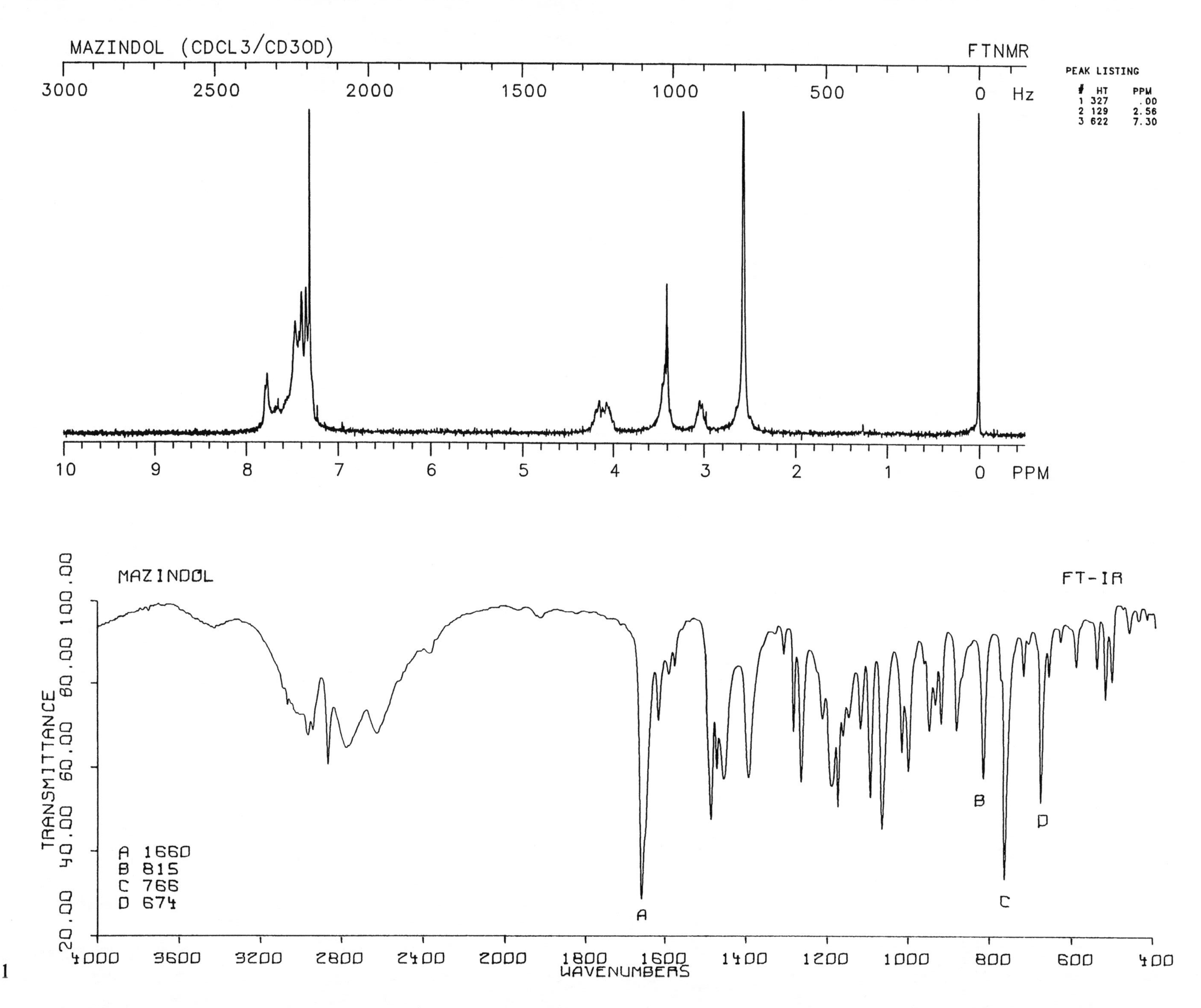
MAZINDOL (CDCL3/CD3OD)
FTNMR
PEAK LISTING
# HT PPM
1 327 .00
2 129 2.56
3 622 7.30
Hz
PPM
MAZINDOL
FT-IR
TRANSMITTANCE
WAVENUMBERS
A 1660
B 815
C 766
D 674

# MEBENDAZOLE

$C_{16}H_{13}N_3O_3$

**Molecular weight:** 295.30 (295.10)
**Synonyms:** (5-Benzoyl-1H-benzimidazol-2-yl)-carbamic acid methyl ester

**Trade names:** Vermox

**Use:** Anthelmintic
**HPLC:**
**GC:**

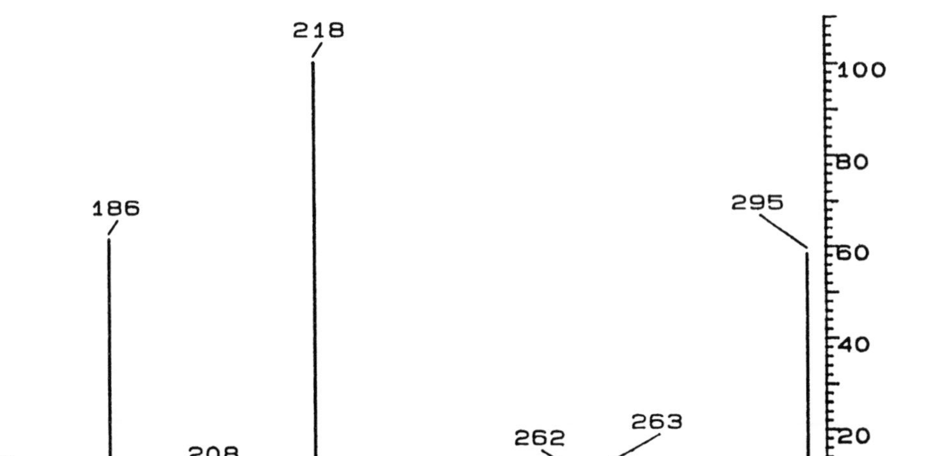


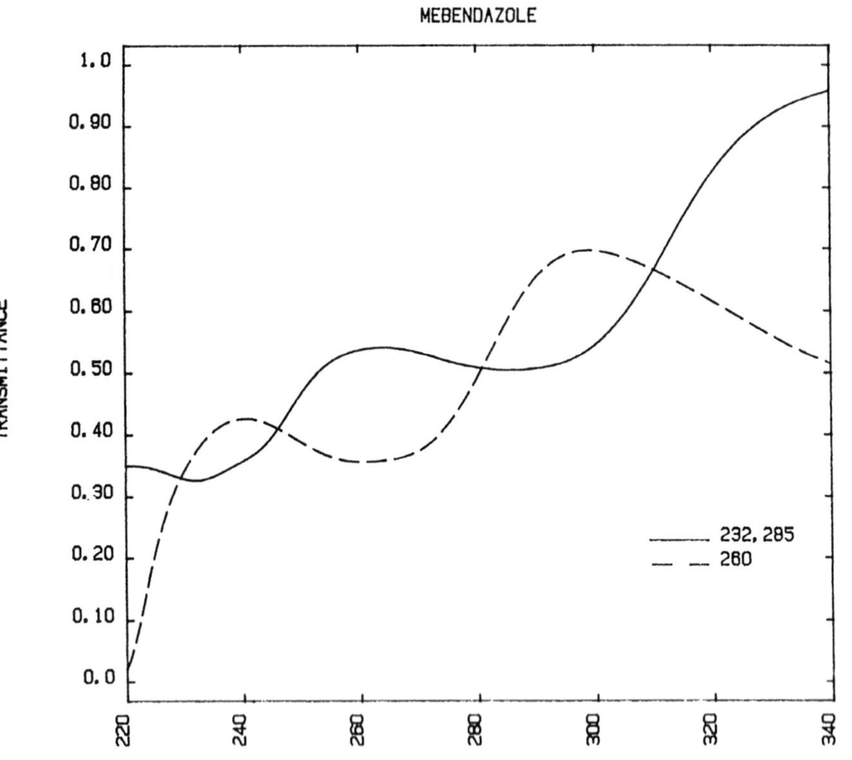


## MEBENDAZOLE--DIP

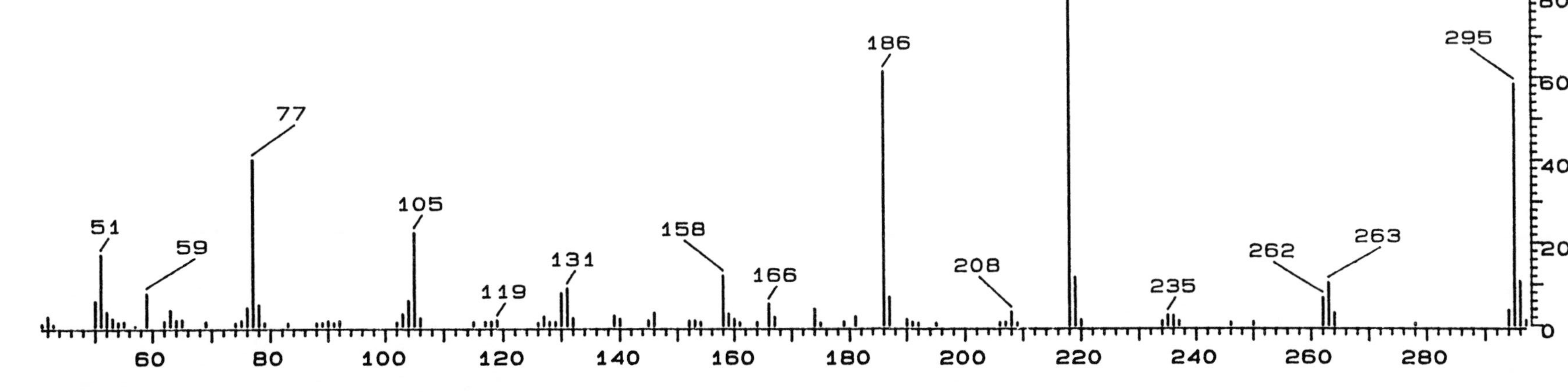

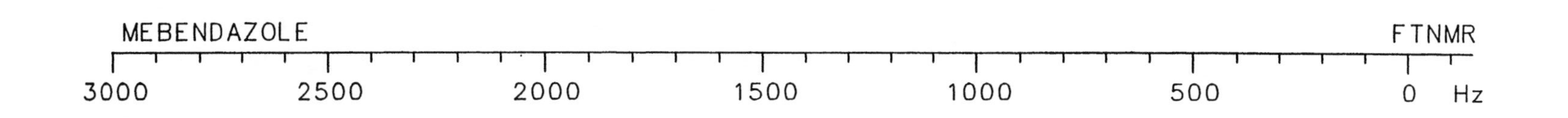


INSUFFICIENT SOLUBILITY

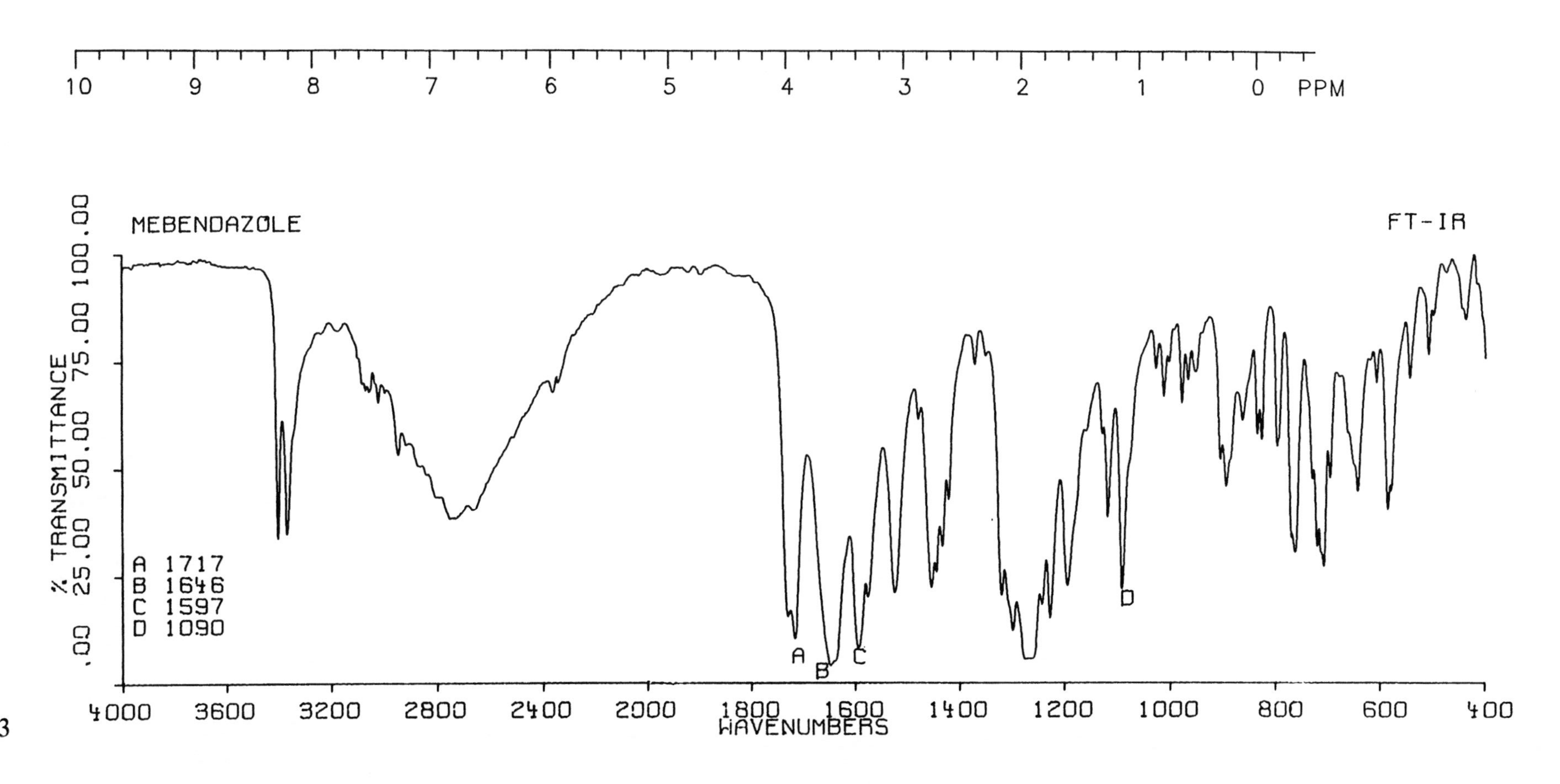

# MECAMYLAMINE

$C_{11}H_{21}N$

**Molecular weight:** 167.30 (167.17)
**Synonyms:** N,2,3,3-Tetramethylbicyclo[2.2.1]heptan-2-amine;
2-methylaminoisocamphane
**Trade names:** Inversine

**Use:** Antihypertensive
**HPLC:**
**GC:** 1821; 200°C

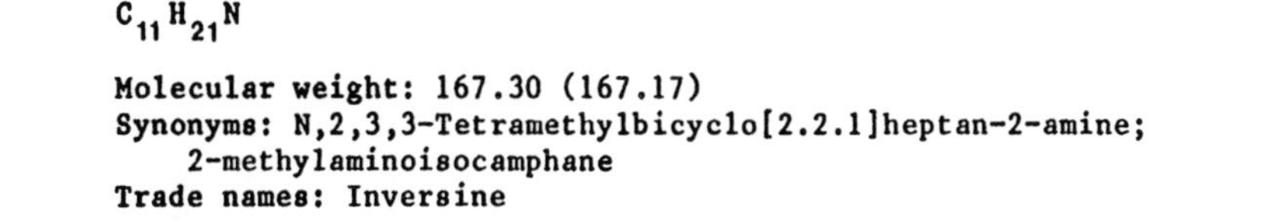


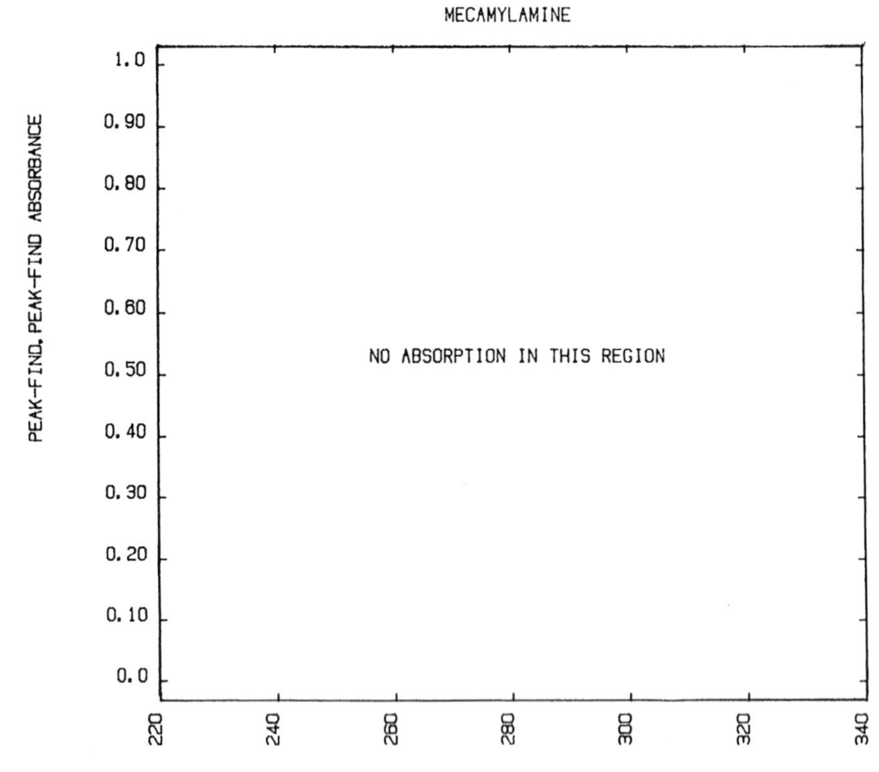

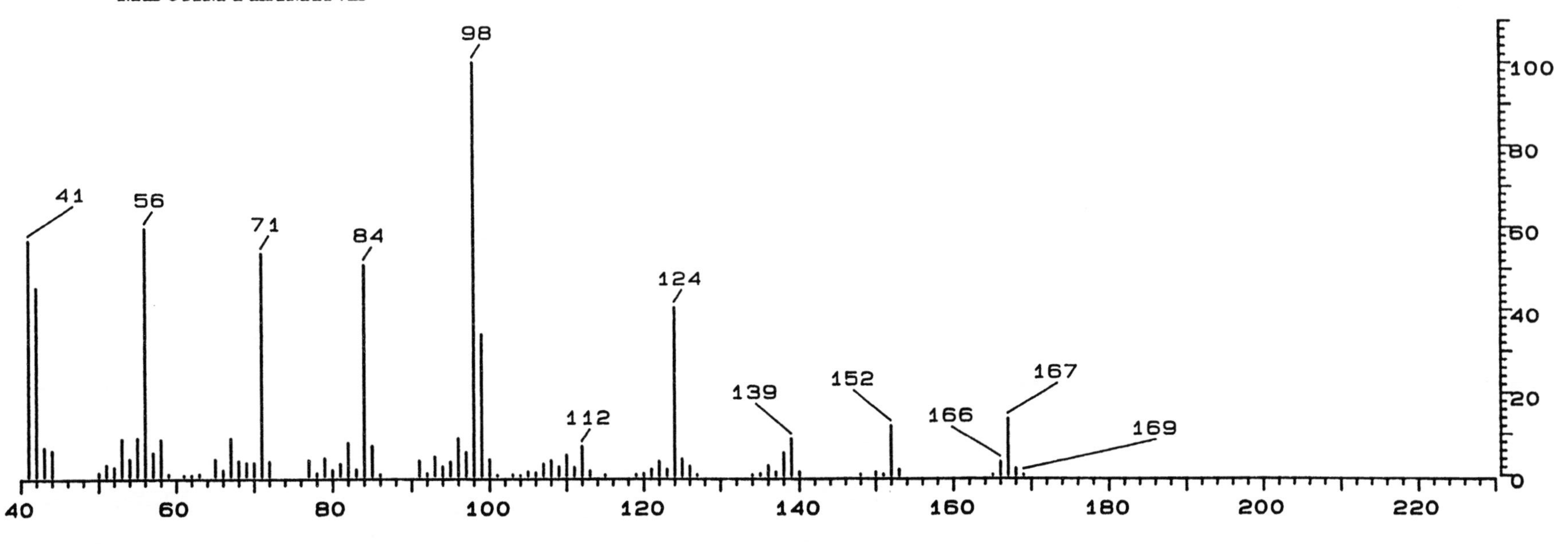

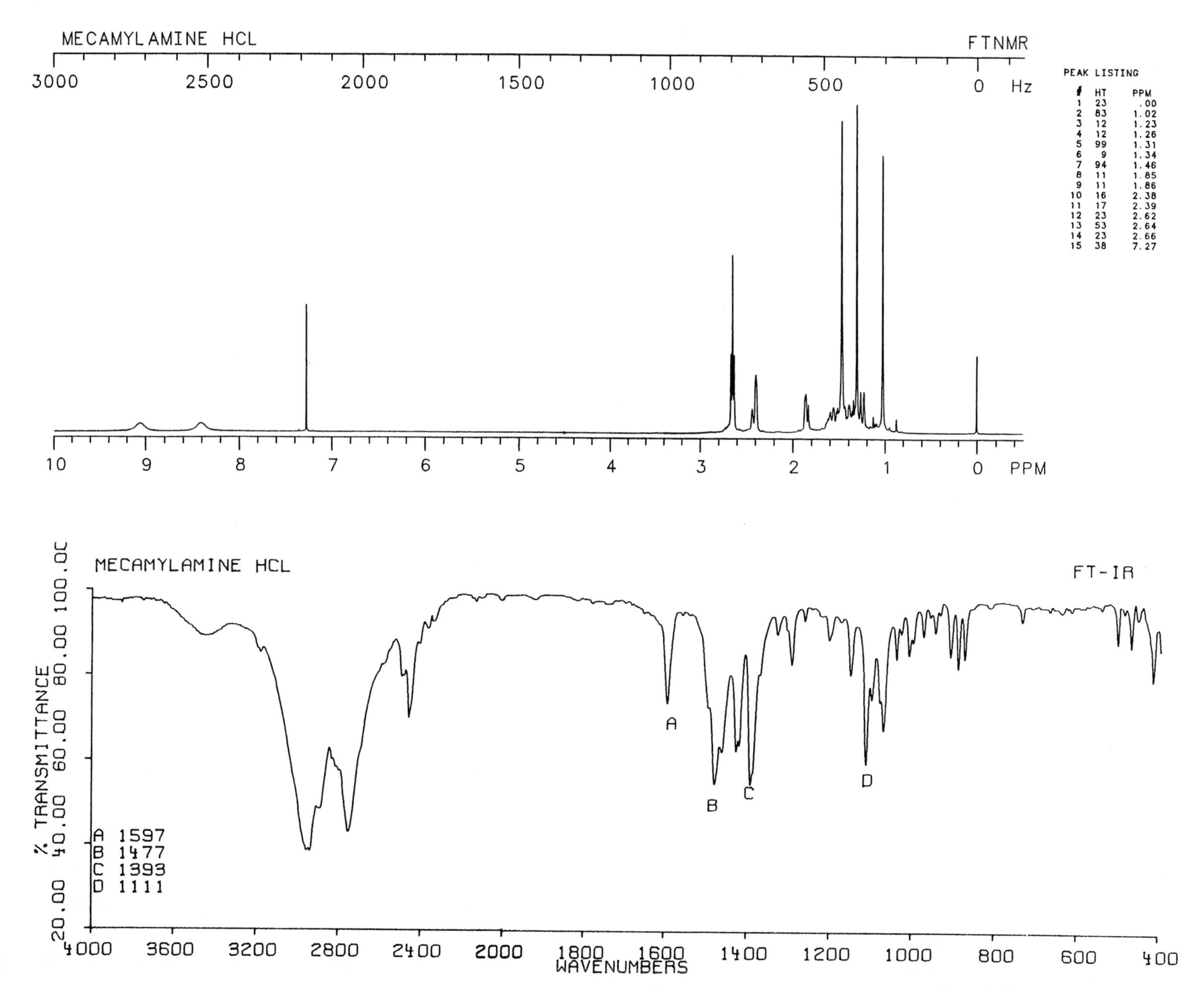
MECAMYLAMINE HCL
FTNMR
3000 2500 2000 1500 1000 500 0 Hz
10 9 8 7 6 5 4 3 2 1 0 PPM
PEAK LISTING
# HT PPM
1 23 .00
2 83 1.02
3 12 1.23
4 12 1.26
5 99 1.31
6 9 1.34
7 94 1.46
8 11 1.85
9 11 1.86
10 16 2.38
11 17 2.39
12 23 2.62
13 53 2.64
14 23 2.66
15 38 7.27
MECAMYLAMINE HCL
FT-IR
% TRANSMITTANCE
20.00 40.00 60.00 80.00 100.0C
4000 3600 3200 2800 2400 2000 1800 1600 1400 1200 1000 800 600 400
WAVENUMBERS
A B C D
A 1597
B 1477
C 1393
D 1111

# MECLIZINE

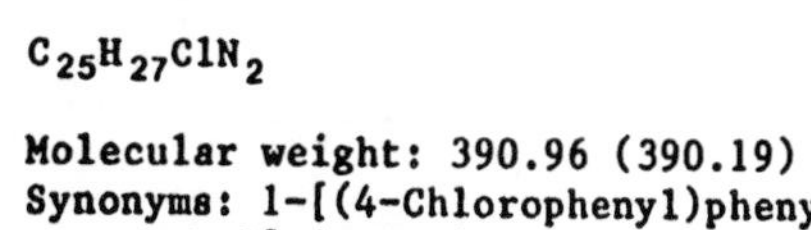

$C_{25}H_{27}ClN_2$

**Molecular weight:** 390.96 (390.19)
**Synonyms:** 1-[(4-Chlorophenyl)phenylmethyl]-4-[(3-methylphenyl)-methyl]piperazine; meclozine
**Trade names:** Antivert, Bonine, Meclizine

**Use:** Antihistamine, antinauseant
**HPLC:** Si-10; 2A:98B; 7.0
**GC:** 3081; 280°C

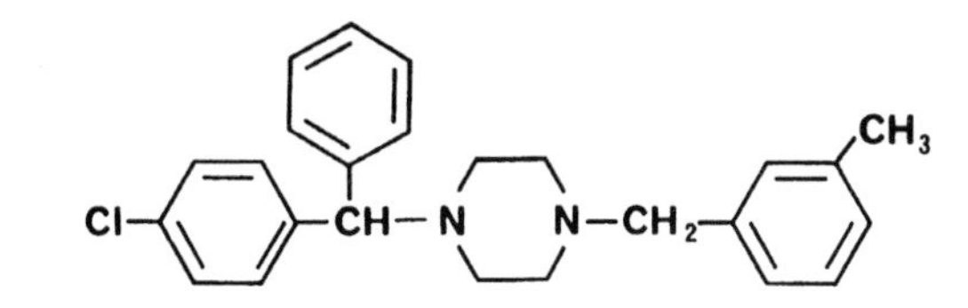


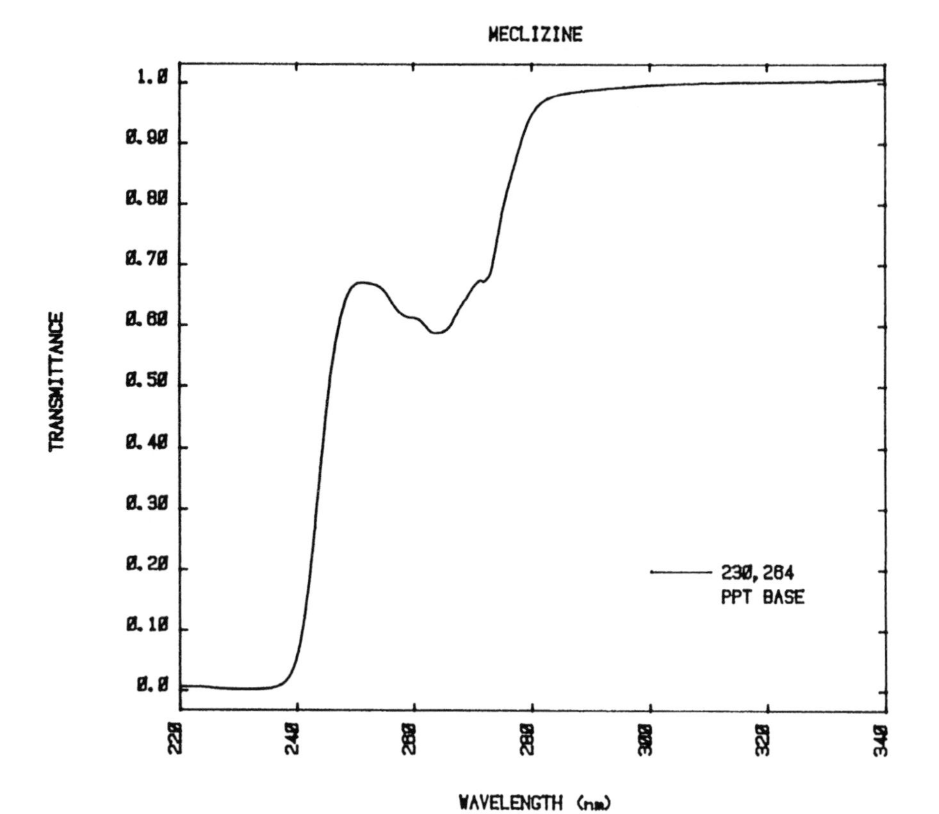


MECLIZINE

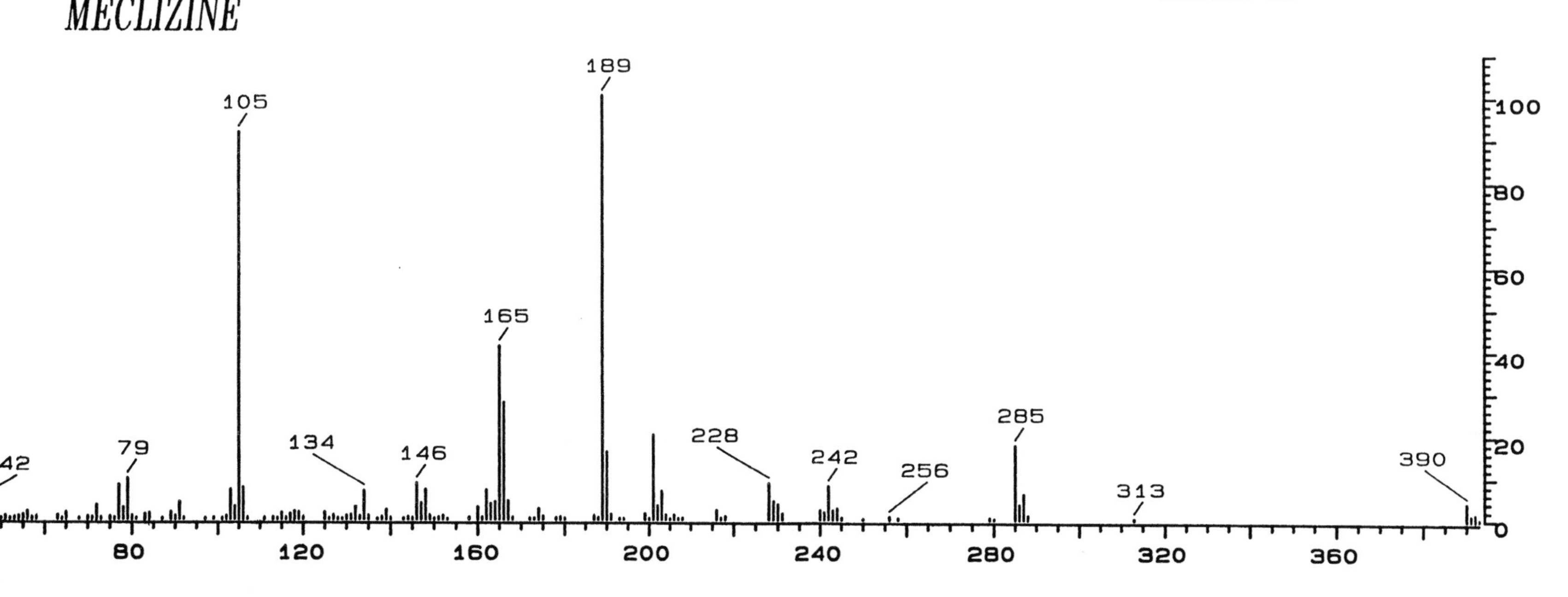

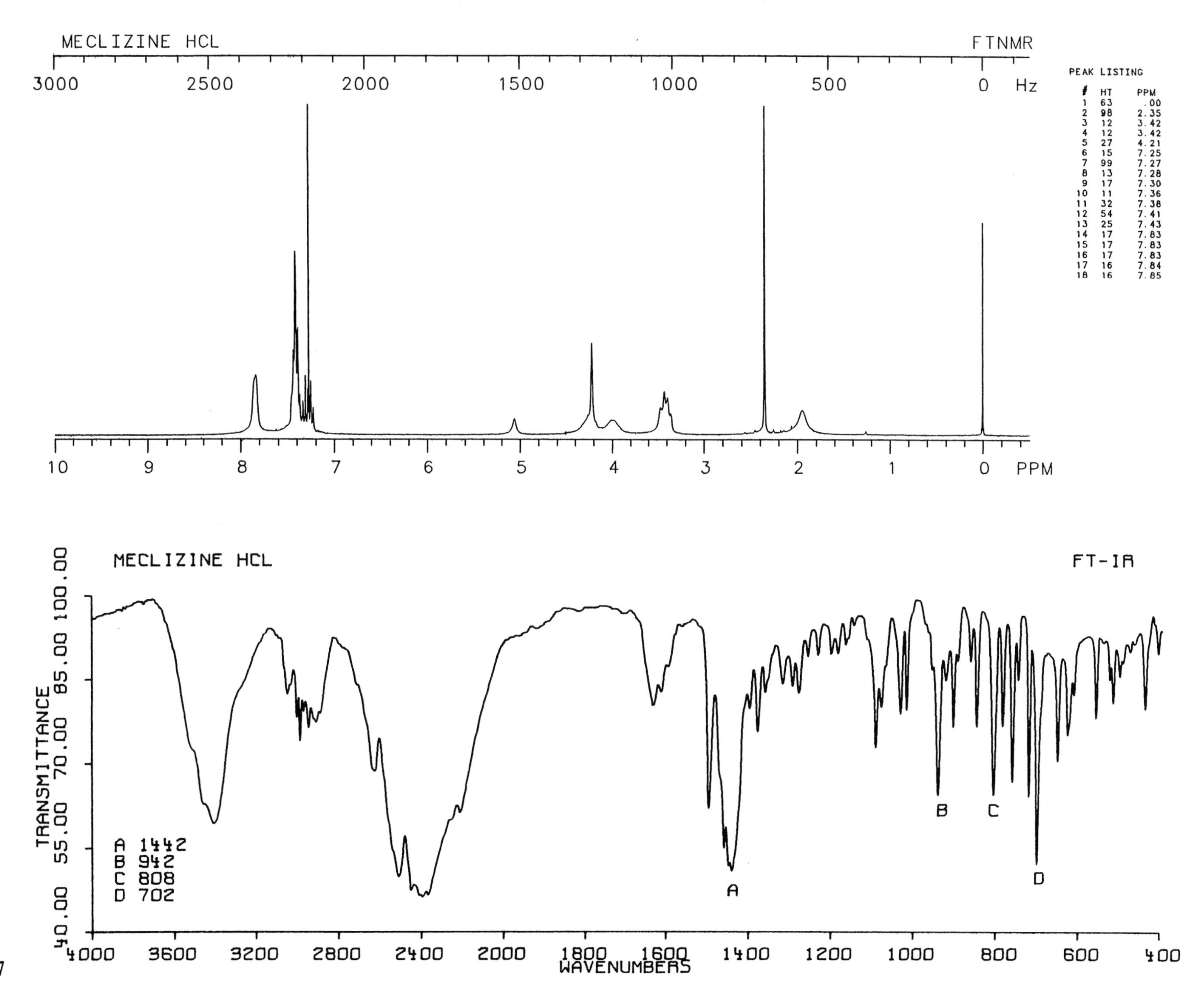
MECLIZINE HCL
FTNMR
Hz
PPM
PEAK LISTING
# HT PPM
1 63 .00
2 98 2.35
3 12 3.42
4 12 3.42
5 27 4.21
6 15 7.25
7 99 7.27
8 13 7.28
9 17 7.30
10 11 7.36
11 32 7.38
12 54 7.41
13 25 7.43
14 17 7.83
15 17 7.83
16 17 7.83
17 16 7.84
18 16 7.85
MECLIZINE HCL
FT-IR
TRANSMITTANCE
WAVENUMBERS
A 1442
B 942
C 808
D 702

# MECLOCYCLINE

$C_{22}H_{21}ClN_2O_8$

**Molecular weight:** 476.87 (476.10)
**Synonyms:** 7-Chloro-4-(dimethylamino)-1,4,4a,5,5a,6,11,11a-octahydro-3,5,10,12,12a-pentahydroxy-6-methylene-1,11-dioxo-2-naphthacene-carboxamide; 7-chloro-6-methylene-5-hydroxytetracycline
**Trade names:** Meclan, Mecloderm, Meclutin, Meclosorb, Traumatociclina

**Use:** Antibacterial
**HPLC:**
**GC:**

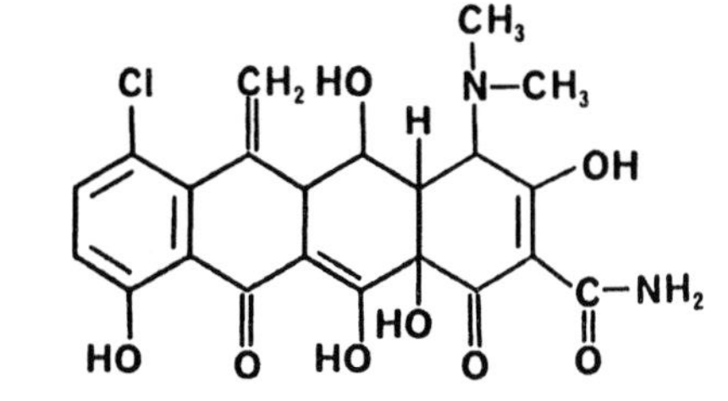


MECLOCYCLINE

TRANSMITTANCE

WAVELENGTH (nm)

244
233,251

MECLOCYCLINE--DIP

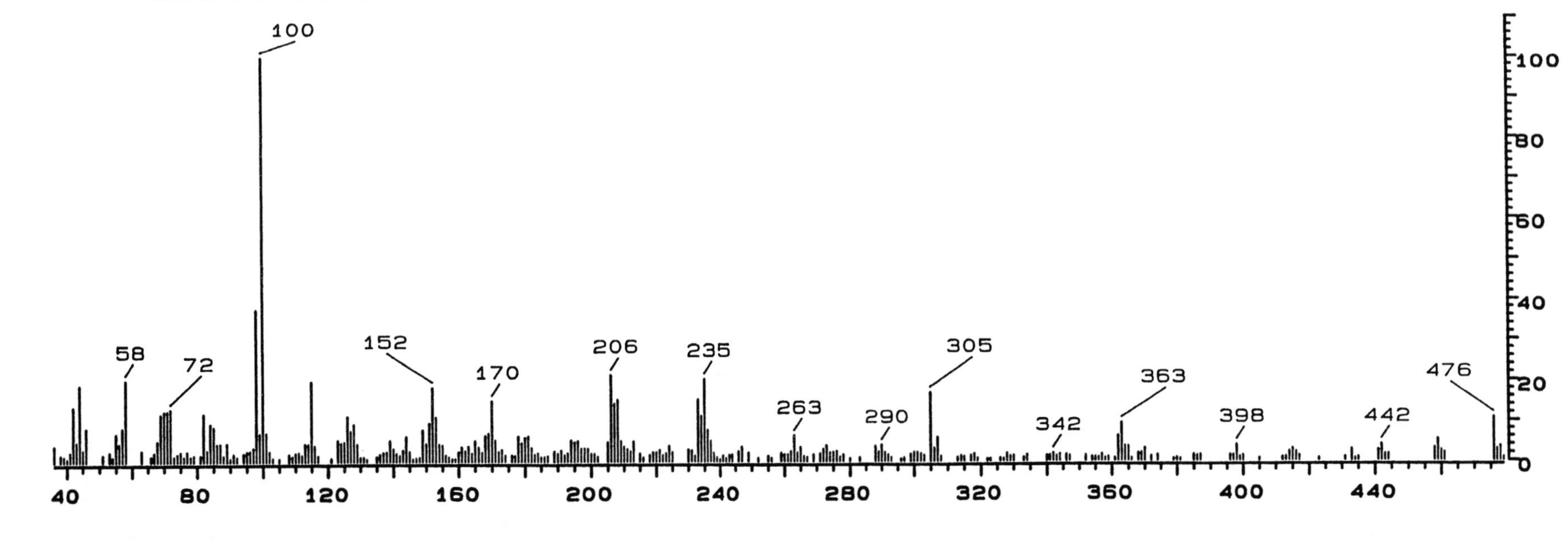

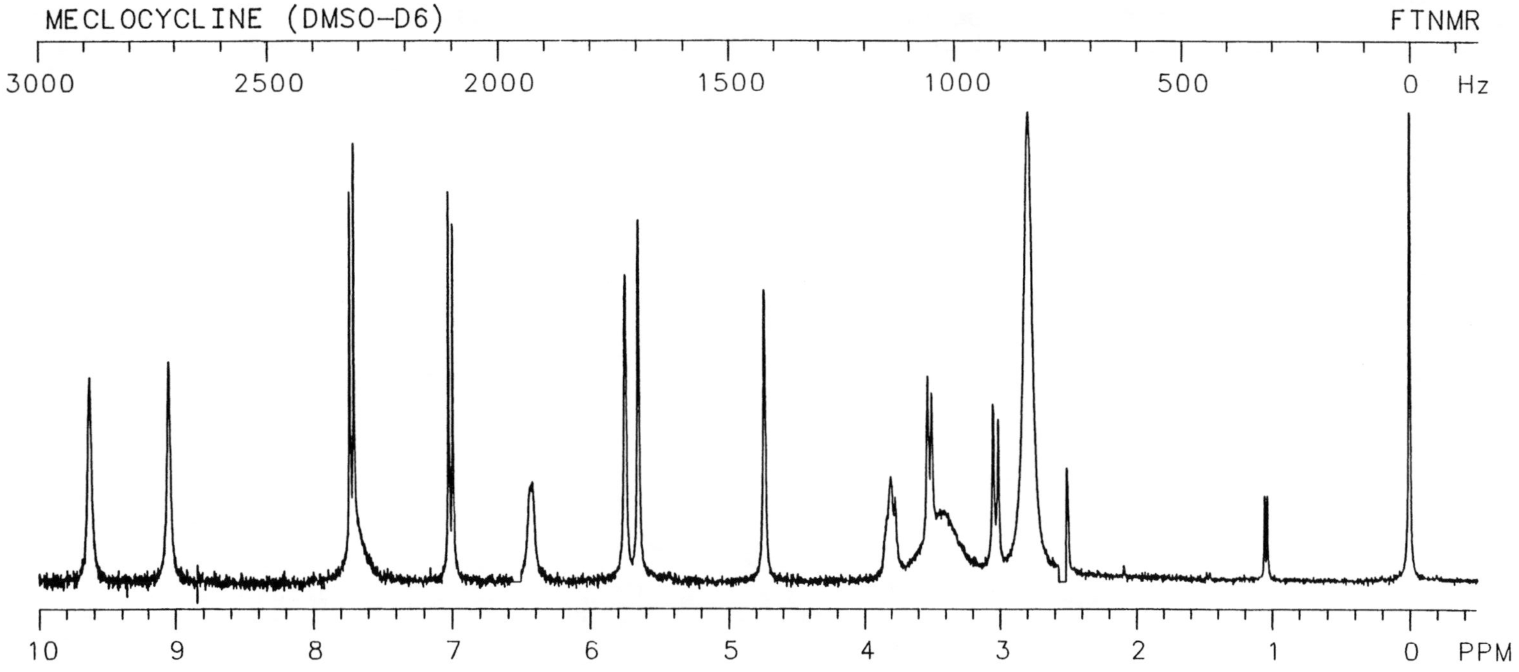

MECLOCYCLINE (DMSO-D6)
FTNMR
3000
2500
2000
1500
1000
500
0
Hz
10
9
8
7
6
5
4
3
2
1
0
PPM


PEAK LISTING

| # | HT | PPM |
|---|---|---|
| 1 | 200 | .00 |
| 2 | 23 | 2.50 |
| 3 | 95 | 2.79 |
| 4 | 95 | 2.79 |
| 5 | 94 | 2.79 |
| 6 | 33 | 3.01 |
| 7 | 36 | 3.05 |
| 8 | 38 | 3.50 |
| 9 | 42 | 3.53 |
| 10 | 59 | 4.73 |
| 11 | 73 | 5.65 |
| 12 | 62 | 5.74 |
| 13 | 72 | 6.99 |
| 14 | 79 | 7.02 |
| 15 | 88 | 7.70 |
| 16 | 79 | 7.73 |
| 17 | 44 | 9.04 |
| 18 | 43 | 9.04 |
| 19 | 39 | 9.62 |
| 20 | 41 | 9.62 |

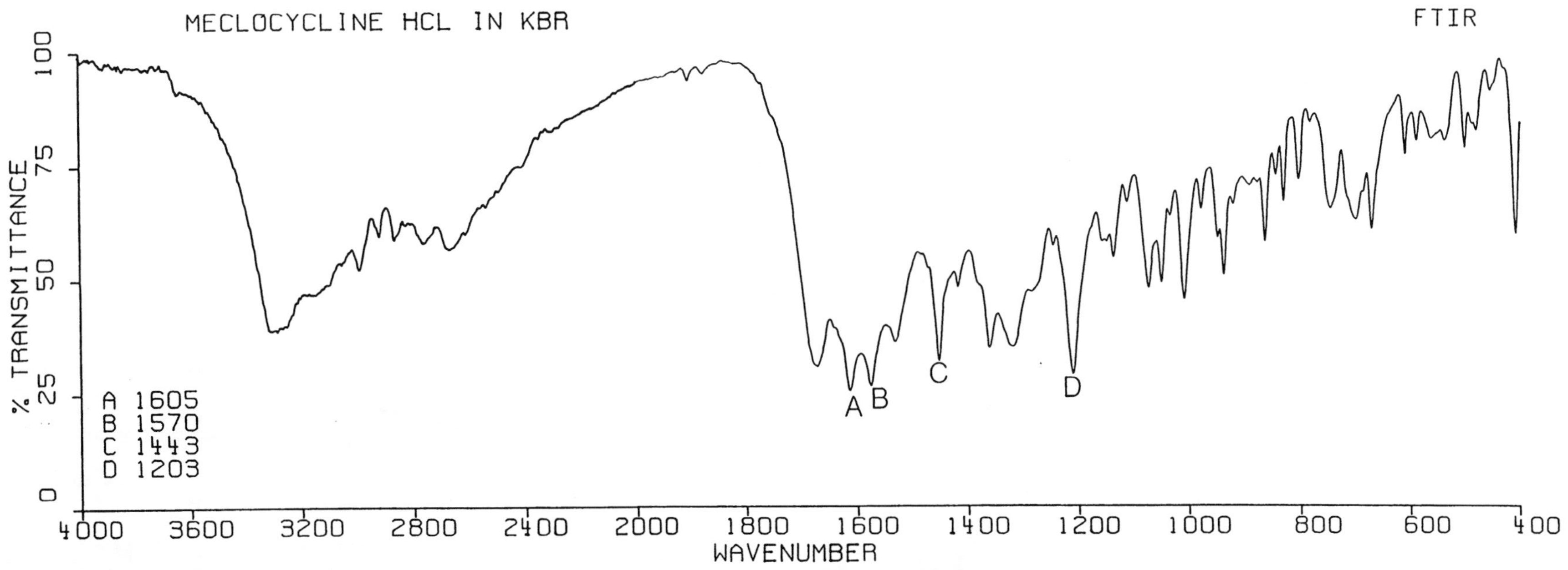

MECLOCYCLINE HCL IN KBR
FTIR
% TRANSMITTANCE
100
75
50
25
0
A 1605
B 1570
C 1443
D 1203
A
B
C
D
4000
3600
3200
2800
2400
2000
1800
1600
1400
1200
1000
800
600
400
WAVENUMBER

# MECLOFENAMIC ACID

$C_{14}H_{11}Cl_2NO_2$

Molecular weight: 296.15 (295.02)
Synonyms: 2-[(2,6-Dichloro-3-methylphenyl)amino]benzoic acid;
meclophenamic acid
Trade names: Meclomen

Use: Anti-inflammatory
HPLC: Si-10; 5A:95B; 6.0
GC:

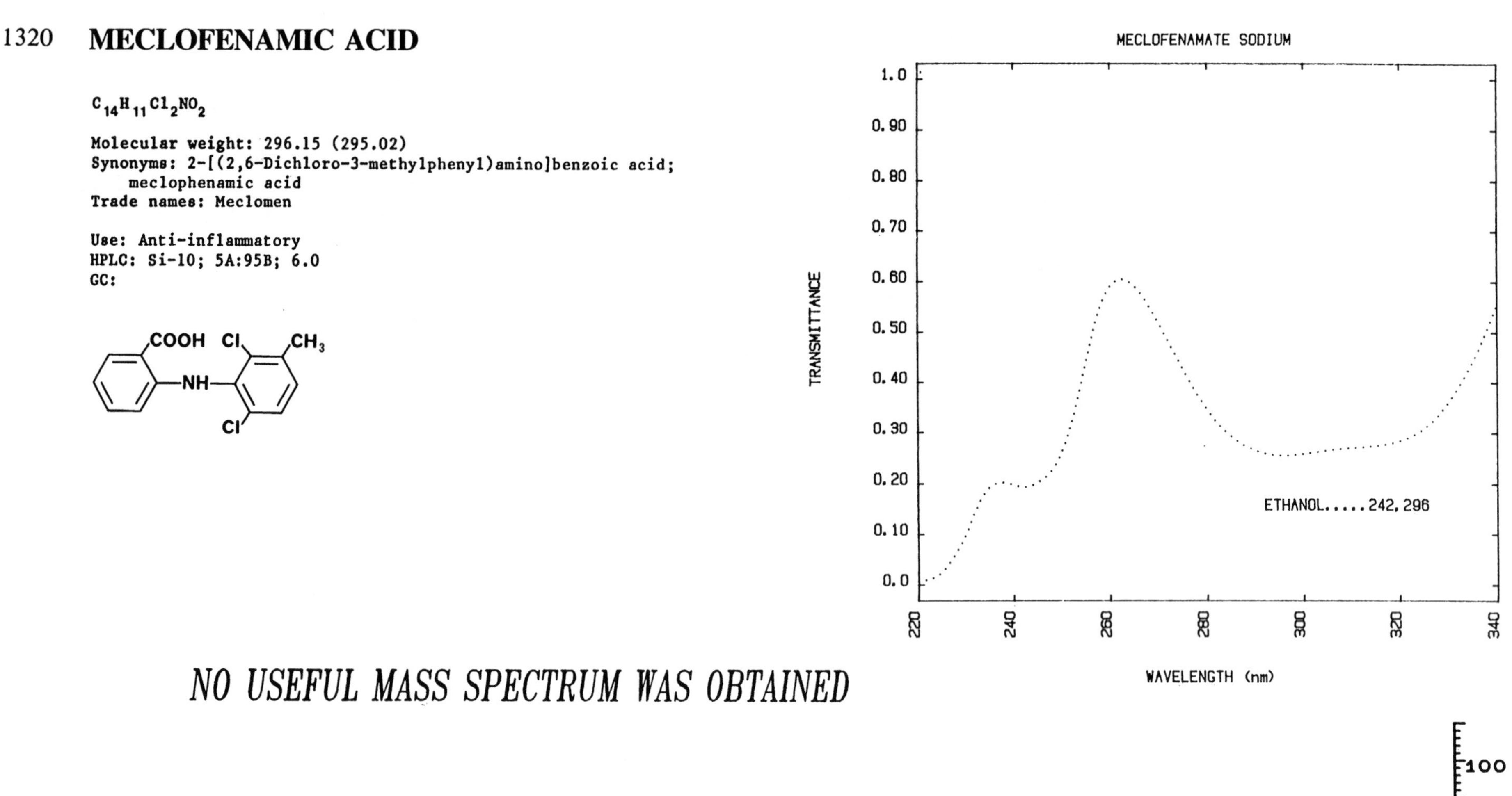


NO USEFUL MASS SPECTRUM WAS OBTAINED

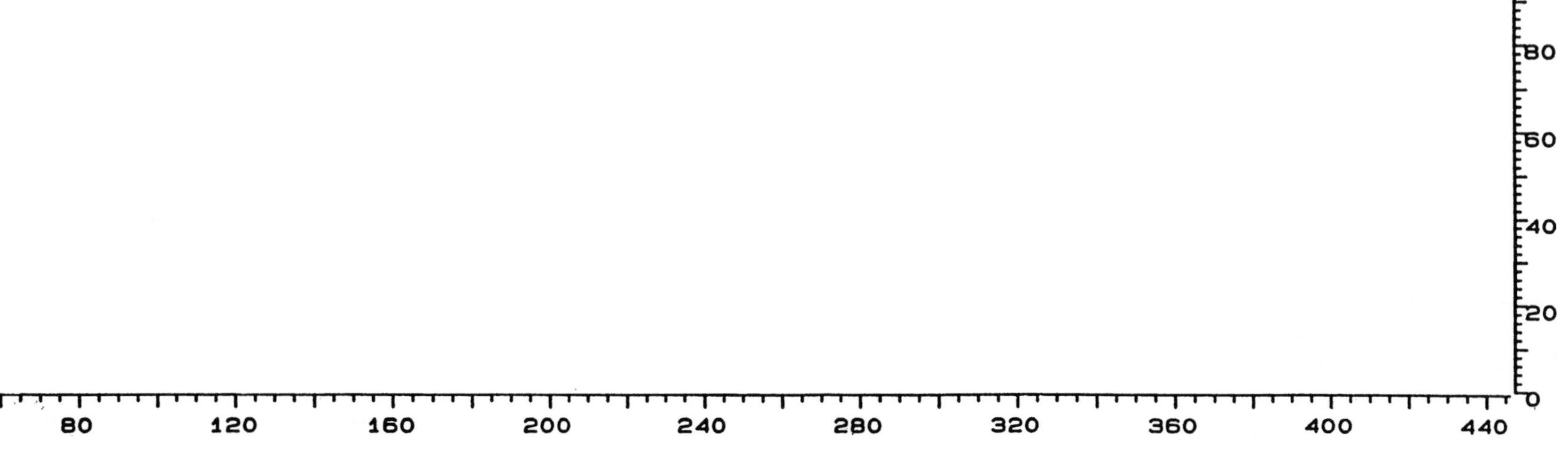

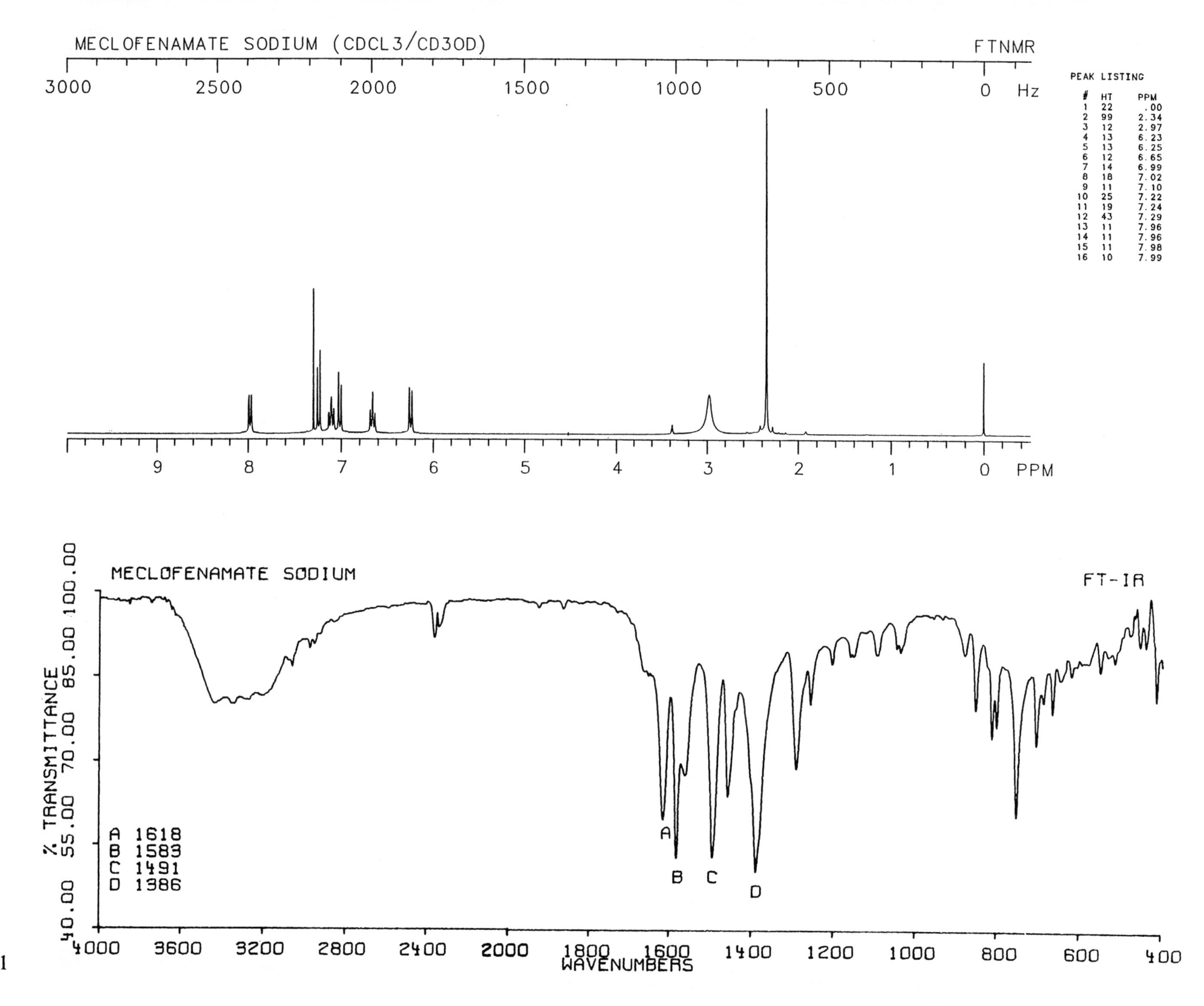
MECLOFENAMATE SODIUM (CDCL3/CD3OD)
FTNMR
Hz
PPM
PEAK LISTING
# HT PPM
1 22 .00
2 99 2.34
3 12 2.97
4 13 6.23
5 13 6.25
6 12 6.65
7 14 6.99
8 18 7.02
9 11 7.10
10 25 7.22
11 19 7.24
12 43 7.29
13 11 7.96
14 11 7.96
15 11 7.98
16 10 7.99
MECLOFENAMATE SODIUM
FT-IR
% TRANSMITTANCE
WAVENUMBERS
A 1618
B 1583
C 1491
D 1386

# MECLOFENOXATE

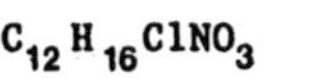
$C_{12}H_{16}ClNO_3$

Molecular weight: 257.73 (257.08)
Synonyms: (4-Chlorophenoxy)acetic acid 2-(dimethylamino)ethyl ester; centrophenoxine; meclofenoxane; acephen
Trade names: Analux, Cerebon, Cetrexin, Helfergin, Lucidril, Proseryl

Use: Central stimulant
HPLC:
GC:

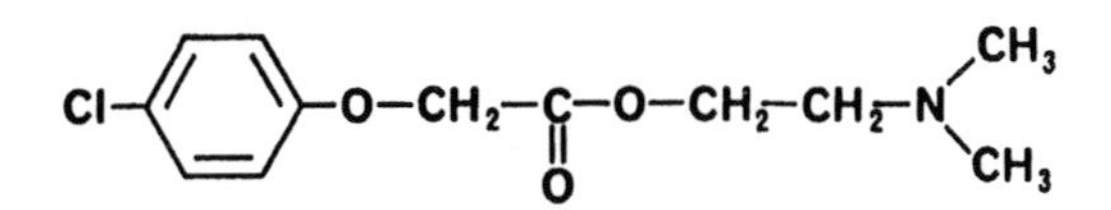


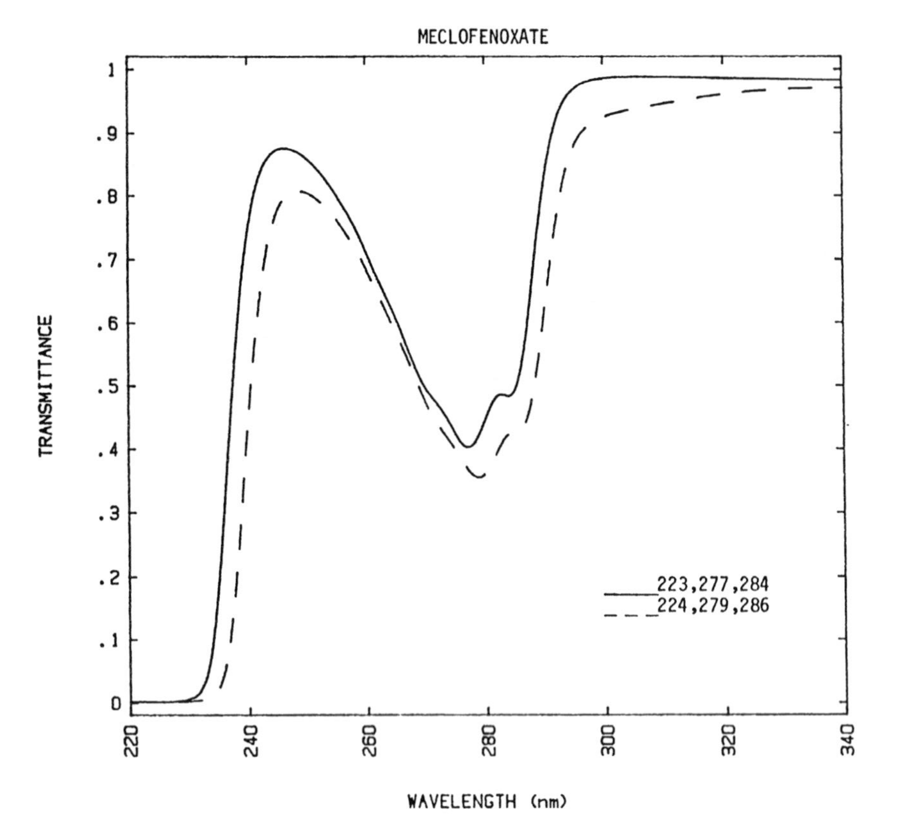

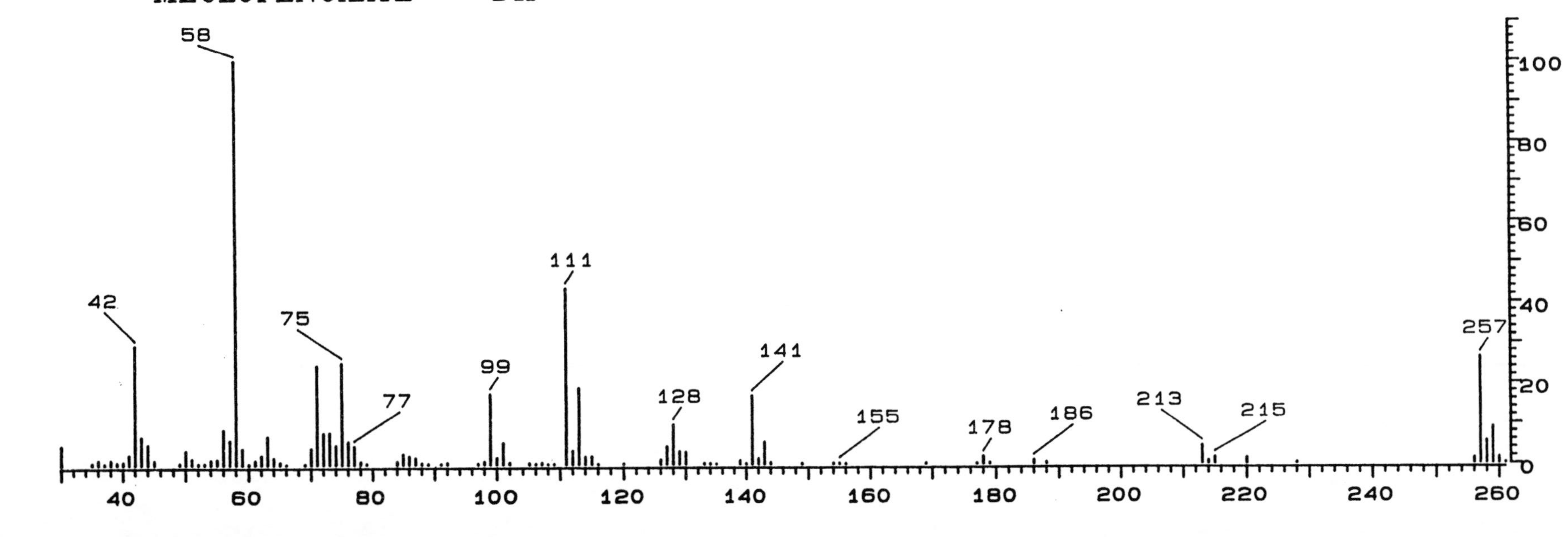

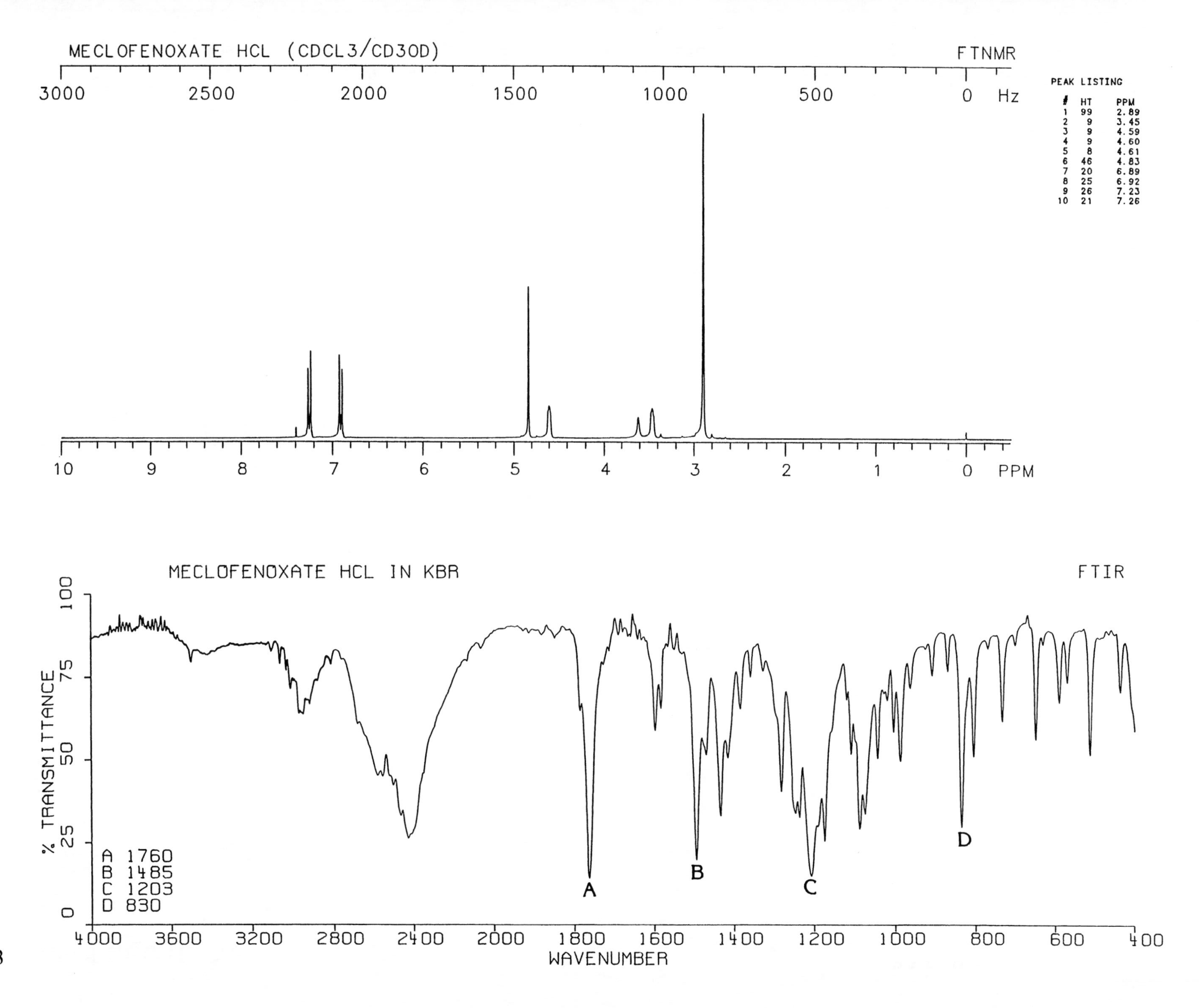

MECLOFENOXATE HCL (CDCL3/CD3OD)
FTNMR
3000 2500 2000 1500 1000 500 0 Hz
PEAK LISTING
# HT PPM
1 99 2.89
2 9 3.45
3 9 4.59
4 9 4.60
5 8 4.61
6 46 4.83
7 20 6.89
8 25 6.92
9 26 7.23
10 21 7.26
10 9 8 7 6 5 4 3 2 1 0 PPM
MECLOFENOXATE HCL IN KBR
FTIR
% TRANSMITTANCE
100 75 50 25 0
A 1760
B 1485
C 1203
D 830
A B C D
4000 3600 3200 2800 2400 2000 1800 1600 1400 1200 1000 800 600 400
WAVENUMBER

# MECLOQUALONE

$C_{15}H_{11}ClN_2O$

Molecular weight: 270.72 (270.06)
Synonyms: 3-(2-Chlorophenyl)-2-methyl-4(3H)-quinazolinone

Trade names: Nubarene

Use: Sedative, hypnotic
HPLC: Si-10; 1A:99B; 3.5
GC: 2318; 250°C

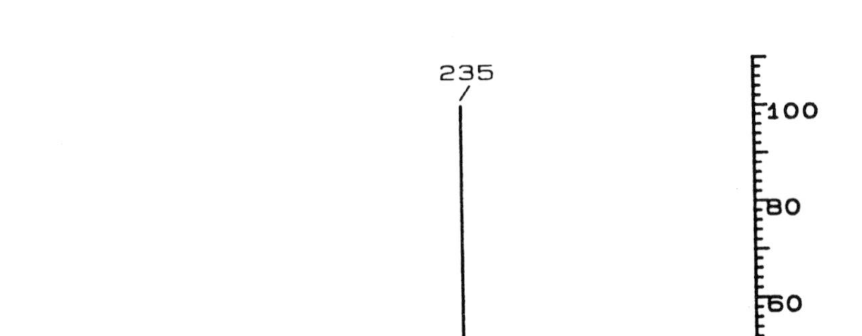


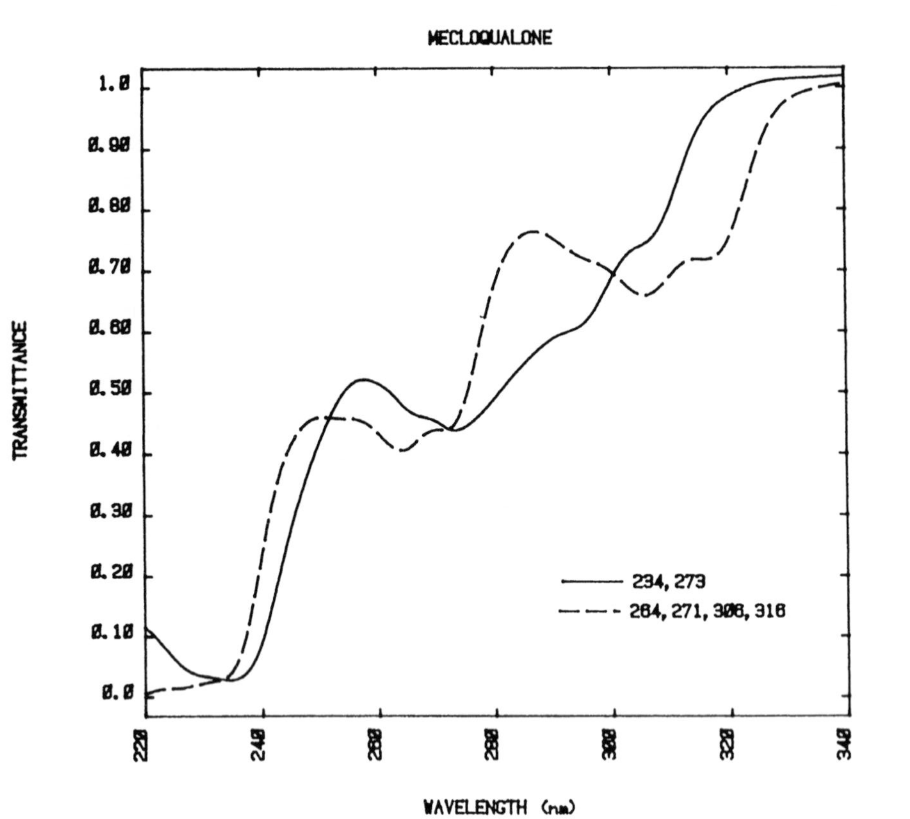


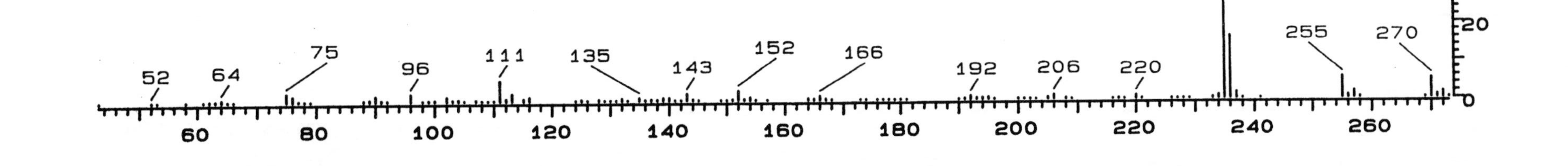

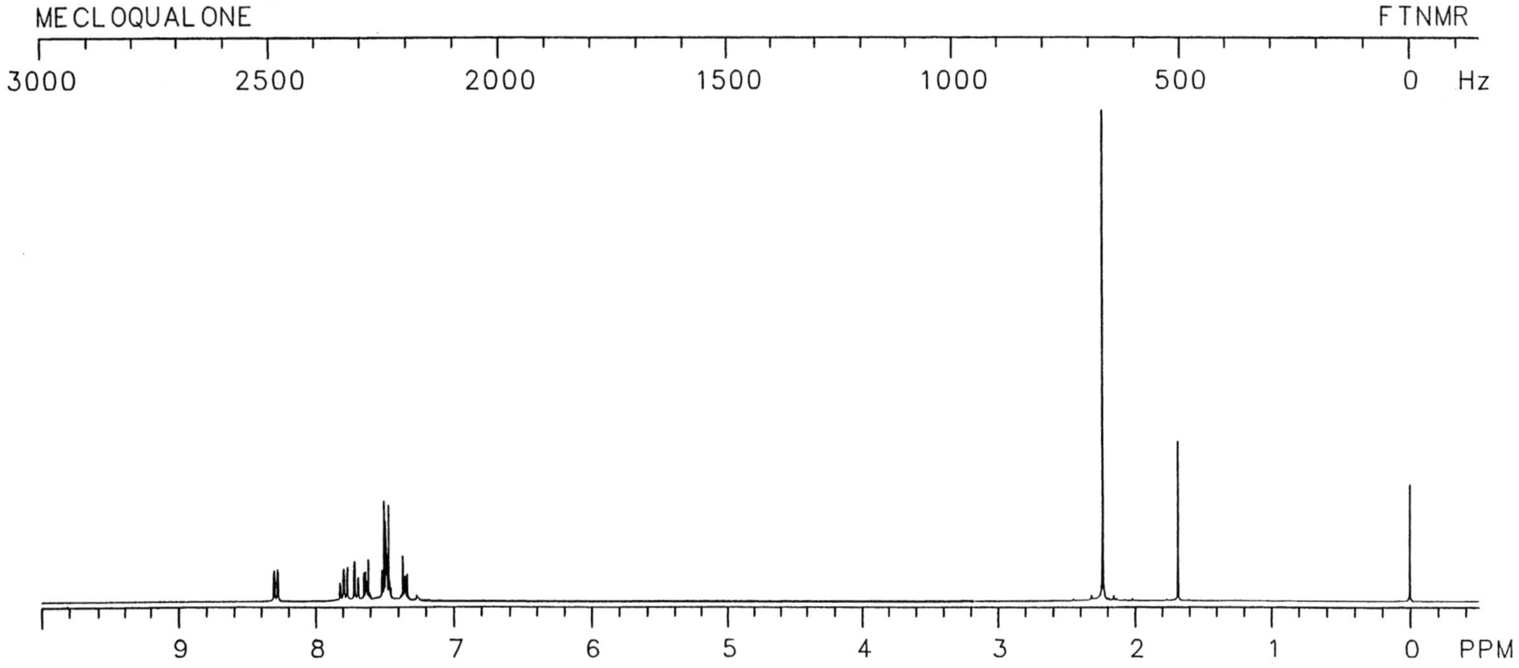

MECLOQUALONE

FTNMR

3000 2500 2000 1500 1000 500 0 Hz

PEAK LISTING

| # | HT | PPM |
|---|---|---|
| 1 | 23 | .00 |
| 2 | 32 | 1.68 |
| 3 | 99 | 2.23 |
| 4 | 29 | 7.26 |
| 5 | 9 | 7.36 |
| 6 | 19 | 7.46 |
| 7 | 9 | 7.47 |
| 8 | 8 | 7.48 |
| 9 | 13 | 7.48 |
| 10 | 16 | 7.48 |
| 11 | 20 | 7.49 |
| 12 | 8 | 7.61 |

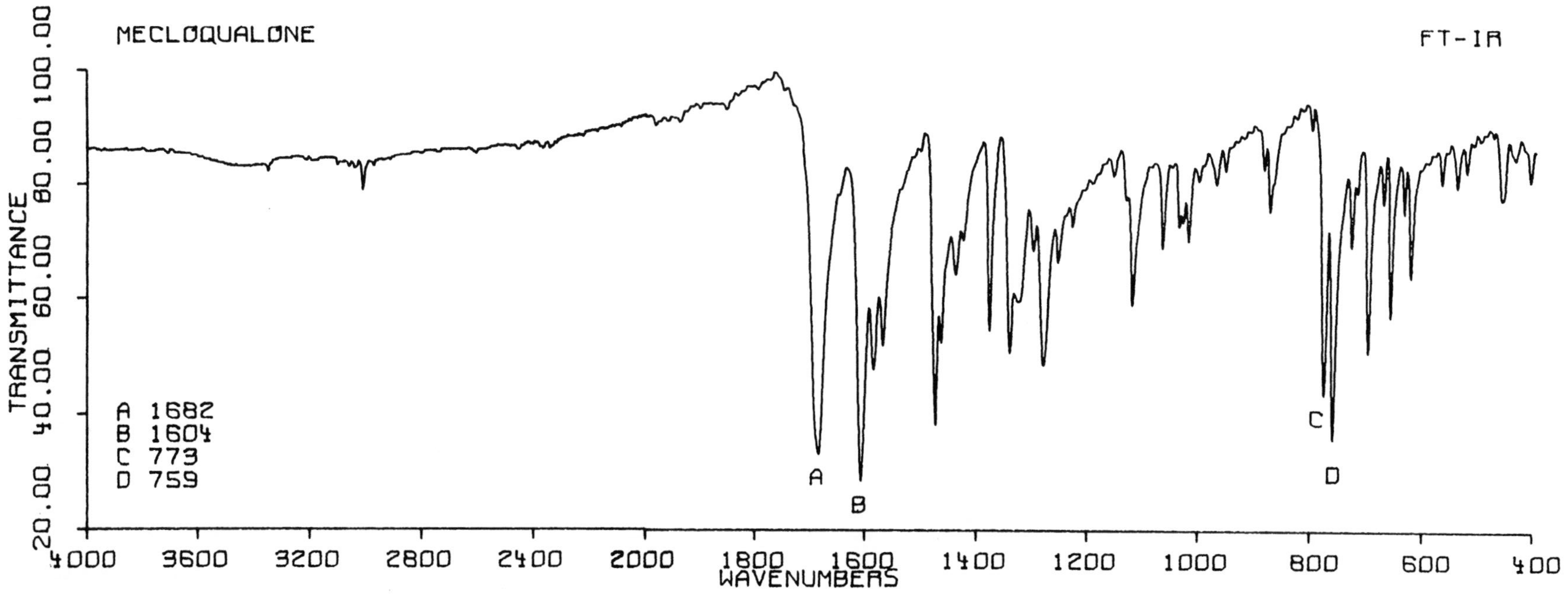

9 8 7 6 5 4 3 2 1 0 PPM

MECLOQUALONE

FT-IR

TRANSMITTANCE

20.00 40.00 60.00 80.00 100.00

A 1682
B 1604
C 773
D 759

4000 3600 3200 2800 2400 2000 1800 1600 1400 1200 1000 800 600 400

WAVENUMBERS

# MEDAZEPAM

$C_{16}H_{15}ClN_2$

Molecular weight: 270.76 (270.09)
Synonyms: 7-Chloro-2,3-dihydro-1-methyl-5-phenyl-1H-1,4-benzodiazepine
Trade names: Ansilan, Diepin, Elbrus, Esmail, Medazepol, Mezepan, Marsis, Megasedan, Nobrium, Pazital, Psiquium, Resmit, Rudotet, Serenium
Use: Tranquilizer
HPLC: Si-10; 100B; 14.0
GC: 2284; 250°C

CH3
N
N
Cl

MEDAZEPAM
TRANSMITTANCE
1
.9
.8
.7
.6
.5
.4
.3
.2
.1
0
220
240
260
280
300
320
340
WAVELENGTH (nm)
253
253

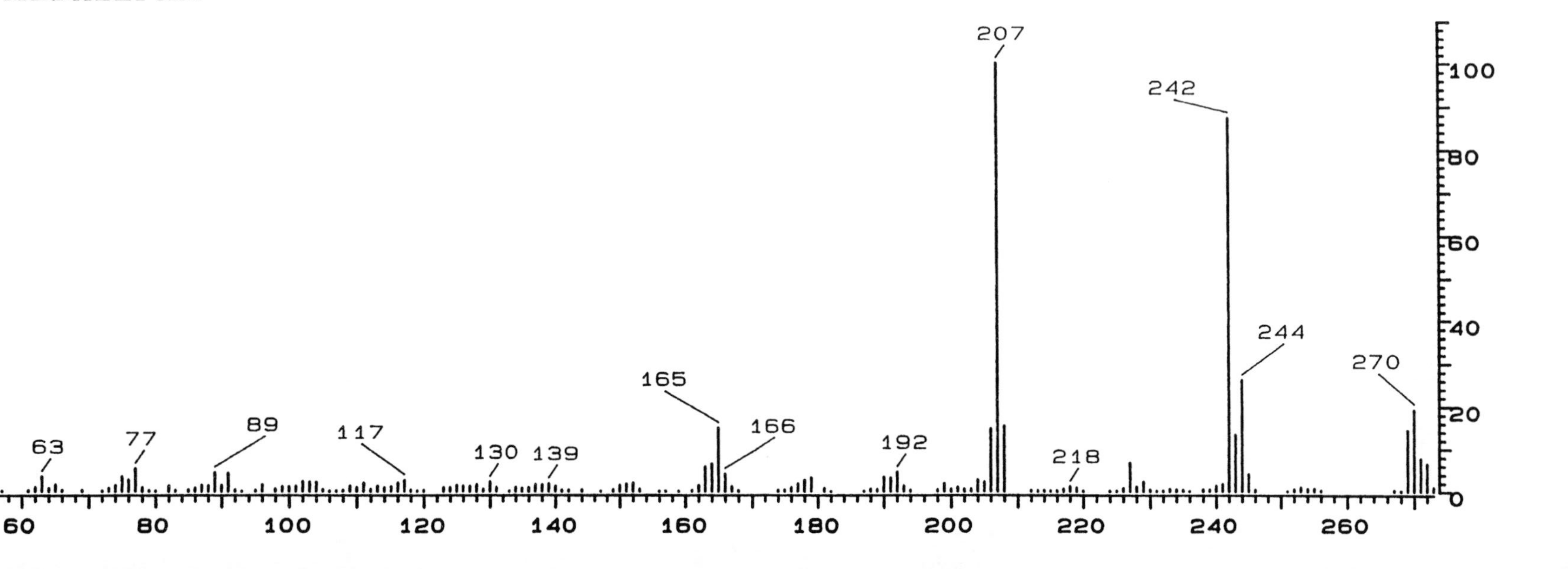

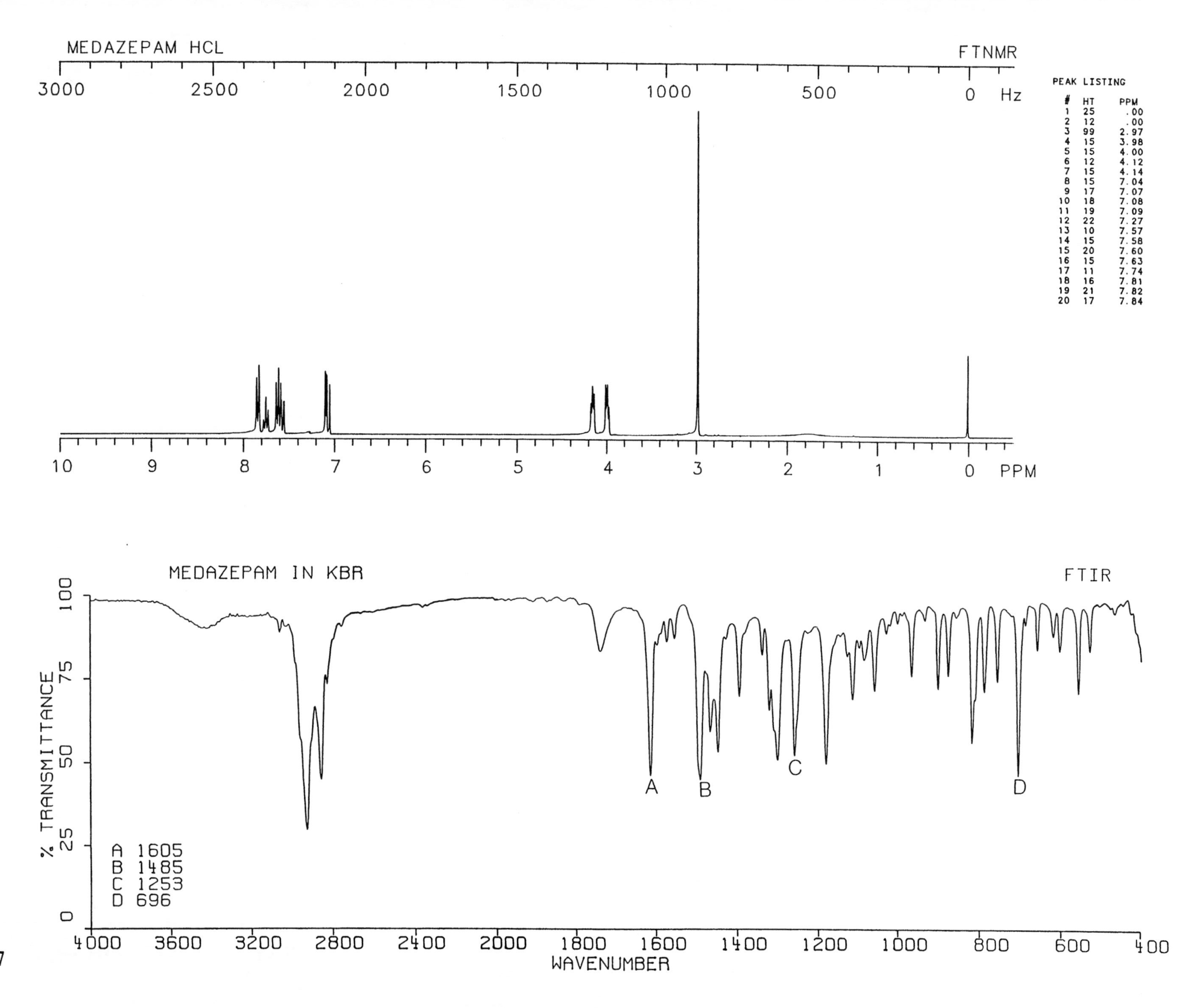
MEDAZEPAM HCL
FTNMR
3000
2500
2000
1500
1000
500
0 Hz
PEAK LISTING
# HT PPM
1 25 .00
2 12 .00
3 99 2.97
4 15 3.98
5 15 4.00
6 12 4.12
7 15 4.14
8 15 7.04
9 17 7.07
10 18 7.08
11 19 7.09
12 22 7.27
13 10 7.57
14 15 7.58
15 20 7.60
16 15 7.63
17 11 7.74
18 16 7.81
19 21 7.82
20 17 7.84
10 9 8 7 6 5 4 3 2 1 0 PPM
MEDAZEPAM IN KBR
FTIR
% TRANSMITTANCE
100
75
50
25
0
A
B
C
D
A 1605
B 1485
C 1253
D 696
4000 3600 3200 2800 2400 2000 1800 1600 1400 1200 1000 800 600 400
WAVENUMBER

# MEDROXYPROGESTERONE

$C_{22}H_{32}O_3$

**Molecular weight:** 344.50 (344.24)
**Synonyms:** 17-Hydroxy-6α-methylpregn-4-ene-3,20-dione

**Trade names:** Amen, Depo-Provera, Provera

**Use:** Oral contraceptive
**HPLC:** Si-10; 1A:99B; 6.0
**GC:** 3047; 280°C

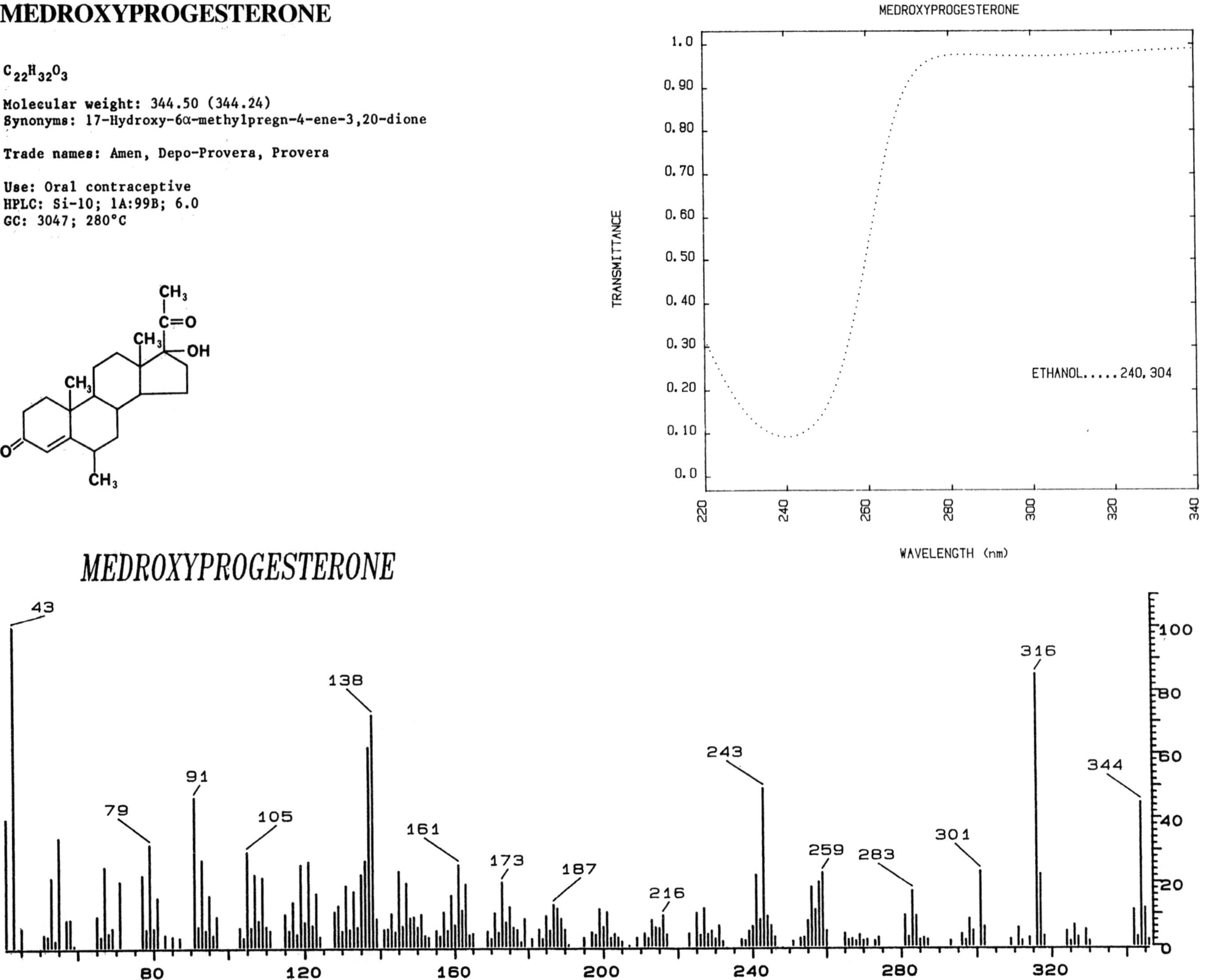

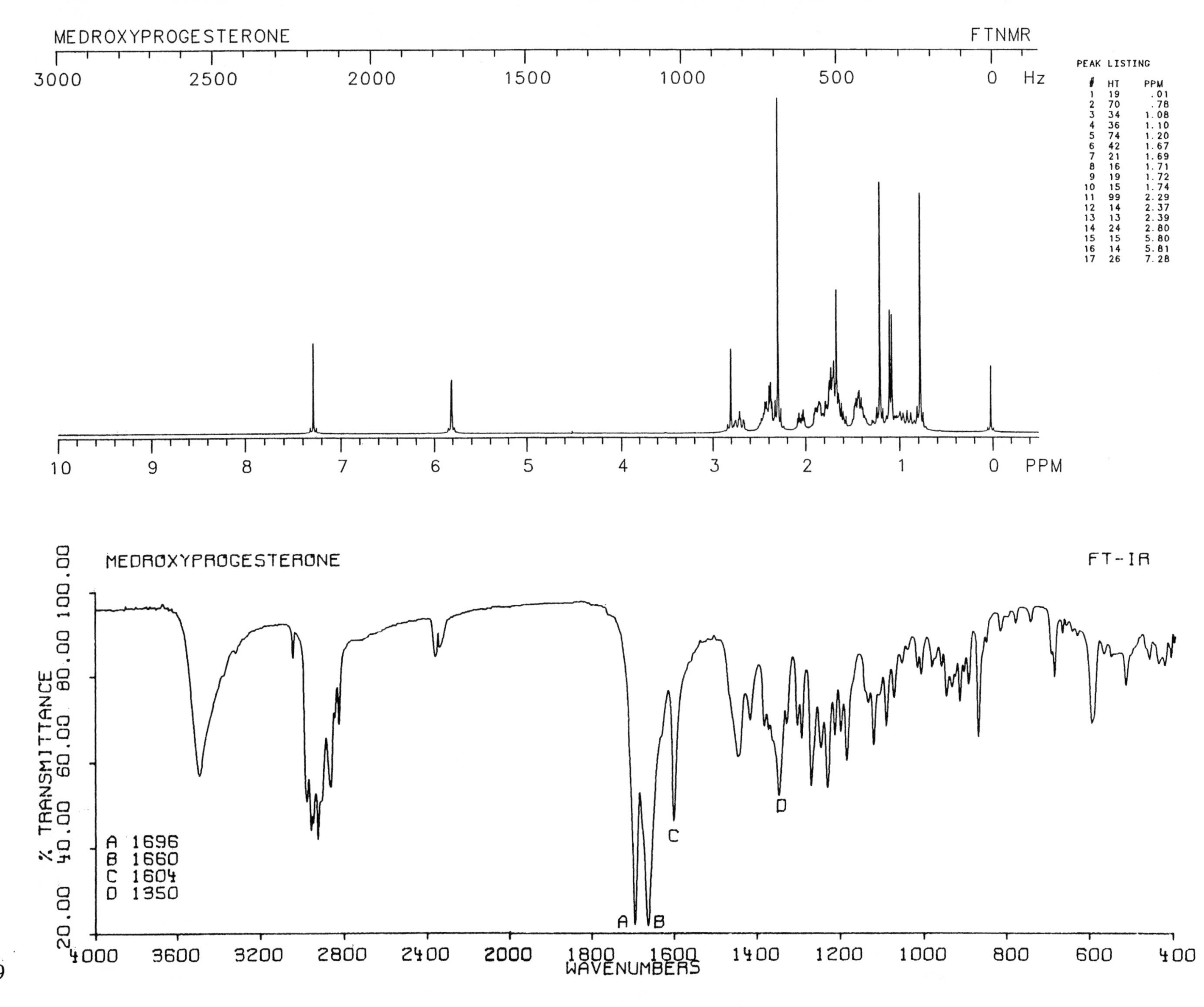

MEDROXYPROGESTERONE
FTNMR
3000 2500 2000 1500 1000 500 0 Hz
PEAK LISTING
# HT PPM
1 19 .01
2 70 .78
3 34 1.08
4 36 1.10
5 74 1.20
6 42 1.67
7 21 1.69
8 16 1.71
9 19 1.72
10 15 1.74
11 99 2.29
12 14 2.37
13 13 2.39
14 24 2.80
15 15 5.80
16 14 5.81
17 26 7.28
10 9 8 7 6 5 4 3 2 1 0 PPM
MEDROXYPROGESTERONE
FT-IR
% TRANSMITTANCE
20.00 40.00 60.00 80.00 100.00
A 1696
B 1660
C 1604
D 1350
A B C D
4000 3600 3200 2800 2400 2000 1800 1600 1400 1200 1000 800 600 400
WAVENUMBERS

# MEFENAMIC ACID

$C_{15}H_{15}NO_2$

Molecular weight: 241.29 (241.11)
Synonyms: 2-[(2,3-Dimethylphenyl)amino]benzoic acid

Trade names: Ponstel

Use: Anti-inflammatory
HPLC: Si-10; 5A:95B; 4.5
GC: 2209; 250°C

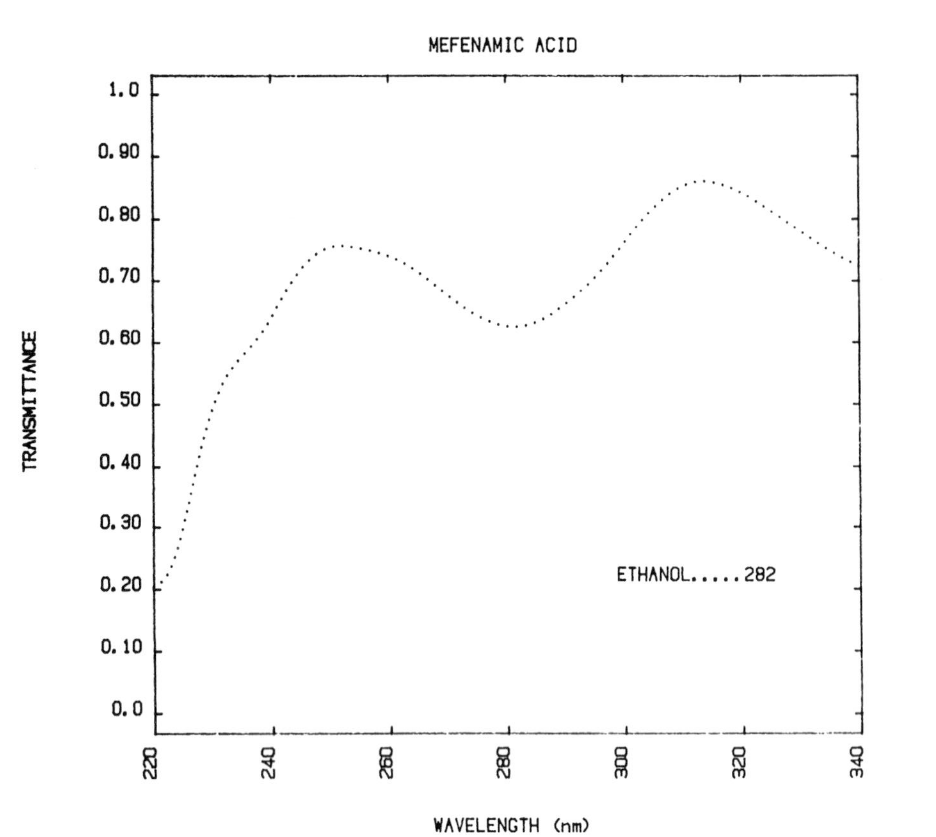

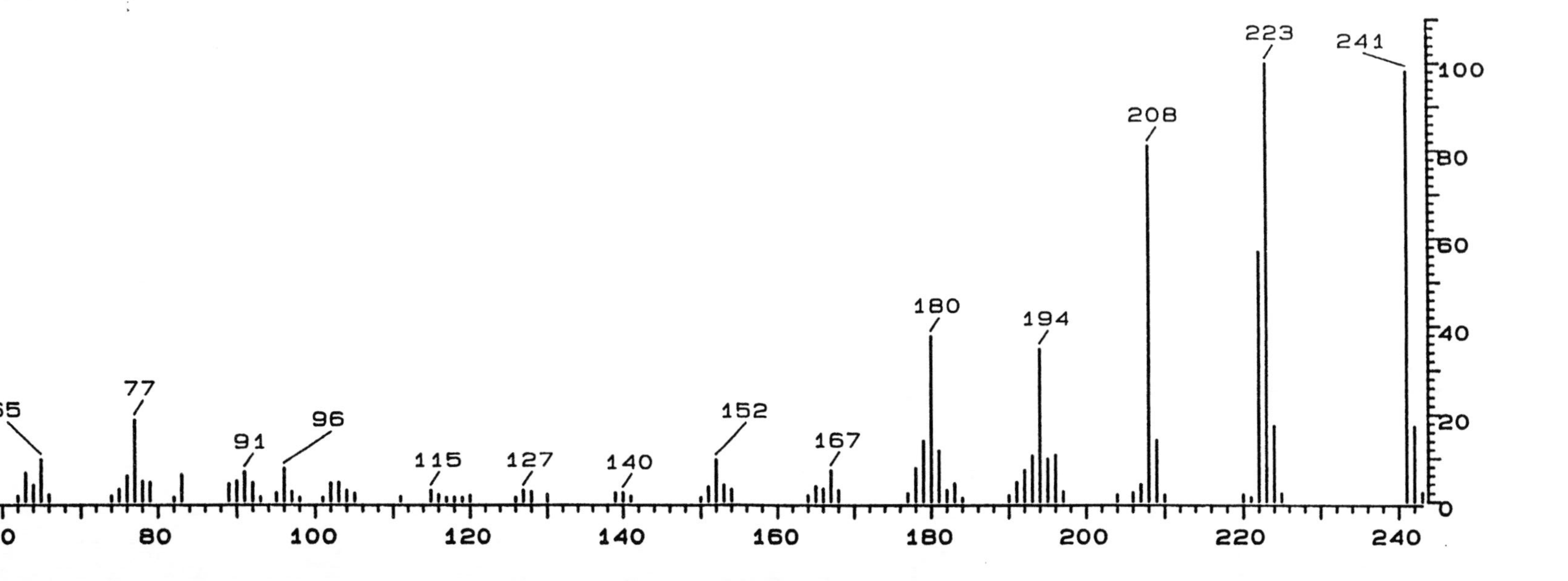

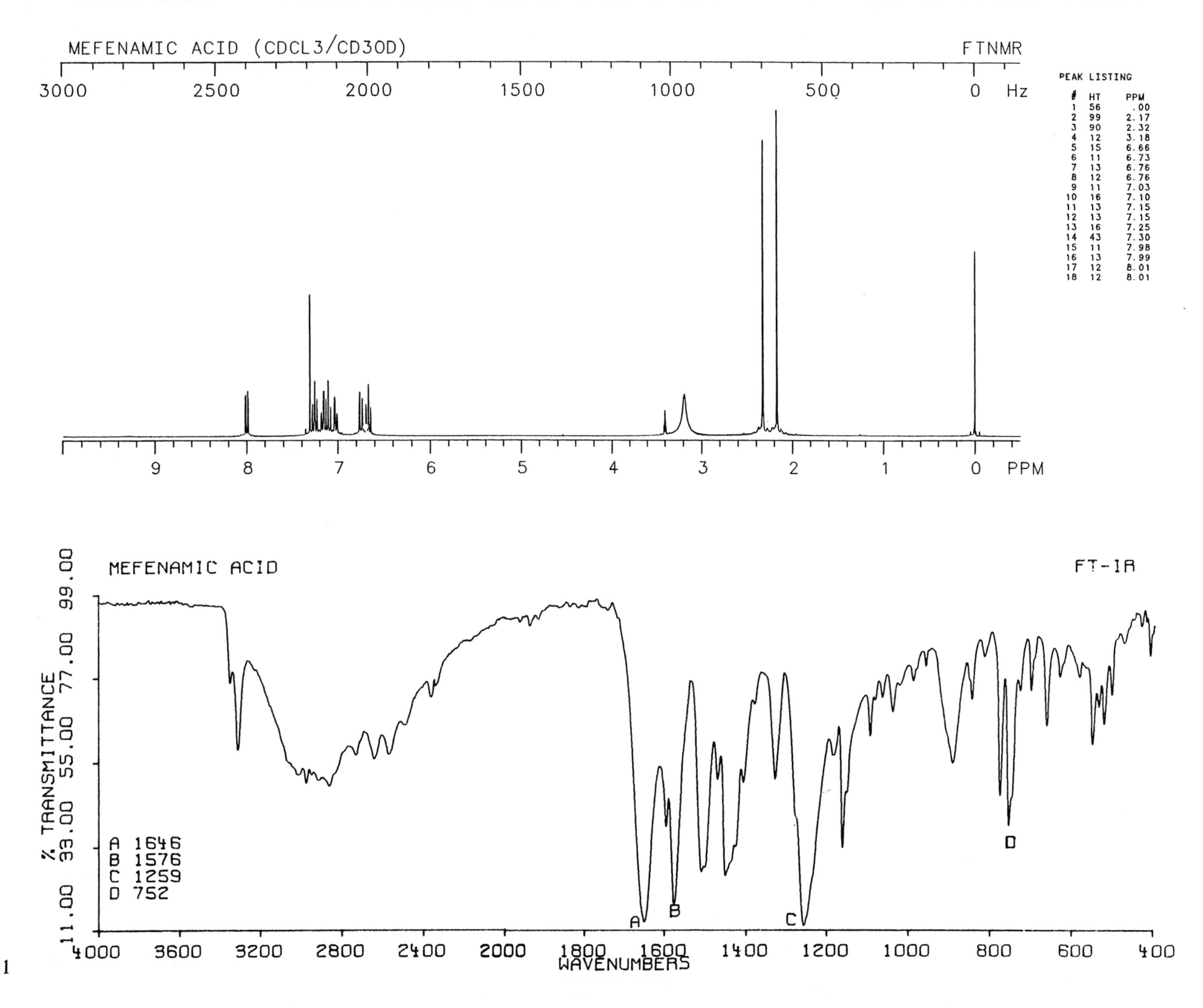
MEFENAMIC ACID (CDCL3/CD3OD)
FTNMR
3000 2500 2000 1500 1000 500 0 Hz
PEAK LISTING
# HT PPM
1 56 .00
2 99 2.17
3 90 2.32
4 12 3.18
5 15 6.66
6 11 6.73
7 13 6.76
8 12 6.76
9 11 7.03
10 16 7.10
11 13 7.15
12 13 7.15
13 16 7.25
14 43 7.30
15 11 7.98
16 13 7.99
17 12 8.01
18 12 8.01
9 8 7 6 5 4 3 2 1 0 PPM
MEFENAMIC ACID
FT-IR
% TRANSMITTANCE
99.00 77.00 55.00 33.00 11.00
A 1646
B 1576
C 1259
D 752
A B C D
4000 3600 3200 2800 2400 2000 1800 1600 1400 1200 1000 800 600 400
WAVENUMBERS

# MEFENOREX

$C_{12}H_{18}ClN$

Molecular weight: 211.74 (211.11)
Synonyms: N-(3-Chloropropyl)-α-methylphenethylamine; 1-phenyl-2-(3-chloropropylamino)propane
Trade names: Anexate, Doracil, Pondinil, Pondinol, Rondimen

Use: Anorexic
HPLC: Si-10; 2A:98B; 3.3
GC: 1615; 200°C

$C_6H_5-CH_2-CH(CH_3)-NH-CH_2-CH_2-CH_2-Cl$

MEFENOREX

TRANSMITTANCE

WAVELENGTH (nm)

251,257,263
253,258,267

*MEFENOREX*

44 56 77 84 91 103 120 123 144 148 160 174 198 210

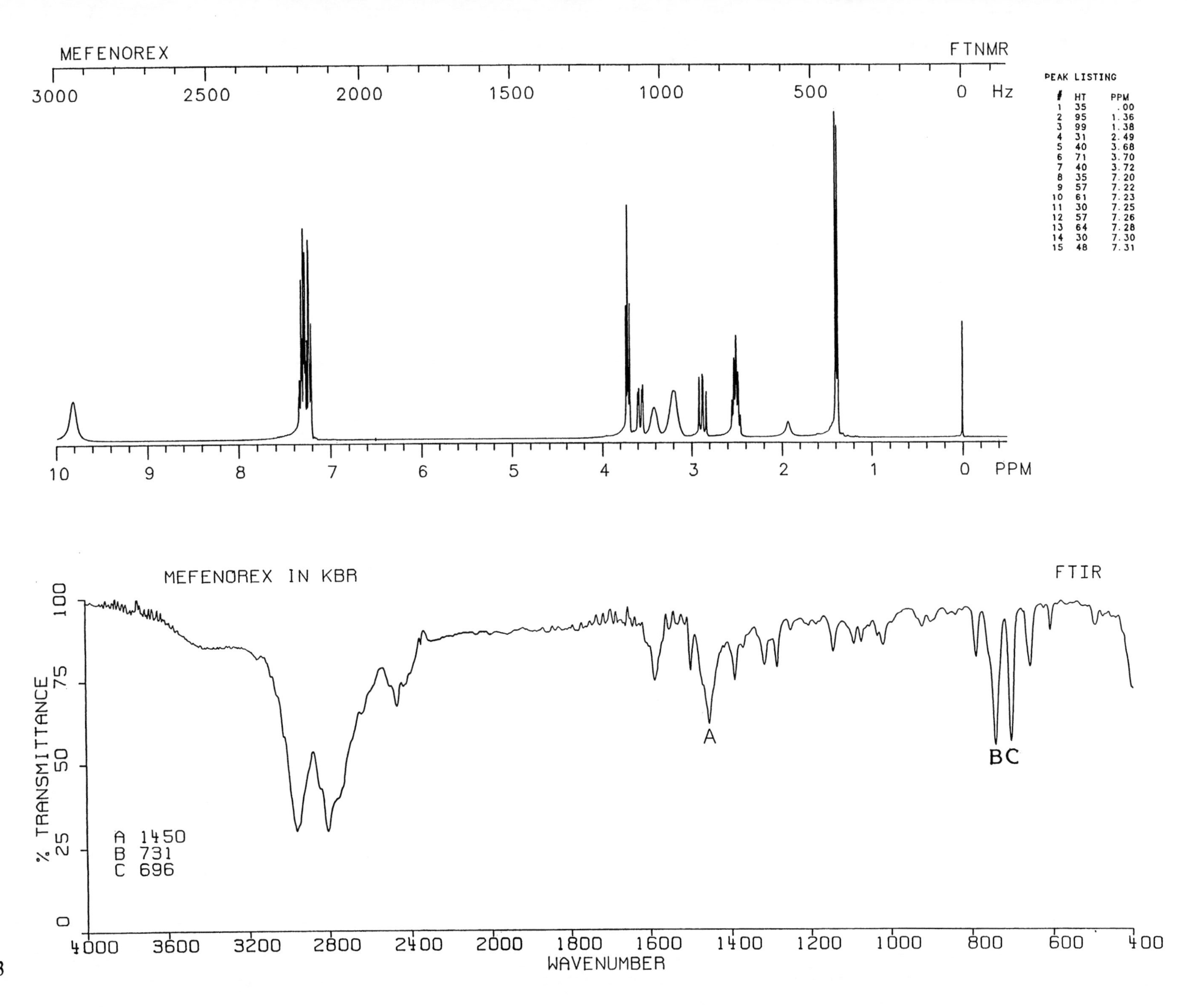
MEFENOREX
FTNMR
3000 2500 2000 1500 1000 500 0 Hz
10 9 8 7 6 5 4 3 2 1 0 PPM
PEAK LISTING
# HT PPM
1 35 .00
2 95 1.36
3 99 1.38
4 31 2.49
5 40 3.68
6 71 3.70
7 40 3.72
8 35 7.20
9 57 7.22
10 61 7.23
11 30 7.25
12 57 7.26
13 64 7.28
14 30 7.30
15 48 7.31
MEFENOREX IN KBR
FTIR
% TRANSMITTANCE
100 75 50 25 0
A
BC
A 1450
B 731
C 696
4000 3600 3200 2800 2400 2000 1800 1600 1400 1200 1000 800 600 400
WAVENUMBER

# MEGESTROL ACETATE

$C_{24}H_{32}O_4$

**Molecular weight:** 384.52 (384.23)
**Synonyms:** 17α-Hydroxy-6-methylpregna-4,6-diene-3,20-dione acetate

**Trade names:** Megace

**Use:** Antineoplastic
HPLC: Si-10; 1A;99B; 4.0
GC: 3145; 280°C

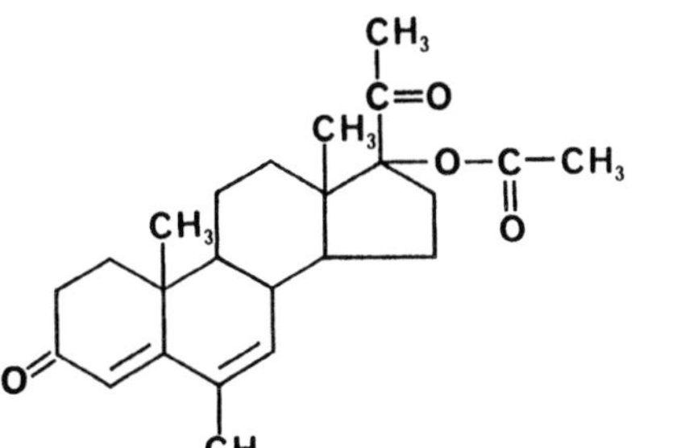


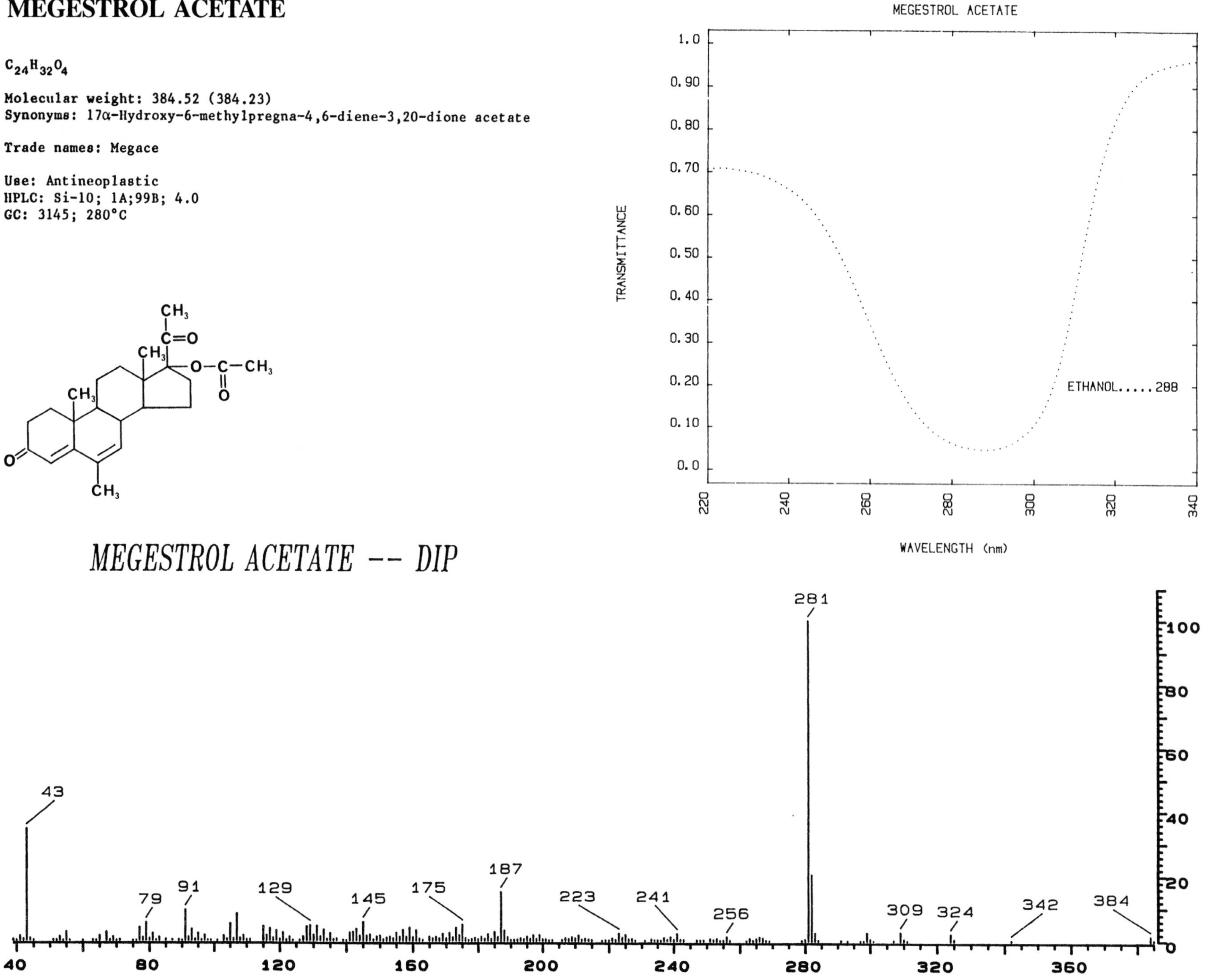

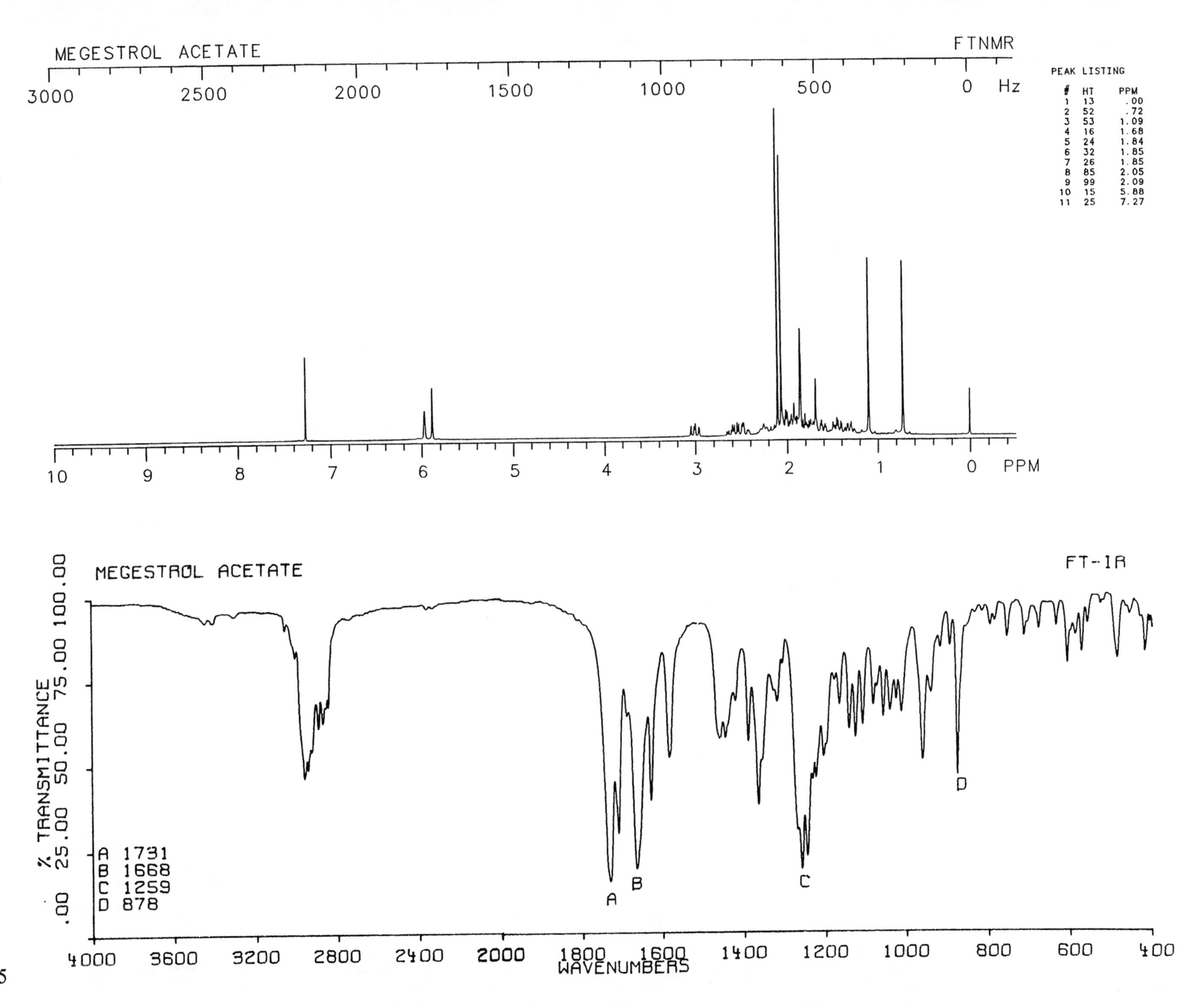
MEGESTROL ACETATE
FTNMR
3000
2500
2000
1500
1000
500
0 Hz
PEAK LISTING
# HT PPM
1 13 .00
2 52 .72
3 53 1.09
4 16 1.68
5 24 1.84
6 32 1.85
7 26 1.85
8 85 2.05
9 99 2.09
10 15 5.88
11 25 7.27
10
9
8
7
6
5
4
3
2
1
0 PPM
MEGESTROL ACETATE
FT-IR
% TRANSMITTANCE
.00
25.00
50.00
75.00
100.00
A 1731
B 1668
C 1259
D 878
A
B
C
D
4000
3600
3200
2800
2400
2000
1800
1600
1400
1200
1000
800
600
400
WAVENUMBERS

# MELPHALAN

$C_{13}H_{18}Cl_2N_2O_2$

Molecular weight: 305.20 (304.08)
Synonyms: 4-[Bis(2-chloroethyl)amino]-L-phenylalanine; L-phenylalanine mustard; alanine nitrogen mustard, PAM; phenylalanine nitrogen mustard
Trade names: Alkeran, Medphalan, Merphalan, Sarcolysine

Use: Antineoplastic
HPLC: Si-10; 30A:70B; 5.0
GC:

$HOOC{-}CH(NH_2){-}CH_2{-}C_6H_4{-}N(CH_2{-}CH_2{-}Cl)_2$

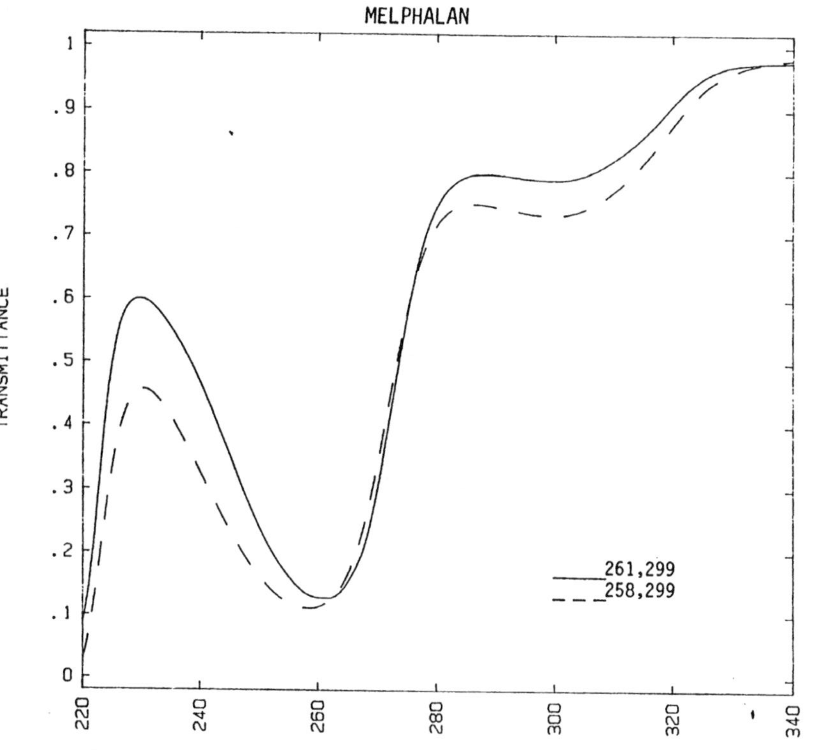

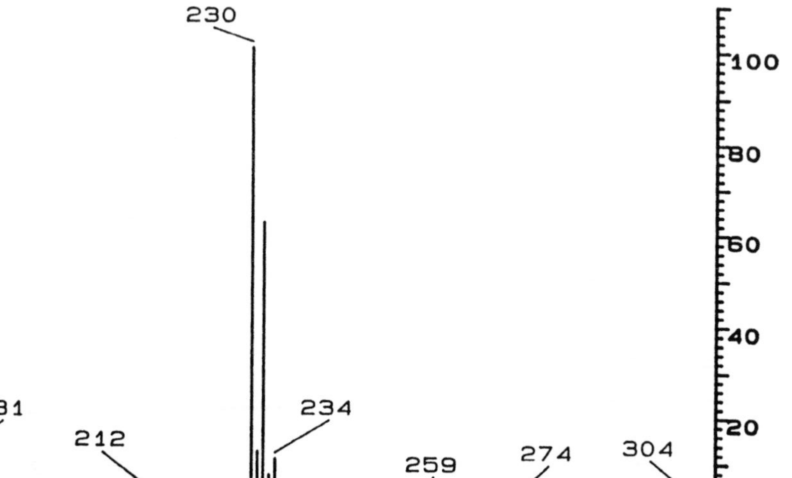

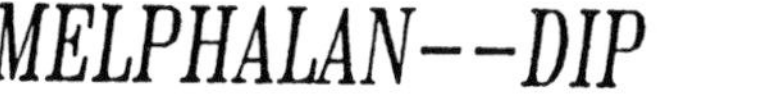


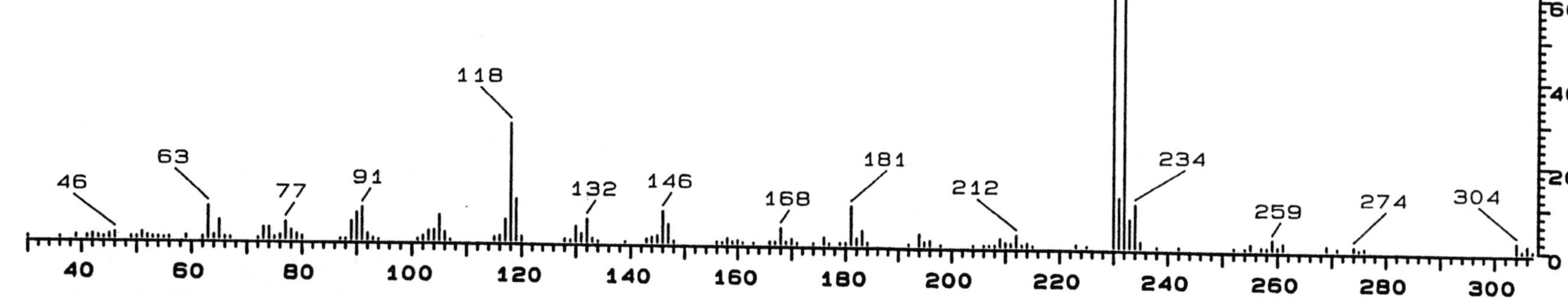

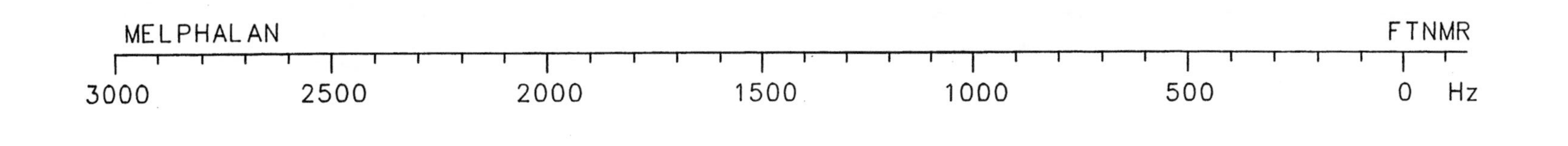


INSUFFICIENT SOLUBILITY

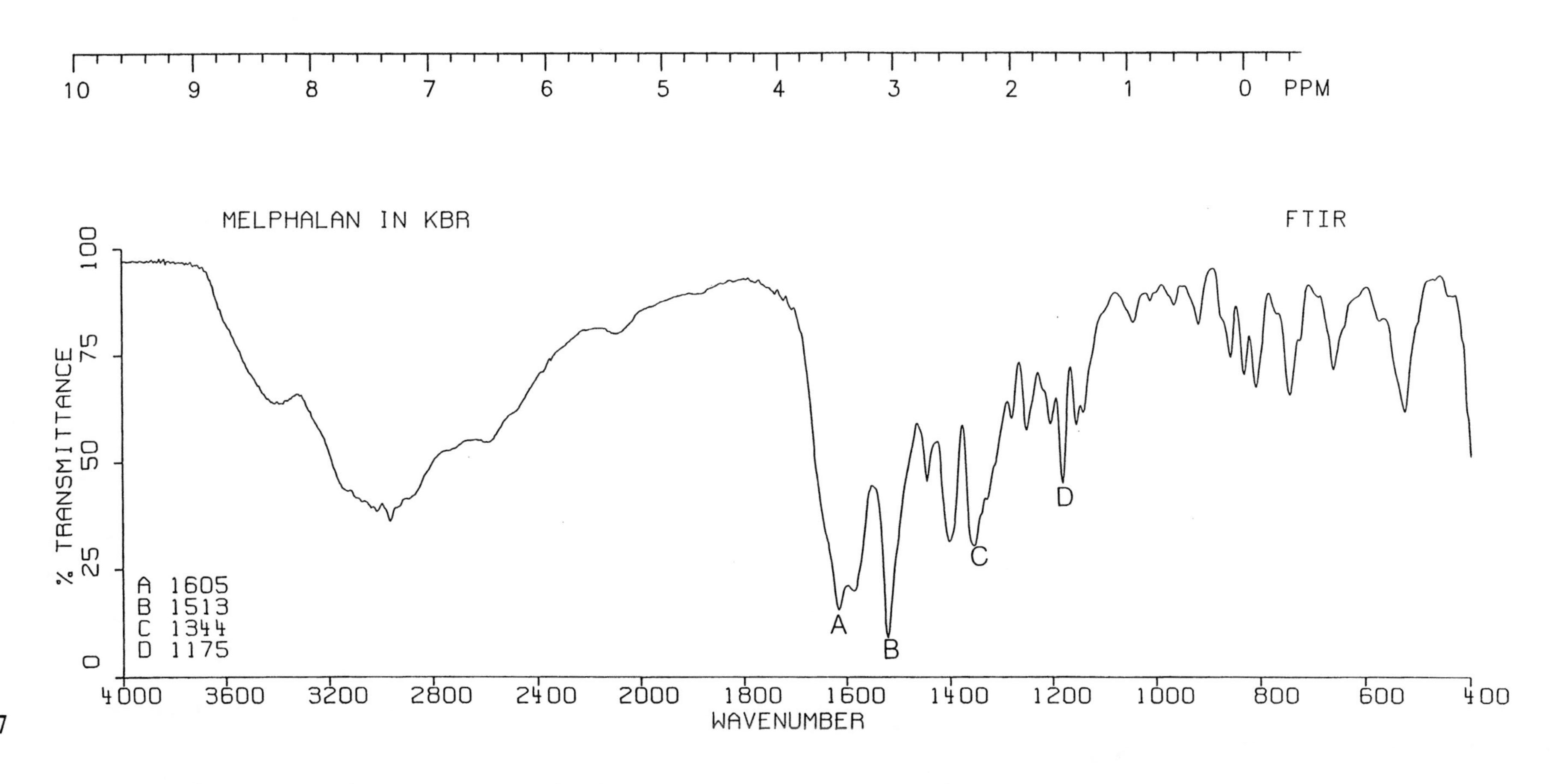

# MENTHOL

$C_{10}H_{20}O$

**Molecular weight:** 156.26 (156.15)
**Synonyms:** 5-Methyl-2-(1-methylethyl)cyclohexanol; ℓ-methol; peppermint camphor; 3-p-menthanol
**Trade names:** Decongestant Elixir, Denorex Medicated Shampoo, Panalgesic, Derma Medicone-HC, Rectal Medicone-HC
**Use:** Topical antipruritic
**HPLC:**
**GC:** 1173; 140°C

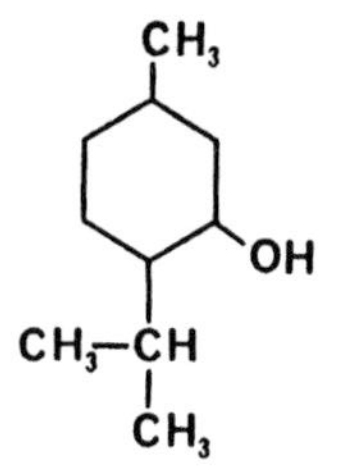

CH3
OH
CH3–CH
CH3

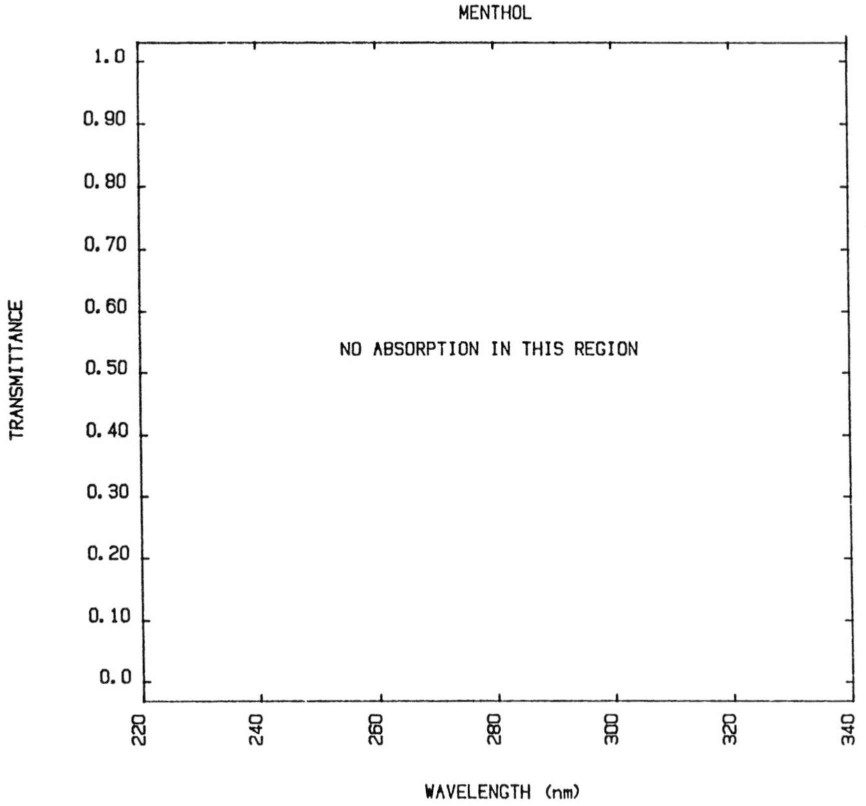

MENTHOL

TRANSMITTANCE
1.0
0.90
0.80
0.70
0.60
0.50
0.40
0.30
0.20
0.10
0.0

NO ABSORPTION IN THIS REGION

220 240 260 280 300 320 340
WAVELENGTH (nm)

*MENTHOL*

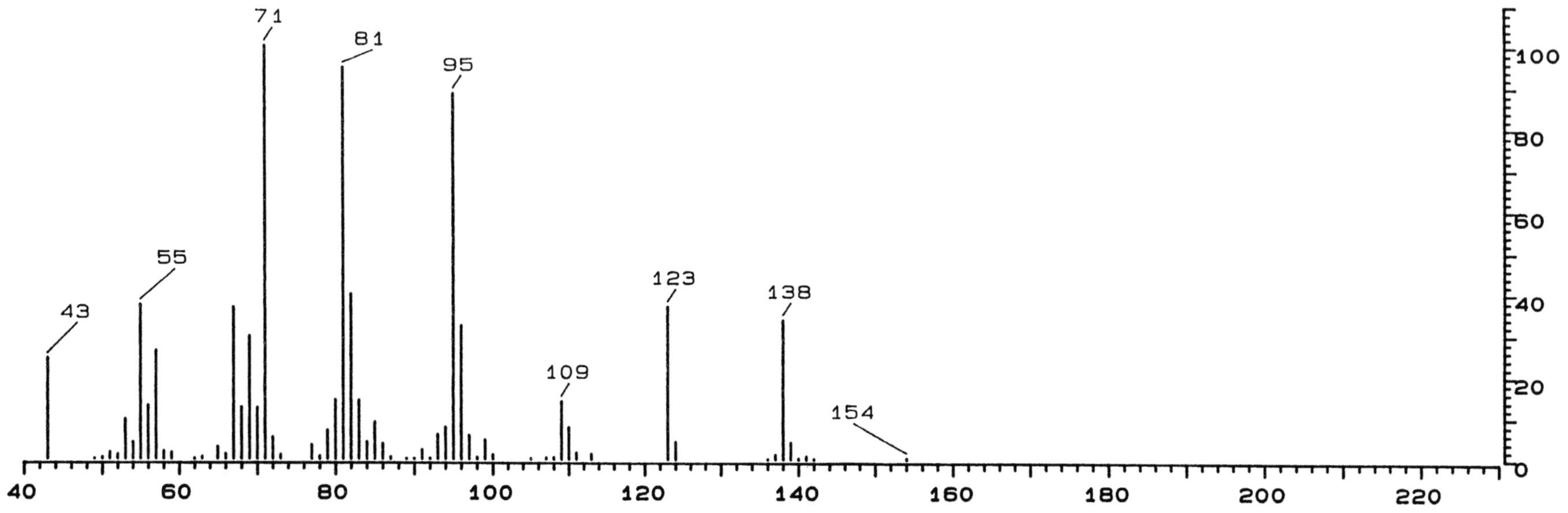

43 55 71 81 95 109 123 138 154

40 60 80 100 120 140 160 180 200 220

0 20 40 60 80 100

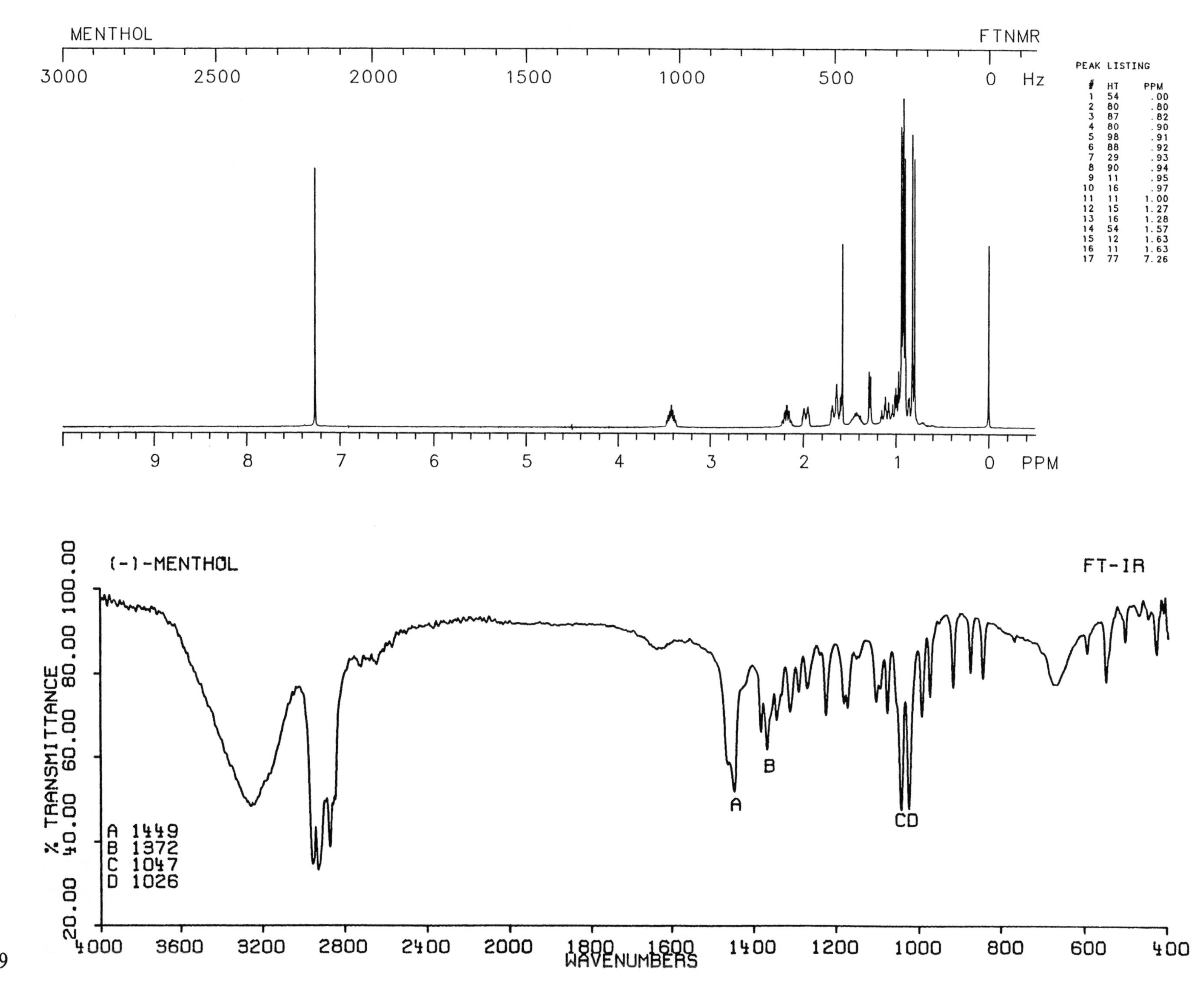
MENTHOL
FTNMR
3000
2500
2000
1500
1000
500
0 Hz
PEAK LISTING
# HT PPM
1 54 .00
2 80 .80
3 87 .82
4 80 .90
5 98 .91
6 88 .92
7 29 .93
8 90 .94
9 11 .95
10 16 .97
11 11 1.00
12 15 1.27
13 16 1.28
14 54 1.57
15 12 1.63
16 11 1.63
17 77 7.26
9
8
7
6
5
4
3
2
1
0 PPM
(-)-MENTHOL
FT-IR
% TRANSMITTANCE
100.00
80.00
60.00
40.00
20.00
A 1449
B 1372
C 1047
D 1026
A
B
CD
4000
3600
3200
2800
2400
2000
1800
1600
1400
1200
1000
800
600
400
WAVENUMBERS

# MEPENZOLATE BROMIDE

$C_{21}H_{26}BrNO_3$

Molecular weight: 420.35 (419.11)
Synonyms: 3-[(Hydroxydiphenylacetyl)oxy]-1,1-dimethyl-piperidinium bromide
Trade names: Cantil

Use: Anticholinergic
HPLC:
GC:

MEPENZOLATE BROMIDE

TRANSMITTANCE

WAVELENGTH (nm)

—— 251, 257, 263
— — 225, 252, 258, 264

NO USEFUL MASS SPECTRUM WAS OBTAINED

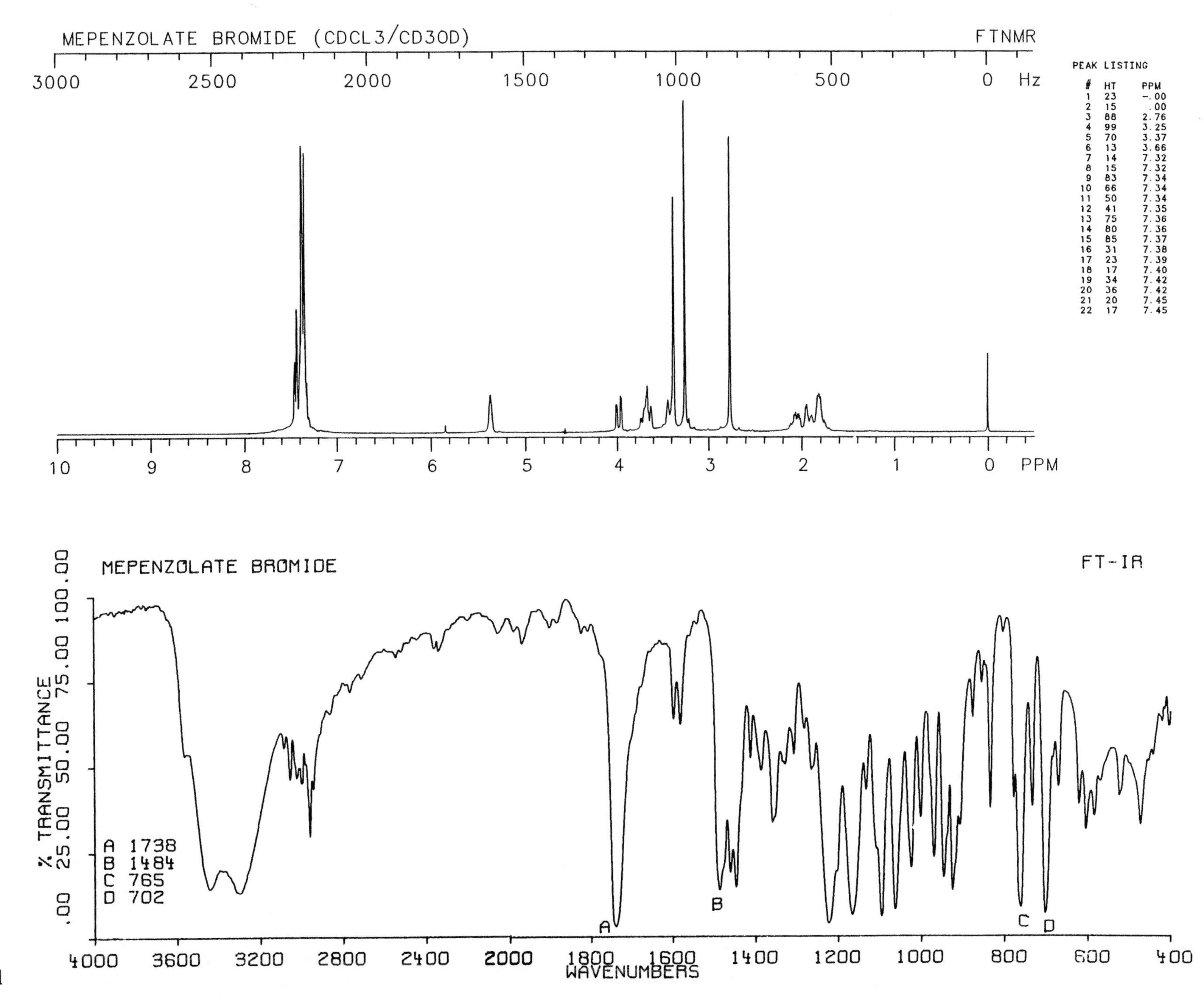
MEPENZOLATE BROMIDE (CDCL3/CD3OD)
FTNMR
3000 2500 2000 1500 1000 500 0 Hz
10 9 8 7 6 5 4 3 2 1 0 PPM
PEAK LISTING
# HT PPM
1 23 -.00
2 15 .00
3 88 2.76
4 99 3.25
5 70 3.37
6 13 3.66
7 14 7.32
8 15 7.32
9 83 7.34
10 66 7.34
11 50 7.34
12 41 7.35
13 75 7.36
14 80 7.36
15 85 7.37
16 31 7.38
17 23 7.39
18 17 7.40
19 34 7.42
20 36 7.42
21 20 7.45
22 17 7.45
MEPENZOLATE BROMIDE
FT-IR
% TRANSMITTANCE
.00 25.00 50.00 75.00 100.00
A 1738
B 1484
C 765
D 702
A B C D
4000 3600 3200 2800 2400 2000 1800 1600 1400 1200 1000 800 600 400
WAVENUMBERS

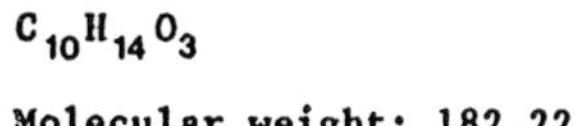

# MEPHENESIN

$C_{10}H_{14}O_3$

Molecular weight: 182.22 (182.09)
Synonyms: 3-(2-Methylphenoxy)-1,2-propanediol

Trade names:

Use: Muscle relaxant
HPLC: Si-10; 2A:98B; 4.0
GC: 1588; 200°C

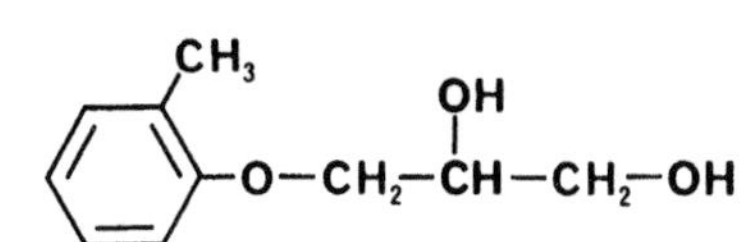


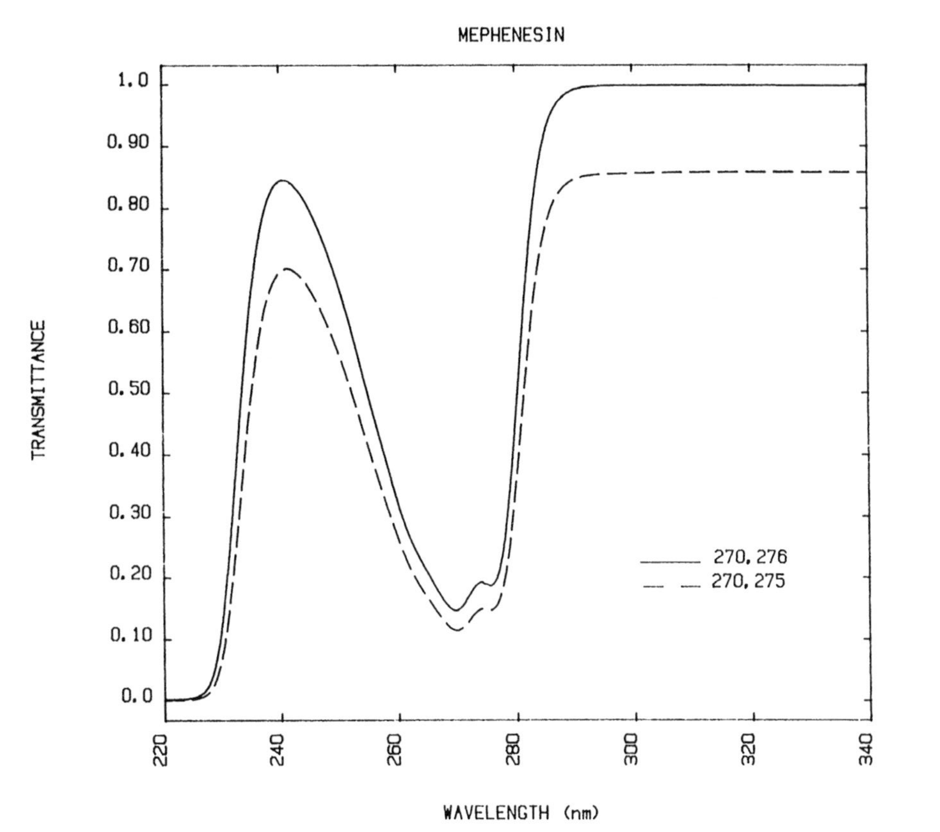

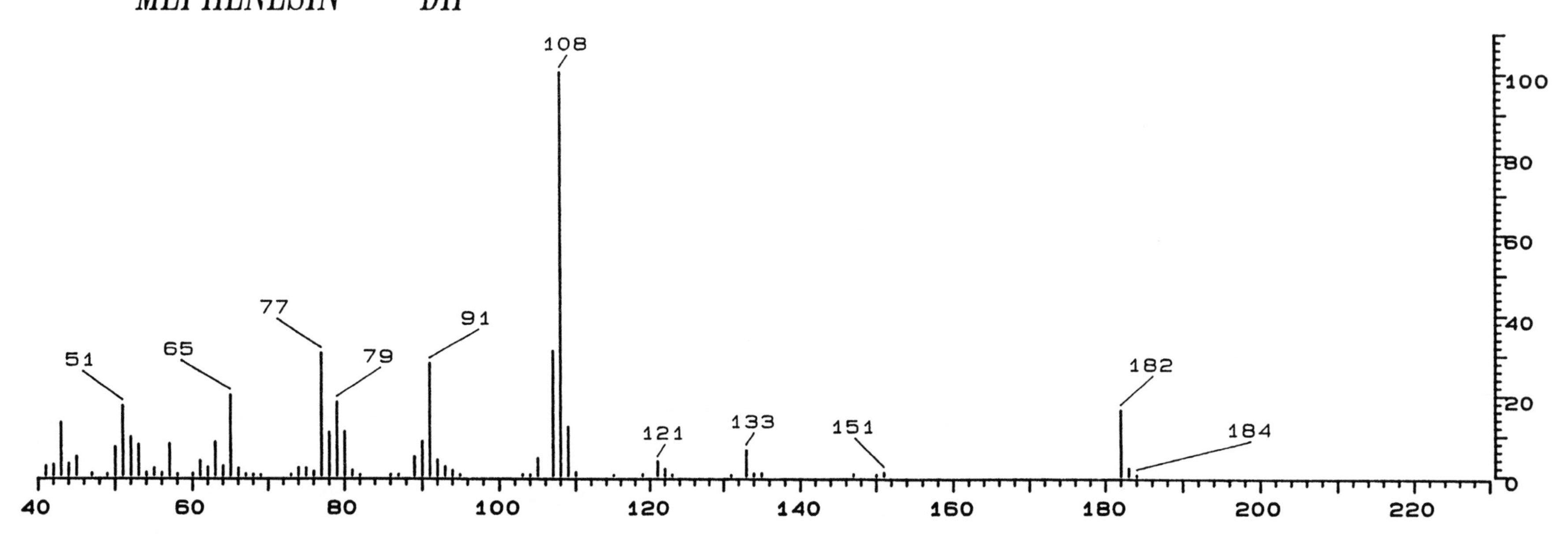

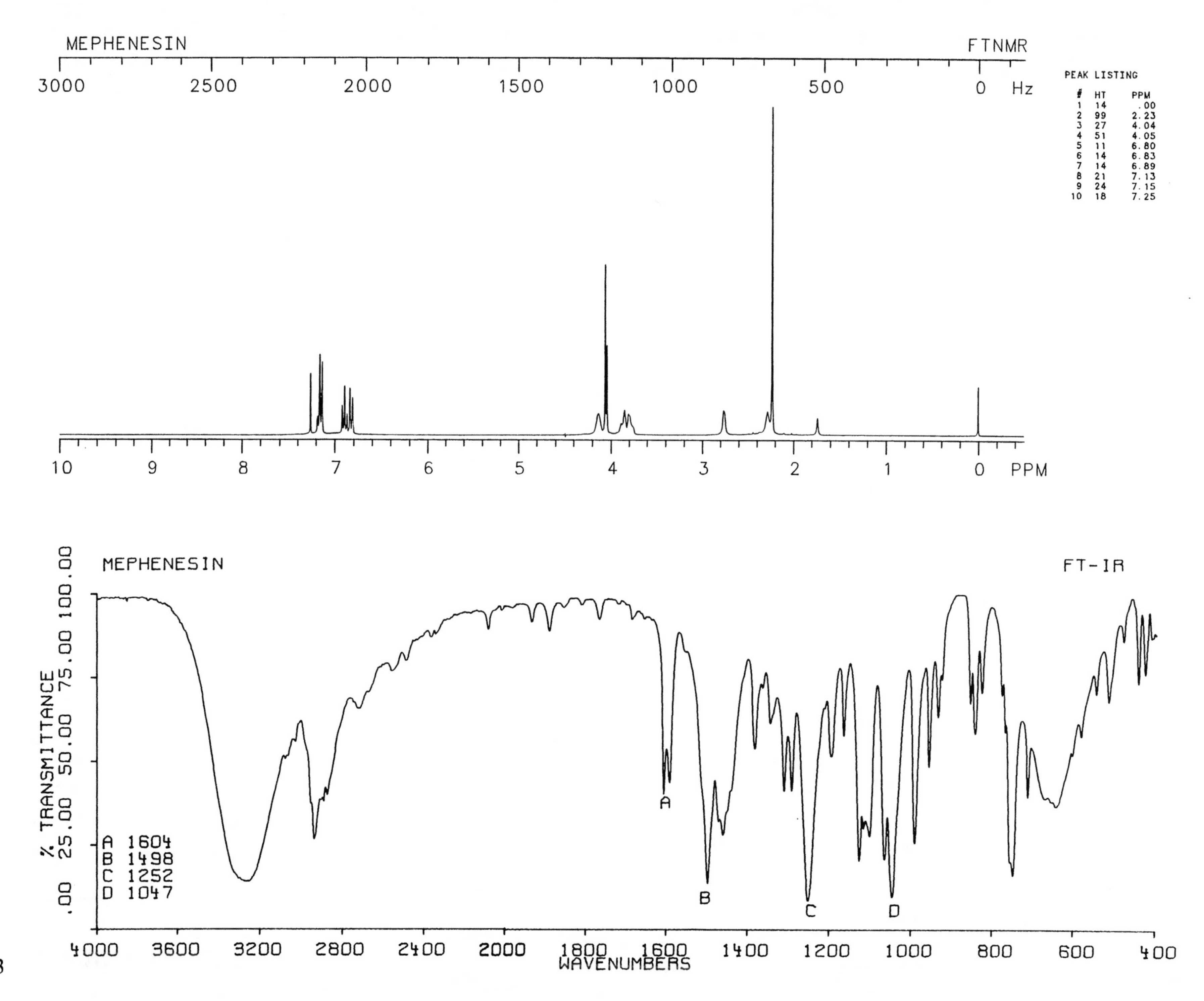
MEPHENESIN
FTNMR
3000 2500 2000 1500 1000 500 0 Hz
PEAK LISTING
# HT PPM
1 14 .00
2 99 2.23
3 27 4.04
4 51 4.05
5 11 6.80
6 14 6.83
7 14 6.89
8 21 7.13
9 24 7.15
10 18 7.25
10 9 8 7 6 5 4 3 2 1 0 PPM
MEPHENESIN
FT-IR
% TRANSMITTANCE
.00 25.00 50.00 75.00 100.00
A 1604
B 1498
C 1252
D 1047
A
B
C
D
4000 3600 3200 2800 2400 2000 1800 1600 1400 1200 1000 800 600 400
WAVENUMBERS

# MEPHENTERMINE

$C_{11}H_{17}N$

Molecular weight: 163.26 (163.14)
Synonyms: N-α,α-Trimethylbenzeneethanamine; N-α,α-trimethyl-phenethylamine
Trade names: Wyamine

Use: Adrenergic
HPLC:
GC: 1257; 140°C

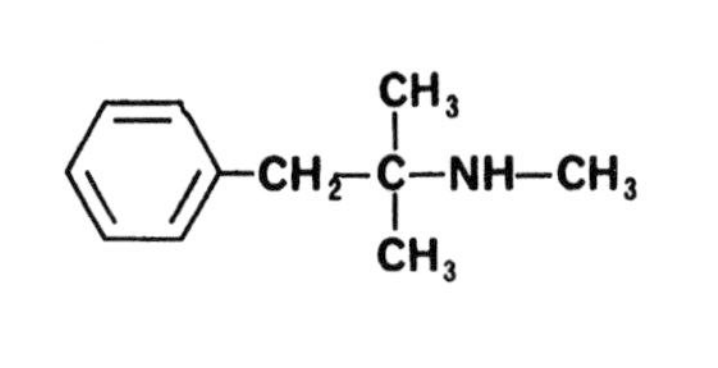


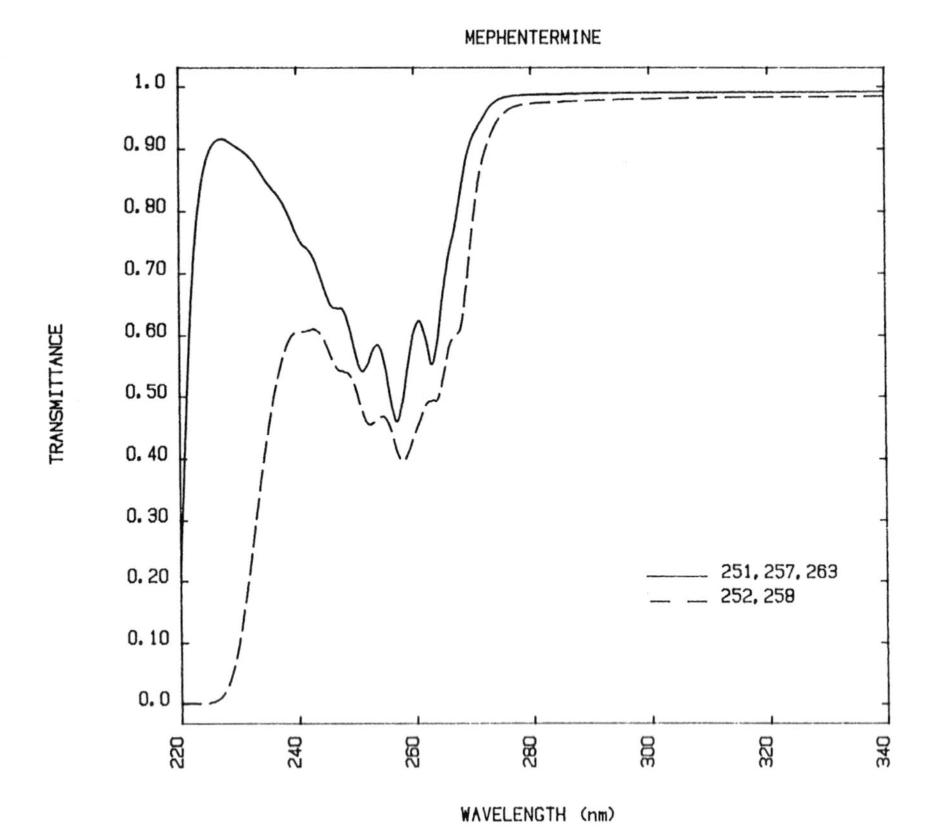

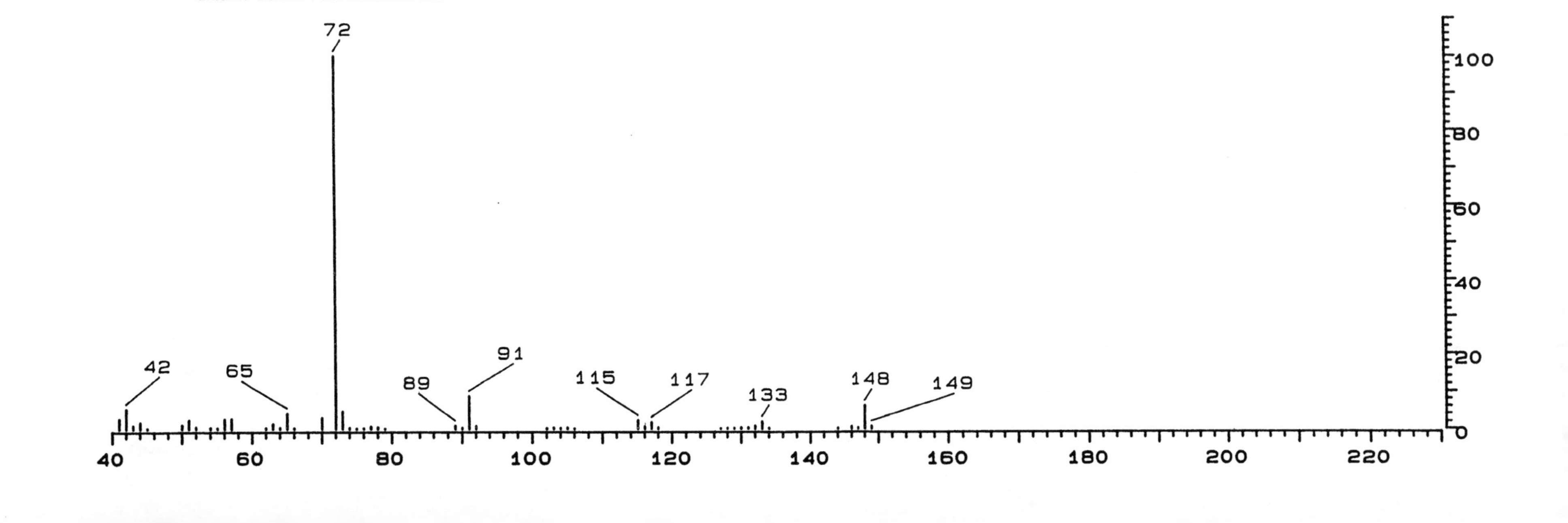

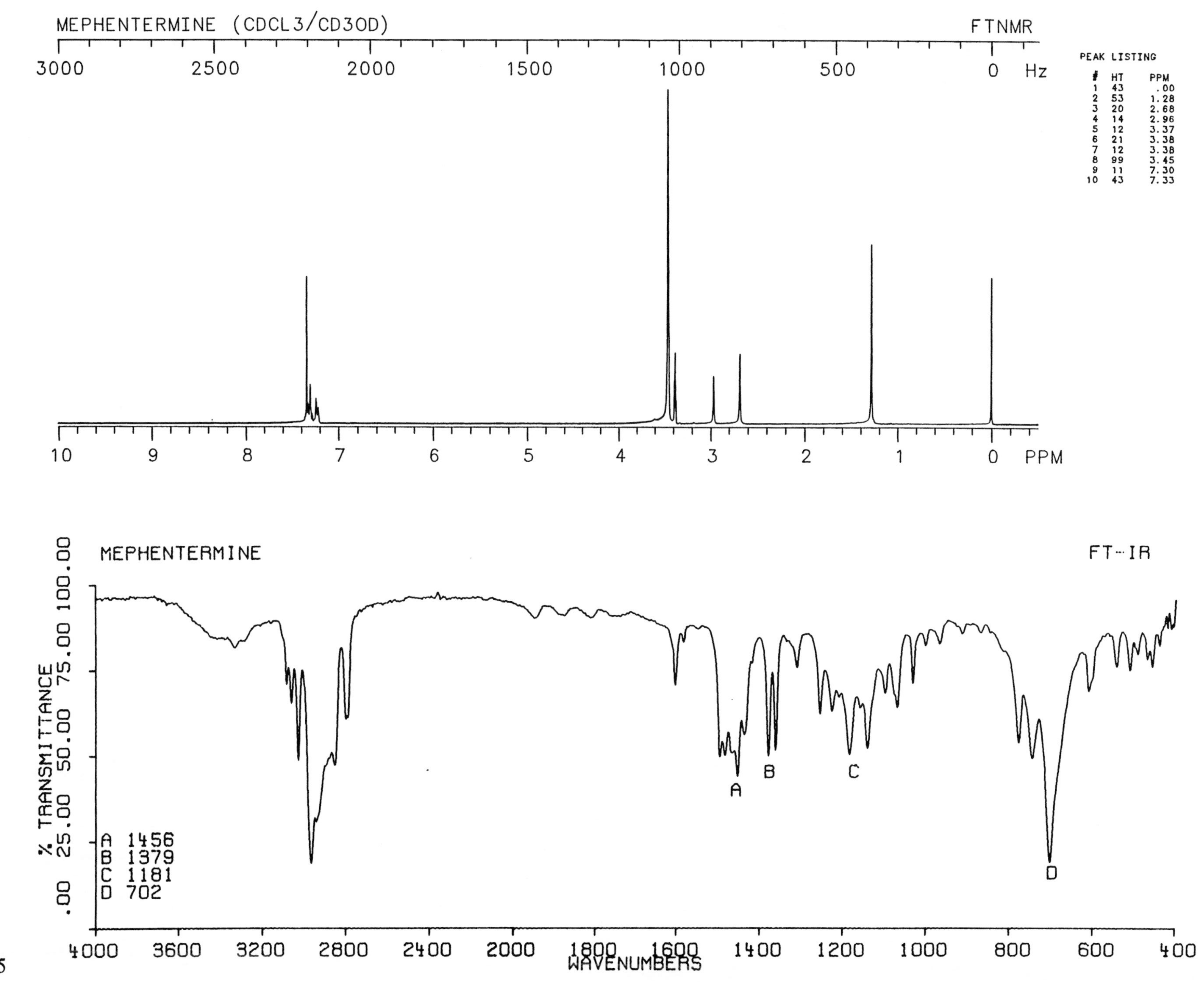
MEPHENTERMINE (CDCL3/CD3OD)
FTNMR
Hz
PPM
PEAK LISTING
# HT PPM
1 43 .00
2 53 1.28
3 20 2.68
4 14 2.96
5 12 3.37
6 21 3.38
7 12 3.38
8 99 3.45
9 11 7.30
10 43 7.33
MEPHENTERMINE
FT-IR
% TRANSMITTANCE
A 1456
B 1379
C 1181
D 702
WAVENUMBERS

# MEPHENYTOIN

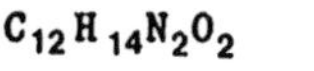

$C_{12}H_{14}N_2O_2$

Molecular weight: 218.25 (218.11)
Synonyms: 5-Ethyl-3-methyl-5-phenyl-2,4-imidazolidinedione;
5-ethyl-3-methyl-5-phenylhydantoin
Trade names: Mesantoin

Use: Anticonvulsant
HPLC: Si-10; 100B; 3.3
GC: 1808; 200°C

$C_6H_5$, $C_2H_5$, H, N, O, N, $CH_3$, O

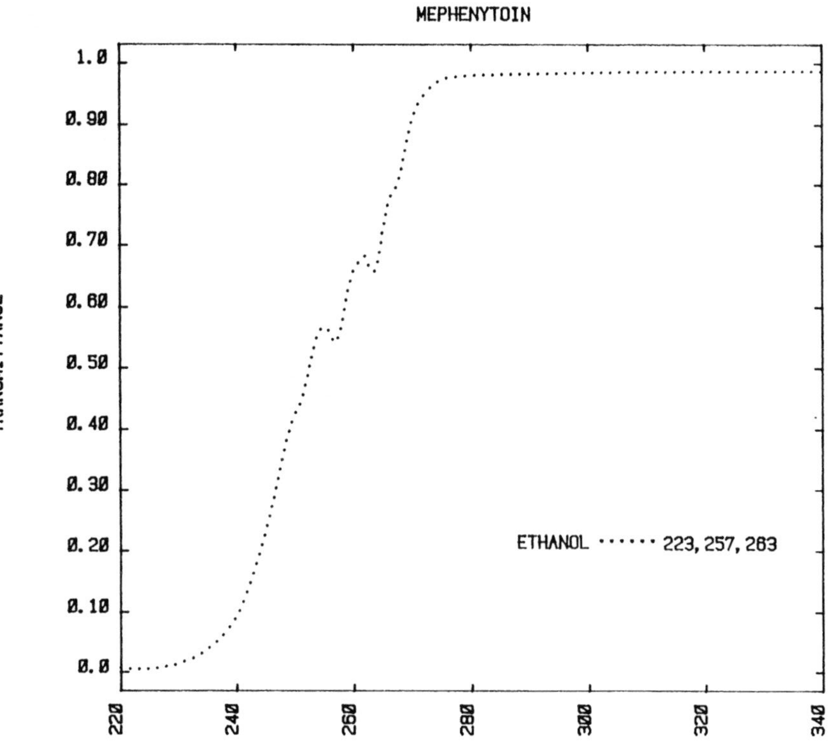

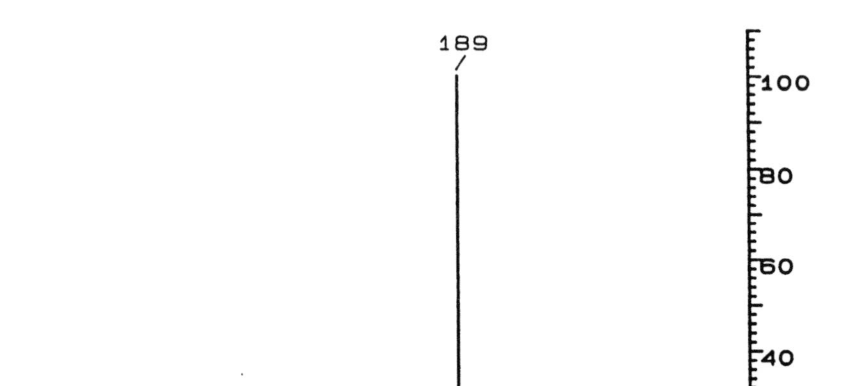

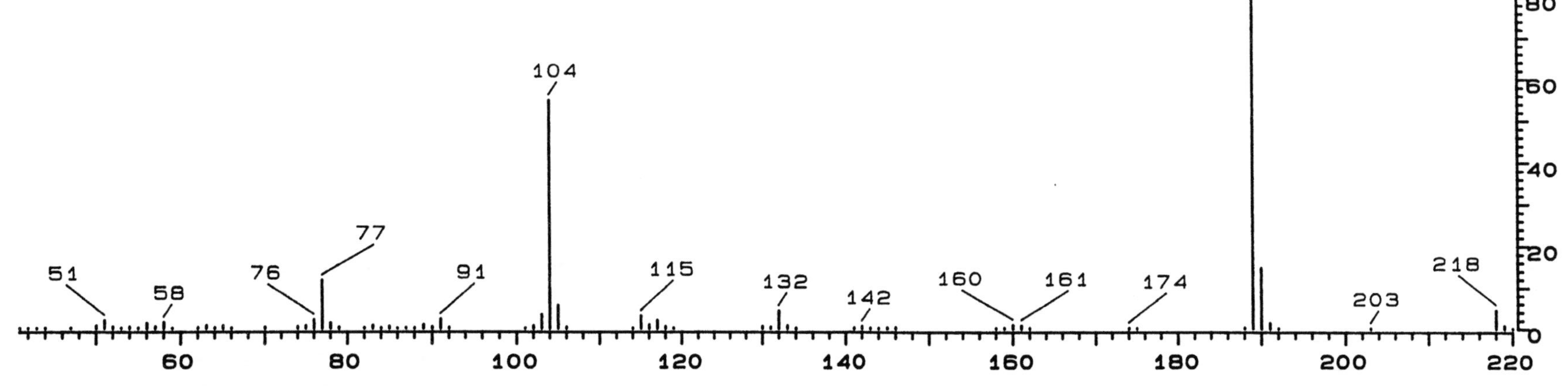

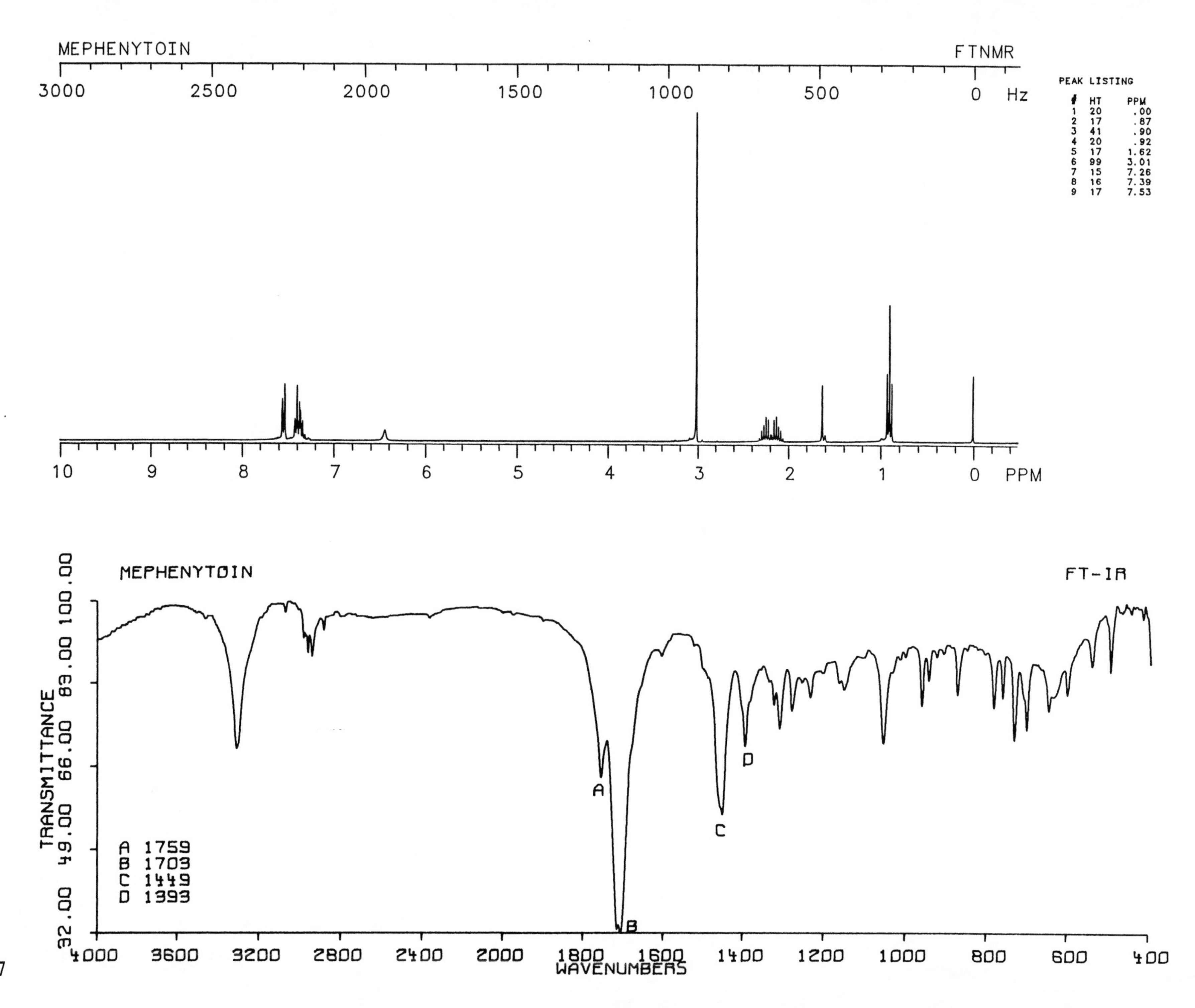
MEPHENYTOIN
FTNMR
3000 2500 2000 1500 1000 500 0 Hz
PEAK LISTING
# HT PPM
1 20 .00
2 17 .87
3 41 .90
4 20 .92
5 17 1.62
6 99 3.01
7 15 7.26
8 16 7.39
9 17 7.53
10 9 8 7 6 5 4 3 2 1 0 PPM
MEPHENYTOIN
FT-IR
TRANSMITTANCE
32.00 49.00 66.00 83.00 100.00
A 1759
B 1703
C 1449
D 1393
A B C D
4000 3600 3200 2800 2400 2000 1800 1600 1400 1200 1000 800 600 400
WAVENUMBERS

# MEPHOBARBITAL

$C_{13}H_{14}N_2O_3$

**Molecular weight:** 246.26 (246.10)
**Synonyms:** 5-Ethyl-1-methyl-5-phenyl-2,4,6(1H,3H,5H)pyrimidinetrione; 5-ethyl-1-methyl-5-phenylbarbituric acid; methylphenobarbitone
**Trade names:** Mebaral

**Use:** Sedative, anticonvulsant
**HPLC:** Si-10; 100B; 6.0
**GC:** 1896; 200°C

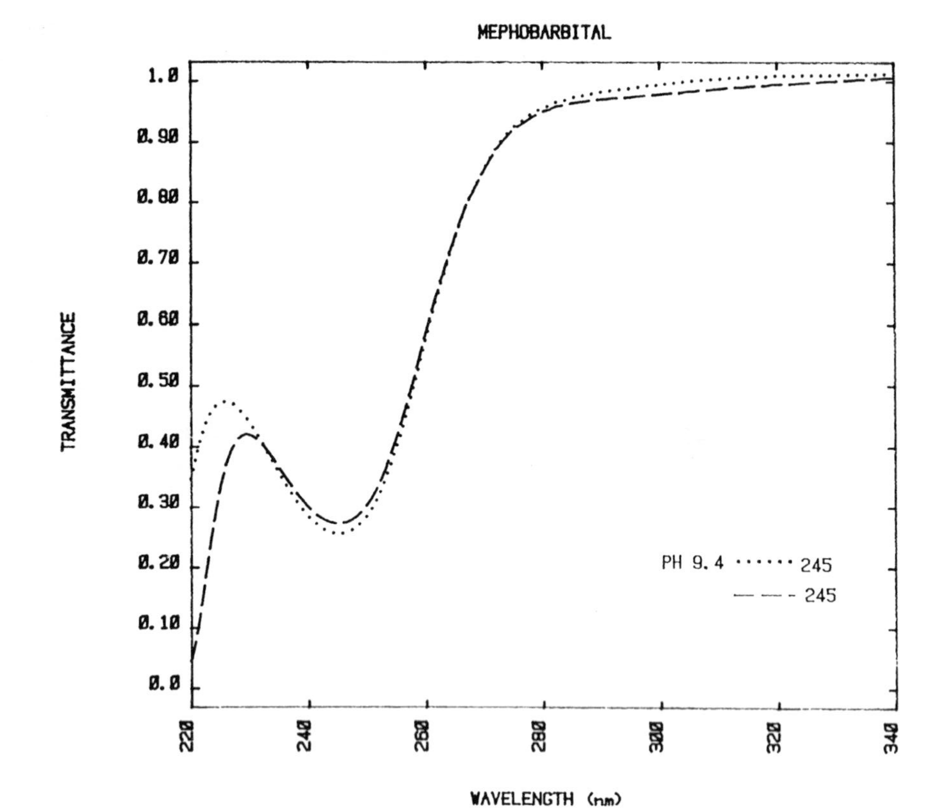

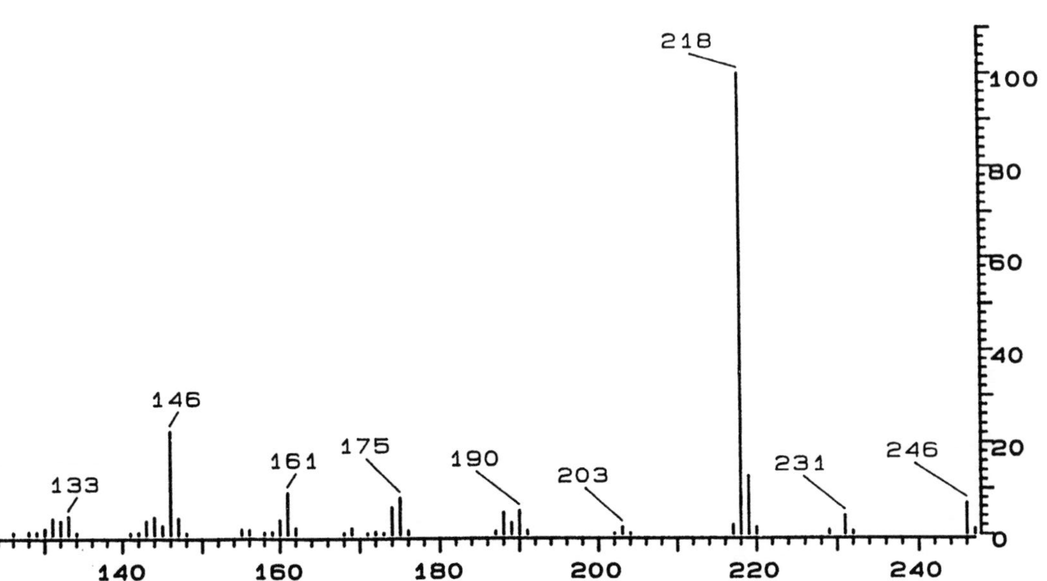

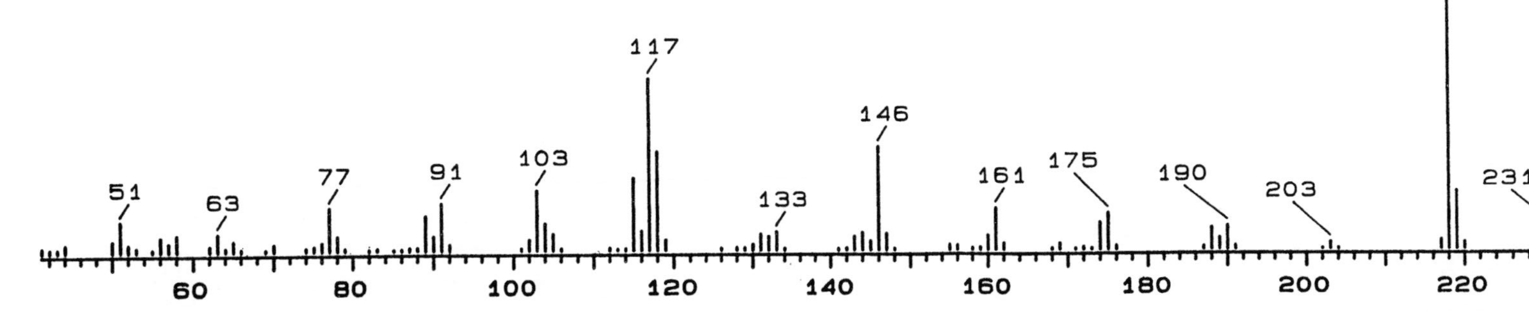

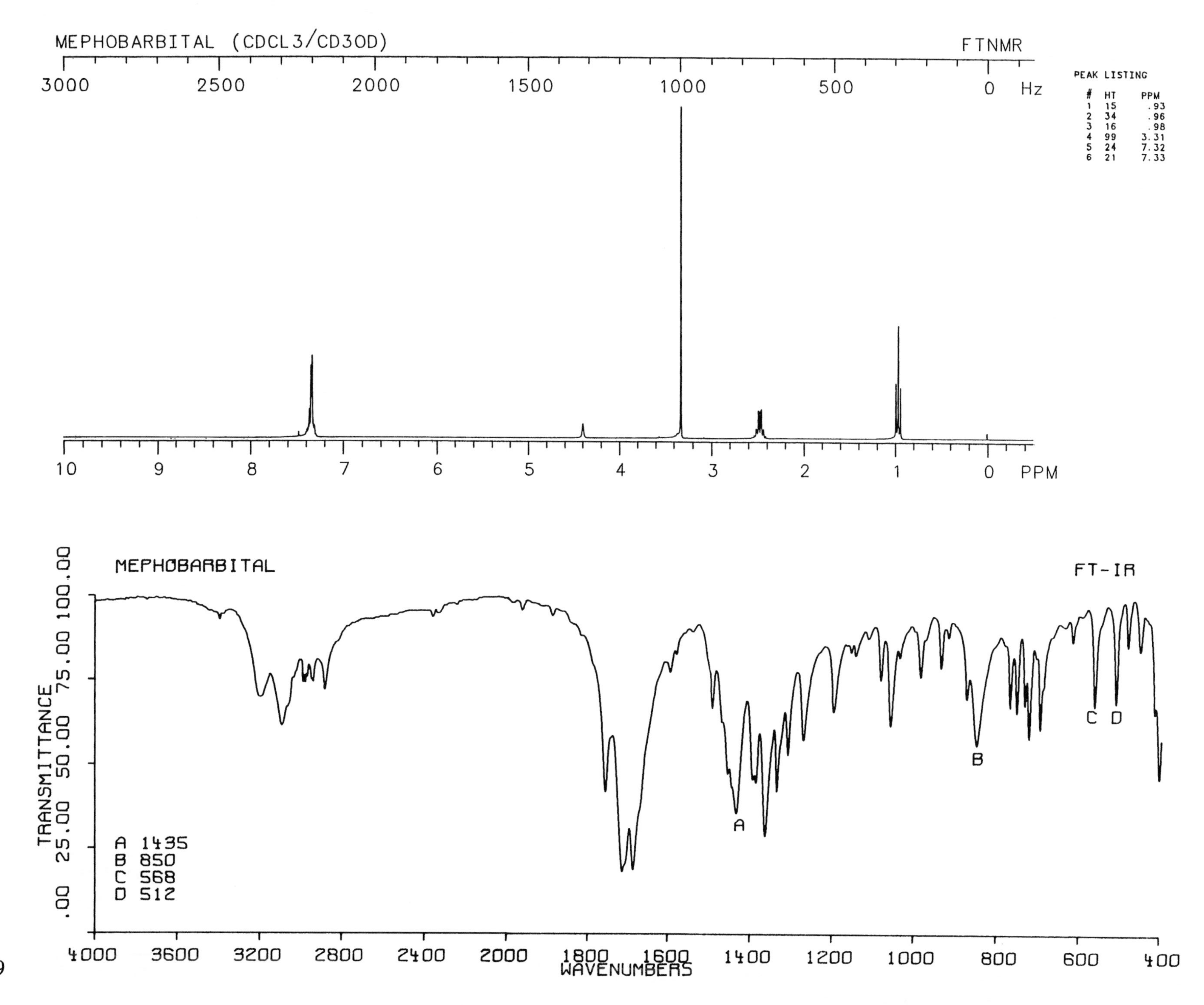
MEPHOBARBITAL (CDCL3/CD3OD)
FTNMR
3000 2500 2000 1500 1000 500 0 Hz
PEAK LISTING
# HT PPM
1 15 .93
2 34 .96
3 16 .98
4 99 3.31
5 24 7.32
6 21 7.33
10 9 8 7 6 5 4 3 2 1 0 PPM
MEPHOBARBITAL
FT-IR
TRANSMITTANCE
.00 25.00 50.00 75.00 100.00
A 1435
B 850
C 568
D 512
A
B
C D
4000 3600 3200 2800 2400 2000 1800 1600 1400 1200 1000 800 600 400
WAVENUMBERS

# MEPIRIZOLE

$C_{11}H_{14}N_4O_2$

Molecular weight: 234.26 (234.11)
Synonyms: 4-Methoxy-2-(5-methoxy-3-1H-methylpyrazol-1-yl)-6-methylpyrimidine; epirizole
Trade names: Mebron

Use: Analgesic, antipyretic, anti-inflammatory
HPLC: Si-10; 1A:99B; 4.0
GC: 1853; 200°C

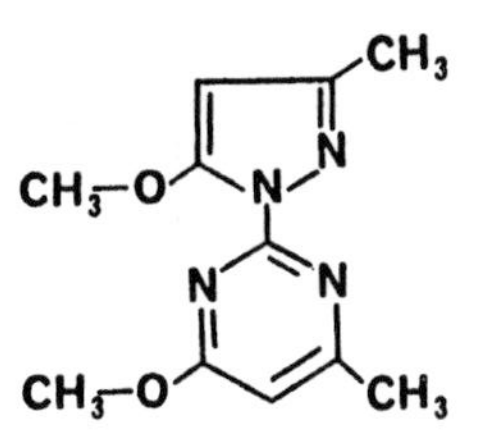


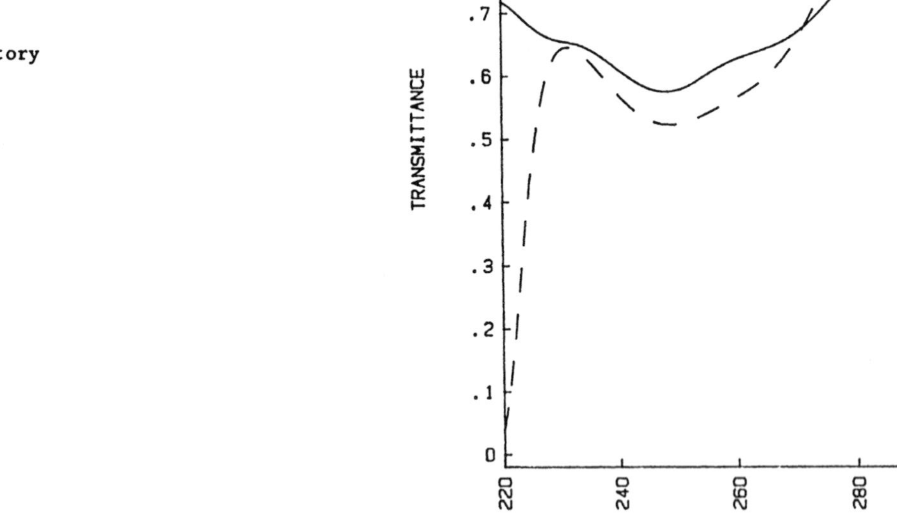


*MEPIRIZOLE*

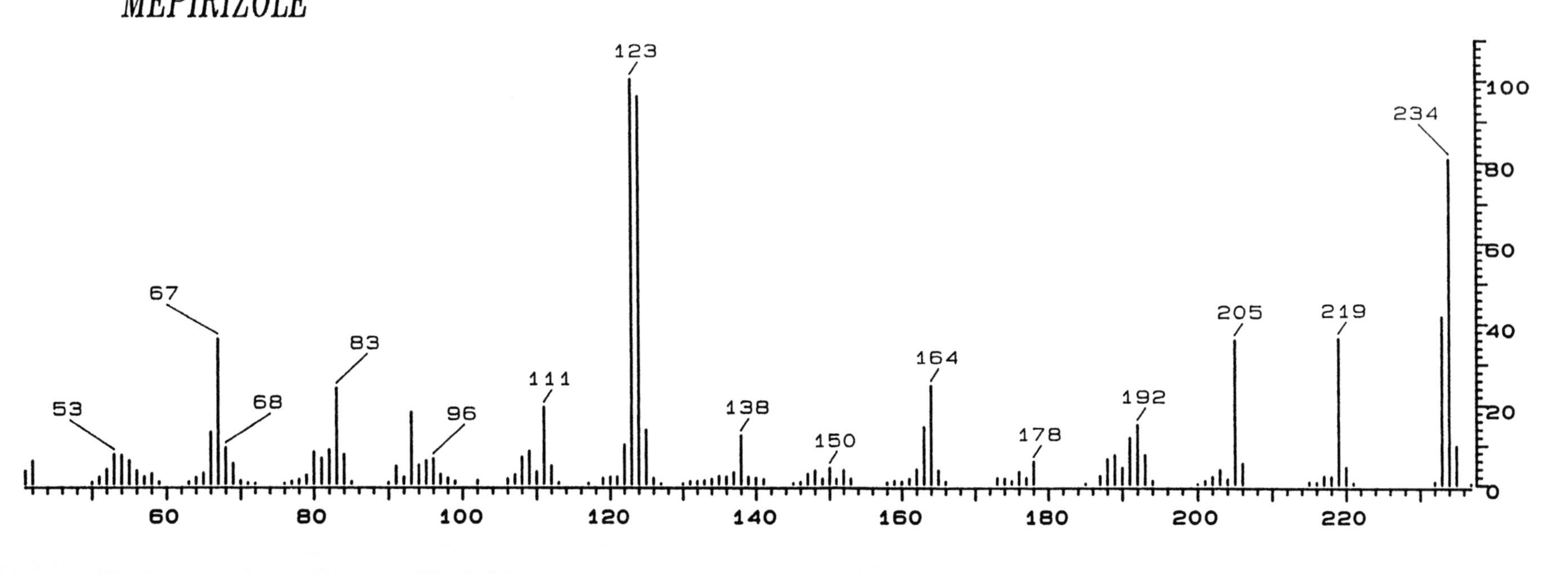

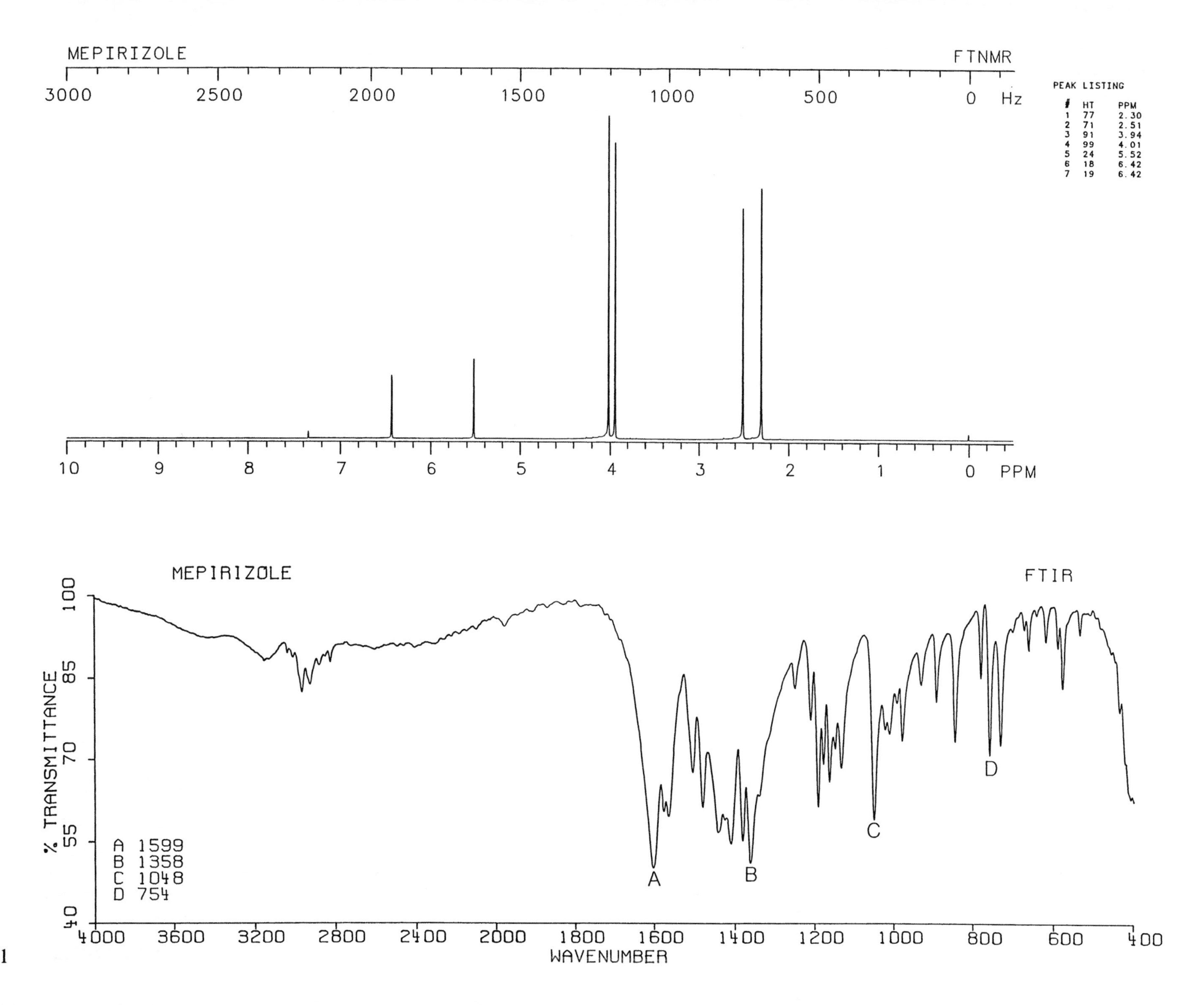
MEPIRIZOLE
FTNMR
PEAK LISTING
# HT PPM
1 77 2.30
2 71 2.51
3 91 3.94
4 99 4.01
5 24 5.52
6 18 6.42
7 19 6.42
Hz
PPM
MEPIRIZOLE
FTIR
% TRANSMITTANCE
WAVENUMBER
A 1599
B 1358
C 1048
D 754

# MEPIVACAINE

$C_{15}H_{22}N_2O$

Molecular weight: 246.35 (246.17)
Synonyms: N-(2,6-Dimethylphenyl)-1-methyl-2-piperidinecarboxamide

Trade names: Carbocaine, Scandicain

Use: Local anesthetic
HPLC: Si-10; 1A:99B; 2.7
GC: 2134; 250°C

MEPIVACAINE

*MEPIVACAINE*

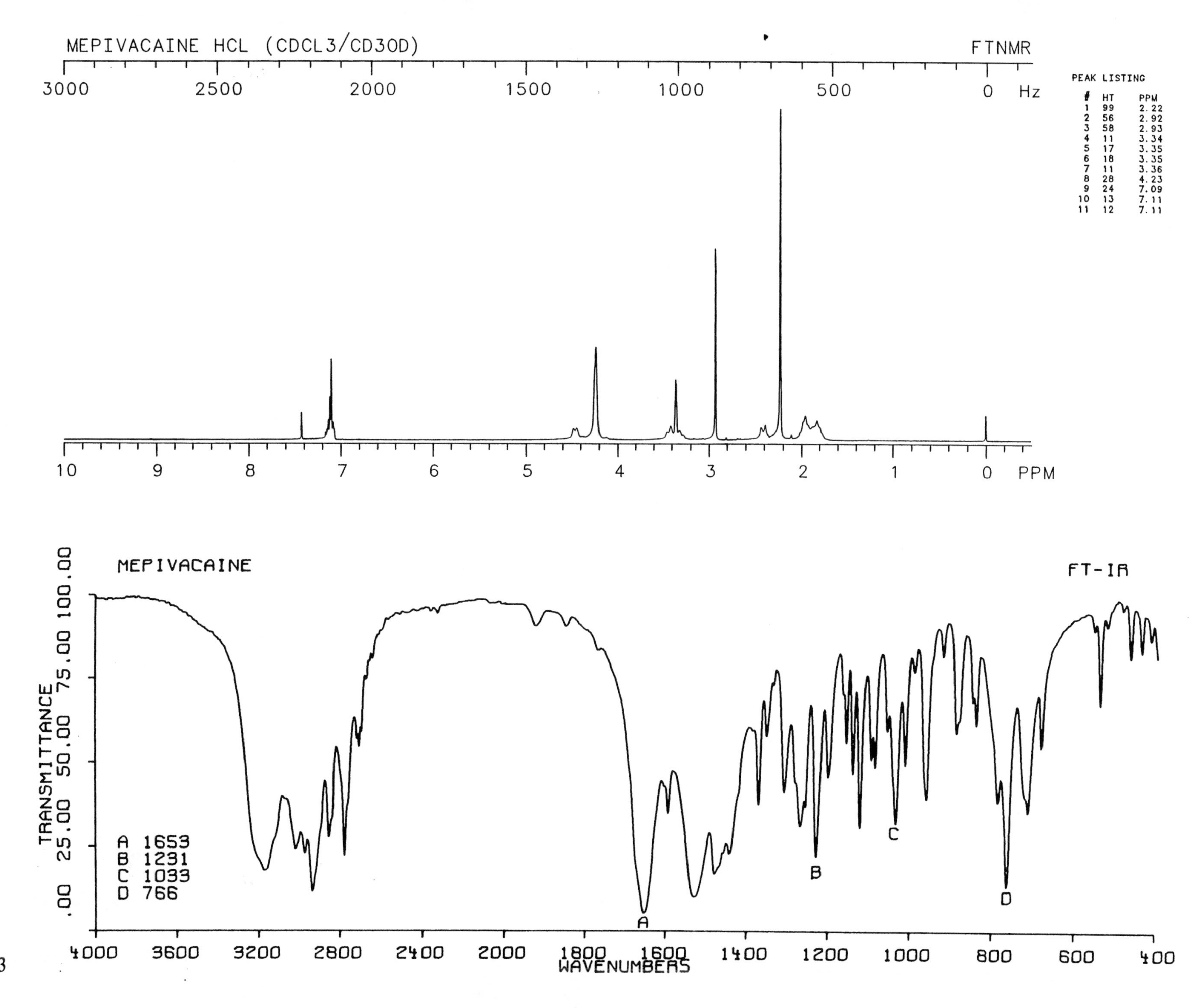

MEPIVACAINE HCL (CDCL3/CD3OD)
FTNMR
3000 2500 2000 1500 1000 500 0 Hz
PEAK LISTING
# HT PPM
1 99 2.22
2 56 2.92
3 58 2.93
4 11 3.34
5 17 3.35
6 18 3.35
7 11 3.36
8 28 4.23
9 24 7.09
10 13 7.11
11 12 7.11
10 9 8 7 6 5 4 3 2 1 0 PPM
MEPIVACAINE
FT-IR
TRANSMITTANCE
.00 25.00 50.00 75.00 100.00
A 1653
B 1231
C 1033
D 766
4000 3600 3200 2800 2400 2000 1800 1600 1400 1200 1000 800 600 400
WAVENUMBERS

# MEPROBAMATE

$C_9H_{18}N_2O_4$

**Molecular weight:** 218.25 (218.13)
**Synonyms:** 2-Methyl-2-propyl-1,3-propanediol dicarbamate; 2,2-di(carbamoyloxymethyl)pentane
**Trade names:** Deprol, Equagesic, Equanil, Meprogesic, Milpath Milprem, Miltown, Pathibamate
**Use:** Tranquilizer
**HPLC:**
**GC:** 1804; 200°C

$$NH_2-C(=O)-O-CH_2-C(CH_3)(CH_2-CH_2-CH_3)-CH_2-O-C(=O)-NH_2$$

MEPROBAMATE

FTNMR

3000 2500 2000 1500 1000 500 0 Hz

PEAK LISTING

| # | HT | PPM |
|---|---|---|
| 1 | 29 | -.00 |
| 2 | 17 | .00 |
| 3 | 28 | .90 |
| 4 | 28 | .91 |
| 5 | 89 | .91 |
| 6 | 67 | 1.25 |
| 7 | 23 | 1.27 |
| 8 | 49 | 1.28 |
| 9 | 27 | 1.29 |
| 10 | 99 | 3.90 |
| 11 | 16 | 4.75 |
| 12 | 36 | 7.26 |
| 13 | 23 | 7.26 |

10 9 8 7 6 5 4 3 2 1 0 PPM

MEPROBAMATE

FT-IR

TRANSMITTANCE .00 25.00 50.00 75.00 100.00

A 1689
B 1590
C 1407
D 1336

A B C D

4000 3600 3200 2800 2400 2000 1800 1600 1400 1200 1000 800 600 400
WAVENUMBERS

# 6-MERCAPTOPURINE

$C_5H_4N_4S$

Molecular weight: 152.18 (152.02)
Synonyms: 6-Purinethiol; 6MP

Trade names: Purinethol

Use: Antineoplastic
HPLC: Si-10; 5A:95B; 4.5
GC:

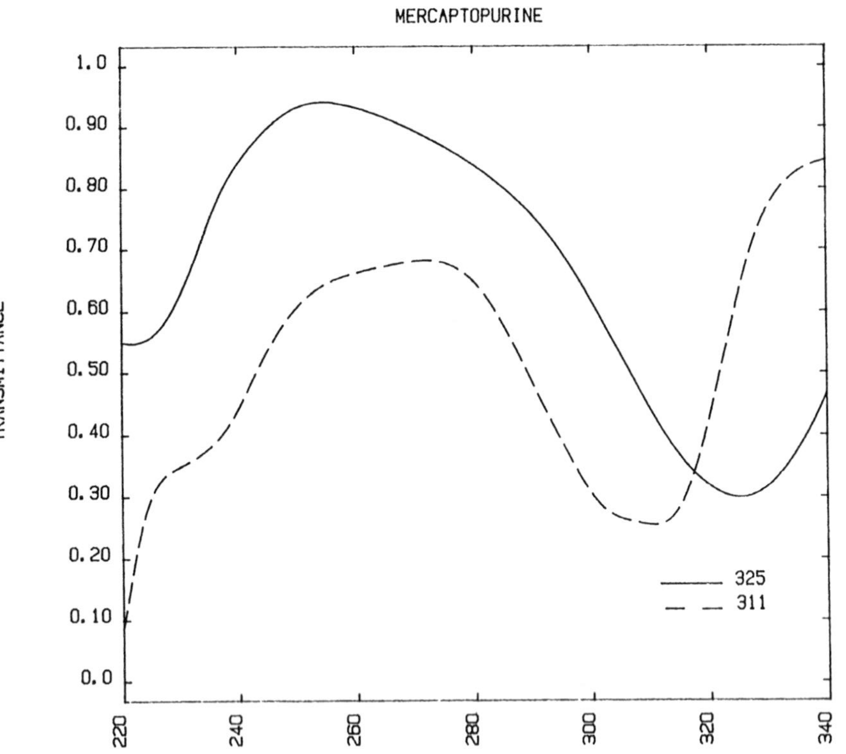

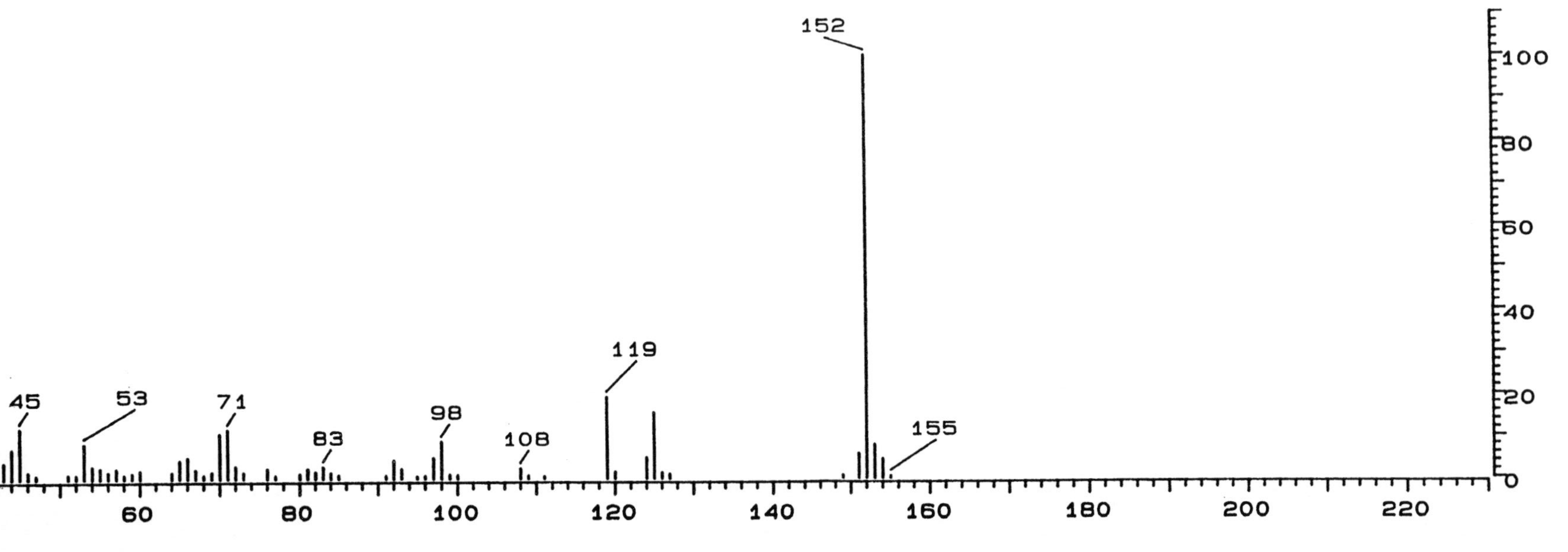

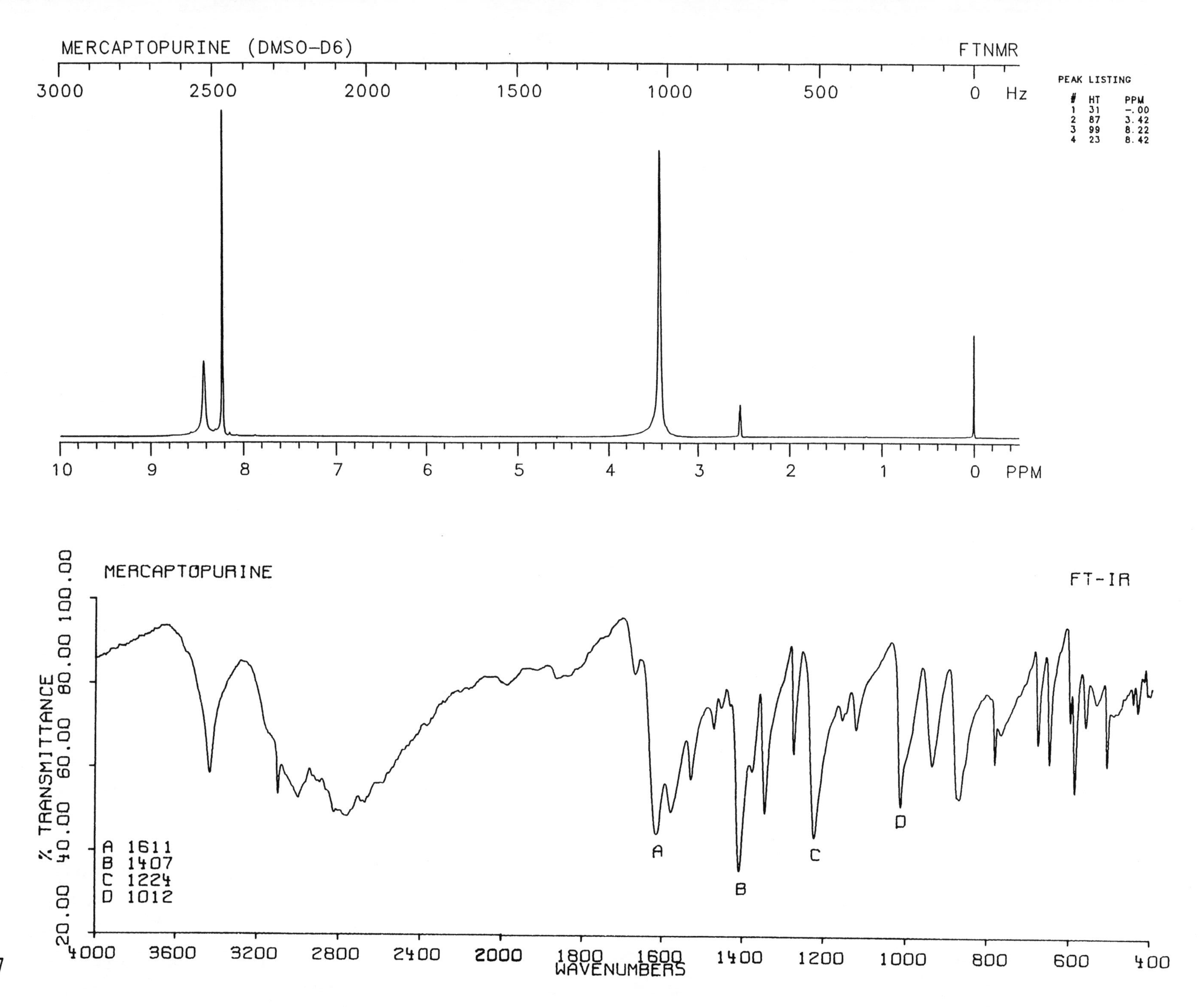

MERCAPTOPURINE (DMSO-D6)
FTNMR
3000
2500
2000
1500
1000
500
0 Hz
PEAK LISTING
# HT PPM
1 31 -.00
2 87 3.42
3 99 8.22
4 23 8.42
10
9
8
7
6
5
4
3
2
1
0 PPM
MERCAPTOPURINE
FT-IR
% TRANSMITTANCE
20.00
40.00
60.00
80.00
100.00
A 1611
B 1407
C 1224
D 1012
A
B
C
D
4000
3600
3200
2800
2400
2000
1800
1600
1400
1200
1000
800
600
400
WAVENUMBERS

# MESCALINE

$C_{11}H_{17}NO_3$

**Molecular weight:** 211.26 (211.12)
**Synonyms:** 3,4,5-Trimethoxybenzeneethanamine; mezcaline

**Trade names:**

**Use:** Hallucinogen
**HPLC:** Si-10; 20A:80B; 5.0
**GC:** 1694; 200°C

$CH_2-CH_2-NH_2$

$CH_3O$ $OCH_3$ $OCH_3$

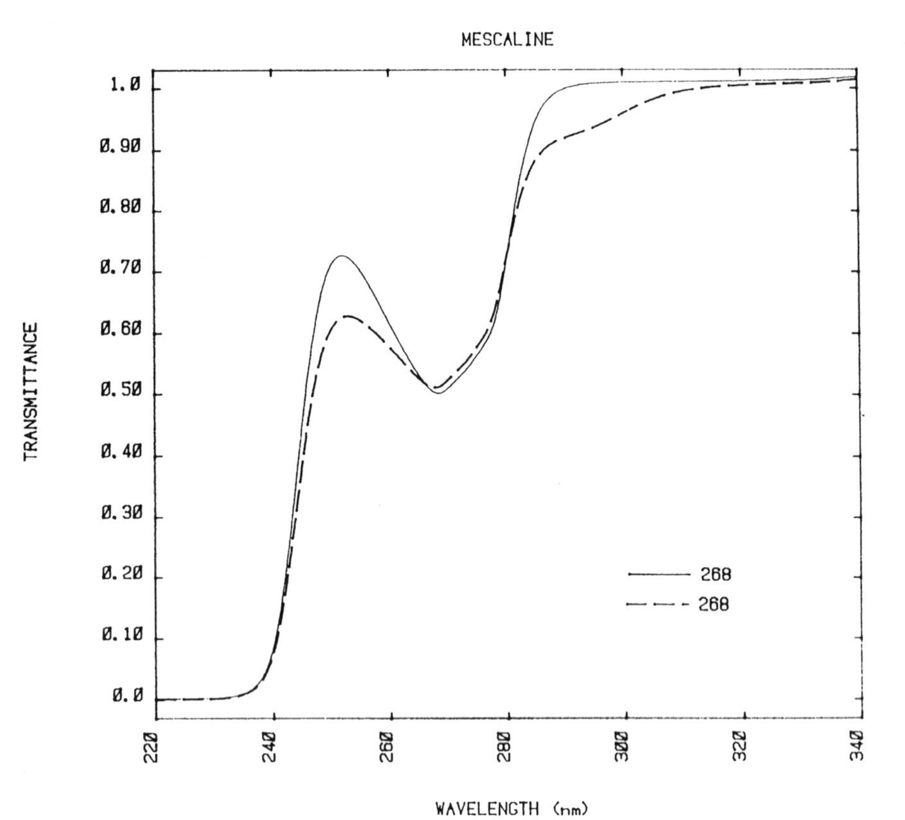


*MESCALINE*

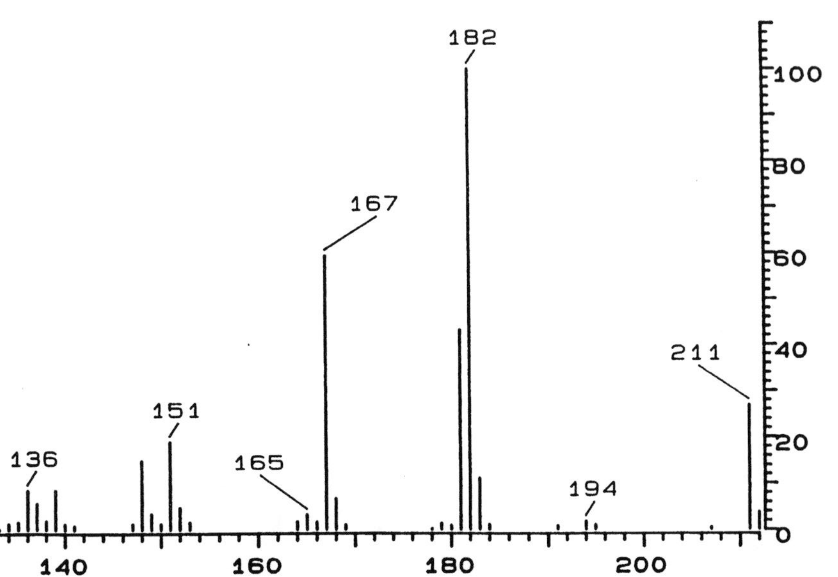


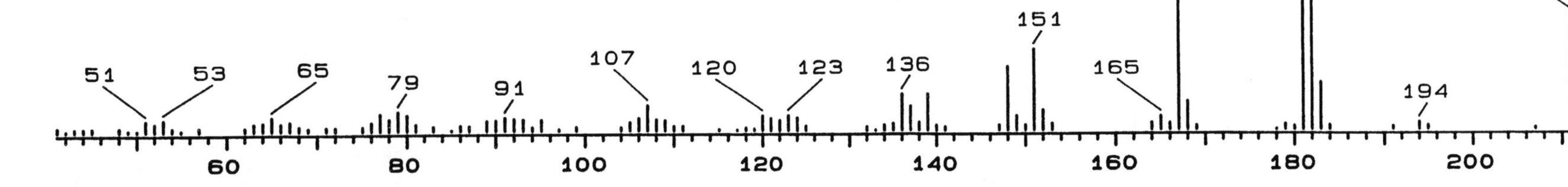

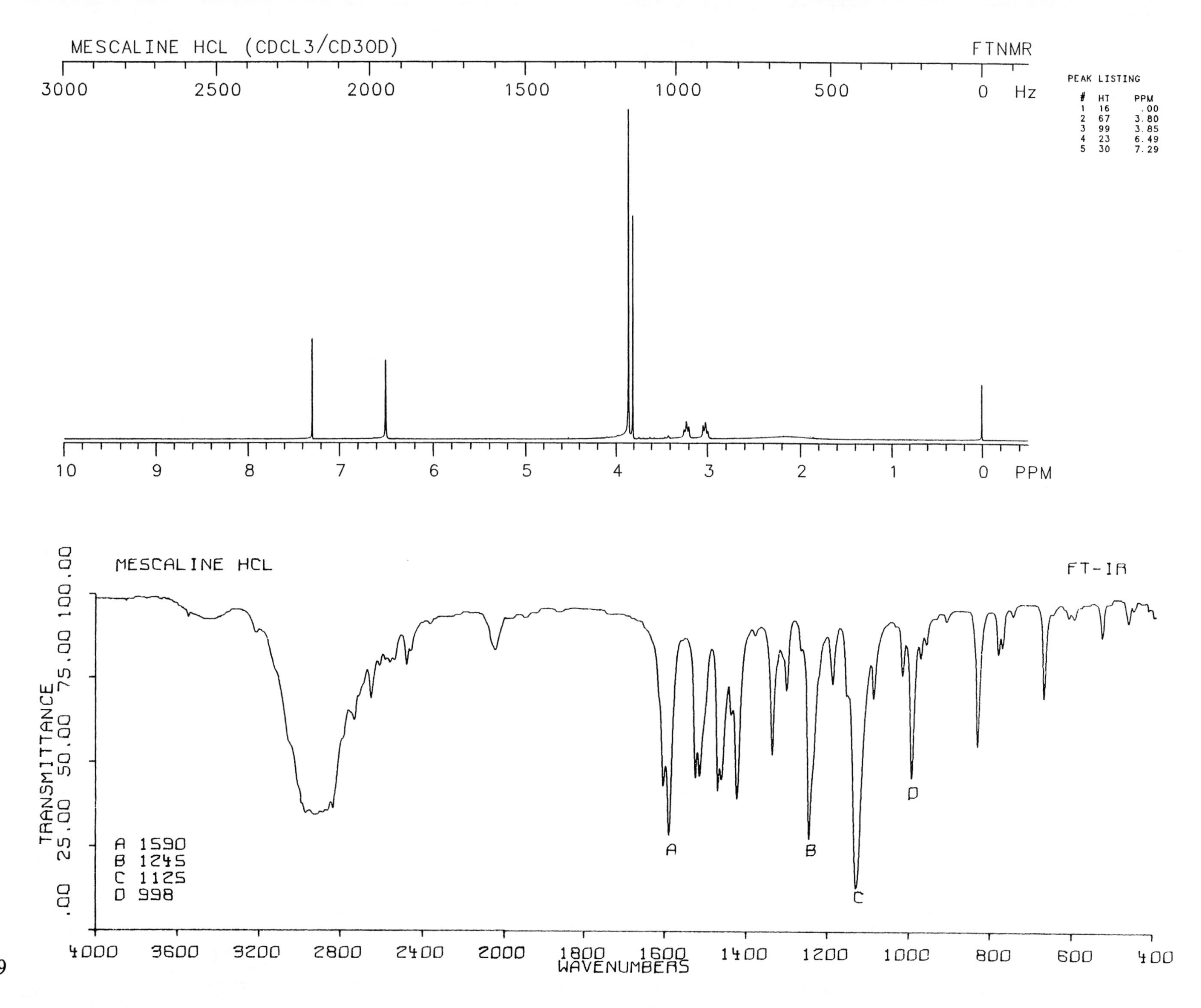
MESCALINE HCL (CDCL3/CD3OD)
FTNMR
3000 2500 2000 1500 1000 500 0 Hz
PEAK LISTING
# HT PPM
1 16 .00
2 67 3.80
3 99 3.85
4 23 6.49
5 30 7.29
10 9 8 7 6 5 4 3 2 1 0 PPM
MESCALINE HCL
FT-IR
TRANSMITTANCE
.00 25.00 50.00 75.00 100.00
A 1590
B 1245
C 1125
D 998
A
B
C
D
4000 3600 3200 2800 2400 2000 1800 1600 1400 1200 1000 800 600 400
WAVENUMBERS

# MESORIDAZINE

$C_{21}H_{26}N_2OS_2$

Molecular weight: 386.57 (386.15)
Synonyms: 10-[2-(1-Methyl-2-piperidinyl)ethyl]-2-(methyl-sulfinyl)-10H-phenothiazine
Trade names: Lidanil, Serentil

Use: Tranquilizer
HPLC: Si-10; 20A:80B; 6.0
GC:

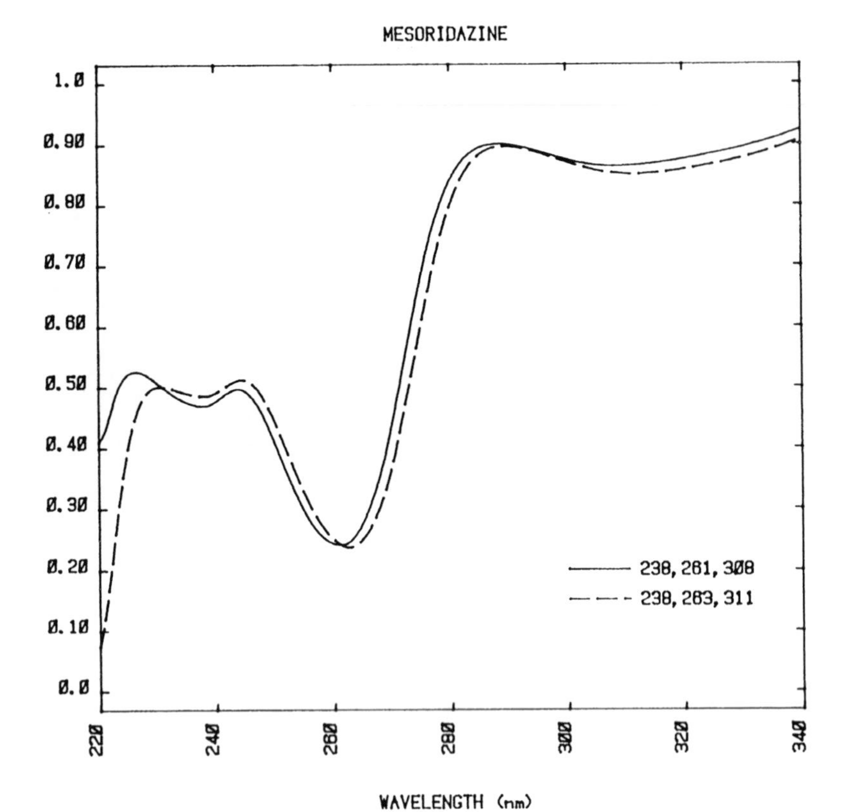


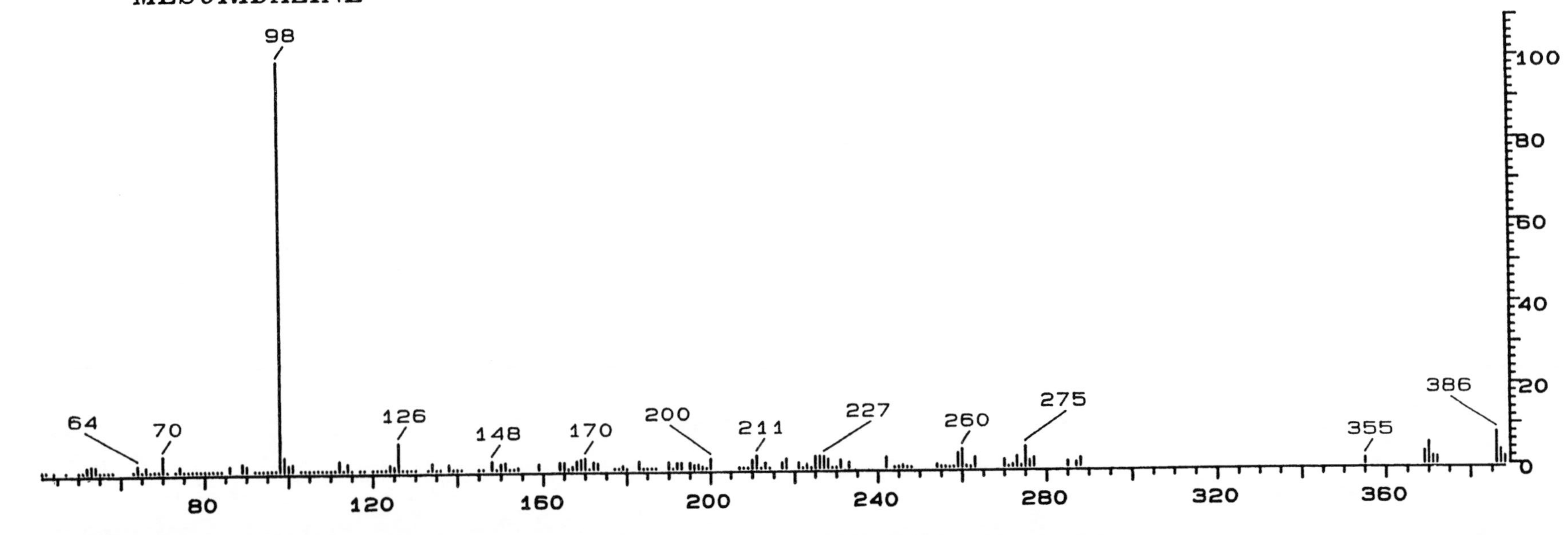

MESORIDAZINE BESYLATE

FTNMR

PEAK LISTING

| # | HT | PPM |
|---|---|---|
| 1 | 18 | .00 |
| 2 | 31 | 2.59 |
| 3 | 35 | 2.61 |
| 4 | 12 | 2.67 |
| 5 | 99 | 2.69 |
| 6 | 41 | 2.70 |
| 7 | 17 | 2.71 |
| 8 | 13 | 7.18 |
| 9 | 15 | 7.26 |
| 10 | 19 | 7.26 |
| 11 | 32 | 7.27 |
| 12 | 15 | 7.29 |
| 13 | 22 | 7.36 |
| 14 | 13 | 7.37 |
| 15 | 28 | 7.37 |
| 16 | 35 | 7.38 |
| 17 | 13 | 7.86 |
| 18 | 15 | 7.87 |
| 19 | 13 | 7.88 |
| 20 | 15 | 7.89 |

3000 2500 2000 1500 1000 500 0 Hz

10 9 8 7 6 5 4 3 2 1 0 PPM

MESORIDAZINE

FT-IR

A 1456
B 1407
C 1040
D 751

TRANSMITTANCE

40.00 55.00 70.00 85.00 100.00

4000 3600 3200 2800 2400 2000 1800 1600 1400 1200 1000 800 600 400

WAVENUMBERS

# MESTRANOL

$C_{21}H_{26}O_2$

Molecular weight: 310.42 (310.19)
Synonyms: 3-Methoxy-19-nor-17α-pregna-1,3-5(10)-trien-20-yn-17-ol; 17α-ethynylestradiol 3-methyl ether
Trade names: Enovid, Metrulen, Norinyl, Ovulen, Ortho-Novum

Use: Estrogen
HPLC: Si-10; 100B; 2.9
GC: 2680; 280°C

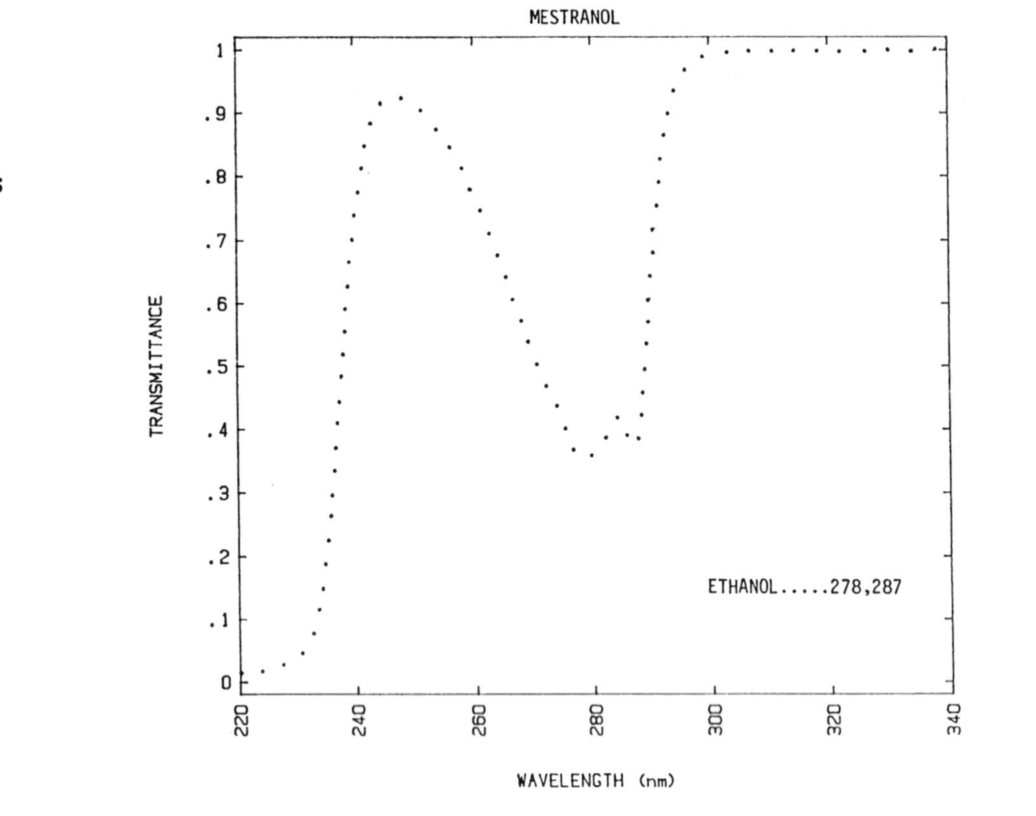


*MESTRANOL*

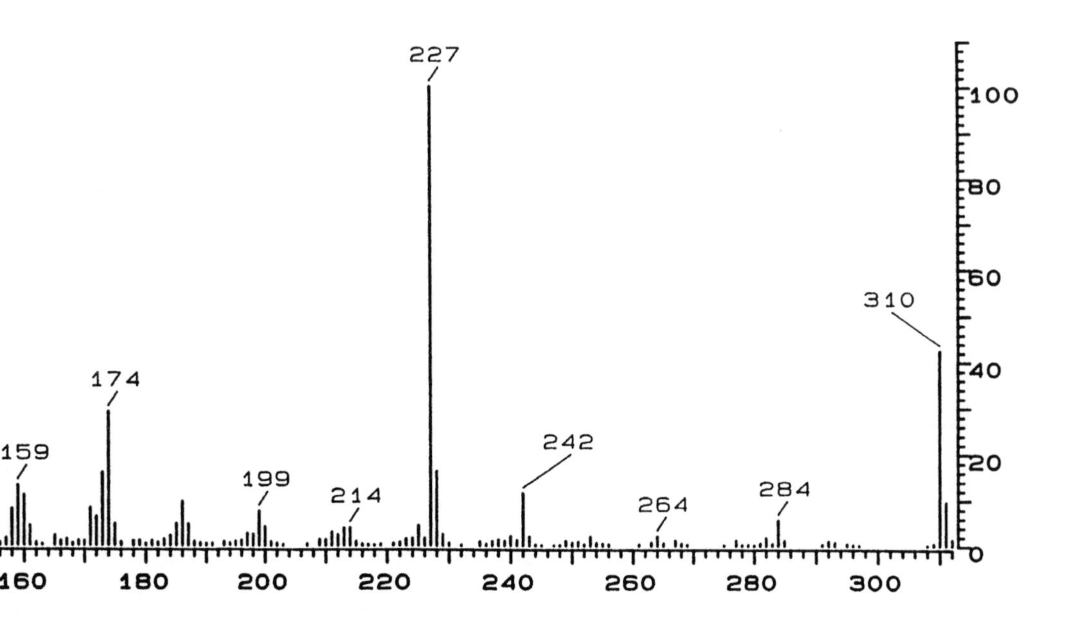

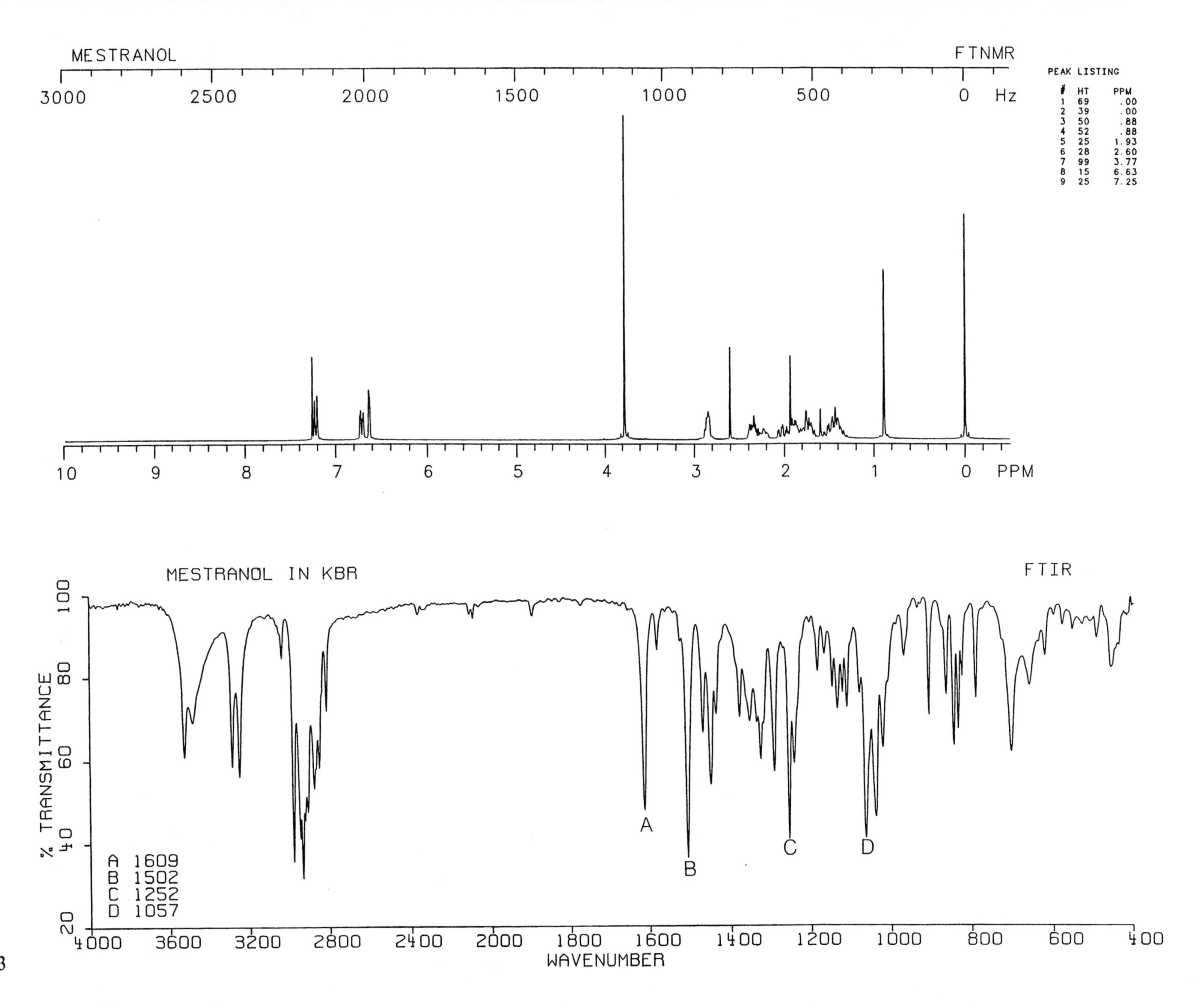
MESTRANOL
FTNMR
3000 2500 2000 1500 1000 500 0 Hz
PEAK LISTING
# HT PPM
1 69 .00
2 39 .00
3 50 .88
4 52 .88
5 25 1.93
6 28 2.60
7 99 3.77
8 15 6.63
9 25 7.25
10 9 8 7 6 5 4 3 2 1 0 PPM
MESTRANOL IN KBR
FTIR
% TRANSMITTANCE
100 80 60 40 20
A 1609
B 1502
C 1252
D 1057
A B C D
4000 3600 3200 2800 2400 2000 1800 1600 1400 1200 1000 800 600 400
WAVENUMBER

# METANEPHRINE

$C_{10}H_{15}NO_3$

Molecular weight: 197.23 (197.11)
Synonyms: 4-Hydroxy-3-methoxy-α-(methylaminomethyl)benzenemethanol; 3-O-methylepinephrine; metadrenaline
Trade names:

Use: Metabolite of adrenaline
HPLC: Si-10; 50A:50B; 3.6
GC:

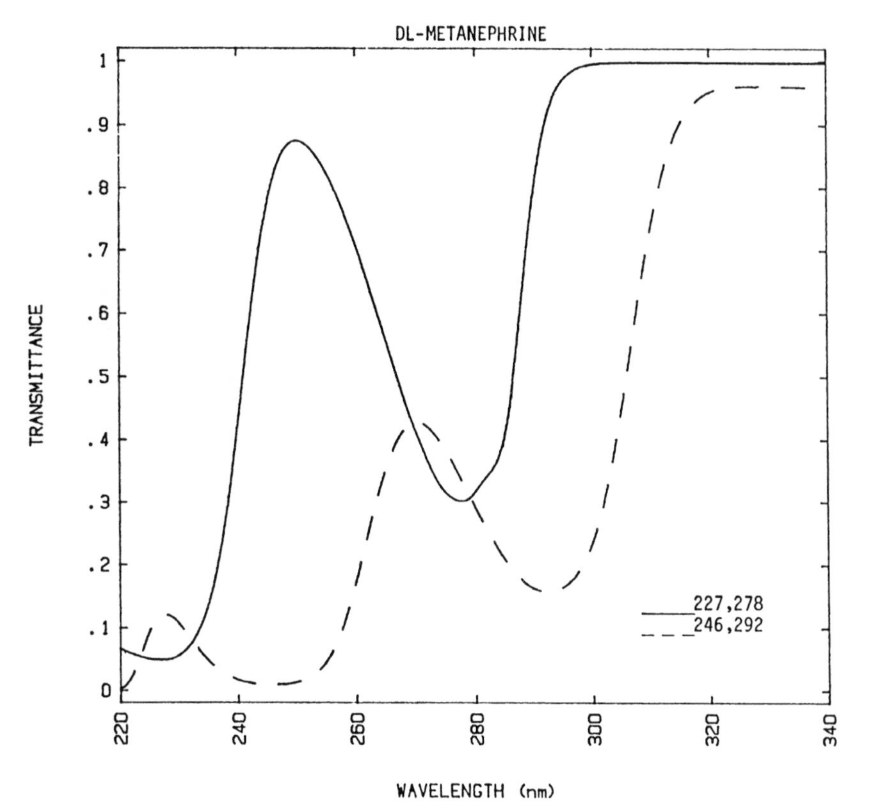


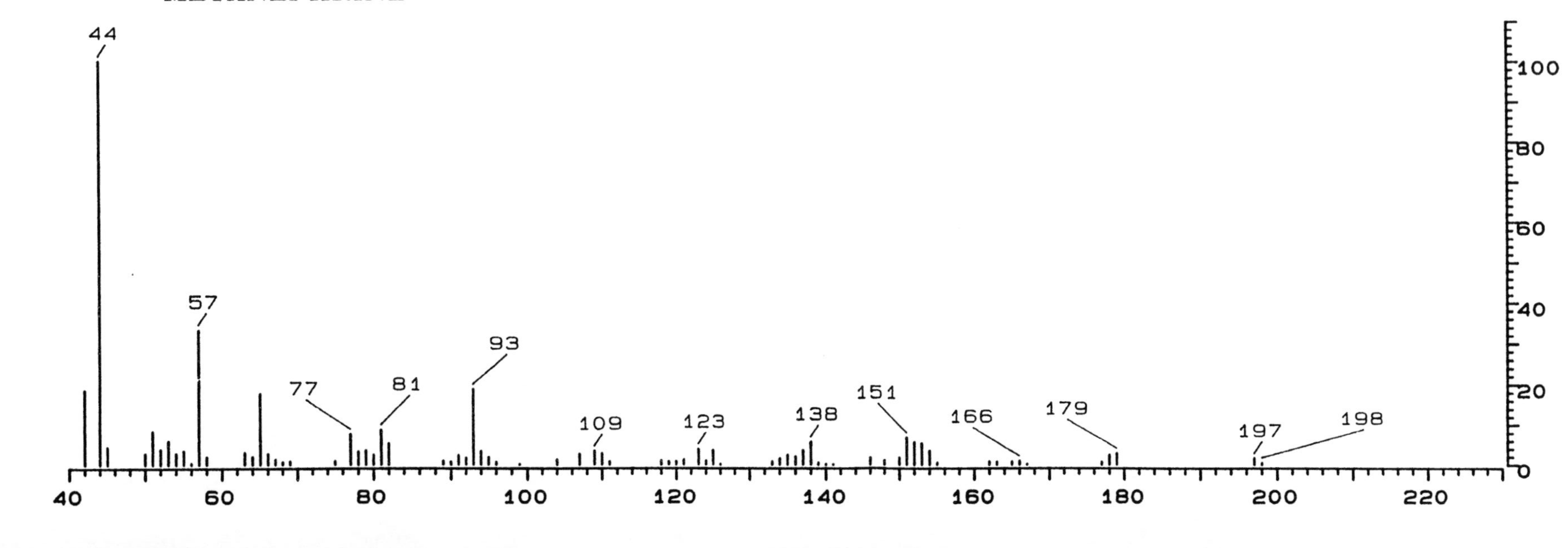

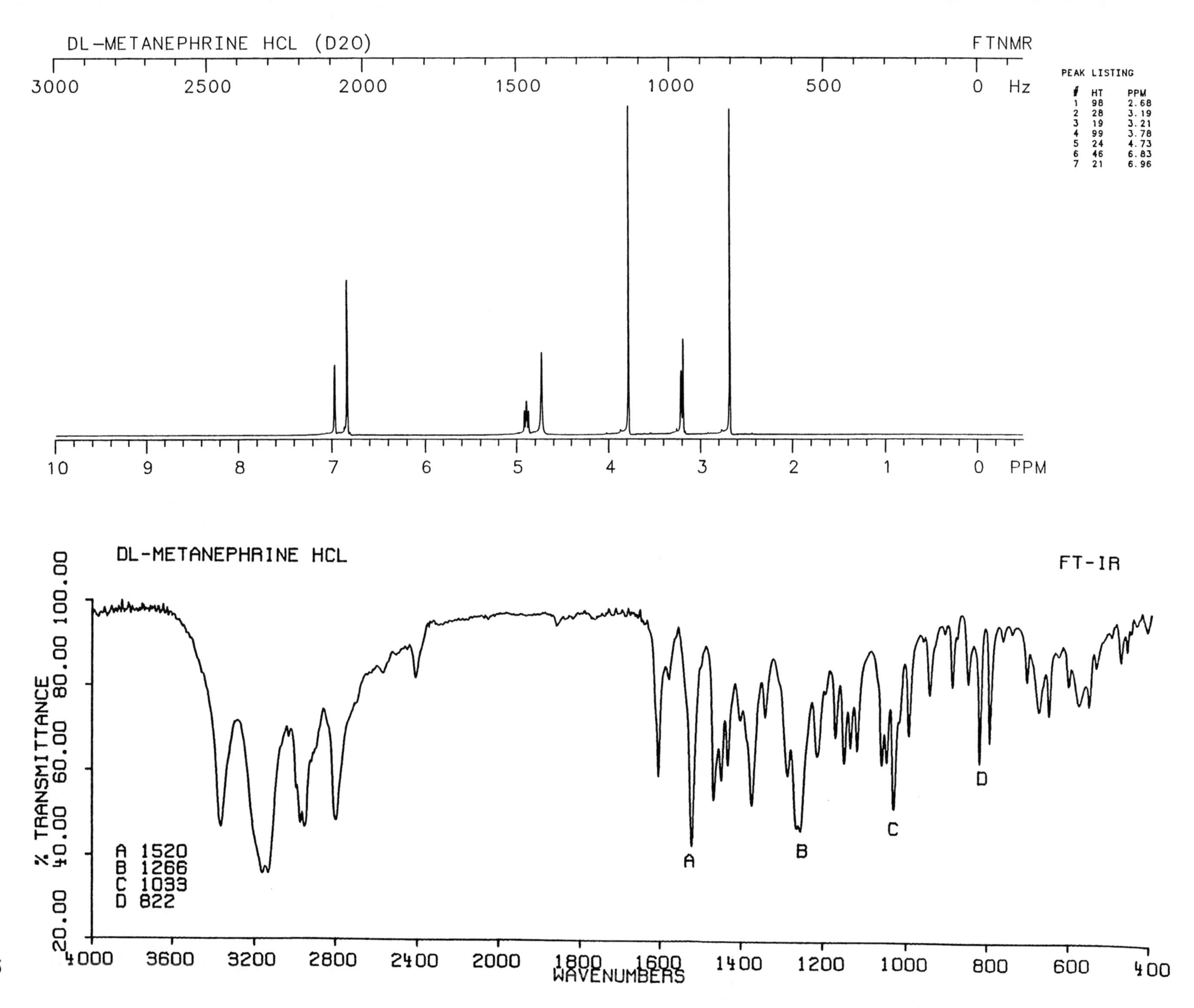
DL-METANEPHRINE HCL (D2O)
FTNMR
3000 2500 2000 1500 1000 500 0 Hz
PEAK LISTING
# HT PPM
1 98 2.68
2 28 3.19
3 19 3.21
4 99 3.78
5 24 4.73
6 46 6.83
7 21 6.96
10 9 8 7 6 5 4 3 2 1 0 PPM
DL-METANEPHRINE HCL
FT-IR
% TRANSMITTANCE
20.00 40.00 60.00 80.00 100.00
A 1520
B 1266
C 1033
D 822
A
B
C
D
4000 3600 3200 2800 2400 2000 1800 1600 1400 1200 1000 800 600 400
WAVENUMBERS

# METAPROTERENOL

$C_{11}H_{17}NO_3$

Molecular weight: 211.26 (211.12)
Synonyms: 5-[1-Hydroxy-2-[(1-methylethyl)amino]ethyl]-
1,3-benzenediol
Trade names: Alupent, Metaprel

Use: Bronchodilator
HPLC:
GC:

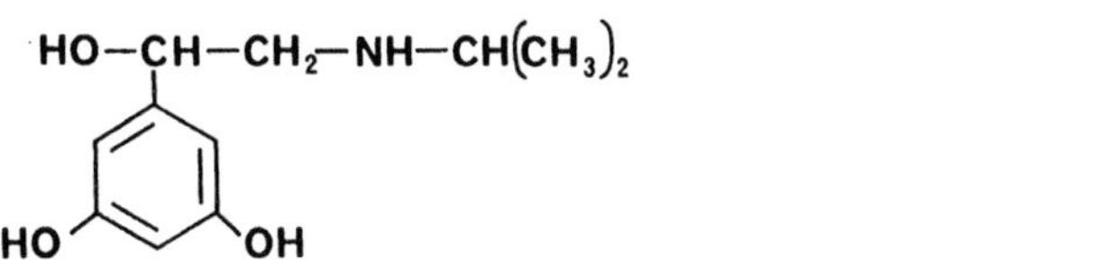


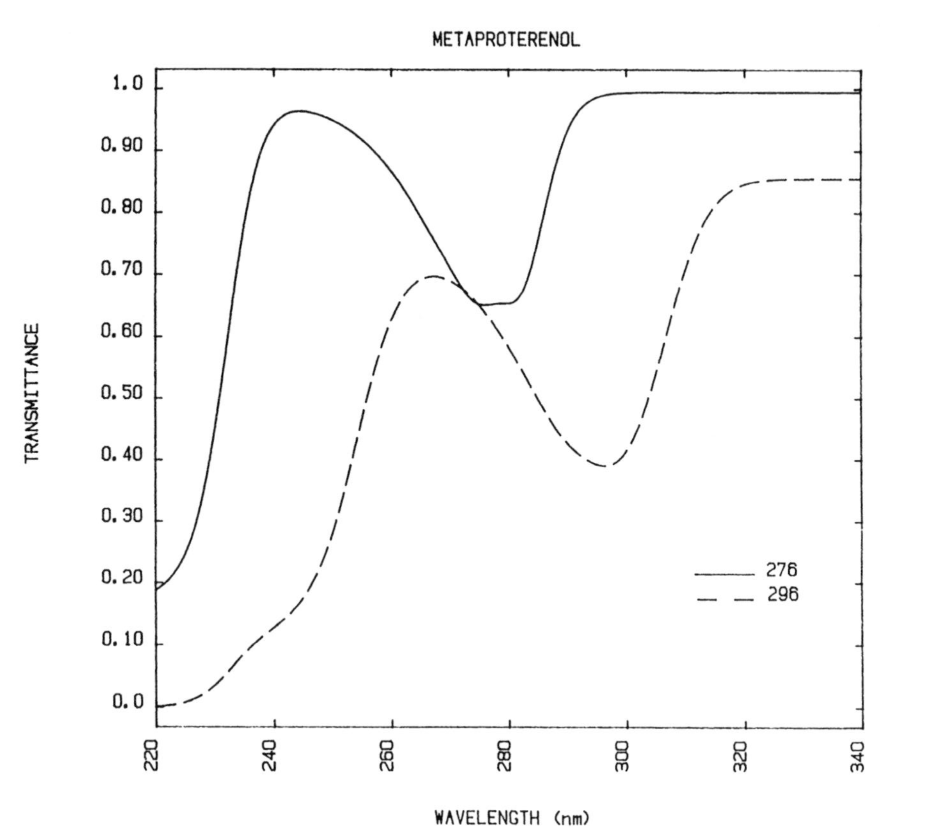

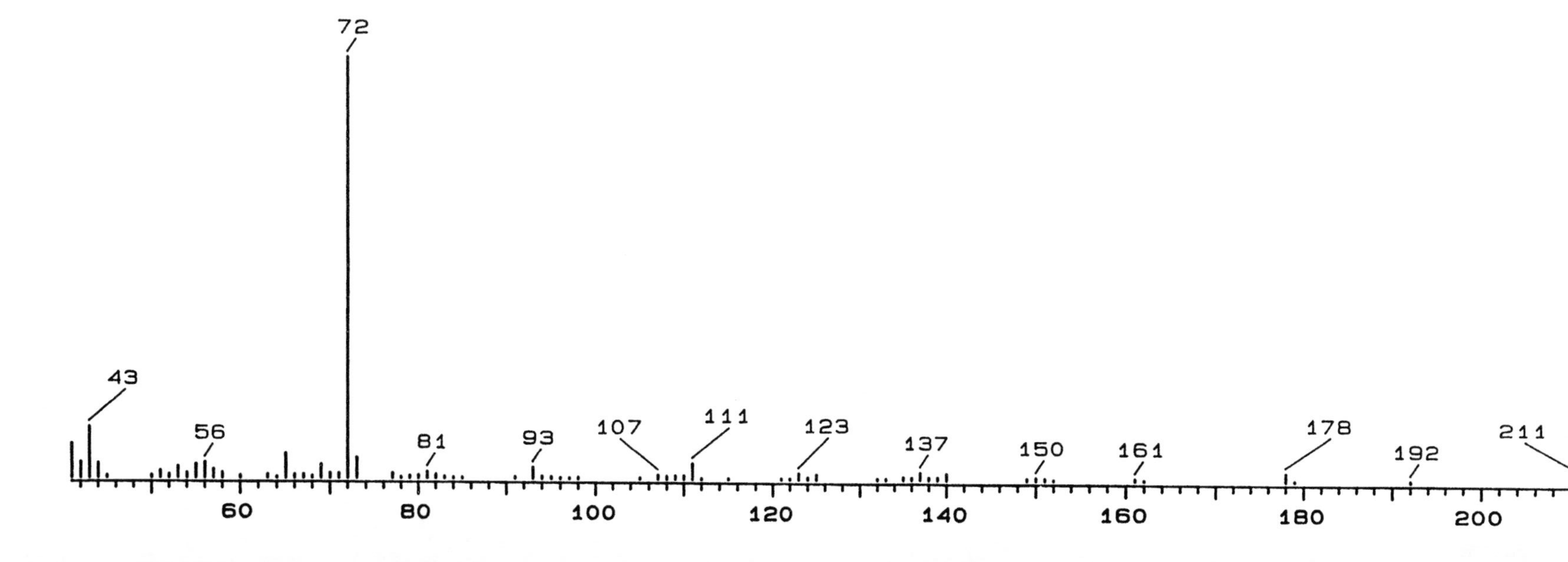

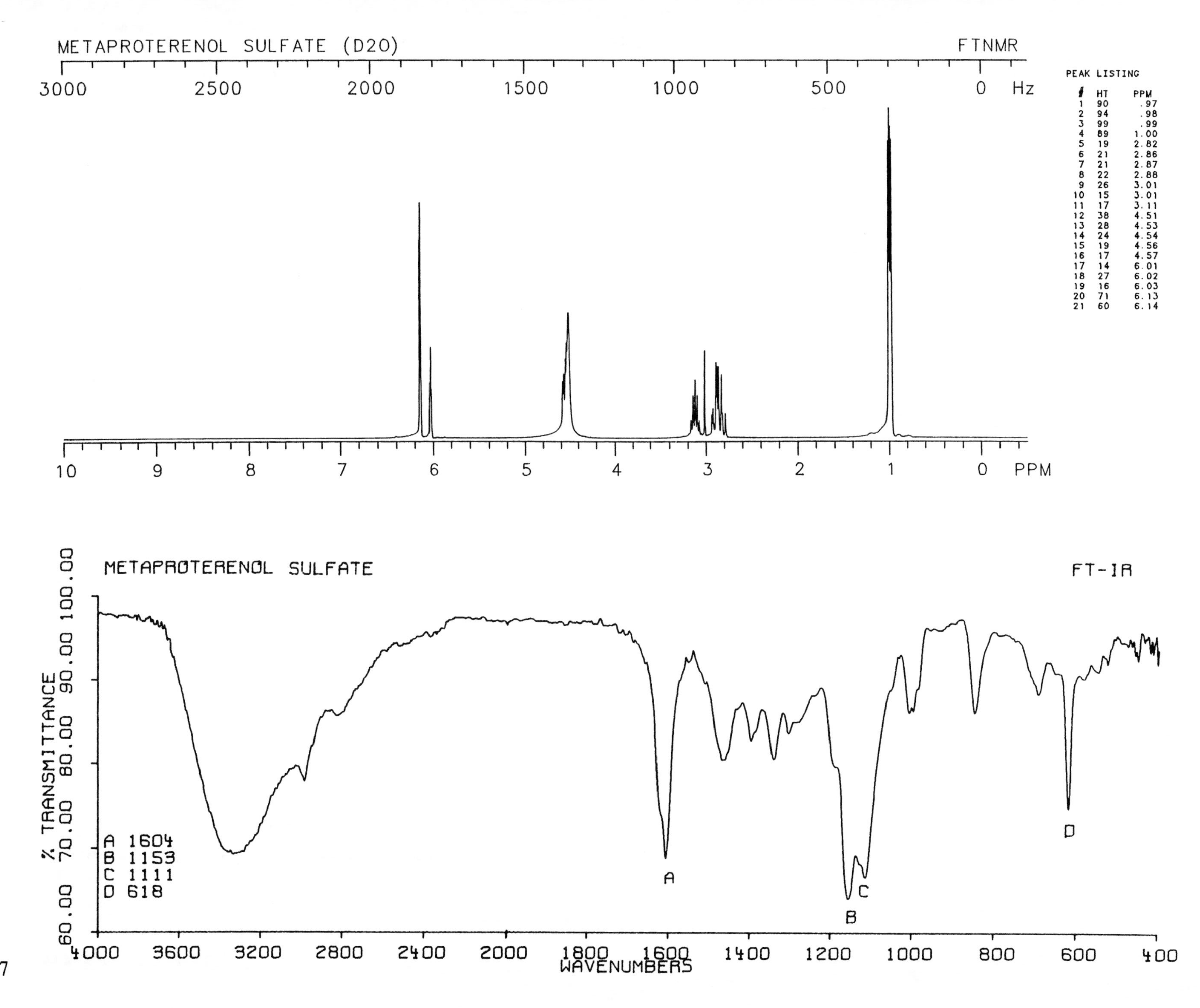
METAPROTERENOL SULFATE (D2O)
FTNMR
3000 2500 2000 1500 1000 500 0 Hz
PEAK LISTING
# HT PPM
1 90 .97
2 94 .98
3 99 .99
4 89 1.00
5 19 2.82
6 21 2.86
7 21 2.87
8 22 2.88
9 26 3.01
10 15 3.01
11 17 3.11
12 38 4.51
13 28 4.53
14 24 4.54
15 19 4.56
16 17 4.57
17 14 6.01
18 27 6.02
19 16 6.03
20 71 6.13
21 60 6.14
10 9 8 7 6 5 4 3 2 1 0 PPM
METAPROTERENOL SULFATE
FT-IR
% TRANSMITTANCE
60.00 70.00 80.00 90.00 100.00
A 1604
B 1153
C 1111
D 618
A
B
C
D
4000 3600 3200 2800 2400 2000 1800 1600 1400 1200 1000 800 600 400
WAVENUMBERS

# METARAMINOL

$C_9H_{13}NO_2$

**Molecular weight:** 167.20 (167.10)
**Synonyms:** α-(1-Aminoethyl)-3-hydroxybenzenemethanol; m-hydroxynorephedrine
**Trade names:** Aramine, Metaraminol

**Use:** Adrenergic
**HPLC:** Si-10; 20A:80B; 5.8
**GC:**

HO–$C_6H_4$–CH(OH)–CH($NH_2$)–$CH_3$

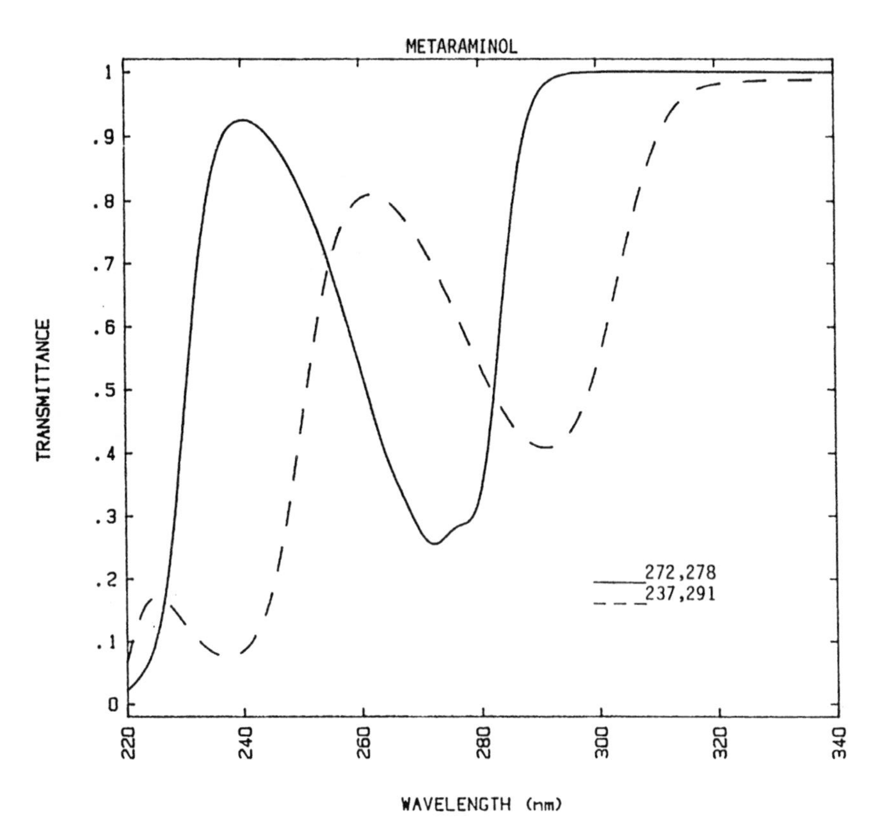


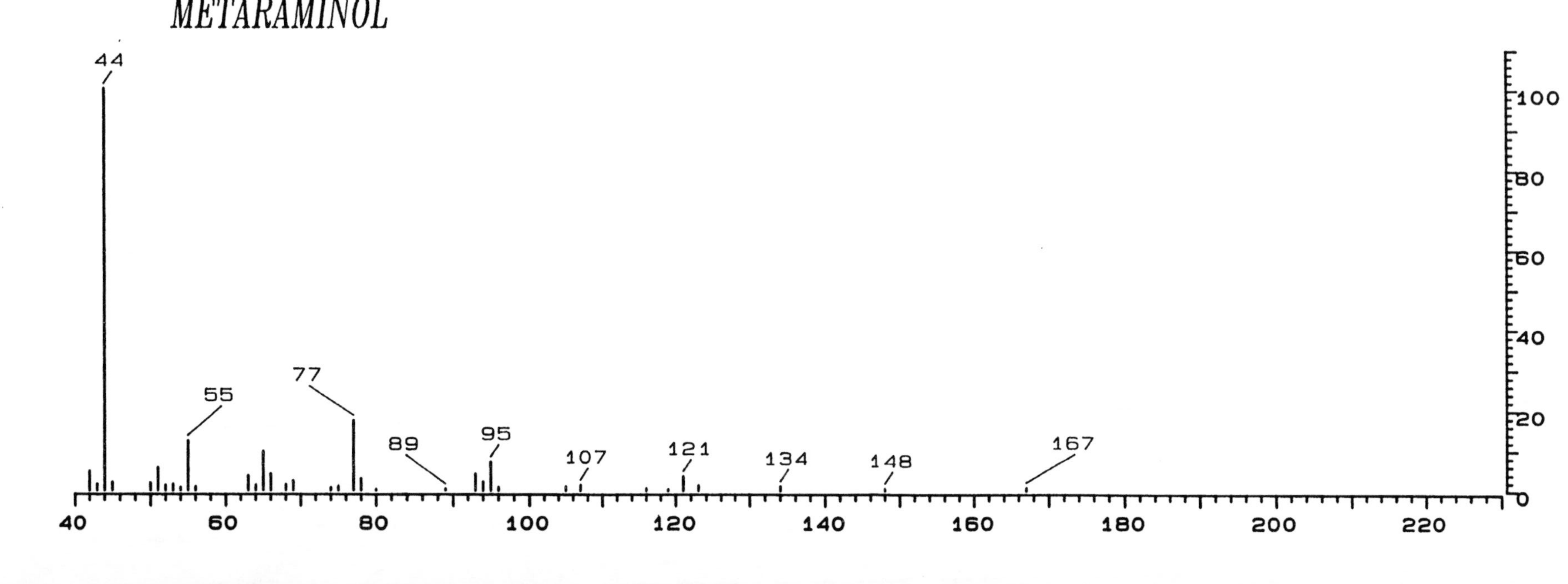

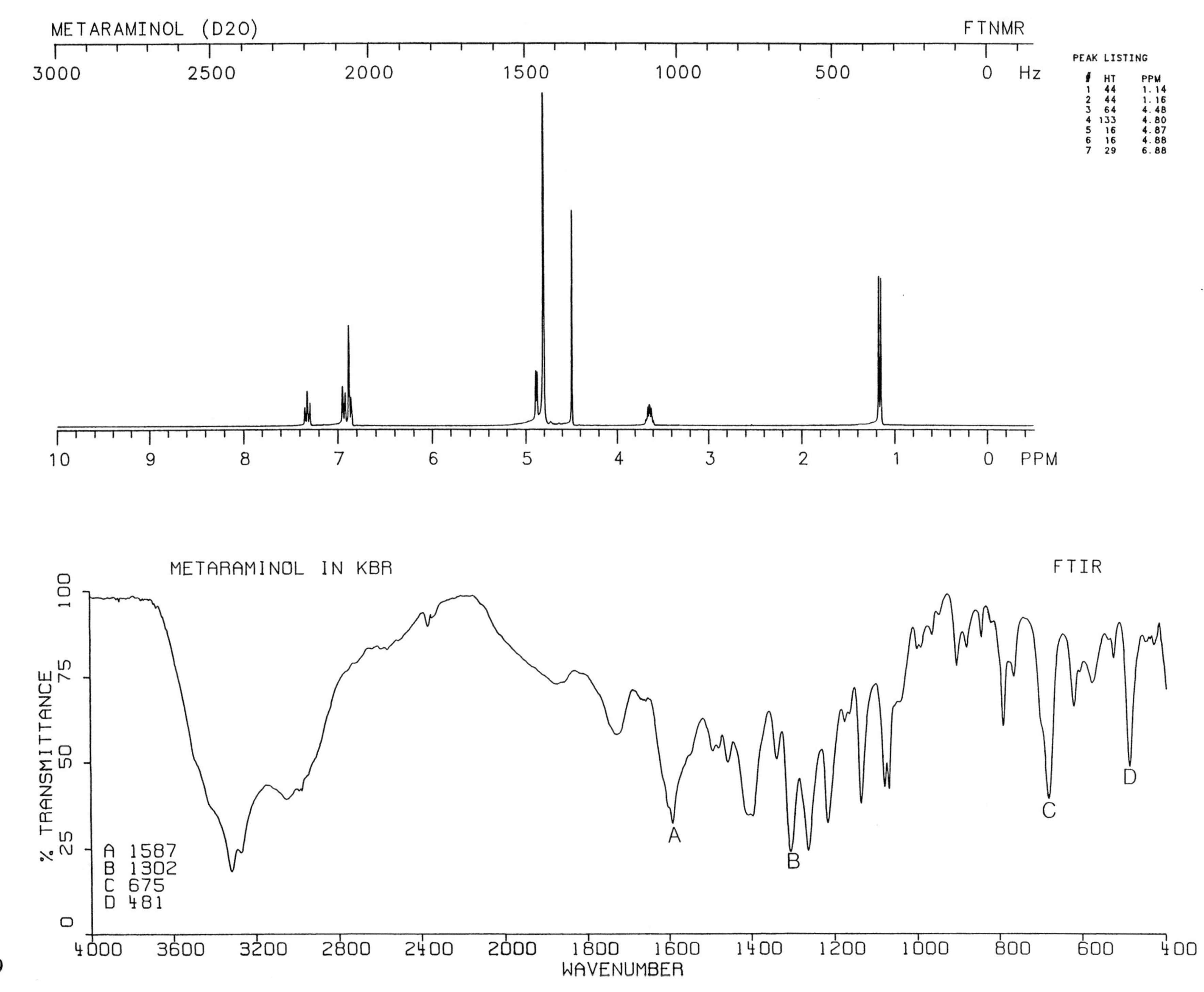
METARAMINOL (D2O)
FTNMR
3000 2500 2000 1500 1000 500 0 Hz
PEAK LISTING
# HT PPM
1 44 1.14
2 44 1.16
3 64 4.48
4 133 4.80
5 16 4.87
6 16 4.88
7 29 6.88
10 9 8 7 6 5 4 3 2 1 0 PPM
METARAMINOL IN KBR
FTIR
% TRANSMITTANCE
100 75 50 25 0
A 1587
B 1302
C 675
D 481
A
B
C
D
4000 3600 3200 2800 2400 2000 1800 1600 1400 1200 1000 800 600 400
WAVENUMBER

# METAXALONE

$C_{12}H_{15}NO_3$

**Molecular weight:** 221.26 (221.10)
**Synonyms:** 5-(3,5-Dimethylphenoxymethyl)-2-oxazolidinone

**Trade names:** Skelaxin

**Use:** Skeletal muscle relaxant
**HPLC:**
**GC:** 2215; 250°C

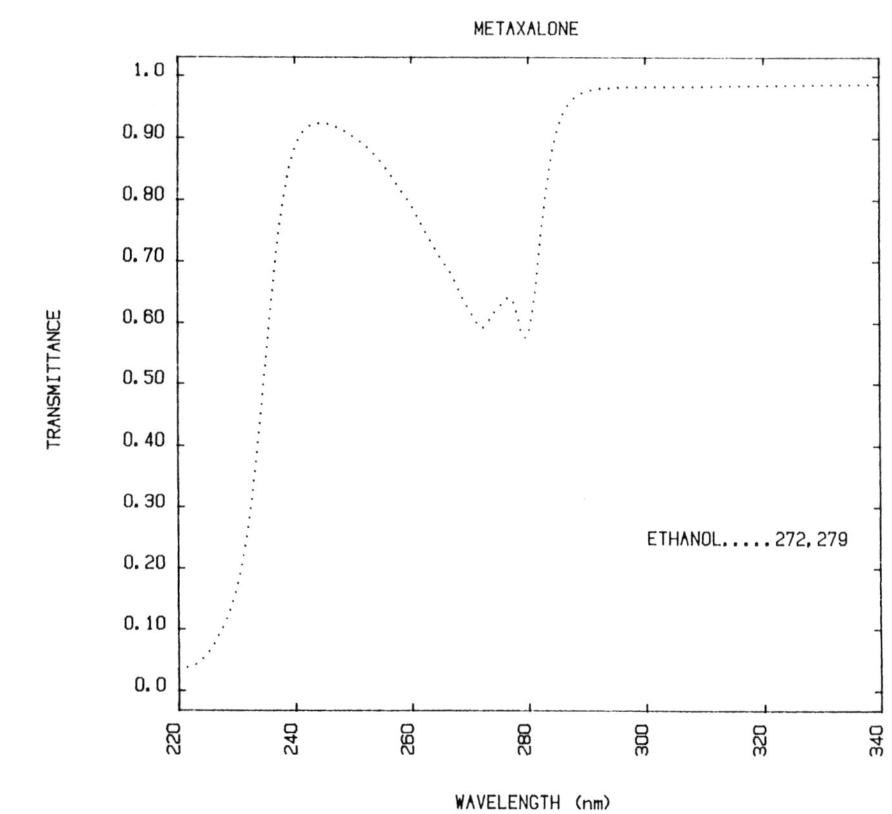


*METAXALONE*

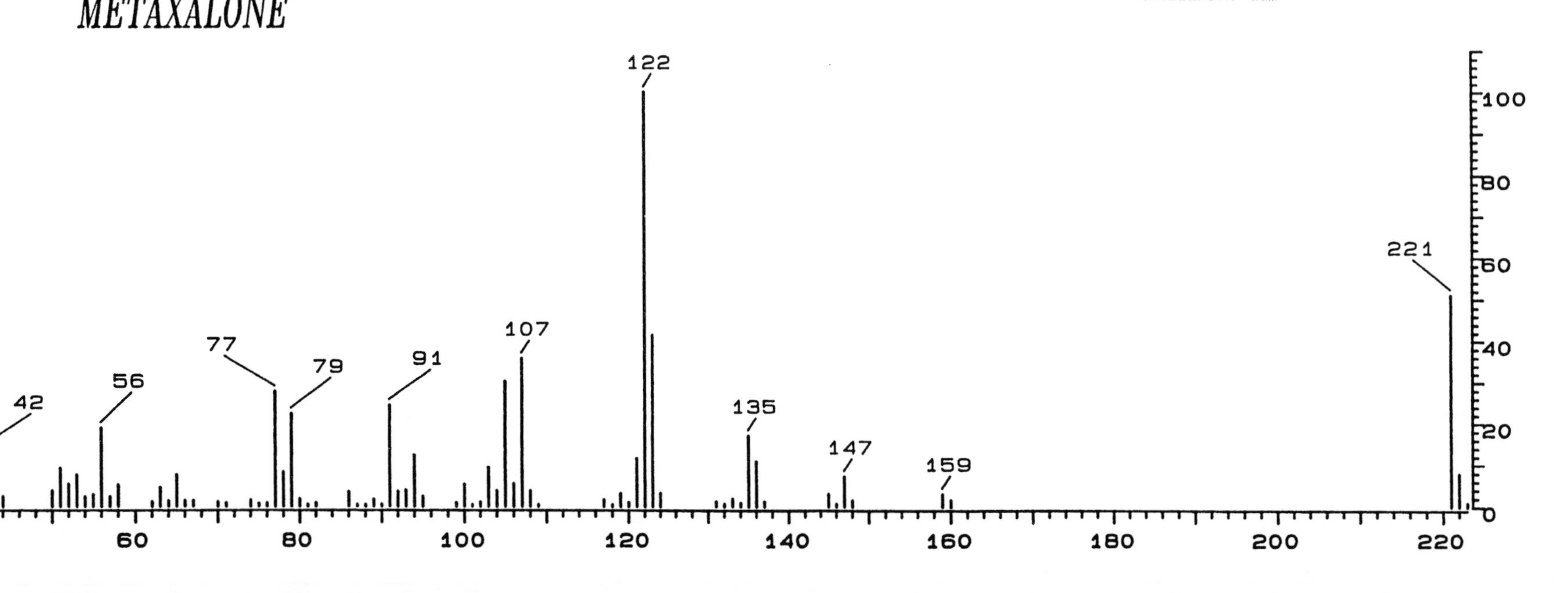

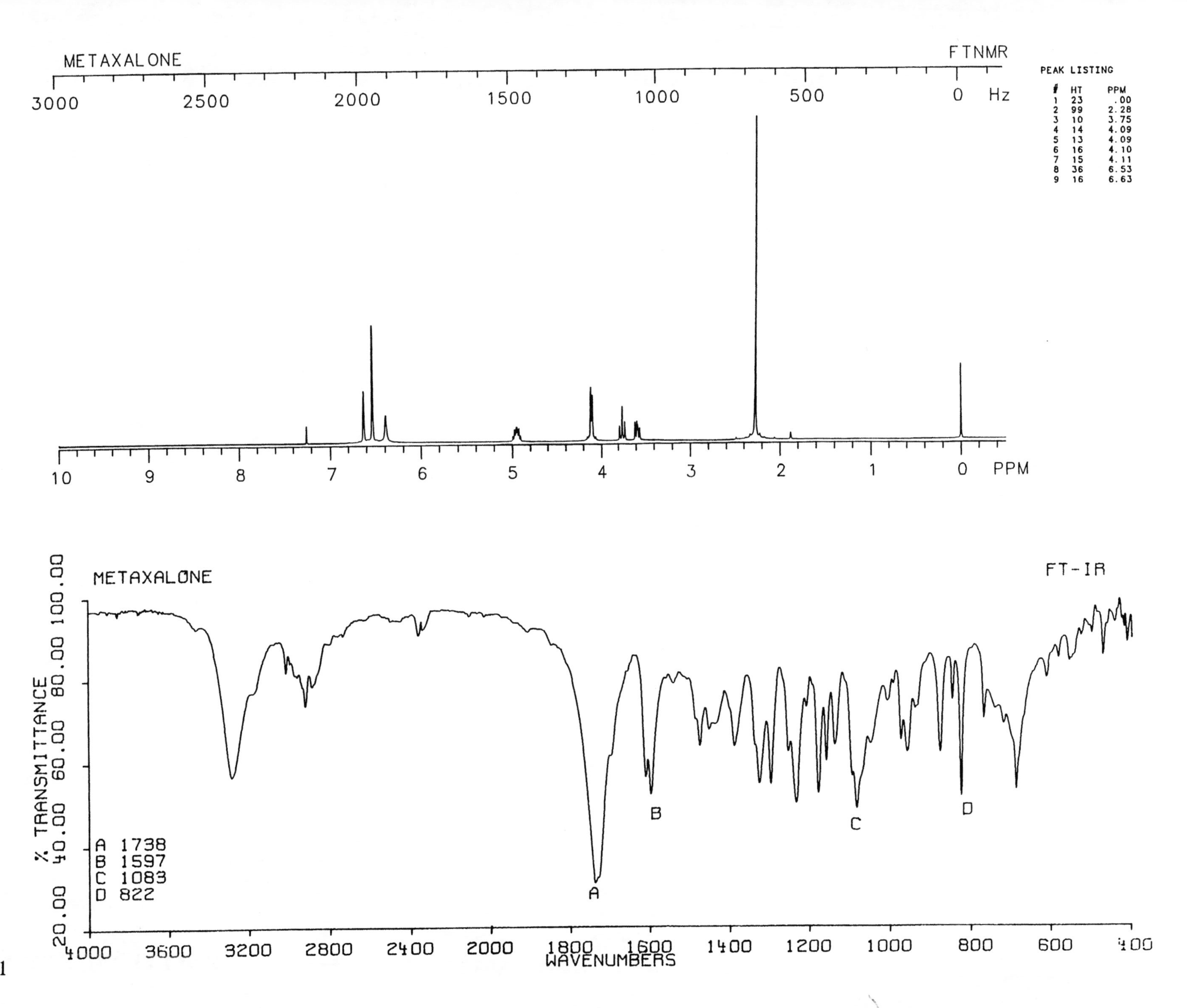
METAXALONE
FTNMR
3000
2500
2000
1500
1000
500
0 Hz
PEAK LISTING
# HT PPM
1 23 .00
2 99 2.28
3 10 3.75
4 14 4.09
5 13 4.09
6 16 4.10
7 15 4.11
8 36 6.53
9 16 6.63
10 9 8 7 6 5 4 3 2 1 0 PPM
METAXALONE
FT-IR
% TRANSMITTANCE
20.00 40.00 60.00 80.00 100.00
A 1738
B 1597
C 1083
D 822
A
B
C
D
4000 3600 3200 2800 2400 2000 1800 1600 1400 1200 1000 800 600 400
WAVENUMBERS

# METHACYCLINE

$C_{22}H_{22}N_2O_8$

**Molecular weight:** 442.43 (442.14)
**Synonyms:** 4-Dimethylamino-1,4,4α,5,5α,6,11,12a-octahydro-3,5,10,-12,12a-pentahydroxy-6-methylene-1,11-dioxo-2-naphthacene-carboxamide
**Trade names:** Rondomycin

**Use:** Antibiotic
**HPLC:**
**GC:**

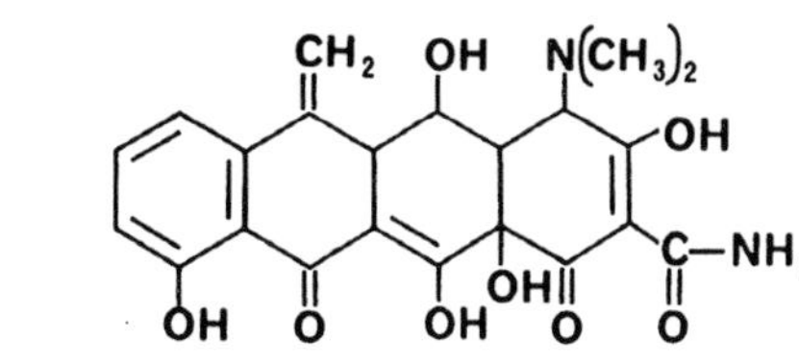

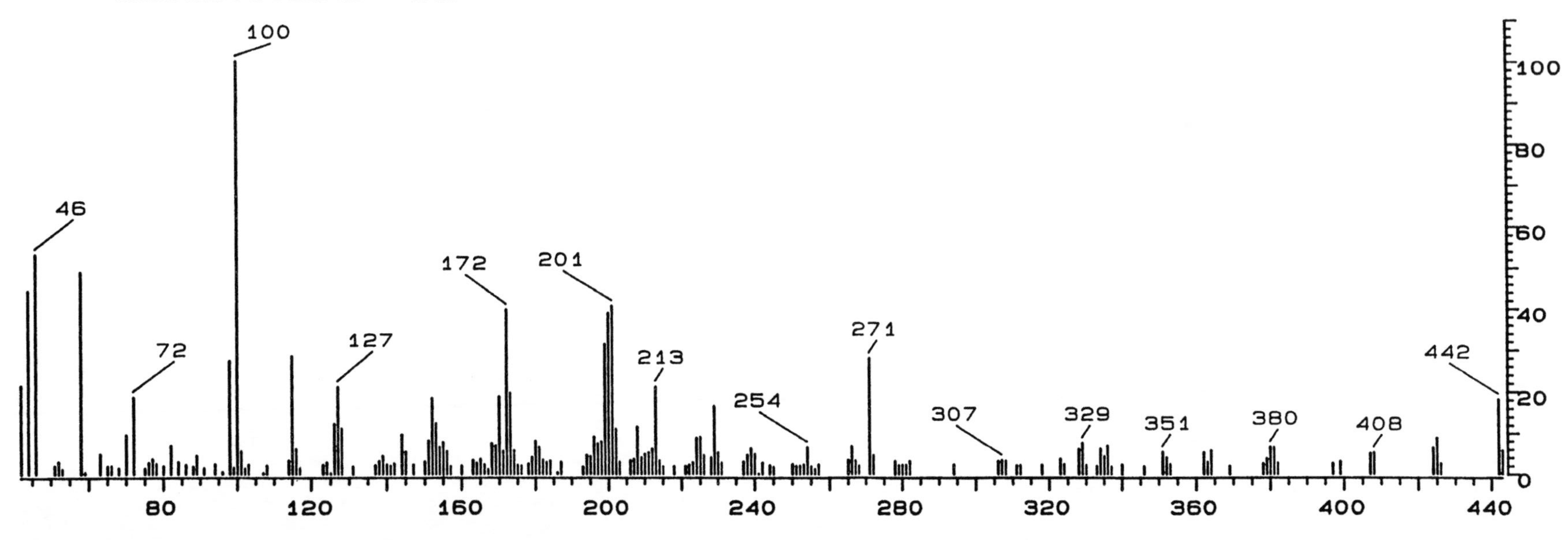

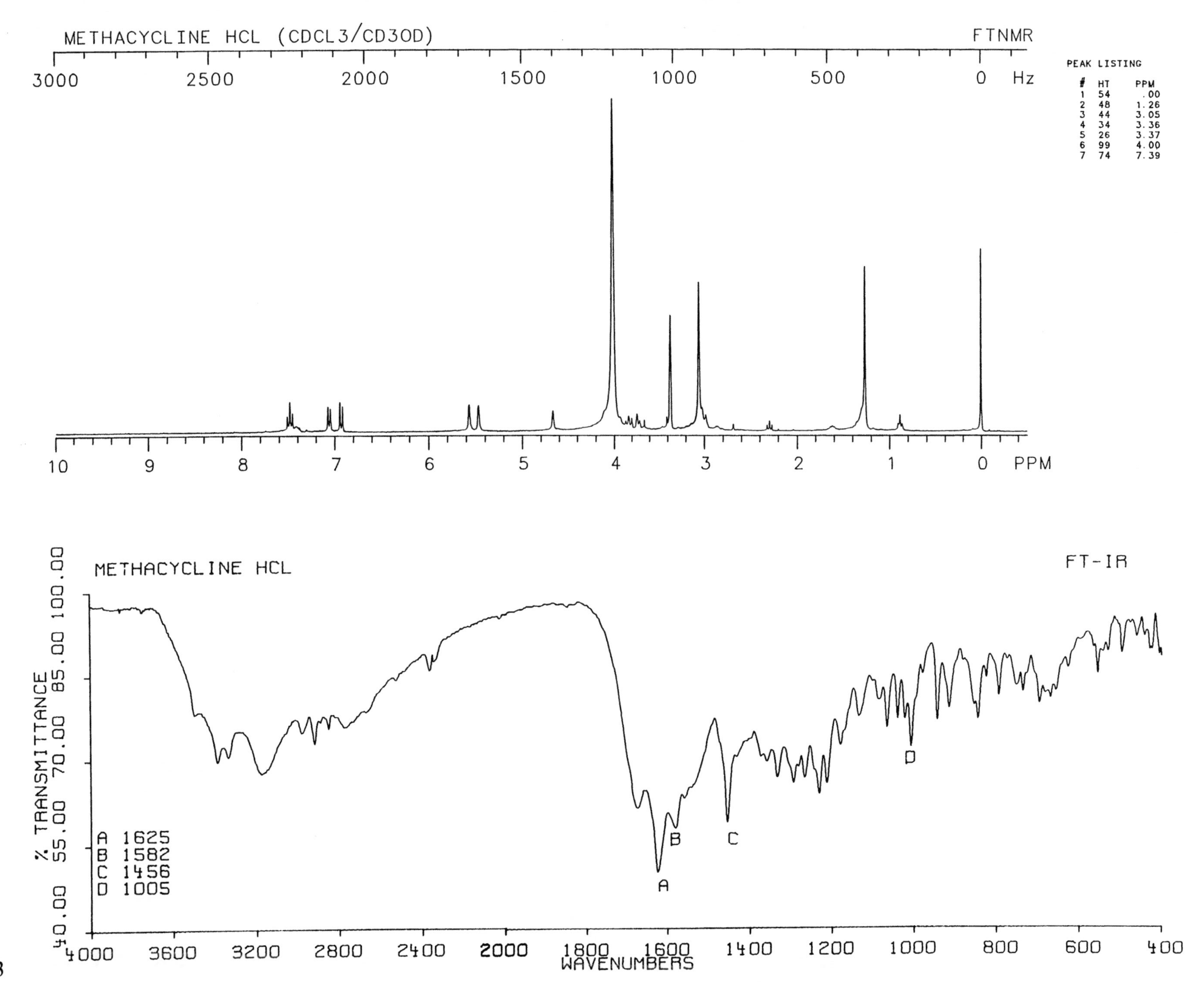
METHACYCLINE HCL (CDCL3/CD3OD)
FTNMR
3000
2500
2000
1500
1000
500
0 Hz
PEAK LISTING
# HT PPM
1 54 .00
2 48 1.26
3 44 3.05
4 34 3.36
5 26 3.37
6 99 4.00
7 74 7.39
10
9
8
7
6
5
4
3
2
1
0 PPM
METHACYCLINE HCL
FT-IR
% TRANSMITTANCE
40.00
55.00
70.00
85.00
100.00
A 1625
B 1582
C 1456
D 1005
A
B
C
D
4000
3600
3200
2800
2400
2000
1800
1600
1400
1200
1000
800
600
400
WAVENUMBERS

# α-METHADOL

$C_{21}H_{29}NO$

Molecular weight: 311.47 (311.23)
Synonyms: β-[2-(Dimethylamino)propyl]-α-ethyl-β-phenylbenzeneethanol; dimepheptanol
Trade names: Methadol, Pangerin

Use: Narcotic analgesic
HPLC: Si-10; 5A:95B; 3.5
GC: 2229; 250°C

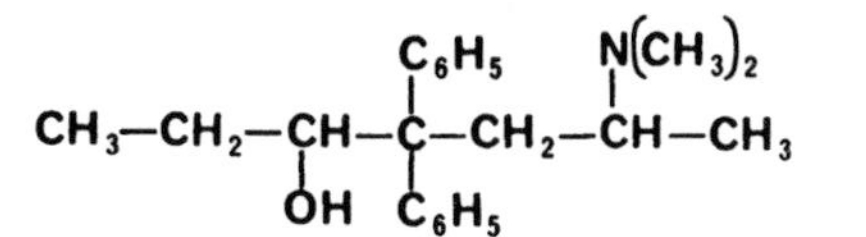

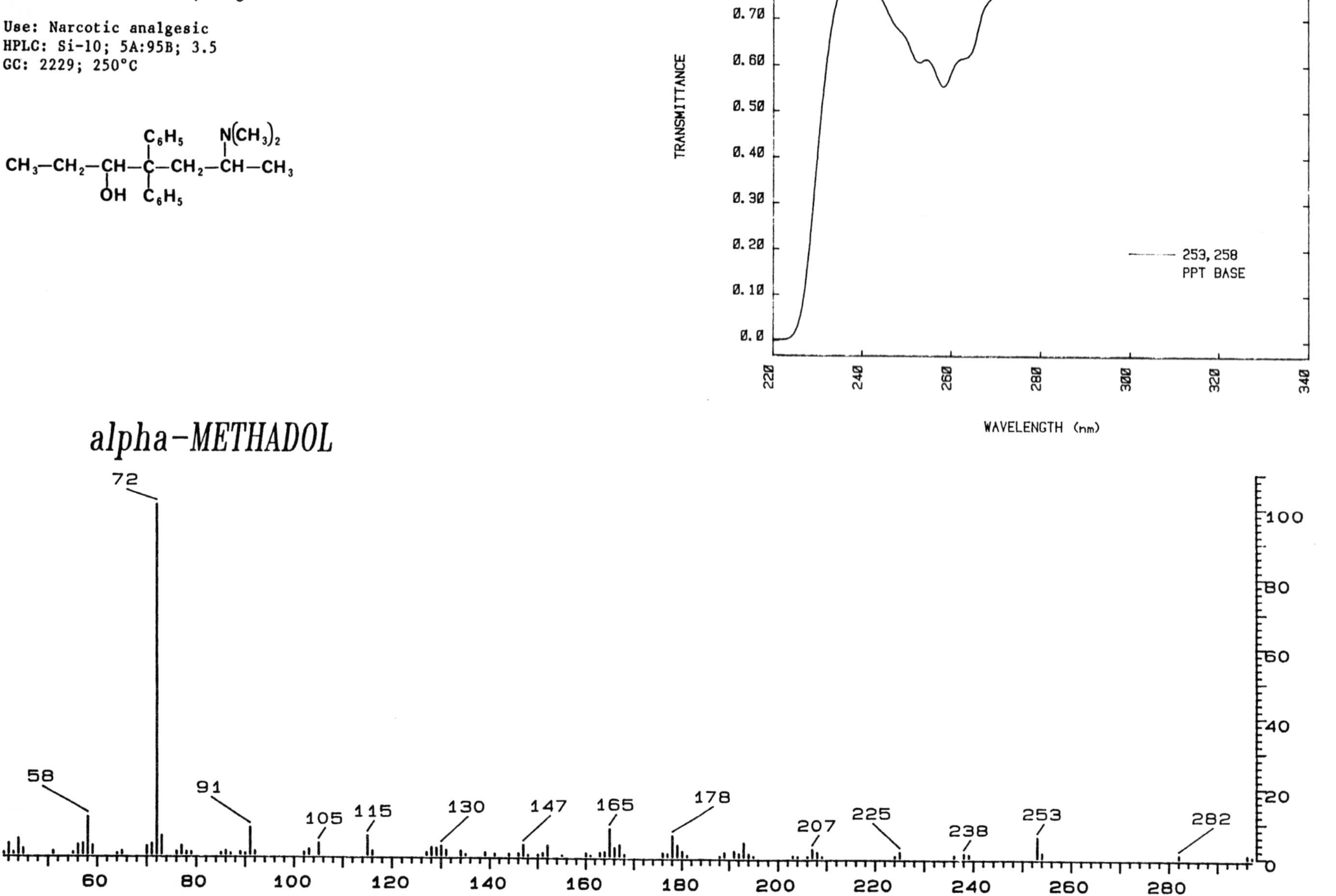


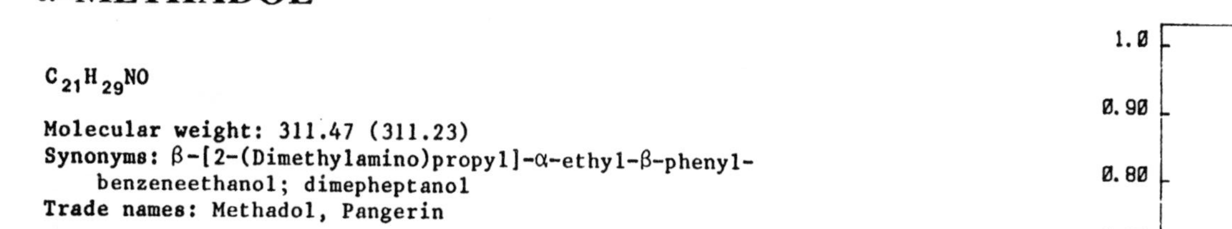

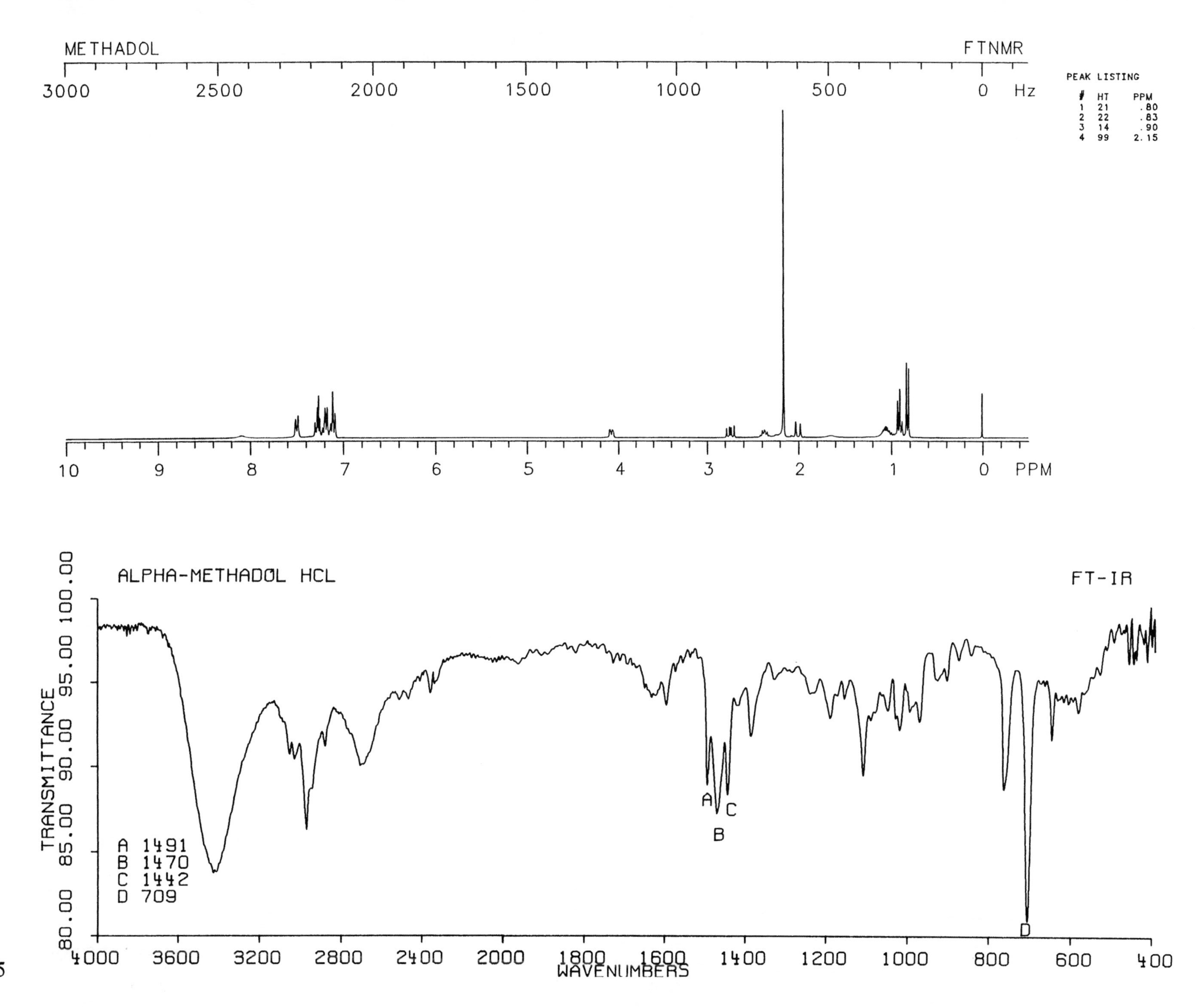
METHADOL
FTNMR
3000 2500 2000 1500 1000 500 0 Hz
PEAK LISTING
# HT PPM
1 21 .80
2 22 .83
3 14 .90
4 99 2.15
10 9 8 7 6 5 4 3 2 1 0 PPM
ALPHA-METHADOL HCL
FT-IR
TRANSMITTANCE
80.00 85.00 90.00 95.00 100.00
A 1491
B 1470
C 1442
D 709
A B C D
4000 3600 3200 2800 2400 2000 1800 1600 1400 1200 1000 800 600 400
WAVENUMBERS

# β-METHADOL

$C_{21}H_{29}NO$

**Molecular weight:** 311.47 (311.23)
**Synonyms:** β-[2-(Dimethylamino)propyl]-α-ethyl-β-phenylbenzene-ethanol; dimepheptanol
**Trade names:**

**Use:**
HPLC: Si-10; 4A:96B; 4.0
GC: 2218; 250°C

C6H5 N(CH3)2
C6H5—C—CH2—CH
HO—CH CH3
C2H5

BETA-METHADOL

TRANSMITTANCE

WAVELENGTH (nm)

258, 263, 269
PPT BASE

*beta–METHADOL*

44 72 91 105 115 147 165 178 193 207 225 253 282 296

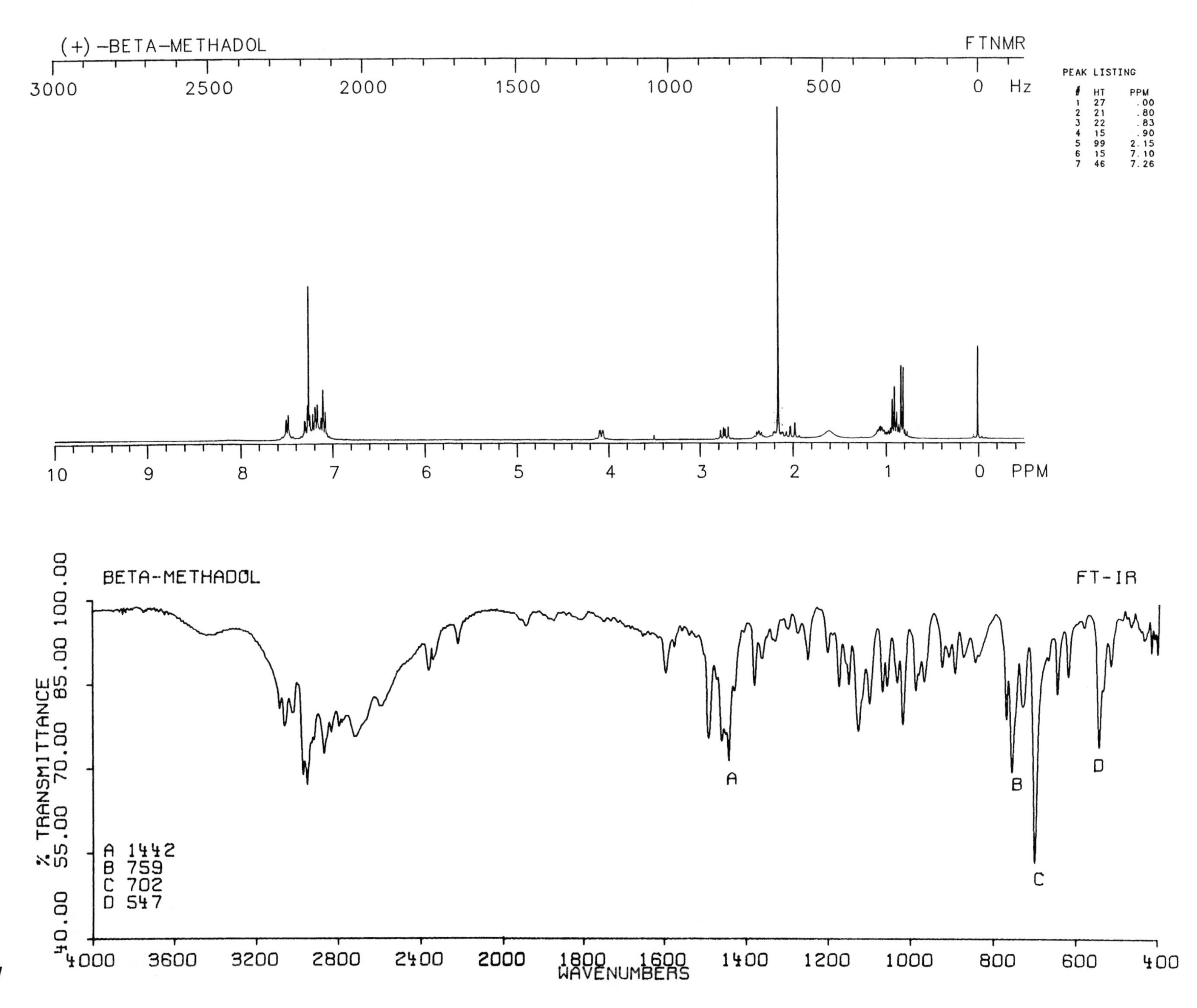
(+) -BETA-METHADOL
FTNMR
3000 2500 2000 1500 1000 500 0 Hz
PEAK LISTING
# HT PPM
1 27 .00
2 21 .80
3 22 .83
4 15 .90
5 99 2.15
6 15 7.10
7 46 7.26
10 9 8 7 6 5 4 3 2 1 0 PPM
BETA-METHADOL
FT-IR
% TRANSMITTANCE
40.00 55.00 70.00 85.00 100.00
A 1442
B 759
C 702
D 547
A
B
C
D
4000 3600 3200 2800 2400 2000 1800 1600 1400 1200 1000 800 600 400
WAVENUMBERS

# METHADONE

$C_{21}H_{27}NO$

**Molecular weight:** 309.45 (309.21)
**Synonyms:** 6-Dimethylamino-4,4-diphenyl-3-heptanone

**Trade names:** Dolophine, Methadone Diskets

**Use:** Narcotic analgesic
**HPLC:** Si-10; 10A:90B; 5.3
**GC:** 2196; 250°C

$CH_3{-}CH_2{-}C({=}O){-}C(C_6H_5)_2{-}CH_2{-}CH(N(CH_3)_2){-}CH_3$

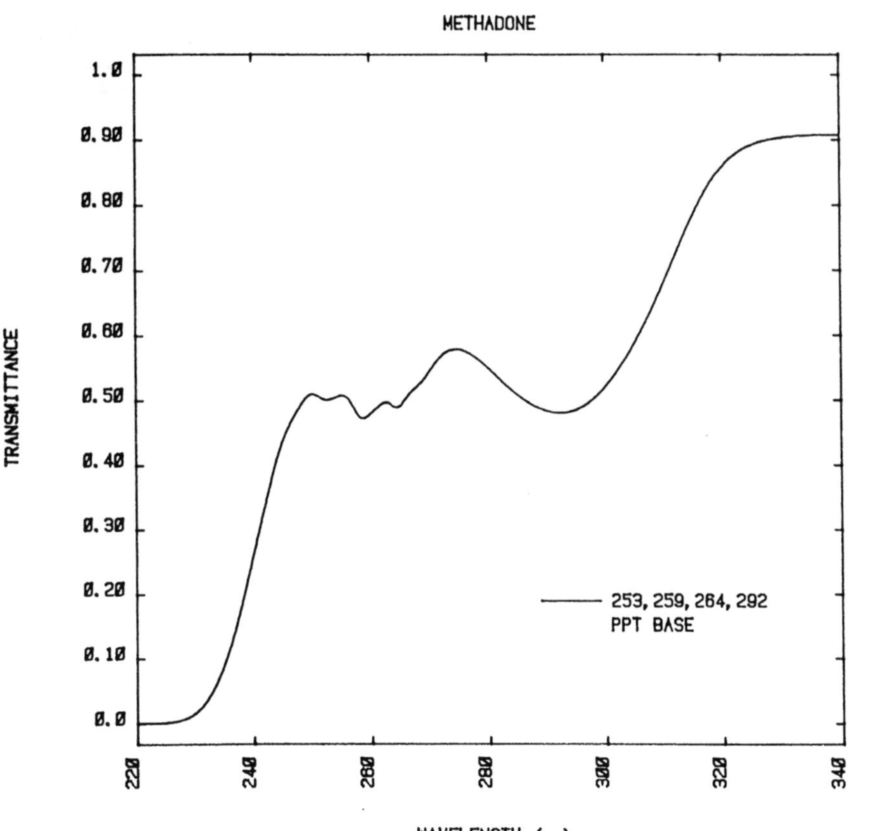


*METHADONE*

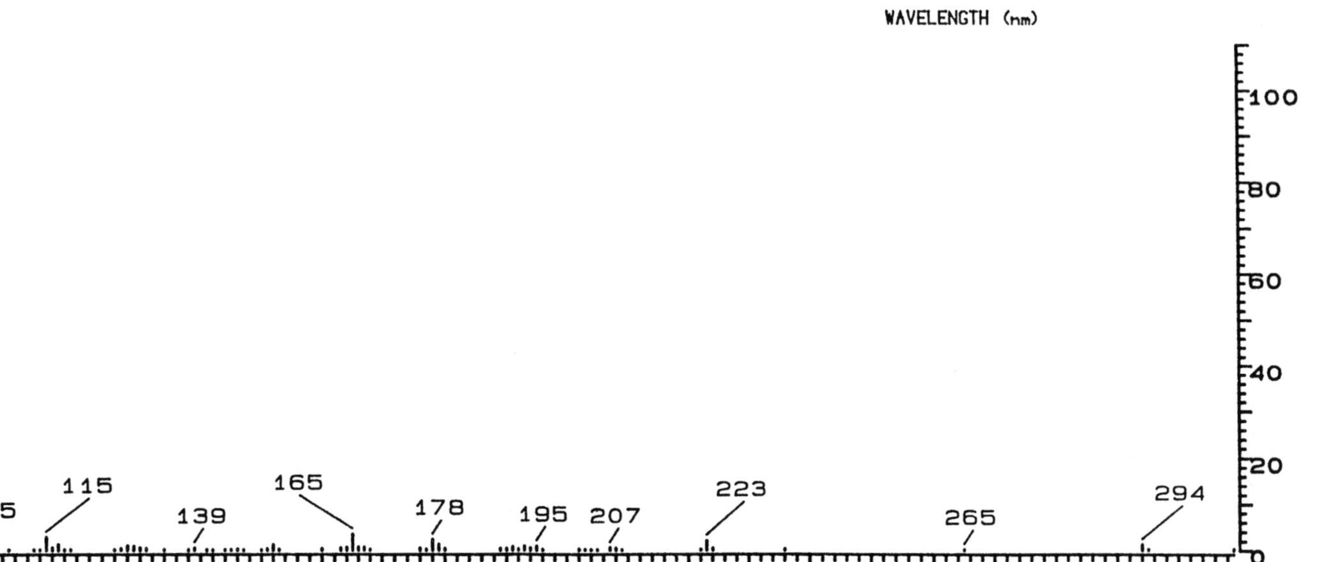
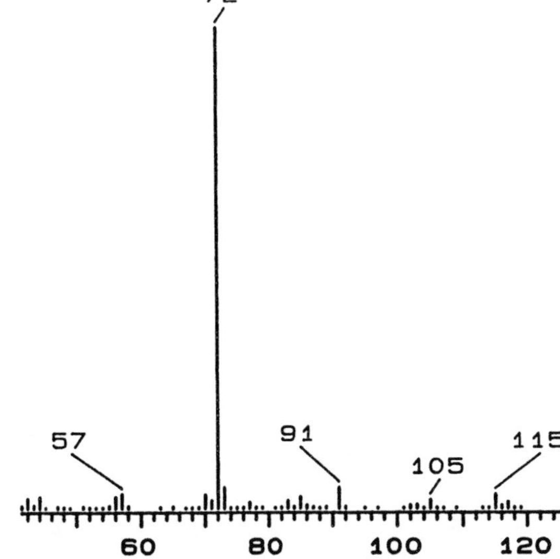

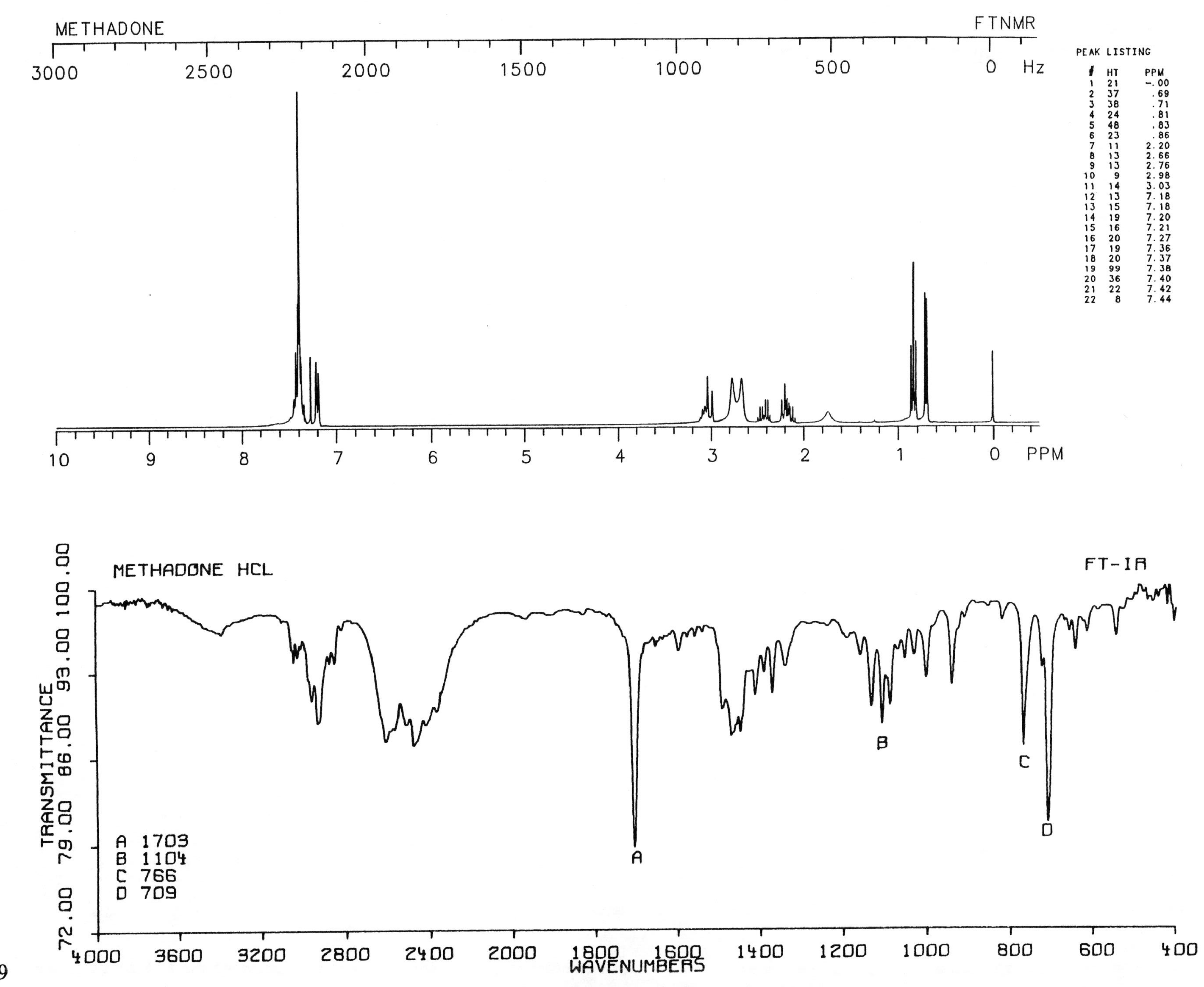
METHADONE
FTNMR
3000
2500
2000
1500
1000
500
0
Hz
PEAK LISTING
# HT PPM
1 21 -.00
2 37 .69
3 38 .71
4 24 .81
5 48 .83
6 23 .86
7 11 2.20
8 13 2.66
9 13 2.76
10 9 2.98
11 14 3.03
12 13 7.18
13 15 7.18
14 19 7.20
15 16 7.21
16 20 7.27
17 19 7.36
18 20 7.37
19 99 7.38
20 36 7.40
21 22 7.42
22 8 7.44
10
9
8
7
6
5
4
3
2
1
0
PPM
METHADONE HCL
FT-IR
100.00
93.00
86.00
79.00
72.00
TRANSMITTANCE
A 1703
B 1104
C 766
D 709
A
B
C
D
4000
3600
3200
2800
2400
2000
1800
1600
1400
1200
1000
800
600
400
WAVENUMBERS

# METHALLATAL

$C_{10}H_{14}N_2O_2S$

**Molecular weight:** 226.29 (226.08)
**Synonyms:** 5-Ethyl-2,3-dihydro-5-(2-methyl-2-propenyl)-2-thioxo-4,6-(1H,5H)-pyrimidinedione; 5-ethyl-5-(2-methylallyl)-2-thiobarbituric acid
**Trade names:** Mosidal

**Use:** Antinauseant
**HPLC:** Si-10; 1A:99B; 3.5
**GC:** 1746; 200°C

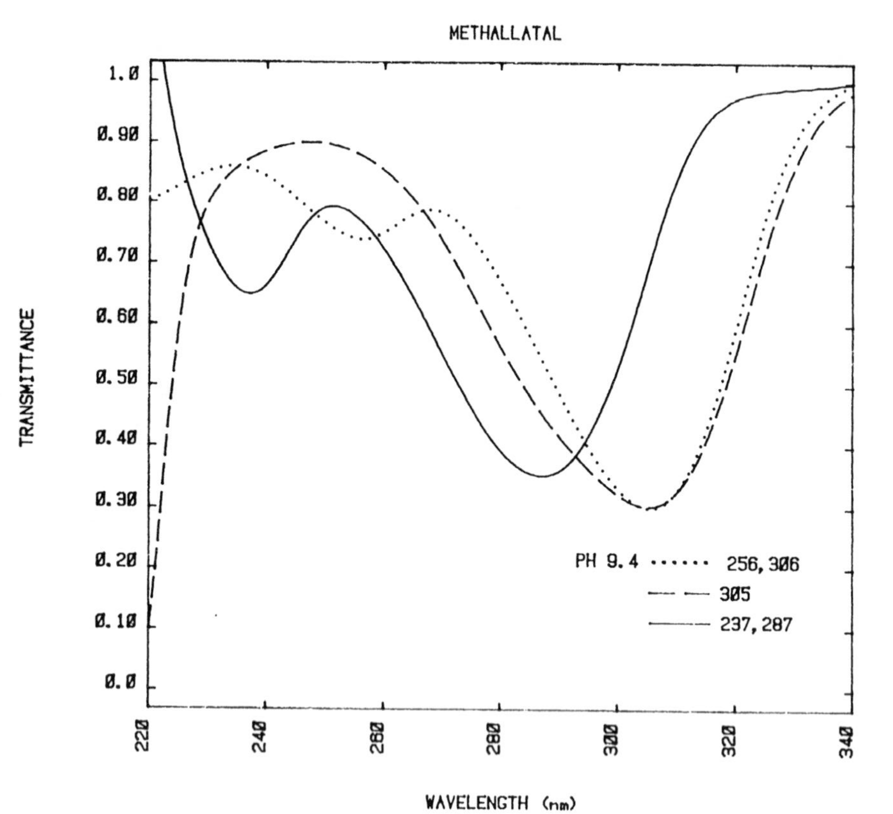

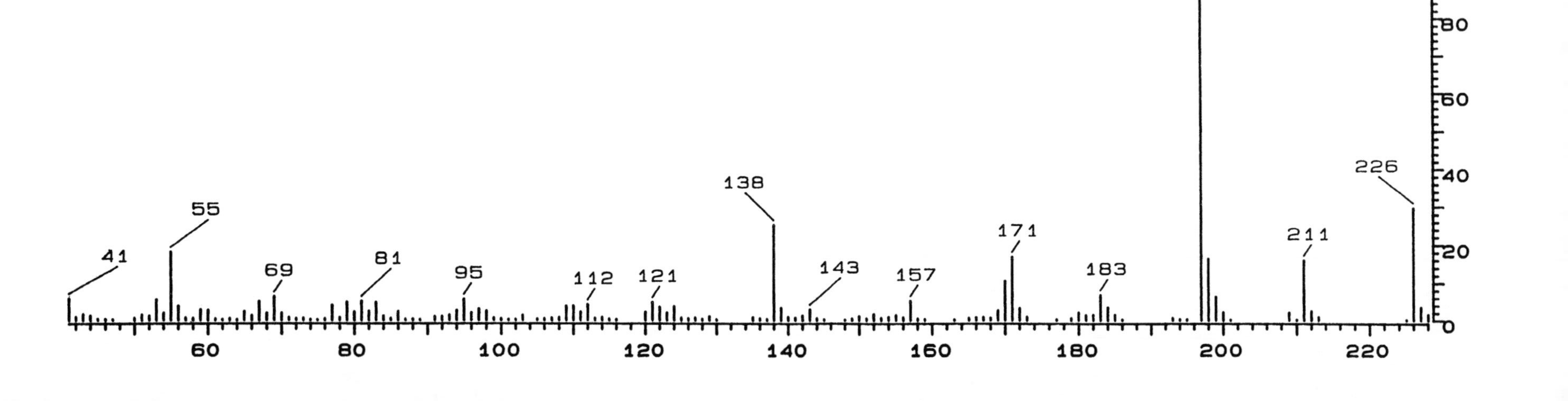

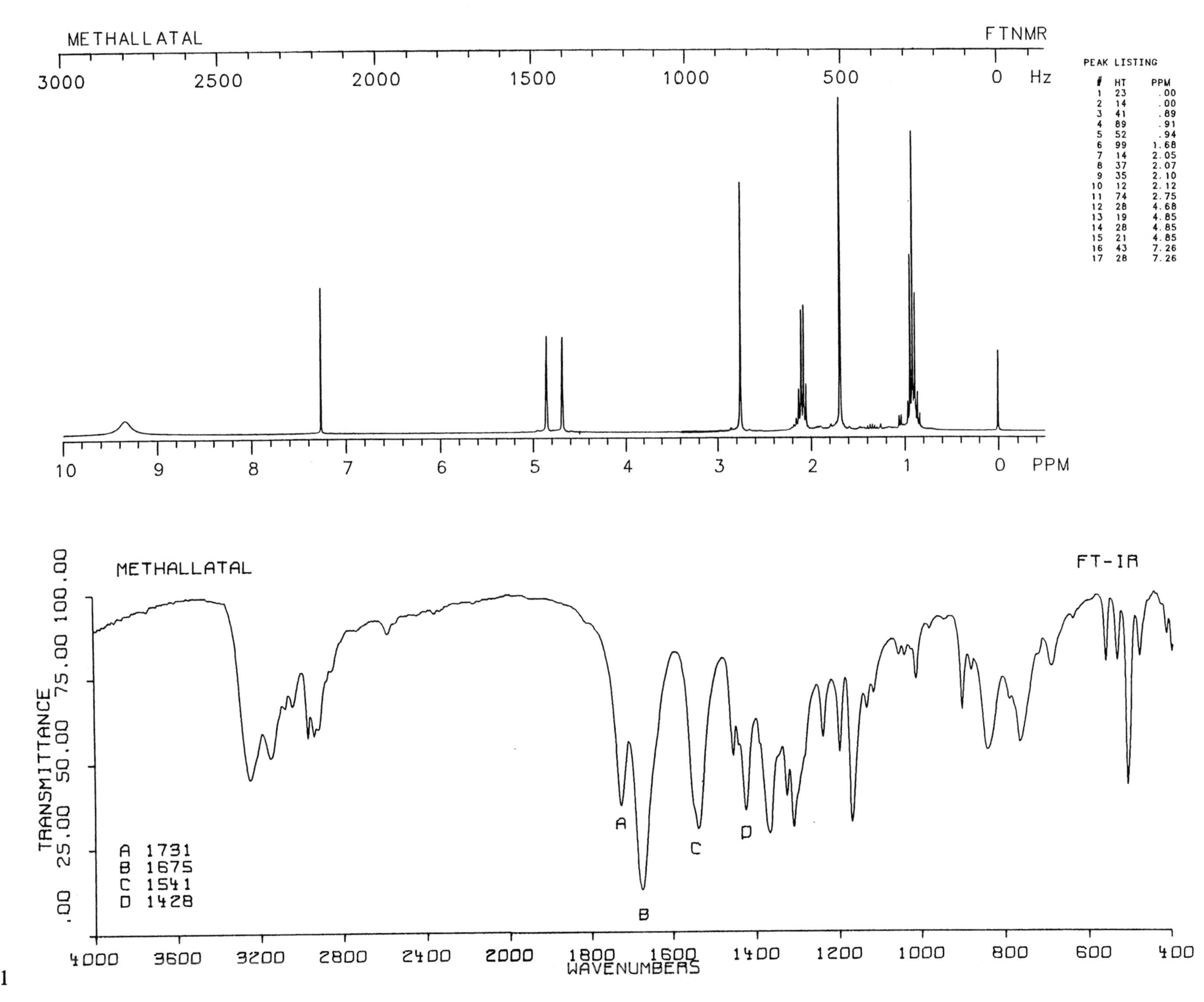

METHALLATAL
FTNMR
3000
2500
2000
1500
1000
500
0 Hz
PEAK LISTING
# HT PPM
1 23 .00
2 14 .00
3 41 .89
4 89 .91
5 52 .94
6 99 1.68
7 14 2.05
8 37 2.07
9 35 2.10
10 12 2.12
11 74 2.75
12 28 4.68
13 19 4.85
14 28 4.85
15 21 4.85
16 43 7.26
17 28 7.26
10
9
8
7
6
5
4
3
2
1
0 PPM
METHALLATAL
FT-IR
TRANSMITTANCE
.00
25.00
50.00
75.00
100.00
A
B
C
D
A 1731
B 1675
C 1541
D 1428
4000
3600
3200
2800
2400
2000
1800
1600
1400
1200
1000
800
600
400
WAVENUMBERS

# METHAMPHETAMINE

$C_{10}H_{15}N$

Molecular weight: 149.24 (149.12)
Synonyms: N,α-Dimethylbenzeneethanamine; deoxyephedrine; desoxyephedrine; speed
Trade names: Desoxyn, Methadrine

Use: Central stimulant
HPLC: Si-10; 20A:80B; 5.7
GC: 1180; 140°C

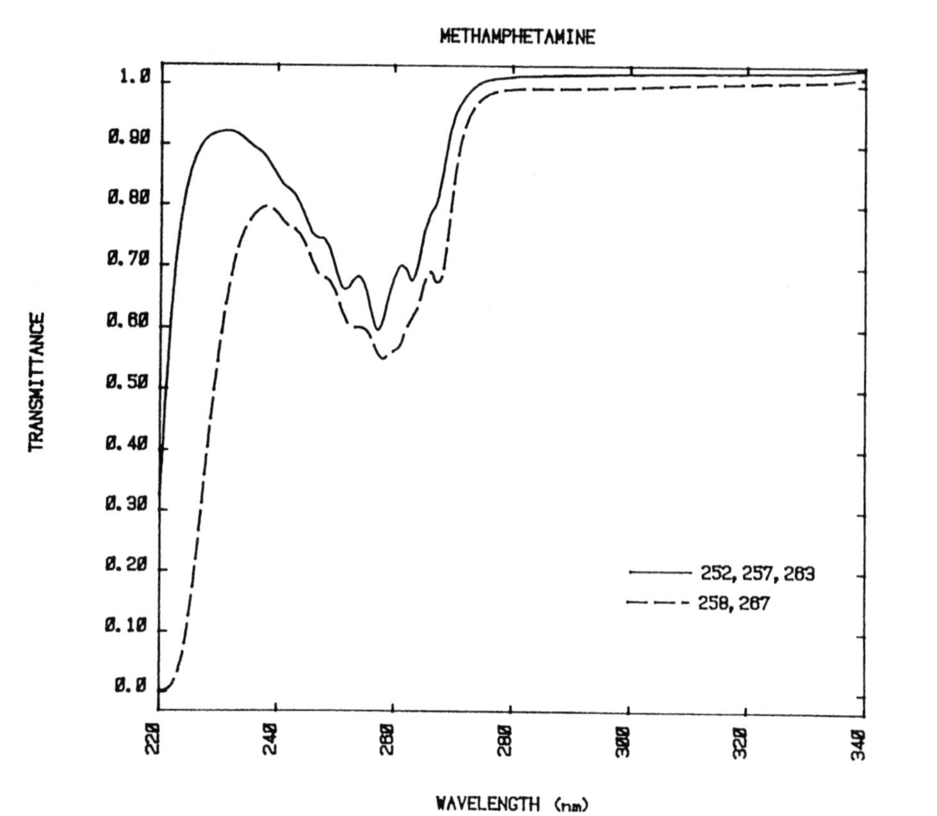


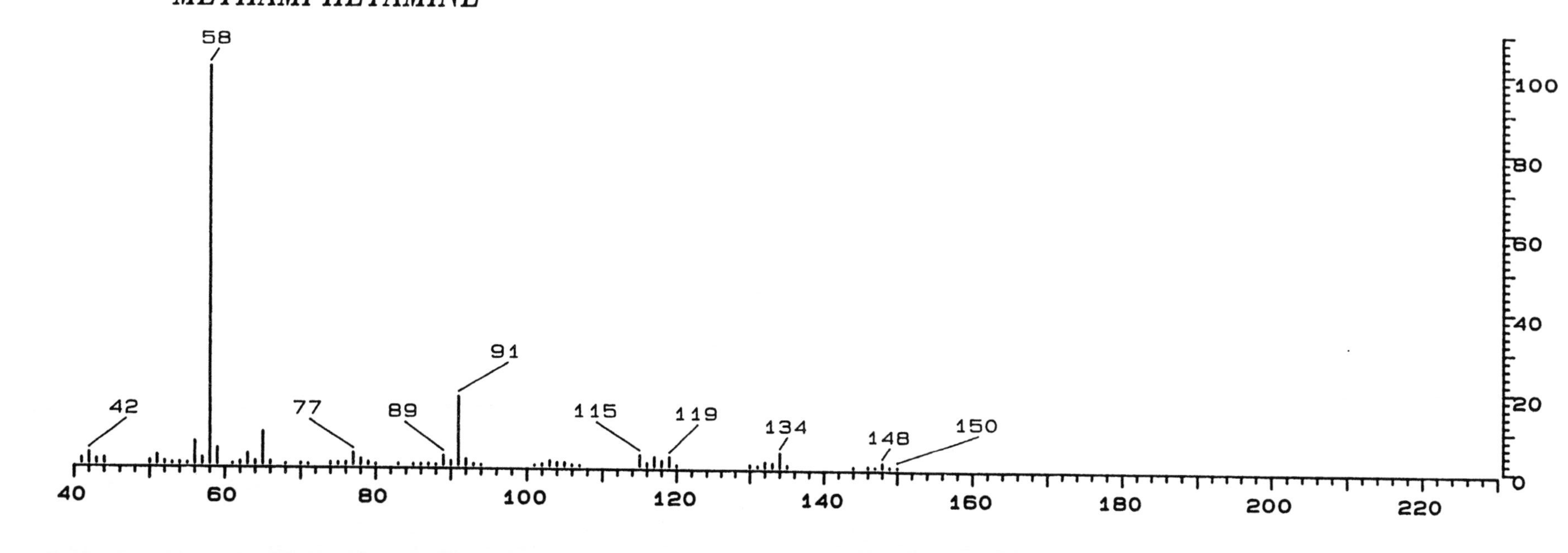

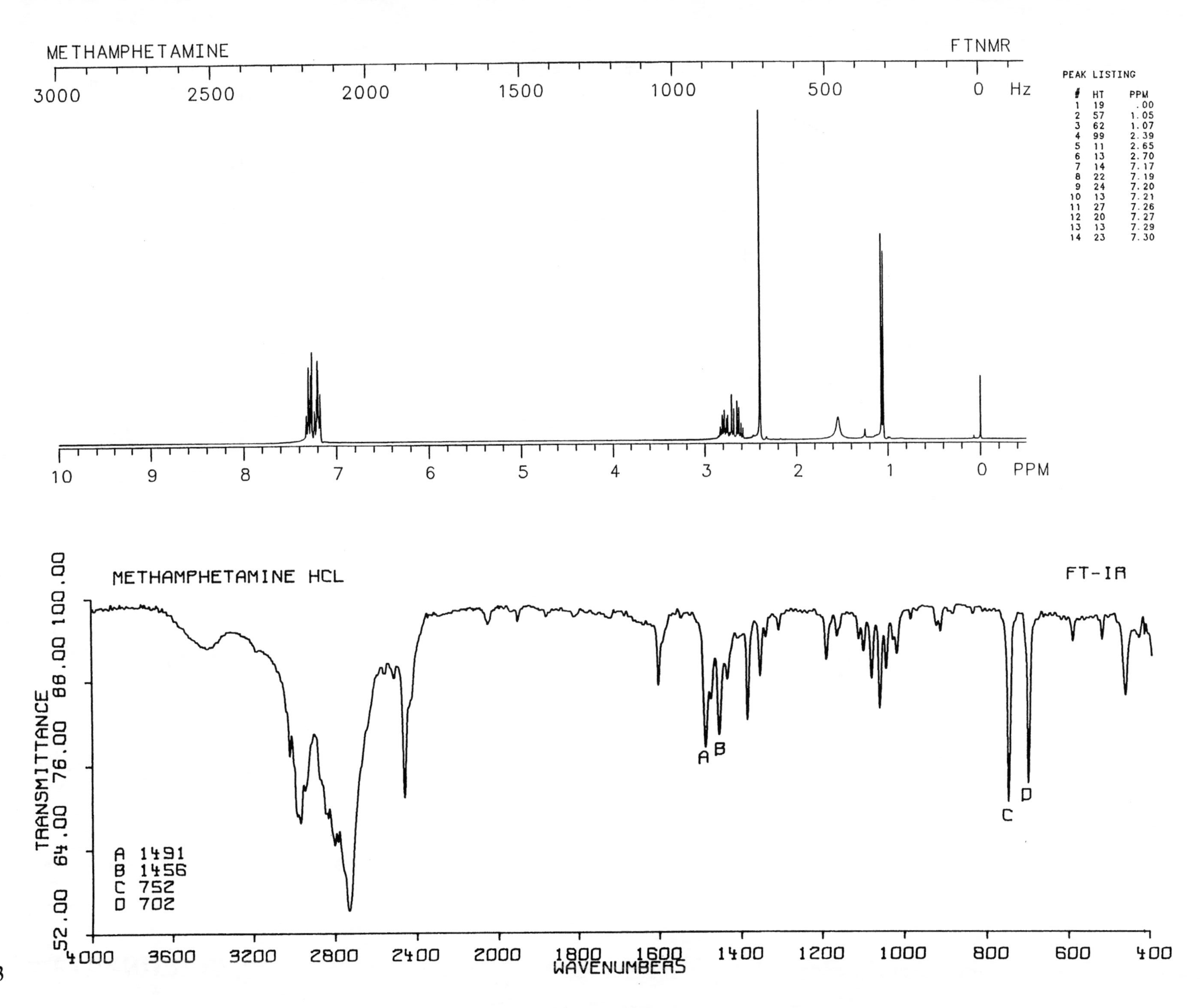
METHAMPHETAMINE
FTNMR
3000
2500
2000
1500
1000
500
0
Hz
PEAK LISTING
# HT PPM
1 19 .00
2 57 1.05
3 62 1.07
4 99 2.39
5 11 2.65
6 13 2.70
7 14 7.17
8 22 7.19
9 24 7.20
10 13 7.21
11 27 7.26
12 20 7.27
13 13 7.29
14 23 7.30
10
9
8
7
6
5
4
3
2
1
0
PPM
METHAMPHETAMINE HCL
FT-IR
TRANSMITTANCE
52.00
64.00
76.00
88.00
100.00
A 1491
B 1456
C 752
D 702
A
B
C
D
4000
3600
3200
2800
2400
2000
1800
1600
1400
1200
1000
800
600
400
WAVENUMBERS

# METHANDRIOL

$C_{20}H_{32}O_2$

**Molecular weight:** 304.46 (304.24)
**Synonyms:** 17α-Methyl-5-androstene-3β-17β-diol; methylandrostenediol; mestenediol; MAD
**Trade names:** Androdiol, Masdiol, Metendiol, Methandiol, Nabolial, Notandron, Neosteron, Probolin, Stenediol
**Use:** Antibiotic
**HPLC:**
**GC:** 2637; 280°C

OH
$CH_3$
$CH_3$
$CH_3$
O

METHANDRIOL

TRANSMITTANCE

1.0 0.90 0.80 0.70 0.60 0.50 0.40 0.30 0.20 0.10 0.0

NO ABSORPTION IN THIS REGION

220 240 260 280 300 320 340

WAVELENGTH (nm)

*METHANDRIOL*

43 77 91 105 119 145 159 171 197 213 228 253 254 271 304

60 80 100 120 140 160 180 200 220 240 260 280 300

0 20 40 60 80 100

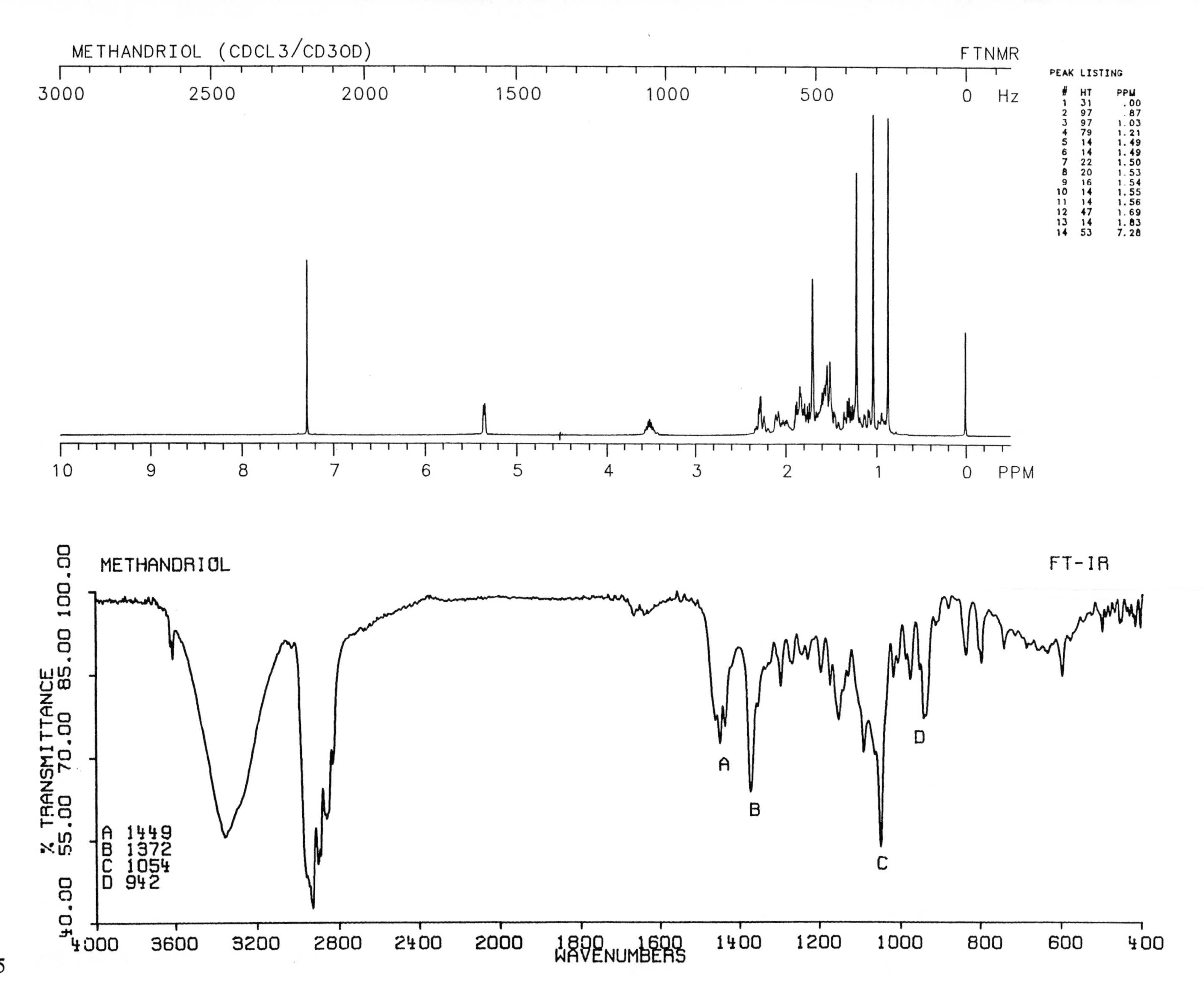
METHANDRIOL (CDCL3/CD3OD)
FTNMR
3000 2500 2000 1500 1000 500 0 Hz
PEAK LISTING
# HT PPM
1 31 .00
2 97 .87
3 97 1.03
4 79 1.21
5 14 1.49
6 14 1.49
7 22 1.50
8 20 1.53
9 16 1.54
10 14 1.55
11 14 1.56
12 47 1.69
13 14 1.83
14 53 7.28
10 9 8 7 6 5 4 3 2 1 0 PPM
METHANDRIOL
FT-IR
% TRANSMITTANCE
40.00 55.00 70.00 85.00 100.00
A 1449
B 1372
C 1054
D 942
A
B
C
D
4000 3600 3200 2800 2400 2000 1800 1600 1400 1200 1000 800 600 400
WAVENUMBERS

# METHANDROSTENOLONE

$C_{20}H_{28}O_2$

**Molecular weight:** 300.44 (300.21)
**Synonyms:** 17β-Hydroxy-17-methylandrosta-1,4-dien-3-one; methandienone
**Trade names:** Dianabol

**Use:** Androgen, anabolic
**HPLC:**
**GC:** 2789; 280°C

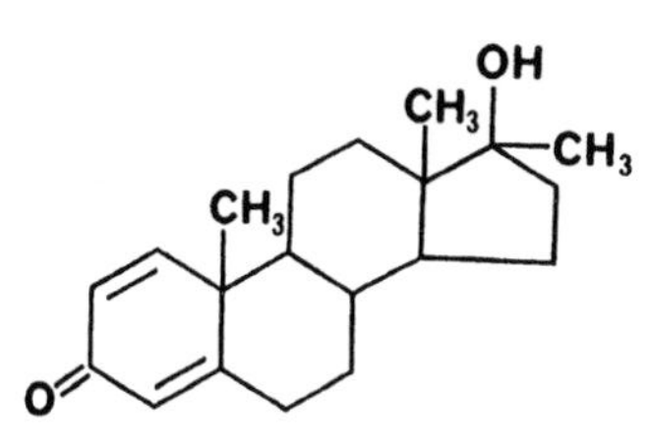


METHANDROSTENOLONE

TRANSMITTANCE

1.0
0.90
0.80
0.70
0.60
0.50
0.40
0.30
0.20
0.10
0.0

ETHANOL.....244

220
240
260
280
300
320
340

WAVELENGTH (nm)

*METHANDROSTENOLONE*

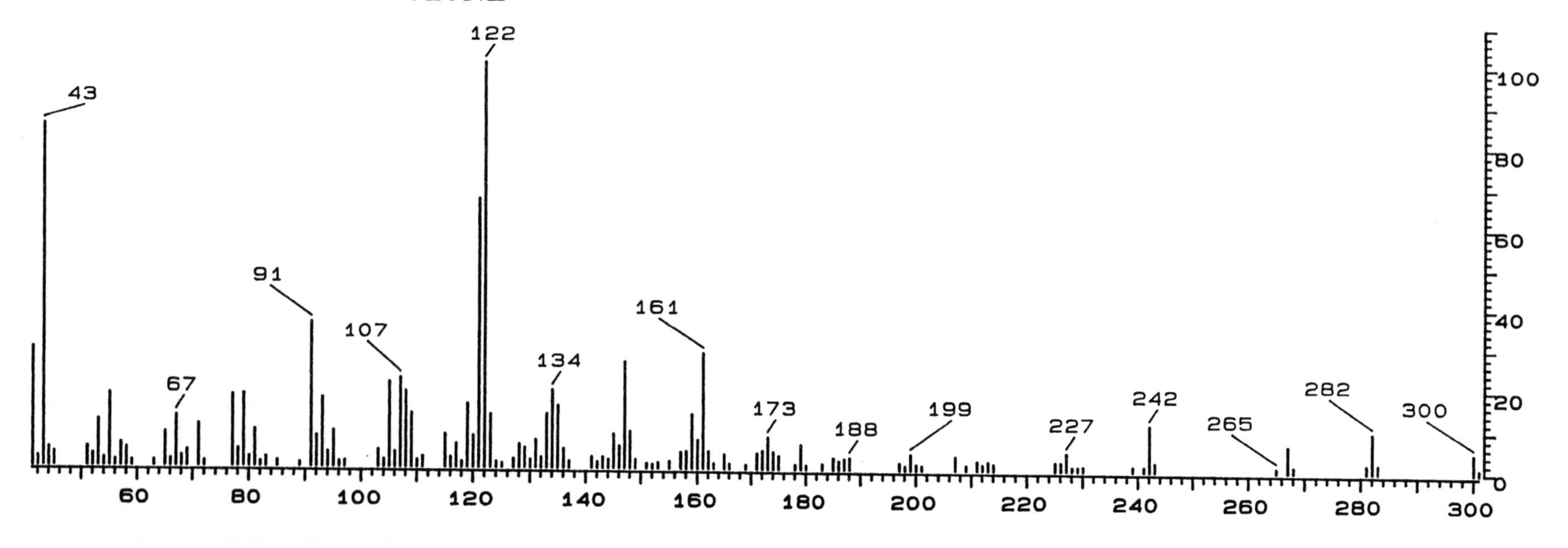

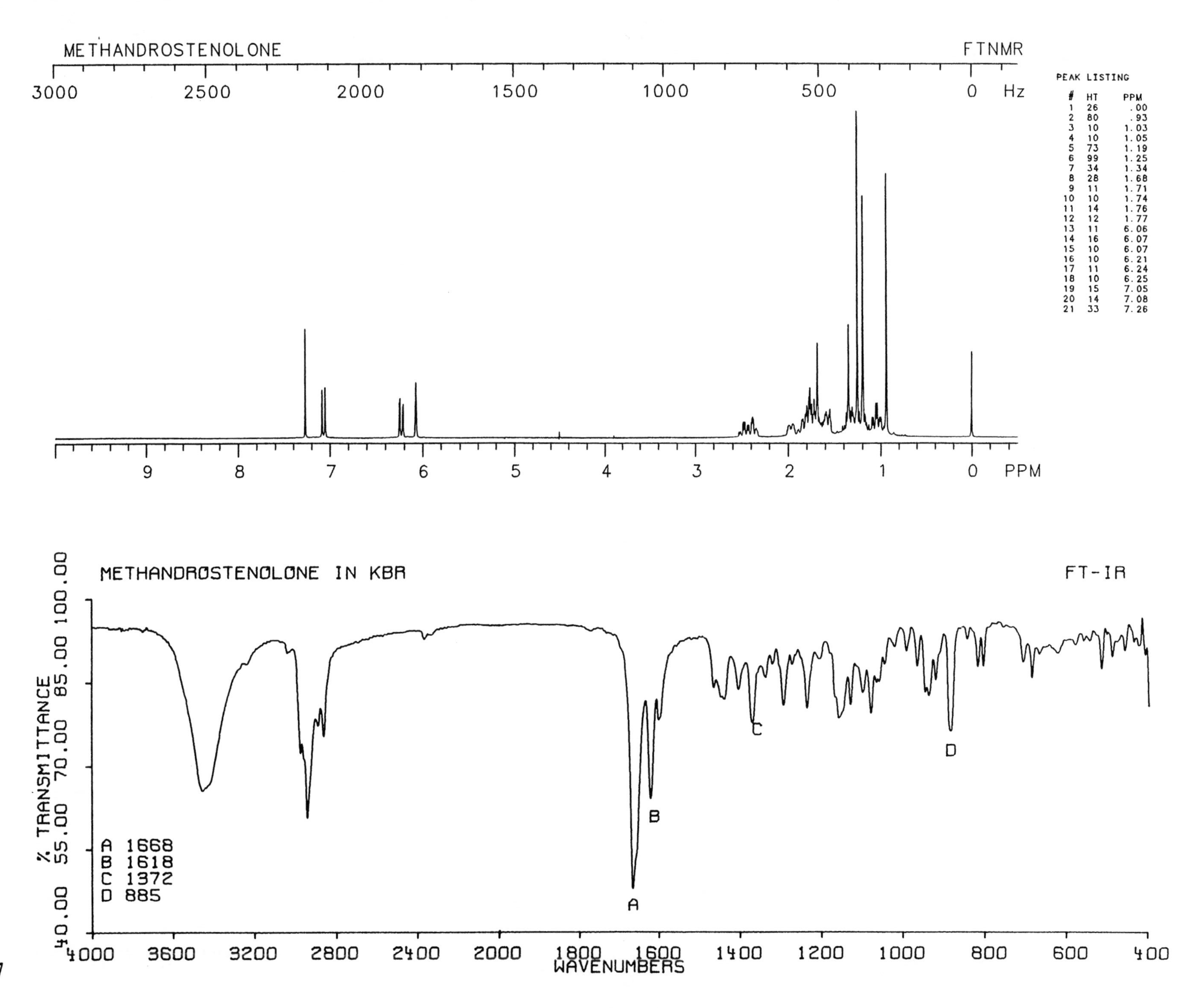

METHANDROSTENOLONE
FTNMR
3000
2500
2000
1500
1000
500
0
Hz
PEAK LISTING
# HT PPM
1 26 .00
2 80 .93
3 10 1.03
4 10 1.05
5 73 1.19
6 99 1.25
7 34 1.34
8 28 1.68
9 11 1.71
10 10 1.74
11 14 1.76
12 12 1.77
13 11 6.06
14 16 6.07
15 10 6.07
16 10 6.21
17 11 6.24
18 10 6.25
19 15 7.05
20 14 7.08
21 33 7.26
9 8 7 6 5 4 3 2 1 0 PPM
METHANDROSTENOLONE IN KBR
FT-IR
% TRANSMITTANCE
40.00 55.00 70.00 85.00 100.00
A 1668
B 1618
C 1372
D 885
A B C D
4000 3600 3200 2800 2400 2000 1800 1600 1400 1200 1000 800 600 400
WAVENUMBERS

# METHAPYRILENE

$C_{14}H_{19}N_3S$

**Molecular weight:** 261.38 (261.13)
**Synonyms:** N,N-Dimethyl-N'-2-pyridinyl-N'-(2-thienylmethyl)-1,2-ethanediamine
**Trade names:**

**Use:** Antihistamine
**HPLC:** Si-10; 10A:90B; 4.1
**GC:** 2039; 250°C

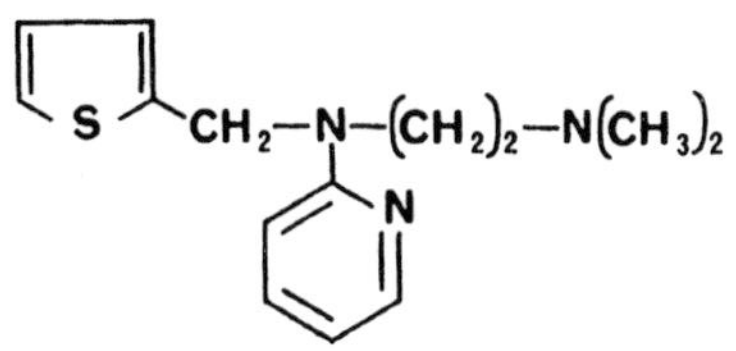

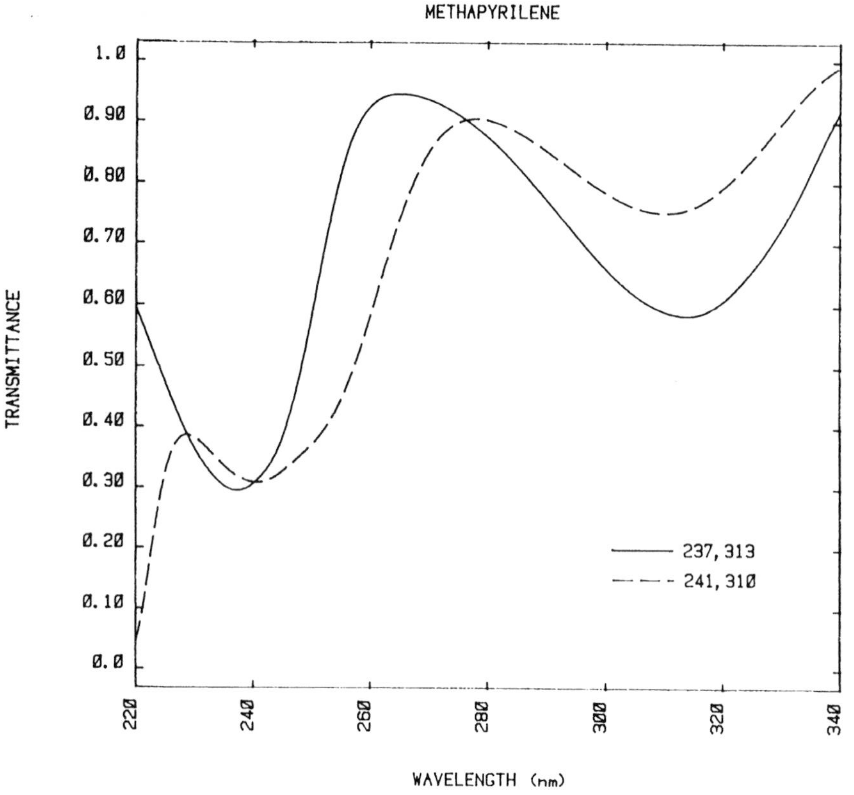

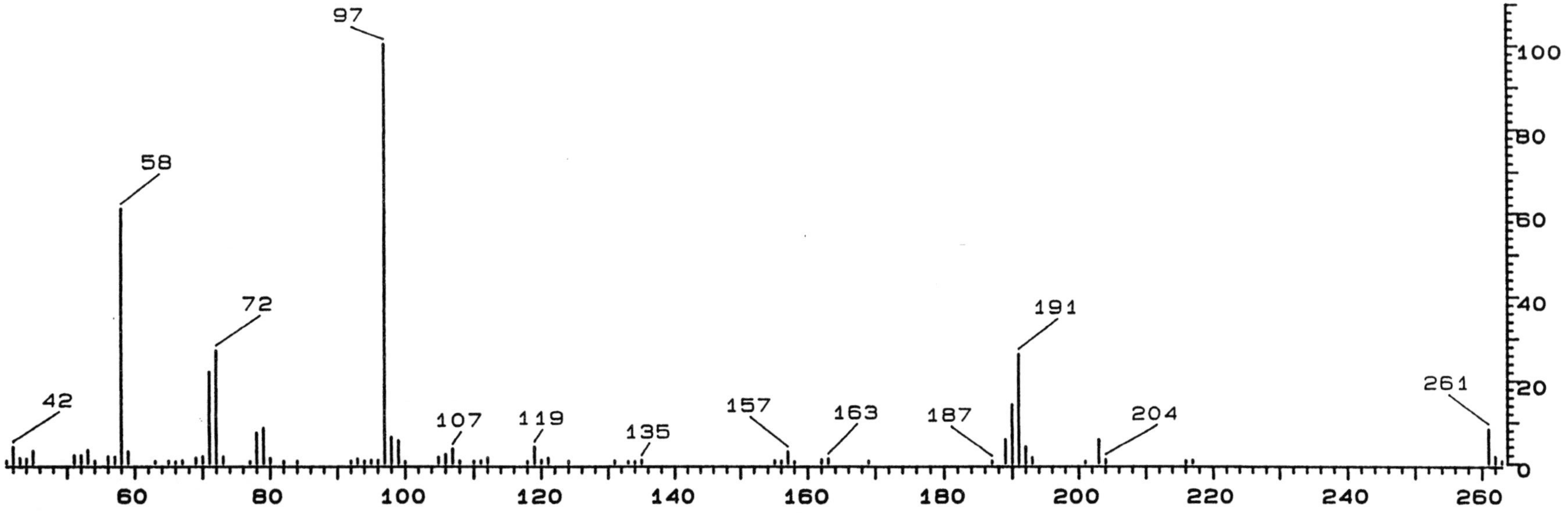

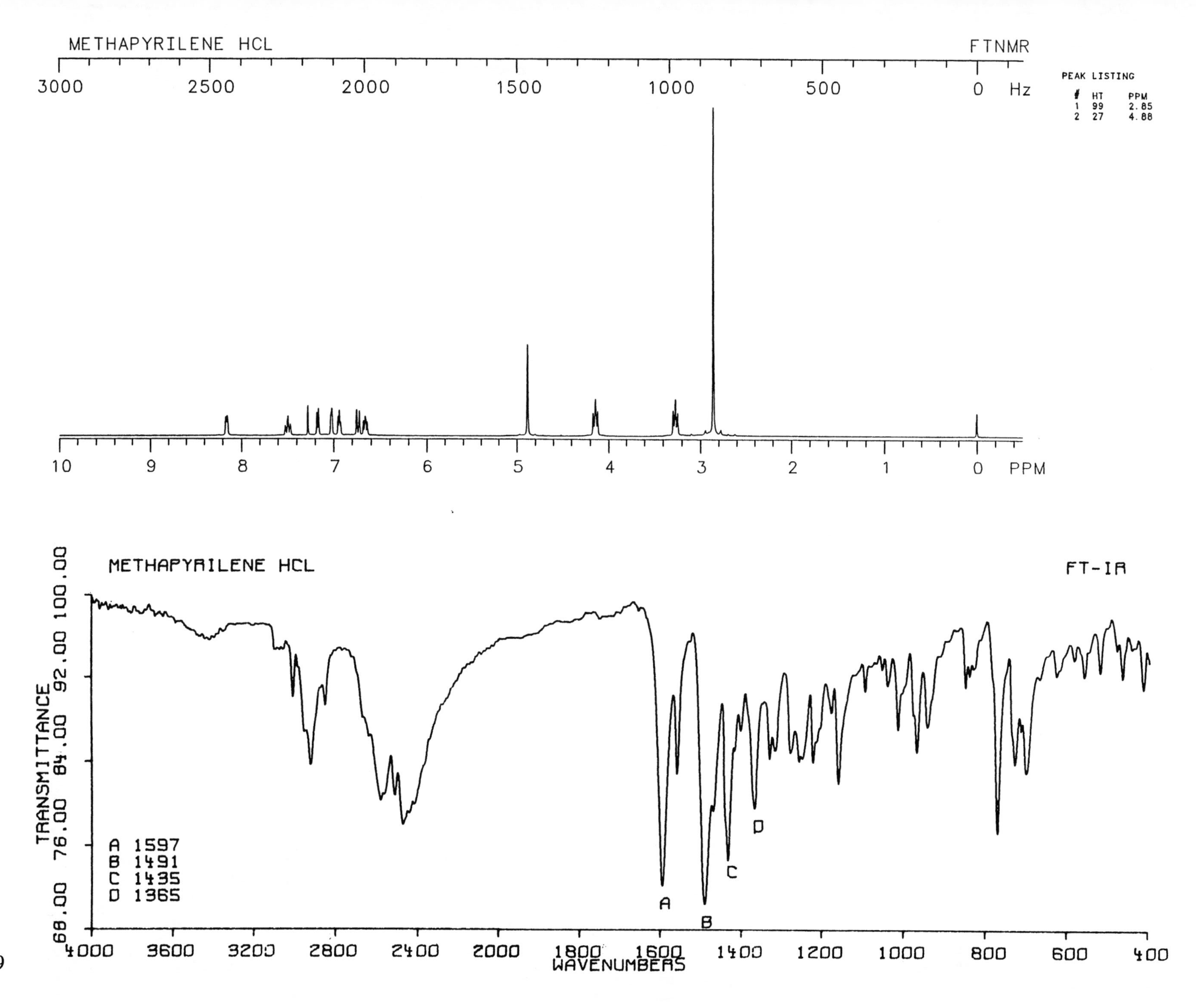
METHAPYRILENE HCL
FTNMR
3000
2500
2000
1500
1000
500
0 Hz
PEAK LISTING
# HT PPM
1 99 2.85
2 27 4.88
10
9
8
7
6
5
4
3
2
1
0 PPM
METHAPYRILENE HCL
FT-IR
TRANSMITTANCE
68.00
76.00
84.00
92.00
100.00
A 1597
B 1491
C 1435
D 1365
A
B
C
D
4000
3600
3200
2800
2400
2000
1800
1600
1400
1200
1000
800
600
400
WAVENUMBERS

# METHAQUALONE

$C_{16}H_{14}N_2O$

**Molecular weight:** 250.30 (250.11)
**Synonyms:** 2-Methyl-3-o-tolyl-4(3H)-quinazolinone; MTQ; ludes

**Trade names:** Mequin, Quaalude, Parest, Sopor

**Use:** Hypnotic
**HPLC:** Si-10; 1A:99B; 6.7
**GC:** 2218; 250°C

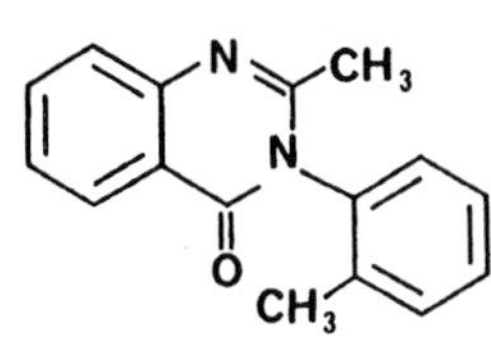


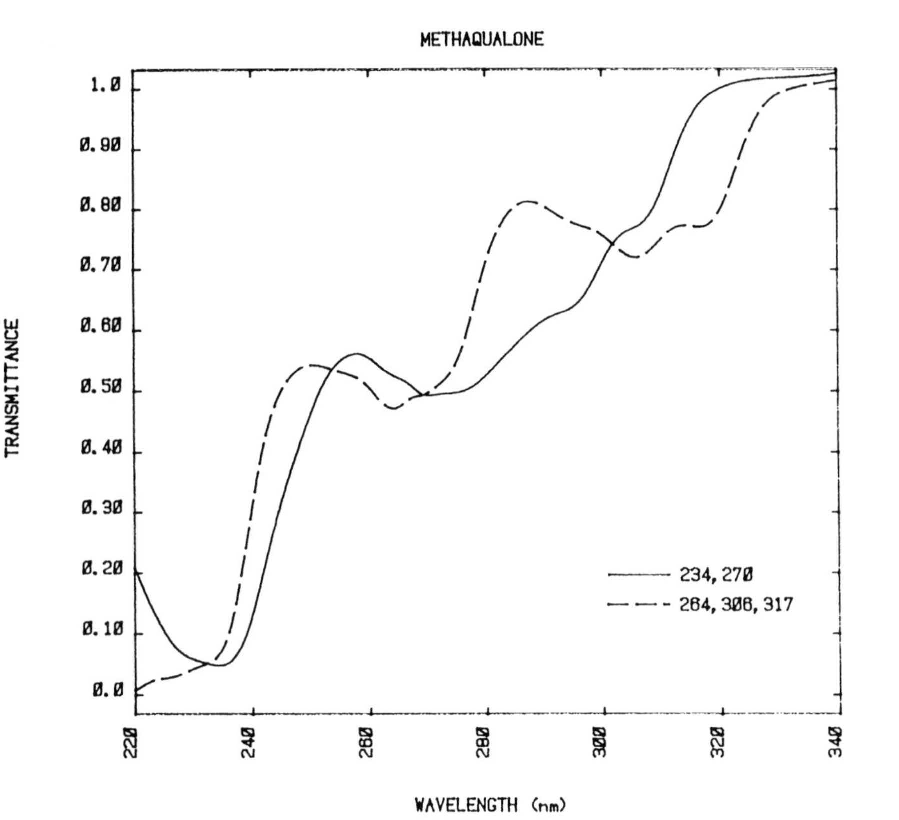

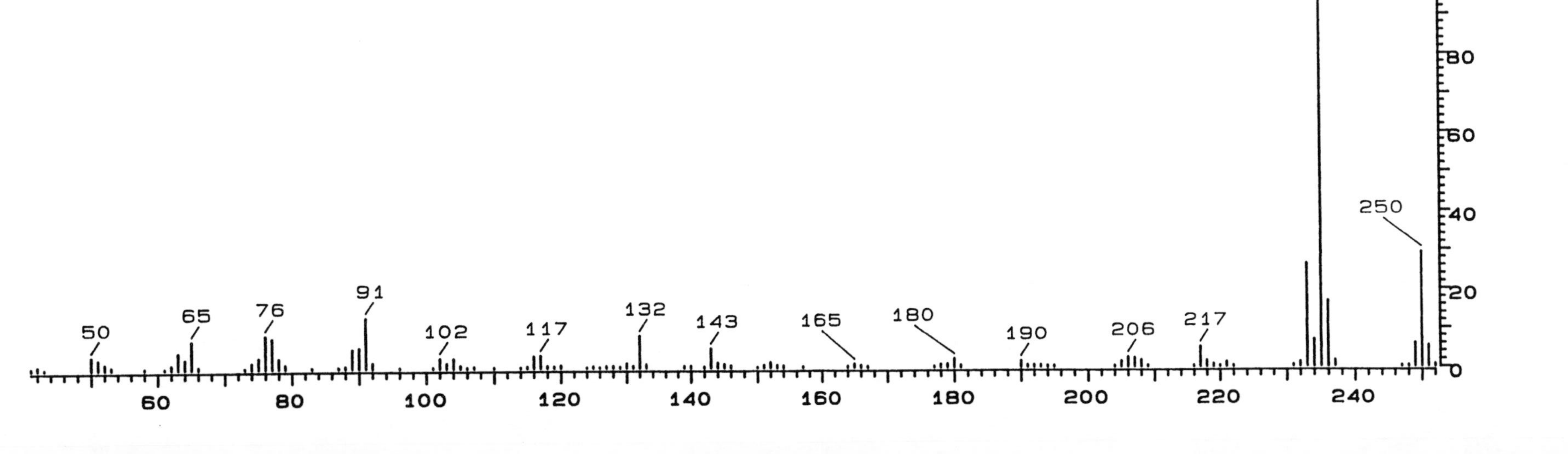

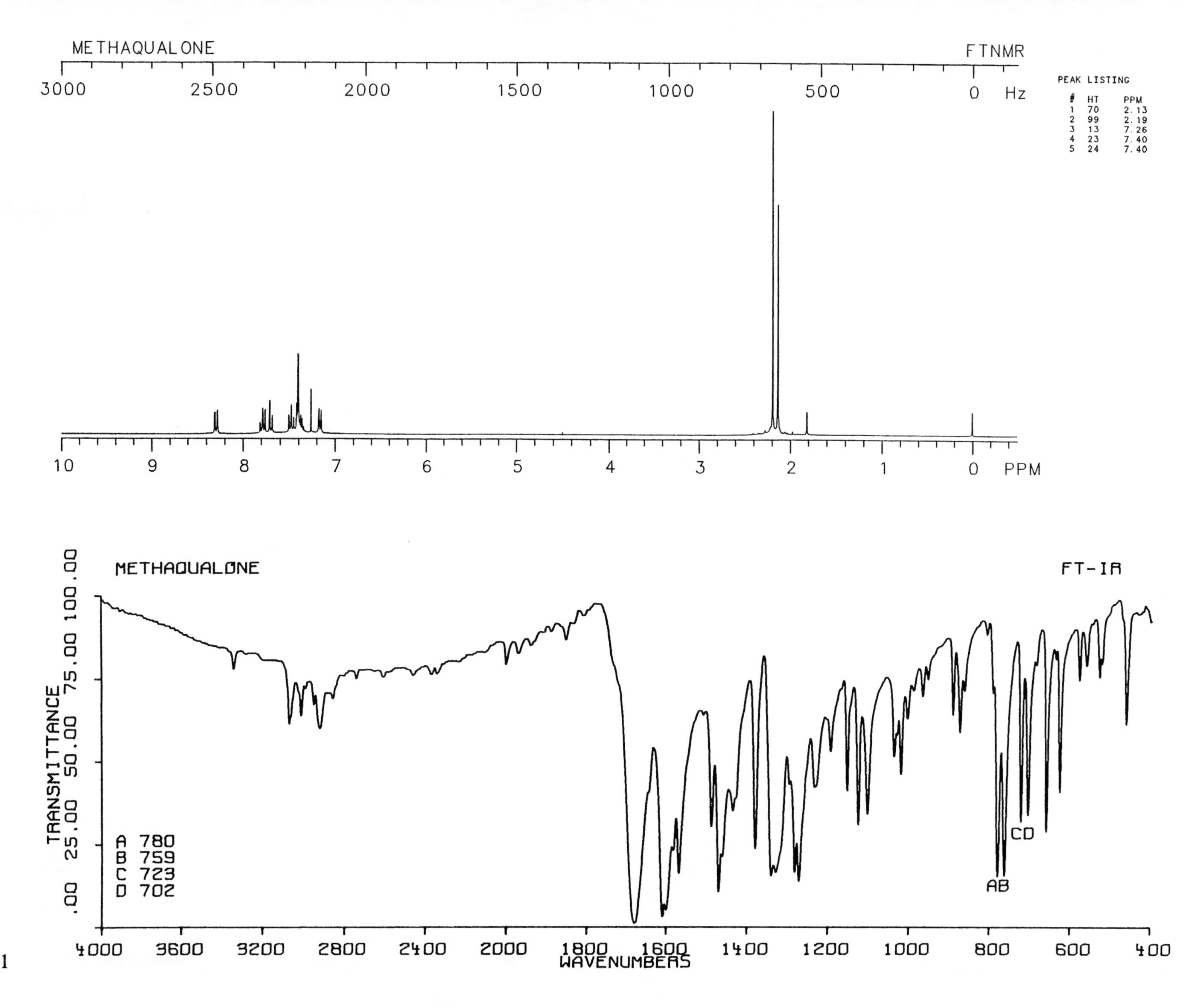
METHAQUALONE
FTNMR
3000 2500 2000 1500 1000 500 0 Hz
PEAK LISTING
# HT PPM
1 70 2.13
2 99 2.19
3 13 7.26
4 23 7.40
5 24 7.40
10 9 8 7 6 5 4 3 2 1 0 PPM
METHAQUALONE
FT-IR
TRANSMITTANCE
.00 25.00 50.00 75.00 100.00
A 780
B 759
C 723
D 702
CD
AB
4000 3600 3200 2800 2400 2000 1800 1600 1400 1200 1000 800 600 400
WAVENUMBERS

# METHARBITAL

$C_9H_{14}N_2O_3$

**Molecular weight:** 198.22 (198.10)
**Synonyms:** 5,5-Diethyl-1-methyl-2,4,6(1H,3H,5H)-pyrimidinetrione; 5,5-diethyl-1-methylbarbituric acid; metharbitone
**Trade names:** Gemonil

**Use:** Anticonvulsant
**HPLC:** Si-10; 1A:99B; 3.7
**GC:** 1450; 200°C

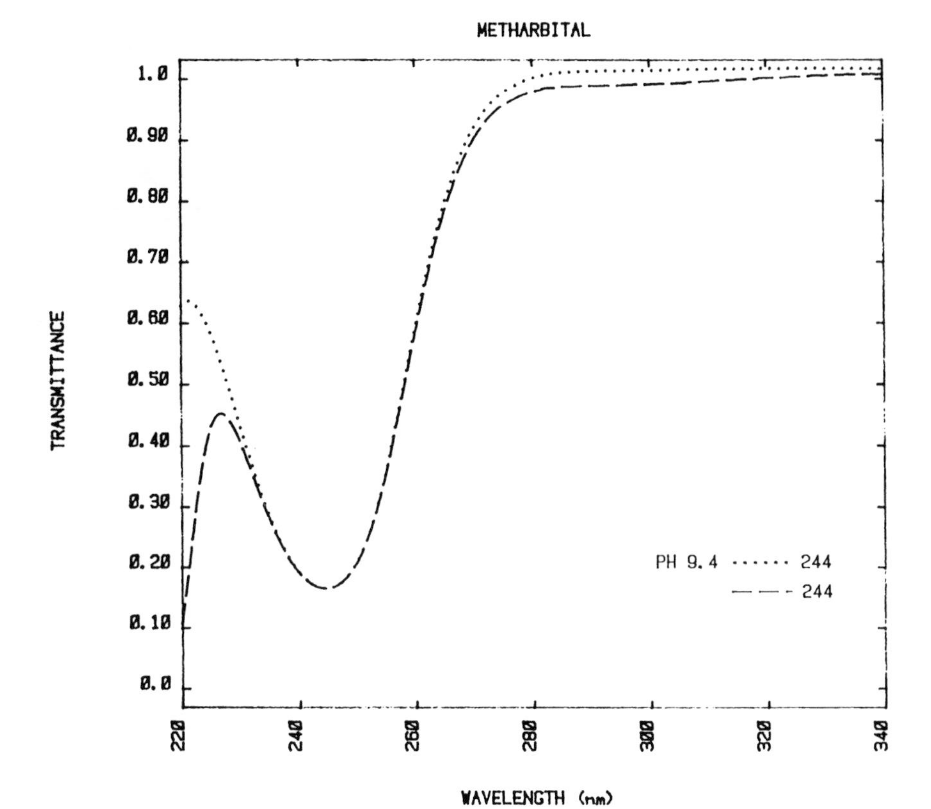

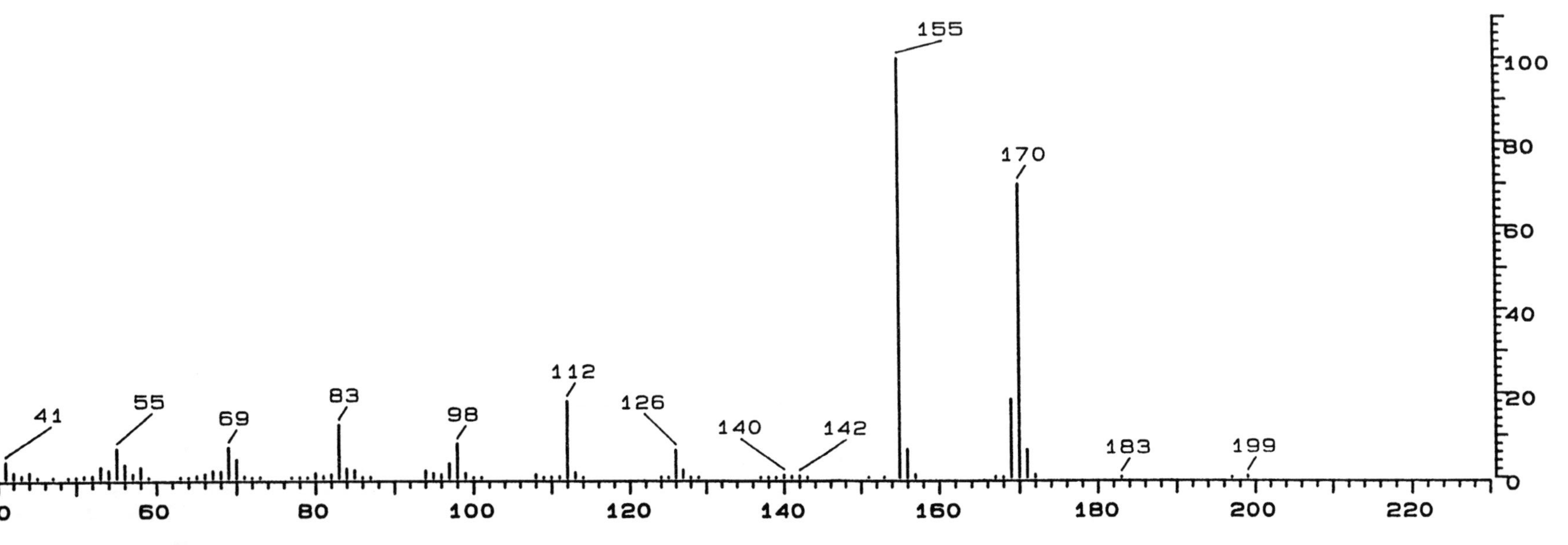

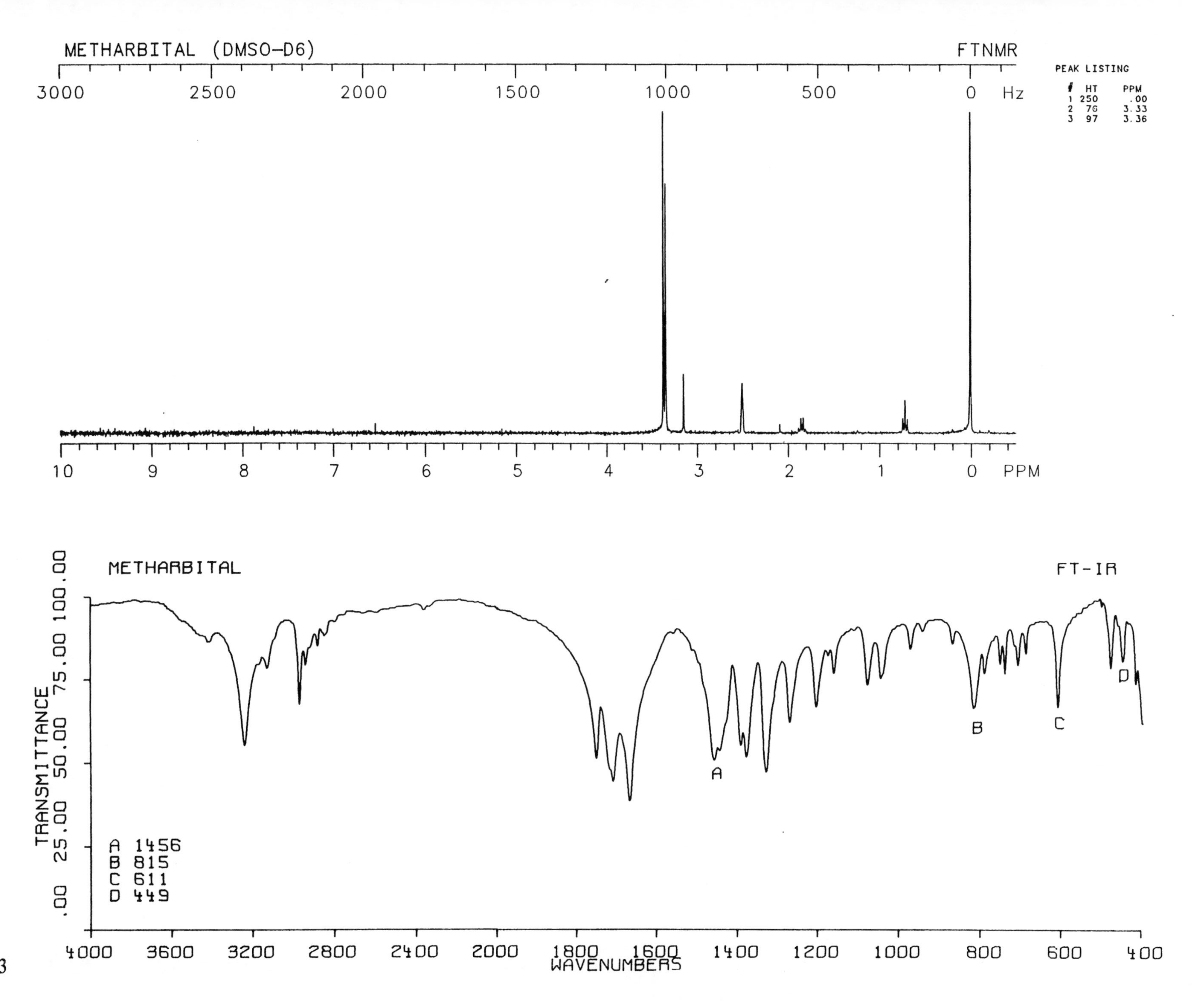
METHARBITAL (DMSO-D6)
FTNMR
3000 2500 2000 1500 1000 500 0 Hz
PEAK LISTING
# HT PPM
1 250 .00
2 76 3.33
3 97 3.36
10 9 8 7 6 5 4 3 2 1 0 PPM
METHARBITAL
FT-IR
TRANSMITTANCE
.00 25.00 50.00 75.00 100.00
A 1456
B 815
C 611
D 449
4000 3600 3200 2800 2400 2000 1800 1600 1400 1200 1000 800 600 400
WAVENUMBERS

# METHAZOLAMIDE

$C_5H_8N_4O_3S_2$

**Molecular weight:** 236.27 (236.00)
**Synonyms:** N-[(5-Aminosulfonyl)-3-methyl-1,3,4-triadiázol-2(3H)-ylidene]acetamide
**Trade names:** Neptazane

**Use:** Carbonic anhydrase inhibitor
**HPLC:**
**GC:**

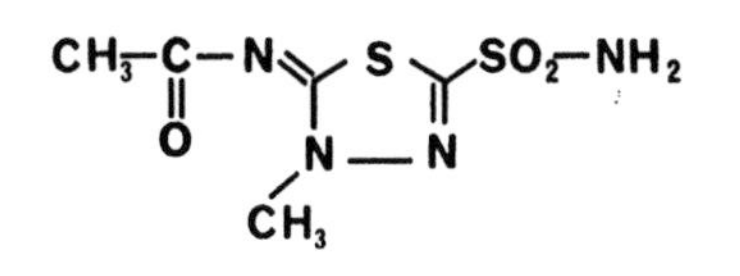


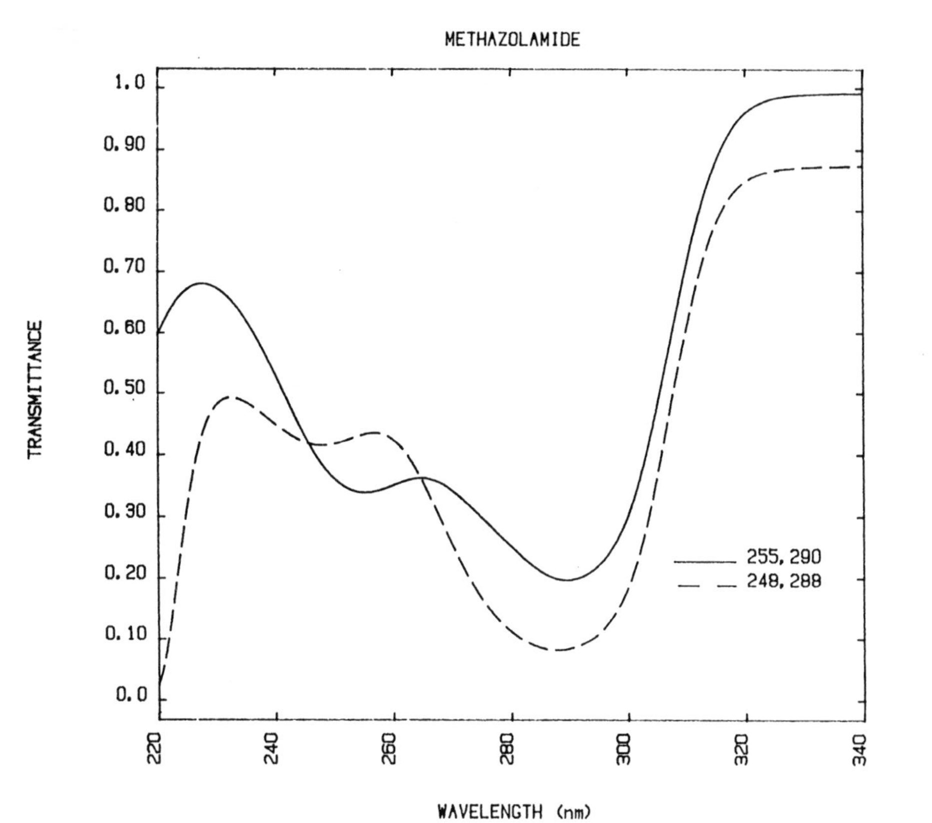

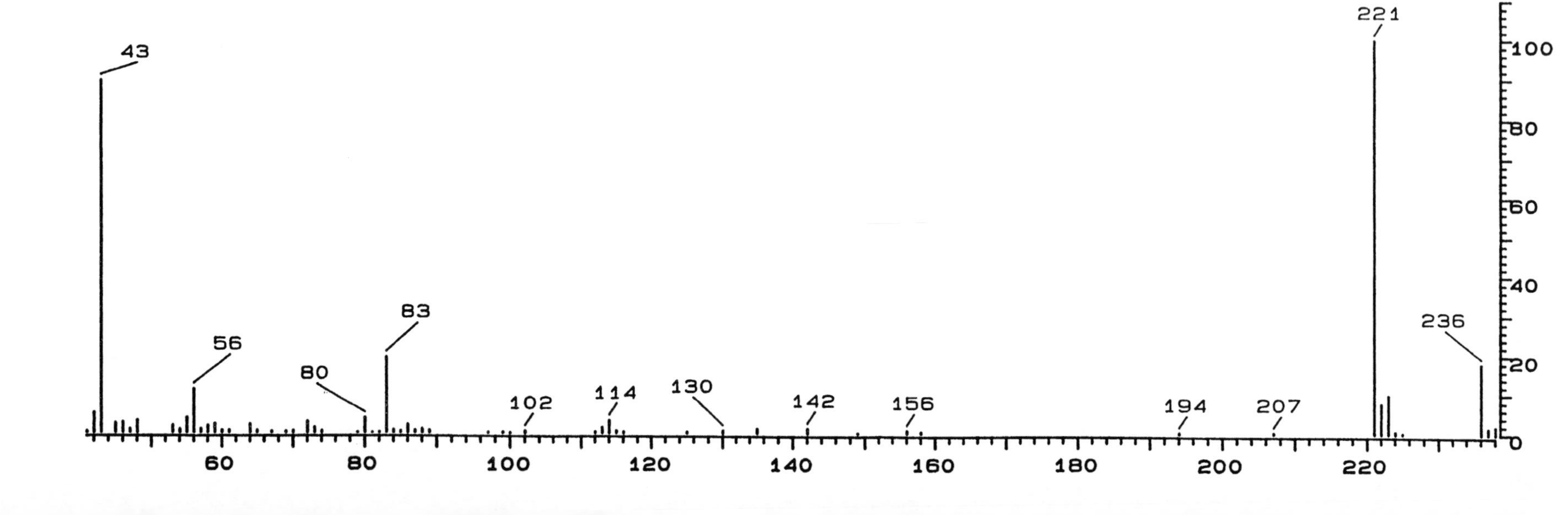

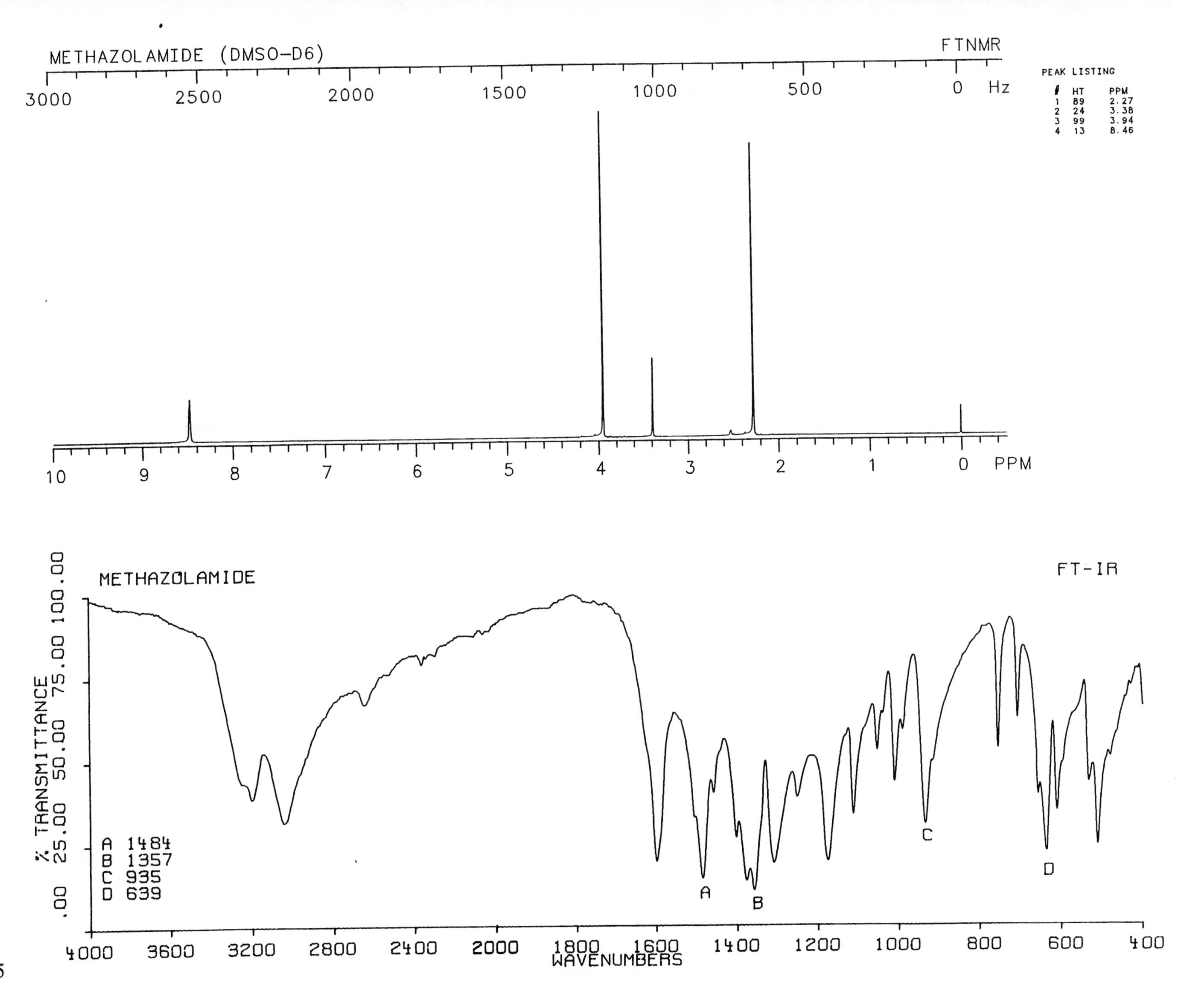
METHAZOLAMIDE (DMSO-D6)
FTNMR
Hz
PPM
PEAK LISTING
# HT PPM
1 89 2.27
2 24 3.38
3 99 3.94
4 13 8.46
METHAZOLAMIDE
FT-IR
% TRANSMITTANCE
WAVENUMBERS
A 1484
B 1357
C 935
D 639

# METHDILAZINE

$C_{18}H_{20}N_2S$

Molecular weight: 296.44 (296.14)
Synonyms: 10-[(1-Methyl-3-pyrrolidinyl)methyl]phenothiazine

Trade names: Tacaryl

Use: Antipruritic
HPLC: Si-10; 20A:80B; 4.0
GC: 2504; 250°C

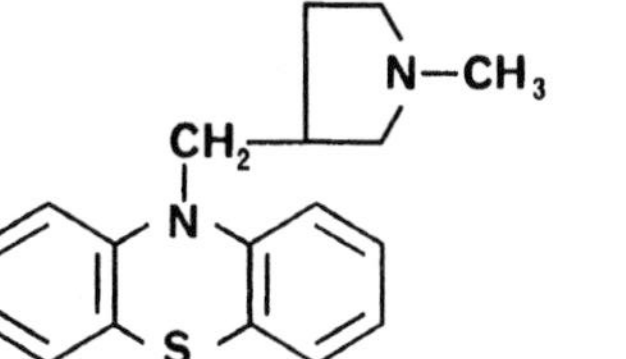
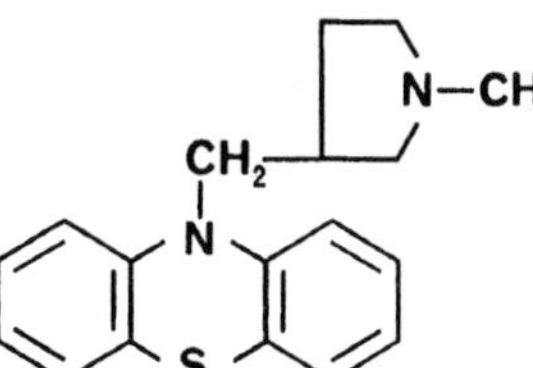


METHDILAZINE

TRANSMITTANCE

1.0
0.90
0.80
0.70
0.60
0.50
0.40
0.30
0.20
0.10
0.0

220 240 260 280 300 320 340

WAVELENGTH (nm)

——— 252, 301
— — 253, 304

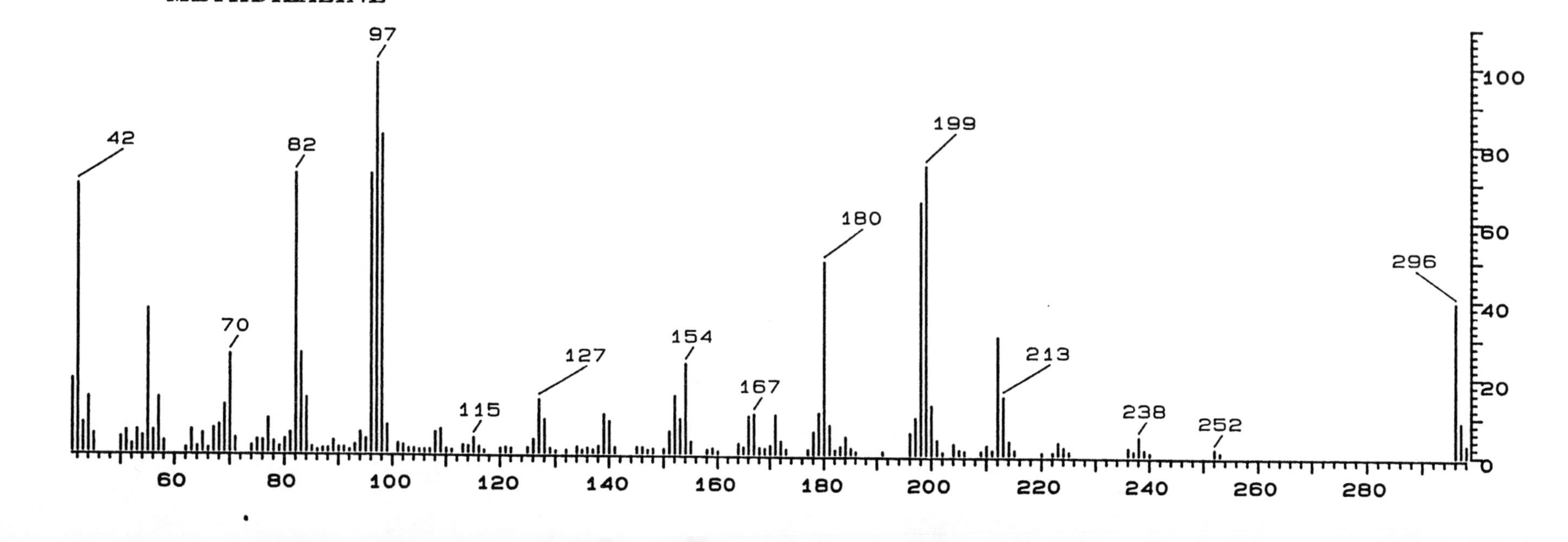

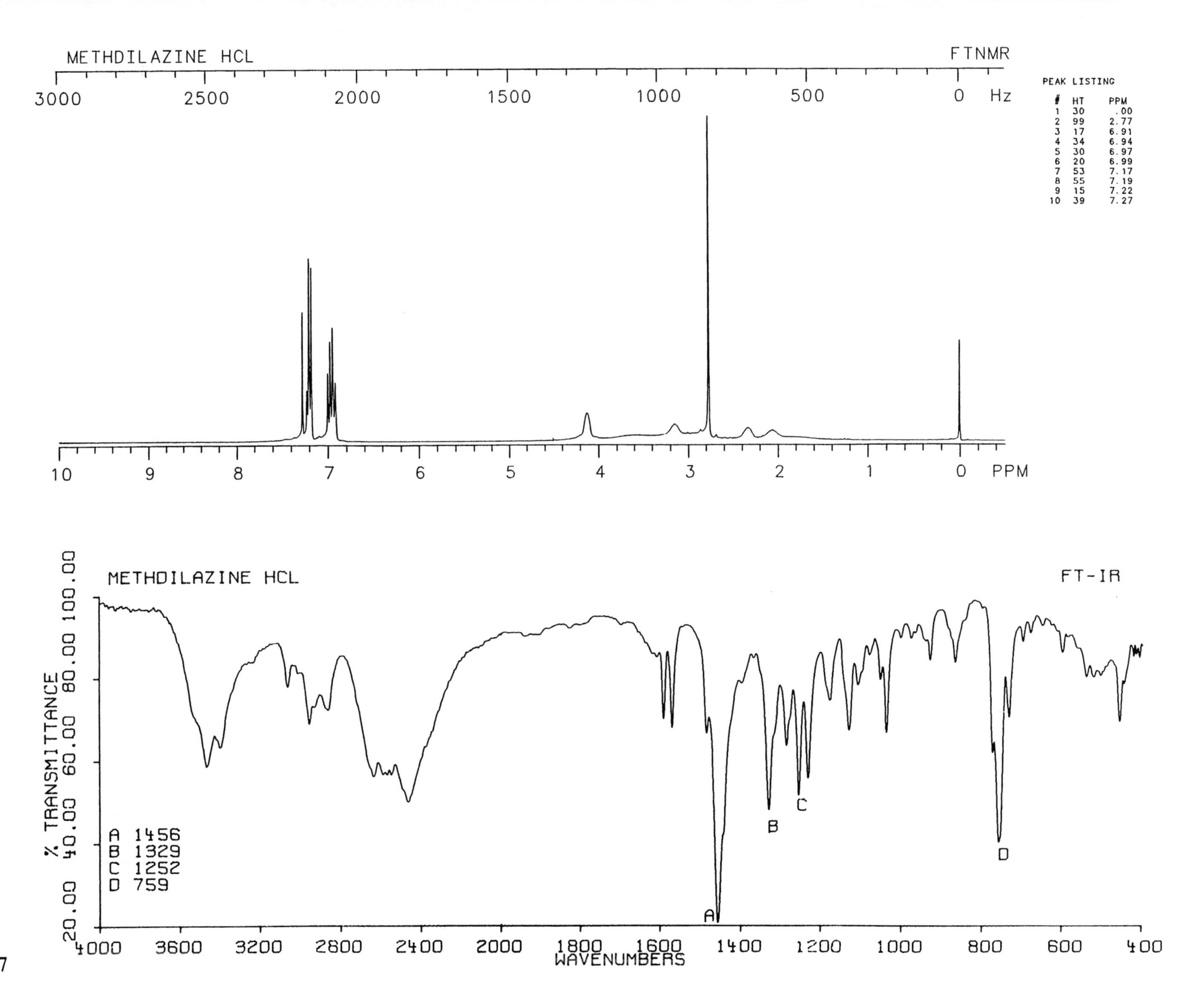

METHDILAZINE HCL
FTNMR
3000 2500 2000 1500 1000 500 0 Hz
PEAK LISTING
# HT PPM
1 30 .00
2 99 2.77
3 17 6.91
4 34 6.94
5 30 6.97
6 20 6.99
7 53 7.17
8 55 7.19
9 15 7.22
10 39 7.27
10 9 8 7 6 5 4 3 2 1 0 PPM
METHDILAZINE HCL
FT-IR
% TRANSMITTANCE
20.00 40.00 60.00 80.00 100.00
A 1456
B 1329
C 1252
D 759
A
B
C
D
4000 3600 3200 2800 2400 2000 1800 1600 1400 1200 1000 800 600 400
WAVENUMBERS

# METHENAMINE

$C_6H_{12}N_4$

**Molecular weight:** 140.19 (140.11)
**Synonyms:** 1,3,5,7-Tetrazatricyclo[3.3.1.1$^{3,7}$]-decane; hexamine; hexamethylenamine
**Trade names:** Thiacide, Trac, Urised, Uroblue, Uro-Phosphate, Uroqid-Acid

**Use:** Antibacterial
**HPLC:**
**GC:** 1223; 140°C

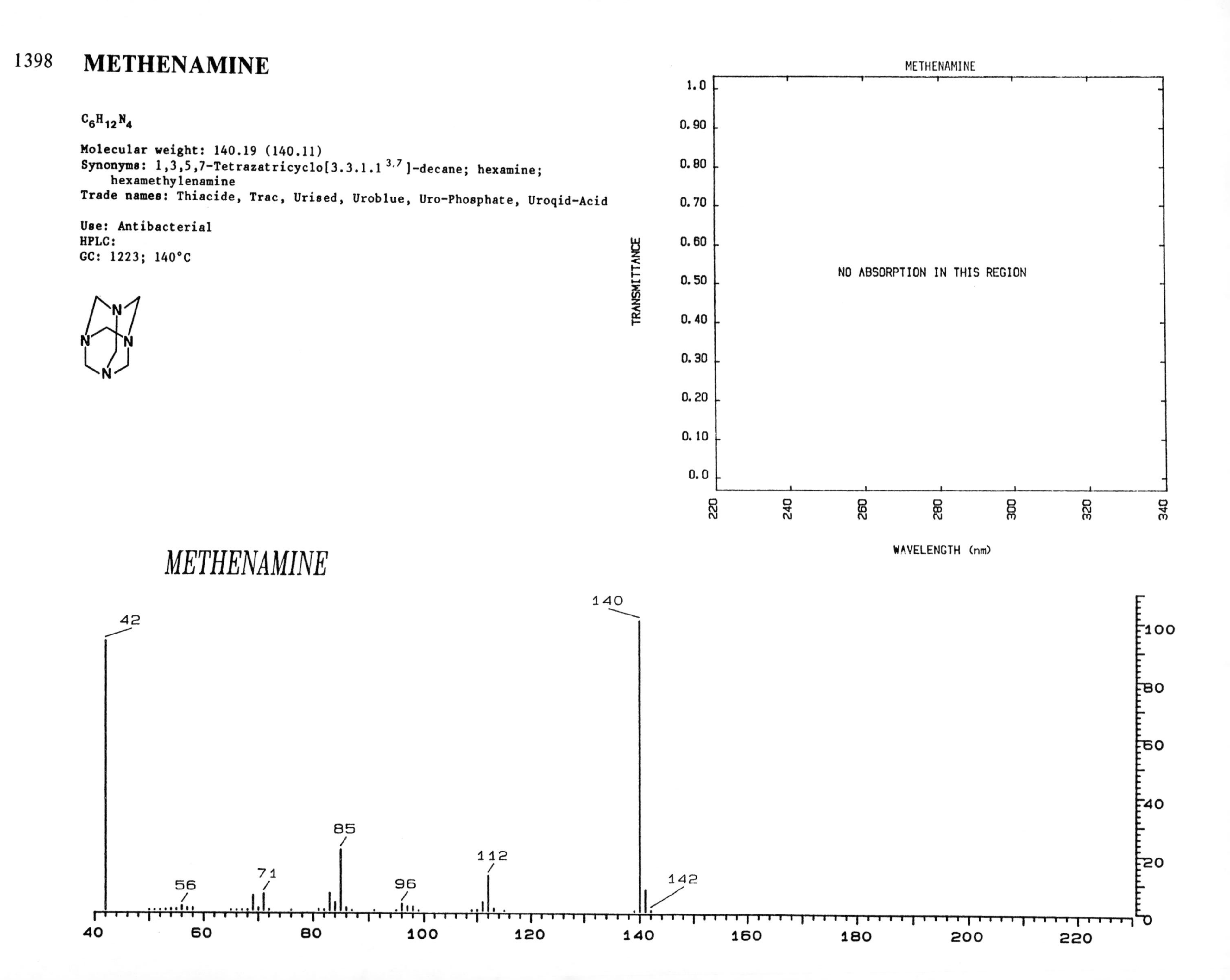

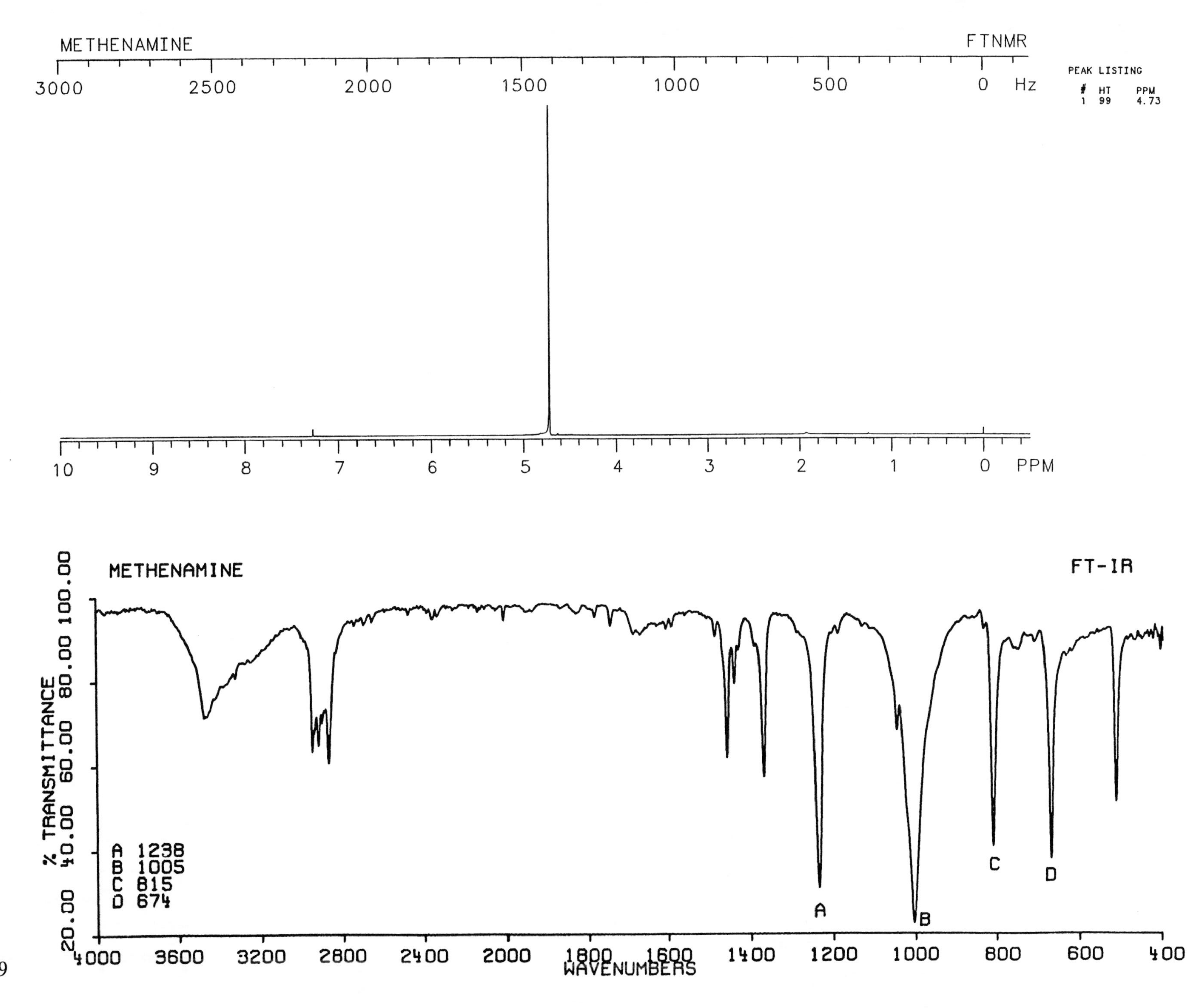
METHENAMINE
FTNMR
3000
2500
2000
1500
1000
500
0 Hz
PEAK LISTING
# HT PPM
1 99 4.73
10
9
8
7
6
5
4
3
2
1
0 PPM
METHENAMINE
FT-IR
% TRANSMITTANCE
20.00
40.00
60.00
80.00
100.00
A 1238
B 1005
C 815
D 674
A
B
C
D
4000
3600
3200
2800
2400
2000
1800
1600
1400
1200
1000
800
600
400
WAVENUMBERS

# METHICILLIN

$C_{17}H_{20}N_2O_6S$

**Molecular weight:** 380.42 (380.10)
**Synonyms:** 6-(2,6-Dimethoxybenzamido)-3,3-dimethyl-7-oxo-4-thia-1-azabicyclo[3.2.0]heptane-2-carboxylic acid; 2,6-dimethoxyphenylpenicillin
**Trade names:** Staphcillin

**Use:** Antimicrobial
**HPLC:**
**GC:**

O—CH3 O COOH CH3 CH3 C—NH N O—CH3

METHICILLIN

TRANSMITTANCE

1.0 0.90 0.80 0.70 0.60 0.50 0.40 0.30 0.20 0.10 0.0

278
280

220 240 260 280 300 320 340

WAVELENGTH (nm)

*NO USEFUL MASS SPECTRUM WAS OBTAINED*

100 80 60 40 20 0

80 120 160 200 240 280 320 360 400 440

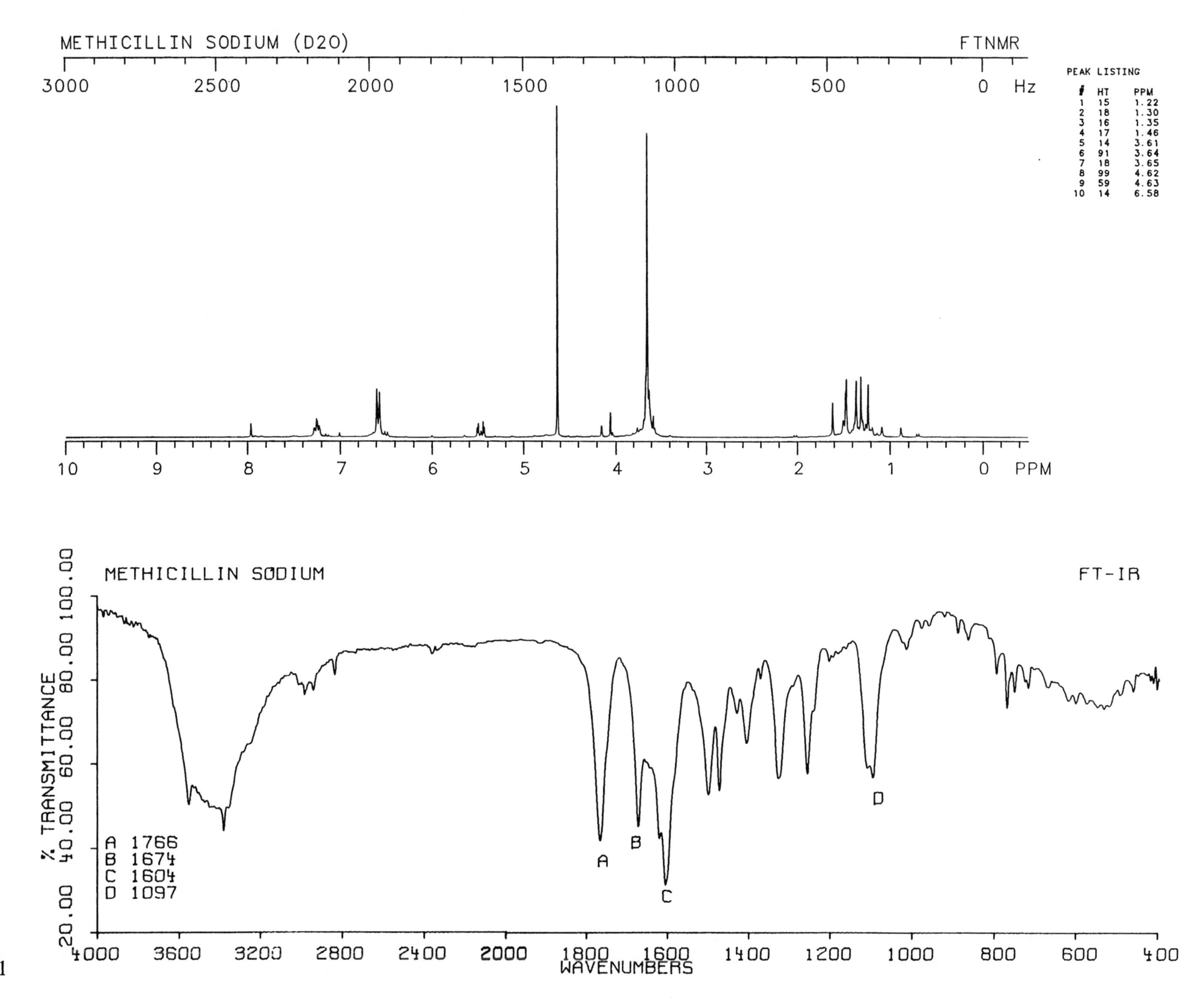

METHICILLIN SODIUM (D2O)
FTNMR
3000
2500
2000
1500
1000
500
0 Hz
PEAK LISTING
# HT PPM
1 15 1.22
2 18 1.30
3 16 1.35
4 17 1.46
5 14 3.61
6 91 3.64
7 18 3.65
8 99 4.62
9 59 4.63
10 14 6.58
10
9
8
7
6
5
4
3
2
1
0 PPM
METHICILLIN SODIUM
FT-IR
% TRANSMITTANCE
100.00
80.00
60.00
40.00
20.00
A 1766
B 1674
C 1604
D 1097
A
B
C
D
4000
3600
3200
2800
2400
2000
1800
1600
1400
1200
1000
800
600
400
WAVENUMBERS

# METHIMAZOLE

$C_4H_6N_2S$

**Molecular weight:** 114.17 (114.03)
**Synonyms:** 1-Methylimidazole-2-thiol; thiamazole

**Trade names:** Tapazole

**Use:** Thyroid inhibitor
**HPLC:** Si-10; 2A:98B; 4.2
**GC:** 1600; 200°C

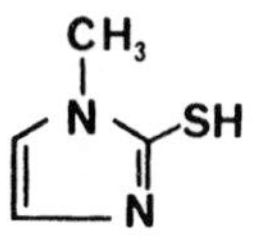

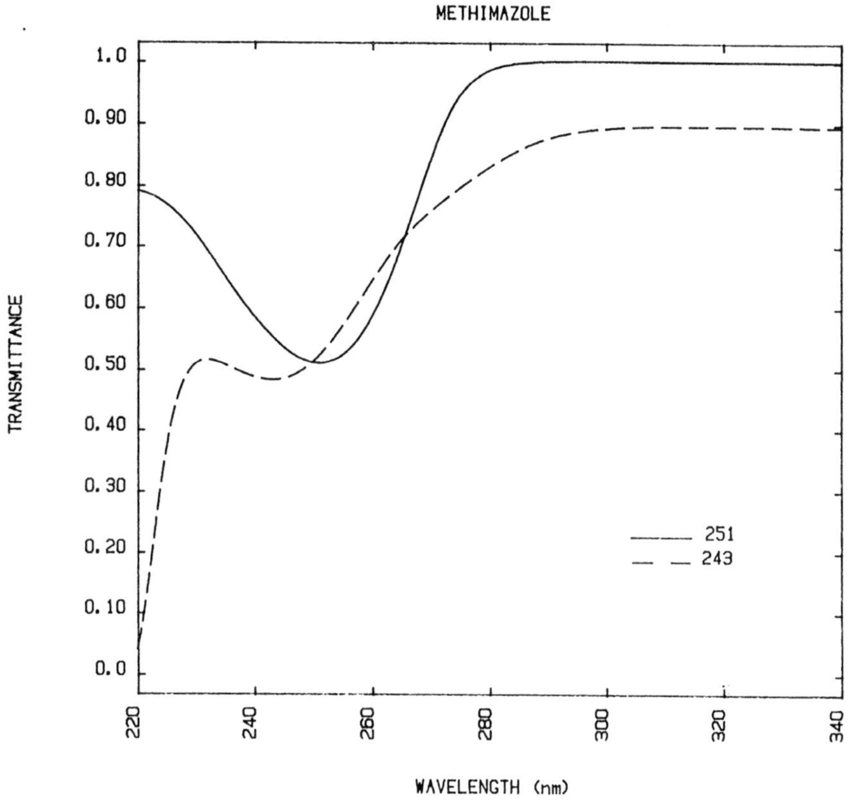

METHIMAZOLE

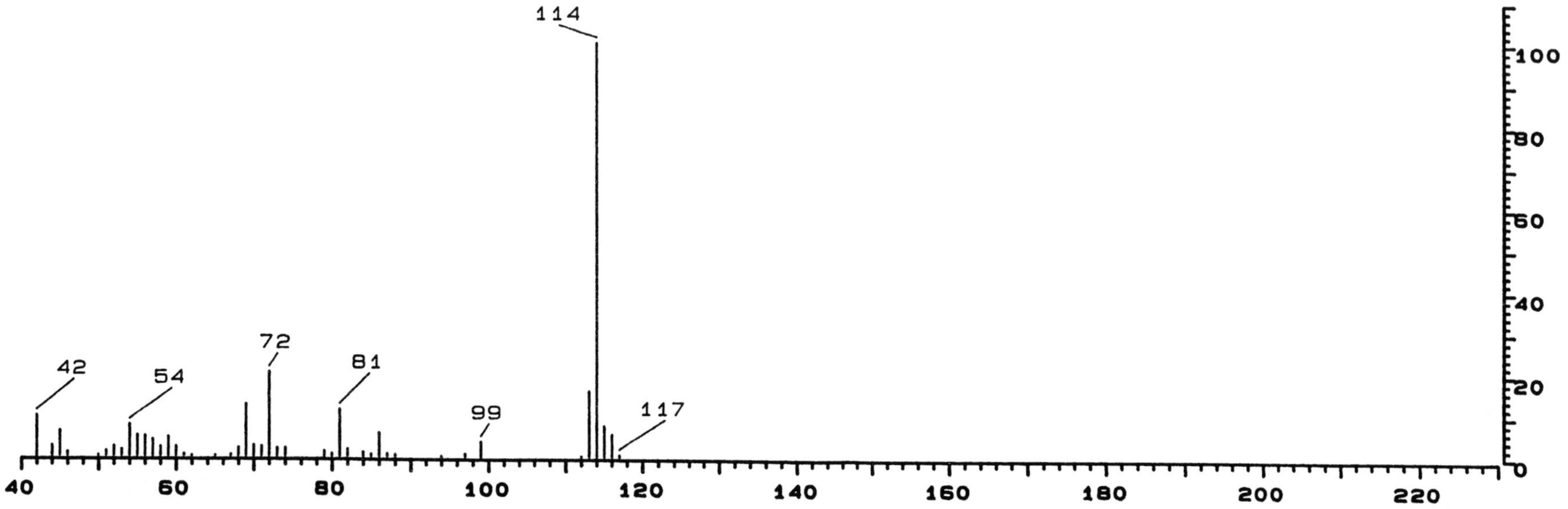

TRANSMITTANCE

WAVELENGTH (nm)

—— 251
— — 243

*METHIMAZOLE*

114, 117, 99, 81, 72, 54, 42

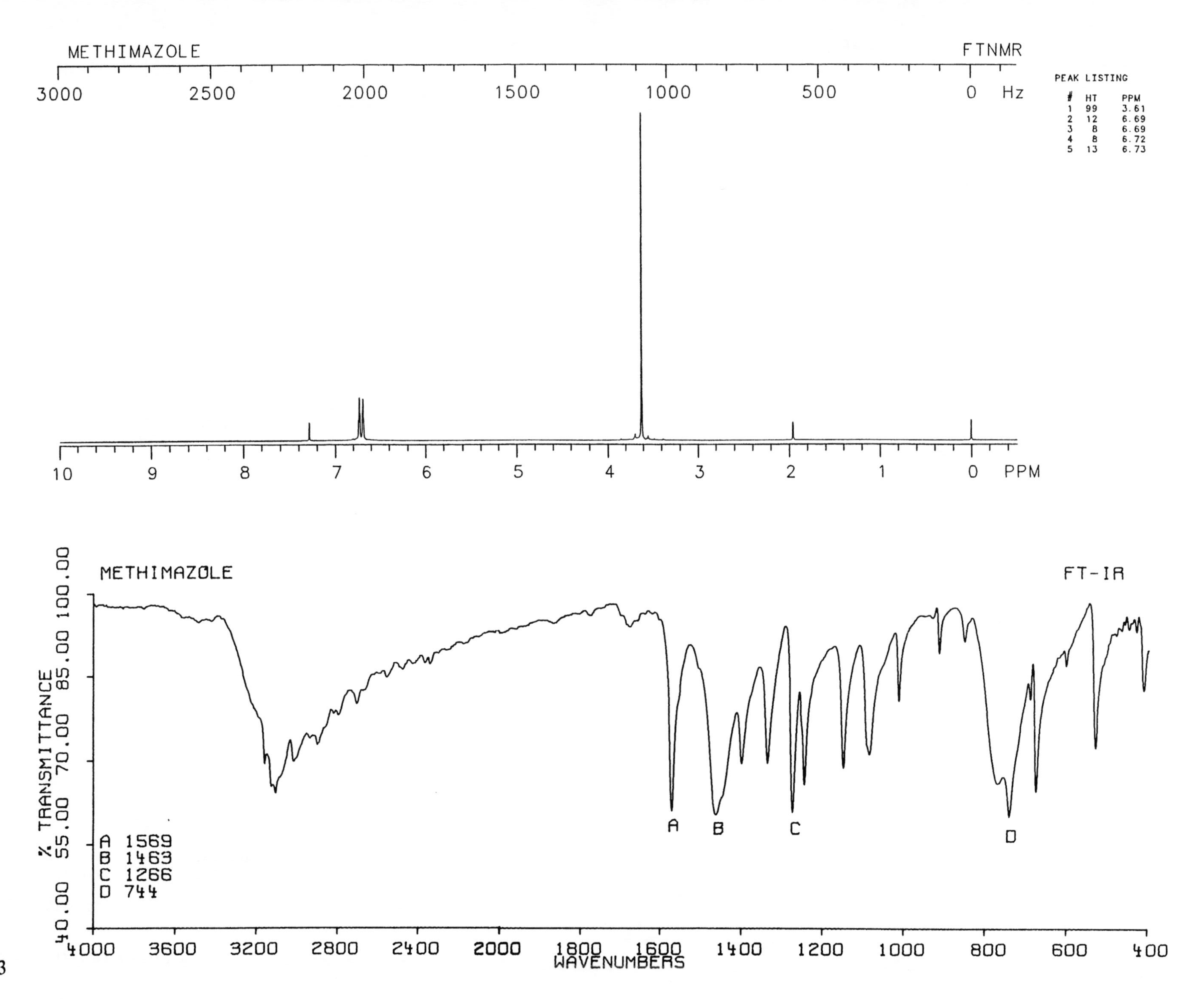
METHIMAZOLE
FTNMR
3000 2500 2000 1500 1000 500 0 Hz
PEAK LISTING
# HT PPM
1 99 3.61
2 12 6.69
3 8 6.69
4 8 6.72
5 13 6.73
10 9 8 7 6 5 4 3 2 1 0 PPM
METHIMAZOLE
FT-IR
% TRANSMITTANCE
40.00 55.00 70.00 85.00 100.00
A 1569
B 1463
C 1266
D 744
A B C D
4000 3600 3200 2800 2400 2000 1800 1600 1400 1200 1000 800 600 400
WAVENUMBERS

# METHIONINE

$C_5H_{11}NO_2S$

Molecular weight: 149.21 (149.05)
Synonyms: 2-Amino-4-(methylthio)butyric acid; MET

Trade names: Pedameth

Use: Lipotropic
HPLC:
GC:

$CH_3-S-CH_2-CH_2-CH(NH_2)-COOH$

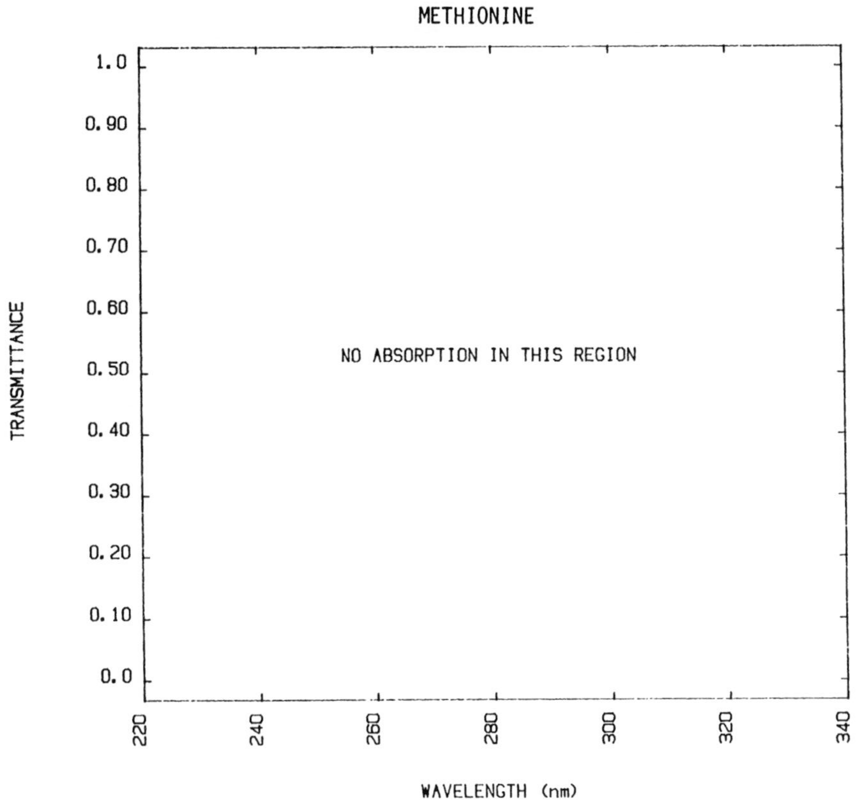

METHIONINE

TRANSMITTANCE

1.0 0.90 0.80 0.70 0.60 0.50 0.40 0.30 0.20 0.10 0.0

NO ABSORPTION IN THIS REGION

220 240 260 280 300 320 340

WAVELENGTH (nm)

*METHIONINE--DIP*

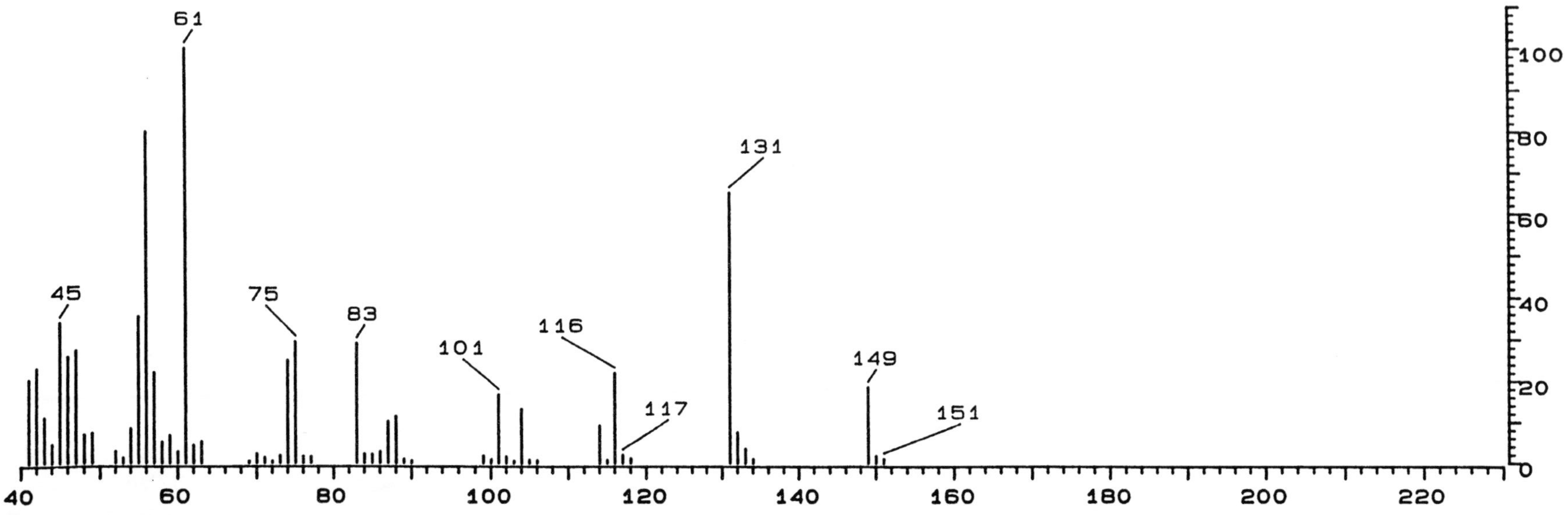

61 45 75 83 101 116 117 131 149 151

100 80 60 40 20 0

40 60 80 100 120 140 160 180 200 220

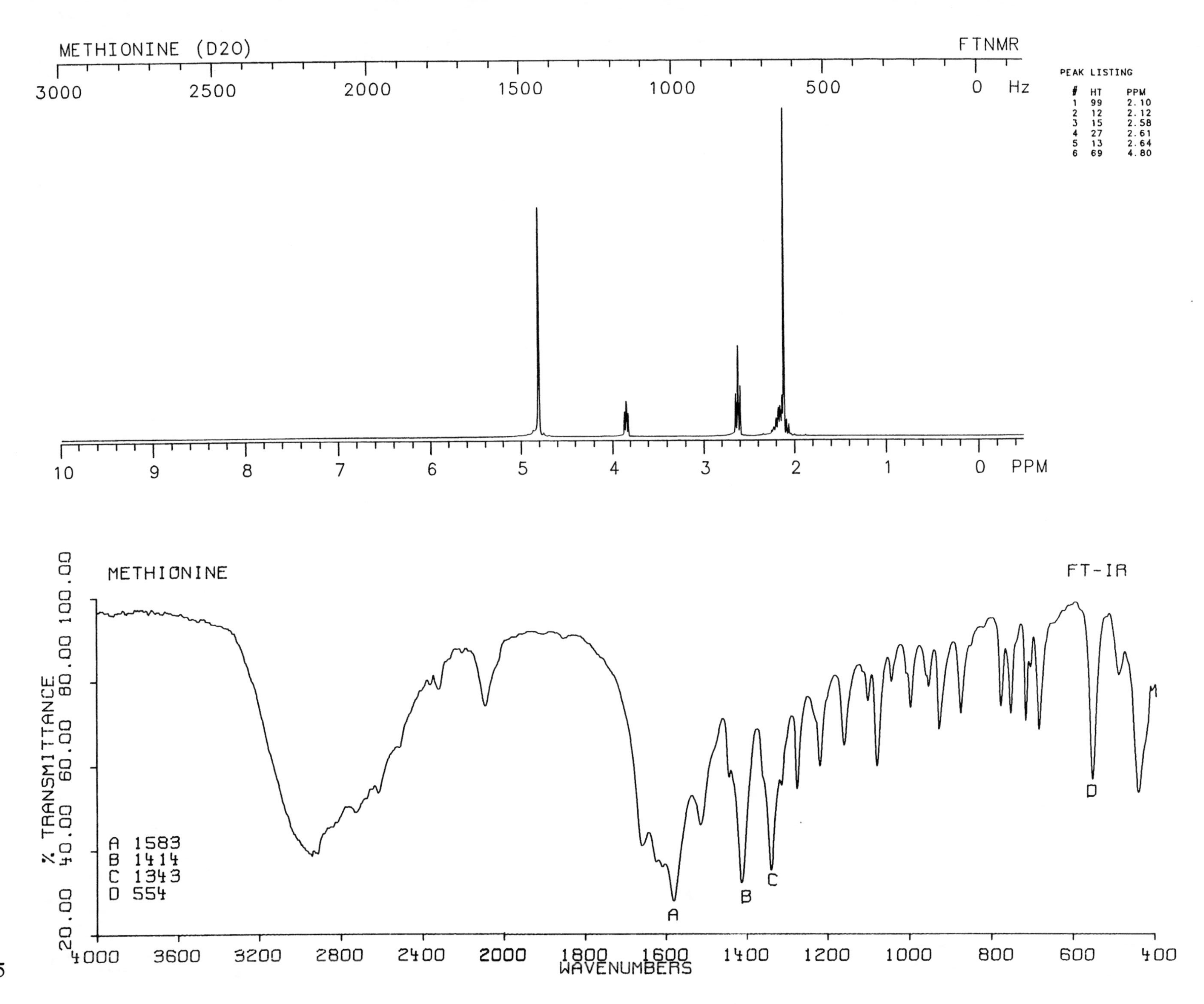
METHIONINE (D2O)
FTNMR
Hz
PPM
PEAK LISTING
# HT PPM
1 99 2.10
2 12 2.12
3 15 2.58
4 27 2.61
5 13 2.64
6 69 4.80
METHIONINE
FT-IR
% TRANSMITTANCE
WAVENUMBERS
A 1583
B 1414
C 1343
D 554

# METHIXENE

$C_{20}H_{23}NS$

Molecular weight: 309.48 (309.16)
Synonyms: 1-Methyl-3-(9H-thioxanthen-9-yl-methyl)piperidine

Trade names: Trest

Use: Smooth muscle relaxant
HPLC: Si-10; 10A:90B; 4.0
GC: 2520; 250°C

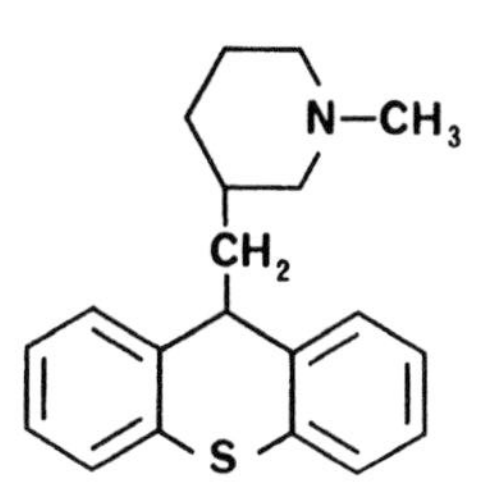


METHIXENE

TRANSMITTANCE

1.0
0.90
0.80
0.70
0.60
0.50
0.40
0.30
0.20
0.10
0.0

220 240 260 280 300 320 340

WAVELENGTH (nm)

——— 267
— — 276

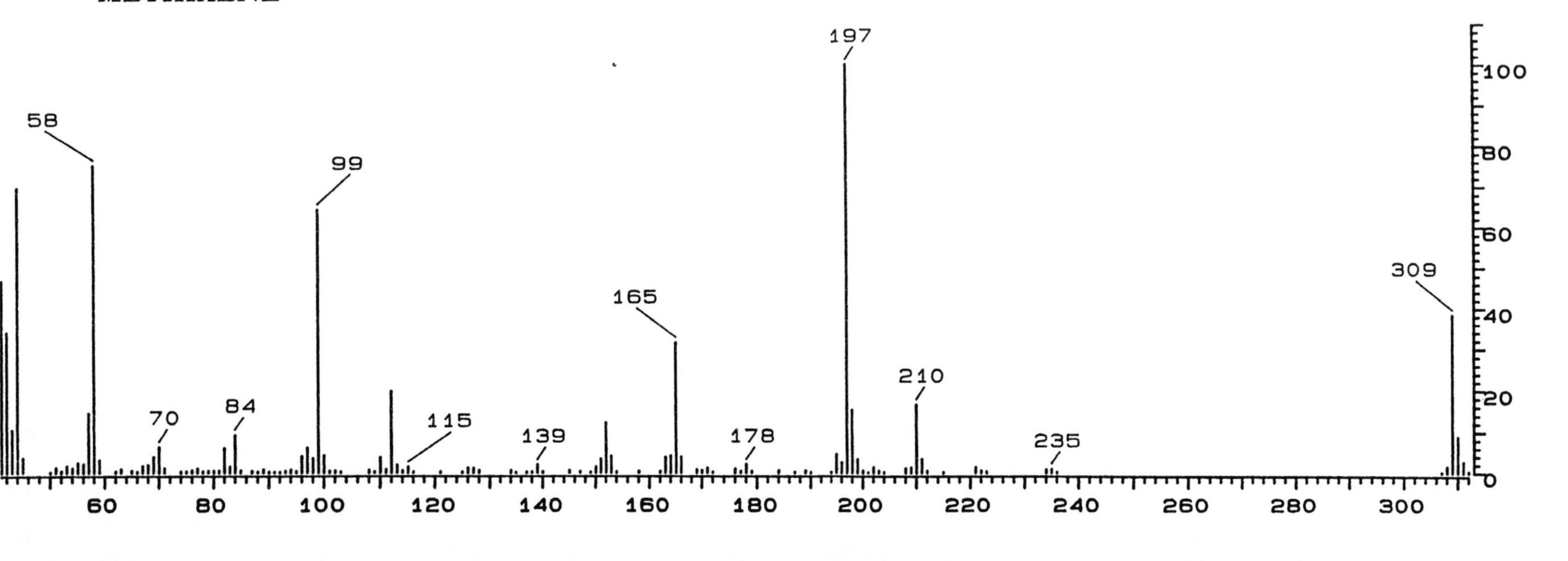

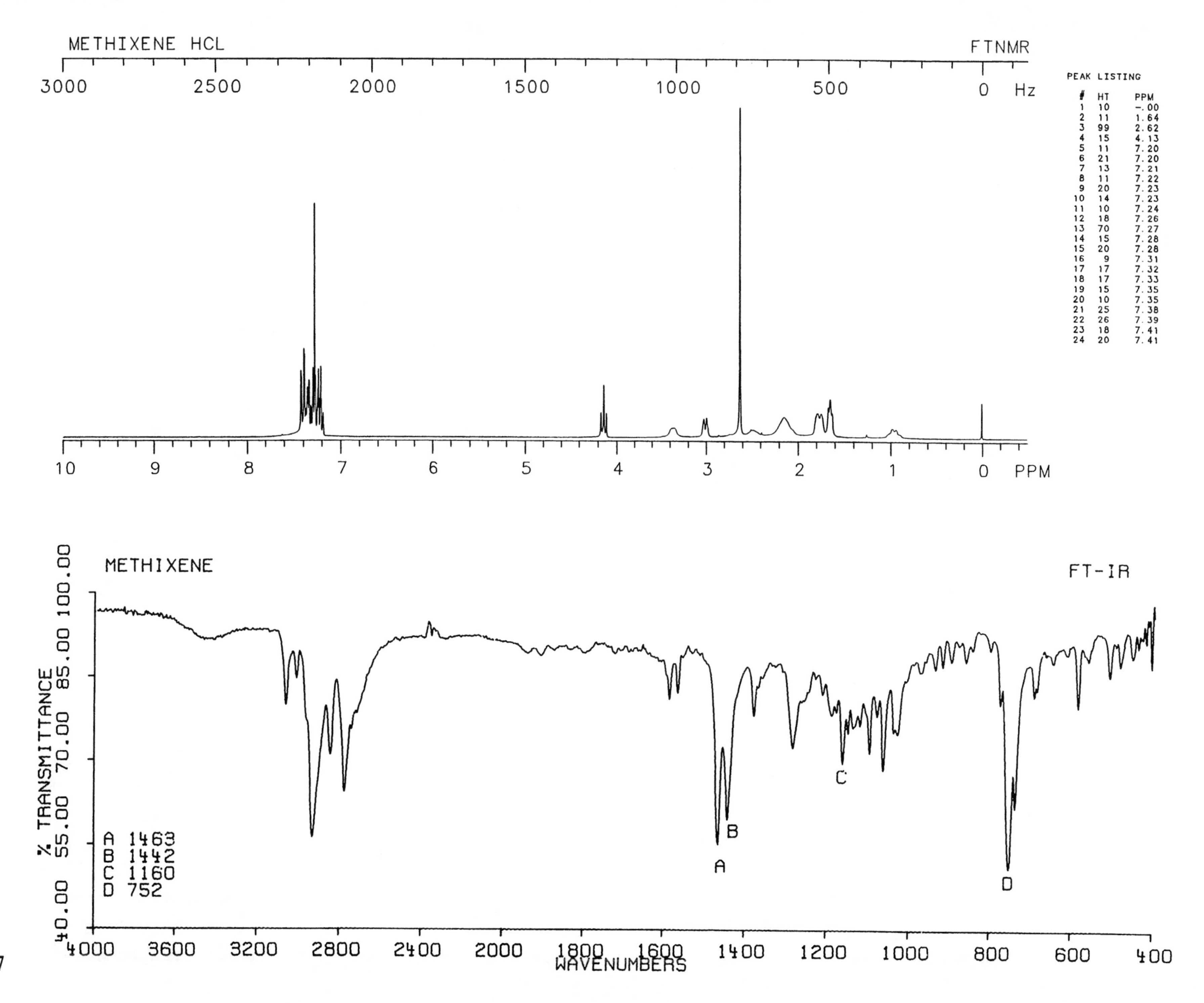

METHIXENE HCL
FTNMR
Hz
PPM
PEAK LISTING
# HT PPM
1 10 -.00
2 11 1.64
3 99 2.62
4 15 4.13
5 11 7.20
6 21 7.20
7 13 7.21
8 11 7.22
9 20 7.23
10 14 7.23
11 10 7.24
12 18 7.26
13 70 7.27
14 15 7.28
15 20 7.28
16 9 7.31
17 17 7.32
18 17 7.33
19 15 7.35
20 10 7.35
21 25 7.38
22 26 7.39
23 18 7.41
24 20 7.41
METHIXENE
FT-IR
% TRANSMITTANCE
WAVENUMBERS
A 1463
B 1442
C 1160
D 752

# METHOCARBAMOL

$C_{11}H_{15}NO_5$

Molecular weight: 241.24 (241.10)
Synonyms: 3-(o-Methoxyphenoxy)-1,2-propanediol 1-carbamate; guaiacol glyceryl ether carbamate
Trade names: Robaxin

Use: Skeletal muscle relaxant
HPLC: Si-10; 2A:98B; 9.0
GC: 1685; 200°C

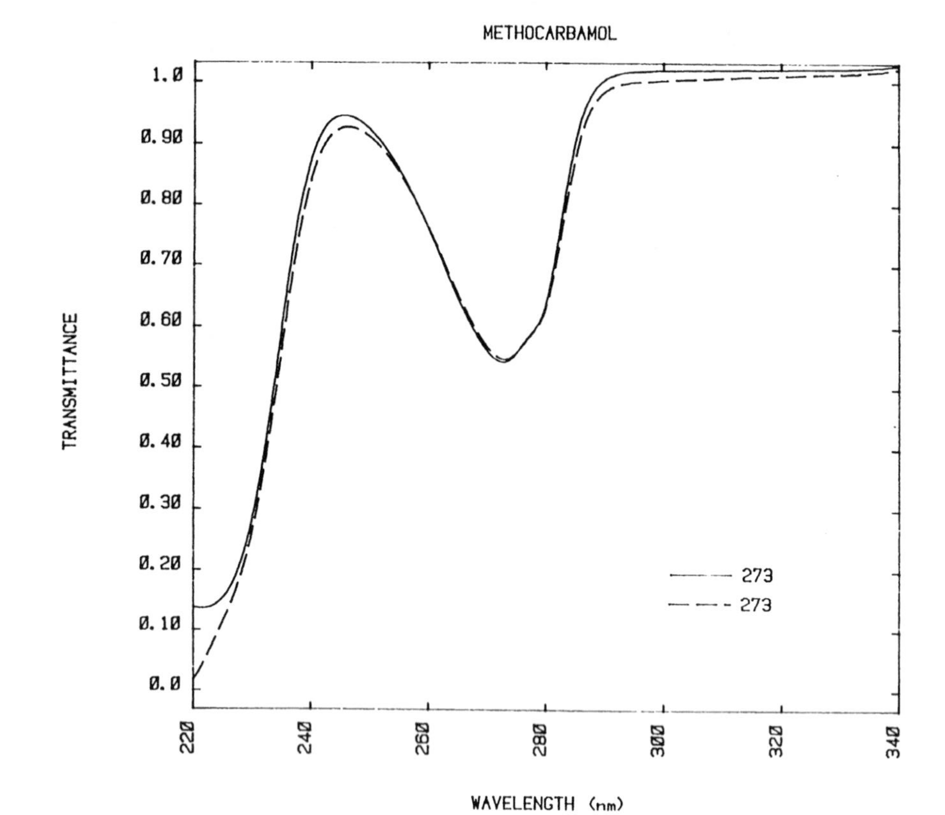

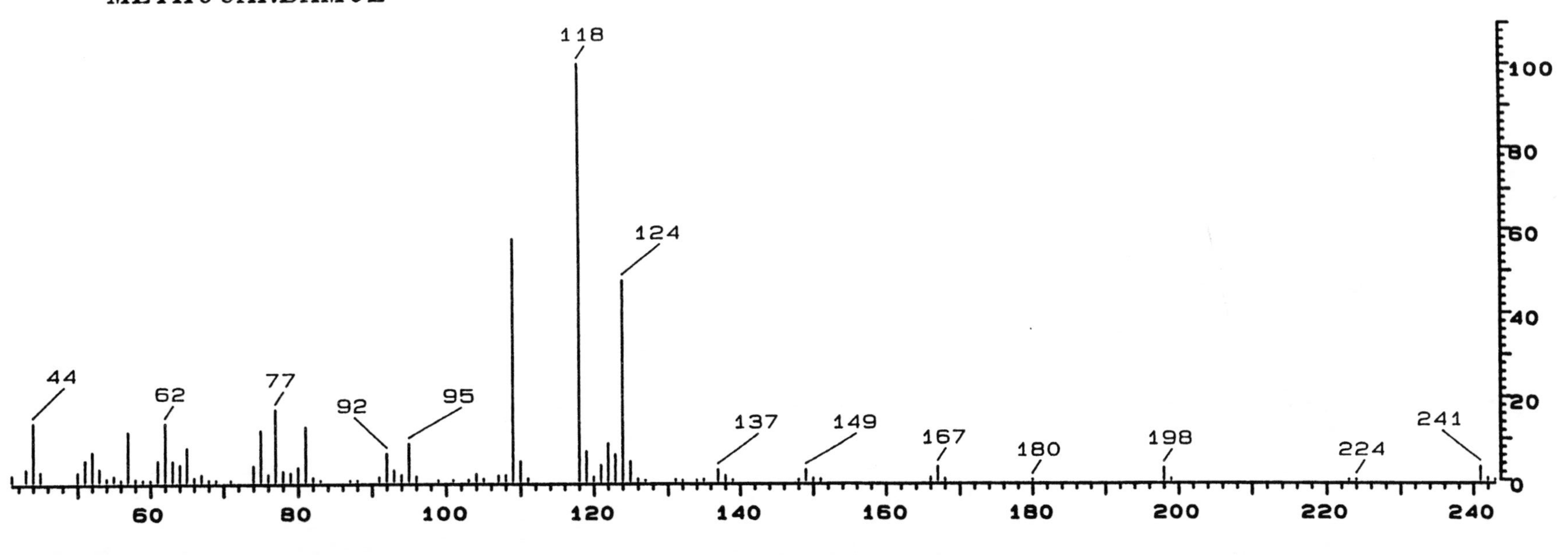

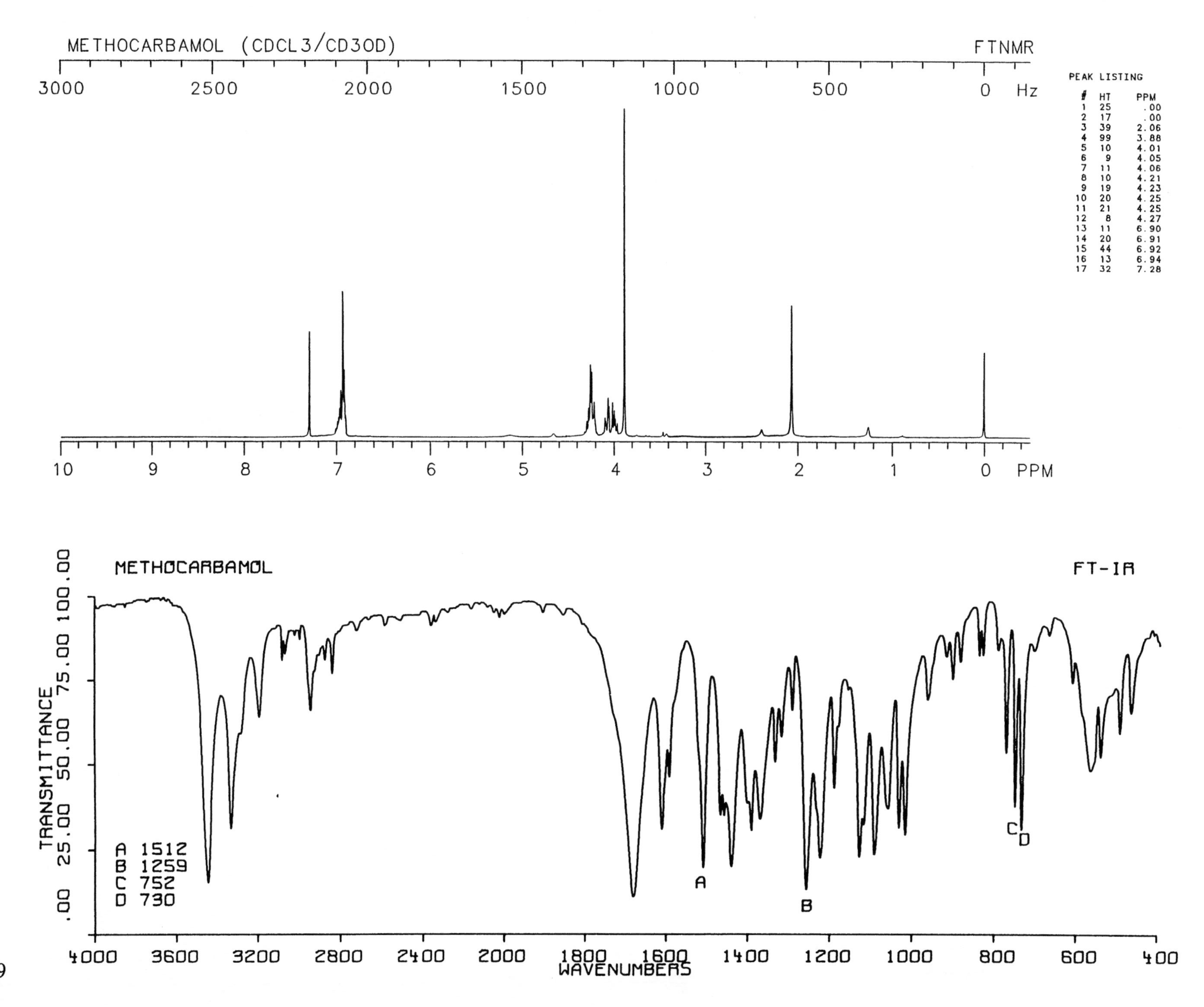
METHOCARBAMOL (CDCL3/CD3OD)
FTNMR
Hz
PPM
PEAK LISTING
# HT PPM
1 25 .00
2 17 .00
3 39 2.06
4 99 3.88
5 10 4.01
6 9 4.05
7 11 4.06
8 10 4.21
9 19 4.23
10 20 4.25
11 21 4.25
12 8 4.27
13 11 6.90
14 20 6.91
15 44 6.92
16 13 6.94
17 32 7.28
METHOCARBAMOL
FT-IR
TRANSMITTANCE
WAVENUMBERS
A 1512
B 1259
C 752
D 730

# METHOHEXITAL

$C_{14}H_{18}N_2O_3$

Molecular weight: 262.31 (262.13)
Synonyms: 1-Methyl-5-(1-methyl-2-pentynyl)-5-(2-propenyl)-2,4,6-(1H,3H,5H)pyrimidinetrione; 5-allyl-1-methyl-5-(1-methyl-2-pentynyl)barbituric acid; methohexitone
Trade names: Brevital

Use: Anesthetic
HPLC: Si-10; 1A:99B; 3.2
GC: 1772; 200°C

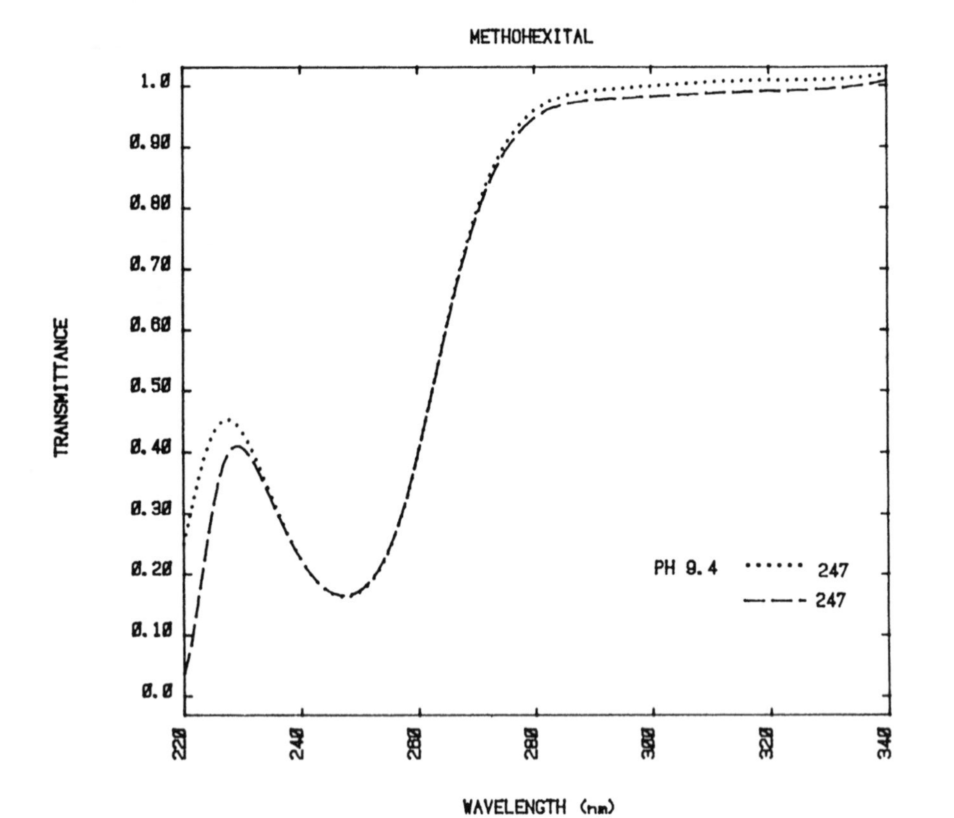


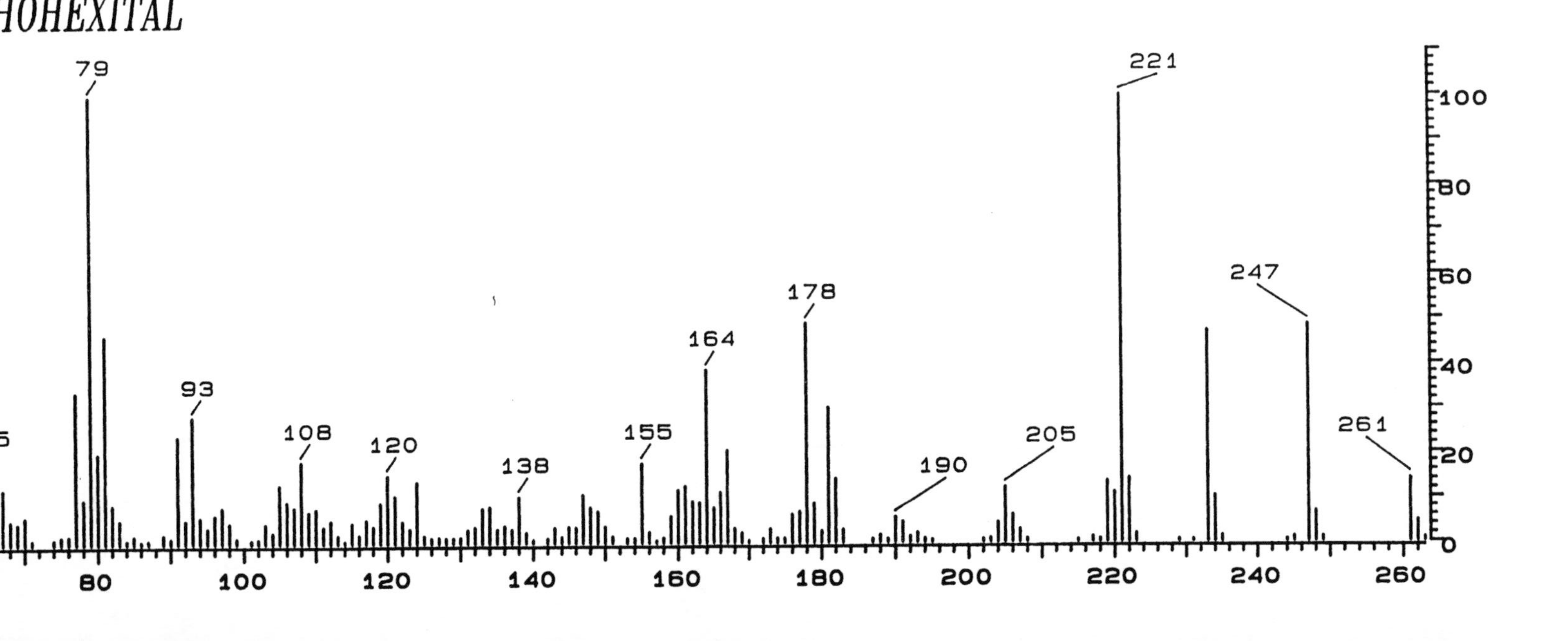

METHOHEXITAL

FTNMR

PEAK LISTING

| # | HT | PPM |
|---|---|---|
| 1 | 17 | .00 |
| 2 | 19 | 1.00 |
| 3 | 44 | 1.03 |
| 4 | 19 | 1.05 |
| 5 | 34 | 1.28 |
| 6 | 33 | 1.31 |
| 7 | 99 | 3.29 |
| 8 | 20 | 7.26 |

METHOHEXITAL

FT-IR

A 1710
B 1682
C 652
D 596

# METHOTREXATE

$C_{20}H_{22}N_8O_5$

Molecular weight: 454.44 (454.17)
Synonyms: N-[p-[[2,4-Diamino-6-pteridinyl)methyl]methylamino]-benzoyl]glutamic acid; amethopterin
Trade names: Mexate

Use: Antineoplastic, antimetabolite
HPLC: Si-10; 50A:50B; 9.0
GC:

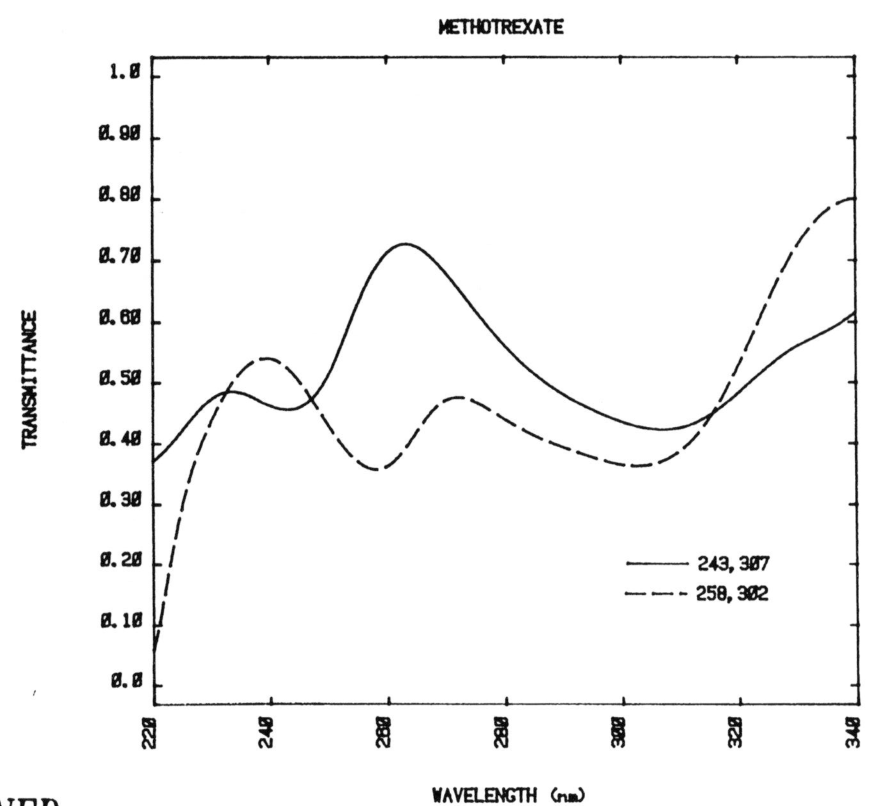


NO USEFUL MASS SPECTRUM WAS OBTAINED

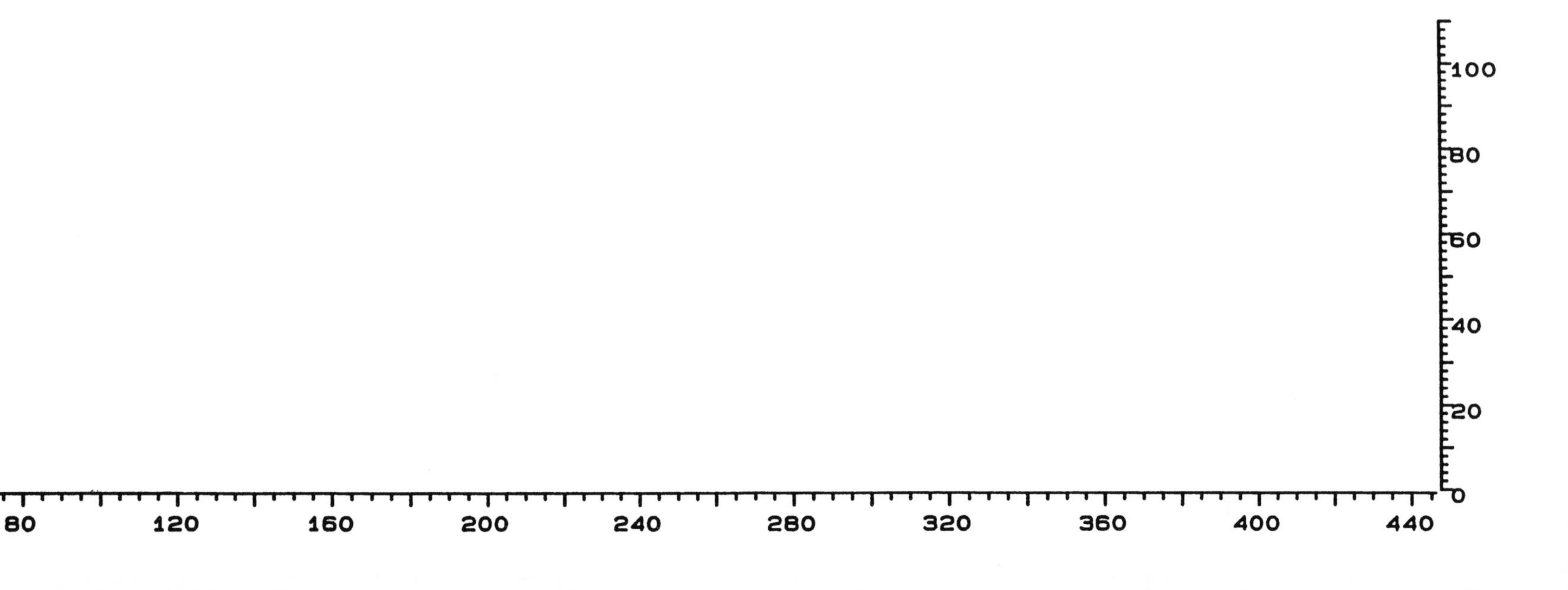

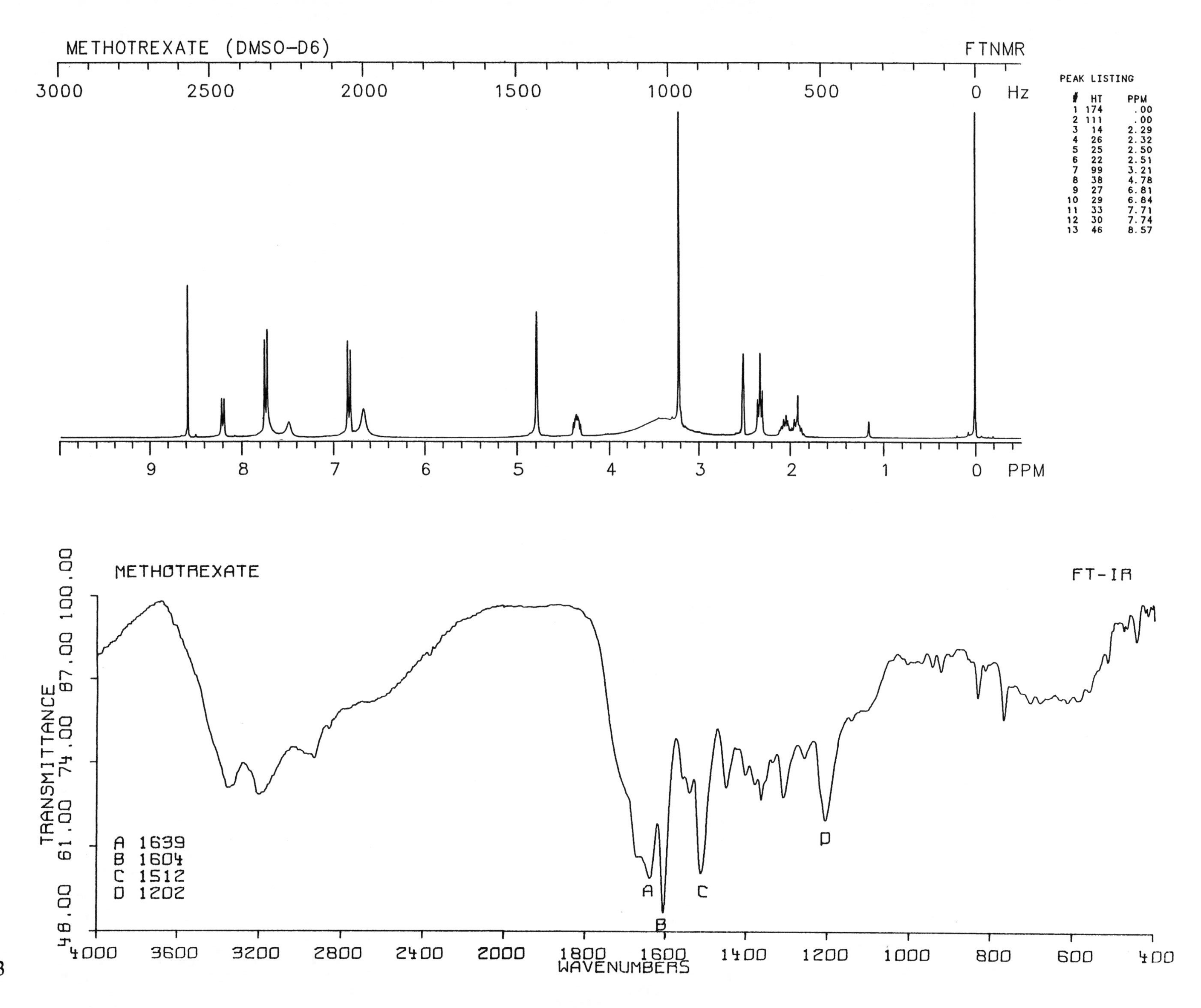

METHOTREXATE (DMSO-D6)
FTNMR
3000 2500 2000 1500 1000 500 0 Hz
PEAK LISTING
# HT PPM
1 174 .00
2 111 .00
3 14 2.29
4 26 2.32
5 25 2.50
6 22 2.51
7 99 3.21
8 38 4.78
9 27 6.81
10 29 6.84
11 33 7.71
12 30 7.74
13 46 8.57
9 8 7 6 5 4 3 2 1 0 PPM
METHOTREXATE
FT-IR
TRANSMITTANCE
48.00 61.00 74.00 87.00 100.00
A 1639
B 1604
C 1512
D 1202
A B C D
4000 3600 3200 2800 2400 2000 1800 1600 1400 1200 1000 800 600 400
WAVENUMBERS

# METHOTRIMEPRAZINE

$C_{19}H_{24}N_2OS$

Molecular weight: 328.47 (328.16)
Synonyms: 2-Methoxy-N,N,β-trimethyl-10H-phenothiazine-10-propanamine; 2-methoxytrimeprazine
Trade names: Levoprome

Use: Analgesic, tranquilizer
HPLC: Si-10; 5A:95B; 4.8
GC: 2563; 250°C

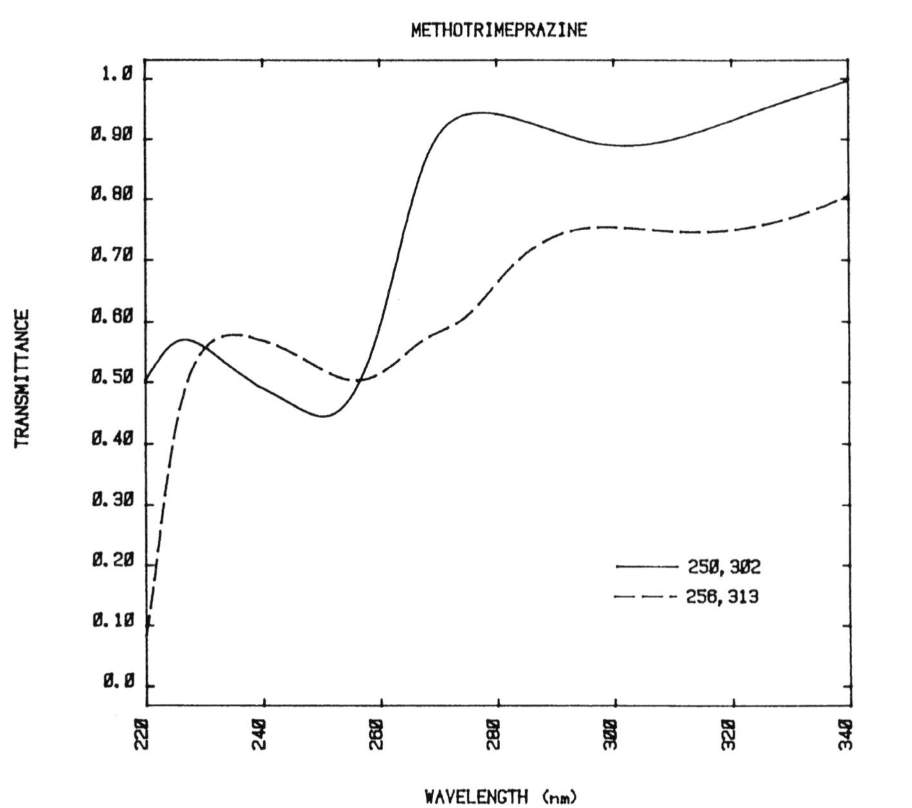

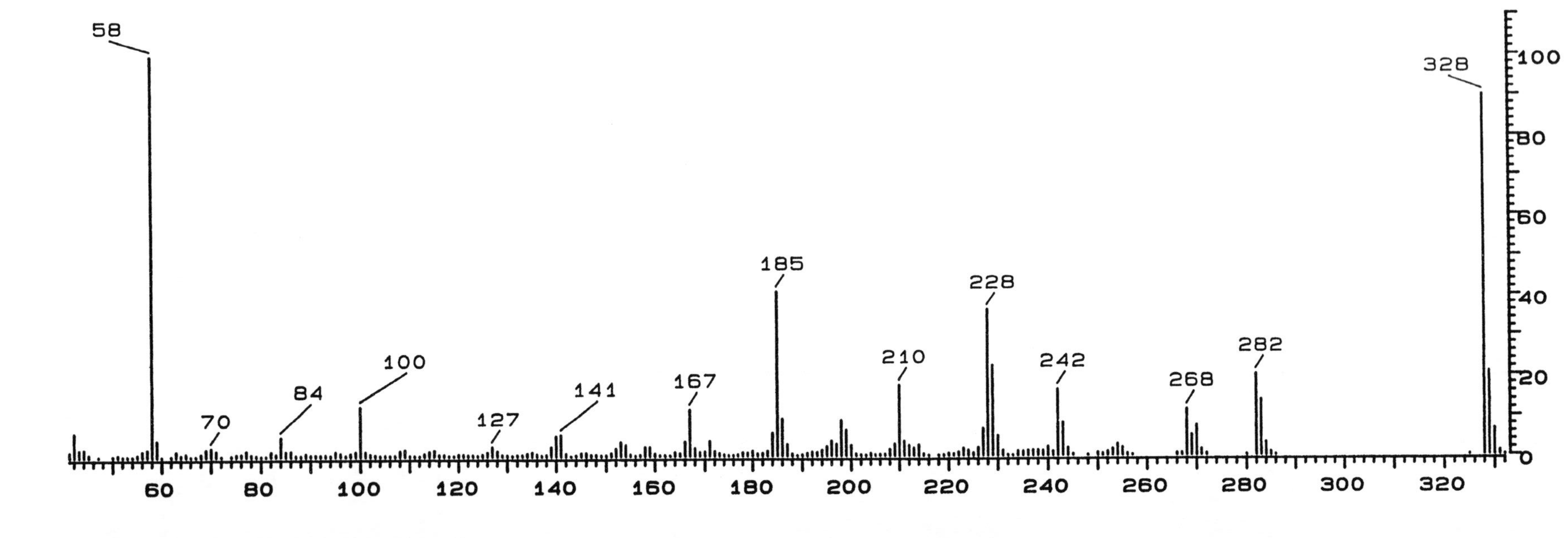

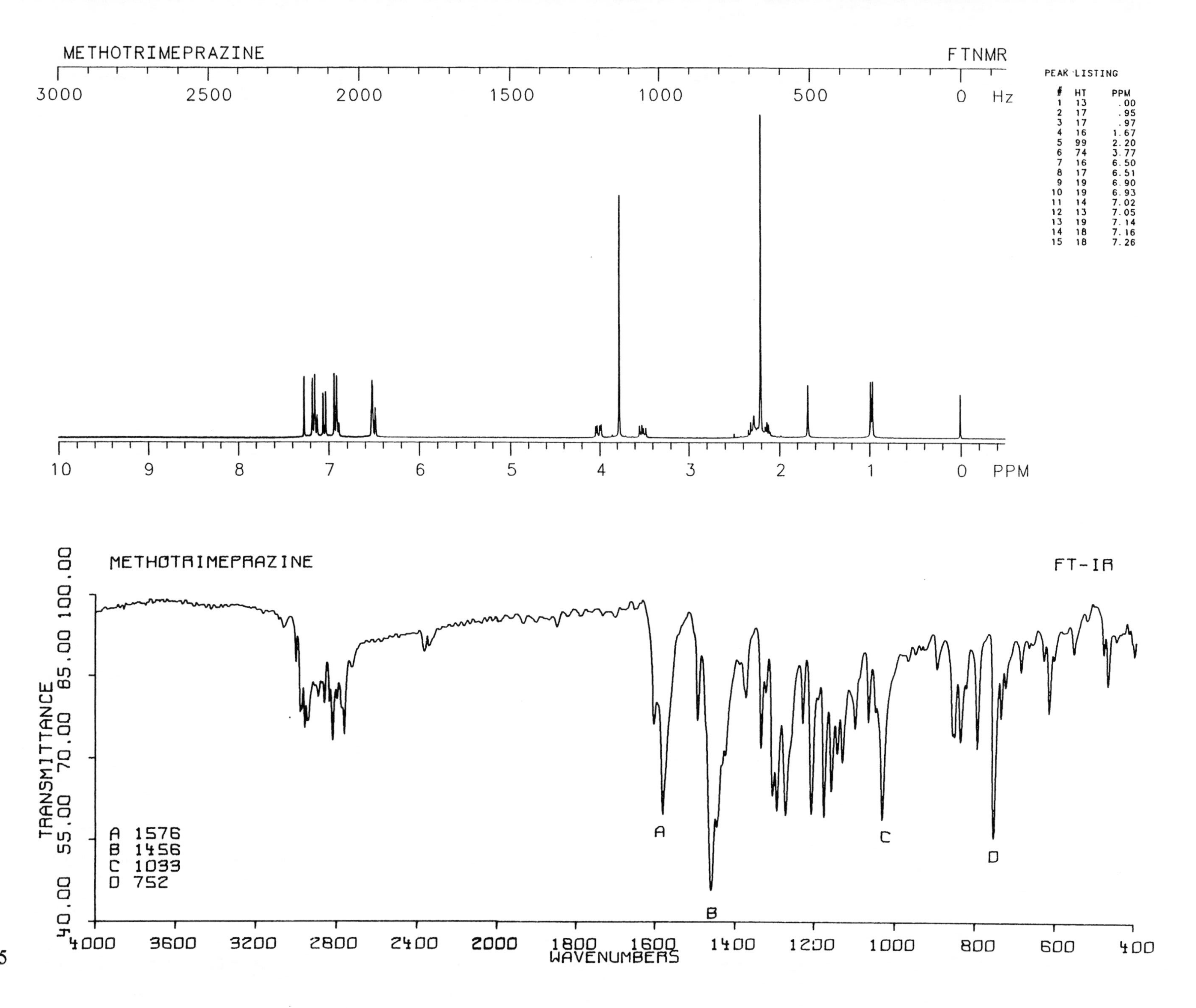
METHOTRIMEPRAZINE
FTNMR
3000 2500 2000 1500 1000 500 0 Hz
PEAK LISTING
# HT PPM
1 13 .00
2 17 .95
3 17 .97
4 16 1.67
5 99 2.20
6 74 3.77
7 16 6.50
8 17 6.51
9 19 6.90
10 19 6.93
11 14 7.02
12 13 7.05
13 19 7.14
14 18 7.16
15 18 7.26
10 9 8 7 6 5 4 3 2 1 0 PPM
METHOTRIMEPRAZINE
FT-IR
TRANSMITTANCE
40.00 55.00 70.00 85.00 100.00
A 1576
B 1456
C 1033
D 752
A
B
C
D
4000 3600 3200 2800 2400 2000 1800 1600 1400 1200 1000 800 600 400
WAVENUMBERS

# METHOXSALEN

$C_{12}H_8O_4$

**Molecular weight:** 216.20 (216.04)
**Synonyms:** 9-Methoxy-7H-furo[3,2-g][1]benzopyran-7-one; xanthotoxin; 8-methoxypsoralen
**Trade names:** Oxsoralen

**Use:** Dermal pigmentation
**HPLC:**
**GC:** 2075; 250°C

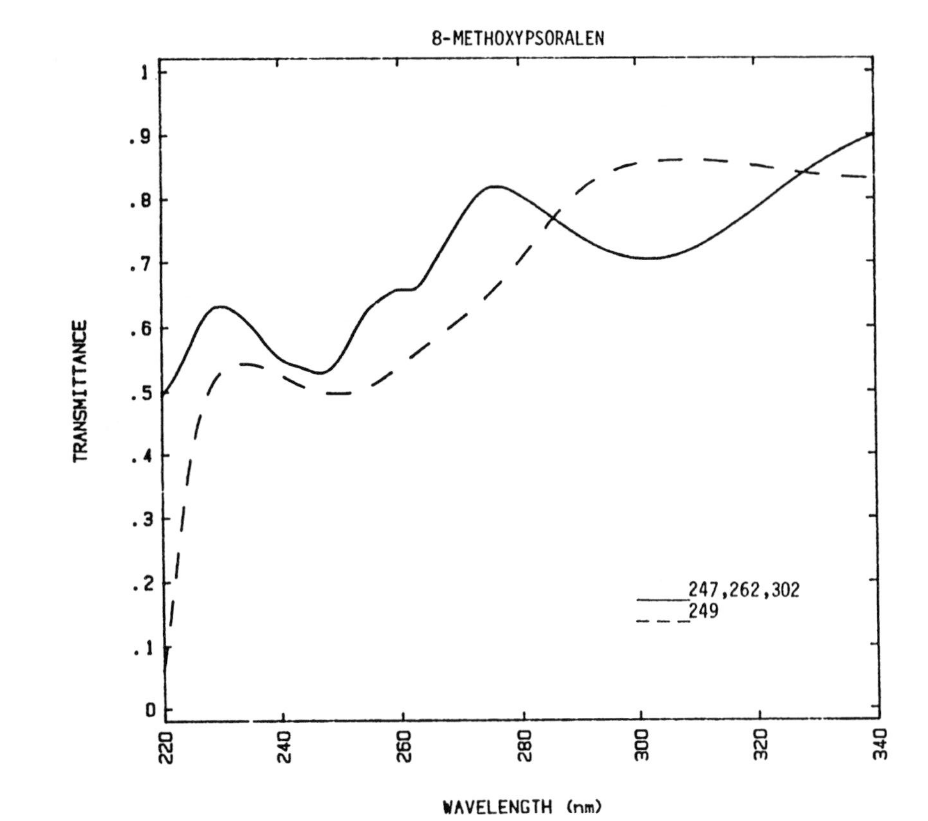

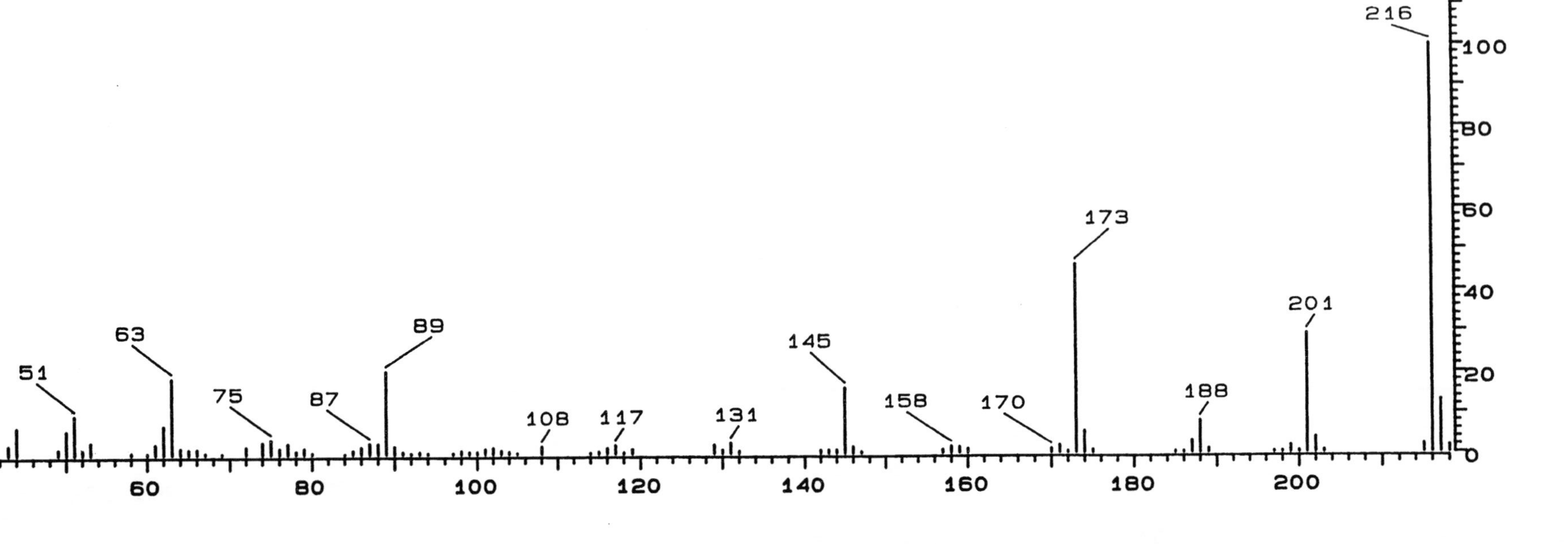

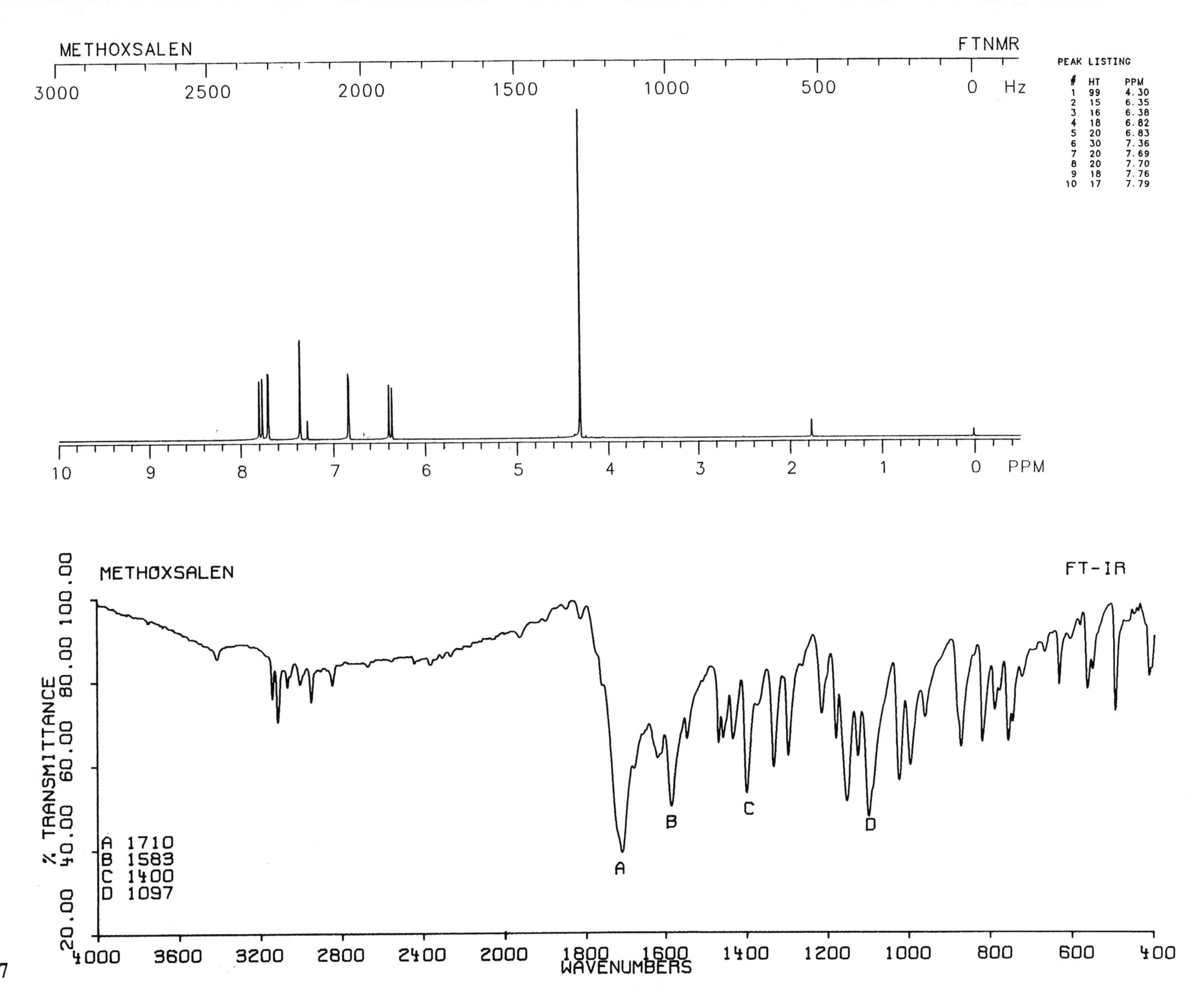
METHOXSALEN
FTNMR
3000
2500
2000
1500
1000
500
0
Hz
PEAK LISTING
# HT PPM
1 99 4.30
2 15 6.35
3 16 6.38
4 18 6.82
5 20 6.83
6 30 7.36
7 20 7.69
8 20 7.70
9 18 7.76
10 17 7.79
10
9
8
7
6
5
4
3
2
1
0
PPM
METHOXSALEN
FT-IR
% TRANSMITTANCE
20.00
40.00
60.00
80.00
100.00
A 1710
B 1583
C 1400
D 1097
A
B
C
D
4000
3600
3200
2800
2400
2000
1800
1600
1400
1200
1000
800
600
400
WAVENUMBERS

# METHOXYAMINE

$CH_5NO$

Molecular weight: 47.05 (47.04)
Synonyms: O-Methylhydroxylamine; methoxylamine; α-methylhydroxylamine; hydroxylamine methyl ether
Trade names:

Use: Synthesis, strong irritant
HPLC:
GC:

$CH_3-O-NH_2$

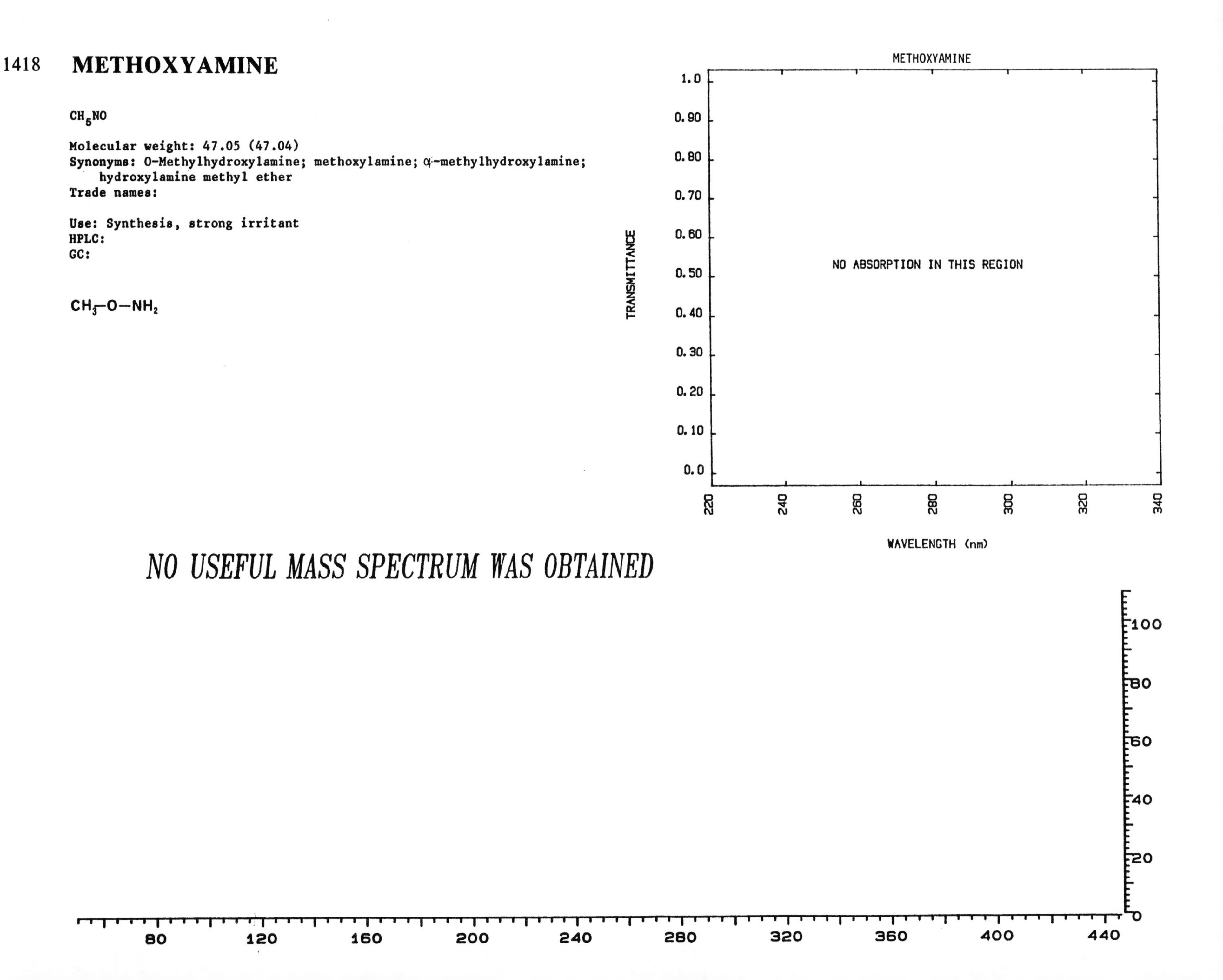

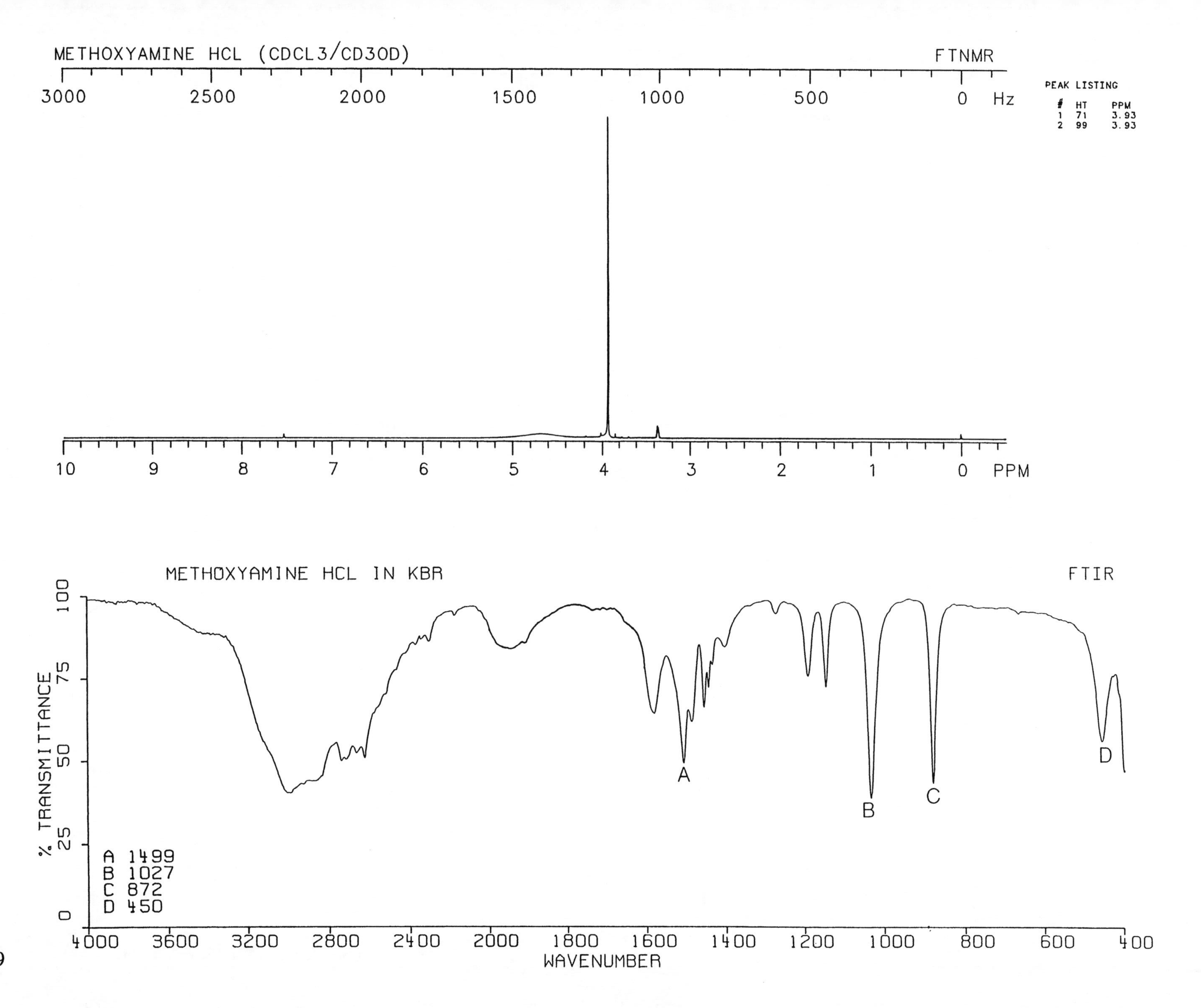

METHOXYAMINE HCL (CDCL3/CD3OD)
FTNMR
3000 2500 2000 1500 1000 500 0 Hz
PEAK LISTING
# HT PPM
1 71 3.93
2 99 3.93
10 9 8 7 6 5 4 3 2 1 0 PPM
METHOXYAMINE HCL IN KBR
FTIR
% TRANSMITTANCE
100 75 50 25 0
A 1499
B 1027
C 872
D 450
A B C D
4000 3600 3200 2800 2400 2000 1800 1600 1400 1200 1000 800 600 400
WAVENUMBER

# 4-METHOXYAMPHETAMINE

$C_{10}H_{15}NO$

**Molecular weight:** 165.24 (165.12)
**Synonyms:** 4-Methoxy-α-methylbenzeneethanamine; p-methoxy-α-methylphenethylamine; PMA; p-methoxyamphetamine
**Trade names:**

**Use:** Hallucinogen
**HPLC:** Si-10; 20A:80B; 4.2
**GC:** 1374; 140°C

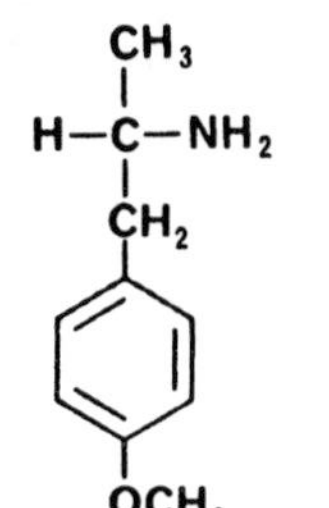


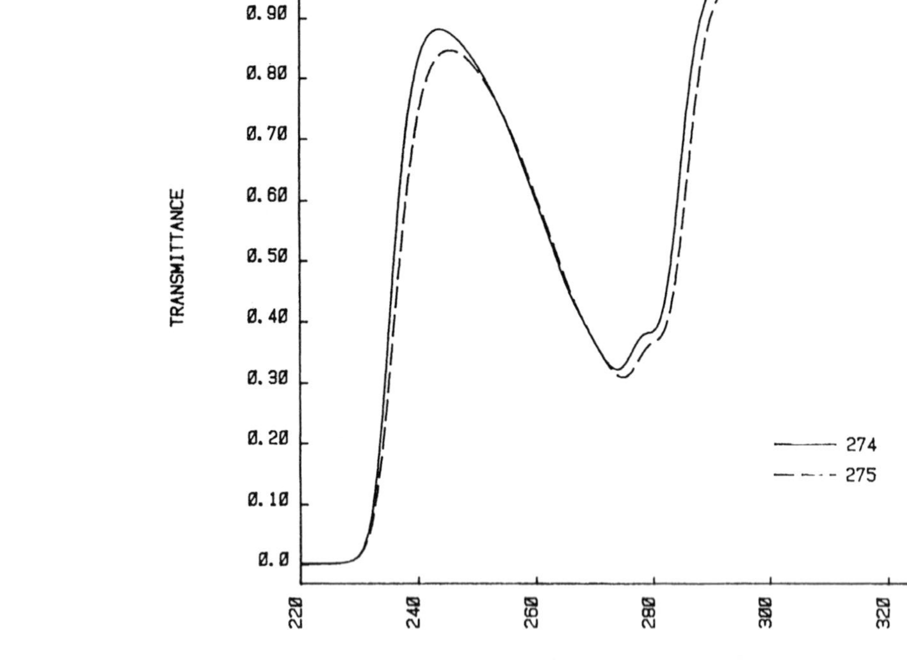


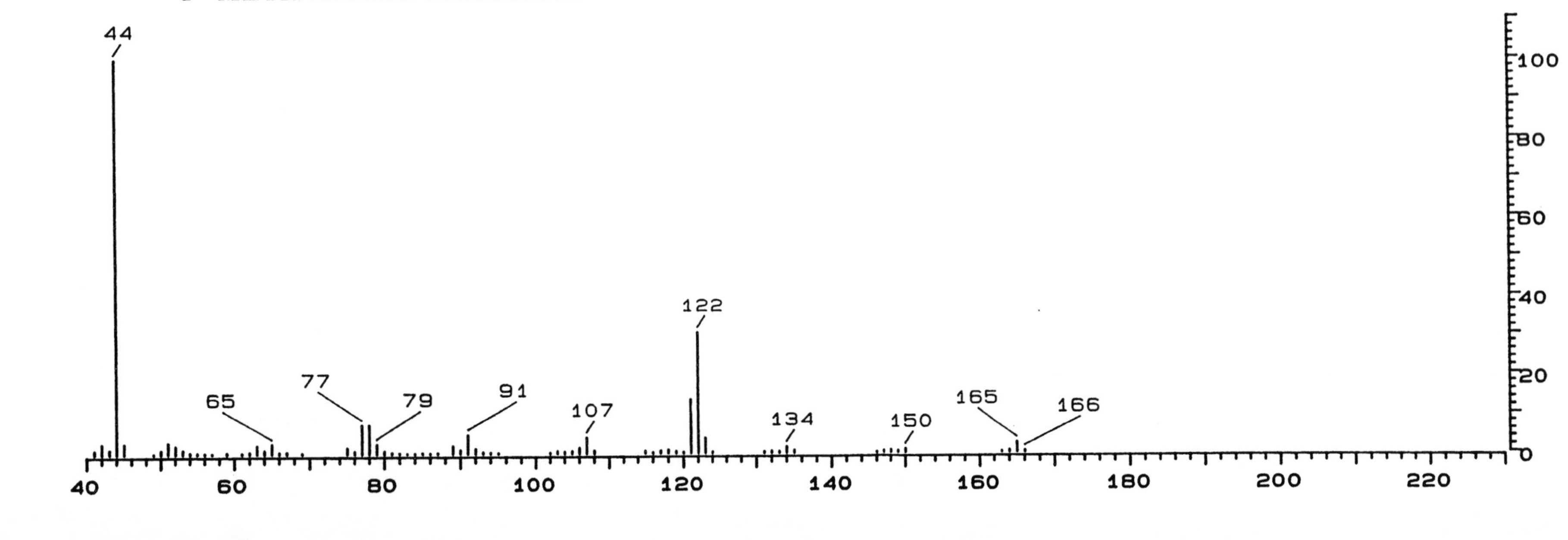

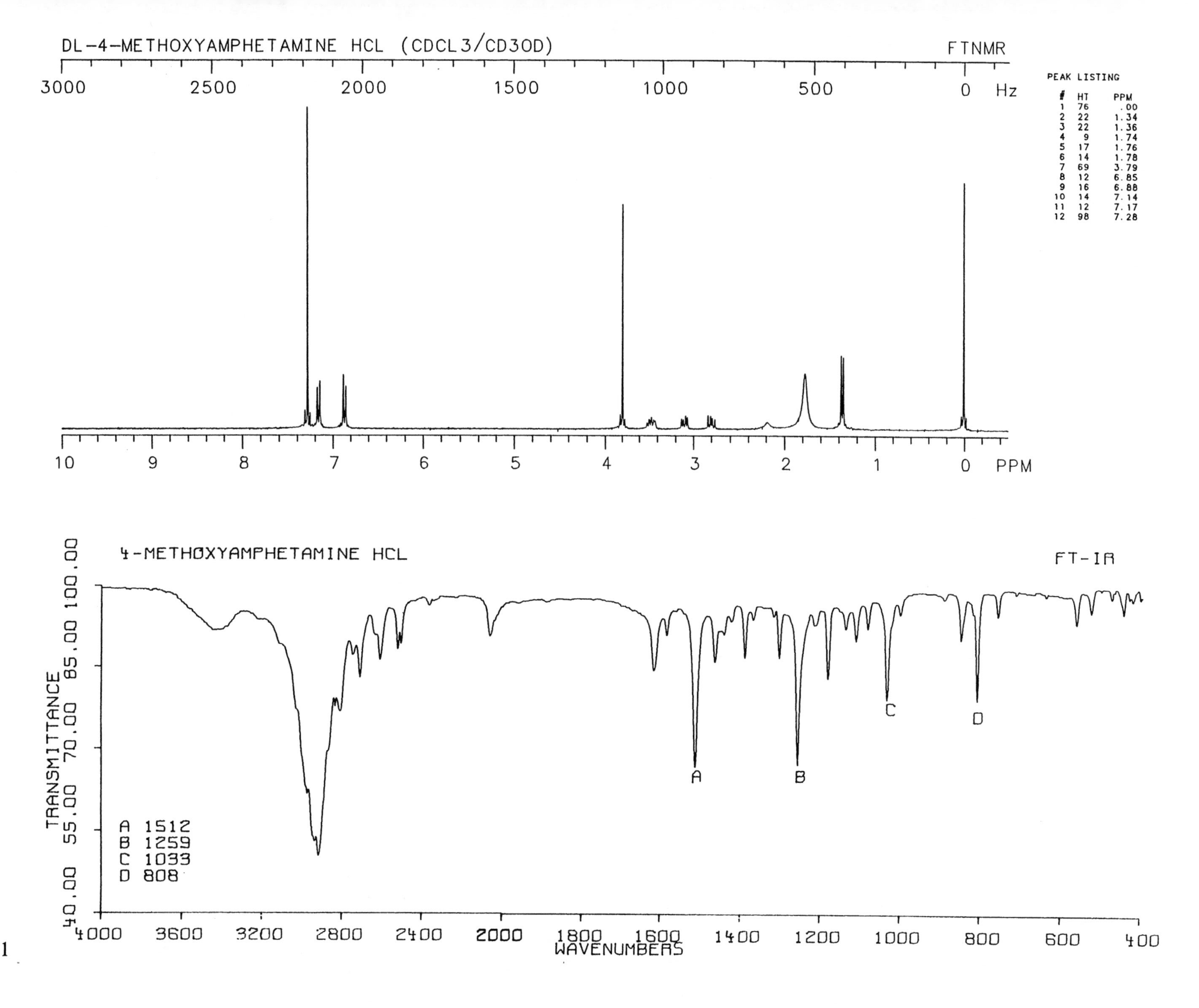

DL-4-METHOXYAMPHETAMINE HCL (CDCL3/CD3OD)
FTNMR
3000
2500
2000
1500
1000
500
0 Hz
PEAK LISTING
# HT PPM
1 76 .00
2 22 1.34
3 22 1.36
4 9 1.74
5 17 1.76
6 14 1.78
7 69 3.79
8 12 6.85
9 16 6.88
10 14 7.14
11 12 7.17
12 98 7.28
10
9
8
7
6
5
4
3
2
1
0 PPM
4-METHOXYAMPHETAMINE HCL
FT-IR
TRANSMITTANCE
40.00
55.00
70.00
85.00
100.00
A 1512
B 1259
C 1033
D 808
A
B
C
D
4000
3600
3200
2800
2400
2000
1800
1600
1400
1200
1000
800
600
400
WAVENUMBERS

# 5-METHOXY-*N,N*-DIMETHYLTRYPTAMINE

$C_{11}H_{17}NO_2$

**Molecular weight:** 195.25 (195.13)
**Synonyms:** 5-Methoxy-4-[2-(dimethylamino)ethyl]phenol; (3-[2-dimethylaminoethyl]-5-methoxyindole)
**Trade names:**

**Use:**
HPLC: Si-10; 20A:80B; 4.0
GC: 2033; 250°C

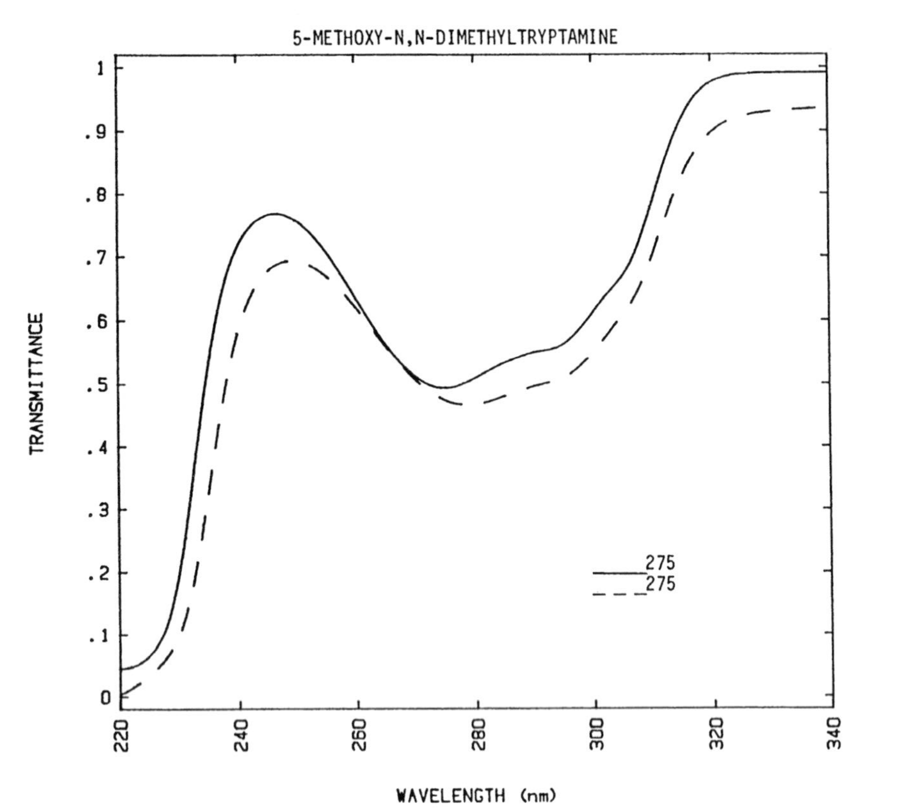


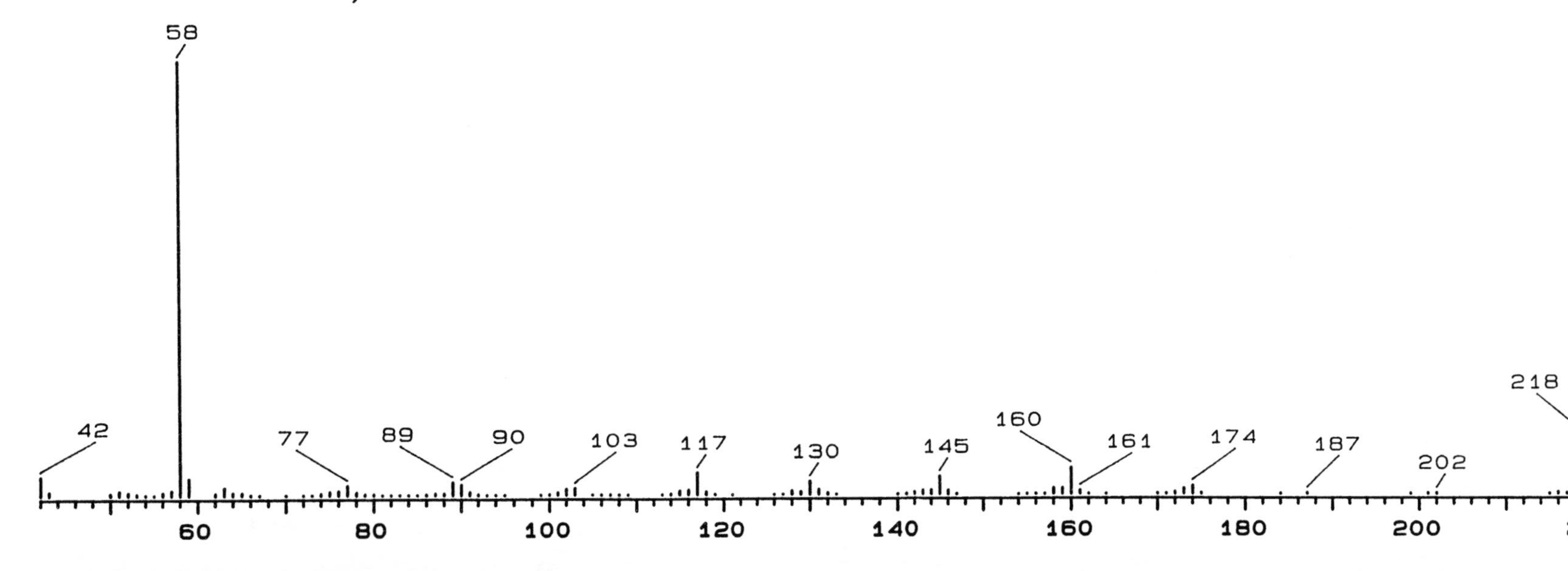

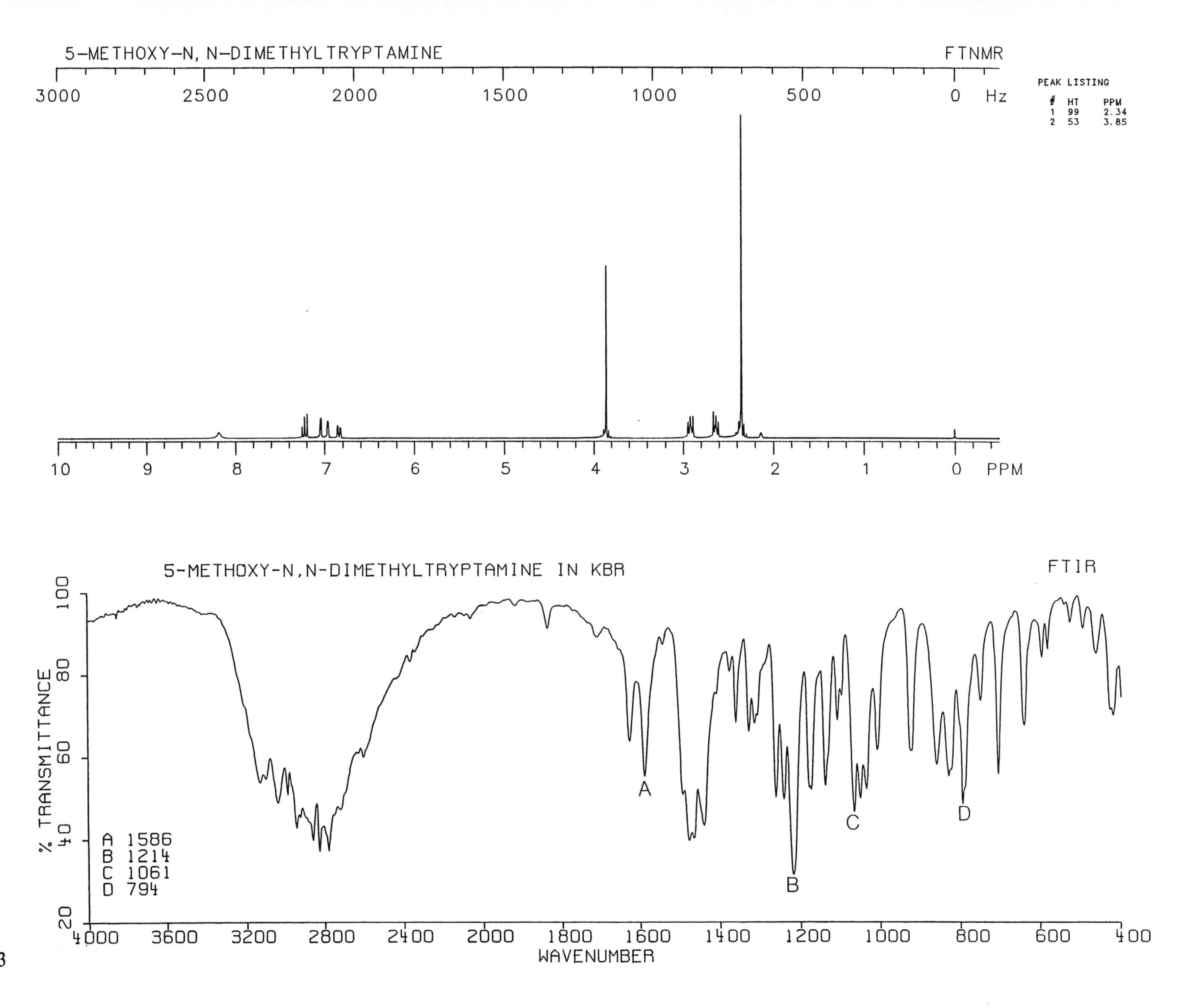

5-METHOXY-N, N-DIMETHYLTRYPTAMINE
FTNMR
3000 2500 2000 1500 1000 500 0 Hz
PEAK LISTING
# HT PPM
1 99 2.34
2 53 3.85
10 9 8 7 6 5 4 3 2 1 0 PPM
5-METHOXY-N,N-DIMETHYLTRYPTAMINE IN KBR
FTIR
% TRANSMITTANCE
100 80 60 40 20
A 1586
B 1214
C 1061
D 794
A
B
C
D
4000 3600 3200 2800 2400 2000 1800 1600 1400 1200 1000 800 600 400
WAVENUMBER

# METHOXYESTRONE

$C_{19}H_{24}O_3$

Molecular weight: 300.38 (300.17)
Synonyms: 2,3-Dihydroxy-1,3,5[10]-estratrien-17-one-2-methyl ether; 2-hydroxyestrone-2-methyl ether; 3-hydroxy-2-methoxy-1,3,5[10]-estratien-17-one
Trade names:

Use: Estrogen
HPLC:
GC: 2765; 280°C

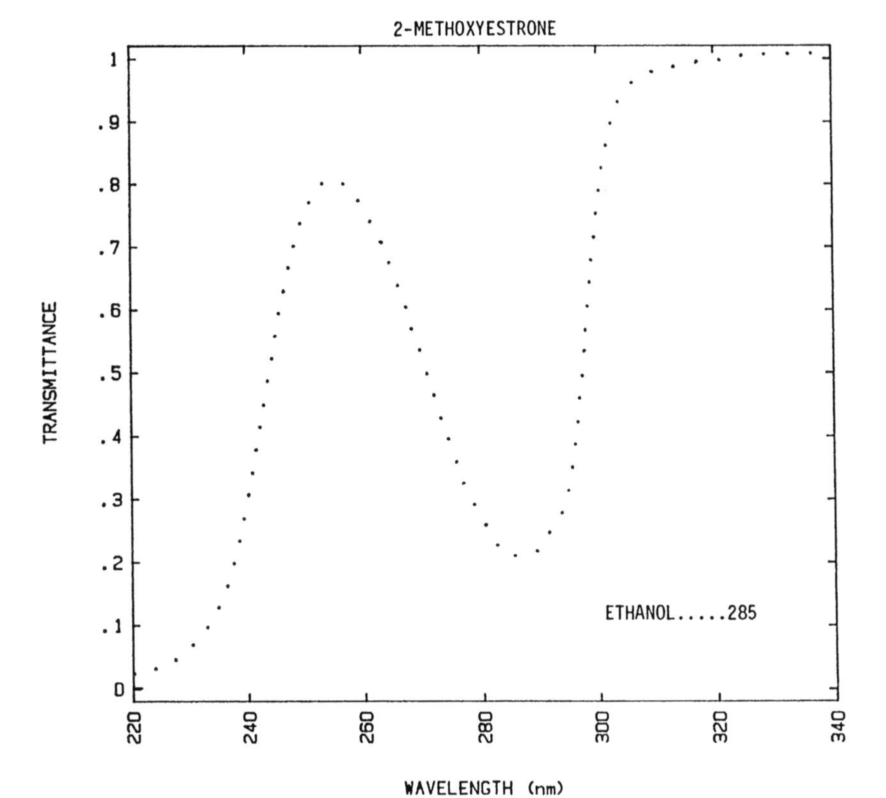

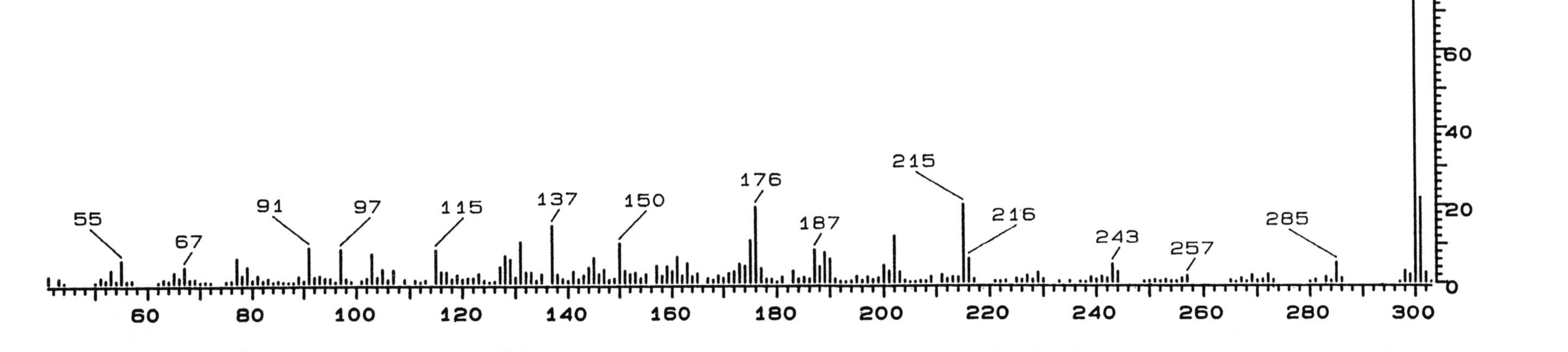

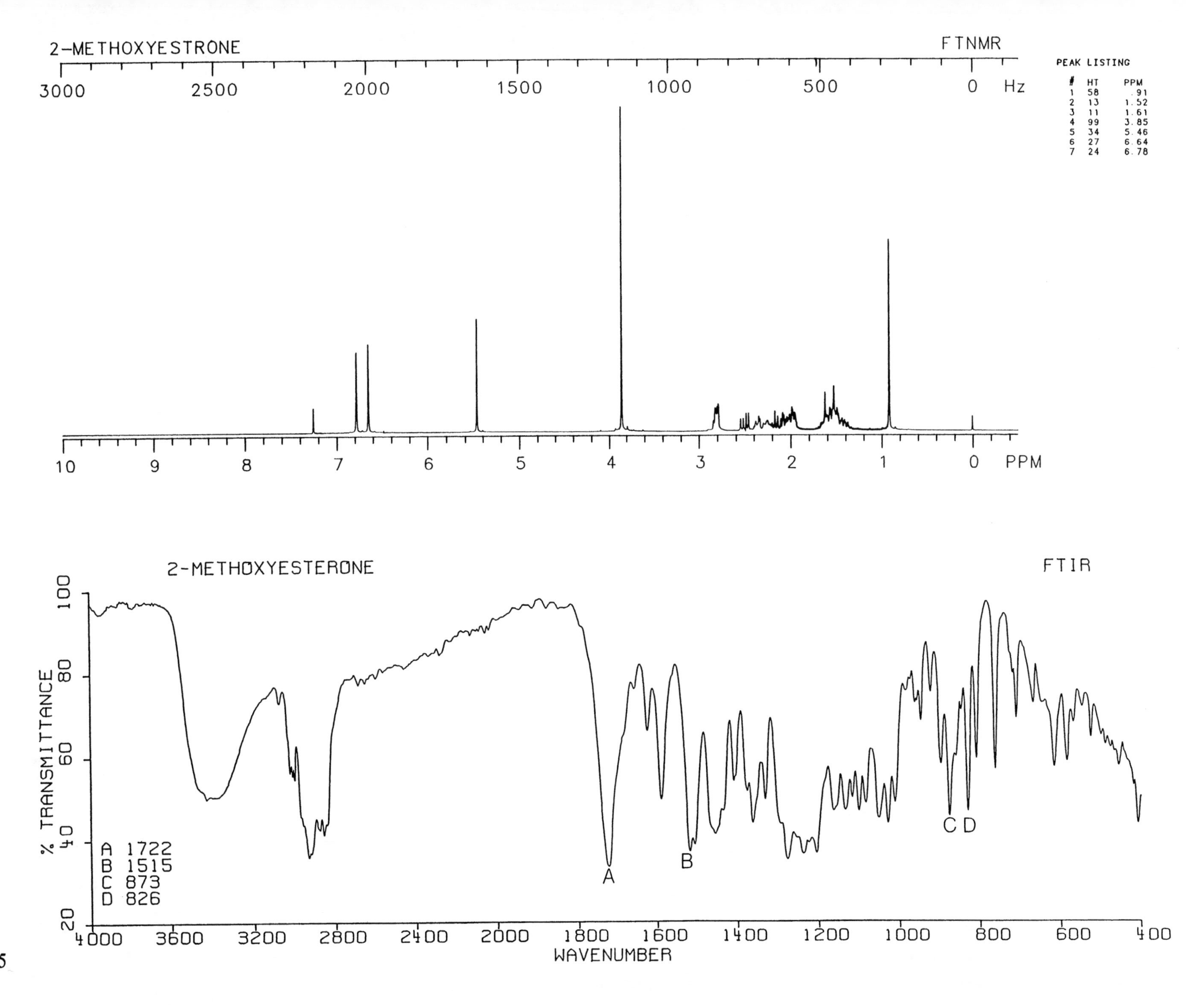
2-METHOXYESTRONE
FTNMR
Hz
PPM
PEAK LISTING
# HT PPM
1 58 .91
2 13 1.52
3 11 1.61
4 99 3.85
5 34 5.46
6 27 6.64
7 24 6.78
2-METHOXYESTERONE
FTIR
% TRANSMITTANCE
WAVENUMBER
A 1722
B 1515
C 873
D 826

# METHOXYFLURANE

$C_3H_4Cl_2F_2O$

Molecular weight: 164.97 (163.96)
Synonyms: 2,2-Dichloro-1,1-difluoro-1-methoxyethane

Trade names: Penthrane

Use: Anesthetic
HPLC:
GC:

$CH_3—O—CF_2—CHCl_2$

METHOXYFLURANE

TRANSMITTANCE

1.0 0.90 0.80 0.70 0.60 0.50 0.40 0.30 0.20 0.10 0.0

NO ABSORPTION IN THIS REGION

220 240 260 280 300 320 340

WAVELENGTH (nm)

*METHOXYFLURANE*

81 47 63 67 98 114 118 133 145 163 167

100 80 60 40 20 0

40 60 80 100 120 140 160 180 200 220

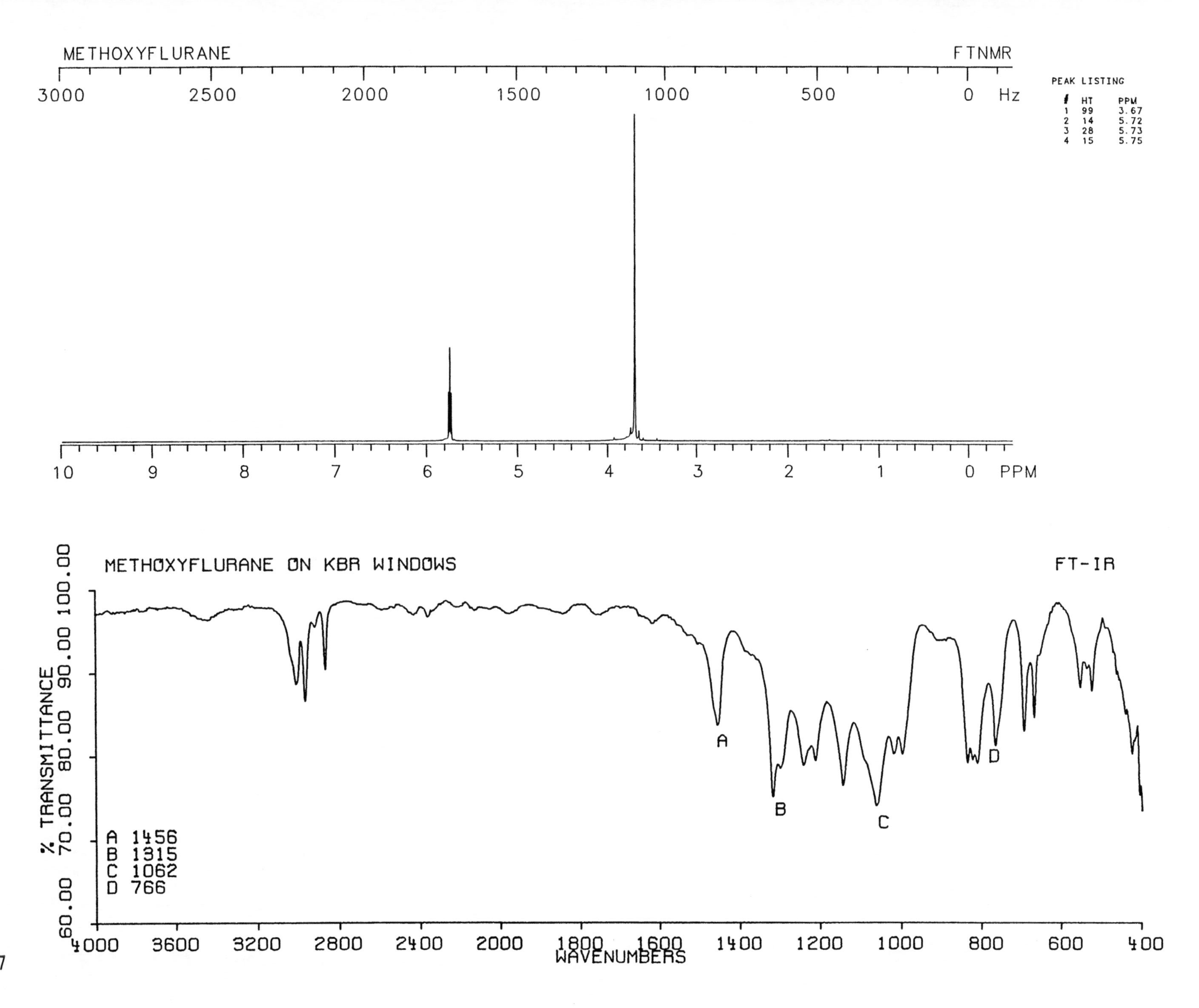
METHOXYFLURANE
FTNMR
3000 2500 2000 1500 1000 500 0 Hz
PEAK LISTING
# HT PPM
1 99 3.67
2 14 5.72
3 28 5.73
4 15 5.75
10 9 8 7 6 5 4 3 2 1 0 PPM
METHOXYFLURANE ON KBR WINDOWS
FT-IR
% TRANSMITTANCE
60.00 70.00 80.00 90.00 100.00
A 1456
B 1315
C 1062
D 766
A
B
C
D
4000 3600 3200 2800 2400 2000 1800 1600 1400 1200 1000 800 600 400
WAVENUMBERS

# 6-METHOXYHARMALAN

$C_{13}H_{14}N_2O$

**Molecular weight:** 214.26 (214.11)
**Synonyms:** 4,9-Dihydro-6-methoxy-1-methyl-3H-pyridol[3,4-b]indole; 10-methoxyharmalan; 3,4-dihydromethoxyharman
**Trade names:**

**Use:** Serotonin inhibitor
**HPLC:** Si-10; 20A:80B; 3.8
**GC:** 2228; 250°C

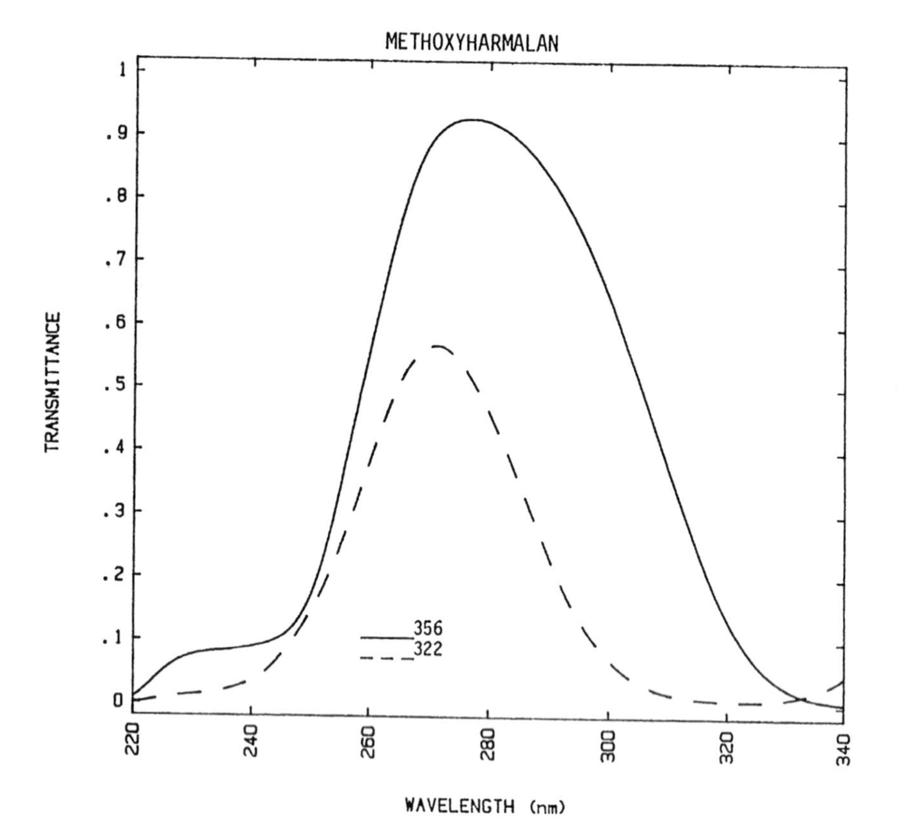

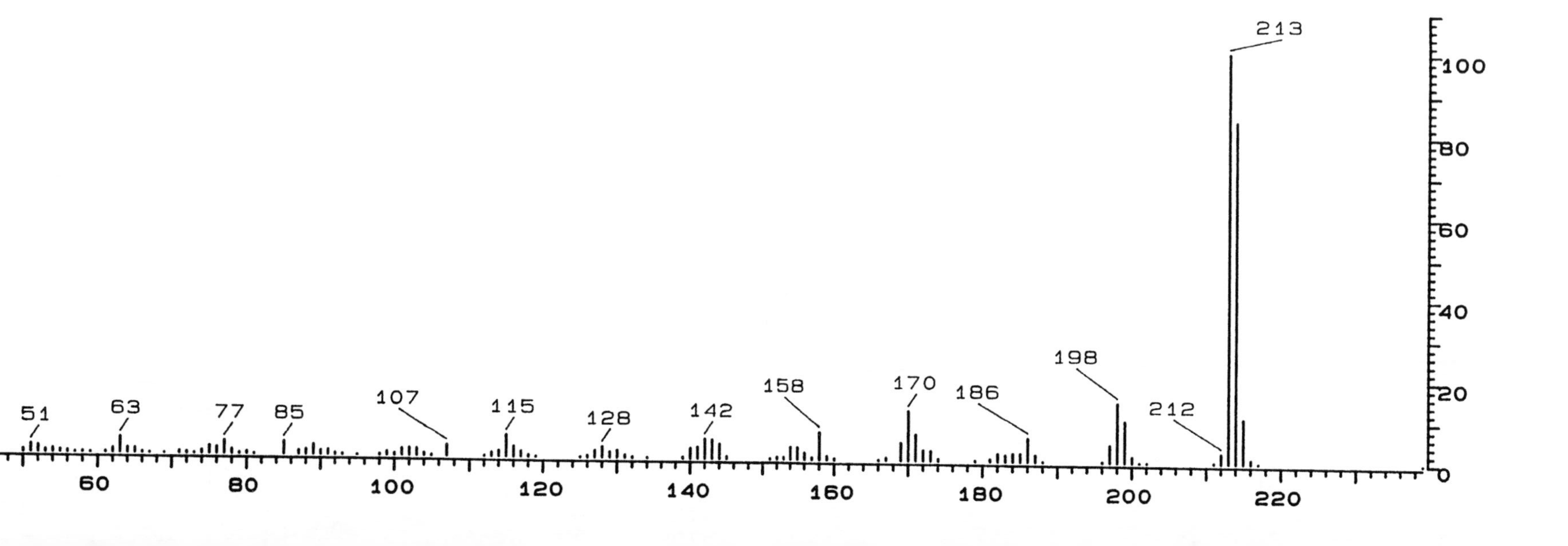

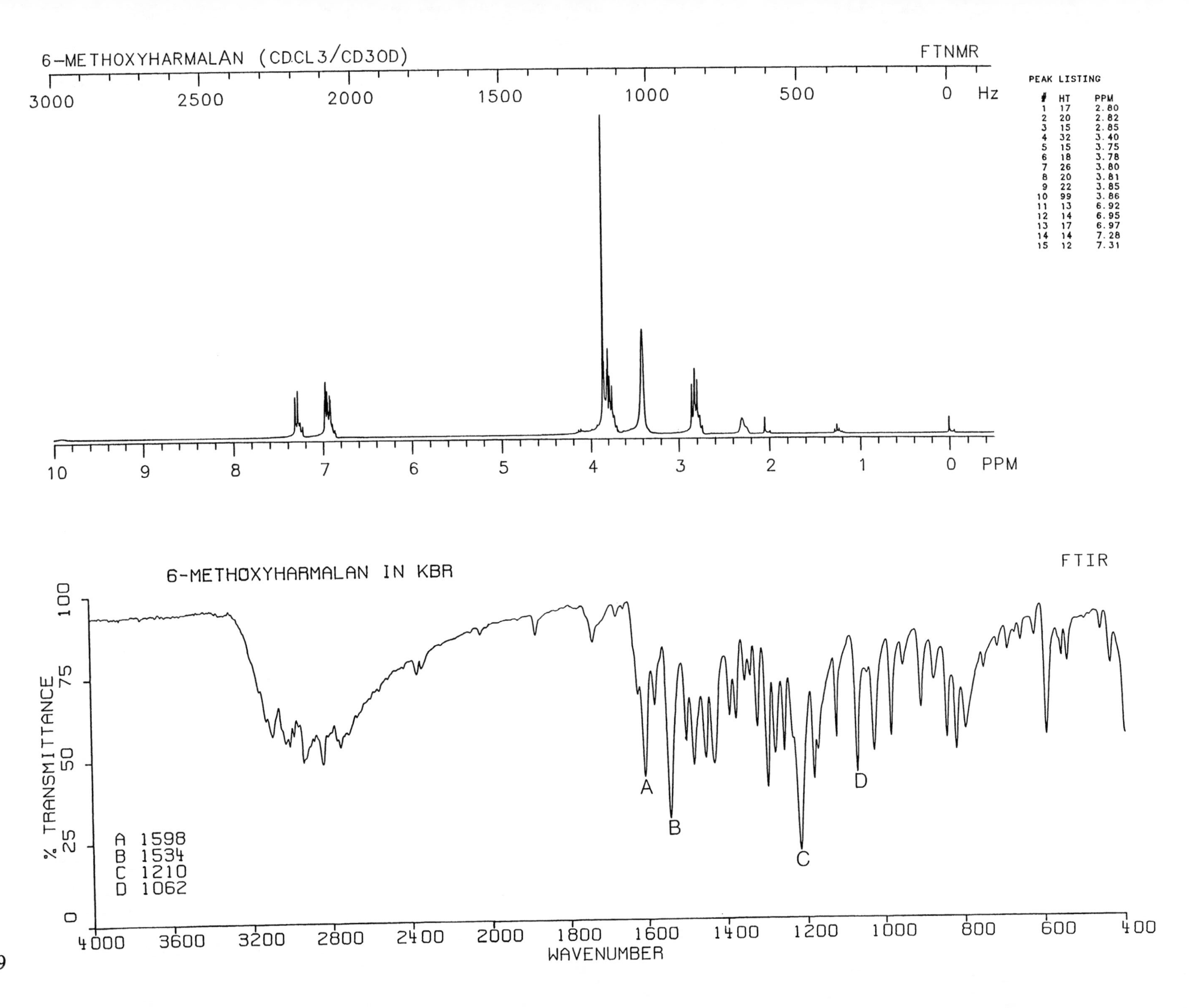
6-METHOXYHARMALAN (CDCL3/CD3OD)
FTNMR
3000
2500
2000
1500
1000
500
0 Hz
PEAK LISTING
# HT PPM
1 17 2.80
2 20 2.82
3 15 2.85
4 32 3.40
5 15 3.75
6 18 3.78
7 26 3.80
8 20 3.81
9 22 3.85
10 99 3.86
11 13 6.92
12 14 6.95
13 17 6.97
14 14 7.28
15 12 7.31
10 9 8 7 6 5 4 3 2 1 0 PPM
6-METHOXYHARMALAN IN KBR
FTIR
% TRANSMITTANCE
100
75
50
25
0
A 1598
B 1534
C 1210
D 1062
A
B
C
D
4000 3600 3200 2800 2400 2000 1800 1600 1400 1200 1000 800 600 400
WAVENUMBER

# 8-METHOXYLOXAPINE

$C_{19}H_{20}ClN_3O_2$

**Molecular weight:** 357.84 (357.12)
**Synonyms:** 2-Chloro-8-methoxy-11-(4-methyl-1-piperazinyl)-dibenz[b,f][1,4]oxazepine
**Trade names:**

**Use:**
**HPLC:** Si-10; 20A:80B; 4.2
**GC:** 2897; 280°C

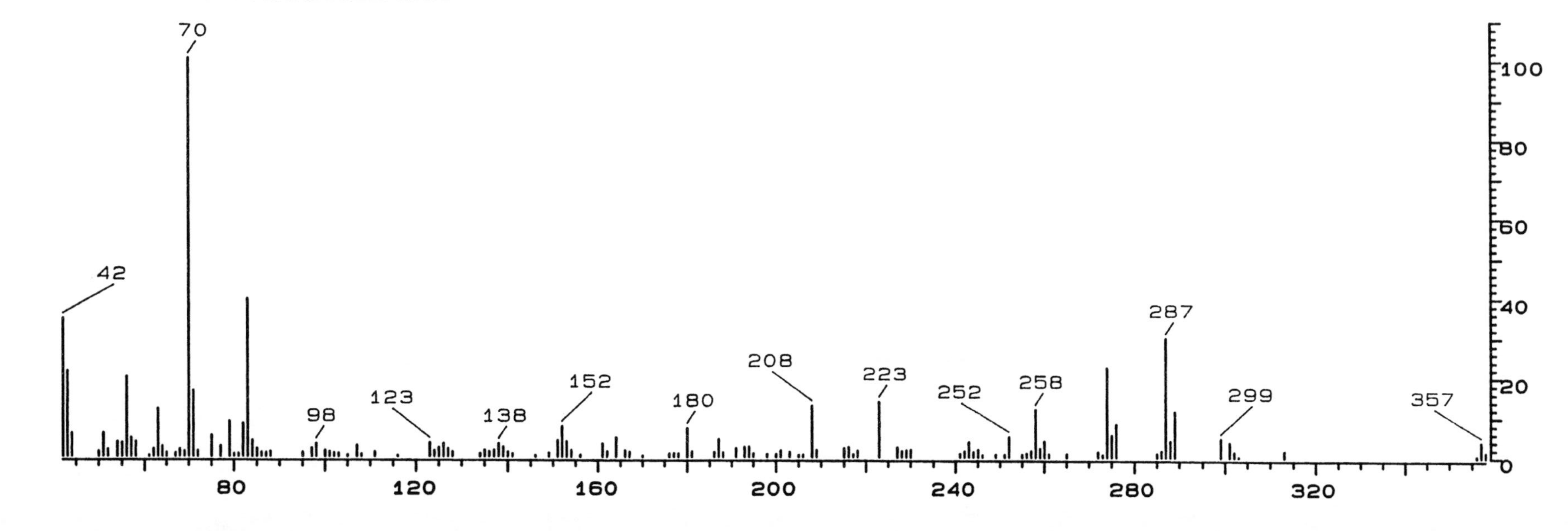

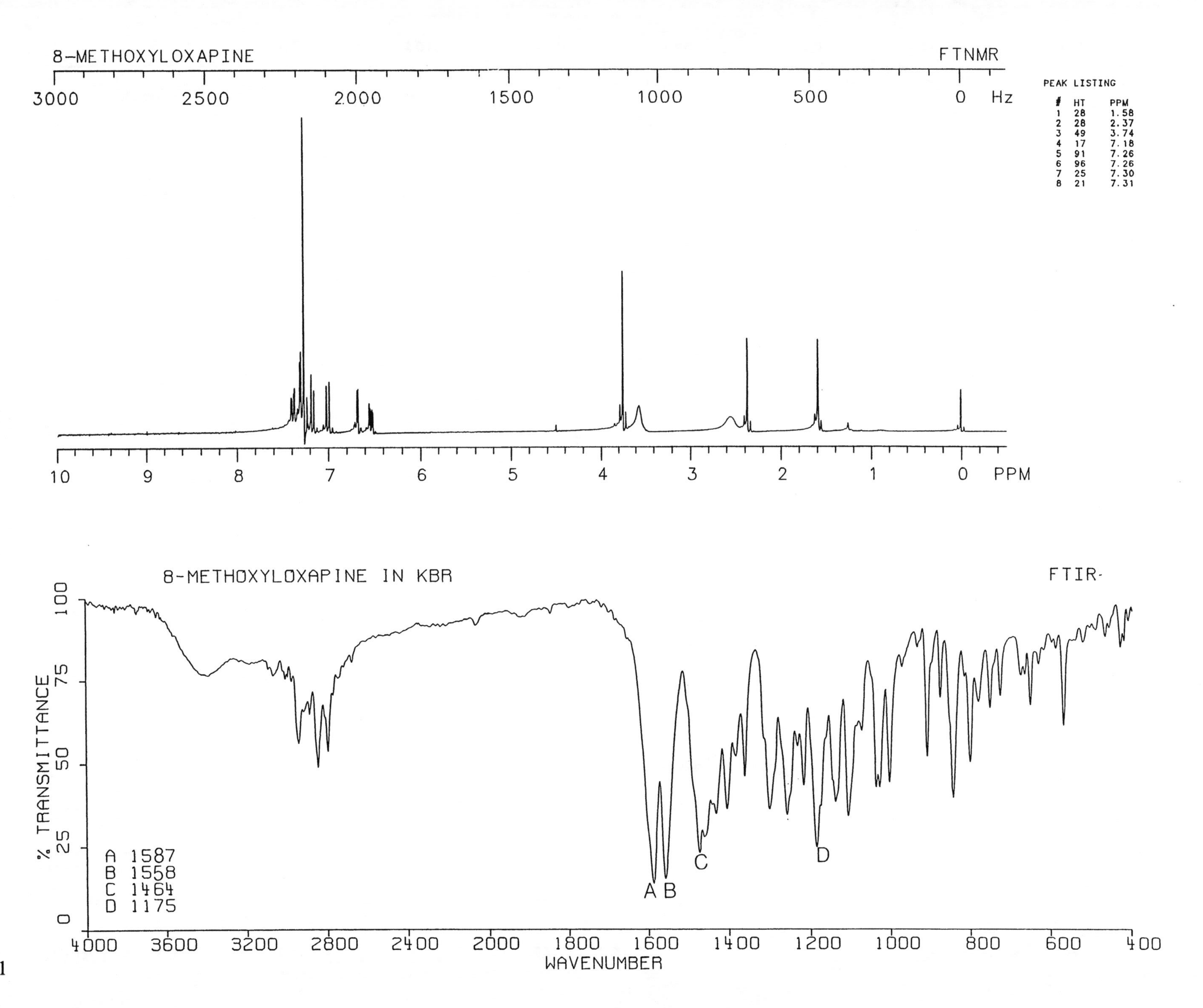
8-METHOXYLOXAPINE
FTNMR
3000 2500 2000 1500 1000 500 0 Hz
PEAK LISTING
# HT PPM
1 28 1.58
2 28 2.37
3 49 3.74
4 17 7.18
5 91 7.26
6 96 7.26
7 25 7.30
8 21 7.31
10 9 8 7 6 5 4 3 2 1 0 PPM
8-METHOXYLOXAPINE IN KBR
FTIR-
% TRANSMITTANCE
100 75 50 25 0
A 1587
B 1558
C 1464
D 1175
A B C D
4000 3600 3200 2800 2400 2000 1800 1600 1400 1200 1000 800 600 400
WAVENUMBER

# 2-METHOXY-4,5-METHYLENEDIOXYAMPHETAMINE

$C_{11}H_{15}NO_3$

**Molecular weight:** 209.24 (209.11)
**Synonyms:** 5-(α-Methyl)ethanamine-6-methoxy-1,3-benzodioxole; MMDA

**Trade names:**

**Use:** Hallucinogen
**HPLC:** Si-10; 20A:80B; 5.4
**GC:** 1686; 200°C

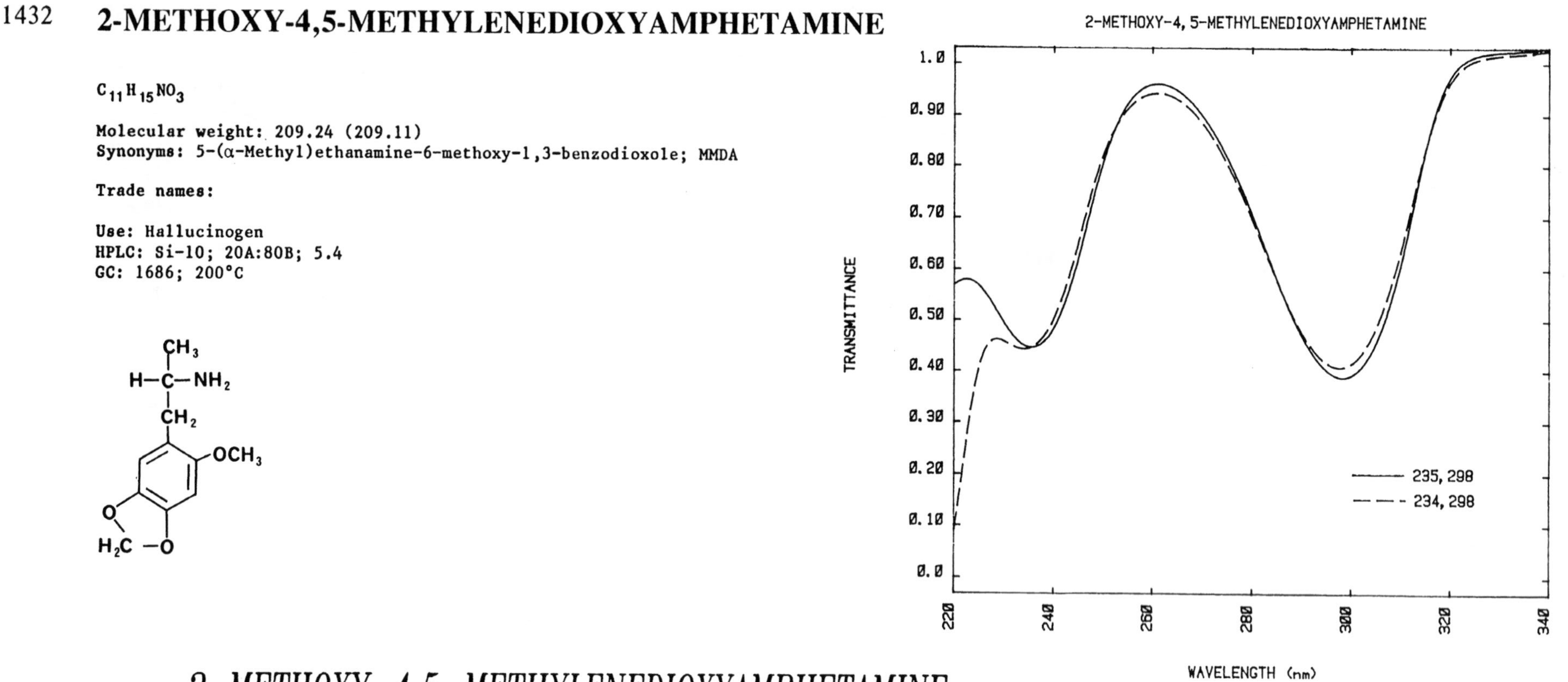


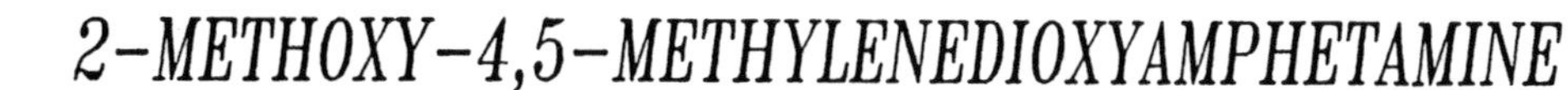


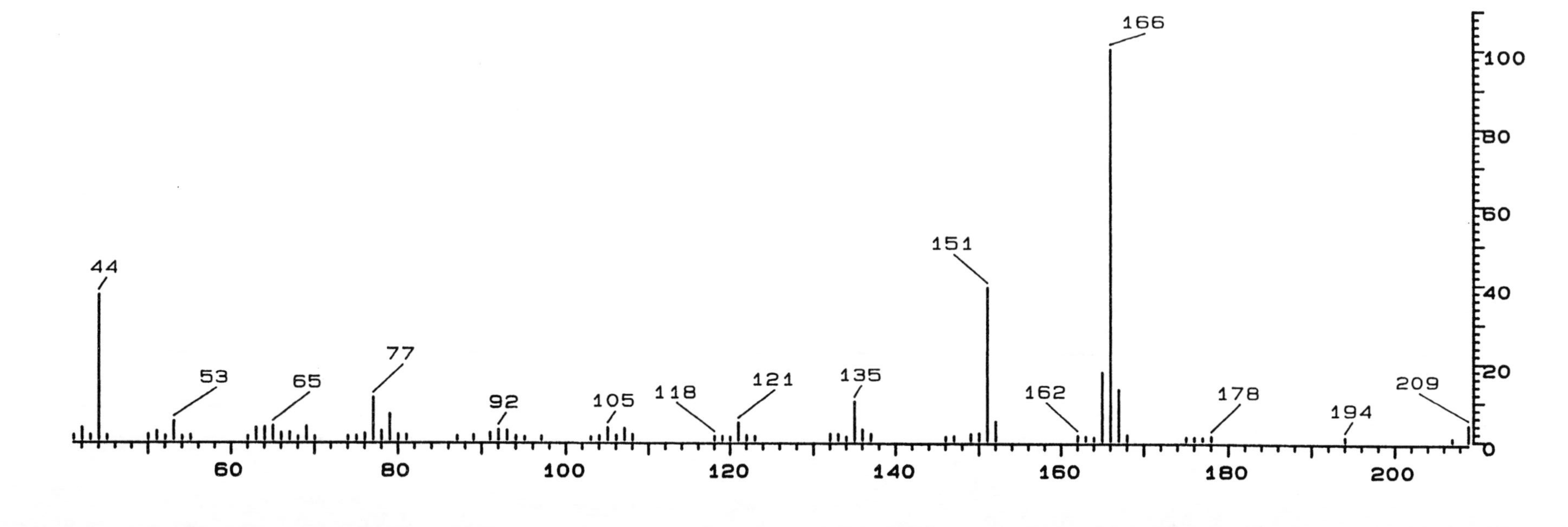

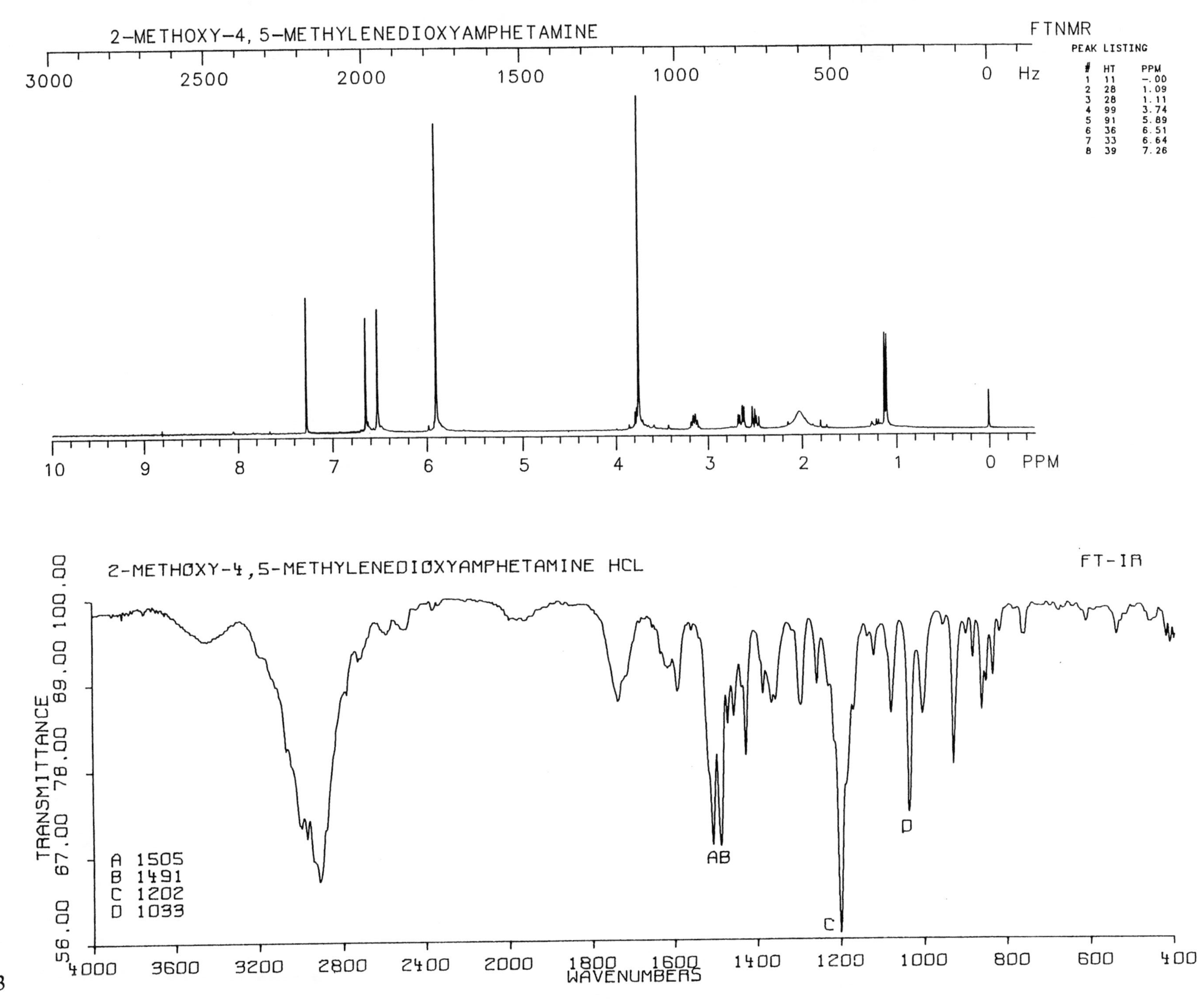

2-METHOXY-4, 5-METHYLENEDIOXYAMPHETAMINE
FTNMR
PEAK LISTING
# HT PPM
1 11 -.00
2 28 1.09
3 28 1.11
4 99 3.74
5 91 5.89
6 36 6.51
7 33 6.64
8 39 7.26
3000 2500 2000 1500 1000 500 0 Hz
10 9 8 7 6 5 4 3 2 1 0 PPM
2-METHØXY-4,5-METHYLENEDIØXYAMPHETAMINE HCL
FT-IR
TRANSMITTANCE
56.00 67.00 78.00 89.00 100.00
A 1505
B 1491
C 1202
D 1033
AB
C
D
4000 3600 3200 2800 2400 2000 1800 1600 1400 1200 1000 800 600 400
WAVENUMBERS

# METHOXYPHENAMINE

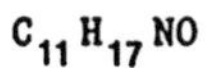


**Molecular weight:** 179.25 (179.13)
**Synonyms:** 2-Methoxy-N,α-dimethylbenzeneethanamine; β-(o-methoxyphenyl)-isopropylmethylamine; methoxiphenadrin; mexyphamine
**Trade names:** Orthoxicol, Orthoxine, Ortodrinex, Proasma

**Use:** Adrenergic, sympathomimetic
**HPLC:** Si-10; 20A:80B; 5.8
**GC:** 1368; 140°C

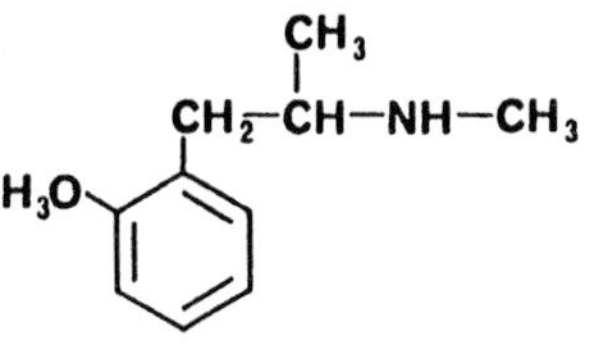


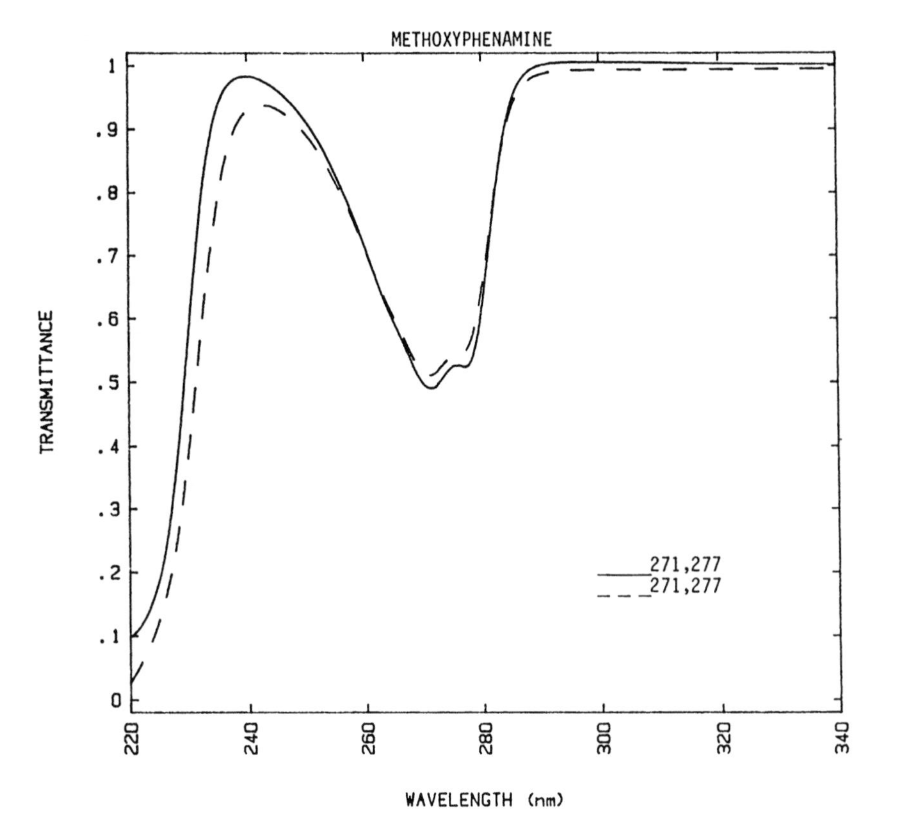

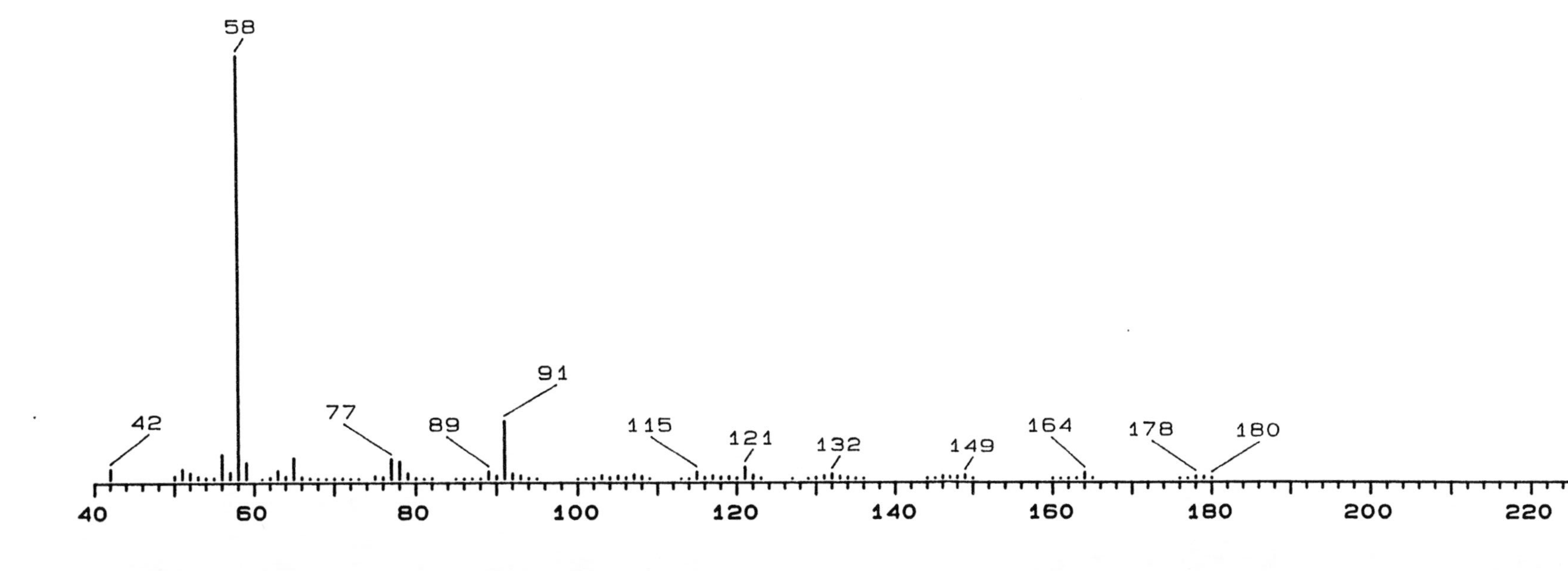

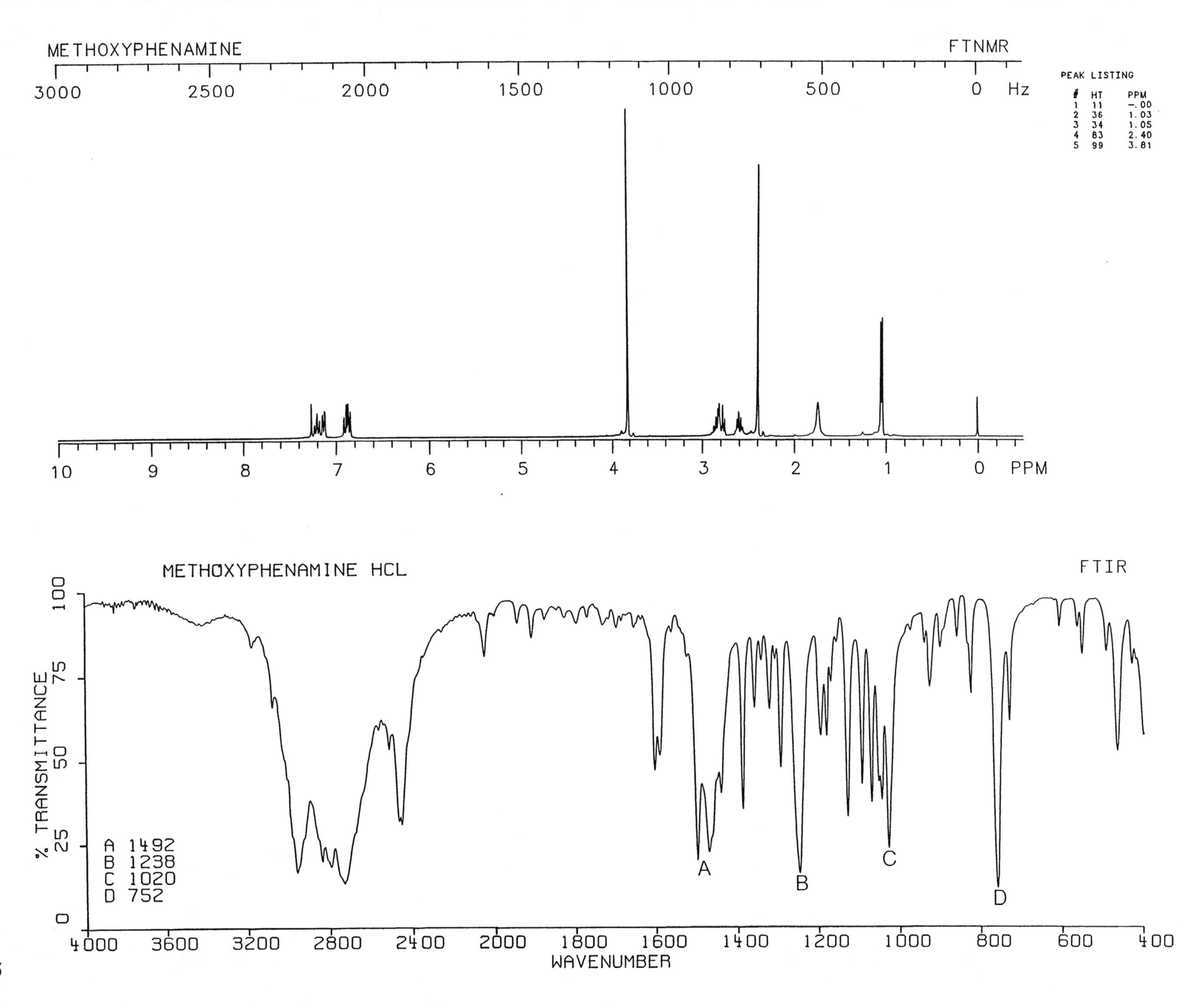
METHOXYPHENAMINE
FTNMR
3000 2500 2000 1500 1000 500 0 Hz
PEAK LISTING
# HT PPM
1 11 -.00
2 36 1.03
3 34 1.05
4 83 2.40
5 99 3.81
10 9 8 7 6 5 4 3 2 1 0 PPM
METHOXYPHENAMINE HCL
FTIR
100 75 50 25 0
% TRANSMITTANCE
A 1492
B 1238
C 1020
D 752
A
B
C
D
4000 3600 3200 2800 2400 2000 1800 1600 1400 1200 1000 800 600 400
WAVENUMBER

# METHOXYVERAPAMIL

$C_{28}H_{40}N_2O_5$

Molecular weight: 484.61 (484.29)
Synonyms: α -[3-[[2-(3,4-Dimethoxyphenyl)ethyl]methylamino]propyl]-3,4,5-trimethoxy-α-(1-methylethyl)benzeneacetonitrile
Trade names:

Use:
HPLC:
GC: 2070; 250°C

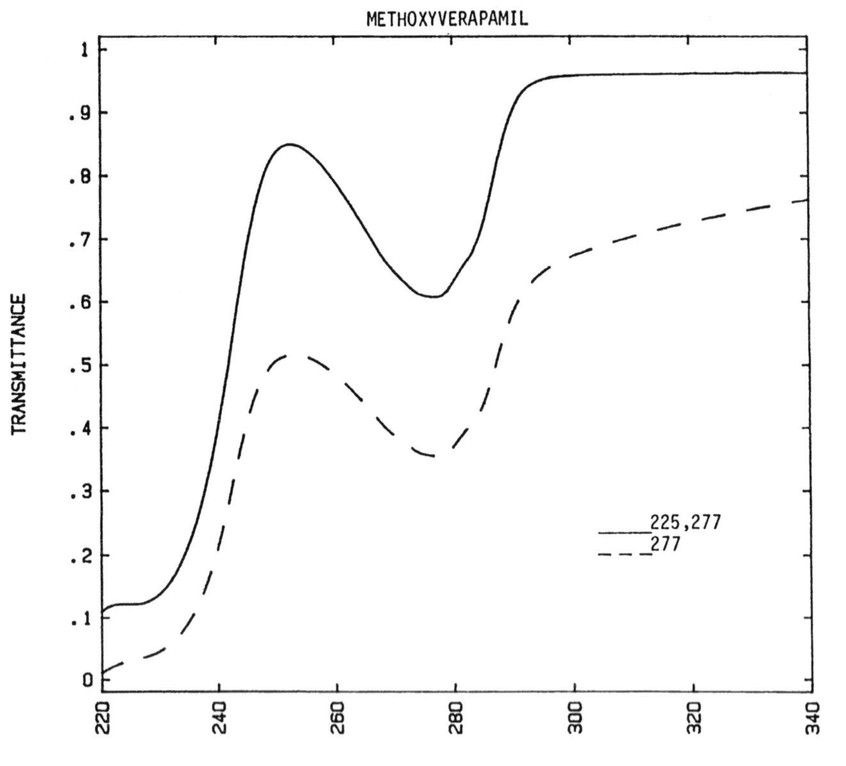

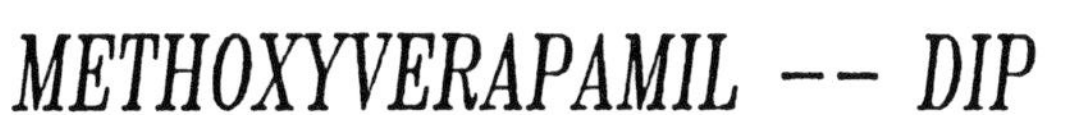


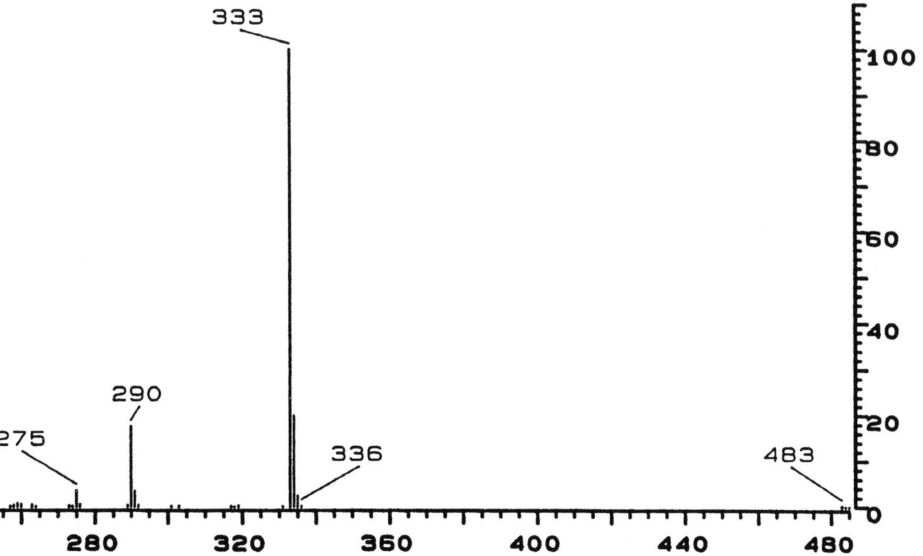

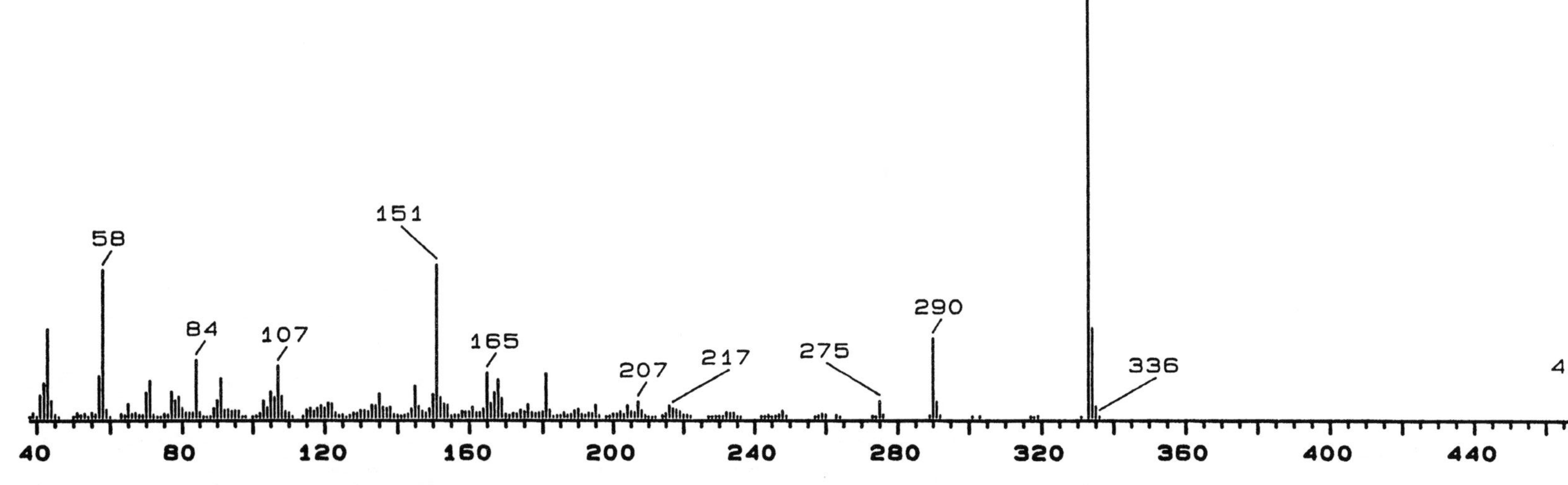

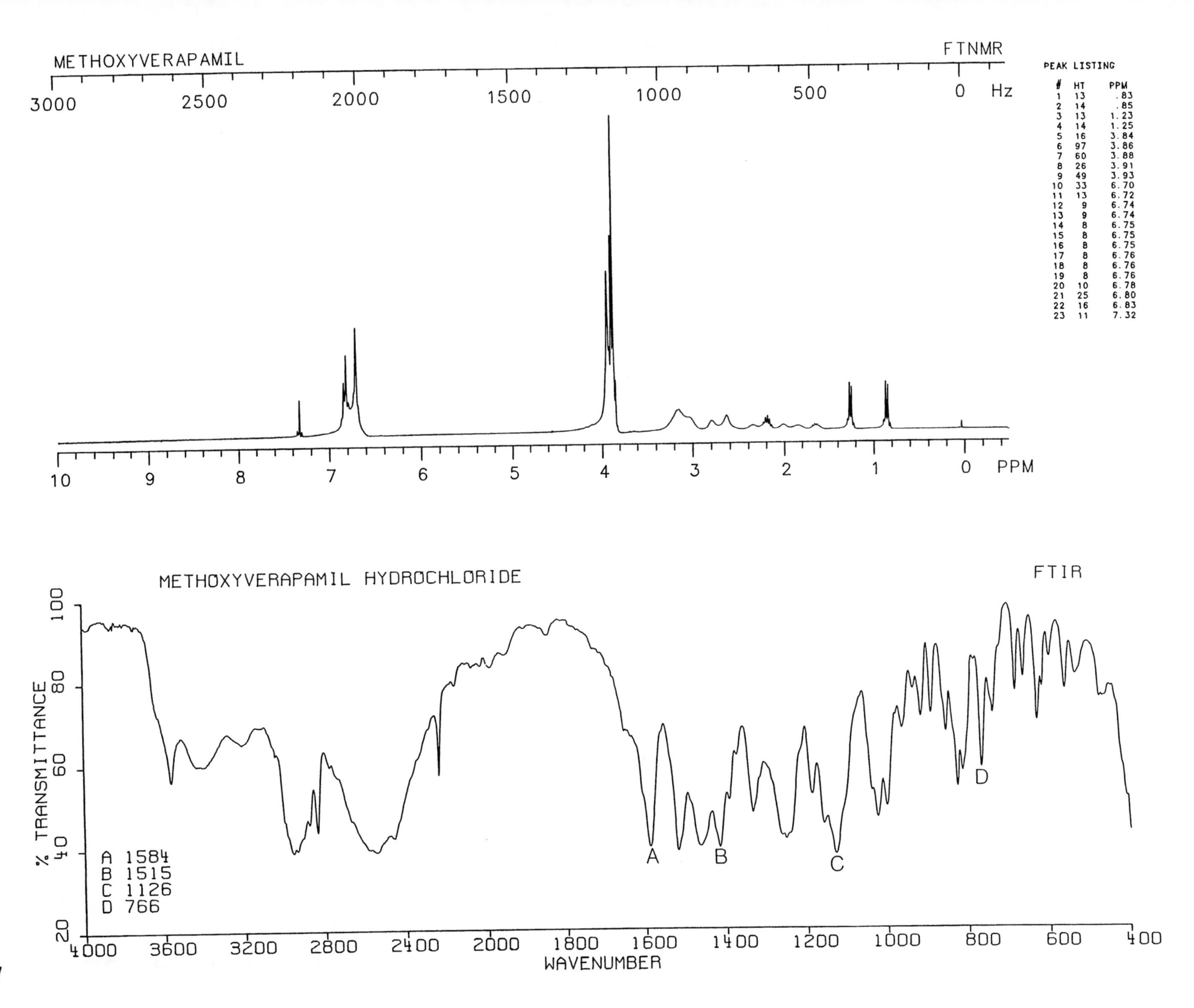
METHOXYVERAPAMIL
FTNMR
3000
2500
2000
1500
1000
500
0
Hz
10
9
8
7
6
5
4
3
2
1
0
PPM
PEAK LISTING
# HT PPM
1 13 .83
2 14 .85
3 13 1.23
4 14 1.25
5 16 3.84
6 97 3.86
7 60 3.88
8 26 3.91
9 49 3.93
10 33 6.70
11 13 6.72
12 9 6.74
13 9 6.74
14 8 6.75
15 8 6.75
16 8 6.75
17 8 6.76
18 8 6.76
19 8 6.76
20 10 6.78
21 25 6.80
22 16 6.83
23 11 7.32
METHOXYVERAPAMIL HYDROCHLORIDE
FTIR
% TRANSMITTANCE
100
80
60
40
20
A 1584
B 1515
C 1126
D 766
A
B
C
D
4000
3600
3200
2800
2400
2000
1800
1600
1400
1200
1000
800
600
400
WAVENUMBER

# METHSCOPOLAMINE BROMIDE

$C_{18}H_{24}BrNO_4$

Molecular weight: 398.30 (397.09)
Synonyms: 7-(3-Hydroxy-1-oxo-2-phenylpropoxy)-9,9-dimethyl-3-oxa-9-azoniatricyclo[3.3.1.0$^{2,4}$]nonane bromide; scopolamine methylbromide
Trade names: Pamine

Use: Anticholinergic
HPLC:
GC:

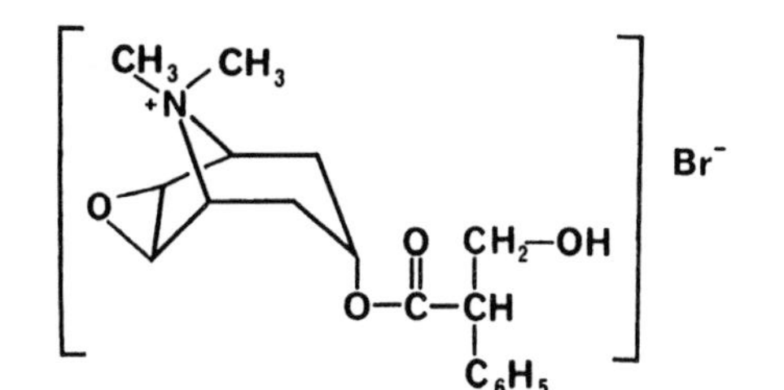


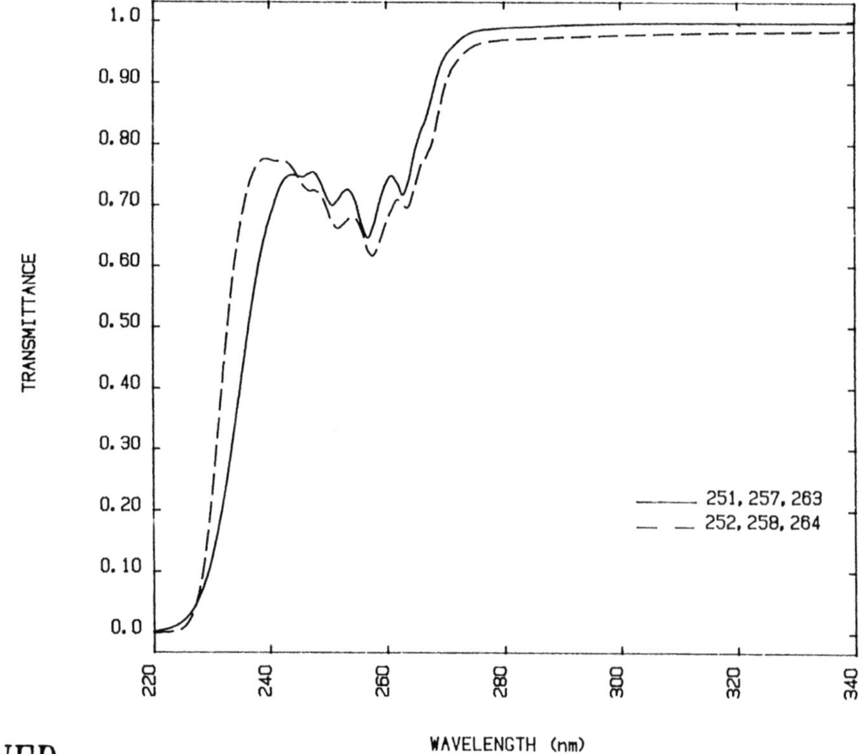

NO USEFUL MASS SPECTRUM WAS OBTAINED

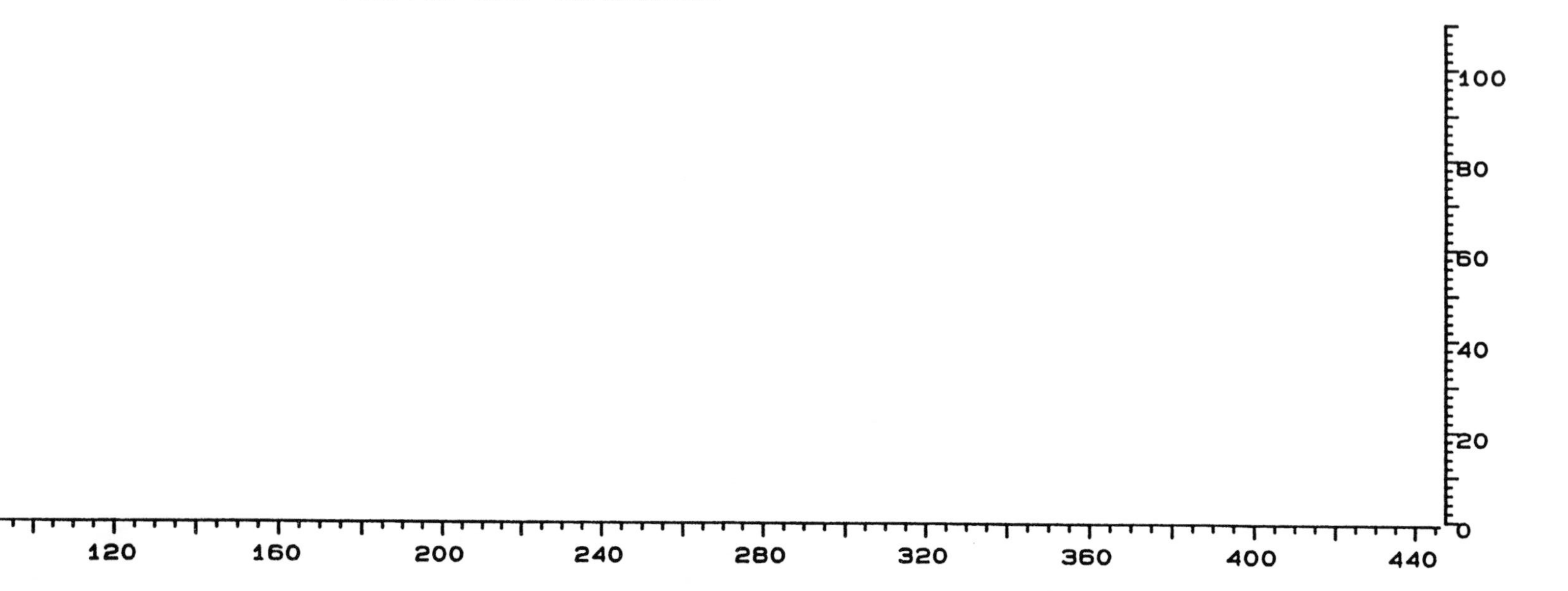

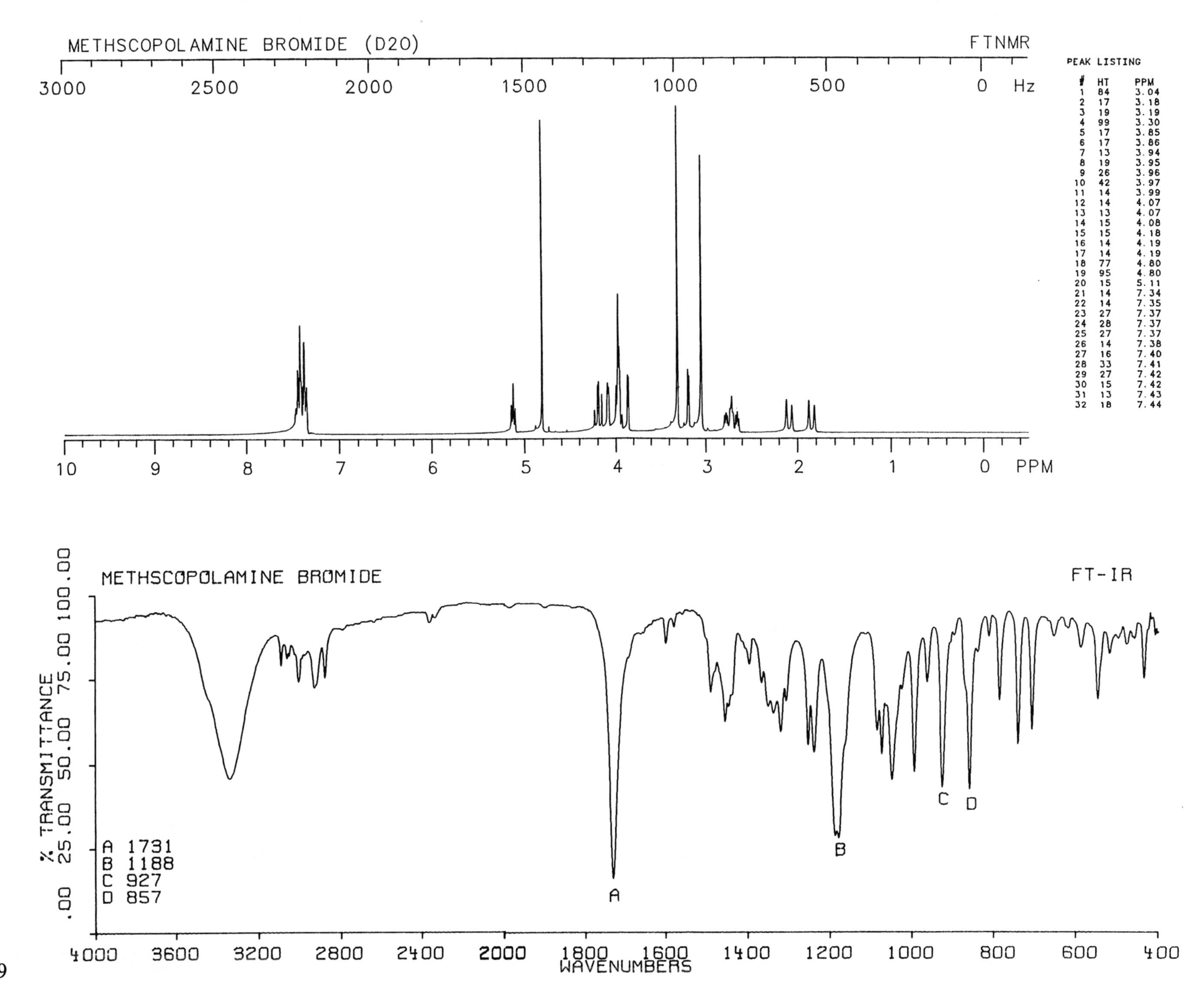

METHSCOPOLAMINE BROMIDE (D2O)
FTNMR
3000 2500 2000 1500 1000 500 0 Hz
10 9 8 7 6 5 4 3 2 1 0 PPM
PEAK LISTING
# HT PPM
1 84 3.04
2 17 3.18
3 19 3.19
4 99 3.30
5 17 3.85
6 17 3.86
7 13 3.94
8 19 3.95
9 26 3.96
10 42 3.97
11 14 3.99
12 14 4.07
13 13 4.07
14 15 4.08
15 15 4.18
16 14 4.19
17 14 4.19
18 77 4.80
19 95 4.80
20 15 5.11
21 14 7.34
22 14 7.35
23 27 7.37
24 28 7.37
25 27 7.37
26 14 7.38
27 16 7.40
28 33 7.41
29 27 7.42
30 15 7.42
31 13 7.43
32 18 7.44
METHSCOPOLAMINE BROMIDE
FT-IR
% TRANSMITTANCE
.00 25.00 50.00 75.00 100.00
A 1731
B 1188
C 927
D 857
A
B
C
D
4000 3600 3200 2800 2400 2000 1800 1600 1400 1200 1000 800 600 400
WAVENUMBERS

# METHSUXIMIDE

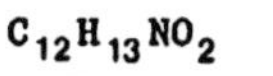

$C_{12}H_{13}NO_2$

Molecular weight: 203.24 (203.10)
Synonyms: 1,3-Dimethyl-3-phenyl-2,5-pyrrolidinedione

Trade names: Celontin

Use: Anticonvulsant
HPLC: Si-10; 100B; 3.3
GC: 1644; 200°C

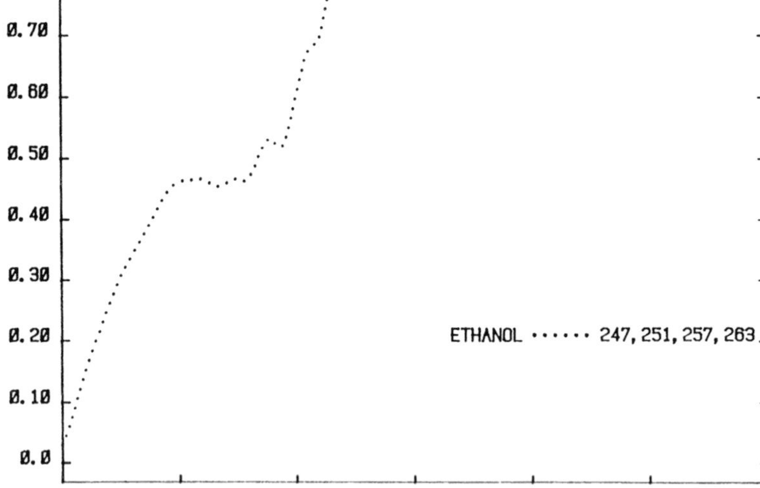


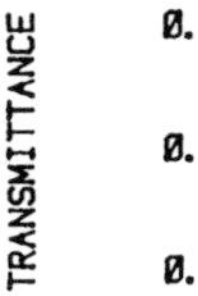

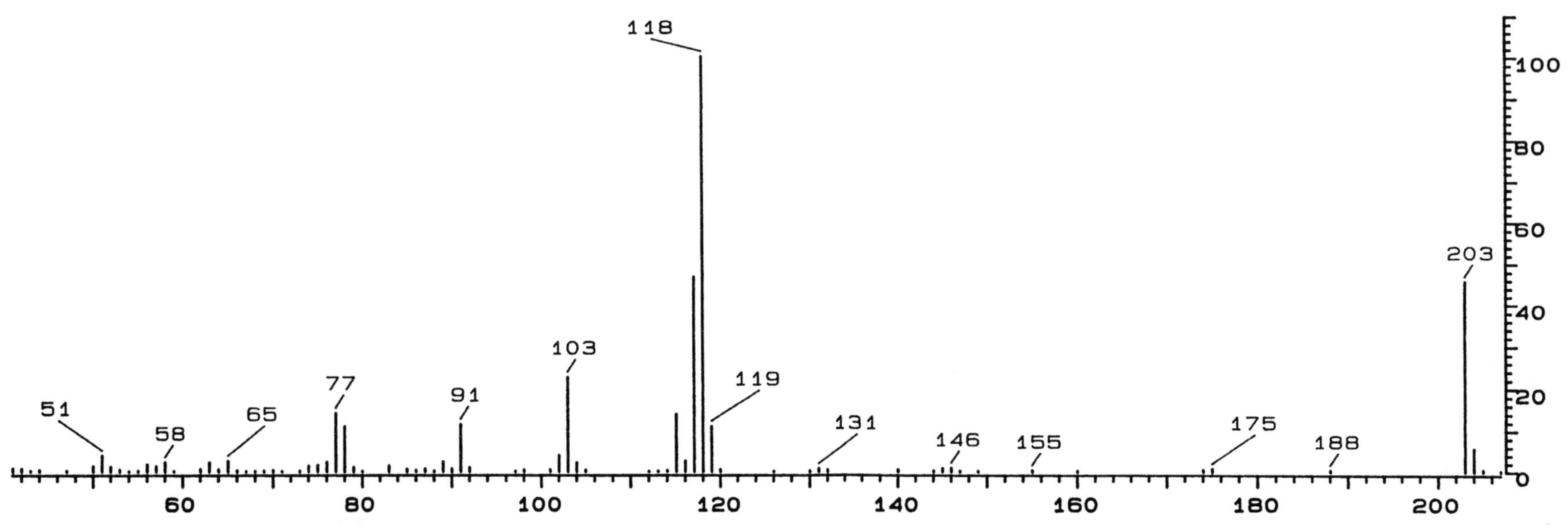

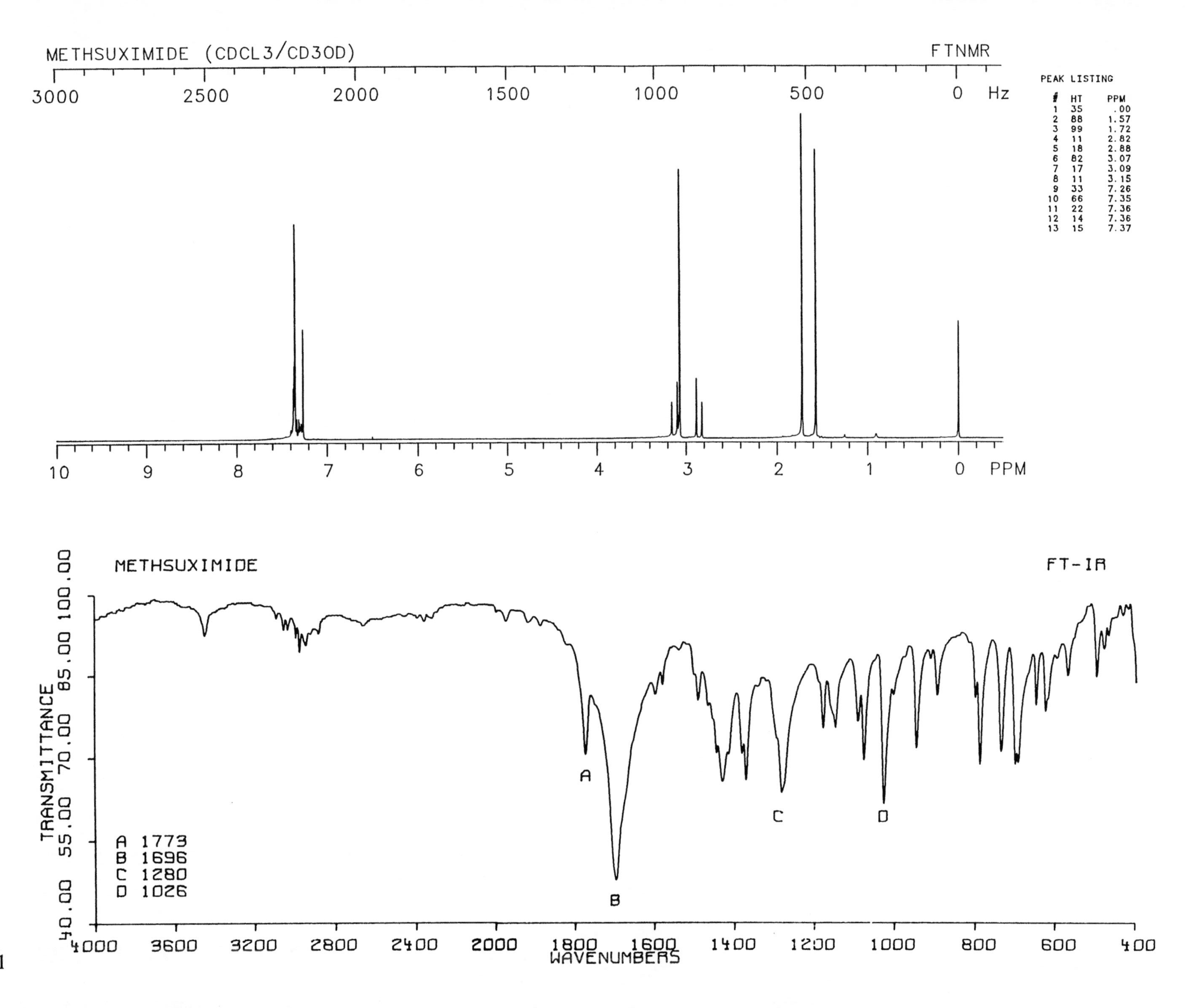
METHSUXIMIDE (CDCL3/CD3OD)
FTNMR
3000 2500 2000 1500 1000 500 0 Hz
10 9 8 7 6 5 4 3 2 1 0 PPM
PEAK LISTING
# HT PPM
1 35 .00
2 88 1.57
3 99 1.72
4 11 2.82
5 18 2.88
6 82 3.07
7 17 3.09
8 11 3.15
9 33 7.26
10 66 7.35
11 22 7.36
12 14 7.36
13 15 7.37
METHSUXIMIDE
FT-IR
TRANSMITTANCE
40.00 55.00 70.00 85.00 100.00
A 1773
B 1696
C 1280
D 1026
A B C D
4000 3600 3200 2800 2400 2000 1800 1600 1400 1200 1000 800 600 400
WAVENUMBERS

# METHYCLOTHIAZIDE

$C_9H_{11}Cl_2N_3O_4S_2$

Molecular weight: 360.24 (258.96)
Synonyms: 6-Chloro-3-(chloromethyl)-3,4-dihydro-2-méthyl-2H-1,2,4-benzothiadiazine-7-sulfonamide-1,1-dioxide
Trade names: Aquatensen, Diutensen, Enduron, Enduronyl, Eutron

Use: Diuretic, antihypertensive
HPLC:
GC:

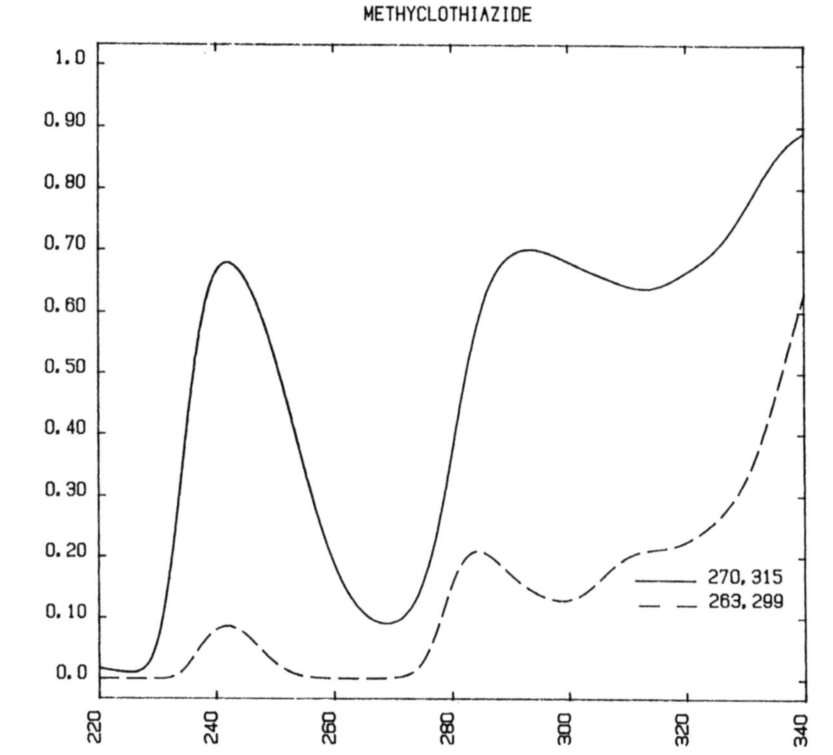


METHYCLOTHIAZIDE -- DIP

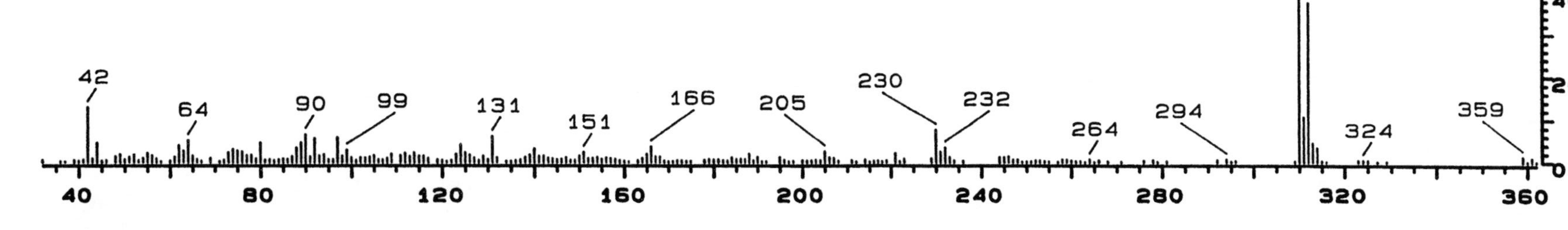


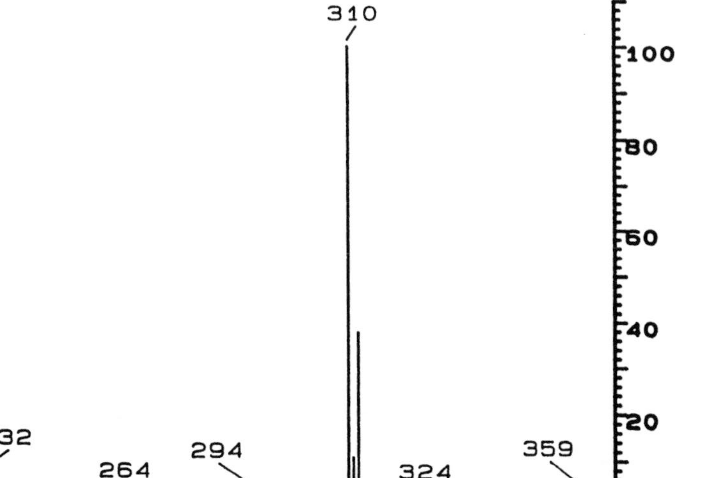

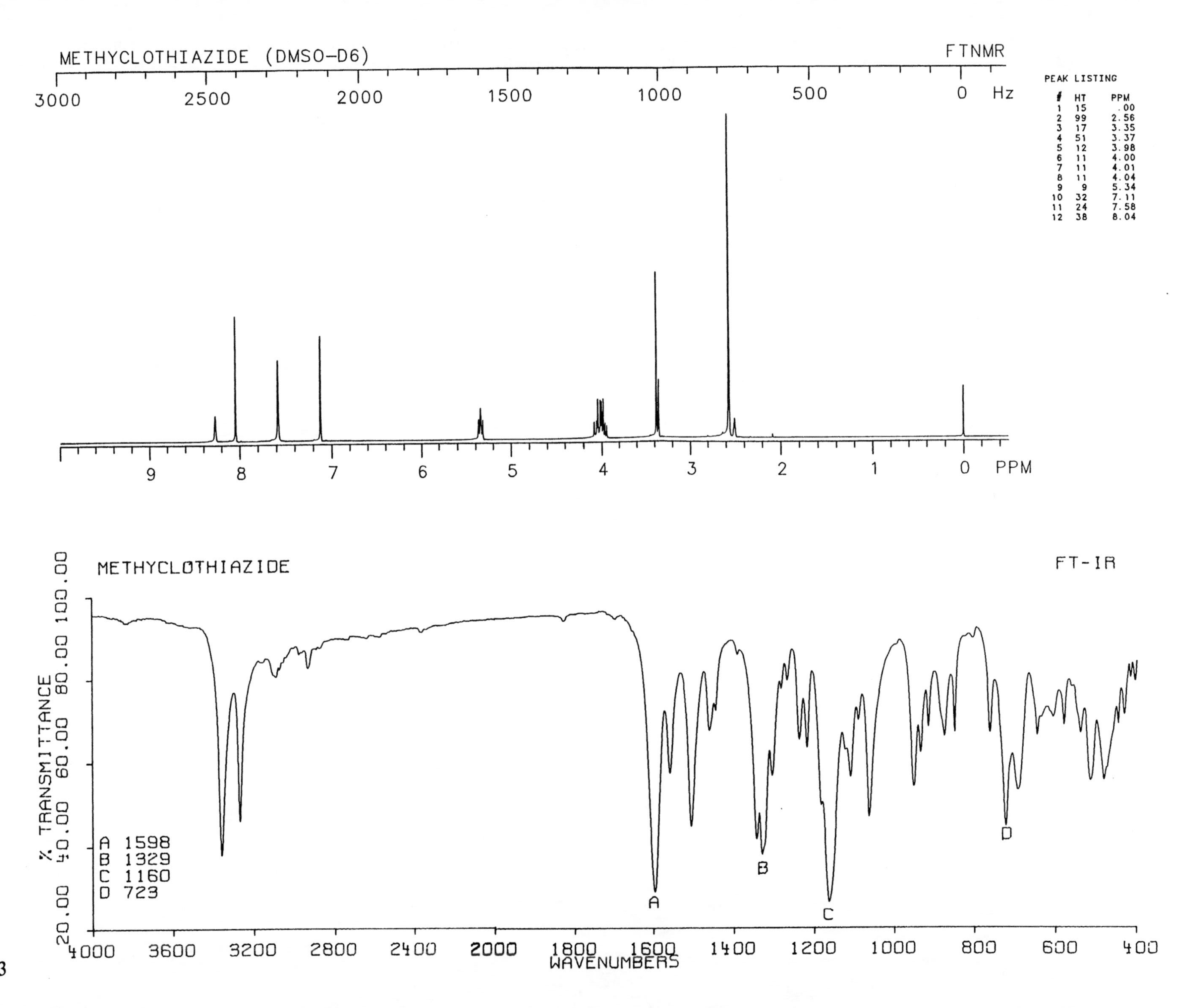
METHYCLOTHIAZIDE (DMSO-D6)
FTNMR
3000 2500 2000 1500 1000 500 0 Hz
PEAK LISTING
# HT PPM
1 15 .00
2 99 2.56
3 17 3.35
4 51 3.37
5 12 3.98
6 11 4.00
7 11 4.01
8 11 4.04
9 9 5.34
10 32 7.11
11 24 7.58
12 38 8.04
9 8 7 6 5 4 3 2 1 0 PPM
METHYCLOTHIAZIDE
FT-IR
% TRANSMITTANCE
20.00 40.00 60.00 80.00 100.00
A 1598
B 1329
C 1160
D 723
A B C D
4000 3600 3200 2800 2400 2000 1800 1600 1400 1200 1000 800 600 400
WAVENUMBERS

# METHYLAMINE

$CH_5N$

Molecular weight: 31.06 (31.04)
Synonyms: Methanamine; monomethylamine

Trade names:

Use: Synthesis
HPLC:
GC: 592; 80°C

$CH_3-NH_2$

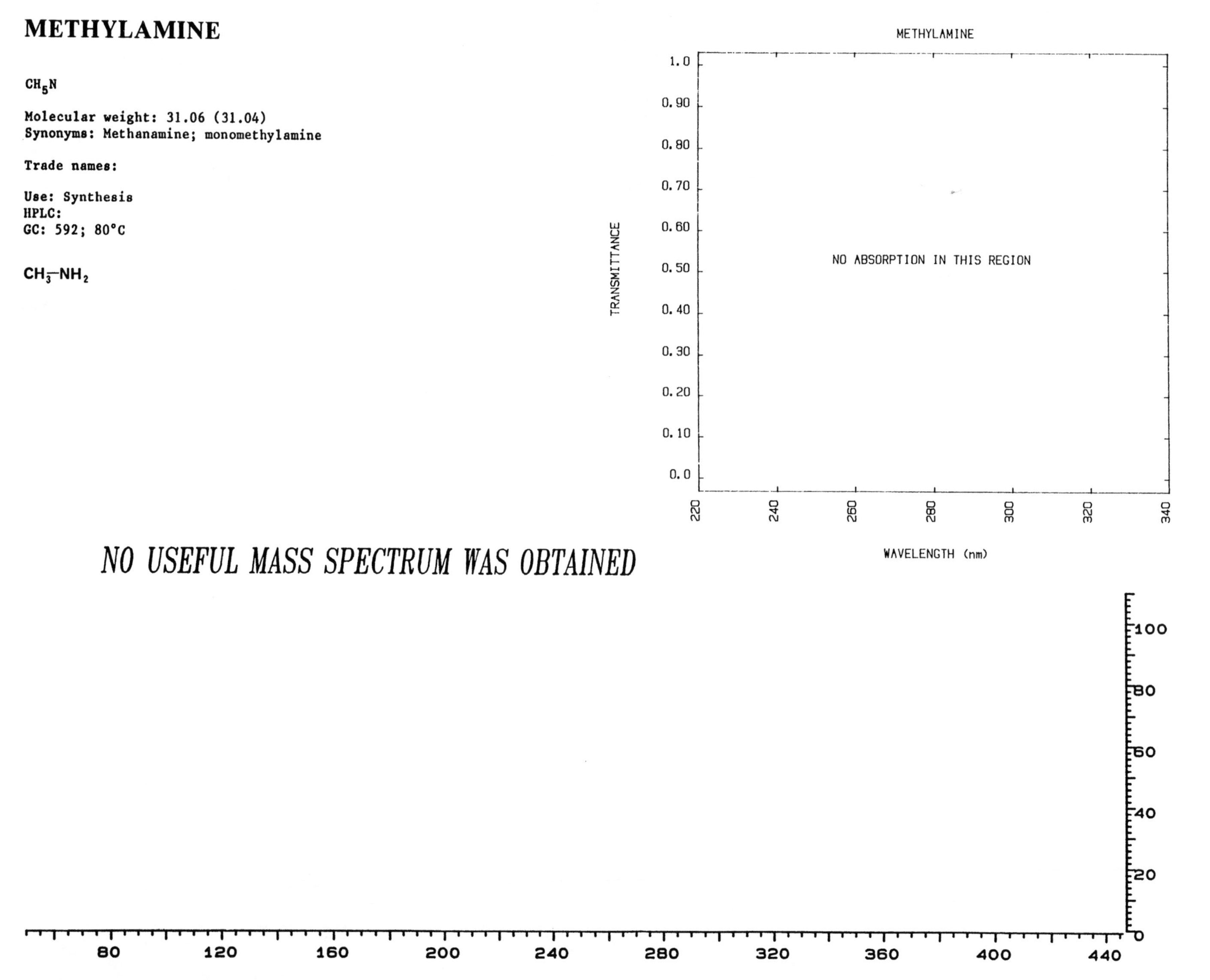

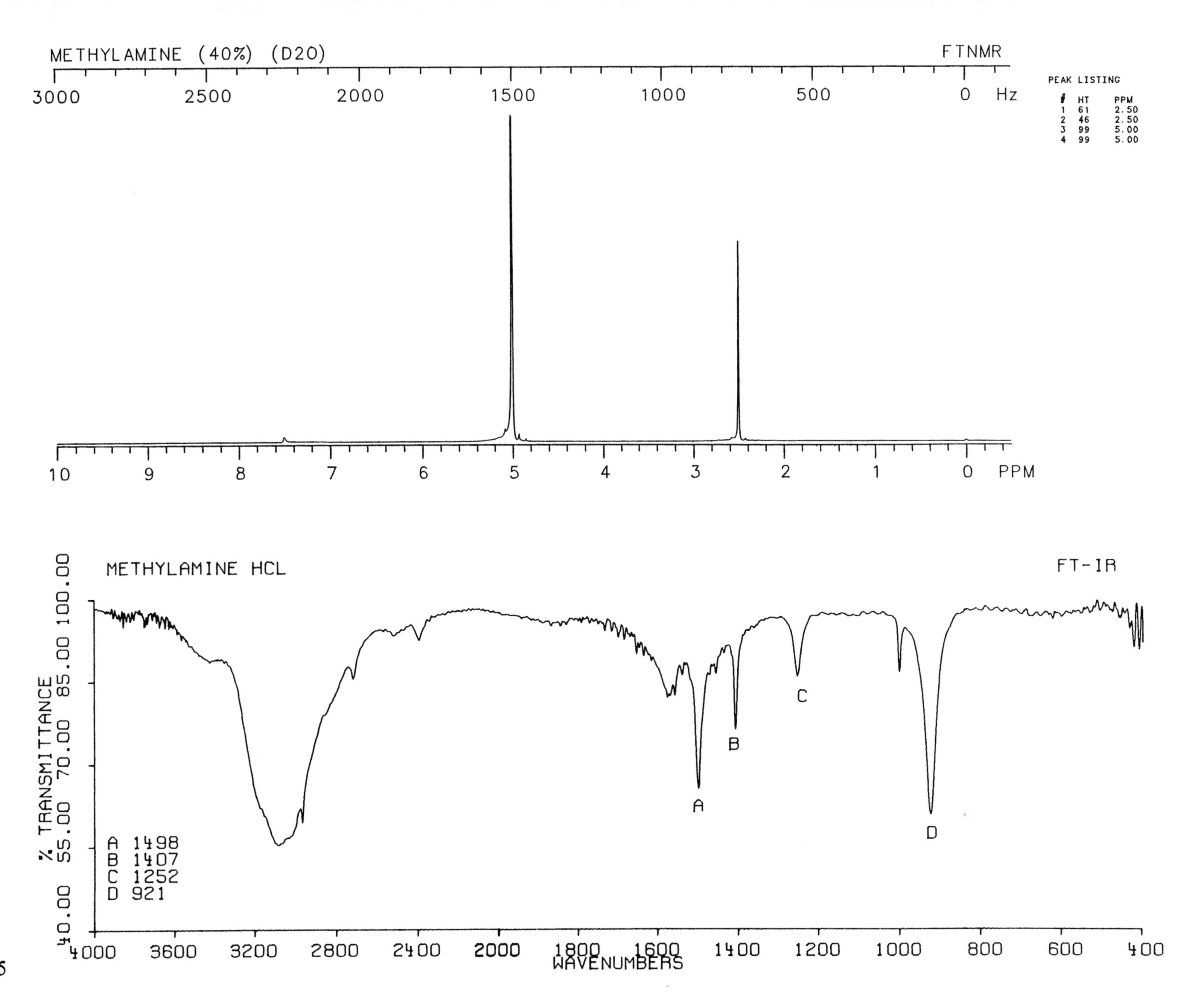
METHYLAMINE (40%) (D2O)
FTNMR
3000
2500
2000
1500
1000
500
0 Hz
PEAK LISTING
# HT PPM
1 61 2.50
2 46 2.50
3 99 5.00
4 99 5.00
10
9
8
7
6
5
4
3
2
1
0 PPM
METHYLAMINE HCL
FT-IR
% TRANSMITTANCE
40.00
55.00
70.00
85.00
100.00
A 1498
B 1407
C 1252
D 921
A
B
C
D
4000
3600
3200
2800
2400
2000
1800
1600
1400
1200
1000
800
600
400
WAVENUMBERS

# 4-METHYLAMPHETAMINE

$C_{10}H_{15}N$

Molecular weight: 149.24 (149.12)
Synonyms: 4-Methyl-α-methylbenzeneethanamine

Trade names:

Use: Hallucinogen
HPLC:
GC:

$CH_3-C_6H_4-CH_2-CH(NH_2)-CH_3$

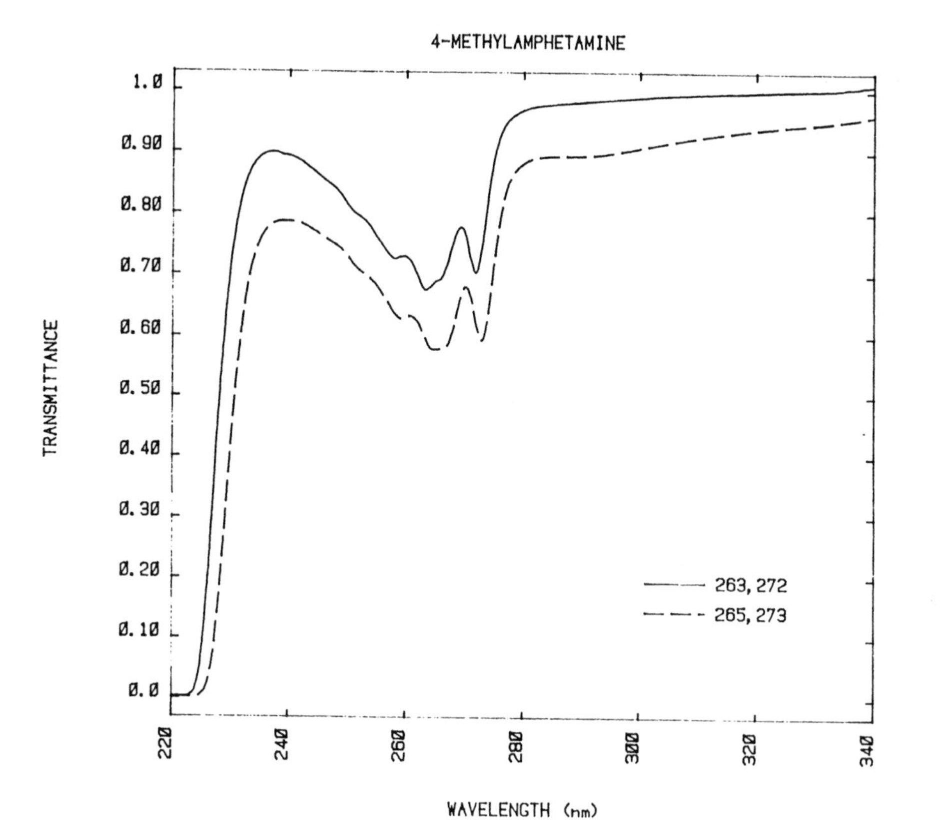

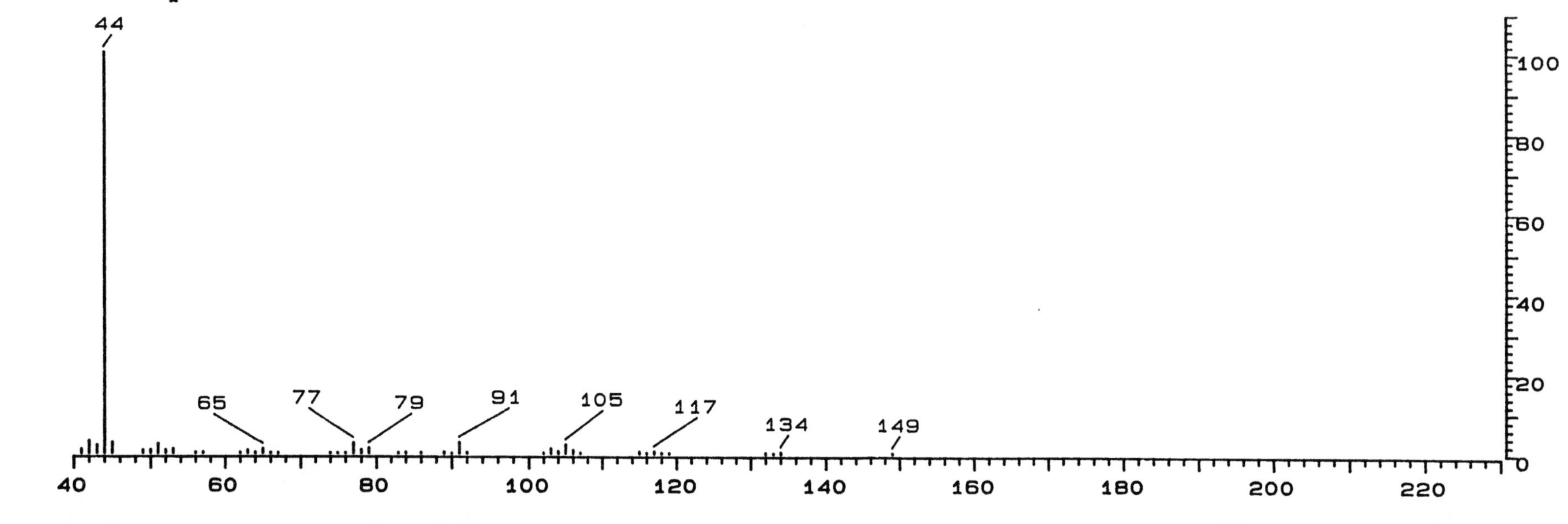

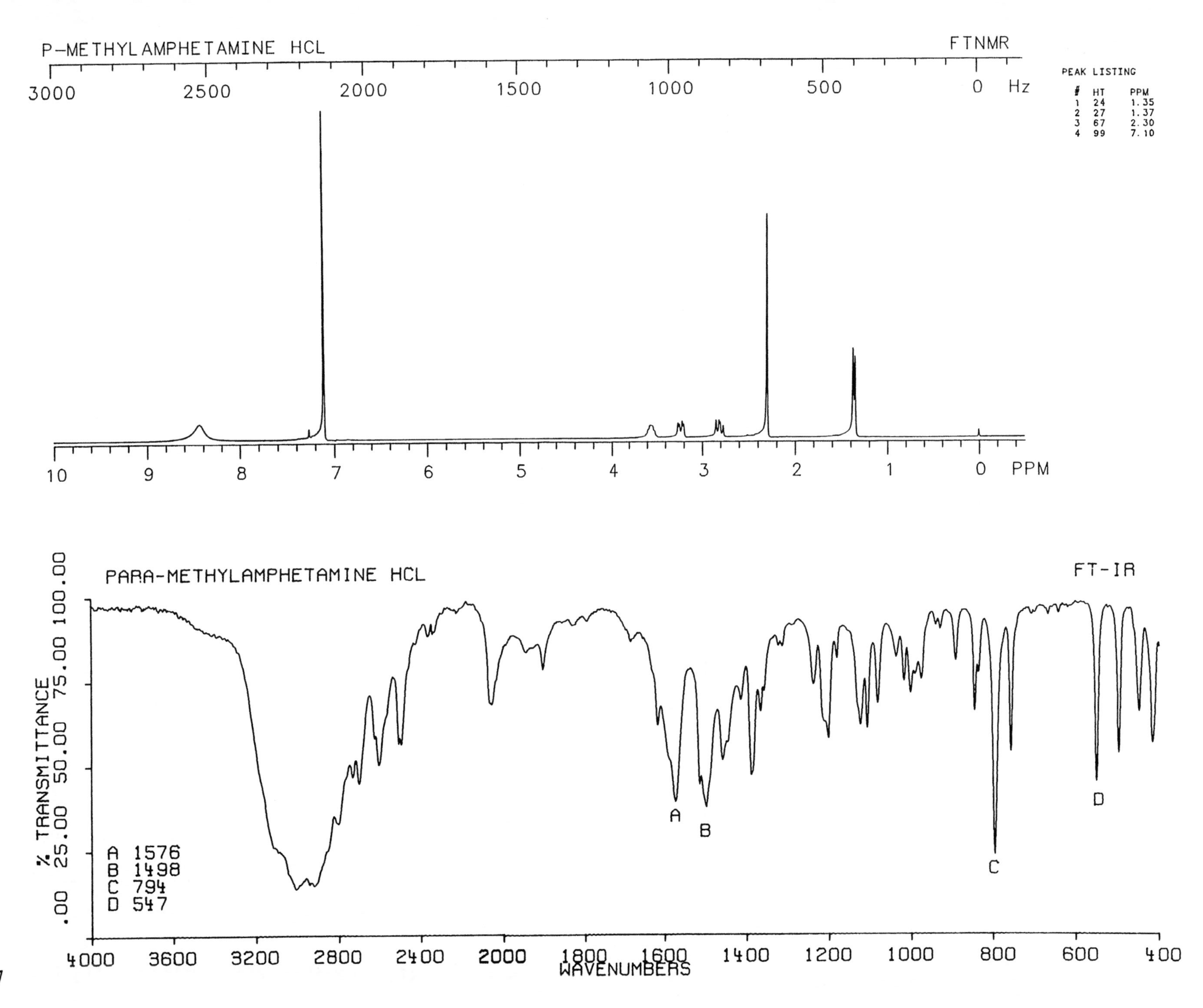
P-METHYLAMPHETAMINE HCL
FTNMR
3000
2500
2000
1500
1000
500
0 Hz
PEAK LISTING
# HT PPM
1 24 1.35
2 27 1.37
3 67 2.30
4 99 7.10
10
9
8
7
6
5
4
3
2
1
0 PPM
PARA-METHYLAMPHETAMINE HCL
FT-IR
% TRANSMITTANCE
.00
25.00
50.00
75.00
100.00
A 1576
B 1498
C 794
D 547
A
B
C
D
4000
3600
3200
2800
2400
2000
1800
1600
1400
1200
1000
800
600
400
WAVENUMBERS

# METHYL-α-L-DAUNOSAMINIDE

$C_7H_{15}NO_3$

Molecular weight: 161.18 (161.11)
Synonyms: Daunosamine methyl glycoside

Trade names:

Use:
HPLC:
GC:

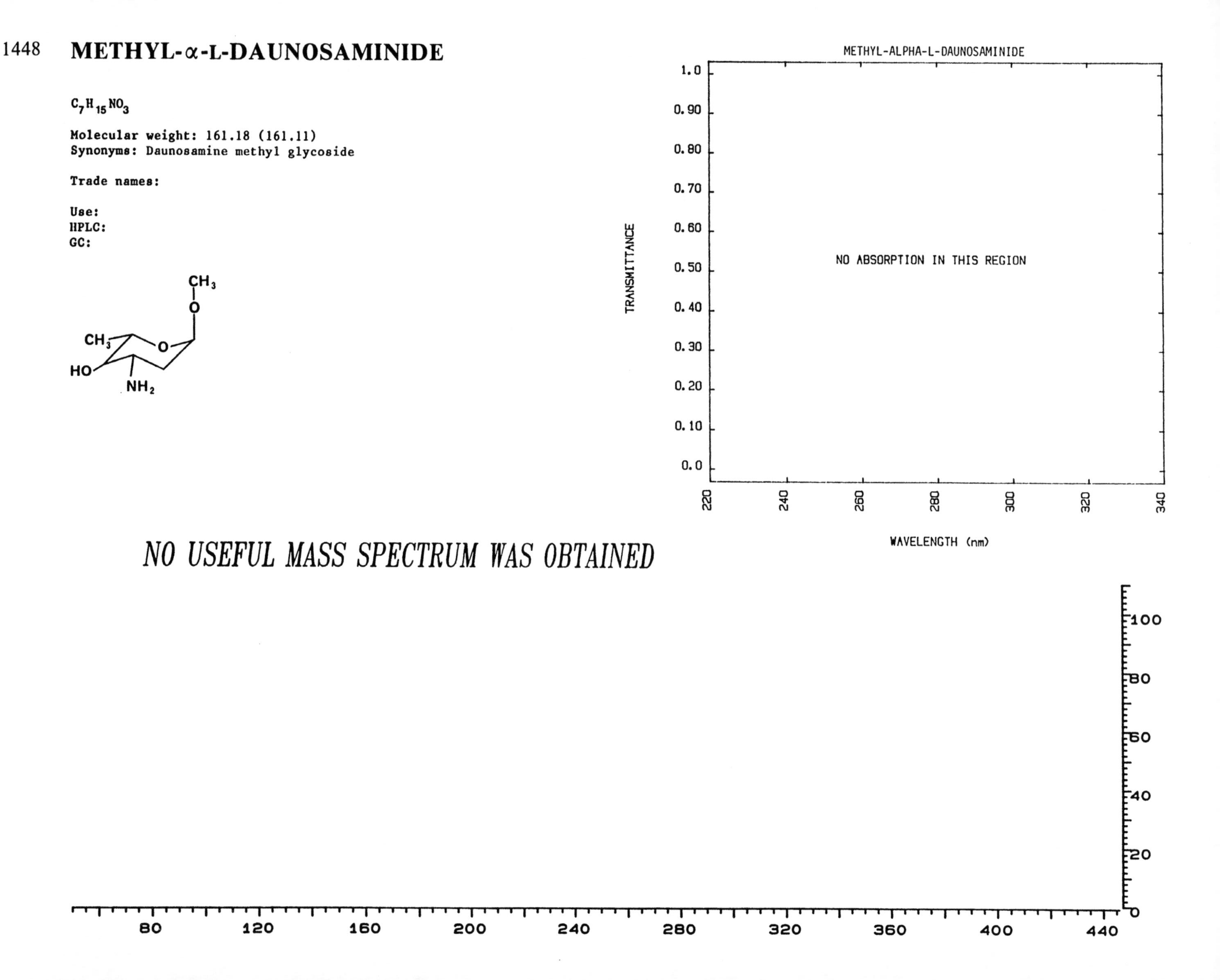

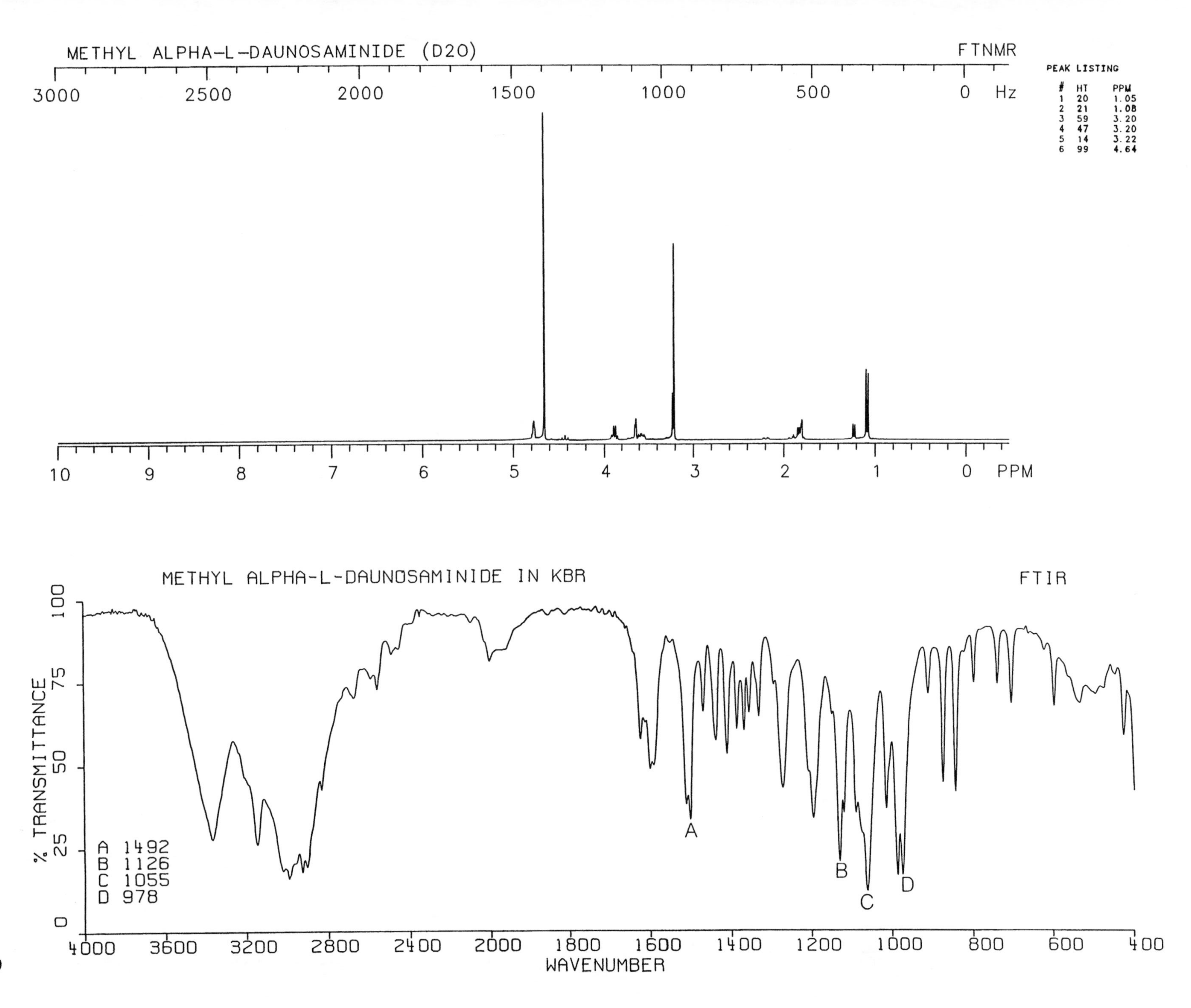
METHYL ALPHA-L-DAUNOSAMINIDE (D2O)
FTNMR
3000 2500 2000 1500 1000 500 0 Hz
PEAK LISTING
# HT PPM
1 20 1.05
2 21 1.08
3 59 3.20
4 47 3.20
5 14 3.22
6 99 4.64
10 9 8 7 6 5 4 3 2 1 0 PPM
METHYL ALPHA-L-DAUNOSAMINIDE IN KBR
FTIR
% TRANSMITTANCE
100 75 50 25 0
A 1492
B 1126
C 1055
D 978
A
B
C
D
4000 3600 3200 2800 2400 2000 1800 1600 1400 1200 1000 800 600 400
WAVENUMBER

# 5-METHYL-*N, N*-DIMETHYLTRYPTAMINE

$C_{13}H_{19}N_2O$

**Molecular weight:** 219.30 (219.15)
**Synonyms:** N,N-Dimethyl-1H-indole-3-ethanamine

**Trade names:**

**Use:** Hallucinogen
**HPLC:**
**GC:**

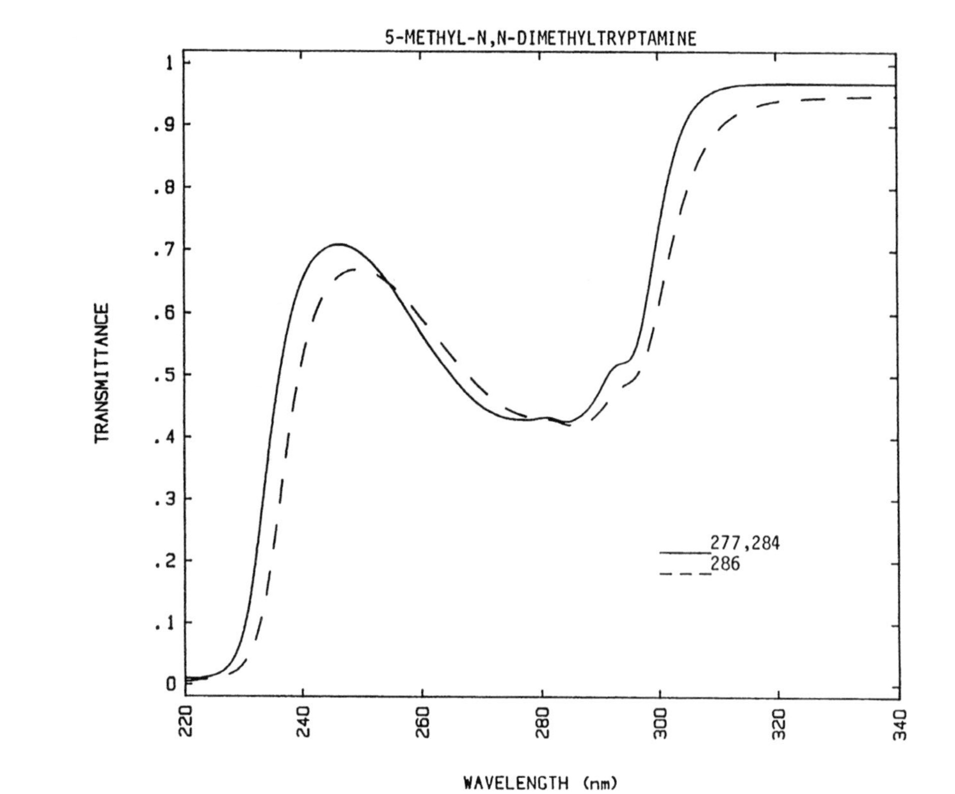


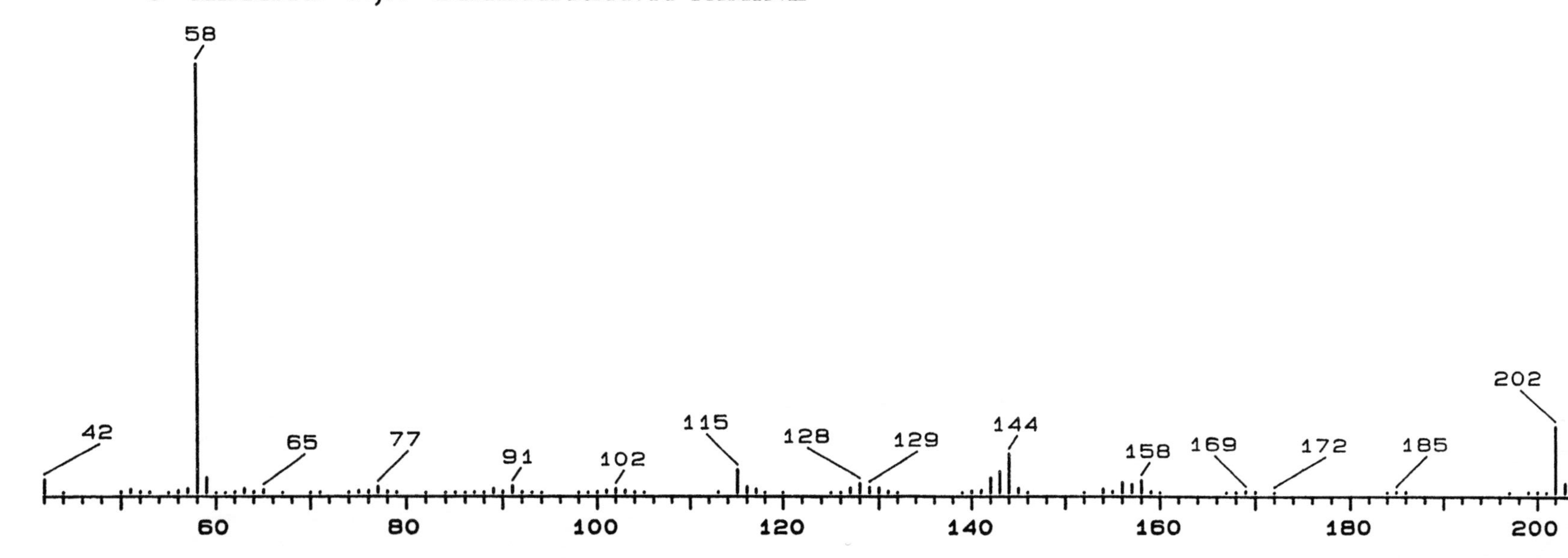

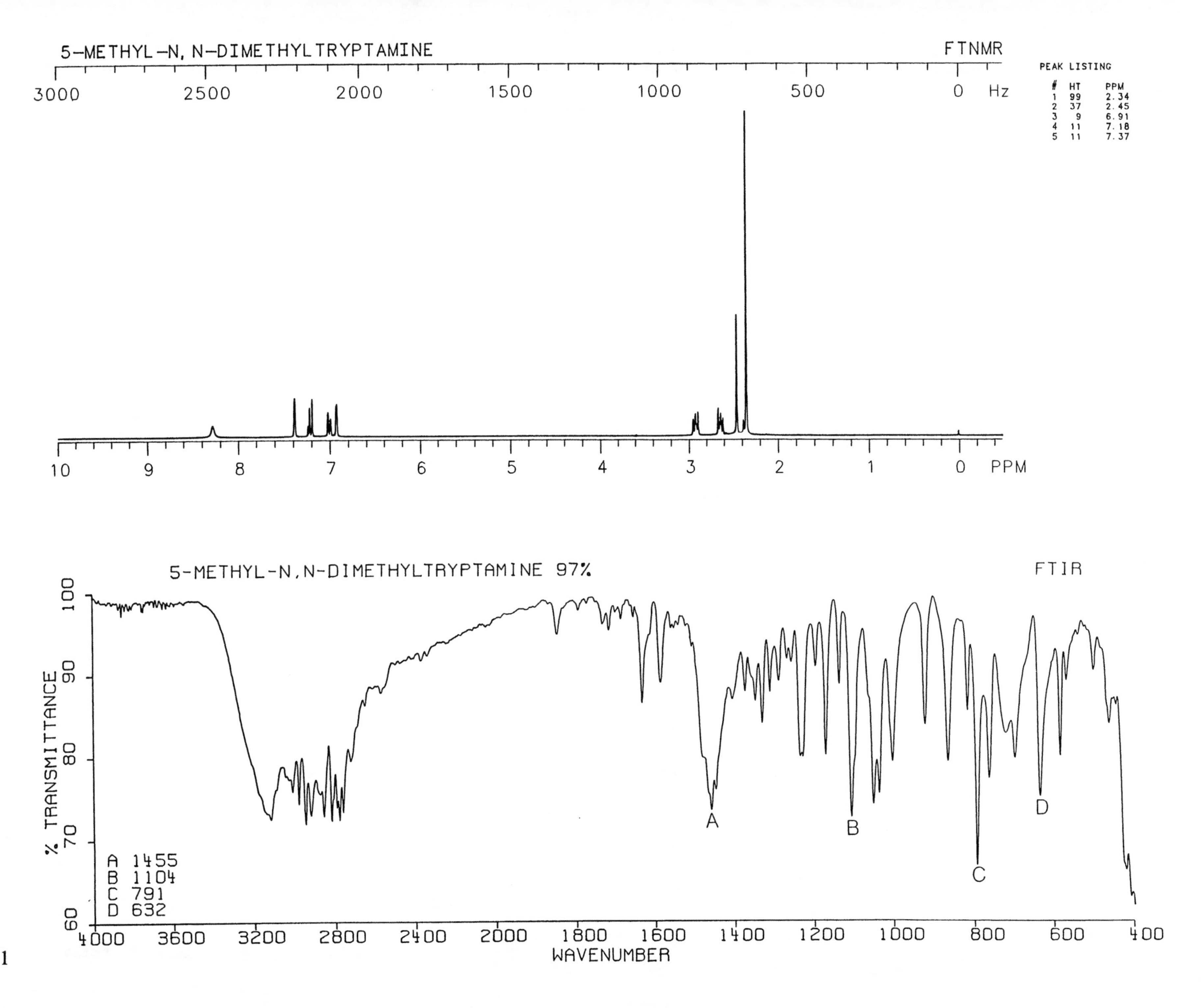
5-METHYL-N, N-DIMETHYLTRYPTAMINE
FTNMR
PEAK LISTING
# HT PPM
1 99 2.34
2 37 2.45
3 9 6.91
4 11 7.18
5 11 7.37
Hz
PPM
5-METHYL-N,N-DIMETHYLTRYPTAMINE 97%
FTIR
% TRANSMITTANCE
WAVENUMBER
A 1455
B 1104
C 791
D 632

# METHYLDOPA

$C_{10}H_{13}NO_4$

Molecular weight: 211.22 (211.08)
Synonyms: 3-Hydroxy-α-methyl-ℓ-tyrosine; α-methyldopa

Trade names: Aldomet, Aldoclor, Aldoril

Use: Antihypertensive
HPLC: Si-10; 50A;50B; 4.5
GC:

HO, HO–C6H3–$CH_2$–C($CH_3$)($NH_2$)–COOH

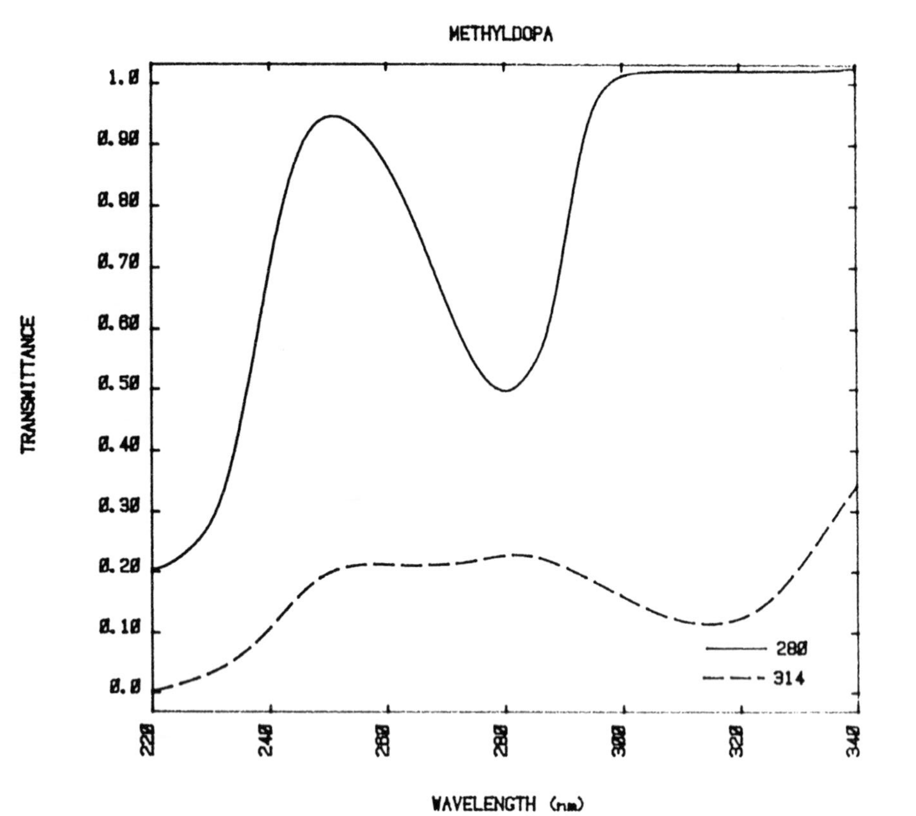


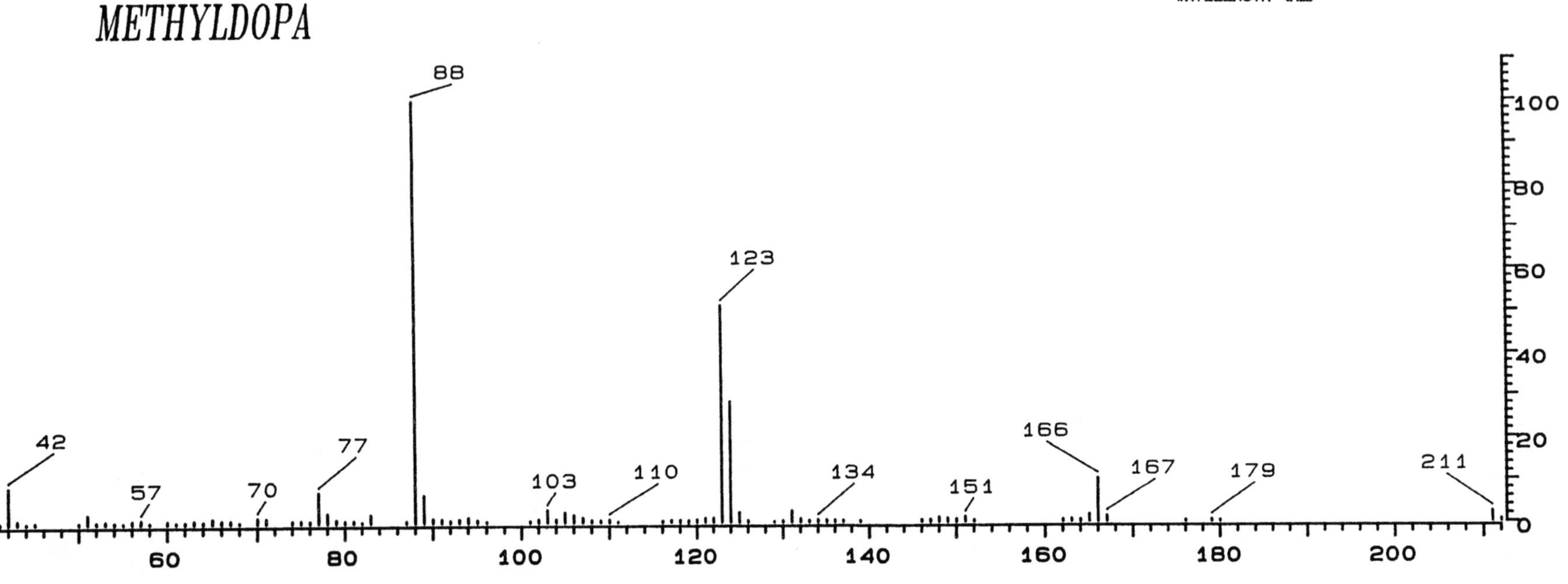

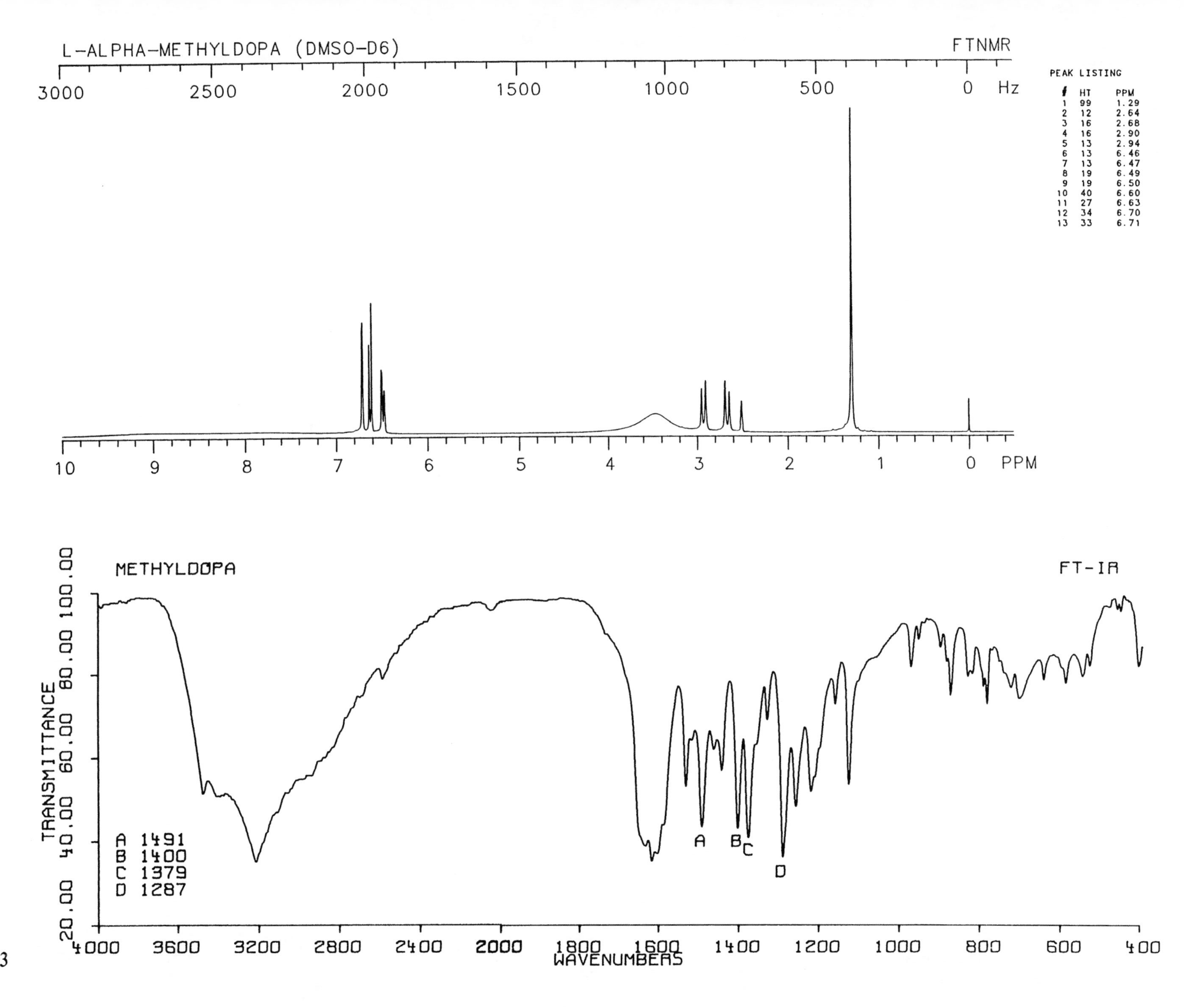

L-ALPHA-METHYLDOPA (DMSO-D6)
FTNMR
Hz
PPM
PEAK LISTING
# HT PPM
1 99 1.29
2 12 2.64
3 16 2.68
4 16 2.90
5 13 2.94
6 13 6.46
7 13 6.47
8 19 6.49
9 19 6.50
10 40 6.60
11 27 6.63
12 34 6.70
13 33 6.71
METHYLDOPA
FT-IR
TRANSMITTANCE
WAVENUMBERS
A 1491
B 1400
C 1379
D 1287

# 0-3-METHYLDOPAMINE

$C_9H_{13}NO_2$

Molecular weight: 167.21 (167.10)
Synonyms: 3-Methoxy-4-(2-aminoethyl)-1,2-benzenediol

Trade names:

Use: Metabolite of L-DOPA
HPLC:
GC: 1651; 200°C

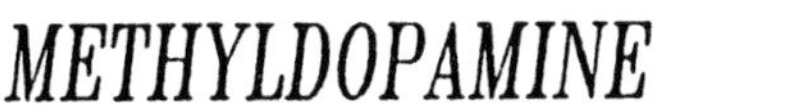


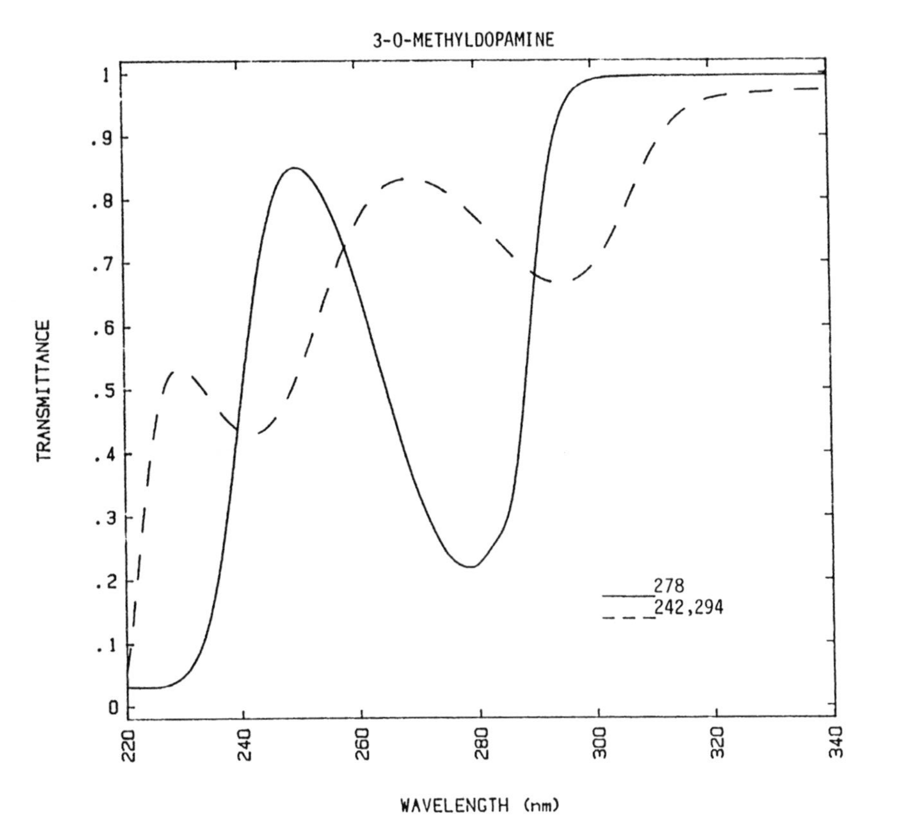


METHYLDOPAMINE

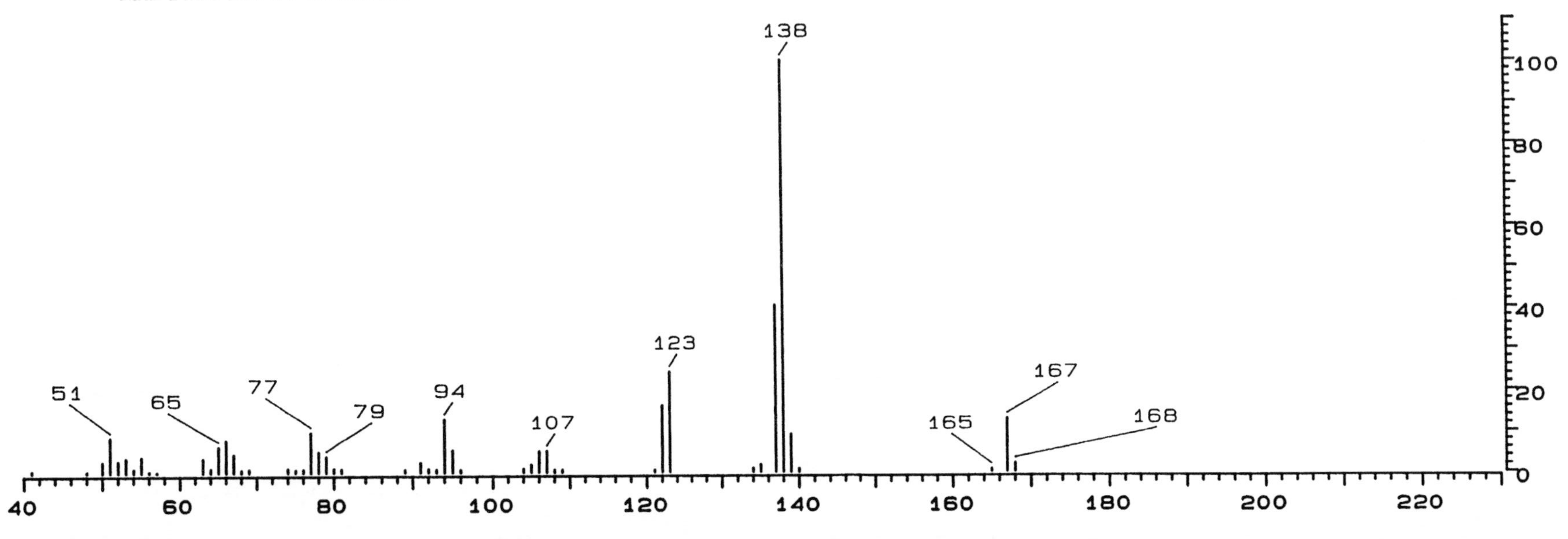

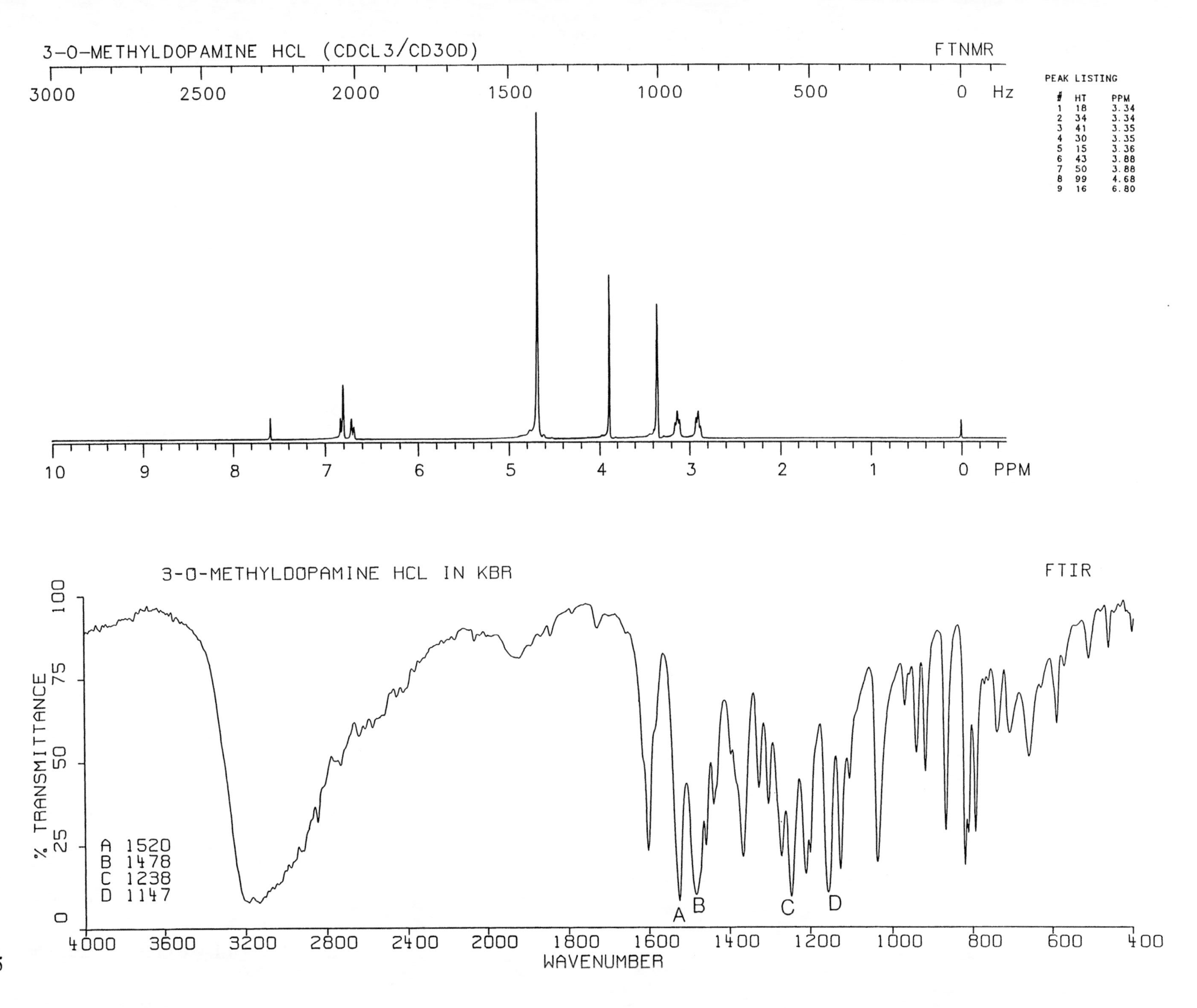
3-O-METHYLDOPAMINE HCL (CDCL3/CD3OD)
FTNMR
PEAK LISTING
# HT PPM
1 18 3.34
2 34 3.34
3 41 3.35
4 30 3.35
5 15 3.36
6 43 3.88
7 50 3.88
8 99 4.68
9 16 6.80
3-O-METHYLDOPAMINE HCL IN KBR
FTIR
A 1520
B 1478
C 1238
D 1147
% TRANSMITTANCE
WAVENUMBER
PPM
Hz

# METHYLDOPATE

$C_{12}H_{17}NO_4$

Molecular weight: 239.27 (239.12)
Synonyms: 3-Hydroxy-α-methyl-L-tyrosine ethyl ester

Trade names: Aldomet Ester HCl Injection

Use: Antihypertensive
HPLC:
GC:

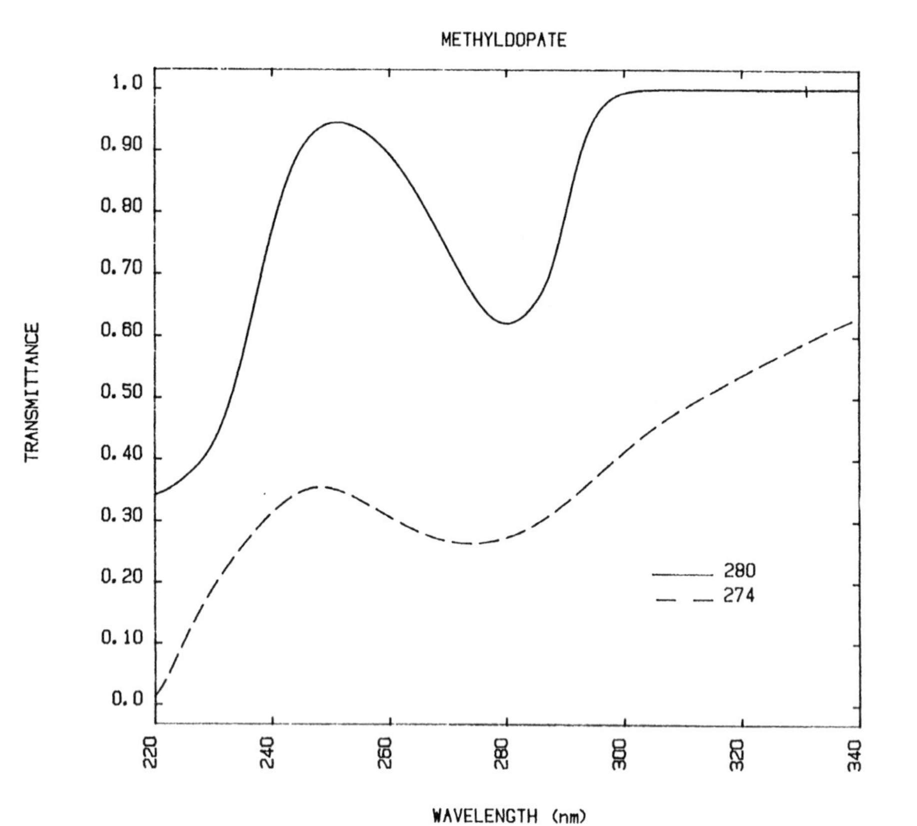


METHYLDOPATE--DIP

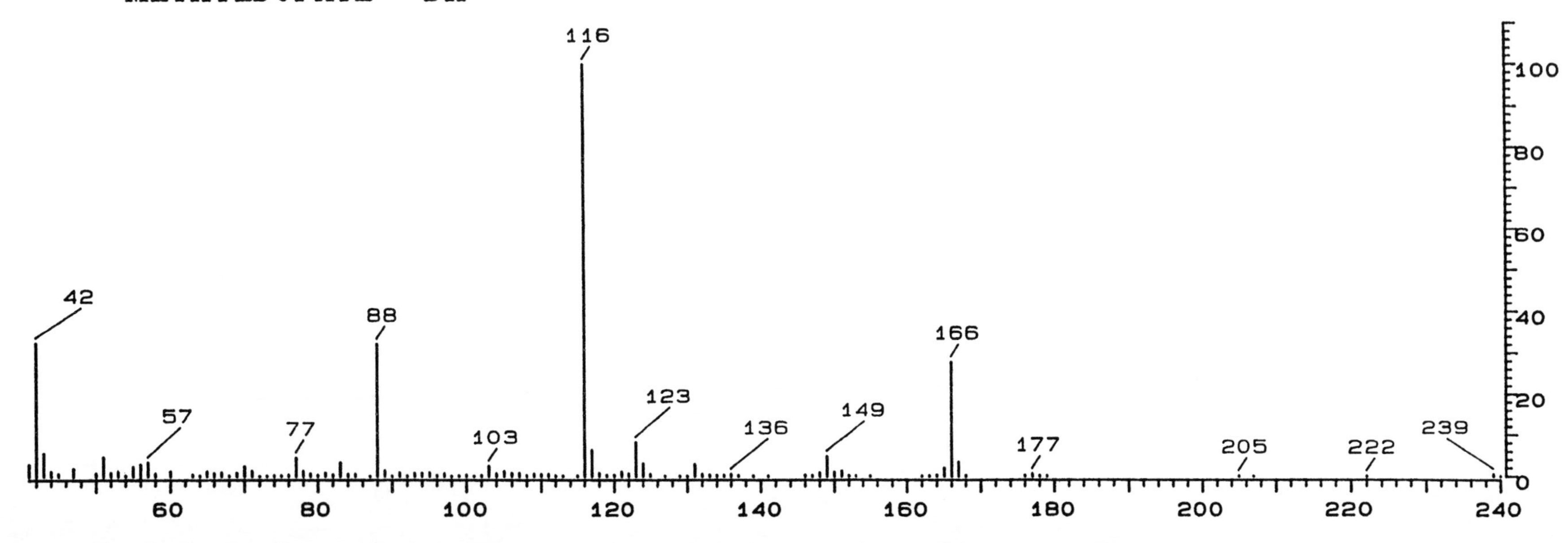

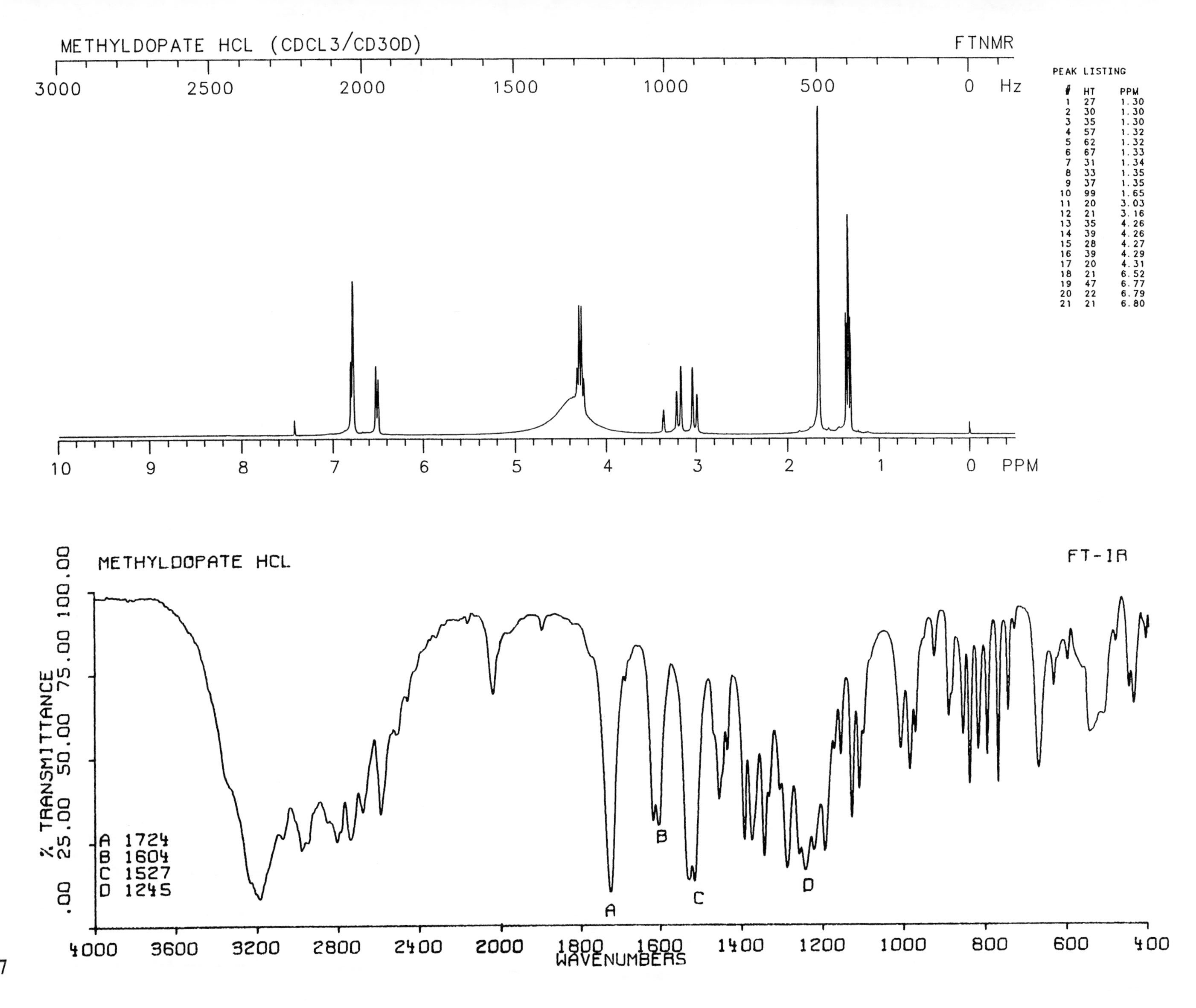
METHYLDOPATE HCL (CDCL3/CD3OD)
FTNMR
PEAK LISTING
# HT PPM
1 27 1.30
2 30 1.30
3 35 1.30
4 57 1.32
5 62 1.32
6 67 1.33
7 31 1.34
8 33 1.35
9 37 1.35
10 99 1.65
11 20 3.03
12 21 3.16
13 35 4.26
14 39 4.26
15 28 4.27
16 39 4.29
17 20 4.31
18 21 6.52
19 47 6.77
20 22 6.79
21 21 6.80
Hz
PPM
METHYLDOPATE HCL
FT-IR
% TRANSMITTANCE
WAVENUMBERS
A 1724
B 1604
C 1527
D 1245

# 4,4-METHYLENEDIANILINE

$C_{13}H_{14}N_2$

Molecular weight: 198.26 (198.12)
Synonyms: 4,4'-Methylenebis[benzenamine]; p,p'-diamino-diphenylmethane
Trade names:

Use: Synthesis
HPLC: Si-10; 100B; 14.0
GC: 2160; 250°C

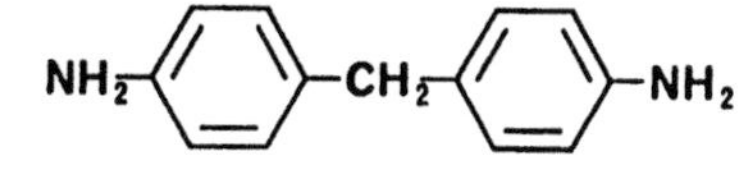


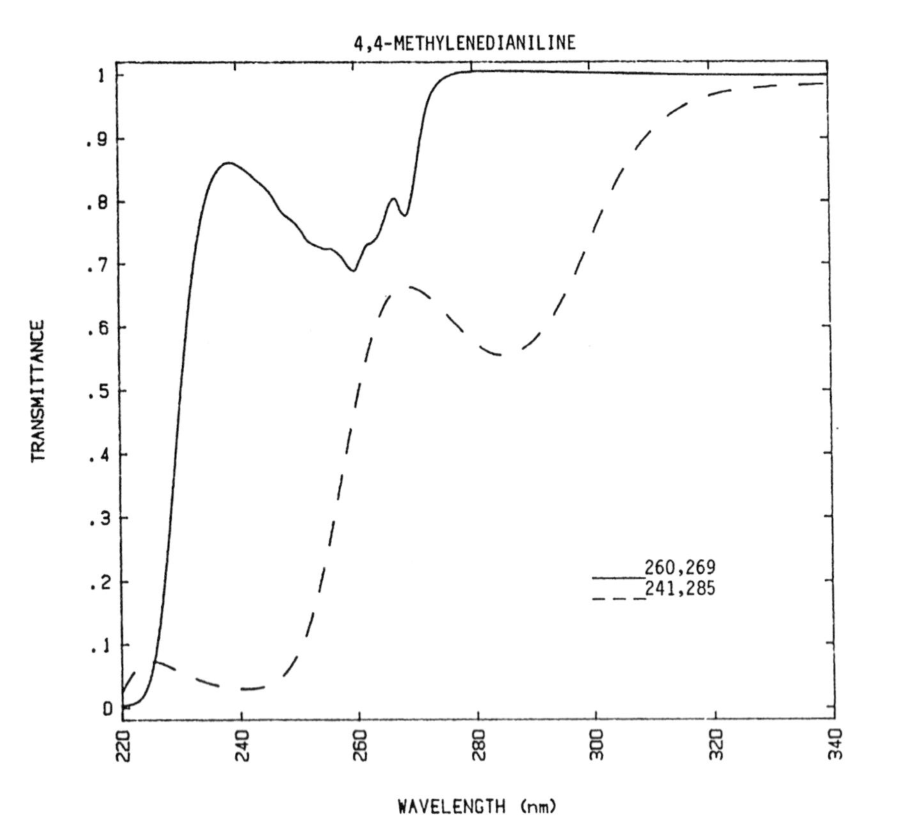

METHYLENEDIANILINE

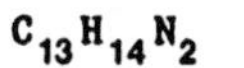

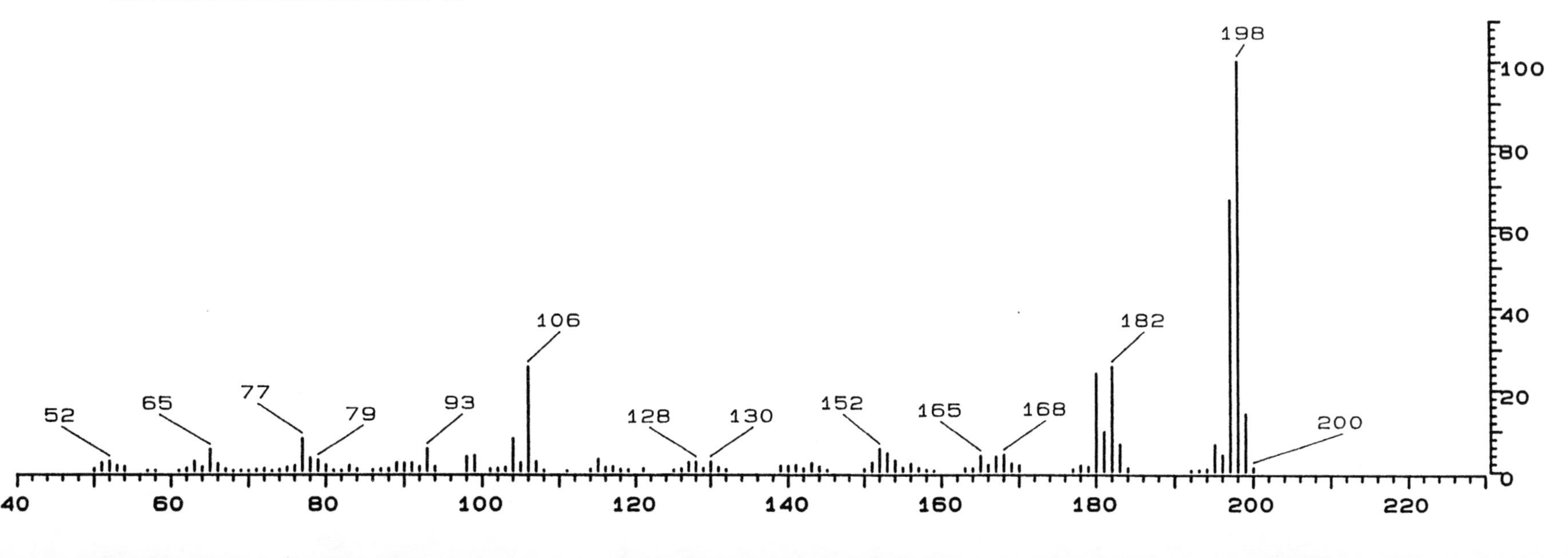

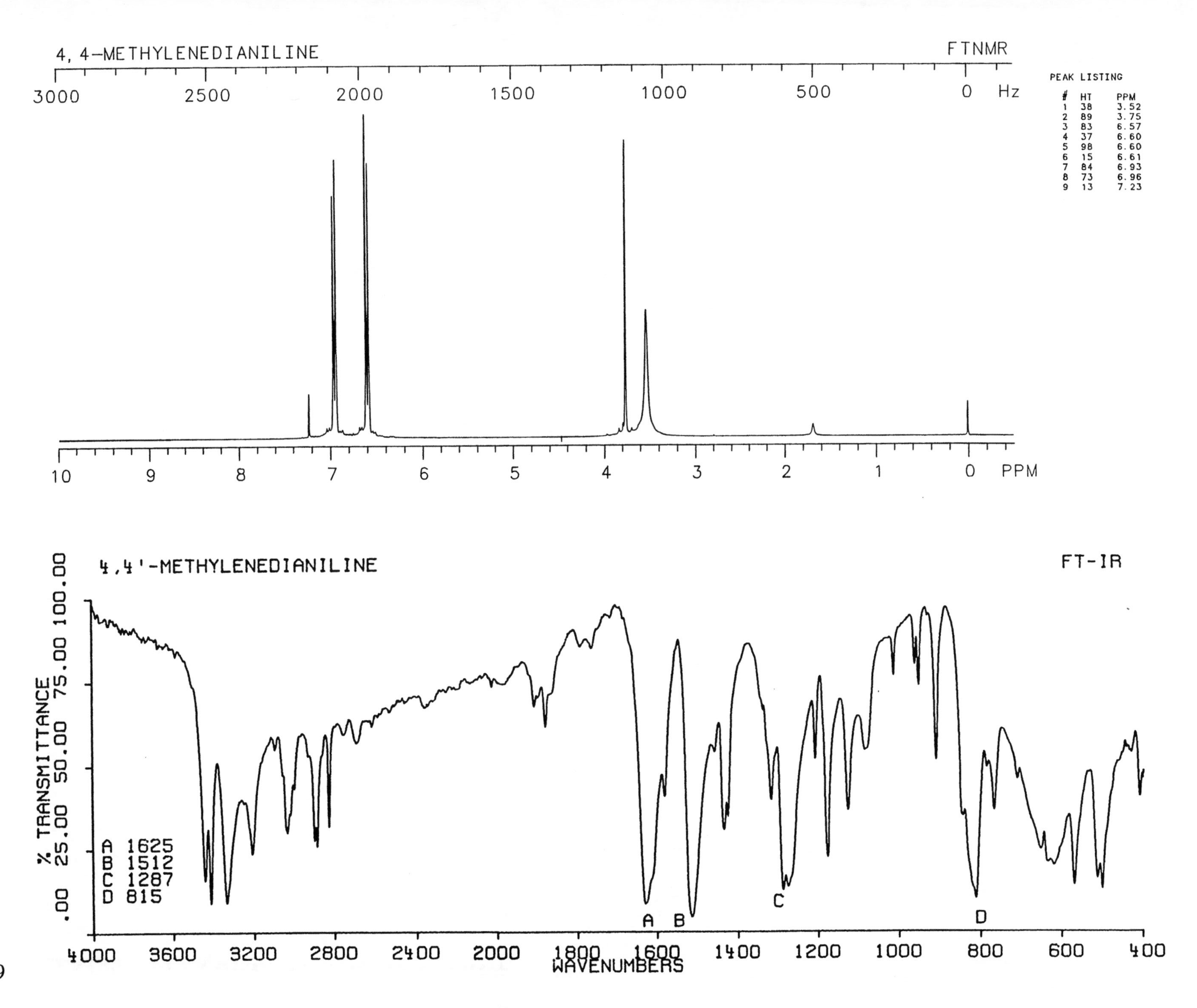

4, 4-METHYLENEDIANILINE
FTNMR
3000
2500
2000
1500
1000
500
0 Hz
PEAK LISTING
# HT PPM
1 38 3.52
2 89 3.75
3 83 6.57
4 37 6.60
5 98 6.60
6 15 6.61
7 84 6.93
8 73 6.96
9 13 7.23
10
9
8
7
6
5
4
3
2
1
0 PPM
4,4'-METHYLENEDIANILINE
FT-IR
% TRANSMITTANCE
.00
25.00
50.00
75.00
100.00
A 1625
B 1512
C 1287
D 815
A
B
C
D
4000
3600
3200
2800
2400
2000
1800
1600
1400
1200
1000
800
600
400
WAVENUMBERS

# 3,4-METHYLENEDIOXYAMPHETAMINE

$C_{10}H_{13}NO_2$

Molecular weight: 179.22 (179.10)
Synonyms: 5-(α-Methyl)ethanamine-1,3-benzodioxole; MDA

Trade names:

Use: Hallucinogen
HPLC: Si-10; 20A:80B; 4.2
GC: 1506; 200°C

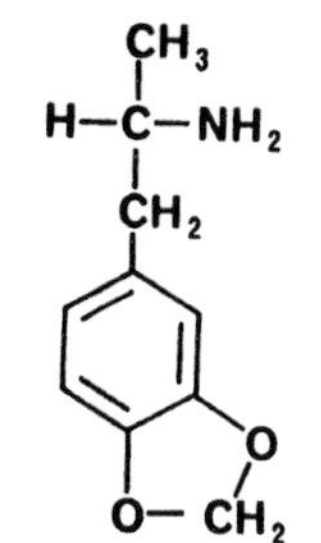


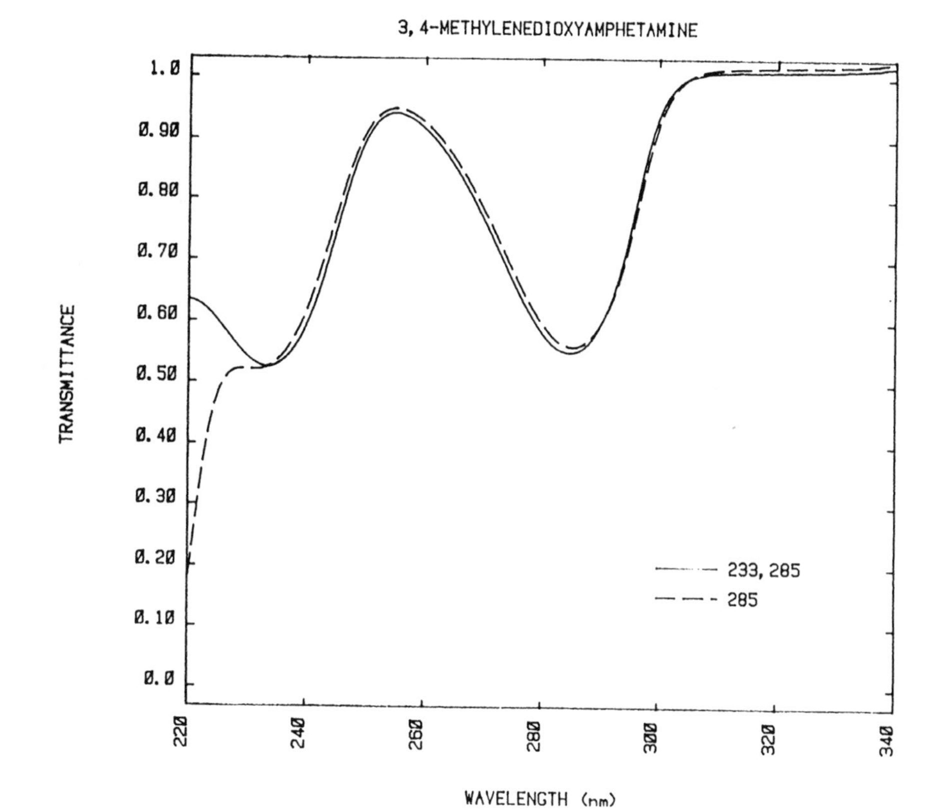

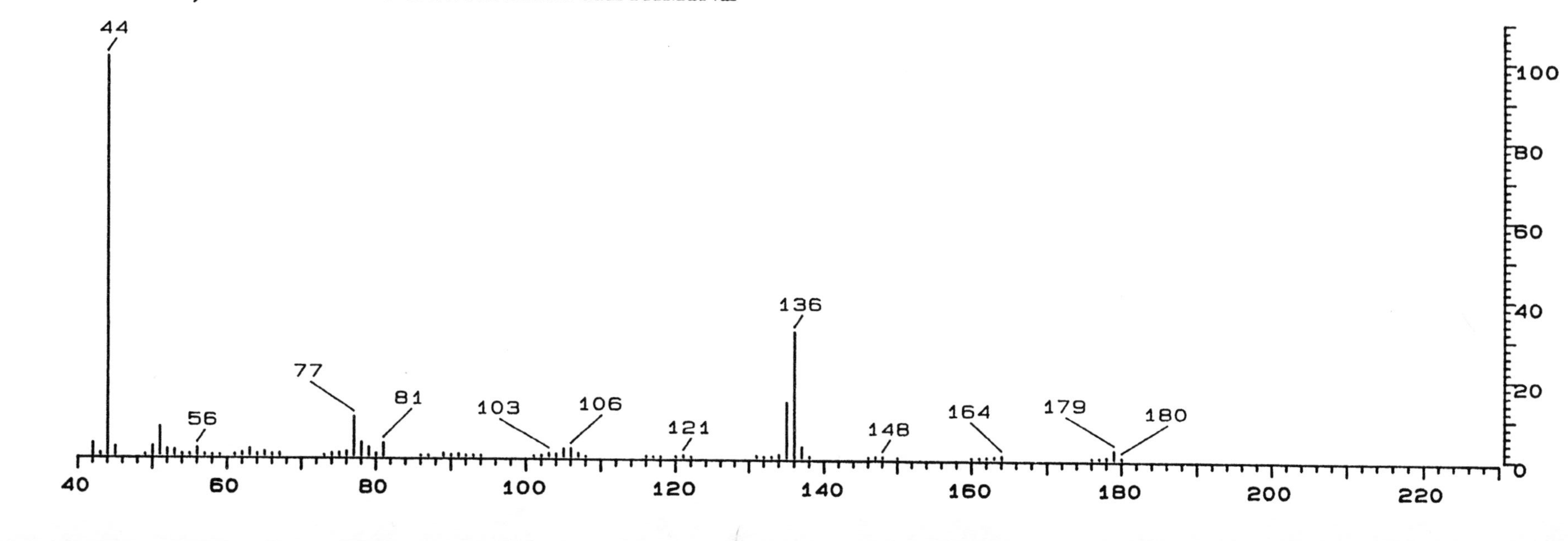

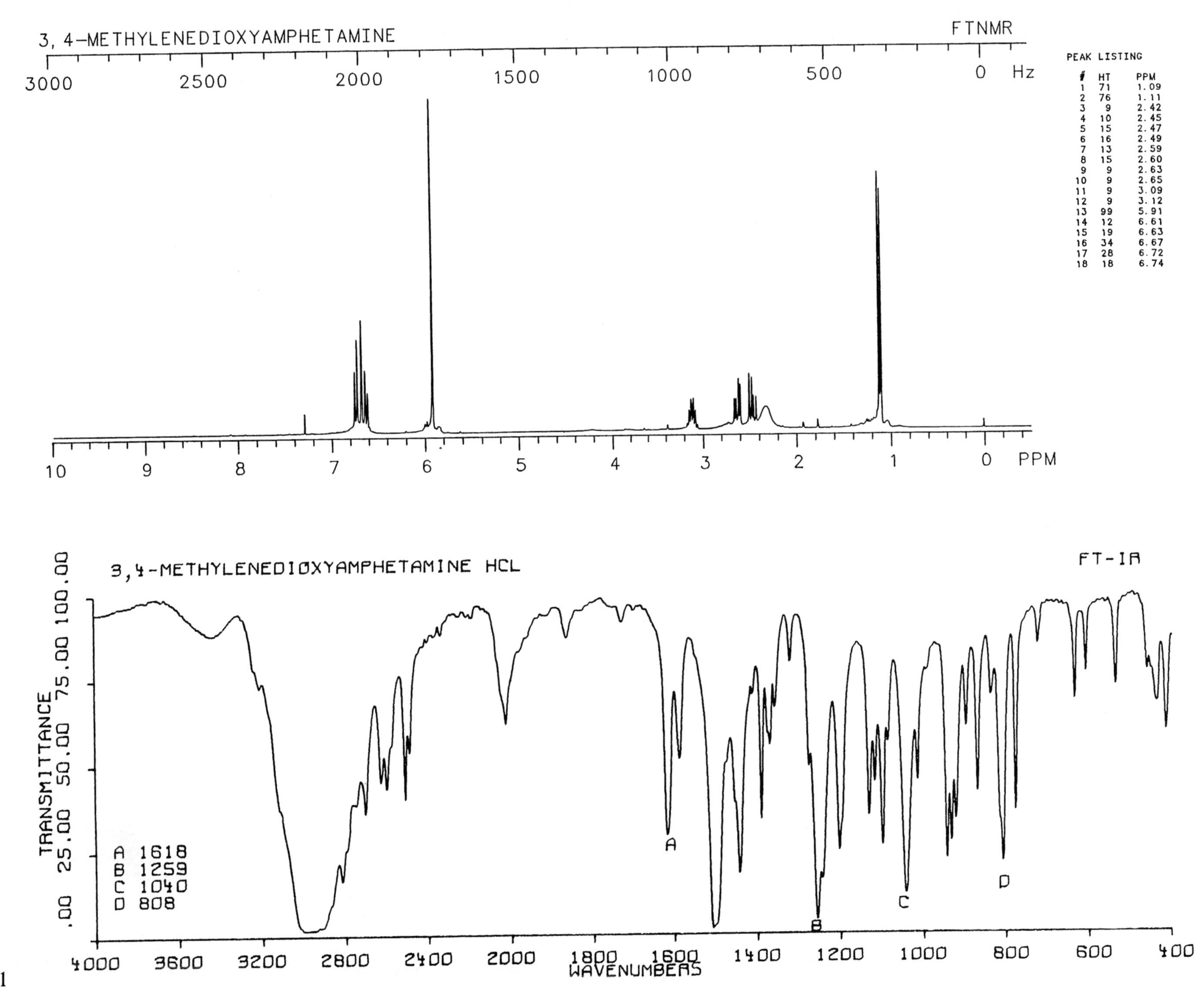
3, 4-METHYLENEDIOXYAMPHETAMINE
FTNMR
3000 2500 2000 1500 1000 500 0 Hz
PEAK LISTING
# HT PPM
1 71 1.09
2 76 1.11
3 9 2.42
4 10 2.45
5 15 2.47
6 16 2.49
7 13 2.59
8 15 2.60
9 9 2.63
10 9 2.65
11 9 3.09
12 9 3.12
13 99 5.91
14 12 6.61
15 19 6.63
16 34 6.67
17 28 6.72
18 18 6.74
10 9 8 7 6 5 4 3 2 1 0 PPM
3,4-METHYLENEDIOXYAMPHETAMINE HCL
FT-IR
TRANSMITTANCE
.00 25.00 50.00 75.00 100.00
A 1618
B 1259
C 1040
D 808
A
B
C
D
4000 3600 3200 2800 2400 2000 1800 1600 1400 1200 1000 800 600 400
WAVENUMBERS

# 3,4-METHYLENEDIOXYETHYLAMPHETAMINE

$C_{12}H_{18}NO_2$

**Molecular weight:** 208.27 (208.13)
**Synonyms:** 5-(α-ethyl)ethanamine-1,3-benzodioxole; N-ethyl MDA analog; MDEA
**Trade names:**

**Use:** Hallucinogen
**HPLC:** Si-10; 10A:90B; 3.3
**GC:** 1637; 200°C

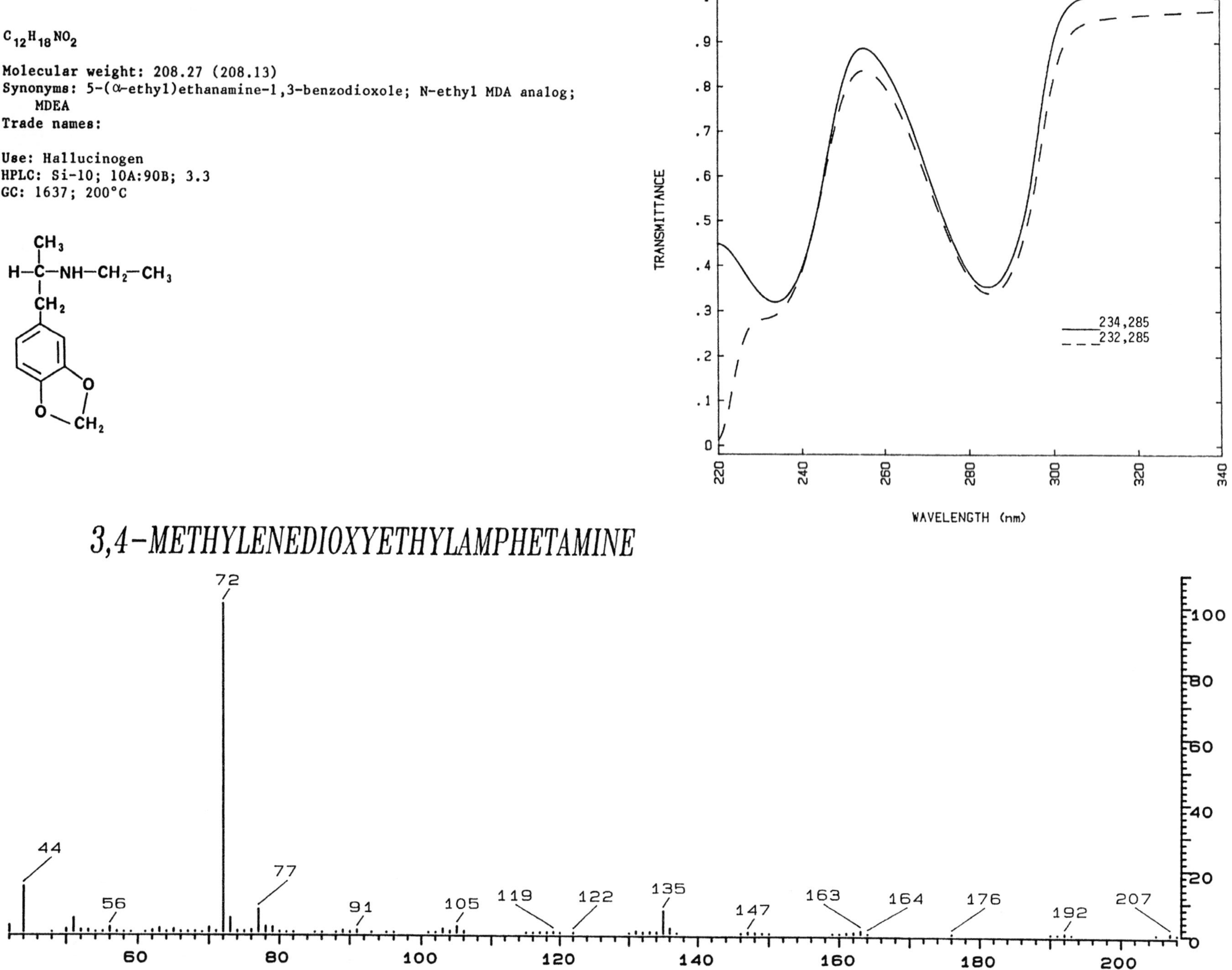

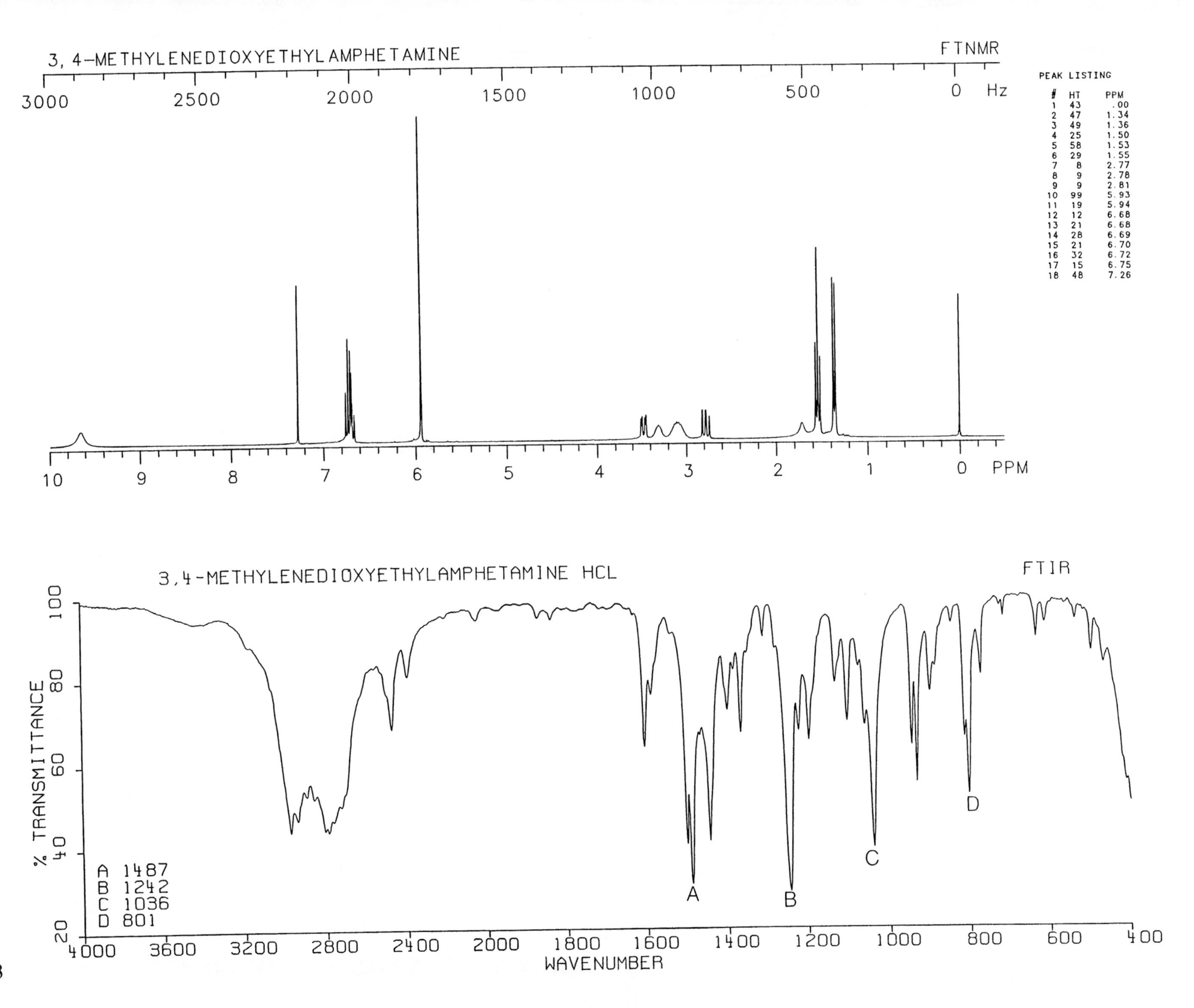
3, 4-METHYLENEDIOXYETHYLAMPHETAMINE
FTNMR
3000
2500
2000
1500
1000
500
0 Hz
10
9
8
7
6
5
4
3
2
1
0 PPM
PEAK LISTING
# HT PPM
1 43 .00
2 47 1.34
3 49 1.36
4 25 1.50
5 58 1.53
6 29 1.55
7 8 2.77
8 9 2.78
9 9 2.81
10 99 5.93
11 19 5.94
12 12 6.68
13 21 6.68
14 28 6.69
15 21 6.70
16 32 6.72
17 15 6.75
18 48 7.26
3,4-METHYLENEDIOXYETHYLAMPHETAMINE HCL
FTIR
% TRANSMITTANCE
100
80
60
40
20
A 1487
B 1242
C 1036
D 801
A
B
C
D
4000
3600
3200
2800
2400
2000
1800
1600
1400
1200
1000
800
600
400
WAVENUMBER

# 3,4-METHYLENEDIOXYMETHAMPHETAMINE

$C_{11}H_{15}NO_2$

Molecular weight: 193.23 (193.11)
Synonyms: 5-(N,α-Dimethyl)ethanamine-1,3-benzodioxole; N-methyl MDA analog; MDMA; Ecstasy; XTC
Trade names:

Use: Hallucinogen
HPLC: Si-10; 10A:90B; 4.4
GC: 1559; 200°C

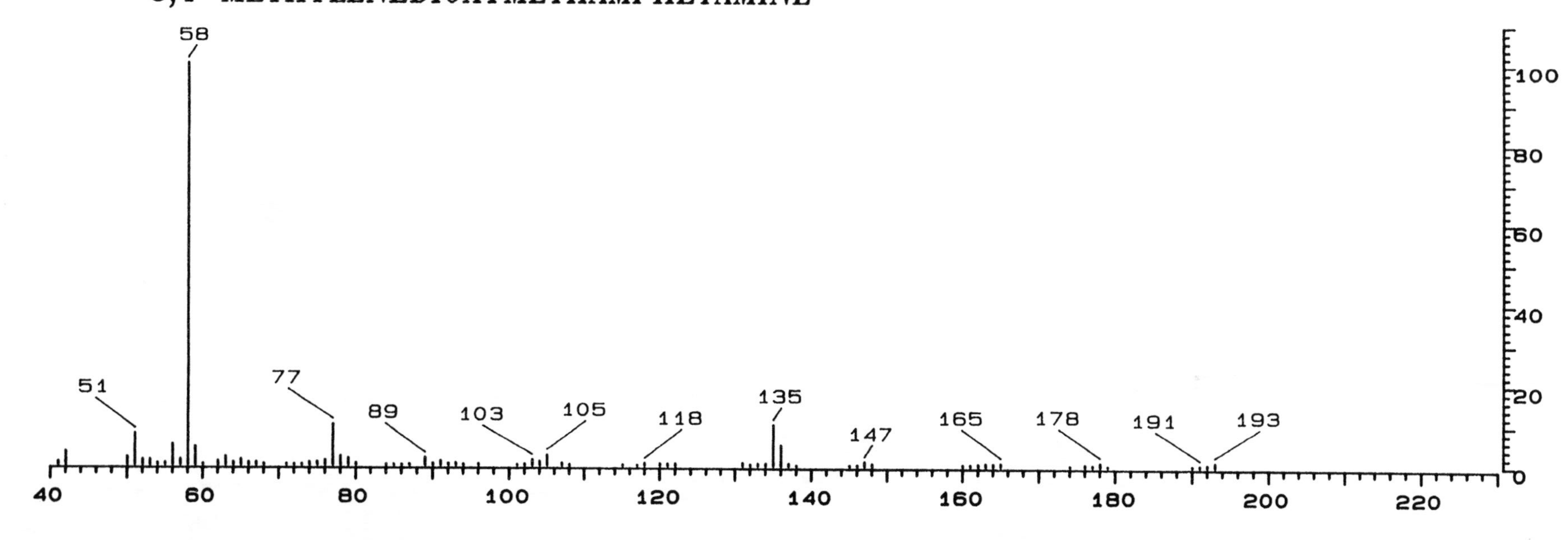


## 3,4-METHYLENEDIOXYMETHAMPHETAMINE

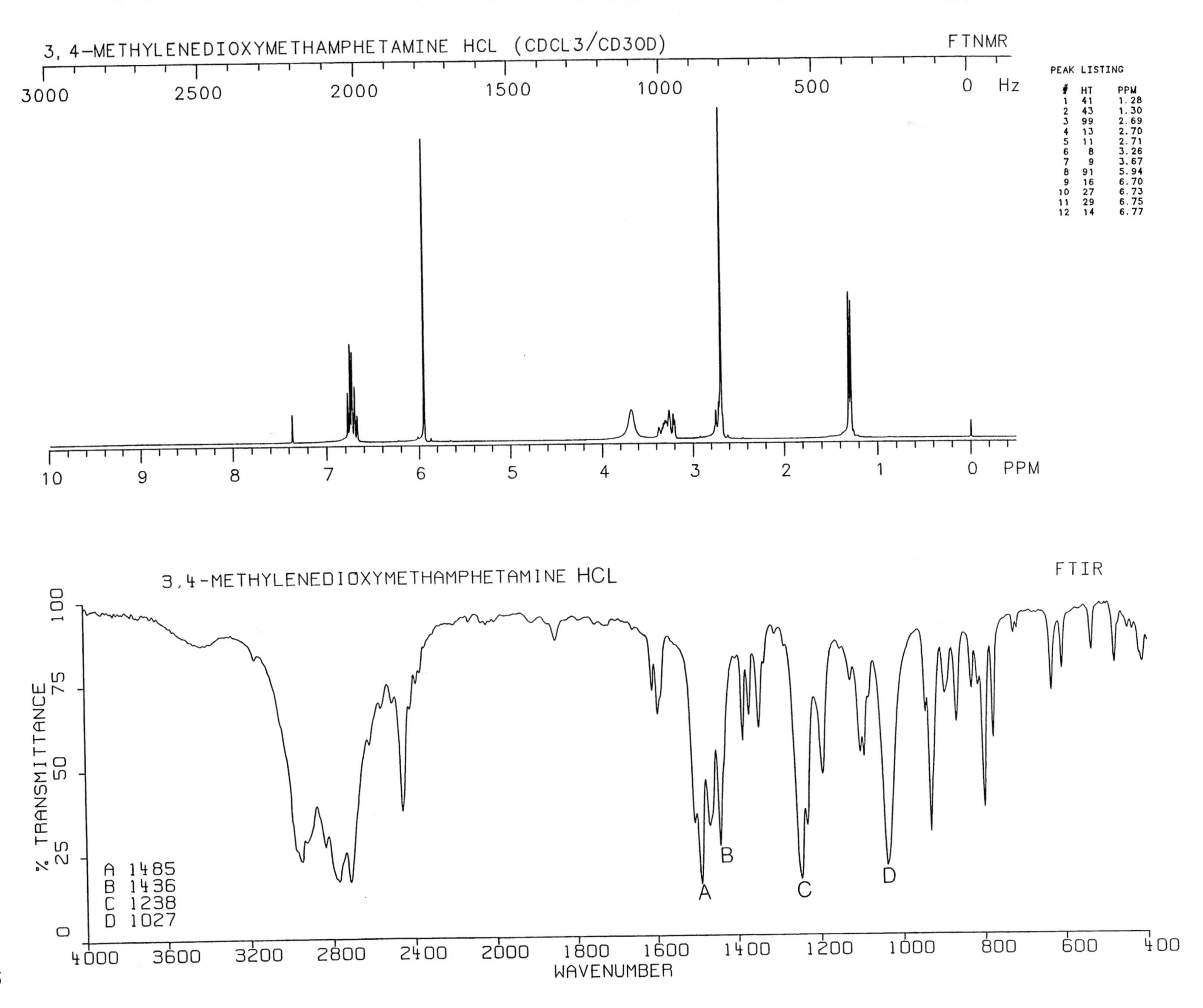
3, 4-METHYLENEDIOXYMETHAMPHETAMINE HCL (CDCL3/CD3OD)
FTNMR
3000
2500
2000
1500
1000
500
0 Hz
PEAK LISTING
# HT PPM
1 41 1.28
2 43 1.30
3 99 2.69
4 13 2.70
5 11 2.71
6 8 3.26
7 9 3.67
8 91 5.94
9 16 6.70
10 27 6.73
11 29 6.75
12 14 6.77
10
9
8
7
6
5
4
3
2
1
0 PPM
3,4-METHYLENEDIOXYMETHAMPHETAMINE HCL
FTIR
% TRANSMITTANCE
100
75
50
25
0
A 1485
B 1436
C 1238
D 1027
A
B
C
D
4000
3600
3200
2800
2400
2000
1800
1600
1400
1200
1000
800
600
400
WAVENUMBER

# 3,4-METHYLENEDIOXYPHENOL ACETONE

$C_{10}H_{10}O_3$

Molecular weight: 178.17 (178.06)
Synonyms:

Trade names:

Use: Synthesis
HPLC: Si-10; 20B:80C; 5.8
GC: 1499; 200°C

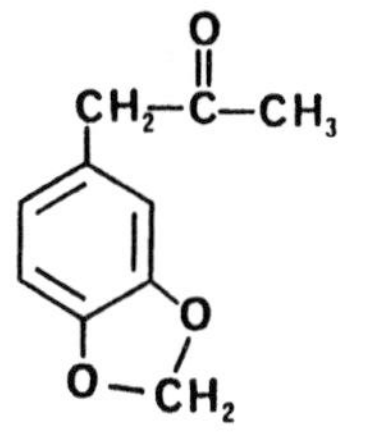


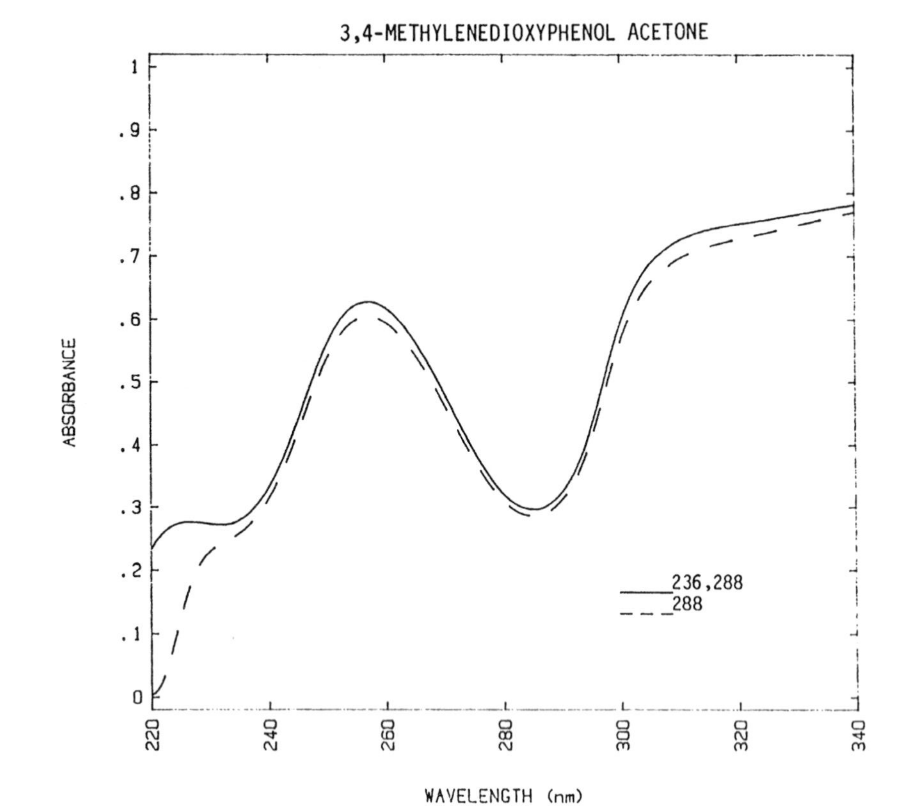

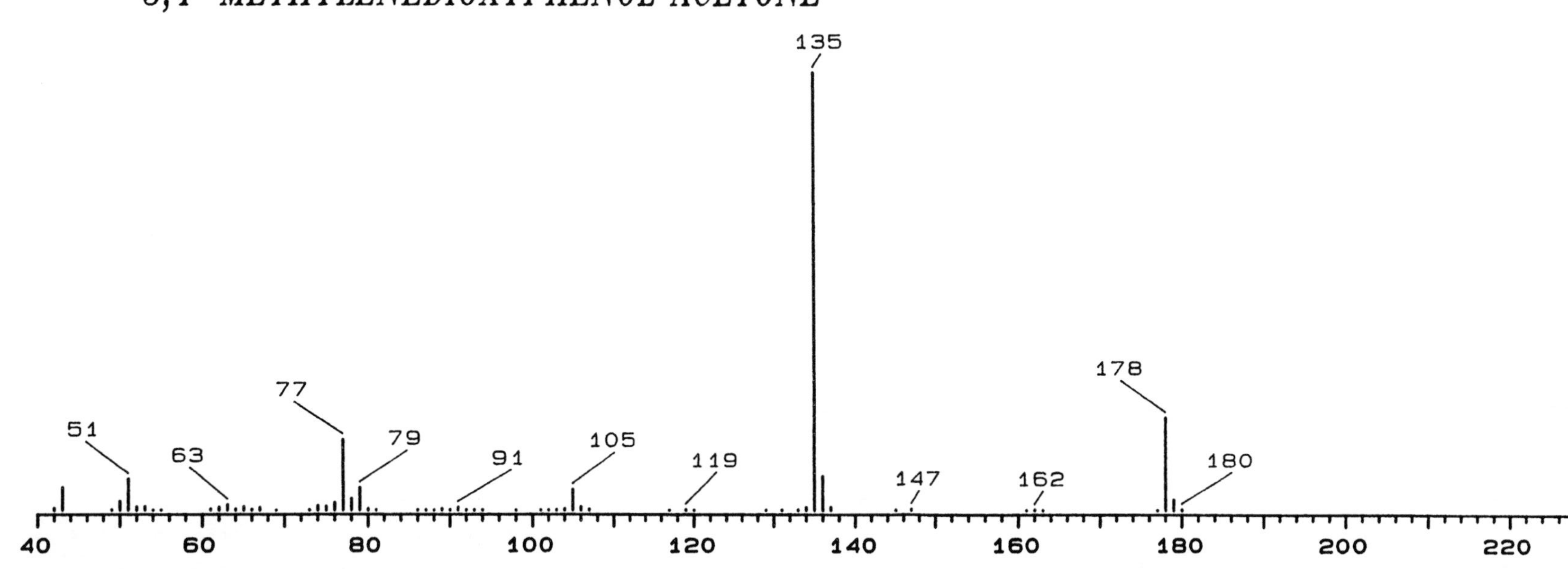

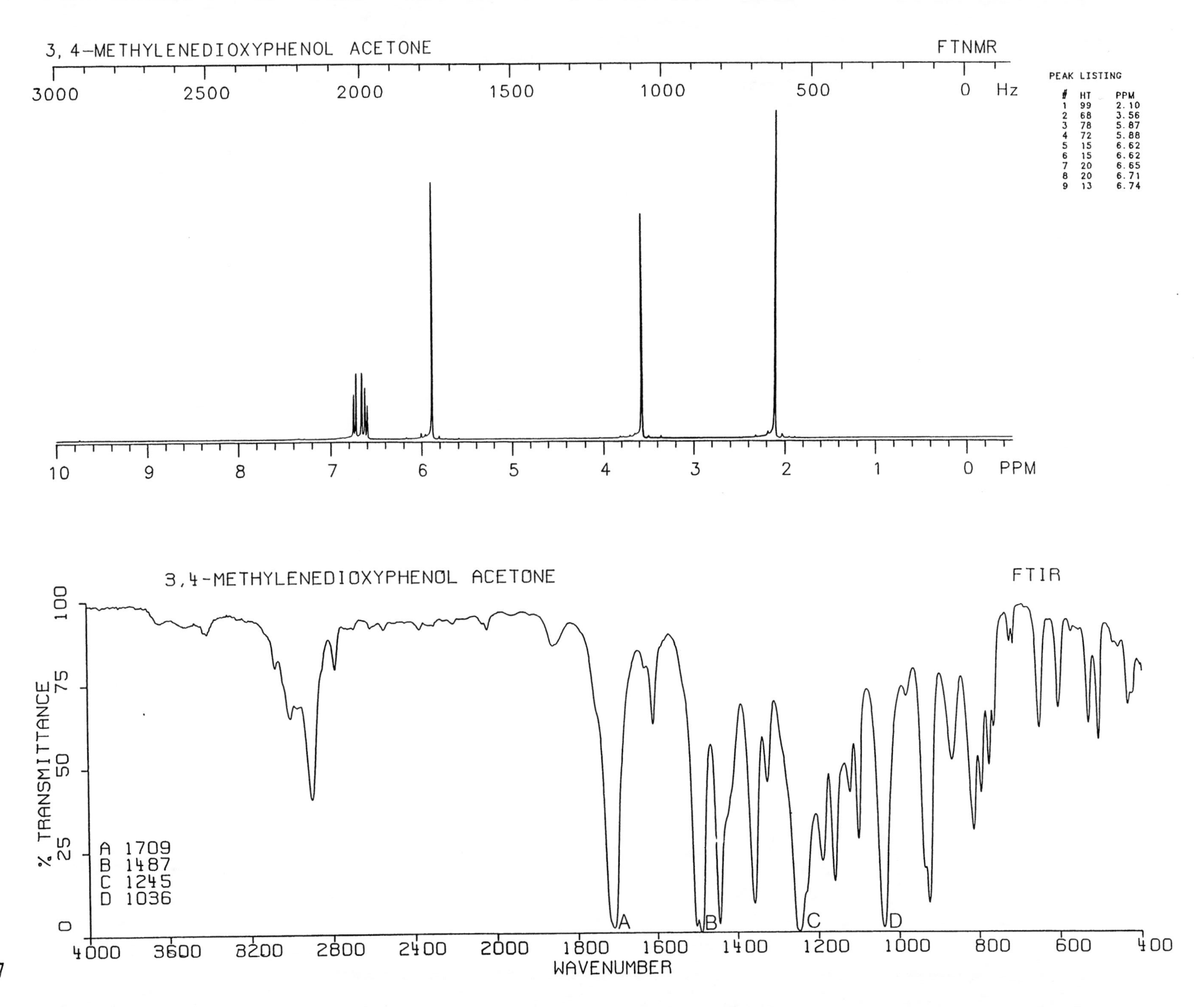
3, 4-METHYLENEDIOXYPHENOL ACETONE
FTNMR
3000 2500 2000 1500 1000 500 0 Hz
PEAK LISTING
# HT PPM
1 99 2.10
2 68 3.56
3 78 5.87
4 72 5.88
5 15 6.62
6 15 6.62
7 20 6.65
8 20 6.71
9 13 6.74
10 9 8 7 6 5 4 3 2 1 0 PPM
3,4-METHYLENEDIOXYPHENOL ACETONE
FTIR
% TRANSMITTANCE
0 25 50 75 100
A 1709
B 1487
C 1245
D 1036
4000 3600 3200 2800 2400 2000 1800 1600 1400 1200 1000 800 600 400
WAVENUMBER

# *N*-METHYLEPHEDRINE

$C_{11}H_{17}NO$

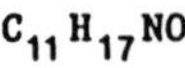

**Molecular weight:** 179.25 (179.13)
**Synonyms:** α-[1-(Dimethylamino)ethyl]benzenemethanol; N,N-dimethyl-norephrine; 2-dimethylamino-1-phenylpropanol
**Trade names:** Metheph, Pholcomed, Tybraine

**Use:** Analeptic
**HPLC:** Si-10; 20A:80B; 4.0
**GC:** 1418; 140°C

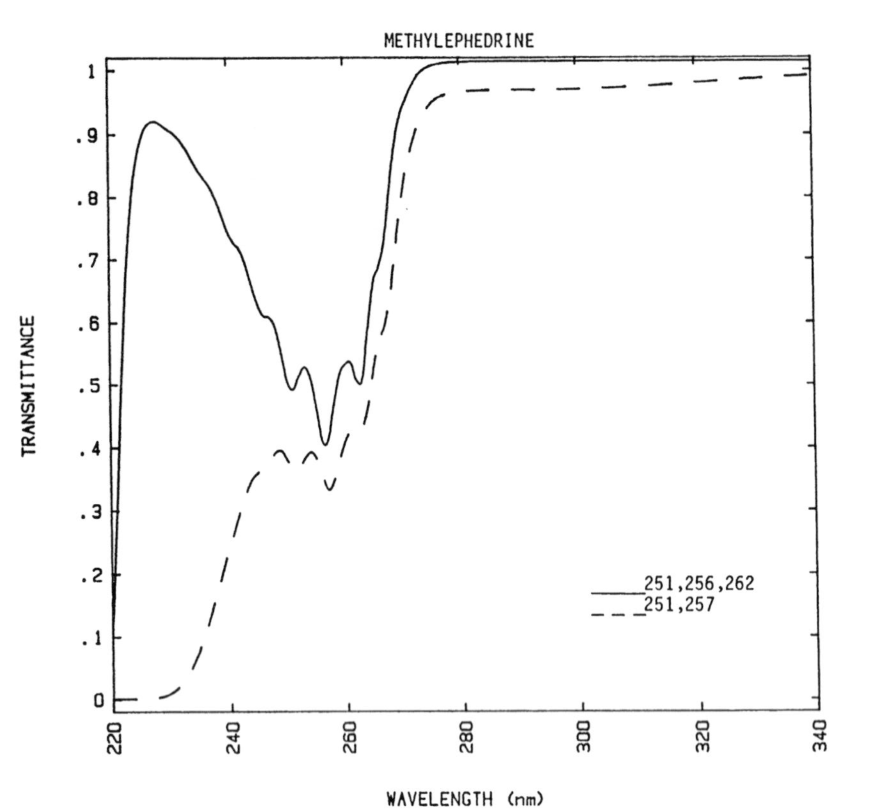


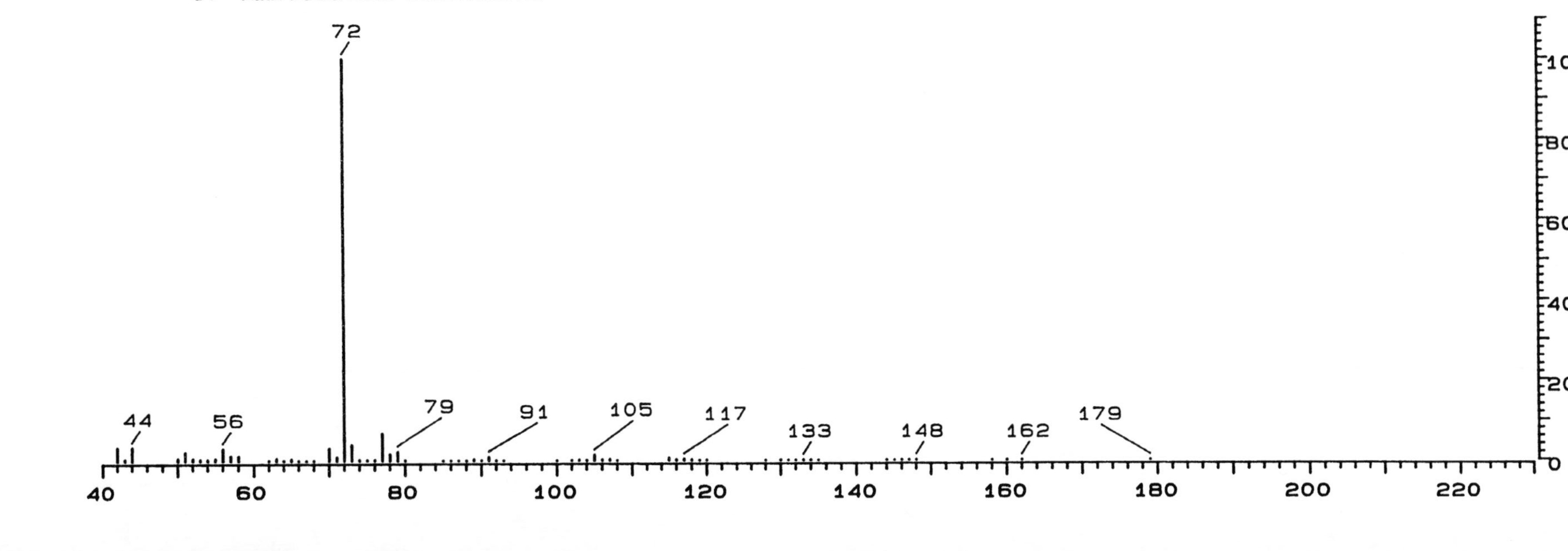

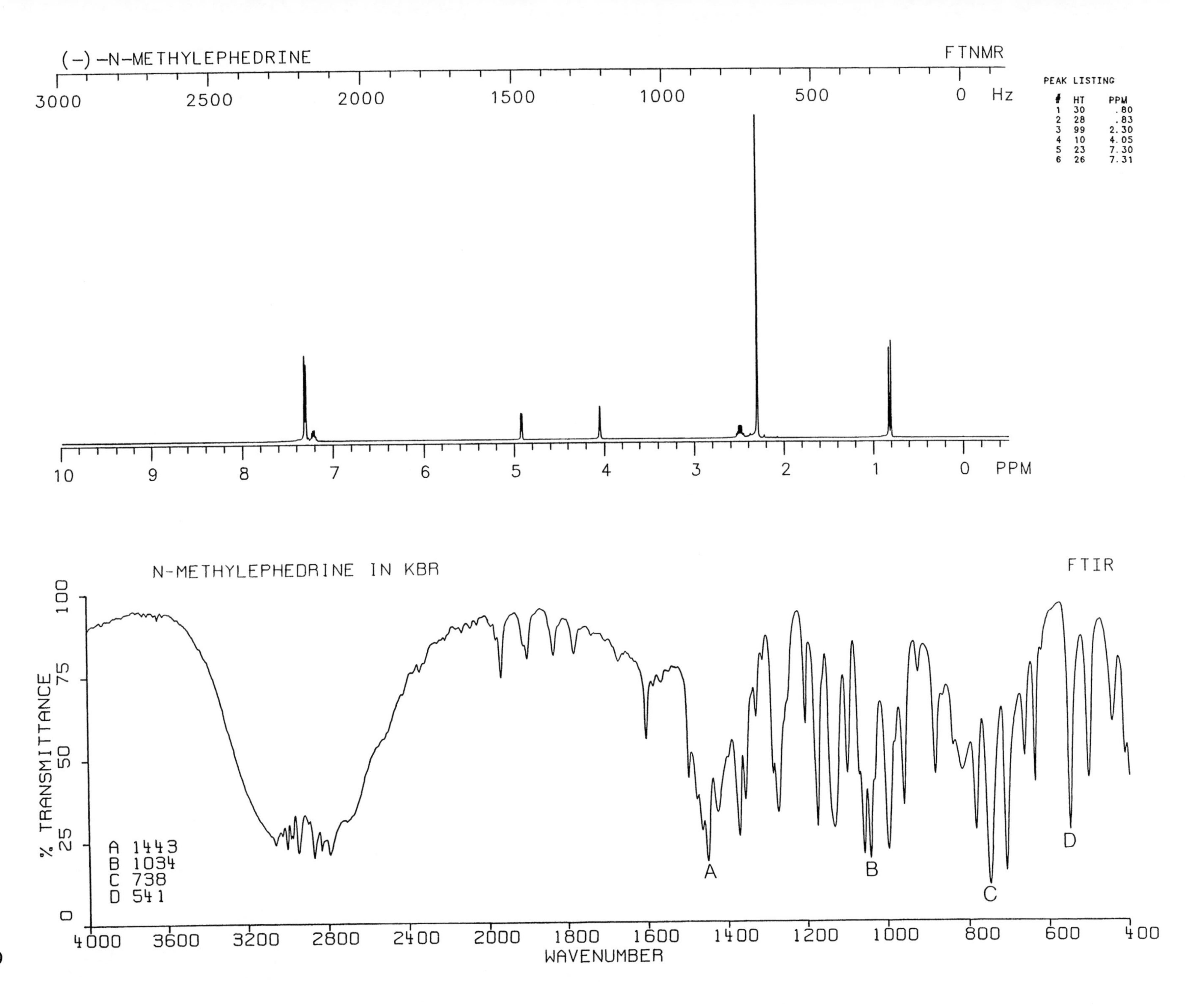

(−)-N-METHYLEPHEDRINE
FTNMR
3000 2500 2000 1500 1000 500 0 Hz
PEAK LISTING
# HT PPM
1 30 .80
2 28 .83
3 99 2.30
4 10 4.05
5 23 7.30
6 26 7.31
10 9 8 7 6 5 4 3 2 1 0 PPM
N-METHYLEPHEDRINE IN KBR
FTIR
% TRANSMITTANCE
0 25 50 75 100
A 1443
B 1034
C 738
D 541
A
B
C
D
4000 3600 3200 2800 2400 2000 1800 1600 1400 1200 1000 800 600 400
WAVENUMBER

# METHYLERGONOVINE

$C_{20}H_{25}N_3O_2$

**Molecular weight:** 339.42 (339.20)
**Synonyms:** 9,10-Didehydro-N-[1-(hydroxymethyl)propyl]-6-methylergoline-8-carboxamide; methylergometrine; methylergobasine
**Trade names:** Methergine

**Use:** Oxytocic
**HPLC:** Si-10; 5A:95B; 5.6
**GC:**

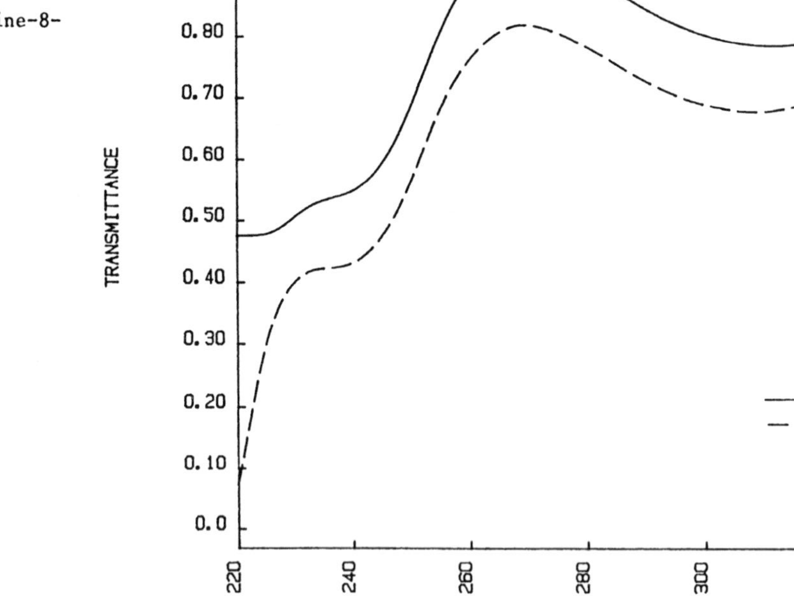

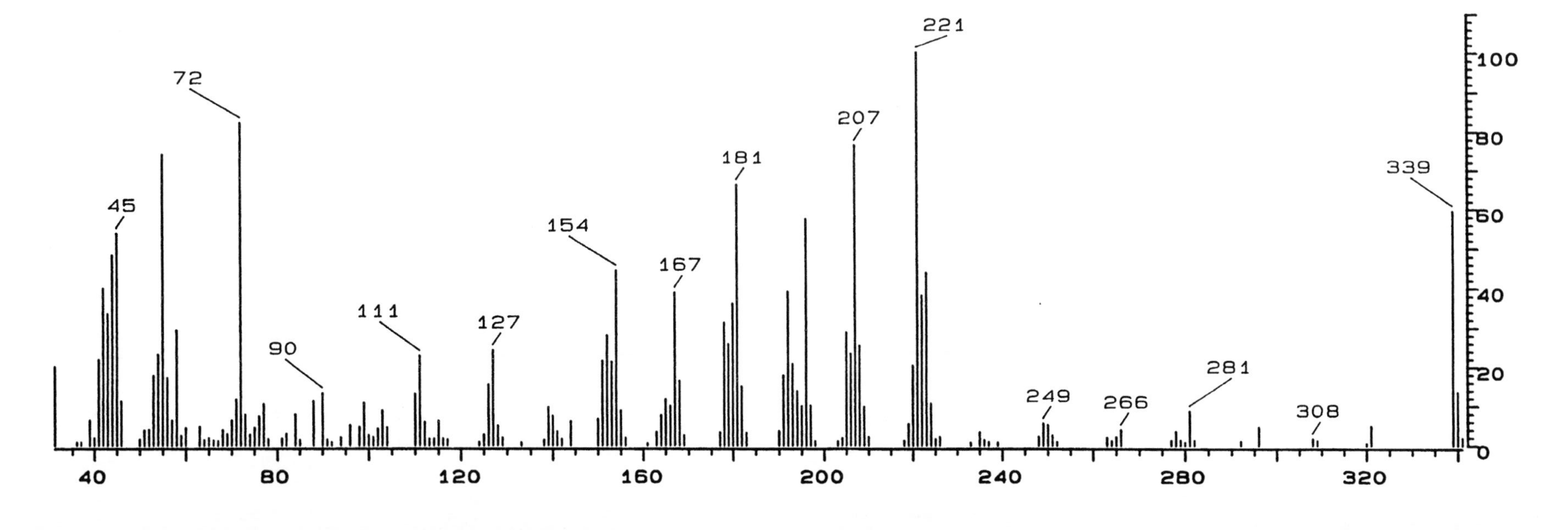

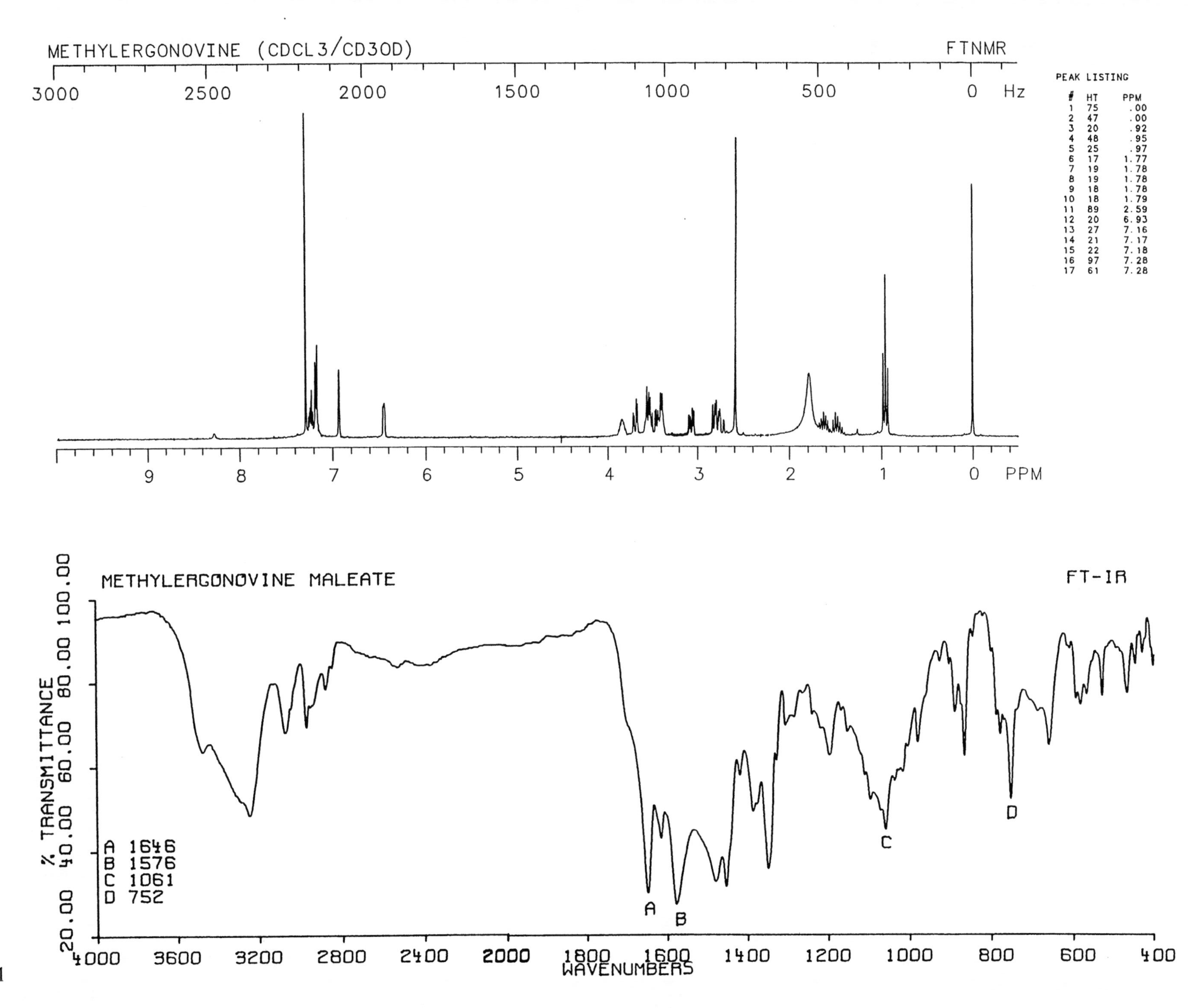
METHYLERGONOVINE (CDCL3/CD3OD)
FTNMR
PEAK LISTING
# HT PPM
1 75 .00
2 47 .00
3 20 .92
4 48 .95
5 25 .97
6 17 1.77
7 19 1.78
8 19 1.78
9 18 1.78
10 18 1.79
11 89 2.59
12 20 6.93
13 27 7.16
14 21 7.17
15 22 7.18
16 97 7.28
17 61 7.28
Hz
PPM
METHYLERGONOVINE MALEATE
FT-IR
A 1646
B 1576
C 1061
D 752
% TRANSMITTANCE
WAVENUMBERS

# α-METHYLFENTANYL

$C_{23}H_{30}N\ O_2$

**Molecular weight:** 350.50 (350.24)
**Synonyms:** N-Phenyl-N-[1-(α-methyl-2-phenylethyl)-4-piperidinyl]-propanamide; China White
**Trade names:**

**Use:** Hypnotic
**HPLC:** Si-10; 2A:98B; 4.5
**GC:** 2831; 280°C

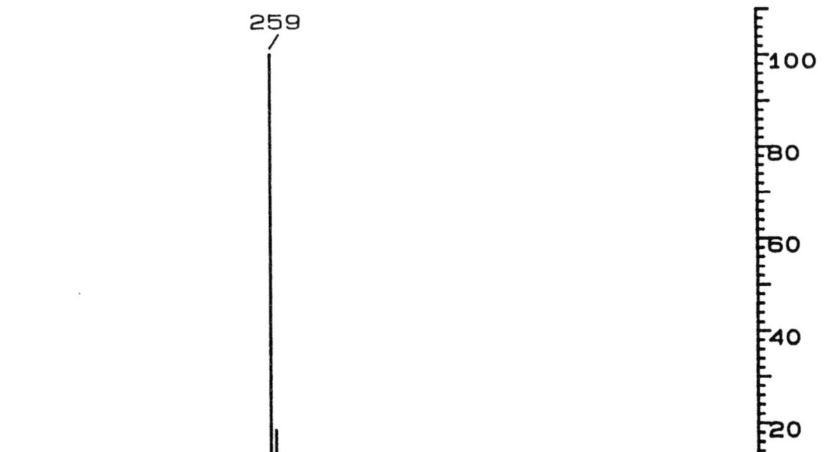

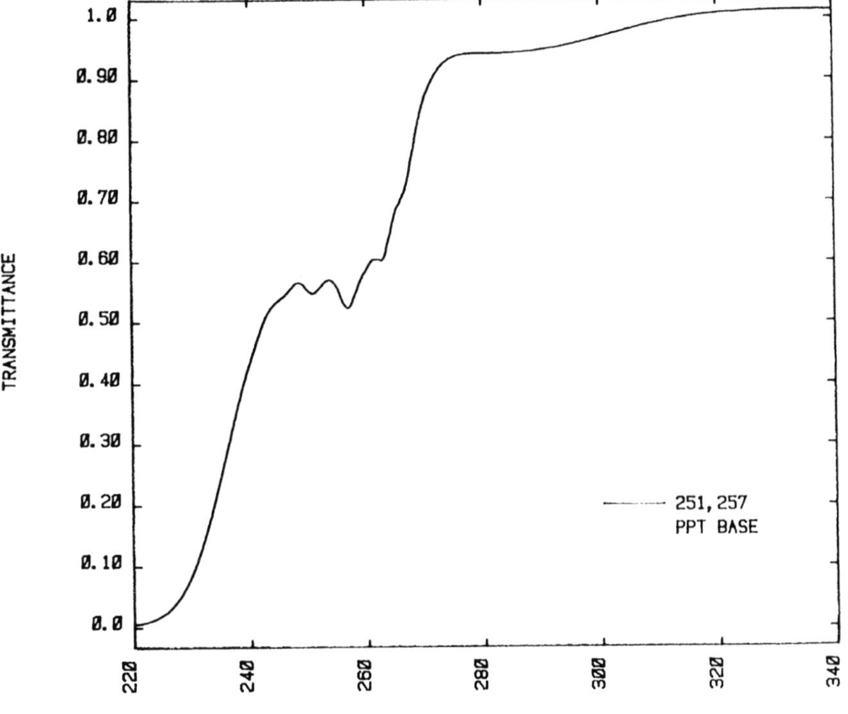

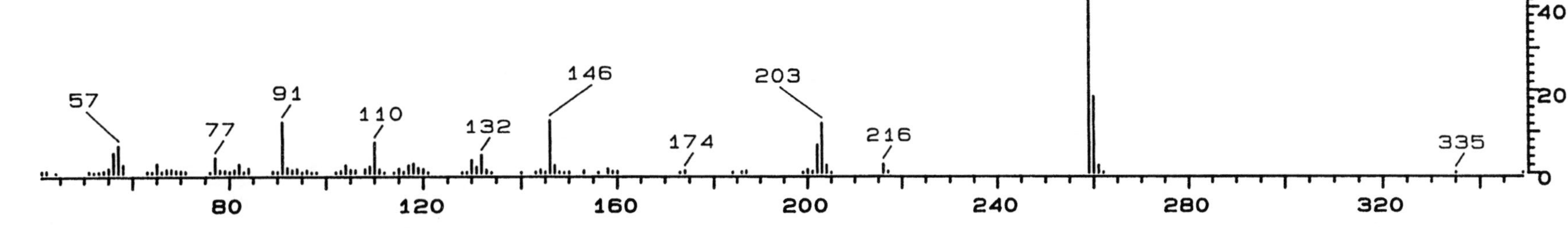

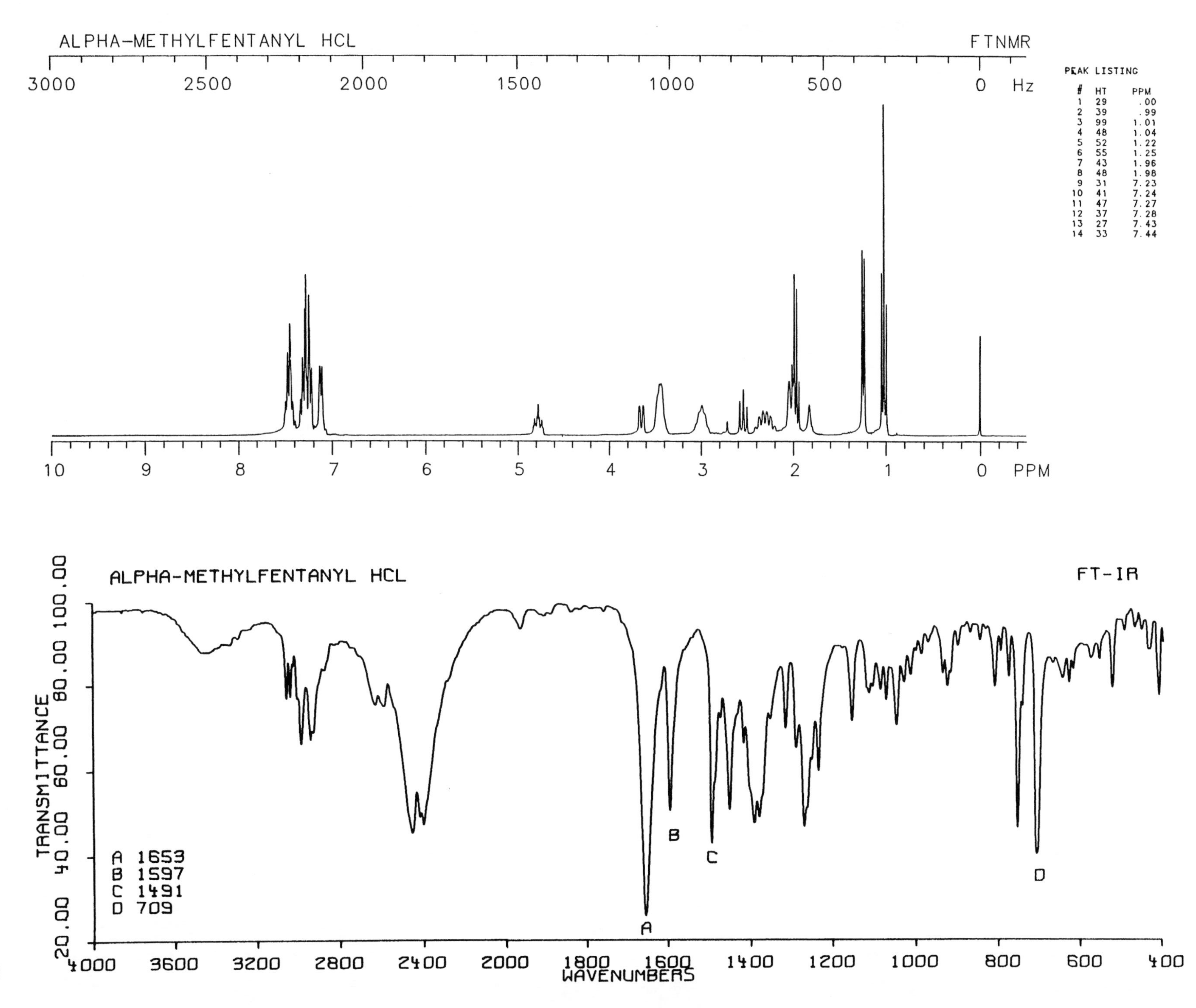
ALPHA-METHYLFENTANYL HCL
FTNMR
3000
2500
2000
1500
1000
500
0
Hz
PEAK LISTING
# HT PPM
1 29 .00
2 39 .99
3 99 1.01
4 48 1.04
5 52 1.22
6 55 1.25
7 43 1.96
8 48 1.98
9 31 7.23
10 41 7.24
11 47 7.27
12 37 7.28
13 27 7.43
14 33 7.44
10
9
8
7
6
5
4
3
2
1
0
PPM
ALPHA-METHYLFENTANYL HCL
FT-IR
TRANSMITTANCE
20.00
40.00
60.00
80.00
100.00
A 1653
B 1597
C 1491
D 709
A
B
C
D
4000
3600
3200
2800
2400
2000
1800
1600
1400
1200
1000
800
600
400
WAVENUMBERS

# 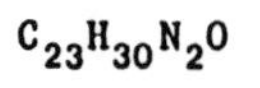-METHYLFENTANYL

$C_{23}H_{30}N_2O$

**Molecular weight:** 350.50 (350.24)
**Synonyms:** N-Phenyl-N[1-(β-methyl-2-phenylethyl)-4-piperidinyl]-propanamide
**Trade names:**

**Use:**
**HPLC:**
**GC:**

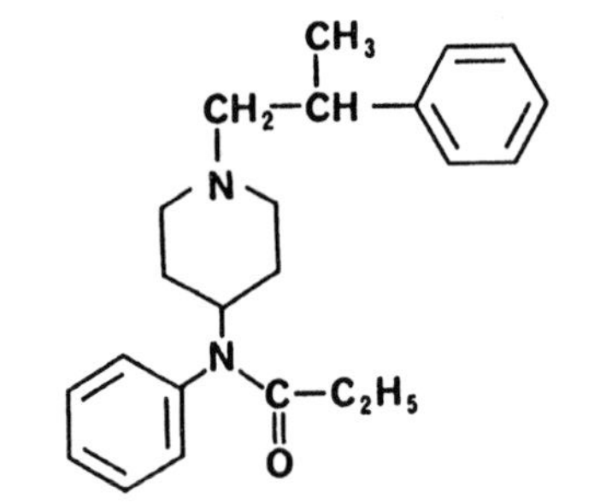


beta-METHYLFENTANYL

ETHANOL.....251,257,263

TRANSMITTANCE

WAVELENGTH (nm)

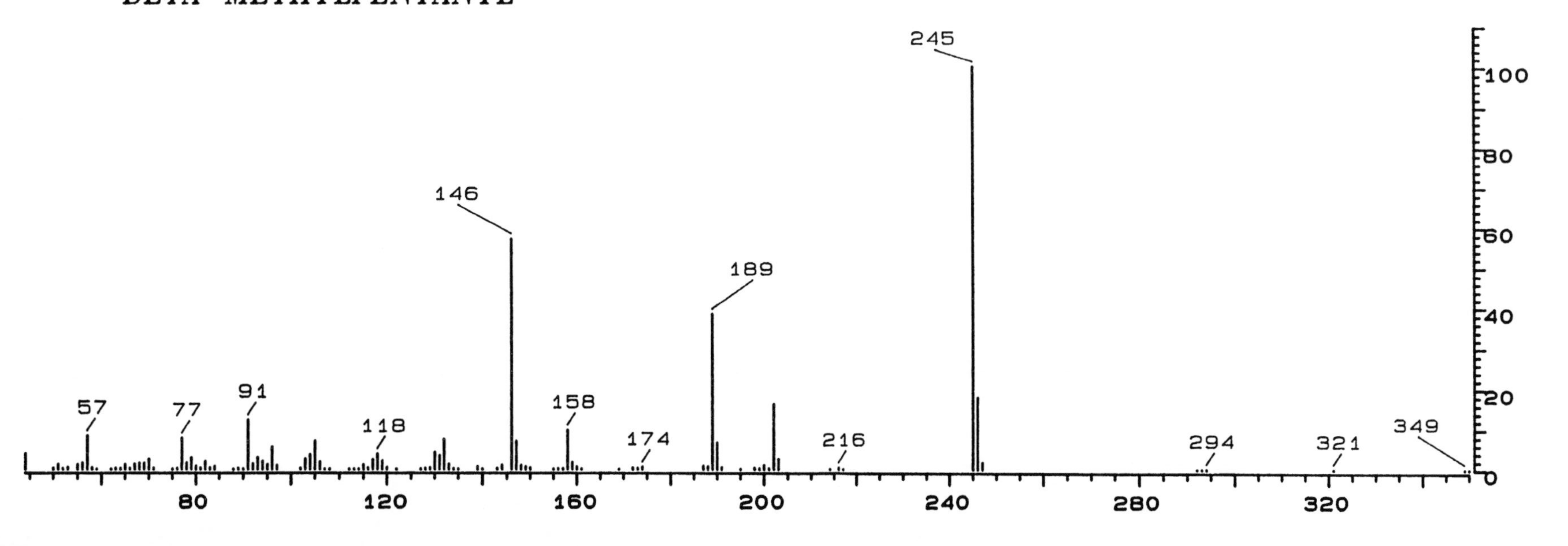

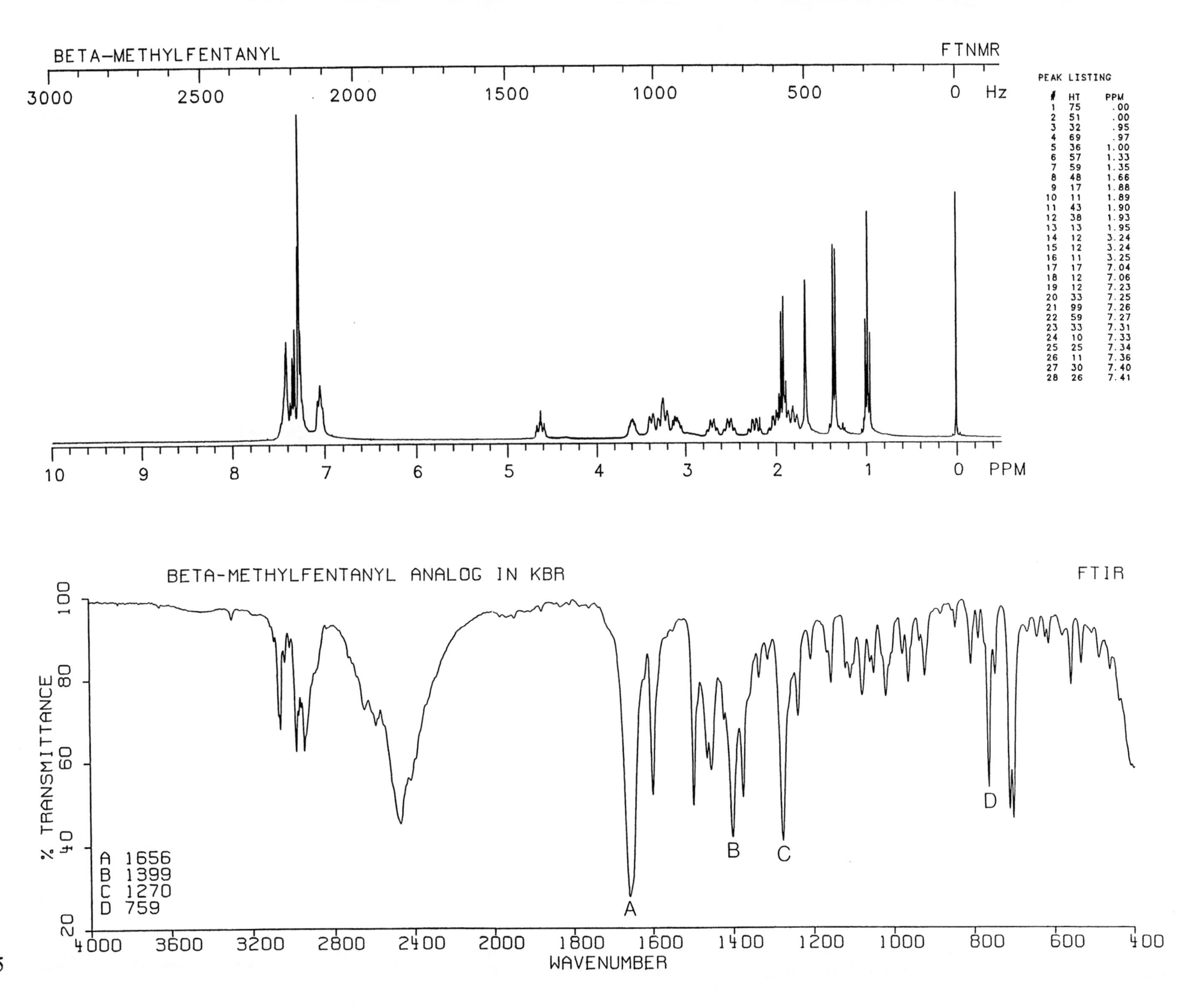

BETA-METHYLFENTANYL
FTNMR
3000 2500 2000 1500 1000 500 0 Hz
10 9 8 7 6 5 4 3 2 1 0 PPM
PEAK LISTING
# HT PPM
1 75 .00
2 51 .00
3 32 .95
4 69 .97
5 36 1.00
6 57 1.33
7 59 1.35
8 48 1.66
9 17 1.88
10 11 1.89
11 43 1.90
12 38 1.93
13 13 1.95
14 12 3.24
15 12 3.24
16 11 3.25
17 17 7.04
18 12 7.06
19 12 7.23
20 33 7.25
21 99 7.26
22 59 7.27
23 33 7.31
24 10 7.33
25 25 7.34
26 11 7.36
27 30 7.40
28 26 7.41
BETA-METHYLFENTANYL ANALOG IN KBR
FTIR
% TRANSMITTANCE
100 80 60 40 20
A 1656
B 1399
C 1270
D 759
A
B
C
D
4000 3600 3200 2800 2400 2000 1800 1600 1400 1200 1000 800 600 400
WAVENUMBER

# α-METHYLFENTANYL ACETYL ANALOG

$C_{22}H_{28}N_2O$

**Molecular weight:** 336.47 (336.22)
**Synonyms:** N-Phenyl-N[1-(α-methyl-2-phenylethyl)-4-piperidinyl]-acetamide
**Trade names:**

**Use:**
**HPLC:**
**GC:**

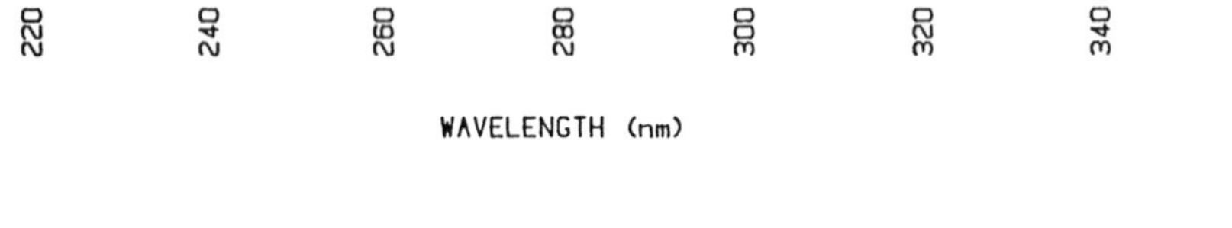


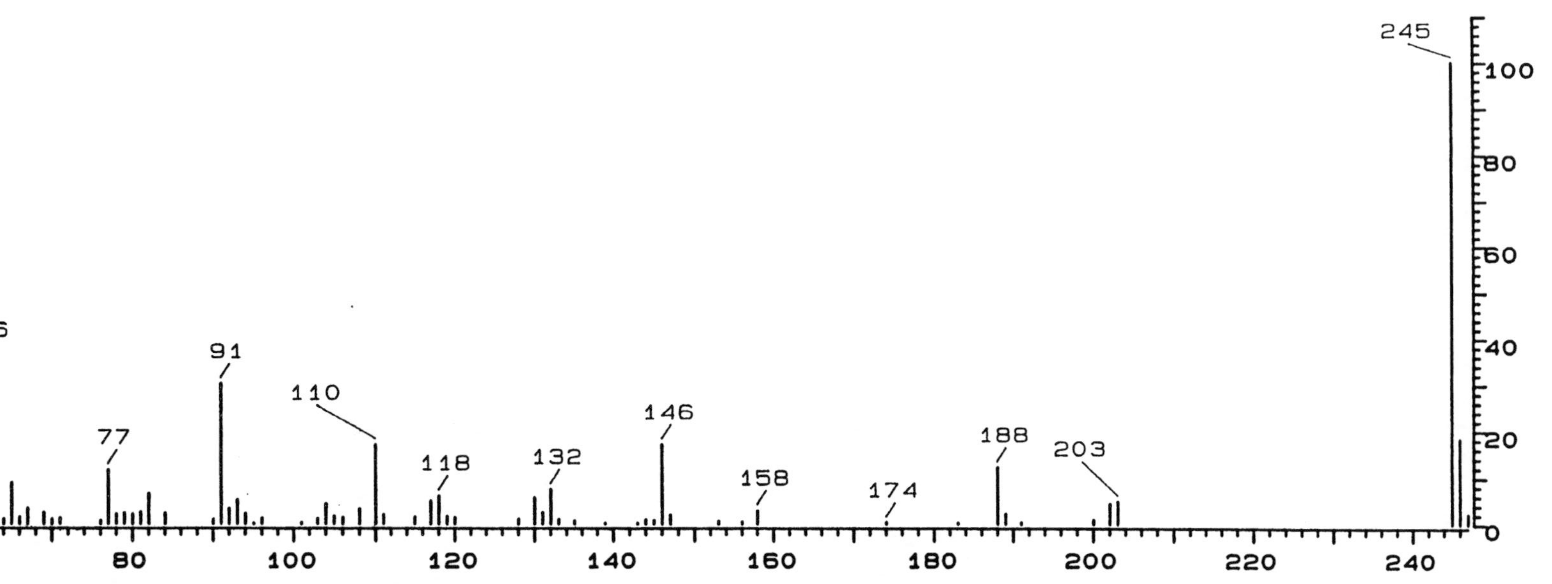

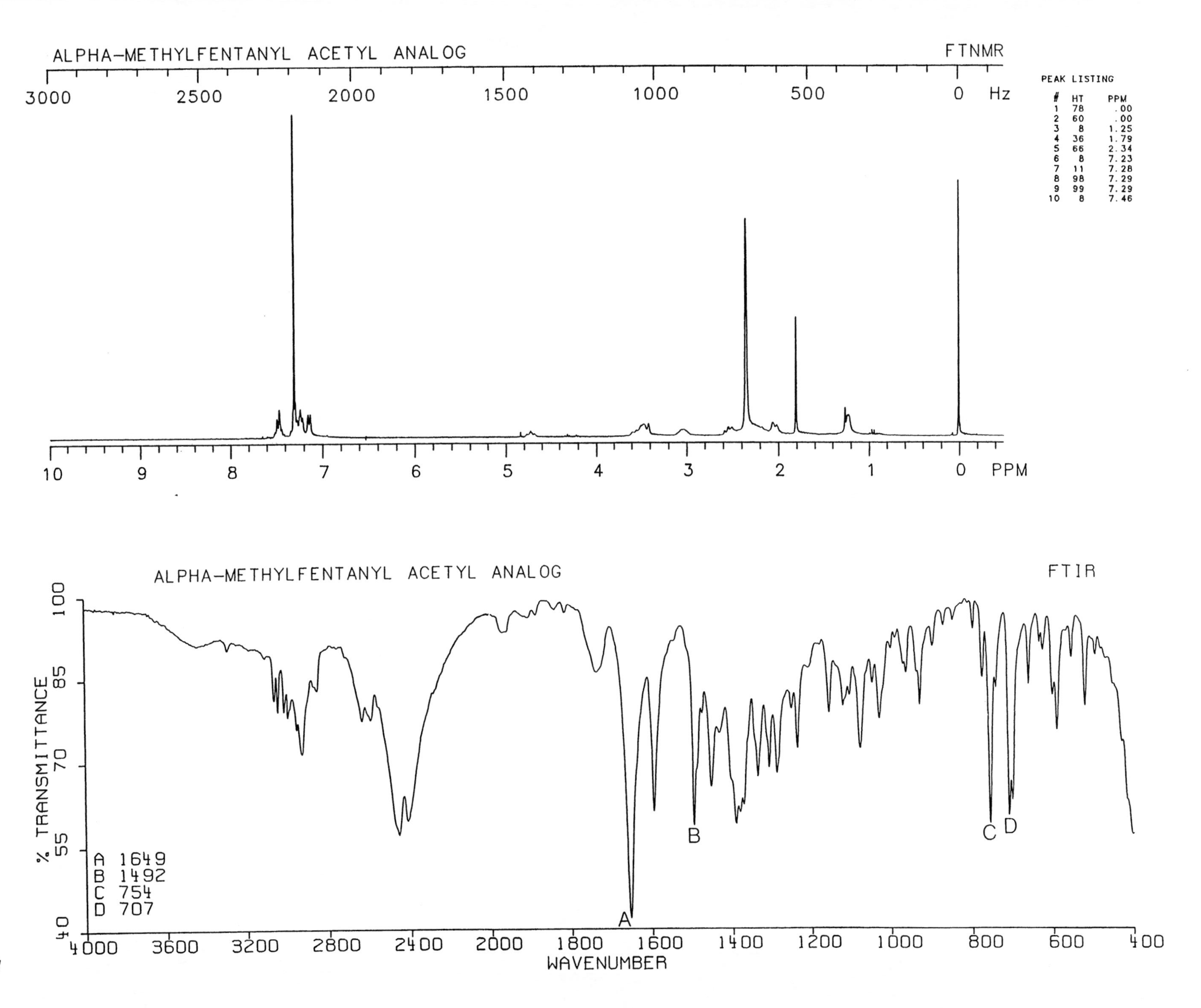
ALPHA-METHYLFENTANYL ACETYL ANALOG
FTNMR
Hz
PPM
PEAK LISTING
# HT PPM
1 78 .00
2 60 .00
3 8 1.25
4 36 1.79
5 66 2.34
6 8 7.23
7 11 7.28
8 98 7.29
9 99 7.29
10 8 7.46
ALPHA-METHYLFENTANYL ACETYL ANALOG
FTIR
% TRANSMITTANCE
WAVENUMBER
A 1649
B 1492
C 754
D 707

# METHYLFORMAMIDE

$C_2H_5NO$

Molecular weight: 59.06 (59.04)
Synonyms: Methylcarbamaldehyde

Trade names:

Use: Synthesis
HPLC:
GC: 842; 80°C

HC(=O)—NH—$CH_3$

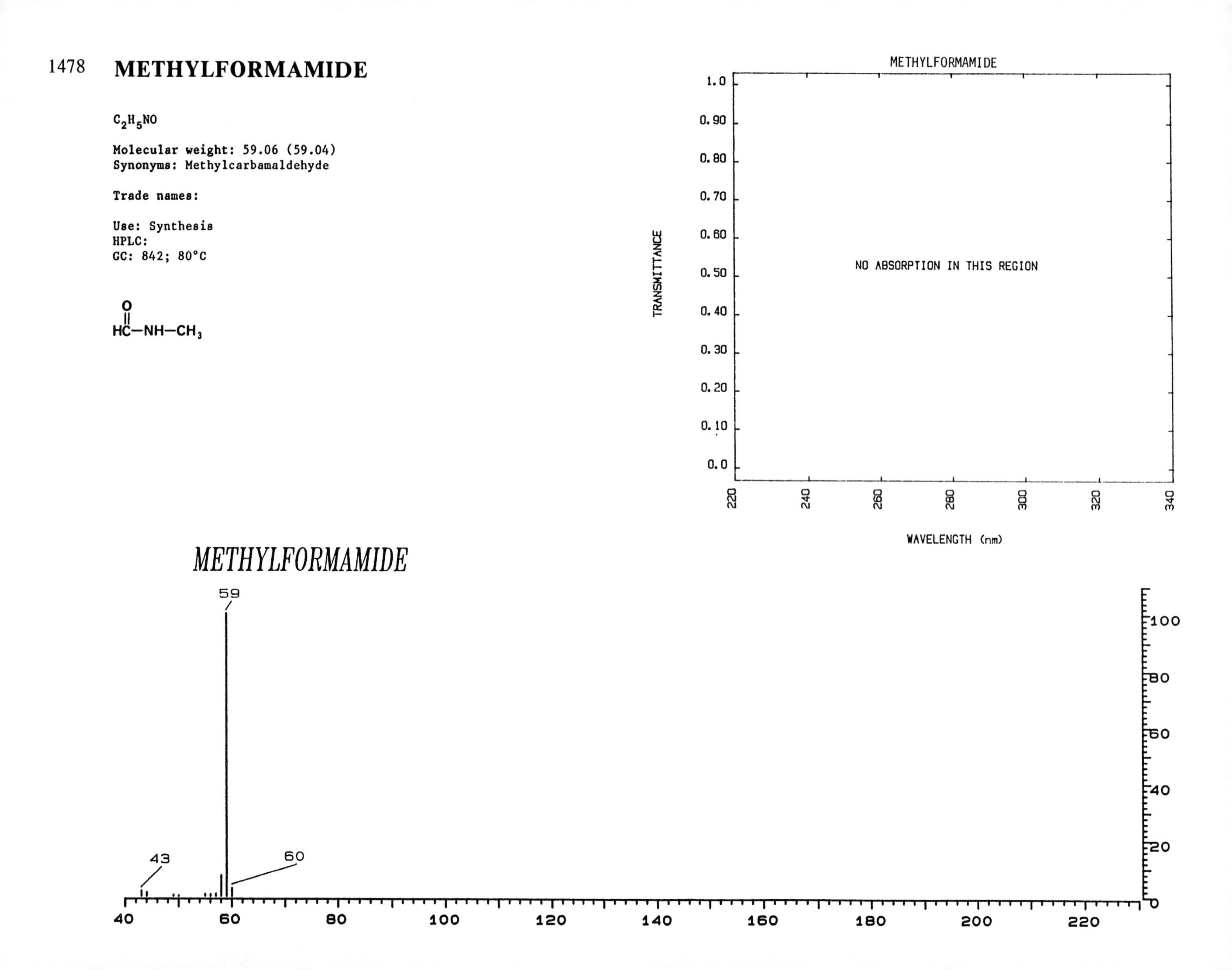

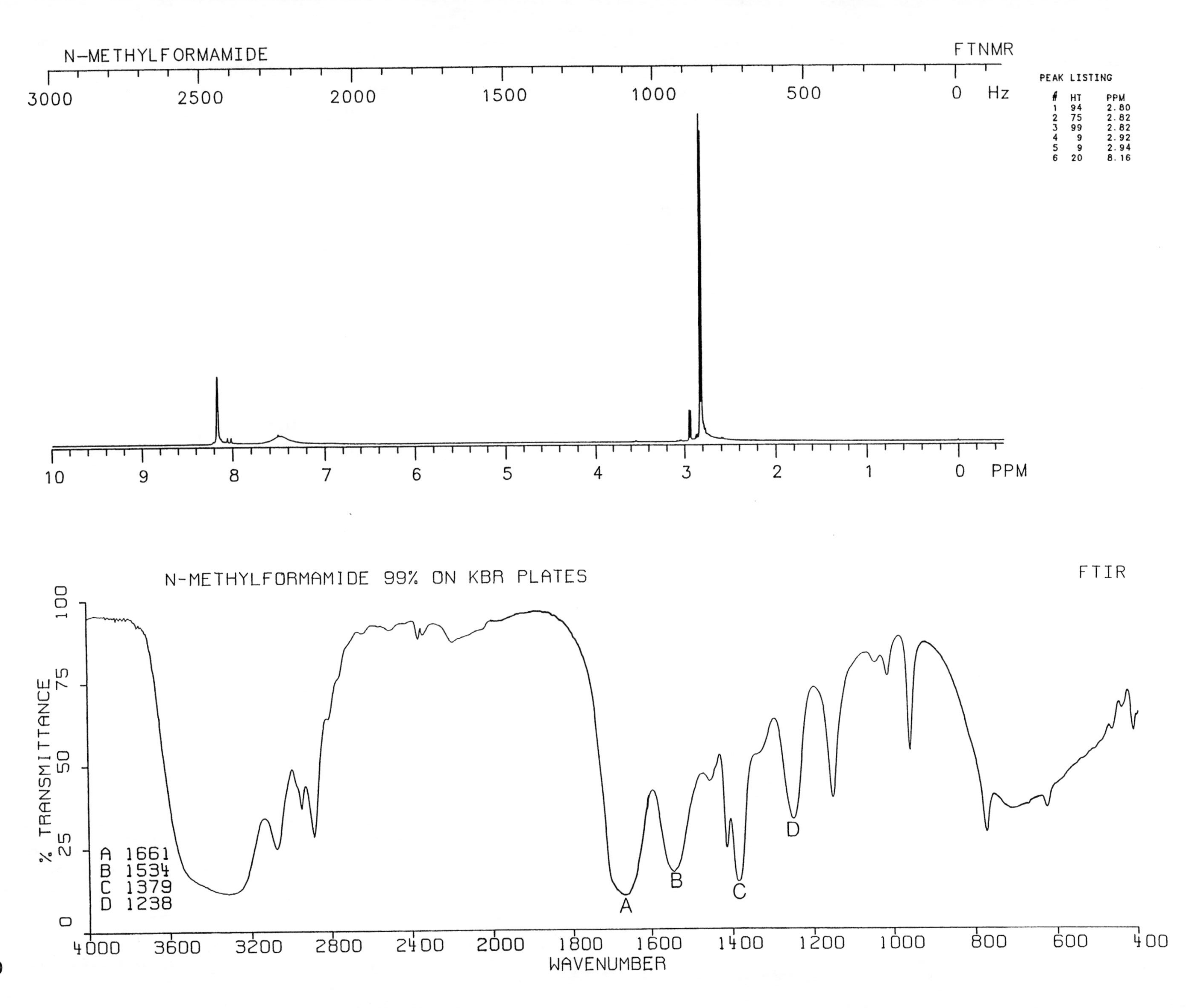
N-METHYLFORMAMIDE
FTNMR
PEAK LISTING
# HT PPM
1 94 2.80
2 75 2.82
3 99 2.82
4 9 2.92
5 9 2.94
6 20 8.16
Hz
PPM
N-METHYLFORMAMIDE 99% ON KBR PLATES
FTIR
% TRANSMITTANCE
A 1661
B 1534
C 1379
D 1238
WAVENUMBER

# METHYLPARABEN

$C_8H_8O_3$

Molecular weight: 152.14 (152.05)
Synonyms: 4-Hydroxybenzoic acid methylester; methyl p-hydroxybenzoate; metagin; methyl parahydroxybenzoate; methylis oxybenzoas
Trade names: Methyl Chemosept, Methyl Parasept, Nipagin, Togosept

Use: Preservative
HPLC: Si-10; 2A:98B; 3.5
GC: 1468; 200°C

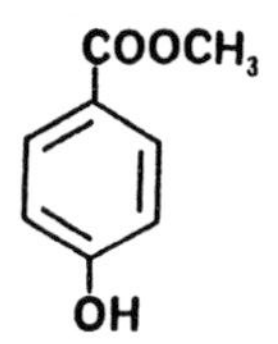


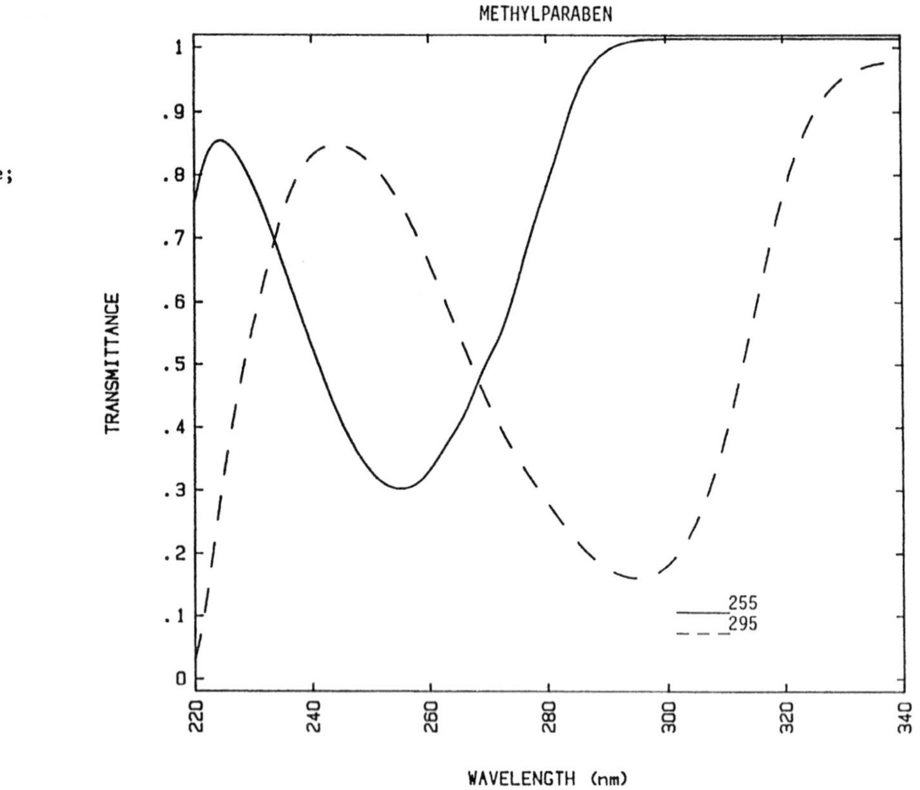

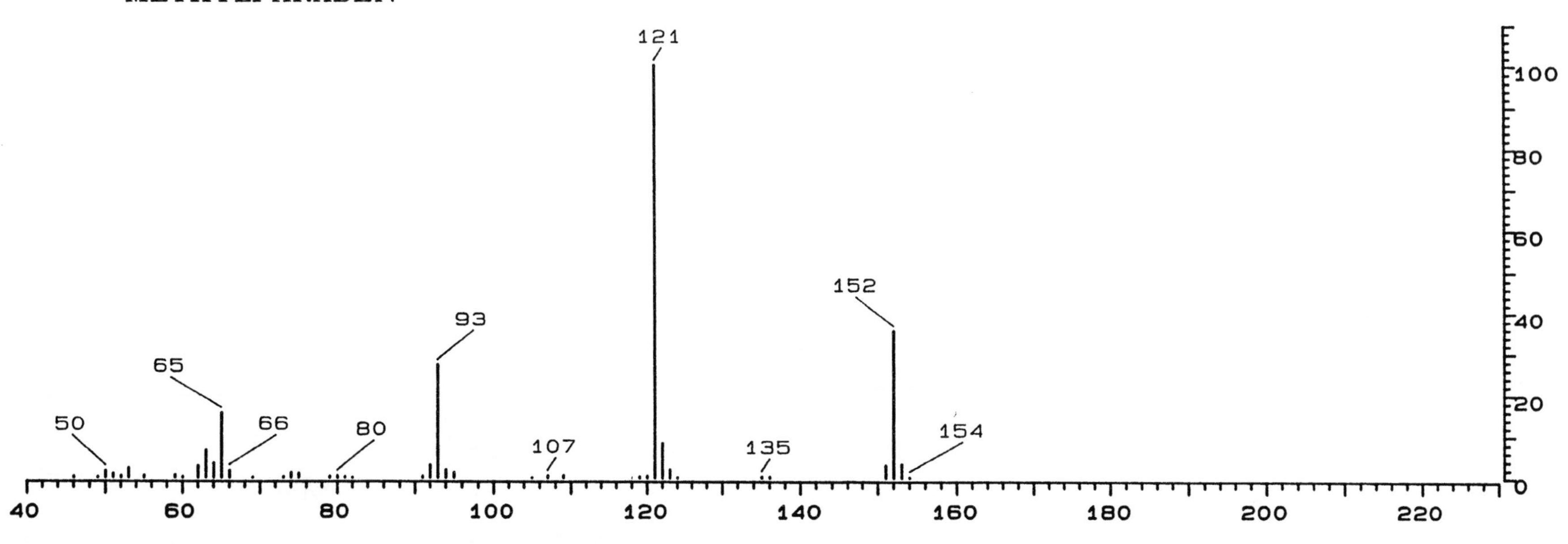

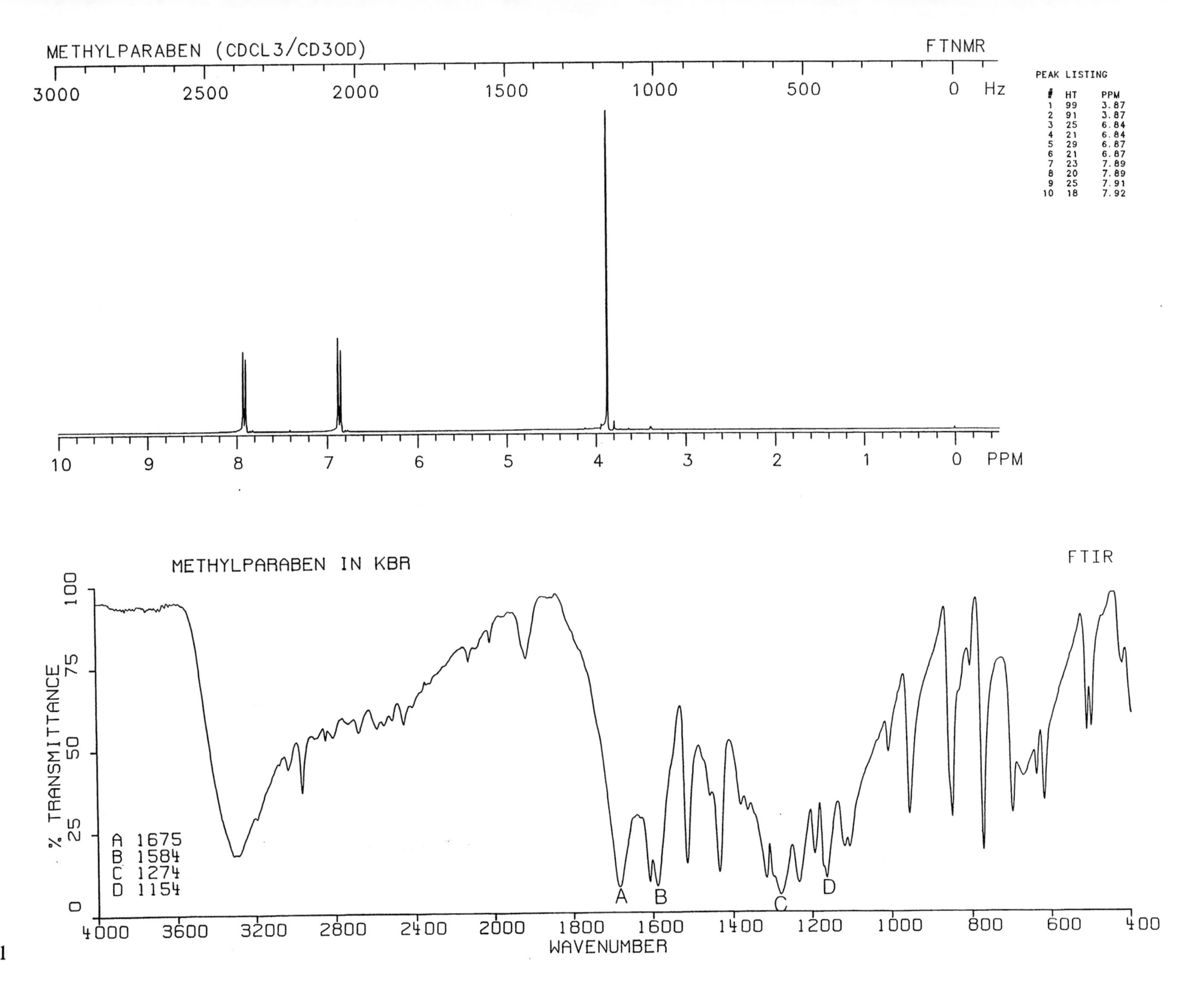
METHYLPARABEN (CDCL3/CD3OD)
FTNMR
3000 2500 2000 1500 1000 500 0 Hz
PEAK LISTING
# HT PPM
1 99 3.87
2 91 3.87
3 25 6.84
4 21 6.84
5 29 6.87
6 21 6.87
7 23 7.89
8 20 7.89
9 25 7.91
10 18 7.92
10 9 8 7 6 5 4 3 2 1 0 PPM
METHYLPARABEN IN KBR
FTIR
% TRANSMITTANCE
100 75 50 25 0
A 1675
B 1584
C 1274
D 1154
A B C D
4000 3600 3200 2800 2400 2000 1800 1600 1400 1200 1000 800 600 400
WAVENUMBER

# 4-METHYLPHENAZONE

$C_{12}H_{14}N_2O$

**Molecular weight:** 202.25 (202.11)
**Synonyms:** 1,2-Dihydro-1,4,5-trimethyl-2-phenyl-3H-pyrazol-3-one

**Trade names:**

**Use:**
HPLC: Si-10; 2A:98B; 5.6
GC:

4-METHYLPHENAZONE

TRANSMITTANCE

.....244,266

WAVELENGTH (nm)

*METHYLPHENAZONE*

56, 65, 77, 91, 96, 107, 119, 132, 142, 144, 159, 173, 174, 187, 202

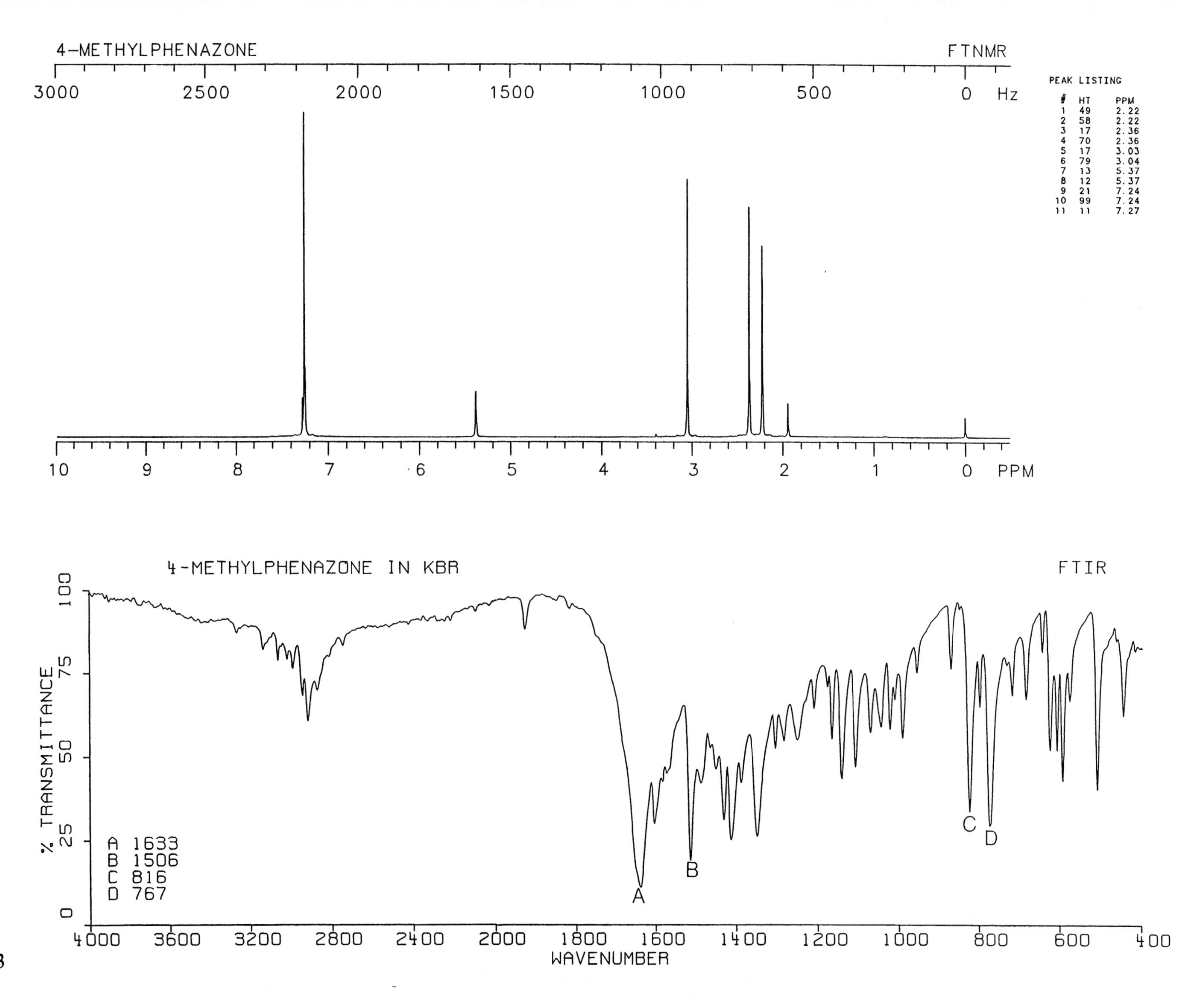
4-METHYLPHENAZONE
FTNMR
3000 2500 2000 1500 1000 500 0 Hz
PEAK LISTING
# HT PPM
1 49 2.22
2 58 2.22
3 17 2.36
4 70 2.36
5 17 3.03
6 79 3.04
7 13 5.37
8 12 5.37
9 21 7.24
10 99 7.24
11 11 7.27
10 9 8 7 6 5 4 3 2 1 0 PPM
4-METHYLPHENAZONE IN KBR
FTIR
% TRANSMITTANCE
100 75 50 25 0
A 1633
B 1506
C 816
D 767
4000 3600 3200 2800 2400 2000 1800 1600 1400 1200 1000 800 600 400
WAVENUMBER

# METHYLPHENIDATE

$C_{14}H_{19}NO_2$

**Molecular weight:** 233.31 (233.14)
**Synonyms:** α-Phenyl-2-piperidineacetic acid methyl ester

**Trade names:** Ritalin

**Use:** Central stimulant
**HPLC:** Si-10; 5A:95B; 4.0
**GC:** 1739; 200°C

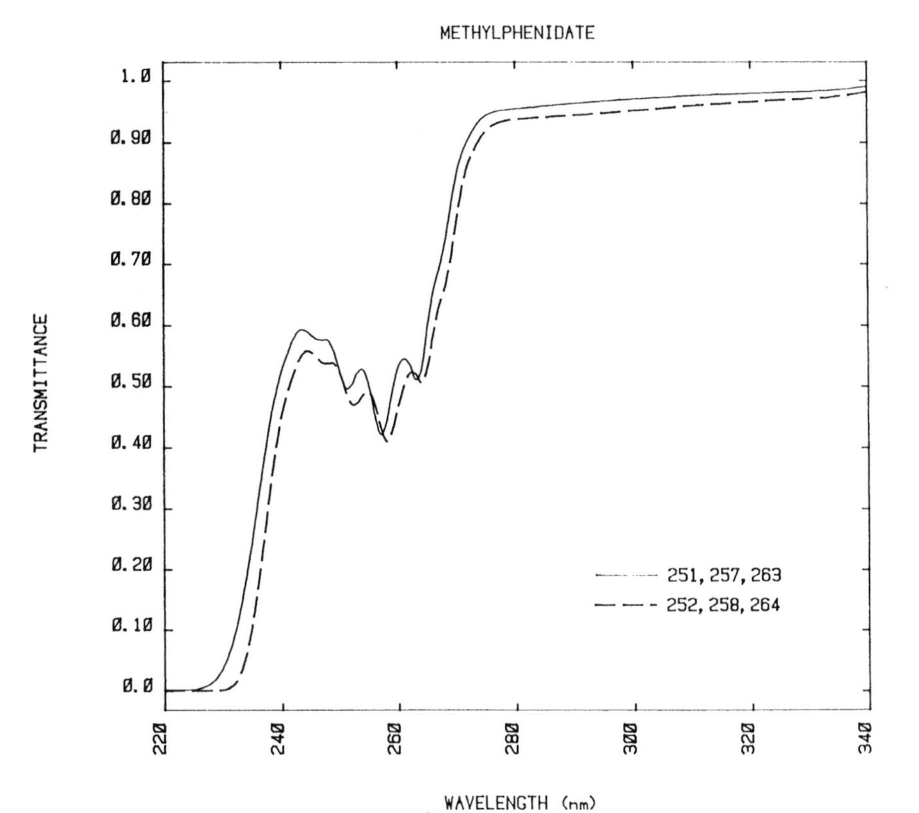


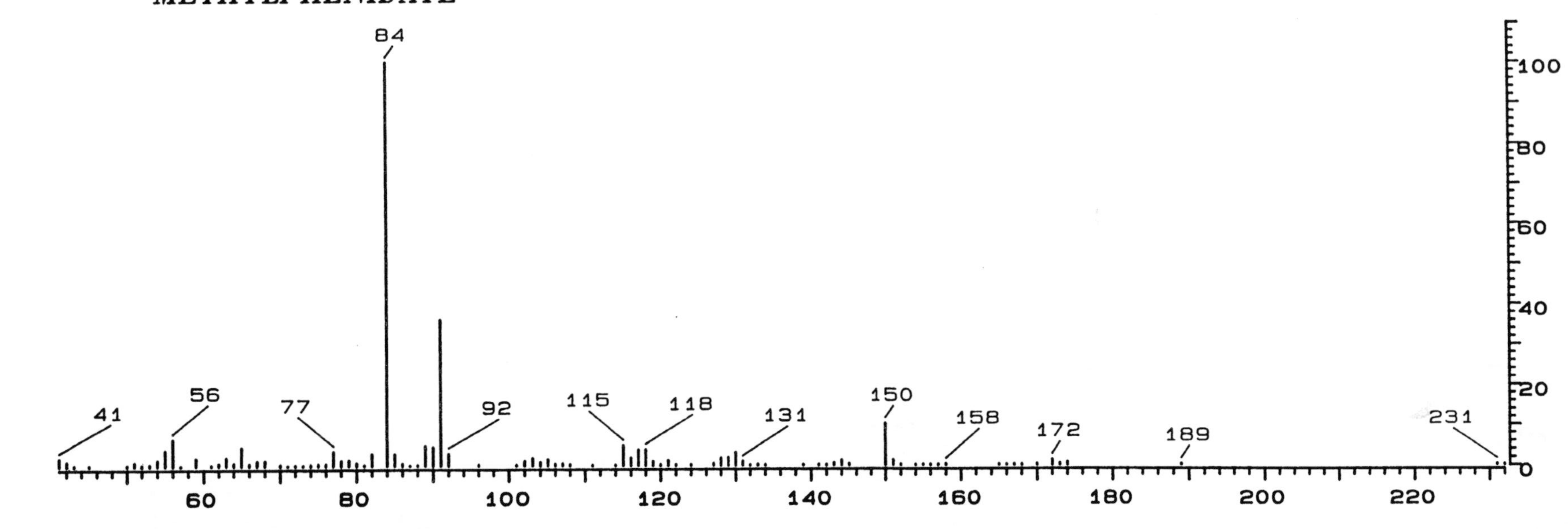

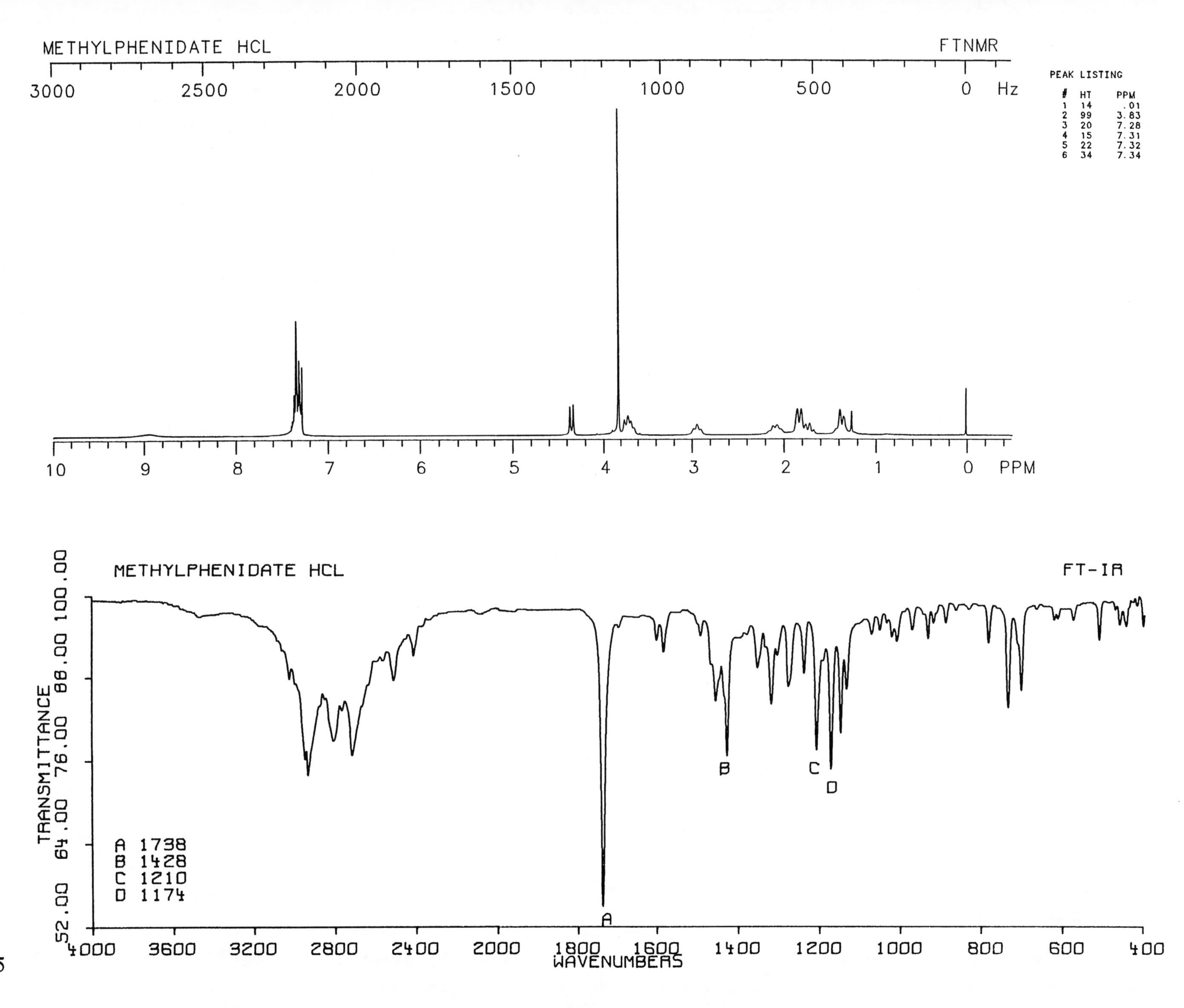

METHYLPHENIDATE HCL
FTNMR
3000
2500
2000
1500
1000
500
0
Hz
PEAK LISTING
# HT PPM
1 14 .01
2 99 3.83
3 20 7.28
4 15 7.31
5 22 7.32
6 34 7.34
10
9
8
7
6
5
4
3
2
1
0
PPM
METHYLPHENIDATE HCL
FT-IR
TRANSMITTANCE
52.00
64.00
76.00
88.00
100.00
A 1738
B 1428
C 1210
D 1174
A
B
C
D
4000
3600
3200
2800
2400
2000
1800
1600
1400
1200
1000
800
600
400
WAVENUMBERS

# N-METHYL-1-PHENYLCYCLOHEXYLAMINE

$C_{13}H_{19}N$

**Molecular weight:** 189.30 (189.15)
**Synonyms:** N-Methyl-1-phenylcyclohexanamine; N-methyl-1-phenylcyclohexamine
**Trade names:**

**Use:** Hallucinogen
**HPLC:** Si-10; 20A:80B; 5.2
**GC:** 1567; 200°C

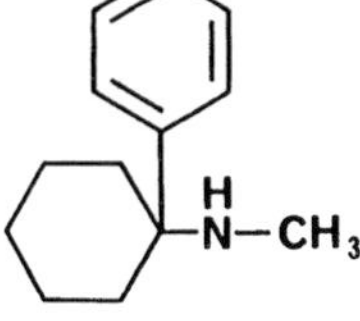


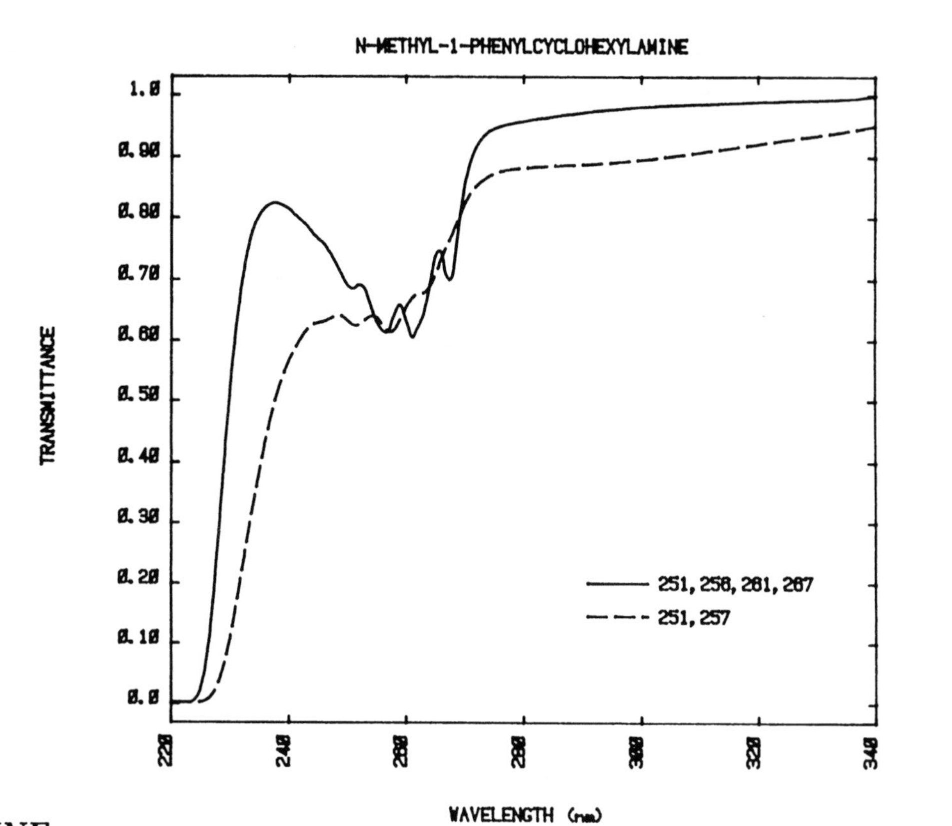

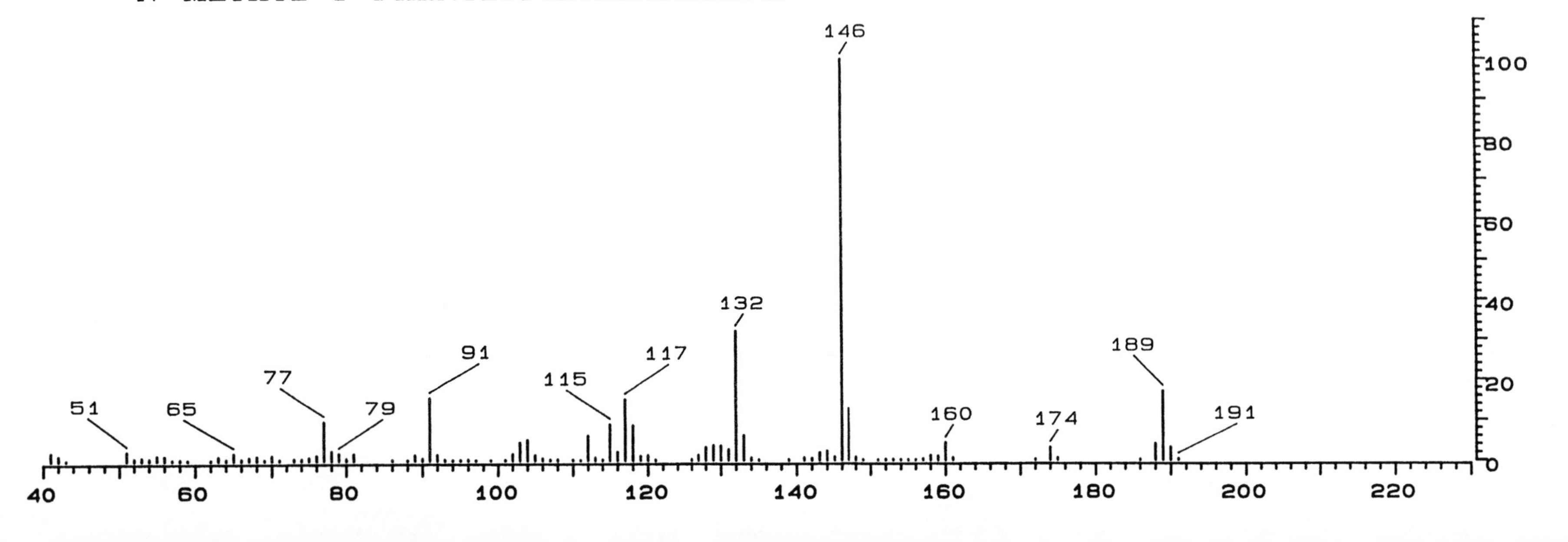

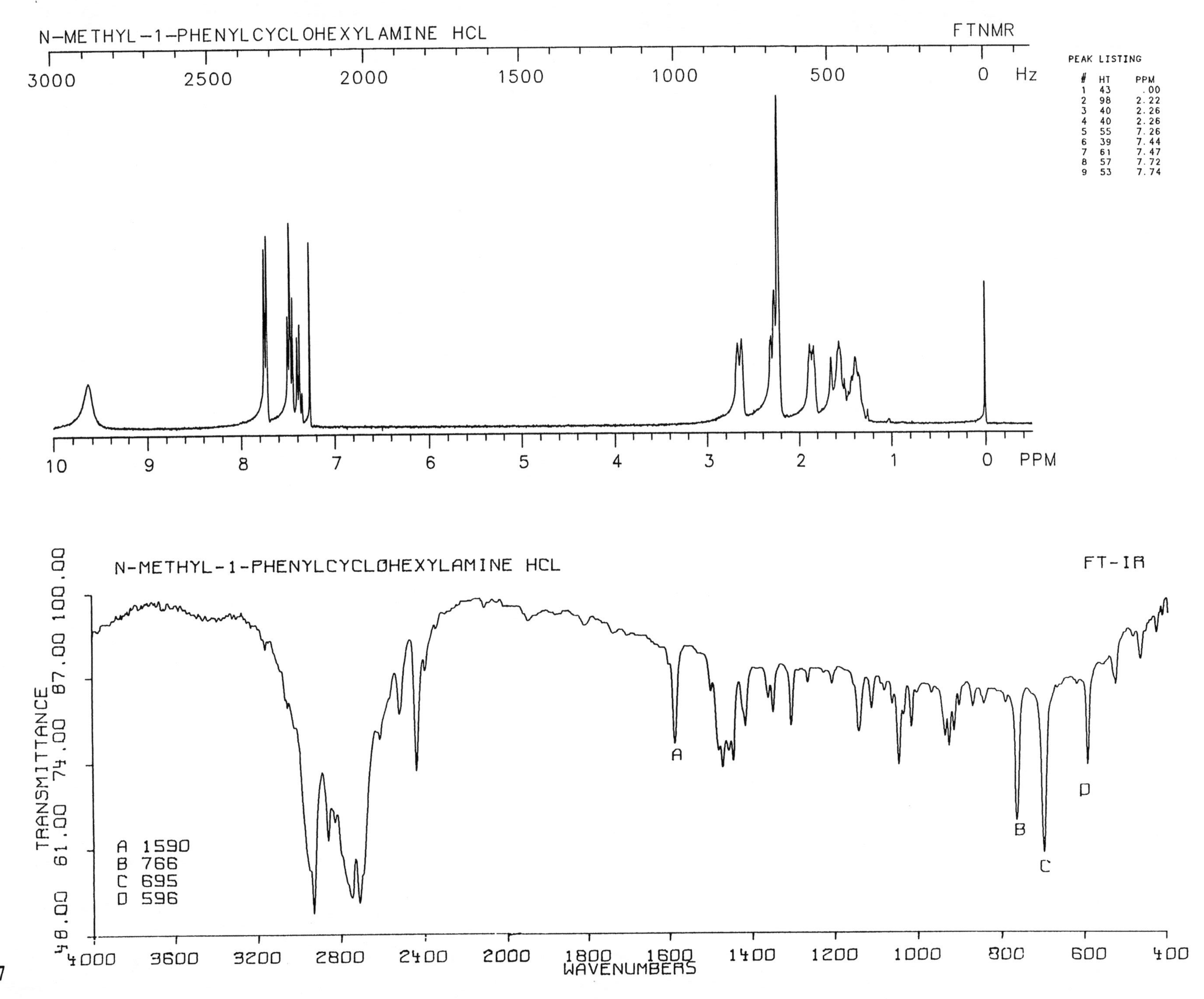
N-METHYL-1-PHENYLCYCLOHEXYLAMINE HCL
FTNMR
3000
2500
2000
1500
1000
500
0 Hz
PEAK LISTING
# HT PPM
1 43 .00
2 98 2.22
3 40 2.26
4 40 2.26
5 55 7.26
6 39 7.44
7 61 7.47
8 57 7.72
9 53 7.74
10
9
8
7
6
5
4
3
2
1
0 PPM
N-METHYL-1-PHENYLCYCLØHEXYLAMINE HCL
FT-IR
TRANSMITTANCE
48.00
61.00
74.00
87.00
100.00
A 1590
B 766
C 695
D 596
A
B
C
D
4000
3600
3200
2800
2400
2000
1800
1600
1400
1200
1000
800
600
400
WAVENUMBERS

# 5-METHYL-5-PHENYLHYDANTOIN

$C_{10}H_{10}N_2O_2$

Molecular weight: 190.20 (190.07)
Synonyms: 5-Methyl-5-phenyl-2,4-imidazolidinedione

Trade names:

Use: Metabolite
HPLC: Si-10; 50A:50B; 4.0
GC: 1908; 200°C

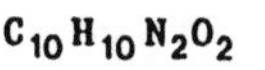


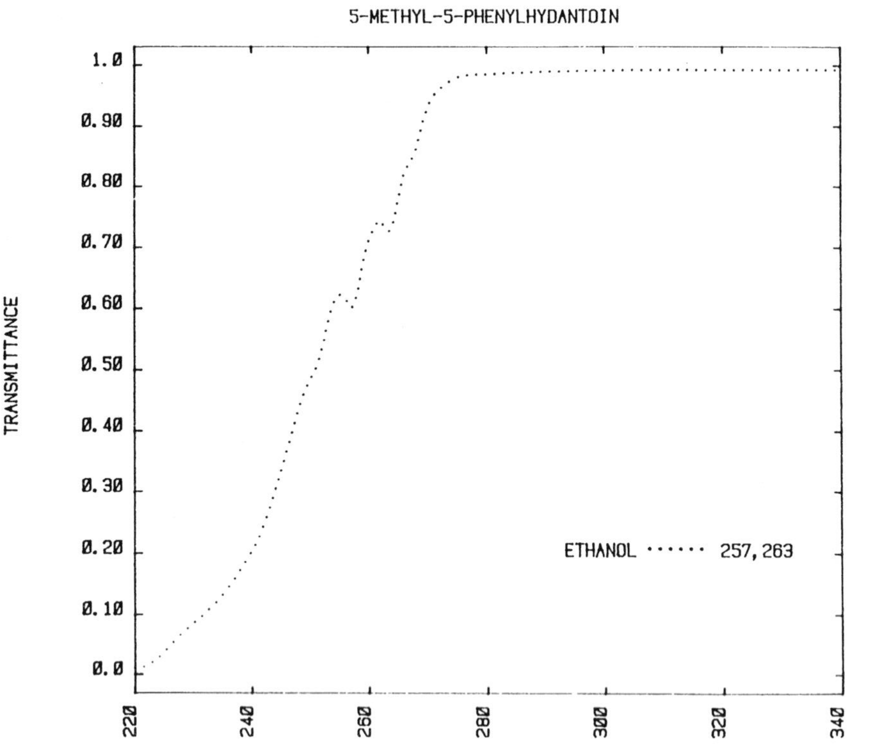

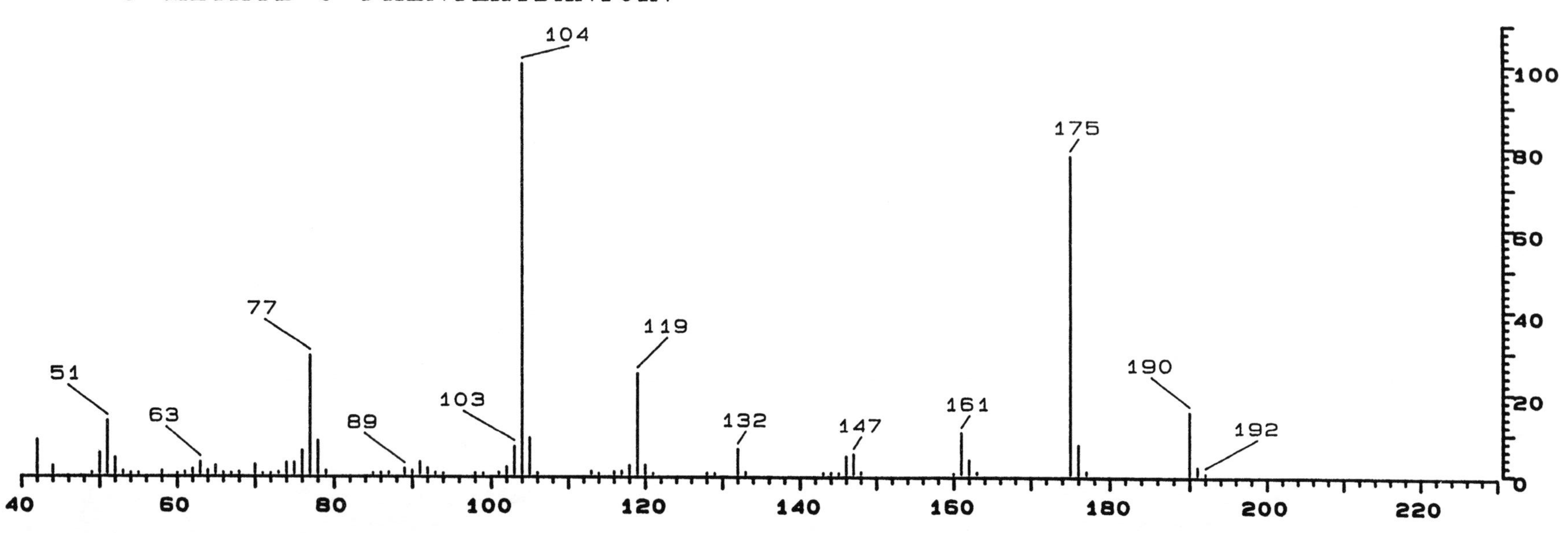

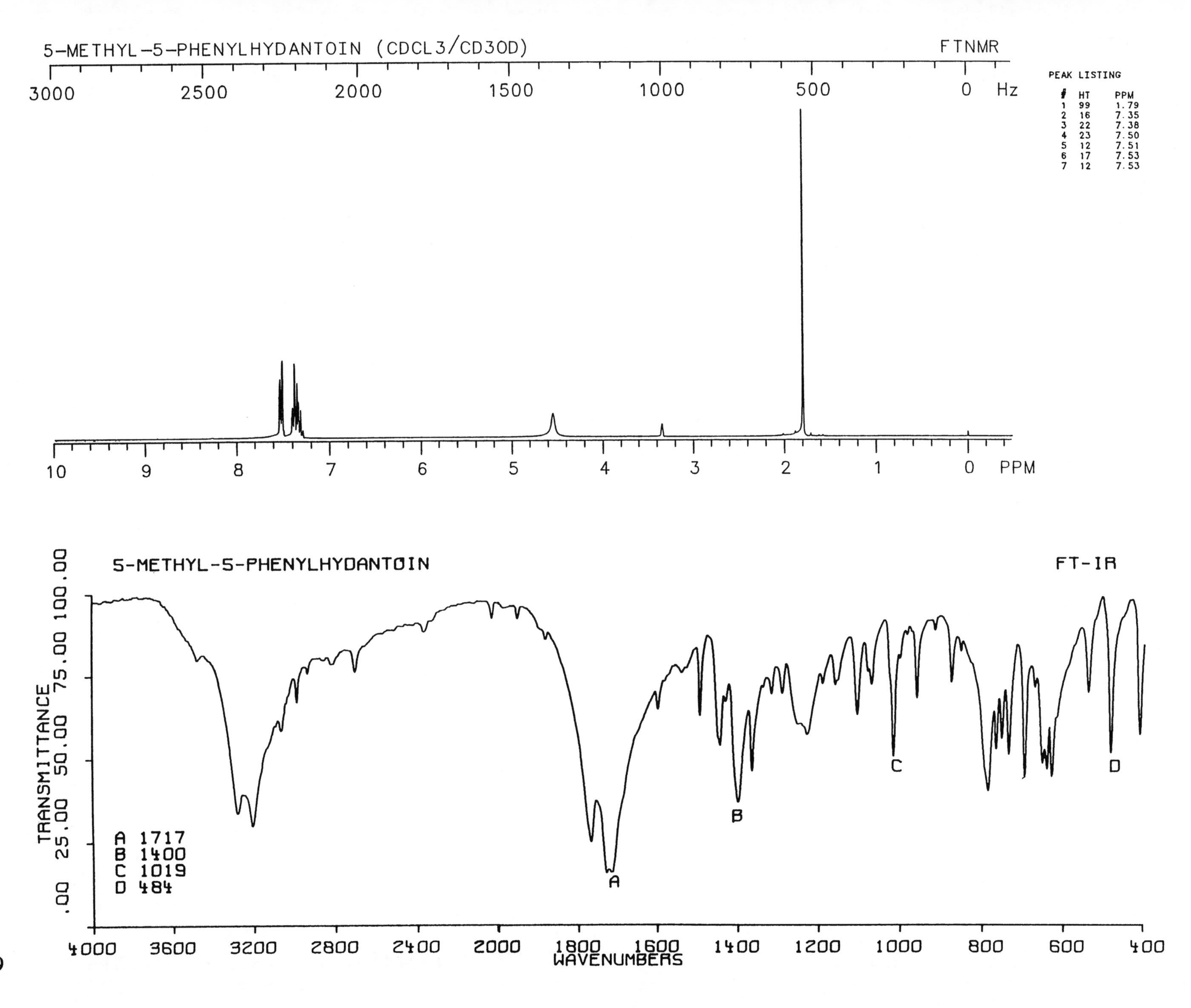
5-METHYL-5-PHENYLHYDANTOIN (CDCL3/CD3OD)
FTNMR
3000
2500
2000
1500
1000
500
0 Hz
PEAK LISTING
# HT PPM
1 99 1.79
2 16 7.35
3 22 7.38
4 23 7.50
5 12 7.51
6 17 7.53
7 12 7.53
10
9
8
7
6
5
4
3
2
1
0 PPM
5-METHYL-5-PHENYLHYDANTOIN
FT-IR
TRANSMITTANCE
.00 25.00 50.00 75.00 100.00
A 1717
B 1400
C 1019
D 484
A
B
C
D
4000 3600 3200 2800 2400 2000 1800 1600 1400 1200 1000 800 600 400
WAVENUMBERS

# 5- (*P*-METHYLPHENYL)-5- PHENYLHYDANTOIN

$C_{16}H_{14}N_2O_2$

**Molecular weight:** 266.30 (266.11)
**Synonyms:** 5-(4-Methylphenyl)-5-phenyl-2,4-imidazolidinedione

**Trade names:**

**Use:** Metabolite
**HPLC:** Si-10; 50A:50B; 3.6
**GC:** 2529; 250°C

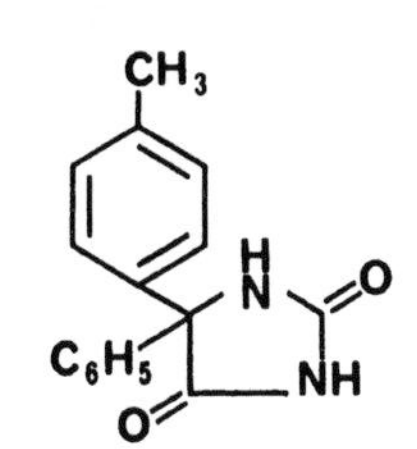


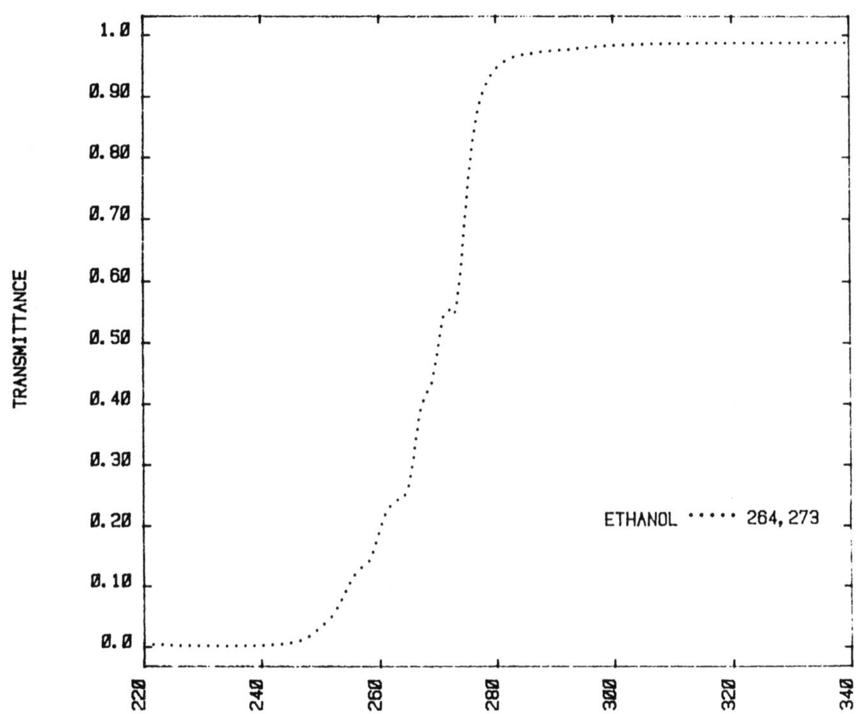


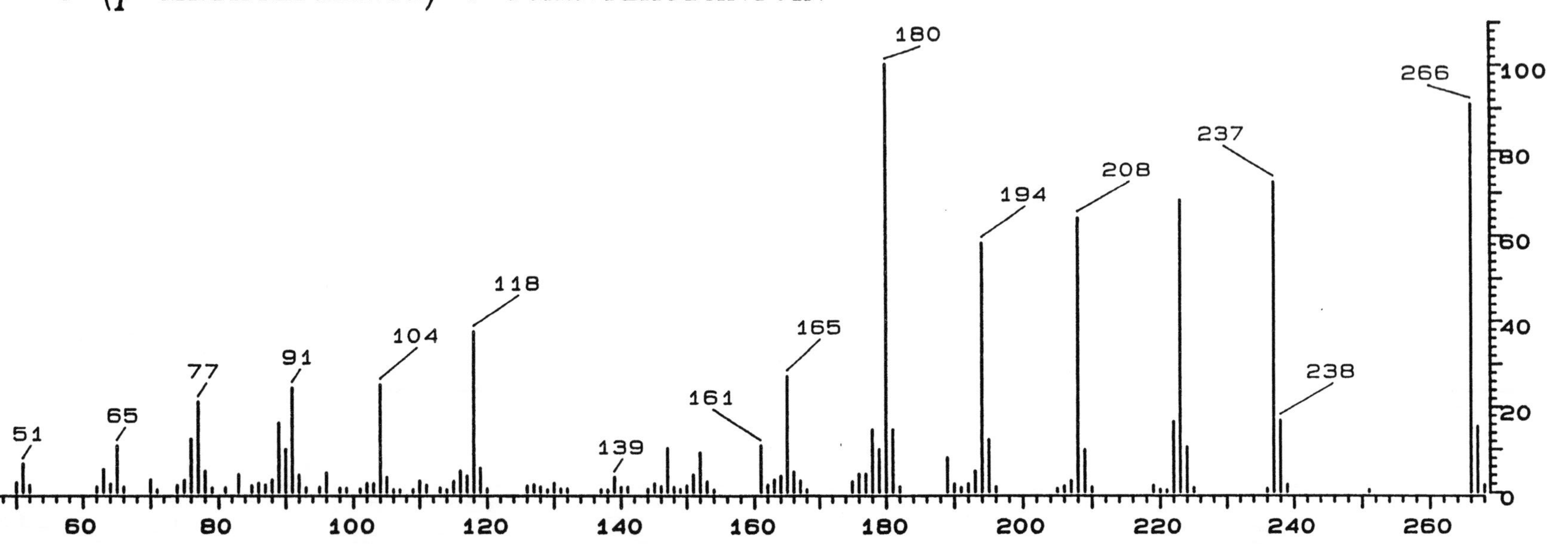

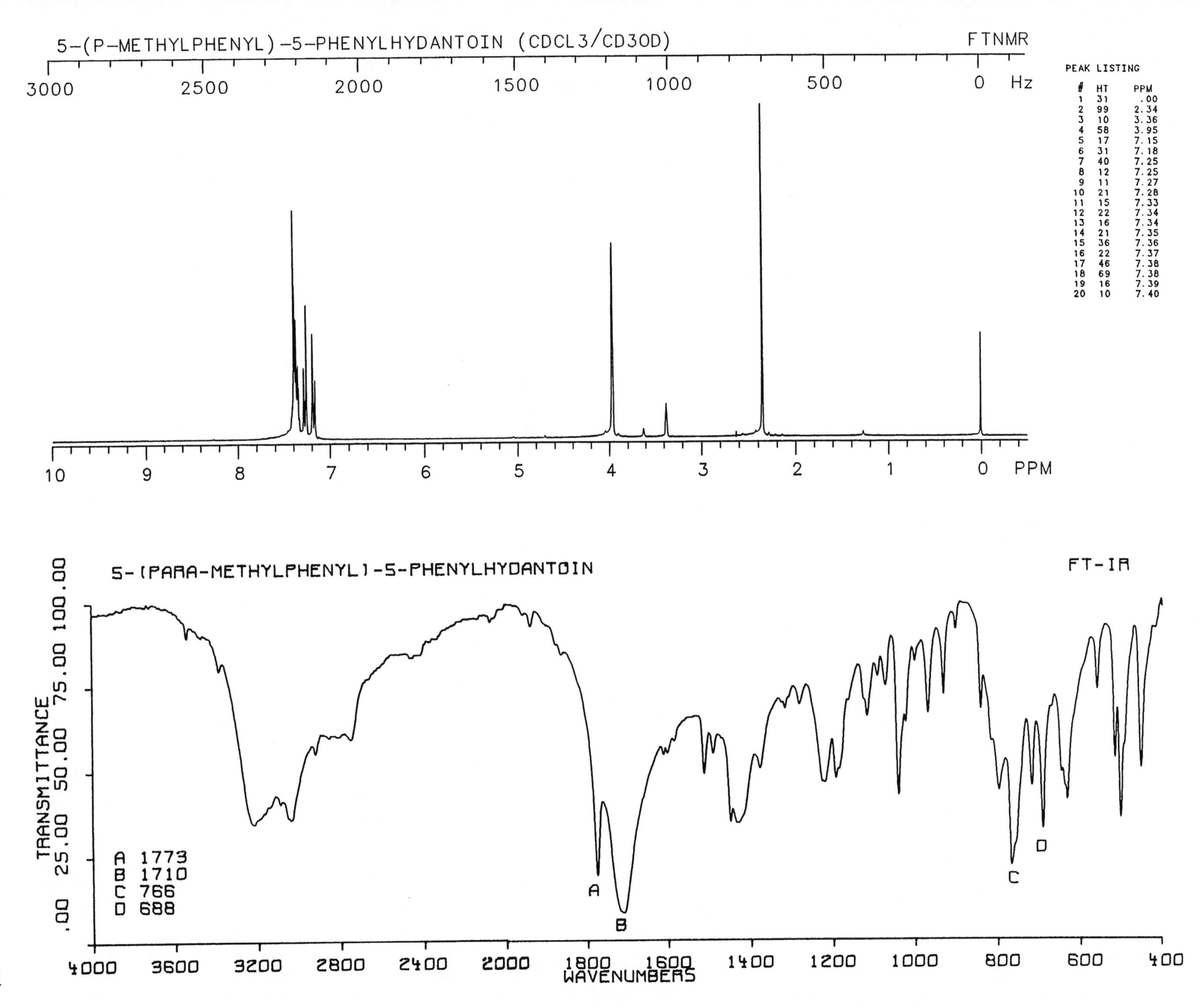

5-(P-METHYLPHENYL)-5-PHENYLHYDANTOIN (CDCL3/CD3OD)
FTNMR
3000
2500
2000
1500
1000
500
0
Hz
PEAK LISTING
# HT PPM
1 31 .00
2 99 2.34
3 10 3.36
4 58 3.95
5 17 7.15
6 31 7.18
7 40 7.25
8 12 7.25
9 11 7.27
10 21 7.28
11 15 7.33
12 22 7.34
13 16 7.34
14 21 7.35
15 36 7.36
16 22 7.37
17 46 7.38
18 69 7.38
19 16 7.39
20 10 7.40
10
9
8
7
6
5
4
3
2
1
0
PPM
5-(PARA-METHYLPHENYL)-5-PHENYLHYDANTOIN
FT-IR
TRANSMITTANCE
.00
25.00
50.00
75.00
100.00
A 1773
B 1710
C 766
D 688
A
B
C
D
4000
3600
3200
2800
2400
2000
1800
1600
1400
1200
1000
800
600
400
WAVENUMBERS

# 1-METHYL-4-PHENYL-1,2,3,6-TETRAHYDROPYRIDINE

$C_{12}H_{15}N$

Molecular weight: 173.26 (173.12)
Synonyms: MPTP

Trade names: Causes severe and irreversible Parkinsonium condition

Use: Synthesis
HPLC:
GC: 1255; 140°C

N
CH₃

METHYL-4-PHENYL-1,2,3,6-TETRAHYDROPYRIDINE

TRANSMITTANCE

WAVELENGTH (nm)

242
248

1-METHYL-4-PHENYL-1,2,3,6-TETRAHYDROPYRIDINE

42 63 77 82 96 115 128 129 144 157 173 175

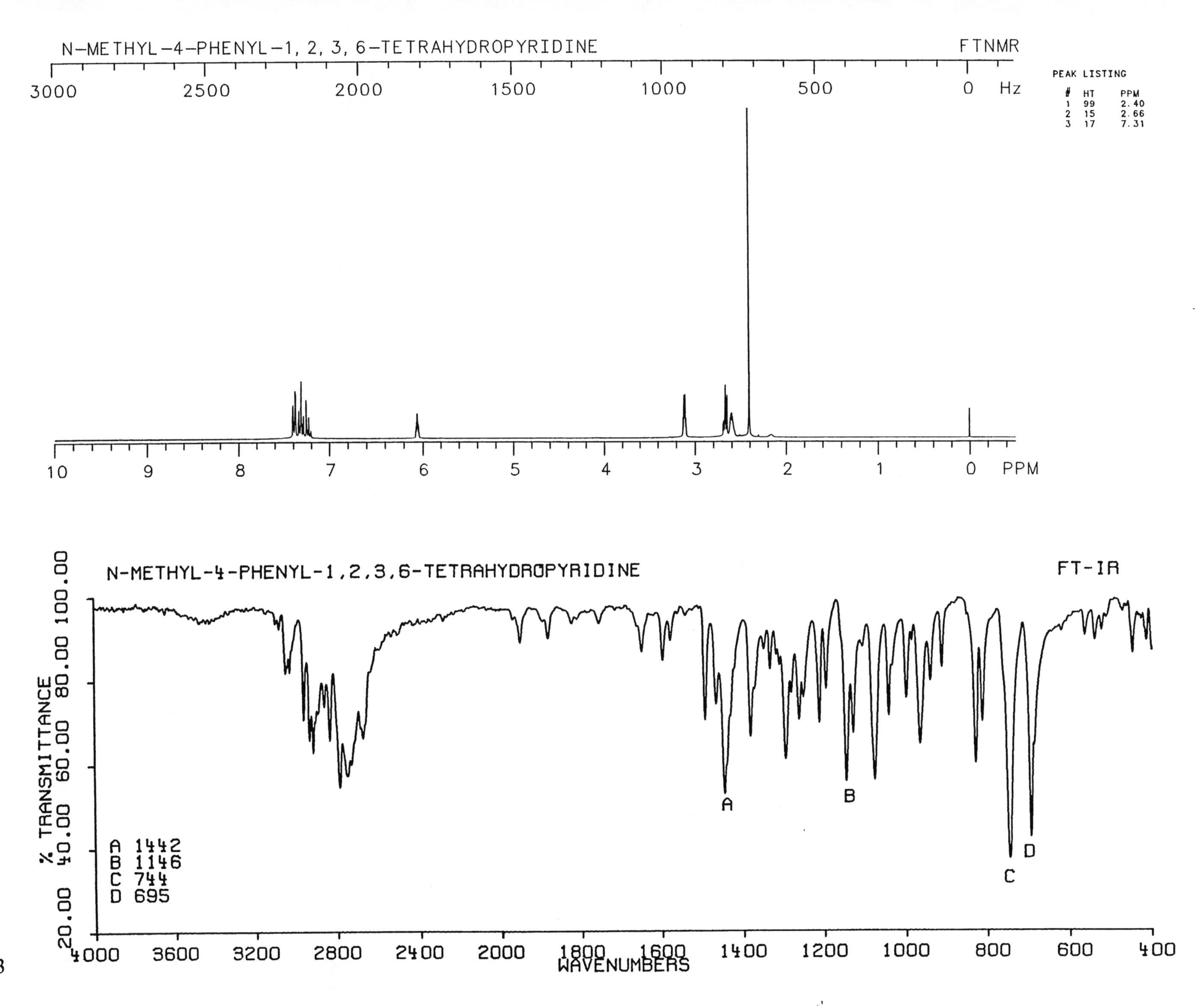

N-METHYL-4-PHENYL-1, 2, 3, 6-TETRAHYDROPYRIDINE
FTNMR
3000
2500
2000
1500
1000
500
0 Hz
PEAK LISTING
# HT PPM
1 99 2.40
2 15 2.66
3 17 7.31
10
9
8
7
6
5
4
3
2
1
0 PPM
N-METHYL-4-PHENYL-1,2,3,6-TETRAHYDROPYRIDINE
FT-IR
% TRANSMITTANCE
20.00
40.00
60.00
80.00
100.00
A 1442
B 1146
C 744
D 695
A
B
C
D
4000
3600
3200
2800
2400
2000
1800
1600
1400
1200
1000
800
600
400
WAVENUMBERS

# METHYLPREDNISOLONE

$C_{22}H_{30}O_5$

Molecular weight: 374.48 (374.21)
Synonyms: 11β,17α,21-Trihydroxy-6α-methyl-1,4-pregnadiene-3,20-dione

Trade names: Medrol, Methylprednisolone

Use: Glucocorticoid, steroid
HPLC:
GC:

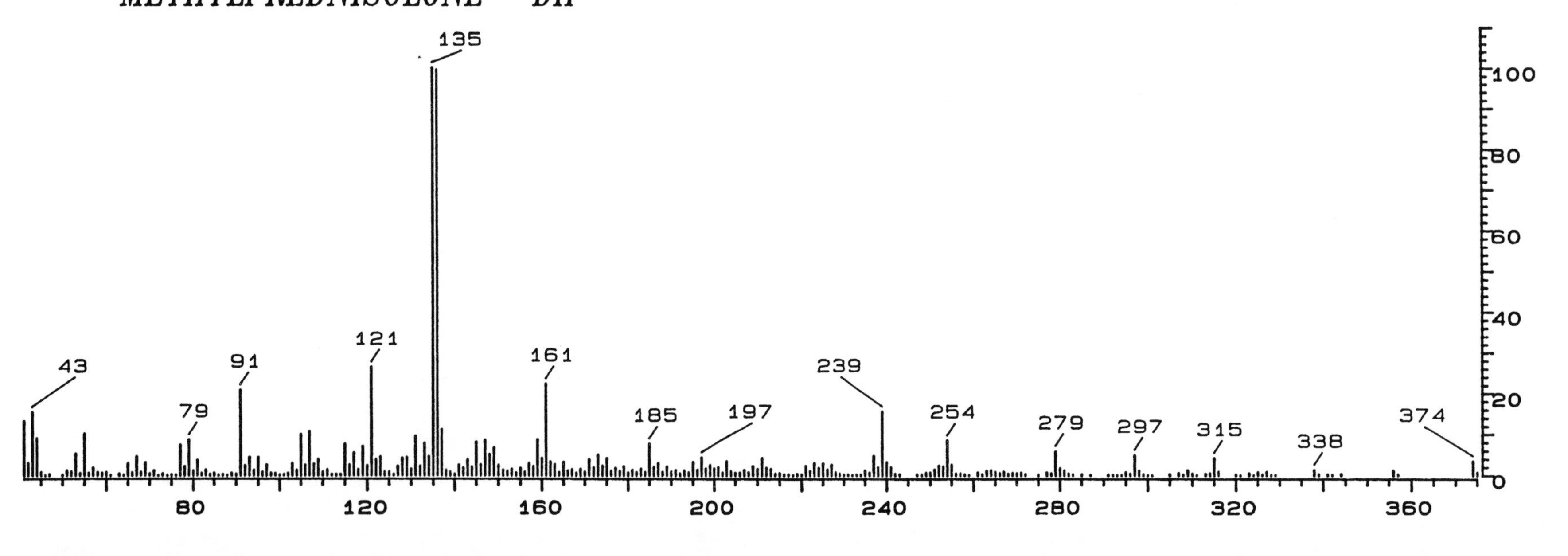

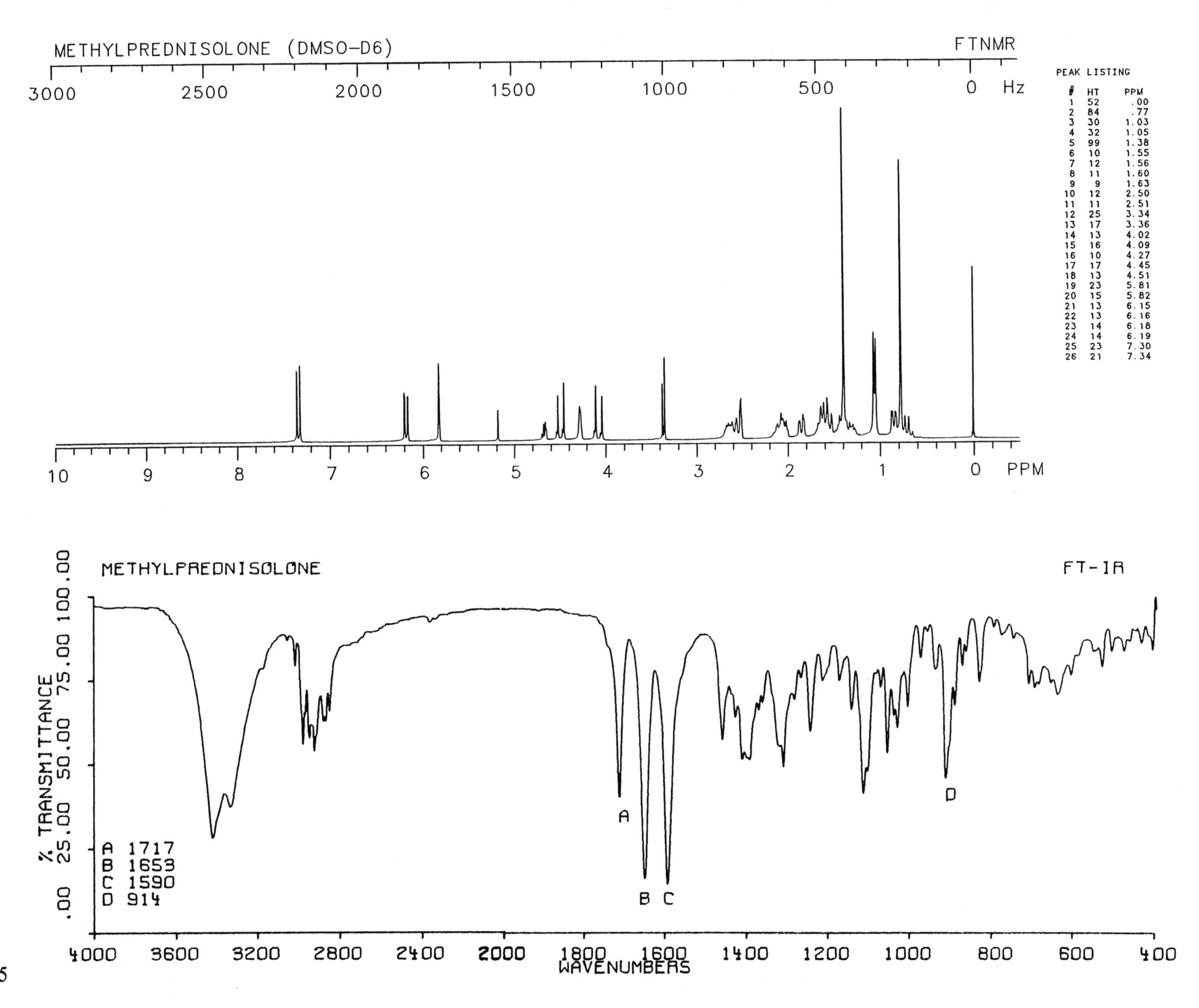
METHYLPREDNISOLONE (DMSO-D6)
FTNMR
PEAK LISTING
# HT PPM
1 52 .00
2 84 .77
3 30 1.03
4 32 1.05
5 99 1.38
6 10 1.55
7 12 1.56
8 11 1.60
9 9 1.63
10 12 2.50
11 11 2.51
12 25 3.34
13 17 3.36
14 13 4.02
15 16 4.09
16 10 4.27
17 17 4.45
18 13 4.51
19 23 5.81
20 15 5.82
21 13 6.15
22 13 6.16
23 14 6.18
24 14 6.19
25 23 7.30
26 21 7.34
Hz
PPM
METHYLPREDNISOLONE
FT-IR
A 1717
B 1653
C 1590
D 914
% TRANSMITTANCE
WAVENUMBERS

# METHYLPRIMIDONE

$C_{13}H_{16}N_2O_2$

Molecular weight: 232.28 (232.12)
Synonyms: 5-Ethyl-5(4-methylphenyl)hexahydropyrimidine-4,6-dione

Trade names:

Use:
HPLC: Si-10; 2A:98B; 6.4
GC: 2389; 250°C

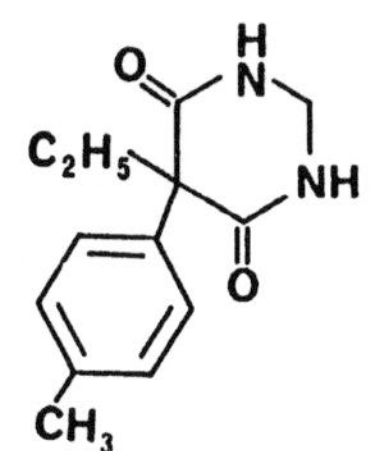


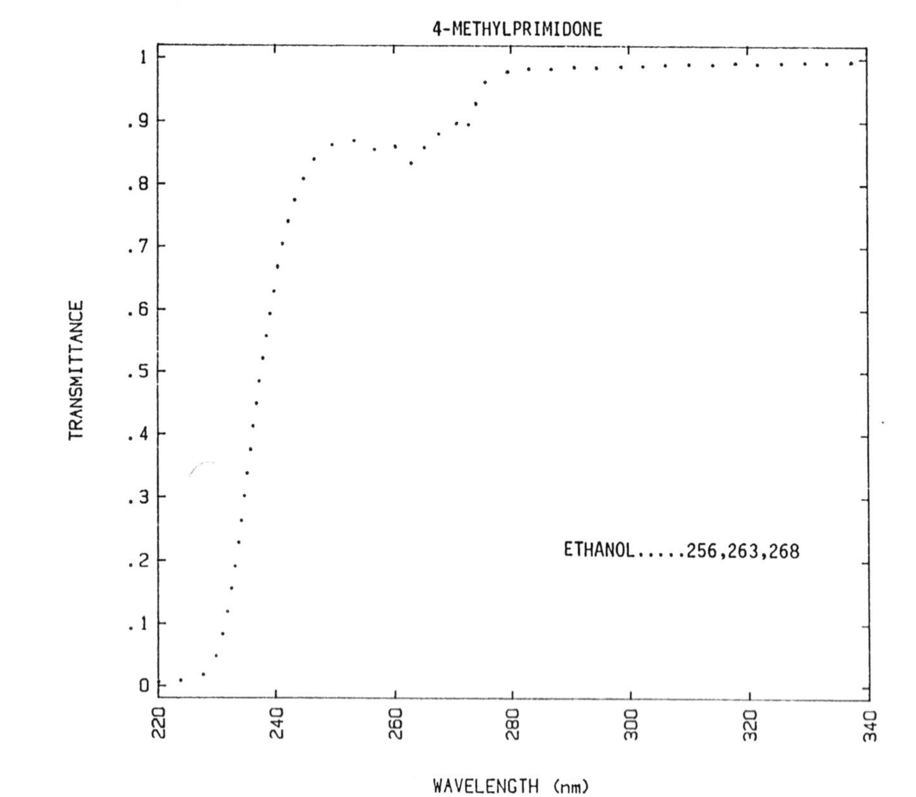

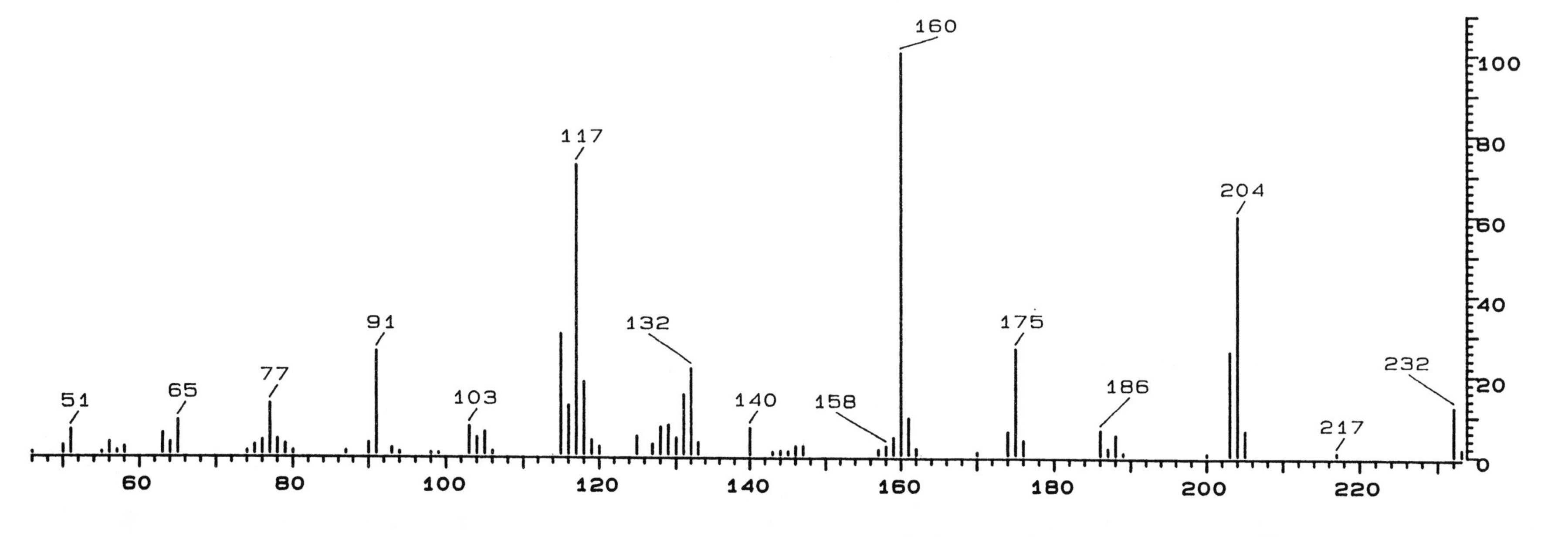

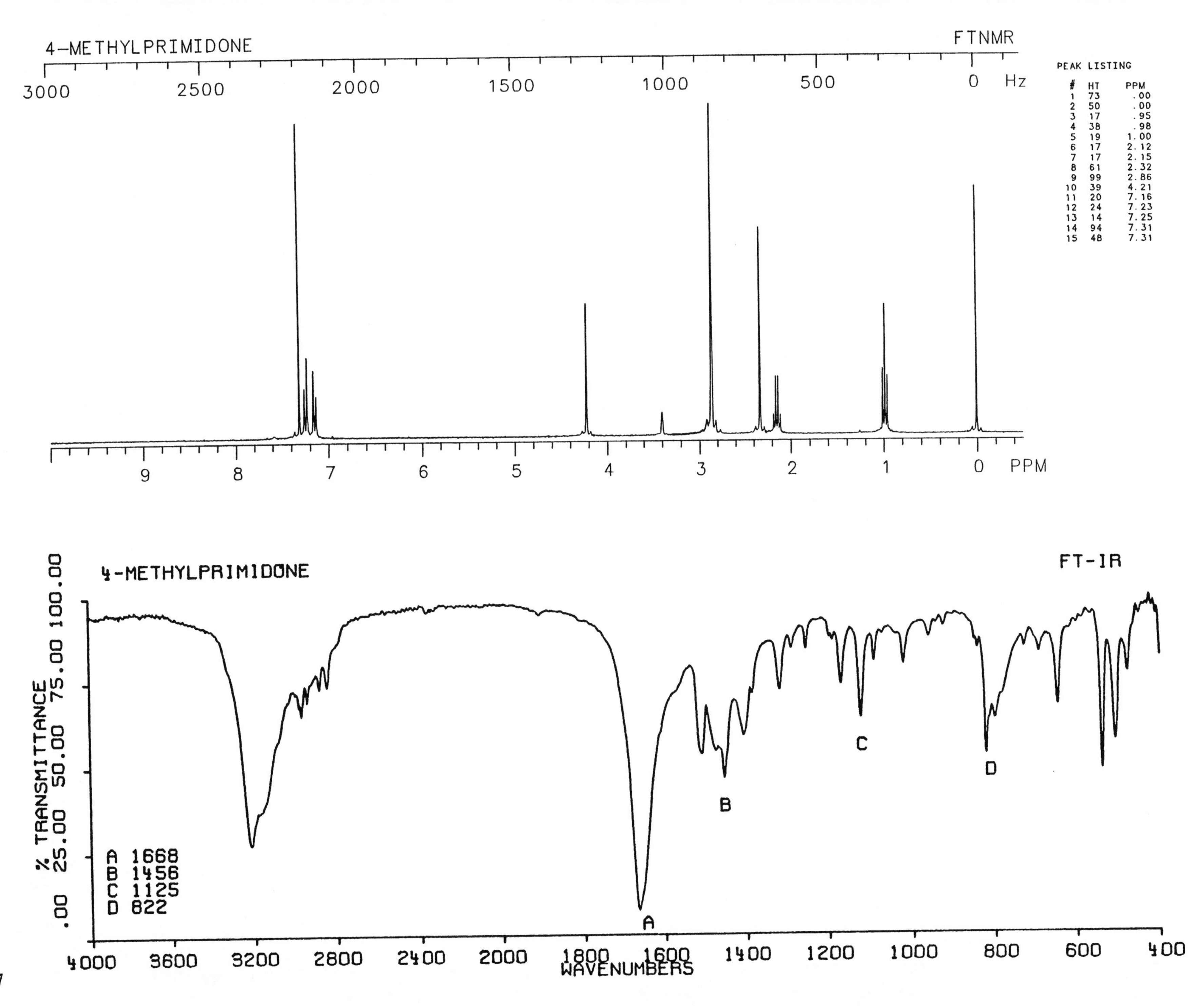

4-METHYLPRIMIDONE
FTNMR
Hz
PPM
PEAK LISTING
# HT PPM
1 73 .00
2 50 .00
3 17 .95
4 38 .98
5 19 1.00
6 17 2.12
7 17 2.15
8 61 2.32
9 99 2.86
10 39 4.21
11 20 7.16
12 24 7.23
13 14 7.25
14 94 7.31
15 48 7.31
4-METHYLPRIMIDONE
FT-IR
% TRANSMITTANCE
WAVENUMBERS
A 1668
B 1456
C 1125
D 822

# α-METHYL-α-PROPYLSUCCINIMIDE

$C_8H_{14}NO_2$

Molecular weight: 156.21 (156.10)
Synonyms: 3-Methyl-3-propyl-pyrrolidinedione

Trade names:

Use:
HPLC:
GC: 1343; 140°C

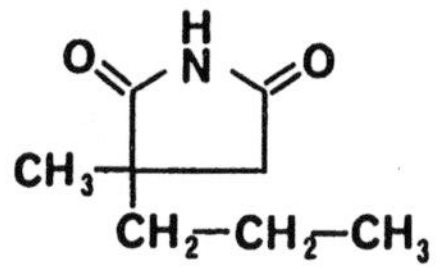

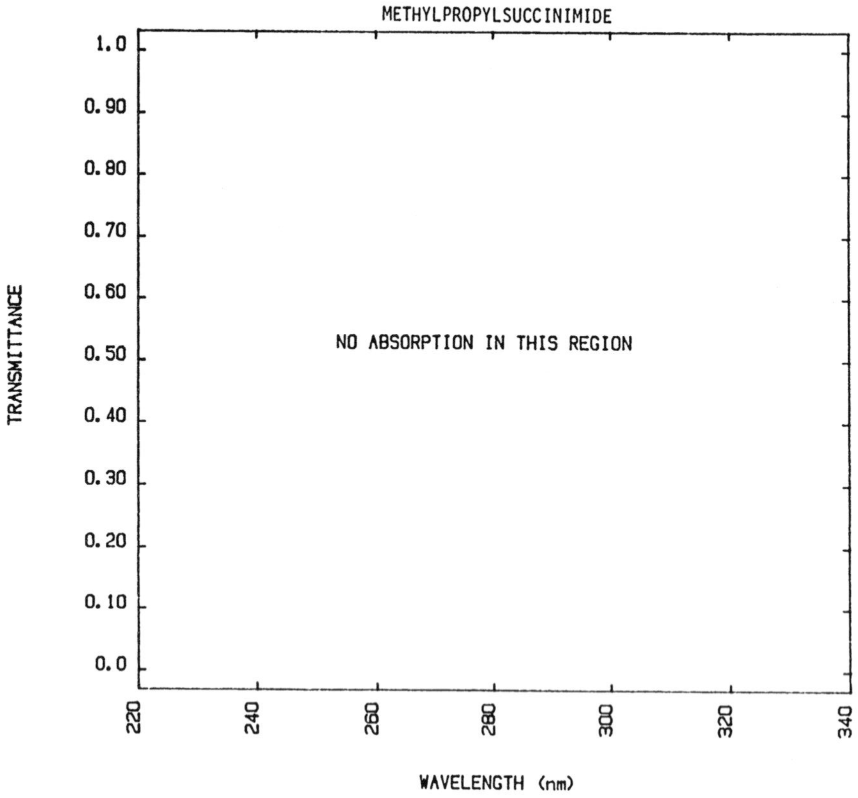

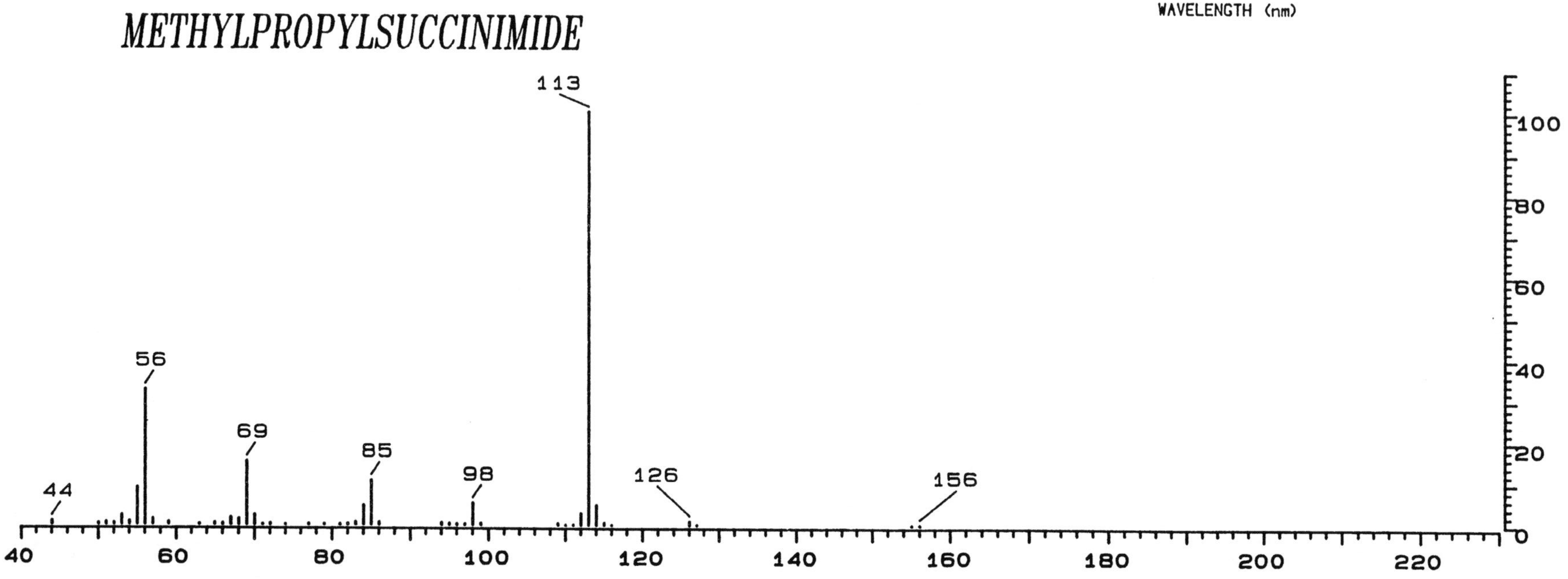

METHYLPROPYLSUCCINIMIDE

TRANSMITTANCE

NO ABSORPTION IN THIS REGION

WAVELENGTH (nm)

METHYLPROPYLSUCCINIMIDE

44 56 69 85 98 113 126 156

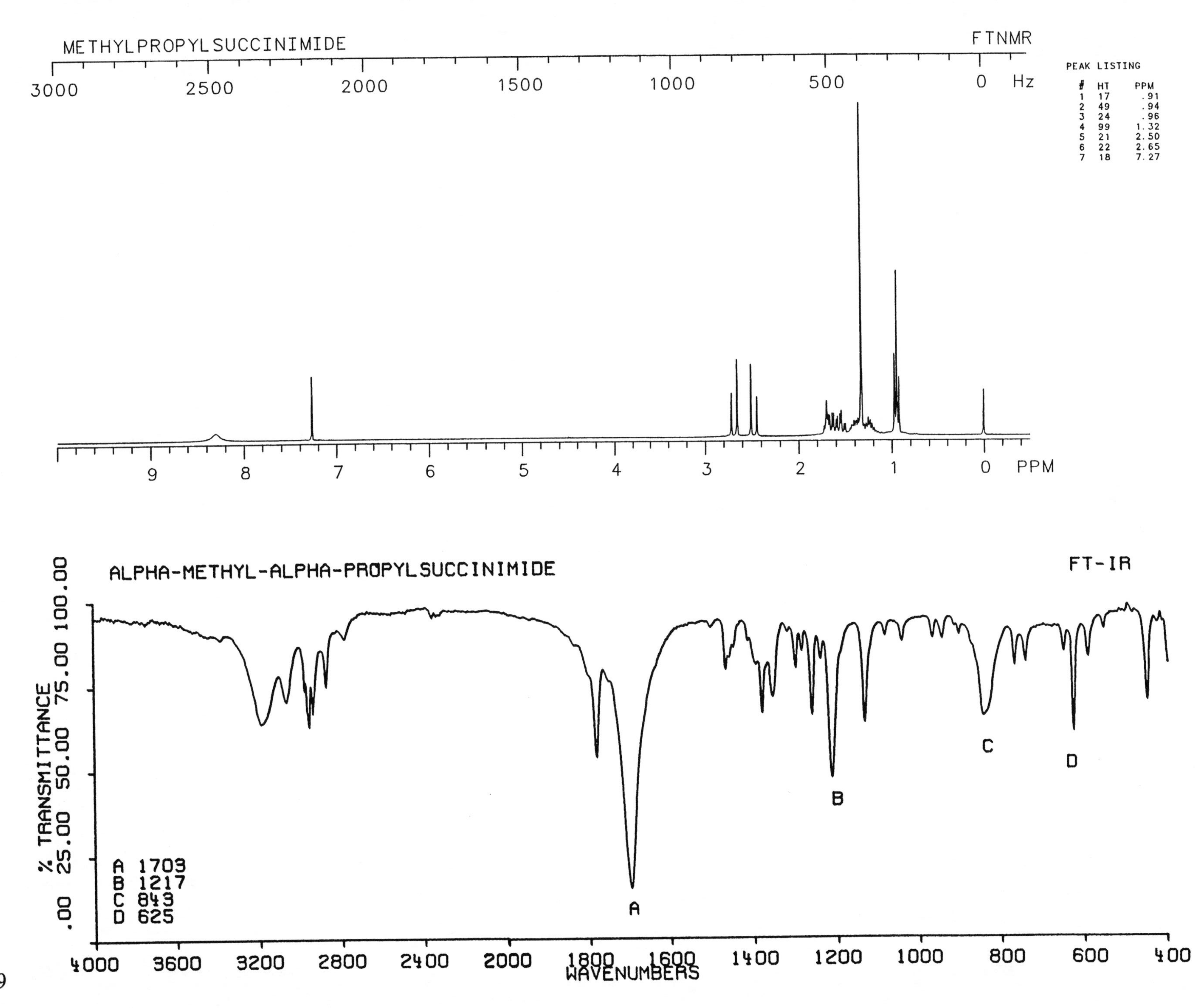
METHYLPROPYLSUCCINIMIDE
FTNMR
3000
2500
2000
1500
1000
500
0 Hz
PEAK LISTING
# HT PPM
1 17 .91
2 49 .94
3 24 .96
4 99 1.32
5 21 2.50
6 22 2.65
7 18 7.27
9
8
7
6
5
4
3
2
1
0 PPM
ALPHA-METHYL-ALPHA-PROPYLSUCCINIMIDE
FT-IR
% TRANSMITTANCE
.00
25.00
50.00
75.00
100.00
A 1703
B 1217
C 843
D 625
A
B
C
D
4000
3600
3200
2800
2400
2000
1800
1600
1400
1200
1000
800
600
400
WAVENUMBERS

# METHYL SALICYLATE

$C_8H_8O_3$

Molecular weight: 152.15 (152.05)
Synonyms: 2-Hydroxybenzoic acid methyl ester; wintergreen oil; teaberry oil
Trade names: Banalg Liniment, Ger-O-Foam, Panalgesic, Sinex Thera-Gesic, Vicks Inhaler
Use: Flavoring, counterirritant
HPLC: Si-10; 2A:98B; 6.0
GC: 1199; 140°C

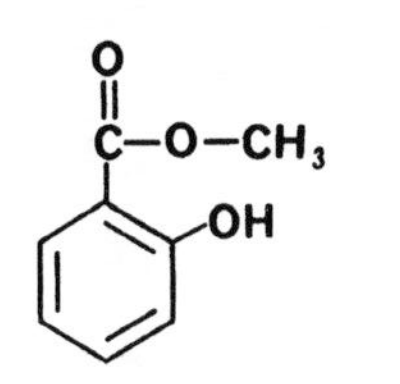


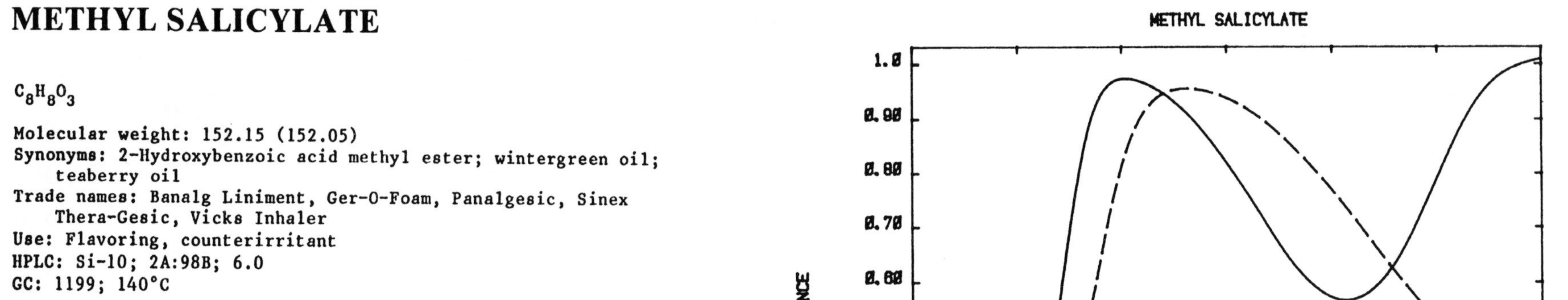


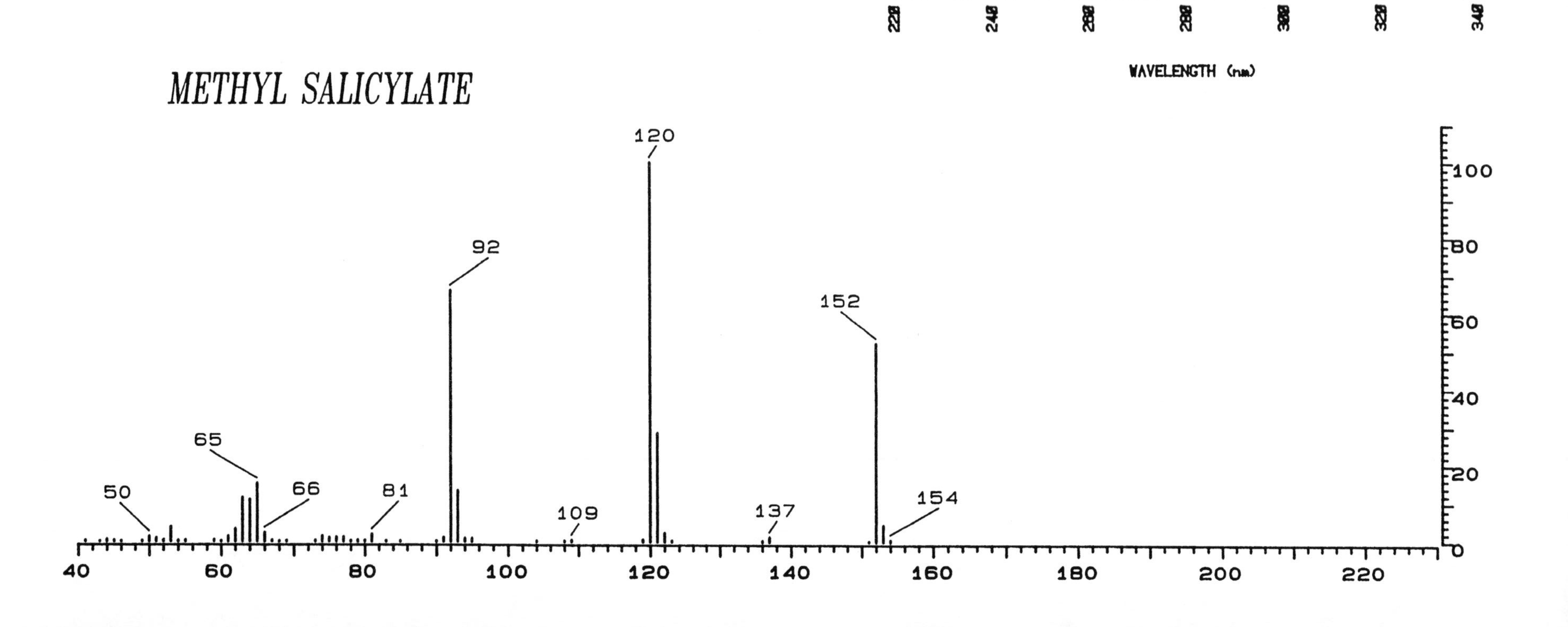

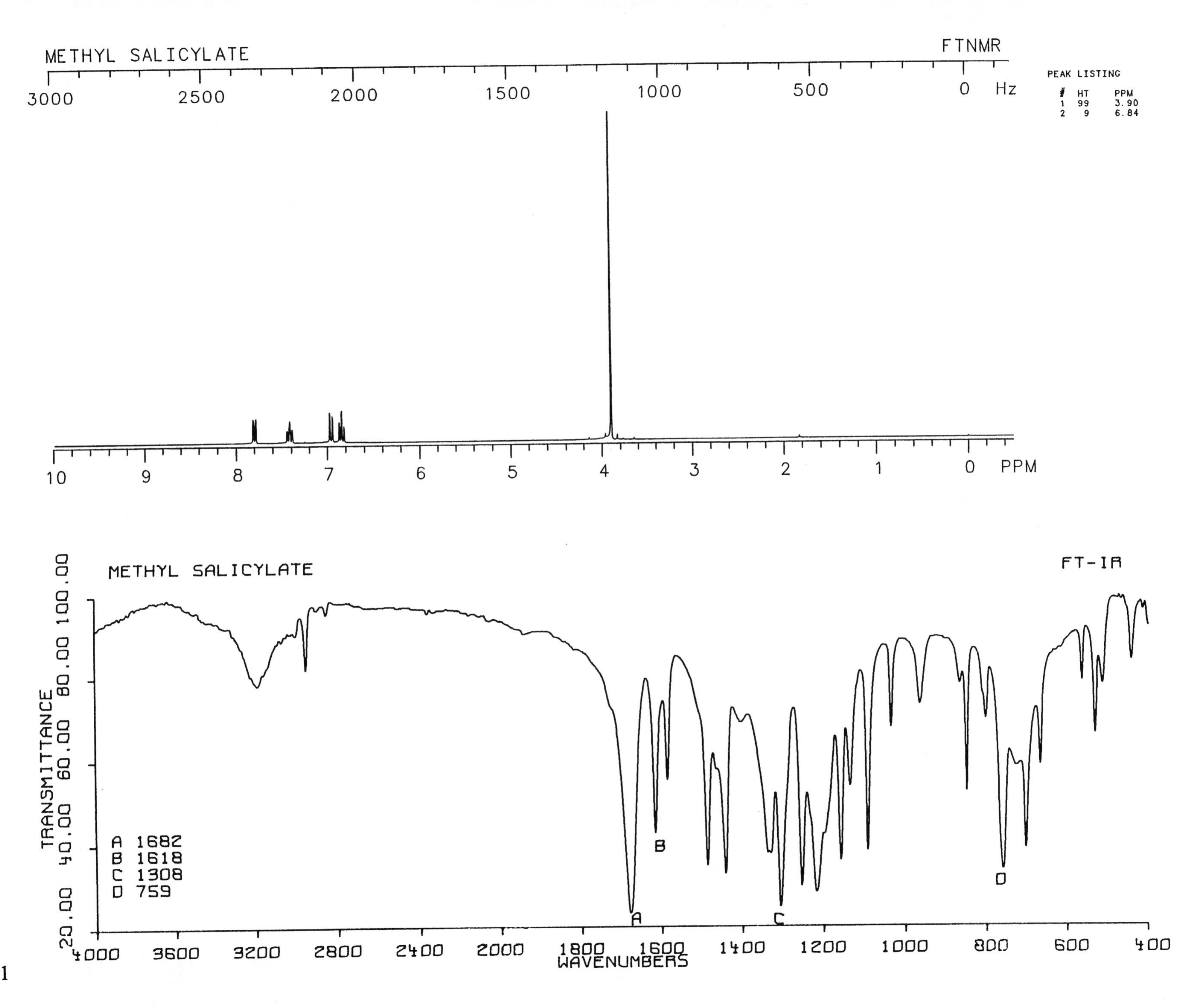
METHYL SALICYLATE
FTNMR
3000
2500
2000
1500
1000
500
0 Hz
PEAK LISTING
# HT PPM
1 99 3.90
2 9 6.84
10
9
8
7
6
5
4
3
2
1
0 PPM
METHYL SALICYLATE
FT-IR
TRANSMITTANCE
20.00
40.00
60.00
80.00
100.00
A 1682
B 1618
C 1308
D 759
A
B
C
D
4000
3600
3200
2800
2400
2000
1800
1600
1400
1200
1000
800
600
400
WAVENUMBERS

# METHYLTESTOSTERONE

$C_{20}H_{30}O_2$

**Molecular weight:** 302.46 (302.22)
**Synonyms:** 17β-Hydroxy-17-methylandrost-4-en-3-one; 17-methyltestosterone
**Trade names:** Android, Estratest, Gynetone, Metandren, Oreton, Premarin, Testred, Virilon
**Use:** Androgen
**HPLC:** Si-10; 2A:98B; 3.7
**GC:** 2703; 250°C

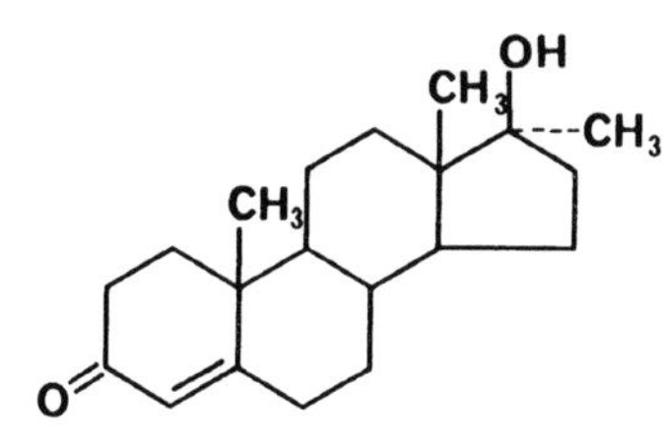


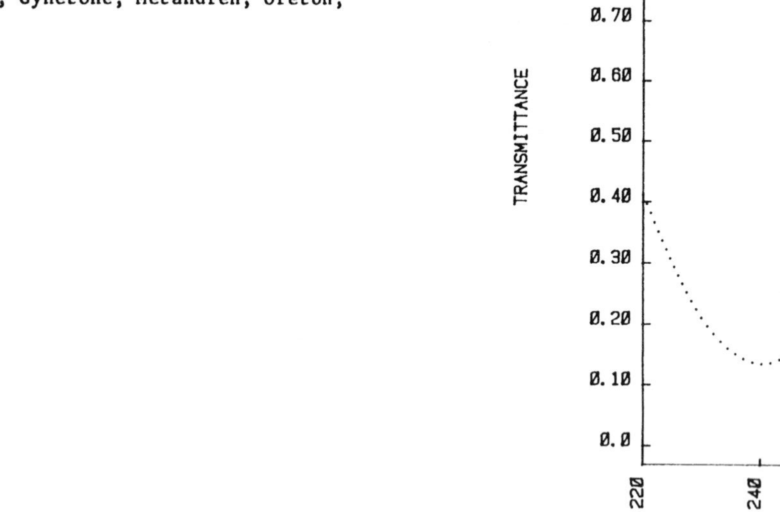

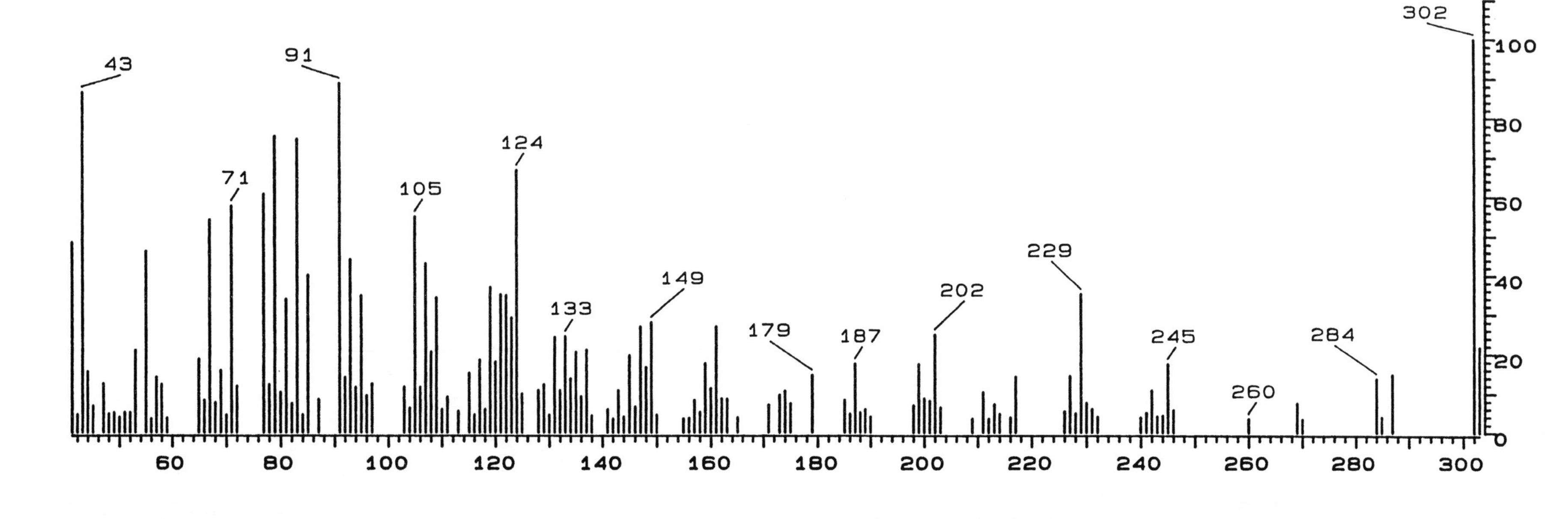

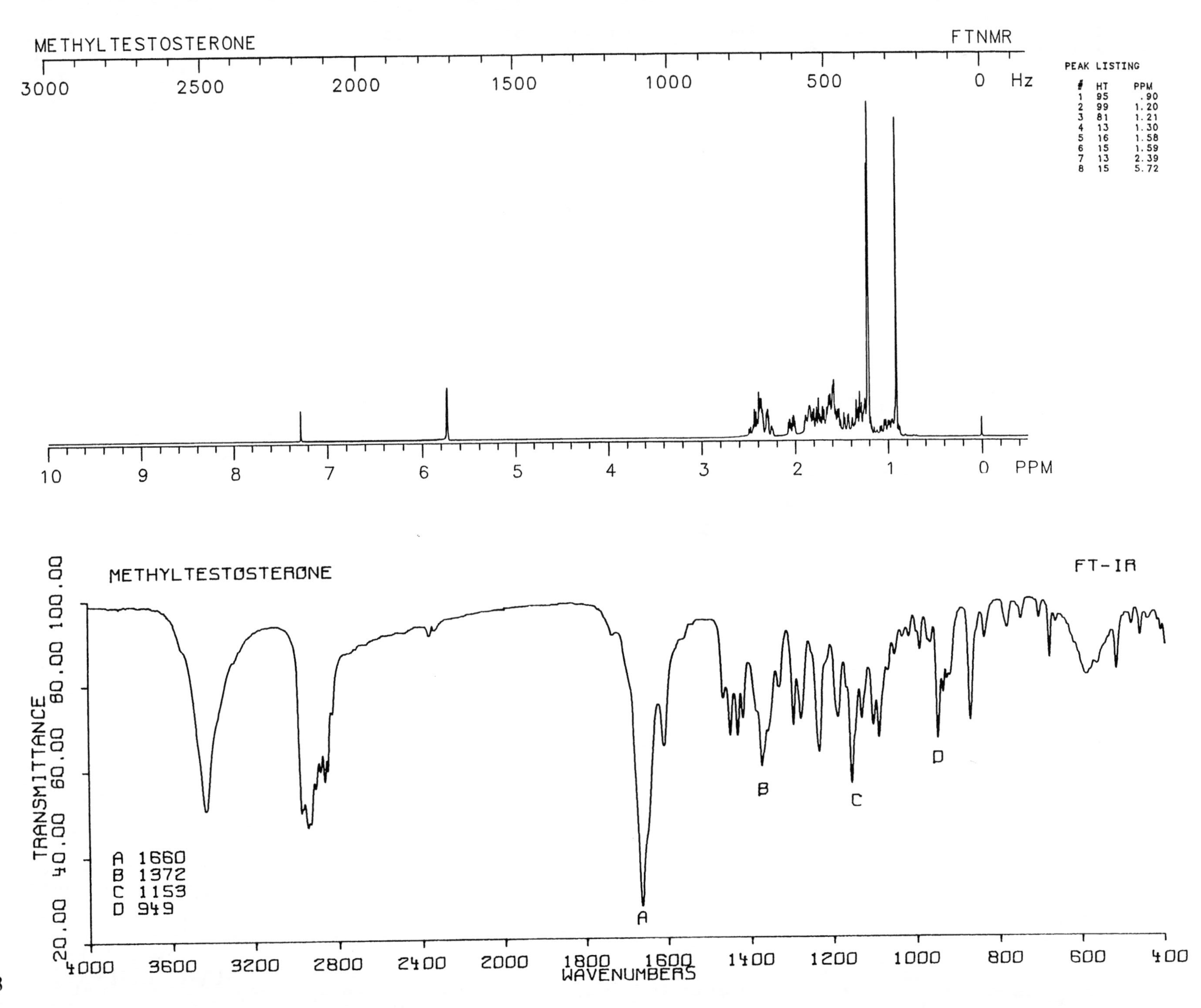
METHYLTESTOSTERONE
FTNMR
3000
2500
2000
1500
1000
500
0
Hz
PEAK LISTING
# HT PPM
1 95 .90
2 99 1.20
3 81 1.21
4 13 1.30
5 16 1.58
6 15 1.59
7 13 2.39
8 15 5.72
10
9
8
7
6
5
4
3
2
1
0
PPM
METHYLTESTOSTERONE
FT-IR
TRANSMITTANCE
20.00
40.00
60.00
80.00
100.00
A 1660
B 1372
C 1153
D 949
A
B
C
D
4000
3600
3200
2800
2400
2000
1800
1600
1400
1200
1000
800
600
400
WAVENUMBERS

# *N*-METHYLTRYPTAMINE

$C_{11}H_{14}N_2$

**Molecular weight:** 174.25 (174.12)
**Synonyms:** 3-(2-Methylaminoethyl)indole

**Trade names:**

**Use:** Hallucinogen
HPLC:
GC:

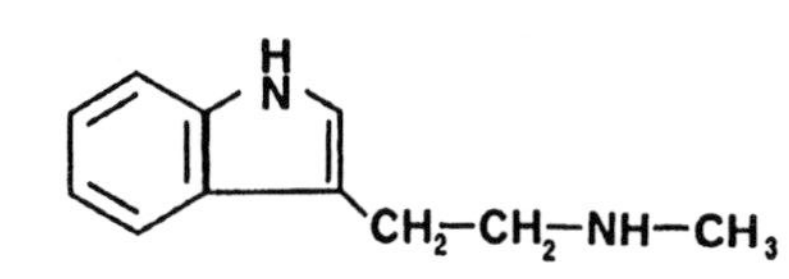


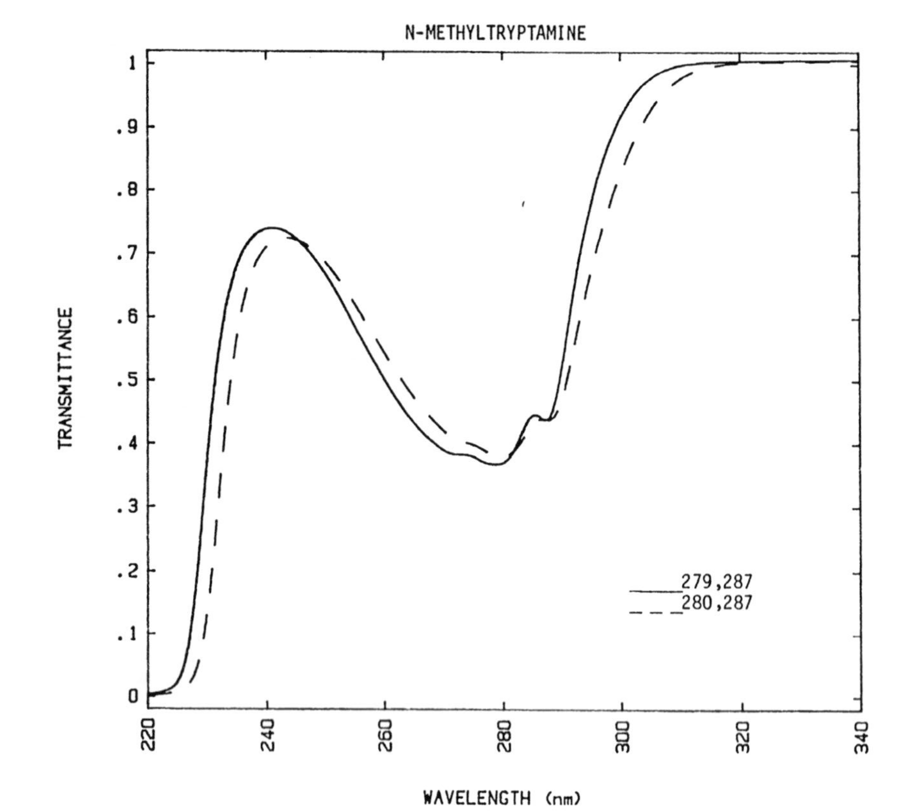

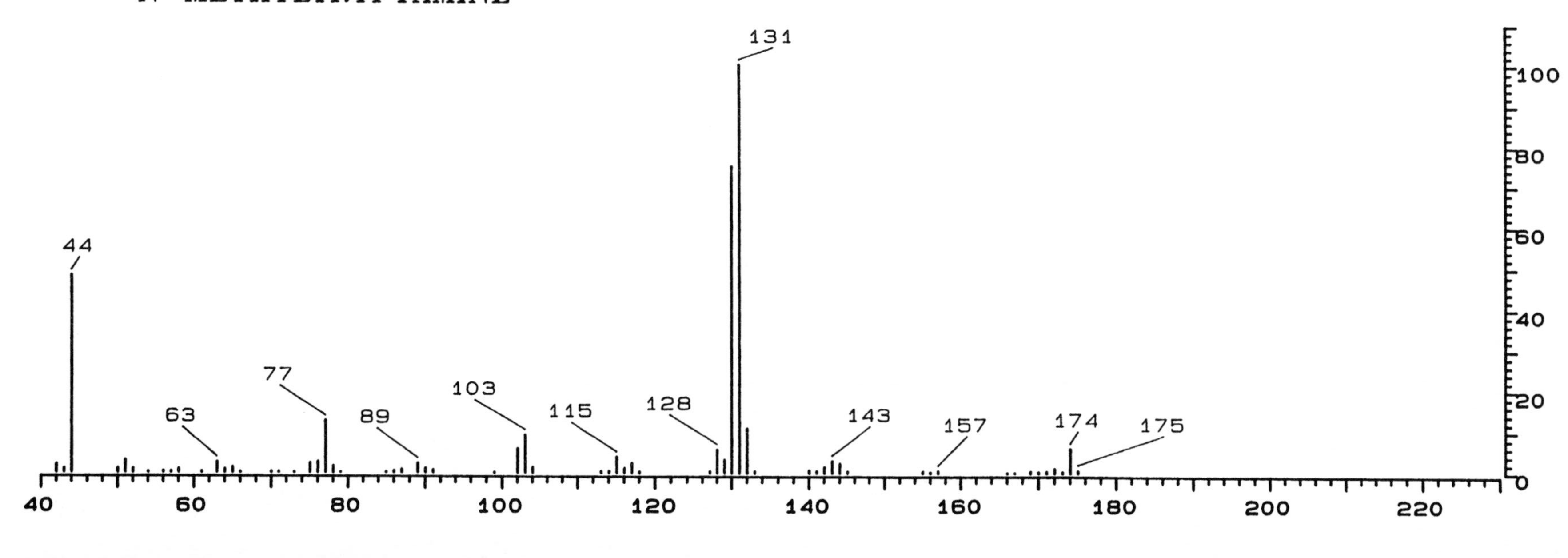

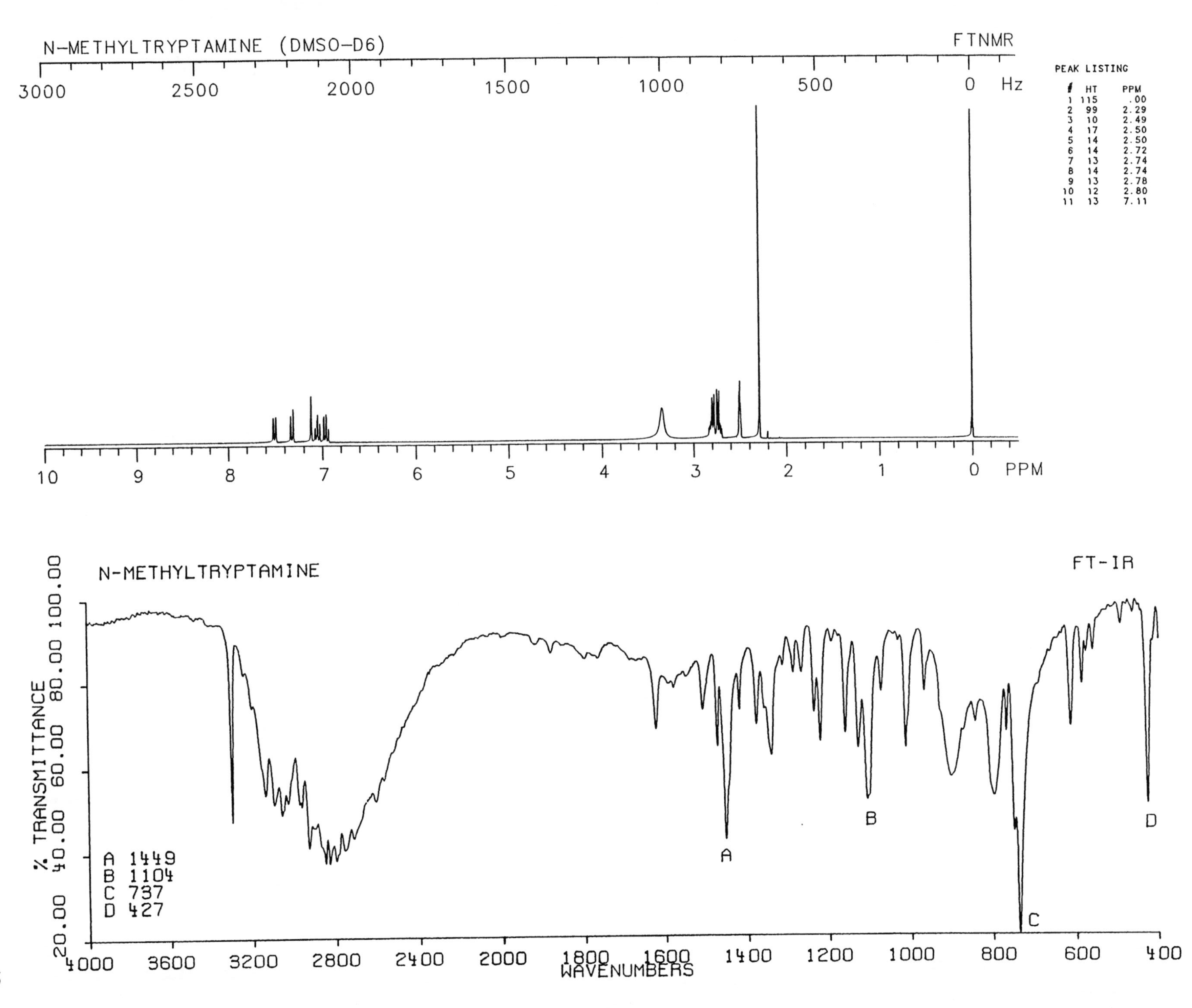
N-METHYLTRYPTAMINE (DMSO-D6)
FTNMR
Hz
PPM
PEAK LISTING
# HT PPM
1 115 .00
2 99 2.29
3 10 2.49
4 17 2.50
5 14 2.50
6 14 2.72
7 13 2.74
8 14 2.74
9 13 2.78
10 12 2.80
11 13 7.11
N-METHYLTRYPTAMINE
FT-IR
% TRANSMITTANCE
WAVENUMBERS
A 1449
B 1104
C 737
D 427

# 5-METHYLTRYPTAMINE

$C_{11}H_{14}N_2$

Molecular weight: 174.25 (174.12)
Synonyms: 3-(2-Aminoethyl)-5-methylindole

Trade names:

Use: Hallucinogen
HPLC: Si-10; 20A:80B; 4.1
GC: 1775; 200°C

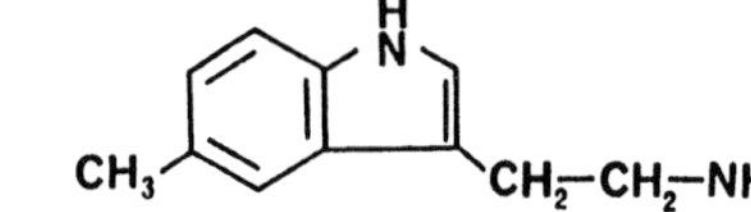


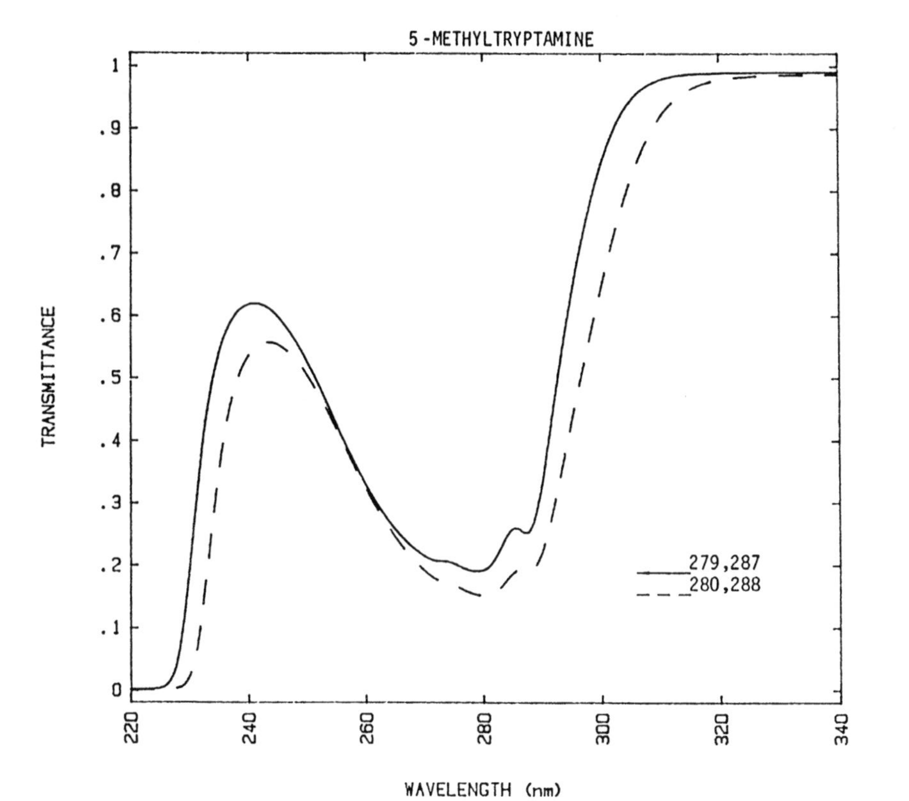

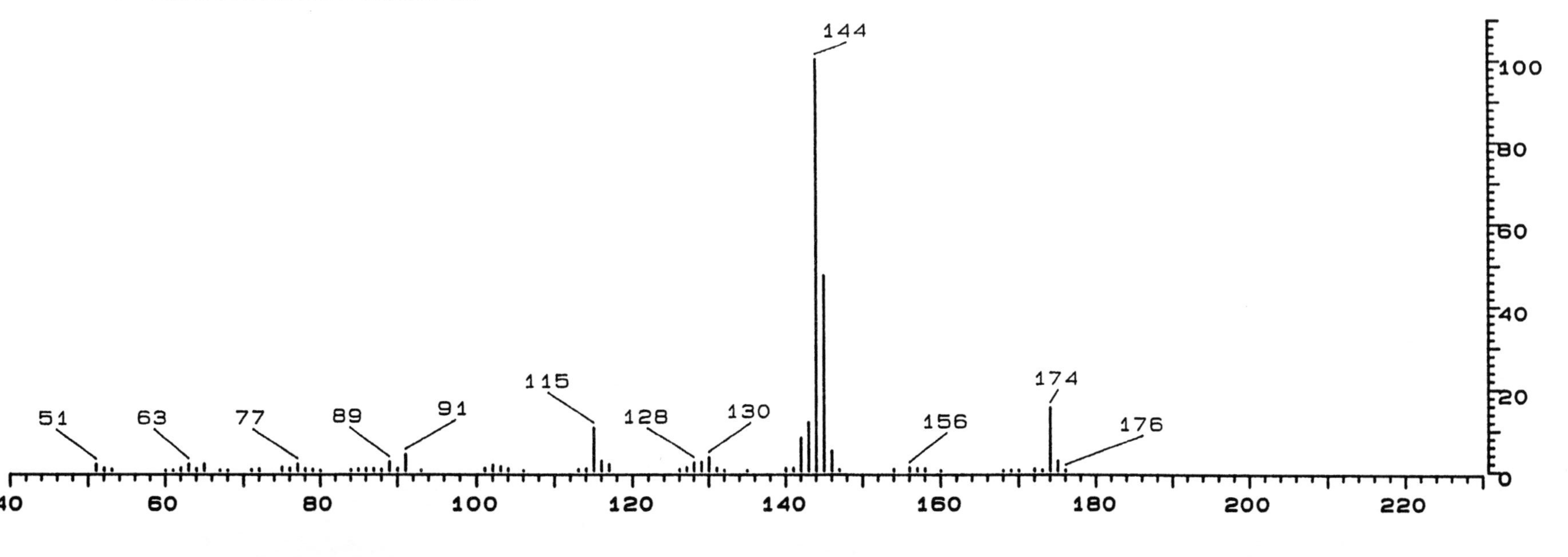

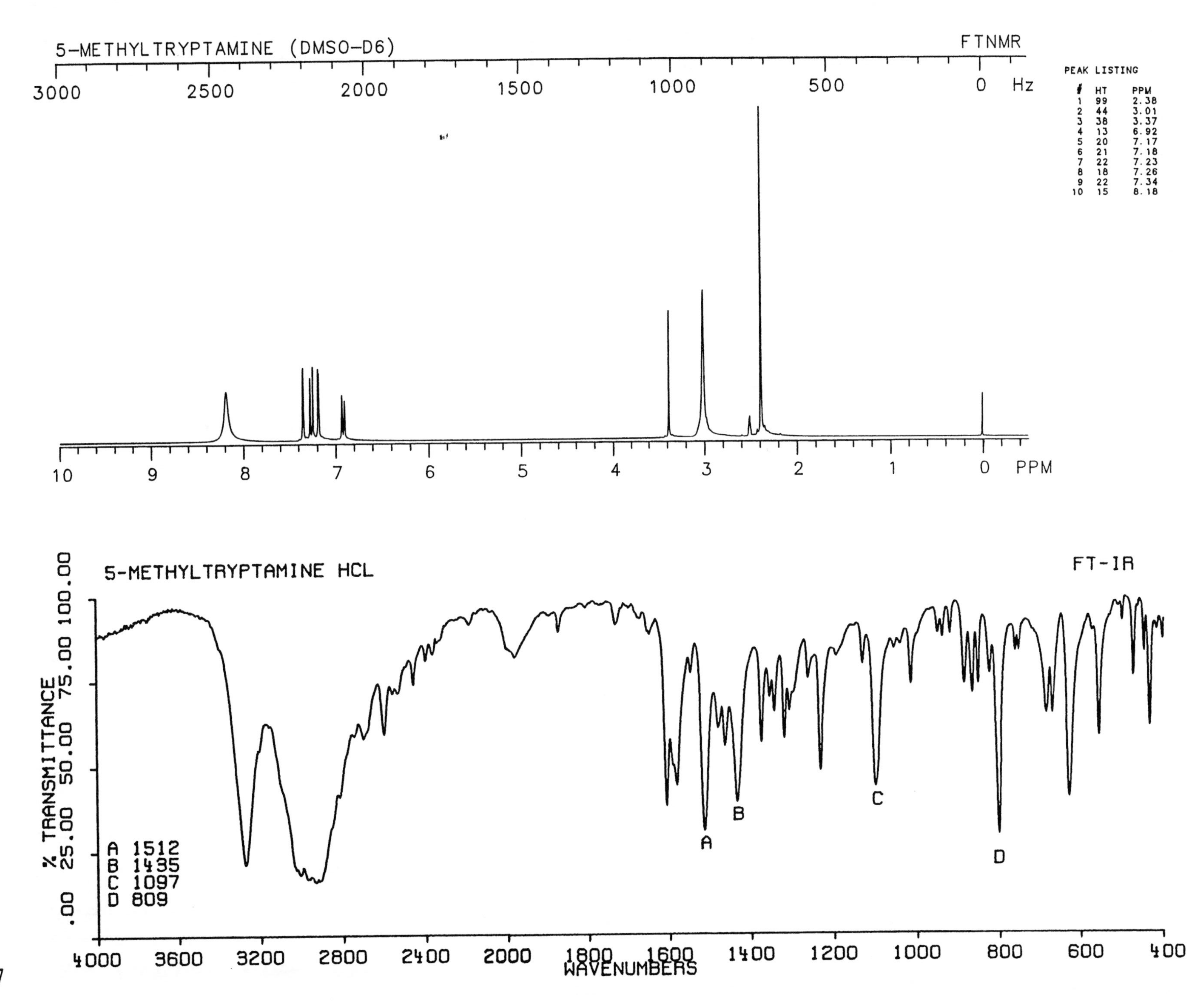
5-METHYLTRYPTAMINE (DMSO-D6)
FTNMR
Hz
PPM
PEAK LISTING
# HT PPM
1 99 2.38
2 44 3.01
3 38 3.37
4 13 6.92
5 20 7.17
6 21 7.18
7 22 7.23
8 18 7.26
9 22 7.34
10 15 8.18
5-METHYLTRYPTAMINE HCL
FT-IR
% TRANSMITTANCE
WAVENUMBERS
A 1512
B 1435
C 1097
D 809

# 3-METHYLXANTHINE

$C_6H_6N_4O_2$

**Molecular weight:** 166.14 (166.05)
**Synonyms:** 3,7-Dihydro-3-methyl-1H-purine-2,6-dione

**Trade names:**

**Use:**
**HPLC:** Si-10; 10A:90B; 4.0
**GC:**

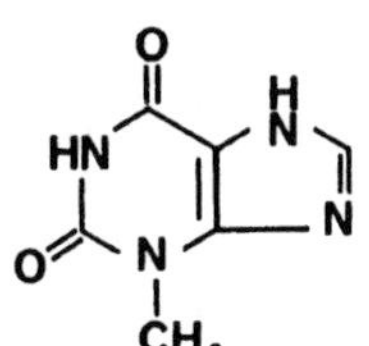


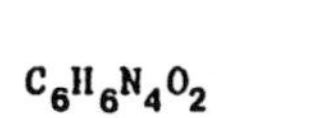

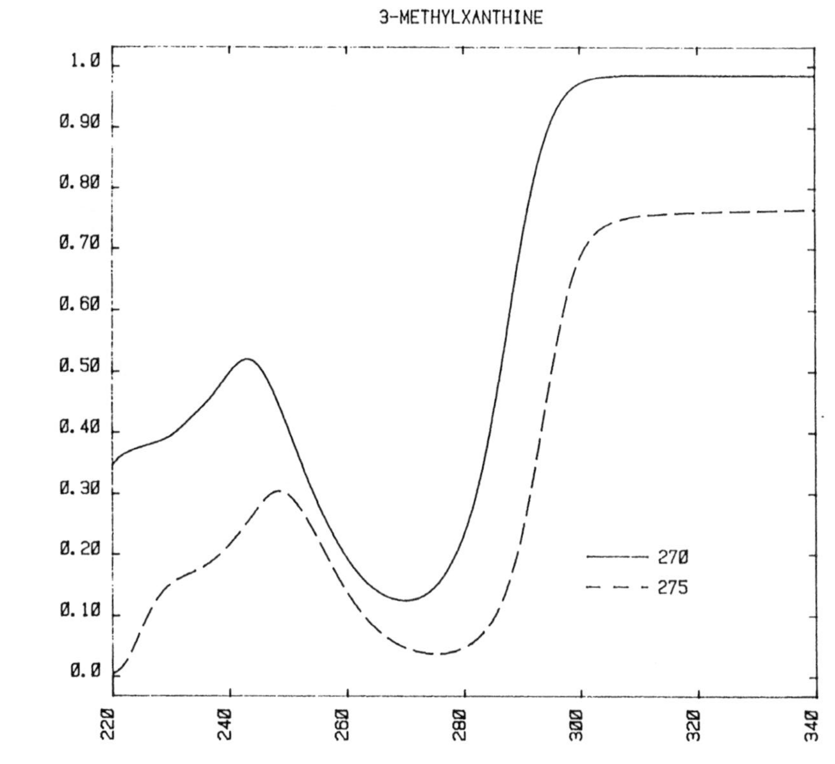


## 3-METHYLXANTHINE--DIP

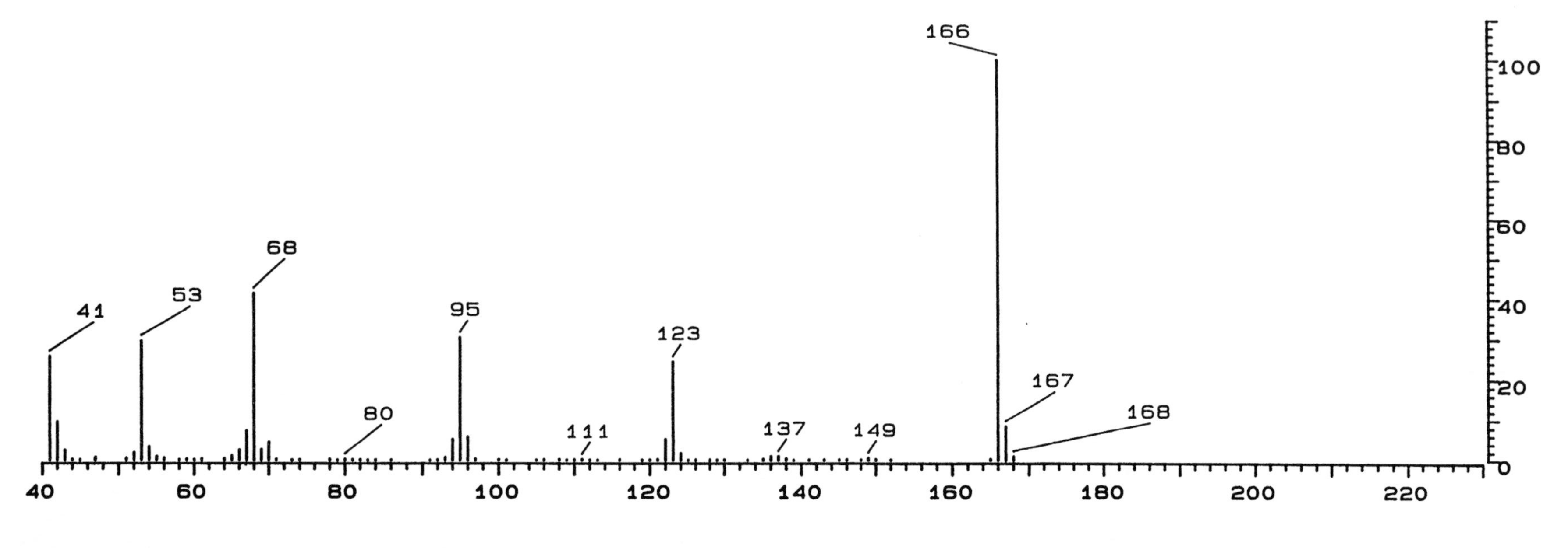

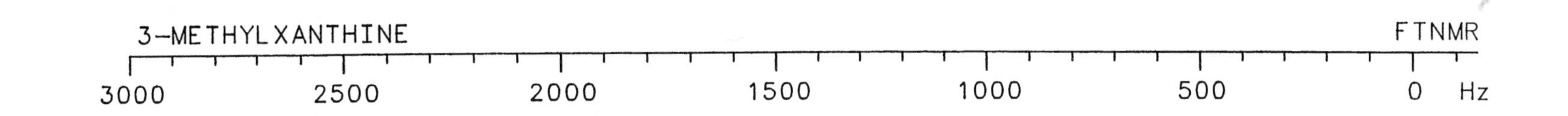


INSUFFICIENT SOLUBILITY

10 9 8 7 6 5 4 3 2 1 0 PPM

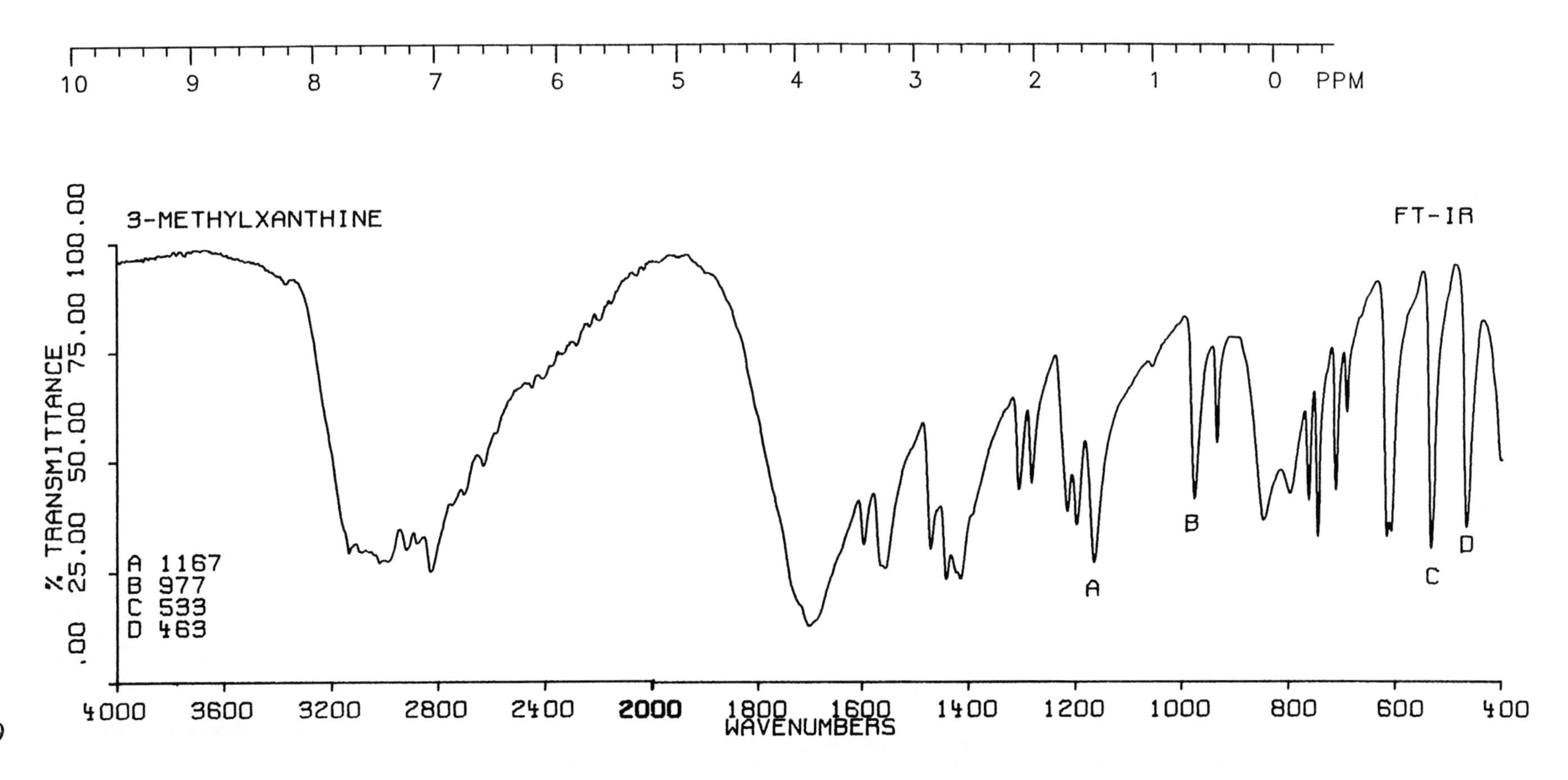

# METHYPRYLON

$C_{10}H_{17}NO_2$

Molecular weight: 183.25 (183.13)
Synonyms: 3,3-Diethyl-5-methyl-2,4-piperidinedione

Trade names: Noludar

Use: Sedative, hypnotic
HPLC: Si-10; 2A:98B; 3.8
GC: 1553; 200°C

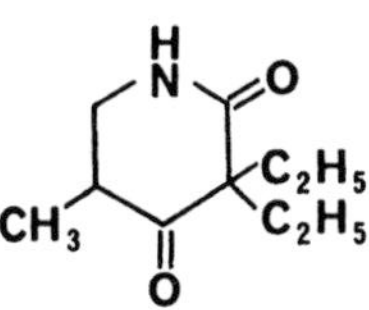

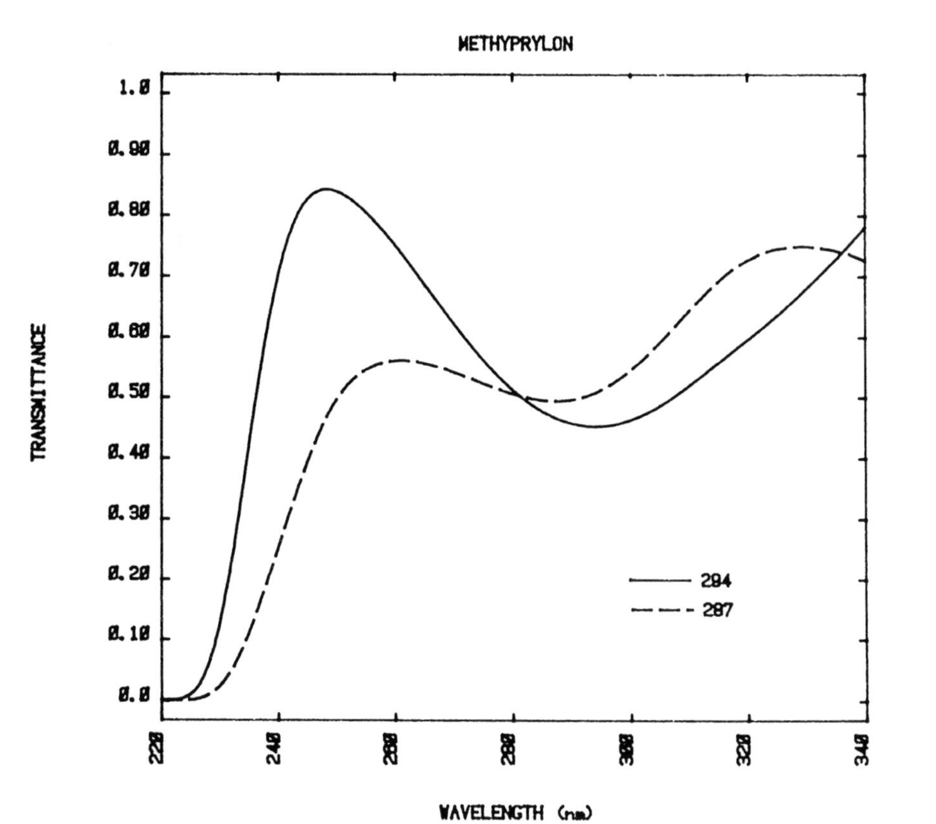


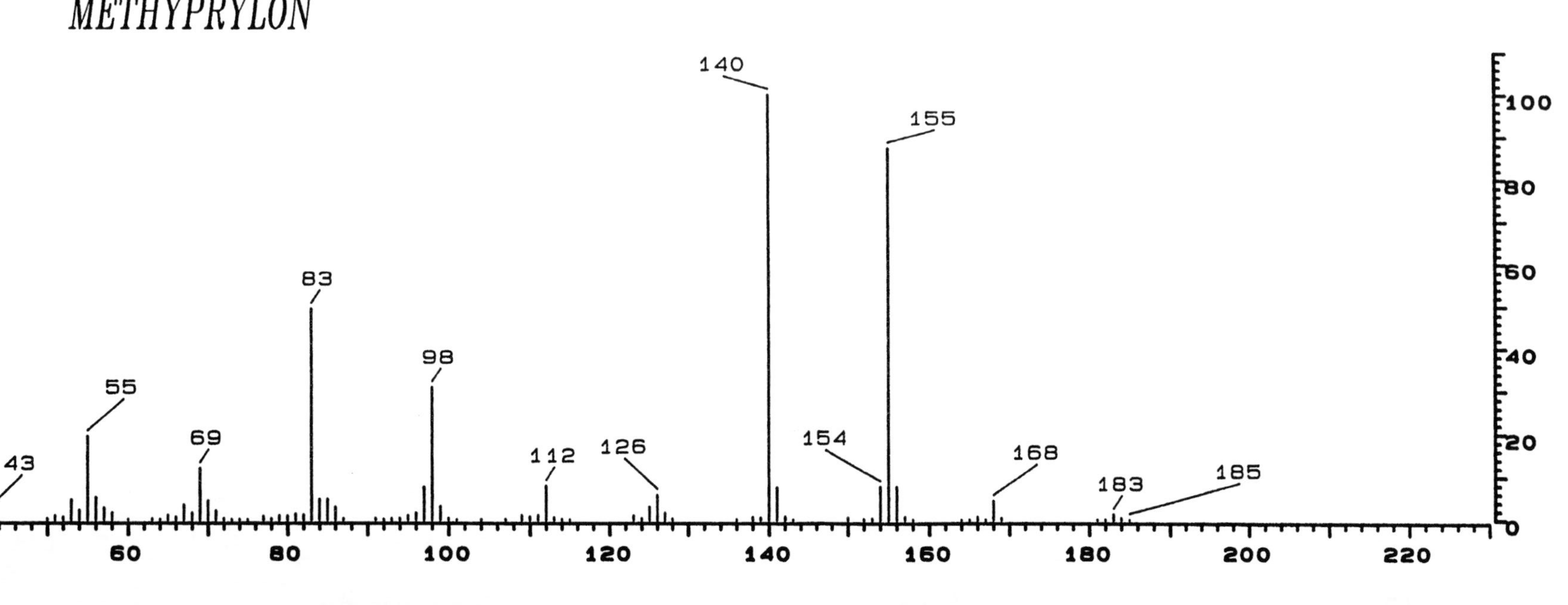

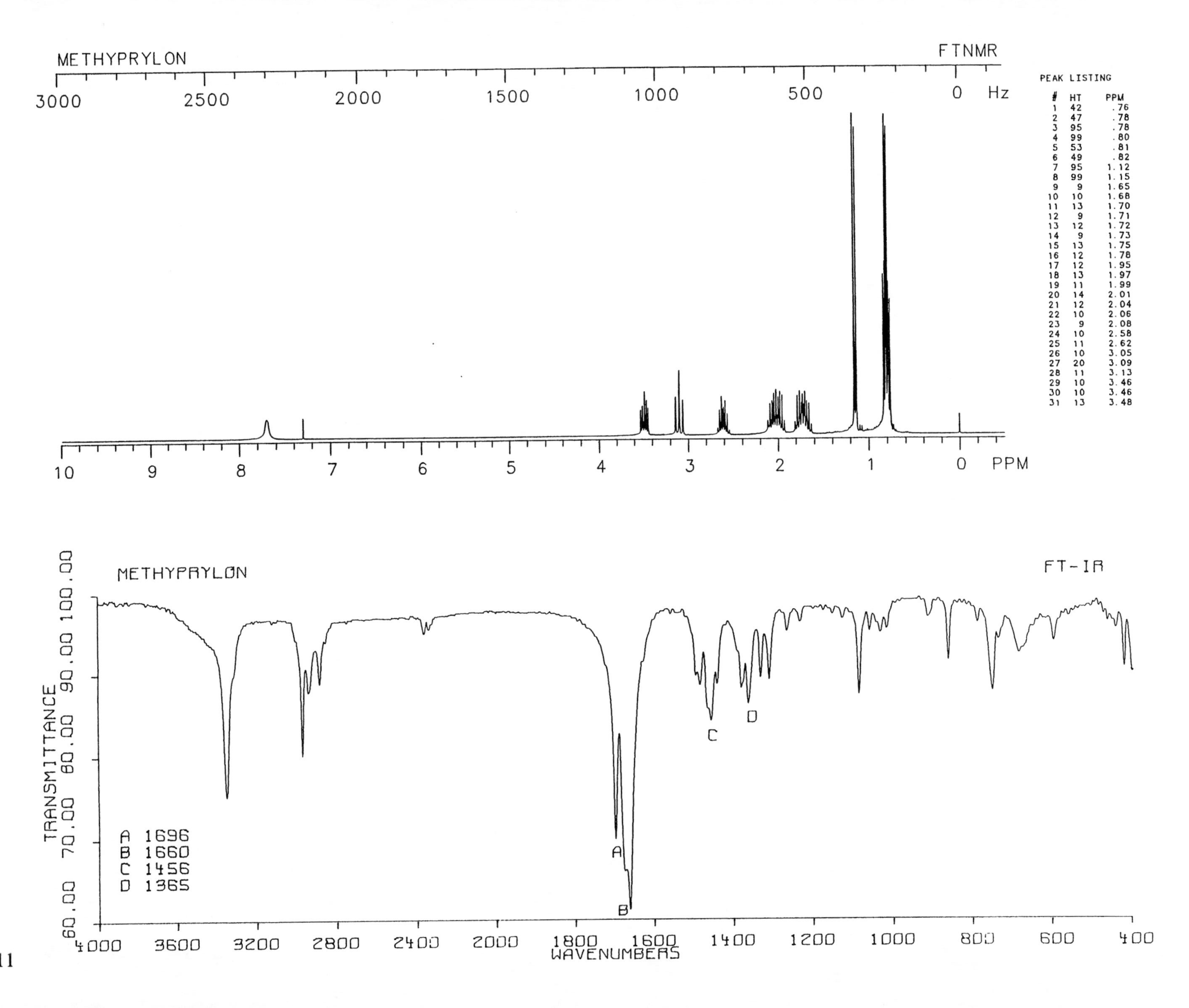
METHYPRYLON
FTNMR
3000
2500
2000
1500
1000
500
0 Hz
10
9
8
7
6
5
4
3
2
1
0 PPM
PEAK LISTING
# HT PPM
1 42 .76
2 47 .78
3 95 .78
4 99 .80
5 53 .81
6 49 .82
7 95 1.12
8 99 1.15
9 9 1.65
10 10 1.68
11 13 1.70
12 9 1.71
13 12 1.72
14 9 1.73
15 13 1.75
16 12 1.78
17 12 1.95
18 13 1.97
19 11 1.99
20 14 2.01
21 12 2.04
22 10 2.06
23 9 2.08
24 10 2.58
25 11 2.62
26 10 3.05
27 20 3.09
28 11 3.13
29 10 3.46
30 10 3.46
31 13 3.48
METHYPRYLON
FT-IR
TRANSMITTANCE
60.00
70.00
80.00
90.00
100.00
A 1696
B 1660
C 1456
D 1365
A
B
C
D
4000
3600
3200
2800
2400
2000
1800
1600
1400
1200
1000
800
600
400
WAVENUMBERS

# METHYSERGIDE

$C_{21}H_{27}N_3O_2$

Molecular weight: 353.47 (353.21)
Synonyms: 9,10-Didehydro-N-[1-(hydroxymethyl)propyl]-1,6-dimethylergoline-8-carboxamide; 1-methylmethylergonovine
Trade names: Sansert

Use: Migraine prophylactic
HPLC:
GC:

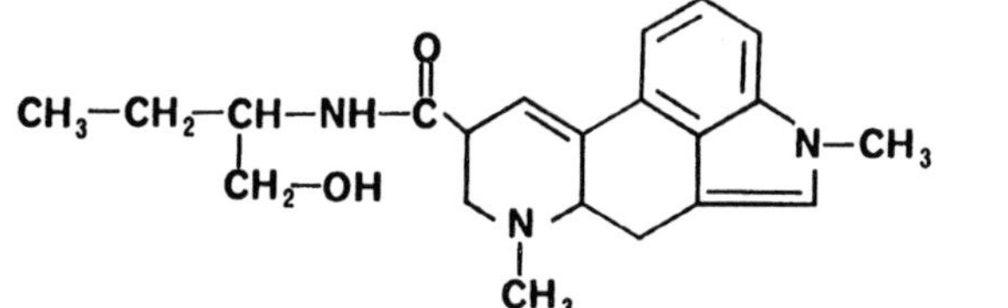


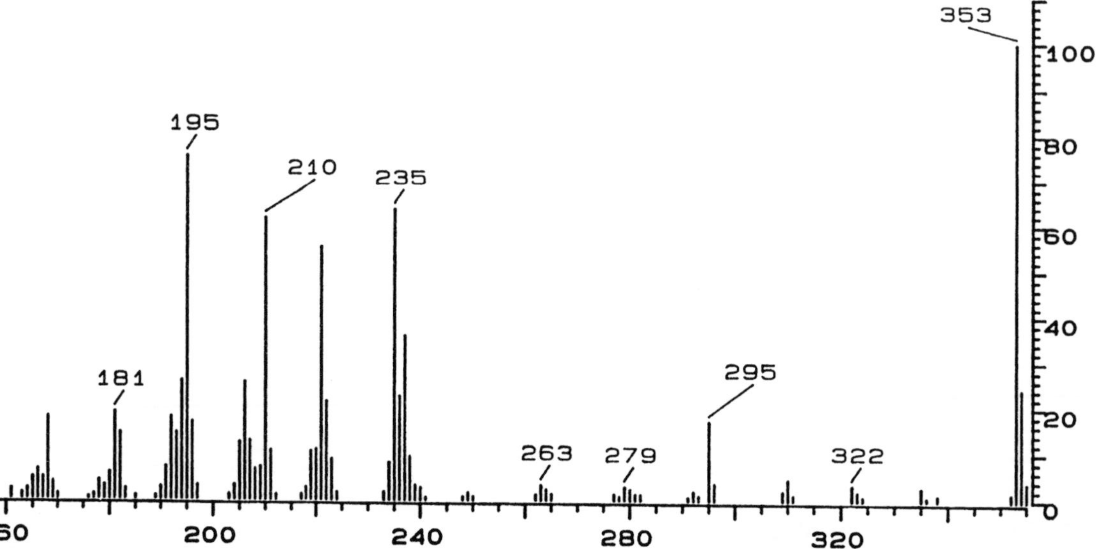


*METHYSERGIDE--DIP*

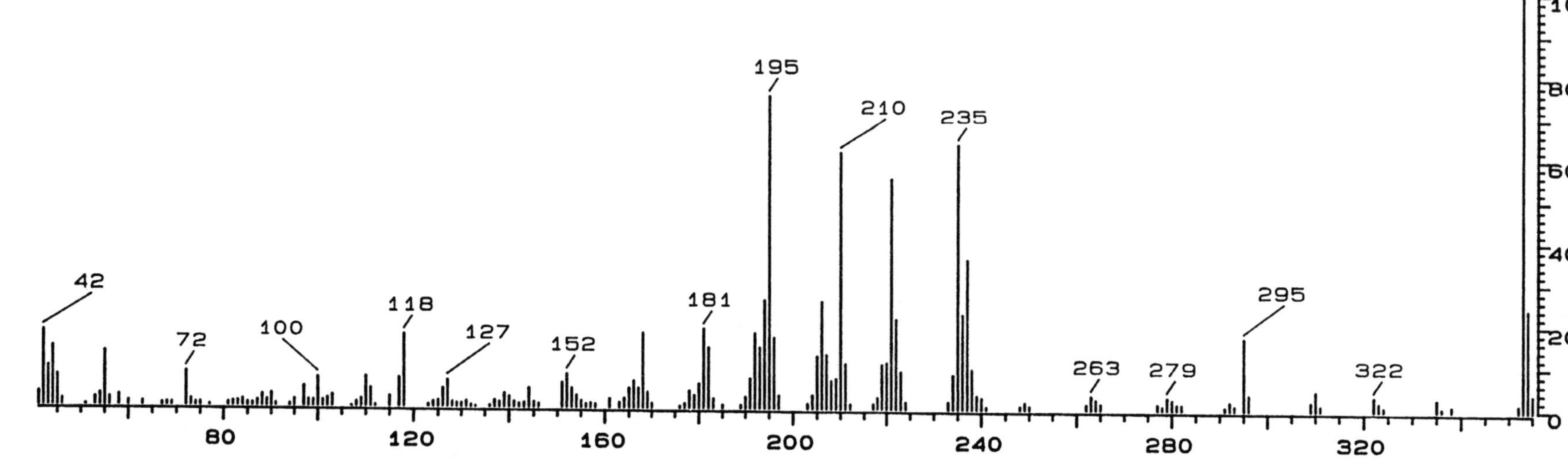


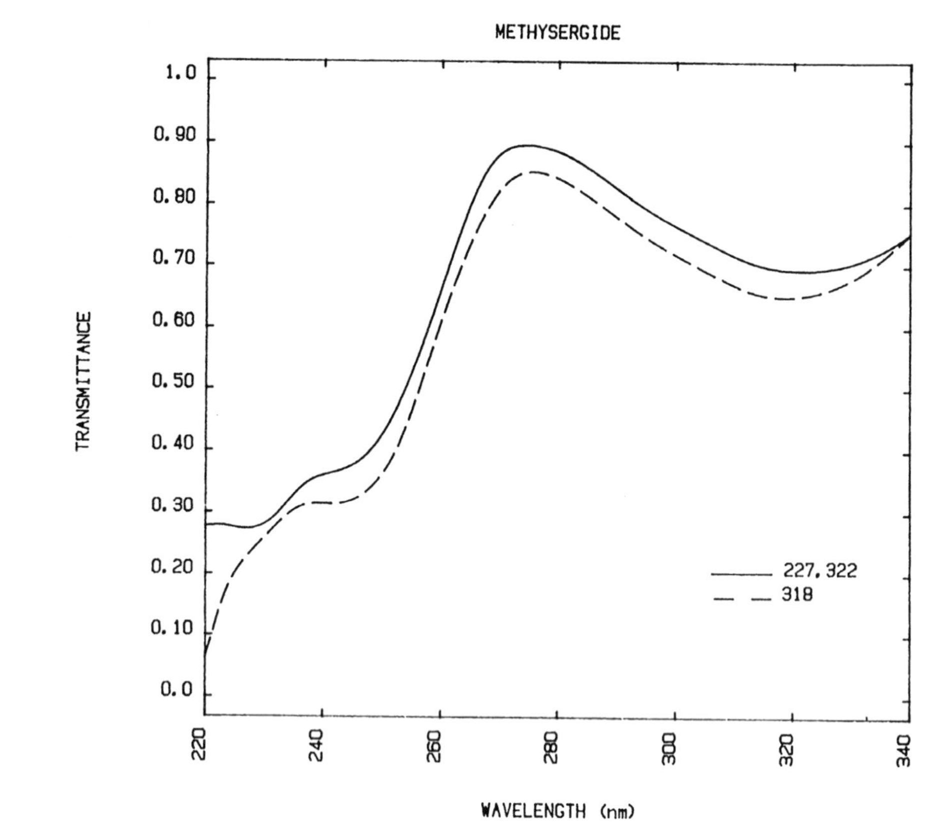

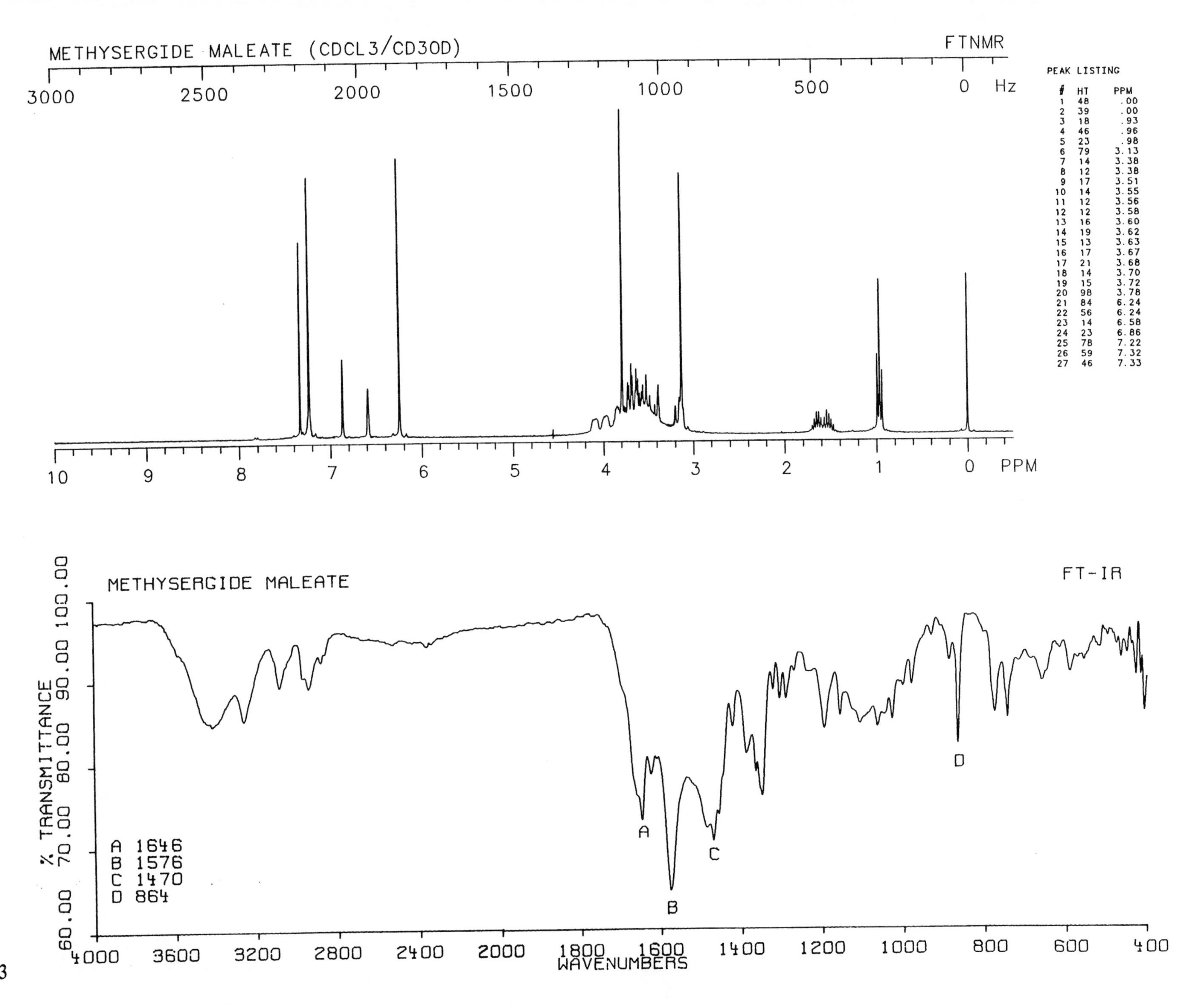
METHYSERGIDE MALEATE (CDCL3/CD3OD)
FTNMR
3000
2500
2000
1500
1000
500
0 Hz
10
9
8
7
6
5
4
3
2
1
0 PPM
PEAK LISTING
# HT PPM
1 48 .00
2 39 .00
3 18 .93
4 46 .96
5 23 .98
6 79 3.13
7 14 3.38
8 12 3.38
9 17 3.51
10 14 3.55
11 12 3.56
12 12 3.58
13 16 3.60
14 19 3.62
15 13 3.63
16 17 3.67
17 21 3.68
18 14 3.70
19 15 3.72
20 98 3.78
21 84 6.24
22 56 6.24
23 14 6.58
24 23 6.86
25 78 7.22
26 59 7.32
27 46 7.33
METHYSERGIDE MALEATE
FT-IR
% TRANSMITTANCE
60.00
70.00
80.00
90.00
100.00
A 1646
B 1576
C 1470
D 864
A
B
C
D
4000
3600
3200
2800
2400
2000
1800
1600
1400
1200
1000
800
600
400
WAVENUMBERS

# METOCLOPRAMIDE

$C_{14}H_{22}ClN_3O_2$

Molecular weight: 299.80 (299.14)
Synonyms: 4-Amino-5-chloro-N-[(2-diethylamino)ethyl]-2-methoxybenzamide
Trade names: Reglan

Use: Antiemetic
HPLC: Si-10; 20A:80B; 4.3
GC:



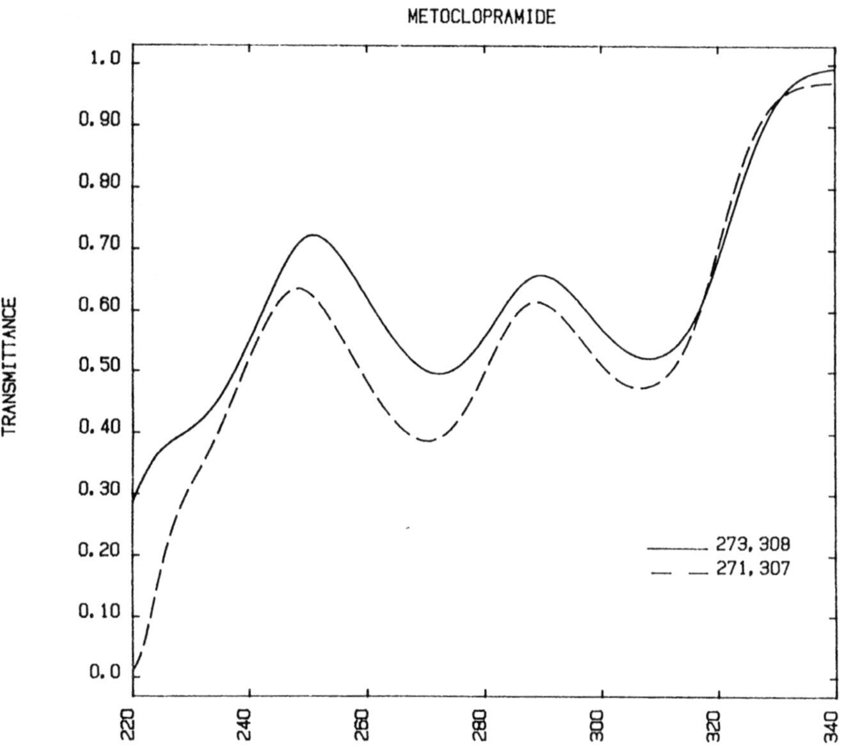


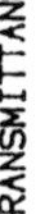

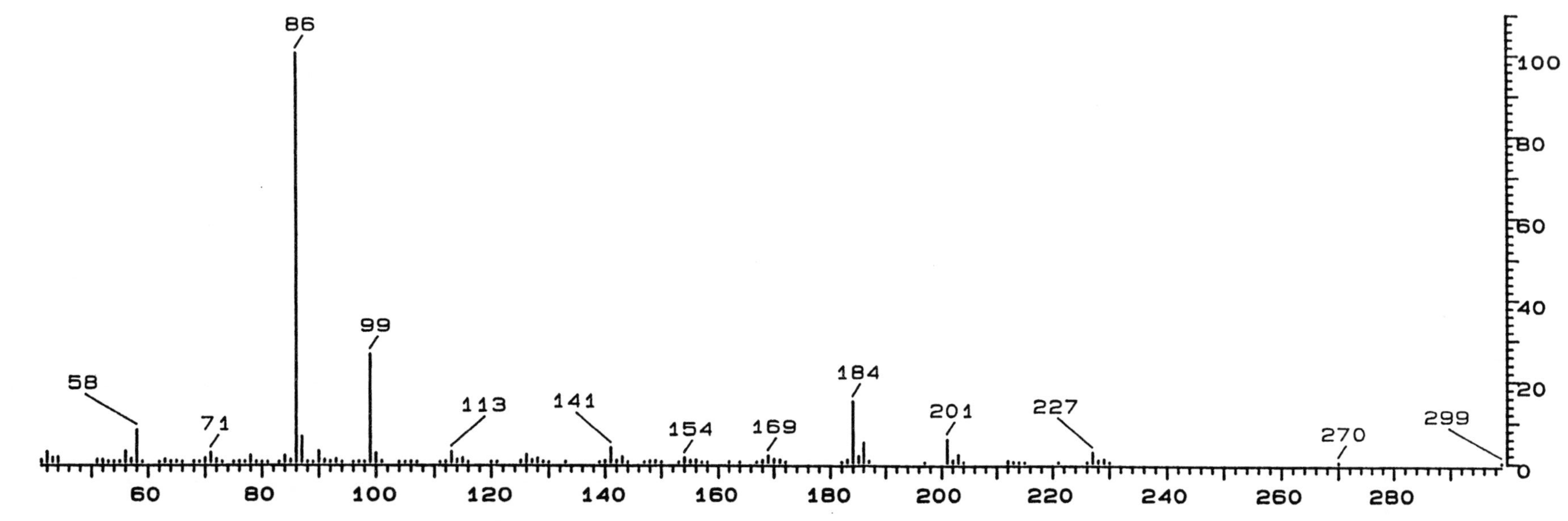

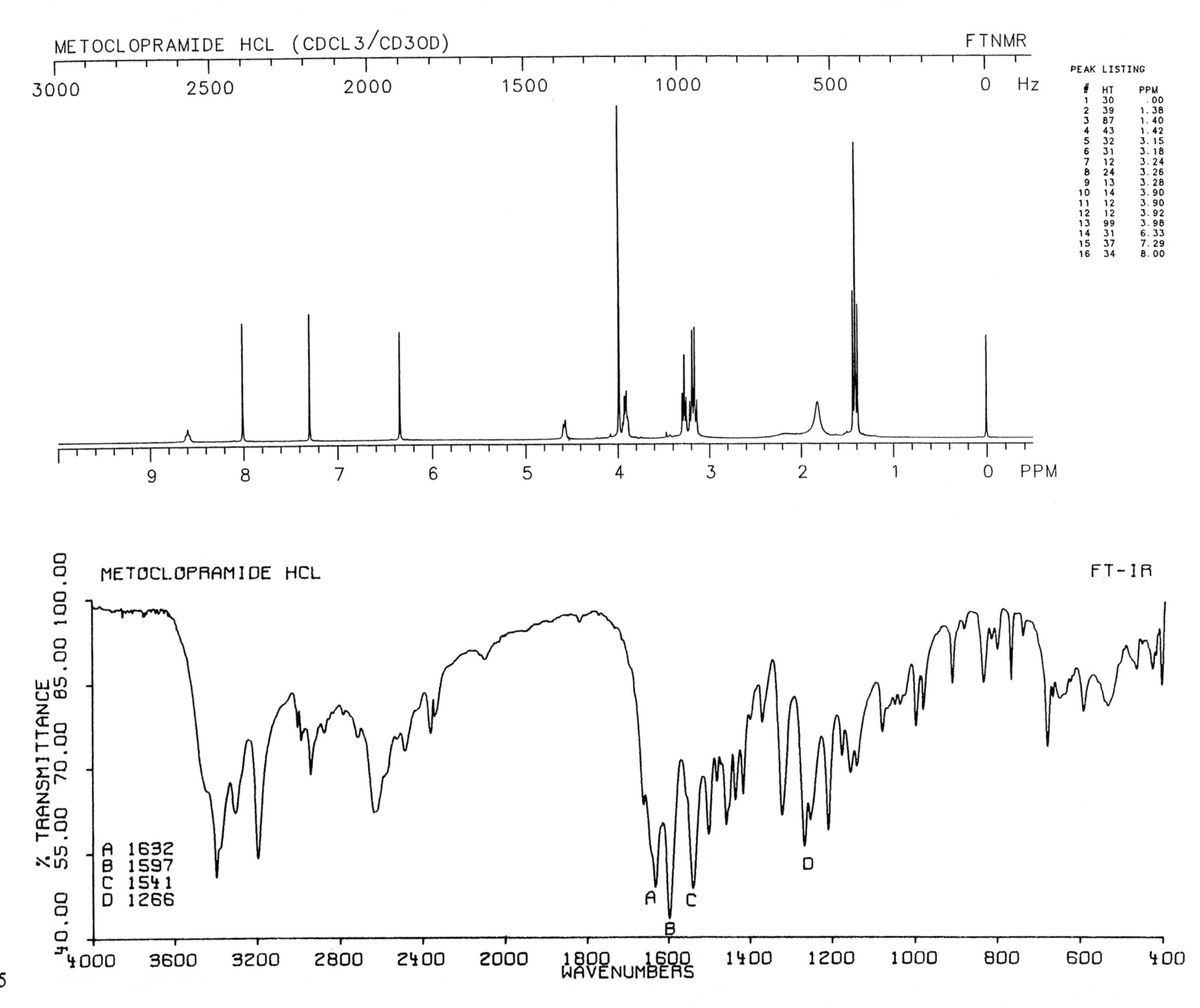
METOCLOPRAMIDE HCL (CDCL3/CD3OD)
FTNMR
3000 2500 2000 1500 1000 500 0 Hz
9 8 7 6 5 4 3 2 1 0 PPM
PEAK LISTING
# HT PPM
1 30 .00
2 39 1.38
3 87 1.40
4 43 1.42
5 32 3.15
6 31 3.18
7 12 3.24
8 24 3.26
9 13 3.28
10 14 3.90
11 12 3.90
12 12 3.92
13 99 3.98
14 31 6.33
15 37 7.29
16 34 8.00
METOCLOPRAMIDE HCL
FT-IR
% TRANSMITTANCE
40.00 55.00 70.00 85.00 100.00
A 1632
B 1597
C 1541
D 1266
A B C D
4000 3600 3200 2800 2400 2000 1800 1600 1400 1200 1000 800 600 400
WAVENUMBERS

# METOLAZONE

$C_{16}H_{16}ClN_3O_3S$

Molecular weight: 365.84 (365.06)
Synonyms: 7-Chloro-1,2,3,4-tetrahydro-2-methyl-3-(2-methylphenyl)-4-oxo-6-quinolinesulfonamide
Trade names: Diulo, Zaroxolyn

Use: Diuretic
HPLC: Si-10; 2A:98B; 4.0
GC:

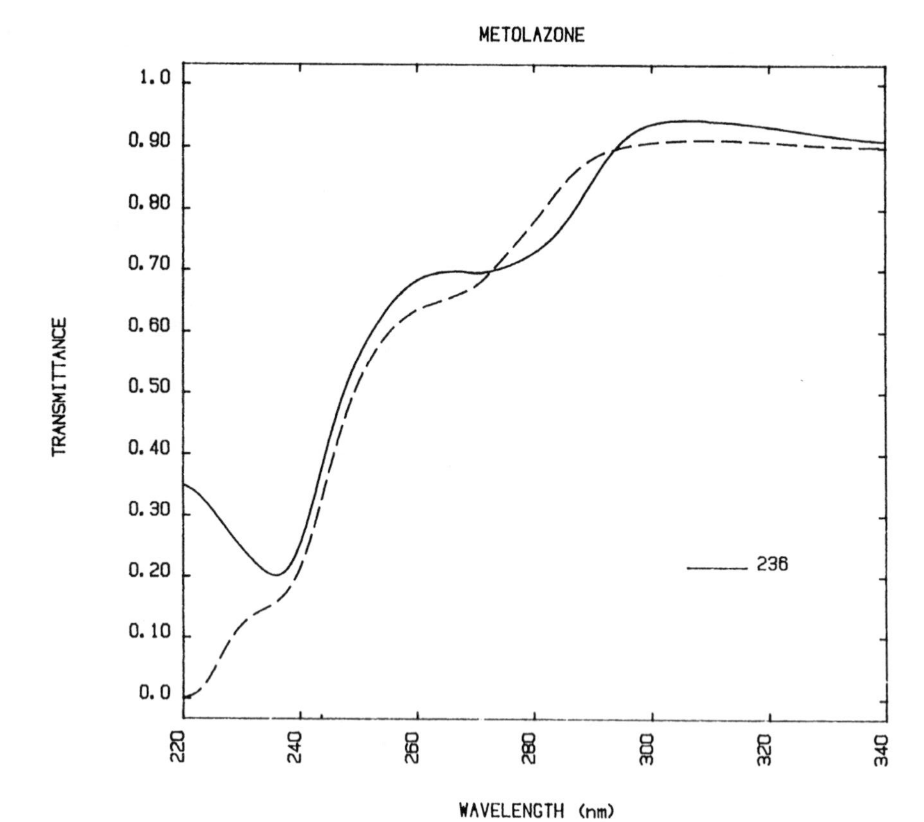

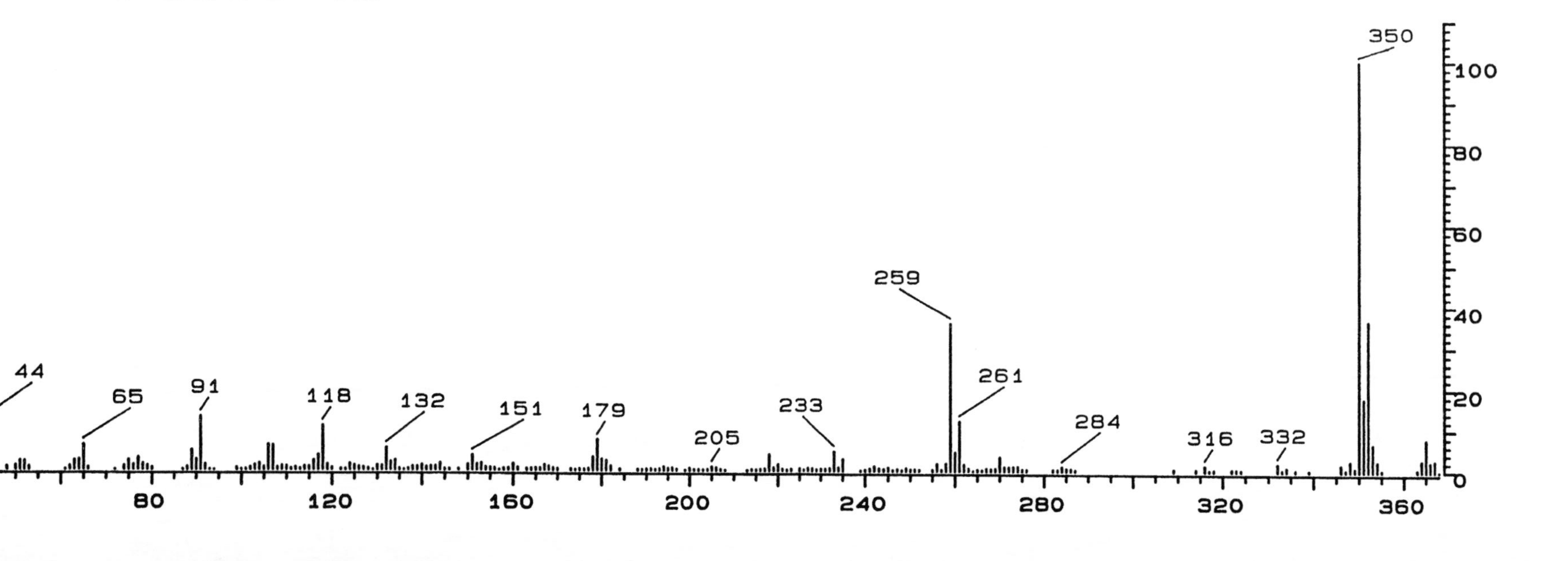

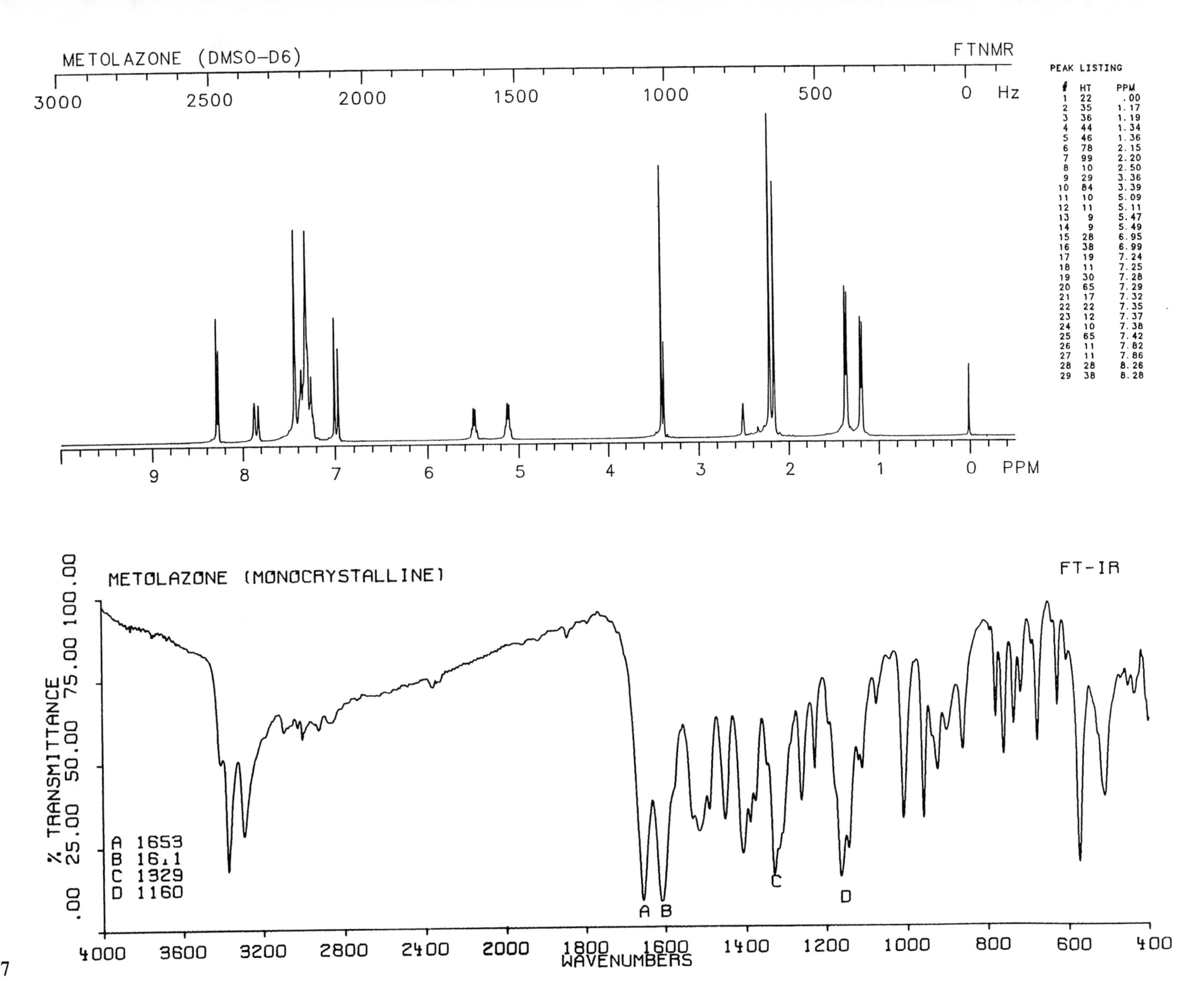
METOLAZONE (DMSO-D6)
FTNMR
3000 2500 2000 1500 1000 500 0 Hz
9 8 7 6 5 4 3 2 1 0 PPM
PEAK LISTING
# HT PPM
1 22 .00
2 35 1.17
3 36 1.19
4 44 1.34
5 46 1.36
6 78 2.15
7 99 2.20
8 10 2.50
9 29 3.36
10 84 3.39
11 10 5.09
12 11 5.11
13 9 5.47
14 9 5.49
15 28 6.95
16 38 6.99
17 19 7.24
18 11 7.25
19 30 7.28
20 65 7.29
21 17 7.32
22 22 7.35
23 12 7.37
24 10 7.38
25 65 7.42
26 11 7.82
27 11 7.86
28 28 8.26
29 38 8.28
METOLAZONE (MONOCRYSTALLINE)
FT-IR
% TRANSMITTANCE
.00 25.00 50.00 75.00 100.00
A 1653
B 16.1
C 1329
D 1160
A B C D
4000 3600 3200 2800 2400 2000 1800 1600 1400 1200 1000 800 600 400
WAVENUMBERS

# METOPROLOL

$C_{15}H_{25}NO_3$

Molecular weight: 267.37 (267.18)
Synonyms: 1-[4-(2-Methoxyethyl)phenoxy]-3-(1-methylethyl)amino]-2-propanol
Trade names: Lopressor

Use: Beta-Adrenergic blocker
HPLC: Si-10; 20A:80B; 5.4
GC:

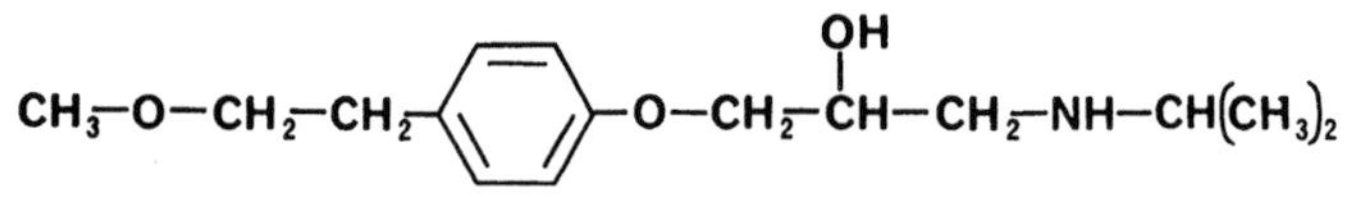
OH
CH3-O-CH2-CH2
O-CH2-CH-CH2-NH-CH(CH3)2

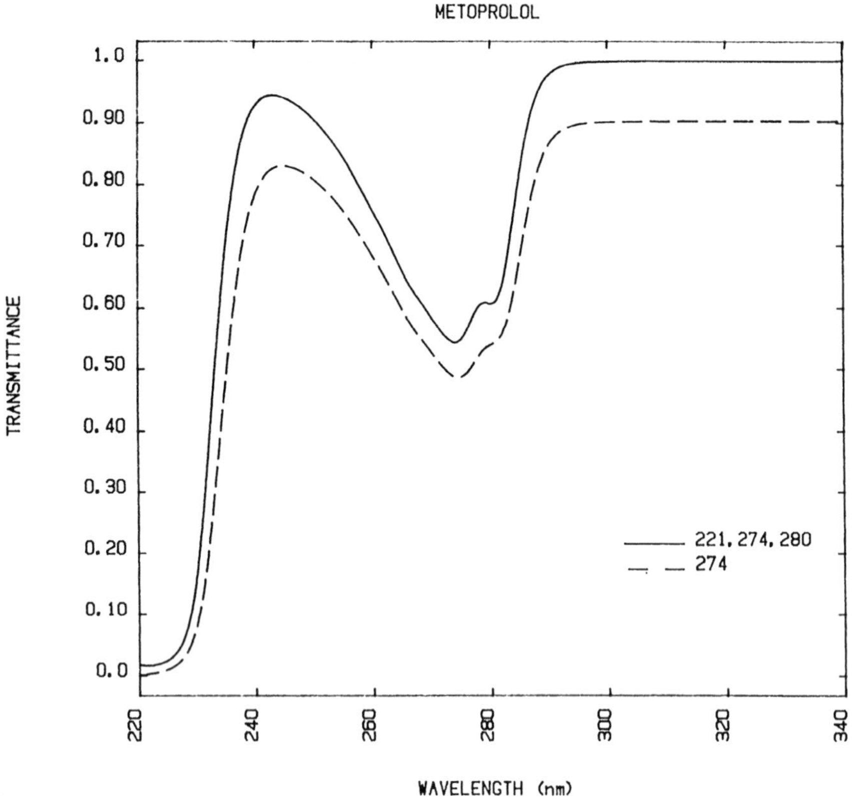
METOPROLOL
TRANSMITTANCE
1.0
0.90
0.80
0.70
0.60
0.50
0.40
0.30
0.20
0.10
0.0
221, 274, 280
274
220
240
260
280
300
320
340
WAVELENGTH (nm)

METOPROLOL--DIP

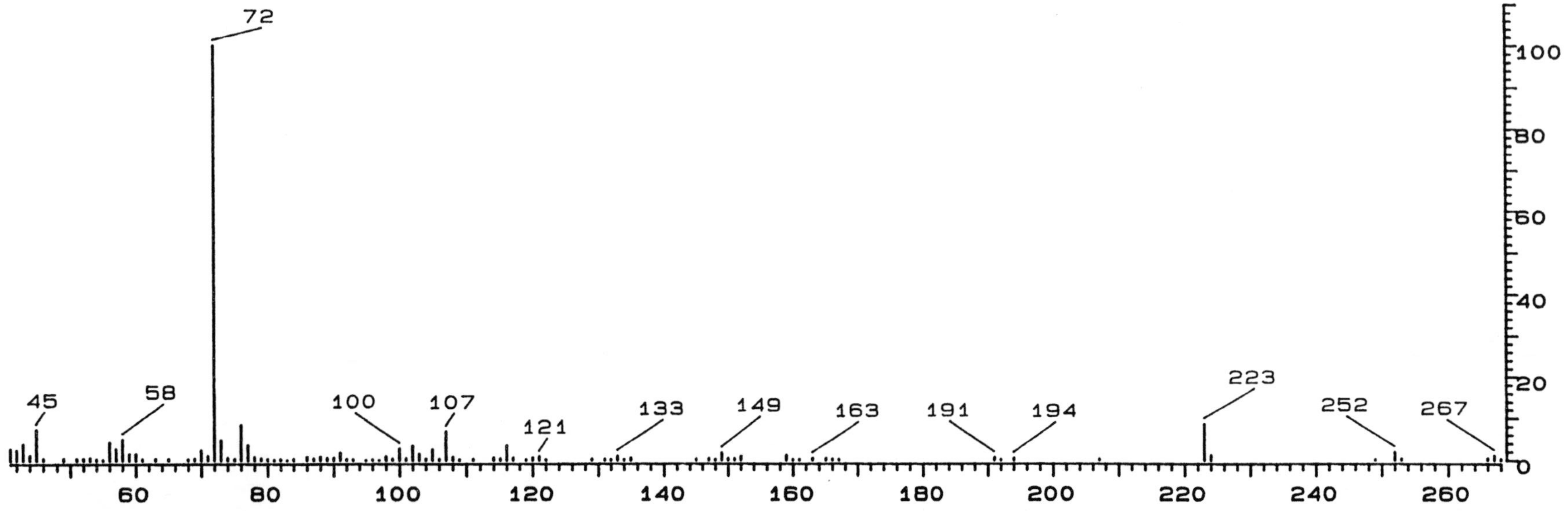
72
45
58
100
107
121
133
149
163
191
194
223
252
267
60
80
100
120
140
160
180
200
220
240
260
100
80
60
40
20
0

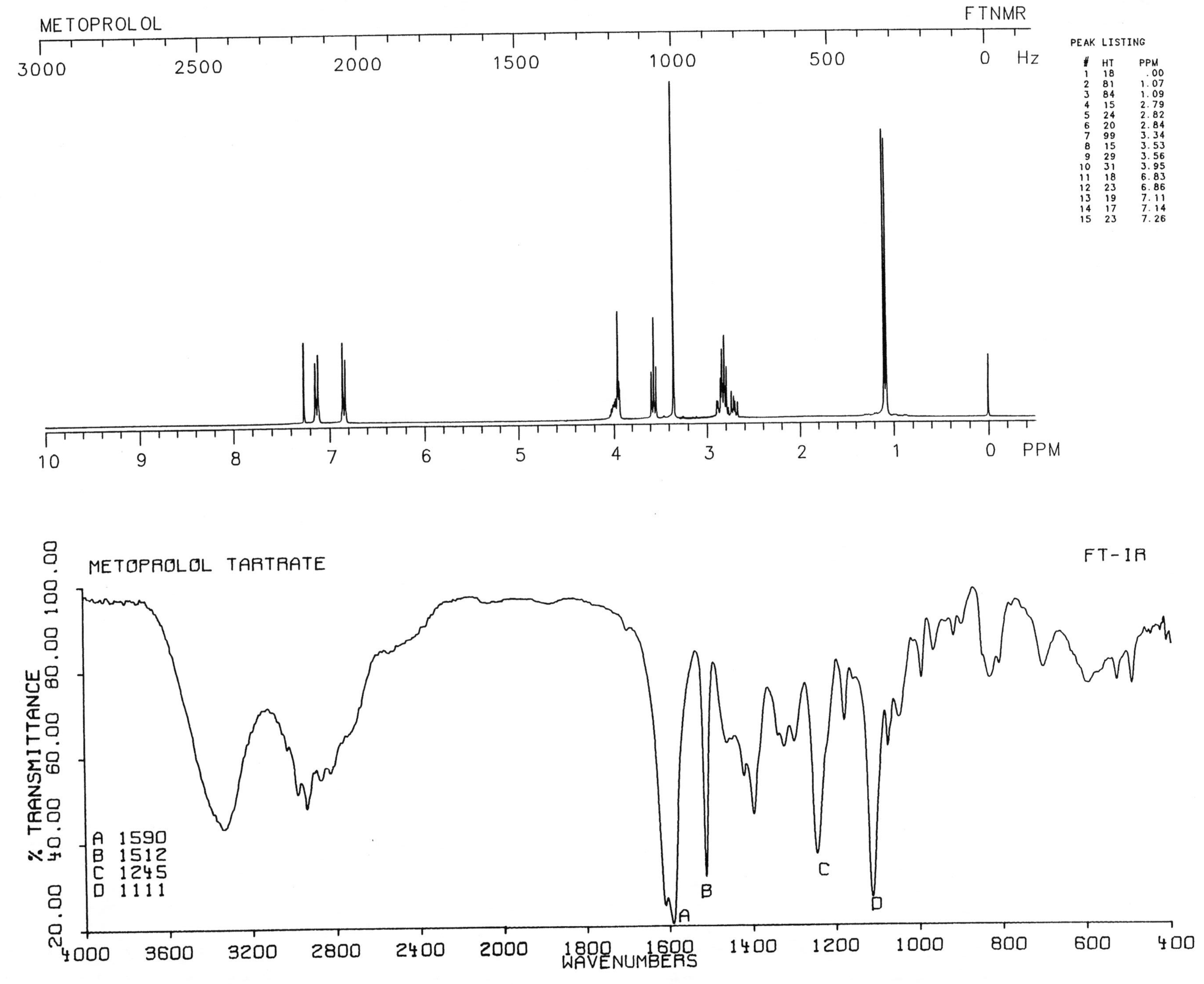
METOPROLOL
FTNMR
3000 2500 2000 1500 1000 500 0 Hz
10 9 8 7 6 5 4 3 2 1 0 PPM
PEAK LISTING
# HT PPM
1 18 .00
2 81 1.07
3 84 1.09
4 15 2.79
5 24 2.82
6 20 2.84
7 99 3.34
8 15 3.53
9 29 3.56
10 31 3.95
11 18 6.83
12 23 6.86
13 19 7.11
14 17 7.14
15 23 7.26
METOPROLOL TARTRATE
FT-IR
% TRANSMITTANCE
20.00 40.00 60.00 80.00 100.00
4000 3600 3200 2800 2400 2000 1800 1600 1400 1200 1000 800 600 400
WAVENUMBERS
A 1590
B 1512
C 1245
D 1111
A
B
C
D

# METRIZAMIDE

$C_{18}H_{22}I_3N_3O_8$

**Molecular weight:** 789.10 (788.85)
**Synonyms:** 2-[[3-(Acetylamino)-5-(acetylmethylamino)-2,4,6-triiodo-benzoyl]amino]-2-deoxy-D-glucose
**Trade names:** Amipaque

**Use:** Diagnostic aid
**HPLC:** Si-10; 20A:80B; 8.2
**GC:**

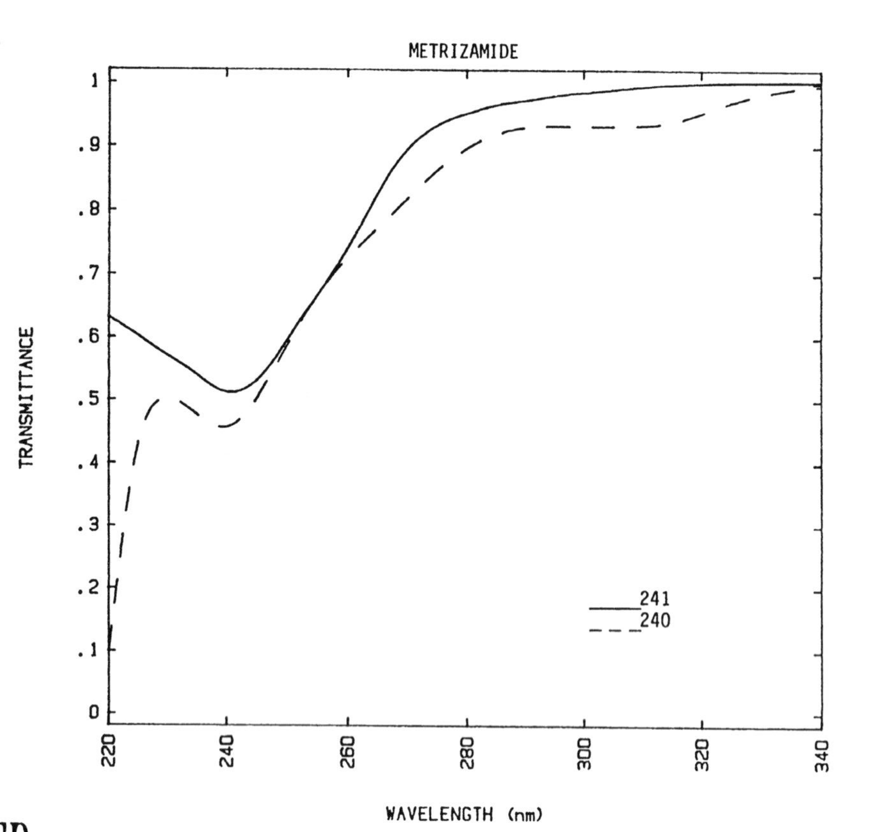


NO USEFUL MASS SPECTRUM WAS OBTAINED

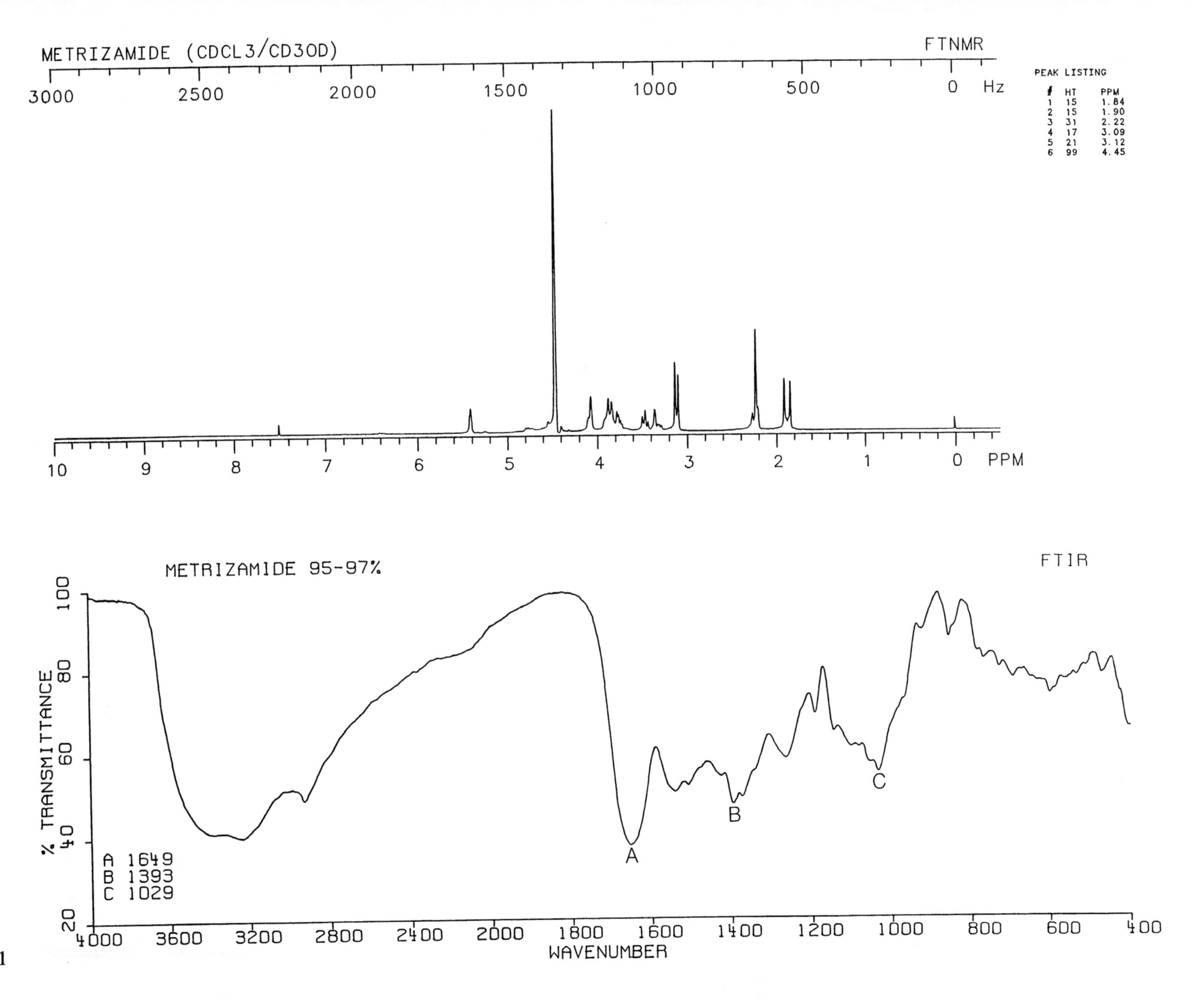
METRIZAMIDE (CDCL3/CD3OD)
FTNMR
3000
2500
2000
1500
1000
500
0
Hz
PEAK LISTING
# HT PPM
1 15 1.84
2 15 1.90
3 31 2.22
4 17 3.09
5 21 3.12
6 99 4.45
10
9
8
7
6
5
4
3
2
1
0
PPM
METRIZAMIDE 95-97%
FTIR
% TRANSMITTANCE
100
80
60
40
20
A 1649
B 1393
C 1029
A
B
C
4000
3600
3200
2800
2400
2000
1800
1600
1400
1200
1000
800
600
400
WAVENUMBER

# METRONIDAZOLE

$C_6H_9N_3O_3$

Molecular weight: 171.16 (171.06)
Synonyms: 2-Methyl-5-nitroimidazole-1-ethanol

Trade names: Flagyl

Use: Trichomonacide
HPLC: Si-10; 5A:95B; 4.0
GC: 1699; 200°C

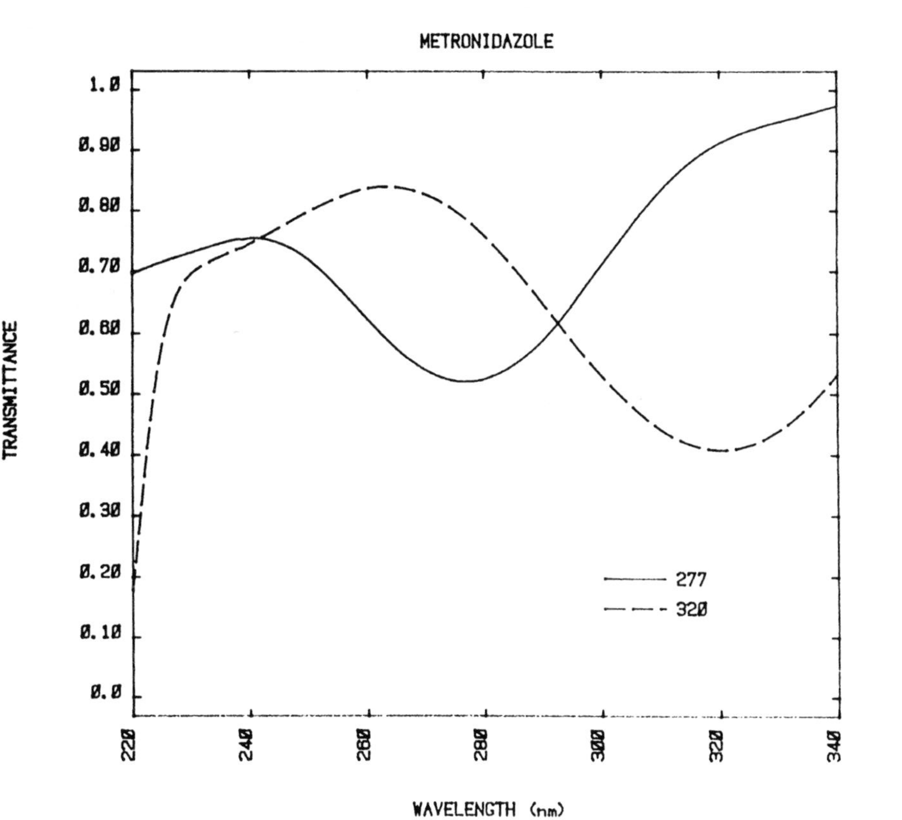


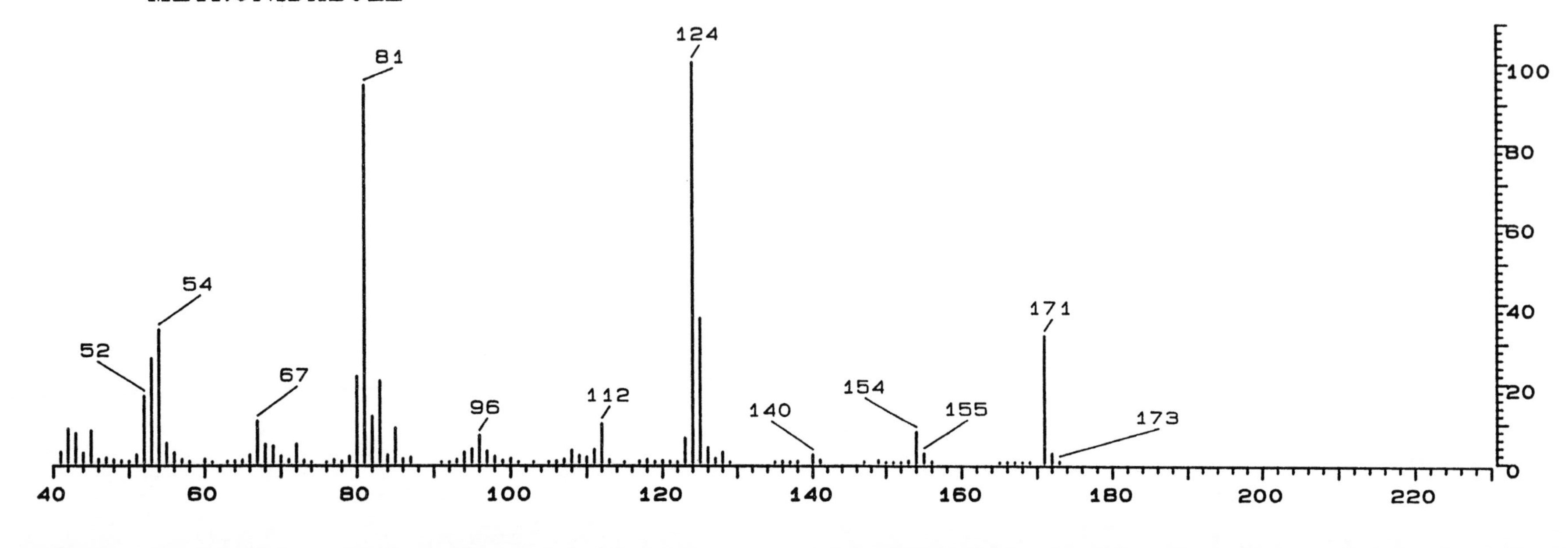

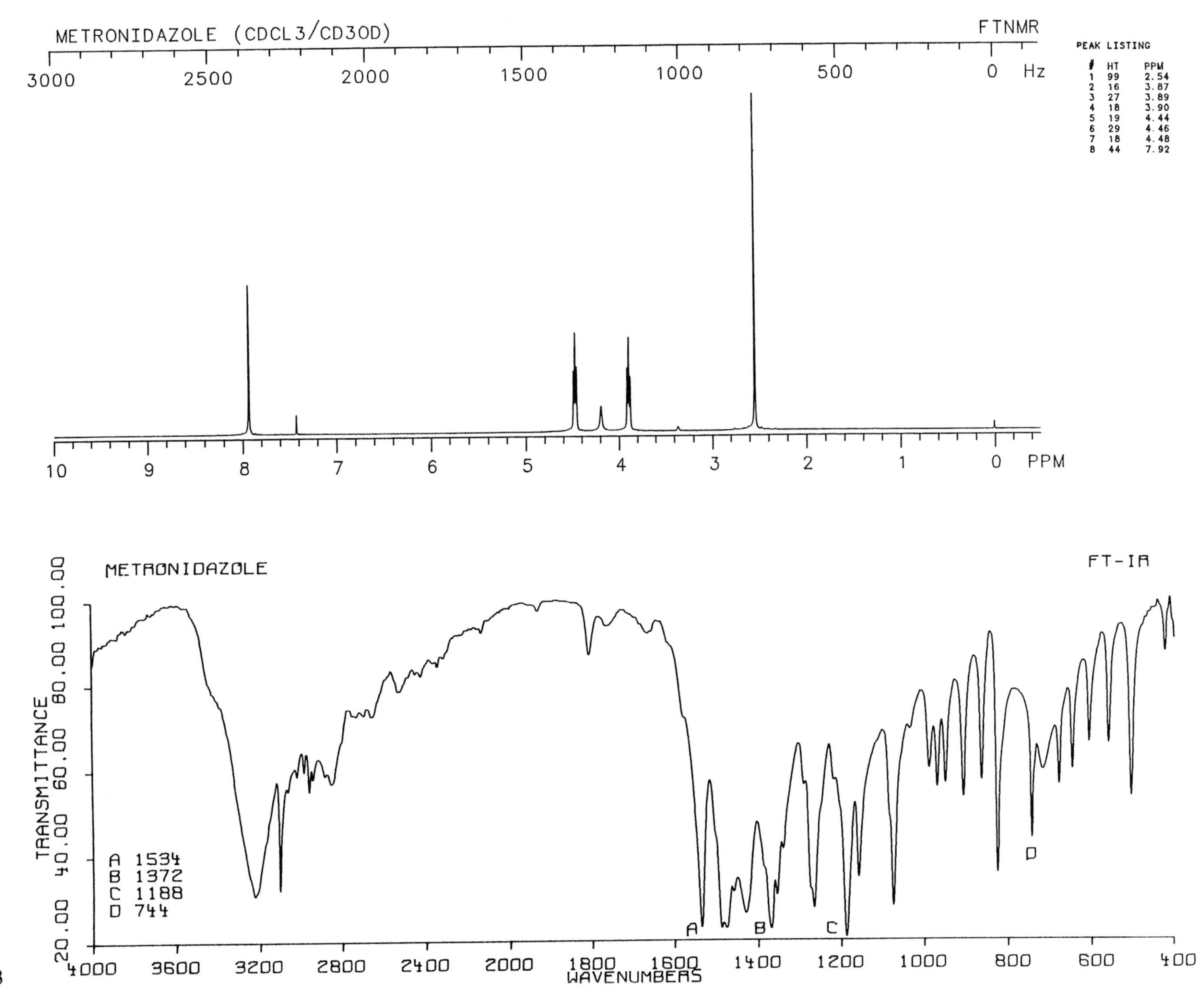
METRONIDAZOLE (CDCL3/CD3OD)
FTNMR
3000 2500 2000 1500 1000 500 0 Hz
PEAK LISTING
# HT PPM
1 99 2.54
2 16 3.87
3 27 3.89
4 18 3.90
5 19 4.44
6 29 4.46
7 18 4.48
8 44 7.92
10 9 8 7 6 5 4 3 2 1 0 PPM
METRONIDAZOLE
FT-IR
TRANSMITTANCE
20.00 40.00 60.00 80.00 100.00
A 1534
B 1372
C 1188
D 744
4000 3600 3200 2800 2400 2000 1800 1600 1400 1200 1000 800 600 400
WAVENUMBERS

# METYRAPONE

$C_{14}H_{14}N_2O$

Molecular weight: 226.28 (226.11)
Synonyms: 2-Methyl-1,2-di-3-pyridyl-1-propanone; mepyrapone; metopyrone
Trade names: Metopirone

Use: Diagnostic aid
HPLC: Si-10; 2A:98B; 6.0
GC: 1898; 200°C

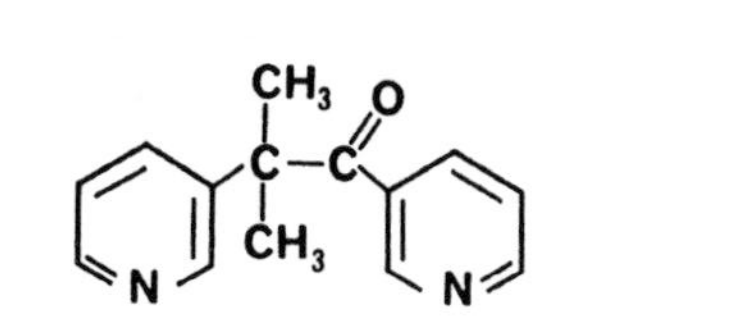

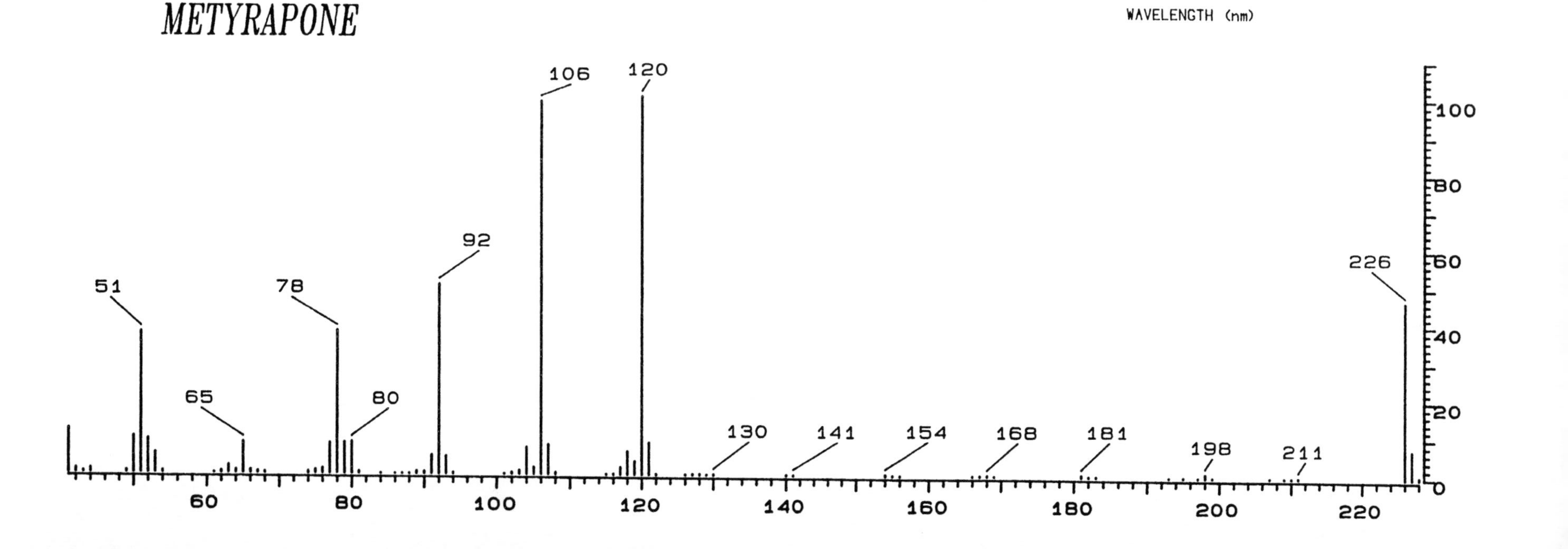

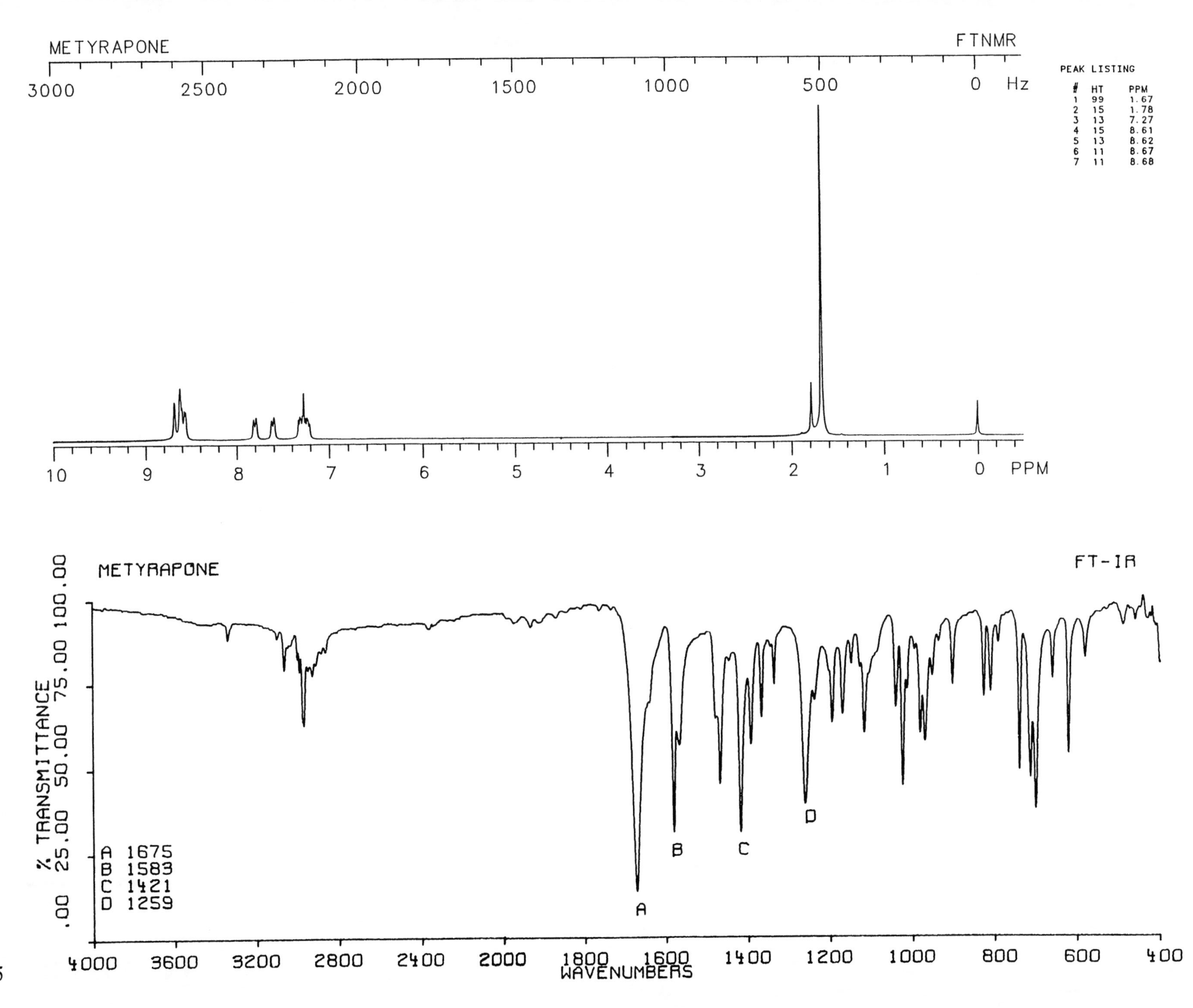
METYRAPONE
FTNMR
Hz
PPM
PEAK LISTING
# HT PPM
1 99 1.67
2 15 1.78
3 13 7.27
4 15 8.61
5 13 8.62
6 11 8.67
7 11 8.68
METYRAPONE
FT-IR
% TRANSMITTANCE
WAVENUMBERS
A 1675
B 1583
C 1421
D 1259

# METYROSINE

$C_{10}H_{13}NO_3$

**Molecular weight:** 195.21 (195.09)
**Synonyms:** α-Methyltyrosine; α-methyl-p-tyrosine; α-methyl-L-tyrosine; metirosine; 4-hydroxy-α-methylphenyl-alanine
**Trade names:** Demser

**Use:** Antihypertensive
**HPLC:**
**GC:**

HO–C6H4–CH2–C(CH3)(NH2)–COOH

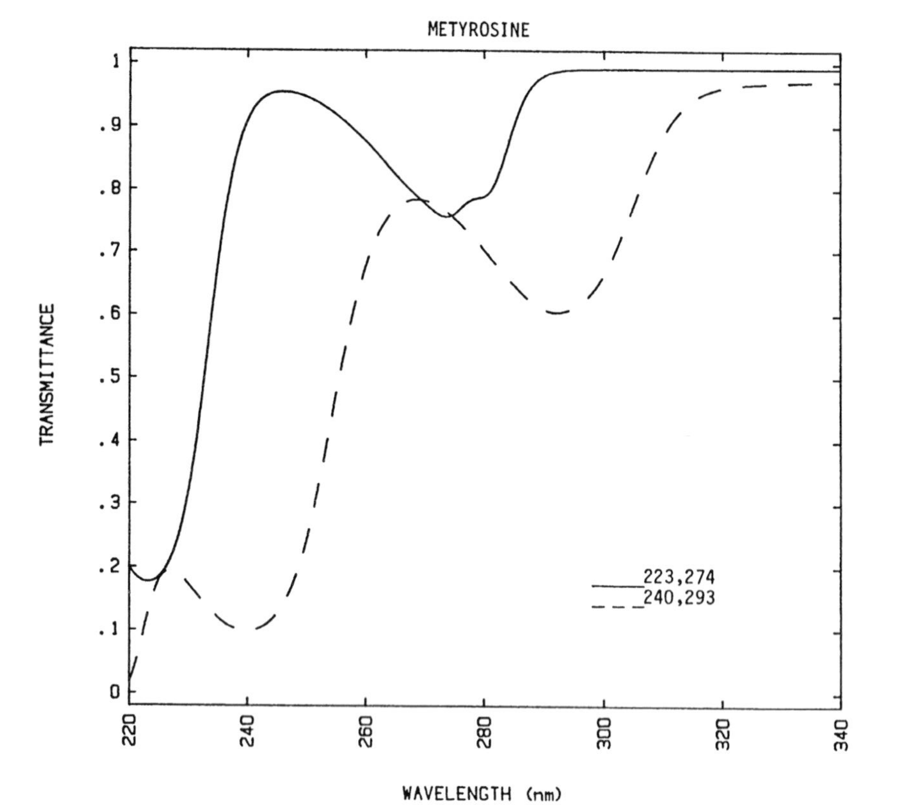


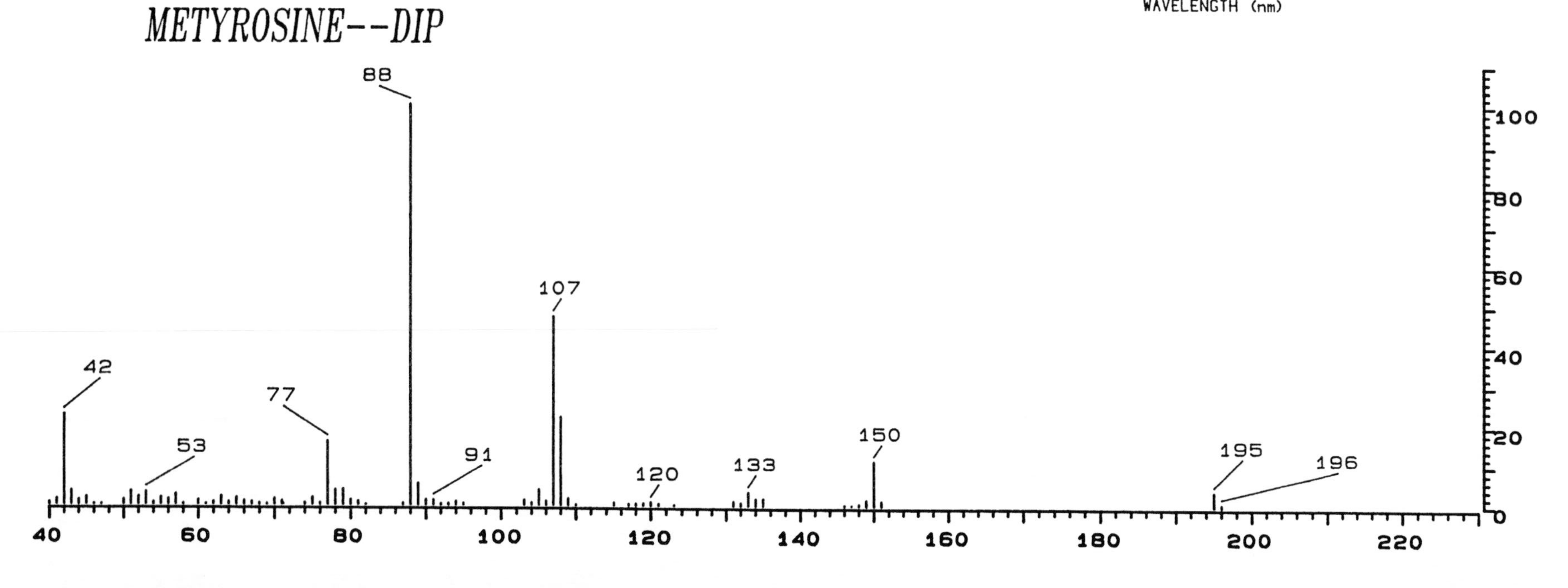

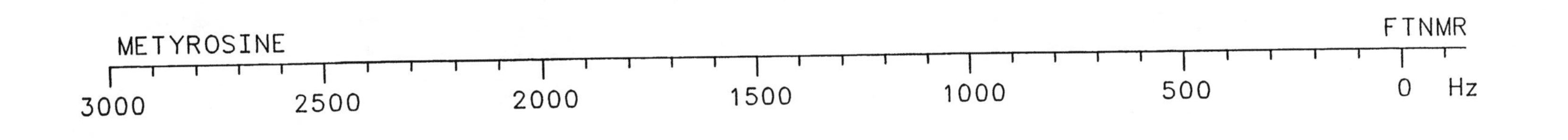


INSUFFICIENT SOLUBILITY

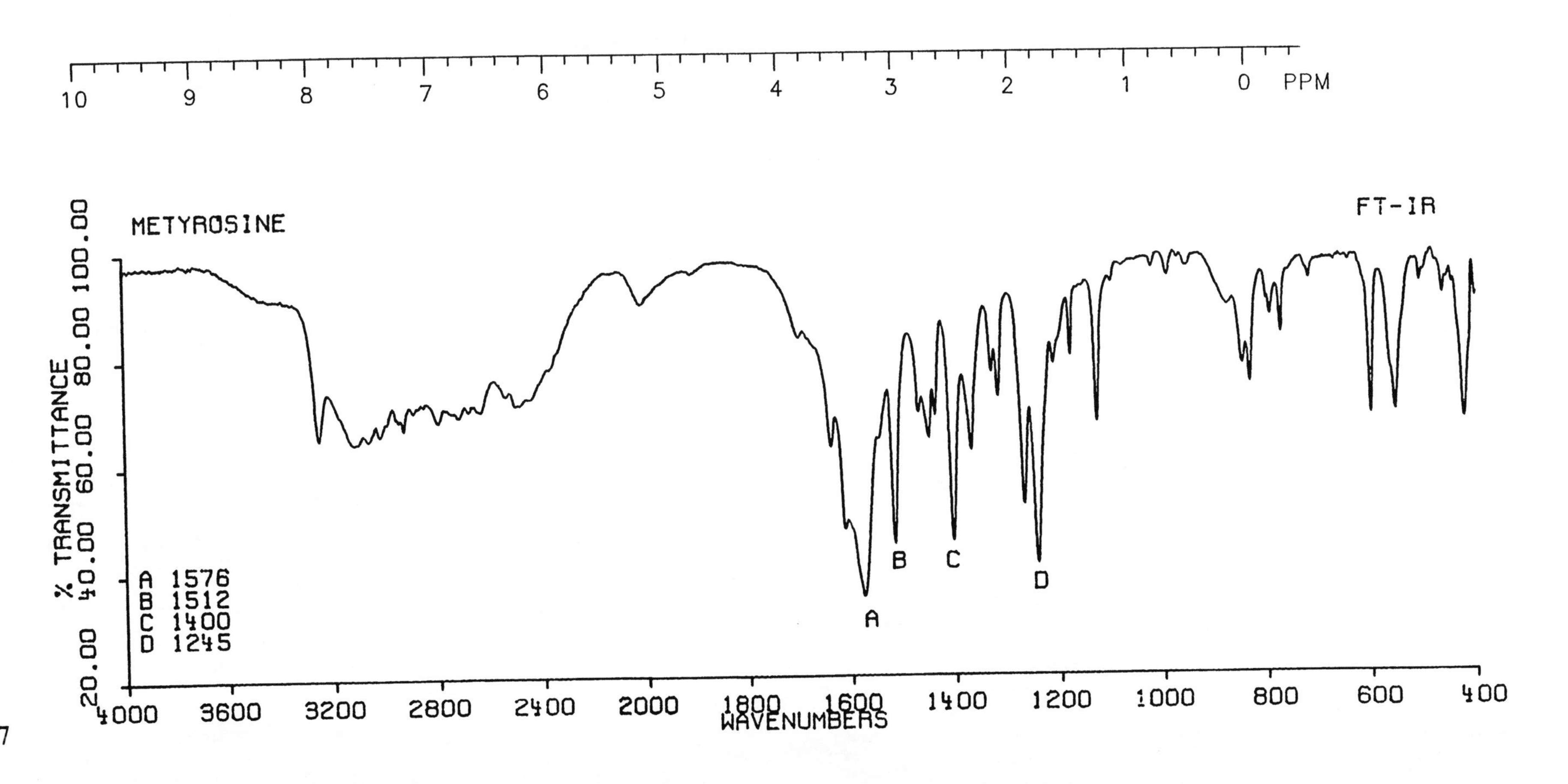

# MEZLOCILLIN

$C_{21}H_{25}N_5O_8S_2$

**Molecular weight:** 539.59 (539.11)
**Synonyms:** 6-[D-2[3-(Methyl-sulfonyl)-2-oxo-imidazolidine-1-carboxamido]-2-phenyl-acetamido]penicillanic acid
**Trade names:** Baycipen, Baypen, Mezlin

**Use:** Antibiotic, antibacterial
**HPLC:** Si-10; 10A:90B; 5.0
**GC:**

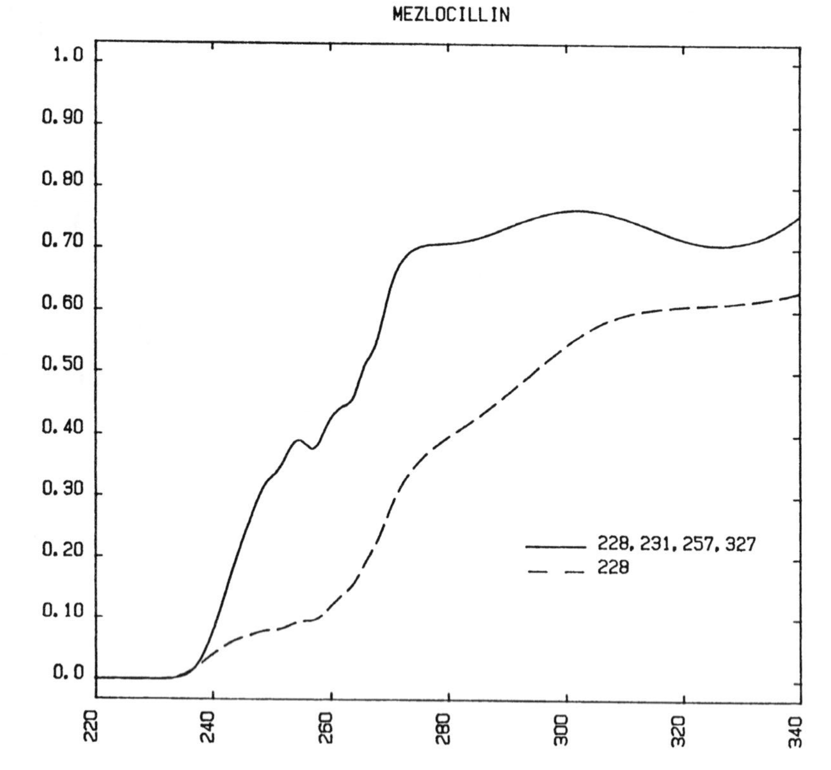


*NO USEFUL MASS SPECTRUM WAS OBTAINED*

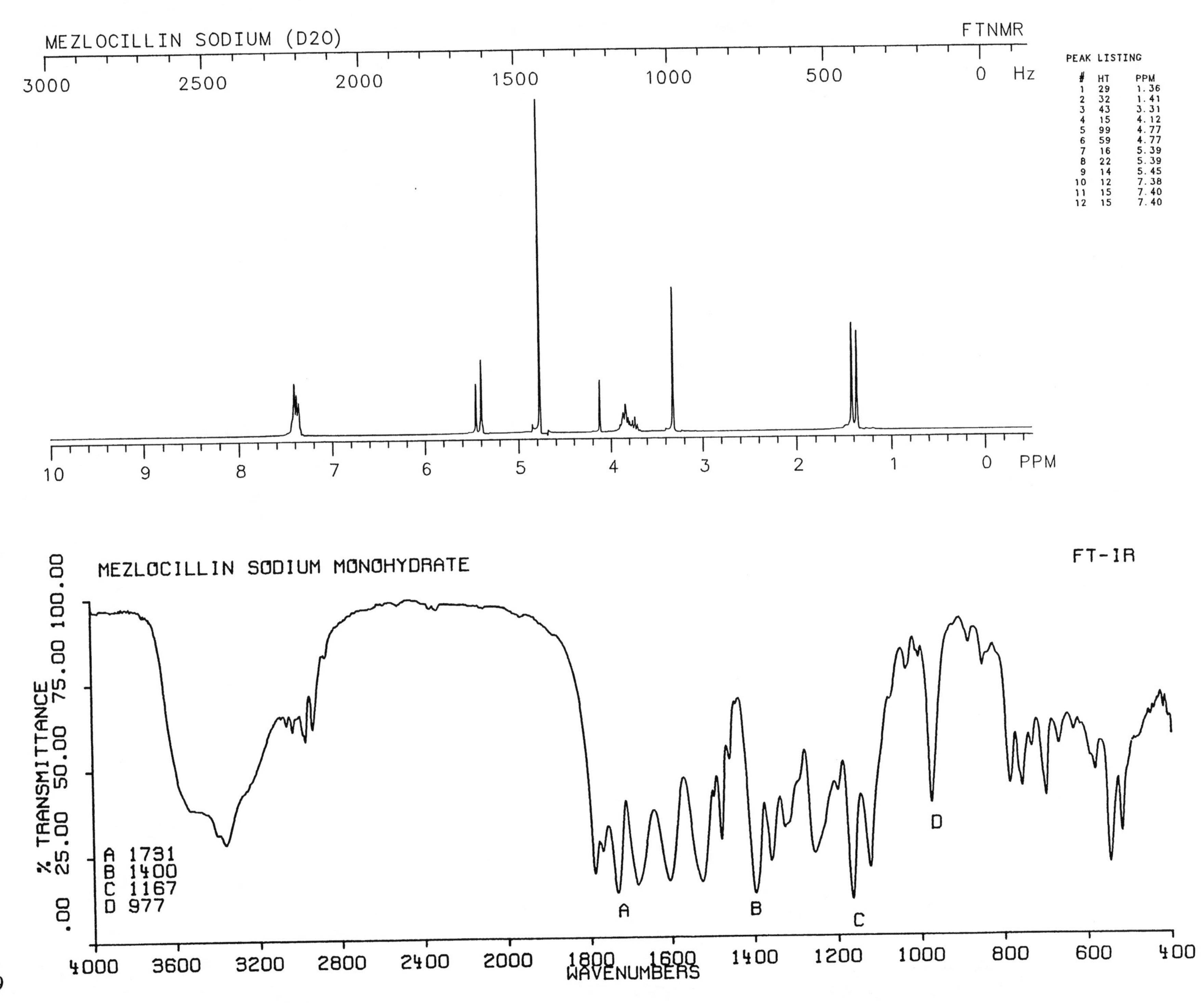
MEZLOCILLIN SODIUM (D2O)
FTNMR
3000 2500 2000 1500 1000 500 0 Hz
PEAK LISTING
# HT PPM
1 29 1.36
2 32 1.41
3 43 3.31
4 15 4.12
5 99 4.77
6 59 4.77
7 16 5.39
8 22 5.39
9 14 5.45
10 12 7.38
11 15 7.40
12 15 7.40
10 9 8 7 6 5 4 3 2 1 0 PPM
MEZLOCILLIN SODIUM MONOHYDRATE
FT-IR
% TRANSMITTANCE
.00 25.00 50.00 75.00 100.00
A 1731
B 1400
C 1167
D 977
A B C D
4000 3600 3200 2800 2400 2000 1800 1600 1400 1200 1000 800 600 400
WAVENUMBERS

# MICONAZOLE

$C_{18}H_{14}Cl_4N_2O$

**Molecular weight:** 416.21 (413.99)
**Synonyms:** 1-[2-(2,4-Dichlorophenyl)-2-[(2,4-dichlorophenyl)methoxy]-ethyl]-1H-imidazole
**Trade names:** Albistat, Daktarin, Dermonistat, Micatin, Monistat IV

**Use:** Antifungal
HPLC: Si-10; 1A:99B; 3.5
GC: 3095; 280°C

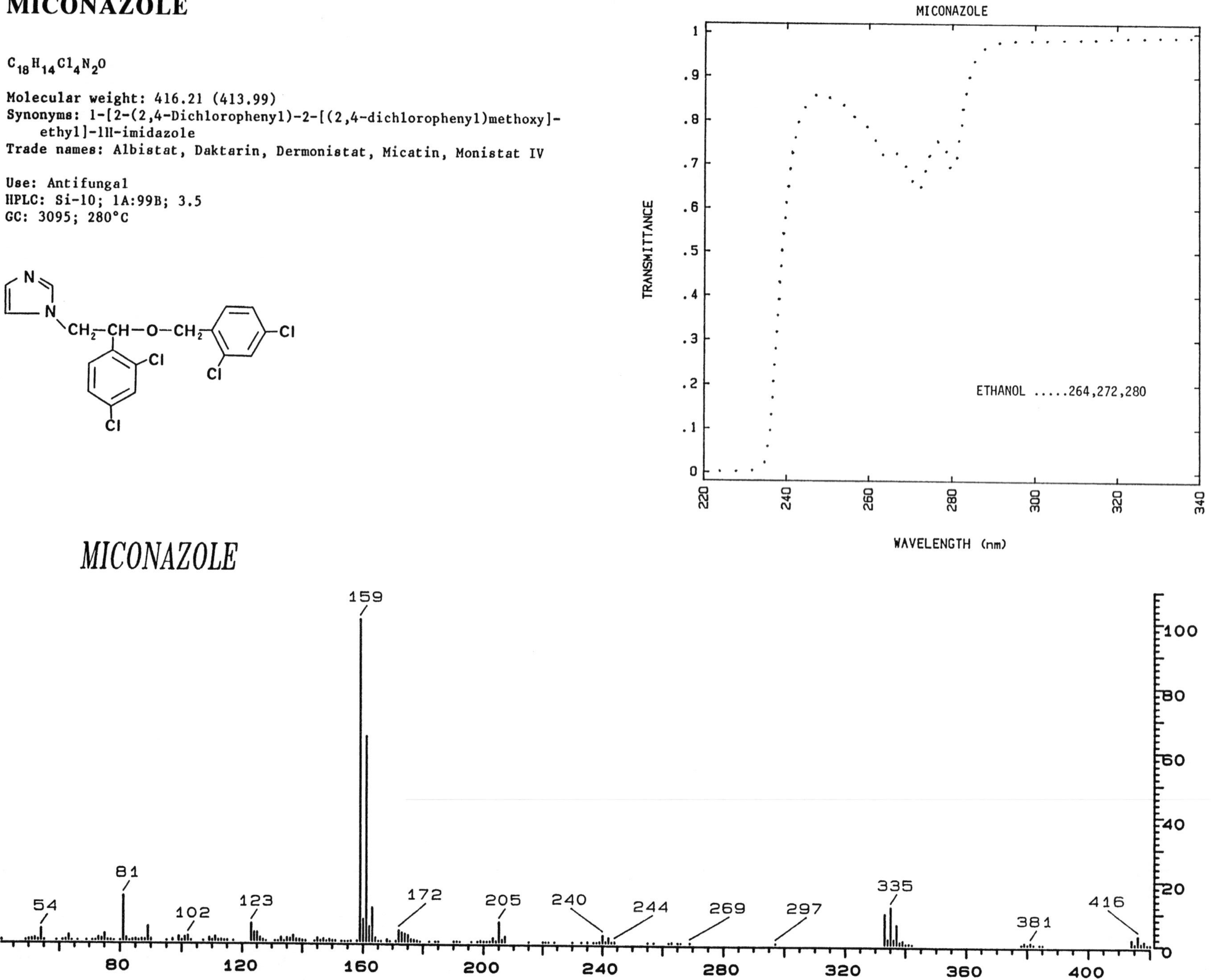

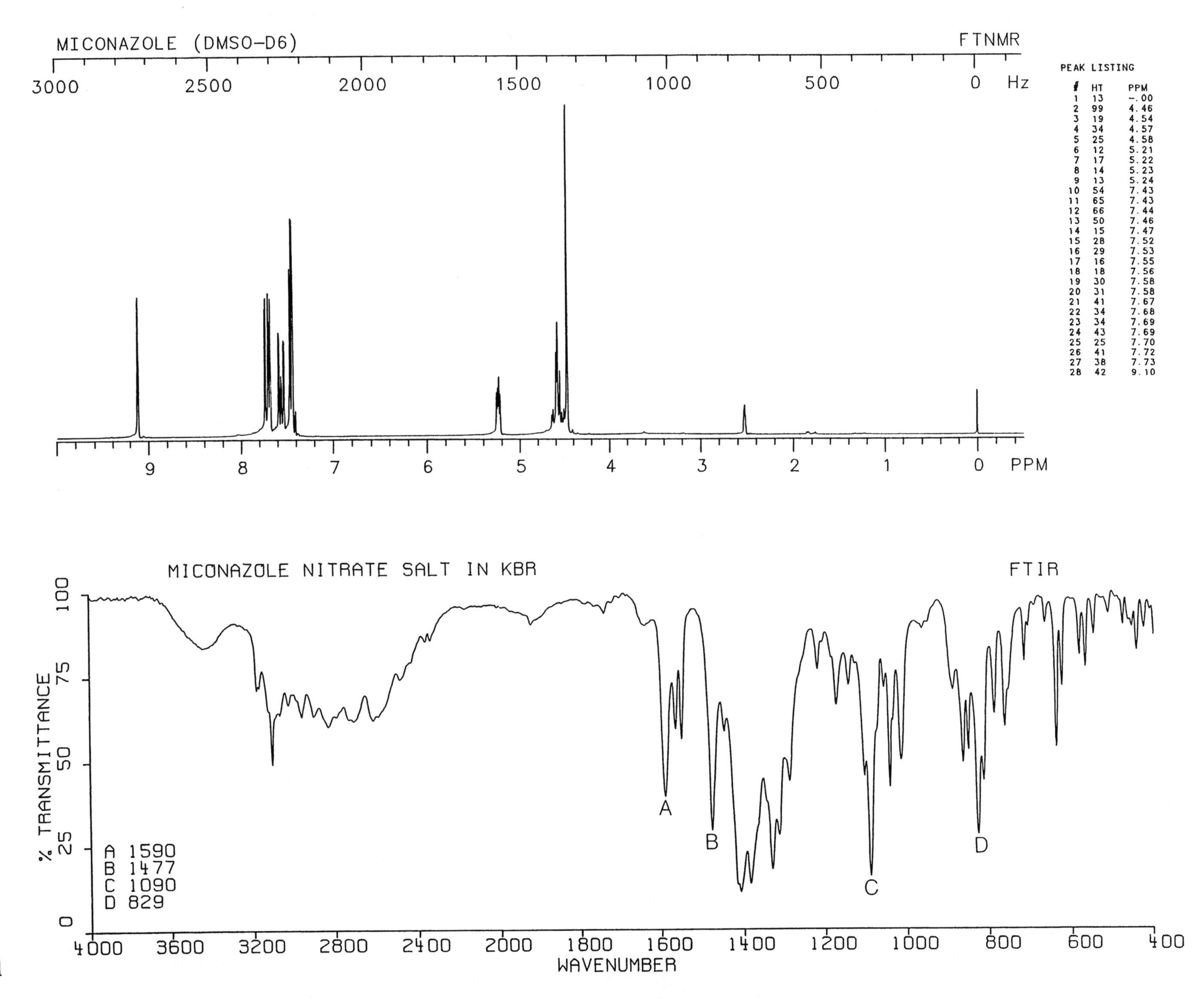
MICONAZOLE (DMSO-D6)
FTNMR
3000 2500 2000 1500 1000 500 0 Hz
9 8 7 6 5 4 3 2 1 0 PPM
PEAK LISTING
# HT PPM
1 13 -.00
2 99 4.46
3 19 4.54
4 34 4.57
5 25 4.58
6 12 5.21
7 17 5.22
8 14 5.23
9 13 5.24
10 54 7.43
11 65 7.43
12 66 7.44
13 50 7.46
14 15 7.47
15 28 7.52
16 29 7.53
17 16 7.55
18 18 7.56
19 30 7.58
20 31 7.58
21 41 7.67
22 34 7.68
23 34 7.69
24 43 7.69
25 25 7.70
26 41 7.72
27 38 7.73
28 42 9.10
MICONAZOLE NITRATE SALT IN KBR
FTIR
% TRANSMITTANCE
100 75 50 25 0
A 1590
B 1477
C 1090
D 829
A B C D
4000 3600 3200 2800 2400 2000 1800 1600 1400 1200 1000 800 600 400
WAVENUMBER

# MIDECAMYCIN

$C_{41}H_{67}NO_{15}$ $A_1$

$C_{41}H_{65}NO_{15}$ $A_3$

Molecular weight: 813.36 (813.45) [$A_1$]; 811.36 (811.44) [$A_3$]
Synonyms: $A_1$= leucomycin V 3,4$^b$-dipropanate; mydecamycin
$A_3$= 9-deoxy-9-oxoleucomycin V 3,4$^b$-dipropanoate
Trade names: Aboren, Medemycin, Midecacine, Rubimycin

Use: Antibacterial
HPLC:
GC:

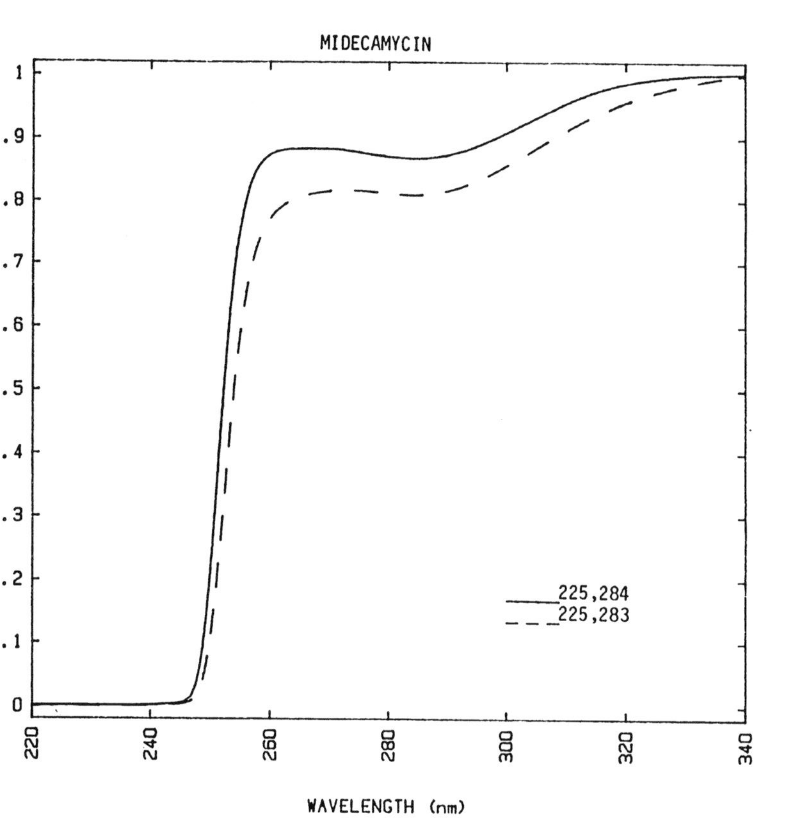


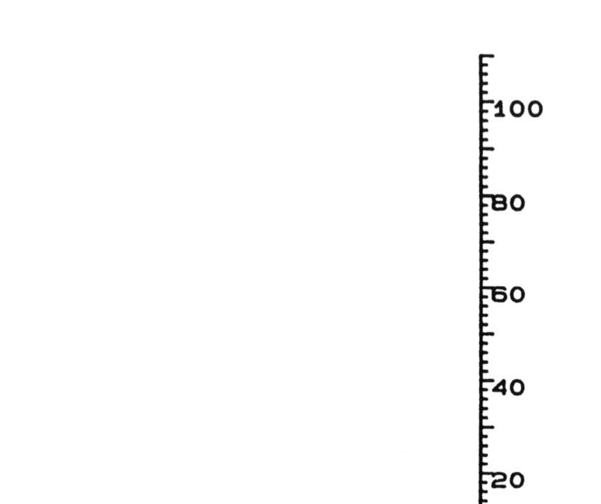


NO USEFUL MASS SPECTRUM WAS OBTAINED

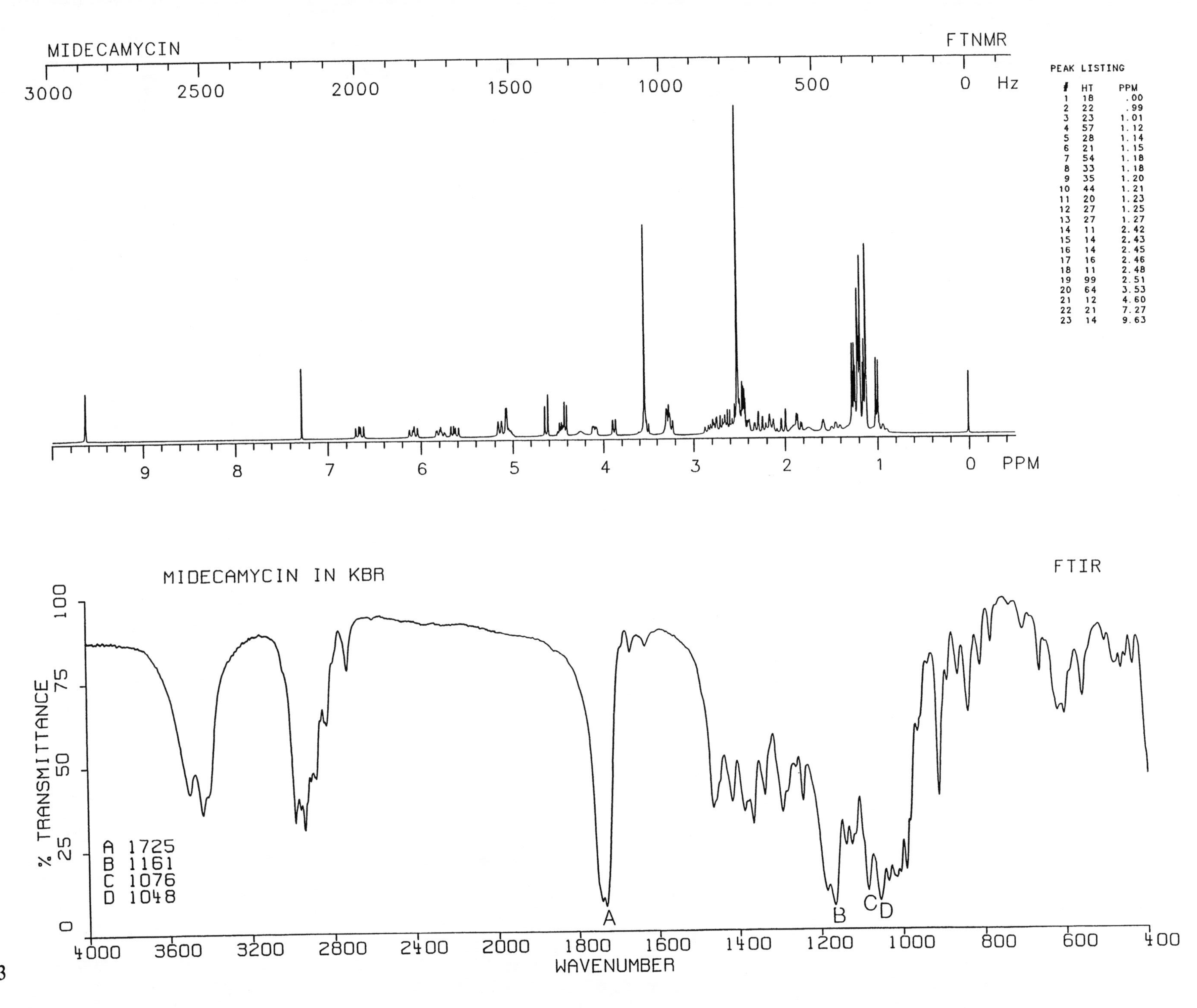
MIDECAMYCIN
FTNMR
3000 2500 2000 1500 1000 500 0 Hz
9 8 7 6 5 4 3 2 1 0 PPM
PEAK LISTING
# HT PPM
1 18 .00
2 22 .99
3 23 1.01
4 57 1.12
5 28 1.14
6 21 1.15
7 54 1.18
8 33 1.18
9 35 1.20
10 44 1.21
11 20 1.23
12 27 1.25
13 27 1.27
14 11 2.42
15 14 2.43
16 14 2.45
17 16 2.46
18 11 2.48
19 99 2.51
20 64 3.53
21 12 4.60
22 21 7.27
23 14 9.63
MIDECAMYCIN IN KBR
FTIR
% TRANSMITTANCE
0 25 50 75 100
A 1725
B 1161
C 1076
D 1048
A B C D
4000 3600 3200 2800 2400 2000 1800 1600 1400 1200 1000 800 600 400
WAVENUMBER

# MINOCYCLINE

$C_{23}H_{27}N_3O_7$

Molecular weight: 457.49 (457.18)
Synonyms: 4,7-Bis(dimethylamino)-1,4,4a,5,5a,6,11,12a-octahydro-3,10,12,12a-tetrahydroxy-1,11-dioxo-2-naphthacenecarboxamide
Trade names: Minocin

Use: Antibiotic
HPLC: Si-10; 5A:95B; 5.5
GC:

MINOCYCLINE--DIP

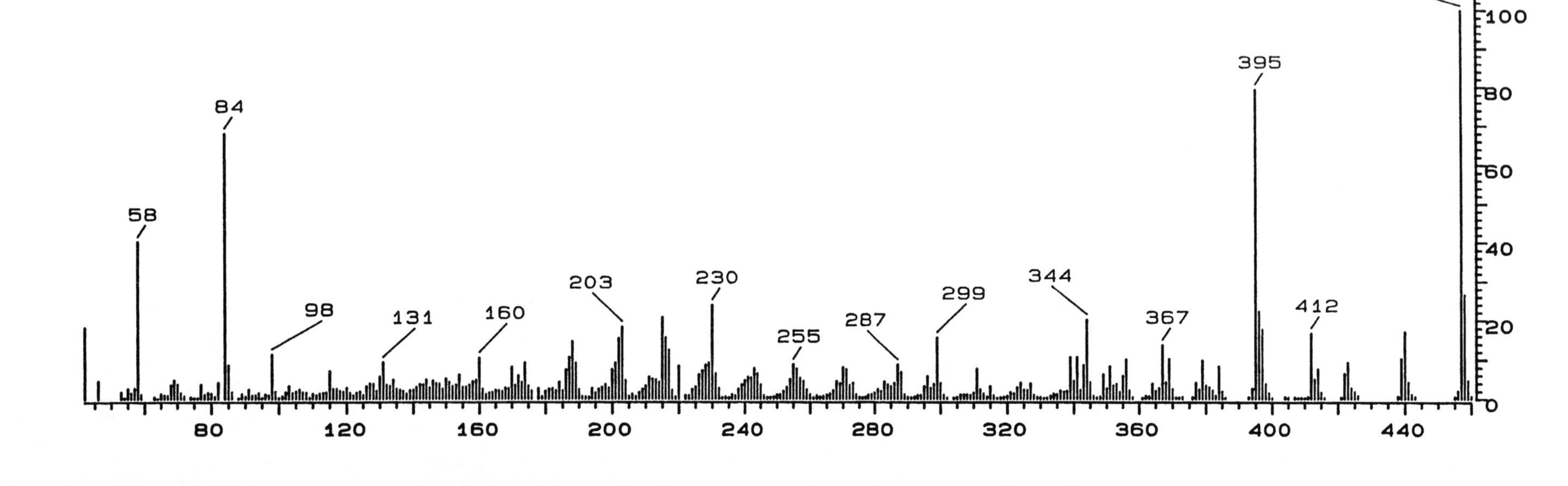

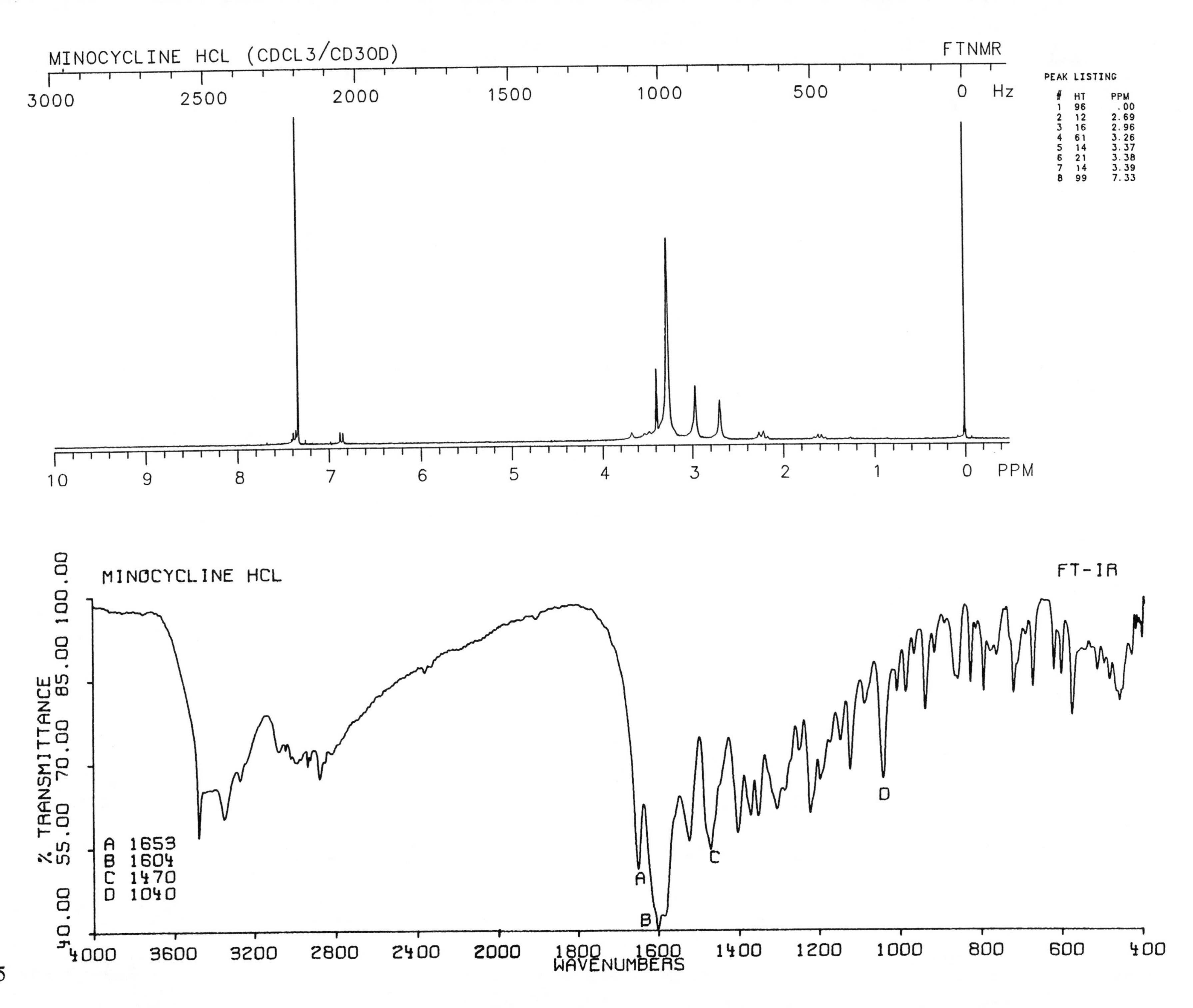

MINOCYCLINE HCL (CDCL3/CD3OD)
FTNMR
3000
2500
2000
1500
1000
500
0 Hz
PEAK LISTING
# HT PPM
1 96 .00
2 12 2.69
3 16 2.96
4 61 3.26
5 14 3.37
6 21 3.38
7 14 3.39
8 99 7.33
10
9
8
7
6
5
4
3
2
1
0 PPM
MINOCYCLINE HCL
FT-IR
% TRANSMITTANCE
40.00
55.00
70.00
85.00
100.00
A 1653
B 1604
C 1470
D 1040
A
B
C
D
4000
3600
3200
2800
2400
2000
1800
1600
1400
1200
1000
800
600
400
WAVENUMBERS

# MINOXIDIL

$C_9H_{15}N_5O$

Molecular weight: 209.25 (209.13)
Synonyms: 2,4-Diamino-6-piperidinopyrimidine-3-oxide

Trade names: Loniten

Use: Antihypertensive
HPLC: Si-10; 20A;80B; 4.8
GC:

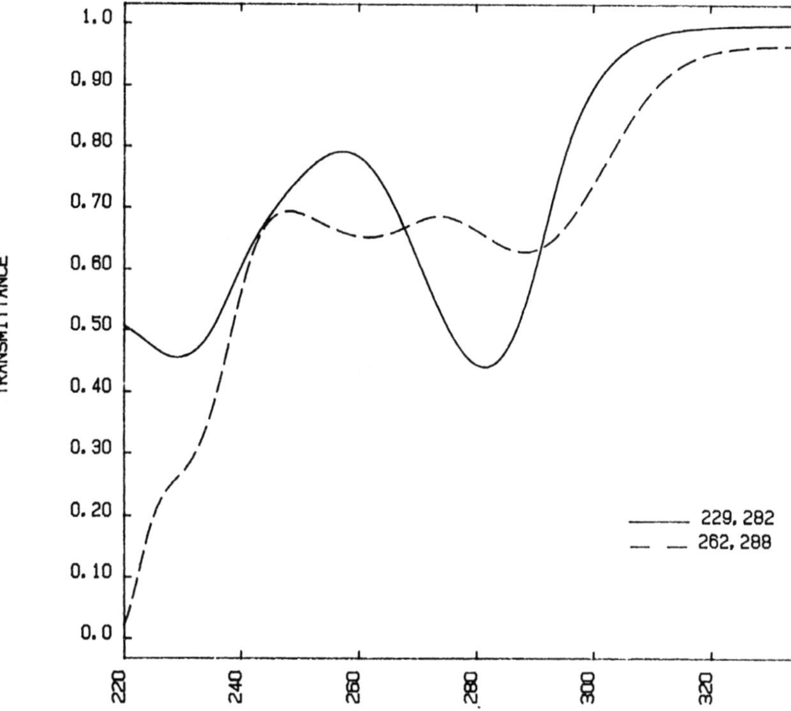


*MINOXIDIL*

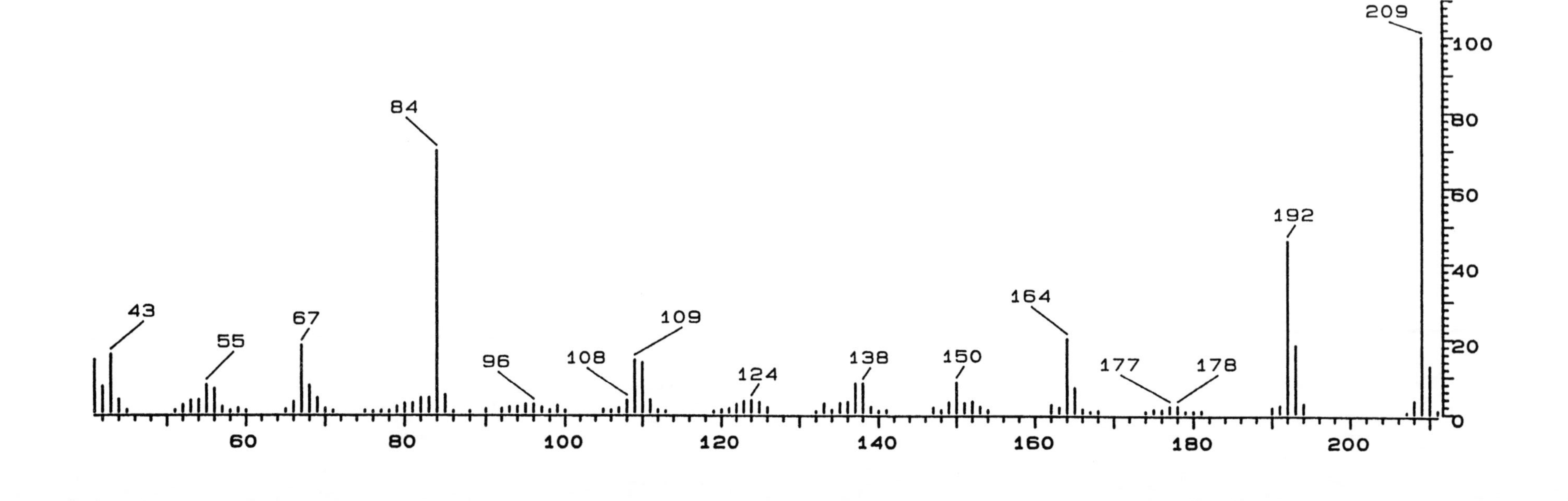

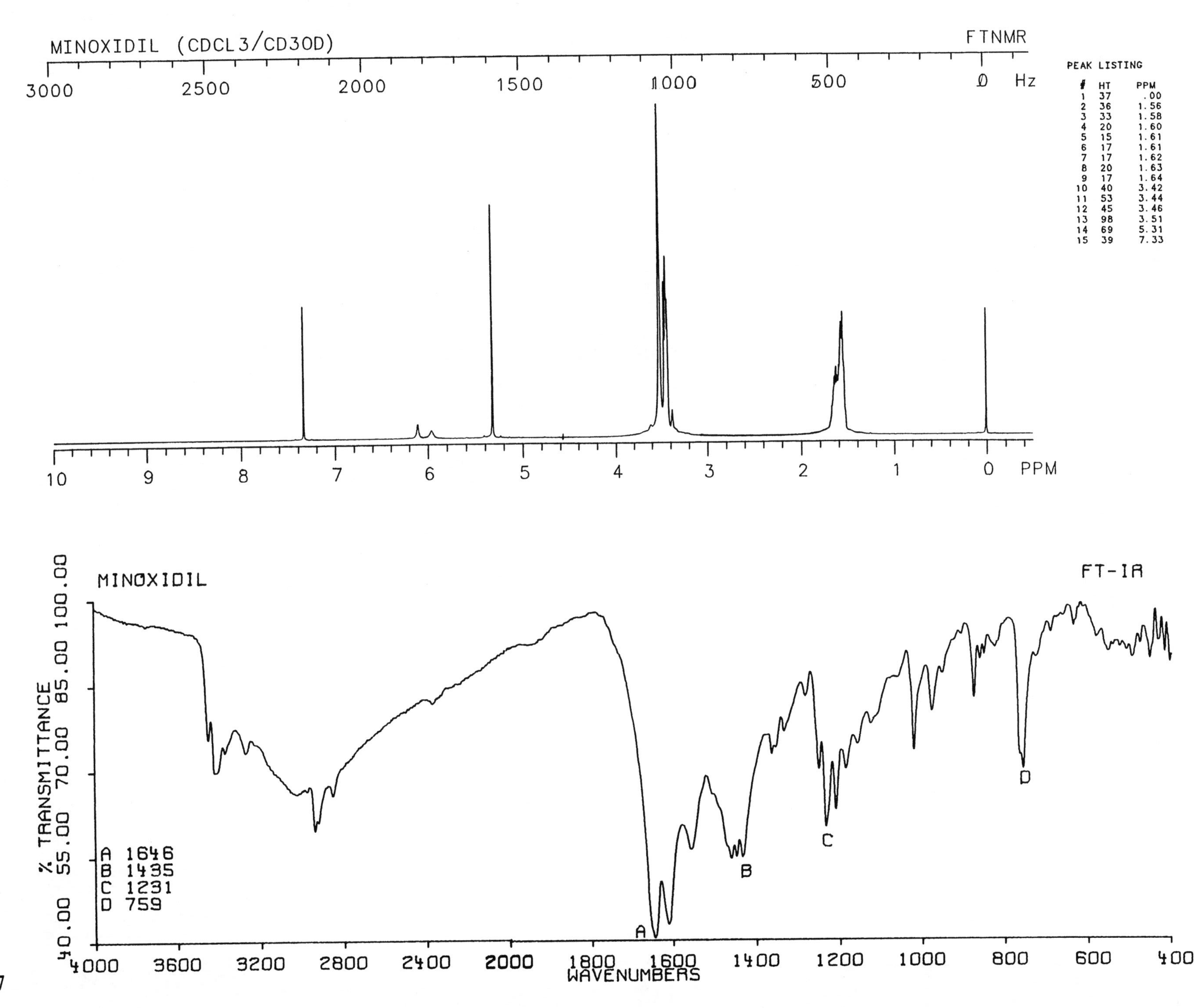
MINOXIDIL (CDCL3/CD3OD)
FTNMR
Hz
PPM
PEAK LISTING
# HT PPM
1 37 .00
2 36 1.56
3 33 1.58
4 20 1.60
5 15 1.61
6 17 1.61
7 17 1.62
8 20 1.63
9 17 1.64
10 40 3.42
11 53 3.44
12 45 3.46
13 98 3.51
14 69 5.31
15 39 7.33
MINOXIDIL
FT-IR
% TRANSMITTANCE
WAVENUMBERS
A 1646
B 1435
C 1231
D 759

# MITHRAMYCIN

$C_{52}H_{75}O_{24}$

Molecular weight: 1085.18 (1083.46)
Synonyms: Aureolic acid; mitramycin, aurelic acid

Trade names: Mithracin

Use: Antineoplastic
HPLC:
GC:

MITHRAMYCIN

TRANSMITTANCE

WAVELENGTH (nm)

229,276,316,330
279

*NO USEFUL MASS SPECTRUM WAS OBTAINED*

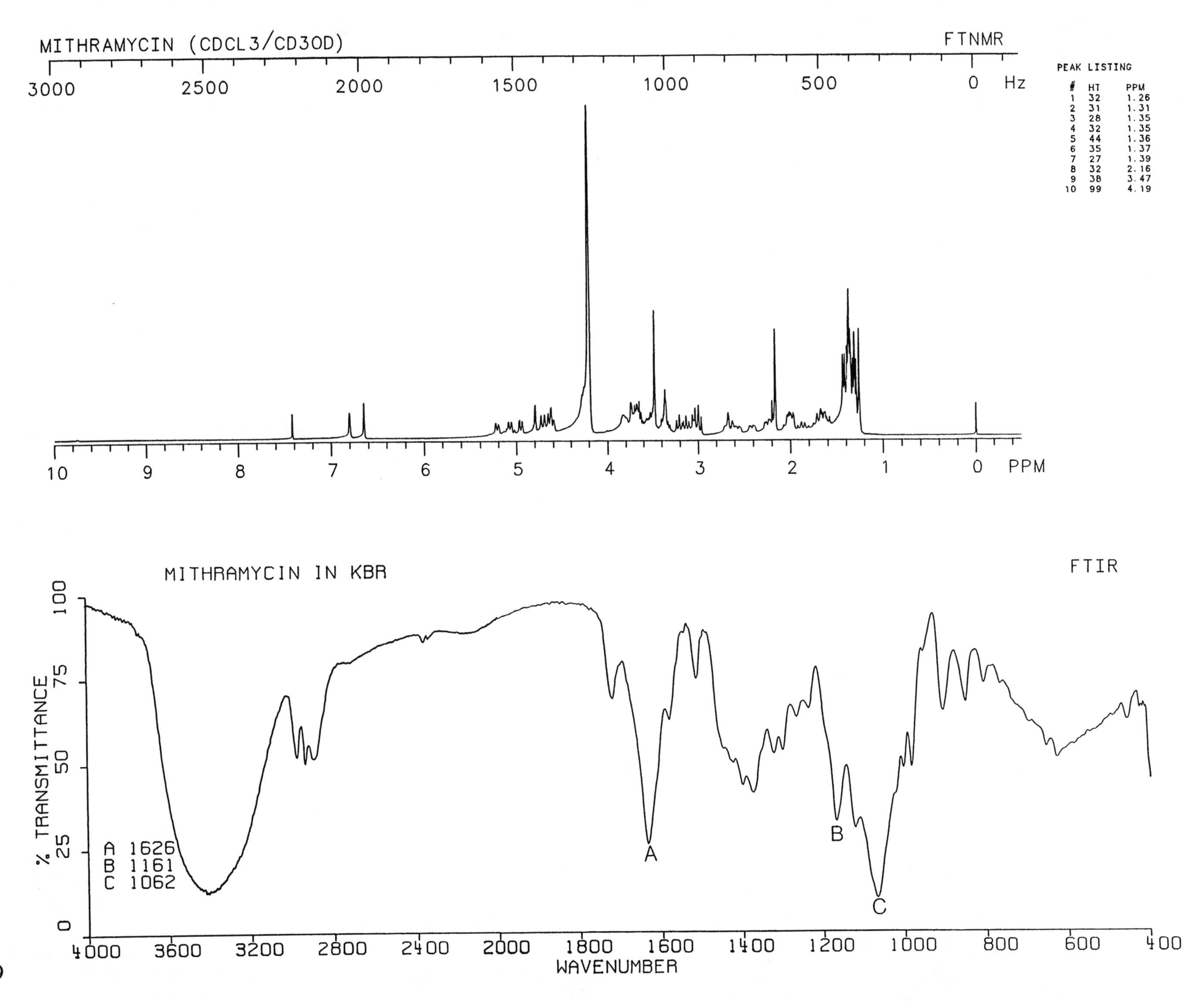
MITHRAMYCIN (CDCL3/CD3OD)
FTNMR
PEAK LISTING
# HT PPM
1 32 1.26
2 31 1.31
3 28 1.35
4 32 1.35
5 44 1.36
6 35 1.37
7 27 1.39
8 32 2.16
9 38 3.47
10 99 4.19
Hz
PPM
MITHRAMYCIN IN KBR
FTIR
% TRANSMITTANCE
A 1626
B 1161
C 1062
WAVENUMBER

# MITOTANE

$C_{14}H_{10}Cl_4$

**Molecular weight:** 320.05 (317.95)
**Synonyms:** 1-Chloro-2-[2,2-dichloro-1-(4-chlorophenyl)ethyl]benzene

**Trade names:** Lysodren

**Use:** Antineoplastic
**HPLC:** Si-10; 100C; 5.0
**GC:**

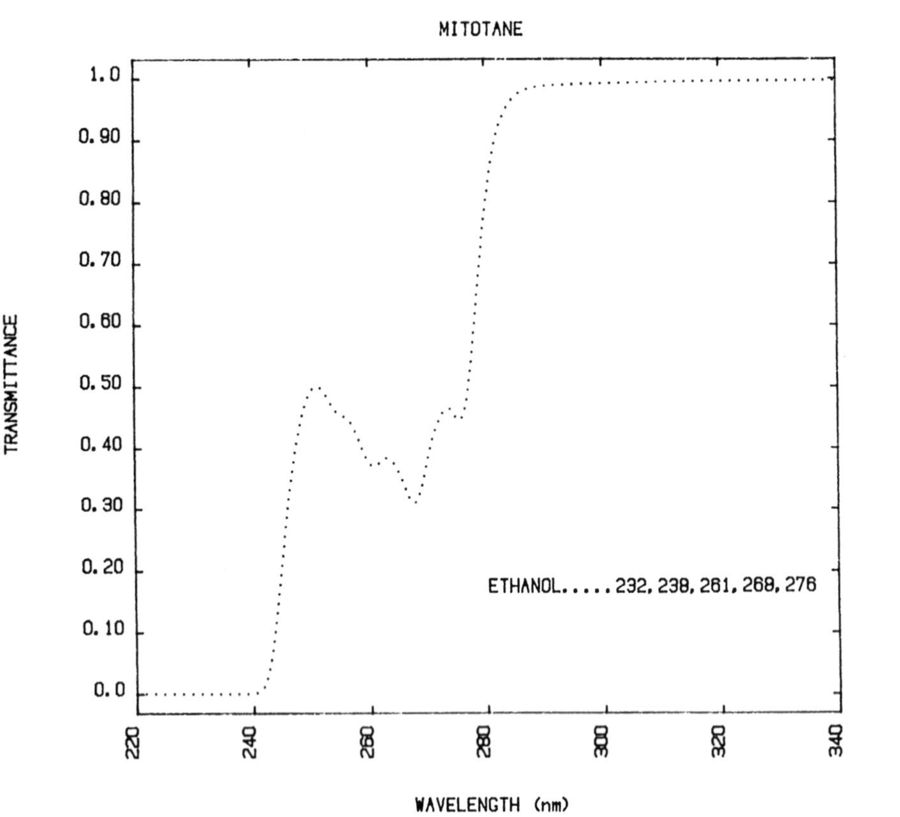

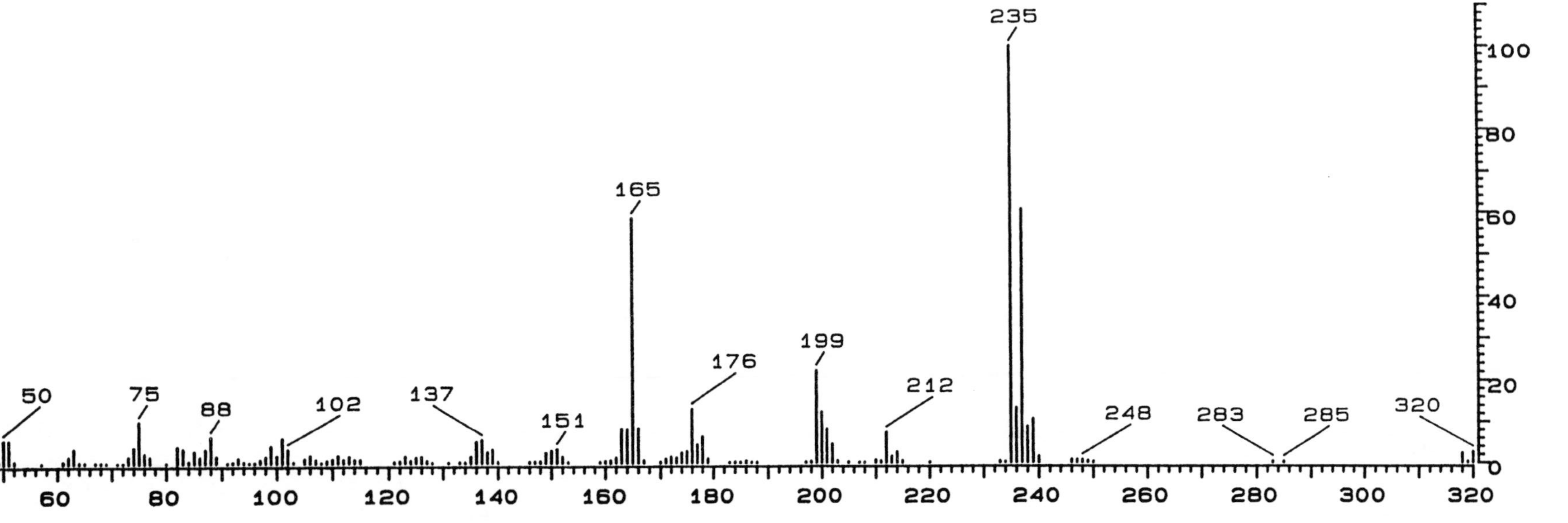

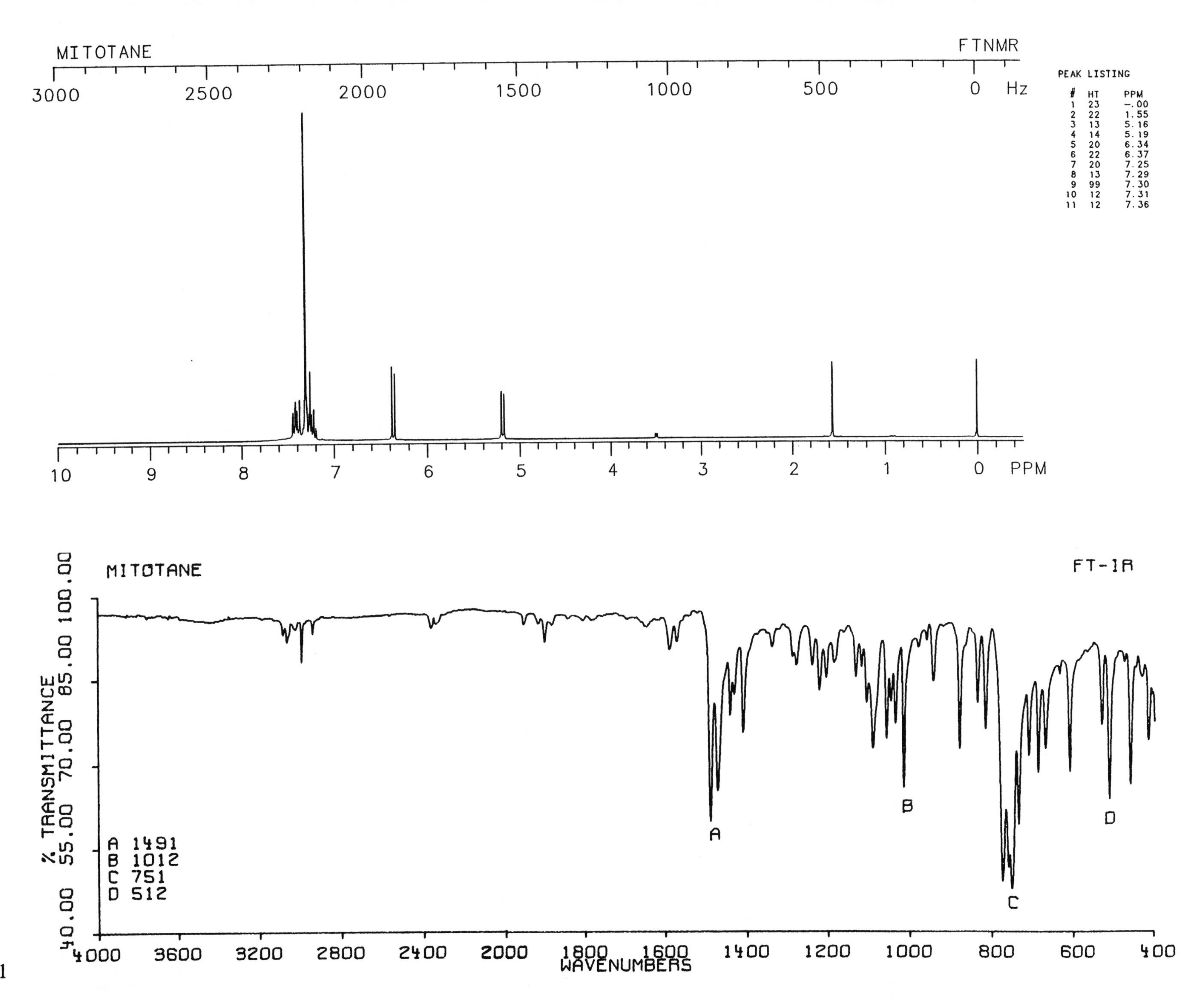
MITOTANE
FTNMR
3000
2500
2000
1500
1000
500
0 Hz
PEAK LISTING
# HT PPM
1 23 -.00
2 22 1.55
3 13 5.16
4 14 5.19
5 20 6.34
6 22 6.37
7 20 7.25
8 13 7.29
9 99 7.30
10 12 7.31
11 12 7.36
10 9 8 7 6 5 4 3 2 1 0 PPM
MITOTANE
FT-IR
% TRANSMITTANCE
40.00 55.00 70.00 85.00 100.00
A 1491
B 1012
C 751
D 512
A B C D
4000 3600 3200 2800 2400 2000 1800 1600 1400 1200 1000 800 600 400
WAVENUMBERS

# MOLINDONE

$C_{16}H_{24}N_2O_2$

Molecular weight: 276.38 (276.18)
Synonyms: 3-Ethyl-1,5,6,7-tetrahydro-2-methyl-5-(4-morpholinymethyl)-4H-indol-4-one
Trade names: Moban

Use: Sedative, tranquilizer
HPLC: Si-10; 2A:98B; 8.0
GC: 1907; 200°C

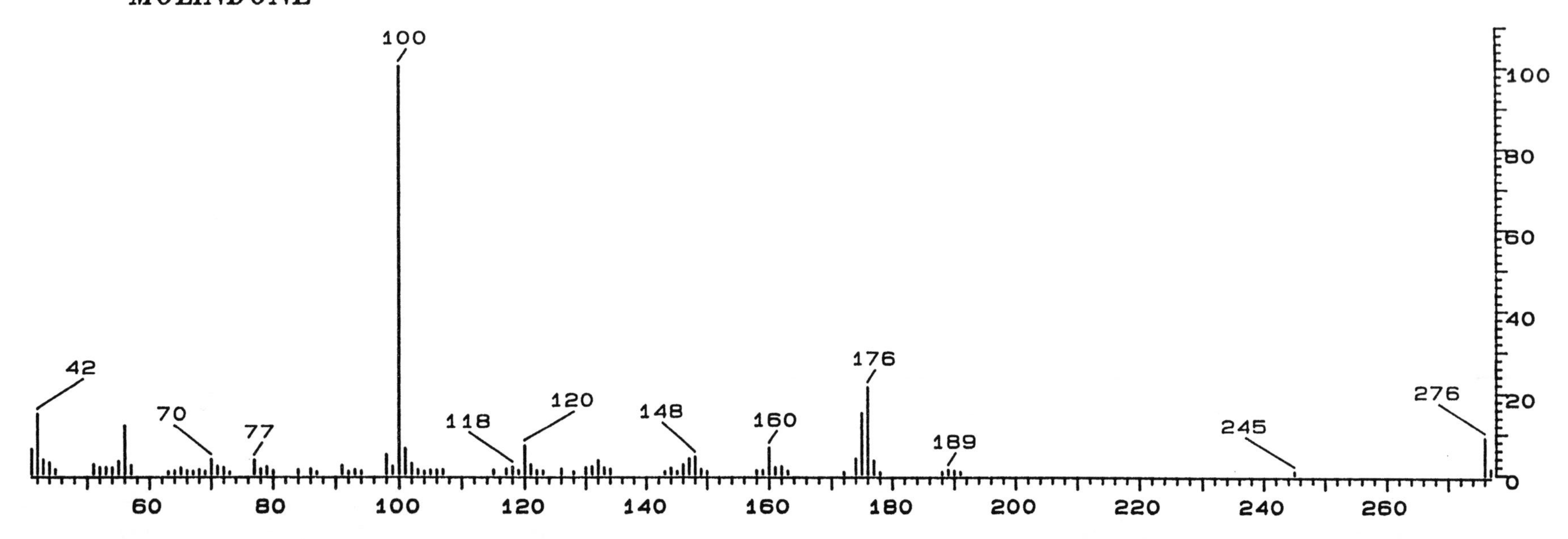

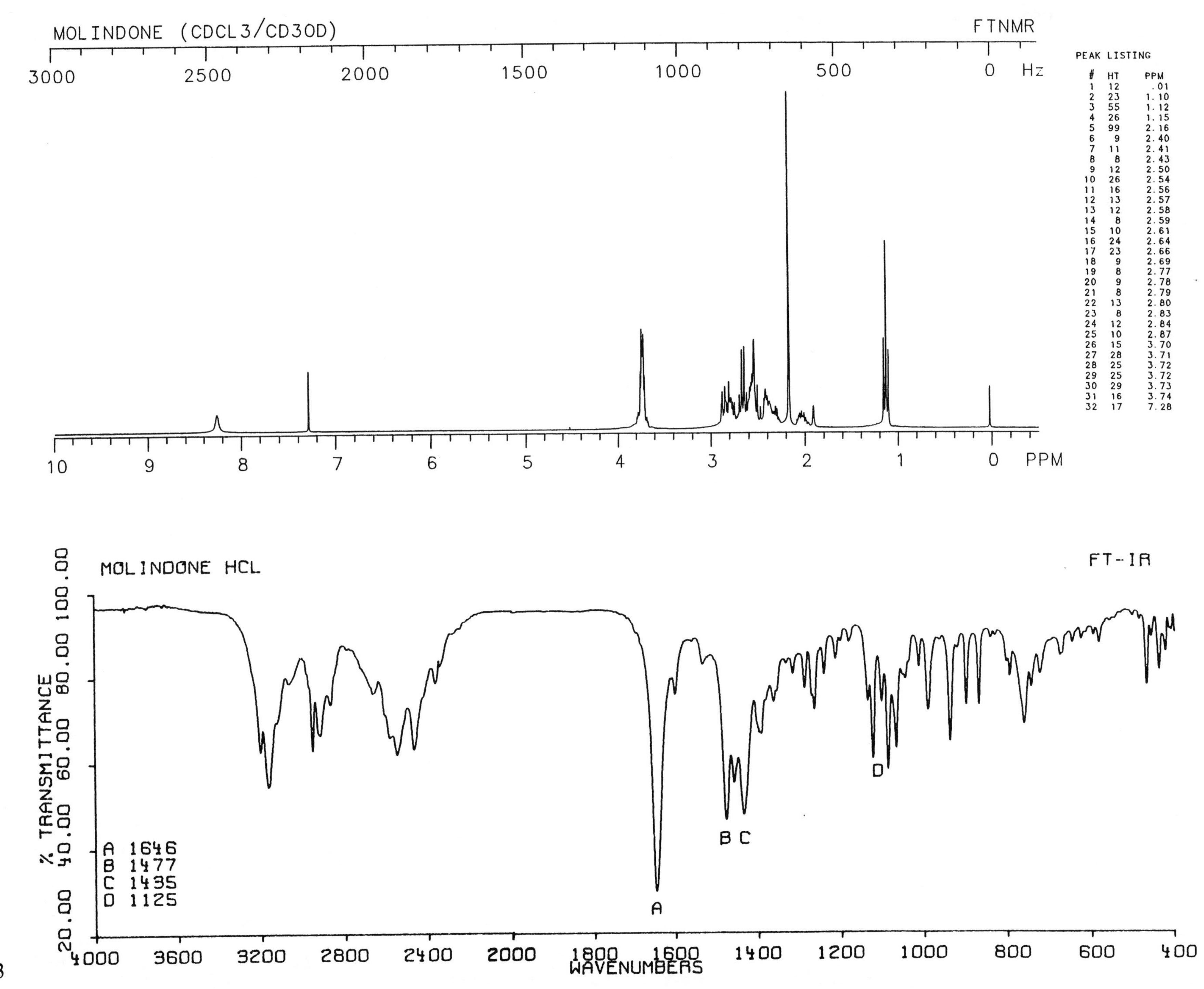

MOLINDONE (CDCL3/CD3OD)
FTNMR
3000 2500 2000 1500 1000 500 0 Hz
10 9 8 7 6 5 4 3 2 1 0 PPM
PEAK LISTING
# HT PPM
1 12 .01
2 23 1.10
3 55 1.12
4 26 1.15
5 99 2.16
6 9 2.40
7 11 2.41
8 8 2.43
9 12 2.50
10 26 2.54
11 16 2.56
12 13 2.57
13 12 2.58
14 8 2.59
15 10 2.61
16 24 2.64
17 23 2.66
18 9 2.69
19 8 2.77
20 9 2.78
21 8 2.79
22 13 2.80
23 8 2.83
24 12 2.84
25 10 2.87
26 15 3.70
27 28 3.71
28 25 3.72
29 25 3.72
30 29 3.73
31 16 3.74
32 17 7.28
MOLINDONE HCL
FT-IR
% TRANSMITTANCE
20.00 40.00 60.00 80.00 100.00
4000 3600 3200 2800 2400 2000 1800 1600 1400 1200 1000 800 600 400
WAVENUMBERS
A 1646
B 1477
C 1435
D 1125
A
B C
D

# 6-MONOACETYLMORPHINE

$C_{19}H_{21}NO_4$

Molecular weight: 327.38 (327.15)
Synonyms: 7,8-Didehydro-4,5-epoxy-17-methylmorphinan-3-ol-6-acetate (ester); 6-acetoxy-7,8-dehydro-4,5-epoxy-3-hydroxy-N-methylmorphinan; MAM
Trade names:

Use: Narcotic
HPLC: Si-10; 10A:90B; 3.5
GC: 2579; 250°C

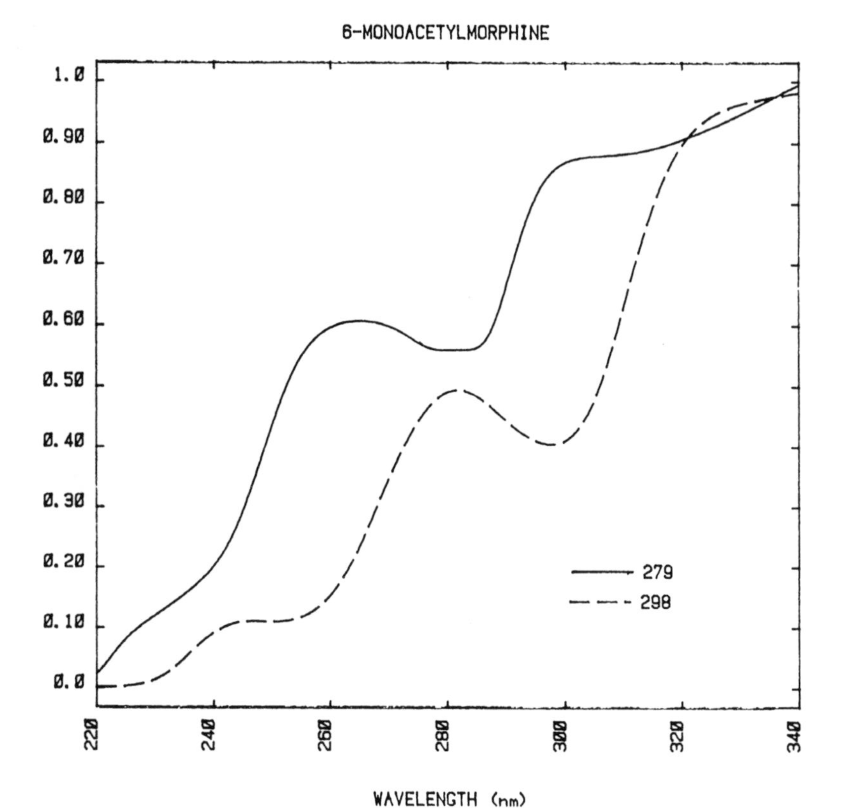


*6-MONOACETYLMORPHINE*

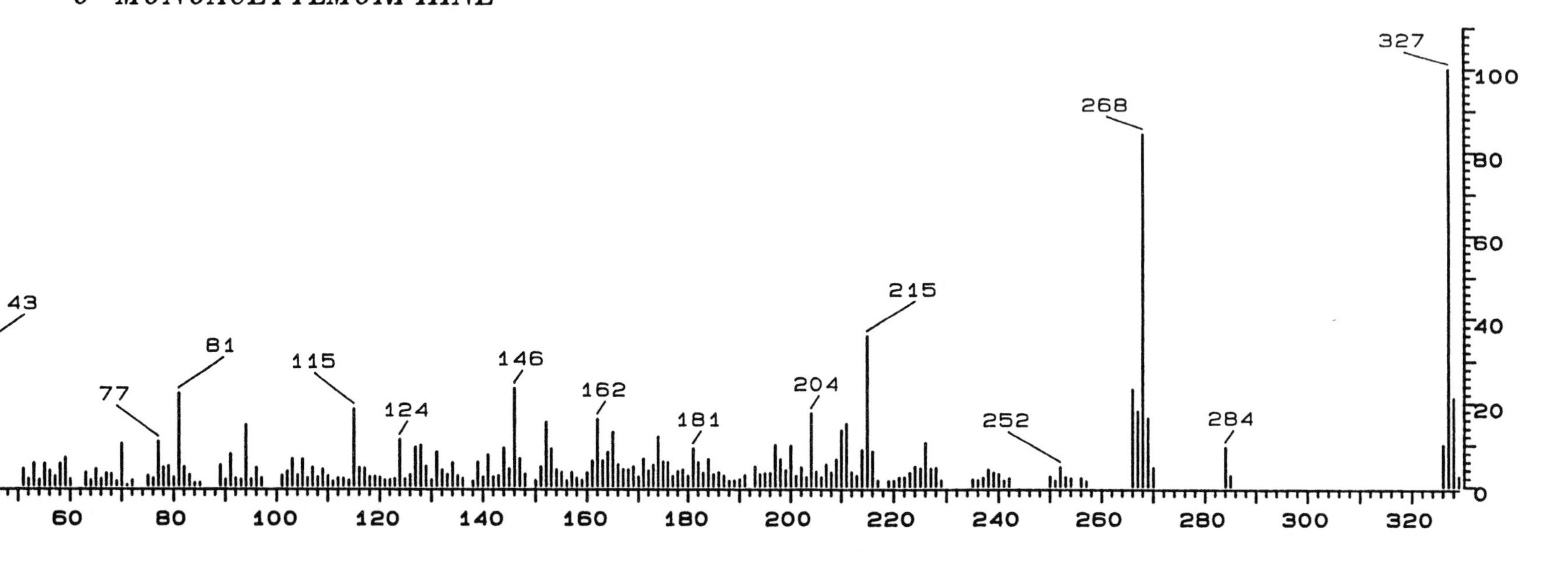

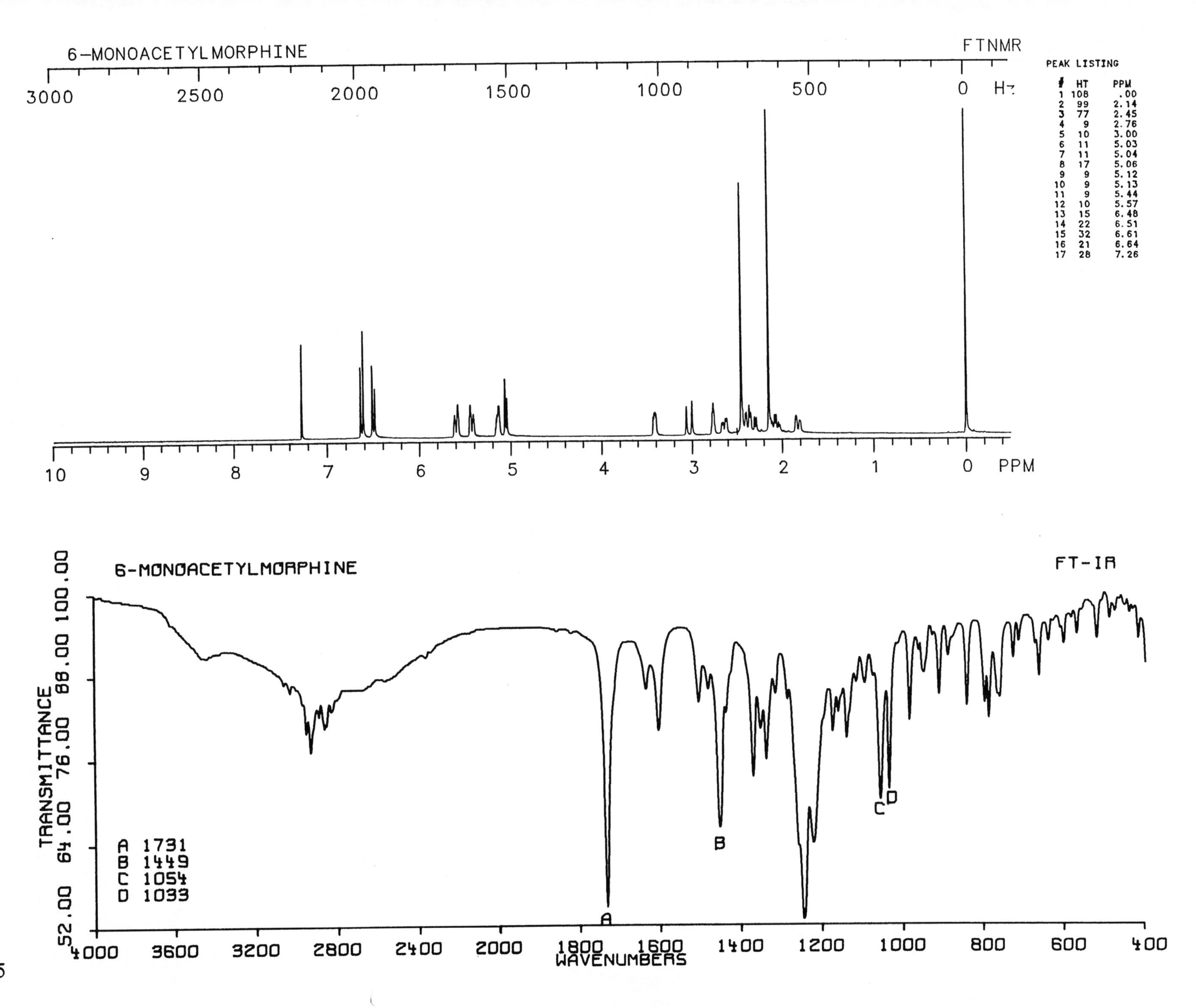
6-MONOACETYLMORPHINE
FTNMR
PEAK LISTING
# HT PPM
1 108 .00
2 99 2.14
3 77 2.45
4 9 2.76
5 10 3.00
6 11 5.03
7 11 5.04
8 17 5.06
9 9 5.12
10 9 5.13
11 9 5.44
12 10 5.57
13 15 6.48
14 22 6.51
15 32 6.61
16 21 6.64
17 28 7.26
Hz
PPM
6-MONOACETYLMORPHINE
FT-IR
TRANSMITTANCE
WAVENUMBERS
A 1731
B 1449
C 1054
D 1033

# MONOBENZONE

$C_{13}H_{12}O_2$

**Molecular weight:** 200.24 (200.08)
**Synonyms:** 4-(Phenylmethoxy)phenol; p-(benzyloxy)phenol

**Trade names:** Benoquin

**Use:** Depigmentor
**HPLC:** Si-10; 100B; 4.5
**GC:** 1852; 200°C

O—CH₂—C₆H₅ / OH

MONOBENZONE

TRANSMITTANCE

1.0, 0.90, 0.80, 0.70, 0.60, 0.50, 0.40, 0.30, 0.20, 0.10, 0.0

220, 240, 260, 280, 300, 320, 340

WAVELENGTH (nm)

—— 287
— — 236, 305

*MONOBENZONE*

51, 53, 65, 81, 91, 109, 128, 141, 152, 200

60, 80, 100, 120, 140, 160, 180, 200

0, 20, 40, 60, 80, 100

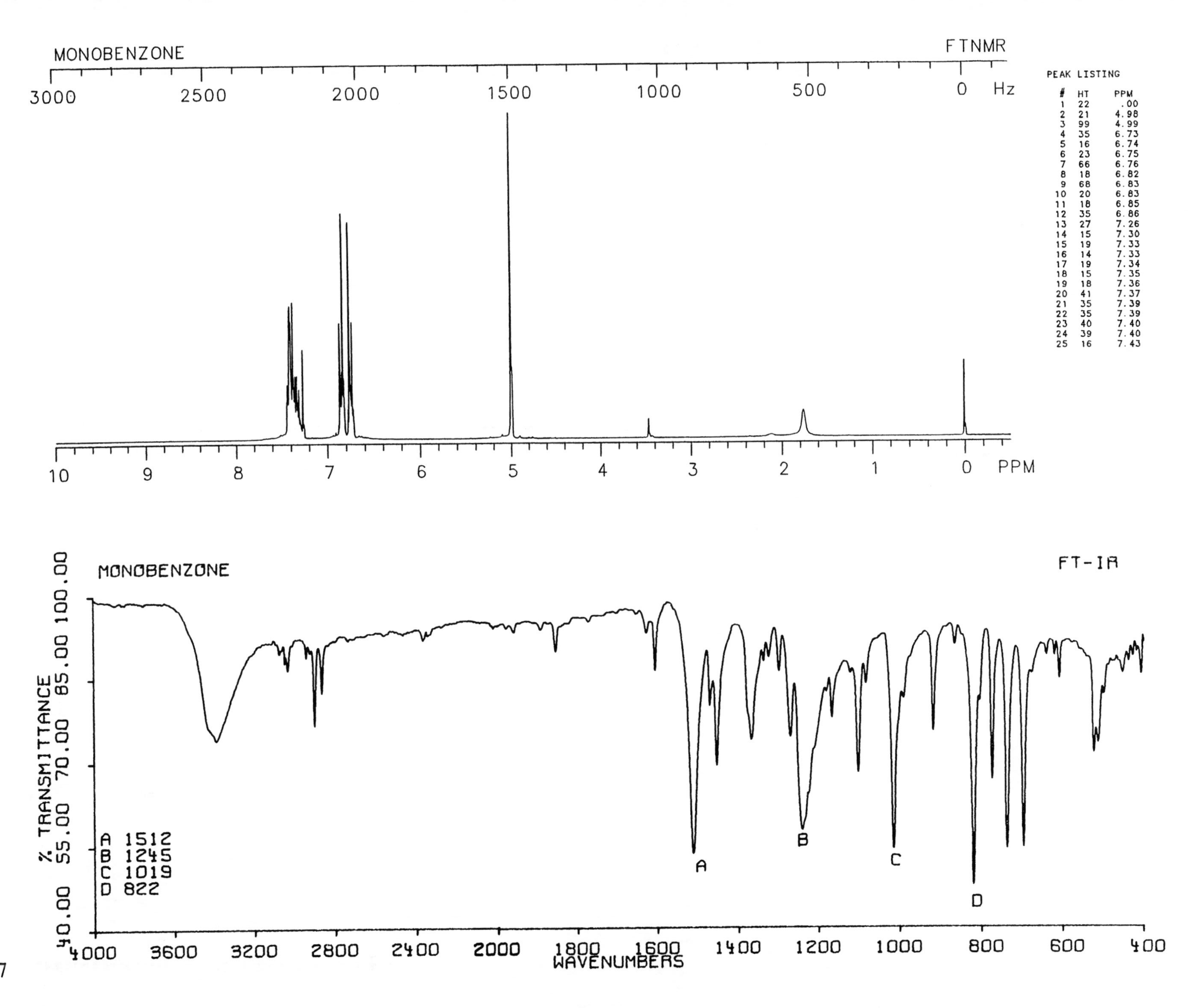
MONOBENZONE
FTNMR
3000 2500 2000 1500 1000 500 0 Hz
10 9 8 7 6 5 4 3 2 1 0 PPM
PEAK LISTING
# HT PPM
1 22 .00
2 21 4.98
3 99 4.99
4 35 6.73
5 16 6.74
6 23 6.75
7 66 6.76
8 18 6.82
9 68 6.83
10 20 6.83
11 18 6.85
12 35 6.86
13 27 7.26
14 15 7.30
15 19 7.33
16 14 7.33
17 19 7.34
18 15 7.35
19 18 7.36
20 41 7.37
21 35 7.39
22 35 7.39
23 40 7.40
24 39 7.40
25 16 7.43
MONOBENZONE
FT-IR
% TRANSMITTANCE
40.00 55.00 70.00 85.00 100.00
A 1512
B 1245
C 1019
D 822
A B C D
4000 3600 3200 2800 2400 2000 1800 1600 1400 1200 1000 800 600 400
WAVENUMBERS

# MONOSODIUM GLUTAMATE

$C_5H_8NNaO_4 \cdot H_2O$

**Molecular weight:** 187.13 (187.06)
**Synonyms:** Glutamic acid monosodium salt monohydrate; monosodium L-glutamate monohydrate; chinese seasoning
**Trade names:** Accent, Ajinomoto, Glutacyl, Glutavene, Vetsin

**Use:** Flavor enhancer
**HPLC:**
**GC:**

$$HOOC-CH(NH_2)-CH_2-CH_2-COONa \cdot H_2O$$

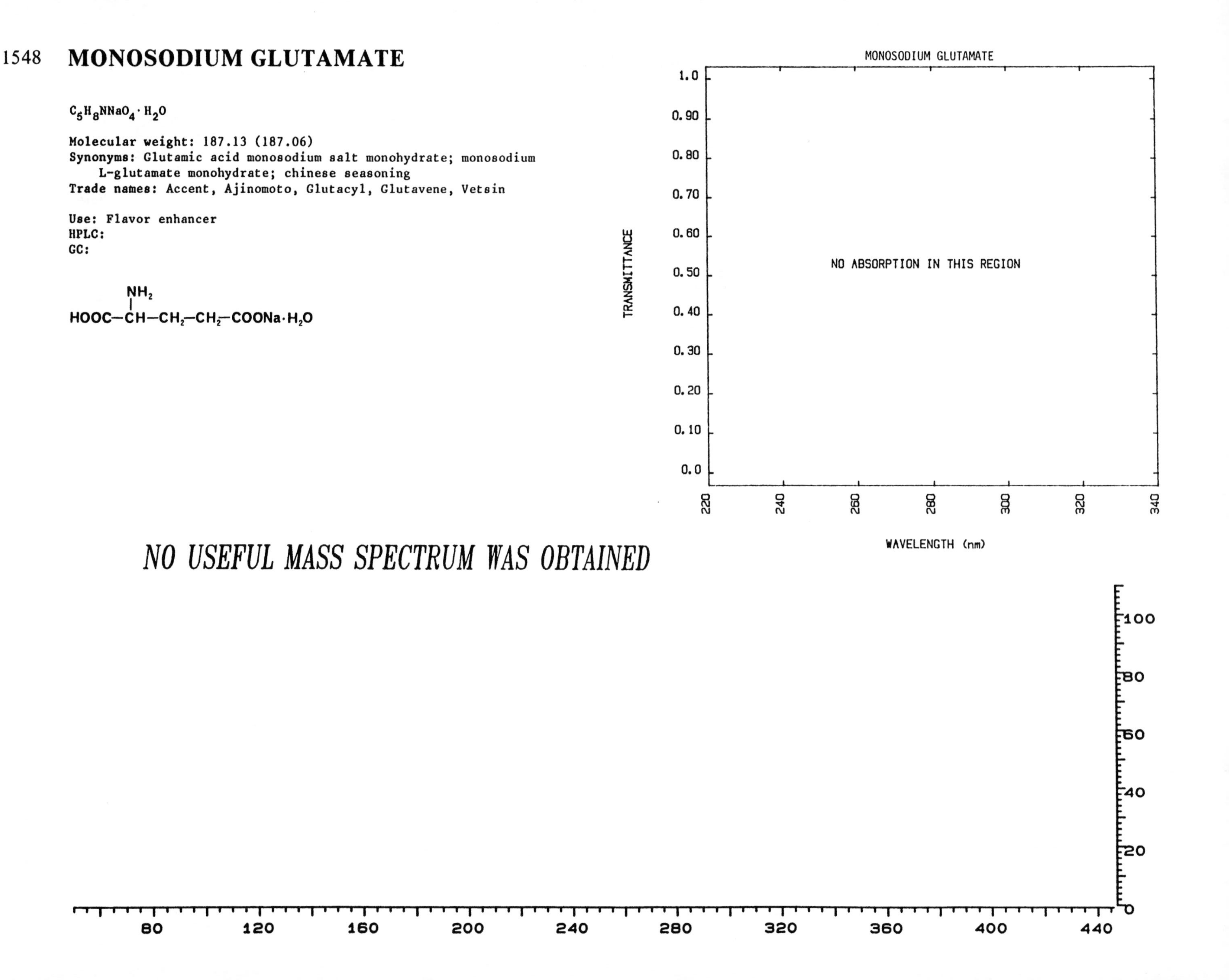

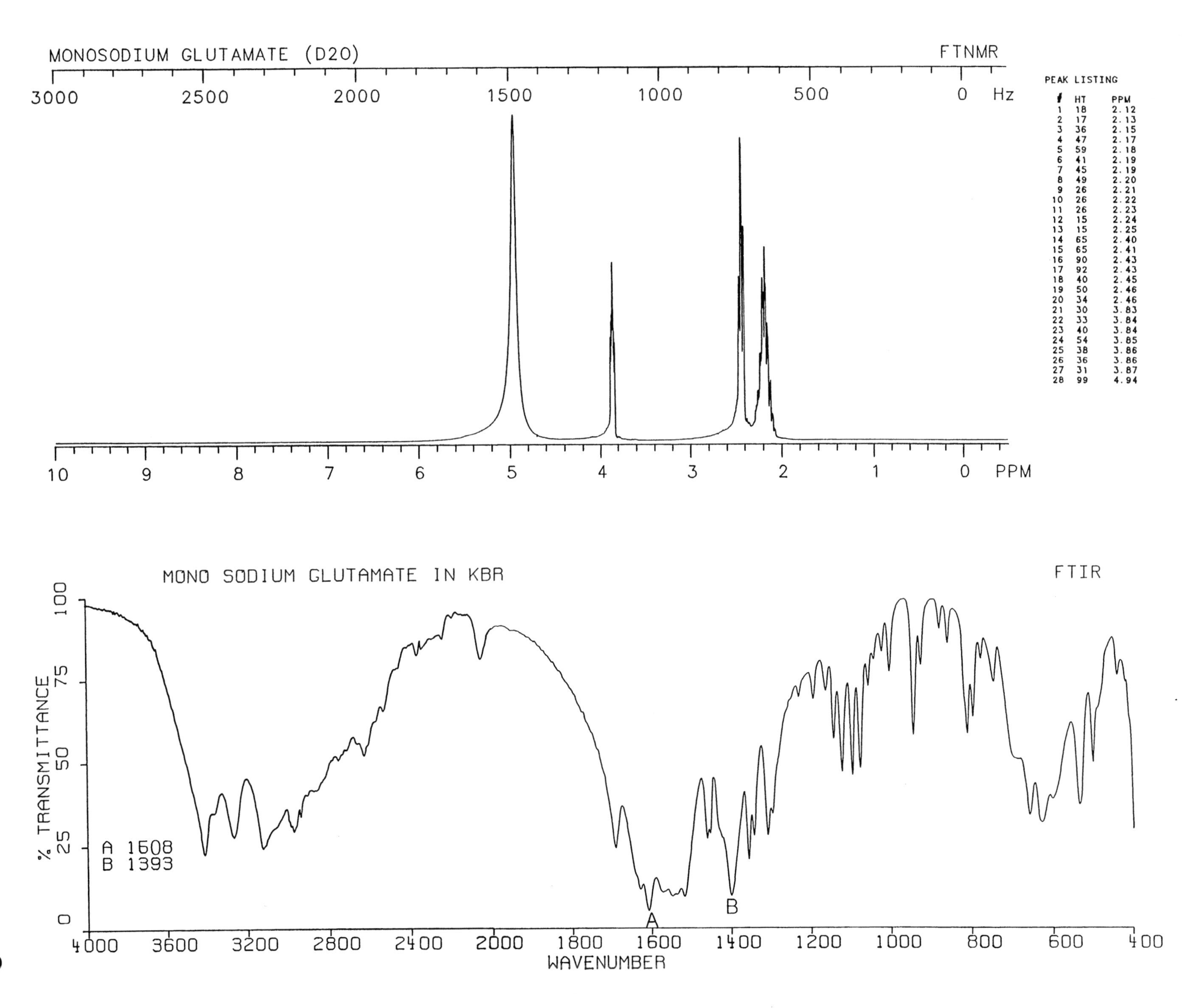
MONOSODIUM GLUTAMATE (D2O)
FTNMR
3000
2500
2000
1500
1000
500
0 Hz
10
9
8
7
6
5
4
3
2
1
0 PPM
PEAK LISTING
# HT PPM
1 18 2.12
2 17 2.13
3 36 2.15
4 47 2.17
5 59 2.18
6 41 2.19
7 45 2.19
8 49 2.20
9 26 2.21
10 26 2.22
11 26 2.23
12 15 2.24
13 15 2.25
14 65 2.40
15 65 2.41
16 90 2.43
17 92 2.43
18 40 2.45
19 50 2.46
20 34 2.46
21 30 3.83
22 33 3.84
23 40 3.84
24 54 3.85
25 38 3.86
26 36 3.86
27 31 3.87
28 99 4.94
MONO SODIUM GLUTAMATE IN KBR
FTIR
% TRANSMITTANCE
100
75
50
25
0
A 1608
B 1393
A
B
4000
3600
3200
2800
2400
2000
1800
1600
1400
1200
1000
800
600
400
WAVENUMBER

# MOPERONE

$C_{22}H_{26}FNO_2$

**Molecular weight:** 355.46 (355.20)
**Synonyms:** 1-(4-Fluorophenyl)-4-[4-hydroxy-4-(4-methylphenyl)-1-piperidinyl]-1-butanone; methylperidol
**Trade names:** Luvatren

**Use:** Tranquilizer
**HPLC:** Si-10; 2A:98B; 7.6
**GC:** 3018; 280°C

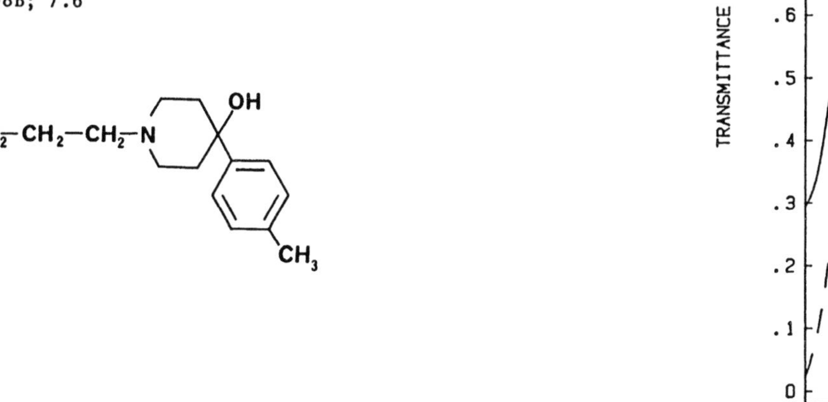

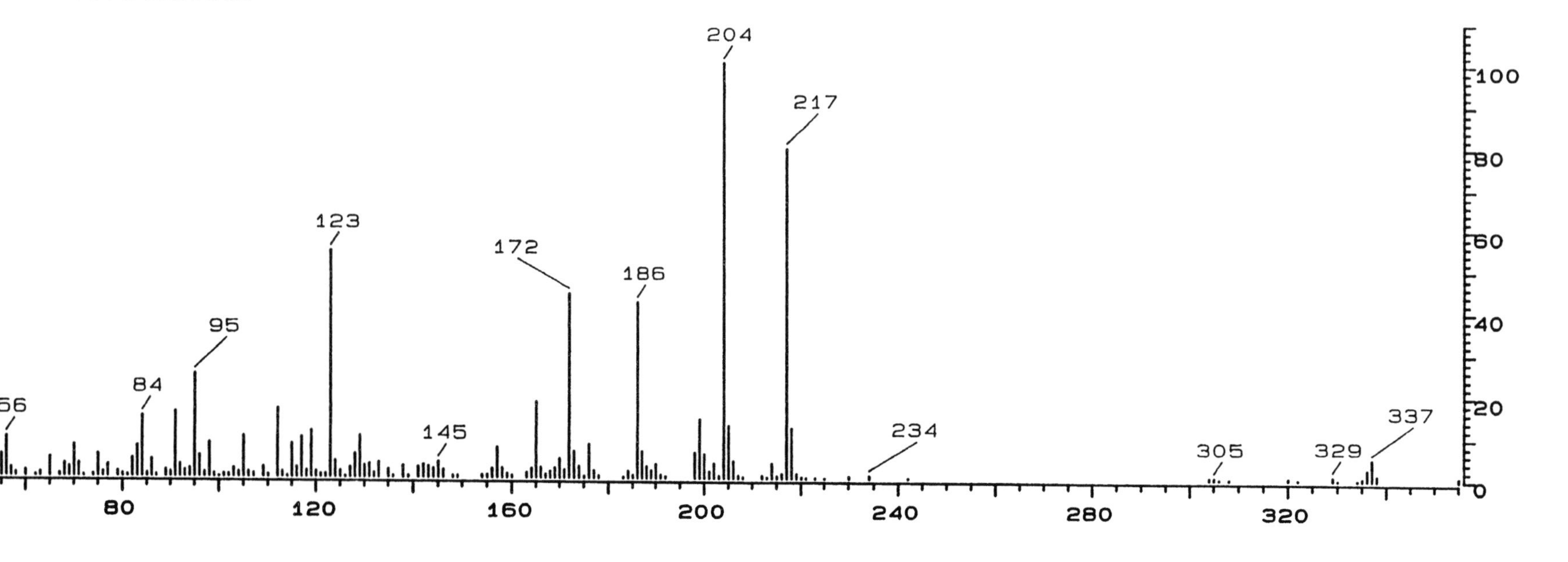

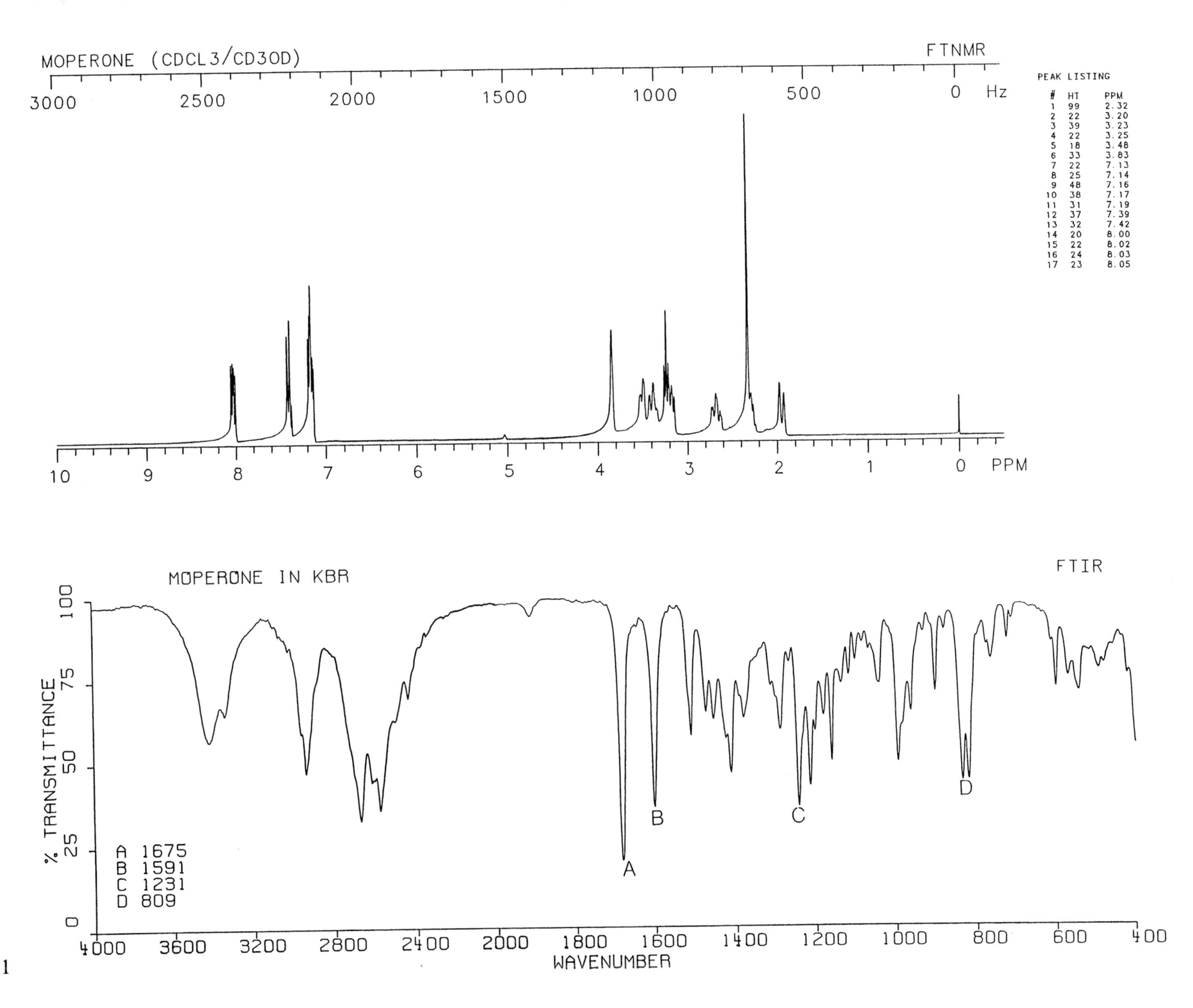
MOPERONE (CDCL3/CD3OD)
FTNMR
3000 2500 2000 1500 1000 500 0 Hz
10 9 8 7 6 5 4 3 2 1 0 PPM
PEAK LISTING
# HT PPM
1 99 2.32
2 22 3.20
3 39 3.23
4 22 3.25
5 18 3.48
6 33 3.83
7 22 7.13
8 25 7.14
9 48 7.16
10 38 7.17
11 31 7.19
12 37 7.39
13 32 7.42
14 20 8.00
15 22 8.02
16 24 8.03
17 23 8.05
MOPERONE IN KBR
FTIR
% TRANSMITTANCE
100 75 50 25 0
A 1675
B 1591
C 1231
D 809
A B C D
4000 3600 3200 2800 2400 2000 1800 1600 1400 1200 1000 800 600 400
WAVENUMBER

# MORPHINE

$C_{17}H_{19}NO_3$

Molecular weight: 285.34 (285.14)
Synonyms: 7,8-Didehydro-4,5-epoxy-17-methylmorphinan-3,6-diol

Trade names: Morphine Sulfate, Morphina

Use: Narcotic analgesic
HPLC: Si-10; 10A:90B; 9.0
GC: 2523; 250°C

$CH_3$ N $CH_2$ $CH_2$ OH O OH

MORPHINE

TRANSMITTANCE

1.0 0.90 0.80 0.70 0.60 0.50 0.40 0.30 0.20 0.10 0.0

220 240 260 280 300 320 340

WAVELENGTH (nm)

285
298

*MORPHINE*

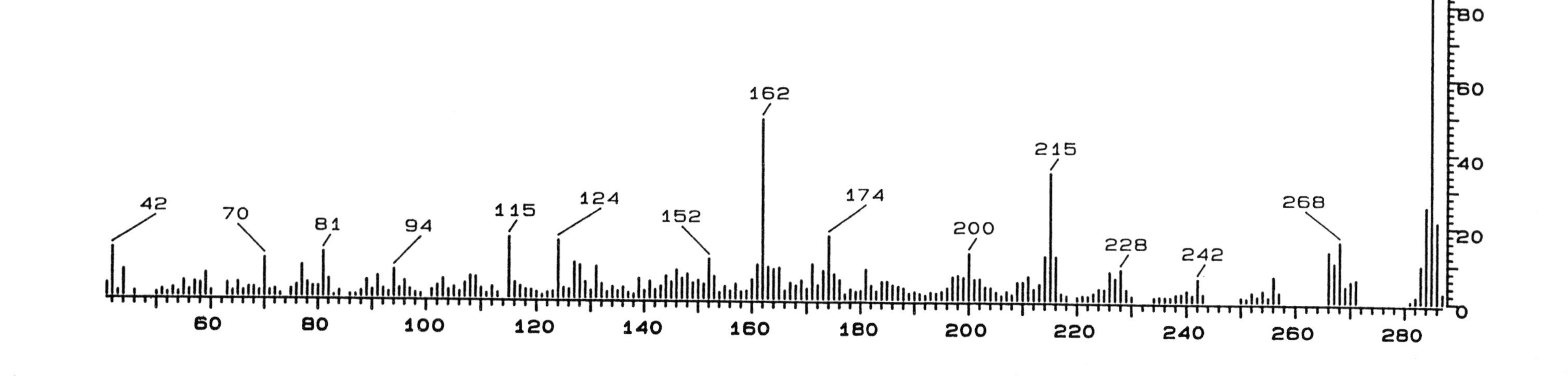

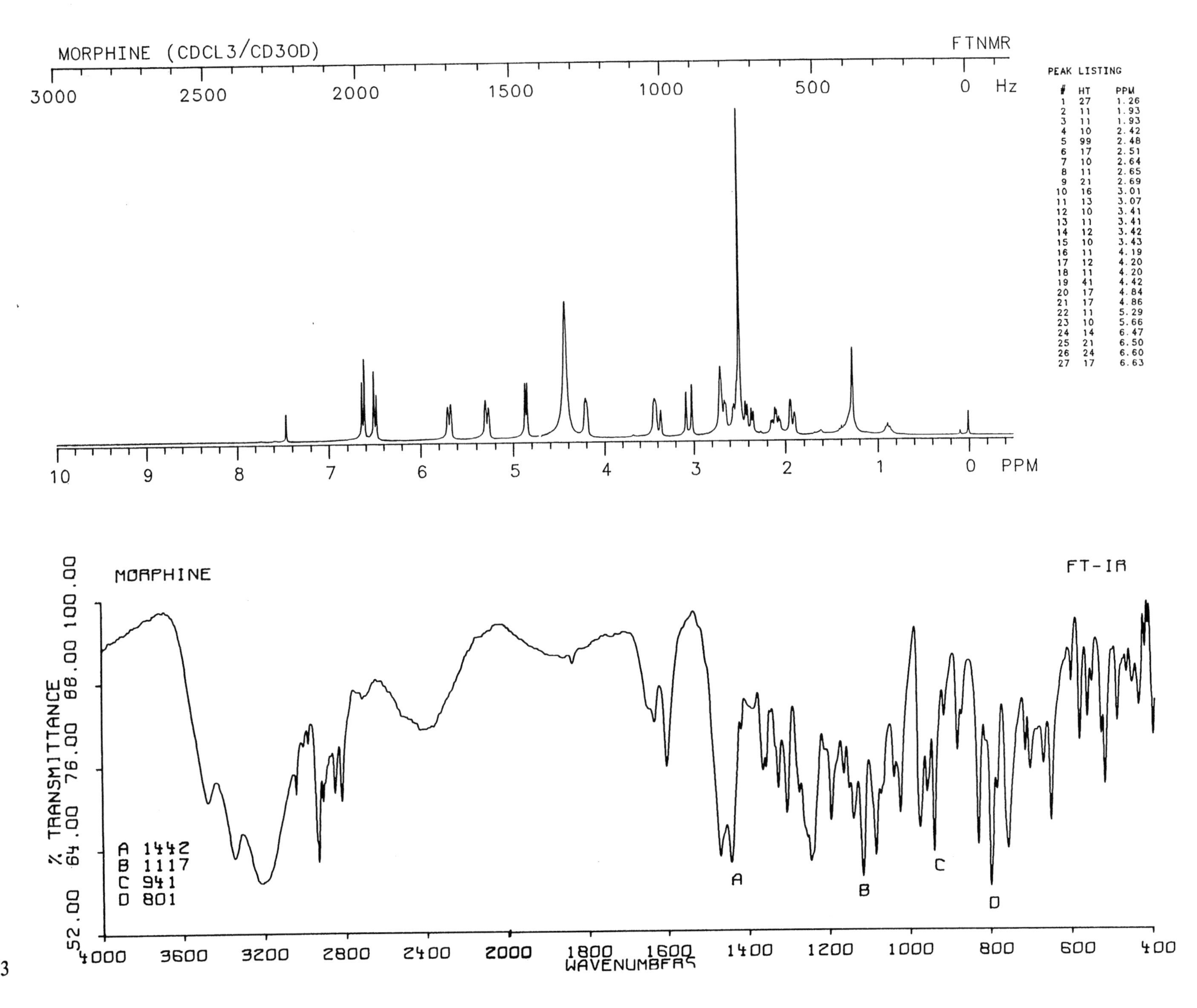
MORPHINE (CDCL3/CD3OD)
FTNMR
Hz
PPM
PEAK LISTING
# HT PPM
1 27 1.26
2 11 1.93
3 11 1.93
4 10 2.42
5 99 2.48
6 17 2.51
7 10 2.64
8 11 2.65
9 21 2.69
10 16 3.01
11 13 3.07
12 10 3.41
13 11 3.41
14 12 3.42
15 10 3.43
16 11 4.19
17 12 4.20
18 11 4.20
19 41 4.42
20 17 4.84
21 17 4.86
22 11 5.29
23 10 5.66
24 14 6.47
25 21 6.50
26 24 6.60
27 17 6.63
MORPHINE
FT-IR
% TRANSMITTANCE
WAVENUMBERS
A 1442
B 1117
C 941
D 801

# MORPHINE-3-GLUCURONIDE

$C_{23}H_{27}NO_9$

Molecular weight: 461.47 (461.17)
Synonyms: 7,8-Didehydro-4,5-epoxy-17-methylmorphinan-6-ol-3-d-glucuronic acid
Trade names:

Use: Metabolite
HPLC:
GC:

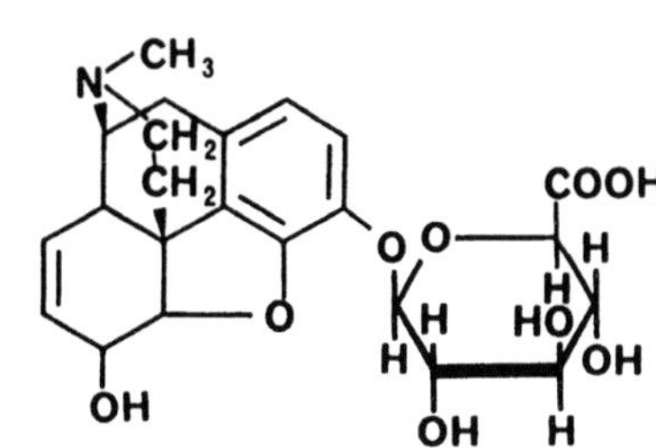


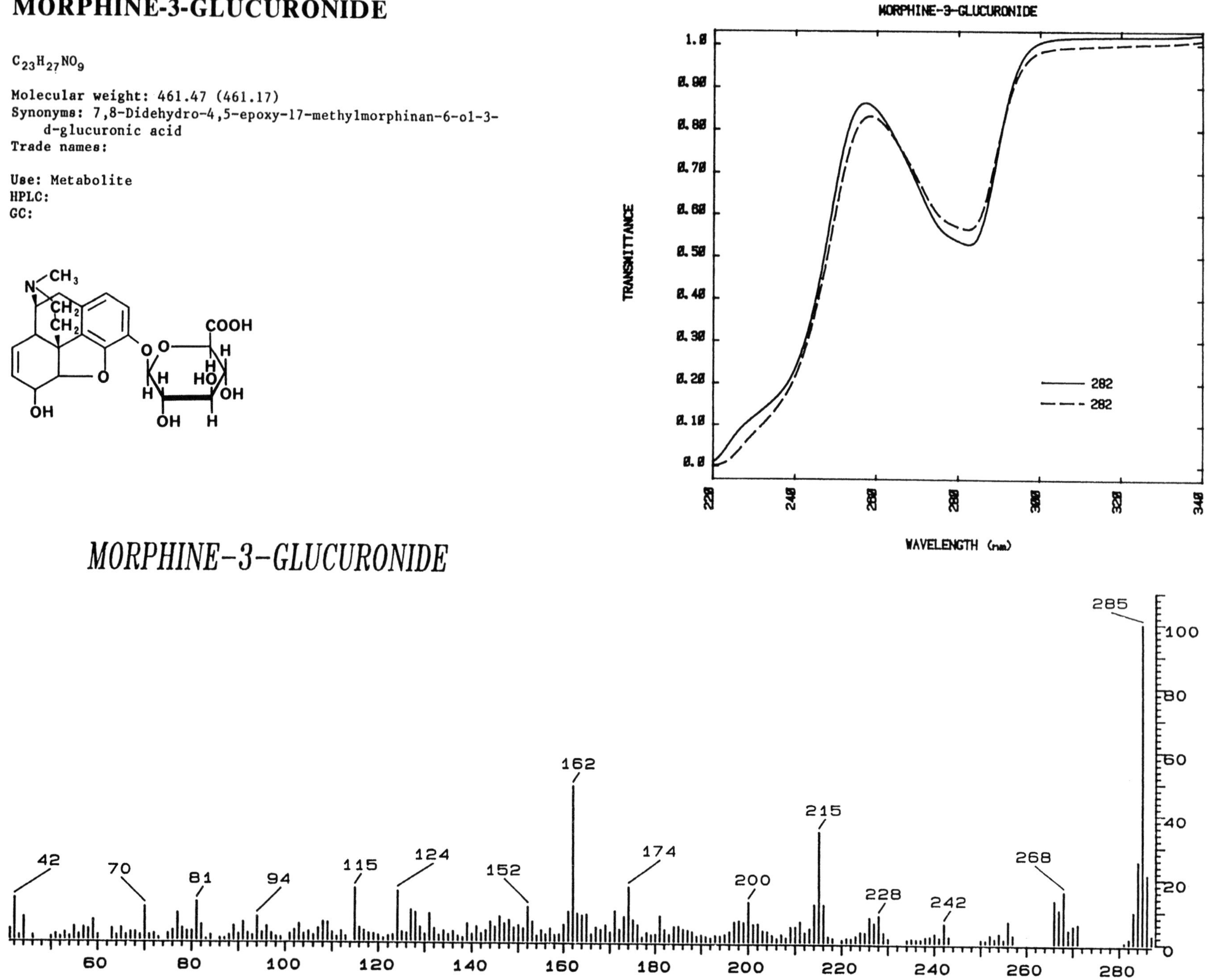

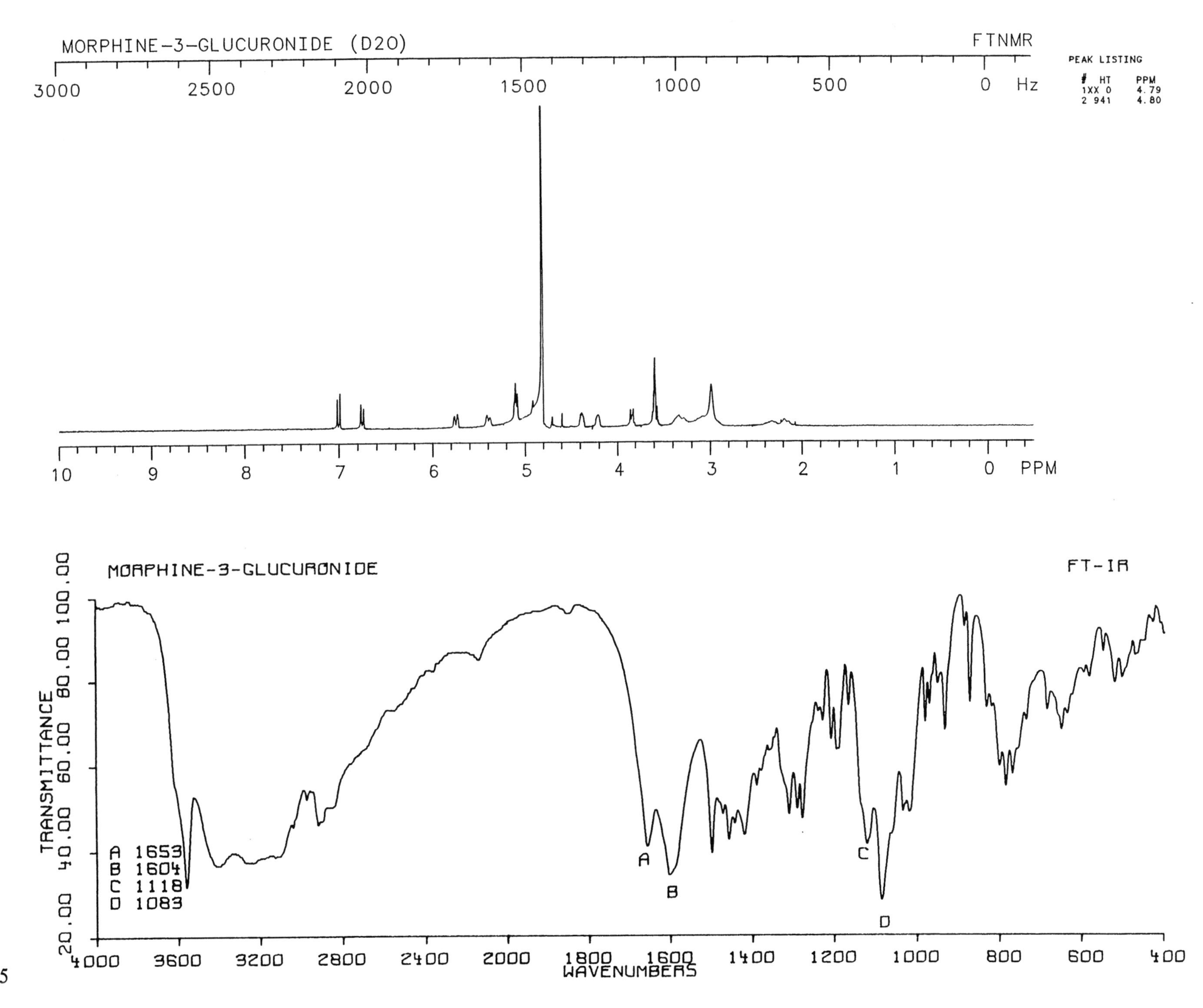

MORPHINE-3-GLUCURONIDE (D2O)
FTNMR
PEAK LISTING
# HT PPM
1XX 0 4.79
2 941 4.80
Hz
PPM
MORPHINE-3-GLUCURONIDE
FT-IR
TRANSMITTANCE
WAVENUMBERS
A 1653
B 1604
C 1118
D 1083

# MORPHINE-*N*-OXIDE

$C_{17}H_{19}NO_4$

**Molecular weight:** 301.33 (301.13)
**Synonyms:** Genomorphine; morphine oxide

**Trade names:**

**Use:** Narcotic analgesic
**HPLC:** Si-10; 20A:80B; 6.2
**GC:**

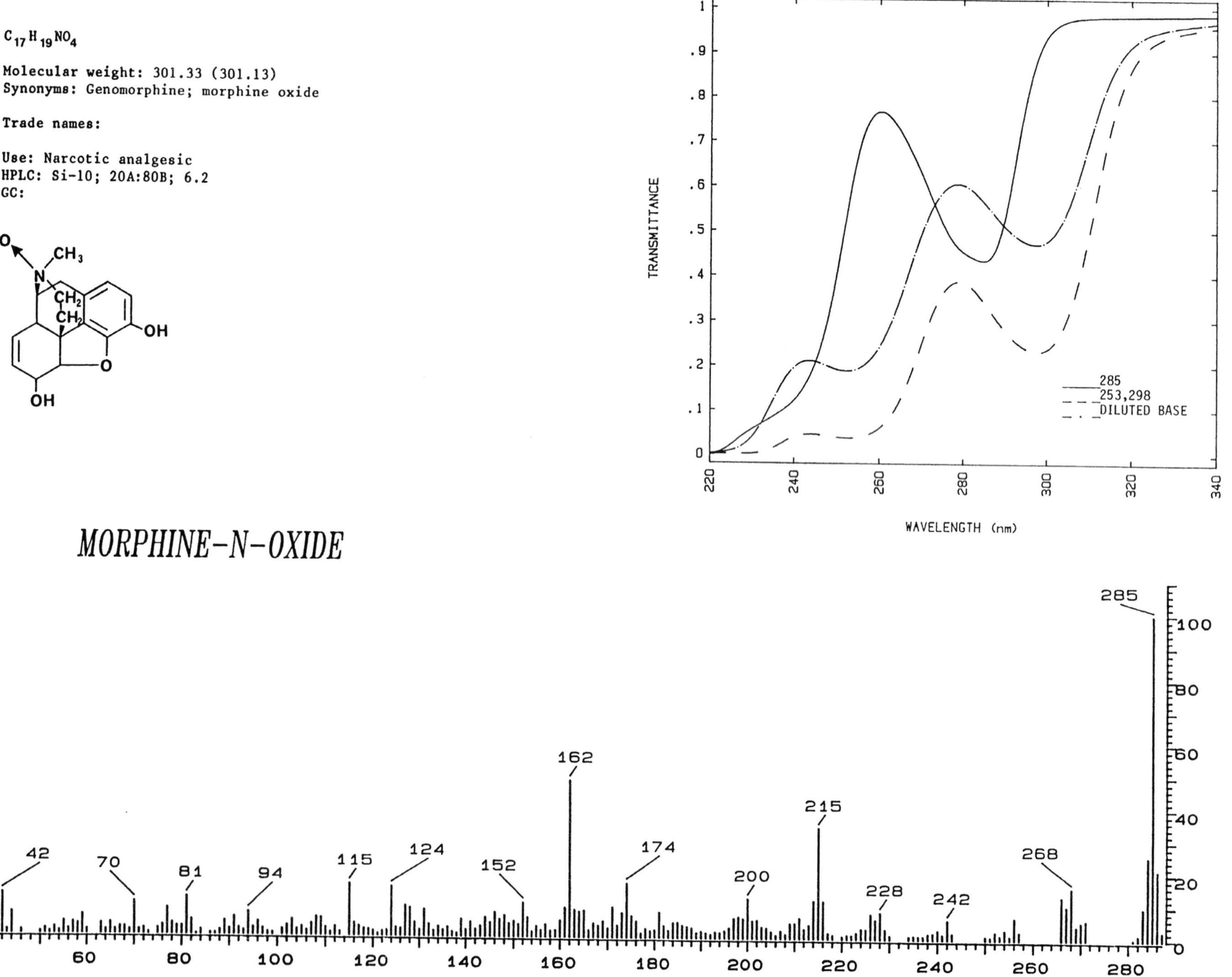

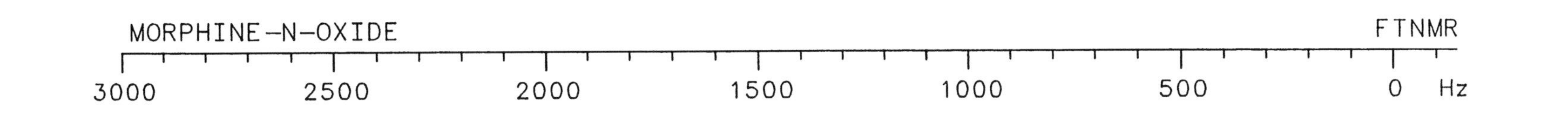
MORPHINE-N-OXIDE
FTNMR
3000
2500
2000
1500
1000
500
0
Hz
INSUFFICIENT SOLUBILITY
10
9
8
7
6
5
4
3
2
1
0
PPM

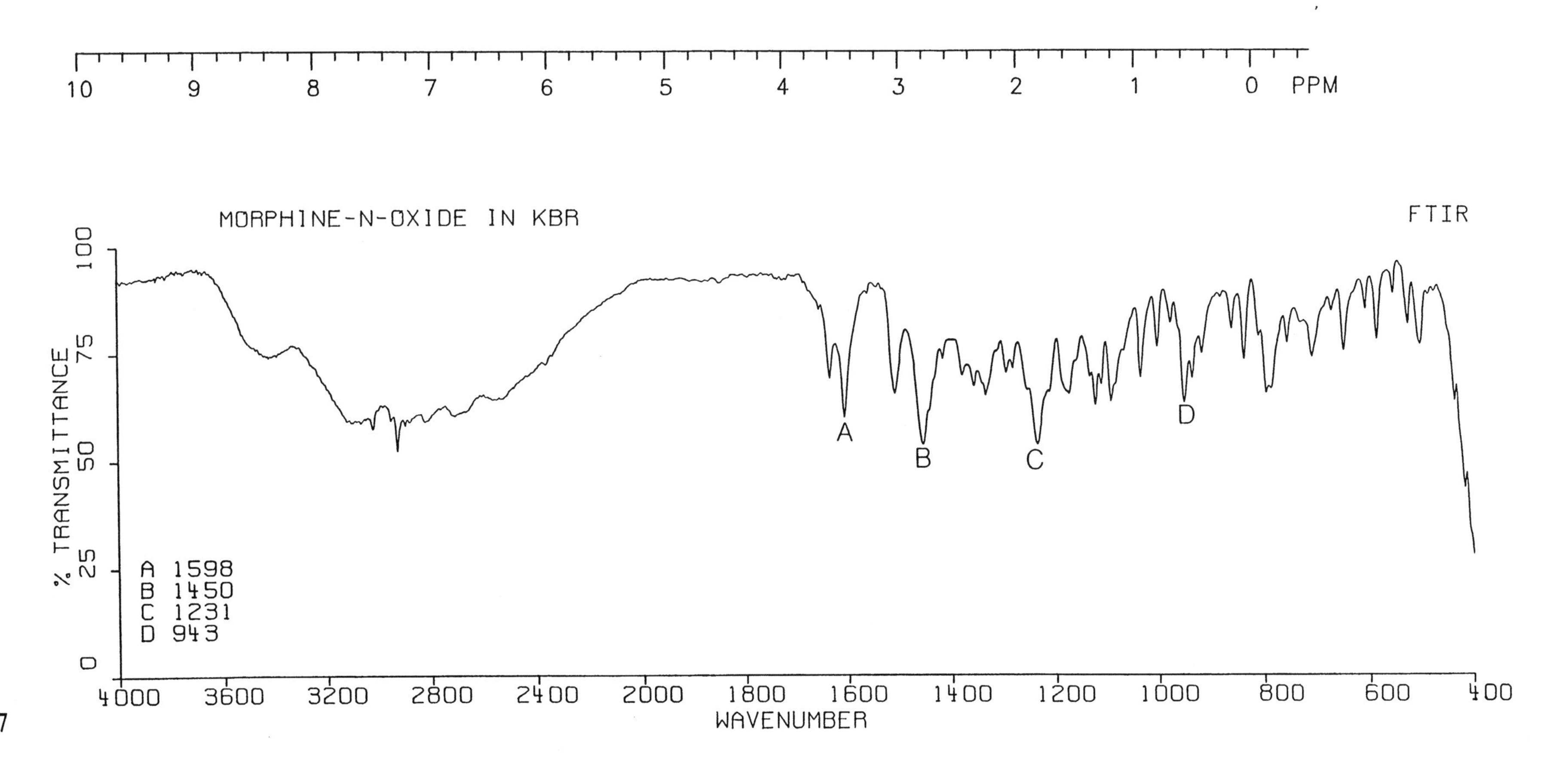
MORPHINE-N-OXIDE IN KBR
FTIR
% TRANSMITTANCE
100
75
50
25
0
A 1598
B 1450
C 1231
D 943
A
B
C
D
4000
3600
3200
2800
2400
2000
1800
1600
1400
1200
1000
800
600
400
WAVENUMBER

# MOXALACTAM

$C_{20}H_{20}N_6O_9S$

**Molecular weight:** 520.48 (520.10)
**Synonyms:** (6R,7R)-7-[[Carboxy(4-hydroxyphenyl)acetyl]amino]-7-methoxy-3-[[(1-methyl-1H-tetrazol-5-yl)thia]methyl]-8-oxo-5-oxa-1-azabicyclo[4.2.0]oct-2-ene-2-carboxylic acid; lamoxactam; latamoxef
**Trade names:** Festamoxin, Moxam, Moxalactam, Shiomarin

**Use:** Antibiotic
**HPLC:**
**GC:**

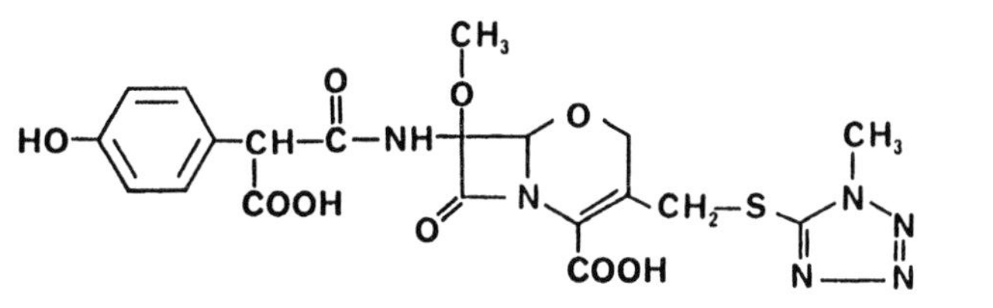


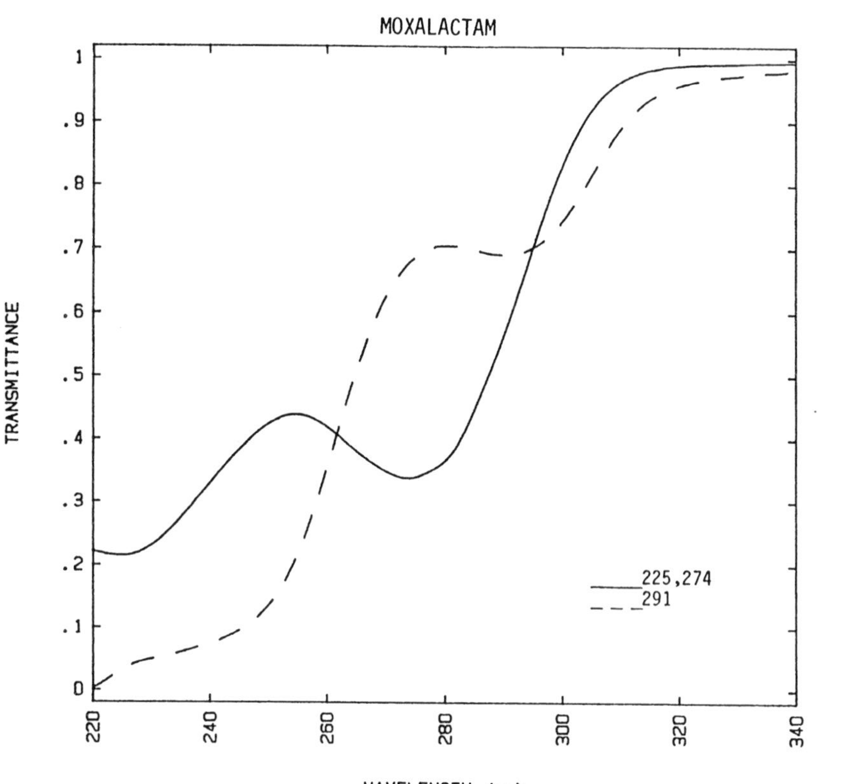

*NO USEFUL MASS SPECTRUM WAS OBTAINED*

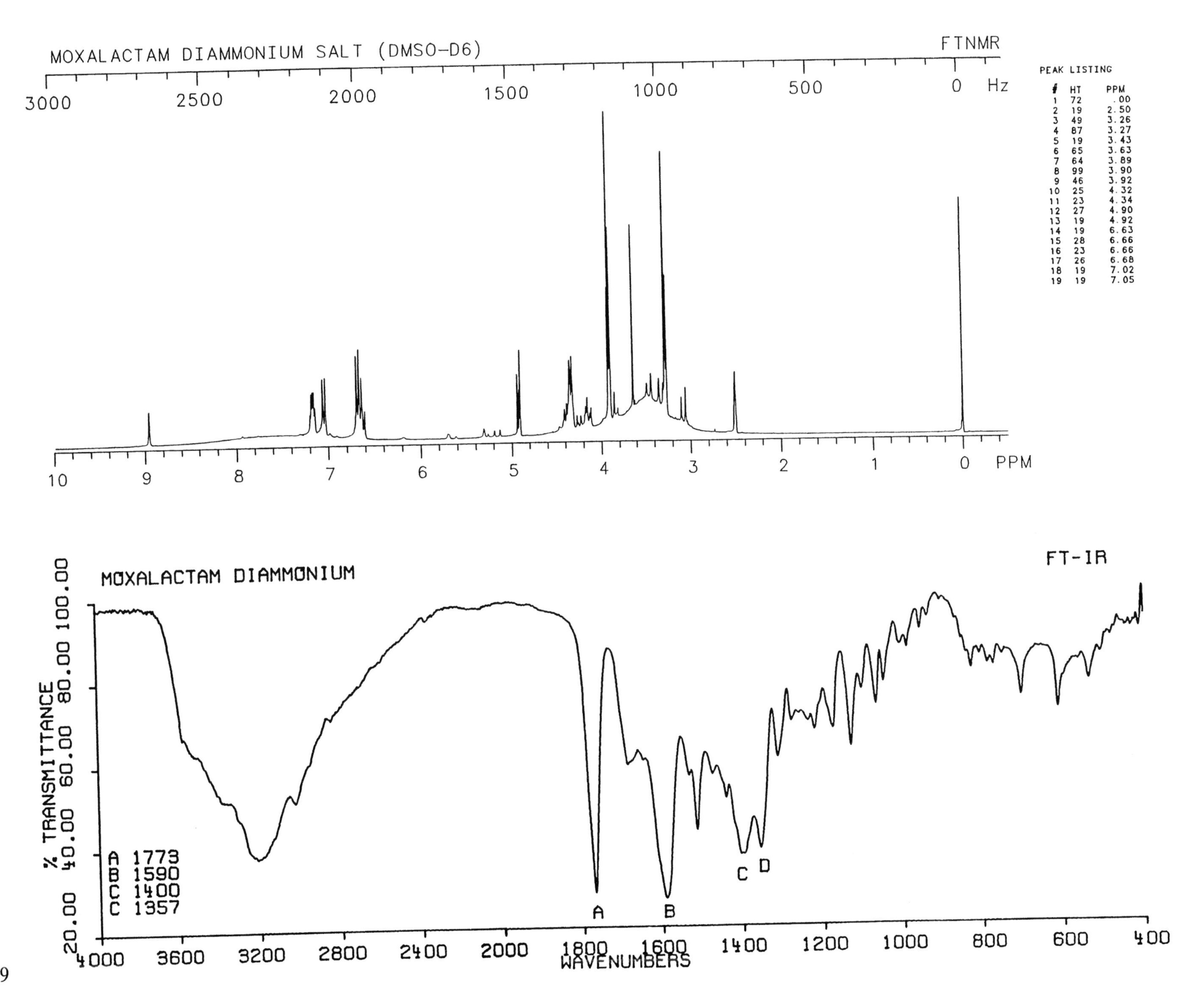

MOXALACTAM DIAMMONIUM SALT (DMSO-D6)
FTNMR
Hz
PPM
PEAK LISTING
# HT PPM
1 72 .00
2 19 2.50
3 49 3.26
4 87 3.27
5 19 3.43
6 65 3.63
7 64 3.89
8 99 3.90
9 46 3.92
10 25 4.32
11 23 4.34
12 27 4.90
13 19 4.92
14 19 6.63
15 28 6.66
16 23 6.66
17 26 6.68
18 19 7.02
19 19 7.05
MOXALACTAM DIAMMONIUM
FT-IR
% TRANSMITTANCE
WAVENUMBERS
A 1773
B 1590
C 1400
C 1357

# NADOLOL

$C_{17}H_{27}NO_4$

**Molecular weight:** 309.41 (309.19)
**Synonyms:** Cis-5-[3-[(1,1-Dimethylethyl)amino]-2-hydroxypropoxy]-1,2,3,4-tetrahydro-2,3-naphthalenediol
**Trade names:** Corgard

**Use:** Beta-adrenergic receptor blocking agent
**HPLC:** Si-10; 20A:80B; 7.5
**GC:**

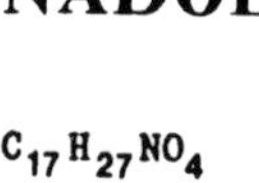

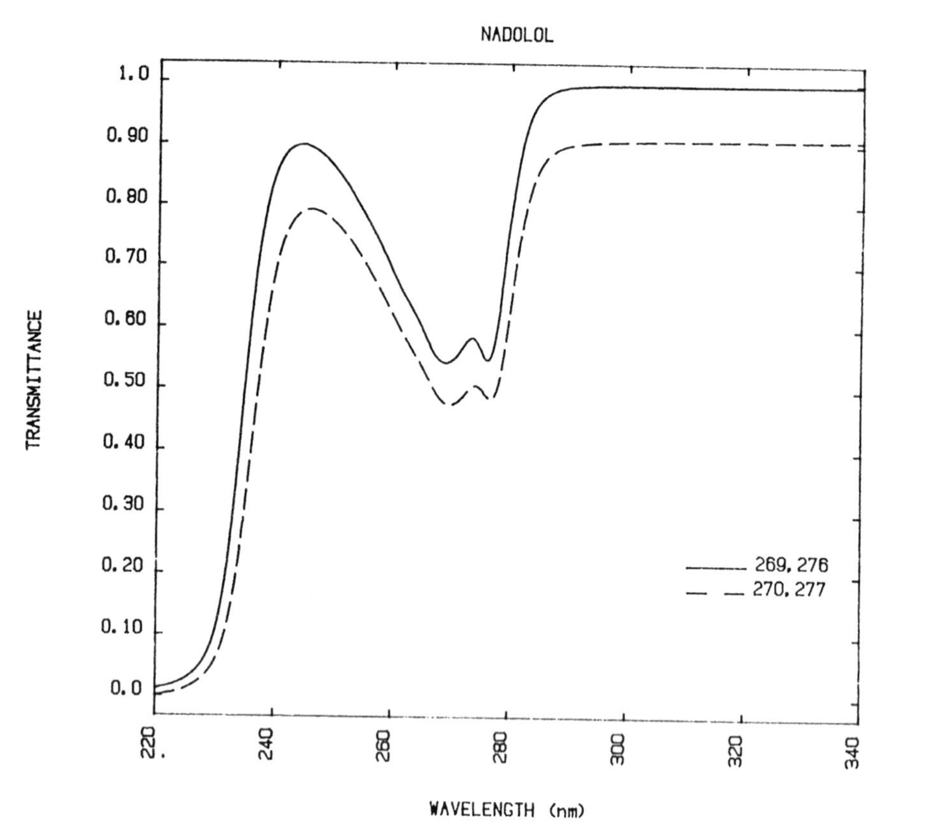


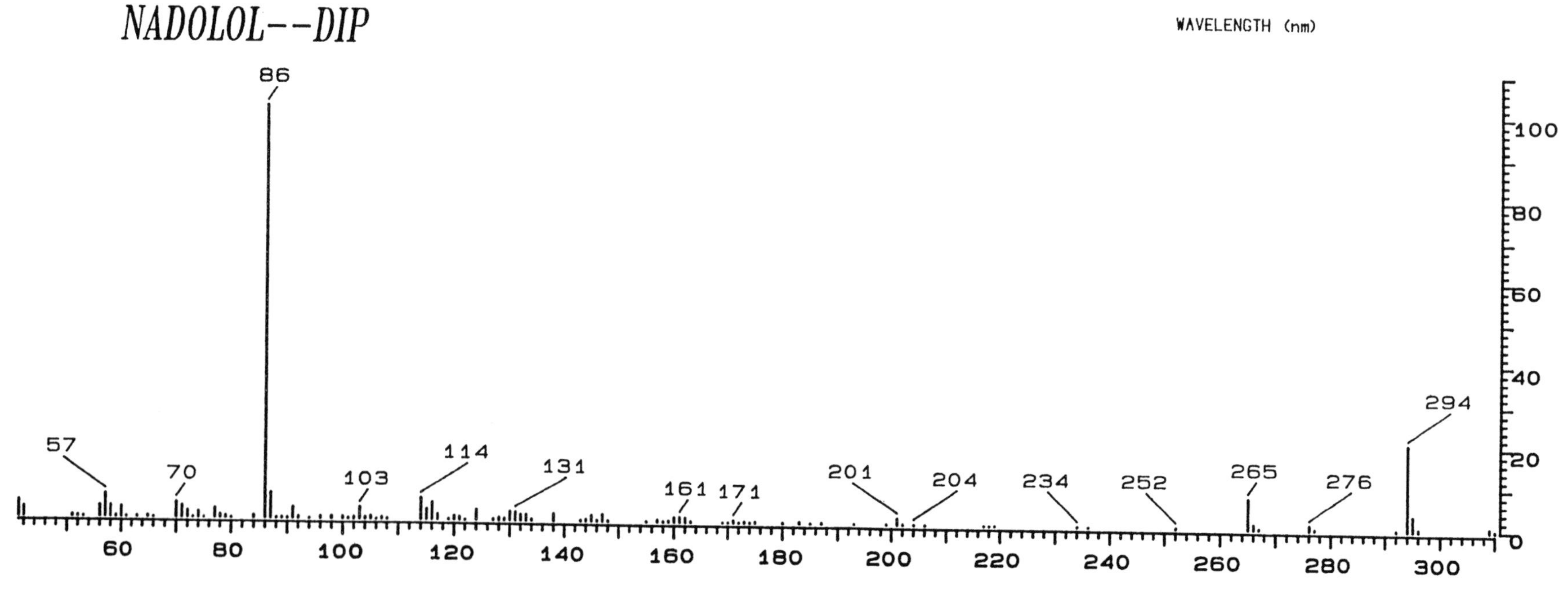

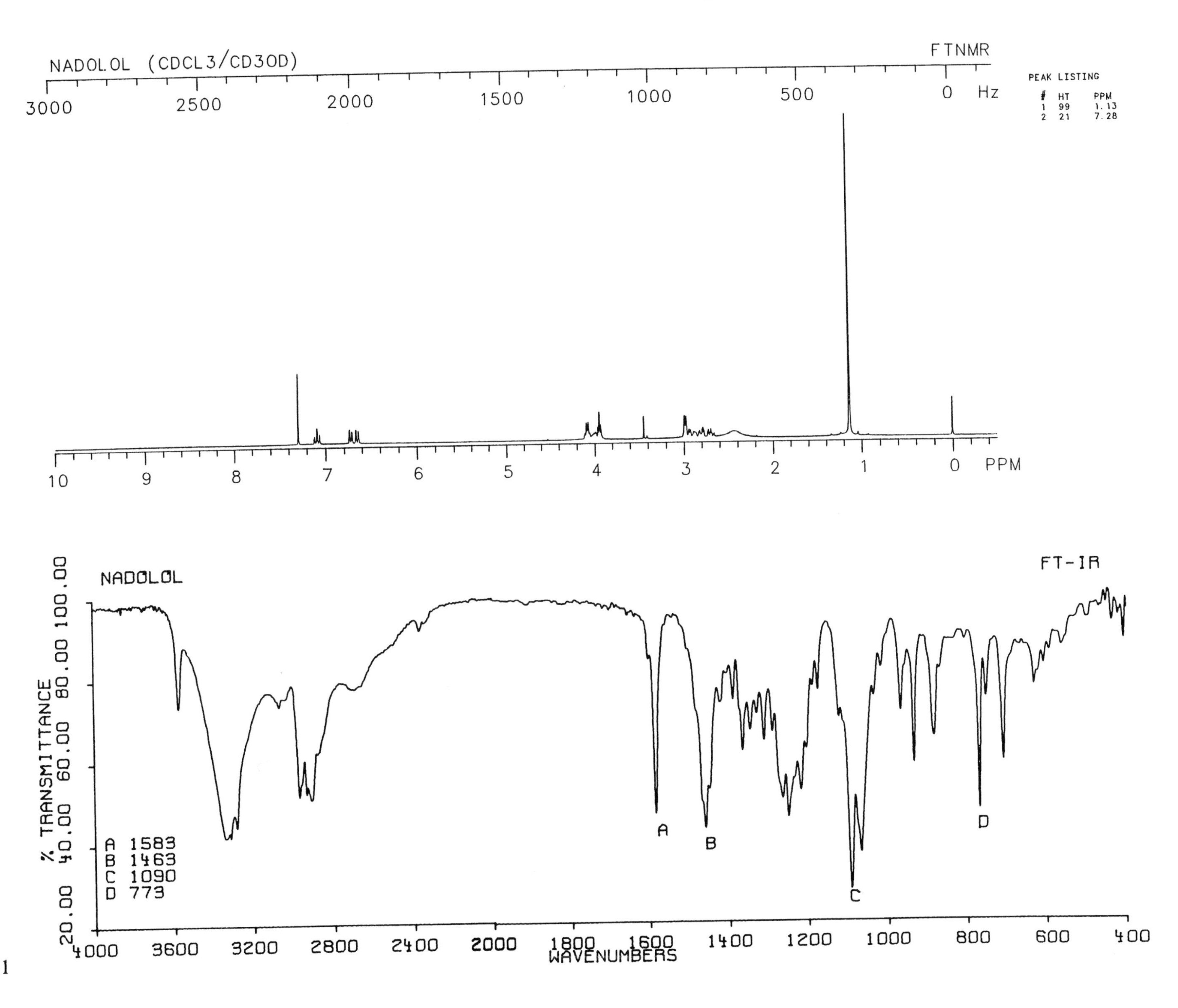
NADOLOL (CDCL3/CD3OD)
FTNMR
3000 2500 2000 1500 1000 500 0 Hz
PEAK LISTING
# HT PPM
1 99 1.13
2 21 7.28
10 9 8 7 6 5 4 3 2 1 0 PPM
NADOLOL
FT-IR
% TRANSMITTANCE
20.00 40.00 60.00 80.00 100.00
A 1583
B 1463
C 1090
D 773
A
B
C
D
4000 3600 3200 2800 2400 2000 1800 1600 1400 1200 1000 800 600 400
WAVENUMBERS

# NAFCILLIN

$C_{21}H_{22}N_2O_5S$

**Molecular weight:** 414.48 (414.13)
**Synonyms:** 6-(2-Ethoxy-1-naphthamido)-3,3-dimethyl-7-oxo-4-thia-1-azabicyclo[3.2.0]heptane-2-carboxylic acid
**Trade names:** Nafcil, Unipen

**Use:** Antibacterial
**HPLC:** Si-10; 10A:90B; 5.2
**GC:**

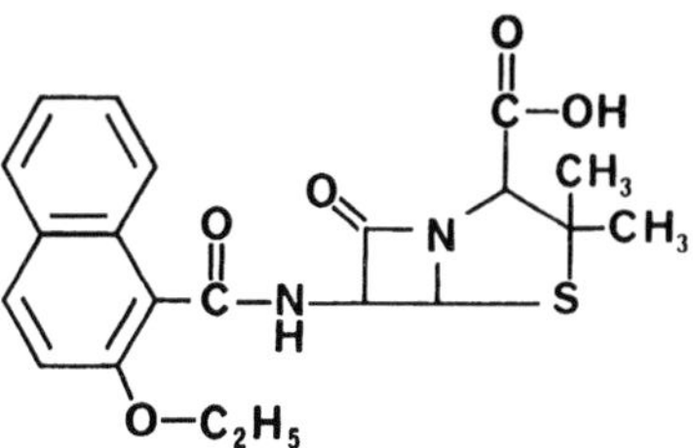
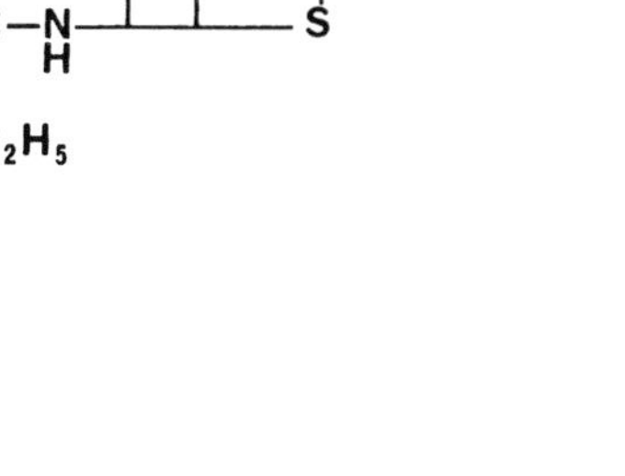

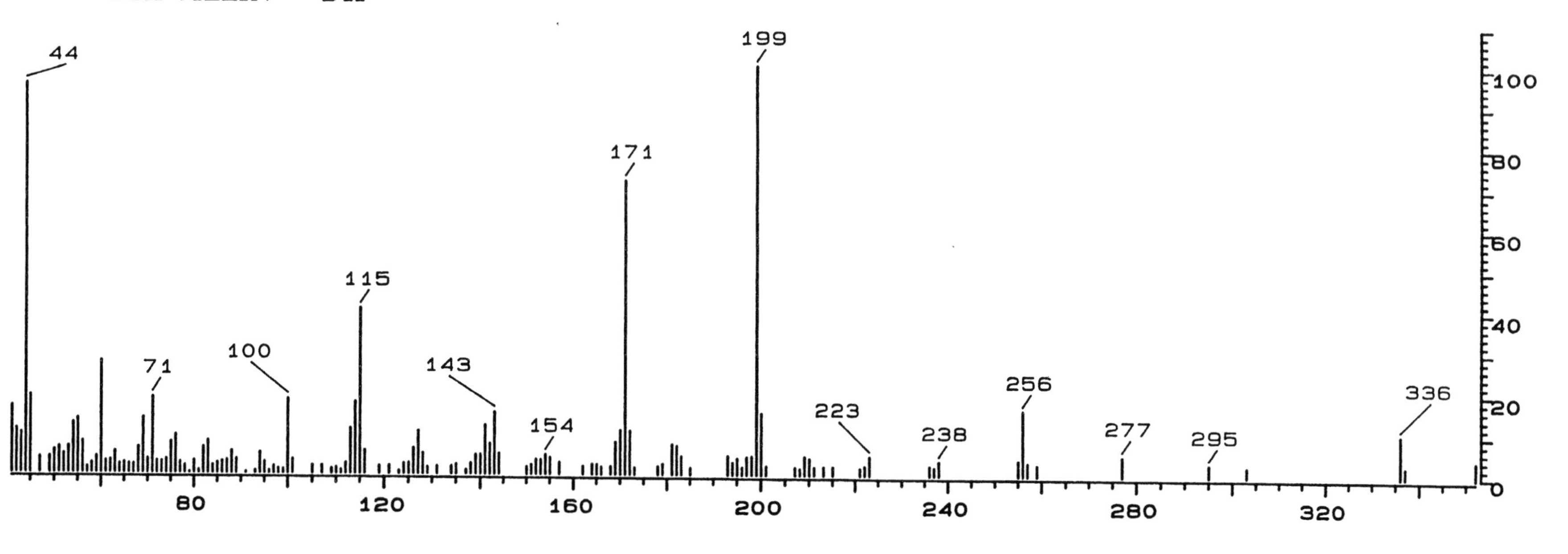


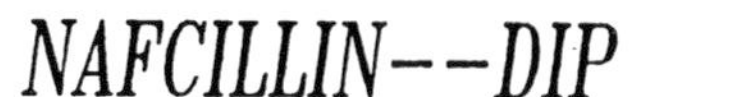

*NAFCILLIN--DIP*

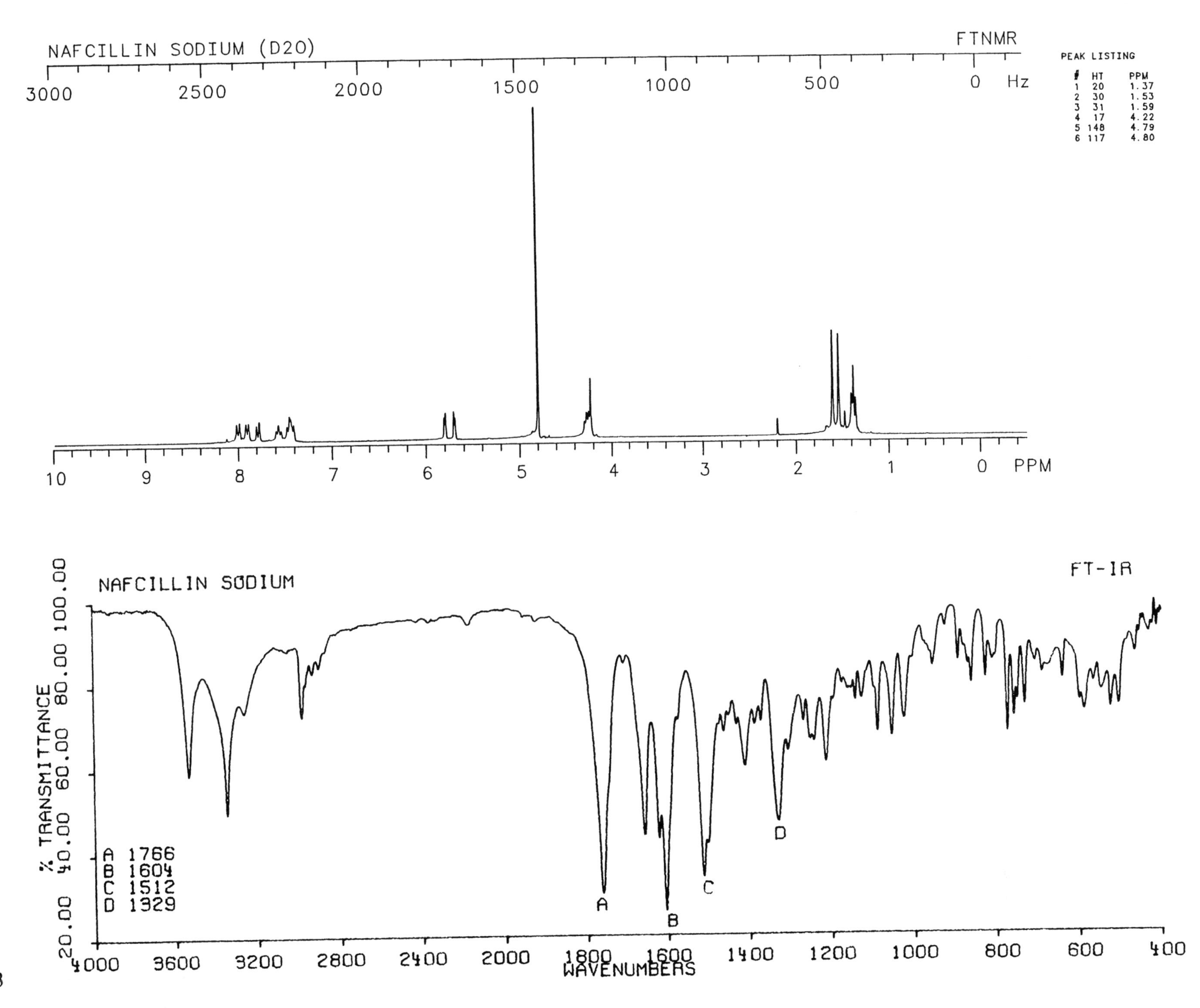

NAFCILLIN SODIUM (D2O)
FTNMR
3000
2500
2000
1500
1000
500
0
Hz
PEAK LISTING
# HT PPM
1 20 1.37
2 30 1.53
3 31 1.59
4 17 4.22
5 148 4.79
6 117 4.80
10
9
8
7
6
5
4
3
2
1
0
PPM
NAFCILLIN SODIUM
FT-IR
% TRANSMITTANCE
20.00
40.00
60.00
80.00
100.00
A 1766
B 1604
C 1512
D 1329
A
B
C
D
4000
3600
3200
2800
2400
2000
1800
1600
1400
1200
1000
800
600
400
WAVENUMBERS

# NAFOXIDINE

$C_{29}H_{31}NO_2$

**Molecular weight:** 425.56 (425.24)
**Synonyms:** 4-[(3,4-Dihydro-6-methoxy-2-phenyl-1-naphthyl)-phenoxy]ethyl pyrrolidine
**Trade names:**

**Use:**
**HPLC:** Si-10; 2A:98B; 5.7
**GC:**

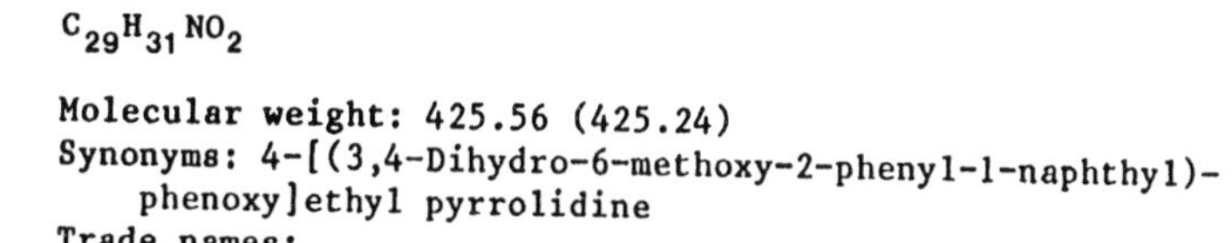


NAFOXIDINE

TRANSMITTANCE

WAVELENGTH (nm)

ETHANOL.....252,304

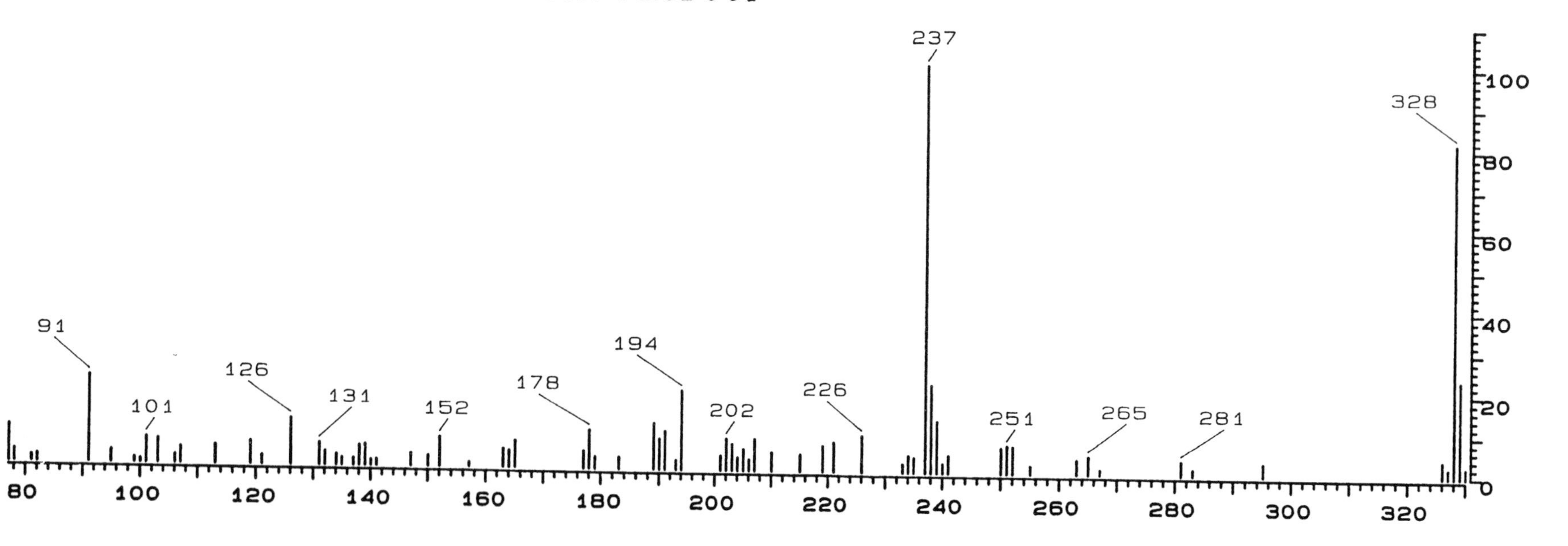

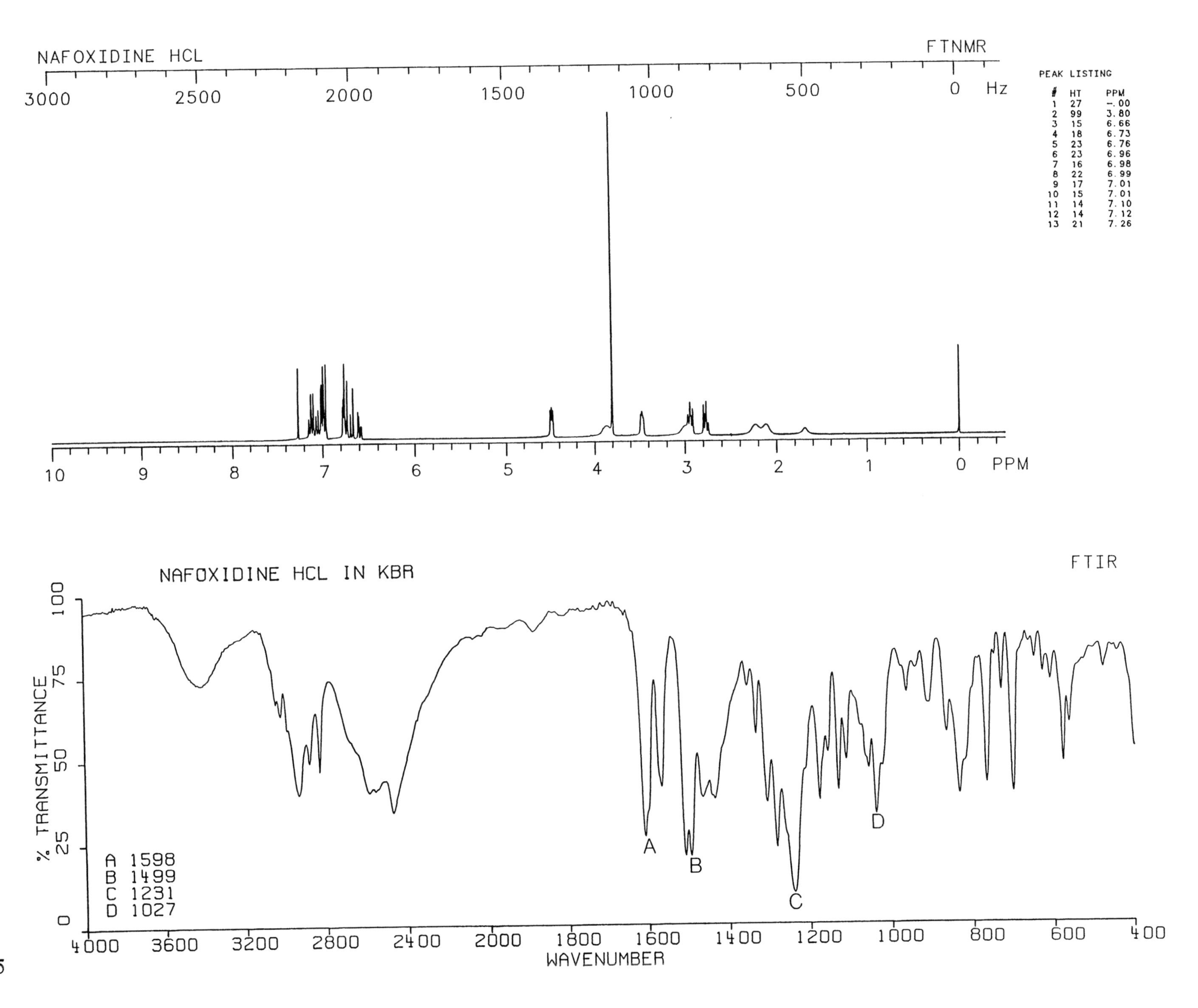

NAFOXIDINE HCL
FTNMR
3000
2500
2000
1500
1000
500
0 Hz
PEAK LISTING
# HT PPM
1 27 -.00
2 99 3.80
3 15 6.66
4 18 6.73
5 23 6.76
6 23 6.96
7 16 6.98
8 22 6.99
9 17 7.01
10 15 7.01
11 14 7.10
12 14 7.12
13 21 7.26
10 9 8 7 6 5 4 3 2 1 0 PPM
NAFOXIDINE HCL IN KBR
FTIR
% TRANSMITTANCE
0 25 50 75 100
A 1598
B 1499
C 1231
D 1027
A
B
C
D
4000 3600 3200 2800 2400 2000 1800 1600 1400 1200 1000 800 600 400
WAVENUMBER

# NALBUPHINE

$C_{21}H_{25}NO_4$

**Molecular weight:** 357.46 (357.19)
**Synonyms:** 17-(Cyclobutylmethyl)-4,5α -epoxymorphinan-3,6α,14-triol
**Trade names:** Nubain

**Use:** Narcotic analgesic
**HPLC:** Si-10; 2A:98B; 8.0
**GC:**

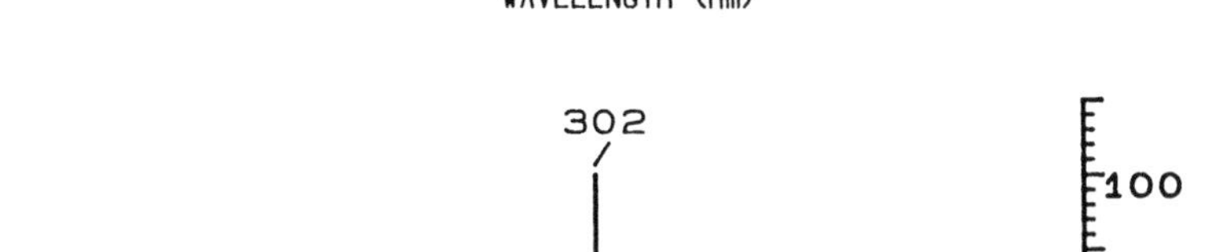


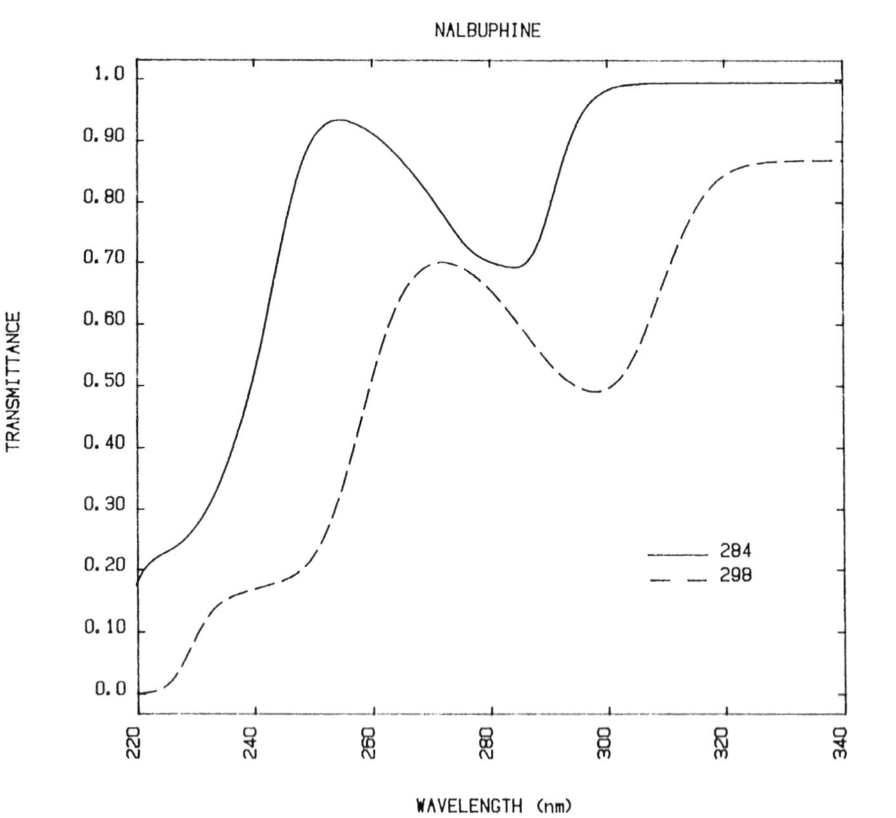


NALBUPHINE--DIP

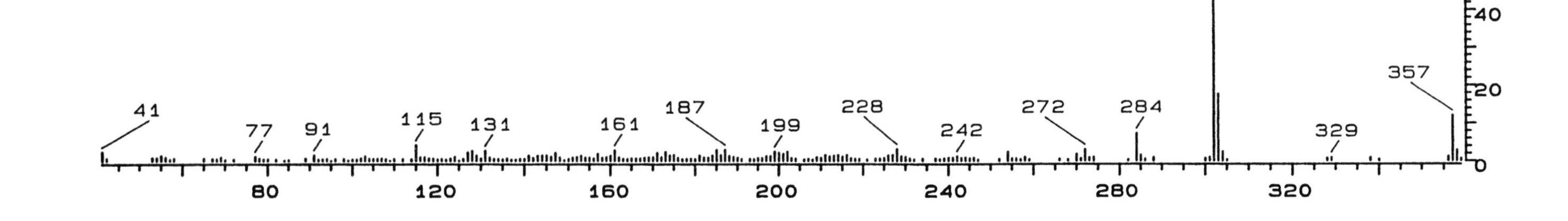

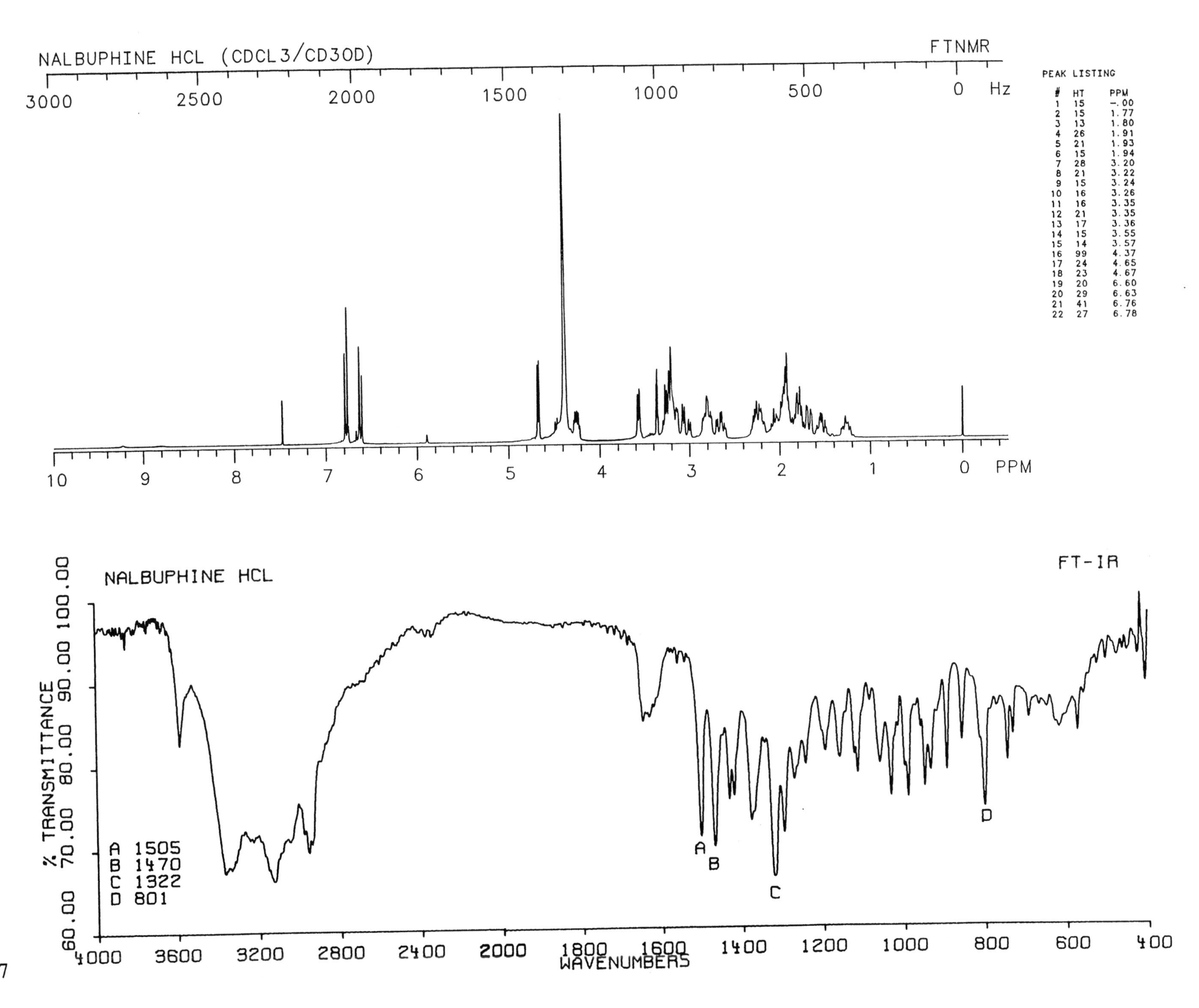
NALBUPHINE HCL (CDCL3/CD3OD)
FTNMR
3000 2500 2000 1500 1000 500 0 Hz
10 9 8 7 6 5 4 3 2 1 0 PPM
PEAK LISTING
# HT PPM
1 15 -.00
2 15 1.77
3 13 1.80
4 26 1.91
5 21 1.93
6 15 1.94
7 28 3.20
8 21 3.22
9 15 3.24
10 16 3.26
11 16 3.35
12 21 3.35
13 17 3.36
14 15 3.55
15 14 3.57
16 99 4.37
17 24 4.65
18 23 4.67
19 20 6.60
20 29 6.63
21 41 6.76
22 27 6.78
NALBUPHINE HCL
FT-IR
% TRANSMITTANCE
60.00 70.00 80.00 90.00 100.00
A 1505
B 1470
C 1322
D 801
A
B
C
D
4000 3600 3200 2800 2400 2000 1800 1600 1400 1200 1000 800 600 400
WAVENUMBERS

# NALIDIXIC ACID

$C_{12}H_{12}N_2O_3$

Molecular weight: 232.23 (232.09)
Synonyms: 1-Ethyl-1,4-dihydro-7-methyl-4-oxo-1,8-naphthyridine-3-carboxylic acid; nalidixinic acid
Trade names: Mictral, NegGram Caplets, Uriben

Use: Antibacterial
HPLC: Si-10; 20A:80B; 3.6
GC: 2484; 250°C

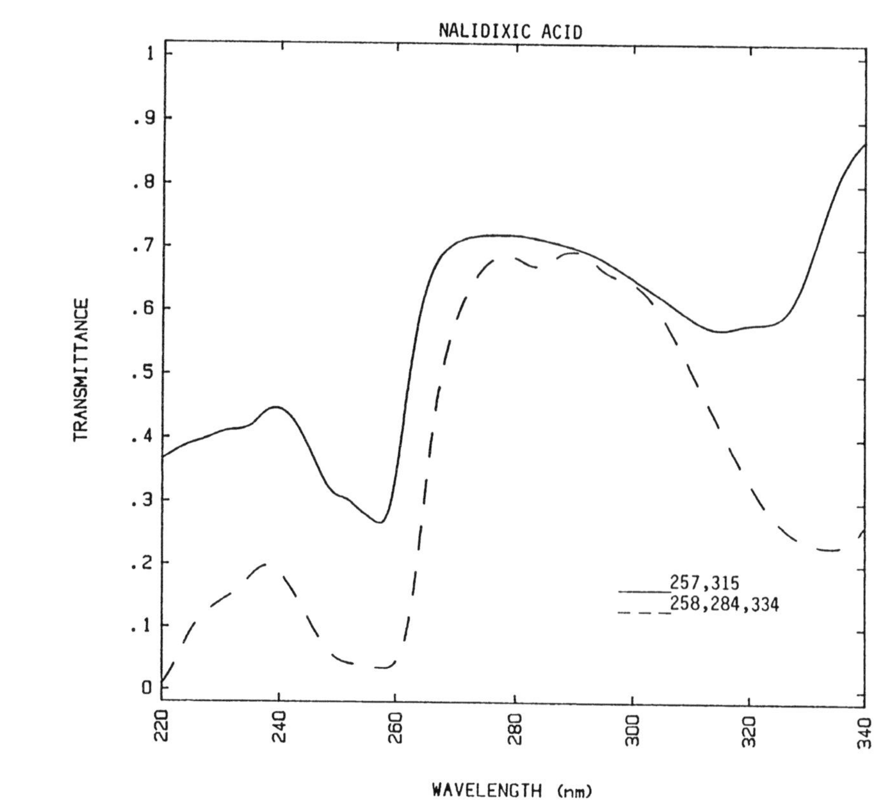


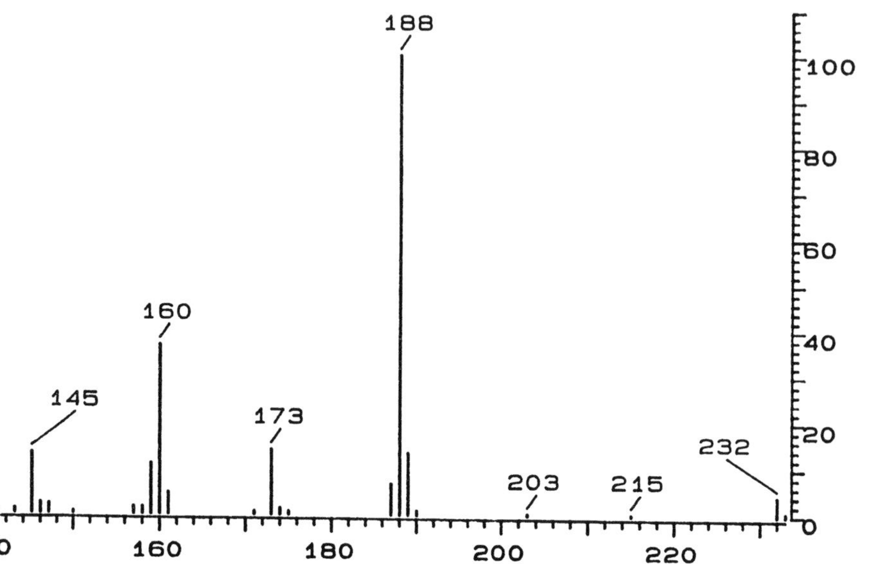


*NALIDIXIC ACID*

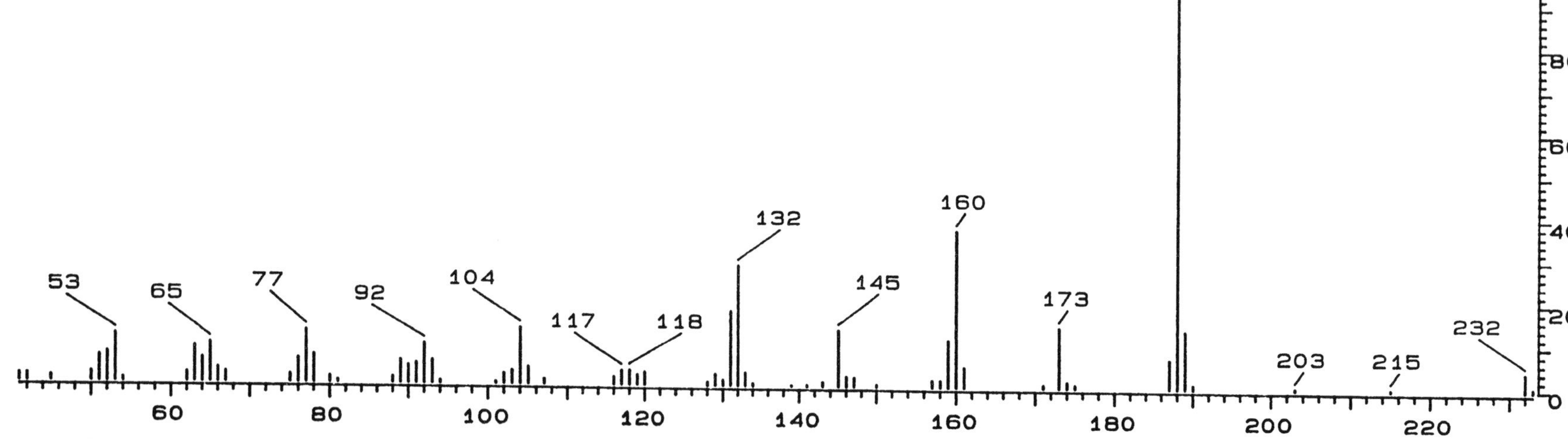

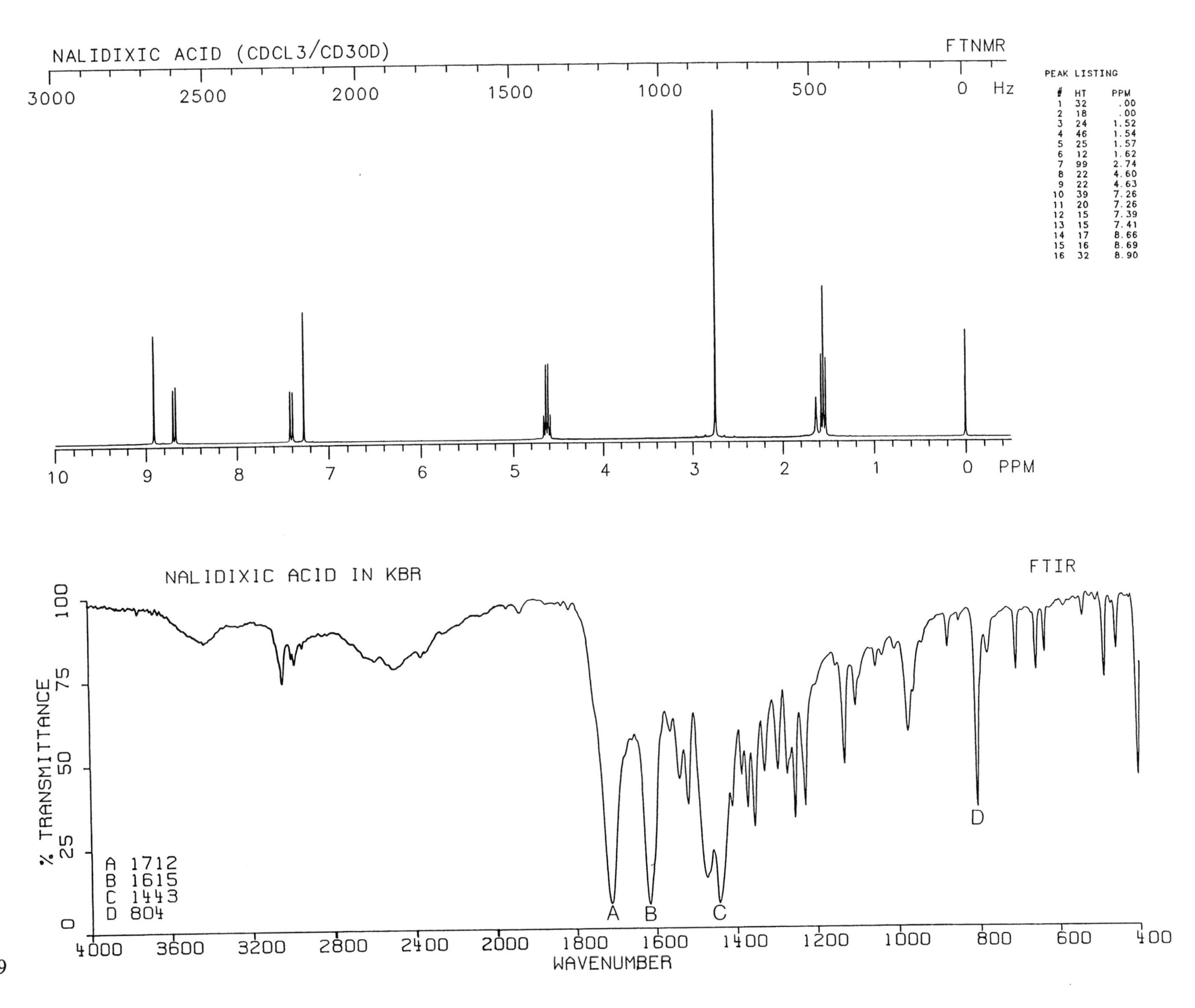
NALIDIXIC ACID (CDCL3/CD3OD)
FTNMR
3000 2500 2000 1500 1000 500 0 Hz
10 9 8 7 6 5 4 3 2 1 0 PPM
PEAK LISTING
# HT PPM
1 32 .00
2 18 .00
3 24 1.52
4 46 1.54
5 25 1.57
6 12 1.62
7 99 2.74
8 22 4.60
9 22 4.63
10 39 7.26
11 20 7.26
12 15 7.39
13 15 7.41
14 17 8.66
15 16 8.69
16 32 8.90
NALIDIXIC ACID IN KBR
FTIR
% TRANSMITTANCE
100 75 50 25 0
A 1712
B 1615
C 1443
D 804
A B C D
4000 3600 3200 2800 2400 2000 1800 1600 1400 1200 1000 800 600 400
WAVENUMBER

# NALORPHINE

$C_{19}H_{21}NO_3$

Molecular weight: 311.38 (311.15)
Synonyms: 7,8-Didehydro-4,5-epoxy-17-(2-propenyl)morphinan-3,6-diol; allorphine; N-allylnormorphine
Trade names: Nalline, Norfin

Use: Narcotic antagonist
HPLC: Si-10; 10A:90B; 4.0
': 2615; 250°C

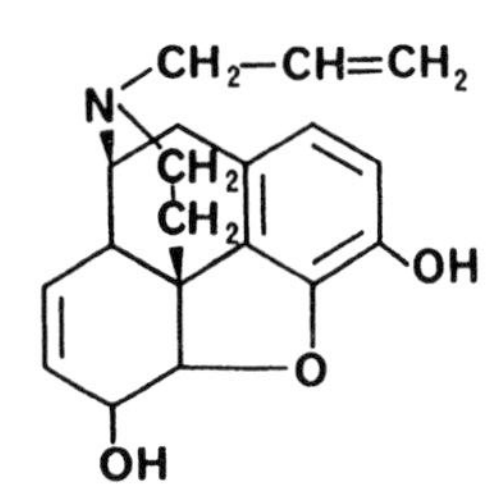


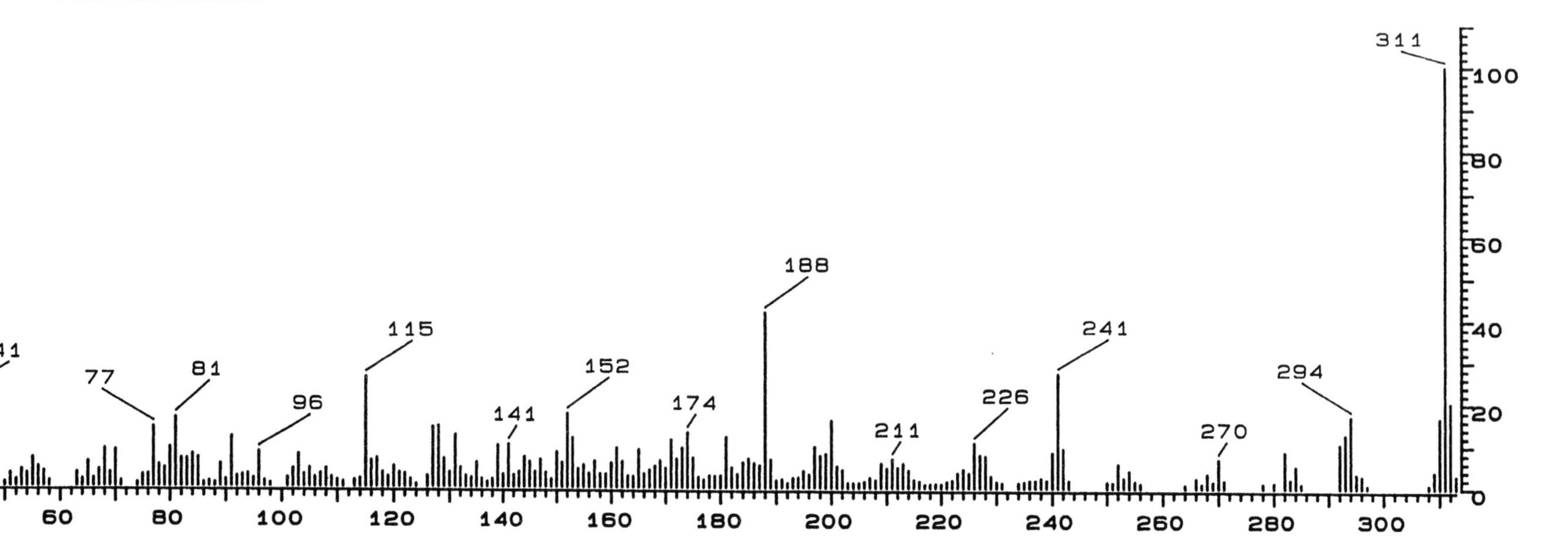

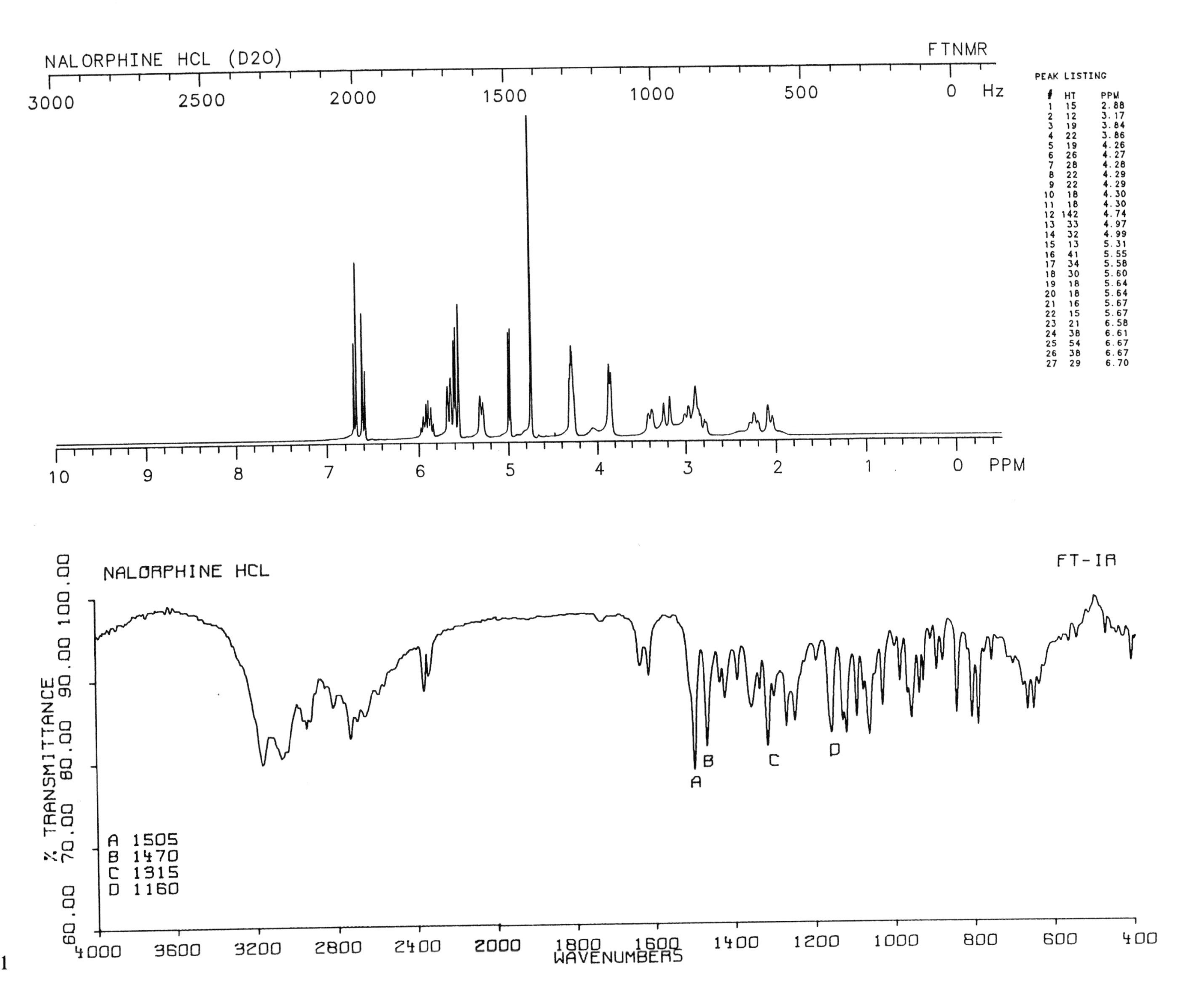
NALORPHINE HCL (D2O)
FTNMR
3000 2500 2000 1500 1000 500 0 Hz
10 9 8 7 6 5 4 3 2 1 0 PPM
PEAK LISTING
# HT PPM
1 15 2.88
2 12 3.17
3 19 3.84
4 22 3.86
5 19 4.26
6 26 4.27
7 28 4.28
8 22 4.29
9 22 4.29
10 18 4.30
11 18 4.30
12 142 4.74
13 33 4.97
14 32 4.99
15 13 5.31
16 41 5.55
17 34 5.58
18 30 5.60
19 18 5.64
20 18 5.64
21 16 5.67
22 15 5.67
23 21 6.58
24 38 6.61
25 54 6.67
26 38 6.67
27 29 6.70
NALORPHINE HCL
FT-IR
% TRANSMITTANCE
60.00 70.00 80.00 90.00 100.00
A 1505
B 1470
C 1315
D 1160
A B C D
4000 3600 3200 2800 2400 2000 1800 1600 1400 1200 1000 800 600 400
WAVENUMBERS

# NALOXONE

$C_{19}H_{21}NO_4$

**Molecular weight:** 327.38 (327.15)
**Synonyms:** 17-Allyl-4,5α-epoxy-3,14-dihydroxy-morphinan-6-one; N-allyl-noroxymorphone
**Trade names:** Narcan, Narcan Neonatal

**Use:** Narcotic antagonist
**HPLC:** Si-10; 100B; 3.8
**GC:** 2820; 280°C

*NALOXONE*

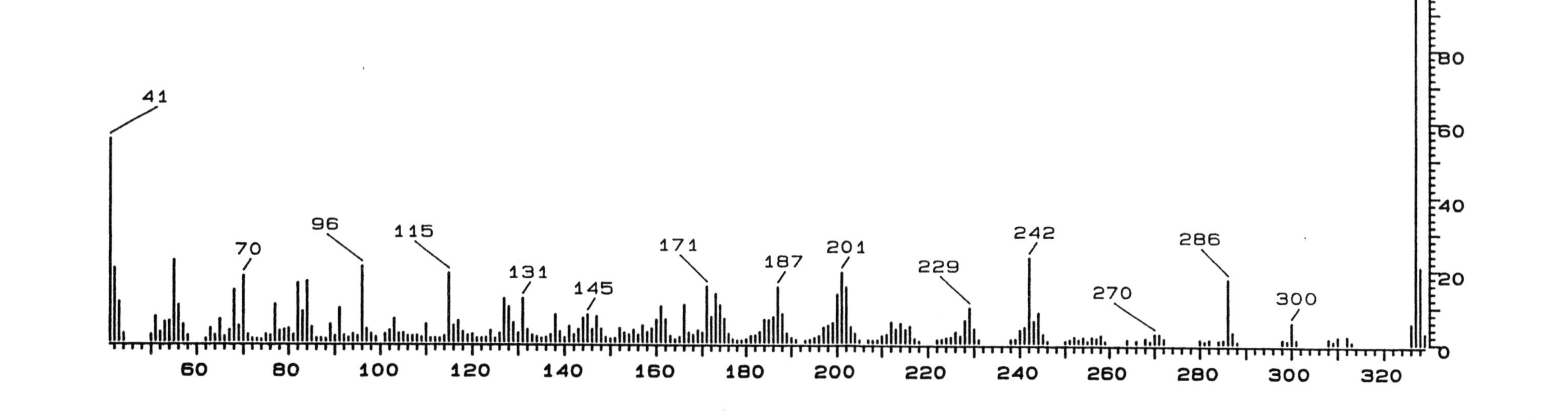

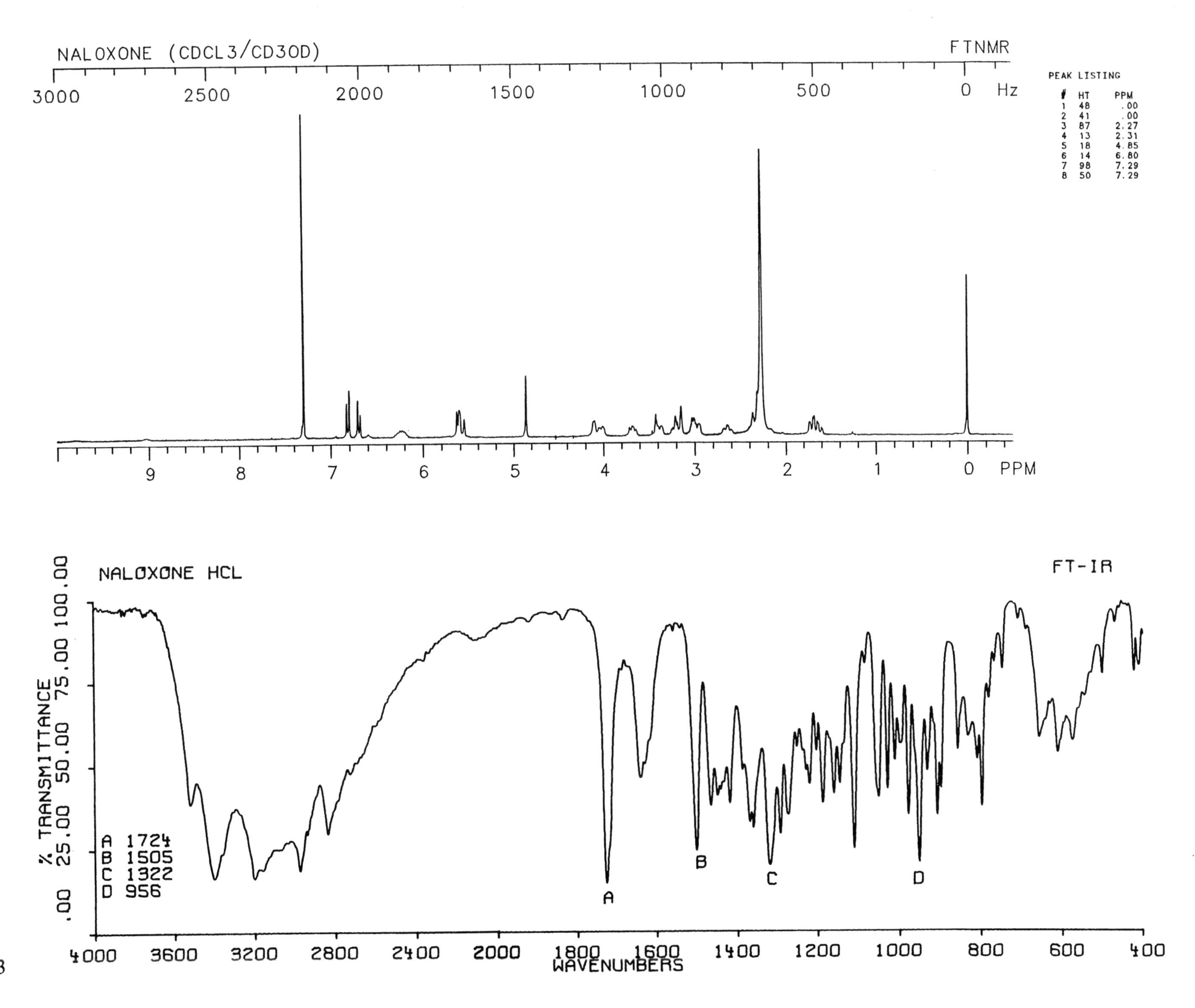
NALOXONE (CDCL3/CD3OD)
FTNMR
Hz
PPM
PEAK LISTING
# HT PPM
1 48 .00
2 41 .00
3 87 2.27
4 13 2.31
5 18 4.85
6 14 6.80
7 98 7.29
8 50 7.29
NALOXONE HCL
FT-IR
% TRANSMITTANCE
WAVENUMBERS
A 1724
B 1505
C 1322
D 956

# 6-β-NALTREXOL

$C_{20}H_{25}NO_4$

Molecular weight: 343.42 (343.18)
Synonyms: 5α-17-(Cyclopropylmethyl)-4,5-epoxy-3,6,14-trihydroxymorphinan
Trade names:

Use: Metabolite of naltrexone
HPLC: Si-10; 20A:80B; 4.0
GC: 2971; 280°C

N
$CH_2$
$CH_2$
$CH_2$
HO
OH
O
H
OH

6-B-NALTREXOL

TRANSMITTANCE

1.0
0.90
0.80
0.70
0.60
0.50
0.40
0.30
0.20
0.10
0.0

220
240
260
280
300
320
340

WAVELENGTH (nm)

284
297

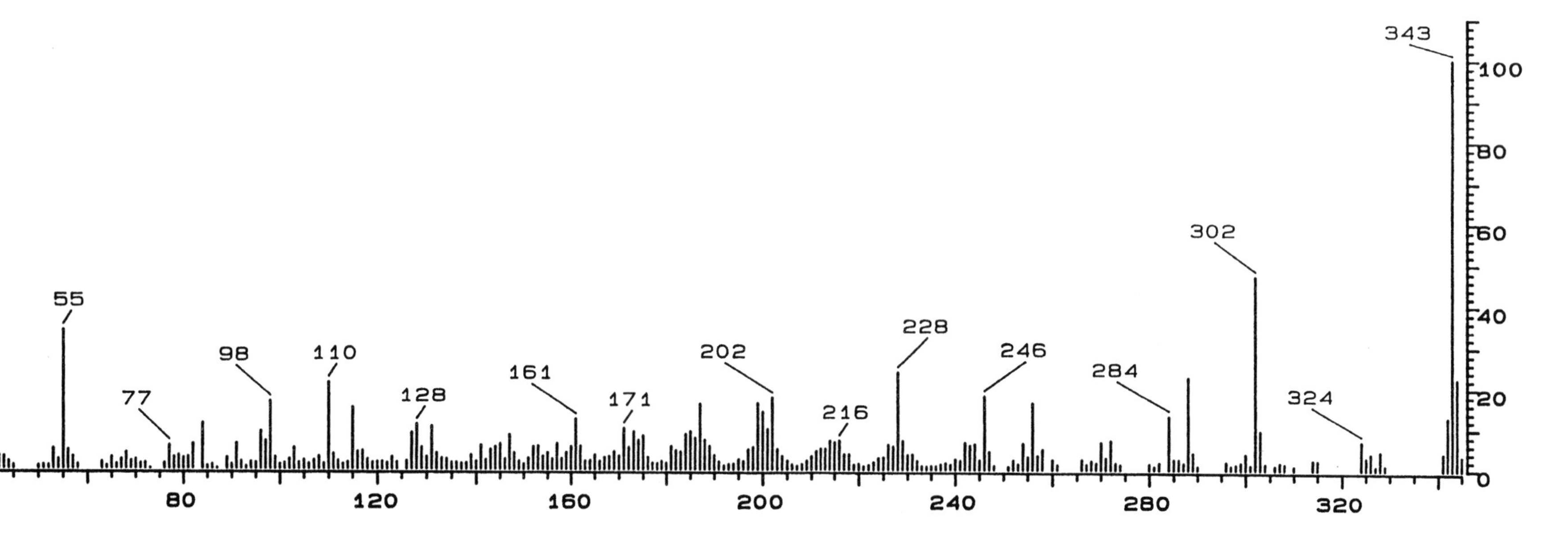

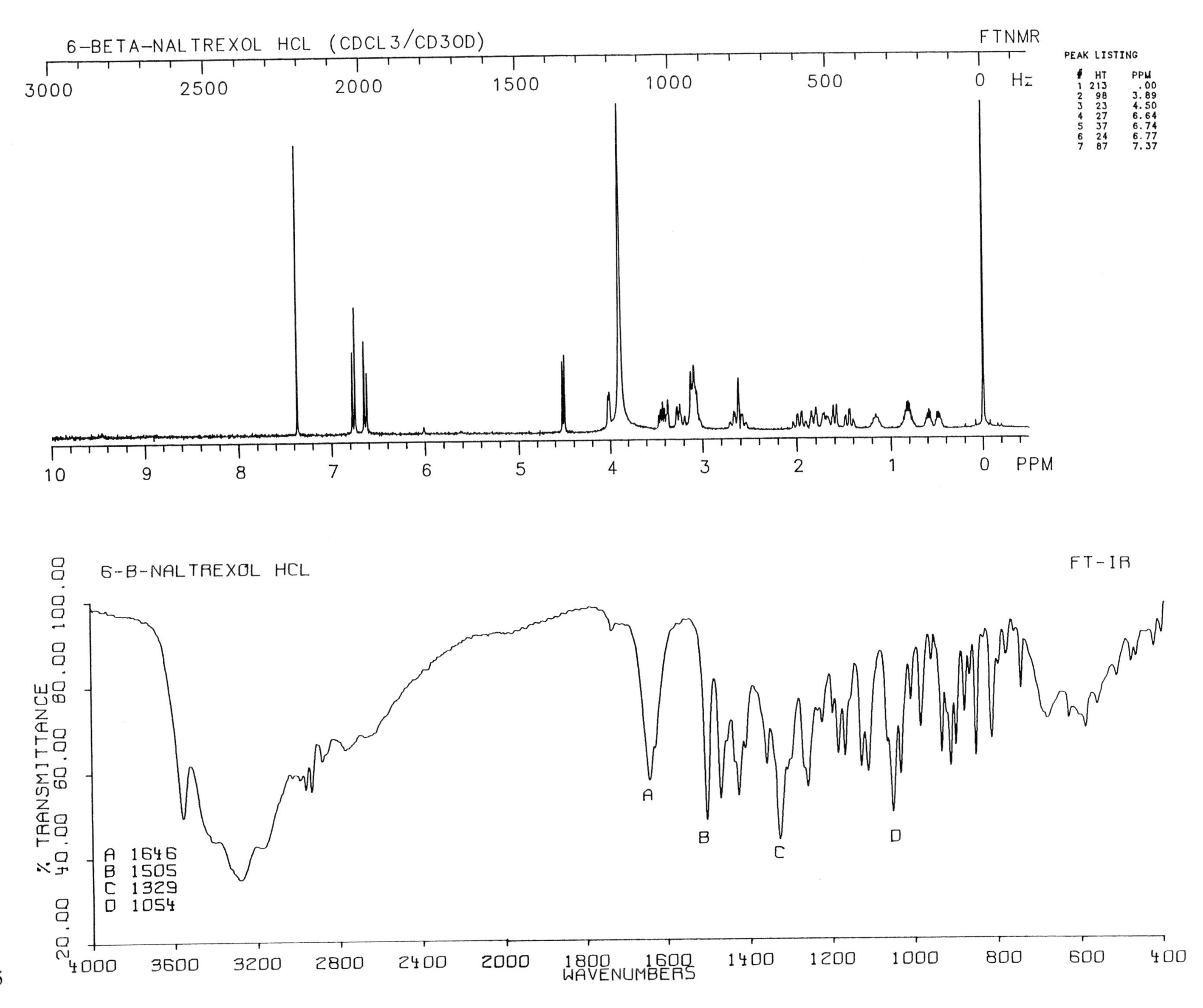
6-BETA-NALTREXOL HCL (CDCL3/CD3OD)
FTNMR
PEAK LISTING
# HT PPM
1 213 .00
2 98 3.89
3 23 4.50
4 27 6.64
5 37 6.74
6 24 6.77
7 87 7.37
Hz
PPM
6-B-NALTREXOL HCL
FT-IR
% TRANSMITTANCE
WAVENUMBERS
A 1646
B 1505
C 1329
D 1054

# NALTREXONE

$C_{20}H_{23}NO_4$

Molecular weight: 341.41 (341.16)
Synonyms: 5α-17-(Cyclopropylmethyl)-4,5-epoxy-3,14-dihydroxymorphinan-6-one
Trade names:

Use: Narcotic antagonist
HPLC: Si-10; 5A:95B; 4.0
GC: 2775; 280°C

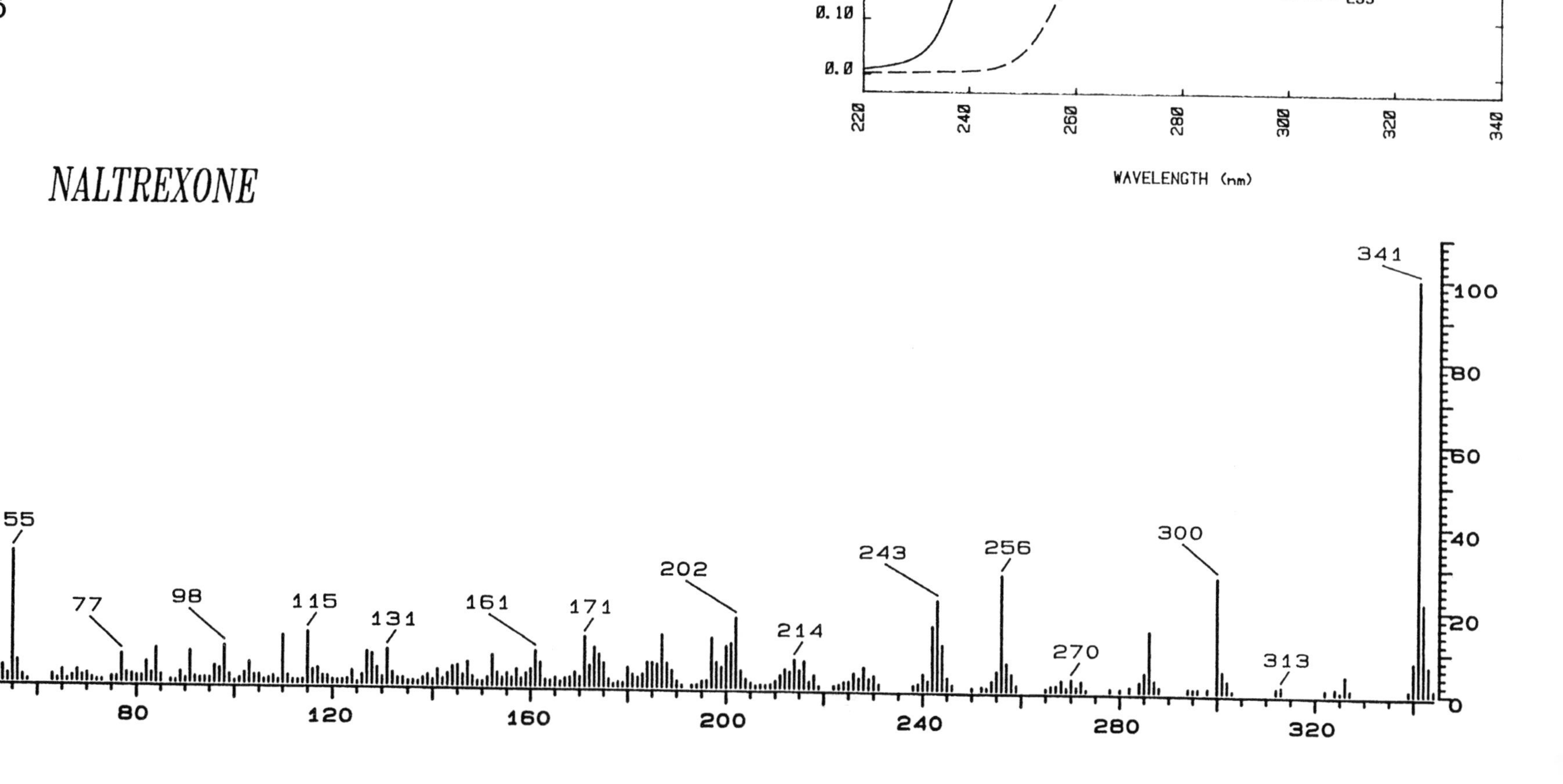

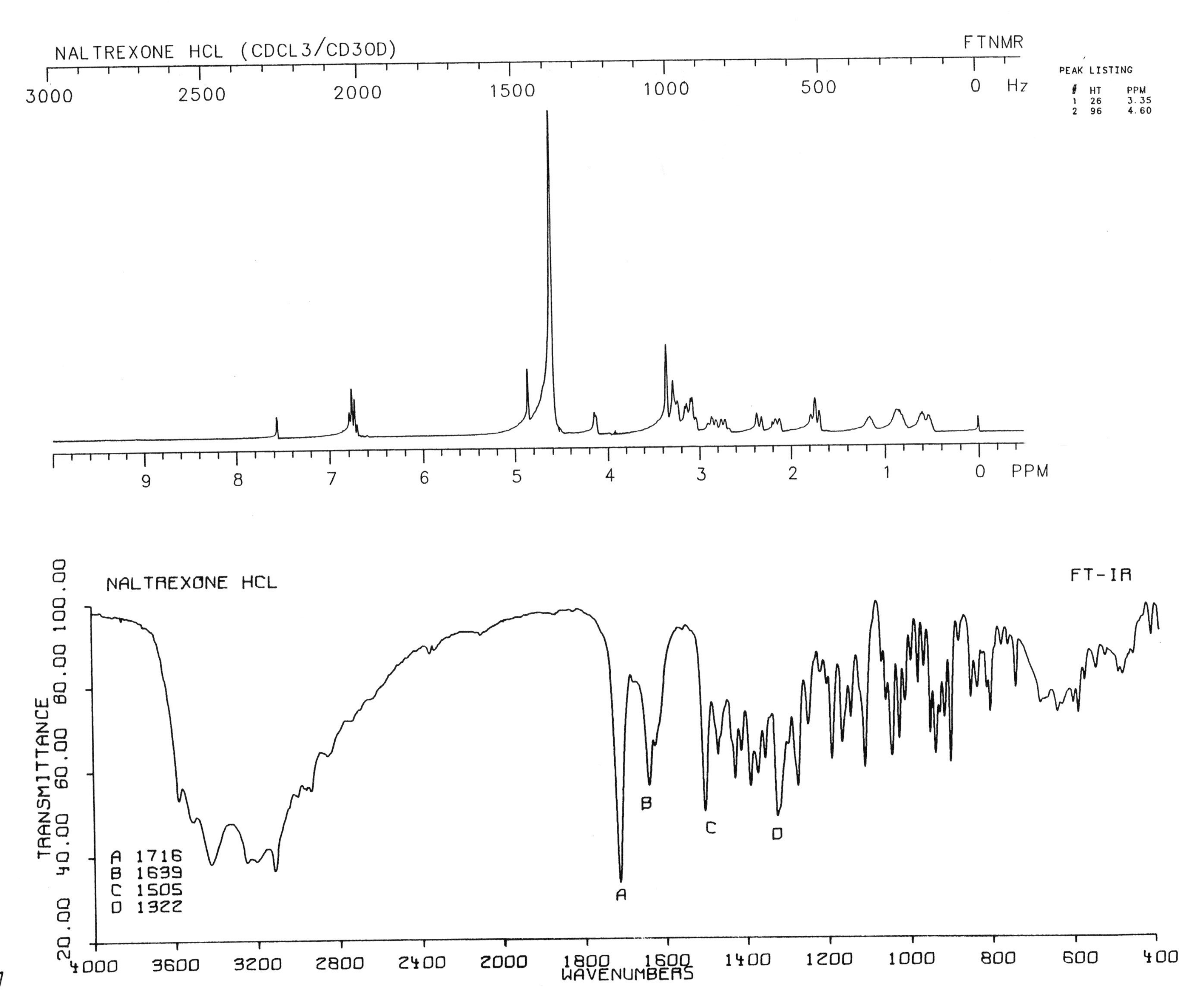
NALTREXONE HCL (CDCL3/CD3OD)
FTNMR
3000
2500
2000
1500
1000
500
0 Hz
PEAK LISTING
# HT PPM
1 26 3.35
2 96 4.60
9
8
7
6
5
4
3
2
1
0 PPM
NALTREXONE HCL
FT-IR
TRANSMITTANCE
20.00
40.00
60.00
80.00
100.00
A 1716
B 1639
C 1505
D 1322
A
B
C
D
4000
3600
3200
2800
2400
2000
1800
1600
1400
1200
1000
800
600
400
WAVENUMBERS